Plant Biotechnology and Agriculture
Prospects for the 21st Century

植物生物技术与农业
展望 21 世纪

Edited by
Arie Altman
Robert H. Smith Institute of Plant Sciences and Genetics in Agriculture
Hebrew University of Jerusalem
Rehovot, Israel
Paul Michael Hasegawa
Bruno C. Moser Distinguished Professor
Horticulture and Landscape Architecture Department
Purdue University
West Lafayette, Indiana, USA

科学出版社
北京

图字：01-2012-2726

This is an annotated version of
Plant Biotechnology and Agriculture：Prospects for the 21st Century
Edited by Arie Altman and Paul Michael Hasegawa.

ISBN：978-0-12-381466-1.

图书在版编目(CIP)数据

植物生物技术与农业：展望21世纪＝Plant Biotechnology and Agriculture Prospects for the 21st Century：英文／（以）奥尔特曼（Altman，A.）等编著.—北京：科学出版社，2012

ISBN 978-7-03-034291-1

Ⅰ.①植… Ⅱ.①奥… Ⅲ.①植物－生物技术－介绍－英文②农业技术－生物技术－介绍－英文 Ⅳ.①Q94②S188

中国版本图书馆CIP数据核字（2012）第093545号

责任编辑：孙红梅／责任印制：钱玉芬
封面设计：耕者设计工作室

科学出版社 出版
北京东黄城根北街16号
邮政编码：100717
http://www.sciencep.com
双青印刷厂 印刷
科学出版社发行 各地新华书店经销
*
2012年7月第 一 版 开本：787×1092 1/16
2012年7月第一次印刷 印张：42 3/4
字数：1 014 000

定价：180.00元

（如有印装质量问题，我社负责调换）

前　言

早在20世纪70年代，伴随着对于细菌和酵母的基因克隆和DNA测序研究的突破，有人就曾预言后续的应用前景不可限量，并将改变人类的生活。迄今为止大量的研究突破验证了这一预言，其中包括通过蛋白合成技术生产治疗人类和动物疾病的药物或疫苗，在食品、饮料和其他工业产品生产中使用的酶类，改良的作物新品种，以及生产生物燃料等生物制品的新途径等等。以上这些成果只是35年前大量研究发现的一部分，因此说这些研究发现带来了科技的巨大进步毫不为过。

20世纪70年代和80年代在科学技术领域的投资在其后的20年间显现出了明显效果，尤其在农业等领域。例如，在玉米遗传和育种领域的投资促进了后来在超级杂交种和骨干种质改良中发挥了重要作用的遗传工程（GE）和其他一些生物技术的研究，并已经取得很大进展。同样由投资支持的由 Mary Dell Chilton，Josef Schell，Marc Van Montague，Rob Schilperoot 等人领导的团队与孟山都公司（St. Louis，MO）Robb Fraley，Steve Rogers，Rob Horsch 领导的团队相互合作，从研究根癌农杆菌的作用机理开始，直到获得转基因农作物植株。上述科学家的研究引领了全球科学界通过基因工程技术培育和改良农作物和非农作物植物性状的研究潮流。

在后面的论述中，将向读者介绍植物生物技术的背景知识，当今遗传学、基因组学以及其他各种组学的研究状况，以及目前对于遗传工程的理解。其后的章节将介绍种质资源的改良和保存、植物育种、种子改良，以及孤雌生殖等方面的科技进展，这些内容都与农业和农业生物技术的短期和长期的成功密切相关，同时为读者深入理解后面关于新科技应用前景的章节提供背景知识。如果能够满足相关安全条例并被消费者接受，这些前景应用将继续改变全球农业的未来。

包括笔者在内的许多科学家投身于遗传学研究，我们希望：1）取得新发现；2）改善人类生存条件，科学家通常在这些目标中的某一个方面有所作为，而难以兼顾。取得新发现固然令人兴奋，当多数科学家们得知其新发现能够转化为现实应用时会感到十分欣慰。自从1996年首个产品进入市场以来，农业生物技术确实对我们的生活产生了重大影响，同时使得人们对科技的未来充满遐想。那些抗虫、抗除草剂、抗病毒的农作物的产生和应用，使人们相信生物技术在其他方面同样可以获得成功。而农民、农业经济、环境、消费者、种子公司都能从中获益。事实上，我们对于这些研究的前景很有信心：研究工作将取得更快的进展，而且将培育出更加抗病、抗虫、需要较少化学保护剂的农作物，能够更好地适应高温、低温、干旱和盐碱的种植条件；更加富含蛋白质、维生素、矿物质，培育出的某些特殊作物将满足不断增长的世界人口在制药、化工以及能源等方面的需要。

当我们的新发现被某些人认为可怕，被认为对人类、动物和环境有害时，我们感到十分吃惊和失望。这种失落现在仍在继续，尤其是在一些刚开始推广基因工程作物的国家。对于一项尚未得到公正评价的技术来说，这曾经并仍然是一个出乎意料的结果。认为这项技术不安全和可怕的人做出这种评价所基于的原因五花八门，以至于无法在此详细讨论。比如有些人无视科学界对于其安全性的广泛共识，认为基因工程的安全性会带来法律问题，有人从哲学角度考虑反对在动植物身上使用基因工程技术，还有人认为除非基因工程产品能够拥有迎合消费者的特性，否则他们不会购买含有基因工程原料的产品。

在生物技术产品生产过程的很多环节已经受到了这种恐惧带来的影响，笔者就曾经遇到过大学教授反对学生参加能够促进农业生物技术发展的研究；有些研究者对于他们的研究成果能否被承认缺乏信心，担心其研究工作会受到基因工程产品反对者的阻挠；对于基因工程产品批准的程序和代价的担心可能使得一些潜在的投资人放弃投资；负责产品上市的管理官员发现会有对相关科技一无所知的地方法官来评价新产品或是新技术的安全性并压制管理机关的决策；种子公司出于对上市程序及消费者反应的担心而放弃采用基因工程技术，转而选用突变和高级育种来得到需要的性状。这些对于创新的负面作用已经影响了不少产品的研发工作。

本书介绍了一些新技术和新产品，我们相信科技发展不会停止，因为这些科技对环境、对发明者、对消费者和生产者都是有利的。笔者认为自己有时是个谨慎的乐观主义者，有时是个轻微的悲观主义者，其原因可能与农业在解决人类要实现可持续发展所面临的重大挑战中的角色有关。这些挑战包括在气候变化背景下的世界粮食安全，生物能源的发展，以及人类的健康和繁衍。随着人口的不断增长，要满足需求的增加需要有新科学技术的参与，这不仅是科学界的共识，也是人类的共识。

从后面的内容中，有人看到的是战胜挑战的机会，有人看到的则只有挑战和困难，不论是满怀希望还是愤世嫉俗，笔者都希望读者能够从本书中有所收获，经过思考，在相比阅读之前有了更全面的了解之后再决定你应该如何行动。面对挑战，我们需要行动。

Roger N. Beachy

President Emeritus, Donald Danforth Plant Science Center, and First Director, National Institute of Food and Nutrition, U. S. Department of Agriculture

（王斌　译）

序

当今世界人口增长的速度已经超出了食物生产增长的速度，因此必须加快新的农业技术尤其是植物生物技术的应用，以尽量减小人类需求和生产能力之间的差距。据预测世界人口将很快达到70亿，到2050年将超过100亿，而目前农业产量的年均增长只有1.8%～2%。农业为人类提供动物饲料，植物纤维，而且是人类食物的来源，既为人类提供热量和常量营养元素又为人类提供微量营养元素。生物技术为农作物的增产和食品生物安全性的提高提供新方法，还能生产植物来源的生物新材料，包括潜在的生物燃料作物。农业同时从经济学的角度对人类生活质量发挥着巨大影响。毫无疑问，单靠传统农业将无法满足人类的需要，人类必须依靠植物生物技术。

尽快广泛地推广农业生物技术和推广医学生物技术同样重要，或者说更加重要，世界范围内死于饥荒和营养不良相关疾病的人数远超过死于“现代”“西方式”疾病的人数。粮食紧缺是巨大的灾难，但与其他自然灾害不同的是，我们可以针对这场灾难进行准备，甚至是预防。

农业起源于对野生动植物的驯养和培育以及长期对产量和质量性状的逐渐选择。在其后的食物存储过程中，伴随着对于微生物的应用，代表着最早期生物技术的传统食物发酵技术产生了，并应用在食物生产和保存过程中。现今传统农业显示出3个方面的局限性：1）市场的局限性（在全球化的背景下，商品的自由流通使得商品的区域定价已经失效，取而代之的是对于全球化贸易策略的需求）；2）自然资源的局限性（以沙漠化、干旱和盐碱化为代表的全球气候变化以及伴随的工业化和城市化进程已经造成了土地和水资源的短缺及土壤、水质、空气质量的恶化）；3）固有生物遗传资源的局限性（通过传统育种提供改良基因型进展缓慢，无法满足要求，而且由于缺少合适的天然基因和等位基因，传统的杂交育种受到很大限制）。

目前在植物生物技术领域取得的进展已经超出了之前的所有预期，而且未来必将更加美好。植物和农业生物技术革命的完全实现有赖于创新的研发工作能够不断取得成果，对产业有利的监督管理环境和大众接受程度的提高。过去十年间在植物分子生物学和生物技术方面取得的新进展使得这些科学与经典的生理学、生物化学和育种等科学有可能有效结合，产生如下结果：1）更好地控制植物的发育（生长和繁育）；2）使植物能够更好地抵抗不同的生物的和非生物的胁迫；3）对传统育种提供帮助；4）生物工程改造生物；5）促进特殊种类食品和新的生物材料的生产。

因此，本专著希望能够向读者提供在植物生物技术本身及其成果方面权威的和最新的知识、技术方法和进展，内容涵盖从基本的生物学发现到其在农业和其他生命科学中的应用。本书内容由相关领域专家精心撰写，包括对取得成果的评述，新的植物生物技术方法、产品在促进经典植物科学和农业科学的发展和在作物及其产品改良上的应用前景，植物生物技术和实用农业技术之间彼此配合、相互促进。

本书还可以作为植物生物学、农业科技、植物分子遗传学、植物育种、食品科学、生物材料科学等领域的研究者的参考用书，以及用作农业、植物、食品和生物技术领域研究生的教学辅助用书。

本书的第一部分（植物生物技术方法简介）主要介绍与生物技术相关的重要生物学进展，内容包括基因组学，蛋白质组学，代谢物组学，组织培养和遗传转化。第二部分的内容将介绍育种生物技术，包括体细胞培养，分子标记辅助选择，杂交种生产，自发变异的鉴定和开发，表观遗传学和表观基因组学。第三部分介绍与种质相关的内容，包括无菌苗的大孢子培养和再生，孤雌生殖的调控，种质资源的收集和保存。之后的两部分内容介绍控制植物对环境中非生物和生物胁迫的应答反应的控制，非生物胁迫包括干旱、盐碱、极端温度条件、污染等，还将介绍植物修复相关的内容，生物胁迫包括真菌、细菌、病毒引起的疾病以及虫害，还将介绍杂草控制方面的内容。第六部分内容介绍用生物技术对产量和品质性状进行改良，将讨论与根系、茎和果实发育，控制开花，以及采收后存储和控制质量性状的次级代谢产物相关的修饰。最后一部分内容介绍与转基因的漂流、知识产权、管理制度相关内容以及作物生物技术在发展中国家的现状。

笔者希望对所有参与编写的来自全球各地的在植物学和农业生物技术领域做出瞩目贡献的杰出先锋科学家表示感谢。书中这些作者表达了各自的观点，我们已经尽力保持风格一致，各章中所列举的结果和观点可能反映了作者各自的偏好。

笔者希望对在本书编纂过程中提供了帮助和宝贵意见的朋友和同事表示感谢，并感谢Elsevier 的工作人员尤其是 Rogue Shindler 的耐心和所提供的帮助。

最后，笔者希望感谢编写者的家庭为他们提供的理解和支持。

Arie Altman

Paul Michael Hasegawa

（王斌　译）

目　　录

第三部分 植物种质资源

第四部分 控制植物对环境的反应：非生物和生物胁迫

（王斌　译）

植物生物技术简介 2011：概况和在农业上的应用

Arie Altman，Hebrew University of Jerusalem，Iseael
Paul Michael Hasegawa，Purdue University，Indiana

1. 植物的作用及农业：食物安全，环境变化及新的生物材料

现在的农作物是植物经过长期驯化和对质量、数量性状的长期自然选择和人为定向选择的产物。动、植物的某种最初的改良可能是源自单一的偶然机会，早期的农民或科学家的眼力和大脑在优良性状选择上起到了决定性作用。因此，农业和植物科学的发展主要是科学的发现和判断，以及长期创新的结果，有些起到了革命性的作用，如食品发酵，无机肥料的使用，和近期的遗传转化技术的应用。

人工技术和生物技术，包括食物和纤维，自古以来就造就了人类的生活。发酵生产的动、植物产品（面包，奶酪，葡萄酒），农业生产出现以来的常规育种，后来的绿色革命，分子标记辅助选择以及重组 DNA 技术等都属于生物技术范畴（Meiri and Altman，1998；Altman，1999；Chrispeels and Sadava，2003；Science，2010）。人们迫切需要寻找新的生物技术，从 1982～1983 年最初获得转基因植物的重大突破开始，实际上加速了植物分子生物技术的应用，主要有以下四个原因：

（1）世界人口的增长需要更多的食物。

（2）人们认识到病原菌侵染和食品营养质量（特别是维生素和矿物质）会影响人类的健康。

（3）不利的气候变化及有害生物和非生物危害威胁着农作物和生态系统。

（4）人类社会要寻求新的植物产品，如生物材料，治疗药物和生物燃料。

在人口增长速度超出了食物生产增长（包括产量和质量）速度的当今世界，需要农业和生物技术的迅速发展。

1.1 世界人口和食物供给

根据联合国粮农组织（FAO）2009 年的估计，到 2050 年世界人口总数将达到 91 亿，比现在增加 34%，而且主要集中在发展中国家。它还预测由于城市化的加速，随着收入水平的提高，大约 70%的人口将为城市人口（目前是 49%）。为了供养这样巨大的城市人口，食物供给必须增加 70%，例如谷物的年产量必须从目前的 2.1 亿吨提高到 3 亿吨。FAO 进一步指出只有对农业生产必要的投资、政策、法规都能得到实施，上述供应的提高才可以实现。据估计在发展中国家，80%所需产品的增加来自产量的提高和种植密度的增加，20%来自种植面积的扩大。完成上述任务对改进农业技术和生物技术提出了更大的需求。应当指出的是讨论食物供给不只是指热量需求，还包括维生素、矿物质，和其他营养因素的摄入，通过生产生物强化食品改善营养不良人群（尤其是儿童和贫穷国家的人

们）的健康。这方面内容在后面还要进一步讨论。

1.2 气候和环境变化

气候变化和沙漠化通过影响植物生理特性和增加土壤侵蚀加重了食物保障的风险。二氧化碳浓度大大提高，从 1750 年的 270μmol mol^{-1}，增加到目前的 385μmol mol^{-1}（Solomon et al.，2007；Le Quéré et al.，2009）。同时由于加剧燃烧矿物油和其他人为活动，使甲烷、臭氧和一氧化氮等温室气体也随之增加，到 2050 年将超过 550 μmol mol^{-1}（Raven and Karley，2006；Brouder and Volenec，2008）。由于二氧化碳浓度提高造成的温室效应是导致全球气候变暖的主要原因，有人预计温度年均升高，在今后 50～100 年会升高 3～5 ℃（Solomon et al.，2007）。气候变化、全球变暖、干旱、盐碱等的综合作用已经使地球上许多地方发生了荒漠化。另外由于当地居民的过度开发引起的荒漠化，也导致土壤侵蚀，使耕地减少和农业生产下降（Altman，1999；Reynolds et al.，2007）。干旱地区占全球面积的 41%，目前 65 亿世界总人口的 38%居住在这些地区。其中 10%～20%的地区土地严重退化，直接影响到发展中国家大约 2.5 亿人口，预计气候变化和人口增长的威胁还会不断加剧（UNCCD，1994；GLP，2005；MEA，2005；Reynolds et al.，2007）。气候变化和荒漠化不仅导致土地干旱，还会导致盐碱化。估计今后 25 年这一地区将上升到 30%耕地丧失，到 2050 年将上升到 50%的耕地丧失。

1.3 新的生物材料

采用植物和农业生物技术增加粮食生产是主要的目标，但是在过去 20 年中，农产品生产已经开始从低价的、大宗商品逐步走向生产高价的、专业化的植物副产品，以及新的非食物产品，这些产品统称为生物材料。这种转移是与消费市场对新的生物技术开发出来的工业产品（如生物塑料，生物胶）、药品、医药用品（疫苗，治疗佐剂，胶原蛋白）以及生物燃料的需求相符的。用粮食作物生产生物燃料不仅价格贵而且影响食物供应，因此，目前用于生产生物燃料的第二代植物不再用农作物了。

除了对传统农业继续进行改革，仅有两种途径可能提高食物供应和农业供给：（1）寻找替代的食物源，例如海洋或地球外的产品，单细胞蛋白；（2）利用转基因作物和分子标记辅助选择等新的生物技术来提高植物育种效率，加强作物改良。目前为止，有可靠的替代食品的期望还未能实现。与此相反，农业生物技术和 DNA 重组技术的产品，从 1996 年开始进入市场以来已经发挥了重要作用。全球 29 个国家约 1540 万农民种植了超过十亿公顷的遗传工程改良作物（James，2010），种植者主要是小型的，资源匮乏的发展中国家的农民，目前种植的作物主要有四种，不久就会有更多种类投入商业化生产。值得一提的是 2010 年发展中国家种植的生物技术作物占全球的 48%，2015 年将超过发达国家。生物技术作物只经历 30 年就对农业产生了革命性影响，这是前所未有的。

2. 需要配合：综合了解植物机理和先进技术与生物技术互相配合

传统农业已经不能充分满足世界上发达国家和发展中国家对粮食和新植物产品的需

求。为了使农业生产力连续增长，人类生存要依赖于传统育种和现代生物技术有效的配合。以绿色革命为例，它使印度和其他东南亚国家小麦产量提高了10倍，可供养的人口数量提高了两倍。随后矮杆水稻品种在农业生产中的应用对于增加粮食生产起到了相同的作用。绿色革命的小麦和矮杆水稻的优点都已被充分开发利用了，农作物品种改良急需新的替代方法。现在传统育种和现代生物技术的配合可能就接近了创造“常绿革命”的边沿。

在过去的十年，对于植物生物学的认识有了很大的长进，这为以后从植物生产食物、纤维、生物聚合物、生物燃料和代谢物等产物打下了坚实的基础。对分子遗传学和组学的了解和运用能力推动着农业生物技术革命往前发展。进一步的发展取决于目前的基础，持续的科技投入以及科技的新发展，这是推动力。一场重大的变革正在酝酿中，就像当年人类从狩猎和采集到耕种的过度一样，这种变革是社会结构变化的最高形式，它使人口巨增，并发展成为地球上的统治物种（Doeley et al.，2006）。我们的祖先通过对自然存在的遗传变异进行农艺性状的选择，对植物和作物进行驯化培育（见第1章）。人类的选择逐步改变了未种植作物的遗传结构，开始了驯化过程，大幅度提高了以植物为基础的食品、纤维和饲料的产量。

在我们的祖先开始驯化千年以后的今天，我们正处在一个新的纪元，组学技术将为技术提升和资源改良做出贡献，最终获得对农作物做出我们现在想象不到的重大改良。这将使得对遗传变异的开发利用更适合未来的社会对食物（作为热量和营养来源）、生物聚合物、生物燃料以及植物代谢物等的需求。在过去的十年中，对基因组学范例的了解有了长足的进展，使得人们对生长和发育、对生物和非生物因子的反应在遗传和表观遗传上是如何调控的有了更根本的、深入的、综合的了解和认识（见第2章）。新的基因组测序技术、同线性连锁数据采集和代谢途径整合等新技术对于深入了解生物体内和生物体间复杂的过程（如高产机理，对生物胁迫和非生物胁迫的耐受机理）起到了巨大的推动作用。组学（包括基因组学、转录组学、表观遗传组学和蛋白质组学）正在促进农作物的生物技术改良，特别是那些受到复杂遗传和表观遗传机制和过程控制的表型性状。

表观遗传组学对植物育种和品种培育具有重要影响（见第14章），许多发育的或对环境的反应是受表观遗传控制的，预计表观基因组的重新编程是通过生物技术改良农作物的重要因素。基因组技术正在从模式植物转向农作物（见第6章）。高通量的测序和数据分析技术使研究人员完全可以获得某种新的作物品种（即使其基因组很大）及其相关品种的基因组信息，因此，建立它们的比较图谱、确定它们间的共线性关系、鉴定重要性状间的等位性都是可行的。

植物基因组学和转录组学的重大进展促进了人们对蛋白质组学和代谢组学的兴趣。当今，现代先进的方法可以对蛋白质组进行结构和数量上的详细分析，还可以对任何一个对代谢和生物功能具有重要作用的翻译后修饰进行分析研究（见第4章）。蛋白质组学可以对植物代谢系统提供详细的了解，从而通过改变代谢途径的流向生产目标产物。蛋白质组学研究将有助于实现对那些能够增强重要农艺性状表型表达的代谢网络（无论在遗传学上是多么复杂）进行生物技术改良。生物技术也将促进利用植物作为生物工厂生产各种蛋白质产物（见第3章），诸如疫苗、抗原、抗体、肽类激素、酶、试剂以及储存蛋白。这种能力是几十年来，分析机理和过程得来的，涉及对细胞内蛋白跟踪和定位，对细胞、组织、器官研究的积累，也包括开发出能把有用信息转化成大规模和小规模的、在经济上可行、对社会有用产品的研究。估计植物代谢组包含 200，000－1，000，000 个不同的分子（见第 5 章）。这些代谢物代表了细胞的化学能力。一些特异的代谢物及其浓度就形成了指导植物对发育和环境条件做出相应反应的特定代谢信号。代谢图就是植物所处状态和可能的调节过程的明确指示。代谢产物是生成食物、营养品、药品、纤维、生物燃料及其他一级和二级产物的原料。依据调控产品前体物和改变代谢流向来调控代谢图可以影响生物分子的形成，从而改变应对发育和环境反应的途径。

通常提到的组织培养技术是最突出的、最早发展起来的植物生物技术。最初的组织培养是用作细胞和组织分化研究的平台，也用于形态建成、激素鉴定和功能研究（见第 9 章）。在组织培养基础上衍生发展起了生物技术的许多新分支学科，包括快速无性系扩增，体细胞育种（原生质体融合、通过花粉培养产生单倍体、细胞筛选等），疾病剔除，转基因植物以及种质资源保护（见第 10，15，17 章）。

植物生物学的一项重要成果就是分子遗传资源和细胞基因转移技术的建立，细胞基因转移包括转移基因成功地整合到受体的基因组，并能稳定地复制（见第 7，8 章）。结合使用组织培养的知识，从那些导入基因能在后代稳定表达并遗传下去的转染的细胞可以培育出植株，这个过程就被称为遗传转化。根癌农杆菌介导的遗传转化是在生物技术中最常用的一种方法。有一种“解甲”的根癌农杆菌，在自然界它是一种植物病原菌，它含有能识别宿主、导入基因复制、导入宿主细胞并整合到宿主基因组中等的多种遗传决定因子（见第 7 章）。基因枪和其他非根癌农杆菌介导的遗传转化方法也用于植物转基因（见第 8 章）。

与组学技术和有效遗传转化相匹配的是对植物生物过程的深入了解和分子应用大大推动了生物技术革命。孤雌生殖是从母本胚珠或未经减数分裂的配子中形成体细胞胚的现象（见第 16 章）。这一自然发生的过程能使优良性状在杂种中固定下来，并且在遗传上是稳定的，可以通过种子繁殖遗传下去。通过这一过程可以把进化上需要保留的某个特定植物品种所含的某种遗传变异类型保留下来。孤雌生殖中的关键遗传决定因子已经清楚，有人认为孤雌生殖可能促进杂交种子的生产。DNA 分子标记辅助选择简化了在分离群体中进行表型筛选的程序，大大加速了品种选育（见第 11 章）。技术的快速发展正在推动把这些技术应用到目前分子遗传鉴定还处于初始阶段的作物中。雄性不育是很多重要农作物杂交种子生产的基础（见第 12 章）。细胞质雄性不育和育性恢复的机理在分子水平上得到了很好的阐述，分子设计的应用也是很有效的。把生物技术应用于自然遗传资源的开发也在进行中（见第 13 章）。从模式植物（如拟南芥）得到的认识和发展的技术使得人们可以从作物的亲缘种中开发优良等位基因。

3. 对生物胁迫和非生物胁迫的耐受性

植物对环境胁迫的适应性受到众多基因和分子网络的控制，环境胁迫分为生物胁迫和非生物胁迫，前者包括细菌病害、真菌病害、病毒病害、昆虫以及其他有害生物，后者包括干旱、盐碱和极端温度。环境胁迫激活相关的反应机制，重建新的平衡以及保护和修复损伤的蛋白和膜结构。使植物能耐受环境胁迫的生物工程策略依赖于某些基因的表达，这些基因或者与信号和调控代谢途径有关，或者是与环境胁迫耐受蛋白和合成代谢物生化途径关键酶的编码基因有关（见第 2，18 章）。目前通过遗传转化改良耐胁迫的努力已获得了几项重要成果，但是由于耐胁迫的遗传机理的复杂性，要获得真正耐胁迫又没有负作用的作物品种是很困难的。因此，生物技术必须与传统的生理学和育种学密切配合，才能有所作为。

由于极端的气候、土壤等非生物胁迫和病、虫害等生物胁迫共同造成的农作物损失大于其他所有影响作物生长、发育和产量的弊端所造成的损失。上述这些胁迫在过去十年中一直是研究重点，特别是那些对作物生长发育起负作用的胁迫发生的机理以及植物对胁迫做出的反应。由于分子遗传学技术的高度发展，在鉴定耐受性遗传决定因子方面已经取得了实质性的进展。本章将集中关注胁迫的成因和植物在对付胁迫条件提高生存能力所产生的适应反应两个方面取得的重要进展。第一方面是植物对某种生物胁迫或非生物胁迫产生的必须的而且是足够的抗性或耐性的遗传决定因子。采用功能的得、失方法已经鉴定出了多个遗传决定因子和多种机制能使植物在有害条件的胁迫下做出反应。此外，对自然变异的研究，通过固定选择已经鉴定出了能够适应一种或多种环境胁迫的适应机制。

干旱和盐碱是限制作物产量的两种最主要的与水有关的胁迫。减少农业用水和人类生活用水标志着淡水补给的生物地球化学循环已经失去平衡（Yoo et al.，2009）。此外，高盐对农业用水起着负面影响。地球人口增加和气候变化是导致用水恶化，补给不平衡的两个主要因素。将经典方法和高通量技术（组学）结合起来是培育节水作物和耐盐碱胁迫作物的有效方法（见第 18 章）。通过对节水和耐盐相关基因十多年的研究，现在导入远缘植物的耐胁迫基因是可行的。把多基因性状的多个基因位点或等位基因聚合起来是完全可能的，但这些多基因性状必须与耐胁迫性状连锁，这样得到的耐胁迫新种质才能保持稳定。

这个时期对于植物遇到低温胁迫做出的反应和耐低温胁迫也有了较多的了解（见第 19 章）。最突出的成果是对耐低温所必需的控制调节子的低温信号网络的成功鉴定。这些网络通过转录和转录后两种机制调控着植物对低温的反应。高温调控网络在这个时期也被鉴定出来了。有趣的是发现和低温信号网络有重叠的特性。我们已经有了一些材料，可以用于分子育种和遗传转化，以便改良作物的耐极端低温和耐极端高温特性。植物和它们的相关微生物有能力通过对有机和无机污染物的降解、挥发分散或积累从而减轻其受污染的影响（见第 20 章）。通过对植物修复过程遗传决定因子的鉴定和功能分析，现在通过使功能恢复的生物技术可以增强植物的修复能力。

一个经典的例子是从基础研究中（包括利用高通量分子遗传技术）获得的知识是如何促进农业防治策略的，突出研究了植物对生物胁迫的应答。其中基本的研究发现是在将经典的遗传学与生物技术的努力结合时得到的。

目前已经可以设计出精细的技术路线对植物进行改良，提高植物对真菌、细菌病害的持久（遗传学上的可持续）抗性（见第 21 章）。持久抗性是由基因的功能决定的，或者是促进植物的免疫系统，或者是干扰病原菌的侵染机制。病原菌的应答启动子和组织特异的表达会大大降低转基因植物从导入基因得到的抗病能力。抗病毒农作物已经在生产中应用，其他的还在培育阶段（见第 22 章）。在抗病毒作物上的成功归功于过去十多年对病毒毒性的认识，即病毒毒性是由 RNA 沉默机制决定的，由于 RNA 沉默导致病毒毒性丧失，植物就表现出抗病毒特性。抗病毒研究的多数策略都是以病毒复制为目标的。

生物技术也可以为控制昆虫等害虫提供经济有效、环境许可、持久耐用的方案（见第 23 章）。主要的例子是转细菌 *Bt* 基因的抗虫棉花和玉米。*Bt* 内毒素能抑制昆虫的消化，从而破坏其生存和繁殖。

4. 控制生长和发育：改良产量和质量性状

生长、发育和代谢的遗传程序控制着产量（总生物量、果实或种子）。根、茎的发育和结构，花和果实的发育是农作物和林木生长、产量和质量的主要决定因子。各器官与环境相互作用，受到环境的影响；当植物处于干旱、盐碱、营养不良以及有毒离子的环境下，根是最直接的传感器和反应组织；茎、花及果实的发育受到多种环境因素（诸如光照、温度、湿度以及收后储存条件）的控制并要分别做出相应的反应。对那些在控制根、茎、花、果实以及营养产物发育方面具有重要作用的基因和基因产物进行仔细的分子分析和生物技术的操作，揭示出的有价值的信息可以在模式植物和农作物中实际应用。

现今至少在模式植物拟南芥中对根的组织结构和生长模型有了详细的描述（见第 24 章）。目前研究焦点集中在土壤环境因子是如何影响根系结构的，根系结构影响水和养分的通透性。对根系结构和发育及其受到环境影响后作出的调整等方面的基本了解是设计未来能更有效获得水分和养料的新型农作物所需要的基本知识。开花时间是结实的一项最主要的时间指标，它受到发育、激素和环境等诸多因素的调节（见第 25 章）。影响开花时间的诸多因素通过受表观遗传学调控的信号和感知决定子控制开花时间，过去十年中对许多控制开花时间的信号和感知决定子已经进行了很好的确定，其中包括难找的成花素。到目前为止，植物育种者都是通过传统方法选择理想的开花性状。以后就将采用生物技术方法研究开花特性。

开花标志着植物发育从幼年到了成年期，随后将成熟并完成繁育后代。花含有雌性繁育结构，包括子房或种子，子房是含有种子的胚珠坐落的地方。人类培育植物种子作为食物利用它的营养和味觉价值。人类对控制果实大小、形状和市场吸引力等性状的基因进行了选择，降低了毒素的积累，提高了营养价值，提高了生长期和成熟期的一致性。对上述性状的一些主要决定因子已经被鉴定出来了，这为生物技术在果树发展中的应用打下了良好基础（见第 26 章）。新鲜果品的采收后储存期是一个重要参数（见第 27 章）。衰老所涉及的生理、生化过程是受遗传控制的，在品种选育过程中都经过了选择。对上述主要过程的遗传决定因子的认知为今后通过分子育种或生物技术对收后质量性状进行改良提供了重要帮助。

如前面提到的，对农业产品的需求是全球面临的威胁。随着人口增长，不仅需求量增大，而且人们要求高营养价值的农产品。质量性状的改良涉及低分子量的代谢物，例如一些必需营养物、促进健康的植物化合物和挥发物（见第 28 章）。次生代谢产物的生物合成和生产方面的新进展为用生物技术对农作物进行改良，使之含有高度理想的代谢物（即使他们在植物中的功能尚不清楚）开发出了有用的资源。

5. 扩大研究领域：利用植物作为工厂生产工业产品，医药品、生物材料和生物能源

传统农业和林业一直试图对来自植物的食物产品的数量和质量、木材生产、纤维结构以及燃料进行改良。实际上过去几千年来人类一直用植物为原料生产常用生物材料、特别是聚合物。因此，上述任务作为生物技术的主要目标一点也不奇怪。尽管每种植物都有固定数目的基因，但是大量的基因突变产生了更大数量的植物代谢物，其中的绝大多数都尚未得到鉴定。植物产生 100，000 种以上次生代谢物，包括许多聚合物，估计总数超过 500，000 种。这还不包括那几千种主要的代谢物（Singer，2003）。植物产生的代谢物的价值在于所有自然产生的植物代谢物都是通过植物光合作用吸收太阳的能量，而不依赖其他能源，另外值得强调的是用植物作为生物技术工厂生产新生物材料具有重要意义。

自然代谢物主要包括根据工程植物的生物合成体系区分的两大类新生物材料。第一类来自对植物结构的直接改良和对代谢物的修饰；第二类是在植物中生产出的非植物化合物（即植物工厂）。植物生物工程的这种潜能由于以下三个原因现在已经成为完全可能的：（1）新基因的发现；（2）代谢物平台的付诸使用；（3）遗传工程的强大能力（包括基因叠加）能够创造出新代谢工程途径。事实上，植物充当了极致的有机生物化学家的角色。

此外，植物还和动物、微生物一样，执行着许多已知的代谢反应（例如：氧化、羟基化、甲基化、水解、还原作用等），植物在生产次生代谢产物和生物聚合物过程中其共轭系统表现出超强的多功能多用途性能。和生产次生代谢产物有关的大多数代谢途径的遗传工程是对某种或某一组化合物的数量起到提高或降低的调整作用。现在可以通过多种途径实现这个目标，包括：（1）对一个或几个基因的表达进行修饰，从而克服代谢途径中特定的限速步骤，或者关闭竞争性代谢途径；（2）敲除代谢途径的某个酶学步骤（例如通过反义 RNA 降低相关 mRNA 的数量、共抑制或 RNA 干扰等技术；或超表达能拮抗该酶的某种抗体）；（3）分流到一个有竞争性的代谢途径中；（4）提高目标化合物的分解代谢作用。此外，通过对生物合成基因的调节基因表达的调控，也成功地实现了对代谢物生物合成的控制（Verpoorte and Memelik，2002）。

生物技术专家们正在对用于食物、化学药品以及能源工业的一些植物组分进行修饰，采用的主要方法就是代谢途径工程。研究的目标化合物除了次生代谢物外还有许多基本的营养代谢物，如：碳水化合物（淀粉产量和分配，高直链淀粉或高支链淀粉的生产以及果聚糖的生产等）；蛋白质（改进氨基酸组成和蛋白质含量）；油和脂肪（饱和脂肪酸和不饱和脂肪酸的比例，提高特定脂肪酸的含量以及其他内容；Altman，1999）。

近来，以下两个原因引起人们增加了对植物生物聚合物的关注：（1）全球的化石燃料（作为能源和以石油为基础的塑料及其他化学工业的原料）供应能力的下滑；（2）气候变化和需要加强碳的回收和减少空气污染。上述原因促进了人们开发植物合成各种化合物和材料的兴趣，特别是生产可再生能源如生物燃料和生物柴油。美国联邦政府已经确定了到 2022 年在运输部门用植物生产的可再生的生物燃料替代目前美国的石油消费 360 亿加仑。

中国、新西兰、欧共体、英国、巴西、加拿大和澳大利亚都已确定了类似的可再生生物能源的生产目标。目前，生产的生物能源主要是来自用淀粉和谷物（如玉米、甘蔗）做原料生产的乙醇。生物燃料的原料主要来自林业和农业资源。美国每年可用做生物燃料的原料总量约为14亿吨干重，其中10亿吨来自农业，4亿吨来自林业（主要有短期轮作的杂交杨树、桉树、火炬松和柳树）和其他生长在边际地的非粮食能源作物（如：高梁和柳枝稷、五节芒等草本植物）（见第30，31章；J. S. Yuan et al.，2008；Simmons et al.，2008；Boernke and Broer，2010）。在发展可以部分替代化石燃料的低碳、可再生能源方面，植物的生物量起到关键作用，但何时能生产出达到数量可观水平的可再生燃料还值得怀疑。

除了用于生产可再生能源外，植物新材料被认为是石油材料的很有吸引力的替代品（如生物降解的热塑性塑料和人造橡胶），也可以用转基因植物生产出各种生物聚合物，如聚羟基脂肪酸酯，聚氨基酸，具有热滞后特性的多聚糖可用于生物亲和纯化，食品工业用的耐温耐盐的酶，用转基因植物生产的新生物聚合物（如聚羟基脂肪酸酯，环式糊精，化学工业用的聚氨基酸等），具有重要特性的纤维蛋白（胶原蛋白，弹性蛋白，丝）（见第30章；Boernke and Broer，2010）。总起来看，在植物中生产的化合物都是一些在植物中生产可更有效、而且对植物产量没有负影响的，或在植物中生产可具有更好性能的新的化合物，或者是通过植物光合作用生产大量的成本更低的原材料。

用植物作为生物反应器生产外来的非植物生物药品在过去十年发展势头汹涌，包括生产活性肽、疫苗、抗体，以及制药工业所用的各种各样的药用蛋白和酶制剂。最初提出的“分子制药”概念是在植物中生产药物，人们吃下植物生产的这些药物作为疫苗（Mason and Arntzen，1995；Twyman et al.，2003）；近来又提出的用植物作为生物反应器生产药品，人们不直接食用这些植物，而从中提取出有效成分。这两种概念正在相互融合形成新的分子制药新概念（见第29章；Ma et al.，2005）。

植物很快成为公认的生物制药的生产者，这主要依赖于稳定的细胞核遗传转化技术，最近的质体遗传转化技术以及各种瞬时表达技术。过去几年中，用植物为基础表达外源蛋白成为几种工业产品和生产多种临床实验用蛋白（包括疫苗、抗原和抗体）制造工艺的基础。目前已有三种药用蛋白获准生产上市，它们是抗龋齿抗体、新城病毒亚组合和葡糖脑苷脂酶。这样，植物除了用于传统农业生产，还被用作工厂生产工业产品，医药品，生物材料，生物能源，最终形成新时代的新型农业。

6. 植物和农业生物技术：社会和公共的领域

以前传统农业的科技成果没有伴随着消极的公众反应，它们通常会受到欢迎和被采用（如使用无机肥料、杀虫剂和精准灌溉）。然而，当前的农业革命（即转基因植物）和分子工具的使用引起了社会公众和监管的关注，及社会的争议。遗传改良作物的公众接受和充分商业化除了科技本身的突破外，还受到适当的管理、经济、环境、伦理及社会因素的影响。

生物技术不仅影响到我们对生物学的理解，还影响到社会和经济环境以及道德评价，因此产生了很多疑问。公众最关注的是哪些问题？最合理的（科学的）解答是什么？首先我们应该意识到对植物遗传改良的争论在科学争论上不是第一次。科学发现上的争论是科学史的一部分，通常是与无知、虚伪的争论，并混杂着与宗教信仰的争论。哥白尼（1473～1543）曾被攻击为魔鬼。布鲁诺（1548～1600）因为他在天文学和宇宙学上的革命性思想，被罗马教皇处罚并被烧死在火刑柱上。对宇宙学科技革命起了重要作用的伽利略（1564～1642）其工作被指责为支持了宇宙以太阳为中心的观点，而遭到谴责和拘禁在家并被警告停止其研究。在此 300 年后，达尔文（1809～1882）被指责为反基督者。但是，所有科学的发现和革命，都不会因受到阻挡而停止前进的，目前的生物技术也一样。它们的发展应该受到监测和调控，但它们不能被阻挡和排除。对这些反对意见是回避不了的，应该听取和了解，但要适当约束。

遗传工程和其他分子技术的公众接受需要其安全和透明使用。因此，生物技术产物和遗传工程植物的进一步释放，要有充足科学依据，积极跟踪，充分透明，有以下 5 个主要的安全议题和预警原则是应该认真考虑到的：食品安全，保持生态和环境平衡，社会经济学考虑，监管和知识产权以及伦理学考虑。

6.1 食品安全

有两方面可能的危险会涉及。第一方面是工程作物中的基因在人体中的整合和表达的问题。这种担心得不到科学支持，因为人类吃下去的重组 DNA 的量是可以忽略不计的（即使吃下 50%的来自转基因食物），而且在动物和人类的胃肠中它们很快被降解成很短的 DNA 或多肽片段。在动物试验中从未发现过完整的重组 DNA（Schubbert et al.，1998；European Food Safety Authority，2007；Lemaux，2008 ）。但有人声称植物基因组中的 DNA 插入具有高度突变性（ Latham et al.，2005)。科学证据表明远亲（界）类群（如植物和人类）间的基因转移和整合是极其罕见和不太可能的（见第 32 章）。第二方面可能的危险是吃下的转基因食物中的有毒或容易引起过敏反应的产物可能进入食物链。尽管这种可能性极其罕见，但这种事情一直是由 USDA，EPA，FDA 和 NIH 严格管理的（见第 33，34，36 章）。正因为这种原因，获得批准的监管转基因食品不管是新鲜的还是加工的都比一般市场的非转基因产品对人类更安全。

6.2 生态平衡和环境

重组 DNA/基因从转基因植物向别的植物和生长在同一生态系统中的其他生物的漂流潜在着会干扰生态平衡的可能性。但我们知道：一种生态平衡的建立是一个十分长期的过程，短期的实验室或田间实验一般不足以预测这种结果，必须依靠历史证据和科学分析的密切配合。

考古学和历史学的证据表明，农业上快速的动植物驯化发生在 10000～14000 年前（Pringle，1988；Gupta，2004 ），早期的驯化最后导致植物从其起源的原产地迁移到别的区域，引起近缘植物间的天然杂交（前孟德尔时代）产生了新的农业植物。这样看来，基因流动是一种自然过程，通常是对社会有利的。在后孟德尔时代发生了新品种控制育种，培育出了许多具有重要经济意义的新品种，其中包括绿色革命的新品种。自然的基因流动和孟德尔育种一直应用到现在，已被过度开发了。

这就要求用新的生物技术/分子工具包括遗传工程和分子标记辅助选择的方法实现无害、快速、全控制的育种。在植物品种间和向生态系统的基因漂流，不论是水平的还是垂直的，或者是不同属间的，都是非常少见的（见第 32 章）。但确有实例报导导入基因（如耐除草剂基因）从转基因作物漂流到农田和路边杂草中（见第 32 章；Gressel，2008；Warwick et al.，2009）。

某些自然界不存在的新性状（如在转基因植物中生产新药品和新生物材料）的基因的整合存在潜在的危险性，它可能对该转基因植物和邻近植物具有未知的好处和坏处。这种情况下，在田间种植时对这些植物要仔细地进行监测，避免导入基因进入到相关品种中去。

如前面谈到的，从转基因植物产生的基因漂流的有害效应是很少见的、不会影响环境。另外，限制基因漂流向其他植物已有了成熟的技术（转基因遏制和缓解技术），还在不断发展新技术。转基因遏制技术是从遗传上遏制基因漂流的发生，例如，把目标基因放在细胞质基因组中，雄性不育，使植物变成无性生殖的，采用限制技术创造种子不育。如果基因漂流已经发生了，转基因缓解技术可以抑制基因漂流的固定和在群体中的蔓延，例如已有多种方法可以使得到的杂交种对化学药物高度敏感，通过种子易毁坏和矮化的方法抑制杂草（见第 32 章；Oliver and Li，2011）。

转基因植物（如 *Bt* 作物）中的有毒基因产物可能是影响生态系统的另一种潜在危险。*Bt* 内毒素被昆虫和别的生物消化降解，一方面可能影响生态系统的生物平衡，另一方面可能引起抗性昆虫群体的发展。有人声称用转 *Bt* 作物饲养的皇冠蝶的死亡率明显增加，经过严格控制下的研究表明这种情况只发生在非常特异的人工控制条件下，在开放环境下不会发生。经过两年的综合研究表明：*Bt* 转基因玉米的花粉对皇冠蝶群体的影响可以忽略不计（Sears et al.，2001）。

综观大量近期的研究，作者们普遍认为尽管在有益节肢动物中有一些负效应的报导，但使用杀虫生物技术农作物比使用杀虫剂控制病虫害更有益于环境（Gatehouse et al.，2011）。这类事件普遍发生的可能性很低，不太可能影响已建立起来的自然生态系统。

6.3 社会经济因素

近来对在农业可持续发展中，植物生物技术对社会经济方面的作用也有些讨论（D. Yuan et al.，2011）。有几个关键问题，如农业投入（如：灌溉，土壤管理，矿物营养平衡）中的技术改进是全社会受益的。而受雇的植物生物技术学家通过对农业可持续发展的贡献，在促进社会经济，消除饥饿、与营养不良有关的疾病、和贫困方面可发挥最大的潜能，特别是在发展中国家和较贫困国家是这样。例如，在很多情况下，把高产的遗传工程作物种在不良的条件下，受到非生物的（干旱、盐碱、高温、污染物）和生物的（各种昆虫、病害、和其他有害生物）胁迫。相对来说，种植工程作物会减少投入成本，减少使用农药、除草剂和其他化学药品。

饥饿、与营养不良是一个复杂的社会经济课题，相关资源不足的穷人购买农业投入来生产他们自己的农作物，或从其他来源购买食品。显然国家间食品分配不均是个大问题，能充分满足全世界人口需要的粮食供应是根本不存在的（见第 35 章；FAO，2009；DFID，2010）。

由于复杂的政治和经济状况不可能改变，至关重要的是发展中国家和较贫困国家要采用新型的生物技术，用较少的投入和走一条可持续发展的道路，来生产他们自己的改良作物，达到自给自足。有几个主要因素阻碍在发展中国家中使用转基因作物，包括：公共研究机构的能力和资源有限；农场主和消费者对转基因作物的认知水平低；昂贵的监管过程，使这项技术难以到达小国家，特别是那些地方小作物（见第 36 章）。

6.4 监管、法律和知识产权

"金稻"（Golden Rice）是遗传工程作物监管和知识产权（IP）方面最好的例子。"金稻"是由 Ingo（一个国际非政府组织）Potrykus 培育的一个维生素 A 的前体物β-胡萝卜素含量大大提高的水稻品种（Ye et al.，2000）。1999 年实验室工作完成；2000 年得到知识产权；2004 年通过维生素 A 的前体物含量足够的检测，2007 年完成整合到目的品系的工作。目前正进行大田试验，在孟加拉、印度、菲律宾和其他国家，与当地品种一起试验。看来在 2012 年可通过监管，并转到农民手中。正如 Potrykus 所说（2010）：遗传工程作物如果可以从极严格的监管中释放出来，将可挽救无数饥饿和营养不良者。

上述的是一个相当正式的过程，应该发展到释放为新的植物品种，它是通过基因转移和相关的分子生物技术得来的，直至到农民手中，使消费者和农场主均获效益。这些品种需要：（1）受到高度的监管和知识产权保护（见第 33，34 章；Lemaux，2009）；（2）评估和监管要根据它们的创新点和推定的危险性，而不是根据培育它们所用的育种技术（常规育种，分子标记辅助选择或重组 DNA 技术）。Hood 等人（见第 34 章），Harfouche 等人（见第 33 章）和另一些人（例如 Pretty，2008；Fedoroff et al.，2010）都谈到，监管应该更好地反映测量的结果，而不是觉察的危险性。要让植物生物技术实现获益，每个遗传工程事件都要强调科学证据。保护知识产权有利于加快经济、社会和农业的发展；改进专利法和许可证发放合同，使之要能反映各方的不断增长的需要，以上做法是为了农业生物技术不断加强，为农业可持续发展做出成功的贡献。

6.5 伦理道德方面

植物和农业生物技术的批评者坚信生物技术是没有科学价值的，他们提出下列问题：我们可以允许以实用和盈利的名义去改变植物和动物的遗传结构吗？生命是神圣的，生物技术学家是自然的主宰者吗？所有生物的遗传结构都是全人类的财富，它可以被公司占用吗？

伦理学的主题是有关道德行为的问题，这一问题首先由亚里斯多德（公元前 384～322 年）创造和讨论的，他的出发点是"人类做的每一件事情都瞄准做些好事，但有的人目标比别人更高些，目标最高的人达到以做好事为最幸福的事、最好的生活"。约翰·穆勒（1806～1873）是亚里斯多德的追随者，在他的实用主义理论中他建立了"最伟大幸福的原则"，其中涉及"要为最多的人做最伟大的好事"。农业和植物生物技术就是以他的实用主义和道德哲学为根据的。以前也已经讨论过，农业和植物生物技术必须是以负责任的方式来操作，必须有管理，必须对公众是公开透明的（Meiri and Altman，1998；Chrispeels and Sadava，2003）。

在极端宗教批判的概念里，生物技术科学家正在“扮演上帝”。宗教信仰是超越目前我们讨论范围的，争论有困难。但请关注：有文章（Genesis 6：18～20）提到，自古以来记载与人的繁育有关的是在诺亚方舟中，那里第一次提到有“育种机构”的记载，那里男人和女人的生命都被给予机会生存，繁殖，产生后代，用于未来世代的选择。在近期一本有关伦理学的刊物的一篇评论中还提到：全世界成千上万的糖尿病患者在使用胰岛素，而现在胰岛素是一种专门的由重组 DNA 技术生产的产品。我们难道可以剥夺这些患者的转基因药物吗？这种药物能治疗他，挽救他的生命。然而生物技术的反对者却企图剥夺另一些人的食物，仅仅因为它是遗传修饰过的食物。伦理道德学家将同意，吃转基因食物和使用转基因药物实际上是一样的。

7. 结论：我们将走向何方？

农业是整个世界经济和环境的基础，是地球上所有居住者的福祉。300 多年前，乔纳森·思维福特（1667～1745）曾写道：“如果谁能使两穗玉米或两片草叶长在一个地方，而以前那个地方只能长一穗玉米或一片草叶，那么他将比整个种族的政客放在一起都能更有利于人类，更多地服务于他的国家。”事实上，今天植物生物技术的成就已经超过以往所有的期望，未来更是充满了希望。此外，利用新的分子遗传学的资源，“组学”技术和数据分析平台，使人们更好地了解植物生物学，并已将它成功地转化到农业实践中去。为了改良农作物连续地使用先进的技术，改善作物的生长和发育的控制（包括营养和繁殖），实现了作物产量和质量的双增长，也改善了植物对生物和非生物胁迫的抵抗；同时，为了支持传统的育种学，也在越来越多地应用分子标记辅助选择工具。先进的生物技术工具可造就新一代的作物，使之具有改进的农业效益，扩大的植物和农作物使用范围，可用来生产特殊的加强食品，各种新生物材料和产生生物能源。

很明显，也有一些针对生物技术和转基因植物有潜在风险的观点，要全面实现植物和农业生物技术革命，一方面要依赖不断地成功和创新研究，开展各种活动；另一方面要依赖一个良好的监管环境和大众的接受。有一些生物技术方法确实是与传统的育种技术有所不同的。然而方法是可以而且也应该仔细评估的（就像对任何其他新技术一样）以避免或实际地说是减少风险。已有很多关于对生物安全性质疑的解答，对设想的转基因食品的危险，及对环境的影响做出了解答；另有一些解答工作正在进行中；其他的还有声称转基因食品有风险的观点，但是没有科学上的支持。因此，我们应该不断地关注社会因素对转基因植物的公众接受程度，商业化的监管框架，以及关注允诺的长期政策上对研究提供的投资是否兑现，及转基因植物商业化的进程。

我们相信，全面和立即着手开展植物和农业生物技术的工作，特别是使用 DNA 重组技术和分子标记辅助选择是至关重要的。它和医学生物技术一样——有时甚至更为重要——因为世界范围内死于饥荒和营养不良疾病的人，要多于因患“现代”西方世界的疾病而死亡的人。重视农业就需要更加强大的，更有效的新的植物育种生物技术，才能释放出经济的，高回报的，有专利权的植物和植物的产品。没有先进的研究和发达的生物技术，生理学，基因组学和对农业植物实用的生物技术，以上目的是无法实现的。通过这样做生物技术将使农业彻底变革，这是满足食品生产（数量和质量）要求的需要，也是开发利用新的以植物为基础的代谢产物和生物聚合物的需要。在短期内生物技术的发展对发展中国家的影响将是非常巨大的，这些国家应该参与进来，制定出工程作物使用的准则。

（王　斌　译）

This book is dedicated to the memory of my dear wife, Yehudit (1947–2009).

Arie Altman.

Contents

Contributors

Arie Altman
Professor of Horticulture and Plant Biotechnology, Robert H. Smith Institute of Plant Sciences and Genetics in Agriculture, Hebrew University of Jerusalem, Rehovot, Israel Introduction

Ute Baumann
Australian Centre for Plant Functional Genomics ACPFG, University of Adelaide, Glen Osmond, Australia

Diane M. Beckles
Department of Plant Sciences, University of California, Davis, California, USA

Giora Ben-Ari
Institute of Horticulture, the Volcani Center ARO, Bet-Dagan, Israel

Philip N. Benfey
Department of Biology and Institute for Genome Science and Policy, Center for Systems Biology, Duke University Durham, North Carolina, USA

Sylvain Bischof
Department of Biology, Plant Biotechnology, ETH Zurich, Zurich, Switzerland

Eduardo Blumwald
Department of Plant Sciences, University of California, Davis, California, USA

J. Hans Bohnert
Department of Plant Biology and Department of Crop Sciences, University of Illinois at Urbana Champaign, Urbana, Illinois, USA; Division of Applied Life Science (BK21 program) and Environmental Biotechnology, National Core Research Center, Graduate School of Gyeongsang National University, Korea

Ray A. Bressan
Division of Applied Life Science (BK21 program) and Environmental Biotechnology, National Core Research Center, Graduate School of Gyeongsang National University, Korea; Department of Horticulture and Landscape Architecture, Purdue University, West Lafayette, Indiana, USA; Center for Plant Stress Genomics, King Abdallah University of Science and Technology, Thuwal, Saudi Arabia

Rafael Catalá
Departamento de Biología Medioambiental, Centro de Investigaciones Biológicas (CIB-CSIC), Madrid, Spain

John M. Cheeseman
Department of Plant Biology, University of Illinois at Urbana Champaign, Urbana, Illinois, USA

Chiang Shiong Loh
Department of Biological Sciences, National University of Singapore, Singapore

Joann A. Conner
Department of Horticulture, The University of Georgia, Tifton, Georgia, USA

Carole Cramer
Arkansas Biosciences Institute, Arkansas State University, Jonesboro, Arkansas, USA

Nikos Darzentas
Institute of Agrobiotechnology, CERTH, Thermi-Thessaloniki, Greece

Maheshi Dassanayake
Department of Plant Biology, University of Illinois at Urbana Champaign, Urbana, Illinois, USA

Hugo De Groote
International Maize and Wheat Improvement Centre (CIMMYT), Nairobi, Kenya

Om Parkash Dhankher
Plant, Soil, and Insect Sciences, University of Massachusetts, Amherst, Massachusetts, USA

Aurora Díaz
Departamento de Biología Medioambiental, Centro de Investigaciones Biológicas (CIB-CSIC), Madrid, Spain
19 Molecular responses to extreme temperatures

Dong-Ha Oh
Department of Horticulture and Landscape Architecture, Purdue University, West Lafayette, Indiana, USA; Division of Applied Life Science (BK21 program) and Environmental Biotechnology, National Core Research Center, Graduate School of Gyeongsang National University, Korea

Sharon Doty
School of Forest Resources, College of the Environment, University of Washington, Seattle, Washington, USA
20 Biotechnological approaches for phytoremediation

Florent Engelmann
IRD, UMR DIADE, Montpellier, France; Bioversity International, Maccarese (Fiumicino), Rome, Italy
17 Germplasm collection, storage, and conservation

Kellye A. Eversole
Eversole Associates, Bethesda, Maryland, USA

Delphine Fleury
Australian Centre for Plant Functional Genomics ACPFG, University of Adelaide, Glen Osmond, Australia

Abed Gera
Department of Plant Pathology, Agricultural Research Organization, the Volcani Center, Bet Dagan, Israel

Anatoli Giritch
Nomad Bioscience GmbH, Halle, Germany

Yuri Y. Gleba
Nomad Bioscience GmbH, Halle, Germany

Dennis Gonsalves
USDA, ARS, Pacific Basin Agricultural Research Center, Hilo, HI, USA

Kannan Grant
Office of Technology Commercialization, University of Alabama Huntsville, Huntsville, Alabama, USA

Jonathan Gressel
Plant Sciences, Weizmann Institute of Science, Rehovot, Israel

Jonas Grossmann
Department of Biology, Plant Biotechnology, ETH Zurich, Zurich, Switzerland

Wilhelm Gruissem
Department of Biology, Plant Biotechnology, ETH Zurich, Zurich, Switzerland

Avtar K. Handa
Department of Horticulture and Landscape Architecture, Purdue University, West Lafayette, Indiana, USA

Christian S. Hardtke
Department of Plant Molecular Biology, University of Lausanne, Lausanne, Switzerland

Antoine Harfouche
Department for Innovation in Biological, Agro food and Forest systems, University of Tuscia, Viterbo, Italy

Paul Michael Hasagawa
Bruno C. Moser Distinguished Professor, Horticulture and Landscape Architecture, Purdue University, West Lafayette, Indiana, USA

J.S. (Pat) Heslop-Harrison
Department of Biology, University of Leicester, Leicester, UK

T.J.V. Higgins
CSIRO, Plant Industry, Canberra, Australia

Elizabeth E. Hood
Arkansas Biosciences Institute, Arkansas State University, Jonesboro, Arkansas, USA

Joseph E. Huesing
Department of Entomology, Purdue University, West Lafayette, Indiana, USA

Takashi Ikeda
School of Agriculture, Meiji University, Japan

Jianfeng Xu
Arkansas Biosciences Institute, Arkansas State University, Jonesboro Arkansas, USA

Aliki Kapazoglou
Institute of Agrobiotechnology, CERTH, Thermi-Thessaloniki, Greece

Masayo Kawaguchi
School of Agriculture, Meiji University, Japan

Eli Khayat
Scientific Director, Rahan Meristem Ltd., Rosh Hanikra, Israel

Prakash P. Kumar
Department of Biological Sciences and Temasek Life Sciences Laboratory, National University of Singapore, Singapore

Peter Langridge
Australian Centre for Plant Functional Genomics ACPFG, University of Adelaide, Glen Osmond, Australia

Uri Lavi
Institute of Horticulture, the Volcani Center ARO, Bet-Dagan, Israel

Amnon Lers
Department of Postharvest Science of Fresh Produce, Agricultural Research Organization, the Volcani Center, Bet-Dagan, Israel

Chiang-Shiong Loh
Department of Biological Sciences, National University of Singapore

Sally Mackenzie
Center for Plant Science Innovation, University of Nebraska, Lincoln, Nebraska, USA

Venu Margam
Department of Entomology, Purdue University, West Lafayette, Indiana, USA

Kenji Matsui
Department of Biological Chemistry, Faculty of Agriculture, and Department of Applied Molecular Bioscience, Graduate School of Medicine, Yamaguchi University, Yamaguchi, Japan

Tracie K. Matsumoto
USDA, ARS, Pacific Basin Agricultural Research Center, Hilo, HI, USA

Autar K. Mattoo
Centro de Investigación en Alimentación y Desarrollo, Sonora, México; USDA-ARS, Sustainable Agricultural Systems Laboratory, Beltsville Agricultural Research Center, Beltsville, Maryland, USA

Munir Mawassi
Department of Plant Pathology, Agricultural Research Organization, The Volcani Center, Bet Dagan, Israel

Richard B. Meagher
Department of Genetics, University of Georgia, Athens, Georgia, USA

Giuliana Medrano
Arkansas Biosciences Institute, Arkansas State University, Jonesboro, Arkansans, USA

Richard Meilan
Department of Forestry and Natural Resources, Purdue University, West Lafayette, Indiana, USA

Larry L. Murdock
Department of Entomology, Purdue University, West Lafayette, Indiana, USA

Martina Newell-McGloughlin
Director, University of California Systemwide Biotechnology Research and Education Program (UCBREP); Adjunct Professor, Plant Pathology, University of California Davis, Davis, California, USA

Kaisa Nieminen
Department of Plant Molecular Biology, University of Lausanne, Lausanne, Switzerland

Peggy Ozias-Akins
Department of Horticulture, The University of Georgia, Tifton, Georgia, USA

Zvi Peleg
Department of Plant Sciences, University of California, Davis, California, USA

Elizabeth A.H. Pilon-Smits
Biology Department, Colorado State University, Fort Collins, Colorado, USA

Yves Poirier
Département de Biologie Moléculaire Végétale, Université de Lausanne, Lausanne, Switzerland

Ute Roessner
Australian Centre for Plant Functional Genomics and Metabolomics Australia, School of Botany, University of Melbourne, Australia

Julio Salinas
Departamento de Biología Medioambiental, Centro de Investigaciones Biológicas (CIB-CSIC), Madrid, Spain

Dor Salomon
Department of Molecular Biology and Ecology of Plants, Tel-Aviv University, Tel-Aviv, Israel

Alon Samach
The R.H. Smith Institute for Plant Sciences and Genetics in Agriculture, The R.H. Smith Faculty of Agricultural, Food and Environment, the Hebrew University of Jerusalem, Rehovot, Israel

Fumihiko Sato
Graduate School of Biostudies, Kyoto University, Sakyo, Kyoto, Japan

Trude Schwarzacher
Department of Biology, University of Leicester, Leicester, UK

Guido Sessa
Department of Molecular Biology and Ecology of Plants, Tel-Aviv University, Tel-Aviv, Israel

Yoel Shiboleth
Department of Molecular, Cellular and Developmental Biology, University of Michigan, Ann Arbor, Michigan, USA

Vincent K Shier
Oblon, Spivak, McClelland, Maier, & Neustadt, L.L.P., Alexandria, Virginia, USA

Blake A. Simmons
Vice-President, Deconstruction Division, Joint BioEnergy Institute, Emeryville, California, USA; Senior Manager, Biological and Materials Science Center, Biofuels and Biomaterials Science and Technology, Sandia National Laboratories, Livermore, California USA

Acram Taji
Queensland University of Technology, Brisbane, Australia

Tanya Tapingkae
Faculty of Agricultural Technology, Chiang Mai Rajabhat University, Thailand

Martín-Ernesto Tiznado-Hernández
Department of Horticulture and Landscape Architecture, Purdue University, West Lafayette, Indiana, USA; Centro de Investigación en Alimentación y Desarrollo, Sonora, México; USDA-ARS, Sustainable Agricultural Systems Laboratory, Beltsville Agricultural Research Center, Beltsville, Maryland, USA

Christopher N. Topp
Department of Biology and Institute for Genome Science and Policy, Center for System, Biology, Duke University, Durham, North Carolina, USA

Athanasios Tsaftaris
Department of Genetics and Plant Breeding, AUTH, Thessaloniki, Greece; Institute of Agrobiotechnology, CERTH, Thermi-Thessaloniki, Greece

Tzvi Tzfira
Department of Molecular, Cellular and Developmental Biology, University of Michigan, Ann Arbor, Michigan, USA

Jan B. van Beilen
Département de Biologie Moléculaire Végétale, Université de Lausanne, Lausanne, Switzerland

Deborah Vicuna Requesens
Arkansas Biosciences Institute, Arkansas State University, Jonesboro, Arkansas, USA

Harkamal Walia
Department of Plant Sciences, University of California, Davis, California, USA

Philip R. Watkins
CSIRO, Plant Industry, Canberra, Australia

Dae-Jin Yun
Division of Applied Life Science (BK21 program) and Environmental Biotechnology, National Core Research Center, Graduate School of Gyeongsang National University, Korea; Department of Horticulture and Landscape Architecture, Purdue University, West Lafayette, Indiana, USA

Zul Zulkarnain
Agricultural Faculty, University of Jambi, Indonesia

Foreword

In the 1970s when breakthroughs in cloning and sequencing of bacterial and yeast DNA were reported, visionaries predicted that there would be a plethora of future applications that would follow. They also predicted that the lives of humanity would be changed as a consequence. In the interim between the cloning of bacterial DNA and today there have been many breakthroughs that confirmed the predictions including: production of protein-based therapies and vaccines that treat human and animal diseases; enzymes that are used as industrial products and in fermentation of foods and drinks; improved varieties of crops; and new ways of producing biofuels and other bioproducts. These are some of the outcomes from discoveries made more than 35 years ago. The list of successes is long. Progress in research and outcomes of technologies have been remarkable, to say the least.

The results of investments made in science and technology in the 1970s and 1980s were evident in the 1990s and in the first decade of the new millennium, including in agriculture. For example, investments made in maize genetics and breeding paid dividends in superior hybrids and elite germplasm that were poised for subsequent improvement through genetic engineering (GE) and other types of biotechnology. Likewise, investments made in understanding how *Agrobacterium tumefaciens* acted as a pathogen served as the starting point through which leaders in the field, including Mary Dell Chilton, Josef Schell, Marc Van Montague, and Rob Schilperoot, along with the research team at Monsanto Company (St. Louis, MO) led by Robb Fraley, Steve Rogers, and Rob Horsch, developed transgenic crop plants. These leaders were responsible for the rush of activity by scientists and technicians around the globe to apply plant genetic engineering to develop new traits in crop and non-crop plants.

In the text that follows the reader is provided with background information on plant biotechnology; on the status of studies in genetics, genomics, and cousin "omics" platforms; and on the current understanding of how genetic engineering is accomplished. These are followed by chapters that present the state of the art in advances in the improvement and preservation of germplasm, plant breeding, and seed improvement, including apomixis. Each of these is essential to the short- and long-term success of agriculture and agricultural biotechnology. They also represent a background for the reader to dive into the chapters that describe the applications of current science and technology. The applications will, if successfully completed and moved through the maze of "regulatory safety processes" and accepted by consumers, continue to shape the future of global agriculture.

Many scientists, including me, became engaged in the genetic sciences because we hoped to; (1) contribute to discovery of new knowledge; and (2) contribute to improvement of the human condition. Most of us are more successful in achieving one, but usually not both, of these goals. While discoveries of new knowledge have their own rewards, for many of us additional satisfaction comes from knowing that knowledge has translated to applications that have impact. Agricultural biotechnology has indeed had impact since the first products came to the marketplace in 1996, and those impacts color the future of the technologies. The successful introduction of crops that exhibit resistance to certain insects, tolerance to selected herbicides, and increased resistance to selected virus diseases gave many of us hope that other applications would be equally well received. The farm producers and farm economy, the environment, the consumer, and the seed companies have all benefited. Indeed, there was great optimism that researchers would continue to make rapid advances and would produce crops with increased resistance to pests and pathogens and requiring fewer chemical protectants; crops able to withstand high and low temperatures, and droughty and saline soils; crops with increased levels of nutrient rich proteins, vitamins, and minerals; and special applications that would yield value-added products to serve the pharmaceutical, industrial chemical, and energy needs of a demanding world population.

Imagine our surprise and disappointment when the new discoveries were branded by some as products to be feared, and as damaging to human and animal health and to the environment. That disappointment continues today in many parts of the globe, and is of increasing concern in countries where GE crops were first introduced. This was and remains an unexpected outcome of the technologies that remain to be adequately addressed. Those who have branded the technologies as unsafe and to be avoided/feared have done so for a variety of reasons so disparate as to be impossible to analyze in any detail here. Some consider that there are legitimate reasons to be concerned about the safety of the product, notwithstanding that widely held scientific consensus uncovered no significant concerns: others are categorically, often philosophically, opposed to the use of advanced genetic technologies on agricultural plants and animals. Still others consider that until a trait that has consumer appeal is offered, they will not purchase products with GE materials.

The effects of the "fear branding" exercise have been felt in many different segments of the pathway to a product of biotechnology. I have encountered university professors who discouraged their students from engaging in science that would enhance agricultural biotechnology; potential innovators who are unsure that their technology would be accepted or worse, that those opposed to GE products would block progress; potential investors who are unsure of the pathway, and eventual cost, of regulatory approval and choose not to invest in a GE product; regulatory officials who work to streamline product approval only to find themselves in court defending a process or product before a magistrate who has no knowledge of the science that evaluates the actual risk of a new product and overrides decisions made by regulatory authorities; and seed company officials who decide to develop traits achieved through mutagenesis and advanced breeding rather than GE technologies because of concern about regulatory approval and consumer acceptance of a GE product. The negative impacts on innovation are stifling and have reduced development of many different types of products.

Some of the new processes and products are described in this volume because there is great hope, even expectation, that new technologies will find their way to approval and application because of their value to the environment, the inventor, and the consumer, as well as to the producer. I characterize myself as cautiously optimistic on one day and narrowly pessimistic on other days. Perhaps it is because of the role that agriculture plays in addressing many of the "grand challenges" that are required for a sustainable future, including global food security in a changing climate, biological solutions for energy needs, and a healthy population, that we continue. With the growing demands brought by a growing population, new technologies will be required as part of the solution: this is the consensus of many, not only of the scientists.

In the following pages the hopeful will see opportunities to meet the grand challenges; the less hopeful, even cynical, will point out the many challenges or pitfalls. Regardless of where you are in this casting, I hope that you learn a bit, discuss a bit, and then make decisions of how you will engage based on a more complete understanding than you had before reading. The grand challenges command it.

Roger N. Beachy
President Emeritus, Donald Danforth Plant Science Center, and First Director, National Institute of Food and Nutrition, U.S. Department of Agriculture

Preface

In a world where population growth is outstripping food supply, new agricultural technologies — especially plant-biotechnology — must be swiftly implemented to narrow the gap between production and human need. The world population will soon reach 7 billion, and over 10 billion in the year 2050, while agricultural production is growing at a slower rate of about 1.8–2% annually. All humans depend on agriculture for animal feed, fiber, and food production; the latter is required as caloric and macronutrient staples to reduce under-nutrition, but also as an important source of micronutrients to alleviate malnutrition. Biotechnology offers novel approaches to enhance crop productivity and food biofortification, and produces novel plant-based biomaterials, including potential energy resources. One also cannot understate the economic aspects of agriculture in human quality of life. It is irrefutable that traditional agricultural methods alone will be unable to meet population demands; consequently, applying plant biotechnology procedures is a necessity.

Massive and immediate implementation of plant and agricultural biotechnology is as crucial as that of medical biotechnology — sometimes even more so — since more people worldwide die from famine and malnutrition-related diseases than from "modern," west-world diseases. Crop product and food shortages are the ultimate disaster. However, unlike most natural disasters, this is one that we can prepare for and sometimes even prevent.

Domestication of plants and animals found in the wild combined with gradual long-term selection for productivity and quality traits were the first attributes of agriculture. Domestication, followed by food storage, coincided with the use of microorganisms. Thus was born classical food fermentation, the earliest known application of biotechnology for the generation and preservation of food products. This traditional agriculture now faces three major limitations: (1) market limitations (the world is becoming a global village where free-market rules negate the effectiveness of local pricing policies, and where a dictate of international trade and policies exists); (2) limitations of natural resources (global climatic changes — mainly desertification, drought, and salinization — as well as industrialization and urbanization, have reduced land and water availability and caused alarming deterioration of soil, water, and air quality); and (3) inherent biological genetic limitations (the release of new improved genotypes by classical breeding is now too slow to cope with the demands, and is considerably limited by the lack of appropriate "natural" genes or alleles that can be introgressed by traditional genetic crosses).

Achievements to date in plant biotechnology have already surpassed all previous expectations, and the future is even more promising. The full realization of the plant and agricultural biotechnology revolution depends on both continued successful and innovative research and development activities, as well as on a favorable regulatory climate and public acceptance. The new achievements of plant molecular biology and biotechnology in the last decade make it possible to be fully integrated with classical physiology, biochemistry, and breeding, thus resulting in: (1) better growth and development control (vegetative and reproductive); (2) improved plant protection against the ever-increasing threats of abiotic and biotic stress; (3) aid to classical breeding; (4) generation of engineered organisms; and (5) expansion of agriculture toward the production of specialty foods and novel biomaterials.

In this respect, the present comprehensive treatise aims at providing an authoritative update of plant biotechnology techniques, procedures, and accomplishments, from basic biological discovery science to applications in agriculture and other life sciences. This book contains expert evaluations of achievements and of the potential that novel plant biotechnological approaches and products will add to "classical" plant science and agriculture for improvement of crops and plant-based agricultural products. Specific emphasis is on the interrelationships between basic plant biotechnologies and applied agricultural applications, and the way they contribute to each other.

This book is also intended as a reference for plant biologists, agricultural scientists, and plant molecular geneticists, as well as for those involved in plant breeding, food science, and biomaterial product development. In addition, it will serve as a comprehensive text for graduate-level students in departments of agricultural, plant, food, and biotechnological sciences.

The first section of the book (Introduction to basic procedures in plant biotechnology) highlights the most significant biological advances that are relevant to biotechnology, including chapters on genomics, proteomics, metabolomics, tissue culture, and genetic transformation. This is followed by a section on breeding biotechnologies, including somatic cell procedures, marker-assisted selection, hybrid seed production, identifying and exploiting natural variation, epigenetics, and epigenomics. The third section deals with germplasm-related topics such as micropropagation and generation of pathogen-free plants, regulation of apomixis, and germplasm collection and preservation. This is followed by two sections with chapters on the control of plant response to the environment, both abiotic and biotic stress, including drought, salinity, extreme temperatures, pollutants, and phytoremediation, as well as fungal, microbial, and viral diseases, and insect and weed control. The sixth section is devoted to the application of biotechnologies to improve and control yield and quality traits, including chapters discussing root, shoot, and fruit development and flowering control, as well as postharvest storage and modification of secondary metabolites for quality traits. Next is a section on the use of plants as factories for pharmaceuticals, biomaterials, and bioenergy. The last section of this book deals with transgene flow, intellectual property rights, regulatory issues, and crop biotechnology in developing countries.

We wish to express our appreciation to all contributing authors. Distinguished scientists worldwide, who have made significant scientific contributions and are pioneers in the field of plant and agricultural biotechnology, wrote each chapter. The various articles present the opinion of the authors and their points of view. While every effort was made to reach uniformity in style, the presented results and ideas and organizational details of the chapters reflect the preferences of the respective authors.

We are grateful to many of our colleagues and friends for their advice and collaboration, and to Elsevier — and especially to Rogue Shindler — for their assistance and patience.

Finally, with love and affection, we are indebted to our families for their patience and understanding.

Arie Altman
Paul Michael Hasegawa

Introduction to plant biotechnology 2011: Basic aspects and agricultural implications

Arie Altman[1] Paul Michael Hasegawa[2]
[1]*Hebrew University of Jerusalem, Israel,* [2]*Purdue University, Indiana*

The Role of Plants and Agriculture: Food Security, the Changing Environment and Novel Biomaterials

Current agricultural crops are products of plant domestication, along with gradual long-term changes in their qualitative and quantitative traits because of continuous natural and human-directed selection. Whereas some of the first improvements in plants and animals could have resulted from chance alone, the eye and brain of the "primitive" scientist-farmer were crucial in selecting the good from the bad, and the productive and quality crop from the less worthy. Thus, advancements in agriculture and plant science are, primarily, the result of scientific discoveries and judgments, historic innovations, and are sometimes revolutionary, such as fermentation of food products, use of inorganic fertilizers, and more recently, plant genetic transformation.

Man-made technologies and biotechnologies, including food and fiber, have shaped human life since time immemorial. Fermented plant and animal products (e.g., bread, cheese, and wine), conventional crop breeding since the birth of agricultural communities, the "Green Revolution" of later years, and molecular marker-assisted selection and recombinant DNA techniques are all biotechnologies by definition (Meiri and Altman, 1998; Altman, 1999; Chrispeels and Sadava, 2003; *Science*, 2010). The urgent need to look for alternative biotechnologies and the actual accelerated rate of adopting plant molecular biotechnologies since the breakthrough report of the first transgenic plant in 1982–1983 (Otten et al., 1981; Barton et al., 1983) is due to four major causes:

1. Increase in world population and the need for more food
2. Recognition that human health is affected by disease-causing pathogenic organisms and by the nutritional quality of foods, especially vitamins and minerals
3. Adverse global climatic changes accompanied by detrimental biotic and abiotic hazards (stresses) to crops and ecosystems
4. Human societies searching for novel, non-food plant products such as biomaterials, therapeutics, and biofuels.

Thus, in a world where population growth is outstripping food production (both for quantity and quality), agricultural and plant biotechnologies need to be swiftly implemented.

World population and food

According to recent reports and learned estimations (FAO, 2009), the world's population will reach 9.1 billion by 2050, 34% higher than today, and nearly all of this increase will occur in developing countries. It is also predicted that due to the accelerated rate of urbanization, about 70% of the world's population will be urban (compared to 49% today) accompanied probably by higher income levels. To feed this larger urban population, food production will have to increase by 70%; for example, cereal production will need to reach about 3 billion tons, compared with 2.1 billion today. The Food and Agriculture Organization (FAO) report further argues that the required increase in food production can be achieved if the necessary investments, policies, and regulations for agricultural production are undertaken. It is estimated that in developing countries 80% of the necessary production increases would come from yield increase and cropping intensity, and only 20% from expansion of arable land, which emphasizes an even greater need for improved agricultural technologies and biotechnologies. It should be mentioned that the discussion of food supply refers not only to caloric demands, but also to increased intake of vitamins, essential minerals, and other nutritional factors via production of biofortified food to improve the health of undernourished people, especially in children and in poorer countries. This aspect is further discussed in the next section.

Climate and environmental changes

Climate changes and desertification pose additional risks to long-term food security by detrimentally affecting plant physiological processes and increasing soil erosion. CO_2 concentrations have increased significantly, rising from about 270 μmol mol^{-1} in 1750 to over 385 μmol mol^{-1} currently (Solomon et al., 2007; Le Quéré et al., 2009). The accompanying increase in methane, ozone, and nitrous oxide (greenhouse gases) due to intensified burning of fossil oils and other man-made activities will exceed concentrations of 550 μmol mol^{-1} by 2050 (Raven and Karley, 2006; Brouder and Volenec, 2008). The resulting increase in the greenhouse effect is the key factor in current global warming, with a projected average annual mean warming increase of 3–5°C in the next 50–100 years (Solomon et al., 2007). The combined effects of climatic changes, global warming, drought, and salinity have led to increased desertification in many parts of the world. Desertification, caused by overexploitation by local inhabitants, results in increased soil erosion and decreased land and agricultural productivity (Altman, 1999; Reynolds et al., 2007). Dry lands cover about 41% of Earth's land surface and are home to more than 38% of the total current global population of 6.5 billion. Some form of severe land degradation is present on 10–20% of these lands, the consequences of which are estimated to affect directly more than 250 million people in the developing world, an estimate likely to increase substantially in the face of climate change and population growth (UNCCD, 1994; GLP, 2005; MEA, 2005; Reynolds et al., 2007). The combined climatic changes and desertification cause not only an alarming increase in dry lands, but also result in soil salinization. This is estimated to result in a 30% arable land loss over the next 25 years, and up to 50% land loss by 2050.

Novel biomaterials

Increasing food production through plant and agricultural biotechnologies is still the major target; however, the last two decades have witnessed a shift from the production of low-priced food and bulk commodities to high-priced, specialized, plant-derived, and novel non-food products, collectively termed biomaterials. This move coincided also with the increasing market demand for novel biotechnology-derived industrial products (e.g., bio-plastics, bio-glues), pharmaceuticals, medical devices (such as vaccines and other therapeutics, and collagen), and plant resource-derived bioenergy production. As for the latter, increased biofuel production from food crops is likely to affect food supply and costs. Therefore, there is currently a need to move from first-generation biofuel crop plants to second-generation plants.

In addition to continuously improving traditional agricultural practices, only two major potential solutions exist to increase food supply and agricultural commodities: (1) a search for alternative food sources (e.g., marine or extraterrestrial products, single-cell proteins); or (2) enhanced plant breeding efficiency for crop improvement by utilizing the new biotechnological tools, including production of transgenic crops and molecular marker-assisted selection. Expectations for reliable sources of alternative foods have not been realized so far. On the other hand, the products of agricultural biotechnology and recombinant DNA technologies, which first came to the marketplace in 1996, have increased impressively since their introduction. Over one billion hectares of genetically modified (GM) crops (currently planted with four major crops, but many more will soon be commercially produced) are now grown by 15.4 million farmers (most of them small-scale, resource-poor farmers in developing countries) in 29 countries around the globe (James, 2010). It is noteworthy that developing countries grew 48% of the global biotech crops in 2010 and will exceed industrialized nations in their plantings of biotech crops by 2015. The revolutionary impact to agriculture in only 30 years is unprecedented.

The Need for Integration: Merging Understanding of Plant Mechanisms and Advancing Technologies to Biotechnology

Classical agriculture is no longer sufficient to supply the demand of developed and developing countries worldwide for food production, as well as for new plant commodities. Human survival, *vis-à-vis* a continuous increase in agricultural productivity, is dependent on an effective merging of classical breeding and plant biotechnology. The Green Revolution, for example, led to a 10-fold increase in wheat production in India and in several other countries of Southeast Asia, feeding three times as many people. This was followed by the adoption of dwarf rice varieties for use in agriculture, which led to comparable increases in food production. However, their benefits have been thoroughly exploited, and alternative solutions are required to breed improved crops. Now biotechnology, integrated with classical breeding, is on the verge of creating the "evergreen revolution."

The last decade has led to a phenomenally greater understanding of plant biology, which is the underpinning foundation for substantial translational potential to enhance plant-derived food, fiber, biopolymer, biofuel, and metabolite production. Technical understanding and capacities in molecular genetics and omics technologies drive this biotechnological revolution in agriculture. Further advances will be made based on this current foundation, continued scientific input, and the technological advances that will be derived. A dynamic paradigm is in motion similar to when humans went from being hunters and gatherers to being farmers, which was paramount to the development of societal structures that led to substantial population increase and the pre-eminence of humans as the dominant species on the planet (Doebley et al., 2006). Plant and crop domestication resulted from naturally existing genetic variation and selection for desirable agricultural traits made by our ancestral progenitors (see Chapter 1). Human selection changed the genetic composition of uncultivated crops, initiating the process of domestication and leading to substantially increased production of plant-based food, fiber, and forage materials.

Millennia after the onset of domestication, we are now in an era where omics technologies are contributing to technical advancements and resources that will lead to improvement of crops at a heretofore unimaginable pace. These will enable exploitation of genetic variation to meet current and future societal needs for food as caloric and nutrient sources, biopolymers and biofuels, and plant metabolites. The understanding of genomics paradigms has advanced so much in the last decade that it has resulted in a fundamentally more integrative and in-depth comprehension about how genetic and epigenetic processes regulate plant growth and development, and responses to abiotic and biotic factors (see Chapter 2). This new age is driven by novel technologies for genomic sequence and synteny linkage data acquisition and mining and pathway integration for a much more robust understanding of intra- and inter-organismal complex processes, such as yield and abiotic and biotic stress tolerance mechanisms. The era of omics including genomics, transcriptomics, epigenomics, and proteomics is poised to facilitate biotechnological improvement of crops, particularly for phenotypes that are controlled by complex genetic and epigenetic mechanisms and processes.

Epigenomics has substantial implications on plant breeding and cultivar development (see Chapter 14). Many developmental and environmental responses are regulated by epigenetics, and it is predicted that reprogramming of the epigenome is a substantial factor in crop development through biotechnology. Genomics technologies are moving from models to crop systems (see Chapter 6). High-throughput sequencing technologies and dataset analyses make it feasible to obtain genomic sequence information of new crop species, even those with large genomes, and related species make it feasible to build comparative maps, define syntenic relationships, and identify allelic variation for important adaptive traits.

Substantial advancements in plant genomics and transcriptomics spurred interest in proteomics and metabolomics. Today, modern methodologies can structurally characterize and quantify the proteome, as well as any post-translational modifications that may be important for metabolic and therefore biological function (see Chapter 4). Proteomics provides a detailed understanding of plant metabolic networks that will lead to predictions about altering pathway fluxes to achieve targeted end-product formation. Biotechnologically, the future will entail protein modifications that will modify metabolic networks to enhance agriculturally important crop phenotypes, regardless of genetic complexity. Biotechnology also will promote using plants for production of protein products, such as using plants as bio-factories (see Chapter 3). Products will include vaccines, antigens, antibodies, peptide hormones, enzymes, reagents, and storage proteins. This capacity is the result of decades of dissecting mechanisms and processes involved in intracellular protein trafficking and targeting; cell, tissue, or organ accumulation; and developing engineering systems that can translate this information into large- and small-scale production that is economically viable and societally useful. It is estimated that the plant metabolome consists of 200,000 to 1,000,000 different molecules (see Chapter 5). These metabolites represent the chemical capacity of cells. Specific individual metabolites and their concentrations make up the metabolic signature(s) that direct plant responses to developmental and environmental cues. The metabolic profile is the most definitive indicator of plant status and a likely regulator of processes. Metabolites are also sources of food, nutraceuticals, pharmaceuticals, fiber and other primary and secondary products, and biofuels. Profile modification based on precursor-production modulation and changes in metabolic flux will substantially affect production of biomolecular resources and alterations in developmental and environmental response pathways.

Tissue culture, as it is generically referred to, is the most prominent first plant biotechnology. It was developed primarily as a platform for basic research on cellular and tissue differentiation, and morphogenesis and hormone identification and function (see Chapter 9). Tissue culture has led to numerous biotechnological spinoffs. These include rapid clonal multiplication, somatic breeding methodologies (e.g., proptoplast fusion, generation of haploids by pollen culture, cell selection, etc.), disease eradication, transgenic plants, and germplasm preservation (see Chapters 10, 15, and 17).

A major achievement in plant biology was the development of the molecular genetic resources and the techniques for cellular gene transfer, which include integration into the host genome and stable replication (see Chapters 7 and 8). Combined with the knowledge derived from tissue culture research, plants can be reconstituted from transfected cells, which are capable of stable expression and inheritance by progeny of transgenes; a process referred to as genetic transformation. *Agrobacterium*-mediated transformation is the predominant method used in biotechnology. "Disarmed" forms of *Agrobacterium*, which is a natural plant pathogen, possess the genetic determinants involved in host recognition, transgene replication, and transfer into host cells and integration into the host genome (see Chapter 7). Microprojectile bombardment and other non-*Agrobacterium* methods of plant transformation are used to generate transgenic plants of species outside the *Agrobacterium* host range (see Chapter 8).

Coincident with the development of omics technologies and genetic transformation efficacy, a new basic understanding of plant biological processes and molecular applications is also driving the biotechnology revolution. Apomixis is the formation of somatic embryos from maternal tissue of the ovule or unreduced gametes (see Chapter 16). This naturally occurring process allows superior traits to be fixed in hybrids and genetically stabilized and inherited through seed propagation. This process is thought to maintain a type of genetic variation that is evolutionarily required for the maintenance of certain plant species. Since key genetic determinants of apomixis are known, it may be feasible to co-opt apomixis to facilitate hybrid seed production. DNA marker-assisted selection refined the process of screening for phenotypes in segregating populations, greatly accelerating cultivar development (see Chapter 11). Rapid technological advances are accelerating the utility of the technique for application to crops whose molecular genetic characterization is rudimentary. Male sterility is the basis for hybrid seed production of many important crops (see Chapter 12). The molecular genetic basis of cytoplasmic male sterility and fertility restoration mechanisms is well-characterized at the molecular level to the extent that effective biotechnology strategies can be designed. Biotechnology is also poised to exploit natural genetic variation (see Chapter 13). Understanding and technology development gained by using model species such as *Arabidopsis* have led to the feasibility of accessing superior alleles from crop relatives.

The Need to Protect Against Abiotic and Biotic Stresses

Plant adaptation to environmental stresses, both abiotic (e.g., drought, salinity, and temperature extremes) and biotic (e.g., bacterial, fungal, and viral diseases, and insect and other pests), is controlled by cascades of genes and molecular networks. These activate numerous stress responsive mechanisms necessary to re-establish homeostasis and to protect and repair damaged proteins and membranes. Plant engineering strategies for environmental stress tolerance rely on the expression of genes that are involved in signaling and regulatory pathways or genes that encode proteins conferring stress tolerance or enzymes present in pathways leading to the synthesis of functional and structural metabolites (see Chapters 2 and 18). Present efforts to improve plant stress tolerance by genetic transformation resulted in several important achievements; however, the genetically complex mechanisms involved in abiotic stress tolerance have made it difficult to achieve substantive effects in crops without side effects on yield. For this reason, biotechnology should be fully integrated with classical physiology and breeding.

Crop losses due to climatic and edaphic (abiotic) extremes, and biotic pathogens and pests are greater than for all other maladies that affect plant growth, development, and yield. These stresses have been the focus of intensive research efforts over the last decade, particularly to determine the mechanisms by which stresses negatively affect plant growth and development, and plants' responses to stresses. Substantial inroads have been made in the identification of genetic determinants that are necessary or sufficient for resistance or tolerance due to the substantial advances in molecular genetic technologies. This section of chapters focuses on critical advances made in the understanding of stress etiology and acclimation, and adaptation responses of plants that enhance the capacity of the organism to cope with these stresses. The primary focus is on the genetic determinants that are necessary and, in some instances, sufficient for plant tolerance/resistance to a particular abiotic or biotic stress. Loss- and gain-of-function approaches have identified determinants and mechanisms for plant responses that help a plant cope with the deleterious effects of stresses. In addition, natural variation studies identified adaptive mechanisms fixed by selection for fitness in an environment afflicted with one or more of these stresses.

Drought and salinity are two of the most significant water-stress-related abiotic stresses that limit crop productivity. Diminishing supplies of water for agriculture and humans are indicative that the biogeochemical cycle for replenishment is no longer in balance to sustain fresh water production (Yoo et al., 2009). In addition, high salinity negatively influences water available for agriculture. Predicted global population increases and climate changes are major factors exacerbating the water use versus replenishment imbalance. Combining and synergizing classical and high-throughput (omics) technologies is an efficient and effective way to develop water and salt stress tolerant crops (see Chapter 18). After studying genes involved in water and salt stress tolerance for more than 10 years, it is now feasible to introgress better stress tolerance alleles from genetically distant relatives. Systems approaches will facilitate pyramiding loci and alleles for multiple traits that must be linked in combination with stress tolerance to maintain yield stability of stress-tolerant germplasm.

This period has also brought greater understanding of plant responses critical for low temperature acclimation and tolerance (see Chapter 19). Striking results have been achieved in the identification of low-temperature signaling networks that control regulons, which are necessary for tolerance. These networks regulate the plant response to low temperature by both transcriptional and post-transcriptional regulatory mechanisms. High-temperature regulatory networks have also been characterized. Interestingly, there are overlapping characteristics of low-temperature signaling pathways. The basic resources appear in hand for molecular breeding and genetic transformation to impact crop tolerance to low- and high-temperature extremes. Plants, and their associated microbes, possess an inherent capacity to mitigate the effects of inorganic and organic pollutants by degradation, volatilization, or accumulation (see Chapter 20). Recent identification of genetic determinants and characterization of their function in phytoremediation is now translated through gain-of-function biotechnology to enhance phytoremediation capacity of plants.

An exemplary example of how knowledge is obtained from basic research, including the use of high-throughput molecular genetic technologies, has contributed to agricultural control strategies highlighted by research in the area of plant responses to biotic stress. Here translation of basic research discoveries is occurring by amalgamating both classical breeding and biotechnology efforts.

Detailed strategies can be devised to enhance durable (genetically sustainable) resistance to fungal and bacterial pathogens (see Chapter 21). Durable resistance is based on gene function that either increases the plant immune system or interferes with the pathogen infection mechanism. Pathogen responsive promoters and tissue-specific expression substantially reduce the yield drag that can occur in conjunction with pathogen resistance achieved by differential expression of a transgene. Virus resistance crops are already under cultivation and others are in the development phase (see Chapter 22). This is attributable to more than a decade long understanding that pathogen-derived viral resistance is due to RNA silencing mechanisms, which result in viral disease resistance of plants. Most strategies target different mechanisms involved mainly in viral replication.

Biotechnology also offers an economically efficacious and environmentally durable and sustainable approach to control of insects and other plant pests (see Chapter 23). The prime example is insect-resistant crops that are obtained by gain-of-function expression of the bacterial *Bt* endotoxin gene in transgenic plants of cotton and maize. The *Bt* endotoxin inhibits insect digestion, which impairs survival and fecundity.

Growth and Development Control: Improving Yield and Quality Traits

Growth and development and metabolic genetic programs govern yield (e.g., biomass, fruit, or seed). Root and shoot development and architecture, and flowering and fruit development are the major determinants of growth, yield, and quality of crop plants and forest trees. These organs interact with and are affected by the environment; roots are the immediate sensors and reacting tissues when plants are confronted with soil drought, salinity, and nutritional or toxic ions; and development of shoots, flowers, and fruits are controlled by and respond in many ways to light, temperature, and humidity signals, as well as to postharvest conditions. Careful molecular analysis and biotechnological manipulations of genes and gene products that play a central role in shoot, root, flower, and fruit development, as well as in the nutritive value of the plant products, revealed a wealth of information that can now be practically exploited (implemented) in both model and crop plants.

Today, at least for the model *Arabidopsis*, the critical genetic determinants of root organization and growth patterning are well-characterized (see Chapter 24). Currently the focus is on how root architecture is influenced by factors in the soil environment, which directs the capacity of these organs to access water and essential nutrients. This basic understanding of root organization and development, and their modulation by the environment, is likely to be baseline information for the design of agricultural crops in the future that are more efficient in nutrient and water acquisition. Flowering time is fundamental to the timing of yield and is regulated by developmental, hormonal, and environmental factors (see Chapter 25). These influence flowering time through signaling and effector determinants that are under epigenetic regulation, many of which have been well-defined in this last decade, including the elusive florigen. To date, plant breeders have selected for desired flowering traits by classical methods. The near future likely will involve research that utilizes current understanding in biotechnological approaches.

Flowering signifies the transition from juvenile to adult stages of plant growth, which ultimately leads to reproductive maturity and reproduction. The flower contains the female reproductive structures, including the ovary or fruit, which is the receptacle for seed-bearing ovules. Fruit has been cultivated as a food source for its nutritive value and organoleptic properties. Humans have applied selection pressure for genes that regulate fruit size, shape, and attractiveness; that reduce accumulation of toxins; that enhance nutraceutical value; and that control ripening to enhance uniformity and shelf life. The critical determinants have been identified, opening a vast potential for biotechnology to contribute to the development of fruit crops (see Chapter 26). Postharvest storage life of fresh market produce is a critical factor (see Chapter 27). The biochemical and physiological processes involved in senescence are genetically determined and have often been selected for during cultivation. Knowledge of critical determinants of focal processes offers substantial promise for the development of crops with improved postharvest qualities through molecular breeding or biotechnology.

As previously indicated, the demand for plant agricultural products is on the rise worldwide. Not only is there greater demand because of an increase in population number, but also because people need crop products with higher nutritive value. Quality-trait improvement involves low molecular weight metabolites such as essential nutrients, health-promoting phytochemicals, and volatiles (see Chapter 28). Recent advances in secondary product biosynthesis and production have developed the resources for biotechnological improvement of crops for highly desirable metabolites, even though their function in plants is not well understood. Nonetheless, these are valued products for improving the quality of crop produce.

Expanding the Horizons: Plants as Factories for Industrial Products, Pharmaceuticals, Biomaterials, and Bioenergy

Traditionally, agriculture and forestry have tried to improve the quantity and quality of plant-derived food products, as well as production of timber and fiber for construction, fabrics, and fuel. In fact, humankind has used plants as a renewable source of common biomaterials, especially polymers, for thousands of years, so it is not surprising these are also the primary targets of plant biotechnology. Even though there are a finite number of genes for each plant species, a large variety of the resulting gene permutations yield a much larger number of plant metabolites, many of them still unidentified. Plants produce a diverse array of more than 100,000 secondary metabolites, including many polymers, and it is estimated that the total number exceeds 500,000. This is in addition to several thousand primary metabolites (Singer, 2003). The wealth of plant-derived metabolites, in addition to the fact that all naturally produced plant metabolites are not dependent on additional energy inputs other than harvesting the sun's energy via photosynthesis, further emphasizes the importance of plants as biotechnological factories for new biomaterials.

Natural metabolites include two major categories of new biomaterials based on engineering plant biosynthetic machinery. These occur from: (1) direct improvement and metabolic modification of plant constituents; or (2) the manufacture in plants of non-plant compounds such as pharmaceuticals (i.e., "plant factories"). This potential of plant biotechnology is now possible because of novel gene discovery, and the availability of metabolomic platforms and the power of genetic engineering (including gene stacking) — resulting in novel metabolic engineering pathways. In fact, plants have emerged as the ultimate organic biochemists.

Moreover, while plants, like microorganisms and animals, perform many of the known metabolic reactions (e.g., oxidation, hydroxylation, methylation, hydrolysis, reduction), they are unique in the versatility of their conjugation systems in the production of secondary metabolites and biopolymers. Genetic engineering of a metabolic pathway, most often those responsible for the production of secondary metabolites, results in modulating the increase or decrease in the quantity of a certain compound or group of compounds. This can now be achieved by a variety of approaches. These include modifying the expression of one or a few genes, thereby overcoming specific rate-limiting steps in the pathway and shutting down competitive pathways; knocking out an enzymatic step in the pathway (e.g., by reducing the level of the corresponding mRNA via antisense, co-suppression, or RNA interference technologies, or by over-expressing an antibody against the enzyme), and diversion of the flux into a competitive pathway, or an increase in the catabolism of the target compound. Control of metabolite biosynthesis has also been achieved by regulating the expression of regulatory genes that control biosynthetic genes (Verpoorte and Memelink, 2002).

Biotechnologies, mostly based on the engineering of metabolic pathways, are now available for modifying many plant constituents that are used in the food, chemical, and energy industries. This includes, in addition to secondary metabolites, many "primary" nutritional metabolites: carbohydrates (starch yield and allocation, production of high-amylose or high-amylopectin starch, fructan production, etc.); proteins (improved amino acid composition and protein content); and oils and fats (ratio of saturated to non-saturated fatty acids, increased content of specific valuable fatty acids and others; Altman, 1999).

Two major factors have recently increased the focus for supply of plant-based biopolymers: (1) the declining global availability of fossil fuels for both energy resources and the petroleum-based plastics and other chemicals; and (2) climatic changes combined with the need for enhanced carbon sequestration and diminishing air pollution. These have led to a major interest in exploiting plants for the sustainable synthesis of a variety of chemicals and materials, especially for the production of renewable energy, such as biofuels and biodiesel. The U.S. federal government has set a target of replacing 36 billion gallons of current U.S. petroleum consumption within the transportation sector with plant-based renewable biofuels by 2022. China, New Zealand, the European Union, UK, Brazil, Canada, and Australia are examples of other countries that have established similar renewable bioenergy production targets. Currently, the majority of biofuel production is ethanol derived from starch- or grain-based feedstocks such as corn and sugarcane. The biofuel feedstocks are primarily distributed among forestry and agricultural resources: total biomass available for conversion into biofuels within the United States is estimated at about 1.4 billion dry tonnes per year, about 1 billion of it from agricultural resources and 400 million from forest resources (mainly short-rotation hybrid poplar, eucalyptus, loblolly pine, and willow) and other non-food energy crops (grasses such as switchgrass, *Miscanthus*, and sorghum) grown on marginal lands (see Chapters 30 and 31; J. S. Yuan et al., 2008; Simmons et al., 2008; Boernke and Broer, 2010). Plant biomass will play a central role in the development of low carbon, renewable energy supplies capable of partially displacing fossil fuels, but the time required to achieve significant levels of sustainable production is still questionable.

In addition to their use for renewable energy, plant-based novel biomaterials are now recognized as attractive substitutes for petrochemical-based materials (such as biodegradable thermoplastics and elastomers) and a variety of novel biopolymers produced in transgenic plants. These biopolymers include polyhydroxyalkanoates; poly-amino acids; polysaccharides with heat hysteresis properties and for bioaffinity purification; temperature and salt-resistant enzymes for the food industry; novel polymers produced in transgenic plants such as polyhydroxyalkanoates, cyclodextrins, and poly-amino acids for the chemical industry; and fibrous proteins, such as collagen, elastin, and silk, which have important material properties (see Chapter 30; Boernke and Broer, 2010). All in all, appropriate targets for novel compound production in plants are compounds that can either be produced more efficiently in plants, can have stable production without negatively affecting crop yield, can have better properties when produced in plants, or are needed as a bulk material at low cost via photosynthesis.

The use of plants as bioreactors for the production of "foreign," non-plant biopharmaceuticals has gained considerable momentum during the last decade. These include production of bioactive peptides, vaccines, antibodies, and a variety of pharmaceutical proteins and enzymes for the pharmaceutical industry. The initial concept of "molecular pharming" — production of edible vaccines where the pharmaceutical product is expressed in the plant, which is consumed by people (Mason and Arntzen, 1995; Twyman et al., 2003) — has been recently combined with efficient pharmaceutical production platforms in plants that serve as bioreactors for extracting the product rather than directly consuming the plant (see Chapter 29; Ma et al., 2005).

Plants are rapidly becoming accepted manufacturing hosts for biopharmaceuticals based on technologies involving the use of stable nuclear genetic transformation, and more recently the use of plastid transformation, as well as various transient expression technologies. Over the last several years, plant-based expression of heterologous proteins has become the basis of several manufacturing processes used for the industrial production and pilot manufacturing of many clinical proteins, including vaccines, antigens, and antibodies. There are currently three pharmaceutical proteins approved for market manufacturing: anti-caries antibody; Newcastle virus subunit; and glucocerebrosidase. Thus, the new era where plants are used, in addition to traditional agriculture, as factories for industrial products, pharmaceuticals, biomaterials, and bioenergy, may eventually lead to alternative types of agriculture.

Plant and Agricultural Biotechnology: A Societal and Public Approach

Former scientific achievements and technical advancements in "traditional" agriculture were not accompanied by negative public reactions; on the contrary, they were usually welcomed and adopted (e.g., use of inorganic fertilizers and pesticides, precision irrigation). However, the very recent agricultural revolution (i.e., generation of transgenic plants) and use of molecular tools has evoked public and regulatory concerns and sociological issues. It is now clear that public acceptance and full commercialization of genetically enhanced crop plants and forest trees depend, in addition to breakthrough science, on proper regulatory, economic, environmental, ethical, and societal factors.

Biotechnology affects not only our biological understanding, but also social and economic conditions and moral values, giving rise to many questions. What are the major public concerns and what are the proper (scientific) answers? First, we should be aware that the plant GM debate is not the only one that has occurred in science. Debates on scientific discoveries are part of our history and result most often from ignorance and hypocrisy, mingled with religious belief. Nicolaus Copernicus (1473–1543) was attacked as being "the devil." Giordano Bruno (1548–1600) was charged by the Roman Inquisition and was burned at the stake because of his revolutionary ideas about astronomy and the universe. Galileo Galilei (1564–1642), who played a major role in the scientific revolution about the universe, was condemned and subjected to home arrest and warned to abandon his work after being charged with supporting the heliocentric view of the universe. Three hundred years later, Charles Darwin (1809–1882) was attacked and charged with being "the antichrist." It is clear, however, that scientific discoveries and revolutions, such as the new biotechnology, cannot be stopped or prevented. These should be monitored and controlled, but they cannot be banished.

Public acceptance of the molecular techniques, both GM and others, calls for their safe and transparent use. Therefore, further distribution of biotechnology products and genetically engineered plants is concerned with five major safety issues and precautionary principles that should be scientifically based, actively followed, and fully transparent: food safety; preserving the ecological balance and the environment; socioeconomic considerations; regulatory aspects and intellectual property rights; and finally, ethical concerns.

Food safety

Two possible risks are involved. The first risk is integration and expression of genes from engineered crops in humans. This concern is not scientifically supported since the negligible amount of the consumed recombinant DNA (even when assuming that 50% of the diet is from GM foods) is quickly degraded into short DNA or peptide fragments in the gastrointestinal tract of animals and humans. Recombinant DNA was never found intact in animal experiments (Schubbert et al., 1998; European Food Safety Authority, 2007; Lemaux, 2008). It was claimed, however, that DNA insertion in the plant genome could be highly mutagenic (Latham et al., 2005). Scientific evidence shows that gene transfer and integration between distant genera, let alone kingdoms (i.e., plants and humans), are extremely rare and unlikely (see Chapter 32). Another risk is penetration of new toxic or allergenic products, which may be present in the consumed GM food, into the food chain. Even though harmful consequences from such residues have been rare, this issue is continuously monitored and assessed at all production stages and is strictly regulated by the USDA, EPA, FDA, and NIH (see Chapters 33, 34, and 36). In view of this, regulated and certified transgenic food, both fresh and processed, is safer for human consumption than many non-transgenic products found regularly in the markets.

Ecological balance and the environment

Detrimental effects of recombinant DNA/gene flow from a transgenic plant to other plants, as well as to other organisms inhabiting the same ecosystems, might potentially disturb the ecological balance. We should be aware, however, that the establishment of an ecological balance is a very long-term process and short-term laboratory/field experiments are not always efficient enough to predict results. Thus it is necessary to rely on a combination of historical evidence and scientifically based analysis.

Archaeological and historical evidence suggests that rapid plant and animal domestication for agricultural use occurred about 10,000–14,000 years ago (Pringle, 1988; Gupta, 2004). Early domestication eventually led to migration of plants from their natural habitat of origin to other regions, resulting in natural (pre-Mendelian) crossing between adjacent plants, and the production of new agricultural plants. Thus, gene flow is a natural process that usually benefits society. Controlled breeding of new crops occurred in post-Mendelian years, resulting in many new varieties of economic importance, including the new varieties of the Green Revolution. However, natural gene flow and Mendelian breeding has been overexploited by now.

This calls for risk-free, rapid, and fully controlled implementation of the new biotechnological/molecular tools, including genetic engineering and marker-assisted selection. Gene flow among plant species and to the ecosystem, whether vertical or horizontal or between genera, is very rare (see Chapter 32). However, while there is little risk of transgene flow from an engineered crop to wild/relative species, cases of transgene flow to weeds in agricultural fields and roadsides have been reported, for example, for herbicide tolerance related genes (see Chapter 32; Gressel, 2008; Warwick et al., 2009).

There are some potential risks of introgression of transgenes for "new traits" that do not exist in nature (e.g., production of pharmaceuticals and novel biomaterials in GM plants), which may have unknown advantages and disadvantages to the engineered plant and its neighbors. In such cases, field plots of these plants should be carefully monitored to avoid introgression into related species and varieties.

As discussed earlier, detrimental effects of gene flow from transgenic plants are very rare and do not seem to affect the environment. In addition, technologies for restricting gene flow to other plants, such as transgene containment and transgene mitigation, are available and are being continuously developed. Transgene containment inherently prevents transgene flow to occur, such as targeting genes to the cytoplasmic genome, male sterility, rendering crops asexual, and creating seed sterility by genetic use restriction technologies. Transgene mitigation prevents the transgene — if flow eventually happened — from establishing and spreading within the population, such as several strategies for rendering the resulting hybrids hypersensitive to chemicals or by suppressing weedy traits like seed shattering and dwarfing (see Chapter 32; Oliver and Li, 2011).

Another potential risk that may affect ecosystems is the biosynthesis of toxic gene products in GM plants, such as is the case with *Bt* crops. The *Bt* endotoxin is digested by insects and other living organisms, thus potentially affecting ecosystem biotic balance on one side, and development of resistant insect populations on the other. It was claimed that the mortality rate of the monarch butterfly was significantly increased following feeding on *Bt*-engineered crops. Carefully controlled studies indicated that it could occur only in very specific, artificial conditions and not in the open environment. A comprehensive two-year study suggested that the impact of *Bt* corn pollen on monarch butterfly populations is negligible (Sears et al., 2001).

In a recent extensive review the authors of the study concluded that despite some reports of negative effects on beneficial arthropods, insecticidal biotech crops have the potential to be much more environmentally benign in this respect than insecticide-based pest management approaches (Gatehouse et al., 2011). Such events, if they occur at all, seem very rare, and it is unlikely that they will affect natural established ecosystems.

Socioeconomic considerations

Some of the socioeconomic aspects of plant biotechnology for sustainable agricultural development have been recently discussed (D. Yuan et al., 2011). A few key issues such as technical improvements in the use of agricultural inputs (e.g., irrigation, soil management, balanced mineral nutrition) are beneficial to society. However, controlled employment of plant biotechnologies has the greatest potential to improve socioeconomic factors and eradicate hunger and malnutrition-associated diseases and poverty — especially in developing and poorer countries — by contributing to sustainable agriculture. Examples, in an increasing number of cases, include high yielding and biofortified genetically engineered crop plants that produce under conditions afflicted by abiotic (drought, salinity, high temperatures, pollution) and biotic (insects, disease, and other pests) stresses. The latter occurs with reduced inputs of costly and polluting pesticides, herbicides, and other chemicals.

Hunger and malnutrition is a complex socioeconomic issue related to lack of resources of poor people to buy agricultural inputs for growing their own crops or to buy food from other sources. It is clear that unequal food distribution among countries is the major problem, not existing global food supplies that could otherwise adequately feed the world population (see Chapter 35; FAO, 2009; DFID, 2010). Since the complex political and economic situation is not likely to change, it is essential that developing and poorer countries employ novel biotechnologies to be self-sufficient by growing their own improved crops with less costly inputs and in a sustainable way. Several major factors hinder the use of GM crops in developing countries, including limited capacity and resources for public research institution, low level of awareness of GM food by farmers and consumers, and an expensive regulatory process that keeps this technology out of the reach of small countries, especially with respect to minor local crops (see Chapter 36).

Regulation, legal aspects, and intellectual property rights

One of the prime examples of regulatory and intellectual property (IP) rights aspects of genetically engineered crops is the case of "Golden Rice" for the vitamin A precursor beta carotene enhancement, which was first developed by Ingo Potrykus (Ye et al., 2000). It was ready in the lab by 1999, IP rights granted in 2000, regulatory-clean rice line with enough vitamin A precursor approved in 2004, and introgression into target varieties achieved in 2007. It is currently under field experiments with local varieties in Bangladesh, India, the Philippines, and other countries, and is likely to pass regulation and transfer to farmers in 2012. As stated by Potrykus (2010), "Genetically-engineered crops could save many millions from starvation and malnutrition — if they can be freed from excessive regulation."

It is substantial that formal protocols should be developed to release new plant varieties derived through gene transfer and related molecular biotechnologies to the farmers, benefiting both consumers and farmers. These varieties need to be: (1) subjected to high levels of regulation and IP protection (see Chapters 33 and 34; Lemaux, 2009); and (2) evaluated and regulated according to their novelty and putative risks, and not according to the breeding technique used to produce them ("conventional," molecular marker-assisted selection, or recombinant DNA technology).

As stated by Hood et al. (Chapter 34), Harfouche et al. (Chapter 33), and others (e.g., Pretty, 2008; Fedoroff et al., 2010), regulation should preferably reflect the measured rather than the perceived risks. To allow the benefits of plant biotechnology to be realized, scientific evidence should underline each GE event. Protecting IP rights helps accelerate economic, social, and agricultural development for the benefit of humanity, and improvements in patent laws and licensing agreements should reflect the ongoing need for public–private sector partnership for continuing and intensifying the successful contribution of plant biotechnology to sustainable agriculture.

Ethical aspects

Critics of plant and agricultural biotechnology maintain that biotechnological science is value-free, posing the following questions: Are we allowed to alter the genetic structure of plants and animals in the name of utility and profit? Is there something sacred about life and do biotechnologists assume to be masters of nature? Is the genetic makeup of all living organisms the heritage of all humanity, or it can be appropriated by corporations?

The subject of ethics, which is concerned with the question of virtue, was coined and discussed first by Aristotle (384–322 BC) whose starting point was that "everything humans do is aimed at some good, with some good higher than others … the highest human good that people aim at is referred to as happiness or living well." John Stuart Mill (1806–1873) followed Aristotle, and established in his utilitarian theory the "greatest-happiness principle," which deals with "doing the greatest good for the most amount of people…." Agricultural and plant biotechnology, and in fact all biotechnologies, are based on his utilitarian and ethical philosophy. As discussed earlier and by others (Meiri and Altman, 1998; Chrispeels and Sadava, 2003), it has to be performed in a responsible way, be regulated, and be transparent to the public.

At the very heart of the extreme religious critique is the notion that scientists are "playing God." Religious beliefs are beyond the present discussion and are difficult to argue. One can only mention that man intervened in breeding for time immemorial, and that Noah's Ark (Genesis 6:18–20) was perhaps the first documented "breeding institution," on which males and females of all living creatures were given the chance to survive, breed, and produce offspring for selection of future generations. One last comment on the issue of ethics: millions of diabetic people around the world consume insulin, which is exclusively produced now by recombinant DNA technologies. Should we deprive a patient of a GM drug that can cure him and save his life? Nevertheless, some opponents of biotechnology want to deprive other groups of people of food because it is genetically modified. Ethical moralists will agree that having GM food is as good as a GM drug.

Concluding Remarks: Where Do We Go from Here?

Agriculture is fundamental to the economies and environments of the entire world, and to the well-being of all its inhabitants. More than 300 years ago Jonathan Swift (1667–1745) wrote, "Whoever could make two ears of corn or two blades of grass to grow upon a spot of ground where only one grew before, would deserve better of mankind, and do more essential service to his country, than the whole race of politicians put together." Indeed, achievements today in plant biotechnology have already surpassed all previous expectations and the future is even more promising. Moreover, a better understanding of plant biology using novel molecular genetic resources, omics technologies, and data analysis platforms have been successfully translated to practical agriculture. The advanced technologies that are continuously implemented for improving crop plants resulted in better growth and development control (both vegetative and reproductive) achieving increases in yields and quality of the crop, resulted in improved plant protection against the ever-increasing threats of abiotic and biotic stress, and supported classical breeding by increasingly applying marker-assisted selection tools. Moreover, the advanced biotechnological tools resulted in the generation of engineered crop plants with improved agricultural benefits, and expanded the use of plants and agricultural crops for the production of specialty biofortified foods, a variety of novel biomaterials, and for bioenergy generation.

It is clear, however, that, in view of some claimed potential risks of biotechnology and GM crops, the full realization of the plant and agricultural biotechnology revolution depends on both continued successful and innovative research and development activities, and on a favorable regulatory climate and public acceptance. Indeed, some of the tools of biotechnology are different from traditional breeding techniques. However, tools can and should be carefully assessed — like any new technique — in order to avoid or, more practically, to minimize risks. Many of the biosafety-associated solutions for the envisaged risks of GM food and effects on the environment are already available and additional solutions are being developed, and other claimed risks are not supported scientifically. Thus, we should continuously take care of societal factors such as public acceptance of transgenic plants, the regulatory framework for their commercialization, and a long-term political commitment to provide impetus into research and commercialization.

We believe that full-scale and immediate implementation of plant and agricultural biotechnologies, especially the use of recombinant DNA technology and molecular-marker assisted selection, is as crucial as that of medical biotechnology — sometimes even more so — since more people worldwide die from famine and malnutrition-related diseases than from "modern," western-world diseases. The intensification of agriculture thus requires enhanced and more efficient novel plant breeding biotechnologies for the release of economical, high-return, and patentable plants and plant-derived products. This cannot be achieved without advanced research and development in biochemistry, physiology, genomics, and practical biotechnology of agricultural plants. By doing so, biotechnology will revolutionize agriculture, which is required to meet food production (quantitative and qualitative) demands and to develop uses for novel plant-based metabolites and biopolymers. In the near term, biotechnology will have the greatest impact on developing countries, and these countries should be involved in developing guidelines for use of engineered crops.

References

Altman, A. (1999). Plant biotechnology in the 21st century: The challenges ahead. *Electronic Journal of Biotechnology*, *2*, 52–55. <http://www.ejb.org/content/vol2/issue2/full/1/>

Barton, K. A., Binns, A. N., Matzke, A. J. M., & Chilton, M.-D. (1983). Regeneration of intact tobacco plants containing full length copies of genetically engineered T-DNA, and transmission of T-DNA to R1 progeny. *Cell*, *32*, 1033–1043.

Beckles and Roessner. (2011). this book

Ben-Ari and Lavi. (2011). this book

Bischof et al. (2011). this book

Boernke, F., & Broer, I. (2010). Tailoring plant metabolism for the production of novel polymers and platform chemicals. *Current Opinion in Plant Biology*, *13*, 353–361.

Brouder, S. M., & Volenec, J. J. (2008). Impact of climate change on crop nutrient and water use efficiencies. *Physiologia Plantarum*, *133*, 705–724.

Catala et al. (2011). this book

Chrispeels, M. J., & Sadava, D. E. (Eds.). (2003). *Plants, Genes, and Crop Biotechnology* (2nd ed.). Jones and Bartlett/American Society of Plant Biologists. ISBN 0-7637-1586-7.

De Groote. (2011). this book

DFID, *The politics of poverty: Elites, citizens and states. Findings from 10 years DFID-funded research on governance and fragile states 2001–2010*. London, UK: DFID.

Dhankher et al. (2011). this book

Doebley, J., Gaut, B. S., & Smith, B. D. (2006). The molecular genetics of domestication. *Cell, 127*, 1309–1321.

Englemann. (2011). this book

European Food Safety Authority. (2007). EFSA statement of the fate of recombinant DNA or proteins in meat, milk and eggs from animals, <http://www.efsa.europa.eu/EFSA/Statement/gmo_EFSA_statement_DNA_proteins_gastroint,0.pdf/>

FAO, *The state of food security in the world 2009. Economic crises, impacts and lessons learned*. Rome: FAO.

Fedoroff, N. V., Battisti, D. S., Beachy, R. N., Cooper, P. J. M., & Fischhoff, D. A., et al. (2010). Radically rethinking agriculture for the 21st century. *Science, 327*, 833–834.

Feeding the Future – Special issue, *Science, 327*, 797–834.

Fluery et al. (2011). this book

Gatehouse, A. M. R., Ferry, N., Edwards, M. G., & Bell, H. A. (2011). Insect-resistant biotech crops and their impacts on beneficial arthropods. *Philosophical Transactions of the Royal Society B, 366*, 1438–1452.

Gleba. (2011). this book

GLP, Global Land Project—Science Plan and Implementation Strategy. 2005. [IGBP (International Geosphere Biosphere Program) Report No. 53/International Human Dimensions Programme Report No. 19, IGBP Secretariat, Stockholm, <www.globallandproject.org/documents.shtml/>

Gressel, J. (2008). *Genetic glass ceiling: Transgenic for crop biodiversity*. John Hopkins University Press.

Gressel. (2011). this book

Gupta, A. K. (2004). Origin and domestication of plants and animals linked to early Holocene climate amelioration. *Current Science, 87*, 54–59.

Handa. et al. (2011). this book

Hardtke and Nieminen. (2011). this book

Harfouche, A. (2011). this book

Heslop-Harrison and Schwarzacher. (2011). this book

Hood. (2011). this book

James, C. (2010). ISAAA Brief 42-2010: Executive summary, global status of commercialized Biotech/GM Crops: 2010. <http://www.isaaa.org/resources/publications/briefs/42/executivesummary/default.asp/>

Khayat. (2011). this book

Kumar and Loh. (2011). this book

Latham, J. R., Wilson, A. K., & Steinbrecher, R. A. (2005). Mutational consequences of plant transformation. *Journal of Biomedicine & Biotechnology, 2006*, 1–7.

Le Qeure, C., Raupach, M. R, Canadell, J. G., Marland, G., & Bopp, L., et al. (2009). Trends in the sources and sinks of carbon dioxide. *Nature Geoscience, 2*, 831–836.

Lemaux, P. G. (2008). Genetically engineered plants and foods: A scientist analysis of the issue I. *Annual Review of Plant Biology, 59*, 771–812.

Lemaux, P. G. (2009). Genetically engineered plants and foods: A scientist analysis of the issue II. *Annual Review of Plant Physiology, 60*, 511–559.

Lers. (2011). this book

Ma, J. K.-C., Barros, E., Bock, R., Christou, P., & Dale, P. J. (2005). Molecular farming for new drugs and vaccines: Current perspectives on the production of pharmaceuticals in transgenic plants. *EMBO Reports, 6*, 593–599.

Mackenzie. (2011). this book

Maheshi et al. (2011). this book

Mason, H. S., & Arntzen, C. J. (1995). Transgene plants as vaccine production systems. *Trends in Biotechnology, 13*, 388–392.

Matsumoto and Gonzalves. (2011). this book

Mawassi and Gera. (2011). this book

MEA, Millennium Ecosystem Assessment, *Ecosystems and Human Well-Being: Desertification Synthesis*. Washington, DC: World Resources Institute.

Meiri, H., & Altman, A. (1998). Agriculture and agricultural biotechnology: Development trends toward the 21st century. In A. Altman (Ed.), *Agricultural Biotechnology*. New York: Marcel Dekker Inc.

Newell McGloughlin. (2011). this book

Oliver, M. J., & Li, Y. (Eds.). (2011). *Gene containment*. Ames: Wiley.

Otten, L. A., De Greve., H., Hernalsteens, J. P., Van Montagu, M., Scheider, O., Straub., J., & Schell, J. (1981). Mendelian transmission of genes introduced into plants by the Ti plasmids of *Agrobacterium tumefaciens*. *Molecular & General Genetics, 783*, 209–213.

Ozias-Akins and Conner. (2011). this book

Paleg et al. (2011). this book

Poirier. (2011). this book

Potrykus, I. (2010). Regulation must be revolutionized. *Nature, 466*, 561.

Pretty, J. (2008). Agricultural sustainability: Concepts, principles and evidence. *Philosophical Transactions of the Royal Society B, 363*, 447–466.

Pringle, H. (1998). The slow birth of agriculture. *Science, 282*, 1446.

Raven, J. A., & Karley, A. J. (2006). Carbon sequestration: Photosynthesis and subsequent processes. *Current Biology, 16*, R165–67.

Reynolds, J. F., Stafford Smith, D. M., Lambin, E. F., Turner, B. L., II, & Mortimore, M., et al. (2007). Global desertification: Building a science for dryland development. *Science, 316*, 847–851.

Salomon and Sessa. (2011). this book

Samach. (2011). this book

Sati abd Matsyu. (2011). this book

Schubbert, R., Hohlweg, U., Renz, D., & Doerfler, W. (1998). On the fate of orally ingested foreign DNA in mice: Chromosomal association and placental transmission in the fetus. *Molecular & General Genetics, 259*, 569–576.

Sears, M. K., Hellmich, R. L., Stanley-Horn, D. E., Oberhauser, K. S., & Pleasants, J. M. (2001). Impact of Bt corn pollen on monarch butterfly populations: A risk assessment. *Proceedings of the National Academy of Sciences, 98*, 11937–11942.

Shiboleth and Tzfira. (2011). this book

Simmons. (2011). this book

Simmons, B. A., Loque, D., & Blanch, H. W. (2008). Next-generation biomass feedstocks for biofuel. *Genome Biology, 9*, 242.

Singer, A. C., Crawley, D. E., & Thompson, I. P. (2003). Secondary plant metabolites in phytoremediation and biotransformation. *Trends in Biotechnology, 21*, 123–129.

Solomon, S., Qin, D., Manning, M., Chen, Z., & Arguis, M. M. (Eds.), (2007). *Intergov. Panel Clim. Change. Climate Change 2007: The Physical Science Basis. Contrib. Work. Group I to the Fourth Assess. Rep. Intergov. Panel Clim. Change, ed*. Cambridge, UK: Cambridge Univ. Press.

Tapingkae. (2011). this book

Topp and Benfy. (2011). this book

Tsaftaris et al. (2011). this book

Twyman, R. M., Stoger, E., Schillberg, S., Christou, P., & Fischer, R. (2003). Molecular farming in plants: Host systems and expression technology. *Trends in Biotechnology, 21*, 570–578.

UNCCD. 1994. United Nations Convention to Combat Desertification, Elaboration of an International Convention to Combat Desertification in Countries Experiencing Serious Drought and/or Desertification, Particularly in Africa" (U.N. Doc. A/AC.241/27, 33 I.L.M. 1328, United Nations)

Verpoorte, R., & Memelink, J. (2002). Engineering secondary metabolite production in plants. *Current Opinion in Biotechnology, 13*, 181–187.

Warwick, S. I., Beckie, H. J., & Hall, L. M. (2009). Gene flow, invasiveness, and ecological impact of genetically modified crops. *Annals of the New York Academy of Sciences, 1168*, 72–99.

Watkins et al. (2011). this book

Ye, X., Al-Babili, S., Klöti, A., Zhang, J., & Lucca, P., et al. (2000). Engineering provitamin A (b-carotene) biosynthetic pathway into (carotenoid-free) rice endosperm. *Science, 287*, 303–305.

Yoo, C. Y., Pence, H. E, Hasegawa, P. M., & Mickelbart, M. V. (2009). Regulation of transpiration to improve crop water use. *Critical Reviews in Plant Sciences, 28*, 410–431.

Yuan, D., et al. (2011). The potential impact of plant biotechnology on the Millennium Development Goals. *Plant Cell Reports, 30*, 249–265.

Yuan, J. S., Tiller, K. H., Al-Ahmad, H., Stewart, N. R., & Stewart, C. N, Jr. (2008). Plants to power: Bioenergy to fuel the future. *Trends in Plant Science, 13*, 421–429.

Section A

Introduction to basic procedures in plant biotechnology

Genetics and genomics of crop domestication

1

J.S. (Pat) Heslop-Harrison Trude Schwarzacher
University of Leicester, Leicester, UK

TABLE OF CONTENTS

Plants and Domestication

Scope

In this review of genetics and genomics related to plant biotechnology and agriculture, we consider the nature of species that are grown as crops and used by mankind, or otherwise associated with people. We will then review aspects of the genetics and genome changes that have been associated with crop plants and their domestication from their wild relatives before speculating about some of the new opportunities for plant biotechnology to meet the challenges faced in the twenty-first century.

Domesticated crops

Domesticated crops are a subset of all plants. Domesticated species, whether plants or animals, are considered as those grown by people for economic or other reasons, and that differ from their closest wild relatives. Domesticated species are reliant on human intervention for their reproduction, nutrition, health, planting, and dispersal. They are harvested with the possibility that a different species will be planted in their place. Additional characteristics selected for domestication include size of harvested parts, yield or yield stability, and quality for the use of the product. There are extensive genetic differences in all of these characteristics between individuals within a species, as well as between species, and multiple characteristics are selected at the time of domestication that make the crop worth growing by farmers for millennia and now by today's plant breeders.

DOI: 10.1016/B978-0-12-381466-1.00001-8

Genomic techniques allow the underlying selection processes to be understood, exploited, and refined for crop improvement. Genomic scientists can now understand and improve the efficiency of exploitation of genes, genetic diversity, and controls present in crop species and their wild relatives. Domestication of plants, including selection of appropriate species and genetic changes, is one of the features of agriculture, but agriculture also requires knowledge beyond suitable genotypes (Janick, 2005), such as the planting, growing, protection, and harvest of the plants and the accurate timing of the various farming operations.

Domesticated plants are grown by the human population to meet a range of needs that can be summarized by the six "Fs": food, feed, fuel, fibers (and chemicals), flowers, and pharmaceuticals. Plants within each of these classes have substantial economic impact. Nevertheless, out of 400,000 species of flowering plants, less than 200 have been domesticated as food and feed plants, and just 12 species provide 75% of the food eaten (FAOStat, 2010). Very few of the 1000 gymnosperms, and arguably none of the 15,000 ferns and allies, have been domesticated. New knowledge of genetics and improved techniques of selection, hybridization, or gene transfer have the potential to enable more species to be domesticated.

As well as domesticated crop species, there are many spices, pharmaceutical (and medicinal), horticultural, and garden ("flowers") plants collected over the last millennia from the wild and cultivated on a small-scale. These plants may be genetically similar and as diverse as their wild relatives, although one or a small number of genes may have been selected. Many of the selections require human intervention to survive, often because they are grown outside their natural climate range or have abnormalities that are regarded as attractive or useful but reduce plant fitness. However, with the exception of some hybrids, the limited changes mean they are not normally considered as domesticated.

Weeds

Weeds and invasive species are associated with human farming and habitation, although they are not normally considered as domesticated species. There has been limited genomic and genetic work on most of these species with notable exceptions, in particular *Arabidopsis thaliana*. Harlan and deWet (1965) defined a weed as "a generally unwanted organism that thrives in habitats disturbed by man"; like crops, weed species are extremely diverse, and have different strategies for survival. The effect of weeds on agriculture can be devastating, such as taking nutrition from the crop, making harvest difficult, or reducing the value and quality of the harvest.

Most crop plants will not establish themselves in an environment where weeds thrive and active intervention is needed to remove competition. In an extensive study of feral oil seed rape (canola or *Brassica napus*), Crawley and Brown (1995) showed the very high level of turnover of site occupancy on highway verges, with local extinction occurring within three years in the absence of new seeding and soil disturbance. In contrast, weeds can be notably persistent, with; for example, nettles (*Urtica dioica*) remaining as markers of sites of habitation after hundreds of years in northern Europe in the absence of further habitation or evidence of crops. There are strong selection pressures on weeds to benefit from the human-created habitat at the farm (rather than plant breeder) level, working with potentially much larger and more widely distributed populations than breeders use. Weeds may mimic the growth forms or seeds of crops and are distributed or grow along with them. The selection is not applied to yield and quality characteristics, but on survival and population distribution or expansion, with key genes such as those for seed dormancy or dehiscence (see the section Genes of Domestication) potentially selected in the opposite direction from the seeds of a crop.

Crops can become weeds. In the *Brassicas*, for example, the same genotype may be a weed with low yield and poor characteristics in one environment, but a robust crop with desirable properties in another. Volunteers — plants from a previous crop on the same land — are a major challenge in growing many field crops. They thrive in the crop conditions; the economic damage from these weeds includes acting as reservoirs of crop-specific diseases over several seasons in a rotation.

Weeds have no harvest value in a crop, reducing yields, and making crop management difficult, so farmers have been improving their weed control methods since the start of agriculture. In advanced commercial farms weed control is an expensive part of the agronomy, while for smallholders and subsistence farmers, the continuous labor required can be one of the most tedious and demeaning operations for the people, usually women and children, that are involved. The removal and control of weeds is environmentally costly and involves burning, herbicides, deep plowing and multiple soil cultivations, processes leading to erosion, poor soil moisture conservation, use of large amounts of energy, loss of soil structure, uncontrolled fires, and smoke or pollution. Approaches to weed control have changed continuously over millennia, including use of fire, planting methods, and plowing. As well as the application of agronomic and technological approaches to limit weed spread, breeders must also consider the genetic characteristics of weeds and both the potential of a crop to become a weed and the ease of control of weeds within a new variety. This work interacts with making models of population biology based on the understanding of weed characteristics such as developmental plasticity or seed dormancy.

Invasive species

Another group of plants associated with humans are the invasive species. Along with habitat destruction, invasive species are often considered to be the major threat to biodiversity worldwide, although Gurevitch and Padilla (2004) pointed out that the cause and effect data are generally weak. Genetics and genomic research is required to understand the biology of invasives, so that the characteristics that led to uncontrolled displacement of native species can be avoided in the breeding of crops. The requirements of crops including high partitioning of the plant's resources to the harvestable product, non-distribution of seeds, and uniformity of growth tend to mean that few domesticated crops have invasive characteristics. However, a number of horticultural plants and those introduced for their novelty value have caused problems both in agriculture and the wild in very diverse environments ranging from temperate and tropical, through fresh water, grasslands, and woods to uplands, with the species taking advantage of man-made or man-influenced habitats. Examples of invasive species causing significant problems include water hyacinth (*Eichhornia* species), *Rhododendron*, knotweed (*Fallopia japonica*; Bailey et al., 2007), kudzu (*Pueraria* spp.), and some ferns (bracken, *Pteridium*, and *Azolla* spp.). This is notable since ferns have not been domesticated as crops.

Model species and crop sciences

The diversity in growth forms, reproduction, and uses between the crops means that most crop scientists have focused their work on a single species, while fundamental studies adopted a small number of convenient models. During much of the twentieth century, major research or model species were crops because they could be easily obtained and grown worldwide, and laboratory protocols, resources, and background information were extensive. Spinach was used for many studies of photosynthesis (e.g., Bassham and Calvin, 1955), maize was used for genetics (e.g., McClintock et al., 1981), and carrot or tobacco was used for tissue culture. However, for genetics, a fast generation time, small plant size, and the ability to mutagenize populations were major advantages. Researchers including Kranz, Redei, and Koornneef (e.g., Koornneef et al., 1983) established *A. thaliana* as a model species in the 1970s, and, because of its small genome size (165 Mbp), *Arabidopsis* was chosen to be the first plant to have its DNA sequenced (Arabidopsis Genome Initiative, 2000). The ease of growing large numbers under controlled conditions and extensive scientific resources led to it becoming the model for plant research in many laboratories. Rice became the second plant genome to be sequenced, because of its status as one of the world's two major crops, relatively small genome size of 435 Mbp, and contrasting taxonomic position to *Arabidopsis* (e.g., Sasaki et al., 2002). A major justification of these sequencing projects was the suggestion that the gene content of all plants would be similar, a prediction that has largely held true (e.g., Figure 3 in Argout et al., 2011), although sequencing led to some surprises including the low total number of genes — typically 30,000 — found in all organisms.

With the advent of plant biotechnologies, genomics, mathematical modeling, and informatics, a large number of tools and results of general nature can be applied across most crops and potential crops (see review by Moose and Mumm, 2008). Few crop scientists are now restricted to work on one species and need to exploit approaches and results with other crops and model species. In the genetics and genomics field there are many parallels between species, making it essential to integrate information. Throughout history and prehistory, humans have been classifying plants, assessing their similarity to use as food or medicines, and avoiding or processing toxic plants long before the advent of agriculture. A succession of techniques including morphological study, crossing, karyotype analysis, DNA sequence comparisons, and now whole genome sequences has established plant relationships. The Angiosperm Phylogeny Group (2009) presents a robust, monophyletic phylogeny showing relationships between all angiosperms; better understanding of the evolution and phylogeny is important for crop genetics because it shows the most closely related species to use to find valuable characteristics.

Understanding Domestication Processes

Evidence of relatives and processes of early domestication

The early processes of domestication can be inferred from examination of wild crop relatives and comparison with existing crops at the morphological, physiological, genetic, or DNA levels. Since farming and domestication is less than 10,000 years old, the archaeological record of the introduction of species into agriculture is rich (Zeder, 2006) and documents some aspects of the transition from hunter–gatherer societies to sedentary, farming-based communities. Indeed, the earliest hunter–gatherer cave paintings date from 32,000 years ago (Clottes, 2010) and in combination with archaeological evidence they show the pre-agricultural period. The domestication process happened independently in Southeast Asia and the Middle East, and soon after it is found in Asia, Europe, Africa, and the Americas after the retreat of the Pleistocene ice around 12,000 years ago. The domestication of all of the major crops now grown started at about the same time. Pictures of domesticated plants appear in Chinese and Arabic manuscripts up to 2000 years ago (Paris et al., 2009; Wang et al., 2008; Janick, 2005) and can be correlated with archaeological evidence. With the use of genetic markers to genotype crops and their relatives found in various locations, Salamini et al. (2002) reviewed how genetic markers traced the sites of domestication of cereals to wild populations of grasses in the Near East, and Gross and Olsen (2010) discussed that genetic inferences about geographical origins of crops and the number of independent domestication events are compatible with archaeological data.

Domestication of particular species, and the genetic characteristics that make them different from their wild relatives, are also associated with technology used in agriculture societies for planting, harvesting, threshing, transport, and storage; or long-lasting infrastructure like roads, habitations, and field organization; and domestic arrangements including specialized storage and preparation premises or cooking processes. All of these give additional information about the genetic changes from wild species since genotypes must complement the societal practices. In the first decade of the twenty-first century, genetic and genomic methods enabled examination of the processes of crop domestication, including both the identification of the genetic basis and its origin and the duration of domestication (Papa et al., 2007).

Genes of domestication

The "suite of traits" including seed dispersal, seed dormancy, gigantism in the harvested parts, determinate and synchronized growth, increased harvest index, and change in sweetness or bitterness have been called the "domestication syndrome" after Hammer (1984). These characteristics make a crop worth growing, and without them the difficulties of planting, cultivation, and poor harvest make them unrewarding to grow. It is likely that a combination of all of the characteristics must be present together for a species to reach the first stage of domestication, since most of these traits in some form are present in all domesticated crops. Doebley (2004) and Doebley et al. (2006) reviewed data showing that the differences in cultivars mean that wild progenitors of crops are not easily recognizable. Furthermore, many of these characters are so disadvantageous in non-cultivated situations that the crop will not establish in the wild (Crawley and Brown, 1995): indehiscent plants will not distribute seeds, whereas an annual plant bearing seeds without dormancy means the species would not survive one bad season.

Further evidence, at least in the cereals, for the importance of the small number of domestication syndrome genes comes from the similarity of changes in several domesticates known as convergent evolution. Paterson et al. (1995) showed that the same genes and gene pathways were involved in domestication of sorghum, rice, and maize. As with other genetic effects, many domestication characteristics are regulated by quantitative trait loci (QTLs) where several genes have effects (Varshney et al., 2006), and transcriptional regulators (rather than enzymatic or structural) genes (Doebley et al., 2006; Martin et al., 2010) are often involved.

Genetic variation and domestication

Genetically, any requirement for change of multiple characters simultaneously requires either an extremely unusual conjunction of genetic mutations or recombination, or selection and intercrossing to bring characters together over many generations. Clearly, the latter did not happen to any great extent, and genetic and genomic data collected over the last decade do suggest that the diversity of alleles present in domesticated species is lower than in their wild progenitors. This supports the domestication syndrome concept with a number of characteristics coming together at one time. This selection has left a "genomic signature" in all current crops, present thousands of generations later, and the loss of diversity compared to the wild species is seen as a "genetic bottleneck" (Doebley, 2004). Genetic analysis has shown that many of the gene alleles involved in the domestication syndrome are present within the gene pool of wild progenitors of crops, although with a low frequency, whereas other traits are apparently new mutations (Doebley et al., 2006; Huang et al., 2007; and see Chapter 13). One important approach to identifying genetic bottlenecks has been comparison of genomic regions neighboring key domestication traits with selectively neutral regions; reduced variation in linked genes suggests that the number of domestication syndrome genes is limited. The "selective sweep" of the genome (Clark et al., 2004) with directional selection leads to reduced variation and linkage disequilibria (Anhalt et al., 2008, 2009) in the selected regions.

Whereas only a few plants have carried critical traits related to domestication and have been used for most subsequent breeding, the genetic bottleneck or "founder effect" will have reduced the diversity to a small number of gene alleles present in the original selected population (changing gene allele frequencies, eliminating rare alleles, and introducing linkage disequilibrium). It has widely been considered, especially on theoretical grounds, that genetic drift will have further reduced the diversity after domestication, given that the selection of a few hundred varieties at most for use in breeding represents a tiny population size. In many cultivated crops, the level of genetic polymorphism has been reduced by 60 to 90% in passing through the genetic bottleneck in cultivars compared to wild relatives (e.g., Buckler et al., 2001 in maize). Similarly, rice cultivars may include only 10 to 20% of the diversity present in the wild relatives (Zhu et al., 2007). Even with extensive data, it remains challenging to distinguish between the monophyletic and polyphyletic origin of a crop using molecular markers. As noted by Zhang et al. (2009), genetic marker data can indicate that the two cultivated rice subspecies, *indica* and *japonica*, either evolved independently at different times and sites (Tang et al., 2006), or had a monophyletic origin from a common wild rice that subsequently separated. The diversity restriction is not universal, and the polyphyletic origin of some polyploid crops has probably reduced the bottleneck effect: hexaploid bread wheat (AABBDD genome constitution) has much of the genetic diversity present in its progenitors (Dubcovsky and Dvorak, 2007) and originated recurrently with ancestral D-genomes (Caldwell et al., 2004), even if all the D-genome variation is not represented (Saeidi et al., 2008). Cifuentes et al. (2010) discussed the polyphyletic origin of canola (oilseed rape, *B. napus*), which incorporates variation from both the duplication of loci as a polyploid and from several ancestors of the tetraploid cultivars.

In domesticated species, artificial selection is the main evolutionary force because humans — farmers and more recently plant breeders — exert strong selection pressure compared to that from the environment where a species is established (Innan and Kim, 2008). These authors pointed out that artificial selection may act on alleles that may have been neutral variants before domestication, and the fixation of these may not remove DNA variation in the surrounding region, depending on the initial frequency of the beneficial alleles. The number of alleles selected during domestication, the population sizes, and the number of independent selection events will all affect the intensity of the selection bottleneck.

Genetic control related to diversity and speciation

While geographical isolation of populations stops gene flow within a species, it is far from the only effect that leads to separation of genotypes. Rieseberg and Blackman (2010) have identified no less than 41 different genes that can lead to reproductive isolation of populations. Genetics related to plant evolution and isolation is a relatively recent research area, and it is clear that the identification of genes that effect reproductive behavior — recombination and interact with effects on fertility, leading to isolation and speciation (Heslop-Harrison, 2010) — may well show how some of the selective sweeps (Nielsen et al., 2005) have been driven during crop domestication. Understanding the genetic effects and genes that enable these processes may show how levels of diversity can be maintained within species, whether in wild ecosystems or crops.

Domestication of maize

One of the best understood examples of genetic and genomic changes during domestication comes from maize (*Zea mays* or corn in the Americas), where the seminal work of Doebley and colleagues (Doebley et al., 2006; Wright et al., 2005) identified the relatively few genes giving rise to the major physiological and morphological differences between maize and its closest wild ancestor, teosinte (represented by several *Zea* species). Maize, with naked grains in multiple rows and 10 to 100 times more kernels per ear, has a very different appearance from the branched teosinte, which has grains with a hard seed coat on inflorescences that shatter (disarticulate) when ripe and carried on multiple stalks. Among the first genes identified was *teosinte branched 1* (*tb1*), a transcriptional regulator that represses the branching (Doebley, 2004). The gene *teosinte glume architecture*, *tga1* (Wang et al., 2005), is a key single-gene that controls development of the hard coat around the kernel in teosinte. It was identified by high-resolution genetic mapping and map-based cloning. Doust (2007) more generally studied the developmental genetics of grass plant-architecture in genetic, evolutionary, and ecological contexts. He concluded that exploring the phylogenetic context of the crop grasses suggests new ways to identify and create combinations of morphological traits that will best suit future needs: knowledge of past events shows how future breeding can proceed.

Technically, works such as those previously mentioned have focused on making experimental hybrid populations for genetically mapping traits that can be identified as domestication related. Another group of researchers took a large-scale approach to characterizing how bottlenecks and artificial selection have altered genetic variation during domestication of teosinte to form maize using an unbiased, genome-wide approach. Wang et al. (2005; see also Vigouroux et al., 2005) measured single nucleotide polymorphism (SNP) levels in 774 genes, and found that the maize inbred lines had only 57% of the variation present in the teosinte sample, showing evidence for the genetic bottleneck. The genes could be divided into two classes based on the variation signatures at single nucleotides (SNPs): 2–5% of the genes were under selection during domestication and have been selected with 10 times the intensity of the selectively neutral genes where limited population size alone has reduced the variation. Yamasaki et al. (2005, 2007) sequenced 1095 maize genes from various lines and identified eight genes with no variation between inbred maize lines, but with SNP variation in teosinte; six showed selection throughout the DNA sequence of the gene, while two had signatures of selection in the 3′ portion of each gene. The functions of the genes, examined after the analysis, were "consistent with agronomic selection for nutritional quality, maturity, and productivity," although most had not been identified previously as being associated with their selection in the crop.

Domestication of legumes

Weeden (2007) examined the domestication of the pea (*Pisum sativum*), and identified approximately 20 genes or QTLs responsible for the domestication of it. Because of the availability of a range of germplasm from the pea, a time line for the "domestication syndrome" genes could be established. Domestication syndrome characters including indehiscent pods, seed dormancy, gigantism as seed weight, and earliness were seen in the most primitive lines, while dwarfing, harvest index, photoperiod-sensitivity and white flowering, along with additional seed weight traits, appeared much more recently. This is evidence for the model shown by Gross and Olsen (2010) that domestication is a two-stage process. First, is a rapid process that makes the crop worthwhile to grow, including the domestication syndrome traits that allow a crop to be reliably sown, cultivated, and harvested such as uniform seed germination and fruit ripening. This is then followed by a stage acquiring traits over a longer period that improves the crop.

A second finding of Weeden (2007) showed that, although the phenotypic characters are similar, the genes involved in pea domestication are different from those in the common bean, *Phaseolus*, contrasting with the conclusion showing convergent evolution in rice, maize, and sorghum (Paterson et al., 1995). Weeden is optimistic that the presence of multiple genes means that there are several ways for breeders to modify unwanted characters and avoid detrimental effects associated with some otherwise valuable alleles.

Several studies have investigated the genetic diversity and signatures of domestication in soybean, a species with a center of origin and domestication in South China. Guo et al. (2010) proposed a single origin with a moderately severe genetic bottleneck during domestication, showing that wild soybeans in South China have an unexploited and valuable gene pool for future breeding. However, Hyten et al. (2006) examined other populations, finding that there were several rounds of reduction of genetic diversity, following domestication in Asia to produce numerous Asian landraces and introduction of a few genotypes to North America. Notably, they found modern cultivars retained 72% of the sequence diversity present in the Asian landraces but lost 79% of rare alleles, with the major constrictions of diversity coming first from the domestication event, and secondly from the introduction of a small number of races to North America, while later breeding has had less effect.

Grasses tend to have inflorescences where all individuals flower together and the seeds reach maturity at a similar time, which is certainly an advantage for agriculture. However, other wild plants flower and set seed over a long part of the crop season, making growing and harvest of the ripe seed difficult. In species such as soybean (*Glycine max*), determinacy of growth through the character of a terminal flower is an agronomically important trait associated with the domestication. Most soybean cultivars are classifiable into indeterminate and determinate growth habit, whereas G. *soja*, the wild progenitor of soybean, is indeterminate. Tian et al. (2010) took a candidate-gene approach to demonstrate that the determinate growth habit in soybean is controlled by a single gene homologous to *TFL1* (terminal flower) in *Arabidopsis*, which is a reasonable expectation. The genetics of the determinate habit has been known since the 1970s, and mapped more recently. There are, as expected from the known genetic background of soybean, four homologous copies for the determinate genes.

Yield traits

Yield, affected by gigantism and number of harvested units, is normally a quantitative trait with continuous variation and complex heritability. However, analysis and partitioning of yield components, combined with use of well-designed test crosses and large populations, is allowing key regions of the genome — in some cases now correlated with genes — to be identified. Genes increasing harvestable yield have been extensively studied using genetic and genomic approaches. Measurements of yield components, starting long before extensive use of genomic approaches, showed that, for example, rice yield includes traits such as grain number and grain weight, or duration and rate of grain-filling, and is regulated by multiple QTLs (Yano, 2001). Use of appropriate hybrid populations segregating for yield characteristics, such as biomass in forage grasses (ryegrass, *Lolium perenne*; Anhalt et al., 2009) or fruit yield in tomato, *Solanum lycopersicoides* (*Lycopersicon esculentum*; Cong et al., 2002), is showing that genetic regions on the map are responsible for a large part of the variation in yield observed. However, often a large number of genetic regions are identified: in tomato, no less than 28 different QTLs affecting fruit weight have been identified (Cong et al., 2002). QTL analysis is also of potential importance when identifying characters where the same gene affects different traits; this could indicate selection in opposite directions is unlikely to succeed (e.g., grain protein and yield or palatability/sweetness and insect resistance).

Hybrid Species and New Polyploids in Domestication

Most of the species previously discussed have a genetic structure similar to their wild relatives, such as fertility and reproduction through seeds. However, a group of crop species have a different genomic constitution from wild species, bringing together copies of genomes from different ancestral species that are not found normally in nature. This includes species that have different chromosome numbers from their relatives or that are hybrids (see Molnár et al., 2010).

Among early domesticates, the banana is an interesting example. Wild, fertile, diploid bananas have small fruits and large seeds with very little fruit pulp that is eaten. However, almost all of the cultivated lines are sterile and parthenocarpic, producing fruits in the absence of seeds. This is a characteristic associated with triploidy, so cultivars have a chromosome constitution of $2n = 3x = 33$, while the fertile wild species are $2n = 2x = 22$. The ultimate origin of the accessions that have become cultivars is unclear, although there are many independent parthenocarpic cultivars selected by early farmers in South East Asia that are vegetatively propagated (Heslop-Harrison and Schwarzacher, 2007). De Langhe et al. (2010) presented evidence indicating the complex origin of the cultivars, involving intermediate hybrids and backcrossing, with additional clonal variation. They note that this gives extensive variation, but makes development of artificial breeding schemes difficult.

Polyploid cytotypes can be larger than their diploid progenitors, and this has lead to their selection and cultivation in blueberries and strawberries (the octaploid hybrid is cultivated, and small wild species are mostly diploid; Schulze et al., 2011). This advantage is not universal and fruit gigantism can be under the control of several genes. Cultivated grapes are much larger than their wild progenitors, but there is no recent genome duplication in their ancestry (French-Italian Public Consortium, 2007).

Post-Domestication Selection

Modifications in crop characteristics

Since early agriculture, crop improvement has been a continuous process over thousands of years, driven by the need for disease resistance, adaptation to new and changing climates, quality, ability to propagate and grow, and yield. Systematic breeding since the mid-twentieth century — required by the need for increased food production, new crop uses, a different socioeconomic environment, climate, and water use changes — and new quality requirements have accelerated breeding with unprecedented speed.

As well as the continuing increase in the world's population, people worldwide have moved into conurbations. In Europe, only 2% of the population is engaged in agriculture, a change seen in global trends, with 2008 being the first year when more than half the human population lived in cities.

Changes in demand have certainly lead to substantial changes in the nature of crops. This divorce of populations from local food production requires new infrastructure to bring agricultural production to the people, and perhaps new genotypes appropriate to the transport chain. Increased meat consumption, particularly of chicken and pork, is seen as a significant global trend, requiring not only farming of the animals, but also production of the crops to feed these animals. Another significant global trend is the increase in plant oil production driven by consumer demand. Major crops such as oil palm, oilseed rape, and soybean show the greatest increase in production over the last 50 years. The increase in animal and oil consumption is widely considered to have negative effects on human health, although the use of biotechnology allows modification of the fatty acid content of oilseed plants for health purposes or to produce nutritional fatty acids not normally found in crop plants (Damude and Kinney, 2008).

Over a similar period, several crops have shown substantial declines either globally or regionally. Notable examples would be fiber crops including hemp, jute, and flax that have been replaced by petrochemical-based ropes and textiles, while oats in northern Europe are no longer grown as feed for horses.

Mechanization of agriculture in the twentieth century in developed countries has directly led to landscape-wide effects such as larger fields, more land in cultivation, and changed water or erosion management. Improved crop protection and artificial nitrogenous fertilizer application has also led to substantial changes in landscape and crop mixes, largely removing the need for fallowing or cover crops. Along with labor costs (which are very high in crops where mechanization has been limited), farming practices have certainly changed both the mix of crops grown and the requirements from the genetics of the varieties in a complex relationship between consumer income, tastes, import availability, and price.

Is biotechnology affecting the species and acreage of crops? Two crops showing substantial production declines regionally have been restored to production with transgenic varieties. Lines of cotton carrying insect resistance with the Bt-toxin have lead directly to an increase (or stability from a declining trend) in acreage of cotton. Papaya, where Hawaiian production fell by 40% and moved from the Oahu island because of devastation by the papaya ringspot virus, has been restored to production by the introduction of the resistance transgenic variety from 1998 (http://www.hawaiipapaya.com/rainbow.htm). It is predictable that intensive production of other crops will follow similar patterns (King et al., 2003, 2004) in the future with either stoppage of production or introduction of transgenic varieties.

New Domestication

Domesticated species

About 10% of all plants species are suitable for food, and a higher proportion has been used as species or flavors, or has medicinal value. However, most species have not met the first requirements for successful domestication. The small number of plants that have been domesticated, along with their repeated domestication in a number of localities, might suggest that no other species are suitable for domestication, at least for production of substantial amounts of food. Are there other plants that have appropriate genetical attributes to become a crop? It may be that the genetics was complex or allelic diversity was restricted so that it never become worthwhile to cultivate the plant at the early period of domestication, and later in history, refinement of genetic traits had already given the major crops a significant advantage so others could not compete in terms of ease of husbandry, harvest, nutritional value, or some other quality. There are prospects for domestication of new species, although in practice the number of new species introduced to agriculture in the last century on a significant scale is very limited. On the FAO list of crops, only kiwi fruit and the hybrid cereal triticale are significant additions in the last 50 years. As well as the new crop classes mentioned earlier, there are species currently collected from the wild that are being prospected for domestication. Overexploitation is threatening the survival of several medicinal plants and spices, providing an incentive for increasing the scope of domestication.

Our understanding of the domestication processes in the major crops can now be applied to advancing domestication in species that have not been domesticated (Heslop-Harrison, 2002). Although the farmer and consumer demands are higher than ever, the use of molecular methods and genomics covered in this volume are likely to mean that relatively rapid selection of unimproved wild species is possible, including some fruits and nuts with limited breeding (Heslop-Harrison, 2004). Appropriate diversity can be found within germplasm collections, and multiple DNA markers can combine traits and select from huge populations of plants or transformation can introduce single-genes.

Lost crops

Major crops were domesticated early in the history of agriculture, and the rarity of introductions was discussed earlier. It is also valuable to consider crops that were cultivated more extensively in prehistory — meeting the early requirements for being worthwhile crops, but not being as useful, or improved at the rate of other species more recently. Proso millet (*Panicum miliaceum*) was, like wheat, widely grown, as is demonstrated by archaeological evidence (Hunt et al., 2008). However, since this early period, its production has reduced to less than 1% of the production of wheat. It is notable for being the most water-efficient grass (Heyduck et al., 2008), but there is minimal genetic work (not even the ancestors of the tetraploid have been defined clearly), so it is one example where further work on its genetics will be valuable. A number of legumes, now minor, may also be interesting "lost crops."

Trees and biofuels

Trees provide fuel and fiber for construction, utensils, and paper. Planting and coppicing of trees has long been practiced, with selection of trees for regeneration and yield. However, the selection and characterization of trees for construction timber or fiber has been less systematic because of the availability of timber trees in native forests, the long timescales involved in cultivation, lack of continuous revenue, and large capital costs involved. The selection methods and genetics being applied to trees are rapidly changing, and application of genomics and marker technology has potential to improve tree characteristics (Neale, 2007; Gailing et al., 2009)

For biofuels, the twenty-first century has seen the establishment of several genomics-based research programs that are looking at both the improvement of existing crops and introduction of new crops with high biomass yields suitable for fuel production as well as new ways of processing using microorganisms to break down biomass (Heaton et al., 2008; Rubin, 2008; Somerville et al., 2010)

Genetics and breeding for new uses: Ecosystem services

Most plants planted and grown on a large scale by man provide products that are traded and used or have horticultural or amenity value. However, as well as conservation management of wild environments, it is probable that there will be increased selective breeding for plants that improve the local or wider environment. During the twentieth century, plants have been used for habitat restoration of mining sites, and the selection of appropriate species and genotypes of trees, shrubs, and grasses for colonization of these sites has, along with improvements in earth handling and planting methods, made enormous improvements in the landscapes of derelict areas (Richardson, 1975). Elsewhere, plantings are used to stabilize soils or sands and prevent erosion or drift. As well as the poor substrates, mining wastes may be contaminated by heavy metals, and Bradshaw and colleagues carried out extensive work on the selection of genotypes for land remediation, which are now widely applied (Antonovics et al., 1971). Beyond use for

Table 1.1 Key features of selected model species and major and minor crops related to the organization of their genomes

Crop	Species	Family	1C Genome size (Mb)[c]	2n Chromosome number	Ploidy level	Life form	Life span	Climatic range
Maize	*Zea mays*[a]	Poaceae	2670	20	2x (4x)	Herb	Annual	Temperate
Wheat	*Triticum aestivum*	Poaceae	17,000	42	2x	Herb	Annual	Temperate
Rice	*Oryza sativa* ssp. *indica*[a]	Poaceae	420	24	2x	Herb	Annual	Tropical
Rice	*Oryza sativa* ssp. *japonica*[a]	Poaceae	466	24	2x	Herb	Annual	Temperate to tropical
Sorghum	*Sorghum bicolor*[a]	Poaceae	730	20	2x	Herb	Annual or perennial	Tropical to subtropical
Barley	*Hordeum vulgare*	Poaceae	5400	14	2x	Herb	Annual	Temperate
Rye	*Secale cereale*	Poaceae	8100	14	2x	Herb	Annual	Temperate
Pearl millet	*Pennisitum glaucum*	Poaceae	2620	14	2x	Herb	Annual	Tropical
Foxtail millet	*Setaria italica*	Poaceae	513	18	2x	Herb	Annual	Temperate to subtropical
Sugar cane	*Saccharum* sp.	Poaceae	3960 (80 chr)	80–128	8x or more	Giant herb	Perennial	Warm temperate to tropical
Potato	*Solanum tuberosum*[a]	Solanaceae	2050	48	4x	Herb	Perennial	Temperate
Tomato	*Solanum lycopersicum*	Solanaceae	1000	24	2x	Herb	Perennial, grown as annual crop	Temperate to subtropical
Cassava	*Manihot esculenta*	Euphorbiaceae	807	36	2x	Woody shrub	Perennial	Tropical to subtropical
Soybean	*Glycine max*[a]	Fabaceae	1100	40	2x	Annual herb	Annual	Temperate to subtropical
Groundnut or peanut	*Arachis hypogaea*	Fabaceae	2807	40	4x	Herb	Annual	Tropical to warm temperate

land remediation, other plant species provide "ecosystem services" such as waste decomposition, water purification, hydrology improvement through root systems, fencing, or hedging. It is certain that the uses of plants to provide these services will increase as their value is recognized through economic methods (e.g., the Millenium Ecosystem Assessment, 2010, undertaken in connection with the United Nations system). New ecosystem values are likely to be introduced, including carbon capture (e.g., through roots; Kell, 2011).

The human uses of plants to provide ecosystem services have currently involved selection of appropriate genotypes from the wild. However, there is a genetic basis for the properties needed, and systematic breeding, including use of biotechnology (Chory et al., 2000), can improve their performance enough so there is enough improvement to balance the research and breeding costs.

Features of Domesticated Genomes

Table 1.1 summarizes the remarkable diversity in fundamental characteristics of some major and minor crops and some other comparator species including features of their genome organization and size (see discussion in Heslop-Harrison and Schwarzacher, 2011). The crops have mostly been domesticated and then selected by farmers and breeders over several

Predominant breeding system	Propagation	Parts used	Commodity	Nutritional use	World production 2009 (Million tonnes)	Storage
Cross-pollinating/monoecious although self-fertile	Seed	Grain endosperm/ leaves	Cereal/forage	Starch, protein, oil	817	Dry/years
Self-pollinating	Seed	Grain endosperm	Cereal	Starch, protein	682	Dry/years
Self-pollinating	Seed	Grain endosperm	Cereal	Starch, protein	679	Dry/years
Self-pollinating	Seed	Grain endosperm	Cereal	Starch, protein		Dry/years
Self-pollinating occasionally out-crossing	Seed	Grain endosperm/ leaves	Cereal/forage	Starch, protein	62	Dry/years
Self-pollinating	Seed	Grain endosperm	Cereal	Starch, protein	150	Dry/years
Out-crossing	Seed	Grain endosperm	Cereal	Starch, protein	18	Dry/years
Out-crossing	Seed	Grain endosperm	Cereal	Starch, protein	32 (millet)	Dry/years
Self-pollinating occasionally out-crossing	Seed	Grain endosperm/ leaves	Cereal/forage	Starch, protein	32 (millet)	Dry/years
Cross-pollinating	Stem cutting	Stalks	Sugar	Sugar	1683	Processed
Not true breeding	Vegetatively (tuber)	Tuber	Vegetable	Starch	330	Months
Self-incompatible, self-fertile in some cultivars	Seed	Fruit	Vegetable	Dietary fiber, antioxidants[b]	141	Fresh
Out-crossing	Vegetatively (stem)	Root	Vegetable	Starch	241	Days/only in ground
Self-pollinating	Seed	Seed (cotyledon)	Protein and oil	Protein, oil	222	One year
Self-pollinating	Seed	Pods with seed (cotyledon)	Vegetable, oil	Protein, fat, nutrient rich	36	One year

(*Continued*)

Table 1.1 (Continued)

Crop	Species	Family	1C Genome size (Mb)[c]	2n Chromosome number	Ploidy level	Life form	Life span	Climatic range
Alfalfa	*Medicago sativa*	Fabaceae	841	16/32	2x/4x	Herb	Annual	Warm temperate
Oil palm	*Elaeis guineensis*	Arecaceae	1800	32	2x	Tree	Perennial	Tropical
Date palm	*Phoenix dactylifera*[a]	Arecaceae	929	36	2x	Tree	Perennial	Tropical
Coconuts	*Cocos nucifera*	Arecaceae	3472	32	2x	Tree	Perennial	Tropical
Sugar beet	*Beta vulgaris*	Amaranthaceae	1223	18 or 36	2x or 4x	Herb	Biennial	Temperate
Banana and plantain	Banana cultivars *Musa*	Musaceae		33	3x	Giant herb	Perennial	Tropical
Wild banana	*Musa acuminata*	Musaceae	550	22	2x	Giant herb	Perennial	Tropical
Sweet potato	*Ipomoea batatas*	Convolvulaceae	1467	60		Herb/vine	Perennial	Tropical to warm temperate
Onion	*Allium cepa*	Alliaceae	16,382	16	2x	Herb	Biennial	Temperate
Rapeseed	*Brassica napus*	Brassicaceae	1125	38	4x	Herb	Annual	Temperate
Cabbage	*Brassica oleracea*	Brassicaceae	758	18	2x	Herb	Annual	Temperate
Arabidopsis	*Arabidopsis thaliana*[a]	Brassicaceae	165	10	2x	Ephemeral	Annual	Temperate
Oranges (citrus)	*Citrus × sinensis*	Rutaceae	611	18	2x	Tree	Perennial	Warm temperate to subtropical
Apples	*Malus domestical*[a]	Rosaceae	327	34/51	2x/3x	Tree	Perennial	Temperate
Strawberry	*Fragaria × ananassa*	Rosaceae	597	56	8x	Herb	Perennial	Temperate
Wild strawberry	*Fragaria vesca*[a]	Rosaceae	240	14	2x	Herb	Perennial	Temperate
Grape	*Vitis vinifera*[a]	Vitaceae	490	38	2x	Shrub	Perennial	Temperate to warm temperate
Cucumber	*Cucumis sativus*[a]	Cucurbitaceae	367	14	2x	Herb	Annual	Temperate
Olive	*Olea europaea*	Oleaceae	1907	46	2x	Tree	Perennial	Warm temperate
Lettuce	*Lactuca sativa*	Asteraceae	2590	18	2x	Herb	Annual or biannual	Temperate

Predominant breeding system	Propagation	Parts used	Commodity	Nutritional use	World production 2009 (Million tonnes)	Storage
Out-crossing and self-pollinating	Seed	Leaves	Forage			Fresh/ processed
Out-crossing/monoecious	F1 seed	Fruit (mesocarp and kernel)	Oil	Oil	207	Months
Out-crossing/dioecious	Seed/cuttings	Fruit (mesocarp)	Fruit	Fruit	7	One year
Out-crossing	Seed	Seed (endosperm)	Fruit	Fruit, fiber	4	
Out-crossing, occasionally self-fertile	Seed	Root	Root vegetable	Sugar	229	Month/ processed
Sterile	Suckers/tissue culture	Fruit	Fruit	Starch tropics, temperate[b]	130	Weeks
Cross-pollinating but self-fertile	Suckers	Fruit/leaves	Fruit/fiber	Starch		Months
Out-crossing	Seed	Tuber	Vegetable	Starch	108	Week
Cross-pollinating but self-fertile	Seed/bulb	Bulb	Vegetable	Flavoring, starch	72	Months
Out-crossing/ self-incompatible	Seed	Seed	Oil	Oil	62	One year
Out-crossing/ self-incompatible	Seed	Leaves	Vegetable	Antioxidants/vitamins[b]	71	Fresh
Self-pollinating	Seed					
Self-fertile, some self-pollinating	Grafting	Fruit	Fruit	Dietary fiber, vitamins[b]	68 (124 Citrus)	Weeks
Out-crossing/ self-incompatible	Grafting	Fruit	Fruit	Dietary fiber, vitamins[b]	72	Months
Hybrid polyploid, vegetative	Runners	Fruit	Fruit	Dietary fiber, vitamins[b]	4	Fresh
Out-crossing	Seeds, runners	Fruit	Fruit	Dietary fiber, vitamins[b]		
Out-crossing, self-incompatible	Grafting	Berry	Fruit, wine	Drink, antioxidants, vitamins	67	Fresh/dried/ processed
Self-fertile, some self-pollinating	Seed	Fruit	Vegetable	Dietary fiber, vitamins[b]	39	Fresh
Out-crossing/ self-incompatible		Fruit	Vegetable/oil	Oil	18	Months
Self-fertile	Seed	Leaf	Leaf vegetable	Dietary fiber, vitamins[b]	24	Fresh

(*Continued*)

Table 1.1 (Continued)

Crop	Species	Family	1C Genome size (Mb)[c]	2n Chromosome number	Ploidy level	Life form	Life span	Climatic range
Celery	*Apium graveolens*	Apiaceae	1050	22	2x	Herb	Biennial	Temperate
Papaya	*Carica papaya*[a]	Caricaceae	367	18	2x	Tree	Perennial	Tropical
Saffron	*Crocus sativus*	Iridaceae	5770	24	3x	Herb	Perennial	Temperate to mediterranean
Cotton	*Gossypium hirsutum*	Malvaceae	2347	52	4x	Woody shrub	Annual	Warm temperate
Poplar	*Populus trichocarpa*[a]	Salicaceae	550	38	2x	Tree	Perennial	Temperate
Human	*Homo sapiens*[a]	Hominidae	3200	46	2x		Perennial	

The crops have been selected intensively by farmers at both the level of choice of species (including new species or hybrids hardly known outside agriculture) and for characteristics including harvestable yield and propagation from diversity within each species. However, few common features related to genome size, chromosome number, and ploidy emerge from the table.

[a]Genome sequenced, public and published by 2011.

[b]Five-a-day: fruit or vegetable with range of properties making it a healthy food, not normally eaten for energy or protein.

[c]1C is the unreplicated haploid DNA content; most DNA contents from angiosperm genome size database, Bennett and Leitch (2011); some from sequencing consortia.

thousand years from the approximately 400,000 plant species. Cereals dominate the list of production figures, and it is clear that the exploitation of the seed has been very important. As a high-energy, harvestable, desiccated, storable, transportable, and robust part of the plant, people have been able to exploit the requirements of the plant for propagation.

Comparative analysis is extremely informative in most of biology. All of the domesticated species in Table 1.1 share at least some key characteristics related to domestication and selection: an imbalance of parts compared to the wild forms with the harvested part being larger; selection against the dispersal mechanisms most common in the wild (including shattering of pods or inflorescences, dropping of fruits, continuous fruiting, and delayed germination of seeds); ability to establish quickly in single-species stands; or reduction in bitter or other compounds in the harvested part.

However, the data in Table 1.1 suggest, perhaps surprisingly, few features of large-scale genome organization that have evidence of selection (Heslop-Harrison and Schwarzacher, 2011), despite the intensive selection of genic characteristics as discussed previously. Of the top three cereals, rice has a very small genome and wheat a very large genome; wheat is a hexaploid, rice is diploid, and maize is an ancient tetraploid. Related to selection, it is also notable that the breeding systems or propagation methods (Dwivedi et al., 2010; Charlesworth, 2006), affecting heterozygosity and gene allele population genetics in agriculture, are diverse. For example, among the Poaceae, maize is out-crossing, rye is self-incompatible, and sugarcane is vegetatively propagated, while other major cereals are self-fertile. Thus it seems there are no "rules" about genome structure — size, number of chromosomes, or ploidy — for plants selected as crops.

Superdomestication

Breeding of new plant varieties requires genetic variation. This can come from wild collections of germplasm (see Heslop-Harrison, 2002), where extensive seed or plant collections are available for most major crops with allelic variation present in most genes. Many genes with the same function are present in different species, and transgenic approaches mean that genes can be transferred from one species to another. Individual genes that are desirable in a crop can be transferred between species: the gene making the Cry toxin from *Bacillus thuringiensis* giving resistance to many lepidopteran pests in Bt crops is not found in any plants but has been transferred from the bacterium. It is also possible to engineer entire biosynthetic pathways that are missing in one species and desirable for cultivation (e.g., "golden rice" includes the pathway for beta-carotene synthesis in the endosperm, Ye et al., 2000; or to alter oil properties, Damude and Kinney, 2008). New mutations can also be identified as beneficial for crop plants; some of the first genetic changes in domestication were selected by farmers from new mutations. Radiation or chemical mutagenesis has also been helpful in the generation of new genetic variation, and the FAO/IAEA mutant variety database (2010) shows that more than 3000 plant mutant cultivars have been released

Predominant breeding system	Propagation	Parts used	Commodity	Nutritional use	World production 2009 (Million tonnes)	Storage
Self-fertile	Seed	Petiole, root or seed	Vegetable	Dietary fiber, vitamins, spice[b]		Fresh, weeks
cross-pollinating, self-pollinating or parthenocarpic		Fruit	Fruit	Dietary fiber, vitamins[b]	10	Week
Sterile	Bulb	Stigma	Spice	Flavoring		One year
Self-pollinating, but out-crossing possible	Seed	Seed	Fiber/textiles		64	Decades
Out-crossing/dioecious	Seed	Trunk	Timber/fiber			Decades
Dioecious				Eats and uses the rest		Decades

commercially worldwide by 2010. The range of characteristics covers nearly all breeding traits and has proved useful for correcting weaknesses in existing varieties or generating new characters where there is no accessible variation in germplasm.

We can expect that synthetic gene construction, random or site-directed mutagenesis, outside the plant cell may increase further availability of gene alleles for specific crop requirements. The current use of green fluorescent proteins (GFP) in plants may provide an indication of the power of this method. The GFP gene used in plant research as a marker for gene expression is modified from that in its source, the jellyfish *Aequorea victoria*, to make it more stable in plants, and several modifications to the coding sequence give variants with different colors and much increased brightness (Chiu et al., 1996). Better understanding of the genetic pathways involved in crops through systems biology (e.g., Kim et al., 2008, 2010) will also be valuable for identifying improvement targets.

With the understanding of both the genetics and the genomics of crop species, we are now able to develop new crop genotypes incorporating designed characteristics (Vaughan et al., 2007). Farmers will be able to deliver appropriate crops to a growing population by exploitation of appropriate technology and use of the gene pool (Tanksley and McCouch, 1997) — the range of genes present in organisms — and perhaps beyond through synthetic biology approaches. Biotechnology and understanding the behavior of the plant genome provides a range of tools and options that allow crop "superdomestication" (the planning of requirements of new characteristics in our crops).

In most plants, conventional crossing programs have followed the paradigm of intercrossing pairs of optimum varieties and then selecting progeny, following inbreeding for several generations, that performed better than either parent. This is summed up in the mantra of "cross the best with the best and hope for the best." In the twenty-first century, crop improvement is accelerating through the use of genetic maps and DNA markers to identify useful variant alleles of genes, to plan recombination between desirable traits, to combine different resistance genes, and accelerate selection, particularly for quantitative traits. Plant breeding is an increasingly targeted and quantitative process.

An important meta-study by van de Wouw et al. (2010) addressed whether there is a continuing reduction in genetic diversity in crop species. Jarvis and Hodgkin (1999) recognized hybridization with undomesticated lines in many species, thus increasing the diversity in the variation available to plant breeders. Analyzing a large amount of research reported in many papers, with a range of cereal (e.g., Huang et al., 2007), leguminous, and other crops, van de Wouw et al. (2010) showed that in the last century there has been no overall decline in genetic diversity in varieties released over each decade, suggesting that introduction of new germplasm has kept pace with the loss of diversity through inbreeding.

This volume cannot be divorced from social, economic, and political areas, not least because research can foresee future challenges or problems, and can indicate options for their solution. Farming, whether for food, fuel, or fibers, never assists biodiversity, uses water, leads to erosion, uses crop protection chemicals, and uses fertilizers. Many of the most pressing problems of mankind are related to plants and the environment, whether for health, food security, or response to climate change. Based on socioeconomic factors, including changes in national and global trade patterns, recognition of requirements for sustainability, nutritional and health needs, and developing crops suitable for changed climates, targets can be set for new crop varieties and occasionally introduction of new crops. Appropriate technologies can then be applied to deliver solutions.

Acknowledgments

We thank IAEA-Coordinated Research Project "Molecular Tools for Quality Improvement in Vegetatively Propagated Crops Including Banana and Cassava," the Generation Challenge programme, and the EU Agri Gen Res project 018 Crocusbank for supporting a portion of our work. We are grateful to our many collaborators and fellows who have worked in our laboratory.

References

Angiosperm Phylogeny Group, An update of the Angiosperm Phylogeny Group classification for the orders and families of flowering plants: APG III. *Botanical Journal of the Linnean Society*, *161*(2), 105–121.

Anhalt, U. C. M., Heslop-Harrison, J. S., Byrne, S., Guillard, A., & Barth, S. (2008). Segregation distortion in *Lolium*: Evidence for genetic effects. *Theoretical Applied Genetics*, *117*, 297–306. doi: 10.1007/s00122-008-0774-7

Anhalt, U. C. M., Heslop-Harrison, J. S., Piepho, H. P., Byrne, S., & Barth, S. (2009). Quantitative trait loci mapping for biomass yield traits in a *Lolium* inbred line derived F2 population. *Euphytica*, *170*, 99–107. doi: 10.1007/s10681-009-9957-9

Antonovics, J., Bradshaw, A. D., & Turner, R. G. (1971). Heavy metal tolerance in plants. In J. B. Cragg (Ed.), *Advances in ecological research* (pp. 2–86). London: Academic Press.

Arabidopsis Genome Initiative, Analysis of the genome sequence of the flowering plant *Arabidopsis thaliana*. *Nature*, *408*(6814), 796–815. Available from: <http://dx.doi.org/10.1038/35048692/>

Argout, X., Salse, J., Aury, J. M., Guiltinan, M. J., Droc, G., & Gouzy, J., et al. (2011, February). The genome of *Theobroma cacao*. *Nature Genetics*, *43*(2), 101–108. Available from: <http://dx.doi.org/10.1038/ng.736/>

Bailey, J. P., Bímová, K., & Mandák, B. (2007). The potential role of polyploidy and hybridisation in the further evolution of the highly invasive *Fallopia taxa* in Europe. *Ecological Research*, *22*, 920–928.

Bassham, James A., & Calvin, M. (1955). *Photosynthesis*. Lawrence Berkeley National Laboratory. LBNL Paper UCRL-2853. Retrieved from: <http://escholarship.org/uc/item/0j6008b4/>

Buckler, E. S., IV, Thornsberry, J. M., & Kresovich, S. (2001, June). Molecular diversity, structure and domestication of grasses. *Genetical Research*, *77*(3), 213–218. Available from: <http://view.ncbi.nlm.nih.gov/pubmed/11486504/>

Caldwell, K. S., Dvorak, J., Lagudah, E. S., Akhunov, E., Luo, M. C. C., & Wolters, P., et al. (2004, June). Sequence polymorphism in polyploid wheat and their D-genome diploid ancestor. *Genetics*, *167*(2), 941–947. Available from: <http://dx.doi.org/10.1534/genetics.103.016303/>

Charlesworth, D. (2006, September 5). Evolution of plant breeding systems. *Current Biology*, *16*, R726–R735. doi: 10.1016/j.cub.2006.07.068

Chiu, W. I., Niwa, Y, Zeng, W., Hirano, T., Kobayashi, H., & Sheen, J. (1996, March). Engineered GFP as a vital reporter in plants. *Current Biology*, *6*(3), 325–330. Available from: <http://www.cell.com/current-biology/abstract/S0960-9822(02)00483-9/>

Chory, J., Ecker, J. R., Briggs, S., Caboche, M., Coruzzi, G. M., & Cook, D., et al. (2000). National Science Foundation-sponsored workshop report: "The 2010 project functional genomics and the virtual plant. A blueprint for understanding how plants are built and how to improve them". *Plant Physiology*, *123*, 423–426. doi:10.1104/pp.123.2.423

Cifuentes, M., Eber, F., Lucas, M. O., Lode, M., Chevre, A. M., & Jenczewski, E. (2010, July). Repeated polyploidy drove different levels of crossover suppression between homoeologous chromosomes in *Brassica napus* allohaploids. *The Plant Cell*, *22*(7), 2265–2276. Available from: <http://dx.doi.org/10.1105/tpc.109.072991/>

Clark, R. M., Linton, E., Messing, J., & Doebley, J. F. (2004, January). Pattern of diversity in the genomic region near the maize domestication gene *tb1*. *Proceedings of the National Academy of Sciences of the United States of America*, *101*(3), 700–707. Available from: <http://dx.doi.org/10.1073/pnas.2237049100/>

Clottes, J. 2010. Chauvet Cave (ca. 30,000 B.C.). (2000). In *Heilbrunn Timeline of Art History*. New York: The Metropolitan Museum of Art. <http://www.metmuseum.org/toah/hd/chav/hd_chav.htm/> Accessed 09.10.

Cong, B., Liu, J., & Tanksley, S. D. (2002, October). Natural alleles at a tomato fruit size quantitative trait locus differ by heterochronic regulatory mutations. *Proceedings of the National Academy of Sciences of the United States of America*, *99*(21), 13606–13611. Available from: <http://dx.doi.org/10.1073/pnas.172520999/>

Crawley, M. J., & Brown, S. L. (1995). Seed limitation and the dynamics of feral Oilseed Rape on the M25 motorway. *Proceedings of the Royal Society of London Series B: Biological Sciences*, *259*(1354), 49–54. Available from: <http://dx.doi.org/10.1098/rspb.1995.0008/>

Damude, H. G., & Kinney, A. J. (2008). Enhancing plant seed oils for human nutrition. *Plant Physiology*, *147*, 962–968. doi:10.1104/pp.108.121681

De Langhe, E., Hřibová, E., Carpentier, S., Doležel, J., & Swennen, R. (2010, December). Did backcrossing contribute to the origin of hybrid edible bananas? *Annals of Botany*, *106*(6), 849–857. Available from: <http://dx.doi.org/10.1093/aob/mcq187/>

Doebley, J. (2004). The genetics of maize evolution. *Annual Review of Genetics*, *38*(1), 37–59. Available from: <http://dx.doi.org/10.1146/annurev.genet.38.072902.092425/>

Doebley, J. F., Gaut, B. S., & Smith, B. D. (2006, December). The molecular genetics of crop domestication. *Cell*, *127*(7), 1309–1321. Available from: <http://dx.doi.org/10.1016/j.cell.2006.12.006/>

Doust, A. (2007, October). Architectural evolution and its implications for domestication in grasses. *Annals of Botany*, *100*(5), 941–950. Available from: <http://dx.doi.org/10.1093/aob/mcm040/>

Dubcovsky, J., & Dvorak, J. (2007, June). Genome plasticity a key factor in the success of polyploid wheat under domestication. *Science*, *316*(5833), 1862–1866. Available from: <http://dx.doi.org/10.1126/science.1143986/>

Dwivedi, S., Perotti, E., Upadhyaya, H., & Ortiz, R. (2010, December). Sexual and apomictic plant reproduction in the genomics era: Exploring the mechanisms potentially useful in crop plants. *Sexual Plant Reproduction*, *23*(4), 265–279. Available from: <http://dx.doi.org/10.1007/s00497-010-0144-x/>

FAOstat (2010). *Production data relating to food and agriculture*. <http://faostat.fao.org/>

FAO/IAEA mutant variety database (2010). <http://mvgs.iaea.org/aboutMutantVarieties.aspx/>

French-Italian Public Consortium for Grapevine Genome Characterization. The grapevine genome sequence suggests ancestral hexaploidization in major angiosperm phyla. *Nature*, *449*, 463–468. doi:10.1038/nature06148

Gailing, O., Vornam, B., Leinemann, L., & Finkeldey, R. (2009, December). Genetic and genomic approaches to assess adaptive genetic variation in plants: Forest trees as a model. *Physiologia Plantarum*, *137*(4), 509–519.

Gross, B. L., & Olsen, K. M. (2010, September). Genetic perspectives on crop domestication. *Trends in Plant Science*, *15*(9), 529–537. Available from: <http://dx.doi.org/10.1016/j.tplants.2010.05.008/>

Guo, J., Wang, Y., Song, C., Zhou, J., Qiu, L., & Huang, H., et al. (2010, September). A single origin and moderate bottleneck during domestication of soybean (*Glycine max*): Implications from microsatellites and nucleotide sequences. *Annals of Botany*, *106*(3), 505–514. Available from: <http://dx.doi.org/10.1093/aob/mcq125/>

Gurevitch, J., & Padilla, D. (2004, September). Are invasive species a major cause of extinctions? *Trends in Ecology & Evolution*, *19*(9), 470–474. Available from: <http://dx.doi.org/10.1016/j.tree.2004.07.005/>

Hammer, K. (1984). Das Domestikations-syndrom. *Kulturpflanze*, *32*, 11–34.

Harlan, J., & de Wet, J. (1965, January). Some thoughts about weeds. *Economic Botany*, *19*(1), 16–24. Available from: <http://dx.doi.org/10.1007/BF02971181/>

Heaton, E. A., Dohleman, F. G., & Long, S. P. (2008). Meeting US biofuel goals with less land: The potential of *Miscanthus*. *Global Change Biology*, *14*, 2000–2014.

Heslop-Harrison, J. S. (2002). Exploiting novel germplasm. *Australian Journal of Agricultural Research*, *53*(8), 873–879.

Heslop-Harrison, J. S. (2004). Biotechnology of fruit and nut crops: Introduction. In R. E. Litz (Ed.), *Biotechnology of Fruit and Nut Crops* (pp. xix–xxiv). CABI.

Heslop-Harrison, J. S. (2010). Genes in evolution: The control of diversity and speciation. *Annals of Botany*, *106*(3), 437–438. Available from: <http://dx.doi.org/10.1093/aob/mcq168/>

Heslop-Harrison, J. S., & Schwarzacher, T. (2007). Domestication, genomics and the future for banana. *Annals of Botany*, *100*(5), 1073–1084. doi: 10.1093/aob/mcm191

Heslop-Harrison, J. S., & Schwarzacher, T. (2011). Organisation of the plant genome in chromosomes. *The Plant Journal*, *66*(1), 18–33. Available from: <http://dx.doi.org/10.1111/j.1365-313X.2011.04544.x/>

Heyduck, R. F., Baltensperger, D. D., Nelson, L. A., & Graybosch, R. A. (2008). Yield and agronomic traits of waxy proso in the central Great Plains. *Crop Science*, *48*(2), 741–748. Available from: <http://dx.doi.org/10.2135/cropsci2007.02.0081/>

Huang, X. Q., Wolf, M., Ganal, M. W., Orford, S., Koebner, R. M. D., & Röder, M. S. (2007). Did modern plant breeding lead to genetic erosion in european winter wheat varieties? *Crop Science*, *47*(1), 343–349. Available from: <http://dx.doi.org/10.2135/cropsci2006.04.0261/>

Hunt, H. V., Vander Linden, M., Liu, X., Motuzaite-Matuzeviciute, G., Colleridge, S., & Jones, M. K. (2008). Millets across Eurasia: Chronology and context of early records of the genera *Panicum* and *Setaria* from archaeological sites in the Old World. *Vegetation History and Archaeobotany*, *17*, 5–18.

Hyten, D. L., Song, Q., Zhu, Y., Choi, I. Y., Nelson, R. L., & Costa, J. M., et al. (2006 November). Impacts of genetic bottlenecks on soybean genome diversity. *Proceedings of the National Academy of Sciences*, *103*(45), 16666–16671. Available from: <http://dx.doi.org/10.1073/pnas.0604379103/>

Innan, H., & Kim, Y. (2008 July). Detecting local adaptation using the joint sampling of polymorphism data in the parental and derived populations. *Genetics*, *179*(3), 1713–1720. Available from: <http://dx.doi.org/10.1534/genetics.108.086835/>

Janick, J. (2005). The origin of fruits, fruit growing, and fruit breeding. *Plant Breeding Reviews*, *25*, 255–320.

Jarvis, D. I., & Hodgkin, T. (1999). Wild relatives and crop cultivars: Detecting natural introgression and farmer selection of new genetic combinations in agroecosystems. *Molecular Ecology*, *8*, S159–S173.

Kell, D. B. (2011). Breeding crop plants with deep roots: Their role in sustainable carbon, nutrient and water sequestration. *Annals of Botany*, Sept. in press. <http://dx.doi.org/10.1093/aob/mcr175/>

Kim, J., Kim, T. -G., Jung, S. H., Kim, J. -R., Park, T., & Heslop-Harrison, P., et al. (2008). Evolutionary design principles of modules that control cellular differentiation: Consequences for hysteresis and multistationarity. *Bioinformatics*, *24*(13), 1516–1522. doi: 10.1093/bioinformatics/btn229

Kim, T. H., Kim, J., Heslop-Harrison, P., & Cho, K. H. (2010). Evolutionary design principles and functional characteristics based on kingdom-specific network motifs. *Bioinformatics* <http://dx.doi.org/10.1093/bioinformatics/btq633/>

King D., Heslop-Harrison J. S., & 23 other members of the GM Science Review Panel. (2003). GM science review: An open review of the science relevant to GM crops and food based on the interests and concerns of the public. <www.gmsciencedebate.org.uk/>. 296pp.

King D., Heslop-Harrison J. S., & 25 other members of the GM Science Review Panel. (2004). GM science review: Second Report. An open review of the science relevant to GM crops and food based on the interests and concerns of the public. Full text from: <http://gmsciencedebate.org.uk/>. 116pp.

Koornneef, M., van Eden, J., Hanhart, C. J., Stam, P., Braaksma, F. J., & Feenstra, W. J. (1983, July). Linkage map of *Arabidopsis thaliana*. *Journal of Heredity*, *74*(4), 265–272. Available from: <http://jhered.oxfordjournals.org/content/74/4/265.abstract/>

Martin, C., Ellis, N., & Rook, F. (2010, October). Do transcription factors play special roles in adaptive variation? *Plant Physiology*, *154*(2), 506–511. Available from: <http://dx.doi.org/10.1104/pp.110.161331/>

McClintock, B., Kato, T. A., & Blumenschein, A. (1981). Chromosome constitution of races of maize. Its significance in the interpretation of relationships between races and varieties in the Americas. Chapingo, Mexico: Colegio de Postgraduados.

Millenium Ecosystem Assessment. (2010). Guide to the Millennium Assessment Reports. <http://www.maweb.org/en/Index.aspx/>

Molnár, I., Cifuentes, M., Schneider, A., Benavente, E., & Molnar-Lang, M. (2011). Association between simple sequence repeat-rich chromosome regions and intergenomic translocation breakpoints in natural populations of allopolyploid wild wheats. *Annals of Botany*, *107*(1), 65–76. <http://dx.doi.org/10.1093/aob/mcq215/>

Moose, S. P., & Mumm, R. H. (2008). Molecular plant breeding as the foundation for 21st century crop improvement. *Plant Physiology*, *147*, 969–977. doi: 10.1104/pp.108.118232

Neale, D. B. (2007, December). Genomics to tree breeding and forest health. *Current Opinion in Genetics & Development*, *17*(6), 539–544.

Nielsen, R., Williamson, S., Kim, Y., Hubisz, M. J., Clark, A. G., & Bustamante, C. (2005, November). Genomic scans for selective sweeps using SNP data. *Genome Research*, *15*(11), 1566–1575. Available from: <http://dx.doi.org/10.1101/gr.4252305/>

Papa, R., Bellucci, E., Rossi, M., Leonardi, S., Rau, D., & Gepts, P., et al. (2007, October). Tagging the signatures of domestication in common bean (*Phaseolus vulgaris*) by means of pooled DNA samples. *Annals of Botany*, *100*(5), 1039–1051. Available from: <http://dx.doi.org/10.1093/aob/mcm151/>

Paris, H. S., Daunay, M. C., & Janick, J. (2009, June). The Cucurbitaceae and Solanaceae illustrated in medieval manuscripts known as the Tacuinum Sanitatis. *Annals of Botany*, *103*(8), 1187–1205. Available from: <http://dx.doi.org/10.1093/aob/mcp055/>

Paterson, A. H., Lin, Y. R., Li, Z., Schertz, K. F., Doebley, J. F., & Pinson, S. R. M., et al. (1995). Convergent domestication of cereal crops by independent mutations. *Science*, *269*, 1714–1718.

Richardson, J. A. (1975). Physical problems of growing plants on colliery waste. M. J. Chadwick & G. T. Goodman (Eds.), *The Ecology of Resource Degradation and Renewal. The Fifteenth Symposium of the British Ecological Society*, Leeds, 10-12 July 1973 (pp. 275-286). Oxford: Blackwell Scientific Publications.

Rieseberg, L. H., & Blackman, B. K. (2010, September). Speciation genes in plants. *Annals of Botany*, *106*(3), 439–455. Available from: <http://dx.doi.org/10.1093/aob/mcq126/>

Rubin, E. M. (2008, August 14). Genomics of cellulosic biofuels. *Nature*, *454*, 841–845. doi: 10.1038/nature07190

Saeidi, H., Rahiminejad, M. R., & Heslop-Harrison, J. S. (2008). Retroelement insertional polymorphisms, diversity and phylogeography within diploid, D-Genome *Aegilops tauschii* (Triticeae, Poaceae) sub-taxa in Iran. *Annals of Botany*, *101*(6), 855–861. doi:10.1093/aob/mcn042

Salamini, F., Ozkan, H., Brandolini, A., Scafer-Pregl, R., & Martin, W. (2002). Genetics and geography of wild cereal domestication in the near East. *Nature Reviews Genetics*, *3*, 429.

Sasaki, T., Matsumoto, T., Yamamoto, K., Sakata, K., Baba, T., & Katayose, Y., et al. (2002, November). The genome sequence and structure of rice chromosome 1. *Nature*, *420*(6913), 312–316. Available from: <http://dx.doi.org/10.1038/nature01184/>

Schulze, J., Stoll, P., Widmer, A., & Erhardt, A. (2011). Searching for gene flow from cultivated to wild strawberries in Central Europe. *Annals of Botany*, *107*(4), 699–707. Available from: <http://dx.doi.org/10.1093/aob/mcr018/>

Somerville, C. R., Youngs, H., Taylor, C., Davis, S. C., & Long, S. P. (2010, August 13). Feedstocks for lignocellulosic biofuels. *Science*, 790–792. doi: 10.1126/science.1189268

Tang, T., Lu, J., Huang, J., He, J., McCouch, S. R., & Shen, Y., et al. (2006, November). Genomic variation in rice: Genesis of highly polymorphic linkage blocks during domestication. *PLoS Genetics*, *2*(11), e199. Available from: <http://dx.doi.org/10.1371/journal.pgen.0020199/>

Tanksley, S. D., & McCouch, S. R. (1997, August). Seed banks and molecular maps: Unlocking genetic potential from the wild. *Science*, *277*(5329), 1063–1066. Available from: <http://dx.doi.org/10.1126/science.277.5329.1063/>

Tian, Z., Wang, X., Lee, R., Li, Y., Specht, J. E., & Nelson, R. L., et al. (2010, May). Artificial selection for determinate growth habit in soybean. *Proceedings of the National Academy of Sciences*, *107*(19), 8563–8568. Available from: <http://dx.doi.org/10.1073/pnas.1000088107/>

Varshney, R. K., Hoisington, D. A., & Tyagi, A. K. (2006). Advances in cereal genomics and applications in crop breeding. *Trends in Biotechnology*, *24*(11), 490–499. doi: 10.1016/j.tibtech.2006.08.006

Vaughan, D. A., Balázs, E., & Heslop-Harrison, J. S. (2007). From crop domestication to super-domestication. *Annals of Botany*, *100*(5), 893–901. doi: 10.1093/aob/mcm224

Vigouroux, Y., Mitchell, S., Matsuoka, Y., Hamblin, M., Kresovich, S., & Smith, J. S. C., et al. (2005, March). An analysis of genetic diversity across the maize genome using microsatellites. *Genetics*, *169*(3), 1617–1630. Available from: <http://dx.doi.org/10.1534/genetics.104.032086/>

Wang, H., Nussbaum-Wagler, T., Li, B., Zhao, Q., Vigouroux, Y., & Faller, M., et al. (2005, August). The origin of the naked grains of maize. *Nature*, *436*(7051), 714–719. Available from: <http://dx.doi.org/10.1038/nature03863/>

Wang, J. X., Gao, T. G., & Knapp, S. (2008, December). Ancient Chinese literature reveals pathways of eggplant domestication. *Annals of Botany*, *102*(6), 891–897. Available from: <http://dx.doi.org/10.1093/aob/mcn179/>

Weeden, N. F. (2007, October). Genetic changes accompanying the domestication of *pisum sativum*: Is there a common genetic basis to the "Domestication Syndrome" for Legumes? *Annals of Botany*, *100*(5), 1017–1025. Available from: <http://dx.doi.org/10.1093/aob/mcm122/>

Wouw, M. van de, van Hintum, T., Kik, C., van Treuren, R., & Visser, B. (2010, April). Genetic diversity trends in twentieth century crop cultivars: A meta analysis. *TAG Theoretical and Applied Genetics Theoretische und angxewandte Genetik*, *120*(6), 1241–1252. Available from: <http://dx.doi.org/10.1007/s00122-009-1252-6/>

Wright, S. I., Bi, I. V., Schroeder, S. G., Yamasaki, M., Doebley, J. F., & McMullen, M. D., et al. (2005, May). The effects of artificial selection on the maize genome. *Science*, *308*(5726), 1310–1314. Available from: <http://dx.doi.org/10.1126/science.1107891/>

Yamasaki, M., Tenaillon, M. I., Vroh, Bi I, Schroeder, S. G., Sanchez-Villeda, H., & Doebley, J. F., et al. (2005, November). A large-scale screen for artificial selection in maize identifies candidate agronomic loci for domestication and crop improvement. *The Plant Cell*, *17*(11), 2859–2872. Available from: <http://dx.doi.org/10.1105/tpc.105.037242/>

Yamasaki, M., Wright, S. I., & McMullen, M. D. (2007, October). Genomic screening for artificial selection during domestication and improvement in maize. *Annals of Botany*, *100*(5), 967–973. Available from: <http://dx.doi.org/10.1093/aob/mcm173/>

Yano, M. (2001). Genetic and molecular dissection of naturally occurring variation. *Current Opinion in Plant Biology*, *4*, 130–135.

Ye, X., Al-Babili, S., Klöti, A., Zhang, J., Lucca, P., & Beyer, P., et al. (2000). Engineering the provitamin A (beta-carotene) biosynthetic pathway into (carotenoid-free) rice endosperm. *Science*, *287*(5451), 303–305. doi:10.1126/science.287.5451.303

Zeder, M. A. (2006). *Documenting domestication: New genetic and archaeological paradigms*. University of California Press.

Zhang, Y., Wang, J., Zhang, X., Chen, J. Q., Tian, D., & Yang, S. (2009, April). Genetic signature of rice domestication shown by a variety of genes. *Journal of Molecular Evolution*, *68*(4), 393–402. Available from: <http://dx.doi.org/10.1007/s00239-009-9217-6/>

Zhu, Q., Zheng, X., Luo, J., Gaut, B. S., & Ge, S. (2007, March). Multilocus analysis of nucleotide variation of *Oryza sativa* and its wild relatives: Severe bottleneck during domestication of rice. *Molecular Biology and Evolution*, *24*(3), 875–888. Available from: <http://dx.doi.org/10.1093/molbev/msm005/>

The scope of things to come: New paradigms in biotechnology

2

Maheshi Dassanayake[1] Dong-Ha Oh[1] Dae-Jin Yun[2]
Ray A. Bressan[2,3,4] John M. Cheeseman[1] J. Hans Bohnert[1,2,5]

[1]*Department of Plant Biology, University of Illinois at Urbana Champaign, Urbana, Illinois,* [2]*Division of Applied Life Science (BK21 program) and Environmental Biotechnology, National Core Research Center, Graduate School of Gyeongsang National University, Korea,* [3]*Department of Horticulture and Landscape Architecture, Purdue University, West Lafayette, Indiana,* [4]*Center for Plant Stress Genomics, King Abdallah University of Science and Technology, Thuwal, Saudi Arabia,* [5]*Department of Crop Sciences, University of Illinois at Urbana Champaign, Urbana, Illinois*

TABLE OF CONTENTS

Introduction

For more than 8000 years after the transition from hunter–gatherer societies to sessile, farming communities, the domestication and improvement of crops proceeded in an informal, yet purposeful way. This trajectory is obvious from archaeological excavations as well as from descriptions of agricultural practices left behind by early travelers (e.g., Columella, ~60 CE; Pliny the Elder, ~100CE). Although all cultures that developed agriculture participated in the process, particularly complete records were kept in China (Bray, 1986). Later records of the progress of crop improvement appeared in the Middle East during Islam's golden age, and were subsequently expanded by the monastic culture of medieval Europe (Postan, 1973; Watson, 1983; Astill and Langdon, 1997; Stone, 2005). New World cultures also contributed significantly. Their crops, among them maize, tomato, and potato, have since marched triumphantly across the world (Doebley et al., 2006). As in other fields of science, the Age of Enlightenment in the mid-seventeenth century accelerated agricultural innovation by initiating fundamental changes in the scientific thought process.

Fundamentally new insights, however, had to wait until the lucid outline of evolution by Darwin (1868), and the introduction of the gene concept by Mendel's studies on heritable quanta (Anonymous, 1951). Ultimately, these have inspired new experimental approaches, usually accompanied and supported by novel technologies, culminating in the age of molecular biology and molecular genetics.

The last two decades, however, have seen yet another sea change. Enabled initially by high-throughput sequencing technologies and advances in bioinformatics that seem quaint by now, a much altered view of the genetic structures that shape organisms has begun to emerge. The concept of a gene as an isolated unit of inheritance is rapidly being replaced by the idea of a genetic continuum or field along chromosome segments that encompasses a memory of the past. In such domains, units can change their character, expression, and function in a flexible, environmentally responsive, way. These concepts are now the focus of both theoretical considerations and experimental enquiry (Jablonka and Lamb, 2006; Pfluger and Wagner, 2007; Swiezewski et al., 2009).

In parallel, fostered by the ease with which primary sequence information can be acquired, biology is heading into a "post-modelian" era. No longer are ideas of how plants operate dependent on a few convenient but often inappropriate "model" systems. Now, each species, genotype, or ecotype can serve as a model.

Each one of the technological advances that broadened our horizon has brought us closer to a new, and equally massive,

DOI: 10.1016/B978-0-12-381466-1.00002-X

barrier — data overload. Desktop computers are no longer able to deal with datasets measured in terabytes or petabytes, or with memory usage in the hundreds of gigabyte ranges, and most biologists need computer science to contribute to computational solutions. Assuming that competent programmers exist to solve these (and there are certainly many who are trying) issues, the real goals still remain over the horizon — the synthesis of genetic and molecular information with phenotypic, physiological, and even ecological data through integrative and responsive databases (Rhee et al., 2003; Stein, 2008).

The progress of the last decade has brought us to a point at which knowledge-based biotechnology in the agricultural arena will enable unprecedented precision in the application of genetic knowledge for crop improvement. In paraphrasing a concept increasingly used in medical literature, plant biotechnology is poised to enter an era of "personalized breeding." Breeders have always selected combinations of alleles approaching optimal adaptation to a specific climate and local soil conditions, but integrated genomics can provide not only the how, but the why, when, and where with respect to which allele combinations should be put together.

This chapter will discuss new technologies in DNA sequence acquisition in monitoring the dynamic behavior of transcript populations on a global scale, recording protein composition and modification, and assembling metabolite profiles in unprecedented complexity and in plant response pathways to the physical and biotic environments. The advances, filtered and organized by bioinformatics, signify nothing less than the emergence of a new organismic biology with immense potential for biotechnology.

Progress Enabled by Next-Generation DNA Sequencing

We will briefly discuss how new technologies — primarily "next-generation" (next-gen) sequencing tools — have made possible the acquisition of genome sequence information for a rapidly increasing number of species and the recording of the dynamic pattern of gene expression in response to external manipulations (Wilhelm and Landry, 2009; Kircher and Kelso, 2010; Lu et al., 2010). Figure 2.1 and Table 2.1 provide a graphic and condensed view of the processes required for a complete description of the genetic makeup of an organism. Figure 2.1 introduces terms and organizes the flow of information for a complete picture, with more detailed

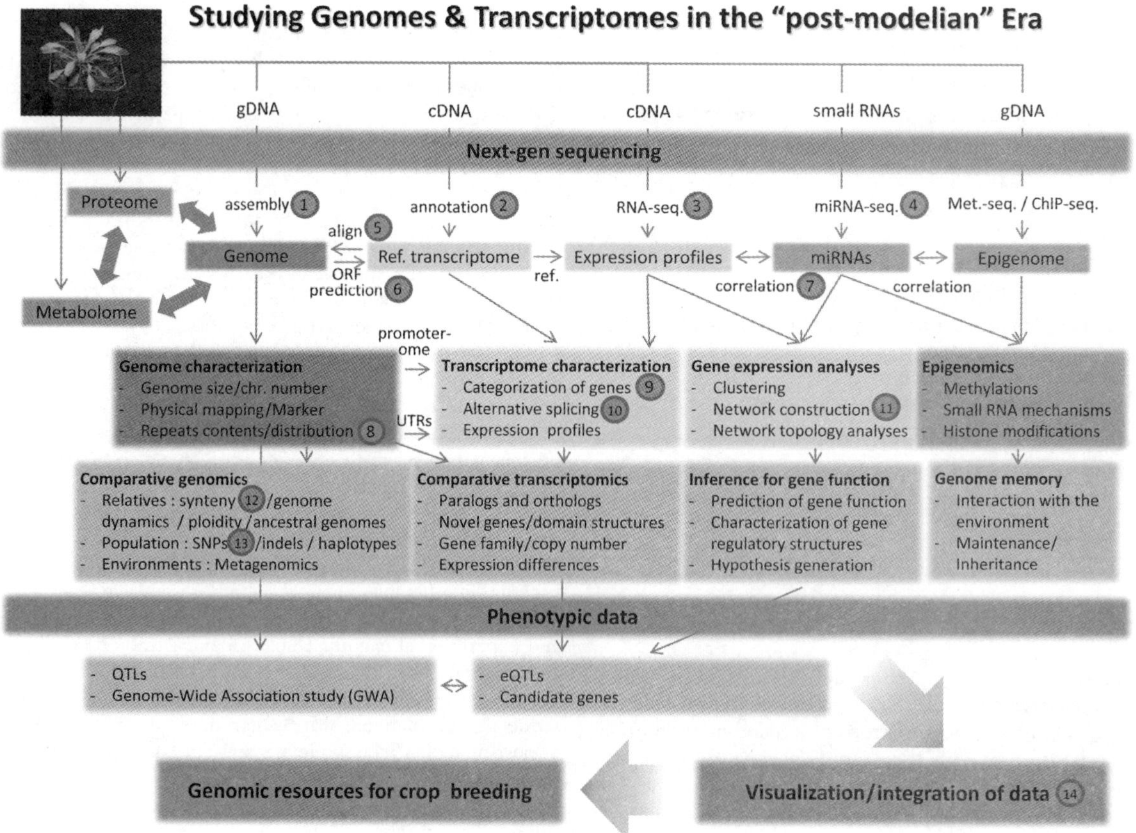

Figure 2.1 • Studying genomes and transcriptomes in the "post-modelian" era. Please see color plate section at the end of the book.

Table 2.1 A compilation of tools for next-gen DNA sequencing and analysis

1. Genome assembly

i. De novo

a. Short reads

1) Velvet — http://www.ebi.ac.uk/~zerbino/velvet/
2) ABySS — http://www.bcgsc.ca/platform/bioinfo/software/abyss/
3) SOAPdenovo — http://soap.genomics.org.cn/soapdenovo.html
4) Newbler — Roche/454
5) Contrail — http://sourceforge.net/apps/mediawiki/contrail-bio/index.php?title = Contrail

b. Sanger-type reads

1) Phred/Phrap — http://www.phrap.org/phredphrapconsed.html
2) CAP3 — http://deepc2.psi.iastate.edu/aat/cap/capdoc.html
3) Celera assembler — http://sourceforge.net/apps/mediawiki/wgs-assembler/index.php?title = Main_Page
4) Arachne — http://www.broadinstitute.org/crd/wiki/index.php/Arachne_Main_Page

ii. Reference aided

1) AMOScmp — http://sourceforge.net/apps/mediawiki/amos/index.php?title = AMOScmp
2) Bambus — http://sourceforge.net/apps/mediawiki/amos/index.php?title = Bambus

iii. Hybrid assembly

1) MIRA — http://www.chevreux.org/projects_mira.html
2) Minimus2 — http://sourceforge.net/apps/mediawiki/amos/index.php?title = Minimus2

2. Transcriptome annotation

1) BLAST — http://www.ncbi.nlm.nih.gov/staff/tao/URLAPI/netblast.html
2) BLAT (**B**last-**L**ike **A**lignment **T**ool) — http://genome.ucsc.edu/goldenPath/help/blatSpec.html

3. RNA-seq processing

1) Qseq by DNASTAR — http://www.dnastar.com/t-nextgen-qseq.aspx
2) Myrna — http://bowtie-bio.sourceforge.net/myrna/index.shtml
3) Cufflinks — http://cufflinks.cbcb.umd.edu/
4) Custom pipelines using bowtie for alignment of RNA-seq data
5) DEGseq — http://bioconductor.org/packages/2.6/bioc/html/DEGseq.html

4. miRNA identification

1) Rfam — http://rfam.sanger.ac.uk/
2) miRBase — http://www.mirbase.org/
3) eumiR — http://miracle.igib.res.in/eumir/
4) miRNAFinder — http://bioinfo3.noble.org/mirna/
5) MiRscan — http://genes.mit.edu/mirscan/index.html

5. Align transcriptome/EST to genome

1) GMAP — http://research-pub.gene.com/gmap/
2) Exonerate — http://www.genome.iastate.edu/bioinfo/resources/manuals/exonerate/
3) ssahaEST — http://www.sanger.ac.uk/resources/software/ssahaest/
4) EXALIN — http://blast.wustl.edu/exalin/
5) Spidey — http://www.ncbi.nlm.nih.gov/IEB/Research/Ostell/Spidey/

6. ORF prediction

1) FGENESH — http://linux1.softberry.com/berry.phtml
2) GeneMark — http://exon.biology.gatech.edu/
3) GENSCAN — http://genes.mit.edu/GENSCAN.html
4) ORF Finder — http://www.ncbi.nlm.nih.gov/gorf/gorf.html
5) Glimmer — www.cbcb.umd.edu/software/glimmer/
6) GenomeThreader — http://www.genomethreader.org

(*Continued*)

Table 2.1 (Continued)

7) Gismo — http://www.cebitec.uni-bielefeld.de/groups/brf/software/gismo/
8) mGene — www.fml.tuebingen.mpg.de/raetsch/suppl/mgene

7. Correlating miRNA and mRNA expression and target prediction

1) Custom pipelines
2) microRNA.org — http://www.microrna.org/microrna/home.do
3) miRU — http://bioinfo3.noble.org/miRNA/miRU.htm
4) RNAhybrid — http://bibiserv.techfak.uni-bielefeld.de/rnahybrid/

8. Repeat identification and distribution

1) RepeatMasker — http://www.repeatmasker.org/
2) SSRIT — http://www.gramene.org/db/markers/ssrtool
3) RECON — http://selab.janelia.org/recon.html
4) REPET — http://urgi.versailles.inra.fr/Tools/REPET

9. Categorization of gene function

1) GO — http://www.geneontology.org/
2) KOG — http://www.ncbi.nlm.nih.gov/COG/
3) KEGG — http://www.genome.jp/kegg/
4) Blast2GO — http://www.blast2go.org

10. Alternative splicing recognition

1) Tophat — http://tophat.cbcb.umd.edu
2) ASmodeler — http://genome.ewha.ac.kr/ECgene/ASmodeler/
3) AceView — http://www.ncbi.nlm.nih.gov/IEB/Research/Acembly/

11. Network

1) ATTED-II — http://atted.jp/
2) AraNet — http://www.functionalnet.org/aranet/about.html
3) GGM (Ma et al., 2007).

12. Viewing synteny

1) Circos — http://mkweb.bcgsc.ca/circos/
2) Mauve — http://gel.ahabs.wisc.edu/mauve/
3) 3 OSLay — http://www-ab.informatik.uni-tuebingen.de/software/oslay
4) Synbrowse — http://www.synbrowse.org/
5) Sybil — http://sybil.sourceforge.net/
6) SyMAP — http://www.symapdb.org/

13. Identification of SNPs

1) Crossbow — http://bowtie-bio.sourceforge.net/crossbow/index.shtml
2) SOAPsnp — http://soap.genomics.org.cn/soapsnp.html

14. Visualization and integration of data

1) gBrowse 2 — http://gmod.org/wiki/Ggb/
2) EagleView — http://bioinformatics.bc.edu/marthlab/EagleView
3) Hawkeye — http://sourceforge.net/apps/mediawiki/amos/index.php?title = Hawkeye
4) ABySS-Explorer — http://www.bcgsc.ca/platform/bioinfo/software/abyss-explorer/releases/1.0

This table does not contain an exhaustive list of programs and tools. Many tools have multiple and overlapping uses, whereas a particular tool may have been included under just one task. Programs are constantly evolving and new programs are introduced. Participating in an online discussion forum like http://seqanswers.com/ is suggested for an idea about current programs. Generally, for a given task, any one tool does not provide a complete answer; more often it requires a combination of tools, simple in-house scripts to customize the general output of the program used, and/or use of custom or publicly available pipelines to efficiently adjust programs for the specific project. Numbers for categories 1–14 are indicated in Figure 2.1.

explanations of the necessary activities in Table 2.1. Examples of the computational and bioinformatics-type tools required for analysis of the data are provided, although we wish to again point to the transitory nature of these tools.

Mapping of comprehensive, genome-wide, treatment-specific transcript profiles

For the last quarter of the twentieth century, the piecemeal acquisition and isolation of genes and transcripts, beginning with chloroplast functions and photosynthesis-specific molecular biochemistry, eventually birthed a molecular interpretation of developmental and physiological parameters of plant life. By the 1990s, it was becoming increasingly recognized that the random use of over- (CaMV35S- or actin promoter-driven) or under-expressed (antisense) coding regions was producing experimental results of questionable value (Thomas et al., 1995). While replacement of these approaches by condition-, cell-, tissue-, or developmental stage-specific promoters or the use of gene knockout mutations might improve specificity (Rivero et al., 2007), the majority of transgenic approaches at plant improvement have shown marginal, albeit reproducible, effects while confirmations through studies in the field have until recently been missing (Tarczynski et al., 1993; Apse et al., 1999; Nelson et al., 2007).

There are still few examples, notwithstanding the generation of crops that are temporarily transgenically protected against root-feeding organisms or herbicide action, where additions or removals of coding regions have had the biotechnological effects hoped for at the start of the experiments. Given this, any scientist would argue that the reasons for the shortcomings are largely centered in incomplete data. A major emphasis for the last 20 years has been overcoming this limitation. The pace of data acquisition, at least with respect to gene, genome, and transcriptome sequences, has been rapidly accelerating during the first decade of the twenty-first century.

Current next-gen sequencing

DNA sequencing based on Sanger chemistry (Sanger et al., 1977) was the prevalent technology from the late 1970s to the mid-2000s. Although the seminal contributions in genome sequencing of primary model organisms from bacteria to humans by Sanger sequencing are real, it is a laborious and expensive technique. Nevertheless, it evolved into an initial high-throughput sequencing phase with the introduction of fluorescently labeled terminators (Prober et al., 1987), parallelized high-resolution capillary-based polymer gel electrophoresis (Hunkapiller et al., 1991; Dolnik, 1999), and automated fluid handling robotic systems (Wada et al., 1989; Meldrum, 2000).

The next-gen sequencing began around 2006 with Roche 454 pyrosequencing. This new technique reduced sequencing costs and increased high-throughput yields by several orders of magnitude, and similarly increased the pool of investigators who could enter the arena. This 454 sequencing was the first technology other than Sanger to sequence and assemble de novo bacterial (Hiller et al., 2007; Tauch et al., 2008) and eukaryotic genomes (Wheeler et al., 2008) and transcriptomes (Vera et al., 2008; Maher et al., 2009).

Since its introduction, the 454 approach has continued to evolve rapidly, spurred on by competition from several other technologies and sequencing strategies (Bentley, 2006; Shendure and Ji, 2008). Two major alternatives include the Illumina/Solexa Illumina® Genome Analyzer, and the ABI SOLiD® sequencing platform. Although it will upset both aficionados and manufacturers, the major differences between the platforms are with respect to error types, sequence lengths, and total outputs. Sequencing errors in particular are likely to be a major concern for the near future. The major drawback of 454 sequencing, for example, is a high error rate associated with homopolymers. Because the length of the homopolymers are determined by a signal intensity that is not directly proportional to the number of bases added, the dominant error type in 454 sequencing is insertion–deletion rather than substitution (Wicker et al., 2006). However, currently, the Roche 454 platform generates the longest read lengths, several hundred bases, among second generation sequencing platforms. An improved platform — with less error, higher numbers of reads, and longer reads — should be available in the summer of 2010.

With the Illumina sequencing platform, in contrast, the dominant error type is substitution, and a major difference and limitation is read length, which is below one hundred. However, massive throughput can be gained with the current Illumina Genome Analyzer series; 2 × 100 bp reads up to 200 GB in total sequences are returned in eight days. This is sufficient to enable de novo sequencing of large genomes (Li et al., 2010). The ABI SOLiD® system also returns massive numbers of short sequences. To increase overall accuracy, it uses a two base encoding system that enables unique internal error checking. The result, however, is error rates that are not significantly lower than other second generation platforms (Harismendy et al., 2009).

Behold the third generation

Although second generation sequencing tools have led to an explosion of data of record-shattering proportions, their time is unquestionably limited. The third generation is in the wings. Third generation sequencing platforms involve completely different approaches compared to the earlier sequencing platforms. Generally they target real-time single molecule sequencing. At present the third generation sequencing technologies have shown proof of concept, some instruments are in a beta-testing loop, but further development is expected to improve their performance. There are at least two approaches close to being marketed. NanoPore® sequencing offers single-molecule detection via electrophoretically driven DNA (or RNA) molecules in solution through a membrane of nano-scale pores at rates ranging from micro- to milliseconds per molecule. Alternately, single molecule real-time (SMRT) DNA sequencing introduced by Pacific Biosciences (PacBio; Menlo Park, CA) involves the real-time monitoring of DNA polymerase activity. With either of these, it is to be expected that DNA sequencing rates exceeding a thousand bases per second can be achieved at very low costs enabling the target of "one genome per $1000" set by NIH in 2004 (Deamer and Akeson, 2000; Rhee and Burns, 2006; Bayley, 2006; Branton et al., 2008).

This century, however, is still young. Second generation sequencing will enjoy a period of dominance that will last less than 10 years, and third generation sequencing is unlikely to be the ultimate answer. We expect continued rapid advancement with competition based on improved speed, lower cost, and acceptable (very high) accuracy in single reads.

Modifications of sequencing tools and protocols that deal with the structure of the emerging epigenome will not be far behind. It is conceivable that we will be able to have simultaneous knowledge of the sequence and the methylations or other modifications in both the DNA sequence and the histone modification dynamics that indicate the epigenetic state. First results have been reported. Large-scale DNA methylation analysis via bisulfite sequencing of human fibroblasts and stem cells has revealed patterns in nuclear reprogramming (Deng et al., 2009). Short read sequences from non-model plants have also been used in comparative genomics without going through an assembly phase. Studies looking into transcription regulation via small RNAs are beginning to become widely explored with pyrosequencing technologies (Ha et al., 2009; Zhou et al., 2009; Fahlgren et al., 2009). This has also been applied to metagenomic ocean water column microbial populations that revealed unique small RNAs (Shi et al., 2009).

Instruments may be conceived that accomplish disentanglement of protein–DNA complexes in chromatin, which would then unravel the next level of complexity.

The Elephant in the Laboratory: Data Handling

An outstanding characteristic of genomics, in fact of all "omics" approaches, is that they produce immense amounts of data that require statistical and bioinformatics tools for handling (Table 2.1). Required are: (1) programs for sorting and assembling the raw data by specific criteria, (2) tools for visualizing and curating intermediate and final results and, finally, (3) ways for generating neutral hypotheses, which list, explain, and signify the important aspects of the omics approaches that result from the immense flood of data. Examples of the tools for raw data processing are sequence assembly pipelines for next-gen DNA sequencing tools. The relevant algorithms and programs are still being updated, often in short succession. They are indispensible factors enhancing the quality of the sequencing. In addition, basically every step of genomics required pipelines involving bioinformatics (Table 2.1). For the visualization, curation, and integration tools, one example is Gbrowse, which — with constant upgrades since 2002 (Stein et al., 2002; Donlin, 2009) — now includes functions not only in browsing and data retrieving, but also for multi-species comparisons.

The financial driver for sequencing technology has been medical applications, and bioinformatics has equally profited from the prospect of personalized medicine based on DNA sequences. The tools are most likely already available, but hidden in the Babylonian towers erected by companies, administrations, and the pet algorithms of individual researchers. To describe the many tools in detail is beyond the scope of this chapter, but an example may outline one of the problems that characterize the difficulties (Le Novere et al., 2009). This article suggests that the community accepts a graphical notation system to represent diagrams that delineate processes and pathways. Underlying is the idea that a graphical language must be developed. This language, comparable to Internet protocols, should be universal. The authors present a "Systems Biology Graphical Notation" language with defined symbols, whereas in the past inconsistency and ambiguity characterized notations of genes, pathways, and processes.

To truly harness the wealth of the data, simply to sort, categorize, and statistically buttress their degree of significance is not sufficient. Needed are mathematical tools that establish correlations or partial correlations that arrange genes, proteins, metabolites, and activities into networks in dependence on development or environmental influences. Examples of bioinformatics tools for hypothesis generation are found among the attempts to draw inferences from large sets of microarray results. Clustering and visualization have already revealed the structure of co-expressed genes and have been powerful tools when combined with ingenious ways of sampling (Gong et al., 2005; Brady et al., 2007; Ma and Bohnert, 2007; Dinneny et al., 2008) and the attempts lead to networks based on graphic theory, depicting either correlations or partial correlations based on the gene expression (Ma et al., 2007; Obayashi et al., 2009; Lee et al., 2010). The new hypothesis and prediction can be generated from the constructed network by using "guilt by association" of genes of unknown functions with known ones or by analyzing the topology of the network and identifying its modules (Mao et al., 2009; Bonnet et al., 2010; MacLean et al., 2010). Co-expression tools are integrated with data from proteomics and metabolomics data to infer the gene functions (Hirai et al., 2007; Popescu et al., 2009).

Future developments will include networks that follow timescales, for example, predictive networks. An example for such a network is with yeast, where changes over time are recorded in strains with defined deletions (Yip et al., 2010). Also, the integration of gene expression with protein:protein interactions, protein modifications, and metabolite dynamics will become more widespread. While most attention has been devoted to *Arabidopsis* (and rudimentary data for some crops), networks will be extended to other species (for reviews, see Fukushima et al., 2009; Moreno-Risueno et al., 2010).

From Sequences to Comparative Genomics

At present, the genomic sequences of nine angiosperm species have been published; the latest addition is the genome of the diploid monocot *Brachypodium distachyon*, which provides a glimpse of the grass ancestral genome (The International *Brachypodium* Initiative, 2010). Driven by the availability of second generation sequencing, however, new genome sequences are emerging at a very fast pace (Paterson et al., 2010). To our knowledge, in July 2010, at least eight more higher plant genomes are now in the publication pipeline or the sequences are through the final assembly pipeline (*Arabidopsis lyrata*, *Thellungiella salsuginea*, *T. parvula*, cacao, castor bean, soybean, and tomato). In addition, genome sequences for a number of algal species in phylogenetically widely separated lineages, from diatoms to brown algae and Chlorophytes, and the moss *Physcomitrella patens* are published (Merchant et al., 2007; Rensing et al., 2008). Within a year's time, this list will be significantly longer.

As limited as it is, with the current collection of completed genomes comparisons are underway to define the angiosperm ancestral gene complement. With this yardstick, we can then trace how different orders, families, clades, and individual species modified the basal set of genes and have enlarged specific characters. We can then identify orthologs and distinguish paralogous genes in families that precede the monocot/dicot divide, and those genetic changes that define angiosperm orders (Wicker and Keller, 2007; Sunkar and Jagadeeswaran, 2008). While it is not an absolute requirement for analyses of transcriptomes, genome sequences are more informative because they allow for easier evolutionary analysis due to the presence of introns. To correlate gene family number, structure, and expression characteristics across species with phenotypic markers in these species is considered a means for understanding physiological and developmental differences. For example, orthologous gene identification of *SOS1*, *HKT1*, *NHX1*, *NHX5*, *AVP1*, and *AVP2* in *Lophopyrum elongatum*, a wild relative of wheat, was established by comparisons to the *Arabidopsis* and rice genomes in an effort to improve salt tolerance in wheat (Mullan et al., 2007). With a complement estimated as ~12,000 ancestral genes (Sterck et al., 2007), it is already possible to address questions of speciation. Expansions and contractions of gene numbers and modifications of functions can be traced against the background of, often repeated, ploidy changes and elimination of genes for which many extant plants carry markers (Tang et al., 2008). Many of these will be found to contribute to biotechnologically relevant genes or identify novel pathway repertoires. Important in this respect will be lineage-specific genes. These will undoubtedly be instrumental to elucidate both distant relationships relevant to environmentally associated lifestyles, features, and processes that distinguish closely related species or genotypes. *Arabidopsis thaliana* accessions undoubtedly have contributed most to comparative genomics within ecotypes (Alonso-Blanco et al., 2005). Even in a compact genome like that of *A. thaliana*, Clark et al. (2007) showed that 9.4% of protein coding genes in the reference genome are altered or deleted by naturally existing polymorphisms in wild accessions. The significance of epigenetic variation among ecotypes of *Arabidopsis* has also been demonstrated in several studies (Vaughn et al., 2007; Zhai et al., 2008). Extending such studies and the identification of differences to crop species will make it possible to trace allele structures in breeding lines and to correlate them to quality traits. Quality trait loci (QTLs) and association mapping to explore genotypic variation in crop improvement are being complemented by next-gen sequencing (Alonso-Blanco et al., 2005; Myles et al., 2009).

Among these are questions about how chromosome structure influences gene expression by the distribution and dynamic behavior of chromatin condensation processes. Epigenomics has gained a central role in physiological processes from plant development to adaptations to different environmental conditions. Although significant recognition was given to DNA methylation, small non-coding RNAs, and histone post-translational modifications in epigenomics in the past decade, next-gen sequencing technologies have begun to open floodgates to large datasets enabling global mapping of epigenetic modification at single-base resolution. Further epigenetic analysis is explored with chromatin immunoprecipitation followed by sequencing (ChIP–seq). This allows genome-wide profiling of DNA-binding proteins, histone modifications, or nucleosomes (Park, 2009). We will finally be able to analyze in depth the functions associated with non-coding but highly conserved stretches of sequences in different clades (Inada et al., 2003). Genetic structures, represented by the type, number, and domain structure of genes in diverse pathways, can be analyzed for traces that guide major plant developmental decisions, such as vegetative growth; circadian rhythmicity; flowering time; and flowering, seed set, and maturation; and dormancy. Likewise, the gene complements and chromosomal complexities underlying innate immunity and pathogen or abiotic stress response pathways can be approached holistically rather than in single-gene by single-gene mutant fashion that has characterized the last several decades. Such a holistic view replaces attempts to find "silver bullets" that define a phenotype, generally without any knowledge of the context in which individual genes are embedded.

Transcriptome profiling

Beyond sequencing of genomes alone, next-gen sequencing has also accelerated our knowledge of plant transcriptomes and their dynamics (Bellin et al., 2009; Wall et al., 2009; Buggs et al., 2010; Dassanayake et al., 2010; Parchman et al., 2010). It is now possible to acquire an essentially complete picture of a plant's transcriptome by a single or very few sequence runs with cDNAs from RNAs isolated from plant tissues and developmental stages that have been subjected to a number of different conditions. Preliminary large-scale transcriptome sequencing efforts have already attempted to investigate physiological processes in non-model plant systems. For example, transcriptome characterization and expression was examined in the healthy and infected tissues of the American chestnut and blight-resistant Chinese chestnut to explore candidate genes and pathways important in biotic stress (Barakat et al., 2009). Because this is not limited to model plants, the twenty-first century will, we hope, see a transition from addressing important plant related problems using model systems inappropriate to the questions at hand, to addressing them by selecting new models that have already solved the problem through evolution. These include extremophiles, plants that are known for their exquisite adaptation to almost any of the most highly stressful environments on Earth (Bressan et al., 2001; Amtmann et al., 2005; Amtmann, 2009).

Next-gen-based recordings of plant transcriptomes provide other advantages. Until now we had to resort to the analysis of microarray hybridizations to shed light on regulated processes that affect multiple genes and pathways (Bohnert et al., 2006; Deyholos, 2010; Schnable et al., 2004). However, comparative chromosome-based genomics and complete transcriptome dynamics based on next-gen sequencing are now within the means of many labs. Such studies promise to truly elucidate the connections in genetic response networks in depth. In the past, transcript changes have been measured through array hybridizations in time series that have typically been based on a few points, whereas next-gen sequencing allows for time series to be more complete, more accurate, specific, and faster to assemble and to analyze once a genome sequence for a species exists. Faster assemblies based on constantly improving programs also contribute.

As a final example, it is now also possible to consider how genome information contributes to our understanding that comes from transcription-related behavior of gene families and duplicated genes with apparently identical functions. How and to what degree such duplicated genes, which we now will know for certain in a species, have undergone functional specialization in space, time, or biochemistry can be determined. The contribution of duplicated genes to morphological complexity and the phenotype can be studied by comparison in many species (Freeling and Thomas, 2006).

Broadening the Genomics Toolbox: Proteins and Metabolites

Proteomics advances

Progress in proteomics has been equally impressive. It is irrefutable that gazing at genomes and transcriptomes alone cannot explain a biological system at the cellular level. Certainly, the proteome is as dynamic as the transcriptome. The chemical complexity — the primary sequence of amino acids, secondary aspects of protein folding, and post-translational modifications — and the large dynamic range of proteins precludes single approaches to complete proteome analysis, even disregarding problems originating from secondary structure changes and complex tertiary or quaternary structures. To understand the scope of current methodologies in studying plant proteomes, see the reviews by Jorrin et al. (2007) and Carpentier et al. (2008a, b).

The effects of mechanisms that lead to post-transcriptional and -translational modifications are apparent only when a transcript's final cause, a protein or the product of a pathway, is known. In turn, the dynamic of proteins, protein modifications, and intra- and intercellular protein trafficking exert deterministic influences on the phenotype (Sullivan and Green, 1993; Boston et al., 1996; Zivy and de Vienne, 2000; Mazzucotelli et al., 2006; Inaba and Schnell et al., 2008). Of particular importance are studies that trace connections between post-translational modifications to proteins and the dynamics of protein:protein interactions related to developmental states, tissues, and environmental conditions.

The biotechnological advances in genomics have had a reciprocal effect on the advancement of proteomics. Mechanisms in transcription are targets of modifications and adjustments; processes like transcription initiation, splicing, or transcript destination are highly regulated and exert a determining influence that is as fundamental as the mere presence of a gene or transcript. Mechanisms that control translation preference and efficiency, as well as stability and decay of mRNAs, can be traced in combinations of transcript profiles and mutants that generate deficiencies in the protein folding, distribution, and turnover machineries (Wei et al., 2008; Murphy et al., 2009; Lois, 2010; Li and Chiu, 2010). To understand the scope of current methodologies in studying plant proteomes, several reviews provide an excellent basis (Jorrin et al., 2007; Carpentier et al., 2008b; Rose et al., 2004). In the past proteomics has been characterized by a two-dimensional separation of proteins or peptides or subsequently by gel-free techniques combined with mass spectrometry (Baerenfaller et al., 2008); the future will belong to methods based on mass spectroscopy (MRI) that can resolve large, complex mixtures of oligopeptides (Gevaert and Vandekerckhove, 2000; Junqueira et al., 2008).

Many recent studies in proteomics have attempted to address the gap between transcriptome knowledge and functional protein interactions. For example, in systemic acquired resistance (SAR), NPR1 is a coactivator responsible for transcription regulation. A recent study showed that proteosome-mediated turnover of phosphorylated NPR1 is required for full induction of target genes and establishment of SAR (Spoel et al., 2009). Similarly, identification of specific protein complexes and their physical interaction with amylo-pectin biosynthesis enzymes in developing wheat endosperm has been explored (Tetlow et al., 2008).

Databases such as GabiPD (http://www.gabipd.org/) attempt to integrate, analyze, and visualize heterogeneous data from different omics platforms containing high-throughput experiments of different plant species (Riano-Pachon et al., 2009).

Metabolomics highlights

The metabolomics node in post-genome analysis assumes eminent importance in functional genomics. It traces the effects of genes and proteins in how metabolites appear and change and the reciprocal roles of metabolites in orchestrating gene and protein action. The complement of metabolites, the metabolome, with its dynamic fluctuations within and between cells defines the biochemical phenotype of cells and tissues. As an emerging tool in functional genomics it is different from the biochemical analysis of a metabolite because it strives to become a high-throughput analysis tool that complements the other omics technologies (Fiehn, 2001; Bino et al., 2004) by providing qualitative and quantitative identification of all metabolites present in an organism under a given set of conditions (Allwood et al., 2008).

Even more than for the proteome, it is impractical to identify the complete metabolome of any organism. As with proteins, there is no single extraction technique or analytical method that can isolate and detect every metabolite in any biological sample (Sekiyama et al., 2010). This is further compounded by the diversity of compounds which, in plants, are estimated to include 1–2 million different metabolites (Oksman-Caldentey and Inze, 2004). To reduce additional ambiguities related to sample processing, analysis, and reporting, and to standardize workflows, the Metabolomics Standards Initiative (http://msi-workgroups.sourceforge.net/) has recommended that protocols be kept to report, in detail, the growth of the material, analysis details, and the ways of data processing in the framework of a controlled vocabulary for data reporting (Fiehn, 2002; Jenkins et al., 2004; Fiehn, et al., 2006). This is certainly an essential first step in making some sense of the mind-boggling large datasets these analyses can produce.

Central to metabolomics are two complementary approaches of metabolite analysis: metabolite fingerprinting and profiling. Metabolic fingerprinting involves identification of metabolite signatures in samples via techniques such as NMR, Fourier transform ion cyclotron resonance mass spectrometry, Fourier transform (FT)-IR spectroscopy, Raman spectroscopy, and electrospray ionization (ESI)-MS. This approach is usually limited to a small number of metabolites, and is at least relatively fast and low cost. Metabolic profiling, in contrast, involves the detection, quantification, and identification of much larger numbers of metabolites within a sample via techniques such as gas or liquid chromatography coupled to mass spectroscopy, and comparison of outputs to pre-analyzed "libraries." A major limitation, hindering metabolic profiling from taking on its full role in the genomic toolbox, is the lack of availability of pure compounds to make those libraries. A number of reviews discuss benefits and drawbacks of individual methods and critical bioinformatics-type analyses used in these approaches (Kopka et al., 2004; Allwood et al., 2008; Shulaev et al., 2008; Hegeman, 2010).

Given that these problems will be solved, probably with much of the twenty-first century to go, a goal for the integration of genomic, proteomic, and metabolomic analyses will be to provide the correlations needed to clarify the many, poorly understood metabolic networks and pathways. It will be possible to explain their broader biological significance with respect to the environmental, developmental, or other test conditions, and then, to assign gene functions, for example, by the "clique metabolite matrices" method (Kose et al., 2001). The algorithm searches for linear metabolite correlations (2-cliques), defines maximal cliques of metabolites that are separated from others, and establishes hierarchical network models.

This will provide important insight into plant metabolic regulation and gene-to-metabolite relationships (Bohnert et al., 2006). The predictive power of combined omics data is highly enhanced when coupled with mathematical modeling. Although such experiments are still limited, with the most complete analysis available for bacteria and yeasts (Yamada and Bork, 2009), a pioneering study conducted by Morioka et al. (2007) showed successful modeling of gene expression linked to metabolic accumulation in sulfur-starved *Arabidopsis* plants. Similarly, spearheaded by groups at the Max Planck Institute for Plant Physiology in Golm, Germany, major advances have been made into the description of the *Arabidopsis* metabolome, in many cases combined with detailed enzymatic measurement (Steinhauser et al., 2010). One promising aspect for the future is the integration of metabolite and enzyme activity data with transcript analyses as they are mirrored in the MapMan program. Although originally developed for *Arabidopsis*, this program has sufficient flexibility to be adapted to other plants (Urbanczyk-Wochniak et al., 2006).

Genomics Unlimited: Getting Beyond Mere Genes

For decades, plant biologists have discussed the best approaches to finding and understanding pathways, proteins, and genes that confer abiotic stress tolerance. The goal is to develop methods to utilize this information in crop improvement as climate change, degradation of agricultural lands, and an increasing population loom large.

Within species, recording of single nucleotide polymorphisms (SNPs) and QTL information that uses molecular markers for increasingly finer resolution have at least partially brought progress. By utilizing the power of next-gen sequencing and statistics, associating genetic variations (mostly SNPs) and QTLs in a genome-wide scale, genome-wide association (GWA) studies become feasible. The GWA was initiated by human geneticists to discover the genetic basis of human disease (Hirschhorn and Daly, 2005). It was adopted rapidly to plants starting with the model plant *Arabidopsis* for varieties of phenotypes (Atwell et al., 2010; Nemri et al., 2010) and crop plants (Waugh et al., 2009; Cirulli and Goldstein, 2010). Although the GWA has its own limitations — that each single genetic polymorphism usually exerts very little effect on the phenotype, which often is a result of complex interaction of multiple genetic variation (Galvan et al., 2010) — approaches to identify QTLs more targeted to more specific traits were also becoming fruitful. In rice, for example, it has been possible to definitively associate a strong QTL for traits such as germination at low temperature with grain yield and identify highly likely candidate genes (Fujino et al., 2008; Wang et al., 2008; Huang et al., 2009). Concepts and databases are emerging that correlate QTL information, transcript expression, and genome structure, as well as for pyramiding beneficial alleles (Ashikari and Matsuoka, 2006; Zeng et al., 2007). Due to public concerns and policy limitation in the use of genetically modified plants, genomics assisted QTL mapping provides a desirable alternative to transgenic modifications in crops. This is explored for many crop species exploiting the plethora of draft genome sequences that are emerging (Barone et al., 2009).

Also poised to become a defining field for the twenty-first century is epigenomics (Katsnelson, 2010). Again, progress in understanding this major influence on organismal behavior will be enabled by whole genome sequencing capacities, but also by the ability to compare diverse genomes. For efficiency it may be necessary to recruit ever more computing power to the problem, which will require large research project units and may ask for commercialization, comparative genomics, and transcriptomics on an increasingly higher scale and throughput. The problem and challenge in epigenomics is its superimposition on the classical genetic landscape in multiple layers, including ATP-dependent chromatin remodeling, post-translational histone modifications, variances in histone combination, and RNA-based epigenetic control mechanisms and DNA methylation, both of which are engrained by the organism's history and previous lifestyle as well as by recent environmental that define the organism's future. Omics-scale characterization of DNA methylation and small RNAs are among those expected to capitalize on the most recent technical advances including direct single molecule sequencing of methylated DNA (Flusberg et al., 2010) and deep sequencing of small RNA species (Creighton et al., 2009). A genome-wide methylome map was developed for *Arabidopsis* (Zhang et al., 2006; Zilberman et al., 2007), and deep sequencing was widely used to identify both species and expression of plant miRNAs (Zhan and Lukens, 2010).

Irrefutable evidence documents the immense importance of the epigenetic machinery for the functioning of eukaryotic organisms. The dynamically imprinted state includes first a memory of the past of an individuum or a line and also carries the possibility of responding to future adverse environmental challenges. Both ways have, at least in plants, been recognized long ago, although analysis and understanding has been lacking until recently. Epigenetics constitutes a set of mechanisms that establishes a memory of past experiences and challenges to the genome that have been successfully counteracted and incorporated into an appropriate gene expression program through which the phenotype adapted to the prevailing environment (Jablonka and Lamb, 2006; Matzke et al., 2009; Turner et al., 2010; Iorio et al., 2010). However, the process is fluid, not a dead end, for which terms such as acclimation, priming, or somaclonal variation provide examples of recurrent epigenetic events.

Into the Future: Genomics-Based Biotechnology and Agriculture

We can live without oil but we cannot live without water. Every successful high culture has had access to ample water, has modified the use of water through irrigation schemes and, in the process, established organized public interactions. Every culture that let slip its control of water has disappeared. At present 2+ billion people are without reliable access to clean water (UN World Water Assessment Report, http://www.unesco.org/water/wwap/wwdr). We can live without computer chips but we cannot live without food that is produced utilizing approximately two-thirds of the renewable fresh water on Earth. Water shortages constitute the biggest barrier to enhancing food production. In the near future, crops will have to be grown on less land, or less suitable land, and with less water. The climatic conditions of this future are such that most crops do not have experience and memory, and lack an appropriate allele structure.

This outlook, considered realistic by many (Foley et al., 2005; Vorosmarty et al., 2000; Wood, 2003; Piao et al., 2007; Nord and Lynch, 2009; Godfray et al., 2010), will require accelerated breeding efforts that begin the adaptation of extant crops to a variety of new environments by focusing on fundamental plant functions. While breeding will remain the backbone of genetic adaptation, solutions will also need transgenic approaches. Next-gen sequencing and the omics tools are already supporting both strategies. For example, utilizing the more economical and efficient sequencing of genomes and transcriptomes enables development of molecular markers and identification of genomic variations by extensive re-sequencing of plant ecotypes, and enhanced mapping and association of QTLs of species for which until recently no genomics-type data were available (Varshney et al., 2009; Barone et al., 2009). There remain as yet various barriers before connections can be made between laboratory findings and field data. We conclude this review by highlighting a few of those barriers and the potentials of genomics tools to solving problems.

From models to crops, from labs to fields

The knowledge and data accumulated for model species must be transferred to crop species. Next-gen sequencing is already providing the basis for such a change, and gradually and inevitably making the concept of "model" obsolete. The continuing reduction of sequencing cost will enable genome sequencing on an immense scale such that breeders may obtain whatever DNA sequence is necessary from hundreds to thousands of lines during a growing season for comparison with last year's lines. Not to belittle the tasks ahead, accumulation of phenotypic information of crop plants modified by either transgenic or breeding approaches will require time and effort, in particular in how information is analyzed and statistically prepared. Also, sequencing and analyzing the large and, often, polyploid genomes of many crop species will be challenging and will drive further improvements in DNA sequencing, assembly methods, and self-learning algorithms that draw conclusions from the immense data flood. Along this process, we will find the tools that will allow us to define the roles of repetitive sequences and non-coding yet transcribed regions, seemingly junk DNA. Their cryptic functions, if any, for phenotypic characters can only be unraveled on a genomic scale and view.

A second aspect is realism in experimentation — the transfer of laboratory knowledge to knowledge about field performance. It is already known that a large number of gene expression characteristics that laboratory scientists term "stress-induced" are constitutively expressed in the field (Miyazaki et al., 2004). Over the last two decades a number of engineering attempts have been made. In many instances a marginal effect could be documented, but to our knowledge genuinely improved performances under field or field stress conditions have not been reported. For example, numerous numbers of transgenic plants were developed with genes for ion transport and compatible solutes accumulation, to enhance salinity stress tolerance, while performance tests used greenhouse conditions (Ashraf and Akram, 2009).

Only few recent experiments have shown improvement of transgenic plants in the field or, in addition, attempted to monitor unexpected interactions with the environment

(Zeller et al., 2010). Expression of a transcription factor gene under the control of a stress condition-specific promoter resulted in maize plants that could withstand moderate drought conditions better than the control line (Nelson et al., 2007). Another transgenic attempt was made to improve the resistance of bread wheat against powdery mildew. Here, the transgenic lines did show the expected improved resistance. However, reactions of the plants to other prevailing environmental factors indicated differences between greenhouse and field, in part revealing adverse effects by the transgene (Zeller et al., 2010). We are clearly not able to predict the unintended consequences caused by transgenes and transformation. Instead, many more experiments will be required in which field conditions and genomic sequences, multifactorial transcriptome profiles, phenotypes, and yield characteristics in the field are monitored. Very likely, better success will have to rely on a more controlled and specific expression of transgenes. Attempts are being made to utilize or develop promoters that are tissue-, condition-, or signal-specific (Venter, 2007; Lv et al., 2009; Kovalchuk et al., 2010). Also, transfer of multiple genes in the pathway will have a better chance of providing more viable solutions than a single transgene. As one example, the transfer of four genes that had been associated with water deficit and salt tolerance in *Arabidopsis*, which is certainly not a stress-tolerant species, resulted in considerable tolerance to drought in cotton (Z. Gong, R.A. Bressan, J-K Zhu et al., personal communication). Accumulation of genomics data and better ways of networking similar experiments and solutions will help in identifying regulatory elements as well as genetic resources in specific pathways and the interactions between pathways. Next-gen sequencing will play a crucial role for assessing interactions of transgenes with the allele structure of the target species and with the environment.

Genetic resources from extremophile species

Extant crops have generally been adapted for growth in the most productive environments, which, at least to some degree, was at the expense of traits that characterize less ideal, more stressful conditions. For the future species that show natural adaptations to heat, cold, freezing, salinity, flooding, and, especially, low water availability are of particular importance. It will be important that their genome will be sequenced. These extremophile species include genes and proteins, protein modification characteristics, biochemical and metabolic pathways, and epigenetic characters that support their protection. Most important will be those extremophile species that are close relatives of established models and/or crops. The importance of work with extremophile species, and in particular extremophile relatives of *Arabidopsis*, has been pointed out before (Bressan et al., 2001; Amtmann et al., 2005; Amtmann, 2009). Another alternative strategy will be to analyze species from different orders or families that grow under similar stressful conditions. As a point in favor of such an approach, an analysis of two mangroves in different orders, their transcriptomes (Dassanayake et al., 2009, 2010) have been instrumental in defining how an extremophile lifestyle leads to transcript populations distinguished from those of glycophytes.

Exploring "unknown unknowns"

A powerful aspect of the holistic approach underlying genomics is to begin without preconceived views and let the data create hypotheses. Up to now gene additions, subtractions (knockouts), or modifications of genes in existing pathways rely on the chance discovery of a desired mutant phenotype or, worse, begin by relying on what we seem to know about a known gene. The years of work using *Arabidopsis* tagged mutants and work on protein:protein interactions revealed functional complexity for many individual proteins. The multiple functions may affect different biochemistries, and removing a protein can have quite unexpected consequences.

Adding to protein multifunctionality, the hand-waving associated with the term "position effect" in explaining unexpected results of transgenes will be another unknown on which genomics approaches can focus. Next-gen DNA sequencing can monitor the place of a transgene in the vicinity of transcribed regions that may be translated or may be functional by carrying epigenetic signals. Not that we know all about how promoter regions are controlled during development and under environmental challenges, about transcriptome to proteome dynamic interactions, or about the influence of genomic memory — but these are known unknowns. Only by integrating the wealth of available and emerging data, and predominantly genomics-type data, can the remaining unknown unknowns be found. Predominantly among those are the elucidation of epigenetic effects governing genome memory, maintenance of this memory, or interactions by which the environment generates non-mutational genome alterations. Also related are questions about the cryptic functions of repetitive elements and non-coding sequences, and about the evolution of mechanisms for silencing invading genetic elements and how this necessity became incorporated into chromatin correction, chromosome composition, and genome complexity control mechanisms.

The importance of stress "tolerance" engineering

Whether by breeding or transgenic engineering, abiotic stress tolerance is a most important character to achieve. Increased stress tolerance and increased ability to grow and produce in a variety of stressful environments is sought after even when productivity becomes partially compromised. The character most desirable is water productivity, and typically, but in a more narrow sense, labeled water use efficiency.

Over the last two decades a number of genetic engineering attempts have been made. In many instances a marginal effect could be documented, but to our knowledge improved performance under field or field stress conditions has not been shown. One recent experiment, however, has shown just such an improvement in the field (Nelson et al., 2004). The expression of a transcription factor gene under the control of a stress condition-specific promoter resulted in maize plants that could withstand some drought stress better than the control line. An increased amount of genomics data and better ways of networking similar experiments and solutions will likely result in similar successes. It must be questioned whether marginal improvements from single-gene transfers will provide viable solutions; it seems that the accumulation of several genes will be superior. Also, the transfer of multiple genes into a transgenic crop line may protect against stress parameters that affect more than one resistance or recovery pathway.

For example, a comparison of genome sequences from *A. thaliana* and the stress-tolerant, halophytic close relative *T. parvula* indicated high sequence identity in translated regions of genes. In contrast low sequence homology was observed in the promoter regions of salt-tolerance determinant genes, such as SOS1, to a degree that suggests expression strength or transcript stability as a major component of *T. parvula*'s extraordinarily high salinity tolerance (Orsini et al., 2010; Oh et al., 2010). The availability of the *T. parvula* nuclear genome — at approximately 163 Mbp only slightly larger than the *A. thaliana* genome — promises to provide an exceptional insight into the differences between a glycophyte and an extremophile relative (Dassanayake et al., 2011). Our recent and ongoing analysis of the genome of the extremely salt-tolerant *Arabidopsis* relative *T. parvula* reveals surprising differences that affect how both species deal with saline water.

Genomics and its related cousins have already generated and will generate unprecedented deeper understanding of plant life. Among the species whose genome will be sequenced, those that show natural adaptations to heat, cold, freezing, salinity, flooding and, especially, low water availability are of particular importance. In this century, studies will not be limited to a few — much less a single — model species. The enduring value of model species will be to use what we have learned in the struggle to assemble them and in providing archetypical gene models and the understanding of gene complexity against which other species can be measured. For the first time we can hope to view plants as an integrated system — we are no longer like blind men touching an elephant — with the prospect of transferring this knowledge into crop improvement that will, we hope, let us meet the challenges of feeding the additional 1.5 to 2 billion people expected to challenge the food chain in this century. The enabling technologies to which next-gen sequencing has recently significantly contributed have been developed, but overall success does not depend on science as solving the problems does not have a simple science-based solution.

Acknowledgments

The work on this article has been made possible by funds from the World Class University (WCU grant number R32 and 8211; 10148) and BK 21 programs, Gyeongsang National University, by the Government of Korea, and by the King Abdallah University for Science and Technology (KAUST) in Thuwal, Saudi Arabia.

References

Allwood, J. W., Ellis, D. I., & Goodacre, R. (2008). Metabolomic technologies and their application to the study of plants and plant-host interactions. *Physiologia Plantarum, 132*, 117–135.

Alonso-Blanco, C., Gomez-Mena, C., Llorente, F., Koornneef, M., Salinas, J., & Martinez-Zapater, J. M. (2005). Genetic and molecular analyses of natural variation indicate CBF2 as a candidate gene for underlying a freezing tolerance quantitative trait locus in *Arabidopsis*. *Plant Physiology, 139*, 1304–1312.

Amtmann, A. (2009). Learning from evolution: *Thellungiella* generates new knowledge on essential and critical components of abiotic stress tolerance in plants. *Molecular Plant, 2*, 3–12.

Amtmann, A., Bohnert, H. J., & Bressan, R. A. (2005). Abiotic stress and plant genome evolution. Search for new models. *Plant Physiology, 138*, 127–130.

Anonymous, (1951). FACSIMILE of Gregor Mendel's "Experiments in plant hybridation." *The Journal of Heredity, 42*, 3–47.

Apse, M. P., Aharon, G. S., Snedden, W. A., & Blumwald, E. (1999). Salt tolerance conferred by overexpression of a vacuolar Na^+/H^+ antiport in *Arabidopsis*. *Science, 285*, 1256–1258.

Ashikari, M., & Matsuoka, M. (2006). Identification, isolation and pyramiding of quantitative trait loci for rice breeding. *Trends in Plant Science, 11*, 344–350.

Ashraf, M., & Akram, N. A. (2009). Improving salinity tolerance of plants through conventional breeding and genetic engineering: An analytical comparison. *Biotechnology Advances, 27*, 744–752.

Astill, G., & Langdon, J. (Eds.), (1997). *Medieval farming and technology – the impact of agricultural change in northwest europe*. Leiden and Boston: Brill.

Atwell, S., Huang, Y. S., Vilhjalmsson, B. J., Willems, G., Horton, M., & Li, Y., et al. (2010). Genome-wide association study of 107 phenotypes in *Arabidopsis thaliana* inbred lines. *Nature, 465*(7298), 627–631.

Baerenfaller, K., Grossmann, J., Grobei, M. A., Hull, R., Hirsch-Hoffmann, M., & Yalovsky, S., et al. (2008). Genome-scale proteomics reveals *Arabidopsis thaliana* gene models and proteome dynamics. *Science, 320*, 938–941.

Barakat, A., DiLoreto, D. S., Zhang, Y., Smith, C., Baier, K., & Powell, W. A., et al. (2009). Comparison of the transcriptomes of American chestnut (*Castanea dentata*) and Chinese chestnut (*Castanea mollissima*) in response to the chestnut blight infection. *BMC Plant Biology, 9*, 51.

Barone, A., Di Matteo, A., Carputo, D., & Frusciante, L. (2009). High-throughput genomics enhances tomato breeding efficiency. *Current Genomics, 10*, 1–9.

Bayley, H. (2006). Sequencing single molecules of DNA. *Current Opinion in Chemical Biology, 10*, 628–637.

Bellin, D., Ferrarini, A., Chimento, A., Kaiser, O., Levenkova, N., & Bouffard, P., et al. (2009). Combining next-generation pyrosequencing with microarray for large scale expression analysis in non-model species. *BMC Genomics, 10*, 555.

Bentley, D. R. (2006). Whole-genome re-sequencing. *Current Opinion in Genetics & Development, 16*, 545–552.

Bino, R. J., Hall, R. D., Fiehn, O., Kopka, J., Saito, K., & Draper, J., et al. (2004). Potential of metabolomics as a functional genomics tool. *Trends in Plant Science, 9*, 418–425.

Bohnert, H. J., Gong, Q., Li, P., & Ma, S. (2006). Unraveling abiotic stress tolerance mechanisms – getting genomics going. *Current Opinion in Plant Biology, 9*, 180–188.

Bonnet, E., Tatari, M., Joshi, A., Michoel, T., Marchal, K., & Berx, G., et al. (2010). Module network inference from a cancer gene expression data set identifies microRNA regulated modules. *PLoS ONE, 5*, e10162.

Boston, R. S., Viitanen, P. V., & Vierling, E. (1996). Molecular chaperones and protein folding in plants. *Plant Molecular Biology, 32*, 191–222.

Brady, S. M., Orlando, D. A., Lee, J. Y., Wang, J. Y., Koch, J., & Dinneny, J. R., et al. (2007). A high-resolution root spatiotemporal map reveals dominant expression patterns. *Science*, *318*, 801–806.

Branton, D., Deamer, D. W., Marziali, A., Bayley, H., Benner, S. A., & Butler, T., et al. (2008). The potential and challenges of nanopore sequencing. *Nature Biotechnology*, *26*, 1146–1153.

Bray, F. (1986). *Agriculture. Science and Civilization in China*. Cambridge University Press.

Bressan, R. A., Zhang, C., Zhang, H., Hasegawa, P. M., Bohnert, H. J., & Zhu, J. K. (2001). Learning from the *Arabidopsis* experience. The next gene search paradigm. *Plant Physiology*, *127*, 1354–1360.

Buggs, R. J., Chamala, S., Wu, W., Gao, L., May, G. D., & Schnable, P. S., et al. (2010). Characterization of duplicate gene evolution in the recent natural allopolyploid *Tragopogon miscellus* by next-generation sequencing and Sequenom iPLEX MassARRAY genotyping. *Molecular Ecology*, *19*(Suppl 1), 132–146.

Carpentier, S. C., Coemans, B., Podevin, N., Laukens, K., Witters, E., & Matsumura, H., et al. (2008a). Functional genomics in a non-model crop: Transcriptomics or proteomics? *Physiologia Plantarum*, *133*, 117–130.

Carpentier, S. C., Panis, B., Vertommen, A., Swennen, R., Sergeant, K., & Renaut, J., et al. (2008b). Proteome analysis of non-model plants: A challenging but powerful approach. *Mass Spectrometry Reviews*, *27*, 354–377.

Cirulli, E. T., & Goldstein, D. B. (2010). Uncovering the roles of rare variants in common disease through whole-genome sequencing. *Nature Reviews Genetics*, *11*, 415–425.

Clark, R. M., Schweikert, G., Toomajian, C., Ossowski, S., Zeller, G., & Shinn, P., et al. (2007). Common sequence polymorphisms shaping genetic diversity in *Arabidopsis thaliana*. *Science*, *317*, 338–342.

Creighton, C. J., Reid, J. G., & Gunaratne, P. H. (2009). Expression profiling of microRNAs by deep sequencing. *Briefings in Bioinformatics*, *10*, 490–497.

Darwin, C. (1868). *The variation of animals and plants under domestication*. London: John Murray.

Dassanayake, M., Haas, J. S., Bohnert, H. J., & Cheeseman, J. M. (2010). Comparative transcriptomics for mangrove species: An expanding resource. *Functional & Integrative Genomics*, *10*(4), 523–532.

Dassanayake, M., Haas, J. S., Bohnert, H. J., & Cheeseman, J. M. (2009). Shedding light on an extremophile lifestyle through transcriptomics. *The New Phytologist*, *183*, 764–775.

Dassanayake, M., Oh, D.-H., Haas, J.S., Hernandez, A., Hong, H., Shahjhan Ali, S., et al., (2011). The genome of the extremophile crucifer *Thellungiella purvula*. *Nature Genetics*. (AOP, 07 August 2011; DOI 10.1038/ng.889.

Deamer, D. W., & Akeson, M. (2000). Nanopores and nucleic acids: Prospects for ultrarapid sequencing. *Trends in Biotechnology*, *18*, 147–151.

Deng, J., Shoemaker, R., Xie, B., Gore, A., LeProust, E. M., & Antosiewicz-Bourget, J., et al. (2009). Targeted bisulfite sequencing reveals changes in DNA methylation associated with nuclear reprogramming. *Nature Biotechnology*, *27*, 353–360.

Deyholos, M. K. (2010). Making the most of drought and salinity transcriptomics. *Plant Cell And Environment*, *33*, 648–654.

Dinneny, J. R., Long, T. A., Wang, J. Y., Jung, J. W., Mace, D., & Pointer, S., et al. (2008). Cell identity mediates the response of *Arabidopsis* roots to abiotic stress. *Science*, *320*, 942–945.

Doebley, J. F., Gaut, B. S., & Smith, B. D. (2006). The molecular genetics of crop domestication. *Cell*, *127*, 1309–1321.

Dolnik, V. (1999). DNA sequencing by capillary electrophoresis (review). *Journal of Biochemical and Biophysical Methods*, *41*, 103–119.

Donlin, M. J. (2009). Using the Generic Genome Browser (GBrowse). *Current Protocols in Bioinformatics* (**Chapter 9**, Unit 9 9).

Fahlgren, N., Sullivan, C. M., Kasschau, K. D., Chapman, E. J., Cumbie, J. S., & Montgomery, T. A., et al. (2009). Computational and analytical framework for small RNA profiling by high-throughput sequencing. *RNA*, *15*, 992–1002.

Fiehn, O. (2001). Combining genomics, metabolome analysis, and biochemical modelling to understand metabolic networks. *Comparative and Functional Genomics*, *2*, 155–168.

Fiehn, O. (2002). Metabolomics – the link between genotypes and phenotypes. *Plant Molecular Biology*, *48*, 155–171.

Fiehn, O., Kristal, B., van Ommen, B., Sumner, L. W., Sansone, S. A., & Taylor, C., et al. (2006). Establishing reporting standards for metabolomic and metabonomic studies: A call for participation. *Omics*, *10*, 158–163.

Flusberg, B. A., Webster, D. R., Lee, J. H., Travers, K. J., Olivares, E. C., & Clark, T. A., et al. (2010). Direct detection of DNA methylation during single-molecule, real-time sequencing. *Nature Methods*, *7*, 461–465.

Foley, J. A., Defries, R., Asner, G. P., Barford, C., Bonan, G., & Carpenter, S. R., et al. (2005). Global consequences of land use. *Science*, *309*, 570–574.

Freeling, M., & Thomas, B. C. (2006). Gene-balanced duplications, like tetraploidy, provide predictable drive to increase morphological complexity. *Genome Research*, *16*, 805–814.

Fujino, K., Sekiguchi, H., Matsuda, Y., Sugimoto, K., Ono, K., & Yano, M. (2008). Molecular identification of a major quantitative trait locus, qLTG3-1, controlling low-temperature germinability in rice. *Proceedings of the National Academy of Sciences of the United States of America*, *105*, 12623–12628.

Fukushima, A., Kusano, M., Redestig, H., Arita, M., & Saito, K. (2009). Integrated omics approaches in plant systems biology. *Current Opinion in Chemical Biology*, *13*, 532–538.

Galvan, A., Ioannidis, J. P., & Dragani, T. A. (2010). Beyond genome-wide association studies: Genetic heterogeneity and individual predisposition to cancer. *Trends in Genetics: TIG*, *26*, 132–141.

Gevaert, K., & Vandekerckhove, J. (2000). Protein identification methods in proteomics. *Electrophoresis*, *21*, 1145–1154.

Godfray, H. C., Beddington, J. R., Crute, I. R., Haddad, L., Lawrence, D., & Muir, J. F., et al. (2010). Food security: The challenge of feeding 9 billion people. *Science*, *327*, 812–818.

Gong, Q., Li, P., Ma, S., Indu Rupassara, S., & Bohnert, H. J. (2005). Salinity stress adaptation competence in the extremophile *Thellungiella halophila* in comparison with its relative *Arabidopsis thaliana*. *Plant Journal*, *44*, 826–839.

Ha, M., Lu, J., Tian, L., Ramachandran, V., Kasschau, K. D., & Chapman, E. J., et al. (2009). Small RNAs serve as a genetic buffer against genomic shock in *Arabidopsis* interspecific hybrids and allopolyploids. *Proceedings of the National Academy of Sciences of the United States of America*, *106*, 17835–17840.

Harismendy, O., Ng, P. C., Strausberg, R. L., Wang, X., Stockwell, T. B., & Beeson, K. Y., et al. (2009). Evaluation of next generation sequencing platforms for population targeted sequencing studies. *Genome Biology*, *10*, R32.

Hegeman, A. D. (2010). Plant metabolomics – meeting the analytical challenges of comprehensive metabolite analysis. *Briefing in Functional Genomics*, *9*, 139–148.

Hiller, N. L., Janto, B., Hogg, J. S., Boissy, R., Yu, S., & Powell, E., et al. (2007). Comparative genomic analyses of seventeen *Streptococcus pneumoniae* strains: Insights into the pneumococcal supragenome. *Journal of Bacteriology*, *189*, 8186–8195.

Hirai, M. Y., Sugiyama, K., Sawada, Y., Tohge, T., Obayashi, T., & Suzuki, A., et al. (2007). Omics-based identification of *Arabidopsis* Myb transcription factors regulating aliphatic glucosinolate biosynthesis. *Proceedings of the National Academy of Sciences of the United States of America*, *104*, 6478–6483.

Hirschhorn, J. N., & Daly, M. J. (2005). Genome-wide association studies for common diseases and complex traits. *Nature Reviews Genetics*, *6*, 95–108.

Huang, X., Qian, Q., Liu, Z., Sun, H., He, S., & Luo, D., et al. (2009). Natural variation at the DEP1 locus enhances grain yield in rice. *Nature Genetics*, *41*, 494–497.

Hunkapiller, T., Kaiser, R. J., Koop, B. F., & Hood, L. (1991). Large-scale and automated DNA sequence determination. *Science*, *254*, 59–67.

Inaba, T., & Schnell, D. J. (2008). Protein trafficking to plastids: One theme, many variations. *The Biochemical Journal*, *413*, 15–28.

Inada, D. C., Bashir, A., Lee, C., Thomas, B. C., Ko, C., & Goff, S. A., et al. (2003). Conserved noncoding sequences in the grasses. *Genome Research*, *13*, 2030–2041.

Iorio, M. V., Piovan, C., & Croce, C. M. (2010). Interplay between microRNAs and the epigenetic machinery: An intricate network. *Biochimica et Biophysica Acta, 1799*(10–12), 694–701.

Jablonka, E., & Lamb, M. J. (2006). The evolution of information in the major transitions. *Journal of Theoretical Biology, 239*, 236–246.

Jenkins, H., Hardy, N., Beckmann, M., Draper, J., Smith, A. R., & Taylor, J., et al. (2004). A proposed framework for the description of plant metabolomics experiments and their results. *Nature Biotechnology, 22*, 1601–1606.

Jorrin, J. V., Maldonado, A. M., & Castillejo, M. A. (2007). Plant proteome analysis: A 2006 update. *Proteomics, 7*, 2947–2962.

Junqueira, M., Spirin, V., Balbuena, T. S., Thomas, H., Adzhubei, I., & Sunyaev, S., et al. (2008). Protein identification pipeline for the homology-driven proteomics. *Journal of Proteomics, 71*, 346–356.

Katsnelson, A. (2010). Genomics goes beyond DNA sequence. *Nature, 465*, 145.

Kircher, M., & Kelso, J. (2010). High-throughput DNA sequencing – concepts and limitations. *Bioessays, 32*, 524–536.

Kopka, J., Fernie, A., Weckwerth, W., Gibon, Y., & Stitt, M. (2004). Metabolite profiling in plant biology: Platforms and destinations. *Genome Biology, 5*, 109.

Kose, F., Weckwerth, W., Linke, T., & Fiehn, O. (2001). Visualizing plant metabolomic correlation networks using clique-metabolite matrices. *Bioinformatics, 17*, 1198–1208.

Kovalchuk, N., Li, M., Wittek, F., Reid, N., Singh, R., & Shirley, N., et al. (2010). Defensin promoters as potential tools for engineering disease resistance in cereal grains. *Plant Biotechnology Journal, 8*, 47–64.

Le Novere, N., Hucka, M., Mi, H., Moodie, S., Schreiber, F., & Sorokin, A., et al. (2009). The systems biology graphical notation. *Nature Biotechnology, 27*, 735–741.

Lee, I., Ambaru, B., Thakkar, P., Marcotte, E. M., & Rhee, S. Y. (2010). Rational association of genes with traits using a genome-scale gene network for *Arabidopsis thaliana*. *Nature Biotechnology, 28*(2), 149–156.

Li, H. M., & Chiu, C. C. (2010). Protein transport into chloroplasts. *Annual Review of Plant Biology, 61*, 157–180.

Li, R., Zhu, H., Ruan, J., Qian, W., Fang, X., & Shi, Z., et al. (2010). *De novo* assembly of human genomes with massively parallel short read sequencing. *Genome Research, 20*, 265–272.

Lois, L. M. (2010). Diversity of the SUMOylation machinery in plants. *Biochemical Society Transactions, 38*, 60–64.

Lu, T., Lu, G., Fan, D., Zhu, C., Li, W., & Zhao, Q., et al. (2010). Function annotation of rice transcriptome at single nucleotide resolution by RNA-seq. *Genome Research, 20*, 1238–1249.

Lv, X., Song, X., Rao, G., Pan, X., Guan, L., & Jiang, X., et al. (2009). Construction vascular-specific expression bi-directional promoters in plants. *Journal of Biotechnology, 141*, 104–108.

Ma, S., & Bohnert, H. J. (2007). Integration of *Arabidopsis thaliana* stress-related transcript profiles, promoter structures, and cell-specific expression. *Genome Biology, 8*, R49.

Ma, S., Gong, Q., & Bohnert, H. J. (2007). An *Arabidopsis* gene network based on the graphical Gaussian model. *Genome Research, 17*, 1614–1625.

MacLean, D., Elina, N., Havecker, E. R., Heimstaedt, S. B., Studholme, D. J., & Baulcombe, D. C. (2010). Evidence for large complex networks of plant short silencing RNAs. *PLoS ONE, 5*, e9901.

Maher, C. A., Kumar-Sinha, C., Cao, X., Kalyana-Sundaram, S., Han, B., & Jing, X., et al. (2009). Transcriptome sequencing to detect gene fusions in cancer. *Nature, 458*, 97–101.

Mao, L., Van Hemert, J. L., Dash, S., & Dickerson, J. A. (2009). *Arabidopsis* gene co-expression network and its functional modules. *BMC Bioinformatics, 10*, 346.

Matzke, M., Weiger, T. M., Papp, I., & Matzke, A. J. (2009). Nuclear membrane ion channels mediate root nodule development. *Trends in Plant Science, 14*, 295–298.

Mazzucotelli, E., Belloni, S., Marone, D., De Leonardis, A., Guerra, D., & Di Fonzo, N., et al. (2006). The e3 ubiquitin ligase gene family in plants: Regulation by degradation. *Current Genomics, 7*, 509–522.

Meldrum, D. (2000). Automation for genomics, part two: Sequencers, microarrays, and future trends. *Genome Research, 10*, 1288–1303.

Merchant, S. S., Prochnik, S. E., Vallon, O., Harris, E. H., Karpowicz, S. J., & Witman, G. B., et al. (2007). The *Chlamydomonas* genome reveals the evolution of key animal and plant functions. *Science, 318*, 245–250.

Miyazaki, S., Fredricksen, M., Hollis, K. C., Poroyko, V., Shepley, D., & Galbraith, D. W., et al. (2004). Expression profiles for *Arabidopsis thaliana* grown at UIUC SoyFACE. *Field Crop Research, 90*, 47–59.

Moreno-Risueno, M. A., Busch, W., & Benfey, P. N. (2010). Omics meet networks – using systems approaches to infer regulatory networks in plants. *Current Opinion in Plant Biology, 13*, 126–131.

Morioka, R., Kanaya, S., Hirai, M. Y., Yano, M., Ogasawara, N., & Saito, K. (2007). Predicting state transitions in the transcriptome and metabolome using a linear dynamical system model. *BMC Bioinformatics, 8*, 343.

Mullan, D. J., Colmer, T. D., & Francki, M. G. (2007). *Arabidopsis*-rice-wheat gene orthologues for Na^+ transport and transcript analysis in wheat-L. elongatum aneuploids under salt stress. *Molecular Genetics and Genomics: MGG, 277*, 199–212.

Murphy, J. E., Padilla, B. E., Hasdemir, B., Cottrell, G. S., & Bunnett, N. W. (2009). Endosomes: A legitimate platform for the signaling train. *Proceedings of the National Academy of Sciences of the United States of America, 106*, 17615–17622.

Myles, S., Peiffer, J., Brown, P. J., Ersoz, E. S., Zhang, Z., & Costich, D. E., et al. (2009). Association mapping: Critical considerations shift from genotyping to experimental design. *Plant Cell, 21*, 2194–2202.

Nelson, D. E., Repetti, P. P., Adams, T. R., Creelman, R. A., Wu, J., & Warner, D. C., et al. (2007). Plant nuclear factor Y (NF-Y) B subunits confer drought tolerance and lead to improved corn yields on water-limited acres. *Proceedings of the National Academy of Sciences of the United States of America, 104*, 16450–16455.

Nemri, A., Atwell, S., Tarone, A. M., Huang, Y. S., Zhao, K., & Studholme, D. J., et al. (2010). Genome-wide survey of *Arabidopsis* natural variation in downy mildew resistance using combined association and linkage mapping. *Proceedings of the National Academy of Sciences of the United States of America, 107*, 10302–10307.

Nord, E. A., & Lynch, J. P. (2009). Plant phenology: A critical controller of soil resource acquisition. *Journal of Experimental Botany, 60*, 1927–1937.

Obayashi, T., Hayashi, S., Saeki, M., Ohta, H., & Kinoshita, K. (2009). ATTED-II provides coexpressed gene networks for *Arabidopsis*. *Nucleic Acids Research, 37*, D987–991.

Oh, D. H., Dassanayake, M., Haas, J. S., Kropornika, A., Wright, C., & d'Urzo, M. P., et al. (2010). Genome structures and halophyte-specific gene expression of the extremophile *Thellungiella parvula* in comparison to *Thellungiella salsuginea* (*T. halophila*) and *Arabidopsis thaliana*. *Plant Physiology, 154*, 1040–1052.

Oksman-Caldentey, K. M., & Inze, D. (2004). Plant cell factories in the post-genomic era: New ways to produce designer secondary metabolites. *Trends in Plant Science, 9*, 433–440.

Orsini, F., D'Urzo, M. P., Inan, G., Serra, S., Oh, D. H., & Mickelbart, M. V., et al. (2010). A comparative study of salt tolerance parameters in 11 wild relatives of *Arabidopsis thaliana*. *Journal of Experimental Botany, 61*, 3787–3798.

Parchman, T. L., Geist, K. S., Grahnen, J. A., Benkman, C. W., & Buerkle, C. A. (2010). Transcriptome sequencing in an ecologically important tree species: Assembly, annotation, and marker discovery. *BMC Genomics, 11*, 180.

Park, P. J. (2009). ChIP-seq: Advantages and challenges of a maturing technology. *Nature Reviews Genetics, 10*, 669–680.

Paterson, A. H., Freeling, M., Tang, H., & Wang, X. (2010). Insights from the comparison of plant genome sequences. *Annual Review of Plant Biology, 61*, 349–372.

Pfluger, J., & Wagner, D. (2007). Histone modifications and dynamic regulation of genome accessibility in plants. *Current Opinion in Plant Biology, 10*, 645–652.

Piao, S., Friedlingstein, P., Ciais, P., de Noblet-Ducoudre, N., Labat, D., & Zaehle, S. (2007). Changes in climate and land use have a larger direct impact than rising CO_2 on global river runoff trends. *Proceedings of the National Academy of Sciences of the United States of America, 104*, 15242–15247.

Popescu, S. C., Popescu, G. V., Bachan, S., Zhang, Z., Gerstein, M., & Snyder, M., et al. (2009). MAPK target networks in *Arabidopsis thaliana* revealed using functional protein microarrays. *Genes & Development, 23*, 80–92.

Postan, M. M. (1973). *Essays on medieval agriculture and general problems of the medieval economy*. Cambridge University Press.

Prober, J. M., Trainor, G. L., Dam, R. J., Hobbs, F. W., Robertson, C. W., & Zagursky, R. J., et al. (1987). A system for rapid DNA sequencing with fluorescent chain-terminating dideoxynucleotides. *Science, 238*, 336–341.

Rensing, S. A., Lang, D., Zimmer, A. D., Terry, A., Salamov, A., & Shapiro, H., et al. (2008). The physcomitrella genome reveals evolutionary insights into the conquest of land by plants. *Science, 319*, 64–69.

Rhee, M., & Burns, M. A. (2006). Nanopore sequencing technology: Research trends and applications. *Trends in Biotechnology, 24*, 580–586.

Rhee, S. Y., Beavis, W., Berardini, T. Z., Chen, G., Dixon, D., & Doyle, A., et al. (2003). The Arabidopsis Information Resource (TAIR): A model organism database providing a centralized, curated gateway to *Arabidopsis* biology, research materials and community. *Nucleic Acids Research, 31*, 224–228.

Riano-Pachon, D. M., Nagel, A., Neigenfind, J., Wagner, R., Basekow, R., & Weber, E., et al. (2009). GabiPD: The GABI primary database – a plant integrative 'omics' database. *Nucleic Acids Research, 37*, D954–959.

Rivero, R. M., Kojima, M., Gepstein, A., Sakakibara, H., Mittler, R., & Gepstein, S., et al. (2007). Delayed leaf senescence induces extreme drought tolerance in a flowering plant. *Proceedings of the National Academy of Sciences of the United States of America, 104*, 19631–19636.

Rose, J. K., Bashir, S., Giovannoni, J. J., Jahn, M. M., & Saravanan, R. S. (2004). Tackling the plant proteome: Practical approaches, hurdles and experimental tools. *Plant Journal, 39*, 715–733.

Sanger, F., Nicklen, S., & Coulson, A. R. (1977). DNA sequencing with chain-terminating inhibitors. *Proceedings of the National Academy of Sciences of the United States of America, 74*, 5463–5467.

Schnable, P. S., Hochholdinger, F., & Nakazono, M. (2004). Global expression profiling applied to plant development. *Current Opinion in Plant Biology, 7*, 50–56.

Sekiyama, Y., Chikayama, E., & Kikuchi, J. (2010). Profiling polar and semipolar plant metabolites throughout extraction processes using a combined solution-state and high-resolution magic angle spinning NMR approach. *Analytical Chemistry, 82*, 1643–1652.

Shendure, J., & Ji, H. (2008). Next-generation DNA sequencing. *Nature Biotechnology, 26*, 1135–1145.

Shi, Y., Tyson, G. W., & DeLong, E. F. (2009). Metatranscriptomics reveals unique microbial small RNAs in the ocean's water column. *Nature, 459*, 266–269.

Shulaev, V., Cortes, D., Miller, G., & Mittler, R. (2008). Metabolomics for plant stress response. *Physiologia Plantarum, 132*, 199–208.

Spoel, S. H., Mou, Z., Tada, Y., Spivey, N. W., Genschik, P., & Dong, X. (2009). Proteasome-mediated turnover of the transcription coactivator NPR1 plays dual roles in regulating plant immunity. *Cell, 137*, 860–872.

Stein, L. D. (2008). Towards a cyberinfrastructure for the biological sciences: Progress, visions and challenges. *Nature Reviews Genetics, 9*, 678–688.

Stein, L. D., Mungall, C., Shu, S., Caudy, M., Mangone, M., & Day, A., et al. (2002). The generic genome browser: A building block for a model organism system database. *Genome Research, 12*, 1599–1610.

Steinhauser, M. C., Steinhauser, D., Koehl, K., Carrari, F., Gibon, Y., & Fernie, A. R., et al. (2010). Enzyme activity profiles during fruit development in tomato cultivars and *Solanum pennellii*. *Plant Physiology, 153*, 80–98.

Sterck, L., Rombauts, S., Vandepoele, K., Rouze, P., & Van de Peer, Y. (2007). How many genes are there in plants (... and why are they there)? *Current Opinion in Plant Biology, 10*, 199–203.

Stone, D. (2005). *Decision-making in medieval agriculture*. Oxford, UK: Oxford University Press.

Sullivan, M. L., & Green, P. J. (1993). Post-transcriptional regulation of nuclear-encoded genes in higher plants: The roles of mRNA stability and translation. *Plant Molecular Biology, 23*, 1091–1104.

Sunkar, R., & Jagadeeswaran, G. (2008). *In silico* identification of conserved microRNAs in large number of diverse plant species. *BMC Plant Biology, 8*, 37.

Swiezewski, S., Liu, F., Magusin, A., & Dean, C. (2009). Cold-induced silencing by long antisense transcripts of an *Arabidopsis* Polycomb target. *Nature, 462*, 799–802.

Tang, H., Wang, X., Bowers, J. E., Ming, R., Alam, M., & Paterson, A. H. (2008). Unraveling ancient hexaploidy through multiply-aligned angiosperm gene maps. *Genome Research, 18*, 1944–1954.

Tarczynski, M. C., Jensen, R. G., & Bohnert, H. J. (1993). Stress protection of transgenic tobacco by production of the osmolyte mannitol. *Science, 259*, 508–510.

Tauch, A., Schneider, J., Szczepanowski, R., Tilker, A., Viehoever, P., & Gartemann, K. H., et al. (2008). Ultrafast pyrosequencing of *Corynebacterium kroppenstedtii* DSM44385 revealed insights into the physiology of a lipophilic corynebacterium that lacks mycolic acids. *Journal of Biotechnology, 136*, 22–30.

Tetlow, I. J., Beisel, K. G., Cameron, S., Makhmoudova, A., Liu, F., & Bresolin, N. S., et al. (2008). Analysis of protein complexes in wheat amyloplasts reveals functional interactions among starch biosynthetic enzymes. *Plant Physiology, 146*, 1878–1891.

The International Brachypodium Initiative. (2010). Genome sequencing and analysis of the model grass *Brachypodium distachyon*. *Nature, 463*, 763–768.

Thomas, J. C., Smigocki, A. C., & Bohnert, H. J. (1995). Light-induced expression of ipt from *Agrobacterium tumefaciens* results in cytokinin accumulation and osmotic stress symptoms in transgenic tobacco. *Plant Molecular Biology, 27*, 225–235.

Turner, T. L., Bourne, E. C., Von Wettberg, E. J., Hu, T. T., & Nuzhdin, S. V. (2010). Population resequencing reveals local adaptation of *Arabidopsis lyrata* to serpentine soils. *Nature Genetics, 42*, 260–263.

Urbanczyk-Wochniak, E., Usadel, B., Thimm, O., Nunes-Nesi, A., Carrari, F., & Davy, M., et al. (2006). Conversion of MapMan to allow the analysis of transcript data from Solanaceous species: Effects of genetic and environmental alterations in energy metabolism in the leaf. *Plant Molecular Biology, 60*, 773–792.

Varshney, R. K., Nayak, S. N., May, G. D., & Jackson, S. A. (2009). Next-generation sequencing technologies and their implications for crop genetics and breeding. *Trends in Biotechnology, 27*, 522–530.

Vaughn, M. W., Tanurd, Ic, M., Lippman, Z., Jiang, H., Carrasquillo, R., & Rabinowicz, P. D., et al. (2007). Epigenetic natural variation in *Arabidopsis thaliana*. *PLoS Biology, 5*, e174.

Venter, M. (2007). Synthetic promoters: Genetic control through *cis* engineering. *Trends in Plant Science, 12*, 118–124.

Vera, J. C., Wheat, C. W., Fescemyer, H. W., Frilander, M. J., Crawford, D. L., & Hanski, I., et al. (2008). Rapid transcriptome characterization for a nonmodel organism using 454 pyrosequencing. *Molecular Ecology, 17*, 1636–1647.

Vorosmarty, C. J., Green, P., Salisbury, J., & Lammers, R. B. (2000). Global water resources: Vulnerability from climate change and population growth. *Science, 289*, 284–288.

Wada, A., Kidokoro, S., & Endo, S. (1989). Expanding roles of computers and robotics in biological macromolecular research. *Annual Review of Biophysics and Biophysical Chemistry, 18*, 1–24.

Wall, P. K., Leebens-Mack, J., Chanderbali, A. S., Barakat, A., Wolcott, E., & Liang, H., et al. (2009). Comparison of next generation sequencing technologies for transcriptome characterization. *BMC Genomics, 10*, 347.

Wang, E., Wang, J., Zhu, X., Hao, W., Wang, L., & Li, Q., et al. (2008). Control of rice grain-filling and yield by a gene with a potential signature of domestication. *Nature Genetics, 40*, 1370–1374.

Watson, A. M. (1983). *Agricultural innovation in the early Islamic world*. Cambridge University Press.

Waugh, R., Jannink, J. L., Muehlbauer, G. J., & Ramsay, L. (2009). The emergence of whole genome association scans in barley. *Current Opinion in Plant Biology, 12*, 218–222.

Wei, C., Peng, J., Xiong, Z., Yang, J., Wang, J., & Jin, Q. (2008). Subproteomic tools to increase genome annotation complexity. *Proteomics, 8*, 4209–4213.

Wheeler, D. A., Srinivasan, M., Egholm, M., Shen, Y., Chen, L., & McGuire, A., et al. (2008). The complete genome of an individual by massively parallel DNA sequencing. *Nature, 452*, 872–876.

Wicker, T., & Keller, B. (2007). Genome-wide comparative analysis of copia retrotransposons in Triticeae, rice, and *Arabidopsis* reveals conserved ancient evolutionary lineages and distinct dynamics of individual copia families. *Genome Research*, *17*, 1072–1081.

Wicker, T., Schlagenhauf, E., Graner, A., Close, T. J., Keller, B., & Stein, N. (2006). 454 sequencing put to the test using the complex genome of barley. *BMC Genomics*, *7*, 275.

Wilhelm, B. T., & Landry, J. R. (2009). RNA-Seq-quantitative measurement of expression through massively parallel RNA-sequencing. *Methods*, *48*, 249–257.

Wood, W. W. (2003). A fresh water odyssey: Some observations on the global resource. *Ground Water*, *41*, 300–305.

Yamada, T., & Bork, P. (2009). Evolution of biomolecular networks: Lessons from metabolic and protein interactions. *Nature Reviews Molecular Cell Biology*, *10*, 791–803.

Yip, K. Y., Alexander, R. P., Yan, K. K., & Gerstein, M. (2010). Improved reconstruction of *in silico* gene regulatory networks by integrating knockout and perturbation data. *PLoS ONE*, *5*, e8121.

Zeller, S. L., Kalinina, O., Brunner, S., Keller, B., & Schmid, B. (2010). Transgene x environment interactions in genetically modified wheat. *PLoS ONE*, *5*, e11405.

Zeng, H., Luo, L., Zhang, W., Zhou, J., Li, Z., & Liu, H., et al. (2007). PlantQTL-GE: A database system for identifying candidate genes in rice and *Arabidopsis* by gene expression and QTL information. *Nucleic Acids Research*, *35*, D879–882.

Zhai, J., Liu, J., Liu, B., Li, P., Meyers, B. C., & Chen, X., et al. (2008). Small RNA-directed epigenetic natural variation in *Arabidopsis thaliana*. *PLoS Genetics*, *4*, e1000056.

Zhan, S., & Lukens, L. (2010). Identification of novel miRNAs and miRNA dependent developmental shifts of gene expression in *Arabidopsis thaliana*. *PLoS ONE*, *5*, e10157.

Zhang, X., Yazaki, J., Sundaresan, A., Cokus, S., Chan, S. W., & Chen, H., et al. (2006). Genome-wide high-resolution mapping and functional analysis of DNA methylation in *Arabidopsis*. *Cell*, *126*, 1189–1201.

Zhou, X., Sunkar, R., Jin, H., Zhu, J. K., & Zhang, W. (2009). Genome-wide identification and analysis of small RNAs originated from natural antisense transcripts in *Oryza sativa*. *Genome Research*, *19*, 70–78.

Zilberman, D., Gehring, M., Tran, R. K., Ballinger, T., & Henikoff, S. (2007). Genome-wide analysis of *Arabidopsis thaliana* DNA methylation uncovers an interdependence between methylation and transcription. *Nature Genetics*, *39*, 61–69.

Zivy, M., & de Vienne, D. (2000). Proteomics: A link between genomics, genetics and physiology. *Plant Molecular Biology*, *44*, 575–580.

Protein targeting: Strategic planning for optimizing protein products through plant biotechnology

3

Elizabeth Hood Carole Cramer Giuliana Medrano Jianfeng Xu
Arkansas State University, Jonesboro, Arkansas

TABLE OF CONTENTS

Introduction: Strategic Decisions about How to Express an Output Trait

Plant-based bioproduction has been explored and documented for several types of proteins in several plant systems by many academic groups and companies (Faye and Gomord, 2010). This chapter reviews various types of protein expression systems and how protein targeting is incorporated into those systems. We do not focus on building a comprehensive catalog of examples of recombinant proteins as this has been recently reviewed (Daniell et al., 2009; Khan et al., 2010). Rather, we are discussing the process of choosing a plant system and the mechanisms of how to direct a protein to a particular site in a tissue or cell.

How does one determine which plant system to use to produce an output trait? The system of choice depends on several factors (Howard and Hood, 2005). For example, should one choose a wild species versus a domesticated species? Much is known about domesticated species (how to grow and process these crops). They are also high-yielding. In contrast, a wild plant species may be clearly distinguishable from a cultivated crop, but little is known about the safety of the plant and if the yield will be low. What about food versus non-food crops? Food crops have the advantage of high safety for human health products, but have the potential of becoming intermixed with foods (Howard and Hood, 2005). However, non-food or wild crops can also become intermixed with food if poorly handled, even if they are very different in appearance. The situation of food intermixing can be prevented through good stewardship, and models for how to safely grow transgenic crops with recombinant proteins have been published (Howard and Hood, 2007). Finally, the characteristics of the product often determine the choice of crop, as discussed in the following sections.

Why is it important to consider these mechanistic issues within the context of the system chosen? Each protein that has merit as a product — whether for pharmaceutical, vaccine, or industrial applications — has a unique set of circumstances that will allow its most efficient production. The ability to make the most protein in the best system requires some decisions about the nature of the final product use, the protein being produced, and the circumstances necessary for efficient product formulation and quality (Howard and Hood, 2007). These factors must be strategically considered to be able to commercialize such a trait, including the type of crop to be used for production. We have suggested a decision matrix to drive those considerations (Table 3.1).

Most output traits — pharmaceuticals, vaccines, and industrial products — are produced for markets rather than being

DOI: 10.1016/B978-0-12-381466-1.00003-1

Table 3.1 Decision matrix for a plant bioproduction system — plant, tissue, subcellular location

Production parameter	Production characteristic	Comments
Protein product	Toxic	Full containment — *in vitro* such as hairy roots
	Pharmaceutical	Consider hazard — plant or culture system
	Industrial	Commodity crop — large volume, deregulated
	Vaccine	Food quality — oral delivery
Market volume	Large	Need stability of storage — seed
	Medium	Other issues may drive
	Small	Not driven by production cost — leaf, root okay
Margin recovered	Large	Can tolerate high cost of goods
	Small	Need large volume and product stability to justify. Seeds most efficient
Yield per target tissue	High	Ability to increase event expression through site-directed insertion or breeding and selection; industrial products; seeds most efficient
	Medium	Vaccines — for direct delivery
	Low	Pharmaceuticals with high margin; *in vitro* or leaf — but requires immediate extraction
Formulation	Purified	Requires high margin to offset cost
	Extract	Value of product drives system
	Direct delivery	Industrial — compatibility with use; vaccine/pharmaceutical — compatibility with recipient
Metabolic impact	High	Test best location; not cytoplasmic, chloroplast, mitochondrial
	Medium	Test best location
	None	Only issue is high accumulation
Glycosylation or disulfide bonds	Necessary	ER/secretion pathway to Golgi
	No effect	Location in cell does not affect activity
	Inhibitory to activity	Non-secretory pathway
Other post-translational modification		As needed
		In correct compartment to accomplish
Regulatory issues	Food/non-food	USDA; FDA
	Wild relatives	Containment required
	Pesticidal	EPA

With thousands of species to select from and a wide variety of possible products, it is unlikely that any one system will work best for all types of products. Thus, depending on the protein product, the need for high- or low-volume production and the need for low cost versus high integrity of the protein are some of the issues that need to be considered when choosing a production system for recombinant proteins for industry or pharmaceuticals. Molecular product considerations aside from market and cost considerations drive decisions about production systems as well. These include post-translational modifications including glycosylation, purification requirements, metabolic impact on the host, and regulatory needs. All issues taken together drive the decision for each protein production system.

studied for their scientific interest or the basic research principles that they address. Thus, commercial considerations drive the system of choice. One of the first considerations is whether a market is known for a particular product in which the investigator has interest. An example of a pharmaceutical product that has recently been in the news is glucocerebrosidase, an enzyme used to treat Gaucher disease (Cramer et al., 1996; Radin et al., 1997). Twenty years ago, this enzyme was considered the most expensive drug on the market, and one could make a lot of money. However, a second consideration is how many customers the market supplies — only about 4000 patients are currently treated for this lysosomal disease — thus the market is fairly small in volume (i.e., a small number of units to be sold). For this type of pharmaceutical, the real story is the margin on each unit sold, which is quite large. Producing this protein in plants has allowed this market to be addressed with less expense and greater safety than producing the enzyme from an animal system (Ratner, 2010). This product has been approved for sales in France and is under consideration for approval at the FDA in the United States.

How much protein is required per tissue extracted to make the product profitable? Often for a high-margin pharmaceutical, there is much flexibility in the answer to this question. The higher the yield per tissue, the less extraction buffer (i.e., water) is required per dry weight and the better the economics. Leaf, cell culture, and root systems work well in the case of pharmaceuticals but require immediate attention upon harvest. Seed systems work as well, but if a commodity, the protein product may be too small a volume to make the seed a viable system when considering containment issues. For a product with a low margin recovered per unit sold, a stable, low-cost system is required.

Product delivery requires that the protein be in a particular formulation. For example, is the protein a pharmaceutical that must be injected to be useful? In that case it will need to be purified and the expression system should offer good purification options — hairy roots or cell culture could work well here. On the other end of the spectrum is direct delivery of the product utilizing plant tissue directly with minimal processing and no extraction. If an industrial enzyme, the tissue must be compatible with the process. If a vaccine or pharmaceutical, the system must be a food product to ensure safety.

Many proteins expressed in plants have an impact on the plant's metabolism (Table 3.1). Examples of this are avidin (Hood et al., 1997) and laccase in maize (Hood et al., 2003). In the avidin case, targeting the protein to the apoplast solved the problem. For laccase, the apoplast was the best location, but the enzyme still had detrimental effects on the plant, including inhibiting germination and browning the leaves. Perhaps this problem can be solved by using a different form of the enzyme, such as a zymogen (Woodard et al., 2003) or activatable enzyme (http://www.agrivida.com/).

The next thing to consider is what types of expression systems are necessary to ensure the protein's integrity. What type of post-translational modifications does the protein have? Are they required for activity or stability? Will plant glycosylation patterns have an impact (Ma et al., 2003)? Does the protein require disulfide bonds to fold properly? Are other modifications present or required? Many post-translational modifications happen in the endomembrane system (Buchanan et al., 2000) and thus the protein should be targeted there. If glycosylation is not required, the cytoplasm or the chloroplast can be the final destination, but each location must be balanced with yield and cost of goods considerations.

Finally, regulatory questions must often be considered. Will the plants be grown in the field or under greenhouse or *in vitro* conditions? How much material is needed? Answers to these questions may depend on the hazard of the product or the size of the market (Howard and Donnelly, 2004). At any rate, if field grown, permits will be required unless the crop is deregulated. If deregulation is the goal, then many other issues become important, such as whether the crop is a food or feed crop or if it has wild relatives in the growth zone. All of these issues can be resolved, but the costs associated with the resolution can vary widely (Chapter 34).

Once a system is chosen, the next thing to consider is how to make the protein express the best in that system. We have reviewed general subcellular targeting strategies as well as seed, leaf, and hairy root expression systems.

Approaches

As described previously, selecting a site for transgene product accumulation must integrate the desired characteristics of the product or trait as well as practical requirements for yield, stability, and recovery. The most common sites for recombinant protein accumulation utilize trafficking through the endomembrane system and include the endoplasmic reticulum (ER), the vacuole, and protein bodies derived from ER or vacuoles, and the apoplast (secretion pathway; Table 3.2). The endomembrane system contains the cellular machinery for protein glycosylation (N- or O-linked) and disulfide bond formation, as well as an elaborate protein folding/quality control system that may be important for biological function of the target protein (e.g., see review by Vitale and Pedrazzini, 2005). Accumulation of heterologous proteins in a non-endomembrane compartment has also been very successful for some proteins. These compartments include the plastid (as products of either the nuclear or the plastid genome) and specialized organelles such as oil bodies prevalent in oilseed crops (Table 3.2). The molecular signals required to direct proteins to each of these compartments are discussed in the next section.

Routing proteins to the endomembrane system

An N-terminal signal peptide is the pivotal signal for delivery of proteins into the ER, the entry point for secreted proteins and all luminal proteins that are subsequently trafficked to various endomembrane compartments. An analogous signal peptide is also utilized by transmembrane proteins whose amino terminus is positioned in the lumen (and may ultimately reside at the extracellular surface of the plasma membrane). This signal peptide is typically 20–30 amino acids long that comprises the amino terminus of the translation product of cytosolic ribosomes and is generally followed by a "stop translation" signal. The newly emerged peptide interacts directly with the signal recognition particle (SRP), which directs docking to the ER membrane via interaction with the membrane-bound SRP receptor. This interaction is required for directing the nascent peptide to a protein translocon complex, a prerequisite for resumption of protein translation. Thus, insertion of proteins into the ER is a co-translational process. The machinery for signal peptide cleavage (signal peptidase), N-linked glycosylation (oligosaccharide-protein, transferase glucose transferases, glucosidases, and α-mannosidase I), and disulfide bond formation (protein disulfide isomerase, PDI) reside in the ER lumen often associated with the inner surface of the ER membrane. Thus initial post-translational processing events occur as the protein is being translated/inserted into the ER lumen. An elaborate chaperone-mediated protein folding and editing process is then instituted to ensure appropriate processing of the protein, and those proteins that do not meet the compartment's quality control requirement (misfolded or incorrectly glycosylated) are shunted into the ERAD pathway for degradation (ER-associated degradation typically involving

Table 3.2 Protein targeting strategies to subcellular compartments

Target organelle	Location in protein	Nature of signal	Signal removed	Comments
Endomembrane system (lumen)	N-term	1–3 basic aa followed by 6–12 hydrophobic aa	Yes	Entry point for secreted proteins and proteins residing in ER, Golgi, vacuoles, PBs, and PSV
ER retention	C-term	KDEL; HDEL; SEKDEL	No	Retention/retrieval for luminal proteins; transmembrane proteins traffic by distinct signals
Vacuole	N-term	NPIR conserved domain	Yes	NTPPs and CTPP interact with distinct receptors for sorting to the vacuole
	C-term	No consensus identified	Yes	
	Internal	No consensus identified	Varies	
Chloroplast	N-term	Varied, often rich in S, T, small hydrophobic aa; lacking D, E	Yes	Additional signals direct to specific locations within thylakoid membranes

Multiple locations are available in the cell to accumulate proteins. However, for over-expressed pharmaceutical or industrial proteins, not all locations are equally viable. Because the "rules" of expression and accumulation are not known, often proteins must be empirically tested in a number of compartments to determine their best location. Some basic characteristics are described here in Table 3.2. Abbreviations: aa, amino acid; PBs, protein bodies; PSV, protein storage vacuoles; NTPP, N-terminal pro-peptide; CTPP, C-terminal pro-peptide; D, E, H, K, L, S, T, amino acid single letter code.

ubiquitinization and delivery to the cytosolic proteosome complex; reviewed in Vitale and Boston, 2008). Proteins that clear the ER quality control are packaged into vesicles for delivery to the Golgi for additional processing (e.g., modification of glycans) and sorting to vacuoles or other membrane-bound organelles. The "default pathway" in plants (i.e., where a protein goes if no additional targeting information is presented), is secretion to the extracellular space (Denecke et al., 1999). For Type I (single pass) and Type IIIa transmembrane proteins (those with N-terminus localized to the ER lumen), the signal peptide is typically cleaved and the hydrophobic transmembrane domains are set within the membrane during protein synthesis by a series of "stop transfer" and "start transfer" signals.

Signal peptides are "necessary and sufficient" for delivery of proteins to the ER and are generally "interchangeable." The signal sequence is not a defined sequence but a pattern or motif that typically comprises one or more positively charged amino acid residues at the N-terminus followed by a stretch of 6–12 hydrophobic residues and a cleavage site. Signal peptide prediction tools and databases have been developed (e.g., http://www.cbs.dtu.dk/services/SignalP/; links cited in http://www.signalpeptide.de/index.php?m=links) to identify putative signal peptides and the signal peptide cleavage site. For expression of heterologous proteins, plant-specific signal peptides (SPs) are often used. Common plant signal sequences include the signal peptides from tobacco extensin (De Loose et al., 1991; Xu et al., 2005), PR-S (Sijmons et al., 1990), and osmotin, (Dieryck et al., 1997), the barley α-amylase SP (Rogers, 1985) and the potato patatin SP (Medrano et al., 2009). In many cases, the signal peptide from the heterologous protein (e.g., a human or fungal sequence) efficiently targets its protein to the plant ER and is recognized by the plant signal peptidase to create the precise N-terminus of the mature product seen in its native organism, such as human IL-2, interferon-β, and β-casein (Edelbaum et al., 1991; Chong et al., 1997; Magnuson et al., 1998); fungal phytase (Verwoerd et al., 1995); and xylanase (Herbers et al., 1995). However, there are other examples of proteins whose expression levels were enhanced by the use of a plant SP (e.g., Sojikul et al., 2003) and ones where plants inefficiently processed more complex "pre- pro-" sequences from mammalian proteins (human serum albumin, Sijmons et al., 1990). For researchers wanting to optimize product yield, often multiple SPs are tested for initial assessment of production strategies. For pharmaceutical applications, precise cleavage of the SP — whether from animal, fungal, or plant sources — is critical and generally confirmed by N-terminal sequencing of the final purified product.

Accumulating proteins in the ER

A specific protein motif is typically required to retain proteins within the ER. The most widely used motif is the KDEL ER retention motif. Proteins having C-terminal KDEL (lys-asp-glu-leu) or HDEL (his-asp-glu-leu) interact with the KDEL receptor, a transmembrane protein that functions in vesicular trafficking primarily between the ER and the Golgi. KDEL-containing proteins are typically released in the ER environment and bound in the Golgi environment leading to ER retention or retrieval, respectively. There are many examples of plant-produced recombinant proteins that have been engineered to contain the KDEL retrieval signal, and in many cases this has led to higher levels of protein accumulation (reviewed in Boothe et al., 2010). However, there are also examples where addition of the KDEL motif did not enhance protein yields in plant systems so each protein of interest must be tested. It should also be noted that retention of the protein of interest in the ER may have consequences with respect to glycan composition, extractability, and "non-native" sequences of the final product. In situations where addition of a KDEL sequence has resulted in significantly higher product levels, it is possible that the greater protein stability is linked with development of "protein bodies" that partition products from proteinases and other destabilizing plant proteins. Systems to specifically engineer product accumulation within ER-derived protein bodies have been developed and are discussed in the next section.

Accumulating proteins in ER-derived protein bodies

Plant seeds function to store all of the resources required to support germination and initial growth prior to onset of photosynthesis. Thus, they have evolved unique mechanisms to accumulate and densely pack proteins in a way that retains function. This typically takes the form of membrane delimited storage organelles that are derived either from the ER, termed "protein bodies" (PBs), or from post-Golgi organelles termed "protein storage vacuoles" (PSVs; reviewed by Herman and Larkins, 1999). By understanding the mechanisms that trigger PB formation during seed development (Galili, 2004), technologies have been developed that induce ER-derived PBs in plant leaves and other tissues (Kogan et al., 2001; Mainieri et al., 2004). The storage proteins aggregate within the ER and either remain within this organelle or are budded off into discrete organelles. Products stored in ER-derived versus vacuole-derived PBs often differ in their glycan composition (routing through the Golgi facilitates processing of high-mannose N-glycans to complex glycans), and both types of protein bodies are considered stable environments for protein storage. It should be noted that plant-derived protein bodies do not seem to display the negative aspects of bacterial protein inclusion bodies that result from high-level expression of heterologous proteins in bacterial systems. Accumulation of foreign proteins in bacterial inclusion bodies often yields insoluble product requiring harsh processes of resolubilization and renaturation to recover an active product. Analogous issues have not been common in plant systems (with the possible exception of plastid localized proteins).

The proline-rich N-terminal domain of the γ-zein (maize storage protein) includes a highly repetitive sequence $(VHLPPP)_n$ that forms an amphipathic polyproline helix and is critical for zein protein aggregation at the ER membrane (Kogan et al., 2001). Mainieri et al. (2004) demonstrated that the fusion of 89 amino acid residues of γ-zein is sufficient to mediate the assembly of a target protein into PBs. A synthetic sequence consisting of $(PPPVHL)_8$ has been developed as a targeting tag (termed Zera®) to facilitate assembly and recovery of recombinant proteins (Torrent et al., 2009). Although PBs are seed storage organelles, the zein-derived sequences enable PB formation in leaf tissue (e.g., in *Arabidopsis* and tobacco) and interestingly, also direct formation of ER-derived protein bodies in non-plant species including fungal, insect, and mammalian cells (Torrent et al., 2009; Llompart et al., 2010). Accumulation of proteins in PBs may facilitate higher levels of accumulation as PBs typically sequester proteins away from proteinases and other destabilizing factors. In addition, the high density of PBs may provide a purification advantage (Llompart et al., 2010).

Accumulating proteins in the vacuole or vacuolar protein bodies

The plant vacuole is one of the largest subcellular compartments in a leaf, stem, or root, particularly if the cell is expanding. They are lytic compartments, function as reservoirs for ions and metabolites, and participate in cellular detoxification and general cell metabolism (Marty, 1999). Because of their lytic nature, vacuoles can be hostile environments for foreign protein accumulation. Nevertheless, some proteins are stable and accumulate to high levels in this acidic environment. In seeds, vacuoles store the protein reserves destined to feed the growing embryo upon germination and thus have the potential for the efficient packing and long-term storage that may be optimal for certain applications. It has been suggested that protein storage vacuoles are distinct organelles with distinct resident biomarkers and biogenesis (Paris et al., 1996; Vitale and Raikhel, 1999; Frigerio et al., 2008). Targeting domains from seed storage proteins have been used to specifically localize recombinant proteins to these types of vacuoles (discussed further in the next section). It should also be noted that vacuoles may possess additional glycan-modifying enzymes and thus, targeting to vacuoles may provide a glycoprotein with more defined or desirable N-linked glycan compositions.

Vacuolar targeting of a protein encoded by a nuclear gene requires dual targeting signals (Table 3.2). First, an ER signal sequence is required for entry into the endomembrane system. A second signal is active after the protein has progressed through the ER and Golgi network where it is carried in vesicles to the vacuole. Receptors for these sequences allow binding and delivery to the organelle. Vacuolar targeting signals are less tightly defined compared to the N-terminal ER-signal peptides and have been identified at the C-terminus (C-terminal pro-peptide, CTPP; e.g., barley lectin, phaseolin, tobacco chitinase) and the N-terminal region of the "mature" protein (N-terminal pro-peptide, NTPP, located immediately upstream of the ER signal sequence; e.g., sporamin, aleurain) as well as internal domains that direct vacuolar targeting (e.g., phytohemagglutinin, legumin, ricin). The NTPP and CTPP are typically removed by proteases within the vacuole. In some cases (e.g., the A-B plant toxins such as ricin and abrin), the internal vacuolar targeting sequences are also removed within the vacuole as part of protein processing to the mature active toxin. All three types of vacuolar targeting signals (C, N, and internal) have been shown to be necessary and sufficient to sort model proteins from the default secretion route into vacuoles (reviewed in Marty, 1999; Neuhaus and Rogers, 1998). However, specific vacuolar targeting signals may not function for a particular heterologous protein of interest. Choices for vacuole signal selection may reflect the target host plant (monocot or dicot), a desire to have the signal cleaved, potential impacts of vacuolar signal on secondary structure and recombinant protein activity, and interest in directing product to the lytic versus storage vacuoles.

Several examples of vacuolar accumulation of recombinant proteins are available including those that exhibit protein accumulation in the seed. In maize, the barley aleurain vacuolar targeting sequence (Holwerda et al., 1992) provides both ER and vacuolar targeting signals and directs the arrival of the foreign protein to the storage vacuole in the seed (embryo). Using the aleurain targeting motif, Streatfield et al. (2001) demonstrated high levels of vacuolar accumulation of the *Escherichia coli* heat labile toxin, LT-B. The β-1,4-endoglucanase from *Acidothermus cellulolyticus* is also highly stable in the vacuole using this sequence, although a cellobiohydrolase from *Trichoderma reesei* did not accumulate within vacuoles but was active when secreted to the cell wall, suggesting that the vacuolar environment was inhospitable for this enzyme (Hood et al., 2007). A recombinant dog gastric lipase (active in a highly acidic environment) was produced in tobacco plants comparing enzymes targeted to the vacuole versus the apoplast for yield and activity (Gruber et al., 2001). Expression levels were 5% (vacuole) and 7% (secreted) of total acid-extractable protein with equivalent activity suggesting that both compartments support this acid-stable protein. These examples illustrate the utility of the compartment as well as the empirics of the process — we need to know more to predict the stability of the protein of interest.

The first plant-made pharmaceutical protein to complete clinical trials (currently under FDA review in the United States but approved for human use in several European countries) is a human enzyme targeted to the plant cell vacuole. Glucocerebrosidase is a lysosomal enzyme and thus was considered likely to be stable in the more acidic vacuolar compartment. As described by Shaaltiel et al. (2007), the human acid β-glucosidase, glucocerebrosidase (enzyme replacement therapeutic for Gaucher disease), was modified to contain the SP from the *Arabidopsis* basic endochitinase gene (Samac et al., 1990) and the C-terminus storage vacuole targeting motif from tobacco chitinase A (DLLVDTM; Neuhaus et al., 1991). Additionally, the mannose-terminated glycan, thought to be the dominant complex glycan of vacuolar glycoproteins (Faye and Gomord, 2010), is considered advantageous for uptake and lysosomal delivery of the protein administered to patients.

Accumulating proteins in the apoplast

Secretion is the default pathway of the plant endomembrane system and without addition of specific signals for sorting or retention, proteins that traffic through the endomembrane system will typically be secreted to the extracellular space. Most large recombinant proteins accumulate within the apoplast — the region between the plasma membrane and the cell wall — as diffusion through the cell wall matrix is size delimiting. However, recombinant protein strategies using plant cell cultures are often employed to recover the target protein in the culture medium, which decreases the complexity of the initial purification stream and minimizes exposure to vacuolar/intracellular proteinases. Strategies to direct and enhance recovery of secreted proteins in plant cell culture systems are discussed in the next section.

Accumulating proteins in the chloroplast

Plastids (primarily chloroplasts but potentially tuber amyloplasts or fruit/flower chromoplasts) represent a relatively unique site for recombinant protein accumulation with several key advantages for high level protein production as well as distinct limitations. Proteins can be routed to plastids by several mechanisms. Proteins encoded by the nuclear genome (or proteins synthesized in the cytosol from viral or transfected DNAs) can be targeted to plastids by plastid transit peptides that direct proteins across the chloroplast membrane and into specific compartments within the chloroplast (reviewed in Bruce, 2000). However, the most compelling approach for using plastids involves direct transformation of the plastid genome (reviewed in Daniell et al., 2009; Bock and Warzecha, 2010). This technology provides significant gene copy number advantages since a plant leaf cell can contain up to 100 chloroplasts and each chloroplast has many copies of the plastid genome, supporting very high levels of recombinant protein production. For example, recombinant protein yields of more than 70% of total soluble proteins have been reported (Oey et al., 2009) and, as reviewed by Bock and Warzecha, 2010, vaccine antigens from numerous groups have been accumulated in tobacco leaves at levels ranging from 1% to more than 30% of total soluble proteins. In addition: (1) plastid transformation supports strong homologous recombination so that efficient sites for transcription can be reliably targeted, (2) the plastid genome appears resistant to gene silencing, and (3) plastids are generally maternally inherited providing strategies to minimize gene-flow issues. However, currently relatively few plants support efficient plastid transformation (solanaceous plants such as tobacco and tomato have been most successful), and plastids do not support the more complex post-translational modifications (e.g., glycosylation, disulfide bonding) that are required by many eukaryotic proteins for bioactivity or stability. In addition, to exploit the high level of protein accumulation in chloroplasts, the tissue of choice is typically leaves (i.e., tissues actively involved in photosynthesis), which elevates plastid numbers and synthetic capacity. Leaf tissues often have greater post-harvest challenges for short- and long-term storage compared to seeds and may have greater variations in production levels in the field and processing recoveries compared to seed-based production strategies. In addition, like bacterial systems, some heterologous proteins are difficult to recover in active form suggesting that inclusion bodies similar to those common in bacterial expression systems may come into play in plastid-based expression (Gleba et al., 2005).

Accumulating proteins on the surface of oil bodies

Plants, especially oilseed crops, typically store seed oils in oil bodies. These are organelles that encompass oils (e.g., triglycerides) in a single layer phospholipid membrane that contains the highly hydrophobic protein oleosin. Heterologous proteins have been expressed as oleosin fusions (Boothe et al., 2010; van Rooijen and Maloney, 1995). Oleosins, low molecular mass (Mr 16–24 kDa) polypeptides, consist of a hydrophobic domain flanked by two hydrophilic domains. Oleosins are initially targeted to the ER membranes although both C- and N-termini remain in the cytosol and the proteins are subsequently transferred to the oil bodies (Napier et al., 1996; Abell et al., 2004). Thus, the fused protein of interest essentially coats the oil bodies and is positioned on the cytosolic face. Advantages of this protein production strategy can include very stable long-term storage in dry seeds and efficient purification based on "floating" the oil bodies. The recombinant protein-decorated oil bodies can be used directly in some applications such as topical applications, vaccines, or *in vitro* catalysis. Additional strategies include the incorporation of a proteolytic cleavage site between the oleosin and the protein of interest enabling release of the target protein after purification of the oil bodies. Applications of this technology could be limiting for some proteins that require post-translational modifications localized to the endomembrane system or are unstable in the cytosolic milieu. Researchers at SemBioSys have cleverly overcome this issue by targeting the protein of interest through the endomembrane systems, often with an ER retrieval domain as well, then retrieving the protein onto oil body surfaces through binding with an anti-oleosin single chain antibody (scFv). Thus, the product is trafficked and accumulated within the endomembrane system for post-translational processing, but associates with the oil bodies upon cell breakage, providing the advantages of oil-body-based flotation centrifugation (see Boothe et al., 2010), combining the benefits of both systems.

Examples of pharmaceutical proteins produced using this oil body system include human insulin (Nykiforuk et al., 2005), human growth factor (Boothe et al., 2010), and Apolipoprotein $A1_{Milano}$ (Nykiforuk et al., 2010). Apo-$A1_{Milano}$, a protein with significant cardioprotective potential, was produced in safflower seeds at levels of 7 g recombinant protein per kilogram of seeds. Human growth factor was expressed at levels up to 1.58% of total seed protein and, using the oil body flotation centrifugation process, more than 90% of soluble plant proteins were removed by the liquid–liquid separation with full retention of the hGF:oleosin fusions (Boothe et al., 2010).

Seed-Based Expression Systems

Seeds are one of the likely locations for accumulating a protein for utilization as a pharmaceutical or vaccine product, or for application to an industrial process. What are some of the major reasons that one would choose seed-based expression for an output trait? Seeds are easy to store and the processes for storage are well-developed. As long as the seeds are from a well-known crop, whether for food/feed or other use, the handling processes are also well-established. Seeds are very stable environments for proteins and many examples show stable long-term storage of recombinant proteins (Kusnadi et al., 1997; Tackaberry et al., 1999; Howard, 2005), with many protease inhibitors and carbohydrates that assist in long-term storage, and stabilization of protein integrity.

A number of proteins have been expressed as output traits specifically in seeds with excellent results (Table 3.3; Daniell et al., 2009; Khan et al., 2010). Proteins from viruses, bacteria, fungi, plants, animals, and humans have been transformed into plants with successful seed-based expression, although relatively few examples from each group are represented. The following discussion focuses on proteins from groups of organisms.

Although few viral proteins have been expressed in seed (Table 3.3; Daniell et al., 2009; Khan et al., 2010), one can observe that the examples to date of viral proteins in seed show a tendency toward relatively low expression levels compared to proteins from other organisms. The glycoprotein B (gB) from cytomegalovirus, a human herpes virus, was expressed in tobacco from the rice glutelin promoter (Tackaberry et al., 1999). gB is a 150 kDa membrane-bound glycoprotein that also has disulfide bonds (i.e., has many post-translational modifications). The protein was correctly processed and showed activity although the accumulation level was only 0.015% of total soluble protein (TSP). This monocot seed storage protein promoter functioned well in tobacco, a dicot, to produce the mRNA for this complex protein. In contrast, the hepatitis E virus (HEV) E2 antigen from the second viral open reading frame (ORF2) was expressed in tobacco plastids using the rice psbA promoter (Zhou et al., 2006). Although significant amounts of the protein were expressed in leaves (0.1% TSP), 100-fold less protein (0.001% TSP) accumulated in seeds using this system (Table 3.3). This is likely due to the plastids being less prevalent or less developed in tobacco seeds than in leaves. In the maize seed system, two viral proteins have been accumulated to several-fold higher amounts than those expressed in tobacco. The hepatitis B virus (HepB) surface antigen was expressed from the maize globulin-1 promoter and targeted to the endomembrane system. This membrane protein likely is localized to the plasma membrane and may form virus-like particles as has been shown for HepB in tobacco leaves (Mason et al., 1992), although the observation of this phenomenon in seeds has not been confirmed. The S (spike) protein from the transmissible gastroenteritis virus (TGEV), a porcine virus, was also expressed in maize seed (Streatfield et al., 2001). A maize constitutive promoter was used and the protein moved through the endomembrane system with the barley alpha amylase signal sequence (Rogers, 1985). The gene was completely re-synthesized with maize codon usage bias and showed expression levels of 0.4% TSP (Table 3.3). Thus, the best viral expression was shown with genes expressed from strong promoters, targeted to the membrane system, and codon optimized.

The heat labile antigen from *E. coli* (LT-B) is the best example of a bacterial protein expressed in plants (Streatfield and Howard, 2003; Chikwamba et al., 2002). Chikwamba et al. (2002) used an LT-B gene that was synthesized with averaged maize and potato optimized codons (Mason et al., 1992). The

Table 3.3 Examples of seed-based expression of recombinant proteins

Plant system	Protein	Promoter	Subcellular location	Expression level	Comments	Reference
Tobacco	Cytomegalovirus glycoprotein B (viral)	Rice glutelin	Apoplast	0.015% TSP 686 μg/g DW	Monocot promoter in dicot; active in cellular assay	Tackaberry et al. (1999)
Tobacco	HEV E2 (viral)	Rice psbA	Plastids	Lvs 0.1% TSP Seeds 0.001% TSP	Plastid accumulation in seed is low with CP transformation	Zhou et al. (2006)
Maize	HepB surface antigen (viral)	Maize globulin	Plasma membrane	0.1% TSP	First 40 codons maize-optimized	J. Howard, unpublished
Maize	TGEV S protein (viral)	Maize ubiquitin 1	Apoplast	0.004% DW (~0.4% TSP)	Gene completely optimized; active in preventing disease	Streatfield et al. (2001)
Maize	*E. coli* LT-B (bacterial)	CaMV 35S γ-zein	Apoplast	0.01% TSP (T1) 0.07% TSP (T1)	Mistargeted to amyloplast; stable over three generations	Chikwamba et al. (2002, 2003)
Maize	*E. coli* LT-B (bacterial)	Maize globulin	Vacuolar	10% TSP	Highest T1 seed; better than ER or apoplast	Streatfield and Howard (2003)
Rice	Lipoprotein (Ag473) (bacterial)	CaMV 35S	Not discussed	0.8% TSP	Also expressed in leaves; better than Rubisco SS in seed	Yiu et al. (2008)
Tobacco	Growth hormone (human)	Sorghum γ-kafirin	Apoplast	0.16% TSP	Monocot promoter in dicot; no protein in leaves or roots	Leite et al. (2000)
Tobacco	hIGFBP-3 (human)	Phaseolin	Protein storage vacuoles	0.08% DW	Correctly targeted; codon optimized	Cheung et al. (2009)
Rice	Insulin-like growth factor (human)	Luminal binding protein (BiP)	ER	6.8% TSP	BiP may be transported out of Golgi with saturated HDEL receptor	Xie et al. (2008)
Arabidopsis	Insulin (human)	Phaseolin	Oleosin fusion	0.13% TSP protein	Visible on stained gel	Nykiforuk et al. (2006)
Tobacco	Acid β-glucosidase (human)	Soybean 7S globulin	Protein storage bodies	200–500 U/kg seed	Glycosylation essential; no fucose or xylose; leaf expression failed; inhibitory to seed viability	Reggi et al. (2005)
Rice	hGM-CSF (human)	Rice glutelin	Protein storage bodies	1.3% TSP	Biologically active	Sardana et al. (2007)
Rice	Type II collagen-4X peptide (human)	Glutelin	Glutelin-fusion protein bodies	1 μg per seed	In protein bodies — correctly targeted	Hashizume et al. (2008)
Maize	Avidin (bovine)	Maize ubiquitin	Apoplast	0.2% DW in mixed population	Toxic in cytoplasm; seventh generation backcrosses; 0.4% DW in positive seed	Hood et al. (1997); Masarik et al. (2003)

Table 3.3 (Continued)

Plant system	Protein	Promoter	Subcellular location	Expression level	Comments	Reference
Maize	Trypsin (bovine)	Maize globulin	Apoplast	0.3% TSP	Requires zymogen form to be expressed	Woodard et al. (2003)
Maize	Brazzein (plant)	Maize globulin	Apoplast	10% TSP	Small, stable sweet protein	Lamphear et al. (2005)
Maize	Laccase (fungal)	Maize globulin	Apoplast	0.8% TSP extractable	Inhibits germination; 90% non-extractable	Hood et al. (2003)
Maize	Mn peroxidase (fungal)	Maize globulin	Apoplast	14%TSP	Plant damage with constitutive promoter	Clough et al. (2006)
Maize	Endo-cellulase E1 (bacterial)	Maize globulin	ER and vacuole	0.1% DW	Seventh generation backcrosses	Hood et al. (2007) and Hood, unpublished
Maize	Exo-cellulase CBH I (fungal)	Maize globulin	CW and ER	0.2% DW	Seventh generation backcrosses	Hood et al. (2007)

Abbreviations: TSP, total soluble protein; DW, dry weight; CW, cell wall; ER, endoplasmic reticulum; CBH I, cellobiohydrolase I; hGM-CSF, human granulocyte macrophage colony stimulating factor.

promoter and subcellular location had a dramatic effect on protein accumulation — γ-zein promoter driven LT-B accumulated to seven times higher levels than CaMV 35S promoted gene expression in maize seed (Chikwamba et al., 2002). In this case, the signal sequence used with the gene was the native bacterial signal peptide plus or minus the SEKDEL for ER retention. The protein with just the signal peptide accumulated in the amyloplast, which was a totally different compartment than expected (Chikwamba et al., 2003). This phenomenon indicates that not all parameters of targeting are completely understood (Hood, 2004a). Streatfield and Howard (2003) expressed the LT-B from a constitutive promoter, maize ubiquitin, and targeted the protein to various compartments. In this case, the targeting sites included the nucleus, the chloroplast, the apoplast, the ER, the cytoplasm, and the vacuole (Streatfield and Howard, 2003). The top three locations were the vacuole, the apoplast, and the ER, although the vacuole was the best for this maize codon optimized gene. The cytoplasm was the worst by over four orders of magnitude.

Human proteins have been expressed in numerous systems, numerous tissues, and with numerous targeting sequences (Giddings, 2001; Ma et al., 2003; Stoger et al., 2005; Daniell et al., 2009). Among the examples in Table 3.3, many expression levels have been achieved, although all are expressed from strong seed-specific promoters. One of the obvious conclusions from the data (Table 3.3) is that the proteins represent extremely divergent character from growth factors, to collagen, to a lysosomal enzyme. Thus, comparing their expression characteristics is not highly relevant. However, two interesting examples stand out. First, the human acid β-glucosidase (glycocerebrosidase) was originally designed to be expressed in leaves but no plants with expression were recovered (Reggi et al., 2005). The authors switched to a seed-specific promoter and recovered plants with high activity. Because glycosylation is critical to protein function, this protein was targeted to the endomembrane system. Although good expression could be recovered, the seeds showed symptoms of physiological damage at the higher expression levels. Nevertheless, this same enzyme was successfully expressed by the Cramer group in tobacco leaves in the early 1990s, but targeted to the vacuole (Cramer et al., 1996; Radin et al., 1997). The promoters and targeting sequences differed from those of the Reggi group, reinforcing the concept of exploring several plant/organ/subcellular combinations to achieve success.

An interesting application of seed-based expression is the oil body fusion (Boothe et al., 2010). Several enzymes and pharmaceutical proteins have been expressed using this technology including insulin, xylanase, and Apolipoprotein B. These protein fusions accumulate to high levels on oil bodies in oil seeds and are easily purified by flotation.

The maize seed expression system has been employed to express for a number of proteins from viral, bacterial, fungal, plant, and animal sources (Table 3.3). Several subcellular locations in the embryo have been tested for the best location(s) for foreign protein accumulation, as described earlier for LT-B (Streatfield and Howard, 2003; Hood, 2004b). Laccase accumulation was also tested in numerous subcellular locations, with the apoplast yielding the best results (Hood et al., 2003). All plants expressing laccase showed damage, suggesting that the tissue-specific promoter was leaky. Interestingly, the ER was predicted to be an additional high-expressing location, but this was true only when expressed from a constitutive promoter (Hood et al., 2003). When the ER was the target location using the high-expressing, embryo-specific maize globulin-1 promoter, no plants were recovered. This suggests that the expression may have been high in the ER, but the enzyme damaged the cells and they could not survive.

Thus, in the maize studies looking at multiple subcellular locations and with tissue-specific as well as constitutive promoters, three locations stand out as the winning locations for high-level accumulation of proteins: the apoplast, the ER, and the vacuole (Table 3.3). This information allows one to narrow the systems to test when a new protein of interest is in the queue. These three locations were used in all subsequent maize protein expression studies. Almost all examples of proteins expressed in maize accumulated to the highest levels in the apoplast with either the tissue-specific or the constitutive promoter. Two exceptions are the endo- and exo-cellulases (Hood et al., 2007). The E1 bacterial endocellulase accumulated to the highest levels in the ER and vacuole, whereas the fungal CBH I exo-cellulase accumulated to the highest levels in the ER. Unfortunately, the ER-localized protein, although enzymatically active, was truncated and not useable for industrial applications. Neither gene was codon optimized but was fused to a maize-optimized barley alpha amylase signal sequence.

In summary, the seed-based expression system offers the advantages of protein storage stability, high volume production, and oral delivery when in a food crop. Seed systems are amenable to breeding for better quality germplasm and higher protein accumulation levels (Hood et al., 2003, 2007). There are numerous organs, tissues, and subcellular locations that allow the accumulation of high levels of different protein classes. The details are still empirical, although a great deal of information is now available to guide our choices.

Leaf Systems

Plant-based bioproducts from plants with large foliage volume (i.e., tobacco, alfalfa, lettuce, etc.) allow high biomass yield and easy scale-up of a target protein. Examples abound of recombinant proteins in tobacco leaves (Table 3.4). *Nicotiana tabacum* and *N. benthamiana* have served as model species for plant sciences for almost two decades, having succeeded as a crop system for molecular farming. They are strong candidates as hosts for production of recombinant proteins (Cramer et al., 1999; Schillberg et al., 2002; Stoger et al., 2002; Commandeur et al., 2003).

The pharmaceutical industry needs an efficient expression system for therapeutic proteins. The advantages of tobacco leaves as a bioproduction system include: (1) transformation procedures are well established; (2) biomass yield is high in a leafy plant; (3) scale-up is easy; (4) high soluble protein levels can be achieved; (5) flexibility is available in the expression methods (transient-based via *Agrobacterium* or viral vectors, and stable nuclear or chloroplast genome transformation); (6) it is not a food or feed crop, therefore it has a low risk for transgenic products to enter the food or feed chain; (7) plant growth requirements are inexpensive and simple; (8) it has the ability to perform eukaryotic post-translational modifications; and (9) it is possible to co-express several genes at the same time. Among the disadvantages of using *Nicotiana* leaves are: (1) high levels of toxic alkaloids exist, although some low-alkaloid varieties are available (such as cultivar 81V9), of particular importance for oral delivery; (2) the protein product may be unstable and subject to degradation; and (3) highly efficient purification is expensive and time-consuming because of the high protein background in leaves (Ma et al., 2003; Clemente, 2006; Goodin et al., 2008; Tremblay et al., 2010).

Antibodies represent one of the major classes of proteins that have been produced in plants. Several plants have been hosts, but the overwhelming majority of them have been produced in tobacco leaves (De Muynck et.al., 2010). They are targeted to the endomembrane system since they require glycosylation, disulfide bonds, and assembly of multiple subunits. Plants are a great system for antibody production because the capacity of mammalian cell culture is limiting. Plants, particularly tobacco, are a rapid system that produces a diverse array of product (De Muynck et al., 2010).

Besides tobacco, alfalfa (*Medicago sativa* L.), cherry tomatillo (*Physalis ixocarpa* Brot), collard (*Brassica oleracea*), lettuce (*Lactuca sativa*), potato leaves (*Solanum tuberosum*), rice leaves (*Oryza sativa*), spinach (*Spinacia oleracea*), soybean (*Glycine max*), and tomato leaves (*Solanum lycopersicum*) are other leafy crops that have been used for production of recombinant proteins (Table 3.4). Alfalfa has ample biomass suitable for the production of high-value and -volume recombinant proteins, and has no nitrogen fertilizer requirements because it fixes atmospheric nitrogen through symbiotic association with root-invading bacteria. Alfalfa is easily propagated by stem cuttings (Vlahova et al., 2005).

One of the major limitations of the expression of recombinant antigens in transgenic plants is achievement of a yield that is sufficient to meet commercial needs. Yield remains largely unpredictable and there is great variation of expression of different proteins even using the same promoter, vector, and expression system. To achieve higher yields, different stages of protein expression in plants can be optimized. Our research results using a tobacco transient transformation system for three homologous genes: mouse, chicken, and human interleukin-12 (IL-12), have shown completely different levels of expression. For example, mouse IL-12 had the highest expression (89 μg/g FW) and can meet commercial levels when expressed in plants, but human IL-12 is very poorly expressed (0.12 μg/g FW; Figure 3.1), whereas the chicken IL-12 counterpart had intermediate expression (4.5 μg/g FW: Medrano et al., 2009, 2010). All three genes had the same promoter and targeting site, illustrating again the empirical state of the science and the principles yet to be elucidated.

Stable versus transient leaf expression systems

Research using plants for recombinant protein expression originally was accomplished by stably transforming the nuclear genome of *N. tabacum*. DNA is first introduced into the target cell. After passage through the plant cell wall and membrane, the introduced DNA then proceeds to the nucleus, passes through the nuclear membrane, and becomes integrated into

Table 3.4 Leaf-based expression of recombinant proteins

Plant system	Protein	Promoter	Subcellular location	Comments	Reference
Alfalfa leaves	Zera-*Yersinia pestis* F1-V antigen	35S CaMV and CsVMV	Cytoplasm, ER	0.015–0.023% TSP	Alvarez et al. (2010)
Alfalfa leaves	Human granulocyte colony stimulating factor (hG-CSF)				Aguilar et al. (2009)
Cherry tomatillo leaves	Hepatitis B surface antigen HBsAg	35S CaMV		90–300 ng/ FW	Gao et al. (2003)
Collard leaves	Smallpox vaccine virus B5 antigenic domain	35S CaMV and RbcS	Apoplast and subcellular outer membrane anchor	100 mg/kg FW	Golovkin et al. (2007)
Lettuce leaves	Malaria (CTB-ama1 and CTB msp1)		Chloroplast	7.3% (CTB-ama1) and 6.1% (CTB msp1) TSP	Davoodi-Semiromi et al. (2010)
Lettuce leaves	Heat-labile enterotoxin B subunit (LTB)	35S CaMV		1–2 % TSP	Kim et al. (2007)
Lettuce leaves	Cholera toxin B subunit-human proinsulin (CTB-Pins)	*psbA*	Chloroplast	Up to ~2.5% TSP	Ruhlman et al. (2007)
Tobacco N. benthamiana	Zera-human growth hormone (hGH)	35S CaMV	ER	3 mg/kg	Llompart et al. (2010)
Tobacco N. benthamiana	Zera-*Yersinia pestis* F1-V antigen	35S CaMV and CsVMV promoter	Cytoplasm, ER	0.08–0.11% TSP	Alvarez et al. (2010)
Tobacco N. benthamiana	Chicken IL-12	Double-enhanced 35S CaMV		0.4% TSP	Medrano et al. (2010)
Tobacco N. benthamiana	Mouse IL-12	Double-enhanced 35S CaMV		3–7% TSP	Medrano et al. (2009)
Tobacco N. benthamiana	Elastin-like polypeptide-GFP fusion	35S CaMV	Cytoplasm, chloroplasts, apoplast, and ER	11% TSP in the ER	Conley et al. (2009)
Tobacco N. benthamiana	Resveratrol synthase (RS)	Double-enhanced 35S CaMV	Cytoplasm	ND	Condori et al. (2009)
Tobacco N. benthamiana	HIV type 1 mucosal vaccine (CTB-MPR) membrane proximal (ectodomain) region of gp41		ER	0.01–0.2% TSP	Matoba et al. (2009)
Tobacco N. benthamiana	Recombinant hemagglutinin (HA) from pathogenic avian influenza virus (H5N1) subtype	35S CaMV	ER	400 mg/kg FW	Shoji et al. (2009a)
Tobacco N. benthamiana	Pathogenic avian influenza virus (H5N1 subtype)	35S CaMV	ER	60 mg/kg FW	Shoji et al. (2009b)
Tobacco N. tabacum	Malaria (CTB-ama1 and CTB msp1)		Chloroplast	13.17% (CTB-ama1) and 10.11% (CTB msp1) TSP	Davoodi-Semiromi et al. (2010)
Tobacco N. tabacum	Zera-Ct and Zera-EGF	Enhanced 35S CaMV	ER	Zera-EGF 0.5 g/kg FW and Zera-hGH 3.2 g/kg FW	Torrent et al. (2009)
Tobacco N. tabacum	Cholera toxin B subunit-human proinsulin (CTB-Pins)	*psbA*	Chloroplast	16% TSP	Ruhlman et al. (2007)

(*Continued*)

Table 3.4 (Continued)

Plant system	Protein	Promoter	Subcellular location	Comments	Reference
Tobacco leaves N. tabacum	Zein and phaseolin	35S CaMV	ER	Zeolin 3.5% TSP and phaseolin-KDEL 0.5% TSP	Mainieri et al. (2004)
Tobacco leaves N. benthamiana and *N. tabacum*	HA1 antigenic domain of the H5N1	Ribulose bisphosphate carboxylase (RBC)	Apoplast secretion and ER	Stable 4 mg/kg of FW	Spitsin et al. (2009)
Potato leaves	Rotavirus caprid protein (VP6)	35S CaMV		0.006% TSP	Matsumura et al. (2002)
Rice leaves	Antigenic lipoprotein (Ag473)	35S CaMV		0.18–0.6% TSP	Yiu et al. (2008)
Spinach leaves	Rabies virus glycoprotein and nucleoprotein	Subgenomic mRNA promoters specific for AlMV and TMV CP		60 μg/g FW	Yusibov et al. (2002)
Soybean leaves	Bean pod mottle virus (BPMV)-GFP			1% TSP	Ghabrial et al. (2009)
Tomato leaves	Accessory colonization factor subunit A (ACFA)-CTB	Double-enhanced 35S CaMV		0.0003% to 0.06% TSP (ACFA) 0.006% to 0.02% TSP (CTB-ACFA)	Sharma et al. (2008)
Tomato leaves	Mouse IL-12	35S CaMV		2.7–7.3 μg/g FW	Gutierrez-Ortega et al. (2005)

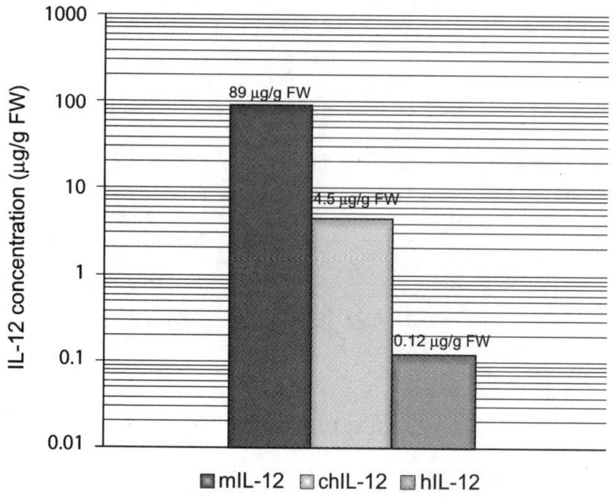

Figure 3.1 • Three different IL-12 homologous genes have different expression levels in the same transient expression system. • Three different sandwich ELISA methods were used to determine protein quantity of each IL-12 species (mouse, chicken, and human IL-12) expressed in a transient system in *N. benthamiana* leaves. Transient expressed empty vector was used as a negative control for all three ELISAs.

the genome. It is believed that the introduced DNA can function for a short time in the nucleus as an extrachromosomal entity, but integration into the genetic material of the target cell is necessary for long-term functionality and expression. Nuclear transformation for tobacco plants has a number of advantages including that transgene expression is more stable and it is useful when the objective is to study protein distribution within cell types other than leaf epidermis.

In contrast, transient expression is typically carried out in tobacco leaves of *N. benthamiana* (Thomas et al., 2003; Voinnet et al., 2003). Transient expression systems are vectored either by *Agrobacterium*, viral vectors, or a combination of these, and are rapid techniques to assess the expression of different targeted proteins whose respective precursor genes are transcribed extrachromosomally. Stably transformed plants are not generated (Wydro et al., 2006). Transient expression usually takes from 2 to 4 days, whereas stable lines are generated after approximately 2 to 6 months — a more time-consuming method. The transient expression method is generally used to facilitate rapid analysis of the transformation construct integrity and to validate activity of new recombinant proteins (Sheen, 2001; Voinnet et al., 2003; Dhillon et al., 2009). Development of this agro-infiltration process for large-scale recombinant protein production is actively being explored by both academic and industry researchers (Fischer et al., 2004; Wydro et al., 2006). In contrast to stable nuclear transformation, the transient infiltration system yields high levels of expression, does not require sterile conditions after transformation, and is a rapid protein production system. One disadvantage of the transient system is that harvesting at a particular time is necessary.

We have developed an *Agrobacterium*-mediated transient expression system in tobacco that provides high yields of protein and a significant amount of leaf tissue. This system provides a rapid source of transgene product for assessing post-translational modifications, purification strategies, and bioactivity, as well as an effective system for optimizing

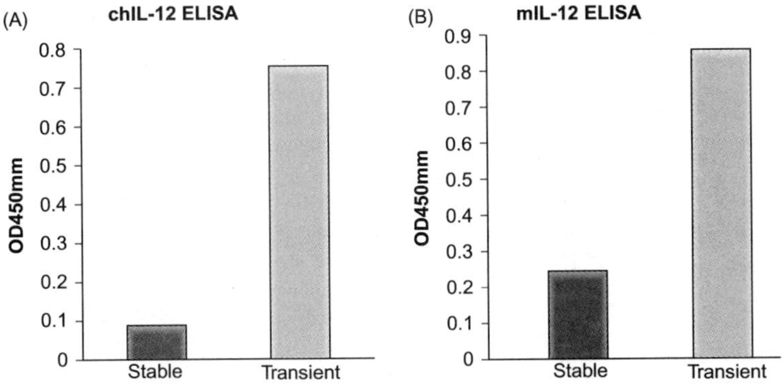

Figure 3.2 • Comparison of the expression levels of IL-12 by two methods, stable and transient transformation. • The highest expressing line obtained by stable transformation in *N. tabacum* was used to compare levels with transient expression system in *N. bethamiana* using the same construct (A) chIL-12 or (B) mIL-12.

construct elements. Using the immunomodulator mouse IL-12 as a model pharmaceutical product, we obtained bioactive recombinant protein at levels exceeding 5% of total soluble leaf protein (Medrano et al., 2009). What we have learned from our experience using IL-12 as a model protein is that higher yields are possible using the transient method compared with stable transformation using the same construct.

In Figure 3.2, IL-12 serves as the model gene of interest for discussion of yields in stable versus transient-based protein expression methods. This 70 kDa heterodimeric glycosylated cytokine is an immunomodulator composed of an α-chain (p35 subunit) and a β-chain (p40 subunit) with promising therapeutic and vaccine applications due to its role in directing cell-mediated immunity (Adorini, 1999; Fieschi and Casanova, 2003). We have expressed a single chain form of chicken (Medrano et al., 2010) and mouse IL-12 (Medrano et al., 2009) with a glycine-serine linker. Both forms retain the full biological function of the native heterodimer (Lieschke et al., 1997). In addition, we analyzed the highest expressor from 26 stable independent chIL-12 clones from *N. tabacum* and compared them with chIL-12 derived from transient transformation of *N. benthamiana*. Higher yield was obtained from the transient expression host, 7 times more for chIL-12 and 3.4 times more for mIL-12 compared with stable expression in *N. tabacum*.

Protein bodies in leaves

The ER seems to be a compartment that efficiently accumulates proteins, where stability, folding, and post-translational modifications are enhanced. This organelle appears to alleviate downstream issues and facilitate production of difficult proteins, making the production of biopharmaceuticals at commercial levels possible. A proprietary technology called Zera® (originally named γ-zein) is being developed by ERA Biotech (Barcelona, Spain). The technology, discussed in the previous section, is based on *in vivo* protein accumulation in ER-derived artificial storage organelles called StorPro bodies. It simplifies purification and enhances productivity of known cell lines by *in vivo* encapsulation and one-step concentration of protein by density followed by Zera®-affinity capture (Herman and Larkins, 1999; Llompart et al., 2010).

The Zera® peptide is a proline-rich domain of a plant storage protein, that when fused to a target protein has two functions: (1) self-assembling of non-secretory storage organelles, and (2) protein body formation properties in the ER. Zera® technology has a number of advantages: (1) streamlines protein recovery by density, delivering pre-purified protein that has an impact on cost-savings, time consumption, and process development; (2) protects proteins inside of StorPro bodies from proteolytic degradation; (3) is suitable for expression of difficult-to-express proteins, such as toxic-to-cell molecules, unstable or labile products, and membrane proteins; (4) is compatible with transient and stable transformation; (5) increases yield and quality due to *in vivo* encapsulation that improves the capacity of the cell factory to assemble and accumulate complex products; and (6) simplifies/accelerates downstream processing.

Hairy Root Cultures

In addressing increased concerns about regulatory compliance and product safety, there has been renewed interest in "molecular farming" with *in vitro* cultured plant cells, tissues, and organs (Huang and McDonald, 2009; Shih and Doran, 2009). Hairy roots appear as one of the most attractive *in vitro* expression systems with several advantages when compared with field-cultured plants (Guillon et al., 2006; Georgiev et al., 2007). Hairy roots are generated by infection of plants with the gram negative soil bacterium *Agrobacterium rhizogenes*, which contains a large root-inducing (Ri) plasmid (Shanks and Morgan, 1999). Integration into the plant genome of the T-DNA carried on the Ri plasmid results in differentiation and growth of neoplastic (oncogenic) roots, called hairy roots, at the infection site (Guillon et al., 2006; Shanks and Morgan, 1999). These induced root tissues can be excised from the host plants and grown indefinitely *in vitro*. For the past three decades, hairy roots have been successfully induced from more than 116 plant species (Rigano and Walmsley, 2005) and have become an essential tool in a wide range of fundamental studies as well as for large-scale tissue culture purposes (Guillon et al., 2006). Except for the natural genes introduced into hairy roots by *A. rhizogenes*, other genes of interest, such as those encoding animal proteins, can be readily introduced using the same process as that used to create transgenic plants in other systems. This makes hairy root cultures a viable alternative production platform for heterologous proteins from which the first product is now on the verge of commercialization.

Advantages of the hairy root culture system

Several general features of hairy root cultures confer significant regulatory and technical advantages compared with the culture of whole plants or dedifferentiated plant suspension cells. Since hairy roots are cultured in sterile and controlled environmental conditions, the cultures are independent of the climate, soil quality, season, and weather (Hellwig et al., 2004). No issues with pathogen and herbicide contaminants, transgene dissemination, or other environmental concerns linked with whole plant systems are present (Guillon et al., 2006a; Putalun et al., 2003). Thus Good Manufacturing Practices (GMP) procedures can be readily implemented throughout the production pipeline in alleviating a number of regulatory concerns regarding plant-made pharmaceuticals. Another important advantage of the hairy root culture system is the possible extracellular secretion of expressed proteins. This Rhizosecretion (Guillon et al., 2006b) offers a simplified method for the recovery of foreign proteins from an inexpensive and well-defined medium that lacks exogenous proteins. Therefore, the cost for downstream processing could be significantly reduced. In addition, hairy roots are fast growing cultures that reach large biomass volumes within a short time (approximately one month), so they are especially useful when the product is required more quickly than that obtained from agriculture (Putalun et al., 2003). Although these advantages are shared by suspension cell culture systems, the genetic and biochemical stability of hairy roots and their efficient productivity offer substantial advantages over suspension cells where somaclonal variations and production instability can occur with relatively high frequency (Martinez et al., 2005). In addition, due to altered auxin metabolism, hairy roots are able to grow on plant hormone-free media (Georgiev et al., 2007), thus offering another attractive advantage over suspension cell cultures. Hairy root cultures integrate the merits of whole-plant cultivation with those of a suspension cell culture system, holding immense potential for the pharmaceutical industry.

Recombinant proteins expressed with hairy root cultures

Due to the rapid growth, long-term genetic stability, and high productivity properties of hairy root cultures, they have been explored for decades for their potential to produce valuable metabolites, particularly secondary metabolites (Kim et al., 2002). The interest in production of pharmaceutical proteins with the hairy root system started in 1997 when Wongsamuth and Doran (1997) reported the first application of hairy roots for synthesis of a full-length murine IgG_1 monoclonal antibody. Fully functional antibody with a maximum yield of 18 mg/L was achieved in their research, and up to 43% of the antibody was secreted when the culture medium was supplemented with polyvinylpyrrolidone (PVP) and gelatin. Since then, production of 15–20 recombinant proteins have been reported from hairy root cultures, as shown in Table 3.5. These include reporter proteins (e.g., GUS and GFP; Medina-Bolivar and Cramer, 2004; Lee et al., 2007), enzymes (Gaume et al., 2003; Woods et al., 2008), antibodies (Sharp and Doran, 2001a; Martinez et al., 2005), antigens (Ko et al., 2006; Kumar et al., 2006; Rukavtsova et al., 2007), growth hormone and growth factor (Komarnytsky et al., 2006; Parsons et al., 2010), and immunomodulators such as ricin B (the non-toxic lectin subunit; Medina-Bolivar et al., 2003) and murine interleukin-12 (mIL-12; Liu et al., 2008; 2009). A high yield of protein was observed in the expression of acetylcholinesterase that accumulated to levels of up to 3.3% of TSP (Woods et al., 2008). Because foreign proteins expressed in hairy roots are often secreted into the culture medium, this feature has been exploited recently for expression of human alkaline phosphatase (Gaume et al., 2003), ricin B (Medina-Bolivar et al., 2003), and IgG antibodies (Wongsamuth and Doran, 1997; Sharp and Doran, 2001a,b; Komarnytsky et al., 2006). These proteins contain a signal sequence for secretion through the endomembrane system. However, the secreted proteins were found to be vulnerable to protease degradation in the medium (Sharp and Doran, 2001a,b; Zhang et al., 2005). To overcome this problem, Zhang et al. (2005) developed an aqueous two-phase extraction (ATPE) system comprising PEG/sodium sulfate for fast recovery of ricin B, not only to partially purify the protein but also to improve its stability. In addition, due to long-term genetic stability of root tissues compared with suspension cells, the production of antibody IgG_1 with tobacco hairy root cultures could sustain a constant level for a period of three years in comparison with severely declining yields in transgenic tobacco cell cultures (Sharp and Doran, 2001a). The biosynthetic capability of hairy root cultures for molecular farming thus has been largely recognized.

Hairy root cultures in bioreactors and scale-up

Scaling up hairy root cultures in a bioreactor is a critical step toward commercial exploitation of this culture system (Mishra and Ranjan, 2008). A number of bioreactor configuration designs have been examined for the growth of hairy roots ranging from conventional airlift (Liu et al., 1999a), bubble column (Suresh et al., 2004; Rudrappa et al., 2005), and stirred tank (Mehrotra et al., 2008) to specially designed trickle-bed (Ramakrishnan and Curtis, 2004), airlift balloon (Sivakumar et al., 2008), and nutrient mist bioreactors (Liu et al., 2009). These bioreactors can be roughly divided into three types: liquid-phase, gas-phase, or hybrid reactors that are a combination of both (Kim et al., 2002). Due to special morphological and physiological characteristics of hairy roots, conventional bioreactors without appropriate modifications are usually inefficient to support hairy roots growing to high-density as the cultured roots tend to form clumps that

Table 3.5 Recombinant proteins expression in hairy root cultures

Recombinant proteins	Plant species	Culture approaches	Protein yields	Reference
Murine IgG_1	*N. tabacum cv* NT-1	Shake flask	18.0 mg/L, 14.0% secreted	Wongsamuth and Doran (1997)
		Airlift bioreactor (2L)	19.8 mg/L, 21.5% secreted.	
	N. tabacum cv NT-1	Shake flask	7.0 mg/L, 26% secreted	Sharp and Doran (2001a)
	N. tabacum cv NT-1	Shake flask	3.6 mg/L or 12% of TSP	Sharp and Doran (2001a)
14D9 murine IgG_1	*N. tabacum*	Shake flask	64.03 mg/L	Martinez et al. (2005)
Human single-chain IgG_1	*N. tabacum*	Shake flask	9.7 μg/gFW/day, secreted	Komarnytsky et al. (2006)
Human IgG_4			21.8 μg/gFW/day, secreted	
Murine single chain interlukin-12 (mIL-12)	*N. tabacum* cv Xanthi	Shake flask	0.5% of TSP	Liu et al. (2008)
	N. tabacum cv Xanthi	Shake flask	434.8 μg/L, 21% secreted	Liu et al. (2009)
		Mist bioreactor (4L)	5.3 μg/gFW	
		Airlift bioreactor (2L)	3.5 μg/gFW	
Hepatitis B surface antigens (HBsAg)	*Potato* (var. Kufri Bahar)	Shake flask	97.1 ng/(gFW)	Kumar et al. (2006)
	N. tabacum L.	Shake flask	0.01% of TSP	Rukavtsova et al. (2007)
Cholera toxin B-surface protective antigen (CTB-spaA)	*N. plumbaginifolia*	Petri dish	N/A	Ko et al. (2006)
Ricin-B	*N. tabacum* cv Xanthi	Shake flask	N/A	Medina-Bolivar et al. (2003), Zhang et al. (2005)
Human epidermal growth factor (hEGF)	*N. tabacum*	Shake flask	2 μg/gFW	Parsons et al. (2010)
Human growth hormone (hGH)	*N. benthamiana*	Shake flask	N/A	Skarjinskaia et al. (2008)
Human acetylcholinesterase	*N. benthamiana*	Shake flask	3.3% of TSP	Woods et al. (2008)
Rabbit cytochrome P450 2E1	*Atropa belladonna*	Shake flask	N/A	Banerjee et al. (2002)
Human secreted alkaline phosphatase (SEAP)	*N. tabacum*	Plastic chamber	280 μg/gDW, secreted	Gaume et al. (2003)
β-glucuronidase (GUS)	*N. tabacum*	Shake flask	N/A	Lee et al. (2007)
Green fluorescence protein (GFP)	*N. tabacum* cv Xanthi	Plastic sleeve bioreactor (5L)	820 μg/L, 20% secreted	Skarjinskaia et al. (2008), Medina-Bolivar and Cramer (2004)
	N. benthamiana	Shake flask	50 μg/gFW	Skarjinskaia et al. (2008)
	Catharanthus roseus	Shake flask	N/A	Hughes et al. (2002)

Abbreviations: FW, fresh weight; DW, dry weight; TSP, total soluble protein; NA: date not available.

resist the percolation of oxygen (Ramakrishnan and Curtis, 2004; Kino-Oka et al., 1999), and are prone to damage when exposed to high-shear environments (Kim and Yoo, 1993). It is therefore necessary to have bioreactors that can maintain low hydrodynamic stress and high volumetric oxygen transfer to achieve high-density cultures of hairy roots (Kim and Yoo, 1993). In this regard, various modifications to conventional bioreactors were attempted such as a stirred tank in which the impellers were separated by a mesh from roots to reduce the shear (Jung and Tepfer, 1987), a stirred tank in which internal supporting matrices were installed to immobilize the roots (Mehrotra et al., 2008; Abbasi et al., 2009), and a bubble column bioreactor with multiple spargers for better aeration (Kwok and Doran, 1995). While these modifications are simple in form, they introduce large differences in operation (Mehrotra et al., 2008).

So far, two types of bioreactor systems are regarded as the most successful for hairy root cultures. One is based on the airlift and bubble column concept. This type of bioreactor has been used for the scale-up of many species of hairy roots such as *Artemisia annua* (Liu et al., 1998), *N. tabacum* (Wongsamuth and Doran, 1997), *Solanum chrysotrichum*

(Caspeta et al., 2005), red beet (Neelwarne and Thimmaraju, 2009), and *Beta vulgaris* (Rudrappa et al., 2004), either for synthesis of secondary metabolites or recombinant proteins. The second culture system, called a mist reactor, was specially designed for growing plant tissues and organs such as shoots and roots (Towler et al., 2006; Weathers et al., 2009). This is a gas-phase bioreactor in which the plant tissues are exposed to humidified air or a gas mixture and nutrients are delivered as droplets by spray nozzles or ultrasonic transducers (Eibl and Eibl, 2008). The mist reactor offers advantages in that the oxygen transfer limitation found in other systems can be reduced or eliminated even at high root-bed densities, and hydrodynamic stress imparted on root biomass is totally eliminated (Weathers and Giles, 1988; Kim et al., 2001; Towler et al., 2006). However, due to the continuous gas phase in mist reactors, internal root-anchor matrices such as horizontal mesh trays and cylindrical stainless steel meshes are required (Eibl and Eibl, 2008). Various species of hairy roots have been grown in mist reactors, mainly for the purpose of production of secondary metabolites (Liu et al., 1999b; Kim et al., 2001; Souret et al., 2003; Huang et al., 2004; Ramakrishnan and Curtis, 2004; Suresh et al., 2005). In a recent report of scaling up tobacco root cultures for expression of mIL-12, better protein production was observed in a mist reactor than in an airlift reactor (Liu et al., 2009). This is actually the first demonstration of successful production of a pharmaceutical protein in a mist bioreactor with potential for large-scale applications. Finally, there has been a trend toward the use of disposable bioreactors for large-scale cultures of hairy roots to reduce production costs and minimize validation efforts under GMP regulations (Ducos et al., 2010; Eibl et al., 2010). These disposable bioreactors consist of a sterile plastic chamber or bag that is partially filled with liquid medium or nutrient mist, inoculated with root tissues, and discarded after harvest (Eibl and Eibl, 2008). The disposable bioreactor has been successfully exploited for growing tobacco hairy roots to produce functional mIL-12 (Liu et al., 2009) and GFP (Medina-Bolivar and Cramer, 2004). Results have proved useful for further optimization and scale-up studies.

Hairy root cultures have shown promising biosynthetic capability for recombinant proteins. Their substantial advantages over field-cultured plants or suspension cells include fast growth, freedom from pathogen and herbicide contamination, genotype and phenotype stability, and autotrophy for plant hormones. However, they provide challenges for culture scale-up due to the special morphological characteristics of the differentiated tissues. Future research should focus on establishing effective and economical bioreactor culture systems for industrial production. If such a goal is achieved, the application of hairy root cultures for molecular farming will become reality.

Summary and Conclusions

Protein targeting within the context of an appropriate plant, organ, tissue, and subcellular location is still a somewhat empirical science. However, due to the intensity of research in plant-made pharmaceutical, vaccine, and industrial proteins, a great deal of information has been gathered on the utility of various combinations. Leaves, seeds, and hairy roots each have their advantages and disadvantages, as do food/feed versus non-food plant systems. The cytoplasm is often an inhospitable compartment for high-level protein accumulation but works well for selectable marker proteins such as for kanamycin or bialaphos resistance. The endomembrane system is necessary for proteins that require post-translational modifications. Many choices exist within the endomembrane system for storage and degree of modification, and mechanisms to reach these specific compartments are well known. However, what is not known in many cases is whether the particular protein of interest will perform well in that location. The chloroplast has worked well for many proteins, particularly if they are from chloroplast transformation methods. However, these proteins are not post-translationally modified.

As new information is learned about protein targeting, better predictions of the most likely location for the best expression can be made. Nevertheless, empirical tests with multiple sites will still be needed because biology is not always predictable.

References

Abbasi, B. H., Liu, R., Saxena, P. K., & Liu, C. Z. (2009). Cichoric acid production from hairy root cultures of *Echinacea purpurea* grown in a modified airlift bioreactor. *Journal of Chemical Technology and Biotechnology*, *84*, 1697–1701.

Abell, B., Hahn, M., Holbrook, L., & Moloney, M. (2004). Membrane topology and sequence requirements for oil body targeting of oleosin. *The Plant Journal*, *37*, 461–470.

Adorini, L. (1999). Interleukin-12, a key cytokine in Th1-mediated autoimmune diseases. *Cellular and Molecular Life Science*, *55*, 1610–1625.

Aguilar, O., Glatz, C. E., & Rito-Palomares, M. (2009). Characterization of green-tissue protein extract from alfalfa (*Medicago sativa*) exploiting a 3-D technique. *Journal of Separation Science*, *32*, 3223–3231.

Alvarez, M. L., Topal, E., Martin, F., & Cardineau, G. A. (2010). Higher accumulation of F1-V fusion recombinant protein in plants after induction of protein body formation. *Plant Molecular Biology*, *72*, 75–89.

Banerjee, S., Shang, T. Q., Wilson, A. M., Moore, A. L., Strand, S. E., & Gordon, M. P., et al. (2002). Expression of functional mammalian P450 2E1 in hairy root cultures. *Biotechnology and Bioengineering*, *77*, 462–466.

Bock, R., & Warzecha, H. (2010). Solar-powered factories for new vaccines and antibiotics. *Trends in Biotechnology*, *28*, 246–252.

Boothe, J., Nykiforuk, C., Shen, Y., Zaplachinski, S., & Szarka, S., et al. (2010). Seed-based expression systems for plant molecular farming. *Plant Biotechnology Journal*, *8*, 588–606.

Bruce, B. (2000). Chloroplast transit peptides: Structure, function, and evolution. *Trends in Cell Biology*, *10*, 440–447.

Buchanan, B., Gruissem, W., & Jones, R. (2000). *Biochemistry and molecular biology of plants*. Somerset, NJ: Wiley and Sons.

Caspeta, L., Quintero, R., & Villarreal, M. L. (2005). Novel airlift reactor fitting for hairy root cultures: Developmental and performance studies. *Biotechnology Progress*, *21*, 735–740.

Cheung, S. C. K., Sun, S. S. M., Chan, J. C. N., & Tong, P. C. Y. (2009). Expression and subcellular targeting of human insulin-like growth factor binding protein-3 in transgenic tobacco plants. *Transgenic Research, 18*, 943–951.

Chikwamba, R. K., Scott, M. P., Mejia, L. B., Mason, H. S., & Wang, K. (2003). Localization of a bacterial protein in starch granules of transgenic maize kernels. *Proceedings of the National Academy of Sciences of the United States of America, 100*, 11127–11132.

Chikwamba, R., Cunnick, J., Hathaway, D., McMurray, J., Mason, H., & Wang, K. (2002). A functional antigen in a practical crop: LT-B producing maize protects mice against *Escherichia coli* heat labile enterotoxin (LT) and cholera toxin (CT). *Transgenic Research, 11*, 479–493.

Chong, D., Roberts, W., Arakawa, T., & Illes, K., et al. (1997). Expression of the human milk protein beta-casein in transgenic potato plants. *Transgenic Research, 6*, 289–296.

Clemente, T. (2006). Nicotiana (*Nicotiana tobaccum, Nicotiana benthamiana*). *Methods in Molecular Biology, 343*, 143–154.

Clough, R. C., Pappu, K., Thompson, K., Beifuss, K., Lane, J., & Delaney, D. E., et al. (2006). Manganese peroxidase from the white-rot fungus *Phanerochaete chrysosporium* is enzymatically active and accumulates to high levels in transgenic maize seed. *Plant Biotechnology Journal, 4*, 53–62.

Commandeur, U., Twyman, R. M., & Fischer, R. (2003). The biosafety of molecular farming in plants. *AgBiotechNet, 110*, 1–9.

Condori, J., Medrano, G., Sivakumar, G., Nair, V., Cramer, C., & Medina-Bolivar, F. (2009). Functional characterization of a stilbene synthase gene using a transient expression system in planta. *Plant Cell Reports, 28*, 589–599.

Conley, A. J., Joensuu, J. J., Menassa, R., & Brandle, J. E. (2009). Induction of protein body formation in plant leaves by elastin-like polypeptide fusions. *BMC Biology, 7*, 48.

Cramer, C. L., Boothe, J. G., & Oishi, K. K. (1999). Transgenic plants for therapeutic proteins: Linking upstream and downstream strategies. *Current Topics in Microbiology and Immunology, 240*, 95–118.

Cramer, C., Weissenborn, D., Oishi, K., & Grabau, E., et al. (1996). Bioproduction of human enzymes in transgenic tobacco. *Annual New York Academy of Science, 792*, 62–71.

Daniell, H., Singh, N. D., Mason, H., & Streatfield, S. J. (2009). Plant-made vaccine antigens and biopharmaceuticals. *Trends in Plant Science, 14*, 669–679.

Davoodi-Semiromi, A., Schreiber, M., Nalapalli, S., Verma, D., Singh, N. D., & Banks, R. K., et al. (2010). Chloroplast-derived vaccine antigens confer dual immunity against cholera and malaria by oral or injectable delivery. *Plant Biotechnology Journal, 8*, 223–242.

De Loose, M., Gheysen, G., Tire, C., Gielen, J., & Villarroel, R., et al. (1991). The extensin signal peptide allows secretion of a heterologous protein from protoplasts. *Gene, 99*, 95–100.

De Muynck, B., Navarre, C., & Boutry, M. (2010). Producton of antibodies in plants: Status after twenty years. *Plant Biotechnology Journal, 8*, 529–563.

Denecke, J., Botterman, J., & Deblaere, R. (1999). Protein secretion in plant cells can occur via a default pathway. *The Plant Cell, 2*, 51–59.

Dhillon, T., Chiera, J., Lindbo, J., & Finer, J. (2009). Quantitative evaluation of six different viral suppressors of silencing using image analysis of transient GFP expression. *Plant Cell Reports, 28*, 639–647.

Dieryck, W., Pagnier, J., Poyart, C., & Marden, M., et al. (1997). Human haemoglobin from transgenic tobacco. *Nature, 386*, 29–30.

Ducos, J. P., Terrier, B., & Courtois, D. (2010). Disposable bioreactors for plant micropropagation and mass plant cell culture. *Advances in Biochemical Engineering/Biotechnology, 115*, 89–115.

Edelbaum, O., Sher, N., Rubinstein, M., Novick, D., & Tal, N., et al. (1991). Two antiviral proteins, gp35 and gp22, correspond to beta-1,3-glucanase and an isoform of PR-5. *Plant Molecular Biology, 17*, 171–173.

Eibl, R., & Eibl, D. (2008). Design of bioreactors suitable for plant cell and tissue cultures. *Phytochemistry Reviews, 7*, 593–598.

Eibl, R., Kaiser, S., Lombriser, R., & Eibl, D. (2010). Disposable bioreactors: The current state-of-the-art and recommended applications in biotechnology. *Applied Microbiology and Biotechnology, 86*, 41–49.

Faye, L., & Gomord, V. (2010). Success stories in molecular farming – a brief overview. *Plant Biotechnology Journal, 8*, 525–528.

Fieschi, C., & Casanova, J. (2003). The role of interleukin-12 in human infectious diseases: Only a faint signature. *European Journal of Immunology, 33*, 1461–1464.

Fischer, R., Stoger, E., Schillberg, S., Christou, P., & Twyman, R. M. (2004). Plant-based production of biopharmaceuticals. *Current Opinion In Plant Biology, 7*, 152–158.

Frigerio, L., Hinz, G., & Robinson, D. (2008). Multiple vacuoles in plant cells: Rule or exception? *Traffic, 9*, 1564–1570.

Galili, G. (2004). ER-derived compartments are formed by highly regulated processes and have special functions in plants. *Plant Physiology, 136*, 3411–3413.

Gao, Y., Ma, Y., Li, M., Cheng, T., Li, S. W., & Zhang, J., et al. (2003). Oral immunization of animals with transgenic cherry tomatillo expressing HBsAg. *World Journal of Gastroenterology: WJG, 9*, 996–1002.

Gaume, A., Komarnytsky, S., Borisjuk, N., & Raskin, I. (2003). Rhizosecretion of recombinant proteins from plant hairy roots. *Plant Cell Reports, 21*, 1188–1193.

Georgiev, M. I., Pavlov, A. I., & Bley, T. (2007). Hairy root type plant *in vitro* systems as sources of bioactive substances. *Applied Microbiology and Biotechnology, 74*, 1175–1185.

Ghabrial, S. A., Zhang C., & Gu, H. (2009). Viral vector useful in soybean and methods of use. In States, U., (ed.).

Giddings, G. (2001). Transgenic plants as protein factories. *Current Opinion in Biotechnology, 12*, 450–454.

Gleba, Y., Klimyuk, V., & Marillonnet, S. (2005). Magnifection – a new platform for expressing recombinant vaccines in plants. *Vaccine, 23*, 2042–2048.

Golovkin, M., Spitsin, S., Andrianov, V., Smirnov, Y., Xiao, Y., & Pogrebnyak, N., et al. (2007). Smallpox subunit vaccine produced in Planta confers protection in mice. *Proceedings of the National Academy of Sciences of the United States of America, 104*, 6864–6869.

Goodin, M. M., Zaitlin, D., Naidu, R. A., & Lommel, S. A. (2008). *Nicotiana benthamiana*: Its history and future as a model for plant-pathogen interactions. *Molecular Plant–Microbe Interactions, 21*, 1015–1026.

Gruber, V., Berna, P., Arnaud, T., Bournat, P., Clement, C., & Mison, D., et al. (2001). Large-scale production of a therapeutic protein in transgenic tobacco plants: Effect of subcellular targeting on quality of a recombinant dog gastric lipase. *Molecular Biology, 7*, 329–340.

Guillon, S., Tremouillaux-Guiller, J., Pati, P. K., Rideau, M., & Gantet, P. (2006). Harnessing the potential of hairy roots: Dawn of a new era. *Trends in Biotechnology., 24*, 403–409.

Guillon, S., Tremouillaux-Guiller, J., Pati, P. K., Rideau, M., & Gantet, P. (2006). Hairy root research: Recent scenario and exciting prospects -- Commentary. *Current Opinion in Plant Biology, 9*, 341–346.

Gutierrez-Ortega, A., Sandoval-Montes, C., de Olivera-Flores, T. J., Santos-Argumedo, L., & Gomez-Lim, M. A. (2005). Expression of functional interleukin-12 from mouse in transgenic tomato plants. *Transgenic Research, 14*, 877–885.

Hashizume, F., Hino, S., Kakehashi, M., & Okajima, T., et al. (2008). Development and evalutation of transgenic rice seeds accumulating a type II – collagen tolergenic peptide. *Transgenic Research, 17*, 1117–1129.

Hellwig, S., Drossard, J., Twyman, R. M., & Fischer, R. (2004). Plant cell cultures for the production of recombinant proteins. *Nature Biotechnology, 22*, 1415–1422.

Herbers, K., Wilke, I., & Sonnewald, U. (1995). A thermostable xylanase from *Clostridium thermocellum* expressed at high levels in the apoplast of transgenic tobacco has no detrimental effects and is easily purified. *Nature Biotechnology, 13*, 63–66.

Herman, E. M., & Larkins, B. A. (1999). Protein storage bodies and vacuoles. *Plant Cell, 11*, 601–614.

Holwerda, C., Padgett, H., & Rogers, J. (1992). Proaleurain vacuolar targeting is mediated by short contiguous peptide interactions. *The Plant Cell, 4*, 307–318.

Hood, E. (2004). Where, oh where has my protein gone? *Trends in Biotechnology, 22*, 53–55.

Hood, E. E. (2004) Bioindustrial and biopharmaceutical products from plants. In *New directions for a diverse planet: Proceedings of the 4th International Crop Science Congress*.

Hood, E. E., Bailey, M. R., Beifuss, K., Magallanes-Lundback, M., Horn, M. E.,

& Callaway, E., et al. (2003). Criteria for high-level expression of a fungal laccase gene in transgenic maize. *Plant Biotechnology Journal*, *1*, 129–140.

Hood, E., Love, R., Lane, J., Bray, J., Clough, R., & Pappu, K., et al. (2007). Subcellular targeting is a key condition for high-level accumulation of cellulase protein in transgenic maize seed. *Plant Biotechnology Journal*, *5*, 709–719.

Hood, E., Witcher, D., Maddock, S., Meyer, T., & Baszczynski, C., et al. (1997). Commercial production of avidin from transgenic maize: Characterization of transformant, production, processing, extracting, and purification. *Molecular Breeding*, *3*, 291–306.

Howard, J. A. (2005). Commercialization of biopharmaceutical and bioindustrial proteins from plants. *Crop Science*, *45*, 468–472.

Howard, J. A., & Donnelly, K. C. (2004). A quantitative safety assessment model for transgenic protein products produced in agricultural crops. *Journal of Agricultural and Environmental Ethics*, *17*, 545–558.

Howard, J. A., & Hood, E. (2005). Bioindustrial and biopharmaceutical products produced in plants. *Advances in Agronomy*, *85*, 91–124.

Howard, J. A., & Hood, E. E. (2007). Methods for growing nonfood products in transgenic plants. *Crop Science*, *47*, 1255–1262.

Huang, S. Y., Hung, C. H., & Chou, S. N. (2004). Innovative strategies for operation of mist trickling reactors for enhanced hairy root proliferation and secondary metabolite productivity. *Enzyme and Microbial Technology*, *35*, 22–32.

Huang, T. K., & McDonald, K. A. (2009). Bioreactor engineering for recombinant protein production in plant cell suspension cultures. *Biochemical Engineering Journal*, *45*, 168–184.

Hughes, E. H., Hong, S. B., Shanks, J. V., San, K. Y., & Gibson, S. I. (2002). Characterization of an inducible promoter system in *Catharanthus roseus* hairy roots. *Biotechnology Progress*, *18*, 1183–1186.

Jung, G., & Tepfer, D. (1987). Use of genetic transformation by the RiT-DNA of Agrobacterium rhizogenes to stimulate biomass and tropane alkaloid production in *Atropa belladonna* and *Calystegia sepium* roots grown *in vitro*. *Plant Science*, *50*, 141–151.

Khan, S., Rajan, V., & Howard, J. (2010). Plant molecular pharming – industrial enzymes. In R. Meyers (Ed.), *Encyclopedia of sustainability science and technology*. New York, NY: Springer Science + Business Media. (In press)

Kim, T. G., Kim, M. Y., Kim, B. G., Kang, T. J., Kim, Y. S., & Jang, Y. S., et al. (2007). Synthesis and assembly of *Escherichia coli* heat-labile enterotoxin B subunit in transgenic lettuce (*Lactuca sativa*). *Protein Expression and Purification*, *51*, 22–27.

Kim, Y. H., & Yoo, Y. J. (1993). Development of a bioreactor for high-density culture of hairy roots. *Biotechnology Techniques*, *7*, 859–862.

Kim, Y., Wyslouzil, B. E., & Weathers, P. J. (2001). A comparative study of mist and bubble column reactors in the *in vitro* production of artemisinin. *Plant Cell Reports*, *20*, 451–455.

Kim, Y., Wyslouzil, B. E., & Weathers, P. J. (2002). Invited review: Secondary metabolism of hairy root cultures in bioreactors. *In Vitro Cellular & Developmental Biology-Plant*, *38*, 1–10.

Kino-Oka, R., Hitaka, Y., Taya, M., & Tone, S. (1999). High-density culture of red beet hairy roots by considering medium flow conditions in a bioreactor. *Chemical Engineering Science*, *54*, 3179–3186.

Ko, S., Liu, J. R., Yamakawa, T., & Matsumoto, Y. (2006). Expression of the protective antigen (SpaA) in transgenic hairy roots of tobacco. *Plant Molecular Biology Reporter*, *24* 251a–251g

Kogan, M., Dalcol, I., & Gorostiza, P., et al. (2001). Self-assembly of the amphipathic helix (VHLPPP): A mechanism for Zein protein body formation. *Journal of Molecular Biology*, *312*, 907–913.

Komarnytsky, S., Borisjuk, N., Yakoby, N., Garvey, A., & Raskin, I. (2006). Cosecretion of protease inhibitor stabilizes antibodies produced by plant roots. *Plant Physiology*, *141*, 1185–1193.

Kumar, G. B. S., Ganapathi, T. R., Srinivas, L., Revathi, C. J., & Bapat, V. A. (2006). Expression of hepatitis B surface antigen in potato hairy roots. *Plant Science*, *170*, 918–925.

Kusnadi, A. R., Nikolov, Z. L., & Howard, J. A. (1997). Production of recombinant proteins in transgenic plants: Practical considerations. *Biotechnology and Bioengineering*, *56*, 473–484.

Kwok, K. H., & Doran, P. M. (1995). Kinetic and stoichiometric analysis of hairy roots in a segmented bubble column bioreactor. *Biotechnology Progress*, *11*, 429–435.

Lamphear, B., Barker, D., Brooks, C., Delaney, D., Lane, J., & Beijuss, K., et al. (2005). Expression of the sweet protein brazzein in maize for production of a new sweetener. *Plant Biotechnology Journal*, *3*, 103–114.

Lee, K. T., Chen, S. C., Chiang, B. L., & Yamakawa, T. (2007). Heat-inducible production of beta-glucuronidase in tobacco hairy root cultures. *Applied Microbiology and Biotechnology*, *73*, 1047–1053.

Leite, A., Kemper, E. L., da Silva, M. J., & Luchessi, A. D., et al. (2000). Expression of correctly processed human growth hormone in seeds of transgenic tobacco plants. *Molecular Breeding*, 47–53.

Lieschke, G., Rao, P., Gately, M., & Mulligan, R. (1997). Bioactive murine and human interleukin-12 fusion proteins which retain antitumor activity *in vivo*. *Nature Biotechnology*, *15*, 35–40.

Liu, C. Z., Towler, M. J., Medrano, G., Cramer, C. L., & Weathers, P. J. (2009). Production of mouse interleukin-12 is greater in tobacco hairy roots grown in a mist reactor than in an airlift reactor. *Biotechnology and Bioengineering*, *102*, 1074–1086.

Liu, C. Z., Wang, Y. C., Ouyang, F., Ye, H. C., & Li, G. F. (1999). Production of artemisinin by *Artemisia annua* hairy root culture in a internal loop airlift bioreactor. *Acta Botanica Sinica*, *41*, 181–183.

Liu, C. Z., Wang, Y. C., Zhao, B., Guo, C., Ouyang, F., & Ye, H. C., et al. (1999). Development of a nutrient mist bioreactor for growth of hairy roots. *In Vitro Cellular & Developmental Biology-Plant*, *35*, 271–274.

Liu, C., Wang, Y., Guo, C., Ouyang, F., Ye, H., & Li, G. (1998). Enhanced production of artemisinin by *Artemisia annua* L. hairy root cultures in a modified inner-loop airlift bioreactor. *Bioprocess Engineering*, *19*, 389–392.

Liu, J., Dolan, M. C., Reidy, M., & Cramer, C. L. (2008). Expression of bioactive single-chain murine IL-12 in transgenic plants. *Journal of Interferon and Cytokine Research*, *28*, 381–392.

Llompart, B., Llop-Tous, I., Marzabal, P., & Torrent, M., et al. (2010). Protein production from recombinant protein bodies. *Process Biochemistry* (in press)

Ma, J. K. -C., Drake, P., & Christou, P. (2003). The production of recombinant pharmaceutical proteins in plants. *Nature Reviews Genetics*, *4*, 794–805.

Magnuson, N., Linzmaier, P., Reeves, R., & An, G., et al. (1998). Secretion of biologically active human interleukin-2 and interleukin-4 from genetically modified tobacco cells in suspension culture. *Protein Expression and Purification*, *13*, 45–52.

Mainieri, D., Rossi, M., Archinti, M., Bellucci, M., De Marchis, F., & Vavassori, S., et al. (2004). Zeolin. A new recombinant storage protein constructed using maize gamma-zein and bean phaseolin. *Plant Physiology*, *136*, 3447–3456.

Martinez, C., Petruccelli, S., Giulietti, A. M. A., & Alvarez, M. A. (2005). Expression of the antibody 14D9 in *Nicotiana tabacum* hairy roots. *Electronic Journal of Biotechnology*, *8*, 170–176.

Marty, F. (1999). Plant vacuoles. *The Plant Cell*, *11*, 587–600.

Masarik, M., Kizek, R., Kramer, K., Billova, S., Brazdova, M., & Vacek, J., et al. (2003). Application of Avidin–Biotin technology transfer stripping square-wave voltammetry for detection of DNA hybridization and avidin in transgenic avidin maize. *Analytical Chemistry*, *75*, 2663–2669.

Mason, H., Lam, D. and Arntzen, C. (1992) Expression of hepatitis B surface antigen in transgenic plants. *PNAS USA*, *89*, 11745–11749.

Matoba, N., Kajiura, H., Cherni, I., Doran, J. D., Bomsel, M., & Fujiyama, K., et al. (2009). Biochemical and immunological characterization of the plant-derived candidate human immunodeficiency virus type 1 mucosal vaccine CTB-MPR. *Plant Biotechnology Journal*, *7*, 129–145.

Matsumura, T., Itchoda, N., & Tsunemitsu, H. (2002). Production of immunogenic VP6 protein of bovine group A rotavirus in transgenic potato plants. *Archives of Virology*, *147*, 1263–1270.

Medina-Bolivar, F., & Cramer, C. (2004). Production of recombinant proteins by hairy roots cultured in plastic sleeve bioreactors. *Methods in Molecular Biology*, *267*, 351–363.

Medina-Bolivar, F., Wright, R., Funk, V., Sentz, D., Barroso, L., & Wilkins, T. D., et al. (2003).

A non-toxic lectin for antigen delivery of plant-based mucosal vaccines. *Vaccine, 21*, 997–1005.

Medrano, G., Dolan, M. C., Stephens, N. T., McMickle, A., Erf, G., & Radin, D., et al. (2010). Efficient plant-based production of chicken interleukin-12 yields a strong immunostimulatory cytokine. *Journal of Interferon and Cytokine Research, 30*, 143–154.

Medrano, G., Reidy, M. J., Liu, J., Ayala, J., Dolan, M. C., & Cramer, C. L. (2009). Rapid system for evaluating bioproduction capacity of complex pharmaceutical proteins in plants. *Methods in Molecular Biology, 483*, 51–67.

Mehrotra, S., Kukreja, A. K., Khanuja, S. P. S., & Mishra, B. N. (2008). Genetic transformation studies and scale up of hairy root culture of *Glycyrrhiza glabra* in bioreactor. *Electronic Journal of Biotechnology, 11*, 6.

Mishra, B. N., & Ranjan, R. (2008). Growth of hairy-root cultures in various bioreactors for the production of secondary metabolites. *Biotechnology and Applied Biochemistry, 49*, 1–10.

Napier, J., Stobart, A., & Shewry, P. (1996). The structure and biogenesis of plant oil bodies: The role of the ER membrane and the oleosin class of proteins. *Plant Molecular Biology, 31*, 945–956.

Neelwarne, B., & Thimmaraju, R. (2009). Bioreactor for cultivation of red beet hairy roots and *in situ* recovery of primary and secondary metabolites. *Engineering in Life Sciences, 9*, 227–238.

Neuhaus, J., & Rogers, J. (1998). Sorting of proteins to vacuoles in plant cells. *Plant Molecular Biology, 38*, 127–144.

Neuhaus, J., Stitcher, L., Meins, F., Jr., & Boller, T. (1991). A short C-terminal sequence is necessary and sufficient for the targeting of chitinase to the plant vacuole. *Proceeding from the National Academy of Science, 88*, 10362–10366.

Nykiforuk, C., Boothe, J., Murray, E., & Keon, R., et al. (2005). Transgenic expression and recovery of biologically active recombinant-human insulin from Arabidopsis thaliana seeds. *Plant Biotechnology Journal, 4*, 77–85.

Nykiforuk, C., Shen, Y., Murray, E., & Boothe, J., et al. (2010). Expression and recovery of biologically active recombinant Apolipoprotein AI-milano from transgenic safflower (*Carthamus tinctorius*) seeds. *Plant Biotechnology Journal, 9*, 250–263.

Oey, M., Lohse, M., Kreikemeyer, B., & Bock, R. (2009). Exhaustion of the chloroplast protein synthesis capacity by massive expression of a highly stable protein antibiotic. *The Plant Journal, 57*, 436–445.

Paris, N., Stanley, C., Jones, R., & Rogers, J. (1996). Plant cells contain two functionally distinct vacuolar compartments. *Cell, 85*, 563–572.

Parsons, J., Wirth, S., Dominguez, M., Bravo-Almonacid, F., Giulietti, A. M., & Rodriguez Talou, J. (2010). Production of human epidermal growth factor (hEGF) by *in vitro* cultures of *Nicotiana tabacum*: Effect of tissue differentiation and sodium nitroprusside addition. *International Journal of Biotechnology and Biochemistry, 6*, 131–138.

Putalun, W., Taura, F., Qing, W., Matsushita, H., Tanaka, H., & Shoyama, Y. (2003). Anti-solasodine glycoside single-chain Fv antibody stimulates biosynthesis of solasodine glycoside in plants. *Plant Cell Reports, 22*, 344–349.

Radin, D., Cramer, C., Oishim, K., & Weissenborn, D. (1997). Production of Lysosomal Enzymes in Plant-based Expression Systems. United States.

Ramakrishnan, D., & Curtis, W. (2004). Trickle-bed root culture bioreactor design and scale-up: Growth, fluid-dynamics, and oxygen mass transfer. *Biotechnology and Bioengineering, 88*, 248–260.

Ratner, M. (2010). Pfizer stakes a claim in plant cell-made biopharmaceuticals. *Nature Biotechnology, 28*, 107–108.

Reggi, S., Marchetti, S., Patti, T., & de Amicis, F., et al. (2005). Recombinant human acid b-glucosidase stored in tobacco seed is stable, active and taken up by human fibroblasts. *Plant Molecular Biology*, 101–113.

Rigano, M. M., & Walmsley, A. M. (2005). Expression systems and developments in plant-made vaccines. *Immunology and Cell Biology, 83*, 271–277.

Rogers, J. C. (1985). Two barley alpha-amylase gene families are regulated differently in aleurone cells. *Journal of Biological Chemistry, 260*, 3731–3738.

Rudrappa, T., Neelwarne, B., & Aswathanarayana, R. G. (2004). *In situ* and *ex situ* adsorption and recovery of betalains from hairy root cultures of *Beta vulgaris*. *Biotechnology Progress, 20*, 777–785.

Rudrappa, T., Neelwarne, B., Kumar, V., Lakshmanan, V., Venkataramareddy, S. R., & Aswathanarayana, R. G. (2005). Peroxidase production from hairy root cultures of red beet (*Beta vulgaris*). *Electronic Journal of Biotechnology, 8*, 185–196.

Ruhlman, T., Ahangari, R., Devine, A., Samsam, M., & Daniell, H. (2007). Expression of cholera toxin B-proinsulin fusion protein in lettuce and tobacco chloroplasts – oral administration protects against development of insulitis in non-obese diabetic mice. *Plant Biotechnology Journal, 5*, 495–510.

Rukavtsova, E. B., Abramikhina, T. V., Shulga, N. Y., Bykov, V. A., & Bur'yanov, Y. I. (2007). Tissue specific expression of hepatitis B virus surface antigen in transgenic plant cells and tissue culture. *Russian Journal of Plant Physiology, 54*, 770–775.

Samac, D., Hironaka, C., Yallaly, P., & Shah, D. (1990). Isolation and characterization of the genes encoding basic and acidic chitinase in *Arabidopsis thaliana*. *Plant Physiology, 93*, 907–914.

Sardana, R., Dudani, A. K., Tackaberry, E., Alli, Z., et al. (2007) Biologically active human GM-CSF produced in the seeds of transgenic rice plants. In *Transgenic research* (pp. 713-721). Springer.

Schillberg, S., et al. (2002). Antibody molecular farming in plants and plant cells. *Phytochemistry Reviews, 1*, 45–54.

Shaaltiel, Y., Bartfed, D., Hashmueli, S., Baum, G., Brill-Almon, E., & Galili, G., et al. (2007). Production of glucoceebrosidase with terminal mannose glycans for enzyme replacement therapy of Gaucher's disease using a plant cell system. *Plant Biotechnology Journal, 5*, 579–590.

Shanks, J. V., & Morgan, J. (1999). Plant 'hairy root' culture. *Current Opinion in Biotechnology, 10*, 151–155.

Sharma, M. K., Jani, D., Thungapathra, M., Gautam, J. K., Meena, L. S., & Singh, Y., et al. (2008). Expression of accessory colonization factor subunit A (ACFA) of *Vibrio cholerae* and ACFA fused to cholera toxin B subunit in transgenic tomato (*Solanum lycopersicum*). *Journal of Biotechnology, 135*, 22–27.

Sharp, J., & Doran, P. (2001). Characterization of monoclonal antibody fragments produced by plant cells. *Biotechnology and Bioengineering, 73*, 338–346.

Sharp, J. M., & Doran, P. M. (2001). Strategies for enhancing monoclonal antibody accumulation in plant cell and organ cultures. *Biotechnology Progress, 17*, 979–992.

Sheen, J. (2001). Signal transduction in maize and *Arabidopsis* mesophyll protoplasts. *Plant Physiology, 127*, 1466–1475.

Shih, S. M. H., & Doran, P. M. (2009). Foreign protein production using plant cell and organ cultures: Advantages and limitations. *Biotechnology Advances, 27*, 1036–1042.

Shoji, Y., Bi, H., Musiychuk, K., Rhee, A., Horsey, A., & Roy, G., et al. (2009). Plant-derived hemagglutinin protects ferrets against challenge infection with the A/ Indonesia/05/05 strain of avian influenza. *Vaccine, 27*, 1087–1092.

Shoji, Y., Farrance, C. E., Bi, H., Shamloul, M., Green, B., & Manceva, S., et al. (2009). Immunogenicity of hemagglutinin from A/ Bar-headed Goose/Qinghai/1A/05 and A/ Anhui/1/05 strains of H5N1 influenza viruses produced in *Nicotiana benthamiana* plants. *Vaccine, 27*, 3467–3470.

Sijmons, P., Dekker, B., Schrammeijier, B., & Verwoerd, T., et al. (1990). Production of correctly processed human serum albumin in transgenic plants. *Nature Biotechnology, 8*, 217–221.

Sivakumar, G., Dolan, M. C., Condori, J., Bennett, S., Bacchetta, L., & Medina-Bolivar, F. (2008). Air lift balloon type bioreactor: Platform for commercial production of plant based small molecules and tissues. *In Vitro Cellular & Developmental Biology-Animal, 44*, S4–S5.

Skarjinskaia, M., Karl, J., Araujo, A., Ruby, K., Rabindran, S., & Streatfield, S. J., et al. (2008). Production of recombinant proteins in clonal root cultures using episomal expression vectors. *Biotechnology and Bioengineering., 100*, 814–819.

Sojikul, P., Buehner, N., & Mason, H. (2003). A plant signal peptide–hepatitis B surface antigen fusion protein with enhanced stability and immunogenicity expressed in plant cells. *Proceedings of the National Academy of Sciences USA, 100*, 2209–2214.

Souret, F. F., Kim, Y., Wysiouzil, B. E., Wobbe, K. K., & Weathers, P. J. (2003). Scale-up of *Artemisia annua* L. hairy root cultures produces complex patterns of terpenoid gene expression. *Biotechnology and Bioengineering., 83*, 653–667.

Spitsin, S., Andrianov, V., Pogrebnyak, N., Smirnov, Y., Borisjuk, N., & Portocarrero, C., et al. (2009). Immunological assessment of plant-derived avian flu H5/HA1 variants. *Vaccine*, *27*, 1289–1292.

Stoger, E., Ma, J. K. C., Fischer, R., & Christou, P. (2005). Sowing the seeds of success: Pharmaceutical proteins from plants. *Current Opinion in Biotechnology*, *16*, 167–173.

Stoger, E., Sack, M., Perrin, Y., Vaquero, C., & Torres, E., et al. (2002). Practical considerations for pharmaceutical antibody production in different crop systems. *Molecular Breeding*, *9*, 149–158.

Streatfield, S. J., & Howard, J. A. (2003). Plant-based vaccines. *International Journal for Parasitology*, *33*, 479–493.

Streatfield, S., Jilka, J., Hood, E., Turner, D., Bailey, M., & Mayor, J., et al. (2001). Plant-based vaccines: Unique advantages. *Vaccine*, *19*, 2742–2748.

Suresh, B., Bais, H. P., Raghavarao, K. S. M. S., Ravishankar, G. A., & Ghildyal, N. P. (2005). Comparative evaluation of bioreactor design using *Tagetes patula* L. hairy roots as a model system. *Process Biochemistry*, *40*, 1509–1515.

Suresh, B., Thimmaraju, R., Bhagyalakshmi, N., & Ravishankar, G. A. (2004). Polyamine and methyl jasmonate-influenced enhancement of betalaine production in hairy root cultures of *Beta vulgaris* grown in a bubble column reactor and studies on efflux of pigments. *Process Biochemistry*, *39*, 2091–2096.

Tackaberry, E. S., Dudani, A., Prior, F., Tocchi, M., Sardana, R., & Alotsaar, I., et al. (1999). Development of biopharmaceuticals in plant expression systems: Cloning, expression, and immunological reactivity of human cytomegalovirus glycoprotein B (UL55) in seeds of transgenic tobacco. *Vaccine*, *17*, 3020–3029.

Thomas, C. L., Leh, V., Lederer, C., & Maule, A. J. (2003). Turnip crinkle virus coat protein mediates suppression of RNA silencing in *Nicotiana benthamiana*. *Virology*, *306*, 33–41.

Torrent, M., Llompart, B., Lasserre-Ramassamy, S., Llop-Tous, I., Bastida, M., & Marzabal, P., et al. (2009). Eukaryotic protein production in designed storage organelles. *BMC Biology*, *7*, 5.

Towler, M., Kim, Y. J., Correll, M., Wyslouzil, B., & Weathers, P. J. (2006). Design, development, and applications of mist bioreactors for micropropagation and hairy root culture. In S. D. Gupta & Y. Ibaraki (Eds.), *Plant tissue culture engineering* (pp. 119–134). Dordrecht: Springer.

Tremblay, R., Wang, D., Jevnikar, A. M., & Ma, S. (2010). Tobacco, a highly efficient green bioreactor for production of therapeutic proteins. *Biotechnology Advances*, *28*, 214–221.

van Rooijen, G., & Maloney, M. (1995). Plant seed oil-bodies as carriers for foreign proteins. *Nature Biotechnology*, *13*, 72–77.

Verwoerd, T., van Paridon, P., van Ooyen, A., van Lent, J., Hoekema, A., & Pen, J. (1995). Stable accumulation of *Aspergillus niger* phytase in transgenic tobacco leaves. *Plant Physiology*, *109*, 1199–1205.

Vitale, A., & Boston, R. (2008). Endoplasmic reticulum quality control and the unfolded protein response: Insights from plants. *Traffic*, *9*, 1581–1588.

Vitale, A., & Pedrazzini, E. (2005). Recombinant pharmaceuticals from plants: The plant endomembrane system as bioreactor. *Molecular Interventions*, *5*, 216–225.

Vitale, A., & Raikhel, N. (1999). What do proteins need to reach different vacuoles? *Trends in Plant Science*, *4*, 149–155.

Vlahova, M., Stefanova, P., Petkov, A., & Barbulova, D., et al. (2005). Genetic modification of alfalfa (*Medicago Sativa* L.) for quality improvement and production of novel compounds. *Biotechnology & Biotechnological Equipment*, *19*, 56–62.

Voinnet, O., Rivas, S., Mestre, P., & Baulcombe, D. (2003). An enhanced transient expression system in plants based on suppression of gene silencing by the p19 protein of tomato bushy stunt virus. *The Plant Journal*, *33*, 949–956.

Weathers, P., & Giles, K. (1988). Regeneration of plants using nutrient mist culture. *In Vitro Cellular & Developmental Biology-Plant*, *24*, 727–732.

Weathers, P. J., Sivakumar, G., Liu, C., Towler, M. J., & Medina-Bolivar, F. (2009). Scaling up a new version of the mist bioreactor for hairy root cultures. *In Vitro Cellular & Developmental Biology-Animal*, *45*, S69–S70.

Wongsamuth, R., & Doran, P. (1997). Hairy roots as an expression system for production of antibodies. In P. M. Doran (Ed.), *Hairy roots: Culture and applications* (pp. 89–97). Amsterdam: Harwood Academic.

Wongsamuth, R., & Doran, P. M. (1997). Production of monoclonal antibodies by tobacco hairy roots. *Biotechnology and Bioengineering*, *54*, 401–415.

Woodard, S., Mayor, J., Bailey, M., Barker, D., Love, R., & Lane, J., et al. (2003). Maize (*Zea mays*)-derived bovine trypsin: Characterization of the first large-scale, commercial protein product from transgenic plants. *Biotechnology and Applied Biochemistry*, *38*, 123–130.

Woods, R. R., Geyer, B. C., & Mor, T. S. (2008). Hairy-root organ cultures for the production of human acetylcholinesterase. *BMC Biotechnology*, *8*

Wydro, M., Kozubek, E., & Lehmann, P. (2006). Optimization of transient Agrobacterium-mediated gene expression system in leaves of *Nicotiana benthamiana*. *Acta Biochimica Polonica*, *53*, 289–298.

Xie, T., Qui, Q., Zhang, W., Ning, T., Yang, W., & Zheng, C., et al. (2008). A biologically active rhIGF-1 fusion accumulated in transgenic rice seeds can reduce blood glucose in diabetic mice via oral delivery. *Peptides*, 1862–1870.

Xu, J., Shpak, E., & Gu, T., et al. (2005). Production of recombinant plant gum with tobacco cell culture in bioreactor and gum characterization. *Biotechnology and Bioengineering*, *90*, 578–588.

Yiu, J., Tseng, M., Yang, C., & Liu, C., et al. (2008). Transgenic rice expresses an antigenic lipoprotein of *Neisseria gonorrhoeae*. *Journal of the Science of Food and Agriculture*, *88*, 1603–1613.

Yusibov, V., Hooper, D. C., Spitsin, S. V., Fleysh, N., Kean, R. B., & Mikheeva, T., et al. (2002). Expression in plants and immunogenicity of plant virus-based experimental rabies vaccine. *Vaccine*, *20*, 3155–3164.

Zhang, C. M., Medina-Bolivar, F., Buswell, S., & Cramer, C. L. (2005). Purification and stabilization of ricin B from tobacco hairy root culture medium by aqueous two-phase extraction. *Journal of Biotechnology*, *117*, 39–48.

Zhou, Y., Lee, M., Ng, J., Chye, M., Yip, W., & Zee, S., et al. (2006). A truncated hepatitis E virus ORF2 protein expressed in tobacco plastids is immunogenic in mice. *World Journal of Gastroenterology.*, *12*, 306–312.

Proteomics and its application in plant biotechnology

4

Sylvain Bischof Jonas Grossmann Wilhelm Gruissem

Department of Biology, Plant Biotechnology, and Functional Genomics Center, Zurich, Switzerland

TABLE OF CONTENTS

Introduction

Modern day crop plants are essential as a food source for humans and animals, as biofuel, and as feedstock for industrial use. At the same time, the world population is growing rapidly and with it the need for plant-based foods. Forty years from now there will be 9.2 billion people living on Earth, 50% more than today. Not only has the daily food demand per capita increased from an average of 2360 kcal in the mid-1960s to currently 2800 kcal, but demands for high-quality food like fruit, vegetables, and meat is also rising. Meat production over the last 15 years has doubled as well. Today, nearly half of the food produced worldwide is fed to farm animals to satisfy the growing demand for meat. More recently, crop plants that have traditionally been used to feed humans and animals are now also used increasingly as feedstock for chemical and biofuel production, placing previously unknown demands on agricultural production. The available arable land is limited, however, which necessitates cultivation strategies to increase crop yield per acre. Therefore it is important to understand the regulation of plant functions and growth, and how a plant allocates biomass to its edible and usable parts to secure sustainable crop yields in a changing environment for feeding a growing world population and meeting demands for chemical feedstocks. The central goal of plant biotechnology is the targeted introduction of desired traits into a crop plant via genetic engineering to produce plants that are better adapted to biotic and abiotic stress, and with higher yield per acre of cultivated land.

Plants can also be used as biochemical factories that catalyze many different complex biochemical reactions. Plant biotechnology exploits this potential to use plants as bioreactors for the production of food and feed additives, antibodies, vaccines, pharmaceuticals, enzymes, or biodegradable polymers like polyhydroxyalkanoates (PHA). Plants are a cost-effective production system because they are easy and inexpensive to cultivate in large quantities. Scale-up and processing of plant material is efficient because the existing agronomic infrastructure can be used for this purpose. In some instances, plant-based products such as vaccines and pharmaceuticals can be synthesized in leaves, seeds, and fruits, and delivered with the normal diet, therefore making further processing or isolation of active ingredients unnecessary (De Wilde et al., 2000).

The increased pressure on plant productivity for food, diverse derivatives, or as raw material has greatly accelerated plant research during the last decade. Especially in the field of plant biotechnology, substantial progress was made that notably lead to the possible genetic transformation of a large number of different plant species. In this context, the establishment of novel analysis techniques and genome-wide profiling technologies plays an important role that allows plant metabolism not only to be better understood, but also to be improved. An essential milestone for plant sciences was the completion of the first sequenced plant genome, *Arabidopsis thaliana*, in 2000 (Arabidopsis Genome Initiative, 2000). This resource laid the foundation for the development and application of high-throughput technologies such as transcriptomics and proteomics. These large-scale approaches offer the

DOI: 10.1016/B978-0-12-381466-1.00004-3

opportunity to monitor a considerable number of components of a system in a single experiment and to investigate how these components vary upon different stresses or genetic perturbations. Integrative analysis of such large-scale datasets allows understanding complex metabolic networks and the interplay of multiple components of a biological system.

In this chapter, we review mass spectrometry-based proteomics as a powerful tool to identify and quantify proteins in complex biological samples. We discuss the current state of the art of the technology, the techniques in use, and address the challenges that should guide the future applications of proteomics in the rapidly evolving field of plant biotechnology.

Mass Spectrometry-Based Proteomics

The term "proteomics" was coined in the late 1990s (James, 1997). The goal of proteomics is a comprehensive, quantitative description of protein expression and its changes under the influence of biological perturbations (Anderson and Anderson, 1998). Various approaches such as large-scale proteome profiling, enrichment, and identification of post-translational modifications (PTMs), subcellular localization, protein–protein interaction, or activity-based protein profiling can be undertaken to understand protein dynamics (Pandey and Mann, 2000; Aebersold and Mann, 2003; Phizicky et al., 2003; Barglow and Cravatt, 2007; Gstaiger and Aebersold, 2009). Identification and quantification of all proteins in an organism or a specific tissue at a certain time point are very ambitious goals. The high degree of complexity of cellular proteomes and the low abundance of several proteins necessitate highly sensitive analytical techniques. Mass spectrometry (MS) has increasingly become the method of choice for analysis of complex protein samples. MS-based proteomics was made possible by the availability of genome sequence databases and various technical advances, most notably the discovery and development of protein and peptide ionization methods (Fenn et al., 1989).

Proteomics is becoming increasingly important as a tool to integrate genome information and transcription activity with translation efficiency and enzyme activity (Gstaiger and Aebersold, 2009; Baginsky et al., 2010). Quantitative protein data are a new source of information about the complexity and function of biological systems, especially when combined with information from large RNA profiling databases such as Genevestigator (Zimmermann et al., 2004; Hruz et al., 2008). Although RNA accumulation is often used as a proxy for protein abundance and complexity, it is often difficult to predict protein concentrations from mRNA levels because post-transcriptional regulatory mechanisms adjust protein concentrations to prevailing conditions. Recent large-scale analyses confirmed that in many cases no apparent correlation exists between transcript, protein, and metabolite levels, suggesting that regulation of gene activity occurs at different nodes in biological networks (Gibon et al., 2004; Kleffmann et al., 2004; Baerenfaller et al., 2008). Tools for high-throughput quantitative determination of proteins are now available and have been reviewed extensively in the current literature (Yan and Chen, 2005; Wienkoop et al., 2010; Domon and Aebersold, 2010; Schulze and Usadel, 2010). Quantitative protein information complements data from transcriptional profiling and metabolomics. It represents a key link between different levels of gene expression regulation and provides insights into their causal relationships. Unlike transcriptional profiling, however, comprehensive proteome analysis remains challenging, and information about proteome complexity and dynamics is far from complete (Cox and Mann, 2007).

In contrast to transcriptomics, different approaches are needed to achieve a high coverage of the proteome (Yates et al., 2009). A typical MS-based proteomics workflow is composed of several steps. Sample preparation, including subcellular fractionation and protein extraction methods, diverse enzymatic digestions, chromatographic separation, or enrichment of peptides have to be adapted depending on the sample complexity and the objective of each analysis. The composition of the extraction buffer and especially the type of detergents used play key roles for the solubilization of proteins of interest and also for downstream analysis. Samples, in most cases trypsin-digested peptides, are then analyzed by MS. Peptide, respectively, protein identification is achieved by comparing measured with theoretical spectra from available proteomic databases that are based on genomic sequences. The output of MS-based proteomics experiments therefore relies on the availability of high-quality full genome sequences or at least a collection of expressed sequence tags (ESTs). Fast improvements of DNA sequencing techniques (Metzker, 2009) led to the release of many sequence databases of plant genomes, including important crop and economically relevant species such as barley, cassava, cucumber, grapevine, maize, papaya, poplar, potato, rice, soybean, tomato, or wheat. In the following section, we highlight important aspects of a typical proteomics workflow (Figure 4.1) and discuss potential problems in proteome analysis. Also, an essential aspect of MS-based proteomics deals with accurate protein quantification strategies, such that changes in protein abundance across different samples and conditions can be monitored with precision. Several methods involving labeling techniques but also label-free approaches were developed and will be discussed here.

Sample preparation prior to mass spectrometry

Bottom-up proteomics is usually applied to identify and characterize many peptides in a complex mixture and then infer protein identity. Very often, protein extraction and the reduction of sample complexity by fractionation techniques are key steps for achieving an increased sensitivity. Furthermore, detection of low abundant proteins can be hindered by one or several very abundant proteins such as ribulose-1,5-bisphosphate carboxylase (RuBisCO), which makes up more than 50% of the total amount of protein in green, photosynthetically active plant tissues. To handle this problem, depletion strategies such as selective removal by immunoaffinity have been developed (Cellar et al., 2007). But caution has to be taken for quantitative analyses because immunodepletion changes the overall proteome composition and

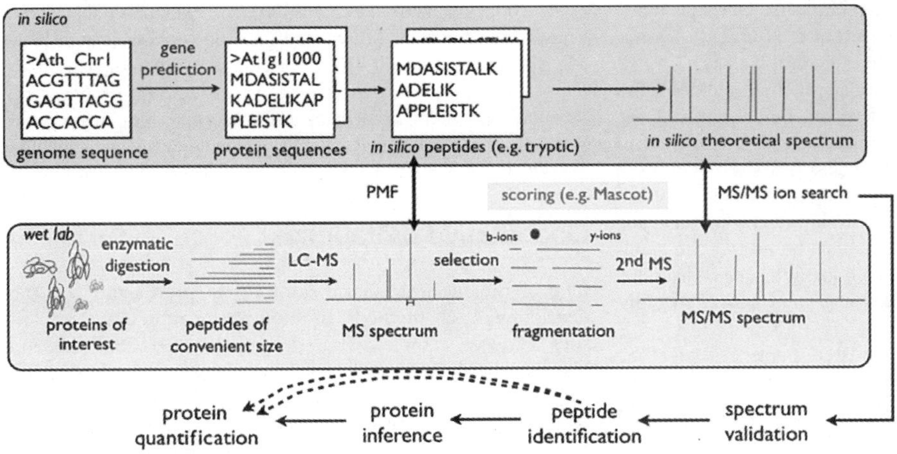

Figure 4.1 • Typical proteomics workflow for protein identification and quantification in complex sample mixtures. • Two parallel procedures constitute a typical proteomics workflow. Protein identification from measured spectra in the wet lab (in blue) directly relay on genomic sequences processed *in silico* (in green). *In silico*, protein sequence databases are generated from available genomic sequences. In this process, a gene prediction algorithm is used to predict open reading frames from which protein sequences are subsequently extracted. Protein sequences are then digested *in silico* into small polypeptides (peptides) following the same enzymatic rules as used in the wet lab. From these *in silico* peptides, fragment ion masses are calculated according to established fragmentation rules (e.g., b- and y-ions for collision-induced dissociation activation) resulting in the theoretical MS/MS spectrum. In the wet lab, protein extracts are enzymatically digested (in most cases with trypsin) and analyzed by tandem mass spectrometry. MS spectra of eluting peptides and MS/MS spectra obtained after fragmentation of selected peptides are recorded. For protein identification, measured MS and MS/MS spectra are compared with *in silico* spectra by using specific algorithms such as Mascot or Sequest. Spectra matching results in peptide and subsequent protein identification if the peptide is not ambiguous to several different proteins. Quantitative information is usually extracted at peptide level and then propagated to the identified proteins. PMF: Peptide mass fingerprint. Please see color plate section at the end of the book.

complexity and also, additional proteins might be removed unintentionally.

Protein extraction and digestion

Protein extraction is a crucial step for MS-based proteomics experiments. The chosen extraction procedure determines the population of proteins that will be analyzed and therefore sample complexity. Stringency of extraction ranging from soluble proteins to hydrophobic membrane proteins highly depends on the type of detergents used. Several excellent studies dedicated to protein extraction procedures are available and will not be discussed further in this chapter (e.g., Saravanan and Rose, 2004; Isaacson et al., 2006; Maldonado et al., 2008; Weiss and Görg, 2009). Nevertheless, detergents used for protein extraction have to be chosen carefully, especially for their compatibility with downstream fractionation or enzymatic digestion.

Enzymatic digestion usually precedes MS analysis. Trypsin is the most widely used enzyme and cleaves amino acid sequences after lysine and arginine to generate smaller peptides (Shevchenko et al., 1996). The size of such peptides is appropriate for ionization and further detection by MS. Enzymatic digestion using multiple proteases such as LysC, ArgC, AspN, and GluC in addition to trypsin have also been explored in order to improve protein identification (Swaney et al., 2010). Enzymatic digestion can either be done prior to or after fractionation steps.

Protein and peptide fractionation

Sample fractionation leads to the reduction of sample complexity. Higher sensitivity can therefore be achieved for the identification of low abundant proteins. Various fractionation techniques have been developed that are based on different properties of proteins or peptides such as size, isoelectric point, hydrophobicity, or charge. Protein fractionation methods include, notably, one-dimensional sodium-dodecyl sulfate polyacrylamide gel electrophoresis (SDS-PAGE), two-dimensional (2D) PAGE, native PAGE, gel filtration, or isoelectric focusing. In contrast, fewer peptide fractionation techniques are available and of these, strong cation exchange is the most widely used (Bodenmiller et al., 2007). In addition, peptide populations obtained after enzymatic digestion are usually separated online by high-pressure liquid chromatography (LC) prior to the MS/MS analysis (Yates et al., 2009).

Mass spectrometry

During the last three decades, MS was established as the method of choice to identify peptides and proteins. MS of peptides is more sensitive than MS of whole proteins and the mass of an intact protein by itself is not sufficient for identification. The high sensitivity and mass accuracy of modern mass spectrometers today allows the confident identification and quantification of a high number of proteins in a complex sample. Also, MS analyses are highly reproducible and compatible with high-throughput approaches. These robust characteristics made MS-based proteomics, together with transcriptomics, a leading domain of the rapidly emerging field of system biology.

A typical mass spectrometer consists of three elements: an ion source, a mass analyzer, and a detector. Before entering the mass spectrometer, peptides are ionized, either by an electrospray ion source (ESI) or by matrix-assisted laser disruption ionization (MALDI; Aebersold and Mann, 2003). ESI ionization is often coupled to a high-pressure liquid chromatography (LC) in very fine capillaries, in which at the end of the capillary peptides are nebulized into small, highly charged droplets. In MALDI, peptides are embedded in a solid matrix and ionized by a laser pulse.

After ionisation, multiple protonated peptides enter the mass spectrometer and a mass spectrum of the peptides eluting at this time point is taken (MS^1 spectrum). A computer generates a prioritized list of these peptides for fragmentation and subsequent tandem mass spectrometric or MS/MS measurements (MS^2 spectra; Figure 4.1). In these measurements a selected peptide ion is isolated, fragmented by energetic collision with gas (usually helium), and the fragment ions are recorded as the tandem mass spectrum. Coupling of two mass spectrometers allows measurement of MS^2 spectra at the same time as MS^1 spectra are recorded. Usually, a high mass accuracy instrument records precursor ions in the MS^1. Three to seven of the most intensive precursor ions are then selected for fragmentation and detection (MS^2) in the second ion trap mass spectrometer. During MS^2 acquisition, the first mass spectrometer already records the successive MS^1. Acquired MS^2 and the associated mass from the MS^1 spectra are then used for matching against protein sequence databases.

Spectra assignment for peptide and protein identification

For peptide identification, measured tandem mass spectra are compared with an *in silico* digested protein database, which is generated (if available) from the genome sequence of the organism under investigation (Figure 4.1). Several computational programs such as Sequest or Mascot (and many others, e.g., Phenyx, Omssa, X!Tandem, Pepsplice) allow comparisons and scoring of spectra that result in peptide identification that is diagnostic for specific proteins (Pappin et al., 1993; Eng et al., 1994). At present, however, full genome sequences are not available for all economically relevant crops, which makes their proteome analysis more challenging. A number of strategies have been developed to facilitate proteomics studies in non-sequenced organisms. First, EST collections are available for most of the major crops. These databases can be searched with MS/MS data for peptide identification. Other strategies include modified MS/MS data analysis workflows such as peptide or protein *de novo* sequencing, which extracts amino acid sequence information directly from an MS/MS spectrum without relying on existing protein databases. A number of *de novo* sequencing tools are available and some of them achieve a good quality *de novo* sequencing result when applied to high-quality spectra (Chen et al., 2001; Johnson and Taylor, 2002; Ma et al., 2003; Searle et al., 2004; Zhang, 2004; Fischer et al., 2005; Frank and Pevzner, 2005; Grossmann et al., 2005). These software tools, however, do not perform well with MS/MS spectra that result from inaccurate measurements, missing peaks (gaps), and chemical or instrument noise. Therefore, *de novo* sequencing is most often not compatible with a high-throughput workflow because all sequences must be carefully examined for correctness, which requires time and expertise.

Quantitative proteomics

One of the most significant advances in proteomics was the development of methods to perform quantitative comparisons between samples (Baginsky et al., 2010; Schulze and Usadel, 2010). For many years, comparative proteome studies were done using 2D gels, in which the staining intensity of protein spots is usually correlated with the abundance of proteins. However, this method suffers limitations in sensitivity and dynamic range (Gygi et al., 2000; Wu and Han, 2006). Improvement in sensitivity, speed, and accuracy of modern mass spectrometers, and progress in statistical data evaluation, however, have now provided a good foundation for quantification in shotgun-type experiments (Lu et al., 2007).

LC-MS data of high mass resolution can be directly used for label-free quantification using appropriate computational methods. Using software such as SuperHirn or Progenesis, peptide signals from LC-MS data can be accurately extracted and these features can be tracked across measurements of different samples (Mueller et al., 2007; www.nonlinear.com/products/progenesis/lc-ms/overview). Other relative quantification methods make use of stable isotope labeling such as isotope-coded affinity tag (ICAT) technology, or metabolically ^{15}N-labeled cell cultures (Gygi et al., 1999; Kierszniowska et al., 2009).

A more recent method, multiple reaction monitoring (MRM), uses the special feature of quadrupole mass spectrometers to specifically select and quantify known peptides (Lange et al., 2008). Three inline quadrupoles allow: (1) the selection for a defined precursor as MS^1, (2) its fragmentation, and (3) a second selection of a specific fragment. This approach enhances sensitivity and therefore the dynamic range for protein detection in a complex protein mixture (Picotti et al., 2009). In combination with the introduction of internal isotopically labeled signature peptides often referred to as proteotypic peptides (PTPs), this method can be used for absolute quantification of peptides. Quantitative measurements can also be achieved by the addition of PTPs into analytical protein samples, where native peptides can then be quantified relative to the added PTPs (Kuster et al., 2005).

Post-translational modifications

Post-translational modifications (PTMs) of proteins often have critical roles in regulating the activity of proteins and therefore the biology of the cell. PTMs are enzymatic, covalent chemical modifications of proteins that typically occur after the translation of mRNAs. These modifications are relevant because they can potentially change the physical or chemical properties, the activity, the localization, or the stability of a protein (Farley and Link, 2009). Modification-specific enrichment techniques combined with advanced MS/MS methods and computational data analysis have revealed a surprisingly large extent of PTMs in proteins, including multi-site, cooperative modifications in individual proteins (Zhao and Jensen, 2009).

Phosphorylation is a prominent modification that regulates a large number of cellular processes. It is a reversible modification catalyzed by protein kinases and can occur on serine, threonine, or tyrosine. Despite the often low abundance of phosphorylated peptides within a complex peptide mixture, precise assignment of the phosphorylation sites can now be achieved using a combination of enrichment methods, such as immobilized metal-affinity chromatography (IMAC) or titanium dioxide (TiO_2), and the sensitivity and high mass accuracy of modern mass spectrometers (Li and Dass, 1999; Larsen et al., 2005; Bodenmiller et al., 2007). In plants, several large-scale studies successfully started to map the phosphoproteome of *Arabidopsis* (Sugiyama et al., 2008; Jones et al., 2009; Reiland et al., 2009), of rice (Whiteman et al., 2008; Nakagami et al., 2010), or of *Medicago* (Grimsrud et al., 2010). Such data are accessible in different databases such as PhosPhAt (Baginsky and Gruissem, 2009; Kersten et al., 2009; Durek et al., 2010).

So far, other PTMs have been studied on a large scale using different enrichment methods. For example, N-linked glycocapture enabled the study of protein N-glycosylation (Zhang et al., 2003). Lysine and N-terminal acetylation were mapped by applying immunoaffinity strategies or combined fractional diagonal chromatography (Gevaert et al., 2003; Choudhary et al., 2009a). But to date, most of these techniques have not been applied to plants on a larger scale.

Proteomics in Plant Biotechnology

Although genetically engineered crops are now widely used in agriculture — in particular insect- and herbicide-resistant maize, soybean, canola, and cotton — very few reports are available on the proteome of transgenic crops. This is not surprising, because crop genome sequence information is only now becoming available on a broader scale and the engineered traits used in crop plants today are not expected to influence the physiological and metabolic activities of the plant. Concerns about the safety of food from genetically engineered crops have been addressed in detail and novel proteins that are expressed in a crop plant are subject to a rigorous evaluation to exclude potential allergenic effects (Metcalfe, 2005; Goodman et al., 2005) or effects on the environment and non-target organisms (Bigler, 2006). Most studies reported that analyzed the substantial equivalence of genetically engineered crops typically found more differences in transcripts, proteins, and metabolites between crop varieties than differences between transgenic plants and their non-transgenic siblings (e.g., Catchpole et al., 2005; Lehesranta et al., 2005; Baker et al., 2006; Baudo et al., 2006; Ioset et al., 2007). This will change, however, as crop plants are engineered to improve their nutritional properties and tolerance to abiotic and biotic stress to address the nutritional health of humans and animals, and to achieve yield stability or increase yield under changing climatic conditions (e.g., Ye et al., 2000; Butelli et al., 2008; Wirth et al., 2009). Therefore it will be increasingly important to understand the complexity of genetic and protein networks in crop plants and how their function relates to physiology and performance.

What has been achieved so far in crop proteomics?

Proteomics research in the agricultural field — so-called crop proteomics — has a history of about two decades. Although only very limited protein sequence information was first available for most crop plants, one common approach was to search the measured spectra using a database of all non-redundant protein sequence databases or to search a subset of this database containing only plant sequences, commonly called the *viridiplantae* protein sequence database. Clearly, the lack of detailed genomic information was limiting these approaches with respect to the completeness of such studies. Furthermore, if a protein of a related organism was identified based on some peptides of the related protein, it could not be taken for granted that the protein also existed in exactly this form in the investigated organism. Nevertheless, many pioneer plant proteomics studies of crop plants succeeded in identifying homologous protein sequences using such databases.

Non-redundant protein sequence databases are still often used in crop proteomics because only a small number of plant genomes have been sequenced to date, although this is changing quickly with next-generation sequencing technologies that are now available. Protein separation by 2D PAGE followed by identification using MALDI-TOF is the most widely used method to investigate the proteome composition of various intracellular compartments in crop plants, but also to investigate quantitative protein changes upon application of various stresses or genetic perturbations. MS-based studies were already reported for a large number of unsequenced plant species including: apple, chickpea, cotton, eucalyptus, ginseng, olive tree, pumpkin, strawberry, sugar beet, or sugarcane. Their discussion goes beyond the scope of this chapter.

Despite the reduced throughput and identification in comparison with shotgun proteomics, 2D PAGE approaches remain powerful because they allow high-confidence identification of differentially regulated proteins between two experimental conditions by quantifying intensities of specific protein spots (Weiss and Görg, 2009). Protein spots on highly resolved 2D gels are mainly composed of one unique protein, which increases the chances for successful protein identification by MALDI-TOF. Mass spectra obtained by MALDI-TOF are highly accurate, therefore, *de novo* sequencing methods can be applied in addition to searches using non-redundant protein sequence databases to achieve an increased peptide and protein identification for unsequenced organisms (Grossmann et al., 2007). Such an approach was undertaken to analyze the proteome of hairy root from ginseng (Kim et al., 2003).

Arabidopsis thaliana as plant model organism

Establishment of novel methods and technologies in plant MS-based proteomics was mainly driven by studies using the

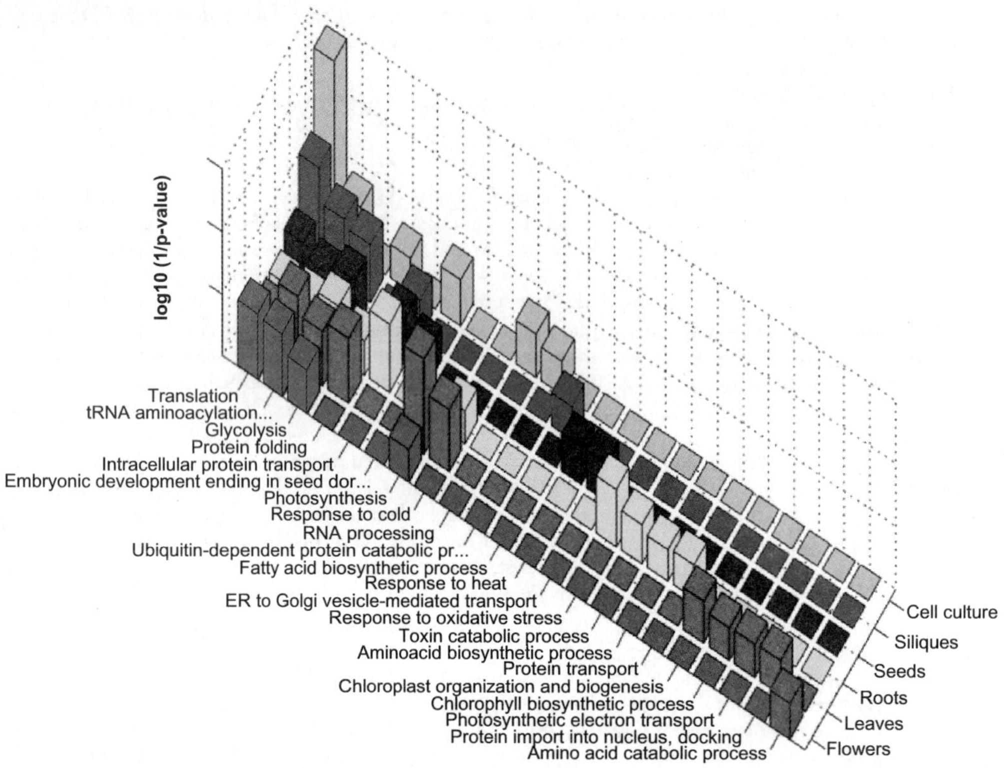

Figure 4.2 • Protein landscape of identified proteins in *Arabidopsis* tissues. • Functional classification of proteins identified in a large-scale proteome study of various *Arabidopsis* tissues. Functional classification is based on TAIR GO categories from the aspect "biological process." Please see color plate section at the end of the book. Figure and legend are adapted from Baerenfaller et al. (2008).

model plant *A. thaliana*. Recently, label-free spectral counting (Lu et al., 2007) has been successfully applied to quantify more than 13,000 proteins of the *Arabidopsis* proteome (Baerenfaller et al., 2008). This study provided an important resource for the plant research community as quantitative information became available for proteins of several tissues including leaves, roots, cotyledons, flowers, seeds, or siliques (Figure 4.2). Furthermore, the quantitative protein inventory of many organelles has been achieved, thus providing valuable information on intracellular protein localization (Lunn, 2007; Lilley and Dupree, 2007; Baginsky, 2009). Information of such large-scale proteomics studies were compiled into publicly available databases highlighted in Table 4.1. Also, MS-based proteomics was used to investigate responses to biotic and abiotic stress (Peck et al., 2001). In addition, regulatory networks have been studied by analyzing PTMs such as phosphorylation (Sugiyama et al., 2008; Jones et al., 2009; Reiland et al., 2009; Baginsky and Gruissem, 2009). Altogether, all of this acquired knowledge is greatly contributing to our understanding of crop and economical relevant species (Rensink and Buell, 2004).

Crops and other economically relevant plant species

Following completion of the map-based genome sequencing in 2005, rice has emerged as another plant model species (International Rice Genome Sequencing Project, 2005). Functional genomics approaches, including proteomics, have therefore become more accessible and powerful for this species (Yazaki et al., 2004). For example, quantitative label-free shotgun proteomic analysis was performed to analyze the proteome of cultured cells exposed to high- and low-temperature stresses (Gammulla et al., 2010). Albeit most rice MS-based proteomics studies remain based on 2D PAGE, these cover a wide range of specific biological questions such as the identification of gibberellic acid-responsive proteins in leaf sheath or drought-induced changes in the anther proteome (Gu et al., 2010; Liu and Bennett, 2010). The proteome composition of various tissues and intracellular compartments such as the nucleus, the plasma membrane, the plastid, or the mitochondria were described (Khan and Komatsu, 2004; Tanaka et al., 2010; von Zychlinski et al., 2005; Komatsu, 2008;

Table 4.1 Available proteome databases for *Arabidopsis*

Compartement	Abbreviation	Name	Web site	No. of proteins	Reference
All	SUBA	SUB-Cellular Location Database for Arabidopsis Proteins	suba.plantenergy.uwa.edu.au	5699	Heazlewood et al. (2007)
All	atProteome	The Arabidopsis Proteome Map	www.atproteome.ethz.ch	13,029	Baerenfaller et al. (2008)
Chloroplast	PlProt	Plastid Proteome Database	www.plprot.ethz.ch	2043	Kleffmann et al. (2006)
Chloroplast	PPDB	The Plant Proteome Database	ppdb.tc.cornell.edu	>5000	Sun et al. (2009)
Chloroplast	AT_CHLORO	Chloroplast Proteome Database	http://www.grenoble.prabi.fr/at_chloro	1323	Ferro et al. (2010)
Mitochondria	AMPDB	The Arabidopsis Mitochondrial Protein DataBase	suba.planetary.uwa.edu.au	416	Heazlewood et al. (2004)
Nucleolus	AtNoPDB	The Arabidopsis Nucleolar Protein Database	bioinf.scri.sari.ac.uk/cgi-bin/atnopdb/home	217	Brown et al. (2005)

Proteome databases for the plant model organism *A. thaliana*. Proteins were identified by LC-MS/MS either in total extracts or in purified compartments such as the chloroplast, the mitochondria, or the nucleolus.

Aki and Yanagisawa, 2009; Choudhary et al., 2009b; Huang et al., 2009). Also the consequences of various stress conditions were investigated by comparative proteomics. These include notable responses to high temperatures (Lee et al., 2007; Han et al., 2009), drought (Salekdeh et al., 2002; Liu and Bennett, 2010), or leaf wounding (Dafoe et al., 2010).

MS-based proteomics studies were not only restricted to model species. For example, the phloem protein composition has been analyzed in rice, pumpkin, and poplar (Dafoe et al., 2009; Lin et al., 2009; Aki et al., 2008; Cho et al., 2010). Because of its role as a "model tree" with economic implications for the wood industry, the proteome composition of poplar and its modulation upon cadmium or heat stress were investigated (Ferreira et al., 2006; Plomion et al., 2006; Kieffer et al., 2009). Also for economic reasons, the proteome of grapevine has recently received much attention. Studies include responses to stresses such as salt (Jellouli et al., 2008), fungal infections (Marsh et al., 2010), hormone treatment (Giribaldi et al., 2010), or herbicide treatment (Castro et al., 2005).

The proteomes of several crop species, in particular maize, wheat, and cassava, were also subjected to many proteome studies. Seed and starch-bound storage proteins, proteins of storage roots, and seed development received particular attention (Sheffield et al., 2006; Grimaud et al., 2008; Sergeant et al., 2009; Shen et al., 2009; Li et al., 2010; Pechanova et al., 2010; Wall et al., 2010). Of interest is also the analysis of potential allergens in maize kernels using proteomic tools (Fasoli et al., 2009) or the investigation of C4 photosynthesis in isolated chloroplasts of bundle sheath cells (Friso et al., 2010).

MS-based proteomics was also recently used to understand the consequences of genetic perturbations or of genetic transformation (Kim et al., 2008; Barros et al., 2010). Taken together, the constantly growing numbers of studies related to crop physiology and more recently also to plant biotechnology reinforce the view that proteomics has emerged as a technology of choice to understand the proteome composition of various crops, tissues, intracellular compartments, and its changes upon application of many different biotic or abiotic stresses.

Future applications and perspectives

Considering the rapid advances in next-generation sequencing technologies, the numbers of sequenced plant genomes will quickly grow in the future. For example, the genome of papaya has recently been released (Yu et al., 2009) and the establishment of adequate protein extraction procedures for 2D PAGE has already started (Rodrigues et al., 2009). Together with the increased number of sequenced and annotated plant genomes, the analytical power of MS-based proteomics approaches will increase. It can therefore be expected that shotgun label-free approaches coupled with statistical downstream analysis using bioinformatics tools will be used more often. In addition, the standardization of quantitative proteomics workflows such as MRM or visual proteomics and various enrichment techniques (e.g., for phosphopeptides), should allow a deeper insight into specific molecular mechanisms (Bodenmiller et al., 2007; Lange et al., 2008; Picotti et al., 2009; Beck et al., 2009).

Despite their expected importance for biofuel or hydrogen production, the proteomes of algae remain poorly analyzed. As a model organism with its genome sequenced at the end of 2007, *Chlamydomonas* is excellently suited for MS-based proteomics analyses and different studies have contributed thus far to describe the overall proteome composition and its response to sulfur-deprivation or metal-rich water (Rolland et al., 2009; Chen et al., 2010; Cid et al., 2010).

MS-based proteomics approaches and functional genomics studies such as transcriptomics or metabolomics simultaneously monitor a large number of components of a system at one time point. An integrative analysis of such large-scale datasets generates a broad overview of the complex plant metabolic network. Subsequent modeling of these data is also expected to allow predictive functions of different components of a metabolic network. Future strategies and applications in plant biotechnology will certainly take advantage of a broader systems view for the selection and enhancement of specific metabolic functions by classical breading strategies or by direct engineering of genetically modified organisms.

References

Aebersold, R., & Mann, M. (2003). Mass spectrometry-based proteomics. *Nature, 422*, 198–207.

Aki, T., Shigyo, M., Nakano, R., Yoneyama, T., & Yanagisawa, S. (2008). Nano scale proteomics revealed the presence of regulatory proteins including three FT-Like proteins in phloem and xylem saps from rice. *Plant and Cell Physiology, 49*, 767–790.

Aki, T., & Yanagisawa, S. (2009). Application of rice nuclear proteome analysis to the identification of evolutionarily conserved and glucose-responsive nuclear proteins. *Journal of Proteome Research, 8*, 3912–3924.

Anderson, N. L., & Anderson, N. G. (1998). Proteome and proteomics: New technologies, new concepts, and new words. *Electrophoresis, 19*, 1853–1861.

Arabidopsis Genome Initiative, (200). Analysis of the genome sequence of the flowering plant *Arabidopsis thaliana*. *Nature, 408*, 796–815.

Baerenfaller, K., Grossmann, J., Grobei, M. A., Hull, R., Hirsch-Hoffmann, M., & Yalovsky, S., et al. (2008). Genome-scale proteomics reveals *Arabidopsis thaliana* gene models and proteome dynamics. *Science, 320*, 938–941.

Baginsky, S., & Gruissem, W. (2009). The chloroplast kinase network: New insights from large-scale phosphoproteome profiling. *Molecular Plant, 2*, 1141–1153.

Baginsky, S., Hennig, L., Zimmermann, P., & Gruissem, W. (2010). Gene expression analysis, proteomics, and network discovery. *Plant Physiology, 152*, 402–410.

Baginsky, S. (2009). Plant proteomics: Concepts, applications, and novel strategies for data interpretation. *Mass Spectrometry Reviews, 28*, 93–120.

Baker, J. M., Hawkins, N. D., Ward, J. L., Lovegrove, A., Napier, J. A., & Shewry, P. R., et al. (2006). A metabolomic study of substantial equivalence of field-grown genetically modified wheat. *Plant Biotechnology Journal, 4*, 381–392.

Barglow, K. T., & Cravatt, B. F. (2007). Activity-based protein profiling for the functional annotation of enzymes. *Nature Methods, 4*, 822–827.

Barros, E., Lezar, S., Anttonen, M. J., van Dijk, J. P., Röhlig, R. M., & Kok, E. J., et al. (2010). Comparison of two GM maize varieties with a near-isogenic non-GM variety using transcriptomics, proteomics and metabolomics. *Plant Biotechnology Journal, 8*, 436–451.

Baudo, M. M., Lyons, R., Powers, S., Pastori, G. M., Edwards, K. J., & Holdsworth, M. J., et al. (2006). Transgenesis has less impact on the transcriptome of wheat grain than conventional breeding. *Plant Biotechnology Journal, 4*, 369–380.

Beck, M., Malmström, J. A., Lange, V., Schmidt, A., Deutsch, E. W., & Aebersold, R. (2009). Visual proteomics of the human pathogen *Leptospira interrogans*. *Nature Methods, 6*, 817–823.

Bigler, F. (2006). Ninth international symposium on the biosafety of genetically modified organisms. Session II: Identifying and defining hazards and potential consequences I: Concepts for problem formulation and non-target risk assessment. *Environmental Biosafety Research, 5*, 183–186.

Bodenmiller, B., Mueller, L. N., Mueller, M., Domon, B., & Aebersold, R. (2007). Reproducible isolation of distinct, overlapping segments of the phosphoproteome. *Nature Methods, 4*, 231–237.

Brown, J. W., Shaw, P. J., Shaw, P., & Marshall, D. F. (2005). *Arabidopsis* nucleolar protein database (AtNoPDB). *Nucleic Acids Research, 1*, 633–636.

Castro, A. J., Carapito, C., Zorn, N., Magné, C., Leize, E., & Van Dorsselaer, A., et al. (2005). Proteomic analysis of grapevine (*Vitis vinifera* L.) tissues subjected to herbicide stress. *Journal of Experimental Botany, 56*, 2783–2795.

Catchpole, G. S., Beckmann, M., Enot, D. P., Mondhe, M., Zywicki, B., & Taylor, J., et al. (2005). Hierarchical metabolomics demonstrates substantial compositional similarity between genetically modified and conventional potato crops. *Proceedings of the National Academy of Sciences of the United States of America, 102*, 14458–14462.

Cellar, N. A., Kuppannan, K., Langhorst, M. L., Ni, W., Xu, P., & Young, S. A. (2007). Cross species applicability of abundant protein depletion columns for ribulose-1,5-bisphosphate carboxylase/oxygenase. *Journal of Chromatography B, Analytical Technologies in the Biomedical and Life Sciences, 1*(861), 29–30..

Chen, M., Zhao, L., Sun, Y. L., Cui, S. X., Zhang, L. F., & Yang, B., et al. (2010). Proteomic analysis of hydrogen photoproduction in sulfur-deprived Chlamydomonas cells. *Journal of Proteome Research, 9*, 3854–3866.

Chen, T., Kao, M. Y., Tepel, M., Rush, J., & Church, G. M. (2001). A dynamic programming approach to *de novo* peptide sequencing via tandem mass spectrometry. *Journal of Computational Biology, 8*, 325–337.

Cho, W. K., Chen, X. Y., Rim, Y., Chu, H., Kim, S., & Kim, S. W., et al. (2010). Proteome study of the phloem sap of pumpkin using multidimensional protein identification technology. *Journal of Plant Physiology, 167*, 771–778.

Choudhary, C., Kumar, C., Gnad, F., Nielsen, M. L., & Rehman, M., et al. (2009). Lysine acetylation targets protein complexes and co-regulates major cellular functions. *Science, 325*, 834–840.

Choudhary, M. K., Basu, D., Datta, A., Chakraborty, N., & Chakraborty, S. (2009). Dehydration-responsive nuclear proteome of rice (*Oryza sativa* L.) illustrates protein network, novel regulators of cellular adaptation, and evolutionary perspective. *Molecular & Cellular Proteomics: MCP, 8*, 1579–1598.

Cid, C., Garcia-Descalzo, L., Casado-Lafuente, V., Amils, R., & Aguilera, A. (2010). Proteomic analysis of the response of an acidophilic strain of Chlamydomonas sp. (*Chlorophyta*) to natural metal-rich water. *Proteomics, 10*, 2026–2036.

Cox, J., & Mann, M. (2007). Is proteomics the new genomics? *Cell, 130*, 395–398.

Dafoe, N. J., Gowen, B. E., & Constabel, C. P. (2010). Thaumatin-like proteins are differentially expressed and localized in phloem tissues of hybrid poplar. *BMC Plant Biology, 10*, 191.

Dafoe, N. J., Zamani, A., Ekramoddoullah, A. K., Lippert, D., Bohlmann, J., & Constabel, C. P. (2009). Analysis of the poplar phloem proteome and its response to leaf wounding. *Journal of Proteome Research, 5*, 2341–2350.

De Wilde, C., Van Houdt, H., De Buck, S., Angenon, G., De Jaeger, G., & Depicker, A. (2000). Plants as bioreactors for protein production: Avoiding the problem of transgene silencing. *Plant Molecular Biology, 43*, 347–359.

Domon, B., & Aebersold, R. (2010). Options and considerations when selecting a quantitative proteomics strategy. *Nature Biotechnology, 28*, 710–721.

Durek, P., Schmidt, R., Heazlewood, J. L., Jones, A., & MacLean, D., et al. (2010). PhosPhAt: The *Arabidopsis thaliana* phosphorylation site database. An update. *Nucleic Acids Research, 38*, D828–834.

Eng, J. K., McCormack, A. L., & III., Yates, J. R. (1994). An approach to correlate tandem mass spectral data of peptides with amino acid sequences in a protein database. *Journal of the American Society for Mass Spectrometry*, *11*, 976–989.

Farley, A. R., & Link, A. J. (2009). Identification and quantification of protein posttranslational modifications. *Methods in Enzymology*, *463*, 725–763.

Fasoli, E., Pastorello, E. A., Farioli, L., Scibilia, J., Aldini, G., & Carini, M., et al. (2009). Searching for allergens in maize kernels via proteomic tools. *Journal of Proteomics*, *13*, 501–510.

Fenn, J. B., Mann, M., Meng, C. K., Wong, S. F., & Whitehouse, C. M. (1989). Electrospray ionization for mass spectrometry of large biomolecules. *Science*, *246*, 64–71.

Ferreira, S., Hjernø, K., Larsen, M., Wingsle, G., Larsen, P., & Fey, S., et al. (2006). Proteome profiling of *Populus euphratica* Oliv. upon heat stress. *Annals of Botany*, *98*, 361–377.

Ferro, M., Brugiere, S., Salvi, D., Seigneurin-Berny, D., Court, M., & Moyet, L., et al. (2010). AT_CHLORO: A comprehensive chloroplast proteome database with subplastidial localization and curated information on envelope proteins. *Molecular Cell Proteomics*, *9*, 1063–1084.

Fischer, B., Roth, V., Roos, F., Grossmann, J., Baginsky, S., & Widmayer, P., et al. (2005). NovoHMM: A hidden Markov model for *de novo* peptide sequencing. *Analytical Chemistry*, *77*, 7265–7273.

Frank, A., & Pevzner, P. (2005). PepNovo: *De novo* peptide sequencing via probabilistic network modeling. *Analytical Chemistry*, *77*, 964–973.

Friso, G., Majeran, W., Huang, M., Sun, Q., & van Wijk, K. J. (2010). Reconstruction of metabolic pathways, protein expression, and homeostasis machineries across maize bundle sheath and mesophyll chloroplasts: Large-scale quantitative proteomics using the first maize genome assembly. *Plant Physiology*, *152*, 1219–1250.

Gammulla, C. G., Pascovici, D., Atwell, B. J., & Haynes, P. A. (2010). Differential metabolic response of cultured rice (*Oryza sativa*) cells exposed to high- and low-temperature stress. *Proteomics*, *10*, 3001–3019.

Gevaert, K., Goethals, M., Martens, L., Van Damme, J., & Staes, A., et al. (2003). Exploring proteomes and analyzing protein processing by mass spectrometric identification of sorted N-terminal peptides. *Nature Biotechnology*, *21*, 566–569.

Gibon, Y., Blaesing, O. E., Hannemann, J., Carillo, P., Höhne, M., & Hendriks, J. H., et al. (2004). A Robot-based platform to measure multiple enzyme activities in *Arabidopsis* using a set of cycling assays: Comparison of changes of enzyme activities and transcript levels during diurnal cycles and in prolonged darkness. *Plant Cell*, *16*, 3304–3325.

Giribaldi, M., Gény, L., Delrot, S., & Schubert, A. (2010). Proteomic analysis of the effects of ABA treatments on ripening *Vitis vinifera* berries. *Journal of Experimental Botany*, *61*, 2447–2458.

Goodman, R. E., Hefle, S. L., Taylor, S. L., & van Ree, R. (2005). Assessing genetically modified crops to minimize the risk of increased food allergy: A review. *International Archives of Allergy and Immunology*, *137*, 153–156.

Grimaud, F., Rogniaux, H., James, M. G., Myers, A. M., & Planchot, V. (2008). Proteome and phosphoproteome analysis of starch granule-associated proteins from normal maize and mutants affected in starch biosynthesis. *Journal of Experimental Botany*, *59*, 3395–3406.

Grimsrud, P. A., den Os, D., Wenger, C. D., Swaney, D. L., Schwartz, D., & Sussman, M. R., et al. (2010). Large-scale phosphoprotein analysis in *Medicago truncatula* roots provides insight into *in vivo* kinase activity in legumes. *Plant Physiology*, *152*, 19–28.

Grossmann, J., Fischer, B., Baerenfaller, K., Owiti, J., Buhmann, J. M., & Gruissem, W., et al. (2007). A workflow to increase the detection rate of proteins from unsequenced organisms in high-throughput proteomics experiments. *Proteomics*, *7*, 4245–4254.

Grossmann, J., Roos, F. F., Cieliebak, M., Lipták, Z., Mathis, L. K., & Müller, M., et al. (2005). AUDENS: A tool for automated peptide *de novo* sequencing. *Journal of Proteome Research*, *4*, 1768–1774.

Gstaiger, M., & Aebersold, R. (2009). Applying mass spectrometry-based proteomics to genetics, genomics and network biology. *Nature Reviews Genetics*, *10*, 617–627.

Gu, J. Y., Wang, Y., Zhang, X., Zhang, S. H., Gao, Y., & An, C. C. (2010). Identification of gibberellin acid-responsive proteins in rice leaf sheath using proteomics. *Frontiers in Bioscience*, *15*, 826–839.

Gygi, S. P., Corthals, G. L., Zhang, Y., Rochon, Y., & Aebersold, R. (2000). Evaluation of two-dimensional gel electrophoresis-based proteome analysis technology. *Proceedings of the National Academy of Sciences of the United States of America*, *97*, 9390–9395.

Gygi, S. P., Rist, B., Gerber, S. A., Turecek, F., & Gelb, M. H., et al. (1999). Quantitative analysis of complex protein mixtures using isotope-coded affinity tags. *Nature Biotechnology*, *17*, 994–999.

Han, F., Chen, H., Li, X. J., Yang, M. F., Liu, G. S., & Shen, S. H. (2009). A comparative proteomic analysis of rice seedlings under various high-temperature stresses. *Biochimica et Biophysica Acta*, *1794*, 1625–1634.

Heazlewood, J. L., Tonti-Filippini, J. S., Gout, A., Day, D. A., Whelan, J., & Millar, A. H. (2004). Experimental analysis of the *Arabidopsis* mitochondrial proteome highlights signalling and regulatory components, provides assessment of targeting prediction programs and points to plant specific mitochondrial proteins. *Plant Cell*, *16*, 241–256.

Heazlewood, J. L., Verboom, R. E., Tonti-Filippini, J., Small, I., & Millar, A. H. (2007). SUBA: The Arabidopsis Subcellular Database. *Nucleic Acids Research*, *35*, D213–218.

Hruz, T., Laule, O., Szabo, G., Wessendrop, F., Bleuler, S., & Oertle, L., et al. (2008). Genevestigator V3: A reference expression database for the meta-analysis of transcriptomes. *Advances in Bioinformatics*, *1* (5), 851–857.

Huang, S., Taylor, N. L., Narsai, R., Eubel, H., Whelan, J., & Millar, A. H. (2009). Experimental analysis of the rice mitochondrial proteome, its biogenesis, and heterogeneity. *Plant Physiology*, *149*, 719–734.

International Rice Genome Sequencing Project. (2005). The map-based sequence of the rice genome. *Nature*, *436*, 793–800.

Ioset, J. R., Urbaniak, B., Ndjoko-Ioset, K., Wirth, J., Martin, F., & Gruissem, W., et al. (2007). Flavonoid profiling among wild type and related GM wheat varieties. *Plant Molecular Biology*, *65*, 645–654.

Isaacson, T., Damasceno, C. M., Saravanan, R. S., He, Y., Catalá, C., & Saladié, M., et al. (2006). Sample extraction techniques for enhanced proteomic analysis of plant tissues. *Nature Protocols*, *1*, 769–774.

James, P. (1997). Protein identification in the post-genome era: The rapid rise of proteomics. *Quarterly Reviews of Biophysics*, *30*, 279–331.

Jellouli, N., Ben Jouira, H., Skouri, H., Ghorbel, A., Gourgouri, A., & Mliki, A. (2008). Proteomic analysis of Tunisian grapevine cultivar *Razegui* under salt stress. *Journal of Plant Physiology*, *165*, 471–481.

Johnson, R. S., & Taylor, J. A. (2002). Searching sequence databases via *de novo* peptide sequencing by tandem mass spectrometry. *Molecular Biotechnology*, *22*, 301–315.

Jones, A. M., MacLean, D., Studholme, D. J., Serna-Sanz, A., Andreasson, E., & Rathjen, J. P., et al. (2009). Phosphoproteomic analysis of nuclei-enriched fractions from *Arabidopsis thaliana*. *Journal of Proteomics*, *72*, 439–451.

Kersten, B., Agrawal, G. K., Durek, P., Neigenfind, J., Schulze, W., & Walther, D., et al. (2009). Plant phosphoproteomics: An update. *Proteomics*, *9*, 964–988.

Khan, M. M., & Komatsu, S. (2004). Rice proteomics: Recent developments and analysis of nuclear proteins. *Phytochemistry*, *65*, 1671–1681.

Kieffer, P., Schröder, P., Dommes, J., Hoffmann, L., Renaut, J., & Hausman, J. F. (2009). Proteomic and enzymatic response of poplar to cadmium stress. *Journal of Proteomics*, *72*, 379–396.

Kierszniowska, S., Seiwert, B., & Schulze, W. X. (2009). Definition of *Arabidopsis* sterol-rich membrane microdomains by differential treatment with methyl-beta-cyclodextrin and quantitative proteomics. *Molecular Cellular Proteomics: MCP*, *8*, 612–623.

Kim, S. I., Kim, J. Y., Kim, E. A., Kwon, K. H., Kim, K. W., & Cho, K., et al. (2003). Proteome analysis of hairy root from Panax ginseng C.A. Meyer using peptide fingerprinting, internal sequencing and expressed sequence tag data. *Proteomics*, *3*, 2379–2392.

Kim, S. T., Kim, S. G., Kang, Y. H., Wang, Y., Kim, J. Y., & Yi, N., et al. (2008). Proteomics analysis of rice lesion mimic mutant (spl1) reveals tightly localized probenazole-induced protein (PBZ1) in cells undergoing programmed cell death. *Journal of Proteome Research*, *7*, 1750–1760.

Kleffmann, T., Hirsch-Hoffmann, M., Gruissem, W., & Baginsky, S. (2006). Plprot: A comprehensive proteome database for different plastid types. *Plant and Cell Physiology*, *47*, 432–436.

Kleffmann, T., Russenberger, D., von Zychlinski, A., Christopher, W., Sjolander, K., & Gruissem, W., et al. (2004). The *Arabidopsis thaliana* chloroplast proteome reveals pathway abundance and novel protein functions. *Current Biology : CB*, *14*, 354–362.

Komatsu, S. (2008). Plasma membrane proteome in *Arabidopsis* and rice. *Proteomics*, *8*, 4137–4145.

Kuster, B., Schirle, M., Mallick, P., & Aebersold, R. (2005). Scoring proteomes with proteotypic peptide probes. *Nature Reviews Molecular Cell Biology*, *6*, 577–583.

Lange, V., Picotti, P., Domon, B., & Aebersold, R. (2008). Selected reaction monitoring for quantitative proteomics: A tutorial. *Molecular Systems Biology*, *4*, 222.

Larsen, M. R., Thingholm, T. E., Jensen, O. N., Roepstorff, P., & Jorgensen, T. J. (2005). Highly selective enrichment of phosphorylated peptides from peptide mixtures using titanium dioxide microcolumns. *Molecular Cellular Proteomics*, *4*, 873–886.

Lee, D. G., Ahsan, N., Lee, S. H., Kang, K. Y., Bahk, J. D., & Lee, I. J., et al. (2007). A proteomic approach in analyzing heat-responsive proteins in rice leaves. *Proteomics*, *7*, 3369–3383.

Lehesranta, S. J., Davies, H. V., Shepherd, L. V., Nunan, N., McNicol, J. W., & Auriola, S., et al. (2005). Comparison of tuber proteomes of potato varieties, landraces, and genetically modified lines. *Plant Physiology*, *138*, 1690–1699.

Li, K., Zhu, W., Zeng, K., Zhang, Z., Ye, J., & Ou, W., et al. (2010). Proteome characterization of cassava (*Manihot esculenta Crantz*) somatic embryos, plantlets and tuberous roots. *Proteome Science*, *8*, 10.

Li, S., & Dass, C. (1999). Iron(III)-immobilized metal ion affinity chromatography and mass spectrometry for the purification and characterization of synthetic phosphopeptides. *Analytical Biochemistry*, *270*, 9–14.

Lilley, K. S., & Dupree, P. (2007). Plant organelle proteomics. *Current Opinion in Plant Biology*, *10*, 594–599.

Lin, M. K., Lee, Y. J., Lough, T. J., Phinney, B. S., & Lucas, W. J. (2009). Analysis of the pumpkin phloem proteome provides insights into angiosperm sieve tube function. *Molecular & Cellular Proteomics : MCP*, *8*, 343–356.

Liu, J. X., & Bennett, J. (2010). Reversible and irreversible drought-induced changes in the Anther Proteome of rice (*Oryza sativa* L.) Genotypes IR64 and Moroberekan. *Molecular Plant 4* (1), 59–69.

Lu, P., Vogel, C., Wang, R., Yao, X., & Marcotte, E. M. (2007). Absolute protein expression profiling estimates the relative contributions of transcriptional and translational regulation. *Nature Biotechnology*, *25*, 117–124.

Lunn, J. E. (2007). Compartmentation in plant metabolism. *Journal of Experimental Botany*, *58*, 35–47.

Ma, B., Zhang, K., Hendrie, C., Liang, C., Li, M., & Doherty-Kirby, A., et al. (2003). PEAKS: Powerful software for peptide *de novo* sequencing by tandem mass spectrometry. *Rapid Communications in Mass Spectrometry*, *17*, 2337–2342.

Maldonado, A. M., Echevarría-Zomeño, S., Jean-Baptiste, S., Hernández, M., & Jorrín-Novo, J. V. (2008). Evaluation of three different protocols of protein extraction for Arabidopsis thaliana leaf proteome analysis by two-dimensional electrophoresis. *Journal of Proteomics*, *71*, 461–472.

Marsh, E., Alvarez, S., Hicks, L. M., Barbazuk, W. B., Qiu, W., & Kovacs, L., et al. (2010). Changes in protein abundance during powdery mildew infection of leaf tissues of Cabernet Sauvignon grapevine (*Vitis vinifera* L.). *Proteomics*, *10*, 2057–2064.

Metcalfe, D. D. (2005). Genetically modified crops and allergenicity. *Nature Immunology*, *6*, 857–860.

Mueller, L. N., Rinner, O., Schmidt, A., Letarte, S., & Bodenmiller, B., et al. (2007). SuperHirn – A novel tool for high resolution LC-MS-based peptide/protein profiling. *Proteomics*, *7*, 3470–3480.

Nakagami, H., Sugiyama, N., Mochida, K., Daudi, A., Yoshida, Y., & Toyoda, T., et al. (2010). Large-scale comparative phosphoproteomics identifies conserved phosphorylation sites in plants. *Plant Physiology*, *153*, 1161–1174.

Pandey, A., & Mann, M. (2000). Proteomics to study genes and genomes. *Nature*, *405*, 837–846.

Pappin, D. J., Hojrup, P., & Bleasby, A. J. (1993). Rapid identification of proteins by peptide-mass fingerprinting. *Current Biology*, *3*, 327–332.

Pechanova, O., Pechan, T., Ozkan, S., McCarthy, F. M., Williams, W. P., & Luthe, D. S. (2010). Proteome profile of the developing maize (*Zea mays* L.) rachis. *Proteomics*, *10*, 3051.

Peck, S., Nühse, T. S., Hess, D., Iglesias, A., Meins, S., & Boller, T. (2001). Directed proteomics identifies a plant-specific protein rapidly phosphorylated in response to bacterial and fungal elicitors. *The Plant Cell*, *13*, 1467–1475.

Phizicky, E., Bastiaens, P. I., Zhu, H., Snyder, M., & Fields, S. (2003). Protein analysis on a proteomic scale. *Nature*, *422*, 208–215.

Picotti, P., Bodenmiller, B., Mueller, L. N., Domon, B., & Aebersold, R. (2009). Full dynamic range proteome analysis of *S. cerevisiae* by targeted proteomics. *Cell*, *138*, 795–806.

Plomion, C., Lalanne, C., Claverol, S., Meddour, H., Kohler, A., & Bogeat-Triboulot, M. B., et al. (2006). Mapping the proteome of poplar and application to the discovery of drought-stress responsive proteins. *Proteomics*, *6*, 6509–6527.

Reiland, S., Messerli, G., Baerenfaller, K., Gerrits, B., & Endler, A., et al. (2009). Large-scale *Arabidopsis* phosphoproteome profiling reveals novel chloroplast kinase substrates and phosphorylation networks. *Plant Physiology*, *150*, 889–903.

Rensink, W. A., & Buell, C. R. (2004). *Arabidopsis* to rice. Applying knowledge from a weed to enhance our understanding of a crop species. *Plant Physiology*, *135*, 622–629.

Rodrigues, S. P., Ventura, J. A., Zingali, R. B., & Fernandes, P. M. (2009). Evaluation of sample preparation methods for the analysis of papaya leaf proteins through two-dimensional gel electrophoresis. *Phytochemical Analysis*, *20*, 456–464.

Rolland, N., Atteia, A., Decottignies, P., Garin, J., Hippler, M., & Kreimer, G., et al. (2009). Chlamydomonas proteomics. *Current Opinion in Microbiology*, *12*, 285–291.

Salekdeh, G. H., Siopongco, J., Wade, L. J., Ghareyazie, B., & Bennett, J. (2002). Proteomic analysis of rice leaves during drought stress and recovery. *Proteomics*, *2*, 1131–1145.

Saravanan, R. S., & Rose, J. K. (2004). A critical evaluation of sample extraction techniques for enhanced proteomic analysis of recalcitrant plant tissues. *Proteomics*, *4*, 2522–2532.

Schulze, W. X., & Usadel, B. (2010). Quantitation in mass-spectrometry-based proteomics. *Annual Review of Plant Biology*, *61*, 491–516.

Searle, B. C., Dasari, S., Turner, M., Reddy, A. P., Choi, D., & Wilmarth, P. A., et al. (2004). High-throughput identification of proteins and unanticipated sequence modifications using a mass-based alignment algorithm for MS/MS *de novo* sequencing results. *Analytical Chemistry*, *76*, 2220–2230.

Sergeant, K., Pinheiro, C., Hausman, J. F., Ricardo, C. P., & Renaut, J. (2009). Taking advantage of nonspecific trypsin cleavages for the identification of seed storage proteins in cereals. *Journal of Proteome Research*, *8*, 3182–3190.

Sheffield, J., Taylor, N., Fauquet, C., & Chen, S. (2006). The cassava (*Manihot esculenta Crantz*) root proteome: Protein identification and differential expression. *Proteomics*, *6*, 1588–1598.

Shen, Z., Li, P., Ni, R. J., Ritchie, M., Yang, C. P., & Liu, G. F., et al. (2009). Label-free quantitative proteomics analysis of etiolated maize seedling leaves during greening. *Molecular & Cellular Proteomics: MCP*, *8*, 2443–2460.

Shevchenko, A., Wilm, M., Vorm, O., & Mann, M. (1996). Mass spectrometric sequencing of proteins silver-stained polyacrylamide gels. *Analytical Chemistry*, *68*, 850–858.

Sugiyama, N., Nakagami, H., Mochida, K., Daudi, A., & Tomita, M., et al. (2008). Large-scale phosphorylation mapping reveals the extent of tyrosine phosphorylation in *Arabidopsis*. *Molecular Systems Biology*, *4*, 193.

Sun, Q., Zybailov, B., Majeran, W., Friso, G., Olinares, P. D., & van Wijk, K. J. (2009). PPDB, the plant proteomics database at Cornell. *Nucleic Acids Research*, *37*, D969–974.

Swaney, D. L., Wenger, C. D., & Coon, J. J. (2010). Value of using multiple proteases for large-scale mass spectrometry-based proteomics. *Journal of Proteome Research*, *9*, 1323–1329.

Tanaka, N., Konishi, H., Khan, M. M., & Komatsu, S. (2010). Proteome analysis of rice tissues by two-dimensional electrophoresis: An approach to the investigation of gibberellin regulated proteins. *Molecular Genetics and Genomics*, *270*, 485–496.

von Zychlinski, A., Kleffmann, T., Krishnamurthy, N., Sjolander, K., Baginsky, S., & Gruissem, W. (2005). Proteome analysis of the rice etioplast: Metabolic and regulatory networks and novel protein functions. *Molecular Cellular Proteomics*, *4*, 1072–1084.

Wall, M. L., Wheeler, H. L., Smith, J., Figeys, D., & Altosaar, I. (2010). Mass spectrometric analysis reveals remnants of host-pathogen molecular interactions at the starch granule surface in wheat endosperm. *Phytopathology*, *100*, 848–854.

Weiss, W., & Görg, A. (2009). High-resolution two-dimensional electrophoresis. *Methods in Molecular Biology*, *564*, 13–32.

Whiteman, S. A., Nühse, T. S., Ashford, D. A., Sanders, D., & Frans Maathuis, J. M. (2008). A proteomic and phosphoproteomic analysis of *Oryza sativa* plasma membrane and vacuolar membrane. *The Plant Journal*, *56*, 146–156.

Wienkoop, S., Baginsky, S., & Weckwerth, W. (2010). *Arabidopsis thaliana* as a model organism for plant proteome research. *Journal of Proteomics*, *73*, 2239–2248.

Wirth, J., Poletti, S., Aeschlimann, B., Yakandawala, N., Drosse, B., & Osorio, S., et al. (2009). Rice endosperm iron biofontification by targeted and synergistic action of nicotianamine synthase and ferritin. *Plant Biotechnology Journal*, *7*, 631–644.

Wu, L., & Han, D. K. (2006). Overcoming the dynamic range problem in mass spectrometry-based shotgun proteomics. *Expert Review of Proteomics*, *3*, 611–619.

Yan, W., & Chen, S. S. (2005). Mass spectrometry-based quantitative proteomic profiling. *Briefings in Functional Genomics & Proteomics*, *4*, 27–38.

Yates, J. R., Ruse, C. I., & Nakorchevsky, A. (2009). Proteomics by mass spectrometry: Approaches, advances, and applications. *Annual Review of Biomedical Engineering*, *11*, 49–79.

Yazaki, J., Kojima, K., Suzuki, K., Kishimoto, N., & Kikuchi, S. (2004). The Rice PIPELINE: A unification tool for plant functional genomics. *Nucleic Acids Research*, *1*, D383–387.

Ye, X., Al-Babili, S., Klöti, A., Zhang, J., Lucca, P., & Beyer, P., et al. (2000). Engineering the provitamine A (beta-carotene) biosynthetic pathway into (carotenoid-free) rice endosperm. *Science*, *287*, 303–305.

Yu, Q., Tong, E., Skelton, R. L., Bowers, J. E., Jones, M. R., & Murray, J. E., et al. (2009). A physical map of the papaya genome with integrated genetic map and genome sequence. *BMC Genomics*, *10*, 371.

Zhang, H., Li, X. J., Martin, D. B., & Aebersold, R. (2003). Identification and quantification of N-linked glycoproteins using hydrazide chemistry, stable isotope labeling and mass spectrometry. *Nature Biotechnology*, *21*, 660–666.

Zhao, Y., & Jensen, O. N. (2009). Modification-specific proteomics: Strategies for characterization of post-translational modifications using enrichment techniques. *Proteomics*, *9*, 4632–4641.

Zimmermann, P., Hirsch-Hoffmann, M., Hennig, L., & Gruissem, W. (2004). GENEVESTIGATOR. *Arabidopsis* microarray database and analysis toolbox. *Plant Physiology*, *136*, 2621–2632

Plant metabolomics: Applications and opportunities for agricultural biotechnology

5

Diane M. Beckles[1] Ute Roessner[2]

[1]*Department of Plant Sciences, University of California, Davis California,* [2]*Australian Centre for Plant Functional Genomics and Metabolomics Australia, University of Melbourne, Melbourne, Australia*

TABLE OF CONTENTS

Introduction

Plants produce a spectacular array of chemical compounds and are some of the most efficient and elegant chemical synthesizing machines known. A single plant has the capacity to produce 5–25,000 compounds at any given time (Verpoorte, 1998; Weckwerth and Fiehn, 2002; Oksman-Caldentey and Inzé, 2004), and there are estimates that as many as 200,000–1,000,000 different metabolites may be produced in the plant kingdom (Saito and Matsuda, 2010). The chemical structures synthesized by plants are our source of food (directly and indirectly), fuel (both fossilized, ethanolic, and oil), polymers, fiber, adhesives, waxes, dyes, fragrances, medicinal compounds, and feedstocks for animals. It is not an over-exaggeration to suggest that our existence on Earth depends on plant metabolic capacity.

Advances in analytical technology now make it possible to take a comprehensive and unbiased look at the chemical inventory of a plant cell. Each cell unit is an individual processing facility responsible for the production of the metabolic intermediates or end products valuable to man. These compounds share some basic functional groups — hydroxyls, alcohols, steroids, alkyls, benzyl rings, and so forth — but it is the chemical diversity that arises from the infinite combination of these structures that leads to unique compounds with different solubilities, melting points, and reactivities typical of plant metabolism (Trethewey, 2004; Roessner and Beckles, 2009). Current profiling technologies permit the simultaneous detection of hundreds of compounds within a single plant extract. This has led to the emergence of the field of "metabolomics," that is, the description of the metabolic state of a biological system in response to environmental and genetic perturbations (Oliver et al., 1998; Fiehn, 2001; Villas-Boas et al., 2005).

Although the metabolome in many plants represents a catalog of thousands of different chemical species, for simplicity they are classified as belonging to one of two groups — primary or secondary — depending on their roles in plant survival. Secondary metabolites are not directly needed for normal growth and development, but their production can enhance the organism's survival under certain conditions of biotic or abiotic stress. Based on DNA sequence analysis of many plant genomes, we know that there are thousands of genes that potentially encode enzymes that catalyze exotic reactions (Saito and Matsuda, 2010), but we have not been able to experimentally resolve the functions of these putative enzymes. Primary

DOI: 10.1016/B978-0-12-381466-1.00005-5

metabolites are immediately necessary for plants to complete their life cycles under normal growth conditions. Quantitatively, they comprise the highest proportion of the metabolome and they tend to be simpler compounds (e.g., sugars and amino and fatty acids), which often act as immediate substrates for compounds related to agronomic yield (e.g., starch, protein, lipids, and cell walls). The human intervention 10,000 years ago that led to the domestication of naturally growing plant species also led to massive changes in plant composition. Many secondary metabolites that enabled plants to tolerate unfavorable conditions in the wild have been diminished in various crop plants, whereas the primary storage compounds have been coaxed for maximal production through breeding and by providing optimal conditions of culture. Although it is pragmatic to make this arbitrary distinction, secondary metabolites may regulate signaling networks that are integral to the successful completion of the plant growth cycle acting as pseudohormones (Pourcel and Grotewold, 2009), and we anticipate that further studies will elucidate these interrelationships.

At its outset biotechnology held great promise for rational and targeted metabolic engineering, but its potential has not yet been fully realized. Although there have been some promising breakthroughs (Itkin and Aharoni, 2009), metabolite engineering, by and large, has been unsuccessful due in large part to the complexity and robustness of plant metabolism and the extent to which the environment influences plant phenotype. So, can metabolomics be applied to enhance the chances of successful biotechnological manipulation? We argue that the answer is yes, and that this can be achieved in three main ways. Metabolomics can:

1. Identify metabolite markers that underscore valuable phenotypes.
2. Assess the outcomes of engineering steps.
3. Guide future metabolic engineering strategies.

Metabolomics will not be a magic bullet for solving biotechnological problems, but may prove to be a valuable technology by providing a more complete picture of the biological system under study. In this chapter we will outline what is known about the general organization of metabolite networks and use it to illustrate broad concepts of metabolic control and regulation, information necessary for metabolite engineering. We will then survey the analytical and data-mining tools available as this will allow us to ascertain what can be measured and how biological information can be extracted from these measurements. We will then outline how metabolomics is being used to support biotechnology. There has literally been an explosion in the number, scope, and depth of applications of metabolomics in recent years and we will not be able to comprehensively address them all. Finally, we will look at the challenges ahead and the key issues that must be addressed if the full potential of this discipline is to be realized for improving agriculture.

Metabolite Networks: The Basics

Network analysis is rather complex and there are many evolving theorems. For the purposes of this chapter, we will explore this subject superficially, so the reader is referred to the recent literature (Ma and Zeng, 2003a,b; Sweetlove and Fernie, 2005; Zhao et al., 2006; Rios-Estepa and Lange, 2007; Kose et al., 2007; Rodriguez and Infante, 2009; Grafahrend-Belau et al., 2009;) for more details. Some commonly constructed plant metabolite networks display stoichiometric relationships between metabolites that imply functionality. The relationship between metabolite x_1 and all others is determined in several samples and metabolites that correlate strongly, that is, maintain the same relationship under different perturbations, are inferred to be connected by common regulatory mechanisms or in an enzymatic step (Weckwerth et al., 2004; Camacho et al., 2005; Stamova et al., 2009). In such networks each metabolite may be considered to be a *node*. When two or more metabolites (nodes) are connected (presumably by an enzymatic step that links them), this is described as an *edge*. Metabolites that are the substrates of many reactions, for example, glucose-6-phosphate, may have many edges highly interlinked to other nodes (metabolites). Where there is a high frequency of such metabolites, these regions of the network are described as *hubs*. Several hubs may form individual *modules*, which can connect to others leading to a supra-hierarchical structure.

Metabolic networks have been extensively characterized in unicellular prokaryotes, which are much less complex than plants. However, some basic patterns have emerged that may be broadly applicable. A "bow-tie" structure is often evident when the nodes of a metabolite network are viewed (Ma and Zeng, 2003a,b; Zhao et al., 2006; Grafahrend-Belau et al., 2009; Figure 5.1). A giant strong component (GSC) that constitutes many reactions of central metabolism can be discerned. Inputs to the GSC are those needed for assimilation so in plants these would be carbon dioxide fixation and N-, S-, and P-assimilation reactions that have high energy consumption but are needed for the synthesis of more complicated compounds such as amino acids, nucleic acids, sugars, and so forth. Outputs from the GSC in plants, which would include reactions that constitute a secondary metabolism, are typically specialized and are less interconnected.

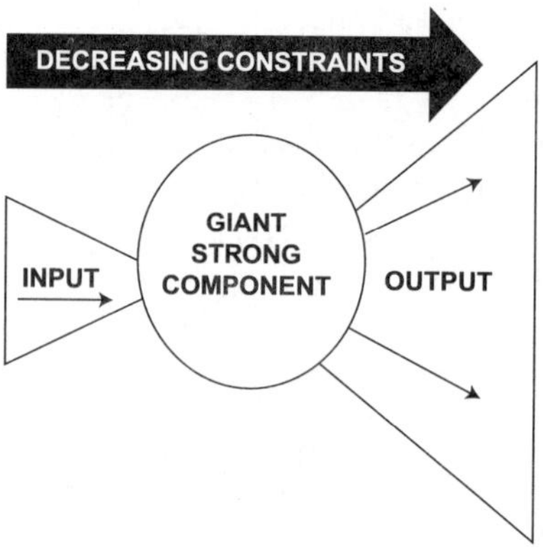

Figure 5.1 • Bow-tie structure illustrating the global properties of metabolic networks. Adapted from Grafahrend-Belau et al. (2009) and Ma and Zeng (2003).

As carbon and energy flow through this system there may be decreasing constraints, such as less regulation of the reactions. Considering plant metabolism in this way can provide a framework relevant to metabolic engineering. Reactions that are part of the GSC may be more difficult to change, or changes may impact normal growth and development, whereas reactions described as outputs may be easier to manipulate since they are less intertwined with core processes needed for survival. Reactions at the juncture of the GSC and the output modules may play a key role controlling fluxes between reactions needed for basal functioning (GSC), and those that can readily react to swift changes in external conditions and boost survival (output) (Modolo et al., 2009). The model we have presented here is based on data for low-complexity organisms; however, with data amassed from metabolomics of plant organs, networks that accurately reflect the metabolomes of higher plants may be built.

Metabolomics: Technologies for Analyses

The metabolome consists of a vast number of different chemical structures hugely diverse in chemical and physical properties such as molecular weight and size, polarity, volatility, solubility, stability, and pk_a (Villas-Bôas et al., 2007). Therefore, a number of analytical technologies have to be employed to enable the separation, detection, and quantification of these chemicals. The most common platforms are liquid and gas chromatography both coupled with mass spectrometry (LC-MS and GC-MS) and nuclear magnetic resonance spectroscopy (NMR). How these instruments function and why they are used in the metabolomics field is extensively described in a number of reviews and books (Villas-Bôas et al., 2007; Roessner and Beckles, 2009, and references therein). There are two major approaches in metabolomics, targeted and untargeted (Figure 5.2). The attempt to measure as many metabolites as possible is defined as untargeted analysis and the selection and accurate quantification of a subset of metabolites of interest is called the targeted approach. There are advantages and disadvantages for both approaches, as outlined in Figure 5.2. Thousands of metabolites can now be analyzed simultaneously by using LC-MS (Callahan et al., 2009). The resulting data matrices are highly multidimensional and require sophisticated statistical and multivariate analysis techniques to make comparisons between stages and identify the differences and similarities between those stages (see the section Data Analysis and Interpretation). One difficulty in the interpretation of data resulting from untargeted analysis is the sheer number of peaks determined that cannot be related to a particular chemical structure, and therefore cannot be related to a particular biochemical pathway. The major advantage of targeted analysis, on the other hand, is that the user knows what is being measured which gives better interpretable results in a biological context. The actual concentrations of the metabolites targeted are determined, which may indicate importance and function of the metabolites in a rather biochemical context. Targeted analysis is generally characterized by a much higher accuracy, greater selectivity and, most important, a greater sensitivity compared to untargeted analysis (e.g., Kitteringham et al., 2009; Wei and Seymour, 2010) since the metabolites of interest are separated from the crude extract using specialized extraction procedures or via specific targeting in the mass spectrometer. A common methodology for specific targeting of the ions of a particular compound utilizes the so-called multiple reaction monitoring (MRM), a method well known from proteomics studies designed to quantify peptides (Kitteringham et al., 2009; Yocum and Chinnaiyan, 2009). In metabolomics, the technique is very comparable, the molecular ion of a compound is first selected and further fragmented. One specific fragment is then selected and monitored (a process called transition) and the number of both the mother and fragment ion is counted. When compared to a standard calibration curve it is then possible to estimate the concentration of the compound in the extract. Using this physical instrumentation developments thus far have allowed the quantification of up to 250 metabolites simultaneously (Wei and Seymour, 2010); the greatest limitation, however, is the availability of authentic standards required for the establishment of calibration curves.

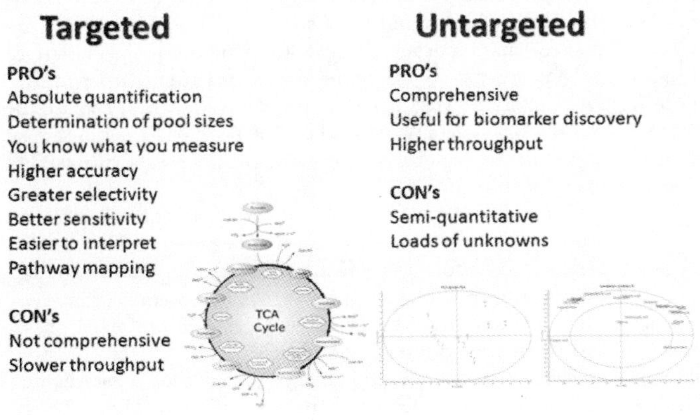

Targeted
PRO's
Absolute quantification
Determination of pool sizes
You know what you measure
Higher accuracy
Greater selectivity
Better sensitivity
Easier to interpret
Pathway mapping
CON's
Not comprehensive
Slower throughput
TCA Cycle
Untargeted
PRO's
Comprehensive
Useful for biomarker discovery
Higher throughput
CON's
Semi-quantitative
Loads of unknowns

Figure 5.2 • Schematics of the advantages and disadvantages of targeted versus untargeted metabolite analysis. Please see color plate section at the end of the book.

Analytical platforms

A number of analytical platforms have to be used in a complementary manner to measure the huge chemical diversity. The most common metabolite detection technologies, often referred to as "work-horses," are mass spectrometry (MS) and NMR. Complex metabolite mixtures extracted from biological fluids and tissues often require a separation step prior to detection, especially when using MS as the detection system. Gas chromatography (GC), liquid chromatography (LC), and capillary electrophoresis (CE) are the most common separation techniques coupled with MS. These separation and detection methods have been extensively described in the literature (Villas-Bôas et al., 2007; Roessner and Beckles, 2009). Here we only present a short summary of the technologies and refer the reader to the extensive literature.

GC-MS

Gas chromatography coupled with mass spectrometry (GC-MS) is characterized by its great separation power and reproducibility. GC-MS allows the analysis of compounds that are either volatile or can be made volatile through derivatization. After injection, the compounds are separated based on their volatility temperature and interaction with the column stationary phase (e.g., polarity). Eluted compounds enter the ion source, usually an electron impact ionization (EI) source, where they are ionized prior to entering the MS. Since the conditions of electron impact ionization are highly standardized and reproducible, comprehensive GC-MS mass spectral libraries for compound identifications have been established that are both commercially (e.g., NIST, http://www.nist.gov/srd/nist1a.cfm; Agilent's FiehnLib; Kind et al., 2009) and publicly available (GMD; Kopka et al., 2005). Ideally, for increased confidence and reproducibility of the data, the libraries do not only offer searching based on mass spectra, but also provide retention times or retention time indices under standardized conditions for each compound (Kopka et al., 2005). Due to the comprehensiveness and validity of the existing GC-MS libraries, a typical GC-MS analysis of a biological sample today can separate up to 400 compounds, including amino and organic acids, sugars, amines, fatty acids, and sterols, of which about 100 to 150 can be unambiguously identified with respect to their chemical nature. Much work must still be done to identify remaining unknown peaks. The bottleneck comes from the lack of availability of authentic standards required to confirm a peak's identity before it can be stored in a mass spectral library. Although GC-MS is currently limited to identification of compounds smaller than 1000 Da, it remains a major analytical platform for metabolomics approaches since it is robust, easy to establish, and requires relatively low capital investments for setup.

LC-MS

The targeted and quantitative approach and untargeted but semi-quantitative analysis using LC-MS have been widely used in the proteomics field and have garnered substantially greater interest in the metabolomics community during the past few years (Griffiths and Wang, 2009). There are a number of advantages to using LC-MS for metabolite analyses. First, compounds do not require derivatization prior to analysis so the chemical structures of the compounds analyzed are therefore not altered. Secondly, higher molecular weight compounds and metabolites with lower thermostability are amendable for analysis.

Table 5.1 List of mass analyzers, their principle of mass separation, and indication of mass resolution

Type of analyzer	Symbol	Principle of mass separation	Resolution
Quadrupole	Q	*m/z* (trajectory stability)	Low
Ion trap	IT	*m/z* (resonance frequency)	Low
Time of flight	TOF	Velocity (flight time)	High
Fourier transform ion cyclotron resonance	FTICR	*m/z* (resonance frequency)	Very high
Fourier transform orbitrap	FT-OT	*m/z* (resonance frequency)	Very high

Source: Modified from Hoffmann and Stroobant (2007).

Eluting compounds must be converted into the gas phase via simultaneous ionization to be analyzed using MS. The most common ionization techniques include electrospray ionization (ESI) and atmospheric pressure ionization (API). Once ionized compounds are separated they are then counted based on their mass to charge ratio. This process takes place in a mass analyzer. The most common analyzers and their modes of ion separation are listed in Table 5.1. A more detailed description of how the different analyzers function is given in Roessner and Beckles (2009).

Modern MS often involves the use of two or more mass analyzers in tandem (e.g., QqTOF, QqQ, QqLT; for graphic description see Roessner and Beckles, 2009). For tandem MS a number of triggered mass selections are carried out, followed by subsequent fragmentation of the selected ions. Potentially, two (MS/MS) or more rounds (MS^n) of this process are conducted, an important feature for structural elucidation of unknown peaks to determine specific fragmentation patterns. Once a target's fragmentation pattern is known, it can also be utilized to assign a name to the same peak observed in several experimental biological samples through the establishment of mass spectral libraries.

There are a great number of different column matrices available for the separation of complex metabolite extracts and it is therefore important to choose the appropriate stationary as well as mobile phase for the compounds of interest. These column matrices include ion exchange, reversed phase, and hydrophobic interaction chromatography. Unfortunately, no single separation methodology is capable of separating the vast number of different chemistries of metabolites. This is

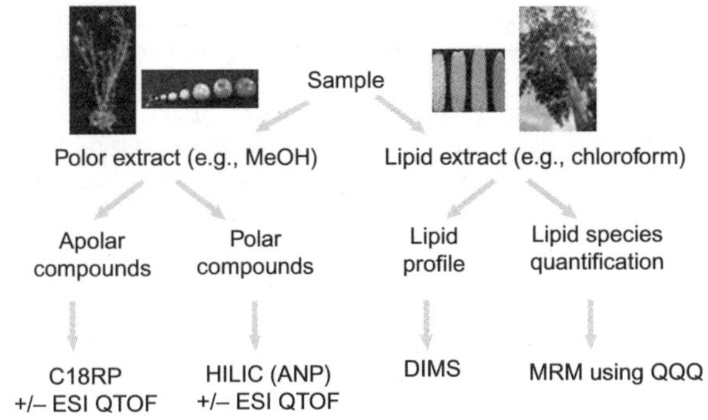

Figure 5.3 • Schematic of the potential methodologies required for untargeted and highly comprehensive metabolite analyses. Please see color plate section at the end of the book.

especially problematic if the aim is to analyze as many metabolites as possible in an untargeted manner. Figure 5.3 illustrates the complexity of methodologies required to determine as many compounds as possible, ranging from highly hydrophobic (such as lipids) to highly hydrophilic (such as amino acids or sugar phosphates) metabolites, from a single sample.

To analyze both hydrophilic and hydrophobic compounds, these classes are ideally separated through an extraction process using a bi-phasic solvent mixture (e.g., methanol, water, chloroform). The chloroform phase, which contains the lipophilic components such as lipids and sterols, can then be analyzed using lipidomics methodologies either to screen for as many lipid classes present as possible (Ivanova et al., 2009) or to quantify individual lipid species using MRM (Ståhlman et al., 2009).

The methanol/water phase must then be analyzed with two different stationary phases. C18 reversed phase is used for the more polar components in the extract (e.g., flavonoids, alkaloids, non-polar amino acids such as tryptophan or phenylalanine, and certain plant hormones) and aqueous normal phase (ANP) or hydrophilic interaction chromatography (HILIC) is used for polar components (e.g., sugars, sugar phosphates, polar amino acids such as alanine and glycine; Callahan et al., 2009). Two types of ionization modes (positive and negative) have to be used to cover a greater complement of metabolites in the extract.

A number of mass spectral libraries are commercially or publicly available (see Table 5.2); however, none of them record retention times of the compounds. Unfortunately, retention times in LC application are often inconsistent, as major changes in the constitution of the mobile phase and temperature or pH of the solvents (and extracts) can have dramatic effects on the retention times of the eluting compounds or even, in part, on the order of elution. Another difficulty for unambiguous peak identification is that a number of metabolites have the exact same accurate mass and therefore the same sum formular, which can be generated from the accurate mass. New initiatives around the world are supporting attempts to establish highly standardized and controlled conditions for LC (e.g., stationary and mobile phase, temperature, gradient, flow, pH) as well as MS (e.g., ionization mode, collision energies, MS scan modes and speeds) with the longer term goal of developing LC-MS libraries that record accurate masses, retention times, and MS^2 and MS^3 fragmentation patterns.

Table 5.2 Examples of mass spectral databases

Name	Web site	Reference
Metlin	http://metlin.scripps.edu/	Smith et al. (2005)
Chemspider	www.chemspider.com/	Williams (2008)
MassBank	www.massbank.jp/	Horai et al. (2010)
HMD	www.hmdb.ca/	Forsythe and Wishart (2009)

In summary, there are two important developments required to improve the utility of LC-MS-based metabolomics. First, high-throughput methodologies to help identify more and more of the unknown mass features in chromatograms need to be established, both by building comprehensive libraries and by selective purification and enrichment technique of those unknown peaks for subsequent NMR-based structural elucidations. The second important development is to extend our current targeted quantitative MRM-based methods toward more comprehensive methods that would allow accurate quantification of hundreds of metabolites simultaneously.

NMR

NMR-based metabolite profiling has been used extensively for the analysis of biological fluids such as blood, serum, or urine and has become a great success in a number of medical applications and biomarker discovery studies (Bollard et al., 2005). There are a number of advantages to using NMR in metabolomics; for instance, it is a non-destructive measurement allowing the sample to be used for other analyses. Also, NMR-based metabolite detection can quantitatively capture a number of different classes of metabolites with different chemical and physical properties in one single analysis, substantially reducing the measurement time per sample (Ward et al., 2007). Compared to other analytical platforms, NMR represents a fast and relatively cheap (per sample base) method; however, NMR requires a substantial capital investment and, therefore, careful consideration. NMR is also characterized by its relatively low sensitivity, meaning that larger sample volumes are required to detect sufficient numbers of metabolites.

Data analysis and interpretation

A single metabolomics experiment can generate hundreds of data points that require cautious treatment and analysis to remove noise from a "real" signal and allow biological interpretation of these data. There are a number of steps involved in this process, from data extraction from the instrument to the ability to visualize them with respect to the biological/biochemical question. Most often there are three stages of data

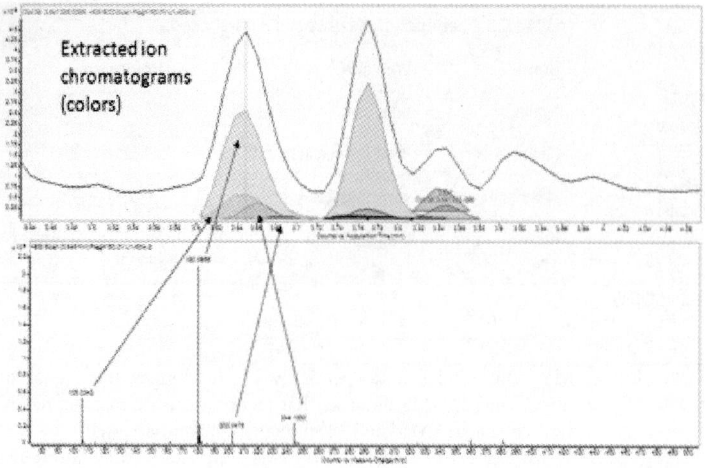

Figure 5.4 • Example of deconvolution. • Upper panel shows the total ion chromatogram (in black) and the deconvoluted extracted ion chromatograms of each compound detected (in color). The lower panel shows the respective accurate mass peaks for the deconvoluted extracted ion chromatograms for the leftmost peak in the upper panel only. Please see color plate section at the end of the book.

processing applied. The first, called pre-processing, includes steps to reduce raw data by removing noise. These include deconvolution, filtering, peak detection or feature finding, and alignment of the peaks found from different samples. The main purpose of these steps is to extract the characteristics of each observed ion (e.g., m/z ratio, retention time, and ion intensity) from the raw data file into an easy-to-interpret form. The second step involves normalization and transformation procedures followed by the third step, the actual statistical analysis using either univariate (e.g., Student's *t*-test) or multivariate (e.g., principal component analysis; PCA) methodologies.

Data pre-processing

The first step for the data pre-processing procedure is to deconvolute the signals and detect the peaks under each deconvoluted signal. Figure 5.4 shows an example of a deconvoluted total ion chromatogram (TIC). Each colored peak underneath the TIC peak represents an individual ion detected at a similar time that can now be related to its accurate mass in the mass spectrum shown in Figure 5.4. A similar approach is taken when GC-MS data are deconvoluted; however, in that situation it is important to consider that each compound is represented by more than one ion due to the electron impact ionization process used in GC-MS.

The next step is to distinguish between noise and the "real" signal from the sample. It is important to always run blank samples with each experiment (e.g., an empty tube treated the same as all other samples), because they will provide estimates of background noise that will be subtracted from experimental samples. The next process is to align all detected signals across all samples within an experiment to account for retention time shifts between runs. Most instrument vendors offer this capability in their instrument-specific software; however, this operation can also be performed using open-source software packages (e.g., XCMS, Smith et al., 2006; or MZmine, Katajamaa et al., 2006; for review see Katajamaa and Oresic, 2007).

Normalization and data transformation

Deconvolution, peak identification, and alignment result in a data matrix in table format, which then undergoes adequate normalization and subsequent transformation prior to statistical analysis. Depending on the properties of the data matrix different normalization procedures are applied for metabolomics data. Most biological data are characterized by heteroscedasticity (i.e., where the standard deviation of each metabolite determined in replicate samples changes with the mean of the signal; e.g., Arneberg et al., 2007; Herberich et al., 2010), which can substantially affect results (Kvalheim et al., 1994). Therefore, it is important to normalize data. This is done by choosing the appropriate normalization procedure, because it is important to consider that every dataset is different. Common normalization procedures include Z-scores, media, sum of selected or all metabolites, and internal and external standard normalizations. These procedures and their advantages and disadvantages are described in detail in Roessner et al. (2011).

Depending on the type of statistical analysis required, different transformation procedures are also applied. For instance, most biological data are either right or left skewed but the Student's *t*-test can only be applied to normally distributed data; therefore, transformation is required prior to performing *t*-tests. Transformation also changes heteroscedasticity into homoscedasticity, so the method used depends on the properties of the heteroscedasticity. For example, if the standard deviation is directly proportional to the mean of each signal, log-transformation is optimal to change heteroscedasticity. However, if the standard deviation is proportional to the root of the mean, then square-root transformation can provide a homoscedastic noise pattern (Roessner et al., 2011).

Statistical analysis

Once the data are appropriately normalized and transformed they can be statistically analyzed. A number of different univariate (e.g., Student's *t*-test) and multivariate statistical analysis tools (e.g., Principal Component Analysis (PCA) linear discriminant analysis, LDA; hierarchical cluster analysis, HCA; partial least squares discriminant analysis, PLS-DA; and independent component analysis, ICA) are available to determine statistical differences between samples and metabolites. Again, the method of choice depends on the biological question as well as on the experimental design and data acquisition. There is extensive literature available describing the functionality of different statistical methodologies and their applicability and usefulness to the analysis of metabolomics data (see Roessner and Beckles, 2009; Roessner et al., 2011, and references therein).

Data visualization

The previously mentioned data analysis tools are great for representing the changes, similarities, and dissimilarities between groups. However, in a biological context metabolomics data need to be mapped onto biochemical pathways to allow interpretation of the influence of the treatment (genetic or environmental) on the levels of metabolites. The low percentage of signals detected that are known with respect to chemical structure and therefore in relation to a metabolite presents a bottleneck in this approach. That is why targeted metabolite analysis has gained more interest in the past few years; the metabolites detected and quantified as a result of targeted analysis are known and can be mapped onto a biochemical pathway. In any untargeted approach the only metabolites that are further investigated are those with statistically significant changes, which may be useful if one's aim is to identify a particular marker characteristic of a sample group; other metabolites are not further investigated (e.g., identified). We want to stress that from a biochemical and biological point of view it is important to know which metabolites do and do not change in their pool sizes.

A number of software tools are available that allow mapping of metabolite changes onto biochemical pathways (for a detailed review see Roessner and Beckles, 2009). Again, each has its advantages and disadvantages. There are tools such as the Omics Viewer (Pathway Tools Omics Viewer – PTOV), which allows mapping of metabolite level changes onto a given, static pathway, and maps gene and protein expression and metabolic flux analysis data (Paley and Karp, 2006). Only a small portion of metabolites are identified in routine metabolomics experiments and mapping those onto the whole biochemical pathway "dilutes" the visualization of the changes. Using tools such as VANTED (Junker et al., 2006), the pathways with which the measured metabolites are involved can be drawn in a customized fashion. Another advantage of the VANTED tool is that it allows mapping of the results of more than two experiments on the same metabolic pathway rather than limiting only one comparison per visualized map.

Metabolomics: Applications in Agricultural Biotechnology

The primary aim of agricultural biotechnology is to maximize the synthesis of a product deemed desirable, to reduce the level of undesirable products or to introduce a novel non-native product to the plant in the hopes of delivering value-added, reasonably priced commodities to market. So how can metabolomics enable this?

Metabolomics can be used to tell: (1) if there is a pathway step that is regulated differently due to a perturbation, (2) the extent to which genetic or environmental changes alter metabolism, and (3) which gene(s) underlie the accumulation of a metabolite. In addition, metabolomics is used to assist with cultivar classification and phenotyping. The attention we give to each of the next sections is not necessarily indicative of its importance; rather we highlight areas that are more often than not overlooked and those studies that best exemplify an application.

Metabolite profiling to test substantial equivalence

Metabolomics plays a major role in the determination of substantial equivalence of GMO crops (Beale et al., 2009). Genetic alteration affects the visible phenotype as well as the biochemical composition of the cells leading to unexpected effects on the basis of current genetic or biochemical knowledge (Cellini et al., 2004; Kuiper et al., 2003). To determine substantial equivalence, a concept initially introduced by the Organization for Economic Co-operation and Development (OECD, 1991; Kuiper et al., 2002) as a framework for safety evaluations where existing crops and foods serve as a baseline for safety and the properties of any new foods and crops are compared to their counterparts is used. The metabolomes of genetically engineered plants are monitored and compared to the natural variation of metabolomes of their wild-type counterparts. This framework is not a safety assessment *per se*, but presents a starting point for a comprehensive comparison on all biochemical and chemical levels of the new product (Kuiper et al., 2001). Metabolomics is only one of the tools utilized in support of this concept (Rischer and Oksman-Caldentey, 2006). There have been examples presented where the introduction (or deletion) of a gene has resulted in unexpected changes in metabolism and metabolite levels, even if the gene of interest is not actively involved in metabolic processes. For instance, the introduction of a gene encoding a plasma membrane Na^+-pumping ATPase, originally from *Physcomitrella patens*, into rice and barley under the control of a constitutive promoter resulted in dramatic effects on metabolite levels compared to the control plants even under controlled conditions (Jacobs et al., 2007). Interestingly, some metabolites were differently changed in either rice or barley, indicating the importance of applying the substantial equivalence concept, not only to every gene used for transformation, but also in every species and most probably also every variety into which it was transferred. Large metabolomics studies have been undertaken on transgenic crops grown in the field (Catchpole et al., 2005; Baker et al., 2006; Zywicki et al., 2005; Venneria et al., 2008), and the consensus from those studies is that, while results have to be treated with caution, strong indications are that traditional breeding or growth in different environments may produce inherently more genetic variation in plants than that from some transgenic manipulations.

Phytochemical diversity, phenotyping, and classification

A metabolite profile provides, to some extent, a biochemical readout of the genetic blueprint of an organism. It can therefore be used to classify and phenotype individuals and to look at evolutionary and phylogenetic relationships. Hundreds of data points can be generated from a single GC run; even if these compounds are unknown, their occurrence and relative content are a signature of the genetic background and environmental conditions under which that individual was cultured.

Metabolomics has already been exploited to survey phytochemical diversity among herbs used in traditional medicine (Jiang et al., 2006), to phenotype varieties based on chemical composition, allowing growers to implement standards for quality control (Zeng et al., 2008; Vekiari et al., 2010). Metabolite profiles also act as a signature to identify cultivars and varieties by the region in which they are cultivated (Anastasiadi et al., 2009). The profiles of secondary metabolites in eucalyptus portrayed important taxonomic and evolutionary relationships leading to the suggestion that it can be a powerful tool for phylogenetic studies (Merchant et al., 2006). DNA sequence analysis is still the standard used for fingerprinting individuals, but metabolomics is set to be part of the growing arsenal for performing these types of studies.

Postharvest quality of horticultural crops

The postharvest quality of fruit and vegetables is a prime determinant of consumer acceptability and producer profitability. Good flavor, appearance, and low costs are important to the public whereas reducing losses along the production chain from harvest to table are of concern to the producer (Beckles, 2011). The quality of fruit and vegetables after harvest are maintained by manipulating their storage conditions — temperature, humidity, and atmospheric composition — to delay senescence and retain nutrient content. Sometimes, regardless of environmental manipulation, there is still loss of product. Metabolomics is being increasingly used to understand the biochemical basis of these traits and to identify relevant steps in the pathways for manipulation. Treatments for preserving shelf life (e.g., exposure to UV treatment; Rudell et al., 2008) and modifying CO_2 and O_2 levels (Franck et al., 2004) also affect the metabolome. Singling out the altered metabolites that change in a manner coincident with the treatment may present targeted genetic strategies to augment environmental manipulation. In addition, analyses of volatiles have been used to detect postharvest disease during storage of onions (Prithiviraj et al., 2004), mangoes (Moalemiyan et al., 2006), and apples (Li et al., 2007). This information could help shippers remove bad batches and preserve the core of the crop.

Chemicals also underscore fruit and vegetable flavor and their complex interaction may make one fruit more desirable than another. Metabolomics will feature prominently when identifying these compounds, allowing biologists to establish a baseline for understanding flavor. It may then be possible to develop tools to definitively tag fruit and vegetables in terms of quality and, with functional genomics approaches, identify genes that underscore the latter (Hall, 2007). For example, broad surveys of aromas and secondary metabolites of a tomato mapping population have been undertaken (Bermudez et al., 2008). Correlations with the levels of specific compounds and fruits classified as desirable by sensory panels were found (Zanor et al., 2009). This kind of study may make it possible to rely less heavily on sensory panels in evaluating horticultural crops. In addition, cheap, user-friendly, "electronic noses," such as gas sensor devices that can detect volatiles (Kende et al., 2010), are being used to indicate peak ripeness (Zhang et al., 2008).

Stress responses

Plants have evolved very elaborate basal and acquired mechanisms to deal with biotic (insect, bacteria, fungi, and viral attack) and abiotic stress (cold, frost, heat, drought, salinity). Abiotic stress alone costs millions of dollars in agricultural losses each year (Mittler, 2006), and problems with heat and drought stress are forecast to be amplified due to global warming in a report from a workshop held by the U.S. Department of Energy Office of Science (http://genomicscience.energy.gov/carboncycle/response.pdf). This topic has been extensively reviewed and is the subject of Chapter 18, so it will be treated only superficially here.

Exposure to stress causes, among other things, metabolic dysfunction in plants. Several soluble, neutral organic compounds, such as polyols, amino acids, quaternary ammonium compounds, and/or tertiary sulfonium compounds (compatible solutes), are altered when plants are placed in unfavorable conditions (Bohnert et al., 1995). Changes in these compounds can be quantified through metabolite profiling (Bowne et al., 2010).

Plant responses to stress may be divided into primary and secondary events (Orr and Raison, 1990). Primary events are those that are the direct result of the initial signals propagated from the stress condition, and may be protective or adaptive. Secondary events are time-dependent, downstream symptoms of the dysfunction. The knowledge derived from studying both periods can be informative. In our lab we studied the primary metabolic events during postharvest chilling stress in fruit from *Solanum lycopersicum* cv. Micro-Tom. Tomato fruits stored below 10°C can be damaged and unsalable (Luengwilai and Beckles, 2010). We measured changes in membrane permeability (a sensitive way to detect response to chilling) before, during, and after the introduction of cold stress and, at the time points when membrane ion leakage was initiated, GC-MS sampling was used to correlate metabolite changes with the trait of interest. Approximately 15% of the metabolome was altered, even though morphological effect of the stress was not manifested in the fruit until 14 days later (Luengwilai et al., 2011).

In the real world plants may face an onslaught of unfavorable conditions with each stress occurring with variable timing and intensity. For example, high solar irradiation often accompanies high temperatures, which together reduce water availability and increase soil electrical conductivity. However, metabolomics has been very useful in understanding the physiology of the plant stress response. Using this approach we have learned that in *Arabidopsis* heat and cold stress cause some similar, overlapping responses (Kaplan et al., 2004), but the effects of two or more stresses on the levels of pathway components cannot be predicted by studying each stress individually (Mittler, 2006; Mittler and Blumwald, 2010). We have further learned that landraces may tolerate abiotic stress better than commercial cultivars by accumulating a different spectrum of compatible solutes (Widodo et al., 2009), although how widespread this mechanism is will need to be examined further. Plant scientists have even been able to isolate marker metabolites associated with freezing tolerance in *Arabidopsis* (Korn et al., 2010).

Less is known about the metabolic responses of plants to biotic stress. The general approach is to separately analyze the metabolomes of the tissue to be infected (host) and the pathogen of interest and then analyze the metabolomes of both in a time-course series after infection. In this way metabolites deployed for plant defense and pathogen virulence may be identified. As proof of concept ESI-MS profiling of a resistant and a sensitive cultivar of *Brachypodium* before and after *Magnaporthe grisea* infection was performed to identify metabolites that underscore the differential response. Phospholipid levels correlated to the differences in symptomology and were implicated in defense (in resistant cultivar) or progression (in the susceptible cultivar) of the disease (Allwood et al., 2010). There is also evidence that different pathogenic agents, such as fungus, bacteria, and viroids, will each induce a specific catalog of metabolites upon host infection (Lopez-Gresa et al., 2009).

Functional genomics

Functional genomics is the study of the physiological and biochemical consequences of gene function and interactions. It is well accepted that metabolomics can inform on changes in gene expression in more direct ways than methods that gauge levels of mRNA or polypeptides (Fiehn, 2004). Transcripts and proteins are modified in response to a myriad of regulatory mechanisms so that their accumulated levels may not reflect their relative importance at that time and place. Metabolite levels are determined by a culmination of high-order regulatory signals and are closer to the phenotype than DNA, RNA, or protein. To identify genes that encode enzymes or regulatory molecules that determine metabolite levels, gene-to-metabolite networks can be drawn (Moreno-Risueno et al., 2010). From such a dataset, transcript changes that correlate with metabolite changes implicate a causal relationship. Several important papers in which genes that control commercially important compounds have been identified have been published recently (Rischer and Oksman-Caldentey, 2006; Rischer et al., 2007; Saito et al., 2008). More detailed studies have been reviewed by Saito and Matsuda (2010).

The genome of *Arabidopsis* was the first of any flowering plant to be fully sequenced (Arabidopsis Genome Initiative, 2000). However, 40–60% of genes have not been functionally annotated. To tackle this, a global program has been established to undertake a systematic genome-wide functional screen to identify and validate the function of all of the genes in the plant. The Arabidopsis Metabolomics Consortium (http://tht.vrac.iastate.edu:81/index.php), supported by the National Science Foundation (NSF), is part of this global program and has brought together a number of expert laboratories in plant metabolomics, utilizing as many analytical platforms as possible, with the goal of elucidating the entire metabolic complement of *Arabidopsis*. The aim is to analyze in great detail as many *Arabidopsis* lines as possible that carry mutations in genes of which the functions are not fully understood. The resulting metabolomics data will then be integrated with the results of transcript, protein, and phenotypic analyses, enhancing the generation of hypotheses about gene functions. All of the data generated are stored in a publicly available database freely accessible by researchers for utility in their research areas (Bais et al., 2010).

Breeding and metabolite quantitative trait loci

Variations in an agronomic trait between related species are often due to differences in the cumulative actions of multiple gene products. The genes encoding these quantitative traits or quantitative trait loci (QTLs), may reside on different chromosomes, but their products work together as part of a regulatory, perception, metabolic, or transduction pathway to determine a phenotype. Metabolomics, however, may provide a useful strategy for linking phenotypes to QTLs (Kliebenstein, 2007; Fernie and Schauer, 2009; Keurentjes, 2009). Specific regions of the genome that co-localize with changes in metabolite levels can be classified as metabolite QTLs (mQTLs). Traditionally differences in easily scored phenotypes such as seed mass, plant height, and fruit lycopene were used as markers and now, DNA sequence is more commonly used. Using metabolites as phenotypes presents many advantages. Metabolomics approaches are not constrained by the need for the sequence of an entire genome and, at the present, are a cheaper platform than DNA sequencing technologies (Fernie and Schauer, 2009). Perhaps the best example comes from work by Schauer et al. (2006). A mapping population consisting of 76 individuals was produced by a cross between the wild *Solanum pennellii* and *S. lycopersicum* cv. "M82." Segments of the M82 genome were introgressed into homologous regions of the wild species, which made it possible to assign loci underlying contrasting agronomic, physiological, and molecular traits (Lippman and Zamir, 2007). Fruit and leaves from each individual and the parental lines were profiled using GC-MS and data on different agronomic traits were collected. Schauer et al. (2006) detected 889 mQTLs, with approximately half of them influencing yield-related traits and several of them influencing entire metabolic networks. Single mQTLs that may regulate or control malate levels, which influence tomato fruit quality, were also found (Schauer et al., 2006). This intersection of quantitative genetics and metabolomics is very powerful and presents such unique possibilities for plant breeding that we anticipate rapid growth in the coming years.

Metabolomics: Challenges and Future Perspectives

Our understanding of metabolism has benefited tremendously from progress made via the field of metabolomics. However, there are several issues that will have to be adequately resolved to definitively characterize plant metabolomes and apply that knowledge for precise genetic manipulation of plants. These limitations are not insurmountable and can be overcome provided that adequate resources are available. Indeed, recent publications indicate that steps are being taken to redress these issues.

From model organisms to crop plants

Our food comes from approximately 3000 different species with 8 — namely rice, wheat, barley, maize, potato, tomato, soybean, and pea — providing 90% of our calories and nutrients (Bowne et al., 2010), yet it is the weedy lab model *Arabidopsis* that has been most extensively characterized with respect to the metabolome (Okuda et al., 2008; Matsuda et al., 2010). Even among related organisms, such as cereal grains, GC-MS profiles of polar metabolites (primary compounds) can be easily differentiated to the point where it has been suggested that metabolomics can be used to fingerprint individual crops or even cultivars (Arbona et al., 2009; Roessner and Bowne, 2009). The comparative genomics approach that leverages information directly from model species to others of economic value may be of limited value in understanding varied metabolomes. Metabolic atlases for plant species of economic importance may need to be developed. Knowledge on how to efficiently optimize extractions, profiling technologies, data-mining techniques, and metabolic databases such as AraCyc (Mueller et al., 2003) gained from studies in *Arabidopsis*, and other model species will undoubtedly be applied to less studied but agriculturally important species.

Compartmentation of plant metabolism

Plant metabolism is highly compartmented. The presence of several organelles, each performing specific physiological and metabolic roles, shows developmental plasticity, which means that their *in situ* metabolic pool sizes will vary within a cell and be dependent on the developmental stage. Knowing the amount of metabolites in the compartment helps to determine if its content is sufficient to participate in the metabolic process of that organelle or cell. This information is critical for accurate pathway engineering. It is not trivial to obtain quantities of relatively pure fractions of subcellular organelles adequate for analysis. Non-aqueous subcellular fractionation can be used to separate organelles and is technically easier with non-bulky tissues such as leaves (Winter et al., 1993; Benkeblia et al., 2009), but great progress has also been made with hitherto difficult organs such as tubers (Farre et al., 2001, 2008). Use of this technique in combination with GC-MS revealed that most organic and amino acids and sugars are sequestered in the tuber vacuoles, but that their concentrations are highest in the cytosol (Farre et al., 2008). This information can be used to determine the potential activation state of enzymes in each compartment and help to develop transgenic approaches for modifying enzyme activity.

High-resolution sampling

Most metabolite profiling is done on crude extracts of a single type of tissue, but often these tissues consist of a heterogeneous collection of different cell types with varying functions and presumably, with different catalogs of metabolites. Homogenization therefore damps metabolite signals associated with specialized cell types within the tissue. Two broad approaches are being taken to map the distribution of cellular metabolites (Moco et al., 2009). One uses laser microdissection for precision harvesting of a population of specific cell types the unique chemical composition of which is then determined using conventional approaches (Schad et al., 2005). This "micro" view of metabolites was used to study secretory cavities in *Dilatris* and led to the identification of a "novel" class of compounds called methoxyphenylphenalenones (Schneider and Holscher, 2007). The other approach uses MS to image metabolites *in situ* in different sections of the tissue (Svatos, 2010); for example, waxes in *Arabidopsis* leaf epidermis have been profiled (Jun et al., 2010) as well as the epicuticular layer of *Arabidopsis* leaf cuticle and the secondary compounds in herbal species (Ng et al., 2007).

Individual cells within a tissue may also have diverse metabolomes. In addition to having different functions, they may be at different stages of the cell cycle and can even be exposed to distinct microenvironments, each producing unique metabolic consequences, as compared to adjacent cells (Kehr, 2003). Single-cell metabolite profiling would offer even richer information, but is hindered by the relatively small size of plant cells and the rigidity of the cell wall. Still, such high-resolution measurements are now possible. Vertes and colleagues recently obtained metabolite profiles of individual cells from onion and lily bulb epidermis and maize roots showing that each had distinct profiles (Shrestha and Vertes, 2009; Day et al., 2010).

Primary and secondary metabolism pose different challenges

Primary metabolism may be subjected to complex regulatory controls to ensure adequate growth and metabolism. This is consistent with the idea of primary metabolism as the GSC consisting of major hubs (Figure 5.1). Metabolites in this hub include simple sugars and nitrogen compounds that are potent regulators of gene expression (Koch, 1996; Coruzzi and Zhou, 2001; Rolland et al., 2006). If linkages within this hub are altered by genetic or environmental perturbation then one of two outcomes may be envisaged. The system may be able to buffer against the perturbation and maintain homeostasis or, if a "vulnerable" step has been targeted in the hub, then other connections will be altered, many of which may directly affect normal growth, likely leading to reduced fitness. Therefore while compounds of primary metabolism are more abundant and as such easier to assay, it is not trivial to purposefully manipulate their levels. When all of the evidence of the past 10 years is taken together, there are more examples of genetic engineering experiments designed to increase sugars, starch, proteins, and amino acids without negatively changing yield that have failed, than have succeeded (Smith, 2008; Morandini, 2009; Stitt et al., 2010). Metabolomics, however, has made it possible to refine our understanding of the control and regulation of metabolism. Organisms with "silent phenotypes," those that have been genetically modified but show no visible symptoms, can be probed to discover how the perturbation was buffered (Weckwerth et al., 2004; Stamova et al., 2009). However, only when all of the regulatory mechanisms that act on chromatin, DNA, RNA, protein and enzyme activity, cooperative binding, and post-translational modifications are identified, will we have an accurate portrayal of the regulation of metabolism.

In contrast to primary metabolism, there has been considerable success in changing the levels of secondary metabolites. The nutraceutical profiles of foods have been improved, volatiles and pigmentation in fruit and flowers have been altered, chemicals have been introduced that provide protection against pathogen attack, and the potency of medicinal agents have been amplified (Itkin and Aharoni, 2009). The bow-tie structure predicts fewer constraints from the GSC on secondary metabolites (Figure 5.1). This does not mean that engineering the levels of secondary metabolites is easy. The genes that encode many of the enzymatic steps are unknown; the metabolites are structurally complex and occur at low levels and, often, only transiently. There is a high energetic cost for the construction of intricate chemical compounds *in planta* and production must provide a clear cost-benefit ratio for the plant (Kliebenstein, 2009). Tampering with these mechanisms that have evolved over millions of years does not always produce the anticipated response under field conditions. Metabolomics can be a useful tool for improving the success of these experiments. Many of the functional genomics approaches combining transcriptomics and targeted metabolomics have led to the identification of new gene targets (see the section Functional Genomics).

Identifying the metabolome

The sheer number of different metabolites analyzed with better instrumentation that has greater sensitivity and detection speed has resulted in datasets containing numerous metabolites that are unknown with respect to their chemical nature. Still, about 60–80% of all signals detected in a profiling experiment, using both GC-MS and LC-MS platforms, cannot be assigned a name. Although the increased mass resolution of newer mass spectrometers (e.g., QTOF, Orbitrap, or FT-ICR technologies; see Table 5.1) offers the opportunity to estimate sum formulars from a detected molecular ion, those sum formulars can still be translated into a number of potential structures. The development of mass spectral libraries is crucial for the deciphering of unknown structures; a number of these libraries are now available either commercially or in the public domain (Table 5.2).

Additionally, to accurately record the compounds in a mass spectral library, researchers and vendors have started to record retention times under well-defined chromatographic and ionization conditions that will further enhance peak identifications. The utilization of mass spectral fragmentation patterns of a particular chemical structure (tandem MS) is also critical for attaining unambiguous peak identifications. However, it will be some time before libraries will be comprehensive enough for compound identifications based on accurate mass, retention time, and mass spectral fragmentation pattern. Currently there are enormous efforts underway to establish high-throughput approaches to enhance identifications of unknown peaks in various biological systems. Also, new software tools, designed to predict the chromatographic behavior and fragmentation pattern of any metabolite predicted to occur in a biological extract, are under development (e.g., see Hummel et al., 2010 for GC-MS and Schulz-Trieglaff et al., 2008 for LC-MS). It is now time to coordinate these efforts and collate and structure the information for best utilization by any metabolomics researcher. The first attempt at such a collaborative effort is MassBank (Horai et al., 2010), a Japanese initiative inviting other participants from around the world to contribute to a publicly available metabolite mass spectral library.

Measurements of metabolic flux

Metabolism may be seen as consisting of two separate entities — concentration and flux (Wiechert et al., 2007). However, steady-state levels of metabolites generated by metabolomics tell nothing of metabolic flux. This is perhaps the most serious hindrance to a biochemist's ability to understand metabolism in plant organs. We have already discussed how difficult it is to accurately measure intracellular concentrations of metabolites (see sections Compartmentation of Plant Metabolism and High-Resolution Sampling) and that at best we are surveying the average *contents* of different metabolite pools. Real-time momentary changes in fluxes are even harder to gauge, especially in plants, because of compartmentation, enzyme multiplicity, and high cellular water content (Morgan and Rhodes, 2002; Weitzel et al., 2007; Wiechert, 2007; Wiechert et al., 2007; Droste et al., 2008; Kruger and Ratcliffe, 2008; Allen et al., 2009; Schryer et al., 2009). In spite of the challenges associated with making flux measurements, a recent advancement was made in looking at metabolic flux in cell wall biosynthesis, a process that is very poorly understood but is of great interest for the generation of cellulosic biofuels. MRM-based quantitative LC-MS was used to separate compounds that are intermediates of cell wall synthesis, and this was coupled with dynamic labeling using ^{13}C-sucrose to monitor the flux of the label from sucrose to cell wall (Alonso et al., 2010).

A further complication is that flux may also be controlled by metabolic channeling. This occurs when the product of one enzymatic reaction is quickly transferred or *channeled* to the active site of the next enzyme in the pathway. Enzyme participants are organized as macromolecular complexes described as *metabolons* (Srere, 1985) wherein they are physically but temporarily associated with each other. These enzyme aggregates result in quasi-compartmentalization in the cytosol (Jorgensen et al., 2005) and conceivably would facilitate the rapid movement of metabolites by reducing free diffusion. The association/dissociation constants of these macromolecular complexes would then be important regulatory factors in determining pathway flux (Srere, 1985). There are several examples of this phenomenon in plants, such as the Calvin cycle (Winkel, 2004) and tryptophan, phenylpropanoid, cyanogenic glucoside, isoprenoid, flavonoid, and alkaloid biosynthesis (Jorgensen et al., 2005; Winkel, 2009). Judging from the number of recent publications on this subject, we may be underestimating the extent to which metabolism is regulated in this manner.

Outlook

Metabolomics is being established as an indispensable tool for all facets of biology. It is driving knowledge-creation and the invention of new technologies to meet the multitude of challenges that face agricultural production. Progress in the use of metabolic QTLs and functional genomics approaches undoubtedly will generate new targets that could drive future breeding strategies and create solutions for preventing loss of crops postharvest. The widespread adoption of efficient and smart technologies for high-resolution sampling and imaging of metabolites coupled with the ability to monitor metabolic fluxes will undoubtedly lead to better information generated from metabolomics. The use of multiple analytical platforms, better compound identification, and new enhanced data-mining strategies will permit the full exploitation of metabolite profiles. We have outlined current uses of metabolomics in support of the development of agricultural biotechnology and some of the new advancements in these areas. We argue that collaborations between researchers around the globe will be crucial to enable the seamless integration of information obtained from metabolomics experiments in each of the research areas (e.g., same crop, same stress condition, similar gene transformation), and to coordinate research efforts for the most successful outcomes in developing stronger crops and better quality plant products.

Acknowledgments

Ute Roessner is thankful for support from both the Australian Centre for Plant Functional Genomics and Metabolomics Australia. Work in Diane Beckles' lab is supported by the National Science Foundation Grant MCB-06211, University of California Hatch projects CA-D*-PLS-7198-H, and CA-D*-PLS-7821-H and grants from the Metabolomic Core University of California Davis Genome Center. Both authors are indebted to Dr. Belinda Martineau for her helpful comments.

References

Allen, D. K., Libourel, I. G. L., & Shachar-Hill, Y. (2009). Metabolic flux analysis in plants: Coping with complexity. *Plant Cell and Environment*, *32*, 1241–1257.

Allwood, J. W., Clarke, A., Goodacre, R., & Mur, L. A. J. (2010). Dual metabolomics: A novel approach to understanding plant-pathogen interactions. *Phytochemistry*, *71*, 590–597.

Alonso, A. P., Piasecki, R. J., Wang, Y., LaClair, R. W., & Shachar-Hill, Y. (2010). Quantifying the labeling and the levels of plant cell wall precursors using ion chromatography tandem mass spectrometry. *Plant Physiology*, *153*, 915–924.

Anastasiadi, M., Zira, A., Magiatis, P., Haroutounian, S. A., Skaltsounis, A. L., & Mikros, E. (2009). H-1 NMR-based metabolomics for the classification of greek wines according to variety, region and vintage. Comparison with HPLC data. *Journal of Agricultural and Food Chemistry*, *57*, 11067–11074.

Arabidopsis Genome Initiative. (2000). Analysis of the genome sequence of the flowering plant *Arabidopsis thaliana*. *Nature*, *408*, 796–815.

Arbona, V., Iglesias, D. J., Talon, M., & Gomez-Cadenas, A. (2009). Plant phenotype demarcation using nontargeted LC-MS and GC-MS metabolite profiling. *Journal of Agricultural and Food Chemistry*, *57*, 7338–7347.

Arneberg, R., Rajalahti, T., Flikka, K., Berven, F. S., Kroksveen, A. C., & Berle, M., et al. (2007). Pretreatment of mass spectral profiles: Application to proteomic data. *Analytical Chemistry*, *79*, 7014–7026.

Bais, P., Moon, S. M., He, K., Leitao, R., Dreher, K., & Walk, T., et al. (2010). PlantMetabolomics.org: A webportal for plant metabolomics experiments. *Plant Physiology*, *152*, 1807–1816.

Baker, J. M., Hawkins, N. D., Ward, J. L., Lovegrove, A., Napier, J. A., & Shewry, P. R., et al. (2006). A metabolomic study of substantial equivalence of field-grown genetically modified wheat. *Plant Biotechnology Journal*, *4*, 381–392.

Beale, M. H., Ward, J. L., & Baker, J. M. (2009). Establishing substantial equivalence: Metabolomics. *Methods Molecular Biology*, *478*, 289–303.

Beckles, D. M. (2011). Factors affecting the postharvest sugar content of tomato *Solanum lycopersicum* L. fruit. *Postharvest Biology and Technology*. In press. doi: 10.1016/j.postharvbio.2011.05.016.

Beckles, D. M., Stamova, L., & Luengwilai, K. (2011). Factors controlling tomato *Solanum lycopersicum* L. fruit sugar accumulation. *Postharvest Biology and Technology* (Submitted)

Benkeblia, N., Shinano, T., & Osaki, M. (2009). Metabolite profiling and assessment of metabolome compartmentation of soybean leaves using non-aqueous fractionation and GC-MS analysis. *Metabolomics*, *3*, 297–305.

Bermudez, L., Urias, U., Milstein, D., Kamenetzky, L., Asis, R., & Fernie, A. R., et al. (2008). A candidate gene survey of quantitative trait loci affecting chemical composition in tomato fruit. *Journal of Experimental Botany*, *59*, 2875–2890.

Bohnert, H. J., Nelson, D. E., & Jensen, R. G. (1995). Adaptations to environmental stresses. *The Plant Cell*, *7*, 1099–1111.

Bollard, M. E., Stanley, E. G., Lindon, J. C., Nicholson, J. K., & Holmes, E. (2005). NMR-based metabolomic approaches for evaluating physiological influences on biofluid composition. *NMR Biomedical*, *18*, 143–162.

Bowne, J., Bacic, A., Tester, M., & Roessner, U. (2010). Abiotic Stress and Metabolomics. In R. D. Hall (Ed.), *Annual plant reviews: The biology of plant metabolomics*. Wiley-Blackwell. (In Press)

Callahan, D. L., De Souza, D., Bacic, A., & Roessner, U. (2009). Profiling of polar metabolites in biological extracts using diamond hydride-based aqueous normal phase chromatography. *Journal of Separation Science*, *32*, 2273–2280.

Camacho, D., de la Fuente, A., & Mendes, P. (2005). The origin of correlations in metabolomics data. *Metabolomics*, *1*, 53–63.

Catchpole, G. S., Beckmann, M., Enot, D. P., Mondhe, M., Zywicki, B., & Taylor, J., et al. (2005). Hierarchical metabolomics demonstrates substantial compositional similarity between genetically modified and conventional potato crops. *Proceedings of the National Academy of Sciences of the United States America, 102*, 11458–14462.

Cellini, F., Chesson, A., Colquhoun, I., Constable, A., Davies, H. V., & Engel, K. H., et al. (2004). Unintended effects and their detection in genetically modified crops. *Food and Chemical Toxicology, 42*, 1089–1125.

Coruzzi, G. M., & Zhou, L. (2001). Carbon and nitrogen sensing and signaling in plants: Emerging "matrix effects". *Current Opinion in Plant Biology, 4*, 247–253.

Day, J.A., Shrestha, B., Nemes, P. and Vertes, A. (2010) Metabolic profiling of corn (*Zea mays*) roots by midinfrared laser ablation electrospray ionization mass spectrometry. Proceedings of the 58th ASMS Conference on Mass Spectrometry and Allied Topics, Salt Lake City, Utah.

Droste, P., Weitzel, M., & Wiechert, W. (2008). Visual exploration of isotope labeling networks in 3D. *Bioprocess and Biosystems Engineering, 31*, 227–239.

Farre, E. M., Fernie, A. R., & Willmitzer, L. (2008). Analysis of subcellular metabolite levels of potato tubers (*Solanum tuberosum*) displaying alterations in cellular or extracellular sucrose metabolism. *Metabolomics, 4*, 161–170.

Farre, E. M., Tiessen, A., Roessner, U., Geigenberger, P., Trethewey, R. N., & Willmitzer, L. (2001). Analysis of the compartmentation of glycolytic intermediates, nucleotides, sugars, organic acids, amino acids, and sugar alcohols in potato tubers using a nonaqueous fractionation method. *Plant Physiology, 127*, 685–700.

Fernie, A., & Schauer, N. (2009). Metabolomics-assisted breeding: A viable option for crop improvement? *Trends in Genetics, 25*, 39–48.

Fiehn, O. (2001). Combining genomics, metabolome analysis, and biochemical modelling to understand metabolic networks. *Comparative and Functional Genomics, 2*, 155–168.

Fiehn, O. (2004). High throughput metabolite profiling for functional genomics. *Plant and Cell Physiology, 45* S7-S7

Forsythe, I. J., & Wishart, D. S. (2009). Exploring human metabolites using the human metabolome database. *Current Protocols in Bioinformatics* (Chapter 14:Unit14.8)

Franck, C., Lammertyn, J., & Nicolai, B. (2004). Metabolic profiling using GC-MS to study biochemical changes during long-term storage of pears. *Acta Horticulturae, 682*, 1991–1998.

Grafahrend-Belau, E., Junker, B. H., Klukas, C., Koschutzki, D., Schreiber, F., & Schwobbermeyer, H. (2009). Topology of Plant Metabolic Networks. In J. Schwender (Ed.), *Plant metabolic networks* (pp. 173–210). Heidelberg: Springer.

Griffiths, W. J., & Wang, Y. (2009). Mass spectrometry: From proteomics to metabolomics and lipidomics. *Chemical Society Reviews, 38*, 1882–1896.

Hall, R. D. (2007). Food metabolomic: META-PHOR a new European research initiative. *AgroFood Industry Hi-Tech, 18*, 14–16.

Herberich, E., Sikorski, J., & Hothorn, T. (2010). A robust procedure for comparing multiple means under heteroscedasticity in unbalanced designs. *Public Library of Science One, 5*, e9788.

Hoffmann, E., & Stroobant, V. (2007). *Mass Spectrometry, principles and applications*. Wiley.

Horai, H., Arita, M., Kanaya, S., Nihei, Y., Ikeda, T., & Suwa, K., et al. (2010). MassBank: A public repository for sharing mass spectral data for life sciences. *Journal of Mass Spectrometry, 45*, 703–714.

Hummel, J., Strehmel, N., Selbig, J., Walther, D., & Kopka, J. (2010). Decision tree supported substructure prediction of metabolites from GC-MS profiles. *Metabolomics, 6*, 322–333.

Itkin, M., & Aharoni, A. (2009). Bioengineering. In A. E. Osbourne & V. Lanzotti (Eds.), *Plant-derived natural products* (pp. 435–473). Heidelberg: Springer.

Ivanova, P. T., Milne, S. B., Myers, D. S., & Brown, H. A. (2009). Lipidomics: A mass spectrometry based systems level analysis of cellular lipids. *Current Opinion of Chemical Biology, 13*, 526–531.

Jacobs, A., Lunde, C., Bacic, A., Tester, M., & Roessner, U. (2007). The impact of constitutive expression of a moss Na^+ transporter on the metabolomes of rice and barley. *Metabolomics, 3*, 307–317.

Jiang, H. L., Xie, Z. Z., Koo, H. J., McLaughlin, S. P., Timmermann, B. N., & Gang, D. R. (2006). Metabolic profiling and phylogenetic analysis of medicinal Zingiber species: Tools for authentication of ginger (*Zingiber officinale Rosc.*). *Phytochemistry, 67*, 1673–1685.

Jorgensen, K., Rasmussen, A. V., Morant, M., Nielsen, A. H., Bjarnholt, N., & Zagrobelny, M., et al. (2005). Metabolon formation and metabolic channeling in the biosynthesis of plant natural products. *Current Opinion in Plant Biology, 8*, 280–291.

Jun, J. H., Song, Z. H., Liu, Z. J., Nikolau, B. J., Yeung, E. S., & Lee, Y. J. (2010). High-spatial and high-mass resolution imaging of surface metabolites of *Arabidopsis thaliana* by laser desorption-ionization mass spectrometry using colloidal silver. *Analytical Chemistry, 82*, 3255–3265.

Junker, B. H., Klukas, C., & Schreiber, F. (2006). VANTED: A system for advanced data analysis and visualization in the context of biological networks. *BMC Bioinformatics, 7*, 109.

Kaplan, F., Kopka, J., Haskell, D. W., Zhao, W., Schiller, K. C., & Gatzke, N., et al. (2004). Exploring the temperature-stress metabolome of *Arabidopsis*. *Plant Physiology, 136*, 4159–4168.

Katajamaa, M., Miettinen, J., & Oresic, M. (2006). MZmine: Toolbox for processing and visualization of mass spectrometry based molecular profile data. *Bioinformatics, 22*, 634–636.

Katajamaa, M., & Oresic, M. (2007). Data processing for mass spectrometry-based metabolomics. *Journal of Chromatography A, 27*, 318–328.

Kehr, J. (2003). Single cell technology. *Current Opinion in Plant Biology, 6*, 617–621.

Kende, A., Portwood, D., Senior, A., Earll, M., Bolygo, E., & Seymour, M. (2010). Target list building for volatile metabolite profiling of fruit. *Journal of Chromatography A* (Ahead of publication)

Keurentjes, J. J. B. (2009). Genetical metabolomics: Closing in on phenotypes. *Current Opinion in Plant Biology, 12*, 223–230.

Kind, T., Wohlgemuth, G., Lee do, Y., Lu, Y., Palazoglu, M., & Shahbaz, O., et al. (2009). FiehnLib: Mass spectral and retention index libraries for metabolomics based on quadrupole and time-of-flight gas chromatography/mass spectrometry. *Analytical Chemistry, 15*, 10038–10048.

Kitteringham, N. R., Jenkins, R. E., Lane, C. S., Elliott, V. L., & Park, B. K. (2009). Multiple reaction monitoring for quantitative biomarker analysis in proteomics and metabolomics. *Journal of Chromatography B Analytic Technology Biomedical Life Science, 877*, 1229–1239.

Kliebenstein, D. J. (2007). Metabolomics and plant quantitative trait locus analysis – The optimal genetical genomics platform? In B. Nikolau & E. S. Wurtele (Eds.), *Concepts in plant metabolomics* (pp. 29–44). Netherlands: Springer.

Kliebenstein, D. J. (2009). Use of Secondary Metabolite Variation in Crop Improvement. In A. E. Osbourne & V. Lanzotti (Eds.), *Plant-derived Natural Products* (pp. 83–95). Heidelberg: Springer.

Koch, K. E. (1996). Carbohydrate-modulated gene expression in plants. *Annual Review of Plant Physiology and Plant Molecular Biology, 47*, 509–540.

Kopka, J., Schauer, N., Krueger, S., Birkemeyer, C., Usadel, B., & Bergmüller, E., et al. (2005). GMD@CSB.DB: The Golm metabolome database. *Bioinformatics, 21*, 1635–1638.

Korn, M., Gartner, T., Erban, A., Kopka, J., Selbig, J., & Hincha, D. K. (2010). Predicting *Arabidopsis* freezing tolerance and heterosis in freezing tolerance from metabolite composition. *Molecular Plant, 3*, 224–235.

Kose, F., Budczies, J., Holschneider, M., & Fiehn, O. (2007). Robust detection and verification of linear relationships to generate metabolic networks using estimates of technical errors. *BMC Bioinformatics, 8*, 162.

Kruger, N. J., & Ratcliffe, G. R. (2008). Metabolic organization in plants: A challenge for the metabolic engineer. *Advances in Plant Biochemistry and Molecular Biology, 1*, 1–27.

Kuiper, H. A., Kleter, G. A., Noteborn, H. P., & Kok, E. J. (2001). Assessment of the food safety issues related to genetically modified foods. *Plant Journal, 27*, 503–528.

Kuiper, H. A., Kleter, G. A., Noteborn, H. P., & Kok, E. J. (2002). Substantial equivalence – An appropriate paradigm for the safety assessment of genetically modified foods? *Toxicology, 181-182*, 427–431.

Kuiper, H. A., Kok, E. J., & Engel, K. H. (2003). Exploitation of molecular profiling techniques for GM food safety assessment. *Current Opinion in Biotechnology*, *14*, 238–243.

Kvalheim, O. M., Brakestad, F., & Liang, Y. (1994). Preprocessing of analytical profiles in the presence of homoscedastic or heteroscedastic noise. *Analytical Chemistry*, *66*, 43–51.

Li, C., Krewer, G., & Kays, S. J. (2007). Detection of apple deterioration using an electronic nose and zNosetm. *Transactions of the ASABE*, *50*, 1417–1425.

Lippman, Z. B., & Zamir, D. (2007). *Natural variation as a tool for gene identification in plants*. Chichester: John Wiley & Sons Ltd.

Lopez-Gresa, M. P., Maltese, F., Belles, J. M., Conejero, V., Kim, H. K., & Choi, Y. H., et al. (2009). Metabolic response of tomato leaves upon different plant-pathogen interactions. *Phytochemical Analysis*, *21*, 89–94.

Luengwilai, K., & Beckles, D. M. (2010). Climacteric ethylene is not essential for initiating chilling injury in tomato (*Solanum lycopersicum*) cv. Ailsa Craig. *Journal of Stored Products and Postharvest Research*, *1*, 1.

Luengwilai, K., Salveit, M. E., & Beckles, D. M. (2011). Metabolite content of harvested Micro-Tom tomato (*Solanum lycopersieum* L.) fruit is altered by chilling and protective heat-shock treatments as shown by GC-Ms metabolic profiling. *Postharvest Biology and Technology*. In press. doi: 10.1016/j.postharvbio.2011.05.014.

Ma, H. W., & Zeng, A. P. (2003). Reconstruction of metabolic networks from genome data and analysis of their global structure for various organisms. *Bioinformatics*, *19*, 270–277.

Ma, H. W., & Zeng, A. P. (2003). The connectivity structure, giant strong component and centrality of metabolic networks. *Bioinformatics*, *19*, 1423–1430.

Matsuda, F., Hirai, M. Y., Sasaki, E., Akiyama, K., Yonekura-Sakakibara, K., & Provart, N. J., et al. (2010). AtMetExpress development: A phytochemical atlas of *Arabidopsis* development. *Plant Physiology*, *152*, 566–578.

Merchant, A., Richter, A., Popp, M., & Adams, M. (2006). Targeted metabolite profiling provides a functional link among eucalypt taxonomy, physiology and evolution. *Phytochemistry*, *67*, 403–408.

Mittler, R. (2006). Abiotic stress, the field environment and stress combination. *Trends in Plant Science*, *11*, 15–19.

Mittler, R., & Blumwald, E. (2010). Genetic engineering for modern agriculture: Challenges and perspectives. *Annual Review of Plant Biology*, *61*, 13.1–13.20.

Moalemiyan, M., Vkram, A., Kushalappa, A. C., & Yaylayan, V. (2006). Volatile metabolite profiling to detect and discriminate stem-end rot and anthracnose diseases of mango fruits. *Plant Pathology*, *55*, 792–802.

Moco, S., schneider, B., & Vervoort, J. (2009). Plant micrometabolomics: The analysis of endogenous metabolites present in a plant cell or tissue. *Journal of Proteome Research*, *8*, 1694–1703.

Modolo, M. V., Reichert, A. I., & Dixon, R. A. (2009). Introduction to the different classes of biosynthetic enzymes. In A. E. Osbourne & V. Lanzotti (Eds.), *Plant-derived natural products* (pp. 143–163). Heidelberg: Springer.

Morandini, P. (2009). Rethinking metabolic control. *Plant Science*, *176*, 441–451.

Moreno-Risueno, M. A., Busch, W., & Benfey, P. N. (2010). Omics meet networks – using systems approaches to infer regulatory networks in plants. *Current Opinion of Plant Biology*, *13*, 126–131.

Morgan, J. A., & Rhodes, D. (2002). Mathematical modeling of plant metabolic pathways. *Metabolic Engineering*, *4*, 80–89.

Mueller, L. A., Zhang, P., & Rhee, S. Y. (2003). AraCyc: A biochemical pathway database for *Arabidopsis*. *Plant Physiology*, *132*, 453–460.

Ng, K. M., Liang, Z. T., Lu, W., Tang, H. W., Zhao, Z. Z., & Che, C. M., et al. (2007). *In vivo* analysis and spatial profiling of phytochemicals in herbal tissue by matrix-assisted laser desorption/ionization mass spectrometry. *Analytical Chemistry*, *79*, 2745–2755.

Oksman-Caldentey, K. -M., & Inzé, D. (2004). Plant cell factories in the post-genomic era: New ways to produce designer secondary metabolites. *Trends in Plant Science*, *9*, 433–440.

Okuda, S., Yamada, T., Hamajima, M., Itoh, M., Katayama, T., & Bork, P., et al. (2008). KEGG Atlas mapping for global analysis of metabolic pathways. *Nucleic Acids Research*, *36*, W423–W426.

Oliver, S. G., Winson, M. K., Kell, D. B., & Baganz, F. (1998). Systematic functional analysis of the yeast genome. *Trends in Biotechnology*, *16*, 373–378.

Orr, G. R., & Raison, J. K. (1990). The effect of changing the composition of phosphatidylglycerol from thylakoid polar lipids of oleander and cucumber on the temperature of the transition related to chilling injury. *Planta*, *181*, 137–143.

Paley, S. M., & Karp, P. D. (2006). The pathway tools cellular overview diagram and Omics viewer. *Nucleic Acid Research*, *34*, 3771–3778.

Pourcel, L., & Grotewold, E. (2009). Participation of phytochemicals in plant development and growth. In A. E. Osbourne & V. Lanzotti (Eds.), *Plant-derived natural products: Synthesis, function and application* (pp. 269–283). Dordrecht, Heidelberg: Springer.

Prithiviraj, B., Vikram, A., Kushalappa, A. C., & Yaylayan, V. (2004). Volatile metabolite profiling for the discrimination of onion bulbs infected by *Erwinia carotovora* ssp. carotovora, *Fusarium oxysporum* and *Botrytis allii*. *European Journal of Plant Pathology*, *110*, 371–377.

Rios-Estepa, R., & Lange, B. M. (2007). Experimental and mathematical approaches to modeling plant metabolic networks. *Phytochemistry*, *68*, 2351–2374.

Rischer, H., & Oksman-Caldentey, K. -M. (2006). Unintended effects in genetically modified crops: Revealed by metabolomics? *Trends in Biotechnology*, *24*, 102–104.

Rischer, H., Oresic, M., Goossens, A., Inze, D., & Oksman-Caldentey, K. M. (2007). Integrating transcriptional and metabolic profiling to unravel secondary metabolite biosynthesis in plants. In Z. Xu, J. Li, Y. Xue & W. Yang (Eds.), *Biotechnology and sustainable agriculture 2006 and beyond* (pp. 135–138). Netherlands: Springer.

Rodriguez, A., & Infante, D. (2009). Network models in the study of metabolism. *Electronic Journal of Biotechnology*, *12*, 2.

Roessner, U., & Beckles, D. M. (2009). Metabolite measurements. In J. Schwender (Ed.), *Plant metabolic networks* (pp. 39–69). New York, New York: Springer Verlag.

Roessner, U., & Bowne, J. (2009). What is metabolomics all about? *Biotechniques*, *46*, 363–365.

Roessner, U., Nahid, A., Hunter, A. and Bellgard, M. (2011) Metabolomics – the combination of analytical chemistry, biology and informatics. In Moo-Young, M. (ed.) *Comprehensive Biotechnology* 2nd ed., vol.1, pp. 447–459, Elsevier.

Rolland, F., Baena-Gonzalez, E., & Sheen, J. (2006). Sugar sensing and signaling in plants: Conserved and novel mechanisms. *Annual Review of Plant Biology*, *57*, 675–709.

Rudell, D. R., Mattheis, J. P., & Curry, F. A. (2008). Prestorage ultraviolet-white light irradation alters apple peel metabolome. *Journal of Agricultural and Food Chemistry*, *56*, 1138–1147.

Saito, K., Hirai, M. Y., & Yonekura-Sakakibara, K. (2008). Decoding genes with coexpression networks and metabolomics – "majority report by precogs." *Trends in Plant Science*, *13*, 36–43.

Saito, K., & Matsuda, F. (2010). Metabolomics for functional genomics, systems biology, and biotechnology. *Annual Review of Plant Biology*, *61*, 463–489.

Schad, M., Mungur, R., Fiehn, O., & Kehr, J. (2005). Metabolic profiling of laser microdissected vascular bundles of *Arabidopsis thaliana*. *Plant Methods*, *1*, 1–10.

Schauer, N., Semel, Y., Roessner, U., Gur, A., Balbo, I., & Carrari, F., et al. (2006). Comprehensive metabolic profiling and phenotyping of interspecific introgression lines for tomato improvement. *Nature Biotechnology*, *24*, 447–454.

Schneider, B., & Holscher, D. (2007). Laser microdissection and cryogenic nuclear magnetic resonance spectroscopy: An alliance for cell type-specific metabolite profiling. *Planta*, *225*, 763–770.

Schulz-Trieglaff, O., Pfeifer, N., Gröpl, C., Kohlbacher, O., & Reinert, K. (2008). LC-MSsim – a simulation software for liquid chromatography mass spectrometry data. *BMC Bioinformatics*, *9*, 423.

Schryer, D. W., Peterson, P., Paalme, T., & Vendelin, M. (2009). Bidirectionality and compartmentation of metabolic fluxes are revealed in the dynamics of isotopomer networks. *International Journal of Molecular Sciences*, *10*, 1697–1718.

Shrestha, B., & Vertes, A. (2009). *In situ* metabolic profiling of single cells by laser ablation electrospray ionization mass spectrometry. *Analytical Chemistry, 81*, 8265–8271.

Smith, A. M. (2008). Prospects for increasing starch and sucrose yields for bioethanol production. *Plant Journal, 54*, 546–558.

Smith, C. A., O'Maille, G., Want, E. J., Qin, C., Trauger, S. A., & Brandon, T. R., et al. (2005). METLIN: A metabolite mass spectral database. *Therapeutic Drug Monitoring, 27*, 747–751.

Smith, C. A., Want, E. J., O'Maille, G., Abagyan, R., & Siuzdak, G. (2006). XCMS: Processing mass spectrometry data for metabolite profiling using nonlinear peak alignment, matching, and identification. *Analytical Chemistry, 78*, 779–787.

Srere, P. A. (1985). The Metabolon. *Trends in Biochemical Sciences, 10*, 109–110.

Stamova, B. S., Roessner, U., Suren, S., Laudencia-Chingcuanco, D., Bacic, A., & Beckles, D. M. (2009). Metabolic profiling of transgenic wheat over-expressing the high-molecular-weight Dx5 glutenin subunit. *Metabolomics, 5*, 239–252.

Ståhlman, M., Ejsing, C. S., Tarasov, K., Perman, J., Borén, J., & Ekroos, K. (2009). High-throughput shotgun lipidomics by quadrupole time-of-flight mass spectrometry. *Journal of Chromatography B, 15*, 2664–2672.

Stitt, M., Sulpice, R., & Keurentjes, J. (2010). Metabolic networks: How to identify key components in the regulation of metabolism and growth. *Plant Physiology, 152*, 428–444.

Svatos, A. (2010). Mass spectrometric imaging of small molecules. *Trends in Biotechnology, 28*, 425–434.

Sweetlove, L. J., & Fernie, A. (2005). Regulation of metabolic networks: Understanding metabolic complexity in the systems biology era. *New Phytologist, 168*, 9–24.

Trethewey, R. N. (2004). Metabolite profiling as an aid to metabolic engineering in plants. *Current Opinion in Plant Biology, 7*, 196–201.

Vekiari, S. A., Oreopoulou, V., Kourkoutas, Y., Kamoun, N., Msallem, M., & Psimouli, V., et al. (2010). Characterization and seasonal variation of the quality of virgin olive oil of the Throumbolia and Koroneiki varieties from Southern Greece. *Grasas Y Aceites, 61*, 221–231.

Venneria, E., Fanasca, S., Monastra, G., Finotti, E., Ambra, R., & Azzini, E., et al. (2008). Assessment of the nutritional values of genetically modified wheat, corn, and tomato crops. *Journal of Agriculture and Food Chemistry, 56*, 9206–9214.

Verpoorte, R. (1998). Exploration of nature's chemodiversity: The role of secondary metabolites as leads in drug development. *Drug Discovery Today, 3*, 232–238.

Villas-Boas, S. G., Rasmussen, S., & Lane, G. A. (2005). Metabolomics or metabolite profiles? *Trends in Biotechnology, 23*, 385–386.

Villas-Bôas, S. G., Roessner, U., & Hansen, M., et al. (2007). *Metabolome analysis: An introduction*. New Jersey, NJ, USA: John Wiley & Sons, Inc..

Ward, J. L., Baker, J. M., & Beale, M. H. (2007). Recent applications of NMR spectroscopy in plant metabolomics. *FEBS Journal, 274*, 1126–1131.

Weckwerth, W., & Fiehn, O. (2002). Can we discover novel pathways using metabolomic analysis? *Current Opinion in Biotechnology, 13*, 156–160.

Weckwerth, W., Loureiro, M. E., Wenzel, K., & Fiehn, O. (2004). Differential metabolic networks unravel the effects of silent plant phenotypes. *Proceedings of the National Academy of Sciences of the United States of America, 101*, 7809–7814.

Wei, R., Li, G., & Seymour, A. B. (2010). High-throughput and multiplexed LC/MS/MRM method for targeted metabolomics. *Analytical Chemistry, 82*, 5527–5533.

Weitzel, M., Wiechert, W., & Noh, K. (2007). The topology of metabolic isotope labeling networks. *BMC Bioinformatics, 8*, 315.

Widodo, , Patterson, J. H., Newbigin, E., Tester, M., Bacic, A., & Roessner, U. (2009). Metabolic responses to salt stress of barley (*Hordeum vulgare* L.) cultivars, Sahara and Clipper, which differ in salinity tolerance. *Journal of Experimental Botany, 60*, 4089–4103.

Wiechert, W. (2007). The thermodynamic meaning of metabolic exchange fluxes. *Biophysical Journal, 93*, 2255–2264.

Wiechert, W., Schweissgut, O., Takanaga, H., & Frommer, W. B. (2007). Fluxomics: Mass spectrometry versus quantitative imaging. *Current Opinion in Plant Biology, 10*, 323–330.

Williams, A. J. (2008). Public chemical compound databases. *Current Opinion in Drug Discovery and Development, 11*, 393–404.

Winkel, B. S. J. (2004). Metabolic channeling in plants. *Annual Review of Plant Biology, 55*, 85–107.

Winkel, B. S. J. (2009). Metabolite channeling and multi-enzyme complexes. In A. E. Osbourne & V. Lanzotti (Eds.), *Plant-derived natural products* (pp. 195–208). Heidleberg: Springer.

Winter, H., Robinson, D. G., & Heldt, H. W. (1993). Subcellular volumes and metabolite concentrations in barley leaves. *Planta, 191*, 180–190.

Yocum, A. K., & Chinnaiyan, A. M. (2009). Current affairs in quantitative targeted proteomics: Multiple reaction monitoring-mass spectrometry. *Brief Functional Genomics and Proteomics, 8*, 145–157.

Zanor, M. I., Rambla, J. L., Chaib, J., Steppa, A., Medina, A., & Granell, A., et al. (2009). Metabolic characterization of loci affecting sensory attributes in tomato allows an assessment of the influence of the levels of primary metabolites and volatile organic contents. *Journal of Experimental Botany, 60*, 2139–2154.

Zeng, Z., Chau, F. -T., Chan, H. -y., Cheung, C. -y., Lau, T. -y., & Wei, S., et al. (2008). Recent advances in the compound-oriented and pattern-oriented approaches to the quality control of herbal medicines. *Chinese Medicine, 3*, 9.

Zhang, H., Chang, M., Wang, J., & Ye, S. (2008). Evaluation of peach quality indices using an electronic nose by MLR, QPST and BP network. *Sensors and Actuators B: Chemical, 134*, 332–338.

Zhao, J., Yu, H., Luo, H. -J., Cao, Z. W., & Li, Y. -X. (2006). Hierarchical modularity of nested bow-ties in metabolic networks. *BMC Bioinformatics, 7*, 386.

Zywicki, B., Catchpole, G., Draper, J., & Fiehn, O. (2005). Comparison of rapid liquid chromatography-electrospray ionization-tandem mass spectrometry methods for determination of glycoalkaloids in transgenic field-grown potatoes. *Analytical Biochemistry, 336*, 178–186.

Plant genome sequencing: Models for developing synteny maps and association mapping

6

Delphine Fleury Ute Baumann Peter Langridge

Australian Centre for Plant Functional Genomics, University of Adelaide Glen Osmond, Australia

TABLE OF CONTENTS

Introduction

Plant genomes are characterized by large variations of genome size and level of ploidy. C-value is the amount of nuclear DNA in the unreplicated gametic nucleus, irrespective of the ploidy level of the species. The genome size of flowering plants varies almost 2000-fold from small genome plants such as *Genlisea margaretae* (a carnivorous herb) with 2C = 0.129 pg (63 Mbp) to the monocot *Trillium hagae* with 2C = 264.9 pg (for comparison *Arabidopsis thaliana* has 2C = 0.321 pg; Bennett and Leitch, 2005; Zonneveld et al., 2005; Greilhuber et al., 2006). C-value information on over 5000 plant species are found in the plant C-values Database (http://data.kew.org/cvalues/introduction.html; Greilhuber et al., 2006; Swift 1950).

Intriguingly large genome size is not associated with evolutionary advancement or ecological competitiveness. Quite the contrary, plants with large genomes appear to have reduced photosynthetic efficiency and are underrepresented in extreme environments. Furthermore genera with large genomes show low species diversity, indicating that they evolve more slowly (Gaut and Ross-Ibarra, 2008). However, there are some interesting correlations between genome size and plant characteristics. For example, guard cell length appears to positively correlate with genome size across a wide range of major taxa with the exception of the Poeae (Hodgson et al., 2010). Other phenotypical characteristics of large genomes, are an increased cell size and slow mitotic activity relative to small genome species. Variation in genome size is found not only within genera, but also within a species. For instance, variation within the nuclear DNA content were found in F1 hybrids of maize (Rayburn et al., 1993) and among different accessions of *A. thaliana* (Schmuths et al., 2004).

Major contributors to plant genome size are repetitive sequences, which may be located at a few defined chromosomal sites or widely dispersed. Among the repetitive sequences of well-defined functions are the telomeric sequences at the ends of the chromosomes, the centromeric sequences, and the ribosomal RNA encoding regions. Telomeres are composed of simple G-rich repeats, which can reach lengths of up to 160 kbp (Richards and Ausubel, 1988; Adams et al., 2001). The centromeric region of a chromosome is usually associated with blocks of highly repetitive tandemly repeated DNA (also called satellite DNA) and retroelements. The highest proportion of repetitive DNA consists of mobile elements, the transposons and retrotransposons, which can make up to >50% of a genome

DOI: 10.1016/B978-0-12-381466-1.00006-7

(San Miguel et al., 1996). In addition to the various blocks of repetitive DNA, many plant genomes contain accessory chromosomes, so-called B-chromosomes. These are highly condensed chromosomes usually devoid of genes and have non-Mendelian modes of inheritance (Puertas, 2002).

Whole genome duplication events provide another means for increasing genome size leading to a change in ploidy level of the plant. There are different types of polyploidy in plants. Autopolyploidy can result from a fusion of two unreduced gametes, i.e., gametes that show non-disjunction of all homologous chromosomes during meiosis I. Combining two or more genomes from different plants leads to allopolyploidy (Chen, 2007). Whole genome duplication events are often followed by chromosomal rearrangements and extensive gene loss leading to a quasi re-diploidization. Interestingly there appears to be a pattern to the loss of genes following the polyploidization events, and among the retained gene pairs more than half showed significant differences in gene expression.

Several studies have provided evidence for widespread occurrence of paleopolyploidy in the angiosperms (Masterson, 1994; Blanc and Wolfe, 2004; Cui et al., 2006), and thus polyploidy seems to have been a major force in plant evolution. Most crop species appear to be polyploid, either due to ancient or relatively recent duplication events. For example, wheat, cotton, tobacco, coffee, sugarcane, peanut, oat, and canola are allopolyploids, whereas potato, watermelon, banana, and alfalfa are autopolyploids. There are also examples, such as soybean, where research indicates that the genome has both allo- and autopolyploid origins (Udall and Wendel, 2006). The preponderance of polyploidization in crop plants is thought to be associated with the increased vigor and larger organ size in polyploids relative to their diploid progenitors (Gepts, 2003). The interactions between the genomes may also result in essentially fixing hybrid vigor (Hilu, 1993). The polyploidy nature of many crop species has provided an important tool for plant breeders since it allows exploration of wild, diploid progenitors as sources of novel genes or alleles for crop improvement. For example, the diploid and tetraploid progenitors of hexaploid bread wheat have provided a critical source for disease, abiotic stress, and even quality genes (Feuillet et al., 2008).

Diversity of plant genome size and the complex nature of crop genome due to polyploidy and repetitive sequences are problematic in gene identification and understanding gene function for their application in plant breeding. In this chapter, we will cover three aspects of plant genomics that help tackle these issues.

Genome Sequencing

Strategies for plant genome sequencing

Since the 1970s, the chain termination method (also called the Sanger method) has been the most broadly used sequencing technology. This technique is based on the neo-synthesis of the target fragment using DNA polymerase, radioactively labeled nucleotides, and dideoxynucleotide triphosphates (ddNTPs), which terminate the reaction. Since its early application the method has been improved with the use of fluorescent labels and capillary electrophoresis, which allow automation and sequence reads of 700–1000 bp. This technology has been used for the first plant genomes sequenced, *A. thaliana* (2000), rice (2005), and poplar (2006).

The first strategy for plant genome sequencing used a map-based approach, also referred to as the hierarchical clone-by-clone method (Figure 6.1). The genomic DNA was fragmented and isolated by cloning into large-insert vectors such as bacterial artificial chromosomes (BACs; Figure 6.2). The BAC clones are 50–200 kbp long and highly redundant to cover six- to ten-fold of the entire genome. The clones are fingerprinted based on their restriction sites and assembled into physical contigs. The contigs are ordered in a physical map by anchoring them onto a genetic map. The BACs are either randomly chosen or selected based on their position on the physical map or their gene content for sequencing. Large-insert clones, such as BACs, are then fragmented into smaller pieces, which are cloned into plasmids. These small fragments are Sanger sequenced and the original BAC DNA sequence is deduced from them. For the *Arabidopsis* project, clones showing minimum overlap at their ends were selected for sequencing. The first map of the Columbia accession covered 115 Mbp of the 125 Mbp genome (The Arabidopsis Genome Initiative, 2000). A similar strategy was used for maize (*Zea mays* cv. B73). The minimal tiling path (MTP) is the minimum number of clones covering the maximum length of the genome. An MTP set of 19,000 BAC clones was selected from the physical map and Sanger sequenced to approximately six-fold coverage.

Rice was the first crop species to have its genome sequenced. The method employed was a combination of hierarchical clone-by-clone and whole genome shotgun sequencing (WGS; Figure 6.1). For WGS sequencing, genomic DNA is sheared into small fragments, cloned into plasmids, and sequenced (Figure 6.2). The construction of large-insert libraries is not necessary. Overlapping sequences are merged by computer algorithms, which assemble the millions of sequenced fragments into contiguous stretches of DNA. Two subspecies of rice were sequenced, *Oryza sativa* L. ssp. *japonica* by the International Rice Genome Sequencing project (2005) and Syngenta (Goff et al., 2002), while the spp. *indica* was sequenced by the Beijing Genomics Institute (Yu et al., 2002). Since then, WGS sequencing has been successfully used for bigger genomes, such as *Populus trichocarpa* (485 Mbp; Tuskan et al., 2006) and soybean (1.1 Gbp; Schmutz et al., 2010).

Because the assembly of large and polyploid genomes would be difficult without a framework sequence or process for deconvoluting the contributions for the different genomes, an intermediate approach has been chosen for the bread wheat genome (Figure 6.1). Initially, individual chromosomes or chromosome arms were isolated by flow cytometry and used for constructing chromosome-specific BAC libraries. The approach was validated with the physical map of chromosome 3B. An MTP of 7440 chromosome 3B-specific BAC clones was assembled in 1036 contigs covering 82% of the 811 Mbp of the chromosome (Paux et al., 2008). This methodology was

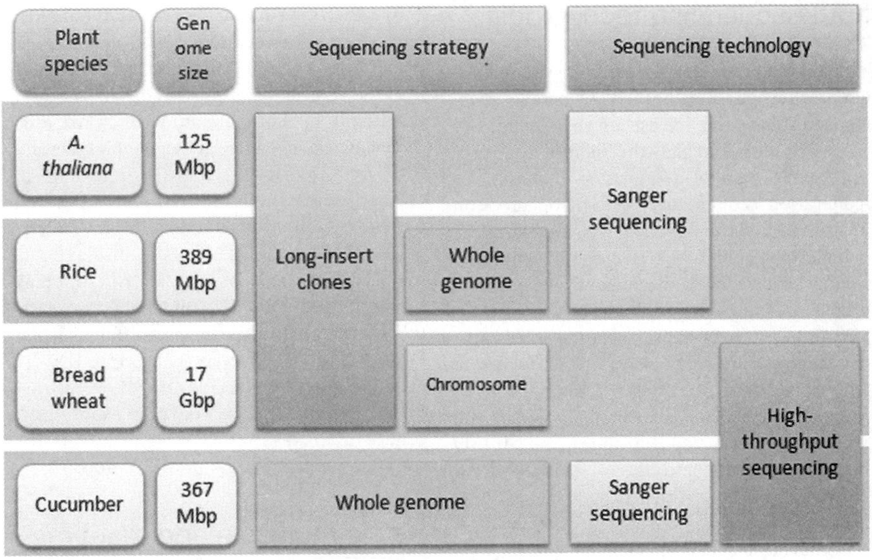

Figure 6.1 • Strategies used for plant genome sequencing • Depending on the genome size, different strategies have been deployed in plants, using long-insert clones such as bacterial artificial chromosomes (BAC) and shotgun sequencing of the whole genome or of specific chromosomes. Sanger sequencing technology, high-throughput sequencing (or next-generation sequencing), and combinations of both technologies were utilized. Please see color plate section at the end of the book.

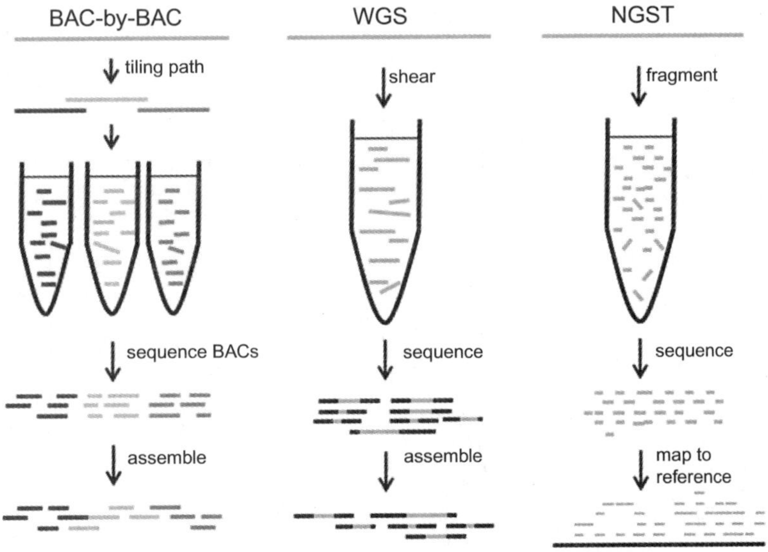

Figure 6.2 • Principles of BAC-by-BAC, whole genome sequencing (WGS), and next-generation sequencing technologies (NGST) approaches of genome sequencing. The BAC-by-BAC approach starts by creating a tilling path of overlapping BACs followed by shotgun sequencing and assembly of the BACs. For WGS, genomic DNA is sheared into random fragments that are cloned into a plasmid vector and sequenced from both ends. Cloning is not necessary for NGST: genomic DNA is fragmented and sequenced, and sequence reads are eventually mapped to a reference genome. Please see color plate section at the end of the book.

also employed for the sorting of each of the 42 wheat chromosome arms from a collection of aneuploid wheat lines (Safar et al., 2010). This strategy greatly reduces the complexity by decreasing the work unit from 17 Gbp to 224–580 Mbp, depending on the considered chromosome arm. The chromosome-specific BAC libraries for wheat are now under construction under the auspices of the International Wheat Genome Sequencing Consortium (www.wheatgenome.org).

High-throughput sequencing technologies

Recent progress in micro-fluidics and nanotechnology is revolutionizing DNA sequencing by enabling millions of parallel reactions in miniature space. Since the release of the Roche 454 GS20 sequencer in 2006, different high-throughput sequencers (HTS) using massively parallel sequencing technologies (also called next-generation sequencing, NGS) appeared on the market. A characteristic of the first HTS is the template preparation: each single-molecule DNA is clonally amplified without vector cloning (Figure 6.2). Because the expensive (~1% of the total cost of sequencing; Scheibye-Alsing et al., 2009) and time-consuming step of vector cloning is circumvented, the NGS technologies are particularly attractive for *de novo* genome sequencing and genome re-sequencing.

The current market-dominant platforms among the NGS — the Solexa Genome Analyzer (Illumina), the SOLiD technology (Applied Biosystems), and the 454 FLXsystem (Roche) — all perform sequencing by synthesis using a DNA polymerase or a ligase as the core biochemistry. Incorporated nucleotides or oligonucleotides, in the case of SoLiD, are detected as fluorescent or chemiluminescent signals. The HTS generates large amounts of sequence data in one run (e.g., ~50 Gbp for Illumina GAIIx; Simpson et al., 2009), which greatly reduces the cost per base. The current read lengths of SOLiD (50 bp) and 454 GSFLX (400 bp) are much shorter than those generated by the Sanger method. However, the sequencing length is compensated for by the large number of sequenced fragments and the consequent high redundancy of reads per fragment. Moreover, the technologies are improving for increasing read length.

The genome sequence of the cucumber (*Cucumis sativus* var. *sativus*) was obtained by combining data from Sanger and Illumina GA sequencing runs, generating a 72.2-fold genome coverage (Figure 6.1; Huang et al., 2009). Combinations of traditional and NGS technologies are also used for investigating the genomes of cacao, tomato, apple, and banana, among others (Edwards and Batley, 2010). With the decreasing cost of sequencing and improvement of the technologies leading to longer sequence read lengths and throughput, we can expect that small genome crops will soon be entirely sequenced, and orphan crops will also become amenable to sequencing.

Although short-read technologies can be applied to *de novo* sequencing, they are not yet suitable for complex genomes with a high content of repeat sequences, duplicated regions, and homologous chromosomes. Therefore, a physical map-based strategy in combination with whole genome shotgun NGS has been chosen for barley (Schulte et al., 2009) and wheat (www.wheatgenome.org) projects. Both chromosome and BAC-based approaches considerably reduce the size and the complexity of the sequencing unit enabling the use of HTS. However, even at the BAC scale, a comparison of clone sequences generated by 454 GS20 and Sanger technologies in barley showed that high-quality assemblies could be generated from 454 GS20 sequences for genic region but not for repetitive DNA (Wicker et al., 2006). Complex genomes with 80% of repeat elements like barley and wheat still require longer reads or read-pairs longer than the longest repeat sequence in the target DNA.

On the other hand, short-read technologies allow rapid and cost-effective re-sequencing of genomes from diverse germplasm, thus enabling the detection of sequence variants (single nucleotide polymorphisms, copy number variation, insertion/deletions) by comparison with a reference genome. These data can be used to generate high-resolution genetic maps (i.e., for association studies), or to map the introgression of chromosome regions from one germplasm into another one in a breeding scheme. Technical improvement such as bar-coding of the DNA samples allows multiplexing of different genetic material in a single sequencing run, opening the area of genotyping by sequencing. This concept already has been applied to 238 rice recombinant inbred lines (RILs; Xie et al., 2010). Despite the low coverage per line (0.05×), Illumina GA reads enable single nucleotide polymorphism (SNP) genotyping of the population. In the rice example, the high-resolution genetic map was used to map a quantitative trait locus (QTL) for grain width in a region of 200 kbp.

Single molecule and real-time sequencing

The weaknesses of the NGS technologies may be overcome by the so-called third generation of sequencers. These platforms work on a single molecule DNA template, avoiding the DNA amplification step, which introduces base errors in NGS. Instruments have been developed by Helicos Biosciences (Harris et al., 2008), Pacific Bioscience (Eid et al., 2009), and Oxford Nanopore. However, many other companies have also joined the race to develop fast, cheap, and reliable sequencing technologies (Gupta, 2008; Pettersson et al., 2009; Ansorge, 2009).

The Helicos system performs sequencing by synthesis of bound DNA molecules at the nanoscale level and presently achieves a read length of up to 30 bp. At the core of the technology developed by Pacific Bioscience are zero-mode waveguides containing a single immobilized DNA polymerase, fluorescently labeled dNTPs, and a DNA template. Sequencing occurs in real time because no interruptions are required in the synthesis process for imaging. This feature greatly increases the speed of sequencing. Read lengths of around 1500 bp have been achieved. The concept behind Oxford Nanopore Technology is to use α-hemolysin nanopores to immobilize single DNA strands, which are then sequentially cleaved by an exonuclease. The bases are identified as they pass through the nanopore (Deamer, 2010). The technology has the potential of sequencing long strands of DNA. Single molecule sequencing could greatly help with the completion of genome sequences by filling gaps in draft sequences and correcting ambiguous reads. Because the preparation of single molecule templates is much simpler than for the first generation of HTS and requires less DNA, it is likely to become the method of choice for sequencing rare DNA, such as fossil DNA in paleogenomics studies.

Assembly and alignment programs

The emergence of the NGS has obvious implications for the way sequences are assembled. Traditional assembly programs designed for long-read Sanger derived sequences are not suitable for the large number of short-read sequences produced by the NGS. Generally speaking the choice of sequence assemblers is dependent on the strategy chosen for genome sequencing, as well as the sequencing platform used for the task.

For plant genome sequencing, particularly for the large genome species, the key issues are related to *de novo* sequencing where no framework is available. Sequence data from any of the platforms come with quality scores that describe the probability that a base call is incorrect. For Sanger reads the base-calling program PHRED (Ewing and Green, 1998) assigns a quality value to each base from the trace data, that is, $q = -10\log_{10}(p_e)$, with p_e being the error probability of a particular base. Similarly, quality scores also come with Illumina (Cock et al., 2010), SoLiD, and Roche data. However, they are only partially convertible (Brockman et al., 2008).

These quality scores are also used to identify sequence overlap. Sequence assembly is based on the assumption that two reads that share a same string of nucleotides are derived from the same region in the genome. Assemblers use such overlaps between sequences to generate larger contiguous pieces of DNA. If the genome being sequenced is very large, then the number of overlaps is also very large and the analysis problem becomes complex. Furthermore, genomic regions that share perfect repeats can be indistinguishable and easily cause misassembly or even repeat collapse, especially if the sequence reads are shorter than the repeats. Clearly the otherwise labor-intensive hierarchical (BAC by BAC) Sanger sequencing approach avoids many of the complexity problems compared to other approaches, especially for large genomes.

In WGS projects genomic DNA is randomly sheared and subcloned, in the case of the Sanger method, for sequencing (Figure 6.2). Therefore, for full coverage and successful assembly there is a necessity to over sample the genome (Lander and Waterman, 1988). It is therefore not surprising that, for instance, the draft genome sequence for *O. sativa* ssp. *indica* was derived from a 4.2-fold coverage (Yu et al., 2002), which represents about 92% of the genome. The genome of poplar had a 7.5-fold coverage (Tuskan et al., 2006) and that of *Brachypodium* was obtained from a 9.4-fold genome coverage (The International Brachypodium Initiative, 2010). In a typical WGS approach, the sheared DNA is sequenced from both ends (double-barreled sequencing step). The resulting read pairs, also called mate pairs, are separated by a known distance.

The first step in any assembly pipeline is the removal of vector or adapter sequences and the removal of low-quality sequences based on quality scores. In the next step, overlap computation is performed. In other words, all pair-wise alignments between the reads are computed. Assemblers assess the quality of an overlap by measures such as length of the overlap and sequence identity. PHRED scores can also be taken into account at this step. During this process, assemblers may also take note of potential repeat fragments; for instance, the TIGR assembler algorithm uses the distribution of potential overlaps for each fragment to identify and label a fragment as repetitive (Sutton et al., 1995). Clusters of fragments that consistently align with one another are merged into contiguous sequences (contigs). In the final step, using information from mate pairs, the contigs are ordered, oriented, and assembled into super-contigs, also called scaffolds. During a process of refinement that can take several years and includes validation of the assembly, closing gaps and resequencing of controversial regions, the scaffolds will become the actual chromosome sequences.

In addition to the TIGR assembler mentioned earlier, several algorithms, including ARACHNE (Batzoglou et al., 2002), Celera (Myers et al., 2000), CAP3 (Huang and Madan, 1999), Phrap (www.phrap.org), Phusion (Mullikin and Ning, 2003), and Atlas (Havlak et al., 2004) have been employed for WGS read assembly from Sanger reads. Common to these programs is the overlap-layout-consensus approach (Pevzner et al., 2001); this involves starting with a pairwise comparison of all reads and then using an overlap graph for the assembly (Pop, 2009). For *de novo* genome assembly from 454 data there is software such as Newbler (Roche); CABOG, which is a revised version of the Celera Assembler (Miller et al., 2008); and MIRA (Chevreux et al., 2004). Chaisson et al. (2009) recently benchmarked five *de novo* short-read assembly algorithms and found that Velvet (Zerbino and Birney, 2008) performed best. However, none of the benchmarked algorithms have been able to successfully assemble anything longer than a bacterial or fungal genome. On the other hand, SOAPdenovo used for Illumina data has been successfully used for the *de novo* assembly of the human and panda genomes (Miller et al., 2010). A mixed strategy for *de novo* genome sequencing would combine long- and short-read data, which can be assembled using MIRA, for instance. Because different types of errors are produced by the different platforms, this strategy would alleviate the weaknesses of both technologies. At last, genome assemblies are validated in conjunction with genome finishing efforts aimed at generating the complete sequence of organisms. Validation is either manually or by using software like Consed (Gordon et al., 1998), Staden package (Staden et al., 2000), or TIGR Editor (Phillippy et al., 2008).

Genome browsers

Once a genome is assembled, genome browsers provide a graphical user interface to access and view the sequence data together with the genome annotation. There are a number of software products available for browsing genomes, including stand-alone tools, such as Artemis (http://www.sanger.ac.uk/resources/software/artemis/; Carver et al., 2008) and Apollo (http://apollo.berkeleybop.org/current/index.html; Lewis et al., 2002). The most common web-based browsers are EnsEMBL, GBrowse, the University of California at Santa Cruz Genome Browser, and the NCBI Map Viewer. Typically the genome sequence is organized into chromosomes or scaffolds and displayed diagrammatically in a chromosome overview panel. Below the diagram, genome annotation tracks that contain information on genes and gene models, repetitive regions, mRNA and EST data, expression data, and comparative genomic and other information are displayed. The Phytozome Web site (http://www.phytozome.net) currently provides an access point to 25 plant genomes (v.7.0) and employs GBrowse to present the genome annotation. Sequence similarity searches (Basic Local Alignment Search tool for Nucleotides, BLASTN) can be performed and information about gene homologs and gene ancestry can be accessed.

Comparative genomics information for the major grass crops species and also *Arabidopsis*, poplar, and grape can be obtained at the Gramene Web site (http://www.gramene.org; Liang et al., 2008). Gramene uses EnsEMBL as its genome viewer and allows the user to configure the display to switch between chromosomal and gene-based views, transcript and protein sequence information and domain annotation, single and multiple species view, and a graphical synteny display. Gramene also provides information about genetic and physical maps, QTLs and markers, germplasm resources, SNPs, phenotypes, pathways, and evolutionary information.

NCBI Map Viewer displays genome information quite differently from other genome browsers. The display is vertically oriented and different maps, typically BAC or chromosome contigs, genetic markers, and genes are displayed side by side. The user can choose which map feature to see and can zoom in and out of a genomic region for more or less detailed views. The map viewer is closely integrated with NCBI's other resources, such as UniGene, RefSeq, and LocusLink (see http://www.ncbi.nlm.nih.gov/genomes/PLANTS/PlantList.html for information on plant genomes). Other Web sites that provide access to a range of plant genomes and their annotations are the PlantGDB (www.plantgdb.org) and the MIPS plant group's sites (http://mips.helmholtz-muenchen.de/plant/genomes.jsp). PlantGDB employs the xGDB (eXtensible Genome Data Broker; Schlueter et al., 2006) genome browser, which allows customization of views, includes a range of tools for sequence retrieval and downloads, and links to other online databases and tools.

Several Web sites have been established that are specific to one plant family; for example, the genomes of *Glycine max*, *Medicago truncatula*, and *Lotus japonicus* can be browsed via the Legume Information System (http://www.comparative-legumes.org/index.php/Home), whereas the peach genome is accessible at the Genome Database for Rosaceae (http://www.rosaceae.org/).

A range of Web sites are dedicated to individual species, their genomes, and related resources. For instance, the Arabidopsis Information Resource (TAIR; www.arabidopsis.org) provides a comprehensive resource for plant scientists working with *Arabidopsis*. It includes the genome sequence, data on gene structure, gene expression and product information, metabolism, DNA and seed stocks, genetic maps, and molecular markers for the research community. Much less centralized are the resources for the crop model species, such as rice. Four independent annotated genome assemblies are available, three for the O. *japonica* (rice.plantbiology.msu.edu, Ouyang et al., 2007; rapdb.dna.affrc.go.jp, Tanaka, 2008; rice.genomics.org.cn/rice/index2.jsp, Zhao et al., 2004) and one for the O. *indica* genome (http://rice.genomics.org.cn/rice/index2.jsp, Zhao et al., 2004). Michigan State University Rice Genome Annotation Project (rice.plantbiology.msu.edu) provides more extensive annotation and also links to gene expression information. Further information on individual genome sequencing projects, their current status, release date, and links to the sequence data can be obtained from the genome project Web sites at NCBI (www.ncbi.nlm.nih.gov/genomes/leuks.cgi).

Models for Developing Syntenic Maps

Definitions

Comparative genomics provides a method to unravel the relationship between genomes by describing conserved (or homologous) chromosomes or chromosomal regions between related species. Shared markers or genes between chromosomes define syntenic regions. If the order of markers is conserved, the region is described as collinear or syntenic (conserved gene or marker order) between chromosomes. Synteny is revealed by building and comparing genetic and physical maps. Genetic maps group and order loci into linkage groups based on recombination rates between loci, and each group represents a chromosome. Genetic distance is the recombination rate between loci in centimorgan (cM). Two loci are linked if the recombination frequency is lower than expected (50%). A physical map assembles genomic fragments into contigs measured in base pairs (bp). The identification of common loci between related species enables comparison of genome structure and the definition of genome changes or evolution from ancestor genomes. Organisms that are closely related show a higher number of syntenic blocks than genetically distant species.

Intraspecies comparison

Genetic maps are constructed using populations that segregate for variants between loci. Consequently, the marker density of genetic maps is limited by the level of polymorphism measurable in each considered population. Whenever shared markers are identified, genetic maps can be compared and potentially merged into a consensus map. The CMap program enables common molecular markers between genetic maps to be readily identified and linked (Fang et al., 2003; Duran et al., 2010). JoinMap has been developed to construct consensus maps from multiple genetic maps (Stam, 1993). Both programs were successfully used for the construction of the first consensus maps of wheat (Somers et al., 2004) and barley (Varshney et al., 2007). By bringing together collections of molecular markers, consensus maps increase the overall number of markers available for molecular breeding and genetic analysis. These maps also linked the first Restriction Fragment Length Polymorphism (RFLP) genetic maps to the more recent PCR marker maps, which are more suitable for high-throughput screening of germplasm in breeding programs. Consensus maps can also be used for the analysis of QTL. The MetaQTL method uses consensus models to merge distinct genetic maps and QTL datasets (Veyrieras et al., 2007). The intraspecies synteny increases the probability of identifying true QTL by integrating information from various genetic sources and multiple sets of phenotypic data.

Cytogenetics for interspecies comparison

Prior to the development of molecular markers and genetic maps, cytogenetics was the method of choice for comparing closely related species, such as species of the same tribe. Using the ability of some plant species to cross-hybridize, synteny was studied by the observation of chromosome pairing during meiosis in interspecific hybrids. Chromosome painting by means of N/C banding enables identification of each chromosome at metaphase I. For example, the chromosome pairing of *ph1a* mutant wheat × rye hybrids revealed a homologous relationship between the two species (Naranjo et al., 1991). Fluorescent/genomic *in situ* hybridization (FISH/GISH) using DNA from different but related species as a probe reveals chromosomal rearrangements at somatic metaphase. The combination of N/C banding with the GISH method in different wheat species showed specific cyclic translocations between hexaploid wheat and its ancestral genomes (Naranjo et al., 1987; Jiang and Gill, 1994). These types of observations provided insights into the phylogeny and evolution of related species.

Sequence comparison

A common way to generate comparative maps between related species is using probes, which can cross-hybridize across species such as cDNA-based probes. Massive projects of EST sequencing and mapping have considerably increased the amount of data for comparative mapping. For example, over 1 million ESTs have been produced in wheat and over 7000 have been physically mapped by hybridization onto lines carrying deletions in specific chromosomal regions (Qi et al., 2004). With the development of a full genome sequence, extensive EST maps, and computational methods for sequence comparison, genome comparison can now be extended to evolutionary distant species.

Because large chromosomal blocks are collinear between genomes of the same family, species of small size genomes have been selected as models for particular families of plants. To be relevant for agriculture, model species should be selected for each key node of phylogenetic trees of crop species. The gene order along a chromosomal region in a model species is indicative of the gene order and content along the syntenic region of a related species. Extensive synteny has been described between model species and crop families such as *Arabidopsis* and Brassicaceae (Schranz et al., 2007), *Brachypodium*, rice, and Triticeae (The International Brachypodium Initiative, 2010), *M. truncatula* and Fabaceae (Young and Udvardi, 2009), and tomato and Solanaceae (Mueller et al., 2005). Despite the large genome size difference between rice and bread wheat (330 Mbp and 16 Gbp, respectively), 13 blocks of collinearity have been identified (Salse et al., 2008), which makes the rice genome a good model for wheat. Fewer rearrangements and duplications were observed in the Solanaceae genomes compared to other crops. Comparative maps show that relative positions of QTLs and major genes are often also conserved, which increases the value of comparative resources such as the SOL Genomics Network (Mueller et al., 2005).

One of the most comprehensive comparative maps covers the Poaceae family in which the level of gene synteny is high. Relationships between the genomes of eight grass species have been described in a consensus map called the "Crop Circle" (Devos, 2005). This map shows the gradual increase in genome size represented by the concentric circles from the smallest genome at the center (rice) to the largest genome on the outside (Triticeae).

Macro- versus micro-synteny

The scale of synteny depends upon the tools that are used to describe it. Chromosome painting (with FISH/GISH), pairing analysis, and non-sequence-based genetic maps limit the analysis to the macro level where only major structural changes are revealed. Genomic sequence provides the resolution for a wide range of comparisons, from the whole genome to specific genes. Although macrosynteny can be very high between related species, the accumulation of mutations such as base changes, insertions, deletions, duplications, and local rearrangements may break the micro-synteny at the sequence level.

In general, microcollinearity is greater in euchromatic (mostly constituted of genes or coding sequences) than in heterochromatic regions, which is illustrated by the comparison of the two sequenced genomes of rice and sorghum (Bowers et al., 2005). However, evolution of certain gene families is lineage specific. Extensive study of the *Ha* locus controlling grain hardness showed significant loss of micro-synteny between highly collinear genomes of the Triticeae. Although the three genes composing the locus were present in diploid wheat, the tetraploid wheat *Triticum turgidum* lost two genes, which were subsequently recovered by acquisition of the D genome in hexaploid wheat (Chantret et al., 2005). Comparison of intron composition between several wheat genomes, barley, and rice also revealed that sequence conservation of the plastic region such as *Ha* is not always linked to the evolutionary distance between species (Chantret et al., 2008).

Nature of the differences

Disruption of synteny is due to chromosomal rearrangement processes such as translocation, inversion, chromosome fusion, and breakage; gene, segment, and chromosomal duplication and loss; polyploidization and return to diploidy; proliferation of repetitive sequences; sequence conversion; unequal homologous recombination; and illegitimate recombination (Figure 6.3).

As noted earlier, polyploidy has been a major mechanism of generating variation and supporting evolutionary change. Intraspecific synteny reveals that some species have been subject to several rounds of duplication with gradual return to a diploid state. For instance, there is strong evidence that the *Arabidopsis* genome has undergone several rounds of polyploidization: the first event predates the monocot-dicot split, the second event took place before *Arabidopsis* diverged from the other dicots studied, and the third event took place

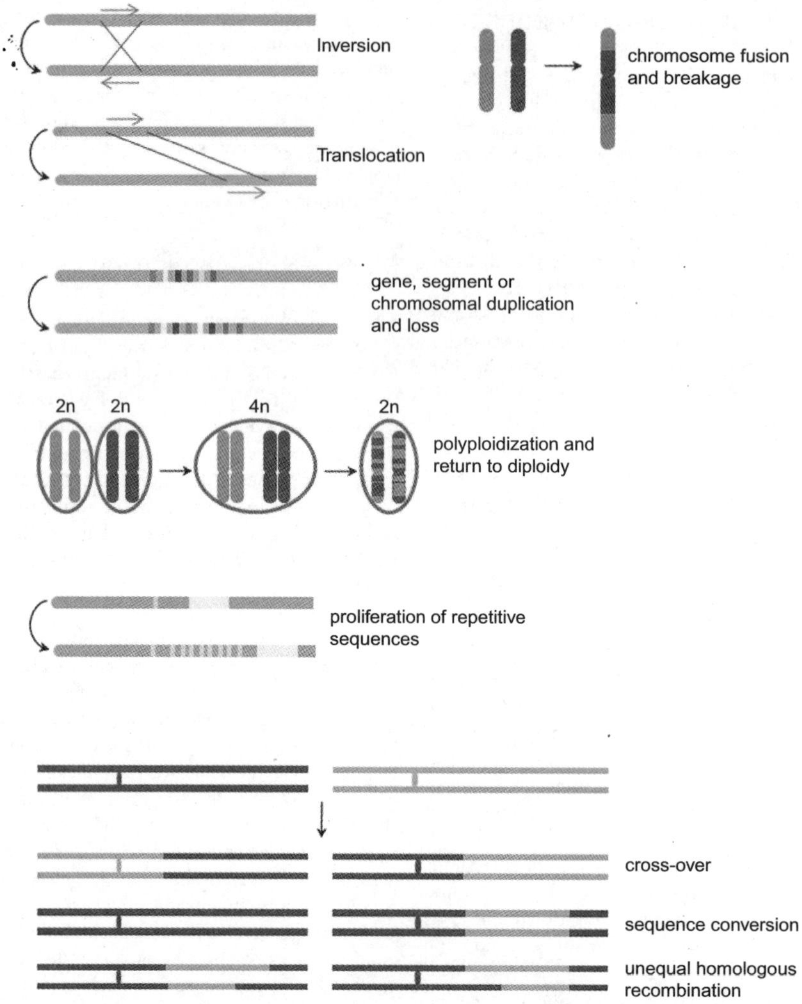

Figure 6.3 • Chromosomal rearrangement processes leading to disruption of synteny • Disruptions of synteny commonly observed involve translocation, inversion, duplication, and loss of DNA segments. A change of chromosome number between species might be due to chromosome fusion and breakage. A common phenomenon in plant evolution involves an increase of ploidy level followed by gradual loss of DNA and return to diploidy. Some rearrangements involve proliferation of repetitive elements such as transposons. Sequence conversion, unequal homologous recombination, and illegitimate recombination might also modify the synteny between homologous chromosomes. Please see color plate section at the end of the book.

sometime after the divergence of *Arabidopsis* and *Brassica* from the Malvaceae, represented by cotton (Bowers et al., 2003).

Polyploidization increases the number of genes available for change and selection, and induces transposable element activity. A study of newly synthesized allopolyploid wheat showed that the polyploidization event is rapidly followed by genome downsizing through elimination of DNA sequences (Eilam et al., 2008). After duplication events, gene copies might be relocated or deleted, leading to disruption of synteny. Some families of genes and some specific chromosomal regions are subject to rapid structural change in the frequency of gene duplication and DNA deletion. The analysis of the genome duplication event in *Arabidopsis* revealed signs of differential retention of duplicated genes, with conservation of genes involved in transcription and signal transduction, and preferential loss of those involved in DNA repair and those encoding organellar proteins (Blanc and Wolfe, 2004a, b). Moreover, as shown in *Arabidopsis* (Thomas et al., 2006) and maize (Woodhouse et al., 2010), the choice of which homolog is removed after polyploidization is also not random. In the case of *Arabidopsis* one homolog was preferentially deleted for 85% of the genome. Gene copies might also acquire different functions. For example, the domestication gene *tb1*, which determines the branching of maize, has a minor effect in foxtail millet (Doust et al., 2004).

Activity of repeat elements such as transposons and retrotransposons is also a major force for evolutionary change and can affect the level of synteny. Although the number of genes shared between haploid genomes of the grasses is about the same, the genome size difference is mainly due to differences in repetitive DNA. Bursts of intense activity of retrotransposon elements increased the genome size and composition in repetitive DNA between close species of the same genera, such as *Oryza* and *Zea* (Devos, 2010). Some transposable elements such as Pack-MULE (Hanada et al., 2009) and Helitron (Yang and Bennetzen, 2009) are able to duplicate fragments of DNA and therefore multiply gene copies.

Synteny is also affected by recombination activity. Synteny between grass genomes decreases along the chromosome from centromere to telomere, and the recombination rate increases along this axis. Insertions and deletions were mainly located in high recombinational regions of chromosomes (Akhunov et al., 2003). It has been observed that clusters of disease resistance genes were preferentially located in high recombination regions where gene duplication and recombination may be frequent and evolutionary processes can be rapid (Friedman and Baker, 2007).

Applications of syntenic maps

Breeding strategies are based on the ability of chromosomal regions to recombine in the progeny. The efficiency of recombining genes of interest in the same germplasm depends upon the structure and behavior of the genome. Synteny studies provide useful information on genome structure and genome evolution.

Identifying chromosomal blocks where synteny is high is very useful for positional cloning. Cloning genes underlying QTL require fine genetic maps, which are difficult to generate for species with low sequencing resources. Model genomes are used to develop cross-species molecular markers; for example, SSR markers were developed by using *M. truncatula* sequences for genetic mapping in alfalfa, pulses, and clover (Gutierrez et al., 2005; Sledge et al., 2005; Zhang et al., 2007). A QTL identified in one species can be transposed onto the syntenic region of model species. The model genomic sequence is then used to identify a candidate gene that may underlie the trait, which can then be isolated from the large and more complex genome (Figure 6.4). For example, loci of flowering time genes in canola have been located to a syntenic region containing major flowering time genes in *Arabidopsis* (Okazaki et al., 2007). Microsynteny between *Arabidopsis* and tomato has been used in fine-mapping the *ovate* gene and *Diageotropica* loci in tomato (Ku et al., 2001; Oh et al., 2002).

Since gene order is often conserved during evolution, synteny gives information about the phylogenetic relationships between species. Paleogenomics describes the identification and characterization of the ancestral genome structure and evolutionary mechanisms that have structured genome species (Salse et al., 2009). For example, the Crop Circle has been used to reconstruct the cereals ancestral genome, with either 5 or 7 chromosomes. In both hypotheses, whole genome duplications are followed by chromosome fusion and diploidization leading to the current rice, sorghum, maize, and Triticeae genomes (Devos, 2010; Bolot et al., 2009).

Tools and limitations

A major problem of syntenic analysis is the identification of orthologous genes, which have evolved from a common ancestor, and paralogous genes, which were duplicated in the same genome. Chromosomal homology is difficult to assess in plants because plant species evolution involves extensive rearrangements, particularly gene and genome duplications. Gradual evolution to a diploid state after genome duplication may make the identification of orthologous genes more difficult. Misidentification of orthologous sequences can lead to misinterpretation of the synteny. Therefore, robust and accurate sequence alignment is essential. Several computational programs have been written for comparative mapping: LineUp (Hampson et al., 2003), CloseUp (Hampson et al., 2005), Compass (Liu et al., 2004), and SyntenyTracker (Donthu et al., 2009). Some of these computer programs use statistical tests to infer the significance of the chromosomal homology. Programs of sequence similarity search use BLASTN. Because the BLASTN default parameters do not discriminate between paralogs and orthologs, two parameters (cumulative identity percentage and cumulative alignment length percentage) have been defined to statistically select the sequences with the highest identity over the longest alignment (Salse et al., 2009).

Disruption of the micro-synteny is a major limitation for positional cloning in crops by using model species. In some cases, genes have been lost by accumulation of a small deletion or relocated to another region. Comparison of wheat genomic sequence with *Brachypodium* and rice shows a high number of non-collinear genes interspersed in a highly conserved region (Choulet et al., 2010). For example, the map-based approach to isolate the 4H boron-tolerance gene in barley identified a syntenic region of 11.2 kb on rice chromosome Os3. Despite the high level of barley/rice gene collinearity between 4H and Os3, the BOR1 ortholog could not be found in rice chromosome 3. The cloning of the barley *Bot1* gene was achieved by a candidate gene approach by mapping barley ESTs with homology to boron transporters.

Association Mapping

Definitions

Gene and genetic mapping of a trait of interest is based on measuring the rate of meiotic recombination between the phenotype (trait) and genes or markers. In its simplest and most common form this is achieved through biparental crosses where two parents differing for the trait of interest are mated and the segregation of the trait and linked markers is measured. For positional cloning of a gene underlying the trait of interest, several rounds of mapping with large segregating populations may be needed to zero in on the target gene. Positional cloning usually requires analysis of several thousand lines to achieve sufficient genetic resolution to identify a small number of candidate genes underlying the target trait.

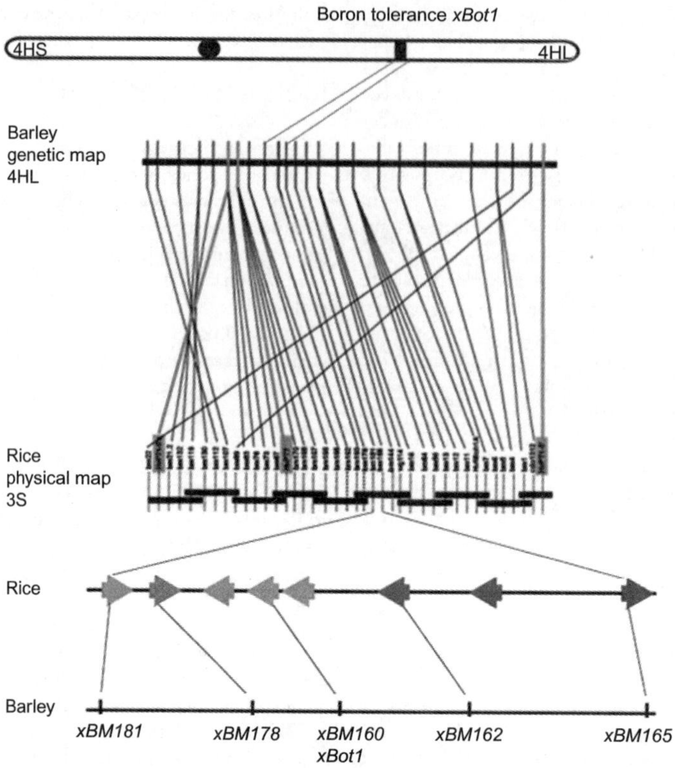

Figure 6.4 • Positional cloning of QTL: example of the boron tolerance *Bot1* locus in barley (modified from Sutton et al., 2007) • Synteny between barley chromosome 4HL and rice chromosome 3S enabled the identification of new molecular markers in barley using rice sequences that were collinear to the boron tolerance locus *xBot1*. Please see color plate section at the end of the book.

The key challenges in this process are finding sufficient polymorphic markers to allow the fine mapping and the phenotyping of the segregating lines. For species with a sequenced genome, marker development is relatively straightforward and for easily measured traits, phenotyping may not be an issue. However, for many traits associated with agronomic performance, such as yield or stress tolerance, and for species without a sequenced genome, positional cloning is problematic.

Although the use of biparental crosses has been very successful, there are alternative population structures that can be used to identify marker trait associations and lead to positional cloning. One of the most important has been association mapping. This approach uses a series of lines, or association mapping panels, that differ in expression of the trait being studied. All of the lines expressing the trait are expected to carry the same or similar gene or gene variant. If the lines in the mapping panels are related either through their breeding and selection history or due to a common area of origin, then one would expect the genomic region surrounding the target trait to also be similar between the lines expressing the trait; such regions are in linkage disequilibrium (LD). Molecular markers in regions of the genome not linked or associated with the trait should show completely random association with the trait or be in linkage equilibrium. A similar approach is to track genomic regions through a pedigree. If the breeding program has been selecting for one or more specific traits, regions of the genome will be seen, identifying by common molecular marker alleles, maintained in those lines expressing the trait (Paull et al., 1998; Jannink et al., 2001).

The principle advantage of association mapping is that it can use existing phenotypic data. For example, considerable phenotypic data will accumulate on individual lines in a breeding program. Genotyping of these lines with molecular markers allows the association of particular genomic regions with trait expression to be identified. The source material for an association mapping study can be from independent breeding programs, large international programs, wild species, or landrace germplasm collections. For example, the CIMMYT collects data for thousands of advanced lines that are generated each year and evaluated for phenology (among other traits) at hundreds of international locations (Reynolds et al., 2009). Such a database collected by international nursery networks are highly valuable for assembling suitable populations for association studies.

Population size and structure

A key advantage of association mapping has been the opportunity to investigate diverse genetic material and potentially identify multiple alleles and mechanisms of underlying traits. Association mapping uses recombination events that have occurred over an extended period of time. The resolution of the mapping will depend on the extent of LD, or non-random association of markers, that has occurred across the genome. If association mapping is conducted within a breeding program, the extent of LD will depend on the breeding system and the size, structure, and age of the germplasm within the breeding program (Mackay and Powell, 2007). In a small population, genetic drift leads to the loss of rare combinations of alleles, which increases LD (Flint-Garcia et al., 2003). Because lines are more likely to be heterozygous in out-crossing species than in selfing species, LD decays more rapidly in out-crossing species. For inbreeding species, LD can extend considerable distances and association mapping is unlikely to give high genetic resolution (Caldwell et al., 2006; Figure 6.5). This problem can be overcome by using different germplasm pools. For example, in barley LD extends over long distances within modern cultivars but is quite narrow in landraces and extremely short in wild barley (Caldwell et al., 2006). Consequently, association mapping in cultivated barleys will only give rough genetic positions, but positional cloning of genes underlying traits would be feasible for some regions if the trait can be successfully phenotyped in landraces or wild barley.

A further limitation of association mapping relates to the structure of mapping panels or germplasm collections used. The difficulty here is that there may be factors other than the trait being assessed that impose a structure on the germplasm selected for analysis. LD between markers and traits might be due to genetic drift, selection, and admixture. Selection for a phenotype controlled by unlinked loci in epistasis may increase LD, although the loci are not physically linked (Flint-Garcia et al., 2003). For example, LD extent fluctuates from 1 cM to over 20 cM depending on the considered region in durum and bread wheat (Breseghello and Sorrells, 2006; Maccaferri et al., 2005). Wheat breeding programs often use locally adapted elite germplasm into which specific loci/traits were introgressed from novel germplasm or other species to address specific problems. Consequently, confounding factors such as phenology, particularly maturity and plant height, make the interpretation of association studies particularly difficult for traits such as yield or drought tolerance (Reynolds et al., 2009). There are now some very useful statistical tools to deal with the problems of population structure (Yu and Buckler, 2006), but strong underlying structure and multiple structures within an association mapping panel can significantly reduce the power of the analysis.

Several alternative population structures have been proposed and evaluated to overcome these limitations. These include the development of complex populations involving multiple parents: Multiparental Advanced Generation InterCross (MAGIC) populations (Cavanagh et al., 2008) and Nested Association Mapping (NAM) populations (McMullen et al., 2009). However, both of these approaches require the development of new, large, and quite complex populations. In MAGIC, multiple parental lines are intercrossed and RILs are generated for each intercross, while in the NAM scheme, parental lines are crossed with the same reference line to develop sets of RILs. For example, 25 parental lines were crossed with a tester line of maize to produce 25 populations of ~200 RILs each, generating 136,000 recombination events (McMullen et al., 2009). These large populations also need to be phenotyped for the trait(s) of interest. The NAM populations are particularly attractive since they offer the genetic power of simple crosses while allowing the investigation of diverse germplasms.

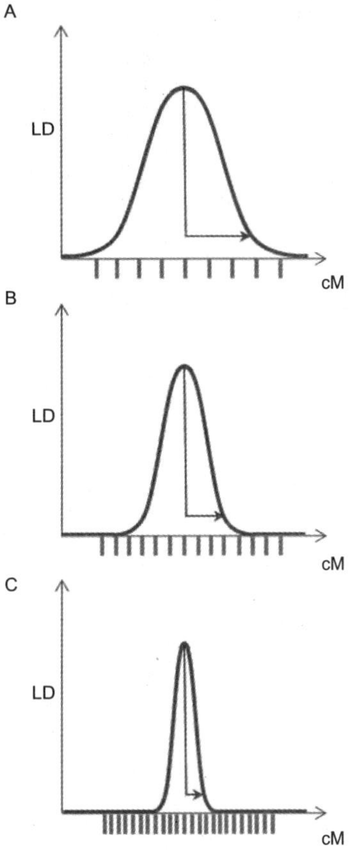

Figure 6.5 • Decay of linkage disequilibrium (LD) with genetic distance along a chromosome • Red arrows represent the LD extent in centimorgan (cM) in different species (ex: A: wheat and C: maize) and germplasm collections (ex: in barley A: cultivars, B: landraces, C: wild). Green bars represent the density of markers necessary to detect a peak of LD. Please see color plate section at the end of the book.

Markers and marker density

The advances in DNA sequencing and genome analysis are opening up new opportunities for association mapping. SNP platforms are now being developed for many crop species including crops that had previously been neglected (Varshney et al., 2009). These platforms offer high-density and low-cost genotyping. Consequently, it is now becoming feasible to genotype very large association mapping panels of several thousand lines with several thousand marker loci. Large mapping panels increase the ability of researchers to deal with complications due to population structure, and the high marker density gives good genetic resolution. However, the problem for inbreeding species remains, due to low levels of recombination seen in most modern breeding programs. This is likely to limit the applicability of association mapping for map-based cloning unless the traits can be tracked in landraces or wild relatives where the recombinational resolution is much higher.

The density of markers needed in association mapping is related to the extent of LD in the mapping panel and the purpose of the mapping study. If LD is extensive then a high marker density is not required, because the marker-trait association would be detected over an extended region of the genome (Figure 6.5A). Nevertheless, even in species where LD extends over several centimorgans at least 1000 markers are likely to be required (assuming the genetic size of the genome is 1000–2000 cM). This density will provide markers sufficiently close for use in breeding and selection programs but not close enough for positional cloning. For species such as maize where LD extends over only short distances, several thousand markers are likely to be needed to identify useful associations (Figure 6.5C; Belo et al., 2008). Gene conversion (Figure 6.3) also affects the LD between markers. When two nearby markers are in strong LD, it can be assumed that all markers in between, defining haplotype blocks, are also in strong LD (Wall et al., 2004). However, some loci might not be in strong LD with the markers delineating the block if the gene conversion is high and the marker density is low. These hot spots of recombination would decrease the power of association studies.

Implications

The past few years have seen major changes in DNA sequencing and molecular marker technologies. The key impact of these changes has been to shift the bottleneck for gene discovery and analysis away from model species to crops. Access to small and fully sequenced genomes was crucial in developing a strong base for plant molecular biology. These resources helped build large research communities able to tackle key questions of plant development, disease, and stress tolerance. However, it was always clear that many traits critical for agriculture could not be effectively addressed through the use of non-crop models. Our most important crops have been through a long period of selection for adaptation to the cropping environment. In this environment, plants are grown at high density and in a monoculture. It is also possible that many of the adaptive traits examined in non-crop models have already been optimized in crops as a result of their long history of selection. Rice has been the exception, and there have been several notable discoveries in key genes associated with yield, disease resistance, and crop performance (Xing and Zhang, 2010).

For major crops there are extensive germplasm collections and large phenotypic datasets available to support molecular biology research. However, these resources were difficult to use for gene discovery and analysis. The identification and characterization of the syntenic relationship between the genomes of plants in the same family provided a tool for translating research outcomes in models to crops and was also a valuable tool for identifying markers in target regions, which is a key step in positional cloning. These tools have underpinned the discovery of important genes in the complex genomes of many of our crop species. The advances in gene and genome sequencing are now opening a new phase in plant molecular biology. The genomes of several crops are now available and more will follow over the next few years. Whole genome re-sequencing is now feasible and is being applied to multiple cultivars, landraces, and wild relatives for some crops. This information will help us understand some of the selective processes that have operated during crop domestication and cultivation. The new sequencing capabilities also mean that SNP discovery is becoming routine for most plant species. Consequently, we expect to see high-density marker platforms available for most major crops.

Taken together, these developments provide a base for a revolution in crop breeding and genetics. High-density marker platforms open the opportunities for the use of new population structures for both gene discovery and for practical plant breeding. Association mapping has been difficult and unreliable while marker densities were low and met with limited success. However, with high marker densities and the opportunity to screen large diversity panels of germplasm this mapping technique is likely to be revitalized. Importantly, association mapping has the huge advantage of allowing the use of existing phenotypic datasets. The germplasm collections and associated phenotypic data are available for many crop species. Essentially genotyping has ceased to be the limitation, but populations and phenotypic datasets are limiting, which pushes the emphasis back to using existing collections. More sophisticated population structures, such as NAM populations, and germplasm collections built to help tackle specific complex traits, such as wheat drought panel (Salekdeh et al., 2009), will become increasingly valuable.

These developments support gene discovery and genetic analysis. What impact will they have on breeding technologies? Detailed genotypic datasets and haplotype definition around key loci can lead to new breeding methodologies. The problems outlined earlier for genetic analysis in large and complex genomes and the generally low marker densities have meant that research has focused on genes or loci of major effect. This approach has been highly successful and many breeding programs include marker-based screening for such loci in their programs. However, the new marker platforms have reduced the cost and increased the marker density so that strategies based around whole genome marker surveys become viable. One such approach, Genome Wide Selection (GWS), requires no information about marker-trait associations, but allows an estimation of breeding values associated with each marker locus (Meuwissen et al., 2001; Bernardo and Yu, 2007). In this way multiple genomic regions of small genetic effect can be selected for. This type of breeding strategy offers the potential for major changes in population structure and screening, and is expected to allow selection for variation that may have been largely overlooked in the past.

These new approaches are expected to open new paths for gene discovery and breeding, but there are also several issues likely to limit the application of these new methods, particularly for public sector breeding programs. The key limitations are the lack of resources, training, and capabilities by most of the world's crop improvement programs. Molecular markers have been available for over a decade and are routine in some, particularly private, breeding programs (Eathington et al., 2007). However, they have only been effectively applied in a small number of public sector breeding programs. Most plant breeders have neither the time nor the resources to deploy new breeding strategies. These limitations will need to be addressed.

The concerns with food security and likely impact of climate change on food production have injected a new urgency into accelerating the rates of genetic gain in breeding programs. The major challenge now will be to ensure that the technological advances are effectively deployed.

References

Adams, S. P., Hartman, T. P., Lim, K. Y., Chase, M. W., Bennett, M. D., & Leitch, I. J., et al. (2001). Loss and recovery of Arabidopsis-type telomere repeat sequences 5′-(TTTAGGG)(n)-3′ in the evolution of a major radiation of flowering plants. *Proceedings Biological Sciences*, *268*, 1541–1546.

Akhunov, E. D., Akhunova, A. R., Linkiewicz, A. M., Dubcovsky, J., Hummel, D., & Lazo, G., et al. (2003). Synteny perturbations between wheat homoeologous chromosomes caused by locus duplications and deletions correlate with recombination rates. *Proceedings of the National Academy of Sciences of the United States of America*, *100*, 10836–10841.

Ansorge, W. J. (2009). Next-generation DNA sequencing techniques. *New Biotechnology*, *25*, 195–203.

Batzoglou, S., Jaffe, D. B., Stanley, K., Butler, J., Gnerre, S., & Mauceli, E., et al. (2002). ARACHNE: A whole-genome shotgun assembler. *Genome Research*, *12*, 177–189.

Belo, A., Zheng, P. Z., Luck, S., Shen, B., Meyer, D. J., & Li, B. L., et al. (2008). Whole genome scan detects an allelic variant of fad2 associated with increased oleic acid levels in maize. *Molecular Genetics and Genomics*, *279*, 1–10.

Bennett, M. & Leitch, I. (2005). Plant DNA C-values database.

Bernardo, R., & Yu, J. M. (2007). Prospects for genome-wide selection for quantitative traits in maize. *Crop Science*, *47*, 1082–1090.

Blanc, G., & Wolfe, K. H. (2004a). Functional divergence of duplicated genes formed by polyploidy during *Arabidopsis* evolution. *Plant Cell*, *16*, 1679–1691.

Blanc, G., & Wolfe, K. H. (2004b). Widespread paleopolyploidy in model plant species inferred from age distributions of duplicate genes. *Plant Cell*, *16*, 1667–1678.

Bolot, S., Abrouk, M., Masood-Quraishi, U., Stein, N., Messing, J., & Feuillet, C., et al. (2009). The "inner circle" of the cereal genomes. *Current Opinion in Plant Biology*, *12*, 119–125.

Bowers, J. E., Arias, M. A., Asher, R., Avise, J. A., Ball, R. T., & Brewer, G. A., et al. (2005). Comparative physical mapping links conservation of microsynteny to chromosome structure and recombination in grasses. *Proceedings of the National Academy of Sciences of the United States of America*, *102*, 13206–13211.

Bowers, J. E., Chapman, B. A., Rong, J., & Paterson, A. H. (2003). Unravelling angiosperm genome evolution by phylogenetic analysis of chromosomal duplication events. *Nature*, *422*, 433–438.

Breseghello, F., & Sorrells, M. E. (2006). Association mapping of kernel size and milling quality in wheat (*Triticum aestivum* L.) cultivars. *Genetics*, *172*, 1165–1177.

Brockman, W., Alvarez, P., Young, S., Garber, M., Giannoukos, G., & Lee, W. L., et al. (2008). Quality scores and SNP detection in sequencing-by-synthesis systems. *Genome Research*, *18*, 763–770.

Caldwell, K. S., Russell, J., Langridge, P., & Powell, W. (2006). Extreme population-dependent linkage disequilibrium detected in an inbreeding plant species, *Hordeum vulgare*. *Genetics*, *172*, 557–567.

Carver, T., Berriman, M., Tivey, A., Patel, C., Bohme, U., & Barrell, B. G., et al. (2008). Artemis and ACT: Viewing, annotating and comparing sequences stored in a relational database. *Bioinformatics*, *24*, 2672–2676.

Cavanagh, C., Morell, M., Mackay, I., & Powell, W. (2008). From mutations to MAGIC: Resources for gene discovery, validation and delivery in crop plants. *Current Opinion in Plant Biology*, *11*, 215–221.

Chaisson, M. J., Brinza, D., & Pevzner, P. A. (2009). *De novo* fragment assembly with short mate-paired reads: Does the read length matter? *Genome Research*, *19*, 336–346.

Chantret, N., Salse, J., Sabot, F., Bellec, A., Laubin, B., & Dubois, I., et al. (2008). Contrasted microcollinearity and gene evolution within a homoeologous region of wheat and barley species. *Journal of Molecular Evolution*, *66*, 138–150.

Chantret, N., Salse, J., Sabot, F., Rahman, S., Bellec, A., & Laubin, B., et al. (2005). Molecular basis of evolutionary events that shaped the hardness locus in diploid and polyploid wheat species (Triticum and Aegilops). *Plant Cell*, *17*, 1033–1045.

Chen, Z. J. (2007). Genetic and epigenetic mechanisms for gene expression and phenotypic variation in plant polyploids. *Annual Review of Plant Biology*, *58*, 377–406.

Chevreux, B., Pfisterer, T., Drescher, B., Driesel, A. J., Muller, W. E., & Wetter, T., et al. (2004). Using the miraEST assembler for reliable and automated mRNA transcript assembly and SNP detection in sequenced ESTs. *Genome Research*, *14*, 1147–1159.

Choulet, F., Wicker, T., Rustenholz, C., Paux, E., Salse, J., & Leroy, P., et al. (2010). Megabase level sequencing reveals contrasted organization and evolution patterns of the wheat gene and transposable element spaces. *Plant Cell*, *22*, 1686–1701.

Cock, P. J., Fields, C. J., Goto, N., Heuer, M. L., & Rice, P. M. (2010). The Sanger FASTQ file format for sequences with quality scores, and the Solexa/Illumina FASTQ variants. *Nucleic Acids Research*, *38*, 1767–1771.

Cui, L., Wall, P. K., Leebens-Mack, J. H., Lindsay, B. G., Soltis, D. E., & Doyle, J. J., et al. (2006). Widespread genome duplications throughout the history of flowering plants. *Genome Research*, *16*, 738–749.

Deamer, D. (2010). Nanopore analysis of nucleic acids bound to exonucleases and polymerases. *Annual Review of Biophysics*, *39*, 79–90.

Devos, K. M. (2005). Updating the "crop circle". *Current Opinion in Plant Biology*, *8*, 155–162.

Devos, K. M. (2010). Grass genome organization and evolution. *Current Opinion in Plant Biology*, *13*, 139–145.

Donthu, R., Lewin, H. A., & Larkin, D. M. (2009). SyntenyTracker: A tool for defining homologous synteny blocks using radiation hybrid maps and whole-genome sequence. *BMC Research Notes*, *2*, 148.

Doust, A. N., Devos, K. M., Gadberry, M. D., Gale, M. D., & Kellogg, E. A. (2004). Genetic control of branching in foxtail millet. *Proceedings of the National Academy of Sciences of the United States of America*, *101*, 9045–9050.

Eathington, S. R., Crosbie, T. M., Edwards, M. D., Reiter, R., & Bull, J. K. (2007). Molecular markers in a commercial breeding program. *Crop Science*, *47*, S154–S163.

Edwards, D., & Batley, J. (2010). Plant genome sequencing: Applications for crop improvement. *Plant Biotechnology Journal*, *8*, 2–9.

Eid, J., Fehr, A., Gray, J., Luong, K., Lyle, J., & Otto, G., et al. (2009). Real-time DNA sequencing from single polymerase molecules. *Science*, *323*, 133–138.

Eilam, T., Anikster, Y., Millet, E., Manisterski, J., & Feldman, M. (2008). Nuclear DNA amount and genome downsizing in natural and synthetic allopolyploids of the genera Aegilops and Triticum. *Genome*, *51*, 616–627.

Ewing, B., & Green, P. (1998). Base-calling of automated sequencer traces using phred. II. Error probabilities. *Genome Research*, *8*, 186–194.

Feuillet, C., Langridge, P., & Waugh, R. (2008). Cereal breeding takes a walk on the wild side. *Trends in Genetics*, *24*, 24–32.

Flint-Garcia, S. A., Thornsberry, J. M., & Buckler, E. S. (2003). Structure of linkage disequilibrium in plants. *Annual Review of Plant Biology*, *54*, 357–374.

Friedman, A. R., & Baker, B. J. (2007). The evolution of resistance genes in multi-protein plant resistance systems. *Current Opinion in Genetics & Development*, *17*, 493–499.

Gaut, B. S., & Ross-Ibarra, J. (2008). Selection on major components of angiosperm genomes. *Science*, *320*, 484–486.

Gepts, P. (2003). Ten thousand years of crop evolution. In M. J. C. and D. E. Sadava (Eds.), *Plants, genes, and crop biotechnology* (pp. 328–359). Sudbury, MA: Jones and Bartlett.

Goff, S. A., Ricke, D., Lan, T. H., Presting, G., Wang, R., & Dunn, M., et al. (2002). A draft sequence of the rice genome (*Oryza sativa* L. ssp. *japonica*). *Science*, *296*, 92–100.

Gordon, D., Abajian, C., & Green, P. (1998). Consed: A graphical tool for sequence finishing. *Genome Research*, *8*, 195–202.

Greilhuber, J., Borsch, T., Muller, K., Worberg, A., Porembski, S., & Barthlott, W. (2006). Smallest angiosperm genomes found in lentibulariaceae, with chromosomes of bacterial size. *Plant Biology (Stuttgart)*, *8*, 770–777.

Gupta, P. K. (2008). Single-molecule DNA sequencing technologies for future genomics research. *Trends in Biotechnology*, *26*, 602–611.

Gutierrez, M. V., Vaz Patto, M. C., Huguet, T., Cubero, J. I., Moreno, M. T., & Torres, A. M. (2005). Cross-species amplification of *Medicago truncatula* microsatellites across three major pulse crops. *Theoretical and Applied Genetics*, *110*, 1210–1217.

Hampson, S. E., Gaut, B. S., & Baldi, P. (2005). Statistical detection of chromosomal homology using shared-gene density alone. *Bioinformatics*, *21*, 1339–1348.

Hampson, S., McLysaght, A., Gaut, B., & Baldi, P. (2003). LineUp: Statistical detection of chromosomal homology with application to plant comparative genomics. *Genome Research*, *13*, 999–1010.

Hanada, K., Vallejo, V., Nobuta, K., Slotkin, R. K., Lisch, D., & Meyers, B. C., et al. (2009). The functional role of pack-MULEs in rice inferred from purifying selection and expression profile. *Plant Cell*, *21*, 25–38.

Harris, T. D., Buzby, P. R., Babcock, H., Beer, E., Bowers, J., & Braslavsky, I., et al. (2008). Single-molecule DNA sequencing of a viral genome. *Science*, *320*, 106–109.

Havlak, P., Chen, R., Durbin, K. J., Egan, A., Ren, Y., & Song, X. Z., et al. (2004). The Atlas genome assembly system. *Genome Research*, *14*, 721–732.

Hilu, K. W. (1993). Polyploidy and the evolution of domesticated plants. *American Journal of Botany*, *80*, 1494–1499.

Hodgson, J. G., Sharafi, M., Jalili, A., Diaz, S., Montserrat-Marti, G., & Palmer, C., et al. (2010). Stomatal vs. genome size in angiosperms: The somatic tail wagging the genomic dog? *Annals of Botany*, *105*, 573–584.

Huang, S., Li, R., Zhang, Z., Li, L., Gu, X., & Fan, W., et al. (2009). The genome of the cucumber, *Cucumis sativus* L. *Nature Genetics*, *41*, 1275–1281.

Huang, X., & Madan, A. (1999). CAP3: A DNA sequence assembly program. *Genome Research*, *9*, 868–877.

International Rice Genome Sequencing Project, The map-based sequence of the rice genome. *Nature*, *436*, 793–800.

Jannink, J. L., Bink, M. C. A. M., & Jansen, R. C. (2001). Using complex plant pedigrees to map valuable genes. *Trends in Plant Science*, *6*, 337–342.

Jiang, J., & Gill, B. S. (1994). Different species-specific chromosome translocations in *Triticum timopheevii* and *T. turgidum* support the diphyletic origin of polyploid wheats. *Chromosome Research*, *2*, 59–64.

Ku, H. M., Liu, J., Doganlar, S., & Tanksley, S. D. (2001). Exploitation of *Arabidopsis*-tomato synteny to construct a high-resolution map of the ovate-containing region in tomato chromosome 2. *Genome*, *44*, 470–475.

Lander, E. S., & Waterman, M. S. (1988). Genomic mapping by fingerprinting random clones: A mathematical analysis. *Genomics*, *2*, 231–239.

Lewis, S. E., Searle, S. M., Harris, N., Gibson, M., Lyer, V., & Richter, J., et al. (2002). Apollo: A sequence annotation editor. *Genome Biology*, *3* (RESEARCH0082).

Liang, C., Jaiswal, P., Hebbard, C., Avraham, S., Buckler, E. S., & Casstevens, T., et al. (2008). Gramene: A growing plant comparative genomics resource. *Nucleic Acids Research*, *36*, D947–953.

Liu, L., Gong, G., Liu, Y., Natarajan, S., Larkin, D. M., & Everts-van der Wind, A., et al. (2004). Multi-species comparative mapping *in silico* using the COMPASS strategy. *Bioinformatics*, *20*, 148–154.

Maccaferri, M., Sanguineti, M. C., Noli, E., & Tuberosa, R. (2005). Population structure and long-range linkage disequilibrium in a durum wheat elite collection. *Molecular Breeding*, *15*, 271–289.

Mackay, I., & Powell, W. (2007). Methods for linkage disequilibrium mapping in crops. *Trends in Plant Science*, *12*, 57–63.

Masterson, J. (1994). Stomatal size in fossil plants: Evidence for polyploidy in majority of angiosperms. *Science*, *264*, 421–424.

McMullen, M. D., Kresovich, S., Villeda, H. S., Bradbury, P., Li, H., & Sun, Q., et al. (2009). Genetic properties of the maize nested association mapping population. *Science*, *325*, 737–740.

Meuwissen, T. H., Hayes, B. J., & Goddard, M. E. (2001). Prediction of total genetic value using genome-wide dense marker maps. *Genetics*, *157*, 1819–1829.

Miller, J. R., Delcher, A. L., Koren, S., Venter, E., Walenz, B. P., & Brownley, A., et al. (2008). Aggressive assembly of pyrosequencing reads with mates. *Bioinformatics*, *24*, 2818–2824.

Miller, J. R., Koren, S., & Sutton, G. (2010). Assembly algorithms for next-generation sequencing data. *Genomics*, *95*, 315–327.

Mueller, L. A., Solow, T. H., Taylor, N., Skwarecki, B., Buels, R., & Binns, J., et al. (2005). The SOL Genomics Network: A comparative resource for Solanaceae biology and beyond. *Plant Physiology*, *138*, 1310–1317.

Mullikin, J. C., & Ning, Z. (2003). The phusion assembler. *Genome Research*, *13*, 81–90.

Myers, E. W., Sutton, G. G., Delcher, A. L., Dew, I. M., Fasulo, D. P., & Flanigan, M. J., et al. (2000). A whole-genome assembly of *Drosophila*. *Science*, *287*, 2196–2204.

Naranjo, T., & Fernandez-Rueda, P. (1991). Homeology of rye chromosome arms to wheat. *Theoretical and Applied Genetics*, *82*, 577–586.

Naranjo, T., Roca, A., Goicoechea, P. G., & Giraldez, R. (1987). Arm homoeology of wheat and rye chromosomes. *Genome*, *29*, 873–882.

Oh, K., Hardeman, K., Ivanchenko, M. G., Ellard-Ivey, M., Nebenfuhr, A., & White, T. J., et al. (2002). Fine mapping in tomato using microsynteny with the *Arabidopsis* genome: The Diageotropica (Dgt) locus. *Genome Biology*, *3* (research0049).

Okazaki, K., Sakamoto, K., Kikuchi, R., Saito, A., Togashi, E., & Kuginuki, Y., et al. (2007). Mapping and characterization of FLC homologs and QTL analysis of flowering time in *Brassica oleracea*. *Theoretical and Applied Genetics*, *114*, 595–608.

Ouyang, S., Zhu, W., Hamilton, J., Lin, H., Campbell, M., & Childs, K., et al. (2007). The TIGR Rice Genome Annotation Resource: Improvements and new features. *Nucleic Acids Research*, *35*, D883–887.

Paull, J. G., Chalmers, K. J., Karakousis, A., Kretschmer, J. M., Manning, S., & Langridge, P. (1998). Genetic diversity in Australian wheat varieties and breeding material based on RFLP data. *Theoretical and Applied Genetics*, *96*, 435–446.

Paux, E., Sourdille, P., Salse, J., Saintenac, C., Choulet, F., & Leroy, P., et al. (2008). A physical map of the 1-gigabase bread wheat chromosome 3B. *Science*, *322*, 101–104.

Pettersson, E., Lundeberg, J., & Ahmadian, A. (2009). Generations of sequencing technologies. *Genomics*, *93*, 105–111.

Pevzner, P. A., Tang, H., & Waterman, M. S. (2001). An Eulerian path approach to DNA fragment assembly. *Proceedings of the National Academy of Sciences of the United States of America*, *98*, 9748–9753.

Phillippy, A. M., Schatz, M. C., & Pop, M. (2008). Genome assembly forensics: Finding the elusive mis-assembly. *Genome Biology*, *9*, R55.

Pop, M. (2009). Genome assembly reborn: Recent computational challenges. *Briefings in Bioinformatics*, *10*, 354–366.

Puertas, M. J. (2002). Nature and evolution of B chromosomes in plants: A non-coding but information-rich part of plant genomes. *Cytogenetic and Genome Research*, *96*, 198–205.

Qi, L. L., Echalier, B., Chao, S., Lazo, G. R., Butler, G. E., & Anderson, O. D., et al. (2004). A chromosome bin map of 16,000 expressed sequence tag loci and distribution of genes among the three genomes of polyploid wheat. *Genetics*, *168*, 701–712.

Rayburn, A. L., Biradar, D. P., Bullock, D. G., & Mcmurphy, L. M. (1993). Nuclear-DNA content in F1 hybrids of maize. *Hermathena*, *70*, 294–300.

Reynolds, M., Manes, Y., Izanloo, A., & Langridge, P. (2009). Phenotyping approaches for physiological breeding and gene discovery in wheat. *Annals of Applied Biology, 155*, 309–320.

Richards, E. J., & Ausubel, F. M. (1988). Isolation of a higher eukaryotic telomere from *Arabidopsis thaliana*. *Cell, 53*, 127–136.

Safar, J., Simkova, H., Kubalakova, M., Cihalikova, J., Suchankova, P., & Bartos, J., et al. (2010). Development of chromosome-specific BAC resources for genomics of bread wheat. *Cytogenetic and Genome Research, 129*

Salekdeh, G. H., Reynolds, M., Bennett, J., & Boyer, J. (2009). Conceptual framework for drought phenotyping during molecular breeding. *Trends in Plant Science, 14*, 488–496.

Salse, J., Abrouk, M., Murat, F., Quraishi, U. M., & Feuillet, C. (2009). Improved criteria and comparative genomics tool provide new insights into grass paleogenomics. *Briefings in Bioinformatics, 10*, 619–630.

Salse, J., Chague, V., Bolot, S., Magdelenat, G., Huneau, C., & Pont, C., et al. (2008). New insights into the origin of the B genome of hexaploid wheat: Evolutionary relationships at the SPA genomic region with the S genome of the diploid relative *Aegilops speltoides*. *BMC Genomics, 9*, 555.

San Miguel, P., Tikhonov, A., Jin, Y. K., Motchoulskaia, N., Zakharov, D., & Melake-Berhan, A (1996). Nested retrotransposons in the intergenic regions of the maize genome. *Science, 274*, 765–768.

Scheibye-Alsing, K., Hoffmann, S., Frankel, A., Jensen, P., Stadler, P. F., & Mang, Y., et al. (2009). Sequence assembly. *Computational Biology and Chemistry, 33*, 121–136.

Schlueter, S. D., Wilkerson, M. D., Dong, Q., & Brendel, V. (2006). xGDB: Open-source computational infrastructure for the integrated evaluation and analysis of genome features. *Genome Biology, 7*, R111.

Schmuths, H., Meister, A., Horres, R., & Bachmann, K. (2004). Genome size variation among accessions of *Arabidopsis thaliana*. *Annals of Botany, 93*, 317–321.

Schmutz, J., Cannon, S. B., Schlueter, J., Ma, J., Mitros, T., & Nelson, W., et al. (2010). Genome sequence of the palaeopolyploid soybean. *Nature, 463*, 178–183.

Schranz, M. E., Song, B. H., Windsor, A. J., & Mitchell-Olds, T. (2007). Comparative genomics in the Brassicaceae: A family-wide perspective. *Current Opinion in Plant Biology, 10*, 168–175.

Schulte, D., Close, T. J., Graner, A., Langridge, P., Matsumoto, T., & Muehlbauer, G., et al. (2009). The International Barley Sequencing Consortium - at the threshold of efficient access to the barley genome. *Plant Physiology, 149*, 142–147.

Simpson, J. T., Wong, K., Jackman, S. D., Schein, J. E., Jones, S. J., & Birol, I. (2009). ABySS: A parallel assembler for short read sequence data. *Genome Research, 19*, 1117–1123.

Sledge, M. K., Ray, I. M., & Jiang, G. (2005). An expressed sequence tag SSR map of tetraploid alfalfa (*Medicago sativa L.*). *Theoretical and Applied Genetics, 111*, 980–992.

Somers, D., Isaac, P., & Edwards, K. (2004). A high-density microsatellite consensus map for bread wheat (*Triticum aestivum* L.). *Theoretical and Applied Genetics, 109*, 1105–1114.

Staden, R., Beal, K. F., & Bonfield, J. K. (2000). The Staden package, 1998. *Methods in Molecular Biology, 132*, 115–130.

Stam, P. (1993). Construction of integrated genetic linkage maps by means of a new computer package: Join map. *The Plant Journal, 3*, 739–744.

Sutton, G., White, O., Adams, M., & Kerlavage, A. (1995). TIGR Assembler: A new tool for assembling large shotgun sequencing projects. *Genome Science & Technology, 1*, 9–19.

Sutton, T., Baumann, U., Hayes, J., Collins, N., Shi, B., & Schnurbusch, T., et al. (2007). Boron-toxicity tolerance in barley arising from efflux transporter amplification. *Science, 318*, 1446–1449.

Swift, H. (1950). The constancy of deoxyribose nucleic acid in plant nuclei. *Proceedings of the National Academy of Sciences of the United States of America, 36*, 643–654.

Tanaka, T., Antonio, B. A., Kikuchi, S., Matsumoto, T., Nagamura, Y., & Numa, H., et al. (2008). The rice annotation project database (RAP-DB): 2008 update. *Nucleic Acids Research, 36*, D1028–1033.

The International Brachypodium Initiative, Genome sequencing and analysis of the model grass *Brachypodium distachyon*. *Nature, 463*, 763–768.

Thomas, B. C., Pedersen, B., & Freeling, M. (2006). Following tetraploidy in an *Arabidopsis* ancestor, genes were removed preferentially from one homeolog leaving clusters enriched in dose-sensitive genes. *Genome Research, 16*, 934–946.

Tuskan, G. A., Difazio, S., Jansson, S., Bohlmann, J., Grigoriev, I., & Hellsten, U., et al. (2006). The genome of black cottonwood, *populus trichocarpa* (Torr. & Gray). *Science, 313*, 1596–1604.

Udall, J. A., & Wendel, J. F. (2006). Polyploidy and crop improvement. *Crop Science, 46*, S3–S14.

Varshney, R. K., Close, T. J., Singh, N. K., Hoisington, D. A., & Cook, D. R. (2009). Orphan legume crops enter the genomics era! *Current Opinion in Plant Biology, 12*, 202–210.

Varshney, R. K., Marcel, T. C., Ramsay, L., Russell, J., Roder, M. S., & Stein, N., et al. (2007). A high density barley microsatellite consensus map with 775 SSR loci. *Theoretical and Applied Genetics, 114*, 1091–1103.

Veyrieras, J. B., Goffinet, B., & Charcosset, A. (2007). MetaQTL: A package of new computational methods for the meta-analysis of QTL mapping experiments. *BMC Bioinformatics, 8*, 49.

Wall, J. D. (2004). Close look at gene conversion hot spots. *Nature Genetics, 36*, 114–115.

Wicker, T., Schlagenhauf, E., Graner, A., Close, T. J., Keller, B., & Stein, N. (2006). 454 sequencing put to the test using the complex genome of barley. *BMC Genomics, 7*, 275.

Woodhouse, M. R., Schnable, J. C., Pedersen, B. S., Lyons, E., Lisch, D., & Subramaniam, S., et al. (2010). Following tetraploidy in maize, a short deletion mechanism removed genes preferentially from one of the two homologs. *PLoS Biology, 8*, e1000409.

Xie, W., Feng, Q., Yu, H., Huang, X., Zhao, Q., & Xing, Y., et al. (2010). Parent-independent genotyping for constructing an ultrahigh-density linkage map based on population sequencing. *Proceedings of the National Academy of Sciences of the United States of America, 107*, 10578–10583.

Xing, Y., & Zhang, Q. (2010). Genetic and molecular bases of rice yield. *Annual Review of Plant Biology, 61*, 421–442.

Yang, L., & Bennetzen, J. L. (2009). Distribution, diversity, evolution, and survival of Helitrons in the maize genome. *Proceedings of the National Academy of Sciences of the United States of America, 106*, 19922–19927.

Young, N. D., & Udvardi, M. (2009). Translating *Medicago truncatula* genomics to crop legumes. *Current Opinion in Plant Biology, 12*, 193–201.

Yu, J., & Buckler, E. S. (2006). Genetic association mapping and genome organization of maize. *Current Opinion in Biotechnology, 17*, 155–160.

Yu, J., Hu, S., Wang, J., Wong, G. K., Li, S., & Liu, B., et al. (2002). A draft sequence of the rice genome (*Oryza sativa* L. ssp. *indica*). *Science, 296*, 79–92.

Zerbino, D. R., & Birney, E. (2008). Velvet: Algorithms for *de novo* short read assembly using de Bruijn graphs. *Genome Research, 18*, 821–829.

Zhang, Y., Sledge, M. K., & Bouton, J. H. (2007). Genome mapping of white clover (*Trifolium repens* L.) and comparative analysis within the Trifolieae using cross-species SSR markers. *Theoretical and Applied Genetics, 114*, 1367–1378.

Zhao, W., Wang, J., He, X., Huang, X., Jiao, Y., & Dai, M., et al. (2004). BGI-RIS: An integrated information resource and comparative analysis workbench for rice genomics. *Nucleic Acids Research, 32*, D377–382.

Zonneveld, B. J., Leitch, I. J., & Bennett, M. D. (2005). First nuclear DNA amounts in more than 300 angiosperms. *Annals of Botany, 96*, 229–244.

Agrobacterium-mediated plant genetic transformation

7

Yoel Shiboleth Tzvi Tzfira
University of Michigan, Ann Arbor, Michigan

TABLE OF CONTENTS

Introduction

Agrobacterium-mediated plant genetic transformation is a multistep process in which the bacterium delivers a specific DNA molecule from its own genome into the host-cell genome (Tzfira and Citovsky, 2002, 2006; Gelvin, 2003b, 2010; Citovsky et al., 2007; Banta and Montenegro, 2008). The transferred DNA molecule (T-DNA) resides on one of the bacterium's genetic elements, designated the tumor-inducing (Ti) plasmid, and encodes a set of oncogenes and opine-catabolism genes (Britton et al., 2008). Integration of the native T-DNA molecule into the plant genome and expression of the oncogenes leads to the formation of tumor-like structures (galls) from the transformed tissues (Aloni and Ullrich, 2008). The transformed gall tissues produce opines, which are amino acid derivatives that serve as a nitrogen source for the growing bacteria (Britton et al., 2008; Aloni and Ullrich, 2008). Interestingly, replacing the native T-DNA molecule with any other DNA sequence does not affect the transformation process, and recombinant *Agrobacterium* strains have therefore been developed as an efficient tool for the delivery of genes of interest into plant species (Banta and Montenegro, 2008). Under controlled conditions, *Agrobacterium* is also capable of transforming yeast, fungi, and even human cells (reviewed by Lacroix et al., 2006; Soltani et al., 2008), which led to the suggestion that *Agrobacterium* may also be used as a tool for functional genomics in fungal species (Michielse et al., 2005). Owing to its unique biology and its important role as a biotechnological tool, *Agrobacterium* has been the subject of numerous studies aimed at understanding its biology and harnessing its power for plant genetic transformation (e.g., Hooykaas and Beijersbergen, 1994; Tzfira and Citovsky, 2002, 2006; Gelvin, 2003b, 2009, 2010; Christie et al., 2005; Christie and Cascales, 2005; Citovsky et al., 2007; Banta and Montenegro, 2008; Pitzschke and Hirt, 2010). In this chapter, we outline the biology of the transformation process and introduce the reader to the basic requirements for using *Agrobacterium* as a biotechnological tool for plant genetic engineering. Prospects for enhancing and controlling the transformation process are also discussed.

The Genetic Transformation Process

Agrobacterium uses a set of chromosomal (*chv*) and Ti-plasmid virulence (*vir*) proteins to sense the presence of wounded plant tissues and to produce the transported form of the T-DNA (a single-stranded copy of its T-DNA region, also known as the T-strand), and to deliver it to the plant cell (reviewed by Citovsky et al., 2007; Tzfira and Citovsky, 2002, 2006; Gelvin, 2003b, 2010). The T-DNA, which on native Ti plasmids resides in *cis* to the *vir* region (Figure 7.1A), is defined by two direct 25 bp repeats, designated left and right borders (Wang et al., 1987). Unveiling the molecular mechanism by which the T-strand is excised from the Ti plasmid by *vir* proteins acting in *trans* to the T-DNA region enabled the development of recombinant and autonomous plasmids (designated "binary plasmids," Figure 7.1B), which were engineered to carry a well-defined synthetic DNA (Hoekema et al., 1983). Furthermore, by removing the native T-DNA region from Ti plasmids, scientists were able to produce disarmed Ti-plasmid strains (Figure 7.1B): these were still capable of responding to wounded plant tissues and shuttle a

DOI: 10.1016/B978-0-12-381466-1.00007-9

recombinant T-DNA molecule from a binary vector, but could not induce the formation of galls in infected plants. Naturally, elimination of the native T-DNA genes required the development of novel constructs and methods for the regeneration and selection of transformed tissues and whole transgenic plants from infected cells, as we describe later in the chapter.

The transformation process has been the subject of numerous reviews (Hooykaas and Beijersbergen, 1994; Tzfira and Citovsky, 2002, 2006; Gelvin, 2003b, 2009, 2010; Christie et al., 2005; Christie and Cascales, 2005; Citovsky et al., 2007; Banta and Montenegro, 2008; Pitzschke and Hirt, 2010) and is typically divided into several general steps (Figure 7.2). Regardless of whether the T-DNA is launched from a native Ti plasmid, engineered Ti plasmid, or binary plasmid, the transformation process begins with sensing of signals from wounded plant tissues by the bacterial VirA-VirG two-component signal-transduction system (step 1) and attachment of the *Agrobacterium* cells to the infected plant tissues, and induction of the virulence(*vir*) region (step 2). The plant molecule acetosyringone, for example, has been identified as a specific activator of the *Agrobacterium vir* region (Stachel et al., 1985) and is now used for activation of *Agrobacterium* cells in many transformation protocols (Sheikholeslam and Weeks, 1987).

Once expressed, the VirD2/VirD1 heterodimer functions as an ssDNA nuclease, which nicks and releases a single-stranded (ss) portion of the T-DNA molecule (step 3). A single VirD2 molecule remains attached to the ssT-DNA (the T-strand), producing the immature T-complex that is then transported via a VirB/VirD4 type four secretion system (T4SS) into the plant cell (step 4). The exact structure of the secretion system, the precise point(s) of contact between components of the T4SS and the plant cell wall, and/or membrane, and the mechanism by which the secretion apparatus delivers the immature T-complex, as well as several other Vir proteins (e.g., VirF, VirE2, and VirE3; step 4) into the plant cells are still mostly unknown (Christie et al., 2005; Christie and Cascales, 2005). A mature T-complex is likely to be formed inside the host-cell cytoplasm (step 5) by coating of the T-strand with numerous VirE2, proteins. VirE2 molecules are thought to provide the T-strand with structure and protection (Citovsky et al., 1989, 1997; Abu-Arish et al., 2004) during its travel through the host-cell cytoplasm and into the host-cell nucleus (step 6). Once inside the nucleus, the T-complex is delivered to the host-cell chromatin, stripped from its escorting proteins (step 7) and, finally, randomly integrated into the host-cell genome as either a single- or double-stranded (ds) T-DNA molecule (step 8) most likely into genomic double-stand breaks.

In addition to Vir and Chv proteins, a growing number of host factors and proteins have been identified as participants in various steps of the transformation process, in particular the last steps (see Table 7.1 and Tzfira and Citovsky, 2002, 2006; Gelvin, 2003b, 2010; Citovsky et al., 2007). More specifically, once assembled in the cytoplasm (Figure 7.2, step 5), the T-complex interacts with a wide range of host factors on its way to the host-cell genome and during its integration there. Biophysical studies indicate that the T-complex may be directed toward the host-cell nucleus along the microtubule system by interactions of VirE2 with putative molecular motors (Salman et al., 2005). VirE2 and VirD2 (both of which have been shown to possess nuclear-localization activity) have also been reported to function in T-strand nuclear import (Shurvinton et al., 1992; Citovsky et al., 1992, 1994; Narasimhulu et al., 1996; Zupan et al., 1996; Ziemienowicz et al., 1999, 2001; Tzfira and Citovsky, 2001). T-complex nuclear import (Figure 7.2, step 6) may also depend on the functions of a wide range of host factors, such as VIP1 (Tzfira et al., 2001), several members of the importin α proteins (Ballas and Citovsky, 1997; Bhattacharjee et al., 2008), cyclophilins (Deng et al., 1998), protein phosphatase 2C (PP2C; Tao et al., 2004), and the bacterial protein VirE3 (Lacroix et al., 2005).

Targeting the T-strand to points of integration and stripping it of its escorting proteins represents the second to last step of the transformation process (Figure 7.2, step 7), and these processes are mediated by the combined action of bacterial and host proteins. The cyclin-dependent kinase-activating kinase (CAK2M) and a TATA-box binding protein (Bako et al., 2003), for example, have been proposed to participate in guiding the T-complex to preintegration sites by interaction with VirD2. Similarly the interaction of VIP1 with VirE2 and with histone H2A (Li et al., 2005), the association of DNA-VirE2-VIP1 with the mononucleosome (Lacroix et al., 2008b), and the function of H2A in T-DNA integration (Mysore et al., 2000) suggest a role for core histone modifications in T-complex targeting the host DNA (Lacroix and Citovsky, 2009). Furthermore, the binding of the nucleosome-associated DNA-VirE2-VIP1 complexes with VirF, a bacterial factor involved in targeted proteolysis of VIP1 and VirE2 (Tzfira et al., 2004), suggests that stripping of the T-complex from its escorting proteins may take place near the integration site. Interestingly, *Agrobacterium* can also induce the expression of VBF, a plant factor that can compensate for the absence of VirF in targeted proteolysis of VIP1 and VirE2 (Zaltsman et al., 2010). These studies (and many others, which have been extensively reviewed by Hooykaas and Beijersbergen, 1994; Tzfira and Citovsky, 2002, 2006; Gelvin, 2003b, 2009, 2010; Christie et al., 2005; Christie and Cascales, 2005; Citovsky et al., 2007; Banta and Montenegro, 2008; Pitzschke and Hirt, 2010) further reveal the complexity of the last steps of the transformation process inside living plant cells.

The final step (step 8) is the integration of presumably naked T-DNA molecules into the plant genome. The T-strand's bacterial chaperones VirE2 and VirD2 are not likely to actively participate in the integration process, because they do not possess the enzymatic activity needed for T-DNA integration in plant cells. Several studies, using yeast and plant cells as hosts for *Agrobacterium* transformation, revealed that various DNA-repair and maintenance proteins are required for the integration process (reviewed by Tzfira et al., 2004; Ziemienowicz et al., 2008). The plant H2A (Mysore et al., 2000), the VirE2-interacting protein 2 (VIP2; Anand et al., 2007), the chromatin-assembly factor complex (CAF-1; Kirik et al., 2006) and the genomic double-strand break (DSB) repair proteins from plants (i.e., Ku80) and from yeast (i.e., Ku70 and Rad50, see van Attikum et al., 2001; van Attikum and Hooykaas, 2003) are just a few examples of plant proteins that have been proposed to function in the integration process (for more reviews see Tzfira

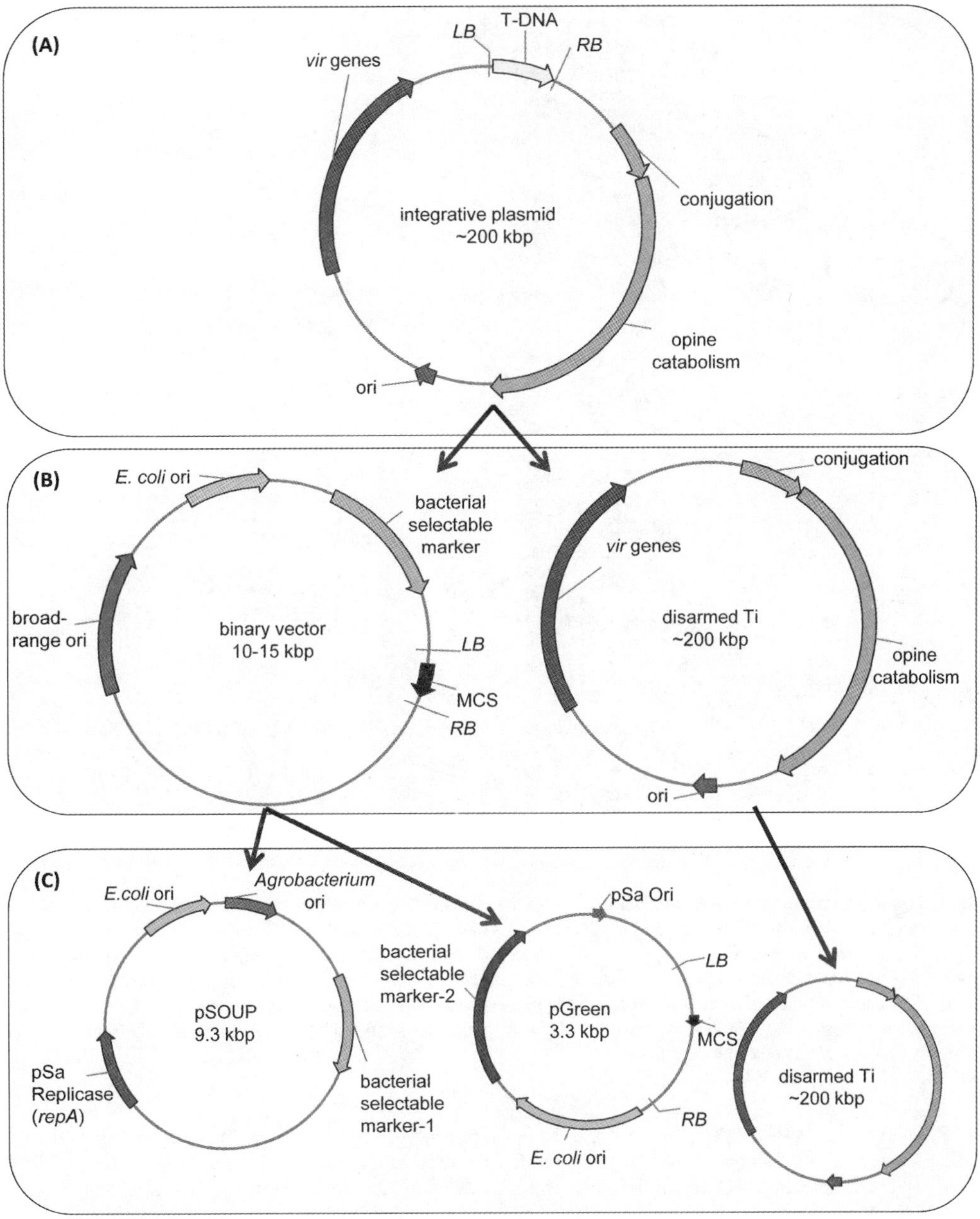

Figure 7.1 • Structure of three types of *Agrobacterium* vector systems • (A) Integrative Ti plasmid: the native T-DNA region was replaced by recombinant T-DNA and is present in *cis* to the *vir* region. Integrative Ti plasmids also typically carry opine catabolism and conjugation genes. (B) Binary system: composed of a disarmed Ti plasmid (which is essentially similar to an integrative Ti plasmid, only without a T-DNA region) and a small, autonomous binary vector. The binary plasmid is engineered to carry an MCS in which genes of interest can be cloned to produce a T-DNA, an *E. coli* ori, and a bacterial selectable marker. Binary plasmids are significantly smaller than Ti plasmids and are easier to manipulate in *E. coli* cells. A broad-range ori facilitates the use of binary vectors in various *Agrobacterium* strains. (C) Split binary system: composed of a disarmed Ti plasmid and a split binary vector, in which the minimized pGreen vector carries the T-DNA region and the pSoup vector is engineered to support the replication and maintenance of pGreen in *Agrobacterium* cells (the *repA* gene on pSoup is required to support the function of pSa ori on pGreen). Abbreviations: LB, left border; RB, right border; ori, origin of replication. Please see color plate section at the end of the book.

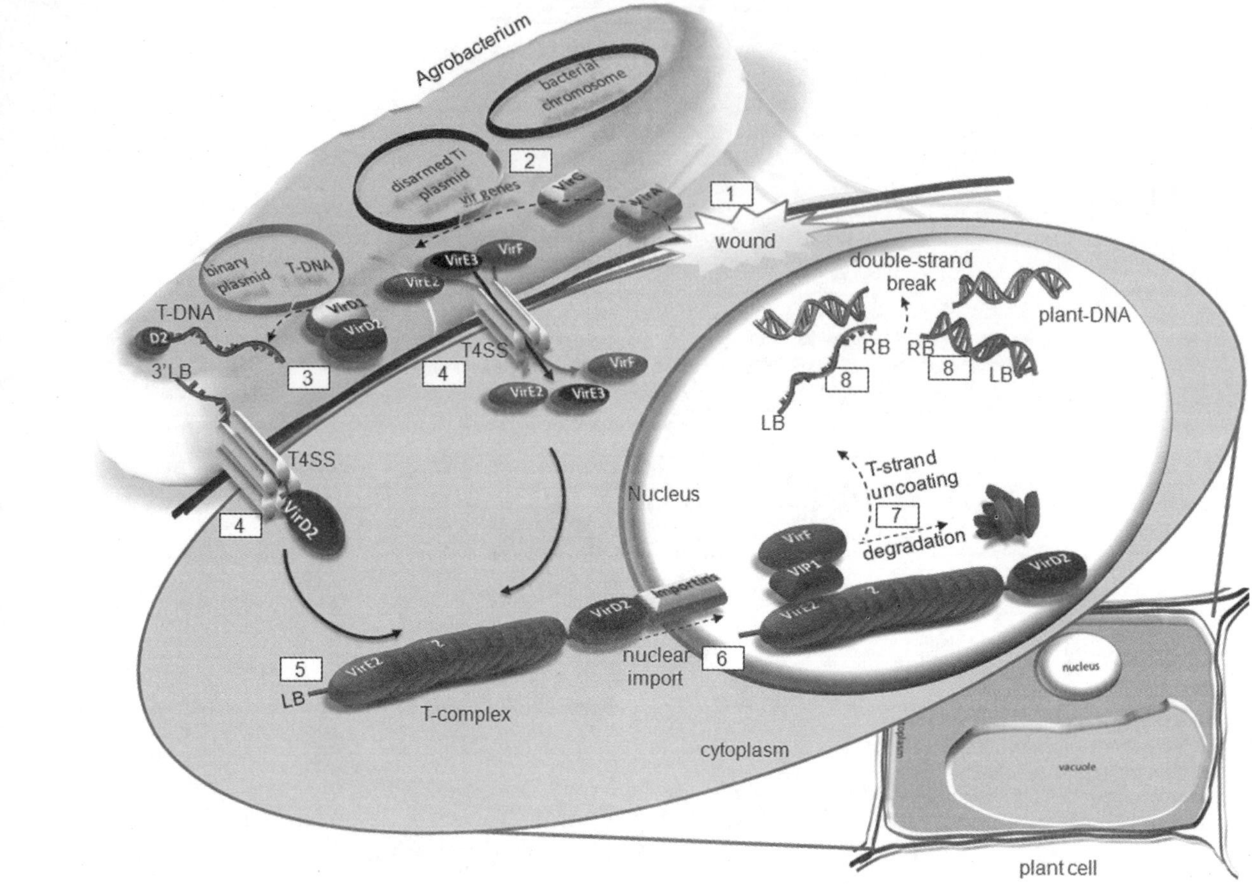

Figure 7.2 • Model for *Agrobacterium*-mediated plant genetic transformation • A typical engineered *Agrobacterium* cell carries, in addition to its own natural chromosome, a disarmed Ti plasmid and a binary plasmid with a recombinant T-DNA region. We divide the transformation process into eight general steps: (1) sensing of wounded plant cells by the VirA/VirG signal-transduction system; (2) activation of the virulence (*vir*) region and production of virulence proteins; (3) production of immature T-complex; (4) export of immature T-complex and Vir proteins via the T4SS; (5) formation of mature T-complex and its cytoplasmic transport; (6) T-complex nuclear import; (7) uncoating of the T-complex and intranuclear transport; (8) T-DNA integration as a single- or double-stranded intermediate, presumably into genomic DSBs. Please see color plate section at the end of the book.

et al., 2004; Ziemienowicz et al., 2008; Gelvin and Kim, 2007). In addition, a special role has been assigned to genomic DSBs in attracting and capturing T-DNA molecules. More specifically, the induction of genomic DSBs at specific locations by the expression of restriction enzymes leads to the preferred integration of dsT-DNA intermediates (and possibly also of T-strands) into the break sites (Salomon and Puchta, 1998; Chilton and Que, 2003; Tzfira et al., 2003). This process is mediated by the plant non-homologous end joining (NHEJ) DNA-repair machinery. It should be noted that T-DNA molecules can be directed to site-specific integration in yeast cells, a process that depends on the sequence of the T-DNA molecule and the function of homologous recombination (HR) DNA-repair proteins (van Attikum et al., 2001; van Attikum and Hooykaas, 2003). Interestingly, altering the plant DNA-repair machinery by expressing a yeast chromatin-remodeling protein, Rad54 (Shaked et al., 2005), resulted in a high frequency of HR-mediated T-DNA integration in *Arabidopsis* plants, an observation further supporting the idea that T-DNA integration is highly dependent on the function of plant factors.

Various models have been proposed to explain the random integration of T-strands and dsT-DNA intermediates into the plant genome (reviewed by Tzfira et al., 2004; Ziemienowicz et al., 2008). However, our lack of knowledge on the precise mechanism of T-DNA integration poses a great challenge for the use of *Agrobacterium* as a tool for site-specific genome engineering in plant cells. Recent advances in the development of zinc finger nucleases (ZFNs) as a tool for the induction of site-specific DSBs (Porteus and Carroll, 2005; Kumar et al., 2006; Wu et al., 2007; Camenisch et al., 2008; Weinthal et al., 2010) and understanding the important role

Table 7.1 Examples of host proteins involved in various steps of the transformation process

Protein	Function	Reference
AtAGP17	Lysine-rich arabinogalactan protein, may be involved in reducing plant resistance response to *Agrobacterium*	Gaspar et al. (2004)
BTI1	Interacts with VirB2, may be involved in initial steps of *Agrobacterium*-plant interactions	Hwang and Gelvin (2004)
Microtubules	Cytoplasmic transport of the T-complex toward the nucleus	Salman et al. (2005)
RocA, Roc4, CypA	Binding to VirD2, possibly involved in maintaining its conformation in the plant cytoplasm	Deng et al. (1998)
AtKapα	Interacts with VirD2, nuclear import of VirD2	Ballas and Citovsky (1997)
VIP1	Interacts with VirE2, nuclear import of VirE2	Tzfira et al. (2001, 2002); Djamei et al. (2007)
IMPa-4	Interacts with VirE2, nuclear import of VirE2	Bhattacharjee et al. (2008)
CAK2Ms	Interacts with VirD2, nuclear import of VirD2	Bako et al. (2003)
PP2C	Interacts with the C-terminal region of VirD2, involved in nuclear import of VirD2	Tao et al. (2004)
VIP2	Interacts with VirE2, involved in T-DNA integration	Anand et al. (2007)
ASK1	Interacts with VirF, involved in T-DNA uncoating	Tzfira et al. (2004); Schrammeijer et al. (2001)
VBF	Interacts with VIP1, involved in T-DNA uncoating	Zaltsman et al. (2010)
H2A	T-DNA integration	Mysore et al. (2000)
Ku70, Rad50	T-DNA integration, discovered in yeast	van Attikum et al. (2001); van Attikum and Hooykaas (2003)
Ku80	T-DNA integration	Li et al. (2005); Friesner and Britt (2003)
CAF-1	T-DNA integration	Endo et al. (2006)
DNA ligase IV	T-DNA integration	Cai et al. (2009)

BTI1, VirB2-interacting protein 1; Roc and CypA, *Arabidopsis* cyclophilin chaperone family; AtKapα, *Arabidopsis thaliana* importin α; VIP1, VirE2-interacting protein 1; IMPa-4, *Arabidopsis* importin α isoform CAK2M, cyclin-dependent kinase-activating kinase; PP2C, protein phosphatase 2C; VIP2, VirE2 interacting protein 2; ASK1, *Arabidopsis* sKp1-like protein; VBF, VIP1-binding F-box protein, CAF-1, chromatin-assembly factor 1.

played by DSBs in inducing the HR-mediated pathway in plant cells (Puchta et al., 1996; Siebert and Puchta, 2002; Orel et al., 2003), may lead to the development of novel strategies for controlling the integration process, as will be described later in the chapter.

Plant factors may also function at earlier stages of the transformation process. Indeed, the interactions of several host proteins (e.g., BTI1) with VirB2 (a component of the bacterial T-pilus) have been shown to be important for the transformation process (Hoekema et al., 1983), and several other proteins have been postulated to affect the bacterial attachment and/or T-DNA transfer to plants (Zhu et al., 2003). *Agrobacterium* has also been shown to alter the plant cell's defense response (Ditt et al., 2001, 2005, 2006; Veena et al., 2003). Veena et al. (2003), for example, showed that the transfer of T-DNA and proteins can suppress host defense response genes, and Ditt et al. (2005) reported on an inverse correlation between the expression level of several defense genes and transformation efficiency. This negative correlation may be attributed to the function of salicylic acid, which was shown to attenuate the function of the VirA kinase domain and was thus proposed to function in the repression of *vir* expression (Yuan et al., 2007). That salicylic acid inhibits the induction of *vir* genes (Anand et al., 2008), and that salicylic acid-over-expressing mutant plants were recalcitrant to tumor formation (Yuan et al., 2007) further supports the notion that attenuating the plant defense response is a critical step in the transformation process.

The precise molecular mechanism by which *Agrobacterium* can alter the plant defense response is still unknown; nevertheless, the *Arabidopsis* EFR1 protein, which is the receptor for the bacterial elongation factor Tu (EF-Tu), was shown to restrict *Agrobacterium*-mediated genetic transformation of *Arabidopsis* plants (Zipfel et al., 2006). Furthermore, a molecular link between *Agrobacterium*-mediated activation of the MAPK signaling pathway and the transformation process were recently discovered (Djamei et al., 2007). More specifically, Djamei et al. (2007) discovered that

MAPK-mediated phosphorylation of VIP1, a host protein involved in T-complex nuclear import, is required for VIP1 nuclear import and for activation of defense response genes. It is intriguing to speculate that *Agrobacterium* developed novel ways of not only suppressing the host defense system, but also using specific components of the defense response for its own needs (Dafny-Yelin et al., 2008a; Pitzschke and Hirt, 2010).

The set of events and the functions that each of the previously mentioned plant and bacterial proteins play in T-complex cytoplasmic trafficking, nuclear import, and integration have been the subject of several recent reviews (Tzfira et al., 2005; Lacroix et al., 2008a; Gelvin, 2010). Understanding their role in the transformation process may lead to the development of methods for enhancing the transformation of recalcitrant species and controlling T-DNA integration, as will be described in the next section.

Agrobacterium as a Tool for Plant Transformation

Agrobacterium can genetically transform a wide range of plant and non-plant species. It is now used as a practical and robust tool for the genetic engineering of many scientifically and agriculturally important plant species, including *Arabidopsis*, tobacco, tomato, poplar, soybean, rice, and wheat (reviewed by Herrera-Estrella et al., 2005; Lacroix et al., 2006; Banta and Montenegro, 2008; Soltani et al., 2008). An understanding of the bacterium's biology and the development of plant tissue-culture techniques were crucial for early adaptation and engineering of this unique bacterium for plant genetic transformation. We focus our discussion on the key factors needed for successful transformation and regeneration of transgenic plants, and readers are referred to the *Agrobacterium Protocals* (Wang 2006a,b) and the *Compendium of Transgenic Crop Plants* (Kole and Hall, 2008) for detailed transformation protocols of scientifically and commercially important plant species and for an in-depth discussion on commercially important transgenic crops and traits.

A prerequisite for the development of a new transformation protocol is the ability to regenerate a whole plant from a single, transformed cell. The regeneration of plants from single cells relies on the totipotency of many plant cells and can be technically achieved under sterile conditions, using plant tissue-culture techniques and the application of exogenous growth regulators. With the exception of *Arabidopsis* plants, which are typically transformed by dipping their flowers in an *Agrobacterium* culture (Zhang et al., 2006), all transformation protocols rely on the *in vitro* co-cultivation of plant explants with aseptic cultures of *Agrobacterium* cells (Wang, 2006a,b; Kole and Hall, 2008). Since different *Agrobacterium* strains' cells vary in their transformation efficiency toward various plant species, cultivars, and even explants of a given plant, establishing a new transformation protocol often involves a long process of mix and match between different *Agrobacterium* strains, plant explants, and co-cultivation and culturing conditions.

In the early years of *Agrobacterium* research, scientists developed integrative plasmids in which the genes of interest were integrated into the T-DNA region of existing Ti plasmids to deliver them into plant cells (Figure 7.1A). In such systems, the genes of interest are first cloned into small, *Escherichia coli*-based cloning plasmids. The plasmids are next transferred into *Agrobacterium* cells where they integrate by HR into the T-DNA region of disarmed or native Ti plasmids. pGV3850, for example, was modified from a native Ti plasmid by replacing its T-DNA region with the sequence of the cloning vector pBR322 (Zambryski et al., 1983). Recombinant pBR322 plasmids, carrying the genes of interest, are next transferred into *Agrobacterium* cells and recombine into the pGV3850 T-DNA region. While useful for plant genetic engineering, the manipulation and engineering of large Ti plasmids can be tedious and difficult to perform. Thus, most modern transformation protocols rely on the use of binary vectors (Figure 7.1B) for gene delivery from disarmed *Agrobacterium* strains into plant cells (reviewed by Hellens et al., 2000; Komari et al., 2006; Lee and Gelvin, 2008). A binary vector is typically designed to carry the T-DNA region and an *E. coli* origin of replication and selection marker (Figure 7.1B). The latter allows the binary plasmid to be engineered using standard molecular biology techniques and common *E. coli* strains. Several examples of early and modern binary plasmids and sets of plasmids can be found in Table 7.2. The T-DNA region is defined by its left and right borders and is usually composed of a plant selection marker and a multiple cloning site (MCS), which typically facilitate the cloning of one or two genes of interest (Figure 7.3). Since the T-strand's right border is protected by VirD2, integration at the right border is usually more accurate than at the left border of the T-strand. Thus, most modern binary plasmids carry the plant selection marker gene near the left border (e.g., Becker et al., 1992; van Engelen et al., 1995), assuring that the gene(s) of interest, which is cloned near the right border, will be present in the selected cells.

Various genes can be used for the selection of transformed plant cells. Genes encoding antibiotic or herbicide resistance (e.g., *nptII*, *hpt*, *ppt*, or *bar*) can be found in many binary plasmids and have been successfully used for positive selection of a wide range of plant species. pBIN19, for example (Bevan, 1984), was one of the first commercialized binary vectors engineered to replicate in *E. coli* and to carry the kanamycin plant selection marker. Interestingly, despite its extensive use for plant genetic engineering, the complete sequence of pBIN19 was only reported in 1995 (Frisch et al., 1995), the same year in which a modified and improved version of this vector (designated pBINPLUS), in which the kanamycin selection marker was located close to the left border, was also revealed (van Engelen et al., 1995). In another example, the pGPTV set of vectors was developed to facilitate the exchange of several plant selection markers (i.e., resistance to bleomycin, basta, hygromycin, kanamycin, and dihydrofolate reductase) between them (Becker, 1990). Other selection genes that rely on negative (e.g., *codA*, thymidine kinase, and cytochrome P450; Koprek et al., 1999; Gardiner and Howlett, 2004), metabolic (e.g., *dao1* and alcohol dehydrogenase; Xiaohui Wang et al., 2001; Erikson et al., 2004), and phenotypical (e.g., *ipt*; Ebinuma et al., 1997) selection, for example, have also been deployed for the production of

Table 7.2 Examples of *Agrobacterium* binary vectors

Plasmid(s)	Key features	Reference
pBIN19	Plant selection marker (Kan) next to the right border; supports blue/white cloning; limited MCS; commercialized by Clontech	Bevan (1984)
pBINPLUS	Plant selection marker (Kan) next to the left border; supports blue/white cloning; limited MCS; MCS also contains two rare restriction sites; a higher copy number in *E. coli*, and two rare restriction sites	van Engelen et al. (1995)
pBI	A set of vectors; plant selection marker (Kan) next to the right border; promoterless gene; commercialized by Clontech	Jefferson et al. (1987)
pBIB/pBIG	Set of vectors; exchangeable plant selection markers (Kan, Hyg) next to the right border; with or without promoterless *uidA* gene	Becker (1990)
pGPTV	A set of vectors; plant selection genes (Kan, Hyg, Dhfr, Bleo, and Bar) next to the left border; promoterless *uidA* gene, limited MCS	Becker et al. (1992)
pCAMBIA	Large set of vectors; constantly being improved and developed by Cambia (an independent non-profit institution); supports several plant selection genes (e.g., Kan, Hyg, and Bar); supports translational fusion to various tags	http://www.cambia.org
pPZP	Set of small binary plasmids; some vectors assembled with plant selection genes (Kan, Gent) next to the left border; some vectors support blue/white cloning; vectors carry bacterial chloramphenicol- or spectinomycin-resistant genes	Hajdukiewicz et al. (1994)
BIBAC	Binary bacterial artificial chromosome; designed to transfer very large T-DNA molecules	Hamilton et al. (1996); Hamilton (1997)
pSAT/pRCS	A set of vectors; supports construction of multigene T-DNAs by rare cutters; facilitates fluorescence protein fusion; based on pPZP	Goderis et al. (2002); Tzfira et al. (2005); Dafny-Yelin et al. (2007)
pSITE	A set of vectors; based on pPZP; facilitates fluorescence protein fusion by Gateway	Chakrabarty et al. (2007)
pMDC	A set of vectors; based on pCAMBIA; supports *gfp* and *uidA* protein fusion by Gateway; supports construction of promoter-reporter and genomic DNA by Gateway	Curtis and Grossniklaus (2003)
pEarleyGate	A set of vectors; based on pCAMBIA; supports translational fusion to various epitope tags (e.g., FLAG, HA, cMyc, and AcV5)	Earley et al. (2006)
pGreen/pSoup	Very small T-DNA-carrying binary (pGreen) vector; pSoup provides replication functions for pGreen; supports several plant selection genes (Kan, Hyg, Sul and Bar); constantly being improved and further developed (see http://www.pgreen.ac.uk)	Hellens et al. (2000)
pMSP	Carries super promoter; supports several plant selection genes (Kan, Hyg, and Bar)	Lee et al. (2007)

Abbreviations: MCS; multiple cloning site; Kan, kanamycin; Hyg, hygromycin; Dhfr, dihydrofolate reductase; Bleo, bleomycin; Bar, basta; Gent, gentamycin; *uidA*, β-D-glucuronidase; Sul, sulfonylurea; *gfp*, green fluorescent protein.

transgenic plants (for an in-depth discussion on plant selection markers, see reviews by Miki and McHugh, 2004; Sundar and Sakthivel, 2008).

In addition to selection genes, plant scientists often use the *uidA* reporter gene (which encodes the β-glucuronidase enzyme) as a visual marker for transformation events. β-glucuronidase activity can be easily monitored in infected plant tissues by enzymatic conversion of the colorless X-Gluc substrate to blue precipitates (Jefferson et al., 1987). The *uidA* reporter gene is thus especially powerful for detecting early transformation events, which are typically monitored 1 to 2 days post co-cultivation. The pBIN19-derivative pBI121 (Jefferson et al., 1987; Chen et al., 2003), for example, was engineered to carry a functional *uidA* expression cassette in addition to the kanamycin plant selection marker, and has been widely used in the development of many transformation protocols. It is important to note that while transient *uidA* expression is a clear indicator of successful T-DNA nuclear import and its complementation to the double-stranded intermediate, production of transgenic plants depends on the integration of the T-DNA molecule into the plant cell genome and the regeneration of a whole plant under selective conditions.

Interestingly, only a small number of *Agrobacterium* strains have been disarmed and are currently used in a large number of transformation protocols. Because many of these strains show a certain degree of resistance to various antibiotics (Lee and Gelvin, 2008), most of the binary vectors have been

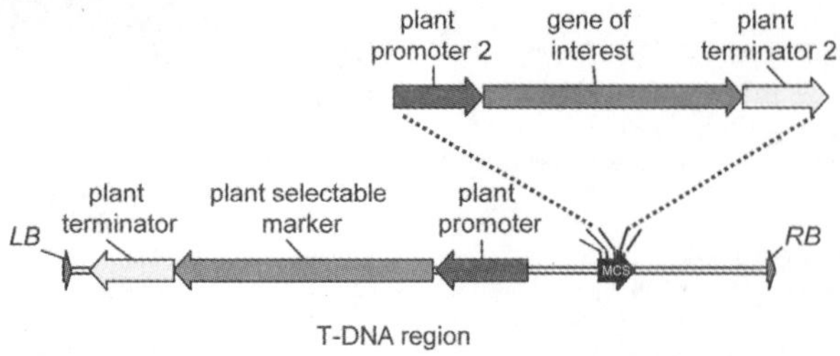

Figure 7.3 • General structure of T-DNA region • Current practice calls for placing the plant selectable marker expression cassette close to the T-DNA's left border (LB) and constructing a versatile MCS next to the T-DNA's right border (RB). Independent plant expression cassette(s) can next be cloned into the MCS. This structure ensures that a high proportion of the selected plants will carry both the selection marker and the gene(s) of interest. Please see color plate section at the end of the book.

designed around a limited number of *Agrobacterium* origins of replication and these vectors carry a relatively small range of *Agrobacterium* selection markers. Thus, for example, the use of *Agrobacterium* strain EHA101 (Hood et al., 1986) was limited by its natural resistance to kanamycin, which was the preferred antibiotic for selection of several early binary vectors (e.g., pBIN, pBIN19, pGA, and pGPTV). EHA101 was later engineered to produce EHA105, which was no longer resistant to kanamycin (Hood et al., 1993), thus expanding the repertoire of *Agrobacterium* strains that can be used in conjunction with kanamycin-resistant binary vectors. The natural resistance of *Agrobacterium* strains adds another layer of complexity to their use in plant genetic engineering. Following a co-cultivation step, it is necessary to eliminate the bacterium from the culture and to allow the infected explants to recover and regenerate into transformed plants. Since the application of antibiotics to plant culture can have a negative effect on plant-cell recovery and regeneration, it is important to evaluate the use of several antibiotics at various concentrations for *Agrobacterium* elimination.

Developing a new transformation system relies on the careful orchestration of a wide range of biological components and procedural steps. These include proper selection of the plant explants, binary vector, selection marker, antibiotics, and *Agrobacterium* strain, and establishing the procedures for regeneration, co-cultivation, bacterium elimination, and selection and recovery of transgenic plants. Even after reporting on the successful transformation of a given plant species, scientists continue to improve existing procedures by changing various parameters of the transformation protocols. Husaini (2009), for example, reported that eliminating *Agrobacterium* cells by a combination of timentin and cefotaxime at low concentrations leads to higher transformation and regeneration efficiency than using just one of these antibiotics at high concentration. This improvement in the transformation efficiency was attributed to lowering of the phytotoxic effect of these antibiotics on the leaf disks (Husaini and Abdin, 2008; Husaini, 2009). Similarly, Alsheikh et al. (2002) reported that combining the use of specific *Agrobacterium* strains with long exposure to low concentrations of *Agrobacterium*-eliminating antibiotics could improve the transformation efficiency of antibiotic-sensitive *Fragaria vesca* and *F. v. semperflorens* plants. Interestingly, even fine-tuning the early steps of the transformation process (e.g., the activation of *Agrobacterium* cells), can affect the transformation efficiency. Thus, for example, while acetosyringone was dispensable for the transformation of wheat cell suspension cultures when using the combination of *A. tumefaciens* strain C58 and the binary vector pMON18365 (Cheng et al., 1997), doubling its concentration in the co-cultivation medium led to improved transformation efficiency of inoculated durum wheat embryos when the combination of *Agrobacterium* strain AGL1 and the split binary system pGreen/pSoup (Figure 7.1C) was used (He et al., 2010). It should be noted that no direct comparison between the use of the different *Agrobacterium* strains and binary vectors was reported in those studies (Cheng et al., 1997; He et al., 2010); nevertheless, other studies clearly indicate that different combinations of *Agrobacterium* strains and binary plasmids vary in their transformation efficiencies (e.g., Przetakiewicz et al., 2004; Gasparis et al., 2008).

Many other factors can potentially affect the transformation efficiency of various plant species. However, there is no clear and straightforward set of instructions for discovering and applying such factors for improvement and control of the transformation process. Thus, much of the technical progress made in plant genetic engineering has occurred through the design of new, and the manipulation of existing, procedures and through the development of novel sets of binary vectors.

Novel and Specialized Vectors for Plant Transformation

Binary vectors have been instrumental in the delivery of a wide range of genes into plant cells (reviewed by Herrera-Estrella et al., 2005; Dafny-Yelin and Tzfira, 2007; Banta and Montenegro, 2008; Lee and Gelvin, 2008). Most binary vectors have been engineered to facilitate the cloning and transfer of a single gene of interest within their MCS (in addition to the selection marker; Figures 7.1B and 7.3), a design that often hinders the cloning of novel plant transformation constructs. The growing use of *Agrobacterium* for basic plant research and biotechnology, and in particular the growing interest in facilitating the expression of fluorescent protein fusions, the delivery of multiple genes into plants, and the introduction of recombination-based cloning systems, has led to the development of a growing number of specialized binary vectors and families of vectors (exemplified in Table 7.2).

The importance of developing specialized binary vectors was realized in the early days of vector development. One of the first specialized binary vectors, for example, was pBI101 (Jefferson et al., 1987), which was designed to facilitate the fusion of plant promoters to the *uidA* reporter gene to enable gene-expression studies in plant cells. The vector, which was available as a commercial product from Clontech (as part of a set of several binary vectors), carried a relatively small MCS, relied on kanamycin as the sole plant selection marker, and was not fully sequenced. Thus, while pBI101 was used in many plant transformation studies, its cloning capacity was limited and its modification was tedious.

The growing interest in plant gene expression and protein subcellular localization studies and the advances in fluorescent proteins (e.g., GFP) and confocal microscopy technologies led to the development of a wide range of specialized binary vectors that could facilitate the cloning and transformation of fluorescent fusion genes into plant cells. The pSITE vector system (Chakrabarty et al., 2007), for example, was designed to facilitate the stable and transient expression of fluorescent protein fusions (e.g., GFP, YFP, and RFP) in plant cells. Not only did this family of vectors provide the user with a wide range of fluorescent proteins, it was also designed for Gateway® cloning (a restriction-enzyme free, recombination-based cloning system) which can significantly reduce the effort needed to fuse genes of interest into these binary vectors (Karimi et al., 2007).

Realizing the power of Gateway-based cloning, many other binary vectors have been designed to support this new technology for plant genetic transformation (e.g., Curtis and Grossniklaus, 2003; Helliwell and Waterhouse, 2003; Cheo et al., 2004; Sasaki et al., 2004; Karimi et al., 2005; Earley et al., 2006; Gehl et al., 2009). The pMDC set of vectors (Curtis and Grossniklaus, 2003), for example, has been designed to facilitate constitutive or inducible gene expression, *uidA* and GFP protein fusion, promoter analysis, and more. Specialized Gateway-based binary vectors have also been developed to facilitate epitope-tag translational fusion (Earley et al., 2006), analysis of protein–protein interactions by bimolecular fluorescence complementation (BiFC) assay (Gehl et al., 2009), and for hairpin RNA-mediated gene-silencing experiments in plants (Helliwell and Waterhouse, 2003).

Further advances in Gateway technology led to additional progress in binary vector construction technology and, in particular, to the introduction of modular binary vector assembly and construction of multigene vectors. More specifically, the development of MultiSite Gateway technology, which enables the recombination of several DNA segments in a single reaction step (Cheo et al., 2004; Sasaki et al., 2004), enabled the assembly of novel T-DNA vectors by simultaneous recombination of multiple DNA fragments (Karimi et al., 2005). MultiSite Gateway also enabled the construction of multigene binary vectors (Wakasa et al., 2006). Chen et al. (2006), for example, introduced MultiRound Gateway as a novel strategy for the construction of multigene binary vectors by altering between different Gateway Entry vectors. In another example, a multiple-round *in vivo* site-specific assembly (designated MISSA) Gateway recombination system (Chen et al., 2010) was developed as a novel *in vivo* assembly method for the construction of multigene binary vectors from a collection of DNA modules that could be designed to code for virtually any type of DNA sequence (e.g., promoters, terminators, genes of interest, selection markers, fluorescent genes, and even complete gene-expression cassettes).

Other designs for the construction of multigene plasmids have also been reported. Lin et al. (2003) developed a novel plant binary assembly system that takes advantage of combining the Cre/loxP recombination system and the use of two rare-cutting endonucleases for the sequential assembly of multiple plant expression cassettes onto a transformation-competent artificial chromosome (TAC) based vector (Liu et al., 1999). The authors used this system to construct T-DNAs with 40- to 80kb genomic DNA fragments and reported on their successful transfer into *Arabidopsis* plants (Liu et al., 1999). Other systems, such as the pPZP-RCS2/pAUX (Goderis et al., 2002) and the pPZP-RCS2/pSAT (Tzfira et al., 2005; Chung et al., 2005), rely exclusively on the use of rare-cutting enzymes for sequential assembly of up to 7 gene-long transformation vectors. While these systems rely on a very limited number of commercially available rare-cutting enzymes (Goderis et al., 2002; Chung et al., 2005; Tzfira et al., 2005), their capacity can potentially be increased by using artificial restriction enzymes (i.e., ZFNs), which have been proven useful for the construction of plant binary vectors (Zeevi et al., 2008, 2010).

The relatively large backbone of many binary vectors may pose a technical obstacle for their genetic manipulation, so scientists developed sets of vectors with minimal backbone sequences. The pCB series of mini binary vectors, for example, are about half the size of pBIN19 and were constructed by eliminating over 5kb that were deemed non-essential for their function in plant transformation from the pBIN19 vector backbone (Xiang et al., 1999). It should be noted that by minimizing pBIN19, the authors eliminated the *traF* region from the vector backbone, and thus limited the transfer of pCB vectors into *Agrobacterium* to direct transformation methods. The pPZP family of small binary vectors, on the other hand, has been designed to maintain the pBR322*bom* site, which enables their conjugal transfer from *E. coli* to *Agrobacterium* cells (Hajdukiewicz et al., 1994). Further minimization of the binary vector backbone was achieved by Hellens et al. (2000), who constructed the pGreen series of vectors by removing the RepA and Mob functions from the vector backbone and supporting the replication of pGreen vectors in *Agrobacterium* cells by a second plasmid, pSoup (Figure 7.1C).

Overcoming the technical challenge of constructing large and complex T-DNA molecules is just one of the concerns when planning their transfer into plant cells. Hamilton et al. (1996, 1997), for example, reported that reinforcing *Agrobacterium* with additional copies of the *virG* and *virE* genes was necessary to transfer a 150kb T-DNA molecule using the binary-BAC (BIBAC) vector system, which they developed, in plant cells. While the BIBAC binary vector system was successfully used for delivery of very long T-DNA molecules into several plant species (e.g., Feng et al., 2006; He et al., 2006; Li, et al., 2007; Vega et al., 2008). Song et al. (2003) reported that BIBAC potato genomic fragments larger than 100kb were not stable in several *Agrobacterium* strains, including two *recA*-deficient ones. Similarly, Liu et al. (1999), who developed the TAC binary vector and used it to construct an *Arabidopsis* genomic DNA fragment (40–80kb) library of approximately 10,000 clones, recognized that large binary plasmid clones may not be stable in *Agrobacterium* cells and recommended testing individual constructs before using them in plant transformation experiments. Interestingly, Chen et al. (2010) reported that using conjugal DNA transfer instead of electroporation for the delivery of a multigene binary vector into a *recA*-deficient *Agrobacterium* strain was crucial for obtaining *Agrobacterium* clones with stable binary vectors. Thus, while Tao and Zhang (1998) demonstrated that even though over 300 kb fragments can be cloned into BIBAC vectors and stably maintained in *E. coli* cells, their transfer and maintenance in *Agrobacterium* cells should be regarded with caution.

Overall, significant progress has been made in the last two decades in the design and construction of a wide range of binary vectors and new vectors are constantly being developed (e.g., Dafny-Yelin et al., 2007; Lee et al., 2007; Thole et al., 2007; Zaltsman et al., 2007; Koizumi and Iwata, 2008). Nevertheless, while it seems that there is a vector for every task (exemplified in Table 7.2), plant scientists and biotechnologists still face the challenges of improving the transformation rate of recalcitrant species, extending the range of *Agrobacterium* cells to new plant species and cultivars, and controlling the outcome of the transformation process. These difficulties can potentially be overcome by manipulating the host and by interfering with the T-DNA integration.

Manipulating the Plant Genome to Improve and Control Transformation

Developing an *Agrobacterium*-mediated genetic transformation procedure is an empirical process in which scientists examine various combinations of bacteria and binary plasmids and inoculation, regeneration, and selection techniques. Remarkably, many of the transformation procedures that have been developed in the last several decades have relied on a relatively small number of *Agrobacterium* strains and binary vectors, while many plant species, varieties, and cultivars still remain recalcitrant to *Agrobacterium* infection. It seems that further genetic manipulation of the bacterium and the development of additional binary vectors may not be sufficient to expand the bacterium host range and to improve existing, albeit low-efficiency, transformation procedures of certain plant species. It was thus suggested that further improvements in transformation technologies may arise by using plant factors known to function during the transformation process (Gelvin, 2003a; Tzfira and Citovsky, 2006).

We previously noted that various host factors have been assigned functional roles during different steps of the transformation process (Table 7.1). In addition, through extensive genetic screens for plant genes that are involved in the transformation process, a large number of *rat* (resistant to *Agrobacterium* infection) and *hat* (hypersensitive to *Agrobacterium* infection) *Arabidopsis* mutants have been identified (e.g., Nam et al., 1999; Zhu et al., 2003). It is thus possible that manipulating these genes, and their putative homologs, in recalcitrant plant species may provide a unique pathway for improving their *Agrobacterium*-mediated genetic transformation efficiencies. Indeed, over-expression of several host proteins in transgenic plants rendered them more susceptible to *Agrobacterium* infection. Thus, for example, over-expression of VIP1, a plant protein involved in T-DNA nuclear import, T-DNA targeting to the plant genome, and T-DNA uncoating (Tzfira et al., 2001, 2004; Citovsky et al., 2004; Li et al., 2005; Lacroix et al., 2008b) in transgenic tobacco plants, increased its susceptibility to *Agrobacterium*-mediated genetic transformation, most likely due to the increased rate of T-DNA nuclear import in the transgenic plants (Tzfira et al., 2002). In another example, over-expression of KU80, a plant DNA-repair protein, increased the susceptibility of transgenic *Arabidopsis* to *Agrobacterium* infection (Li et al., 2005). Since KU80, involved in DSB repair (West et al., 2002), was suggested to mediate the integration of dsT-DNA intermediates into genomic DSBs (Tzfira et al., 2004; Li et al., 2005; Ziemienowicz et al., 2008), the increased susceptibility may be attributed to increased T-DNA integration rate in the transgenic plants. Increased transformation efficiency was also reported in transgenic BTI1-over-expressing *Arabidopsis* plants (Hwang and Gelvin, 2004). The precise mechanism by which this VirB2-interacting protein functions during the transformation process is still unknown, but it exemplifies how over-expression of plant proteins, which presumably function during early stages of the transformation process, can increase plant transformation efficiency.

Enhancing plant genetic transformation by genetic modification is still mostly limited to model plants (i.e., *Arabidopsis* and tobacco). Nevertheless, in a recent report, Zheng et al. (2009) showed that over-expression of the *Arabidopsis* HTA1 in rice leads to higher transformation efficiency of the transgenic plants. The histone H2A protein was originally described as an *Arabidopsis* protein involved in T-DNA integration (Mysore et al., 2000) and its expression, as well as the expression of several other *Arabidopsis* histone-encoding genes, increased the efficiency of *Agrobacterium*-mediated transformation of H2A-overexpressing transgenic *Arabidopsis* plants (Mysore et al., 2000; Tenea et al., 2009). It is important to note that not only stable, but also transient expression of HTA1 was sufficient to increase *Agrobacterium*-mediated transformation in rice, as evidenced by transformation experiments in which both *HTA1* expression and the target gene vectors were co-transformed into rice tissues (Zheng et al., 2009). Thus transient delivery of other plant proteins may be sufficient to improve the transformation of other agronomically important and hard-to-transform plant species and cultivars. Naturally, other gene delivery methods (e.g., microbombardment) may be used for the initial plant gene expression in plant species that are highly recalcitrant to *Agrobacterium* infection. Alternatively, *Agrobacterium* cells, which are capable of exporting several Vir proteins independently of the T-DNA molecule (Vergunst et al., 2000), can potentially be engineered to export recombinant plant proteins during the infection process. Indeed, Vergunst et al. (2005) showed that Cre recombinase protein, which was tagged with a VirF export signal, could be exported into plant cells, as determined by Cre-mediated activation of the GFP reporter gene in the infected plants. Similarly, Lacroix et al. (2005) showed that *Agrobacterium* could export the mGAL4-VP16 chimeric transcription activator, when fused to the VirE3 protein, into plant cells, and that the fusion was capable of activating the *uidA* reporter gene in transgenic plants. It is thus possible that future technologies will rely on fusing plant proteins to full-length Vir proteins or to short Vir export signals to export them, in parallel (or prior) to infection by *Agrobacterium* cells that carry the genes of interest on their T-DNA.

Manipulating the host genome can also affect the efficiency of the overall process, as well as that of T-DNA integration. van Attikum et al. (2001) used a set of DNA-repair mutants to investigate T-DNA integration in yeast cells. The authors discovered that the yeast proteins Ku70, Rad50, Mre11, Xrs2, Lig4, and Sir4 were all required for NHEJ-mediated T-DNA integration, whereas Rad51 and Rad52 were required for HR-mediated T-DNA integration in yeast cells. They also reported that a Ku70/Rad52 double mutant was blocked in T-DNA integration. These observations support the notion that genetic manipulation of the host (in that case yeast cells) can affect the route of T-DNA integration and the final outcome of the transformation process.

In the past several years, various attempts have been made to alter the route of T-DNA integration in plant cells by expressing heterologous genes or by downregulating native DNA-repair genes. Reiss et al. (2000), for example, expressed the *recA* gene from *E. coli* in tobacco in an attempt to induce HR-mediated T-DNA integration in plant cells. Nevertheless, while RecA indeed functions in plant cells (as evidenced by an increase in sister chromatid exchanges), no increase in HR-mediated gene targeting was observed. In another report, Shalev et al. (1999) over-expressed RuvC (a bacterial protein involved in resolving Holliday junctions) and demonstrated that it can increase somatic crossovers, and intra- and extra-chromosomal recombination. The authors did not determine whether it can also be used to enhance T-DNA integration via HR. Impressive enhancement of T-DNA integration via HR was observed in *Arabidopsis* plants over-expressing Rad54 (a member of the SWI2/SNF2 chromatin-remodeling gene family; Shaked et al., 2005). More specifically, the authors reported that HR-mediated T-DNA integration in Rad54-transgenic plants was, on average, 27-fold higher than HR-mediated T-DNA integration in non-transgenic, wild-type plants. The authors thus suggested that the lack of Rad54-like activity in wild-type plants may be a limiting factor for T-DNA integration via HR and proposed that further improvements in such integration might be achieved by stacking *rad54* with additional HR-related genes or by using plants mutated in NHEJ genes (e.g., Gherbi et al., 2001; Molinier et al., 2004; Takeda et al., 2004). Indeed, several reports have suggested that altering plant DNA-repair genes may affect DNA-repair and maintenance responses. Thus, for example, decreased intrachromosomal HR was observed in MIM mutant plants (Hanin et al., 2000), whereas increased intrachromosomal HR was observed in MIM-over-expressing plants (Hanin et al., 2000), *rad50* (Gherbi et al., 2001), and chromatin-assembly factor 1 (*caf-1*; Endo et al., 2006) mutants. Whether T-DNA integration via HR and/or NHEJ can also be affected in these plants still needs to be studied.

Using Novel Selection Methods and Restriction Enzymes to Control T-DNA Integration

T-DNA molecules can integrate via HR even without host-genome manipulation. However, since HR-mediated T-DNA integration occurs at a much lower frequency than NHEJ-mediated T-DNA integration, scientists suggested that using novel selection and screening methods may assist in indentifying and selecting for these rare HR-mediated integration events (reviewed by Hohn and Puchta, 1999; Kumar and Fladung, 2001; Puchta, 2002, 2003; Britt and May, 2003; Iida and Terada, 2005). Paszkowski et al. (1988) were the first to demonstrate that foreign DNA molecules can integrate into a predetermined site via HR. In their experimental approach, a defective selection marker in transgenic plants was repaired by retransformation with a linear plasmid construct that was engineered to carry regions that were complementary and homologous to the defective selection marker. This report was later followed by the work of Offringa et al. (1990) who demonstrated that T-DNA molecules can also function as substrates for HR-mediated selection marker repair in transgenic plants, by Lee et al. (1990) who used T-DNA molecules to target the native selectable gene coding for ALS (acetolactate synthase) in tobacco, and by Miao and Lam (1995) who targeted the TGA3 (a bZIP-like transcription factor) locus in *Arabidopsis*, again using T-DNA molecules. In the latter case, targeting of the TGA3-encoding gene was detected by replacing part of the gene with a kanamycin-resistance-encoding sequence. Since kanamycin-resistant cells could also be obtained via random T-DNA integration, the authors included the GUS-encoding gene outside the region of homology on the incoming T-DNA molecule, which allowed them to distinguish between NHEJ- and HR-mediated T-DNA integration. These studies, and those of others (e.g., Hrouda and Paszkowski, 1994; Risseeuw et al., 1995; Hanin et al., 2001; Endo et al., 2006, 2007), clearly showed that T-DNA and dsDNA molecules (i.e., plasmid DNA) can be directed to HR-mediated integration in plant cells, albeit at a low frequency (for further reading see Hohn and Puchta, 1999; Kumar and Fladung, 2001; Puchta, 2002, 2003; Britt and May, 2003; Iida and Terada, 2005).

To overcome the problem of selecting for rare HR-mediated T-DNA integration events in non-selectable genes, scientists developed T-DNA molecules that facilitate the use of positive-negative selection schemes (Mansour et al., 1988). Positive-negative selection schemes are useful for discriminating between NHEJ- and HR-mediated integration events. From a technical point of view, a positive-negative selection approach calls for the construction of a novel T-DNA molecule in which a positive selection marker is flanked by regions of homology to the target gene, and a negative selection gene is cloned outside of the target sequence. Both NHEJ- and HR-mediated T-DNA integration events can be selected for by positive selection, whereas negative selection enables the elimination of NHEJ-mediated integration events from the population of transgenic events. The positive-negative selection approach was used to drive T-DNA integration into the *Gln1* and *Pzf* genes in *Lotus japonicus* (Thykjaer et al., 1997) and into the chalcone synthase (CHS) encoding gene and an artificial *htp* gene in *Arabidopsis* cell-suspension culture (Gallego et al., 1999). In neither case was there any conclusive evidence supporting the notion that the T-DNA molecules indeed integrated into the target genes. Yet, molecular analysis revealed that the alcohol dehydrogenase (ADH) encoding gene in *Arabidopsis* and *Waxy*, *Adh2*, and the methyltransferase (MET1a) encoding genes in rice (Terada et al., 2002, 2007; Johzuka-Hisatomi et al., 2008; Yamauchi et al., 2009; Xiaohni Wang et al., 2001) were all targeted via HR-mediated T-DNA integration using positive-negative selection schemes. These observations further support the notion that T-DNA molecules can indeed integrate via HR into the plant genome. Nevertheless, achieving HR-mediated T-DNA integration in crop and model plants remains a challenge (reviewed by Hohn and Puchta, 1999; Kumar and Fladung, 2001; Puchta, 2002, 2003; Britt and May, 2003; Iida and Terada, 2005; Puchta and Hohn, 2005; Porteus, 2009).

Deploying novel selection schemes does not alter the pathways for T-DNA integration, but merely assists in selecting specific and rare HR-mediated integration events. Studies in various eukaryotic cells showed that induction of genomic DSBs can stimulate HR-mediated DNA repair (reviewed by Carroll, 2004; Puchta, 2005; Porteus, 2009), and it was thus suggested that induction of DSBs may enhance T-DNA integration via HR. Indeed, intrachromosomal HR in plant cells has been enhanced by the excision of transposable elements (Xiao and Peterson, 2000) or by X-ray (Tovar and Lichtenstein, 1992), both of which cause DSBs in plant cells. Similarly, expression of site-specific endonucleases (e.g., HO and I-*Sce*I) also enhanced intrachromosomal HR in plant cells (Chiurazzi et al., 1996; Puchta et al., 1996; Siebert and Puchta, 2002; Orel et al., 2003). It should still be noted (as previously mentioned) that NHEJ is the preferred mode of DSB repair in plant cells (Britt and May, 2003; Puchta, 2005), and that T-DNA molecules are preferentially captured via NHEJ in DSBs (Salomon and Puchta, 1998; Chilton and Que, 2003; Tzfira et al., 2003). Nevertheless, combining the use of artificial rare-cutting restriction enzymes (e.g., ZFNs) with T-DNA molecules, which can be engineered to carry regions of homology to genomic sequences, holds great promise for enhancing site-specific T-DNA integration via HR (Carroll, 2004; Paques and Duchateau, 2007; Weinthal et al., 2010).

ZFNs are synthetic restriction enzymes that can induce DSBs in a wide range of eukaryotic cells, including plant cells (for recent reviews see Porteus and Carroll, 2005; Durai et al., 2005; Carroll et al., 2006). Expression of ZFNs in plants often results in NHEJ-mediated, site-specific mutagenesis (e.g., Lloyd et al., 2005; de Pater et al., 2009; Tovkach et al., 2009; Osakabe et al., 2010; Tanaka et al., 2010). ZFNs are also useful for inducing HR-mediated integration of plasmid and T-DNA molecules in plants. Wright et al. (2006), for example, showed that plasmid dsDNA molecules were directed to HR-mediated, site-specific integration in tobacco protoplasts upon expression of ZFNs. More specifically, the authors produced several transgenic tobacco plants that carried a non-functional *gus::nptII* fusion marker and a ZFN-recognition site. They next co-transformed transgenic protoplasts with a ZFN-expressing construct and a plasmid DNA molecule that carried homologous regions and repair sequences to the non-functional *gus::nptII* fusion marker. Kanamycin-resistant and GUS-expressing calluses were recovered, indicating the nature of the HR-mediated integration and *gus::nptII* repair by the plasmid DNA molecule. The same research group later reported that ZFN and HR-mediated gene replacement (delivered by plasmid DNA) can be used to target and mutate native tobacco genes (i.e., *SuR* genes' coding sequences; Townsend et al., 2009).

ZFN and HR-mediated integration are not limited to plasmid DNA. Cai et al. (2009) showed that ZFN expression combined with engineered T-DNA molecules can lead to site-specific integration and replacement of not only pre-integrated, non-functional reporter genes (i.e., a disabled *gfp* and *pat* genes) but also an endogenous locus (i.e., the endochitinase gene *CHN50*). In addition, Shukla et al. (2009) demonstrated that ZFNs can direct T-DNA molecules to HR-mediated, site-specific integration in corn plants by targeting and replacing the *IPK1* gene (which encodes the enzyme that catalyzes the final step in phytate biosynthesis). Collectively, these studies clearly indicate that by combining selection methods, novel constructs, and expression of restriction enzymes, T-DNA molecules can potentially be directed to HR-mediated integration and be selected and used for gene targeting in plant cells.

Conclusions and Future Prospects

Over the past several decades, *Agrobacterium* has been established as a novel and critical tool for plant research and biotechnology. Extensive research has revealed much of the bacterium's unique and fascinating biology and a wide range of transformation protocols have been developed for an impressive number of plant species, varieties, and cultivars. Furthermore, extensive progress has been made in the development and construction of both versatile and specialized *Agrobacterium* binary vectors that have further facilitated the use of *Agrobacterium* for various tasks in basic plant cell research. Nevertheless, many commercially important plant species and cultivars still remain recalcitrant to *Agrobacterium* transformation and plant scientists are still limited in their ability to control the transformation process, even in susceptible plant species. Unveiling the role of plant factors and cellular mechanisms in the transformation process and developing novel tools for the induction of site-specific genomic DSBs has allowed scientists to interfere with the transformation process and to improve and even control the outcome of transformation events. Still, more research is needed to systematically dissect the route of the T-complex and other proteins inside the plant cytoplasm and nucleus and to determine the role of DNA-repair machineries in T-DNA integration. Some of the questions that still remain open are: (1) how the T-complex and Vir proteins pass through the plant cell wall barrier, (2) the precise structure and function of the bacterial T4SS, (3) how the mature T-complex is assembled and transported through the plant-cell cytoplasm, (4) which plant proteins are responsible for complementing the T-strands into double-stranded intermediates, and (5) how the T-complex is delivered into genomic DSBs (or other genomic locations) for integration. In addition, further development in ZFN technology and vector design is needed to allow scientists and plant biotechnologists to harness this powerful tool for controlling T-DNA and for genome engineering in plant cells.

Acknowledgments

We apologize to colleagues whose work was not cited due to lack of space. Our lab is supported by grants from The Consortium for Plant Biotechnology Research (CPBR), Dow Agrosciences, Biotechnology Research and Development Corporation (BRDC), the U.S.-Israel Binational Agricultural Research and Development (BARD), and the U.S.-Israel Binational Science Foundation (BSF).

References

Abu-Arish, A., Frenkiel-Krispin, D., Fricke, T., Tzfira, T., Citovsky, V., & Grayer Wolf, S., et al. (2004). Three-dimensional reconstruction of *Agrobacterium* VirE2 protein with single-stranded DNA. *Journal of Biological Chemistry, 279*, 25359–25363.

Aloni, R., & Ullrich, C. I. (2008). Biology of crown gall tumors. In T. Tzfira & V. Citovsky (Eds.), *Agrobacterium* (pp. 565–591). New York: Springer.

Alsheikh, M. K., Suso, H. -P., Robson, M., Battey, N. H., & Wetten, A. (2002). Appropriate choice of antibiotic and *Agrobacterium* strain improves transformation of antibiotic-sensitive *Fragaria vesca* and *F. v. semperflorens*. *Plant Cell Reports, 20*, 1173–1180.

Anand, A., Krichevsky, A., Schornack, S., Lahaye, T., Tzfira, T., & Tang, Y., et al. (2007). *Arabidopsis* VIRE2 INTERACTING PROTEIN2 is required for *Agrobacterium* T-DNA integration in plants. *Plant Cell, 19*, 1695–1708.

Anand, A., Uppalapati, S. R., Ryu, C. M., Allen, S. N., Kang, L., & Tang, Y., et al. (2008). Salicylic acid and systemic acquired resistance play a role in attenuating crown gall disease caused by *Agrobacterium tumefaciens*. *Plant Physiology, 146*, 703–715.

Bako, L., Umeda, M., Tiburcio, A. F., Schell, J., & Koncz, C. (2003). The VirD2 pilot protein of *Agrobacterium*-transferred DNA interacts with the TATA box-binding protein and a nuclear protein kinase in plants. *Proceedings of the National Academy of Sciences of the United States of America, 100*, 10108–10113.

Ballas, N., & Citovsky, V. (1997). Nuclear localization signal binding protein from *Arabidopsis* mediates nuclear import of *Agrobacterium* VirD2 protein. *Proceedings of the National Academy of Sciences of the United States of America, 94*, 10723–10728.

Banta, L. M., & Montenegro, M. (2008). *Agrobacterium* and plant biotechnology. In T. Tzfira & V. Citovsky (Eds.), *Agrobacterium* (pp. 73–147). New York: Springer.

Becker, D. (1990). Binary vectors which allow the exchange of plant selectable markers and reporter genes. *Nucleic Acids Research, 18*, 203.

Becker, D., Kemper, E., Schell, J., & Masterson, R. (1992). New plant binary vectors with selectable markers located proximal to the left T-DNA border. *Plant Molecular Biology, 20*, 1195–1197.

Bevan, M. W. (1984). Binary *Agrobacterium* vectors for plant transformation. *Nucleic Acids Research, 12*, 1811–1821.

Bhattacharjee, S., Lee, L. Y., Oltmanns, H., Cao, H., Veena, Cuperus, J., & Gelvin, S. B. (2008). IMPa-4, an *Arabidopsis* importin alpha isoform, is preferentially involved in *Agrobacterium*-mediated plant transformation. *Plant Cell, 20*, 2661–2680.

Britt, A. B., & May, G. D. (2003). Re-engineering plant gene targeting. *Trends in Plant Science, 8*, 90–95.

Britton, M. T., Escobar, M. A., & Dandekar, A. M. (2008). The oncogenes of *Agrobacterium tumefaciens* and *Agrobacterium rhizogenes*. In T. Tzfira & V. Citovsky (Eds.), *Agrobacterium* (pp. 524–563). New York: Springer.

Cai, C. Q., Doyon, Y., Ainley, W. M., Miller, J. C., Dekelver, R. C., & Moehle, E. A., et al. (2009). Targeted transgene integration in plant cells using designed zinc finger nucleases. *Plant Molecular Biology, 69*, 699–709.

Camenisch, T. D., Brilliant, M. H., & Segal, D. J. (2008). Critical parameters for genome editing using zinc finger nucleases. *Mini Reviews in Medicinal Chemistry, 8*, 669–676.

Carroll, D. (2004). Using nucleases to stimulate homologous recombination. *Methods in Molecular Biology, 262*, 195–207.

Carroll, D., Morton, J. J., Beumer, K. J., & Segal, D. J. (2006). Design, construction and *in vitro* testing of zinc finger nucleases. *Nature Protocols, 1*, 1329–1341.

Chakrabarty, R., Banerjee, R., Chung, S. -M., Farman, M., Citovsky, V., & Hogenhout, S. A., et al. (2007). pSITE vectors for stable integration or transient expression of autofluorescent protein fusions in plants: Probing *Nicotiana benthamiana*-virus interactions. *Molecular Plant-Microbe Interactions, 20*, 740–750.

Chen, P. -Y., Wang, C. -K., Soong, S. -C., & To, K. -Y. (2003). Complete sequence of the binary vector pBI121 and its application in cloning T-DNA insertion from transgenic plants. *Molecular Breeding, 11*, 287–293.

Chen, Q. J., Xie, M., Ma, X. X., Dong, L., Chen, J., Wang, X. C. (2010) MISSA is a highly efficient *in vivo* DNA assembly method for plant multiple-gene transformation. *Plant Physiology 153*, 41–51.

Chen, Q. J., Zhou, H. M., Chen, J., & Wang, X. C. (2006). A Gateway-based platform for multigene plant transformation. *Plant Molecular Biology, 62*, 927–936.

Cheng, M., Fry, J. E., Pang, S., Zhou, H., Hironaka, C. M., & Duncan, D. R., et al. (1997). Genetic transformation of wheat mediated by *Agrobacterium tumefaciens*. *Plant Physiology, 115*, 971–980.

Cheo, D. L., Titus, S. A., Byrd, D. R., Hartley, J. L., Temple, G. F., & Brasch, M. A. (2004). Concerted assembly and cloning of multiple DNA segments using *in vitro* site-specific recombination: Functional analysis of multi-segment expression clones. *Genome Research, 14*, 2111–2120.

Chilton, M. -D., & Que, Q. (2003). Targeted integration of T-DNA into the tobacco genome at double-strand breaks: New insights on the mechanism of T-DNA integration. *Plant Physiology, 133*, 956–965.

Chiurazzi, M., Ray, A., Viret, J. F., Perera, R., Wang, X. H., & Lloyd, A. M., et al. (1996). Enhancement of somatic intrachromosomal homologous recombination in *Arabidopsis* by the HO endonuclease. *Plant Cell, 8*, 2057–2066.

Christie, P. J., Atmakuri, K., Krishnamoorthy, V., Jakubowski, S., & Cascales, E. (2005). Biogenesis, architecture, and function of bacterial type IV secretion systems. *Annual Review of Microbiology, 59*, 451–485.

Christie, P. J., & Cascales, E. (2005). Structural and dynamic properties of bacterial type IV secretion systems. *Molecular Membrane Biology, 22*, 51–61.

Chung, S. M., Frankman, E. L., & Tzfira, T. (2005). A versatile vector system for multiple gene expression in plants. *Trends in Plant Science, 10*, 357–361.

Citovsky, V., Guralnick, B., Simon, M. N., & Wall, J. S. (1997). The molecular structure of *Agrobacterium* VirE2-single stranded DNA complexes involved in nuclear import. *Journal of Molecular Biology, 271*, 718–727.

Citovsky, V., Kapelnikov, A., Oliel, S., Zakai, N., Rojas, M. R., & Gilbertson, R. L., et al. (2004). Protein interactions involved in nuclear import of the *Agrobacterium* VirE2 protein *in vivo* and *in vitro*. *Journal of Biological Chemistry, 279*, 29528–29533.

Citovsky, V., Kozlovsky, S. V., Lacroix, B., Zaltsman, A., Dafny-Yelin, M., & Vyas, S., et al. (2007). Biological systems of the host cell involved in *Agrobacterium* infection. *Cellular Microbiology, 9*, 9–20.

Citovsky, V., Warnick, D., & Zambryski, P. C. (1994). Nuclear import of *Agrobacterium* VirD2 and VirE2 proteins in maize and tobacco. *Proceedings of the National Academy of Sciences of the United States of America, 91*, 3210–3214.

Citovsky, V., Wong, M. L., & Zambryski, P. C. (1989). Cooperative interaction of *Agrobacterium* VirE2 protein with single stranded DNA: Implications for the T-DNA transfer process. *Proceedings of the National Academy of Sciences of the United States of America, 86*, 1193–1197.

Citovsky, V., Zupan, J., Warnick, D., & Zambryski, P. C. (1992). Nuclear localization of *Agrobacterium* VirE2 protein in plant cells. *Science*, *256*, 1802–1805.

Curtis, M. D., & Grossniklaus, U. (2003). A Gateway cloning vector set for high-throughput functional analysis of genes *in planta*. *Plant Physiology*, *133*, 462–469.

Dafny-Yelin, M., Chung, S. -M., Frankman, E. L., & Tzfira, T. (2007). pSAT RNA interference vectors: A modular series for multiple gene down-regulation in plants. *Plant Physiology*, *145*, 1272–1281.

Dafny-Yelin, M., Levy, A., & Tzfira, T. (2008). The ongoing saga of *Agrobacterium*-host interactions. *Trends in Plant Science*, *13*, 102–105.

Dafny-Yelin, M., & Tzfira, T. (2007). Delivery of multiple transgenes to plant cells. *Plant Physiology*, *145*, 1118–1128.

de Pater, S., Neuteboom, L. W., Pinas, J. E., Hooykaas, P. J., & van der Zaal, B. J. (2009). ZFN-induced mutagenesis and gene-targeting in *Arabidopsis* through *Agrobacterium*-mediated floral dip transformation. *Plant Biotechnology Journal*, *7*, 821–835.

Deng, W., Chen, L., Wood, D. W., Metcalfe, T., Liang, X., & Gordon, M. P., et al. (1998). *Agrobacterium* VirD2 protein interacts with plant host cyclophilins. *Proceedings of the National Academy of Sciences of the United States of America*, *95*, 7040–7045.

Ditt, R. F., Kerr, K. F., de Figueiredo, P., Delrow, J., Comai, L., & Nester, E. W. (2006). The *Arabidopsis thaliana* transcriptome in response to *Agrobacterium tumefaciens*. *Molecular Plant-Microbe Interactions*, *19*, 665–681.

Ditt, R. F., Nester, E. W., & Comai, L. (2001). Plant gene expression response to *Agrobacterium tumefaciens*. *Proceedings of the National Academy of Sciences of the United States of America*, 10954–10959.

Ditt, R. F., Nester, E., & Comai, L. (2005). The plant cell defense and *Agrobacterium tumefaciens*. *FEMS Microbiology Letters*, *247*, 207–213.

Djamei, A., Pitzschke, A., Nakagami, H., Rajh, I., & Hirt, H. (2007). Trojan horse strategy in *Agrobacterium* transformation: Abusing MAPK defense signaling. *Science*, *318*, 453–456.

Durai, S., Mani, M., Kandavelou, K., Wu, J., Porteus, M. H., & Chandrasegaran, S. (2005). Zinc finger nucleases: Custom-designed molecular scissors for genome engineering of plant and mammalian cells. *Nucleic Acids Research*, *33*, 5978–5990.

Earley, K. W., Haag, J. R., Pontes, O., Opper, K., Juehne, T., & Song, K., et al. (2006). Gateway-compatible vectors for plant functional genomics and proteomics. *Plant Journal*, *45*, 616–629.

Ebinuma, H., Sugita, K., Matsunaga, E., & Yamakado, M. (1997). Selection of marker-free transgenic plants using the isopentenyl transferase gene. *Proceedings of the National Academy of Sciences of the United States of America*, *94*, 2117–2121.

Endo, M., Ishikawa, Y., Osakabe, K., Nakayama, S., Kaya, H., & Araki, T., et al. (2006). Increased frequency of homologous recombination and T-DNA integration in *Arabidopsis* CAF-1 mutants. *The EMBO Journal*, *25*, 5579–5590.

Endo, M., Osakabe, K., Ichikawa, H., & Toki, S. (2006). Molecular characterization of true and ectopic gene targeting events at the acetolactate synthase gene in *Arabidopsis*. *Plant and Cell Physiology*, *47*, 372–379.

Endo, M., Osakabe, K., Ono, K., Handa, H., Shimizu, T., & Toki, S. (2007). Molecular breeding of a novel herbicide-tolerant rice by gene targeting. *Plant Journal*, *52*, 157–166.

Erikson, O., Hertzberg, M., & Nasholm, T. (2004). A conditional marker gene allowing both positive and negative selection in plants. *Nature Biotechnology*, *22*, 455–458.

Feng, J., Vick, B. A., Lee, M. K., Zhang, H. B., & Jan, C. C. (2006). Construction of BAC and BIBAC libraries from sunflower and identification of linkage group-specific clones by overgo hybridization. *Theoretical And Applied Genetics*, *113*, 23–32.

Friesner, J., & Britt, A. B. (2003). *Ku80*- and *DNA ligase IV*-deficient plants are sensitive to ionizing radiation and defective in T-DNA integration. *Plant Journal*, *34*, 427–440.

Frisch, D. A., Harris-Haller, L. W., Yokubaitis, N. T., Thomas, T. L., Hardin, S. H., & Hall, T. C. (1995). Complete sequence of the binary vector Bin 19. *Plant Molecular Biology*, *27*, 405–409.

Gallego, M. E., Sirand-Pugnet, P., & White, C. I. (1999). Positive-negative selection and T-DNA stability in *Arabidopsis* transformation. *Plant Molecular Biology*, *39*, 83–93.

Gardiner, D. M., & Howlett, B. J. (2004). Negative selection using thymidine kinase increases the efficiency of recovery of transformants with targeted genes in the filamentous fungus *Leptosphaeria maculans*. *Current Genetics*, *45*, 249–255.

Gaspar, Y. M., Nam, J., Schultz, C. J., Lee, L. Y., Gilson, P. R., & Gelvin, S. B., et al. (2004). Characterization of the *Arabidopsis* lysine-rich arabinogalactan-protein AtAGP17 mutant (*rat1*) that results in a decreased efficiency of *Agrobacterium* transformation. *Plant Physiology*, *135*, 2162–2171.

Gasparis, S., Bregier, C., Orczyk, W., & Nadolska-Orczyk, A. (2008). *Agrobacterium*-mediated transformation of oat (*Avena sativa* L.) cultivars via immature embryo and leaf explants. *Plant Cell Reports*, *27*, 1721–1729.

Gehl, C., Waadt, R., Kudla, J., Mendel, R. R., & Hansch, R. (2009). New GATEWAY vectors for high throughput analyses of protein–protein interactions by bimolecular fluorescence complementation. *Molecular Plant*, *2*, 1051–1058.

Gelvin, S. B. (2003a). Improving plant genetic engineering by manipulating the host. *Trends in Biotechnology*, *21*, 95–98.

Gelvin, S. B. (2003b). *Agrobacterium*-mediated plant transformation: The biology behind the "gene-jockeying" tool. *Microbiology and Molecular Biology Reviews*, *67*, 16–37.

Gelvin, S. B. (2009). *Agrobacterium* in the genomics age. *Plant Physiology*, *150*, 1665–1676.

Gelvin, S. B. (2010). Plant proteins involved in *Agrobacterium*-mediated genetic transformation. *Annual Review of Phytopathology*, *48*, 45–68.

Gelvin, S. B., & Kim, S. I. (2007). Effect of chromatin upon *Agrobacterium* T-DNA integration and transgene expression. *Biochimica et Biophysica Acta*, *1769*, 410–421.

Gherbi, H., Gallego, M. E., Jalut, N., Lucht, J. M., Hohn, B., & White, C. I. (2001). Homologous recombination *in planta* is stimulated in the absence of Rad50. *EMBO Reports*, *2*, 287–291.

Goderis, I. J., De Bolle, M. F., Francois, I. E., Wouters, P. F., Broekaert, W. F., & Cammue, B. P. (2002). A set of modular plant transformation vectors allowing flexible insertion of up to six expression units. *Plant Molecular Biology*, *50*, 17–27.

Hajdukiewicz, P., Svab, Z., & Maliga, P. (1994). The small, versatile pPZP family of *Agrobacterium* binary vectors for plant transformation. *Plant Molecular Biology*, *25*, 989–994.

Hamilton, C. M. (1997). A binary-BAC system for plant transformation with high-molecular-weight DNA. *Gene*, *200*, 107–116.

Hamilton, C. M., Frary, A., Lewis, C., & Tanksley, S. D. (1996). Stable transfer of intact high molecular weight DNA into plant chromosomes. *Proceedings of the National Academy of Sciences of the United States of America*, *93*, 9975–9979.

Hanin, M., Mengiste, T., Bogucki, A., & Paszkowski, J. (2000). Elevated levels of intrachromosomal homologous recombination in *Arabidopsis* overexpressing the MIM gene. *Plant Journal*, *24*, 183–189.

Hanin, M., Volrath, S., Bogucki, A., Briker, M., Ward, E., & Paszkowski, J. (2001). Gene targeting in *Arabidopsis*. *Plant Journal*, *28*, 671–677.

He, R. F., Wang, Y. Y., Du, B., Tang, M., You, A. Q., & Zhu, L. L., et al. (2006). Development of transformation system of rice based on binary bacterial artificial chromosome (BIBAC) vector. *Yi Chuan Xue Bao*, *33*, 269–276.

He, Y., Jones, H. D., Chen, S., Chen, X. M., Wang, D. W., & Li, K. X., et al. (2010). *Agrobacterium*-mediated transformation of durum wheat (*Triticum turgidum* L. var. durum cv Stewart) with improved efficiency. *Journal of Experimental Botany*, *61*, 1567–1581.

Hellens, R. P., Edwards, E. A., Leyland, N. R., Bean, S., & Mullineaux, P. M. (2000). pGreen: A versatile and flexible binary Ti vector for *Agrobacterium*-mediated plant transformation. *Plant Molecular Biology*, *42*, 819–832.

Hellens, R., Mullineaux, P., & Klee, H. (2000). Technical Focus: A guide to *Agrobacterium* binary Ti vectors. *Trends in Plant Science*, *5*, 446–451.

Helliwell, C., & Waterhouse, P. (2003). Constructs and methods for high-throughput gene silencing in plants. *Methods*, *30*, 289–295.

Herrera-Estrella, L., Simpson, J., & Martinez-Trujillo, M. (2005). Transgenic plants: An historical perspective. *Methods in Molecular Biology*, *286*, 3–32.

Hoekema, A., Hirsch, P. R., Hooykaas, P. J. J., & Schilperoort, R. A. (1983). A binary plant vector strategy based on separation of *vir*- and T-region of the *Agrobacterium tumefaciens* Ti-plasmid. *Nature, 303*, 179–180.

Hohn, B., & Puchta, H. (1999). Gene therapy in plants. *Proceedings of the National Academy of Sciences of the United States of America, 96*, 8321–8323.

Hood, E. E., Gelvin, S. B., Melchers, L. S., & Hoekema, A. (1993). New *Agrobacterium* helper plasmids for gene transfer to plants. *Transgenic Research, 2*, 33–50.

Hood, E. E., Helmer, G. L., Fraley, R. T., & Chilton, M. D. (1986). The hypervirulence of *Agrobacterium tumefaciens* A281 is encoded in a region of pTiBo542 outside of T-DNA. *Journal of Bacteriology, 168*, 1291–1301.

Hooykaas, P. J. J., & Beijersbergen, A. G. M. (1994). The virulence system of *Agrobacterium tumefaciens*. *Annual Review of Phytopathology, 32*, 157–179.

Hrouda, M., & Paszkowski, J. (1994). High fidelity extrachromosomal recombination and gene targeting in plants. *Molecular & General Genetics, 243*, 106–111.

Husaini, A. M. (2009). Pre- and post-agroinfection strategies for efficient leaf disk transformation and regeneration of transgenic strawberry plants. *Plant Cell Reports, 29*, 97–110.

Husaini, A. M., & Abdin, M. Z. (2008). Development of transgenic strawberry (*Fragaria* × *ananassa* Duch.) plants tolerant to salt stress. *Plant Science, 174*, 446–455.

Hwang, H. H., & Gelvin, S. B. (2004). Plant proteins that interact with VirB2, the *Agrobacterium tumefaciens* pilin protein, mediate plant transformation. *Plant Cell, 16*, 3148–3167.

Iida, S., & Terada, R. (2005). Modification of endogenous natural genes by gene targeting in rice and other higher plants. *Plant Molecular Biology, 59*, 205–219.

Jefferson, R. A., Kavanagh, T. A., & Bevan, M. W. (1987). GUS fusions: Beta-glucuronidase as a sensitive and versatile gene fusion marker in higher plants. *The EMBO Journal, 6*, 3901–3907.

Johzuka-Hisatomi, Y., Terada, R., & Iida, S. (2008). Efficient transfer of base changes from a vector to the rice genome by homologous recombination: Involvement of heteroduplex formation and mismatch correction. *Nucleic Acids Research, 36*, 4727–4735.

Karimi, M., De Meyer, B., & Hilson, P. (2005). Modular cloning in plant cells. *Trends in Plant Science, 10*, 103–105.

Karimi, M., Depicker, A., & Hilson, P. (2007). Recombinational cloning with plant gateway vectors. *Plant Physiology, 145*, 1144–1154.

Kirik, A., Pecinka, A., Wendeler, E., & Reiss, B. (2006). The chromatin assembly factor subunit FASCIATA1 is involved in homologous recombination in plants. *Plant Cell, 18*, 2431–2442.

Koizumi, N., & Iwata, Y. (2008). Construction of a binary vector for transformation of *Arabidopsis thaliana* with a new selection marker. *Bioscience, Biotechnology, and Biochemistry, 72*, 3041–3043.

Kole, C., & Hall, T. C. (2008). *Compendium of transgenic crop plants, 10 volume set*. Wiley-Blackwell.

Komari, T., Takakura, Y., Ueki, J., Kato, N., Ishida, Y., & Hiei, Y. (2006). Binary vectors and super-binary vectors. *Methods in Molecular Biology, 343*, 15–41.

Koprek, T., McElroy, D., Louwerse, J., Williams-Carrier, R., & Lemaux, P. G. (1999). Negative selection systems for transgenic barley (*Hordeum vulgare* L.): Comparison of bacterial *codA*- and cytochrome P450 gene-mediated selection. *Plant Journal, 19*, 719–726.

Kumar, S., Allen, G. C., & Thompson, W. F. (2006). Gene targeting in plants: Fingers on the move. *Trends in Plant Science, 11*, 159–161.

Kumar, S., & Fladung, M. (2001). Controlling transgene integration in plants. *Trends in Plant Science, 6*, 155–159.

Lacroix, B., & Citovsky, V. (2009). *Agrobacterium* aiming for the host chromatin: Host and bacterial proteins involved in interactions between T-DNA and plant nucleosomes. *Communicative & Integrative Biology, 2*, 42–45.

Lacroix, B., Elbaum, M., Citovsky, V., & Tzfira, T. (2008a). Intracellular transport of *Agrobacterium* T-DNA. In T. Tzfira & V. Citovsky (Eds.), *Agrobacterium* (pp. 365–394). New York: Springer.

Lacroix, B., Loyter, A., & Citovsky, V. (2008b). Association of the *Agrobacterium* T-DNA-protein complex with plant nucleosomes. *Proceedings of the National Academy of Sciences of the United States of America, 105*, 15429–15434.

Lacroix, B., Tzfira, T., Vainstein, A., & Citovsky, V. (2006). A case of promiscuity: *Agrobacterium*'s endless hunt for new partners. *Trends in Genetics, 22*, 29–37.

Lacroix, B., Vaidya, M., Tzfira, T., & Citovsky, V. (2005). The VirE3 protein of *Agrobacterium* mimics a host cell function required for plant genetic transformation. *The EMBO Journal, 24*, 428–437.

Lee, K. Y., Lund, P., Lowe, K., & Dunsmuir, P. (1990). Homologous recombination in plant cells after *Agrobacterium*-mediated transformation. *Plant Cell, 2*, 415–425.

Lee, L. Y., & Gelvin, S. B. (2008). T-DNA binary vectors and systems. *Plant Physiology, 146*, 325–332.

Lee, L. Y., Kononov, M. E., Bassuner, B., Frame, B. R., Wang, K., & Gelvin, S. B. (2007). Novel plant transformation vectors containing the superpromoter. *Plant Physiology, 145*, 1294–1300.

Li, J., Krichevsky, A., Vaidya, M., Tzfira, T., & Citovsky, V. (2005). Uncoupling of the functions of the *Arabidopsis* VIP1 protein in transient and stable plant genetic transformation by *Agrobacterium*. *Proceedings of the National Academy of Sciences of the United States of America, 102*, 5733–5738.

Li, J., Vaidya, M., White, C., Vainstein, A., Citovsky, V., & Tzfira, T. (2005). Involvement of KU80 in T-DNA integration in plant cells. *Proceedings of the National Academy of Sciences of the United States of America, 102*, 19231–19236.

Li, Y., Uhm, T., Ren, C., Wu, C., Santos, T. S., & Lee, M. K., et al. (2007). A plant-transformation-competent BIBAC/BAC-based map of rice for functional analysis and genetic engineering of its genomic sequence. *Genome, 50*, 278–288.

Lin, L., Liu, Y. G., Xu, X., & Li, B. (2003). Efficient linking and transfer of multiple genes by a multigene assembly and transformation vector system. *Proceedings of the National Academy of Sciences of the United States of America, 100*, 5962–5967.

Liu, Y. G., Shirano, Y., Fukaki, H., Yanai, Y., Tasaka, M., & Tabata, S., et al. (1999). Complementation of plant mutants with large genomic DNA fragments by a transformation-competent artificial chromosome vector accelerates positional cloning. *Proceedings of the National Academy of Sciences of the United States of America, 96*, 6535–6540.

Lloyd, A., Plaisier, C. L., Carroll, D., & Drews, G. N. (2005). Targeted mutagenesis using zinc-finger nucleases in *Arabidopsis*. *Proceedings of the National Academy of Sciences of the United States of America, 102*, 2232–2237.

Mansour, S. L., Thomas, K. R., & Capecchi, M. R. (1988). Disruption of the proto-oncogene int-2 in mouse embryo-derived stem cells: A general strategy for targeting mutations to non-selectable genes. *Nature, 336*, 348–352.

Miao, Z. H., & Lam, E. (1995). Targeted disruption of the TGA3 locus in *Arabidopsis thaliana*. *Plant Journal, 7*, 359–365.

Michielse, C. B., Hooykaas, P. J., van den Hondel, C. A., & Ram, A. F. (2005). *Agrobacterium*-mediated transformation as a tool for functional genomics in fungi. *Current Genetics, 48*, 1–17.

Miki, B., & McHugh, S. (2004). Selectable marker genes in transgenic plants: Applications, alternatives and biosafety. *Journal of Biotechnology, 107*, 193–232.

Molinier, J., Ramos, C., Fritsch, O., & Hohn, B. (2004). CENTRIN2 modulates homologous recombination and nucleotide excision repair in *Arabidopsis*. *Plant Cell, 16*, 1633–1643.

Mysore, K. S., Nam, J., & Gelvin, S. B. (2000). An *Arabidopsis* histone H2A mutant is deficient in *Agrobacterium* T-DNA integration. *Proceedings of the National Academy of Sciences of the United States of America, 97*, 948–953.

Nam, J., Mysore, K. S., Zheng, C., Knue, M. K., Matthysse, A. G., & Gelvin, S. B. (1999). Identification of T-DNA tagged *Arabidopsis* mutants that are resistant to transformation by *Agrobacterium*. *Molecular & General Genetics, 261*, 429–438.

Narasimhulu, S. B., Deng, X. -B., Sarria, R., & Gelvin, S. B. (1996). Early transcription of *Agrobacterium* T-DNA genes in tobacco and maize. *Plant Cell, 8*, 873–886.

Offringa, R., de Groot, M. J., Haagsman, H. J., Does, M. P., van den Elzen, P. J., & Hooykaas, P. J. (1990). Extrachromosomal homologous recombination and gene targeting in plant cells after *Agrobacterium* mediated transformation. *The EMBO Journal*, *9*, 3077–3084.

Orel, N., Kyryk, A., & Puchta, H. (2003). Different pathways of homologous recombination are used for the repair of double-strand breaks within tandemly arranged sequences in the plant genome. *Plant Journal*, *35*, 604–612.

Osakabe, K., Osakabe, Y., & Toki, S. (2010). Site-directed mutagenesis in *Arabidopsis* using custom-designed zinc finger nucleases. *Proceedings of the National Academy of Sciences of the United States of America*, *107*, 12034–12039.

Paques, F., & Duchateau, P. (2007). Meganucleases and DNA double-strand break-induced recombination: Perspectives for gene therapy. *Current Gene Therapy*, *7*, 49–66.

Paszkowski, J., Baur, M., Bogucki, A., & Potrykus, I. (1988). Gene targeting in plants. *The EMBO Journal*, *7*, 4021–4026.

Pitzschke, A., & Hirt, H. (2010). New insights into an old story: *Agrobacterium*-induced tumour formation in plants by plant transformation. *The EMBO Journal*, *29*, 1021–1032.

Porteus, M. H. (2009). Plant biotechnology: Zinc fingers on target. *Nature*, *459*, 337–338.

Porteus, M. H., & Carroll, D. (2005). Gene targeting using zinc finger nucleases. *Nature Biotechnology*, *23*, 967–973.

Przetakiewicz, A., Karas, A., Orczyk, W., & Nadolska-Orczyk, A. (2004). *Agrobacterium*-mediated transformation of polyploid cereals. The efficiency of selection and transgene expression in wheat. *Cellular & Molecular Biology Letters*, *9*, 903–917.

Puchta, H. (2002). Gene replacement by homologous recombination in plants. *Plant Molecular Biology*, *48*, 173–182.

Puchta, H. (2003). Towards the ideal GMP: Homologous recombination and marker gene excision. *Journal of Plant Physiology*, *160*, 743–754.

Puchta, H. (2005). The repair of double-strand breaks in plants: Mechanisms and consequences for genome evolution. *Journal of Experimental Botany*, *56*, 1–14.

Puchta, H., Dujon, B., & Hohn, B. (1996). Two different but related mechanisms are used in plants for the repair of genomic double-strand breaks by homologous recombination. *Proceedings of the National Academy of Sciences of the United States of America*, *93*, 5055–5060.

Puchta, H., & Hohn, B. (2005). Green light for gene targeting in plants. *Proceedings of the National Academy of Sciences of the United States of America*, *102*, 11961–11962.

Reiss, B., Schubert, I., Kopchen, K., Wendeler, E., Schell, J., & Puchta, H. (2000). RecA stimulates sister chromatid exchange and the fidelity of double-strand break repair, but not gene targeting, in plants transformed by *Agrobacterium*. *Proceedings of the National Academy of Sciences of the United States of America*, *97*, 3358–3363.

Risseeuw, E., Offringa, R., Franke-van Dijk, M. E., & Hooykaas, P. J. (1995). Targeted recombination in plants using *Agrobacterium* coincides with additional rearrangements at the target locus. *Plant Journal*, *7*, 109–119.

Salman, H., Abu-Arish, A., Oliel, S., Loyter, A., Klafter, J., & Granel, R., et al. (2005). Nuclear localization signal peptides induce molecular delivery along microtubules. *Biophysical Journal*, *89*, 2134–2145.

Salomon, S., & Puchta, H. (1998). Capture of genomic and T-DNA sequences during double-strand break repair in somatic plant cells. *The EMBO Journal*, *17*, 6086–6095.

Sasaki, Y., Sone, T., Yoshida, S., Yahata, K., Hotta, J., & Chesnut, J. D., et al. (2004). Evidence for high specificity and efficiency of multiple recombination signals in mixed DNA cloning by the Multisite Gateway system. *Journal of Biotechnology*, *107*, 233–243.

Schrammeijer, B., Risseeuw, E., Pansegrau, W., Regensburg-Tuïnk, T. J. G., Crosby, W. L., & Hooykaas, P. J. J. (2001). Interaction of the virulence protein VirF of *Agrobacterium tumefaciens* with plant homologs of the yeast Skp1 protein. *Current Biology*, *11*, 258–262.

Shaked, H., Melamed-Bessudo, C., & Levy, A. A. (2005). High frequency gene targeting in *Arabidopsis* plants expressing the yeast RAD54 gene. *Proceedings of the National Academy of Sciences of the United States of America*, *102*, 12265–12269.

Shalev, G., Sitrit, Y., Avivi-Ragolski, N., Lichtenstein, C., & Levy, A. A. (1999). Stimulation of homologous recombination in plants by expression of the bacterial resolvase *ruvC*. *Proceedings of the National Academy of Sciences of the United States of America*, *96*, 7398–7402.

Sheikholeslam, S. N., & Weeks, D. P. (1987). Acetosyringone promotes high efficiency transformation of *Arabidopsis thaliana* explants by *Agrobacterium tumefaciens*. *Plant Cell Reports*, *8*, 291–298.

Shukla, V. K., Doyon, Y., Miller, J. C., Dekelver, R. C., Moehle, E. A., & Worden, S. E., et al. (2009). Precise genome modification in the crop species *Zea mays* using zinc-finger nucleases. *Nature*, *459*, 437–441.

Shurvinton, C. E., Hodges, L., & Ream, L. W. (1992). A nuclear localization signal and the C-terminal omega sequence in the *Agrobacterium tumefaciens* VirD2 endonuclease are important for tumor formation. *Proceedings of the National Academy of Sciences of the United States of America*, *89*, 11837–11841.

Siebert, R., & Puchta, H. (2002). Efficient repair of genomic double-strand breaks by homologous recombination between directly repeated sequences in the plant genome. *Plant Cell*, *14*, 1121–1131.

Soltani, J., van Heusden, G. P. H., & Hooykaas, P. J. (2008). *Agrobacterium*-mediated transformation of non-plant organisms. In T. Tzfira & V. Citovsky (Eds.), *Agrobacterium* (pp. 650–674). New York: Springer.

Song, J., Bradeen, J. M., Naess, S. K., Helgeson, J. P., & Jiang, J. (2003). BIBAC and TAC clones containing potato genomic DNA fragments larger than 100 kb are not stable in *Agrobacterium*. *Theoretical and Applied Genetics*, *107*, 958–964.

Stachel, S. E., Messens, E., Van Montagu, M., & Zambryski, P. C. (1985). Identification of the signal molecules produced by wounded plant cell that activate T-DNA transfer in *Agrobacterium tumefaciens*. *Nature*, *318*, 624–629.

Sundar, I. K., & Sakthivel, N. (2008). Advances in selectable marker genes for plant transformation. *Journal of Plant Physiology*, *165*, 1698–1716.

Takeda, S., Tadele, Z., Hofmann, I., Probst, A. V., Angelis, K. J., & Kaya, H., et al. (2004). BRU1, a novel link between responses to DNA damage and epigenetic gene silencing in *Arabidopsis*. *Genes & Development*, *18*, 782–793.

Tanaka, S., Ishii, C., Hatakeyama, S., & Inoue, H. (2010). High efficient gene targeting on the AGAMOUS gene in an *Arabidopsis* AtLIG4 mutant. *Biochemical and Biophysical Research Communications*, *396*, 289–293.

Tao, Q., & Zhang, H. B. (1998). Cloning and stable maintenance of DNA fragments over 300 kb in *Escherichia coli* with conventional plasmid-based vectors. *Nucleic Acids Research*, *26*, 4901–4909.

Tao, Y., Rao, P. K., Bhattacharjee, S., & Gelvin, S. B. (2004). Expression of plant protein phosphatase 2C interferes with nuclear import of the *Agrobacterium* T-complex protein VirD2. *Proceedings of the National Academy of Sciences of the United States of America*, *101*, 5164–5169.

Tenea, G. N., Spantzel, J., Lee, L. Y., Zhu, Y., Lin, K., & Johnson, S. J., et al. (2009). Overexpression of several *Arabidopsis* histone genes increases *Agrobacterium*-mediated transformation and transgene expression in plants. *Plant Cell*, *21*, 3350–3367.

Terada, R., Johzuka-Hisatomi, Y., Saitoh, M., Asao, H., & Iida, S. (2007). Gene targeting by homologous recombination as a biotechnological tool for rice functional genomics. *Plant Physiology*, *144*, 846–856.

Terada, R., Urawa, H., Inagaki, Y., Tsugane, K., & Iida, S. (2002). Efficient gene targeting by homologous recombination in rice. *Nature Biotechnology*, *20*, 1030–1034.

Thole, V., Worland, B., Snape, J. W., & Vain, P. (2007). The pCLEAN dual binary vector system for *Agrobacterium*-mediated plant transformation. *Plant Physiology*, *145*, 1211–1219.

Thykjaer, T., Finnemann, J., Schauser, L., Christensen, L., Poulsen, C., & Stougaard, J. (1997). Gene targeting approaches using positive-negative selection and large flanking regions. *Plant Molecular Biology*, *35*, 523–530.

Tovar, J., & Lichtenstein, C. (1992). Somatic and meiotic chromosomal recombination between inverted duplications in transgenic tobacco plants. *Plant Cell*, *4*, 319–332.

Tovkach, A., Zeevi, V., & Tzfira, T. (2009). A toolbox and procedural notes for characterizing novel zinc finger nucleases for genome editing in plant cells. *Plant Journal*, *57*, 747–757.

Townsend, J. A., Wright, D. A., Winfrey, R. J., Fu, F., Maeder, M. L., & Joung, J. K., et al. (2009). High-frequency modification of plant genes using engineered zinc-finger nucleases. *Nature*, *459*, 442–445.

Tzfira, T., & Citovsky, V. (2001). Comparison between nuclear import of nopaline- and octopine-specific VirE2 protein of *Agrobacterium* in plant and animal cells. *Molecular Plant Pathology, 2*, 171–176.

Tzfira, T., & Citovsky, V. (2002). Partners-in-infection: Host proteins involved in the transformation of plant cells by *Agrobacterium. Trends in Cell Biology, 12*, 121–129.

Tzfira, T., & Citovsky, V. (2006). *Agrobacterium*-mediated genetic transformation of plants: Biology and biotechnology. *Current Opinion in Biotechnology, 17*, 147–154.

Tzfira, T., Frankmen, L., Vaidya, M., & Citovsky, V. (2003). Site-specific integration of *Agrobacterium tumefaciens* T-DNA *via* double-stranded intermediates. *Plant Physiology, 133*, 1011–1023.

Tzfira, T., Lacroix, B., & Citovsky, V. (2005). Nuclear Import of *Agrobacterium* T-DNA. In T. Tzfira & V. Citovsky (Eds.), *Nuclear Import and Export in Plants and Animals* (pp. 83–99). New York: Landes Bioscience/ Kluwer Academic/Plenum Publishers.

Tzfira, T., Li, J., Lacroix, B., & Citovsky, V. (2004). *Agrobacterium* T-DNA integration: Molecules and models. *Trends in Genetics, 20*, 375–383.

Tzfira, T., Tian, G. -W., Lacroix, B., Vyas, S., Li, J., & Leitner-Dagan, Y., et al. (2005). pSAT vectors: A modular series of plasmids for autofluorescent protein tagging and expression of multiple genes in plants. *Plant Molecular Biology, 57*, 503–516.

Tzfira, T., Vaidya, M., & Citovsky, V. (2001). VIP1, an *Arabidopsis* protein that interacts with *Agrobacterium* VirE2, is involved in VirE2 nuclear import and *Agrobacterium* infectivity. *The EMBO Journal, 20*, 3596–3607.

Tzfira, T., Vaidya, M., & Citovsky, V. (2002). Increasing plant susceptibility to *Agrobacterium* infection by overexpression of the *Arabidopsis* nuclear protein VIP1. *Proceedings of the National Academy of Sciences of the United States of America, 99*, 10435–10440.

Tzfira, T., Vaidya, M., & Citovsky, V. (2004). Involvement of targeted proteolysis in plant genetic transformation by *Agrobacterium. Nature, 431*, 87–92.

van Attikum, H., Bundock, P., & Hooykaas, P. J. J. (2001). Non-homologous end-joining proteins are required for *Agrobacterium* T-DNA integration. *The EMBO Journal, 20*, 6550–6558.

van Attikum, H., & Hooykaas, P. J. J. (2003). Genetic requirements for the targeted integration of *Agrobacterium* T-DNA in *Saccharomyces cerevisiae. Nucleic Acids Research, 31*, 826–832.

van Engelen, F. A., Molthoff, J. W., Conner, A. J., Nap, J. P., Pereira, A., & Stiekema, W. J. (1995). pBINPLUS: An improved plant transformation vector based on pBIN19. *Transgenic Research, 4*, 288–290.

Veena, , Jiang, H., Doerge, R. W., & Gelvin, S. B. (2003). Transfer of T-DNA and Vir proteins to plant cells by *Agrobacterium tumefaciens* induces expression of host genes involved in mediating transformation and suppresses host defense gene expression. *Plant Journal, 35*, 219–236.

Vega, J. M., Yu, W., Han, F., Kato, A., Peters, E. M., & Zhang, Z. J., et al. (2008). *Agrobacterium*-mediated transformation of maize (*Zea mays*) with Cre-lox site specific recombination cassettes in BIBAC vectors. *Plant Molecular Biology, 66*, 587–598.

Vergunst, A. C., Schrammeijer, B., den Dulk-Ras, A., de Vlaam, C. M. T., Regensburg-Tuink, T. J., & Hooykaas, P. J. J. (2000). VirB/D4-dependent protein translocation from *Agrobacterium* into plant cells. *Science, 290*, 979–982.

Vergunst, A. C., van Lier, M. C., den Dulk-Ras, A., Grosse Stuve, T. A., Ouwehand, A., & Hooykaas, P. J. (2005). Positive charge is an important feature of the C-terminal transport signal of the VirB/D4-translocated proteins of *Agrobacterium. Proceedings of the National Academy of Sciences of the United States of America, 102*, 832–837.

Wakasa, Y., Yasuda, H., & Takaiwa, F. (2006). High accumulation of bioactive peptide in transgenic rice seeds by expression of introduced multiple genes. *Plant Biotechnology Journal, 4*, 499–510.

Wang, K. (2006a). *Agrobacterium protocols* (Vol. 1). Humana Press.

Wang, K. (2006b). *Agrobacterium protocols* (Vol. 2). Humana Press.

Wang, K., Stachel, S. E., Timmerman, B., Van Montagu, M., & Zambryski, P. C. (1987). Site-specific nick occurs within the 25 bp transfer promoting border sequence following induction of vir gene expression in *Agrobacterium tumefaciens. Science, 235*, 587–591.

Weinthal, D., Tovkach, A., Zeevi, V., & Tzfira, T. (2010). Genome editing in plant cells by zinc finger nucleases. *Trends in Plant Science, 15*, 308–321.

West, C. E., Waterworth, W. M., Story, G. W., Sunderland, P. A., Jiang, Q., & Bray, C. M. (2002). Disruption of the *Arabidopsis AtKu80* gene demonstrates an essential role for AtKu80 protein in efficient repair of DNA double-strand breaks *in vivo. Plant Journal, 31*, 517–528.

Wright, D. A., Thibodeau-Beganny, S., Sander, J. D., Winfrey, R. J., Hirsh, A. S., & Eichtinger, M., et al. (2006). Standardized reagents and protocols for engineering zinc finger nucleases by modular assembly. *Nature Protocols, 1*, 1637–1652.

Wu, J., Kandavelou, K., & Chandrasegaran, S. (2007). Custom-designed zinc finger nucleases: What is next? *Cellular and Molecular Life Sciences, 64*, 2933–2944.

Xiang, C., Han, P., Lutziger, I., Wang, K., & Oliver, D. J. (1999). A mini binary vector series for plant transformation. *Plant Molecular Biology, 40*, 711–717.

Xiao, Y. L., & Peterson, T. (2000). Intrachromosomal homologous recombination in *Arabidopsis* induced by a maize transposon. *Molecular & General Genetics, 263*, 22–29.

Xiaohui Wang, H., Viret, J. F., Eldridge, A., Perera, R., Signer, E. R., & Chiurazzi, M. (2001). Positive-negative selection for homologous recombination in *Arabidopsis. Gene, 272*, 249–255.

Yamauchi, T., Johzuka-Hisatomi, Y., Fukada-Tanaka, S., Terada, R., Nakamura, I., & Iida, S. (2009). Homologous recombination-mediated knock-in targeting of the MET1a gene for a maintenance DNA methyltransferase reproducibly reveals dosage-dependent spatiotemporal gene expression in rice. *Plant Journal, 60*, 386–396.

Yuan, Z. C., Edlind, M. P., Liu, P., Saenkham, P., Banta, L. M., & Wise, A. A., et al. (2007). The plant signal salicylic acid shuts down expression of the *vir* regulon and activates quormone-quenching genes in *Agrobacterium. Proceedings of the National Academy of Sciences of the United States of America, 104*, 11790–11795.

Zaltsman, A., Krichevsky, A., Loyter, A., & Citovsky, V. (2010). *Agrobacterium* induces expression of a host F-box protein required for tumorigenicity. *Cell Host & Microbe, 7*, 197–209.

Zaltsman, A., Yi, B. Y., Krichevsky, A., Gafni, Y., & Citovsky, V. (2007). Yeast-plant coupled vector system for identification of nuclear proteins. *Plant Physiology, 145*, 1264–1271.

Zambryski, P. C., Joos, H., Genetello, C., Leemans, J., Van Montagu, M., & Schell, J. (1983). Ti plasmid vector for the introduction of DNA into plant cells without alteration of their normal regeneration capacity. *The EMBO Journal, 2*, 2143–2150.

Zeevi, V., Tovkach, A., & Tzfira, T. (2008). Increasing cloning possibilities using artificial zinc finger nucleases. *Proceedings of the National Academy of Sciences of the United States of America, 105*, 12785–12790.

Zeevi, V., Tovkach, A., & Tzfira, T. (2010). Artificial zinc finger nucleases for DNA cloning. *Methods in Molecular Biology, 649*, 209–225.

Zhang, X., Henriques, R., Lin, S. S., Niu, Q. W., & Chua, N. H. (2006). *Agrobacterium*-mediated transformation of *Arabidopsis thaliana* using the floral dip method. *Nature Protocols, 1*, 641–646.

Zheng, Y., He, X. W., Ying, Y. H., Lu, J. F., Gelvin, S. B., & Shou, H. X. (2009). Expression of the *Arabidopsis thaliana* histone gene AtHTA1 enhances rice transformation efficiency. *Molecular Plant, 2*, 832–837.

Zhu, Y., Nam, J., Humara, J. M., Mysore, K. S., Lee, L. Y., & Cao, H., et al. (2003). Identification of *Arabidopsis rat* mutants. *Plant Physiology, 132*, 494–505.

Ziemienowicz, A., Görlich, D., Lanka, E., Hohn, B., & Rossi, L. (1999). Import of DNA into mammalian nuclei by proteins originating from a plant pathogenic bacterium. *Proceedings of the National Academy of Sciences of the United States of America, 96*, 3729–3733.

Ziemienowicz, A., Merkle, T., Schoumacher, F., Hohn, B., & Rossi, L. (2001). Import of *Agrobacterium* T-DNA into plant nuclei: Two distinct functions of VirD2 and VirE2 proteins. *Plant Cell, 13*, 369–384.

Ziemienowicz, A., Tzfira, T., & Hohn, B. (2008). Mechanisms of T-DNA integration. In T. Tzfira & V. Citovsky (Eds.),

Agrobacterium (pp. 395–440). New York: Springer.

Zipfel, C., Kunze, G., Chinchilla, D., Caniard, A., Jones, J. D., & Boller, T., et al. (2006). Perception of the bacterial PAMP EF-Tu by the receptor EFR restricts *Agrobacterium*-mediated transformation. *Cell*, *125*, 749–760.

Zupan, J., Citovsky, V., & Zambryski, P. C. (1996). *Agrobacterium* VirE2 protein mediates nuclear uptake of ssDNA in plant cells. *Proceedings of the National Academy of Sciences of the United States of America*, *93*, 2392–2397.

Biolistic and other non-*Agrobacterium* technologies of plant transformation

8

Tracie K. Matsumoto Dennis Gonsalves
USDA, ARS, Pacific Basin Agricultural Research Center, Hilo, Hawaii

TABLE OF CONTENTS

Introduction

The ability to insert foreign DNA into plants is viewed as one of the major advances in modern agriculture. The elegant and classical studies on *Agrobacterium tumefaciens* provided the foundation that has made *Agrobacterium* the dominant manner by which plants are transformed. However, due to the initial restrictions of host specificity and other restraints of *Agrobacterium*-mediated transformation, other methods for transforming plants have been developed and have proven useful for fundamental and practical purposes. Although recently reclassified as *Rhizobium radiobacter* (Young, 2001), due to the history and universal use of the term *Agrobacterium*, we will continue to refer to this method of transformation as *Agrobacterium*-mediated. Transformation via *Agrobacterium* has been reviewed in Chapter 7 by Dr. Tzvi Tzfira. This chapter deals with an overview of the non-*Agrobacterium* transformation approaches.

Many methods have been described to introduce foreign DNA into plant cells. Here we will cover some of the methods that have been successfully used to create transgenic plants. Methods such as electrophoretic transfection, electroporation of intact plant cells and tissues, bioactive-bead-mediated gene transfer, microinjection, pollen-tube pathway, and silica carbide-mediated transformation will briefly be discussed first. These methods have been successful in the laboratory, but there are very few examples where these techniques have been applied on a commercial basis. Biolistics has served as an excellent approach for transforming monocots and recalcitrant agronomic crops such as soybeans. Indeed, a number of the high acreage transgenic crops grown in the world has been transformed by biolistics (James, 2009). Since biolistics is the most dominant non-*Agrobacterium* alternative approach, it will be the primary focus of this chapter.

Other Non-*Agrobacterium* Transformation

Transformation of plants by *A. tumefaciens* and biolistics are, by far, the most widely used methods. However, a number of novel transformation approaches that do not use *Agrobacterium* or biolistics have been developed. Some of them are useful in specific cases. These alternative approaches have been nicely summarized in a recent review by Rao et al. (2009).

Electrophoretic transfection

Theoretically, stable integration of foreign DNA into the meristematic tissue of plants could be a very easy and inexpensive way to transform plants. Indeed, Ahokas (1989) described a process, which he coined electrophoretic transfection, to electrophoretically introduce foreign DNA into germinating

DOI: 10.1016/B978-0-12-381466-1.00008-0

barley seeds. They showed transient expression of the GUS gene in the seeds, but stable expression was not demonstrated. Perhaps the only successful case for this approach has been with transformation of protocorms of the orchid *Calanthe sedenii* (Griesbach, 1994). This method was reproduced largely as described by Griesbach (1994). The details provide the rationale for the approach and at the same time reveal what makes this approach successful. To prepare material, immature ovules are excised and placed in tissue culture media to develop into protocorms, which contain the apical meristem that eventually will develop into plants. Protocorms approximately 2 mm in diameter were subjected to electrophoresis with plasmid DNA containing the GUS gene and control elements. The DNA is dissolved in agarose and filled in the tip of a 1 ml plastic pipette. The tip of a second pipette is filled with agarose without DNA. The pipettes are then filled with conductivity buffer. The DNA pipette is placed over the apical meristem of the protocorm and the buffer pipette is placed on the base of the protocorm. Throughout the procedure a constant electrical current was applied. After treatment, the protocorms were replated in fresh media and subsequently raised to plants growing under greenhouse conditions. In their study, they achieved a transformation frequency of about 50%, and transformed plants did not show chimeric patterns. Despite these early breakthroughs, other successful efforts have not been reported (Rao, 2009).

Electroporation

Electroporation involves the use of an electrical pulse to create pores in cells through which DNA can enter the cell. Tissues to be electroporated are suspended in a cuvette containing DNA and subjected to an electrical pulse. The steps after this procedure are similar to other transformation procedures: selection, regeneration, and analysis of putative transgenic plants. This approach is widely used to introduce DNA into protoplasts, since these cells do not have walls. Importantly, this approach has been applied for successful transformation via electroporation of calli, which are generally much easier to regenerate than protoplast. In this section, we briefly describe the transfer of DNA into embryogenic calli of maize and the subsequent development of transgenic plants that expressed the introduced gene (D'Halluin et al., 1992). Unlike protoplasts, initial work showed that electroporation of embryogenic calli in the absence of wounding was not successful. Embryogenic calli were wounded by enzymatic (Macrozyme) treatment to partially degrade cell wall or by finely chopping the calli tissue prior to electroporation with DNA containing the octopine synthase gene. Electroporated calli were selected in kanamycin-containing medium, and plants were subsequently regenerated from kanamycin-resistant calli. Over a 4 month period 148 R0 plants were regenerated from 31 kanamycin-resistant calli. The transgenes were stably expressed through three generations, which conclusively showed the stability of the inserted gene. Moreover, these experiments were done with type I calli or those derived from preconditioned immature zygotic embryos. This procedure removed the prerequisite of only being able to use embryogenic, friable type II cell cultures for transformation via biolistic transformation. A later work showed that maize type II calli could be transformed via electroporation (Pescitelli and Sukhapinda, 1995).

Bioactive-beads-mediated gene transfer

This approach focuses on using bioactive beads (calcium alginate microbeads) to entrap and thus protect the foreign DNA which is then introduced into protoplasts (Sone et al., 2008). The entrapped DNA can be introduced into protoplasts via electroporation (Murakawa et al., 2008a) or polyethylene glycol (PEG). A major advantage is that very large DNA can be introduced into cells without shearing. The bioactive beads are prepared mixing sodium alginate with isoamyl alcohol and the solution emulsified by sonication. The resulting beads are about 1.6 μm in diameter. The DNA is then added to the emulsion, which results in the entrapment of the DNA by the bioactive beads. This method has been improved by linking the DNA with liposomes derived from synthetic cationic lipid (Murakawa et al., 2008b). The resulting DNA–liposome complex is then entrapped by the bioactive beads. Introduction of the bioactive beads into the protoplasts was done via the PEG method. This approach resulted in a fourfold increase in transformation efficiency of protoplasts. The maximum transformation efficiency (the number of GFP expressing protoplasts divided by the total number of protoplasts) was 0.75%. This approach provides a way to introduce DNA protoplasts, but the protoplasts have not been regenerated into plants to evaluate the stability of gene expression in whole plants and subsequent generations. Thus, its use for the development of transgenics remains to be demonstrated.

Microinjection

Transformation of plants via microinjection of foreign DNA has been successful in only a few cases. In this section we briefly mention two reports that successfully introduced foreign DNA by microinjecting the DNA into young floral tillers of rye plants (De La Pena, 1987), and into isolated zygote protoplasts of barley (Holm et al., 2000). De La Pena et al. (1987) introduced *Agrobacterium* or DNA containing the aminoglycoside II gene (confers kanamycin resistance) into the target plants that were at a stage that is most receptive for DNA. In their case, injection was made into young floral tillers. They postulated that the DNA is transported through the vascular system to germ cells, which can take up the DNA only when they are in a competent stage, such as two weeks after meiosis when the germ cells are not surrounded by a callose wall that presumably allows the DNA to be taken up by the germ cell. Although the transformation rate was very low (e.g., in a DNA injection procedure), in one experiment 3 seeds out of 1000 from 37 plants expressed the target DNA. The great advantage of this approach is that tissue culture techniques are not necessary to obtain transformed plants.

Direct injection of protoplast is another approach that has shown marginal success. This approach requires technical skills for isolation of protoplasts, injection of protoplast, and then subsequent regeneration of the protoplast into plants. With barley, Holm et al. (2000) showed that 62% of the cells injected with the GUS gene survived and 55% of the survivors developed into plants with 21% of the plants expressing the gene of interest. However, only two plants had significant GUS expression. Their conclusions were that transformation can be achieved by introducing the DNA into protoplast via microinjection, but that gene expression was rare. The latter observation was surmised to be due to gene shearing.

Microinjection has also been used to achieve "chromosome-mediated" transformation of petunia hybrids (Griesbach, 1987). Although success with this approach has been reported in only this case, it is briefly described because it shows that chromosomal fragments can be introduced into plants and be stably integrated, although at a very low frequency. The rationale for this approach was to introduce properties (i.e., genes that govern the flavonoid biosynthesis pathways) that require the expression of multiple genes or polygenes, which might be clustered in segments of the chromosome. The flavonoid biosynthetic pathway was chosen because it is well characterized, and the author had developed a rapid high-performance liquid chromatography (HPLC) method to assay for the intermediates in this pathway. Another requirement was to microinject evacuolated protoplasts to avoid interference of toxic substances that might be released from vacuoles when protoplasts are injected. Isolated chromosomes were injected (0.1 pL) into evacuolated protoplasts that were obtained following centrifugation in Percoll™ gradients. The injected protoplasts were cultured on feeder plants, and subsequently grew into small calli. Samples of microcalli were analyzed for protein differences in polyacrylamide gels, and those showing differences were regenerated into plants and subsequently subjected to flavanol analysis. Out of the 500 injected protoplasts, 23% divided. Further analysis showed that plants regenerated from 2 of 60 calli showed properties that were different from the non-transformed plants. One was sterile and the other only slightly fertile. The properties of the single plant that was obtained between the cross of the slightly fertile transformant and a non-transformed plant suggested that genes from the introduced chromosome segments were being expressed. Other successes have not been reported using this technique (Rao et al., 2009).

Pollen-tube pathway

The pollen-tube pathway transformation approach is ingenious in the way that transformation is effected by inserting foreign DNA in the pathway of a naturally occurring physical phenomenon of pollination. During the pollination process of higher plants, after the pollen germinates on the stigmatic surface, a pollen tube is formed from the pollen and extends down the style into the ovule; nuclei from the pollen travels through the pollen tube to fertilize the ovule. The pollen-tube transformation approach takes advantage of this natural phenomenon by introducing DNA into the pollen tube channel. A few hours after pollination, the style is cut and the DNA is placed on the cut surface. The DNA then enters the pollen tube and moves in the tube into the ovary, and subsequently is incorporated into the nuclei of the receiving ovary. In other words, the foreign DNA "fertilizes" the ovule analogously like the pollen grain DNA would. The obvious advantage of this approach is that in successful integration cases some of the harvested seeds would have the foreign DNA incorporated into its genome, and the plants from the germinated seed would be transgenic. Furthermore, this approach eliminates tissue culture steps and is not limited by the ability to regenerate plants from transformed protoplast, calli, or leaf pieces. Some shortcomings are that this approach would work only with plants that flower and readily produce seeds, and that it can be done only during the blooming period of the plant.

Apparently, the method was first reported by Duan and Chen (1985) who transformed rice. A later report by Luo and Wu (1989)provided the first detailed molecular data that critically established this approach for transforming rice. In their case, 20% of the recovered seeds expressed the introduced foreign DNA. These authors suggested that the two critical points in this procedure were: (1) to determine the optimal time after pollination for the application of the foreign DNA to the excised florets, and (2) to cut the upper part of the floret so that the ovary is not damaged. This approach has been used in crops such as cotton (Zhang et al., 2009), corn (Yang et al., 2009), soybean (Liu et al., 2009), wheat, and petunia as cited in Rao (2009). With soybean, Liu et al. (2009), investigated the various parameters of the method and found their best results were obtained when the whole style was removed 6–8 hours after pollination and by incorporating a surfactant into the DNA solution.

Silica carbide whisker-mediated transformation

The silica carbide whisker-mediated transformation approach (Kaeppler et al., 1990) is arguably the simplest, most direct, and economical way to introduce foreign DNA into cells that results in stable transgenic plants that express the foreign DNA. Silica carbide whiskers are needle-like microfibers about 0.5 μm in diameter and 10–80 μm long that puncture the plant cells to form an entrance portal for the foreign DNA. Regeneration and selection of cells can then result in transgenic plants that express the foreign DNA. This approach is briefly described for cotton (Asad et al., 2008). Embryogenic calli are produced in tissue culture from embryos of cotton seeds; sterile silica carbide whiskers are mixed with a solution of plasmid DNA in a tube and the tube is vortexed to create wounds in the cell walls of the callus tissue through which the plasmid DNA can enter the cells. The treated calli are placed on selection media and transformed cells regenerated into plants. Extensive characterization of the transformed cotton plants over three generations showed that transformation was stable and the salt-tolerant property was expressed. Simplicity is the elegance of this approach and it has been successfully used for other crops such as rice and maize (Thompson et al., 1995; Mizuno et al., 2005).

Biolistic Transformation

The history of biolistic transformation and the different biolistic approaches are discussed in this section, along with the current state of the invention. We look into the impact this technology has had on agricultural plants and provide a background on the development of a commercial crop using biolistics as the transformation method.

John Sanford was investigating DNA transfer through the use of microlasers to cut holes in living pollen tubes as a method to introduce foreign DNA into embryos (Sanford, 2000). While looking into the feasibility of using ion and electron beams to cut holes in plant cell walls, the concept of biolistics or microprojectile bombardment was first conceived by John Sanford and Ed Wolf. The term biolistics was derived from the words *bio*logy and bal*listics* and has become a prevalent method of plant transformation, second only to *Agrobacterium*-mediated transformation. Although multiple reviews have been published on this technique, our goal here is to present a general background of this concept and describe cases of the impact it has had on modern agriculture.

The invention

What began as a meeting between Cornell scientists looking into the feasibility of using microbeams to cut holes in cell walls to transfer DNA into a plant cell eventually became the origin of the modern biolistic microprojectile bombardment system (Sanford et al., 1987; Sanford, 2000). This system revolutionized the field of genetic engineering by allowing gene transfer to *Agrobacterium* recalcitrant crops such as corn (Klein et al., 1987, 1988), as well as reliable organelle transformation such as chloroplast in algae (Boynton et al., 1988) and tobacco (Daniell et al., 1990), and mitochondria in yeast (Johnston et al., 1988).

The origin of the modern biolistic microprojectile bombardment system started from humble beginnings as a BB gun (a store-bought air gun that propels spherical shaped pellets) with a modified barrel to facilitate loading of the tungsten powder to accelerate DNA into onion epidermal cells (Sanford et al., 1987; Sanford, 2000). Gas-generated devices were found to be too destructive when used in close proximity to the plant material, and could not propel the particles with a high enough velocity at longer distances. Thus, modifications of this initial model relied upon a gunpowder-driven nylon projectile (Sanford et al., 1987). This nylon projectile carried the RNA- and DNA-coated tungsten particles until it reached a steel stopping plate. The particles were allowed to pass through a small aperture in the steel plate and penetrate the plant tissue. Using this apparatus, the expression of both a plasmid containing a 35s promoter driving the chloramphenicol acetyltransferase (CAT) reporter gene and RNA from the tobacco mosaic virus (TMV) were expressed in onion epidermal cells (Klein et al., 1987), demonstrating the first biolistic transformation of plant cells.

The next version of the gene gun consisted of an airtight vacuum chamber to house the gun and target tissue to reduce air resistance that slows the velocity of the particles and subsequently penetration of the particles into the plant tissue (Sanford, 2000). With this innovative gun system the first stably transformed plants were produced by insertion of the neomycin phosphotransferase (NPTII) gene into tobacco leaves and suspension cells (Klein et al., 1988b). This system produced the first transformation of chloroplasts (Boynton et al., 1988), the first transformation of mitochondria (Johnston et al., 1988), and the first transformation of corn (Klein et al., 1988a). In addition, this system was also used for the stable transformation of papaya for papaya ringspot virus (PRSV) resistance, which resulted in the first commercial transgenic fruit crop in the United States (Fitch et al., 1990; Gonsalves et al., 2004).

Although the technology rights of Biolistics Inc. were sold to DuPont in 1989, with the exception of the ornamental rights, research was conducted to improve the biolistic gene gun system (Sanford, 2000). High pressure helium replaced the gunpowder-driven systems and generated pressure to produce an instantaneous burst or gas shock from membranes designed to rupture at specific pressures (Sanford, 1991). This gas shock was enclosed under partial or full vacuum to protect the operator's hearing and allowed to vent the expanding gas to prevent back pressure that can slow the gas shock. The gas shock accelerates the microprojectiles and deflects the remaining gas through the "throat" of the system, which is made of interchangeable inserts to allow flexibility in transforming different tissues including yeast, bacteria, plants, cultured animal cells, and the skin of live mice. This helium-driven system increased transformation efficiency by a 4- to 331-fold increase over the gunpowder-driven system (Sanford, 1991).

Electric discharge particle acceleration

An alternative method to the biolistic system developed by Sanford and colleagues was developed by Christou (1988) and colleagues at Agracetus. Using a similar concept to the biolistic system of propelling DNA-coated heavy metal particles into the plant cell for transformation, this system uses electrical discharge as the method for particle acceleration (McCabe and Christou, 1993). This system consists of two chambers separated by a 10 μL drop of water, and the initial electrical shock wave is generated in the spark chamber. This wave is prevented from directly interacting with the carrier sheet, instead secondary waves produced by reflection off the chamber floor and walls provide a uniform force across the carrier sheet containing the DNA-coated gold particles. The accelerated carrier is stopped by a retaining screen, but the gold particles continue to travel toward the target tissue (McCabe and Christou et al., 1993). Similar to the biolistic system, this entire assembly may be placed under vacuum to reduce the air resistance that slows the particles.

Using this ACCELL™ technology, Agracetus was able to develop genotype-independent transformation of soybeans while *Agrobacterium*-mediated transformation was limited to one non-commercial variety (Christou et al., 1988, 1989; Christou, 1990). Although the heritability of the transgene was limited to a few transformed lines, using GUS expression in stem, leaf, and petiole/midrib of the initial transformants, plants with transformed germlines could be identified (Christou et al., 1989). In addition, the ACCELL™ gene gun was used to transform cotton lines that were difficult to regenerate in tissue culture by using meristematic tissue from excised embryonic axes of cotton (McCabe and Martinelli, 1993). This technology was not marketed commercially as a transformation unit; instead Agracetus offered this technology as in "in house" service (Batty and Evans, 1992). In 1996, the plant transformation technology was purchased by Monsanto, while the mammalian DNA gene delivery technology was retained as Auragen Inc.

Current status of the "invention" hardware

Currently, the biolistic technology is exclusively licensed by DuPont to Bio-Rad Laboratories (Hercules, California) and is available in two devices: the PDS-1000/He and the Helios gene gun. Based upon the helium-driven device described by Sanford (1991), the PDS-1000/He consists of an acceleration mechanism and a main chamber (Kikkert, 1993). Helium gas fills the acceleration tube to pressures between 450 and 2200 psi, and the bursting pressure is determined by a user-selected Kapton membrane (rupture disk) laminated to burst at the desired pressure. When the critical pressure is reached in the acceleration chamber, the rupture disk bursts and releases a gas shock wave that propels the macrocarrier that holds the microscopic gold or tungsten particles (microcarrier) toward the target tissue. The macrocarrier impacts the stopping screen, which allows passage of the microcarriers to proceed downward toward the target tissue that is held under vacuum in the main chamber. Transformation efficiency is determined by changing several parameters including the: (1) helium pressure by using different rupture disks, (2) distance between the rupture disk and macrocarrier, (3) distance between the macrocarrier and stopping screen, (4) position of the target tissue from the stopping screen, and (5) the level of vacuum used in the main chamber (Kikkert, 1993).

The Helios gene gun is a handheld device that can be used to transform larger target tissues that may not fit in the main chamber of the PDS-1000/He, such as epidermal cells of the skin for vaccination studies (Tang et al., 1992). Similar to the PDS-1000/He system, the Helios uses pressurized helium to accelerate the microcarriers coated with the nucleic acid from the cartridge through the acceleration channel. The microcarriers spread while traveling down the barrel and reduce the shock wave to the target tissue (Bio-Rad, Bulletin 9541) In addition, this technology has been used for inoculation of leaves for virus infection (Yamagishi et al., 2006) and transient gene expression (Ueki et al., 2009).

Advantages of Biolistic Transformation

Biolistics, as mentioned earlier, is a physical method of accelerating DNA into the plant tissue that is not dependent on host specificity or species limitation such as with *Agrobacterium*-mediated transformation. Until the advent of biolistic technologies, transformation of important agronomic crops such as maize (Klein et al., 1988a), soybean (Christou et al., 1988, 1989; Christou, 1990), and cotton (McCabe and Christou, 1993) were not attainable.

Diverse cell types can be transformed with biolistics, whereas embryogenic callus is often preferable, and organized tissues such as embryos or meristems can successfully be transformed (Christou, 1996). This reduces the potential somaclonal variation that arises during tissue culture. Transformation of organized tissue also allows regeneration of genotypes that may be recalcitrant to plant regeneration from callus, such as in the case of cotton (McCabe and Martinell, 1993). In addition to being used as a sole transformation protocol for some plant species, biolistics has also been used to wound sunflower meristems prior to *Agrobacterium* infection to improve transformation efficiency (Bidney et al., 1992).

A perceived disadvantage of biolistic transformation is that the resulting plants contain multiple highly complex inserts prone to recombination and silencing. Comparison of rice and barley plants transformed with *Agrobacterium* and biolistics revealed that particle bombardment results in plants with a greater number of inserts: 2.7 versus 1.8 in rice, and >8 versus 1–3 in barley (Dai et al., 2001; Travella et al., 2005). All *Agrobacterium* transformed barley lines resulted in the co-expression of the two transformed genes, however, only 70% of those transformed with biolistics contained intact copies (Travella et al., 2005). Similarly, in the case of rice, GUS activity was more stable in *Agrobacterium*-transformed lines versus those transformed with biolistics (Dai et al., 2001). Although some instability was observed in the inheritance of *Agrobacterium*-transformed traits in barley, only 3 or 9 plants followed an expected Mendelian segregation ratio (Travella et al., 2005). In contrast, inheritance in rice followed expected Mendelian segregation ratios. Multiple copies of the transgene in *Agrobacterium*-transformed lines were located in different loci compared to co-integration of multiple transgenes in the same loci in the biolistic-derived lines (Dai et al., 2001).

However, as reviewed by Altpeter et al. (2005), the integration of multiple genes into a common loci with biolistics may be advantageous when multiple gene transformation or transformation of high molecular weight DNA is desired. Multiple gene transformation is often limited by the amount of DNA that can be cloned within the border regions of the T-DNA during *Agrobacterium* transformation. As many as 14 different transgenes have been co-bombarded into rice and resulted in 125 independent lines of which 3 contained 11, 10, and 9 transgenes at one or two genetic loci (Chen et al., 1998). Multiple gene transformation has been used for virus resistance, metabolic pathway engineering, and multimeric protein production (Altpeter et al., 2005).

In a survey of gene expression in rice, ryegrass, fescue, rye, and wheat, the transgene copy number was a critical factor effecting expression, as long as the integration of the transgene was intact and did not contain deleterious rearrangements (Altpeter et al., 2005). Often the undesirable effects, such as transgene rearrangements and adverse effects on endogenous gene expression, are often associated with the vector backbone. Linearization of the transformation vector to exclude the unnecessary vector sequences resulted in low-copy-number transgene integrations with low frequency of transgene rearrangements relative to supercoiled or linear plasmids (Fu et al., 2000). Lowering the amount of DNA used for each bombardment also decreased the number of insertions. In corn, 46% of 1600 transformation events contained single insertions when 2.5 ng of DNA was used per bombardment versus 25% single insertion events observed when 25 ng of DNA was used over 18 independent experiments with the helium gun (Lowe et al., 2009).

A key advantage to biolistic transformation has been in the area of organelle transformation. The gunpowder-driven microprojectile system developed by Sanford produced the first transformed mitochondria in the yeast *Saccharomyces cerevisiae* (Johnston et al., 1988), and currently is the only method for reliable transformation of mitochondria (Bonnefoy and Fox, 2007). In addition, this device was able to produce the first transformation of chloroplasts in *Chamydomonas* (Boynton et al., 1988) and the first stable transformation of tobacco chloroplast (Daniell et al., 1990). Currently, biolistics is still the preferred method to transform chloroplast due to the high transformation efficiency (Wang et al., 2009). Chloroplast transformation offers the advantages of high expression level, uniparental transgene inheritance, which can be used in homologous recombination for transgene targeting and is not limited by gene silencing and position effects (reviewed in Altpeter et al., 2005). Chloroplast transformed plants have been produced to improve disease and insect resistance, increased salt tolerance, production of biopharmaceuticals and edible vaccines, metabolic pathway engineering, and research on RNA editing (Altpeter et al., 2005; Wang et al., 2009).

Implications of Biolistics in Agricultural Biotechnology

As the predominant alternative to *Agrobacterium*-mediated transformation, biolistics was a contemporary transformation method during the initial push to deregulate genetically engineered crops. Here we will use the current deregulated crops as a gauge to measure the impact biolistics has on agriculture. In addition, we will provide a background and rationale behind a commercial crop (papaya) that was developed using the biolistic transformation method.

Application of biolistics in agriculture crops

Table 8.1 contains a list of the petitions for non-regulated status that were granted by USDA APHIS in the United States as of July 29, 2011, for dicotyledonous plants and Table 8.2 for monocotyledonous plants. These tables consist of the transformed crop, the APHIS petition number, institution applying for the petition, transgenic phenotype, and transformation event or line. The transformation method was obtained from the individual petitions. The biolistic transformation designation includes microprojectile bombardment or particle acceleration method and particle gun process.

This table is not an accurate depiction of the exact commercial status of transgenic crops in the United States, since many of the institutions filing the petitions no longer exist or not all crops were commercialized. However, it does give a general idea on the history and relative importance of each transformation technique for each crop. From this table it is evident that *Agrobacterium*-mediated transformation is the predominant method for gene transfer for most crops. Biolistics transformation was used to produce transgenic lines of monocot crops; corn and rice and dicot crops; and cotton, papaya, and soybean. Two petitions included direct gene uptake and protoplast direct uptake, and one petition listed immature embryo electroporation as the transformation method and another Confidential Business Information (CBI). Interestingly, while early releases (year of petition filed is indicated by the first two digits of petition number) of the monocot corn were largely dependent on microprojectile bombardment, newer releases are reliant upon *Agrobacterium*-mediated transformation; the dicot crop soybean is predominantly transformed with biolistics and not *Agrobacterium*-mediated transformation. As the number of transformed crops increases and more research is conducted to refine the insertion events, it will be interesting to see what new innovations will be developed in the transformation field.

The majority of the applicant institutions are large agricultural companies, but four of the applicants are public sector institutions, including the sulfonyl urea herbicide resistant flax from the University of Saskatchewan, plum pox virus resistant plum from the USDA, and ARS- and PRSV-resistant papaya from Cornell University and the University of Florida. The PRSV papaya from Cornell University was the first of these institutions to achieve non-regulated status in the United States. In the next section we provide the reader with a first-hand perspective on the development of papaya as a commercial crop using biolistics as the transformation method.

Papaya: A case study of biolistic transformation

The papaya (*Carica papaya*) serves as a good case to examine the biolistic process and impact of the procedure in a transgenic dicot plant, because it provides intriguing molecular biology and a commercial application. In this section, we describe the Hawaiian papaya case as an interrelated series of events, of which the transformation of papaya via the biolistic approach played a crucial role.

Papaya and papaya ringspot virus

Papaya is a large herbaceous plant with a single stem that grows in tropical and subtropical lowland regions. In Hawaii, the dominate papaya is the small Hawaiian solo type papaya that weighs an average of 1–2.5 pounds. Although papaya is polygamous, with male, female, and hermaphrodite plants, only the hermaphrodite is grown commercially (Gonsalves et al., 2008) in Hawaii.

Table 8.1 USDA, APHIS non regulated dicot crops as of July 29, 2011

Regulated article	APHIS petition #	Applicant institution	Transgenic phenotype	Transformation event or line	Transformation method
Cichorium intybus	97-148-01p	Bejo	Male sterile	RM3-3, RM3-4, RM3-6	*Agrobacterium*-mediated transformation
Cotton	93-196-01p	Calgene	Bromoxynil tolerant	BXN	*Agrobacterium*-mediated transformation
	94-308-01p	Monsanto	Lepidopteran resistant	531, 757, 1076	*Agrobacterium*-mediated transformation
	95-045-01p	Monsanto	Glyphosate tolerant	1445, 1698	*Agrobacterium*-mediated transformation
	95-256-01p	DuPont	Sulfonylurea tolerant	19-51a	*Agrobacterium*-mediated transformation
	97-013-01p	Calgene	Bromoxynil tolerant and Lepidopteran resistant	Events 31807 and 31808	*Agrobacterium*-mediated transformation
	00-342-01p	Monsanto	Lepidopteran resistant	Cotton Event 15985	Biolistic transformation
	02-042-01p	Aventis	Phosphinothericin tolerant	LLCotton25	*Agrobacterium*-mediated transformation
	03-036-01p	Mycogen/Dow	Lepidopteran resistant	281-24-236	*Agrobacterium*-mediated transformation
	03-036-02p	Mycogen/Dow	Lepidopteran resistant	3006-210-23	*Agrobacterium*-mediated transformation
	03-155-01p	Syngenta	Lepidopteran resistant	COT 102	*Agrobacterium*-mediated transformation
	04-086-01p	Monsanto	Glyphosate tolerant	MON 88913	*Agrobacterium*-mediated transformation
	06-332-01p	Bayer CropScience	Glyphosate tolerant	GHB614	*Agrobacterium*-mediated transformation
Flax	98-335-01p	University of Saskatchewan	Tolerant to soil residues of sulfonyl urea herbicide	CDC Triffid	*Agrobacterium*-mediated transformation
Papaya	96-051-01p	Cornell University	PRSV resistant	55-1, 63-1	Biolistic transformation
	04-337-01p	University of Florida	Papaya ringspot virus resistant	X17-2	*Agrobacterium*-mediated transformation
Plum	04-264-01p	ARS	Plum pox virus resistant	C5	*Agrobacterium*-mediated transformation
Potato	94-257-01p	Monsanto	Coleopteran resistant	BT6, BT10, BT12, BT16, BT17, BT18, BT23	*Agrobacterium*-mediated transformation
	95-338-01p	Monsanto	CPB resistant	SBT02-5 and -7, ATBT04-6 and -27, -30, -31, -36	*Agrobacterium*-mediated transformation
	97-204-01p	Monsanto	CPB and PLRV resistant	RBMT21-129 & RBMT21-350	*Agrobacterium*-mediated transformation
	97-339-01p	Monsanto	CPB and PVY resistant	RBMT15-101, SEMT15-02, SEMT15-15	*Agrobacterium*-mediated transformation
	99-173-01p/97-204-01p	Monsanto	PLRV and CPB resistant	RBMT22-82	*Agrobacterium*-mediated transformation
Rapeseed	94-090-01p	Calgene	Oil profile altered	pCGN3828-212/86-18 and -23	*Agrobacterium*-mediated transformation
	97-205-01p	AgrEvo	Phosphinothricin tolerant	T45	*Agrobacterium*-mediated transformation

(*Continued*)

Table 8.1 (Continued)

Regulated article	APHIS petition #	Applicant institution	Transgenic phenotype	Transformation event or line	Transformation method
	98-216-01p	Monsanto	Glyphosate tolerant	RT73	*Agrobacterium*-mediated transformation
	98-278-01p	AgrEvo	Phosphinothricin tolerant and pollination control	MS8 and RF3	*Agrobacterium*-mediated transformation
Soybean	01-324v01p/98-216-01p	Monsanto	Glyphosate tolerant	RT200	*Agrobacterium*-mediated transformation
	01-206-01p/98-278-01p	Aventis	Phosphinothricin tolerant and pollination control	MS1 and RF1/RF2	*Agrobacterium*-mediated transformation
	01-206-02p/97-205-01p	Aventis	Phosphinothricin tolerant	Topas 19/2	*Agrobacterium*-mediated transformation
	93-258-01p	Monsanto	Glyphosate tolerant	40-3-2	Biolistic transformation
	96-068-01p	AgrEvo	Phosphinothricin tolerant	W62, W98, A2704-12, A2704-21, A5547-35	Biolistic transformation
	97-008-01p	DuPont	Oil profile altered	G94-1, G94-19, G-168	Biolistic transformation
	98-014-01p/96-068-01p	AgrEvo	Phosphinothricin tolerant	A5547-127	Biolistic transformation
	98-238-01p	AgrEvo	Phosphinothricin tolerant	GU262	Biolistic transformation
	06-271-01p	Pioneer	Glyphosate and acetolactate synthase tolerant	356043	Biolistic transformation
	06-178-01p	Monsanto	Glyphosate tolerant	MON 89788	*Agrobacterium*-mediated transformation
	06-354-01p	Pioneer	High oleic acid	DP-3054	Biolistic transformation
Squash	92-204-01p	Upjohn	WMV2 and ZYMV resistant	ZW-20	*Agrobacterium*-mediated transformation
	95-352-01p	Asgrow	CMV, ZYMV, WMV2 resistant	CZW-3	*Agrobacterium*-mediated transformation
Sugar beet	97-336-01p	AgrEvo	Phosphinothricin tolerant	T-120-7	Agrobacterium-mediated transformation
	98-173-01p	Novartis Seeds and Monsanto	Glyphosate tolerant	GTSB77	Agrobacterium-mediated transformation
	03-323-01p	Monsanto	Glyphosate tolerant	H7-1	*Agrobacterium*-mediated transformation
Tobacco	01-121-01p	Vector	Reduced nicotine	Vector 21-41	*Agrobacterium*-mediated transformation
Tomato	92-196-01p	Calgene	Fruit ripening altered	FLAVR SAVR	*Agrobacterium*-mediated transformation
	94-227-01p/92-196-01p	Calgene	Fruit ripening altered	Line N73 1436-111	*Agrobacterium*-mediated transformation
	94-228-01p	DNA Plant Tech	Fruit ripening altered	1345-4	*Agrobacterium*-mediated transformation
	94-230-01p/92-196-01p	Calgene	Fruit ripening altered	9 additional FLAVRSAVR lines	*Agrobacterium*-mediated transformation
	94-290-01p	Zeneca and Petoseed Fruit	Polygalacturonase level decreased	B, Da, F	*Agrobacterium*-mediated transformation
	95-030-01p/92-196-01p	Calgene	Fruit ripening altered	20 additional FLAVRSAVR lines	*Agrobacterium*-mediated transformation

(*Continued*)

Table 8.1 (Continued)

Regulated article	APHIS petition #	Applicant institution	Transgenic phenotype	Transformation event or line	Transformation method
	95-053-01p	Monsanto	Fruit ripening altered	8338	*Agrobacterium*-mediated transformation
	95-179-01p/92-196-01p	Calgene	Fruit ripening altered	2 additional FLAVRSAVR lines	*Agrobacterium*-mediated transformation
	95-324-01p	Agritope	Fruit ripening altered	35 1 N	*Agrobacterium*-mediated transformation
	96-248-01p/92-196-01p	Calgene	Fruit ripening altered	1 additional FLAVRSAVR line	*Agrobacterium*-mediated transformation
	97-287-01p	Monsanto	Lepidopteran resistant	5345	*Agrobacterium*-mediated transformation

Papaya ringspot virus (PRSV) is the most important virus affecting papaya worldwide. The virus belongs to the potyvirus genus, a very large and economically important group of plant viruses (Gonsalves et al., 2008). These viruses are vectored by aphids in a non-persistent manner; that is, the aphids can acquire the virus quickly and transmit it quickly, but the virus is not retained by the aphid vector for long periods of time. The severity of PRSV infection on papaya is manifested by the stunting of the papaya plant, mosaic and distortion of leaves, and loss of crop productivity. Unfortunately, resistance to PRSV has not been found in C. *papaya* so conventional breeding has not been successful in developing resistance (Gonsalves et al., 2006).

Genes that confer resistance to PRSV have been identified in the genus *Vasconcellea* (Gonsalves et al., 2006). However, crosses between C. *papaya* and *Vasconcellea* species are sterile. Progress is being made, however, in rescuing embryos of these crosses and using various approaches to bridge C. *papaya* and these other plants so that later generations would be more fertile when crossed to C. *papaya* and also contain the resistant gene (Gonsalves et al., 2006). However, commercial lines have not yet been obtained.

Tolerance to PRSV has been identified in C. *papaya* and the process of recurrent selection has been used to integrate tolerance into selected lines, some of which have been sold commercially, and others that have been released by the government for use by homeowners and farmers. The latter includes the release of Tha Pra 2 (Prasartsee et al., 1998; Gonsalves et al., 2006). Although these lines get infected by PRSV, they continue to produce marketable fruit. Nevertheless, tolerant cultivars of papaya are not commonly grown.

Papaya and PRSV in Hawaii

The entire papaya story has been summarized in a number of reviews (Gonsalves, 1998, 2004; Gonsalves et al., 1998, 2004). The Hawaiian solo papaya has been grown in Hawaii for over a hundred years, but was free of PRSV until the 1940s when Jensen observed the disease and actually coined the term papaya ringspot virus (Gonsalves, 1998). The virus became especially severe in the 1950s and virtually decimated papaya production on Oahu island, which forced the papaya industry to be relocated in the Puna district of Hawaii island. Puna was an ideal region because an excellent variety "Kapoho" had been selected and was very adaptable to the Puna area where there was well-drained lava rock type soil and lots of rainfall and sunshine. Most important, PRSV was not present in Puna. By 1978, however, PRSV was in Hilo, only 19 miles away and thus the potential danger of PRSV causing large-scale damage in Puna was real (Gonsalves, 1998).

Biolistic approach to transform papaya for resistance to PRSV

With the potential danger of PRSV to the papaya in Puna, a proactive research program was started in 1985 to use the "parasite-derived resistance" approach to develop transgenic papaya with resistance to PRSV. Parasite-derived resistance is a phenomenon where transgenic plants containing a gene of the parasite, or pathogen, are protected against detrimental effects of the same pathogen (Sanford and Johnston, 1985). A case for this approach was made in 1986 when transgenic tobacco and tomato expressing the coat protein gene of tobacco mosaic virus showed resistance or tolerance to the virus (Powell-Abel et al., 1986). We followed the same lead and embarked on transforming papaya such that it would express the coat protein gene of PRSV and hopefully be resistant to PRSV. There are many aspects to the effort, but for this purpose we will focus on the transgenic effort and the rationale on how we came to use the biolistic approach. Details on the complete story are given in a number of reviews (Gonsalves, 1998; Gonsalves et al., 2006).

Agrobacterium transformation of papaya leaf pieces was initially tried. Numerous efforts to transform papaya via *Agrobacterium* failed and the project was at standstill. Two factors helped to move the project forward. The first was the report that embryogenic calli of walnut could be transformed by *Agrobacterium*, which provided a somatic embryogenic calli approach rather than organogensis (McGranahan et al., 1988). The second factor was that the gene gun had been recently

Table 8.2 USDA, APHIS non-regulated monocot crops as of July 29, 2011

Crop	APHIS petition #	Applicant institution	Transgenic phenotype	Transformation event or line	Transformation method
Corn	94-319-01p	Ciba Seeds	Lepidopteran resistant	Event 176	Biolistic transformation
	94-357-01p	AgrEvo	Phosphinothricin tolerant	T14, T25	Protoplast direct uptake
	95-093-01p	Monsanto	Lepidopteran resistant	MON 80100	Biolistic transformation
	95-145-01p	DeKalb	Phosphinothricin tolerant	B16	Biolistic transformation
	95-195-01p	Northrup King	European Corn Borer resistant	Bt11	CBI
	95-228-01p	Plant Genetic Systems	Male sterile	MS3	Immature embryo electroporation
	96-017-01p/95-093-01p	Monsanto	European Corn Borer resistant	MON809 and MON810	Biolistic transformation
	96-291-01p	DeKalb	European Corn Borer resistant	DBT418	Biolistic transformation
	96-317-01p	Monsanto	Glyphosate tolerant and ECB resistant	MON802	Biolistic transformation
	97-099-01p	Monsanto	Glyphosate tolerant	GA21	Biolistic transformation
	97-265-01p	AgrEvo	Phosphinothricin tolerant and lepidopteran resistant	CBH 351	Biolistic transformation
	97-342-01p	Pioneer	Male sterile and phosphinothricin tolerant	676, 678, 680	Biolistic transformation
	98-349-01p/95-228-01p	AgrEvo	Phosphinothricin tolerant and male sterile	MS6	Biolistic transformation
	00-011-01p/97-099-01p	Monsanto	Glyphosate tolerant	NK603	Biolistic transformation
	00-136-01p	Mycogen c/o Dow and Pioneer	Lepidopteran resistant and phosphinothricin tolerant	Line 1507	Biolistic transformation
	01-137-01p	Monsanto	Corn rootworm resistant	MON 863	Biolistic transformation
	03-181-01p/00-136-01p	Dow	Lepidopteran resistant and phosphinothricin tolerant	TC-6275	*Agrobacterium*-mediated transformation
	03-353-01p	Dow	Corn rootworm resistant	59122	*Agrobacterium*-mediated transformation
	04-125-01p	Monsanto	Corn rootworm resistant	88017	*Agrobacterium*-mediated transformation
	04-229-01p	Monsanto	High lysine	LY038	Biolistic transformation
	04-362-01p	Syngenta	Corn rootworm protected	MIR604	*Agrobacterium*-mediated transformation
	05-280-01p	Syngenta	Thermostable alpha-amylase	3272	*Agrobacterium*-mediated transformation
	06-298-01p	Monsanto	European Corn Borer resistant	MON 89034	*Agrobacterium*-mediated transformation
	07-152-01p	Pioneer	Glyphosate and imidazolinone tolerant	DP-098140-6	*Agrobacterium*-mediated transformation
	07-253-01p	Syngenta	Lepidopteran resistant	MIR-162 Maize	*Agrobacterium*-mediated transformation
	08-338-01p	Pioneer	Male sterility fertility restored, visual marker	OP-32138-1	*Agrobacterium*-mediated transformation
Rice	98-329-01p	AgrEvo	Phosphinothricin tolerant	LLRICE06, LLRICE62	Direct gene transfer
	06-234-01p/98-329-01p	Bayer CropScience	Phosphinothricin tolerant	LLRICE601	*Agrobacterium*-mediated transformation

invented by John Sanford who was a colleague of Gonsalves at Cornell University, which gave him timely access to the gene gun equipment and expertise of John Sanford (Klein et al., 1987; Sanford et al., 1992). Maureen Fitch successfully developed an embryogenic calli system for papaya, and transformation was initiated with the gene gun from 1988 to 1989 (Fitch et al., 1990; Fitch and Manshardt, 1990, 1992). By 1991, genetically engineered transgenic papaya with resistance to PRSV had been developed and tests in the greenhouse showed that it was resistant to PRSV (Fitch et al., 1992).

Testing, deregulation, commercialization, and impact of the transgenic papaya in Hawaii

As might be expected, initial efforts at transforming embryogenic calli via the biolistic approach were not highly efficient, but they did result in a transgenic line that showed resistance to PRSV (Gonsalves, 1998). Fifteen lines of transgenic Sunset and Kapoho papaya were tested for resistance, and one of them, designated line 55-1, showed resistance under greenhouse conditions. Clones of line 55-1 also showed excellent resistance in a 1992 field trial on Oahu island in Hawaii. Ironically, in 1992 PRSV was discovered in Puna on Hawaii island where 95% of Hawaii's papaya was grown. As expected, PRSV moved rapidly throughout Puna and by October 1994, efforts to suppress PRSV in Puna were abandoned. By 1995, PRSV was widespread in Puna and severely reduced papaya production. Fortunately, efforts to develop a control method had been undertaken proactively and, as noted earlier, by 1992 a PRSV-resistant papaya had been identified. Efforts were focused on testing the papaya under field conditions in Puna and on moving toward deregulation and commercialization of the transgenic papaya.

The Puna field trial was started in 1995 and resulted in a transgenic papaya that was resistant under field conditions (Ferreira et al., 2002). Using plants in the 1992 field trial on Oahu island as a source for improving line 55-1, Richard Manshardt (1998) successfully developed two commercial quality transgenic cultivars called SunUp and Rainbow. Rainbow would go on to be the most popular transgenic papaya cultivar in Hawaii. Line 55-1 was deregulated by the Animal Plant Health Inspection Service (APHIS) and the Environmental Protection Agency (EPA), and consultation with the Food and Drug Administration (FDA) was completed by the end of 1997. The transgenic papaya was commercialized in May 1998 and seeds were released to the growers starting that same month (Gonsalves et al., 2004).

The impact of the transgenic papaya transformed via the biolistic approach has been documented elsewhere (Gonsalves, 2006). In summary, since its release in 1998 the resistance of the transgenic papaya has been highly effective, and the transgenic papaya currently accounts for about 80% of the papaya production. One could safely say that without the PRSV-resistant transgenic papaya, the papaya industry in Puna would not have survived the PRSV. Efforts are now focused on obtaining regulatory approval of the Hawaii transgenic papaya (line 55-1 and its derivatives) in Japan, and to thus expand Hawaii's papaya market. In 1992, when PRSV was first discovered in Puna, Hawaii exported about 25% of its papaya to Japan, but the amount being exported to Japan is currently less than 10% of Hawaii's production. The latter is due to the low production of non-transgenic papaya in Hawaii and the current prohibition of the transgenic papaya in Japan. Progress has been made in this effort and we are hopeful to have regulatory approval in 2011.

Characteristics of transgene inserts in biolistically transformed line 55-1 and its derivatives

Unlike *Agrobacterium* transformation, the biolistic approach for DNA insertion is expected to be a rather random process that could result in numerous inserts of DNA. Interestingly, a thorough study of the transgenic line 55-1 revealed only three inserts (Suzuki et al., 2008). The inserts included a single contiguous 9789 bp fragment encoding the coat protein, *uida* and *nptII* genes, a 290 bp fragment of the 822 bp nptII gene, and a 222 bp tetA gene fragment with flanking plasmid vector DNA 1533 bp in length. Furthermore, genetic studies also showed that these inserts segregate as a unit, suggesting that they are at the same loci. Lastly, five of the six flanking genomic DNA sequences are nuclear plastid sequences. We have no evidence that these nuclear plastid sequences are preferred integration sites.

The biolistically transgenic papaya line 55-1, which was first tested for resistance in 1991 and has been commercially planted since 1998, has been remarkably stable, and its resistance to PRSV has held up under a variety of field conditions and virus pressure. Studies have shown that the mechanism of resistance is via post-transcriptional gene silencing (Tennant et al., 2001). These observations under long-term field conditions reinforce the value and impact of the biolistic transformation approach for developing effective transgenic virus-resistant crops using the "pathogen-derived concept."

References

Ahokas, H. (1989). Transfection of germinating barley seed electrophoretically with exogenous DNA. *Theoretical and Applied Genetics*, *77*, 469–472.

Altpeter, F., Baisakh, N., Beachy, R., Bock, R., Capell, T., & Christou, P., et al. (2005). Particle bombardment and the genetic enhancement of crops: Myths and realities. *Molecular Breeding*, *15*, 305–327.

Asad, S., Mukhtar, Z., Nazir, F., Hashmi, J. A., Mansoor, S., & Zafar, Y., et al. (2008). Silicon carbide whisker-mediated embryogenic callus transformation of cotton (*Gossypium hirsutum* L.) and regeneration of salt tolerant plants. *Molecular Biotechnology*, *40*, 161–169.

Batty, N., & Evans, J. (1992). Biological ballistics–no longer a shot in the dark. *Transgenic Research*, *1*, 107–113.

Bidney, D., Scelonge, C., Martich, J., Burrus, M., Sims, L., & Huffman, G. (1992). Microprojectile bombardment of plant tissues increases transformation frequency by *Agrobacterium tumefaciens*. *Plant Molecular Biology*, *18*, 301–313.

Bonnefoy, N., & Fox, T. D. (2007). Directed alteration of *Saccharomyces cerevisiae* mitochondrial DNA by biolistic transformation and homologous recombination. *Methods in Molecular Biology (Clifton, N.J.)*, *372*, 153–166.

Boynton, J., Gillham, N., Harris, E., Hosler, J., Johnson, A., & Jones, A., et al. (1988). Chloroplast transformation in Chlamydomonas with high velocity microprojectiles. *Science*, *240*, 1534–1538.

Chen, L., Marmey, P., Taylor, N. J., Brizard, J. -P., Espinoza, C., & D'Cruz, P., et al. (1998). Expression and inheritance of multiple transgenes in rice plants. *Nature Biotechnology*, *16*, 1060–1064.

Christou, P. (1990). Morphological description of transgenic soybean chimeras created by the delivery, integration and expression of foreign DNA using electric discharge particle acceleration. *Annals of Botany*, *66*, 379–386.

Christou, P. (1996). Transformation technology. *Trends in Plant Science*, *1*, 423–431.

Christou, P., McCabe, D. E., & Swain, W. F. (1988). Stable transformation of soybean callus by DNA-coated gold particles. *Plant Physiology*, *87*, 671–674.

Christou, P., Swain, W. F., Yang, N. -S., & McCabe, D. E. (1989). Inheritance and expression of foreign genes in transgenic soybean plants. *Proceedings of the National Academy of Sciences of the United States of America*, *86*, 7500–7504.

Dai, S., Zheng, P., Marmey, P., Zhang, S., Tian, W., & Chen, S., et al. (2001). Comparative analysis of transgenic rice plants obtained by *Agrobacterium*-mediated transformation and particle bombardment. *Molecular Breeding*, *7*, 25–33.

Daniell, H., Vivekananda, J., Nielsen, B. L., Ye, G. N., Tewari, K. K., & Sanford, J. C. (1990). Transient foreign gene expression in chloroplasts of cultured tobacco cells after biolistic delivery of chloroplast vectors. *Proceedings of the National Academy of Sciences of the United States of America*, *87*, 88–92.

De La Pena, A., Loerz, H., & Schell, J. (1987). Transgenic rye plants obtained by injecting DNA into young floral tillers. *Nature (London)*, *325*, 274–276.

Duan, X., & Chen, S. (1985). Variation of the characters of rice (*Oryza sativa*) induced by foreign DNA uptake. *China Agricultural Science*, *3*, 6–9.

D'Halluin, K., Bonne, E., Bossut, M., De Beuckeleer, M., & Leemans, J. (1992). Transgenic maize plants by tissue electroporation. *Plant Cell*, *4*, 1495–1505.

Ferreira, S. A., Pitz, K. Y., Manshardt, R., Zee, F., Fitch, M., & Gonsalves, D. (2002). Virus coat protein transgenic papaya provides practical control of papaya ringspot virus in Hawaii. *Plant Disease*, *86*, 101–105.

Fitch, M., & Manshardt, R. (1990). Somatic embryogenesis and plant regeneration from immature zygotic embryos of papaya (*Carica papaya* L.). *Plant Cell Reports*, *9*, 320–324.

Fitch, M. M. M., Manshardt, R. M., Gonsalves, D., Slightom, J. L., & Sanford, J. C. (1990). Stable transformation of papaya via microprojectile bombardment. *Plant Cell Reports*, *9*, 189–194.

Fitch, M. M. M., Manshardt, R. M., Gonsalves, D., Slightom, J. L., & Sanford, J. C. (1992). Virus resistant papaya derived from tissues bombarded with the coat protein gene of papaya ringspot virus. *BioTechnology*, *10*, 1466–1472.

Fu, X., Duc, L. T., Fontana, S., Bong, B. B., Tinjuangjun, P., & Sudhakar, D., et al. (2000). Linear transgene constructs lacking vector backbone sequences generate low-copy-number transgenic plants with simple integration patterns. *Transgenic Research*, *9*, 11–19.

Gonsalves, C., Lee, D. & Gonsalves, D. (2004). Transgenic virus resistant papaya: The Hawaiian "Rainbow" was rapidly adopted by farmers and is of major importance in Hawaii today. *Online. APSnet Feature, American Phytopathological Society, August-September* <http://www.apsnet.org/online/feature/rainbow/>

Gonsalves, D. (1998). Control of papaya ringspot virus in papaya: A case study. *Annual Review of Phytopathology*, *36*, 415–437.

Gonsalves, D., Ed. (2004). Virus Resistant Transgenic Papaya in Hawaii: A Case for Technology Transfer to Lesser Developed Countries. OECD/USAID/ARS Conference on Virus Resistant Transgenic Papaya in Hawaii: A Case for Technology Transfer to Lesser Developed Countries, held in Hilo, HI October 22–24, 2003. Edited by Gonsalves, D. Hilo, Hawaii: Petroglyph Press, LTD, Hilo, Hawaii 96720.

Gonsalves, D. (2006). Transgenic papaya: Development, release, impact, and challenges. *Advances in Virus Research*, *67*, 317–354.

Gonsalves, D., Ferreira, S., Manshardt, R., Fitch, M. & Slightom, J. (1998). Transgenic virus resistant papaya: New hope for control of papaya ringspot virus in Hawaii. *APSnet feature story for September 1998 on world wide web. Address is: <http://www.apsnet.org/education/feature/papaya/Top.htm/>*.

Gonsalves, D., Gonsalves, C., Ferreira, S., Pitz, K., Fitch, M., Manshardt, R. & et al. (2004). Transgenic virus resistant papaya: From hope to reality for controlling of papaya ringspot virus in Hawaii. *Online. APSnet Feature, American Phytopathological Society, August-September, <http://www.apsnet.org/online/feature/ringspot/>* July 2004.

Gonsalves, D., Suzuki, J., Tripathi, S. & Ferreira, S. (2008). Papaya ringspot virus (Potyviridae). In Mahy, B. & Van Regenmortel, M., (eds.), *Encyclopedia of virology* (5 vols, 3rd Edition Vol. 4, pp. vol. 4, pp. 1–8. 5 vols). Elsevier, Oxford.

Gonsalves, D., Vegas, A., Prasartsee, V., Drew, R., Suzuki, J., & Tripathi, S. (2006). Developing papaya to control papaya ringspot virus by transgenic resistance, intergeneric hybridization, and tolerance breeding. *Plant Breeding Reviews*, *26*, 35–78.

Griesbach, R. J. (1987). Chromosome-mediated transformation via microinjection. *Plant Science (Shannon)*, *50*, 69–78.

Griesbach, R. J. (1994). An improved method for transforming plants through electrophoresis. *Plant Science (Limerick)*, *102*, 81–89.

Holm, P. B., Olsen, O., Schnorf, M., Brinch-Pedersen, H., & Knudsen, S. (2000). Transformation of barley by microinjection into isolated zygote protoplasts. *Transgenic Research*, *9*, 21–32.

James, C. (2009). Global status of commercialized biotech/GM crops: 2009. *ISAAA Brief 41-2009: Executive Summary*.

Johnston, S., Anziano, P., Shark, K., Sanford, J., & Butow, R. (1988). Mitochondrial transformation in yeast by bombardment with microprojectiles. *Science*, *240*, 1538–1541.

Kaeppler, H. E., Gu, W., Somers, D. A., Rines, H. W., & Cockburn, A. E. (1990). Silicon carbide fiber-mediated DNA delivery into plant cells. *Plant Cell Reports*, *9*, 415–418.

Kikkert, J. (1993). The Biolistic® PDS-1000/He device. *Plant Cell, Tissue and Organ Culture*, *33*, 221–226.

Klein, T., Wolf, E., Wu, R., & Sanford, J. (1987). High-velocity microprojectiles for delivery of nucleic acids in living cells. *Nature*, *326*, 70–73.

Klein, T. M., Fromm, M., Weissinger, A., Tomes, D., Schaaf, S., & Sletten, M., et al. (1988). Transfer of foreign genes into intact maize cells with high-velocity microprojectiles. *Proceedings of the National Academy of Sciences of the United States of America*, *85*, 4305–4309.

Klein, T. M., Harper, E. C., Svab, Z., Sanford, J. C., Fromm, M. E., & Maliga, P. (1988). Stable genetic transformation of intact *Nicotiana* cells by the particle bombardment process. *Proceedings of the National Academy of Sciences of the United States of America*, *85*, 8502–8505.

Klein, T. M., Wolf, E. D., Wu, R., & Sanford, J. C. (1987). High-velocity microprojectiles for delivering nucleic acids into living cells. *Nature*, *327*, 70–73.

Liu, J., Su, Q., Au, L., & Yang, A. (2009). Transfer of a minimal linear marker-free and vector-free smGFP cassette into soybean via ovary-drip transformation. *Biotechnology Letters*, *31*, 295–303.

Lowe, B., Shiva Prakash, N., Way, M., Mann, M., Spencer, T., & Boddupalli, R. (2009). Enhanced single copy integration events in corn via particle bombardment using low quantities of DNA. *Transgenic Research*, *18*, 831–840.

Luo, Z. -x., & Wu, R. (1989). A simple method for the transformation of rice via the pollen-tube pathway. *Plant Molecular Biology Reporter*, *7*, 69–77.

Manshardt, R. M. (1998). *"UH Rainbow" papaya*. University of Hawaii College of Tropical Agriculture and Human Resources New Plants for Hawaii-1. (2pp).

McCabe, D., & Christou, P. (1993). Direct DNA transfer using electric discharge particle acceleration (ACCELL™ technology). *Plant Cell, Tissue and Organ Culture*, *33*, 227–236.

McCabe, D. E., & Martinell, B. J. (1993). Transformation of elite cotton cultivars via particle bombardment of meristems. *Nat. Biotech.*, *11*, 596–598.

McGranahan, G. H., Leslie, C. A., Uratsu, S. L., Martin, L. A., & Dandekar, A. M. (1988). *Agrobacterium*-mediated transformation of walnut somatic embryos and regeneration of transgenic plants. *BioTechnology*, *6*, 800–804.

Mizuno, K., Takahashi, W., Beppu, T., Shimada, T., & Tanaka, O. (2005). Aluminum borate whisker-mediated production of transgenic tobacco plants. *Plant Cell Tissue and Organ Culture*, *80*, 163–169.

Murakawa, T., Kajiyama, S. i., & Fukui, K. (2008). Improvement of bioactive bead-mediated transformation by concomitant application of electroporation. *Plant Biotechnology*, *25*, 387–390.

Murakawa, T., Kajiyama, S. i., Ikeuchi, T., Kawakami, S., & Fukui, K. (2008). Improvement of transformation efficiency by bioactive-beads-mediated gene transfer using DNA-lipofectin complex as entrapped genetic material. *Journal of Bioscience and Bioengineering*, *105*, 77–80.

Pescitelli, S. M., & Sukhapinda, K. (1995). Stable transformation via electroporation into maize type-II callus AND regeneration of fertile transgenic plants. *Plant Cell Reports*, *14*, 712–716.

Powell-Abel, P., Nelson, R. S., De, B., Hoffmann, N., Rogers, S. G., & Fraley, R. T., et al. (1986). Delay of disease development in transgenic plants that express the tobacco mosaic virus coat protein gene. *Science*, *232*, 738–743.

Prasartsee, V., Chikiatiyos, S., Palakorn, K., Chaeuychoom, P., Kongpolprom, W., Wichainum, S., et al. (1998). *The New Technologies for the Development of Sustainable Farming in Northeast. Proceedings of JIRCAS-ITCAD Seminar, March 24, 1998, Khon Kaen, Thailand.*

Rao, A. Q., Bakhsh, A., Kiani, S., Shahzad, K., Shahid, A. A., & Husnain, T., et al. (2009). The myth of plant transformation. *Biotechnology Advances*, *27*, 753–763.

Sanford, J. (1991). An improved, helium-driven biolistic device. *Technique*, *3*, 3–16.

Sanford, J. C. (2000). Turning point article: The development of the biolistic process. *In Vitro Cellular and Developmental Biology-Plant*, *36*, 303–308.

Sanford, J. C., & Johnston, S. A. (1985). The concept of parasite-derived resistance–Deriving resistance genes from the parasite's own genome. *Journal of Theoretical Biology*, *113*, 395–405.

Sanford, J. C., Klein, T. M., Wolf, E. D., & Allen, N. (1987). Delivery of substances into cells and tissues using a particle bombardment process. *Particulate Science and Technology*, *5*, 27–37.

Sanford, J. C., Smith, F. D., & Russell, J. A. (1992). Optimizing the biolistic process for different biological applications. *Methods in Enzymology*, *217*, 483–509.

Sone, T., Nagamori, E., Ikeuchi, T., Mizukami, A., Takakura, Y., & Kajiyama, S. I., et al. (2008). A novel gene delivery system in plants with calcium alginate micro-beads. *Journal of Bioscience and Bioengineering*, *94*, 87–91.

Suzuki, J. Y., Tripathi, S., Fermin, G. A., Jan, F. -J., Hou, S., & Saw, J. H., et al. (2008). Characterization of insertion sites in rainbow papaya, the first commercialized transgenic papaya fruit crop. *Tropical Plant Biology*, *1*, 293–309.

Tang, D. -c., DeVit, M., & Johnston, S. A. (1992). Genetic immunization is a simple method for eliciting an immune response. *Nature*, *356*, 152–154.

Tennant, P., Fermin, G., Fitch, M. M., Manshardt, R. M., Slightom, J. L., & Gonsalves, D. (2001). Papaya ringspot virus resistance of transgenic rainbow and sunup is affected by gene dosage, plant development, and coat protein homology. *European Journal of Plant Pathology*, *107*, 645–653.

Thompson, J. A., Drayton, P. R., Frame, B. R., Wang, K., & Dunwell, J. M. (1995). Maize transformation utilizing silicon carbide whiskers: A review. *Euphytica*, *85*, 75–80.

Travella, S., Ross, S. M., Harden, J., Everett, C., Snape, J. W., & Harwood, W. A. (2005). A comparison of transgenic barley lines produced by particle bombardment and *Agrobacterium*-mediated techniques. *Plant Cell Reports*, *23*, 780–789.

Ueki, S., Lacroix, B., Krichevsky, A., Lazarowitz, S. G., & Citovsky, V. (2009). Functional transient genetic transformation of *Arabidopsis* leaves by biolistic bombardment. *Nature Protocols*, *4*, 71–77.

Wang, H. -H., Yin, W. -B., & Hu, Z. -M. (2009). Advances in chloroplast engineering. *Journal of Genetics and Genomics*, *36*, 387–398.

Yamagishi, N., Terauchi, H., Kanematsu, S., & Hidaka, S. (2006). Biolistic inoculation of soybean plants with soybean dwarf virus. *Journal of Virological Methods*, *137*, 164–167.

Yang, A., Su, Q., & An, L. (2009). Ovary-drip transformation: a simple method for directly generating vector- and marker-free transgenic maize (*Zea mays* L.) with a linear GFP cassette transformation. *Planta (Berlin)*, *229*, 793–801.

Young, J., Kuykendall, L., Martinez-Romero, E., Kerr, A., & Sawada, H. (2001). A revision of *Rhizobium Frank* 1889, with an amended description of the genus, and the inclusion of all species of *Agrobacterium* Conn 1942 and Allorhizobium undicola de Lajudie et al. 1998 as new combinations: Rhizobium radiobacter, R. rhizogenes, R. rubi, R. undicola and R. vitis. *International Journal of Systematic and Evolutionary Microbiology*, *51*, 89–103.

Zhang, H., Zhao, F., Zhao, Y., Guo, C., Li, C., & Xiao, K. (2009). Establishment of transgenic cotton lines with high efficiency via pollen-tube pathway. *Frontiers of Agriculture in China*, *3*, 359–365.

Plant tissue culture for biotechnology

9

Prakash P. Kumar Chiang Shiong Loh
National University of Singapore, Singapore

TABLE OF CONTENTS

Introduction

Regeneration of plants via tissue culture is based on the principle of totipotency originally proposed by Haberlandt in 1902 (see Vasil, 2008). Early attempts to culture excised plant parts included root culture (White, 1934) and shoot tip or axillary bud culture for micropropagation (Morel, 1960). Subsequently, regenerating whole plants via somatic embryogenesis from cultured callus tissue of carrot (Steward et al., 1958; Reinert, 1959) as well as regeneration of a whole plant starting from a single cell of tobacco (Vasil and Hildebrandt, 1965) were accomplished. Around the same time Miller et al. (1955) reported that using the appropriate relative ratio of auxin to cytokinin in the nutrient medium can induce plant regeneration in culture. Murashige and Skoog (1962) developed a well-defined mineral nutrient formulation based on their analysis of tobacco leaf composition, which helped to sustain growth and division of tobacco cells and tissues. The MS medium, as it is now known, has indeed become a noteworthy contribution to plant tissue culture. More recently, the technique of *Agrobacterium*-mediated gene transfer and regeneration of transgenic plants was discovered (Herrera-Estrella et al., 1983; De Block et al., 1984), which has proven to be useful when introducing desirable agronomic traits to transgenic crop plants (Shah et al., 1986). These were seminal findings that firmly established plant tissue culture as a valuable tool for research, as well as crop improvement by biotechnology.

As the name indicates, plant tissue culture involves excising plant tissues and growing them on nutrient media. The term "tissue culture" is used rather broadly to include several variations. such as meristem culture for propagation of virus-free plants (e.g., orchids, strawberry, grape, and potato), protoplast culture, cell suspension culture, tissue and organ culture (Gamborg, 2002), and anther or pollen culture for producing haploid plants (Guha and Maheshwari, 1966). In general, the goal for crop improvement via tissue culture could be achieved by:

1. Initiation and establishment of cultures.
2. Performing the specific modifications, which include clonal propagation, virus elimination, selection of variants, genetic transformation, and so forth.
3. Regeneration of plants with the desired modifications.

In this chapter we focus on various technical aspects of plant tissue culture. The basic laboratory setup, handling of explant tissue, nutrient medium and establishing the culture, and incubation of cultures will be discussed. This will be followed by a brief discussion of the implications of plant tissue culture technology for agricultural biotechnology (see also Chapter 10).

Plant Tissue Culture Technology

The basic laboratory setup

A laboratory that can handle plant biochemistry or physiology-type experiments will meet most of the general requirements

DOI: 10.1016/B978-0-12-381466-1.00009-2

of plant tissue culture. Some of the major items needed include general glassware and disposable plasticware, chemicals that serve as mineral nutrients for media preparation, a supply of distilled water or high-quality deionized water, pH meter, magnetic stirrer, weighing balance, autoclave, filter sterilization units, microwave oven, refrigerator, and freezer. Specialized equipment such as laminar flow, a sterile air workstation to handle aseptic explants, media, and so forth will be essential. Incubation of the cultures may be done on shelves fitted with lighting. Cultures may also be incubated in growth chambers or rotary shakers (for cell suspensions). For data recording and image capture, cameras and microscopes with photography attachments will be required.

Preparation of tissue for culturing

The part of the plant that is excised and put in culture is referred to as the explant. This may be leaf disks, cotyledons, hypocotyls, roots, shoot tips, axillary buds, or zygotic embryos. The selected explants have to be rendered aseptic, which often involves surface sterilization using various dilutions (10–30% v/v) of Clorox® or commercial bleach that has sodium hypochlorite (5.25% w/v) as the active ingredient. Other chemicals such as ethanol (1 to 5 min rinses in 70% ethanol) or silver nitrate solution may also be used for surface sterilization. Prior to inoculating the explants onto the nutrient medium, they have to be rinsed thoroughly with autoclaved distilled water to remove traces of the chemical used and then trimmed to get rid of the cells along the edges that might have been killed by the harsh chemicals used for sterilizing.

There are several simplified, modified methods of surface sterilization where the explant donor tissues, such as seed pods and floral buds, may be directly dipped in ethanol (70%) and, in some cases, flamed briefly. The seeds or anthers are then excised aseptically for culture. With this procedure the explants never come in contact with the chemical surface-sterilization agent, giving a better survival rate for the explants. This method is useful for intact seed pods of orchids or anthers in unopened floral buds of several species.

Certain materials, especially field-grown plants, are extremely difficult to surface sterilize. In addition, selected field-grown plants with desired characteristics for clonal propagation might be located too far from the laboratory. Thus, pre-surface sterilization treatments have to be devised. For example, in guava, scions were obtained from selected field-grown plants and grafted to seedling root-stocks (Loh and Rao, 1989). Grafted plants were kept in the laboratory for harvesting nodal explants. To remove apical dominance and encourage sprouting of the axillary buds, healthy scion branches were decapitated 5 to 8 days prior to excision of the nodal explants. The nodal segments were then surface sterilized in 80% ethanol followed by 5% and 3% Clorox solutions before successful establishment of cultures.

The age and physiological state of the explant donor plant may have significant influence on the success of regeneration of plants. For example, various studies have shown that cotyledons from 3- to 6-day-old seedlings of *Brassica* species serve as the best source of explants for adventitious shoot regeneration and *Agrobacterium*-mediated genetic transformation (Sharma et al., 1990; Gasic and Korban, 2006). When leaves are the sources of explant, for example in petunia, the first fully expanded leaf is the preferred material. For mangosteen, a woody tree species, only young red leaves produced shoot buds in culture (Goh et al., 1990). When mature, green leaves were used the frequency of shoot bud formation decreased (Goh et al., 1994). Furthermore, leaf segments (3 mm transverse sections of mangosteen leaves) showed a strong polarity of regeneration with shoot buds arising from the midrib near the distal (apical) cut end of leaf segments (Goh et al., 1994). Seedling roots and hypocotyls are also used in several species as the explant. In some of the cereals such as rice and corn, and various grasses (Hiei et al., 1994; Frame et al., 2002), as well as in several of the coniferous tree species (Gupta and Durzan, 1986; Lu et al., 1991), the zygotic embryo serves as the preferred explant for culture initiation.

Once a suitable explant is selected and prepared for culture, it should be incubated on an appropriate nutrient medium for growth and differentiation.

Nutrient media

Various mineral formulations are available to culture plant tissues. The major media include MS medium (Murashige and Skoog, 1962) and Gamborg's B5 medium (Gamborg et al., 1968). Generally, the plant tissue culture media are made up of macro- and micronutrients, vitamins, phytohormones, and other adjuvants (such as coconut water), as well as sucrose (2–3% w/v, it can be commercial grade refined sugar for routine propagation, but should be analytical grade sucrose for research purposes). The nutrient media can be prepared by mixing stock solutions of various chemical ingredients (Gamborg et al., 1976) or from commercially available ready mixed powder (e.g., www.plantmedia.com) as recommended by the manufacturers. Other formulations for specific species may be used based on earlier publications. Adjuvants such as ascorbic acid, polyvinyl pyrrolidone, and activated charcoal may be required for some species that show extreme cases of tissue browning on excision and secretion of polyphenolic substances from the damaged cells.

The pH of the medium is adjusted to 5.8 ±0.2 using dilute NaOH or HCl and a suitable gelling agent is added when a semisolid matrix is desired. Agar (8–10 g/l) and the gellan gum known as Phytagel® or Gelrite® (2–3 g/l) are two of the most common gelling agents used for plant tissue culture media. The medium is sterilized by autoclaving for about 20 minutes. Care should be taken to fill the containers to no more than half the total volume for effective autoclaving.

Cell culture can be in the form of a liquid suspension of small clusters of cells, apart from callus culture on semisolid nutrient medium as previously indicated. When liquid nutrient medium is required it may be similarly prepared (without the gelling agent) and sterilized by autoclaving or by filter sterilization using filters with 0.22 μm pore size. Also, heat labile components of the media such as some phytohormones (indoleacetic acid, gibberellic acid) and antibiotics, if used, should be filter sterilized and added to autoclaved medium that is cooled to about 60°C before aliquoting the medium to aseptic culture vessels.

Selection of the appropriate nutrient medium for a given species or tissue is generally arrived at by empirical trials. This can be done by a systematic screen of various concentrations of the media components such as the broad spectrum screen using stock solutions of the various groups of mineral components (De Fossard; ronalddefossard.com/author.htm). Thus the medium components would be grouped under four categories and use three concentrations for each (low, medium, high) category and prepare various combinations of the components. Alternatively, one can start from the standard MS medium and vary the composition of the various macro- and micronutrients, vitamins, and phytohormones.

Perhaps one of the major components that has a significant effect on regeneration is the type and concentration of phytohormones in the medium. It is worth remembering that the discovery of the phytohormone cytokinins led to the central dogma of tissue culture, namely, endogenous ratio of cytokinin to auxin determining the nature of regeneration (Skoog and Miller, 1957). According to this, a relatively high ratio of cytokinin to auxin within the explant favors shoot regeneration, a relatively high auxin to cytokinin ratio leads to root regeneration, and the intermediate ratio causes callus proliferation. Often the concentration of the phytohormones in the medium is higher (normally 10^{-7}–10^{-5} M), because the endogenous concentration depends on the efficiency of uptake of the compound by the explant from the external medium. Hence, optimization of the appropriate phytohormone concentrations in the medium can also be empirically determined in the earlier set of exploratory experiments. Once an optimum medium combination is identified, it can be used as a defined medium for the species, and often for closely related species of plants.

Types of culture

We can identify various types of plant tissue culture. Based on the scale of operations, one can classify laboratory- and industrial-scale cell cultures. The laboratory-scale cultures can be handled in relatively small growth chambers. However, industrial-scale cultures (i.e., commercial micropropagation programs) will require sufficiently large facilities, the details of which are beyond the scope of this discussion.

Based on the nature and exposure of the explants to the nutrient medium, we can identify continuous immersion or batch cultures and periodic immersion, as well as culture on semisolid medium (solidified with agar or Gelrite™). The major types of cultures that can be established are summarized in Figure 9.1. The type of culture to be established will depend on the purpose of culturing, such as micropropagation, secondary metabolite production, *in vitro* flowering, and so forth. In all cases, periodic subculturing has to be conducted to replenish nutrients and to eliminate potentially toxic exudates. Subculture intervals for cell suspension cultures can be generally between 10 to 14 days. Cultures on semisolid nutrient media need to be transferred to fresh media once in 4 to 6 weeks.

Protoplast isolation and culture has been a valuable tool for early attempts at transgenic plant production for species that were not amenable for *Agrobacterium*-mediated transformation. When plasmid DNA is incubated with protoplasts in the presence of polyethylene glycol and calcium, they can be induced to take up the DNA and subsequently, a small percentage of the resulting cells may have the DNA incorporated into their genomes. This process is inefficient for transgenic plant production with most of the species. However, in more recent years, this has become a valuable research tool to study transient expression of proteins *in vivo* and subcellular localization of tagged proteins (e.g., fusion protein with green fluorescent protein), or for testing the activation and compartmentalization of such tagged proteins (Yoo et al., 2007). This offers a relatively quick experimental system to test hypotheses and obtain valuable preliminary data that can form the basis of more detailed studies of complex regulatory networks and whole plant physiology.

Environmental aspects of tissue culture

The choice of culture vessels can vary and Erlenmeyer flasks, culture tubes, petri dishes, and specialized containers may be used to establish plant tissue cultures. Cell cultures for extraction of metabolites (e.g., Taxol or recombinant proteins) are generally established in bioreactor vessels of desirable scale.

The influence of the headspace gas composition (or the gas space above the culture medium within the culture container) on growth and differentiation is well recorded (Kumar et al., 1987, 1996, 1998; Kumar and Thorpe, 1991; Kozai et al., 1995). One of the major components in the headspace gas is ethylene, the gaseous phytohormone. It can exert an influence on regeneration even after it is released from the explant tissues (Kumar et al., 1987, 1996, 1998). The partial pressures of carbon dioxide and oxygen in the headspace gas can also exert some influence on regeneration (Nguyen et al., 2001). The use of vented lids, specialty vented plastic film containers, or incorporation of chemicals that act as inhibitors of ethylene action (e.g., $AgNO_3$, silver thiosulfate) into the culture medium help in optimizing regeneration.

Once the explants are inoculated onto the nutrient medium under aseptic conditions working in a laminar flow workstation, they are incubated at 25 ±2°C. Depending on the needs of the culture, incubation may be under a photoperiod with 12 to 26 hours of light or in darkness for a specific initial culture period. Incubation of the cultures may be done on shelves fitted with lighting or in growth chambers. Although the most common type of light source is the fluorescent bulb, more recently, light emitting diode (LED) light sources are becoming available for this purpose. The LED light panels consume less power, generate less heat, and can be configured to deliver lights of specific wavelengths (e.g., 460 nm blue and 630 nm red) so that plants can be grown under light suitable for absorption by chlorophyll.

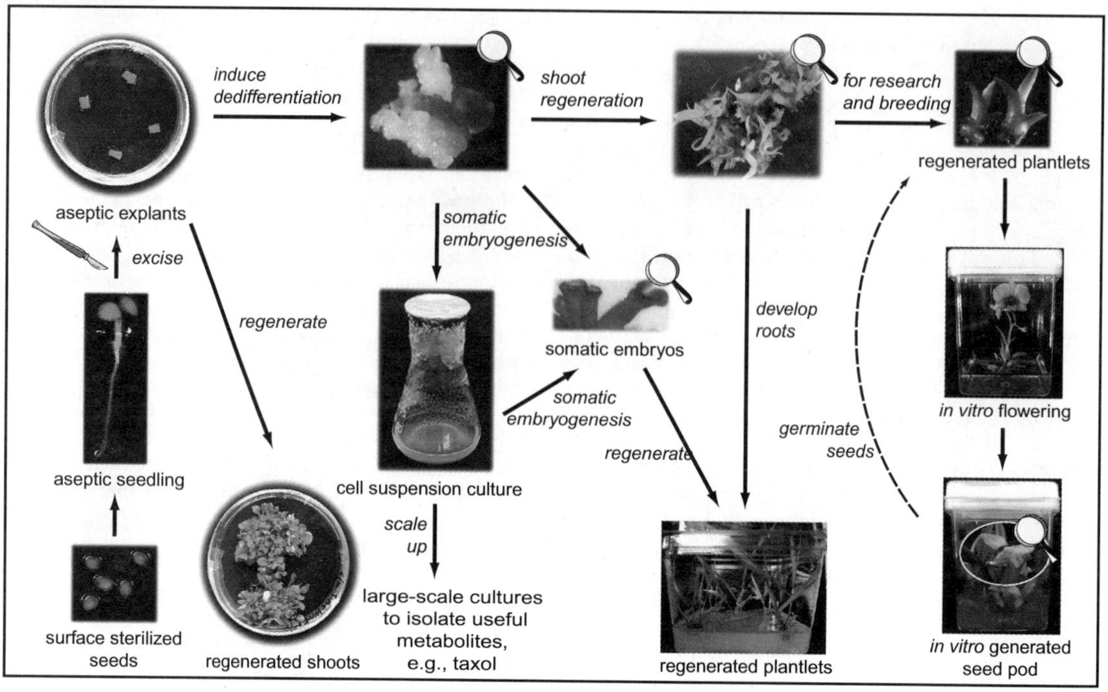

Figure 9.1 • Schematic representation of various types of plant tissue cultures • Photographs corresponding to the different stages of culture (from the left) leading to plant regeneration from excised leaf segments or establishment of callus and cell suspension cultures with somatic embryogenesis are arranged sequentially to illustrate the process. Regeneration of switchgrass plantlets from callus and orchid plant regeneration via protocorm-like bodies as well as *in vitro* flower induction and seed pod formation in *Dendrobium* sp. are illustrated on the right half of the collage. Photographs with a magnifying lens indicate that they are close-up views of a larger field of culture. Please see color plate section at the end of the book.

Modes of regeneration

As indicated earlier, the concept of totipotency was realized in plant cells with the ability to regenerate whole new plants starting from single cells. Such a process involves dedifferentiation from partially differentiated cell/tissue type (e.g., parenchyma) to a meristem-like state followed by redifferentiation (regeneration) into well-organized structures. The totipotency of cells in plant meristem was the basis of the concept of "stem cells" in animals that has become such an intense research area in the past decade or so. *De novo* organogenesis (Figure 9.2A, B) and somatic embryogenesis (Figure 9.2C–E) are the two major routes of regeneration of plants in tissue culture. Organogenesis can occur directly as shoot buds or roots from the cultured tissues or via an intermediate callus stage. Direct regeneration of adventitious shoots is generally preferred for clonal propagation applications, because chromosomal aberrations may be introduced during some callus induction treatments. However, the choice of the pathway of regeneration is genetically determined and, depending on the species, protocols for only one route of regeneration may be available; for example, in cereals such as rice and corn regeneration is via callus. In either case, generally a small number of cells are involved in establishing the organ or somatic embryo.

The plantlets developed from shoot buds or by germination of somatic embryos can be grown to a sufficiently large size before they are put through a hardening process and acclimation to be grown in pots.

Implications for Agricultural Biotechnology

As indicated previously, plant tissue culture has contributed enormously to agricultural biotechnology. For example, many plants grown from seed show considerable variation in growth, flower characteristics, yield, disease resistance, resistance to environmental stress, and so forth (Altman, 2003; Thorpe, 2007). It would therefore be valuable to select those that possess desirable characters for vegetative multiplication. Tissue culture allows large-scale micropropagation applications to recover large numbers of plants with a comparable genotype (clones). This is also desirable in horticulture (i.e., slow-growing ornamental orchids; Figure 9.2G, H) and for fruit crops such as strawberry and grape. As is well documented, introduction of the transgene occurs in a limited number of cells after co-cultivation with *Agrobacterium* or biolistic bombardment of transgenes into explants. To recover

Figure 9.2 • Examples of various types of tissue culture • (A, B) Adventitious shoot regeneration from excised leaves of *Paulownia fortunei*. (B) Well-developed individual rooted plantlets from the leaf cultures in (A). (C) Cell suspension culture of *Brassica napus* ssp. *oleifera* in liquid medium. (D) Cultured cell clusters from the suspension. (E) Somatic embryogenesis in cell suspension culture. (F) Tobacco anther cultures with germinated haploid embryos. (G) Orchid propagation starting from seeds (inset) and multiplication of protocorm-like bodies resulting from the seed germination *in vitro*. (H) Well-established orchid plantlets arising from the cultures in (G). This method is routinely used for large-scale propagation of various ornamental orchid species. Please see color plate section at the end of the book.

transgenic plants, plants from only those populations of cells need to be regenerated. Therefore, plant tissue culture has been the key tool to recovering transgenic plants in crop biotechnology. The thousands of hectares of currently planted transgenic crops in over 40 countries would not have been feasible without tissue culture tools.

Besides genetic transformation for crop improvement, other applications of tissue culture include selection of variants/mutants (somaclonal variation, mutation), cryopreservation, dihaploid production and embryo rescue, and production of useful chemicals (e.g., Taxol). Liquid cultures have been used for detecting trace elements and determining their toxicities to crop plants (Kopittke et al., 2010). Some of the specific examples of tissue culture technology applied for crop improvement are discussed in the next section.

Haploid tissue culture

The importance of haploid tissue culture merits special mention. Haploid tissue cultures are most often obtained by culturing microspores, pollen grains, and anthers (Figure 9.2F). Haploids in plant breeding programs are useful as a means of shortening the time required to complete backcross programs, as a means of helping to retain the character under transfer, and as a means of stabilizing the transferred genetic materials in a homozygous form. The use of haploids in a breeding program also allows the possibility of isolating an array of individual genomes whether dominant or recessive for selection, study, and recombination. The individual carrying the lethal genes would theoretically be eliminated from the population.

Somatic embryogenesis

In somatic embryogenesis, vegetative cells develop into plants through embryogenetic stages without the fusion of gametes (Williams and Maheswaran, 1986). Since the first reports of somatic embryogenesis in the tissue culture of carrot (Steward et al., 1958; Reinert, 1959), the importance of somatic embryogenesis in combining efficient cloning of desired genotypes has been realized (Sharp et al., 1980). The key rationale is that plants regenerated from direct somatic embryogenesis are often more uniform than plants regenerated via callus tissues (Maheswaran and Williams, 1984). Somatic embryos could also generate secondary somatic embryos from their surfaces. Secondary embryogenesis (also called recurrent, repetitive, cyclic, accessory, or proliferative embryogenesis) occurs when primary somatic embryos give rise to successive cycles of embryos (Figure 9.3). Secondary embryogenesis systems provide a way to produce large populations of vegetative propagules in a short span of time (Lee et al., 1997). Such secondary embryos may be useful for recovering a large number of plants from clonal propagation, genetic transformation, and induced mutation. Developing embryos or embryogenic cells could be exposed to microprojectile bombardment or other means of genetic transformation, and the transformed cells could be selected and regenerated into plants (Chen and Beversdorf, 1994). Thus, secondary embryos of cassava were used for induction of mutation *in vitro* through γ-irradiation and mutant plants with altered starch composition were obtained (Joseph et al., 2004).

Artificial seeds

Production of artificial seeds by embedding somatic embryos in a suitable matrix such as agarose or calcium alginate can be a useful tool for large-scale multiplication projects (Figure 9.3). As early as 1992, successful field planting of alfalfa artificial seeds derived from somatic embryos encapsulated in calcium alginate and subsequent conversion to plants had been reported (Fuji et al., 1992). The main objective for developing artificial seeds has been for the production of "clonal seeds" (Redenbaugh, 1990), so somatic embryos are appropriate for this purpose. However, the concept is now extended to include encapsulation of other tissue cultured or *in vitro* prepared materials (e.g., protocorm-like bodies, rooted shoot buds), so the objective for the making of artificial seeds may need to be expanded (Khor and Loh, 2005). It is interesting to note that the concept was extended to include the encapsulation of non-endospermic seeds or protocorms of commercially high-value species, such as orchids (Khor et al., 1998).

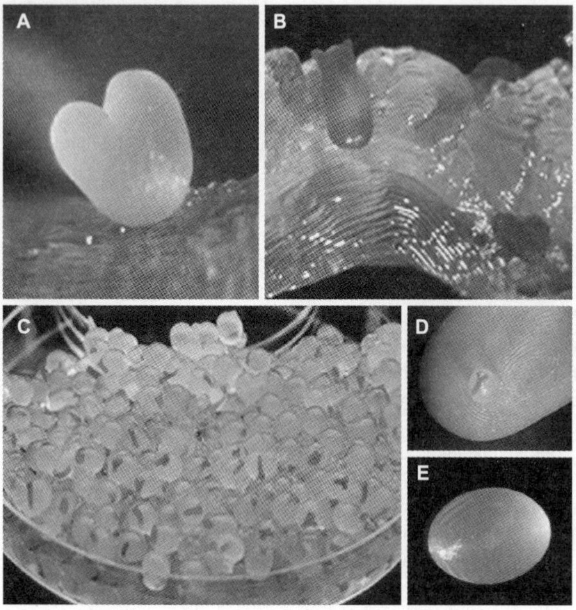

Figure 9.3 • (A) Heart-shaped somatic embryo of *Brassica napus* ssp. *oleifera*. (B) Secondary somatic embryogenesis from the hypocotyls of *B. napus* ssp. *oleifera*. (C, D, E) Artificial seeds prepared by embedding somatic embryos of *B. napus* ssp. *oleifera* in calcium alginate. Please see color plate section at the end of the book.

They demonstrated aseptic encapsulation of orchid embryos in a two-coat (alginate-chitosan and alginate-gelatin) "artificial seed" of about 4 mm diameter with simulated endosperm and seed coat in the final product. The aim was to allow such naturally non-endospermic seeds to germinate like seeds of other angiosperms without having to go through elaborate *in vitro* handling at the site of planting and reduce labor cost.

In vitro flowering

One of the more recent applications of tissue culture is to hasten the breeding cycle of ornamental species that have long juvenile phases; for example, orchids that require over three years of vegetative growth before flowering. Experience with *in vitro* flowering of *Dendrobium* hybrid seedlings showed that the flowering time could be significantly shortened. Under optimal conditions, *in vitro* flowering could be observed 5 months after seed germination instead of the over 30 months required in field-grown plants (Sim et al., 2007). Segregation of flower colors was observed in these *in vitro* flowers, hence early assessment of the flower characteristics, such as color, shape, and size is possible using the *in vitro* flowering system. This will, in turn, reduce labor costs and optimize the space required for normal orchid breeding (Sim et al., 2007). In addition, flowers induced in culture could be self-pollinated (Hee et al., 2007) *in vitro* or pollinated with pollen grains harvested from field-grown plants (Sim et al., 2007) resulting in seed pod formation in culture. Therefore, *in vitro* flower formation followed by *in vitro* pollination and seed production for evaluation of the flower phenotypes in the next generation could be achieved within 11 months for *Dendrobium* hybrid, which would have taken over 6 years under the normal breeding cycle (Hee et al., 2007).

Future Perspectives

As can be seen from the previous discussion, plant tissue culture promises to continue to be a valuable tool for research on morphogenesis, cell signaling, physiology, and molecular biology, as well as crop improvement by biotechnology. The benefits of biotechnologically improved crops are apparent with the wide introduction of genetically modified cotton (Bt cotton) in China and India. Based on the 2009 estimation, there was a net reduction of 60% in pesticide use (reduction of 43 kg of pesticide application per hectare) in China, and an increase in yield of about 50% (from about 300 kg/ha in 2002–2003 to about 567 kg/ha in 2007–2008) in India (Paarlberg, 2010). Also, apart from the resulting socio-economic benefits for farmers, such a remarkable increase in yield has made India a major exporter of cotton (1.2 million bales in 2003–2004 to about 8.5 million bales in 2007–2008). Furthermore, China has already approved GM rice for cultivation, and this is expected to result in improvements in yield similar to that which has been realized with other food crops such as corn, soybean, and canola. The biofuel industry is also poised for rapid expansion in the coming decade. The approximately 12 billion gallons of American bioethanol production in 2010 is largely from corn starch and first generation cellulosic ethanol plants such as switchgrass, with about 30% of the annual corn crop being used for ethanol production (Pueppke, 2010). Bioethanol production is predicted to increase to about 30 billion gallons in 2020, with corn starch and advanced biofuels contributing equally to this (i.e., 15 billion gallons from genetically improved energy crops such as switchgrass, *Miscanthus*, sugarcane, and microalgae along with another 15 billion gallons from corn starch).

With the predicted need of crop yield to be doubled by 2050 in order to sustain the food, fiber, and fuel needs of the ever-increasing human population, the need to use technology for crop improvement cannot be overemphasized. It is clear that development of improved crops such as those with high water use efficiency and tolerance to multiple biotic and abiotic stresses combined with high yield are currently progressing with judicious use of plant biotechnology. Hence, it is safe to predict that the firmly established technology of plant tissue culture will continue to contribute significantly to agricultural biotechnology in the decades to come.

Acknowledgments

We would like to thank Ms. Petra Stamm for helping to prepare the figures and Mr. Koh Teng Seah for providing the orchid culture photographs. The current research in Kumar Lab is funded by grants from the National University of Singapore (R154-000-407-112) and the Science and Engineering Research Council (SERC), Singapore (R154-000-441-305).

References

Altman, A. (2003). From plant tissue culture to biotechnology: Scientific revolutions, abiotic stress tolerance, and forestry. *In Vitro Cellular and Developmental Biology-Plant, 39*, 75–84.

Chen, K. L., & Beversdorf, W. D. (1994). A combined use of microprojectile and DNA imbibitions enhances transformation frequency of canola (*Brassica napus* L.). *Theoretical and Applied Genetics, 88*, 187–192.

De Block, M., Herrera-Estrella, L., Van Montagu, M., Schell, J., & Zambryski, P. (1984). Expression of foreign gene in regenerated plants and their progeny. *EMBO Journal, 3*, 1681–1689.

Frame, B. R., Shou, H., Chikwamba, R. K., Zhang, Z., Xiang, C., & Fonger, T. M., et al. (2002). *Agrobacterium tumefaciens*-mediated transformation of maize embryos using a standard binary vector system. *Plant Physiology, 129*, 13–22.

Fuji, J. A., Slade, D., Aguirre-Rascon, J., & Redenbaugh, K. (1992). Field planting of alfalfa artificial seeds. *In Vitro Cellular and Developmental Biology-Plant, 28P*, 73–80.

Gamborg, O. L. (2002). Plant tissue culture. Biotechnology. Milestones. *In Vitro Cellular and Developmental Biology-Plant, 38*, 84–98.

Gamborg, O. L., Miller, R. A., & Ojima, K. (1968). Nutrient requirements of suspension cultures of soybean root cells. *Experimental Cell Research, 50*, 151–158.

Gamborg, O. L., Murashige, T., Thorpe, T. A., & Vasil, I. K. (1976). Plant tissue culture media. *In Vitro Cellular and Developmental Biology-Plant, 12*, 473–478.

Gasic, K., & Korban, S. S. (2006). Indian Mustard [*Brassica juncea* (L.) Czern.]. *Methods in Molecular Biology, 343*, 281–290.

Goh, C. J., Lakshmanan, P., & Loh, C. S. (1994). High frequency direct shoot bud regeneration from excised leaves of mangosteen (*Garcinia mangostana* L.). *Plant Science, 101*, 173–180.

Goh, K. L., Rao, A. N., & Loh, C. S. (1990). Direct shoot bud formation from leaf explants of seedlings and mature mangosteen (*Garcinia mangostana* L.) trees. *Plant Science, 68*, 113–121.

Guha, S., & Maheshwari, S. C. (1966). Cell division and differentiation of embryos in the pollen grains of Datura *in vitro*. *Nature, 212*, 97–98.

Gupta, P. K., & Durzan, D. J. (1986). Somatic polyembryogenesis from callus of mature sugar pine embryos. *Bio/Technology, 4*, 643–645.

Hee, K. H., Loh, C. S., & Yeoh, H. H. (2007). Early *in vitro* flowering and seed production in culture in *Dendrobium* Chao Praya Smile (Orchidaceae). *Plant Cell Reports, 26*, 2055–2062.

Herrera-Estrella, L., Depicker, A., Van Montagu, M., & Schell, J. (1983). Expression of chimaeric genes transferred in to plant cells using a Ti plasmid-derived vector. *Nature, 303*, 209–213.

Hiei, Y., Ohta, S., Komari, T., & Kumashiro, T. (1994). Efficient transformation of rice (*Oryza sativa* L.) mediated by *Agrobacterium* and sequence analysis of the boundaries of the T-DNA. *Plant Journal, 6*, 271–282.

Joseph, R., Yeoh, H. H., & Loh, C. S. (2004). Induced mutations in cassava using somatic embryos and the identification of mutant plants with altered starch yield and composition. *Plant Cell Reports, 23*, 91–98.

Khor, E., & Loh, C. S. (2005). Artificial seeds. In V. Nedovi & R. Willaert (Eds.), *Application of cell immobilisation biotechnology* (pp. 527–537). Berlin: Springer.

Khor, E., Ng, W. F., & Loh, C. S. (1998). Two-coat systems for encapsulation of *Spathoglottis plicate* (Orhcidaceae) seeds and protocorms. *Biotechnology and Bioengineering, 59*, 635–639.

Kopittke, P. M., Blamey, F. P. C., Asher, C. J., & Menzies, N. W. (2010). Trace metal phytotoxicity in solution culture: A review. *Journal of Experimental Botany, 61*, 945–954.

Kozai, T., Kitaya, Y., Fujiwara, K., Smith, M. A. L., & Aitken-Christie, J. (1995). Environmental measurement and control systems. In J. Aitken-Chistie (Ed.), *Automation and environmental control in plant tissue culture* (pp. 539–574). Dordrecht, The Netherlands: Kluwer Academic Publishers.

Kumar, P. P., Lakshmanan, P., & Thorpe, T. A. (1998). Regulation of morphogenesis in plant tissue culture by ethylene. *In Vitro Cellular and Developmental Biology-Plant, 34*, 94–103.

Kumar, P. P., Rao, C. D., & Goh, C. J. (1996). Ethylene and CO_2 affect direct shoot regeneration from the petiolar ends of *Paulownia kawakamii* leaves cultured *in vitro*. *Plant Growth Regulation, 20*, 237–243.

Kumar, P. P., Reid, D. M., & Thorpe, T. A. (1987). The role of ethylene and carbon dioxide in differentiation of shoot buds in excised cotyledons of *Pinus radiata in vitro*. *Physiologia Plantarum, 69*, 244–252.

Kumar, P. P., & Thorpe, T. A. (1991). A setup for incubating plant cultures under continuous flow of gases. *In Vitro Cellular and Developmental Biology-Plant, 27P*, 43–44.

Lee, K. S., Van Duren, M., & Mopurgo, R. (1997). Somatic embryogenesis in cassava: A tool for mutation breeding. In B. S. Ahloowalia (Ed.), *Improvement of basic food crops in Africa through plant breeding, including the use of use of induced mutations* (pp. 55–60). International Atomic Energy Agency. (Vienna-TECDOC-951).

Loh, C. S., & Rao, A. N. (1989). Clonal propagation of guava (*Psidium guajava* L.) from seedlings and grafted plants and adventitious shoot formation *in vitro*. *Scientia Horticulturae, 39*, 31–39.

Lu, C. -Y., Harry, I. S., Thompson, M. R., & Thorpe, T. A. (1991). Plantlet regeneration from cultured embryos and seedling parts of red spruce (*Picea rubens* Sarg.). *Botanical Gazette, 152*, 42–50.

Maheswaran, G., & Williams, E. G. (1984). Direct somatic embryoid formation in immature embryos of *Trifolium repens, T. pretense* and *Medicago sativa*, and rapid clonal propagation of *T. repens*. *Annals of Botany, 54*, 201–211.

Miller, C. O., Skoog, F., Von Saltza, M. H., & Strong, F. M. (1955). Kinetin, a cell division factor from deoxyribonucleic acid. *Journal of the American Chemical Society, 77*, 1392.

Morel, G. M. (1960). Producing virus-free cymbidium orchids. *American Orchid Society Bulletin, 29*, 495–497.

Murashige, T., & Skoog, F. (1962). A revised medium for rapid growth and bioassays with tobacco tissue cultures. *Physiologia Plantarum, 15*, 473–497.

Nguyen, Q. T., Kozai, T., Heo, J., & Thai, D. X. (2001). Photoautotrophic growth responses of *in vitro* cultured coffee plantlets to ventilation methods and photosynthetic photon fluxes under carbon dioxide enriched conditions. *Plant Cell Tissue and Organ Culture, 66*, 217–225.

Paarlberg, R. (2010). How agricultural biotechnology is being kept out of Africa. Plenary lecture at the 12th IAPB Congress, St. Louis, MO, USA, June 6–11, 2010.

Pueppke, S. (2010). The global rush to biofuels and what we are learning about food production, sustainability and the carbon economy. Plenary lecture at the 12th IAPB Congress, St. Louis, MO, USA, June 6–11, 2010.

Redenbaugh, K. (1990). Application of artificial seeds to tropical crops. *HortScience, 25*, 251–255.

Reinert, J. (1959). Uber die controlle der morphogenese und die induction von adventive-embryonem am gevebeculturen aus carotten. *Planta, 53*, 318–333.

Shah, D. M., Horsch, R. B., Klee, H. J., Kishore, G. M., Winter, J. A., & Tumer, N. E., et al. (1986). Engineering herbicide tolerance in transgenic plants. *Science, 233*, 478–481.

Sharma, K. K., Bhojwani, S. S., & Thorpe, T. A. (1990). Factors affecting high frequency differentiation of shoots and roots from cotyledon explants of *Brassica juncea* (L.) Czern. *Plant Science, 66*, 247–253.

Sharp, W. R., Sondahl, M. R., Caldas, L. S., & Maraffa, S. B. (1980). The physiology of *in vitro* asexual embryogenesis. *Horticulture Review, 2*, 268–310.

Sim, G. E., Loh, C. S., & Goh, C. J. (2007). High frequency early *in vitro* flowering of *Dendrobium* Madame Thong-In (Orchidaceae). *Plant Cell Reports, 26*, 383–393.

Skoog, F., & Miller, C. O. (1957). Chemical regulation of growth and organ formation in plant tissues cultured *in vitro*. *Symposium of the Society of Experimental Biology, 11*, 118–130.

Steward, F. C., Mapes, M. O., & Mears, K. (1958). Growth and organized development of cultured cells. II. Organization in cultures grown from freely suspended cells. *American Journal of Botany*, *45*, 705–708.

Thorpe, T. A. (2007). History of plant tissue culture. *Molecular Biotechnology*, *37*, 169–180.

Vasil, I. K. (2008). A history of plant biotechnology: From the cell theory of Schleiden and Schwann to biotech crops. *Plant Cell Reports*, *27*, 1423–1440.

Vasil, V., & Hildebrandt, A. C. (1965). Differentiation of tobacco plants from single, isolated cells in microcultures. *Science*, *150*, 889–892.

White, P. R. (1934). Potentially unlimited growth of excised tomato root tips in a liquid medium. *Plant Physiology*, *9*, 585–600.

Williams, E. G., & Maheswaran, G. A. (1986). Somatic embryogenesis: Factors influencing coordinated behaviour of cells in embryogenic groups. *Annals of Botany*, *57*, 443–462.

Yoo, S. -D., Cho, Y. H., & Sheen, J. (2007). *Arabidopsis* mesophyll protoplasts: A versatile cell system for transient gene expression analysis. *Nature Protocols*, *2*, 1565–1572.

Section B

Breeding biotechnologies

Somatic (asexual) procedures (haploids, protoplasts, cell selection) and their applications

10

Tanya Tapingkae[1] Zul Zulkarnain[2] Masayo Kawaguchi[3]
Takashi Ikeda[3] Acram Taji[4]

[1]*Chiang Mai Rajabhat University, Thailand*
[2]*University of Jambi, Indonesia*
[3]*Meiji University, Japan*
[4]*Queensland University of Technology, Brisbane, Australia*

TABLE OF CONTENTS

General Introduction

In angiosperms, double fertilization generates the embryo and the endosperm simultaneously, the joint development of which leads to a viable seed. Somatic embryos, however, are formed from plant cells that are not normally involved in the development of embryos (fertilization or gamete fusion). The term somatic refers to embryos developing asexually from vegetative (or somatic) tissue. Somatic plant cells are terminally differentiated and can regain totipotency and initiate embryo development under appropriate conditions. The development of techniques and protocols to produce plant embryos asexually has had a huge technological and economical impact on agricultural systems, and these biotechnologies represent an integral part of the breeding programs of important crops.

In this chapter, recent advances in somatic (asexual) procedures and their applications are reviewed, including somatic embryogenesis, haploid technologies, protoplast and somatic hybridization, and use of somatic procedures in screening and development of stress-resistant plants. This method is now a well-established technology that has made significant contributions to plant improvement and mass propagation in horticulture, agriculture, and to some extent in forestry, as a means of rapidly multiplying elite varieties or clones.

Somatic Embryogenesis

Introduction

Over the past fifty years, somatic embryogenesis has been the subject of increasing research in a large number of monocotyledonous and dicotyledonous plant species. Somatic

DOI: 10.1016/B978-0-12-381466-1.00010-9

embryogenesis has become one of the most desired pathways in the regeneration of plants via tissue culture, because it bypasses the necessity of time-consuming and costly manipulation of individual explants, which is a problem with organogenesis (Folta and Dhingra, 2006; Carneros et al., 2009). Somatic embryogenesis might also overcome difficulties with micropropagation procedures in species that are difficult to root, such as chickpea (reviewed by Anwar et al., 2010) and rubber (reviewed by Nayanakantha and Seneviratne, 2007).

Patterns of somatic embryogenesis

Two patterns of somatic embryogenesis are recognized: direct embryogenesis, where the embryo develops directly on the explant, and indirect embryogenesis, in which the embryo arises from a callus. Direct somatic embryogenesis is the desirable approach to obtaining genetically stable regenerated plants, because callus formation could cause somaclonal variation (Mizukami et al., 2008).

Direct somatic embryogenesis has been developed for numerous species, including alfalfa (Rudus et al., 2009), *Arabidopsis* (Kurczynska et al., 2007), broccoli (Yang et al., 2010), camphor tree (Du et al., 2007; Shi et al., 2009), carrot (Mizukami et al., 2008), chickpea (Kiran et al., 2005; Kiran Ghanti et al., 2010), coffee (reviewed by Ducos et al., 2007; Santana-Buzzy et al., 2007), cotton (Zhang et al., 2009), eucalyptus (Prakash and Gurumurthi, 2010), legumes (reviewed by Lakshmanan and Taji, 2000), millet (Ceasar and Ignacimuthu, 2010), rice (reviewed by Silva, 2010), rubber (Lardet et al., 2009), saffron (Demeter et al., 2010), and sugarcane (Snyman et al., 2006). It has been used to obtain genetically stable regenerated plants and has been reported to be useful in the regeneration of recombinant plants (Yang et al., 2009). Most studies applied direct somatic embryogenesis to the regeneration of valuable plants.

Factors affecting somatic embryo induction

Explant and genotype

The choice of explant is considered to be an important factor in the induction of somatic embryogenesis. Genotype, explant source, and developmental stage (age) of explant all affect somatic embryogenesis (Choudhury et al., 2008). In guava, most workers have used immature zygotic embryos as the primary explants for the induction of somatic embryogenesis. Other explants, such as leaf, node, internode, petal, and mesocarp, have also been tested. These explants failed to induce somatic embryos, with the exception of mesocarp, where there was some success (reviewed by Rai et al., 2010). In kodo millet, somatic embryogenesis was obtained from explants, such as immature inflorescences, immature and mature zygotic embryos, mature caryopses, immature inflorescences, leaf segments, root, mesocotyl, and shoot apex (reviewed by Ceasar and Ignacimuthu, 2010). Shoot apical meristems have been used effectively to develop regeneration systems across the other cereals, and apical meristem explants have been successfully employed as starting material to recover stably transformed maize, wheat, rice, oat, barley, sorghum, and millet (reviewed by Sticklen and Orabya, 2005).

Somatic embryogenesis is highly species- or genotype-dependent. A recent study by Binott et al. (2008) confirmed that callus induction, somatic embryo formation, and plant regeneration in maize were found to be genetically linked, and hybrids showed better tissue culture response compared with their inbred lines counterparts. Somatic embryogenesis in conifers is often highly genotype-dependent; consequently, an increased number of genotypes capable of somatic embryogenesis can be made available through breeding (reviewed by Bonga et al., 2010). Fu et al. (2008) found that the highest levels of embryogenic callus induction of *Dianthus chinensis* were achieved when the donor buds had been cold pretreated and the subsequent anther culture was maintained in darkness. Furthermore, there appeared to be a correlation between genotype and culture conditions.

Chemical factors

Somatic embryogenesis is controlled by a number of different chemical factors. The agents used to induce *in vitro* somatic embryogenesis are highly variable, ranging from various plant hormones to stress treatments (Fehér et al., 2003). Involvement of plant hormones and plant growth regulators with *in vitro* somatic embryogenesis has been intensively reviewed by Jiménez (2001, 2005), Fehér et al. (2003), and Jiménez and Thomas (2005), while molecular and genetic studies of hormonal regulation were reviewed by Rose and Nolan (2006). Molecular development studies with plant hormones and apical meristems might also be expected to aid in the interpretation of the role of auxin and cytokinins in somatic embryogenesis induction (Ikeda et al., 2006).

In most species studied in which growth regulators are needed for induction of somatic embryogenesis, auxins and cytokinins are key factors determining the embryogenic response. This is probably due to their participation in the regulation of cell cycling, division, and differentiation (Fehér et al., 2003). Auxin 2,4-dichlorophenoxyacetic acid (2,4-D), in isolation or combined with other growth regulators, mainly cytokinins, has been used for the induction of somatic embryogenesis via tissue culture of seeds and zygotic embryos for several species (Fehér et al., 2003; Ikeda et al., 2006; Raghavan, 2006; Fehér, 2008). Somatic embryogenesis may proceed on a single medium or may require medium alterations specific for a specific developmental stage (reviewed by Anwar et al., 2010). Thidiazuron (TDZ), a synthetic phenylurea, has been reported to be the most active cytokinin-like substance for shoot induction in plant tissue culture (reviewed by Lakshmanan and Taji, 2000). TDZ induces high-frequency somatic embryogenesis in some plant species when used alone or in combination with other growth regulators (reviewed by Anwar et al., 2010).

In addition, somatic embryogenesis can be induced by various stresses, specific hormone treatments, and over-expression of specific genes (Rose and Nolan, 2006). Rudus et al. (2009) demonstrated that stress-related phytohormones, abscisic acid

(ABA), and jasmonic acid (JA) are synthesized in tissues of *Medicago sativa* cultured *in vitro* during the whole process of indirect somatic embryogenesis, but the biosynthetic capacity changes substantially in distinct somatic embryogenesis phases.

Other inductive factors

The mineral composition of the media, type, and concentration of carbon sources, amino acids, heavy metal ions, light-dark regime, and temperature can play vital roles in somatic embryogenesis (Mohamed et al., 2004). The type and concentration of carbon source has been found to affect the induction of somatic embryos. The superiority of maltose over sucrose on somatic embryogenesis was proved in kodo millet, rice, wheat, barley, and rubber (reviewed by Ceasar and Ignacimuthu, 2010). An increase in sucrose concentrations in the embryogenesis medium enhanced the frequency of somatic embryogenesis in sugarcane (Gandonou et al., 2005) and melon (Nakagawa et al., 2001). Zhang et al. (2008) suggested that sugars function as both a carbon source and as an osmotic regulator in culture media.

There are a number of reports indicating that the three main aliphatic polyamines (PAs) — putrescine (put-diamine), spermidine (spd-triamine), and spermine (spm-tetraamine) — which are organic polycations, particularly in their free form, play a crucial role in somatic embryo formation, development, and conversion to plant (Nakagawa et al., 2006; De la Peña et al., 2008). High concentrations of PAs are often observed in tissues undergoing somatic embryogenesis (Monteiro et al., 2002). The specific role of PAs during embryogenesis has been reviewed by Baron and Stasolla (2008) and Mattoo et al. (2010). Genotypic differences in cellular PA concentrations or ratios serve as indicators of regeneration — embryogenic potential in both angiosperm and gymnosperm species and transgenic manipulation of polyamine biosynthesis has proven successful in enhancing somatic embryogenesis (reviewed by Baron and Stasolla, 2008).

Somatic embryos are very sensitive to toxic substances and volatiles produced during *in vitro* culturing (da Silva et al., 2009). Several anti-browning compounds (ascorbic acid, charcoal, DTE, DTT, PVP, PVPP, and silver nitrate) were tested in *Eucalyptus globulus* by adding them to the expression medium (MS), but all decreased somatic embryogenesis potential and only DTE, charcoal, and silver nitrate reduced explant browning. When added only during the induction period, anti-browning agents reduced accumulation of phenolics, but also severely reduced somatic embryogenesis potential (Pinto et al., 2008b). The number of somatic embryos of *Myrciaria aureana* was significantly reduced in activated charcoal-supplemented medium (Motoike et al., 2007). Conversely, Bahgat et al. (2008) reported that the addition of ascorbic and citric acids to media enhanced the production of somatic embryos of *Vicia faba*. Positive results in induction of somatic embryogenesis by supplementation of the culture medium with activated charcoal were reported for *Bixa orellana* (Paiva Neto et al., 2003) and *Bactris gasipaes* (Steinmacher et al., 2007).

The presence of light promoted somatic embryogenesis in some species, such as finger millet (Corrêa et al., 2009) and camphor tree (Shi et al., 2009). However, induction of direct somatic embryogenesis from *Phalaenopsis* leaf explants required a 60 day induction period in darkness (Gow et al., 2009). In general, red light promotes and blue light inhibits the induction of somatic embryogenesis (Bach and Krol, 2001; Takanori and Cuello, 2005; Hoshino and Cuello, 2006). However, blue light stimulated the germination of mature somatic embryos of *Freesia refracta* (Bach et al., 2000). The effect of light quality on the growth of embryogenic tissue can be genotype-dependent (Latkowska et al., 2000). Until now, the literature relating to the relationship between light and somatic embryogenesis provided inadequate and/or conflicting information, preventing a clear understanding of the role of light in somatic embryogenesis.

Histodifferentiation

Auxins are generally involved in cell division and expansion and differentiation of the vascular system. Auxins are also associated with regulation of the embryonic patterns of histodifferentation (Bassuner et al., 2007). Cangahuala-Inocente et al. (2009) studied the accumulation of protein, total sugars, starch, amino acids, polyamines, indole-3-acetic acid (IAA), and ABA in different stages of *Acca sellowiana* zygotic embryogenesis. Starch is the predominant storage compound during zygotic embryo development. All amino acids are synthesized during embryogenesis, with an increased accumulation in the heart and cotyledonary stages. Asparagine is the major amino acid observed. Free polyamine synthesis during early developmental stages, as well as the accumulation of polyamine conjugates in the cotyledonary stage, may be considered as reliable biochemical markers of embryonal development in *A. sellowiana*. IAA and ABA levels were inversely proportional in the heart-shaped and cotyledonary stages, suggesting their involvement with histodifferentiation patterns, mainly the establishment of embryonal symmetries.

Plant maturation

The morphology of somatic embryogenesis is affected by a number of factors, such as plant growth regulators, medium composition, subculture time, or frequency. The media of somatic embryo development are different from the media used for somatic embryo induction. Auxins in any medium play a key role during the induction stage, but inhibit plantlet formation. Hence, somatic embryos should be transferred into a medium without plant growth regulator during plantlet formation. The use of hormone-free medium was very fruitful in maturing somatic embryos and obtaining plantlets in many species, such as *V. faba* (Bahgat et al., 2008), triticale (Atak et al., 2008), eucalyptus (Pinto et al., 2008a), and waxflowers (Ratanasanobon and Seaton, 2010). However, it is reported that the addition of L-proline, maltose, and TDZ to the maturation medium increased the frequency of somatic embryo maturation and plantlet formation (Ceasar and Ignacimuthu, 2010). Several studies report that decreasing sucrose content induces maturation of somatic embryos in plant tissue cultures (Lee et al., 2001; Demeter et al., 2010).

Plant regeneration

Establishment of an efficient somatic embryogenesis and regeneration system has become an integral part of plant biotechnology, as regeneration of transgenic plants in most of the important crops is dependent on the formation of somatic embryos (reviewed by Vasil, 2008a,b). An efficient, rapid, and dependable regeneration system is required for transformation and other genetic improvement. Plant regeneration through somatic embryogenesis is generally preferred over other *in vitro* developmental processes, such as organogenesis, because it offers more target cells (Ogita et al., 2002) to develop into whole plants. Furthermore, somatic embryos are believed to be of single cell origin, any regenerated transformed cell would give rise to a transformed plant without unwanted chimeric events often associated with transformed meristems, which results in organogenesis (Sakhanokho et al., 2009).

Regeneration efficiency varies among species and often among cultivars of the same species, and genotype-specific regeneration protocols will need to be established for best results (Jauhar, 2006). Bahgat et al. (2008) reported that cultivar differences and explant differences play important roles in determining both callus formation and regeneration rate in the faba bean (*Vicia faba* L.). Some of the most important factors for successful plant regeneration are the culture medium and environmental incubation conditions. Recently, Ceasar and Ignacimuthu (2008, 2010) reported a highly efficient somatic embryogenesis and plant regeneration system for kodo millet and finger millet by using a two-step culture. They suggested that this two-step culture procedure might be adopted in the future for efficient somatic embryogenesis and regeneration of the rest of the millets. A two-step culture process has also been proved successful in carrot (Mizukami et al., 2008) and potato (Seabrook and Douglass, 2001) for efficient somatic embryogenesis and regeneration. Feng et al. (2010) found that 70% of embryos with abnormal cotyledons developed into plantlets in *Koelreuteria bipinnata*. This indicates that somatic embryos with abnormalities during early developmental stages do not always affect plantlet formation.

Gene expression during somatic embryogenesis

Somatic embryogenesis is the developmental restructuring of somatic cells toward the embryogenic pathway that forms the basis of cellular totipotency in higher plants. Various models have been widely investigated to gain an understanding of the mechanisms of gene regulation during this developmental process, and an array of genes activated or differentially expressed during somatic embryogenesis has been isolated employing various molecular techniques (reviewed by Ikeda et al., 2006; Raghavan, 2006; Rose and Nolan, 2006; Jenik et al., 2007; Bonga et al., 2010). Somatic embryogenesis involves many molecular events encompassing differential gene expression, as well as various signal transduction pathways for activating/repressing numerous gene sets, many of which are yet to be identified and characterized (reviewed by Berleth and Chatfield, 2002; Chugh and Khurana, 2002; Quiroz-Figueroa et al., 2006; Rose and Nolan, 2006; Namasivayam, 2007; Curtis and Grossniklaus, 2008; Fehér, 2008; Karami et al., 2009; Aquea et al., 2010). Thus, future trends involve characterization of development-specific genes during somatic embryogenesis to provide a deeper insight into the mechanisms involved during differentiation of somatic cells and phenotypic expression of cellular totipotency in higher plants.

Mass propagation and somaclonal variation

In vitro somatic embryogenesis is an important prerequisite for the use of many biotechnological tools for mass propagation. Large-scale liquid cultures have been used for micropropagation through organogenesis or somatic embryogenesis pathways in both agitated vessels and in bioreactors (reviewed by Ziv, 2005; Akin-Idowu et al., 2009; Ducos et al., 2009). Research directed at improving bioreactors for somatic embryogenesis has been reported for several plant species (reviewed by Ziv, 2000; Ducos et al., 2008). For example, coffee somatic embryogenesis is used in industry for large-scale, rapid propagation of selected heterozygous varieties, such as Robusta clones or F1 hybrid Arabica (Ducos et al., 2007). Pilot bioreactor units producing up to a million somatic seedlings per year are currently being implemented by various teams (reviewed by Ducos et al., 2008).

Genetic variation is very important in crop improvement, because it forms the basis of development of new varieties. Somaclonal variation is a valuable tool in plant breeding, wherein variation in tissue culture regenerated plants from somatic cells can be used in the development of crops with novel traits. The molecular basis of somaclonal variation is not precisely known; however, both genetic and epigenetic mechanisms are thought to play a role (Gao et al., 2009). Somaclonal variation has been associated with changes in chromosome number and structure, point mutations, and DNA methylation (reviewed by Cardoza and Steward, 2004; Baenziger et al., 2006; Bartoszewski et al., 2007; Santana-Buzzy et al., 2007).

Haploid Technology

Introduction

Haploid technology is of significant interest for developmental and genetic research, as well as for plant breeding and biotechnology. Haploid plants are useful in understanding cellular totipotency because they develop from single male or female gametes without fertilization (Powell, 1990). Haploid individuals also provide an excellent example when studying induced mutagenesis, where recessive traits can be easily detected (Seguí-Simarro and Nuez, 2008b). Tolerance to unfavorable conditions, such as drought, cold, heavy metals, or low nutrients, are among recessive traits that can be detected promptly in haploid plants. Homozygous lines can be generated more rapidly via haploid development compared to conventional means (Tomasi et al., 1999). Taji et al. (2002) claimed that such lines could be generated in only one generation, while conventional methods need at least five generations.

Since it was first demonstrated by Guha and Maheshwari (1964) on *Datura inoxia* and Nitsch and Nitsch (1969) on tobacco, haploids have been used to produce homozygous genotypes in a number of economically important species: tobacco (*Nicotiana tabacum*), wheat (*Triticum aestivum*), barley (*Hordeum vulgare*), canola (*Brassica napus*), rice (*Oryza sativa*), and maize (*Zea mays*; Aryan, 2002; Maluszynsky et al., 2003; Maraschin et al., 2005b). Success has also been reported on legumes: pigeon pea (*Cajanus cajan*), soybean (*Glycine max*), cowpea (*Vigna unguiculata*), winged bean (*Psophocarpus tetragonolobus*), East Indian Walnut (*Albizia lebbeck*) and Yellow Flame Tree (*Peltophorum pterocarpum*; Crosser et al., 2006).

However, haploid individuals tend to be smaller in size, less vigorous, more sensitive to disease and environmental stress and, most important, they are sterile (Seguí-Simarro and Nuez, 2008b). Therefore, for practical purposes it is usually desirable to obtain doubled haploids. Regardless of the mode of development, doubling of haploid plants either spontaneously via endoreduplication (DNA duplication without mitosis), nuclear fusion (merging of coalescing nuclei into a larger nuclei, mixing both DNA contents), endomitosis (mitosis in the absence of mitotic spindle and nuclear envelope breakdown), or by chemical means such as colchicine or oryzalin, leads to a homozygous doubled haploid individual with two identical copies of each chromosome (Crosser et al., 2006). For many families of angiosperms, doubled haploidy has been used as a routine tool in breeding programs because this technique provides pure lines in a single generation, which may save considerable time in the breeding of new cultivars (Tuvesson et al., 2000; Maluszynsky et al., 2003).

Cytological basis underlying haploid plant induction

The term androgenesis refers to plant regeneration directly from microspore culture under *in vitro* conditions. The underlying principle of androgenesis is to stop the development of pollen cells, which normally become sexual cells, and to force their development directly into a complete plant (Nitsch, 1981). This process inhibits typical gametophytic differentiation and instead allows cell division and regeneration to occur (Dunwell, 1986).

As gametogenesis (microspore development) takes place, mature pollen grains are formed via mitosis. Since the developmental route is not yet determined during microgametogenesis, there is a chance to interrupt the normal gametophytic pathway and to induce sporophytic development. As the result of sporophytic divisions, multicellular microspores develop within the anthers. Differentiation of these multicellular units may result in pollen embryos, which then develop into haploid plants.

In certain conditions, microspores may undergo sporophytic development instead of entering the gamete-producing pathway. Touraev et al. (2001) suggested that there are a variety of microspore developmental fates upon exposure to inductive conditions. Many of the microspores arrest and/or die, some develop pollen-like structures prior to death or arrest, and others develop a multinucleate, haploid callus-like structure. Meanwhile, other microspores are directly committed to embryogenesis and undergo numerous changes at different levels to become microspore-derived embryos. From this, the microspores enlarge significantly, the nucleus repositions to the cell center, the cytoplasm clears, and the large vacuole breaks apart into smaller fragments. Maraschin et al. (2005a) suggested that these cytological rearrangements provide a transient early morphological marker of embryogenic commitment, a star-like microspore. In addition to morphological alterations, the microspore also undergoes numerous biochemical changes (Seguí-Simarro and Nuez, 2008a). Pauls et al. (2006) proposed that these changes include cytoplasmic pH alteration, wherein alkalinization occurs upon embryogenesis induction, as opposed to acidification during pollen-like development.

Changes in the microspore development pathway are also shown by the alteration of cytokinesis pattern. Gervais et al. (2000) suggested that during the induction of microspore embryogenesis, phragmoplast microtubules take a more parallel orientation and a preprophase band that is normally absent during pollen mitosis is now developed. Additionally, there is a shift in the division plane from asymmetrical and lens-shaped (typical of first pollen division) to asymmetrical and equatorial, similar to that of somatic cell development (Seguí-Simarro and Nuez, 2008a). All of these coincide with upregulation of a number of cytoskeleton and vesicle trafficking related genes, as proposed by Pauls et al. (2006) and Hosp et al. (2007b).

Other cytological markers of microspore embryogenic development are shown by the nucleus. Typically, the gametophytic pathway leads to the differentiation of generative and vegetative nuclei, which are significantly different in size, shape, status of chromatin condensation, and transcriptional activity. Meanwhile, sporophytic development of microspores is similar to that found in vacuolated microspores and significantly different from that found in generative and vegetative nuclei. Testillano et al. (2000) found that in canola (*B. napus*), pepper (*Capsicum annuum*), and tobacco (*N. tabacum*) nuclei, two- to four-celled embryogenic microspores all exhibited features of cycling. Further, the chromatin was also similar to that of cycling cells, and the immunogold labeling of embryogenic nuclei revealed a clustered pattern of the proliferating cell nuclear antigen (PCNA), a typical pattern of replicative status (Testillano et al., 2000). In addition to this, the nucleolus was highly active in ribosome biogenesis, the typical architecture of proliferating cells (Testillano et al., 2005). Other ultrastructural characteristics of microspore embryogenic development include the presence of high levels of perichromatin fibrils in the interchromatin region and condensed chromatin borders, as well as an increase in number per cell and the percentage of cells containing Cajal bodies, together with a decrease in Cajal body size (Seguí-Simarro et al., 2006; Seguí-Simarro and Nuez, 2008a).

Factors affecting the induction of microspore embryos

Microspore embryogenesis has considerable potential in plant breeding due to its potential to produce haploid plants from immature microspores, which can be doubled to obtain fertile and true-to-type doubled-haploid individuals (Tuvesson et al., 2000; Zamani et al., 2000; Crosser et al., 2006; Guzy-Wróbelska et al., 2007). For it to be successfully applied in plant breeding, a high yield of plants must be obtained consistently from anther or microspore culture. There are a number of factors that influence the embryogenic response of cultured microspores (Smykal and Pechan, 2000; Wang et al., 2000; Datta, 2005; Seguí-Simarro et al., 2006; Seguí-Simarro and Nuez, 2008b), as reviewed below.

Plant genotype

It is obvious that the response of anthers during *in vitro* culture is dependent upon the genotype of the donor plants. It is not only different species, but also different cultivars within species and, even individuals of the same cultivar that may show differences in embryogenic responses (Seguí-Simarro and Nuez, 2008a). The genotype of the donor plant affects the frequency of embryogenesis, as well as the quality of the embryo produced (Palmer and Keller, 1997). The significant genotype dependence in androgenic response is reported by Mitykó et al. (1996) on pepper anther culture. They found that large-fruited genotypes produced the highest number of plantlets per cultured anther compared to small- and medium-fruited genotypes, which proved to be poor or non-responsive. Genotypic-dependence is also reported in *Raphanus sativus* (Takahata et al., 1996), *T. aestivum* (Bitsch et al., 1998), *Linum usitatissimum* (Chen et al., 1998), *Scabiosa columbaria* (Romeijn and Lammeren, 1999), and *Phleum pratense* (Guo et al., 1999). Although the basis of genetic control remains unexplained, it is clear that genetic factors interact with other factors to control pollen embryogenesis (Palmer and Keller, 1997; Rudolf et al., 1999; Smykal and Pechan, 2000).

Developmental stage of the microspore

The particular stage of microspore development at the time of culture initiation plays an important role in achieving success with microspore embryogenesis. This is because only microspores at certain developmental stages can be redirected to undergo embryogenesis. Once maturation gene products begin to accumulate after the first mitosis, the gametophytic development of the microspore has been determined and embryogenesis cannot take place (Pechan and Smykal, 2001).

The exact stage of microspore development, which is more readily diverted to a sporophytic pathway, seems to vary with species. For most species, the optimal developmental stage for induction of embryogenesis is around the first mitosis, from vacuolate microspore to early bicellular pollen (Pechan and Smykal, 2001; Touraev et al., 2001). However, in some species, narrower time frames, including vacuolate stage or early bicellular, are usually reported (Crosser et al., 2006; Seguí-Simarro and Nuez, 2008a). Although many workers proposed that the optimum period for pollen response lies in between the first pollen mitosis and early bicellular, an exact determination of microspore stage before being introduced to culture requires a cytological analysis. As such, for practical programs many workers rely on morphological parameters such as the length of floral buds and petals (Mitykó et al., 1996; Takahata et al., 1996; Tomasi et al., 1999; Laura et al., 2006; Bal et al., 2009). Furthermore, da Silva Lauxen et al. (2003) demonstrated that microspores within the anther of the same flower were at different developmental stages; and also, different bud sizes were found to contain microspores at different stages depending on plant genotype (Crosser et al., 2006). Although the optimum bud size varies among genotypes, the microspore populations represented in these buds are mostly in the range of uninucleate and binucleate stages.

In conjunction with the best response in culture, Dodds and Roberts (1985) proposed three categories of microspore developmental stages: premitotic, mitotic, and postmitotic. Plants in the premitotic category show best embryoid formation when microspores have completed meiosis but have not yet started the first pollen division (e.g., *Hyoscyamus niger* and *H. vulgare*). Plants belonging to the mitotic group respond best about the time of the first pollen division (e.g., *N. tabacum*, *Datura innoxia*, and *Paeonia* spp.). Meanwhile, plants in the postmitotic category show the best response when microspores are in the early binucleate stage (e.g., *Atropa belladonna*).

Stress pre-treatment

Culture conditions, particularly stress pre-treatment applied to microspores prior to culture initiation, play an important role in the success of embryogenesis induction. It is well known that cultures must be subjected to a stress pre-treatment to be committed to embryogenesis. Various types of stress pre-treatment have been found to improve microspore embryogenesis in a wide range of species. These stresses include low temperature (Bishnoi et al., 2000; Pauk et al., 2000; Tuvesson et al., 2000), high temperature (Kiviharju and Pehu, 1998; Pechan and Smykal, 2001), carbon starvation (Caredda et al., 2000; Kasha et al., 2001; Cistué et al., 2006; Labbani et al., 2007), colchicine (Smykal and Pechan, 2000; Seguí-Simarro et al., 2003; Obert and Barnabas, 2004), and combinations of these (reviewed by Shariatpanahi et al., 2006).

These stress pre-treatments of anthers are critically important for blocking gametophytic development and for triggering pollen embryogenesis in competent microspores. Under stress conditions, microspores enlarge and cytoplasm undergoes structural reorganization (Touraev et al., 1997). Scant amyloplasts and liposomes and abundant vacuoles are found (Hofer et al., 1999; Wojciechowicz and Kikowska, 2009). In addition, Hosp et al. (2007a) claimed that microspores exposed to stressful conditions often exhibit signs of cellular component degradation, such as a decrease in the number of amyloplasts, lipid bodies, and ribosomes. These microspores are capable of switching from their gametophytic to sporophytic developmental pathways, giving rise to embryos or embryogenic callus.

Culture medium

Successful embryoid production of different species is achieved on media of different compositions. The common medium employed in most tissue culture systems including anther culture is MS medium (Murashige and Skoog, 1962). However, other media, such as WH (White, 1962), LS (Linsmaier and Skoog, 1965), B5 (Gamborg et al., 1968), Nitsch's (Nitsch, 1969), SH (Schenk and Hildebrandt, 1972), and Woody Plant Medium (Lloyd and McCown, 1980), are also used. The nutritional requirements of isolated microspores are more complex than those of excised anthers. In isolated microspores, certain factors responsible for the induction of androgenesis, which might have been provided by the anther, are missing, and these have to be provided through the culture medium (Reinert and Bajaj, 1977). For example, tobacco-excised anthers can be successfully cultured on a simple basal medium, whereas the isolated microspores require a higher amount of nitrogen in the form of amino acids (Reinert et al., 1975).

In addition to the nutrient composition, the physical properties of the medium have to be considered before starting the culture. Agar-solidified media have been extensively used in many anther culture systems, but the use of liquid medium is also becoming popular, particularly in microspore culture. Androgenesis in *T. aestivum* (Trottier et al., 1993) and *H. vulgare* (Trottier et al., 1993; Cistué et al., 1998) involved the use of liquid media. Meanwhile, microspore embryogenesis in anther culture of *O. sativa* (Bishnoi et al., 2000; Aryan, 2002) was successfully induced on a solid medium. A new approach in anther culture is the use of a double-phase medium. The anther is grown in a thin layer of liquid medium, which is on top of solid agar in the agar plate. This method has proven to be useful in the anther culture of triticale (× *Triticosecale*; Immonen and Robinson, 2000). Another way of improving the yield of microspore-derived plants has included culture in a liquid medium supplemented with Ficoll 400™, a non-ionic copolymer of sucrose and epichlorohydrin, which functions as a buoyancy-increasing component to allow anthers and calli to float. The detrimental anaerobic conditions of the liquid media are believed to cause the reduction in plant regeneration because anthers and calli develop under the medium surface. However, the results have not been consistent among studies or with different species or cultivars.

Growth regulators

The application of plant growth regulators to the culture medium is necessary for the successful induction of microspore embryogenesis. Auxin and cytokinin are the two most extensively used growth regulators in the anther culture of a wide range of plant species. In the anther culture of Poaceae and Brassicaceae, 2,4-D is usually applied (Bishnoi et al., 2000). Mitykó et al. (1996) used 2,4-D and kinetin to induce haploid embryo formation in the anther culture of *C. annuum*, and the same growth regulators were also used by Metwally et al. (1998b) to produce haploid plantlets from the anther culture of *Cucurbita pepo*. Vigorous green plantlets are regenerated from the microspore culture of *H. vulgare* in the presence of auxins such as IAA or naphthaleneacetic acid (NAA; Castillo et al., 2000). Although the application of growth regulators is found to enhance microspore embryogenesis in most species, their removal from the culture medium, coupled with simultaneous lowering of the sucrose concentration, resulted in the initiation of embryogenesis or shoot organogenesis in the anther culture of *Speciosa columbaria* (Romeijn and Lammeren, 1999). These observations support the assumption that the requirement of exogenous auxin and cytokinin depends on their endogenous level within the anther (Reinert and Bajaj, 1977). An adequate auxin balance is important for microspore embryogenesis to take place, particularly during the transition period from the radial symmetry of globular embryos to the bilateral symmetry of later stages (Ramesar-Fortner and Yeung, 2006). Although it is crucial for the transition, a high level of auxin may inhibit embryos from further elongation (Seguí-Simarro and Nuez, 2008a).

Many plants and microspores exposed to environmental stress are known to produce abscisic acid (ABA). Seguí-Simarro and Nuez (2008a) suggested that ABA signaling caused activation in some genes. They suggested that the internalization of stress signals by ABA signaling pathways would be an intermediate step in the activation of gene expression programs leading, directly or indirectly, to microspore embryogenesis (Maraschin et al., 2005b; Tsuwamoto et al., 2007). Besides its previously mentioned role, ABA has a morphogenetic importance in maintaining morphological integrity during early embryo growth, especially in the development of shoot apical meristem (Seguí-Simarro and Nuez, 2008a).

Haploid induction via ovary and ovule cultures

Ovaries and unfertilized ovules are also possible alternative sources for haploid or doubled-haploid production (Miyoshi and Asakura, 1996; Metwally et al., 1998a; Sato et al., 2000; Thomas, 2004; Shalaby, 2007). Thus far, the success of ovary and ovule culture in the production of haploid and doubled-haploids has been reported on a laboratory scale for a number of taxa (Miyoshi and Asakura, 1996; Alan et al., 2003, 2004 Tang et al., 2006; Bhat and Murthy, 2008). No commercial application of ovary or ovule cultures has been found thus far, therefore research on the use of ovary and ovule culture in induction of haploids should be encouraged in the future. A number of papers have revealed factors associated with the response of ovaries or ovules cultured *in vitro*. Plant genotype was found to be important in ovule culture of *Cucumis sativus* (Suprunova and Shmykova, 2008), while in the *Morus alba* ovule culture, Thomas (2004) found that the stage at which the explants were cultured was critical for the gynogenic response. In addition, the type of basal medium and plant growth regulators used also play important roles, as reported by Kouassi et al. (2008) in ovule culture of *Hevea brasiliensis*. Despite progress, this approach is faced with many constraints impeding its widespread application; consequently the gynogenic response of ovaries and ovules under various *in vitro* conditions needs further exploration.

Protoplast and Somatic Hybridization

Introduction

Somatic hybridization and cybridization have great potential in plant improvement. Somatic hybridization through protoplast fusion provides the ability to combine parent genes in higher plants to overcome sexual incompatibility among plant species or genera. Protoplast fusion enables transfer of desirable qualities, for example, resistance to diseases (bacterial, fungal, viral), pests, herbicides, and other stress factors.

Hundreds of reports on somatic hybridization have been published during the past four decades. A recent review of somatic hybrids was provided by Liu et al. (2005a). Grosser and Gmitter (2005) and Singh and Rajam (2009) reviewed applications of somatic hybridization and cybridization in citrus improvement. Wang et al. (2009) reviewed progress in somatic hybridization in banana (*Musa* spp.). Application to genetics and breeding in somatic hybrids of potato (*Solanum tuberosum*) was reviewed by Orczyk et al. (2003). Somatic hybridization in the family Brassicaceae was reviewed by Navratilova (2004). Reddy et al. (2008) reviewed developments in seaweed protoplast research and their potential in genetic improvement. This section primarily focuses on literature published in the past decade, with particular emphasis on types of somatic hybrids, protoplast fusion methods, selection and identification of somatic hybrids, and the recent efforts in somatic hybridization are reviewed.

Types of somatic hybrids

Somatic hybrids can be classified into three types: symmetric somatic hybrids, asymmetric somatic hybrids, and cybrids (cytoplasmic hybrids). Symmetric somatic hybridization is defined as the combination of nuclear and cytoplasmic genetic information from both parental species. Asymmetric somatic hybridization is incomplete, with the loss of some cytoplasmic or nuclear DNA, and this type of hybridization has been used to introduce fragments of the nuclear genome from one donor species into the intact genome of another recipient species. Cybrids harbor only one parental nuclear genome and either the cytoplasmic genome of the other (non-nuclear) parent or that of a combination of both parental species (see Guo et al., 2004 for a review). Asymmetric hybrids and cybrids are widely used for crop improvement because one or more traits can be added while maintaining cultivar integrity (Wang et al., 2008).

Protoplast fusion methods

Protoplast surfaces bear strong negative charges, and intact protoplasts in suspension repel each other, hence fusion is accomplished by the addition of calcium ions or polyethylene glycol (PEG), or by using electric fields. Electrofusion is preferred over chemical fusion, as fusion conditions are much easier to control (Duquenne et al., 2007). According to work by Trigiano and Gray (2000), the yield of fusion products increases 20-fold when electrofusion is used. Even though binary chemical fusion is more efficient in some plants, it allows more efficient cell division and plant regeneration afterwards (Assani et al., 2005).

In recent years, somatic hybridization has been accomplished by electrofusion of protoplast, and characterizing the regenerated plantlets by flow cytometry and isozyme or DNA marker analysis. Electrochemical protoplast fusion is a process that combines the merits of both somatic hybridization and chemical methods (Olivares-Fuster et al., 2005). Olivares-Fuster et al. (2005) reported production of citrus somatic hybrids and cybrids via electrochemical protoplast fusion, where protoplasts of sweet orange and Mexican lime were induced to undergo fusion in the presence of PEG and electric impulses of direct current. High rates of embryogenesis were exhibited.

Since most of the hybrids obtained via symmetric protoplast fusion may contain numerous undesirable genes, repeated backcrossing and selection are required for elimination of undesirable traits. The obtained somatic hybrids often show chromosome loss, weakness, or sterility, probably due to somatic incompatibility (Fu et al., 2009). Chromosome elimination is an important issue and a complicated process influenced by many factors, such as genotype, type of irradiation rays, irradiation dose, and phylogenetic relatedness (Liu et al., 2005a; Yang et al., 2007).

In asymmetric fusion, metabolism inhibitors inactivate recipient protoplasts and simplify the selection of the regenerants. Iodoacetate (IOA) has been used to inactivate protoplasts of parents for fusion and has been found to be efficient in selectively producing somatic hybrids (Shimonaka et al., 2002). For transferring a part of the genome or cytoplasm, the donor protoplasts are usually irradiated with X- or γ-rays prior to fusion (Fu et al., 2009), but these two kinds of irradiating rays are dangerous and inconvenient to use. Therefore, UV has been increasingly used to break and fragment chromosomes of donors for production of asymmetric hybrids (Yang et al., 2007; Wang et al., 2008).

Effects of UV on donor chromosome elimination and fragmentation are dose dependent (Xiang et al., 2003). Using the viability, division percentage, and plating efficiency of UV-irradiated protoplasts as indicators, Fu et al. (2009) were able to enhance chromosome elimination by increasing the irradiation dose. Xiao et al. (2009) found that when UV dosage increased, the differentiation ability of colony formation and the frequency of plant regeneration decreased in protoplast fusion in banana. In addition, the UV treatment given to the donor protoplast influenced the growth and development of fused products, suggesting that optimal UV irradiation is a key factor for asymmetric protoplast fusion.

Many plant species have been used in intergeneric somatic hybridization, as shown in Table 10.1. Some important economic characteristics transferred through protoplast fusion, such as tolerance to disease, herbicides, and salt are shown in Table 10.2.

Table 10.1 Intergeneric symmetric and asymmetric fusions

Parents		Fusions	
A	B	Symmetric fusions	Asymmetric fusions or microfusions
Atropa	*Nicotiana*	Yemets et al. (2000)	Yemets et al. (2000)
Brassica	*Arabidopsis*		Yamagishi et al. (2002)
	Camelina		Jiang et al. (2009)
	Crambe		Wang et al. (2003)
	Orychophragmus	Hu et al. (2002b)	Hu et al. (2002b)
			Zhao et al. (2008)
	Isatis	Du et al. (2009)	Du et al. (2009)
		Tu et al. (2008)	
	Sinapis		Hu et al. (2002a)
Citrus	*Fortunella*	Takami et al. (2004)	
	Poncirus	Guo et al. (2002)	Liu and Deng (2000)
	Clausena	Fu et al. (2003)	
	Microcitrus	Liu and Deng (2002)	
		Xu et al. (2004)	
	Severinia	Grosser and Chandler (2000)	
Dendranthema	*Artemisia*	Furuta et al. (2004)	
Daucus	*Panax*		Han et al. (2009)
Helianthus	*Cichorium*		Varotto et al. (2001)
Hyoscyamus	*Nicotiana*		Zubko et al. (2002)
Lathyrus	*Pisum*	Durieu and Ochatt (2000)	
Raphanus	*Isatis*	Tu et al. (2008)	
Triticum	*Avena*		Xiang et al. (2003)
			Xiang et al. (2010)
	Aeleuropus		Yue et al. (2001)
	Agropyron	Cui et al. (2009)	Xia et al. (2003)
		Gao et al. (2010)	Liu et al. (2009)
	Bupleurum		Zhou et al. (2006)
	Lolium		Cheng and Xia (2004)
	Psathyrostachys	Xing et al. (2001)	
	Setaria		Xiang et al. (2004)
	Haynaldia	Zhou et al. (2001)	Zhou et al. (2001)
	Zea	Szarka et al. (2002)	Xu et al. (2003)

Table 10.2 Important economic characteristics transferred through protoplast fusion

Somatic hybrids	Characters	References
Musa acuminata cv. Mas + *M. silk* cv. Guoshanxiang	Disease resistance (*Fusarium*)	Xiao et al. (2009)
Brassica napus + *Isatis indigotica*	Disease resistance	Du et al. (2009)
Helianthus annuus L. + *H. maximiliani*	Disease resistance	Taski-Ajdukovic et al. (2006)
Triticum aestivum + *Aeleuropus littorulis T. aestivum* + *Agropyron elongatum*	Salt tolerance	Yue et al. (2001)
		Xia et al. (2003)
B. napus + *Crambe abyssinica*	Higher erucic acid content	Wang et al. (2003)
Oryza meyeriana + *O. sativa* ssp. Japonica	Bacterial blight resistance	Yan et al. (2004)
Dendranthema × *grandiflorum* + *Artemisia sieversiana*	Rust resistance	Furuta et al. (2004)
Sinapsis arvensis + *Arabidopsis thaliana*	Stem canker resistance	Hu et al. (2002a)
Nicotiana plumbaginifolia + *Atropa belladonna*	Amiprophos methyl resistance	Yemets et al. (2000)
Citrus reticulata cv. Red Tangerine + *Poncirus trifoliata*	Tolerant to CTV (citrus tristeza virus) and CEV (citrus exocortis virus)	Guo et al. (2002)
T. aestivum + *A. elongatum*	Higher protein content	Gao et al. (2010)
		Cui et al. (2009)
B. napus + *Camelina sativa*	Higher level of linolenic and eicosanoic acids	Jiang et al. (2009)
Citrus reticulata + *C. limon*	Carotenoid compounds	Bassene et al. (2009)
R. sativus + *B. rapa*	Medicinal components	Tu et al. (2008)
Daucus carota + *Panax quinquefolius*	Ginsenoside	Han et al. (2009)
Ipomoea batatas + *I. triloba*	Storage root-bearing	Yang et al. (2009)
Murcott tangor (*Citrus reticulata* Blanco × *C. sinensis* (L.) Osbeck) + Hirado Buntan Pink pummelo (HBP) (*C. grandis* (L.) Osbeck)	Seedless	Cai et al. (2010)
Citrus sinensis Osbeck cv. Yoshida navel orange + *Citrus unshiu* Marc cv. Okitsu satsuma mandarin	Seedless	An et al. (2008)
Bingtang orange (*Citrus sinensis* (L.) Osbeck) + Calamondin (*C. microcarpa* Bunge)	Asiatic citrus canker-tolerant and ornamental citrus breeding	Cai et al. (2010)

Selection of somatic hybrids

It is necessary to select the hybrid products from the unfused and homo-fused protoplast populations. Without a strategy for identification and selection of hybrid cells, a very time-consuming and tedious identification of somatic hybrids among large numbers of regenerated calli or plants must occur. Several techniques have been used for selecting hybrid cell lines, including manual selection, selective media, metabolic inhibitors, and complementation systems, such as chlorophyll deficiency complementation, auxotroph complementation, resistance markers and double mutants, and culture and application of the green fluorescent protein (GFP) marker gene (reviewed by Guo et al., 2004; Navratilova, 2004).

Identification of somatic hybrids

Hybrid plants developed from post-fused protoplast cultures can be identified based on morphological, cytological (chromosome counting, flow cytometry analysis, and isozyme analysis), and molecular analysis (DNA markers). Genetic analysis can be undertaken only if the hybrid plants are fertile. Many hybrid plants of distant related species are sterile. Thus there has been a constant effort to exploit molecular markers, especially to characterize the organellar genomes of somatic hybrids and cybrids. Modern molecular technologies used to characterize the nature of somatic hybrids include random amplified polymorphic DNA (RAPD), restriction fragment length polymorphism (RFLP), amplified fragment length polymorphism (AFLP), simple sequence repeat (SSR), inter-simple sequence repeat (ISSR), genomic *in situ* hybridization (GISH), chloroplast simple sequence repeat (cpSSR), and cleaved amplified polymorphism sequence (CAPS) analyses. A number of excellent comprehensive reviews are available (Guo et al., 2004; Navratilova, 2004; Liu et al., 2005a). Multiple techniques are often used because results from one method are not sufficient to draw unequivocal conclusions about hybridity. For example, RAPD and ISSR analysis were used to identify somatic hybrids of disease-resistant banana (Xiao et al., 2009). Two new citrus somatic hybrids identified by flow cytometry analysis showed that all plants from pre-selected lines of the two combinations were tetraploid. Using SSR analysis confirmed their hybrid nature, with nuclear DNA from both fusion parents and an absence of parental-specific bands. Furthermore, cytoplasmic genomes of the recovered plants were further revealed by CAPS and cpSSR analysis (Cai et al., 2010).

Factors affecting regeneration of hybrid plants

Successful gene transfer via protoplast fusion depends on the ability to regenerate a mature plant from the fusion product. Many factors affecting regeneration of plants, such as the nature of protoplast (Nassour and Dorion, 2002), culture method (Thomas, 2009), culture medium (Sonntag et al., 2009), protoplast density (Khentry et al., 2006), and external conditions (Qin et al., 2005) have been reported. Selection of parents is very important in using hybridization to create new variation. Despite the obvious importance of this issue, more research has been done on methods of selection in breeding populations than on selection of parents to create these populations (Baenziger et al., 2006). The somatic hybrids between remotely related species were, in general, more difficult to root and culture to mature plants outside *in vitro* conditions (Tu et al., 2008). Du et al. (2009) reported that most of the hybrids between *B. napus* and *Isatis indigotica* developed shoots and roots very slowly, and even some rooted plants failed to survive after being transferred to soil or stopped growing and died during the flowering stage, although care was taken with soil and humidity conditions. The poor performance of these hybrids may be related to the incorporation of excessive genomes of the two very distantly related species and strong genetic incompatibilities. Somatic hybrids from parents of a widely divergent genetic background may have low fertility and viability (Navratilova, 2004).

Efficient protoplast isolation, fusion, and regeneration of fusion products are all indispensable steps toward the creation of a somatic hybrid and therefore demand an integrated approach of techniques (Duquenne et al., 2007; Pati et al., 2008). To date, no standardized method has been proposed that is suitable for a wide range of species. Using a specific crop as a testing model will have wider implications. Citrus is one of a few plants where somatic hybridization via protoplast fusion is reaching its predicted potential. Citrus somatic hybrid plants have been produced from more than 250 parental combinations, covering a wide range of germplasm, and an extensive review of the subject has recently been published (Grosser and Gmitter, 2005; Peña et al., 2008; Singh and Rajam, 2009; Castle, 2010). Somatic hybridization has been successful within the Brassicaceae family, which is very amenable to protoplast fusion. A wide range of interspecific, intergeneric, and intertribal hybrids have been synthesized with *B. napus* as one fusion parent (Du et al., 2009).

Screening and Development of Stress-Resistant Plants Using *in vitro* Selection Techniques

Introduction

Screening of plant germplasms to obtain desired characteristics and to create novel plants has been undertaken for millennia. Since *in vitro* selection techniques became available, they have played important roles in plant screening and breeding programs. Compared to traditional methods, handling of larger plant populations, increased mutation induction efficiency, more precisely controlled environmental conditions, and higher frequency of mutant recovery are possible by using *in vitro* selection techniques (Predieri and Di Virgilio, 2007). These techniques have been intensively used especially for obtaining stress-resistant plants (Suprasanna et al., 2009). In this section, biotic and abiotic stress-resistant plants screened and developed using *in vitro* selection techniques in the last decade are discussed.

General methods of screening and breeding using *in vitro* selection techniques

Screening and breeding methods using *in vitro* selection techniques are described here. These methods include induction of variants or mutants, *in vitro* selection of plants with desired traits, and assessment of the plants. Tissue culture generates a wide range of somaclonal variation, which can express useful traits for breeders (Larkin and Scowcroft, 1981). Although the frequency of genetic variation is reasonably high (Hammerschlag et al., 2006), it can be increased by using physical and chemical mutagens, such as γ-ray and ethylmethane sulfonate. To screen for plants with target traits from a large number of variants, mutants or germplasm under *in vitro* conditions, efficient selection agents need to be employed (Kumar et al., 2008a,b). The agents differ depending on the target traits. If the target trait is resistance to a certain environmental stress, the stress must be added to the *in vitro* conditions at the time of selection. For example, to select salt-tolerant plants, media containing salt can be used. Generally, a sublethal dose of the particular selective agent is used, and plants that survive under these conditions are selected as resistant plants. If calli, cells, or protoplasts are used for selection, they need to be regenerated through organogenesis or somatic embryogenesis after the selection. Finally, *in vivo* assessment of selected plants, as well as their progenies, is required to examine the stability of the obtained traits.

Biotic stress resistance

Table 10.3 shows disease-resistant plants recently obtained using *in vitro* selection systems. Although physical or chemical mutagens may be used, it was revealed that mutation depends mainly on somaclonal variations to select and develop disease-resistant plants (Table 10.3). As a selective agent, culture filtrates of pathogens are often used (Table 10.3). Culture filtrates contain the metabolites secreted from pathogen to medium, including toxins. The kind of toxins may vary depending on the type of diseases and plant pathogens. For many plant diseases well-characterized toxins may be unavailable, hence culture filtrates can be used effectively (Suprasanna et al., 2009). If a specific toxin is known to be involved in the cause of disease, the toxin can be employed as a selective agent. Several fungi of the genus *Fusarium* produce fusaric acid (5-*n*-butylpicolinic acid), causing wilt of some plant species. This toxin has been successfully used for the development of disease-resistant plants (Table 10.3). Oxalic acid was also employed for the development of *Sclerotinia sclerotiorum* resistant plant in *B. napus* (Table 10.3).

Selection is popularly carried out with callus, with great success in development of disease-resistant plants (Table 10.3). In general, callus surviving on the selection media will develop into resistant plants. This is because for many host-specific toxins, a good correlation has been demonstrated between resistance to the pathogen at the plant level and insensitivity to the toxin at the cellular level (van den Bulk, 1991). However, Gonzalez et al. (2006) reported that a resistant bean cultivar of halo-blight pathogen showed a significantly smaller percentage of callus formation than a susceptible cultivar on a medium with the pathogen culture filtrates including a certain toxin. The authors assumed that the resistant bean cultivars triggered a hypersensitivity reaction including cell death. When the *in vitro* grown bud shoots of the same bean cultivars were subjected to the selection agent, the resistant cultivar showed higher growth than the susceptible cultivars (Gonzalez et al., 2006).

Resistance is a quantitative characteristic and, consequently, through selection, more resistant varieties can be obtained (Schiva et al., 1985; Thakur et al., 2002). Hammerschlag et al. (2006) confirmed this by assessing the degree of anthracnose resistance in *in vitro* grown strawberry plants that were regenerated from the explants of various cultivars with different degrees of resistance. It was concluded that increased levels of resistance can be generated from a susceptible cultivar, but to obtain the highest levels of resistance, it is best to start with germplasm already exhibiting some degree of resistance (Hammerschlag et al., 2006).

Abiotic stress tolerance

In vitro screening and development of salt-tolerant plants recently obtained is shown in Table 10.4. Various mutagens, as well as somaclonal variations, have been employed to obtain salt-tolerant plants. Sodium chloride with the concentrations of 100–200 mM is often used as a selective agent. The selection is usually applied to callus, although in some cases it is applied to *in vitro* shoots.

It is difficult to assess whether the salt-tolerant plants obtained from *in vitro* selection techniques have adapted to high salt concentration (i.e., a temporary or epigenetic change vs. a permanent or heritable change). Hasegawa et al. (1994) suggested that virtually all plant species could adapt to salt stress if the stress is imposed gradually. The salt-stress adapted plants are of great use if the plants are propagated

Table 10.3 Recently obtained disease-resistant plants using *in vitro* selection systems

Plant species	Aim	Mutagens	Type of culture and selective agents	Assessment	References
B. napus	Development of *Sclerotinia sclerotiorum* resistant plant.	Somaclonal variation or haploid calli treated with ethylmethane sulfonate.	Calli were grown on media containing 3–8 mM of oxalic acid.	Plants regenerated from selected calli showed resistance to the pathogen in both field and greenhouse.	Liu et al. (2005b)
Cauliflower (*B. oleracea*)	Development of black rot disease resistant plant.	Calli treated with ethyl methane sulfonate or gamma ray.	Calli were grown on media containing culture filtrate of *Xanthomonas campestris* pv. *campestris*.	*In vivo* plants regenerated from selected calli showed resistance to the pathogen.	Mangal and Sharma (2002)
Safflower (*Carthamus tinctorius*)	Development of *Alternaria carthami* resistant plant.	Somaclonal variation.	Embryogenic calli were grown on medium containing culture filtrate of *Alternaria carthami*.	Regenerated plants and their progenies showed resistance to the pathogen in the field.	Kumar et al. (2008a)
Turmeric (*Curcuma longa*)	Development of root rot disease resistant.	Somaclonal variation.	Calli were treated with culture filtrate of *Pythium graminicolum*.	Plants regenerated from selected calli showed resistance to the pathogen under *in vivo* conditions.	Gayatri et al. (2005)

(*Continued*)

Table 10.3 (Continued)

Plant species	Aim	Mutagens	Type of culture and selective agents	Assessment	References
Chrysanthemum (*Dendranthema grandiflorum*)	Development of leaf spot disease resistant plant.	Somaclonal variation.	Callus was grown on medium containing culture filtrate of *Septoria obesa*.	Plants regenerated from selected cell lines and cuttings of the generated plants acquired disease resistance in the field.	Kumar et al. (2008b)
Carnation (*Dianthus caryophyllus*)	Development of *Fusarium* resistant plant.	Somaclonal variation.	Calli were grown on media containing culture filtrate of *Fusarium oxysporum* f. sp. *dianthi*.	Plants regenerated from the resistant calli had acquired considerable resistance against the pathogen in the field.	Thakur et al. (2002)
Gladiolus	Development of *Fusarium* yellows resistant plant.	*In vitro* grown shoot clumps irradiated by gamma ray.	Shoots were cultured on media containing fusaric acid (1–1.5 mM).	*In vitro* selected mutants acquired resistance to the pathogen under *in vivo* conditions.	Pathania and Misra (2003)
Gladiolus	Development of *Fusarium* yellows resistant plant.	*In vitro* grown shoot clumps irradiated by gamma ray.	Shoots were cultured on media containing culture filtrate of *Fusarium oxysporum* f. sp. *gladioli*	*In vitro* selected mutants acquired resistance to the pathogen under *in vivo* conditions.	Pathania and Misra (2003)
Sugarcane (*Saccharum* sp.)	Development of red rot disease resistant plant.	Somaclonal variation.	Callus was grown on medium containing culture filtrate of *Colletotrichum falcatum* Went.	Somaclones regenerated from selected calli exhibited resistance to the pathogen in the field.	Sengar et al. (2009)
Grapevine (*Vitis vinifera*)	Development of anthracnose disease resistant plant.	Somaclonal variation.	Proembryogenic masses were grown in media containing culture filtrate of *Elsinoe ampelina* (deBary) Shear.	Plants regenerated from selected lines exhibited resistance to the pathogen in a greenhouse.	Jayasankar et al. (2000)
Pineapple (*Ananas comosus*)	Screening of fusariosis resistant plant.	—	Leaf segments of *in vitro* plantlets were spotted with culture filtrate of *Fusarium subglutinas*.	Plants exhibiting various levels of pathogen resistance were obtained.	Borras et al. (2001)
Pineapple (*Ananas comosus*)	Screening of fusariosis resistant plant.	—	Leaf segments of *in vitro* plantlets were spotted with 0.75 g/L of fusaric acid.	Plants exhibiting various levels of pathogen resistance were obtained.	Borras et al. (2001)
Yam (*Dioscorea* spp.)	Screening of resistant clones of anthracnose diseases.	—	Excised leaves were spotted with extracted metabolites of culture filtrate of *Colletotrichum gloeosporioides*.	Plants exhibiting various levels of pathogen resistance were obtained.	Amusa (2000)
Strawberry (*Fragaria* × *ananassa*)	Screening of anthracnose disease resistant plant.	—	*In vitro* leaves were incubated in a spore suspension of *Colletotrichum acutatum* isolate Goff.	Plants exhibiting various levels of pathogen resistance were obtained.	Hammerschlag et al. (2006)
Cassava (*Manihot* spp.)	Screening of resistant clones of anthracnose diseases.	—	Stem cuttings and excised leaves were spotted with extracted metabolites of culture filtrate of *Colletotrichum gloeosporioides* f. sp. *Manihotis*.	Plants exhibiting various levels of pathogen resistance were obtained.	Amusa (2000)

(*Continued*)

Table 10.4 Recently obtained NaCl tolerant plants using *in vitro* selection systems

Plant species	Mutagens	Type of culture and selective agents	Assessment
Sugar beet (*Beta vulgaris*)	*In vitro* grown bud clumps irradiated by gamma ray.	Bud clumps were grown on media supplemented with 171, 257, 342, or 428 mM of NaCl.	Progenies obtained from self-pollination of regenerated selected lines showed resistance to NaCl in the field.
Cauliflower (*B. oleracea*)	Microshoots treated with *N*-nitroso-*N*-ethylurea and *N*-nitroso-*N*-methyl-urea.	Microshoots were grown with media containing 3 mM of hydroxyproline.	Regenerated *in vitro* and *in vivo* plants obtained resistance to NaCl, however, the resistance was not heritable.
Greater yam (*Dioscorea alata*)	Adaptation or somaclonal variation.	Nodal segments from *in vitro* plantlets were placed on media containing 100 mM of NaCl.	Selected *in vitro* plantlets showed enzyme activity profiles reflecting the biochemical adjustments of the plantlets to cope with the saline conditions.
Strawberry (*Fragaria* × *ananassa* Duch.)	Somaclonal variation.	Medium containing 200 mM of NaCl was used at the seed germination and early growth phase of seedlings.	Progenies of regenerated selected lines showed resistance to NaCl under *in vitro* conditions.
Sweet potato (*Ipomoea batatas*)	Calli treated with ethyl methanesulfonate.	Calli were cultured on medium supplemented with 200 mM of NaCl.	Plants regenerated from selected calli showed resistance to NaCl under *in vitro* conditions.
Sweet potato (*I. batatas*)	Cell aggregates from embryogenic suspension cultures were irradiated with gamma ray.	Cell aggregates were cultured in a medium containing 342 mM of NaCl.	Plants regenerated from selected cell lines showed resistance to NaCl under *in vitro* and *in vivo* conditions.
Rice (*Oryza sativa*)	Somaclonal variation.	Calli were cultured on medium supplemented with 300 mM of NaCl.	Progenies of regenerated selected lines showed resistance to NaCl under *in vivo* conditions.
Robinia pseudoacacia	Somaclonal variation.	Calli were grown on media including 0.15 or 0.2 M of NaCl.	Plants regenerated from selected calli showed resistance to NaCl under *in vitro* conditions.
Sugarcane (*Saccharum* sp.)	Calli irradiated with gamma ray.	Calli were grown on media with 85.6 or 171.1 mM NaCl.	Radiated calli were more tolerant to NaCl than non-radiated calli.
Sugarcane (*Saccharum* sp.)	Calli irradiated with gamma ray or somaclonal variants.	Calli were grown on media containing 42.8, 85.6, 128.3, and 171.1 mM of NaCl.	The RAPD profiles of putatively salt-tolerant regenerants differed from those of others under *in vitro* conditions.
Sweet potato (*I. batatas*)	—	Shoot apexes were inoculated in test tubes containing media supplemented with 0.5 or 1.0% (w/v) of NaCl.	Based on the levels of tolerance to NaCl under *in vitro* conditions, 15 genotypes were separated into 3 different categories.
Mulberry (*Morus* spp.)	—	Axillary buds were cultured on medium supplemented with 0.25–1.00% of NaCl.	A wide variation in the response to NaCl was observed among 63 genotypes under *in vitro* conditions. Correlations were found in the NaCl tolerance levels between the *in vitro* test and *ex vitro* test.
Mulberry (*Morus* spp.)	—	Seeds were germinated on medium containing 0.25, 0.5, 0.75, 1.0, and 1.25% of NaCl.	Seeds from 43 different genotypes showed wide variability of tolerance to salinity under *in vitro* conditions, some of which showed considerable tolerance to salinity.
Mulberry (*Morus* spp.)	—	Seeds were germinated on medium containing 1.25% of NaCl.	Selected plants showed resistance to NaCl in the field.
Sugarcane (*Saccharum* sp.)	—	Calli were grown on media including 17, 34, 68, and 102 mM of NaCl.	Two genotypes of GT54-9 and NCo310 were found to be more tolerant to NaCl than a genotype of Co413 under *in vitro* conditions.

asexually (Wheatley et al., 2003). However, if the plants go through sexual propagation, assessment of the traits in their progenies is necessary. Table 10.4 summarizes both heritable traits of salt tolerance and salt tolerance developed as a result of adaptation. Biswas et al. (2002) simulated drought conditions *in vitro* by using PEG, which induces water stress in the medium by acting as a non-penetrating osmotic agent that lowers the water potential of the medium. Rice selected with this chemical proved to have drought tolerance in the field, and the traits showed a monogenic inheritance pattern (Biswas et al., 2002). Drought-tolerant plants of *Tagetes minuta* were obtained when 60 mM mannitol was used as a selective agent (Mohamed et al., 2000).

McClinchey and Kott (2008) successfully selected cold-tolerant spring canola plants by using salicylic acid, jasmonic acid, 3,4-dehydro-D,L-proline, azetidine-2-carboxylate, and hydroxyproline. In winter barley, embryogenic calli derived from anther cultures were placed on media including 10–20 mM of hydroxyproline (Tantau et al., 2004). Regenerated plants from the selected calli proved to have frost resistance, and the resistance was confirmed to be heritable (Tantau et al., 2004). A heritable frost resistance was also obtained in cauliflowers by selecting microshoots with media containing hydroxyproline before the shoots were stored at 5°C for 2 years (Fuller et al., 2006). High temperature tolerant potato plants were obtained when microtubers were produced at a high temperature of 28°C from γ-irradiated shoots (Das et al., 2000).

In vitro selection techniques are also used to produce herbicide-tolerant plants with a combination of gene transformation techniques. Herbicides have also been used as selective agents (Guosheng et al., 2002; Venkataiah et al., 2005; Shizukawa and Mii, 2008).

Future perspective of screening and breeding using *in vitro* selection techniques

In vitro selection techniques are now taking the place of traditional methods in plant screening and breeding programs because of their convenience. However, the lack of well-established *in vitro* techniques often prevents it from practical use. A complete operation of *in vitro* screening and breeding involves variant or mutant induction, *in vitro* selection, *in vitro* plant regeneration, acclimatization of *in vitro* plants, and assessment of *in vivo* plants. In chili pepper, cell and tissue cultures have been applied successfully to the selection of variant cells exhibiting increased resistance to abiotic stress, but no plants exhibiting the selected traits have been regenerated (Ochoa-Alejo and Ramirez-Malagon, 2001). Ben-Hayyim and Moore (2007) also reported that attempts to regenerate salt-tolerant citrus plants via *in vitro* production of salt-tolerant callus or mutagenesis have been rather limited and are not in use. Extensive research in the establishment of each *in vitro* technique is required in the future.

Conclusions and Future Directions

Much has been achieved using the technologies described in this section for the selection of useful traits in plants. While some problems remain, techniques such as somatic embryogenesis, haploid techniques, protoplast, and cell and callus culture are now routinely used in the selection of traits important in agriculture. For example, somatic hybridization is applied to increase the capacity of plants to cope better with biotic and abiotic stresses, which may become a key focus in the future for sustainable agriculture. These techniques can be applied to improving plants for ornamental purposes and nutritional quality, creating unique and highly desirable plants that can meet market demand. Future research needs to focus on development of somatic hybridization techniques to improve plant capabilities for use in biofuel and bioremediation. *In vitro* selection techniques are now complementing traditional methods in plant screening and breeding programs because of their convenience and because of the potential they have for reducing the duration of the breeding cycle. However, the lack of well-established *in vitro* techniques often prevents its practical use. A complete operation of *in vitro* screening and breeding involves variant or mutant induction, selection, plant regeneration, acclimatization, and assessment of *in vivo* plants. Extensive research into the establishment of each *in vitro* technique protocol for the success of plant screening and breeding programs is required. Recently, climate change has been affecting the growth and yield of plants all over the world. Growers may be forced to change to a new crop that is suitable for new environmental conditions such as extreme temperatures, drought, or chemical accumulated/contaminated soil. The endless possibilities for developing improvements in plants through technologies described in this chapter have the potential to help solve world hunger and problems faced in agriculture as a result of climate change.

Acknowledgments

The authors wish to thank Mr. Youske Sato from Meiji University for his help in retrieving many of the references for this chapter.

References

Akin-Idowu, P. E., Ibitoye, D. O., & Ademoyegun, O. T. (2009). Tissue culture as a plant production technique for horticultural crops. *African Journal of Biotechnology*, *8*, 3782–3788.

Alan, A. R., Brants, A., Cobb, E., Goldschmied, P. A., Mutschler, M. A., & Earle, E. D. (2004). Fecund gynogenic lines from onion (*Allium cepa* L.) breeding materials. *Plant Science*, *167*, 1055–1066.

Alan, A. R., Mutschler, M. A., Brants, A., Cobb, E., & Earle, E. D. (2003). Production of gynogenic plants from hybrids of *Allium cepa* L. and *A. roylei* Stern. *Plant Science*, *165*, 1201–1211.

Amusa, N. A. (2000). Screening of cassava and yam cultivars for resistance to anthracnose using toxic metabolites of colletotrichum species. *Mycopathologia*, *150*, 137–142.

An, H., Jin, S. -B., Kang, B., & Park, H. (2008). Production of somatic hybrids between satsuma mandarin (*Citrus unshiu*) and navel orange (*Citrus sinensis*) by protoplast fusion. *Journal of Plant Biology*, *51*, 186–191.

Anwar, F., Sharmila, P., & Saradhi, P. P. (2010). No more recalcitrant: Chickpea regeneration and genetic transformation. *African Journal of Biotechnology*, *9*, 782–797.

Aquea, F., Johnston, A. J., Canon, P., Grossniklaus, U. Arce-Johnson, P. (2010). *TRAUCO*, a *Trithorax*-group gene homologue, is required for early embryogenesis in *Arabidopsis thaliana*. *Journal of Experimental Botany*, erp396.

Aryan, A. P. (2002). Production of double haploids in rice: Anther vs. microspore culture. *Proceedings of the importance of plant tissue culture and biotechnology in plant sciences* (pp. 201–208). Armidale.

Assani, A., Chabane, D., & Haicour, R., et al. (2005). Protoplast fusion in banana (*Musa* spp): Comparison of chemical (PEG: Polyethylene glycol) and electrical procedure. *Plant Cell, Tissue and Organ Culture*, *83*, 145–151.

Atak, M., Kaya, M., Khawar, M. M., Saglam, S., Özcan, S., & Ciftci, Y. (2008). Effect of age on somatic embryogenesis from immature zygotic embryos of 5 Turkish triticale genotypes. *African Journal of Biotechnology*, *7*, 1765–1768.

Bach, A., & Krol, A. (2001). Effect of light quality on somatic embryogenesis in *Hyacinthus orientalis* L. "Delft's Blue." *Biological Bulletin Poznan*, *38*, 103–107.

Bach, A., Malik, M., Ptak, A., & Kedra, M. (2000). Light effects on ornamental microplant shoots and bulbs quality. *Acta Horticulturae*, *530*, 173–179.

Baenziger, P. S., Russell, W. K., Graef, G. L., & Campbell, B. T. (2006). Improving lives: 50 years of crop breeding, genetics, and cytology (C-1). *Crop Science*, *46*, 2230–2244.

Bahgat, S., Shabban, O. A., El-Shihy, O., Lightfoot, D. A., & El-Shemy, H. A. (2008). Establishment of the regeneration system for *Vicia faba* L. *Current Issues in Molecular Biology*, *11*, i47–54.

Bal, U., Ellialtioglu, S., & Abak, K. (2009). Induction of symetrical nucleus division and multi-nucleate structures in microspores of eggplant (*Solanum melongena* L.) cultured *in vitro*. *Scientia Agricultura*, *66*, 535–539.

Baron, K., & Stasolla, C. (2008). The role of polyamines during *in vivo* and *in vitro* development. *In Vitro Cellular & Developmental Biology–Plant*, *44*, 384–395.

Bartoszewski, G., Havey, M. J., Ziółkowska, A., Dlugosz, M., & Malepszy, S. (2007). The selection of mosaic (MSC) phenotype after passage of cucumber (*Cucumis sativus* L.) through cell culture – a method to obtain plant mitochondrial mutants. *Journal of Applied Genetics*, *48*, 1–9.

Bassene, J., Froelicher, Y., Dhuique-Mayer, C., Mouhaya, W., Ferrer, R., & Ancillo, G., et al. (2009). Non-additive phenotypic and transcriptomic inheritance in a citrus allotetraploid somatic hybrid between C. *reticulata* and C. *limon*: The case of pulp carotenoid biosynthesis pathway. *Plant Cell Reports*, *28*, 1689–1697.

Bassuner, B. M., Lam, R., Lukowitz, W., & Yeung, E. C. (2007). Auxin and root initiation in somatic embryos of *Arabidopsis*. *Plant Cell Reports*, *26*, 1–11.

Ben-Hayyim, G., & Moore, G. A. (2007). Recent advances in breeding citrus for drought and saline stress tolerance. In M. A. Jenks (Ed.), *Advances in molecular breeding toward drought and salt tolerant crops* (pp. 627–642). Dordrecht, Netherlands: Springer.

Berleth, T., & Chatfield, S. (2002). Embryogenesis: Pattern formation from a single cell. In *The arabidopsis book*: American Society of Plant Biologists, ed.

Bhat, J. G., & Murthy, H. N. (2008). Haploid plant regeneration from unpollinated ovule cultures of niger (*Guizotia abyssinica* (L.f.) Cass). *Russian Journal of Plant Physiology*, *55*, 241–245.

Binott, J. J., Songa, J. M., Ininda, J., Njagi, E. M., & Machuka, J. (2008). Plant regeneration from immature embryos of Kenyan maize inbred lines and their respective single cross hybrids through somatic embryogenesis. *African Journal of Biotechnology*, *7*, 981–987.

Bishnoi, U., Jain, R. K., Rohilla, J. S., Chowdhury, V. K., Gupta, K. R., & Chowdury, J. B. (2000). Anther culture of recalcitrant *indica* x Basmati rice hybrids. *Euphytica*, *114*, 93–101.

Biswas, J., Chowdhury, B., Bhattacharya, A., & Mandal, A. B. (2002). *In vitro* screening for increased drought tolerance in rice. *In Vitro Cellular Development Biology- Plant*, *38*, 525–530.

Bitsch, C., Gröger, S., & Lelley, T. (1998). Effect of parental genotypes on haploid embryo and plantlet formation in wheat×maize crosses. *Euphytica*, *103*, 319–323.

Bonga, J. M., Klimaszewska, K. K., & von Aderkas, P. (2010). Recalcitrance in clonal propagation, in particular of conifers. *Plant Cell, Tissue and Organ Culture*, *100*, 241–254.

Borras, O., Santos, R., Matos, A. P., Cabral, R. S., & Arzola, M. (2001). A first attempt to use a *Fusarium subglutinans* culture filtrate for the selection of pineapple cultivars resistant to fusariose disease. *Plant Breeding*, *120*, 435–438.

Cai, X., Duan, Y., Fu, J., & Guo, W. (2010). Production and molecular characterization of two new Citrus somatic hybrids for scion improvement. *Acta Physiologiae Plantarum*, *32*(1), 215–221.

Cangahuala-Inocente, G., Silveira, V., Caprestano, C., Ducroquet, J., Floh, E., & Guerra, M. (2009). Dynamics of biochemical and morphophysiological changes during zygotic embryogenesis in *Acca sellowiana* (Berg.) Burr. *Plant Growth Regulation*, *59*, 103–115.

Cardoza, V., & Steward, C. N. (2004). *Brassica* biotechnology: Progress in cellular and molecular biology. *In Vitro Cellular & Developmental Biology–Plant*, *40*, 542–551.

Caredda, S., Doncoeur, C., Devaux, P., Sangwan, R. S., & Clément, C. (2000). Plastid differentiation during androgenesis in albino and non-albino producing cultivars of barley (*Hordeum vulgare* L.). *Sexual Plant Reproduction*, *13*, 95–104.

Carneros, E., Celestino, C., Klimaszewska, K., Park, Y. S., Toribio, M., & Bonga, J. (2009). Plant regeneration in Stone pine (*Pinus pinea* L.) by somatic embryogenesis. *Plant Cell, Tissue and Organ Culture*, *98*, 165–178.

Castillo, A. M., Vallés, M. P., & Cistué, L. (2000). Comparison of anther and isolated microspore cultures in barley. Effects of culture density and regeneration medium. *Euphytica*, *113*, 1–8.

Castle, W. S. (2010). A career perspective on *Citrus* rootstocks, their development, and commercialization. *HortScience*, *45*, 11–15.

Ceasar, S., & Ignacimuthu, S. (2010). Effects of cytokinins, carbohydrates and amino acids on induction and maturation of somatic embryos in kodo millet (*Paspalum scorbiculatum* Linn.). *Plant Cell, Tissue and Organ Culture*, *102*, 153–162.

Ceasar, S. A., & Ignacimuthu, S. (2008). Efficient somatic embryogenesis and plant regeneration from shoot apex explants of different Indian genotypes of finger millet (*Eleusine coracana* (L.) Gaertn.). *In Vitro Cellular & Developmental Biology-Plant*, *44*, 427–435.

Chen, Y., Kenaschuk, E. O., and Procunier, D. J. (1998). Plant regeneration from anther culture in Canadian cultivars of flax (*Linum usitatissimum* L.). *Euphytica*, 102.

Cheng, A. X., & Xia, G. M. (2004). Somatic hybridization between common wheat and Italian ryegrass. *Plant Science*, *166*, 1219–1226.

Choudhury, H., Kumaria, S., & Tandon, P. (2008). Induction and maturation of somatic embryos from intact megagametophyte explants in Khasi pine (*Pinus kesiya Royle* ex. Gord.). *Current Science*, *95*, 1433–1438.

Chugh, A., & Khurana, P. (2002). Gene expression during somatic embryogenesis – recent advances. *Current Science*, *83*, 715–730.

Cistué, L., Ramos, A., & Castillo, A. M. (1998). Influence of anther pretreatment and culture medium composition on the production of barley doubled haploids from model and low responding cultivars. *Plant Cell, Tissue and Organ Culture*, *55*, 159–166.

Cistué, L., Soriano, M., Castillo, A. M., Vallés, M. P., Sanz, J. M., & Echavarri, B. (2006). Production of doubled haploid in *durum* wheat (*Triticum turgidum* L.) through isolated microspore culture. *Plant Cell Reports*, *25*, 257–264.

Corrêa, C. M., de Oliveira, G. N., Astarita, L. V., & Santarém, E. R. (2009). Plant regeneration through somatic embryogenesis of yacón [*Smallanthus sonchifolius* (Poepp. and Endl.) H. Robinson]. *Brazilian Archives of Biology and Technology*, *52*, 549–554.

Crosser, J. S., Lülsdorf, L. L., Davies, P. A., Clarke, H. J., Bayliss, K. L., & Mallikarjuna, N., et al. (2006). Toward doubled haploid production in the Fabaceae: Progress, constraints, and opportunities. *Critical Review in Plant Sciences*, *25*, 139–157.

Cui, H., Yu, Z., Deng, J., Gao, X., Sun, Y., & Xia, G. (2009). Introgression of bread wheat chromatin into tall wheatgrass via somatic hybridization. *Planta*, *229*(2), 323–330.

Curtis, M., & Grossniklaus, U. (2008). Molecular control of autonomous embryo and endosperm development. *Sexual Plant Reproduction*, *21*, 79–88.

da Silva, M., Pinto, D., Guerra, M., Floh, E., Bruckner, C., & Otoni, W. (2009). A novel regeneration system for a wild passion fruit species (*Passiflora cincinnata* Mast.) based on somatic embryogenesis from mature zygotic embryos. *Plant Cell, Tissue and Organ Culture*, *99*, 47–54.

da Silva Lauxen, M., Kaltchuk-Santos, E., Hu, C., Callegari-Jaques, S. M., & Bodanese-Zanetti, M. H. (2003). Association between floral loud size and developmental stage in soybean microspores. *Brazilian Archieves of Biology and Technology*, *46*, 515–520.

Das, A., Gosal, S. S., Sidhu, J. S., & Dhaliwal, H. S. (2000). Induction of mutations for heat tolerance in potato by using *in vitro* culture and radiation. *Euphytica*, *114*, 205–209.

Datta, S. (2005). Androgenic haploids: Factors controlling development and its application in crop improvement. *Current Science*, *89*, 1870–1878.

De la Peña, C., Galaz-Ávalos, R., & Loyola-Vargas, V. (2008). Possible role of light and polyamines in the onset of somatic embryogenesis of *Coffea canephora*. *Molecular Biotechnology*, *39*, 215–224.

Demeter, Z., Surányi, G., Molnár, V., Sramkó, G., Beyer, D., & Kónya, Z., et al. (2010). Somatic embryogenesis and regeneration from shoot primordia of *Crocus heuffelianus*. *Plant Cell, Tissue and Organ Culture*, *100*, 349–353.

Dodds, J. H., & Roberts, L. W. (1985). *Experiments in plant tissue culture*. Cambridge: Cambridge University Press.

Du, L., Zhou, S., & Bao, M. Z. (2007). Effect of plant growth regulators on direct somatic embryogenesis in camphor tree (*Cinnamomum camphora* L.) from immature zygotic embryos and embryogenic calli induction. *Forestry Studies in China*, *9*, 267–271.

Du, X. Z., Ge, X. H., Yao, X. C., Zhao, Z.G., & Li, Z. Y. (2009). Production and cytogenetic characterization of intertribal somatic hybrids between *Brassica napus* and *Isatis indigotica* and backcross progenies. *Plant Cell Reports*, *28*(7), 1105–1113.

Ducos, J., Terrier, B., Courtois, D., & Pétiard, V. (2008). Improvement of plastic-based disposable bioreactors for plant science needs. *Phytochemistry Reviews*, *7*, 607–613.

Ducos, J. P., Lambot, C., & Petiard, V. (2007). Bioreactors for coffee mass propagation by somatic embryogenesis. *International Journal of Plant Developmental Biology*, *1*, 1–12.

Ducos, J. P., Terrier, B. & Courtois, D. (2009). Disposable bioreactors for plant micropropagation and mass plant cell culture. In *Disposable Bioreactors*, Vol. 115: Advances in biochemical engineering/Biotechnology (pp. 89–115).

Dunwell, J. M. (1986). Pollen, ovule and embryo culture as tools in plant breeding. In P. G. Alderson (Ed.), *Plant Tissue Culture and its Agricultural Application* (pp. 375–404). London: Butterworths.

Duquenne, B., Eeckhaut, T., Werbrouck, S., & Van Huylenbroeck, J. (2007). Effect of enzyme concentrations on protoplast isolation and protoplast culture of *Spathiphyllum* and *Anthurium*. *Plant Cell, Tissue and Organ Culture*, *91*(2), 165–173.

Durieu, P., & Ochatt, S. J. (2000). Efficient intergeneric fusion of pea (*Pisum sativum* L.) and grass pea (*Lathyrus sativus* L.) protoplasts. *Journal of Experimental Botany*, *51*, 1237–1242.

Fehér, A. (2008). The initiation phase of somatic embryogenesis: What we know and what we don't. *Acta Biologica Szegediensis*, *52*, 53–56.

Fehér, A., Pasternak, T. P., & Dudits, D. (2003). Transition of somatic plant cells to an embryogenic state. *Plant Cell, Tissue and Organ Culture*, *74*, 201–228.

Feng, D. L., Meng, Q. R., Li, W. P., Hu, Y. H., Li, M., & Gu, A.X. (2010). Morphology of somatic embryogenesis and plantlet formation in tissue cultures of lantern tree (*Koelreuteria bipinnata* var. *integrifoliola*). *Forestry Studies in China*, *12*, 31–36.

Folta, K., & Dhingra, A. (2006). Transformation of strawberry: The basis for translational genomics in Rosaceae. *In Vitro Cellular & Developmental Biology–Plant*, *42*, 482–490.

Fu, C. H., Guo, W. W., Liu, J. H., & Deng, X. X. (2003). Regeneration of *Citrus sinensis* + *Clausena lansium* intergeneric triploid and tetraploid somatic hybrids and their molecular identification. *In Vitro Cellular & Developmental Biology Plant*, *39*, 360–364.

Fu, L. L., Yang, X. Y., Zhang, X. L., Wang, Z. W., Feng, C. H., & Liu, C. X., et al. (2009). Regeneration and identification of interspecific asymmetric somatic hybrids obtained by donor-recipient fusion in cotton. *Chinese Science Bulletin*, *54*(17), 3035–3044.

Fu, X., Yang, S., & Bao, M. (2008). Factors affecting somatic embryogenesis in anther cultures of Chinese pink (*Dianthus chinensis* L.). *In Vitro Cellular & Developmental Biology–Plant*, *44*, 194–202.

Fuller, M. P., Metwali, E. M. R., Eed, M. H., & Jellings, A. J. (2006). Evaluation of abiotic stress resistance in mutated populations of cauliflower (*Brassica oleracea* var. *botrytis*). *Plant Cell, Tissue and Organ Culture*, *86*, 239–248.

Furuta, H., Shinoyama, H., Nomura, Y., Maeda, M., & Makara, K. (2004). Production of intergeneric somatic hybrids of chrysanthemum [*Dendranthema* × *grandiflorum* (Ramat.) Kitamura] and wormwood (*Artemisia sieversiana* J. F. Ehrh. ex. Willd) with rust (*Puccinia horiana* Henning) resistance by electrofusion of protoplasts. *Plant Science*, *166*, 695–702.

Gamborg, O. L., Millers, R. A., & Ojima, K. (1968). Nutrient requirements of suspension cultures of soybean root cells. *Experimental Cell Research*, *50*, 151–158.

Gandonou, C., Errabii, T., Abrini, J., Idaomar, M., Chibi, F., & Senhaji, N. S. (2005). Effect of genotype on callus induction and plant regeneration from leaf explants of sugarcane (*Saccharum* spp.). *African Journal of Biotechnology*, *4*, 1250–1255.

Gao, X., Liu, S., Sun, Q., & Xia, G. (2010). High frequency of HMW-GS sequence variation through somatic hybridization between *Agropyron elongatum* and common wheat. *Planta*, *231*(2), 245–250.

Gao, X., Yang, D., Cao, D., Ao, M., Sui, X., Wang, Q., et al. (2010) *In vitro* micropropagation of *Freesia hybrida* and the assessment of genetic and epigenetic stability in regenerated plantlets. *Journal of Plant Growth Regulation*, 29, 257–267.

Gayatri, M. C., Roopa-darshini, V. R., & Kavyashree., R. (2005). Selection of turmeric callus for tolerant to culture filtrate of *Pythium graminicolum* and regeneration of plants. *Plant Cell, Tissue and Organ Culture*, *83*, 33–40.

Gervais, C., Newcomb, W., & Simmonds, D. H. (2000). Rearrangement of the actin filament and microtubule cytoskeleton during induction of microspore embryogenesis in *Brassica napus* L. cv. Topas. *Protoplasma*, *213*, 194–202.

Gonzalez, A. I., Polanco, C., & Ruiz, M. L. (2006). *In vitro* culture response of common bean explants to filtrate from *Pseudomonas syringae* pv. *phaseolicola* and correlation with disease resistance. *In Vitro Cellular Development Biology–Plant*, *42*, 160–164.

Gow, W.P., Chen, J. T. Chang, W. C. (2009. Enhancement of direct somatic embryogenesis and plantlet growth from leaf explants of *Phalaenopsis* by adjusting culture period and explant length. *Acta Physiologiae Plantarum*.

Grosser, J. W., & Chandler, J. L. (2000). Somatic hybridization of high yield, cold-hardy and disease resistant parents for citrus rootstock improvement. *Journal of Horticultural Science and Biotechnology*, *75*, 641–644.

Grosser, J. W., & Gmitter, F. G., Jr. (2005). 2004 SIVB Congress symposium proceedings "Thinking outside the cell:" Applications of somatic hybridization and cybridization in crop improvement, with citrus as a model. *In Vitro Cellular & Developmental Biology–Plant*, *41*(3), 220–225.

Guha, S., & Maheshwari, S. C. (1964). *In vitro* production of embryos from anthers of *Datura*. *Nature*, *204*, 497.

Guo, W. W., Cai, X. D., & Grosser, J. W. (2004). Somatic cell cybrids and hybrids in plant improvement. In H. Daniell & C. D. Chase (Eds.), *Molecular Biology and Biotechnology of Plant Organelles* (pp. 635–659). The Netherlands: Springer.

Guo, W. W., Cheng, Y. J., & Deng, X. X. (2002). Regeneration and molecular characterization of intergeneric somatic hybrids between *Citrus reticulata* and *Poncirus trifoliata*. *Plant Cell Reports*, *20*, 829–834.

Guo, Y. -D., Sewón, P., & Pulli, S. (1999). Improved embryogenesis from anther culture and plant regeneration in timothy. *Plant Cell, Tissue and Organ Culture*, *57*, 85–93.

Guosheng, L., Qingwei, Z., Juren, Z., Yuping, B., & Lei, S. (2002). Establishment of multiple shoot clumps from maize (*Zea mays* L.) and regeneration of herbicide-resistant transgenic plantlets. Science in China Series C. *Life Sciences*, *45*, 40–49.

Guzy-Wróbelska, J., Labocha-Pawlowska, A., Kwasniewski, M., & Szarejko, I. (2007). Different recombination frequencies in wheat doubled haploid population obtained through maize pollination and anther culture. *Euphytica*, *156*, 173–183.

Hammerschlag, F., Garces, S., Koch-Dean, M., Ray, S., Lewers, K., & J. Maas, J., et al. (2006). *In vitro* response of strawberry cultivars and regenerants to Colletotrichum acutatum. *Plant Cell, Tissue and Organ Culture*, *84*, 255–261.

Han, L., Zhou, C., Shi, J., Zhi, D., & Xia, G. (2009). Ginsenoside Rb_1 in asymmetric somatic hybrid calli of *Daucus carota* with *Panax quinquefolius*. *Plant Cell Reports*, *28*(4), 627–638.

Hasegawa, P. M., Bressan, R. A., Nelson, D. E., Samaras, Y., & Rhodes, D. (1994). Tissue culture in the improvement of salt tolerance in plants. In A. R. Yeo & T. J. Flower (Eds.), *Soil mineral stresses: Approaches to crop improvement*. (pp. 83–125). Berlin: Springer-Verlag.

Hofer, M., Touraev, A., & Haberle-Bors, E. (1999). Induction of embryogenesis from isolated apple microspores. *Plant Cell Reports*, *18*, 1012–1017.

Hoshino, T., & Cuello, J. (2006). Environmental design considerations for somatic embryogenesis. In A. Mujib & J. Samaj (Eds.), *Somatic embryogenesis* (pp. 25–34). Berlin, Heidelberg: Springer.

Hosp, J., Maraschin, S. F., Touraev, A., & Boutilier, K. (2007a). Functional genomic of microspore embryogenesis. *Euphytica*, *158*, 275–285.

Hosp, J., Tashpulatov, A., Rossner, U., Barsova, E., Katholnigg, H., & Steinborn, R., et al. (2007b). Transcriptional and metabolic profiles of stress-induced, embryogenic tobacco microspores. *Plant Molecular Biology*, *63*, 137–149.

Hu, Q., Anderson, S. B., Dixelius, C., & Hansen, L. N. (2002a). Production of fertile intergeneric somatic hybrids between *Brassica napus* and *Sinapis arvensis* for the enrichment of the rapeseed gene pool. *Plant Cell Reports*, *21*, 147–152.

Hu, Q., Hansen, L. N., Laursen, J., Dixelius, C., & Andersen, S. B. (2002b). Intergeneric hybrids between *Brassica napus* and *Orychophragmus violaceus* containing traits of agronomic importance for oilseed rape breeding. *Theoretical and Applied Genetics*, *105*, 834–840.

Ikeda, M., Umehara, M., & Kamada, H. (2006). Embryogenesis-related genes: Its expression and roles during somatic and zygotic embryogenesis in carrot and *Arabidopsis*. *Plant Biotechnology*, *23*, 153–161.

Immonen, S., & Robinson, J. (2000). Stress treatment and ficoll for improving green plant regeneration in triticale anther culture. *Plant Science*, *150*, 77–84.

Jauhar, P. P. (2006). Modern biotechnology as an integral supplement to conventional plant breeding: The prospects and challenges. *Crop Science*, *46*, 1841–1859.

Jayasankar, S., Li, Z., & Gray, D. J. (2000). *In vitro* selection of *Vitis vinifera* "Chardonnay" with *Elsinoe ampelina* culture filtrate is accompanied by fungal resistance and enhanced secretion of chitinase. *Planta*, *211*, 200–208.

Jenik, P. D., Gillmor, C. S., & Lukowitz, W. (2007). Embryonic patterning in *Arabidopsis thaliana*. *Annual Review in Cell and Developmental Biology*, *23*, 207–236.

Jiang, J. J., Zhao, X. X., Tian, W., Li, T. B., & Wang, Y. P. (2009). Intertribal somatic hybrids between *Brassica napus* and *Camelina sativa* with high linolenic acid content. *Plant Cell, Tissue and Organ Culture*, *99*(1), 91–95.

Jiménez, V. M. (2001). Regulation of *in vitro* somatic embryogenesis with emphasis on the role of endogenous hormones. *Revista Brasileira de Fisiologia Vegetale*, *13*, 196–223.

Jiménez, V. M. (2005). Involvement of plant hormones and plant growth regulators on *in vitro* somatic embryogenesis. *Plant Growth Regulation*, *47*, 91–110.

Jiménez, V. M., & Thomas, C. (2005). Participation of plant hormones in determination and progression of somatic embryogenesis. *Plant Cell Monographs*, *2*, 103–118.

Karami, O., Aghavaisi, B., & Mahmoudi Pour, A. (2009). Molecular aspects of somatic-to-embryogenic transition in plants. *Journal of Chemical Biology*, *2*, 177–190.

Kasha, K. J., Simion, E., Oro, R., Yao, Q. A., Hu, T. C., & Carlson, A. R. (2001). An improved *in vitro* technique for isolated microspore culture of barley. *Euphytica*, *120*, 379–385.

Khentry, Y., Paradornuvat, A., Tantiwiwat, S., Phansiri, S., & Thaveechai, N. (2006). Protoplast isolation and culture of *Dendrobium* Sonia "Bom 17." *Kasetsart Journal (Natural Science)*, *40*, 361–369.

Kiran, G., Kaviraj, C. P., Jogeswar, G., Kishor, K. V. K., & Rao, S. (2005). Direct and high frequency somatic embryogenesis and plant regeneration from hypocotyls of chickpea (*Cicer arietinum* L.), a grain legume. *Current Science*, *89*, 1012–1018.

Kiran Ghanti, S., Sujata, K., Srinath Rao, M., & Kavi Kishor, P. (2010). Direct somatic embryogenesis and plant regeneration from immature explants of chickpea. *Biologia Plantarum*, *54*, 121–125.

Kiviharju, E., & Pehu, E. (1998). The effect of cold and heat pretreatment on anther culture response of *Avena sativa* and *A. sterilis*. *Plant Cell, Tissue and Organ Culture*, *54*, 97–104.

Kouassi, K. M., Koffi, K. E., Gnagne, Y. M., O. N'nan, Y., Coulibaly, , & Sangare, A. (2008). Production of *Hevea brassiliensis* embryos from *in vitro* culture of unpollinated ovules. *Biotechnology*, *7*, 793–797.

Kumar, J. V., Kumari, B. D. R., Sujatha, G., & Castano, E. (2008a). Production of plants resistant to *Alternaria carthami* via organogenesis and somatic embryogenesis of safflower cv. NARI-6 treated with fungal culture filtrates. *Plant Cell, Tissue and Organ Culture*, *93*, 85–96.

Kumar, S., Kumer, S., Negi, S. P., & Kanwar, J. K. (2008b). *In vitro* selection and regeneration of chrysanthemum (*Dendranthema grandiflorum* Tzelev) plants resistant to culture filtrate of *Septoria obesa* Syd. *In Vitro Cellular Development Biology-Plant*, *44*, 474–479.

Kurczynska, E. U., Gaj, M. D., Ujczak, A., & Mazur, E. (2007). Histological analysis of direct somatic embryogenesis in *Arabidopsis thaliana* (L.) Heynh. *Planta*, *226*, 619–628.

Labbani, Z., De Buyser, J., & Picard, E. (2007). Effect of mannitol pretreatment to improve green plant regeneration on isolated microspore culture in *Triticum turgidum* ssp. *durum* cv. "Jennah Khetifa." *Plant Breeding*, *126*, 565–568.

Lakshmanan, P., & Taji, A. (2000). Somatic embryogenesis in leguminous plants. *Plant Biology*, *2*, 136–148.

Lardet, L., Dessailly, F., Carron, M. P., Rio, M. A., Ferriere, N., & Montoro, P. (2009). Secondary somatic embryogenesis in *Hevea brasiliensis* (Mull. Arg.): An alternative process for long-term somatic embryogenesis. *Journal of Rubber Research*, *12*, 215–228.

Larkin, P. J., & Scowcroft, W. R. (1981). Somaclonal variation – a novel source of variability from cell cultures for plant improvement. *Theoretical and Applied Genetics*, *60*, 197–214.

Latkowska, M. J., Kvaalen, H., & Appelgren, M. (2000). Genotype dependent blue and red light inhibition of the proliferation of the embryogenic tissue or Norway spruce. *In Vitro Cellular and Developmental Biology-Plant*, *36*, 57–60.

Laura, M., Safaverdi, G., & Allavena, A. (2006). Androgenetic plants of *Anemone coronaria* derived through anther culture. *Plant Breeding*, *125*, 629–634.

Lee, E. K., Cho, D. Y., & Soh, W. Y. (2001). Enhanced production and germination of somatic embryos by temporary starvation in tissue cultures of *Daucus carota*. *Plant Cell Reports*, *20*, 408–415.

Linsmaier, E. M., & Skoog, F. (1965). Organic growth factor requirements of tobacco tissue cultures. *Physiologia Plantarum*, *18*, 100–127.

Liu, H., Liu, S., & Xia, G. (2009). Generation of high frequency of novel alleles of the high molecular weight glutenin in somatic hybridization between bread wheat and tall wheatgrass. *Theoretical and Applied Genetics*, *118*(6), 1193–1198.

Liu, J., Xu, X., & Deng, X. (2005a). Intergeneric somatic hybridization and its application to crop genetic improvement. *Plant Cell, Tissue and Organ Culture*, *82*(1), 19–44.

Liu, J. H., & Deng, X. X. (2000). Regeneration of intergeneric hybrid embryoids via protoplast asymmetric fusion (in Chinese). *Acta Hortic Sin*, *27*, 207–209.

Liu, J. H., & Deng, X. X. (2002). Regeneration and analysis of citrus interspecific mixoploid hybrid plants from asymmetric somatic hybridization. *Euphytica*, *125*, 13–20.

Liu, S., Wang, H., Zhang, J., Fitt, B. D. L., Xu, Z., & Evans, N., et al. (2005b). *In vitro* mutation and selection of doubled-haploid *Brassica napus* lines with improved resistance to *Sclerotinia sclerotiorum*. *Plant Cell Reports*, *24*, 133–144.

Lloyd, G. B., & McCown, B. H. (1980). Commercially feasible micropropagation of mountain laurel (*Kalmia latifolia*) by use of shoot tip culture. *Proceedings of the International Plant Propagators' Society*, *30*, 412–427.

Maluszynsky, M., Kasha, K. J., Forster, B. P., & Szarejko, I. (2003). *Doubled Haploid Production in Crop Plants*. Dordrecht, The Netherlands: Kluwer Academic Publishers.

Mangal, M., & Sharma, D. R. (2002). *In vitro* mutagenesis and cell selection for the induction of black rot resistance in cauliflower. *Journal of Horticultural Science and Biotechnology*, *77*, 268–272.

Maraschin, S. D., Vennik, M., Lamers, G. E. M., Spaink, H. P., & Wang, M. (2005a). Time-lapse tracking of barley androgenesis reveals position-determined cell death within pro-embryos. *Planta*, *220*, 531–540.

Maraschin, S. F., de Priester, W., Spaink, H. P., & Wang, M. (2005b). Androgenic switch: An example of plant embryogenesis from the male gametophyte perspective. *Journal of Bxperimental Botany*, *56*, 1711–1726.

Mattoo, A., Minocha, S., Minocha, R., & Handa, A. (2010). Polyamines and cellular metabolism in plants: Transgenic approaches reveal different responses to diamine putrescine versus higher polyamines spermidine and spermine. *Amino Acids*, *38*, 405–413.

McClinchey, S. L., & Kott, L. S. (2008). Production of mutants with high cold tolerance in spring canola (*Brassica napus*). *Euphytica*, *162*, 51–67.

Metwally, E. I., Moustafa, S. A., El-Sawy, B. I., Haroun, S. A., & Shalaby, T. A. (1998a). Production of haploid plants from *in vitro* culture of unpollinated ovules of *Cucurbita pepo*. *Plant Cell, Tissue and Organ Culture*, *52*, 117–121.

Metwally, E. I., Moustafa, S. A., El-Sawy, B. I., & Shalaby, T. A. (1998b). Haploid plantlets derived by anther culture of *Cucurbita pepo*. *Plant Cell, Tissue and Organ Culture*, *52*, 171–176.

Mitykó, J., Andrásfalvy, A., & Fári, M. (1996). Hungarian pepper cultivar breaks through traditional barriers in breeding practices. *Hungarian Agricultural Research*, *1*, 18–22.

Miyoshi, K., & Asakura, N. (1996). Callus induction, regeneration of haploid plants and chromosome doubling in ovule cultures of pot gerbera (*Gerberra jamesonii*). *Plant Cell Reports*, *16*, 1–5.

Mizukami, M., Takeda, T., Satonaka, H., & Matsuoka, H. (2008). Improvement of propagation frequency with two-step direct somatic embryogenesis from carrot hypocotyls. *Biochemical Engineering Journal*, *38*, 55–60.

Mohamed, M. A. H., Harris, P. J. C., & Henderson, J. (2000). *In vitro* selection and characterisation of a drought tolerant clone of *Tagetes minuta*. *Plant Science*, *159*, 213–222.

Mohamed, S. V., Wang, C. S., Thiruvengadam, M., & Jayabalan, N. (2004). *In vitro* plant regeneration via somatic embryogenesis through cell suspension cultures of horsegram [*Macrotyloma uniflorum* (Lam.) Verdc.]. *In Vitro Cellular & Developmental Biology–Plant*, *40*, 284–289.

Monteiro, M., Kevers, C., Dommes, J., & Gasper, T. (2002). A specific role for spermidine in the initiation phase of somatic embryogenesis in *Panax ginseng* CA meyer. *Plant Cell, Tissue and Organ Culture*, *68*, 225–232.

Motoike, S. Y., Saraiva, E. S., Ventrella, M. C., Silva, C. V., & Salomaõ, L. C. C. (2007). Somatic embryogenesis of *Myrciaria aureana* (Brazilian grape tree). *Plant Cell, Tissue and Organ Culture*, *89*, 75–81.

Murashige, T., & Skoog, F. (1962). A revised medium for rapid growth and bio assays with tobacco tissue cultures. *Physiologia Plantarum*, *15*, 473–497.

Nakagawa, H., Saijyo, T., Yamauchi, N., Shigyo, M., Kako, S., & Ito, A. (2001). Effects of sugars and abscisic acid on somatic embryogenesis from melon (*Cucumis melo* L.) expanded cotyledon. *Scientia Horticulturae*, *90*, 85–92.

Nakagawa, R., Ogita, S., Kubo, T., & Funada, R. (2006). Effect of polyamines and L-ornithine on the development of proembryogenic masses of *Cryptomeria japonica*. *Plant Cell, Tissue and Organ Culture*, *85*, 229–234.

Namasivayam, P. (2007). Acquisition of embryogenic competence during somatic embryogenesis. *Plant Cell, Tissue and Organ Culture*, *90*, 1–8.

Nassour, M., & Dorion, N. (2002). Plant regeneration from protoplasts of micropropagated *Pelargonium* × hortorum "Alain:" Effect of some environmental and medium factors on protoplast system efficiency. *Plant Science*, *163*, 169–176.

Navratilova, B. (2004). Protoplast cultures and protoplast fusion focused on *Brassicaceae*–A review. *HortScience (Prague)*, *31*, 140–157.

Nayanakantha, N. M. C., & Seneviratne, P. (2007). Tissue culture of rubber: Past, present and future prospects. *Ceylon Journal of Science (Biological Sciences)*, *36*, 116–125.

Nitsch, C. (1981). Production of isogenic lines: Basic technical aspects of androgenesis. In T. A. Thorpe (Ed.), *Plant tissue culture: Methods and application in agriculture* (pp. 241–252). New York: Academic Press, Inc.

Nitsch, J. P. (1969). Experimental androgenesis in *Nicotiana*. *Phytomorphology*, *19*, 389–404.

Nitsch, J. P., & Nitsch, C. (1969). Haploid plants from pollen grains. *Science*, *163*, 85–87.

Obert, B., & Barnabas, B. (2004). Colchicine induced embryogenesis in maize. *Plant Cell, Tissue and Organ Culture*, *77*, 283–285.

Ochoa-Alejo, N., & Ramirez-Malagon, R. (2001). *In vitro* chili pepper biotechnology. *In Vitro Cellular Development Biology-Plant*, *37*, 701–729.

Ogita, S., Uefuji, H., Choi, Y., Hatanaka, T., Ogawa, M., & Yamaguchi, Y., et al. (2002). Genetic modification of coffee plants. *Journal of Plant Biotechnology*, *3*, 91–94.

Olivares-Fuster, O., Duran-Vila, N., & Navarro, L. (2005). Electrochemical protoplast fusion in citrus. *Plant Cell Reports*, *24*, 112–119.

Orczyk, W., Przetakiewicz, J., & Nadolska-Orczyk, A. (2003). Somatic hybrids of *Solanum tuberosum* – application to genetics and breeding. *Plant Cell, Tissue and Organ Culture*, *74*(1), 1–13.

Paiva Neto, V. B., Botelho, M. N., Aguiar, R., Silva, E. A. M., & Otoni, W. C. (2003). Somatic embryogenesis from immature zygotic embryos of Annato (*Bixa orellana* L.). *In Vitro Cellular & Developmental Biology–Plant*, *39*, 629–634.

Palmer, C. E., & Keller, W. A. (1997). Pollen Embryos. In V. K. Sawhney (Ed.), *Pollen Biotechnology for Crop Production and Improvement* (pp. 392–422). Cambridge: Cambridge University Press.

Pathania, N. S., & Misra, R. L. (2003). *In vitro* mutagenesis studies in Gladiolus for induction of resistance to *Fusarium oxysporum* f. sp. *gladioli*. *Acta Horticulturae*, *624*, 487–494.

Pati, P. K., Sharma, M., & Ahuja, P. S. (2008). Rose protoplast isolation and culture and heterokaryon selection by immobilization in extra thin alginate film. *Protoplasma*, *233*(1), 165–171.

Pauk, J., Paulimatka, M., Toth, K. L., & Monostori, T. (2000). *In vitro* androgenesis of triticale in isolated microspore culture. *Plant Cell, Tissue and Organ Culture*, *61*, 221–229.

Pauls, K. P., Chan, J., Woronuk, G., Schulze, D., & Brazolot, J. (2006). When microspores decide to become embryos – cellular and molecular changes. *Canadian Journal of Botany*, *84*, 668–678.

Pechan, P. M., & Smykal, P. (2001). Androgenesis: Affecting the fate of the male gametophyte. *Physiologia Plantarum*, *111*, 1–8.

Peña, L., Cervera, M., Fagoaga, C., Romero, J., Ballester, A., & Soler, N., et al. (2008). Citrus. In C. Kole & T. C. Hall (Eds.), *Compendium of transgenic crop plants: Transgenic tropical and subtropical fruits and nuts*. Blackwell Publishing Ltd..

Pinto, G., Park, Y. S., Silva, S., Neves, L., Araújo, C., & Santos, C. (2008a). Factors affecting maintenance, proliferation, and germination of secondary somatic embryos of *Eucalyptus globulus* Labill. *Plant Cell, Tissue and Organ Culture*, *95*, 69–78.

Pinto, G., Silva, S., Park, Y. S., Neves, L., Araújo, C., & Santos, C. (2008b). Factors influencing somatic embryogenesis induction in Eucalyptus globulus Labill.: Basal medium and anti-browning agents. *Plant Cell, Tissue and Organ Culture*, *95*, 79–88.

Powell, W. (1990). Environmental and Genetical Aspects of Pollen Embryogenesis. In Y. P. S. Bajay (Ed.), *Haploids in crop improvement* (pp. 45–65). Berlin: Springer-Verlag.

Prakash, M. G., & Gurumurthi, K. (2010). Effects of type of explant and age, plant growth regulators and medium strength on somatic embryogenesis and plant regeneration in *Eucalyptus camaldulensis*. *Plant Cell, Tissue and Organ Culture*, *100*, 13–20.

Predieri, S., & Dii Virgilio, N. (2007). *In vitro* mutagenesis and mutant multiplication. In S. M. Jain & H. Haggman (Eds.), *Protocols for micropropagation of woody trees and fruits* (pp. 323–333). Dordrecht, Netherlands: Springer.

Qin, Y. H., Zhang, S. L., Asghar, S., Zhang, L. X., Qin, Q. P., & Chen, K. S., et al. (2005). Regeneration mechanism of Toyonoka strawberry under different color plastic films. *Plant Science*, *168*, 1425–1431.

Quiroz-Figueroa, F. R., Rojas-Herrera, R., Galaz-Avalos, R. M., & Loyola-Vargas, V. M. (2006). Embryo production through somatic embryogenesis can be used to study cell differentiation in plants. *Plant Cell, Tissue and Organ Culture*, *86*, 285–301.

Raghavan, V. (2006). Can carrot and *Arabidopsis* serve as model systems to study the molecular biology of somatic embryogenesis? *Current Science*, *90*, 1336–1343.

Rai, M. K., Asthana, P., Jaiswal, V. S., & Jaiswal, U. (2010). Biotechnological advances in guava (*Psidium guajava* L.): Recent developments and prospects for further research. *Trees–Structure and Function*, *24*, 1–12.

Ramesar-Fortner, N. S., & Yeung, E. C. (2006). Physiological influences in the development and function of the shoot apical meristem of microspore-derived embryos of *Brassica napus* "Topas". *Canadian Journal of Botany*, *84*, 371–383.

Ratanasanobon, K., & Seaton, K. A. (2010). Development of *in vitro* plant regeneration of Australian native waxflowers (*Chamelaucium* spp.) via somatic embryogenesis. *Plant Cell, Tissue and Organ Culture*, *100*, 59–64.

Reddy, C., Gupta, M., Mantri, V., & Jha, B. (2008). Seaweed protoplasts: Status, biotechnological perspectives and needs. *Journal of Applied Phycology*, *20*(5), 619–632.

Reinert, J., & Bajaj, Y. P. S. (1977). Anther culture: Haploid production and its significance. In Y. P. S. Bajaj (Ed.), *Applied and fundamental aspects of plant cell, tissue and organ culture* (pp. 251–267). Berlin: Springer-Verlag.

Reinert, J., Bajaj, Y. P. S., & Haberle-Bors, E. (1975). Induction of haploid tobacco plants from isolated pollen. *Protoplasma*, *84*, 191–196.

Romeijn, G., & Van Lammeren, A. A. (1999). Plant regeneration through callus initiation from anthers and ovules of *Scabiosa columbaria*. *Plant Cell, Tissue and Organ Culture*, *56*, 169–177.

Rose, R., & Nolan, K. (2006). Invited review: Genetic regulation of somatic embryogenesis with particular reference to *Arabidopsis thaliana* and *Medicago truncatula*. *In Vitro Cellular & Developmental Biology–Plant*, *42*, 473–481.

Rudolf, K., Bohanec, B., & Hansen, M. (1999). Microspore culture of white cabbage, *Brassica oleracea* var. capitata L.: Genetic improvement of non-responsive cultivars and effect of genome doubling agents. *Plant Breeding*, *118*, 237–241.

Rudus, I., Weiler, E., & Kepczynska, E. (2009). Do stress-related phytohormones, abscisic acid and jasmonic acid play a role in the regulation of *Medicago sativa* L. somatic embryogenesis? *Plant Growth Regulation*, *59*, 159–169.

Sakhanokho, H. F., Rajasekaran, K., & Kelley, R. Y. (2009). Somatic embryogenesis in *Hedychium bousigonianum*. *HortScience*, *44*, 1487–1490.

Santana-Buzzy, N., Rojas-Herrera, R., Galaz-Ávalos, R., Ku-Cauich, J., Mijangos-Cortés, J., & Gutiérrez-Pacheco, L., et al. (2007). Advances in coffee tissue culture and its practical applications. *In Vitro Cellular & Developmental Biology–Plant*, *43*, 507–520.

Sato, S., Katoh, N., Yoshida, H., Iwai, S., & Hagimori, M. (2000). Production of doubled haploid plants of carnation (*Dianthus caryophyllus* L.) by pseudofertilized ovule culture. *Scientia Horticulturae*, *83*, 301–310.

Schenk, A. V., & Hildebrandt, A. C. (1972). Medium and techniques for induction and growth of monocotyledonous and dicotyledonous plant cell cultures. *Canadian Journal of Botany*, *50*, 199–204.

Schiva, T., Dalla-Goda, C., & Mercuri, A. (1985). Analisi genetica della resistenza a *Fusarium oxysporum* f.sp. *dianthi* (Prill et Del) Snyd et Hans nel garotano "Ecotipo Mediterraneo." *Annali Istituto Sperimentale Floricoltura*, *16*, 23–38.

Seabrook, J. E. A., & Douglass, L. K. (2001). Somatic embryogenesis on various potato tissues from a range of genotypes and ploidy levels. *Plant Cell Reports*, *20*, 175–182.

Seguí-Simarro, J. M., Bárány, I., Suárez, R., Fadón, B., Testillano, P. S., & Risueño, M. C. (2006). Nuclear bodies domain changes with microspore reprogramming to embryogenesis. *European Journal of Histochemistry*, *50*, 35–44.

Seguí-Simarro, J. M., & Nuez, F. (2008a). How microspores transform into haploid embryos: Changes associated with embryogenesis induction and microspore-derived embryogenesis. *Physiologia Plantarum*, *134*, 1–12.

Seguí-Simarro, J. M., & Nuez, F. (2008b). Pathways to doubled haploidy: Chromosome doubling during androgenesis. *Cytogenetic and Genome Research*, *120*, 358–369.

Seguí-Simarro, J. M., Testillano, P. S., & Risueño, M. C. (2003). Hsp70 and Hsp90 change their expression and subcellular localization after microspore embryogenesis induction in *Brassica napus* L. cv. Topas. *Journal of Structural Biology*, *142*, 379–391.

Sengar, A. S., Thind, K. S., Kumar, B., Pallavi, M., & Gosal., S. S. (2009). *In vitro* selection at cellular level for red rot resistance in sugarcane (*Saccharum* sp.). *Plant Growth Regulator*, *58*, 201–209.

Shalaby, T. A. (2007). Factors affecting haploid induction through *in vitro* gynogenesis in summer squash (*Cucurbita pepo* L.). *Scientia Horticulturae*, *115*, 1–6.

Shariatpanahi, M. E., Bal, U., Haberle-Bors, E., & Touraev, A. (2006). Stresses applied for the re-programming of plant microspores towards *in vitro* embryogenesis. *Physiologia Plantarum*, *127*, 519–534.

Shi, X., Dai, X., Liu, G., & Bao, M. (2009). Enhancement of somatic embryogenesis in camphor tree (*Cinnamomum camphora* L.): Osmotic stress and other factors affecting somatic embryo formation on hormone-free medium. *Trees – Structure and Function*, *23*, 1033–1042.

Shimonaka, M., Hosoki, T., Tomita, M., & Yasumuro, Y. (2002). Production of somatic hybrid plants between Japanese bunching onion (*Allium fistulosuml*) and bulb onion (*A. cepal.*) via electrofusion. *Japanese Society for Horticultural Science*, *71*, 623–631.

Shizukawa, Y., & Mii, M. (2008). Production of bialaphos-resistant *Nierembergia repens* by electroporation. *Plant Biotechnology Reports*, *2*, 219–226.

Silva, T. (2010). Indica rice anther culture: Can the impasse be surpassed? *Plant Cell, Tissue and Organ Culture*, *100*, 1–11.

Singh, S., & Rajam, M. V. (2009). Citrus biotechnology: Achievements, limitations and future directions. *Physiology and Molecular Biology of Plants*, *15*, 3–22.

Smykal, P., & Pechan, P. M. (2000). Stress, as assessed by the appearance of sHsp transcripts, is required but not sufficient to initiate androgenesis. *Physiologia Plantarum*, *110*, 135–143.

Snyman, S. J., Meyer, G. M., Richards, J. M., Haricharan, N., Ramgareeb, S., & Huckett, B. I. (2006). Refining the application of direct embryogenesis in sugarcane: Effect of the developmental phase of leaf disc explants and the timing of DNA transfer on transformation efficiency. *Plant Cell Reports*, *25*, 1016–1023.

Sonntag, K., Ruge-Wehling, B., & Wehling, P. (2009). Protoplast isolation and culture for somatic hybridization of *Lupinus angustifolius* and *L. subcarnosus*. *Plant Cell, Tissue and Organ Culture*, *96*, 297–305.

Steinmacher, D. A., Cangahuala-Inocente, G. C., Clement, C. R., & Guerra, M. P. (2007). Somatic embryogenesis from peach palm zygotic embryos. *In Vitro Cellular & Developmental Biology–Plant*, *43*, 124–132.

Sticklen, M. B., & Orabya, H. F. (2005). Shoot apical meristem: A sustainable explant for genetic transformation of cereal crops. *In Vitro Cellular & Developmental Biology*, *41*, 187–200.

Suprasanna, P., Sidha, M., & Bapat, V. A. (2009). Integrated approaches of mutagenesis and *in vitro* selection for crop improvement. In A. Kumar & N. S. Shekhawat (Eds.), *Plant tissue culture and molecular markers: Their role in improving crop productivity* (pp. 73–92). New Delhi: I. K. International Publishing House Pvt. Ltd..

Suprunova, T. and N. Shmykova. 2008. *In vitro* induction of haploid plants in umpollinated ovules, anther and microspore culture of *Cucumis sativus*. In M. Pitrat (Ed.), *Proceedings of the nineth EUCARPIA meeting on genetics and breeding of Cucurbitaceae*, May 21–24th, 2008, (pp. 371–374). INRA, Avignon (France).

Szarka, B., Gonter, L., Mlonar-lang, M., Morocz, S., & Dudits, D. (2002). Mixing of maize and wheat genomic DNA by somatic hybridization in regenerated sterile maize plants. *Theoretical and Applied Genetics*, *105*, 1–7.

Taji, A., Kumar, P., & Lakshmanan, P. (2002). *In vitro plant breeding*. New York: Haworth Press, Inc..

Takahata, Y., Komatsu, H., & Kaizuma, N. (1996). Microspore culture of radish (*Raphanus sativus* L.): Influence of genotype and culture conditions on embryogenesis. *Plant Cell Reports, 16*, 163–166.

Takami, K., Matsumara, A., Yahata, M., Imayama, T., Kunitake, H., & Komatsu, H. (2004). Production of intergeneric somatic hybrids between round kumquat (*Fortunella japonica* Swingle) and "Morita navel" orange (*Citrus sinensis* Osbeck). *Plant Cell Reports, 23*, 39–45.

Takanori, T. & Cuello, J. (2005). Regulating radiation quality and intensity using narrow-band LEDs for optimization of somatic embryogenesis. In *Proceedings of the 2005 annual meeting of the American society of agricultural engineers* (pp. 1–12).

Tang, F., Tao, Y., Zhao, T., & Wang, G. (2006). *In vitro* production of haploid and double-haploid plants from pollinated ovaries of maize (*Zea mays*). *Plant Cell, Tissue and Organ Culture, 84*, 233–237.

Tantau, H., Balko, C., Brettschneider, B., Melz, G., & Dorffling, K. (2004). Improved frost tolerance and winter survival in winter barley (*Hordeum vulgare* L.) by *in vitro* selection of proline overaccumulating lines. *Euphytica, 139*, 19–32.

Taski-Ajdukovic, K., Vasic, D., & Nagl, N. (2006). Regeneration of interspecific somatic hybrids between *Helianthus annuus* L. and *Helianthus maximiliani* (Schrader) via protoplast electrofusion. *Plant Cell Reports, 25*(7), 698–704.

Testillano, P. S., Coronado, M. J., Seguí-Simarro, J. M., Domenech, J., Gonzáles-Melendi, P., & Raska, I., et al. (2000). Defined nuclear changes accompany the reprogramming of the microspores to embryogenesis. *Journal of Structural Biology, 129*, 223–232.

Testillano, P. S., Gonzáles-Melendi, P., Coronado, M. J., Seguí-Simarro, J. M., Moreno, M. A., & Risueño, M. C. (2005). Differentiating plant cells switched to proliferation remodel the functional organization of nuclear domains. *Cytogenetic and Genome Research, 109*, 166–174.

Thakur, M., Sharma, D., & Sharma, S. (2002). *In vitro* selection and regeneration of carnation (*Dianthus caryophyllus* L.) plants resistant to culture filtrate of *Fusarium oxysporum* f. sp. *dianthi*. *Plant Cell Reports, 20*, 825–828.

Thomas, T. D. (2004). Embryological observations on unpollinated ovary culture of mulberry (*Morus alba* L.). *Acta Biologica Cracoviensia Series Botanica, 46*, 87–94.

Thomas, T. D. (2009). Isolation, callus formation and plantlet regeneration from mesophyll protoplasts of *Tylophora indica* (Burm. f.) Merrill: An important medicinal plant. *In Vitro Cellular & Developmental Biology–Plant, 45*(5), 591–598.

Tomasi, P., Dierig, D. A., Backhaus, R. A., & Pigg, K. B. (1999). Floral bud and mean petal length as morphological predictors of microspore cytological stage in Lasquerella. *HortScience, 34*, 1269–1270.

Touraev, A., Pfosser, M., & Heberle-Bors, E. (2001). The microspores: A haploid multipurpose cell. *Advances in Botanical Research, 35*, 53–109.

Touraev, A., Vicente, O., & Heberle-Bors, E. (1997). Initiation of microspore embryogenesis by stress. *Trends in Plant Science, 2*, 285–323.

Trigiano, R., & Gray, D. (2000). *Plant tissue culture concepts and laboratory exercises*. Washington DC: CRC Press.

Trottier, M. C., Collin, J., & Comeau, A. (1993). Comparison of media for their aptitude in wheat anther culture. *Plant Cell, Tissue and Organ Culture, 35*, 59–67.

Tsuwamoto, R., Fukuoka, H., & Takahata, Y. (2007). Identification and characterization of genes expressed in early embryogenesis from microspores of *Brassica napus*. *Planta, 225*, 641–652.

Tu, Y., Sun, J., Liu, Y., Ge, X., Zhao, Z., & Yao, X., et al. (2008). Production and characterization of intertribal somatic hybrids of *Raphanus sativus* and *Brassica rapa* with dye and medicinal plant *Isatis indigotica*. *Plant Cell Reports, 27*(5), 873–883.

Tuvesson, S., Ljungberg, A., Johansson, N., Karlsson, K. E., Suijs, L. W., & Josset, J. P. (2000). Large-scale production of wheat and triticale double haploids through the use of single-anther culture method. *Plant Breeding, 119*, 455–459.

van den Bulk, R. W. (1991). Application of cell and tissue culture and *in vitro* selection for disease resistance breeding – a review. *Euphytica, 56*, 269–285.

Varotto, S., Nenz, E., Lucchin, M., & Parrini, P. (2001). Production of asymmetric somatic hybrid plants between *Cichorium intybus* L. and *Helianthus annuus* L. *Theoretical and Applied Genetics, 102*, 950–956.

Vasil, I. (2008). A history of plant biotechnology: From the Cell Theory of Schleiden and Schwann to biotech crops. *Plant Cell Reports, 27*, 1423–1440.

Vasil, I. (2008). A short history of plant biotechnology. *Phytochemistry Reviews, 7*, 387–394.

Venkataiah, P., Christopher, T., & Karampuri, S. (2005). Selection of atrazine-resistant plants by *in vitro* mutagenesis in pepper (*Capsicum annuum*). *Plant Cell, Tissue and Organ Culture, 83*, 75–82.

Wang, M., van Bergen, S., & Van Duijn, B. (2000). Insights into a key developmental switch and its importance for efficient plant breeding. *Plant Physiology, 124*, 523–530.

Wang, M. Q., Zhao, J. S., & Peng, Z. Y., et al. (2008). Chromosomes are eliminated in the symmetric fusion between *Arabidopsis thaliana* L. and *Bupleurum scorzonerifolium* Willd. *Plant Cell, Tissue and Organ Culture, 92*, 121–130.

Wang, X., Xia, H., & Yue-rong, W. (2009). Progress in protoplast culture and somatic hybridization in banana (*Musa* spp.). *Journal of Fruit Science*

Wang, Y. P., Sonntag, K., & Rudloff, E. (2003). Development of rapeseed with high erucic acid content by asymmetric somatic hybridization between *Brassica napus* and *Crambe abyssinica*. *Theoretical and Applied Genetics, 106*, 1147–1155.

Wheatley, A. O., Ahmad, M. H., & Asemota, H. N. (2003). Development of salt adaptationin *in vitro* greater yam (*Dioscorea alata*) plantlets. *In Vitro Cellular Development Biology–Plant, 39*, 346–353.

White, P. R. (1962). *The Cultivation of Animal and Plant Cells*. New York: Ronald Press.

Wojciechowicz, M. K., & Kikowska, M. A. (2009). Induction of multi-nucleate microspores in anther culture of *Salix viminalis* L. *Dendrobiology, 61*, 55–64.

Xia, G. M., Xiang, F. N., Zhou, A. F., Wang, H., He, S. X., & Chen, H. M. (2003). Asymmetric somatic hybridization between wheat (*Triticum aestivum* L.) and *Agropyron elongatum* (Host) Nevski. *Theoretical and Applied Genetics, 107*, 299–305.

Xiang, F., Wang, J., Xu, C., & Xia, G. (2010). The chromosome content and genotype of two wheat cell lines and of their somatic fusion product with oat. *Planta, 231*(5), 1201–1210.

Xiang, F., Xia, G., & Chen, H. (2003). Effect of UV dosage on somatic hybridisation between common wheat (*Triticum aestivum* L.) and *Avena sativa* L. *Plant Science, 164*, 697–707.

Xiang, F. N., Xia, G. M., Zhi, D. Y., Wang, J., Nie, H., & Chen, H. M. (2004). Regeneration of somatic hybrids in relation to the nuclear and cytoplasmic genomes of wheat and *Setaria italica*. *Genome, 47*, 680–688.

Xiao, W., Huang, X., Gong, Q., Dai, X. M., Zhao, J. T., & Wei, Y. R., et al. (2009). Somatic hybrids obtained by asymmetric protoplast fusion between *Musa* Silk cv. Guoshanxiang (AAB) and *Musa acuminata* cv. Mas (AA). *Plant Cell, Tissue and Organ Culture, 97*(3), 313–321.

Xing, H. Q., Xia, G. M., & Chen, H. M. (2001). Preliminary study on hybrid chromosome composition and relationship in symmetric hybridization between *Triticum aestivum* and intergeneric grasses (in Chinese). *Bulletin of Botanical Research, 21*, 74–78.

Xu, C. H., Xia, G. M., Zhi, D. Y., Xiang, F. N., & Chen, H. M. (2003). Integration of maize nuclear and mitochondrial DNA into the wheat genome through somatic hybridization. *Plant Science, 165*, 1001–1008.

Xu, X. Y., Liu, J. H., & Deng, X. X. (2004). Production and characterization of intergeneric diploid cybrids derived from symmetric fusion between *Microcitrus papuana* Swingle and sour orange (*Citrus aurantium*). *Euphytica, 136*, 115–123.

Yamagishi, H., Landgren, M., Forsberg, J., & Glimelius, K. (2002). Production of asymmetric hybrids between *Arabidopsis thaliana* and *Brassica napus* utilizing an efficient protoplast culture system. *Theoretical and Applied Genetics, 104*, 959–964.

Yan, C. Q., Qian, K. X., Yan, Q. S., Zhang, X. Q., Xue, G. P., & Huang, W. G., et al. (2004). Use of asymmetric somatic hybridization for transfer of the bacterial blight resistance trait from *Oryza meyeriana* L. to *O. sativa* L. ssp. *japonica*. *Plant Cell Reports, 22*, 569–575.

Yang, J. L., Seong, E. S., Kim, M. J., Ghimire, B. K., Kang, W. H., & Yu, C. Y., et al. (2010). Direct somatic embryogenesis from pericycle cells of broccoli (*Brassica oleracea* L. var. *italica*) root explants. *Plant Cell, Tissue and Organ Culture, 100*, 49–58.

Yang, X. Y., Zhang, X. L., & Jin, S. X., et al. (2007). Production and characterization of asymmetric hybrids between upland cotton Coker 201 (*Gossypium hirsutum*) and wild cotton (*G. klozschianum* Anderss). *Plant Cell, Tissue and Organ Culture, 89*, 225–235.

Yang, Y., Guan, S., Zhai, H., He, S., & Liu, Q. (2009). Development and evaluation of a storage root-bearing sweet potato somatic hybrid between *Ipomoea batatas* (L.) Lam. and *I. triloba* L. *Plant Cell, Tissue and Organ Culture, 99*, 83–89.

Yemets, A. I., Kundel'chuk, O. P., Smertenko, A. P., Solodushko, V. G., Rudas, V. A., & Gleba, Y. Y., et al. (2000). Transfer of amiprophosmethyl resistance from a *Nicotiana plumbaginifolia* mutant by somatic hybridization. *Theoretical and Applied Genetics, 100*, 847–857.

Yue, W., Xia, G. M., Zhi, D. Y., & Chen, H. M. (2001). Transfer of salt tolerance from *Aeleuropus littorulis* Sinensis to wheat (*Triticum aestivum*) via asymmetric somatic hybridization. *Plant Science, 161*, 259–263.

Zamani, I., Kovacks, G., Gouli-Vavdinoudi, E., Roupakias, D. G., & Barnabas, B. (2000). Regeneration of fertile doubled haploid plants from colchicine-supplemented media in wheat anther culture. *Plant Breeding, 119*, 461–465.

Zhang, B. H., Wang, Q. L., Liu, F., Wang, K. B., & Frazier, T. P. (2009). Highly efficient plant regeneration through somatic embryogenesis in 20 elite commercial cotton (*Gossypium hirsutum* L.) cultivars. *Plant Omics, 2*, 259–268.

Zhang, C. L., Chen, D. F., Kubalakova, M., Zhang, J., Scott, N. W., & Elliott, M. C., et al. (2008). Efficient somatic embryogenesis in sugar beet (*Beta vulgaris* L.) breeding lines. *Plant Cell, Tissue and Organ Culture, 93*, 209–221.

Zhao, Z. G., Hu, T. T., Ge, X. H., Du, X. Z., Ding, L., & Li, Z. Y. (2008). Production and characterization of intergeneric somatic hybrids between *Brassica napus* and *Orychophragmus violaceus* and their backcrossing progenies. *Plant Cell Reports, 27*(10), 1611–1621.

Zhou, A. F., Xia, G. M., & Chen, H. M. (2001). Comparative study of symmetric and asymmetric somatic hybridization between common wheat and *Haynaldia villosa*. *Science China (Series C), 44*, 294–304.

Zhou, C. E., Xia, G. M., Zhi, D. Y., & Chen, Y. (2006). Genetic characterization of asymmetric somatic hybrids between *Bupleurum scorzonerifolium* Willd and *Triticum aestivum* L.: Potential application to the study of the wheat genome. *Planta, 223*, 714–724.

Ziv, M. (2000). Bioreactor technology for plant micropropagation. *Horticultural Review, 24*, 1–30.

Ziv, M. (2005). Simple bioreactors for mass propagation of plants. *Plant Cell, Tissue and Organ Culture, 81*, 277–285.

Zubko, M. K., Zubko, E. I., & Gleba, Y. Y. (2002). Self-fertile cybrids *Nicotiana tabacum* (+ *Hyoscyamus aureus*) with a nucleo–plastome incompatibility. *Theoretical Applied Genetics, 105*, 822–828.

Marker-assisted selection in plant breeding

11

Giora Ben-Ari Uri Lavi
Institute of Horticulture, The Volcani Center ARO, Bet-Dagan, Israel

TABLE OF CONTENTS

Background

The concept of marker-assisted selection

The concept of marker-assisted selection (MAS) was suggested 25 years ago by C. Smith and P. Simpson (1986) and by Soller and Beckmann (1983; see more details in the next section). The idea was that: "It is unlikely that many of the polymorphisms identified by the new laboratory techniques will be the QTL themselves. However, many of them may be linked to the QTL and so will allow indirect selection." Following the discovery of restriction fragment length polymorphism (RFLP; Botstein et al., 1980) and other kinds of DNA markers (see the following section), it was assumed that classical plant breeding (as well as animal breeding) would experience a major revolution, in that most of the selection would not be carried out on the basis of the phenotypes (the performance in the field), but would depend significantly on markers genetically linked to the genes of interest. This concept is also known in the literature as marker-assisted breeding (MAB). In this chapter we will use the term MAS. The practice of MAS relies upon a situation of linkage disequilibrium (LD; for explanation see the section Classical Methods of Gene Identification), which exists between the DNA marker and a specific gene (quantitative trait locus; QTL). LD can be exploited by selection as if the effects are caused by the marker.

The advantages of MAS results from the fact that many of the traits of interest to breeders are not easily assessed. Thus, selection, which is based on linked DNA markers, is much more efficient. Selection based on markers can be carried out at an early age (plantlets); therefore, it has the potential to significantly reduce the number of individuals assessed by the breeder, thus reducing costs. MAS has greater potential for efficient gene pyramiding; namely, combining several important genes in one cultivar. At the same time MAS does not reduce the time of the breeding project because the selected plants need to be tested and evaluated in the field. The length of the evaluation process depends mainly on the length of the juvenile period of each species.

The prerequisites for the classical procedure of MAS are the DNA markers (see the following sections) and linkage analysis, which will identify markers that are linked to the genes controlling the trait(s) of interest. It is noteworthy that the "quality" and the number of markers have a major impact on the success of MAS. The quality of markers relates to their characteristics and to the cost and the efficiency of the genotyping process. The number of markers affects the reliability of the linkage between them and the gene(s). In other words, screening a large number of markers has the potential to identify close and reliable linkage between the marker and the gene of interest.

DOI: 10.1016/B978-0-12-381466-1.00011-0

Linkage analysis is basically aimed at determining the linkage between sequence variation (specific DNA markers) and a specific phenotype. The details of such analyses can be found elsewhere (Martin et al., 1993, 1994; Okada et al., 2010; Udagawa et al., 2010; Yang et al., 2010) and will not be discussed here. Cost-benefit analysis is required, because for some traits it may be cheaper and more efficient to rely upon phenotypic selection rather than on MAS. It is essential to emphasize that MAS must be carried out in family structures (progeny of known crosses) where there is significant LD in one or two generations. It is essential to emphasize that any identification of linkage between markers and genes of interest must be verified before application in a breeding project. We will distinguish between applications of MAS to traits that are controlled by single genes versus quantitative traits that are controlled by a number of genes. Due to the similarities between the search for markers for human diseases and the search for markers associated with agricultural traits, we will include examples from human genetics. Many reviews have described MAS principles and applications (Collard et al., 2008; Collins et al., 2008; Neil et al., 2009; Varshney and Dubey, 2009; Xu et al., 2009; Varshney et al., 2009a; Thomson et al., 2010; Varshney, 2010; Xu, 2010). In this review we would like to discuss MAS in a very broad sense; namely to discuss various technologies that are based on DNA variations and that have the potential to improve classical breeding.

The topics to be covered in this chapter include: a historical review of MAS; identification of genes controlling the important agricultural traits; a description of the major technological tools used in MAS (emphasizing their pros and cons); a description of modern genotyping technologies; various applications of DNA markers in plant breeding; modern methods for identification of genes controlling important agricultural traits; actual applications of MAS in commercial plant breeding programs; difficulties in the application of MAS to plant breeding programs; and finally, future prospects of new technologies that identify sequence variation.

Historical review

Intuitive phenotypic selection of domesticated species has been practiced by farmers for thousands of years with cumulatively rewarding results. In recent years, however, MAS has emerged as a strategy for dramatically increasing selection gains. Sax (1923) showed in the early twentieth century that an observable gene having simple Mendelian inheritance could act as a marker for the segregation of a gene involved in the expression of a quantitative trait. About 60 years later, following the discovery of RFLPs (Botstein et al., 1980), the first DNA marker map consisting of 57 RFLP markers was constructed for tomatoes (Bernatzky and Tanksley, 1986).

The use of DNA markers for the purposes of selection in breeding programs was first suggested by Neimann-Sorensen and Robertson (1961), Smith (1967), Soller (1978), and Stuber et al. (1982). Tanksley et al. (1981) published the first MAS experimental study with plants and suggested that isozyme selection could precede morphological selection in tomatoes, but not replace it. The number of studies aimed at establishing genetic markers linked to important phenotypic traits has increased rapidly, and has virtually exploded since the publication of the first use of DNA markers (RFLP) for the analysis of quantitative traits in tomatoes (Paterson et al., 1988). These studies dealt with both theoretical and applied aspects of the application of DNA markers (Lander and Botstein, 1989; Lande and Thompson, 1990; Zhang and Smith, 1992; Dudley, 1993; Gimelfarb and Lande, 1994; Lee, 1995; Hospital and Charcosset, 1997; Whittaker et al., 1997). as well as experimental studies of MAS (Fatokun et al., 1992; Meksem et al., 1995; Austin and Lee, 1996; Peterhansel et al., 1997). A recent search of the ISI Web of Knowledge (www.isiwebofknowledge.com) yielded 3385 references to "Marker Assisted Selection." The number of publications per year on this subject expanded from two publications in 1989 to 496 papers in 2010.

Young (1999) expressed some caution in his optimistic vision of MAS by saying: "Even though marker-assisted selection now plays a prominent role in the field of plant breeding, examples of successful, practical outcomes are rare." Limited availability of markers and efficient genotyping methodologies were among the main limitations in the practical application of QTL studies. Technical developments such as high-throughput sequencing technologies have overcome some of these limitations, and the last decade has witnessed numerous QTL detection experiments in various species. The first fully described plant gene responsible for a quantitative trait was the QTL *fw2.2* controlling fruit size in tomato. This gene was mapped by analysis of recombinants, cosmid complementation, and genomic sequencing, and identified the *OFRX* gene as the underlying gene (Frary et al., 2000). To date, more than 2500 studies on QTL mapping in plants have been published. Yet, Young's (1999) optimistic vision has still not become a reality, and breeding programs based on DNA markers for improving quantitative traits in plants are rare.

Plant Traits, DNA Markers, Technologies, and Applications

Genes controlling important traits

The interest of breeders in MAS is focused mainly on the important traits including tolerance to biotic and abiotic stresses as well as agronomic (or horticultural) traits.

MAS for biotic stresses

Numerous pathogens attack plants, including viruses, bacteria, nematodes, and fungi. Considerable effort is required to determine pathogenic mechanisms and to identify host plant genes that confer resistance. These mechanisms are under the control of dominant or recessive major genes or QTLs. DNA markers are used to identify resistance genes and for their introgression and pyramiding into new cultivars. Plant disease resistance genes (R-genes) play a key role in recognizing proteins expressed by specific avirulence (Avr) genes in pathogens. Different techniques have been developed to identify R-genes, and to date a total of 73 have been isolated from

22 plant species, most of them from the Solanaceae family. These have been found to interact with 31 different pathogens (Sanseverino et al., 2010). Examples of characterized R-genes are the *ASC1* gene conferring resistance to the pathogen *Alternaria alternata*, which causes the disease *Alternaria* stem canker (Brandwagt et al., 2000), and the broad-spectrum mildew resistance gene *RPW8* in *Arabidopsis thaliana* (Xiao et al., 2001).

However, not all disease resistance traits are monogenic. Some types of resistance behave as quantitative traits and are controlled by multiple genes. An example of this type of disease is caused by filamentous fungi of the *Fusarium* species that are widely distributed in soil. Numerous *Fusarium* species have been associated with *Fusarium* head blight (FHB) in wheat (*Triticum aestivum*), barley (*Hordeum vulgare*), and other small-grain cereals (Parry et al., 1995). The most prevalent species of this pathogen are *Fusarium graminearum*, *F. culmorum*, and *F. avenaceum* (Walter et al., 2010). Apart from losses in grain yield and seed quality, the major damage caused by FHB is the contamination of the crop with toxic secondary metabolites of the fungus, known as mycotoxins. More than 100 QTLs for FHB resistance have been reported in wheat (Buerstmayr et al., 2009), but few of them have been validated. Among those validated QTLs, *Fhb1* has been found to have a major effect on FHB resistance and can be mapped as a single Mendelian gene with high precision. Thus, for the purposes of MAS, *Fhb1* is the diagnostic marker (Pumphrey et al., 2007).

MAS for abiotic stress

Abiotic stress is defined as the negative impact of non-living factors on living organisms in a specific environment. The stresses include drought, salinity, low or high temperatures, and other environmental extremes. Abiotic stresses, especially hypersalinity and drought, are the primary causes of crop loss worldwide. In contrast to plant resistance to biotic stresses, which mostly depends on monogenic traits, the genetically complex responses to abiotic stresses are multigenic and thus more difficult to identify, control, and manipulate.

To date, 27 drought-resistant cultivars/lines of different crops have been registered in Crop Science or reported in other sources. These crops included peanut, common bean, sunflower, chickpea, wheat, tall fescue, wheatgrass, barley, and maize (Ashraf, 2010). QTL mapping of drought tolerance has been carried out for various crops including maize, wheat, barley, cotton, sorghum, and rice (Quarrie et al., 1994; Teulat et al., 1997; Sari-Gorla et al., 1999; Saranga et al., 2001; Sanchez et al., 2002; Bernier et al., 2008; Chin et al., 2010; Nevo and Chen, 2010).

Due to inappropriate irrigation regimes and the wider use of brackish water for irrigation, areas devastated by secondary salinization are increasing. To cope with this escalating problem, efforts are being made to enhance salt tolerance of economically important crops. QTLs for salt tolerance or related physiological traits have been reported in *A. thaliana* (Quesada et al., 2002), barley (Ellis et al., 1997, 2002; Mano and Takeda, 1997), citrus (Tozlu et al., 1999), rice (Flowers et al., 2000; Prasad et al., 2000; Koyama et al., 2001; Lin et al., 2004; Ren et al., 2005; Lee et al., 2006), soybean (Lee et al., 2004), tomato (Foolad et al., 1998, 2001; Villalta et al., 2007, 2008), wheat (Ma et al., 2007; Quarrie et al., 2005), and white clover (Wang et al., 2010). However, despite the extensive research on the salt response in various crops, genetically engineered salt-tolerant cultivars (with few exceptions like the transgenic wheat expressing a vacuolar Na^+/H^+ antiporter gene, *AtNHX1* from *A. thaliana*, with improved grain yields in saline soils; Xue et al., 2004) have not been successfully tested under natural field conditions (Ashraf and Akram, 2009).

MAS for agronomic traits

Nearly 100 genes and functional polymorphisms underlying natural variation in plant development and physiology have been identified. In crop plants these include genes involved in domestication traits, such as those related to the timing of germination and flowering, plant architecture, and fruit and seed structure and morphology, as well as yield and quality traits (Alonso-Blanco et al., 2009). For example: *DELAY OF GERMINATION 1* (*DOG1*) induces seed dormancy and is specifically expressed during seed development (Alonso-Blanco et al., 2003). In addition, it has been shown that a *DOG1* functional nucleotide polymorphism affects its expression, and dormant seeds have higher RNA levels than non-dormant ones (Bentsink et al., 2006). It was shown that *DOG1* plays a role in the germination of non-dormant seeds such as *Lepidium sativum* and *Brassica rapa* of the Brassicaceae family (Graeber et al., 2010).

Nine genes contributing to the variation in the flowering time of *A. thaliana* have been isolated (Alonso-Blanco et al., 2009). Among these are: *FRIGIDA* (*FRI*), *FRIGIDA LIKE1* (*FRL1*), *FRIGIDA LIKE2* (*FRL2*), *FLOWERING LOCUS C* (*FLC*), and others. In contrast, analysis of variability in flowering time in maize showed no evidence of any QTL having a major effect on this trait. Instead, numerous small-effect QTLs were identified (Buckler et al., 2009). More than 20 loci having a major effect on plant architecture have been identified and, in many instances, linked to genes encoding transcription factors (Alonso-Blanco et al., 2009).

A large number of studies have focused on the genetic basis of positive heterosis or hybrid vigor. Heterosis is observed in many hybrids and is a major plant breeding objective. Numerous QTLs with different levels of dominance, overdominance, and epistatic effects have been mapped for heterosis in maize (Frascaroli et al., 2007), rice (Li et al., 2001), tomato (Semel et al., 2006), rapeseed (*B. napus;* Radoev et al., 2008), and *A. thaliana* (Kusterer et al., 2007; Melchinger et al., 2007). Heterosis in tomato yield and Brix value (sugar content) has recently been nicely demonstrated by Krieger et al. (2010). They reported that heterozygosity for tomato loss-of-function alleles of *SINGLE FLOWER TRUSS* (*SFT*) increases yields by up to 60% and Brix by about 30%. This effect was repeated in distinct genetic backgrounds and environments.

DNA markers

Genetic variation (and mainly phenotypic variation) has been a focus of interest and research for a long time. The discovery of

RFLP by Botstein et al. (1980) revolutionized this field since it first allowed a quantitative, accurate, genome-wide assessment of polymorphisms at the DNA level. Ever since, numerous classes of DNA markers have been developed and have been the subject of many reviews (Masojc, 2002; Collard et al., 2008; Appleby et al., 2009; Barone et al., 2009; Gailing et al., 2009; Varshney et al., 2009b; Edwards and Batley, 2010). DNA markers are the tools for MAS in plants, animals, humans, and other organisms. In this chapter we will discuss only a few important classes of these markers that have been applied to plants, emphasizing their advantages and disadvantages.

It is noteworthy that many of these DNA markers were discovered and developed in humans and then applied to plants based on their general characteristics. Development of tens of classes of markers resulted from the desire to improve these tools using new technologies, as well as from the creativity of various scientists.

The ideal DNA marker should be: (1) highly polymorphic (as assessed by the number of alleles in the population) to have the potential to identify the analyzed phenotypes; (2) highly abundant and uniformly distributed in the genome to have the potential to identify tight linkage between markers and the genes of interest and to construct saturated genetic maps; (3) co-dominant (the ability to distinguish the heterozygote) to link the marker allele to a specific allele of the gene controlling the phenotype of interest; and (4) easy to genotype, which is the ability to genotype large numbers of individuals in a large number of marker loci at a reasonable price (see the section Genome-wide association studies).

DNA markers are indirect alternatives for whole genome sequence since full sequencing of many genotypes in each study is not yet feasible. However, major developments have been made and seven green plant species (*A. thaliana, Glycine max, Oryza sativa, Physcomitrella patens, Populus trichocarpa, Sorghum bicolor,* and *Vitis vinifera*) have been fully sequenced and quite a number of other species are in the process of being sequenced. Moreover, efforts are being invested in the development and improvement of sequencing methodologies, thus, it is expected that the DNA sequence of many other species will soon be available. The characteristics of the various DNA markers are summarized in Table 11.1.

Restriction fragment length polymorphism

RFLP was developed by Botstein et al. (1980).

Genotyping technology: DNA is cut with a restriction enzyme, the resulting fragments are size separated on an agarose gel, blotted onto a membrane, hybridized, and exposed to a labeled probe. Specific probes are usually generated from genomic or c-DNA libraries.

Genotyping: Each band represents an allele having a specific size in base pairs (bp).

Source of polymorphism: Existence or absence of a restriction site.

Characteristics: An RFLP marker usually has two alleles at each locus and thus has a low level of polymorphism.

The number of available RFLP loci in plants varies between tens to a few thousands. RFLPs are single locus markers and their mode of inheritance is co-dominant. The genotyping technology is rather complicated (including restriction, gel electrophoresis, blotting, hybridization, and exposure of X-ray film); therefore, the actual number of analyzed RFLP loci varies between tens to a few hundreds. RFLP was the first DNA marker discovered and thus the drive for the development

Table 11.1 Major characteristics of the various DNA markers

DNA marker	Laboratory technique	Number of detected loci	Source of polymorphism	Level of polymorphism	Dominance	Abundance	References
RFLP	Southern blot; agarose gel	Single locus	Point mutation (in a restriction site)	Low	Co-dominant	Medium	Botstein et al. (1980)
RAPD	PCR; agarose gel	Multiloci	Point mutation (affecting the annealing of the primer)	Low	Dominant	Low	Welsh and McClelland(1990); Williams et al. (1990)
AFLP	PCR; acrylamide gel	Multiloci	Point mutation (mainly in a restriction site)	Low (for each band) but one reaction detects many loci	Dominant	Medium	Vos et al. (1995)
SSR	PCR; acrylamide gel	Single locus	Variation in the number of the repeats	Very high	Co-dominant	Medium	Litt and Luty (1989); Edwards et al. (1991)
SNP	Primer extension; chips	Single locus	Point mutation (with sequence information)	Low for each locus, but high for high-throughput genotyping	Co-dominant	Very high	Rafalski (2002)

of the MAS concept. Many studies have applied RFLP over the years (Paterson et al., 1988, 1991; Tanksley et al., 1992; Martin et al., 1993; Wing et al., 1994). However, lately, due to the aforementioned disadvantages of this marker, other DNA markers have become more popular. It is noteworthy that RFLP is still applied to studies of synteny due to its ability to genotype various species that are genetically close (Moore et al., 1995a,b; Devos and Gale, 1997, 2000; Gebhardt et al., 2003; Srinivasachary et al., 2007).

Random amplified polymorphic DNA

Random amplified polymorphic DNA (RAPD) markers were developed and applied in plants (Welsh and McClelland, 1990; Williams et al., 1990).

Genotyping technology: Total DNA is amplified using short single (10 nucleotides) random primers in PCR and the resulting fragments are size separated on an ethidium bromide stained agarose gel.

Genotyping: The banding pattern is transformed onto a 1/0 matrix to determine the existence or absence of each band.

Source of polymorphism: Homology between the primer and the template DNA results in the existence of a certain band. Any mutation that prevents the hybridization of a primer to the template DNA at a certain locus results in the absence of this band.

Characteristics: RAPD usually has two alleles at each locus (existence or absence of a certain band).

The number of RAPD loci used for the analyses of plant genomes varies between several tens to several hundreds. (Note: although the number of short, random, 10 base primers is high, most of them are not polymorphic). RAPD are multilocus markers and their mode of inheritance is dominant. The genotyping technology is very simple (the major advantage of this system), therefore, RAPD has become very popular in many laboratories. The main disadvantages of this system are low level of polymorphism, dominant mode of inheritance that is less suitable for MAS, and low reliability. RAPD primers are shorter than regular PCR primers (which are about 16–22 nucleotides) in order to detect polymorphism. On the other hand, the short (10 nucleotide) primers result in very "sensitive" PCRs and as a result repeated PCR results are not always identical.. Many studies have applied RAPD for various goals (Paran et al., 1991; Reiter et al., 1992; Adamblondon et al., 1994; Xue et al., 2010).

Amplified fragment length polymorphism

Amplified fragment length polymorphism (AFLP) markers were developed by Keygene (Vos et al., 1995).

Genotyping technology: Total DNA (or c-DNA) is restricted by both a "4 cutter" (a restriction enzyme whose restriction site contains 4 bp) and a "6 cutter." Specific universal adaptors homologous to the specific restriction sites are then ligated to the resulting DNA fragments. Restriction and ligation are carried out simultaneously in one reaction making the procedure easy to perform. The resulting fragments flanked by a pair of primers are amplified by PCR using primers homologous to the universal primers. To decrease the number of fragments, a second PCR with primers elongated at their 3′ end by 1–3 random nucleotides is carried out. During the latter PCR the primers of the amplified fragments are labeled radioactively or by fluorescent dye. The amplified fragments are size separated on a sequencing acrylamide gel. The AFLP banding pattern can also be generated on a DNA sequencer.

Genotyping: The band pattern is transformed onto a 1/0 matrix to determine the existence or absence of each band.

Source of polymorphism: Existence or absence of a restriction site (and/or nucleotide polymorphism detected in the second PCR).

Characteristics: An AFLP band usually has two alleles in each locus (existence or absence of a certain band).

The number of bands in a typical run of AFLP is several tens. Since one can use various combinations of restriction endonucleases and several combinations of nucleotides to elongate the primers for the second PCR, band patterns consisting of several hundred bands and more for each tested individual can be generated. AFLP are multilocus markers and their mode of inheritance is dominant. The genotyping technology is rather simple. The main advantages of this system are the relative ease of the genotyping, the relative high number of loci detected in each reaction, and the reliability of the system. The major disadvantages are the low level of polymorphism and the dominant mode of inheritance, which is less suitable for MAS. It is noteworthy that AFLP markers are applied to the detection of DNA methylation by using pairs of restriction enzymes (methylation sensitive and methylation resistant). Many studies have applied AFLP for various goals (Mueller and Wolfenbarger, 1999; Meudt and Clarke, 2007).

Simple sequence repeats (also referred to as microsatellites)

Simple sequence repeats (SSRs) were discovered and developed by Litt and Luty (1989) and by Edwards et al. (1991) in humans and were first applied to plants by Akkaya et al. (1992).

Genotyping technology: Specific loci are PCR amplified by specific primers flanking a simple repeat that consists of 1–5 nucleotides. Primers are generated by screening genomic libraries with probes consisting of the various combinations of 1–5 nucleotides (when the genome sequence is available, the sequence for microsatellites and synthesized PCR primers homologous to the flanking sequences can be screened). The number of available SSR primers in various plant species varies from a few to several hundred (the SSR human map consists of about 20,000 SSRs loci). During this PCR, the primers are labeled radioactively or by fluorescent dye. The amplified fragments are size separated on a sequencing acrylamide gel. The SSR banding pattern can also be generated and assessed on a DNA sequencer.

Genotyping: Each band represents an allele having a specific size in base pairs.

Source of polymorphism: SSRs are one of several sequence variations named variable number of tandem repeats (VNTR), and are distinguished from the macrosatellites by the size of the core sequence (few nucleotides vs. several tens). Thus, the polymorphism is the number of the tandem repeats of a specific microsatellite at a specific locus.

Characteristics: SSRs are single locus markers and have a few to even more than ten alleles in each locus; thus, SSRs are highly polymorphic.

SSRs are co-dominant markers so they can distinguish heterozygotes from homozygotes. The main advantages of SSRs are their high level of polymorphism and their reliability. Many studies have applied SSRs to various goals (Zhao and Kochert, 1993; Provan et al., 1998; Kim et al., 2002; Guo et al., 2006; Scascitelli et al., 2010).

Single nucleotide polymorphism

Single nucleotide polymorphism (SNP) is a point mutation (a change in one nucleotide) for which a short flanking sequence is known. The system is used in humans, animal husbandry, and to a lesser extent in plants.

Genotyping technology: There are various technologies to genotype SNP including the use of DNA sequencers, mass spectrometry, and more. Modern technologies are automated, requiring expensive machinery, but they are also offered by commercial companies as a service (see section Modern genotyping technologies).

Genotyping: Each allele is defined by the actual nucleotide in the sequence.

Source of polymorphism: SNPs are generated by point mutations; namely, any change of a single nucleotide at any place in the genome. Changes can be the replacement of one nucleotide with another and/or deletion or addition (Indels) of a single nucleotide.

Characteristics: SNPs are single locus markers and mostly have two alleles in each locus (low level of polymorphism).

Their mode of inheritance is co-dominant. SNPs have become the marker of choice due to two main advantages: they are highly abundant (there are about 10 million SNPs in the human genome; http://www.ncbi.nlm.nih.gov/snp); and high-throughput technologies of genotyping SNPs (allowing the genotyping of hundreds of thousands SNP in each individual) are available. These advantages compromise their low level of polymorphism (see also the section Genome-wide Association). SNPs have been applied to only a few studies in plants (Nordborg and Weigel, 2008; Ganal et al., 2009; Krause et al., 2010), although, based on the work in humans and farm animals, this number is expected to rise.

Modern genotyping technologies

DNA sequencing technologies and DNA markers were developed separately. However, when sequence data began to yield information on sequence variation among different accessions of the same species, these two avenues merged. A significant advancement was the development of the high-throughput genotyping of SNPs. Although the efficiency of the sequencing technologies has increased dramatically and the cost has decreased, they are mainly used in humans and to a lesser degree in farm animals. The genotyping of various agricultural plant species is still a bottleneck. Some of the high-throughput technologies that have been applied to plants for genotyping SNPs are described in the following sections.

Mass spectrometry

Kwon et al (2001) pioneered the use of mass spectrometry (MS) for DNA sequencing. They demonstrated that matrix-assisted laser disruption/ionization (MALDI) time of flight (TOF) was very efficient (in terms of high-throughput, accuracy, and cost-efficiency) for analysis of complex mixtures of molecules that are separated by their mass/charge ratios. An automated system is currently offered by Sequenom, Inc. (www.sequenom.com) and is used for high-throughput SNP analysis (Jurinke et al., 2001). The short sequence flanking the SNP is amplified in PCR and serves as a template for DNA synthesis in the presence of specific dideoxy nucleotides. The SNP alleles differ in the location of the first chosen dideoxy nucleotide, and thus the primer extension reaction results in alleles with different mass/charge ratios. These fragments are separated, as described previously, and the automated system can generate tens of thousands (multiplexing and analysis of DNA pools are available) of alleles at various loci/various individuals, each given a number equivalent to its size in base pairs (Figure 11.1). The MassARRAY system can be either purchased or supplied as a service. It should be noted that in addition to SNP genotyping, the system can be applied to studies of DNA methylation, analyses of copy number variation (CNV), and gene expression (Cullinan and Cantor, 2008).

Diversity arrays technology

The diversity arrays technology (DArT) was developed and first demonstrated in barley (having a large genome) by Wenzl et al. (2004). It is based on microarray hybridizations as described by Jaccoud et al. (2001). DArT arrays are generated from genomic libraries by amplification of either random or candidate clones that are spotted on polylysine covered slides. The genotyped samples are genomic DNA cut with *Pst1* and a 4 cutter to which suitable adapters were ligated. The resulting fragments are amplified by PCR (similar to AFLP). The samples are denatured and hybridized to the microarrays, and washed and scanned on a confocal laser scanner such as Affymetrix 428 (Santa Clara, CA) or Tecan LS300 (Grödig, Austria). Several articles describe the application of DArT to various plants (Bolibok-Bragoszewska et al., 2009; Kopecky et al., 2009; Hippolyte et al., 2010).

SNP arrays

Two commercial platforms are currently available for very high throughput genotyping of SNPs (500,000 SNPs). These platforms are mainly used in humans in genome-wide association (GWA) studies. In these studies the association between specific SNPs and human diseases is analyzed by comparing thousands, and even tens of thousands, of "cases" (individuals

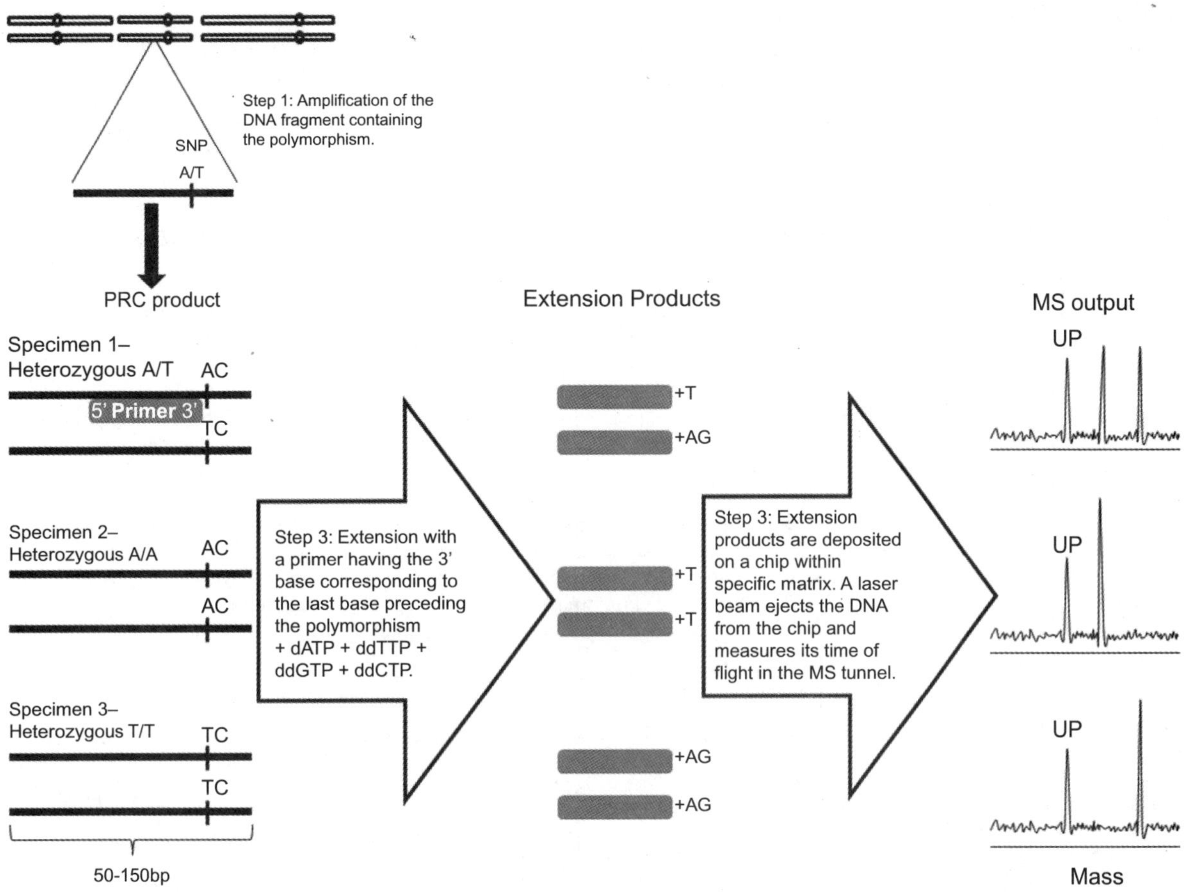

Figure 11.1 • Mass spectrometry • High-throughput single nucleotide polymorphism (SNP) analysis by matrix-assisted laser disruption/ionization (MALDI) time of flight (TOF). Genotyping of three DNA samples containing a specific SNP locus (A/T) by the MALDI-TOF-MS. The MS output contains either two or three peaks, the first one represents the unextended primer (UP), followed by one peak (of a homozygote), or two peaks (of a heterozygote). The peaks differ in their TOF on the basis of the ratio mass/charge affected by the sequence.

having a certain disease) and a similar number of controls, each genotyped by hundreds of thousands of SNPs. A number of studies using these platforms have been carried out (Ge et al., 2009; Thomas et al., 2009; Ku et al., 2010). In addition, the technology is now starting to be applied to farm animals (M. Ron, personal communication). These platforms are offered by Affymetrix, Inc. (www.affymetrix.com) and Illumina, Inc. (www.Illumina.com).

Affymetrix offers a GWA Human SNP array featuring over 900,000 SNPs, and the same number of probes for the detection of CNV. Illumina offers the Genome Analyzer, which is based on a different technology but is applied to the same purposes. Illumina also offers "custom-made" chips that can be generated for any organism.

Modern sequencing technologies

As stated earlier, only seven green plant species have been fully sequenced due to the very high cost of these projects. However, the importance of sequencing large numbers of human individuals for the detection of genes controlling complex diseases has encouraged the development of new high-throughput technologies at a moderate price. Some of these technologies are already commercially available and others are close to being launched. It is reasonable to think that these new technologies will be applied to generating the full sequence of various plant species. Today, these technologies can be applied to sequencing c-DNA libraries of various tissues under various conditions. Thus, it is an important tool for the assessment of patterns of gene expression and the identification of specific genes controlling important agricultural traits. The description of three major modern sequencing technologies is presented next.

Solexa-Illumina

The technology is described on the Illumina Web site (www.illumina.com). This technology relies on the attachment of randomly fragmented genomic DNA to a planar, optically transparent surface. Attached DNA fragments are extended and amplified to generate an ultra-high density flow cell with hundreds of millions of clusters each containing ~1000 copies of the same template. These templates are sequenced using a four-color technology that employs reversible terminators with removable fluorescent dyes. High-sensitivity fluorescent detection is achieved using laser excitation and total internal reflection optics. Sequence reads are aligned against a reference genome and genetic differences are called using specific software. After completion of the first read, the templates can be regenerated *in situ* to enable a second 75+ bp read from the opposite end of the fragments. The Paired-End Module directs the regeneration and amplification operations to prepare the templates for the second round of sequencing. First, the newly sequenced strands are stripped off and the complementary strands are amplified to form clusters. Once the original templates are cleaved and removed, the reverse strands undergo sequencing by synthesis. The second round of sequencing occurs at the opposite end of the templates, generating 75+ bp reads for a total of >20 Gbp of paired-end data per run. Alternative sample preparation methods allow the same system to be applied for analysis of gene expression, small RNA discovery, and protein–nucleic acid interactions.

454 (now Roche)

This technology is based on miniaturizing a pyrosequencing reaction and carrying it out on a solid phase (Margulies et al., 2005; Rothberg and Leamon, 2008). Genomic DNA (or c-DNA) is isolated, fragmented, ligated to adapters, and separated into single strands. Fragments are bound to beads under conditions that favor one fragment per bead. The beads are isolated and compartmentalized in the droplets of a "PCR-reaction-mixture-in-oil emulsion," and PCR amplification occurs within each droplet. Ten million copies of a unique DNA template are generated in each bead. The emulsion is broken down, the DNA strands are denatured, and beads are deposited into wells of a fiber-optic slide. Smaller beads carrying immobilized required enzymes are deposited into each well. CCD camera-based imaging is used to image the fiber-optic slide.

Pacific biosciences

Pacific Biosciences has developed the single molecule real-time sequencing technology (SMRT™) (www.pacificbiosciences.com). The technology operates at single molecule resolution and its main features are: (1) an SMRT Cell, which enables observation of individual fluorophores by maintaining a high signal-to-noise ratio; (2) phospho-linked nucleotides serving as the building blocks for the fast accurate synthesis of natural DNA; and (3) a detection platform that enables single molecule, real-time detection. Thus, the prospect is an accurate and very high-throughput DNA sequencing at a low cost.

Identification of genes controlling commercially important traits

The phenotypes of plants are determined by the interaction between their genome and the environment. A successful breeding program manipulates the plant genome with the aim of developing new cultivars with desirable traits. Genetic control of physiological traits is traditionally studied by analyzing the segregation of the trait of interest in conjunction with molecular markers in a mapping population. The identification of plant genes, which determines economically important traits, is an important step toward further manipulation of the plant phenotype. This knowledge can be applied to MAS using the gene sequence as a marker for the trait, or through direct manipulation of the gene to create a transgenic cultivar of improved phenotype (see the section Classical Methods of Gene Identification).

Most target traits, such as yield, seed dormancy, flowering time, fruit production, fruit quality, resistance to some diseases and insects, and stress tolerance, are quantitative in nature and have a complex mode of inheritance (Alonso-Blanco et al., 2005). Understanding the genetic control of quantitative traits is necessary and relevant in exploring the theory, methods, and strategies of cultivar development.

The ultimate aim of molecular genetic studies of quantitative genetic variation is to identify the genes that collectively control the desired trait. The use of MAS enables the breeder to manipulate the trait of interest with the help of linked markers, even without identification of the genes involved (Dekkers and Hospital, 2002).

Classical methods of gene identification

Map-based positional cloning has been a popular tool for identifying genes of interest without prior knowledge of the molecular mechanism responsible for the desired phenotype (Tanksley et al., 1995; Glazier et al., 2002; Peters et al., 2003). Positional cloning has been used successfully for gene isolation in numerous studies of human, animal, and plant genetics (Martin et al., 1993; Yano, 2001; Korstanje and Paigen, 2002; Andersson and Georges, 2004; Salvi and Tuberosa, 2005; Varshney et al., 2006) and is considered a routine technique, particularly in plants (Lukowitz et al., 2000; Jander et al., 2002). Saturated genetic maps have already been established for many crops, and modern genotyping technologies are capable of providing comprehensive genome coverage, even in organisms in which DNA sequence information is not available (Tester and Langridge, 2010).

Linkage disequilibrium (LD) is the non-random association of alleles at separate loci located on the same chromosome (Mackay and Powell, 2007). Family-based linkage analysis is generated by establishing a population from a small number of founders in the recent past in which LD is still maintained over short genetic distances and during few generations. Two genetically divergent parents, who show clear genetic differences for one or more traits of interest, need to be selected and crossed. Crosses are made to generate segregating populations (e.g., F1 populations in naturally out-crossing plants; or F2 populations in inbreeding species and/or backcross populations). Existence of LD allows the detection of linkage between a DNA marker and a gene of interest.

Introgression lines (ILs) are a series of lines each having most of its genetic material derived from a commercial cultivar and each having a defined fragment of a certain chromosome derived from a genetically close wild species. A complete collection of ILs (the IL library) covers the complete genome of the donor by having a representative of each chromosomal fragment of the wild species. Such IL libraries have been created for tomato (Eshed and Zamir, 1995), rice (Li et al., 2005), and wheat (Pestsova et al., 2006; Maccaferri et al., 2008). IL libraries are generated by crossing a commercial cultivar with a wild-type donor followed by a series of backcrosses with the recipient parent and several generations of selfing to obtain a single homozygous fragment of the donor in each line (Figure 11.2). IL libraries serve as an efficient tool for the identification of chromosomal segments controlling traits of interest.

A sufficient number of markers that reveal polymorphism between parents must be identified and applied for genotyping the entire population. Once the phenotypic and genotypic data are determined for each individual, significant linkage between the markers and the relevant trait are established through various statistical analyses ranging from simple techniques, such as analysis of variance (ANOVA), to models that deal with multiple markers and interactions (Doerge, 2002). We will not elaborate here on these analyses but mention a simple method called "tail analysis" (Hillel et al., 1989; Plotsky et al., 1990) or "bulked segregant analysis" (Michelmore et al., 1991), which is used to rapidly identify markers linked to gene(s) of interest. The two bulked DNA samples are generated from a segregating population (from a single cross). The two bulks differ in the trait of interest or represent the two tails of the progeny distribution regarding the trait of interest, and then they are screened by the available DNA markers. The assumption is that alleles of DNA markers, which are statistically different between the two bulks, are associated with the trait of interest. Michelmore et al. (1991) applied this method to lettuce and identified RAPD markers linked to a gene conferring resistance to downy mildew.

At a second stage, fine-scale mapping is applied to narrow down the candidate region until sufficient resolution is achieved to anchor the target gene to the physical map (Collins, 1992; Glazier et al., 2002). Chromosome walking is used to zoom in on a targeted gene from a very tightly linked marker(s), ultimately leading to isolation of the target gene (Collins, 1992; Tanksley et al., 1995). This procedure involves construction of large insert libraries composed of overlapping clones to generate contigs. Identification of the clone(s) of interest is based on screening the libraries with the markers found to flank the gene of interest. Large DNA clones, in the order of several megabases, include mainly bacterial artificial chromosomes (BAC), and yeast artificial chromosomes (YAC; Kumar, 1999).

The candidate gene is usually sequenced in the various phenotypes to identify the causative DNA variant. Such variants include non-synonymous and deletion polymorphisms in coding regions (Mackay et al., 2009), as well as differences in gene expression level (Doebley et al., 1997; Kliebenstein et al., 2001; Liu et al., 2003), polymorphism in promoters, and introns affecting the binding of transcription factors and mRNA splicing (Mackay et al., 2009). The last step of gene identification is validation, which can be performed in various ways, such as introducing specific alleles followed by a change of a specific phenotype, testing outlier populations, and more (Borevitz and Chory, 2004).

An alternative approach for using DNA markers that are evenly spaced throughout the genome is the use of candidate genes that are selected because of *a priori* hypotheses about their role in the trait of interest. Identification of QTLs is based on genotype–phenotype association within this set of genes (Tabor et al., 2002; Ben-Ari et al., 2006).

Modern methods for gene identification

The concept of DNA markers results from the fact that in most cases the gene(s) controlling the trait(s) of interest are unknown. However, it is obvious that when the gene(s) controlling the phenotypic difference is known, it is the best tool for MAS. Recently, a number of methodologies aimed toward identification of genes have been developed. The following are some examples based on modern high-throughput technologies.

Targeting-induced local lesions in genomes

Targeting-induced local lesions in genomes (TILLING) is a powerful reverse-genetic strategy that employs a mismatch-specific endonuclease (an enzyme that identifies a mismatch and cuts the double-stranded DNA at this specific site) to detect induced or natural DNA polymorphisms. High-throughput TILLING allows the rapid and cost-effective detection of induced point mutations in populations of chemically mutagenized individuals. The strategy of TILLING was first reported by McCallum et al. (2000a,b) ten years ago. They used HPLC to detect mismatches in heteroduplex DNAs that had been generated by PCR amplification of specific genes on a DNA template from a pooled population of wild-type and mutant *Arabidopsis* plants. A high-throughput TILLING protocol was published a year later and employed a mismatch-specific celery nuclease, CEL1, to identify SNPs (Colbert et al., 2001). The advantages of TILLING are that it is a simple procedure, with high sensitivity, high efficiency, and non-transgenic modifications. Thus, TILLING provides a powerful approach for gene discovery, assessment of DNA polymorphism, and plant improvement (Wang et al., 2006). TILLING has been successfully applied to various crops including maize (Till et al., 2004), rice (Till et al., 2007), potato (Muth et al., 2008), and soybean (Cooper et al., 2008), as well as tetraploid and hexaploid wheat (Uauy et al., 2009).

Genome-wide association

The genome-wide association (GWA) is increasingly being adopted as the method of choice for identifying genes. Association analysis involves searching for genotype–phenotype correlations in unrelated individuals (Myles et al., 2009). This method can reveal genes or QTLs by examining the marker-trait associations that can be attributed to the strength of LD between markers and functional polymorphisms across a set of diverse germplasm. GWA was originally pioneered by human geneticists as a potential solution to the challenging problem of finding the genetic basis of common complex human diseases such as diabetes, high blood pressure, and diseases of the heart and the nervous system

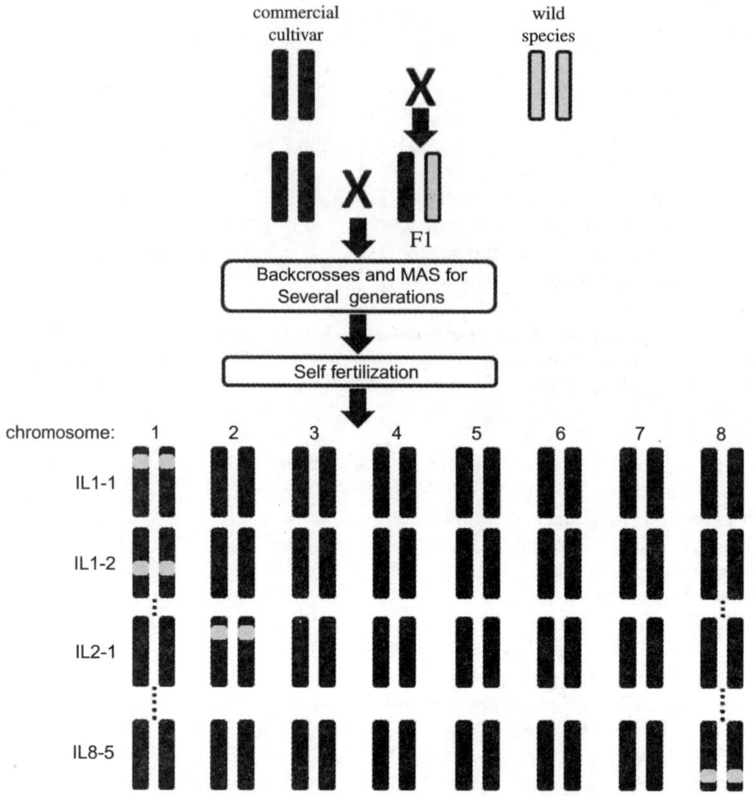

Figure 11.2 • Introgression lines (ILs) library • A wild-type parent (red genome) is crossed with the commercial cultivar (blue genome) to generate the F1 hybrid, which is backcrossed for several generations with the commercial cultivar. The progenies are then selfed for several generations. This procedure results in a series of plants, each of which are carrying a known, different, homozygous segment of the wild species genome. These chromosomal segments are identified by genotyping the library with genome-wide DNA markers distinguishing between the wild and commercial parental strains. The whole IL library covers all of the wild genome. For example, IL1-1 consists of the entire commercial cultivar genome with the exception of a single segment donated by the wild species located at the north of chromosome 1, whereas IL8-5 contains the wild segment on the south of chromosome 8. Please see color plate section at the end of the book.

(Hirschhorn and Daly, 2005; Burton et al., 2007). Application of GWA depends on the scale of LD, which in turn determines the degree to which molecular markers may be associated with the relevant phenotype (Mitchell-Olds, 2010). The main advantage of GWA over linkage analysis is that it exploits all of the recombination events that have occurred in the evolutionary history of a sample, which results in much higher resolution. In addition, the number of mapped QTLs for a given phenotype is not limited to those that are significantly different between the parents of a particular cross, but rather by the number of real QTL underlying the trait (Zhu et al., 2008). Due to higher genome density, lower mutation rate, and better amenability to high-throughput detection systems, SNPs are rapidly becoming the marker of choice for GWA studies. High-capacity DNA sequencing instruments or high-density oligonucleotide arrays efficiently identify SNPs at a density that accurately reflects genome-wide LD structure and haplotype diversity. A proper statistical analysis is needed to identify genotype–phenotype associations (Zhu et al., 2008). Recently, several studies used GWA to identify QTLs in *Arabidopsis* (Chan et al., 2009; Atwell et al., 2010; Brachi et al., 2010), barley (Stracke et al., 2009), maize (Sherry et al., 2005), pearl millet (Saidou et al., 2009), and teosinte (Weber et al., 2008). This high-density genotyping becomes increasingly accessible and was recently used to generate genetic linkage maps in several other plants, including soybean (Hyten et al., 2010) and wheat (Akhunov et al., 2009). In cases where relevant alleles are present at high frequencies, association studies can disclose causal polymorphism affecting agronomic traits in crop plants (Rafalski, 2010). The detection power of association mapping depends not only on the magnitude of the effect that can be ascribed to a locus relative to other loci present in the population, but also on the allele frequency. The limitations of this method are that rare alleles cannot be detected with confidence unless their effect is considerable, and that population structure may lead to wrong results. Therefore, linkage analysis in segregating families is more appropriate for the identification of rare alleles (Rafalski, 2010). It is reasonable to assume that the new high-throughput genotyping methods will be applied to segregating populations and thus overcome some of the difficulties of GWA (Mitchell-Olds, 2010).

RNA interference

The functions of many genes have been elucidated by selective gene inactivation and subsequent phenotypic analysis. For example, specific mutations, gene knockouts, and RNA interference can all result in decreased production of a specific protein, yielding informative phenotypes (Raina and Crews, 2010). Small RNAs are riboregulators that have critical roles in most eukaryotes. They repress gene expression by acting either on DNA to guide sequence modifications and chromatin remodeling or on RNA to guide degradation and/or prevent translation. Small RNA-mediated regulation is often referred to as RNA silencing, gene silencing, or RNA interference (RNAi). RNA silencing was discovered in plants about 20 years ago during the course of transgenic experiments that eventually led to gene silencing (Napoli et al., 1990; Smith et al., 1990; Van Der Krol et al., 1990). At the post-transcriptional level, RNAi has emerged as a useful tool for gene silencing due to its ability to knock down levels of any gene for which a rather short sequence is available. Since its discovery, RNAi has been shown to play an essential role in plants (Vaucheret, 2006), fungi (Nakayashiki et al., 2006), and mammals (McCaffrey et al., 2002). RNA-silencing can be induced by short nucleotide double-stranded RNA (dsRNA). Significantly, the production of large-scale small interfering RNA and short hairpin RNA libraries has made genome-wide RNAi analysis possible (Ruiz-Ferrer and Voinnet, 2009).

Expression QTLs

Expression QTLs (eQTLs) are identified by applying QTL analysis to data assessing gene expression in a segregating population. Transcript abundance of each gene is the analyzed phenotype.

eQTL analysis requires genetic markers that can be genotyped in all individuals in the analyzed population. The outcome of this analysis is a statistical association between genetic markers located at specific regions of the genome and the expression level of the assayed gene. The resulting eQTL plot indicates the likely genetic location of DNA sequence variation (eQTL) that causes the observed variation in transcript abundance. Furthermore, eQTL data enable the modeling of genetic regulatory networks and provide a better understanding of the underlying phenotypic variation.

Expression QTLs are empirically divided into two classes: *cis* and *trans*. In the former, the sequence variation controlling transcript levels is assumed to be determined by the sequence variation that lies within or in close proximity to the gene. A *cis*-eQTL coincides with the location of the underlying gene. In case of *trans*-eQTL, the observed location of the eQTL does not coincide with the location of the gene. This implies that the observed eQTL represents the position of a locus that controls the expression of the target gene. Target gene expression can be controlled by a combination of both *cis*- and *trans*-acting elements (Arnis et al., 2010). Expression QTL analysis has been demonstrated in *Arabidopsis*, maize, barley, wheat, poplar, and eucalyptus (Kirst et al., 2005; Street et al., 2006; Jordan et al., 2007; Keurentjes et al., 2007; Shi et al., 2007; West et al., 2007; Druka et al., 2008; Potokina et al., 2008).

Chemical genetics

Chemical genetics is a powerful new discipline in plant science. Bioactive small molecules can be used to identify novel signaling pathways and unravel redundant networks. Small molecules are used to alter protein function and explore biological roles of these target proteins. In principle, small molecules allow rapid, conditional, and reversible alteration of biological functions; therefore they can generate new and often complementary information compared to classical genetic studies. Small molecules can also overcome limitations of genetic approaches such as lethality, genetic redundancy, and pleiotropic effects observed in genetic mutants (Tóth and van der Hoorn, 2010). A chemical genetic approach helped reveal the hormone signaling pathways of auxin (Hayashi et al., 2008), brassinosteroid (De Rybel et al., 2009), and abscisic acid (Park et al., 2009).

Application of DNA markers to breeding

DNA markers have been applied in plant breeding programs for various purposes in addition to the identification of genes of interest and the development of MAS. These applications include identification of individual plants and plant populations, assessment of genetic distances (phylogeny), identification of parents for the production of hybrid seeds, and significant decreases of the number of backcross generations needed to achieve gene introgression.

Identification

The ability to identify individual plants is at the core of many applications. In many cases, it is necessary to individually characterize the sampled genotypes. The use of DNA markers in cultivar identification is particularly important for protection of proprietary germplasm. DNA markers have also been useful in pedigree analysis and parental identification (Joshi et al, 1999).

Most forest tree species exhibit high levels of genetic diversity that can be used to trace the origin of living plants or their products such as timber and processed wood. Molecular tools are applied to test the origin of forest reproductive material used for the establishment of plantations and in international trade of timber and wood products (Finkeldey et al., 2010). DNA markers are also used to confirm purity, especially in hybrid cultivars where the maintenance of high levels of genetic purity is essential (Collard and Mackill, 2008).

Improving classical breeding projects

The major applications of DNA markers for the improvement of classical breeding projects include conserving the genetic diversity of parental stocks, selecting the suitable parents for the generation of heterosis, decreasing the number of backcross generations in gene introgression projects, and pyramiding various genes in the same cultivar.

Conserving diversity

To achieve progress in a breeding program, breeders need to carefully select the appropriate parents. In addition to the performance of the parental lines, their genetic diversity should be taken into account as well. Using diverse cultivars as the core breeding material increases the chances of achieving new combinations and new elite strains excelling in various agricultural traits. Therefore, the assessment of biodiversity is crucial for parental selection in breeding programs (Collard and Mackill, 2008).

Understanding the critical role of genetic information has played an important part in the development of conservation biology. Genetic drift generates loss of genetic variation and dramatically affects small populations. The potential of future adaptation to a changing environment depends on the scope of genetic variation. A second consequence of the loss of genetic variation is that the number of homozygous individuals within a population necessarily increases. Such inbreeding may be associated with a reduction in individual fitness. Thus, maintenance of genetic variation is regarded as a primary goal in conservation efforts and breeding projects (Milligan et al., 1994).

Selection of parents for the generation of heterosis

Heterosis refers to the superior phenotypes observed in hybrids relative to their inbred parents with respect to traits such as growth rate, reproductive success, and yield. Heterosis was discovered in maize about a century ago and has subsequently been found to occur in many crop species. The increase in yield as a result of the cultivation of hybrid offspring ranges from 15 to 50%, depending on the crop (Lippman and Zamir, 2007).

The theory of quantitative genetics predicts a positive correlation between parental divergence and the estimated degree of heterosis (Falconer and Mackay, 1996). However, the data obtained to date in support of this correlation are not conclusive, and the ability to predict levels of heterosis based on the genetic distance between parents varies with different traits and crops (Flint-Garcia et al., 2009).

Introgression

Commercial elite cultivars may often be improved by a desirable allele (like resistance to a specific disease), which exists in a distantly related (and even a wild-type) genotype. Gene introgression can be achieved by crossing the elite cultivar with the donor plant, and then by repeated backcrossing of the progeny with the recipient line, while selecting the desirable allele in each generation. The problem is that about six (and more) generations are needed to achieve this goal. The use of DNA markers can improve and shorten the introgression program by minimizing linkage drag and reducing the number of necessary backcrosses. In this context, markers distinguishing between the two parents are used to select the desired genotype even after one or two backcross generations (Figure 11.3).

The objective of application of DNA markers is to recover, as rapidly as possible, the maximum proportion of recurrent regions of the parent genome at non-target loci through many markers that are distributed evenly throughout the genome. Data from simulation studies suggest that two or three backcross generations may be sufficient (Hillel et al., 1990; Visscher et al., 1996; Hospital and Charcosset, 1997; Frisch et al., 1999).

Pyramiding

Pyramiding is the accumulation of several desired alleles in a single line or cultivar (Pedersen and Leath, 1988). Marker-based selection is a very useful approach to maximal utilization of the existent gene resources. Genes controlling various agronomic traits can be theoretically combined together in a single plant. Furthermore, genes responsible for resistance to different diseases or insect pests can be pyramided to generate a variety/cultivar exhibiting multiresistance to pathogens.

When several favorable genes are hosted by two different parents, the simplest strategy involves the production of a segregating population and subsequent screening of the progeny on the basis of molecular markers for individuals homozygous at the targeted loci. Pyramiding of multiple genes from various parental lines is more complex and can be achieved by several generations of pairwise crossing (Servin et al., 2004). In the past, it has been difficult to pyramid multiple resistance genes because they generally exhibit identical phenotypes and progeny testing is required to determine which plants carry more than one desired gene. DNA markers may facilitate the pyramiding process because DNA markers for multiple specific genes or QTLs can be tested using a single DNA sample without the need for phenotyping. Such pyramids have been developed in rice against bacterial blight and blast. Kottapalli et al. (2010) introgressed three bacterial blight resistance genes, *xa5*, *xa13*, and *Xa21*, into a fine rice grain variety. Another successful pyramiding was achieved in wheat by combining several QTLs conferring resistance to *F. pseudograminearum* (Bovill et al., 2010).

MAS in breeding programs

It has been predicted for over two decades that molecular marker technology would reshape breeding programs and facilitate rapid gains from selection. About 14 years ago, Concibido et al. (1996) described the application of MAS for breeding of cyst nematode (*Heterodera glycines*) resistance in soybean. However, while MAS is effectively used in breeding of various monogenic traits, it has not been significantly successful in the breeding of polygenic traits, especially in cases where many alleles of small effect are involved in producing a specific phenotype (Jannink et al., 2010).

MAS has been widely used in breeding programs for gene introgression and gene pyramiding, particularly for disease resistance genes in the major crops, but also in crops of lesser economic importance. For obvious reasons, most breeders do not reveal their efforts to apply MAS in their breeding projects. Thus, references are not available for many of these activities. The information we have gained from breeders at both research institutes and commercial companies (see the Acknowledgments section) is that most efforts are invested in

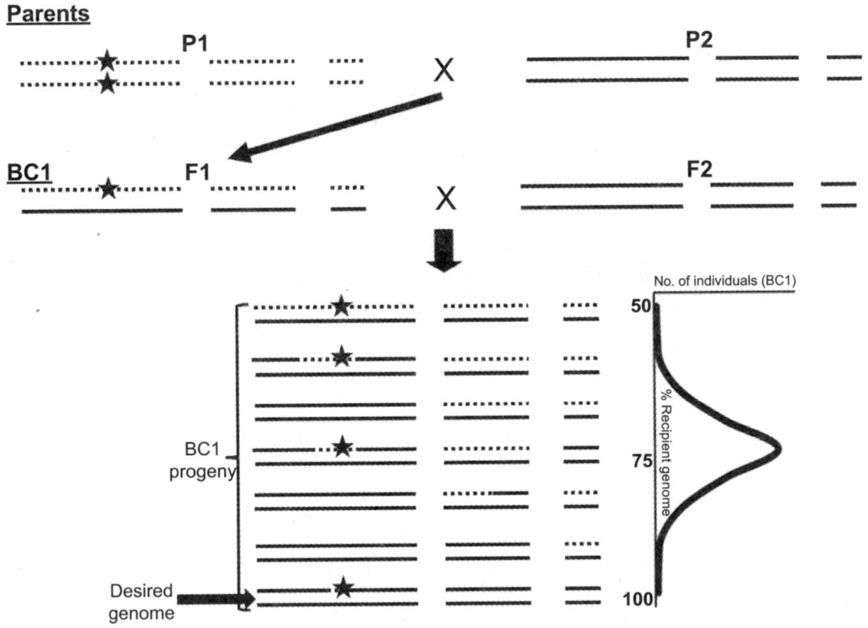

Figure 11.3 • Gene introgression • Carried out by crossing a donor parent (P1; represented by a red line) and a recipient parent (P2; represented by a black line) followed by repeated backcrosses (BC) of the F1 with the recipient parent to recover the recipient genome. The target gene (star) is selected for in each of the backcross generations. The average content of the recipient genome (black and red continuous line) in BC1 progeny is 75%. However, the content varies among the progeny from 50 to 100% and is characterized by a normal distribution curve (right graph). Recovery of the recipient genome can be enhanced by using DNA markers distributed across the entire genome and distinguishing the two parents. Thus, by the proper use of DNA markers one can significantly decrease the number of required back crosses. Please see color plate section at the end of the book.

applying MAS to the improvement of commercially important crops such as wheat, rice, corn, soybean, and some vegetables like tomato and pepper. Public funding to support validation and refinement of MAS techniques and its applications in field breeding has been limited (Xu and Crouch, 2008), and most successes are with traits controlled by single genes.

A large number of monogenic traits or major QTLs have been targeted for wheat improvement through MAS. These traits include resistance to various rusts, FHB, barley yellow dwarf virus (BYDV), nematodes, and Hessian fly/Russian wheat aphid. Also included are some quality traits such as grain protein content, grain hardness, tolerance to pre-harvest sprouting, grain color, bread-making quality, grain texture, and gluten strength (Gupta et al., 2010).

Among the abiotic stresses, drought and salinity are the major ones. Certain progress has been achieved in developing drought- and salt-tolerant cultivars using conventional breeding approaches.

Numerous QTLs affecting drought and salt tolerance have been identified. For example, the AtNHX1 antiporter is involved in the sub-cellular compartmentalization of potassium, which in turn affects potassium nutrition and sodium (Na) tolerance (Leidi et al., 2010). The phosphate transporter PHT4;6 was identified to be a determinant of salt tolerance localized in the Golgi apparatus of *Arabidopsis* (Cubero et al., 2009). Zhang et al. (2005) showed the effect of a change in the membrane fatty acid on tolerance to drought and salinity. However, to our knowledge, the degree of application of these findings to plant breeding projects is limited.

Some progress has been made at the International Rice Research Institute (IRRI) by the development of tolerant introgression lines in elite genetic backgrounds (Li et al., 2007). A number of QTLs conferring drought tolerance were mapped in barley. These include several physiological/biochemical traits such as osmotic adjustment capacity, proline content, stomatal conductance, water soluble carbohydrates, relative water content, leaf turgor, ABA content, and various morphological traits, but the challenges of application of MAS still remain (Li et al., 2007). Despite considerable traditional and molecular studies in tomato, there is no report of any commercial tomato cultivar with salt or drought tolerance (Foolad, 2007a). Attempts to generate tolerant potato cultivars by introducing genes such as osmotin-like protein, GPD, and trehalose synthesis protein have not been successful (Byun et al., 2007), and QTL mapping for salt tolerance in cotton is moving very slowly (Lubbers et al., 2007). It is believed that the complexity of these traits slows down the progress along this line.

Next are some examples of the application of MAS in various crops. In tomatoes, P. Lindhout (Bai and Lindhout, 2007) reviewed the tools used in tomato breeding. The most prominent application of MAS in tomato breeding is for the introduction of resistance genes for diseases and pests such as bacterial canker, *Fusarium* wilt, tobacco mosaic virus, and *Verticillium* wilt. In addition, MAS has been used for selection of self-pruning (determinant vs. indeterminant) cultivars. Mutant genes that delay maturation (*rin* and *nor*) have been incorporated into local cultivars and the abscission related gene *JOINTLESS-2* has been used for selection of low fruit removal force (FRF) cultivars (Foolad, 2007b). Zamir et al. (Krieger et al., 2010) showed that the flowering gene *SINGLE FLOWER TRUSS* (*SFT*) drives heterosis for yield in tomato.

Attempts to find markers for fruit color, taste, and quality are in their infancy. One difficulty is that environmental factors such as light and temperature tend to have an impact on the color, so that plants carrying the lycopene cyclase gene (generating beta carotene from lycopene) vary in their pigmentation (I. Levin, personal communication).

Regarding fruit quality, the major locus, *Fgr* (Fructose to glucose *r*atio), was mapped to chromosome 4, while a second locus was mapped to chromosome 6 near the loci for a fructokinase (*FK2*) and a hexokinase (Levin et al., 2006). The two loci show an epistatic relationship, with $FK2^{hab}$ further increasing the *Fgr* only in the presence of the wild species allele for *Fgr* (Pedersen and Leath, 1988). Another gene that proved to be useful for increasing Brix without decreasing yield parameters is the *AGPase-LS1* (Petreikov et al., 2006, 2009).

MAS is better applied in pre-breeding, where on the basis of information derived from ILs (see details in the section RNA Interference), one can generate cultivars carrying specific chromosomal segments known to have major alleles that affect important quantitative traits like sugar content (Brix) and yield (E. Kopilevits, personal communication).

In pepper, MAS is applied to confer resistance to Tobamoviruses (Kim et al., 2008), tomato spotted wilt virus (Moury et al., 2000), root knot nematode (Wang et al., 2009), and for potyvirus (Yeam et al., 2005). In addition, there are markers for fruit color (yellow, red, and brown) and for fruit cracking.

Examples from crop plants are reviewed in this section. Dr. D.J. Makill of IRRI has been successful in the development of submergence — tolerant rice cultivars using the *Sub1* locus (Septiningsih et al., 2009). Two varieties with the submergence tolerance gene *SUB1* were released recently. MAS was applied to decrease the number of backcross (BC) generations (to two to three generations) and reduce the linkage drag. "Swarna-Sub1" is grown in India and Bangladesh and IR64-Sub1 in the Philippines and Indonesia where submergence is a problem. Dr. R.K. Varshney of the International Crops Research Institute for the Semi-Arid Tropics (ICRISAT) and Generation Challenge Programme (GCP) stated that they are in the process of introgressing a major QTL (controlling 30% of the phenotypic variation for root traits) to chickpea. Dr. Y. Xu of CIMMYT reported the identification of loci conferring drought resistance by using introgression lines of maize (Hao et al., 2009) and the use of the Opaque2 mutant for the improvement of protein quality (Xu et al., 2009).

Although quite a number of marker loci have been reported to be linked to various genes of interest, it is our feeling that only a few of these markers are currently used routinely in breeding projects. At the same time, it needs to be emphasized that application of MAS to breeding is on the rise.

Discussion

Bottlenecks and difficulties in the application of MAS

The goal of breeding is to develop cultivars with improved agricultural characteristics. In general this goal can be achieved either by incorporation of traits that do not exist in the current commercial cultivars or by pyramiding traits available in the primary gene pool into a specified cultivar. For the application of MAS, whenever new traits are considered, one needs first to identify markers that are linked to these traits. Thus, linkage or association analysis based on phenotypic assessment must be carried out because breeders need to continue to assess plants in the field. Once linkage has been identified, it is obvious that MAS has the potential to be an efficient tool for the breeder. When pyramiding is considered and when linkage has already been identified, efficient MAS would be a very useful tool for the breeder. As noted, breeders will still need to assess the plants in the field and make sure that the pyramiding is successful, and that no negative effects are associated with the new cultivars.

Efficient MAS is based on three requisites: (1) the ability to identify DNA variations by DNA markers or the genes themselves (see the section DNA Markers); (2) the analyzed population (see the section Application of DNA Markers to Breeding); and (3) the statistical procedure for identifying linkage or association between a DNA marker and a gene of interest.

Successful MAS depends upon the availability of a very large number of DNA markers (to cover the whole genome) as well as the availability of efficient genotyping technologies to allow the genotyping of a large number of individuals, each with numerous markers. If the human genome is to be the landmark for future developments in this area, SNPs are the marker of choice, both due to their high abundance and the availability of high-throughput genotyping technologies for SNPs. CNV, (Iafrate et al., 2004; Sebat et al., 2004), which has not been applied to plants, represents another kind of DNA variation that is associated with human disorders (Lee et al., 2007; Cook and Scherer, 2008) and may have a wider significance in the future. It is important to state that the high abundance of SNPs and their efficient genotyping technologies compensate for their low level of polymorphism. Thus, it is expected that availability of greater amounts of sequence information will increase the number of available SNPs in various plant species. Obviously, availability of full sequences of the analyzed individuals will be the ultimate solution of this bottleneck.

One of the major advantages of plants is the relative ease (compared to other biological systems) of generating large segregating populations. F2 and F1 (in the case of heterozygous plants) as well as BCs are rather easy to generate in very large numbers (hundreds, thousands, and more). The major bottleneck is the long juvenile period of some plant species, especially fruit trees. In addition, in some species like mango and avocado, it is not feasible to carry out artificial pollination to generate large populations, and alternative procedures such as caging the two parents under a net in the presence of a beehive must be used.

Once the population is generated, the phenotypes need to be assessed. This assessment may be a major bottleneck in developing efficient MAS. Dr. A. Blum (personal communication and http://www.plantstress.com/), strongly emphasized the need for the correct assessment of the phenotypes. It is clear that errors in the phenotypic assessment lead to either false negative or false positive identification of markers. A major reason for errors in the phenotypic assessment is the wrong choice of traits. For example, since yield is a very complex trait, one should not search for DNA markers for this trait but rather identify its major components such as head number, grains per head, and grain size in cereals, and then search for DNA markers that are linked or associated with these components. It should be noted that phenotyping, especially of agronomic and physiological traits in the field, is often difficult and sometimes even inaccurate.

In this chapter we have not elaborated on the statistical tools for the identification of DNA markers genetically linked or associated with the gene(s) controlling important agricultural traits. However, a major bottleneck in the development of MAS is the ability to distinguish between false positive and true significant linkage or association between markers and genes of interest. The large numbers of analyzed markers and individuals necessary for the development of MAS creates a major challenge for this analysis and a need to develop suitable tools. Identification of statistically significant results is a major task due to the multi-testing (i.e., the large number of comparisons). Finally, it is essential to be able to identify interactions, such as considering the possibility that the desired phenotypes may result from the interaction between different genes and/or alleles and identifying the DNA markers that reflect these interactions. Although bioinformatic and statistical tools for these analyses are developing at a rapid pace, better, more efficient and more accurate methods are needed.

The GWA has been applied in the last several years to the analyses of human disorders and has identified a large number of DNA variants associated with various diseases (Rosenberg et al., 2010). However, there are also some difficulties; for example, GWA aimed at identification of SNPs associated with human height were carried out on tens of thousands of individuals genotyping 100–500,000 SNPs in each individual. These studies resulted in the identification of 40 DNA variants associated with height. However, these variants altogether accounted for only about 5% of the genetic variation, even though heritability of human height is estimated to be 80–90%. Similar results characterize the situation in various complex diseases in which the identified variants explain only a small part of the genetic variation (Maher, 2008). There can be several explanations for these results: (1) the causative DNA variants were not the analyzed SNPs, but rather other DNA variants; (2) the desired phenotypes resulted from rare variants or variants of low penetrance; (3) the heritability estimates were wrong and the actual heritability estimates are rather low; (4) phenotyping (in humans, the diagnosis of some disorders) was not accurate enough (see earlier information); (5) the analyzed phenotypes are controlled by dominant or epistatic interactions of several genes/alleles; and (6) the phenotypes are controlled by epigenetic factors (like DNA methylation).

It is our view that there are ways to overcome these bottlenecks, such as widening the types of genetic variants and increasing their numbers (eventually analyzing full sequences), increasing the number of analyzed individuals, and developing better statistical tools that will enable the detection of dominant and epistatic interactions between several loci. Regarding the future of MAS in plants, we believe that the genotyping analyses of a high number of DNA variants (SNPs) in a large number of individuals originating from segregated populations (F1, F2, and BC) have the potential to generate effective MAS and more efficient breeding projects.

Future prospects of application of genetic variations to breeding

The ultimate knowledge of genetic variation will be generated from comparisons of whole genome sequences of individuals from the same species. It is assumed that having the full sequence of the various phenotypes will allow the identification of the genes controlling these phenotypes. It can be argued that in addition to the sequence information, epigenetics (such as methylation) plays a major role in controlling phenotypes and that due to the complexity of the control of quantitative traits, having even the full sequence will not be sufficient for the understanding of the molecular mechanisms behind the important agricultural traits. However, most scientists agree that full genome sequencing is highly desirable. The major technical developments (described in the section Modern sequencing technologies) suggest that it is probable that in the near future the genome sequences of more agricultural plant species will be available. The next step will then be the full sequencing of individuals of the same species having various phenotypes.

We expect that before the era of full sequencing, there will be other developments related to genetic variation of plant species. On the one hand, it is expected that an approach similar to the application of high-throughput SNP analysis in animal husbandry (the analysis of tens of thousands of SNPS in both cattle and chickens) will be seen in some of the commercially important plant species. On the other hand, there are already several examples of plant transcriptome analyses, and we believe that with the availability of various technologies the transcriptome of more plant species will be analyzed. Another approach is knockout (either by transposons, T-DNA, or RNAi). The problem with this approach is that it is based on the availability of a transformation methodology (usually based on regeneration capability), which is not available to all plant species. However, massive efforts have been invested in developing transformation systems for a wider range of plant species and to make the available ones more efficient in terms of percent of successful events of transformation. The most obvious approach that will widen and deepen in terms of more genes and more plant species is the identification of important genes in various commercial plant species on the basis of homology with known genes from "model" plants (the species in which these genes are currently known), such as the identification of orthologs using simple molecular techniques.

These approaches will identify important genes that are of interest to the breeder who will then be able to make the breeding projects much more efficient by looking for the desirable alleles of these genes.

Finally, we would like to emphasize that in our opinion, plants have some built-in advantages, which in view of the difficulties mentioned in the section Bottlenecks and difficulties in the application of MAS, make us rather optimistic regarding the future prospects of MAS in plants. These advantages include the:

- Rather short juvenile period in some plant species (although this is not so for many fruit trees).
- Ability (in some cases) to carry out self-pollinations or to generate dihaploids to generate homozygous plants. This advantage was applied to the citrus sequencing project (Nir Carmi, personal communication).
- Ability to generate large populations in a "family structure." It is noteworthy that one of the explanations for the limited "success" of human GWA projects is that the same phenotype could be generated by various genotypes (see section Genome-wide association). Analysis of the progeny of a certain cross should overcome this difficulty.

Acknowledgment

The authors are very thankful to those who contributed their knowledge and critical views on the subject in long discussions and in e-mails: Professor G. Ben-Haim, ARO-Volcani Center, Israel; Dr. A. Blum, plantstress.com; Professor A. Cahaner, the Faculty of Agriculture at the Hebrew University of Jerusalem, Israel; Dr. P. Cregan, the ARS-USDA, Beltsville, MD; Dr. Y. Elkind, the Faculty of Agriculture at the Hebrew University of Jerusalem, Israel; Dr. E. Kopilevits, the Faculty of Agriculture at the Hebrew University of Jerusalem, Israel; Dr. I. Levin, ARO-Volcani Center, Israel; Dr. D.J. Mackill, IRRI; Dr. M. Moshelion, the Faculty of Agriculture at the Hebrew University of Jerusalem, Israel; Dr. I. Paran, ARO-Volcani Center, Israel; Dr. M. Ron, ARO-Volcani Center, Israel; Dr. R. K. Varshney, ICRISAT and GCP; Dr. N. Yonash, Hazera-Genetics Ltd, Israel; Professor D. Zamir, the Faculty of Agriculture at the Hebrew University of Jerusalem, Israel; and Dr. Y. Xu, CIMMYT. Special thanks to Dr. P. Cregan, the ARS-USDA, Beltsville, MD, who also made valuable comments on the manuscript.

References

Adamblondon, A. F., Sevignac, M., Dron, M., & Bannerot, H. (1994). A genetic-map of common bean to localize specific resistance genes against anthracnose. *Genome*, *37*, 915–924.

Akhunov, E., Nicolet, C., & Dvorak, J. (2009). Single nucleotide polymorphism genotyping in polyploid wheat with the Illumina GoldenGate assay. *TAG Theoretical and Applied Genetics*, *119*, 507–517.

Akkaya, M. S., Bhagwat, A. A., & Cregan, P. B. (1992). Length polymorphisms of Simple Sequence Repeat DNA in soybean. *Genetics*, *132*, 1131–1139.

Alonso-Blanco, C., Aarts, M. G. M., Bentsink, L., Keurentjes, J. J. B., Reymond, M., & Vreugdenhil, D., et al. (2009). What has natural variation taught us about plant development, physiology, and adaptation? *Plant Cell*, *21*, 1877–1896.

Alonso-Blanco, C., Bentsink, L., Hanhart, C. J., Vries, H. B. E., & Koornneef, M. (2003). Analysis of natural allelic variation at seed dormancy loci of *Arabidopsis thaliana*. *Genetics*, *164*, 711–729.

Alonso-Blanco, C., Mendez-Vigo, B., & Koornneef, M. (2005). From phenotypic to molecular polymorphisms involved in naturally occurring variation of plant development. *International Journal of Developmental Biology*, *49*, 717–732.

Andersson, L., & Georges, M. (2004). Domestic-animal genomics: Deciphering the genetics of complex traits. *Nature Reviews Genetics*, *5*, 202–212.

Appleby, N., Edwards, D., & Batley, J. (2009). New technologies for ultra-high throughput genotyping in plants. In D. J. Somers, P. Langridge & J. P. Gustafson (Eds.), *Plant genomics* (pp. 19–39). New-York: Humana Press.

Arnis, D., Elena, P., Zewei, L., Ning, J., Xinwei, C., & Mike, K., et al. (2010). Expression quantitative trait loci analysis in plants. *Plant Biotechnology Journal*, *8*, 10–27.

Ashraf, M., & Akram, N. A. (2009). Improving salinity tolerance of plants through conventional breeding and genetic engineering: An analytical comparison. *Biotechnology Advances*, *27*, 744.

Ashraf, M. (2010). Inducing drought tolerance in plants: Recent advances. *Biotechnology Advances*, *28*, 169–183.

Atwell, S., Huang, Y. S., Vilhjalmsson, B. J., Willems, G., Horton, M., & Li, Y., et al. (2010). Genome-wide association study of 107 phenotypes in *Arabidopsis thaliana* inbred lines. *Nature*, *465*, 627–631.

Austin, D. F., & Lee, M. (1996). Genetic resolution and verification of quantitative trait loci for flowering and plant height with recombinant inbred lines of maize. *Genome*, *39*, 957–968.

Bai, Y., & Lindhout, P. (2007). Domestication and breeding of tomatoes: What have we gained and what can we gain in the future? *Annals of Botany*, *100*, 1085–1094.

Barone, A., Di Matteo, A., Carputo, D., & Frusciante, L. (2009). High-throughput genomics enhances tomato breeding efficiency. *Current Genomics*, *10*, 1–9.

Ben-Ari, G., Zenvirth, D., Sherman, A., David, L., Klutstein, M., & Lavi, U., et al. (2006). Four linked genes participate in controlling sporulation efficiency in budding yeast. *PLoS Genetics*, *2*, e195.

Bentsink, L. n., Jowett, J., Hanhart, C. J., & Koornneef, M. (2006). Cloning of DOG1, a quantitative trait locus controlling seed dormancy in *Arabidopsis*. *Proceedings of the National Academy of Sciences*, *103*, 17042–17047.

Bernatzky, R., & Tanksley, S. D. (1986). Toward a saturated linkage map in tomato based on isozymes and random cDNA sequences. *Genetics*, *112*, 887–898.

Bernier, J., Atlin, G. N., Serraj, R., Kumar, A., & Spaner, D. (2008). Breeding upland rice for drought resistance. *Journal of the Science of Food and Agriculture*, *88*, 927–939.

Bolibok-Bragoszewska, H., Heller-Uszynska, K., Wenzl, P., Uszynski, G., Kilian, A., & Rakoczy-Trojanowska, M. (2009). DArT markers for the rye genome – genetic diversity and mapping. *BMC Genomics*, *10*, 578.

Borevitz, J. O., & Chory, J. (2004). Genomics tools for QTL analysis and gene discovery. *Current Opinion in Plant Biology*, *7*, 132–136.

Botstein, D., White, R. L., Skolnick, M., & Davis, R. W. (1980). Construction of a genetic-linkage map in man using restriction fragment length polymorphisms. *American Journal of Human Genetics*, *32*, 314–331.

Bovill, W., Horne, M., Herde, D., Davis, M., Wildermuth, G., & Sutherland, M. (2010). Pyramiding QTL increases seedling resistance to crown rot (*Fusarium pseudograminearum*) of wheat (*Triticum aestivum*). *TAG Theoretical and Applied Genetics*, *121*, 127–136.

Brachi, B., Faure, N., Horton, M., Flahauw, E., Vazquez, A., & Nordborg, M., et al. (2010). Linkage and association mapping of *Arabidopsis thaliana* flowering time in nature. *PLoS Genetics*, *6*, e1000940.

Brandwagt, B. F., Mesbah, L. A., Takken, F. L., Laurent, P. L., Kneppers, T. J., & Hille, J., et al. (2000). A longevity assurance gene homolog of tomato mediates resistance to *Alternaria alternata* f. sp. *lycopersici* toxins and fumonisin B1. *Proceedings of the*

National Academy of Sciences of the United States of America, *97*, 4961–4966.

Buckler, E. S., Holland, J. B., Bradbury, P. J., Acharya, C. B., Brown, P. J., & Browne, C., et al. (2009). The genetic architecture of maize flowering time. *Science*, *325*, 714–718.

Buerstmayr, H., Ban, T., & Anderson, J. A. (2009). QTL mapping and marker-assisted selection for *Fusarium* head blight resistance in wheat: A review. *Plant Breeding*, *128*, 1–26.

Burton, P. R., Cardon, L. R., Craddock, N., Deloukas, P., Duncanson, A., & Kwiatowski, D. P., et al. (2007). Genome-wide association study of 14,000 cases of seven common diseases and 3,000 shaded controls. *Nature*, *447*, 661–678.

Byun, M. -O., Kwon, H. -B., & Park, S. -C. (2007). Recent advances in genetic engineering of potato crops for drought and saline stress tolerance. In M. A. Jenks, P. M. Hasegawa & S. M. Jain (Eds.), *Advances in molecular breeding toward drought and salt solerant crops* (pp. 713–737). Dordrecht: Springer.

Chan, E. K. F., Rowe, H. C., & Kliebenstein, D. J. (2009). Understanding the evolution of defense metabolites in *Arabidopsis thaliana* using Genome-Wide Association mapping. *Genetics*, *109*, 108522.

Chin, J., Lu, X., Haefele, S., Gamuyao, R., Ismail, A., & Wissuwa, M., et al. (2010). Development and application of gene-based markers for the major rice QTL phosphorus uptake 1. *TAG Theoretical and Applied Genetics*, *120*, 1073–1086.

Colbert, T., Till, B. J., Tompa, R., Reynolds, S., Steine, M. N., & Yeung, A. T., et al. (2001). High-throughput screening for induced point mutations. *Plant Physiology*, *126*, 480–484.

Collard, B. C. Y., & Mackill, D. J. (2008). Marker-assisted selection: An approach for precision plant breeding in the twenty-first century. *Philosophical Transactions of the Royal Society B-Biological Sciences*, *363*, 557–572.

Collins, F. S. (1992). Positional cloning: Let's not call it reverse anymore. *Nature Genetics*, *1*, 3–6.

Collins, N. C., Tardieu, F., & Tuberosa, R. (2008). Quantitative Trait Loci and crop performance under abiotic stress: Where do we stand? *Plant Physiology*, *147*, 469–486.

Concibido, V. C., Denny, R. L., Lange, D. A., Orf, J. H., & Young, N. D. (1996). RFLP mapping and Marker-Assisted Selection of Soybean cyst nematode resistance in PI 209332. *Crop Science*, *36*, 1643–1650.

Cook, E. H., Jr, & Scherer, S. W. (2008). Copy-number variations associated with neuropsychiatric conditions. *Nature*, *455*, 919–923.

Cooper, J., Till, B., Laport, R., Darlow, M., Kleffner, J., & Jamai, A., et al. (2008). TILLING to detect induced mutations in soybean. *BMC Plant Biology*, *8*, 9.

Cubero, B., Nakagawa, Y., Jiang, X. -Y., Miura, K. -J., Li, F., & Raghothama, K. G., et al. (2009). The phosphate transporter PHT4;6 is a determinant of salt tolerance that is localized to the golgi apparatus of *Arabidopsis*. *Molecular Plant*, *2*, 535–552.

Cullinan, A., & Cantor, C. (2008). Sequenom, Inc. *Pharmacogenomics*, *9*, 1211–1215.

De Rybel, B., Audenaert, D., Vert, G., Rozhon, W., Mayerhofer, J., & Peelman, F., et al. (2009). Chemical inhibition of a subset of *Arabidopsis thaliana* GSK3-like kinases activates brassinosteroid signaling. *Chemistry & Biology*, *16*, 594–604.

Dekkers, J. C. M., & Hospital, F. (2002). The use of molecular genetics in the improvement of agricultural populations. *Nature Reviews Genetics*, *3*, 22–32.

Devos, K. M., & Gale, M. D. (1997). Comparative genetics in the grasses. *Plant Molecular Biology*, *35*, 3–15.

Devos, K. M., & Gale, M. D. (2000). Genome relationships: The grass model in current research. *Plant Cell*, *12*, 637–646.

Doebley, J., Stec, A., & Hubbard, L. (1997). The evolution of apical dominance in maize. *Nature*, *386*, 485–488.

Doerge, R. W. (2002). Mapping and analysis of quantitative trait loci in experimental populations. *Nature Reviews Genetics*, *3*, 43–52.

Druka, A., Druka, I., Centeno, A. G., Li, H. Q., Sun, Z. H., & Thomas, W. T. B., et al. (2008). Towards systems genetic analyses in barley: Integration of phenotypic, expression and genotype data into GeneNetwork. *BMC Genetics*, *9*, 73.

Dudley, J. W. (1993). Molecular markers in plant improvement – Manipulation of genes affecting quantitative traits. *Crop Science*, *33*, 660–668.

Edwards, A., Civitello, A., Hammond, H. A., & Caskey, C. T. (1991). DNA typing and genetic-mapping with trimeric and tetrameric tandem repeats. *American Journal of Human Genetics*, *49*, 746–756.

Edwards, D., & Batley, J. (2010). Plant genome sequencing: Applications for crop improvement. *Plant Biotechnology Journal*, *8*, 2–9.

Ellis, R. P., Forster, B. P., Gordon, D. C., Handley, L. L., Keith, R. P., & Lawrence, P., et al. (2002). Phenotype/genotype associations for yield and salt tolerance in a barley mapping population segregating for two dwarfing genes. *Journal of Experimental Botany*, *53*, 1163–1176.

Ellis, R. P., Forster, B. P., Waugh, R., Bonar, N., Handley, L. L., & Robinson, D., et al. (1997). Mapping physiological traits in barley. *New Phytologist*, *137*, 149–157.

Eshed, Y., & Zamir, D. (1995). An introgression line population of *Lycopersicon pennellii* in the cultivated tomato enables the identification and fine mapping of yield-associated QTL. *Genetics*, *141*, 1147–1162.

Falconer, D. S., & Mackay, T. F. C. (1996). *Introduction to quantitative genetics* (4th ed.). pp. 253–261 Harlow, Essex, UK: Longmans Green.

Fatokun, C. A., Menanciohautea, D. I., Danesh, D., & Young, N. D. (1992). Evidence for orthologous seed weight genes in cowpea and mung bean based on RFLP mapping. *Genetics*, *132*, 841–846.

Finkeldey, R., Leinemann, L., & Gailing, O. (2010). Molecular genetic tools to infer the origin of forest plants and wood. *Applied Microbiology and Biotechnology*, *85*, 1251–1258.

Flint-Garcia, S. A., Buckler, E. S., Tiffin, P., Ersoz, E., & Springer, N. M. (2009). Heterosis is prevalent for multiple traits in diverse maize germplasm. *PLoS ONE*, *4*, e7433.

Flowers, T. J., Koyama, M. L., Flowers, S. A., Sudhakar, C., Singh, K. P., & Yeo, A. R. (2000). QTL: Their place in engineering tolerance of rice to salinity. *Journal of Experimental Botany*, *51*, 99–106.

Foolad, M. R. (2007a). Current status of breeding tomatoes for salt and drought tolerance. In M. A. Jenks, P. M. Hasegawa & S. M. Jain (Eds.), *Advances in molecular breeding toward drought and salt solerant crops* (pp. 669–700). Dordrecht: Springer.

Foolad, M. R. (2007b). Genome mapping and molecular breeding of tomato. *International Journal of Plant Genomics*, *2007*, 64358.

Foolad, M. R., Chen, F. Q., & Lin, G. Y. (1998). RFLP mapping of QTLs conferring salt tolerance during germination in an interspecific cross of tomato. *Theoretical and Applied Genetics*, *97*, 1133–1144.

Foolad, M. R., Zhang, L. P., & Lin, G. Y. (2001). Identification and validation of QTLs for salt tolerance during vegetative growth in tomato by selective genotyping. *Genome*, *44*, 444–454.

Frary, A., Nesbitt, T. C., Frary, A., Grandillo, S., van der Knaap, E., & Cong, B., et al. (2000). fw2.2: A quantitative trait locus key to the evolution of tomato fruit size. *Science*, *289*, 85–88.

Frascaroli, E., Cane, M. A., Landi, P., Pea, G., Gianfranceschi, L., & Villa, M., et al. (2007). Classical genetic and quantitative trait loci analyses of heterosis in a maize hybrid between two elite inbred lines. *Genetics*, *176*, 625–644.

Frisch, M., Bohn, M., & Melchinger, A. E. (1999). Comparison of selection strategies for marker-assisted backcrossing of a gene. *Crop Science*, *39*, 1295–1301.

Gailing, O., Vornam, B., Leinemann, L., & Finkeldey, R. (2009). Genetic and genomic approaches to assess adaptive genetic variation in plants: Forest trees as a model. *Physiologia Plantarum*, *137*, 509–519.

Ganal, M. W., Altmann, T., & Roder, M. S. (2009). SNP identification in crop plants. *Current Opinion in Plant Biology*, *12*, 211–217.

Ge, D., Fellay, J., Thompson, A. J., Simon, J. S., Shianna, K. V., & Urban, T. J., et al. (2009). Genetic variation in IL28B predicts hepatitis C treatment-induced viral clearance. *Nature*, *461*, 399–401.

Gebhardt, C., Walkemeier, B., Henselewski, H., Barakat, A., Delseny, M., & Stuber, K. (2003). Comparative mapping between potato (*Solanum tuberosum*) and *Arabidopsis thaliana* reveals structurally conserved domains and ancient duplications in the potato genome. *Plant Journal*, *34*, 529–541.

Gimelfarb, A., & Lande, R. (1994). Simulation of Marker Assisted Selection for nonadditive traits. *Genetical Research*, *64*, 127–136.

Glazier, A. M., Nadeau, J. H., & Aitman, T. J. (2002). Finding genes that underlie complex traits. *Science*, *298*, 2345–2349.

Graeber, K., Linkies, A., Muller, K., Wunchova, A., Rott, A., & Leubner-Metzger, G. (2010). Cross-species approaches to seed dormancy and germination: Conservation and biodiversity of ABA-regulated mechanisms and the Brassicaceae DOG1 genes. *Plant Molecular Biology*, *73*, 67–87.

Guo, W. Z., Wang, W., Zhou, B. L., & Zhang, T. Z. (2006). Cross-species transferability of G-arboreum-derived EST-SSRs in the diploid species of Gossypium. *Theoretical and Applied Genetics*, *112*, 1573–1581.

Gupta, P., Langridge, P. & Mir, R. (2010). Marker-assisted wheat breeding: Present status and future possibilities. *Molecular Breeding*, 10.1007/s11032-009-9359-7.

Hao, Z., Liu, X., Li, X., Xie, C., Li, M., & Zhang, D., et al. (2009). Identification of quantitative trait loci for drought tolerance at seedling stage by screening a large number of introgression lines in maize. *Plant Breeding*, *128*, 337–341.

Hayashi, K., Tan, X., Zheng, N., Hatate, T., Kimura, Y., & Kepinski, S., et al. (2008). Small-molecule agonists and antagonists of F-box protein-substrate interactions in auxin perception and signaling. *Proceedings of the National Academy of Sciences of the United States of America*, *105*, 5632–5637.

Hillel, J., Plotzky, Y., Gal, O., Haberfeld, A., Lavi, U., Dunnington, E. A., et al. (1989). DNA fingerprints in chickens. In *Proceedings of the thirty-first British poultry breeders' roundtable reading* (pp. 1–11). England.

Hillel, J., Schaap, T., Haberfeld, A., Jeffreys, A. J., Plotzky, Y., & Cahaner, A., et al. (1990). DNA fingerprints applied to gene introgression in breeding programs. *Genetics*, *124*, 783–789.

Hippolyte, I., Bakry, F., Seguin, M., Gardes, L., Rivallan, R., & Risterucci, A. -M., et al. (2010). A saturated SSR/DArT linkage map of *Musa acuminata* addressing genome rearrangements among bananas. *BMC Plant Biology*, *10*, 65.

Hirschhorn, J. N., & Daly, M. J. (2005). Genome-wide association studies for common diseases and complex traits. *Nature Reviews Genetics*, *6*, 95–108.

Hospital, F., & Charcosset, A. (1997). Marker-assisted introgression of quantitative trait loci. *Genetics*, *147*, 1469–1485.

Hyten, D. L., Choi, I. -Y., Song, Q., Specht, J. E., Carter, T. E., Jr., & Shoemaker, R. C., et al. (2010). A high density integrated genetic linkage map of Soybean and the development of a 1536 universal soy linkage panel for Quantitative Trait Locus mapping. *Crop Science*, *50*, 960–968.

Iafrate, A. J., Feuk, L., Rivera, M. N., Listewnik, M. L., Donahoe, P. K., & Qi, Y., et al. (2004). Detection of large-scale variation in the human genome. *Nature Genetics*, *36*, 949–951.

Jaccoud, D., Peng, K., Feinstein, D., & Kilian, A. (2001). Diversity arrays: A solid state technology for sequence information independent genotyping. *Nucleic Acids Research*, *29*, E25.

Jander, G., Norris, S. R., Rounsley, S. D., Bush, D. F., Levin, I. M., & Last, R. L. (2002). *Arabidopsis* map-based cloning in the Post-Genome Era. *Plant Physiology*, *129*, 440–450.

Jannink, J. -L., Lorenz, A. J., & Iwata, H. (2010). Genomic selection in plant breeding: From theory to practice. *Briefings in Functional Genomics*, *9*, 166–177.

Jordan, M. C., Somers, D. J., & Banks, T. W. (2007). Identifying regions of the wheat genome controlling seed development by mapping expression quantitative trait loci. *Plant Biotechnology Journal*, *5*, 442–453.

Joshi, S. P., Ranjekar, P. K., & Gupta, V. S. (1999). Molecular markers in plant genome analysis. *Current Science*, *77*, 230–240.

Jurinke, C., Boom, D., Cantor, C. R., & Köster, H. (2001). Automated genotyping using the DNA MassArray™ technology. In J. B. (2001). Rampal (Ed.), *DNA arrays: Methods and protocols* (Vol. 170, pp. 103–116). New-York: Humana Press.

Keurentjes, J. J. B., Fu, J. Y., Terpstra, I. R., Garcia, J. M., van den Ackerveken, G., & Snoek, L. B., et al. (2007). Regulatory network construction in *Arabidopsis* by using genome-wide gene expression quantitative trait loci. *Proceedings of the National Academy of Sciences of the United States of America*, *104*, 1708–1713.

Kim, H. J., Han, J. H., Yoo, J. H., Cho, H. J., & Kim, B. D. (2008). Development of a sequence characteristic amplified region marker linked to the L-4 locus conferring broad spectrum resistance to tobamoviruses in pepper plants. *Molecules and Cells*, *25*, 205–210.

Kim, N. S., Armstrong, K. C., Fedak, G., Ho, K., & Park, N. I. (2002). A microsatellite sequence from the rice blast fungus (*Magnaporthe grisea*) distinguishes between the centromeres of Hordeum vulgare and H-bulbosum in hybrid plants. *Genome*, *45*, 165–174.

Kirst, M., Basten, C. J., Myburg, A. A., Zeng, Z. B., & Sederoff, R. R. (2005). Genetic architecture of transcript-level variation in differentiating xylem of a eucalyptus hybrid. *Genetics*, *169*, 2295–2303.

Kliebenstein, D. J., Lambrix, V. M., Reichelt, M., Gershenzon, J., & Mitchell-Olds, T. (2001). Gene duplication in the diversification of secondary metabolism: Tandem 2-oxoglutarate-dependent dioxygenases control glucosinolate biosynthesis in *Arabidopsis*. *Plant Cell*, *13*, 681–693.

Kopecky, D., Bartos, J., Lukaszewski, A., Baird, J., Cernoch, V., & Kolliker, R., et al. (2009). Development and mapping of DArT markers within the Festuca–Lolium complex. *BMC Genomics*, *10*, 473.

Korstanje, R., & Paigen, B. (2002). From QTL to gene: The harvest begins. *Nature Genetics*, *31*, 235–236.

Kottapalli, K., Lakshmi Narasu, M., & Jena, K. (2010). Effective strategy for pyramiding three bacterial blight resistance genes into fine grain rice cultivar, Samba Mahsuri, using sequence tagged site markers. *Biotechnology Letters*, *32*, 989–996.

Koyama, M. L., Levesley, A., Koebner, R. M. D., Flowers, T. J., & Yeo, A. R. (2001). Quantitative Trait Loci for component physiological traits determining salt tolerance in rice. *Plant Physiology*, *125*, 406–422.

Krause, J., Fu, Q., Good, J. M., Viola, B., Shunkov, M. V., & Derevianko, A. P., et al. (2010). The complete mitochondrial DNA genome of an unknown hominin from southern Siberia. *Nature*, *464*, 894–897.

Krieger, U., Lippman, Z. B., & Zamir, D. (2010). The flowering gene SINGLE FLOWER TRUSS drives heterosis for yield in tomato. *Nature Genetics*, *42*, 459–463.

Ku, C. S., Loy, E. Y., Pawitan, Y., & Chia, K. S. (2010). The pursuit of genome-wide association studies: Where are we now? *Journal of Human Genetics*, *55*, 195–206.

Kumar, L. S. (1999). DNA markers in plant improvement: An overview. *Biotechnology Advances*, *17*, 143–182.

Kusterer, B., Muminovic, J., Utz, H. F., Piepho, H. P., Barth, S., & Heckenberger, M., et al. (2007). Analysis of a triple testcross design with recombinant inbred lines reveals a significant role of epistasis in heterosis for biomass-related traits in *Arabidopsis*. *Genetics*, *175*, 2009–2017.

Kwon, Y. -S., Tang, K., Cantor, C. R., Koster, H., & Kang, C. (2001). DNA sequencing and genotyping by transcriptional synthesis of chain-terminated RNA ladders and MALDI-TOF mass spectrometry. *Nucleic Acids Research*, *29*, e11.

Lande, R., & Thompson, R. (1990). Efficiency of Marker-Assisted Selection in the improvement of quantitative traits. *Genetics*, *124*, 743.

Lander, E. S., & Botstein, D. (1989). Mapping Mendelian factors underlying quantitative traits using RFLP linkage maps. *Genetics*, *121*, 185–199.

Lee, G. J., Boerma, H. R., Villagarcia, M. R., Zhou, X., Carter, T. E., & Li, Z., et al. (2004). A major QTL conditioning salt tolerance in S-100 soybean and descendent cultivars. *Theoretical and Applied Genetics*, *109*, 1610–1619.

Lee, J. A., Carvalho, C. M. B., & Lupski, J. R. (2007). A DNA replication mechanism for generating nonrecurrent rearrangements associated with genomic disorders. *Cell*, *131*, 1235–1247.

Lee, M. (1995). DNA markers and plant breeding programs. In D. L. Sparks (Ed.), *Advances in agronomy* (Vol. 55, pp. 265–344). San Diego: Academic Press Inc.

Lee, S. Y., Ahn, J. H., Cha, Y. S., Yun, D. W., Lee, M. C., & Ko, J. C., et al. (2006). Mapping of quantitative trait loci for salt tolerance at the seedling stage in rice. *Molecules and Cells*, *21*, 192–196.

Leidi, E. O., Barragan, V., Rubio, L., El-Hamdaoui, A., Ruiz, M. T., & Cubero, B., et al. (2010). The AtNHX1 exchanger mediates potassium compartmentation in vacuoles of transgenic tomato. *Plant Journal*, *61*, 495–506.

Levin, I., Gilboa, N., Cincarevsky, F., Oguz, I., Petreikov, M., & Yeselson, Y., et al. (2006). Epistatic interaction between the Fgr and FK2 genes determines the fructose to glucose ratio in mature tomato fruit. *Israel Journal of Plant Sciences*, *54*, 215–222.

Li, C., Zhang, G., & Lance, R. (2007). Recent advances in breeding barley for drought and saline stress tolerance. In M. A. Jenks, P. M. Hasegawa & S. M. Jain (Eds.), *Advances in molecular breeding toward drought and salt solerant crops* (pp. 603–626). Dordrecht: Springer.

Li, Z. K., Fu, B. Y., Gao, Y. M., Xu, J. L., Ali, J., & Lafitte, H. R., et al. (2005).). Genome-wide introgression lines and their use in genetic and molecular dissection of complex phenotypes in rice (*Oryza sativa* L.). *Plant Molecular Biology*, *59*, 33–52.

Li, Z. K., Luo, L. J., Mei, H. W., Wang, D. L., Shu, Q. Y., & Tabien, R., et al. (2001). Overdominant epistatic loci are the primary genetic basis of inbreeding depression and heterosis in rice. I. Biomass and grain yield. *Genetics*, *158*, 1737–1753.

Lin, H. X., Zhu, M. Z., Yano, M., Gao, J. P., Liang, Z. W., & Su, W. A., et al. (2004). QTLs for Na^+ and K^+ uptake of the shoots and roots controlling rice salt tolerance. *Theoretical and Applied Genetics*, *108*, 253–260.

Lippman, Z. B., & Zamir, D. (2007). Heterosis: Revisiting the magic. *Trends in Genetics*, *23*, 60–66.

Litt, M., & Luty, J. A. (1989). A hypervariable microsatellite revealed by *in vitro* amplification of a dinucleotide repeat within the cardiac-muscle actin gene. *American Journal of Human Genetics*, *44*, 397–401.

Liu, J., Cong, B., & Tanksley, S. D. (2003). Generation and analysis of an artificial gene dosage series in tomato to study the mechanisms by which the cloned Quantitative Trait Locus fw2.2 controls fruit size. *Plant Physiology*, *132*, 292–299.

Lubbers, E. L., Chee, P. W., Saranga, Y., & Paterson, A. H. (2007). Recent advances and future prospective in molecular breeding of cotton for drought and salinity stress tolerance. In M. A. Jenks, P. M. Hasegawa & S. M. Jain (Eds.), *Advances in molecular breeding toward drought and salt solerant crops* (pp. 775–796). Dordrecht: Springer.

Lukowitz, W., Gillmor, C. S., & Scheible, W. R. (2000). Positional cloning in *Arabidopsis*. Why it feels good to have a genome initiative working for you. *Plant Physiology*, *123*, 795–805.

Ma, L. Q., Zhou, E. F., Huo, N. X., Zhou, R. H., Wang, G. Y., & Jia, J. Z. (2007). Genetic analysis of salt tolerance in a recombinant inbred population of wheat (*Triticum aestivum* L.). *Euphytica*, *153*, 109–117.

Maccaferri, M., Sanguineti, M. C., Corneti, S., Ortega, J. L. A., Salem, M. B., & Bort, J., et al. (2008). Quantitative Trait Loci for grain yield and adaptation of Durum wheat (*Triticum durum* Desf.) across a wide range of water availability. *Genetics*, *178*, 489–511.

Mackay, I., & Powell, W. (2007). Methods for linkage disequilibrium mapping in crops. *Trends in Plant Science*, *12*, 57–63.

Mackay, T. F. C., Stone, E. A., & Ayroles, J. F. (2009). The genetics of quantitative traits: Challenges and prospects. *Nature Reviews Genetics*, *10*, 565–577.

Maher, B. (2008). Personal genomes: The case of the missing heritability. *Nature*, *456*, 18–21.

Mano, Y., & Takeda, K. (1997). Mapping quantitative trait loci for salt tolerance at germination and the seedling stage in barley (*Hordeum vulgare* L). *Euphytica*, *94*, 263–272.

Margulies, M., Egholm, M., Altman, W. E., Attiya, S., Bader, J. S., & Bemben, L. A., et al. (2005). Genome sequencing in microfabricated high-density picolitre reactors. *Nature*, *437*, 376–380.

Martin, G. B., Brommonschenkel, S. H., Chunwongse, J., Frary, A., Ganal, M. W., & Spivey, R., et al. (1993). Map-based cloning of a protein kinase gene conferring disease resistance in tomato. *Science*, *262*, 1432–1436.

Martin, G. B., Frary, A., Wu, T., Brommonschenkel, S., Chunwongse, J., & Earle, E. D., et al. (1994). A member of the tomato Pto gene family confers sensitivity to fenthion resulting in rapid cell death. *Plant Cell*, *6*, 1543–1552.

Masojc, P. (2002). The application of molecular markers in the process of selection. *Cellular & Molecular Biology Letters*, *7*, 499–509.

McCaffrey, A. P., Meuse, L., Pham, T. T. T., Conklin, D. S., Hannon, G. J., & Kay, M. A. (2002). Gene expression – RNA interference in adult mice. *Nature*, *418*, 38–39.

McCallum, C. M., Comai, L., Greene, E. A., & Henikoff, S. (2000a). Targeted screening for induced mutations. *Nature Biotechnology*, *18*, 455–457.

McCallum, C. M., Comai, L., Greene, E. A., & Henikoff, S. (2000b). Targeting Induced Local Lesions IN Genomes (TILLING) for plant functional genomics. *Plant Physiology*, *123*, 439–442.

Meksem, K., Leister, D., Peleman, J., Zabeau, M., Salamini, F., & Gebhardt, C. (1995). A high-resolution map of the vicinity of the R1 locus on chromosome-V of potato based on RFLP and AFLP markers. *Molecular & General Genetics*, *249*, 74–81.

Melchinger, A. E., Piepho, H. P., Utz, H. F., Muminovic, J., Wegenast, T., & Torjek, O., et al. (2007). Genetic basis of heterosis for growth-related traits in *Arabidopsis* investigated by testcross progenies of near-isogenic lines reveals a significant role of epistasis. *Genetics*, *177*, 1827–1837.

Meudt, H. M., & Clarke, A. C. (2007). Almost forgotten or latest practice? AFLP applications, analyses and advances. *Trends in Plant Science*, *12*, 106–117.

Michelmore, R. W., Paran, I., & Kesseli, R. V. (1991). Identification of markers linked to disease-resistance genes by bulked segregant analysis – a rapid method to detect markers in specific genomic regions by using segregating populations. *Proceedings of the National Academy of Sciences of the United States of America*, *88*, 9828–9832.

Milligan, B. G., Leebensmack, J., & Strand, A. E. (1994). Conservation genetics – beyond the maintenance of marker diversity. *Molecular Ecology*, *3*, 423–435.

Mitchell-Olds, T. (2010). Complex-trait analysis in plants. *Genome Biology*, *11*, 113.

Moore, G., Devos, K. M., Wang, Z., & Gale, M. D. (1995). Cereal genome evolution – grasses, line up and form a circle. *Current Biology*, *5*, 737–739.

Moore, G., Foote, T., Helentjaris, T., Devos, K., Kurata, N., & Gale, M. (1995). Was there a single ancestral cereal chromosome? *Trends in Genetics*, *11*, 81–82.

Moury, B., Pflieger, S., Blattes, A., Lefebvre, V., & Palloix, A. (2000). A CAPS marker to assist selection of tomato spotted wilt virus (TSWV) resistance in pepper. *Genome*, *43*, 137–142.

Mueller, U. G., & Wolfenbarger, L. L. (1999). AFLP genotyping and fingerprinting. *Trends in Ecology & Evolution*, *14*, 389–394.

Muth, J., Hartje, S., Twyman, R. M., Hofferbert, H. R., Tacke, E., & Prufer, D. (2008). Precision breeding for novel starch variants in potato. *Plant Biotechnology Journal*, *6*, 576–584.

Myles, S., Peiffer, J., Brown, P. J., Ersoz, E. S., Zhang, Z., & Costich, D. E., et al. (2009). Association mapping: Critical considerations shift from genotyping to experimental design. *Plant Cell*, *21*, 2194–2202.

Nakayashiki, H., Kadotani, N., & Mayama, S. (2006). Evolution and diversification of RNA silencing proteins in fungi. *Journal of Molecular Evolution*, *63*, 127–135.

Napoli, C., Lemieux, C., & Jorgensen, R. (1990). Introduction of a chimeric chalcone synthase gene into Petunia results in reversible co-suppression of homologous genes in trans. *Plant Cell*, *2*, 279–289.

Neil, J., Helen, O., Howard, T., & Izolda, P. A. (2009). Markers and mapping revisited: Finding your gene. *New Phytologist*, *183*, 935–966.

Neimann-Sorensen, A., & Robertson, A. (1961). The association between blood groups and several production characteristics in three danish cattle breeds. *Acta Agriculturae Scandinavica*, *11*, 163–196.

Nevo, E., & Chen, G. X. (2010). Drought and salt tolerances in wild relatives for wheat and barley improvement. *Plant Cell and Environment*, *33*, 670–685.

Nordborg, M., & Weigel, D. (2008). Next-generation genetics in plants. *Nature*, *456*, 720–723.

Okada, M., Lanzatella, C., Saha, M. C., Bouton, J., Wu, R., & Tobias, C. M. (2010). Complete switchgrass genetic maps reveal subgenome collinearity, preferential pairing, and multilocus interactions. *Genetics*, *genetics*, *110*, 113910.

Paran, I., Kesseli, R., & Michelmore, R. (1991). Identification of Restriction-Fragment-Length-Polymorphism and Random Amplified Polymorphic DNA markers linked to downy mildew resistance genes in lettuce, using near-isogenic lines. *Genome*, *34*, 1021–1027.

Park, S. Y., Fung, P., Nishimura, N., Jensen, D. R., Fujii, H., & Zhao, Y., et al. (2009). Abscisic acid inhibits type 2C protein phosphatases via the PYR/PYL family of START proteins. *Science*, *324*, 1068–1071.

Parry, D. W., Jenkinson, P., & McLeod, L. (1995). Fusarium ear blight (Scab) in small-grain cereals – A review. *Plant Pathology*, *44*, 207–238.

Paterson, A. H., Damon, S., Hewitt, J. D., Zamir, D., Rabinowitch, H. D., & Lincoln, S. E., et al. (1991). Mendelian factors underlying quantitative traits in tomato – comparison across species, generations, and environments. *Genetics*, *127*, 181–197.

Paterson, A. H., Lander, E. S., Hewitt, J. D., Peterson, S., Lincoln, S. E., & Tanksley, S. D. (1988). Resolution of quantitative traits into Mendelian factors by using a complete linkage map of restriction fragment length polymorphisms. *Nature*, *335*, 721–726.

Pedersen, W. L., & Leath, S. (1988). Pyramiding major genes for resistance to maintain residual effects. *Annual Review of Phytopathology*, *26*, 369–378.

Pestsova, E. G., Borner, A., & Roder, M. S. (2006). Development and QTL assessment of *Triticum aestivum–Aegilops tauschii* introgression lines. *Theoretical and Applied Genetics*, *112*, 634–647.

Peterhansel, C., Freialdenhoven, A., Kurth, J., Kolsch, R., & Schulze-Lefert, P. (1997). Interaction analyses of genes required for resistance responses to powdery mildew in barley reveal distinct pathways leading to leaf cell death. *Plant Cell*, *9*, 1397–1409.

Peters, J. L., Cnudde, F., & Gerats, T. (2003). Forward genetics and map-based cloning approaches. *Trends in Plant Science*, *8*, 484–491.

Petreikov, M., Shen, S., Yeselson, Y., Levin, I., Bar, M., & Schaffer, A. (2006). Temporally extended gene expression of the ADP-Glc pyrophosphorylase large subunit (AgpL1) leads to increased enzyme activity in developing tomato fruit. *Planta*, *224*, 1465–1479.

Petreikov, M., Yeselson, L., Shen, S., Levin, I., Schaffer, A. A., & Efrati, A., et al. (2009). Carbohydrate balance and accumulation during development of near-isogenic tomato lines differing in the AGPase-L1 allele. *Journal of the American Society for Horticultural Science*, *134*, 134–140.

Plotsky, Y., Cahaner, A., Haberfeld, A., Lavi, U. & Hillel, J. (1990). Analysis of genetic association between DNA fingerprint bands and quantitative traits by DNA mixes. In *Proceedings of the fourth world congress on genetics applied to livestock production* (pp. 133–136), Edinburgh, Scotland.

Potokina, E., Druka, A., Luo, Z. W., Wise, R., Waugh, R., & Kearsey, M. (2008). Gene expression quantitative trait locus analysis of 16,000 barley genes reveals a complex pattern of genome-wide transcriptional regulation. *Plant Journal*, *53*, 90–101.

Prasad, S. R., Bagali, P. G., Hittalmani, S., & Shashidhar, H. E. (2000). Molecular mapping of quantitative trait loci associated with seedling tolerance to salt stress in rice (*Oryza sativa* L.). *Current Science*, *78*, 162–164.

Provan, J., Soranzo, N., Wilson, N. J., McNicol, J. W., Forrest, G. I., & Cottrell, J., et al. (1998). Gene-pool variation in Caledonian and European Scots pine (*Pinus sylvestris* L.) revealed by chloroplast simple-sequence repeats. *Proceedings of the Royal Society of London Series B-Biological Sciences*, *265*, 1697–1705.

Pumphrey, M. O., Bernardo, R., & Anderson, J. A. (2007). Validating the Fhb1 QTL for Fusarium head blight resistance in near-isogenic wheat lines developed from breeding populations. *Crop Science*, *47*, 200–206.

Quarrie, S. A., Gulli, M., Calestani, C., Steed, A., & Marmiroli, N. (1994). Location of a gene regulating drought-induced abscisic-acid production on the long arm of chromosome 5a of wheat. *Theoretical and Applied Genetics*, *89*, 794–800.

Quarrie, S. A., Steed, A., Calestani, C., Semikhodskii, A., Lebreton, C., & Chinoy, C., et al. (2005). A high-density genetic map of hexaploid wheat (*Triticum aestivum* L.) from the cross Chinese Spring X SQ1 and its use to compare QTLs for grain yield across a range of environments. *Theoretical and Applied Genetics*, *110*, 865–880.

Quesada, V., Garcia-Martinez, S., Piqueras, P., Ponce, M. R., & Micol, J. L. (2002). Genetic architecture of NaCl tolerance in *Arabidopsis*. *Plant Physiology*, *130*, 951–963.

Radoev, M., Becker, H. C., & Ecke, W. (2008). Genetic analysis of heterosis for yield and yield components in rapeseed (*Brassica napus* L.) by quantitative trait locus mapping. *Genetics*, *179*, 1547–1558.

Rafalski, A. (2002). Applications of single nucleotide polymorphisms in crop genetics. *Current Opinion in Plant Biology*, *5*, 94–100.

Rafalski, J. A. (2010). Association genetics in crop improvement. *Current Opinion in Plant Biology*, *13*, 174–180.

Raina, K., & Crews, C. M. (2010). Chemical inducers of targeted protein degradation. *Journal of Biological Chemistry*, *285*, 11057–11060.

Reiter, R. S., Williams, J. G. K., Feldmann, K. A., Rafalski, J. A., Tingey, S. V., & Scolnik, P. A. (1992). Global and local genome mapping in *Arabidopsis-thaliana* by using recombinant inbred lines and Random Amplified Polymorphic DNAs. *Proceedings of the National Academy of Sciences of the United States of America*, *89*, 1477–1481.

Ren, Z. H., Gao, J. P., Li, L. G., Cai, X. L., Huang, W., & Chao, D. Y., et al. (2005). A rice quantitative trait locus for salt tolerance encodes a sodium transporter. *Nature Genetics*, *37*, 1141–1146.

Rosenberg, N. A., Huang, L., Jewett, E. M., Szpiech, Z. A., Jankovic, I., & Boehnke, M. (2010). Genome-wide association studies in diverse populations. *Nature Reviews Genetics*, *11*, 356–366.

Rothberg, J. M., & Leamon, J. H. (2008). The development and impact of 454 sequencing. *Nature Biotechnology*, *26*, 1117–1124.

Ruiz-Ferrer, V., & Voinnet, O. (2009). Roles of plant small RNAs in biotic stress responses. *Annual Review of Plant Biology*, *60*, 485–510.

Saidou, A. -A., Mariac, C., Luong, V., Pham, J. -L., Bezancon, G., & Vigouroux, Y. (2009). Association studies identify natural variation at PHYC linked to flowering time and morphological variation in pearl millet. *Genetics*, *182*, 899–910.

Salvi, S., & Tuberosa, R. (2005). To clone or not to clone plant QTLs: Present and future challenges. *Trends in Plant Science*, *10*, 297–304.

Sanchez, A. C., Subudhi, P. K., Rosenow, D. T., & Nguyen, H. T. (2002). Mapping QTLs associated with drought resistance in sorghum (*Sorghum bicolor* L. Moench). *Plant Molecular Biology*, *48*, 713–726.

Sanseverino, W., Roma, G., De Simone, M., Faino, L., Melito, S., & Stupka, E., et al. (2010). PRGdb: A bioinformatics platform for plant resistance gene analysis. *Nucleic Acids Research*, *38*, D814–21.

Saranga, Y., Menz, M., Jiang, C. X., Wright, R. J., Yakir, D., & Paterson, A. H. (2001). Genomic dissection of genotype x environment interactions conferring adaptation of cotton to arid conditions. *Genome Research*, *11*, 1988–1995.

Sari-Gorla, M., Krajewski, P., Di Fonzo, N., Villa, M., & Frova, C. (1999). Genetic analysis of drought tolerance in maize by molecular markers. II. Plant height and flowering. *Theoretical and Applied Genetics*, *99*, 289–295.

Sax, K. (1923). The association of size differences with seed-coat pattern and pigmentation in *Phaseolus vulgaris*. *Genetics*, *8*, 552–560.

Scascitelli, M., Whitney, K. D., Randell, R. A., King, M., Buerkle, C. A., & Rieseberg, L. H. (2010). Genome scan of hybridizing sunflowers from Texas (*Helianthus annuus* and *H. debilis*) reveals asymmetric patterns of introgression and small islands of genomic differentiation. *Molecular Ecology*, *19*, 521–541.

Sebat, J., Lakshmi, B., Troge, J., Alexander, J., Young, J., & Lundin, P., et al. (2004). Large-scale copy number polymorphism in the human genome. *Science*, *305*, 525–528.

Semel, Y., Nissenbaum, J., Menda, N., Zinder, M., Krieger, U., & Issman, N., et al. (2006). Overdominant quantitative trait loci for yield and fitness in tomato. *Proceedings of the National Academy of Sciences*, *103*, 12981–12986.

Septiningsih, E. M., Pamplona, A. M., Sanchez, D. L., Neeraja, C. N., Vergara, G. V., & Heuer, S., et al. (2009). Development of submergence-tolerant rice cultivars: The Sub1 locus and beyond. *Annals of Botany*, *103*, 151–160.

Servin, B., Martin, O. C., Mezard, M., & Hospital, F. (2004). Toward a theory of marker-assisted gene pyramiding. *Genetics*, *168*, 513–523.

Sherry, A. F. -G., Anne-Céline, T., Jianming, Y., Gael, P., Susan, M. R., & Sharon, E. M., et al. (2005). Maize association population: A high-resolution platform for quantitative trait locus dissection. *The Plant Journal*, *44*, 1054–1064.

Shi, C., Uzarowska, A., Ouzunova, M., Landbeck, M., Wenzel, G., & Lubberstedt, T. (2007). Identification of candidate genes associated with cell wall digestibility and eQTL (expression quantitative trait loci) analysis in a Flint x Flint maize recombinant inbred line population. *BMC Genomics*, *8*, 22.

Smith, C., & Simpson, S. P. (1986). The use of genetic polymorphisms in livestock improvement. *Journal of Animal Breeding and Genetics*, *103*, 205–217.

Smith, C. (1967). Improvement of metric traits through specific genetic loci. *Animal Production*, *9*, 349–358.

Smith, C. J. S., Watson, C. F., Bird, C. R., Ray, J., Schuch, W., & Grierson, D.

(1990). Expression of a truncated tomato polygalacturonase gene inhibits expression of the endogenous gene in transgenic plants. *Molecular & General Genetics, 224*, 477–481.

Soller, M., & Beckmann, J. S. (1983). Genetic-polymorphism in varietal identification and genetic-improvement. *Theoretical and Applied Genetics, 67*, 25–33.

Soller, M. (1978). The use of loci associated with quantitative effects in dairy cattle improvement. *Animal Production, 27*, 133–139.

Srinivasachary, , Dida, M. M., Gale, M. D., & Devos, K. M. (2007). Comparative analyses reveal high levels of conserved colinearity between the finger millet and rice genomes. *Theoretical and Applied Genetics, 115*, 489–499.

Stracke, S., Haseneyer, G., Veyrieras, J. -B., Geiger, H., Sauer, S., & Graner, A., et al. (2009). Association mapping reveals gene action and interactions in the determination of flowering time in barley. *TAG Theoretical and Applied Genetics, 118*, 259–273.

Street, N. R., Skogstrom, O., Sjodin, A., Tucker, J., Rodriguez-Acosta, M., & Nilsson, P., et al. (2006). The genetics and genomics of the drought response in Populus. *Plant Journal, 48*, 321–341.

Stuber, C. W., Goodman, M. M., & Moll, R. H. (1982). Improvement of yield and ear number resulting from selection at allozyme loci in a maize population. *Crop Science, 22*, 737–740.

Tabor, H. K., Risch, N. J., & Myers, R. M. (2002). Candidate-gene approaches for studying complex genetic traits: Practical considerations. *Nature Reviews Genetics, 3*, 391–397.

Tanksley, D., Medina-Filho, H., & Rick, C. M. (1981). The effect of isozyme selection on metric characters in an interspecific backcross of tomato — basis of an early screening procedure. *TAG Theoretical and Applied Genetics, 60*, 291.

Tanksley, S. D., Ganal, M. W., & Martin, G. B. (1995). Chromosome landing – a paradigm for map-based gene cloning in plants with large genomes. *Trends in Genetics, 11*, 63–68.

Tanksley, S. D., Ganal, M. W., Prince, J. P., Devicente, M. C., Bonierbale, M. W., & Broun, P., et al. (1992). High-density molecular linkage maps of the tomato and potato genomes. *Genetics, 132*, 1141–1160.

Tester, M., & Langridge, P. (2010). Breeding technologies to increase crop production in a changing world. *Science, 327*, 818–822.

Teulat, B., Monneveux, P., Wery, J., Borries, C., Souyris, I., & Charrier, A., et al. (1997). Relationships between relative water content and growth parameters under water stress in barley: A QTL study. *New Phytologist, 137*, 99–107.

Thomas, D. L., Thio, C. L., Martin, M. P., Qi, Y., Ge, D., & O'hUigin, C., et al. (2009). Genetic variation in IL28B and spontaneous clearance of hepatitis C virus. *Nature, 461*, 798–801.

Thomson, M. J., Ismail, A. M., McCouch, S. R., & Mackill, D. J. (2010). Marker Assisted Breeding. In A. Pareek, S. K. Sopory, H. J. Bohnert Govindjee (Eds.), *Abiotic stress adaptation in plants: Physiological, molecular and genomic foundation* (pp. 451–469). Dordrecht: Springer.

Till, B. J., Reynolds, S. H., Weil, C., Springer, N., Burtner, C., & Young, K., et al. (2004). Discovery of induced point mutations in maize genes by TILLING. *BMC Plant Biology, 4*, 12.

Till, B., Cooper, J., Tai, T., Colowit, P., Greene, E., & Henikoff, S., et al. (2007). Discovery of chemically induced mutations in rice by TILLING. *BMC Plant Biology, 7*, 19.

Tóth, R., & van der Hoorn, R. A. L. (2010). Emerging principles in plant chemical genetics. *Trends in Plant Science, 15*, 81–88.

Tozlu, I., Guy, C. L., & Moore, G. A. (1999). QTL analysis of morphological traits in an intergeneric BC1 progeny of Citrus and Poncirus under saline and non-saline environments. *Genome, 42*, 1020–1029.

Uauy, C., Paraiso, F., Colasuonno, P., Tran, R. K., Tsai, H., & Berardi, S., et al. (2009). A modified TILLING approach to detect induced mutations in tetraploid and hexaploid wheat. *BMC Plant Biology, 9*, 115.

Udagawa, H., Ishimaru, Y., Li, F., Sato, Y., Kitashiba, H. & Nishio, T. (2010). Genetic analysis of interspecific incompatibility in *Brassica rapa*. *Theoretical and Applied Genetics*, 121(4), 689–696.

Van Der Krol, A. R., Mur, L. A., Beld, M., Mol, J. N. M., & Stuitje, A. R. (1990). Flavonoid genes in Petunia: Addition of a limited number of gene copies may lead to a suppression of gene expression. *Plant Cell, 2*, 291–299.

Varshney, R. K., & Dubey, A. (2009). Novel genomic tools and modern genetic and breeding approaches for crop improvement. *Journal of Plant Biochemistry & Biotechnology, 18*, 127–138.

Varshney, R. K. (2010). Gene-based marker systems in plants: High throughput approaches for marker discovery and genotyping. In S. M. Jain & D. S. Brar (Eds.), *Molecular techniques in crop improvement* (pp. 119–142). Dordrecht: Springer.

Varshney, R. K., Hoisington, D. A., & Tyagi, A. K. (2006). Advances in cereal genomics and applications in crop breeding. *Trends in Biotechnology, 24*, 490–499.

Varshney, R. K., Hoisington, D. A., Nayak, S. N., & Graner, A. (2009). Molecular plant breeding: Methodology and achievements. In D. J. Somers, P. Langridge & J. P. Gustafson (Eds.), *Plant Genomics* (pp. 283–304). New-York: Humana Press.

Varshney, R. K., Nayak, S. N., May, G. D., & Jackson, S. A. (2009). Next-generation sequencing technologies and their implications for crop genetics and breeding. *Trends in Biotechnology, 27*, 522–530.

Vaucheret, H. (2006). Post-transcriptional small RNA pathways in plants: Mechanisms and regulations. *Genes & Development, 20*, 759–771.

Villalta, I., Bernet, G. P., Carbonell, E. A., & Asins, M. J. (2007). Comparative QTL analysis of salinity tolerance in terms of fruit yield using two solanum populations of F-7 lines. *Theoretical and Applied Genetics, 114*, 1001–1017.

Villalta, I., Reina-Sanchez, A., Bolarin, M. C., Cuartero, J., Belver, A., & Venema, K., et al. (2008). Genetic analysis of Na^+ and K^+ concentrations in leaf and stem as physiological components of salt tolerance in Tomato. *Theoretical and Applied Genetics, 116*, 869–880.

Visscher, P. M., Haley, C. S., & Thompson, R. (1996). Marker-assisted introgression in backcross breeding programs. *Genetics, 144*, 1923–1932.

Vos, P., Hogers, R., Bleeker, M., Reijans, M., Vandelee, T., & Hornes, M., et al. (1995). AFLP – A new technique for DNA-fingerprinting. *Nucleic Acids Research, 23*, 4407–4414.

Walter, S., Nicholson, P., & Doohan, F. M. (2010). Action and reaction of host and pathogen during Fusarium head blight disease. *New Phytologist, 185*, 54–66.

Wang, D. -K., Sun, Z. -X., & Tao, Y. -Z. (2006). Application of TILLING in plant improvement. *Acta Genetica Sinica, 33*, 957–964.

Wang, J. P., Drayton, M. C., George, J., Cogan, N. O. I., Baillie, R. C., & Hand, M. L., et al. (2010). Identification of genetic factors influencing salt stress tolerance in white clover (*Trifolium repens* L.) by QTL analysis. *Theoretical and Applied Genetics, 120*, 607–619.

Wang, L. H., Gu, X. H., Hua, M. Y., Mao, S. L., Zhang, Z. H., & Peng, D. L., et al. (2009). A SCAR marker linked to the N gene for resistance to root knot nematodes (*Meloidogyne* spp.) in pepper (*Capsicum annuum* L.). *Scientia Horticulturae, 122*, 318–322.

Weber, A. L., Briggs, W. H., Rucker, J., Baltazar, B. M., Sanchez-Gonzalez, J. D., & Feng, P., et al. (2008). The genetic architecture of complex traits in Teosinte (*Zea mays* ssp *parviglumis*): New evidence from association mapping. *Genetics, 180*, 1221–1232.

Welsh, J., & McClelland, M. (1990). Fingerprinting genomes using PCR with arbitrary primers. *Nucleic Acids Research, 18*, 7213–7218.

Wenzl, P., Carling, J., Kudrna, D., Jaccoud, D., Huttner, E., & Kleinhofs, A., et al. (2004). Diversity Arrays Technology (DArT) for whole-genome profiling of barley. *Proceedings of the National Academy of Sciences of the United States of America, 101*, 9915–9920.

West, M. A. L., Kim, K., Kliebenstein, D. J., van Leeuwen, H., Michelmore, R. W., & Doerge, R. W., et al. (2007). Global eQTL mapping reveals the complex genetic architecture of transcript-level variation in *Arabidopsis*. *Genetics, 175*, 1441–1450.

Whittaker, J. C., Haley, C. S., & Thompson, R. (1997). Optimal weighting of information in marker-assisted selection. *Genetical Research, 69*, 137–144.

Williams, J. G. K., Kubelik, A. R., Livak, K. J., Rafalski, J. A., & Tingey, S. V. (1990). DNA polymorphisms amplified by arbitrary primers are useful as genetic-markers. *Nucleic Acids Research, 18*, 6531–6535.

Wing, R. A., Zhang, H. B., & Tanksley, S. D. (1994). Map-based cloning in crop plants – tomato as a model system .1. Genetic and physical mapping of *JOINTLESS*. *Molecular & General Genetics*, *242*, 681–688.

Xiao, S., Ellwood, S., Calis, O., Patrick, E., Li, T., & Coleman, M., et al. (2001). Broad-spectrum mildew resistance in *Arabidopsis thaliana* mediated by RPW8. *Science*, *291*, 118–120.

Xu, Y., & Crouch, J. H. (2008). Marker-Assisted Selection in plant breeding: From publications to practice. *Crop Science*, *48*, 391–407.

Xu, Y. (2010). *Molecular plant breeding*. pp. 292–389 Wallingford: CABI publication.

Xu, Y., Skinner, D. J., Wu, H., Palacios-Rojas, N., Araus, J. L., & Yan, J., et al. (2009). Advances in maize genomics and their value for enhancing genetic gains from breeding. *International Journal of Plant Genomics*, *2009*, 957602.

Xue, D. W., Feng, S. G., Zhao, H. Y., Jiang, H., Shen, B., & Shi, N. N., et al. (2010). The linkage maps of Dendrobium species based on RAPD and SRAP markers. *Journal of Genetics and Genomics*, *37*, 197–204.

Xue, Z. -Y., Zhi, D. -Y., Xue, G. -P., Zhang, H., Zhao, Y. -X., & Xia, G. -M. (2004). Enhanced salt tolerance of transgenic wheat (*Tritivum aestivum* L.) expressing a vacuolar Na^+/H^+ antiporter gene with improved grain yields in saline soils in the field and a reduced level of leaf Na^+. *Plant Science*, *167*, 849–859.

Yang, X., Yan, J., Shah, T., Warburton, M., Li, Q., Li, L., et al. (2010). Genetic analysis and characterization of a new maize association mapping panel for quantitative trait loci dissection. *Theoretical and Applied Genetics*, *121*(3), 417–431.

Yano, M. (2001). Genetic and molecular dissection of naturally occurring variation. *Current Opinion in Plant Biology*, *4*, 130–135.

Yeam, I., Kang, B. -C., Lindeman, W., Frantz, J., Faber, N., & Jahn, M. (2005). Allele-specific CAPS markers based on point mutations in resistance alleles at the pvr1 locus encoding eIF4E in Capsicum. *TAG Theoretical and Applied Genetics*, *112*, 178–186.

Young, N. D. (1999). A cautiously optimistic vision for marker-assisted breeding. *Molecular Breeding*, *5*, 505–510.

Zhang, M., Barg, R., Yin, M., Gueta-Dahan, Y., Leikin-Frenkel, A., & Salts, Y., et al. (2005). Modulated fatty acid desaturation via overexpression of two distinct omega-3 desaturases differentially alters tolerance to various abiotic stresses in transgenic tobacco cells and plants. *The Plant Journal*, *44*, 361–371.

Zhang, W., & Smith, C. (1992). Computer-simulation of Marker-Assisted Selection utilizing linkage disequilibrium. *Theoretical and Applied Genetics*, *83*, 813–820.

Zhao, X. P., & Kochert, G. (1993). Phylogenetic distribution and genetic-mapping of a (GGC) (N) microsatellite from rice (*Oryza-sativa* l). *Plant Molecular Biology*, *21*, 607–614.

Zhu, C., Gore, M., Buckler, E. S., & Yu, J. (2008). Status and prospects of association mapping in plants. *The Plant Genome*, *1*, 5–20.

12

Male sterility and hybrid seed production

Sally Mackenzie

Center for Plant Science Innovation, University of Nebraska, Lincoln, Nebraska

TABLE OF CONTENTS

Introduction

Plant reproduction represents one of the most highly coordinated and complex developmental processes to be elucidated biologically. Over the past several years, the angiosperm flower has been resolved to a multicellular integration of genetically definable whorls (Bowman et al., 1989) programmed metabolically to present volatile and pigment attractants (Dudareva et al., 2004), synchronously modulating cellular behaviors in response to environmental cues, to genetically coordinate two distinct generations within a single functional organ (Ursin et al., 1989). One of the most stunning features of the higher plant reproductive structure is the seemingly near-autonomy of its individual components, permitting interconversion of numerous mating strategies. Dioecy, gynodioecy, self-incompatibility, cleistogamy, insect, and wind or animal pollination capabilities all derive from variations on this set of developmental programs.

Because so much of agricultural production depends on the success of plant reproduction, extensive research has been carried out to understand the developmental process. Identifying the genes that participate in stamen development has provided targets for gene manipulation and tools for precisely controlling gene expression during male gametogenesis. This, in turn, has allowed the genetic engineering of male sterility by a variety of strategies. In this chapter, pollen development and some of the most recent advances in the understanding of male reproductive biology as it pertains to the natural and artificial induction of male sterility for agricultural hybrid seed production are reviewed.

Male Gametogenesis

Pollen mitosis I

In the process of male gamete formation, a pollen mother cell undergoes meiosis to form a tetrad of microspores enclosed in callose (1,3-β glucan; Figure 12.1). Dissolution of the callose occurs by enzymes released from the tapetal cells, dispersing the individual 1N microspores. The microspores increase their size and undergo a programmed pattern of division to produce a three-celled gametophyte that comprises the pollen grain. The first of these divisions is necessarily asymmetric, giving rise to a larger vegetative cell that provides essential products to the pollen grain and gives rise to the pollen tube, and a smaller generative cell contained within the cytoplasm of the vegetative cell. The generative cell divides again to produce two sperm cells. Without this divisional asymmetry,

DOI: 10.1016/B978-0-12-381466-1.00012-2

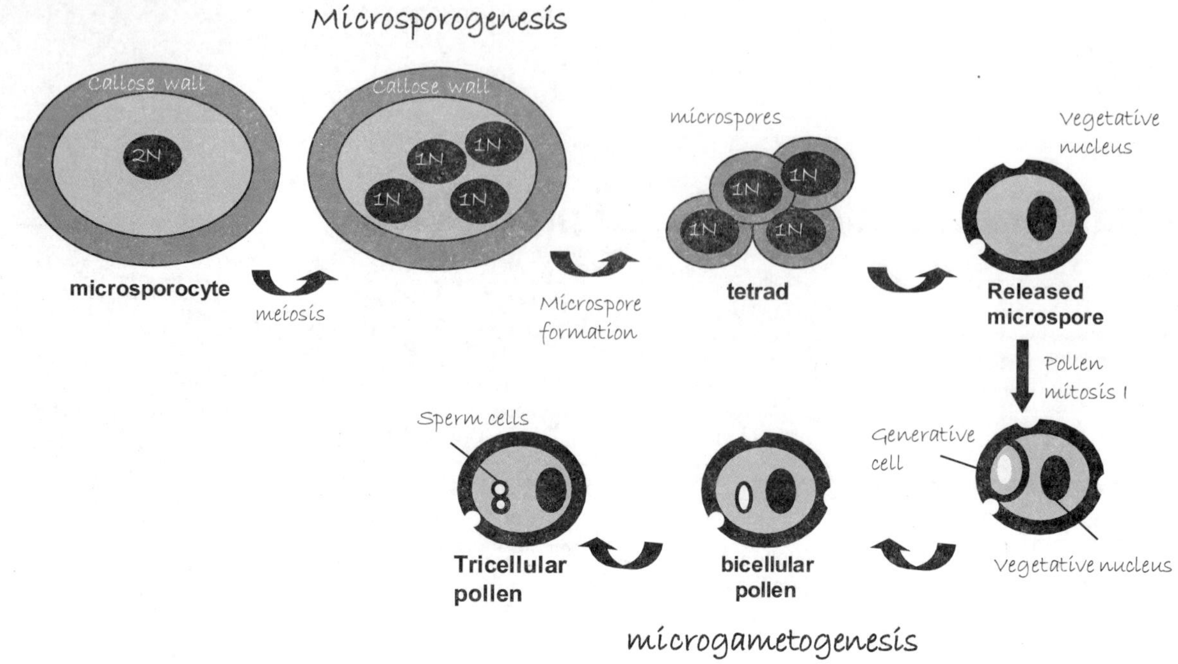

Figure 12.1 • The process of microgametogenesis • Please see color plate section at the end of the book.

both daughter cells would exhibit features of vegetative cells (Eady et al., 1994). The first gametophyte mutant identified, SIDECAR POLLEN (SDP) showed alterations in this division pattern (Chen and McCormick, 1996), undergoing a symmetrical division, followed by asymmetric division of one of the daughter cells. What are shed are pollen grains with an additional vegetative cell attached. Other genes have since been identified to influence this gametophytic cell division pattern, including GEMINI POLLEN 1 (GEM1), a MAP 215 microtubule-associated protein that influences microspore polarity and cytokinesis by acting on the interphase spindle and phragmoplast microtubules (Twell et al., 2002) and the TWO-IN-ONE gene, a member of the FUSED Ser/Thr protein kinase family, which associates with the phragmoplast ring and is essential for proper cytokinesis (Oh et al., 2005). Mutations at these loci and several others that influence cell division have served to underscore the importance of asymmetric division to the proper determination of cell fate in the male gametophyte.

Pollen mitosis II

Following asymmetric division for the first pollen mitosis (PM I), the vegetative cell exits the cell cycle at G1 while the generative cell continues through a second round of mitosis (PM II). This differential control of the cell cycle is essential for the production of identical sperm cells that will carry out double fertilization. In eukaryotes, the protein kinase cdc2 acts as a key cell cycle regulator. In *Arabidopsis*, a homolog of cdc2, A-type cyclin-dependent kinase (CDKA), is present as a single gene. A *cdka;1* mutant produces mature bicellular pollen grains, with a single sperm-like cell and a vegetative cell, due to failure of PM II (Iwakawa et al., 2006). During fertilization, the sperm-like cell fuses with the egg cell to initiate embryogenesis, but the central cell nucleus remains unfertilized, resulting in arrest of embryo development at the globular stage because of the lack of developed endosperm. A second mutant, F-BOX-LIKE 17 (FBL17), produces a strikingly similar phenotype to the *cdka;1* mutant (Kim et al., 2008; Gusti et al., 2009). FBL17 is a substrate-specific component of an E3 ubiquitin protein ligase complex, targeting KRP6 and KRP7 for proteasome-dependent degradation. KRP6 and KRP7 are inhibitors of CDKA. The expression of FBL17 specifically in the pollen germ cell following asymmetric microspore division retains CDKA activity and allows progression through PM II, while KRP6 and KRP7 stability in the vegetative cell maintains the cell in G1 phase. Interestingly, both the *cdka;1* and *fbl17* mutant phenotypes, associated with incomplete endosperm development, can be partially suppressed by a mutation in any of the FIS-class polycomb group chromatin modifying genes that participate in genome imprinting (Nowack et al., 2007); thus releasing embryo development from the requirement of double fertilization. With this second mutation, seeds are able to develop, albeit smaller in size, without differentiation of an endosperm.

More recent integration of genetic mutation data with transcript profiling analysis targeting male gametogenesis has allowed elucidation of intricate genetic networks that direct pollen formation, germination, and fertilization (Borg et al., 2009). These studies have provided a wealth of promoter and regulatory sequence information that allows the targeting of transgene expression specifically to tapetal or gamete cells (McCormick et al., 1991), as well as the opportunity to manipulate naturally occurring events during gametogenesis for precise disruption. The influences of the epigenome and organelle metabolic functions on these pollen processes have not been fully explored.

Male Sterility Mutants Elucidate Anther Development

The floral reproductive structure is organized into four genetically distinguishable whorls (Smyth et al., 1990). The outermost third whorl gives rise to the male reproductive structures of the plant, with the stamen comprised of several tissues types. Stamen development must be coordinated with the pistil to allow for successful reproduction. Genetic variation in this coordination has been associated with enhancement of out-crossing and, ultimately, speciation, while other mutations render the plant male sterile. These mutations can include alterations in filament elongation, pollen maturation, or anther dehiscence (Chaudhury, 1993; Taylor et al., 1998).

Fairly comprehensive genetic screens for pollen mutants have been carried out by several labs, with outcomes that underscore the complexity of microsporogenesis (McCormick, 2004). While many of the mutations arising in these screens involved genes that also function elsewhere in development, a number of male sterility mutants were identified in otherwise phenotypically normal plants. This observation implies not only that pollen development involves numerous distinct processes, but that the opportunity exists to identify male sterility mutants that may be useful agriculturally with relatively few pleiotropic effects.

Several components of the cellular program for anther and pollen development have been elucidated by detailed mutant analysis (Figure 12.2; Ma, 2005; Wijeratne et al., 2007). In the ABC model for floral organ identity, *Arabidopsis* B genes APETALA3 (AP3) and PISTILLATA (PI) and C genes AGAMOUS (AG) and SEPALLATA (SEP) are essential to stamen differentiation. Downstream to these is Sporocyteless (SPL, also known as NOZZLE, NZZ), which is activated by AG (Ito et al., 2004) and participates in the formation of archesporial cells and cell proliferation (Yang et al., 1999; Balasubramanian and Schneitz, 2000). The number of archesporial cells that form and the initiation of surrounding tapetal cell formation are controlled by EXCESS MICROSPOROCYTES 1 (EMS1)/EXS (Canales et al., 2002; Zhao et al., 2002), a leucine-rich repeat receptor protein kinase. Whereas meiotic nuclear division in the *ems1/exs* mutant occurs normally, microsporocytes do not undergo cytokinesis, producing a male sterility phenotype. Acting downstream to EMS1/EXS is the gene DISFUNCTIONAL TAPETUM 1 (DYT1), which is essential for tapetal development. DYT1 encodes a basic helix-loop-helix (bHLH) transcription factor that is highly expressed in the tapetum and, to a lesser extent, in meiocytes, and which appears to be important to controlling expression of tapetum-preferential genes (Zhang et al., 2006).

Another key component to tapetal function, acting downstream to DYT1, is Defective in Tapetal Development and Function 1 (TDF1). TDF1 encodes an R2R3 MYB transcription factor, and expresses in tapetum, meiocytes, and microspore cells (Zhu et al., 2008). This gene is thought to be an important component for controlling callose dissolution, and may participate in exine formation of the pollen wall. Downstream to TDF1 are two additional genes regulating tapetal development, ABORTED MICROSPORES (AMS) and AtMYB103. AMS encodes an MYC transcription factor that interacts with two additional bHLH proteins (AtbHLH089 and AtbHLH091) and a tapetal-specific gene ATA20 (Xu et al., 2010). These express in tapetum and microspores and participate in microspore mitosis (Sorensen et al., 2003). AtMYB103 is an R2R3 MYB gene family member that expresses in tapetum to influence tapetal development and microsporogenesis, and in the trichomes to regulate their pattern of branching (Higginson et al., 2003).

Downstream to AtAMYB103, the PHD transcriptional activator MALE STERILITY 1 (MS1) functions to regulate postmeiotic male gametogenesis by influencing exine development and cytosolic features of developing microspores (Wilson et al., 2001; Ito et al., 2007). Following this process, the MALE STERILITY 2 (MS2) gene appears to act at the point of tetrad release to influence pollen wall development, altering sporopollenin synthesis and exine patterning (Aarts et al., 1997).

In parallel with identification of the stamen development regulatory network, efforts have begun to identify many of the downstream genes that participate coordinately in developmental processes. For example, the AMS transcription factor influences nearly 550 genes that express in the anther (Xu et al., 2010). Identified genes, many shown to contain the 6 bp consensus motif CANNTG, are associated with tapetal functions and pollen wall formation. It was feasible to confirm several of these genes by chromatin immunoprecipitation. They include functions involved, for example, in lipid transport, fatty acid biosynthesis, flavonol accumulation, and pectin dynamics. Functional analysis of at least one by insertional mutation also confirms a role in pollen development and provides information that may prove useful in the engineering of non-pleiotropic male sterility in crops. Understanding the precise role and expression of both regulatory and functional loci in pollen development permits the design of male-sterility-inducing strategies that involve plant-derived gene information rather than non-plant, cytotoxic sources, which are more likely to find disfavor in agricultural deployment.

Hormonal Influences on Male Reproduction in Plants

Gibberellic acid

Floral development and reproduction processes are controlled by numerous hormonal influences. Some of these hormone effects have been identified by mutations. Gibberellic acid (GA) plays an important role in plant reproduction. Mutants

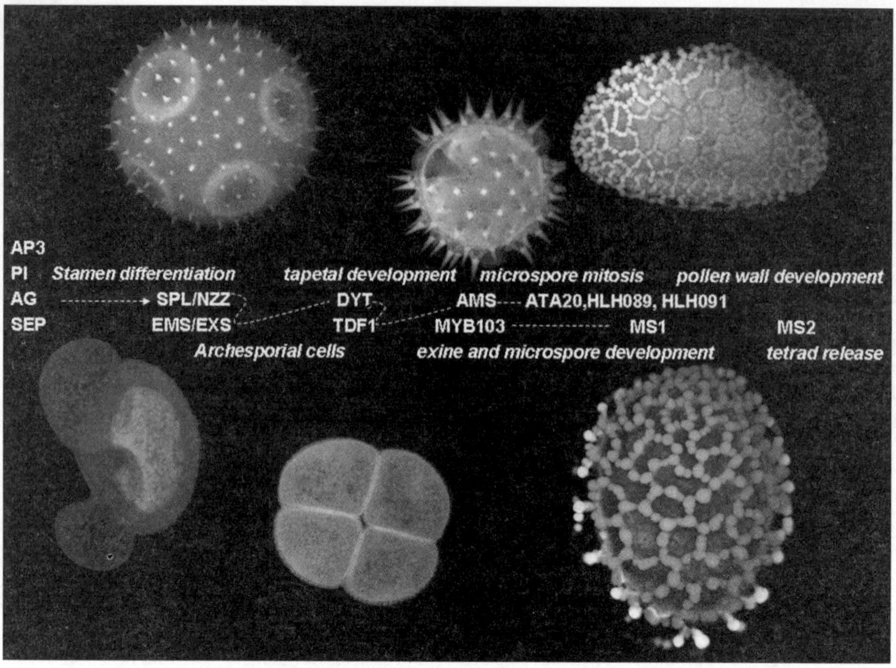

Figure 12.2 • Laser scanning confocal microscopy reveals the extensive diversity and remarkable beauty of plant pollen • Mutant analysis in *Arabidopsis* has allowed the identification of several genetic components regulating the development of these structures, shown here in their predicted functional order relative to the developmental stages at which they act. Mutation at many of these loci results in a male-sterile plant phenotype. Please see color plate section at the end of the book.

such as *gal-3* in *Arabidopsis* have allowed the elucidation of GA influence, displaying the arrest of stamen filament and petal elongation (Cheng et al., 2004; Tyler et al., 2004). GA regulates cell elongation during stamen development, as well as the developmental pathway from microspore to mature pollen grain. DELLA is a transcriptional regulator that operates to negatively control GA response in plants. DELLA proteins RGA, RGA1, and RGL2 together repress petal, stamen, and anther development. Transcript levels of floral homeotic genes APETALA3 (AP3), PISTILLATA (PI), and AGAMOUS (AG) are detectably upregulated in young flowers of *ga1-3* when treated with GA (Yu et al., 2004). Application of GA reverses the effects of DELLA in suppressing floral development to restore pollen fertility.

Gibberellins also impact pollen viability and pollen tube elongation (Singh et al., 2002). In rice, mutations influencing GA biosynthesis are reduced in their transmission to the subsequent generation as a consequence of reduced pollen tube growth. These genes are generally expressed postmeiotically and, therefore, mutants display gametophytic transmission patterns (Chhun et al., 2007).

GA regulates jasmonic acid biosynthesis

Research has begun to clarify the mode of GA regulation. GAMYB is a class of GA-regulated transcription factors (Woodger et al., 2003). Two closely related genes in *Arabidopsis*, MYB33 and MYB65, result in defective tapetal development and premeiotic pollen abortion when dually mutated (Millar and Gubler, 2005). Likewise, GA upregulates the expression of jasmonic acid (JA) biosynthesis genes DAD1 and LOX1 to promote JA production. This activity by GA to regulate JA promotes expression of three factors, MYB21, MYB24, and MYB57, which are necessary for stamen development (Cheng et al., 2009). The JA-deficient mutant *opr3* and JA signaling mutant *coiI* (Xie et al., 1998) also show evidence of incomplete stamen development and male sterility. Exogenous application of JA, but not its precursor, restores fertility to the *opr3* mutant, demonstrating that JA is the active signaling molecule required for male gametophyte development (Stintzi and Browse, 2000). Consequently, the process of normal stamen development appears to involve a complex interplay between GA, DELLA proteins, JA, and several MYB factors.

Brassinosteroids

Brassinosteroids (BR) have been shown to regulate anther and pollen development. For example, the exogenous application of BRs in some crops can induce pollen tube elongation (Hewitt et al., 1985). Likewise, BR mutants such as the BR-deficient mutant *cpd* (Szekeres et al., 1996), the brassinosteroid-insensitive mutant *bin2* (Li et al., 2001), and the BR receptor mutant *bri1-201* (Bouquin et al., 2001) in

Arabidopsis all show evidence of male sterility. Interestingly, the *bri1-201* allele also reveals cross-talk between GA and BR pathways, with *bri1-201* demonstrating altered expression of *GA5*, a GA biosynthesis gene, and the GA-responsive locus *GASA1* (Bouquin et al., 2001). Perhaps the most direct evidence of BR influence on anther and pollen development comes from recent ChIP analysis demonstrating that the transcription factor BES1, important for BR signaling, can directly bind to the promoter regions of microsporogenesis regulators *SPL/NZZ*, *TDF1*, *AMS*, *MS1*, and *MS2* (Ye et al., 2010; see Figure 12.2). Consequently, BRs appear to have the capacity to regulate male fertility both indirectly, by an intersection with GA processes, and directly by controlling expression of key genes involved in multiple steps of stamen development.

Auxins

Evidence indicates that auxins also influence plant male reproductive development. The auxin biosynthetic double mutant *yuc2 yuc6* and the quadruple mutant *tir1 afb1 afb2 afb3*, involving auxin signaling genes, demonstrate loss in male fertility (Dharmasiri et al., 2005; Cheng et al., 2006). Use of auxin-sensitive reporter genes allows detection of auxin effects in microspores and in anther somatic tissues during dehiscence and in the anther/filament junction (Cecchetti et al., 2008). Auxin synthesis occurs in anthers, and TIR1, AFB1, AFB2, and AFB3 auxin receptor genes are transcribed in late stages of anther development following meiosis. Combined mutations at three and four of the auxin receptor loci result in premature anther dehiscence and pollen maturation, with pollen release occurring prior to filament elongation, suggesting that auxins are essential to coordinating pollen maturation with their release.

Cytoplasmic Male Sterility Systems in Agriculture

Cytoplasmic male sterility (CMS) is the maternally inherited inability of a plant to produce or shed viable pollen. CMS is conditioned by gain-of-function mitochondrial genome mutations, giving rise to novel mitochondrial proteins and facilitating the use of CMS in hybrid crop seed production. A mitochondrial mutation is transmitted to all progeny of a cross, which allows the breeder to maintain the male sterility trait without the inconvenience of gene segregation each generation. Implementation of a CMS system for hybrid seed production requires the breeder to identify not only a stable CMS mutation, and a crossing strategy to maintain the male sterile trait, but also nuclear suppressor gene(s) that will restore pollen fertility. Hybrid seed production involves crossing an inbred, maintained on the CMS cytoplasm, with complementary inbred lines that contain the nuclear fertility restorer gene(s) to produce a fertile hybrid. Such a genetic system, when properly implemented, can greatly reduce the cost of seed production by eliminating the need for manual emasculation, and provide incentive for seed purchase each season to avoid segregation of male sterile plants.

Plant mitochondrial mutations

Plant and animal mitochondrial genomes differ in their large and variable sizes and in their tendency to undergo recombination (Arrieta-Montiel and Mackenzie, 2010). In plants, this recombination activity can subdivide the genome to a highly redundant, multi-partite structure, with different submolecules maintained at varying levels. Sites with the potential to undergo recombination can be numerous within the mitochondrial genome, with 47 repeat pairs functionally identified in the *Arabidopsis* mitochondrial genome (Arrieta-Montiel et al., 2009; Davila et al., 2011). This type of recombination provides an opportunity for accelerated genome evolution and capacity for extensive genomic variation. Two forms of variation are particularly prevalent in the mitochondrial genomes of plants, sequence chimeras from non-homologous end joining, and rapid intermolecular genomic copy number changes within the mitochondrial population, produced by recombination and termed substoichiometric shifting (SSS). Both phenomena are key components of CMS systems.

All CMS mitochondrial mutations identified to date have been comprised of sequence chimeras, generally fusing identifiable mitochondrial gene sequences with segments of unknown origin (Schnable and Wise, 1998; Hanson and Bentolila, 2004). Yet, no two CMS mutations have been the same. These rearranged sequences give rise to new gene products that generally display some degree of hydrophobicity and likely interact with the mitochondrial membrane. Many CMS sequences contain segments of genes involved in the ATPase complex, implying that energy-associated alterations might participate in the CMS phenotype.

CMS mutations differ not only in their sequence origins, but often in their phenotype, from alterations in microspore development (Datta et al., 2002), to callose dissolution (Abad et al., 1995), to premature tapetal breakdown patterns (González-Melendi et al., 2008). Whether these phenotypic differences are the consequence of distinct mechanisms of action or simply a reflection of variation in spatial and temporal expression patterns of the sterility lesion is not known.

Fertility restoration

The male sterile phenotype is reversibly suppressed by nuclear fertility restorer genes. Identification and cloning of these genes has provided valuable insight into the nature of mitochondrial gene regulation (Schmitz-Linneweber and Small, 2008). Several fertility restorer genes have been cloned, the vast majority of them have been found to contain pentatricopeptide repeats (PPR) within the protein. PPR proteins are known to participate in mRNA splicing, editing, and translation initiation processes (Delannoy et al., 2007). In plants, this family of genes functions in both mitochondria and plastids and is unusually large, numbering over 400 in *Arabidopsis* (Small and Peeters, 2000). Although essential for the proper processing and translation of native organellar transcripts, PPR-mediated processing can serve to inactivate a rearranged, CMS-associated gene transcript. Interestingly, regions of a genome may find PPR protein-coding genes organized in large duplicated gene clusters, representing fertility restorer loci (Barr and Fishman, 2010). Organization of these genes within plant genomes suggests that they are subject to intense evolutionary selection, somewhat similar in pattern to disease-resistance genes (Geddy and Brown, 2007). In naturally occurring CMS hybrid incompatibility systems, where the male sterility phenotype can arise by particular hybridizations, evidence from PPR protein gene organization and conservation across angiosperms suggest co-evolution of these nuclear and cytoplasmic factors to influence cross compatibility.

Two other types of nuclear fertility restorers have also been reported. The first is the *Rf2* restorer of maize CMS-T cytoplasm, which encodes an aldehyde dehydrogenase that is suggested to play a detoxification role (Cui et al., 1996; Liu et al., 2001). Another type of nuclear restorer was reported in the common bean *Fr* gene, which effects specific stoichiometric changes in mitochondrial genome organization and results in dramatic reduction of the mitochondrial male sterility sequence, *pvs-orf239*, to near undetectable levels (Mackenzie and Chase, 1990). This latter fertility restoring mechanism is non-reversible in subsequent generations; therefore, it does not fall under the formal definition of a fertility restorer gene. However, such nuclear-directed influences on mitochondrial genome organization likely play an important role in controlling mitochondrial genetic variation in natural plant populations.

Stability of the CMS trait

An essential feature of the CMS mutation for use in crop breeding is its stability, with little or no spontaneous reversion to fertility. Reversion to fertility is associated with SSS activity (Arrieta-Montiel and Mackenzie, 2010). Recombination-mediated genomic shifting can result in rapid copy number reduction of the DNA molecule on which the CMS mutation resides, leaving the remainder of the mitochondrial genome essentially unchanged (Janska et al., 1998; Bellaoui et al., 1998). In this manner, the CMS mutation is silenced and the plant displays a male fertile phenotype. Reversion frequency in most CMS systems that have been studied appears to be influenced by nuclear genetic background (Mackenzie and McIntosh, 1999). Therefore, a breeder wishing to implement a CMS system must test the stability of the system on a range of germplasm in the program.

Male Sterility: Metabolic and Evolutionary Implications

CMS is a naturally found condition

CMS is observed in over 150 plant species (Kaul, 1988), suggesting that the capacity to convert to male sterility is prevalent in plants. In CMS common bean (Arrieta-Montiel et al., 2001) and *Brassica* spp. (Murayama et al., 2004; Giancola et al., 2007), fertile plants from wild populations have been shown to contain the male sterility gene at substoichiometric levels. These observations imply that it is feasible to interconvert male fertile and sterile types under particular, not yet defined conditions. At least two nuclear genes have been identified that control the mitochondrial SSS process, MSH1 (Abdelnoor et al., 2003) and RECA3 (Shedge et al., 2007). Disruption or RNAi suppression of the nuclear gene MSH1 results in a marked increase in the frequency of mitochondrial recombination (Sandhu et al., 2007) and, interestingly, results in a CMS phenotype in six different plant species tested to date (Shedge et al., 2010; Sandhu and Mackenzie, unpublished). This raises the intriguing possibility that CMS induction in nature can occur under conditions that suppress MSH1 expression. Interestingly, transcription of MSH1 is markedly downregulated under environmental stresses, including heat, drought, cold, and high light (Zimmermann et al., 2004; Shedge et al., 2010).

While male sterility occurs by numerous distinct mechanisms in plants, the phenotype is presumably adaptive within a plant population under certain conditions. Male sterility has been associated with increases in seed production and vigor as the consequences of out-crossing (Charlesworth, 2002). The frequency of gynodioecy, or the mixture within a plant population of both hermaphrodites and females (male steriles), has been estimated to occur in over 7% of all plant species (McCauley and Olson, 2008). In many of these cases, the male sterility appears to be cytoplasmic, likely a consequence of the dynamic nature of the mitochondrial genome. Important questions remaining in this field revolve around the nature of environmental and/or cellular cues that trigger interconversion of male fertile and male sterile phenotypes in natural populations and under agricultural field conditions.

Organelle metabolism influences pollen development

Aside from the CMS trait, there is other evidence to suggest that organelle metabolism, and overall cellular metabolic state, influences pollen development. Mutation of the plastidial glyceraldehyde-3-phosphate dehydrogenase, in the form of a *gapcp1 gapcp2* double mutant, renders the plant male sterile (Muñoz-Bertomeu et al., 2010). Pollen from this mutant displays unusual collapsed shapes due to alterations in the tapetum and perhaps changes in carbohydrate pools and/or trehalose levels. Likewise, disruption of cytosolic phosphoglucomutase prevents both male and female gametogenesis, suggesting that proper carbohydrate distribution is critical at these stages (Egli et al., 2010). Temperature has a major impact on pollen development and viability. Cold temperature stress can impede pollen development, and male-gametophyte-specific classes of transcription factors appear to influence pollen response to adverse temperature effects (Zou et al., 2010).

The tapetum is naturally programmed for an apoptosis-like process (Wang et al., 1999; Wu and Cheung, 2000; Love et al., 2008). The timing of this programmed tapetal degeneration is critical for proper pollen development (Parish and Li, 2010). It remains largely speculation whether mitochondrial cues signal this PCD response, but there appears to be indication in some CMS systems that mitochondria influence this process (Chase, 2007). Likewise, interference with the process can provide a means of engineering male sterility.

Genetic Engineering of Male Sterility

Agricultural systems can benefit from the availability of reliable male sterility, useful in hybrid seed production, transgene containment, and reduction of overall airborne allergens. The deployment of transgenic crops that are male sterile provides greater assurance that the transgenic genotype will remain confined to the area in which it is planted, without concern for transport of pollen to neighboring fields. The abundance of biological detail now available for pollen pathways facilitates the engineering of transgenic approaches for male sterility induction, albeit with mixed success to date. A number of anther-specific promoters are readily obtained from genes identified to specifically express during anther and pollen stages. As a recent example of this approach, five anther-specific promoters were isolated from *Brassica oleracea* and *B. rapa* and shown to express in tapetal cells during meiocyte development. Fusion of these promoters to candidate male sterility genes such as cysteine proteases BoCysP1 and BoCP3, and Gibberellin-insensitive (*gai*) genes for introduction to *Arabidopsis* (Konagaya et al., 2008) produced mixed results. Expression of BoCysP1 controlled by two specific promoters produced plants with complete male sterility. Tapetal and middle layer cells of anthers expressing BoCysP1 showed swelling and vacuolation, perhaps interference with the normal pattern of programmed cell death. Similar approaches have been taken expressing the barnase gene for cell ablation during pollen development (Mariani et al., 1990; Gardner et al., 2009), DNA adenine methylase (Cigan and Albertsen, 2000), phosphonate monoesterase (Dotson et al. 1996), ribosome inactivating protein (Cho et al., 2001), Avadin (Albertsen et al., 1999), Stilbene synthase (Hofig et al., 2006), and endo-β-1,3,glucanase (Tsuchiya et al., 1995). It has also been feasible to capitalize on the discovery of genes known to function distinctly during pollen development to manipulate expression and alter pollen processes. The BECLIN I gene is an ortholog of the yeast ATG6 gene involved in autophagy, and disruption of this locus in *Arabidopsis* results in male sterility. Transgenic over-expression of the gene in tobacco tapetum during early stages of anther development can also alter the timing of programmed cell death and tapetal degeneration (Singh et al., 2010).

Some of the limitations existing in engineering male sterility systems have been the occurrence of an incomplete sterility phenotype, difficulties with restoring fertility, or complications associated with segregation of fertile progeny in subsequent generations. Some challenges have been addressed by fusion of the male sterility gene to an inducible promoter, and some by the design of a fertility restoration strategy. Transgenic repression of the anther-specific cell wall invertase or interference with its activity blocks early stages of pollen development, causing male sterility without having additional pleiotropic effects (Engelke et al., 2010). Restoration of fertility occurs by substituting the downregulated endogenous plant invertase activity by a yeast invertase fused to the N-terminus of a potato vacuolar proteinase II (PiII-ScSuc2), under control of an anther-specific invertase promoter Nin88 from tobacco. The chimeric PiII-ScSuc2 is N-glycosylated and efficiently secreted to the apoplast, a necessary property of the native form. In addition, the Nin88::PiII-ScSuc2 fusion does not affect pollen development in a normal background, allowing this genotype to be used as a fertile restorer line.

Implementation of Male Sterility in Agricultural Systems

Attempts at genetic engineering of male sterility, while often effective in suppressing or disrupting normal pollen development patterns, have generally suffered from their infeasibility to scale up for agricultural implementation. An important motivation for identifying an effective genetic system for male sterility is the opportunity to cultivate hybrids. An F_1 hybrid provides more uniform crop performance in the field and, more importantly, the significant advantage of heterosis. Heterosis, or hybrid vigor, refers to the ability of the F_1 hybrid to outperform either of the parents used in the cross. Consequently, development of hybrids has, in the case of many crops, accounted for significant gains in overall crop yield. For example, hybrid rice in China can be credited with reducing the total amount of land planted to rice from 36 Mha in 1975 to 30 Mha in 2000, and increasing total production from 128 to 189 million tons, a yield increase of up to 6.2 tons/ha (http://www.fao.org/rice2004).

There have been several different genetic models proposed to account for heterosis, including dominance, overdominance, and pseudo-overdominance, all appearing to have some merit (Chen, 2010). It has also been suggested that epigenetic factors, influenced by circadian-controlled regulatory mechanisms, may underlie much of the heterosis phenomenon. However, what has not been investigated sufficiently is the influence of maternal effects on hybrid vigor. Often reciprocal differences in cross-compatibility and vigor outcomes will be observed in a cross. In some cases a CMS plant crossed to its sterility maintainer line will produce a more vigorous progeny than self-pollination of the same maintainer line (S. Mackenzie, unpublished). In these crosses, the two hybridizing plants are nearly identical in nuclear genotype, implying that the vigor may result from cytoplasmic factors. In fact, very little research has been conducted to investigate the possible influence of mitochondria or plastids, the two essential generators of ATP, on heterosis. It is intriguing to speculate that organellar metabolic factors, together with epigenetic signals, may integrate to enhance vigor in the hybrid. Organellar changes arising as a consequence of mutation or suppression of MSH1 in *Arabidopsis* and other crop species give rise to dramatic changes in growth patterns and transcript profiles, suggesting that organellar functions can directly influence several developmental pathways in plants (Shedge et al., 2007, 2010; Sandhu et al., 2007; Sandhu and Mackenzie, unpublished).

The natural system of cytoplasmic male sterility remains the most broadly utilized agriculturally. CMS has been successfully implemented in a number of crops, including rice, sorghum, millet, onion, sugar beet, and carrot, but limitations exist with this genetic system as well. Following the Southern Corn Leaf Blight of 1970, when a vast area of the Midwestern United States was planted with CMS-T type maize that proved susceptible to the T-toxin of *Bipolaris maydis* (Ullstrup, 1970), CMS in maize is no longer used to such a great extent. What was learned from that episode was the need for multiple distinct, stable sources of CMS for any given crop to avoid overplanting to a single cytoplasm type. Furthermore, a useful source of CMS has not yet been identified in many crops in which hybrid vigor is well documented. With the recent demonstration that CMS can be transgenically induced (Sandhu et al., 2007), the opportunity may now exist for identification of additional CMS lines. Presumably, this type of cytoplasmic genetic manipulation can be carried out in multiple species and in several distinct genotypes. If MSH1 modulation occurs naturally in plant populations, and the CMS arising in these manipulations also occurs in nature, we can assume that fertility restoration mechanisms are also relatively prevalent.

The wealth of fundamental information available on the genetic and physiological pathways regulating stamen development provides well-documented opportunities to influence these processes transgenically. More practical strategies for introduction of male sterility traits to large-scale agricultural production are needed now.

References

Aarts, M. G., Aarts, M. G., Hodge, R., Kalantidis, K., Florack, D., & Wilson, Z. A., et al. (1997). The *Arabidopsis* MALE STERILITY 2 protein shares similarity with reductases in elongation/condensation complexes. *Plant Journal, 12*, 615–623.

Abad, A., Mehrtens, B., & Mackenzie, S. (1995). Specific expression and fate of a mitochondrial sterility-associated mitochondrial protein in cms common bean. *Plant Cell, 7*, 271–28557.

Abdelnoor, R. V., Yule, R., Elo, A., Christensen, A., Meyer-Gauen, G., & Mackenzie, S. (2003). Substoichiometric shifting in the plant mitochondrial genome is influenced by a gene homologous to MutS. *Proceedings of the National Academy of Sciences of the United States of America, 100*, 5968–5973.

Albertsen, M. C., Howard, J. A., & Maddock, S. (1999). Induction of male sterility in plants by expression of high levels of avidin. US Patent 5962769.

Arrieta-Montiel, M., Lyznik, A., Woloszynska, M., Janska, H., Tohme, J., & Mackenzie, S. (2001). Tracing evolutionary and developmental implications of mitochondrial stoichiometric shifting in the common bean. *Genetics, 158*, 851–864.

Arrieta-Montiel, M., & Mackenzie, S. A. (2010). Plant Mitochondrial Genomes and Recombination. In F. Kempken (Ed.), *Advances in plant biology: Plant mitochondria* (pp. 65–84). Springer Publ.

Arrieta-Montiel, M. P., Shedge, V., Davila, J., Christensen, A. C., & Mackenzie, S. A. (2009). Diversity of the *Arabidopsis* mitochondrial genome occurs via nuclear-controlled recombination activity. *Genetics, 183*, 1261–1268.

Balasubramanian, S., & Schneitz, K. (2000). NOZZLE regulates proximal-distal pattern formation, cell proliferation and early sporogenesis during ovule development in *Arabidopsis thaliana*. *Development, 127*, 4227–4238.

Barr, C. M., & Fishman, L. (2010). The nuclear component of a cytonuclear hybrid incompatibility in *Mimulus* maps to a cluster of pentatricopeptide repeat genes. *Genetics, 184*, 455–465.

Bellaoui, M., Martin-Canadell, A., Pelletier, G., & Budar, F. (1998). Low-copy-number molecules are produced by recombination, actively maintained and can be amplified in the mitochondrial genome of Brassicaceae: Relationship to reversion of the male sterile phenotype in some cybrids. *Molecular & General Genetics, 257*, 177–185.

Borg, M., Brownfield, L., & Twell, D. (2009). Male gametophyte development: A molecular perspective. *Journal of Experimental Botany, 60*, 1465–1478.

Bouquin, T., Meier, C., Foster, R., Nielsen, M., & Mundy, J. (2001). Control of specific gene expression by gibberellin and brassinosteroid. *Plant Physiology, 127*, 450–458.

Bowman, J. L., Smyth, D. R., & Meyerowitz, E. M. (1989). Genes directing flower development in *Arabidopsis*. *Plant Cell, 1*, 37–52.

Canales, C., Bhatt, A. M., Scott, R., & Dickinson, H. (2002). EXS, a putative LRR receptor kinase, regulates male germline cell number and tapetal identity and promotes seed development in *Arabidopsis*. *Current Biology, 12*, 1718–1727.

Cecchetti, V., Altamura, M., Falasca, G., Costantino, P., & Cardarelli, M. (2008). Auxin regulates *Arabidopsis* anther dehiscence, pollen maturation, and filament elongation. *Plant Cell, 20*, 1760–1774.

Charlesworth, D. (2002). What maintains male-sterility factors in plant populations? *Heredity, 89*, 408–409.

Chase, C. D. (2007). Cytoplasmic male sterility: A window to the world of plant mitochondrial–nuclear interactions. *Trends in Genetics, 23*, 81–90.

Chaudhury, A. M. (1993). Nuclear genes controlling male fertility. *Plant Cell, 5*, 1277–1283.

Chen, Y. C., & McCormick, S. (1996). sidecar pollen, an *Arabidopsis thaliana* male gametophytic mutant with aberrant cell divisions during pollen development. *Development, 122*, 3243–3253.

Chen, Z. J. (2010). Molecular mechanisms of polyploidy and hybrid vigor. *Trends in Plant Science, 15*, 57–71.

Cheng, H., Qin, L., Lee, S., Fu, X., Richards, D. E., & Cao, D., et al. (2004). Gibberellin regulates *Arabidopsis* floral development via suppression of DELLA protein function. *Development, 131*, 1055–1064.

Cheng, H., Song, S., Xiao, L., Soo, H. M., Cheng, Z., & Xie, D., et al. (2009). Gibberellin acts through jasmonate to control the expression of MYB21, MYB24, and MYB57 to promote stamen filament growth in *Arabidopsis*. *PLoS Genetics, 5*, e1000440.

Cheng, Y., Dai, X., & Zhao, Y. (2006). Auxin biosynthesis by the YUCCA flavin monooxygenases controls the formation of floral organs and vascular tissues in *Arabidopsis*. *Genes & Development, 20*, 1790–1799.

Chhun, T., Aya, K., Asano, K., Yamamoto, E., Morinaka, Y., & Watanabe, M., et al. (2007). Gibberellin regulates pollen viability and pollen tube growth in rice. *Plant Cell, 19*, 3876–3888.

Cho, H. J., Kim, S., Kim, M., & Kim, B. D. (2001). Production of transgenic male sterile tobacco plants with the cDNA encoding a ribosome inactivating protein in *Dianthus sinensis* L. *Molecules and Cells, 11*, 326–333.

Cigan, A. M., & Albertsen, M. C. (2000). Reversible nuclear genetic system for male sterility in transgenic plants. US Patent 6072102.

Cui, X., Wise, R. P., & Schnable, P. S. (1996). The rf2 nuclear restorer gene of male-sterile T-cytoplasm maize. *Science, 272*, 1334–1336.

Datta, R., Chamusco, K. C., & Chourey, P. S. (2002). Starch biosynthesis during pollen maturation is associated with altered patterns of gene expression in maize. *Plant Physiology, 130*, 1645–1656.

Davila, J. I., Arrieta-Montiel M. P., Wamboldt Y., Shedge,V., Xu Y.-Z., Mackenzie S. A. (2011) Double-strand break repair processes drive evolution of the mitochondrial genome in *Arabidopsis*. Submitted.

Delannoy, E., Stanley, W. A., Bond, C. S., & Small, I. D. (2007). Pentatricopeptide repeat (PPR) proteins as sequence-specificity factors in post-transcriptional processes in organelles. *Biochemical Society Transactions, 35*, 1643–1647.

Dharmasiri, N., Dharmasiri, S., Weijers, D., Lechner, E., Yamada, M., & Hobbie, L., et al. (2005). Plant development is regulated by a family of auxin receptor F box proteins. *Developmental Cell*, *9*, 109–119.

Dotson, S. B., Lanahan, M. B., Smith, A. G., & Kishore, G. M. (1996). A phosphonate monoester hydrolase from *Burkholderia caryophilli* PG2982 is useful as a conditional lethal gene in plants. *Plant Journal*, *10*, 383–392.

Dudareva, N., Pichersky, E., & Gershenzon, J. (2004). Biochemistry of plant volatiles. *Plant Physiology*, *135*, 1893–1902.

Eady, C., Lindsey, K., & Twell, D. (1994). Differential activation and conserved vegetative cell-specific activity of a late pollen promoter in species with bicellular and tricellular pollen. *Plant Journal*, *5*, 543–550.

Egli, B., Kolling, K., Köhler, C., Zeeman, S. C., & Streb, S. (2010). Loss of cytosolic phosphoglucomutase is compromises for gametophyte development in *Arabidopsis*. *Plant Physiology*, *154*, 1659–1671.

Engelke, T., Hirsche, J., & Roitsch, T. (2010). Anther-specific carbohydrate supply and restoration of metabolically engineered male sterility. *Journal of Experimental Botany*, *61*, 2693–2706.

Gardner, N., Felsheim, R., & Smith, A. G. (2009). Production of male- and female-sterile plants through reproductive tissue ablation. *Journal of Plant Physiology*, *166*, 871–881.

Geddy, R., & Brown, G. G. (2007). Genes encoding pentatricopeptide repeat (PPR) proteins are not conserved in location in plant genomes and may be subject to diversifying selection. *BMC Genomics*, *8*, 130.

Giancola, S., Rao, Y., Chaillou, S., Hiard, S., Martin-Canadell, A., & Pelletier, G., et al. (2007). Cytoplasmic suppression of Ogura cytoplasmic male sterility in European natural populations of *Raphanus raphanistrum*. *Theoretical and Applied Genetics*, *114*, 1333–1343.

González-Melendi, P., Uyttewaal, M., Morcillo, C. N., Hernández Mora, J. R., Fajardo, S., & Budar, F., et al. (2008). A light and electron microscopy analysis of the events leading to male sterility in Ogu-INRA CMS of rapeseed (*Brassica napus*). *Journal of Experimental Botany*, *59*, 827–838.

Gusti, A., Baumberger, N., Nowack, M., Pusch, S., Eisler, H., & Potuschak, T., et al. (2009). The *Arabidopsis thaliana* F-box protein FBL17 is essential for progression through the second mitosis during pollen development. *PLoS ONE*, *4*, e4780.

Hanson, M., & Bentolila, S. (2004). Interactions of mitochondrial and nuclear genes that affect male gametophyte development. *Plant Cell*, *16*(suppl), S154–S169.

Hewitt, F. R., Hough, T., O'Neill, P., Sasse, J. M., Williams, E. G., & Rowan, K. S. (1985). Effect of brassinolide and other growth regulators on the germination and growth of pollen tubes of *Prunus avium* using a multiple hangingdrop assay. *Australian Journal of Plant Physiology*, *1*, 201–211.

Higginson, T., Li, S. F., & Parish, R. W. (2003). AtMYB103 regulates tapetum and trichome development in *Arabidopsis thaliana*. *Plant Journal*, *35*, 177–192.

Hofig, K. P., Moller, R., Donaldson, L., Putterill, J., & Walter, C. (2006). Towards male sterility in *Pinus radiata*, a stilbene synthase approach to genetically engineer nuclear male sterility. *Plant Biotechnology Journal*, *4*, 333–343.

http://www.fao.org/rice2004/

Ito, T., Nagata, N., Yoshiba, Y., Ohme-Takagi, M., Ma, H., & Shinozaki, K. (2007). *Arabidopsis* MALE STERILITY1 encodes a PHD-Type transcription factor and regulates pollen and tapetum development. *Plant Cell*, *19*, 3549–3562.

Ito, T., Wellmer, F., Yu, H., Das, P., Ito, N., & Alves-Ferreira, M., et al. (2004). The homeotic protein AGAMOUS controls microsporogenesis by regulation of SPOROCYTELESS. *Nature*, *430*, 356–360.

Iwakawa, H., Shinmyo, A., & Sekine, M. (2006). *Arabidopsis* CDKA;1, a cdc2 homologue, controls proliferation of generative cells in male gametogenesis. *Plant Journal*, *45*, 819–831.

Janska, H., Sarria, R., Woloszynska, M., Arrieta-Montiel, M., & Mackenzie, S. A. (1998). Stoichiometric shifts in the common bean mitochondrial genome leading to male sterility and spontaneous reversion to fertility. *The Plant Cell*, *10*, 1163–1180.

Kaul, M. (1988). Male sterility in higher plants. In R. Frankel, M. Grossman, & P. Maliga (Eds.), *Monographs on theoretical and applied genetics* (Vol. 10, pp. 775–795). Berlin: Springer-Verlag.

Kim, H. J., Oh, S. -A., Brownfield, L., Hong, S. H., Ryu, H., & Hwang, I., et al. (2008). Control of plant germline proliferation by SCF (FBL17) degradation of cell cycle inhibitors. *Nature*, *455*, 1134–1137.

Konagaya, K., Ando, S., Kamachi, S., Tsuda, M., & Tabei, Y. (2008). Efficient production of genetically engineered, male-sterile *Arabidopsis thaliana* using anther-specific promoters and genes derived from *Brassica oleracea* and *B. rapa*. *Plant Cell Reports*, *27*, 1741–1754.

Li, J., Nam, K., Vafeados, D., & Chory, J. (2001). BIN2, a new brassinosteroid-insensitive locus in *Arabidopsis*. *Plant Physiology*, *127*, 14–22.

Liu, F., Cui, X., Horner, H. T., Weiner, H., & Schnable, P. S. (2001). Mitochondrial aldehyde dehydrogenase activity is required for male fertility in maize. *Plant Cell*, *13*, 1063–1078.

Love, A. J., Milner, J. J., & Sadanandom, A. (2008). Timing is everything: Regulatory overlap in plant cell death. *Trends in Plant Science*, *13*, 589–595.

Ma, H. (2005). Molecular genetic analyses of microsporogenesis and microgametogenesis in flowering plants. *Annual Review of Plant Biology*, *56*, 393–434.

Mackenzie, S., & McIntosh, L. (1999). Higher plant mitochondria. *Plant Cell*, *11*, 571–585.

Mackenzie, S. A., & Chase, C. D. (1990). Fertility restoration is associated with loss of a portion of the mitochondrial genome in cytoplasmic male-sterile common bean. *Plant Cell*, *2*, 905–912.

Mariani, C., Beuckeleer, M. D., Truettner, J., Leemans, J., & Goldberg, R. B. (1990). Induction of male sterility in plants by a chimeric ribonuclease gene. *Nature*, *347*, 737–741.

McCauley, D. E., & Olson, M. S. (2008). Do recent findings in plant mitochondrial molecular and population genetics have implications for the study of gynodioecy and cytonuclear conflict? *Evolution; International Journal of Organic Evolution*, *62*, 1013–1025.

McCormick, S. (2004). Control of male gametophyte development. *Plant Cell*, *16*, S142–S153.

McCormick, S., Twell, D., Vancanneyt, G., & Yamaguchi, J. (1991). Molecular analysis of gene regulation and function during male gametophyte development. *Symposia of the Society for Experimental Biology*, *45*, 229–244.

Millar, A. A., & Gubler, F. (2005). The *Arabidopsis* GAMYB-like genes, MYB33 and MYB65, are microRNA-regulated genes that redundantly facilitate anther development. *Plant Cell*, *17*, 705–721.

Muñoz-Bertomeu, J., Cascales-Miñana, B., Irles-Segura, A., Mateu, I., Nunes-Nesi, A., & Fernie, A. R., et al. (2010). The plastidial glyceraldehyde-3-phosphate dehydrogenase is critical for viable pollen development in *Arabidopsis*. *Plant Physiology*, *152*, 1830–1841.

Murayama, K., Yahara, T., & Terachi, T. (2004). Variation of female frequency and cytoplasmic male-sterility gene frequency among natural gynodioecious populations of wild radish (*Raphanus sativus* L.). *Molecular Ecology*, *13*, 2459–2464.

Nowack, M. K., Shirzadi, R., Dissmeyer, N., Dolf, A., Endl, E., & Grini, P. E., et al. (2007). Bypassing genomic imprinting allows seed development. *Nature*, *447*, 312–315.

Oh, S. A., Johnson, A., Smertenko, A., Rahman, D., Park, S. K., & Hussey, P. J., et al. (2005). A divergent cellular role for the FUSED kinase family in the plant-specific cytokinetic phragmoplast. *Current Biology*, *15*, 2107–2111.

Parish, R. W., & Li, S. F. (2010). Death of a tapetum: A programme of developmental altruism. *Plant Science*, *178*, 73–89.

Sandhu, A. S., Abdelnoor, R. V., & Mackenzie, S. A. (2007). Transgenic induction of mitochondrial rearrangements for cytoplasmic male sterility in crop plants. *Proceedings of the National Academy of Sciences of the United States of America*, *104*, 1766–1770.

Schmitz-Linneweber, C., & Small, I. (2008). Pentatricopeptide repeat proteins: A socket set for organelle gene expression. *Trends in Plant Science*, *13*, 663–670.

Schnable, P., & Wise, R. (1998). The molecular basis of cytoplasmic male sterility and fertility restoration. *Trends in Plant Science*, *3*, 175–180.

Shedge, V., Arrieta-Montiel, M., Christensen, A. C., & Mackenzie, S. A. (2007). Plant mitochondrial recombination surveillance requires novel *RecA* and *MutS* homologs. *Plant Cell*, *19*, 1251–1264.

Shedge, V., Davila, J., Arrieta-Montiel, M. P., Mohammed, S., & Mackenzie, S. A. (2010). Extensive rearrangement of the *Arabidopsis* mitochondrial genome elicits cellular conditions for thermotolerance. *Plant Physiology, 152*, 1960–1970.

Singh, D., Jermakow, A., & Swain, S. (2002). Gibberellins are required for seed development and pollen tube growth in *Arabidopsis*. *Plant Cell, 14*, 3133–3147.

Singh, S. P., Pandey, T., Srivastava, R., Verma, P. C., Singh, P. K., & Tuli, R., et al. (2010). BECLIN1 from *Arabidopsis thaliana* under the generic control of regulated expression systems, a strategy for developing male sterile plants. *Plant Biotechnology Journal, 8*(9), 1005–1022.

Small, I. D., & Peeters, N. (2000). The PPR motif – a TPR-related motif prevalent in plant organellar proteins. *Trends in Biochemical Sciences, 25*, 46–47.

Smyth, D. R., Bowman, J. L., & Meyerowitz, E. M. (1990). Early flower development in *Arabidopsis*. *Plant Cell, 2*, 755–767.

Sorensen, A., Kröber, S., Unte, U. S., Huijser, P., Dekker, K., & Saedler, H. (2003). The *Arabidopsis* ABORTED MICROSPORES (AMS) gene encodes an MYC class transcription factor. *Plant Journal, 33*, 413–423.

Stintzi, A., & Browse, J. (2000). The *Arabidopsis* male-sterile mutant, *opr3*, lacks the 12-oxophytodienoic acid reductase required for jasmonate synthesis. *Proceedings of the National Academy of Sciences of the United States of America, 97*, 10625–10630.

Szekeres, M., Németh, K., Koncz-Kálmán, Z., Mathur, J., Kauschmann, A., & Altmann, T., et al. (1996). Brassinosteroids rescue the deficiency of CYP90, a cytochrome P450, controlling cell elongation and de-etiolation in *Arabidopsis*. *Cell, 85*, 171–182.

Taylor, P. E., Glover, J. A., Lavithis, M., Craig, S., & Singh, M. B., et al. (1998). Genetic control of male fertility in *Arabidopsis thaliana*: Structural analyses of postmeiotic developmental mutants. *Planta, 205*, 492–505.

Tsuchiya, T., Toriyama, K., Yoshikawa, M., Ejiri, S., & Hinata, K. (1995). Tapetum-specific expression of the gene for an endobeta-1,3-glucanase causes male sterility in transgenic tobacco. *Plant and Cell Physiology, 36*, 487–494.

Twell, D., Park, S. K., Hawkins, T. J., Schubert, D., Schmidt, R., & Smertenko, A., et al. (2002). Mor1/Gem1 has an essential role in the plant-specific cytokinetic phragmoplast. *Nature Cell Biology, 4*, 711–714.

Tyler, L., Thomas, S. G., Hu, J., Dill, A., Alonso, J. M., & Ecker, J. R., et al. (2004). Della proteins and gibberellin-regulated seed germination and floral development in *Arabidopsis*. *Plant Physiology, 135*, 1008–1019.

Ullstrup, A. J. (1970). The impacts of the Southern Corn Leaf Blight epidemics of 1970–1971. *Annual Review of Phytopathology, 10*, 37–50.

Ursin, V. M., Yamaguchi, J., & McCormick, S. (1989). Gametophytic and sporophytic expression of anther-specific genes in developing tomato anthers. *Plant Cell, 1*, 727–736.

Wang, M., Hoekstra, S., van Bergen, S., Lamers, G. E. M., Oppedijk, B. J., & van der Heiden, M. W., et al. (1999). Apoptosis in developing anthers and the role of ABA in this process during androgenesis in *Hordeum vulgare* L. *Plant Molecular Biology, 39*, 489–501.

Wijeratne, A. J., Zhang, W., Sun, Y., Liu, W., Albert, R., & Zheng, Z., et al. (2007). Differential gene expression in *Arabidopsis* wild-type and mutant anther: Insights into anther cell differentiation and regulatory networks. *Plant Journal, 52*, 14–29.

Wilson, Z., Morroll, S., Dawson, J., Swarup, R., & Tighe, P. (2001). The *Arabidopsis* MALE STERILITY1 (MS1) gene is a transcriptional regulator of male gametogenesis, with homology to the PHD-finger family of transcription factors. *Plant Journal, 28*, 27–39.

Woodger, F. J., Millar, A., Murray, F., Jacobsen, J. V., & Gubler, F. (2003). The role of GAMYB transcription factors in GA-regulated gene expression. *Journal of Plant Growth Regulation, 22*, 176–184.

Wu, H. -M., & Cheung, A. Y. (2000). Programmed cell death in plant reproduction. *Plant Molecular Biology, 44*, 267–281.

Xie, D. X., Feys, B. F., James, S., Nieto-Rostro, M., & Turner, J. G. (1998). COI1: An *Arabidopsis* gene required for jasmonate-regulated defense and fertility. *Science, 280*, 1091–1094.

Xu, J., Yang, C., Yuan, Z., Zhang, D., Gondwe, M. Y., & Ding, Z., et al. (2010). The ABORTED MICROSPORES regulatory network is required for postmeiotic male reproductive development in *Arabidopsis thaliana*. *Plant Cell, 22*, 91–107.

Yang, W. C., Ye, D., Xu, J., & Sundaresan, V. (1999). The SPOROCYTELESS gene of *Arabidopsis* is required for initiation of sporogenesis and encodes a novel nuclear protein. *Genes & Development, 13*, 2108–2117.

Ye, Q., Zhu, W., Li, L., Zhang, S., Yin, Y., & Ma, H., et al. (2010). Brassinosteroids control male fertility by regulating the expression of key genes involved in *Arabidopsis* anther and pollen development. *Proceedings of the National Academy of Sciences of the United States of America, 107*, 6100–6105.

Yu, H., Ito, T., Zhao, Y., Peng, J., Kumar, P., & Meyerowitz, E. M. (2004). Floral homeotic genes are targets of gibberellin signaling in flower development. *Proceedings of the National Academy of Sciences of the United States of America, 101*, 7827–7832.

Zhang, W., Sun, Y., Timofejeva, L., Chen, C., Grossniklaus, U., & Ma, H. (2006). Regulation of *Arabidopsis* tapetum development and function by DYSFUNCTIONAL TAPETUM1 (DYT1) encoding a putative bHLH transcription factor. *Development, 133*, 3085–3095.

Zhao, D. Z., Wang, G. F., Speal, B., & Ma, H. (2002). The EXCESS MICROSPOROCYTES1 gene encodes a putative leucine-rich repeat receptor protein kinase that controls somatic and reproductive cell fates in the *Arabidopsis* anther. *Genes & Development, 16*, 2021–2031.

Zhu, J., Chen, H., Li, H., Gao, J. F., Jiang, H., & Wang, C., et al. (2008). Defective in Tapetal Development and Function 1 is essential for anther development and tapetal function for microspore maturation in *Arabidopsis*. *Plant Journal, 55*, 266–277.

Zimmermann, P., Hirsch-Hoffmann, M., Hennig, L., & Gruissem, W. (2004). GENEVESTIGATOR. Arabidopsis Microarray Database and Analysis Toolbox. *Plant Physiology, 136*, 2621–2632.

Zou, C., Jiang, W., & Yu, D. (2010). Male gametophyte-specific WRKY34 transcription factor mediates cold sensitivity of mature pollen in *Arabidopsis*. *Journal of Experimental Botany, 61*, 3901–3914. (Epub 2010 Jul 19).

Advances in identifying and exploiting natural genetic variation

13

Christian S. Hardtke Kaisa Nieminen
University of Lausanne, Lausanne, Switzerland

TABLE OF CONTENTS

Natural Genetic Variation in Crop Breeding: From Prehistory to the Green Revolution

The creation of crops adapted for human use through plant breeding marks a major transition in human history from a sedentary to a mainly sessile lifestyle. This was achieved by domestication of wild plants through selection of phenotypic characteristics that are beneficial for human consumption (Purugganan and Fuller, 2009). The selection process relied on the natural genetic variation present in the ancestors of domesticated crop species, and on occasional spontaneous mutations that contributed to within-species natural genotypic variation. Today, the natural variation present in domesticated crop species is much lower than in their wild ancestor populations (Tanksley and McCouch, 1997; Wright et al., 2005), as it has been largely eliminated over centuries of active selection for agriculturally adaptive traits and exclusive mating within a domesticated population of a limited size (Buckler et al., 2001; Doebley et al., 2006). The out-crossing of the domesticated cultivars with their wild varieties was deliberately avoided by breeders, as this would have diluted many of the beneficial features accumulated during the domestication process. Natural variation available for selective breeding was thus limited, even if in some cases domestication might have occurred multiple times independently (Allaby et al., 2008; Brown et al., 2009), and most of the crop species varieties stayed relatively inert from the ancient times up to the twentieth century. By then, more artificial variation was introduced by induced mutations through deliberate human activity, such as chemical or radiation mutagenesis (Waugh et al., 2006). This artificial increase in the mutation spectrum contributed to breeding

DOI: 10.1016/B978-0-12-381466-1.00013-4

efforts; however, the so-called Green Revolution was initially driven by natural germplasm. The key event that dramatically increased worldwide agricultural productivity during the Green Revolution was the generation of semi-dwarfed wheat varieties with increased resistance to disease and lodging, as well as proportionally higher grain yield by Norman Borlaug and co-workers (Huang et al., 2002). Notably, the dwarfing alleles were contributed by ancient germplasm from East Asia and thus were supposedly derived from spontaneous mutation events (Borojevic, 2005). Similar, existing natural dwarfing alleles were also found in other crops, such as rice (Jodon and Beachell, 1943), or later induced through mutagenesis (Waugh et al., 2006).

Although a useful tool, mutagenesis of domesticated crops is limited in its ability to produce novel genetic variation. Chemical mutagenesis mostly introduces point mutations that can result in amino acid changes in the encoded proteins or premature stop codons leading to truncated proteins. By contrast, radiation generally induces larger size mutations, such as gene deletions or other structural genome rearrangements. Both methods can also result in gene expression (pattern) changes if regulatory sequences are affected. Thus, mutagenesis has the potential to produce loss-of-function and gain-of-function alleles, including hypo- and hyperactive ones. However, in the short term mutagenesis is only able to modify existing genetic material, not to introduce novel genes into the genome. Given the emerging picture of significant copy numbers and gene variation between strains (see the next section), relying exclusively on mutagenesis would leave the large potential of natural genetic variation untapped. This chapter will focus on presenting an overview of the work on natural genetic variation, including selected examples, and on outlining how this variation could be efficiently exploited for crop breeding.

The Genetic Limits of Evolving Domesticated Crops

The yield improvement through the Green Revolution reached its peak in the 1960s with the creation of novel dwarf varieties for various crops and the improvement of agricultural traits through mutagenesis approaches. Since then, the success of crop breeding has been more modest, although significant yield increases were still obtained throughout the remainder of the twentieth century (Huang et al., 2002; Wollenweber et al., 2005). However, the latter mainly represent the success of optimized farming practice and chemical inputs, such as fertilizers, herbicides, and insecticides. The success of the Green Revolution has to some degree become a bane for today's agricultural environment, as on the one hand, the zeitgeist demands more sustainable, less intensive agricultural schemes, while on the other, the global population growth driven by the Green Revolution creates pressure to ever improve and globally level crop productivity to feed the growing human population (Huang et al., 2002; Wollenweber et al., 2005). This is aggravated by the continuous race against the evolution of insects and diseases, and the deterioration of farm land, which will all decrease the yield of crop species if these are not continuously improved.

The far-reaching consequences of the reduction of genetic diversity in domesticated crops as compared to their wild progenitors are only now being recognized. Beyond the avoidance of out-crossing of domesticated cultivars to wild relatives to preserve their beneficial features, this reduction is also a consequence of the presumably small effective population size involved in individual domestication events (Buckler et al., 2001; Doebley et al., 2006; Allaby et al., 2008; Brown et al., 2009). Genetic diversity was further constrained by selection for the desired agronomic traits that distinguish crops from their wild ancestors. This is because selection does not only reduce diversity at selected loci as favorable alleles are increasing in frequency, but it also reduces diversity at genetically linked loci ("genetic hitchhiking"), partly by increasing the extent of linkage disequilibrium (LD). This can, for example, result in the linkage and fixation of a selected, favorable allele (e.g., for yield increase) of one gene with a hypoactive, eventually detrimental allele of another (e.g., of a disease resistance gene).

Tapping the natural genetic variation present in wild ancestors

The populations of the wild counterparts have been evolving without the pressure of human selection, thus harboring a larger natural variation than the domesticated crop species (Tanksley and McCouch, 1997). Plant breeders have been aware of the lack of genetic diversity in crops for some time; however, various constraints have prevented the exploitation of natural genetic variation available in the wild ancestors for the improvement of domesticated crop species. The most important one is the general loss of the beneficial, selected characteristics of a crop in the progeny from a cross to a wild ancestor. This is due to the additive effects of many QTL alleles that have accumulated during the improvement of a crop over centuries, a genetic constellation that is re-created at very low frequency in the progeny from the out-cross. For instance, to improve a parental crop whose performance hypothetically depends on a combination of ten homozygous loci by introducing one additional homozygous locus of interest from the wild ancestor is a daunting task. The desired variety carrying all 11 loci homozygously would statistically only be recovered once in more than 4 million F2 progeny (4^{11}), under the idealized assumption of free Mendelian segregation of all loci and no epistatic masking by other loci from the wild ancestor. Thus, this probability is even lower if problems arise, such as LD between one of the crop QTLs and the allele of interest from the wild ancestor. Therefore, in many ways the problem faced by plant breeders in bringing in more natural variation is a purely quantitative one that could be solved by marker-assisted breeding (i.e., the selection of progeny according to the genotypes of molecular markers known to be linked to the QTL), ideally combined with high-throughput phenotyping. The complexity can be further reduced by a multistep strategy, which identifies progeny that are homozygous for some, but heterozygous for the remaining QTL, and the stepwise identification of progeny fully homozygous for all QTL in the successive generations. However, this strategy assumes that all QTL contributing to the parental crop's performance are known and genetically tractable. Since this is not the case, even more time-consuming introgression of loci through repeated backcrossing to the parental elite variety is the best existing, and thus practiced, alternative.

In an ideal situation the markers linked to the QTL and the identity of the genes underlying those QTL and their allelic variants of interest would be known, which would allow direct genotyping of quantitative trait nucleotides (QTNs). In such a scenario, the isolation of any desired genotype derived from a cross would only depend on inherent biological limits, such as recombination or genetic incompatibilities and technical limits, such as the number of progeny that can be analyzed by high-throughput technology. To arrive at such a situation, the major technical obstacles to routinely identify QTNs of interest have to be overcome. Progress in model organisms in this respect over the last decade suggests that we might be getting close.

Natural Genetic Variation in *Arabidopsis*

In the 1980s, *Arabidopsis thaliana* (*Arabidopsis*) was chosen to be developed into a powerful model system for the investigation of the molecular basis of plant growth and development (Koornneef and Meinke, 2010). The focus of basic plant research on this model organism proved to be highly productive, leading to the isolation of molecular players in nearly every aspect of plant development and physiology through the mutagenesis approach. Thus, a wealth of knowledge on *Arabidopsis* genes and their role in various developmental and physiological processes was accumulated, together with excellent biological and molecular genetic resources for their investigation (e.g., Sussman et al., 2000; Initiative, 2000; Alonso et al., 2003; Rhee et al., 2003).

The need to develop *Arabidopsis* as a model system initially focused efforts toward a few isogenized standard genotypes, so-called accessions, which had been selected for their growth habit and short generation time, such as Col-0, Ws-0, or Ler-0. Mutagenesis approaches were usually performed in one of these backgrounds to enable direct comparisons of mutant phenotypes without additional, confounding natural genetic variation (i.e., through genetic modifiers). The use of other accessions was mostly restricted to their role as crossing partners to allow map-based cloning of mutated loci of interest (Lukowitz et al., 2000). However, the phenotypes that could be investigated in the standard backgrounds were inherently limited because of the characteristics they were chosen for, such as early flowering or absence of vernalization requirement (Johanson et al., 2000). Interestingly, these characteristics arise from partly known natural mutations in important modifier loci, such as a *phytochrome D* loss-of-function allele in Ws-0 that affects shade avoidance, elongation growth, and flowering time (Devlin et al., 1999), or a loss-of-function, mutation in the pleiotropic regulator of plant growth and architecture, *ERECTA*, in Ler (Torii et al., 1996; Godiard et al., 2003).

The utility of natural allelic *Arabidopsis* variants for molecular genetics was first convincingly demonstrated by the isolation of key regulators of vernalization, *FRIGIDA* (*FRI*) and *FLOWERING LOCUS C* (*FLC*) (Michaels and Amasino, 1999; Johanson et al., 2000). This was not possible in the standard backgrounds, which all carry null alleles of *fri* or its target, *FLC*, and do not require vernalization (Johanson et al., 2000; Caicedo et al., 2004; Shindo et al., 2005, 2006). However, the natural phenotypic variation of *Arabidopsis* had been a major interest of early *Arabidopsis* researchers, leaving a legacy of collections of isogenized natural accessions as well as crosses (Alonso-Blanco and Koornneef, 2000), such as the one leading to the identification of *FRI* (Napp-Zinn, 1957). These successes, as well as an influential review (Alonso-Blanco and Koornneef, 2000), fueled a renewed interest in the natural variation of *Arabidopsis*. This coincided with a key achievement in developing *Arabidopsis* into a model system, that is the deciphering of the complete genome sequence of the *Arabidopsis* reference strain Col-0 (Initiative, 2000), the first genome of a multicellular eukaryote to be sequenced. Availability of the genome sequence opened up the possibility of rapid identification of polymorphisms in other *Arabidopsis* accessions, and thus easy generation of molecular markers for segregation analyses of progeny derived from crosses between accessions. Systematic surveys of disparate genomes did indeed yield hundreds of novel molecular markers, greatly accelerating, for example, map-based cloning efforts or molecular evolutionary analyses (Borevitz et al., 2003; Schmid et al., 2003; Nordborg et al., 2005; West et al., 2006; Clark et al., 2007; Zhang et al., 2008; Zeller et al., 2008).

QTL Analyses in *Arabidopsis*

The easy access to molecular markers also rendered QTL analyses in *Arabidopsis* much more amenable than before. The theoretically limiting factors in QTL analysis are principally the extent of LD, the marker density, and the number of recombination events. In practice, the latter limitation could be overcome by creating a sufficiently large population of, ideally, recombinant inbred lines (RILs) through repeated selfing of F2 progeny and successive single seed descent generations up to near isogenization (typically from the seventh selfed generation onward; Alonso-Blanco et al., 1998; Singer et al., 2006). The new marker sets overcame the hurdle of sufficiently high marker density, and were also instrumental in estimating the extent of LD on a genome-wide scale (Nordborg et al., 2002; Plagnol et al., 2006; Kim et al., 2007). Thus, the isolation of QTL and eventually QTNs appeared to become feasible in *Arabidopsis*, coinciding with the renewed interest in its natural genetic variation (Alonso-Blanco and Koornneef, 2000; Koornneef et al., 2004). Numerous QTL studies were thus initiated with many of them focusing on traits of clearly adaptive value, such as flowering time or seed dormancy. This led to the isolation of mostly novel alleles of known genes and, in a few cases, it led to novel genes that had been missed in mutagenesis approaches. Next we will highlight a few examples of interest, since an excellent up-to-date summary of

QTL identified both in model organisms and crops has been recently published (Alonso-Blanco et al., 2009). We will also focus on hardwired natural genetic variation as opposed to epigenetic natural variation, which likely contributes significantly to phenotypic variety (Shindo et al., 2006; Vaughn et al., 2007; Zhai et al., 2008; Johannes et al., 2009) and is discussed in chapter 14 of this volume.

Novel *Arabidopsis* genes isolated through the natural variation approach

A remarkable practical aspect of QTL analyses is the quasi-immortal conservation of the biological material (i.e., the RILs) due to their homozygosity, which is maintained over subsequent generations (Alonso-Blanco et al., 1998). Thus, an RIL population that has been established and genotyped once can be screened repeatedly for multiple phenotypes to identify the respective underlying allelic variation. For instance, an RIL population derived from a cross between the Ler and Cvi accessions has been analyzed for dozens of phenotypes, eventually leading to the isolation of several QTLs (Alonso-Blanco et al., 2009). The fact that the Cvi accession from the Cape Verde Islands might represent a somewhat isolated accession adapted to a rather unusual latitude for *Arabidopsis* might have helped in this respect. The Ler × Cvi RIL population also led to the identification of a novel gene underlying a germination QTL, the *DELAY OF GERMINATION 1* (*DOG1*) gene, based on the low seed dormancy of Cvi and the higher dormancy of Ler (Bentsink et al., 2006). Genetic variation for seed dormancy in nature is a typical quantitative trait controlled by multiple loci on which environmental factors have a prominent effect. *DOG1* delays germination by imposing a requirement for a longer period of seed after-ripening, while at the same time increasing seed viability. The *DOG1* alleles appear to differ in the expression levels of their mRNA, which encodes a novel protein of unknown biochemical function. Decreased *DOG1* activity in Cvi is consistent with an adaptation to a mild, stable climate, since dormancy has mainly evolved in temperate climates to prevent germination when conditions are suitable, but the probability of survival of the seedling until reproduction is low (e.g., in warm days of autumn in contrast to spring).

Successful QTL analyses were also performed for various physiological and developmental traits, such as salt tolerance or root system architecture (Alonso-Blanco et al., 2009). Generally, this again revealed novel alleles of known genes, but in a few cases also previously uncharacterized genes. For instance, a screen for root system architecture variation identified a loss-of-function allele of the previously uncharacterized *BREVIS RADIX* (*BRX*) gene as a major QTL for primary root growth in an RIL population derived from the Uk-1 and Sav-0 parents (Mouchel et al., 2004). *BRX* encodes the founding member of a novel family of highly conserved plant-specific proteins found in both the mono- and dicotyledons, with a role in plant hormone signaling pathways (Mouchel et al., 2006). Moreover, a medium frequency hyperactive *BRX* allele of potentially adaptive value that leads to increased root system growth has also been identified (Beuchat et al., 2010). This is particularly interesting because such cases are still very rare, that is, most QTL alleles identified so far represent loss-of-function variants (Alonso-Blanco et al., 2009).

What to Expect: Intraspecific Variation in Gene Structure and Content

In summary, by now numerous QTLs for a whole suite of phenotypes have been identified in *Arabidopsis*. The majority of those loci have large effects, but some of them explain only a smaller portion of the phenotypic variance, proving the feasibility of the isolation of small effect QTLs (e.g., Kroymann and Mitchell-Olds, 2005; Bentsink et al., 2006; Sergeeva et al., 2006; Filiault et al., 2008; Beuchat et al., 2010). However, certain biases appear to prevail. For instance, one feature shared by the large majority of QTLs isolated so far is that compared to the Col-0 reference allele, they mostly represent rare loss-of-function alleles (Alonso-Blanco et al., 2009). Only a few examples of differentially active, functional alleles that have been experimentally confirmed exist (e.g., El-Din El-Assal et al. 2001; Maloof et al., 2001; Balasubramanian et al., 2006; Filiault et al., 2008; Beuchat et al., 2010). Moreover, amino acid polymorphisms appear to prevail over regulatory polymorphisms, since only few QTLs found are due to *bona fide* expression differences, such as mutations that affect regulatory regions but leave the transcript intact (Koornneef et al. 2004; Alonso-Blanco et al., 2009). In general, hardwired expression differences in *Arabidopsis* appear to be rare and are often likely to represent copy number variation or partial gene deletions rather than QTNs in regulatory motifs (Plantegenet et al., 2009). From a basic research perspective, this is somewhat disappointing when it comes to the ecological-evolutionary aspects of QTL. Although it is conceivable that loss of gene function can be adaptive, for instance, the loss of *FRI* function in the Cvi accession (Calcedo et al., 2004), this is expected to be the exception rather than the norm.

Structural genome variation: Higher than expected?

A straightforward explanation for this bias in isolated QTLs would be a selection bias by the investigators, who often focus on accessions that display the opposite extremes of a phenotype of interest. This increases the phenotypic variance, facilitating QTL isolation but possibly selecting against, or masking, alleles with subtle effects. Moreover, it appears possible that the genome evolution of *Arabidopsis* introduces a systemic bias toward loss-of-function alleles. The latter is suggested by the first glimpses at genetic variation between accessions at the whole genome sequence level. Only a few accessions have been re-sequenced by high coverage shotgun sequencing so far; however, the picture emerging from these data already indicates a highly dynamic *Arabidopsis* genome (Ossowski et al., 2008; Santuari et al., 2010). While single nucleotide polymorphisms (SNPs) are prevalent at the frequencies already predicted by SNP searches using marker fragments, whole genome sequencing also indicates a surprising number of sizeable deletions and insertions (indels). These partially reflect common ancestry, since there is substantial overlap between the indels observed in individual accessions (Santuari et al., 2010). Frequently, deletions affect large parts of coding regions. Thus, compared to the Col-0 reference sequence, divergent accessions carry numerous null alleles. Given the genome duplication history of *Arabidopsis*, this may not be surprising and might be a species-specific feature. Although the underlying molecular causes might differ, substantial structural variation has also been observed in crop species (Huang et al., 2008; Gore et al., 2009; Springer et al., 2009).

QTL Analysis and Sequence Variation in Crops

Domestication genes of maize

Isolation of QTL alleles in crops has been equally boosted by the elucidation of the whole genome sequences of various grasses, most prominently rice and maize (Project, 2005; Schnable et al., 2009). In maize, however, QTL for several agricultural traits have been isolated through the pioneering work of Doebley and co-workers. Their studies aimed to isolate alleles that were selected during maize domestication by analyses of crosses with the wild maize ancestor, the Mexican grass teosinte (Doebley et al., 2006). During the domestication process, profound differences between the maize and teosinte architecture have been fixed through human selection. For instance, while teosinte shoots have several elongated branches with male inflorescences (tassels) at their tips and female inflorescences (ears) in secondary axillary branches next to the leaves, maize shoots instead produce only short branches with seed-bearing ears at the ends. The suppression of branching and branch elongation, as well as the change in the inflorescence sex at the tip of the branch in maize, resulted in production of a lower number of ears per plant, but allowed each ear to grow larger (Doebley et al., 1995). Through QTL analysis, a major locus regulating these morphological differences, *teosinte branched 1 (tb1)*, was identified (Doebley et al., 1997). *tb1* encodes a TCP class transcription factor and acts as a repressor of lateral organ growth and a promoter of the development of female inflorescences. In teosinte, *tb1* is expressed in the axillary meristems of the main branches, whereas in maize its expression is high in the main lateral branch meristems. The domestication process has therefore altered the expression pattern of this gene. Indeed, during maize domestication, polymorphisms in the promoter region of *tb1* were drastically reduced as the selective pressure by the breeders, aimed to optimize maize morphology for human use, unconsciously modified the level and pattern of *tb1* expression. Thus, the *tb1* QTL represents a unique example of regulatory variation in that both expression pattern and level are affected, probably because of an indel affecting an upstream enhancer (Clark et al., 2006). Interestingly, part of this change is likely due to another, epistatic QTL, which might be identical to the *barren stalk 1* (*ba1*) gene, since *ba1* is required for normal expression levels of *tb1*(Gallavotti et al., 2004). *ba1* encodes a non-canonical member of the bHLH transcription factor class. A natural *ba1* allele, which carries a large insertion in the 5′ regulatory regions, has been identified. Interestingly, this allele appears to have been selected by breeders relatively recently in attempts to enhance the *tb1* phenotype. Thus, even in domesticated crops, residual natural variation that can be selected for exists. It is possible, however, that this is more prevalent in maize than in other crops because of its highly dynamic genome (Gore et al., 2009; Schnable et al., 2009), which also suggests that the *ba1* mutation might have arisen within the domesticated germplasm.

Examples from rice

Interestingly, isolation of the *ba1* gene was aided by a candidate gene approach, since a homolog, the *LAX PANICLE* (*LAX*) gene, had already been identified in rice (Komatsu et al., 2003). *LAX* loss-of-function results in a rice phenotype reminiscent of the *ba1* phenotype. Rice shoots consist of tillers, which are panicle (a grain-bearing organ) forming shoots that emerge from the basal internode of the mother stem (the culm). Similar to other monocotyledon crops, rice architecture is crucial for grain yield and is determined by plant height, tiller number and angle, number of panicles per tiller, and branching of the panicles. Identification of genes and allelic variants that regulate these traits would help us to further understand the molecular mechanisms that regulate productivity and would facilitate breeding of new varieties that increase grain yield. For instance, the wild ancestor of domesticated rice shows a prostate growth habit with a wide tiller angle and a short stature with many tillers. To allow high-density cultivation, ancient humans selected for plants with a more erect growth habit, due to a narrower tiller angle. The transition from the prostate growth of ancestral wild rice (*Oryza rufipogon*) to the erect growth of *O. sativa* was one of the most critical events in rice domestication. Through genetic studies of a cross between domesticated and wild rice, the gene *PROSTATE GROWTH 1* (*PROG1*) was identified as an important regulator for this change in growth habit during rice domestication (Jin et al., 2008; Tan et al., 2008). *PROG1* encodes a C_2H_2 zinc finger protein, which likely acts as a transcription factor and is predominantly expressed in axillary meristems of the culm, where the tiller buds form. Several fixed mutations have been identified within the *PROG1* coding region of domesticated varieties and it is thought that one of them, a threonine to serine change, inactivates the protein (Jin et al., 2008).

While *PROG1* behaves as a single Mendelian, large effect locus, two QTLs for grain yield were mapped in a recent rice breeding approach (Miura et al., 2010). One of them, the *SQUAMOSA PROMOTER BINDING PROTEIN-LIKE 14* (*OsSPL14*) locus, showed up as a single, semi-dominant Mendelian locus in a parallel study (Jiao et al., 2010), emphasizing the advantage of analyzing independent populations with different tester parents in parallel. Both studies compared classic *japonica* rice varieties with different lines that exhibit less tillers but more branched and productive panicles. Strikingly, the studies revealed that the lines with the more productive panicles carry a point mutation, which affects the *OsSPL14* transcript stability by rendering the transcript less susceptible to microRNA-mediated decay. Consequently, more OsSPL14 protein is expressed, which both represses tiller formation during the vegetative stage and promotes panicle branching during the reproductive stage. This results in rice plants with fewer tillers, but increased lodging resistance and more productive panicles. Since the reduced tillering can be compensated by a higher planting density, such varieties consequently display higher grain yields.

Examples from other cereals

A trait that is historically important for optimizing yield in both seed and vegetative crops is the timing of the transition between the vegetative and reproductive phases of the life cycle, for example, flowering. This transition is stimulated by both environmental and developmental signals, and has been characterized in great detail in *Arabidopsis*, revealing the major, conserved molecular players (Baurle and Dean, 2006). The timing of flowering is a major factor in regional climatic adaptation of crop species. To maximize yield, it is essential that a crop's life cycle is adapted to the agro-environments in which it is grown. In vegetative crops, the aim is usually to delay flowering, while in seed crops, the exact timing of flowering is a major determinant of grain yield. Vernalization accelerates flowering by promoting the switch from vegetative to reproductive development, depending on the length of the cold period (Baurle and Dean, 2006). The longer the cold period, the faster flowering will occur, up to a certain threshold when the vernalization response becomes saturated. Winter varieties require vernalization and are sown in late summer or autumn, so that they can be vernalized during winter and then flower in spring. Spring cereals do not require vernalization and are sown in early spring and harvested in late summer. Natural variation studies have identified the major vernalization response genes in wheat and barley, *VERNALIZATION 1* (*VRN1*), *VRN2*, and *VRN3* (note that these loci are not homologous to the *VERNALIZATION* loci of *Arabidopsis*; Trevaskis et al., 2007). These genes regulate the response to cold period and promotion of flowering by long days. *VRN1* encodes a transcription factor that promotes inflorescence meristem identity (Yan et al., 2003). In winter varieties, expression of *VRN1* is strongly induced by a period of low temperature. The extent of this induction depends on the length of the vernalization period, resulting in a quantitative effect on the timing of inflorescence initiation. The activity of *VRN1* is enhanced by *VRN3*, the homolog of the *Arabidopsis* florigen, *FLOWERING LOCUS T* (*FT*), whose expression is induced by long days and further accelerates reproductive apex development (Yan et al., 2006). *VRN2*, a repressor of flowering, integrates vernalization and day-length responses by repressing *VRN3* until plants are vernalized (Yan et al., 2004, 2006). *VRN2* is repressed by *VRN1* when days start to lengthen in the spring, thus allowing the *VRN3*-mediated flowering response to long days. The *VRN1*, *VRN2*, and *VRN3* genes should be useful to cereal breeding programs. An increased understanding of how these genes interact to control flowering time should also help us to predict how environmental factors affect flowering in different genetic backgrounds. This will aid in the development of varieties of major food crops, such as wheat, that are able to cope with climate change.

Toward Prediction of Variation in Molecular Function: Why Model Organisms are here to Stay

An interesting aspect of the findings highlighted earlier, as well as other QTL studies, is the importance of architectural changes in domestication and yield increase (Doust, 2007), which might also explain the prevalence of regulatory factors identified in those studies. Another example is an important locus selected during maize domestication, *teosinte glume architecture* (*tga1*), a SQUAMOSA PROMOTER BINDING PROTEIN (SBP) class transcription factor that affects seed case development (Wang et al., 2005). A *tga1* allele carrying several mutations, most notably a likely causal amino acid change, led to the naked grain phenotype of maize.

The presumed causal polymorphism of *tga1* has not been directly verified in experiments, which illustrates a common problem in the analysis of crop QTL alleles. With very few exceptions, QTL alleles isolated from maize, rice, or wheat thus far have not been characterized with respect to molecular function. While in some cases, rearrangements in the 5′ regulatory region and associated expression level differences suggest a loss-of-function, in most cases the causal polymorphisms can only be guessed at. For example, a maize QTL for flowering time selected during domestication, *Dwarf8*, carries several polymorphisms, including indels in the regulatory region, as well as amino acid deletions (Thornsberry et al., 2001). These polymorphisms are genetically linked and found together in all varieties investigated. Thus, which of these polymorphisms is causal can only be a best guess and has not been formally determined. This is true for most crop QTLs isolated so far, principally because tracking down the causal polymorphism would require laborious and lengthy transgenic analyses.

Crucial support from model organism candidate genes

Notably, when a putative causal polymorphism has been identified, it is mostly by inference from mutant analyses in model organisms. In particular, the analysis of *Arabidopsis* growth and development has furnished a huge amount of data on gene function that can be exploited to understand the molecular basis of allelic variation in the crop homologs. Identification of many crop QTL profited largely from candidate approaches driven by known genes from model organisms, significantly facilitating and accelerating QTL identification once coarse mapping had been achieved. This was possible because of the general evolutionary proximity of flowering plants to each other, and therefore the often conceptually similar role of the homologous genes in different species.

Maybe the most prominent example is the *Arabidopsis* gene *GIBBERELLIC ACID INSENSITIVE* (*GAI*), a negative regulator of GA response in *Arabidopsis*. In the dwarfed *Arabidopsis gai* mutant, the gai protein misses an amino acid stretch, resulting in its stabilization and constitutive activity, and thus the inhibition of stem elongation (Peng et al., 1997). Strikingly, the dwarfing Green Revolution gene *Reduced height 1* of wheat turned out to be a *GAI* homolog, and the dwarfing alleles are equally affected by a deletion in the same amino acid motif (Peng et al., 1999). This is also true for dwarfing alleles of the maize homolog, *Dwarf8*. Notably, these deletions are not found in the *dwarf8* alleles that have been selected for flowering time mentioned earlier, providing an example of a pleiotropic locus that can affect various traits, as is frequently observed in *Arabidopsis* (e.g., *FLC* or *ERECTA*). Similarly, isolation of many other crop QTLs was aided by the preceding detailed analysis of their model organism homologs, for example, *tb1* (a homolog of *Antirrhinum* *CYCLOIDEA*) or *VRN3* (a homolog of *Arabidopsis FT*). In some cases however, *Arabidopsis* or generally dicotyledon homologs do not appear to exist, such as for the *LAX* gene of rice. *LAX* was instrumental in isolating *ba1* from maize (Gallavotti et al., 2004), emphasizing the need for a well-developed monocotyledon model system, such as *Brachypodium distachyon* (Draper et al., 2001; Initiative, 2010).

Model systems as references to characterize allele activities

Advantages of the model systems include short generation time, well-established methodology, and biological resources. This permits routine and quick generation of transgenic lines. Combined with null mutants for nearly each gene available (e.g., Alonso et al., 2003), direct testing of the activity of allelic variants if the mutants display a quantifiable phenotype is possible. In *Arabidopsis*, this has been used to evaluate the importance of amino acid residues or regulatory elements, or to measure the activity of different functional (i.e., not null) alleles in the context of natural variation. Such analyses might be extended to include homologous genes from other species (e.g., Beuchat et al., 2010), first to prove their equivalent biochemical role, and secondly to determine the impact of mutations observed in nature. If done at high throughput, this strategy could be applied to survey the activity of the natural polymorphisms in the alleles of a given gene (e.g., Filiault et al., 2008; Beuchat et al., 2010). Subsequently, this information could then be used to pre-select varieties that carry alleles of interest for breeding efforts, without *a priori* knowledge of the varieties' phenotypic characteristics. Thus, the results from complementation assays in a model organism could be exploited to predict the activity and phenotypic impact of natural crop alleles to drive crop improvement.

Beyond Simple Traits: Epigenetics, Heterosis, Genetic Incompatibility, and Trade-offs

Incompatibility between natural accessions

Another aspect in which model systems will remain invaluable is to understand genetic phenomena that are relevant for plant breeding. This includes the molecular basis of heterosis (e.g., Rosas et al., 2010), epigenetic phenomena (e.g., Shindo et al., 2006; Vaughn et al., 2007), or the limits of breeding due to incompatibility. The latter phenomenon occurs frequently enough to pose a practical problem, but its potential molecular causes are only beginning to be understood very recently thanks to analyses of *Arabidopsis*. These studies demonstrated that certain combinations of natural alleles in one genome are either not viable or result in severely decreased fitness, thus limiting the genetic constitutions that can be recovered from the respective cross between natural accessions. For instance, in a pioneering study by Weigel and co-workers, decreased fitness due to hybrid necrosis was demonstrated to result from an incompatible interaction between alleles of certain disease-resistance loci (Bomblies et al., 2007). This was first discovered in a cross between the Uk-1 and Uk-3 accessions, but eventually shown to occur in several other crosses, suggesting that such "autoimmune response" might be a common cause for incompatibility.

Other studies demonstrated that synthetic lethality can arise from genetic interaction between paralogous loci, highlighting the dynamics of structural genome rearrangements and mutation rate in *Arabidopsis* (Ossowski et al., 2008, 2010; Santuari et al., 2010). For example, an essential gene in amino acid biosynthesis was shown to be duplicated on two different chromosomes. Differential evolutionary trajectories of those paralogs in the Col-0 and Cvi accessions have resulted in their differential expression levels and the deletion of one paralog in Cvi (Bikard et al., 2009). As a consequence, the functional (expressed) paralog in the Col-0 accession is located in the position of the deleted Cvi allele, whereas the remaining functional Cvi paralog on a different chromosome is located in the position of the hypoactive (i.e., lowly expressed) Col-0 allele. Consequently, only certain combinations between the four alleles result in sufficient overall gene activity, whereas other combinations impair growth rate or are not even viable. Notably, independent non-functionalization events of one, but not the other, paralog were observed in other accessions, suggesting that this type of incompatibility could be widespread. A conceptually similar case, involving a transposition event, was also identified for reduced shoot growth (Vlad et al., 2010).

The previously mentioned studies illustrate the very dynamic rate of duplication and transposition events, even over evolutionary short periods of time, resulting in various divergent although still functional allele combinations fixed in different accessions. This variation remains largely hidden in wild populations of *Arabidopsis* because it is mainly selfing, but it restricts the options for out-crossing, thereby contributing to genetic drift and eventually speciation (Bomblies and Weigel, 2007). Because *Arabidopsis* is selfing, these incompatibilities might not be as strongly selected against as in obligate out-crossing species, but they might occur more frequently in crops because of their inbreeding history. Thus, understanding of these processes in model organisms might help us to understand, and eventually even predict, incompatibilities between crop varieties.

Trade-offs between different beneficial traits

Another phenomenon that could limit crop improvement are trade-offs between different traits that have been shaped by selective pressures. The molecular basis of these trade-offs is not very well understood, although there are some general notions that there is a constant balancing of growth versus defense. A recent study has confirmed this concept for *Arabidopsis* accessions, demonstrating that allelic diversity at the *ACCELERATED CELL DEATH 6* (*ACD6*) locus determines pleiotropic differences in both vegetative growth and resistance to microbial infection, as well as herbivory (Todesco et al., 2010). A hyperactive *ACD6* allele strongly enhances pathogen resistance while at the same time slowing biomass accumulation. This allele is found at intermediate frequency throughout the worldwide range of *Arabidopsis*, suggesting that it provides a substantial fitness benefit despite its marked impact on growth. Similar genetic scenarios are conceivable in crops and might occur there even more frequently because of the largely unidirectional selection pressures imposed by breeders.

Extending the Toolbox: Genome-wide Association Mapping

In summary, the characterization of natural genetic variation has made big strides during the last decade, thanks to the accumulating wealth of biological data from model systems and the quantum leaps in high-throughput technologies. The technological progress also fuels more ambitious projects that are relevant for plant biotechnology, such as association mapping. In *Arabidopsis*, a first attempt at large-scale genome-wide association (GWA) mapping was published recently, presenting a set of 95 accessions that have been genotyped for about 250,000 SNPs on average for 1 SNP per 500 bp (Atwell et al., 2010). These accessions were phenotyped for more than 100 traits, and several strong associations were obtained after correction for possible confounding effects of population structure. Among them, previously identified large effect loci for traits with a wide quantitative range, such as flowering time, were confirmed. For many other traits, significant associations were found. It thus appears likely that GWA studies will become a standard tool for dissecting natural variation and identifying candidate genes. Independent experimental verification will remain necessary, however, as GWA alone will often not allow accurate estimation of allelic effects. Moreover, GWA can only detect alleles that explain a sufficient fraction of the variation in the mapping population and therefore have to be reasonably common. The resolution of GWA can thus be improved by increasing the sample size of accessions or varieties investigated, as well as by characterizing the complete polymorphism set of all lines by whole genome sequencing like those attempted by the 1001 Genomes Project (Weigel and Mott, 2009).

However, some inherent limits to GWA cannot be overcome, such as distortions due to co-evolution of physically unlinked loci seen in the examples of incompatibility presented earlier in this chapter. To what degree GWA will drive marker-assisted crop breeding remains to be seen, but so far only small-scale studies with limited success, in the sense that no new loci were identified, exist (Aranzana et al., 2005; Zhao et al., 2007; Ehrenreich et al., 2009; Weber et al., 2009). Nevertheless, it appears likely that GWA combined with dedicated QTL analyses of RILs and experimental verification of allelic variant activities will be a complementing and very powerful approach to dissect the allelic contributions to traits of interest. The availability of more sophisticated, multi-parental RIL populations and RIL populations characterized at the whole genome sequence level will greatly contribute to these efforts (e.g., Sibout et al., 2008; Kover et al., 2009; Santuari et al., 2010).

The Route to Effectively Exploit Natural Variation for Plant Biotechnology

Through technology developed in studies of natural variation of model species, we will eventually be able to routinely associate natural variation with adaptive agricultural features in crops and identify novel genes or adaptive alleles from natural variation present in the wild. As an increasing number of genes regulating agronomical traits in crop species are identified, their sequences can be used by breeders as functional markers for selecting favorable genotypes.

A major question in plant biotechnology will be how to best exploit the identified natural variation of interest. This aspect is indivisible from developments in legislation and public perception concerning such issues as patenting or genetically modified crops. The most inoffensive approach would be classic marker-assisted breeding that aims to recover a desired allelic combination in the progeny of a given cross or introgression between varieties carrying the alleles of interest. In principal, the new technologies enable us to effectively breed crop varieties even from crosses with their wild counterparts through a limited number of generations. Identification of alleles responsible for the desired features enables us to confirm their presence in the genome even in the absence of any visible phenotypic traits. Thus, we do not have to rely exclusively on phenotypic traits any longer; we can directly identify individuals carrying the desirable alleles through the use of genetic markers and select them for further breeding. As outlined in the introduction, in practice, this approach is mainly limited by the available genotyping throughput and the number of individuals that can be analyzed. Another important constraint, however, is the limited knowledge about all of the alleles responsible for the additive beneficial features of the parental crop varieties. Conservation of these beneficial features in the derived, desired variety would thus require time-consuming, precise introgression of the alleles of interest from the other parent. Therefore, the rational and straightforward solution would be the introduction of desired alleles into elite varieties by transgenic means. Beyond simple transgenesis, the recent breakthroughs in gene targeting in plants would even enable targeted replacement of alleles (Shukla et al., 2009; Townsend et al., 2009; Osakabe et al., 2010; Zhang et al., 2010). Arguably, this would be the optimal way to bring beneficial natural variation back into crop germplasms and pave the way for true designer crops.

References

Allaby, R. G., Fuller, D. Q., & Brown, T. A. (2008). The genetic expectations of a protracted model for the origins of domesticated crops. *Proceedings of the National Academy of Sciences of the United States of America*, *105*, 13982–13986.

Alonso, J. M., Stepanova, A. N., Leisse, T. J., Kim, C. J., Chen, H., & Shinn, P., et al. (2003). Genome-wide insertional mutagenesis of *Arabidopsis thaliana*. *Science*, *301*, 653–657.

Alonso-Blanco, C., Aarts, M. G., Bentsink, L., Keurentjes, J. J., Reymond, M., & Vreugdenhil, D., et al. (2009). What has natural variation taught us about plant development, physiology, and adaptation? *Plant Cell*, *21*, 1877–1896.

Alonso-Blanco, C., & Koornneef, M. (2000). Naturally occurring variation in *Arabidopsis*: An underexploited resource for plant genetics. *Trends in Plant Science*, *5*, 22–29.

Alonso-Blanco, C., Koornneef, M., & Stam, P. (1998). The use of recombinant inbred lines (RILs) for genetic mapping. *Methods in Molecular Biology*, *82*, 137–146.

Aranzana, M. J., Kim, S., Zhao, K., Bakker, E., Horton, M., & Jakob, K., et al. (2005). Genome-wide association mapping in *Arabidopsis* identifies previously known flowering time and pathogen resistance genes. *PLoS Genetics*, *1*, e60.

Atwell, S., Huang, Y. S., Vilhjalmsson, B. J., Willems, G., Horton, M., & Li, Y., et al. (2010). Genome-wide association study of 107 phenotypes in *Arabidopsis thaliana* inbred lines. *Nature*, *465*, 627–631.

Balasubramanian, S., Sureshkumar, S., Agrawal, M., Michael, T. P., Wessinger, C., & Maloof, J. N., et al. (2006). The *PHYTOCHROME* C photoreceptor gene mediates natural variation in flowering and growth responses of *Arabidopsis thaliana*. *Nature Genetics*, *38*, 711–715.

Baurle, I., & Dean, C. (2006). The timing of developmental transitions in plants. *Cell*, *125*, 655–664.

Bentsink, L., Jowett, J., Hanhart, C. J., & Koornneef, M. (2006). Cloning of DOG1, a quantitative trait locus controlling seed dormancy in *Arabidopsis*. *Proceedings of the National Academy of Sciences of the United States of America*, *103*, 17042–17047.

Beuchat, J., Li, S., Ragni, L., Shindo, C., Kohn, M. H., & Hardtke, C. S. (2010). A hyperactive quantitative trait locus allele of *Arabidopsis* BRX contributes to natural variation in root growth vigor. *Proceedings of the National Academy of Sciences of the United States of America*, *107*, 8475–8480.

Bikard, D., Patel, D., Le Mette, C., Giorgi, V., Camilleri, C., & Bennett, M. J., et al. (2009). Divergent evolution of duplicate genes leads to genetic incompatibilities within *A. thaliana*. *Science*, *323*, 623–626.

Bomblies, K., Lempe, J., Epple, P., Warthmann, N., Lanz, C., & Dangl, J. L., et al. (2007). Autoimmune response as a mechanism for a Dobzhansky-Muller-type incompatibility syndrome in plants. *PLoS Biology*, *5*, e236.

Bomblies, K., & Weigel, D. (2007). *Arabidopsis*: A model genus for speciation. *Current Opinion in Genetics & Development*, *17*, 500–504.

Borevitz, J. O., Liang, D., Plouffe, D., Chang, H. S., Zhu, T., & Weigel, D., et al. (2003). Large-scale identification of single-feature polymorphisms in complex genomes. *Genome Research*, *13*, 513–523.

Borojevic, K. (2005). The transfer and history of "reduced height genes" (Rht) in wheat from Japan to Europe. *The Journal of Heredity*, *96*, 455–459.

Brown, T. A., Jones, M. K., Powell, W., & Allaby, R. G. (2009). The complex origins of domesticated crops in the Fertile Crescent. *Trends in Ecology & Evolution*, *24*, 103–109.

Buckler, E. S., Thornsberry, J. M., & Kresovich, S. (2001). Molecular diversity, structure and domestication of grasses. *Genetical Research*, *77*, 213–218.

Caicedo, A. L., Stinchcombe, J. R., Olsen, K. M., Schmitt, J., & Purugganan, M. D. (2004). Epistatic interaction between *Arabidopsis* FRI and FLC flowering time genes generates a latitudinal cline in a life history trait. *Proceedings of the National Academy of Sciences of the United States of America*, *101*, 15670–15675.

Clark, R. M., Schweikert, G., Toomajian, C., Ossowski, S., Zeller, G., & Shinn, P., et al. (2007). Common sequence polymorphisms shaping genetic diversity in *Arabidopsis thaliana*. *Science*, *317*, 338–342.

Clark, R. M., Wagler, T. N., Quijada, P., & Doebley, J. (2006). A distant upstream enhancer at the maize domestication gene *tb1* has pleiotropic effects on plant and inflorescent architecture. *Nature Genetics*, *38*, 594–597.

Devlin, P. F., Robson, P. R., Patel, S. R., Goosey, L., Sharrock, R. A., & Whitelam, G. C. (1999). *Phytochrome D* acts in the shade-avoidance syndrome in *Arabidopsis* by controlling elongation growth and flowering time. *Plant Physiology*, *119*, 909–915.

Doebley, J., Stec, A., & Gustus, C. (1995). *teosinte branched 1* and the origin of maize: Evidence for epistasis and the evolution of dominance. *Genetics*, *141*, 333–346.

Doebley, J., Stec, A., & Hubbard, L. (1997). The evolution of apical dominance in maize. *Nature*, *386*, 485–488.

Doebley, J. F., Gaut, B. S., & Smith, B. D. (2006). The molecular genetics of crop domestication. *Cell*, *127*, 1309–1321.

Doust, A. (2007). Architectural evolution and its implications for domestication in grasses. *Annals of Botany*, *100*, 941–950.

Draper, J., Mur, L. A., Jenkins, G., Ghosh-Biswas, G. C., Bablak, P., & Hasterok, R., et al. (2001). *Brachypodium distachyon*. A new model system for functional genomics in grasses. *Plant Physiology*, *127*, 1539–1555.

Ehrenreich, I. M., Hanzawa, Y., Chou, L., Roe, J. L., Kover, P. X., & Purugganan, M. D. (2009). Candidate gene association mapping of *Arabidopsis* flowering time. *Genetics*, *183*, 325–335.

El-Din El-Assal, S., Alonso-Blanco, C., Peeters, A. J., Raz, V., & Koornneef, M. (2001). A QTL for flowering time in *Arabidopsis* reveals a novel allele of CRY2. *Nature Genetics*, *29*, 435–440.

Filiault, D. L., Wessinger, C. A., Dinneny, J. R., Lutes, J., Borevitz, J. O., & Weigel, D., et al. (2008). Amino acid polymorphisms in *Arabidopsis phytochrome B* cause differential responses to light. *Proceedings of the National Academy of Sciences of the United States of America*, *105*, 3157–3162.

Gallavotti, A., Zhao, Q., Kyozuka, J., Meeley, R. B., Ritter, M. K., & Doebley, J. F., et al. (2004). The role of *barren stalk1* in the architecture of maize. *Nature*, *432*, 630–635.

Godiard, L., Sauviac, L., Torii, K. U., Grenon, O., Mangin, B., & Grimsley, N. H., et al. (2003). *ERECTA*, an LRR receptor-like kinase protein controlling development pleiotropically affects resistance to bacterial wilt. *Plant Journal*, *36*, 353–365.

Gore, M. A., Chia, J. M., Elshire, R. J., Sun, Q., Ersoz, E. S., & Hurwitz, B. L., et al. (2009). A first-generation haplotype map of maize. *Science*, *326*, 1115–1117.

Huang, J., Pray, C., & Rozelle, S. (2002). Enhancing the crops to feed the poor. *Nature*, *418*, 678–684.

Huang, X., Lu, G., Zhao, Q., Liu, X., & Han, B. (2008). Genome-wide analysis of transposon insertion polymorphisms reveals intraspecific variation in cultivated rice. *Plant Physiology*, *148*, 25–40.

Initiative, I. B. (2010). Genome sequencing and analysis of the model grass *Brachypodium distachyon*. *Nature*, *463*, 763–768.

Initiative, T. A. G. (2000). Analysis of the genome sequence of the flowering plant *Arabidopsis thaliana*. *Nature*, *408*, 796–815.

Jiao, Y., Wang, Y., Xue, D., Wang, J., Yan, M., & Liu, G., et al. (2010). Regulation of *OsSPL14* by *OsmiR156* defines ideal plant architecture in rice. *Nature Genetics*, *42*, 541–544.

Jin, J., Huang, W., Gao, J. P., Yang, J., Shi, M., & Zhu, M. Z., et al. (2008). Genetic control of rice plant architecture under domestication. *Nature Genetics*, *40*, 1365–1369.

Jodon, N. E., & Beachell, M. H. (1943). Rice dwarf mutations and their inheritance. *Journal of Heredity*, *34*, 155–160.

Johannes, F., Porcher, E., Teixeira, F. K., Saliba-Colombani, V., Simon, M., & Agier, N., et al. (2009). Assessing the impact of transgenerational epigenetic variation on complex traits. *PLoS Genetics*, *5*, e1000530.

Johanson, U., West, J., Lister, C., Michaels, S., Amasino, R., & Dean, C. (2000). Molecular analysis of *FRIGIDA*, a major determinant of natural variation in *Arabidopsis* flowering time. *Science*, *290*, 344–347.

Kim, S., Plagnol, V., Hu, T. T., Toomajian, C., Clark, R. M., & Ossowski, S., et al. (2007). Recombination and linkage disequilibrium in *Arabidopsis thaliana*. *Nature Genetics*, *39*, 1151–1155.

Komatsu, K., Maekawa, M., Ujiie, S., Satake, Y., Furutani, I., & Okamoto, H., et al. (2003). LAX and SPA: Major regulators of shoot branching in rice. *Proceedings of the National Academy of Sciences of the United States of America*, *100*, 11765–11770.

Koornneef, M., Alonso-Blanco, C., & Vreugdenhil, D. (2004). Naturally occurring genetic variation in *Arabidopsis thaliana*. *Annual Review of Plant Biology*, *55*, 141–172.

Koornneef, M., & Meinke, D. (2010). The development of *Arabidopsis* as a model plant. *Plant Journal*, *61*, 909–921.

Kover, P. X., Valdar, W., Trakalo, J., Scarcelli, N., Ehrenreich, I. M., & Purugganan, M. D., et al. (2009). A Multiparent Advanced Generation Inter-Cross to fine-map quantitative traits in *Arabidopsis thaliana*. *PLoS Genetics*, *5*, e1000551.

Kroymann, J., & Mitchell-Olds, T. (2005). Epistasis and balanced polymorphism influencing complex trait variation. *Nature*, *435*, 95–98.

Lukowitz, W., Gillmor, C. S., & Scheible, W. R. (2000). Positional cloning in *Arabidopsis*. Why it feels good to have a genome initiative working for you. *Plant Physiology*, *123*, 795–805.

Maloof, J. N., Borevitz, J. O., Dabi, T., Lutes, J., Nehring, R. B., & Redfern, J. L., et al. (2001). Natural variation in light sensitivity of *Arabidopsis*. *Nature Genetics*, *29*, 441–446.

Michaels, S. D., & Amasino, R. M. (1999). FLOWERING LOCUS C encodes a novel MADS domain protein that acts as a repressor of flowering. *Plant Cell*, *11*, 949–956.

Miura, K., Ikeda, M., Matsubara, A., Song, X. J., Ito, M., & Asano, K., et al. (2010). *OsSPL14* promotes panicle branching and higher grain productivity in rice. *Nature Genetics*, *42*, 545–549.

Mouchel, C. F., Briggs, G. C., & Hardtke, C. S. (2004). Natural genetic variation in *Arabidopsis* identifies BREVIS RADIX, a novel regulator of cell proliferation and elongation in the root. *Genes & Development*, *18*, 700–714.

Mouchel, C. F., Osmont, K. S., & Hardtke, C. S. (2006). BRX mediates feedback between brassinosteroid levels and auxin signalling in root growth. *Nature*, *443*, 458–461.

Napp-Zinn, K. (1957). Untersuchungen zur Genetik des Kältebedürfnisses bei *Arabidopsis thaliana* (L.) Heynh. *Z. Indukt. Abstamm Vererbungsl*, *88*, 253–285.

Nordborg, M., Borevitz, J. O., Bergelson, J., Berry, C. C., Chory, J., & Hagenblad, J., et al. (2002). The extent of linkage disequilibrium in *Arabidopsis thaliana*. *Nature Genetics*, *30*, 190–193.

Nordborg, M., Hu, T. T., Ishino, Y., Jhaveri, J., Toomajian, C., & Zheng, H., et al. (2005). The pattern of polymorphism in *Arabidopsis thaliana*. *PLoS Biology*, *3*, e196.

Osakabe, K., Osakabe, Y., & Toki, S. (2010). Site-directed mutagenesis in *Arabidopsis* using custom-designed zinc finger nucleases. *Proceedings of the National Academy of Sciences of the United States of America*, *107*, 12034–12039.

Ossowski, S., Schneeberger, K., Clark, R. M., Lanz, C., Warthmann, N., & Weigel, D. (2008). Sequencing of natural strains of *Arabidopsis thaliana* with short reads. *Genome Research*, *18*, 2024–2033.

Ossowski, S., Schneeberger, K., Lucas-Lledo, J. I., Warthmann, N., Clark, R. M., & Shaw, R. G., et al. (2010). The rate and molecular spectrum of spontaneous mutations in *Arabidopsis thaliana*. *Science*, *327*, 92–94.

Peng, J., Carol, P., Richards, D. E., King, K. E., Cowling, R. J., & Murphy, G. P., et al. (1997). The *Arabidopsis GAI* gene defines a signaling pathway that negatively regulates gibberellin responses. *Genes & Development*, *11*, 3194–3205.

Peng, J., Richards, D. E., Hartley, N. M., Murphy, G. P., Devos, K. M., & Flintham, J. E., et al. (1999). "Green revolution" genes encode mutant gibberellin response modulators. *Nature*, *400*, 256–261.

Plagnol, V., Padhukasahasram, B., Wall, J. D., Marjoram, P., & Nordborg, M. (2006). Relative influences of crossing over and gene conversion on the pattern of linkage disequilibrium in *Arabidopsis thaliana*. *Genetics*, *172*, 2441–2448.

Plantegenet, S., Weber, J., Goldstein, D. R., Zeller, G., Nussbaumer, C., & Thomas, J., et al. (2009). Comprehensive analysis of *Arabidopsis* expression level polymorphisms with simple inheritance. *Molecular Systems Biology*, *5*, 242.

Project, I. R. G. S. (2005). The map-based sequence of the rice genome. *Nature*, *436*, 793–800.

Purugganan, M. D., & Fuller, D. Q. (2009). The nature of selection during plant domestication. *Nature*, *457*, 843–848.

Rhee, S. Y., Beavis, W., Berardini, T. Z., Chen, G., Dixon, D., & Doyle, A., et al. (2003). The *Arabidopsis* Information Resource (TAIR): A model organism database providing a centralized, curated gateway to *Arabidopsis* biology, research materials and community. *Nucleic Acids Research*, *31*, 224–228.

Rosas, U., Barton, N. H., Copsey, L., Barbier de Reuille, P., & Coen, E. (2010). Cryptic variation between species and the basis of hybrid performance. *PLoS Biology*, *8*, e1000429.

Santuari, L., Pradervand, S., Amiguet-Vercher, A. M., Thomas, J., Dorcey, E., & Harshman, K., et al. (2010). Substantial deletion overlap among divergent *Arabidopsis* genomes revealed by intersection of short reads and tiling arrays. *Genome Biology*, *11*, R4.

Schmid, K. J., Sorensen, T. R., Stracke, R., Torjek, O., Altmann, T., & Mitchell-Olds, T., et al. (2003). Large-scale identification and analysis of genome-wide single-nucleotide polymorphisms for mapping in *Arabidopsis thaliana*. *Genome Research*, *13*, 1250–1257.

Schnable, P. S., Ware, D., Fulton, R. S., Stein, J. C., Wei, F., & Pasternak, S., et al. (2009). The B73 maize genome: complexity, diversity, and dynamics. *Science*, *326*, 1112–1115.

Sergeeva, L. I., Keurentjes, J. J., Bentsink, L., Vonk, J., van der Plas, L. H., & Koornneef, M., et al. (2006). Vacuolar invertase regulates elongation of *Arabidopsis thaliana* roots as revealed by QTL and mutant analysis. *Proceedings of the National Academy of Sciences of the United States of America*, *103*, 2994–2999.

Shindo, C., Aranzana, M. J., Lister, C., Baxter, C., Nicholls, C., & Nordborg, M., et al. (2005). Role of *FRIGIDA* and *FLOWERING LOCUS C* in determining variation in flowering time of *Arabidopsis*. *Plant Physiology*, *138*, 1163–1173.

Shindo, C., Lister, C., Crevillen, P., Nordborg, M., & Dean, C. (2006). Variation in the epigenetic silencing of FLC contributes to natural variation in *Arabidopsis* vernalization response. *Genes & Development*, *20*, 3079–3083.

Shukla, V. K., Doyon, Y., Miller, J. C., DeKelver, R. C., Moehle, E. A., & Worden, S. E., et al. (2009). Precise genome modification in the crop species *Zea mays* using zinc-finger nucleases. *Nature*, *459*, 437–441.

Sibout, R., Plantegenet, S., & Hardtke, C. S. (2008). Flowering as a condition for xylem expansion in *Arabidopsis* hypocotyl and root. *Current Biology*, *18*, 458–463.

Singer, T., Fan, Y., Chang, H. S., Zhu, T., Hazen, S. P., & Briggs, S. P. (2006). A high-resolution map of *Arabidopsis* recombinant inbred lines by whole-genome exon array hybridization. *PLoS Genetics*, *2*, e144.

Springer, N. M., Ying, K., Fu, Y., Ji, T., Yeh, C. T., & Jia, Y., et al. (2009). Maize inbreds exhibit high levels of copy number variation (CNV) and presence/absence variation (PAV) in genome content. *PLoS Genetics*, *5*, e1000734.

Sussman, M. R., Amasino, R. M., Young, J. C., Krysan, P. J., & Austin-Phillips, S. (2000). The *Arabidopsis* knockout facility at the University of Wisconsin-Madison. *Plant Physiology*, *124*, 1465–1467.

Tan, L., Li, X., Liu, F., Sun, X., Li, C., & Zhu, Z., et al. (2008). Control of a key transition from prostrate to erect growth in rice domestication. *Nature Genetics*, *40*, 1360–1364.

Tanksley, S. D., & McCouch, S. R. (1997). Seed banks and molecular maps: Unlocking genetic potential from the wild. *Science*, *277*, 1063–1066.

Thornsberry, J. M., Goodman, M. M., Doebley, J., Kresovich, S., Nielsen, D., & Buckler, E. S. (2001). Dwarf8 polymorphisms associate with variation in flowering time. *Nature Genetics*, *28*, 286–289.

Todesco, M., Balasubramanian, S., Hu, T. T., Traw, M. B., Horton, M., & Epple, P., et al. (2010). Natural allelic variation underlying a major fitness trade-off in *Arabidopsis thaliana*. *Nature*, *465*, 632–636.

Torii, K. U., Mitsukawa, N., Oosumi, T., Matsuura, Y., Yokoyama, R., & Whittier, R. F., et al. (1996). The *Arabidopsis ERECTA* gene encodes a putative receptor protein kinase with extracellular leucine-rich repeats. *Plant Cell*, *8*, 735–746.

Townsend, J. A., Wright, D. A., Winfrey, R. J., Fu, F., Maeder, M. L., & Joung, J. K., et al. (2009). High-frequency modification of plant genes using engineered zinc-finger nucleases. *Nature*, *459*, 442–445.

Trevaskis, B., Hemming, M. N., Dennis, E. S., & Peacock, W. J. (2007). The molecular basis of vernalization-induced flowering in cereals. *Trends in Plant Science*, *12*, 352–357.

Vaughn, M. W., Tanurdzic, M., Lippman, Z., Jiang, H., Carrasquillo, R., & Rabinowicz, P. D., et al. (2007). Epigenetic natural variation in *Arabidopsis thaliana*. *PLoS Biology*, *5*, e174.

Vlad, D., Rappaport, F., Simon, M., & Loudet, O. (2010). Gene transposition causing natural variation for growth in *Arabidopsis thaliana*. *PLoS Genetics*, *6*, e1000945.

Wang, H., Nussbaum-Wagler, T., Li, B., Zhao, Q., Vigouroux, Y., & Faller, M., et al. (2005). The origin of the naked grains of maize. *Nature*, *436*, 714–719.

Waugh, R., Leader, D. J., McCallum, N., & Caldwell, D. (2006). Harvesting the potential of induced biological diversity. *Trends in Plant Science*, *11*, 71–79.

Weber, A. L., Zhao, Q., McMullen, M. D., & Doebley, J. F. (2009). Using association mapping in teosinte to investigate the function of maize selection-candidate genes. *PLoS One*, *4*, e8227.

Weigel, D., & Mott, R. (2009). The 1001 genomes project for *Arabidopsis thaliana*. *Genome Biology*, *10*, 107.

West, M. A., van Leeuwen, H., Kozik, A., Kliebenstein, D. J., Doerge, R. W. & Clair, D. A., et al. (2006). High-density haplotyping with microarray-based expression and single feature polymorphism markers in *Arabidopsis*. *Genome Research*, *16*, 787–795.

Wollenweber, B., Porter, J. R., & Lubberstedt, T. (2005). Need for multidisciplinary research towards a second green revolution. *Current Opinion in Plant Biology*, *8*, 337–341.

Wright, S. I., Bi, I. V., Schroeder, S. G., Yamasaki, M., Doebley, J. F., & McMullen, M. D., et al. (2005). The effects of artificial selection on the maize genome. *Science*, *308*, 1310–1314.

Yan, L., Fu, D., Li, C., Blechl, A., Tranquilli, G., & Bonafede, M., et al. (2006). The wheat and barley vernalization gene *VRN3* is an orthologue of FT. *Proceedings of the National Academy of Sciences of the United States of America*, *103*, 19581–19586.

Yan, L., Loukoianov, A., Blechl, A., Tranquilli, G., Ramakrishna, W., & SanMiguel, P., et al. (2004). The wheat *VRN2* gene is a flowering repressor down-regulated by vernalization. *Science*, *303*, 1640–1644.

Yan, L., Loukoianov, A., Tranquilli, G., Helguera, M., Fahima, T., & Dubcovsky, J. (2003). Positional cloning of the wheat vernalization gene VRN1. *Proceedings of the National Academy of Sciences of the United States of America*, *100*, 6263–6268.

Zeller, G., Clark, R. M., Schneeberger, K., Bohlen, A., Weigel, D., & Ratsch, G. (2008). Detecting polymorphic regions in *Arabidopsis thaliana* with resequencing microarrays. *Genome Research*, *18*, 918–929.

Zhai, J., Liu, J., Liu, B., Li, P., Meyers, B. C., & Chen, X., et al. (2008). Small RNA-directed epigenetic natural variation in *Arabidopsis thaliana*. *PLoS Genetics*, *4*, e1000056.

Zhang, F., Maeder, M. L., Unger-Wallace, E., Hoshaw, J. P., Reyon, D., & Christian, M., et al. (2010). High frequency targeted mutagenesis in *Arabidopsis thaliana* using zinc finger nucleases. *Proceedings of the National Academy of Sciences of the United States of America*, *107*, 12028–12033.

Zhang, X., Shiu, S. H., Cal, A., & Borevitz, J. O. (2008). Global analysis of genetic, epigenetic and transcriptional polymorphisms in *Arabidopsis thaliana* using whole genome tiling arrays. *PLoS Genetics*, *4*, e1000032.

Zhao, J., Paulo, M. J., Jamar, D., Lou, P., van Eeuwijk, F., & Bonnema, G., et al. (2007). Association mapping of leaf traits, flowering time, and phytate content in *Brassica rapa*. *Genome*, *50*, 963–973.

From epigenetics to epigenomics and their implications in plant breeding

14

Athanasios Tsaftaris Aliki Kapazoglou Nikos Darzentas
Institute of Agrobiotechnology, CERTH, Thermi-Thessaloniki, Greece

TABLE OF CONTENTS

Mechanisms of Epigenetic Inheritance and their Interactions

Introduction

After having different meanings in the last few decades, epigenetics is now defined as the mitotically (from cell to cell during development) and meiotically (from one generation to the next) heritable variation that does not entail a change in DNA.

Indicative to the stability of epigenetic variation (epialleles) is perhaps the peloric mutant phenotype described by Linnaeus 250 years ago. Peloric characterizing, a change from bilateral to radial flower symmetry both in *Antirrhinum majus* and *Linaria vulgaris* still showing the same phenotype today, is due to an epigenetic change in the cycloidea gene, causing it to be silenced.

Almost 35 years have passed since it was proposed that DNA methylation at cytosine residues in eukaryotes could act as a stably inherited epigenetic modification controlling gene expression and be involved in cellular differentiation. Since then intense research efforts have expanded the field of epigenetics by expanding better understanding of diverse aspects of DNA methylation deciphering the involvement of methylation in multivarious biological processes, and expanding the epigenetic field with the discovery of extra novel mechanisms of epigenetic inheritance. Thus, the ever-expanding list of epigenetic mechanisms includes, beyond DNA methylation, histone protein variants for all four histone proteins participating in nucleosome core formation and their numerous covalent modifications, as well as a cloud of small interfering RNAs of varying sizes, modes of biogenesis, amplification, and targeting.

The current explosion of different "omics" technologies, that marked not only the transition from the study of individual genes and their products to the study of whole genomes, but also allowed a more integrated and holistic view of the interrelation of different genomic and epigenomic (genome-wide map of epigenetic modifications) systems. It was known, for instance, that the different epigenetic mechanisms mentioned do not operate independently of each other. Histone deacetylation and other modifications, particularly the methylation of lysine 9 within histone H3 (H3K9) of histone tails, cause chromatin condensation and block transcription. In addition, histone modification can also attract DNA MTases to initiate 5 methyl cytosine (5mC) formation, which in turn can reinforce histone modification patterns conducive to silencing (Tariq and Paszkowski, 2004; Fuks, 2005). The

DOI: 10.1016/B978-0-12-381466-1.00014-6

opposite is also true; that is, DNA methylation reinforces silencing by the attraction of histone modifications (methylation and deacetylation), leading to a more compact and thus inactive chromatin (Tamaru and Selker, 2001; Tariq and Paszkowski, 2004). With the recent advent of RNA interference in the epigenetic field, it became clear almost from the start that among their other roles, sRNAs also provide a common denominator of both DNA methylation and/or histone modifications. Experiments in plants have clearly shown the involvement of sRNAs in the establishment of heterochromatic states and silencing (Wassenegger et al., 1994;Matzke et al., 2001). Disruption of one of these interacting systems can lead to expression or silencing of genes, resulting in epigenetically controlled phenotypes.

Work in epigenomics is now reinforcing the view for the whole genome providing significant outputs. The very recent sequence of the maize genome opens new avenues to this research. Maize is the largest plant genome to be fully sequenced and characterized. It contains a high percentage of transposable elements (TEs) of different types, providing an excellent opportunity to study and understand the role of different TEs in eukaryotic large genome organization and function, but most important it provides an understanding of the function of epigenomic mechanisms as TE regulators at a genomic level. Keeping in mind that TEs were discovered in maize many years ago, thanks to the pioneering work of B. McClintock and P. Peterson, and that significant biological phenomena involving epigenetic phenomena like paramutation and parental imprinting of genes were discovered and studied for a long time in maize, the opportunities now offered by maize genomic fields are obvious. Maize development is well studied, but most important is that maize went through a domestication process for more than 10,000 years, dramatically changing its phenotype, and through intense breeding in the last 100 years increasing its yields almost 100-fold. This was done mainly by discovering and exploiting heterosis in this crop.

In this chapter, after a brief description of the different epigenetic mechanisms, this transition from epigenetics to epigenomics will be described in the section From Epigenetics to Epigenomics, together with the genome-wide methods used and the data, resources, and tools available in exploring, integrating, and extracting useful information. Finally, the role of epigenomics in plant development, phenotypic variation, response to environmental cues and stresses, and the implication of different epigenetic phenomena in plant breeding will be discussed in the section Epigenetic Phenomena and their Implications in Plant Breeding. Due to space limitations, only indicative phenomena like parental imprinting involved in endosperm size and thus yield in many crops, or the epigenetic control of flowering time with implications in breeding early or late varieties for many annual crops, will be described.

For different topics related to this chapter, there are excellent recent reviews that will be mentioned accordingly, but the integrative approach, based on recent findings on epigenomic mechanisms and their implication in plant breeding, is the main objective of this chapter.

Epigenetic mechanisms and their interactions

DNA cytosine methylation

Cytosine-5-methylation is a critical epigenetic modification that is established and maintained by multiple interacting cellular mechanisms. Similar to animals, cytosine methylation in plants is found predominately in a symmetrical CG dinucleotide site. However, unlike animals, it also occurs at CHG and asymmetric CHH sites (where H is A, C, or T). A dynamic interplay between methylation and demethylation accomplished through specific enzymes is critical for proper cellular regulation during plant development. Distinct DNA methylation mechanisms operate in higher plants involving *de novo* and "maintenance" DNA methyltransferases (MTases), together with DNA demethylases, histone-modifying enzymes, chromatin remodeling factors, and small interfering RNAs (siRNAs; Law and Jacobsen, 2010). The mammalian DNMT1 homolog, METHYLTRANSFERASE 1 (MET1), is responsible for maintenance of primarily CG methylation, whereas the plant-specific CHROMOMETHYLASE3 (CMT3) interacts with H3K9 histone methyltransferase KRYPTONITE (SUVH4) and maintains DNA methylation at CHG sites. The DOMAINS REARRANGED METHYLTRANSFERASE 1 and 2 (DRM1/2; homologs of the mammalian DNMT3a/b) catalyze *de novo* methylation and maintain CHH methylation (Chan et al., 2005).

In plants, as in other higher eukaryotes, DNA methylation has been thought to protect genomes from endogenous selfish DNA elements (TEs) and exogenous viral invasions (Zilberman, 2008). These elements are usually heavily methylated and, consequently, transcriptionally inactive. In addition, cytosine methylation has a prominent role in regulating gene expression during plant development and in response to conditions of stress (Zilberman et al., 2007; Zhu, 2008).Processes such as vernalization, flower and seed development, and abiotic and biotic stress tolerance, are largely regulated by cytosine methylation (Chinnusamy and Zhu, 2009a).

Although for several years it has been believed that DNA methylation resided in the promoter regions of genes, recent studies based on large-scale, genome-wide analysis of the *Arabidopsis* "methylome" (Zhu, 2008; Cokus et al., 2008; Lister et al., 2008) have shown that at least in this model plant cytosine methylation covers both regulatory regions and the "bodies" of genes (exons, introns, and 5′ and 3′ untranslated regions). Of all expressed genes, about 5% are methylated in the promoters and 33% in transcribed regions. In plants, promoter methylation usually leads to transcriptional repression. Surprisingly, body-methylated genes showed no or only a moderate effect in gene expression, and were transcribed in a less tissue-specific manner than unmethylated genes. However, there are exceptions to these rules. For example, cytosine methylation in the transcribed regions of the *Arabidopsis* floral genes *SUPERMAN* (*SUP*) and *AGAMOUS* (*AG*) results in substantial reduction in gene expression (Ito et al., 2003). Furthermore, methylation in a long distance element 3′ downstream of the gene *PHERES1*

(*PHE1*), important in seed development in *Arabidopsis*, results in gene activation (Makarevich et al., 2008), and a recent study of a *Petunia* floral gene, *pMADS3*, showed that methylation in a particular CG was associated with upregulated ectopic gene expression (Shibuya et al., 2009).

Even though DNA methylation is a relatively stable epigenetic mark, it is subject to passive or active demethylation during plant development (Zhu et al., 2007). Passive demethylation can occur when methylated cytosines are replaced by non-modified cytosines during DNA replication (Tsaftaris et al., 2008). Active DNA demethylation is achieved by specific DNA glycosylases through the base excision repair pathway, which recognizes and removes methylated cytosines (Zhu, 2009). The DNA glycosylase *DEMETER* (*DME*) functions only in the central cell of the female gametophyte to establish parental genomic imprinting crucial for seed viability (Choi et al., 2002; Huh et al., 2008; see the section Epigenetic Phenomena and their Implications in Plant Breeding). Its homologs *REPRESSOR OF SILENCING* (*ROS1*), *DML2*, and *DML3* are expressed in vegetative tissues and target transposons, repetitive elements, and small RNA-generating loci (Agius et al., 2006; Penterman et al., 2007a). In addition, they seem to target loci that reside at the boundaries of euchromatin and heterochromatin, where they might protect genes from deleterious methylation (Penterman et al., 2007b). Very recent reports on global DNA methylation profiling of endosperm and embryo genomes demonstrated widespread reduction of DNA methylation in the endosperm, particularly at regions corresponding to TEs and small RNAs (Gehring et al., 2009; Hsieh et al., 2009). These were largely due to *DME* action in the central cell, the progenitor of the endosperm, which develops after fusion of the central cell with one of the sperm cells of the pollen (while the other sperm cell fuses to the egg cell to produce the embryo). Although *DME* action is restricted to the central cell, global demethylation persists in the developing endosperm post-fertilization. These data suggested that DME carries out less specific and more global demethylation than previously thought (see the section Endosperm Development and Parental Imprinting).

Unlike mammals, which erase and reset DNA methylation in embryogenesis, plants can inherit epigenetic changes through meiosis. "Epialleles" — phenotypic variants that are epigenetically rather than genetically different from their parents — are frequently encountered in plants and in certain cases found to be inherited by the next progeny generation. A striking example of the inherited stability of epialleles is the *peloric* mutant (a change from bilateral to radial flower symmetry) found both in *Antirrhinum* and *Linaria*, which was originally described by Linnaeus and whose phenotype remained stable for 250 years. Cubas et al. (1999) demonstrated that in the *Linaria* mutant, the *CYCLOIDEA* (*CYC*) gene (encoding for a TCP transcription factor) is extensively methylated and transcriptionally silenced.

Finally, it should be noted that DNA methylation, in addition to being the cause of epigenetic variations, is the cause of mutations and genetic variation. The C methylation pathway is inherently mutagenic, particularly under conditions of limiting supply of the methyl donor S-Ade-Met (Matzke et al., 2001). Such a low in the S-Ade-Met environment permits the accumulation of an intermediate in the C methylation pathway (5,6-dihydrocytosine), which has a 10^4-fold higher rate of deamination than 5-methylcytosine.

Histone modifications

Modifications on nucleosomal histone tails include methylation, acetylation, phosphorylation, ubiquitylation, biotinylation, and sumoylation on specific lysines and arginines. A complex pattern of site-specific combinations of histone modifications on different residues known as the "epigenetic histone code" leads to specific chromatin states that regulate transcriptional activity and are inherited by daughter cells (Strahl and Allis, 2000). In plants, the best characterized modifications are histone methylation and histone acetylation. Generally, di-methylation of histone 3 on lysine 9 (H3K9me2) and tri-methylation of histone 3 on lysine 27 (H3K27me3) leads to a repressive chromatin state correlating with gene silencing; whereas tri-methylation of histone 3 on lysine 4 (H3K4me3) leads to a permissive chromatin state correlating with gene activation (Kouzarides, 2007). In plants H3K9me2 is mostly associated with transcriptional silencing of transposons and repeats in centromeric/pericentromeric heterochromatin regions. Genome-wide analysis in *Arabidopsis* (Turck et al., 2007; Zhang et al., 2007a, 2009) has determined the global H3 methylation patterns along chromatin (see the section Epigenetic Mechanisms and their Interactions) and showed an agreement between the localization of H3 methylation marks and their proposed functions based on previous locus-specific studies. H3K4me3, which is associated with gene activation, was found mostly in endogenous genes but not in transposons, whereas H3K9me2 coincides almost exclusively with transposons and other repeats. H3K27me3 was correlated with a large number of tissue-specific expressed genes transcriptionally silent in the tissue studied (young seedlings; Turck et al., 2007; Zhang et al., 2007a). However, inactivation through H3K27me3 methylation is a reversible process, but when accompanied by H3K9 methylation along with histone deacetylation and DNA methylation, chromatin becomes progressively more and more condensed and heterochromatic, which leads to more permanent (cemented) inactivation that is almost irreversible during development (Tariq and Paszkowski, 2004). This type of compact silenced chromatin characterizes the majority of the heterochromatic genome comprised of TEs and other repeats.

The concerted and/or antagonizing actions of different histone methyltransferases (HMTs) and histone demethylases (HDMs) with different specificities determine the histone methylation pattern of chromatin in different cell types at different times during plant development. Similarly, dynamic histone acetylation and deacetylation patterns established by histone acetyltransferases (HATs) and histone deacetylases (HDACs) result in active and silent chromatin forms, respectively (Chen and Tian, 2007). Interplay between the two mechanisms allows for dynamic chromatin remodeling during development and in response to environmental changes.

Like cytosine methylation, histone modifications function in various aspects of plant growth and development including flower and seed formation and stress responses (Kohler and Villar, 2008; Chinnusamy and Zhu, 2009a; Amasino, 2010). Extensive analysis of *Arabidopsis* mutants has suggested that there is significant cross-talk between histone modifications and DNA methylation (Vaillant and Paszkowski, 2007). For example, in *kyp* mutants where the action of an H3K9me2 HMT, KRYPTONITE (KYP) is perturbed, severe reduction of CHG DNA methylation was observed (Jackson et al., 2002; Malagnac et al., 2002). In addition, genome-wide profiling of H3K9me2 and CHG methylation revealed a high correlation between these two marks (Bernatavichute et al., 2008). It was proposed that CHG methylation catalyzed by CMT3 is recognized by KYP, which is recruited at those chromatin sites. In turn, CMT3 binds these methylated histone marks through its chromodomain, thus reinforcing CHG methylation (Law and Jacobsen, 2010). This kind of interdependent cooperative action between two silencing marks results in further chromatin compaction in heterochromatic regions, most likely to ensure transcriptional silencing of transposons and other repetitive elements.

Specific HMT homologs belonging to the Polycomb group (PcG) complex variants perform H3K27me3 in a spatiotemporal-specific manner leading to specific transcriptional repression at different developmental stages (Schatlowski et al., 2008). Three different PcG complexes have been described in *Arabidopsis*, operating during the processes of seed development, vernalization, and flower development, respectively (Kohler and Villar, 2008). The FIS2-FIE-MEA-MSI1 complex functions during seed development, and the EMF2-CLF/SWN-FIE-MSI1 complex has been suggested to play a role in suppressing the transition from vegetative development to flowering and flower organ formation (Chanvivattana et al., 2004; Schubert et al., 2006) and the VRN2-CLF/SWN-FIE-MSI1 complex has been implicated in the regulation of the process of vernalization (Sung and Amasino, 2004; Wood et al., 2006). In all three complexes FIE and MSI1 (WD40 proteins) remain constant, whereas the homologs MEA/CLF/SWN (encoding H3K27me3 HMTs) and FIS2/EMF2/VRN2 (encoding *Drosophila* Su(Z)12 homologs) change depending on the PcG variant (Kohler and Villar, 2008).

Trithorax group (TrxG) proteins antagonize the action of PcG proteins and lead to activation of PcG-silenced genes. These include SET-domain histone methyltransferases with H3K4me3 specificity, as well as ATP-dependent chromatin remodeling factors such as the chromodomain/helicase/DNA-binding domain CHD3 protein PICKLE (PKL). PKL targets PcG genes like *SWN* and *EMF2* known to silence embryonic-activating regulators, and uplifts the suppressive PcG-effect on embryonic development (Aichinger et al., 2009).

Histone demethylases are a recent discovery (Shi et al., 2004). Human lysine-specific demethylase1 (LSD1), a nuclear amine oxidase, specifically demethylates H3K4me1 and H3K4me2, and a number of JmjC domain-containing proteins have been demonstrated to be H3K4, H3K9, H3K27, H3K36, H3R2, and H4R3 demethylases. In *Arabidopsis*, 21 genes encoding JmjC family proteins were identified (Hong et al., 2009). *AtJmj1* and *AtJmj2* were shown to be involved in photoperiodic flowering and *FLOWERING LOCUS C* (*FLC*) regulation, respectively. *AtJmj15* represses genic cytosine methylation and *AtJmj1* and *AtJmj4* were shown to directly repress *FLOWERING LOCUS T* (*FT*) expression via demethylation of H3K4me3 (Jeong et al., 2009).

Two HDAC families, the RPD3/HDA1 and the plant-specific HD2, have been well characterized in *Arabidopsis*, and have been associated with distinct functions. For example, the *AtHDAC6* gene (RPD3/HDA1-Class I) was shown to play a role in transgenic and repetitive DNA silencing and in the nuclear organization of rDNA loci (Aufsatz et al., 2002; Probst et al., 2004). *AtHDA6* and At*HDA19* have been implicated in repression of embryonic-specific gene functions during germination (Tanaka et al., 2008) and At*HDA6* is required for jasmonate response, senescence, and flowering (Wu et al., 2008). Similarly, *HD2a* has been associated with seed development and maturation (Wu et al., 2000; Zhou et al., 2004; Sridha and Wu, 2006). In recent years, histone modifiers such as HMTs and HDACs have also been identified and analyzed in agriculturally important crop plants. For example, the genes encoding PcG complexes and HDAC proteins have been characterized in rice (Fu et al., 2007; Luo et al., 2009). Likewise, in maize a number of *HDAC* genes have been identified (Lusser et al., 2001; Rossi et al., 2003) and the *HDA101* gene (RPD3/HDA1-Class I) has been shown to be involved in vegetative to reproductive transitions and in other chromatin modifications such as histone methylations and phosphorylations (Rossi et al., 2007). In barley, our group has identified, characterized, and mapped *PcG* and *HDAC* genes, and has examined their expression in different barley varieties and upon exposure to stress-related hormones (Demetriou et al., 2009, 2010; Kapazoglou et al., 2010). Differential gene expression for both *HDAC* and *PcG* genes was evidenced during seed development in different barley varieties. Most *HDAC* genes were induced by exposure to jasmonate, and the barley *PcG* genes *E(Z)* and FIE were strongly induced by ABA. Similar studies were performed for different members of barley *HAT* genes, which showed that differential expression during seed development and induction by ABA has also characterized (Papaefthimiou et al., 2010).

In addition to modifications of the so-called canonical histones participating in nucleosomal formation ($2H_2A$, $2H_2B$, $2H_3$, and $2H_4$), histone variants can also be incorporated into nucleosomes during plant development. The canonical histones comprise the major part of histone proteins, are encoded by multiple genes, and are expressed predominately during the S phase of the cell cycle. Thus, their deposition into chromatin is strictly coupled to DNA replication. Even though histones are the slowest evolving proteins, allelic variants with differences in amino acid sequence do exist. Unlike canonical histones, histone variants are often encoded by a single gene and are expressed throughout the cell cycle; hence, they can be incorporated into nucleosomes during the entire cell cycle. Extensive studies during the last decade have pointed to a role for histone variants in the structural and functional aspects of chromatin. The $H_2A.Z$ is involved in transcription and plays an important role in plant development. Recent studies on *Arabidopsis* have shown that $H_2A.Z$ deposition acts against DNA methylation, implying that

$H_2A.Z$ has a role in protecting genes from silencing (March-Diaz and Reyes, 2009). Genome-wide profiling of histone variants using the ever-improving high-throughput technologies is possible. Such a large-scale analysis would help to unravel the histone variant landscape and its association with other histone modifications and DNA methylation patterns and transcription factor activities.

Small RNAs

Small RNAs were found to interfere with many aspects of gene activity and genome organization; consequently, they control crucial processes in the cell cycle and the development of eukaryotic organisms. Four major pathways producing endogenous RNAi have been identified in plants. The micro RNA (miRNA), transacting siRNA (ta-siRNA), and natural-antisense siRNA (nat-siRNA) function predominately at the post-transcriptional level through mRNA degradation and/or translational inhibition (Chapman and Carrington, 2007). The heterochromatic (hc-RNAs) siRNA pathway is involved mainly in transcriptional gene silencing by directing DNA methylation (Henderson and Jacobsen, 2007). miRNAs constitute only 5% of the total small RNA mass and less than 0.1% of sequence complexity in *Arabidopsis*. The rest represents more than 100,000 different siRNAs that originate from thousands of loci (Kasschau, 2007; Mosher et al., 2008).

Growing evidence in recent years has demonstrated that siRNAs direct methylation of DNA sequences complementary to the siRNAs; a process known as RNA-directed DNA methylation (RdDM; Chinnusamy and Zhu, 2009b). The majority of siRNAs are 24nt RNAs corresponding to TEs and other repetitive elements and cause epigenetic gene silencing by directing *de novo* cytosine methylation of DNA (Zhang et al., 2007b). They originate from double-stranded (ds) RNA precursors generated by the RNA polymerase isoform IVa (Pol IVa) and RNA-dependent RNA polymerase 2 (RDR2). The dsRNA precursors are then processed by DCL3 to 24-nt siRNAs and are loaded to ARGONAUTE4 containing RNA-induced silencing complex (RISC). The complex interacts with PolIVb to guide DRM1/2-mediated DNA methylation (Zhang et al., 2007b; Mosher et al,. 2008). Genome-wide DNA methylation profiling by microarray and shotgun bisulfate sequencing in *Arabidopsis* revealed that DNA methylation is most abundant in heterochromatic regions (Zilberman et al., 2007; Cokus et al., 2008). Comparison of the global DNA methylation pattern and siRNA profiles demonstrated that about 37% of the methylated loci are associated with siRNA clusters.

Apart from silencing transposons, RdDM has a role in gene regulation and proper plant development. For example, a 3′ non-coding region of the *Arabidopsis BONSAI* gene was found to harbor a long interspersed nuclear element (LINE) retrotransposon insertion that gives rise to siRNAs required for hypermethylation (Saze and Kakutani, 2007). A mutant in the chromatin remodeling factor DDM1 caused a developmental defect that was associated with siRNA-mediated hypermethylation of the *BONSAI* 3′ region. Likewise, RdDM-mediated silencing of the *SUPPRESSOR OF DRM1 DRM2 CMT3* (*SDC*) gene was shown to be important for normal leaf development. Over-expression of *SDC* results in a curled-leaf phenotype. In wild-type plants siRNAs originating from tandem repeats on the *SDC* promoter direct DNA methylation of the repeats and the SDC gene is silenced (Henderson and Jacobsen, 2008).

The regulation of the *FWA* (*Flower Wageningen*) gene is another example where siRNA, as well as other epigenetic mechanisms such as DNA and histone modifications, are involved in gene suppression. *FWA* encodes a homeobox transcription factor and *fwa* mutants cause a late-flowering phenotype. The gene is only expressed from the maternal allele in the endosperm, whereas it is silenced biparentally in the other tissues. Ectopic expression of *FWA* in adult tissue in the *fwa* mutant results in delayed flowering. The *FWA* silencing is associated with methylation of tandem repeats around the transcriptional start site, which are hypomethylated in the endosperm or in the mutant where the gene is expressed (Soppe et al., 2000; Kinoshita et al., 2004, 2007). It was demonstrated that promoter methylation is under RNA-directed methylation as the tandem repeats produced siRNAs, which in turn direct DNA methylation at those sites (Chan et al., 2005; Kinoshita et al., 2007). In addition to RNA-directed DNA methylation, it was recently shown that sporophytic silencing of *FWA* is also achieved through H3K4 histone demethylation by the *Arabidopsis* LDL1and LDL2 homologs (Jiang et al., 2007).

MiRNAs also play crucial roles in a wide range of plant developmental and stress-related processes. In the last decade a plethora of information concerning the diverse functions of miRNAs in both monocots (rice, maize, *Brachypodium*) and dicots (*Arabidopsis*, *Populus*) has been reported (reviewed in Martin et al., 2010). Aside from their post-transcriptional mode of action, miRNAs act at the transcriptional level and regulate gene expression through DNA methylation, as current studies have revealed (Wu et al., 2009; Khraiwesh et al., 2010).

Small RNAs can follow non-cell autonomous travel throughout the plant organism using different proteins as carriers. The effects of post-transcriptional RNA silencing can extend beyond its sites of initiation, owing to the movement of signal molecules. Breakthrough grafting experiments demonstrated the transmission of the *Nia* (nitrate reductase) co-suppressed state from silenced root-stocks to non-silenced transgenic scions (Palauqui et al., 1997). Shortly after this it was found that leaf infiltration into transgenic tobacco plants expressing the green fluorescent protein (GFP) triggered a systemic sequence-specific loss of GFP expression (Voinnet et al., 1998). The nature of the mobile signal had been unclear until very recently when it was demonstrated that mobile siRNAs and miRNAs can migrate in grafted plants from roots to shoots and *vice versa* and direct epigenetic modifications in recipient cells (Dunoyer et al., 2010; Martienssen, 2010; Molnar et al., 2010). Understanding the small RNA-mediated mechanisms of systemic signaling is of major importance given the growing interest of using grafted plants in vegetable production (Tsaftaris et al., 2008).

Finally, in the same context of small RNA action, the phenomenon of transgene silencing should not go unmentioned. In the past decade the production of transgenic plants was made possible through the development of techniques for introducing and expressing foreign genes. However, often transgene silencing was observed where the foreign gene was not expressed as expected. Intensive research discovered small RNA-mediated silencing mechanisms underlying this phenomenon. Stabilizing transgene expression is a priority, especially to the agricultural biotechnology industry that avidly pursued transgenic approaches to modulate gene expression and produce superior plants. For a detailed description of the implications of small RNAs to breeding, in relation to systemic signaling and stability of transgenes, refer to Tsaftaris et al. (2008).

The functional relationships between the different epigenetic mechanisms mentioned earlier and their mutual reinforcement are unexpectedly complex. This is best exemplified by their involvement in the formation of large complex structures engaged in chromatin remodeling. In addition to DNA methylases/demethylases and the histone-modifying enzymes, these complexes also contain histone chaperones and ATP-dependent chromatin remodeling enzymes. Thanks to their individual epigenetic mechanisms, these complexes could have direct interaction with each other, reinforcing each other's activity. Furthermore, due to their modular structure, by exchanging parts, complexes could lead to gene activation or suppression. Different types of such complexes have been described in more detail in yeast and animal systems (see Ho and Crabtree, 2010), but this knowledge is now expanding to plant chromatin remodeling systems too (Jarillo et al., 2009). As described by these authors, chromatin remodeling complexes control gene expression both in shoot and root apical meristems by keeping actively dividing cells undifferentiated, thus totipotent, or by keeping fate-determined cells committed to later giving rise to specific organs. The specific chromatin remodeling complexes involved in pluripotent stem cells able to both self-renew and generate undifferentiated cells committed to the formation of new tissues and organs were also described recently (Shen and Xu, 2009).

Finally, recent findings indicated the relevance of the spatial distribution of genes within the nucleus to their transcriptional control (Sexton et al., 2007; Singh et al., 2008). The topology of the chromatin at the position where a particular gene is located could determine its proper expression in the context of a developmental program, revealing a mechanistic link between gene position and gene expression.

From Epigenetics to Epigenomics

Deciphering epigenomes: A matter of scale and complexity

Life sciences have entered an era of immense data generation capabilities, enabled by significant steps forward in relevant technologies. Soon, one or more epigenomes will be available — a genome-wide map of epigenetic modifications — for every plant genome, focusing on different cultivars, hybrids, tissues, developmental stages, environmental conditions, abiotic and biotic stresses, and so forth. Such a development is now seen as a prerequisite to explaining the diversity of life, since genomes have proved too similar to do so (Slack, 2010).

Arabidopsis and rice have been relatively simple case studies compared to the ever clearer complexity of maize, which is a massive genome, 85% of which consists of TEs (Baucom et al., 2009; Schnable et al., 2009), and thus is very dynamic and capable of great plasticity. It has been the center of much attention (also in this chapter). The Panzea consortium (http://panzea.org) has been working on a large number of landraces, hybrids, lines, and so forth, of maize and its wild relative, the teosinte. The goal is to capture their diversity on a genetic, phenotypic, and, consequently, functional basis, making an annotated DNA mosaic available with which the additional layers of large-scale "omic" data can be integrated for further research. Such data integration requires expertise from the fields of bioinformatics and computational biology for gathering, organizing, storing, and connecting data in databases; data mining; knowledge and association discovery; and modeling of reactions, processes, pathways, cells, and visualization.

Epigenomic methods and the type of data collected

ChIP-chip

Chromatin immunoprecipitation (ChIP) is a technique used for assaying protein–DNA binding *in vivo*. In principle, genomic DNA is sheared by sonication into small fragments, preferentially in the range of a 200 to 800 base pairs. The proteins are then linked to the DNA by gentle formaldehyde treatment. An antibody specific to the protein of interest is used to immunoprecipitate the protein–DNA assembly and enrich the DNA fragments associated with the protein. The cross-links are then reversed by acid treatment to release the DNA, the DNA is purified, and then the DNA fragments are amplified by PCR. For ChIP-chip experiments PCR products are then labeled, hybridized on microarrays, and fluorescence read by a scanner; this enables a genome-scale view of DNA–protein interactions. On high-density tiling arrays, oligonucleotide probes can be placed across an entire genome or across selected regions of a genome (e.g., promoters) at a preferred resolution. Usually when the epigenetic landscape of histone modifications are to be determined, antibodies raised against the particular modification (e.g., H327K3me or H34Kme3) are used. The quality of ChIP-chip data depends on a variety of factors, the most important one is the specificity of the antibody. Another factor is the development of the platform. To precisely map the locations of binding or modification, sufficient probe resolution is crucial, and this might be compromised by the total number of probes that can be fitted on an array. Striking progress has been made in array manufacturing technology with the use of oligonucleotide probes. Whole genome tiling at low cost has been feasible for small genomes. For example, a single "tiling" array manufactured by Affymetrix covers the entire genome of *Drosophila melanogaster* with 25 mer probes at 38 bp resolution (Park, 2008). A variety of tools is available for processing the data and reducing noise (Johnson et al., 2008).

ChIP-seq

Progress in next-generation sequencing (NGS) technology has allowed the combination of ChIP and NGS known as ChIP-seq. In ChIP-seq experiments the enriched DNA fragments of interest are sequenced directly, instead of hybridized on an array. Currently there are four sequencing platforms: the Illumina (Solexa) Genome Analyzer; the Applied Biosystems SOLiD; the Roche 454 pyrosequencing; and the Helicos platform. The Illumina and SOLiD technologies can sequence 35–50bp fragments, and a single lane out of eight on a flow cell can generate 8–12 million reads. Most ChIP-seq data have been generated with the Illumina Genome Analyzer. In this platform a library of adapter-ligated ChIP DNA fragments is constructed and loaded onto a solid substrate. Cluster amplification ensues that generates many clonal copies of each fragment by bridge PCR. The sequencing step involves the enzyme-driven extension of all templates in parallel with fluorescently labeled reversible terminator nucleotides. After each extension, the fluorescent labels that have been incorporated are detected through high-resolution imaging. There are two main advantages of ChIP-seq compared to ChIP-chip. First, it provides single base pair resolution of direct sequencing. Secondly, it suffers from much less noise than ChIP-chip. The microarray approach relies on hybridization chemistry between the probe and the DNA target. Nucleic acid hybridization depends on many factors such as GC content, length, concentration, and secondary structure of the two sequences. Nonspecific hybridization between imperfect matches frequently occurs and contributes to the noise. In addition to signal quantification, the amount of signal captured through a scanner might not be linear over its entire range (Park, 2008, 2009).

Genome-wide DNA methylation profiling

Methylated DNA immunoprecipitation (MeDIP) is a technique for performing large-scale analysis of whole genome DNA methylation. It utilizes a monoclonal antibody that recognizes 5-methyl-cytosine to immunoprecipitate methylated DNA sequences. High molecular weight genomic DNA is sheared to 300–500bp fragments by sonication and is subsequently immunoprecipitated using the anti-5-methyl-cytosine antibody. The enriched fragments are then analyzed using genomic tiling arrays. The only drawback of this method is the CpG-density factor, that is, the efficiency by which the DNA is immunoprecipitated depends on the degree of CpG methylation within the fragment. This complicates DNA methylation quantification and cross-comparisons between samples with different profiles. However, by adding external control DNA the data can be normalized and used reliably. MeDIP was used for genome-wide DNA methylation mapping of the *Arabidopsis* genome (Zhang and Jacobsen, 2006; Zilberman et al., 2007). The combination of MeDIP with massive parallel sequencing, MeDIP-seq, could be a powerful approach for genome-wide DNA methylation profiling (Beck and Rakyan, 2008). However, the presence of bulk quantities of heavily methylated repeat DNA poses a problem for this method. These fragments will be recovered with very high efficiencies and cannot be assigned unequivocally to specific genomic regions. This represents a major disadvantage for the MeDIP-seq method. MeDIP-ChIP does not have this problem, because genomic tiling microarray probes are repeat-masked. Another very powerful technique, sodium bisulfate (BS)-seq, has been used to generate a DNA methylation map at one-nucleotide resolutions across the genome, through the coupling of bisulfate treatment of DNA and NGS technology. Sodium bisulfate converts unmethylated cytosines to uracils, whereas methylated cytosines remain unconverted. Consequently, after amplification by PCR, unmethylated cytosines are displayed as thymines and methylated cytosines as cytosines. Genomic DNA is first sheared, linked to adapters, subjected to bisulfate treatment, and then shotgun sequenced. This technique was first employed by using the Illumina GA to sequence 3.8Gbp of *Arabidopsis* sulfate-converted DNA, providing the first methylome at single base resolution of the entire *Arabidopsis* genome (Cokus et al., 2008). Owing to its high sensitivity, BS-seq proved superior to previously used methylome-profiling techniques based on tiling arrays. First, it identified particular methylated CGs that had been previously missed. Secondly, it allowed distinction between symmetric (CG and CHG) and asymmetric (CHH) DNA methylation. Thirdly, it enabled analysis of repetitive sequences including telomeric repeats, which are difficult to analyze by microarray techniques. Finally, it allowed identification of precise boundaries between methylated and unmethylated regions such as tandem repeats and flanking regions (Cokus et al., 2008).

Epigenomic resources

Plant species have been relatively slow to become part of the omics era, mainly because of their complexity and genome size. This is even more obvious in the field of plant epigenomics, which had to follow efforts in plant genomics, transcriptomics, and so forth, to have data to lie upon and interconnect. Nevertheless, epigenomic-related data and methods are now becoming increasingly abundant, and Table 14.1, although by no means comprehensive, provides an overview of publicly available plant epigenomics resources.

Access to information on the molecular machinery behind chromatin modifications can be critical in understanding these processes. Protein families playing major roles include DNA methyltransferases, and histone acetyltransferases and deacetylases. ChromDB (1; first resource in Table 14.1; Gendler et al., 2008) is a resource offering such information, including on RNA interference, based on plant transcription products (ESTs and cDNAs) and genome sequences (where available). Snapshots of the effects of these processes on the genome are becoming readily available for key plant species. The Jacobsen Laboratory at the University of California in Los Angeles (3; in Table 14.1) provides links to diverse data from large-scale *Arabidopsis* epigenomics investigations, covering histone modifications, methylation, nucleosomes, and small RNAs. Rice and maize have seen similar efforts from front-running research groups in the United States and China, producing extensive datasets on epigenetic modifications and their relationships to small RNA and mRNA transcriptomes in maize (10 in Table 14.1; Wang et al., 2009), and on global mapping of epigenetic modifications in rice (11 in Table 14.1; Li et al., 2008).

Table 14.1 Resources table

#	Date	Method(s)	Educational	Name	Description	Chromatin/ histone modification	Methylation	Small RNAs/RNAI	Plant(s)	Country	URL	PMID
1	X			ChromBB	display chromatin-associated proteins, including RANi-associated proteins	X		X	various	US	http://www.chromdb.org/	17942414
2	X			miRBase	Published miRNA sequences and annotation			X	various	UK	http://www.mirbase.org/	17991681
3	X			Jacobsen lab	diverse epigenomics data	X	X	X	*Arabidopsis*	US	http://epigenomics.mcdb.ucla.edu/	various
4	X			*Arabidopsis* Small RNA Project	microRNA, endogenous siRNA and other small RNA-related data			X	*Arabidopsis*	US	http://asrp.cgrb.oregonstate.edu/	17999994
5		X		RNA22	pattern-based method for the identification of microRna-target sites and their corresponding RNA/RNA complexes			X	generic	US	http://cbcsrv.watson.ibm.com/ma22.html	16990141
6		X		UEA sRNA tools	analysis of high-throughput small RNA data			X	generic	UK	http://sma-tools.cmp.uea.ac.uk/plant/cgi-bin/sma-tools.cgi	18713789
7	X			SiLoDB	genomic loci that produce sRNAs			X	various	UK	http://silodb.cmp.uea.ac.uk/	n/a
8		X		miRanda	finding genomic targets for microRNAs			X	generic	US/UK	http://www.microma.org/microma.org/microma/getDownloads.do	15502875

9		X		psRNATarget	plant small RNA regulators and targets prediction			X	generic	US	http://bioinfo3.noble.org/psRNATarget/	15980567*
10	X			NCBI GEO dataset	epigenetic modifications and their relationships to smRNA and nRNA transcriptomes in maize	X	X	X	*Zea mays*	US/China	http://www.ncbi.nlm.nih.gov/geo/query/acc.cgi?acc = GSE15286	19376930
11	X			NCBI GEO dataset	global mapping of histone modifications and methylation in rice	X	X	X	*Oryza sativa*	China	http://www.ncbi.nlm.nih.gov/geo/query/acc.cgi?acc = GSE9925	18263775
12	X			Compartaive Sequencing of Plant small RNAs	comparative sequencing of plant small RNAs			X	various	US	http://smallrna.udel.edu/	n/a
13	X			Massively Parallel Signature Sequencing (MPSS)	different projects for many plant species, including multiple sRNA collections			X	various	US	http://mpss.udel.edu/	16381968
14		X	X	Teaching Tools in Plant Biology	online feature of The Plant Cell consisting of teaching materrials**	X	X	X	generic	n/a	http://www.plantcell.org/teachingtools/	n/a

*the publication refers to the previous version of the systim
**first six sections free, require subscription to Plant Cell

The majority of available resources and methods on plant epigenomics are concentrating on RNA interference and small RNAs such as siRNAs and microRNAs (miRNAs). A well-known computational method to check whether miRNAs can potentially bind to DNA sequences is miRanda (8; in Table 14.1), based on established sequence alignment and thermodynamic principles, and actually validated on mammalian data — both inputs are user-provided, with miRNAs taken from specialized repositories of published data such as the miRBase (2 in Table 14.1; John et al., 2004). Another excellent example of how bioinformatics acts as a horizontal platform for research, is the RNA22 algorithm (5 in Table 14.1; Miranda et al., 2006), developed at IBM Research. Based on a generic pattern discovery algorithm (TEIRESIAS; Rigoutsos and Floratos, 1998), it is able to comprehensively capture the characteristics of a known miRNA set (e.g., by a resource like miRBase) and search for potential targets in DNA sequences in a very sensitive manner, thus providing unprecedented depths of prediction. Providing a more plant-orientated methodology, the Plant Small RNA Regulator Target (psRNA-Target) Analysis Server (9 in Table 14.1; Zhang et al., 2005) allows users to search for targets of small RNAs and to annotate potential regulators based on published ones.

As already mentioned, technological advancements have allowed the production of many millions of short sequences from biological samples, potentially of great importance in understanding their regulatory role and effect on plants. To mine useful information from this mountain of data, state-of-the-art bioinformatics tools are necessary, and a useful collection is found at the University of East Anglia (UEA) sRNA toolkit (6 in Table 14.1; Moxon et al., 2008). It has a plant-specific section based on large-scale sRNA datasets, so it can assist in filtering sequences, identify miRNAs and create their expression profiles, predict targets and regulatory patterns, and visualization. Consequently, placing large sets of such sequences on genomes can lead to the identification of potential sRNA-producing genomic loci — see SiLoDB (7 in Table 14.1; http://silodb.cmp.uea.ac.uk/).

Data for such tools exist due to the effort of a number of U.S.-based (and -funded) research groups. The *Arabidopsis* Small RNA Project at Oregon State University (4 in Table 14.1; Backman et al., 2008) provides access to miRNAs, endogenous siRNAs, and other small RNA-related data on a genome-wide level for *A. thaliana* in different mutants, developmental stages, and treatments — such a wide range of data provide the necessary context and depth to understand the impact of small RNA pathways. The Meyers Laboratory (http://ag.udel.edu/meyers_lab/) at the University of Delaware has been a key player in the large-scale epigenomics data generation and analysis partnered with the collaborating laboratories of Pam Green at the University of Delaware and (http://ag.udel.edu/plsc/faculty/pamgreen/) and Xing Deng at Yale (http://www.yale.edu/denglab/). They first spearheaded the Massively Parallel Signature Sequencing (MPSS) of several plant species and the creation of signature-based transcriptional resources for analyses of mRNA and small RNA (13 in Table 14.1; Nakano et al., 2006); Illumina's SBS method has recently replaced the original MPSS/454 providing considerably better quality data. The labs of Blake Meyers and Pamela Green at the University of Delaware are also behind the Comparative Sequencing of Plant Small RNAs (12 in Table 14.1; http://smallrna.udel.edu/), an extensive sequence resource of plant small RNAs (21 to 24 nucleotides) from representative species across the plant kingdom.

Finally, it is important to highlight efforts in educating and involving students and the general public. An initiative by the journal *The Plant Cell* provides online educational material, which includes two parts on epigenetics and small RNAs (14 in Table 14.1; http://www.plantcell.org/teachingtools/) that provide an accessible, comprehensive, current, and illustrated view. The aforementioned Panzea project also has an outreach program on evolution, genetics, and agriculture, recognizing the value of maize as a showcase for the effect of domestication on genetic and phenotypic variability, and the value of research-driven breeding programs.

Transposable elements on the emerging epigenomic landscape(s)

Transposable elements (TEs) are highly active and mobile genetic elements that can be found within the genomes of virtually all living organisms. Initially considered as junk or selfish DNA, their profound influence in shaping eukaryotic genomes is now widely accepted. Long terminal repeat (LTR) retrotransposons form the most abundant TE type in plants, largely accounting for the vast differences in plant genome sizes (Vitte and Panaud, 2005). Their "copy and paste" replication mechanism has the potential to greatly amplify their numbers, even over short evolutionary timescales, enabling them to become a major component of plant genomes. Characteristically, plant genome size positively correlates with LTR retrotransposon abundance. Small genome plants like *Arabidopsis* and rice are sparsely populated by LTR retrotransposons, 5.6% (Pereira, 2004) and 17% (McCarthy et al., 2002), respectively, in contrast to medium and large genome plants, where the LTR retrotransposon fraction may reach up to 70% (barley; Vicient et al., 1999) or 75% (maize) (Baucom et al., 2009; Schnable et al., 2009).

The ability of TEs to insert in various areas of the genome, including gene-rich regions, can lead to deleterious mutations detrimental to the host plant. Consequently, it comes as no surprise that their activity is tightly regulated both at the transcriptional and post-transcriptional levels (Matzke and Birchler, 2005). The advent of technologies that enable the precise positional determination of epigenetic marks added either to DNA or chromatin/histone proteins has already revealed, and in the future will elucidate even further, how areas occupied by TEs carry chemical marks that may even vary according to the respective TE family. Overall, the concept that DNA methylation is primarily a mechanism of genome defense against TE is now robustly supported (Suzuki and Bird, 2008). DNA methylation of genomic regions rich in TE insertions induces heterochromatin formation and subsequent TE silencing. In contrast, repression of DNA methylation, by DNA methyltransferase gene mutants, is concomitant to an increase in TE activity. Plants have evolved a TE methylation strategy guided by siRNAs; for example, an RNA-dependent DNA methylation system (Vaucheret, 2008). Centromeric and TE regions have a similar repertoire of both DNA and chromatin modifications that usually associate with silent chromatin formation. However, different classes of TEs interact with different histone modification marks. The near-gene heterochromatin insertion patterns of the maize Sireviruses Opie and Ji correlate with the modified histone H3K4me2 (Lamb et al., 2007); whereas other LTR retrotransposons prefer accumulating into large gene-free heterochromatic blocks, and the centromere-specific retrotransposons of maize (CRM) may target for insertion centromere histone H3 (CENH3) or its methylation variants (Wolfgruber et al., 2009).

Along with guiding DNA methylation, siRNAs also silence TE expression at the post-transcriptional level by degradation of homologous mRNA TE targets (Hollick, 2008). Massive parallel sequencing of small RNAs and mRNAs has shown the complicated interactions between siRNAs and TE regulation in plants. In principle, interruption of the RNA interference (RNAi) pathway results in the over-expression of most TE families (Jia et al., 2009). Intriguingly, analysis of siRNA libraries produced from B73 wild-type or mop1-mutant maize plants, or from different maize tissues (young leaves, immature ears, and immature tassels) showed that siRNA species of different lengths (22 bp vs. 24 bp) appear to originate from or target TEs under these different circumstances and tissues (Nobuta et al., 2008; Wang et al., 2009).

Our insights into the complex interactions between TEs and plant hosts have been greatly facilitated by the development of technologies that can conduct large-scale, genome-wide epigenomic analyses. Crucially, all of the components of the network — TE regulation and activity, small RNAs abundance, DNA methylation and histone modification patterns — are dynamic entities that are greatly influenced by plant biotic and abiotic stresses. The genome and epigenome fluidity found among different maize genotypes that underlies phenotypic variability can provide valuable new information to breeders in the quest for ever more tolerant and productive varieties.

An illustrative and practical example of data and resources integration

In an attempt to highlight a few aspects of bioinformatics research, we decided to visualize datasets from the Wang et al. (2009) publication of in *The Plant Cell* (10; in Table 14.1). Such an exercise involves accessing online data repositories (in this case NCBI's Gene Expression Omnibus), downloading and organizing data, converting sequence coordinates data from *Zea mays* BACs (as provided) to chromosomes, transforming raw data to binned frequency counts on the maize genome, and finally visualizing all of the information using Circos (Krzywinski et al., 2009), a popular software package for visualizing data and information in a circular layout, making it ideal for exploring relationships between objects or positions.

Figure 14.1 depicts the 10 chromosomes of maize as ideograms in a clockwise manner, starting with chromosome 1 at 12 o'clock. The centromere is highlighted with a red-filled band. All lengths are relative; that is, chromosome 1 has the longest ideogram since it is the longest chromosome. Concentric heat map tracks represent different epigenomic datasets from the Wang et al. (2009) paper, as well as the distributions of a specific retrotransposon class (Bousios et al., 2010) of genes as found by the sequencing consortium (http://www.maizesequence.org/), and high-quality haplotype SNP data from the Panzea consortium (http://panzea.org/index.html; Gore et al., 2009). Heat map colors begin with white for the lowest frequency and reach very dark shades of the perspective color for the highest frequencies.

In particular, and from outside to inside, the first five tracks in Figure 14.1 in shades of red are H3K27me3, H3K36me3, H3K4me3, H3K9ac, and methylation data (respectively) in maize roots, the next five in shades of green are the equivalent data but in shoots. The track shaded in blue represents the Sirevirus retrotransposons, and the track shaded in gray provides the distribution of genes. A glimpse of the variability at the nucleotide level between the two well-studied varieties of maize, B73 and MO17, can be seen in the next inward track containing 275,000 high-quality SNPs (that differ between the two) in the form of a kaleidoscope of colors representing the four bases and deletions (A, red; C, blue; G, green; T, black; deletion, yellow; n/a, white). In the remaining space in the circle, known miRNA identified in the high-throughput sequencing data are mapped on the chromosomes with lines connecting occurrences of the same sequence — in green the sequences are identified in shoots and in red they are identified in roots.

Although the purpose of this exercise was to stress the need for and power of data integration (and the tools that make it possible), and therefore no biological insight is intended, it is interesting to note visual biases such as the high density in almost all tracks at around three quarters along chromosome four, or shoot-only (all green) or root-only (all red) instances of miRNAs.

Epigenetic Phenomena and their Implications in Plant Breeding

Epigenetic controls during vegetative development and the role of the environment

Flowering plants undergo major developmental transitions, such as the transition from the embryonic to the vegetative stage, and finally to the reproductive stage.

Epigenetic mechanisms are at the core of these endogenous developmental regulatory mechanisms responsible for materializing a certain developmental stage and transit to the next one, while at the same time blocking the genes responsible for the one further down. For example, for a plant to get from the juvenile vegetative (where the genes controlling this stage are active) to the next adult vegetative phase requires the suppression of the juvenile vegetative stage genes, the activation of the adult vegetative stage genes, and the blockage of the genes of the ensuing reproductive stage for allowing the period to materialize the adult vegetative stage. The involvement of

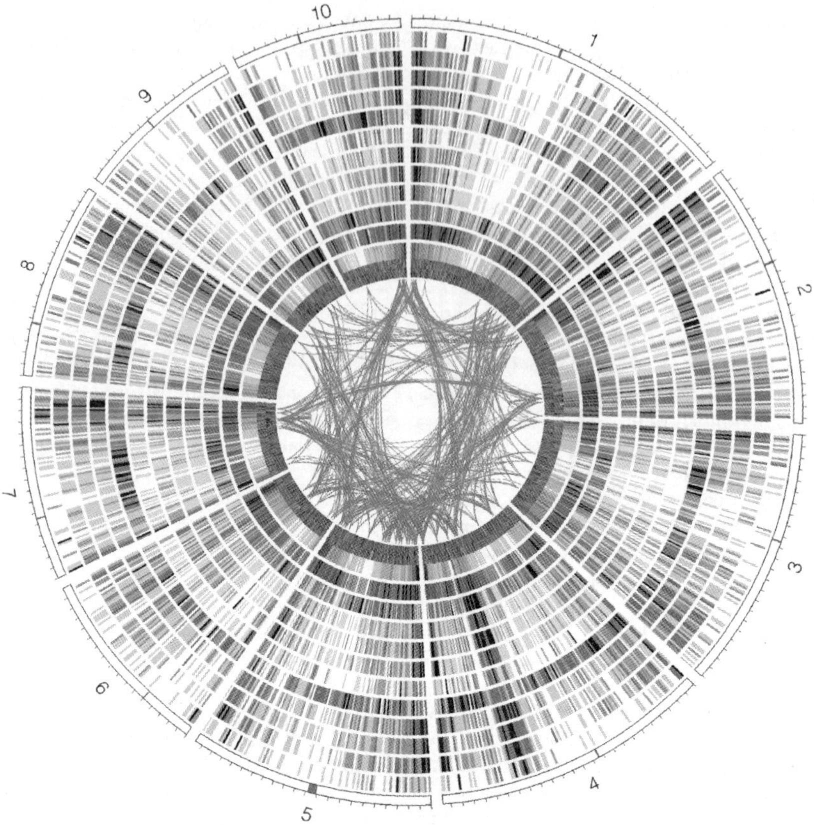

Figure 14.1 • The ten chromosomes of *Zea mays* B73, encircling layers of epigenomic information (red and green tracks), specific retrotransposable elements (blue track), genes (gray track), haplotype SNPs between the B73 and MO17 maize lines (multicolor double-track), and miRNAs (red and green links in the center). All tracks are heat maps of frequencies, from white to the darkest shade of the perspective color, apart from the SNP data – see details in the text. Visualization was created with Circos and in-house PERL scripts. Please see color plate section at the end of the book.

epigenetic chromatin remodeling complexes in controlling SAM and RAM during early and late embryonic stages were mentioned in the section Epigenetic Mechanisms and their Interactions. The role of the complexes in later stages, like the transition to reproductive stage and seed development, will be described in the sections Epigenetic Control of Flowering and Endosperm Development and Parental Imprinting. It is worth mentioning that small RNAs, and particularly miRNAs originally discovered as master regulators of developmental timing in *Caenorhabditis elegans*, regulate developmental timing in plants too. As described in a recent review (Poethig, 2009), small RNAs play a central role in the timing of developmental transitions. Focusing on the aforementioned transition from the juvenile vegetative stage to the adult vegetative stage, the review specifically describes two miRNAs: miR156, targeting an SBP-box transcription factor; and miR172, targeting the AP2-LIKE transcription factor that suppresses flowering. miRNA156 is strongly expressed during the juvenile vegetative stage and it is mainly responsible for the manifestation of this stage. SBP-box transcription factors are important for controlling the expression of genes for the adult vegetative stage. Thus, by targeting SBP-transcription factors, miR156 prevents the transition to the next stage, that is, the adult vegetative stage. In addition, miR156 targets an SBP transcription factor that negatively controls the expression of miR172. In this way miR156 allows for the expression of miR172, which in turn targets and cleaves *AP2-LIKE* transcripts. Destruction of the AP2-LIKE suppressor promotes the transition to flowering.

Numerous other examples of different miRNAs controlling other developmental transitions have been reported. The importance of this epigenetic mechanism on imposing heterochronic changes prolonging or shortening developmental stages, and the implications of this to breeding different plant varieties, is obvious. Furthermore, by their involvement in controlling the developmental programs during the different developmental stages, epigenetic mechanisms can operate as integrators of external environmental signals affecting development. Reprogramming of cell differentiation in response to environmental stress leads to phenological and developmental plasticity, both important mechanisms of plant resistance. Phenotypic plasticity helps adjust the durations of various phenological phases in plants, and thus allows plants to avoid exposure of critical growth phases, and especially reproductive development, to stress. Further, adjustment of growth and development is critical for effective use of resources under stress (Chinnusamy and Zhu, 2009a).

A number of studies, including our own, have indicated that epigenetic changes are taking place as a result of changing environmental cues/stresses, such as light (circadian clock, photoperiod, light quality, temperature (high or low), water availability (drought or flooding), and so forth. For example, in *Arabidopsis*, *HDAC* genes from both the RPD3/HDA1 and the HD2 families of HDACs have been associated with the response to biotic and abiotic stress. *AtHDA19* was suggested to mediate jasmonic acid (JA) and ethylene signaling during pathogen defense (Tian and Chen, 2001; Tian et al., 2005; Zhou et al., 2005), and *HDA6* was shown to be required for jasmonate response, senescence, and flowering (Wu et al., 2008). In addition, histone modifications have been suggested to play a major role in ABA-mediated response to abiotic stress (Chinnusamy et al., 2008). ABA treatment caused severe reduction in expression of the histone deacetylase, *AtHD2C*, whereas over-expression of *AtHD2C* resulted in enhanced abiotic stress tolerance to salt and drought stress, and both the repression of several ABA-responsive genes and the induction of others (Sridha and Wu, 2006).

In rice, cold, osmotic and salt stress, and external application of hormones such as JA, ABA, and SA were shown to increase the expression of certain *HDA1* genes and reduce the expression of others (Fu et al., 2007). Recent work from our group in barley has identified and characterized *HDAC*, *HAT*, and *PcG* genes (Demetriou et al., 2009, 2010; Kapazoglou et al., 2010; Papaefthimiou et al., 2010). Members of the *RPD3/HDA1* family of HDACs displayed differential expression in two barley cultivars differing in various characteristics, such as seed size and tolerance to stress, implying a possible association of these genes with different traits. In addition, *HDA1* genes representing each of the four classes of the *HDA1* family were found to be induced by exogenous JA application of seedlings. Similarly, two representative members of the *HD2* family were shown to respond both to exogenous JA and ABA. *HAT* genes from barley were also shown to be induced by treatment with ABA. Furthermore, additional studies with two members of a barley Polycomb group, *E(Z)* and *FIE*, demonstrated a marked induction of these genes as a response to exogenous ABA. The observed changes in expression of barley epigenetic genes upon treatment with stress-related hormones implied an association of these genes with biotic and abiotic stress in barley.

Accordingly, as was previously mentioned, mechanisms engaging chromatin remodeling complexes are involved in responses to environmental cues. These provide a cell memory for past environmental events such as vernalization or stabilization of specific developmental phases and patterns of differentiation. Chromatin remodeling mechanisms also participate in the regulation of gene expression in differentiated cells in response to environmental cues, either when they represent expected regular variations, such as light–dark cycles, or unpredictable stress conditions (Jarillo et al., 2009).

Most of these stress-induced modifications are reset to the basal level once the stress is relieved, while some of the modifications may be stable, that is, may be carried forward as "stress memory" and may be inherited across mitotic or even meiotic cell divisions. Epigenetic stress memory may help plants to cope more effectively with subsequent stresses. Developmental and environmental signals can induce epigenetic modifications in the genome, and thus, the single genome in a plant cell gives rise to multiple epigenomes in response to developmental and environmental cues. Understanding stress-induced epigenetic processes in plant stress tolerance requires further clarification of these questions: How much of the stress-induced gene expression changes are associated with alterations in DNA methylation and histone modification marks? Are stress-induced DNA and histone modifications during acclimation, or during the first experience of stress, memorized and inherited mitotically and meiotically? What are the adaptive values of epigenetic stress memory (Chinnusamy and Zhu, 2009a)?

Epigenetic control of flowering

The transition from the vegetative to the reproduction phase in plants is a major developmental process, and proper flowering time in different environments is crucial for reproductive success. Plants have evolved several ways to control flowering time effectively, such as perception of day length, vernalization — during which exposure to prolonged cold renders meristems competent to flower and prevents flowering in the winter — and the juvenile to adult phase transition (Amasino, 2010). The central integrator of flowering regulation in *Arabidopsis* is the gene *FLOWERING LOCUS C* (*FLC*). *FLC* is a MADS-box transcription factor that inhibits floral transition by reducing the expression of three master regulators of flowering time, *FLOWERING LOCUS T* (*FT*), *SUPPRESSION OF OVEREXPRESSION OF CONSTANS 1* (*SOC1*), and *FD*. *FT* acts as a floral promoter and is controlled by *CONSTANS* (CO). *SOC1* is also a floral promoter and is partly induced by *FT*. In turn, *FT*, *FD*, and *SOC1* activate floral identity genes, *APETALA 1* (*AP1*) and *LEAFY*, leading to flower formation (Turck et al., 2008). The expression of *FLC* is under the control of a multitude of both positive and negative chromatin regulatory pathways, both endogenous (developmental) and exogenous (environmental).

Upon vernalization in *Arabidopsis*, *FLC* is repressed by a series of histone modifications that drive *FLC* chromatin to a repressive state. These include H3K27me2, H3K27me3, and H3K9me histone methylation, H3K4 histone demethylation, and H4R3sme2 (histone 4 arginine 3 symmetric dimethylation (Sung and Amasino, 2006; Finnegan and Dennis, 2007; Greb et al., 2007; Wang et al., 2007; Schmitz et al., 2008). Prolonged exposure to cold induces expression of the gene *VERNALIZATION INSENSITIVE 3* (*VIN3*) encoding a protein with a plant homeodomain (PHD) finger. VIN3 interacts physically with VIL1/VERNALIZATION 5 (VRN5), a VIN3 homolog, and both are required for *FLC* repression (Sung and Amasino, 2004; Greb et al., 2007). It was recently shown that VIN3 and VIL1/VRN5 belong to a PRC2-like PcG repressive complex consisting of four PRC2 subunits, SWN, VERNALIZATION 2 (VRN2), FIE, and

MSI1 (homologs of the *Drosophila* E(Z), ESC, SU(Z)12, and p55, respectively) along with one more PHD-domain protein, VERNALIZATION5/VIN3 LIKE 1 (VEL1; Wood et al., 2006; De Lucia et al., 2008).

Repressive histone modifications such as PRC2-mediated H3K72me3 have been suggested to be involved in *FLC* silencing, also in the absence of vernalization. *Arabidopsis* mutants defective in *SWN* or *CLF* (*EZ* homologs), *EMF2* (*VRN2* homolog), *FIE*, and *MSI1*, displayed *FLC* derepression in young seedlings, implying the action of a CLF-SWN/EMF2/FIE/MSI1 complex in *FLC* suppression during vegetative development (Wood et al., 2006; Jiang et al., 2008). Similarly, other repressive mechanisms such as H3K4me demethylation were shown to take place in *FLC* regulation. Loss-of-function of *FLD*, a plant homolog of the human *LSD1*, resulted in decreased levels of H3K4me2 and H3K4me3 in *FLC* chromatin and subsequent *FLC* repression (Jiang et al., 2007).

Histone modifications leading to active chromatin have also been reported for the *FLC* locus. *ARABIDOPSIS TRITHORAX 1* (*ATX1*), encoding a histone methyltransferase responsible for the H3K4me3 chromatin activation mark, was shown to be associated with H3K4me3 on *FLC* chromatin and consequently *FLC* expression (Pien et al., 2008). Recently, four genes responsible for histone H2B monoubiquitination were found to promote *FLC* activation and subsequent repression of flowering in *Arabidopsis* (Cao et al., 2008; Gu et al., 2009). Moreover, a recently identified *Arabidopsis* homolog of the yeast SWR1c complex, which deposits the histone H2A.Z variant in chromatin, has been implicated in H2A.Z deposition into *FLC* chromatin, and subsequent *FLC* activation and floral repression (Choi et al., 2002; Deal et al., 2007).

Along with *FLC*, other flowering genes are under epigenetic control. A CLF-containing PRC2-like complex was proposed to mediate H3K27me3 marking at both *AGAMOUS-LIKE 19* (a MADS-box transcription factor upregulated upon vernalization and promoting flowering) and *FT* chromatin, and cause repression of both genes (Schonrock et al., 2006; Jiang et al., 2008). In addition a plant-specific JmjC domain histone demethylase was found to mediate H3K4me3 in *FT* and an *FT* homolog, *TWIN SISTER OF FT* (*TSF*), chromatin (Yang et al., 2010). Furthermore, miRNAs have been mentioned already as important regulators of flowering (Poethig, 2009), whereas other epigenetic mechanisms such as RdDM and cold-induced long antisense transcripts have been reported to silence chromatin at the *FLC* locus in *Arabidopsis* (Swiezewski et al., 2007, 2009).

Interestingly, natural variation in flowering time has been associated with an epiallele of *FLC*, which was created by the action of siRNA-dependent DNA methylation (Zhai et al., 2008). In the *Arabidopsis Landsberg erecta* (*Ler*) ecotype an MPF (Methylated region near Promoter of FLC) element was found adjacent to the promoter of *FLC*. This produced a high abundance of 24 nt siRNAs that directed DNA methylation at the MPF in the *Ler* ecotype and resulted in FLC suppression. However, in the Columbia (Col) ecotype the corresponding region produces low levels of matching siRNAs that cannot direct DNA methylation. In the same study, a large number of additional loci were identified with complementarity to siRNAs, which showed a more significant increase in DNA methylation in *Ler* than Col. Furthermore, a previous report revealed that DNA methylation in transcribed regions of genes is highly polymorphic between the two ecotypes (Vaughn et al., 2007). These observations suggested that epigenetic changes are likely to be a major contribution to phenotypic variation, and this could be exploited for understanding the molecular basis determining agricultural traits.

Integrating the responses to developmental and environmental cues to ensure flowering under the most optimal conditions is critical for reproductive success. A variety of epigenetic molecular mechanisms participate in the regulation of such a vital step in plant development. Additional genes and regulatory networks involved in the flowering process have been described (He, 2009; Amasino, 2010).

Along with the wealth of information on regulatory pathways and epigenetic molecular mechanisms governing flowering in the model-plant *Arabidopsis*, a great deal of progress has been made in the past few years in delineating the genetic and epigenetic molecular mechanisms of the floral transition in crop species including rice, wheat, and barley. Floral transition was proven to hide behind the early or late cultivars identified in these crops; cultivars were bred alleviating the need for vernalization allowing dispersal into new environments. For example, the genes and regulatory pathways involved in the process of photoperiod and vernalization have been described in detail in cereals (Distelfeld et al., 2009). One of the key genes in cereal flowering, *VRN1* (different from *Arabidopsis VRN1*) encoding a MADS-box transcription factor was found to be in an active chromatin state upon vernalization, which was induced by increased levels of H3K4me3 and loss of H3K27me3 (Oliver et al., 2009). In barley and wheat cultivars without the need for vernalization, the *VRN1* gene was inactivated. Thorough molecular understanding of the floral transition can have important implications in plant breeding, as proper flowering time has an important impact on crop yield. Epigenetic changes can lead to the formation of epialleles associated with optimal flowering under different environmental conditions.

Endosperm development and parental imprinting

In flowering plants, the seed is formed through the process of double fertilization (Olsen, 2004). Bypass of double fertilization is observed in natural apomicts or after mutagenesis of specific gene loci. In *Arabidopsis* mutations of the *FIS* (fertilization independent seed) class of genes — *FERTILIZATION INDEPENDENT SEED* (*FIE*), *MEDEA* (*MEA*), *FERTILIZATION INDEPENDENT SEED2* (*FIS2*), *MULTICOPY SUPRESSOR OF IRA1* (*MSI1*) — encoding members of a PcG-histone methyltransferase complex (described earlier) result in the formation of seed-like structures in the absence of double fertilization that eventually collapse (Gehring et al., 2004; Huh et al., 2008). The phenotype of *fie*, *mea*, *fis2*, and *msi1* mutants is accompanied by another impressive characteristic, that is, endosperm overproliferation (Kohler et al., 2003a; Guitton et al., 2004). Endosperm overproduction and seed size increase are also observed in crosses where extra paternal genomes or hypomethylated maternal genomes are used (Adams et al., 2000). Another feature of the *fis* phenotype is that when the female gametophyte carrying a mutation in a *fis* locus is fertilized by a wild-type plant, the seed finally collapses indicative of a lethal maternal effect of the *fis* mutation and a gametophytic maternal control over seed development by *FIS* genes (Gehring et al., 2004; Huh et al., 2008). This is partially attributed to parental imprinting of *FIE*, *MEA*, and *FIS2*. Parental genomic imprinting refers to preferential expression of a gene from only the maternal or paternal allele (uniparental gene expression). So far, a number of imprinted genes have been identified in *Arabidopsis* and maize (Berger and Chaudhury, 2009; Mosher et al., 2009). Most of these are preferentially expressed in the endosperm and are controlled by one or more epigenetic mechanisms. The combined action of DNA methylation/demethylation and histone modifications is exemplified in the phenomenon of parental imprinting of the *PHERES1* (*PHE1*) and the *MEDEA* (*MEA*) genes in the *Arabidopsis* endosperm during seed development. Highly regulated spatiotemporal expression of only the paternal allele of *PHE1*, a Type-I MADS-box gene, is crucial for proper seed development (Kohler et al., 2003b). Strict repression of *PHE1* prior to fertilization and gene activation after fertilization is under the control of both DNA cytosine methylation and histone modifications. Specifically, the requirement for *PHE1* repression prior to fertilization is accomplished, in part, by the unmethylated status of both paternal and maternal alleles in the sperm cell and in the central cell of the female gametophyte, respectively. Post-fertilization methylation of a distantly located 3′ downstream region of the *PHE1* paternal allele is associated with gene activation after fertilization, whereas the unmethylated maternal *PHE1* allele remains transcriptionally silent (Makarevich et al., 2008). Notably, in this case DNA methylation is associated with gene activation and not repression as is usually the case, pointing to different mechanisms underlying this phenomenon. In addition to DNA methylation there is H3K27me3 at *PHE1* histones, established by the gene *MEA*, an H3K27me3 methyltransferase and subunit of a PcG complex. H3K27me3 marking keeps both paternal and maternal *PHE1* alleles silent until fertilization when an unknown mechanism removes H3K27me3 suppression from only the paternal *PHE1* allele. *MEA* and *FIS2* are also parentally imprinted, but unlike *PHE1* it is the maternal allele that is expressed. Prior to fertilization, in the central cell of the female gametophyte, both *MEA* and *FIS2* maternal alleles are repressed by methylation at 5′ upstream and 3′ downstream elements, most likely established by *MET1*. Likewise, in the sperm cell both *MEA* and *FIS2* paternal alleles are repressed (Huh et al., 2008; Kohler and Villar, 2008). The maternal allele of the DNA glycosylase *DME* only expressed in the central cell and not in the sperm cell, and acting antagonistically to *MET1*, demethylates maternal *MEA* and *FIS2* and allows for the expression of the maternal *MEA* and *FIS2* alleles and production of the MEA/FIE/FIS2/MSI1 complex (Gehring et al., 2006; Jullien et al., 2006; Huh et al., 2008). After fertilization, this complex establishes H3K27me3 marks at paternal *MEA* histones within the zygote, further repressing the paternal *MEA* allele (as well as the maternal *PHE1* allele mentioned previously) and ensuring genomic imprinting at the *MEA* and *PHE1* loci, a condition crucial for seed viability (Huh et al., 2008). In conclusion, imprinting at these loci is regulated by a combination of DNA methylation by MET1, DNA demethylation via transcriptional repression of MET1 through the RBR1/MSI1 pathway, by active DNA demethylation via the DNA glycosylase DME, and by histone modifications, in particular H3K27me3 (Berger and Chaudhury, 2009).

Up to now DME-mediated DNA demethylation appeared to target individual genes of the maternal genome, but two recent reports have demonstrated DME-dependent, genome-wide DNA demethylation during female gametogenesis and seed development. These studies mapped the global DNA methylation pattern of both the embryo and endosperm genomes in *Arabidopsis* through deep sequencing of methylcytosine immunoprecipitated DNA and high-throughput bisulfite DNA sequencing (Gehring et al., 2009; Hsieh et al., 2009). Bulk methylation decrease in all sequence contexts (CG, CHG, CHH) was observed in the endosperm as compared to the embryo. CG-methylation was reduced in both gene bodies and repeats, and was partially restored in *dme* mutants. The most prominent loss of genic methylation occurred at 5′ of the transcriptional start site and genes with reduced 5′ upstream methylation were preferentially expressed in the endosperm. Furthermore, methylation levels at sequences of known imprinted genes were also calculated, specifically the *MEA* 3′ repeats, the *FWA* promoter and transcriptional start site, the *FIS2* promoter and the *PHE1* 3′ repeats, and significant methylation reduction was evidenced in the endosperm. Moreover, CG-methylation was largely restored for all genes in *dme* endosperm. It was proposed that imprinting is not confined to specific genes but rather affects and reprograms the entire maternal genome in the endosperm (Hsieh et al., 2009). Unlike mammals, which have to establish imprinting by differential methylation of specific sequences in their gametes, plants seem to carry out global demethylation of the central cell leading to a hypomethylated maternal endosperm genome, regardless of massive transposon derepression. The endosperm genome is not transmitted to the next generation; therefore, transient transposon activation may have little negative impact, and resetting the DNA methylation pattern of the endosperm genome is not required.

It seems that DNA methylation and parental imprinting through DME demethylation in the central cell and endosperm have critical roles in seed development and seed viability. This is most likely because maintaining a specific ratio between maternal and paternal genomes is vital for proper seed development. According to the parental conflict hypothesis, maternal and paternal genomes have conflicting roles. The maternal genome has evolved to distribute resources equally to all offspring, whereas the paternal genome has evolved to maximize resource allocation to the best-fit offspring (Haig and Westoby, 1991). In support of this theory, interploidy studies in *Arabidopsis* have shown that crosses between a tetraploid mother and a diploid father produce smaller than wild-type seeds, and crosses between a tetraploid father and a diploid mother produced larger than wild-type seeds (Scott et al., 1998). Thus excess of parental genomes has an impact on seed size. Similar seed-size increase was demonstrated when hypomethylated maternal genomes were used (Adams et al., 2000). It has been shown that loss-of-function mutations in *MET1* and *CMT3* lead to *Arabidopsis* embryos with abnormal development and reduced viability (Xiao et al., 2006a). It has also been shown that global demethylation in *met1* pollen mutants inhibits endosperm growth and results in smaller seeds (Xiao et al., 2006b). On the other hand, crosses between wild-type pollen and ovules from *met1/met1* plants or ovules carrying the *MET1* antisense gene produce plants with larger seeds. This suggested that *MET1* dosage in the maternal vegetative tissue has an impact on seed size.

Whereas in dicots the endosperm is consumed by the embryo during seed development, in monocots such as cereals, the endosperm persists after embryo development is completed and constitutes the major portion of the mature kernel. Since the endosperm of cereals provides the world's largest supply of caloric intake, understanding seed-dependent epigenetic processes such as imprinting in monocot plants is of great importance. The epigenetic regulation of imprinting of genes associated with seed development and seed size and yield could have important uses in plant breeding, for example, by driving mono-allelic expression of desirable genes in agriculturally important species. Recently genes such as HDACs, HATs, and PcG genes, involved in epigenetic regulation in seed development have been identified in barley (Demetriou et al., 2009, 2010; Kapazoglou et al., 2010; Papaefthimiou et al., 2010). Genes encoding members of a PcG complex exhibit significant differences in expression during seed development in barley cultivars with different seed size. In addition, they have been mapped on barley chromosomes and have been associated with seed-related traits such as malting quality.

Further understanding and manipulating of imprinting and other seed developmental epigenetic mechanisms in economically important monocots could have significant implications in plant breeding for improved varieties.

Conclusions and Prospects

The availability of the maize genome opens new horizons not only in further understanding a number of biological phenomena first described in maize and having an epigenetic basis like paramutation, parental imprinting, cycling of TEs, and so forth, but to also understand genetic diversity and variation from thousands of years during the domestication process, from teosinte to different maize races. Equally relevant are the substantial changes made by breeders the last hundred years or so, including the exploitation of heterosis. As was stressed in a recent review (Hollick and Springer, 2009), the maize epigenome can perceive and respond in heritable ways to stressful conditions, but the mode of this perception is completely unknown. It is anticipated that further understanding of the maize epigenome and its control system, as well as in other crop species, will eventually illuminate this connection between environmental perception and the heritable "genetic" changes.

Acknowledgments

We would like to acknowledge the members of our institute for critically reading the review and the General Secretariat for Research and Technology, Greece, for continuous support of our institute.

Abbreviations

5mC	5 methyl cytosine
MTase	Methyltransferase
CLF	Curly Leaf
CMT	Chromomethyltransferase
DCL	Dicer-like protein
DME	Demeter
E(Z)	Enhancer of Zeste
FIE	Fertilization-independent endosperm
FIS	Fertilization-independent seed
FLC	Flowering Locus C
HAT	Histone acetyltransferase
HDAC	Histone deacetylase
HDM	Histone demethylase
HMT	Histone methyltransferase
SU(Z)12	Suppressor of Zeste (12)
SAM	Shoot apical meristem
RAM	Root apical meristem
TE	Transposable Element
TCP	TEOSINTE BRANCHED1, CYCLOIDEA, and PCF

References

Adams, S., Vinkenoog, R., Spielman, M., Dickinson, H. G., & Scott, R. J. (2000). Parent-of-origin effects on seed development in *Arabidopsis thaliana* require DNA methylation. *Development*, *127*, 2493–2502.

Agius, F., Kapoor, A., & Zhu, J. -K. (2006). Role of the *Arabidopsis* DNA glycosylase/lyase ROS1 in active DNA demethylation. *Proceedings of the National Academy of Sciences*, *103*, 11796–11801.

Aichinger, E., Villar, C. B., Farrona, S., Reyes, J. C., Hennig, L., & Kohler, C. (2009). CHD3 proteins and polycomb group proteins antagonistically determine cell identity in *Arabidopsis*. *PLoS Genetics*, *5*, e1000605.

Amasino, R. M. (2010). Seasonal and developmental timing of flowering. *Plant Journal*, *61*, 1001–1013.

Aufsatz, W., Mette, M. F., van der Winden, J., Matzke, M., & Matzke, A. J. (2002). HDA6, a putative histone deacetylase needed to enhance DNA methylation induced by double-stranded RNA. *Embo Journal*, *21*, 6832–6841.

Backman, T. W. H., Sullivan, C. M., Cumbie, J. S., Miller, Z. A., Chapman, E. J., & Fahlgren, N., et al. (2008). Update of ASRP: The *Arabidopsis* small RNA project database. *Nucleic Acids Research*, *36*, D982–D985.

Baucom, R. S., Estill, J. C., Chaparro, C., Upshaw, N., Jogi, A., & Deragon, J. M., et al. (2009). Exceptional diversity, non-random distribution, and rapid evolution of retroelements in the B73 maize genome. *PloS Genetics*, *5*(11), e1000732.

Beck, S., & Rakyan, V. K. (2008). The methylome: Approaches for global DNA methylation profiling. *Trends in Genetics*, *24*, 231–237.

Berger, F., & Chaudhury, A. (2009). Parental memories shape seeds. *Trends in Plant Science*, *14*, 550–556.

Bernatavichute, Y. V., Zhang, X., Cokus, S., Pellegrini, M., & Jacobsen, S. E. (2008). Genome-wide association of histone H3 lysine nine methylation with CHG DNA methylation in *Arabidopsis thaliana*. *PLoS One*, *3*, e3156.

Bousios, A., Darzentas, N., Tsaftaris, A., & Pearce, S. R. (2010). Highly conserved motifs in non-coding regions of Sirevirus retrotransposons: The key for their pattern of distribution within and across plants? *BMC Genomics*, *11*, 89.

Cao, Y., Dai, Y., Cui, S., & Ma, L. (2008). Histone H2B monoubiquitination in the chromatin of FLOWERING LOCUS C regulates flowering time in *Arabidopsis*. *Plant Cell*, *20*, 2586.

Chan, S. W. L., Henderson, I. R., & Jacobsen, S. E. (2005). Gardening the genome: DNA methylation in *Arabidopsis thaliana*. *Nature Reviews Genetics*, *6*, 351–360.

Chanvivattana, Y., Bishopp, A., Schubert, D., Stock, C., Moon, Y. -H., & Sung, Z. R., et al. (2004). Interaction of Polycomb-group proteins controlling flowering in *Arabidopsis*. *Development*, *131*, 5263–5276.

Chapman, E. J., & Carrington, J. C. (2007). Specialization and evolution of endogenous small RNA pathways. *Nature Reviews Genetics*, *8*, 884–896.

Chen, Z. J., & Tian, L. (2007). Roles of dynamic and reversible histone acetylation in plant development and polyploidy. *Biochimica et Biophysica Acta*, *1769*, 295–307.

Chinnusamy, V., & Zhu, J. K. (2009a). Epigenetic regulation of stress responses in plants. *Current Opinion in Plant Biology*, *12*, 133–139.

Chinnusamy, V., & Zhu, J. -K. (2009b). RNA-directed DNA methylation and demethylation in plants. *Science in China Series C: Life Sciences*, *52*, 331–343.

Chinnusamy, V., Gong, Z., & Zhu, J. K. (2008). Abscisic acid-mediated epigenetic processes in plant development and stress responses. *Journal of Integrative Plant Biology*, *50*, 1187–1195.

Choi, Y., Gehring, M., Johnson, L., Hannon, M., Harada, J. J., & Goldberg, R. B., et al. (2002). DEMETER, a DNA glycosylase domain protein, is required for endosperm gene imprinting and seed viability in *Arabidopsis*. *Cell*, *110*, 33–42.

Cokus, S. J., Feng, S., Zhang, X., Chen, Z., Merriman, B., & Haudenschild, C. D., et al. (2008). Shotgun bisulphite sequencing of the *Arabidopsis* genome reveals DNA methylation patterning. *Nature*, *452*, 215–219.

Cubas, P., Vincent, C., & Coen, E. (1999). An epigenetic mutation responsible for natural variation in floral symmetry. *Nature*, *401*, 157–161.

De Lucia, F., Crevillen, P., Jones, A. M., Greb, T., & Dean, C. (2008). A PHD-polycomb repressive complex 2 triggers the epigenetic silencing of FLC during vernalization. *Proceedings of the National Academy of Sciences of the United States of America*, *105*, 16831–16836.

Deal, R. B., Topp, C. N., McKinney, E. C., & Meagher, R. B. (2007). Repression of flowering in *Arabidopsis* requires activation of FLOWERING LOCUS C expression by the histone variant H2A.Z. *Plant Cell*, *19*, 74.

Demetriou, K., Kapazoglou, A., Bladenopoulos, K., & Tsaftaris, A. (2010). Epigenetic chromatin modifiers in barley: II. Characterization and expression analysis of the HDA1 family of barley histone deacetylases during development and in response to Jasmonic acid. *Plant Molecular Biology Reporter*, *28*, 9–21.

Demetriou, K., Kapazolgou, A., Tondelli, A., Francia, E., Stanca, M. S., & Bladenopoulos, K., et al. (2009). Epigenetic chromatin modifiers in barley: I. Cloning, mapping and expression analysis of the plant specific HD2 family of histone deacetylases from barley, during seed development and after hormonal treatment. *Physiologia Plantarum*, *136*, 358–368.

Distelfeld, A., Li, C., & Dubcovsky, J. (2009). Regulation of flowering in temperate cereals. *Current Opinion in Plant Biology*, *12*, 178–184.

Dunoyer, P., Schott, G., Himber, C., Meyer, D., Takeda, A., & Carrington, J. C., et al. (2010). Small RNA duplexes function as mobile silencing signals between plant cells. *Science*, *328*, 912–916.

Finnegan, E. J., & Dennis, E. S. (2007). Vernalization-induced trimethylation of histone H3 lysine 27 at FLC is not maintained in mitotically quiescent cells. *Current Biology*, *17*, 1978–1983.

Fu, W., Wu, K., & Duan, J. (2007). Sequence and expression analysis of histone deacetylases in rice. *Biochemical and Biophysical Research Communications*, *356*, 843–850.

Fuks, F. (2005). DNA methylation and histone modifications: Teaming up to silence genes. *Current Opinion in Genetics & Development*, *15*, 490–495.

Gehring, M., Bubb, K. L., & Henikoff, S. (2009). Extensive demethylation of repetitive elements during seed development underlies gene imprinting. *Science*, *324*, 1447–1451.

Gehring, M., Choi, Y., & Fischer, R. L. (2004). Imprinting and seed development. *Plant Cell*, *16*(Suppl), S203–213.

Gehring, M., Huh, J. H., Hsieh, T. F., Penterman, J., Choi, Y., & Harada, J. J., et al. (2006). DEMETER DNA glycosylase establishes MEDEA polycomb gene self-imprinting by allele-specific demethylation. *Cell*, *124*, 495–506.

Gendler, K., Paulsen, T., & Napoli, C. (2008). ChromDB: The chromatin database. *Nucleic Acids Research*, *36*, D298–D302.

Gore, M. A., Chia, J. M., Elshire, R. J., Sun, Q., Ersoz, E. S., & Hurwitz, B. L., et al. (2009). A first-generation haplotype map of maize. *Science*, *326*, 1115–1117.

Greb, T., Mylne, J. S., Crevillen, P., Geraldo, N., An, H., & Gendall, A. R., et al. (2007). The PHD finger protein VRN5 functions in the epigenetic silencing of *Arabidopsis* FLC. *Current Biology*, *17*, 73–78.

Gu, X., Jiang, D., Wang, Y., Bachmair, A., & He, Y. (2009). Repression of the floral transition via histone H2B monoubiquitination. *Plant Journal*, *57*, 422.

Guitton, A. E., Page, D. R., Chambrier, P., Lionnet, C., Faure, J. E., & Grossniklaus, U., et al. (2004). Identification of new members of fertilisation independent seed polycomb group pathway involved in the control of seed development in *Arabidopsis thaliana*. *Development*, *131*, 2971–2981.

Haig, D., & Westoby, M. (1991). Seed size, pollination costs and angiosperm success. *Evolutionary Ecology*, *5*, 231–247.

He, Y. (2009). Control of the transition to flowering by chromatin modifications. *Molecular Plant*, *2*, 554–564.

Henderson, I. R., & Jacobsen, S. E. (2007). Epigenetic inheritance in plants. *Nature*, *447*, 418–424.

Henderson, I. R., & Jacobsen, S. E. (2008). Tandem repeats upstream of the *Arabidopsis* endogene SDC recruit non-CG DNA methylation and initiate siRNA spreading. *Genes & Development*, *22*, 1597–1606.

Ho, L., & Crabtree, G. R. (2010). Chromatin remodelling during development. *Nature*, *463*, 474–484.

Hollick, J. B. & Springer, N. (2009). Epigenetic Phenomena and Epigenomics in Maize. In *Epigenomics* (pp. 119–147).

Hollick, J. B. (2008). Sensing the epigenome. *Trends in Plant Science*, *13*, 398–404.

Hong, E. -H., Jeong, Y. -M., Ryu, J. -Y., Amasino, R., Noh, B., & Noh, Y. -S. (2009). Temporal and spatial expression patterns of nine *Arabidopsis* genes encoding Jumonji C-domain proteins. *Molecules and Cells*, *27*, 481–490.

Hsieh, T. F., Ibarra, C. A., Silva, P., Zemach, A., Eshed-Williams, L., & Fischer, R. L., et al. (2009). Genome-wide demethylation of *Arabidopsis* endosperm. *Science*, *324*, 1451–1454.

Huh, J. H., Bauer, M. J., Hsieh, T. F., & Fischer, R. L. (2008). Cellular programming of plant gene imprinting. *Cell*, *132*, 735–744.

Ito, T., Sakai, H., & Meyerowitz, E. M. (2003). Whorl-specific expression of the SUPERMAN gene of *Arabidopsis* is mediated by cis elements in the transcribed region. *Current Biology*, *13*, 1524–1530.

Jackson, J. P., Lindroth, A. M., Cao, X., & Jacobsen, S. E. (2002). Control of CpNpG DNA methylation by the KRYPTONITE histone H3 methyltransferase. *Nature*, *416*, 556–560.

Jarillo, J. A., Piñeiro, M., Cubas, P., & Martínez-Zapater, J. M. (2009). Chromatin remodeling in plant development. *The International Journal of Developmental Biology*, *53*, 1581–1596.

Jeong, J. -H., Song, H. -R., Ko, J. -H., Jeong, Y. -M., Kwon, Y. E., & Seol, J. H., et al.

(2009). Repression of *FLOWERING LOCUS T* Chromatin by functionally redundant histone H3 Lysine 4 Demethylases in *Arabidopsis*. *PLoS One*, *4*, e8033.

Jia, Y., Lisch, D. R., Ohtsu, K., Scanlon, M. J., Nettleton, D., & Schnable, P. S. (2009). Loss of RNA-dependent RNA polymerase 2 (RDR2) function causes widespread and unexpected changes in the expression of transposons, genes, and 24-nt small RNAs. *Plos Genetics*, *5*

Jiang, D., Wang, Y., Wang, Y., & He, Y. (2008). Repression of FLOWERING LOCUS C and FLOWERING LOCUS T by the *Arabidopsis* Polycomb repressive complex 2 components. *PLoS One*, *3*, e3404.

Jiang, D., Yang, W., He, Y., & Amasino, R. M. (2007). *Arabidopsis* relatives of the human lysine-specific Demethylase1 repress the expression of FWA and FLOWERING LOCUS C and thus promote the floral transition. *Plant Cell*, *19*, 2975.

John, B., Enright, A. J., Aravin, A., Tuschl, T., Sander, C., & Marks, D. S. (2004). Human MicroRNA targets. *Plos Biology*, *2*, 1862–1879.

Johnson, D. S., Li, W., Gordon, D. B., Bhattacharjee, A., Curry, B., & Ghosh, J., et al. (2008). Systematic evaluation of variability in ChIP-chip experiments using predefined DNA targets. *Genome Research*, *18*, 393–403.

Jullien, P. E., Kinoshita, T., Ohad, N., & Berger, F. (2006). Maintenance of DNA methylation during the *Arabidopsis* life cycle is essential for parental imprinting. *Plant Cell*, *18*, 1360–1372.

Kapazoglou, A., Tondelli, A., Papaefthimiou, D., Ampatzidou, H., Francia, E., & Stanca, M., et al. (2010). Epigenetic chromatin modifiers in barley: IV. The study of barley Polycomb group (PcG) genes during seed development and in response to external ABA. *BMC Plant Biology*, *10*, 73.

Kasschau, K. D. e. a. (2007). Genome-wide profiling and analysis of *Arabidopsis* siRNAs. *PLoS Biology*, *5*, 0479–0493.

Khraiwesh, B., Arif, M. A., Seumel, G. I., Ossowski, S., Weigel, D., & Reski, R., et al. (2010). Transcriptional control of gene expression by MicroRNAs. *Cell*, *140*, 111–122.

Kinoshita, T., Miura, A., Choi, Y., Kinoshita, Y., Cao, X., & Jacobsen, S. E., et al. (2004). One-way control of FWA imprinting in *Arabidopsis* endosperm by DNA methylation. *Science*, *303*, 521–523.

Kinoshita, Y., Saze, H., Kinoshita, T., Miura, A., Soppe, W. J., & Koornneef, M., et al. (2007). Control of FWA gene silencing in *Arabidopsis thaliana* by SINE-related direct repeats. *Plant Journal*, *49*, 38–45.

Kohler, C., & Villar, C. B. R. (2008). Programming of gene expression by Polycomb group proteins. *Trends in Cell Biology*, *18*, 236–243.

Kohler, C., Hennig, L., Bouveret, R., Gheyselinck, J., Grossniklaus, U., & Gruissem, W. (2003a). *Arabidopsis* MSI1 is a component of the E(Z)/FIE Polycomb group complex and required for seed development. *Embo Journal*, *22*, 4804–4814.

Kohler, C., Hennig, L., Spillane, C., Pien, S., Gruissem, W., & Grossniklaus, U. (2003b). The Polycomb-group protein MEDEA regulates seed development by controlling expression of the MADS-box gene PHERES1. *Genes & Development*, *17*, 1540–1553.

Kouzarides, T. (2007). Chromatin modifications and their function. *Cell*, *128*, 693–705.

Krzywinski, M., Schein, J., Birol, I., Connors, J., Gascoyne, R., & Horsman, D., et al. (2009). Circos: An information aesthetic for comparative genomics. *Genome Research*, *19*, 1639–1645.

Lamb, J. C., Meyer, J. M., Corcoran, B., Kato, A., Han, F. P., & Birchler, J. A. (2007). Distinct chromosomal distributions of highly repetitive sequences in maize. *Chromosome Research*, *15*, 33–49.

Law, J. A., & Jacobsen, S. E. (2010). Establishing, maintaining and modifying DNA methylation patterns in plants and animals. *Nature Reviews Genetics*, *11*, 204–220.

Li, X. Y., Wang, X. F., He, K., Ma, Y. Q., Su, N., & He, H., et al. (2008). High-resolution mapping of epigenetic modifications of the rice genome uncovers interplay between DNA methylation, histone methylation, and gene expression. *Plant Cell*, *20*, 259–276.

Lister, R., O'Malley, R. C., Tonti-Filippini, J., Gregory, B. D., Berry, C. C., & Millar, A. H., et al. (2008). Highly integrated single-base resolution maps of the epigenome in *Arabidopsis*. *Cell*, *133*, 523–536.

Luo, M., Platten, D., Chaudhury, A., Peacock, W. J., & Dennis, E. S. (2009). Expression, imprinting, and evolution of rice homologues of the Polycomb group genes. *Molecular Plant*, *2*, 711–723.

Lusser, A., Kolle, D., & Loidl, P. (2001). Histone acetylation: Lessons from the plant kingdom. *Trends in Plant Science*, *6*, 59–65.

Makarevich, G., Villar, C. B. R., Erilova, A., & Kohler, C. (2008). Mechanism of PHERES1 imprinting in *Arabidopsis*. *Journal of Cell Science*, *121*, 906–912.

Malagnac, F., Bartee, L., & Bender, J. (2002). An *Arabidopsis* SET domain protein required for maintenance but not establishment of DNA methylation. *The EMBO Journal*, *21*, 6842–6852.

March-Diaz, R., & Reyes, J. C. (2009). The beauty of being a variant: H2A.Z and the SWR1 complex in plants. *Molecular Plant*, *2*, 565–577.

Martienssen, R. (2010). Small RNA makes its move. *Science*, *328*, 834–835.

Martin, R. C., Liu, P. -P., Goloviznina, N. A., & Nonogaki, H. (2010). microRNA, seeds, and Darwin? Diverse function of miRNA in seed biology and plant responses to stress. *Journal of Experimental Botany*, *61*, 2229–2234.

Matzke, M. A., & Birchler, J. A. (2005). RNAi-mediated pathways in the nucleus. *Nature Reviews Genetics*, *6*, 24–35.

Matzke, M. A., Matzke, A. J. M., Pruss, G. J., & Vance, V. B. (2001). RNA-based silencing strategies in plants. *Current Opinion in Genetics & Development*, *11*, 221–227.

McCarthy, E. M., Liu, J. D., Lizhi, G., & McDonald, J. F. (2002). Long terminal repeat retrotransposons of *Oryza sativa*. *Genome Biology*, *3*(10), 0053.

Miranda, K. C., Huynh, T., Tay, Y., Ang, Y. S., Tam, W. L., & Thomson, A. M., et al. (2006). A pattern-based method for the identification of microRNA binding sites and their corresponding heteroduplexes. *Cell*, *126*, 1203–1217.

Molnar, A., Melnyk, C. W., Bassett, A., Hardcastle, T. J., Dunn, R., & Baulcombe, D. C. (2010). Small silencing RNAs in plants are mobile and direct epigenetic modification in recipient cells. *Science*, *328*, 872–875.

Mosher, R. A., Melnyk, C. W., Kelly, K. A., Dunn, R. M., Studholme, D. J., & Baulcombe, D. C. (2009). Uniparental expression of PolIV-dependent siRNAs in developing endosperm of *Arabidopsis*. *Nature*, *460*, 283–286.

Mosher, R. A., Schwach, F., Studholme, D., & Baulcombe, D. C. (2008). PolIVb influences RNA-directed DNA methylation independently of its role in siRNA biogenesis. *Proceedings of the National Academy of Sciences of the United States of America*, *105*, 3145–3150.

Moxon, S., Schwach, F., Dalmay, T., MacLean, D., Studholme, D. J., & Moulton, V. (2008). A toolkit for analysing large-scale plant small RNA datasets. *Bioinformatics*, *24*, 2252–2253.

Nakano, M., Nobuta, K., Vemaraju, K., Tej, S. S., Skogen, J. W., & Meyers, B. C. (2006). Plant MPSS databases: Signature-based transcriptional resources for analyses of mRNA and small RNA. *Nucleic Acids Research*, *34*, D731–D735.

Nobuta, K., Lu, C., Shrivastava, R., Pillay, M., De Paoli, E., & Accerbi, M., et al. (2008). Distinct size distribution of endogenous siRNAs in maize: Evidence from deep sequencing in the mop1-1 mutant. *Proceedings of the National Academy of Sciences of the United States of America*, *105*, 14958–14963.

Oliver, S. N., Finnegan, E. J., Dennis, E. S., Peacock, W. J., & Trevaskis, B. (2009). Vernalization-induced flowering in cereals is associated with changes in histone methylation at the VERNALIZATION1 gene. *Proceedings of the National Academy of Sciences*, *106*, 8386–8391.

Olsen, O. A. (2004). Nuclear endosperm development in cereals and *Arabidopsis thaliana*. *Plant Cell*, *16*, S214–S227.

Palauqui, J. -C., Elmayan, T., Pollien, J. -M., & Vaucheret, H. (1997). Systemic acquired silencing: Transgene-specific post-transcriptional silencing is transmitted by grafting from silenced stocks to non-silenced scions. *The EMBO Journal*, *16*, 4738–4745.

Papaefthimiou, D., Likotrafiti, E., Kapazoglou, A., Bladenopoulos, K., & Tsaftaris, A. (2010). Epigenetic chromatin modifiers in barley: III. Isolation and characterization of the barley GNAT-MYST family of histone acetyltransferases and responses to exogenous ABA. *Plant Physiology and Biochemistry*, *48*, 98–107.

Park, P. J. (2008). Epigenetics meets next-generation sequencing. *Epigenetics*, *3*, 318–321.

Park, P. J. (2009). ChIP-seq: Advantages and challenges of a maturing technology. *Nature Reviews Genetics, 10*, 669–680.

Penterman, J., Uzawa, R., & Fischer, R. L. (2007a). Genetic interactions between DNA demethylation and methylation in *Arabidopsis*. *Plant Physiology, 145*, 1549–1557.

Penterman, J., Zilberman, D., Huh, J. H., Ballinger, T., Henikoff, S., & Fischer, R. L. (2007b). DNA demethylation in the *Arabidopsis* genome. *Proceedings of the National Academy of Sciences, 104*, 6752–6757.

Pereira, V. (2004). Insertion bias and purifying selection of retrotransposons in the *Arabidopsis thaliana* genome. *Genome Biology, 5*(10), R79.

Pien, S., Fleury, D., Mylne, J. S., Crevillen, P., Inze, D., & Avramova, Z., et al. (2008). ARABIDOPSIS TRITHORAX1 dynamically regulates FLOWERING LOCUS C activation via histone 3 lysine 4 trimethylation. *Plant Cell, 20*, 580–588.

Poethig, R. S. (2009). Small RNAs and developmental timing in plants. *Current Opinion in Genetics & Development, 19*, 374–378.

Probst, A. V., Fagard, M., Proux, F., Mourrain, P., Boutet, S., & Earley, K., et al. (2004). *Arabidopsis* histone deacetylase HDA6 is required for maintenance of transcriptional gene silencing and determines nuclear organization of rDNA repeats. *Plant Cell, 16*, 1021–1034.

Rigoutsos, I., & Floratos, A. (1998). Combinatorial pattern discovery in biological sequences: The TEIRESIAS algorithm. *Bioinformatics, 14*, 55–67.

Rossi, V., Locatelli, S., Lanzanova, C., Boniotti, M. B., Varotto, S., & Pipal, A., et al. (2003). A maize histone deacetylase and retinoblastoma-related protein physically interact and cooperate in repressing gene transcription. *Plant Molecular Biology, 51*, 401–413.

Rossi, V., Locatelli, S., Varotto, S., Donn, G., Pirona, R., & Henderson, D. A., et al. (2007). Maize histone deacetylase hda101 is involved in plant development, gene transcription, and sequence-specific modulation of histone modification of genes and repeats. *Plant Cell, 19*, 1145–1162.

Saze, H., & Kakutani, T. (2007). Heritable epigenetic mutation of a transposon-flanked *Arabidopsis* gene due to lack of the chromatin-remodeling factor DDM1. *Embo Journal, 26*, 3641–3652.

Schatlowski, N., Creasy, K., Goodrich, J., & Schubert, D. (2008). Keeping plants in shape: Polycomb-group genes and histone methylation. *Seminars in Cell & Developmental Biology, 19*, 547–553.

Schmitz, R. J., Sung, S., & Amasino, R. M. (2008). Histone arginine methylation is required for vernalization-induced epigenetic silencing of FLC in winter-annual *Arabidopsis thaliana*. *Proceedings of the National Academy of Sciences of the United States of America, 105*, 411–416.

Schnable, P. S., Ware, D., Fulton, R. S., Stein, J. C., Wei, F. S., & Pasternak, S., et al. (2009). The B73 maize genome: Complexity, diversity, and dynamics. *Science, 326*, 1112–1115.

Schonrock, N., Bouveret, R., Leroy, O., Borghi, L., Kohler, C., & Gruissem, W., et al. (2006). Polycomb-group proteins repress the floral activator AGL19 in the FLC-independent vernalization pathway. *Genes & Development, 20*, 1667–1678.

Schubert, D., Primavesi, L., Bishopp, A., Roberts, G., Doonan, J., & Jenuwein, T., et al. (2006). Silencing by plant Polycomb-group genes requires dispersed trimethylation of histone H3 at lysine 27. *The EMBO Journal, 25*, 4638–4649.

Scott, R. J., Spielman, M., Bailey, J., & Dickinson, H. G. (1998). Parent-of-origin effects on seed development in *Arabidopsis thaliana*. *Development, 125*, 3329–3341.

Sexton, T., Schober, H., Fraser, P., & Gasser, S. M. (2007). Gene regulation through nuclear organization. *Nature Structural & Molecular Biology, 14*, 1049–1055.

Shen, W. -H., & Xu, L. (2009). Chromatin remodeling in stem cell maintenance in *Arabidopsis thaliana*. *Molecular Plant, 2*, 600–609.

Shi, Y., Lan, F., Matson, C., Mulligan, P., Whetstine, J. R., & Cole, P. A., et al. (2004). Histone demethylation mediated by the nuclear amine oxidase homolog LSD1. *Cell, 119*, 941–953.

Shibuya, K., Fukushima, S., & Takatsuji, H. (2009). RNA-directed DNA methylation induces transcriptional activation in plants. *Proceedings of the National Academy of Sciences, 106*, 1660–1665.

Singh, J., Freeling, M., & Lisch, D. (2008). A position effect on the heritability of epigenetic silencing. *PLoS Genetics, 4*, e1000216.

Slack, F. J. (2010). Time for the epigenome. *Nature, 463*, 587.

Soppe, W. J. J., Jacobsen, S. E., Alonso-Blanco, C., Jackson, J. P., Kakutani, T., & Koornneef, M., et al. (2000). The late flowering phenotype of fwa mutants is caused by gain-of-function epigenetic alleles of a homeodomain gene. *Molecular Cell, 6*, 791–802.

Sridha, S., & Wu, K. (2006). Identification of AtHD2C as a novel regulator of abscisic acid responses in *Arabidopsis*. *Plant Journal, 46*, 124–133.

Strahl, B. D., & Allis, C. D. (2000). The language of covalent histone modifications. *Nature, 403*, 41–45.

Sung, S., & Amasino, R. M. (2004). Vernalization in *Arabidopsis thaliana* is mediated by the PhD finger protein VIN3. *Nature, 427*, 159–164.

Sung, S., & Amasino, R. M. (2006). Molecular genetic studies of the memory of winter. *Journal of Experimental Botany, 57*, 3369–3377.

Suzuki, M. M., & Bird, A. (2008). DNA methylation landscapes: Provocative insights from epigenomics. *Nature Reviews Genetics, 9*, 465–476.

Swiezewski, S., Crevillen, P., Liu, F., Ecker, J. R., Jerzmanowski, A., & Dean, C. (2007). Small RNA-mediated chromatin silencing directed to the 3′ region of the *Arabidopsis* gene encoding the developmental regulator, FLC. *Proceedings of the National Academy of Sciences of the United States of America, 104*, 3633–3638.

Swiezewski, S., Liu, F., Magusin, A., & Dean, C. (2009). Cold-induced silencing by long antisense transcripts of an *Arabidopsis* Polycomb target. *Nature, 462*, 799–802.

Tamaru, H., & Selker, E. U. (2001). A histone H3 methyltransferase controls DNA methylation in *Neurospora crassa*. *Nature, 414*, 277–283.

Tanaka, M., Kikuchi, A., & Kamada, H. (2008). The *Arabidopsis* histone deacetylases HDA6 and HDA19 contribute to the repression of embryonic properties after germination. *Plant Physiology, 146*, 149–161.

Tariq, M., & Paszkowski, J. (2004). DNA and histone methylation in plants. *Trends in Genetics, 20*, 244–251.

Tian, L., & Chen, Z. J. (2001). Blocking histone deacetylation in *Arabidopsis* induces pleiotropic effects on plant gene regulation and development. *Proceedings of the National Academy of Sciences of the United States of America, 98*, 200–205.

Tian, L., Fong, M. P., Wang, J. J., Wei, N. E., Jiang, H., & Doerge, R. W., et al. (2005). Reversible histone acetylation and deacetylation mediate genome-wide, promoter-dependent and locus-specific changes in gene expression during plant development. *Genetics, 169*, 337–345.

Tsaftaris, A., Polidoros, A., Kapazoglou, A., Tani, E., Kovacevic, N. (2008). Epigenetics and plant breeding In *Plant breeding reviews* (Vol. 30, pp. 49–179).

Turck, F., Fornara, F., & Coupland, G. (2008). Regulation and identity of Florigen: FLOWERING LOCUS T moves center stage. *Annual Review of Plant Biology, 59*, 573–594.

Turck, F., Roudier, F. o., Farrona, S., Martin-Magniette, M. -L., Guillaume, E., & Buisine, N., et al. (2007). *Arabidopsis* TFL2/LHP1 specifically associates with genes marked by trimethylation of histone H3 lysine 27. *PLoS Genetics, 3*, e86.

Vaillant, I., & Paszkowski, J. (2007). Role of histone and DNA methylation in gene regulation. *Current Opinion in Plant Biology, 10*, 528–533.

Vaucheret, H. (2008). Plant ARGONAUTES. *Trends in Plant Science, 13*, 350–358.

Vaughn, M. W., Tanurdzic, M., Lippman, Z., Jiang, H., Carrasquillo, R., & Rabinowicz, P. D., et al. (2007). Epigenetic natural variation in *Arabidopsis thaliana*. *PLoS Biology, 5*, e174.

Vicient, C. M., Suoniemi, A., Anamthawat-Jonsson, K., Tanskanen, J., Beharav, A., & Nevo, E., et al. (1999). Retrotransposon BARE-1 and its role in genome evolution in the genus Hordeum. *Plant Cell, 11*, 1769–1784.

Vitte, C., & Panaud, O. (2005). LTR retrotransposons and flowering plant genome size: Emergence of the increase/decrease model. *Cytogenetic and Genome Research, 110*, 91–107.

Voinnet, O., Vain, P., Angell, S., & Baulcombe, D. C. (1998). Systemic spread of sequence-specific transgene RNA degradation in plants is initiated by localized introduction of ectopic promoterless DNA. *Cell*, *95*, 177–187.

Wang, X. F., Elling, A. A., Li, X. Y., Li, N., Peng, Z. Y., & He, G. M., et al. (2009). Genome-wide and organ-specific landscapes of epigenetic modifications and their relationships to mRNA and Small RNA Transcriptomes in Maize. *Plant Cell*, *21*, 1053–1069.

Wang, X., Zhang, Y., Ma, Q., Zhang, Z., Xue, Y., & Bao, S., et al. (2007). SKB1-mediated symmetric dimethylation of histone H4R3 controls flowering time in *Arabidopsis*. *Embo Journal*, *26*, 1934–1941.

Wassenegger, M., Heimes, S., Riedel, L., & Sбnger, H. L. (1994). RNA-directed *de novo* methylation of genomic sequences in plants. *Cell*, *76*, 567–576.

Wolfgruber, T. K., Sharma, A., Schneider, K. L., Albert, P. S., Koo, D. H., & Shi, J. H., et al. (2009). Maize centromere structure and evolution: Sequence analysis of centromeres 2 and 5 reveals dynamic loci shaped primarily by retrotransposons. *Plos Genetics*, *5*

Wood, C. C., Robertson, M., Tanner, G., Peacock, W. J., Dennis, E. S., & Helliwell, C. A. (2006). The *Arabidopsis thaliana* vernalization response requires a polycomb-like protein complex that also includes VERNALIZATION INSENSITIVE 3. *Proceedings of the National Academy of Sciences*, *103*, 14631–14636.

Wu, G., Park, M. Y., Conway, S. R., Wang, J. -W., Weigel, D., & Poethig, R. S. (2009). The sequential action of miR156 and miR172 regulates developmental timing in *Arabidopsis*. *Cell*, *138*, 750–759.

Wu, K., Tian, L., Malik, K., Brown, D., & Miki, B. (2000). Functional analysis of HD2 histone deacetylase homologues in *Arabidopsis thaliana*. *Plant Journal*, *22*, 19–27.

Wu, K., Zhang, L., Zhou, C., Yu, C. W., & Chaikam, V. (2008). HDA6 is required for jasmonate response, senescence and flowering in *Arabidopsis*. *Journal of Experimental Botany*, *59*, 225–234.

Xiao, W., Brown, R. C., Lemmon, B. E., Harada, J. J., Goldberg, R. B., & Fischer, R. L. (2006b). Regulation of seed size by hypomethylation of maternal and paternal genomes. *Plant Physiology*, *142*, 1160–1168.

Xiao, W., Custard, K. D., Brown, R. C., Lemmon, B. E., Harada, J. J., & Goldberg, R. B., et al. (2006a). DNA methylation is critical for *Arabidopsis* embryogenesis and seed viability. *Plant Cell*, *18*, 805–814.

Yang, W., Jiang, D., Jiang, J., & He, Y. (2010). A plant-specific histone H3 lysine 4 demethylase represses the floral transition in *Arabidopsis*. *Plant Journal*, *62*, 663–673.

Zhai, J., Liu, J., Liu, B., Li, P., Meyers, B. C., & Chen, X., et al. (2008). Small RNA-directed epigenetic natural variation in *Arabidopsis thaliana*. *PLoS Genetics*, *4*, e10000056.

Zhang, X., & Jacobsen, S. E. (2006). Genetic analyses of DNA methyltransferases in *Arabidopsis thaliana*. *Cold Spring Harbor Symposia on Quantitative Biology*, *71*, 439–447.

Zhang, X., Bernatavichute, Y., Cokus, S., Pellegrini, M., & Jacobsen, S. (2009). Genome-wide analysis of mono-, di- and trimethylation of histone H3 lysine 4 in *Arabidopsis thaliana*. *Genome Biology*, *10*, R62.

Zhang, X., Clarenz, O., Cokus, S., Bernatavichute, Y. V., Pellegrini, M., & Goodrich, J., et al. (2007a). Whole-genome analysis of histone H3 lysine 27 trimethylation in *Arabidopsis*. *PLoS Biology*, *5*, e129.

Zhang, X., Henderson, I. R., Lu, C., Green, P. J., & Jacobsen, S. E. (2007b). Role of RNA polymerase IV in plant small RNA metabolism. *Proceedings of the National Academy of Sciences of the United States of America*, *104*, 4536–4541.

Zhang, Y. J. (2005). miRU: An automated plant miRNA target prediction server. *Nucleic Acids Research*, *33*, W701–W704.

Zhou, C., Labbe, H., Sridha, S., Wang, L., Tian, L., & Latoszek-Green, M., et al. (2004). Expression and function of HD2-type histone deacetylases in *Arabidopsis* development. *Plant Journal*, *38*, 715–724.

Zhou, C., Zhang, L., Duan, J., Miki, B., & Wu, K. (2005). HISTONE DEACETYLASE19 is involved in jasmonic acid and ethylene signaling of pathogen response in *Arabidopsis*. *Plant Cell*, *17*, 1196–1204.

Zhu, J., Kapoor, A., Sridhar, V. V., Agius, F., & Zhu, J. -K. (2007). The DNA Glycosylase/Lyase ROS1 functions in pruning DNA methylation patterns in *Arabidopsis*. *Current Biology*, *17*, 54–59.

Zhu, J. -K. (2008). Epigenome sequencing comes of age. *Cell*, *133*, 395–397.

Zhu, J. -K. (2009). Active DNA demethylation mediated by DNA glycosylases. *Annual Review of Genetics*, *43*, 143–166.

Zilberman, D. (2008). The evolving functions of DNA methylation. *Current Opinion in Plant Biology*, *11*, 554–559.

Zilberman, D., Gehring, M., Tran, R. K., Ballinger, T., & Henikoff, S. (2007). Genome-wide analysis of *Arabidopsis thaliana* DNA methylation uncovers an interdependence between methylation and transcription. *Nature Genetics*, *39*, 61–69.

Section C

Plant germplasm

An engineering view to micropropagation and generation of true to type and pathogen-free plants

15

Eli Khayat
Rahan Meristem Ltd., Israel

TABLE OF CONTENTS

Preface

The premise that *in vitro* propagation will replace seed propagation has yet to become reality. Despite the great advantages of micropropagated plants, tissue-culture technologists have not succeeded in adapting this mode of propagation to a large industrialized scale. Pursuing micropropagation in four stages was first suggested by Murashige and Skoog (1974). The four stages include: explant preparation (stage I); multiplication (stage II); rooting and elongation (stage III); and hardening (stage IV). This partitioning is still employed by most production laboratories. Tens of thousands of protocols have been developed for *in vitro* propagation of a wide range of crop species. Regardless of the enormous advances in the biological aspects, in almost four decades of existence, the major obstacles to modernizing the micropropagation industry remain unresolved. High labor costs in developed countries play a major economic factor when considering tissue culture over alternative modes of propagation. The expenditure on labor in micropropagation laboratories situated in developing countries is 50–60% of the total production cost (Anderson et al., 1977; Ilan and Khayat, 1997; Standaert de Metsenaere, 1991). Consequently, in recent years the commercial micropropagation industry migrated from developed to underdeveloped countries. Other industrial constraints such as loss due to microbial contamination, failure to sustain phenotypic conformity, and high operational costs contributed to confining the commercial application of micropropagation to a narrow range of crop species.

The increase in volume of micropropagated plants is confined to a short list of species that greatly gain from the technology including banana, date palm, orchid, ferns, ornamental greens, and *Prunus* and *Malus* root-stocks. Other alternatives for propagating these species are too complex and entail high cost. The biological technology for propagating these species has been perfected, mainly in commercial laboratories, and consequently has not been disclosed publicly.

On the other hand, protocols for explantation, multiplication, rooting, and hardening have been published for a broad range of crops. Unfortunately, most published protocols solely address developmental aspects, leaving the industrial aspects largely ignored. It would be fair to say that in the last 15–20 years the engineering aspects of the micropropagation industry have regressed rather than advanced.

This chapter will highlight commercial micropropagation from an engineering point of view rather than the biological side.

Shoot Multiplication Through Meristem Culture

Meristem culture, when possible, is by far the safest, most rapid, and most effective mode of micropropagation. At

DOI: 10.1016/B978-0-12-381466-1.00015-8

the apex of all shoots, whether terminal or lateral, there are dome-shaped masses of cells known as meristems. Cells of apical meristems have large nuclei, a compact cytoplasm, and small vacuoles. While the cells in the shoot tip meristem are actively dividing, the axillary meristems in dormant buds are quiescent. When conditions are favorable, cells in axillary meristems divide rapidly and the daughter cells increase in size, becoming in most instances larger than the cells from which they originated.

The efficiency of micropropagation by subculturing of *de novo* generated meristems (culture of apical buds or nodal segments) is influenced by several genetic and external factors including *ex vitro* (e.g., genotype and physiological state) and *in vitro* conditions (e.g., media constituents, temperature, and light; Dobránszki and Teixeira da Silva, 2010).

Stage 0: Disinfection and start of axenic culture

Initially stage 0 was set up as a means for introducing uncontaminated and true to type initials. To assure cultures start with minimal contamination, mother stock plantations are often used. Using mother stock collections having a high level of sanitation reduces the risk of introducing infected material into the culture. Disinfecting the initials could start in the mother stock greenhouses or in the field collection. In contained facilities, mother plants are sprayed with pesticides two to three weeks prior to the introduction of explants to the tissue-culture laboratory (Nasib et al., 2008).

Mortality of initials can also occur due to secretion of oxidizing compounds from the wounded tissue. The oxidation of polyphenols to quinones is common in many species, and is well documented in the literature (Romano et al., 1992; Titov et al., 2006; North et al., 2010).

Although almost any tissue could regenerate into a complete plant, shoot tips and axillary buds are most often used as explant material. In a few species, leaves are used as explants (*Anthurium* spp., Pierik and Steegmans, 1976; *Ficus lyrata*, Debergh and De Wael, 1977; *Saintpaulia*, Margara and Poillat, 1985; *Stevia*, Ferreira and Handro, 1988; *Cryptanthus*, Koh and Davies, 1997; Atak and Celik, 2009).

Stage I: Initiation of culture

Following disinfection, initials are placed in axenic conditions for monitoring microbial growth from weeks 4–12. The concept of microbial indexing (reviewed by Leifert and Cassells, 2001) is particularly practical in cases when a vast number of plants is propagated from a single initial (e.g., ornamental ferns), or in cases in which the value of each initial is very high (e.g., date palms).

Typically, the medium of choice for shoot branching and elongation in stage I is MS (Murashige and Skoog, 1962) supplemented with both auxins and cytokinins. These media are often supplemented with antioxidation reagents such as PVP (Rabha Abdelwahd et al., 2008), 8-hydroxy quinoline citrate (8-HQC; Alizadeh et al., 2010) at a concentration of 200 mg l^{-1}, and ascorbic and citric acids. Dark cultivation is helpful for inhibiting oxidation, as well as preventing degradation of the auxin component in the medium.

Stage II: Multiplication

In stage II micropropagators aim at high proliferation of unrooted propagules. A high cytokinin to auxin ratio is often applied to enhance shoot proliferation. Depending on the genotype and mode of propagation, propagators provide from 0.5 to 10 mg l^{-1}. However, since the total number of transfers, as well as the rate of multiplication in each transfer cycle, induces somaclonal variations, it is customary to normalize both parameters by regulating the levels of cytokinins, planning the total number of transfers per explant and setting up an optimal time of incubation between transfer cycles.

There is a debate among researchers regarding the effect of excess cytokinins on off-types. Bairu et al. (2006) reported that higher levels of BAP and IBA elevated the percentages of somaclonal variants in "Cavendish" banana. On the other hand, Reuveni and Israeli (1990) showed opposite results. These discrepancies could stem from differences in the sensitivity to BAP of different genotypes. Regardless of the cause, the correlation between the number of propagules per explant and the frequency of somaclonal variation is decisive. Certain plant species are more prone to somaclonal variation. In this regard strawberries are a good example (Boxus et al., 2000).

The genetic mechanisms for somaclonal variation were extensively studied in micropropagated banana plants (Hwang, 1986; Cote et al., 1993; Krikorian et al., 1993; Khayat et al., 2004). Plant architecture is greatly affected by somaclonal variation. Plant size greatly fluctuates in Cavendish populations (*Musa acuminata* Cavendish AAA) produced by tissue culture. "Dwarf" or "tall" mutants are very common. In general, the dwarf mutants develop a small deformed bunch (Reuveni et al., 1996; Khayat et al., 2004). For practical reasons, lower stature plants are considered advantageous. However, in most cultivars, taller plants are positively correlated with a larger bunch. A somaclonal mutant of "Zelig," (a Cavendish selection), registered under the name "Adi," was selected for a combined low stature and large bunch. The average plant height of the selected clone is approximately 230 cm, compared to 315 cm of its originator Zelig. The average bunch weight and finger size of Adi are both larger than Zelig (Khayat, 2010). The dwarf and tall mutants in Cavendish have been characterized as under- or oversensitive to the hormone Gibberellic acid (GA), respectively. An *in vitro* bioassay was developed where a low concentration of GA3 was included in the growth medium during the elongation stage of the tissue culture. Wild-type plants in the GA3-containing medium elongated above the plants that were placed on a substrate devoid of the hormone (control treatment). Dwarf mutants on the hormone medium grew below the height of the control, while oversensitive mutants grew above (Khayat et al., 2004). A mutant that was depicted as "Long and Narrow Leaves" (LNL) was created by exhaustive cycling of tissue culture leading to induction of retrotransposon expression and transposition. These experiments (Khayat et al., 2004) led to the conclusion that the long internodes mutation is caused by insertions of retroelements in a chromosomal locus having an effect on plant height. An extensive number of cycles in tissue culture enhance the activity of retrotransposable elements in the genome. Similar results were reported in rice where specific genes were mutated by transposition of Tos 17, a rice retrotransposon activated by tissue culture (Hirochika et al., 1996). Interestingly, in rice, at least 50% of the Tos 17 landing sites were detected as coding regions. The "Landing Pad" hypothesis of retrotransposable elements may explain the high frequency of dwarf mutants in *Musa*. The conserved *Ty1-copia* group of retrotransposons, highly expressed in banana (Khayat et al., 2004), is ubiquitous to various dicotyledonous and monocotyledonous plant species including potatoes, onions, fava beans, barley, and rye (Kumar et al., 2004).

The process of tissue culture also induces mutations mediated by DNA methylation (Peraza-Echeverria et al., 2001). It was pointed out that the source of explants had an effect on the degree of polymorphism. *In vitro* plants produced from floral meristems showed greater frequencies of methylation in comparison to sucker meristems.

Stage III: Elongation and promotion of shoot and root development

Given that rooting is the final *in vitro* stage, the efficacy of the entire process often relies on the success of root and shoot formation. Commercial laboratories consider stage III the most expensive stage, since it is the most labor intensive part of the *in vitro* process.

It is widely accepted that roots and leaves formed *in vitro* have a limited functionality in *ex vitro* conditions (Wardle et al., 1979; Grout et al., 1985; Fila et al., 1998). Water conductivity is impaired due to discontinuous vessels between the stem and adventitious roots formed *in vitro*. However, the process of *in vitro* root induction is crucial for formation of secondary functional roots outside of the *in vitro* conditions.

Given that for most species rooting and elongation require a high ratio of auxin/cytokinins; for recalcitrant species removal of excess cytokinins demands an intermediate medium containing activated charcoal (Ouma et al., 2004; Dobránszki and Teixeira da Silva, 2010; North et al., 2010). Various soil substrates have been experimentally used in stage III, but none have been utilized by commercial micropropagators. The main advantage of non-agar substrates in stage III is avoiding damaging roots upon transplanting. Since *in vitro* produced roots do not function during hardening, producers often prefer short and stubby roots rather than long roots that would be prone to entanglement and ripping.

Stage IV: Acclimatization and hardening

The eco-physiological parameters for hardening *ex vitro* plantlets differ greatly between plant species. For many species this is the most crucial stage, in which losses are very significant. The anatomy and physiology of leaves developed *in vitro* are not tolerant of normal ambient conditions. As a result of the indulging *in vitro* conditions, the plants develop an abnormal leaf anatomy that includes underdeveloped stomata and lack of a cuticle (Wetzstein and Sommer, 1982; Gilly et al., 1997). Consequently, the unhardened plants are susceptible to rapid water loss. In addition, the photosynthetic machinery is impaired. The transition to autotrophic existence is often possible only in photosynthetic organs that were developed in *ex vitro* conditions. In many cases, the success of hardening depends on the survival capability of the *in vitro* developed tissues until *ex vitro* organs (mainly roots and leaves) differentiate (reviewed by Pospisilova et al., 1999).

Susceptibility of unacclimatized plants to airborne, as well as soil-borne, pathogens could cause rapid death in the hardening greenhouses. Due to the sterility of the tissue-cultured plants they tend to be more vulnerable to pathogens that benefit from the lack of competing microbes on the host. In recent years commercial producers have realized the benefit of inoculating *ex vitro* plants with beneficial microorganisms.

Finally, the choices of soil substrate and plug parameters are highly important. Susceptibility to waterlogging is often overlooked. Due to the small size of the plug, there is a tendency to keep the plants under continuous flooding, which severely impairs the rate of survival. Especially in warm climate zones, due to the small plug size, water loss through the plug could be rapid and difficult to handle.

Automation

The process of micropropagation requires a mesh of integrated processes extending from explantation, media preparation, sterilization of media and vessels, media pouring, cutting the propagules in the multiplication stage, planting in a new vessel, growing the plants in environmentally controlled growth rooms, screening out contaminated vessels, size sorting, and hardening. Several attempts at limiting manual labor in micropropagation were reported (Deleplanque et al., 1985; Kutz, 1986; Watake and Kinase, 1988; Kinase and Watake, 1991; Peleg et al., 1993; Alchanatis et al., 1994; Sobey et al., 1997; Sluis, 2006). The main constraint for robotics is the inverse relationship between the volume of production versus the cost per propagule.

A detailed description of the Vitron™ 501 (developed for ForBio) was described by Sobey et al.(1997). The instrument comprises a box-loader that unloads a stack of vessels. The lids of the vessels are individually opened by a pneumatic cap opener, and a set of grippers and scissors cut the tip and lower nodes of the plant and place them in media-containing vessels. A CCD camera visualizes the plant and calculates the exact position of cutting. Color-based artificial visualization has been proposed by Alchanatis et al. (1993) and Peleg et al. (1993). Colors and shades were analyzed by a computer program that determines the exact chopping position of the micropropagated plants. The processing rate of the Vitron 501 apparatus is approximately six million plants per year, a moderate volume considering the capital investment for the apparatus. In addition, the Vitron 501 is merely suitable for excision of internodes.

Wang et al. (1998) developed a robust computer algorithm for identification, selection and separation of clustered sugarcane shoots in stage II meristem culture.

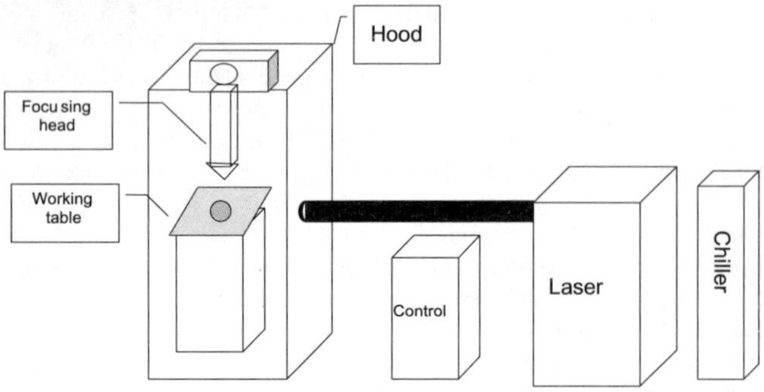

Figure 15.1 • Schematic drawing of the laser beam cutter • Schematic presentation of a laser apparatus for mechanical cutting of meristems. The apparatus is designed to fit a laminar flow hood under aseptic conditions. Please see color plate section at the end of the book.

Other geometric structures such as globular meristems (i.e., banana meristem culture, bulbous plants, etc.) require a different cutting apparatus. The multiple meristems are clustered into randomly arranged, compact spheres that are easily separated manually. However, random cutting with a mechanical instrument requires passage through very dense tissue. We have tried a CO_2 laser-beam cutting apparatus (Rahan Meristem, unpublished data) with a beam diameter of 20 mm. Surprisingly, the required beam intensity for cutting submeristems of a banana is in the magnitude of 2 kW. Consequently, the apparatus (described in Figure 15.1), requires a chiller of 25 kW. To utilize the cutting tool in a practical manner and efficiently, the culture would require morphological and physiological adjustments that combine vigorous growth with compact clamps comprised of numerous meristems.

Automated tissue-culture systems developed for plants without distinguishable nodes have also been investigated by other researchers. Miwa and Mitsubishi Heavy Industries developed a tissue-culture robot that handles bulbous plants (Takayama et al., 1991). Cooper and Grant (1992) developed an automated system of cutting and placing clustered propagules. The system was designed for the compact multimeristematic form of growth in conjunction with a range of conventional plastic vessels. McFarlane (1993) described an image-guided robotic harvesting system of micropropagation plants. The vision system used a stem location algorithm and directed a Cartesian robot to perform the harvesting.

Energy and Lights

A significant expense in micropropagation laboratories is related to energy. Energy consumption is mainly due to air conditioning, air filtration, and lights. The energy component of the production cost depends mainly on the climate in which the laboratory is situated. As a rule, micropropagation laboratories require low humidity (<60%) and moderate temperatures (20–24°C). High humidity in the ambient air is conducive to fungal contaminations. In tropical regions, under warm and humid climate conditions, a large portion of the air-conditioning wattage is used for air drying.

Most stage II and III protocols require lighting in the growth rooms. Although various types of lamps were tested in tissue-culture laboratories, fluorescent bulbs are the most common light source in the industry (Tanaka et al., 2009). Until recently, when compared to other light sources, for example, incandescent and luminescent bulbs (high pressure sodium), fluorescent bulbs are relatively efficient with respect to PPF and chlorophyll absorption spectrum. However, even with electronic ballasts and starters, cool white fluorescent bulbs emit significant excess heat. In Israeli laboratories approximately 60% of air-conditioning BTUs are used to remove excess heat load produced by the light bulbs (unpublished data). In addition to the high cost of air cooling, the heat emitted by the bulbs damages the plants and induces the growth of endogenous bacteria that otherwise would not grow on the growth medium.

In Mediterranean type climates, the expenditure of energy is in the range of 15 to 20% of the total cost of production (Ilan and Khayat, 1997). With increasing energy cost the relative expenditure on energy is proportionally higher.

Photoautotrophic Cultures

Sucrose or hexoses have traditionally been included in the medium of micropropagated plants as a source of carbon and energy. Inclusion of sugars alleviates the need to provide environmental conditions that support photosynthesis. This tradition of photoheterotrophic or photomixotrophic cultures has been challenged since 1988 (Kozai et al., 1988). It is argued that for many plant species, including Calla lilies (*Zantedeschia elliottiana*) and China firs (*Cunninghamia lanceolata*), plant growth and development is inhibited in photomixotrophic conditions due to the lack of fixed carbon, low evapotranspiration, low rate of nutrient uptake, and ethylene accumulation in the vessels (Kozai et al., 2005). The advantages of autotrophic cultures include reduced losses due to microbial contamination, easier rooting and hardening, and the possibility of growing plants more economically in larger vessels with reduced risk of contamination (Kozai et al., 2005). In fact, several reports show that under forced ventilation, growth and multiplication of photoautotrophic cultures is faster than with mixotrophic cultures (Nguyen et al., 1999; Nguyen and Kozai, 2001; Zobayed et al., 2004; Kozai et al., 2005).

Kozai and co-workers (1988) showed that the limiting factors for autotrophic cultures are mainly environmental. Chlorophyllus *in vitro* plantlets are capable of supplying their carbon and energy needs under a photon flux of 50–200 μmoles. $m^{-2}.sec^{-1}$ (compared to 30 μmoles.$m^{-2}.sec^{-1}$), air humidity below 70%, and ambient levels of CO_2. Despite many advantages in using photoautotrophic cultures in large-scale production, the technology was not adopted due to unresolved technical and logistic difficulties. The relatively high photosynthetic photon flux needed for autotrophic cultures generates a high heat load in the laboratory. To supply adequate carbon the vessels require high ventilation, which is hard to achieve in small containers. It was therefore suggested to increase the vessels size. Fujiwara (1988) developed a 20L vessel with forced ventilation for micropropagation of strawberry plants (for a review of culture vessels see Zobayed et al., 2000). A greater impact on growth is achieved in a CO_2 enriched environment (Kozai et al., 1987; Kozai and Iwanami, 1988; Woltering, 1990). This would not be practical when the culture is in small vessels (Kozai et al., 2005). Normally, tissue-culture vessels are capped with an airtight plastic film. Plant aeration is attained by slow diffusion of the gas molecules over a concentration gradient between the outside and inside air. Installing small membranes (with a pore size of 0.2–0.5 microns) on the surface of the lid greatly improved aeration. Alternatively, ambient air could be forced through a filter to the vessel. Yue et al. (1993) developed a forced ventilation system for micropropagated cultures. Aside from the higher levels of CO_2, it was possible to reduce the relative humidity in the cultures from almost 100 to 46%. This concept was later used in photoautotrophic cultures (Armstrong et al., 1997; Heo and Kozai, 1999).

Interestingly, for the case of employing a temporary immersion system, the inclusion of sucrose in the medium of micropropagated plantains was beneficial to the hardening process (*ex vitro* stage) even at high PPF and enriched CO_2 (Aragón et al., 2010).

Micropropagation in Liquid Media

The rates of growth and multiplication determine the efficiency of tissue cultures. These parameters are largely influenced by the ability of the plants to absorb nutrients from the growth medium. Solid media necessitates diffusion of nutrients from the medium to the plants. The gradual depletion of nutrients in the vicinity of the plant requires frequent transfers of plants to new media. Recurrent preparation of media is laborious, costly, and promotes microbial contaminations.

For large-scale plant cell culture, liquid media offer many advantages over solid or semi-solid matrices (Yesil-Celiktas et al., 2010). In liquids the plant cells benefit from intimate contact with the medium, providing a long-lasting continuous supply of nutrients. Consequently, provided that aeration is adequate, growth and multiplication rates are higher. However, liquid media introduce morphological and physiological abnormalities stemming from hyperhydration of the tissue leading to dysfunctional stomata and protocorm-like structures that do not differentiate into complete plants. In addition, super hydrated plants have a "glassy" phenotype (Ziv, 2000). Poor aeration is a severe drawback in liquid cultures. Due to the low solubility of oxygen in water, vigorous agitation or air bubbling does not provide adequate aeration of the growth medium. Consequently, the plantlets often suffer under the stress of anoxia.

The invention of the temporary immersion system (TIS) signifies major progress in the field of micropropagation in liquid media. Similar technologies in which the plant material is intermittently flooded preceded TIS. Weathers and Giles (1988) proposed the use of nutrient mist culture in *Daucus, Lycopersicon, Ficus, Cinchona*, and *Brassica*. All of the plants grew at least as well as they did on an agar-based medium and the rate of multiplication was greater. Another innovative approach introduced propagation of potato plantlets in slowly rotating bioreactors (Akita and Ohta, 1998). TIS was first introduced by Alvard et al. (1993) in micropropagated banana plants. Since then, it has been tested for micropropagation of pineapples (Escalona et al., 2003), plantains (Aragón et al., 2005), *Hevea brasiliensis* (Etienne et al., 1997), *Psidium guajava* (Kosky et al., 2005), and *Cacao* (Niemenak et al., 2008). Small TIS bioreactors are commercialized under the name RITA® (http://plantstissueculture.blogspot.com/2008/11/rita-temporary-immersion-system.html). Apparently, due to proper aeration, the physiology and morphology of the *in vitro* plants growing in TIS bioreactors do not show hyperhydration symptoms (Teisson et al., 1996; Etienne and Berthouly, 2002; Ziv, 2005) and gain better absorption of the nutrients.

Plant—Microbe Interaction During *in vitro* and *ex vitro* Stages of Micropropagation

Commercial tissue-culture laboratories realize that although sanitation is effective for reducing microbial contamination, it does not eliminate contaminants. Experience teaches us that endophytic bacteria could express latently, appearing on the medium in advanced cycles of multiplication (Khayat et al., 1996). Despite a harsh process of disinfection, explants are often introduced to the *in vitro* culture "contaminated" with fungi and bacteria. Indexing the culture for non-fastidious contaminants is laborious and expensive, whereas separation of axenic to contaminated cultures is not always achievable (Reed et al., 1995; Cassells and Tahmatsidou, 1996). Most latent endophytic bacteria are opportunistic but not pathogenic. Given the right conditions they will slow down growth and proliferation of the culture (Leifert et al., 1994). Furthermore, commercial laboratories often fail to detect contaminating bacteria that do not manifest visual symptoms other than slow growth (Debergh and Vanderschaeghe, 1988).

Although some success has been reported using antibiotics (Young et al., 1984; Falkiner, 1990; Leifert et al., 1991; Kneifel and Leonhardt, 1992), it is largely agreed among commercial micropropagators that the damage to the plants outweighs the benefit. When selecting the proper antibiotic, the micropropagator should consider the effect on the plant and the bacteria. Another consideration is the nature of the chosen antibiotic. Pollock et al. (1983) reported that on tobacco cells the family of β-lactam antibiotics (e.g., penicillin and carbenicillin) was least toxic among 20 different antibiotics. Leifert et al. (1992) reported that carbenicillin was toxic to a variety of ornamental plants. Aminoglycosides such as streptomycin, neomycin, and gentamicin undergo complex interactions with

Figure 15.2 • Effect of bacterial strains and *Fusarium oxysporum* on banana plants • (A) control, (B) cocktail combination of bacterial strains and *G. intraradices*, (C) *Glomus intraradices*, and (D) *F. oxysporum*. Please see color plate section at the end of the book.

polyribosomes. But the phytotoxic effects are species-specific. After six days in culture Reed et al. (1995) observed that neomycin caused severe damage to mint shoots, rifampicin slightly inhibited growth, and streptomycin had no effect.

The choice of bacteriostatic compounds is considered safer for use in comparison to most antibiotics. To this end, quaternary ammonium cations are successful against a wide range of gram negative bacteria (unpublished results). Although there is a risk of the contaminants reappearing at a later stage, this class of compounds is relatively effective and safe to use in most plant species.

In many cases, a useful strategy to combat endophytic bacterial contaminations is by modifying the environmental conditions of the cultured plants. This would be effective after the bacteria has been isolated, identified, and characterized. Upon recognition of the contaminating species, arresting its growth could be achieved by changes of pH, ambient temperature, aeration, and so forth.

Inoculation with Beneficial Microorganisms

Micropropagation techniques reduce or even eliminate the population of beneficial microorganisms such as mycorrhizal fungi and plant growth promoting bacteria in the rhizosphere, at least during the early stages of the post *in vitro* acclimation. The lack of beneficial microorganisms can adversely affect the survival and growth of *in vitro* produced plantlets. Soil microorganisms are integral components in agricultural, horticultural, and forest ecosystems, fulfilling critical roles in nutrient cycling, maintenance of soil structure, regulation of plant growth, and combating harmful pathogens. Microbial activity is particularly intense in the rhizosphere; the nutrient rich zone of soil surrounding plant roots that accumulates a variety of organic compounds due to root exudation, secretion, and decomposition. They stimulate plant growth through various mechanisms including enhancing nutrient and water uptake, producing plant growth regulators, acting as biological control agents against root pathogens, helping plants to tolerate various stresses (salt, drought, and toxic chemicals), and are helpful to plants growing on industrial waste and degraded lands (bioremediation; Granger et al., 1983; Chavez and Ferrera-Cerrato, 1990; Vidal et al., 1992; Lucy et al., 2004; Kapoor et al., 2008).

Plant-growth-promoting bacteria enhance plant growth and development by increasing nitrogen and phosphorus uptake, synthesizing plant-growth-promoting substances, and producing antibiotics in the rhizosphere to combat against root pathogens.

Mycorrhizal fungi are soil-borne and form beneficial symbiosis with host plants. The fungal hyphae grow within the plant roots and extend out into the surrounding soil acting as an extension of the roots, which increases the absorptive surface area of the root system. In mycorrhizal associations, both partners (plant and fungus) benefit from each other. The host plants receive mineral nutrition and water supply from the fungus while the fungus obtains photosynthetically derived carbohydrates from the host plant. Because of the mycorrhizal association, plants also become resistant to certain root pathogens and can tolerate stressful conditions such as drought, salt, or toxic chemicals. In general, mycorrhizal fungi help in the development of a stronger root system. Several types of mycorrhizal associations have been recognized, but the two most common types are ecto- and endo- (or arbuscular) mycorrhizae. Ectomycorrhizal fungal hyphae develop externally forming dense mycelial growths or sheaths in the epidermal layer (fungal mantle). The fungal hyphae also grow between the cortical cells (Hartig net) without penetrating them. Ectomycorrhizal associations are common in conifers and several hardwood trees. Endomycorrhizal fungi grow between the cells of plant roots and penetrate the cortical cells forming vesicles and arbuscules (VA). Vesicles serve as storage organs for lipids, whereas arbuscules serve as major metabolic exchange points between the plant root and the fungus. Endomycorrhizal associations are common in certain conifers, hardwood and horticultural trees, and most agricultural crops.

Inoculation of *ex vitro* plants of various species with bacteria and mycorrhizal fungi has been found to have a synergistic beneficial effect on plant growth, mineral nutrition, and disease resistance (see Figure 15.2 and Tables 15.1–15.3).

Table 15.1 Morphological parameters of 8-week-old *ex vitro* banana (*Musa acuminata* AAA) plantlets produced by meristem culture at Rahan Meristem, Rosh Hanikra Israel

Treatment	Survival (%)	Shoot length (cm)	Root length (cm)	Shoot dry wt. (g)	Root dry wt. (g)	Total dry wt. (g)	S:R ratio	Mycorrhizal colonization (%)
Control (A)	100a	5.5dc	8.5a	0.091f	0.042f	0.133f	2.17ab	0.0c
Enterobacter cloacae								
(CAL2) (B)	100a	7.0bcd	8.0a	0.115cd	0.061cd	0.176cd	1.88cd	0.0c
Enterobacter cloacae								
(UW4) (C)	100a	9.0ab	9.0a	0.119bc	0.065c	0.184c	1.83d	0.0c
UWF-ACC deaminase (D)	100a	5.0d	8.0a	0.090f	0.039f	0.129f	2.30a	0.0c
Pseudomonas putida								
ATTC + pRK 415 (E)	100a	8.0abc	8.0a	0.113de	0.057de	0.170d	1.97bcd	0.0c
P. putida + pRK 415 + ACC deaminase from UW4 (F)	100a	8.0abc	8.1a	0.107e	0.052e	0.159e	2.05bc	0.0c
Glomus intraradices (G)	100a	9.0ab	8.5a	0.125b	0.089b	0.214b	1.40e	87.0ab
G + B	100a	10.0a	9.0a	0.120bc	0.090b	0.210b	1.33e	85.0ab
G + C	100a	9.5ab	8.5a	0.123b	0.089b	0.212b	1.38e	88.0a
G + D	100a	9.0ab	8.0a	0.122b	0.088b	0.210b	1.38e	88.0a
G + E	100a	8.5ab	9.0a	0.125b	0.085b	0.210b	1.47e	83.0b
G + F	100a	9.0ab	9.0a	0.123b	0.088b	0.211b	1.39c	88.0a
G + B + C + D + E + F	100a	9.5ab	9.0a	0.198a	0.110a	0.308a	1.79d	87.0ab

Values are the means of eight plants. Means followed by the same letter for a particular parameter are not significantly ($P = <0.001$) different from each other.
The plants were inoculated with different strains of bacteria and fungi prior to transfer to a plastic covered greenhouse for hardening. The bacteria were kindly provided by Professor B. Glick, University of Waterloo, Canada.

Table 15.2 Analysis of macro- and microelements in leaf samples of 8-week-old *ex vitro* banana (*Musa acuminata* AAA) plants

Treatment	N (%)	P (cm)	K (cm)	Ca (%)	Mg (%)	B (ppm)	Cu (ppm)	Fe (ppm)	Mn (ppm)	Zn (ppm)
Control (A)	1.82g	0.170j	3.80f	0.80d	0.35b	60a	10d	80e	300a	75.2d
Enterobacter cloacae										
(CAL2) (B)	3.20e	0.285g	5.10abc	0.99c	0.75a	62a	19bc	200c	302a	101bc
Enterobacter cloacae										
(UW4) (C)	3.50d	0.270h	5.05abcd	0.105e	0.82a	58a	20ab	179d	299a	110b
UWF-ACC deaminase (D)	1.79g	0.180i	3.51g	0.82d	0.38b	59a	11cd	85e	301a	80d
Pseudomonas putida										
ATTC + pRK 415 (E)	2.59f	0.300f	4.98de	0.100e	0.80a	60a	20ab	205c	303a	95c
P. putida + pRK 415 + ACC deaminase from UW4 (F)	2.60f	0.272h	5.00bcde	0.101e	0.78a	62a	21ab	272b	300a	109b
Glomus intraradices (G)	3.62d	0.450b	5.12a	1.57a	0.81a	60a	27ab	305a	298a	129a
G + B	4.00b	0.430d	5.09abcd	1.60a	0.79a	62a	28a	298a	300a	130a
G + C	4.09b	0.440c	4.99cde	1.55ab	0.82a	61a	25ab	300a	303a	132a
G + D	3.60d	0.430d	4.93e	1.47b	0.80a	60a	27ab	302a	299a	132a
G + E	3.80c	0.450b	5.00bcde	1.53ab	0.83a	59a	27ab	300a	300a	130a
G + F	3.75c	0.420e	5.11ab	1.52ab	0.81a	60a	27ab	302a	300a	129a
G + B + C + D + E + F	4.32a	0.460a	5.00bcde	1.53ab	0.80a	62a	25ab	300a	302a	132a

Values are the means of five replicates. Means followed by the same letter for a particular parameter are not significantly ($P = < 0.001$) different from each other.

The plants were produced by meristem culture at Rahan Meristem, Rosh Hanikra Israel. The plants were inoculated with different strains of bacteria and fungi prior to transfer to a plastic covered greenhouse for hardening. The bacteria were kindly provided by Professor B. R. Glick of the University of Waterloo, Canada.

Table 15.3 Analysis of macro- and microelements in leaf samples of 6-month-old *ex vitro* banana (*Musa acuminata* AAA) plants

Treatment	N (%)	P (cm)	K (cm)	Ca (%)	Mg (%)	B (ppm)	Cu (ppm)	Fe (ppm)	Mn (ppm)	Zn (ppm)
Control (A)	1.9ef	0.170e	3.74c	0.804d	0.38e	72.2b	12.4f	91.6c	351.6a	81.8c
Enterobacter cloacae (CAL2) (B)	3.16bcd	0.27de	5.18abc	0.97cd	0.80de	82.4ab	21.4abcdef	210.4b	360.2a	105.2b
Enterobacter cloacae (UW4) (C)	3.54bcd	0.28cde	5.5abc	1.01cd	0.77d	85.2ab	25abcdef	194.2b	371a	115b
UWF-ACC deaminase (D)	1.78f	0.188e	3.80bc	0.85d	0.42e	85.4ab	17.2ef	96c	356.2a	98bc
Pseudomonas putida ATTC + pRK 415 (E)	2.3def	0.32abcde	5.00abc	1.10bcd	0.84bcd	82.8ab	20.6cdef	207.8b	361a	97bc
P. putida + pRK 415 + ACC deaminase from UW4 (F)	2.50def	0.306bcde	5.06abc	1.00cd	0.86abcd	83.2ab	19.2def	236.4b	369.8a	1000bc
Glomus intraradices (G)	4.04abc	0.46ab	6.26a	1.82a	0.97ab	89ab	36abc	324.8a	384.4a	141a
G + B	4.22abc	0.424abcd	6.04a	1.76a	0.92abcd	84.4ab	32.4abcde	316.6a	384.2a	139.2a
G + C	4.46ab	0.444ab	5.92a	1.79a	0.98ab	86.6ab	37.2ab	322.8a	387a	143.2a
G + D	3.82abc	0.398abcd	6.06a	1.56abc	0.90abcd	83ab	34.8abcd	319.2a	379a	134.4a
G + E	4.22abc	0.434abc	5.78a	1.69ab	0.948abc	81.2ab	34abcd	319.2a	380a	136a
G + F	4.08abc	0.418abcd	5.68ab	1.73a	0.90abcd	83.8ab	34abcd	324.8a	381.2a	135a
G + B + C + D + E + F	4.86a	0.476a	6.60a	1.83a	1.02a	93.4a	38.40a	338.8a	397.8a	145a

Values are the means of five replicates. Means followed by the same letter for a particular parameter are not significantly ($P = <0.001$) different from each other.
The plants were produced by meristem culture at Rahan Meristem, Rosh Hanikra Israel. The plants were inoculated with different strains of bacteria and fungi prior to transfer to a plastic covered greenhouse for hardening. The bacteria were kindly provided by Professor B. R. Glick of the University of Waterloo, Canada.

The synergistic effect was due to plant growth substances produced by the bacteria and fungi rather than the soluble phosphate ions released. The author found that co-inoculation with bacteria and mycorrhizal fungi can significantly decrease the incidence of root rot diseases of certain agricultural crops and forest trees compared to single inoculation.

Elimination of Viruses by *in vitro* Techniques

In light of the wide spread of viral diseases in intense cropping systems, combined with tighter governmental regulations imposed on nurseries, virus elimination is a crucial part of vegetative propagation. For some important crops that are propagated vegetatively, stock plants are kept in specialized greenhouses protected from the invasion of virus-transmitting insects and mites. Stock plants are commonly cleaned from viruses using *in vitro* tissue-culture procedures and the clean stock is then transferred to nuclear greenhouses. Nuclear plants are the source of foundation plantations and following labeling and certification, propagation material is distributed to field nurseries. Several *in vitro* techniques have been developed for virus elimination in plant material. Most methods take advantage of the fact that plant viruses use vascular tracts for mobility. Given that apices do not contain vascular differentiated cells, they are virus free. This phenomenon has been realized in a very broad array of plant species including banana and plantains (Helliot et al., 2001), potatoes (Matthews et al., 1992; Faccioli, 2001), grapes (Sahar et al., 2009), laurel trees (Lloyd and McCown, 1980), strawberries (Khanam et al., 1998; Biswas et al., 2007), citrus (Navarro, 1992; Fifaei et al., 2007), and bamboo (Hsu et al., 2000). Shoot tip culture is often combined with thermotherapy. Using this procedure meristematic cells grow faster than the differentiation of vascular cells, reducing the risk of infected excised shoots. In recent years, numerous antiviral compounds have been discovered and effectively used for virus elimination in plants (Borissenko et al., 1985; Wambugu, 1985; Bittner et al., 1989). The most common compound is 1-β-ribofuranosyl-1,2,4-triazole-3-carboxamide (Virazole; Sidwell et al., 1972). Virazole is included in the growth medium of the shoot tips (Cassels, 1987; Khayat et al., 1999). This compound is lethal to viruses due to inhibiting RNA replication as a base analog of cytosine (Elia et al., 2008). The use of Virazole combined with thermotherapy and shoot tip culture was highly effective in elimination of artichoke latent virus (ALV) in comparison to each of the techniques alone (Khayat et al., 1999; see Tables 15.4 and 15.5).

Table 15.4 Effect of thermotherapy and Virazole® on artichoke latent virus (ALV) elimination in Globe artichoke (*Cynara scolymus* L.)

Treatment	Virazole® (ppm)	Thermotherapy	Clean plants after 5 weeks in culture (%)
1	0	+	40
2	0	+	55
3	10	+	71
4	20	+	100
5	0	+	57
6	0	+	40
7	10	+	100
8	20	+	100

Thermotherapy: 31 consecutive days: 10 hours @ 42°C (light); 14 hours @ 10°C (dark).

Table 15.5 Effect of thermotherapy and explant size on artichoke latent virus (ALV) elimination in Globe artichoke (*Cynara scolymus* L.)

Explant size (mm)	Thermotherapy	Clean plants after 5 weeks of culture (%)
1	+	100
2	+	66
3	+	50
4	+	29

Thermotherapy: 31 consecutive days: 10 hours @ 42°C (light); 14 hours @ 10°C (dark).

Concluding Remarks

In light of recent changes in world agriculture and incorporation of genetic engineering as a tool of plant breeding, it behooves us to take a fresh look at micropropagation. The combination of automated systems with modern visual intelligence aided by computer technology would properly counter the high cost of manual labor in micropropagation and open the path for efficient *in vitro* plant production. Thus the emphasis for development should shift from the biological aspects to a new focus on engineering.

Acknowledgments

The author would like to thank Mrs. Lyn Marantz and Ms. Yonat Tal-Or for their helpful comments.

References

Akita, M., & Ohta, Y. (1998). A simple method for mass propagation of potato (*Solanum tuberosum* L.) using a bioreactor without forced aeration. *Plant Cell Reports*, *18*(1), 284–287.

Alchanatis, V., Peleg, K., & Ziv, M. (1993). Classification of tissue culture segments by colour machine vision. *Journal of Agricultural Engineering Research*, *55*(4), 299–311.

Alchanatis, V., Peleg, K., & Ziv, M. (1994). Morphological control and mensuration of potato plantlets from tissue-cultures for automated micropropagation. *Plant Cell Tissue and Organ Culture*, *36*(3), 331–338.

Alizadeh, M., Singh, S. K., & Patel, V. B. (2010). Comparative performance of *in vitro* multiplication in four grape (*Vitis* spp.) rootstock genotypes. *International Journal of Plant Production*, *4*(1), 41–50.

Alvard, D., Cote, F., & Teisson, C. (1993). Comparison of methods of liquid medium culture for banana micropropagation. *Plant Cell, Tissue and Organ Culture*, *32*, 55–60.

Anderson, W. C., Meagher, G. W., & Nelson, A. G. (1977). Cost of propagating broccoli plants through tissue culture. *HortScience*, *12*, 543–544.

Aragón, C. E., Escalona, M., Capote, I., Pina, D., Cejas, I., & Rodriguez, R., et al. (2005). Photosynthesis and carbon metabolism in plantain (*Musa* AAB) plantlets growing in temporary immersion bioreactors and during *ex vitro* acclimatization. *In Vitro Cellular & Developmental Biology-Plant*, *41*(4), 550–554.

Aragón, C. E., Escalona, M., Rodriguez, R., Cañal, M. J., Capote, I., & Pina, D., et al. (2010). Effect of sucrose, light, and carbon dioxide on plantain micropropagation in temporary immersion bioreactors. *In Vitro Cellular & Developmental Biology–Plant*, *46*, 89–94.

Armstrong, J., Lemos, E. E. P., Zobayed, S. M. A., Justin, S. H. F. W., & Armstrong, W. (1997). A humidity induced convective through-flow ventilation system benefits *Annona squamosa* L. explants and coconut calloid. *Annals of Botany*, *79*, 31–40.

Atak, C., & Celik, O. (2009). Micropropagation of *Anthurium andraeanum* from leaf explants. *Pakistan Journal of Botany*, *41*(3), 1155–1161.

Bairu, M. W., Fennell, C. W., & van Staden, I. (2006). The effect of plant growth regulators on somaclonal variation in Cavendish bananas (*Musa* AAA cv. "Zelig").

Biswas, M. K., Hossain, M., & Islam, R. (2007). Virus free plantlets production of strawberry through meristem culture. *World Journal of Agricultural Sciences*, *3*, 757–763.

Bittner, H., Schenk, G., Schuster, G., & Kluge, S. (1989). Elimination by chemotherapy of potato virus S from potato plants grown *in vitro*. *Potato Research*, *32*, 175–179.

Borissenko, S., Schuster, G., & Schmygla, W. (1985). Obtaining a high percentage of explants with negative serological reactions against viruses by combining potato meristem culture with antiphytoviral chemotherapy. *Journal of Phytopathology*, *114*, 185–188.

Boxus, P., Jemmali, A. J., Terzi, M., & Arezki, O. (2000). Drift in genetic stability in micropropagation: The case of strawberry. *Acta Horticulturae*, *530*, 155–161.

Cassells, A., & Tahmatsidou, V. (1996). The influence of local plant growth conditions on non-fastidious bacterial contamination of meristem-tips of Hydrangea cultured *in vitro*. *Plant Cell, Tissue and Organ Culture*, *47*(1), 15–26.

Cassels, A. C. (1987). *In vitro* induction of virus-free potatoes by chemotherapy. In Y. P. S. Bajaj (Ed.), *Biotechnology in agriculture and forestry* (pp. 40–50). Springer Verlag.

Chavez, M. C. G., & Ferrera-Cerrato, F. (1990). Effect of vesicular-arbuscular mycorrhizae on tissue culture-derived plantlets of strawberry. *HortScience*, *25*(8), 903–905.

Cooper, P. A., & Grant, J. E. (1992). Development of prototype automated cutting and placing system for tissue culture multiplication. *Combined Proceedings of the International Plant Propagators Society*, *42*, 309–313.

Cote, F. X., Sandoval, J. A., Marie, Ph., & Auboiron, E. (1993). Variations in micropropagated bananas and plantains: Literary survey. *Fruits*, *48*, 15–22.

Debergh, P. C., & De Wael, J. (1977). Mass propagation of *Ficus lyrata*. *Acta Horticulturae*, *78*, 361–364.

Debergh, P. C., & Vanderschaeghe, A. M. (1988). Some symptoms indicating the presence of bacterial contaminants in plant tissue culture. *Acta Horticulturae*, *225*, 77–81.

Deleplanque, H., Bonnet, P., Postaire, J. G. (1985). An intelligent robotic system for *in vitro* plantlet production. *Proceedings fifth international conference Robot Vision and sensory controls* (pp. 305–314).

Dobránszki, J., & Teixeira da Silva, J. A. (2010). Micropropagation of apple – a review. *Biotechnology Advances*, *28*, 462–488.

Elia, G., Belloli, C., & Cirone, F. (2008). *In vitro* efficacy of ribavirin aganst canine distemper virus. *Antiviral Research*, *77*(2), 108–113.

Escalona, M., Samson, G., Borroto, C., & Desjardins, Y. (2003). Physiology of effects of temporary immersion bioreactors on micropropagated pineapple plantlets. *In Vitro Cellular & Developmental Biology–Plant*, *39*(6), 651–656.

Etienne, H., & Berthouly, M. (2002). Temporary immersion systems in plant micropropagation. *Plant Cell Tissue and Organ Culture*, *69*, 215–231.

Etienne, H., Lartaud, M., MichauxFerriere, N., Carron, M. P., Berthouly, M., & Teisson, C. (1997). Improvement of somatic embryogenesis in *Hevea brasiliensis* (Mull Arg) using the temporary immersion technique. *In Vitro Cellular & Developmental Biology-Plant*, *33*(2), 81–87.

Faccioli, G. (2001). Control of Potato viruses using meristem and stem-cutting cultures, thermotherapy and chemotherapy. In G. Loebenstein (Ed.), *Virus and Virus-like diseases of Potatoes and production of seed-potatoés* (pp. 365–390).

Falkiner, F. R. (1990). The criteria for choosing an antibiotic for control of bacteria in plant tissue culture. *International Association for Plant Tissue Culture, Newsletter*, *60*, 14–21.

Ferreira, C. M., & Handro, W. (1988). Micropropagation of *Stevia rebaudiana* through leaf explants from adult plants. *Planta Medica*, *54*(2), 157–160.

Fifaei, R., Golein, B., Taheri, H., & Tadjvar, Y. (2007). Elimination of Citrus tristeza virus of Washington navel orange (*Citrus sinensis* [L] Osbeck) through shoot-tip grafting. *International Journal of Agriculture and Biology*, *9*(1), 27–30.

Fila, G., Ghashghaie, J., Hoarau, J., & Cornic, G. (1998). Photosynthesis, leaf conductance and water relations of *in vitro* cultured grapevine rootstock in relation to acclimatization. *Physiologia Plantarum*, *102*(3), 411–418.

Fujiwara, K., Kozai, T., & Watanabe, T. (1988). Development of a photoautotrophic tissue culture system for shoots and/or plantlets at rooting and acclimatization stages. *Acta Horticulturae*, *230*, 153–158.

Gilly, C., Rohr, R., & Chamel, A. (1997). Ultrastructure and radiolabelling of leaf cuticles from ivy (*Hedera helix* L.) plants in vitro and during *ex vitro* acclimatization. *Annals of Botany*, *80*(2), 139–145.

Granger, R. L., Plenchette, C., & Fortin, J. A. (1983). Effect of vesicular-arbuscular endomycorrhizal fungus *Glomus epigaeum* on the growth and leaf mineral content of two apple clones propagated *in vitro*. *Canadian Journal of Plant Science*, *63*, 551–555.

Grout, B. W. W., & Millam, S. (1985). Photosynthetic development of micropropagated strawberry plantlets following transplanting. *Annals of Botany*, *55*(1), 129–131.

Helliot, B., Panis, B., Locicero, A., Reyniers, K., Muylle, H., & Vandewalle, M., et al. (2001). Development of *in vitro* techniques for elimination of virus diseases from *Musa*. *Acta Horticulturae*, *560*, 535–538.

Heo, J., & Kozai, T. (1999). Forced ventilation micropropagation system for enhancing photosynthesis, growth and development of sweet potato plantlets. *Environmental Control in Biology*, *37*(1), 83–92.

Hirochika, H., Sugimoto, K., Otsuki, Y., Tsugawa, H., & Kanda, M. (1996). Retrotransposons of rice involved in mutations induced by tissue culture. *Proceedings of the National Academy of Sciences of the United States of America*, *93*(15), 7783–7788.

Hsu, Y. H., Annamalai, P., Lin, C. S., Chen, Y. Y., Chang, W. C., & Lin, N. S. (2000). A sensitive method for detecting bamboo mosaic virus (BaMV) and establishment of BaMV-free meristem-tio cultures. *Plant Pathology*, *49*, 101–107.

Hwang, S. C. (1986). Variations in banana plants propagated through tissue culture.

Journal of the Chinese Society for Horticultural Science, *32*, 117–125.

Ilan, A., & Khayat, E. (1997). An overview of commercial and technological limitations to marketing of micropropagated plants. *Acta Horticulturae*, *447*, 643–648.

Kapoor, R., Sharma, D., & Bhatnagar, A. K. (2008). Arbiscular mycorrhizae in micropropagation systems and their potential applications. *Scientia Horticulturae*, *116*, 227–239.

Khanam, D., Khatun, M. M., Faisal, S. M., & Hoque, M. A. (1998). *In vitro* propagation of strawberry through meristem tip culture. *Plant Tissue Culture*, *8*(1), 35–39.

Khayat, E. (1999). Effective cleaning of globe artichoke and strawberry plants from viral and bacterial contaminants. *Phytoparasitica*, *27*(2), 128–129.

Khayat, E., (2010). Banana plant named "ADI". In USPTO, USPTO, (Eds.), Rahan Meristem (1998) Ltd.: USA.

Khayat, E., Abdul Razek, A., Halevy, D., & Gontmakher, T. (1996). Effective cleaning of strawberry plants from pathogenic fungal and endophytic bacterial contaminants. *Acta Horticulturae*, *439*, 369–372.

Khayat, E., Duvdevani, A., Lahav, E., & Ballesteros, B. A. (2004). Somaclonal variation in banana (*Musa acuminata* cv. Grande Naine). Genetic mechanism, frequency and application as a tool for clonal selection. In M. S. Jain & R. Swennen (Eds.), *Banana improvement: Cellular, molecular biology, and induced mutations* (pp. 97–110). Enfield, USA: Science Publishers Inc.

Kinase, A., Watake, H. (1991). Robot for mass propagation. *Proceedings of the IFAC mathematical and control applications in agriculture and horticulture* (pp. 225–230).

Kneifel, W., & Leonhardt, W. (1992). Testing of different antibiotics against Gram-positive and Gram-negative bacteria isolated from plant tissue culture. *Plant Cell, Tissue and Organ Culture*, *29*(2), 139–144.

Koh, Y. C., & Davies, J. F. T. (1997). Micropropagation of Cryptanthus with leaf explants with attached intercalary meristems excised from greenhouse stock plants. *Scientia Horticulturae*, *70*(4), 301–307.

Kosky, R., Perozo, J., Valero, N., Peñalver, D. (2005). Somatic embryo germination of *Psidium guajava* L. in the Rita® temporary immersion system and on semisolid medium. In *Liquid Culture Systems for in vitro Plant Propagation*, Hvoslef-Eicle, A. K., & Preil, W. (eds), 225–229. The Netherlands: Springer.

Kozai, T., & Iwanami, Y. (1988). Effects of CO_2 enrichment and sucrose concentration under high photon fluxes on plantlet growth of Carnation (*Dianthus caryuphyllus* L.) in tissue culture during the preparation stage. *Journal of the Japanese Society for Horticultural Science*, *57*, 279–288.

Kozai, T., Kubota, C., & Watanebe, I. (1988). Effect of basal medium composition on the growth of carnation plantlets in auto- and mixotrophic tissue culture. *Acta Horticulturae*, *230*, 159–166.

Kozai, T., Oki, H., Fujiwara, K. (1987). Effects of CO_2 enrichment and sucrose concentration under high photosynthetic photonfluxes on growth of tissue-cultured *Cymbidium* plantlet during the preparation stage. *Symposium Florizel on plant micropropagation in horticultural industries*, 135-141.

Kozai, T., Xiao, Y., Nguyen, Q. T., Afreen, F., & Zobayed, S. M. A. (2005). Photoautotrophic (sugar-free medium) micropropagation systems for large-scale commercialization. *Propagation of Ornamental Plants*, *5*(1), 25–34.

Krikorian, A. D., (1993). Banana and plantain breeding program, Annual report 1992.

Kumar, A., Pearce, S. R., McLean, K., Harrison, G., Heslop-Harrison, J. S., & Waugh, R., et al. (1997). The Ty1-copia group of retrotransposons in plants: Genomic organisation, evolution, and use as molecular markers. *Genetica*, *100*(1–3), 205–217.

Kutz, L. J. (1986). Robottic transplanting. ASAE paper 86:1088.

Leifert, C., Camotta, H., & Waites, W. M. (1992). Effect of combinations of antibiotics on micropropagated *Clematis*, *Delphinium*, *Hosta*, *Iris* and *Photinia*. *Plant Cell, Tissue and Organ Culture*, *29*(2), 153–160.

Leifert, C., Camotta, H., Wright, S. M., Waites, B., Cheyne, V. A., & Waites, W. M. (1991). Elimination of *Lactobacillus plantarum*, *Corynebacterium* spp., *Staphylococcus saprophyticus* and *Pseudomonas paucimobillis* from micropropagated *Hemerocallis*, *Choisya* and *Delphinium* cultures using antibiotics. *Journal of Applied Microbiology*, *71*(4), 307–330.

Leifert, C., & Cassells, A. C. (2001). Microbial hazards in plant tissue and cell cultures. *In Vitro Cellular & Developmental Biology–Plant*, *37*(2), 133–138.

Leifert, C., Morris, C. E., & Waites, W. M. (1994). Ecology of microbial saphrophytes and pathogens in tissue-culture and field-grown plants–reasons for contamination problems *in vitro*. *Critical Reviews in Plant Sciences*, *13*, 139–183.

Lloyd, G., & McCown, B. (1980). Commercially feasible micropropagation of mountain laurel, *Kalmia latifolia*, by use of shoot tip culture. *Combined Proceedings of the Plant Propagators Society*, *30*, 421–426.

Lucy, M., Reed, E., & Glick, B. R. (2004). Applications of free living plant growth-promoting rhizobacteria. *Antonie van Leeuwenhoek*, *86*(1), 1–25.

Margara, J., & Poillat, M. T. (1985). Evolution de l'aptitude a l'organogenese *in vitro* a partir de feuilles de *Saintpaulia ionantha* au cours des cultures successives. *Comptes Rendus de l'Academie des sciences Paris*, *301*, 265–268.

Matthews, S., Stuchbury, T., & Thompson, W. J. (1992). The application of micropropagation to seed potato production in Northern Scotland. *Research Investigations and Field Trials*, *1992–1993*, 192–196.

McFarlane, N. J. B. (1993). Image-guidance for robotic harvesting of micropropagated plants. *Computers and Electronics in Agriculture*, *8*(1), 43–56.

Murashige, T., & Skoog, F. (1974). Plant propagation through tissue culture. *Annual Review of Plant Physiology*, *25*, 135–166.

Nasib, A., Ali, K., & Khan, S. (2008). *In vitro* propagation of Croton (*Codiaeum variegatum*). *Pakistan Journal of Botany*, *40*(1), 99–104.

Navarro, L. (1992). Citrus shoot tip grafting *in vitro*. *Biotechnology in Agriculture and Forestry*, *18*, 327–338.

Nguyen, Q. T., & Kozai, T. (2001). Growth of *in vitro* banana (*Musa* spp.) shoots under photomixotrophic and photoautotrophic conditions. *In Vitro Cellular & Developmental Biology–Plant*, *37*, 824–829.

Nguyen, Q. T., Kozai, T., Nguyen, K. L., & Nguyen, U. V. (1999). Effects of sucrose concentration, supporting material and number of air exchanges of the vessel on the growth of *in vitro* coffee plantlets. *Plant Cell, Tissue and Organ Culture*, *58*, 51–57.

Niemenak, N., Saare-Surminski, K., Rohsius, C., Ndoumou, D. O., & Lieberei, R. (2008). Regeneration of somatic embryos in *Theobroma cacao* L. in temporary immersion bioreactor and analyses of free amino acids in different tissues. *Plant Cell Reports*, *27*, 667–676.

North, J. J., Ndakidemi, P. A., & Laubscher, C. P. (2010). The potential of developing an *in vitro* method for propagating Strelitziaceae. *African Journal of Biotechnology*, *9*(45), 7583–7588.

Ouma, J. P., Young, M. M., & Reichert, N. A. (2004). Rooting of *in vitro* regenerated cotton (*Gossypium hirsutum* L.) is influenced by genotype, medium composition, explant type and age. *African Journal of Biotechnology*, *3*(6), 313–318.

Peleg, K., Cohen, O., Ziv, M., & Kimmel, E. (1993). Machine identification of buds in images of plant shoots. *Machine Vision and Applications*, *6*(4), 224–232.

Peraza-Echeverria, S., Herrera-Valencia, V. A., & Kay, A. -J. (2001). Detection of DNA methylation changes in micropropagated banana plants using methylation-sensitive amplification polymorphism (MSAP). *Plant Science*, *161*(2), 359–367.

Pierik, R. L. M., & Steegmans, H. H. M. (1976). Vegetative propagation of *Anthurium scherzerianum* Schott through callus cultures. *Scientia Horticulturae*, *4*(3), 291–292.

Pollock, K., Barfield, D. G., & Shields, L. R. (1983). The toxicity of antibiotics to plant cell cultures. *Plant Cell Reports*, *2*, 36–39.

Posposilova, J., Tichá, I., Kadle ek, P., Haisel, D., & Plzáková, Š. (1999). Acclimatization of micropropagated plants to *ex vitro* conditions. *Biologia Plantarum*, *42*(4), 481–497.

Rabha Abdelwahd, R. N., Hakam., M., Labhilili, , & Udupa, S. M. (2008). Use of an adsorbent and antioxidants to reduce the effects of leached phenolics in *in vitro* plantlet regeneration of faba bean. *African Journal of Biotechnology*, *7*(8), 997–1002.

Reed, B., Buckley, P., & DeWilde, T. (1995). Detection and eradication of endophytic bacteria from micropropagated mint plants. *In Vitro Cellular & Developmental Biology–Plant*, *31*(1), 53–57.

Reuveni, O., & Israeli, Y. (1990). Measures to reduce somaclonal variation in *in vitro* propagated bananas. *Acta Horticulturae, 275*, 307–313.

Reuveni, O., Israely, Y., & Lahav, E. (1996). Somaclonal variation in banana and plantains (*Musa* species). In Y. P. S. Bajaj (Ed.), *Biotechnology in agriculture and forestry, vol 36. Somaclonal varation in crop improvement II* (pp. 174–195). Berlin Heidelberg New York: Springer.

Romano, A., Naroha, C., Martins, L., & Aucao, M. A. (1992). Influence of growth regulators on shoot proliferation in *Quercus ruber* l. *Annals of Botany, 70*(6), 531–536.

Sahar, A. Y., Al-Dhaher, M. M. A., & Shalaby, A. A. (2009). Elimination of Grapevine fanleaf virus (GFLV) and Grapevine leaf roll-associated virus-1 (GLRaV-1) from infected Grapevine plants using meristem tip culture. *International Journal of Virology, 5*(2), 89–99.

Sidwell, R., Huffman, J. H., Khare, G. P., Allen, L. B., Wittkowski, J. T., & Robuns, R. K. (1972). Broad-spectrum antiviral activity of Virazole: 1-β-D-ribofuranosyl-1,2,4-triazole-3-carboxamide. *Science, 177*, 705–706.

Sluis, C. J. (2006). Integrating automation technologies with commercial micropropagation. In *Plant Tissue Culture Engineering*, Dutta Gupta, S & Ibaraki, Y. (eds), pp. 231-251. The Netherlands: Springer.

Sobey, P. J., Harter, B., Hinsch, A. (1997). Automated micro-propagation of plant material. *Proceedings of the fourth annual conference on mechatronics and machine vision in practice*, 60-65.

Standaert de Metsenaere, R. E. A. (Ed.). (1991). *Economic considerations*. Dordrecht: Kluwer Academic Publishers.

Takayama, S., Swedlund, B., & Miwa, Y. (1991). Automated propagation of microbulbs of lilies. In I. K. Vasil (Ed.), *Cell culture and somatic cell genetics of plants* (pp. 112–131). New York: Academic Press Inc..

Tanaka, M., Norikane, A., & Watanabe, T. (2009). Cold cathode fluorescent lamps (CCFL): Revolutionary light source for plant micropropagation. *Biotechnology & Biotechnological Equipment, 23*(4), 1497–1503.

Teisson, C., Alvard, D., Berthouly, B., Côte, F., Escalant, J. V., & Etienne, H., et al. (1996). Simple apparatus to perform plant tissue culture by temporary immersion. *Acta Horticulturae, 440*, 521–526.

Titov, S., Bhowmik, S. K., Mandal, A., Alam, S. M., & Uddin, S. N. (2006). Control of phenolic secretion and effect of growth regulators for organ formation from *Musa* spp. cv. Kanthali Floral Bud Explants. *American Journal of Biochemistry and Biotechnology, 2*(3), 97–104.

Vidal, M. T., Azcon-Aguilar, C., Barea, J. M., & Pliego-Alfaro, F. (1992). Mycorrhizal inoculation enhances growth and development of micropropagated plants of avocado. *HortScience, 27*, 785–787.

Wambugu, F. M. (1985). Eradication of potato virus Y and S from potato by chemotherapy of cultured axillary bud tips. *American Potato Journal, 62*, 667–672.

Wang, Z., Heinemann, P. H., Morrow, C. T., & Heuser, C. (1998). Identification and separation of micropropagated sugarcane shoots based on the Hough transform. *Transactions of the Asae, 41*(5), 1535–1541.

Wardle, K., Quinlan, A., & Simpkins, I. (1979). Abscisic acid and the regulation of water loss in plantlets of *Brassica oleracea* L. var. botrytis regenerated through apical meristem culture. *Annals of Botany, 43*(6), 745–752.

Watake, H., & Kinase, A. (1988). Robot for tissue culture. *Robot, 64*, 74–79.

Weathers, P., & Giles, K. (1988). Regeneration of plants using nutrient mist culture. *In Vitro Cellular & Developmental Biology–Plant, 24*(7), 727–732.

Wetzstein, H. Y., & Sommer, H. E. (1982). Leaf anatomy of tissue-cultured *Liquidambar styraciflua* (Hamamelidaceae) during acclimatization. *American Journal of Botany, 69*, 1579–1586.

Woltering, E. J. (1990). Beneficial effects of carbon dioxide on development of gerbera and rose plantlets grown *in vitro*. *Scientia Horticulturae, 44*, 341–345.

Yesil-Celiktas, O., Gurel, A., Vardar-Sukan, F. (2010). Large scale cultivation of plant cell and tissue culture in bioreactors. *Transworld Research Network Kerala*. 1–54.

Young, P. M., Hutchins, A. S., & Canfield, M. L. (1984). Use of antibiotics to control bacteria in shoot cultures of woody plants. *Plant Science Letters, 34*(1–2), 203–209.

Yue, D., Gosselin, A., & Desjardins, Y. (1993). Effects of forced ventilation at different relative humidities on growth, photosynthesis and transpiration of geranium plantlets *in vitro*. *Canadian Journal of Plant Science, 73*, 249–256.

Ziv, M. (2000). Bioreactor technology for plant micropropagation. In J. Janick (Ed.), *Horticultural reviews* (pp. 1–30). John Wiley & Sons, Inc..

Ziv, M. (2005). Simple bioreactors for mass propagation of plants. In A. K. Hvoslev-Eide & W. Preil (Eds.), *Liquid culture systems for in vitro plant propagation* (pp. 79–94). Berlin: Springer.

Zobayed, S. M. A., Afreen, F., ZXiao, Y., & Kozai, T. (2004). Recent advancement in research on photoautotrophic micropropagation using large vessels with forced ventilation. *In Vitro Cellular and Developmental Biology–Plant, 40*, 450–458.

Zobayed, S. M. A., Afreen-Zobayed, F., Kubota, C., & Kozai, T. (2000). Mass propagation of *Eucalyptus camaldulensis* in a scaled up vessel under *in vitro* photoautotrophic condition. *Annals of Botany, 85*, 587–592.

Regulation of apomixis

16

Peggy Ozias-Akins   Joann A. Conner
The University of Georgia, Tifton, Georgia

TABLE OF CONTENTS

Introduction

Apomixis, asexual reproduction through seeds, is increasingly being viewed as a deregulation of sexual reproduction rather than an independent pathway (Koltunow and Grossniklaus, 2003; Bicknell and Koltunow, 2004; Curtis and Grossniklaus, 2008). The asynchrony hypothesis put forth by Carman (1997) alluded to deregulation through temporally conflicting developmental programs brought about by hybridization, but such global deregulation is difficult to reconcile with data from genetic studies that have shown the transmission of apomixis by specific genomic regions (Ozias-Akins and van Dik, 2007). Nevertheless, these genomic regions could directly or indirectly alter the regulation of sexual reproduction through genetic and/or epigenetic mechanisms. If we work from the premise that apomixis results from deregulation of sexual reproduction, an understanding of the sexual developmental pathway at the molecular level will contribute to the identification of genes influencing key components of apomixis. Since apomixis can incorporate different developmental events to achieve the end result of clonal progeny through seeds, multiple steps in a pathway may have undergone mutation and selection during evolution.

Apomixis includes both sporophytic and gametophytic types (Nogler, 1984). Somatic cells of the ovule directly form embryos during sporophytic apomixis; therefore, sexual reproduction is not perturbed, although competition between embryos derived from both processes is inevitable. Gametophytic apomixis is characterized by the formation of an unreduced embryo sac and parthenogenesis of the unreduced egg. Therefore, it would not be surprising if many of the same genes function during embryo sac maturation and polarization in apomicts as in sexuals. Universal differences between the two modes of reproduction are: (1) the origin or developmental state of cells specified to become embryo sac mother cells; and (2) egg activation in the absence of fertilization. In some apomicts, endosperm formation also is independent of fertilization. The early event of embryo sac formation in an apomict may begin with the megaspore mother cell (MMC) or its meiotic restitution product (diplospory), or a somatic cell of the nucellus (apospory). In the former, meiosis is replaced by mitosis and in the latter, meiosis usually is subverted by competing aposporous initials. Premature initiation of an embryo sac developmental program in permissive cells might heterochronically shift an entire pathway and preclude the completion of certain events such as meiosis. For the late event of parthenogenesis, heterochronic expression of genes responsible for egg activation could prevent fertilization of the egg while still allowing fertilization of the central cell, an essential event for seed development in pseudogamous apomicts.

With the advent of the genomics era, gene discovery has been greatly accelerated, even for relatively inaccessible tissues such as the nucellus and embryo sac. Furthermore a combination of forward and reverse genetics in model plants has contributed to our understanding of gene function and interactions (Mercier and Grelon, 2008). This review will focus

DOI: 10.1016/B978-0-12-381466-1.00016-X

on recent advances in our knowledge of genes and gene functions that may regulate or act to effect displacement of meiosis, heterochronic or ectopic development of embryo sacs, and spontaneous formation of embryos from egg cells, as studied in natural apomicts or predominantly in mutants from sexual species.

Overview of Ovule Development During Sexual Reproduction

The ovule is the organ that forms the seeds of flowering plants. It is borne in the ovary of the flower and consists of nucellus protected by integuments, precursors of embryo/endosperm, and seed coat, respectively. The nucellus is the central, micropylar-oriented tissue bounded by the integuments, that is, the site of female meiosis and female gametophyte (embryo sac) formation (Yadegari and Drews, 2004). Most common among flowering plants, a single hypodermal archesporial cell becomes specified and morphs into the MMC. During sexual reproduction, the single MMC undergoes meiosis, generating four haploid spores. Typically, only the chalazal-most megaspore survives while the other three undergo programmed cell death (Bell, 1996). The surviving megaspore enters a mitotic phase to form the haploid megagametophyte containing the egg and two synergids at the micropylar pole, three antipodals at the chalazal pole, and a binucleate central cell. Double fertilization of the egg and central cell leads to embryo and endosperm development, respectively. Ovule development has been extensively characterized in sexual, diploid *Arabidopsis thaliana* using mutants that affect ovule primordium initiation during carpel development such as *WUSCHEL* (*WUS*) and *SHOOT MERISTEMLESS* (*STM*), to inner and outer integument formation such as *AINTEGUMENTA* (*ANT*; Skinner et al., 2004; Kelley and Gasser, 2009).

Overview of Ovule Development During Apomictic Reproduction

Gametophytic apomicts are broadly classified as either aposporous or diplosporous. Chromosomally unreduced embryo sacs develop from an MMC (diplospory) or a nearby nucellar cell (apospory). Single embryo sacs are typical of diplospory, whereas multiple embryo sacs are more characteristic of apospory. Mature unreduced embryo sacs may be indistinguishable from reduced embryo sacs, each having 7 cells/8 nuclei, with the exception of *Panicum*-type embryo sacs, which probably represent an adaptation for maintaining non-lethal dosages of maternal and paternal genomes in the endosperm (Haig and Westoby, 1991). These 4-nucleate embryo sacs lack antipodal cells and most frequently have a uninucleate rather than a binucleate central cell (Ozias-Akins et al., 2003). Endosperm development in the majority of apomicts requires fertilization of the central cell (pseudogamy), but autonomous endosperm formation is characteristic of apomicts in the Asteraceae. Autonomous development of the egg into an embryo (parthenogenesis) completes the apomixis pathway. Seed-derived progeny of an obligate apomict are therefore clonal in origin and genetically identical to the maternal parent, although epigenetic states may vary (Leblanc et al., 2009).

Germline Specification

Plant life cycles differ from animal life cycles with respect to having a distinct gametophyte generation. Meiosis in animals occurs in predetermined germline cells destined to produce haploid gametes (Dickinson and Grant-Downton, 2009). Meiosis in plants requires induction of a sporogenous state in properly positioned somatic cells (Figure 16.1). Spatial and temporal cues undoubtedly act to switch the fate of certain somatic cells. Mutants affected in this cell fate transition have been identified. A very early acting *SPOROCYTELESS/NOZZLE* (*SPL/NZZ*) gene, studied through mutation in *Arabidopsis*, is required for differentiation of an archesporial cell into an MMC (Yang, 1999). This gene encodes a putative transcription factor also involved in establishing chalazal identity. A role for *SPL/NZZ* in the initiation of aposporous initials is unlikely given that expression of an *AtSPL:GUS* construct in apomictic *Hieracium* led to staining of the MMC but not the aposporous initials (Tucker et al., 2003).

Another well-studied germline mutation in maize (*Zea mays*) whose gene has not been cloned is *multiple archesporial cells 1* (*mac1*; Sheridan et al., 1996). In this mutant, multiple MMCs develop and undergo meiosis with some ovules even forming multiple embryo sacs. Either multiple hypodermal cells are being directly signaled to make the vegetative to sporogenous tissue transition or respond indirectly through the release of suppression. A mutation with a similar phenotype to *mac1*, *multiple sporocyte* (*msp1*), has been characterized in rice and the gene cloned (Nonomura et al., 2003). The Gramene resource for comparative grass genomics (release 30; www.gramene.org) shows *MSP1* at rice locus LOC_Os01g68870 to have a single homolog in maize on chromosome 3 (GRMZM2G447447). This is not the maize chromosome to which *MAC1* maps (short arm of chromosome 10; Sheridan et al., 1996). Furthermore, newly generated *Mutator* (*Mu*)-tagged alleles of *mac1* did not show evidence of *Mu* insertions in *ZmMSP1* (Ma et al., 2007). Therefore, in spite of the almost identical phenotypes, *MAC1* and *MSP1* appear to be different genes. The retrotransposon-tagged rice *MSP1* gene is similar to a leucine-rich repeat receptor-like kinase (Nonomura et al., 2003). Both male and female sporogenesis are affected in *msp1* mutants, but sporogenous cells degenerate only in the male, resulting in male sterility. As in maize *mac1*, *msp1* mutants have multiple ovule-derived sporocytes that can complete meiosis. Functional megaspores can develop into gametophytes and mutant ovules yield a high frequency of twin embryos; in the rice mutant it was not clear whether the twin embryos were more likely to arise from independent embryo sacs or from multiple cells within a single highly abnormal embryo sac, although cytological study supported the latter. Compared with *Arabidopsis* genes, *MSP1* has highest similarity with *EXCESS MICROSPOROCYTES1/EXTRA SPOROGENOUS CELLS* (*EMS1/EXS*), although *ems1/exs* does not show a female

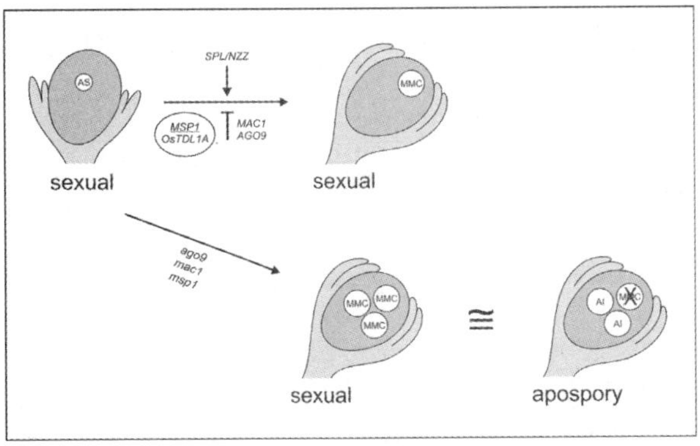

Figure 16.1 • Mutation in genes involved in sexual reproduction can lead to apospory-like phenotypes
• During sexual reproduction, a single archesporial cell (AS) is converted into the megaspore mother cell (MMC) through the expression of *SPOROCYTELESS/NOZZLE* (*SPL/NZZ*). *MULTIPLE ARCHESPORIAL CELLS 1* (*MAC1*), *ARGONAUTE 9* (*AGO9*), and *MULTIPLE SPOROCYTE* (*MSP1*) with *TAPETUM DETERMINANT-LIKE1A* (*OsTDL1A*) act to repress the formation of multiple MMCs. *ago9*, *mac1*, and *msp1* mutants lead to the formation of multiple MMCs, which is similar to the formation of multiple aposporous initials (AI) seen in aposporous apomictic development.

germline phenotype (Canales et al., 2002), but it has significant similarity with other receptor-like kinase genes including the rice disease resistance gene, *XA21*, the brassinosteroid receptor, *BRASSINOSTEROID INSENSITIVE* (*BRI1*), and the developmental gene, *CLAVATA1* (*CLV1*). MSP1 is expressed in cells surrounding the sporogenous tissue, but not in the sporogenous cells, indicating that its expression is suppressing entry into the sporocyte pathway. MSP1 is predicted to be a transmembrane protein whose extracellular LEUCINE-RICH REPEAT (LRR) domain interacts with a protein ligand encoded by *TAPETUM DETERMINANT-LIKE1A* (*OsTDL1A*). MSP1 and OsTDL1A are co-expressed and protein interaction has been demonstrated in a yeast two-hybrid system. Zhao et al. (2008) hypothesized that aposporous initials and extra sporogenous cells develop from the same cell populations; however, the extra sporogenous cells are usually confined to the nucellar hypodermal layer of both maize and rice, a position not always so well defined for aposporous initials that most frequently originate from cells adjacent to meiotic or degenerating meiotic products (Peel et al., 1997; Koltunow et al., 1998; Wen et al., 1998).

Interestingly, a gene with similar expression pattern to *MSP1*, *ARGONAUT9* (*AGO9*), was recently identified through analysis of *Arabidopsis* insertional mutants. The mutant phenotype is characterized by the formation of multiple enlarged subepidermal cells in premeiotic ovules (Olmedo-Monfil et al., 2010). Only one of these typically adopts MMC fate as demonstrated by callose deposition and progression through meiosis, but several of the enlarged cells can express a marker specific to functional megaspores. AGO9 is expressed in the distal region of the premeiotic ovule at the sporophyte–gametophyte boundary, but not in the MMC. AGO9 is a member of the ARGONAUT gene family, and therefore a component of the RNA silencing pathway. The protein was shown to be associated with 24-nucleotide sequences diagnostic of several retrotransposon families, suggesting that transposable element inactivation in companion cells of reproductive tissues restricts MMC specification to a single cell. Another member of the *AGO* gene family from rice, *MEIOSIS ARRESTED AT LEPTOTENE* (*MEL1*), displays a complementary expression pattern compared with *Arabidopsis AGO9*, that is, expression is restricted to sporocytes and disappears once meiosis is initiated (Nonomura et al., 2007). Data suggest that modification of heterochromatin may be one function of MEL1.

The elusive signal for aposporous initial specification in natural apomicts is sought by transcriptional and mutational approaches. For example, *loss of apomeiosis* (*loa*) is a mutant phenotype recovered in *Hieracium aurantiacum* while employing insertional (transposon, T-DNA) and γ-irradiation mutagenesis strategies (Bicknell et al., 2001; Okada et al., 2007). The mutant, *loa1*, was almost completely female sterile and the few progeny were predominantly polyhaploid (Okada et al., 2007). Evidence indicated that the mutant locus was not tagged with T-DNA or the transposon, but may have been generated by somaclonal variation. Nevertheless, the mutant has provided material for detailed cytological analysis of this sterility phenotype. The mutant was in an aneuploid apomictic background where aneuploidy alone may reduce fertility once apomixis becomes defective. Indeed, even though aposporous initials were formed, their position and direction of enlargement with respect to the MMC was altered (more posterior), reducing the displacement of developing megaspores, yet sexual reproduction was rarely completed. The defective aposporous initials also degenerated after assuming partial sexual identity (callose deposition in cell walls). A more extensive collection of *loa* mutants was generated by γ-irradiation (Catanach et al., 2006), but molecular characterization of mutant loci is still in progress.

Transcriptional analysis of ovules at an early stage of development corresponding to MMC specification has shown heterochronic shifts in expression of more than 500 genes between apomictic and sexual *Boechera* (Sharbel et al., 2009). These data support the hybrid-dependent floral asynchrony hypothesis of Carman (1997), since even in diploid *Boechera*, clear evidence of hybrid origin has been shown with molecular cytogenetics (Kantama et al., 2007) and sequence tags (Sharbel et al., 2009). Of the large number of apomict-specific differentially expressed genes, gene ontology terms related to transcriptional regulation were overrepresented (Sharbel et al., 2010). Comparative transcriptome analysis of early-stage ovules from apomictic and sexual genotypes of several species has been conducted using cDNA-amplified fragment length polymorphism (AFLP) and expressed sequence tag (EST) data. These studies include *Paspalum notatum* (Laspina et al., 2008), *P. simplex* (Polegri et al., 2010), *Poa pratensis* (Albertini et al., 2004), *Panicum maximum* (Yamada-Akiyama et al., 2009), *Brachiaria brizantha* (Rodrigues et al., 2003), and *Eragrostis curvula* (Cervigni et al., 2008). As with the more global analysis of Sharbel et al. (2010), many expression differences were observed between apomictic and sexual genotypes at similar stages of development. Many of the apomixis-linked and stage-specific clones in the *P. simplex* study also had putative regulatory functions. Whether substantial heterochronic shifts in gene expression have a convergent regulatory mechanism that would be more consistent with genetic studies awaits further investigation. AFLP mapping in *Hypericum perforatum* L. identified the *Hypericum APOSPORY* (*HAPPY*) locus. A truncated *ARIADNE* ortholog (*HpARI*) resides within *HAPPY* and is speculated to play a role in apospory. ARIADNE proteins, found in yeast, animals, and plants, are involved in the control of ubiquitin-dependent protein degradation (Schallau et al., 2010).

Apomeiosis

More than 50 plant meiosis genes have been cloned and functionally characterized in little more than a decade, mainly due to the broad genetic resources available for the model plant, *A. thaliana* (Mercier and Grelon, 2003; Caryl et al., 2003; Figure 16.2). Several mutants of these genes may be particularly relevant to the regulation of apomixis. Diplosporous apomixis precludes meiosis from initiating or completing; for example, the fate of the MMC is altered such that meiosis is blocked at some point prior to crossing over. Apomeiotic mutants have been described where no phase of meiosis can be observed. For example, mutants of *AMEIOTIC1* (*AM1*) in maize affect microspore and megaspore development, causing mitotic rather than meiotic divisions in mother cells. In the female, MMCs can complete more than one mitotic division to form a linear array of up to eight cells that subsequently degenerate (Golubovskaya et al., 1997). A recently cloned mutant allele of maize *AMEIOTIC1* (*am1-1*) showed no evidence for entry into meiosis in that chromosomes displayed a mitotic prophase arrangement, did not show telomere-specific bouquet formation at the nuclear envelope, lacked homologous chromosome pairing, did not recruit meiosis-specific proteins such as ABSENCE OF FIRST DIVISION1 (AFD1; a cohesin complex Rec8 homolog) or ASYNAPTIC (ASY1; a HOMOLOG PAIRING, HOP1, homolog), and did not show meiotic double-strand breaks (Pawlowski et al., 2009). AM1 is thus required for the transition from mitosis to meiosis in maize. A homolog of maize AM1 in *Arabidopsis* is SWITCH1 (SWI1)/DYAD, although the two proteins may have somewhat different functions since meiosis-specific Rec8 is installed on chromosomes of *Arabidopsis* in *swi1/dyad* mutants indicating entry into meiosis. However, Rec8 also was recruited to chromosomes of one *am1* mutant, *am1-praI*, suggesting that the differences between maize and *Arabidopsis* could be due to allelic diversification. The *dyad* allele of *SWI1* only affects female meiosis wherein development terminates after a single equational division of the MMC, leading to predominantly female sterility (Siddiqi et al., 2000). The mutant is incompletely penetrant, however, such that it is not lethal and a few seeds can be produced. Ploidy level analysis of progeny indicated that occasional mitotic products developed further into unreduced embryo sacs and that unreduced eggs could be fertilized to produce embryos with elevated ploidy levels. These relatively rare events established that functional unreduced eggs where maternal parent heterozygosity was retained could be produced in the *dyad* mutant (Ravi et al., 2008).

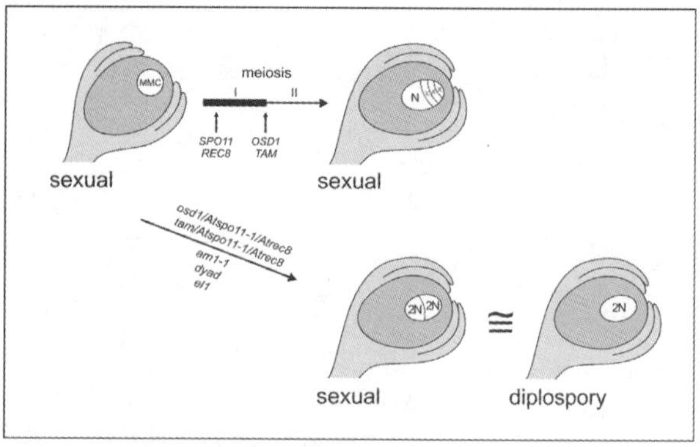

Figure 16.2 • Mutation in genes involved in sexual reproduction can lead to diplospory-like phenotypes • During sexual reproduction, the megaspore mother cell (MMC) undergoes meiosis to produce a chromosomally reduced (N) female megaspore. Genes required for meiosis I include *SPO11* for pairing and recombination and *REC8* for chromatid cohesion. *OMISSION OF SECOND DIVISION* (*OSD1*) and *TARDY ASYNCHRONOUS MEIOSIS* (*TAM*) are required for the transition from meiosis I to meiosis II. Triple mutants of *osd1/Atspo11-1/Atrec8* or *tam/Atspo11-1/Atrec8*, like the outcome of diplospory, lead to 2N megaspores that have not undergone recombination and are genetically identical to the maternal genotype. Mutations in *ameiotic1 (am1-1), dyad*, and *elongate* (*el1*) also lead to 2N megaspores, although they are not genetically identical to the mother.

To produce a relatively fertile mutant *Arabidopsis* phenotype forming unreduced eggs with non-recombined chromosomes required mutations in three genes: *OMISSION OF SECOND DIVISION* (*OSD1*), a plant-specific gene required for entry into the second meiotic division; *SPO-11*, required for pairing and recombination; and *REC8*, participating in chromatid cohesion (d'Erfurth et al., 2009). The double mutant *spo-11/rec8* replaced meiosis I with mitosis, and when combined with *osd1* produced an apomeiosis phenotype (mitosis instead of meiosis; MiMe). A second MiMe phenotype (MiMe-2) was produced by replacing the *osd1* mutation with a mutant allele of *TARDY ASYNCHRONOUS MEIOSIS/CYCA1;2* (*tam*), a cyclin gene (d'Erfurth et al., 2010). As expected, ploidy increased with each generation due to fertilization of unreduced eggs. A phenotype similar to *osd1* has been observed in the *elongate* (*el1*) mutant (Rhoades and Dempsey, 1966) of maize and attributed to the absence of meiosis II (Barrell and Grossniklaus, 2005), but no gene has been cloned to date. Although a triple recessive mutant to derive apomeiosis as generated from sexual *Arabidopsis* is unlikely the evolutionary pathway followed by natural apomicts, this knowledge may nevertheless be useful for synthesizing apomixis *de novo* in sexual plants.

None of these mutations has been shown to play a role in natural apomixis, although intriguing evidence for linkage of apomixis and meiosis loci was published from *Tripsacum dactyloides* (Grimanelli et al., 1998). Comparative mapping between diplosporous *Tripsacum* and sexual maize identified two regions in maize syntenic with the apomixis-linked region in *Tripsacum*. The syntenic chromosomal regions of maize 6L and 8L also contained *afd1* and *el1* loci.

The signals that control degeneration of meiotic products in aposporous apomicts are not known. It is possible that timing of aposporous initial specification, growth rate, and competition for space with the functional megaspore are key mechanisms providing an advantage for the surviving embryo sacs. It is also possible that a signal similar to one responsible for micropylar megaspore degeneration is generated by enlarging aposporous initials, but only sporogenous cells, and not somatic cells of the nucellus, are competent to respond by cell death. Calcium is known to play important signaling and regulatory roles in plant reproduction (Ge et al., 2007a). Degenerating micropylar megaspores in lettuce (*Lactuca sativa* L.) showed a redistribution of loosely bound calcium that preceded cell death (Qiu et al., 2008). Only studying additional plants will determine if calcium flux is a general phenomenon during megaspore degeneration, and whether it plays a role in death of the functional megaspore or its meiotic derivatives upon aposporous initial formation.

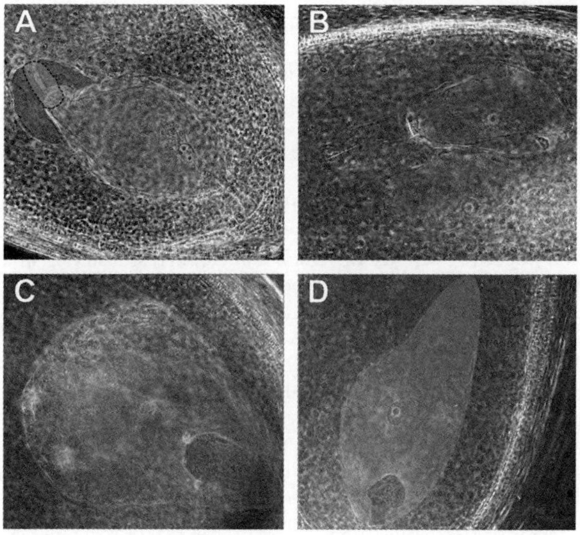

Figure 16.3 • Cleared *Pennisetum* sexual and apomictic ovaries • Ovaries were cleared and mounted using methyl salicylate and photographed using differential interference contrast (DIC) optics. Features of interest are highlighted with color. (A) Sexual *Pennisetum glaucum* ovary with embryo sac consisting of antipodal cells (gray), antipodal cell on a different plane of focus (orange), central cell (yellow) with unfused polar nuclei (blue), and reduced egg and synergid cells (pink). (B) Apomictic *Pennisetum squamulatum* ovary with the formation of multiple aposporous embryo sacs. Central cells (green) with a single 2N polar nucleus (blue) are shown. (C) Developing endosperm (red) and embryo (purple) in an ovary from a BC_8 *P. glaucum* introgression line. (D) Apomictic *P. squamulatum* ovary with parthenogenetic precocious embryo development (purple) in the embryo sac with a uninucleate (blue) central cell (green). Please see color plate section at the end of the book.

Megagametogenesis

The female gametophytic program observed in apomicts typically parallels that observed during sexual reproduction with the most prominent exception being the 4-nucleate *Panicum*-type embryo sac (Figure 16.3). The 4-nucleate pattern actually resembles a more primitive embryo sac structure based on the module theory of quartets of nuclei (Friedman and Ryerson, 2009), but is almost certainly a derived state. In both 4- and 8-nucleate embryo sacs, polarity clearly is established with the egg apparatus most often positioned toward the micropyle, although in aposporous *Pennisetum ciliare* (*Cenchrus ciliaris*), the egg apparatus may be oriented toward the closest boundary of the ovule (Snyder et al., 1955). More deviation in arrangement may occur in aposporous apomicts that frequently initiate multiple competing embryo sacs, some of which may remain immature (Snyder et al., 1955; Koltunow et al., 2000; Guan et al., 2007).

Gamete Specification

Evidence continues to build that auxin plays a key role in establishing developmental gradients in the ovule. Correlation between patterning in the female gametophyte of *Arabidopsis* and auxin gradient was demonstrated with the green fluorescent protein reporter gene under the control of an auxin responsive promoter (Pagnussat et al., 2009). Auxin signaling resulting from auxin biosynthesis at the micropylar pole of the nucellus and developing embryo sac led to establishment of gametic cell identity, although it did not alter the positioning of nuclei or number of cell divisions within the embryo sac. It would be interesting to analyze auxin gradients in ovules of aposporous apomicts. Constitutive expression of *rolB* in apomictic *H. piloselloides* presumably enhanced cell sensitivity to auxin and led to an increase in the number of aposporous embryo sacs, some more distant from the MMC than typical but retaining the capacity to develop embryo and endosperm (Koltunow et al., 2001). Two auxin response factor clones were identified as early-apomixis-stage-specific during cDNA-AFLP analysis in *P. simplex* (Polegri et al., 2010); however, these clones were not shown to be linked to the apomixis-controlling locus (ACL). It is possible that auxin signaling, either directly in aposporous embryo sacs or indirectly from surrounding ovule tissues, may play a role in nuclear division as well as embryo sac patterning.

Three *Arabidopsis* genes, *lachesis* (*lis*), *clotho* (*clo*)/*GAMETOPHYTIC FACTOR 1*(*GFA1*), and *atropos* (*atos*) are involved in gamete specification. In *lis*, *clo*/*gfa1*, and *atos* mutants, ectopic expression of an egg cell marker was observed in synergids and the central cell. LIS is expressed at all stages of female gametophyte development; however, expression of LIS is upregulated in gametic cells and downregulated in accessory cells after cellularization of the embryo sac. *LIS* encodes a homolog of yeast *PRP4*, which is a component of the spliceosome (Gross-Hardt et al., 2007). Interestingly, *CLO/GFA1* and *ATO* also encode genes with sequence similarity to the splicing factor Snu114 and a human pre-mRNA splicing factor SF3a60, respectively (Moll et al., 2008).

The indeterminate gametophyte (*ig1*) mutant of maize (Kermicle, 1969) was initially characterized by its high frequency of haploid production (both maternally and paternally derived), and was shown to extend the free-nuclear stage and affect the polarity of female gametophytes, resulting in a phenotype of increased numbers of eggs, synergids, and polar nuclei (Lin, 1978; Huang and Sheridan, 1996). The *IG1* gene has been cloned and shown to encode a LATERAL ORGAN BOUNDARIES (LOB) domain protein, a member of a plant-specific gene family (Evans, 2007). In the wild-type embryo sac of maize, *IG1* is detectable in the single-cell stage (functional megaspore) and subsequently is expressed in the egg and antipodals, but not the synergids. IG1 may control the switch from cell proliferation to differentiation in the female gametophyte. Ultimately, variable numbers of micropylar cells have cell fates as eggs or synergids, although they are not ultrastructurally distinguishable as such (Huang and Sheridan, 1996).

Genes controlling central cell differentiation have also been isolated. In *diana* (*dia*, *agl61*) mutants, the absence of a normal central cell triggers the degeneration of the embryo sac at female gametophyte (FG) stage FG7. DIA is expressed in the polar nuclei starting at late-stage FG5 and continues through the formation of the secondary nucleus of the central cell (Bemer et al., 2008). DIA forms a heterodimer with AGAMOUS-LIKE80 (AGL80), both Type I MADS-box genes (Bemer et al., 2008; Steffen et al., 2008) that are suspected to control downstream genes essential for central cell and embryo development.

A detailed analysis of cellular transcriptomes of *Arabidopsis* embryo sacs has been made possible by the technological advances of laser capture microdissection and expression arrays based on whole-genome sequence data. Distinct transcriptional profiles for egg, synergid, and central cell implicated epigenetic regulatory mechanisms through small RNA pathways in the egg and transcription factor over-representation in the female gametophyte compared with sporophytic tissues (Wuest et al., 2010). Other studies of female gametophyte transcription and mutation analysis are reviewed by Kagi and Gross-Hardt (2007, 2010).

Parthenogenesis

Parthenogenesis is fertilization independent development of the embryo during seed formation. Parthenogenesis is rare, but well documented, in sexual plants and usually leads to the formation of haploids (Kimber and Riley, 1963; Dunwell, 2010). Induction of gynogenic haploids is one breeding technology used for crops recalcitrant to induction of androgenic haploids (Forster et al., 2007). Spontaneous formation of gynogenic haploids may be accompanied by a diploid embryo in the same seed. The haploid embryo in a twin-embryo seed probably arises most frequently from a synergid cell that assumes egg cell fate (Kimber and Riley, 1963). In the maize *ig1* mutant described earlier, the micropylar-oriented embryo sac cells are competent to form embryos without fertilization. Embryogenesis most likely is triggered post-pollination even though fertilization is avoided.

The strict sense of parthenogenesis as observed among apomicts would entail egg activation in the absence of either pollination or fertilization. Evidence that eggs from the "Salmon" system of wheat are activated comes from their *in vitro* response upon isolation from the surrounding tissues. The Salmon system is characterized by a 1BS/1RS wheat/rye chromosomal translocation that is essential to confer the capacity for parthenogenesis in specific heterocytoplasmic backgrounds. Chromosome 1BS is postulated to carry a parthenogenesis-suppressing gene while 1RS has a parthenogenesis-inducing gene (Matzk et al., 1995). The mechanisms underlying this complex nuclear–cytoplasmic interaction are not understood, but the isogenic lines developed with different cytoplasms provide valuable genetic tools for comparing gene expression differences between parthenogenetic and non-parthenogenetic eggs (Kumlehn et al., 2001). Of particular importance was the demonstration that egg cells isolated from the parthenogenetic (cS and kS) lines initiated embryos in up to 25% of the cases, whereas unfertilized egg cells from the non-parthenogenetic (aS) isogenic line showed no egg cell division *in vitro*. Metabolic differences were indicated by ultrastructural comparison of aS and kS lines three days prior to anthesis where nuclear size, number of nucleoli, and number of ribosomes were greater in the kS than aS eggs (Naumova and Matzk, 1998). The precocious metabolic activity did not lead to egg cell division prior to anthesis, as has been demonstrated for some apomicts.

The factors responsible for egg activation in plants are unknown. In fact, the egg of *Arabidopsis* assumes a quiescent state with an abundance of stored transcripts whose translation predominates for approximately three days after zygote formation and prior to the maternal to zygotic transition (Vielle-Calzada et al., 2000; Pillot et al., 2010). Enhancer trap lines with *β-glucuronidase* (*GUS*) as the reporter gene and representing 19 embryo expressed genes showed GUS expression in seeds only when maternally transmitted. Furthermore, the GUS evidence for maternal allele expression of some genes was validated using allele-specific RT-PCR. Paternal alleles of these genes could not be detected in seeds, either embryo or endosperm, until more than three days after pollination. Data generated to address the timing of maternal to zygotic transition in maize are contrasting, however, perhaps due to different analytical methods and crosses tested (Grimanelli et al., 2005; Meyer and Scholten, 2007). Allele-specific RT-PCR showed no evidence for paternal allele expression of 16 genes in maize seeds three days after pollination (Grimanelli et al., 2005), whereas a single nucleotide polymorphism (SNP)-detection assay identified paternal alleles for 24 active genes in micro-dissected zygotes 1 day after pollination (Meyer and Scholten, 2007). Paternal allele expression was consistent with observation of heterosis early in embryo development, manifested as a higher rate of cell division.

In *Arabidopsis* as well as maize, there is evidence for repression of paternal allele expression in endosperm and embryo, although the two fertilization products were shown to have different transcriptional states. The requirement for transcription during early seed development was tested by knockdown of RNA POLYMERASE II through RNA interference (Pillot et al., 2010). Transcription was shown to be essential for functional megaspore mitosis and endosperm development, but dispensable for embryo initiation. Microarray analysis of seeds with developing embryos but no endosperm (apomictic maize-*Tripsacum* hybrid with fertilization-independent and precocious embryo initiation), and seeds with embryo and endosperm three days after pollination (sexual maize) showed no differential gene expression for the embryo-only sample compared to unfertilized maize ovules. In contrast, more than 2% of the genes were differentially expressed once endosperm began to develop (Grimanelli et al., 2005). These collective data support that zygotic transcription is not required for embryo initiation, nor is there zygote-specific gene expression. Yet there must be similar signals generated during the formation of a zygote that are precociously generated and perceived by unreduced eggs in apomictic plants, or repressive signals that are absent. In apomicts, egg activation and assumption of a zygotic fate does not require a signal from pollination or pollen tube penetration of a synergid, such as generation of a calcium wave. Furthermore, synergids undergo accelerated degeneration compared with those in the egg apparatus of a sexual plant (Vielle et al., 1995). Perhaps a quiescent state is never imposed on the egg of apomicts. The egg of sexual buffelgrass has a large chalazal vacuole, whereas many small vacuoles are more centrally positioned in the egg of apomictic buffelgrass. A similar pattern of vacuolation was observed in aposporous *P. maximum* (Naumova and Willemse, 1995).

How is fertilization of an unreduced egg prevented? Mechanisms could be physical or physiological, although there is little direct evidence for either. The cell wall of the egg in a sexual individual is incomplete, particularly at the chalazal end and where it is in contact with a synergid. The exposed membrane presumably would be necessary for gamete recognition via membrane localized receptor molecules (Peng and Sun, 2008). In apomictic buffelgrass, the cell wall around the egg is completed shortly after pollination, imposing a physical barrier to syngamy (Vielle et al., 1995), but this is not a universal phenomenon in apomicts as demonstrated for *P. maximum* (Naumova and Willemse, 1995). Early pollination of *Pennisetum* and *Paspalum* species has been shown to increase the frequency of fertilization events leading to generation of 2n + n (BIII) hybrids with elevated ploidy levels (Martinez et al., 1994; Burson et al., 2002), implying that the fertilization barrier intensifies shortly before anthesis. Lack of cell cycle synchronization also could influence fertilization if the egg is receptive only at G1, yet early egg activation in an apomict presumably would yield eggs at multiple phases of the cell cycle.

What genes might therefore be responsible for initiating embryogenesis in the absence of fertilization? Positive regulators of embryogenesis have been studied more thoroughly in somatic rather than zygotic embryogenesis. Over-expression of numerous genes provides evidence for a role in embryo morphogenesis. Expression profiling of mutants of some of these genes also has provided evidence for their direct and indirect targets. The embryogenesis gene *SOMATIC EMBRYOGENESIS RECEPTOR-LIKE KINASE* (*SERK*) is a leucine-rich repeat receptor-like kinase that is a marker for embryogenically competent cells, both somatic and zygotic (Schmidt et al., 1997). In *Arabidopsis*, *AtSERK1* is expressed in the ovule primordium, continuing through megasporogenesis in the distal region of the ovule, including the MMC, in all cells of the megagametophyte, and post-fertilization in both early embryo and endosperm (Hecht et al., 2001). Mutants of *AtSERK1*, however, do not show attenuation of embryogenesis suggesting that other *SERK* genes may have overlapping functions. Over-expression of *AtSERK1* in *Arabidopsis* led to enhanced competence for somatic embryo induction, although somatic embryos did not form in the absence of auxin (Hecht et al., 2001). Among natural apomicts, a *SERK* gene differentially expressed in apomictic versus sexual ovules was identified in *Poa pratensis* (Albertini et al., 2005), suggesting that its expression could play a role in acquisition of embryogenic competence by either the MMC or aposporous initials. The pattern of expression of *AtSERK:GUS* in *Hieracium* ovules, however, did not precisely parallel that of *Arabidopsis*; for example, no expression was observed during megagametogenesis or in the egg cell, but globular stage embryos were clearly marked by GUS signal. Furthermore, no differences in *AtSERK:GUS* expression were observed between sexual and apomictic reproduction (Tucker et al., 2003).

Several transcriptional regulators that play a role in embryogenesis have been identified including LEAFY COTYLEDON 1 (LEC1) and LEC2, FUSCA3 (FUS3), AGAMOUS-LIKE 15 (AGL15), WUSCHEL (WUS), and BABY BOOM (BBM). *LEC1* encodes a HAP3 subunit of the CCAAT-box binding factor whereas *LEC2* and *FUS3* encode B3-domain proteins. Over-expression of LEC1 or LEC2 in vegetative tissues causes ectopic embryo formation indicating that either is sufficient for embryo formation under such permissive conditions, and their loss-of-function mutants have impaired embryo development indicating their requirement for embryo morphogenesis. LEC1 and LEC2 both act to target FUS3 as well as each other. Other targets of LEC2 are auxin biosynthesis genes, YUC2 and YUC4 (Stone et al., 2008), a finding that suggests LEC2 may indirectly influence levels or gradients of auxin, a growth regulator known to stimulate somatic embryogenesis and establish polarity in zygotic embryos (Weijers and Jurgens, 2005). LEC2 also upregulates AGL15, a MADS domain transcription factor that activates expression of GA2ox6, an inhibitor of gibberellin biosynthesis. Furthermore, ChIP-chip experiments showed that LEC2 and FUS3 were direct targets of AGL15 (Zheng et al., 2009), suggesting positive feedback regulation. While a role for LEC1 and LEC2 in early embryogenesis is implicated by these experiments, LEC2 has been shown to bind to RY elements found in the promoter of many seed-specific genes, particularly those involved in seed maturation such as seed storage proteins (Braybrook et al., 2006). Similarly, LEC1 promotes expression of fatty acid biosynthetic genes during seed maturation albeit indirectly (Mu et al., 2008). Evidence therefore supports that LEC1 and LEC2 have functions in both early (Lotan et al., 1998) and late (Braybrook and Harada, 2008) embryogenesis.

Another transcriptional regulator, WUS, is necessary for shoot apical meristem development, including the shoot meristem of embryos; however, ectopic expression of WUS is sufficient to induce somatic embryo formation (Zuo et al., 2002). *WUS* is a homeobox gene whose product confers stem cell identity. In addition to being expressed in shoot meristems, it is expressed in the nucellus (thought to have evolved from a shoot meristem; Kenrick and Crane, 1997), but acts non-cell-autonomously through a signaling pathway to stimulate integument initiation (Gross-Hardt et al., 2002). WUS expression is positively regulated by NOZZLE/SPOROCYTELESS (Sieber et al., 2004), a putative transcription factor that affects ovule ontogeny and specification of the MMC (Yang et al., 1999; Schiefthaler et al., 2004). Since WUS has multiple signaling functions, its expression in embryogenically competent cells and response to auxin gradients is probably related to its regulation of stem cell function (Su et al., 2009).

BABY BOOM is an AP2-domain transcription factor, a member of the AP2-ERF gene family (Weigel, 1995; Ohme-Takagi and Shinshi, 1995). BABY BOOM falls in the AINTEGUMENTA (ANT) clade, members of which are involved in root, ovule, and embryo development (Kim et al., 2006). The discovery of *BBM* was as a gene expressed during *Brassica* microspore embryogenesis. Over-expression of BBM leads to ectopic embryo formation in seedlings of *Brassica* and *Arabidopsis* (Boutilier et al., 2002). It is likely that BBM functions through a different pathway than LEC, given the gene targets that have been identified (Passarinho et al., 2008). Interestingly, a novel homolog of BBM has been identified in two apomicts, *Pennisetum squamulatum* and *C. ciliaris*, that is tightly linked with apomixis (Conner et al., 2008). The gene is expressed one to two days prior to anthesis and during embryo development, whereas transcript from the most closely related homolog from a sexual genotype is barely detectable on the day of anthesis (Huo, 2008; Zeng, 2009).

Non-viable parthenogenetic haploid embryos of up to 20 cells have been associated with the loss-of-function of the *Arabidopsis* gene *MULTICOPY SUPPRESSOR IRA 1* (*MSI1*; Guitton and Berger, 2005). Rodrigues et al. (2010) found similar expression of the *Hieracium* homolog *HMSI1* in both apomictic and sexual *Hieracium* ovules, suggesting that autonomous egg cell development is not caused by a lack of *HMS1* expression in apomicts. Furthermore, *HMS1* did not map to the loss of parthenogenesis (LOP) locus, which controls both autonomous embryo and endosperm development in *Hieracium*.

Whether mis-expression or expression of novel forms of any of the previously mentioned transcriptional regulators or signaling pathway components plays a role in parthenogenesis is still open to speculation and hypothesis testing. Regulated expression of several of these transcription factors in the ovule and egg cell has failed to induce parthenogenesis (reported without supporting data in a review by Curtis and Grossniklaus, 2008), although these experiments were likely to have been carried out in sexual, diploid *Arabidopsis*, with genes from sexual diploids, and would not have tested the effect of a novel gene or its regulatory elements from an apomict.

Endosperm Development

A unique evolutionary feature of angiosperms is double fertilization to yield embryo and endosperm through maternal and paternal contributions. Seed development rarely occurs in the absence of endosperm. While fertilization of the egg is prevented in apomixis, the requirement for fertilization of the central cell, termed pseudogamy, remains in most apomicts. Exceptions are largely found in the Asteraceae where autonomous endosperm development is typical. Members of the Poaceae, however, show aberrant and often aborted seed development when maternal to paternal genome ratios in the endosperm are other than 2:1 (Lin, 1984). Apomicts demonstrate adaptations to this requirement, for example, tolerance of imbalanced ratios, such as diplosporous *Tripsacum* (Grimanelli et al., 1997); altered embryo sac structure, such as species with *Panicum*-type embryo sacs (Vielle et al., 1995); or participation of both sperm in fertilization of the central cell (Grossniklaus et al., 2001). Whether apomixis with autonomous endosperm development can be engineered into a sexual grass is open to debate and further experimentation. Nevertheless, considerable progress in identifying genes that normally arrest endosperm formation in the absence of fertilization has been made through the isolation of *fertilization independent seed* (*fis*) class mutants in *Arabidopsis* (see Curtis and Grossniklaus, 2008). The best studied of these mutants are *fis2*, *fertilization independent endosperm* (*fie*), *medea* (*mea*), and *multisuppressor of ira1* (*msi1*). Most of these genes participate in the formation of a Polycomb group (PcG) complex, a type of multimeric complex originally identified in *Drosophila* as binding to chromatin and suppressing target gene expression, thereby suppressing cell proliferation. Mutants for these genes in *Arabidopsis* show proliferation of endosperm and fruit development in the absence of fertilization. Several of these genes are regulated by genomic imprinting, that is, their expression is dependent on maternal or paternal origin. Paternally inherited alleles are not expressed during early seed development. When promoters for *Arabidopsis FIS2*, *FIE*, and *MEA* genes were fused with the coding region for *GUS* and introduced into sexual and apomictic *Hieracium*, expression of the three genes was coordinately regulated, although expression was less localized within ovules of *Hieracium* compared with *Arabidopsis* (Tucker et al., 2003). Furthermore, no differences in expression were observed between sexual and apomictic *Hieracium* genotypes, suggesting that the same molecular pathways were operating during endosperm development even though the genotypes differed in their requirement for central cell fertilization. RNAi knockdown of *HFIE* failed to promote autonomous endosperm formation in sexual *Hieracium* suggesting that the suppressive role of *FIE* in the central cell is not conserved between *Hieracium* and *Arabidopsis* (Rodrigues et al., 2008). Similarly, a T-DNA knockout of *OsFIE1*, which shows endosperm-specific expression, also failed to promote autonomous endosperm formation. In this study, a role for constitutively expressed *OsFIE2* in endosperm suppression could not be ruled out (Luo et al., 2009).

Chromatin Modification and Epigenetic Regulation

DNA methylation, chromatin modifications, and transcriptional gene silencing mediated by small interfering (si) RNA affect gene expression of diverse plant development pathways including vernalization and flowering time, endosperm development, stem cell maintenance, and response to stress (Grant-Downton and Dickinson, 2005, 2006; Berr and Shen, 2010; Mosher and Melnyk, 2010). Examples during sexual plant development include control of parental genomic imprinting of endosperm genes and the transcriptional activation of transposable elements (TEs) in both the pollen vegetative nucleus (Slotkin et al., 2009) and endosperm (Gehring et al., 2009, Hsieh et al., 2009). Production of siRNAs by the transcriptional activation of TEs in pollen vegetative cells and endosperm is hypothesized to protect the developing embryo from TE activation. An epigenetic change of gene expression is also seen during interspecific hybridization and allopolyploidization (Chen, 2007; Martienssen, 2010).

Epigenetics has been hypothesized to play a role in the creation of apomicts through gene expression changes occurring during the stabilization of apomictic genomes derived from polyploidy or interspecific hybridization. Do the mapped genetic loci for apomixis have an epigenetic component? A frequent occurrence among mapped apomixis loci is the suppression of recombination noted through either the identification of multiple linked markers or by a decrease in recombination rates in apomicts among apomixis-linked markers compared with homologous or syntenic regions of sexual relatives, along with hemizygosity of molecular markers (Ozias-Akins and van Dik, 2007). Physical mapping of apomixis loci has been limited to the ASGR (apospory-specific genomic region) in *Pennisetum/Cenchrus* (Goel et al., 2003; Akiyama et al., 2004, 2005), the ACL in *P. simplex* (Pupilli et al., 2001), and the *Beta corolliflora* apomixis carrier chromosome and corresponding chromosomal BACs in the apomictic monosomic addition line Beta M14 ($2n = 2x = 18 + 1$; Ge et al., 2007b). In all four species, hemizygosity was confirmed by BAC-FISH analysis. Quantitative ideograms from *P. squamulatum*/*C. ciliaris* show the ASGR to be highly heterochromatic and abundant in an Opie-2-like retrotransposon for the ASGR in *P. squamulatum* and throughout the genome in *C. ciliaris*. The heterochromatic characteristic of the ASGR does not preclude the identification of non-transposable element coding regions (Conner et al., 2008) or expression of ASGR specific genes. Given the limited data from apomicts, it is easy to speculate a role for epigenetic control of the trait; however, final proof remains elusive.

Conclusions and Future Prospects for Apomixis in Crops

The study of development at the molecular level in both sexual and apomictic genotypes is vastly expanding our knowledge of the obscured female gametophyte generation from spores to gametes. How the key events of apomixis, nonreduction, and parthenogenesis have evolved their regulation remains unknown. Since the function of female expressed genes identified from *Arabidopsis* and maize is not completely orthologous in apomicts, continued investigation of genes expressed in naturally evolved apomicts is an essential complement to the study of mutants in sexual species that express components of apomixis. It is anticipated that knowledge emerging from the study of both sexual and apomictic species will enable the eventual development of apomictic crops where hybrid vigor could be retained through clonal seeds. Furthermore, engineered regulation of apomixis for conditional expression would allow the outcomes of each mode of reproduction to be accessed to develop either new genetic combinations or clonal genotypes as desired. Apomixis then would become an even more valuable tool for plant breeding and seed production.

References

Akiyama, Y., Conner, J. A., Goel, S., Morishige, D. T., Mullet, J. E., & Hanna, W. W., et al. (2004). High-resolution physical mapping in *Pennisetum squamulatum* reveals extensive chromosomal heteromorphism of the genomic region associated with apomixis. *Plant Physiology*, *134*, 1733–1741.

Akiyama, Y., Hanna, W. W., & Ozias-Akins, P. (2005). High-resolution physical mapping reveals that the apospory-specific genomic region (ASGR) in *Cenchrus ciliaris* is located on a heterochromatic and hemizygous region of a single chromosome. *Theoretical and Applied Genetics*, *111*, 1042–1051.

Albertini, E., Marconi, G., Barcaccia, G., Raggi, L., & Falcinelli, M. (2004). Isolation of candidate genes for apomixis in *Poa pratensis* L. *Plant Molecular Biology*, *56*, 879–894.

Albertini, E., Marconi, G., Reale, L., Barcaccia, G., Porceddu, A., & Ferranti, F., et al. (2005). SERK and APOSTART. Candidate genes for apomixis in *Poa pratensis*. *Plant Physiology*, *138*, 2185–2199.

Barrell, P. J., & Grossniklaus, U. (2005). Confocal microscopy of whole ovules for analysis of reproductive development: The *elongate1* mutant affects meiosis II. *Plant Journal*, *43*, 309–320.

Bell, P. R. (1996). Megaspore abortion: A consequence of selective apoptosis? *International Journal of Plant Sciences*, *157*, 1–7.

Bemer, M., Wolters-Arts, M., Grossniklaus, U., & Angenent, G. C. (2008). The MADS domain protein DIANA acts together with AGAMOUS-LIKE80 to specify the central cell in *Arabidopsis* ovules. *Plant Cell*, *20*, 2088–2101.

Berr, A., & Shen, W. (2010). Molecular mechanisms in epigenetic regulation of plant growth and development. *Plant Developmental Biology: Biotechnological Perspectives*, *2*, 325–344.

Bicknell, R., Podivinsky, E., Catanach, S., Erasmuson, S., & Lambie, S. (2001). Strategies for isolating mutants in *Hieracium* with dysfunctional apomixis. *Sexual Plant Reproduction*, *14*, 227–232.

Bicknell, R. A., & Koltunow, A. M. (2004). Understanding apomixis: Recent advances and remaining conundrums. *Plant Cell*, *16*, S228–S245.

Boutilier, K., Offringa, R., Sharma, V. K., Kieft, H., Ouellet, T., & Zhang, L., et al. (2002). Ecotopic expression of BABY BOOM triggers a conversion from vegetative to embryonic growth. *Plant Cell*, *14*, 1737–1749.

Braybrook, S. A., & Harada, J. J. (2008). LECs go crazy in embryo development. *Trends in Plant Science*, *13*, 624–630.

Braybrook, S. A., Stone, S. L., Park, S., Bui, A. Q., Le, B. H., & Fischer, R. L., et al. (2006). Genes directly regulated by LEAFY COTYLEDON2 provide insight into the control of embryo maturation and somatic embryogenesis. *Proceedings of the National Academy of Sciences of the United States of America*, *103*, 3468–3473.

Burson, B. L., Hussey, M. A., Actkinson, J. M., & Shafer, G. S. (2002). Effect of pollination time on the frequency on 2n + n fertilization in apomictic buffelgrass. *Crop Science*, *42*, 1075–1080.

Canales, C., Bhatt, A. M., Scott, R., & Dickinson, H. (2002). EXS, a putative LRR receptor kinase, regulates male germline cell number and tapetal identity and promotes seed development in *Arabidopsis*. *Current Biology*, *12*, 1718–1727.

Carman, J. G. (1997). Asynchronous expression of duplicate genes in angiosperms may cause apomixis, bispory, tetraspory, and polyembryony. *Biological Journal of the Linnean Society*, *61*, 51–94.

Caryl, A. P., Jones, G. H., & Franklin, F. C. H. (2003). Dissecting plant meiosis using *Arabidopsis thaliana* mutants. *Journal of Experimental Botany*, *54*, 25–38.

Catanach, A. S., Erasmuson, S. K., Podivinsky, E., Jordan, B. R., & Bicknell, R. (2006). Deletion mapping of genetic regions associated with apomixis in *Hieracium*. *Proceedings of the National Academy of Sciences of the United States of America*, *103*, 18650–18655.

Cervigni, G. D. L., Paniego, N., Diaz, M., Selva, J. P., Zappacosta, D., & Zanazzi, D., et al. (2008). Expressed sequence tag analysis and development of gene associated markers in a near-isogenic plant system of *Eragrostis curvula*. *Plant Molecular Biology*, *67*, 1–10.

Chen, Z. J. (2007). Genetic and epigenetic mechanisms for gene expression and phenotypic variation in plant polyploids. *Annual Review of Plant Biology*, *58*, 377–406.

Conner, J. A., Goel, S., Gunawan, G., Cordonnier-Pratt, M. M., Johnson, V. E., & Liang, C., et al. (2008). Sequence analysis of bacterial artificial chromosome clones from the apospory-specific genomic region of *Pennisetum* and *Cenchrus*. *Plant Physiology*, *147*, 1396–1411.

Curtis, M. D., & Grossniklaus, U. (2008). Molecular control of autonomous embryo and endosperm development. *Sexual Plant Reproduction*, *21*, 79–88.

Dickinson, H. G., & Grant-Downton, R. (2009). Bridging the generation gap: Flowering plant gametophytes and animal germlines reveal unexpected similarities. *Biological Reviews*, *84*, 589–615.

Dunwell, J. M. (2010). Haploids in flowering plants: Origins and exploitation. *Plant Biotechnology Journal*, *8*, 377–424.

d'Erfurth, I., Cromer, L., Jolivet, S., Girard, C., Horlow, C., & Sun, Y., et al. (2010). The cyclin-A CYCA1;2/TAM is required for the meiosis I to meiosis II transition and cooperates with OSD1 for the prophase to first meiotic division transition. *PLoS Genetics*, *6*, e1000989.

d'Erfurth, I., Jolivet, S., Froger, N., Catrice, O., Novatchkova, M., & Mercier, R. (2009). Turning meiosis into mitosis. *PLoS Biology*, *7*(6), e1000124.

Evans, M. M. S. (2007). The *indeterminate gametophyte1* gene of maize encodes a LOB domain protein required for embryo sac and leaf development. *Plant Cell*, *19*, 46–62.

Forster, B. P., Heberle-Bors, E., Kasha, K. J., & Touraev, A. (2007). The resurgence of haploids in higher plants. *Trends in Plant Science*, *12*, 368–375.

Friedman, W. E., & Ryerson, K. C. (2009). Reconstructing the ancestral female gametophyte of angiosperms: Insights from *Amborella* and other ancient lineages of flowering plants. *American Journal of Botany*, *96*, 129–143.

Ge, L. L., Tian, H. Q., & Russell, S. D. (2007a). Calcium function and distribution during fertilization in angiosperms. *American Journal of Botany*, *94*, 1046–1060.

Ge, Y., He, G. C., Wang, Z. W., Guo, D. D., Qin, R., & Li, R. T. (2007b). GISH and BAC-FISH study of apomictic Beta M14. *Science in China Series C-Life Sciences*, *50*, 242–250.

Gehring, M., Bubb, K. L., & Henikoff, S. (2009). Extensive demethylation of repetitive elements during seed development underlies gene imprinting. *Science*, *324*, 1447–1451.

Goel, S., Chen, Z., Conner, J. A., Akiyama, Y., Hanna, W. W., & Ozias-Akins, P. (2003). Physical evidence that a single hemizygous chromosomal region is sufficient to confer aposporous embryo sac formation in *Pennisetum squamulatum* and *Cenchrus ciliaris*. *Genetics*, *163*, 1069–1082.

Golubovskaya, I., Avalkina, N., & Sheridan, W. F. (1997). New insights into the role of the maize *ameiotic1* locus. *Genetics*, *147*, 1339–1350.

Grant-Downton, R. T., & Dickinson, H. G. (2005). Epigenetics and its implications for plant biology. 1. The epigenetic network in plants. *Annals of Botany*, *96*, 1143–1164.

Grant-Downton, R. T., & Dickinson, H. G. (2006). Epigenetics and its implications for plant biology 2. The "Epigenetic Epiphany": Epigenetics, evolution and beyond. *Annals of Botany*, *97*, 11–27.

Grimanelli, D., Hernandez, M., Perotti, E., & Savidan, Y. (1997). Dosage effects in the endosperm of diplosporous apomictic *Tripsacum* (Poaceae). *Sexual Plant Reproduction*, *10*, 279–282.

Grimanelli, D., Leblanc, O., Espinosa, E., Perotti, E., Gonzales de Leon, D., & Savidan, Y. (1998). Mapping diplosporous apomixis in tetraploid *Tripsacum*: One gene or several genes? *Heredity*, *80*, 33–39.

Grimanelli, D., Perotti, E., Ramirez, J., & Leblanc, O. (2005). Timing of the maternal-to-zygotic transition during early seed development in maize. *Plant Cell*, *17*, 1061–1072.

Gross-Hardt, R., Kagi, C., Baumann, N., Moore, J. M., Baskar, R., & Gagliano, W. B., et al. (2007). LACHESIS restricts gametic cell fate in the female gametophyte of *Arabidopsis*. *PLoS Biology*, *5*, 494–500.

Gross-Hardt, R., Lenhard, M., & Laux, T. (2002). WUSCHEL signaling functions in interregional communication during *Arabidopsis* ovule development. *Genes & Development*, *16*, 1129–1138.

Grossniklaus, U., Spillane, C., Page, D. R., & Kohler, C. (2001). Genomic imprinting and seed development: Endosperm formation with and without sex. *Current Opinion in Plant Biology*, *4*, 21–27.

Guan, L. M., Chen, L. Z., & Terao, H. (2007). Ultrastructural studies of gametophytic apomicts in guinea grass (*Panicum maximum*) II. Characteristics of aposporous initial cell-derived embryo sac. *Cytologia*, *72*, 145–153.

Guitton, A. E., & Berger, F. (2005). Loss of function of MULTICOPY SUPPRESSOR OF IRA 1 produces nonviable parthenogenetic embryos in *Arabidopsis*. *Current Biology*, *15*, 750–754.

Haig, D., & Westoby, M. (1991). Genomic imprinting in endosperm – its effect on seed development in crosses between species, and between different ploidies of the same species, and its implications for the evolution of apomixis. *Philosophical Transactions of the Royal Society of London. Series B, Biological Sciences*, *333*, 1–13.

Hecht, V., Vielle-Calzada, J. -P., Hartog, M. V., Schmidt, E. D. L., Boutilier, K., & Grossniklaus, U., et al. (2001). The *Arabidopsis* Somatic Embryogenesis Receptor Kinase 1 gene is expressed in developing ovules and embryos and enhances embryogenic competence in culture. *Plant Physiology*, *127*, 803–816.

Hsieh, T. F., Ibarra, C. A., Silva, P., Zemach, A., Eshed-Williams, L., & Fischer, R. L., et al. (2009). Genome-wide demethylation of *Arabidopsis* endosperm. *Science*, *324*, 1451–1454.

Huang, B. Q., & Sheridan, W. F. (1996). Embryo sac development in the maize *indeterminate gametophyte1* mutant: Abnormal nuclear behavior and defective microtubule organization. *Plant Cell*, *8*, 1391–1407.

Huo, H. (2008). Genetic analysis of the apospory-specific genomic region (ASGR) in *Pennisetum squamulatum*: From mapping to candidate gene. 1-211. The University of Georgia. (GENERIC)

Kagi, C., & Gross-Hardt, R. (2007). How females become complex: Cell differentiation in the gametophyte. *Current Opinion in Plant Biology*, *10*, 633–638.

Kagi, C., & Gross-Hardt, R. (2010). Analyzing female gametophyte development and function: There is more than one way to crack an egg. *European Journal of Cell Biology*, *89*, 258–261.

Kantama, L., Sharbel, T. F., Schranz, M. E., Mitchell-Olds, T., de Vries, S., & de Jong, H. (2007). Diploid apomicts of the *Boechera holboellii* complex display large-scale chromosome substitutions and aberrant chromosomes. *Proceedings of the National Academy of Sciences of the United States of America*, *104*, 14026–14031.

Kelley, D. R., & Gasser, C. S. (2009). Ovule development: Genetic trends and evolutionary considerations. *Sexual Plant Reproduction*, *22*, 229–234.

Kenrick, P., & Crane, P. R. (1997). The origin and early evolution of plants on land. *Nature, 389*, 33–39.
Kermicle, J. L. (1969). Androgenesis conditioned by a mutation in maize. *Science, 166*, 1422–1424.
Kim, S., Soltis, P., Wall, K., & Soltis, D. (2006). Phylogeny and domain evolution in the *APETALA2*-like gene family. *Molecular Biology and Evolution, 23*, 107–120.
Kimber, G., & Riley, R. (1963). Haploid angiosperms. *Botanical Review, 29*, 480–531.
Koltunow, A. M., & Grossniklaus, U. (2003). Apomixis: A developmental perspective. *Annual Review of Plant Biology, 54*, 547–574.
Koltunow, A. M., Johnson, S. D., & Bicknell, R. A. (1998). Sexual and apomictic development in *Hieracium*. *Sexual Plant Reproduction, 11*, 213–230.
Koltunow, A. M., Johnson, S. D., & Bicknell, R. A. (2000). Apomixis is not developmentally conserved in related, genetically characterized *Hieracium* plants of varying ploidy. *Sexual Plant Reproduction, 12*, 253–266.
Koltunow, A. M., Johnson, S. D., Lynch, M., Yoshihara, T., & Costantino, P. (2001). Expression of *rolB* in apomictic *Hieracium piloselloides* Vill causes ectopic meristems in planta and changes in ovule formation, where apomixis initiates at higher frequency. *Planta, 214*, 196–205.
Kumlehn, J., Kirik, V., Czihal, A., Altschmied, L., Matzk, F., & Lorz, H., et al. (2001). Parthenogenetic egg cells of wheat: Cellular and molecular studies. *Sexual Plant Reproduction, 14*, 239–243.
Laspina, N. V., Vega, T., Seijo, J. G., Gonzalez, A. M., Martelotto, L. G., & Stein, J., et al. (2008). Gene expression analysis at the onset of aposporous apomixis in *Paspalum notatum*. *Plant Molecular Biology, 67*, 615–628.
Leblanc, O., Grimanell, D., Hernandez-Rodriguez, M., Galindo, P. A., Soriano-Martinez, A. M., & Perotti, E. (2009). Seed development and inheritance studies in apomictic maize–*Tripsacum* hybrids reveal barriers for the transfer of apomixis into sexual crops. *The International Journal of Developmental Biology, 53*, 585–596.
Lin, B. Y. (1978). Structural modifications of female gametophyte associated with *indeterminate-gametophyte* (*Ig*) mutant in maize. *Canadian Journal of Genetics and Cytology, 20*, 249–257.
Lin, B. Y. (1984). Ploidy barrier to endosperm development in maize. *Genetics, 107*, 103–115.
Lotan, T., Ohto, M., Yee, K. M., West, M. A. L., Lo, R., & Kwong, R. W., et al. (1998). *Arabidopsis* LEAFY COTYLEDON1 is sufficient to induce embryo development in vegetative cells. *Cell, 93*, 1195–1205.
Luo, M., Platten, D., Chaudhury, A., Peacock, W. J., & Dennis, E. S. (2009). Expression, imprinting, and evolution of rice homologs of the polycomb group genes. *Molecular Plant, 2*, 711–723.
Ma, J., Duncan, D., Morrow, D. J., Fernandes, J., & Walbot, V. (2007). Transcriptome profiling of maize anthers using genetic ablation to analyze pre-meiotic and tapetal cell types. *Plant Journal, 50*, 637–648.
Martienssen, R. A. (2010). Heterochromatin, small RNA and post-fertilization dysgenesis in allopolyploid and interploid hybrids of *Arabidopsis*. *The New Phytologist, 186*, 46–53.
Martinez, E. J., Quarin, C. L., & Espinoza, F. (1994). BIII progeny (2n + n) from apomictic *Paspalum notatum* obtained through early pollination. *The Journal of Heredity, 85*, 295–297.
Matzk, F., Meyer, H. -M., Baumlein, H., Balzer, H. -J., & Schubert, I. (1995). A novel approach to the analysis of the initiation of embryo development in Gramineae. *Sexual Plant Reproduction, 8*, 266–272.
Mercier, R., & Grelon, M. (2008). Meiosis in plants: Ten years of gene discovery. *Cytogenetic and Genome Research, 120*, 281–290.
Meyer, S., & Scholten, S. (2007). Equivalent parental contribution to early plant zygotic development. *Current Biology, 17*, 1686–1691.
Moll, C., von Lyncker, L., Zimmermann, S., Kagi, C., Baumann, N., & Twell, D., et al. (2008). CLO/GFA1 and ATO are novel regulators of gametic cell fate in plants. *Plant Journal, 56*, 913–921.
Mosher, R. A., & Melnyk, C. W. (2010). siRNAs and DNA methylation: Seedy epigenetics. *Trends in Plant Science, 15*, 204–210.
Mu, J. Y., Tan, H. L., Zheng, Q., Fu, F. Y., Liang, Y., & Zhang, J. A., et al. (2008). LEAFY COTYLEDON1 is a key regulator of fatty acid biosynthesis in *Arabidopsis*. *Plant Physiology, 148*, 1042–1054.
Naumova, T. N., & Matzk, F. (1998). Differences in the initiation of the zygotic and parthenogenetic pathway in the Salmon lines of wheat: Ultrastructural studies. *Sexual Plant Reproduction, 11*, 121–130.
Naumova, T. N., & Willemse, M. T. M. (1995). Ultrastructural characterization of apospory in *Panicum maximum*. *Sexual Plant Reproduction, 8*, 197–204.
Nogler, G. A. (1984). Gametophytic apomixis. In B. M. Johri (Ed.), *Embryology of angisoperms* (pp. 475–518). Berlin: Springer.
Nonomura, K. I., Miyoshi, K., Eiguchi, M., Suzuki, T., Miyao, A., & Hirochika, H., et al. (2003). The *MSP1* gene is necessary to restrict the number of cells entering into male and female sporogenesis and to initiate anther wall formation in rice. *Plant Cell, 15*, 1728–1739.
Nonomura, K. I., Morohoshi, A., Nakano, M., Eiguchi, M., Miyao, A., & Hirochika, H., et al. (2007). A germ cell-specific gene of the ARGONAUTE family is essential for the progression of premeiotic mitosis and meiosis during sporogenesis in rice. *Plant Cell, 19*, 2583–2594.
Ohme-Takagi, M., & Shinshi, H. (1995). Ethylene-inducible DNA binding proteins that interact with an ethylene-responsive element. *Plant Cell, 7*, 173–182.
Okada, T., Catanach, A. S., Johnson, S. D., Bicknell, R. A., & Koltunow, A. M. (2007). An *Hieracium* mutant, *loss of apomeiosis 1* (*loa1*) is defective in the initiation of apomixis. *Sexual Plant Reproduction, 20*, 199–211.
Olmedo-Monfil, V., Duran-Figueroa, N., Orteaga-Vazquez, M., mesa-Arevalo, E., Autran, D., & Grimanelli, D., et al. (2010). Control of female gamete formation by a small RNA pathway in *Arabidopsis*. *Nature, 464*, 628–632.
Ozias-Akins, P., Akiyama, Y., & Hanna, W. W. (2003). Molecular characterization of the genomic region linked with apomixis in *Pennisetum/Cenchrus*. *Functional & Integrative Genomics, 3*, 94–104.
Ozias-Akins, P., & van Dik, P. J. (2007). Mendelian genetics of apomixis in plants. *Annual Review of Genetics, 41*, 509–537.
Pagnussat, G. C., Alandete-Saez, M., Bowman, J. L., & Sundaresan, V. (2009). Auxin-dependent patterning and gamete specification in the *Arabidopsis* female gametophyte. *Science, 324*, 1684–1689.
Passarinho, P., Ketelaar, T., Xing, M. Q., van Arkel, J., Maliepaard, C., & Hendriks, M. W., et al. (2008). BABY BOOM target genes provide diverse entry points into cell proliferation and cell growth pathways. *Plant Molecular Biology, 68*, 225–237.
Pawlowski, W. P., Wang, C. J. R., Golubovskaya, I. N., Szymaniak, J. M., Shi, L., & Hamant, O., et al. (2009). Maize AMEIOTIC1 is essential for multiple early meiotic processes and likely required for the initiation of meiosis. *Proceedings of the National Academy of Sciences of the United States of America, 106*, 3603–3608.
Peel, M. D., Carman, J. G., & Leblanc, O. (1997). Megasporocyte callose in apomictic buffelgrass, Kentucky bluegrass, *Pennisetum squamulatum* Fresen, *Tripsacum* L, and weeping lovegrass. *Crop Science, 37*, 724–732.
Peng, X. B., & Sun, M. X. (2008). Gamete recognition in higher plants: An abstruse but charming mystery. *Journal of Integrative Plant Biology, 50*, 868–874.
Pillot, M., Baroux, C., Vazquez, M. A., Autran, D., Leblanc, O., & Vielle-Calzada, J. P., et al. (2010). Embryo and endosperm inherit distinct chromatin and transcriptional states from the female gametes in *Arabidopsis*. *Plant Cell, 22*, 307–320.
Polegri, L., Calderini, O., Arcioni, S., & Pupilli, F. (2010). Specific expression of apomixis-linked alleles revealed by comparative transcriptomic analysis of sexual and apomictic *Paspalum simplex* Morong flowers. *Journal of Experimental Botany, 61*, 1869–1883.
Pupilli, F., Labombarda, P., Caceres, M. E., Quarin, C. L., & Arcioni, S. (2001). The chromosome segment related to apomixis in *Paspalum simplex* is homoeologous to the telomeric region of the long arm of rice chromosome 12. *Molecular Breeding, 8*, 53–61.
Qiu, Y. L., Liu, R. S., Xie, C. T., Russell, S. D., & Tian, H. Q. (2008). Calcium changes during megasporogenesis and megaspore degeneration in lettuce (*Lactuca sativa* L.). *Sexual Plant Reproduction, 21*, 197–204.
Ravi, M., Marimuthu, M. P. A., & Siddiqi, I. (2008). Gamete formation without meiosis in *Arabidopsis*. *Nature, 451*, 1121–1U10.

Rhoades, M. M., & Dempsey, E. (1966). Induction of chromosome doubling at meiosis by elongate gene in maize. *Genetics, 54*, 505–522.

Rodrigues, J. C. M., Cabral, G. B., Dusi, D. M. A., de Mello, L. V., Rigden, D. J., & Carneiro, V. T. C. (2003). Identification of differentially expressed cDNA sequences in ovaries of sexual and apomictic plants of *Brachiaria brizantha*. *Plant Molecular Biology, 53*, 745–757.

Rodrigues, J. C. M., Tucker, M. R., Johnson, S. D., Hrmova, M., & Koltunow, A. M. G. (2008). Sexual and apomictic seed formation in *Hieracium* requires the plant polycomb-group gene *FERTILIZATION INDEPENDENT ENDOSPERM*. *Plant Cell, 20*, 2372–2386.

Rodrigues J. C. M., Okada T., Johnson S. D. and Koltunow A. M. (2010) A MULTICOPY SUPPRESSOR OF IRA1 (MSI1) homologue is not associated with the switch to autonomous seed development in apomictic (asexual) *Hieracium* plants. *Plant Science*, (in press).

Schallau, A., Arzenton, F., Johnston, A. J., Hahnel, U., Koszegi, D., & Blattner, F. R., et al. (2010). Identification and genetic analysis of the APOSPORY locus in *Hypericum perforatum* L. *Plant Journal, 62*, 773–784.

Schiefthaler, U., Balasubramanian, S., Sieber, P., Chevalier, D., Wisman, E., & Schneitz, K. (1999). Molecular analysis of *NOZZLE*, a gene involved in pattern formation and early sporogenesis during sex organ development in *Arabidopsis thaliana*. *Proceedings of the National Academy of Sciences of the United States of America, 96*, 11664–11669.

Schmidt, E. D. L., Guzzo, F., Toonen, M. A. J., & De Vries, S. C. (1997). A leucine-rich repeat containing receptor-like kinase marks somatic plant cells competent to form embryos. *Development, 124*, 2049–2062.

Sharbel, T. F., Voigt, M. L., Corral, J. M., Galla, G., Kumlehn, J., & Klukas, C., et al. (2010). Apomictic and sexual ovules of *Boechera* display heterochronic global gene expression patterns. *Plant Cell, 22*, 655–671.

Sharbel, T. F., Voigt, M. L., Corral, J. M., Thiel, T., Varshney, A., & Kumlehn, J., et al. (2009). Molecular signatures of apomictic and sexual ovules in the *Boechera holboellii* complex. *Plant Journal, 58*, 870–882.

Sheridan, W. F., Avalkina, N. A., Shamrov, I. I., Batygina, T. B., & Golubovskaya, I. N. (1996). The *mac1* gene: Controlling the commitment to the meiotic pathway in maize. *Genetics, 142*, 1009–1020.

Siddiqi, I., Ganesh, G., Grossniklaus, U., & Subbiah, V. (2000). The *dyad* gene is required for progression through female meiosis in *Arabidopsis*. *Development, 127*, 197–207.

Sieber, P., Gheyselinck, J., Gross-Hardt, R., Laux, T., Grossniklaus, U., & Schneitz, K. (2004). Pattern formation during early ovule development in *Arabidopsis thaliana*. *Developmental Biology, 273*, 321–334.

Skinner, D. J., Hill, T. A., & Gasser, C. S. (2004). Regulation of ovule development. *Plant Cell, 16*, S32–S45.

Slotkin, R. K., Vaughn, M., Borges, F., Tanurdzic, M., Becker, J. D., & Feijo, J. A., et al. (2009). Epigenetic reprogramming and small RNA silencing of transposable elements in pollen. *Cell, 136*, 461–472.

Snyder, L. A., Hernandez, A. R., & Warmke, H. E. (1955). The mechanism of apomixis in *Pennisetum ciliare*. *Botanical Gazette, 116*, 209–221.

Steffen, J. G., Kang, I. H., Portereiko, M. F., Lloyd, A., & Drews, G. N. (2008). AGL61 interacts with AGL80 and is required for central cell development in *Arabidopsis*. *Plant Physiology, 148*, 259–268.

Stone, S. L., Braybrook, S. A., Paula, S. L., Kwong, L. W., Meuser, J., & Pelletier, J., et al. (2008). *Arabidopsis* LEAFY COTYLEDON2 induces maturation traits and auxin activity: Implications for somatic embryogenesis. *Proceedings of the National Academy of Sciences of the United States of America, 105*, 3151–3156.

Su, Y. H., Zhao, X. Y., Liu, Y. B., Zhang, C. L., O'Neill, S. D., & Zhang, X. S. (2009). Auxin-induced WUS expression is essential for embryonic stem cell renewal during somatic embryogenesis in *Arabidopsis*. *Plant Journal, 59*, 448–460.

Tucker, M. R., Araujo, A. C. G., Paech, N. A., Hecht, V., Schmidt, E. D. L., & Rossell, J. B., et al. (2003). Sexual and apomictic reproduction in *Hieracium* subgenus *Pilosella* are closely interrelated developmental pathways. *Plant Cell, 15*, 1524–1537.

Vielle, J. P., Burson, B. L., Bashaw, E. C., & Hussey, M. A. (1995). Early fertilization events in the sexual and aposporous egg apparatus of *Pennisetum ciliare* (L.) Link. *Plant Journal, 8*, 309–316.

Vielle-Calzada, J. P., Baskar, R., & Grossniklaus, U. (2000). Delayed activation of the paternal genome during seed development. *Nature, 404*, 91–94.

Weigel, D. (1995). The APETALA2 domain is related to a novel type of DNA binding domain. *Plant Cell, 7*, 388–389.

Weijers, D., & Jurgens, G. (2005). Auxin and embryo axis formation: The end's in sight? *Current Opinion in Plant Biology, 8*, 32–37.

Wen, X. S., Ye, X. L., Li, Y. Q., Chen, Z. L., & Xu, S. X. (1998). Embryological studies on apomixis in *Pennisetum squamulatum*. *Acta Botanica Sinica, 40*, 598–604.

Wuest, S. E., Vijverberg, K., Schmidt, A., Weiss, M., Gheyselinck, J., & Lohr, M., et al. (2010). *Arabidopsis* female gametophyte gene expression map reveals similarities between plant and animal gametes. *Current Biology, 20*, 506–512.

Yadegari, R., & Drews, G. N. (2004). Female gametophyte development. *Plant Cell, 16*, S133–S141.

Yamada-Akiyama, H., Takahara, M., Kikuchi, S., Takamiza, T., Nakagawa, H., & Sugita, S. i., et al. (2009). Analysis of expressed sequence tags in apomictic guineagrass (*Panicum maximum*). *Journal of Plant Physiology, 166*, 750–761.

Yang, W. C., Ye, D., Xu, J., & Sundaresan, V. (1999). The *SPOROCYTELESS* gene of *Arabidopsis* is required for initiation of sporogenesis and encodes a novel nuclear protein. *Genes & Development, 13*, 2108–2117.

Zeng Y. J. (2009). *Identification and characterization of apospory candidate genes in Pennisetum and Cenchrus*. 1–162. Ph.D. Thesis, The University of Georgia. (GENERIC)

Zhao, X. A., de Palma, J., Oane, R., Gamuyao, R., Luo, M., & Chaudhury, A., et al. (2008). OsTDL1A binds to the LRR domain of rice receptor kinase MSP1, and is required to limit sporocyte numbers. *Plant Journal, 54*, 375–387.

Zheng, Y. M., Ren, N., Wang, H., Stromberg, A. J., & Perry, S. E. (2009). Global identification of targets of the *Arabidopsis* MADS domain protein AGAMOUS-Like15. *Plant Cell, 21*, 2563–2577.

Zuo, J. R., Niu, Q. W., Frugis, G., & Chua, N. H. (2002). The *WUSCHEL* gene promotes vegetative-to-embryonic transition in *Arabidopsis*. *Plant Journal, 30*, 349–359.

Germplasm collection, storage, and conservation

17

Florent Engelmann
IRD, UMR DIADE Montpellier, France

TABLE OF CONTENTS

Introduction

Strategies for conserving plant biodiversity

Two basic conservation strategies, *in situ* and *ex situ*, each composed of various techniques, are employed for conservation of plant biodiversity. The Convention on Biological Diversity provides the following definitions for these categories (UNCED, 1992). *Ex situ* conservation means the conservation of components of biological diversity outside their natural habitat. *In situ* conservation means the conservation of ecosystems and natural habitats and the maintenance and recovery of viable populations of species and, in the case of domesticated or cultivated species, in the surroundings where they have developed their distinctive properties. *Ex situ* conservation is appropriate for conserving crops and their wild relatives, whereas *in situ* conservation is especially appropriate for wild species and for landrace material on farms.

Until recently, most conservation efforts, apart from work on forest genetic resources, have concentrated on *ex situ* conservation, particularly seed genebanks. In the 1950s and 1960s, major advances in plant breeding brought about the Green Revolution, which resulted in the widescale adoption of high-yielding varieties and genetically uniform cultivars of staple crops, particularly wheat and rice. Consequently, global concern about the loss of genetic diversity in these crops increased, as farmers abandoned their locally adapted landraces and traditional varieties and replaced them with improved, yet genetically uniform, modern ones. In response to this concern, the International Agricultural Research Centers (IARC) of the Consultative Group on International Agricultural Research (CGIAR) started to assemble germplasm collections of the major crop species within their respective mandates. It is in this context that the International Board for Plant Genetic Resources (IBPGR, today Bioversity International) was established in 1974 to coordinate the global effort to systematically collect and conserve the world's threatened plant genetic diversity (Engelmann and Engels, 2002).

Today, as a result of this global effort, there are over 1750 genebanks worldwide, about 130 of which hold more than 10,000 accessions each. It is estimated that around 7.4 million accessions are maintained *ex situ* globally (FAO, 2009). It should be mentioned that these collecting and conservation activities focused largely on the major food crops, including cereals and some legumes, that is, species that can be conserved easily as seed. This has resulted in overrepresentation of those species in the world's major genebanks, as well as in the fact that conservation strategies and concepts are biased toward such material. It is only more recently that the establishment of field genebanks, allowing the conservation of species for which seed conservation is not appropriate or impossible, as well as the development of new storage technologies, including *in vitro* conservation and cryopreservation, was given due attention by the international community (Engelmann and Engels, 2002).

Until now, most activities on *ex situ* conservation of plant biodiversity have focused on crop species. However, conservation of wild, rare, and endangered plant species has also become an issue of concern. Indeed, as highlighted by Sarasan et al. (2006), the world's biodiversity is declining at an unprecedented rate. During 1996 to 2004, a total of 8321 plant species have been added to the Red List of Threatened

DOI: 10.1016/B978-0-12-381466-1.00017-1

Species (IUCN, 2004), and the number of plants recorded as critically endangered has increased by 60%. For wild species, the traditional conservation approach is *in situ* conservation. However, it is now recognized that *ex situ* techniques can be efficiently used to complement *in situ* methods, and they may represent the only option for conserving certain highly endangered and rare species (Ramsay et al., 2000). It is therefore of paramount importance to develop techniques ensuring optimal storage and rapid multiplication of such species. Botanical gardens play a very important role in *ex situ* conservation of plant biodiversity. UNEP (UNEP, United Nations Environment Programme, 1995) estimated that botanical gardens, of which there are over 2500 around the world, conserve over 80,000 species (one-third of the world's flowering plants), among which Botanic Gardens Conservation International identified over 15,000 threatened species (http://www.bgci.org/ourwork/1977/). Botanical gardens and agricultural genebanks should be seen as playing a complementary role for the conservation of plant biodiversity (Engels and Engelmann, 1998).

Ex situ conservation technologies

Many of the world's major food plants produce seeds that undergo maturation drying, and are thus tolerant to extensive desiccation and can be stored dry at low temperatures. Such seeds are termed orthodox (Roberts, 1973). Storage of orthodox seeds is the most widely practiced method of *ex situ* conservation of plant genetic resources, since 90% of the 7.3 million accessions stored in genebanks are maintained as seed.

In contrast to orthodox seeds, a considerable number of species, predominantly tropical or subtropical in origin, such as coconut, cacao, and many forest and fruit tree species, produce seeds that do not undergo maturation drying and are shed at relatively high moisture content (Chin, 1988). Such seeds are unable to withstand desiccation and are often sensitive to chilling. Therefore, they cannot be maintained under the conventional seed storage conditions previously described, such as storage at low moisture content and low temperature. Seeds of this type are called recalcitrant and have to be kept in moist, relatively warm conditions to maintain viability (Roberts, 1973; Chin and Roberts, 1980). Even when recalcitrant seeds are stored in an optimal manner, their life span is limited to weeks, and occasionally months. Of more than 7000 species for which information on seed storage behavior has been published (Hong et al., 1996), approximately 3% are recorded as recalcitrant, and an additional 4% as possibly recalcitrant.

More recent investigations have identified species exhibiting intermediate storage behavior. While such seeds can tolerate desiccation to fairly low moisture contents, once dried, they become particularly susceptible to injury caused by low temperatures (Ellis et al., 1990, 1991). Even though a continuum in desiccation sensitivity is observed within the intermediate seed storage category, from highly desiccation sensitive to relatively tolerant (Berjak and Pammenter, 1994), the storage life of intermediate seeds can be prolonged by further drying, but it remains impossible to achieve the long-term conservation of orthodox seeds. About 1% of the aforementioned 7000 species studied and included in the Compendium on Seed Storage Behaviour are reported as producing intermediate seeds, and another 1% have been characterized as possibly intermediate (Hong et al., 1996). Included in this category are some economically important species such as coffee, citrus, rubber, oil palm, and many tropical forest tree species.

It should be noted that the percentages of intermediate and recalcitrant seed-producing species previously cited are likely to be largely underestimated (Engelmann and Engels, 2002). These figures are based on scientific and technical publications, which, by default, concern mainly temperate species. In addition, it can be expected that a large proportion of the species for which no information is available, which are predominantly of tropical or subtropical origin, exhibit recalcitrant, or to a lesser extent, intermediate seed storage behavior. As an example, it has been estimated that more than 70% of tree species in humid tropical forest ecosystems have recalcitrant seeds (Ouédraogo et al., 1999).

There are other species for which conservation as seed is problematic. First, there are those that do not produce seeds at all and, consequently, are propagated vegetatively (i.e., banana and plantain; *Musa* spp.). Secondly, there are crops such as potato (*Solanum tuberosum*), other root and tuber crops such as yams (*Dioscorea* spp.), cassava (*Manihot esculenta*) and sweet potato (*Ipomoea batatas*), and sugarcane (*Saccharum* spp.) that have either some sterile genotypes and/or some that produce orthodox seed. However, these seeds are highly heterozygous and, therefore, of limited utility for the conservation of particular genotypes. These crops are usually propagated vegetatively to maintain genotypes as clones (Simmonds, 1982).

Traditionally, the field genebank has been the *ex situ* storage method of choice for the aforementioned problem materials (Engelmann and Engels, 2002). According to the *First Report on the State of the World's Plant Genetic Resources for Food and Agriculture* (FAO, 1996), around 527,000 accessions were maintained in field genebanks at that time. There are no updated data in the second report (FAO, 2009), but this number can only have increased during this period. In some ways, this method offers a satisfactory approach to conservation. The genetic resources under conservation can be readily accessed and observed, thus permitting detailed evaluation. However, there are certain drawbacks that limit its efficiency and threaten its security (Withers and Engels, 1990; Engelmann, 1997a). The genetic resources are exposed to pests, diseases, and other natural hazards such as drought, weather damage, human error, and vandalism. They are also not in a condition that is readily conducive to germplasm exchange because of the great risks of disease transfer through the exchange of vegetative material. Field genebanks are costly to maintain and are thus prone to economic decisions that may limit the level of replication of accessions, the quality of maintenance, and even their very survival in times of economic stringency. Even under the best circumstances, field genebanks require considerable inputs in the form of land (often needing multiple sites to allow for rotation), labor, management, and materials, and as well their capacity to ensure the maintenance of much diversity is limited.

There are other categories of plant material that require the availability of improved storage technologies. The development of biotechnology has led to the production of a new category of germplasm including clones obtained from elite genotypes, cell lines with special attributes, and genetically transformed material (Engelmann, 1991). This new germplasm is often of high added value and very difficult to produce. The development of efficient techniques to ensure its safe conservation is therefore of great importance. Finally, it is of paramount importance to have efficient technologies to ensure the long-term conservation of rare and endangered plant species, the number of which is increasing rapidly.

In the light of the difficulties outlined earlier, efforts have been made to improve the quality and security of conservation offered by field genebanks, and to understand and overcome seed recalcitrance to make seed storage more widely available. It has also been recognized that alternative approaches to genetic conservation were needed for these problem materials and, since the early 1970s, attention has turned to the possibilities offered by biotechnology, specifically *in vitro* or tissue cultures.

Tissue-culture techniques are of great interest for the collecting, multiplication, and storage of plant germplasm (Engelmann, 1991; Bunn et al., 2007). Tissue-culture systems allow propagating of plant material with high multiplication rates in an aseptic environment. Virus-free plants can be obtained through meristem culture in combination with thermotherapy, thus ensuring the production of disease-free stocks and simplifying quarantine procedures for the international exchange of germplasm. The miniaturization of explants reduces space requirements and consequently labor costs for the maintenance of germplasm collections. *In vitro* propagation protocols have been established for several thousands of plant species (George, 1996).

Different *in vitro* conservation methods are employed, depending on the storage duration requested. For short- and medium-term storage, the aim is to reduce growth and to increase the intervals between subcultures. For long-term storage, cryopreservation (i.e., storage at ultra-low temperature, usually with liquid nitrogen at $-196^{\circ}C$), is the only current method. At this temperature, all cellular divisions and metabolic processes are stopped. The plant material can thus be stored without alteration or modification for a theoretically unlimited period of time. Moreover, cultures are stored in a small volume, protected from contamination, requiring very limited maintenance. *In vitro* collecting, slow growth, and cryopreservation techniques are described and analyzed in the following sections.

Applications of Biotechnologies for Conservation

In vitro collecting

Collectors are faced with various problems when collecting germplasm of recalcitrant seed and vegetatively propagated plant species (Engelmann, 2009). Collecting missions often require traveling for relatively long periods in remote areas. It is thus necessary to keep the material collected in a good state for some days and/or weeks before it can be placed in optimal growth or storage conditions. Because of this, there are great risks that recalcitrant seeds will either germinate or deteriorate before they are brought back to the genebank (Allen and Lass, 1983). In addition, many recalcitrant seeds have a sheer weight and bulk, which can be a problem in terms of volume of material to handle and its additional cost, if an adequate sample of the population is to be collected. With vegetatively propagated species, the material collected will consist of stakes, pieces of budwood, tubers, corms, or suckers. Most of these explants will not be adapted to survival once excised from the parent plant, and they also present health risks due to their vegetative nature and contamination with soil-borne pathogens (Withers, 1987). Difficulties can also be encountered when collecting germplasm of orthodox seed-producing species. Even with careful planning during the collecting mission, there may be little or no seed available for all or part of the germplasm to be collected, or seeds might not be at the optimal developmental stage, shed from the plant, or eaten by grazing animals (Guarino et al., 1995). The problems described previously can be overcome if it is realized that the seed is not the only material that can be collected: zygotic embryos or vegetative tissues such as pieces of budwood, shoots, or apices can be sampled, transported, and grown successfully if placed under adequate conditions.

Following an expert meeting organized by IBPGR in 1984 and sponsorship of various research programs, *in vitro* collecting techniques have been developed for different materials including embryos of coconut, cacao, avocado, *Citrus*, vegetative tissues of cacao, *Musa*, coffee, *Prunus*, grape, cotton, and several forage grasses (Pence et al., 2002)). The critical points to consider for the development of *in vitro* collecting techniques have been synthesized and analyzed by Withers (1995). The techniques developed are very simple and flexible, as illustrated next with coconut and cacao.

With coconut, the *in vitro* field-collecting technique developed by Assy-Bah et al. (1987) consists of extracting a plug of endosperm containing the embryo from the nut with a cork borer. After surface sterilization with calcium hypochlorite or commercial bleach, the embryos are dissected on the spot under the shelter of a wooden box and inoculated onto semi-solid medium, or the endosperm plugs are transported to the laboratory where the embryos are dissected and inoculated onto semi-solid medium in aseptic conditions. This technique is very efficient. After approximately 6 to 9 months in culture in the laboratory under standard conditions, an average of 75% of the embryos collected developed into plantlets, which could be successfully transferred to the nursery and then to the field. In addition, embryos inoculated *in vitro* in the field could be kept in the open for two months before being grown in the laboratory without incidence for further development. This protocol is now used on a large scale for the international exchange of coconut germplasm between Côte d'Ivoire, the Philippines, Sri Lanka, and Papua New Guinea in the framework of a project funded by the Global Crop Diversity Trust (Rome, Italy; http://www.croptrust.org/main/).

With cacao, an *in vitro* collecting method was developed for budwood (Yidana et al., 1987). Considering that absolute sterility would be difficult to achieve in the field and would not necessarily be essential for robust, woody material, the aim was to place the samples collected in conditions that would allow suppressing or delaying deterioration. Stem nodal cuttings were disinfected with boiled water that contained drinking water sterilizing tablets and fungicides, then they were inoculated onto semi-solid medium supplemented with fungicide and antibiotics. Explants could be maintained in a relatively clean (although not necessarily completely sterile) condition for up to 6 weeks.

These examples illustrate the flexibility of *in vitro* collecting. There is no one formula to be followed, nor should there be. The approach should be based upon prior knowledge of the requirements of the species and explant in question, combined with the collective experience gained with diverse species in different collecting environments. As in any germplasm transfer operation, particular attention should be given to phytosanitary considerations. *In vitro* collected explants should be treated with the same care and observance of regulations as any other type of collected material (Withers and Engelmann, 1998).

Slow growth storage

Classical techniques

Standard culture conditions can be used for medium-term storage with species which have a naturally slow growing habit. However, for most species, growth reduction is achieved by modifying the environmental conditions and/or the culture medium. The most widely applied technique is temperature reduction, which can be combined with a decrease in light intensity or culture in the dark. Temperatures ranging from 0 to 5°C are employed with cold-tolerant species. Strawberry (*Fragaria* × *ananassa*) plantlets have been stored at 4°C in the dark and kept viable for 6 years with the regular addition of a few drops of liquid medium (Mullin and Schlegel, 1976). Apple (*Malus domestica*) and *Prunus* shoots survived 52 weeks at 2°C (Druart, 1985). Tropical species are often cold-sensitive and have to be stored at higher temperatures. *Musa in vitro* plantlets can be stored at 15°C without transfer for up to 15 months (Banerjee and De Langhe, 1985). Other tropical species such as cassava are much more cold-sensitive, and cassava shoot cultures have to be conserved at temperatures higher than 20°C (Roca et al., 1984).

Various modifications can be made to the culture medium to reduce growth. Embryogenic cultures of carrot could be conserved on a medium without sucrose for two years, and reproliferate if a sucrose solution was supplied (Jones, 1974). Kartha et al. (1981) conserved coffee plantlets on a medium devoid of sugar and with only half of the mineral elements of the standard medium. The addition of osmotic growth inhibitors (e.g., mannitol) or hormonal growth inhibitors (e.g., absicic acid) is also employed successfully to reduce growth (Ng and Ng, 1991). The type of explant, as well as its physiological state when entering storage, can influence the duration of storage achieved. Roxas et al. (1985) indicated that with chrysanthemum, nodal segments showed higher survival than apical buds. The presence of a root system generally increases the storage capacities, as observed by Kartha et al. (1981) with coffee plantlets. Microtubers can be successfully employed as storage propagules, as demonstrated with potato (Kwiatowski et al., 1988; Gopal et al., 2005). Preconditioning the explants by exposing them briefly to temperature and light conditions intermediate between standard and storage conditions was favorable for *Nephrolepis* and *Cordyline* cultures (Hvoslef-Heide, 1992). The type of culture vessel, its volume, and the volume as well as the type of closure of the culture vessel can greatly influence the survival of stored cultures (Engelmann, 1991). Roca et al. (1984) indicated that cassava shoot cultures could be stored for longer periods in better condition by increasing the size of the storage containers. The use of heat-sealable polypropylene bags instead of glass test tubes or plastic boxes was beneficial for the storage of several strawberry varieties (Reed, 1991).

At the end of a storage period, cultures are transferred onto fresh medium and usually placed for a short period in optimal conditions to stimulate regrowth before entering the next storage cycle.

Alternative techniques

Various alternative techniques have been tried for medium-term storage of plant *in vitro* cultures (Engelmann, 1997a; Withers and Engelmann, 1998). They include modification of the gaseous environment of cultures, and desiccation and/or encapsulation of explants. Growth reduction can be achieved by reducing the quantity of oxygen available to the cultures. The simplest method consisted of covering the explants with paraffin or mineral oil or liquid medium. As an example, shoot cultures of several ginger species could be conserved for up to two years under mineral oil with high viability (Dekkers et al., 1991). Reduction of the quantity of oxygen could also be achieved by decreasing the atmospheric pressure of the culture chamber or by using a controlled atmosphere. Desiccation of cultures as a means of achieving medium-term conservation was also employed. Alfalfa somatic embryos desiccated to 10–15% moisture content could be stored for one year, showing only a 5% decrease in their conversion rate at the end of the storage period (Senaratna et al., 1990). Finally, encapsulated explants can be successfully conserved for extended durations. Encapsulated *Valeriana wallichii* shoot tips could be conserved for over 6 months at 4–6°C (Mathur et al., 1989), grape shoot tips were stored for 9 months at 23°C (Guanino et al., 2009), and encapsulated date palm somatic embryos were conserved for 6 months at 4°C (Hamed et al., 2003).

Current development and use of *in vitro* slow growth storage

In vitro slow growth storage techniques are routinely used for medium-term conservation of numerous species, both from temperate and tropical origin, including crops, forest trees, endangered species, and medicinal plants (Ashmore, 1997; Razdan and Cocking, 1997; Engelmann, 1999; Lambardi et al., 2006; Sarasan et al., 2006; Keshavachandran et al., 2007). In 1996, the FAO's *Report on the State of the World's Plant Genetic Resources for Food and Agriculture* (FAO, 1996) estimated that around 38,000 accessions were conserved *in vitro*. Again, there are no updated data available in the new FAO report (FAO, 2009), but this number has certainly increased significantly since that time. However, if *in vitro* conservation appears as a simple and practical option for long-term conservation of numerous species, and has obvious wide medium-term applications, its implementation still needs customizing to any new material, continuous inputs are required, and long-term questions remain regarding the genetic stability of the stored material. Moreover, it is not always possible to apply one single protocol for conserving genetically diverse material. As an example, a storage experiment performed with an *in vitro* collection of African coffee germplasm including 21 diversity groups revealed a large variability in the response of the diversity groups to the storage conditions (Dussert et al., 1997). Some groups showed high genetic erosion during storage, whereas others did not show any erosion.

During the workshop Consultation on the Management of Field and *In vitro* Genebanks held at CIAT, Cali, Colombia on January 15–20, 1996, a panel of experts of *in vitro* conservation identified the gaps in the development of techniques as well as limitations in basic scientific knowledge, which are listed thereafter (Ashmore, 1997). Further information is needed on the nature, as well as the underlying causes and consequences of somaclonal variation. More studies are needed on genetic stability after prolonged storage in culture to establish the relative safety of these forms of storage when compared with other methods such as storage in field genebanks. There is a need for the development and application of characterization systems, including molecular genetic markers for initial identification, as well as monitoring of the genetic stability of stored accessions. Reproducible, simple, and more generally applicable techniques for slow growth storage are needed. *In vitro* conservation techniques need to be developed for additional species, particularly species with recalcitrant seed and minor crop species. There is a need to both test and optimize available *in vitro* conservation technology on a range of species and genotypes in genebank facilities. Technical guidelines have been published recently (Reed et al., 2004) that provide guidance to researchers and genebank managers for the establishment and management of *in vitro* germplasm collections.

Cryopreservation

Cryopreservation is the only technique currently available to ensure the safe and cost-efficient long-term conservation of the germplasm of problem species. In this section, we briefly describe the various cryopreservation techniques available, summarize the achievements made and problems faced with vegetatively propagated and recalcitrant species, and present the current and future utilizations of cryopreservation for plant biodiversity conservation.

Cryopreservation techniques

Some materials, such as orthodox seeds or dormant buds, display natural dehydration processes and can be cryopreserved without any pre-treatment. However, most biological materials employed in cryopreservation (cell suspensions, calluses, shoot tips, embryos) contain high amounts of cellular water, and are thus extremely sensitive to freezing injury, since most of them are not inherently tolerant to freezing. Thus, cells have to be dehydrated artificially to protect them from the damage caused by the crystallization of intracellular water into ice (Mazur, 1984). The techniques employed and the physical mechanisms upon which they are based are different for classical and new cryopreservation techniques (Withers and Engelmann, 1997). In classical techniques, dehydration of samples takes place both before and during freezing (freeze-induced dehydration), whereas in new techniques dehydration takes place only before freezing. In optimal conditions, all freezable water is removed from the cells during dehydration and the highly concentrated internal solutes vitrify upon immersion in liquid nitrogen. Vitrification can be defined as the transition of water directly from the liquid phase into an amorphous phase or glass, while avoiding the formation of crystalline ice (Fahy et al., 1984).

Classical cryopreservation techniques

Classical cryopreservation techniques involve slow cooling down to a defined pre-freezing temperature, followed by rapid immersion in liquid nitrogen. With temperature reduction during slow cooling, the cells and the external medium initially supercool, followed by ice formation in the medium (Mazur, 1984). The cell membrane acts as a physical barrier and prevents the ice from seeding the cell interior, and the cells remain unfrozen but supercooled. As the temperature is further decreased, an increasing amount of the extracellular solution is converted into ice, thus resulting in the concentration of intracellular solutes. Since cells remain supercooled and their aqueous vapor pressure exceeds that of the frozen external compartment, cells equilibrate by loss of water to external ice. Depending upon the rate of cooling and the pre-freezing temperature, different amounts of water will leave the cell before the intracellular contents solidify. In optimal conditions, most or all intracellular freezable water is removed, thus reducing or avoiding detrimental intracellular ice formation upon subsequent immersion of the specimen in liquid nitrogen, during which vitrification of internal solutes occurs. However, freeze-induced dehydration that is too intense can incur different damaging events due to concentration of intracellular salts and changes in the cell membrane (Meryman et al., 1977). Re-warming should be as rapid as possible to avoid the phenomenon of recrystallization in which ice melts and reforms at a thermodynamically favorable, larger, and more damaging crystal size (Mazur, 1984).

Classical freezing procedures include the following successive steps: pre-growth of samples, cryoprotection, slow cooling (0.5–2.0°C/min) to a determined pre-freezing temperature (usually around −40°C), rapid immersion of samples in liquid nitrogen, storage, rapid thawing, and recovery. Classical techniques are generally operationally complex, because they require the use of sophisticated and expensive programmable freezers. In some cases, their use can be avoided by performing the slow freezing step with a domestic or laboratory freezer (Kartha and Engelmann, 1994). Classical cryopreservation techniques have been successfully applied to undifferentiated culture systems such as cell suspensions and calluses (Kartha and Engelmann, 1994; Withers and Engelmann, 1998) and apices of cold-tolerant species (Reed and Uchendu, 2008).

New cryopreservation techniques

In vitrification-based procedures, cell dehydration is performed prior to freezing by exposure of samples to concentrated cryoprotective media and/or air desiccation. This is followed by rapid cooling. As a result, all factors that affect intracellular ice formation are avoided. Glass transitions (changes in the structural conformation of the glass) during cooling and re-warming have been recorded with various materials using thermal analysis (Sakai et al., 1990; Dereuddre et al., 1991; Niino et al., 1992). Vitrification-based procedures offer practical advantages in comparison to classical freezing techniques. Like ultra-rapid freezing, they are more appropriate for complex organs (shoot tips, embryos), which contain a variety of cell types, each with unique requirements under conditions of freeze-induced dehydration. By precluding ice formation in the system, vitrification-based procedures are operationally less complex than classical ones (e.g., they do not require the use of controlled freezers) and have greater potential for broad applicability, requiring only minor modifications for different cell types (Engelmann, 1997a). A common feature of all of these new protocols is that the critical step to achieve survival is the dehydration step, not the freezing step, as in classical protocols (Engelmann, 2000). Therefore, if samples to be frozen can be dehydrated down to sufficiently low water content with little or no decrease in survival in comparison to non-dehydrated controls, limited or no further drop in survival is generally observed after cryopreservation (Engelmann, 2009).

Seven different vitrification-based procedures can be identified: (1) encapsulation-dehydration; (2) vitrification; (3) encapsulation-vitrification; (4) dehydration; (5) pre-growth; (6) pre-growth-dehydration; and (7) droplet-vitrification. The encapsulation-dehydration procedure is based on the technology developed for the production of artificial seeds. Explants are encapsulated in alginate beads, pre-grown in liquid medium enriched with sucrose for 1 to 7 days, partially desiccated in the air current of a laminar airflow cabinet or with silica gel to a water content around 20% (fresh weight basis), and then frozen rapidly. Survival is high and growth recovery of cryopreserved samples is generally rapid and direct without callus formation. This technique has been applied to apices of numerous species from temperate and tropical origins, as well as to cell suspensions and somatic embryos of several species (Gonzalez-Arnao and Engelmann, 2006; Engelmann et al., 2008). A simplification of this technique has been recently proposed by Bonnart and Volk (2010), which involves encapsulation of samples in alginate medium containing 2 M glycerol + 0.5 M sucrose, immediately followed by air-dehydration.

Vitrification involves treatment of samples with cryoprotective substances, such as exposure of samples to a loading solution with intermediate concentration (2 M glycerol + 0.4 M sucrose; Matsumoto et al., 1994), then dehydration with highly concentrated vitrification solutions, rapid freezing and thawing, removal of cryoprotectants in an unloading solution (containing 1.2 M sucrose), and recovery. The most common vitrification solutions are the so-called Plant Vitrification Solutions PVS2 (Sakai et al., 1990) and PVS3 (Nishizawa et al., 1993) developed by Professor Sakai's group in Japan. This procedure has been developed for apices, cell suspensions, and somatic embryos of numerous different species (Sakai and Engelmann, 2007; Sakai et al., 2008). Encapsulation-vitrification is a combination of encapsulation-dehydration and vitrification procedures, where samples are encapsulated in alginate beads, and then subjected to freezing by vitrification. It has been applied to apices of an increasing number of species (Sakai and Engelmann, 2007; Sakai et al., 2008).

Dehydration is the simplest procedure because it consists of dehydrating explants, then freezing them rapidly by direct immersion in liquid nitrogen. This technique is mainly used with zygotic embryos or embryonic axes extracted from seeds. It has been applied to embryos of a large number of recalcitrant and intermediate species (Engelmann, 1997b). Desiccation is usually performed in a laminar airflow cabinet, but more precise and reproducible dehydration conditions are achieved by using a flow of sterile compressed air or silica gel. Ultra-rapid drying in a stream of compressed dry air (a process called flash drying developed by P. Berjak's group in South Africa) allows freezing samples with relatively high water content, thus reducing desiccation injury (Berjak et al., 1989). Optimal survival is generally obtained when samples are frozen with a water content comprised between 10 and 20% (fresh weight basis).

The pre-growth technique consists of cultivating samples in the presence of cryoprotectants, then freezing them rapidly by direct immersion in liquid nitrogen. The pre-growth technique has been developed for *Musa* meristematic cultures (Panis et al., 2002). In a pre-growth dehydration procedure, explants are pre-grown in the presence of cryoprotectants, dehydrated under the laminar airflow cabinet or with silica gel, and then frozen rapidly. This method has been applied notably to asparagus stem segments (Uragami et al., 1990), oil palm somatic embryos (Dumet et al., 1993), coconut zygotic embryos (Assy-Bah and Engelmann, 1992), and more recently to coriander somatic embryos (Popova et al., 2010).

Droplet-vitrification is the latest technique (Sakai and Engelmann, 2007). The number of species to which it has been successfully applied is steadily increasing. Apices are pre-treated with a loading solution, then treated with a vitrification solution, placed on an aluminum foil in minute droplets of vitrification solution, and frozen rapidly in liquid nitrogen. Alternative loading and vitrification solutions have been developed that allow successful application of this technique to plant species sensitive to "classical" loading and vitrification solutions (Kim et al., 2009a,b).

Cryopreservation of vegetatively propagated and recalcitrant seed species

Vegetatively propagated species

A number of publications provide lists of species that have been successfully cryopreserved (Engelmann, 1997a,b; Engelmann and Takagi, 2000; Reed, 2008). For vegetatively propagated species, cryopreservation has a wide applicability. It is used in a large number of genotypes/varieties of many species, including roots, and tubers, fruit and forest trees, ornamentals, and plantation crops, both from temperate and tropical origins. With a few exceptions, vitrification-based protocols have been employed in these cases. It is also interesting to note that in many cases, different protocols can be employed for a given species, resulting in comparable results. Survival is generally high to very high and up to 100% survival could be achieved in some cases, for example, *Allium*, yam, and potato. Regeneration is rapid and direct, and callusing is observed only in cases where the technique is not optimized.

Different reasons can be mentioned to explain these positive results (Engelmann, 2009). The meristematic zone of apices, from which organized growth originates, is composed of a relatively homogenous population of small, actively dividing cells, with few vacuoles and a high nucleocytoplasmic ratio. These characteristics make them more susceptible to withstand desiccation than highly vacuolated and differentiated cells. As mentioned earlier, no ice formation takes place in vitrification-based procedures, thus avoiding the extensive damage caused by ice crystals formed during classical procedures. The whole meristem is generally preserved when vitrification-based techniques are employed, thus allowing direct, organized regrowth. By contrast, classical procedures often lead to destruction of large zones of the meristems, and callusing only or transitory callusing is often observed before organized regrowth starts. Other reasons for the good results obtained are linked with tissue-culture protocols (Engelmann, 2009). Many vegetatively propagated species successfully cryopreserved until now are cultivated crops, often of great commercial importance, for which cultural practices, including *in vitro* micropropagation, are well established. In addition, *in vitro* material is "synchronized" by the tissue culture and pregrowth procedures. As a result, relatively homogenous samples in terms of size, cellular composition, physiological state, and growth response are employed for freezing, thus increasing the chances of positive and uniform response to treatments. Finally, vitrification-based procedures use samples of a relatively large size (shoot tips of 0.5 to 2–3 mm) that can regrow directly without any difficulty.

Freezing techniques are now operational for large-scale experimentation in an increasing number of cases. In view of the wide range of efficient and operationally simple techniques available, any vegetatively propagated species should be amenable to cryopreservation, provided that the tissue culture protocol is sufficiently operational for this species.

Recalcitrant seed species

Some publications present extensive lists of plant species whose embryos and/or embryonic axes have been successfully cryopreserved (e.g., Kartha and Engelmann, 1994; Engelmann et al., 1995; Engelmann, 1997a,b; Reed, 2008; Pence, 2008). However, careful examination of the species mentioned in these papers reveals that only a limited number of truly recalcitrant seed species are included, because research in this area is recent and addressed by very few teams worldwide. However, recalcitrance is a dynamic concept that evolves with research on the biology of species and improvement in classical storage procedures. As a result, some species previously classified as recalcitrant have thus been moved to the intermediate or even sub-orthodox categories and stored using classical or new storage techniques (Engelmann, 2000).

In comparison to the results obtained with vegetatively propagated species, research is still at a very preliminary stage for recalcitrant seeds. The desiccation technique is mainly employed for freezing embryos and embryonic axes (Normah and Makeen, 2008). Survival is extremely variable, regeneration frequently restricted to callusing or incomplete development of plantlets, and the number of accessions tested per species generally very low. There are a number of reasons to explain the current limited development of cryopreservation for recalcitrant seed species (Engelmann, 2000). First, there is a huge number of species with recalcitrant or suspected recalcitrant seeds, and most of them are wild species. As a consequence, little or nothing is known about the biology and the seed storage behavior of many of these species. Where some information on seed storage behavior is available, tissue-culture protocols, including inoculation *in vitro*, germination, and growth of plantlets, propagation, and acclimatization — which are needed for regrowth of embryos and embryonic axes after freezing — are often non-existent or not fully operational. Seeds and embryos of recalcitrant species also display very important variations in moisture content and maturity stage between provenances, between and among seed lots, and between successive harvests, which make their cryopreservation difficult.

Seeds of many species are too large to be frozen directly, and embryos or embryonic axes are thus successfully employed for cryopreservation. However, embryos are often of very complex tissue composition, which displays differential sensitivity to desiccation and freezing, the root pole seeming more resistant than the shoot pole. In some species, embryos are extremely sensitive to desiccation and even minor reduction in their moisture content — down to levels much too high to obtain survival after freezing — leads to irreparable structural damage, as observed notably with cacao (Chandel et al., 1995). Finally, seeds of some species do not contain well-defined embryos.

There are various options to consider for improving storage of non-orthodox seeds. With some species, very precisely controlled desiccation (e.g., using saturated salt solutions) and cooling conditions may freeze whole seeds, as demonstrated recently with various coffee species (Dussert et al., 1997; Dussert and Engelmann, 2006). There is scope for technical improvements in the current cryopreservation protocols for embryos and embryonic axes. Pregrowth on media containing cryoprotective substances confers the tissue's increased tolerance to further desiccation and reduces the heterogeneity of the material. Flash drying, followed by ultra-rapid freezing, has also been very effective for cryopreservation of several species (Berjak et al., 1989; Wesley-Smith et al., 1992). Other cryopreservation techniques including pregrowth-desiccation, encapsulation-dehydration, and vitrification, which have been seldom employed with recalcitrant species, should be tried (Engelmann, 2000; Pence, 2008) selecting embryos at the right developmental stage is of critical importance for the success of any cryopreservation experiment (Engelmann et al., 1995). However, in these cases, basic protocols for disinfection, inoculation *in vitro*, germination of embryos or embryonic axes, plantlet development, and possibly limited propagation will have to be established prior to any cryopreservation experiment. With species for which freezing of whole embryos or embryonic axes have proven unsuccessful, it has been suggested to use alternative explants such as shoot apices of embryos (Varghese et al., 2009), adventitious buds, or somatic embryos induced from the embryonic tissues (Pence, 2008). This might be the only solution for species without well-defined embryos; however, this will require more sophisticated tissue-culture procedures to be developed and mastered.

Finally, analytical tools, which allow better understanding of the biological and physical processes that take place during cryopreservation, are very useful in establishing freezing protocols, especially for problem materials (Engelmann, 2004). Examples can be found with zygotic embryos of *Parkia speciosa*, a tropical tree species (Nadarajan et al., 2008) and several Australian Citrus species (Hamilton et al., 2009), where measuring the thermal events taking place during cooling and re-warming using differential scanning calorimetry (DSC) has been instrumental in establishing efficient cryopreservation protocols for these materials.

Large-scale utilization of cryopreservation for germplasm conservation

Even though its routine use is still limited, there is a growing number of genebanks and botanical gardens where cryopreservation is employed on a large scale for different types of materials, which are or are not tolerant to dehydration. With orthodox seed species, cryopreservation is used mainly for storing seeds with limited longevity and of rare or endangered species. The National Center for Genetic Resources Preservation (NCGRP, Fort Collins, Colorado) conserves 43,400 accessions over the vapors of liquid nitrogen (C. Walters, personal communication). The National Bureau for Plant Genetic Resources (NBPGR, New Delhi, India) stores 1200 accessions from 50 different species, consisting mainly of endangered medicinal plants (Mandal, 2000). This technique is also used in botanical gardens for storing rare and endangered species (Engelmann, 2010). Over 110 accessions of rare or threatened species are stored under cryopreservation at the Kings Park and Botanic Garden in Perth, Australia (Touchell and Dixon, 1994; see also http://www.bgpa.wa.gov.au/). In the United States, the Center for Conservation and Research of Endangered Wildlife (CREW) at the Cincinnati Zoo and Botanical Garden conserves several cryopreserved collections, including collections of seeds of regional endangered species (Pence, 1991), of endangered plant tissues, and of spores and tissues of Bryophytes and pteridophytes (Pence, 2008; see also http://www.cincinnatizoo.org/conservation/crew/crew-plant-research/cryobiobank%E2%84%A2).

It should be noted that recent publications on seed longevity studies showed that seeds of several orthodox species with limited seed longevity (such as *Brassica*) lost viability more rapidly than predicted using classical viability equations, and that the rate of this viability loss increased in line with increasing storage temperature (Walters et al., 2004; 2005). These results, as well as those obtained by Caswell and Kartha (2009), which showed no change in viability of pea and strawberry meristems stored in liquid nitrogen for 28 years, provide compelling evidence that ultra-cold (liquid nitrogen) storage can and does enhance longevity considerably over that at −20°C (Pritchard et al., 2009). Therefore, in the case of orthodox seed species with limited longevity, it might be recommended to systematically store a seed subsample at −196°C, in addition to storage at −20°C to ensure their long-term conservation. Cryopreservation is also applied to intermediate seeds that are tolerant to freezing. Cryopreserved collections of coffee seeds are being established in CATIE (Tropical Agricultural Research and Higher Education Center, Costa Rica) and in IRD Montpellier (France) using a protocol including controlled dehydration and freezing (Dussert and Engelmann, 2006). In France, in the framework of a national grape genetic resource conservation project, seeds of several hundreds of accessions are being cryopreserved after partial desiccation (Chatelet et al., 2009).

With dormant buds, the 2200 accessions of the U.S. apple germplasm field collection are duplicated under cryopreservation (Forsline et al., 1999), as is the case for the 420 accessions of the mulberry field collection maintained at the National Institute of Agrobiological Resources (NIAR, Yamagata, Japan; Niino, 1995). Dormant buds of over 440 European elm accessions are conserved in liquid nitrogen by Afocel (Nangis, France; Harvengt et al., 2004) and research is underway in France (IRD, Montpellier) and the United States (NCGRP, Fort Collins) on the development of a protocol for cryopreservation of grape dormant buds. Breeders routinely store pollen in liquid nitrogen in the framework of their improvement programs (Towill and Walters, 2000). Pollen, which is an interesting material for genetic resource conservation of various species, is stored by several institutes. In India, the NBPGR conserves cryopreserved pollen of 65 accessions belonging to different species (Mandal, 2000), and the Indian Institute for Horticultural Research (IIHR, Bangalore) conserves pollen of 600 accessions belonging to 40 species from 15 different families, some of which have been stored for over 15 years (Ganeshan and Rajashekaran, 2000). In the United States, the NCGRP conserves pollen of 13 pear cultivars and 24 *Pyrus* species (Reed et al., 2000). In China, pollen of over 700 accessions of traditional Chinese flower species is conserved under cryopreservation (Li et al., 2009).

Cryopreservation is also applied to biotechnology products. Over 1000 callus strains of species of pharmaceutical interest are stored at −196°C in the UK (Spencer, 1999), as well as several thousands of conifer embryogenic cell lines employed in large-scale clonal planting programs in Canada (Cyr, 2000). In France, cryopreservation is systematically employed for storing all of the new embryogenic cell lines of coffee and cacao produced by the Biotechnology Laboratory of the Nestlé Company (Florin et al., 1999). Embryogenic cultures of around 80 oil palm accessions have been cryopreserved and stored at IRD (Montpellier, France; Dumet, 1994).

Finally, cryopreservation is being applied in genebanks for long-term storage of genetic resources of vegetatively propagated species, using apices sampled from *in vitro* plantlets. Cryopreservation is the most advanced in potato genebanks;

over 1000 old potato varieties are cryostored in Germany at the Leibnitz Institute of Plant Genetics and Crop Plant Research (IPK; Keller et al., 2005, 2006) and over 200 accessions at the International Potato Center (CIP, Lima, Peru; Golmirzaie and Panta, 2000). Around 100 accessions of the *Pyrus* field collection are cryostored at the National Clonal Germplasm Repository (NCGR, Corvallis, Oregon), and it is duplicated at the NCGRP in Fort Collins, Colorado (Reed et al., 2000). In Korea, two cryopreserved collections of *Allium* have been established, which comprise a total of over 800 accessions (Kim et al., 2009). Finally, cryopreserved collections are under development for long-term storage of tropical plants: 630 banana accessions have been cryopreserved at the INIBAP International Transit Center (Panis et al., 2007) and 540 cassava accessions at the CIAT in Cali, Colombia (Gonzalez-Arnao et al., 2008).

The large-scale utilization of cryopreservation implies a scaling up in the amount of material to handle and to store, from one or a few genotypes in the laboratory to several tens, hundreds or even thousands, which requires the establishment of specific management procedures. For this aim, probabilistic tools have been developed recently to assist genebank curators in the establishment and management of cryopreserved germplasm collections (Dussert et al., 2003). A better understanding of the long-term benefits of cryopreservation and its further integration into general genebank management is also needed (Keller et al., 2008). Recommended approaches include comparative validation of methods between different laboratories, detailed comparisons of crop-based methods, economic analyses, and efficient integration strategies of cryobanks by genebanks, including safe duplication of cryopreserved resources for the limitation of risk of loss.

Cryopreservation imposes a series of stresses to the plant material, which can induce modifications in cryopreserved cultures and regenerated plants. It is thus necessary to verify that the genetic stability of the cryopreserved material is not altered before routinely using this technique for the long-term conservation of plant genetic resources. There is no report so far of modifications at the phenotypical, biochemical, chromosomal, or molecular level that could be attributed to cryopreservation (Engelmann, 1997a; Engelmann, 2004). Recent studies comparing the development of plants originating from control and cryopreserved material performed with several species including sugarcane (Gonzalez-Arnao, 1996), banana (Côte et al., 2000), potato (Mix-Wagner et al., 2003), oil palm (Konan et al., 2007), apple (Liu et al., 2008), oak (Sanchez et al., 2008), yam (Mandal et al., 2008), and gentian (Mikul et al., 2008) did not reveal any differences in the characters studied.

Studies performed on the cost of cryopreservation confirm its low utilization cost. Hummer and Reed (1999) indicated that, at the NCGR, the annual cost of one temperate fruit tree accession is $77 in the field, $23 under *in vitro* slow growth storage, and only $1 under cryopreservation, to which $50–60 should be added for the preparation and cryopreservation of *in vitro* shoot tips of this accession. W.M. Roca (personal communication) evaluated the annual maintenance cost of CIAT's cassava collection, which included 5000 accessions, at around $5000 under cryopreservation, as compared with $30,000 under *in vitro* slow growth storage. More recently, a detailed study compared the costs of maintaining one of the world's largest coffee field collections with those of establishing a coffee cryo-collection at CATIE in Costa Rica (Dulloo et al., 2009). The results indicate that cryopreservation costs less (in perpetuity per accession) than conservation in field genebanks. A comparative analysis of the costs of both methods showed that the more accessions there are in cryopreservation storage, the lower the per accession cost. In addition to costs, the study examined the advantages of cryopreservation over field collection and showed that for species that are difficult to conserve using seeds, and that can only be conserved as live plants, cryopreservation may be the method of choice for long-term conservation of genetic diversity.

Additional uses of cryopreservation

Recently, cryopreservation has been used for cryotherapy, i.e., for eliminating viruses from infected plants, as a substitute or a complement to classical virus eradication techniques such as meristem culture and cryotherapy (Wang et al., 2008). In cryotherapy, plant pathogens such as viruses, phytoplasmas, and bacteria are eradicated from shoot tips by exposing them briefly to liquid nitrogen. Uneven distribution of viruses and obligate vasculature-limited microbes in shoot tips allows elimination of the infected cells by injuring them with the cryo-treatment and regeneration of healthy shoots from the surviving pathogen-free meristematic cells. Thermotherapy followed by cryotherapy of shoot tips can be used to enhance virus eradication. Cryotherapy of shoot tips is easy to implement; it allows treatment of large numbers of samples and results in a high frequency of pathogen-free regenerants. Difficulties related to excision and regeneration of small meristems are largely circumvented. To date, severe pathogens in banana (*Musa* spp.), *Citrus* spp., grapevine (*Vitis vinifera*), *Prunus* spp., raspberry (*Rubus idaeus*), potato (*S. tuberosum*), and sweet potato (*I. batatas*) have been eradicated using cryotherapy. These pathogens include nine viruses (banana streak virus, cucumber mosaic virus, grapevine virus A, plum pox virus, potato leaf roll virus, potato virus Y, raspberry bushy dwarf virus, sweet potato feathery mottle virus, and sweet potato chlorotic stunt virus), sweet potato little leaf phytoplasma, and Huanglongbing bacterium causing "citrus greening" (Wang et al., 2008).

Cryopreservation: progress and prospects

Even though cryopreservation is still routinely employed in a limited number of cases, the development of the new vitrification-based freezing techniques has made its application to a broad range of species possible, especially vegetatively propagated ones (Engelmann, 2004). An important advantage of these new techniques is their operational simplicity, since they will be mainly applied in developing tropical countries where the largest amount of genetic resources of problem species is located. Research is actively performed by various groups in universities, research institutes, botanical gardens, and genebanks worldwide to improve knowledge of biological mechanisms underlying the tolerance of plant tissues to desiccation and cryopreservation. It is hoped that new findings on critical issues such as understanding and control of desiccation sensitivity will contribute significantly to the development of improved cryopreservation techniques for recalcitrant seed species. In this regard, it is interesting to mention that an EU-funded COST (European Cooperation in the Field of Scientific and Technical Research) Action (COST 871: Cryopreservation of Crop Species in Europe) was performed between 2007 and 2010. This action, involving 19 European countries, aimed notably at improving fundamental knowledge about cryoprotection through the determination of physicobiochemical changes associated with tolerance toward cryopreservation, and at developing and applying new cryopreservation protocols. For additional information, see http://www.biw.kuleuven.be/dtp/tro/cost871/Home.htm. It can thus be realistically expected that in the coming years our understanding of the biological mechanisms involved in cryopreservation will increase, and that cryopreservation will become more frequently employed for the long-term conservation of plant genetic resources.

Conclusions

In this chapter, we have presented the new biotechnological possibilities for improving *ex situ* conservation of plant biodiversity in genebanks and botanical gardens. During recent years, dramatic progress has been made with the development of new conservation techniques for non-orthodox and vegetatively propagated species, especially in the area of cryopreservation, and the current *ex situ* conservation concepts should be modified accordingly to accommodate these technological advances. It is now well recognized that an appropriate conservation strategy for a particular plant gene pool requires a holistic approach, combining the different *ex situ* and *in situ* conservation techniques available in a complementary manner. *In situ* and *ex situ* methods, including a range of techniques for the latter, are options available for the different gene pool elements. Selection of the appropriate methods should be based on a range of criteria, including the biological nature of the species in question and practicality and feasibility of the particular methods chosen (which depend on the availability of the necessary infrastructures) as well as the cost-effectiveness and security afforded by their application. As already mentioned in this chapter, the complementarity between genebanks and botanical gardens should be fully recognized and capitalized on to optimize plant biodiversity conservation. Considerations of complementarity with respect to efficiency and cost-effectiveness of the various conservation methods chosen are also important. In many instances, the development of appropriate complementary conservation strategies will still require further research to define the criteria, refine the methods, and test their application for a range of gene pools and situations. In this context, it is important to stress that the new, efficient *in vitro* conservation techniques developed over recent years are not seen as a replacement for conventional *ex situ* approaches. They offer genebank and botanical garden curators additional tools to optimize the conservation of germplasm collections.

References

Allen, J. B., & Lass, R. A. (1983). London cocoa trade Amazon project: Final report, phase 1. *Cocoa Growers Bulletin*, *34*, 1–71.

Ashmore, S. (1997). Status report on the development and application of *in vitro* techniques for the conservation and use of plant genetic resources. In F. Engelmann (Ed.), Rome: International Plant Genetic Resources Institute.

Assy-Bah, B., & Engelmann, F. (1992). Cryopreservation of mature embryos of coconut (*Cocos nucifera* L.) and subsequent regeneration of plantlets. *CryoLetters*, *13*, 117–126.

Assy-Bah, B., Durand-Gasselin, T., & Pannetier, C. (1987). Use of zygotic embryo culture to collect germplasm of coconut (*Cocos nucifera* L.). *FAO/IBPGR Plant Genet. Res. Newsl.*, *71*, 4–10.

Banerjee, N., & De Langhe, E. (1985). A tissue culture technique for rapid clonal propagation and storage under minimal growth conditions of *Musa* (banana and plantain). *Plant Cell Reports*, *4*, 351–354.

Berjak, P., Farrant, J. M., Mycock, D. J., & Pammenter, N. W. (1989). Homoiohydrous (recalcitrant) seeds: The enigma of their desiccation sensitivity and the state of water in axes of *Landolphia kirkii* Dyer. *Planta*, *186*, 249–261.

Berjak, P., & Pammenter, N. W. (1994). Recalcitrance is not an all-or-nothing situation. *Seed Science Research*, *4*, 263–264.

Bonnart, R., & Volk, G. M. (2010). Increased efficiency using the encapsulation-dehydration cryopreservation technique for *Arabidopsis thaliana*. *CryoLetters*, *31*, 95–100.

Bunn, E., Turner, S. R., Panaia, M., & Dixon, K. W. (2007). The contribution of *in vitro* technology and cryogenic storage to conservation of indigenous plants. *Australian Journal of Botany*, *55*, 345–355.

Caswell, K. L., & Kartha, K. K. (2009). Recovery of plants from pea and strawberry meristems cryopreserved for 28 years. *CryoLetters*, *30*, 41–46.

Chandel, K. P. S., Chaudhury, R., Radhamani, J., & Malik, S. K. (1995). Desiccation and freezing sensitivity in recalcitrant seeds of tea, cocoa and jackfruit. *Annals of Botany*, *76*, 443–450.

Chatelet P., Chabrillange, N., Peros, J.P., Ortigosa, P., Peyrière, A., Lacombe, T., et al. (2009). Cryopreservation in the Vitaceae: Seed cryopreservation is effective for wild species accessions. In *Abstwithin first international symposium on cryopreservation in horticultural species*, KULeuven, Belgium, 5–8 April 2009, p. 34.

Chin, H. F. (1988). *Recalcitrant seeds: A status report*. Rome: International Plant Genetic Resources Institute.

Chin, H. F., & Roberts, E. H. (1980). *Recalcitrant Crop Seeds*. Kuala Lumpur: Tropical Press Sdn. Bhd.

Côte, F. X., Goue, O., Domergue, R., Panis, B., & Jenny, C. (2000). In-field behavior of banana plants (*Musa* AA sp.) obtained after regeneration of cryopreserved embryogenic cell suspensions. *CryoLetters*, *21*, 19–24.

Cyr, D. R. (2000). Cryopreservation: Roles in clonal propagation and germplasm conservation of conifers. In F. Engelmann & H. Takagi (Eds.), *Cryopreservation of Tropical Plant Germplasm – Current Research Progress and Applications* (pp. 265–268). Rome: JIRCAS, Tsukuba/IPGRI.

Dekkers, A. J., Rao, A. N., & Goh, C. J. (1991). *In vitro* storage of multiple shoot cultures of gingers at ambient temperature. *Scientia Horticulturae*, *47*, 157–167.

Dereuddre, J., Hassen, M., Blandin, S., & Kaminski, M. (1991). Resistance of alginate-coated somatic embryos of carrot (*Daucus carota* L.) to desiccation and freezing in liquid nitrogen: 2. Thermal analysis. *CryoLetters, 12*, 135–148.

Druart, P. (1985). *In vitro* germplasm preservation techniques for fruit trees. In A. Schäfer-Menuhr (Ed.), *In Vitro Techniques – Propagation and Long-Term Storage* (pp. 167–171). Dordrecht: Nijhoff/Junk for CEC.

Dulloo, M. E., Ebert, A. W., Dussert, S., Gotor, E., Astorga, C., Vasquez, N., et al. (2009). Cost efficiency of cryopreservation as a long term conservation method for coffee genetic resources. *Crop Science, 49*, 2123–2138.

Dumet, D. (1994). Cryoconservation des massifs d'embryons somatiques de palmier à huile (*Elaeis guineensis* Jacq.) par déshydratation-vitrification. Etude du rôle du saccharose pendant le prétraitement. PhD Thesis, Université Paris 6, France.

Dumet, D., Engelmann, F., Chabrillange, N., & Duval, Y. (1993). Cryopreservation of oil palm (*Elaeis guineensis* Jacq.) somatic embryos involving a desiccation step. *Plant Cell Reports, 12*, 352–355.

Dussert, S., & Engelmann, F. (2006). New determinants of coffee (*Coffea arabica* L.) seed tolerance to liquid nitrogen exposure. *CryoLetters, 27*, 169–178.

Dussert, S., Chabrillange, N., Anthony, F., Engelmann, F., Recalt, C., & Hamon, S. (1997). Variability in storage response within a coffee (*Coffea* spp.) core collection under slow growth conditions. *Plant Cell Reports, 16*, 344–348.

Dussert, S., Chabrillange, N., Engelmann, F., Anthony, F., & Hamon, S. (1997). Cryopreservation of coffee (*Coffea arabica* L.) seeds: Importance of the precooling temperature. *CryoLetters, 18*, 269–276.

Dussert, S., Engelmann, F., & Noirot, M. (2003). Development of probalistic tools to assist in the establishment and management of cryopreserved plant germplasm collections. *CryoLetters, 24*, 149–160.

Ellis, R. E., Hong, T., & Roberts, E. H. (1990). An intermediate category of seed storage behaviour? I. Coffee. *Journal of Experimental Botany, 41*, 1167–1174.

Ellis, R. H., Hong, T., Roberts, E. H., & Soetisna, U. (1991). Seed storage behaviour in *Elaeis guineensis*. *Seed Science Research, 1*, 99–104.

Engelmann, F. (1991). *In vitro* conservation of tropical plant germplasm – a review. *Euphytica, 57*, 227–243.

Engelmann, F. (1997). *In vitro* conservation methods. In B. V Ford-Lloyd, J. H. Newburry & J. A. Callow (Eds.), *Biotechnology and plant genetic resources: Conservation and use* (pp. 119–162). Wellingford: CABI.

Engelmann, F. (1997). Importance of desiccation for the cryopreservation of recalcitrant seed and vegetatively propagated species. *Plant Genetic Resources Newsletter, 112*, 9–18.

Engelmann, F. (1999). Management of Field and *In Vitro* Germplasm Collections, Proceedings of a Consultation Meeting, 15-20 January, 1996, CIAT, Cali, Colombia. International Plant Genetic Resources Institute, Rome.

Engelmann, F. (2000). Importance of cryopreservation for the conservation of plant genetic resources. In F. Engelmann & H. Takagi (Eds.), *Cryopreservation of tropical plant germplasm–Current research progress and applications* (pp. 8–20). Rome: JIRCAS, Tsukuba/IPGRI.

Engelmann, F. (2004). Plant cryopreservation: Progress and prospects. *In Vitro Cellular and Developmental Biology–Plant, 40*, 427–433.

Engelmann, F. (2009). Use of biotechnologies for conserving plant biodiversity. *Acta Horticulturae, 812*, 63–82.

Engelmann, F. (2010). Use of biotechnologies for the conservation of plant biodiversity. *In Vitro Cellular and Developmental Biology–Plant*, in press.

Engelmann, F., & Engels, J. M. M. (2002). Technologies and strategies for *ex situ* conservation. In J. M. M. Engels, V. R. Rao, A. H. D. Brown & M. T. Jackson (Eds.), *Managing plant genetic diversity* (pp. 89–104). Wallingford and IPGRI, Rome: CABI.

Engelmann, F., & Takagi, H. (2000). *Cryopreservation of tropical plant germplasm – Current research progress and applications*. Rome: JIRCAS, Tsukuba/IPGRI.

Engelmann, F., Dumet, D., Chabrillange, N., Abdelnour-Esquivel, A., Assy-Bah, B., Dereuddre, J., et al. (1995). Factors affecting the cryopreservation of coffee, coconut and oil palm embryos. *Plant Genetic Resources Newsletter, 103*, 27–31.

Engelmann, F., Gonzalez-Arnao, M. T., Wu, W. J., & Escobar, R. E. (2008). Development of encapsulation-dehydration. In B. M. Reed (Ed.), *Plant cryopreservation: A practical guide* (pp. 59–76). Berlin: Springer Verlag.

Engels, J.M.M. & Engelmann, F. (1998). Botanic gardens and agricultural genebanks: Building on complementary strengths for more effective global conservation of plant genetic resources. In *Proceedings of fifth international botanic gardens conservation Congress*, Kirstenbosch, South Africa, 14–18 Sept. 1998.

Fahy, G. M., MacFarlane, D. R., Angell, C. A., & Meryman, H. T. (1984). Vitrification as an approach to cryopreservation. *Cryobiology, 21*, 407–426.

FAO. (1996). *Report on the state of the World's plant genetic resources for food and agriculture*. Rome: Food and Agriculture Organization of the United Nations.

FAO. (2009). *Draft second report on the World's plant genetic resources for food and agriculture*. Rome: Food and Agriculture Organization of the United Nations.

Florin, B., Brulard, E. & Lepage, B. (1999). Establishment of a cryopreserved coffee germplasm bank. In: *Abststract CRYO'09. annual meeting of the society for Cryobiology*, Tsukuba, Japan, 21–26 July 2009, p. 167.

Forsline, P. L., McFerson, J. R., Lamboy, W. F., & Towill, L. E. (1999). Development of base and active collections of *Malus* germplasm with cryopreserved dormant buds. *Acta Horticulturae, 484*, 75–78.

Ganeshan, S., & Rajashekaran, P. E. (2000). Current status of pollen cryopreservation research: Relevance to tropical agriculture. In F. Engelmann & H. Takagi (Eds.), *Cryopreservation of Tropical Plant Germplasm – Current Research Progress and Applications* (pp. 360–365). Rome: JIRCAS, Tsukuba/IPGRI.

George, E. F. (1996). *Plant propagation by tissue culture. Part 2 – In practice* (2nd ed.). Edington: Exegetics.

Golmirzaie, A., & Panta, A. (2000). Advances in potato cryopreservation at the International Potato Center, Peru. In F. Engelmann & H. Takagi (Eds.), *Cryopreservation of Tropical Plant Germplasm – Current research progress and applications* (pp. 250–254). Rome: JIRCAS, Tsukuba/IPGRI.

Gonzalez-Arnao, M.T. (1996). Desarollo de una tecnica para la crioconservacion de meristemos apicales de caña de azucar. Tesis presentada en opcion al grado de Doctor en ciencias tecnicas. Centro Nacional de Investigaciones Cientificas, La Habana, Cuba.

Gonzalez-Arnao, M. T., & Engelmann, F. (2006). Cryopreservation of plant germplasm using the encapsulation-dehydration technique: Review and case study on sugarcane. *CryoLetters, 27*, 155–168.

Gonzalez-Arnao, M. T., Panta, A., Roca, W. M., Escobar, R. H., & Engelmann, F. (2008). Development and large scale application of cryopreservation techniques for shoot and somatic embryo cultures of tropical crops. *Plant Cell Tissue and Organ Culture, 92*, 1-13.

Gopal, J., Chamali, A., & Debabrata, S. (2005). Use of micro-tubers for slow-growth *in vitro* conservation of potato germplasm. *Plant Genetic Resources Newsletter, 141*, 56–60.

Guanino, T., Silvanini, A., Benelli, C., Beghe, D., & Fabbri, A. (2009). Synthetic seed production and conservation of Kober 5BB grapevine rootstock. *Italian Society for Horticulture, 16*, 267–270.

Guarino, L., Rao, R., & Reid, R. (1995). *Collecting plant genetic diversity, technical guidelines*. Wallingford: CAB International.

Hamed, A. M., Gomma, A. H., Aly, A. S., & Ibrahim, I. A. (2003). *In vitro* storage of encapsuylated somatic embryos of *Phenix dactylifera* L. *Bulletin Faculty of Agriculture, Cairo University, 54*, 585–598.

Hamilton, K. N., Ashmore, S. E., & Pritchard, H. W. (2009). Thermal analysis and cryopreservation of seeds of Australian wild *Citrus* species (Rutaceae): *Citrus australasica, C. inodora* and *C. garrawayi*. *CryoLetters, 30*, 268–279.

Harvengt, L., Meier-Dinkel, A., Dumas, E., & Collin, E. (2004). Establishment of a cryopreserved gene bank of European elms. *Canadian Journal of Forest Research, 34*, 43–55.

Hong, T.D., Linington, S. & Ellis, R.H. (1996). *Seed storage behaviour: A compendium*. Handbooks for Genebanks No. 4. Rome: International Plant Genetic Resources Institute.

Hummer, K.E. & Reed, B.M. (1999). Establishment and operation of a temperate clonal field genebank. In: Management of Field and *In Vitro* Germplasm Collections, Proceedings of a Consultation Meeting, 15–20 January, 1996, CIAT, Cali, Colombia,

(F. Engelmann, Ed.) pp. 29–31. International Plant Genetic Resources Institute, Rome.

Hvoslef-Heide, A. K. (1992). Effect of pre-storage conditions on storage of *in vitro* cultures of *Nephrolepsis exaltata* (L.) Schoot and *Cordyline fruticosa* (L.) A. Chev. *Plant Cell, Tissue and Organ Culture, 28*, 167–174.

IUCN (2004). IUCN Red List of Threatned Species. <www.iucnredlist.org/>.

Jones, L. H. (1974). Long term survival of embryos of carrot (*Daucus carota* L.). *Plant Science Letters, 2*, 221–224.

Kartha, K. K., & Engelmann, F. (1994). Cryopreservation and germplasm storage. In I. K. Vasil & T. A. Thorpe (Eds.), *Plant cell and tissue culture* (pp. 195–230). Dordrecht: Kluwer.

Kartha, K. K., Mroginski, N. L., Pahl, K., & Leung, N. L. (1981). Germplasm preservation of coffee (*Coffea arabica* L.) by *in vitro* culture of shoot apical meristems. *Plant Science Letters, 22*, 301–307.

Keller, E. R. J., Kaczmarczyk, A., & Senula., A. (2008). Cryopreservation for plant genebanks: A matter between high expectations and cautious reservation. *CryoLetters, 29*, 53–62.

Keller, E. R. J., Senula, A., Leunufna, S., & Grübe, M. (2006). Slow growth storage and cryopreservation – tools to facilitate germplasm maintenance of vegetatively propagated crops in living plant collections. *International Journal of Refrigeration, 29*, 411–417.

Keller, E.R.J., Grübe, M. & Senula, A. (2005). Cryopreservation in the Gatersleben genebank – state of the art in potato, garlic and mint. In: Mem. Congr. Int. Biotecnología y Agricultura (Bioveg 2005), Centro de Bioplantas, Ciego de Avila, Cuba, June 2005.

Keshavachandran R., Smitha, M.K. & Vijayaraghavan, R. (2007). *In vitro* conservation of endangered medicinal plants. In *Recent trends in horticultural biotechnology, Vol. I and II. ICAE national symposium on biotechnological interventions for improvement of horticultural crops: Issues and strategies*, Vellanikkara, Kerala, India, 10–12 January, 2005, pp. 439–442.

Kim, H. H., Lee, Y. G., Ko, H. C., Park, S. U., Gwag, J. G., Cho, E. G., et al. (2009). Development of alternative loading solutions in droplet-vitrification procedures. *CryoLetters, 30*, 291–299.

Kim, H. H., Lee, Y. G., Shin, D. J., Kim, T., Cho, E. G., & Engelmann, F. (2009). Development of alternative plant vitrification solutions in droplet-vitrification procedures. *CryoLetters, 30*, 320–334.

Kim, H. H., Shin, D. J., No, N. Y., Yoon, M. K., Choi, H. S., Lee, J. S., et al. (2009). Cryopreservation of garlic germplasm collections using the droplet-vitrification technique. In: *Abststract first international symposium on cryopreservation in horticultural species*, Leuven, Belgium, 5–8 April 2009. p. 39.

Konan, E. K., Durand-Gasselin, T., Koadio, Y. J., Niamké, A. C., Dumet, D., Duval, Y., et al. (2007). Field development of oil palms (*Elaeis guineensis* Jacq.) originating from cryopreserved stabilized polyembryonic cultures (SPCs). *CryoLetters, 28*, 377–386.

Kwiatowski, S., Martin, M. W., Brown, C. R., & Sluis, C. J. (1988). Serial microtuber formation as a long-term conservation method for *in vitro* germplasm. *American Potato Journal, 65*, 369–375.

Lambardi, M., Benelli, C., & de Carlo, A. (2006). Technologies for the *in vitro* conservation of woody plants. *Italian Society for Horticulture, 9*, 58–60.

Li, B.L., Zhang, Y.L., Wang, H., Song, C.H. & Liu, Y. (2009). Pollen cryo-bank establishment and application of traditional Chinese flowers. In: *Abststract CRYO'09. Annual meeting of the society for Cryobiology*, Tsukuba, Japan, 21–26 July 2009, p. 108.

Liu, Y. G., Liu, L. X., Wang, L., & Gao., A. Y. (2008). Determination of genetic stability in surviving apple shoots following cryopreservation by vitrification. *CryoLetters, 29*, 7–14.

Mandal, B. B., Ahuja-Ghosh, S., & Srivastava, P. S. (2008). Cryopreservation of *Dioscorea rotundata* Poir.: A comparative study with two cryogenic procedures and assessment of true-to-type regenerants by RAPD analysis. *CryoLetters, 29*, 399–408.

Mandal, B. B. (2000). Cryopreservation research in India: Current status and future perspectives. In F. Engelmann & H. Takagi (Eds.), *Cryopreservation of Tropical Plant Germplasm – Current Research Progress and Applications* (pp. 282–286). Rome: JIRCAS, Tsukuba/IPGRI.

Mathur, J., Ahuja, P. S., Lal, N., & Mathur, A. K. (1989). Propagation of *Valeriana wallichii* DC. using encapsulated apical and axial shoot buds. *Plant Science, 60*, 111–116.

Matsumoto, T., Sakai, A., & Yamada, K. (1994). Cryopreservation of *in vitro* grown apical meristems of wasabi (*Wasabia japonica*) by vitrification and subsequent high plant regeneration. *Plant Cell Reports, 13*, 442–446.

Mazur, P. (1984). Freezing of living cells: Mechanisms and applications. *American Journal of Physiology, 16*, C125–C142. 247 Cell Physiol.

Meryman, H. T., Williams, R. J., & Douglas, M. S. J. (1977). Freezing injury from solution effects and its prevention by natural or artificial cryoprotection. *Cryobiology, 14*, 287–302.

Mikul, A., Olas, M., Sliwinska, E., & Rybczynki, J. J. (2008). Cryopreservation by encapsulation of *Gentiana* spp. cell suspensions maintains regrowth, embryonic competence and DNA content. *CryoLetters, 29*, 409–418.

Mix-Wagner, G., Schumacher, H. M., & Cross, R. J. (2003). Recovery of potato apices after several years of storage in liquid nitrogen. *CryoLetters, 24*, 33–41.

Mullin, R. H., & Schlegel, D. E. (1976). Cold storage maintenance of strawberry meristem plantlets. *Horticultural Science, 11*, 100–101.

Nadarajan, J., Mansor, M., Krishnapillay, B., Staines, H. J., Benson, E. E., & Harding, K. (2008). Applications of differential scanning calorimetry in developing cryopreservation strategies for *Parkia speciosa*, a tropical tree producing recalcitrant seeds. *CryoLetters, 29*, 95–110.

Ng, S., & Ng, N. Q. (1991). Reduced growth storage of germplasm. In J. H. Doods (Ed.), *In Vitro Methods for Conservation of Plant Genetic Resources* (pp. 11–39). London: Chapman & Hall.

Niino, T. (1995). Cryopreservation of germplasm of mulberry. In Y. P. S. Bajaj (Ed.), *Biotechnology in agriculture and forestry Vol. 32. Cryopreservation of plant Germplasm I* (pp. 102–113). Berlin: Springer Verlag.

Niino, T., Sakai, A., Yakuwa, H., & Nojiri, K. (1992). Cryopreservation of *in vitro*-grown shoot tips of apple and pear by vitrification. *Plant Cell Tissue and Organ Culture, 28*, 261–266.

Nishizawa, S., Sakai, A., Amano, A. Y., & Matsuzawa, T. (1993). Cryopreservation of asparagus (*Asparagus officinalis* L.) embryogenic suspension cells and subsequent plant regeneration by vitrification. *Plant Science, 91*, 67–73.

Normah, M. N., & Makeen, A. M. (2008). Cryopreservation of excised embryos and embryonic axes. In B. M. Reed (Ed.), *Plant cryopreservation: A practical guide* (pp. 211–240). Berlin: Springer Verlag.

Ouédraogo, A.S., Thomsen, K., Engels, J.M.M. & Engelmann, F. (1999). Challenges and opportunities for enhanced use of recalcitrant and intermediate tropical forest tree seeds through improved handling and storage. In: *Proceedings of the IUFRO seed symposium 1998 recalcitrant seeds*, 12–15 Oct. 1998, Kuala Lumpur, Malaysia (M. Marzalina, K.C. Khoo, F.Y. Tsan and B. Krishnapillay, Eds.), pp. 227–234. FRIM, Kepong, Malaysia.

Panis, B., Strosse, H., Van den Henda, S., & Swennen, R. (2002). Sucrose preculture to simplify cryopreservation of banana meristem cultures. *CryoLetters, 23*, 375–384.

Panis, B., Van den houwe, I., Piette, B. & Swennen, R. (2007). Cryopreservation of the banana germplasm collection at the ITC (INIBAP Transit centre). In: *Proceedings of the first meeting of COST 871 working group 2: Technology, application and validation of plant cryopreservation*, Florence, Italy 10–13 May 2007, pp. 34–35.

Pence, V. C. (1991). Cryopreservation of seeds of Ohio native plants and related species. *Seed Science and Technology, 19*, 235–251.

Pence, V. C. (2008). Cryopreservation of bryophytes and ferns. In B. M. Reed (Ed.), *Plant Cryopreservation: A Practical Guide* (pp. 117–140). Berlin: Springer Verlag.

Pence, V. C. (2008). Cryopreservation of recalcitrant seeds. In Y. P. S. Bajaj (Ed.), *Biotechnology in Agriculture and Forestry Vol. 32, Cryopreservation of Plant Germplasm I* (pp. 29–52). Berlin: Springer Verlag.

Pence, V.C., Sandoval, J., Villalobos, V. & Engelmann, F. (2002). *In Vitro* Collecting Techniques for Germplasm Conservation. IPGRI Technical Bulletin N°7. IPGRI, Rome.

Popova, E., Kim, H. H., & Paek, K. Y. (2010). Cryopreservation of coriander (*Coriandrum sativum* L.) somatic embryos using sucrose preculture and air desiccation. *Scientia Horticulturae, 124*, 522–528.

Pritchard H.W., Ashmore, S., Berjak, P., Engelmann, F., González-Benito, E., Li, D.Z., et al. (2009). Storage stability and the biophysics of preservation. In: *Abststract plant conservation for the next decade: A celebration of Kew's 250th Anniversary*, 12–16 October 2009, Royal Botanic Gardens Kew, UK.

Ramsay, M.M., Jackson, A.D. & Porley, R.D. (2000). A pilot study for the *ex situ* conservation of UK bryophytes. In *Eurogard 2000 – II European botanic garden congress, Canary Islands, Spain*: Las Palmas de Gran Canaria, pp. 52–57, (BGCI, Ed.), London, UK.

Razdan, M. K., & Cocking, E. C. (1997). *Conservation of plant genetic resources In Vitro, Vol. 1: General Aspects*. Enfield: Science Publishers Inc.

Reed, B. M. (1991). Application of gas-permeable bags for *in vitro* cold storage of strawberry germplasm. *Plant Cell Reports, 10*, 431–434.

Reed, B. M. (2008). *Plant cryopreservation: A practical guide*. Berlin: Springer Verlag.

Reed, B. M., & Uchendu, E. (2008). Controlled rate cooling. In B. M. Reed (Ed.), *Plant cryopreservation: A practical guide* (pp. 77–92). Berlin: Springer Verlag.

Reed, B. M., DeNoma, J., & Chang, Y. (2000). Application of cryopreservation protocols at a clonal genebank. In F. Engelmann & H. Takagi (Eds.), *Cryopreservation of tropical plant germplasm – current research progress and applications* (pp. 246–249). Rome: JIRCAS, Tsukuba/IPGRI.

Reed, B.M., Engelmann, F., Dulloo, M.E. & Engels, J.M.M. (2004). Technical Guidelines for the Management of Field and *In Vitro* Germplasm Collections. Handbook for Genebanks N°7. IPGRI/SGRP, Rome.

Roberts, H. F. (1973). Predicting the viability of seeds. *Seed Science and Technology, 1*, 499–514.

Roca, W.M., Reyes, R. & Beltran, J. (1984). Effect of various factors on minimal growth in tissue culture storage of cassava germplasm. In *Proceedings of the sixth symposium of the international society for tropical root crops*, Lima, Peru, 21–26 February 1984, pp. 441–446.

Roxas, N. J. L., Tashiro, Y., Miyasaki, S., Isshiki, S., & Takeshita, A. (1985). *In vitro* preservation of Higo chrysanthemum (*Dendratema x grandiflorum* (Ramat.) Kitam.). *Journal of the Japanese Horticulture Society, 63*, 863–870.

Sakai, A., Hirai, D., & Niino, T. (2008). Development of PVS-based vitrification and encapsulation-vitrification protocols. In B. M. Reed (Ed.), *Plant cryopreservation: A practical guide* (pp. 33–58). Berlin: Springer Verlag.

Sakai, A., & Engelmann, F. (2007). Vitrification, encapsulation-vitrification and droplet-vitrification: A review. *CryoLetters, 28*, 151–172.

Sakai, A., Kobayashi, S., & Oiyama, I. (1990). Cryopreservation of nucellar cells of navel orange (*Citrus sinensis* Osb. *var. brasiliensis* Tanaka) by vitrification. *Plant Cell Reports, 9*, 30–33.

Sanchez, C., Martinez, M. T., Vidal, N., San-José, M. C., Valladeres, S., & Vieitez, A. M. (2008). Preservation of *Quercus robur* germplasm by cryostorage of embryogenic cultures derived from mature trees and RAPD analysis of genetic stability. *CryoLetters, 29*, 493–504.

Sarasan, V., Cripps, R., Ramsay, M. M., Atherton, C., McMichen, M., Prendergast, G., et al. (2006). Conservation *in vitro* of threatened plants – progress in the past decade. *In Vitro Developmental Cellullar Biology–Plant, 42*, 206–214.

Senaratna, T., McKersie, B. D., & Bowley, S. R. (1990). Artificial seeds of alfalfa (*Medicago sativa* L.). Induction of desiccation tolerance in somatic embryos. *In Vitro Cellular & Developmental Biology–Plant, 26*, 85–90.

Simmonds, N. W. (1982). The context of the workshop. In L. A. Withers & J. T. Williams (Eds.), *Crop Genetic Resources–the Conservation of Difficult Material, IUBS Series B42* (pp. 1–3). Paris: IUBS/IBPGR/IGF.

Spencer, M. (1999). The challenges of developing cryopreservation strategies to suit the requirements of a large industrial *in vitro* plant cell collection. In: *Abststract Cryo'99, World Congress of Cryobiology*, Marseille, France, 12–15 July 1999, p. 245.

Touchell, D. H., & Dixon, K. W. (1994). Cryopreservation for seedbanking of Australian species. *Annals of Botany, 40*, 541–546.

Towill, L. E., & Walters, C. (2000). Cryopreservation of pollen. In F. Engelmann & H. Takagi (Eds.), *Cryopreservation of tropical plant germplasm – Current research progress and applications* (pp. 115–129). Rome: JIRCAS, Tsukuba/IPGRI.

UNCED. (1992). *Convention on biological diversity*. Geneva: United Nations Conference on Environment and Development.

UNEP. (1995). *Global biodiversity assessment*. Cambridge: Cambridge University Press.

Uragami, A., Sakai, A., & Magai, M. (1990). Cryopreservation of dried axillary buds from plantlets of *Asparagus officinalis* L. grown *in vitro*. *Plant Cell Reports, 9*, 328–331.

Varghese, D., Berjak, P., & Pammenter, N. W. (2009). Cryopreservation of shoot tips of *Trichilia emetica*, a tropical recalcitrant-seeded species. *CryoLetters, 30*, 280–290.

Walters, C., Wheeler, L., & Grotenhuis, J. M. (2005). Longevity of seeds stored in a genebank: Species characteristics. *Seed Science Research, 15*, 1–20.

Walters, C., Wheeler, L., & Stanwood, P. C. (2004). Longevity of cryogenically stored seeds. *Cryobiology, 48*, 229–244.

Wang, Q., Panis, B., Engelmann, F., Lambardi, M. and Valkonen, J.P.T. (2008). Elimination of plant pathogens by cryotherapy of shoot tips: A review. *Annals of Applied Biology.* 154: < http://www3.interscience.wiley.com/journal/119879031/issue/ >.

Wesley-Smith, J., Vertucci, C. W., Berjak, P., Pammenter, N. W., & Crane, J. (1992). Cryopreservation of desiccation-sensitive axes of *Camellia sinensis* in relation to dehydration, freezing rate and the thermal properties of tissue water. *Journal of Plant Physiology, 140*, 596–604.

Withers, L. A. (1987). *In vitro* methods for germplasm collecting in the field. *FAO/IBPGR Plant Genetic Resources Newsletter, 69*, 2–6.

Withers, L. A. (1995). Collecting *in vitro* for genetic resources conservation. In L. Guarino, R. Rao & R. Reid (Eds.), *Collecting plant genetic diversity* (pp. 511–515). Wallingford: CABI.

Withers, L. A., & Engelmann, F. (1997). *In vitro* conservation of plant genetic resources. In A. Altman (Ed.), *Biotechnology in Agriculture* (pp. 57–88). New York: Marcel Dekker Inc.

Withers, L. A., & Engels, J. M. M. (1990). The test tube genebank – a safe alternative to field conservation. *IBPGR Newsletter for Asia and the Pacific, 3*, 1–2.

Yidana, J. A., Withers, L. A., & Ivins, J. D. (1987). Development of a simple method for collecting and propagating cocoa germplasm *in vitro*. *Acta Horticulturae, 212*, 95–98.

Section D

Controlling plant response to the environment: Abiotic and biotic stress

Integrating genomics and genetics to accelerate development of drought and salinity tolerant crops

18

Zvi Peleg Harkamal Walia Eduardo Blumwald
University of California, Davis, California

TABLE OF CONTENTS

Impact of Abiotic Stresses on Crop Plant Productivity

Crop plants are often grown under unfavorable environmental conditions that prevent the full expression of their genetic yield potential. The abiotic stress conditions that most adversely affect crop yield are associated with water deficiency ion imbalance and temperature extremes. The ever-increasing human population, concomitant with loss of farmland (due to urbanization) and diminishing water availability (associated with climate change), poses serious challenges to world agriculture (reviewed by Mittler and Blumwald, 2010). To fulfill the food supply requirements for the projected population by 2050, a significant increase (an estimated 50%) in grain yield of major crops such as rice (*Oryza sativa* L.), wheat (*Triticum aestivum* L.), and maize (*Zea mays* L.) is required (Godfray et al., 2010). Boyer (1982) reported that the average production of major U.S. crops (corn, wheat, soybean, sorghum, oat, barley, potato, sugar beet) is only 21.6% of the highest yields attained under optimal conditions. Diseases, pests, and weed competition losses account for 4.1% and 2.6% yield reductions, respectively, with the remainder of the yield reduction attributed to unfavorable physicochemical (abiotic) environments induced by problematic soils and erratic climate patterns (69.1%). Certainly, some of these losses are caused by inherently unfavorable environments and some by suboptimal management practices by farmers, often due to economic constraints or lack of training. Nevertheless, there is no doubt that a large fraction of potential crop productivity is lost to abiotic stress factors.

Plants respond to environmental stresses at multiple levels such as molecular, cellular, tissue, anatomical, morphological, and whole-plant physiological levels (Bray, 1993, 1997; Munns, 2002; Chaves et al., 2003; Bartels and Sunkar, 2005; Munns and Tester, 2008; Witcombe et al., 2008). These responses to one or more stresses also vary with species and genotype. The plant response to stress also depends on the duration and severity of the event, as well as the age and developmental stage of the plant when the stress is imposed (Bray, 1997). In crop plants the sensitivity/tolerance to abiotic stresses is determined by yield loss rather than survival. Typically, early plant establishment (germination and seedling) and the reproductive stage are the most sensitive in determining yield under stress (Barnabas et al., 2008). However, a large segment of the abiotic stress research in models systems, particularly in *Arabidopsis thaliana*, in the past has focused primarily on the vegetative phase and has strived to identify survival phenotypes. This has hindered our ability to readily translate the discoveries into crop plants.

DOI: 10.1016/B978-0-12-381466-1.00018-3

Water Deficit: A Major Abiotic Stress Factor

Among the various abiotic stress conditions, water deficiency is the most devastating factor (Boyer, 1982; Araus et al., 2008). About 44.9% of U.S. soils are subjected to water deficits due to either drought (25.3%) or shallowness (19.6%), with an additional 43% subjected to other environmental limitations, primarily cold (16.5%) and waterlogging (15.7%; Boyer, 1982). Only 12.1% of U.S. soils are considered free from physicochemical problems. In agreement with this, the total indemnification made to U.S. farmers for crop losses due to drought, flooding, and low temperature accounts for 40.8%, 16.4%, and 13.8%, respectively (Boyer, 1982).

About one-third of the world's arable land suffers from chronically inadequate water availability for agriculture; in virtually all agricultural regions, crop yields are periodically reduced by drought (Bruce et al., 2002). Currently ~80% of the world's useable water resources are consumed by irrigated agriculture (Condon et al., 2004), within a few decades, the expanding world population will require more water for domestic, municipal, industrial, and environmental needs (Hamdy et al., 2003). This trend is expected to grow due to global climatic change and increased aridity (Vorosmarty et al., 2000). Thus, to meet the projected food demands, more crop per drop is required (Condon et al., 2004). Understanding plant responses to water deficiency is important, not only because it is the major stress affecting crop productivity, but also due to other associated stresses such as heat, cold, and salinity, which induce water deficiency (osmotic shock/stress) in plant tissues (Bohnert et al., 1995).

Salinity

Salinity (see definition of saline and sodic soils; Richards, 1954) is a major constraint on crop plant productivity (reviewed by Apse and Blumwald, 2002; Flowers, 2004; Munns and Tester, 2008; Witcombe et al., 2008). More than 800 million hectares of land throughout the world are salt affected, which accounts for 6% of the world total land area (Munns and Tester, 2008). In most cases salinity results from natural causes (salt accumulation over long periods of time). In addition, a significant portion of the cultivated agricultural land is becoming saline due to deforestation or excess irrigation (and fertilization; Shannon, 1997). Current estimates indicate that 20% of the roughly 230 million hectares of irrigated land is affected by salinity. Given that one-third of the food production comes from irrigated agriculture, salinity is becoming a serious problem for crop plant productivity.

Plant Responses to Abiotic Stress

Plant *resistance* to stress conditions may arise from escape, avoidance, or tolerance strategies (Levitt, 1972). *Escape* relies on successful completion of reproduction before the onset of severe stress (i.e., developmental plasticity) achieved by early flowering and/or short growth duration (Mooney et al., 1987). *Avoidance* involves the prevention or decreasing the impact of the stress on the plant, such as minimizing water loss and maximizing water uptake (Chaves et al., 2003), or exclusion of salt ions, a feature observed in halophytes (Munns and Tester, 2008). *Tolerance* relies on the inherent ability of the plant to sustain growth (likely at a reduced rate) even when the conditions are unfavorable for the maintenance of basic plant processes. This strategy involves coordination of physiological and biochemical alterations at the cellular and molecular levels, such as osmotic adjustment (Morgan, 1984), and the sequestration of ion in the plants in the vacuole or leaf sheath and/or older leaves (Mimura et al., 2003). In most cases, plants subjected to stress conditions combine a suite of responses, exhibiting a number of physiological and biochemical responses at the molecular, cellular, and whole plant level (Bray, 1993, 1997; Bohnert et al., 1995; Chaves et al., 2003).

Several physiological and biochemical aspects of the responses of plants to stress have been identified and examined in great detail over a period of several decades. Here we focus on two stress-related physiological mechanisms: *water use efficiency* (WUE) and *osmotic adjustment*(OA). Water use efficiency is defined in agronomic terms as the ratio between dry matter (or yield) production per unit water use (Condon and Hall, 1997). In a physiological context, however, WUE is defined as the ratio between the rate of carbon fixed and the rate of water transpired. Carbon isotope ratio ($^{13}C/^{12}C$, expressed with differential notation as $\delta^{13}C$) is often used as an indirect indicator of WUE (Farquhar et al., 1982; Araus et al., 2008). Osmotic adjustment is the net increase in intercellular solutes in response to water stress (Morgan, 1984). Osmotic adjustment is considered one of the crucial processes in plant adaptation to drought and salinity, because it allows turgor maintenance at lower water potentials, sustains tissue metabolic activity, and enables regrowth upon rewatering or when the high salt levels recede. Plant productivity under arid conditions has been associated with osmotic adjustment in a number of species such as wheat (Morgan, 1984; Peleg et al., 2009), sunflower (Chimenti et al., 2002), cotton (Saranga et al., 2004), chickpea (Moinuddin and Khanna-Chopra, 2004), and sorghum (Tangpremsri et al., 1995). At a physiological level, salinity stress has been considered to have a two-phase effect on plants (Munns, 2002). The initial phase of salinity stress is primarily attributed to an osmotic shock, similar to that of water stress. This initial phase likely has a strong osmotic adjustment component. The second long-term phase involves ion toxicity, which typically takes longer (days or weeks) to accumulate. In most cases, the salt tolerance, which eventually translates to greater yield, is derived from the ability of the plant to recover from the osmotic stress and maintain growth and photosynthesis in the longer second phase. A plant's tolerance to soil salinity is a result of three major components: Na^+ exclusion; tissue tolerance to Na^+; and osmotic tolerance (Munns and Tester, 2008; Plett and Moller, 2010). The impact of each component varies between and even within species. Further, the impact of each component can vary during the life cycle of the plant.

Under field conditions, crop plants are routinely subjected to a combination of different abiotic stresses. Plant responses to one stress can be synergistically or antagonistically modified by the superimposition of other stresses. Water loss from a plant (transpiration) is an unavoidable consequence of photosynthesis (Cowan, 1986), whereby the energy of solar radiation is used for carbon fixation. Although increased transpiration without a corresponding increase in photosynthesis reduces WUE, it is also beneficial for dissipating excess heat. Water stress and heat stress almost invariably co-occur under arid-region field conditions. The resulting need for a balance between plant tolerance to heat and drought complicates strategies for manipulating plant water use to improve productivity under arid conditions. Another example of interaction between two abiotic stresses occurs under salinity. The immediate effect of salinity, reduced soil osmotic potential, is similar to water stress and only its later effect, ion toxicity, is specific to salt (Munns, 2002; Munns and Tester, 2008). Moreover, primary abiotic stresses, such as drought, salinity, cold, heat, and chemical pollution are often interconnected with secondary oxidative stress by inducing the production of reactive oxygen species (ROS) in plants (Vinocur and Altman, 2005).

Breeding for Drought and Salinity Tolerance: "The Conventional Approach"

The primary goal of crop plant breeding is to increase yield and other commercial value traits (e.g., nutritional value, fruit size, color, etc.). On a global scale, plant breeding has been enormously successful in improving yields; for example, the introduction of dwarf wheat and rice varieties after the Green Revolution (Evenson and Gollin, 2003) and the development of hybrid maize (Duvick, 2001). However, the Green Revolution was primarily driven for irrigated agricultural systems. This has often resulted in minimal breeding resources for improved productivity in water limiting or saline environments. Furthermore, the improvement of abiotic tolerance in crops was hindered due to the inherent complexity in phenotyping, polygenic control of traits, low inheritance, and narrow genetic variation.

Effective management of the available water resources through agronomic practices plays a major role in sustainable production under drought (Peterson and Westfall, 2004). Agro-technical practices can have a significant impact on improving the WUE, and yield in marginal environments by using drip systems for saline water (Mizrahi et al., 1988), deficit irrigation (Geerts and Raes, 2009), no tillage (Klein et al., 2002), and precision agriculture (Cassman, 1999). However, these management strategies are too expensive to implement and require infrastructure that is difficult to introduce in both developing and underdeveloped countries. Therefore, genetic improvement in crop plants leading to sustained yield with low inputs is proposed to be a critical factor for sustaining the increasing population food needs.

Genetic improvement of stress tolerance can be achieved through selection; either directly, for a primary trait (such as yield) in a target environment (i.e., empirical breeding), or indirectly, for a secondary trait related to stress adaptation of the crop when grown in a stressful environment (i.e., analytical breeding). Most of the breeding progress in the last 50 years has been derived from empirical (also termed conventional) breeding, which has taken yield as the main trait for selection. However, empirical breeding appears to be reaching a plateau, and complementation of traditional with analytical selection methodologies is required to further improve yields (Araus et al., 2008).

Many physiological traits have been suggested as secondary traits for different species and environments (reviewed by Araus et al., 2008; Cattivelli et al., 2008). Phenology is the most widely used secondary trait because of ease of measurement and high heritability. Flowering data and plant height are the major drought escape traits, and can be used efficiently under terminal drought conditions. However, this approach has several limitations, and earlier varieties may suffer yield penalties in stress-free growing seasons. One of the main limiting factors for increasing productivity, even under non-stress conditions, is photosynthesis capacity (reviewed by Murchie et al., 2009). Stay-green is an important trait for post-flowering drought resistance for many cereal crops such as sorghum, rice, and maize (Campos et al., 2004). Osmoregulation was reported to be an effective selection criterion for drought adaptation under a water deficit environment (Morgan, 1983; Ludlow and Muchow, 1990; Zhang et al., 1999). Recent studies confirmed the role of osmotic adjustment in drought, although a number of contrasting reports were also published (reviewed by Serraj and Sinclair, 2002). In some cases OA was associated with increased yield (Morgan, 1983), and in other studies no such association was founded (Turner et al., 2007). It has been suggested that using OA as selection criteria will be beneficial to improve yields only under severe water stress conditions, where yield potentials tend to be low (Serraj and Sinclair, 2002). The identification and manipulation of candidate genes for such traits could facilitate improved tolerance and increase yields.

WUE is an important determinant of plant adaptation under limited water availability (Condon and Hall, 1997). WUE is often represented by carbon isotope ratio, $\delta^{13}C$ (Farquhar et al., 1982). Increased WUE can be obtained by either increased photosynthetic rate or reduced stomatal conductance (e.g., in response to stress and particularly under water deficit). In both cases it would reflect increase in transpiration efficiency at the leaf scale, although the former is more likely to be associated with increased productivity, while the latter is likely to be critical for survival and sustainable productivity under drought. Accordingly, $\delta^{13}C$ was reported to be negatively, positively or not at all associated with yield (Condon et al., 2004; Araus et al., 2008). An example of a successful breeding program based on WUE was reported by Rebetzke et al. (2002). Using $\delta^{13}C$ they selected bread wheat lines with high WUE under drought conditions that produced greater aerial biomass and grain yield (Rebetzke et al., 2002).

Stomatal density and stomatal aperture are two of the components that can control how efficiently a plant optimizes water loss with photosynthesis during stress. In *Arabidopsis*, QTLs affecting transpiration efficiency have been identified (Juenger et al., 2005; Masle et al., 2005) and one gene, *ERECTA*, has been shown to regulate transpiration efficiency (TE; Masle et al., 2005). *ERECTA* encodes a leucine-rich repeat receptor-like kinase (Torii et al., 1996) and regulates TE via the control of stomatal density, g_s, mesophyll cell proliferation, and photosynthetic capacity (Masle et al., 2005). Recently, the α-subunit of the heterotrimeric G protein in *Arabidopsis*, *GPA1*, was found to regulate stomatal density via the control of epidermal cell size and stomata formation (Nilson and Assmann, 2010). In rice, a C_2H_2 zinc finger transcription factor, DST (drought and salt tolerance), was cloned and reported to control stomatal aperture under drought and salt stress (Huang et al., 2009). DST contributed to stomata movement via regulation of genes involved in ROS homeostasis. It was proposed that DST regulated H_2O_2 homeostasis by inhibiting ROS accumulation during stress, and possibly preventing stomatal closure during drought and salt stress. These results would indicate that the genetic manipulation of the stomatal responses during stress could provide a powerful tool for stress tolerance improvement in crop plants.

Germplasm resources for drought and salinity tolerance

Domestication of wild plants and modern plant breeding imposed a genetic bottleneck on crop plants that has severely eroded the allelic variations of genes originally found in the wild species (Tanksley and McCouch, 1997). Modern crop cultivars are the product of intensive selection to facilitate higher yields under irrigated environments. This selection has narrowed the genetic variability for abiotic stress resistance, making them relatively sensitive to stress. Natural biodiversity is an underexploited sustainable resource that can enrich the genetic basis of cultivated plants with novel alleles that improve productivity and adaptation (McCouch, 2004; Gur and Zamir, 2004). For instance, wild emmer wheat (*Triticum turgidum* ssp. *dicoccoides*), the direct progenitor of domesticated wheat, was found to harbor a rich allelic repertoire for various morphophysiological traits conferring drought and salinity tolerance (Nevo et al., 1992; Peleg et al., 2005, 2008). Likewise, wild species of *Oryza* also showed potential to improve drought tolerance related traits (maintenance of leaf elongation and stomatal conductance under stress, high levels of membrane stability) in rice (Liu et al., 2004). The wild relative of tomato (*Solanum pennellii*) and bean (*Phaseolus* ssp.) were reported to be more resistant to salinity than modern cultivars (Dehan and Tal, 1978; Bayuelo-Jimenez et al., 2002). These studies, and others, highlighted the potential for stress tolerance improvement by introducing novel alleles from wild related species and landraces. A major challenge for breeding is the rapid introgression of desired traits found in landraces and wild relatives, with a minimal drag, to enrich the cultivated gene pools. However, the traditional introgression of novel alleles from wild relatives via conventional breeding is a slow and tedious process. The adoption of new genomics tools will accelerate and improve the introgression of genes into crop plants, and can prevent the drag of undesirable traits (e.g., seed dispersal, flowering time, height, etc.) from the wild. Furthermore, genetic engineering approaches can help overcome species barriers and extend the search beyond the realm of sexually compatible species.

Genetic dissection of plant responses to abiotic stress

One factor that has limited the development of stress-tolerant crops is the low heritability of complex traits such as yield, significant genotype-by-environment interaction (G × E), and QTLs-by-environment interaction (QTL × E). Furthermore, a general lack of understanding of the physiological parameters that reflect genetic potential for improved productivity in a water-limited environment has also contributed. Using genetic mapping to dissect the inheritance of complex traits in the same population is a powerful means to distinguish common heredity from casual associations among such traits. In principle, this can permit a direct test of the role of specific physiological traits in genetic improvement of plant productivity under abiotic stresses, such as those imposed by drought or salinity conditions.

Genetic mapping has been used to identify quantitative trait loci (QTLs) responsible for improved productivity under drought or salinity conditions and for morphophysiological traits associated with stress responses (reviewed by Tuberosa and Salvi, 2006; Collins et al., 2008). However, in only a few studies were productivity and physiological differences genetically mapped in the same populations, enabling researchers to identify possible association between these traits (Saranga et al., 2004; Peleg et al., 2009). When testing genetic maps across various sites/years, QTLs that promote high and stable yield in multiple environment conditions were prioritized (Collins et al., 2008). These QTLs, expressed across most environments, are termed "constitutive QTLs," while QTLs that expressed only under specific environments are called "adaptive QTLs." However, when breeding for drought-prone environments, QTLs that are specific for water-limited conditions, rather than those that are strictly constitutive, could be more beneficial for improving plant performance (Saranga et al., 2004; Peleg et al., 2009) Moreover, since different subsets of QTLs account for productivity under contrasting conditions, a wide environmental adaptation can be obtained by combining constitutive QTLs with various sets of environment-specific QTLs.

QTLs analyses of physiological traits, such as OA, relative water content (RWC), WUE, and cell membrane stability (CMS), have been widely used as stress tolerance indicators, especially drought and salinity. In rice, rooting and OA were found to be associated with improved drought tolerance (Babu et al., 2003). A QTL for root length was found to explain 30% of the variation (Price and Tomos, 1997), however, this was not stable when other environmental constrains were tested (Price et al., 2002). These results suggested that manipulating root morphology would not necessarily improve drought tolerance. QTLs responsible for variation in K^+ and Na^+ content were indentified, and one of these QTLs was cloned, designated as *SALTOL*, and it was found to explain 40% of the variation in salinity tolerance (Lin et al., 2004).

To date, hundreds of QTLs for productivity and abiotic-stress related morphophysiology traits have been reported (Tuberosa and Salvi, 2006; Vij and Tyagi, 2007; Collins et al., 2008). These identified stress-related QTLs might substantially contribute toward a better understanding of the genetic and functional basis of plant responses to abiotic stress (reviewed by Salvi and Tuberosa, 2005), directly or via positional cloning. However, their contribution to crop plant breeding, using molecular-assisted selection (MAS), has been very limited in releasing new cultivars with improved stress tolerance. While MAS was found to be very efficient in biotic stress resistance (reviewed in Collard and Mackill, 2008), only few examples of its use can be found for abiotic stress (Collins et al., 2008). The major weakness of traditional MAS comes from its two components: QTL identification; and the estimation of their effects. Current QTL identification methods are usually based on biparental populations that are not representative, and do not have the same level of allelic diversity and phase as the breeding program as a whole. Generating and genotyping such mapping populations is costly. In many cases, MAS was not efficient due to low heritability, high QTL × E, reliability of QTLs mapping (due to population size and/or number of replicates and environments; Collard and Mackill, 2008).

Introducing new technologies for abiotic stress breeding

Newly developed tools will facilitate the mapping, cloning, and use of stress-related QTLs for plant adaptations to stress conditions:

1. New molecular platforms, for high-throughput marker genotyping such as array based (diversity array technology, DArT), single feature polymorphism (SFP; Rostoks et al., 2005), and single nucleotide polymorphism (SNP) are becoming increasingly accessible.
2. Development of new mapping approaches, such as association mapping and consensus maps with high marker saturation and advance mapping software (Ronin et al., 1999).
3. Development of advance high-throughput phenotyping and statistical models to test environmental effects (Sadok et al., 2007).
4. Increasing availability of sequence information (e.g., "deep" sequencing of DNA and mRNA) and tools for functional analyses of the candidate genes (TILLING; Till et al., 2007, RNA interference; Lukens and Zhan, 2007, and microRNAs).
5. More efficient transformation techniques will enable major stress-related QTLs to be considered for positional cloning, aiming to manipulate the target trait more directly by genetic engineering.

Model plants, such as *Arabidopsis thaliana*, *Brachypodium distachyon*, and *Medicago truncatula*, have been and will continue to offer insights into the genetic and biochemical basis of abiotic stress adaptations (Bohnert et al., 2006; Hirayama and Shinozaki, 2010). Furthermore, the identification of stress-related genes and pathways has been facilitated by introducing new tools and resources developed in these models plants. Microarray technology is a very powerful tool to discover genes related to stress adaptation. In *Arabidopsis*, thousand of transcripts profiles were used to indentify candidate genes for stress responses (Seki et al., 2009; Urano et al., 2010). Recent progress in transcriptome analyses has lead to the routine use of high-density microarrays, tiling arrays (Zeller et al., 2009), and deep sequencing (RNA-seq) in functional genomics laboratories (Nelson et al., 2007). RNA-seq has the potential to not only identify transcript abundance, but also to discern the allelic sequence variant of the transcripts that can be useful in case of hybrids or allopolyploids such as wheat and cotton. However, robust statistical methods are required to interpret large RNA-seq datasets.

Increasing DNA sequencing information will open new options to study the function of stress-related genes. For rice and *Arabidopsis*, various T-DNA or transposome tagged populations are publicly available (Hirayama and Shinkozaki, 2010). TILLING (Targeting Induced Local Lesions IN Genomes) is a reverse genetics technique combining chemical mutagenesis with high-throughput mutation detection to identify point mutations within a specific locus. TILLING resources have been developed for many crop plants, including maize (Till et al., 2004), rice (Till et al., 2007), wheat (Uauy et al., 2009), and tomato (Minoia et al., 2010), which allow the production of targeted knockout genes. Allelic variants generated from TILLING will enable functional analyses of candidate genes, especially those where complete knockout is lethal. Power of detection of allelic mutants is likely to improve considerably as deep-sequencing becomes more accessible.

Engineering-Tolerant Crop Plants: The Transgenic Approach

Response of plants to environmental stress affects all aspects of plant physiology and metabolism, leading to severe yield losses. Thus, tolerance mechanisms depend on the prevention or alleviation of cellular damage, the re-establishment of homeostatic conditions, and the resumption of growth. Discovering and understanding the molecular/genetic basis of these tolerance components has been the focus of crop biotechnology in the past two decades. Despite these enormous research efforts, the role of very few genes in enhancing abiotic stress tolerance has been demonstrated under field conditions. However, this is expected to change, primarily because research is increasingly focused on high yields under stress rather than plant survival. Other factors include better facilities for testing the transgenic materials and the increasing acceptance of genetically engineered plants by the communities. Genetic engineering of candidate genes for abiotic stress was found to be successful in model plants growing under controlled conditions, and provided insights on the role of these genes in key physiological and biochemical processes (reviewed by Vinocur and Altman, 2005; Umezawa et al., 2006; Pardo, 2010). In this review, we have focused on efforts for improving the tolerance of crop plants to abiotic stress.

Genes for osmoregulation

Active accumulation of solute in the vacuoles (i.e., osmotic adjustment) is a common physiological component affecting drought and salinity tolerance (Munns, 2002). Increasing the

Table 18.1 Summary of drought and salt stress-related genes introduced into crop plants

Gene engineered (source)	Promoter	Crop	Growth conditions and stage	Result	References
Osmoregulation associated genes					
P5GS (*Vigna aconitifolia*)	35S	Tobacco	Six-week-old seedlings, greenhouse	10- to 18-fold more proline than control plants, and enhances root growth	Kishor et al. (1995)
P5CS (*V. aconitifolia*)	Act1	Rice	Three-weeks seedling, greenhouse	Faster growth of shoots and roots	Su and Wu (2004)
P5CS (*V. aconitifolia*)	*AIPC*	Wheat	Water withholding at booting stage, greenhouse	Accumulation of proline, which resulted in improved tolerance to water deficit	Vendruscolo et al. (2007)
P5CSF129A (*V. aconitifolia*)	35S	Chickpea	Greenhouse	Higher proline accumulation, however, modest increase in TE	Bhatnagar-Mathur et al. (2009)
ZFP252 (*Oryza sativa*)	35S	Rice	Four-leaf stage, greenhouse	Increase amount of proline, drought and salinity tolerance	Xu et al. (2008)
betA (*Escherichia coli*)	NA	Maize	Drought at reproductive stage, in the field	Increase of 10–23% in yield relative to wild-type, under stress	Quan et al. (2004)
betA (*E. coli*)	35S	Cotton	Withholding water, at first flower stage, greenhouse	Increase glycine betaine level, drought tolerance and increase yield	Lv et al. (2007)
AhCMO (*Atriplex hortensis*)	NA	Cotton	Two-true-leaf stage, pots experiment in greenhouse	Enhancing glycine betaine content, and resistance to salinity stress	Zhang et al. (2009)
BADH (*A. hortensis*)	ubi-1 (maize)	Wheat	Water stress applied post-flowering, greenhouse	Tolerance to drought, improved photosynthesis capacity	Wang et al. (2010)
codA (*A. globiformis*)	35S	Rice		Production of glycine betaine, improved growth	Sakamoto and Murata (1998)
TPSP (*E. coli*)	rcbS and ABA-induced	Rice	Five-week-old seedlings, hydroponic	Accumulation of more trehalose, less photo-oxidative damage, improved mineral balance	Garg et al. (2002)
mtlD (*E. coli*)	*ubi-1* (maize)	Wheat	Three-week-old seedlings, growth chamber	Improved tolerance to water stress and salinity	Abebe et al. (2003)
AVP1	35S	Tomato	Five-week-old plants were subjected to drought, greenhouse	Increased root biomass, and enhanced recovery of plants from an episode of water stress	Park et al. (2005)
AVP1 (*Arabidopsis thaliana*)	35S	Cotton	Twenty-day-old plants were subjected to drought or salinity in greenhouse	Significantly improved tolerance to both drought and salt stresses, with increase of 20% fiber yield	Pasapula et al. (2011)
AVP1 (*A. thaliana*)	35S	Rice	5-week-old seedlings were subjected to salt treatments	Enhanced salt tolerance	Zhao et al. (2006)
AVP1 (*A.thaliana*)	35S	Alfalfa	Growth room	About 215% improvement in shoot dry weight	Bao et al. (2009)
AP2/ERF transcription factor family					
HvCBF4 (*Hordeum vulgare*)	*Ubi1*	Rice	4-week-old seedlings subjected to drought or salinity	Higher survival rate, confirmed improved drought and salinity tolerance	Oh et al. (2007)
TERF1 (*Solanum lycopersicum*)	35S	Rice	2-week-old seedlings were exposed to drought or salinity	Better recovery rate, enhanced tolerance to abiotic stress	Gao et al. (2008)

Table 18.1 (Continued)

Gene engineered (source)	Promoter	Crop	Growth conditions and stage	Result	References
DREB1A (*Arabidopsis*)	*Ubi1*	Rice	Four-week-old plants were subjected to four days of drought following re-watering, greenhouse	Improved photosynthesis, and survival rate	Oh et al. (2005)
WXP1(*Medicago truncatula*)	35S	Alfalfa	Seedling, subjected to cycles drought, greenhouse	Increase in cuticular wax loading on leaves, drought tolerance, and higher recovery rate	Zhang et al. (2005)
OsDBEB1A and *OsDBEB1B* (*O. sativa*)	35S and Ubi1	Rice	Seedling, grown in soil or hydroponic, greenhouse	Tolerance to drought and salinity (survival), reduced growth under normal conditions	Ito et al. (2006)
Nac family					
SNAC1 (*O. sativa*)	35S	Rice	Severe water stress at the reproductive stage in the field	22–34% more seed setting than control	Hu et al. (2006)
NAC10 (*O. sativa*)	GOS2 and RCc3	Rice	4-week-old plants subjected to three days of drought in greenhouse and field experiment	Enlarged roots, improved drought tolerance, with increased yield	Jeong et al. (2010)
OsNAC045 (*O. sativa*)	35S	Rice	Seedlings at three-leaf stage subjected to stress	Enhanced tolerance to drought and salt treatment	Zheng et al. (2009)
Transcription factors					
OsWRKY11 (*O. sativa*)	HSP101	Rice	2-week old seedlings, greenhouse	Slow water loss, and reduced wilting	Wu et al. (2009)
AtNF-YBI, ZmNF-YB2 (*A. thaliana* and *Zea mays*)	35S, OsActin1	Maize	Greenhouse and field evaluation	Improved photosynthesis and polysaccharide metabolism	Nelson et al. (2007)
Ion transporters					
AtNHX1 (*A. thaliana*)	35S	Tomato	Hydroponics in greenhouse	Increased growth, and flower and seed production	Zhang and Blumwald (2001)
AgNHX1 (*A. thaliana*)	35S	Rice	Greenhouse and hydroponics in growth room	50% or 81–100% survival of seedlings	Ohta et al. (2002)
PgNHX1 (*Pennisetum glaucum*)	NA	Rice	10-day-old seedlings, hydroponics in greenhouse	About 81% improvement in shoot and root lengths	Verma et al. (2007)
AtNHX1 (*A. thaliana*)	35S	Wheat	14-day-oldseedlings in greenhouse and field conditions	About 68% increase in shoot dry weight and 26% in root dry weight	Xue et al. (2004)
AtNHX1 (*A. thaliana*)	35S	Tall fescue	Different level of salinity, pots in growth chamber	Transgenic plants thrived well at 200 mM NaCl	Tian et al. (2006)
OsNHX1 (*O. sativa*)	35S	Perennial ryegrass	2-week-old seedlings subjected to salt, in growth room	Transgenic plants survived at 350 mM NaCl when all wild-type plants died	Wu et al. (2005)
AtNHX1 (*A. thaliana*)	35S	Brassica napus	2-week-old seedlings, watered with 200 mM, greenhouse	Improved growth, seed yield and oil quality under high salinity, salt tolerance	Zhang et al. (2001)
SOD2 (yeast)	35S	Rice	5-week-old seedlings subjected to 150 and 300 mM	Higher levels of photosynthesis and root proton exportation capacity	Zhao et al. (2006)

(*Continued*)

Table 18.1 (Continued)

Gene engineered (source)	Promoter	Crop	Growth conditions and stage	Result	References
Hormones related genes					
IPT (*Agrobacterium*)	SARK	Tobacco	Flowering plants, greenhouse and field	Delayed senescence, improved photosynthesis and drought tolerance	Rivero et al. (2007, 2009, 2010)
AtLOS5 (*A. thaliana*)	OsActin1/ HVA22P	Rice	Drought stress applied at booting stage in field experiment	Significantly increased the spikelet fertility and yield of field grown	Xiao et al. (2009)
LeNCED1	35S	Tomato	Five-leaf stage plant subjected to drought	Increased ABA accumulation but had no effect on dehydration tolerance	Thompson et al. (2007)

production of diverse osmolytes was tested in various plants to improve tolerance to abiotic stress (Table 18.1). The manipulation of the mannitol level by over-expression of the mannitol-1-phosphate dehydrogenase (*mtlD*) gene resulted in improved tolerance to salinity and drought in model plants *Arabidopsis* (Thomas et al., 1995) and tobacco (Karakas et al., 1997). In wheat, where mannitol is normally not synthesized, expression of the *mtlD* (from *Escherichia coli*) gene also improved tolerance to water stress and salinity (Abebe et al., 2003). Another *Arabidopsis* gene encoding for a vacuolar pyrophosphatase proton pump (*AVP1*) was over-expressed in *Arabidopsis* (Gaxiola et al., 2001), tomato (Park et al., 2005), cotton (Pasapula et al., 2011), and rice (Zhao et al., 2006), and induced improved growth during drought and salt stress. Interestingly, the overexpressed *AVP1* resulted in a more robust root system that could possibly improve the plant's ability to absorb more water from the soil (Pasapula et al., 2011).

Chickpea (*Cicer arietinum*) was transformed with an osmoregulatory gene *P5CSF129A* (under 35S promoter) encoding the mutagenized Δ^1-pyrroline-5-carboxylate synthetase (*P5CS*) for the overproduction of proline (Bhatnagar-Mathur et al., 2009). The transgenic plants showed significantly higher proline accumulation; however, it resulted only in a modest increase in transpiration efficiency, suggesting that enhanced proline had little bearing on the components of yield in chickpea. Wheat plants over-expressing *P5CS* (under the control of a stress-induced promoter complex; AIPC), showed accumulation of proline, which resulted in improved tolerance to water deficiency (Vendruscolo et al., 2007). Likewise, transgenic rice over-expressing *P5CS* showed significantly higher tolerance to salinity and water stress produced in terms of faster shoot and root growth (Su and Wu, 2004). Rice plants over-expressing the *ZFP252* gene resulted in an increased amount of free proline and soluble sugars, elevated the expression of stress defense genes, and enhanced tolerance to salt and drought stresses (Xu et al., 2008).

Glycine betaine (GB), a fully N-methyl-substituted derivative of glycine, accumulates in the chloroplasts and plastids of many species such as Poaceae, Amaranthaceae, Asteraceae, Malvaceae, and Chenopodiaceae in response to drought and salinity. GB effectively stabilizes the quaternary structures of enzymes and complex proteins, and protects the photosynthetic machineries via ROS scavenging (Chen and Murata, 2008). Transgenic maize expressing *betA*, encoding a choline dehydrogenase, showed more GB accumulation under drought and salinity in the field (Quan et al., 2004). Drought stress, applied at the reproductive stage, resulted in a 10–23% higher yield relative to the wild-type plants. Similarly, cotton plants expressing *betA* were also more drought tolerant (Lv et al., 2007). Under water stress conditions, transgenic cotton had higher RWC, OA increased photosynthesis, reduced ion leakage, and lower lipid membrane peroxidation than wild-type plants. A CMO gene (*AhCMO*) cloned from *Atriplex hortensis* was introduced into cotton, showing enhanced resistance to salinity stress (Zhang et al., 2009). Transgenic potato plants (*S. tuberosum* L.) developed via the introduction of the bacterial choline oxidase (*codA*) gene, expressed under the control of an oxidative stress-inducible *SWPA2* promoter, evidenced enhanced tolerance to NaCl and drought stress at the whole plant level (Ahmad et al., 2008). Recently, wheat plants over-expressing a *BADH* gene, encoding betaine aldehyde dehydrogenase, were shown to be more tolerant to drought and heat by improving the photosynthesis capacity of flag leafs (Wang et al., 2010).

Dehydration-responsive element

In *Arabidopsis*, a dehydration-responsive element (DRE)/C-repeat (CRT) was identified as a *cis*-acting element regulating gene expression in response to dehydration (drought, salinity, and cold stress; Yamaguchi-Shinozaki and Shinozaki, 1994). Several dehydration responsive element binding proteins (DREB) have been identified. The over-expression of these genes activated the expression of many downstream genes with the DRE elements in their promoters, and the resulting transgenic plants showed improved stress tolerance (Agarwal et al., 2006). The over-expression of ERF genes such as *HvCBF4* (from barley), *TERF1* (from tomato), and *HARDY, DREB1A* (from *A. thaliana*) in rice conferred increased tolerance to abiotic stresses (Oh et al., 2005, 2007; Karaba et al., 2007; Gao et al., 2008). Furthermore, expression of the rice ERF gene *AP37* (under the control of a constitutive promoter) enhanced rice yields under stress (Oh et al., 2009). Rice plants over-expressing an ethylene response factor (ERF), a protein TSRF1 that binds to the GCC-box, showed enhanced osmotic and drought tolerance in seedlings (Quan et al., 2010). Transgenic rice lines over-expressing *OsDBEB1A* and *OsDBEB1B* under the control of a constitutive ubiquitin promoter showed more tolerance to drought and salinity conditions; however, under normal conditions, the transgenic lines showed reduced growth (Ito et al., 2006). The DREB1 pathway was also associated with proline accumulation. Tomato and rice plants over-expressing the *AtDREB1B/CBF1* or *OsDBEB1B* showed higher levels of proline as compared with the wild-type plants grown under water stress. Rice plants over-expressing *OsDREB1G* and *OsDREB2B* also showed significantly improved tolerance to water deficit stress at the seedling stage (Chen et al., 2008).

NAC proteins

Several NAC, including (*NAM*, (no apical meristem), *ATAF1-2*, and *CUC2* (cup-shaped cotyledon domains, which are one of the largest plant transcription factor families; Riechmann et al., 2000), have been reported to be associated with abiotic stresses. In rice, over-expression of a stress responsive gene *SNAC1* (*STRESS-RESPONSIVE NAC 1*) significantly enhanced drought resistance (22–34% increased in seed setting) in the field under severe water stress conditions at the reproductive stage (Hu et al., 2006). Over-expression of *OsNAC045* induced enhanced tolerance to drought and salt treatments (Zheng, 2009). Recently, the over-expression of *OsNAC10* in rice, under the control of the constitutive promoter *GOS2* and the root-specific promoter *RCc3*, improved tolerance to drought and salinity. More important, the *RCc3*:*OsNAC10* plants displayed significantly enhanced drought tolerance at the reproductive stage, increasing grain yield (25–42%) in the field under drought conditions (Jeong et al., 2010).

Genes for ionic balance

The accumulation of Na^+ ions into vacuoles through the operation of a vacuolar Na^+/H^+ antiporter provided an efficient strategy to avert the deleterious effect of Na^+ in the cytosol and maintain osmotic balance by using Na^+ (and Cl^-) accumulated in the vacuole to drive water into the cells (Apse et al., 1999; Apse and Blumwald, 2002). Transgenic plants over-expressing an *Arabidopsis* vacuolar Na^+/H^+ antiporter, *AtNHX1*, exhibited improved salt tolerance in *Brassica napus* (Zhang et al., 2001), tomato (Zhang and Blumwald, 2001), cotton (He et al., 2005), and wheat (Xue et al., 2004). The transformation of an ortholog gene (*AgNHX1*) from halophytic plant *A. gmelini* into rice improved salt tolerance of the transgenic rice (Ohta et al., 2002). Transformation of another Na^+/H^+ antiporter family member, *AtNHX3* (from *A. thaliana*), in sugar beet (*Beta vulgaris* L.) resulted in increased salt accumulation in leaves, but not in the storage roots, with enhanced constituent soluble sugar contents under salt stress condition (Liu et al., 2008).

The major focus for salt tolerance gene search specifically in *Arabidopsis* has been on ion exclusion, uptake, and sequestration to limit ion toxicity. The over-expression of SOS1, a plasma membrane-bound Na^+/H^+ antiporter, improved the ability of the transgenic plants to grow in the presence of high NaCl concentrations (Shi et al., 2003). The removal of sodium from the xylem, reducing sodium transfer rate to the shoots, is mediated by the HKT gene family (Plett and Moller, 2010). *AtHKT1;1* in *Arabidopsis*, *OsHKT1;5* in rice, and *HKT1;4* in wheat are all critical in transporting Na^+ into the root stele, decreasing Na^+ in the shoot (Hauser and Horie, 2010). One strategy for improving salinity tolerance is to increase the expression of such genes to further reduce sodium concentrations in the xylem (Plett et al., 2010).

The introduction of genes associated with the maintenance of ion homeostasis from halotolerance into crop plants confirmed salinity tolerance. The yeast gene *HAL1* introduced into tomato (Gisbert et al., 2000), watermelon (*Citrullus lanatus*; Thunb.; Ellul et al., 2003, and melon (*Cucumis melo* L.; Bordas et al., 1997), confirmed a higher level of salt tolerance, with higher cellular K^+ to Na^+ ratio under salt stress. Likewise, introducing the yeast *HAL2* gene into tomato resulted in improved root growth under NaCl conditions, which contributed to improved salt tolerance (Arrillaga et al., 1998). Over-expression of HAL3 (from *Saccharomyces cerevisiae*) homolog *NtHAL3* in tobacco increased proline biosynthesis and the enhancement of salt and osmotic tolerance in cultured tobacco cells (Yonamine et al., 2004).

Genes for redox regulation

Another physiological and biochemical cellular component common to a suite of abiotic stresses including drought and salt stress is oxidative stress. Oxidative stress involves the generation of ROS during stress. The most common ROS are hydrogen peroxide (H_2O_2), superoxide, the hydroxyl radical, and the singlet oxygen. ROS are continuously produced through cellular metabolism and plant cells, and are well equipped with antioxidants and scavenging enzymes to keep their levels low under normal growth conditions (Jaspers and Kangasjärvi, 2010). In rice, transformation of manganese superoxide dismutase (*MnSOD)* into chloroplasts from pea (*Pisum sativum*) under the control of an oxidative stress-inducible *SWPA*2 promoter resulted in improved drought tolerance. This suggested that increased SOD levels can boost the ROS scavenging system in plant chloroplasts (Wang et al., 2005). Recently, a rice gene encoding for a receptor-like kinase (RLK) was reported to improve the drought and salt tolerance of transgenic plants over-expressing the RLK (*OsSIK1*; Ouyang et al., 2010). The transgenic plants had higher activity of the ROS scavengers, such as peroxidases and catalase activity during stress, indicating the ability of the rice plant to limit the damage induced by enhanced ROS production. Over-expression of the *Arabidopsis* gene GF14λ, encoding a 14-3-3 protein that interacts with proteins involved in numerous metabolic processes, including antioxidant activity, demonstrated a "stay-green" phenotype and improved tolerance to moderate water stress in cotton (Yan et al., 2004).

More useful strategies to improve the antioxidant activities in the cell are the transformation of a pyramid of ROS-scavenging enzymes, which could improve tolerance to various stresses. Two genes (from *Suaeda salsa*) coding GST (glutathione S-transferase, EC 2.5.1.18) and CAT (catalase, EC 1.11.1.6) were transformed under the control of a constitutive promoter into rice plants. Transgenic rice seedlings showed a marked enhanced tolerance to salinity and oxidative stresses (Zhao and Zhang, 2006). Expression of three antioxidant enzymes, copper zinc superoxide dismutase (CuZnSOD), ascorbate peroxidase (APX), and dehydroascorbate (DHA) reductase (DHAR) in tobacco chloroplasts resulted in a higher tolerance to oxidative stress induced by salinity stress (Lee et al., 2007). These studies suggested that the simultaneous expression of multiple antioxidant enzymes could be more effective than the expression of single genes for developing transgenic plants with enhanced tolerance to abiotic stresses.

Aquaporins

Aquaporins are considered to be the main channels for the transport of water, as well as small neutral solutes and CO_2, through the plant cell membrane (Tyerman et al., 2002). The aquaporins are small, very hydrophobic, intrinsic membrane proteins. The regulatory role of aquaporins in cellular water transport and been demonstrated (Knepper, 1994). In lowland rice, over-expression of the aquaporin, RWC3, a member of the plasma-membrane intrinsic protein 1 (PIP1) subfamily, under the control of the stress-inducible SWPA2 promoter, resulted in improvement of general water status of the plants under osmotic stress (Lian et al., 2004). The beneficial results found in this study contradict previous studies in tobacco. Transgenic tobacco, constitutively over-expressing the *Arabidopsis* plasma membrane aquaporin PIP1b, showed higher vigor under well-watered conditions, but no beneficial effect under salt stress, and even negative effects causing fast wilting during water stress (Aharon et al., 2003). It may be suggested that the gradual drought treatment applied by Aharon et al. (2003) compared with the osmotic shock used by Lian et al. (2004), affected the results obtained. Furthermore, transgenic rice plants over-expressing barley HvPIP2:1 (a plasma membrane aquaporin), constitutively driven by the 35S-promoter, showed more sensitivity (reduction in growth rate) to salinity stress (Katsuhara et al., 2003).

Recently, tomato plants constitutively over-expressing tonoplast *SlTIP2:2* showed increased osmotic water permeability of the cell and whole-plant transpiration (Sade et al., 2009). The expression of *SlTIP2:2* had multiple functions, increasing transpiration under normal growth conditions, limiting the reduction in transpiration under drought and salt stresses, and also speeding up the revival of transpiration upon recovery from these stresses. Transgenic plants showed significant increases in fruit yield, harvest index, and plant mass relative to the control under both normal and water-stress conditions (Sade et al., 2009). It was postulated that over-expression of the *SlTIP2:2* can bypass the stress-induced down-regulation of the endogenous *aquaporins* genes of the tonoplast, and thus prevent the slowdown of tonoplastic osmotic water permeability (Sade et al., 2009). The results of this study raise the possibility of manipulating *Aquaporins* genes to produce drought- and salinity-tolerant crop-plants.

Other transcription factors

Although multiple transcription factors (TFs) have been well characterized in various plant species, transcriptional reprogramming under drought and stress is not fully understood. Over-expression of the *AtMYB2* gene in rice under the control of an abscisic acid (ABA)-inducible promoter conferred salt stress tolerance, with higher biomass and decreased ion leakage (Malik and Wu, 2005). Over-expression of *OsWRKY11* (encoding a transcription factor comprising a WRKY domain) under the control of an *HSP101 promoter*, conferred heat and drought tolerance at the seedling stage (slower leaf-wilting and less-impaired survival rate of green parts of plants; Wu et al., 2009).

Recently, in was shown that the constitutive over-expression of ubiquitous chaperones (from bacterial origin) in rice and maize resulted in tolerance to various environmental stresses (Castiglioni et al., 2008). These results suggested that chaperone molecules may be good candidates for abiotic stress enhancement.

Hormone Balance and Abiotic Stress

Hormones play a major role in stress signaling. One of the fast responses of plants to soil water stress is the accumulation of ABA in the roots (Thompson et al., 2007), which is transported through the xylem to the shoot (Wilkinson and Davies, 2010) causing stomatal closure to reduce water loss via transpiration (Schroeder et al., 2001) and eventually restricting cellular growth. ABA can also be synthesized in leaf cells and transported through the plant (Wilkinson and Davies, 2010). In *Arabidopsis*, many genes in the ABA metabolism have been characterized, and those for ABA perception and the downstream signal relay are described (reviewed by Huang et al., 2008; Cutler et al., 2010). However, in crop plants, only one ABA-related gene (*LOS5/ABA3*, a key enzyme in the last step of ABA biosynthesis) has been manipulated in rice with enhanced drought tolerance (Xiao et al., 2009). Over-expression of *LOS5* under the control of a constitutive or drought-inducible promoter significantly increased the spikelet fertility and yield of field grown transgenic rice plants (Xiao et al., 2009). In tomato, over-expression of *LeNCED1* (drought-inducible and a rate-limiting enzyme for ABA biosynthesis) resulted in increased ABA accumulation but had no effect on dehydration tolerance (Thompson et al., 2007).

Cytokinins (CK) have been linked to a variety of abiotic stresses (Hare et al., 1997). In *Arabidopsis*, examination of public microarray expression data revealed many genes encoding proteins in the CK signaling pathway that were differentially affected by various abiotic stresses (reviewed by Argueso et al.,

2009). CK is an antagonist to ABA, and the exposure of plants to drought results in a decreased level of CK. The elevated CK level could promote survival under water-stress conditions and inhibit leaf senescence and increased levels of proline (Alvarez et al., 2008). Manipulation of CK was found to be efficient in delaying senescence by spraying (Gan and Amasino, 1997), and more recently in transgenic plants. Isopentenyltransferase (*IPT*, rate-limiting step in CK biosynthesis) has been widely engineered in several plant species (Ma, 2008). However, drought tolerance varied with the type of promoter used to drive *IPT* expression. Transgenic tobacco (*Nicotiana tabacum*) expressing the *IPT* gene under control of a drought-induced promoter (senescence associated receptor kinase; SARK) resulted in increased drought tolerance, photosynthetic capacity, and yield (Rivero et al., 2007, 2009, 2010).

Likewise, transgenic rice plants expressing P_{SARK}::*IPT* showed improved drought tolerance with enhanced grain yield (Peleg et al., 2011).

Transgenic Cassava (*Manihot esculenta* Crantz) expressing *IPT* gene under the control of the senescence induced promoter, *SAG12*, were tested for drought tolerance under field conditions (Zhang et al., 2010). Under drought conditions, the transgenic cassava plants were found to be more tolerant thean the wild-type, due to prohibition of leaf abscission and fast recovery from stress. Creeping bentgrass (*Agrostis stolonifera*) expressing P_{SAG12}::*IPT* was tested in a hydroponic system using different levels of osmotic stress (PEG) (Merewitz et al., 2011). The transgenic plants were able to maintain higher osmotic adjustment, chlorophyll content, water-use efficiency, and greater root viability under osmotic stress compared with the wild-type plants (Merewitz et al., 2011). However, the use of PEG to stimulate osmotic stress is artificial, and not representative of the multi-dimensional response of plants to water-deficit under natural conditions.

Jasmonic acid (JA) is involved in plant development and the defense response. Transgenic rice plants over-expressing the *Arabidopsis* JA carboxyl methyltransferase gene (*AtJMT*) under the *Ubi1* promoter showed an increase in JA levels in panicles (Kim et al., 2009). Plants were grown in a greenhouse, and before panicle ignition were subjected to 2 weeks of drought. The *Ubi1*::*AtJMT* plants resulted in significant grain yield reduction, due to a lower number of spikelets and lower filling rates than wild-type plants (Kim et al., 2009).

Interactions and regulation of various hormone-regulated biosynthetic pathways (see Nemhauser et al., 2006) during water stress play important roles in abiotic stress adaptation. The homeostatic regulation of phytohormones could play significant roles in the regulation of source/sink relationships, and its manipulation could provide a significant avenue for the development of abiotic stress tolerance in plants (reviewed by Peleg and Blumwald, 2011).

Challenges and Prospects

New technologies are providing opportunities to address the challenging problem of maintaining high-yield crop production under stressful and changing climates. The information provided by high-resolution transcript profiling, the identification of large-scale specific protein networks, and their association with the plant responses to environmental perturbations are allowing the application of a systems-level approach to uncover the bases of plant responses to environmental changes. Pursuing a systems-level approach is of paramount importance, because crops of the future are likely to be stacked with multiple traits (water deficit, nitrogen use efficiency, pathogen challenges, etc.). A systems-level approach will contribute to the development of strategies for the combination of an array of traits without affecting the crop performance in the absence of stress. Maintaining normal crop performance when the growth conditions are favorable is likely to be an important factor because of the weather fluctuations predicted due to climate changes.

The manipulation of stress signaling components seems to be the most promising approach, as it has the potential to control a board range of events. More specific promoters are needed to improve the expression of genes of interest under the desirable environmental conditions and/or growth phase. This is particularly important for salt tolerance as regulation of ion homeostasis is highly tissue specific (e.g., xylem loading, ion exclusion in the roots, etc.). The common constitutive promoters are not always efficient, and can have a negative effect on plant growth and development.

Numerous genes related to plant response to abiotic stress have been identified and characterized (Wang et al., 2003; Araus et al., 2008). However, the vast majority of these studies were conducted on the model plant *A. thaliana* (e.g., Masle et al., 2005; Seki et al., 2009) under laboratory conditions (i.e. dehydration), and only a few have been evaluated under field environmental conditions. Moreover, most of these studies showed stress tolerance and/or survival, but not the effects of the different stress conditions on plant productivity (Parry et al., 2005). Thus, more emphasis should be given to the study of the response of crop plants to water deficiency at the reproductive stage and under field conditions. The new powerful molecular and genetic tools, functional genomics, metabolomics, and proteomics will contribute to shed light on key regulators of the plant response to stresses. The use of the indentified candidate genes as targets for genetic engineering will accelerate the developing of crop plants with improved stress tolerance and increase yields.

Acknowledgments

This study was supported by grants from NSF-IOS-0802112, CGIAR GCP#3008.03, UC Discovery #bio06-10627, and the Will W. Lester Endowment of University of California. Z.P. was supported by Vaadia-BARD postdoctoral Fellowship Award No. FI-419-08 from the United States-Israel Binational Agricultural Research and Development Fund (BARD).

References

Abebe, T., Guenzi, A. C., Martin, B., & Cushman, J. C. (2003). Tolerance of mannitol-accumulating transgenic wheat to water stress and salinity. *Plant Physiology, 131*, 1748–1755.

Agarwal, P., Agarwal, P., Reddy, M., & Sopory, S. (2006). Role of DREB transcription factors in abiotic and biotic stress tolerance in plants. *Plant Cell Reports, 25*, 1263–1274.

Aharon, R., Shahak, Y., Wininger, S., Bendov, R., Kapulnik, Y., & Galili, G. (2003). Over-expression of a plasma membrane aquaporin in transgenic tobacco improves plant vigor under favorable growth conditions but not under drought or salt stress. *Plant Cell, 15*, 439–447.

Ahmad, R., Kim, M., Back, K.-H., Kim, H.-S., Lee, H.-S., Kwon, S.-Y., et al. (2008). Stress-induced expression of choline oxidase in potato plant chloroplasts confers enhanced tolerance to oxidative, salt, and drought stresses. *Plant Cell Reports, 27*, 687–698.

Alvarez, S., Marsh, E. L., Schroeder, S. G., & Schachtman, D. P. (2008). Metabolomic and proteomic changes in the xylem sap of maize under drought. *Plant, Cell & Environment, 31*, 325–340.

Apse, M. P., Aharon, G. S., Snedden, W. A., & Blumwald, E. (1999). Salt tolerance conferred by overexpression of a vacuolar Na^+/H^+ antiport in *Arabidopsis*. *Science, 285*, 1256–1258.

Apse, M. P., & Blumwald, E. (2002). Engineering salt tolerance in plants. *Current Opinion in Biotechnology, 13*, 146–150.

Araus, J. L., Slafer, G. A., Royo, C., & Serret, M. D. (2008). Breeding for yield potential and stress adaptation in cereals. *Critical Reviews in Plant Sciences, 27*, 377–412.

Argueso, C. T., Ferreira, F. J., & Kieber, J. J. (2009). Environmental perception avenues: The interaction of cytokinin and environmental response pathways. *Plant, Cell & Environment, 32*, 1147–1160.

Arrillaga, I., Gil-Mascarell, R., Gisbert, C., Sales, E., Montesinos, C., Serrano, R., et al. (1998). Expression of the yeast *HAL2* gene in tomato increases the *in vitro* salt tolerance of transgenic progenies. *Plant Science, 136*, 219–226.

Babu, R. C., Nguyen, B. D., Chamarerk, V., Shanmugasundaram, P., Chezhian, P., Jeyaprakash, P., et al. (2003). Genetic analysis of drought resistance in rice by molecular markers: Association between secondary traits and field performance. *Crop Science, 43*, 1457–1469.

Bao, A.-K., Wang, S.-M., Wu, G.-Q., Xi, J.-J., Zhang, J.-L., & Wang, C.-M. (2009). Overexpression of the *Arabidopsis* H^+-PPase enhanced resistance to salt and drought stress in transgenic alfalfa (*Medicago sativa* L.). *Plant Science, 176*, 232–240.

Barnabas, B., Jager, K., & Feher, A. (2008). The effect of drought and heat stress on reproductive processes in cereals. *Plant, Cell & Environment, 31*, 11–38.

Bartels, D., & Sunkar, R. (2005). Drought and salt tolerance in plants. *Critical Reviews in Plant Sciences, 24*, 23–58.

Bayuelo-Jimenez, J. S., Debouck, D. G., & Lynch, J. P. (2002). Salinity tolerance in Phaseolus species during early vegetative growth. *Crop Science, 42*, 2184–2192.

Bhatnagar-Mathur, P., Vadez, V., Jyostna Devi, M., Lavanya, M., Vani, G., & Sharma, K. (2009). Genetic engineering of chickpea (*Cicer arietinum* L.) with the *P5CSF129A* gene for osmoregulation with implications on drought tolerance. *Molecular Breeding, 23*, 591–606.

Bohnert, H. J., Gong, Q., Li, P., & Ma, S. (2006). Unraveling abiotic stress tolerance mechanisms – getting genomics going. *Current Opinion in Plant Biology, 9*, 180–188.

Bohnert, H. J., Nelson, D. E., & Jensen, R. G. (1995). Adaptations to environmental stresses. *Plant Cell, 7*, 1099–1111.

Bordas, M., Montesinos, C., Dabauza, M., Salvador, A., Roig, L. A., Serrano, R., et al. (1997). Transfer of the yeast salt tolerance gene HAL1 to *Cucumis melo* L. cultivars and *in vitro* evaluation of salt tolerance. *Transgenic Research, 6*, 41–50.

Boyer, J. S. (1982). Plant productivity and environment. *Science, 218*, 443–448.

Bray, E. A. (1993). Molecular responses to water deficit. *Plant Physiology, 103*, 1035–1040.

Bray, E. A. (1997). Plant responses to water deficit. *Trends in Plant Science, 2*, 48–54.

Bruce, W. B., Edmeades, G. O., & Barker, T. C. (2002). Molecular and physiological approaches to maize improvement for drought tolerance. *Journal of Experimental Botany, 53*, 13–25.

Campos, H., Cooper, M., Habben, J. E., Edmeades, G. O., & Schussler, J. R. (2004). Improving drought tolerance in maize: A view from industry. *Field Crops Research, 90*, 19–34.

Cassman, K. G. (1999). Ecological intensification of cereal production systems: Yield potential, soil quality, and precision agriculture. *Proceedings of the National Academy of Sciences of the United States of America, 96*, 5952–5959.

Castiglioni, P., Warner, D., Bensen, R. J., Anstrom, D. C., Harrison, J., Stoecker, M., et al. (2008). Bacterial RNA chaperones confer abiotic stress tolerance in plants and improved grain yield in maize under water-limited conditions. *Plant Physiology, 147*, 446–455.

Cattivelli, L., Rizza, F., Badeck, F.-W., Mazzucotelli, E., Mastrangelo, A. M., Francia, E., et al. (2008). Drought tolerance improvement in crop plants: An integrated view from breeding to genomics. *Field Crops Research, 105*, 1–14.

Chaves, M. M., Maroco, J. P., & Pereira, J. S. (2003). Understanding plant responses to drought – from genes to the whole plant. *Functional Plant Biology, 30*, 239–264.

Chen, J.-Q., Meng, X.-P., Zhang, Y., Xia, M., & Wang, X.-P. (2008). Over-expression of *OsDREB* genes lead to enhanced drought tolerance in rice. *Biotechnology Letters, 30*, 2191–2198.

Chen, T. H. H., & Murata, N. (2008) Glycinebetaine: An effective protectant against abiotic stress in plants, 13, 499–505.

Chimenti, C. A., Pearson, J., & Hall, A. J. (2002). Osmotic adjustment and yield maintenance under drought in sunflower. *Field Crops Research, 75*, 235–246.

Collard, B. C. Y., & Mackill, D. J. (2008). Marker-assisted selection: An approach for precision plant breeding in the twenty-first century. *Philosophical Transactions of the Royal Society B: Biological Sciences, 363*, 557–572.

Collins, N. C., Tardieu, F., & Tuberosa, R. (2008). Quantitative trait loci and crop performance under abiotic stress: Where do we stand? *Plant Physiology, 147*, 469–486.

Condon, A. G., & Hall, A. E. (1997). *Adaptation to diverse environments: Genotypic variation in water-use efficiency within crop species*. San Diego, CA: Academic Press.

Condon, A. G., Richards, R. A., Rebetzke, G. J., & Farquhar, G. D. (2004). Breeding for high water-use efficiency. *Journal of Experimental Botany, 55*, 2447–2460.

Cowan, I. R. (1986). *Economics of carbon fixation in higher plants*. Cambridge: Cambridge University Press.

Cutler, S. R., Rodriguez, P. L., Finkelstein, R. R., & Abrams, S. R. (2010). Abscisic acid: Emergence of a core signaling network. *Annual Review of Plant Biology, 61*.

Dehan, K., & Tal, M. (1978). Salt tolerance in the wild relatives of the cultivated tomato: Responses of *Solanum pennellii* to high salinity. *Irrigation Science, 1*, 71–76.

Duvick, D. N. (2001). Biotechnology in the 1930s: The development of hybrid maize. *Nature Reviews Genetics, 2*, 69–74.

Ellul, P., Ríos, G., Atarés, A., Roig, L. A., Serrano, R., & Moreno, V. (2003). The expression of the Saccharomyces cerevisiae *HAL1* gene increases salt tolerance in transgenic watermelon [*Citrullus lanatus* (Thunb.) Matsun. & Nakai.]. *Theoretical and Applied Genetics, 107*, 462–469.

Evenson, R. E., & Gollin, D. (2003). Assessing the impact of the Green Revolution, 1960 to 2000. *Science, 300*, 758–762.

Farquhar, G. D., O'Leary, M. H., & Berry, J. A. (1982). On the relationship between carbon isotope discrimination and the intercellular carbon dioxide concentration in leaves. *Australian Journal of Plant Physiology, 9*, 121–137.

Flowers, T. J. (2004). Improving crop salt tolerance. *Journal of Experimental Botany, 55*, 307–319.

Gan, S., & Amasino, R. M. (1997). Making sense of senescence. Molecular genetic regulation and manipulation of leaf senescence. *Plant Physiology, 113*, 313–319.

Gao, S., Zhang, H., Tian, Y., Li, F., Zhang, Z., Lu, X., et al. (2008). Expression of *TERF1* in rice regulates expression of stress-responsive genes and enhances tolerance to drought and high-salinity. *Plant Cell Reports, 27*, 1787–1795.

Garg, A. K., Kim, J.-K., Owens, T. G., Ranwala, A. P., Choi, Y. D., Kochian, L. V., et al. (2002). Trehalose accumulation in rice plants confers high tolerance levels to different abiotic stresses. *Proceedings of the National Academy of Sciences of the United States of America, 99*, 15898–15903.

Gaxiola, R. A., Li, J., Undurraga, S., Dang, L. M., Allen, G. J., Alper, S. L., et al. (2001). Drought- and salt-tolerant plants result from overexpression of the *AVP1* H^+-pump. *Proceedings of the National Academy of Sciences of the United States of America, 98*, 11444–11449.

Geerts, S., & Raes, D. (2009). Deficit irrigation as an on-farm strategy to maximize crop water productivity in dry areas. *Agricultural Water Management, 96*, 1275–1284.

Gisbert, C., Rus, A. M., Bolarin, M. C., Lopez-Coronado, J. M., Arrillaga, I., Montesinos, C., et al. (2000). The yeast *HAL1* gene improves salt tolerance of transgenic tomato. *Plant Physiology, 123*, 393–402.

Godfray, H. C. J., Beddington, J. R., Crute, I. R., Haddad, L., Lawrence, D., Muir, J. F., et al. (2010). Food security: The challenge of feeding 9 billion people. *Science, 327*, 812–818.

Gur, A., & Zamir, D. (2004). Unused natural variation can lift yield barriers in plant breeding. *PLoS Biology, 2*, e245.

Hamdy, A., Ragab, R., & Scarascia-Mugnozza, E. (2003). Coping with water scarcity: Water saving and increasing water productivity. *Irrigation and Drainage, 52*, 3–20.

Hare, P. D., Cress, W. A., & van Staden, J. (1997). The involvement of cytokinins in plant responses to environmental stress. *Plant Growth Regulation, 23*, 79–103.

Hauser, F., & Horie, T. (2010). A conserved primary salt tolerance mechanism mediated by *HKT* transporters: A mechanism for sodium exclusion and maintenance of high K^+/Na^+ ratio in leaves during salinity stress. *Plant, Cell & Environment, 33*, 552–565.

He, C., Yan, J., Shen, G., Fu, L., Holaday, A. S., Auld, D., et al. (2005). Expression of an *Arabidopsis* vacuolar sodium/proton antiporter gene in cotton improves photosynthetic performance under salt conditions and increases fiber yield in the field. *Plant and Cell Physiology, 46*, 1848–1854.

Hirayama, T., & Shinozaki, K. (2010). Research on plant abiotic stress responses in the post-genome era: Past, present and future. *The Plant Journal, 61*, 1041–1052.

Hu, H., Dai, M., Yao, J., Xiao, B., Li, X., Zhang, Q., et al. (2006). Overexpressing a *NAM, ATAF*, and *CUC* (NAC) transcription factor enhances drought resistance and salt tolerance in rice. *Proceedings of the National Academy of Sciences, 103*, 12987–12992.

Huang, D., Wu, W., Abrams, S. R., & Cutler, A. J. (2008). The relationship of drought-related gene expression in *Arabidopsis thaliana* to hormonal and environmental factors. *Journal of Experimental Botany, 59*, 2991–3007.

Huang, X.-Y., Chao, D.-Y., Gao, J.-P., Zhu, M.-Z., Shi, M., & Lin, H.-X. (2009). A previously unknown zinc finger protein, *DST*, regulates drought and salt tolerance in rice via stomatal aperture control. *Genes & Development, 23*, 1805–1817.

Ito, Y., Katsura, K., Maruyama, K., Taji, T., Kobayashi, M., Seki, M., et al. (2006). Functional analysis of rice *DREB1/CBF*-type transcription factors involved in cold-responsive gene expression in transgenic rice. *Plant and Cell Physiology, 47*, 141–153.

Jaspers, P., & Kangasjärvi, J. (2010). Reactive oxygen species in abiotic stress signaling. *Physiologia Plantarum, 138*, 405–413.

Jeong, J. S., Kim, Y. S., Baek, K. H., Jung, H., Ha, S.-H., Do Choi, Y., et al. (2010). Root-specific expression of *OsNAC10* improves drought tolerance and grain yield in rice under field drought conditions. *Plant Physiology, 153*, 185–197.

Juenger, T. E., Mckay, J. K., Hausmann, N., Keurentjes, J. J. B., Sen, S., Stowe, K. A., et al. (2005). Identification and characterization of QTL underlying whole-plant physiology in *Arabidopsis thaliana*: $\delta^{13}C$, stomatal conductance and transpiration efficiency. *Plant, Cell & Environment, 28*, 697–708.

Karaba, A., Dixit, S., Greco, R., Aharoni, A., Trijatmiko, K. R., Marsch-Martinez, N., et al. (2007). Improvement of water use efficiency in rice by expression of *HARDY*, an *Arabidopsis* drought and salt tolerance gene. *Proceedings of the National Academy of Sciences, 104*, 15270–15275.

Karakas, B., Ozias-Akins, P., Stushnoff, C., Suefferheld, M., & Rieger, M. (1997). Salinity and drought tolerance of mannitol-accumulating transgenic tobacco. *Plant, Cell and Environment, 20*, 609–616.

Katsuhara, M., Koshio, K., Shibasaka, M., Hayashi, Y., Hayakawa, T., & Kasamo, K. (2003). Over-expression of a barley aquaporin increased the shoot/root ratio and raised salt sensitivity in transgenic rice plants. *Plant & Cell Physiology, 44*, 1378–1383.

Kim, E. H., Kim, Y. S., Park, S.-H., Koo, Y. J., Choi, Y. D., & Chung, Y.-Y., et al. (2009). Methyl jasmonate reduces grain yield by mediating stress signals to alter spikelet development in rice. *Plant Physiology, 149*, 1751–1760.

Kishor, P., Hong, Z., Miao, G. H., Hu, C., & Verma, D. (1995). Overexpression of D^1-pyrroline-5-carboxylate synthetase increases proline production and confers osmotolerance in transgenic plants. *Plant Physiology, 108*, 1387–1394.

Klein, J. D., Mufradi, I., Cohen, S., Hebbe, Y., Asido, S., Dolgin, B., et al. (2002). Establishment of wheat seedlings after early sowing and germination in an arid Mediterranean environment. *Agronomy Journal, 94*, 585–593.

Knepper, M. A. (1994). The aquaporin family of molecular water channels. *Proceedings of the National Academy of Sciences of the United States of America, 91*, 6255–6258.

Lee, Y.-P., Kim, S.-H., Bang, J.-W., Lee, H.-S., Kwak, S.-S., & Kwon, S.-Y. (2007). Enhanced tolerance to oxidative stress in transgenic tobacco plants expressing three antioxidant enzymes in chloroplasts. *Plant Cell Reports, 26*, 591–598.

Levitt, J. (1972). *Responses of plant to environmental stress*. New York: Academic Press.

Lian, H.-L., Yu, X., Ye, Q., Ding, X.-S., Kitagawa, Y., & Kwak, S.-S., et al. (2004). The role of aquaporin RWC3 in drought avoidance in rice. *Plant & Cell Physiology, 45*, 481–489.

Lin, H. X., Zhu, M. Z., Yano, M., Gao, J. P., Liang, Z. W., Su, W. A., et al. (2004). QTLs for Na^+ and K^+ uptake of the shoots and roots controlling rice salt tolerance. *Theoretical and Applied Genetics, 108*, 253–260.

Liu, H., Wang, Q., Yu, M., Zhang, Y., Wu, Y., & Zhang, H. (2008). Transgenic salt-tolerant sugar beet (*Beta vulgaris* L.) constitutively expressing an *Arabidopsis thaliana* vacuolar Na^+/H^+ antiporter gene, *AtNHX3*, accumulates more soluble sugar but less salt in storage roots. *Plant, Cell & Environment, 31*, 1325–1334.

Liu, L., Lafitte, R., & Guan, D. (2004). Wild *Oryza* species as potential sources of drought-adaptive traits. *Euphytica, 138*, 149–161.

Ludlow, M. M., & Muchow, R. C. (1990). A critical evaluation of traits for improving crop yields in water-limited environments. In N. C. Brady (Ed.), *Advances in agronomy* (pp. 107–153). San Diego, CA: Academic Press.

Lukens, L. N., & Zhan, S. (2007). The plant genome's methylation status and response to stress: Implications for plant improvement. *Current Opinion in Plant Biology, 10*, 317–322.

Lv, S., Yang, A., Zhang, K., Wang, L., & Zhang, J. (2007). Increase of glycinebetaine synthesis improves drought tolerance in cotton. *Molecular Breeding, 20*, 233–248.

Ma, Q.-H. (2008). Genetic engineering of cytokinins and their application to agriculture. *Critical Reviews in Biotechnology, 28*, 213–232.

Malik, V., & Wu, R. (2005). Transcription factor *AtMyb2* increased salt-stress tolerance in rice (*Oryza sativa* L.). *Rice Genetics Newsletter, 22*, 63.

Masle, J., Gilmore, S. R., & Farquhar, G. D. (2005). The ERECTA gene regulates plant transpiration efficiency in *Arabidopsis*. *Nature, 436*, 866–870.

McCouch, S. (2004). Diversifying selection in plant breeding. *PLoS Biology, 2*, e347.

Merewitz, E. B., Gianfagna, T., & Huang, B. (2011). Photosynthesis, water use, and root viability under water stress as affected by expression of *SAG12-ipt* controlling cytokinin synthesis in *Agrostis stolonifera*. *Journal of Experimental Botany, 62*, 383–395.

Mimura, T., Kura-Hotta, M., Tsujimura, T., Ohnishi, M., Miura, M., Okazaki, Y., et al. (2003). Rapid increase of vacuolar volume in response to salt stress. *Planta, 216*, 397–402.

Minoia, S., Petrozza, A., D'Onofrio, O., Piron, F., Mosca, G., Sozio, G., et al. (2010). A new mutant genetic resource for tomato crop improvement by TILLING technology. *BMC Research Notes*, *3*, 69.

Mittler, R., & Blumwald, E. (2010). Genetic engineering for modern agriculture: Challenges and perspectives. *Annual Review of Plant Biology*, *61*, 443–462.

Mizrahi, Y., Taleisnik, E., Kagan-Zur, V., Zohar, Y., Offenbach, R., Matan, E., et al. (1988). A saline irrigation regime for improving tomato fruit quality without reducing yield. *Journal of American Society Horticultural Science*, *113*, 202–205.

Moinuddin, , & Khanna-Chopra, R. (2004). Osmotic adjustment in chickpea in relation to seed yield and yield parameters. *Crop Science*, *44*, 449–455.

Mooney, H. A., Pearcy, R. W., & Ehleringer, J. (1987). Plant physiological ecology today. *BioScience*, *37*, 18–20.

Morgan, J. (1983). Osmoregulation as a selection criterion for drought tolerance in wheat. *Australian Journal of Agricultural Research*, *34*, 607–614.

Morgan, J. M. (1984). Osmoregulation and water stress in higher plants. *Annual Review of Plant Physiology*, *35*, 299–319.

Munns, R. (2002). Comparative physiology of salt and water stress. *Plant, Cell & Environment*, *25*, 239–250.

Munns, R., & Tester, M. (2008). Mechanisms of salinity tolerance. *Annual Review of Plant Biology*, *59*, 651–681.

Murchie, E. H., Pinto, M., & Horton, P. (2009). Agriculture and the new challenges for photosynthesis research. *New Phytologist*, *181*, 532–552.

Nelson, D. E., Repetti, P. P., Adams, T. R., Creelman, R. A., Wu, J., Warner, D. C., et al. (2007). Plant nuclear factor Y (NF-Y) B subunits confer drought tolerance and lead to improved corn yields on water-limited acres. *Proceedings of the National Academy of Sciences*, *104*, 16450–16455.

Nemhauser, J. L., Hong, F., & Chory, J. (2006). Different plant hormones regulate similar processes through largely nonoverlapping transcriptional responses. *Cell*, *126*, 467–475.

Nevo, E., Gorham, J., & Beiles, A. (1992). Variation for ^{22}Na uptake in wild emmer wheat, *Triticum dicoccoides* in Israel: Salt tolerance resources for wheat improvement. *Journal of Experimental Botany*, *43*, 511–518.

Nilson, S. E., & Assmann, S. M. (2010). The a-subunit of the *Arabidopsis* heterotrimeric G protein, *GPA1*, is a regulator of transpiration efficiency. *Plant Physiology*, *152*, 2067–2077.

Oh, S.-J., Kim, Y. S., Kwon, C.-W., Park, H. K., Jeong, J. S., & Kim, J.-K. (2009). Overexpression of the transcription factor AP37 in rice improves grain yield under drought conditions. *Plant Physiology*, *150*, 1368–1379.

Oh, S.-J., Kwon, C.-W., Choi, D.-W., Song, S. I., & Kim, J.-K. (2007). Expression of barley *HvCBF4* enhances tolerance to abiotic stress in transgenic rice. *Plant Biotechnology Journal*, *5*, 646–656.

Oh, S.-J., Song, S. I., Kim, Y. S., Jang, H.-J., Kim, S. Y., Kim, M., et al. (2005). *Arabidopsis CBF3/DREB1A* and *ABF3* in transgenic rice increased tolerance to abiotic stress without stunting growth. *Plant Physiology*, *138*, 341–351.

Ohta, M., Hayashi, Y., Nakashima, A., Hamada, A., Tanaka, A., Nakamura, T., et al. (2002). Introduction of a Na^+/H^+ antiporter gene from *Atriplex gmelini* confers salt tolerance to rice. *FEBS Letters*, *532*, 279–282.

Ouyang, S.-Q., Liu, Y.-F., Liu, P., Lei, G., He, S.-J., Ma, B., et al. (2010). Receptor-like kinase *OsSIK1* improves drought and salt stress tolerance in rice *Oryza sativa* plants. *The Plant Journal*, *62*, 316–329.

Pardo, J. M. (2010). Biotechnology of water and salinity stress tolerance. *Current Opinion in Biotechnology*, *21*, 185–196.

Park, S., Li, J., Pittman, J. K., Berkowitz, G. A., Yang, H., Undurraga, S., et al. (2005). Up-regulation of a H^+-pyrophosphatase (H^+-PPase) as a strategy to engineer drought-resistant crop plants. *Proceedings of the National Academy of Sciences of the United States of America*, *102*, 18830–18835.

Parry, M. A. J., Flexas, J., & Medrano, H. (2005). Prospects for crop production under drought: Research priorities and future directions. *Annals of Applied Biology*, *147*, 211–226.

Pasapula, V., Shen, G., Kuppu, S., Paez-Valencia, J., Mendoza, M., Hou, P., et al. (2011) Expression of an *Arabidopsis* vacuolar H^+-pyrophosphatase gene (*AVP1*) in cotton improves drought- and salt tolerance and increases fibre yield in the field conditions, *Plant Biotechnology Journal*, *9*, 88–9.

Peleg, Z., Fahima, T., Abbo, S., Krugman, T., Nevo, E., Yakir, D., et al. (2005). Genetic diversity for drought resistance in wild emmer wheat and its ecogeographical associations. *Plant, Cell & Environment*, *28*, 176–191.

Peleg, Z., Fahima, T., Krugman, T., Abbo, S., Yakir, D., Korol, A. B., et al. (2009). Genomic dissection of drought resistance in durum wheat × wild emmer wheat recombinant inbreed line population. *Plant, Cell & Environment*, *32*, 758–779.

Peleg, Z., Saranga, Y., Krugman, T., Abbo, S., Nevo, E., & Fahima, T. (2008). Allelic diversity associated with aridity gradient in wild emmer wheat populations. *Plant, Cell & Environment*, *31*, 39–49.

Peleg, Z., & Blumwald, E. (2011). Hormone balance and abiotic stress tolerance in crop plants. *Current Opinion in Plant Biology*, *14*, 290–295.

Peleg, Z., Reguera, M., Tumimbang, E., Walia, H., & Blumwald, E. (2011). Cytokinin mediated source-sink modifications improve drought tolerance and increase grain yield in rice under water stress. *Plant Biotechnology Journal*, (DOI: 10.1111/j.1467-7652.2010.00584.x). In press.

Peterson, G. A., & Westfall, D. G. (2004). Managing precipitation use in sustainable dryland agroecosystems. *Annals of Applied Biology*, *144*, 127–138.

Plett, D., Safwat, G., Gilliham, M., Skrumsager Møller, I., Roy, S., Shirley, N., et al. (2010). Improved salinity tolerance of rice through cell type-specific expression of *AtHKT1;1*. *PLoS ONE*, *5*, e12571.

Plett, D. C., & Moller, I. S. (2010). Na^+ transport in glycophytic plants: What we know and would like to know. *Plant, Cell & Environment*, *33*, 612–626.

Price, A. H., Cairns, J. E., Horton, P., Jones, H. G., & Griffiths, H. (2002). Linking drought-resistance mechanisms to drought avoidance in upland rice using a QTL approach: Progress and new opportunities to integrate stomatal and mesophyll responses. *Journal of Experimental Botany*, *53*, 989–1004.

Price, A. H., & Tomos, A. D. (1997). Genetic dissection of root growth in rice (*Oryza sativa* L.). II: Mapping quantitative trait loci using molecular markers. *Theoretical and Applied genetics*, *95*, 143–152.

Quan, R., Hu, S., Zhang, Z., Zhang, H., Zhang, Z., & Huang, R. (2010). Overexpression of an ERF transcription factor *TSRF1* improves rice drought tolerance. *Plant Biotechnology Journal*, *8*, 476–488.

Quan, R., Shang, M., Zhang, H., Zhao, Y., & Zhang, J. (2004). Engineering of enhanced glycine betaine synthesis improves drought tolerance in maize. *Plant Biotechnology Journal*, *2*, 477–486.

Rebetzke, G. J., Condon, A. G., Richards, R. A., & Farquhar, G. D. (2002). Selection for reduced carbon isotope discrimination increases aerial biomass and grain yield of rainfed bread wheat. *Crop Science*, *42*, 739–745.

Richards, L. A. (1954). *Diagnosis and improvements of saline and alkali soils* (Vol. 60). Salinity Laboratory DA: US Department of Agriculture, U.S.

Riechmann, J. L., Heard, J., Martin, G., Reuber, L., Jiang, C., Keddie, J., et al. (2000). *Arabidopsis* transcription factors: Genome-wide comparative analysis among eukaryotes. *Science*, *290*, 2105–2110.

Rivero, R. M., Kojima, M., Gepstein, A., Sakakibara, H., Mittler, R., Gepstein, S., et al. (2007). Delayed leaf senescence induces extreme drought tolerance in a flowering plant. *Proceedings of the National Academy of Sciences*, *104*, 19631–19636.

Rivero, R. M., Shulaev, V., & Blumwald, E. (2009). Cytokinin-dependent photorespiration and the protection of photosynthesis during water deficit. *Plant Physiology*, *150*, 1530–1540.

Rivero, R. M., Gimeno, J., Van Deynze, A., Walia, H., & Blumwald, E. (2010). Enhanced cytokinin synthesis in tobacco plants expressing P_{SARK}::*IPT* prevents the degradation of photosynthetic protein complexes during drought. *Plant and Cell Physiology*, *51*, 1929–1941.

Ronin, Y. I., Korol, A. B., & Nevo, E. (1999). Single- and multiple-trait mapping analysis of linked quantitative trait loci: Some asymptotic analytical approximations. *Genetics*, *151*, 387–396.

Rostoks, N., Borevitz, J., Hedley, P., Russell, J., Mudie, S., Morris, J., et al. (2005). Single-feature polymorphism discovery in the barley transcriptome. *Genome Biology*, *6*, R54.

Sade, N., Vinocur, B. J., Diber, A., Shatil, A., Ronen, G., & Nissan., H., et al. (2009). Improving plant stress tolerance and yield production: Is the tonoplast aquaporin SlTIP2:2 a key to isohydric to anisohydric conversion? *New Phytologist*, *181*, 651–661.

Sadok, W., Naudin, P., Boussuge, B., Muller, B., Welcker, C., & Tardieu, F. (2007). Leaf growth rate per unit thermal time follows QTL-dependent daily patterns in hundreds of maize lines under naturally fluctuating conditions. *Plant, Cell Environment*, *30*, 135–146.

Sakamoto, A., & Murata, A. N. (1998). Metabolic engineering of rice leading to biosynthesis of glycinebetaine and tolerance to salt and cold. *Plant Molecular Biology*, *38*, 1011–1019.

Salvi, S., & Tuberosa, R. (2005). To clone or not to clone plant QTLs: Present and future challenges. *Trends in Plant Science*, *10*, 297–304.

Saranga, Y., Jiang, C.-X., Wright, R. J., Yakir, D., & Paterson, A. H. (2004). Genetic dissection of cotton physiological responses to arid conditions and their inter-relationships with productivity. *Plant, Cell & Environment*, *27*, 263–277.

Schroeder, J. I., Kwak, J. M., & Allen, G. J. (2001). Guard cell abscisic acid signalling and engineering drought hardiness in plants. *Nature*, *410*, 327–330.

Seki, M., Okamoto, M., Matsui, A., Kim, J.-M., Kurihara, Y., Ishida, J., et al. (2009) Microarray analysis for studying the abiotic stress responses in plants, 333-355.

Serraj, R., & Sinclair, T. R. (2002). Osmolyte accumulation: Can it really help increase crop yield under drought conditions? *Plant, Cell & Environment*, *25*, 333–341.

Shannon, M. C. (1997). Adaptation of plants to salinity. In L. S. Donald (Ed.), *Advances in agronomy* (pp. 75–120). San Diego, CA: Academic Press.

Shi, H., Lee, B.-h., Wu, S.-J., & Zhu, J.-K. (2003). Overexpression of a plasma membrane Na^+/H^+ antiporter gene improves salt tolerance in *Arabidopsis thaliana*. *Nature Biotechnology*, *21*, 81–85.

Su, J., & Wu, R. (2004). Stress-inducible synthesis of proline in transgenic rice confers faster growth under stress conditions than that with constitutive synthesis. *Plant Science*, *166*, 941–948.

Tangpremsri, T., Fukai, S., & Fischer, K. S. (1995). Growth and yield of sorghum lines extracted from a population for differences in osmotic adjustment. *Australian Journal of Agricultural Research*, 61–74.

Tanksley, S. D., & McCouch, S. R. (1997). Seed banks and molecular maps: Unlocking genetic potential from the wild. *Science*, *277*, 1063–1066.

Thomas, J. C., Sepahi, M., Arendall, B., & Bohnert, H. J. (1995). Enhancement of seed germination in high salinity by engineering mannitol expression in *Arabidopsis thaliana*. *Plant, Cell Environment*, *18*, 801–806.

Thompson, A. J., Andrews, J., Mulholland, B. J., McKee, J. M. T., Hilton, H. W., Horridge, J. S., et al. (2007). Overproduction of abscisic acid in tomato increases transpiration efficiency and root hydraulic conductivity and influences leaf expansion. *Plant Physiology*, *143*, 1905–1917.

Tian, L., Huang, C., Yu, R., Liang, R., Li, Z., Zhang, L., et al. (2006). Overexpression *AtNHX1* confers salt-tolerance of transgenic tall fescue. *African Journal of Biotechnology*, *5*, 1041–1044.

Till, B., Cooper, J., Tai, T., Colowit, P., Greene, E., Henikoff, S., et al. (2007). Discovery of chemically induced mutations in rice by TILLING. *BMC Plant Biology*, *7*, 19.

Till, B., Reynolds, S., Weil, C., Springer, N., Burtner, C., Young, K., et al. (2004). Discovery of induced point mutations in maize genes by TILLING. *BMC Plant Biology*, *4*, 12.

Till, B. J., Comai, L., & Henikoff, S. (2007). Tilling and Ecotilling for Crop Improvement. In R. K. Varshney & R. Tuberosa (Eds.), *Genomics-assisted crop improvement* (pp. 333–349). Netherlands: Springer.

Torii, K. U., Mitsukawa, N., Oosumi, T., Matsuura, Y., Yokoyama, R., Whittier, R. F., et al. (1996). The *Arabidopsis ERECTA* gene encodes a putative receptor protein kinase with extracellular leucine-rich repeats. *Plant Cell*, *8*, 735–746.

Tuberosa, R., & Salvi, S. (2006). Genomics-based approaches to improve drought tolerance of crops. *Trends in Plant Science*, *11*, 405–412.

Turner, N. C., Abbo, S., Berger, J. D., Chaturvedi, S., French, R. J., Ludwig, C., et al. (2007). Osmotic adjustment in chickpea (*Cicer arietinum* L.) results in no yield benefit under terminal drought. *Journal of Experimental Botany*, *58*, 187–194.

Tyerman, S. D., Niemietz, C. M., & Bramley, H. (2002). Plant aquaporins: Multifunctional water and solute channels with expanding roles. *Plant, Cell & Environment*, *25*, 173–194.

Uauy, C., Paraiso, F., Colasuonno, P., Tran, R., Tsai, H., Berardi, S., et al. (2009). A modified TILLING approach to detect induced mutations in tetraploid and hexaploid wheat. *BMC Plant Biology*, *9*, 115.

Umezawa, T., Fujita, M., Fujita, Y., Yamaguchi-Shinozaki, K., & Shinozaki, K. (2006). Engineering drought tolerance in plants: Discovering and tailoring genes to unlock the future. *Current Opinion in Biotechnology*, *17*, 113–122.

Urano, K., Kurihara, Y., Seki, M., & Shinozaki, K. (2010). "Omics" analyses of regulatory networks in plant abiotic stress responses. *Current Opinion in Plant Biology*, *13*, 132–138.

Vendruscolo, E. C. G., Schuster, I., Pileggi, M., Scapim, C. A., Molinari, H. B. C., Marur, C. J., et al. (2007). Stress-induced synthesis of proline confers tolerance to water deficit in transgenic wheat. *Journal of Plant Physiology*, *164*, 1367–1376.

Verma, D., Singla-Pareek, S., Rajagopal, D., Reddy, M., & Sopory, S. (2007). Functional validation of a novel isoform of Na^+/H^+ antiporter from *Pennisetum glaucum* for enhancing salinity tolerance in rice. *Journal of Biosciences*, *32*, 621–628.

Vij, S., & Tyagi, A. K. (2007). Emerging trends in the functional genomics of the abiotic stress response in crop plants. *Plant Biotechnology Journal*, *5*, 361–380.

Vinocur, B., & Altman, A. (2005). Recent advances in engineering plant tolerance to abiotic stress: Achievements and limitations. *Current Opinion in Biotechnology*, *16*, 123–132.

Vorosmarty, C. J., Green, P., Salisbury, J., & Lammers, R. B. (2000). Global water resources: Vulnerability from climate change and population growth. *Science*, *289*, 284–288.

Wang, F.-Z., Wang, Q.-B., Kwon, S.-Y., Kwak, S.-S., & Su, W.-A. (2005). Enhanced drought tolerance of transgenic rice plants expressing a pea manganese superoxide dismutase. *Journal of Plant Physiology*, *162*, 465–472.

Wang, G.-P., Hui, Z., Li, F., Zhao, M.-R., Zhang, J., & Wang, W. (2010) Improvement of heat and drought photosynthetic tolerance in wheat by overaccumulation of glycinebetaine. *Plant Biotechnology Reports*, *4*, 213–222.

Wang, W. X., Vinocur, B., & Altman, A. (2003). Plant responses to drought, salinity and extreme temperatures: Towards genetic engineering for stress tolerance. *Planta*, *218*, 1–14.

Wilkinson, S., & Davies, W. J. (2010). Drought, ozone, ABA and ethylene: New insights from cell to plant to community. *Plant, Cell & Environment*, *33*, 510–525.

Witcombe, J. R., Hollington, P. A., Howarth, C. J., Reader, S., & Steele, K. A. (2008). Breeding for abiotic stresses for sustainable agriculture. *Philosophical Transactions of the Royal Society B: Biological Sciences*, *363*, 703–716.

Wu, X., Shiroto, Y., Kishitani, S., Ito, Y., & Toriyama, K. (2009). Enhanced heat and drought tolerance in transgenic rice seedlings overexpressing *OsWRKY11* under the control of *HSP101* promoter. *Plant Cell Reports*, *28*, 21–30.

Wu, Y.-Y., Chen, Q.-J., Chen, M., Chen, J., & Wang, X.-C. (2005). Salt-tolerant transgenic perennial ryegrass (*Lolium perenne* L.) obtained by Agrobacterium tumefaciens-mediated transformation of the vacuolar Na^+/H^+ antiporter gene. *Plant Science*, *169*, 65–73.

Xiao, B.-Z., Chen, X., Xiang, C.-B., Tang, N., Zhang, Q.-F., & Xiong, L.-Z. (2009). Evaluation of seven function-known candidate genes for their effects on improving drought resistance of transgenic rice under field conditions. *Molecular Plant*, *2*, 73–83.

Xu, D.-Q., Huang, J., Guo, S.-Q., Yang, X., Bao, Y.-M., Tang, H.-J., et al. (2008). Overexpression of a *TFIIIA*-type zinc finger protein gene *ZFP252* enhances drought and salt tolerance in rice (*Oryza sativa* L.). *FEBS Letters*, *582*, 1037–1043.

Xue, Z.-Y., Zhi, D.-Y., Xue, G.-P., Zhang, H., Zhao, Y.-X., & Xia, G.-M. (2004). Enhanced salt tolerance of transgenic wheat (*Tritivum aestivum* L.) expressing a vacuolar Na^+/H^+ antiporter gene with improved grain yields in saline soils in the field and a reduced level of leaf Na^+. *Plant Science*, *167*, 849–859.

Yamaguchi-Shinozaki, K., & Shinozaki, K. (1994). A novel cis-acting element in an *Arabidopsis* gene is involved in responsiveness to drought, low-temperature, or high-salt stress. *Plant Cell*, *6*, 251–264.

Yan, J. Q., He, C. X., Wang, J., Mao, Z. H., Holaday, S. A., Allen, R. D., et al. (2004). Overexpression of the *Arabidopsis* 14-3-3 protein GF14 λ in cotton leads to a "Stay-Green" phenotype and improves stress tolerance under moderate drought conditions. *Plant and Cell Physiology*, *45*, 1007–1014.

Yonamine, I., Yoshida, K., Kido, K., Nakagawa, A., Nakayama, H., & Shinmyo, A. (2004). Overexpression of *NtHAL3* genes confers increased levels of proline biosynthesis and the enhancement of salt tolerance in cultured tobacco cells. *Journal of Experimental Botany*, *55*, 387–395.

Zeller, G., Henz, S. R., Widmer, C. K., Sachsenberg, T., Rätsch, G., Weigel, D., et al. (2009). Stress-induced changes in the *Arabidopsis thaliana* transcriptome analyzed using whole-genome tiling arrays. *The Plant Journal*, *58*, 1068–1082.

Zhang, H., Dong, H., Li, W., Sun, Y., Chen, S., & Kong, X. (2009). Increased glycine betaine synthesis and salinity tolerance in *AhCMO* transgenic cotton lines. *Molecular Breeding*, *23*, 289–298.

Zhang, H.-X., & Blumwald, E. (2001). Transgenic salt-tolerant tomato plants accumulate salt in foliage but not in fruit. *Nature Biotechnology*, *19*, 765–768.

Zhang, H.-X., Hodson, J. N., Williams, J. P., & Blumwald, E. (2001). Engineering salt-tolerant Brassica plants: Characterization of yield and seed oil quality in transgenic plants with increased vacuolar sodium accumulation. *Proceedings of the National Academy of Sciences of the United States of America*, *98*, 12832–12836.

Zhang, P., Wang, W.-Q., Zhang, G.-L., Kaminek, M., Dobrev, P., & Xu, J., et al. (2010). Senescence-inducible expression of *isopentenyl transferase* extends leaf life, increases drought stress resistance and alters cytokinin metabolism in cassava. *Journal of Integrative Plant Biology*, *52*, 653–669.

Zhang, J., Nguyen, H., & Blum, A. (1999). Genetic analysis of osmotic adjustment in crop plants. *Journal of Experimental Botany*, *50*, 291–302.

Zhang, J.-Y., Broeckling, C. D., Blancaflor, E. B., Sledge, M. K., Sumner, L. W., & Wang, Z.-Y. (2005). Overexpression of *WXP1*, a putative *Medicago truncatula* AP2 domain-containing transcription factor gene, increases cuticular wax accumulation and enhances drought tolerance in transgenic alfalfa (*Medicago sativa*). *The Plant Journal*, *42*, 689–707.

Zhao, F., Guo, S., Zhang, H., & Zhao, Y. (2006). Expression of yeast *SOD2* in transgenic rice results in increased salt tolerance. *Plant Science*, *170*, 216–224.

Zhao, F., & Zhang, H. (2006). Salt and paraquat stress tolerance results from co-expression of the Suaeda salsa glutathione S-transferase and catalase in transgenic rice. *Plant Cell, Tissue and Organ Culture*, *86*, 349–358.

Zhao, F.-Y., Zhang, X.-J., Li, P.-H., Zhao, Y.-X., & Zhang, H. (2006). Co-expression of the Suaeda salsa *SsNHX1* and *Arabidopsis AVP1* confer greater salt tolerance to transgenic rice than the single *SsNHX1*. *Molecular Breeding*, *17*, 341–353.

Zheng, X., Chen, B., Lu, G., & Han, B. (2009). Overexpression of a *NAC* transcription factor enhances rice drought and salt tolerance. *Biochemical and Biophysical Research Communications*, *379*, 985–989.

Molecular responses to extreme temperatures

19

Rafael Catalá[1] Aurora Díaz[2] Julio Salinas[1]

[1]*Centro de Investigaciones Biológicas (CIB-CSIC),Madrid, Spain,*
[2]*Instituto de Biología Molecular y Celular de Plantas (IBMCP), CSIC/Universidad Politécnica de Valencia, Valencia, Spain*

TABLE OF CONTENTS

Introduction

In temperate regions, living organisms have to face an ever-changing environment. For plants, as sessile organisms, extreme temperatures are among the environmental factors that most condition their development and distribution. Learning how plants cope with extreme temperatures is interesting not only to understanding how they develop and reproduce, but also from the more practical aspect of designing molecular tools to improve crop tolerance to these adverse environmental conditions. In the past, different laboratories have devoted their study to unveiling the molecular mechanisms underlying plant response to low temperature; more specifically the process of cold acclimation, the adaptive process where many plants from temperate regions increase their freezing tolerance after a period of exposure to low non-freezing temperature (Levitt, 1980). Cold acclimation constitutes a representative example of plant interaction with the environment and how this interaction has conditioned the evolution of some species. Interesting data have been reported on the characterization of the mechanisms and components that govern this adaptive response, which have revealed that plant response to low temperature is very complex and involves many biochemical and physiological changes (Salinas, 2002; Ruelland et al., 2009). Currently, it seems clear that cold acclimation is mediated by an intricate signaling network subjected to tight regulation. Research determined that low temperature response is mainly controlled at the transcriptional level through an extensive reprogramming of gene expression, although post-transcriptional, translational, and post-translational regulation is also required (Catala and Salinas, 2007; Chinnusamy et al., 2010). In addition, plant response to low temperature is also subjected to epigenetic regulation (Chinnusamy et al., 2010).

Temperatures above the optimum are also highly stressful for plants, disturbing cellular homeostasis and leading to severe retardation in growth and development, and even death. Although it could be assumed that the stresses provoked by high and low temperatures are opposite, they cause many common deleterious effects to plants, including the accumulation of reactive oxygen species (ROS). Accordingly, plants subjected to low temperature increase their tolerance to heat (Fu et al., 1998) and, *vice versa*, heat shock improves the chilling tolerance of cold-sensitive species (Saltveit and Hepler, 2004; Pressman et al., 2006). Heat stress response (HSR) has been extensively studied in plants (Wang et al., 2004); however, little attention has been given to identifying common mechanisms and intermediates implicated in both low and high temperature responses. In this chapter, we present a general view about the current knowledge of how plants perceive and transduce the cold signal, and how low temperature response is regulated. We will also discuss what has been reported on how plant responses to low and high temperatures interact and integrate, and on the potential common mechanisms and components that can mediate such an integration.

Plant Response to Low Temperature

Exposure to low temperature initiates numerous physiological perturbations; the first includes changes in cell membrane

DOI: 10.1016/B978-0-12-381466-1.00019-5

fluidity state and fatty acid composition (Murata and Los, 1997; Suzuki et al., 2001). The alteration of fatty acid composition involves an increase in polyunsaturated lipids content that is essential for plant survival under cold conditions due to the deleterious effects of low temperature on membrane-bound metabolic processes such as respiration and photosynthesis (Cossins, 1994). For example, the combination of light and cold induces a reduction in photosynthetic activity and causes photo-oxidative stress that has been correlated with low temperature sensitivity (Harvaux and Kloppstech, 2001). In response to light and cold, plants increase the accumulation of light-screening pigments (i.e., anthocyanins) to defend photosystems from excess light (Mancinelli, 1984; Christie et al., 1994). In addition, the protein composition of photosystems is altered in response to cold, and some of the changes produced have been described as protecting against photo-oxidative stress (Gray et al., 1997; Capel et al., 1998; Kro et al., 1999). The increase in fatty acid content is not the only modification in cell membrane composition induced by low temperature. It has been reported that levels of 38 plasma membrane proteins in *Arabidopsis* increase or decrease at 4°C (Kawamura and Uemura, 2003). Interestingly, some of these proteins are involved in important processes such as CO_2 fixation, membrane repair, membrane protection against osmotic stress, or proteolysis (Kawamura and Uemura, 2003). Exposure to cold also induces the accumulation of many other proteins that are not located in cellular membranes, some of them playing protective roles (Bae et al., 2003; Gao et al., 2009). Thus, it is well known that low temperature induces the production of antifreeze proteins (AFPs; Venketesh and Dayananda, 2008). The mechanism of AFP action occurs through its adsorption on the ice surface, leading to a curved surface, which prevents further growth of ice by the "Kelvin" effect (Venketesh and Dayananda, 2008). AFPs have been isolated from several plant species including winter rye, carrot, and perennial ryegrass (Antikainen et al., 1996; Worrall et al., 1998; Sidebottom et al., 2000). The content of sugar is also altered in plants in response to low temperature. Several studies have found a high correlation between sugar levels and tolerance to low temperature in different plant species (Guy et al., 1992; Sasaki et al., 1996; Sundblad et al., 2001). Finally, exposure to low temperature also provokes structural changes in the plant, such as modifications in the cell wall composition (Wei et al., 2006).

Data obtained indicate that changes induced by low temperature are subjected to tight control, and many different mechanisms implicated in this regulation have been described. In the following paragraphs, we provide a general overview on how the cold signal is perceived, decoded, and transduced, and about the regulatory mechanisms controlling plant response to low temperature.

Low temperature perception

Perception is the first step in the activation of plant response to cold conditions. Although the molecular mechanisms used by plants to perceive low temperature remain unknown, in some cases changes in membrane fluidity seem to function as a biological thermometer. Orvar and colleagues (2000) found a direct relationship between the rigidification of the plasma membrane, the expression of the alfalfa cold-inducible gene *CAS30*, and a higher capacity of alfalfa cells to cold acclimate at warm temperatures. In contrast, an increase in membrane fluidity provokes a decrease in *CAS30* cold-induction and a reduction in the capacity of alfalfa cells to cold acclimate (Orvar et al., 2000). Recent genetic evidence corroborates these results. Vaultier and colleagues (2006) reported the identification of plants defective in the activity of FAD2 desaturase, which have a steeper membrane rigidification in response to cold. In addition, *Arabidopsis* transgenic plants over-expressing another desaturase, *FAD3*, show decreased membrane rigidification in response to temperature reduction (Vaultier et al., 2006). Interestingly, in *fad2* mutant plants the activation of the cold-inducible diacylglycerol kinase pathway occurs at a higher temperature than in wild-type plants, and a lower temperature is necessary to activate this pathway in *FAD3* over-expessing plants (Vaultier et al., 2006). From these results, the authors inferred that the rigidification of the cell membrane is indeed one of the primary events in plant response to low temperature.

Work in *Synechocystis* has provided some clues on how changes in membrane fluidity could be perceived and subsequently transduced. The characterization of *Synechocystis* PCC6803 insertion mutants allowed the identification of two putative histidine kinases, HIK33 and HIK19, and a protein with typical two-component regulator domains, RER1, that acts as a cold-receptor system (Suzuki et al., 2000). The proposed model suggests that HIK33 would be activated by a decrease in membrane fluidity, inducing its own phosphorylation and subsequent transfer of the phosphorus group to HIK19 and finally to RER1. Expression analysis of *hik* mutants revealed that this system controls the expression of several cold-regulated genes (Suzuki et al., 2000). Although a similar mechanism has not been described in plants, different results support the existence of a related system. In this way, several genes encoding histidine kinases related to the perception of different environmental stimuli have been isolated in plants (Urao et al., 2000).

The implication of light perception in regulating plant response to low temperature has been known for a long time. Many plant species, including alfalfa (Dexter, 1933), ivy (Steponkus, 1971), winter rye (Griffith and McIntyre, 1993), or *Arabidopsis* (Wanner and Junttila, 1999), need light to completely develop the freezing tolerance induced by cold acclimation. Moreover, it has been described that cold-induced gene expression controlled by the C-repeat/dehydration responsive element (CRT/DRE) motif (see next section) is compromised in the *phyB* mutant (Kim et al., 2002a), revealing an important role for the light receptor PHYB as a mediator of gene expression in response to low temperature. Additionally, changes in the red to far-red ratio (R/FR) are sufficient to control the expression of *CBF* genes in a circadian-clock-dependent manner and to increase *Arabidopsis* freezing tolerance (Franklin and Whitelam, 2007). On the other hand, the over-expression of *PHYA* in hybrid aspen reduces the critical day length to which these trees are able to cold acclimate, indicating that PHYA functions as a negative regulator of plant response to cold (Olsen et al., 1997). All of these data stress the importance of light perception in low temperature response.

Transducing the low-temperature signal

Changes in temperature perceived by cellular membranes seem to induce an increase in cytoplasmic calcium concentration ($[Ca^{2+}]_{cyt}$). Afterwards, this Ca^{2+} signal presumably would be transmitted through Ca^{2+}-binding proteins to activate reversible phosphorylation/dephosphorylation cascades. Finally, these phosphorylation/dephosphorylation events would control the accumulation of different signaling molecules, such as plant hormones that, in turn, would modulate the activity of different transcription factors and, therefore, the induction of cold-regulated gene expression.

Ca^{2+} as a second messenger in low-temperature response

Most eukaryotic cells use Ca^{2+} to transduce external signals to cytosolic information and, therefore, to control the response to these stimuli. In plants, Ca^{2+} is involved in the response to several environmental conditions (Reddy, 2001). Low temperature induces a transient increase in $[Ca^{2+}]_{cyt}$ in different species, including tobacco (Knight et al., 1991) and *Arabidopsis* (Knight et al., 1996). It seems that Ca^{2+} influx into the cytoplasm is mediated by Ca^{2+}-permeable channels, which are assumed to be solely dependent on the cooling rate, and Ca^{2+} efflux is mediated by Ca^{2+} pumps, which have been shown to be dependent on absolute temperature (Carpaneto et al., 2007). This hypothesis suggests a role for Ca^{2+}-permeable channels as a primary temperature sensor in plants (Plieth, 1999). However, Sangwan and colleagues (2002) reported pharmacological data revealing that, along with Ca^{2+}-permeable channels, mechanosensitive Ca^{2+} channels are also implicated in cold response. Nonetheless, temperature-sensitive Ca^{2+} channels have not been described. Using tobacco protoplasts, it has been demonstrated that the magnitude of the peak of $[Ca^{2+}]_{cyt}$ depends on the cytoskeleton organization (Mazars et al., 1997). In addition to the influx of extracellular Ca^{2+}, Knight and colleagues (1996) showed that the increase in $[Ca^{2+}]_{cyt}$ induced by cold is also dependent on vacuolar Ca^{2+}. Different results confirm that Ca^{2+} influx into the cytoplasm is essential for the cold acclimation process. Extracellular Ca^{2+} chelators and Ca^{2+} channel inhibitors reduce cold-induced gene expression and the cold acclimation capacity of alfalfa cells and *Arabidopsis*; in addition, they alter protein phosphorylation levels during cold acclimation (Monroy et al., 1993). In this way, it has been described that the activation of SAMK, an alfalfa stress-inducible mitogen-activated protein kinase (MAPK; Jonak et al., 1996), by cold depends on Ca^{2+} influx into the cytoplasm and the activation of Ca^{2+}-dependent protein kinases (CDPK; Mazars et al., 1997).

Due to the role of Ca^{2+} as a second messenger in cold response, the regulation of its levels in the cytoplasm and in the different cellular organelles is essential. The basal levels of $[Ca^{2+}]_{cyt}$ need to be restored, both in the cytoplasm and the organelles, immediately after the increase induced by low temperature (Sanders et al., 2002). This is mainly carried out by Ca^{2+}-ATPases and Ca^{2+}/H^{+} exchangers (Knight, 2000). CAX1 (CAtion eXchanger 1) was the first Ca^{2+}/H^{+} exchanger to be described in plants (Hirschi et al., 1996). Subsequently, a highly homologous, VCAX1, has been described in *Vigna radiate* (Ueoka-Nakanishi et al., 1999). CAX1 and VCAX1 localize in the vacuolar membrane (Ueoka-Nakanishi et al., 1999; Cheng et al., 2003). Interestingly, transgenic tobacco plants over-expressing *CAX1* are more sensitive to low temperature than wild-type plants (Hirschi et al., 1996). Moreover, *cax1* null mutants show higher capacity to cold acclimate than wild-type plants; this phenotype correlates with an enhanced induction of the *CBFs* and of the CBF-regulon under low-temperature conditions (Catala et al., 2003).

Several intermediates involved in the transmission of cold-induced Ca^{2+} signal have been identified. In 1996, Knight and colleagues reported that the alteration of the inositol triphosphate (IP3) metabolism affects the rise in $[Ca^{2+}]_{cyt}$ induced by low temperature in *Arabidopsis* seedlings, denoting a possible role for IP3 in decoding the Ca^{2+} signal. A temperature drop induces the activation of two enzymes of the IP3 biosynthetic pathway, PLC and PLD, leading to an increase in phosphatidic acid (PtdOH) content (Ruelland et al., 2002). This activation is blocked by inhibitors of the phosphoinositide signaling pathway and is Ca^{2+} dependent (Ruelland et al., 2002). The identification and characterization of the *Arabidopsis* mutant *fry1* provided the first genetic evidence that IP3 is implicated in abiotic stress responses. *FRY1* encodes an inositol polyphosphate 1-phosphatase involved in IP3 catabolism (Xiong et al., 2004). *fry1* mutation confers higher cold induction of ABA- and stress-responsive genes and increased sensitivity of germination in ABA. However, tolerance to freezing, drought, and salt stresses is lower in *fry1* than in wild-type plants (Xiong et al., 2004). The *fry1* mutant accumulates higher levels of IP3 than wild-type plants in the presence of ABA, suggesting that IP3 acts as a negative regulator attenuating ABA and stress signaling (Xiong et al., 2004).

Because of their expression patterns under cold conditions, calmodulin-like proteins (CaM) are also good candidates to transduce the Ca^{2+} signal. The expression of *NpCaM-1*, which encodes a tobacco CaM, is induced by cold through a Ca^{2+}-dependent pathway (van der Luit et al., 1999). Other Ca^{2+}-dependent cold-inducible genes encoding CaM are *TCH2* and *TCH3* (Polisensky and Braam, 1996). Yang and Poovaiah (2002) identified a family of nuclear proteins in *Arabidopsis* (AtSR) that interact with CaM and with a DNA sequence present in some promoters. The expression of the corresponding genes is rapidly and differentially induced by cold and other related stresses, which suggests their involvement in abiotic stress signaling (Yang and Poovaiah, 2002). In addition, CBL1, an *Arabidopsis* Ca^{2+} sensor showing homology to the regulatory subunit of the calcineurin, has been implicated in plant response to different stresses (Cheong et al., 2003). Transgenic and genetic analysis revealed that CBL1 has a negative role in the modulation of plant response to low and freezing temperatures, most probably by regulating gene expression (Cheong et al., 2003). However, CBL1 positively regulates plant response to other abiotic stresses, such as drought and high salt, indicating that it has a complex role in the integration of adverse environmental conditions. Kim and colleagues (2003) characterized an *Arabidopsis* mutant defective in *CIPK3* expression, a cold-inducible gene that encodes a protein kinase (PK) that interacts with CBL-like proteins. Data support that CIPK3 also constitutes a point of cross-talk between different Ca^{2+}-dependent signaling pathways when plants are exposed to low temperature (Kim et al., 2003).

Different works reveal a possible role for CDPKs in plant response to cold. Alfalfa cell cultures treated with W7, a CDPK antagonist, showed an inhibition of cold-induced gene expression and a decrease in their cold acclimation capacity (Monroy et al., 1993). In maize protoplasts, the expression of *AtCDPK1* is enough to activate the promoter of the barley cold-inducible gene *HVA1* (Sheen, 1996). In rice, a CDPK is activated by cold, manifesting the implication of a post-translational regulation in the function of this type of PK under low-temperature conditions (Martin and Busconi, 2001).

In addition to CDPKs, MAPKs are other PKs that seem to be implicated in transducing the Ca^{2+} signal in the cold. As mentioned previously, Sangwan and colleagues (2002) described the activation of SAMK as mediated by an increase in $[Ca^{2+}]_{cyt}$ induced by cold, which implicates this MAPK in low-temperature signaling. On the other hand, several cold-inducible genes encoding components of the MAPK cascade have been described. Among them, the alfalfa *MMK4*, which encodes a MAPK (Jonak et al., 1996), and the *Arabidopsis AtMPK3*, *AtMEKK1*, and *AtMAP3Kβ3*, which code for a MAPK, a MAPKK (MAP kinase kinase), and a MAPKKK (MAP kinase kinase kinase), respectively (Mizoguchi et al., 1996; Jouannic et al., 1999). AtMEK3 participates in the phosphorylation cascade initiated by the MAPKKK ANP1 in response to H_2O_2 (Kovtun et al., 2000). Interestingly, transgenic tobacco plants over-expressing *NPK1*, an ortholog of *ANP1*, display increased tolerance to low temperature (Kovtun et al., 2000). Teige and colleagues (2004) reported the characterization of an MAPK cascade involved in *Arabidopsis* cold response. This cascade is activated by low temperature and is composed of MEKK1, a MAPKKK, MKK2, a MAPKK, and MPK2/MPK6, which is a MAPK (Teige et al., 2004). MKK2 controls cold acclimation by regulating the expression of at least 152 cold-inducible genes (Teige et al., 2004). Other genes coding for PKs different than MAPKs, such as *PKABA1* from wheat (Holappa and Walker-Simmons, 1995), and *RPK1* (Hong et al., 1997) and *DBF2* from *Arabidopsis* (Lee et al., 1999), are also induced by low temperature. This suggests a possible implication in cold response; constitutive expression of *AtDBF2* in *Arabidopsis* induces the expression of several cold-inducible genes (Lee et al., 1999).

Transduction of a low-temperature signal depends on the reversible phosphorylation of already present proteins. This denotes that, in addition to PKs, phosphatase activity is required (Monroy et al., 1993, 1997, 1998; Tahtiharju et al., 1997). The expression of *AtPP2CA*, an *Arabidopsis* gene encoding a protein phosphatase (PP), is induced by cold (Tahtiharju and Palva, 2001). Transgenic plants over-expressing an antisense version of this gene show increased levels of cold-induced gene expression when exposed to 4°C, and higher capacity to cold acclimate than wild-type plants, providing evidence that AtPP2CA acts as a negative regulator of *Arabidopsis* response to low temperature (Tahtiharju and Palva, 2001).

Other molecules involved in transducing the cold signal

Some plant hormones, including ABA, polyamines, gibberellic acid (GA), nitric oxide (NO), brassinosteroids (BR), and ethylene (ET), also act as intermediates in cold signaling (Figure 19.1).

ABA levels increase in plants exposed to low temperature and exogenous treatment with ABA enhances freezing tolerance (Chen and Gusta, 1983; Lang et al., 1989). In *Cicer arietinum* and wheat, it has been observed that the increase in freezing tolerance induced by exogenous ABA treatments is mediated by changes in the composition of the plasma membrane (i.e., increasing sterols, unsaturated fatty acid, etc.; Bakht et al., 2006). Moreover, ABA-deficient and insensitive mutants have a decreased capacity to cold acclimate, indicating that ABA is essential for a full cold acclimation (Heino et al., 1990; Gilmour and Thomashow, 1991; Llorente et al., 2000). Additionally, ABA is implicated in the constitutive tolerance of *Arabidopsis* to freezing (Llorente et al., 2000). Some components of the ABA-signaling pathway have also been involved in the regulation of cold response. ABI3, a transcriptional activator that regulates ABA-induced gene expression and different ABA-mediated vegetative processes in *Arabidopsis*, enhances the expression of cold-inducible genes and freezing tolerance (Tamminen et al., 2001). An ABA-hypersensitive wheat mutant, *aba27*, shows increased constitutive freezing tolerance and increased expression of ABA-responsive genes under both control and

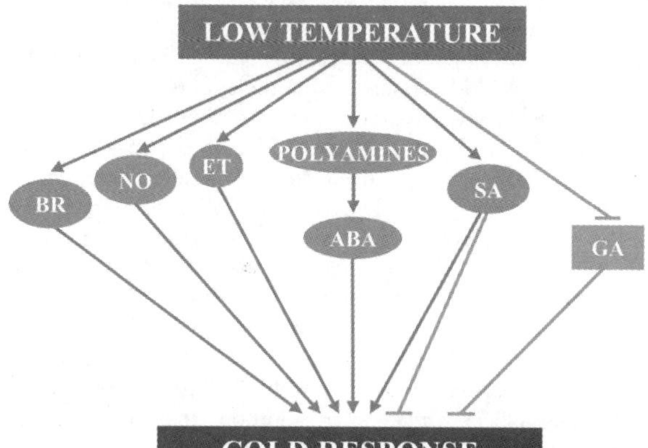

Figure 19.1 • Schematic representation showing the role of plant hormones in low-temperature response. Brassinosteroids (BR), nitric oxide (NO), ethylene (ET), ABA, and polyamines act as positive regulators. Gibberellins (GA) act as negative regulators. Salicylic acid (SA) can act as a positive (cereals, banana) or negative (*Arabidopsis*) regulator. Arrowheads and end lines indicate positive and negative regulation, respectively. Please see color plate section at the end of the book.

low-temperature conditions (Kobayashi et al., 2008a). All of these results manifest a complex role of the ABA-signaling pathway in low-temperature response. Recent works have reported the identification of ABA receptors (Ma et al., 2009; Park et al., 2009a), although their implication in cold response still remains to be determined.

Polyamines accumulate in plants under low-temperature conditions (Lee et al., 1995; Shen et al., 2000; Kim et al., 2002b; Pillai et al., 2004; Cuevas et al., 2008; Kovacs et al., 2010). They are compounds that, because of their biochemical properties, are able to interact with macromolecules such as DNA, RNA, proteins, or phospholipids, regulating their structure and function (Galston and Sawhney, 1990). Kim and colleagues (2002b) reported that cold exposure increases putrescine accumulation in tomato leaves. This increase seems to have a protective role and is negatively controlled by ABA (Kim et al., 2002b). Hummel and colleagues (2004) described that cold tolerance of *Pringlea antiscorbutica* is associated with polyamine accumulation and maintenance. Furthermore, the over-expression of *CaPF1*, a gene encoding a pepper freezing-tolerance-related protein, in pine enhances freezing tolerance due to a high polyamine accumulation (Tang et al., 2007). Recently, Cuevas and co-workers (2008) obtained genetic evidence on the implication of polyamines in the regulation of cold acclimation by modulating ABA biosynthesis and gene expression. They characterized the freezing tolerance of two mutants defective in polyamine biosynthesis, *adc1* and *adc2*, and found that both have a reduced ability to cold acclimate compared to wild-type. Low polyamine levels in *adc1* and *adc2* mutants correlate with low levels of cold-induced ABA accumulation due to a reduced induction of ABA-biosynthetic genes in response to low temperature, pointing out that polyamines positively regulate the accumulation of ABA under low temperature (Cuevas et al., 2008).

The over-expression in tobacco of *GhDREB1*, a cotton gene encoding a transcription factor, increases the tolerance to low temperature, although growth and flowering time were delayed (Shan et al., 2007). *GhDREB1* expression in cotton is induced by low temperature but inhibited by GA, revealing a possible implication of this gene in the regulation of growth via GA (Shan et al., 2007). Recently, genetic evidence has been published demonstrating a reduction in GA function when plants are exposed to low temperature (Achard et al., 2008). It has been proposed that the reduction in plant growth is mainly due to an accumulation of DELLAs, a family of proteins that acts as negative regulators of GA-responsive plant processes (Achard et al., 2008). Furthermore, DELLA accumulation also seems to contribute to the enhanced freezing tolerance that is mediated by *CBF1* expression, although through a pathway that is independent of the CBF-regulon (Achard et al., 2008).

Results revealing an important role for NO in plant response to low temperature have also been described. In pea, low temperature has been reported to increase the activity of NO synthase (NOS) and S-nitrosoglutathione reductase (GSNOR), two enzymes involved in NO biosynthesis (Corpas et al., 2008). In *Arabidopsis*, NO levels increase under cold conditions as a consequence of the induction of nitrate reductase (NR) activity (Zhao et al., 2009). Furthermore, double mutants for NR1 and NR2 (*nia1nia2*), which are unable to accumulate NO after cold exposure, display reduced cold acclimation capacity compared to wild-type plants (Zhao et al., 2009). Molecular analysis of *nia1nia2* mutants revealed that cold induction of *P5CS1* and repression of *ProDH*, and therefore proline accumulation, is much lower in these plants than in wild-type ones, which would account for their decreased cold acclimation capability (Zhao et al., 2009).

Brassinosteroids are a class of steroids implicated in the control of a broad range of plant responses (Sasse, 2003). Results obtained in recent work support a possible role of BR in cold signaling. *Arabidopsis* and *Brassica napus* seedlings treated with exogenous BR display increased tolerance to low temperature that is dependent on changes in cold-induced gene expression (Kagale et al., 2007). Xia and colleagues (2009) found that BRs also positively regulate tolerance to low temperature in cucumber, controlling NADPH oxidase activity and H_2O_2 accumulation.

Exposure to low temperature induces ET production in species such as bean (Field, 1981), tomato (Ciardi et al., 1997), or winter rye (Yu et al., 2001). Moreover, exogenous ET application to rhododendron plants increases their tolerance to freezing (Harber and Fuchigami, 1989), and winter rye plants exposed to cold show higher levels of ET that control the accumulation of antifreeze proteins in leaves and increase plant freezing tolerance (Yu et al., 2001). Constitutive over-expression in tobacco of the tomato gene *JERF3*, which encodes a transcription factor related to ET signaling, increases freezing tolerance by inducing the expression of genes encoding antioxidant enzymes with the consequent reduction of ROS (Wu et al., 2008). In addition, antisense expression in tomato and tobacco of *TERF2/LeERF2*, a tomato gene encoding an ET-inducible transcription factor, inhibits ET production and reduces freezing tolerance. In contrast, *TERF2/LeERF2* over-expression in tobacco and tomato increases freezing tolerance by activating cold-induced gene expression (Zhang and Huang, 2010). As described previously, the implication of ET in cold signaling seems to be clear and very little is still known about the mechanisms that mediate its function.

Substantial work has reported potential valuable effects of salicylic acid (SA) treatment on cold and/or freezing tolerance in cereals (Janda et al., 1999; Kang and Saltveit, 2002; Szalai et al., 2000; Tasgin et al., 2003; Saltveit and Hepler, 2004), bean (Ding et al., 2002), cucumber (Kang and Saltveit, 2002), tomato (Ding et al., 2002), banana (Kang et al., 2007), and Persian lilac (Bernard et al., 2002). The role of SA in plant tolerance to low temperature appears to be related to different protective responses, such as the increase in the activity of antioxidant enzymes (Janda et al., 1999; Kang and Saltveit, 2002) and the enhancement of ice nucleation activity in the apoplast (Tasgin et al., 2003). In contraposition with these data, recent results indicate that SA is a negative regulator of cold acclimation in *Arabidopsis* (Miura and Ohta, 2010). Two SA-accumulating mutants, *siz1* and *acd6*, have reduced cold acclimation capacity and are more sensitive to chilling stress, mainly because of a lower expression of *CBF3* and the corresponding regulon (Miura and Ohta, 2010). The restoration of SA accumulation to wild-type levels suppresses the sensitivity of these mutants to chilling and freezing temperatures (Miura and Ohta, 2010).

Gene expression in response to low temperature

As mentioned previously, most changes induced by low temperature are controlled by extensive reprogramming of gene expression. Global analyses of gene expression in *Arabidopsis* have revealed that about 1400 genes are regulated by low temperature (Matsui et al., 2008; Zeller et al., 2009). Among these genes, one-third are repressed, with only one gene encoding a transcription factor, which denotes that gene expression during cold acclimation is mainly activated (Lee et al., 2005). In addition, many early cold-induced genes encode transcription factors or proteins involved in transcription. Thus, more than 100 genes have been annotated to function in transcription, 95 of them coding for transcription factors (Lee et al., 2005). Interestingly, many cold inducible genes are also induced by other related abiotic stresses such as drought or high salt (Matsui et al., 2008). The use of tiling array technology has allowed the identification of new non-annotated transcriptional units, as well as a high number of non-protein-coding and sense-antisense transcripts whose expression is also controlled by low temperature (Matsui et al., 2008). Different functions for antisense RNAs have been described, including gene silencing, RNA stability, alternative splicing, RNA editing, RNA masking, and RNA methylation (Matsui et al., 2008); however, the role of non-protein-coding transcripts is still poorly understood. Although the majority of the results reported so far have shown that the regulation of gene expression by low temperature is mainly controlled at the transcriptional level, recent data manifest that post-transcriptional regulatory events are also relevant.

Transcriptional control of cold-regulated gene expression

One of the main goals when studying the transcriptional regulation of gene expression in response to cold has been the identification of the *cis* and *trans* elements implicated. To date, several *cis*-acting elements and transcription factors have been related to cold-regulated gene expression.

Cis*-acting elements mediating cold-regulated gene expression*

The first low-temperature response element (LTRE) described in plants was the CRT/DRE motif (CCGAC). It was identified in the promoter regions of *Arabidopsis* cold-inducible genes *COR15A* and *RD29A* (Baker et al., 1994; Yamaguchi-Shinozaki and Shinozaki, 1994), and has also been reported in rapeseed (Jiang et al., 1996), maize (Wang et al., 2008), and wheat (Ouellet et al., 1998; Takumi et al., 2008) cold-inducible genes. In addition, a similar sequence (CCGAAA) identified in the promoter region of the barley gene *blt4.9* functions as an LTRE (Dunn et al., 1998). Interestingly, the CRT/DRE element also mediates the induction of gene expression by drought and ABA (Baker et al., 1994; Yamaguchi-Shinozaki and Shinozaki, 1994; Knight et al., 2004). Deletion analysis in the promoter of the barley cold-inducible gene *blt101.1* allowed the identification of a new LTRE, named CR1, which is also included in the promoters of other components of the *blt101* gene family (Brown et al., 2001). Different *cis*-acting elements that account for the cold-induction of *CBFs* have been described (Chinnusamy et al., 2003; Zarka et al., 2003; Agarwal et al., 2006). The promoter region of *CBF3* contains five MYC-recognition sequences, named MYC1-5, that seem to control the cold-induction of this gene (Chinnusamy et al., 2003). In contrast, the *CBF3* promoter also includes three MYB binding sites that are involved in the negative regulation of *CBF3* expression (Agarwal et al., 2006). Analysis of the *CBF2* promoter region revealed the existence of two sequences, ICEr1 and ICEr2, which contain the consensus recognition sequence for bHLH proteins and could control the low-temperature induction of *CBF2* (Zarka et al., 2003). Moreover, the *CBF2* promoter also contains seven conserved DNA motifs, named CM1-7 (Doherty et al., 2009). While CM4 and CM6 act as negative regulators, CM2 seems to have both negative and positive activity (Doherty et al., 2009). CM2 overlaps with the ICEr1 present in the *CBF2* promoter, which can account for its role as a positive regulator (Doherty et al., 2009). CM motifs have also been detected in the promoters of other early cold-inducible genes, including *ZAT12* that encodes a transcription factor, indicating that they play a relevant role in the control of early-induced gene expression by low temperature (Doherty et al., 2009).

Bioinformatic analysis of the promoters of cold-induced genes unveiled an enrichment of evening elements (EEL; Mikkelsen and Thomashow, 2009), which were previously related to the regulation of gene expression by the circadian clock (Harmer et al., 2000). Mikkelsen and Thomashow (2009) showed that the EEL motif mediates the cold induction of *COL1* and *COR27*, confirming its role in cold signaling. The RSRE box (rapid stress response element, CGCGTT) was identified because of its elevated presence in the promoter regions of *Arabidopsis* genes rapidly induced by wounding (Walley et al., 2007). Intriguingly, around 49% of these genes are also cold induced, suggesting a hypothetical role for the RSRE box in low-temperature response. A synthetic promoter including six copies of RSRE is able to induce the expression of a reporter gene under cold conditions, demonstrating that it is an LTRE (Walley et al., 2007). The fact that some LTREs identified so far are also implicated in the response to other stimuli, such as light (EEL) or wounding (RSRE), points out the importance for plants, as sessile organisms, to integrate different environmental cues to adequately respond to variations in growth conditions.

Trans*-acting elements mediating cold-regulated gene expression*

CBF1, also known as DREB1B, was the first *trans*-acting factor involved in cold signaling to be identified due to its ability to interact with the CRT/DRE motif (Baker et al., 1994; Yamaguchi-Shinozaki and Shinozaki, 1994). It is a 24kDa protein with an AP2 DNA binding domain and an acidic C-terminal region that acts as a transcriptional activator (Stockinger et al., 1997). Subsequently, it was reported that CBF1 belongs to a small family composed of two more proteins, CBF2/DREB1C and CBF3/DREB1A (Gilmour et al., 1998; Liu et al., 1998; Medina et al., 1999). The expression of *CBF* genes is specifically induced by cold in a fast and transient way (Gilmour et al., 1998; Liu et al., 1998; Medina et al., 1999). The CBF proteins seem to regulate the expression of around 12% of the *Arabidopsis* cold-induced genes and its over-expression significantly increases *Arabidopsis* freezing tolerance (Fowler and Thomashow, 2002). Furthermore, plants over-expressing *CBF3* show metabolic changes associated with cold acclimation, such as an increment in the levels of proline and different soluble sugars (Gilmour et al., 2000). Recently, Novillo and co-workers (2004) reported that the absence of CBF2 enhances the tolerance of *Arabidopsis* to freezing temperatures, before and after cold acclimation, drought, and salt stress. This phenotype correlates with a higher accumulation of *CBF1* and *CBF3* transcripts, and consequently with the activation of the CBF-regulon, both under control conditions and after low-temperature exposure (Novillo et al., 2004). These results are evidence that CBF2 negatively controls the expression of *CBF1* and *CBF3*, ensuring an accurate regulation of the CBF-regulon and, therefore, of *Arabidopsis* response to cold stress (Figure 19.2). On the other hand, the characterization of *Arabidopsis* plants with reduced induction of *CBF1* or *CBF3* by low temperature revealed that, contrary to CBF2, CBF1 and CBF3 are not implicated in regulating the expression of other *CBF* genes, confirming that the three CBFs do not have the same function (Novillo et al., 2007). Moreover, data showed that although CBF1 and CBF3 seem to positively control the cold-induction of the same target genes, they are concertedly required to induce the whole CBF-regulon and, therefore, the complete development of the cold acclimation process in *Arabidopsis* (Novillo et al., 2007).

ICE1 is an *Arabidopsis* MYC-like bLHL transcriptional activator that induces *CBF3* expression under low-temperature conditions by interacting with the MYC sequences present in the *CBF3* promoter (as previously mentioned; Figure 19.2) (Chinnusamy et al., 2003). *ice1* mutant plants show an altered induction of *CBF3* and of CBF-target genes and reduced freezing tolerance when compared to wild-type (Chinnusamy et al., 2003). Lee and colleagues (2005) reported that ICE1 controls the expression of around 40% of cold-regulated genes. Unexpectedly, ICE1 expression is not cold induced, indicating that it should be activated by post-transcriptional mechanisms (see the next paragraph). Fursova and co-workers (2009) described another *Arabidopsis* positive *trans*-acting regulator, ICE2, with high sequence homology to ICE1 that is also related to cold acclimation. *ICE2* over-expression enhances *CBF1* expression and *Arabidopsis* freezing tolerance (Figure 19.2; Fursova et al., 2009). However, its ability to bind to the ICE box has not been analyzed yet.

In addition to ICE1 and ICE2, CAMTA3 is another transcription factor that positively controls *CBF* expression by low temperature (Figure 19.2; Doherty et al., 2009). CAMTA proteins are calmodulin-binding transcriptional activators that bind to the CM2 motif (CCGCGT) in the *CBF2* promoter inducing its expression under cold conditions (Doherty et al., 2009). *camta3* null mutants show reduced cold induction of *CBF1*, *CBF2*, and some CBF-target genes, although they are not affected in their capacity to cold acclimate. *camta1/camta3* double mutant, however, has reduced cold acclimation capacity, demonstrating that both CAMTA1 and CAMTA3 are required for full development of this adaptive process (Doherty et al., 2009).

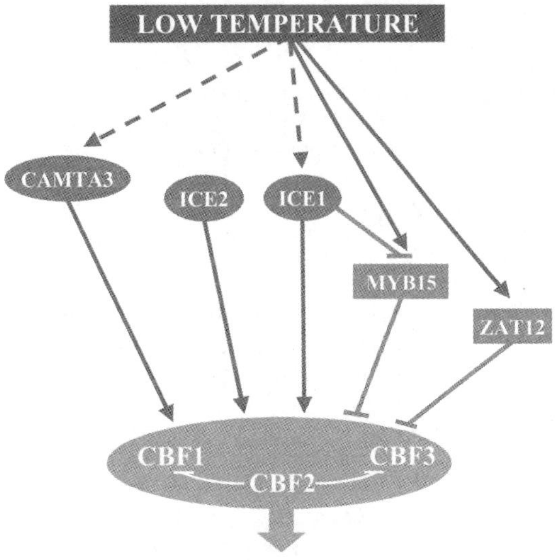

Figure 19.2 • CBFs constitute central intermediates in cold signaling to induce gene expression. ICE1, ICE2, and CAMTA3 positively regulate *CBF* expression. MYB15 and ZAT12 act as negative regulators of *CBFs*. CBF2 negatively regulates the expression of *CBF1* and *CBF3*. ICE1 inhibits *MYB15* expression. Broken arrow means activation by unknown mechanisms. Arrowheads and end lines indicate positive and negative regulation, respectively. Please see color plate section at the end of the book.

Analysis of data obtained from microarray experiments in *Arabidopsis* revealed that some genes have similar expression patterns under cold conditions like the *CBFs*, denoting that they could be coordinately regulated (Vogel e al., 2005). One of these genes is *ZAT12*, whose expression is activated after 1 hour of cold exposure and encodes a zinc-finger protein (Vogel et al., 2005). Over-expression of *ZAT12* in *Arabidopsis* enhances the induction of 24 cold-induced genes, some of them CBF-targets, and produces a slight increase in constitutive freezing tolerance (Vogel et al., 2005). Interestingly, ZAT12 negatively controls the cold induction of *CBFs* (Figure 19.2; Vogel et al., 2005). The mechanism with which ZAT12 regulates gene expression still remains unknown.

HOS9 is a homeodomain transcription factor that has also been implicated in activating gene expression during cold acclimation (Zhu et al., 2004). *hos9* mutant plants display reduced freezing tolerance, both before and after cold acclimation (Zhu et al., 2004). Intriguingly, whereas the cold-induction of *CBFs* in *hos9* mutants is as in the wild-type, the cold-induction of some CBF-target genes is higher, providing evidence that HOS9 controls the cold acclimation process by ensuring a proper cold-inducible gene expression through CBF-independent pathways (Zhu et al., 2004). The identification and characterization of *Arabidopsis lov1-4* mutant allowed the isolation of an NAC-domain transcription factor,

LOV1, which mediates the activation of gene expression in response to low temperature by a CBF-independent pathway (Yoo et al., 2007). *lov1-4* null mutant is hypersensitive to freezing temperatures and has a reduced cold acclimation capacity, confirming that LOV1 is a positive regulator of freezing tolerance and cold acclimation (Yoo et al., 2007). *lov1-4* shows reduced cold induction of CBF-target genes when compared to wild-type plants, but the cold induction of *CBFs* is not affected, which demonstrates that LOV1 function is CBF-independent (Yoo et al., 2007). Although the implication of HOS9 and LOV1 in cold signaling is clear, the DNA motifs to which they bind to remain to be determined.

Several cold-inducible genes, including some CBF targets, contain in their promoters an ABA-responsive element (ABRE, PyACGTGCC; Guiltinan et al., 1990) that might control cold-induced CBF-independent gene expression (Baker et al., 1994; Yamaguchi-Shinozaki and Shinozaki, 1994; Wang and Cutler, 1995). *SCOF1* is a gene whose expression is induced by cold and ABA, and its over-expression in *Arabidopsis* provokes constitutive expression of cold-induced genes and increased tolerance to freezing temperature (Kim et al., 2001). *SCOF1* encodes a zinc finger protein that enhances the binding of SGB1, a bZIP transcription factor, to the ABRE motif (Kim et al., 2001), indicating that it is a positive regulator of cold acclimation through the ABRE-dependent pathway.

LOS2 constitutes a very interesting *trans*-acting element involved in the regulation of cold-induced gene expression. The *Arabidopsis* mutant *los2* displays impaired cold induction of *RD29A* and other cold-inducible genes, and reduced cold acclimation capacity (Lee et al., 2002). However, the cold induction of *CBFs* is as in the wild-type. *LOS2* codes for an enolase that also seems to have a function in transcriptional repression (Lee et al., 2002). It can bind to the c-MYC *cis*-element in the promoter of *STZ/ZAT10*, an *Arabidopsis* cold-inducible gene that encodes a transcriptional repressor (Lee et al., 2002). STZ/ZAT10 is able to bind to the *RD29A* promoter to repress its transcription in transient expression analysis (Lee et al., 2002). The cold induction of *STZ/ZAT10* is increased in *los2* mutant plants, accounting for the reduced induction of *RD29A* by low temperature. Therefore, in response to low temperature, LOS2 would bind to the promoter of *STZ/ZAT10* to induce its expression, which in turn would repress the cold induction of genes such as *RD29A*.

Recent data have established that MYB15, a R2R3-MYB transcription factor that binds to the MYB recognition sites in the promoters of the three *CBFs*, acts as negative regulator of *CBF* expression (Figure 19.2; Agarwal et al., 2006). *myb15* mutant plants show enhanced induction of *CBFs* and their targets under low-temperature conditions, and increased capacity to cold acclimate. On the other hand, transgenic *Arabidopsis* over-expressing *MYB15* display reduced cold induction of *CBFs* and impaired freezing tolerance (Agarwal et al., 2006). *MYB15* expression is induced by cold and negatively regulated by ICE1 (Agarwal et al., 2006). Interestingly, it has been shown that MYB15 can physically interact with ICE1, although the function of this interaction has not been determined (Agarwal et al., 2006). RAP2.1 is another *trans*-acting factor that negatively controls cold-regulated gene expression in *Arabidopsis* (Dong and Liu, 2010). The expression of *RAP2.1* is induced by cold through an ABA-independent pathway (Dong and Liu, 2010). RAP2.1 has an AP2 domain that binds to the CRT/DRE domain, an ERF-associated amphiphilic repression (EAR) motif, and represses the induction of several cold-induced genes, including itself, by binding to the CRT/DRE motifs present in their promoters (Dong and Liu, 2010). The function of RAP2.1 in cold acclimation, if any, needs to be determined. Finally, *MYBC1* encodes an *Arabidopsis* nuclear R3-MYB protein that negatively controls the constitutive freezing tolerance and the cold induction of the *CBFs* and their targets (Zhai et al., 2010). Interestingly, however, the induction of *MYBC1* by low temperature seems to be CBF-independent, supporting the idea that MYBC1 negatively regulates freezing tolerance through a CBF-independent pathway (Zhai et al., 2010). As in the case of RAP2.1, the role of MYBC1 in cold acclimation is still unknown.

In addition to the *Arabidopsis trans*-acting regulators just mentioned, some other factors have been described in different plant species that control gene expression in response to low temperature. Thus, the over-expression of three soybean genes encoding bZIP transcription factors (*GmbZIP44*, *GmbZIP62*, and *GmbZIP78*) in *Arabidopsis* increases freezing tolerance through the activation of ABA-induced gene expression (Liao et al., 2008). In rice, *OsMYB4* encodes an MYB transcription factor and its expression is induced under cold conditions (Vannini et al., 2004). Constitutive expression of *OsMYB4* in *Arabidopsis* increases cold and freezing tolerance by inducing cold-regulated gene expression (Vannini et al., 2004). Also in rice, *SNAC2*, a gene encoding an NAC-type transcription factor, is cold induced and its over-expression in rice increases plant tolerance to cold stress (Hu et al., 2008). *WABI5* is a cold-induced wheat gene encoding an AREB/ABF-type transcription factor that increases freezing tolerance when constitutively expressed in tobacco (Kobayashi et al., 2008b). Recently, Liu and co-workers (2005) reported the identification of an LOS2 homolog in *Capsella bursapastoris*, named CbLOS2, that has the enolase domain as well as the DNA-binding and repression domains. Similarly to *LOS2*, *CbLOS2* expression is induced by low temperature and seems to play a role in the cold acclimation process (Liu et al., 2005). In tomato, transcription factors JERF3 and TERF2/LeERF2 (as previously mentioned) are also implicated in increasing freezing tolerance by activating cold-induced gene expression (Zhang and Huang, 2010). Finally, CBF homologs have been found in different species, including dicots (Xue, 2003; Zhang et al., 2004; Benedict et al., 2006; Huang et al., 2007; Pennycooke et al., 2008a,b; Welling and Palva et al., 2008) and moncots (Dubouzet et al., 2003; Qin et al., 2004; Pellegrineschi et al., 2004; Brautigam et al., 2005; Skinner et al., 2005; Vagujfalvi et al., 2005; Xiong and Fei, 2006; James et al., 2008; Zhao and Bughrara, 2008). The functional characterization of some of these homologs revealed that they function like the *Arabidopsis* factors. Overexpression of *CBF* homologs from rice, perennial ryegrass, maize, barley, and wheat in transgenic tobacco or *Arabidopsis* plants resulted in increased freezing tolerance and expression of cold-regulated genes belonging to the CBF-regulon (Qin et al., 2004; Pellegrineschi et al., 2004; Brautigam et al., 2005; Vagujfalvi et al., 2005; James et al., 2008).

All of the results described earlier confirm the requirement for an accurate regulation of gene expression for the development of a correct plant response to low temperature, including the cold acclimation process, where, in addition to the activation of gene expression, the repression also plays an essential role.

Post-transcriptional control of cold-regulated gene expression

Data reveal that, in addition to the transcriptional regulation, controlling mRNA stability and translation are also pivotal for an accurate cold response. The accumulation of some transcripts during cold acclimation seems to be regulated at the post-transcriptional level (Hajela et al., 1990; Wolfraim et al., 1993; Dunn et al., 1994). Using pharmacological approaches, Phillips and colleagues (1997) showed that an unknown nuclear protein actively stabilizes transcripts corresponding to *blt4.0*, a cold-induced barley gene. In the case of the *Arabidopsis TCH4* gene, cold induction is mediated by a 102 bp 5′-untranscribed region that seems to participate in the stabilization of *TCH4* mRNAs (Iliev et al., 2002). On the other hand, Xiong and colleagues (2002a) identified an *Arabidopsis* gene, *FRY2*, which encodes a protein similar to animal and yeast factors involved in transcription regulation and pre-mRNA processing. Recessive mutations in *FRY2* cause the superinduction of CBF-target genes by cold, indicating that FRY2 is a negative regulator of cold-induced gene expression (Xiong et al., 2002a). In contrast, *fry2* mutants are more sensitive to freezing temperatures than wild-type plants. Taking into consideration the RNA binding domain of FRY2, Xiong and colleagues (2002a) proposed that it could act as an RNA chaperone under low-temperature conditions. The glycine-rich RNA-binding proteins (GR-RBPs) have also been proposed to act as RNA chaperones in plants. In this regard, it has been described that the expression of several *GR-RBP* genes is induced by different abiotic stresses, including cold (Sachetto-Martins et al., 2000). Functional characterization of one *Arabidopsis* GR-RBP, the AtRZ-1, provided interesting data about the role of these proteins in cold acclimation (Kim et al., 2005). *atRZ-1* mutants are more sensitive to low and freezing temperatures, whereas *AtRZ-1* over-expressing plants are more tolerant to these stresses (Kim et al., 2005). On the basis of these results, Kim and colleagues (2005) concluded that AtRZ-1 might have RNA chaperone activity leading to the destabilization of the overstabilized secondary structures, facilitating an efficient translation under low-temperature conditions. *GR-RBPs* genes regulated by cold have also been described in other species. Shinozuka and co-workers (2006) reported the isolation of a *GR-RBP* gene from ryegrass whose expression is induced after low-temperature exposure, suggesting a role in the cold response. Park and colleagues (2009b) reported the identification of *OsDEG10*, a *GR-RBP* gene from rice whose expression is induced by different abiotic stresses, including cold. Interestingly, rice *OsDEG10* RNAi transgenic plants showed a reduction in their tolerance to low temperature compared to wild-type plants (Park et al., 2009b). STA1 is another *Arabidopsis* protein related to pre-mRNA processing and cold signaling. STA1 is similar to the human U5 small ribonucleoprotein-associated 102 kDa protein and to the yeast pre-mRNA splicing factors Prp1p and Prp6p, whose corresponding genes, as *STA1*, are induced by low temperature (Lee et al., 2006). Mutations in *STA1* increase *Arabidopsis* chilling sensitivity, although they do not seem to affect cold-induced gene expression. The stability of some mRNAs and the splicing of the cold-induced gene *COR15A* are altered in *sta1-1* mutants, providing evidence that these processes are important in low-temperature signaling (Lee et al., 2006). Further data demonstrating the implication of alternative splicing in plant response to cold stress have recently been reported. The isolation and characterization of a *DREB2* homolog in wheat, *Wdreb2*, showed that three transcript forms, *Wdreb2*α, *Wdreb2*β, and *Wdreb2*γ, are produced by alternative splicing (Egawa et al., 2006). *Wdreb2* expression is induced by drought, salt, ABA, and cold. Under drought and salt stress conditions only an increase in the accumulation of *Wdreb2*α and *Wdreb2*γ transcripts is observed. In contrast, when plants are exposed to low temperature the levels of all transcript forms are significantly enhanced (Egawa et al., 2006).

The control of mRNA export to the cytoplasm also seems to be influential in cold response. Thus, the implication of the *Arabidopsis* nucleoporin AtNUP160 in cold signaling has been described (Dong et al., 2006a). This protein is involved in mRNA export since a null mutation in the corresponding gene causes mRNA accumulation in the nucleus (Dong et al., 2006a). Furthermore, the *atnup160-1* mutant shows a reduced tolerance to chilling and freezing temperatures, probably due to a decrease in the induction of several cold-responsive genes, including the *CBFs* (Dong et al., 2006a). LOS4, a DEAD-box RNA helicase, has also been implicated in cold acclimation (Gong et al., 2005). *los4* mutants display enhanced cold-induction of *CBF* genes under low-temperature conditions, and are more sensitive to freezing than wild-type plants (Gong et al., 2005). As in the case of AtNUP160, *in situ* poly(A) hybridizations revealed that LOS4 is needed for a correct partition of mRNAs in the cell (Gong et al., 2005). Taken together, all of these results provide a new scenario on the regulation of gene expression during cold acclimation, where mRNA stability and processing, as well as mobilization, seem to be as important as transcriptional activation.

MicroRNAs (miRNAs) are small non-coding RNAs that have emerged as important regulators of gene expression in a multitude of plant processes. Cold-induced miRNAs have been identified by different laboratories. Thus, Sunkar and Zhu (2004) reported that the expression of *miR393*, *miR319*, and *miR402* is induced at different levels in *Arabidopsis* plants exposed to 4°C. Moreover, recent works have described additional miRNAs whose expression is induced by low temperature (Zhou et al., 2008; Jian et al., 2010). In any case, the role of these miRNAs in cold signaling still remains to be deciphered.

Translational and post-translational control of cold-regulated gene expression

Cold-regulated gene expression is also subjected to translational control, indicating that protein synthesis is essential for an adequate response to low temperature in plants. Thus, the translation elongation factor-2-like protein, LOS1, has been shown to be a positive regulator of cold-induced gene expression and cold acclimation (Guo et al., 2002). The *los1-1* mutant displays reduced cold acclimation capacity due to a reduced cold induction of the CBF-target genes (Guo et al., 2002). The induction of the *CBFs* by low temperature, however, is higher in *los1-1* than in wild-type plants (Guo et al., 2002). Synthesis of new proteins under cold conditions is severely impaired in *los1-1*, which may account for the decrease in cold-induced gene expression (Guo et al., 2002). These results suggest that the expression of *CBFs* is independent of new protein synthesis, since they are overinduced in *los1-1* mutants growing under low-temperature conditions; therefore, it is subjected to feedback regulation (Guo et al., 2002). Another protein that might also have a role in protein synthesis under stress conditions is the pea DNA helicase 47 (PDH47). *PDH47* expression is induced by cold in shoots and roots of the pea (Vashisht et al., 2005). Biochemical analysis of purified PDH47 protein revealed that it has helicase activity, both 3′ to 5′ and 5′ to 3′, and promotes protein synthesis *in vitro*. PDH47 is localized into the nucleus and its activity seems to be regulated by phosphorylation events (Vashisht et al., 2005). Further genetic analysis of this protein will help to figure out the function of DNA metabolism in the regulation of plant response to temperature stress.

Protein stability has also been revealed to have important implications in the control of cold-regulated gene expression. In *Arabidopsis*, Ishitani and colleagues (1998) identified HOS1, a negative regulator of the CBF-regulon that contains a RING finger domain characteristic of some E3 ubiquitin ligases. In agreement with its role as negative regulator, *HOS1* expression is transiently inhibited by cold treatment (Ishitani et al., 1998). Although HOS1 negatively regulates cold-induced gene expression, *hos1-1* mutants are chilling- and freezing-sensitive when compared to wild-type plants, indicating that accurate expression of *HOS1* is essential for the correct response to low temperature (Ishitani et al., 1998). Subsequently, it was demonstrated that HOS1 has E3 ubiquitin ligase activity and polyubiquitinates ICE1 (see previous discussion) mediating its inactivation and, therefore, inhibiting the cold-induction of *CBF3* and the CBF-regulon (Dong et al., 2006b). *AtCHIP* also encodes an *Arabidopsis* E3 ubiquitin ligase and its expression is cold induced (Yan et al., 2003). When over-expressed in *Arabidopsis*, it increases the sensitivity to freezing temperature denoting that AtCHIP is a negative regulator of freezing tolerance (Yan et al., 2003). The A subunit of protein phosphatase 2A (PP2A) has been identified as one of the AtCHIP substrates (Luo et al., 2006). PP2A is a negative regulator of the sucrose synthase (SUS) and, therefore, of sugar biosynthesis, which has an important role in chilling and freezing tolerance. *AtCHIP*-over-expressing plants show an increased PP2A activity under cold conditions that could account for the freezing sensitivity of these plants (Luo et al., 2006). AtCHIP seems to monoubiquitinate the A subunit of PP2A, increasing its phosphatase activity, which would, in turn, decrease freezing tolerance (Luo e al., 2006). A different proteolysis pathway is controlled by the cysteine proteases, whose activity is inhibited by cysteine protease inhibitors (cystatins). The characterization of two *Arabidopsis* cystatins, AtCYSa and AtCYSb, revealed that they play a role in cold tolerance (Zhang et al., 2008a). *AtCYSa* and *AtCYSb* genes are induced under low-temperature conditions and constitutively expressed in *CBF3* overexpressing plants, which suggests that they are induced by cold through a CBF-dependent pathway (Zhang et al., 2008a). Accordingly, the promoter region of both genes contains a CRT/DRE element (Zhang et al., 2008a). Transgenic *Arabidopsis* overexpressing *AtCYSa* or *AtCYSb* have increased tolerance to low temperature, confirming that the corresponding cystatins have a function in cold signaling (Zhang et al., 2008a).

SUMOylation is another post-translational modification involved in protein stability and activity. SUMOylation, as ubiquitination, requires the sequential action of three enzymes, E1, E2, and E3 (Kurepa et al., 2003; Colby et al., 2006). The *Arabidopsis* genome only contains one gene, *SIZ1*, that has homology with genes encoding E3 SUMO ligases (Miura et al., 2007). *SIZ1* expression is not induced by cold, although mutations in *SIZ1* decrease *Arabidopsis* tolerance to freezing and chilling temperatures, indicating that SUMOylation also plays an important role in plant response to low temperature (Miura et al., 2007). SIZ1 catalyzes the SUMOylation of ICE1, impeding its ubiquitination (as previously mentioned) and, therefore, the repression of *CBF3* (Miura et al., 2007). Furthermore, the SUMOylation of ICE1 inhibits the expression of *MYB15*, a negative regulator of *CBF* expression (previously mentioned; Miura et al., 2007). It has been proposed that, most probably, SUMOylation of ICE1 would induce its stabilization (Miura et al., 2007). A hypothetical model that would integrate HOS1, SIZ1, and ICE1 activities in the regulation of cold acclimation is presented in Figure 19.3. In this model, under control conditions, HOS1 would be targeting ICE1 for its degradation by the proteasome, maintaining low levels of protein. Temperature decrease would activate SIZ1, by still unknown pathways, and would inhibit *HOS1* expression. SIZ1 activation would induce the SUMOylation of ICE1 and, together with the reduction of *HOS1* expression, would produce a decrease in its ubiquitination levels. The increase in ICE1 accumulation would then induce the expression of the *CBFs*, directly by its interaction with their promoters, and indirectly by the repression of *MYB15* and, therefore, of the CBF-regulon and the cold acclimation process. Results obtained by Chaikam and Karlson (2010) showed that the SUMO pathway could also contribute to the cold response in rice. Exposure of rice plants to low temperature induces an accumulation of SUMO-conjugates (Chaikam and Karlson, 2010). Although the expression of the genes encoding SUMO protein and SUMO activating enzymes is repressed at 4°C, the expression of the conjugating enzymes and the SUMO ligase is activated (Chaikam and Karlson, 2010).

Epigenetic regulation of low-temperature response

Histone acetylation mediated by histone acetyltransferases (HATs) and histone deacetylases (HDAC) has also been implicated in low-temperature signaling. Several reports have connected HATs with transcriptional regulation by remodeling

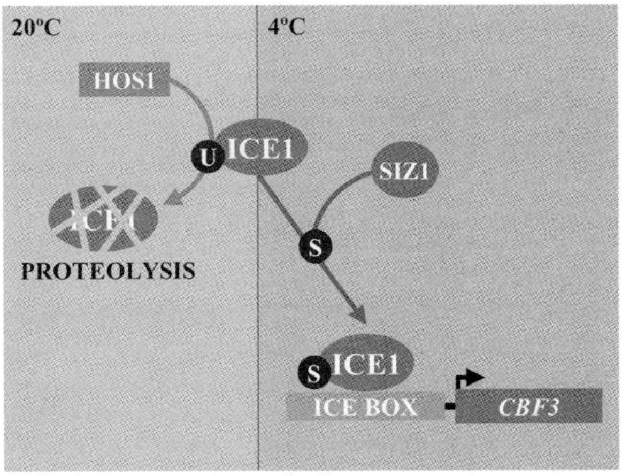

Figure 19.3 • Hypothetical model for post-translational regulation of *CBF3* expression. Ubiquitin (U) and SUMO (S). From Medina et al. (2010). Please see color plate section at the end of the book.

repressive chromatin structures (Sterner and Berger, 2000; Roth et al., 2001). In *Arabidopsis*, different components of the HATs complex (ADA2a, ADA2b, and GCN5) have been described to interact with CBF1 *in vitro* and seem to be needed for CBF1 function (Stockinger et al., 2001). Functional characterization of ADA2b and GCN5 showed that these proteins control cold-induced gene expression (Vlachonasios et al., 2003). *ada2b* and *gcn5* mutant plants display wild-type *CBF* expression, although the cold induction of their targets is reduced (Vlachonasios et al., 2003). Interestingly, only the induction of the CBF-regulon is affected in *ada2b* and *gcn5* mutants, suggesting that the activity of the HATs complex in response to cold is highly specific (Vlachonasios et al., 2003). Reversibility of histone acetylation also seems to be involved in regulating cold-induced gene expression. Thus, Kim and colleagues (2004) identified the *Arabidopsis* mutant *acg1* as a negative regulator of the CBF-regulon. The *acg1* mutant has a constitutive freezing tolerance similar to the wild-type but increased capacity to cold acclimate, which correlates with a higher induction of *CBFs* and their target genes (Kim et al., 2004). *ACG1* resulted to encode FVE, a component of the *Arabidopsis* HDAC complex homolog to the mammalian retinoblastoma associated protein (Kim et al., 2004). These data show that histone acetylation is important for the interaction of CBFs with the CRT/DRE element and the subsequent activation of gene expression under low temperature; in all likelihood preventing inhibitory chromatin structures. Another protein related to the control of histone acetylation under cold conditions is the *Arabidopsis* HOS15, which is very similar to a component of the human SMRT/N-CoR gene repressor complex that associates with HDACs (Zhu et al., 2008). HOS15 specifically interacts with histone H4 and mediates its deacetylation (Zhu et al., 2008). Compared to wild-type plants, *hos15* mutants show increased levels of H4 acetylation that, interestingly, correlate with a hyperinduction of the cold-induced gene *LTI78* at 4°C (Zhu et al., 2008). In contrast, *hos15* plants are more sensitive to freezing temperature than wild types and have reduced capacity to cold acclimate (Zhu et al., 2008). All of these results indicate that the levels of histone acetylation are regulated to ensure the adequate reprogramming of gene expression required for the accurate plant response to low temperature, including the full development of the cold acclimation process.

Sokol and co-workers (2007) reported a nucleosomal analysis in cells of tobacco and *Arabidopsis* under different abiotic stress conditions. After cold exposure, they observed a transient increase in the phosphorylation levels of histone H3-Ser10, followed by the upregulation of the phosphoacetylation of histone H3 and the acetylation of histone H4. All of these changes correlated with the induction of stress-regulated genes, denoting a possible role of the nucleosome in the regulation of cold-induced gene expression (Sokol et al., 2007). On the other hand, HMGB is a family of proteins implicated in nucleosome assembly, replication, and transcription that could also have a function in cold signaling. In this way, the expression of *Arabidopsis HMGB2*, *HMGB3*, and *HMGB4* genes is induced by low temperature (Kwak et al., 2007).

Methylation is another mechanism linked to histone modification and transcriptional regulation that is involved in plant response to low temperature. Kwon and colleagues (2009) reported that cold exposure gradually decreases the trimethylation levels of histone H3 (H3K27me3) in the promoter region of the cold-induced genes *COR15A* and *AtGOLS2*. Interestingly, in plants returned to control temperature the expression of these genes decreases to unstimulated levels, but HE3K27me3 still remains low. It was proposed that reduction of HE3K27me3 after exposure to low temperature could be a mark of recent transcriptional activity (Kwon et al., 2009). In tobacco, it has been reported that the levels of DNA methylation in the cold-inducible gene *NtGPDL* are reduced when plants are exposed to cold (Choi and Sano, 2007). Moreover, the reduction in methylation levels correlates with the induction of *NtGPDL* expression (Choi and Sano, 2007). In the same way, global methylation analysis in the maize genome revealed a general demethylation after low-temperature exposure (Steward et al., 2002). The analysis of a DNA fragment named *ZmMI1*, which includes the coding region of a putative protein and a retrotransposon-like sequence, showed that cold demethylation is followed by transcriptional activation. This suggests that methylation levels may serve to coordinate the cold induction of several stress-inducible genes (Steward et al., 2002).

Cross-talk between Plant Responses to Extreme Temperatures

Cold and heat stress responses have been extensively studied in plants as independent processes, but little is known about their connections and interactions. Strikingly, heat shock seems to improve chilling tolerance in cold-sensitive species (Saltveit and Hepler, 2004; Pressman et al., 2006) and *vice versa*, plants exposed to low temperature develop heat tolerance (Fu et al., 1998), which indicates the existence of a cross-talk between plant responses to extreme temperatures.

As in the case of low temperature, heat shock has direct consequences on plant growth and development that adversely influence the capacity of plants to reproduce. Under high temperature conditions, many physiological and metabolic processes are compromised. At the molecular level, protein synthesis and degradation are negatively affected and the heat-induced oxidative stress can cause important damage leading plant cells ultimately to death. In general, the chances of a plant to survive depend on basal and acquired tolerance mechanisms. Similarly to cold acclimation, when subjected to mild non-lethal temperature, plants can acquire thermotolerance to otherwise lethal high temperature. The heat shock response (HSR) orchestrated by plants is very complex, with a plethora of signaling molecules interlacing their pathways and the related genes tightly regulated at different levels.

The membrane as a node in the perception of temperature oscillations

As previously commented, membrane rigidification seems to be one of the first steps in cold perception in plant cells. In contrast, the maintenance of membrane fluidity has been revealed as one of the earliest responses to high temperature (Hofmann, 2009). Actually, the role of the membrane as a sensor of high temperature has been recently reinforced by Saidi and colleagues (2009) who used different pharmacological compounds that fluidize the membrane to mimic the HSR. Cell membrane fluidity is controlled, at least in part, by transcriptional and post-translational regulation of some fatty acid desaturase (FAD) isoforms. The *Arabidopsis* ω-3 desaturase FAD8 is a good illustration of dual regulation at both transcriptional and post-translational levels. At high temperature, the expression levels of *FAD8* are reduced and the stability of FAD8 is impaired by an autoregulatory post-translational mechanism (Matsuda et al., 2005). The most determining type of fatty acids in plant adaptation to high temperature seem to be the trienoics, and more precisely their content in the plastidial membranes. Accordingly, FAD8 over-expressing *Arabidopsis* plants with altered plastidial trienoic fatty acid content were more compromised in their heat tolerance than FAD3 over-expressing lines in which only the extraplastidic trienoic fatty acid content was affected (Zhang et al., 2008b).

Transducing the signals initiated by temperature variations

Ca^{2+} is a versatile second messenger in plant responses to extreme temperatures

The essential role of Ca^{2+} at the onset of plant response to high temperature has become clear as the addition of Ca^{2+} chelators and/or Ca^{2+} channel blockers and CaM inhibitors during heat shock dramatically *inhibits* the HSR in different species (Gong et al., 1997; Larkindale and Knight, 2002; Saidi et al., 2009). Furthermore, treatment with these pharmacological compounds during the acclimation to high temperature blocks the accumulation of an HSP70 protein, inhibiting plants from acquired thermotolerance (Saidi et al., 2009). Inversely, pre-treatment of maize seeds with $CaCl_2$ has been reported to significantly enhance the extent of the acquired thermotolerance (Gong et al., 1997), highlighting that, as in the cold response, the extracellular Ca^{2+} and its transport through the membrane are necessary for the heat response. Moreover, under cold and heat conditions, the accumulation of the second messenger IP^3, likely due to the hydrolytic activity of the PLC, has been described to intensify the Ca^{+2} influx (Liu et al., 2006). However, in contrast to the cold response, $[Ca^{+2}]_{cyt}$ remains unaltered after high-temperature exposure. Only during the subsequent recovery period are transient elevations in $[Ca^{+2}]_{cyt}$ detected, and these peaks are more pronounced in plants having acquired thermotolerance (Larkindale and Knight, 2002).

Hormones mediate extreme temperature signaling

ABA biosynthesis is essential during the cold acclimation process, as seen earlier, and has been recently reported to be required for the acquisition of thermotolerance. Intriguingly, however, it seems to be unnecessary for basal thermotolerance (Larkindale et al., 2005). Under low and high temperature, ABA biosynthesis is induced (Xiong et al., 2002b; Toh et al., 2008). Furthermore, genetic analysis supports that ABA is a major player in tolerance to high temperature. The *Arabidopsis* ABA-hypersensitive mutant *hat1* is able to survive up to five more days at 42°C and 10–15% humidity than wild-type plants, unveiling an ABA-dependent heat response pathway (Yan et al., 2006). Conversely, ABA-insensitive mutants, like *abi1*, are heat sensitive and the addition of exogenous ABA protects them from heat-induced oxidative damages (Larkindale and Knight, 2002). The implication of ABA in protecting plants from oxidative stress makes this hormone an important element of intersection in the cross-talk among plant responses to low and high temperatures. Nonetheless, the precise mode of action of ABA during HSR is far from clear. It seems that ABA positively regulates the expression of genes encoding critical proteins for plant survival at high temperature, such as HSPs, and more precisely HSP70, which acts as a chaperone ensuring the proper folding of other proteins (Zou et al., 2009; Hu et al., 2010). The HSP70 family is also known to function in protecting cells from the oxidative stress produced under drought and heat stress in maize in an ABA-dependent way (Hu et al., 2010). Members of the HSP70 family that are strongly induced by heat shock are also positively controlled by ABA in other plant species (Zou et al., 2009).

Gibberellin metabolism has a pivotal function in plant development under high temperature conditions. In vegetative tissues of citrus, a rise in the levels of transcripts corresponding to GA20-oxidase, an enzyme involved in GA biosynthesis, has been observed under high temperature conditions (Vidal et al., 2003). Interestingly, the opposite occurs in *Arabidopsis* plants growing under cold conditions, where the induction of genes corresponding to GA catabolism enzymes, like GA2-oxidase, and the repression of genes corresponding to GA biosynthetic enzymes, like GA20-oxidase, have been reported (Lee and Zeevaart, 2005). Recent findings have revealed that CBFs regulate the induction of *GA2-oxidase* expression by low temperature and the subsequent stabilization of DELLAs (Achard et al., 2008), supporting previous observations and offering a coordinated mechanism to explain the growth retardation normally observed when *Arabidopsis* plants are exposed to low temperature. Perhaps, this coordination between cold response and GA degradation could be extrapolated to heat stress conditions. In this way, exogenous treatment with tetcyclacis (an inhibitor of GA biosynthesis) confers some heat-shock protection to moth bean seedlings, avoiding the lipid peroxidation normally caused by high temperature (Upadhyaya et al., 1991). Something similar has been observed in seeds, as barley aleurone layers treated with GA_3 show reduced heat-shock tolerance compared to untreated controls.

In reed, NO levels increase when exposed to heat stress, and seem to act as a molecular intermediate in the acquisition of ABA-induced thermotolerance (Song et al., 2008). In *Arabidopsis*, NO also plays an important role in heat acclimation. The *HOT5* gene encodes a GSNOR, which metabolizes the NO adduct S-nitrosoglutathione (Lee et al., 2008). *Arabidopsis* mutants carrying *hot5* null and missense alleles cannot heat acclimate due to the heat sensitivity associated with a rise in NO species levels (Lee et al., 2008). As expected, *nox1*, an NO-overproducing mutant, turned out to be defective in its thermotolerance capacity (Lee et al., 2008).

The levels of SA increase rapidly under low- and high-temperature conditions, pointing out the importance of this plant hormone in an early phase of the response to extreme temperatures (Kaplan et al., 2004). Salicylic acid seems to be implicated in protecting *Arabidopsis* from heat-induced oxidative damage, since transgenic plants with blocked SA biosynthesis show increased susceptibility to oxidative stress (Larkindale and Knight, 2002). Subsequently, a reduced capacity to acquire thermotolerance in SA deficient mutants was reported, suggesting a function for SA in acquired thermotolerance (Larkindale and Knight, 2002). Nonetheless, there is a lack of consensus on this matter, as SA has been reported to promote basal thermotolerance and not to be essential for acquired thermotolerance (Clarke et al., 2004).

Regulation of gene expression in response to extreme temperatures

Transcriptional regulation of gene expression in response to high and low temperatures

Although HSFs and HSPs were first described to function under high temperature conditions, both of them have been shown to take part in cellular responses induced by abiotic stresses including high and low temperatures. Thus, Swindell and co-workers (2007) proposed the existence of an intense cross-talk between heat and cold stress regulatory networks mediated by HSFs and HSPs in *Arabidopsis*. The expression of *HSF* and *HSP* genes is enhanced by both cold and heat stress, although a different expression pattern was observed in the case of the *HSFs* genes with a late induction under cold conditions and an early response under high temperature (Swindell et al., 2007). This expression pattern under heat shock conditions is consistent with the function of HSFs as transcriptional regulators of HSPs, which act as chaperones preventing heat denatured protein aggregations and assisting proteins to refold properly (Borges and Ramos, 2005). Interestingly, an *Arabidopsis* HSF, AtHsfA1b, seems to play a relevant role in high-temperature-induced chilling tolerance (Li et al., 2003). Tomato transgenic plants over-expressing *AtHsfA1b* exhibit enhanced tolerance to both high and low temperatures, the latter without the need for a previous heat acclimation period (Li et al., 2003). On the other hand, *Arabidopsis* plants over-expressing the yeast *HSP26* gene show increased amounts of free proline and soluble sugars, higher expression levels of stress defense genes and, ultimately, enhanced tolerance to freezing temperature (Xue et al., 2009). *LeHSP23.8* is a gene from tomato that encodes a small HSP and whose expression is responsive to high and low temperatures (Yi et al., 2006). The HSP90 family is also involved in heat and cold tolerance in *Arabidopsis* (Krishna and Gloor, 2001). Recently, the *Arabidopsis* protein AtBAG7, which is a component of the unfolded protein response (UPR), a mechanism triggered by the presence of misfolded proteins under stress conditions to increase the production of molecular chaperones, has been reported to have a protective function during heat shock and under low-temperature conditions (Williams et al., 2010). Consistently, *Atbag7* null mutants are hypersensitive to both extreme temperatures (Williams et al., 2010).

Post-transcriptional regulation of gene expression in response to high and low temperatures

In maize, the DREB2A transcription factor has been reported to be regulated by alternative splicing among other mechanisms (Qin et al., 2007). The longest transcript, coding for the functional protein, is more abundant under cold stress, whereas at high temperature it is underproduced (Qin et al., 2007). Adding complexity to the regulatory networks, the pre-mRNAs of a conserved family of splicing regulators in eukaryotes, the Serine/arginine-Rich (SR) proteins, are subjected to alternative splicing under cold and heat stresses in *Arabidopsis* (Lazar and Goodman, 2000; Palusa et al., 2007). More precisely, extreme temperatures dramatically affect the alternative splicing of pre-mRNAs of most *SR* genes studied; some splicing products are increased or decreased depending on the temperature tested (Lazar and Goodman, 2000; Palusa et al., 2007). These results unveil an enormous transcriptome complexity, as the different SR isoforms produced under extreme temperatures probably have altered functions in pre-mRNA splicing.

As mentioned earlier, miRNAs are known to silence genes at a post-transcriptional level (Baulcombe, 2004). In *Populus trichocarpa*, the expression of a considerable number of miRNAs has been described to be repressed by both low and high temperatures, which suggests that they can also constitute common intermediaries in plant response to extreme temperatures (Lu et al., 2008).

Translational and post-translational regulation of gene expression in response to high and low temperatures

Post-translational modifications like ubiquitination and SUMOylation regulate the activity of proteins under both low and high temperatures. In the case of ubiquitination, Calderon-Villalobos and colleagues (2007) have reported that the expression of *AtFBP7* gene from *Arabidopsis*, which encodes an F-box protein essential for an efficient translation under temperature stress, is induced by low and high temperatures. In addition to function as an E3 ligase (see earlier discussion), AtCHIP also has a role as HSP70 co-chaperone. This role has been described to be temperature sensitive and, as expected, markedly induced by heat stress (Rosser et al., 2007). The accumulation of *AtCHIP* transcripts has been reported under high-temperature conditions, although it has not been correlated with an enhanced stress tolerance (Yan et al., 2003). In contrast, the over-expression of *AtCHIP* in *Arabidopsis* renders plants more sensitive to both low and high temperatures (Yan et al., 2003). This indicates that this E3 ligase functions as a negative regulator during temperature stress, which is necessary for an accurate modulation of plant response to extreme temperatures. The classical function assigned to E3 ligases under heat shock has been to tag the unstable proteins generated for their degradation via proteasome. In light of the results described earlier, however, this could be only one of a range of functions.

Regarding SUMOylation, as in the case of low temperature response, an increase in SUMO-conjugates mediated by SIZ1 has been reported in *Arabidopsis* exposed to high-temperature conditions (Kurepa et al., 2003; Miura et al., 2005; Saracco et al., 2007). Interestingly, *siz1* mutants exhibit reduced basal thermotolerance and wild-type levels of acquired thermotolerance (Yoo et al., 2006). The effect of SIZ1 on basal thermotolerance seems to be SA-independent, although a substantial accumulation of SA was observed in *siz1* mutants subjected to heat shock (Yoo et al., 2006). These results support a role for SA in promoting basal thermotolerance, but are not essential for the acquisition of thermotolerance in agreement with the data reported by Clarke and co-workers (2004). In Figure 19.4, a schematic model that summarizes the regulation levels in which low and high temperature signaling pathways converge and diverge is presented.

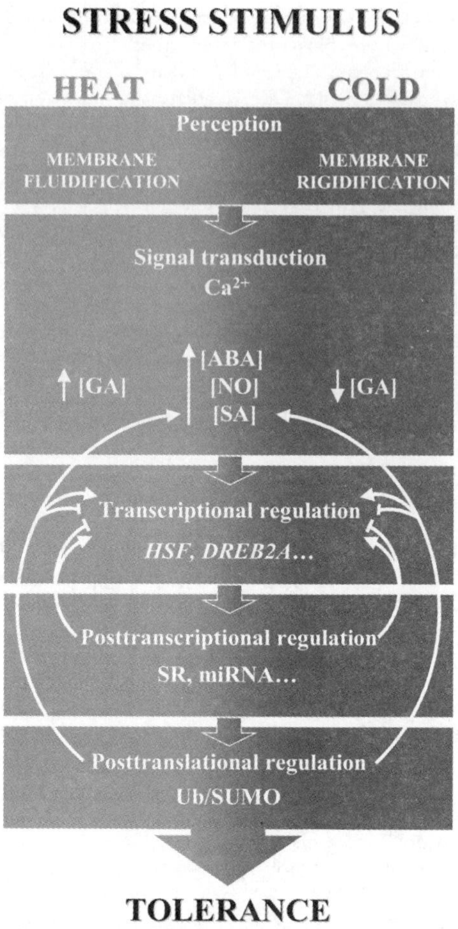

Figure 19.4 • Proposed model for integration of low- and high-temperature signaling pathways. Changes in membrane fluidity are thought to be one of the first steps in the perception of temperature variation. Calcium acts as a second messenger in both low- and high-temperature responses, transducing the signal generated. On the other hand, cold and heat stresses cause an increase in ABA, NO, and SA levels that mediate changes in gene expression and are required for plant tolerance. Gibberellins (GA) are also involved in extreme temperature responses. Finally, plant responses to both low and high temperatures are regulated at the transcriptional, post-transcriptional and post-translational levels. Arrowheads and end lines indicate positive and negative regulation, respectively. Please see color plate section at the end of the book.

Conclusions

As sessile organisms, plants are constantly challenged by a myriad of biotic and abiotic stresses simultaneously. To cope, plants have evolved sophisticated cross-talk strategies between the different response mechanisms. Specifically, from the temperature change sensing to the temperature-responsive gene expression, with the signal transduction pathways in between, there are many levels at which cold and heat responses overlap. This intricate web of interactions is beginning to be revealed and the molecular mechanisms underlying plant responses to low and high temperatures are unveiling. Unfortunately, little is known on how extreme temperatures are perceived and their signals integrated and transmitted to ultimately activate the protective mechanisms. As commented in this chapter, some intermediates have been identified that seem to be particularly relevant and would constitute important nodes of integration between low- and high-temperature responses. In the current scenario of global climate change, steeper temperature variations can have profound effects on crop yield. Understanding the molecular mechanisms controlling plant responses to low and high temperatures, and deciphering the cross-talk nodes governing such responses, should provide new strategies for developing crops with enhanced plasticity against extreme temperatures. These strategies would be of particular interest in temperate regions where dramatic variations in temperature in seasons or even during the same day (i.e., between day and night), can have devastating effects on agriculture.

Acknowledgments

Work in our laboratory is supported by grants GEN2006-27787-E/VEG, BIO2007-65284, CSD2007-00057, and EUI2009-04074 from the Spanish Ministry of Science and Innovation, and grant P2006/GEN-0191 from the Regional Government of Madrid.

References

Achard, P., Gong, F., Cheminant, S., Alioua, M., Hedden, P., & Genschik, P. (2008). The cold-inducible CBF1 factor-dependent signaling pathway modulates the accumulation of the growth-repressing DELLA proteins via its effect on gibberellin metabolism. *Plant Cell, 20*, 2117–2129.

Agarwal, M., Hao, Y., Kapoor, A., Dong, C. H., Fujii, H., & Zheng, X., et al. (2006). A R2R3 type MYB transcription factor is involved in the cold regulation of CBF genes and in acquired freezing tolerance. *Journal of Biological Chemistry, 281*, 37636–37645.

Antikainen, M., Griffith, M., Zhang, J., Hon, W. C., Yang, D., & Pihakaski-Maunsbach, K. (1996). Immunolocalization of antifreeze proteins in winter rye leaves, crowns, and roots by tissue printing. *Plant Physiology, 110*, 845–857.

Bae, M. S., Cho, E. J., Choi, E. Y., & Park, O. K. (2003). Analysis of the *Arabidopsis* nuclear proteome and its response to cold stress. *Plant Journal, 36*, 652–663.

Baker, S. S., Wilhelm, K. S., & Thomashow, M. F. (1994). The 5'-region of *Arabidopsis thaliana cor15a* has *cis*-acting elements that confer cold-, drought- and ABA-regulated gene expression. *Plant Molecular Biology, 24*, 701–713.

Bakht, J., Bano, A., & Dominy, P. (2006). The role of abscisic acid and low temperature in chickpea (*Cicer arietinum*) cold tolerance. II. Effects on plasma membrane structure and function. *Journal of Experimental Botany, 57*, 3707–3715.

Baulcombe, D. (2004). RNA silencing in plants. *Nature, 431*, 356–363.

Benedict, C., Skinner, J. S., Meng, R., Chang, Y., Bhalerao, R., & Huner, N. P., et al. (2006). The CBF1-dependent low temperature signalling pathway, regulon and increase in freeze tolerance are conserved in *Populus* spp. *Plant, Cell & Environment, 29*, 1259–1272.

Bernard, F., Shaker-Bazarnov, H., & Kaviani, B. (2002). Effects of salicylic acid on cold preservation and cryopreservation of encapsulated embryonic axes of Persian lilac (*Melia azedarach* L.). *Euphytica, 123*, 85–88.

Borges, J. C., & Ramos, C. H. (2005). Protein folding assisted by chaperones. *Protein & Peptide Letters, 12*, 257–261.

Brautigam, M., Lindlof, A., Zakhrabekova, S., Gharti-Chhetri, G., Olsson, B., & Olsson, O. (2005). Generation and analysis of 9792 EST sequences from cold acclimated oat, *Avena sativa*. *BMC Plant Biology, 5*, 18.

Brown, A. P., Dunn, M. A., Goddard, N. J., & Hughes, M. A. (2001). Identification of a novel low-temperature-response element in the promoter of the barley (*Hordeum vulgare* L) gene *blt101.1*. *Planta, 213*, 770–780.

Calderon-Villalobos, L. I., Nill, C., Marrocco, K., Kretsch, T., & Schwechheimer, C. (2007). The evolutionarily conserved *Arabidopsis thaliana* F-box protein AtFBP7 is required for efficient translation during temperature stress. *Gene, 392*, 106–116.

Capel, J., Jarillo, J. A., Madueno, F., Jorquera, M. J., Martinez-Zapater, J. M., & Salinas, J. (1998). Low temperature regulates *Arabidopsis* Lhcb gene expression in a light-independent manner. *Plant Journal, 13*, 411–418.

Carpaneto, A., Ivashikina, N., Levchenko, V., Krol, E., Jeworutzki, E., & Zhu, J. K., et al. (2007). Cold transiently activates calcium-permeable channels in *Arabidopsis* mesophyll cells. *Plant Physiology, 143*, 487–494.

Catala, R., & Salinas, J. (2007). Regulatory mechanisms involved in cold acclimation response. *Spanish Journal of Agricultural Research, 6*, 211–220.

Catala, R., Santos, E., Alonso, J. M., Ecker, J. R., Martinez-Zapater, J. M., & Salinas, J. (2003). Mutations in the Ca^{2+}/H^{+} transporter CAX1 increase *CBF/DREB1* expression and the cold-acclimation response in *Arabidopsis*. *Plant Cell, 15*, 2940–2951.

Chaikam, V., & Karlson, D. T. (2010). Response and transcriptional regulation of rice SUMOylation system during development and stress conditions. *BMB Reports, 43*, 103–109.

Chen, T. H., & Gusta, L. V. (1983). Abscisic acid-induced freezing resistance in cultured plant cells. *Plant Physiology, 73*, 71–75.

Cheng, N. H., Pittman, J. K., Barkla, B. J., Shigaki, T., & Hirschi, K. D. (2003). The *Arabidopsis cax1* mutant exhibits impaired ion homeostasis, development, and hormonal responses and reveals interplay among vacuolar transporters. *Plant Cell, 15*, 347–364.

Cheong, Y. H., Kim, K. N., Pandey, G. K., Gupta, R., Grant, J. J., & Luan, S. (2003). CBL1, a calcium sensor that differentially regulates salt, drought, and cold responses in *Arabidopsis*. *Plant Cell, 15*, 1833–1845.

Chinnusamy, V., Ohta, M., Kanrar, S., Lee, B. H., Hong, X., & Agarwal, M., et al. (2003). ICE1: A regulator of cold-induced transcriptome and freezing tolerance in *Arabidopsis*. *Genes and Development, 17*, 1043–1054.

Chinnusamy, V., Zhu, J. K., & Sunkar, R. (2010). Gene regulation during cold stress acclimation in plants. *Methods in Molecular Biology (Totowa, NJ, U. S.), 639*, 39–55.

Choi, C. S., & Sano, H. (2007). Abiotic-stress induces demethylation and transcriptional activation of a gene encoding a glycerophosphodiesterase-like protein in tobacco plants. *Molecular Genetics and Genomics, 277*, 589–600.

Christie, P. J., Alfenito, M. R., & Walbot, V. (1994). Impact of low-temperature stress on general phenylpropanoid and anthocyanin pathways: Enhancement of transcript abundance and anthocyanin pigmentation in maiz seedlings. *Planta, 194*, 541–549.

Ciardi, J. A., Deikman, J., & Orzolek, M. D. (1997). Increased ethylene synthesis enhances chilling tolerance in tomato. *Physiologia Plantarum, 101*, 333–340.

Clarke, S. M., Mur, L. A., Wood, J. E., & Scott, I. M. (2004). Salicylic acid dependent signaling promotes basal thermotolerance but is not essential for acquired thermotolerance in *Arabidopsis thaliana*. *Plant Journal, 38*, 432–447.

Colby, T., Matthai, A., Boeckelmann, A., & Stuible, H. P. (2006). SUMO-conjugating and SUMO-deconjugating enzymes from *Arabidopsis*. *Plant Physiology, 142*, 318–332.

Corpas, F. J., Chaki, M., Fernandez-Ocana, A., Valderrama, R., Palma, J. M., & Carreras, A., et al. (2008). Metabolism of reactive nitrogen species in pea plants under abiotic stress conditions. *Plant and Cell Physiology, 49*, 1711–1722.

Cossins, A. R. (1994). *Temperature adaptation in biological membranes (pp. 63–76)*. London: Portland Press.

Cuevas, J. C., Lopez-Cobollo, R., Alcazar, R., Zarza, X., Koncz, C., & Altabella, T., et al. (2008). Putrescine is involved in *Arabidopsis* freezing tolerance and cold acclimation by regulating abscisic acid levels in response to low temperature. *Plant Physiology, 148*, 1094–1105.

Dexter, S. T. (1933). Effect of several environmental factors on the hardening of plants. *Plant Physiology*, *8*, 123–129.

Ding, C. K., Wang, C. Y., Gross, K. C., & Smith, D. L. (2002). Jasmonate and salicylate induce the expression of pathogenesis-related-protein signal transducers genes and increase resistance to chilling injury in tomato fruit. *Planta*, *214*, 895–901.

Doherty, C. J., Van Buskirk, H. A., Myers, S. J., & Thomashow, M. F. (2009). Roles for *Arabidopsis* CAMTA transcription factors in cold-regulated gene expression and freezing tolerance. *Plant Cell*, *21*, 972–984.

Dong, C. H., Agarwal, M., Zhang, Y., Xie, Q., & Zhu, J. K. (2006b). The negative regulator of plant cold responses, HOS1, is a RING E3 ligase that mediates the ubiquitination and degradation of ICE1. *Proceedings of the National Academy of Sciences of the United States of America*, *103*, 8281–8286.

Dong, C. H., Hu, X., Tang, W., Zheng, X., Kim, Y. S., & Lee, B. H., et al. (2006a). A putative *Arabidopsis* nucleoporin, AtNUP160, is critical for RNA export and required for plant tolerance to cold stress. *Molecular and Cellular Biology*, *26*, 9533–9543.

Dong, C. J., & Liu, J. Y. (2010). The *Arabidopsis* EAR-motif-containing protein RAP2.1 functions as an active transcriptional repressor to keep stress responses under tight control. *BMC Plant Biology*, *10*, 47.

Dubouzet, J. G., Sakuma, Y., Ito, Y., Kasuga, M., Dubouzet, E. G., & Miura, S., et al. (2003). *OsDREB* genes in rice, *Oryza sativa* L., encode transcription activators that function in drought-, high-salt- and cold-responsive gene expression. *Plant Journal*, *33*, 751–763.

Dunn, M. A., Goddard, N. J., Zhang, L., Pearce, R. S., & Hughes, M. A. (1994). Low-temperature-responsive barley genes have different control mechanisms. *Plant Molecular Biology*, *24*, 879–888.

Dunn, M. A., White, A. J., Vural, S., & Hughes, M. A. (1998). Identification of promoter elements in a low-temperature-responsive gene (*blt4.9*) from barley (*Hordeum vulgare* L.). *Plant Molecular Biology*, *38*, 551–564.

Egawa, C., Kobayashi, F., Ishibashi, M., Nakamura, T., Nakamura, C., & Takumi, S. (2006). Differential regulation of transcript accumulation and alternative splicing of a *DREB2* homolog under abiotic stress conditions in common wheat. *Genes & Genetic Systems*, *81*, 77–91.

Field, R. J. (1981). The effect of low temperature on ethylene production by leaf tissues of *Phaseolus vulgaris* L. *Annals of Botany*, *47*, 215–221.

Fowler, S., & Thomashow, M. F. (2002). *Arabidopsis* transcriptome profiling indicates that multiple regulatory pathways are activated during cold acclimation in addition to the CBF cold response pathway. *Plant Cell*, *14*, 1675–1690.

Franklin, K. A., & Whitelam, G. C. (2007). Light-quality regulation of freezing tolerance in *Arabidopsis thaliana*. *Nature Genetics*, *39*, 1410–1413.

Fu, P., Wilen, R. W., Robertson, A. J., Low, N. H., Tyler, R. T., & Gusta, L. V. (1998). Heat tolerance of cold acclimated puma winter rye seedlings and the effect of a heat shock on freezing tolerance. *Plant and Cell Physiology*, *39*, 942–949.

Fursova, O. V., Pogorelko, G. V., & Tarasov, V. A. (2009). Identification of *ICE2*, a gene involved in cold acclimation which determines freezing tolerance in *Arabidopsis thaliana*. *Gene*, *429*, 98–103.

Galston, A. W., & Sawhney, R. K. (1990). Polyamines in plant physiology. *Plant Physiology*, *94*, 406–410.

Gao, F., Zhou, Y., Zhu, W., Li, X., Fan, L., & Zhang, G. (2009). Proteomic analysis of cold stress-responsive proteins in *Thellungiella* rosette leaves. *Planta*, *230*, 1033–1046.

Gilmour, S. J., & Thomashow, M. F. (1991). Cold acclimation and cold-regulated gene expression in ABA mutants of *Arabidopsis thaliana*. *Plant Molecular Biology*, *17*, 1233–1240.

Gilmour, S. J., Sebolt, A. M., Salazar, M. P., Everard, J. D., & Thomashow, M. F. (2000). Overexpression of the *Arabidopsis* CBF3 transcriptional activator mimics multiple biochemical changes associated with cold acclimation. *Plant Physiology*, *124*, 1854–1865.

Gilmour, S. J., Zarka, D. G., Stockinger, E. J., Salazar, M. P., Houghton, J. M., & Thomashow, M. F. (1998). Low temperature regulation of the *Arabidopsis CBF* family of AP2 transcriptional activators as an early step in cold-induced *COR* gene expression. *Plant Journal*, *16*, 433–442.

Gong, M., Li, Y. J., Dai, X., Tian, M., & Li, Z. G. (1997). Involvement of calcium and calmodulin in the acquisition of heat-shock induced thermotolerance in maize seedlings. *Journal of Plant Physiology*, *150*, 615–621.

Gong, Z., Dong, C. H., Lee, H., Zhu, J., Xiong, L., & Gong, D., et al. (2005). A DEAD box RNA helicase is essential for mRNA export and important for development and stress responses in *Arabidopsis*. *Plant Cell*, *17*, 256–267.

Gray, G. R., Chauvin, L. P., Sarhan, F., & Huner, N. (1997). Cold acclimation and freezing tolerance (A complex interaction of light and temperature). *Plant Physiology*, *114*, 467–474.

Griffith, M., & McIntyre, H. C. H. (1993). The interrelationship of growth and frost tolerance in winter rye. *Physiologia Plantarum*, *87*, 335–344.

Guiltinan, M. J., Marcotte, W. R., Jr., & Quatrano, R. S. (1990). A plant leucine zipper protein that recognizes an abscisic acid response element. *Science*, *250*, 267–271.

Guo, Y., Xiong, L., Ishitani, M., & Zhu, J. K. (2002). An *Arabidopsis* mutation in translation elongation factor 2 causes superinduction of CBF/DREB1 transcription factor genes but blocks the induction of their downstream targets under low temperatures. *Proceedings of the National Academy of Sciences of the United States of America*, *99*, 7786–7791.

Guy, C. L., Huber, J. L., & Huber, S. C. (1992). Sucrose phosphate synthase and sucrose accumulation at low temperature. *Plant Physiology*, *100*, 502–508.

Hajela, R. K., Horvath, D. P., Gilmour, S. J., & Thomashow, M. F. (1990). Molecular cloning and expression of *cor* (Cold-Regulated) genes in *Arabidopsis thaliana*. *Plant Physiology*, *93*, 1246–1252.

Harber, R. M., & Fuchigami, L. H. (1989). *Ethylene-induced stress resistance*. Boca Raton: CRC Press.

Harmer, S. L., Hogenesch, J. B., Straume, M., Chang, H. S., Han, B., & Zhu, T., et al. (2000). Orchestrated transcription of key pathways in *Arabidopsis* by the circadian clock. *Science*, *290*, 2110–2113.

Harvaux, M., & Kloppstech, K. (2001). The protective functions of carotenoid and flavonoid pigments against excess visible radiation at chilling temperature investigated in *Arabidopsis npq* and *tt* mutants. *Planta*, *213*, 953–966.

Heino, P., Sandman, G., Lang, V., Nordin, K., & Palva, E. T. (1990). Abscisic acid deficiency prevents development of freezing tolerance in *Arabidopsis thaliana* (L.) Heynh. *Theoretical and Applied Genetics*, *79*, 801–806.

Hirschi, K. D., Zhen, R. G., Cunningham, K. W., Rea, P. A., & Fink, G. R. (1996). CAX1, an $H^+/Ca2^+$ antiporter from *Arabidopsis*. *Proceedings of the National Academy of Sciences of the United States of America*, *93*, 8782–8786.

Hofmann, N. R. (2009). The plasma membrane as first responder to heat stress. *Plant Cell*, *21*, 2544.

Holappa, L. D., & Walker-Simmons, M. K. (1995). The wheat abscisic acid-responsive protein kinase mRNA, PKABA1, is up-regulated by dehydration, cold temperature, and osmotic stress. *Plant Physiology*, *108*, 1203–1210.

Hong, S. W., Jon, J. H., Kwak, J. M., & Nam, H. G. (1997). Identification of a receptor-like protein kinase gene rapidly induced by abscisic acid, dehydration, high salt, and cold treatments in *Arabidopsis thaliana*. *Plant Physiology*, *113*, 1203–1212.

Hu, H., You, J., Fang, Y., Zhu, X., Qi, Z., & Xiong, L. (2008). Characterization of transcription factor gene *SNAC2* conferring cold and salt tolerance in rice. *Plant Molecular Biology*, *67*, 169–181.

Hu, X. L., Liu, R. X., Li, Y. H., Wang, W., Tai, F. J., & Xue, R. L., et al. (2010). Heat shock protein 70 regulates the abscisic acid-induced antioxidant response of maize to combined drought and heat stress. *Plant Growth Regulation*, *60*, 225–235.

Huang, B., Jin, L., & Liu, J. (2007). Molecular cloning and functional characterization of a DREB1/CBF-like gene (*GhDREB1L*) from cotton. *Science in China. Series C, Life sciences*, *50*, 7–14.

Hummel, I., El-Amrani, A., Gouesbet, G., Hennion, F., & Couee, I. (2004). Involvement of polyamines in the interacting effects of low temperature and mineral supply on *Pringlea antiscorbutica* (Kerguelen cabbage) seedlings. *Journal of Experimental Botany*, *55*, 1125–1134.

Iliev, E. A., Xu, W., Polisensky, D. H., Oh, M. H., Torisky, R. S., & Clouse, S. D., et al. (2002). Transcriptional and posttranscriptional regulation of *Arabidopsis*

TCH4 expression by diverse stimuli. Roles of *cis* regions and brassinosteroids. *Plant Physiology, 130*, 770–783.

Ishitani, M., Xiong, L., Lee, H., Stevenson, B., & Zhu, J. K. (1998). *HOS1*, a genetic locus involved in cold-responsive gene expression in *Arabidopsis*. *Plant Cell, 10*, 1151–1161.

James, V. A., Neibaur, I., & Altpeter, F. (2008). Stress inducible expression of the DREB1A transcription factor from xeric, *Hordeum spontaneum* L. in turf and forage grass (*Paspalum notatum Flugge*) enhances abiotic stress tolerance. *Transgenic Research, 17*, 93–104.

Janda, T., Szalai, G., Tari, I., & Páldi, E. (1999). Hydroponic treatment with salicylic acid decreases the effects of chilling injury in maize (*Zea mays* L.) plants. *Planta, 208*, 175–180.

Jian, X., Zhang, L., Li, G., Zhang, L., Wang, X., & Cao, X., et al. (2010). Identification of novel stress-regulated microRNAs from *Oryza sativa* L. *Genomics, 95*, 47–55.

Jiang, C., Iu, B., & Singh, J. (1996). Requirement of a CCGAC cis-acting element for cold induction of the *BN115* gene from winter *Brassica napus*. *Plant Molecular Biology, 30*, 679–684.

Jonak, C., Kiegerl, S., Ligterink, W., Barker, P. J., Huskisson, N. S., & Hirt, H. (1996). Stress signaling in plants: A mitogen-activated protein kinase pathway is activated by cold and drought. *Proceedings of the National Academy of Sciences of the United States of America, 93*, 11274–11279.

Jouannic, S., Hamal, A., Leprince, A. S., Tregear, J. W., Kreis, M., & Henry, Y. (1999). Plant MAP kinase kinase kinases structure, classification and evolution. *Gene, 233*, 1–11.

Kagale, S., Divi, U. K., Krochko, J. E., Keller, W. A., & Krishna, P. (2007). Brassinosteroid confers tolerance in *Arabidopsis thaliana* and *Brassica napus* to a range of abiotic stresses. *Planta, 225*, 353–364.

Kang, G. Z., Wang, Z. X., Xia, K. F., & Sun, G. C. (2007). Protection of ultrastructure in chilling-stressed banana leaves by salicylic acid. *Journal of Zhejiang University. Science, 8*, 277–282.

Kang, H. M., & Saltveit, M. E. (2002). Chilling tolerance of maize, cucumber and rice seedling leaves and roots are differentially affected by salicylic acid. *Physiologia Plantarum, 115*, 571–576.

Kaplan, F., Kopka, J., Haskell, D. W., Schiller, C., Gatzke, N., & Sung, D. Y., et al. (2004). Exploring the temperature-stress metabolome of *Arabidopsis*. *Plant Physiology, 136*, 4159–4168.

Kawamura, Y., & Uemura, M. (2003). Mass spectrometric approach for identifying putative plasma membrane proteins of *Arabidopsis* leaves associated with cold acclimation. *Plant Journal, 36*, 141–154.

Kim, H. J., Hyun, Y., Park, J. Y., Park, M. J., Park, M. K., & Kim, M. D., et al. (2004). A genetic link between cold responses and flowering time through FVE in *Arabidopsis thaliana*. *Nature Genetics, 36*, 167–171.

Kim, H. J., Kim, Y. K., Park, J. Y., & Kim, J. (2002a). Light signalling mediated by phytochrome plays an important role in cold-induced gene expression through the C-repeat/dehydration responsive element (C/DRE) in *Arabidopsis thaliana*. *Plant Journal, 29*, 693–704.

Kim, J. C., Lee, S. H., Cheong, Y. H., Yoo, C. M., Lee, S. I., & Chun, H. J., et al. (2001). A novel cold-inducible zinc finger protein from soybean, SCOF-1, enhances cold tolerance in transgenic plants. *Plant Journal, 25*, 247–259.

Kim, K. N., Cheong, Y. H., Grant, J. J., Pandey, G. K., & Luan, S. (2003). CIPK3, a calcium sensor-associated protein kinase that regulates abscisic acid and cold signal transduction in *Arabidopsis*. *Plant Cell, 15*, 411–423.

Kim, T. E., Kim, S. K., Han, T. J., Lee, J. S., & Chang, S. C. (2002b). ABA and polyamines act independently in primary leaves of cold-stressed tomato (*Lycopersicon esculentum*). *Physiologia Plantarum, 115*, 370–376.

Kim, Y. O., Kim, J. S., & Kang, H. (2005). Cold-inducible zinc finger-containing glycine-rich RNA-binding protein contributes to the enhancement of freezing tolerance in *Arabidopsis thaliana*. *Plant Journal, 42*, 890–900.

Knight, H. (2000). Calcium signaling during abiotic stress in plants. *International Review of Cytology, 195*, 269–324.

Knight, H., Trewavas, A. J., & Knight, M. R. (1996). Cold calcium signaling in *Arabidopsis* involves two cellular pools and a change in calcium signature after acclimation. *Plant Cell, 8*, 489–503.

Knight, H., Zarka, D. G., Okamoto, H., Thomashow, M. F., & Knight, M. R. (2004). Abscisic acid induces *CBF* gene transcription and subsequent induction of cold-regulated genes via the CRT promoter element. *Plant Physiology, 135*, 1710–1717.

Knight, M., Campbell, A. K., Smith, S. M., & Trewavas, A. J. (1991). Transgenic plant aequorin reports the effects or touch and cold-shock and elicitors on cytoplasmic calcium. *Nature, 352*, 524–526.

Kobayashi, F., Maeta, E., Terashima, A., & Takumi, S. (2008b). Positive role of a wheat HvABI5 ortholog in abiotic stress response of seedlings. *Physiologia Plantarum, 134*, 74–86.

Kobayashi, F., Takumi, S., & Nakamura, C. (2008a). Increased freezing tolerance in an ABA-hypersensitive mutant of common wheat. *Journal of Plant Physiology, 165*, 224–232.

Kovacs, Z., Simon-Sarkadi, L., Szucs, A., & Kocsy, G. (2010). Differential effects of cold, osmotic stress and abscisic acid on polyamine accumulation in wheat. *Amino acids, 38*, 623–631.

Kovtun, Y., Chiu, W. L., Tena, G., & Sheen, J. (2000). Functional analysis of oxidative stress-activated mitogen-activated protein kinase cascade in plants. *Proceedings of the National Academy of Sciences of the United States of America, 97*, 2940–2945.

Krishna, P., & Gloor, G. (2001). The Hsp90 family of proteins in *Arabidopsis thaliana*. *Cell Stress & Chaperones, 6*, 238–246.

Krol, M., Ivanov, A. G., Jansson, S., Kloppstech, K., & Huner, N. P. (1999). Greening under high light or cold temperature affects the level of xanthophyll-cycle pigments, early light-inducible proteins, and light-harvesting polypeptides in wild-type barley and the chlorina f2 mutant. *Plant Physiology, 120*, 193–204.

Kurepa, J., Walker, J. M., Smalle, J., Gosink, M. M., Davis, S. J., & Durham, T. L., et al. (2003). The small ubiquitin-like modifier (SUMO) protein modification system in *Arabidopsis*. Accumulation of SUMO1 and -2 conjugates is increased by stress. *Journal of Biological Chemistry, 278*, 6862–6872.

Kwak, K. J., Kim, J. Y., Kim, Y. O., & Kang, H. (2007). Characterization of transgenic *Arabidopsis* plants overexpressing high mobility group B proteins under high salinity, drought or cold stress. *Plant and Cell Physiology, 48*, 221–231.

Kwon, C. S., Lee, D., Choi, G., & Chung, W. I. (2009). Histone occupancy-dependent and -independent removal of H3K27 trimethylation at cold-responsive genes in *Arabidopsis*. *Plant Journal, 60*, 112–121.

Lang, V., Heino, P., & Palva, E. T. (1989). Low temperature acclimation and treatment with exogenous abscisic acid induce common polypeptides in *Arabidopsis thaliana* (L.) Heynh. *Theoretical and Applied Genetics, 77*, 729–734.

Larkindale, J., & Knight, M. R. (2002). Protection against heat stress-induced oxidative damage in *Arabidopsis* involves calcium, abscisic acid, ethylene, and salicylic acid. *Plant Physiology, 128*, 682–695.

Larkindale, J., Hall, J. D., Knight, M. R., & Vierling, E. (2005). Heat stress phenotypes of *Arabidopsis* mutants implicate multiple signaling pathways in the acquisition of thermotolerance. *Plant Physiology, 138*, 882–897.

Lazar, G., & Goodman, H. M. (2000). The *Arabidopsis* splicing factor SR1 is regulated by alternative splicing. *Plant Molecular Biology, 42*, 571–581.

Lee, B. H., Henderson, D. A., & Zhu, J. K. (2005). The *Arabidopsis* cold-responsive transcriptome and its regulation by ICE1. *Plant Cell, 17*, 3155–3175.

Lee, B. H., Kapoor, A., Zhu, J., & Zhu, J. K. (2006). STABILIZED1, a stress-upregulated nuclear protein, is required for pre-mRNA splicing, mRNA turnover, and stress tolerance in *Arabidopsis*. *Plant Cell, 18*, 1736–1749.

Lee, D. J., & Zeevaart, J. A. (2005). Molecular cloning of GA 2-oxidase3 from spinach and its ectopic expression in *Nicotiana sylvestris*. *Plant Physiology, 138*, 243–254.

Lee, H., Guo, Y., Ohta, M., Xiong, L., Stevenson, B., & Zhu, J. K. (2002). *LOS2*, a genetic locus required for cold-responsive gene transcription encodes a bi-functional enolase. *EMBO Journal, 21*, 2692–2702.

Lee, J. H., Van Montagu, M., & Verbruggen, N. (1999). A highly conserved kinase is an essential component for stress tolerance in yeast and plant cells. *Proceedings of the National Academy of Sciences of the United States of America, 96*, 5873–5877.

Lee, T. M., Lur, H. S., & Chu, C. (1995). Abscisic acid and putrecine accumulation in chilling-tolerant rice cultivars. *Crop Science, 35*, 502–508.

Lee, U., Wie, C., Fernandez, B. O., Feelisch, M., & Vierling, E. (2008). Modulation of nitrosative stress by S-nitrosoglutathione reductase is critical for thermotolerance and plant growth in *Arabidopsis*. *Plant Cell, 20*, 786–802.

Levitt, J. (1980). *Responses of plants to environmental stresses: Chilling, freezing and high temperatures stresses*. New York: Academic Press.

Li, H. Y., Chang, C. S., Lu, L. S., Liu, C. A., Chan, M. T., & Charng, Y. Y. (2003). Over-expression of *Arabidopsis thaliana* heat shock factor gene (AtHsf1b) enhances chilling tolerance in transgenic tomato. *Botanical Bulletin of Academia Sinica, 44*, 129–140.

Liao, Y., Zou, H. F., Wei, W., Hao, Y. J., Tian, A. G., & Huang, J., et al. (2008). Soybean *GmbZIP44*, *GmbZIP62* and *GmbZIP78* genes function as negative regulator of ABA signaling and confer salt and freezing tolerance in transgenic *Arabidopsis*. *Planta, 228*, 225–240.

Liu, H. T., Gao, F., Cui, S. J., Han, J. L., Sun, D. Y., & Zhou, R. G. (2006). Primary evidence for involvement of IP3 in heat-shock signal transduction in *Arabidopsis*. *Cell Research, 16*, 394–400.

Liu, Q., Kasuga, M., Sakuma, Y., Abe, H., Miura, S., & Yamaguchi-Shinozaki, K., et al. (1998). Two transcription factors, DREB1 and DREB2, with an EREBP/AP2 DNA binding domain separate two cellular signal transduction pathways in drought- and low-temperature-responsive gene expression, respectively, in *Arabidopsis*. *Plant Cell, 10*, 1391–1406.

Liu, S. X., Wang, X. L., Yang, J. S., Fan, Z. Q., Sun, X. F., & Wang, X. R., et al. (2005). Molecular cloning of a novel LOS2 gene from *Capsella bursa-pastoris*. *Yichuan Xuebao, 32*, 600–607.

Llorente, F., Oliveros, J. C., Martinez-Zapater, J. M., & Salinas, J. (2000). A freezing-sensitive mutant of *Arabidopsis*, *frs1*, is a new *aba3* allele. *Planta, 211*, 648–655.

Lu, S., Sun, Y. H., & Chiang, V. L. (2008). Stress-responsive microRNAs in *Populus*. *Plant Journal, 55*, 131–151.

Luo, J., Shen, G., Yan, J., He, C., & Zhang, H. (2006). AtCHIP functions as an E3 ubiquitin ligase of protein phosphatase 2A subunits and alters plant response to abscisic acid treatment. *Plant Journal, 46*, 649–657.

Ma, Y., Szostkiewicz, I., Korte, A., Moes, D., Yang, Y., & Christmann, A., et al. (2009). Regulators of PP2C phosphatase activity function as abscisic acid sensors. *Science, 324*, 1064–1068.

Mancinelli, A. L. (1984). Photoregulation of anthocyanin synthesis : VIII. Effect of light pretreatments. *Plant Physiology, 75*, 447–453.

Martin, M. L., & Busconi, L. (2001). A rice membrane-bound calcium-dependent protein kinase is activated in response to low temperature. *Plant Physiology, 125*, 1442–1449.

Matsuda, O., Sakamoto, H., Hashimoto, T., & Iba, K. (2005). A temperature-sensitive mechanism that regulates post-translational stability of a plastidial omega-3 fatty acid desaturase (FAD8) in *Arabidopsis* leaf tissues. *Journal of Biological Chemistry, 280*, 3597–3604.

Matsui, A., Ishida, J., Morosawa, T., Mochizuki, Y., Kaminuma, E., & Endo, T. A., et al. (2008). *Arabidopsis* transcriptome analysis under drought, cold, high-salinity and ABA treatment conditions using a tiling array. *Plant and Cell Physiology, 49*, 1135–1149.

Mazars, C., Thion, L., Thuleau, P., Graziana, A., Knight, M. R., & Moreau, M., et al. (1997). Organization of cytoskeleton controls the changes in cytosolic calcium of cold-shocked *Nicotiana plumbaginifolia* protoplasts. *Cell Calcium, 22*, 413–420.

Medina, J., Bargues, M., Terol, J., Perez-Alonso, M., & Salinas, J. (1999). The *Arabidopsis CBF* gene family is composed of three genes encoding AP2 domain-containing proteins whose expression is regulated by low temperature but not by abscisic acid or dehydration. *Plant Physiology, 119*, 463–470.

Mikkelsen, M. D., & Thomashow, M. F. (2009). A role for circadian evening elements in cold-regulated gene expression in *Arabidopsis*. *Plant Journal, 60*, 328–339.

Miura, K., & Ohta, M. (2010). SIZ1, a small ubiquitin-related modifier ligase, controls cold signaling through regulation of salicylic acid accumulation. *Journal of Plant Physiology, 167*, 555–560.

Miura, K., Jin, J. B., Lee, J., Yoo, C. Y., Stirm, V., & Miura, T., et al. (2007). SIZ1-mediated sumoylation of ICE1 controls CBF3/DREB1A expression and freezing tolerance in *Arabidopsis*. *Plant Cell, 19*, 1403–1414.

Miura, K., Rus, A., Sharkhuu, A., Yokoi, S., Karthikeyan, A. S., & Raghothama, K. G., et al. (2005). The *Arabidopsis* SUMO E3 ligase SIZ1 controls phosphate deficiency responses. *Proceedings of the National Academy of Sciences of the United States of America, 102*, 7760–7765.

Mizoguchi, T., Irie, K., Hirayama, T., Hayashida, N., Yamaguchi-Shinozaki, K., & Matsumoto, K., et al. (1996). A gene encoding a mitogen-activated protein kinase kinase kinase is induced simultaneously with genes for a mitogen-activated protein kinase and an S6 ribosomal protein kinase by touch, cold, and water stress in *Arabidopsis thaliana*. *Proceedings of the National Academy of Sciences of the United States of America, 93*, 765–769.

Monroy, A. F., Labbe, E., & Dhindsa, R. S. (1997). Low temperature perception in plants: Effects of cold on protein phosphorylation in cell-free extracts. *FEBS Letters, 410*, 206–209.

Monroy, A. F., Sangwan, V., & Dhindsa, R. S. (1998). Low temperature signal transduction during cold acclimation: Protein phosphatase 2A as an early target for cold-inactivation. *Plant Journal, 13*, 653–660.

Monroy, A. F., Sarhan, F., & Dhindsa, R. S. (1993). Cold-induced changes in freezing tolerance, protein phosphorylation, and gene expression (evidence for a role of calcium). *Plant Physiology, 102*, 1227–1235.

Murata, N., & Los, D. A. (1997). Membrane fluidity and temperature perception. *Plant Physiology, 115*, 875–879.

Novillo, F., Alonso, J. M., Ecker, J. R., & Salinas, J. (2004). CBF2/DREB1C is a negative regulator of CBF1/DREB1B and CBF3/DREB1A expression and plays a central role in stress tolerance in *Arabidopsis*. *Proceedings of the National Academy of Sciences of the United States of America, 101*, 3985–3990.

Novillo, F., Medina, J., & Salinas, J. (2007). *Arabidopsis* CBF1 and CBF3 have a different function than CBF2 in cold acclimation and define different gene classes in the CBF regulon. *Proceedings of the National Academy of Sciences of the United States of America, 104*, 21002–21007.

Olsen, J. E., Junttila, O., Nilsen, J., Eriksson, M. E., Martinussen, I., & Olsson, O., et al. (1997). Ectopic expression of oat phytochrome A in hybrid aspen changes critical day length for growth and prevents cold acclimation. *Plant Journal, 12*, 1339–1350.

Orvar, B. L., Sangwan, V., Omann, F., & Dhindsa, R. S. (2000). Early steps in cold sensing by plant cells: The role of actin cytoskeleton and membrane fluidity. *Plant Journal, 23*, 785–794.

Ouellet, F., Vazquez-Tello, A., & Sarhan, F. (1998). The wheat *wcs120* promoter is cold-inducible in both monocotyledonous and dicotyledonous species. *FEBS Letters, 423*, 324–328.

Palusa, S. G., Ali, G. S., & Reddy, A. S. (2007). Alternative splicing of pre-mRNAs of *Arabidopsis* serine/arginine-rich proteins: Regulation by hormones and stresses. *Plant Journal, 49*, 1091–1107.

Park, H. Y., Kang, I. S., Han, J. S., Lee, C. H., An, G., & Moon, Y. H. (2009b). *OsDEG10* encoding a small RNA-binding protein is involved in abiotic stress signaling. *Biochemical and Biophysical Research Communications, 380*, 597–602.

Park, S. Y., Fung, P., Nishimura, N., Jensen, D. R., Fujii, H., & Zhao, Y., et al. (2009a). Abscisic acid inhibits type 2C protein phosphatases via the PYR/PYL family of START proteins. *Science, 324*, 1068–1071.

Pellegrineschi, A., Reynolds, M., Pacheco, M., Brito, R. M., Almeraya, R., & Yamaguchi-Shinozaki, K., et al. (2004). Stress-induced expression in wheat of the *Arabidopsis thaliana DREB1A* gene delays water stress symptoms under greenhouse conditions. *Genome, 47*, 493–500.

Pennycooke, J. C., Cheng, H., & Stockinger, E. J. (2008a). Comparative genomic sequence and expression analyses of *Medicago truncatula* and alfalfa subspecies falcata COLD-ACCLIMATION-SPECIFIC genes. *Plant Physiology, 146*, 1242–1254.

Pennycooke, J. C., Cheng, H., Roberts, S. M., Yang, Q., Rhee, S. Y., & Stockinger, E. J. (2008b). The low temperature-responsive, *SolanumCBF1* genes maintain high identity in their upstream regions in a genomic environment undergoing gene duplications, deletions, and rearrangements. *Plant Molecular Biology, 67*, 483–497.

Phillips, J. R., Dunn, M. A., & Hughes, M. A. (1997). mRNA stability and localisation of the low-temperature-responsive barley gene family *blt14*. *Plant Molecular Biology, 33*, 1013–1023.

Pillai, M. A., & Akiyama, T. (2004). Differential expression of an S-adenosyl-L-methionine decarboxylase gene involved in polyamine biosynthesis under low temperature stress in japonica and indica rice genotypes. *Molecular Genetics and Genomics*, *271*, 141–149.

Plieth, C. (1999). Temperature sensing by plants: Calcium-permeable channels as primary sensors – a model. *Journal of Membrane Biology*, *172*, 121–127.

Polisensky, D. H., & Braam, J. (1996). Cold-shock regulation of the *Arabidopsis TCH* genes and the effects of modulating intracellular calcium levels. *Plant Physiology*, *111*, 1271–1279.

Pressman, E., Shaked, R., & Firon, N. (2006). Exposing pepper plants to high day temperatures prevents the adverse low night temperature symptoms. *Physiologia Plantarum*, *126*, 618–626.

Qin, F., Kakimoto, M., Sakuma, Y., Maruyama, K., Osakabe, Y., & Tran, L. S., et al. (2007). Regulation and functional analysis of ZmDREB2A in response to drought and heat stresses in *Zea mays* L. *Plant Journal*, *50*, 54–69.

Qin, F., Sakuma, Y., Li, J., Liu, Q., Li, Y. Q., & Shinozaki, K., et al. (2004). Cloning and functional analysis of a novel DREB1/CBF transcription factor involved in cold-responsive gene expression in *Zea mays* L. *Plant and Cell Physiology*, *45*, 1042–1052.

Reddy, A. S. (2001). Calcium: Silver bullet in signaling. *Plant Science*, *160*, 381–404.

Rosser, M. F., Washburn, E., Muchowski, P. J., Patterson, C., & Cyr, D. M. (2007). Chaperone functions of the E3 ubiquitin ligase CHIP. *Journal of Biological Chemistry*, *282*, 22267–22277.

Roth, S. Y., Denu, J. M., & Allis, C. D. (2001). Histone acetyltransferases. *Annual Review of Biochemistry*, *70*, 81–120.

Ruelland, E., Cantrel, C., Gawer, M., Kader, J. C., & Zachowski, A. (2002). Activation of phospholipases C and D is an early response to a cold exposure in *Arabidopsis* suspension cells. *Plant Physiology*, *130*, 999–1007.

Ruelland, E., Vaultier, M. N., Zachowski, A., & Hurry, V. (2009). Cold signalling and cold acclimation in plants. *Advances in Botanical Research*, *49*, 36–126.

Sachetto-Martins, G., Franco, L. O., & de Oliveira, D. E. (2000). Plant glycine-rich proteins: A family or just proteins with a common motif? *Biochimica et Biophysica ACTA*, *1492*, 1–14.

Saidi, Y., Finka, A., Muriset, M., Bromberg, Z., Weiss, Y. G., & Maathuis, F. J., et al. (2009). The heat shock response in moss plants is regulated by specific calcium-permeable channels in the plasma membrane. *Plant Cell*, *21*, 2829–2843.

Salinas, J. (2002). *Molecular mechanisms of signal transduction in cold acclimation.* Oxford University Press.

Saltveit, M. E., & Hepler, P. K. (2004). Effect of heat shock on the chilling sensitivity of trichomes and petioles of African violet (*Saintpaulia ionantha*). *Physiologia Plantarum*, *121*, 35–43.

Sanders, D., Pelloux, J., Brownlee, C., & Harper, J. F. (2002). Calcium at the crossroads of signaling. *Plant Cell*, *14*(Suppl), S401–417.

Sangwan, V., Orvar, B. L., Beyerly, J., Hirt, H., & Dhindsa, R. S. (2002). Opposite changes in membrane fluidity mimic cold and heat stress activation of distinct plant MAP kinase pathways. *Plant Journal*, *31*, 629–638.

Saracco, S. A., Miller, M. J., Kurepa, J., & Vierstra, R. D. (2007). Genetic analysis of SUMOylation in *Arabidopsis*: Conjugation of SUMO1 and SUMO2 to nuclear proteins is essential. *Plant Physiology*, *145*, 119–134.

Sasaki, H., Ichimura, K., & Oda, M. (1996). Changes in sugar content during cold acclimation and deacclimation of cabbage seedlings. *Annals of Botany*, *78*, 365–369.

Sasse, J. M. (2003). Physiological actions of brassinosteroids: An update. *Journal of Plant Growth Regulation*, *22*, 276–288.

Shan, D. P., Huang, J. G., Yang, Y. T., Guo, Y. H., Wu, C. A., & Yang, G. D., et al. (2007). Cotton GhDREB1 increases plant tolerance to low temperature and is negatively regulated by gibberellic acid. *New Phytologist*, *176*, 70–81.

Sheen, J. (1996). Ca^{2+}-dependent protein kinases and stress signal transduction in plants. *Science*, *274*, 1900–1902.

Shen, W., Nada, K., & Tachibana, S. (2000). Involvement of polyamines in the chilling tolerance of cucumber cultivars. *Plant Physiology*, *124*, 431–439.

Shinozuka, H., Hisano, H., Yoneyama, S., Shimamoto, Y., Jones, E. S., & Forster, J. W., et al. (2006). Gene expression and genetic mapping analyses of a perennial ryegrass glycine-rich RNA-binding protein gene suggest a role in cold adaptation. *Molecular Genetics and Genomics*, *275*, 399–408.

Sidebottom, C., Buckley, S., Pudney, P., Twigg, S., Jarman, C., & Holt, C., et al. (2000). Heat-stable antifreeze protein from grass. *Nature*, *406*, 256.

Skinner, J. S., von Zitzewitz, J., Szucs, P., Marquez-Cedillo, L., Filichkin, T., & Amundsen, K., et al. (2005). Structural, functional, and phylogenetic characterization of a large *CBF* gene family in barley. *Plant Molecular Biology*, *59*, 533–551.

Sokol, A., Kwiatkowska, A., Jerzmanowski, A., & Prymakowska-Bosak, M. (2007). Up-regulation of stress-inducible genes in tobacco and *Arabidopsis* cells in response to abiotic stresses and ABA treatment correlates with dynamic changes in histone H3 and H4 modifications. *Planta*, *227*, 245–254.

Song, L. L., Ding, W., Shen, J., Zhang, Z. G., Bi, Y. R., & Zhang, L. X. (2008). Nitric oxide mediates abscisic acid induced thermotolerance in the calluses from two ecotypes of reed under heat stress. *Plant Science*, *175*, 826–832.

Steponkus, P. L. (1971). Cold acclimation of *Hedera helix*: Evidence for a two phase process. *Plant Physiology*, *47*, 175–180.

Sterner, D. E., & Berger, S. L. (2000). Acetylation of histones and transcription-related factors. *Microbiology and Molecular Biology Reviews*, *64*, 435–459.

Steward, N., Ito, M., Yamaguchi, Y., Koizumi, N., & Sano, H. (2002). Periodic DNA methylation in maize nucleosomes and demethylation by environmental stress. *Journal of Biological Chemistry*, *277*, 37741–37746.

Stockinger, E. J., Gilmour, S. J., & Thomashow, M. F. (1997). *Arabidopsis thaliana CBF1* encodes an AP2 domain-containing transcriptional activator that binds to the C-repeat/DRE, a cis-acting DNA regulatory element that stimulates transcription in response to low temperature and water deficit. *Proceedings of the National Academy of Sciences of the United States of America*, *94*, 1035–1040.

Stockinger, E. J., Mao, Y., Regier, M. K., Triezenberg, S. J., & Thomashow, M. F. (2001). Transcriptional adaptor and histone acetyltransferase proteins in *Arabidopsis* and their interactions with CBF1, a transcriptional activator involved in cold-regulated gene expression. *Nucleic Acids Research*, *29*, 1524–1533.

Sundblad, L. G., Andersson, M., Geladi, P., Salomonson, A., & Sjostrom, M. (2001). Fast, nondestructive measurement of frost hardiness in conifer seedlings by VIS + NIR spectroscopy. *Tree Physiology*, *21*, 751–757.

Sunkar, R., & Zhu, J. K. (2004). Novel and stress-regulated microRNAs and other small RNAs from *Arabidopsis*. *Plant Cell*, *16*, 2001–2019.

Suzuki, I., Los, D. A., Kanesaki, Y., Mikami, K., & Murata, N. (2000). The pathway for perception and transduction of low-temperature signals in *Synechocystis*. *EMBO Journal*, *19*, 1327–1334.

Suzuki, R., Noguchi, R., Ota, T., Abe, M., Miyashita, K., & Kawada, T. (2001). Cytotoxic effect of conjugated trienoic fatty acids on mouse tumor and human monocytic leukemia cells. *Lipids*, *36*, 477–482.

Swindell, W. R., Huebner, M., & Weber, A. P. (2007). Transcriptional profiling of *Arabidopsis* heat shock proteins and transcription factors reveals extensive overlap between heat and non-heat stress response pathways. *BMC Genomics*, *8*, 125.

Szalai, G., Tari, I., Janda, T., Pestenácz, A., & Páldi, E. (2000). Effects of cold acclimation and salicylic acid on changes in ACC and MACC contents in maize during chilling. *Biologia Plantarum*, *43*, 637–640.

Tahtiharju, S., & Palva, T. (2001). Antisense inhibition of protein phosphatase 2C accelerates cold acclimation in *Arabidopsis thaliana*. *Plant Journal*, *26*, 461–470.

Tahtiharju, S., Sangwan, V., Monroy, A. F., Dhindsa, R. S., & Borg, M. (1997). The induction of kin genes in cold-acclimating *Arabidopsis thaliana*. Evidence of a role for calcium. *Planta*, *203*, 442–447.

Takumi, S., Shimamura, C., & Kobayashi, F. (2008). Increased freezing tolerance through up-regulation of downstream genes via the wheat *CBF* gene in transgenic tobacco. *Plant Physiology and Biochemistry*, *46*, 205–211.

Tamminen, I., Makela, P., Heino, P., & Palva, E. T. (2001). Ectopic expression of ABI3 gene enhances freezing tolerance in response to abscisic acid and low temperature in *Arabidopsis thaliana*. *Plant Journal*, *25*, 1–8.

Tang, W., Newton, R. J., Li, C., & Charles, T. M. (2007). Enhanced stress tolerance in transgenic pine expressing the pepper *CaPF1* gene is associated with the polyamine biosynthesis. *Plant Cell Reports, 26*, 115–124.

Tasgin, E., Attici, O., & Nalbantogly, B. (2003). Effect of salicylic acid and cold on freezing tolerance in winter wheat leaves. *Plant Growth Regulation, 41*, 231–236.

Teige, M., Scheikl, E., Eulgem, T., Doczi, R., Ichimura, K., & Shinozaki, K., et al. (2004). The MKK2 pathway mediates cold and salt stress signaling in *Arabidopsis*. *Molecular Cell., 15*, 141–152.

Toh, S., Imamura, A., Watanabe, A., Nakabayashi, K., Okamoto, M., & Jikumaru, Y., et al. (2008). High temperature-induced abscisic acid biosynthesis and its role in the inhibition of gibberellin action in *Arabidopsis* seeds. *Plant Physiology, 146*, 1368–1385.

Ueoka-Nakanishi, H., Nakanishi, Y., Tanaka, Y., & Maeshima, M. (1999). Properties and molecular cloning of Ca^{2+}/H^{+} antiporter in the vacuolar membrane of mung bean. *European Journal of Biochemistry, 262*, 417–425.

Upadhyaya, A., Davis, T. D., & Sankhla, N. (1991). Heat-shock tolerance and antioxidant activity in moth bean seedlings treated with tetcyclacis. *Plant Growth Regulation, 10*, 215–222.

Urao, T., Yamaguchi-Shinozaki, K., & Shinozaki, K. (2000). Two-component systems in plant signal transduction. *Trends in Plant Science, 5*, 67–74.

Vagujfalvi, A., Aprile, A., Miller, A., Dubcovsky, J., Delugu, G., & Galiba, G., et al. (2005). The expression of several *CBF* genes at the Fr-A2 locus is linked to frost resistance in wheat. *Molecular Genetics and Genomics, 274*, 506–514.

van der Luit, A. H., Olivari, C., Haley, A., Knight, M. R., & Trewavas, A. J. (1999). Distinct calcium signaling pathways regulate calmodulin gene expression in tobacco. *Plant Physiology, 121*, 705–714.

Vannini, C., Locatelli, F., Bracale, M., Magnani, E., Marsoni, M., & Osnato, M., et al. (2004). Overexpression of the rice *Osmyb4* gene increases chilling and freezing tolerance of *Arabidopsis thaliana* plants. *Plant Journal, 37*, 115–127.

Vashisht, A. A., Pradhan, A., Tuteja, R., & Tuteja, N. (2005). Cold- and salinity stress-induced bipolar pea DNA helicase 47 is involved in protein synthesis and stimulated by phosphorylation with protein kinase C. *Plant Journal, 44*, 76–87.

Vaultier, M. N., Cantrel, C., Vergnolle, C., Justin, A. M., Demandre, C., & Benhassaine-Kesri, G., et al. (2006). Desaturase mutants reveal that membrane rigidification acts as a cold perception mechanism upstream of the diacylglycerol kinase pathway in *Arabidopsis* cells. *FEBS letters, 580*, 4218–4223.

Venketesh, S., & Dayananda, C. (2008). Properties, potentials, and prospects of antifreeze proteins. *Critical Reviews in Biotechnology, 28*, 57–82.

Vidal, A. M., Ben-Cheikh, W., Talon, M., & Garcia-Martinez, J. L. (2003). Regulation of gibberellin 20-oxidase gene expression and gibberellin content in citrus by temperature and citrus exocortis viroid. *Planta, 217*, 442–448.

Vlachonasios, K. E., Thomashow, M. F., & Triezenberg, S. J. (2003). Disruption mutations of ADA2b and GCN5 transcriptional adaptor genes dramatically affect *Arabidopsis* growth, development, and gene expression. *Plant Cell, 15*, 626–638.

Vogel, J. T., Zarka, D. G., Van Buskirk, H. A., Fowler, S. G., & Thomashow, M. F. (2005). Roles of the CBF2 and ZAT12 transcription factors in configuring the low temperature transcriptome of *Arabidopsis*. *Plant Journal, 41*, 195–211.

Walley, J. W., Coughlan, S., Hudson, M. E., Covington, M. F., Kaspi, R., & Banu, G., et al. (2007). Mechanical stress induces biotic and abiotic stress responses via a novel *cis*-element. *PLoS genetics, 3*, 1800–1812.

Wang, H., & Cutler, A. J. (1995). Promoters from *kin1* and *cor6.6*, two *Arabidopsis thaliana* low-temperature- and ABA-inducible genes, direct strong beta-glucuronidase expression in guard cells, pollen and young developing seeds. *Plant Molecular Biology, 28*, 619–634.

Wang, L., Luo, Y., Zhang, L., Zhao, J., Hu, Z., & Fan, Y., et al. (2008). Isolation and characterization of a C-repeat binding transcription factor from maize. *Journal of Integrative Plant Biology, 50*, 965–974.

Wang, W., Vinocur, B., Shoseyov, O., & Altman, A. (2004). Role of plant heat-shock proteins and molecular chaperones in the abiotic stress response. *Trends in Plant Science, 9*, 244–252.

Wanner, L. A., & Junttila, O. (1999). Cold-induced freezing tolerance in *Arabidopsis*. *Plant Physiology, 120*, 391–400.

Wei, H., Dhanaraj, A. L., Arora, R., Rowland, L. J., Fu, Y., & Sun, L. (2006). Identification of cold acclimation-responsive Rhododendron genes for lipid metabolism, membrane transport and lignin biosynthesis: Importance of moderately abundant ESTs in genomic studies. *Plant Cell and Environment, 29*, 558–570.

Welling, A., & Palva, E. T. (2008). Involvement of CBF transcription factors in winter hardiness in birch. *Plant Physiology, 147*, 1199–1211.

Williams, B., Kabbage, M., Britt, R., & Dickman, M. B. (2010). *AtBAG7*, an *Arabidopsis* Bcl-2-associated athanogene, resides in the endoplasmic reticulum and is involved in the unfolded protein response. *Proceedings of the National Academy of Sciences of the United States of America, 107*, 6088–6093.

Wolfraim, L. A., Langis, R., Tyson, H., & Dhindsa, R. S. (1993). cDNA sequence, expression, and transcript stability of a cold acclimation-specific gene, *cas18*, of alfalfa (*Medicago falcata*) cells. *Plant Physiology, 101*, 1275–1282.

Worrall, D., Elias, L., Ashford, D., Smallwood, M., Sidebottom, C., & Lillford, P., et al. (1998). A carrot leucine-rich-repeat protein that inhibits ice recrystallization. *Science, 282*, 115–117.

Wu, L., Zhang, Z., Zhang, H., Wang, X. C., & Huang, R. (2008). Transcriptional modulation of ethylene response factor protein JERF3 in the oxidative stress response enhances tolerance of tobacco seedlings to salt, drought, and freezing. *Plant Physiology, 148*, 1953–1963.

Xia, X. J., Wang, Y. J., Zhou, Y. H., Tao, Y., Mao, W. H., & Shi, K., et al. (2009). Reactive oxygen species are involved in brassinosteroid-induced stress tolerance in cucumber. *Plant Physiology, 150*, 801–814.

Xiong, L., Lee, H., Huang, R., & Zhu, J. K. (2004). A single amino acid substitution in the *Arabidopsis* FIERY1/HOS2 protein confers cold signaling specificity and lithium tolerance. *Plant Journal, 40*, 536–545.

Xiong, L., Lee, H., Ishitani, M., Tanaka, Y., Stevenson, B., & Koiwa, H., et al. (2002a). Repression of stress-responsive genes by FIERY2, a novel transcriptional regulator in Arabidopsis. *Proceedings of the National Academy of Sciences of the United States of America, 99*, 10899–10904.

Xiong, L., Schumaker, K. S., & Zhu, J. K. (2002b). Cell signaling during cold, drought, and salt stress. *Plant Cell, 14*(Suppl), S165–183.

Xiong, Y., & Fei, S. Z. (2006). Functional and phylogenetic analysis of a DREB/CBF-like gene in perennial ryegrass (*Lolium perenne L.*). *Planta, 224*, 878–888.

Xue, G. P. (2003). The DNA-binding activity of an AP2 transcriptional activator HvCBF2 involved in regulation of low-temperature responsive genes in barley is modulated by temperature. *Plant Journal, 33*, 373–383.

Xue, Y., Peng, R., Xiong, A., Li, X., Zha, D., & Yao, Q. (2009). Yeast heat-shock protein gene HSP26 enhances freezing tolerance in *Arabidopsis*. *Journal of Plant Physiology, 166*, 844–850.

Yamaguchi-Shinozaki, K., & Shinozaki, K. (1994). A novel *cis*-acting element in an *Arabidopsis* gene is involved in responsiveness to drought, low-temperature, or high-salt stress. *Plant Cell, 6*, 251–264.

Yan, C. S., Shen, H., Li, Q., & He, Z. H. (2006). A novel ABA-hypersensitive mutant in *Arabidopsis* defines a genetic locus that confers tolerance to xerothermic stress. *Planta, 224*, 889–899.

Yan, J., Wang, J., Li, Q., Hwang, J. R., Patterson, C., & Zhang, H. (2003). AtCHIP, a U-box-containing E3 ubiquitin ligase, plays a critical role in temperature stress tolerance in *Arabidopsis*. *Plant Physiology, 132*, 861–869.

Yang, T., & Poovaiah, B. W. (2002). A calmodulin-binding/CGCG box DNA-binding protein family involved in multiple signaling pathways in plants. *Journal of Biological Chemistry, 277*, 45049–45058.

Yi, S. Y., Sun, A. Q., Sun, Y., Yang, J. Y., Zhao, C. M., & Liu, J. (2006). Differential regulation of *Lehsp23.8* in tomato plants: Analysis of a multiple stress-inducible promoter. *Plant Science, 171*, 398–407.

Yoo, C. Y., Miura, K., Jin, J. B., Lee, J., Park, H. C., & Salt, D. E., et al. (2006). SIZ1 small ubiquitin-like modifier E3 ligase facilitates basal thermotolerance in *Arabidopsis* independent of salicylic acid. *Plant Physiology, 142*, 1548–1558.

Yoo, S. Y., Kim, Y., Kim, S. Y., Lee, J. S., & Ahn, J. H. (2007). Control of flowering time and cold response by a NAC-domain protein in *Arabidopsis*. *PloS One*, *2*, e642.

Yu, X. M., Griffith, M., & Wiseman, S. B. (2001). Ethylene induces antifreeze activity in winter rye leaves. *Plant Physiology*, *126*, 1232–1240.

Zarka, D. G., Vogel, J. T., Cook, D., & Thomashow, M. F. (2003). Cold induction of *Arabidopsis CBF* genes involves multiple ICE (inducer of CBF expression) promoter elements and a cold-regulatory circuit that is desensitized by low temperature. *Plant Physiology*, *133*, 910–918.

Zeller, G., Henz, S. R., Widmer, C. K., Sachsenberg, T., Ratsch, G., & Weigel, D., et al. (2009). Stress-induced changes in the *Arabidopsis thaliana* transcriptome analyzed using whole-genome tiling arrays. *Plant Journal*, *58*, 1068–1082.

Zhai, H., Bai, X., Zhu, Y., Li, Y., Cai, H., & Ji, W., et al. (2010). A single-repeat R3-MYB transcription factor MYBC1 negatively regulates freezing tolerance in *Arabidopsis*. *Biochemical and Biophysical Research Communications*, *394*, 1018–1023.

Zhang, X., Fowler, S. G., Cheng, H., Lou, Y., Rhee, S. Y., & Stockinger, E. J., et al. (2004). Freezing-sensitive tomato has a functional CBF cold response pathway, but a CBF regulon that differs from that of freezing-tolerant *Arabidopsis*. *Plant Journal*, *39*, 905–919.

Zhang, X., Li, M., Wei, D., & Xing, L. (2008b). Identification and characterization of a novel yeast omega3-fatty acid desaturase acting on long-chain n-6 fatty acid substrates from *Pichia pastoris*. *Yeast*, *25*, 21–27.

Zhang, X., Liu, S., & Takano, T. (2008a). Two cysteine proteinase inhibitors from *Arabidopsis thaliana*, AtCYSa and AtCYSb, increasing the salt, drought, oxidation and cold tolerance. *Plant Molecular Biology*, *68*, 131–143.

Zhang, Z., & Huang, R. (2010). Enhanced tolerance to freezing in tobacco and tomato overexpressing transcription factor TERF2/LeERF2 is modulated by ethylene biosynthesis. *Plant Molecular Biology*, *73*, 241–249.

Zhao, H., & Bughrara, S. S. (2008). Isolation and characterization of cold-regulated transcriptional activator *LpCBF3* gene from perennial ryegrass (*Lolium perenne* L.). *Molecular Genetics and Genomics*, *279*, 585–594.

Zhao, M. G., Chen, L., Zhang, L. L., & Zhang, W. H. (2009). Nitric reductase-dependent nitric oxide production is involved in cold acclimation and freezing tolerance in *Arabidopsis*. *Plant Physiology*, *151*, 755–767.

Zhou, X., Wang, G., Sutoh, K., Zhu, J. K., & Zhang, W. (2008). Identification of cold-inducible microRNAs in plants by transcriptome analysis. *Biochimica et Biophysica ACTA*, *1779*, 780–788.

Zhu, J., Jeong, J. C., Zhu, Y., Sokolchik, I., Miyazaki, S., & Zhu, J. K., et al. (2008). Involvement of *Arabidopsis* HOS15 in histone deacetylation and cold tolerance. *Proceedings of the National Academy of Sciences of the United States of America*, *105*, 4945–4950.

Zhu, J., Shi, H., Lee, B. H., Damsz, B., Cheng, S., & Stirm, V., et al. (2004). An *Arabidopsis* homeodomain transcription factor gene, *HOS9*, mediates cold tolerance through a CBF-independent pathway. *Proceedings of the National Academy of Sciences of the United States of America*, *101*, 9873–9878.

Zou, J., Liu, A. L., Chen, X. B., Zhou, X. Y., Gao, G. F., & Wang, W. F., et al. (2009). Expression analysis of nine rice heat shock protein genes under abiotic stresses and ABA treatment. *Journal of Plant Physiology*, *166*, 851–861.

Biotechnological approaches for phytoremediation

20

Om Parkash Dhankher[1] Elizabeth A.H. Pilon-Smits[2]
Richard B. Meagher[3] Sharon Doty[4]

[1]*University of Massachusetts Amherst, Massachusetts*
[2]*Colorado State University, Fort Collins, Colorado*
[3]*University of Georgia, Athens, Georgia*
[4]*Sharon Doty, University of Washington, Seattle, Washington*

TABLE OF CONTENTS

Introduction

Plants can be used in various ways to prevent or remediate environmental pollution (Figure 20.1; for reviews see also Pilon-Smits, 2005; Doty 2008). In some cases plants can degrade pollutants inside their tissues; this method is called phytodegradation and is mostly suitable for organic pollutants, since inorganics can only be moved and not degraded. When degradation happens in the rhizosphere this method is called rhizodegradation, and when it involves microbes it is also called phytostimulation. In some cases the pollutant is immobilized in the root zone, which is called phytostabilization. The combined rhizosphere (root zone) processes contributing to phytoremediation are also termed rhizoremediation. Phytoextraction is the term used for the accumulation of pollutants in harvestable plant tissues, particularly shoot tissues, and this method is mostly used for inorganics. If plant accumulation/adsorption is mainly by plant roots in hydroponic systems the technique is called rhizofiltration. The same principle can be used on a large scale in constructed wetlands. Some pollutants can also be volatilized by plants, such as volatile organic compounds (VOCs) and certain metal(loid)s such as mercury (Hg) and selenium (Se); this is called phytovolatilization.

Different plant processes are important for different remediation strategies; for instance, rhizoremediation may be facilitated by root-released compounds, as well as by

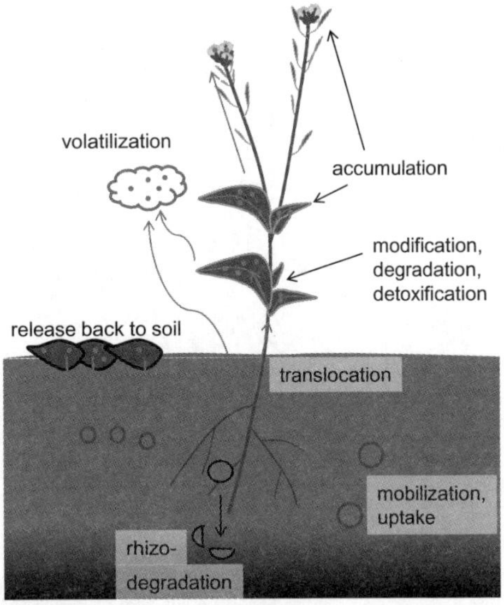

Figure 20.1 • Overview of plant–pollutant interactions and the possible fates of the pollutant (represented as circles) • Degradation of certain organic pollutants in the rhizosphere can be facilitated by root-released compounds. Some root-released compounds can also mobilize pollutants from soil particles and affect uptake. Pollutants can be taken up into the root symplast and translocated via the xylem (apoplast) to the shoot, in the transpiration stream. In the shoot the pollutant can be taken up into the shoot symplast. There it may be further modified (assimilated, degraded, side groups attached, conjugate/chelator attached) and either sequestered in vacuole or cell wall, or volatilized. From leaves pollutants may be remobilized via the phloem to reproductive tissues. Pollutants (or their downstream products) may also be returned to the soil after leaf drop. Please see color plate section at the end of the book.

DOI: 10.1016/B978-0-12-381466-1.00020-1

species-specific, root-associated microbial, flora. When pollutants are taken up into the root symplast, and either accumulated in root cells or exported and translocated to the shoot via the dead xylem in the transpiration stream, this process may involve plant membrane transporters (more so for inorganics than organics). Moreover, root-shoot translocation is driven by transpiration from the plant shoot, as regulated by stomatal opening. In the shoot, the pollutant can be taken up from the xylem into leaf cells, which again may involve membrane transporter proteins. Inside cells the pollutant may be further modified enzymatically (assimilated, degraded, side groups attached, conjugate/chelator attached) and either sequestered (often in vacuoles or cell wall, by means of transporter proteins) or enzymatically mineralized or volatilized. From leaves, pollutants may also be remobilized via the living phloem to young leaves, roots, or reproductive organs, a process that may again involve membrane transporters. Pollutants or their downstream products may also be returned to the soil after leaf drop. As may be clear from the active plant processes involved, plant species differ in their ability to remediate different pollutants, depending on their abundance of transporters and enzymes, their microbial partners, and their transpiration rate. In addition, some general properties of a good phytoremediator species are fast growth and high biomass, hardiness, and tolerance to pollutants. It is an added bonus if a plant species has economic value. All of these biological properties important for phytoremediation may potentially be ameliorated using genetic engineering. In this chapter, we focus on biotechnological approaches to improve plants' ability to tolerate pollutants and phytoremediation efficiency.

As mentioned earlier, different pollutants have different fates in plant-substrate systems, so they have different rate-limiting factors for phytoremediation that may be targeted using genetic engineering (Figure 20.2). For instance, remediation of hydrophobic organics may be limited by their release from soil particles, which may be improved by enhanced production of biosurfactants by roots or root-associated microbes. Similarly, certain metals may be made more bioavailable by root excretion of metal chelators and protons. In the case of rhizodegradation, the secretion of degrading enzymes from roots may be upregulated, as can the secretion of compounds that stimulate microbial density or activity. Uptake and transport into/inside plants may be limited by the abundance of membrane transporters, particularly for inorganics, which depend on uptake on transporter proteins. Organics, when moderately hydrophobic, can often pass membranes passively and do not need transporters. If it is known which transporters mediate pollutant uptake and translocation, these may be overproduced in plants. Plant tolerance, in turn, may be limited by the abundance of enzymes that modify, degrade, or chelate pollutants, or general antioxidant enzymes. Depending

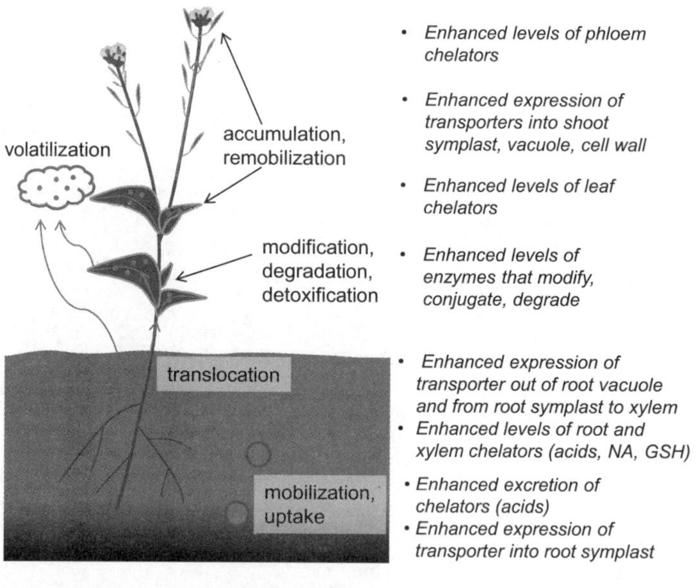

Figure 20.2 • Overview of biotechnological approaches that may enhance various rate-limiting steps in phytoremediation.
• Excreted compounds may facilitate mobilization, and enhanced expression of transporters in the root cell membrane may facilitate import into the root symplast. Enhanced expression of exporters out of the root vacuole and out of the root symplast into the xylem may facilitate translocation to the shoot. Increased levels of root and xylem chelators (acids, GSH) may enhance plant tolerance to the pollutant and pollutant mobilization in the xylem. Uptake into the leaf symplast may be enhanced by increased expression of transporters in the mesophyll cell membrane. Inside leaf cells, enhanced levels of enzymes that modify, conjugate, or degrade pollutants can facilitate tolerance, degradation, sequestration, or volatilization. Tolerance and sequestration are also enhanced by higher levels of leaf chelators or transporter proteins that export pollutants out of the cytosol and into the vacuole or cell wall. Enhanced levels of phloem chelators may facilitate remobilization to reproductive tissues. Please see color plate section at the end of the book.

on the suspected limiting factors, any such enzymes may be over-expressed to enhance phytoremediation capacity. In addition to boosting the expression of existing genes, novel genes may be introduced from other plant species or any organism. In this way, a totally new phytoremediation capacity may be introduced into a suitable plant species for phytoremediation. All of these approaches have been used successfully. In the next section we describe representative cases in more detail.

Overview of results from biotechnological approaches for different pollutants

Inorganic pollutants

Inorganic pollutants include metals/metalloids (e.g., As, Cd, Cu, Hg, Mn, Se, Zn), radionuclides (e.g., Cs, P, U), and plant fertilizers (e.g., nitrate, phosphate). All occur in nature mainly as positively or negatively charged ions and depend on plant transporters for uptake and translocation. Inorganics can be altered (reduced/oxidized), moved into/inside plants, or in some cases volatilized (Hg, Se), but cannot be degraded. Thus, phytoremediation methods available for inorganics include immobilization (phytostabilization), sequestration in harvestable plant tissues (phytoextraction or rhizofiltration) and, in exceptional cases, phytovolatilization. As reviewed by Pilon-Smits (2005) and Doty (2008), biotechnological approaches that have successfully altered the capacity of plants for phytoremediation of inorganics have focused on both tolerance and accumulation. Genes targeted include metal transporter genes, as well as genes that facilitate chelator production. Also, in the case of elements that can be volatilized, genes that facilitate conversion to volatile forms were over-expressed. In the next section we highlight three inorganics As, Hg, and Se.

Arsenic

Arsenic pollution and toxicity

Arsenic (As) contamination in soil and water is a growing problem worldwide, and millions of people face the risk of cancer and poisoning due to As in their drinking water and food supplies. Arsenic is both a carcinogen and a toxin, and is damaging to most human organs (Kaiser, 2001). Arsenic contamination of groundwater used for domestic water supplies has been reported from over 70 countries, affecting the health of an estimated 150 million people (Ravenscroft et al., 2009). This situation is worse in Bangladesh and the West Bengal state of India, and the World Health Organization dubbed this as the "worst mass poisoning" event in human history. Arsenic-contaminated ground waters, apart from use for drinking, are widely used for irrigation of many crops, particularly rice (*Oryza sativa*), adding more than 1000 tons of As per year to the agricultural soils in Bangladesh alone (Ali et al., 2003). Arsenic species are non-biodegradable and they remain in the surface and subsurface of agricultural soils (Juhasz et al., 2003). Significantly high levels of arsenic in the edible crops grown in contaminated soils have been reported in many countries (Larsen et al., 1992; Das et al., 2004; Williams et al., 2005). Tens of thousands of Superfund sites in the United States and other countries are listed as having unacceptably high levels of As and other toxic metals (http://www.epa.gov/superfund/sites/npl/index.htm) and are recommended for clean up.

Inorganic species of As, particularly the oxyanions arsenate (AsO_4^{-3}, referred to as AsV) and arsenite (AsO_3^{-3}, referred to as AsIII), are prevalent in the environment and are often more toxic than its organic forms (Bentley and Chasteen, 2002). AsV is the predominant species in aerobic soils, whereas AsIII predominates in anaerobic environments such as submerged soils. In addition, organic forms of arsenicals such as monomethylarsenate (MMA) and dimethylarsenate (DMA) are also present in the environment. Organic forms of As are generally less toxic than the inorganic As species (Chen et al., 2005). Arsenic toxicity in a cell depends to a large extent on the type of As species. AsIII is highly thiol-reactive and thus has a high affinity to the sulfhydryl group of the amino acid cysteine. Binding of AsIII to Cys residues disrupts protein structure and function, thus affecting many key metabolic processes in the cell, such as oxidative phosphorylation, glutathione production, ATP synthesis, fatty acid metabolism, and gluconeogenesis (Hughes, 2002; Carey et al., 2010). Additionally, the binding ability of AsIII to the non-protein thiol glutathione can deplete this important antioxidant, which leads to increased levels of reactive oxygen species (ROS), and thus oxidative stress (Castlehouse et al., 2003). AsV, being a phosphate analog, can substitute for inorganic phosphate in many biochemical processes, which is evident from the formation of glucose-6-arsenate (Lagunas, 1980) and ADP-arsenate (Gresser, 1981), inhibiting the formation of ATP.

Arsenic in foods and implications for human health

Rice is the most important staple food for over half of the world's population (Fageria, 2007), and it is grown widely in areas where As contamination is widespread. Almost 30–50% of the areas of Bangladesh and West Bengal (India) are irrigated with As-contaminated groundwater to grow paddy rice (Meharg and Rahman, 2003). In the flooded paddy fields that create a reducing environment paddy rice accumulates high levels of AsIII (Meharg, 2004; Williams et al., 2007; Meharg et al., 2009). In U.S.-grown rice, considerably higher levels of total As were found where organic As species constituted the major fraction of arsenic in the grains (Williams et al., 2005; Zavala and Duxbur 2008). Additionally, rice straw is used as forage in many countries, including the United States, India, China, and Bangladesh. High As concentrations in straw may have adverse health effects on cattle and may result in an increased As exposure in humans via the plant–animal–human pathway (Abedin et al., 2002). Therefore, there is a significant concern regarding accumulation of As in meat and dairy products, and agricultural crops and vegetables grown in arsenic-affected areas.

Inorganic As species are phytotoxic and the elevated concentration of As in the soil causes a significant reduction in crop yield (Marin et al., 1993; Meharg, 2004; Zhu et al., 2008). Rice yield decreases by 10% at 25 mg/kg soil As concentrations (Xiong et al., 1987; Marin et al., 1993). Due to severe surface water shortage, more and more farmers are using recycled water and sewage sludge that further contributes to the As buildup in agricultural lands. Additionally, AsV, being a phosphate analog, competes with phosphate uptake and thus causes the inhibition of phosphate and other nutrient uptake (Meharg and Macnair, 1992; Abedin et al., 2002; Dhankher et al., 2006). The phytotoxic effects suffered by crops grown in soil with As residues could be overcome by developing crops resistant to As uptake. Biotechnological approaches may help achieve this goal.

Mechanism of As uptake and detoxification in microbes and plants

Mechanisms of As detoxification have been well characterized in bacteria and yeast. Because of its similarity to phosphate, AsV enters yeast cells via phosphate transporters. A common mechanism by which these microorganisms achieve tolerance to As is by the reduction of AsV to AsIII, and then the exclusion of toxic AsIII oxyanions from the cell by inducible and selective transporters (Rosen, 2002; Mukhopadhyay and Rosen, 2002). In *Escherichia coli*, arsenate reductase, ArsC, reduces AsV to AsIII (Chen et al., 1986) and the latter is subsequently transported out of the cell by an AsIII export pump, ArsAB (Cervantes et al., 1994; Rosen, 2002), thus conferring As resistance. Another important detoxification pathway in prokaryotes involves removal of As by converting inorganic As to volatile organic compounds such as trimethyl arsine through a series of methylation reactions catalyzed by S-adenosylmethionine methyltransferases (Qin et al., 2006).

AsIII has been shown to enter bacterial, yeast, and mammalian cells via aquaglyceroporins (Sanders et al., 1997; Wysocki et al., 2001; Liu et al., 2002), which belong to the major intrinsic protein (MIP) superfamily. In *E. coli*, a glycerol uptake facilitator, GlpF, has been identified as an AsIII transporter (Sanders et al., 1997). Yeast Fps1p, a GlpF homolog, has been shown to facilitate the uptake of metalloids AsIII and antimonite (SbIII) in yeast (Wysocki et al., 2001).

The mechanisms of As uptake and detoxification in plants have recently been reviewed in depth (Tripathi et al., 2007; Zhao et al., 2009, 2010). Several studies support the contention that AsV, being a phosphate analog, is taken up in plants via phosphate uptake systems (Meharg and Macnair, 1992). Phosphate transporter PHT1;1 has been shown to be implicated in AsV uptake in *Arabidopsis thaliana* (Shin et al., 2004; Catarecha et al., 2007). Furthermore, AsV represses genes involved in the phosphate starvation response, suggesting that AsV interferes with phosphate sensing and alters the phosphate signaling mechanism (Catarecha et al., 2007). In *A. thaliana* there are nine high-affinity phosphate transporters (PHT), and different PHTs may vary in their affinity for arsenate. Further studies are needed to identify the relative affinities of various PHTs for AsV and phosphate. While the molecular mechanisms of As detoxification and tolerance in plants remain to be fully determined, it has been shown that plants detoxify As by reducing AsV to AsIII, which is subsequently detoxified via forming complexes with thiol-reactive peptides such as γ-glutamylcysteine (γ-EC), glutathione (GSH), and phytochelatins (PCs; Pickering et al., 2000; Dhankher et al., 2002; Vatamaniuk et al., 2002). The AsIII-thiol complexes are then suggested to be sequestered into vacuoles of both root and shoot cells by glutathione-conjugating pumps and GCPs (Dhankher et al., 2002; Wang et al., 2002), although direct evidence of this remains elusive. Several studies suggested the reduction of AsV to AsIII by endogenous arsenate reductases inside plant cells: genes encoding plant arsenate reductases have recently been isolated and characterized from *Arabidopsis*, rice, *Holcus lanatus*, and *Pteris vittata* (Dhankher et al., 2006; Bleeker et al., 2006; Ellis et al., 2006; Duan et al., 2007). The *Arabidopsis ACR2* gene complemented the function of arsenate reductase in *E. coli* strains deficient in arsenate reductase, ArsC. In addition, *Arabidopsis* lines silenced for ACR2 expression by RNAi showed a clear arsenate-dependent phenotype, which translocated 10- to 15-fold higher levels of As from roots to aboveground tissues (Dhankher et al., 2006).

Recent studies have shown that members of plant aquaporins belonging to the MIP superfamily transport AsIII in rice. Plant MIPs are grouped into four major subfamilies: the plasma membrane intrinsic proteins (PIPs); tonoplast intrinsic proteins (TIPs); nodulin 26-like intrinsic proteins (NIPs); and small and basic intrinsic proteins (SIPs; Weig et al., 1997). Ma et al. (2008) have shown that a silicon (Si) transporter Lsi1 (OsNIP2;1), which is a member of the NIP subfamily, played a major role for the entry of AsIII into rice roots. A mutation in Lsi1 resulted in a nearly 60% reduction in AsIII influx in rice roots. A second Si efflux carrier, Lsi2 (another rice NIP subfamily member) has been shown to be responsible for loading AsIII in the xylem, and a T-DNA insertion in Lsi2 locus resulted in almost 50% reduction in As accumulation in shoot (Ma et al., 2008). Heterologous expression of *Arabidopsis* NIP5;1, NIP6;1, and NIP7;1 in yeast showed AsIII permeability (Bienert et al., 2008; Isayenkov and Maathuis, 2008). Suppression of an *Arabidopsis* NIP7;1 expression in T-DNA insertion lines also resulted in decreased uptake of AsIII (Isayenkov and Maathuis, 2008). Although significant progress has been made in understanding the mechanism of AsIII uptake and transport in plants, there is still more to unravel here. Apart from the role of Si efflux protein Lsi2 in translocation of AsIII from root to shoot tissues, the exact mechanism of xylem or phloem loading of AsIII in root tissues and unloading in the shoot tissues is not known. Further studies are needed to identify transporters that either exclusively transport AsIII or AsIII co-transporters to reduce As uptake without significantly affecting other essential nutrient uptake in plants.

Biotechnological approaches for As remediation and reducing As in food crops

Arsenic phytoremediation

The Chinese brake fern (*P. vittata*) has an exceptional ability to hyperaccumulate very high levels of As (Ma et al., 2001), and thrives in tropical and subtropical places. Thus, *P. vittata* could be highly useful for phytoremediation of As in those regions. In contrast to other land plants, AsIII is the main form of accumulated As in *P. vittata*, where As is transported from rhizome to the frond region and stored as free AsIII (Zhao et al., 2003). A gene, PvACR3, encoding a protein weakly homologous to the yeast ACR3 arsenite effluxer, has been shown to be localized to the vacuolar membrane in the fern gametophyte, indicating that it likely effluxes AsIII into the vacuole for sequestration (Indriolo et al., 2010). Similar genes from higher plants are yet to be identified. Since this fern has such a unique As hyperaccumulation mechanism, it could also serve as an ideal source of genes to biotechnologically enhance As hyperaccumulation in other plants. Although this plant demonstrates considerable potential for As extraction from soil in greenhouse studies, there are several limitations for using *P. vittata* for As remediation in the field. The molecular mechanism of As detoxification in this hyperaccumulating fern is largely unknown, and its perennial growth is restricted to the tropical or subtropical areas. Furthermore, the fern could be invasive and its introduction to non-indigenous areas should be evaluated carefully with regard to potential ecological consequences.

An often explored alternative to *P. vittata* has been the development of genetics-based, cost-effective, portable strategies for cleaning up As polluted soil and water that can be transferred to any plant species adapted to various geo-climatic regions. Different strategies of genetic manipulation may be pursued depending on the desired outcome; for example, increased tolerance to better withstand an As-contaminated environment, increased uptake and tolerance for phytoextraction, decreased uptake, and/or increased methylation for improved food safety (Tripathi et al., 2007, Zhao et al., 2009). A number of transgenic plants have been engineered for increased As tolerance and accumulation. Over-expression of genes involved in the synthesis of PCs or their precursor GSH significantly enhanced As tolerance but failed to significantly enhance As accumulation (Li et al., 2004, 2005; Gasic and Korban, 2007). These studies indicated that increasing GSH and PC synthesis alone is insufficient to achieve enhanced As accumulation in the shoots. Co-expression of both *γ-ECS* and *PCS* in *Arabidopsis* produced a greater effect on As tolerance and accumulation than over-expression of either gene alone (Guo et al., 2008). Therefore, modifying the levels of GSH and PCs in plants is an effective approach for increasing the As tolerance of plants, and could be used for producing novel plants with strong phytoremediation potential.

Transgenic plants with strong tolerance to As and enhanced As accumulation in the shoots were developed by co-expressing two bacterial genes (Dhankher et al., 2002). The *E. coli* arsenate reductase, *ars*C, gene was expressed in leaves as driven by a light-induced soybean RuBisCo small subunit 1 (*SRS1*) promoter. In addition, the *E. coli* γ-glutamylcysteine synthatase, *γ-ECS*, was expressed in both roots and shoots, driven by a strong constitutive *Actin2* promoter (Dhankher et al., 2002). The double transgenic plants were highly tolerant as compared to the plants expressing *γ-ECS* alone. Further, these double transgenic plants attained almost 17-fold higher biomass and hyperaccumulated three-fold more As in the aboveground biomass than wild-type plants when grown on 125 μM sodium arsenate. This work was a significant proof-of-concept for phytoremediation of As-contaminated soil and water by transgenic plants. The leaf-specific expression of *ars*C presumably enhances arsenate reduction, whereas *γ-ECS* over-expression enhanced the biosynthesis of thiol-rich peptides for AsIII complexation. These results imply that plants over-expressing the *γ-ECS* sink peptides can tolerate the increased AsIII generated by ArsC activity in leaves and thus provide increased AsV tolerance, which may have the effect of driving more As accumulation in shoots. This novel strategy demonstrated that stacking traits into transgenic plants with multiple transgenes can be used to get a synergistic effect that transcends what either gene could accomplish on its own. However, in contrast to *P. vittata* where most of the As is hyperaccumulated in the aboveground frond, most angiosperms retained a major fraction (~95%) of As in the roots and only a small amount of total As extracted was translocated to the shoot (Pickering et al., 2000; Dhankher et al., 2002). The speciation of As by X-ray absorption fine structure (EXAFS) of shoots and roots of wild-type plants revealed that most of the As was in the form of arsenite-glutathione complexes, $As(GS)_3$ (Dhankher et al., 2002). Therefore, these plants appeared to have high levels of endogenous arsenate reductase activity that convert AsV into AsIII in roots, immobilizing this toxic AsIII below ground (Dhankher et al., 2002; Dhankher, 2005).

Addressing this proposed problem of endogenous arsenate reductase reducing the mobility of arsenate from roots to shoots, Dhankher et al. (2006) focused their efforts on identifying and blocking arsenate reductase activity. The goal was to further enhance translocation and hyperaccumulation of As in the aboveground tissues. The endogenous arsenate reductase designated as AtACR2 from *Arabidopsis* was cloned and characterized (Dhankher et al., 2006). The knockdown of AtACR2 by RNAi in *Arabidopsis* caused translocation of significantly higher As levels from roots to shoots. Various AtACR2 RNAi knockdown lines translocated 10- to 16-fold more As to shoots and retained slightly less As in their roots than wild-type. These results suggested that blocking AtACR2 function in roots enhances arsenate transport from roots to shoots. Bleeker et al. (2006) also characterized the AtACR2 (*Arath* CDC25) using T-DNA insertion lines and showed that mutant lines were sensitive to AsV, further confirming the functional role of this enzyme. Additionally, an arsenate-activated glutaredoxin from *P. vittata* (PvGRX5) was implicated in As metabolism, and its heterologous over-expression in *Arabidopsis* increased As tolerance and decreased As accumulation in shoots (Sundaram et al., 2008).

So far there have been few studies that indicate the feasibility of manipulating one or two genes for the phytoremediation of As-contaminated environments. In the future, for successful field phytoremediation, efforts should be focused on combining the various genetic elements controlling traits such as As uptake in roots, tolerance, translocation from roots to shoot, thiol complexation of AsIII, sequestration into vacuoles, methylation of As species, and eventually volatilization into high-biomass, non-food fast growing plant species with low agronomic inputs. So far no field trials for As phytoremediation using the genetically modified plants have been conducted. However, with careful selection of non-food and non-invasive plant species and integration of multiple pathways of As uptake and metabolism, As phytoremediation may be effective and acceptable to the public as an environmentally friendly green remediation approach.

Preventing arsenic uptake in food crops

As discussed in previous sections, As accumulation in food crops such as rice and vegetables is a significant health concern. Through a biotechnological approach, it is possible to engineer crops with enhanced tolerance to, and reduced uptake of, As for human consumption. There are several options that may be highly effective in this endeavor. One option, reduced accumulation of As in the aboveground tissues and seeds, could be achieved by enhancing the reduction of AsV to AsIII in roots and increased AsIII-thiol complexation by over-expression of genes encoding arsenate reductases and PC biosynthetic pathway proteins in roots only by means of root-specific promoters. Enhanced PC production in roots could restrict As translocation to shoots through the formation of AsIII-PC complexes and vacuolar sequestration in roots. Secondly, uptake of AsV in roots could be blocked by manipulating the PHTs that have higher affinity for AsV than phosphate. Thirdly, accumulation of AsIII in crops such as rice could be reduced by knocking down the expression of Lsi1 that mediates AsIII uptake in roots and Lsi2, which transports AsIII from roots to shoots. Fourthly, conversion of inorganic As to methylated organic As forms will reduce As toxicity and further volatilization of methylated As species such as MMA and DMA to the gaseous form trimethylarsine (TMA) by over-expression of bacterial or algal As (III-S-adenosyl methyltransferase, ArsM; Qin et al., 2006, 2009), or plant genes coding for As methyltransferases once they have been identified. However, the toxicity of TMA in the paddy field remains to be studied. Fifthly, As accumulation in food crops could be decreased by amendments of soil with phosphate and Se, which are known to inhibit As uptake in plants. This may be achieved by over-expression of PHT genes specialized for only phosphate transport, or transporter genes that mediate Se uptake as described in the selenium section.

Developing novel strategies for phytoremediation as well as preventing As contamination in the food chain will require a comprehensive understanding of As uptake, detoxification, and sequestration in plants and microbes, as well as better knowledge about genes and gene networks involved in As metabolism. Recently, a genome-wide transcriptome analysis identified numerous genes differentially expressed in response to As treatment in *Arabidopsis*, rice, and *Crambe abyssinica* (Abercrombie et al., 2008; Ahsan et al., 2008; Paulose et al., 2010). These genes and gene networks in plants, once characterized, may be transferred to high-biomass, fast-growing plant species to enhance As uptake phytoremediation efficiency or to prevent As accumulation in food crops.

Mercury

Mercury pollution and toxicity

Mercury is a highly toxic pollutant and its widespread contamination in the soil and water is threatening human and environmental health (Dean et al., 1972; Kraemer and Chardonnens, 2001). Mercury is usually released into the environment in inorganic forms, either elemental metallic (Hg(0)) or ionic (Hg(II)) forms. Hg(II) tends to bind strongly to soil components, which reduces its availability and absorption. Organic forms of Hg, particularly methylmercury, dimethylmercury, and phenylmercury, are highly toxic and get accumulated in membrane-bound organelles where these compounds inhibit essential oxidative and photosynthetic pathways. Methylmercury (CH_3Hg) is particularly toxic and the greatest danger to humans and the environment, because of its efficient biomagnifications in the food web (Meagher and Rugh, 1998; Patra and Sharma, 2000). The world first became aware of the extreme dangers of methylmercury in the 1950s after a large, tragic incident of human Hg poisoning at Minamata Bay, Japan (Harada, 1995). Because the various forms of Hg are immutable to biological processes and many bind tightly to organic materials, the vast majority of large Hg-polluted sites remain contaminated indefinitely and remain an environmental threat.

Natural Hg emissions have led to the distribution of Hg throughout the globe. Several natural processes such as volcanoes, fires, and biological processes such as electrochemical reduction to Hg(0) can all serve as the primary vehicles for Hg distribution in the environment (WHO, 2003). In addition to the natural Hg emissions, anthropogenic activities such as burning of fossil fuels; mining of gold, coal, and silver (Nriagu, 1993); and various industrial activities have increased the Hg emission several fold more than the natural emissions (WHO, 2003). Once in the atmosphere, elemental Hg oxidizes into ionic Hg, which is more efficiently deposited in the environment, causing elevated Hg levels. Anaerobic sulfate-reducing bacteria that are particularly active in wetland environments convert inorganic Hg(II) to methylmercury. The most serious Hg pollution involves methylmercury produced by these native bacteria at nearly all Hg-contaminated wetland sites (Kannan et al., 1998). In the aquatic environment, methylmercury-contaminated bacteria are consumed by protozoans, protozoans by small invertebrates, invertebrates by small fish, small fish by big fish, and finally fish by aquatic birds and humans at the top of the food web. Because of methylmercury's relatively high solubility in gut and neural membranes it is more concentrated in each step of the food chain than ionic Hg, which may pass through the digestive system of many of these organisms. Thus, methylmercury is biomagnified in food webs and poses a significant threat to the health of humans and other animals (Boischio and Henshel, 1996; Keating et al., 1997). Consumption of Hg-contaminated fish and other seafood is known to be the major source of Hg in the human diet. Because of the resulting current Hg levels, the U.S. EPA cautions pregnant women and young children against frequent consumption of fish.

Mercury detoxification in bacteria and plants

Bacteria mediate resistance to organomercurial and inorganic mercuric salts by metabolic conversion to non-toxic elemental mercury Hg(0). The bacterial genes responsible for Hg resistance are organized in the *mer* operon. The mer operons among different bacteria vary in structure. In narrow-spectrum Hg resistance the *mer* operon is constituted by genes that encode the functional proteins for regulation (merR), transport (merT, merP and/or merC, merF), and electrochemical reduction (merA; Summers, 1992). Bacteria with broad-spectrum Hg resistance carry the additional gene encoding merB that confers resistance to many organomercurials, such as methylmercury and phenylmercury. MerB, organomercury lyase, catalyzes the protonolytic conversion of R-Hg to Hg(II) and reduced R-H, where R can be a wide variety of organic moieties such as a methyl or phenyl group. MerA, mercuric ion reductase, catalyzes the electrochemical reduction of Hg(II) to Hg(0). Hg(0) has orders of magnitude lower toxicity relative to ionic Hg or to organic forms of Hg (Summers, 1992; Osborn et al., 1997). Metallic Hg is relatively inert, has very low solubility, and is gaseous at standard temperatures, allowing its diffusion from the bacteria that produce it. Under some circumstances metallic Hg will rapidly evaporate from the bacterial habitat and be diluted to apparently harmless levels in the atmosphere.

In plants as in animals, Hg(II) tends to cause problems at the plasma membrane, where it damages membrane transporters such as aquaporins, affecting nutrient and water transport (Zhang and Tyerman, 1999). Because of its extremely high thioreactivity, Hg(II) becomes toxic to numerous enzyme systems when it is at high enough concentrations. Organomercurials have been reported to rapidly enter membrane-rich plastids where they accumulate and disrupt electron transport and oxygen evolution (Bernier and Carpentier, 1995), photophosphorylation, chlorophyll content, and chlorophyll fluorescence (Kupper et al., 1996; Sinha et al., 1996). Plants have no requirement for Hg and typically play a relatively passive role in the biogeochemistry of Hg compounds. To date no naturally occurring plant species with significant capabilities for accumulation, degradation, or removal of Hg have been identified. Several plant species convert modest amounts of Hg(II) to Hg(0) by the activities of several redox enzymes such as catalase and peroxidase (Heaton et al., 1998). Hg(0) is released into the soil from roots or into the atmosphere from shoots. On the other hand, Hg(II) is highly reactive, tends to bind sulfhydryl groups of sulfur containing enzymes, and forms particularly stable chemical products with reduced thiols. Although reaction with thiols of various enzymes and proteins may destroy their activity, proteins and protein complexes with thiol-bound Hg(II) are relatively nontoxic and may be sequestered in vacuoles.

Biotechnological approaches for Hg transformation and phytoremediation

The plants examined cannot successfully detoxify or convert highly toxic methylmercury to less toxic inorganic forms. As discussed previously, the genes encoding bacterial mercury transformations have been well characterized (Smith et al., 1998), laying the molecular genetic groundwork for enhancing Hg tolerance in plants. A strategy to develop plants with improved abilities for Hg removal and detoxification was initiated in the early 1990s by Richard Meagher and co-workers. They made use of the two bacterial genes discussed above from the well-characterized *mer* operon, *merA*, and *merB*, to engineer an Hg transformation and remediation system in plants (Rugh et al., 1996, 1998; Bizily et al., 1999, 2000). Diverse plant species such as *A. thaliana* (Rugh et al., 1996), tobacco (Heaton et al., 2005), yellow poplar (Rugh et al., 1998), cottonwood (Che et al., 2003), and rice (Heaton et al., 2003) constitutively expressing modified *merA* were resistant to at least ten times greater concentrations of Hg(II) than those that kill non-transgenic controls. These transgenic plants showed significant levels of Hg(0) volatilization relative to controls. The ability of genetically engineered deep-rooted yellow poplar and cottonwood to grow on increased concentrations of ionic Hg(II) may demonstrate the potential for phytovolatilization methods of Hg remediation in wetlands. In a pot soil experiment study, transgenic plants expressing *merA* outperformed wild-type plants on Hg-contaminated soil (Heaton et al., 1998). However, the movement of Hg(0) from the roots of these plants to the soil and to the atmosphere has not been examined.

As discussed earlier, methylmercury is more toxic than ionic or metallic mercury and is efficiently biomagnified up the food chain, while ionic and metallic Hg species are not biomagnified. Methylmercury, thus, poses an immediate and serious threat to human populations. Because merA only converts Hg(II) to Hg(0) and thus cannot detoxify and protect plants against the more toxic and environmentally relevant methylmercury, both the *merA* and *merB* genes are needed to protect cells from methylmercury. MerB catalyzes the protonolysis of the carbon-mercury bond, removing the organic ligand and releasing Hg(II), which is a more reactive and less mobile Hg species. Plants were engineered by over-expressing a modified bacterial organomercury lyase gene (*merB*) to transform methylmercury to ionic Hg (Bizily et al., 1999, 2000). Transgenic *A. thaliana* plants expressing *merB* grew vigorously on a wide range of concentrations of monomethylmercuric chloride and phenylmercuric acetate as compared to non-transformed plants, which were severely inhibited or died at the same organomercurial concentrations (Bizily et al., 1999). These results showed that expression of *merB* alone is sufficient to confer methylmercury tolerance, probably because of the extreme toxicity of methylmercury to most eukaryotic cells. These transgenic plants manage to outgrow the environmentally relevant concentrations of methylmercury by converting methylmercury to Hg(II), which accumulates in these plants. In an attempt to more efficiently detoxify methylmercury, plants were engineered to co-express *merA* and *merB*. The resulting transgenic plants carry out the two-step conversion of methylmercury to volatile Hg0 and are tolerant to 50 times greater concentrations of methylmercury than are required to kill control plants, and five times greater than the concentrations that kill *merB* plants (Bizily et al., 2000). Transgenic Eastern cottonwood trees expressing both *merB* and *merA* genes were also highly tolerant to organic mercury (Lyyra et al., 2007). These results demonstrated that plants (trees, shrubs, and grasses) can be engineered to detoxify the most abundant forms of ionic and organic mercury found at polluted sites, and it is likely that a number of phytoremediation strategies that block its flow into the environment can be adopted.

The chloroplast and endoplasmic reticulum (ER) have been shown to be significant targets for Hg poisoning (Bernier and Carpentier, 1995; Bizily et al., 2003). Therefore, engineering Hg detoxification systems in chloroplasts or ER may offer high levels of Hg tolerance and detoxification. Ruiz et al. (2003) used chloroplast engineering for Hg detoxification by integrating the *merA* and *merB* genes into the chloroplast genome. Transgenic tobacco plants exhibited high levels of tolerance to the organomercurial compound phenylmercuric acetate (PMA) and accumulated 100- and 4-fold more Hg in the shoot in the presence of PMA or $HgCl_2$ than untransformed plants, respectively (Hussein et al., 2007). Therefore,

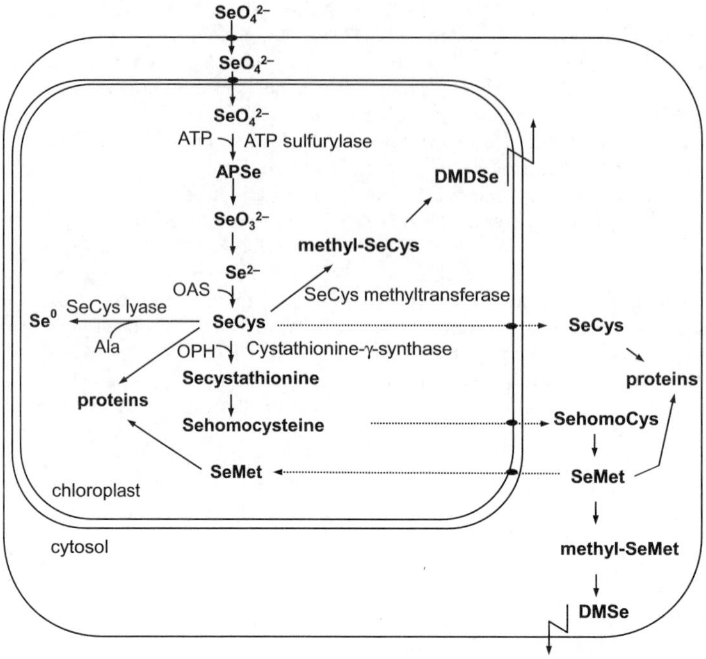

Figure 20.3 • Schematic overview of Se metabolism in plants, showing (in italics) enzymes that were used to alter plant properties with respect to Se tolerance, accumulation, and/or volatilization. APSe: adenosine phosphoselenate; OAS: O-acetylserine; OPH: O-phosphohomoserine; SeCys: selenocysteine; SeMet: selenomethionine; DMSe: dimethylselenide; DMDSe: dimethyldiselenide.

chloroplast engineering may prove a beneficial approach for Hg phytoremediation as well.

Mercury hyperaccumulation

Mercury detoxification and complete volatilization is an ideal strategy in certain situations where immediate removal of Hg is needed. However, there is a certain amount of public resistance and criticism concerning this technology: some people argue that the volatilized Hg(0), although less toxic, will eventually be deposited on the earth's surface. Therefore, additional strategies that can trap and hyperaccumulate Hg(II) aboveground in plant parts for later harvest are needed to prevent the release of Hg(0) in air. In an initial step forward in this direction, plants were engineered that combine the expression of *merB*, which catalyzes the conversion of methylmercury to Hg(II) and enzymes that enhance the synthesis of phytochelatins that bind Hg(II) (Li and Meagher, unpublished data). These plants accumulate more Hg and are more resistant to methylmercury than plants expressing either transgene alone. Future genetic manipulations to improve the phytoremediation of Hg need to include enhanced Hg uptake into roots, transport to shoot, and sequestration into aboveground tissues.

In conclusion, Hg is a neurotoxin and its widespread contamination poses significant human health concerns. Taking advantage of two bacterial genes, *merB* and *merA*, various plant species have been engineered for detoxification of highly toxic methylmercury to the less toxic volatile form of mercury, Hg(0). Although these engineered plants showed significant potential for Hg detoxification in laboratory and greenhouse conditions, the effectiveness of these plants in the natural environment needs to be proven. Future efforts should be focused on engineering plants native to various climatic regions where Hg contamination is widespread. As an alternative to Hg(0) volatilization, there is a need to develop strategies for reduction of methylmercury to Hg(II) and eventual sequestration of Hg(II) in the aboveground tissues. Additionally, engineering the Hg detoxification enzymes in subcellular compartments such as ER and chloroplast will be advantageous.

Selenium

Overview of Se metabolism in plants

Selenium is an essential nutrient for many organisms including humans, but is toxic at elevated levels. Selenium deficiency and toxicity are problems worldwide. There is no evidence that Se is essential for higher plants, but due to its similarity to sulfur Se is readily taken up and assimilated by plants via sulfur transporters and biochemical pathways. Plants accumulate Se in all organs including seeds, and can also volatilize Se into the atmosphere. Some species can even hyperaccumulate Se up to 1% of their dry weight. The ability of plants to accumulate and volatilize Se may be used for phytoremediation. Biotechnology has proven useful to obtain better insight into the genetic and biochemical mechanisms that control Se tolerance, accumulation, and volatilization in plants, and the resulting transgenics with enhanced levels of these processes show promise for use in phytoremediation and as fortified food. Next we present an overview of plant Se metabolism with a focus on biotechnological advances.

As summarized in Figure 20.3, plants readily take up selenate (SeVI) or selenite (SeIV) from their environment and incorporate it into organic compounds using sulfate assimilation enzymes (for reviews see Terry et al., 2000; Sors et al., 2005; Pilon-Smits and Quinn, 2010). The toxicity of Se is thought to be due to non-specific incorporation of the resulting amino acids, selenocysteine (SeCys) and selenomethionine (SeMet), into proteins. To prevent this, plants may break down SeCys into relatively innocuous elemental Se (Se0) or methylate it to relatively non-toxic methyl-SeCys, which may be accumulated or further methylated to volatile dimethyldiselenide (DMDSe). SeMet can also be methylated to form volatile dimethylselenide (DMSe). Methylation of SeCys occurs primarily in Se hyperaccumulators, and is thought to be a key mechanism for their Se tolerance (Neuhierl et al., 1999).

Plant Se accumulation and volatilization are both useful for Se phytoremediation. Moreover, species that accumulate selenocompounds with anticarcinogenic properties, particularly methyl-SeCys, may be useful as fortified foods (Bañuelos, 2009). Examples of such species are broccoli, garlic, and the hyperaccumulators two-grooved milkvetch (*Astragalus bisulcatus*), and prince's plume (*Stanleya pinnata*). Sulfur-loving plants such as *Brassica* and *Allium* species (mustards, cabbages, onion, garlic) typically also accumulate Se well (0.01–0.1% of dry weight) and have been called (secondary) accumulator species. Hyperaccumulators are unique in that they preferentially take up Se over S, can hyperaccumulate and tolerate Se up to 1% of dry weight under field conditions, and accumulate methyl-SeCys (Neuhierl et al., 1999). Selenium hyperaccumulators are only found on seleniferous soils, perhaps because they depend on Se as a defense compound against herbivores or pathogens (Pilon-Smits and Quinn, 2010).

Biotechnological approaches to study and manipulate Se metabolism in plants

In a first approach to manipulate plant Se tolerance, accumulation, and/or volatilization, genes involved in sulfur/selenium assimilation and volatilization were over-expressed. *Brassica juncea* (Indian mustard) over-expressing ATP sulfurylase (APS), involved in selenate-to-selenite conversion, showed enhanced selenate reduction, judged from the finding that transgenic APS plants supplied with selenate accumulated an organic form of Se while wild-type plants accumulated selenate (Pilon-Smits et al., 1999). The APS transgenics accumulated two- to three-fold more Se than wild-type, and 1.5-fold more sulfur. The APS plants tolerated the accumulated Se better than wild-type, perhaps because of the organic form of Se accumulated. Selenium volatilization rate was not affected in the APS transgenics. Indian mustard over-expressing cystathionine gamma synthase (CgS, the first enzyme in the conversion of SeCys to SeMet) showed two- to three-fold higher volatilization rates compared to untransformed plants (Van Huysen et al., 2003). They accumulated 40% less Se in their tissues than wild-type, presumably as a result of their enhanced volatilization. The CgS transgenics were also more Se tolerant than wild-type plants, perhaps due to their lower tissue Se levels.

A second approach to manipulate plant Se metabolism focused on the prevention of SeCys incorporation into proteins. In one strategy, selenocysteine lyase (SL) was expressed in *A. thaliana* and Indian mustard, initially using a mouse SL (Pilon et al., 2003; Garifullina et al., 2003), and subsequently an *A. thaliana* homolog called cpNifS (Van Hoewyk et al., 2005). SeCys lyase breaks down SeCys into alanine and Se(0). The transgenics showed reduced Se incorporation into proteins, enhanced Se tolerance, and about a two-fold enhanced Se accumulation compared to wild-type plants. In another strategy to prevent SeCys incorporation into proteins, SeCys methyltransferase (SMT) from hyperaccumulator *A. bisulcatus* was over-expressed in *A. thaliana* or *B. juncea* (Ellis et al., 2004; LeDuc et al., 2004). The SMT transgenics showed enhanced Se tolerance and enhanced Se accumulation, in the form of methyl-SeCys. The SMT plants also had increased rates of Se volatilization in the form of DMDSe. Double-transgenic Indian mustard plants (over)expressing both APS and SMT even showed up to nine-fold higher Se accumulation compared to wild-type (LeDuc et al., 2006).

Selenium phytoremediation field studies

The results from these various transgenics, including enhanced Se tolerance, up to nine-fold higher Se accumulation and up to three-fold faster Se volatilization, are promising for phytoremediation but all obtained under laboratory conditions. To better determine the transgenics' potential for phytoremediation, they were tested for their capacity to accumulate Se from naturally seleniferous soil and from Se-contaminated sediment. On seleniferous soil in a greenhouse pot experiment, the APS transgenics accumulated Se to three-fold higher levels than wild-type Indian mustard, and the CgS transgenics contained 40% lower Se levels than wild-type (Van Huysen et al., 2004), all in agreement with the laboratory results. Plant growth was the same for all plant types in this experiment. Similarly, when growing in the field on Se (selenate)-contaminated sediment in the San Joaquin Valley in California, the APS transgenics accumulated Se to a four-fold higher level than wild-type Indian mustard (Bañuelos et al., 2005). In a second field experiment on the same Se-polluted sediment, the SL and SMT transgenics showed two-fold higher Se accumulation than wild-type Indian mustard, also in agreement with earlier laboratory experiments (Bañuelos et al., 2007). In both field experiments biomass production was comparable for the different plant types. Therefore, the results obtained using naturally seleniferous or Se-contaminated soils in greenhouse or field are similar to those obtained under controlled laboratory conditions. The various transgenics showed enhanced Se accumulation, volatilization, and/or tolerance, all promising traits for use in phytoremediation or as Se-fortified foods.

Organic pollutants

Phytoremediation of organic pollutants offers the potential for complete degradation of the pollutant if the chemical can be taken up by the plant and if all the necessary biodegradation genes are present. Most of the organic pollutants do have phytotoxic effects that must be overcome for phytoremediation to be effective. Another limiting factor includes the solubility of the pollutant that can hinder the ability of the plant to uptake the chemical. There are several classes of organic pollutants: solvents (i.e., trichloroethylene); explosives such as trinitrotoluene (TNT) and cyclotrimethylenetrinitramine or Research Department Explosive (RDX); polycyclic aromatic hydrocarbons (i.e., naphthalene, pyrene); petroleum products including benzene, toluene, ethylbenzene, and xylene (BTEX); polychlorinated biphenyls (PCBs); and herbicides/pesticides (i.e., atrazine, chlorpyrifos, 2,4-D). In general, plants use a three-step pathway for the detoxification of organic pollutants. In the first phase, a reactive group, such as a hydroxyl, amino, or sulfhydryl group, is added to the xenobiotic. In the second phase, another compound such as a sugar moiety, is conjugated via the reactive group. Finally, in the third phase the conjugated pollutant is sequestered into the vacuole or integrated into cell wall components, thus rendering the compound less toxic. Efforts to increase the effectiveness of phytoremediation of organic pollutants involve either the over-expression of the plant genes involved in any of these steps, introduction of microbial genes known to be involved in pollutant biodegradation, or the inoculation of the plant with pollutant-degrading endophytes (reviewed in Doty, 2008; Dowling and Doty, 2009; Weyens et al., 2009b). In the next section we will summarize the results from these various approaches for different classes of pollutants (Table 20.1).

Solvents

Environmental pollution from solvents is often caused by dumping of the used solvent directly on the ground, eventually leading to contamination of the groundwater. One of the most widespread of the organic pollutants is the solvent trichloroethylene (TCE). Engineering methods for remediation of TCE include air sparging where the contaminated water is pumped through a cylinder into which air is blown, causing the TCE to volatilize and enter the atmosphere. Another common engineering method is the addition of oxidants such as potassium permanganate or hydrogen peroxide that react with the TCE. Bioremediation of TCE using anaerobic bacteria is a popular biological remediation strategy. The site is first made anaerobic by the addition of large quantities of substrates such

Table 20.1 Summary of transgenic plants for phytoremediation of organic pollutants

Compound	Gene	Plant	Reference
Atrazine	*bphC*	Tobacco	Wang et al. (2005)
		Arabidopsis, alfalfa	
Atrazine	hCYP1A1/A2	Tobacco	Bode et al. (2004)
Atrazine	hCYP1A1/2B6/2C19	Rice	Kawahigashi et al. (2006)
Benzene	rCYP2E1	Poplar	Doty et al. (2007)
Benzene	hCYP2E1	Tobacco	James et al. (2008)
PCBs	*bphC*	Tobacco	Francova et al. (2003)
PCBs	*bphA, bphE, bphF, bphG*	Tobacco	Mohammadi et al. (2007)
PCBs	*bphC*	Tobacco	Novakova et al. (2009)
RDX	*xplA*	*Arabidopsis*	Rylott et al. (2006)
RDX	*xplA, xplB*	*Arabidopsis*	Jackson et al. (2007)
TCE	hCYP2E1	Tobacco	Doty et al. (2000)
TCE	rCYP2E1	*Atropa belledonna*	Banerjee et al. (2002)
TCE	rCYP2E1	Poplar	Doty et al. (2007)
TNT	PETNr	Tobacco	French et al. (1999)
TNT	*nfsl*	Tobacco	Hannink et al. (2001)
TNT	*pnrA*	Poplar	van Dillewijn et al. (2008)
TNT	UGTs	*Arabidopsis*	Gandia-Herrero et al. (2008)
Toluene	hCYP2E1	Tobacco	James et al. (2008)

as molasses, vegetable oil, or lactate. The area is then bioaugmented with a consortia of bacteria that degrade TCE to cis-DCE, vinyl chloride, and finally to harmless ethane. All of these engineering methods are effective ways of remediating TCE-contaminated sites, which can usually be achieved within a few years. However, all of these methods are very expensive, challenging, and can sometimes lead to worse environmental situations. Air sparging results in the contamination of the air with the pollutant, which is not an ideal solution for the neighboring communities. The use of potassium permanganate often results in purple residue because of the difficulty in dispersing the oxidant precisely in the required location and at the proper quantity to react with the TCE. Anaerobic biodegradation of TCE frequently stalls at vinyl chloride (VC), a highly toxic metabolite known to cause cancer. Although specific strains of *Dehalococcoides* bacteria are able to degrade the VC to ethane, the strains are sensitive to the low pH that results from the original substrates used to make the site anaerobic. Addition of buffers often leads to fouling of the pipes and increased expense. Although these methods are able to remediate TCE-contaminated sites, the expense and difficulties in practice make them out of the realistic realm for the many sites with low-level or widespread contamination.

Phytoremediation of solvents including TCE is effective for sites with shallow groundwater within the range of tree roots. Poplar trees are especially well suited for phytoremediation of TCE as they are deep-rooted, and a variety of herbaceous species (tobacco, *Leucaena leucocephala*, *Arabidopsis*) also have the genetic capability to degrade TCE (Shang et al., 2001; Doty et al., 2003; Doty, 2008). Plants seem to utilize a TCE degradation pathway similar to mammals, since both result in the metabolite trichloroethanol (Shang et al., 2001). However, phytoremediation of TCE is limited by the apparent low expression of the cytochrome P450 enzyme that activates TCE prior to its degradation. The metabolism of TCE in plants is often considered too slow and may lead to phytovolatilization of the pollutant.

Strategies to improve phytoremediation of TCE include genetic engineering or endophyte-assisted phytoremediation (Doty, 2008; James and Strand, 2009). Over-expression of the mammalian cytochrome P450 CYP2E1 in transgenic tobacco (Doty et al., 2000) and poplar (Doty et al., 2007) led to a strong increase in the metabolism of TCE. There was an increase in TCE removal rate both from the liquid and from air by the transgenic poplar. Although only the first gene in the pathway was over-expressed in the transgenics, dozens of other genes with homology to pollutant degradation genes were also upregulated in response to TCE in the transgenic poplar (Kang et al., 2010). These genes included those involved in pollutant activation, conjugation to sugars, and transport. Field trials of the transgenic poplar are in progress using a simulated pump and treat system (James et al., unpublished). As in the lab studies, the CYP2E1 transgenic plants had more of the TCE metabolite, trichloroethanol, than did the vector-control plants. There was also an elevated level of chloride ions in the test beds with the transgenic plants, indicating dechlorination of the TCE. However, the concentration of TCE in the effluent was not further reduced by the transgenic plants compared to the controls. Therefore, the plants may be more suitable for contaminated water as simulated in the lab studies (Doty et al., 2007) rather than for pump and treat systems that have a continuous source of TCE.

The second approach to improving phytoremediation of TCE is through the use of endophytes that can metabolize TCE. In this strategy, it is thought that the plant and bacteria work together, with the plant effectively taking up the pollutant, and the endophytic bacteria known to colonize the vascular tissues and intercellular spaces degrading the pollutant (Doty, 2008; Weyens et al., 2009b). A well-studied aerobic TCE-degrading bacterium is the *Burkholderia* strain G4 that has a large, self-transmissible, degradative plasmid, TOM (Shields et al., 1995). This bacterium co-metabolizes TCE using a toluene *ortho*-monooxygenase encoded by an operon of *tom* genes, resulting in harmless metabolites. The plasmid, or a mutant version with constitutive expression of the *tom* genes, can be transferred to native endophytes of poplar (Taghavi et al., 2005). Poplar trees were inoculated *in situ* with the modified endophytic strain containing pTOM-Bu61 to improve phytoremediation of TCE (Weyens et al., 2009a). Although no increased uptake of TCE was reported, there was a reduction in the phytovolatilization of TCE from an average of $0.07\,ng\,cm^{-2}h^{-1}$ for the uninoculated plants to an average of about $0.01\,ng\,cm^{-2}h^{-1}$ from the inoculated plants. Since most contaminated sites have both inorganic and organic pollutants, the endophyte approach was then used to determine if inoculation could improve tolerance to both classes of pollutants (Weyens et al., 2010). Yellow lupine plants were inoculated with a *Burkholderia* strain containing both the *tom* genes and the *ncc-nre* Ni resistance genes. Although there was no increase in shoot mass after inoculation, root mass was increased by 30% in the inoculated plants exposed to nickel and TCE relative to unexposed plants. While there was a decrease in phytotranspiration of TCE in the inoculated plants, the difference was not statistically significant. Overall, endophyte-assisted phytoremediation of TCE has led to some improvements, but so far no changes in TCE removal have been reported.

Explosives

At military training ranges there is a need for remediation of the nitroaromatic explosives, TNT and RDX (hexahydro-1,3,5-trinitro-1,3,5-triazine), to prevent the spread into neighboring communities. TNT causes anemia and liver damage, while RDX affects the central nervous system, causing convulsions. Extensive areas are contaminated with these pollutants, approximately 40 million acres in the United States alone (U.S. Defense Science Board Task Force, 1998; Rylott and Bruce, 2008). Nitroaromatic explosives contain an aromatic ring with attached nitro ($-NO_2$) groups. A difficulty in phytoremediation of TNT is the toxicity of this pollutant, and its concentration on sites can be as high as $87{,}000\,mg\,kg^{-1}$ (Talmage et al., 1999). RDX is less toxic but its high solubility in water gives it a higher potential of leaving the site and entering the water table. Hot spots of RDX can be nearly as high as those for TNT.

Some plant species are able to tolerate relatively low levels of TNT, transforming it to an aminodinitrotoluene that is then conjugated to sugars or glutathione, and then probably stored in the vacuole or cell walls, or secreted. Microarray and other gene expression assays have revealed several important classes of enzymes involved in the plant responses to nitroaromatics (Gandia-Herrero et al., 2008). In *A. thaliana*, a small family of oxophytodienoate reductases (OPRs) is upregulated in response to TNT (Mezzari et al., 2005). In *Chlamydomonas reinhardtii* and in *Populus trichocarpa*, glutathione-S-transferases (GSTs) are also upregulated. Comparisons of plant gene expression responses to TNT and RDX revealed little commonality; therefore there seems to be little overlap in the detoxification pathways for these two explosives. Similarly, different bacteria metabolize TNT and RDX using different genes. Although some microorganisms are able to degrade the nitroaromatics, they seem to lack the necessary mass to significantly remediate the contaminated sites (Rylott and Bruce, 2008).

Transgenic model plants expressing these bacterial genes for the degradation of TNT and RDX have successfully overcome the phytotoxic effects, removing more of these pollutants (recently reviewed in Rylott and Bruce, 2008; van Aken, 2009). The pentaerythritol tetranitrate (PETN) reductase gene from *Enterobacter cloacae* strain PB2 encodes an enzyme that removes nitrate from TNT, allowing the bacteria to use TNT as a nitrogen source. Expression of this gene in transgenic tobacco resulted in improved tolerance to TNT (French et al., 1999). Transgenic tobacco containing a bacterial nitroreductase gene (*nfsI*) from the same bacterium metabolized TNT at much faster rates than control plants (Hannink et al., 2001). Not only do the transgenics have improved TNT remediation abilities, they also help restore the rhizospheric community by reducing the soil toxicity (Travis et al., 2007). Expression of a nitroreductase gene (*pnrA*) from a *Pseudomonas* strain in transgenic poplar resulted in increased uptake of TNT from water and soil (van Dillewijn et al., 2008). The transgenic poplar were tolerant to more than five times as high a concentration of TNT in hydroponics, and twice as high in soils, compared to the non-transgenic poplar. Following the first phase of plant detoxification of xenobiotics, plants generally conjugate the activated molecule and sequester it in the vacuole. Since microarray analysis revealed increased expression of genes involved in conjugation (Gandia-Herrero et al., 2008), phytoremediation of TNT may be improved by upregulating genes involved not only in the nitroreductase step, but also in the conjugation step. Indeed, over-expression of two uridine glycosyltransferases from *Arabidopsis* that had been identified by microarray analysis resulted in increased conjugate production and TNT detoxification (Gandia-Herrero et al., 2008).

RDX can be degraded and used as a source of nitrogen by several bacterial strains isolated from contaminated sites (Crocker et al., 2006). The *xplA* gene responsible for RDX biodegradation encodes a novel, fused flavodoxin-cytochrome P450 enzyme (Rylott et al., 2006). Transgenic *Arabidopsis* plants expressing *xplA* (CYP177) from *Rhodococcus rhodochrous* 11Y tolerated and removed high levels of RDX, whereas non-transgenic plants did not take up any significant amount. The *xplA* transgenics grew in soils containing 2000 mg kg^{-1}, a level nearly ten times higher than non-transgenic plants could tolerate. In recent studies, co-expression of both *xplA* and *xplB* in transgenic plants resulted in even greater improvements in RDX removal rates, 30-fold higher than with xplA alone (Jackson et al., 2007). Since military sites are co-contaminated with both TNT and RDX, plants with the ability to detoxify both types of explosives would be desirable. Poplar plants with *nfsI* and *xplA* have increased removal of both TNT and RDX, and triple transformants with *xplA*, *xplB*, and *nfsI* are being constructed (Doty, unpublished).

Although there are no reports of endophyte-assisted phytoremediation of explosives, there are indications that this could be an alternative approach. A natural endophyte of poplar, *Methylobacterium* sp. Strain BJ001, is capable of degrading TNT, RDX, and HMX, mineralizing 60% of the chemicals in about two months (van Aken et al., 2004). Characterization of the xplA/xplB systems of a variety of bacterial strains revealed that the genes are on a plasmid that can be conjugatively transferred (Andeer et al., 2009). Adding the ability to degrade RDX to natural endophytes of high biomass plants such as poplar and willow could result in improved phytoremediation of this class of pollutants.

BTEX, PAHs, and PCBs

Improved phytoremediation using biotechnology is also being pursued for BTEX, PAHs, and PCBs. Petroleum pollutants, including hydrocarbon chains, and the aromatics benzene, toluene, ethylbenzene, and xylene (BTEX) can be remediated using plants if the concentrations are low. Plants growing on sites contaminated with these pollutants often contain petroleum-degrading bacteria in the roots or in the rhizosphere (Siciliano et al., 2003). Poplar trees growing on a BTEX-contaminated site harbored a few dozen endophytes with pollutant-degrading capabilities that may improve phytoremediation (Moore et al., 2006). To increase phytoremediation of BTEX chemicals, the genes for degrading the BTEX component, toluene, were transferred to an endophytic strain and inoculated onto lupine (Barac et al., 2004). The inoculated plants were able to tolerate levels of toluene ten times the normally phytotoxic levels. When the original toluene-degrading strain was inoculated into the more suitable remediation plant, poplar, the strain conjugatively transferred the plasmid *in planta* to the native endophytes, resulting in increased tolerance to toluene (Taghavi et al., 2005). The presence of the endophyte also reduced the phytotranspiration of the chemical. Furthermore, transgenic plants expressing mammalian cytochrome P450 2E1 had greatly increased rates of removal of toluene and benzene (James et al., 2008). Toluene was removed from the hydroponic solution within two days, at a rate ten times faster than the vector-control plants. Benzene was nearly completely removed within three days by CYP2E1 transgenic tobacco, while the vector-control plants removed benzene no better than the unplanted controls.

Phytoremediation of PAHs has had limited success due to the high toxicity of this class of pollutants. Many of these ring-structured compounds are strong carcinogens, formed from the incomplete combustion of fossil fuels. Naphthalene, a low molecular weight PAH, is one of the most common pollutants on the National Priorities List of the United States Environmental Protection Agency (U.S. EPA). It is phytotoxic, causing growth inhibition, reduced transpiration, chlorosis, and wilting (Germaine et al., 2009). Thygesen and Trapp (2002) exposed hydroponically grown willow plants to a variety of PAHs. Although phenanthrene and benzo(a) pyrene did not affect willow growth, naphthalene (325 mg L^{-1}) killed the plants. In a similar study, willow plants readily took up low levels of naphthalene but uptake of pyrene and phenanthrene stalled after three days with phytotoxic effects (Khan and Doty, unpublished). Therefore, different willow species may have different PAH tolerance capacities.

Improving phytoremediation of PAHs using endophytes is a promising approach. Several PAH-degrading endophytes were isolated from poplar and willow that can utilize PAHs as the sole carbon source (Doty, 2008). Experiments to determine if the PAH-degrading endophyte, PD1, can improve phytoremediation in willow are now underway. Using conjugative transfer of a plasmid conferring PAH degradation into an endophytic *Pseudomonas putida* strain, Germaine and colleagues (2009) effectively reduced phytotoxicity of naphthalene. The native endophyte did not reduce phytotoxicity, nor did the non-endophytic *P. putida* PAH-degrading strain. The endophyte with pNAH7, however, protected the inoculated peas, resulting in significantly higher germination rates in soil containing 30 mg naphthalene per kg soil, from 20% germination of the uninoculated plants to 80% in the inoculated seeds. At very high levels of naphthalene up to 100 mg kg^{-1}, none of the uninoculated seeds germinated while 20% of the inoculated ones germinated. Inoculated pea plants had higher transpiration rates, up to 35% higher than the uninoculated controls in hydroponics containing naphthalene. Plant growth was also improved with the PAH-degrading endophyte. Mass of the plants after two weeks of naphthalene treatment was significantly greater in the inoculated plants. In terms of potential phytoremediation improvements, the inoculated plants strongly reduced the amount of naphthalene remaining in soils, removing 37% more of the PAH than the uninoculated pea plants. Therefore, PAH-degrading endophytes have the potential to provide the means for large biomass plants such as poplar and willow to effectively remove this important class of pollutants.

Polychlorinated biphenyls are chlorinated aromatic compounds that are toxic, highly persistent xenobiotics, and are listed as U.S. EPA Priority Pollutants. PCBs were used extensively in a variety of industrial applications due to their thermal stability. Some microorganisms can degrade PCBs aerobically or non-aerobically under a variety of conditions (reviewed in Borja et al., 2005; Pieper and Seeger, 2008). Anaerobic reductive dechlorination of PCBs is usually achieved by a consortia of bacteria including *Dehalococcoides* (Abraham et al., 2002), the genera used extensively in bioremediation of TCE. To date, none of the genes involved in PCB anaerobic dechlorination have been cloned. Aerobic degradation of lower chlorinated PCBs is via co-metabolism by dioxygenases, resulting in ring cleavage and possibly complete mineralization. One of the most effective PCB degraders characterized is *Burkholderia xenovorans* strain LB400. Two operons of genes (*bph*) for aerobic PCB degradation have been identified.

Phytoremediation of PCBs was recently reviewed (van Aken et al., 2010). Plants can assist in microbial remediation in several ways: by releasing root exudates that stimulate microbial growth; by secreting phenolics necessary for PCB co-metabolism; by increasing soil oxygen; and by releasing surfactants that help release soil-bound PCBs. In a test with 7 different plant species, 38% of extractable PCBs remained in the soil compared to 82% remaining in the unplanted control soil (Chekol et al., 2004). In a study with nine different plant species in PCB-contaminated soil, even some of the higher chlorinated congeners of PCBs were translocated within the plant to the shoot tissues (Zeeb et al., 2006). However, in studies with hybrid poplar, only the lower chlorinated PCBs were translocated, with the tetrachlorinated PCBs adsorbed on plant roots (Liu and Schnoor, 2008). Using axenic plant cultures, it was demonstrated that a variety of plants are able to metabolize the PCBs directly (Lee and Fletcher, 1992; Wilken et al., 1995; Mackova et al., 1997; Kucerova et al., 2000; Harms et al., 2003; Rezek et al., 2007). In general, plant metabolism appears limited to tetra chlorinated and lower congeners with slow degradation rates in field trials. Furthermore, the PCB metabolism in plants involves cytochrome P450s that result in toxic intermediates that are not fully degraded, whereas some of the bacterial pathways result in ring cleavage and complete degradation of the PCBs. Therefore, engineering plants with genes for the bacterial pathway may offer a more effective strategy for remediation of this recalcitrant pollutant. To this end, some of the *bph* genes have been introduced into transgenic plants and functional enzymes were produced (Sylvestre et al., 2009; van Aken et al., 2010). The effect of expression of *bph*C in transgenic tobacco plants on PCB remediation was only recently assessed (Novakova et al., 2009). In this study, one of the transgenic plant lines had higher tolerance than the wild-type plants. It may be that transfer of the complete operons for PCB metabolism will be necessary to achieve more effective PCB phytoremediation (Sylvestre et al., 2009). Chloroplast engineering allows for the transfer of entire operons from bacteria into transgenic plants for high expression of enzymes to improve phytoremediation (Ruiz et al., 2003) and may be useful for this particularly challenging pollutant. Another approach is to engineer plant-associated bacteria. By transferring a plasmid with the *bph* operon into *Sinorhizobium meliloti* and inoculating alfalfa with the modified strain, degradation of 2,3,4-trichlorobiphenyl was doubled compared with plants inoculated with the wild-type strain (Chen et al., 2005). In a related study, the *bph* operon was chromosomally inserted into *Pseudomonas fluorescens* strains, and these were inoculated into the rhizosphere of willow plants (de Carcer et al., 2007). After six months, there was significantly more degradation of the PCBs in the rhizosphere containing the modified strains. As of yet, the approach of modifying endophytes with PCB-degrading genes has not been tested (van Aken et al., 2010).

Pesticides

Since pesticides can cause chronic abnormalities in humans and they generally lead to reduced environmental quality, multiple methods including incineration and land filling have been used to remove this class of pollutants; however, these physical methods are expensive and inefficient. Bioremediation using microorganisms capable of degrading the polluting pesticide and enhanced phytoremediation of pesticides using transgenic plants are emerging as more effective solutions (Hussain et al., 2009). The topic of biodegradation of pesticides, including herbicides, is large; therefore, this review will focus primarily on only three: the pesticide chlorpyrifos and the herbicides atrazine, and 2,4-dichlorophenoxyacetic acid (2,4-D). For a more thorough review of remediation of a wide range of pesticides, see the recent review by Hussain and colleagues (2009). Chlorpyrifos, a common organophosphate pesticide, can be degraded by certain strains of bacteria isolated from contaminated environments (Singh et al., 2006). A strain of *Stenotrophonomonas* sp. isolated from contaminated sludge degraded chlorpyrifos at a faster rate than uninoculated soils (Yang et al., 2006). A *Klebsiella* strain, also isolated from sludge from a waste water treatment plant, was shown to biodegrade the pesticide (Ghanum et al., 2007). Chlorpyrifos is also degraded by cultures of plant pathogenic fungi (Al-Mihanna et al., 1998) and other fungi (Bumpus et al., 1993). Atrazine, a non-acidic pesticide that has become a common contaminating herbicide in surface water, can be biodegraded as well (Topp, 2001; Wackett et al., 2002; Satsuma, 2006; Chirnside et al., 2007). In some of these cases, the genes necessary for the biodegradation have been cloned (Shapir et al., 2002; Sajjaphan et al., 2004). Atrazine degradation involves hydrolases, ureases, dehalogenases, and cytochrome P450s encoded by *atz*ABCDEF, *trz*ND, and *psb*A1 (Hussain et al., 2009). The herbicide 2,4-D can be degraded by a strain of *Pseudomonas* (Musarrat et al., 2000). As with atrazine degradation, many of the necessary genes have been identified and cloned (Itoh et al., 2002). One advantage of bioremediation over mechanical methods is that it is a natural system that can cause less of a disturbance to the environment. As described earlier, there have been successful isolations of bacteria and fungi that are capable of degrading these pesticides. However, bioremediation of pesticides as a technology is still under development. Several environmental factors affect the success of bioremediation of pesticides, including soil pH, organic matter content, temperature, aeration, and moisture levels. Although the soil organic content affected degradation of atrazine (Boivin et al., 2005), it did not affect the biodegradation of chlorpyrifos (Singh et al., 2006). Degradation of organophosphate pesticides, including chlorpyrifos, was slower in low pH soils compared to in neutral or alkaline soils (Singh et al., 2006). In addition to maintaining a suitable environment for bioremediation, another problem with bioremediation of pesticides is that some microbes produce toxic intermediates that can be more toxic or persistent than the parent compound (Hussain et al., 2009).

Another approach for remediation of pesticides is to add the purified enzyme responsible for biodegradation. This direct approach to remediation is limited by the ability of the enzyme to function in less than ideal conditions, in pure form without co-factors, and it must have a cheap source of isolation (Sutherland et al., 2004). In one case study, 84,000 L of methyl-parathion-contaminated waste water was treated with purified microbial enzymes with a ten-fold reduction in contaminant level in just 10 minutes (Russell et al., 2001). Chlorpyrifos can be detoxified with mammalian paraoxonase, PON1, an enzyme that can be produced in culture and purified for use in treatment of exposed individuals (Stevens et al., 2008). It can also be produced as a stable foam for emergency treatment of pesticide spills in the environment (C. Furlong, personal communication). Development of inexpensive, large-scale production of PON1 using tobacco chloroplast engineering is underway (Doty, unpublished).

Phytoremediation of pesticides can be effective on several levels including reduction of chemical leaching, aerating the soil, and providing nutrients for microbial biodegradation, as well as direct uptake and phytodegradation of the pollutants (Hussain et al., 2009). A naturally tolerant plant species, *Lolium multiflorum*, was able to germinate and grow in high levels of atrazine (Merini et al., 2009). This ability was strongly inhibited by the P450 inhibitor 1-aminobenzotriazole, suggesting that a cytochrome P450 is responsible for the high tolerance in this species. It was demonstrated that poplar cuttings can take up atrazine and metabolize it through hydrolysis and dealkylation (Chang et al., 2005). Aquatic plants in a constructed wetland were able to remove and retain about 25% of influent chlorpyrifos (Moore et al., 2002). The ability of plants to uptake and translocate pesticides from roots to shoots varies among different plant species, even in hydroponics without the effects of soil binding. For example, atrazine was taken up better by *Juncus effuses* (soft rush) than *Ludwigia peploides* in which it was sequestered in roots rather than translocated to shoots (Bouldin et al., 2006). In a separate study *J. effuses* took up both atrazine and chlorpyrifos; however, it was better able to take up and degrade the chlorpyrifos (Lytle and Lytle, 2000). In this study, chlorpyrifos was rapidly taken up within 24 hours by this prominent wetland species. In recent research with chlorpyrifos, it was shown that poplar and willow trees are able to remove and metabolize this common organophosphate pesticide (Lee et al., 2010). These riparian species have the potential to remove agricultural pollutants before they enter the waterways.

Transgenic plant technology is investigated to improve remediation of pesticides. In research by Wang and colleagues (2005), the *atzA* gene encoding the first enzyme, atrazine chlorohydrolase, of a 6-step pathway was expressed in transgenic plants. The transgenic tobacco, *Arabidopsis*, and alfalfa actively expressed *atzA*, resulting in increased tolerance to a wide range of atrazine concentrations. The pesticide was dechlorinated to hydroxyatrazine in all of the plant organs. In another approach, the mammalian cytochrome P450 genes CYP1A1 and CYP1A2 were expressed in a transgenic tobacco cell culture, resulting in increased metabolism of atrazine (Bode et al., 2004). Profound enhancement of metabolism of a broad range of herbicides including atrazine and metolachlor was achieved in transgenic rice plants co-expressing CYP1A1, CYP2B6, and CYP2C19 (Kawahigashi et al., 2006). The transgenic plants had strong tolerance to eight different herbicides. Whereas control plants were killed with atrazine, which inhibits photosynthesis, the growth of the transgenic plants was unaffected. In terms of remediation of contaminated surface water, the transgenic rice plants removed twice as much of the herbicides after one week than did the control plants. The transgenics also removed significantly more of the atrazine from soil than did the controls. Methods to improve remediation of chlorpyrifos using mammalian CYP2B6 and PON1 in transgenic poplar are currently underway (Lee et al., unpublished).

Endophyte-assisted phytoremediation of pesticides is also showing promising results. In pioneering work by Germaine and colleagues (2006), remediation of the systemic herbicide 2,4-D was improved through inoculation with a GFP-tagged 2,4-D degrading endophyte . The endophyte, a strain of *P. putida*, effectively colonized the pea plants and led to a significant increase in plant growth that correlated with the level of colonization. Uninoculated plants showed severe root toxicity effects in response to the 2,4-D application, while the inoculated plants maintained a healthy root system. The inoculated pea plants removed more of the herbicide and, unlike the control plants, did not accumulate the herbicide in aerial tissues, indicating that the herbicide was degraded within the plant by the *P. putida* endophyte.

Future Prospects

In the past decade there has been a tremendous increase in our knowledge of plant processes involved in, and limiting for phytoremediation of a wide variety of inorganic and organic pollutants. This knowledge has been obtained in part through plant biotechnology and conversely has led to plant biotechnological approaches to enhance the phytoremediation potential of plants. In some cases, as described earlier, natural plant processes involved in uptake, assimilation, or detoxification were manipulated, while in other cases entirely new processes were introduced, often by introducing bacterial genes or even entire bacterial endophytes. Results from field studies are starting to come in and tend to confirm results from initial lab and greenhouse trials. Clearly, plant biotechnological approaches have played an important role in moving the field of phytoremediation forward. For better acceptance in the remediation industry, it is important that new transgenics continue to be tested in the field. In that context it will be helpful if regulatory restrictions can be regularly re-evaluated to make the use of transgenics for phytoremediation less cumbersome.

Acknowledgments

National Science Foundation grant #IOS-0817748 to EAHPS and a grant (#GO12026-273) from the Consortium of Plant Biotech Research (CPBR) to OPD supported the writing of this manuscript.

References

Abedin, M. J., Cotyter-Howells, J., & Meharg, A. A. (2002). Arsenic uptake and accumulation in rice (*Oryza sativa* L.) irrigated with contaminated water. *Plant and Soil*, *240*, 311–319.

Abercrombie, J. M., Halfhill, M. D., Ranjan, P., Rao, M. R., Saxton, A. M., & Yuan, J. S., et al. (2008). Transcriptional responses of *Arabidopsis thaliana* plants to AsV stress. *BMC Plant Biology*, *8*, 87.

Abraham, W. R., Nogales, B., Golyshin, P. N., Pieper, D. H., & Timmis, K. N. (2002). Polychlorinated biphenyl-degrading microbial communities in soils and sediments. *Current Opinion in Microbiology*, *5*, 246–253.

Ahsan, N., Lee, D. -G., & Alam, I. (2008). Comparative proteomic study of arsenic-induced differentially expressed proteins in rice roots reveals glutathione plays a central role during As stress. *Proteomics*, *8*, 3561–3576.

Al-Mihanna, A. A., Salama, A. K., & Abdalla, M. Y. (1998). Biodegradation of chlorpyrifos by either single or combined cultures of some soilborne plant pathogenic fungi. *Journal of Environmental Science and Health*, *33*, 693–704.

Ali, M. A., Badruzzaman, A. B. M., Jalil, M. A., Hossain, M. D., & Ahmed, M. F., et al. (2003). Fate of arsenic extracted with groundwater. In M. F. Ahmed (Ed.), *Fate of arsenic in the environment* (pp. 7–20). Dhaka: ITN Int. Train. Netw.

Andeer, P. F., Stahl, D. A., Bruce, N. C., & Strand, S. E. (2009). Lateral transfer of genes for hexahydro-1,3,5-trinitro-1,3,5-triazine (RDX) degradation. *Applied and Environmental Microbiology*, *75*, 3258–3262.

Banerjee, S., Shang, Q. T., Wilson, A. M., Moore, A. L., Strand, S. E., & Gordon, M. P., et al. (2002). Expression of active mammalian P450 2E1 in hairy root cultures. *Biotechnology and Bioengineering*, *77*, 462–466.

Bañuelos, G., LeDuc, D. L., Pilon-Smits, E. A. H., Tagmount, A., & Terry, N. (2007). Transgenic Indian mustard overexpressing selenocysteine lyase or selenocysteine methyltransferase exhibit enhanced potential for selenium phytoremediation under field conditions. *Environmental Science & Technology*, *41*, 599–605.

Bañuelos, G., Terry, N., LeDuc, D. L., Pilon-Smits, E. A. H., & Mackey, B. (2005). Field trial of transgenic Indian mustard plants shows enhanced phytoremediation of selenium contaminated sediment. *Environmental Science & Technology*, *39*, 1771–1777.

Bañuelos, G. S. (2009). Phytoremediation of selenium-contaminated soil and water produces biofortified products and new agricultural byproducts. In G. S. Banuelos & Z. -Q. Lin (Eds.), *Biofortification and development of new agricultural products* (pp. 57–60). Boca Roca, FL: CRC Press.

Barac, T., Taghavi, S., Borremans, B., Provoost, A., Oeyen, L., & Colpaert, J. V., et al. (2004). Engineered endophytic bacteria improve phytoremediation of water-soluble, volatile, organic pollutants. *Nature Biotechnology*, *22*, 583–588.

Bentley, R., & Chasteen, T. G. (2002). Microbial methylation of metalloids: Arsenic, antimony, and bismuth. *Microbiology and Molecular Biology Reviews*, *66*, 250–271.

Bernier, M., & Carpentier, R. (1995). The action of mercury on the binding of extrinsic polypeptides associated with water oxidizing complex of photosystem II. *FEBS Letters*, *360*, 251–254.

Bienert, G. P., Thorsen, M., Schüssler, M. D., Nilsson, H. R., Wagner, A., & Tamás, M. J., et al. (2008). A subgroup of plant aquaporins facilitate the bi-directional diffusion of As $(OH)_3$ and $Sb(OH)_3$ across membranes. *BMC Biology*, *6*, 26.

Bizily, S., Rugh, C. C., & Meagher, R. B. (2000). Phytoremediation of hazardous organomercurials by genetically engineered plants. *Nature Biotechnology*, *18*, 213–217.

Bizily, S., Rugh, C. C., Summers, A. O., & Meagher, R. B. (1999). Phytoremediation of methylmercury pollution: *MerB* expression in *Arabidopsis thaliana* plants confer resistance to organomercurial. *Proceedings of the National Academy of Sciences of the United States of America*, *96*, 6808–6813.

Bizily, S. P., Kim, T., Kandasamy, M. K., & Meagher, R. B. (2003). Subcellular targeting of methylmercury lyase enhances its specific activity for organic mercury detoxification in plants. *Plant Physiology*, *131*, 463–471.

Bleeker, P. M., Hakvoort, H. W., Bliek, M., Souer, E., & Schat, H. (2006). Enhanced arsenate reduction by a CDC25-like tyrosine phosphatase explains increased phytochelatin accumulation in arsenate-tolerant *Holcus lanatus*. *Plant Journal*, *45*, 917–929.

Bode, M., Stobe, P., Thiede, B., Schuphan, I., & Schmidt, B. (2004). Biotransformation of atrazine in transgenic tobacco cell culture expressing human P450. *Pest Management Science*, *60*, 49–58.

Boischio, A. A., & Henshel, D. S. (1996). Mercury contamination in the Brazilian Amazon. Environmental and occupational aspects. *Water Air Soil Pollution*, *80* 109–107.

Boivin, A., Cherrier, R., & Schiavon, M. (2005). A comparison of five pesticides adsorption and desorption processes in thirteen contrasting field soils. *Chemosphere*, *61*, 668–676.

Borja, J., Taleon, D. M., Auresenia, J., & Gallardo, S. (2005). Polychlorinated biphenyls and their biodegradation. *Process Biochemistry*, *40*, 1999–2013.

Bouldin, J., Farris, J., Moore, M. T., SJr, Smith, & Cooper, C. M. (2006). Hydroponic uptake of atrazine and lambda-dyhalothrin in *Juncus effusus* and *Ludwigia peploides*. *Chemosphere*, *65*, 1049–1057.

Bumpus, J. A., Kakar, S. N., & Coleman, R. D. (1993). Fungal degradation of organophosphorous insecticides. *Applied Biochemistry and Biotechnology*, *39*, 715–726.

Carey, A. -M., Scheckel, K. G., Lombi, E., Newville, M., & Choi, Y., et al. (2010). Grain unloading of arsenic species in rice. *Plant Physiology*, *152*, 309–319.

Castlehouse, H., Smith, C., Raab, A., Deacon, C., Meharg, A. A., & Feldmann, J. (2003). Biotransformation and accumulation of arsenic in soil amended with seaweed. *Environmental Science & Technology*, *37*, 951–957.

Catarecha, P., Segura, M. D., Franco-Zorrilla, J. M., Garcia-Ponce, B., & Lanza, M., et al. (2007). A mutant of the *Arabidopsis* phosphate transporter PHT1;1 displays enhanced arsenic accumulation. *Plant Cell*, *19*, 1123–1133.

Cervantes, C., Ji, G., Ramirez, J. L., & Silver, S. (1994). Resistance to arsenic compounds in microorganisms. *FEMS Microbiology Reviews*, *15*, 355–367.

Chang, S. W., Lee, S. J., & Je, C. H. (2005). Phytoremediation of atrazine by poplar trees: Toxicity, uptake, and transformation. *Journal of Environmental Science and Health*, *40*, 801–811.

Che, D., Meagher, R. B., Heaton, A. C., Lima, A., Rugh, C. L., & Merkle, S. A. (2003). Expression of mercuric ion reductase in Eastern cottonwood (*Populus deltoides*) confers mercuric ion reduction and resistance. *Plant Biotechnology Journal*, *1*, 311–319.

Chekol, T., Vough, L. R., & Chaney, R. L. (2004). Phytoremediation of polychlorinated biphenyl-contaminated soils: The rhizosphere effect. *Environment International*, *30*, 799–804.

Chen, C. M., Misra, T. K., Silver, S., & Rosen, B. P. (1986). Nucleotide sequence of the structural genes for an anion pump. The plasmid-encoded arsenical resistance operon. *Journal of Biological Chemistry*, *261*, 15030–15038.

Chen, Y. Q., Adam, A., Toure, O., & Dutta, S. K. (2005). Molecular evidence of genetic modification of *Sinorhizobium meliloti*: Enhanced PCB bioremediation. *Journal of Industrial Microbiology & Biotechnology*, *32*, 561–566.

Chen, Z., Zhu, Y. G., Liu, W. J., & Meharg, A. (2005). Direct evidence showing the effect of root surface iron plaque on arsenite and arsenate uptake into rice (*Oryza sativa*) roots. *New Phytologist*, *165*, 91–97.

Chirnside, A. E. M., Ritter, W. F., & Radosevich, M. (2007). Isolation of a selected microbial consortium from a pesticide-contaminated mix-load site soil capable of degrading the herbicides atrazine and alachlor. *Soil Biology and Biochemistry*, *39*, 3056–3065.

Crocker, F. H., Indest, K. J., & Fredrickson, H. L. (2006). Biodegradation of the cyclic nitramine explosives RDX, HMX, and CL-20. *Applied Microbiology and Biotechnology*, *73*, 274–290.

Das, H. K., Mitra, A. K., Sengupta, P. K., Hossain, A., Islam, F., & Rabbani, G. H. (2004). Arsenic concentrations in rice, vegetables, and fish in Bangladesh: A preliminary study. *Environmental International*, *30*, 383–387.

de Carcer, D. A., Martin, M., Mackova, M., Macek, T., Karlson, U., & Rivilla, R. (2007). The introduction of genetically modified microorganisms designed for rhizoremediation induces changes on native bacteria in the rhizosphere but not in the surrounding soil. *The ISME Journal*, *1*, 215–223.

Dean, J. G., Bosqui, F. L., & Lanouette, V. H. (1972). Removing heavy metals from waste water. *Environmental Science & Technology*, *6*, 518–522.

Dhankher, O. P. (2005). Arsenic metabolism in plants: An inside story. *New Phytologist*, *168*, 503–505.

Dhankher, O. P., Li, Y., Rosen, B. P., Shi, J., Salt, D., & Senecoff, J. F., et al. (2002). Engineered tolerance and hyperaccumulation of arsenic in plants by combining arsenate reductase and γ-glutamylcysteine synthetase expression. *Nature Biotechnology*, *20*, 1140–1145.

Dhankher, O. P., Rosen, B. P., McKinney, E. C., & Meagher, R. B. (2006). Enhanced arsenic uptake in *Arabidopsis* plants by suppressing endogenous arsenate reductase AtACR2 gene. *Proceedings of the National Academy of Sciences of the United States of America*, *103*, 5413–5418.

Doty, S. L. (2008). Tansley Review: Enhancing phytoremediation through the use of transgenics and endophytes. *New Phytologist*, *179*, 318–333.

Doty, S. L., James, C. A., Moore, A. L., Vajzovic, A., Singleton, G. L., & Ma, C., et al. (2007). Enhanced phytoremediation of volatile environmental pollutants with transgenic trees. *Proceedings of the National Academy of Sciences of the United States of America*, *104*, 16816–16821.

Doty, S. L., Shang, Q. T., Wilson, A. M., Moore, A. L., Newman, L. A., & Strand, S. E., et al. (2003). Metabolism of the soil and groundwater contaminants, ethylene dibromide and trichloroethylene, by the tropical leguminous tree, *Leucaena leucocephala*. *Water Research*, *37*, 441–449.

Doty, S. L., Shang, Q. T., Wilson, A. M., Tangen, J., Westergreen, A., & Newman, L. A., et al. (2000). Enhanced metabolism of halogenated hydrocarbons in transgenic plants containing mammalian P450 2E1. *Proceedings of the National Academy of Sciences of the United States of America*, *97*, 6287–6291.

Dowling, D. N., & Doty, S. L. (2009). Improving phytoremediation through biotechnology. *Current Opinion in Biotechnology*, *20*, 204–206.

Duan, G. -L., Zhou, Y., Tong, Y. P., Mukhopadhyay, R., Rosen, B. P., & Zhu, Y. G. (2007). A CDC25 homologous from rice functions as an arsenate reductase. *New Phytologist*, *174*, 31–321.

Ellis, D. R., Gumaelius, L., Indriolo, E., Pickering, I. J., Banks, J. A., & Salt, D. E. (2006). A novel arsenate reductase from the arsenic hyperaccumulating fern *Pteris vittata*. *Plant Physiology*, *141*, 1544–1554.

Ellis, D. R., Sors, T. G., Brunk, D. G., Albrecht, C., Orser, C., & Lahner, B., et al. (2004). Production of Se-methylselenocysteine in transgenic plants expressing selenocysteine methyltransferase. *BMC Plant Biology*, *4*, 1–11.

Fageria, N. K. (2007). Yield physiology of rice. *Journal of Plant Nutrition*, *30*, 843–879.

Francova, K., Sura, M., Macek, T., Szekeres, M., Bancos, S., & Demnerova, K., et al. (2003). Preparation of plants containing bacterial enzyme for degradation of polychlorinated biphenyls. *Fresenius Environment Bulletin*, *12*, 309–313.

French, C. E., Rosser, S. J., Davies, G. J., Nicklin, S., & Bruce, N. C. (1999). Biodegradation of explosives by transgenic plants expressing pentaerythritol tetranitrate reductase. *Nature Biotechnology*, *17*, 491–494.

Gandia-Herrero, F., Lorenz, A., Larson, T., Graham, I. A., Bowles, D. J., & Rylott, E. L., et al. (2008). Detoxification of the explosive 2,4,6-trinitrotoluene in *Arabidopsis*: Discovery of bifunctional O-and C-glucosyltransferases. *The Plant Journal*, *56*, 963–974.

Garifullina, G. F., Owen, J. D., Lindblom, S. -D., Tufan, H., Pilon, M., & Pilon-Smits, E. A. H. (2003). Expression of a mouse selenocysteine lyase in *Brassica juncea* chloroplasts affects selenium tolerance and accumulation. *Physiologia Plantarum*, *118*, 538–544.

Gasic, K., & Korban, S. S. (2007). Transgenic Indian mustard (*Brassica juncea*) plants expressing an *Arabidopsis* phytochelatin synthase (AtPCS1) exhibit enhanced As and Cd tolerance. *Plant Molecular Biology*, *64*, 361–369.

Germaine, K. J., Keogh, E., Ryan, D., & Dowling, D. N. (2009). Bacterial endophyte-mediated naphthalene phytoprotection and phytoremediation. *FEMS Microbiology Letters*, *296*, 226–234.

Germaine, K. J., Liu, X., Cabellos, G. G., Hogan, J. P., Ryan, D., & Dowling, D. N. (2006). Bacterial endophyte-enhanced phytoremediation of the organochlorine herbicide 2,4-dichlorophenoxyacetic acid. *FEMS Microbiology Ecology*, *57*, 302–310.

Ghanum, I., Orfi, M., & Shamma, M. (2007). Biodegradation of chlorpyrifos by *Klebsiella* sp. isolated from an activated sludge sample of waste water treatment plant in Damascus. *Folia Microbiologica*, *52*, 423–427.

Gresser, M. J. (1981). ADP-arsenate formation by submitochondrial particles under phosphorylating conditions. *Journal of Biological Chemistry*, *256*, 5981–5983.

Guo, J. B., Dai, X. J., Xu, W. Z., & Ma, M. (2008). Overexpressing *GSH1* and *AsPCS1* simultaneously increases the tolerance and accumulation of cadmium and arsenic in *Arabidopsis thaliana*. *Chemosphere*, *72*, 1020–1026.

Hannink, N., Rosser, S. J., French, C. E., Basran, A., Murray, J. A. H., & Nicklin, S., et al. (2001). Phytodetoxification of TNT by transgenic plants expressing a bacterial nitroreductase. *Nature Biotechnology*, *19*, 1168–1172.

Harada, M. (1995). Minamata disease: Methylmercury poisoning in Japan caused by environmental pollution. *Critical Reviews in Toxicology*, *25*, 1–24.

Harms, H., Bokern, M., Kolb, M., & Bock, C. (2003). Transformation of organic contaminants by different plant systems. In S. C. McCutcheon & J. L. Schnoor (Eds.), *Phytoremediation. Transformation and control of contaminants* (pp. 285–316). Hoboken, NJ: John Whiley.

Heaton, A. C., Rugh, C. C., Kim, T., & Meagher, R. B. (2003). Toward detoxifying mercury-polluted aquatic sediments with rice genetically engineered for mercury resistance. *Environmental Toxicology and Chemistry*, *22*, 2940–2947.

Heaton, A. C. P., Rugh, C. L., Wang, N. J., & Meagher, R. B. (2005). Physiological responses of transgenic *merA* tobacco (*Nicotiana tabacum*) to foliar and root mercury exposure. *Water, Air, & Soil Pollution*, *161*, 137–155.

Heaton, A. C. P., Rugh, C. L., Wang, N., & Meagher, R. B. (1998). Phytoremediation of mercury – and methylmercury – polluted soils using genetically engineered plants. *Journal of Soil Contamination*, *7*, 497–510.

Hughes, M. F. (2002). Arsenic toxicity and potential mechanisms of action. *Toxicology Letters*, *133*, 1–16.

Hussain, S., Siddique, T., Arshad, M., & Saleem, M. (2009). Bioremediation and phytoremediation of pesticides: Recent advances. *Critical Reviews in Environmental Science and Technology*, *39*, 843–907.

Hussein, H., Ruiz, O. N., Terry, N., & Daniell, H. (2007). Phytoremediation of mercury and organomercurials in chloroplast transgenic plants: Enhanced root uptake, translocation to shoots and volatilization. *Environmental Science & Technology*, *41*, 8439–8446.

Indriolo, E., Na, G. N., Ellis, D., Salt, D. E., & Banks, J. A. (2010). A vacuolar arsenite transporter necessary for arsenic tolerance in the arsenic hyperaccumulating fern *Pteris vittata* is missing in flowering plants. *Plant Cell*, *22*, 2045–2057.

Isayenkov, S. V., & Maathuis, F. J. M. (2008). The *Arabidopsis thaliana* aquaglyceroporin AtNIP7;1 is a pathway for arsenite uptake. *FEBS Letters*, *582*, 1625–1628.

Itoh, K., Kanda, R., Sumita, Y., Kim, H., Kamagata, Y., & Suyama, K., et al. (2002). tfdA-like genes in 2,4-dichlorophenoxyacetic acid-degrading bacteria belonging to the Bradyrhizobium-Agromonas-Nitrobacter-Afipia cluster in alpha-Proteobacteria. *Applied and Environmental Microbiology*, *68*, 3449–3454.

Jackson, R. G., Rylott, E. L., Fournier, D., Hawari, J., & Bruce, N. C. (2007). Exploring the biochemical properties and remediation applications of the unusual explosive-degrading P450 system XplA/B. *Proceedings of the National Academy of Sciences of the United States of America*, *104*, 16822–16827.

James, C. A., & Strand, S. E. (2009). Phytoremediation of small organic contaminants using transgenic plants. *Current Opinion in Biotechnology*, *20*, 237–241.

James, C. A., Xin, G., Doty, S. L., & Strand, S. E. (2008). Degradation of low molecular weight volatile organic compounds by plants genetically modified with mammalian cytochrome P450 2E1. *Environmental Science & Technology*, *42*, 289–293.

Juhasz, A. L., Naidu, R., Zhu, Y. G., Wang, L. S., Jiang, J. Y., & Cao, Z. H. (2003). Toxicity issues associated with geogenic arsenic in the grondwater–soil–plant–human continuum. *Bulletin of Environmental Contamination and Toxicology*, *71*, 1100–1107.

Kaiser, J. (2001). Second look at arsenic finds higher risk. *Science*, *293*, 2189.

Kang, J. W., Wilkerson, H. W., Farin, F. M., Bammler, T. K., & Beyer, R. P., et al. (2010). Mammalian cytochrome CYP2E1 triggered differential gene regulation in response to trichloroethylene (TCE) in a transgenic poplar. *Functional & Integrative Genomics*, *10*, 417–424.

Kannan, K., Smith, R. G., Lee, R. F., Windom, H. L., Heimuller, P. T., & Macauley, J. M., et al. (1998). Distribution of total mercury and methyl mercury in water, sediment and fish from South Florida estuaries. *Archives of Environmental Contamination and Toxicology*, *34*, 109–118.

Kawahigashi, H., Hirose, S., Ohkawa, H., & Ohkawa, Y. (2006). Phytoremediation of the herbicides atrazine and metolachlor by transgenic rice plants expressing human CYP1A1, CYP2B6, and CYP2C19. *Journal of Agricultural and Food Chemistry*, *54*, 2985–2991.

Keating, M.H., Mahaffey, K.R., Schoeny, R., Rice, G.E., Bullock, O.R., Ambrose, R.B., et al., (1997). *Mercury Study Report to Congress* (1–51). EPA452/R-9-003. US Environmental Protection Agency.

Kraemer, U., & Chardonnens, A. N. (2001). The use of transgenic plants in the bioremediation of soil contaminated with trace elements. *Applied Microbiology and Biotechnology*, *55*, 661–672.

Kucerova, P., Mackova, M., Chroma, L., Burkhard, J., Triska, J., & Demnerova, K., et al. (2000). Metabolism of polychlorinated biphenyls by *Solanum nigrum* hairy root clone SNC-9O and analysis of transformation products. *Plant and Soil*, *225*, 109–115.

Kupper, H., Kupper, F., & Spiller, M. (1996). Environmental relevance of heavy metal substituted chlorophylls using the example of water plants. *Journal of Experimental Botany*, *47*, 259–266.

Lagunas, R. (1980). Sugar-arsenate esters: Thermodynamics and biochemical behavior. *Archives of Biochemistry and Biophysics*, *205*, 67–75.

Larsen, E. H., Moseholm, L., & Nielsen, M. M. (1992). Atmospheric deposition of trace elements around point sources and human healthy risk assessment: II. Uptake of arsenic and chromium by vegetables grown near a wood preservation factory. *Science of the Total Environment*, *126*, 263–275.

LeDuc, D. L., AbdelSamie, M., Montes-Bayón, M., Wu, C. P., Reisinger, S. J., & Terry, N. (2006). Overexpressing both ATP sulfurylase and selenocysteine methyltransferase enhances selenium phytoremediation traits in Indian mustard. *Environmental Pollution*, *144*, 70–76.

LeDuc, D. L., Tarun, A. S., Montes-Bayon, M., Meija, J., Malit, M. F., & Wu, C. P., et al. (2004). Overexpression of selenocysteine methyltransferase in *Arabidopsis* and Indian mustard increases selenium tolerance and accumulation. *Plant Physiology*, *135*, 377–383.

Lee, I., & Fletcher, J. S. (1992). Involvement of mixed-function oxidase systems in polychlorinated biphenyl metabolism by plant cells. *Plant Cell Reports*, *11*, 97–100.

Lee, K.Y., Strand, S.E., Doty, S.L. (2012). Phytoremediation of chlorpyrifos by *Populus* and *Salix*. *International Journal of Phytoremediation*, *14*(1), 48–61. Available online 2 Aug. 2011, DOI: 10.1080/15226514.2011.560213.

Li, Y., Dhankher, O., Carreira, L., Balish, R., & Meagher, R. (2005). Engineered over-expression of γ-glutamylcysteine synthetase in plants confers high levels arsenic and mercury tolerance. *Environmental Toxicology & Chemistry*, *24*, 1376–1386.

Li, Y. J., Dhankher, O. P., Carreira, L., Lee, D., & Chen, A., et al. (2004). Overexpression of phytochelatin synthase in *Arabidopsis* leads to enhanced arsenic tolerance and cadmium hypersensitivity. *Plant and Cell Physiology*, *45*, 1787–1797.

Liu, J. Y., & Schnoor, J. L. (2008). Uptake and translocation of lesser-chlorianted polychlorinated biphenyls (PCBs) in whole hybrid poplar plants after hydroponic exposure. *Chemosphere*, *73*, 1608–1616.

Liu, Z., Shen, J., Carbrey, J. M., Mukhopadhyay, R., Agre, P., & Rosen, B. P. (2002). Arsenite transport by mammalian aquaglyceroporins AQP7 and AQP9. *Proceedings of the National Academy of Sciences of the United States of America*, *99*, 6053–6058.

Lytle, J. S., & Lytle, T. F. (2000). Uptake and loss of chlorpyrifos and atrazine by

Juncus effusus L. in a mesocosm study with a mixture of pesticides. *Environmental Toxicology and Chemistry*, *21*, 1817–1825.

Lyyra, S., Meagher, R. B., Kim, T., Heaton, A., Montello, P., & Balish, R. S., et al. (2007). Coupling two mercury resistance genes in Eastern cottonwood enhances the processing of organomercury. *Plant Biotechnology Journal*, *5*, 254–262.

Ma, J. F., Yamaji, N., Mitani, N., Xu, X. Y., Su, Y. H., & McGrath, S. P., et al. (2008). Transporters of arsenite in rice and their role in arsenic accumulation in rice grain. *Proceedings of the National Academy of Sciences of the United States of America*, *105*, 9931–9935.

Ma, L. Q., Komar, K. M., Tu, C., Zhang, W. H., Cai, Y., & Kennelley, E. D. (2001). A fern that hyperaccumulates arsenic: A hardy versatile fast-growing plant helps to remove arsenic from contaminated soils. *Nature*, *409* 579–579.

Mackova, M., Macek, T., Ocenaskova, J., Burkhard, J., Demnerova, K., & Pazlarova, J. (1997). Biodegradation of polychlorinated biphenyls by plant cells. *International Biodeterioration & Biodegradation*, *39*, 317–325.

Marin, A. R., Masscheleyn, P. H., & Patrik, W. H., Jr. (1993). Soil redox-pH stability of arsenic and its influence on arsenic uptake in rice. *Plant & Soil*, *152*, 245–253.

Meagher, R.B. Rugh, C.L. (1998). Phytoremediation of heavy metal pollution: Ionic and methyl mercury. In Organization for Economic Cooperation and Development Document: Biotechnology for Water Use and Conservation. The Mexico '96 Workshop, Organization for Economic Cooperation and Development, Paris, pp. 305–321.

Meharg, A. A. (2004). Arsenic in rice: Understanding a new disaster for South-East Asia. *Trends in Plant Science*, *9*, 415–417.

Meharg, A. A., & Macnair, M. R. (1992). Suppression of the high-affinity phosphate-uptake system: A mechanism of arsenate tolerance in *Holcus lanatus* L. *Journal of Experimental Botany*, *43*, 519–524.

Meharg, A. A., & Rahman, M. (2003). Arsenic contamination of Bangladesh paddy field soils: Implications for rice contribution to As consumption. *Environmental Science & Technology*, *37*, 229–234.

Meharg, A. A., Williams, P. N., Adomako, E., Lawgali, Y. Y., Deacon, D., & Villada, A., et al. (2009). Geographical variation in total and inorganic As content of polished (white) rice. *Environmental Science & Technology*, *43*, 1612–1617.

Merini, L. J., Bobillo, C., Cuadrado, V., Corach, D., & Giulietti, A. M. (2009). Phytoremediation potential of the novel atrazine tolerant *Lolium multiflorum* and studies on the mechanisms involved. *Environmental Pollution*, *157*, 3059–3063.

Mezzari, M. P., Walters, K., Jelinkova, M., Shih, M. -C., Just, C. L., & Schnoor, J. L. (2005). Gene expression and microscopic analysis of *Arabidopsis* exposed to chloroacetanilide herbicides and explosive compounds. A phytoremediation approach. *Plant Physiology*, *138*, 858–869.

Mohammadi, M., Chalavi, V., Novakova-Sura, M., Laliberte, J. F., & Sylvestre, M. (2007). Expression of bacterial biphenyl-chlorobiphenyl dioxygenase genes in tobacco plants. *Biotechnology and Bioengineering*, *97*, 496–505.

Moore, F.P., Barac, T., Borremans, B., Oeyen, L., Vangronsveld, J., van der, L.D., et al. (2006). Endophytic bacterial diversity in poplar trees growing on a BTEX-contaminated site: The characterisation of isolates with potential to enhance phytoremediation. *Systematic and Applied Microbiology 29*(7), 539–556.

Moore, M. T., Schulz, R., Cooper, C. M., Smith, S., Jr, & Rodgers, J. H., Jr. (2002). Mitigation of chlorpyrifos runoff using constructed wetlands. *Chemosphere*, *46*, 827–835.

Mukhopadhyay, R., & Rosen, B. P. (2002). The phosphatase C(X)5R motif is required for catalytic activity of the *Saccharomyces cerevisiae* Acr2p arsenate reductase. *The Journal of Biological Chemistry*, *28*, 34738–34742.

Musarrat, J., Bano, N., & Rao, R. A. K. (2000). Isolation and characterization of 2,4-dichlorophenoxyacetic acid-catabolizing bacteria and their biodegradation efficiency in soil. *World Journal of Microbiology and Biotechnology*, *16*, 495–497.

Neuhierl, B., Thanbichler, M., Lottspeich, F., & Böck, A. (1999). A family of S-methylmethionine dependent thiol/selenol methyltransferases. Role in selenium tolerance and evolutionary relation. *Journal of Biological Chemistry*, *274*, 5407–5414.

Novakova, M., Mackova, M., Chrastilova, Z., Viktorova, J., Szekeres, M., & Demnerova, K., et al. (2009). Cloning the bacterial bphC gene into *Nicotiana tabacum* to improve the efficiency of PCB-phytoremediation. *Biotechnology and Bioengineering*, *102*, 29–37.

Nriagu, J. O. (1993). Legacy of Mercury. *Nature*, *363*, 589.

Osborn, A. M., Bruce, K. D., Strike, P., & Ritchie, D. A. (1997). Distribution, diversity and evolution of the bacterial mercury resistance (*mer*) operon. *FEMS Microbiology Reviews*, *19*, 239–262.

Patra, M., & Sharma, A. (2000). Mercury toxicity in plants. *Botany Review*, *66*, 379–422.

Paulose, B., Kandasamy, S., & Dhankher, O. P. (2010). Exprerssion profiling of *Crambe abyssinica* under arsenate stress identifies genes and gene networks involved in arsenic metabolism and detoxification. *BMC Plant Biology*, *10*, 108.

Pickering, I. J., Prince, R. C., George, M. J., Smith, R. D., George, G. N., & Salt, D. E. (2000). Reduction and coordination of arsenic in Indian mustard. *Plant Physiology*, *122*, 1171–1177.

Pieper, D. H., & Seeger, M. (2008). Bacterial metabolism of polychlorinated biphenyls. *Journal of Molecular Microbiology and Biotechnology*, *15*, 121–138.

Pilon, M., Owen, J. D., Garifullina, G. F., Kurihara, T., Mihara, H., & Esaki, N., et al. (2003). Enhanced selenium tolerance and accumulation in transgenic *Arabidopsis thaliana* expressing a mouse selenocysteine lyase. *Plant Physiology*, *131*, 1250–1257.

Pilon-Smits, E. A. H. (2005). Phytoremediation. *Annual review of Plant Biology*, *56*, 15–39.

Pilon-Smits, E. A. H., Hwang, S., Lytle, C. M., Zhu, Y., Tai, J. C., & Bravo, R. C., et al. (1999). Overexpression of ATP sulfurylase in Indian mustard leads to increased selenate uptake, reduction, and tolerance. *Plant Physiology*, *119*, 123–132.

Pilon-Smits, E. A. H., & Quinn, C. F. (2010). Selenium Metabolism in Plants. In R. Hell & R. Mendel (Eds.), *Cell biology of metals and nutrients* (pp. 225–241). Berlin Heidelberg: Springer Press.

Qin, J., Lehr, C. R., Yuan, C. G., Le, X. C., McDermott, T. R., & Rosen, B. P. (2009). Biotransformation of arsenic by a Yellowstone thermoacidophilic eukaryotic alga. *Proceedings of the National Academy of Sciences of the United States of America*, *106*, 5213–5217.

Qin, J., Rosen, B. P., Zhang, Y., Wang, G. J., Franke, S., & Rensing, C. (2006). Arsenic detoxification and evolution of trimethylarsine gas by a microbial arsenite S-adenosylmethionine methyltransferase. *Proceedings of the National Academy of Sciences of the United States of America*, *103*, 2075–2080.

Ravenscroft, P., Brammer, H., & Richards, K. S. (2009). *Arsenic pollution: A global synthesis*. U.K: Wiley-Blackwell.

Rezek, J., Macek, T., Mackova, M., & Triska, J. (2007). Plant metabolites of polychlorinated biphenyls in hairy root culture of black nightshade *Solanum nigrum* SNC-90. *Chemosphere*, *69*, 1221–1227.

Rosen, B. P. (2002). Biochemistry of arsenic detoxification. *FEBS Letters*, *529*, 86–92.

Rugh, C. C., Summers, A. O., & Meagher, R. B. (1996). Mercuric ion reductase and resistance in transgenic *Arabidopsis thaliana* expressing modified bacterial merA gene. *Proceedings of the National Academy of Sciences of the United States of America*, *93*, 3182–3187.

Rugh, C. L., Senecoff, J. F., Meagher, R. B., & Merkle, S. A. (1998). Development of transgenic yellow poplar for mercury phytoremediation. *Nature Biotechnology*, *16*, 925–928.

Ruiz, O. N., Hussein, H. S., Terry, N., & Daniell, H. (2003). Phytoremediation of organomercurials via the chloroplast genetic engineering. *Plant Physiology*, *132*, 1344–1352.

Russell, R. J., Sutherland, T. D., Horne, I., Oakeshott, J. G., Zachariou, M., & Nguyen, H. V., et al. (2001). Enzymatic biodegradation of chemical pesticides. *Australasian Biotechnology*, *11*, 24–26.

Rylott, E. L., & Bruce, N. C. (2008). Plants disarm soil: Engineering plants for the phytoremediation of explosives. *Trends in Biotechnology*, *27*, 73–81.

Rylott, E. L., Jackson, R. G., Edwards, J., Womack, G. L., Seth-Smith, H. M., & Rathbone, D. A., et al. (2006). An explosive-degrading cytochrome P450 activity and its targeted application for the phytoremediation of RDX. *Nature Biotechnology*, *24*, 216–219.

Sajjaphan, K., Shapir, N., Wackett, L. P., Palmer, M., Blackmon, B., & Tomkins, J., et al. (2004). *Arthrobacter aurescens* TC1 atrazine catabolism genes trzN, atzB, and atzC are linked on a 160-kilobase region and are functional in *Escherichia coli*. *Applied and Environmental Microbiology*, *70*, 4402–4407.

Sanders, O. I., Rensing, C., Kuroda, M., Mitra, B., & Rosen, B. P. (1997). Antimonite is accumulated by the glycerol facilitator GlpF in *Escherichia coli*. *Journal of Bacteriology*, *179*, 3365–3367.

Satsuma, K. (2006). Characterization of new strains of atrazine-degrading *Nocardiodes* sp. isolated from Japanese riverbed sediment using naturally derived river ecosystem. *Pest Management Science*, *62*, 340–349.

Shang, Q. T., Doty, S. L., Wilson, A. M., Howald, W. N., & Gordon, M. P. (2001). Trichloroethylene oxidative metabolism in plants: The trichloroethanol pathway. *Phytochemistry*, *58*, 1055–1065.

Shapir, N., Osborne, J. P., Johnson, G., Sadowsky, M. J., & Wackett, L. P. (2002). Purification, substrate range, and metal center of AtzC: The N-isopropylammelide aminohydrolase involved in bacterial atrazine metabolism. *Journal of Bacteriology*, *184*, 5376–5384.

Shields, MS, Reagin, MJ, Gerger, RR, Campbell, R, & Somerville, C. (1995). TOM, a new aromatic degradative plasmid from *Burkholderia* (*Pseudomonas*) cepacia G4. *Applied and Environmental Microbiology*, *61*, 1352–1356.

Shin, H., Shin, H. S., Dewbre, G. R., & Harrison, M. J. (2004). Phosphate transport in *Arabidopsis*: Pht1;1 and Pht1;4 play a major role in phosphate acquisition from both low- and high-phosphate environments. *Plant Journal*, *39*, 629–642.

Siciliano, S. D., Germida, J. J., Banks, K., & Greer, C. W. (2003). Changes in microbial community composition and function during a polyaromatic hydrocarbon phytoremediation field trial. *Applied and Environmental Microbiology*, *69*, 483–489.

Singh, B. K., Walker, A., & Wright, D. J. (2006). Bioremedial potential of fenamiphos and chlorpyrifos degrading isolates: Influence of different environmental conditions. *Soil Biology & Biochemistry*, *38*, 2682–2693.

Sinha, S., Gupta, M., & Chandra, P. (1996). Bioaccumulation and biochemical effects of mercury in the plant *Bacopa monnieri* (L). *Environmental Toxicology and Water Quality*, *11*, 105–112.

Sors, T. G., Ellis, D. R., & Salt, D. E. (2005). Selenium uptake, translocation, assimilation and metabolic fate in plants. *Photosynthesis Research*, *86*, 373–389.

Stevens, R. C., Suzuki, S. M., Cole, T. B., Park, S. S., Richter, R. J., & Furlong, C. E. (2008). Engineered recombinant human paraoxonase 1 (rHuPON1) purified from *Escherichia coli* protects against organophosphate poisoning. *Proceedings of the National Academy of Sciences of the United States of America*, *105*, 12780–12784.

Summers, A. (1992). Untwist and shout: A heavy metal-responsive transcriptional regulator. *Journal of Bacteriology*, *174*, 3097–3101.

Sundaram, S., Rathinasabapathi, B., Ma, L. Q., & Rosen, B. P. (2008). An arsenate-activated glutaredoxin from the arsenic hyperaccumulator fern *Pteris vittata* L. regulates intracellular arsenite. *Journal of Biological Chemistry*, *283*, 6095–6101.

Sutherland, T. D., Horne, I., Weir, K. M., Coppin, C. W., Williams, M. R., & Selleck, M., et al. (2004). Enzymatic bioremediation: From enzyme discovery to applications. *Clinical and Experimental Pharmacology & Physiology*, *31*, 817–821.

Sylvestre, M., Macek, T., & Mackova, M. (2009). Transgenic plants to improve rhizoremediation of polychlorinated biphenyls (PCB's). *Current Opinion in Biotechnology*, *20*, 242–247.

Taghavi, S., Barac, T., Greenberg, B., Borremans, B., Vangronsveld, J., & van der Lelie, D. (2005). Horizontal gene transfer to endogenous endophytic bacteria from poplar improves phytoremediation of toluene. *Applied and Environmental Microbiology*, *71*, 8500–8505.

Talmage, S. S., Opresko, D. M., Maxwell, C. J., Welsh, C. J., Cretella, F. M., & Reno, P. H., et al. (1999). Nitroaromatic munition compounds: Environmental effects and screening values. *Reviews of Environment Contamination and Toxicology*, *161*, 1–156.

Terry, N., Zayed, A. M., de Souza, M. P., & Tarun, A. S. (2000). Selenium in higher plants. *Annual Review of Plant Physiology and Plant Molecular Biology*, *51*, 401–432.

Thygesen, R., & Trapp, S. (2002). Phytotoxicity of polycyclic aromatic hydrocarbons to willow trees. *Journal of Soils and Sediments*, *2*, 77–82.

Topp, E. (2001). A comparison of tree atrazine-degrading bacteria for soil bioremediation. *Biology and Fertility of Soils*, *33*, 529–534.

Travis, E. R., Hannink, N. K., Van der Gast, C. J., Thompson, I. P., Rosser, S. J., & Bruce, N. C. (2007). Impact of transgenic tobacco on trinitrotoluene (TNT) contaminated soil community. *Environmental Science & Technology*, *41*, 5854–5861.

Tripathi, R. D., Srivastava, S., Mishra, S., Singh, N., & Tuli, R., et al. (2007). Arsenic hazards: Strategies for tolerance and remediation by plants. *Trends in Biotechnology*, *25*, 158–165.

US Defense Science Board Task Force, *Unexploded ordnance (UXO) clearance, active range UXO clearance, and explosive ordnance disposal programs*. Washington, D.C.: Office of the Undersecretary of Defense for Acquisition and Technology. Ref Type: Report.

van Aken, B. (2009). Transgenic plants for enhanced remediation of toxic explosives. *Current Opinion in Biotechnology*, *20*, 231–236.

van Aken, B., Correa, P. A., & Schnoor, J. L. (2010). Phytoremediation of polychlorinated biphenyls: New trends and promises. *Environmental Science & Technology*, *44*, 2767–2776.

van Aken, B., Yoon, J. M., & Schnoor, J. L. (2004). Biodegradation of nitro-substituted explosives 2,4,6-trinitrotoluene, hexahydro-1,3,5-trinitro-1,3,5-triazine, and octahydro-1,3,5,7-tetranitro-1,3,5-tetrazocine by a phytosymbiotic *Methylobacterium* sp. associated with poplar tissues (*Populus deltoides* x *nigra* DN34). *Applied and Environmental Microbiology*, *70*, 508–517.

van Dillewijn, P., Couselo, J., Corredoira, E., Delgado, A., Wittich, R. -M., & Ballester, A., et al. (2008). Bioremediation of 2,4,6-trinitrotoluene by bacterial nitroreductase expressing transgenic aspen. *Environmental Science & Technology*, *42*, 7405–7410.

Van Hoewyk, D., Garifullina, G. F., Ackley, A. R., Abdel-Ghany, S. E., Marcus, M. A., & Fakra, S., et al. (2005). Overexpression of AtCpNifS enhances selenium tolerance and accumulation in *Arabidopsis*. *Plant Physiology*, *139*, 1518–1528.

Van Huysen, T., Abdel-Ghany, S., Hale, K. L., LeDuc, D., Terry, N., & Pilon-Smits, E. A. H. (2003). Overexpression of cystathionine-γ-synthase in Indian mustard enhances selenium volatilization. *Planta*, *218*, 71–78.

Van Huysen, T., Terry, N., & Pilon-Smits, E. A. H. (2004). Exploring the Selenium phytoremediation potential of transgenic *Brassica juncea* overexpressing ATP sulfurylase or cystathionine γ-synthase. *International Journal of Phytoremediation*, *6*, 111–118.

Vatamaniuk, O. K., Bucher, E. A., & Rea, P. A. (2002). Worms take the "phyto" out of "phytochelatins". *Trends in Biotechnology*, *20*, 61–64.

Wackett, L. P., Sadowsky, M. J., Martinez, B., & Shapir, N. (2002). Biodegradation of atrazine and related s-triazine compounds: From enzymes to field studies. *Applied Microbiology and Biotechnology*, *58*, 39–45.

Wang, J., Zhao, F., Meharg, A. A., Raab, A., Feldmann, J., & McGrath, S. P. (2002). Mechanisms of arsenic hyperaccumulation in *Pteris vittata*. Uptake kinetics interactions with phosphate and arsenic speciation. *Plant Physiology*, *130*, 1552–1561.

Wang, L., Samac, D. A., Shapir, N., Wackett, L. P., Vance, C. P., & Olszewski, N. E., et al. (2005). Biodegradation of atrazine in transgenic plants expressing a modified bacterial atrazine chlorohydrolase (atzA) gene. *Plant Biotechnology Journal*, *3*, 475–486.

Weig, A., Deswarte, C., & Chrispeels, M. J. (1997). The major intrinsic protein family of *Arabidopsis* has 23 members that form three distinct groups with functional aquaporins in each group. *Plant Physiology*, *114*, 1347–1357.

Weyens, N., Croes, S., Dupae, J., Newman, L., van der Lelie, D., & Carleer, R., et al. (2010). Endophytic bacteria improve phytoremediation of Ni and TCE co-contamination. *Environmental Pollution*, *158*, 2422–2427.

Weyens, N., van der, L. D., Artois, T., Smeets, K., Taghavi, S., & Newman, L., et al. (2009). Bioaugmentation with engineered endophytic bacteria improves contaminant fate in phytoremediation. *Environmental Science & Technology*, *43*, 9413–9418.

Weyens, N., van der, L. D., Taghavi, S., & Vangronsveld, J. (2009). Phytoremediation: Plant-endophyte partnerships take the challenge. *Current Opinion in Biotechnology*, *20*, 248–254.

Wilken, A., Bock, C., Bokern, M., & Harms, H. (1995). Metabolism of different PCB congeners in plant cell cultures. *Environmental Toxicology and Chemistry*, *14*, 2017–2022.

Williams, P. N., Price, A. H., Raab, A., Hossain, S. A., Feldmann, J., & Meharg, A. A. (2005). Variation in arsenic speciation and concentration in paddy rice related to dietary exposure. *Environmental Science & Technology*, *39*, 5531–5540.

Williams, P. N., Villada, A., Deacon, C., Raab, A., Figuerola, J., & Green, A. J., et al. (2007). Greatly enhanced As shoot assimilation in rice leads to elevated grain levels compared to wheat and barley. *Environmental Science & Technology*, *41*, 6854–6859.

World Health Organization (WHO, 2003). *Elemental mercury and inorganic mercury compounds: Human health aspects*. Geneva, Switzerland.

Wysocki, R., Chery, C. C., Wawrzycka, D., Hulle, M. V., Cornelis, R., & Thevelein, J. M., et al. (2001). The glycerol channel Fps1p mediates the uptake of arsenite and antimonite in *Saccharomyces cerevisiae*. *Molecular Microbiology*, *40*, 1391–1401.

Xiong, X. Z., Li, P. J., Wang, Y. S., Ten, H., Wang, L. P., & Song, S. H. (1987). Environmental capacity of arsenic in soil and mathematical model. *Huanjing Kexue*, *8*, 8–14.

Yang, C., Liu, N., Guo, X., & Qiao, C. (2006). Cloning of mpd gene from a chlorpyrifos-degrading bacterium and use of this strain in bioremediation of contaminated soil. *FEMS Microbiology Letters*, *265*, 118–125.

Zavala, Y. J., & Duxbury, J. M. (2008). Arsenic in rice: I. Estimating normal levels of total arsenic in rice grain. *Environmental Science & Technology*, *42*, 3856–3860.

Zeeb, B. A., Amphlett, J. S., Rutter, A., & Reimer, K. J. (2006). Potential for phytoremediation of polychlorinated biphenyl-(PCB) contaminated soil. *International Journal of Phytoremediation*, *8*, 199–221.

Zhang, W. H., & Tyerman, S. D. (1999). Inhibition of water channels by HgCl2 in intact wheat root cells. *Plant Physiology*, *120*, 849–858.

Zhao, F. J., Ma, J. F., Meharg, A. A., & McGrath, S. (2009). Arsenic uptake and metabolism in plants. *New Phytologist*, *181*, 777–794.

Zhao, F. J., McGrath, S. P., & Meharg, A. A. (2010). Arsenic as a food chain contaminant: Mechanisms of plant uptake and metabolism and mitigation strategies. *Annual Review of Plant Biology*, *61*, 7–25.

Zhao, F. J., Wang, J. R., Barker, J. H. A., Schat, H., Bleeker, P. M., & McGrath, S. P. (2003). The role of phytochelatinsin arsenic tolerance in the hyperaccumulator *Pteris vittata*. *New Phytologist*, *159*, 403–410.

Zhu, Y. G., Williams, P. N., & Meharg, A. (2008). Exposure to inorganic arsenic from rice: A global health issue? *Environmental Pollution*, *154*, 169–171.

Biotechnological strategies for engineering plants with durable resistance to fungal and bacterial pathogens

21

Dor Salomon Guido Sessa
Tel-Aviv University, Tel-Aviv, Israel

TABLE OF CONTENTS

Introduction

Fungal and bacterial pathogens are a major threat to the yield of crop plants worldwide. Therefore, an important goal in plant science is the production of plants with broad-spectrum and durable resistance to microbial diseases. Potential strategies to reach this goal include classical or marker-assisted breeding, and genetic engineering (Michelmore, 2003). While breeding techniques are time-consuming and have the disadvantage of introducing undesired traits by linkage, genetic engineering is relatively fast and allows transference of individual traits into crops in a calculated manner. In addition, genetic engineering may allow for the rapid introduction of desirable traits from other plant species or organisms into crops, and the precise manipulation of temporal or tissue-specific expression of the trait of interest.

Despite the large variety of bacterial and fungal pathogens present in nature, disease is rare and resistance is prevalent as an outcome of plant–pathogen interactions. The efficient plant immune system relies both on pre-formed structural and chemical barriers, and on inducible defense responses that are activated by complex mechanisms of molecular recognition and cellular signal transduction. However, modern agricultural practices based on cultivation of large areas of genetically identical crops enable pathogens to quickly adapt and defeat plant immunity, causing devastating losses in yield. Attempts to limit pathogen outbreaks by classical breeding procedures have provided only temporary control, due to the ability of pathogens to overcome the resistance genes generally employed in breeding programs (McDowell and Woffenden, 2003). Alternative approaches for achieving durable disease resistance are based on genetic engineering of desired traits in transgenic plants. In many cases, such approaches have been undermined by side effects of transgene expression on plant development, yield, and seed production, necessitating refinement in the design of such strategies (Gurr and Rushton, 2005a,b; Gust et al., 2010).

In recent years, genetic and genomics analyses of plants and interacting pathogens significantly advanced our understanding of the molecular basis of microbial infection and plant immunity, providing the opportunity to select new targets for durable resistance. In addition, discovery of pathogen-responsive and tissue-specific promoters may now allow a tight control of transgene expression. These new molecular tools combined with lessons from previous unsuccessful biotechnological attempts have the potential to lead to successful approaches for the achievement of durable disease resistance.

In this chapter we will review critical issues that should be considered in the design of transgenic technologies for durable and broad-spectrum resistance. We will focus our discussion on the choice of target genes to be expressed, on the number of genes to be employed, and on the control of their expression.

DOI: 10.1016/B978-0-12-381466-1.00021-3

Choosing the Target Gene for Transgenic Expression

Our knowledge of the molecular basis of plant immunity and microbial pathogenicity has greatly increased over the last two decades, providing a better understanding of plant–pathogen interactions, and hence of the available targets for disease control in economically important crop plants. The repertoire of genes that may potentially be manipulated to confer durable disease resistance to a broad spectrum of bacterial and fungal pathogens can be divided into two main groups. The first group is represented by plant or pathogen genes that may induce or potentiate plant immunity. This includes genes that encode plant immune receptors, their corresponding pathogen ligands, and components of signal transduction networks activated by immune receptors. The second group includes plant or exogenous genes that possess antimicrobial activity or counteract pathogen virulence factors. The advantage of employing this second group of genes for transgenic expression is that they act directly against the pathogen without activating plant immune responses and thus avoid energetic burden on the plant.

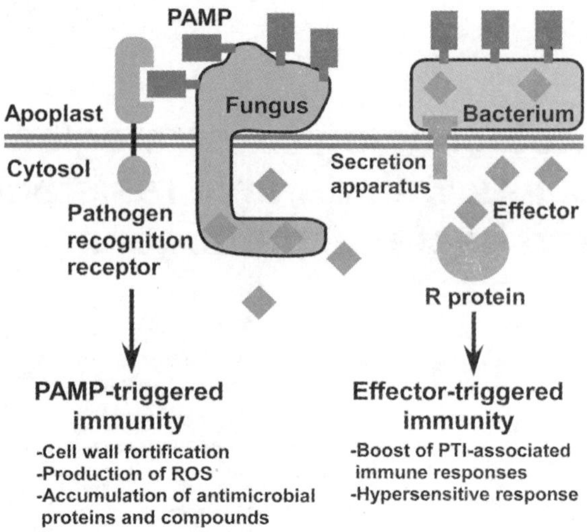

Figure 21.1 • A model illustrating the two main branches of plant immunity. • Plant cells detect invasion by a fungal or bacterial microbe through pathogen recognition receptors (PRRs). Most of the PRRs are embedded in the plasma membrane and recognize conserved pathogen determinants, designated as pathogen- or microbe-associated molecular patterns (PAMPs or MAMPs). This recognition event activates a network of signaling cascades resulting in the onset of a first line of immune response, collectively referred to as PAMP-triggered immunity (PTI). PTI includes fortification of the cell wall, production of reactive oxygen species (ROS), and accumulation of antimicrobial proteins and compounds. A second line of immunity is represented by plant resistance (R) proteins, mostly cytoplasmic, that specifically recognize a pathogen effector protein and mount very efficient immune responses, referred to as effector-triggered immunity (ETI). ETI boosts PTI-associated immune response, often culminating in a localized hypersensitive response (HR) cell death, and inhibits the spread of the pathogen to neighboring healthy tissues. Please see color plate section at the end of the book.

Plant immune receptors mediating pathogen recognition

Plant immune receptors detect the presence of disease-causing bacteria and fungi by recognizing pathogen molecules that are either constitutively expressed or exclusively synthesized during the infection process (Jones and Dangl, 2006). Transmembrane pattern-recognition receptors (PRRs) are a class of plant immune receptors that recognize constitutively expressed and highly conserved molecular structures of microbes, generally referred to as pathogen- or microbe-associated molecular patterns (PAMPs or MAMPs; Boller and Felix, 2009). Immune responses activated by PRRs are collectively referred to as PAMP-triggered immunity (PTI) and represent the basis of broad-spectrum resistance against non-host microbial pathogens (Boller and Felix, 2009; Zipfel, 2009; Figure 21.1).

An important advantage of using PRRs for disease control resides in the conservation of their corresponding PAMPs in different pathogens. Engineering susceptible plants with PRR-mediated recognition of PAMPs, which are present in the majority of the isolates of a certain microbe or in different microbes, has the potential of conferring resistance against a broad range of pathogens. For example, the rice PRR Xa21 recognizes a secreted sulfated peptide that is conserved in diverse strains of the bacterium *Xanthomonas oryzae* pv. *oryzae* (Lee et al., 2009). As predicted, transgenic expression of Xa21 in rice plants missing endogenous Xa21 was shown to confer a broad-range resistance against multiple strains of the pathogen (Wang et al., 1996).

A recently developed strategy to confer disease resistance in plants is the transfer of PAMP recognition capacities across plant families. While some PAMPs are recognized by many plant species, others are recognized only by certain plant families (Boller and Felix, 2009). The bacterial PAMP EF-Tu is known to be recognized naturally only by members of the Brassicaceae family through the EFR receptor (Kunze et al., 2004; Zipfel et al., 2006). It was recently shown that transgenic expression of EFR in the Solanaceous species tomato and *Nicotiana benthamiana* confers EF-Tu responsiveness and resistance to various foliar and vascular bacterial pathogens of different genera (Lacombe et al., 2010). The *BAK1* gene, which encodes a protein structurally similar to several PRRs and is required for the function of multiple transmembrane receptors (Chinchilla et al., 2009), has also been shown to represent a possible target for disease control. In fact, overexpression of BAK1 in transgenic rice enhanced host resistance to blast fungus infection (Hu et al., 2005).

Resistance (R) proteins represent a second class of immune receptors that specifically recognize pathogen virulence effectors (acting under these circumstances as avirulence factors), and trigger a line of efficient immune responses known as effector-triggered immunity (ETI; Jones and Dangl, 2006;

Figure 21.1). Importantly, ETI includes a rapid and localized cell death at the site of infection, known as the hypersensitive response (HR), which is thought to prevent further spread of the pathogen to uninfected tissues (Greenberg and Yao, 2004). Most *R* genes isolated to date encode proteins with an amino-terminal nucleotide binding site (NBS) and a carboxy-terminal leucine rich repeat (LRR; Caplan et al., 2008). *R* genes have been extensively used as a tool to achieve disease control by classical breeding practices as well as by transgenic approaches (Rommens and Kishore, 2000; McDowell and Woffenden, 2003). However, important limitations have compromised most of the attempts to confer durable resistance by the introduction of *R* genes into crop plants. In most cases, *R* genes turn out to lack durability in the field because pathogens are usually able to avoid recognition by R proteins, either by shedding the corresponding effector protein or by introducing mutations in the effector gene. An additional important drawback in the employment of *R* genes for disease control is that their over-expression may result in the constitutive activation of immune responses even in the absence of pathogen, as was described for over-expression of the *R* genes *Pto* and *RPS2* in tomato and *Arabidopsis* plants, respectively (Tang et al., 1999; Tao et al., 2000). Such a constitutive activation of immune responses may have a negative impact on plant growth and yield. Finally, *R* genes may not be able to mount a resistance response when expressed in distant plant species, probably due to the lack of suitable signal transduction mechanisms. This phenomenon is termed "restricted taxonomic functionality" and may hinder the transfer of *R* genes between distantly related crops and from model species to crops (Tai et al., 1999).

Despite these limitations, several reports indicate that certain *R* genes still have great potential for successful employment in disease control. A first example is the pepper *Bs2* gene, whose protein product recognizes the AvrBs2 effector of the gram negative phytopathogenic bacterium *Xanthomonas campestris* (Kearney and Staskawicz, 1990). Because *avrBs2* is widespread among *X. campestris* pathovars and its absence results in bacteria that are much less virulent than wild-type strains (Kearney and Staskawicz, 1990; Swords et al., 1996), the *Bs2 R* gene was selected as a candidate for conferring durable disease resistance against a wide range of *X. campestris* pathovars (Tai et al., 1999). Indeed, transgenic expression of *Bs2* in tomato plants conferred disease resistance against pathovars of *X. campestris* carrying *avrBs2*. In addition, transgenic *N. benthamiana* plants expressing *Bs2* developed an HR when challenged with bacterial strains expressing *avrBs2*. However, the prospected durability of *Bs2* remains to be demonstrated in field trials.

Examples of *R* genes with a proven potential against fungal pathogens include the barley *Rpg1* gene (Brueggeman et al., 2002), the tomato *Ve1* gene (Kawchuk et al., 2001; Fradin et al., 2009), and the *Arabidopsis RPW8.1* and *RPW8.2* genes (Xiao et al., 1997). The barley *Rpg1* gene confers resistance to most pathotypes of the stem rust fungus *Puccinia graminis* f. sp. *tritici* (Brueggeman et al., 2002). Remarkably, this *R* gene was proven to be of extraordinary durability (Brueggeman et al., 2002). North American barley cultivars with *Rpg1* were first released in 1942, and since then there have been no major losses as a result of stem rust (Brueggeman et al., 2002). Resistance provided by the tomato *Ve1* gene against *Verticillium* species, which cause wilt disease in many different crops, has been incorporated into most commercial tomato varieties and has proven to be durable (Kawchuk et al., 2001). Importantly, *Ve1* was also shown to be functional in potato when expressed as a transgene (Kawchuk et al., 2001). The *Arabidopsis RPW8.1* and *RPW8.2* genes represent an example of the potential use of *R* genes for achieving broad-spectrum resistance. These two homologous *Arabidopsis* genes confer resistance to multiple powdery mildew isolates that belong to distinct *Erysiphe* spp. (Xiao et al., 1997). Interestingly, over-expression of *RPW8.1* and *RPW8.2* in transgenic *Arabidopsis* plants conferred resistance to the biotrophic oomycete *Hyaloperonospora parasitica* and to the cauliflower mosaic virus, but renders plants more susceptible to the necrotrophic fungal pathogens *Alternaria* and *Botrytis* spp. (Wang et al., 2007). The differential effect of *RPW8.1* and *RPW8.2* over-expression toward biotrophic and necrotrophic pathogens is reminiscent of the activity of the plant hormone salicylic acid (SA) that activates immune responses against biotrophs while repressing immune responses against necrotrophs (Glazebrook, 2005). These observations raise the possibility that broad-spectrum resistance mediated by *RPW8.1* and *RPW8.2* is the result of constitutive activation of SA-dependent immunity rather than specific recognition of pathogen effectors. Therefore, when manipulating *R* genes for engineering disease resistance it should be taken into account that their over-expression may activate defense pathways that are beneficial against certain pathogens, but detrimental against others. An additional *R*-like gene that when over-expressed induces SA-dependent immune responses is the NBS-LRR *Prf* gene, which is required for tomato resistance against the bacterium *Pseudomonas syringae* (Salmeron et al., 1996). Tomato plants over-expressing *Prf* displayed broad-spectrum resistance against different bacterial pathogens, while they did not differ from wild-type plants with respect to growth and fruit production (Oldroyd and Staskawicz, 1998).

The search for new *R* genes that confer resistance against pathogens for which natural plant defenses are currently unavailable is an important task. An ideal situation would be to identify *R* genes that recognize avirulence determinants indispensable for pathogen virulence. Such *R* genes could potentially provide durable resistance, since the pathogen cannot afford to lose its avirulence determinant to evade plant resistance without hampering its own virulence.

Elicitors of plant immunity

An important strategy to enhance plant disease resistance is to express in plants microbial-derived genes encoding proteins that are recognized by the plant surveillance system and elicit effective immune responses against pathogens (generally referred to as elicitors). Examples of well-characterized elicitors include PAMPs, avirulence effectors, Hrp proteins delivered by the type III secretion system, and fungal or oomycetal proteins, such as cryptogein and cerato-platanins (Gurr and Rushton, 2005b; Table 21.1). Here we will discuss a few

Table 21.1 Main classes of elicitors of plant immunity

Elicitor type	Source	Examples
PAMPs[a]	Bacteria and fungi	Flagellin (Gomez-Gomez and Boller, 2002), EF-Tu[b], chitin[c], β-glucans[d]
DAMPs[e]	Plants	AtPEP1 (Huffaker et al., 2006)
Harpins	Bacteria	Hrf-1 (Shao et al., 2008)
Effector proteins	Bacteria and fungi	AvrPto[f], AvrBs2 (Kearney and Staskawicz, 1990), Avr9[g]
Elicitins	Fungi	Cryptogein (Ricci et al., 1989)

[a]Pathogen-associated molecular patterns.
[b]Kunze et al. (2004).
[c]Baureithel et al. (1994) *J. Biol. Chem.* 269:17931–17938.
[d]Klarzynski et al. (2000) *Plant Physiol.* 124:1027–1038.
[e]Damage-associated molecular patterns.
[f]Ronald et al. (1992) *J. Bacteriol.* 174:1604–1611.
[g]van Kan et al. (1991) *Mol. Plant-Microbe Interact.* 4:52–59.

examples of elicitors whose potential for disease control has been demonstrated.

As detailed previously, PAMPs play a key role in activating PTI, an important branch of plant immunity (Boller and Felix, 2009). Flagellin is a well-characterized PAMP constituting the basic unit of the bacterial flagellum, which is responsible for motility and allows bacteria to respond to stimuli in their environment (Gomez-Gomez and Boller, 2002). In *Arabidopsis* plants, flagellin is recognized by the PRR FLS2 that upon flagellin perception activates signaling networks leading to the induction of immune responses (Gomez-Gomez and Boller, 2002). Interestingly, expression of a flagellin gene from the bacterial pathogen *Acidovorax avenae* in transgenic rice enhanced disease resistance against the fungus *Magnaporthe grisea*, which is the causal agent of blast disease (Takakura et al., 2008). However, the transgenic plants also showed small chlorotic spots on leaves, indicating that the constitutive activation of flagellin-mediated immunity causes detrimental effects on the plant.

Cryptogein, a potent activator of plant defense responses from the oomycete *Phytophthora cryptogea* (Ricci et al., 1989), represents an additional example of the successful implementation of an elicitor to generate enhanced disease resistance in a model plant (Keller et al., 1999). Transgenic expression of cryptogein under a pathogen-inducible promoter caused the appearance of necrotic lesions upon *Phytophthora parasitica* var. *nicotianae* infection and provided enhanced disease resistance against several unrelated pathogenic fungi in tobacco plants (Keller et al., 1999). Harpins constitute another class of pathogen-derived elicitors that can be used to generate disease-resistant transgenic plants (Cornelis and Van Gijsegem, 2000). These are heat-stable and glycine-rich proteins that are produced by gram negative phytopathogenic bacteria and are secreted by their type III secretion system, presumably to the plant apoplast (Cornelis and Van Gijsegem, 2000). When infiltrated at relatively high concentration into the plant intercellular spaces, harpins elicit an HR-like reaction (Cornelis and Van Gijsegem, 2000). Interestingly, over-expression of harpin-encoding genes from different phytopathogenic bacteria in rice and tobacco plants enhanced disease resistance against a number of fungal and bacterial pathogens (Peng et al., 2004; Jang et al., 2006; Sohn et al., 2007). Notably, expression of Hrf1, a harpin from *X. oryzae* pv. *oryzae*, conferred non-specific resistance to the blast fungus *M. grisea* in field conditions (Shao et al., 2008). Hrf1-expressing plants also showed constitutively enhanced expression of genes involved in plant defense pathways (Shao et al., 2008).

Families of small-secreted fungal proteins, for example, hydrophobins and metallothioneins (Gijzen and Nurnberger, 2006; Cox and Hooley, 2009), are widespread among fungi and are likely to be secreted in the host or to have structural roles in the fungal cell wall. Because several of these proteins were shown to act as elicitors of plant immune responses, their ability to confer broad-spectrum resistance has been explored. For example, the *MgSM1* gene from the fungus *M. grisea* encodes a small protein belonging to the cerato-platinin family (Yang et al., 2009). Interestingly, *MgSM1* expression induced immune responses in various plants, and conferred resistance to bacterial and fungal pathogens when expressed in transgenic *Arabidopsis* plants (Yang et al., 2009). Finally, the discovery that immune responses can also be activated by plant-derived elicitors, which are released from damaged plant tissues during infection and termed damage-associated molecular patterns (DAMPs), provides new candidate transgenes for crop improvement (Boller and Felix, 2009). For example, the *Arabidopsis* AtPep1 peptide is encoded by a stress-induced precursor gene and its over-expression induced constitutive transcription of pathogenesis-related (PR) genes and enhanced resistance to a fungal pathogen (Huffaker et al., 2006).

Plant genes involved in signaling networks of plant immunity

The accumulating knowledge on immunity-related signal transduction pathways provides new opportunities to manipulate plant disease resistance (Figure 21.2). The expression of transgenes encoding signaling components that are common intersects of diverse resistance pathways can be used to activate a complete battery of defense responses. This approach could provide a broad-spectrum resistance because activation of immune responses will not be dependent on specific pathogen recognition. In addition, it is expected to be durable because it cannot be defeated by simple mutations or deletions of pathogen genes. However, it should be noted that constitutive activation of immune responses by over-expression of central immunity regulators may result in detrimental effects on growth and development that can eventually reduce crop yields.

Potential candidate genes for this approach are those encoding immunity-related signaling molecules such as protein kinases. In recent years it has become clear that MAP kinase cascades participate in the activation of plant immune

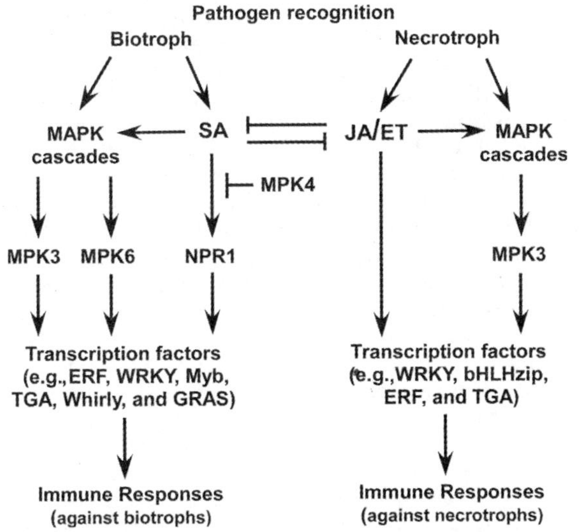

Figure 21.2 • Schematic representation of plant defense signaling networks. • Perception of a pathogen by infected plant tissues triggers activation of multiple signaling cascades that differ according to the lifestyle of the invading microbe (Glazebrook, 2005). Signaling pathways that trigger immune responses against biotrophic organisms are mostly dependent on the plant hormone salicylic acid (SA), include mitogen-activated protein (MAP) kinase modules, and lead to the activation of various classes of defense-related transcription factors. Signaling cascades that induce immune responses against necrotrophic pathogens are dependent on the plant hormones jasmonic acid (JA) and ethylene (ET). These include MAP kinase cascades and lead to the activation of classes of transcription factors that only partially overlap with those induced by biotrophic organisms. Mutual inhibition occurs between SA- and JA/ET-signaling networks (Glazebrook, 2005). *Arabidopsis* NPR1 and MPK6 are examples of signaling molecules playing central roles in SA-dependent networks (Dong, 2004; Pedley and Martin, 2005), while MPK3 represents signaling components that participate both in SA- and JA/ET-dependent pathways (Pedley and Martin, 2005). *Arabidopsis* MPK4 is a negative regulator of SA-dependent immune responses (Petersen et al., 2000).

responses mediated by R and PRR proteins (Pedley and Martin, 2005). An example of the potential use of these proteins for disease control is the enhanced resistance against the bacterium *P. syringae* and the fungus *Botrytis cinerea* that was gained by over-expressing several components of a PAMP-induced MAPK cascade in *Arabidopsis* (Asai et al., 2002). Identification of additional potential candidate genes stems from the observation that a vast transcriptional reprogramming occurs during the activation of immune responses (Wise et al., 2007). Representatives of the ERF, WRKY, Myb, TGA-bZIP, Whirly, and GRAS families of transcription factors have been shown to play central roles in plant immunity, and thus represent strong candidates for engineering enhanced disease resistance (Century et al., 2008). ERF and WRKY family members are excellent examples of well-characterized transcription factors whose potential for disease control has been widely explored (Gutterson and Reuber, 2004; Pandey and Somssich, 2009). In a large number of studies, over-expression of ERF and WRKY transcription factors in transgenic model and crop plants resulted in enhanced disease resistance (Shin et al., 2002). Interestingly, over-expression of defense-related transcription factors may provide disease resistance to multiple unrelated pathogens. For example, over-expression of the ERF Tsi1 protein in transgenic hot pepper enhanced resistance to viral, bacterial, and oomycete pathogens (Shin et al., 2002). Similarly, over-expression of the *Arabidopsis* transcription factor MYB30 in transgenic *Arabidopsis* and tobacco plants enhanced resistance to the bacterium *P. syringae* and the fungus *Cercospora nicotianae* (Vailleau et al., 2002).

Another class of genes that can be used to activate immune responses independent of pathogen recognition are those involved in signaling pathways activated by defense-related plant hormones, such as SA, jasmonic acid (JA), and ethylene (ET; Pieterse et al., 2009; Shah, 2009). A typical example for the potential use of these genes is *NPR1*, which encodes a key regulator of systemic acquired resistance (SAR; Dong, 2004). SAR is a mechanism of induced defense that confers long-lasting protection against a broad spectrum of microorganisms (Durrant and Dong, 2004). *NPR1* represents a particularly relevant candidate for disease control because it is required both for *R*-gene-mediated resistance and SAR (Dong, 2004). The *NPR1* gene product activates expression of defense genes by interacting with members of the TGA family of bZIP transcription factors in the plant cell nucleus (Zhang et al., 1999). Interestingly, *Arabidopsis* plants over-expressing NPR1 displayed enhanced resistance toward bacterial, fungal, and oomycete pathogens (Cao et al., 1998; Friedrich et al., 2001). Notably, enhanced resistance observed in these plants did not result from constitutive activation of immune responses, but from a more efficient response to the SAR-inducing hormone salicylic acid. Although in transgenic *Arabidopsis* plants overexpression of NPR1 did not cause any detrimental effect, in transgenic rice it caused the appearance of necrotic lesions that can negatively affect yield (Fitzgerald et al., 2004). In a more recent report, the introduction of the *Arabidopsis NPR1* gene into carrot plants enhanced resistance against a broad range of pathogens including biotrophic and necrotrophic fungi and a foliar bacterial pathogen, without evidence of any undesirable growth or developmental phenotypes (Wally et al., 2009).

Activation of SAR and concomitant increase of disease resistance was also obtained by manipulating the polyamine metabolism (Moschou et al., 2009). Transgenic tobacco plants over-expressing the polyamine oxidase gene, which catalyzes the oxidative catabolism of spermidine and spermine thus generating hydrogen peroxidase, displayed increased levels of SAR marker genes and of cell wall defense (Moschou et al., 2009). Moreover, the transgenic plants showed pre-induced disease tolerance against the biotrophic bacterium *P. syringae* and the hemibiotrophic oomycete *Ph. parasitica* (Moschou et al., 2009).

An additional avenue for boosting plant immunity is to manipulate biosynthesis of defense-related plant hormones. An example for this approach is represented by over-expression of the pathogen-inducible rice gene *OsAOS2*, which encodes allene oxide synthase, a key enzyme of the JA biosynthetic pathway (Gfeller et al., 2010). Transgenic rice plants expressing *OsAOS2* under the control of a strong pathogen-inducible promoter accumulated high levels of JA, exhibited accumulation of PR genes, and displayed increased resistance to the fungal pathogen *M. grisea* (Mei et al., 2006). However, it should be noted that because of the involvement of defense-related plant hormones in diverse physiological functions other than plant immunity, this type of manipulation is likely to affect plant growth and development, and to cause undesired phenotypes.

Finally, the suppression of plant genes that negatively regulate plant immune responses represents another possible strategy to engineer disease resistance. For example, activation of plant immunity was achieved in *Arabidopsis* by suppressing the expression of the MAP kinase MPK4, which is a negative regulator of SAR (Petersen et al., 2000). *Arabidopsis* plants mutated in *mpk4* exhibited constitutive activation of SAR and increased resistance to the bacterium *P. syringae*, and to the oomycete *Peronospora parasitica* (Petersen et al., 2000). However, similar to other mutants that constitutively express immune responses, *mpk4* mutants displayed developmental phenotypes such as dwarfism, curled leaves, reduced pollen production, and reduced fertility (Petersen et al., 2000).

Antimicrobial genes

The plant immune response culminates in the production of a myriad of antimicrobial agents including reactive oxygen species, secondary metabolites, hydrolytic enzymes, and antimicrobial peptides and proteins. Increasing the accumulation of antimicrobial compounds of different biological origin (plant or microbial) in transgenic plants represents a promising strategy for engineering plant disease resistance against fungal and bacterial pathogens. Antimicrobial agents can be used to counteract the action of specific pathogens without the need to induce complex signaling pathways that lead to a complete array of immune responses. One advantage of this approach is that it allows bypassing the induction of energy-consuming processes in the plant, and reduces the incidence of growth and developmental phenotypes that may negatively affect yields. In early implementations of this strategy, hydrolytic enzymes (e.g., glucanases and chitinases) that target the microbial cell wall were used to enhance disease resistance in important crops, such as rice and wheat (Gomez-Ariza et al., 2007; Shin et al., 2008). In addition, defense-related proteins with protease inhibitor activity were shown to confer disease resistance to bacterial and fungal pathogens when expressed in transgenic tobacco plants (Senthilkumar et al., 2010). Similarly, expression of the grapevine phytoalexin resveratrol in transgenic tobacco plants enhanced resistance to fungal infection (Hain et al., 1993).

More recently, plant-derived genes that encode peptides with antimicrobial activities have emerged as important candidates for generating disease-resistant plants (van der Biezen, 2001; Manners, 2007; Montesinos, 2007; Rossi et al., 2008). Plant antimicrobial peptides are cysteine-rich short amino acid sequences that share general features, such as an overall positive charge and the presence of disulfide bonds, and typically target outer membrane structures (e.g., ion channels). Main families of antimicrobial peptides are represented by α-defensins, thionins, lipid transfer proteins, cyclotides, snakins, and hevein-like (Manners et al., 2007). A database containing structural and functional information on more than 270 antimicrobial plant peptides and their target organisms was recently developed (PhytAMP; http://phytamp.pfba-lab-tun.org; Hammami et al., 2009). This information may assist in isolating and characterizing novel plant antimicrobial peptides, in designing synthetic peptides with higher potency against pathogens or with broad antimicrobial spectra, and in identifying candidate genes for engineering disease resistance against desired pathogens.

Several plant genes that encode peptides with antimicrobial activity were successfully used to generate transgenic plants with enhanced disease resistance. For instance, over-expression of the *snakin-1* gene enhanced disease resistance against the soil-borne fungal pathogen *Rhizoctonia solani* and the bacterial pathogen *Erwinia carotovora* in transgenic potato plants (Segura et al., 1999; Almasia et al., 2008). Likewise, the pepper antimicrobial gene *CaAMP1* conferred broad-spectrum resistance to bacterial, oomycete, and fungal pathogens when over-expressed in *Arabidopsis* plants (Lee et al., 2008). An additional example is the *PnAMP-h2* gene from *Pharbitis nil* (morning glory) plants encoding a hevein-like antimicrobial peptide with potent antifungal activity (Koo et al., 2002). *PnAMP-h2* over-expression in tobacco plants resulted in enhanced resistance against the oomycete *Ph. parasitica* (Koo et al., 2002). Finally, a barley chitinase (*chi-2*) gene and a wheat lipid-transfer-protein (*ltp*), when simultaneously expressed in carrot plants, conferred enhanced resistance against the fungal pathogens *Alternaria radicicola* and *B. cinerea* (Jayaraj and Punja, 2007).

Antimicrobial genes that are efficient against bacterial and fungal pathogens can also be derived from heterologous non-plant systems. Non-phytopathogenic microbes represent a rich source of genes that can be used to engineer disease-resistant plants (Lorito et al., 1998; Punja and Utkhede, 2003). The mycoparasitic fungus *Trichoderma virens* is the source of an endochitinase gene that was introduced into cotton and tobacco plants to enhance disease resistance (Emani et al., 2003). Endochitinases are enzymes that specifically degrade chitin, a major constituent of the cell wall of many fungi (Adams, 2004). Interestingly, transgenic plants over-expressing the *T. virens* endochitinase gene were resistant to both the soil-borne pathogen *R. solani* and the foliar pathogen *A. alternata* (Emani et al., 2003). In another study, fungal-borne viruses were shown to be a source of antimicrobial genes that can be used for crop improvement (Schlaich et al., 2006). KP4 is a virally encoded antifungal toxin secreted by strains of the fungus *Ustilago maydis* (Park et al., 1994; Gage et al., 2001). Interestingly, transgenic wheat plants expressing the *KP4* gene showed a 30% decrease in disease symptoms after infection with the smut fungus *Tilletia caries* in a glasshouse, and a 10% increase in resistance in field experiments (Schlaich et al., 2006).

An original study combined the use of antimicrobial peptides with that of specific antibodies to achieve disease control against the fungal pathogen *Fusarium oxysporum* (Peschen et al., 2004). Three antifungal peptides were each fused to a *Fusarium*-specific antibody and expressed in transgenic *Arabidopsis* plants. By this strategy, the antifungal peptides were specifically targeted to the site of infection where they efficiently interfered with fungal growth and development. Remarkably, transgenic plants expressing the fusion proteins showed significantly enhanced resistance to the pathogen (Peschen et al., 2004). In comparison, transgenic plants that expressed only the antifungal peptides or the antibody displayed lower levels of resistance. It is noteworthy that the morphology of the transgenic plants was similar to that of the non-transformed control plants (Peschen et al., 2004).

Genes targeting pathogen virulence determinants

Virulence determinants on which microbes rely for their pathogenicity can be directly targeted to achieve plant disease resistance. In this section we review a few examples of how virulence mechanisms can be manipulated for disease control. Oxalic acid was found to be associated with pathogenesis of several phytopathogenic fungi and thus represents a target for genetic manipulations (Godoy et al., 1990; Cessna et al., 2000). Based on the hypothesis that introduction of a gene that can specifically degrade oxalic acid in crop plants would confer disease resistance to fungal pathogens, the oxalate decarboxylase gene, whose translation product catabolizes oxalic acid to CO_2 and formic acid, was over-expressed in tobacco and tomato plants (Kesarwani et al., 2000). Interestingly, the obtained transgenic plants showed resistance to the fungal pathogen *Sclerotinia sclerotiorum*, which requires oxalic acid for a successful plant infection (Kesarwani et al., 2000). By a similar strategy, expression in oilseed rape plants of a wheat oxalate oxidase, which oxidizes oxalic acid to CO_2 and H_2O_2, enhanced resistance to *S. sclerotiorum* (Dong et al., 2008).

Programmed cell death is often associated with both susceptible and resistant host responses in plants (Glazebrook, 2005). Because necrotrophic pathogens benefit from host cell death, it was suggested that inhibiting plant programmed cell death by expression of anti-apoptotic genes would result in plants with increased resistance to necrotrophic pathogens. Based on this premise, animal genes that negatively regulate apoptosis were expressed in transgenic tobacco plants (Dickman et al., 2001). Interestingly, expression of the anti-apoptotic human *Bcl-2* and *Bcl-xl*, nematode *CED-9*, and baculovirus *Op-IAP* transgenes conferred heritable resistance to several necrotrophic fungal pathogens including *S. sclerotiorum*, *B. cinerea*, and *C. nicotianae* (Dickman et al., 2001). Similarly, the expression of the baculovirus anti-apoptotic gene *p35* in lupin resulted in a reduction of disease symptoms caused by necrotrophic fungal pathogens (Wijayanto et al., 2009). Together, these studies indicate the potential of anti-apoptotic genes to generate effective disease resistance to necrotrophic fungi in economically important crops.

Recent studies have indicated that pathogens should suppress plant immunity to infect a plant species (Cui et al., 2009; De Wit et al., 2009). To this purpose, some pathogens utilize effector molecules to activate host targets that function as negative regulators of plant immunity. In principle, knocking out such effector targets or susceptibility factors would release the suppression of immune responses and lead to resistance (Pavan et al., 2010). A well-characterized example of such susceptibility factors is the barley plasma membrane protein MLO that is required for susceptibility to powdery mildew infections (Buschges et al., 1997; Piffanelli et al., 2004). Naturally occurring loss-of-function *mlo* alleles in barley and tomato plants provide durable resistance to a variety of powdery mildews without affecting yield (Piffanelli et al., 2004; Bai et al., 2008). As MLO protein families are present in several plant species (including crops), their targeted inactivation bears the potential to confer disease resistance against numerous powdery mildews. Indeed, targeted inactivation of MLO in *Arabidopsis* plants conferred resistance to powdery mildew pathogenesis (Consonni et al., 2006). As discussed for MLO, manipulation of other plant susceptibility factors represents a promising strategy to achieve disease resistance against pathogens (Pavan et al., 2010).

How Many Transgenes Should be Expressed in a Single Plant for Efficient Disease Control?

The introduction of a single gene into susceptible plants, either by breeding or genetic engineering, rarely results in durable disease resistance due to the ability of pathogens to evade its effect and break resistance. A useful strategy to overcome this limitation in the attempt to engineer resistant plants is to "pyramid" or "stack" several transgenes in a single plant (Figure 21.3). This approach is predicted to provide a more durable resistance, since it is unlikely that a pathogen would be able to simultaneously lose or mutate multiple traits. In addition, pyramiding several genes that target multiple races of a pathogen, or even different pathogens, would provide a broader spectrum of resistance compared to a single-gene strategy. In principle, this approach may assist in generating plants that are resistant to both bacteria and fungi, despite the very different virulence mechanisms of these pathogens.

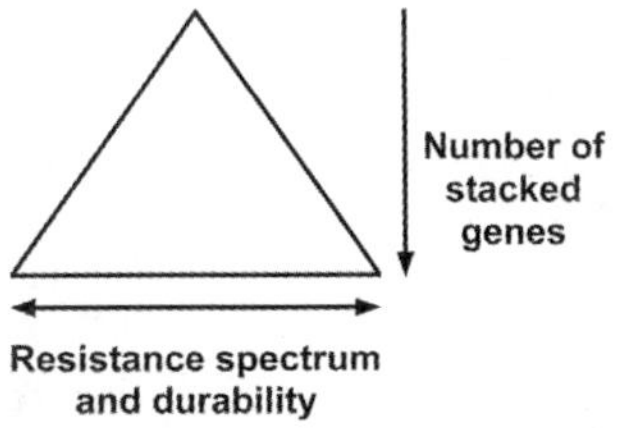

Figure 21.3 • Schematic representation of the pyramiding concept. • Increase in the number of defense transgenes (stacking) in a plant widens the spectrum of pathogens against which the plant is resistant. In addition, stacking of transgenes enhances durability of resistance because pathogens need to develop multiple strategies to evade the effect of different transgenes.

Several reports demonstrate that stacking *R* genes in crop plants represents a successful strategy to confer broad-spectrum and durable disease resistance. For example, multiple *R* genes were introduced by breeding into rice plants to enhance disease resistance to bacterial blight caused by the pathogen *X. oryzae* pv. *oryzae* (Singh et al., 2001). Higher levels of resistance to various *X. oryzae* pv. *oryzae* isolates were obtained by pyramiding into rice different combinations of the *R* genes *Xa*-4, *xa*-5, *xa*-13, and *Xa*-21, as compared to plants expressing only a single *R* gene. Moreover, the combinations of *R* genes tested in this study provided a wider spectrum of resistance to the *X. oryzae* pv. *oryzae* population prevalent in the region in which field experiments were conducted (Singh et al., 2001). Similarly, a strategy of pyramiding *R* genes was successfully applied to confer disease resistance against the blast fungus *M. grisea* (Hittalmani et al., 2000). In this study, rice lines expressing the *R* genes *Pi*1, *Piz*-5, and *Pi*ta displayed enhanced and broad-spectrum resistance against various *M. grisea* isolates compared to lines with a single *R* gene (Hittalmani et al., 2000). Although in these studies the stacking of *R* genes was achieved by classical breeding techniques and not by genetic engineering, the results obtained are supportive of the use of the *R* genes pyramiding strategy. In the future, the faster pace of genetic engineering compared to classical breeding techniques should allow for more efficient pyramiding of *R* genes in crops.

Other transgenes, besides *R* genes, showed promising results when stacked in a single plant. One successful example is the pyramiding of protease-inhibitors in transgenic tobacco plants (Senthilkumar et al., 2010). The protease-inhibitor genes *Sporamin* from sweet potato and *CeCPI* from taro driven by a wound- and pathogen-responsive promoter were expressed in tobacco plants. Transgenic plants over-expressing the stacked genes displayed enhanced resistance against bacterial and oomycete pathogens and insects (Senthilkumar et al., 2010). The pyramiding of PR genes was also demonstrated to be efficient in enhancing plant disease resistance (Jayaraj and Punja, 2007). Stacking of the barley chitinase *chi-2* gene and the wheat lipid transfer protein *ltp* gene in carrot plants resulted in transgenic plants with significantly higher resistance against necrotrophic foliar fungal pathogens compared to plants with a single transgene (Jayaraj and Punja, 2007).

An additional form of pyramiding is manifested in nature as quantitative disease resistance (QDR; Poland et al., 2009). QDR is an incomplete resistance conditioned by multiple genes of partial effect commonly referred to as quantitative resistance loci (QRLs). For certain diseases, particularly those caused by necrotrophic pathogens, QDR is the most important or only form of resistance available. Observations on the performance of crop cultivars with different types of resistance have led to the conclusion that QDR tends to be more durable than typical *R*-gene-mediated resistance (Parlevliet, 2002; Kliebenstein and Rowe, 2009). *R* genes can be rapidly overcome as a result of a strong selection for compatible pathogen variants, whereas breakdown of QDR is less frequent because the partial contribution of QRLs to resistance results in a lower selection pressure on the pathogen. Therefore, a thorough understanding of QDR should contribute to the design and deployment of durable resistance crops. However, the molecular mechanisms that control QDR remain poorly understood, largely due to the incomplete and inconsistent nature of the resistance phenotype, which is usually conditioned by many loci of small effect (Poland et al., 2009). Despite these limitations, several QRLs have been identified in different plant pathosystems in recent years (Poland et al., 2009). An example of the applicability of QDR for conferring durable resistance is the stacking of five QRLs in a rice line susceptible to rice blast (Manosalva et al., 2009). Interestingly, the obtained line exhibited superior resistance to rice blast for over 14 cropping seasons, compared to the susceptible parental line (Manosalva et al., 2009). As more QRLs are isolated and characterized, it will be possible to integrate them in breeding or genetic engineering programs for disease control.

When and Where Should the Transgene(s) be Expressed?

Most of the attempts to enhance disease resistance by constitutive and ectopic expression of immunity-related genes resulted in plants affected in growth and development, and with reduced yield. This is to be ascribed to the vast transcriptional reprogramming occurring in plants during the onset of resistance that re-allocates energetic resources from growth to defense responses (Eulgem et al., 1999). Therefore, when engineering disease resistance in transgenic plants, it is fundamentally important to tightly control the timing and localization of transgene expression to reduce detrimental side effects on plant growth and productivity (Gurr and Rushton, 2005a).

The use of pathogen-responsive promoters to drive transgene expression represents a promising strategy to bypass the undesired effects of constitutive expression. Only a few pathogen-inducible promoters have been employed for the control of bacterial and fungal disease to date (Gurr and Rushton, 2005a), although their implementation for enhancing disease resistance was already proposed in the early 1990s (De Wit, 1992). This is probably related to the many important features that a pathogen-responsive promoter should display to be suitable as a biotechnological tool (Table 21.2): (1) rapid induction upon sensing of the pathogen to allow plants to mount a defense response before the pathogen spreads to neighboring tissues and disease symptoms appear; (2) a minimal degree of constitutive transcriptional activity in the absence of pathogens to ensure that the transgene is expressed only during pathogen attack; (3) ability to respond to various potential pathogens for achieving a wide resistance spectrum; (4) unresponsiveness to processes activated by the transgene to avoid uncontrolled spread of plant responses; and (5) unresponsiveness to stresses other than pathogen attack to ensure specificity to biotic cues. Despite the fact that very few promoters that meet these requirements are available, gene expression profiles and other genomic technologies applied to plant–pathogen interactions are expected to lead to the identification of pathogen-responsive promoters suitable for disease control (Wise et al., 2007).

A complementary approach to limit the energetic cost of transgene constitutive expression is to employ promoters that drive expression to defined tissues or specific stages of the plant life cycle that are most vulnerable to pathogens (Kovalchuk et al., 2010). Spatial regulation of expression could even be combined with pathogen inducibility. Based on the extensive knowledge on virulence mechanisms used by pathogens of many major crops, ideally it should be possible to target transgene expression only to tissues that come in direct contact with the pathogen in the initial infection stages. Such a strategy should allow the plant to arrest the pathogen from spreading from the infection site, while other plant tissues that are not under direct attack continue to function normally without the burden of over-expressing the transgene. Another emerging possibility is to express the transgene at a specific stage of the plant life cycle in which the plant is most vulnerable to pathogen attack. This would allow the plant to resist the infection until endogenous barriers build up, without having to bear the energetic cost of activating defense responses throughout the entire plant life (Gurr and Rushton, 2005a).

An important alternative to native promoters are synthetic promoters that contain single or multiple *cis*-acting elements (Rushton et al., 2002). A combination of different regulatory elements can potentially provide a more tightly controlled temporal expression of the transgene, tissue specificity, and resistance to a wider spectrum of pathogens (bacterial or fungal) with various lifestyles (biotrophs or necrotrophs). Here we describe examples of promising native and synthetic promoters that were characterized to date and were employed to develop disease-resistant transgenic plants.

Pathogen-responsive and tissue-specific promoters

Pathogen-responsive promoters represent a tool for the expression of transgenes in a spatially and temporally controlled manner. In recent years several such promoters have been characterized and their possible employment for engineering disease-resistant plants assessed. Among them is the promoter of the rice *PBZ1* gene, which encodes an intracellular PR protein (Midoh and Iwata, 1996). The PBZ1 protein participates in systemic defense responses activated in rice by the chemical probenazole, which is widely used in agriculture to protect plants against the rice blast fungus *M. grisea* and the bacterial blight pathogen *X. oryzae* pv. *oryzae* (Midoh and Iwata, 1996; Lin et al., 2008). By monitoring *PBZ1* mRNA accumulation, the *PBZ1* promoter was shown to be induced by inoculation with the rice blast fungus, and by treatment with various compounds that induce systemic resistance, but not by wounding (Midoh and Iwata, 1996). Because of its induction characteristics, the *PBZ1* promoter was employed in transgenic rice plants to drive strong pathogen-inducible expression of the *OsAOS2* gene, which encodes a key enzyme in the biosynthetic pathway of the defense-related hormone JA (Mei et al., 2006). Interestingly, upon *M. grisea* infection, the transgenic plants displayed higher levels of JA, enhanced activation of PR genes, and increased resistance to the pathogen (Mei et al., 2006).

Also well characterized is the pathogen-inducible *Pgst1* (formerly *prp1-1*) promoter that drives expression of a potato PR gene (Martini et al., 1993). The *Pgst1* promoter is activated at early stages of potato infection by the late blight fungus *Ph. infestans*, but not in response to abiotic environmental cues (Martini et al., 1993). This promoter was also characterized in apple and shown to maintain its pathogen responsiveness, albeit with relatively low expression levels, when plants were treated with the fungus *Venturia inaequalis* and the bacterium *E. amylovora* (Malnoy et al., 2006).

In an additional significant study, a constitutively active form of a MAP kinase kinase was expressed under the control of a pathogen-inducible promoter to activate immune response upon pathogen attack (Yamamizo et al., 2006). To this end, the promoter of the *PVS3* gene, which encodes a key enzyme of phytoalexin biosynthesis, was first isolated from potato plants and shown to be significantly induced in tubers and leaves by both virulent and avirulent isolates of the late blight fungal pathogen *Ph. infestans*, but not by wounding

Table 21.2 Desired characteristics for a promoter designed to drive defense-related transgenes

Promoter characteristic	Desired	Undesired
Timing of transcription	Only in the presence of a pathogen — minimal degree of transcription in absence of pathogen	Constitutive
Tissue specificity	Restricted to tissues that come in direct contact with the pathogen	Non-specific
Induction kinetics	Rapid — upon sensing the pathogen and before it can spread and cause disease	Slow — after pathogen spread and appearance of disease symptoms
Responsiveness to pathogen spectrum	Wide — transcription is activated upon sensing a wide spectrum of pathogens	Narrow — transcription is activated upon sensing only one or few different pathogens
Responsiveness to transgene	Non-responsive to processes activated by the driven transgene	Responsive to processes activated by the driven transgene
Responsiveness to abiotic stress	Non-responsive	Responsive
Activity spectrum in different plants	Wide — active in various plant species	Narrow — active only in a limited number of plant species

(Yamamizo et al., 2006). Transgenic potato plants carrying a constitutively active form of a MAPK kinase driven by the *PVS3* promoter were generated. Interestingly, these transgenic plants showed high resistance to *Ph. infestans* as well as to the early blight pathogen *A. solani* (Yamamizo et al., 2006).

Recently, the promoter of the barley *GER4c* gene, which belongs to a gene cluster encoding the germin-like protein, has been proposed as a possible biotechnological tool for engineering strictly localized and pathogen-regulated disease resistance in transgenic cereal crops (Himmelbach et al., 2010). This promoter mediates expression specific to epidermal tissues, and it is activated by several host and non-host pathogens in barley (Himmelbach et al., 2010). Additional examples of tissue-specific and stress-inducible promoters are defensin promoters from wheat and rice (Kovalchuk et al., 2010). Their specific spatial and temporal expression patterns in developing grains and germinating seeds make these promoters valuable tools for the targeted expression of transgenes in cereal crops. Several other pathogen-responsive promoters, including *ZmPR4* from maize, *win3.12T* from poplar, and *PmPR10-1.13* from western white pine, were described and even used to engineer disease-resistant plants (Liu et al., 2005; Moreno et al., 2005; Yevtushenko et al., 2005). However, these promoters are induced under both biotic and abiotic stress, and may therefore be less effective as tools to engineer specific disease resistance in plants (Liu et al., 2005; Moreno et al., 2005; Yevtushenko et al., 2005).

Pathogen-responsive elements and synthetic promoters

Transcription factors of multiple families orchestrate the transcriptional reprogramming that takes place in plants challenged by pathogens (Gutterson and Reuber, 2004; Eulgem, 2005; Pandey and Somssich, 2009). Representatives of several families of transcription factors have been shown to bind to promoters of immunity-related genes and regulate their expression (Rushton and Somssich, 1998; Eulgem, 2005). An interesting and promising strategy to control the expression of transgenes for disease control is to isolate pathogen-responsive promoter elements and to use them as building blocks of synthetic promoters (Gurr and Rushton, 2005a). Such promoters could be designed to respond specifically to biotic stress and to recognize a broader spectrum of pathogens than that of native promoters.

Only a few pathogen-responsive promoter elements have been identified and characterized. An example of such elements is the W-box that is found in promoters of many genes activated by WRKY transcription factors (Rushton et al., 1996; Eulgem et al., 1999). Two W-box elements were identified in the promoters of the immediate-early response gene *CMPG1* from parsley and *Arabidopsis*, and represent promising building blocks for synthetic pathogen-responsive promoters (Kirsch et al., 2001; Heise et al., 2002). These elements are rapidly and strongly induced by pathogen elicitors, but not by wounding, in contrast to the *CMPG1* promoter from which they are derived. In addition, they show a strongly synergistic activity in artificial combinations with other *cis*-acting elements (Heise et al., 2002). These characteristics make W-box elements particularly suited for engineering synthetic promoters for disease control. The possibility of employing W-box elements for the enhancement of disease resistance was demonstrated by fusing promoter regions of the DRR206 gene from pea, which contain a W-box and a wound/pathogen-inducible box (W/P-box; Palm et al., 1990), with a DNase elicitor gene from *F. solani* f. sp. *phaseoli* (Choi et al, 2004). Interestingly, this construct conferred resistance against the bacterial pathogen *P. syringae* and the fungus *A. alternata* in tobacco (Choi et al., 2004).

An additional pathogen-responsive element was identified in the tobacco *hsr203J* gene, whose expression is specifically induced during the onset of the HR, several hours before the appearance of necrotic lesions (Pontier et al., 1994, 2001). Remarkably, the *hsr203J* promoter was successfully used to drive pathogen-inducible expression of the cryptogein elicitor in transgenic tobacco plants (Keller et al., 1999). Expression of the transgene in these plants was not detected in the absence of pathogen, whereas it was induced by infection with the virulent oomycete *Ph. parasitica* var. *nicotianae* causing the appearance of localized necrosis (Keller et al., 1999). However, less successful was the attempt to use this promoter to drive expression of the *popA* elicitor gene from the bacterium *Ralstonia solanacearum* (Belbahri et al., 2001). Although tobacco plants carrying the *hsr203J:popA* transgene showed enhanced resistance to *Ph. parasitica* var. *nicotianae*, they occasionally developed an uncontrolled spreading of necrosis that initiated at the site of pathogen infection (Belbahri et al., 2001). Deletion and mutagenesis analysis of the *hsr203J* promoter led to the identification of a 12 bp motif termed HSRE (HSR203J responsive element), which is necessary and sufficient for transcriptional activation of the hsr203J gene in response to pathogen attack (Pontier et al., 2001).

Pathogen-responsive elements can serve as modules for engineering synthetic pathogen-inducible promoters. The potential of this approach has been demonstrated in an elegant study, in which several *cis*-acting elements were tested for their ability to drive pathogen-inducible expression in planta when removed from their respective native promoter context (Rushton et al., 2002). Synthetic promoters containing tetramers of a single type *cis*-acting element were shown to mediate local gene expression in transgenic *Arabidopsis* plants infected by pathogens, although various levels of background expression were also observed in the uninfected plants (Rushton et al., 2002). This study also demonstrated the possibility of constructing improved synthetic promoters that contain elements of different types. Several combinations of pathogen-responsive *cis*-acting elements of different types resulted in promoters with high inducibility by a wide range of pathogens and low background transcriptional activity (Rushton et al., 2002).

An additional promising strategy for engineering synthetic pathogen-responsive promoters stems from the observation that certain plant promoters are directly activated by pathogen virulence factors that act as transcriptional activators. Transcription-activator like (TAL) effectors are virulence proteins found in many phytopathogenic bacteria of the genus

Xanthomonas that are delivered into the plant cell via a type III secretion system (White et al., 2009). TAL effectors have a characteristic central domain comprising a variable number of tandemly arranged copies of a 34/35-amino acid motif that specifically binds to a corresponding *UPA* (upregulated by AvrBs3) box found in promoters of their target host genes (Boch et al., 2009; Moscou and Bogdanove, 2009). In addition, they contain nuclear localization signals and an acidic transcriptional activation domain. Based on these observations, a strategy has been devised to use *UPA* boxes to engineer synthetic promoters by stacking up several different *UPA* boxes in the same promoter (Romer et al., 2009). These synthetic promoters, driving the expression of an *R* gene, were shown to respond and confer disease resistance to various pathogens that possess different TAL effectors (Romer et al., 2009). The recent deciphering of the code of DNA-binding specificity of TAL type III effectors should allow the design of more sophisticated synthetic TAL-responsive promoters in the near future (Boch et al., 2009; Moscou and Bogdanove, 2009).

Conclusions and Prospects

In recent years, the understanding of molecular mechanisms that govern plant immunity and virulence of important pathogens has significantly improved. Thus far engineering of disease resistance in crops remains largely elusive, but a wide array of new genetic tools is now becoming available and may be employed to achieve this goal. Custom-made combinations of promoters and transgenes may be designed and tested for disease control. Type and number of transgene(s), and the native or synthetic promoter(s) to be employed should be carefully selected according to criteria that include the desired spectrum of disease resistance (narrow or broad), the nature (bacteria, fungi, or both), lifestyle (biotrophic or necrotrophic) and virulence strategies of the pathogens(s), and the genetic and physiological characteristics of the engineered crop. Toward the successful engineering of durable resistance in crop plants, several promising novel strategies have proven successful in controlled environments and await confirmation in field trials.

Acknowledgments

Our research on plant disease resistance and susceptibility is supported by grants from the United States-Israel Binational Agricultural Research and Development Fund (BARD; grant no. IS-4159-08C) and from the Israel Science Foundation (ISF; grant no. 326/10).

References

Adams, D. J. (2004). Fungal cell wall chitinases and glucanases. *Microbiology*, *150*, 2029–2035.

Almasia, N. I., Bazzini, A. A., Hopp, H. E., & Vazquez-Rovere, C. (2008). Overexpression of snakin-1 gene enhances resistance to *Rhizoctonia solani* and in transgenic potato plants. *Molecular Plant Pathology*, *9*, 329–338.

Asai, T., Tena, G., Plotnikova, J., Willmann, M. R., Chiu, W. L., & Gomez-Gomez, L., et al. (2002). MAP kinase signalling cascade in *Arabidopsis* innate immunity. *Nature*, *415*, 977–983.

Bai, Y., Pavan, S., Zheng, Z., Zappel, N. F., Reinstadler, A., & Lotti, C., et al. (2008). Naturally occurring broad-spectrum powdery mildew resistance in a Central American tomato accession is caused by loss of *mlo* function. *Molecular Plant-Microbe Interactions*, *21*, 30–39.

Belbahri, L., Boucher, C., Candresse, T., Nicole, M., Ricci, P., & Keller, H. (2001). A local accumulation of the *Ralstonia solanacearum* PopA protein in transgenic tobacco renders a compatible plant–pathogen interaction incompatible. *Plant Journal*, *28*, 419–430.

Boch, J., Scholze, H., Schornack, S., Landgraf, A., Hahn, S., & Kay, S., et al. (2009). Breaking the code of DNA binding specificity of TAL-type III effectors. *Science*, *326*, 1509–1512.

Boller, T., & Felix, G. (2009). A renaissance of elicitors: Perception of microbe-associated molecular patterns and danger signals by pattern-recognition receptors. *Annual Review of Plant Biology*, *60*, 379–406.

Brueggeman, R., Rostoks, N., Kudrna, D., Kilian, A., Han, F., & Chen, J., et al. (2002). The barley stem rust-resistance gene *Rpg1* is a novel disease-resistance gene with homology to receptor kinases. *Proceedings of the National Academy of Sciences of the United States of America*, *99*, 9328–9333.

Buschges, R., Hollricher, K., Panstruga, R., Simons, G., Wolter, M., & Frijters, A., et al. (1997). The barley *Mlo* gene: A novel control element of plant pathogen resistance. *Cell*, *88*, 695–705.

Cao, H., Li, X., & Dong, X. (1998). Generation of broad-spectrum disease resistance by overexpression of an essential regulatory gene in systemic acquired resistance. *Proceedings of the National Academy of Sciences of the United States of America*, *95*, 6531–6536.

Caplan, J., Padmanabhan, M., & Dinesh-Kumar, S. P. (2008). Plant NB-LRR immune receptors: From recognition to transcriptional reprogramming. *Cell Host & Microbe*, *3*, 126–135.

Century, K., Reuber, T. L., & Ratcliffe, O. J. (2008). Regulating the regulators: The future prospects for transcription-factor-based agricultural biotechnology products. *Plant Physiology*, *147*, 20–29.

Cessna, S. G., Sears, V. E., Dickman, M. B., & Low, P. S. (2000). Oxalic acid, a pathogenicity factor for *Sclerotinia sclerotiorum*, suppresses the oxidative burst of the host plant. *Plant Cell*, *12*, 2191–2200.

Chinchilla, D., Shan, L., He, P., de Vries, S., & Kemmerling, B. (2009). One for all: The receptor-associated kinase BAK1. *Trends in Plant Science*, *14*, 535–541.

Choi, J. J., Klosterman, S. J., & Hadwiger, L. A. (2004). A promoter from pea gene *DRR206* is suitable to regulate an elicitor-coding gene and develop disease resistance. *Phytopathology*, *94*, 651–660.

Consonni, C., Humphry, M. E., Hartmann, H. A., Livaja, M., Durner, J., & Westphal, L., et al. (2006). Conserved requirement for a plant host cell protein in powdery mildew pathogenesis. *Nature Genetics*, *38*, 716–720.

Cornelis, G. R., & Van Gijsegem, F. (2000). Assembly and function of type III secretory systems. *Annual Review of Microbiology*, *54*, 735–774.

Cox, P. W., & Hooley, P. (2009). Hydrophobins: New prospects for biotechnology. *Fungal Biology Reviews*, *23*, 40–47.

Cui, H., Xiang, T., & Zhou, J. M. (2009). Plant immunity: A lesson from pathogenic bacterial effector proteins. *Cellular Microbiology*, *11*, 1453–1461.

De Wit, P. J. (1992). Molecular characterization of gene-for-gene systems in plant–fungus interactions and the application of avirulence genes in control of plant pathogens. *Annual Review of Phytopathology*, *30*, 391–418.

De Wit, P. J., Mehrabi, R., Van den Burg, H. A., & Stergiopoulos, I. (2009). Fungal effector proteins: Past, present and future. *Molecular Plant Pathology*, *10*, 735–747.

Dickman, M. B., Park, Y. K., Oltersdorf, T., Li, W., Clemente, T., & French, R. (2001). Abrogation of disease development in plants expressing animal antiapoptotic genes.

Proceedings of the National Academy of Sciences of the United States of America, 98, 6957–6962.

Dong, X. (2004). NPR1, all things considered. *Current Opinion in Plant Biology, 7*, 547–552.

Dong, X., Ji, R., Guo, X., Foster, S. J., Chen, H., & Dong, C., et al. (2008). Expressing a gene encoding wheat oxalate oxidase enhances resistance to *Sclerotinia sclerotiorum* in oilseed rape (*Brassica napus*). *Planta, 228*, 331–340.

Durrant, W. E., & Dong, X. (2004). Systemic acquired resistance. *Annual Review of Phytopathology, 42*, 185–209.

Emani, C., Garcia, J. M., Lopata-Finch, E., Pozo, M. J., Uribe, P., & Kim, D. J., et al. (2003). Enhanced fungal resistance in transgenic cotton expressing an endochitinase gene from *Trichoderma virens*. *Plant Biotechnology Journal, 1*, 321–336.

Eulgem, T. (2005). Regulation of the *Arabidopsis* defense transcriptome. *Trends in Plant Science, 10*, 71–78.

Eulgem, T., Rushton, P. J., Schmelzer, E., Hahlbrock, K., & Somssich, I. E. (1999). Early nuclear events in plant defence signalling: Rapid gene activation by WRKY transcription factors. *EMBO Journal, 18*, 4689–4699.

Fitzgerald, H. A., Chern, M. S., Navarre, R., & Ronald, P. C. (2004). Overexpression of *(At)NPR1* in rice leads to a BTH- and environment-induced lesion-mimic/cell death phenotype. *Molecular Plant-Microbe Interactions, 17*, 140–151.

Fradin, E. F., Zhang, Z., Juarez Ayala, J. C., Castroverde, C. D., Nazar, R. N., & Robb, J., et al. (2009). Genetic dissection of *Verticillium* wilt resistance mediated by tomato Ve1. *Plant Physiology, 150*, 320–332.

Friedrich, L., Lawton, K., Dietrich, R., Willits, M., Cade, R., & Ryals, J. (2001). *NIM1* overexpression in *Arabidopsis* potentiates plant disease resistance and results in enhanced effectiveness of fungicides. *Molecular Plant-Microbe Interactions, 14*, 1114–1124.

Gage, M. J., Bruenn, J., Fischer, M., Sanders, D., & Smith, T. J. (2001). KP4 fungal toxin inhibits growth in *Ustilago maydis* by blocking calcium uptake. *Molecular Microbiology, 41*, 775–785.

Gfeller, A., Dubugnon, L., Liechti, R., & Farmer, E. E. (2010). Jasmonate biochemical pathway. *Science Signaling, 3*, cm3.

Gijzen, M., & Nurnberger, T. (2006). Nep1-like proteins from plant pathogens: Recruitment and diversification of the NPP1 domain across taxa. *Phytochemistry, 67*, 1800–1807.

Glazebrook, J. (2005). Contrasting mechanisms of defense against biotrophic and necrotrophic pathogens. *Annual Review of Phytopathology, 43*, 205–227.

Godoy, G., Steadman, J. R., Dickman, M. B., & Dam, R. (1990). Use of mutants to demonstrate the role of oxalic acid in pathogenicity of *Sclerotinia sclerotiorum* on *Phaseolus vulgaris*. *Physiological and Molecular Plant Pathology, 37*, 179–191.

Gomez-Ariza, J., Campo, S., Rufat, M., Estopa, M., Messeguer, J., & San Segundo, B., et al. (2007). Sucrose-mediated priming of plant defense responses and broad-spectrum disease resistance by overexpression of the maize pathogenesis-related PRms protein in rice plants. *Molecular Plant-Microbe Interactions, 20*, 832–842.

Gomez-Gomez, L., & Boller, T. (2002). Flagellin perception: A paradigm for innate immunity. *Trends in Plant Science, 7*, 251–256.

Greenberg, J. T., & Yao, N. (2004). The role and regulation of programmed cell death in plant-pathogen interactions. *Cellular Microbiology, 6*, 201–211.

Gurr, S. J., & Rushton, P. J. (2005). Engineering plants with increased disease resistance: How are we going to express it? *Trends in Biotechnology, 23*, 283–290.

Gurr, S. J., & Rushton, P. J. (2005). Engineering plants with increased disease resistance: What are we going to express? *Trends in Biotechnology, 23*, 275–282.

Gust, A. A., Brunner, F. and Nurnberger, T. (2010) Biotechnological concepts for improving plant innate immunity. *Current Opinion in Biotechnology, 21*, 204–210.

Gutterson, N., & Reuber, T. L. (2004). Regulation of disease resistance pathways by AP2/ERF transcription factors. *Current Opinion in Plant Biology, 7*, 465–471.

Hain, R., Reif, H. J., Krause, E., Langebartels, R., Kindl, H., & Vornam, B., et al. (1993). Disease resistance results from foreign phytoalexin expression in a novel plant. *Nature, 361*, 153–156.

Hammami, R., Ben Hamida, J., Vergoten, G., & Fliss, I. (2009). PhytAMP: A database dedicated to antimicrobial plant peptides. *Nucleic Acids Research, 37*, D963–D968.

Heise, A., Lippok, B., Kirsch, C., & Hahlbrock, K. (2002). Two immediate-early pathogen-responsive members of the *AtCMPG* gene family in *Arabidopsis thaliana* and the W-box-containing elicitor-response element of *AtCMPG1*. *Proceedings of the National Academy of Sciences of the United States of America, 99*, 9049–9054.

Himmelbach, A., Liu, L., Zierold, U., Altschmied, L., Maucher, H., & Beier, F., et al. (2010). Promoters of the barley germin-like *GER4* gene cluster enable strong transgene expression in response to pathogen attack. *Plant Cell, 22*, 937–952.

Hittalmani, S., Parco, A., Mew, T. V., Zeigler, R. S., & Huang, N. (2000). Fine mapping and DNA marker-assisted pyramiding of the three major genes for blast resistance in rice. *Theoretical and Applied Genetics, 100*, 1121–1128.

Hu, H., Xiong, L., & Yang, Y. (2005). Rice *SERK1* gene positively regulates somatic embryogenesis of cultured cell and host defense response against fungal infection. *Planta, 222*, 107–117.

Huffaker, A., Pearce, G., & Ryan, C. A. (2006). An endogenous peptide signal in *Arabidopsis* activates components of the innate immune response. *Proceedings of the National Academy of Sciences of the United States of America, 103*, 10098–10103.

Jang, Y. S., Sohn, S. I., & Wang, M. H. (2006). The *hrpN* gene of *Erwinia amylovora* stimulates tobacco growth and enhances resistance to *Botrytis cinerea*. *Planta, 223*, 449–456.

Jayaraj, J., & Punja, Z. K. (2007). Combined expression of chitinase and lipid transfer protein genes in transgenic carrot plants enhances resistance to foliar fungal pathogens. *Plant Cell Reports, 26*, 1539–1546.

Jones, J. D., & Dangl, J. L. (2006). The plant immune system. *Nature, 444*, 323–329.

Kawchuk, L. M., Hachey, J., Lynch, D. R., Kulcsar, F., van Rooijen, G., & Waterer, D. R., et al. (2001). Tomato *Ve* disease resistance genes encode cell surface-like receptors. *Proceedings of the National Academy of Sciences of the United States of America, 98*, 6511–6515.

Kearney, B., & Staskawicz, B. J. (1990). Widespread distribution and fitness contribution of *Xanthomonas campestris* avirulence gene *avrBs2*. *Nature, 346*, 385–386.

Keller, H., Pamboukdjian, N., Ponchet, M., Poupet, A., Delon, R., & Verrier, J. L., et al. (1999). Pathogen-induced elicitin production in transgenic tobacco generates a hypersensitive response and nonspecific disease resistance. *Plant Cell, 11*, 223–235.

Kesarwani, M., Azam, M., Natarajan, K., Mehta, A., & Datta, A. (2000). Oxalate decarboxylase from *Collybia velutipes*. Molecular cloning and its overexpression to confer resistance to fungal infection in transgenic tobacco and tomato. *Journal of Biological Chemistry, 275*, 7230–7238.

Kirsch, C., Logemann, E., Lippok, B., Schmelzer, E., & Hahlbrock, K. (2001). A highly specific pathogen-responsive promoter element from the immediate-early activated *CMPG1* gene in *Petroselinum crispum*. *Plant Journal, 26*, 217–227.

Kliebenstein, D. J., & Rowe, H. C. (2009). Plant science. Anti-rust antitrust. *Science, 323*, 1301–1302.

Koo, J. C., Chun, H. J., Park, H. C., Kim, M. C., Koo, Y. D., & Koo, S. C., et al. (2002). Over-expression of a seed specific hevein-like antimicrobial peptide from *Pharbitis nil* enhances resistance to a fungal pathogen in transgenic tobacco plants. *Plant Molecular Biology, 50*, 441–452.

Kovalchuk, N., Li, M., Wittek, F., Reid, N., Singh, R., & Shirley, N., et al. (2010). Defensin promoters as potential tools for engineering disease resistance in cereal grains. *Plant Biotechnology Journal, 8*, 47–64.

Kunze, G., Zipfel, C., Robatzek, S., Niehaus, K., Boller, T., & Felix, G. (2004). The N terminus of bacterial elongation factor Tu elicits innate immunity in *Arabidopsis* plants. *Plant Cell, 16*, 3496–3507.

Lacombe, S., Rougon-Cardoso, A., Sherwood, E., Peeters, N., Dahlbeck, D., & van Esse, H. P., et al. (2010). Interfamily transfer of a plant pattern-recognition receptor confers broad-spectrum bacterial resistance. *Nature Biotechnology, 28*, 365–369.

Lee, S. C., Hwang, I. S., Choi, H. W., & Hwang, B. K. (2008). Involvement of the pepper antimicrobial protein *CaAMP1* gene in broad spectrum disease resistance. *Plant Physiology, 148*, 1004–1020.

Lee, S. W., Han, S. W., Sririyanum, M., Park, C. J., Seo, Y. S., & Ronald, P. C. (2009). A type I-secreted, sulfated peptide triggers XA21-mediated innate immunity. *Science, 326*, 850–853.

Lin, Y. Z., Chen, H. Y., Kao, R., Chang, S. P., Chang, S. J., & Lai, E. M. (2008). Proteomic analysis of rice defense response induced by probenazole. *Phytochemistry, 69*, 715–728.

Liu, J. J., Ekramoddoullah, A. K., Piggott, N., & Zamani, A. (2005). Molecular cloning of a pathogen/wound-inducible *PR10* promoter from *Pinus monticola* and characterization in transgenic *Arabidopsis* plants. *Planta, 221*, 159–169.

Lorito, M., Woo, S. L., Garcia, I., Colucci, G., Harman, G. E., & Pintor-Toro, J. A., et al. (1998). Genes from mycoparasitic fungi as a source for improving plant resistance to fungal pathogens. *Proceedings of the National Academy of Sciences of the United States of America, 95*, 7860–7865.

Malnoy, M., Reynoird, J. P., Borejsza-Wysocka, E. E., & Aldwinckle, H. S. (2006). Activation of the pathogen-inducible *Gst1* promoter of potato after elicitation by *Venturia inaequalis* and *Erwinia amylovora* in transgenic apple (*Malus x domestica*). *Transgenic Research, 15*, 83–93.

Manners, J. M. (2007). Hidden weapons of microbial destruction in plant genomes. *Genome Biology, 8*, 225.

Manosalva, P. M., Davidson, R. M., Liu, B., Zhu, X., Hulbert, S. H., & Leung, H., et al. (2009). A germin-like protein gene family functions as a complex quantitative trait locus conferring broad-spectrum disease resistance in rice. *Plant Physiology, 149*, 286–296.

Martini, N., Egen, M., Runtz, I., & Strittmatter, G. (1993). Promoter sequences of a potato pathogenesis-related gene mediate transcriptional activation selectively upon fungal infection. *Molecular & General Genetics, 236*, 179–186.

McDowell, J. M., & Woffenden, B. J. (2003). Plant disease resistance genes: Recent insights and potential applications. *Trends in Biotechnology, 21*, 178–183.

Mei, C., Qi, M., Sheng, G., & Yang, Y. (2006). Inducible overexpression of a rice allene oxide synthase gene increases the endogenous jasmonic acid level, PR gene expression, and host resistance to fungal infection. *Molecular Plant-Microbe Interactions, 19*, 1127–1137.

Michelmore, R. W. (2003). The impact zone: Genomics and breeding for durable disease resistance. *Current Opinion in Plant Biology, 6*, 397–404.

Midoh, N., & Iwata, M. (1996). Cloning and characterization of a probenazole-inducible gene for an intracellular pathogenesis-related protein in rice. *Plant and Cell Physiology, 37*, 9–18.

Montesinos, E. (2007). Antimicrobial peptides and plant disease control. *FEMS Microbiology Letters, 270*, 1–11.

Moreno, A. B., Penas, G., Rufat, M., Bravo, J. M., Estopa, M., & Messeguer, J., et al. (2005). Pathogen-induced production of the antifungal AFP protein from *Aspergillus giganteus* confers resistance to the blast fungus *Magnaporthe grisea* in transgenic rice. *Molecular Plant-Microbe Interactions, 18*, 960–972.

Moschou, P. N., Sarris, P. F., Skandalis, N., Andriopoulou, A. H., Paschalidis, K. A., & Panopoulos, N. J., et al. (2009). Engineered polyamine catabolism preinduces tolerance of tobacco to bacteria and oomycetes. *Plant Physiology, 149*, 1970–1981.

Moscou, M. J., & Bogdanove, A. J. (2009). A simple cipher governs DNA recognition by TAL effectors. *Science, 326*, 1501.

Oldroyd, G. E., & Staskawicz, B. J. (1998). Genetically engineered broad-spectrum disease resistance in tomato. *Proceedings of the National Academy of Sciences of the United States of America, 95*, 10300–10305.

Palm, C. J., Costa, M. A., An, G., & Ryan, C. A. (1990). Wound-inducible nuclear protein binds DNA fragments that regulate a proteinase inhibitor II gene from potato. *Proceedings of the National Academy of Sciences of the United States of America, 87*, 603–607.

Pandey, S. P., & Somssich, I. E. (2009). The role of WRKY transcription factors in plant immunity. *Plant Physiology, 150*, 1648–1655.

Park, C. M., Bruenn, J. A., Ganesa, C., Flurkey, W. F., Bozarth, R. F., & Koltin, Y. (1994). Structure and heterologous expression of the *Ustilago maydis* viral toxin KP4. *Molecular Microbiology, 11*, 155–164.

Parlevliet, J. (2002). Durability of resistance against fungal, bacterial and viral pathogens; present situation. *Euphytica, 124*, 147–156.

Pavan, S., Jacobsen, E., Visser, R. G., & Bai, Y. (2010). Loss of susceptibility as a novel breeding strategy for durable and broad-spectrum resistance. *Molecular Breeding, 25*, 1–12.

Pedley, K. F., & Martin, G. B. (2005). Role of mitogen-activated protein kinases in plant immunity. *Current Opinion in Plant Biology, 8*, 541–547.

Peng, J. L., Bao, Z. L., Ren, H. Y., Wang, J. S., & Dong, H. S. (2004). Expression of harpin(xoo) in transgenic tobacco induces pathogen defense in the absence of hypersensitive cell death. *Phytopathology, 94*, 1048–1055.

Peschen, D., Li, H. P., Fischer, R., Kreuzaler, F., & Liao, Y. C. (2004). Fusion proteins comprising a *Fusarium*-specific antibody linked to antifungal peptides protect plants against a fungal pathogen. *Nature Biotechnology, 22*, 732–738.

Petersen, M., Brodersen, P., Naested, H., Andreasson, E., Lindhart, U., & Johansen, B., et al. (2000). *Arabidopsis* map kinase 4 negatively regulates systemic acquired resistance. *Cell, 103*, 1111–1120.

Pieterse, C. M., Leon-Reyes, A., Van der Ent, S., & Van Wees, S. C. (2009). Networking by small-molecule hormones in plant immunity. *Nature Chemical Biology, 5*, 308–316.

Piffanelli, P., Ramsay, L., Waugh, R., Benabdelmouna, A., D'Hont, A., & Hollricher, K., et al. (2004). A barley cultivation-associated polymorphism conveys resistance to powdery mildew. *Nature, 430*, 887–891.

Poland, J. A., Balint-Kurti, P. J., Wisser, R. J., Pratt, R. C., & Nelson, R. J. (2009). Shades of gray: The world of quantitative disease resistance. *Trends in Plant Science, 14*, 21–29.

Pontier, D., Balague, C., Bezombes-Marion, I., Tronchet, M., Deslandes, L., & Roby, D. (2001). Identification of a novel pathogen-responsive element in the promoter of the tobacco gene *HSR203J*, a molecular marker of the hypersensitive response. *Plant Journal, 26*, 495–507.

Pontier, D., Godiard, L., Marco, Y., & Roby, D. (1994). *hsr203J*, a tobacco gene whose activation is rapid, highly localized and specific for incompatible plant/pathogen interactions. *Plant Journal, 5*, 507–521.

Punja, Z. K., & Utkhede, R. S. (2003). Using fungi and yeasts to manage vegetable crop diseases. *Trends in Biotechnology, 21*, 400–407.

Ricci, P., Bonnet, P., Huet, J. C., Sallantin, M., Beauvais-Cante, F., & Bruneteau, M., et al. (1989). Structure and activity of proteins from pathogenic fungi *Phytophthora* eliciting necrosis and acquired resistance in tobacco. *European Journal of Biochemistry, 183*, 555–563.

Romer, P., Recht, S., & Lahaye, T. (2009). A single plant resistance gene promoter engineered to recognize multiple TAL effectors from disparate pathogens. *Proceedings of the National Academy of Sciences of the United States of America, 106*, 20526–20531.

Rommens, C. M., & Kishore, G. M. (2000). Exploiting the full potential of disease-resistance genes for agricultural use. *Current Opinion in Biotechnology, 11*, 120–125.

Rossi, L. M., Rangasamy, P., Zhang, J., Qiu, X. Q., & Wu, G. Y. (2008). Research advances in the development of peptide antibiotics. *Journal of Pharmaceutical Sciences, 97*, 1060–1070.

Rushton, P. J., Reinstadler, A., Lipka, V., Lippok, B., & Somssich, I. E. (2002). Synthetic plant promoters containing defined regulatory elements provide novel insights into pathogen- and wound-induced signaling. *Plant Cell, 14*, 749–762.

Rushton, P. J., & Somssich, I. E. (1998). Transcriptional control of plant genes responsive to pathogens. *Current Opinion in Plant Biology, 1*, 311–315.

Rushton, P. J., Torres, J. T., Parniske, M., Wernert, P., Hahlbrock, K., & Somssich, I. E. (1996). Interaction of elicitor-induced DNA-binding proteins with elicitor response elements in the promoters of parsley PR1 genes. *EMBO Journal, 15*, 5690–5700.

Salmeron, J. M., Oldroyd, G. E., Rommens, C. M., Scofield, S. R., Kim, H. S., & Lavelle, D. T., et al. (1996). Tomato *Prf* is a member of the leucine-rich repeat class of plant disease resistance genes and lies embedded within the *Pto* kinase gene cluster. *Cell, 86*, 123–133.

Schlaich, T., Urbaniak, B. M., Malgras, N., Ehler, E., Birrer, C., & Meier, L., et al. (2006). Increased field resistance to *Tilletia caries* provided by a specific antifungal virus gene in genetically engineered wheat. *Plant Biotechnology Journal, 4*, 63–75.

Segura, A., Moreno, M., Madueno, F., Molina, A., & Garcia-Olmedo, F. (1999). Snakin-1, a peptide from potato that is active against plant pathogens. *Molecular Plant-Microbe Interactions*, *12*, 16–23.

Senthilkumar, R., Cheng, C. P., & Yeh, K. W. (2010). Genetically pyramiding protease-inhibitor genes for dual broad-spectrum resistance against insect and phytopathogens in transgenic tobacco. *Plant Biotechnology Journal*, *8*, 65–75.

Shah, J. (2009). Plants under attack: Systemic signals in defence. *Current Opinion in Plant Biology*, *12*, 459–464.

Shao, M., Wang, J., Dean, R. A., Lin, Y., Gao, X., & Hu, S. (2008). Expression of a harpin-encoding gene in rice confers durable nonspecific resistance to *Magnaporthe grisea*. *Plant Biotechnology Journal*, *6*, 73–81.

Shin, R., Park, J. M., An, J. M., & Paek, K. H. (2002). Ectopic expression of *Tsi1* in transgenic hot pepper plants enhances host resistance to viral, bacterial, and oomycete pathogens. *Molecular Plant-Microbe Interactions*, *15*, 983–989.

Shin, S., Mackintosh, C. A., Lewis, J., Heinen, S. J., Radmer, L., & Dill-Macky, R., et al. (2008). Transgenic wheat expressing a barley class II chitinase gene has enhanced resistance against *Fusarium graminearum*. *Journal of Experimental Botany*, *59*, 2371–2378.

Singh, S., Sidhu, J. S., Huang, N., Vikal, Y., Li, Z., & Brar, D. S., et al. (2001). Pyramiding three bacterial blight resistance genes (*xa5*, *xa13* and *Xa21*) using marker-assisted selection into indica rice cultivar PR106. *Theoretical and Applied Genetics*, *102*, 1011–1015.

Sohn, S. I., Kim, Y. H., Kim, B. R., Lee, S. Y., Lim, C. K., & Hur, J. H., et al. (2007). Transgenic tobacco expressing the *hrpN(EP)* gene from *Erwinia pyrifoliae* triggers defense responses against *Botrytis cinerea*. *Molecules and Cells*, *24*, 232–239.

Swords, K. M., Dahlbeck, D., Kearney, B., Roy, M., & Staskawicz, B. J. (1996). Spontaneous and induced mutations in a single open reading frame alter both virulence and avirulence in *Xanthomonas campestris* pv. *vesicatoria avrBs2*. *Journal of Bacteriology*, *178*, 4661–4669.

Tai, T. H., Dahlbeck, D., Clark, E. T., Gajiwala, P., Pasion, R., & Whalen, M. C., et al. (1999). Expression of the *Bs2* pepper gene confers resistance to bacterial spot disease in tomato. *Proceedings of the National Academy of Sciences of the United States of America*, *96*, 14153–14158.

Takakura, Y., Che, F. S., Ishida, Y., Tsutsumi, F., Kurotani, K., & Usami, S., et al. (2008). Expression of a bacterial flagellin gene triggers plant immune responses and confers disease resistance in transgenic rice plants. *Molecular Plant Pathology*, *9*, 525–529.

Tang, X., Xie, M., Kim, Y. J., Zhou, J., Klessig, D. F., & Martin, G. B. (1999). Overexpression of *Pto* activates defense responses and confers broad resistance. *Plant Cell*, *11*, 15–29.

Tao, Y., Yuan, F., Leister, R. T., Ausubel, F. M., & Katagiri, F. (2000). Mutational analysis of the *Arabidopsis* nucleotide binding site-leucine-rich repeat resistance gene *RPS2*. *Plant Cell*, *12*, 2541–2554.

Vailleau, F., Daniel, X., Tronchet, M., Montillet, J. L., Triantaphylides, C., & Roby, D. (2002). A R2R3-MYB gene, *AtMYB30*, acts as a positive regulator of the hypersensitive cell death program in plants in response to pathogen attack. *Proceedings of the National Academy of Sciences of the United States of America*, *99*, 10179–10184.

van der Biezen, E. A. (2001). Quest for antimicrobial genes to engineer disease-resistant crops. *Trends in Plant Science*, *6*, 89–91.

Wally, O., Jayaraj, J., & Punja, Z. K. (2009). Broad-spectrum disease resistance to necrotrophic and biotrophic pathogens in transgenic carrots (*Daucus carota* L.) expressing an *Arabidopsis NPR1 gene*. *Planta*, *231*, 131–141.

Wang, G. L., Song, W. Y., Ruan, D. L., Sideris, S., & Ronald, P. C. (1996). The cloned gene, *Xa21*, confers resistance to multiple *Xanthomonas oryzae* pv. *oryzae* isolates in transgenic plants. *Molecular Plant-Microbe Interactions*, *9*, 850–855.

Wang, W., Devoto, A., Turner, J. G., & Xiao, S. (2007). Expression of the membrane-associated resistance protein RPW8 enhances basal defense against biotrophic pathogens. *Molecular Plant-Microbe Interactions*, *20*, 966–976.

White, F. F., Potnis, N., Jones, J. B., & Koebnik, R. (2009). The type III effectors of *Xanthomonas*. *Molecular Plant Pathology*, *10*, 749–766.

Wijayanto, T., Barker, S. J., Wylie, S. J., Gilchrist, D. G., & Cowling, W. A. (2009). Significant reduction of fungal disease symptoms in transgenic lupin (*Lupinus angustifolius*) expressing the anti-apoptotic baculovirus gene p35. *Plant Biotechnology Journal*, *7*, 778–790.

Wise, R. P., Moscou, M. J., Bogdanove, A. J., & Whitham, S. A. (2007). Transcript profiling in host-pathogen interactions. *Annual Review of Phytopathology*, *45*, 329–369.

Xiao, S., Ellwood, S., Findlay, K., Oliver, R. P., & Turner, J. G. (1997). Characterization of three loci controlling resistance of *Arabidopsis thaliana* accession Ms-0 to two powdery mildew diseases. *Plant Journal*, *12*, 757–768.

Yamamizo, C., Kuchimura, K., Kobayashi, A., Katou, S., Kawakita, K., & Jones, J. D., et al. (2006). Rewiring mitogen-activated protein kinase cascade by positive feedback confers potato blight resistance. *Plant Physiology*, *140*, 681–692.

Yang, Y., Zhang, H., Li, G., Li, W., Wang, X., & Song, F. (2009). Ectopic expression of MgSM1, a Cerato-platanin family protein from *Magnaporthe grisea*, confers broad-spectrum disease resistance in *Arabidopsis*. *Plant Biotechnology Journal*, *7*, 763–777.

Yevtushenko, D. P., Romero, R., Forward, B. S., Hancock, R. E., Kay, W. W., & Misra, S. (2005). Pathogen-induced expression of a cecropin A-melittin antimicrobial peptide gene confers antifungal resistance in transgenic tobacco. *Journal of Experimental Botany*, *56*, 1685–1695.

Zhang, Y., Fan, W., Kinkema, M., Li, X., & Dong, X. (1999). Interaction of NPR1 with basic leucine zipper protein transcription factors that bind sequences required for salicylic acid induction of the *PR-1* gene. *Proceedings of the National Academy of Sciences of the United States of America*, *96*, 6523–6528.

Zipfel, C. (2009). Early molecular events in PAMP-triggered immunity. *Current Opinion in Plant Biology*, *12*, 414–420.

Zipfel, C., Kunze, G., Chinchilla, D., Caniard, A., Jones, J. D., & Boller, T., et al. (2006). Perception of the bacterial PAMP EF-Tu by the receptor EFR restricts *Agrobacterium*-mediated transformation. *Cell*, *125*, 749–760.

Controlling plant response to the environment: Viral diseases

22

Munir Mawassi Abed Gera

Agricultural Research Organization, the Volcani Center, Bet Dagan, Israel

TABLE OF CONTENTS

Introduction

Plant viruses cause significant agricultural losses worldwide, particularly in tropical and subtropical ecosystems. Viruses are very difficult to combat due to the scarcity of effective control measures, placing them among the most important agricultural pathogens. Globalization and the development of international trade of agricultural and horticultural produce are breaking down traditional geographical barriers to pathogen movement (Gera et al., 2010). As trade becomes more international, the potential for importation of non-indigenous plant viruses grows significantly. Despite some successes in eradication, many of these new pathogens are now established, and industry must deal with the consequences of their introduction. Another factor that might influence the spread of plant viruses in the future is, of course, global climate change (Garrett et al., 2006). Although it is difficult to reach firm conclusions about the potential direct influence on virus diseases, global climate change could have a significant, indirect influence on patterns of virus spread. Examples include changes in cropping patterns (e.g., subtropical crops planted in previously more temperate regions), and changes in the distribution of virus vectors, either through vector invasions into new territories or enhanced over-wintering of vectors that allow viruses to establish themselves in areas from which they had previously been restricted. The western flower thrips (WFT, *Frankliniella occidentalis*), the most widespread vector of *Tospoviruses* (Ullman et al., 1997), illustrates such potential impact. Although WFT is currently established only in glasshouses in many areas of northern Europe and North America, global warming and milder winters increase the potential for this insect to become established outdoors as well (Kirk and Terry, 2003). This would introduce the possibility of new and devastating diseases in field crops, as already witnessed in warmer climates.

Most emerging infectious diseases of plants are caused by viruses (Anderson et al., 2004; Gera et al., 2010). This is exemplified by recent widespread epidemics of viruses from the genus *Tospovirus* in the family *Bunyaviridae*, which cause crop losses that account for losses of more than one billion dollars per year (Prins and Goldbach, 1998). Viruses of the genus *Begomovirus* in the family *Geminiviridae* are devastating and cause the loss of millions of tons of food annually (Fargette et al., 2006). Outbreaks of plum pox virus (PPV) in stone fruit trees in the Northern Hemisphere, and of Pepino mosaic virus (PepMV) in tomato crops in Europe and North America also have been significantly devastating.

Viral diseases can affect food quality as well as reduce yields, yet quality is also affected by measures such as spraying with pesticides used to control insect vectors. An alternative control strategy for viruses is the use of crop cultivars or varieties that are genetically resistant. These resistant genotypes carry a heritable trait or set of traits that are responsible for the suppression of virus multiplication and/or spread even under environmental conditions that favor virus infection of the given species. Plant breeding has been used to

DOI: 10.1016/B978-0-12-381466-1.00022-5

incorporate resistance genes into commercially important varieties, but this method suffers from the limitation of sexual compatibility and the species gene pool. Gene transfer to plants abolishes both of these limitations by overcoming species boundaries and therefore extending the gene pool to all life on Earth. In some cases, natural resistance genes have been transferred from one plant to another. In other cases pathogen-related proteins or defensive peptides have been over-expressed. Recombinant antibodies can also be included in the category of heterologous resistance proteins. Various types of virus-resistant antibodies, including single gene Fv fragments (scFvs), have been expressed in plants. The use of genetic resistance is advantageous because it can provide effective protection with no additional cost implications for the producer during the growing season. It is also environmentally friendly and safe for the consumer. Most crop species have virus-resistance traits available, and in many cases cultivars with varying degrees of resistance have achieved commercial status. However, many issues concerning the use of genetic resistance for the sustainable control of viral diseases remain to be investigated. This chapter is restricted to the discussion of development and potential use of natural and transgenic resistant varieties.

Phytosanitation and Quarantine Regulation

Some viruses have been widely distributed around the world and have caused significant damage and economic loss. When trade in plant propagation materials was initiated many years ago, the threat of distributing viruses was minimized because much of the germplasm was exchanged as seed. In the last six decades, the transport of vegetatively propagated clonal material has resulted in the dissemination of many viral diseases. Recently, vegetative propagation of horticultural crops served as virus carriers. In some instances the agents were spread before it was recognized that the material was diseased. In other instances, methods of detection for a specific disease agent were not developed for certification purposes.

Quarantine regulations are designed primarily to exclude the movement of pathogens into geographic areas where they were not previously known to occur. The past two decades have seen the development of an increasingly secure global system of plant health inspection and quarantine (phytosanitary system) that balances agriculture protection against international trade (McGee, 1997). Phytosanitary regulations and policies established by individual countries and increasingly by organizations formed by groups of countries (i.e., the European and Mediterranean Plant Protection Organization, EPPO; North American Plant Protection Organization, NAPPO) are designed to prevent the entry and spread of pests, including viruses, over natural barriers. EPPO is a regional organization operating under guidelines from the International Plant Protection. Today, the organization includes 50 member states in the European and Mediterranean region (http://www.eppo.org). EPPO sets regional standards for phytosanitary protection of plant products. These regulations are enforced by quarantine and other authorized inspection agencies (EPPO, 2009).

Global surveillance and reporting of plant viral diseases require an international coalition that does not yet exist. There is a lack of standardization in testing protocols for seed-borne diseases and significant variation in seed protection treatments among countries. Strengthening the global phytosanitary system (i.e., international agreements on the inspection and quarantining of plants to prevent spread of pathogens and disease) requires improved information, standardized seed testing, improved access to seed treatment protocols, and revised pathology research priorities (McGee, 1997), but funding for adequate diagnostic facilities and programs is often lacking. The economic, environmental, and social costs of incorrect diagnoses are high.

Reducing the threats and impact of emerging plant viral diseases will require novel approaches to integrated research and long-term commitments from the scientific and political communities alike.

Transmission of Plant Viruses

Viruses are unable to pass through the cuticle of plant cells. They have evolved to avoid the cuticle by passive transmission through vegetative propagation and seed transmission. They penetrate the host through wounds that occur in mechanical transmission or transmission by vectors.

Plant viruses are transmitted in many different ways including mechanical rubbing and grafting, and naturally through pollen, seed, arthropods, nematodes, and fungi (Lawson et al., 1995). Knowledge of the factors that influence spread is essential for development of control procedures.

Cultural Strategies of Virus Control

At present no antiviral pesticides that affect the virus without harming the host plant are available. It is, however, possible to reduce the dissemination of viruses by interfering with their spread by vectors. Many cultural procedures used for virus control are aimed at eradicating or altering one or more of the primary participants in the transmission process (vector, virus source plants, and the crop) or preventing their coming together. In this section we describe cultural strategies that were developed to impede the spread of soil- and air-borne viruses.

Management of soil-borne viruses

Soil-borne virus diseases are spread either by soil organisms, like nematodes and fungi, or mechanically transmitted without the assistance of a known vector (Hull, 2002). The latter group of viruses is characterized by high stability and long persistence in soil and drainage water (Mandahar, 1990; Lewandowski, 2000). The use of "intermediating media" such as condensed peat trays or Perlite sleeves can serve as an efficient means to block soil-borne epidemics of tobamoviruses in infested soils (Antignus et al., 2005). This procedure proved highly efficient when tested to protect pepper crops in commercial greenhouses from root infection by pepper mild mottle virus (PMMoV; Y. Antignus, personal communication).

Squash root-stocks were also used successfully to protect cucumbers, watermelon, melon, tomato, and eggplants from soil-borne pathogens (Lee, 1994; Oda, 1999). Pepper root-stocks carrying the L^4 allele of the gene for resistance against PMMoV successfully protected susceptible pepper scions from infection through virus-infested soil in commercial greenhouses (Y. Antignus, personal communication).

Management of airborne viruses

The major group of plant viruses is airborne and transmitted in nature by insects that feed on plants; for example, 192 aphid species are known as vectors of 275 plant viruses (Hull, 2002). Whiteflies and thrips are also known as major virus vectors that are responsible for the spread of virus diseases that may devastate economically important crops. The simplest way to prevent contact between these viruliferous vectors and the crop is by forming a physical barrier to mechanically prevent the invasion of the vector into the crop. Crops grown in the open field can be protected efficiently from virus infection by either floating row covers that lay over the plants without any support, or by the use of "low tunnels" where the protecting cover rests on wire arches. In both cases the protection is time limited because the cover should be removed for cultural practices or bee pollination (Hilje et al., 2001). However, delay in infection may mitigate effects of yield reduction (Antignus and Ben-Yakir, 2004). In general, any material capable of blocking insect invasion on the basis of its hole size can form an efficient barrier. The materials used include polypropylene (agryl), a lightweight material with a texture that allows a certain air flow, and fine mesh nets (50 mesh) that were first implemented in Israeli greenhouses to protect tomato crops from tomato yellow leaf curl virus (TYLCV; Cohen and Berlinger, 1986). Similar protection effects against whiteflies, aphids, and their vectored viruses were reported later by Csizinszky et al. (1997) and Summers et al. (2005).

Vegetable crops are efficiently protected from insect pests (aphids, thrips, whiteflies), and viral diseases transmitted by them, when grown in "walk-in" tunnels or greenhouses covered either with UV-absorbing polyethylene films or with UV-absorbing 50-mesh nets (Antignus et al., 1996, 2001; Antignus, 2001). Tomatoes and cucumbers grown under UV-absorbing greenhouse covers were highly protected against whitefly-borne viruses such as TYLCV and cucurbit yellow stunting disorder virus (CYSDV; Antignus et al., 1996). These results suggest that the elimination of the UV portion of the light spectrum is interfering with the "UV vision" of insects and, as a consequence, may affect their ability to orient themselves to the crop (Antignus, 2001; Raviv and Antignus, 2004). Similar results were reported by Doukas and Payne (2007) and Mutwiwa et al. (2005).

Resistance to Virus Transmission by Insects

In epidemiological terms, approximately 80% of viruses depend on transmission by vectors, mainly insects, but also by nematodes, fungi, and mites (Stacesmith and Hamilton, 1988). For vector transmission, viruses must be transmitted efficiently to a new host when the vector feeds (Hull, 2002). The majority of plant viruses are transmitted by insects through a highly specific process in which interactions among the virus, vector, and plant must occur (Ng and Falk, 2006). Resistance mechanisms acting against vectors have been historically classified as antixenosis if they influence vector behavior in terms of feeding preference (Kogan and Ortman, 1978), or as antibiosis if they increase mortality or reduce fitness or reproductive capacity of the vector (Smith, 1989). Both types of resistance may be due to the presence of non-glandular trichomes or waxy surfaces that act as physical barriers (Sadasivam and Thayumanavan, 2003), or the presence of certain secondary metabolites or deterrent compounds on leaf surfaces, vacuoles, or glandular trichomes that may possess insecticidal or antifeedant properties (Singer et al., 2003). Furthermore, these metabolites may play an active role in resistance, because their synthesis can be enhanced by jasmonic acid (JA) and suppressed by salicylic acid (SA; Traw et al., 2003; Li et al., 2004).

Genes that confer aphid resistance have been identified and mapped at least in lettuce, melon, tomato, soybean, and wheat (Klingler et al., 2005; Li et al., 2007; Wroblewski et al., 2007; Mensah et al., 2008). However, only two of them have been cloned and characterized (Dogimont et al., 2008), both of them belonging to the nucleotide binding site leucine-rich repeat (NBS-LRR) family of resistance genes (Dangl and Jones, 2001; Takken et al., 2006). The first was *Mi-1*, cloned from a wild relative of tomato (*Solanum peruvianum*), which showed high levels of resistance to root-knot nematodes (Rossi et al., 1998), to the whitefly *Bemisia tabaci* (Milligan et al., 1998), and to the aphid *Macrosiphum euphorbiae* (Blackman and Eastop, 2000). *Mi-1* is constitutively expressed in roots and leaves, and the resistance that it confers against nematodes is associated with a hypersensitive response (HR; Lopez-Perez et al., 2006), whereas for whiteflies and aphids it appears to inhibit their feeding, fecundity, and survival on leaves of mature plants (Martinez de Ilarduya et al., 2003).

The second NBS-LRR resistance gene is *Vat*, which confers resistance to *Aphis gossypii* and to viruses that are non-persistently transmitted by this vector in melon (Pitrat and Lecoq, 1980). Using a map-based cloning strategy and functional analysis in transgenic melon plants, Dogimont et al. (2008) demonstrated that a single gene encoding a coiled-coil (CC)-NBS-LRR protein confers both functions. Recognition of an aphid gene product by *Vat* probably triggers the plant's responses, preventing aphid feeding, reducing fecundity and survival, and inhibiting non-specific antiviral responses (Dogimont et al., 2008).

Pathogen-derived resistance

Pathogen-derived resistance (PDR) is a phenomenon where transgenic plants expressing genes or sequences of a pathogen are protected against the cognate or related pathogens. Sanford and Johnston (1985) apply the concept of PDR to the bacteriophage Qβ in *Escherichia coli*. They proposed that pathogens express genes or molecules critical for pathogenesis, and expressing dysfunctional pathogen-derived gene products in a susceptible host results in disrupting pathogen–host interaction and inhibiting the pathogen. The application of the PDR concept to plant viruses was first demonstrated by Powell-Abel et al. (1986), where they showed that transgenic *Nicotiana tabacum* cv. *Xanthi* and cv. *Samsun* expressing the coat protein gene of tobacco mosaic virus (TMV) exhibited resistance following mechanical inoculation of TMV. Resistance of transgenic plants varied as some plants did not express symptoms of disease for the duration of the experiment, whereas other plants exhibited a delay of up to 14 days in symptom development. Nejidat and Beachy (1990) demonstrated that resistance in transgenic plants was strong to tobamoviruses closely related to TMV, but weak or not detectable to distantly related tobamoviruses. Following the discovery by Powell-Abel et al. (1986), the practical effectiveness of PDR for controlling plant viruses has been established for many plant viruses in many crops. Resistance was engineered primarily by using coat protein genes, as well as by other viral proteins such as the movement protein RdRp and proteinase. Yet over the past decade, much work has shown that PDR for plant viruses are RNA-mediated and can occur through expression of non-coding RNA viral sequences of satellite RNA, defective interfering RNA, and 5′ and 3′ non-coding sequences (Table 22.1). It is becoming obvious that almost any genetic element of any virus can have the potential to confer resistance to virus infection in plants.

Table 22.1 Examples of pathogen-derived sequences used to control virus diseases

Pathogen-derived genes or sequences	Examples of resistance against	Hosts
Coat protein	TMV, PVX, TVMV	Tobacco, potato, tomato
Replicase protein	TMV, PVX, PEBV	Tobacco, *Nicotiana benthamiana*
Antisense RNA	TGMV, PLRV	Tobacco, potato
3′-untranslated region	TYMV	*Brassica napus*
Replicon	GVA	N. benthamiana
Defective replicon	TMV, CMV, BMV	Tobacco
Defective interfering RNA or DNA	ACMV, BCTV, BMV	*N. benthamiana*, barley
Movement protein	TMV, PLRV, BMV	Tobacco, potato
Helper component protein	PVY	Tobacco
Satellite RNAs	CMV, TRSV	Tobacco, pepper, tomato

Abbreviations: ACMV, African cassava mosaic virus; BCTV, beet curly top virus; BMV, brome mosaic virus; CMV, cucumber mosaic virus; GVA, grapevine virus A; PEBV, pea early browning virus; PLRV, potato leaf roll virus; PVX, potato virus X; PVY, potato virus Y; TGMV, tobacco golden mosaic virus; TMV, tobacco mosaic virus; TRSV, tobacco ringspot virus; TYMV, turnip yellow mosaic virus.

RNA-mediated resistance

Initially, the mechanism of PDR was poorly understood. It was suggested that expression of the viral coat protein in the transgenic plant would interfere with the uncoating of virus particles in the early event of virus multiplication cycle (Register and Beachy, 1988; Osbourn et al., 1989). It was hypothesized that the coat protein transgene product interacts with a host factor or with the challenge viral RNA, inhibiting or disrupting translation, assembly, or replication of the viral RNA (Asurmendi et al., 2007). However, subsequent work of Lindbo and Dougherty (1992a,b) showed that transgenic tobacco plants expressing untranslatable coat protein gene of the tobacco etch virus (TEV) were immune to TEV infection, providing evidence that the RNA and not the protein of the TEV coat protein gene was responsible for virus resistance in transgenic plants. Since this breakthrough discovery, it soon became apparent that expression of the viral transgene protein product was essentially not needed for obtaining resistance (Dougherty et al., 1994), as it has been shown that most examples of PDR for plant viruses are RNA-mediated and occur through the mechanism of post-transcriptional gene silencing (PTGS), which acts as a defense mechanism against viruses now referred to as RNA silencing (Voinnet, 2001, 2005; Waterhouse et al., 2001; Lindbo and Dougherty, 2005).

The discovery of RNA silencing is arguably the most important biological finding of the last decade. RNA silencing is a mechanism of sequence-specific RNA degradation conserved among multicellular organisms, including plants (Hannon, 2002). It functions in developmental regulation and in host defense against transposable elements and viruses. The pathway of viral RNA silencing is initially triggered by the dsRNA precursors, like those generated from replicating viral RNA intermediates and/or viral RNA secondary structures. This dsRNA is recognized and cleaved into small interfering RNAs (siRNAs) of 21 to 25 nucleotide duplexes by an RNase-III-like enzyme named DICER. The siRNAs are next incorporated and converted to single-stranded RNAs in the Argonaute-containing multi-subunit ribonuclease named RNA-induced silencing complex (RISC). The RISC, consequently, targets and specifically degrades RNAs that share sequence similarity with the siRNA (Waterhouse et al., 1998; Hamilton and Baulcombe, 1999; Hammond et al., 2001; Voinnet, 2001, 2005, 2008; Hannon, 2002; Deleris et al., 2006; Obbard et al., 2009).

Application of the PDR Concept for Developing Transgenic Virus Resistance to Horticultural Crops

Following the first development of TMV resistance in *N. tabacum* cv. *Xanthi* and cv. *Samsun* plants (Powell-Abel, 1986), the concept of PDR was applied in various horticultural crops for the purpose of providing control of viral diseases. Nelson et al. (1988) reported on the development of transgenic tomato plants expressing the coat protein from TMV. In their field trial, most (95%) of the transgenic plants were found to be asymptomatic following mechanical inoculation by TMV. Furthermore, Sanders et al. (1992) showed that such transgenic tomato plants were resistant to distinct TMV strains. Inoculated transgenic tomato plants provided fruit yield that was identical to that obtained from non-inoculated, non-transgenic tomato plants, indicating that the process of transformation and expression of the transgene did not affect the horticultural performance of the transgenic plants (Nelson et al., 1988). Effective resistance was obtained in cucumber transgenic plants expressing the coat protein gene of the CMV. In their work, Gonsalves et al. (1992) reported on a high incidence of asymptomatic cucumber transgenic plants relative to non-transformed plants following CMV inoculation via aphid vectors. This illustrated the efficacy of the PDR concept at providing practical control of aphid-transmitted virus diseases (for review see Fuchs and Gonsalves, 2007).

The development of transgenic papaya resistant to the papaya ringspot virus (PRSV) has been thoroughly reported (for review see Gonsalves, 1998). Transgenic papaya plants expressing the coat protein gene of PRSV have been commercialized, providing a tangible approach to control PSRV in Hawaii (Gonsalves, 2006) and promise for control of the virus worldwide. Potato lines resistant to the potato leafroll virus (PLRV) were also created, field tested, deregulated, and commercialized (Kaniweski and Thomas, 2004). However, resistance to PLRV infection was obtained in potato lines expressing the replicase gene but not the coat protein gene. Another potato cultivar with resistance to PVY was created by expressing its coat protein gene (Kaniewski et al., 1990).

Resistance to multiple viruses for control of virus diseases is worthwhile as mixed infections of viruses are common among horticultural crops. To this extent, the PDR concept was applied for creation of transgenic plants resistant to more than one virus. In early studies performed by Kaniewski et al. (1990) and Lawson et al. (1990), transgenic potato plants expressing the coat protein genes of PVX and PVY were developed and found to be remarkably resistant to infections by both PVX and PVY viruses. Later on, transgenic summer squash plants were developed to express the CP gene of zucchini yellow mosaic virus (ZYMV) and watermelon mosaic virus (WMV). These plants were found to be resistant to two viruses, ZYMV and WMV, using mechanical inoculation or aphid vectors (Fuchs and Gonsalves, 1995). Furthermore, transgenic squash plants expressing the CP genes of CMV, ZYMV, and WMV were found to be resistant to these three viruses (Tricoli et al., 1995; Fuchs et al., 1998b).

RNA silencing-based applications for developing virus resistant plants

A central key to trigger RNA silencing in plants is the formation of dsRNA structures. Many approaches have been applied to produce viral dsRNA for activation of RNA silencing in plants. These included expression of sense and antisense viral sequences, of viral sequences with inverted repeat regions capable to forming hairpin RNA structures, and of virus minireplicon and artificial plant micro RNA (amiRNA; Smith et al., 2000; Prins et al., 2008; Brumin et al., 2009).

Transgenic potato plants, of the commercial "Spunta" cultivar, expressing inverted repeat regions of the 3′-terminal of the coat protein gene of PVY, which is highly conserved in sequences among different PVY isolates, were created (Missiou et al., 2004). Twelve of fifteen transgenic lines produced siRNAs and were highly resistant to three strains of PVY belonging to the subtypes PVY^{N}, PVY^{O}, and PVY^{NTN}. Praveen et al. (2010) developed transgenic tomato plants by transforming RNA inhibition (RNAi) constructs with sequences ranging from 21 to 200 nt of the viral target AC4 gene of various viruses causing the tomato leaf curl disease. The constructs carry the sense and antisense portions of these sequences and are separated by different introns. They found that dsRNA derived from longer arm-length constructs generating a pool of siRNAs were more effective in targeting gene silencing. Tougou et al. (2006) reported on the development of soybean transgenic plants that were transformed with a construct containing inverted repeat-SbDV coat protein gene of the soybean dwarf virus (SbDV) interspersed with β-glucuronidase (*GUS*) sequences. Two months after inoculation of SbDV by aphid, T2 plants contained little SbDV-specific RNA and remained symptomless. These plants contained SbDV-*CP*-specific siRNA, indicating that plants achieved resistance to SbDV by an RNA-silencing-mediated process. Other work evaluated different constructs with intron-spliced hpRNA (ihpRNA), ihpRNA overhang, and ihpRNA spacer for resistance to PVY (Smith et al., 2000; Wesley et al., 2001). The ihpRNA was found to be the most efficient construct as resistance to PVY was obtained in 90% of the plants exhibiting RNA silencing. By arranging viral transgenes as inverted repeats, Bucher et al. (1996) showed that efficient simultaneous targeting of four different tospoviruses can be achieved by using a single small transgene based on the production of minimal sized chimeric cassettes. Due to simultaneous RNA silencing, the transgenic expression of genetic segments of several viruses from a single transgene construct rendered up to 82% of the transformed plant lines with heritable resistance against multiple viruses.

Another approach that has been used to trigger RNA silencing in plants and to create transgenic plant with virus resistance is the expression of artificial plant amiRNAs. Niu et al. (2006) modified an *Arabidopsis thaliana* miR159 precursor to express artificial miRNAs (amiRNAs) targeting viral mRNA sequences encoding two gene silencing suppressors, P69 of TYMV and HC-Pro of turnip mosaic virus (TuMV). Transgenic *A. thaliana* plants expressing amiR-P69159 and amiR-HC-Pro159 were found to be specifically resistant to TYMV and TuMV, respectively. In addition, transgenic plants that express both amiR-P69159 and amiR-HC-Pro159 from a dimeric pre-amiR-P69159/amiR-HC-Pro159 transgene were resistant to both viruses. Interestingly, when inoculated with the specific virus, these amiRNA-based transgenic plants displayed specific resistance even at low temperatures that usually inhibit post-transcriptional gene silencing.

In an attempt to develop resistance to the GVA, Brumin et al. (2009) made a GFP-tagged GVA minireplicon and utilized it as a tool to consistently activate RNA silencing. Transgenic *N. benthamiana* plants expressing the GVA-minireplicon displayed resistance against GVA infection, which was found in 60% and in 90–95%, of T1 and T2 progenies, respectively.

PDR stability and suppression of RNA silencing

Diverse RNA and DNA viruses of plants encode RNA silencing suppressors (RSSs), which in addition to their role in viral counter defense, were implicated in the efficient accumulation of viral RNAs, virus transport, and pathogenesis, as well as determination of the virus host range. Viral RSSs are also known to cause transgene silencing. The Hc-Pro of TEV was among the first RSS shown to suppress PTGS of the *β-glucuronidase* (GUS) transgene (Kasschau and Carrington, 1998). The Hc-Pro of PVY and the 2b protein of CMV were found to suppress PTGS of the GFP transgene in *N. benthamiana* plants (Brigneti et al., 1998). Although a major class of viral RSSs are dsRNA-binding proteins, as revealed first for the tombusviral P19 (Silhavy et al., 2002), RSSs from different plant viruses counteract various steps in the PTGS process (Li and Ding, 2006). The existence of suppressors of PTGS, encoded by plant viruses, poses questions about the stability of PTGS-based resistance in transgenic plants. Mitter et al. (2001, 2003) investigated the suppression of dsRNA-induced PTGS by CMV in a tobacco line expressing a single copy dsRNA-encoding transgene to target PVY. They showed that the immunity to PVY in this transgenic line can break down upon infection with CMV. Similarly, infection of transgenic *N. benthamiana* plants expressing a GVA minireplicon with the GVB or PVY did affect the stability of PTGS-based resistance as these plants turned and became susceptible to GVA (Brumin et al., 2009).

The stability of PTGS-based resistance in transgenic plants was examined for C5 plum trees transformed with the PPV coat protein gene (Zagrai et al., 2008). This transgenic plum displays PTGS, which makes it highly resistant to PPV infection. Yet, the engineered resistance to PPV was stable and was not suppressed by the presence of challenging heterologous viruses.

The stability of engineered resistance is modulated by environmental factors. At low temperature, both virus and transgene triggered RNA silencing are inhibited. Consequently, in cold, plants become more susceptible to viruses, and RNA silencing-based phenotypes of transgenic plants are lost. Szittya et al. (2003) found that siRNA-mediated RNA silencing is temperature dependent in different dicot species, which would suggest that inefficient siRNA generation at low temperature is probably a universal feature of higher plants. In contrast, temperature does not influence the accumulation of micro (mi) RNAs, which play a role in developmental regulation, suggesting that the two classes of small (si and mi) RNAs are generated by different nuclease complexes (Szittya et al., 2003).

Resistance breakage is a critical issue associated with virus-resistant transgenic plants. Consideration of resistance breakdown is important because of the long time required to create transgenic resistant plants, and because one desires to maximize the efficacy of the engineered resistance over time. Engineered resistance tended to be stable in summer squash and papaya transgenic plants as resistance breakage was not reported for more than 10 years (Fuchs and Gonsalves, 2007). However, resistance breakage was reported for wheat transformed with the coat protein gene of wheat streak mosaic virus (WSMV) as transgenic lines that showed WSMV resistance in greenhouse trials did not express resistance under field conditions.

Assessment of Risks Associated with Transgenic Virus Resistance in Plants

In the past, major attention has been paid to potential risks associated with virus-resistant transgenic plants. Potential food safety issues have been extensively reviewed (see König et al., 2004; Fuchs and Gonsalves, 2007; Reddy et al., 2009; Alderborn et al., 2010). Hence, in this section we focus on major potential issues of phenotype- and genotype-mediated impacts, although their occurrence is not as pronounced.

Recombination between the transgene mRNA and the RNA of an infecting virus could occur in the transgenic plant cells during replication of genomic RNA of the challenge virus, resulting in chimeric viruses with a different virus genotype. Recombination between the mRNA encoded by a viral transgene and the RNA of an infecting virus has been documented in several studies (Greene and Allison, 1994; Borja et al., 1999; Adair and Kearney, 2000; Turturo et al., 2008). Thus, recombination raises the issue of whether this phenomenon, if it occurred in transgenic plants in the field, could lead to the creation of novel viral genomes, and consequently to the emergence of new diseases. Despite being reported in several laboratories to occur in transgenic herbaceous plants and by applying conditions of high selective pressure, recombination events have not been proven to occur in transgenic plants in the field. The work carried out by Vigne et al. (2004) addressed the question of whether the recombination event occurs in transgenic grapevines expressing the CP gene of grapevine fanleaf virus (GFLV) in two natural GFLV-infected vineyards. From their results, there was no evidence that transgenic grapevines assisted the creation of viable GFLV recombinants or affected the molecular diversity of GFLV populations. More recently, Turturo et al. (2008) compared the populations of recombinant viruses generated in transgenic and non-transgenic plants using two cucumoviral systems, involving either two strains of CMV or CMV and the related tomato aspermy virus (TAV). Using a highly sensitive RT-PCR assay, which specifically amplifies recombinant viruses, they showed that equivalent populations of recombinant viruses were present in plants expressing a CMV CP transgene and in non-transgenic plants infected simultaneously with both viruses. This suggests that there was no evidence for novel recombination mechanisms in the transgenic plants.

Heteroencapsidation of one virus by the coat protein of a closely related virus could also occur in plants infected with

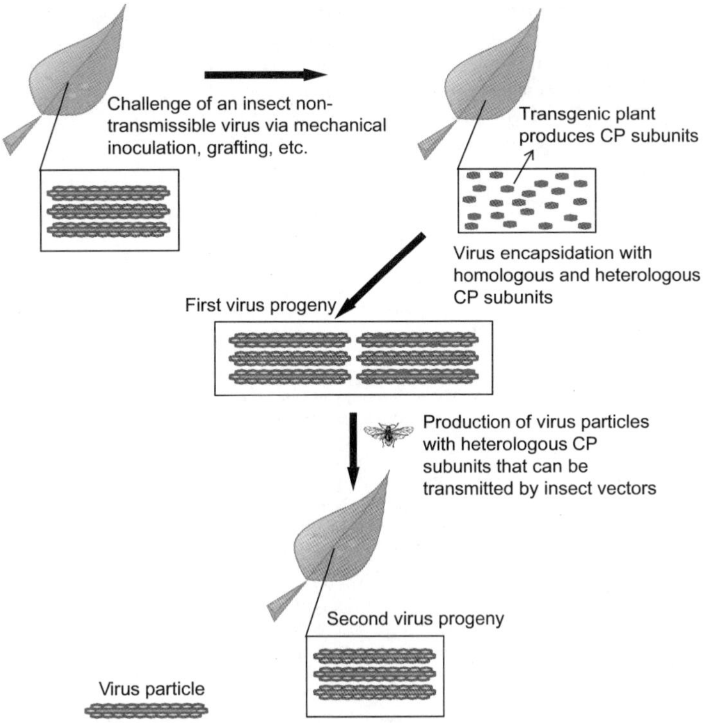

Figure 22.1 • Schematic illustration of heteroencapsidation • A challenge virus, which is non-transmittable by insect vectors, can be transmitted from infected plant to a transgenic plant by mechanical inoculation or grafting. In the transgenic plant, the progeny of the challenge virus can be encapsidated with homologous CP subunits as well as with those heterologous CP subunits encoded by the transgene. Those virus particles with the heterologous CP subunits can be acquired by insect vectors and further transmitted. Please see color plate section at the end of the book.

multiple viruses. This phenomenon could occur in virus-resistant transgenic plants expressing a coat protein, where subunits of the protein encapsidate genomic RNA of a related virus. Because the coat protein can be associated with other features such as pathogeneses or insect-vector specificity, heteroencapsidation might change properties of infecting viruses in transgenic plants. For example, an infecting non-transmissible virus could become transmissible through heteroencapsidation in a transgenic plant when acquiring the coat protein of a transmissible donor virus (Figure 22.1). Consequently, it is theoretically possible that new virus epidemics could result from heteroencapsidation.

Heteroencapsidation has been extensively studied in transgenic herbaceous plants for many viruses such as TMV, CMV, PVY, and ZYMV, (Farinelli et al., 1992; Candelier-Harvey and Hull, 1993; Lecoq et al., 1993; Hammond and Dienelt, 1997). These studies reported that coat proteins expressed in transgenic plants are capable of encapsidation of the RNA genomes of challenge viruses. In transgenic plants expressing the capsid protein of PPV, heteroencapsidation was found to assist aphid transmission and spread of another non-aphid-transmissible strain of ZYMV (Lecoq et al., 1993). In contrast, this has not been found to be the case in the field. For example, transgenic squash and melon expressing the CP gene of an aphid-transmissible strain of CMV have been tested in the field for their capacity to trigger the transmission of an aphid-non-transmissible CMV strain through heteroencapsidation. The results indicated that spread of the aphid-non-transmissible strain did not occur from mechanically inoculated transgenic plants (Fuchs et al., 1998a). In addition, for transgenic papaya and squash, no unexpected emergence of virus species with undesired characteristics was reported 8–10 years post-commercialization (Fuchs and Gonsalves, 2007). However, the only occurrence of heteroencapsidation in the field that has been documented to date was with transgenic squash expressing the CP gene of WMV, where a low rate of transmission of an aphid-non-transmissible ZYMV strain (2%) was reported (Fuchs et al., 1999), although spread of the aphid-non-transmissible strain of ZYMV did not reach epidemic proportions.

Conclusion

Virus diseases cause serious loss of yield and quality in many crop species. Control mechanisms include use of virus-free seed or propagation material, prevention of infection by phytosanitation, prevention of transmission by vectors, and breeding for resistance. Due to the limited effectiveness of physical, chemical, and biological methods in the control of these diseases, the use of genetic or transgenic resistance and RNA silencing has become the best management strategy on a medium-long-term basis. The study of the mechanisms responsible for the resistances found and of their genetic control, as well as the development of new biotechnologies, would enable the pyramiding of different resistance genes, thus making a positive contribution to the development of environmentally friendly and safe durable resistance to viral diseases.

References

Adair, T. L., & Kearney, C. M. (2000). Recombination between a 3-kilobase tobacco mosaic virus transgene and a homologous viral construct in the restoration of viral and nonviral genes. *Archives of Virology*, *145*, 1867–1883.

Alderborn, A., Sundström, J., Soeria-Atmadja, D., Sandberg, M., Andersson, H. C., & Hammerling, U. (2010). Genetically modified plants for non-food or non-feed purposes: Straightforward screening for their appearance in food and feed. *Food and Chemical Toxicology*, *48*, 453–464.

Anderson, P. K., Cunningham, A. A., Patel, N. G., Morales, F. J., Epstein, P. R., & Daszak, P. (2004). Emerging infectious diseases of plants: Pathogen pollution, climate change and agrotechnology drivers. *Trends in Ecology & Evolution*, *19*, 535–544.

Antignus, Y. (2001). Manipulation of wavelength-dependent behaviour of insects: An IPM tool to impede insects and restrict epidemics of insect-borne viruses. *Virus Research*, *71*, 213–220.

Antignus, Y., & Ben-Yakir, D. (2004). Greenhouse photoselective cladding materials serve as an IPM tool to control the spread of insect pests and their vectored viruses. In R. Horowitz & Y. Ishaya (Eds.), *Insect pest management* (pp. 319–333). Berlin: Springer.

Antignus, Y., Lachman, O., Pearlsman, M., & Koren, A. (2005). Containment of cucumber fruit mottle mosaic *Tobamovirus* (CFMMV) infection through roots by planting into a virus-free intermediating medium. *Phytoparasitica*, *33*, 85–87.

Antignus, Y., Lapidot, M., & Cohen, S. (2001). Interference with ultraviolet vision of insects to impede insect pests and insect-borne plant viruses. In K. S. Harris (Ed.), *Virus-Insect-Plant Interactions* (pp. 331–350). U.S.A: Academic Press.

Antignus, Y., Mor, N., Ben-Joseph, R., Lapidot, M., & Cohen, S. (1996). UV-absorbing plastic sheets protect crops from insect pests and from virus diseases vectored by insects. *Environmental Entomology*, *25*, 919–924.

Asurmendi, S., Berg, R. H., Smith, T. J., Bendahmane, M., & Beachy, R. N. (2007). Aggregation of TMV CP plays a role in CP functions and in coat protein-mediated resistance. *Virology*, *366*, 98–106.

Blackman, R. L., & Eastop, V. F. (2000). *Aphids on the World's Crops: An identification and information guide* (2nd ed.), p. 250. New York: Wiley.

Borja, M., Rubio, T., Scholthof, H. B., & Jackson, A. O. (1999). Restoration of wild-type virus by double recombination of tombusvirus mutants with a host transgene. *Molecular Plant–Microbe Interactions*, *12*, 153–162.

Brigneti, G., Voinnet, O., Li, W., Ji, L., Ding, S. W., & Baulcombe, D. (1998). Viral pathogenicity determinants are suppressors of transgene silencing in *Nicotiana benthamiana*. *EMBO Journal*, *17*, 6739–6746.

Brumin, M., Stukalov, S., Haviv, S., Muruganantham, M., Moskovitz, Y., & Batuman, O., et al. (2009). Post-transcriptional gene silencing and virus resistance in *Nicotiana benthamiana* expressing a Grapevine virus a minireplicon. *Transgenic Research*, *18*, 331–345.

Bucher, E., Lohuis, D., Van Poppel, P. M. J. A., Geerts-Dimitriadou, C., Golbach, R., & Prins, M. (1996). Multiple virus resistance at a high frequency using a single transgene construct. *Journal of General Virology*, *87*, 3697–3701.

Candelier-Harvey, P., & Hull, R. (1993). Cucumber mosaic virus genome is encapsidated in alfalfa mosaic virus coat protein expressed in transgenic plants. *Transgenic Research*, *2*, 277–285.

Cohen, S., & Berlinger, M. J. (1986). Transmission and cultural control of whitefly borne viruses. *Agriculture, Ecosystems & Environment*, *17*, 89–97.

Csizinszky, A. A., Schuster, D. J., & Kring, J. B. (1997). Evaluation of color mulches and oil sprays for yield and for the control of silverleaf whitefly, *Bemisia Argentifolii* (Bellows and Perring) on tomatoes. *Crop Protection*, *16*, 475–481.

Dangl, J. L., & Jones, J. D. G. (2001). Plant pathogens and integrated defence responses to infection. *Nature*, *411*, 826–833.

Deleris, A., Gallego-Bartolome, J., Bao, J., Kasschau, K. D., Carrington, J. C., & Voinnet, O. (2006). Hierarchical action and inhibition of plant dicer-like proteins in antiviral defense. *Science*, *313*, 68–71.

Dogimont, C., Chovelon, V., Tual, S., Boissot, N., Rittener-Rüff, V., Giovinazzo, N., et al. (2008). Molecular diversity at the Vat/Pm-W resistance locus in melon. In: M. Pitrat (Ed), *Cucurbitaceae, Proceedings of the nineth EUCARPIA meeting on genetics and breeding of Cucurbitaceae* (pp. 219–228). Avignon (France), May 21–24th, 2008.

Dougherty, W. G., Lindbo, J. A., Smith, H. A., Parks, T. F., Swaney, S., & Proebsting, W. M. (1994). RNA-mediated virus resistance in transgenic plants: Exploitation of a cellular pathway possibly involved in RNA degradation. *Molecular Plant–Microbe Interactions*, *7*, 544–552.

Doukas, D., & Payne, C. (2007). Greenhouse whitefly (Homoptera: Aleyrodidae) dispersal under different UV-light environments. *Journal of Economic Entomology*, *100*, 380–397.

EPPO. (2009). *EPPO summaries of phytosanitary regulations*. EPPO Technical Documents.

Fargette, D., Konate, G., Fauquet, C., Muller, E., Peterschmitt, M., & Thresh, J. M. (2006). Molecular ecology and emergence of tropical plant viruses. *Annual Review of Phytopathology*, *44*, 235–260.

Farinelli, L., Malnöe, P., & Collet, G. F. (1992). Heterologous encapsidation of potato virus Y strain P (PVYO) with the transgenic coat protein of PVY strain N (PVYN) in *Solanum tuberosum* cv. Bintje. *Bio/Technology*, *10*, 1020–1025.

Fuchs, M., Gal-On, A., Raccah, B., & Gonsalves, D. (1999). Epidemiology of an aphid nontransmissible potyvirus in fields of nontransgenic and coat protein transgenic squash. *Transgenic Research*, *99*, 429–439.

Fuchs, M., & Gonsalves, D. (1995). Resistance of transgenic squash Pavo ZW-20 expressing the coat protein genes of zucchini yellow mosaic virus and watermelon mosaic virus 2 to mixed infections by both potyviruses. *Bio/Biotechnology*, *13*, 1466–1473.

Fuchs, M., & Gonsalves, D. (2007). Safety of virus-resistant transgenic plants two decades after their introduction: Lessons from realistic field risk assessment studies. *Annual Review of Phytopathology*, *45*, 173–202.

Fuchs, M., Klas, F. E., McFerson, J. R., & Gonsalves, D. (1998a). Transgenic melon and squash expressing coat protein genes of aphid-borne viruses do not assist the spread of an aphid non-transmissible strain of cucumber mosaic virus in the field. *Transgenic Research*, *7*, 1–14.

Fuchs, M., Tricoli, D. M., McMaster, J. R., Carney, K. J., Schesser, M., & McFerson, J. R., et al. (1998b). Comparative virus resistance and fruit yield of transgenic squash with single and multiple coat protein genes. *Plant Disease*, *82*, 1350–1356.

Garrett, K. A., Dendy, S. P., Frank, E. E., Rouse, M. N., & Travers, S. E. (2006). Climate change effects on plant disease: Genomes to ecosystems. *Annual Review of Phytopathology*, *44*, 489–509.

Gera, A., Katzir, P., Tam, Y., Spiegel, S., and Zeidan, M. (2010). New and emerging viruses in ornamental crops. *Acta Horticulturae* (In press).

Gonsalves, D. (1998). Control of papaya ringspot virus in papaya: A case study. *Annual Review of Phytopathology*, *36*, 415–437.

Gonsalves, D. (2006). Transgenic papaya: Development, release, impact, and challenges. *Advances in Virus Research*, *67*, 317–354.

Gonsalves, D., Chee, P., Provvidenti, R., Seem, R., & Slightom, J. L. (1992). Comparison of coat protein-mediated and genetically-derived resistance in cucumbers to infection by cucumber mosaic virus under field conditions with natural challenge inoculations by vectors. *Biotechnology*, *10*, 1562–1570.

Greene, A. E., & Allison, R. F. (1994). Recombination between viral RNA and transgenic plant transcripts. *Science*, *263*, 1423–1425.

Hamilton, A. J., & Baulcombe, D. C. (1999). A species of small antisense RNA in posttranscriptional gene silencing in plants. *Science*, *286*, 950–952.

Hammond, J., & Dienelt, M. M. (1997). Encapsidation of potyviral RNA in various forms of transgene coat protein is not correlated with resistance in transgenic plants. *Molecular Plant–Microbe Interactions*, *10*, 1023–1027.

Hammond, S. M., Caudy, A. A., & Hannon, G. J. (2001). Post-transcriptional gene silencing by double-stranded RNA. *Nature Reviews Genetics*, *2*, 110–119.

Hannon, G. J. (2002). RNA interference. *Nature*, *418*, 244–251.

Hilje, L., Costa, H. S., & Stansly, P. A. (2001). Cultural practices for managing *Bemisia tabaci* and associated viral diseases. *Crop Protection*, *20*, 801–812.

Hull, R. (2002). *Matthews' plant virology* (4th ed.). San Diego, CA: Academic.

Kaniewski, W., Lawson, C., Sammons, B., Haley, L., Hart, J., & Delannay, X., et al. (1990). Field resistance of transgenic Russet Burbank potato to effects of infection by Potato virus X and Potato virus Y. *Biotechnology*, *8*, 750–754.

Kaniweski, W. K., & Thomas, P. E. (2004). The potato story. *Agricultural and Biological Forum*, *7*, 41–46.

Kasschau, K. D., & Carrington, J. C. (1998). A counterdefensive strategy of plant viruses: Suppression of posttranscriptional gene silencing. *Cell*, *95*, 461–470.

Kirk, W. D. J., & Terry, L. I. (2003). The spread of the western flower thrips *Frankliniella occidentalis* (Pergande). *Agricultural and Forest Entomology*, *5*, 301–310.

Klingler, J., Creasy, R., Gao, L., Nair, R. M., & Calix, A. S. (2005). Aphid resistance in *Medicago truncatula* involves antixenosis and phloem-specific, inducible antibiosis, and maps to a single locus flanked by NBS-LRR resistance gene analogs. *Plant Physiology*, *137*, 1445–1455.

Kogan, M., & Ortman, E. E. (1978). Antixenosis—a new term proposed to replace painter's "nonpreference" modality of resistance. *Bulletin of the Entomological Society of America*, *24*, 175–176.

König, A., Cockburn, A., Crevel, R. W., Debruyne, E., Grafstroem, R., & Hammerling, U., et al. (2004). Assessment of the safety of foods derived from genetically modified (GM) crops. *Food and Chemical Toxicology*, *42*, 1047–1088.

Lawson, C., Kaniewski, W., Haley, L., Rozman, R., Newell, C., & Sanders, P., et al. (1990). Engineering resistance to mixed virus infection in a commercial potato cultivar: Resistance of potato virus X and potato virus Y in transgenic Russet Burban. *Biotechnology*, *8*, 127–134.

Lawson, R., Gera, A., & Hsu, H. T. (1995). Transmission of plant viruses. In G. Loebenstein, R. Lawson & A. Brunt (Eds.), *Viruses and virus-like diseases of bulb and flower crops* (pp. 117–134). Chichester, U.K: Wiley.

Lecoq, H., Ravelonandro, M., Wipf-Scheibel, C., Mansion, M., Raccah, B., & Dunez, J. (1993). Aphid transmission of a non-aphid transmissible strain of zucchini yellow mosaic virus from transgenic plants expressing the capsid protein of plum pox potyvirus. *Molecular Plant–Microbe Interactions*, *6*, 403–406.

Lee, J. M. (1994). Cultivation of grafted vegetables. 1. Current status, grafting methods and benefits. *HortScience*, *29*, 235–239.

Lewandowski, D. (2000). Genus Tobamovirus. In M. H. V. van Regenmortel, C. M. Fauquet, D. H. L. Bishop, E. B. Carstens, M. K. Estes, S. M. Lemon, J. Maniloff, M. A. Mayo, D. J. McGeoch, C. R. Pringle & R. B. Wickner (Eds.), *Virus taxonomy, seventh report of the international committee of taxonomy of viruses* (pp. 889–894). San Diego, CA: Academic Press.

Li, F., & Ding, S. -W. (2006). Virus counterdefense: Diverse strategies for evading the RNA-silencing immunity. *Annual Review of Microbiology*, *60*, 503–531.

Li, L., Zhao, Y., McCaig, B. C., Wingerd, B. A., Wang, J., & Whalon, M. E., et al. (2004). The tomato homolog of CORONATINE-INSENSITIVE1 is required for the maternal control of seed maturation, jasmonate-signaled defense responses, and glandular trichome development. *Plant Cell*, *16*, 126–143.

Li, Y., Hill, C., Carlson, S., Diers, B., & Hartman, G. (2007). Soybean aphid resistance genes in the soybean cultivars Dowling and Jackson map to linkage group M. *Molecular Breeding*, *19*, 25–34.

Lindbo, J. A., & Dougherty, W. G. (1992a). Untranslatable transcripts of the tobacco etch virus coat protein gene sequence can interfere with tobacco etch virus replication in transgenic plants and protoplasts. *Virology*, *189*, 725–733.

Lindbo, J. A., & Dougherty, W. G. (1992b). Pathogen-derived resistance to a potyvirus: Immune and resistant phenotypes in transgenic tobacco expressing altered form of a potyvirus coat protein nucleotide sequence. *Molecular Plant–Microbe Interactions*, *5*, 144–153.

Lindbo, J. A., & Dougherty, W. G. (2005). Plant pathology and RNAi: A brief history. *Annual Review of Phytopathology*, *43*, 191–204.

Lopez-Perez, J. A., Le Strange, M., Kaloshian, I., & Ploeg, A. T. (2006). Differential response of *Mi* gene-resistant tomato rootstocks to root-knot nematodes (*Meloidogyne incognita*). *Crop Protection*, *25*, 382–388.

Mandahar, C. L. (1990). Virus transmission. In C. L. (1990). Mandahar (Ed.), *Plant viruses* (Vol. II, pp. 205–242). Boca Raton, FL: CRC Press.

Martinez de Ilarduya, O., Xie, Q., & Kaloshian, I. (2003). Aphid-induced defense responses in Mi-1-mediated compatible and incompatible tomato interactions. *Molecular Plant–Microbe Interactions*, *16*, 699–708.

McGee, D. C. (1997). *Plant pathogens and the worldwide movement of seeds*. APS Press.

Mensah, C., Di Fonzo, C., & Wang, D. (2008). Inheritance of soybean aphid resistance in PI567541B and PI567598B. *Crop Science*, *48*, 1759–1763.

Milligan, S. B., Bodeau, J., Yaghoobi, J., Kaloshian, I., Zabel, P., & Williamson, V. M. (1998). The root knot nematode resistance gene *Mi* from tomato is a member of the leucine zipper, nucleotide binding, leucine-rich repeat family of plant genes. *Plant Cell*, *10*, 1307–1319.

Missiou, A., Kalantidis, K., Boutla, A., Tzortzakaki, S., Tabler, M., & Tsagris, M. (2004). Generation of transgenic potato plants highly resistant to potato virus Y (PVY) through RNA silencing. *Molecular Breeding*, *14*, 185–197.

Mitter, N., Sulistyowati, E., & Dietzgen, R. G. (2003). Cucumber mosaic virus infection transiently breaks dsRNA-induced transgenic immunity to potato virus Y in tobacco. *Molecular Plant-Microbe Interactions*, *16*, 936–944.

Mitter, N., Sulistyowati, E., Graham, M. W., & Dietzgen, R. G. (2001). Suppression of gene silencing: A threat to virus-resistant transgenic plants? *Trends in Plant Science*, *6*, 246–247.

Mutwiwa, U. N., Brogemeister, C., Von Elsner, B., & Tantu, H. (2005). Effects of UV–absorbing plastic films on greenhouse whitefly (Homoptera: Aleyrodidae). *Journal of Economic Entomology*, *98*, 1221–1228.

Nejidat, A., & Beachy, R. N. (1990). Transgenic tobacco plants expressing a coat protein gene of Tobacco mosaic virus are resistant to some other tobamoviruses. *Molecular Plant-Microbe Interactions*, *3*, 247–251.

Nelson, R. S., McCormick, S. M., Delannay, X., Dube, P., Layton, J., & Anderson, E. J., et al. (1988). Virus tolerance, plant growth and field performance of transgenic tomato plants expressing coat protein from tobacco mosaic virus. *Biotechnology*, *6*, 403–409.

Ng, J. C., & Falk, B. W. (2006). Virus-vector interactions mediating nonpersistent and semipersistent transmission of plant viruses. *Annual Review of Phytopathology*, *44*, 183–212.

Niu, Q. -W., Lin, S. -S., Reyes, J. L., Chen, K. -C., Wu, H. -W., & Yeh, S. -D., et al. (2006). Expression of artificial microRNAs in transgenic *Arabidopsis thaliana* confers virus resistance. *Nature Biotechnology*, *11*, 1420–1428.

Obbard, D. J., Gordon, K. H., Buck, A. H., & Jiggins, F. M. (2009). The evolution of RNAi as a defence against viruses and transposable elements. *Philosophical Transactions of the Royal Society B*, *364*, 99–115.

Oda, M. (1999). Grafting of vegetables to improve greenhouse production. <http://www.agnet.org/library/eb/480/>

Osbourn, J. K., Watts, J. W., Beachy, R. N., & Wilson, T. M. A. (1989). Evidence that nucleocapsid disassembly and a later step in virus replication are inhibited in transgenic tobacco protoplasts expressing TMV coat protein. *Virology, 172*, 370–373.

Pitrat, M., & Lecoq, H. (1980). Inheritance of resistance to *Cucumber mosaic virus* transmission by *Aphis gossypii* in *Cucumis melo*. *Phytopathology, 70*, 958–961.

Powell-Abel, P., Nelson, R. S., De, B., Hoffmann, N., Rogers, S. G., & Fraley, R. T., et al. (1986). Delay of disease development in transgenic plants that express the tobacco mosaic virus coat protein gene. *Science, 232*, 738–743.

Praveen, S., Ramesh, S. V., Mishra, A. K., Koundal, V., & Palukaitis, P. (2010). Silencing potential of viral derived RNAi constructs in Tomato leaf curl virus-AC4 gene suppression in tomato. *Transgenic Research, 19*, 45–55.

Prins, M., & Goldbach, R. (1998). The emerging problem of tospovirus infection and nonconventional methods of control. *Trends in Microbiology, 6*, 31–35.

Prins, M., Laimer, M., Noris, E., Schubert, J., Wassenegger, M., & Tepfer, M. (2008). Strategies for antiviral resistance in transgenic plants. *Molecular Plant Pathology, 9*, 73–83.

Raviv, M., & Antignus, Y. (2004). UV Radiation effects on pathogens and insect pests of greenhouse-grown crops. *Photochemistry and Photobiology, 79*, 219–226.

Reddy, D. V. R., Sudashana, M. R., Fuch, M., Rao, N. C., & Thottappilly, G. (2009). Genetically engineered virus-resistant plants in developing countries: Current status and future prospects. *Advances in Virus Research, 75*, 185–220.

Register, J. C., III, & Beachy, R. N. (1988). Resistance to TMV in transgenic plants results from interference with an early event in infection. *Virology, 166*, 524–532.

Rossi, M., Goggin, F. L., Milligan, S. B., Kaloshian, I., Ullman, D. E., & Williamson, V. W. (1998). The nematode resistance gene *Mi* of tomato confers resistance against the potato aphid. *Proceedings of the National Academy of Sciences of the United States of America, 95*, 9750–9754.

Sadasivam, S., Thayumanavan, B. B. (2003). Molecular host plant resistance to pests (pp. 479). New York; Marcel Dekker.

Sanders, P. R., Sammons, B., Kaniewski, W., Haley, L., Layton, J., & Lavallee, B. J., et al. (1992). Field resistance of transgenic tomatoes expressing the Tobacco mosaic virus or Tomato mosaic virus coat protein genes. *Phytopathology, 82*, 683–690.

Sanford, J. C., & Johnston, S. A. (1985). The concept of parasite-derived resistance — Deriving resistance genes from the parasite's own genome. *Journal of Theoretical Biology, 113*, 395–405.

Silhavy, D., Molnar, A., Lucioli, A., Szittya, G., Hornyik, C., & Tavazza, M., et al. (2002). A viral protein suppresses RNA silencing and binds silencing-generated, 21- to 25-nucleotide doublestranded RNAs. *EMBO Journal, 21*, 3070–3080.

Singer, A. C., Crowley, D. E., & Thompson, I. P. (2003). Secondary plant metabolites in phytoremediation and biotransformation. *Trends in Biotechnology, 21*, 123–130.

Smith, C. M. (1989). *Plant resistance to insects: A fundamental approach*. New York: Wiley.

Smith, N. A., Singh, S. P., Wang, M. B., Stoutjesdijk, P. A., Green, A. G., & Waterhouse, P. M. (2000). Total silencing by intron-spliced hairpin RNAs. *Nature, 407*, 319–320.

Stacesmith, R., & Hamilton, R. I. (1988). Inoculum thresholds of seedborne pathogens-viruses. *Phytopathology, 78*, 875–880.

Summers, C. G., Michell, J. P., & Stapleton, J. J. (2005). Mulches reduce aphid-borne viruses and whiteflies in cantaloupe. *California Agriculture, 59*, 90–94.

Szittya, G., Silhavy, D., Molnar, A., Havelda, Z., Lowas, A., & Lakatos, L., et al. (2003). Low temperature inhibits RNA silencing-mediated defence by the control of siRNA generation. *EMBO Journal, 22*, 633–640.

Takken, F. L. W., Albrech, T. M., & Tameling, W. I. L. (2006). Resistance proteins: Molecular switches of plant defense. *Current Opinion in Plant Biology, 9*, 383–390.

Tougou, M., Furutani, N., Yamagishi, N., Shizukawa, Y., Takahata, Y., & Hidaka, S. (2006). Development of resistant transgenic soybeans with inverted repeat-coat protein genes of soybean dwarf virus. *Plant Cell Reports, 25*, 1213–1218.

Traw, M. B., Kim, J., Enright, S., Cipollini, D. F., & Bergelson, J. (2003). Negative cross-talk between salicylate- and jasmonate-mediated pathways in the Wassilewskija ecotype of *Arabidopsis thaliana*. *Molecular Ecology, 12*, 1125–1135.

Tricoli, D. M., Carney, K. J., Russell, P. F., McMaster, J. R., Groff, D. W., & Hadden, K. C., et al. (1995). Field evaluation of transgenic squash containing single or multiple virus coat protein gene constructs for resistance to Cucumber mosaic virus, Watermelon mosaic virus 2, and Zucchini yellow mosaic virus. *Biotechnology, 13*, 1458–1465.

Turturo, C., Friscina, A., Gaubert, S., Jacquemond, M., Thompson, J. R., & Tepfer, M. (2008). Evaluation of potential risks associated with recombination in transgenic plants expressing viral sequences. *Journal of General Virology, 89*, 327–335.

Ullman, D. E., Sherwood, J. L., & German, T. L. (1997). Thrips as vectors of plant pathogens. In T. L. Lewis (Ed.), *Thrips as crop pests* (pp. 539–565). London: CAB International.

Vigne, E., Komar, V., & Fuchs, M. (2004). Field safety assessment of recombination in transgenic grapevines expressing the coat protein gene of Grapevine fanleaf virus. *Transgenic Research, 13*, 165–179.

Voinnet, O. (2001). RNA silencing as a plant immune system against viruses. *Trends in Genetics, 17*, 449–459.

Voinnet, O. (2005). Induction and suppression of RNA silencing: Insights from viral infections. *Nature Reviews Genetics, 6*, 206–221.

Voinnet, O. (2008). Post-transcriptional RNA silencing in plant–microbe interactions: A touch of robustness and versatility. *Current Opinion in Plant Biology, 11*, 464–470.

Waterhouse, P., Gramham, M., & Wang, M. -B. (1998). Virus resistance and gene silencing in plants can be induced by simultaneous expression of sense and antisense RNA. *Proceedings of the National Academy of Sciences of the United States of America, 95*, 13959–13964.

Waterhouse, P. M., Wang, M. B., & Lough, T. (2001). Gene silencing as an adaptative defense against viruses. *Nature, 411*, 834–842.

Wesley, V., Helliwell, C. A., Smith, N. A., Wang, M. B., Rouse, D. T., & Liu, Q., et al. (2001). Construct design for efficient, effective and high-throughput gene silencing in plants. *Plant Journal, 27*, 581–591.

Wroblewski, T., Piskurewicz, U., Tomczak, A., Ochoa, O., & Michelmore, R. W. (2007). Silencing of the major family of NBS-LRR-encoding genes in lettuce results in the loss of multiple resistance specificities. *Plant Journal, 51*, 803–818.

Zagrai, I., Capote, N., Ravelonandro, M., Ravelonandro, M., Cambra, M., & Zagrai, L., et al. (2008). Plum pox virus silencing of C5 transgenic plums is stable uner challenge inoculation with heterologous viruses. *Journal of Plant Pathology, 90*, S1.63–S1.71.

Insects, nematodes, and other pests

23

Philip R. Watkins[1] Joseph E. Huesing[2] Venu Margam[2]
Larry L. Murdock[2] T.J.V. Higgins[1]

[1]*CSIRO, Plant Industry, Canberra, Australia,*
[2]*Purdue University, West Lafayette, Indiana*

TABLE OF CONTENTS

Introduction — Genetically Modified Crops for Insect Resistance

Presently, all commercialized bioengineered insect-resistant crops (with one exception) contain genes derived from or based on the soil-dwelling bacterium *Bacillus thuringiensis* (Bt) (James and Choudhary, 2010). To guard against the potential for resistance to these Bt toxins new strategies have been developed to maintain transgenic insect protected plants. This involves a range of stewardship principles including rotating the type of modified crop to ensure insects are exposed to different toxins, and the use of Bt free refugia. These refuges allow populations of non-resistant insects to multiply and breed with any insects that may have developed resistance.

Pyramiding resistance genes, modifying the Bt genes and using different promoters to alter the level and site of expression, have also been used to reduce the risk of development of insect resistance. Combined with the fact that not all plant pests are susceptible to Bt, other ways of modifying plants to resist insects have been under investigation for some time. Vegetative insecticidal protein toxins and RNA interference technologies are likely candidates for near to medium-term commercial release. More speculative alternatives include other bacterial toxins (e.g., from *Photorhabdus* and *Xenorhabdus*), digestive enzyme inhibitors, plant lectins, chitinases, invertebrate spider and scorpion toxins, and secondary metabolites (including volatiles).

History of *B. thuringiensis*

Bacillus thuringiensis was first discovered by Shigetane Ishiwatari in 1901. It was isolated as the causative agent in the death of large numbers of silkworms (Ishiwatari, 1901). In 1911, the bacteria were rediscovered in a Mediterranean flour moth and named after the nearby German province of Thuringia (Berliner, 1911). Commercial production of Bt as a spray began in France in 1938, and was introduced to the United States in 1958 (Lambert and Peferoen, 1992). Originally it was not widely used due to the availability of effective synthetic insecticides and its practical limitations; it rapidly washes away in rain and is degraded by sunlight. By the mid-1970s, 13 Bt strains effective against lepidopteran

DOI: 10.1016/B978-0-12-381466-1.00023-7

insects (moths and butterflies) had been identified. In 1977 the *israelensis* subspecies, which is toxic to dipteran insects (flies), was found (Goldberg and Margalit, 1977). This was followed by the discovery of the *tenebrionis* subspecies, which is toxic to coleopterans (beetles; Krieg et al., 1983). Currently, thousands of strains of Bt have been discovered that contain toxins affecting many orders of insects and nematodes.

By the 1980s, insects were well known to develop resistance to synthetic pesticides and there was growing awareness of the environmental effects of chemical overuse. Commercial interest in alternative pest management technologies grew, leading to increased government and private funding for alternatives including Bt research. In 1984, the feasibility of genetically modified plants was established when foreign genetic material was successfully expressed in the tobacco plant without any observed loss of phenotype (Horsch et al., 1984). The first Bt gene modified to express well in plants was developed soon after for tobacco (Vaeck et al., 1987). Bt maize and Bt cotton became the first commercialized transgenic crops to express a Bt gene in 1996. In 2009, 14 years after their initial commercialization, over 50 million hectares of Bt crops were planted in 23 countries (James, 2009).

Bt genes have now been expressed in many experimental systems from a variety of plant species (including various trees, cereals, legumes, and vegetables). The number of commercialized plants is much lower, in part due to intellectual property rights but mostly due to the extremely high cost of developing, testing (including regulatory studies), and releasing the modified plants (Kalaitzandonakes et al., 2007). For these reasons the present commercialized Bt crops consist of crops such as maize and cotton (Table 23.1), with Bt soybean expected soon, which are grown on very large production scales. Field trials and safety tests are being carried out on many other Bt crops, with rice and brinjal closest to commercialization currently over half the worldwide Bt field area consists of crops grown in the USA (see Figure 23.2).

Cry proteins

Bacillus thuringiensis is a gram positive bacterium that produces insecticidal proteins in the vegetative state as well as during sporulation. The δ-endotoxins produced in the sporulation phase are composed of one or more crystal (Cry) and/or cytolytic (Cyt) proteins and are a major source of the bacterium's toxicity (Bravo et al., 2007). The toxicity, mode of action, and specificity of Cry proteins have been experimentally verified (OECD, 2007). *cry* genes constitute nearly all of the anti-insect genes in transgenic insect protected plants. Currently over 80 different types of Cry proteins have been identified. They belong to distinct protein families, and the holoproteins range in size from 50 to 140 kDa (Crickmore et al., 1998). In addition, binary forms of Cry proteins occur in the bacterium and are used in transgenic crops. The best characterized Cry34A/Cry35A binary protein has constituent masses of 14 and 44 kDa, respectively (Schnepf et al., 2005). Most Cry proteins have a distinct specificity and target only a single order or a few species from that order. Some, however, have a broader spectrum of activity that spans two or three orders.

The Cry proteins are organized into three main groups based on structure and function: the three-domain, the mosquitocidal-like, and the binary-like Cry toxins. Three-domain Cry proteins are the largest group and the majority of the Cry toxin genes used to transform plants to impart insect resistance belong to this group. The three-domain group is further divided into more than 40 different types (Cry1 to Cry55) with many different subgroups (Cry1Aa, Cry1Ab, Cry1Bb, Cry2Aa, etc.; Crickmore et al., 1998). New three-domain Cry proteins are assigned to a group based primarily on their sequence. Domain I of the Cry protein is responsible for pore formation and the other two domains determine the insect specificity of the toxin.

The standard model for describing the sequence of events leading to larval intoxication and death is based on the three-domain lepidopteran Cry proteins. A majority of the transgenic Cry proteins currently in use follow this model (see Figure 23.1). Briefly, when ingested by lepidopteran insect larvae the Cry protein, a protoxin, is solubilized by the high pH of the gut lumen. After solubilization the protoxin is activated through cleavage by digestive enzymes into a smaller (~60 kDa) fragment (Hofte and Whiteley, 1989; OECD, 2007). The activated toxic fragment then binds to receptors on the membrane of the insect's midgut epithelial cells (Bravo et al., 1992). In the generally accepted model of Cry action, the toxic fragment forms oligomers that insert into the membrane and form a pore, causing loss of homeostasis. This leads to osmotic shock, cell lysis, septicemia, and insect death (Lorence et al., 1995). Alternatively, it has recently been proposed that toxicity is due to the activation of an apoptotic signal cascade pathway after binding to the midgut receptors (Zhang et al., 2006). In some species enteric bacteria are required for insect death (Broderick et al., 2006, 2009). The narrow target range of the Cry proteins is due to the unique binding receptors and the specificity of the digestive enzymes needed to activate the toxin (OECD, 2007).

Commercially Available Insect Protected Crops

Bt maize

Maize was domesticated around 12,000 years ago in Mesoamerica and is now the largest crop in the Americas. The United States is the largest producer of maize in the world, with China, Brazil, and South Africa also growing significant areas. In the United States only 2.5% of maize is grown for human consumption; the majority being used for livestock feed. An estimated 29% of the total 2007 maize production in the United States was used for biofuels, and this is expected to increase in the future (Dhugga, 2007). Two major pests of maize in the United States and Canada are the lepidopteran European corn borer (ECB; *Ostrinia nubilalis*) and the coleopteran western corn rootworm (WCR; *Diabrotica virgifera virgifera*). It is estimated that the WCR alone is responsible for $1 billion in lost revenue each year in the United States, which includes $800 million in yield loss and $200 million in

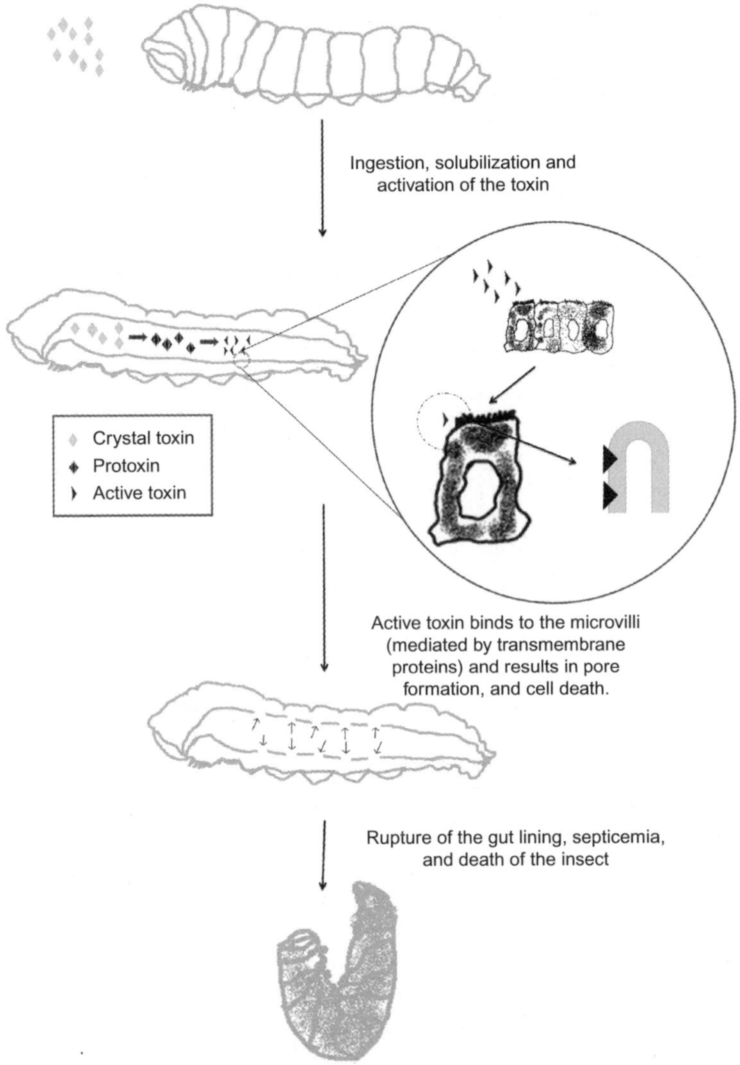

Figure 23.1 • The mode of action of *B. thuringiensis* cry toxins on susceptible insect larvae • The larvae ingest the Cry toxin, which is solubilized in the high pH of the midgut lumen. Digestive enzymes then activate the toxin by cleaving it into smaller fragments. The activated toxin binds to the membrane of the midgut epithelial cells and forms a pore that ruptures the gut lining. This leads to osmotic shock, cell lysis, septicemia, and insect death.

insect control costs. The ECB larvae live hidden within the maize stalk and the WCR larvae bore into the maize roots, making both pests extremely difficult to control with conventional pesticides. Since the release of Bt maize targeted to ECB in 1996, the area planted with genetically modified maize had increased to over 35 million hectares by 2008 (James, 2008).

Currently, there are 21 registered varieties of maize available commercially that contain a *cry* gene. Maize has also been developed to carry multiple (stacked) traits, and many of these varieties contain two or more different *cry* genes and one or more herbicide tolerance genes. Some of the Bt maize lines offer protection from the WCR and other related corn rootworms through the inclusion of modified *Cry3Bb1* or Cry34/Cry35A genes. The maize hybrids are resistant to the major lepidopteran and coleopteran pests.

Some technology developers have collaborated to produce a hybrid maize called SmartStax™, which contains eight transgenes (James, 2009). It produces the Bt toxins Cry2Ab, Cry1A.105, Cry1F, Cry3Bb1, Cry34, and Cry35Ab1, and two herbicide tolerance genes. It gained approval from the U.S. Environmental Protection Agency (EPA) and the Canadian Food Inspection agency and was planted commercially in 2010.

The first versions of Bt maize were protected against lepidopteran species, specifically, ECB. Most studies of the effectiveness of Bt maize have therefore focused on this pest. While Bt maize has been shown to be effective at controlling ECB, the crop's location, climate, time of planting, and how much if any pesticides are used can also affect the level of infestation. The benefit of Bt maize to the farmer is the ability to control the pest with reduced labor and less chemical pesticides (Payne et al., 2003). A recent study estimates that farm income from planting Bt maize has been enhanced by over $8 billion in the first 13 years (1996–2008) since commercialization (Brookes and Barfoot, 2010). Bt maize has been widely accepted by low resource farmers in South Africa where Bt white maize is used as a traditional food (yellow maize is used for feed). First released in 2001, Bt maize now accounts for two-thirds of the total South African white maize hectarage of 1.5 million in 2009 (James, 2009).

Bt maize protected from WCR was first introduced in 2003. These plants are very effective against corn rootworm larvae (Al-Deeb and Wilde, 2005). The yield losses and the control costs of this pest are estimated at $1 billion annually in the United States alone (Metcalf, 1986). Estimates carried out before the commercial release of the modified maize suggested savings to farmers could be $14 to $69 million annually. The recent commercialization of hybrid maizes with multiple insect resistances has the potential to contribute further savings.

An indirect benefit of Bt maize has been a reduction in pathogen infections (Munkvold et al., 1997). Insects wound the plants when they feed, increasing the risk that mycotoxin-producing fungi or other microorganisms will infect the plant. As well as reducing potential yield, the fungal infections pose a risk to the health of humans and animals and decrease the value of the crop. It has been estimated that the reduced fungal damage is worth $17 million annually (Wu et al., 2004).

Bt cotton

Cotton is a major global crop, with 26 million tonnes produced in 2008 at yields just under 800 kg/ha. China, India, and the United States account for two-thirds of all cotton grown, with Pakistan, Brazil, and Uzbekistan being other major cotton growing nations. Cotton is predominately grown for its fiber, although it is also harvested as a source of oil and seed meal. Insects are the major constraint to cotton production; they result in global yield losses of $5 billion annually (Thomson, 2006). About 25% of all insecticides used in agriculture are applied to cotton; in West Africa this may be as high as 80% (James, 2004). The potential for high returns has made genetically modified cotton popular; currently 46% of all cotton grown is either genetically modified for insect resistance, herbicide tolerance, or both (James, 2008).

The major pests of cotton are the lepidopteran tobacco budworm (*Heliothis virescens*), cotton bollworm (*Helicoverpa zea*), and pink bollworm (*Pectinophora gossypiella*). These insects lay their eggs on the cotton boll and after hatching the larvae chew through the cotton lint, causing considerable damage to the plant. Being lepidopterans they are susceptible to control using Bt Cry toxins. Bt cotton (Bollgard®) expressing *cry1Ac* was first released in 1996. In its first year, the crop was attacked by an unusually high number of cotton bollworms, which are less susceptible to Cry1Ac, and some of the planted cotton was lost (Kaiser, 1996). The following years were more successful and the number of Bt cotton hectares subsequently increased in the United States and worldwide. In 2003, Bollgard II was released. This contained the *cry2Ab* gene in addition to *cry1Ac*, providing greater protection from the cotton bollworm and decreasing the likelihood of resistance appearing in bollworm populations.

Bt cotton is also in widespread use in several developing nations. China, the world's largest cotton producer, developed its own Bt cotton using a fused *cry1Ab/cry1Ac*, which has been available commercially since 1997 (James and Choudhary, 2010). In China the cowpea trypsin inhibitor gene has been added to some lines, providing a second mode of insect protection. These lines were released commercially in 2001. In 2002 India commercialized Bt cotton containing the *cry1Ac* gene, and Bt cotton is also grown in Australia, Brazil, Argentina, South Africa, Burkina Faso, Mexico, Columbia, and Costa Rica

Like Bt maize, the economic benefits of Bt cotton are an increase in yield protection due to reduced pest damage and lower pesticide costs. A study carried out by the University of Arizona concluded that in 2001 American and Chinese farmers gained $179 million through the use of Bt cotton, despite a worldwide decline in the price of cotton (Frisvold et al., 2006). A recent study has attributed an increase of $15.6 billion to farm income over 13 years due to the commercialization of Bt cotton (Brookes and Barfoot, 2010). The economic benefits of Bt cotton have been more pronounced in countries outside of the United States. This is likely due to the relatively low insect pressure experienced in the United States compared to other countries. China, India, and Australia sprayed their cotton crops more than ten times a season compared to just two for the United States (James, 2001). South Africa and Australia have grown Bt cotton since the late 1990s, and studies have documented positive economic and/or environmental benefits (Morse et al., 2004; Fitt, 2008). In Burkina Faso, where Bollgard® cotton was first field tested in 2003, trials over 3 years found cotton yields averaged 20% above the non-transgenic controls and reduced insecticide applications by two-thirds (Vitale et al., 2008). Bt cotton was commercialized in Burkina Faso in 2008.

The greatest returns on investments in Bt cotton have been seen in China and India. This appears largely due to higher pest infestations, greater resistance to conventional pesticides, and the majority of growers being small-scale famers (<1 ha), many of whom lack the money to purchase pesticides. Small-scale farmers in China almost doubled their income after adopting Bt cotton, although consumers did not benefit due to governmental control of cotton prices (Pray et al., 2001). In India where pest damage is substantial, the yield of Bt cotton increases have been up to 80% (Qaim and Zilberman, 2003). Even though the initial seed costs for the farmers were higher, the Bt cotton provided greater returns (Bambawale et al., 2004).

The success of Bt cotton in India and China has created problems with stewardship and regulation of the products. In India the spread and adoption of Bt cotton has been rapid and almost universal in many cotton growing areas (Stone, 2007).

Table 23.1 Distribution of Bt and genetically modified crops by country in 2009 (in thousand hectares)

Country	Crops	Bt crop area	Genetically modified crop area
United States	Maize and cotton	27,163	64,000
India	Cotton	8,381	8,400
Brazil	Maize and cotton	5,145	21,400
China	Cotton and poplar	3,672	3,700
Argentina	Maize and cotton	2,205	21,300
South Africa	Maize and cotton	1,614	2,100
Canada	Maize	1,200	8,200
Philippines	Maize	392	490
Australia	Cotton	184	230
Burkina Faso	Cotton	115	115
Uruguay	Maize	90	790
Spain	Maize	76	76
Mexico	Cotton	56	73
Chile	Maize	28	32
Colombia	Cotton	24	24
Honduras	Maize	12	15
Czech Republic	Maize	6	6
Portugal	Maize	5	5
Romania	Maize	3	3
Poland	Maize	3	3
Costa Rica	Cotton	2	2
Egypt	Maize	1	1
Slovakia	Maize	<1	<1
Total		**50,398**	**130,966**

Adapted from James (2009).

This has led to a rise in the number of Bt hybrids being produced, both officially (137 in 2007) and illegally (Jayaraman, 2004). The illegal varieties are more popular among some farmers primarily because they are cheaper, and because they are generally created by hybridizing with local varieties, which means they can be better suited to the local environment (Roy et al., 2007). Further, the high demand and limited supply of quality seed, especially in the early years of commercialization, together with the absence of seed certification, resulted in the sale of fraudulent Bt seeds (Herring, 2008).

Discontinued Bt crops

There are Bt crops that have been discontinued for various reasons. A Bt potato that targeted the Colorado potato beetle (*Letinotarsa decemlineata*) was produced in 1996. This beetle had developed resistance to many chemical pesticides, but remained susceptible to Bt spray containing Cry3A toxins. A released variety of potato expressing the *cry3A* gene was successful at controlling the insect (Perlak et al., 1993). However, many buyers and processors were hesitant to buy the genetically modified potatoes because of perceived public resistance to biotechnology crops. The Bt potato was withdrawn from the market in 2001.

StarLink™ was a maize variety containing the *cry9C* gene. The Cry9C protein is more slowly digested in a simulated stomach than other Cry proteins, leading regulatory authorities to raise concerns about its potential allergenicity in humans; they approved the maize only for livestock feed. In 2000 it was reported that traces of StarLink™ maize were detected in taco shells destined for human consumption (Kaufman, 2000). StarLink™ was withdrawn from registration and an extensive program to remove all traces of the variety from the food chain was initiated. At the time of withdrawal it made up less than 1% of the total maize area in the United States and no positive samples have been detected since 2004. Since the "contamination" was publicized, 28 people have claimed allergic reactions after eating maize products that may have contained StarLink™. After testing by the U.S. Food and Drug Administration and Centers for Disease Control and Prevention, it was concluded that there was no evidence that the reactions reported were consistent with an allergic response (CDC, 2001).

As part of the normal product development cycle, other Bt crops have been withdrawn by developers as they are superseded by new and improved varieties. Knockout™ maize was phased out in 2003, partly because of its relatively higher risk to the monarch butterfly (see next section), but also because other Bt lines were more efficient. Another variety of Bt maize (Bt Xtra™) was removed from the market after company mergers. With the current development of stacked trait crops many of the first generation of single trait crops will or have been replaced. This is necessary to ensure that resistance to the pyramided insect resistance genes is delayed for as long as possible. The Cry1Ac expressing cotton will be completely replaced by Bollgard II, which expresses both the *cry1Ac* and *cry2Ab* genes.

Bt Crops Under Development

Bt brinjal

Brinjal (eggplant/aubergine) is an important food crop in India, Bangladesh, and the Philippines. Its major insect pest is the lepidopteran fruit and shoot borer (FSB; *Leucinodes arbonalis*). The FSB larvae bore into the vegetable and then plug the entry hole with excreta, which shields them from insecticide sprays. In India brinjal is sprayed with chemicals

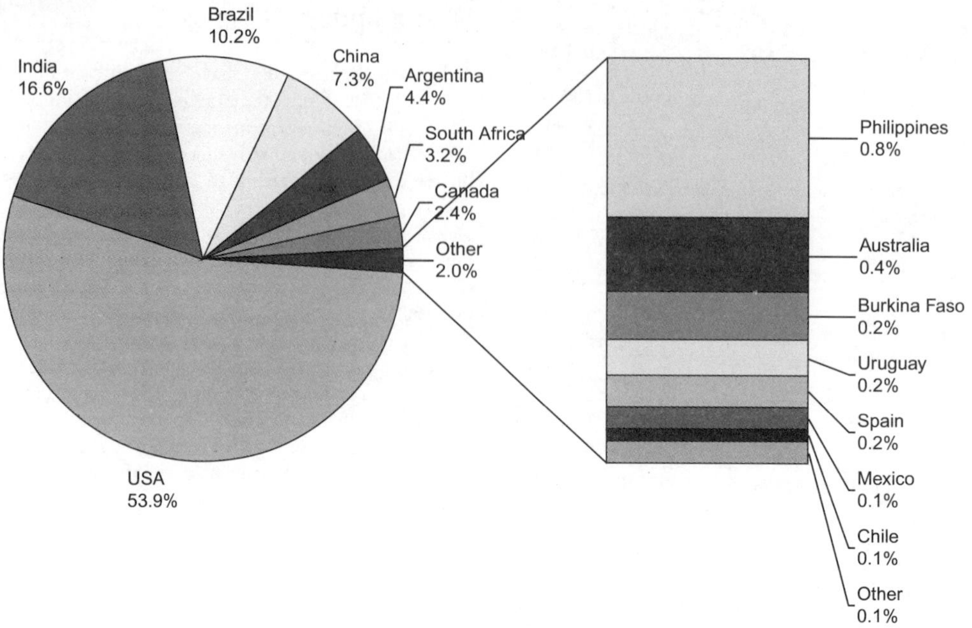

Figure 23.2 • Global distribution of Bt Crops in 2009. (Adapted from James 2009.) Please see color plate section at the end of the book.

15 to 40 times each season. A Bt brinjal was cleared for release in India by the Genetic Engineering Approval Committee in 2009. It is the first food crop for broad-scale, direct human consumption containing a Bt gene that has passed the regulatory phase.

Bt brinjal was developed using a hybrid *Cry1Ac/Cry1Ab* gene and has been very effective at controlling FSB with an average of seven-fold less fruit damage compared to non-Bt lines. Choudhary and Gaur (2008) have predicted that Bt brinjal will result in yields 25% above untreated controls. Safety assessments have been carried out since 2002 and include feeding trials with a range of animals (rats, rabbits, goats, chickens, fish, and lactating cows), measurements of potential gene transfer, weediness, impact on non-target organisms, accumulation of Bt in the soil, and variations in pest and disease susceptibility (Maharashtra Hybrid Seeds Company Ltd., 2009). Some public interest groups strongly oppose the release of the brinjal, and in early 2010 the Indian government imposed a moratorium on the release of Bt brinjal until further independent studies could be carried out.

Bt rice

Rice is one of the most important global food crops and a major source of calories in the human diet. In 2004 Bt rice was commercialized on a small scale in Iran with the help of the International Rice Research Institute (James, 2005). Large-scale trials have been carried out in China since 1999 (Zhang et al., 1999). Two Bt rice varieties have reached the pre-production trial stage, and are being grown under field conditions in selected villages within two of China's provinces. Independent research investigating the financial impact of Bt rice for small-scale farmers in the pre-production trial suggest that the drop in pesticide use and increase in yield will have positive financial benefits for China's rural poor (Huang et al., 2008).

The major hurdle to commercialization of rice in China is concern over the possible ecological consequences of gene flow into wild rice species. China is the center of origin and diversity for rice and contains many wild rice varieties. Gene flow from cultivated rice to wild rice is unavoidable (Chen et al., 2004), and it is possible that wild rice will gain a selective advantage when exposed to lepidopteran pests (Cohen et al., 2008). There are also concerns about the effect of genetically modified rice on trade, in particular to Europe, and a possible unwillingness to become the first nation to commercialize a major genetically modified food for direct human consumption. Nevertheless, in late 2009 China approved preparations for the release of Bt rice, which could occur in two to four years.

Other Bt crops

Other Bt crops expressing *cry* genes are being developed for commercial use. Some are at the field trial stage, while others are still very much in the development stage. Bt soybean containing a *Cry1Ac* gene is in the final stages of development for release into South America (Miklos et al., 2007). Field trials have been conducted in Nigeria using a cowpea transformed with a *cry1Ab* gene. Other crops that have been engineered to produce Cry toxins include wheat, sugarcane, chickpea, canola, tobacco, apples, kiwifruit, cabbage, peanuts, broccoli, pigeon pea, strawberry, poplar, spruce, and eucalyptus.

Impact of Bt

Benefits of Bt crops

As well as the increased yield protection and economic value (see Figure 23.3), other advantages of Bt crops include human safety as a result of reduced handling of insecticides, ease of use, an all-in-one product insect control, time and labor savings, and better pest control (Rice, 2004). The reduction of pesticide use has the advantage of decreasing the environmental contamination risk, increasing insect biodiversity by reducing non-target deaths, lowering the resource cost of producing the pesticides, and lessening the amount of on-farm waste generated. In addition, some traits, such as those that protect the roots of plants from insect pests, can indirectly result in drought tolerance when compared to non-transformed lines (Clark et al., 2006).

Figure 23.3 • Unsprayed, non-transgenic cotton (left) and unsprayed Bt (Cry 1Ac) cotton (right) prior to harvest. (Photo courtesy of C. Mares.) Please see color plate section at the end of the book.

Concerns about Bt crops

Concerns about Bt crops have been raised by some political, environmental, and academic groups (Mendelsohn et al., 2003). Most concerns deal with factors common to all genetically modified crops, in particular human and environmental safety, corporate control of the food supply, and ethical issues over labeling and intellectual property. These will be addressed later in the book, but now we will concentrate on questions raised that are specific to Bt crops. These generally involve: (1) the health impact on humans or animals that consume the cry proteins; (2) non-target effects on beneficial insects; and (3) the development of resistant pests (Shelton et al., 2003).

As a pesticide spray, *B. thuringiensis* has long been considered a safe alternative to chemical pesticides. Despite over 60 years of use, few human health problems have been associated with Bt sprays, and studies conclude that Bt poses very little risk to mammals at field-level dosages (Glare and O'Callaghan, 2000; Mendelsohn et al., 2003; OECD, 2007). A major difference between the Bt spray and Bt engineered crops is that the insecticidal cry proteins are expressed at higher doses and within the plant tissues. This has helped make them successful at controlling insect infestations, but concerns have been raised about how this could affect human health. Currently most crops expressing Bt genes (except Bt sweetcorn) are not directly consumed by humans (the majority of maize is harvested for animal feed), although cottonseed oil is used for frying in the food service industry. However, there are crops in development containing *cry* genes that will directly enter the human food supply (e.g., rice and brinjal), and more are planned for the medium-term future.

Food from genetically modified plants is highly regulated. Before a new genetically modified crop can be approved for commercial release, it must comply with national and international human health safety assessments (Mendelsohn et al., 2003; OECD, 2007; Johnson et al., 2007). To date the vast majority of published studies related to Bt crops reveal similar nutritional performance and growth between animals fed genetically modified and conventional food (Flachowsky et al., 2005; Batista and Oliveira, 2009). The differences that have sometimes been observed at the microscopic and molecular level (Magaña-Gómez and de la Barca, 2008) are difficult to assess and frequently not confirmed by other investigators.

Concerns have been raised about the toxicity of Bt crops to non-target organisms, especially those involved in pollination, decomposition of biomass, and biological control methods (Dale et al., 2002). The potential for adverse effects is evaluated as part of the environmental risk assessment before any crops are commercialized (Nap et al., 2003; Conner et al., 2003). A tiered approach to assessing the potential for harm to beneficial insects has been developed (Romeis et al., 2008). Since Cry proteins are active against specific orders of insects, it is possible to narrow down which beneficial insects might be affected. Also, targeting expression of the *cry* gene to a specific tissue can eliminate the need to evaluate the effects on certain insects (i.e., if the gene is not expressed in the pollen there is no need to evaluate pollen feeding insects).

The most publicized research into the possibility of Bt crops harming beneficial insects relates to the Monarch butterfly (Ferber, 1999). Monarch caterpillars feed on milkweed plants, which often grow near maize fields. In a laboratory situation, only 56% of the Monarch larvae fed milkweed leaves dusted with Bt maize pollen survived, significantly less than a control group fed milkweed that was not dusted (Losey et al., 1999). Additional large-scale studies were commissioned to quantify the risk to the Monarch butterfly in the field. Because the gene promoters used in Yieldgard®CornBorer and Bt11 were almost inactive in pollen, the larvae did not receive lethal amounts of toxin, although the levels expressed in the pollen of one line (Knockout™) were 40-fold higher and did have a deleterious effect (Hellmich et al., 2001). Knockout™ has since been withdrawn from the market.

It was also discovered that the pollen on milkweed leaves in the field was less than had been anticipated and that the exposure rapidly declined with distance from the maize field (Pleasants et al., 2001). The breeding season of the Monarch butterfly had limited overlap with maize pollination and only a small proportion of the butterflies utilized milkweed close to maize crops (Sears et al., 2001). In 2002, despite a 40% increase from 2001 in Bt maize plantings, the Monarch butterfly population of North America increased by 30%, with loss of habitat in central Mexico and California seen as the major threat to their survival (Gatehouse et al., 2002).

Because Bt crops can exhibit 100% lethality to certain pests, there will naturally be effects on the predators of these pests. Meta-analyses comparing Bt plants to their non-sprayed, non-transgenic counterparts have revealed no unexpected negative or positive ecological effects, although predator numbers were reduced on the Bt crops (Wolfenbarger et al., 2008; Duan et al., 2009; Naranjo, 2009). The smaller predator population was attributed to the reduced number of prey and the low nutritional quality of the sick prey that were eaten before they died. In an ecological setting the use of Bt crops can increase biodiversity as pesticide levels drop, and the number of off-target effects will be much less (Cattaneo et al., 2006).

Insects developing resistance to the Cry toxins is a major concern. Over 400 species of insects have already developed resistance to various chemical pesticides and it is expected that they will eventually develop resistance to the Cry toxins if the technology is not properly protected. This could affect farmers who use Bt sprays as well as those who plant Bt crops. Resistance to Bt sprays had already been documented for populations of diamondback moth (*Plutella xylostella*), cabbage looper (*Trichoplusia ni*), and the Indian meal worm (*Plodia interpunctella*; Tabashnik, 1994) before commercial release of Bt crops. In the field, by contrast, resistance to Bt crops has been slow to develop (Fox, 2003). After 14 years of use there is evidence of three pest populations developing some field resistance: the bollworm (*H. zea*); the fall army worm (*Spodoptera frugiperda*); and the stem borer (*Busseola fusa*; Tabashnik, 2008).

The main strategy to delay insect resistance involves refuges. These consist of non-Bt plants grown near the Bt crop allowing susceptible pests to survive and breed. As long as the resistance trait is recessive and enough susceptible insects are available, they should dilute out any resistance genes that have been selected. This strategy has worked well in the past and according to entomologist Bruce Tabashnik, the evolution of *H. zea* resistance could be due to the concentration of Cry toxin not being high enough to kill off the heterozygous hybrids (i.e., resistance is not recessive; Tabashnik et al., 2008). Recently, stacked Bt crops have been commercialized that pyramid two or more *cry* genes. This is expected to further delay the onset of resistance as insects will have to evolve resistance to both toxins at the same time (Zhao et al., 2005). For pyramiding to work efficiently the two toxins must demonstrate no cross-resistance, so if resistance is developed to one toxin they are still completely vulnerable to the other toxin (Tabashnik et al., 2009). Therefore new insecticidal proteins that have a distinctly different mode of action than Cry toxins make attractive stacking options. In addition, both toxins must independently meet the high dose, high lethality requirement of accepted insect resistance management (IRM) plans (Ferré et al., 2008).

Improving Bt

New Cry proteins with the potential to be used as insecticides are still being discovered (Guo et al., 2008). However, with nearly 500 different types of δ-endotoxins annotated from *B. thuringiensis* (Crickmore, 2010) there is no shortage of Cry toxins to investigate. The early *cry* genes used in plants were controlled by the cauliflower mosaic virus 35S gene promoter, resulting in constitutive expression of the toxin at high levels in every cell. With tissue-specific promoters the toxin can be preferentially expressed predominantly where insect damage occurs. Gene promoters are now available that allow expression predominantly in the epidermal cells or the photosynthetic cells. Seed-specific gene promoters can be used to target insects that feed on seeds. The use of gene promoters that express the toxin only in response to insect stress (such as cell damage) would allow even greater control. Subcellular targeting sequences can also be added to the *cry* genes so that the protein can be localized to different cell compartments, including the cell wall or the chloroplast. This often results in higher levels of Cry protein, which also reduces the risk that insect resistance will develop.

The effectiveness of Bt in controlling target insects can also be increased. Stacking Bt genes is one way to do this. As additional new and safe insecticidal proteins are discovered (see the next section), they will inevitably be stacked with the *cry* genes. Deliberate mutation of Cry toxins has been used to determine their mode of action and has also produced some novel proteins with increased activity. Site-directed mutations have been used to increase toxicity to target insects (Rajamohan et al., 1996) and even led to the creation of Cry proteins toxic to mosquitoes from lepidopteran-specific toxins (Liu and Dean, 2006).

An alternative approach is to create hybrid *cry* genes by fusing two *cry* genes together or with other genes, and controlling them from the same gene promoter (Bohorova et al., 2001). Hybrid Cry toxins can increase the range of insects that can be controlled and even produce toxins that affect insects that were previously immune (de Maagd et al., 2000; Naimov et al., 2003; Singh et al., 2004; Karlova et al., 2005). Experiments have been conducted in which a *cry* gene is fused with a gene coding for the non-toxic part of the ricin toxin (Mehlo et al., 2005). This region also binds to the insect midgut, providing a secondary binding receptor for the Cry protein. As well as enhancing the Bt toxin's effectiveness, it is expected to delay resistance as both receptors would need to mutate simultaneously for resistance to develop.

Cowpea Trypsin Inhibitor

The only commercialized crop genetically modified for insect resistance that does not solely use a Cry protein utilizes the cowpea trypsin inhibitor (CpTI). Insects rely on proteases to digest proteins in their diet and trypsin is the major protease used by Lepidoptera. CpTI interferes with the insect's digestion of protein, negatively affecting the organism's nutrition, growth, and metabolism (Gatehouse and Boulter, 1983). Cowpeas are a significant food crop throughout the tropics and are widely consumed by humans and animals in sub-Saharan Africa without any obvious ill effects. Purified CpTI protein fed to rats showed only minimal short-term, anti-nutritional effects (Pusztai et al., 1992), and expression in tobacco and rice provided good resistance against insects (Boulter et al., 1989; Xu et al., 1996). The first commercial crop expressing CpTI was Zhong-41 cotton developed in China. It contains the *Cry1Ac* gene along with *CpTI* and was commercialized in 2001 with the intent of delaying the evolution of resistance in the cotton bollworm (Guo et al., 1999). Rice that expresses the *CpTI* gene is due to be released in China alongside the Bt varieties (James, 2009). The use of CpTI as an IRM tool does not fit the currently accepted IRM models, since the CpTI protein does not independently provide high dose, high lethality performance against target pests. For this reason, use of *CpTI* as a resistance gene in combination with Bt toxins would probably not be approved outside of China.

Novel Insecticidal Protection

VIP genes

In the mid-1990s a new insecticidal protein group was discovered in *B. thuringiensis*, which unlike the Cry proteins are not expressed when the bacterium sporulates. These proteins were found during the bacterium's vegetative state and hence were called vegetative insecticidal proteins (VIPs). Like the Cry proteins, it has been suggested that VIP proteins are activated by proteases, bind to membrane proteins within the insect's midgut epithelium, and form pores (Lee et al., 2003). They are of particular interest as they provided a new range of insecticidal activity (Estruch et al., 1996) and, while they share a similar mechanism of action, they bind different receptors and do not impede Cry toxin activity (Fang et al., 2007). This makes them good candidates for controlling resistance through stacking with *cry* genes.

A maize variety with the *Vip3A* gene, called Viptera™ (Agrisure Viptera 3111 trait stack) gained regulatory approval in 2010. The variety's stacked genes (*cry1Ab, vip3Aa20, cry3A*) will give resistance to the primary maize pests, borers, and rootworms, as well as secondary pests, many of which cannot be controlled by Cry proteins alone. These include corn earworms, armyworms, cutworms, sugarcane borers, and common stalk borers. A *VIP* cotton line, VipCot™, contains the transgenic cotton event COT67B, which produces the Cry1Ab protein and COT102, which features the novel Vip3A protein. It has received U.S. EPA approval for use in natural refuges as well as import approvals for Brazil, Japan, and Mexico.

Microorganism-derived toxins

Other bacteria also contain insecticidal toxins. *Photorhabdus luminescens*, a bacterium symbiotic with nematodes, has been studied and shown to contain numerous insecticidal factors. A gene that produces Toxin A from this bacterium was cloned and modified for expression in *Arabidopsis*, in which it gave almost complete protection against tobacco hornworm (*Manduca sexta*) larvae (Liu et al., 2003). Analysis of the *P. luminescens* genome has revealed more predicted insecticidal toxins than any other bacterium investigated (Duchaud et al., 2003). Other promising genes being cloned and assessed for insecticidal activity are the "Tc toxin complexes" (Tca, Tcb, Tcc, and Tcd), the "*Photorhabdus* insect-related" (Pir) toxins, the "Makes caterpillars floppy" toxins (Mcf1 and Mcf2), the "*Photorhabdus* virulence cassettes (PVCs)," and the "*Photorhabdus* insecticidal toxins" (Pits; ffrench-Constant et al., 2007; Lang et al., 2010). In principle, the toxins from *P. luminescens* and similar ones from another nematode symbiont (*Xenorhabdus nematophilus*) could be promising alternatives or complements to the well-established Cry proteins, but technical difficulties in expressing these large complex proteins will likely delay their development in commercialized products.

Brevibacillus laterosporus is a spore-forming bacterium with insecticidal activity against a wide spectrum of pests including Lepidoptera, Coleoptera, and the Mollusks (de Oliveira et al., 2004). Although no information on transgenic plants expressing *B. laterosporus* genes has been published in the peer reviewed literature, the patents have been submitted for the insecticidal proteins (Floris et al., 2007; Sampson et al., 2009). Cholesterol oxidase from *Streptomyces* is a potent inhibitor of cotton boll weevil larvae by inducing lysis in the midgut epithelium. It has been expressed in transgenic plants, but unless it is targeted to the chloroplast it negatively affects plant development and fertility (Corbin et al., 2001). With the current success of the *cry* genes in controlling the boll weevil, it has not yet been pursued commercially.

Baculoviruses infect insects and have been developed as biological insecticides similar to *B. thuringiensis*, although the cost of manufacturing and their relatively slow onset of action has limited commercial success. However, baculovirus bioactivity can be enhanced dramatically through engineering with other toxins (reviewed in the next section). The virus-enhancing factor (Enhancin) disrupts the peritrophic matrix (PM), making the insects more vulnerable to infection. Baculovirus Enhancin has been expressed in tobacco plants. The levels of expression were generally low, but the development of *T. ni* larvae was slowed on some plants (Cao et al., 2002). Fusolin is a chitin binding protein that also disrupts the PM. When the *fusolin* gene from the armyworm entomopoxvirus was inserted into transgenic rice, it rendered armyworm larvae 42 times more susceptible to Baculovirus infection (Hukuhara et al., 1999).

Plant-derived toxins

Plants have natural defenses against insect attack and transferring plant defense genes between plants could extend the spectrum of resistance. However, as a result of co-evolution, herbivorous insects have adapted to the plant defenses they normally encounter. This limits the prospects of some transgenics, but some show real promise.

CpTI is only one inhibitor of insect digestive enzymes, and there is a wide range of natural plant toxins that make use of this digestive enzyme inhibitor strategy. Plants express digestive inhibitors in their seeds, fruits, tubers, and corms, which appear to protect them from insect attack. They have detrimental effects on insect development and larval survival, both key measures of insecticidal activity. Proteinase inhibitors from insect-resistant plants have been cloned and expressed at high levels in different hosts (Malone et al., 2008), but generally only provide partial protection (Gatehouse, 2008). Insects appear to adapt to the presence of such inhibitors in their diets by elevating the levels of other digestive enzymes. This means that genes for additional inhibitors would need to be stacked for effective insect protection. In addition, the lack of lethality of these proteins will make adhering to the current insect-resistant management regulations difficult.

A digestive inhibitor that has successfully conferred resistance to an insect pest is the bean α-amylase inhibitor (αAI). When expressed in the seeds of pea using a strong promoter it made up to 3% of the seed protein (Shade et al., 1994). Field trials confirmed that the peas were completely resistant to the pea weevil (*Bruchus pisorum*; Morton et al., 2000). However, post-translational processing of the αAI in pea appeared to cause an immune response in mice (Prescott et al., 2005). Following these results, development of the αAI pea was suspended.

Plant lectins are another source of potential insect resistance genes (Murdock et al., 1990; Murdock and Shade, 2002). Lectins are a diverse group of carbohydrate binding proteins that are involved in a range of functions (De Hoff et al., 2009). Some lectins are toxic and are believed to be part of the plant's natural defense against insects. Lectins are attractive as they target many insects not affected by Bt and other toxins. Transgenic plants have been created by introducing genes for plant lectins into new hosts. The snowdrop and garlic lectin have been introduced into rice giving partial resistance to the sap-sucking rice brown planthopper (*Nilaparvata lugens*; Rao et al., 1998; Saha et al., 2006). The same arguments mentioned earlier pertaining to high dose, high lethality and IRM apply here as well. Concerns have been raised about the toxicity of some lectins to mammals, although consumption of transgenic rice containing the snowdrop lectin by rats evoked no detrimental symptoms (Poulsen et al., 2007).

Plant peroxidases (POX) have been implicated in a wide range of physiological processes, including insect resistance. Transgenic plants expressing high levels of tobacco anionic POX have resulted in reduced numbers and feeding by caterpillars, beetles, whiteflies, aphids, and grasshoppers in five different groups (Dowd et al., 1998). However, POX's multiple roles within cells could lead to incompatibility with endogenous and other introduced resistance proteins. Ribosome-inactivating proteins (RIPs) are a group of plant proteins capable of inactivating eukaryotic ribosomes. RIPs are widely distributed around the plant kingdom and, although they can be very toxic (e.g., ricin from the castor bean), most have low or selective toxicities. Transgenic tobacco expressing RIPs have shown reduced damage from insect pests (Shahidi-Noghabi et al., 2009) and hybrid RIP/POX expressing plants do not exhibit any cross interference (Dowd et al., 2006).

Transgenic tobacco, rice, papaya, and potato expressing plant defensins increase resistance to a wide range of plant pathogenic microbes and fungi (Carvalho and Gomes, 2009). A defensin from mung bean has shown *in vitro* lethality to larvae of the bruchid *Callosobruchus chinensis*, suggesting some also have insecticidal activity (Chen et al., 2002; Liu et al., 2006). Patatin is a lipid acyl hydrolase and major protein found in potato tubers, which inhibits the growth of southern and western corn rootworms fed an artificial diet (Strickland et al., 1995). Although no commercial transgenic plants have been developed, patatin and related lipid acyl hydrolases have been patented (Alibhai and Rydel, 2010).

Secondary metabolites

Plants synthesize a range of secondary chemicals as defenses against insects (Fraenkel, 1959). It is possible to increase resistance in plants that do not produce these metabolites through the addition of genes for the required enzymes (Jones et al., 1987). Three genes from sorghum (*Sorghum bicolour*) that express the entire biosynthetic pathway of the tyrosine-derived cyanogenic glucoside dhurrin were transferred into *Arabidopsis* (Kristensen et al., 2005). This resulted in the production of hydrogen cyanide in damaged tissue and increased resistance to the flea beetle (*Phyllotreta nemorum*). Three genes have also been added to tobacco allowing the production of caffeine, a natural insecticide (Kim et al., 2006). A transgenic tobacco plant with an enzyme that converts tryptophan into the alkaloid tryptamine has been created (Thomas et al., 1995). When fed to sweet potato whiteflies (*Bemisia tabaci*) there was a dramatic reduction in fly fertility, although strictly speaking the plants were not insecticidal.

The isopentenyl-transferase (*ipt*) gene from *Agrobacterium tumefaciens* codes for a key enzyme in the cytokinin biosynthetic pathway. Cytokinins have been correlated with the accumulation of insecticidal secondary metabolites in plants. The *ipt* gene was expressed in both tobacco and tomato under a wound-inducible promoter, resulting in cytokinin levels 70 times higher than in the control (Smigocki et al., 2000). The resulting plants exhibited reduced leaf consumption by *M. sexta* and the peach potato aphid (*Myzus persicae*). Unfortunately, the *ipt* gene also had a detrimental effect on plant development as it encodes one of the major plant hormones. Obviously there are risks associated with altering biosynthetic pathways, as the increase of a certain metabolite can disrupt many other pathways.

Another strategy is to modify the volatiles (e.g., scent) expressed by plants to attract insects for pollination. This can be used to deter insect colonization (Wang et al., 2001; Aharoni et al., 2003) or to attract natural enemies of insect pests. *Arabidopsis* has been transformed with a gene for an aphid volatile (*sesquiterpene* (*e*)-*β*-*farnesene*) that is released in response to aphid predators (Beale et al., 2006). It warns other aphids to stay away and this was observed through choice experiments. *Arabidopsis* was also transformed with a maize terpene (*TPS10*) gene that emits volatiles that were attractive to parasitoid wasps (Schnee et al., 2006). When maize is attacked by herbivores it releases this terpene that wasps associate with a nearby source of food. Plants expressing the *TPS10* gene attracted wasps, suggesting it could possibly be used as a pre-emptive defensive mechanism for some crops.

Other toxins

Chitinases digest chitin, a major component of insect cell walls, and their genes are found in species from all kingdoms (Arakane and Muthukrishnan, 2010). Transgenic tobacco expressing an *M. sexta* chitinase prevented growth of *H. virescens*, but did not affect *M. sexta*, probably due to their thicker peritrophic matrix (Ding et al., 1998). Spraying with a sublethal dose of Bt increased the mortality of *M. sexta* larvae, suggesting the use of chitinases with Cry proteins could increase resistance. Papaya engineered with the *M. sexta* chitinase has greater tolerance to spider mites (McCafferty et al., 2006), whereas transgenic cotton containing *M. sexta* chitinase conferred strong insect resistance in the field (Wang et al., 2005). However, expression of baculovirus chitinase in tobacco had no effect on *H. virescens*, although it did provide some resistance to fungal pathogens (Shi et al., 2000). Surprisingly, some examples of transgenic plants expressing chitinase genes have promoted insect growth (Saguez et al., 2005) or reduced toxicity (Gatehouse et al., 1996).

Toxins from animals have also been assessed as insecticides. Scorpion depressant toxins are a group of evolutionarily conserved polypeptides that target sodium channels and induce flaccid paralysis in insects. A toxic gene from the Manchurian scorpion has been expressed in rapeseed along with a *M. sexta* chitinase gene. This resulted in high resistance to diamondback moths (Wang et al., 2005). The scorpion toxin *Androctonus australis* Hector Insect Toxin (AaIT) has been expressed in poplar trees, which showed resistance to gypsy moth larvae (Wu et al., 2000), and in cotton for control of the cotton bollworm (Wu et al., 2008). Spider toxins are primarily targeted toward insects, and spider genes expressed in transgenic rice conferred resistance to striped stem borer (*Chilo suppressalis*) and rice leaffolder (*Cnaphalocrocis medinalis*; Huang et al., 2001). Transgenic tobacco expressing spider toxin was also highly resistant to larvae of the cotton bollworm (*H. armigera*) and the Egyptian armyworm (*Spodoptera littoralis*; Khan et al., 2006), and transgenic tobacco lines stably expressing the toxic teratocyte secretory protein from the parasitic wasp *Microplitis croceipes* caused growth retardation and larval mortality in *M. sexta* and *H. virescens* (Maiti et al., 2003).

Avidin is an egg white glycoprotein that binds to the vitamin biotin with high affinity. Avidin and a related protein, streptavidin, produced by the bacterium *Streptomyces avidinii*, is toxic to insects that require a lot of biotin; including species of Coleoptera, Lepidoptera, Diptera, and Sarcoptiformes. Maize expressing an avidin gene was fully resistant to three coleopteran larvae, but showed no toxicity in mice fed for 21 days (Kramer et al., 2000). Since plants require biotin, expressing the gene in the cytoplasm can lead to developmental abnormalities (Murray et al., 2002). By targeting the protein to vacuoles (Burgess et al., 2002; Markwick et al., 2003) or the seeds (Kramer et al., 2000) this problem is overcome. Because biotin binding proteins do not require a receptor on the insect midgut for activity, they would complement the insecticidal activity of the Cry toxins.

RNAi

RNA interference (RNAi) is an alternative to expressing toxic or inhibitory proteins for insect control (Baum et al., 2007). RNAi is the introduction of double-stranded RNA (dsRNA) into a cell, inhibiting the expression of a gene. RNAi was initially discovered in the nematode *Caenorhabditis elegans* (Fire et al., 1998). It has since proved to be a valuable tool for studying gene function in fungi, plants, insects, and animals (Mello and Conte, 2004). The mechanism for silencing via RNAi has been well investigated (Filipowicz et al., 2008; Carthew and Sontheimer, 2009).

Feeding nematodes and planarians with bacteria containing dsRNA has been shown to initiate RNAi gene silencing (Timmons and Fire, 1998; Newmark et al., 2003). Recent studies have shown that oral delivery of specific dsRNA can induce silencing in ticks (Soares et al., 2005), the light brown apple moth (Turner et al., 2006), triatomine bugs (Araujo et al., 2006), termites (Zhou et al., 2008), the tsetse fly (Walshe et al., 2009), and the beet armyworm (Tian et al., 2009). Plants that are resistant to herbivorous insect attack due to RNAi have now been created. Maize transformed with dsRNA that targeted a vacuolar ATPase of the western corn rootworm, conferred resistance similar to that achieved by the Bt toxin (Baum et al., 2007). In a similar experiment dsRNA was expressed in cotton that targeted a cotton bollworm P450 mRNA encoding an antidote to a natural cotton toxin, creating bollworm-resistant cotton (Mao et al., 2007). They showed that ingestion of either the processed short interfering (siRNA) or long dsRNA was enough to induce gene silencing, although the long dsRNA was more effective.

RNAi may have certain advantages over protein-based toxins. First it degrades much faster in the mammalian digestive system, which will make it safer for human and animal consumption. Also humans and animals already consume large amounts of RNA and DNA in their diets without any ill effects. In addition, there are likely to be fewer regulatory issues associated with RNAi, since the technology has already been used in a number of commercially approved biotechnology products (Auer and Frederick, 2009), and is undergoing regulatory review for use in combating insect diseases such as the colony collapse disorder affecting honey bees worldwide (Hunter et al., 2010). Another advantage is the very high insect species specificity of RNAi molecules, meaning only non-target arthropods closely related to the pest taxonomically would be negatively affected (Huesing et al., 2009). This specificity also means that non-target organism testing can be much more limited to nearest neighbor taxa for this class of molecules (Romeis et al., 2010). RNAi has the potential to target any insect gene, although genes expressed in the midgut make the most attractive targets. As only a short sequence of the insect's RNA needs to be targeted, it is relatively easy to combine multiple gene targets. The plant is only expressing an RNA fragment (plants routinely produce RNAi molecules as part of their normal physiology) so the resource drain should be low. There is a negligible risk that the RNA will show homology to any of the plant's own genes, so there should be no effect on the plant's biological processes.

Nematode-Resistant Crops

Plant parasitic nematodes comprise 15% of the total number of nematode species currently known (Fuller et al., 2008). They can result in yield losses of 20% in a single crop (Bird and Kaloshian, 2003) and cost an estimated $125 billion worth of crop losses globally each year (Chitwood, 2003). The majority of these losses are inflicted by the root-knot nematodes and cyst nematodes (Koenning et al., 1999). Chemical control is expensive and crop rotation does not work against species with wide host ranges (Abad et al., 2003). While natural resistance does exist in some cultivars, genetic engineering allows greater scope, either by introducing resistant genes from other species or using RNAi to target endogenous nematode genes.

Some of the commercialized resistant genes used in insect control can also have a negative effect on nematode populations. CpTI and Bt proteins expressed in transgenic plants have shown varying levels of protection against nematodes. CpTI expressing transgenic *Arabidopsis* affected the sexual development of nematodes (Urwin et al., 1998), and exposure to Bt toxins reduced nematode fertility and viability (Marroquin et al., 2000). Resistance to the root-knot nematode *Meloidogyne incognita* was observed when the Cry6A protein was expressed in tomato roots (Li et al., 2007). Other nematode-resistant genes engineered into transgenic plants include serine protease inhibitors (Cai et al., 2003; Vishnudasan et al., 2005), cystatins (Urwin et al., 1997; Vain et al., 1998; Samac and Smigocki, 2003; Atkinson et al., 2004), the snowdrop lectin (Ripoll et al., 2003), and monoclonal antibodies (Baum et al., 1996). Proteinase inhibitors show the most promise in conferring nematode resistance and field trials of transgenic potatoes expressing cystatins have been successfully completed (Urwin et al., 2001).

Recently more focus has been on the use of RNAi technology to control plant-parasitic nematodes (Gheysen and Vanholme, 2007). Yadav et al. (2006) first reported the development of a transgenic plant conferring nematode resistance by expressing dsRNA. By targeting two root-knot nematode housekeeping genes they produced transgenic tobacco over 95% more resistant to *M. incognita* than wild-type. Huang et al. (2006) expressed dsRNA for a gene involved in nematode parasitism in *Arabidopsis thaliana*. This resulted in effective resistance to four major species of root-knot nematode, more than any other known natural resistant gene. However, inducing plant-delivered RNAi in cyst nematodes has been difficult and resistance has been much lower (Steeves et al., 2006; Sindhu et al., 2009). With the recent sequencing of the plant-parasitic nematode *M. hapla* genome (Opperman et al., 2008) more RNAi targets will be identified.

Recombinant Insecticides

Another related method of insect protection involves the genetic modification of biological control agents to make them more efficient (Inceoglu et al., 2006; Park and Federici, 2009). The *B. thuringiensis* ssp. *israelensis* (Bti) and *B. sphaericus* (Bs) both exhibit mosquitocidal activity. Toxicity to mosquitoes has been increased ten-fold by introducing Bt genes from Bti into Bs (Federici et al., 2003). Toxic genes from *B. thuringiensis* have been inserted into *Photorhabdus* (Jamoussi et al., 2009). However, the most advanced recombinant pathogenic control agents are baculoviruses.

An early recombinant baculovirus was generated by removing the ecdysteroid UDP glucosyltransferase (*egt*) gene (O'Reilly and Miller, 1989). The *egt* deletion viruses killed host larvae 30% faster than the wild-type virus and reduced feeding damage by about 40% (O'Reilly and Miller, 1991). Other early attempts to improve baculoviruses involved addition of genes encoding juvenile hormone esterase (Hammock et al., 1990), a scorpion AaIT toxin (Stewart et al., 1991; Chejanovsky et al., 1995), chitinase (Gopalakrishnan et al., 1995), and spider toxins (Hughes et al., 1997). However, none of these additions increase efficacy beyond the *egt* deletion.

The introduction of a digestive cathepsin L-like protease from the flesh fly (Harrison and Bonning, 2001), a neurotoxin (tox34) from the straw-itch mite (Popham et al., 1997; Burden et al., 2000), LqhIT2 scorpion toxin (Harrison and Bonning, 2000), μ-Aga-IV spider toxin (Prikhodko et al., 1996), and As II and Sh I derived from sea anemones (Prikhodko et al., 1996) into transgenic baculoviruses increased the killing speed by 50–60%. Attempts to introduce Bt endotoxins into baculovirus showed no increase in insecticidal activity (Merryweather et al., 1990). As the Bt protoxin was expressed in the insect cytoplasm it was unable to reach the insect midgut for activation. To overcome this problem the Bt toxin has been fused with polyhedron to produce polyhedra that occlude the BT toxin into the insect midgut, increasing baculovirus killing efficacy by over 60% (Chang et al., 2003).

Field trials using AaIT expressing baculovirus showed 23–29% lower feeding damage compared to wild-type (Cory et al., 1994) and can result in a 20% increase in yield (Sun et al., 2004). The modified baculoviruses offer levels of protection similar to chemical pesticides and have little adverse effect on beneficial insects. Also fitness is reduced as they are unable to propagate as efficiently within the host as wild-type baculoviruses (Zwart et al., 2009). Future modifications could involve removing genes that will reduce fitness without impacting the efficacy, stacking toxic genes to delay insect resistance and increase virulence, and adding genes that enhance the durability of the virus in the field, such as UV resistance.

Conclusion

So far the benefits of genetically modified insect-protected crop plants have far outweighed the risks associated with them. Bt crops expressing Cry proteins are in their second decade of safe use. The use of Bt free refugia and other IRM techniques have delayed the onset of insect resistance to single Cry toxins, but many experts consider it only a matter of time before they lose their effectiveness in controlling their target pests. Crops stacked with multiple protective genes have already been developed and should delay resistance further. Plants with genes for fused hybrid Cry proteins have also been released commercially and should offer the same type of longer term protection. It is important that any stacked or hybrid genes have different modes of action; meet the high dose, high lethality standard; and do not interfere with each other's performance. Ultimately, however, the future of genetically modified insect protected plants relies on the discovery and development of new approaches.

Fortunately promising alternatives to Cry proteins exist and are being developed. The VIP toxins, also from *B. thuringiensis*, are likely to be the next toxins released commercially. The bacterium *P. luminescens* has a whole host of toxic genes, many of which are still being characterized, and could rival or even exceed *B. thuringiensis* in importance, but development of this class of protein toxins has proven difficult. Plants have developed their own defense mechanisms against insects and by swapping systems between different species further protection may be elicited. While plant digestive inhibitors and lectins have shown mixed results at this stage, there are still opportunities to develop them further. Toxins from spiders, wasps, and scorpions have been tested and, although effective against insects, there are concerns about their safety and perception for humans and other mammals. RNAi is a relatively new possibility and if it proves effective against insects in the field it has major potential as an insect control agent.

References

Abad, P., Favery, B., Rosso, M. N., & Castagnone Sereno, P. (2003). Root knot nematode parasitism and host response: Molecular basis of a sophisticated interaction. *Molecular Plant Pathology*, *4*, 217–224.

Aharoni, A., Giri, A. P., Deuerlein, S., Griepink, F., de Kogel, W. J., & Verstappen, F. W. A., et al. (2003). Terpenoid metabolism in wild-type and transgenic *Arabidopsis* plants. *Plant Cell*, *15*, 2866–2884.

Al-Deeb, M. A., & Wilde, G. E. (2005). Effect of Bt corn expressing the Cry3Bb1 toxin on western corn rootworm (Coleoptera: *Chrysomelidae*) biology. *Journal of the Kansas Entomological Society*, *78*, 142–152.

Alibhai, M. F. & Rydel, T. J. (2010). Insect inhibitory lipid acyl hydrolases. United States Patent 7662372.

Arakane, Y., & Muthukrishnan, S. (2010). Insect chitinase and chitinase-like proteins. *Cellular and Molecular Life Science*, *67*, 201–216.

Araujo, R. N., Santos, A., Pinto, F. S., Gontijo, N. F., Lehane, M. J., & Pereira, M. H. (2006). RNA interference of the salivary gland nitrophorin 2 in the triatomine bug *Rhodnius prolixus* (Hemiptera: Reduviidae) by dsRNA ingestion or injection. *Insect Biochemistry and Molecular Biology*, *36*, 683–693.

Atkinson, H. J., Grimwood, S., Johnston, K., & Green, J. (2004). Prototype demonstration of transgenic resistance to the nematode *Radopholus similis* conferred on banana by a cystatin. *Transgenic Research*, *13*, 135–142.

Auer, C., & Frederick, R. (2009). Crop improvement using small RNAs: Applications and predictive ecological risk assessments. *Trends in Biotechnology*, *27*, 644–651.

Bambawale, O. M., Singh, A., Sharma, O. P., Bhosle, B. B., Lavekar, R. C., & Dhandhapani, A., et al. (2004). Performance of Bt cotton (MECH-162) under integrated pest management in farmers' participatory field trial in Nanded district, Central India. *Current Science*, *86*, 1628–1637.

Batista, R., & Oliveira, M. M. (2009). Facts and fiction of genetically engineered food. *Trends in Biotechnology*, *27*, 277–286.

Baum, J. A., Bogaert, T., Clinton, W., Heck, G. R., Feldmann, P., & Ilagan, O., et al. (2007). Control of coleopteran insect pests through RNA interference. *Nature Biotechnology*, *25*, 1322–1326.

Baum, T. J., Hiatt, A., Parrott, W. A., Pratt, L. H., & Hussey, R. S. (1996). Expression in tobacco of a functional monoclonal antibody specific to stylet secretions of the root-knot nematode. *Molecular Plant–Microbe Interactions*, *9*, 382–387.

Beale, M. H., Birkett, M. A., Bruce, T. J. A., Chamberlain, K., Field, L. M., & Huttly, A. K., et al. (2006). Aphid alarm pheromone produced by transgenic plants affects aphid and parasitoid behavior. *Proceedings of the National Academy of Sciences of the United States of America*, *103*, 10509–10513.

Berliner, E. (1911). Uber die Schlaffsucht der Mehlmottenraupe. *Zeitschrift fur das Gesamstadt*, *252*, 3160–3162.

Bird, D. M. K., & Kaloshian, I. (2003). Are roots special? Nematodes have their say. *Physiological and Molecular Plant Pathology*, *62*, 115–123.

Bohorova, N., Frutos, R., Royer, M., Estanol, P., Pacheco, M., & Rascon, Q., et al. (2001). Novel synthetic *Bacillus thuringiensis* cry1B gene and the cry1B-cry1Ab translational fusion confer resistance to southwestern corn borer, sugarcane borer and fall armyworm in transgenic tropical maize. *Theoretical and Applied Genetics*, *103*, 817–826.

Boulter, D., Gatehouse, A. M. R., & Hilder, V. (1989). Use of cowpea trypsin inhibitor (CpTI) to protect plants against insect predation. *Biotechnology Advances*, *7*, 489–497.

Bravo, A., Gill, S. S., & Soberon, M. (2007). Mode of action of *Bacillus thuringiensis* Cry and Cyt toxins and their potential for insect control. *Toxicon*, *49*, 423–435.

Bravo, A., Hendricks, K., Jancens, S., & Peferoen, M. (1992). Immunocytochemical analysis of specific binding of *Bacillus thuringiensis* crystal proteins to lepidopteran and coleopteran midgut membranes. *Journal of Invertebrate Pathology*, *60*, 247–253.

Broderick, N. A., Raffa, K. F., & Handelsman, J. (2006). Midgut bacteria required for *Bacillus thuringiensis* insecticidal activity. *Proceedings of the National Academy of Sciences of the United States of America*, *103*, 15196–15199.

Broderick, N. A., Robinson, C. J., McMahon, M. D., Holt, J., Handelsman, J., & Raffa, K. F. (2009). Contributions of gut bacteria to *Bacillus thuringiensis*-induced mortality vary across a range of Lepidoptera. *BMC Biology*, *7*, 11.

Brookes, G., & Barfoot, P. (2010). *GM crops: Global socio-economic and environmental impacts 1996–2008*. Dorchester, UK: P.G Economics Ltd.

Burden, J. P., Hails, R. S., Windass, J. D., Suner, M. M., & Cory, J. S. (2000). Infectivity, speed of kill, and productivity of a baculovirus expressing the itch mite toxin txp-1 in second and fourth instar larvae of *Trichoplusia ni*. *Journal of Invertebrate Pathology*, *75*, 226–236.

Burgess, E. P. J., Malone, L. A., Christeller, J. T., Lester, M. T., Murray, C., & Philip, B. A., et al. (2002). Avidin expressed in transgenic tobacco leaves confers resistance to two noctuid pests, *Helicoverpa armigera* and *Spodoptera litura*. *Transgenic Research*, *11*, 185–198.

Cai, D., Thurau, T., Tian, Y., Lange, T., Yeh, K. W., & Jung, C. (2003). Sporamin-mediated resistance to beet cyst nematodes (*Heterodera schachtii* Schm.) is dependent on trypsin inhibitory activity in sugar beet (*Beta vulgaris* L.) hairy roots. *Plant Molecular Biology*, *51*, 839–849.

Cao, J., Ibrahim, H., Garcia, J., Mason, H., Granados, R., & Earle, E. (2002). Transgenic tobacco plants carrying a baculovirus enhancin gene slow the development and increase the mortality of *Trichoplusia ni* larvae. *Plant Cell Reports*, *21*, 244–250.

Carthew, R. W., & Sontheimer, E. J. (2009). Origins and mechanisms of miRNAs and siRNAs. *Cell*, *136*, 642–655.

Carvalho, A. O., & Gomes, V. M. (2009). Plant defensins: Prospects for the biological functions and biotechnological properties. *Peptides*, *30*, 1007–1020.

Cattaneo, M. G., Yafuso, C., Schmidt, C., Huang, C., Rahman, M., & Olson, C., et al. (2006). Farm-scale evaluation of the impacts of transgenic cotton on biodiversity, pesticide use, and yield. *Proceedings of the National Academy of Sciences of the United States of America*, *103*, 7571–7576.

CDC, *Investigation of human health effects associated with potential exposure to genetically modified corn: A report to the U.S. Food and Drug Administration from the Centers for Disease Control and Prevention*. National Center for Environmental Health.

Chang, J. H., Choi, J. Y., Jin, B. R., Roh, J. Y., Olszewski, J. A., & Seo, S. J., et al. (2003). An improved baculovirus insecticide producing occlusion bodies that contain *Bacillus thuringiensis* insect toxin. *Journal of Invertebrate Pathology*, *84*, 30–37.

Chejanovsky, N., Zilberberg, N., Rivkin, H., Zlotkin, E., & Gurevitz, M. (1995). Functional expression of an alpha anti-insect scorpion neurotoxin in insect cells and lepidopterous larvae. *FEBS Letters*, *376*, 181–184.

Chen, K. C., Lin, C. Y., Kuan, C. C., Sung, H. Y., & Chen, C. S. (2002). A novel defensin encoded by a mungbean cDNA exhibits insecticidal activity against bruchid. *Journal of Agricultural and Food Chemistry*, *50*, 7258–7263.

Chen, L. J., Lee, D. S., Song, Z. P., Suh, H. S., & Lu, B. R. (2004). Gene flow from cultivated rice (*Oryza sativa*) to its weedy and wild relatives. *Annals of Botany*, *93*, 67–73.

Chitwood, D. J. (2003). Research on plant parasitic nematode biology conducted by the United States Department of Agriculture – Agricultural Research Service. *Pest Management Science*, *59*, 748–753.

Choudhary, B., & Gaur, K. (2008). *The development and regulation of Bt brinjal in India (Eggplant/Aubergine)* (Vol. 38). International Service for the Acquisition of Agri-Biotech Applications.

Clark, P. L., Vaughn, T. T., Meinke, L. J., Molina-ochoa, J., & Foster, J. E. (2006). *Diabrotica virgifera virgifera* (Coleoptera: Chrysomelidae) larval feeding behavior on transgenic maize (MON 863) and its isoline. *Journal of Economic Entomology*, *99*, 722–727.

Cohen, M. B., Arpaia, S., Lan, L. P., Chau, L. M., & Snow, A. A. (2008). Shared flowering phenology, insect pests, and pathogens among wild, weedy, and cultivated rice in the Mekong Delta, Vietnam: Implications for transgenic rice. *Environmental Biosafety Research*, *7*, 73–85.

Conner, A. J., Glare, T. R., & Nap, J. P. (2003). The release of genetically modified crops into the environment. *Plant Journal*, *33*, 19–46.

Corbin, D. R., Grebenok, R. J., Ohnmeiss, T. E., Greenplate, J. T., & Purcell, J. P. (2001). Expression and chloroplast targeting of cholesterol oxidase in transgenic tobacco plants. *Plant Physiology*, *126*, 1116–1128.

Cory, J. S., Hirst, M. L., Williams, T., Hails, R. S., Goulson, D., & Green, B. M., et al. (1994). Field trial of a genetically improved baculovirus insecticide. *Nature*, *370*, 138–140.

Crickmore, N. (2010) Full list of delta-endotoxins. <http://www.lifesci.sussex.ac.uk/home/Neil_Crickmore/Bt/toxins2.html/>

Crickmore, N., Zeigler, D. R., Feitelson, J., Schnepf, E., Van Rie, J., & Lereclus, D., et al. (1998). Revision of the nomenclature for the *Bacillus thuringiensis* pesticidal crystal proteins. *Microbiology and Molecular Biology Reviews*, *62*, 807–813.

Dale, P. J., Clarke, B., & Fontes, E. M. G. (2002). Potential for the environmental impact of transgenic crops. *Nature Biotechnology*, *20*, 567–574.

De Hoff, P. L., Brill, L. M., & Hirsch, A. M. (2009). Plant lectins: The ties that bind in root symbiosis and plant defense. *Molecular Genetics and Genomics*, *282*, 1–15.

de Maagd, R. A., Weemen-Hendriks, M., Stiekema, W., & Bosch, D. (2000). *Bacillus thuringiensis* delta-endotoxin Cry1C domain III can function as a specificity determinant for *Spodoptera exigua* in different, but not all, Cry1-Cry1C hybrids. *Applied and Environmental Microbiology*, *66*, 1559–1563.

de Oliveira, E. J., Rabinovitch, L., Monnerat, R. G., Passos, L. K. J., & Zahner, V. (2004). Molecular characterization of *Brevibacillus laterosporus* and its potential use in biological control. *Applied and Environmental Microbiology*, *70*, 6657.

Dhugga, K. S. (2007). Maize biomass yield and composition for biofuels. *Crop Science*, *47*, 2211–2227.

Ding, X., Gopalakrishnan, B., Johnson, L. B., White, F. F., Wang, X., & Morgan, T. D., et al. (1998). Insect resistance of transgenic tobacco expressing an insect chitinase gene. *Transgenic Research*, *7*, 77–84.

Dowd, P. F., Holmes, R. A., Pinkerton, T. S., Johnson, E. T., Lagrimini, L. M., & Boston, R. S. (2006). Relative activity of a tobacco hybrid expressing high levels of a tobacco anionic peroxidase and maize ribosome-inactivating protein against *Helicoverpa zea* and *Lasioderma serricorne*. *Journal of Agricultural and Food Chemistry*, *54*, 2629–2634.

Dowd, P. F., Lagrimini, L. M., & Nelsen, T. C. (1998). Relative resistance of transgenic tomato tissues expressing high levels of tobacco anionic peroxidase to different insect species. *Natural Toxins*, *6*, 241–249.

Duan, J. J., Lundgren, J. G., Naranjo, S., & Marvier, M. (2009). Extrapolating non-target risk of Bt crops from laboratory to field. *Biology Letters*, *6*, 74–77.

Duchaud, E., Rusniok, C., Frangeul, L., Buchrieser, C., Givaudan, A., & Taourit, S., et al. (2003). The genome sequence of the entomopathogenic bacterium *Photorhabdus luminescens*. *Nature Biotechnology*, *21*, 1307–1313.

Estruch, J. J., Warren, G. W., Mullins, M. A., Nye, G. J., Craig, J. A., & Koziel, M. G. (1996). Vip3A, a novel *Bacillus thuringiensis* vegetative insecticidal protein with a wide spectrum of activities against lepidopteran insects. *Proceedings of the National Academy of Sciences of the United States of America*, *93*, 5389–5394.

Fang, J., Xu, X., Wang, P., Zhao, J. Z., Shelton, A. M., & Cheng, J., et al. (2007). Characterization of chimeric *Bacillus thuringiensis* Vip3 toxins. *Applied and Environmental Microbiology*, *73*, 956–961.

Federici, B. A., Park, H. W., Bideshi, D. K., Wirth, M. C., & Johnson, J. J. (2003). Recombinant bacteria for mosquito control. *Journal of Experimental Biology*, *206*, 3877–3885.

Ferber, D. (1999). GM crops in the cross hairs. *Science*, *286*, 1662–1666.

Ferré, J., Rie, J., & Macintosh, S. C. (2008). Insecticidal genetically modified crops and insect resistance management (IRM). In J. Romeis, A. M. Shelton & G. G. Kennedy (Eds.), *Integration of insect-resistant genetically modified crops within IPM programs* (pp. 41–85). New York: Springer.

ffrench-Constant, R. H., Dowling, A., & Waterfield, N. R. (2007). Insecticidal toxins from Photorhabdus bacteria and their potential use in agriculture. *Toxicon*, *49*, 436–451.

Filipowicz, W., Bhattacharyya, S. N., & Sonenberg, N. (2008). Mechanisms of post-transcriptional regulation by microRNAs: Are the answers in sight? *Nature Reviews Genetics*, *9*, 102–114.

Fire, A., Xu, S. Q., Montgomery, M. K., Kostas, S. A., Driver, S. E., & Mello, C. C. (1998). Potent and specific genetic interference by double-stranded RNA in *Caenorhabditis elegans*. *Nature*, *391*, 806–811.

Fitt, G. P. (2008). Have Bt crops led to changes in insecticide use patterns and impacted IPM? In J. Romeis, A. M. Shelton, & G. G. Kennedy (Eds.), *Integration of insect-resistant genetically modified crops within IPM programs* (Vol. V, pp. 303–328). New York: Springer.

Flachowsky, G., Chesson, A., & Aulrich, K. (2005). Animal nutrition with feeds from genetically modified plants. *Archives of Animal Nutrition*, *59*, 1–40.

Floris, I., Ruiu, L., Satta, A., Delrio, G., Rubino, S., Paglietti, B., et al. (2007). US Patent App. 20,100/003,227.

Fox, J. L. (2003). Resistance to Bt toxin surprisingly absent from pests. *Nature Biotechnology*, *21*, 958–959.

Fraenkel, G. (1959). The *raison d'etre* of secondary plant substances. *Science*, *129*, 1466–1470.

Frisvold, G. B., Tronstad, R., & Reeves, J. M. (2006). International Impacts of Bt Cotton Adoption. In R. E. Evenson & V. Santaniello (Eds.), *International trade and policies for genetically modified products* (pp. 191–201). Cambridge: CAB International.

Fuller, V. L., Lilley, C. J., & Urwin, P. E. (2008). Nematode resistance. *New Phytologist*, *180*, 27–44.

Gatehouse, A. M. R., & Boulter, D. (1983). Assessment of antimetabolic effects of trypsin inhibitors from cowpea (*Vigna unguiculata*) and other legumes on development of the bruchid beetle *Callosobruchus maculatus*. *Journal of the Science of Food and Agriculture*, *34*, 345–350.

Gatehouse, A. M. R., Down, R. E., Powell, K. S., Sauvion, N., Rahbé, Y., & Newell, C. A., et al. (1996). Transgenic potato plants with enhanced resistance to the peach-potato aphid *Myzus persicae*. *Entomologia Experimentalis Et Applicata*, *79*, 295–307.

Gatehouse, A. M. R., Ferry, N., & Raemaekers, R. J. M. (2002). The case of the monarch butterfly: A verdict is returned. *Trends in Genetics*, *18*, 249–251.

Gatehouse, J. A. (2008). Biotechnological prospects for engineering insect-resistant plants. *Plant Physiology, 146*, 881–887.

Gheysen, G., & Vanholme, B. (2007). RNAi from plants to nematodes. *Trends in Biotechnology, 25*, 89–92.

Glare, T. R., & O'Callaghan, M. (2000). *Bacillus thuringiensis: Biology, ecology and safety*. Chichester: John Wiley & Sons.

Goldberg, L. J., & Margalit, J. (1977). A bacterial spore demonstrating rapid larvicidal activity against *Anopheles sergentii, Uranotaenia unguiculata, Culex univitattus, Aedes aegypti* and *Culex pipiens*. *Mosquito News, 37*, 355–358.

Gopalakrishnan, B., Muthukrishnan, S., & Kramer, K. J. (1995). Baculovirus-mediated expression of a *Manduca sexta* chitinase gene: Properties of the recombinant protein. *Insect Biochemistry and Molecular Biology, 25*, 255–265.

Guo, S., Liu, M., Peng, D., Ji, S., Wang, P., & Yu, Z., et al. (2008). New strategy for isolating novel nematicidal crystal protein genes from *Bacillus thuringiensis* strain YBT-1518. *Applied and Environmental Microbiology, 74*, 6997–7001.

Guo, S. D., Cui, H. Z., Xia, L. Q., Wu, D., Ni, W. C., & Zhang, Z. L., et al. (1999). Development of bivalent insect-resistant transgenic cotton plants. *Scientia Agricultura Sinica, 28*, 1–7.

Hammock, B. D., Bonning, B. C., Possee, R. D., Hanzlik, T. N., & Maeda, S. (1990). Expression and effects of the juvenile hormone esterase in a baculovirus vector. *Nature, 344*, 458–461.

Harrison, R. L., & Bonning, B. C. (2000). Use of scorpion neurotoxins to improve the insecticidal activity of *Rachiplusia* ou multicapsid nucleopolyhedrovirus. *Biological Control, 17*, 191–201.

Harrison, R. L., & Bonning, B. C. (2001). Use of proteases to improve the insecticidal activity of baculoviruses. *Biological Control, 20*, 199–209.

Hellmich, R. L., Siegfried, B. D., Sears, M. K., Stanley-Horn, D. E., Daniels, M. J., & Mattila, H. R., et al. (2001). Monarch larvae sensitivity to *Bacillus thuringiensis*-purified proteins and pollen. *Proceedings of the National Academy of Sciences of the United States of America, 98*, 11925–11930.

Herring, R. J. (2008). Whose numbers count? Probing discrepant evidence on transgenic cotton in the Warangal district of India. *International Journal of Multiple Research Approaches, 2*, 145–159.

Hofte, H., & Whiteley, H. R. (1989). Insecticidal crystal proteins of *Bacillus thuringiensis*. *Microbiology and Molecular Biology Reviews, 53*, 242–255.

Horsch, R. B., Fraley, R. T., Rogers, S. G., Sanders, P. R., Lloyd, A., & Hoffmann, N. (1984). Inheritance of functional foreign genes in plants. *Science, 223*, 496–498.

Huang, G., Allen, R., Davis, E. L., Baum, T. J., & Hussey, R. S. (2006). Engineering broad root-knot resistance in transgenic plants by RNAi silencing of a conserved and essential root-knot nematode parasitism gene. *Proceedings of the National Academy of Sciences of the United States of America, 103*, 14302–14306.

Huang, J., Hu, R., Rozelle, S., & Pray, C. (2008). Genetically modified rice, yields, and pesticides: Assessing farm level productivity effects in china. *Economic Development and Cultural Change, 56*, 241–263.

Huang, J. Q., Zhi Ming, W. E. I., Hai Long, A. N., & Zhu, Y. X. (2001). *Agrobacterium tumefaciens*-mediated transformation of rice with the spider insecticidal gene conferring resistance to leaffolder and striped stem borer. *Cell Research, 11*, 149–155.

Huesing, J., Lloyd, F., Levine, S., & Vaughn, T. (2009). Approaches to tier-based NTO testing of RNAi pest control traits. Entomological Society of America. Annual Meeting Tuesday, December 15, 2009 Indianapolis, IN, USA.

Hughes, P. R., Wood, H. A., Breen, J. P., Simpson, S. F., Duggan, A. J., & Dybas, J. A. (1997). Enhanced bioactivity of recombinant Baculoviruses expressing insect-specific spider toxins in Lepidopteran crop pests. *Journal of Invertebrate Pathology, 69*, 112–118.

Hukuhara, T., Hayakawa, T., & Wijonarko, A. (1999). Increased baculovirus susceptibility of armyworm larvae feeding on transgenic rice plants expressing an entomopoxvirus gene. *Nature Biotechnology, 17*, 1122–1124.

Hunter, W., Ellis, J., van Engelsdorp, D., Hayes, J., Westervelt, D., & Glick, E., et al. (2010). Large-scale field application of RNA: technology reducing Israeli acute paralysis virus disease in honey bees (*Apis mellifera*, Hymenoptera: Apidae). *Plos pathogens*, 6(12), e100116.

Inceoglu, A. B., Kamita, S. G., & Hammock, B. D. (2006). Genetically modified baculoviruses: A historical overview and future outlook. *Advances in Virus Research, 68*, 323–360.

Ishiwatari, S. (1901). On a kind of severe flacherie (sotto disease). *Dainihon Sanshi Kaiho, 114*, 1–5.

James, C. (2001). *Global status of commercialized biotech/GM crops: 2001* (Vol. 24). Ithaca, NY: International Service for the Acquisition of Agri-Biotech Applications. (ISAAA Brief No. 24).

James, C. (2004). *Global status of commercialized biotech/GM crops: 2004* (Vol. 32). Ithaca, NY: International Service for the Acquisition of Agri-Biotech Applications. (ISAAA Brief No. 32).

James, C. (2005). *Global status of commercialized biotech/GM crops: 2005* (Vol. 34). Ithaca, NY: International Service for the Acquisition of Agri-Biotech Applications. (ISAAA Brief No. 34).

James, C. (2008). *Global status of commercialized biotech/GM crops: 2008* (Vol. 39). Ithaca, NY: International Service for the Acquisition of Agri-Biotech Applications. (ISAAA Brief No. 39).

James, C. (2009). *Global status of commercialized biotech/GM crops: 2009* (Vol. 41). Ithaca, NY: International Service for the Acquisition of Agri-Biotech Applications. (ISAAA BriefNo. 41).

James, C., & Choudhary, B. (2010). Global adoption of biotech cotton, 1996 to 2007. In J. Widholm, H. Lorz & T. Nagata (Eds.), *Biotechnology in agriculture and forestry* (pp. 177–196). Heidelburg: Springer.

Jamoussi, K., Sellami, S., Abdelkefi-Mesrati, L., Givaudan, A., & Jaoua, S. (2009). Heterologous expression of *Bacillus thuringiensis* vegetative insecticidal protein-encoding gene vip3LB in *Photorhabdus temperata* strain K122 and oral toxicity against the Lepidoptera *Ephestia kuehniella* and *Spodoptera littoralis*. *Molecular Biotechnology, 43*, 97–103.

Jayaraman, K. S. (2004). India produces homegrown GM cotton. *Nature Biotechnology, 22*, 255–256.

Johnson, K. L., Raybould, A. F., Hudson, M. D., & Poppy, G. M. (2007). How does scientific risk assessment of GM crops fit within the wider risk analysis? *Trends in Plant Science, 12*, 1–5.

Jones, D., Huesing, J., Zador, E., & Heim, C. (1987). The tobacco-insect model system for genetically engineering plants for non-protein insect resistance factors. *Molecular Entomology*, 469–478.

Kaiser, J. (1996). Pests overwhelm Bt cotton crop. *Science, 273*, 423.

Kalaitzandonakes, N., Alston, J. M., & Bradford, K. J. (2007). Compliance costs for regulatory approval of new biotech crops. *Nature Biotechnology, 25*, 509–511.

Karlova, R., Weemen-Hendriks, M., Naimov, S., Ceron, J., Dukiandjiev, S., & de Maagd, R. A. (2005). *Bacillus thuringiensis* δ-endotoxin Cry1Ac domain III enhances activity against *Heliothis virescens* in some, but not all Cry1-Cry1Ac hybrids. *Journal of Invertebrate Pathology, 88*, 169–172.

Kaufman, M. (2000, September 18) *Biotech critics cite unapproved corn in taco shells* (pp. A02). Washington Post.

Khan, S. A., Zafar, Y., Briddon, R. W., Malik, K. A., & Mukhtar, Z. (2006). Spider venom toxin protects plants from insect attack. *Transgenic Research, 15*, 349–357.

Kim, Y. S., Uefuji, H., Ogita, S., & Sano, H. (2006). Transgenic tobacco plants producing caffeine: A potential new strategy for insect pest control. *Transgenic Research, 15*, 667–672.

Koenning, S. R., Overstreet, C., Noling, J. W., Donald, P. A., Becker, J. O., & Fortnum, B. A. (1999). Survey of crop losses in response to phytoparasitic nematodes in the United States for 1994. *Journal of Nematology, 31*, 587–618.

Kramer, K. J., Morgan, T. D., Throne, J. E., Dowell, F. E., Bailey, M., & Howard, J. A. (2000). Transgenic avidin maize is resistant to storage insect pests. *Nature Biotechnology, 18*, 670–674.

Krieg, A., Huger, A. M., Langenbruch, G. A., & Schnetter, W. (1983). *Bacillus thuringiensis* var. *tenebrionis*, a new pathotype effective against larvae of Coleoptera. *Zeitschrift für Angewandte Entomologie, 96*, 500–508.

Kristensen, C., Morant, M., Olsen, C. E., Ekstrøm, C. T., Galbraith, D. W., & Lindberg Møller, B., et al. (2005). Metabolic engineering of dhurrin in transgenic *Arabidopsis* plants with marginal inadvertent effects on the metabolome and transcriptome. *Proceedings of the National*

Academy of Sciences of the United States of America, 102, 1779–1784.

Lambert, B., & Peferoen, M. (1992). Insecticidal promise of *Bacillus thuringiensis*. Facts and mysteries about a sucessful biopesticide. *BioScience, 42*, 112–122.

Lang, A. E., Schmidt, G., Schlosser, A., Hey, T. D., Larrinua, I. M., & Sheets, J. J., et al. (2010). *Photorhabdus luminescens* toxins ADP-Ribosylate actin and RhoA to force actin clustering. *Science, 327*, 1139–1142.

Lee, M. K., Walters, F. S., Hart, H., Palekar, N., & Chen, J. S. (2003). The mode of action of the *Bacillus thuringiensis* vegetative insecticidal protein Vip3A differs from that of Cry1Ab δ-endotoxin. *Applied and Environmental Microbiology, 69*, 4648–4657.

Li, X. Q., Wei, J. Z., Tan, A., & Aroian, R. V. (2007). Resistance to root knot nematode in tomato roots expressing a nematicidal *Bacillus thuringiensis* crystal protein. *Plant Biotechnology Journal, 5*, 455–464.

Liu, D., Burton, S., Glancy, T., Li, Z. S., Hampton, R., & Meade, T., et al. (2003). Insect resistance conferred by 283-kDa *Photorhabdus luminescens* protein TcdA in *Arabidopsis thaliana*. *Nature Biotechnology, 21*, 1222–1228.

Liu, X. S., & Dean, D. H. (2006). Redesigning *Bacillus thuringiensis* Cry1Aa toxin into a mosquito toxin. *Protein Engineering, Design & Selection, 19*, 107–111.

Liu, Y. J., Cheng, C. S., Lai, S. M., Hsu, M. P., Chen, C. S., & Lyu, P. C. (2006). Solution structure of the plant defensin VrD1 from mung bean and its possible role in insecticidal activity against bruchids. *Proteins, 63*, 777–786.

Lorence, A., Darszon, A., Díaz, C., Liévano, A., Quintero, R., & Bravo, A. (1995). δ-Endotoxins induce cation channels in *Spodoptera frugiperda* brush border membranes in suspension and in planar lipid bilayers. *FEBS Letters, 360*, 217–222.

Losey, J. E., Rayor, L. S., & Carter, M. E. (1999). Transgenic pollen harms monarch larvae. *Nature, 399*, 214.

Magaña-Gómez, J. A., & de la Barca, A. M. C. (2008). Risk assessment of genetically modified crops for nutrition and health. *Nutrition Reviews, 67*, 1–16.

Maharashtra Hybrid Seeds Company Ltd. (Mahyco), University of Agricultural Sciences (UAS), D. and Tamil Nadu Agricultural University (TNAU), C. (2009) Report of the Expert Committee (EC-II) on Bt Brinjal Event EE-1.

Maiti, I. B., Oey, N., Dahlman, D. L., & Webb, B. A. (2003). Antibiosis-type insect resistance in transgenic plants expressing a teratocyte secretory protein (TSP14) gene from a hymenopteran endoparasite (*Microplitis croceipes*). *Plant Biotechnology Journal, 1*, 209–219.

Malone, L. A., Gatehouse, A. M. R., & Barratt, B. I. P. (2008). Beyond Bt: Alternative strategies for insect-resistant genetically modified crops. In J. Romeis, A. M. Shelton & G. G. Kennedy (Eds.), *Integration of insect-resistant genetically modified crops within IPM programs* (pp. 357–417). New York: Springer.

Mao, Y. B., Cai, W. J., Wang, J. W., Hong, G. J., Tao, X. Y., & Wang, L. J., et al. (2007). Silencing a cotton bollworm P450 monooxygenase gene by plant-mediated RNAi impairs larval tolerance of gossypol. *Nature Biotechnology, 25*, 1307–1313.

Markwick, N. P., Docherty, L. C., Phung, M. M., Lester, M. T., Murray, C., & Yao, J. L., et al. (2003). Transgenic tobacco and apple plants expressing biotin-binding proteins are resistant to two cosmopolitan insect pests, potato tuber moth and lightbrown apple moth, respectively. *Transgenic Research, 12*, 671–681.

Marroquin, L. D., Elyassnia, D., Griffitts, J. S., Feitelson, J. S., & Aroian, R. V. (2000). *Bacillus thuringiensis* (Bt) toxin susceptibility and isolation of resistance mutants in the nematode *Caenorhabditis elegans*. *Genetics, 155*, 1693–1699.

McCafferty, H. R. K., Moore, P. H., & Zhu, Y. J. (2006). Improved Carica papaya tolerance to carmine spider mite by the expression of *Manduca sexta* chitinase transgene. *Transgenic Research, 15*, 337–347.

Mehlo, L., Gahakwa, D., Nghia, P. T., Loc, N. T., Capell, T., & Gatehouse, J. A., et al. (2005). An alternative strategy for sustainable pest resistance in genetically enhanced crops. *Proceedings of the National Academy of Sciences of the United States of America, 102*, 7812–7816.

Mello, C. C., & Conte, D. (2004). Revealing the world of RNA interference. *Nature, 431*, 338–342.

Mendelsohn, M., Kough, J., Vaituzis, Z., & Matthews, K. (2003). Are Bt crops safe? *Nature Biotechnology, 21*, 1003–1009.

Merryweather, A. T., Weyer, U., Harris, M. P., Hirst, M., Booth, T., & Possee, R. D. (1990). Construction of genetically engineered baculovirus insecticides containing the *Bacillus thuringiensis* subsp. *kurstaki* HD-73 delta endotoxin. *Journal of General Virology, 71*, 1535–1544.

Metcalf, R. L. (1986). Forward. In J. L. Krysan, T. A. Miller & J. F. Andersen (Eds.), *Methods for the study of pest Diabrotica*. New York: Springer-Verlag.

Miklos, J. A., Alibhai, M. F., Bledig, S. A., Connor-Ward, D. C., Gao, A. G., & Holmes, B. A., et al. (2007). Characterization of soybean exhibiting high expression of a synthetic *Bacillus thuringiensis* cry1A transgene that confers a high degree of resistance to Lepidopteran pests. *Crop Science, 47*, 148–157.

Morse, S., Bennett, R., & Ismael, Y. (2004). Why Bt cotton pays for small-scale producers in South Africa. *Nature Biotechnology, 22*, 379–380.

Morton, R. L., Schroeder, H. E., Bateman, K. S., Chrispeels, M. J., Armstrong, E., & Higgins, T. J. V. (2000). Bean α-amylase inhibitor 1 in transgenic peas (*Pisum sativum*) provides complete protection from pea weevil (*Bruchus pisorum*) under field conditions. *Proceedings of the National Academy of Sciences of the United States of America, 97*, 3820–3825.

Munkvold, G. P., Hellmich, R. L., & Showers, W. B. (1997). Reduced Fusarium ear rot and symptomless infection in kernels of maize genetically engineered for European corn borer resistance. *Phytopathology, 87*, 1071–1077.

Murdock, L. L., & Shade, R. E. (2002). Lectins and protease inhibitors as plant defenses against insects. *Journal of Agricultural and Food Chemistry, 50*, 6605–6611.

Murdock, L. L., Huesing, J. E., Nielsen, S. S., Pratt, R. C., & Shade, R. E. (1990). Biological effects of plant lectins on the cowpea weevil. *Phytochemistry, 29*, 85–89.

Murray, C., Sutherland, P. W., Phung, M. M., Lester, M. T., Marshall, R. K., & Christeller, J. T. (2002). Expression of biotin-binding proteins, avidin and streptavidin, in plant tissues using plant vacuolar targeting sequences. *Transgenic Research, 11*, 199–214.

Naimov, S., Dukiandjiev, S., & de Maagd, R. A. (2003). A hybrid *Bacillus thuringiensis* delta-endotoxin gives resistance against a coleopteran and a lepidopteran pest in transgenic potato. *Plant Biotechnology Journal, 1*, 51–57.

Nap, J. P., Metz, P. L. J., Escaler, M., & Conner, A. J. (2003). The release of genetically modified crops into the environment. *Plant Journal, 33*, 1–18.

Naranjo, S. E. (2009). Impacts of Bt crops on non-target invertebrates and insecticide use patterns. *CAB Reviews: Perspectives in Agriculture, Veterinary Science, Nutrition and Natural Resources, 4*, 1–23.

Newmark, P. A., Reddien, P. W., Cebria, F., & Alvarado, A. S. (2003). Ingestion of bacterially expressed double-stranded RNA inhibits gene expression in planarians. *Proceedings of the National Academy of Sciences of the United States of America, 100*, 11861–11865.

OECD (2007) Consensus document on safety information on transgenic plants expressing *Bacillus thuringiensis*-derived insect control proteins. No. 42. Paris Environment Directorate, Organisation for Economic Co-operation and Development.

Opperman, C. H., Bird, D. M., Williamson, V. M., Rokhsar, D. S., Burke, M., & Cohn, J., et al. (2008). Sequence and genetic map of *Meloidogyne hapla*: A compact nematode genome for plant parasitism. *Proceedings of the National Academy of Sciences of the United States of America, 105*, 14802–14807.

O'Reilly, D. R., & Miller, L. K. (1989). A baculovirus blocks insect molting by producing ecdysteroid UDP-glucosyl transferase. *Science, 245*, 1110.

O'Reilly, D. R., & Miller, L. K. (1991). Improvement of a baculovirus pesticide by deletion of the egt gene. *Nature Biotechnology, 9*, 1086–1089.

Park, H. W., & Federici, B. A. (2009). Recombinant Bacterial Larvicides for Control of Important Mosquito Vectors of Disease. In S. P. Stock, J. Vandenberg, I. Glazer & N. Boemare (Eds.), *Insect pathogens: Molecular approaches and techniques* (pp. 163–176). Oxford: CAB International.

Payne, J., Fernandez-Cornejo, J., & Daberkow, S. (2003). Factors affecting the likelihood of corn rootworm Bt seed adoption. *AgBioForum*, *6*, 79–82.

Perlak, F. J., Stone, T. B., Muskopf, Y. M., Petersen, L. J., Parker, G. B., & McPherson, S. A., et al. (1993). Genetically improved potatoes: Protection from damage by Colorado potato beetles. *Plant Molecular Biology*, *22*, 313–321.

Pleasants, J. M., Hellmich, R. L., Dively, G. P., Sears, M. K., Stanley-Horn, D. E., & Mattila, H. R., et al. (2001). Corn pollen deposition on milkweeds in and near cornfields. *Proceedings of the National Academy of Sciences of the United States of America*, *98*, 11919–11924.

Popham, H. J. R., Li, Y., & Miller, L. K. (1997). Genetic improvement of *Helicoverpa zea* nuclear polyhedrosis virus as a biopesticide. *Biological Control*, *10*, 83–91.

Poulsen, M., Kroghsbo, S., Schrøder, M., Wilcks, A., Jacobsen, H., & Miller, A., et al. (2007). A 90-day safety study in Wistar rats fed genetically modified rice expressing snowdrop lectin *Galanthus nivalis* (*GNA*). *Food and Chemical Toxicology*, *45*, 350–363.

Pray, C., Ma, D., Jikun, H., & Qiao, F. (2001). Impact of Bt cotton in China. *World Development*, *29*, 813–825.

Prescott, V. E., Campbell, P. M., Moore, A., Mattes, J., Rothenberg, M. E., & Foster, P. S., et al. (2005). Transgenic expression of bean α-amylase inhibitor in peas results in altered structure and immunogenicity. *Journal of Agricultural and Food Chemistry*, *53*, 9023–9030.

Prikhodko, G. G., Robson, M., Warmke, J. W., Cohen, C. J., Smith, M. H. M., & Wang, P., et al. (1996). Properties of three baculovirus-expressing genes that encode insect-selective toxins: Mu-Aga-IV, As II, and SH I. *Biological Control*, *7*, 236–244.

Pusztai, A., Grant, G., Brown, D. J., Stewart, J. C., Bardocz, S., & Ewen, S. W. B., et al. (1992).) Nutritional evaluation of the trypsin (EC 3.4. 21.4) inhibitor from cowpea (*Vigna unguiculata* Walp.). *British Journal of Nutrition*, *68*, 783–791.

Qaim, M., & Zilberman, D. (2003). Yield effects of genetically modified crops in developing countries. *Science*, *299*, 900–902.

Rajamohan, F., Alzate, O., Cotrill, J. A., Curtiss, A., & Dean, D. H. (1996). Protein engineering of *Bacillus thuringiensis* δ-endotoxin: Mutations at domain II of CryIAb enhance receptor affinity and toxicity toward gypsy moth larvae. *Proceedings of the National Academy of Sciences of the United States of America*, *93*, 14338–14343.

Rao, K. V., Rathore, K. S., Hodges, T. K., Fu, X., Stoger, E., & Sudhakar, D., et al. (1998). Expression of snowdrop lectin (*GNA*) in transgenic rice plants confers resistance to rice brown planthopper. *Plant Journal*, *15*, 469–477.

Rice, M.E. (2004) Transgenic rootworm corn: Assessing potential agronomic, economic, and environmental benefits. *Plant Health Program*. <http://www.plantmanagementnetwork.org/pub/php/review/2004/rootworm/>

Ripoll, C., Favery, B., Lecomte, P., Van Damme, E., Peumans, W., & Abad, P., et al. (2003). Evaluation of the ability of lectin from snowdrop (*Galanthus nivalis*) to protect plants against root-knot nematodes. *Plant Science*, *164*, 517–523.

Romeis, J., Bartsch, D., Bigler, F., Candolfi, M. P., Gielkens, M. M. C., & Hartley, S. E., et al. (2008). Assessment of risk of insect-resistant transgenic crops to nontarget arthropods. *Nature Biotechnology*, *26*, 203–208.

Romeis, J., Hellmich, R. L., Candolfi, M. P., Carstens, K., De Schrijver, A., & Gatehouse, A. M. R., et al. (2010). Recommendations for the design of laboratory studies on non-target arthropods for risk assessment of genetically engineered plants. *Transgenic Research*, *19*, 1–22.

Roy, D., Herring, R. J., & Geisler, C. C. (2007). Naturalising transgenics: Official seeds, loose seeds and risk in the decision matrix of Gujarati cotton farmers. *Journal of Development Studies*, *43*, 158–176.

Saguez, J., Hainez, R., Cherqui, A., Van Wuytswinkel, O., Jeanpierre, H., & Lebon, G., et al. (2005). Unexpected effects of chitinases on the peach-potato aphid (*Myzus persicae* Sulzer) when delivered via transgenic potato plants (*Solanum tuberosum* Linne) and *in vitro*. *Transgenic Research*, *14*, 57–67.

Saha, P., Majumder, P., Dutta, I., Ray, T., Roy, S. C., & Das, S. (2006). Transgenic rice expressing *Allium sativum* leaf lectin with enhanced resistance against sap-sucking insect pests. *Planta*, *223*, 1329–1343.

Samac, D. A., & Smigocki, A. C. (2003). Expression of oryzacystatin I and II in alfalfa increases resistance to the root-lesion nematode. *Phytopathology*, *93*, 799–804.

Sampson, K. S., Tomso, D. J. & Guo, R. (2009). Pesticidal genes from *Brevibacillus* and methods for their use. United States Patent Application 12/644,632.

Schnee, C., Köllner, T. G., Held, M., Turlings, T. C. J., Gershenzon, J., & Degenhardt, J. (2006). The products of a single maize sesquiterpene synthase form a volatile defense signal that attracts natural enemies of maize herbivores. *Proceedings of the National Academy of Sciences of the United States of America*, *103*, 1129–1134.

Schnepf, H. E., Lee, S., Dojillo, J. A., Burmeister, P., Fencil, K., & Morera, L., et al. (2005). Characterization of Cry34/Cry35 binary insecticidal proteins from diverse *Bacillus thuringiensis* strain collections. *Applied and Environmental Microbiology*, *71*, 1765–1774.

Sears, M. K., Hellmich, R. L., Stanley-Horn, D. E., Oberhauser, K. S., Pleasants, J. M., & Mattila, H. R., et al. (2001). Impact of Bt corn pollen on monarch butterfly populations: A risk assessment. *Proceedings of the National Academy of Sciences of the United States of America*, *98*, 11937–11942.

Shade, R. E., Schroeder, H. E., Pueyo, J. J., Tabe, L. M., Murdock, L. L., & Higgins, T. J. V., et al. (1994). Transgenic pea seeds expressing the α-amylase inhibitor of the common bean are resistant to bruchid beetles. *Nature Biotechnology*, *12*, 793–796.

Shahidi-Noghabi, S., Van Damme, E. J. M., & Smagghe, G. (2009). Expression of *Sambucus nigra* agglutinin (*SNA*-I) from elderberry bark in transgenic tobacco plants results in enhanced resistance to different insect species. *Transgenic Research*, *18*, 249–259.

Shelton, A. M., Zhao, J. Z., & Roush, R. T. (2003). Economic, ecological, food safety, and social consequences of the deployment of Bt transgenic plants. *Annual Review of Entomology*, *47*, 845–881.

Shi, J., Thomas, C. J., King, L. A., Hawes, C. A., Possee, R. D., & Edwards, M. L., et al. (2000). The expression of a baculovirus-derived chitinase gene increased resistance of tobacco cultivars to brown spot (*Alternaria alternata*). *Annals of Applied Biology*, *136*, 1–8.

Sindhu, A. S., Maier, T. R., Mitchum, M. G., Hussey, R. S., Davis, E. L., & Baum, T. J. (2009). Effective and specific *in planta* RNAi in cyst nematodes: Expression interference of four parasitism genes reduces parasitic success. *Journal of Experimental Botany*, *60*, 315–324.

Singh, P., Kumar, M., Chaturvedi, C., Yadav, D., & Tuli, R. (2004). Development of a hybrid delta-endotoxin and its expression in tobacco and cotton for control of a polyphagous pest *Spodoptera litura*. *Transgenic Research*, *13*, 397–410.

Smigocki, A., Heu, S., & Buta, G. (2000). Analysis of insecticidal activity in transgenic plants carrying the *ipt* plant growth hormone gene. *Acta Physiologiae Plantarum*, *22*, 295–299.

Soares, C. A. G., Lima, C. M. R., Dolan, M. C., Piesman, J., Beard, C. B., & Zeidner, N. S. (2005). Capillary feeding of specific dsRNA induces silencing of the isac gene in nymphal *Ixodes scapularis* ticks. *Insect Molecular Biology*, *14*, 443–452.

Steeves, R. M., Todd, T. C., Essig, J. S., & Trick, H. N. (2006). Transgenic soybeans expressing siRNAs specific to a major sperm protein gene suppress *Heterodera glycines* reproduction. *Functional Plant Biology*, *33*, 991–999.

Stewart, L. M. D., Hirst, M., Ferber, M. L., Merryweather, A. T., Cayley, P. J., & Possee, R. D. (1991). Construction of an improved baculovirus insecticide containing an insect-specific toxin gene. *Nature*, *352*, 85–88.

Stone, G. D. (2007). Agricultural deskilling and the spread of genetically modified cotton in Warangal. *Current Anthropology*, *48*, 67–103.

Strickland, J. A., Orr, G. L., & Walsh, T. A. (1995). Inhibition of Diabrotica larval growth by patatin, the lipid acyl hydrolase from potato tubers. *Plant Physiology*, *109*, 667–674.

Sun, X., Wang, H., Chen, X., Peng, C., Pan, D., & Jehle, J. A., et al. (2004). Biological activity and field efficacy of a genetically modified *Helicoverpa armigera* single-nucleocapsid nucleopolyhedrovirus expressing an insect-selective toxin from a chimeric promoter. *Biological Control*, *29*, 124–137.

Tabashnik, B. E. (1994). Evolution of resistance to *Bacillus thuringiensis*. *Annual Review of Entomology*, *39*, 47–79.

Tabashnik, B. E. (2008). Delaying insect resistance to transgenic crops. *Proceedings of the National Academy of Sciences of the United States of America*, *105*, 19029–19030.

Tabashnik, B. E., Gassmann, A. J., Crowder, D. W., & Carriére, Y. (2008). Insect resistance to Bt crops: Evidence versus theory. *Nature Biotechnology*, *26*, 199–202.

Tabashnik, B. E., Van Rensburg, J. B. J., & Carrière, Y. (2009). Field-evolved insect resistance to Bt crops: Definition, theory, and data. *Journal of Economic Entomology*, *102*, 2011–2025.

Thomas, J. C., Adams, D. G., Nessler, C. L., Brown, J. K., & Bohnert, H. J. (1995). Tryptophan decarboxylase, tryptamine, and reproduction of the whitefly. *Plant Physiology*, *109*, 717–720.

Thomson, J. A. (2006). *GM crops: The impact and the potential*. Collingwood: CSIRO Publishing.

Tian, H., Peng, H., Yao, Q., Chen, H., Xie, Q., & Tang, B., et al. (2009). Developmental control of a Lepidopteran pest *Spodoptera exigua* by Ingestion of bacteria expressing dsRNA of a non-midgut gene. *PLoS ONE*, *4*, 1–13.

Timmons, L., & Fire, A. (1998). Specific interference by ingested dsRNA. *Nature*, *395*, 854.

Turner, C. T., Davy, M. W., MacDiarmid, R. M., Plummer, K. M., Birch, N. P., & Newcomb, R. D. (2006). RNA interference in the light brown apple moth, *Epiphyas postvittana* (Walker) induced by double-stranded RNA feeding. *Insect Molecular Biology*, *15*, 383–391.

Urwin, P. E., Lilley, C. J., McPherson, M. J., & Atkinson, H. J. (1997). Resistance to both cyst and root knot nematodes conferred by transgenic *Arabidopsis* expressing a modified plant cystatin. *Plant Journal*, *12*, 455–461.

Urwin, P. E., McPherson, M. J., & Atkinson, H. J. (1998). Enhanced transgenic plant resistance to nematodes by dual proteinase inhibitor constructs. *Planta*, *204*, 472–479.

Urwin, P. E., Troth, K. M., Zubko, E. I., & Atkinson, H. J. (2001). Effective transgenic resistance to *Globodera pallida* in potato field trials. *Molecular Breeding*, *8*, 95–101.

Vaeck, M., Reynaerts, A., Höfte, H., Jansens, S., De Beuckeleer, M., & Dean, C., et al. (1987). Transgenic plants protected from insect attack. *Nature*, *328*, 33–37.

Vain, P., Worland, B., Clarke, M. C., Richard, G., Beavis, M., & Liu, H., et al. (1998). Expression of an engineered cysteine proteinase inhibitor (Oryzacystatin-IΔD86) for nematode resistance in transgenic rice plants. *Theoretical and Applied Genetics*, *96*, 266–271.

Vishnudasan, D., Tripathi, M. N., Rao, U., & Khurana, P. (2005). Assessment of nematode resistance in wheat transgenic plants expressing potato proteinase inhibitor (*PIN2*) gene. *Transgenic Research*, *14*, 665–675.

Vitale, J., Glick, H., Greenplate, J., & Traore, O. (2008). The economic impacts of second generation Bt cotton in West Africa: Empirical evidence from Burkina Faso. *International Journal of Biotechnology*, *10*, 167–183.

Walshe, D. P., Lehane, S. M., Lehane, M. J., & Haines, L. R. (2009). Prolonged gene knockdown in the tsetse fly *Glossina* by feeding double stranded RNA. *Insect Molecular Biology*, *18*, 11–19.

Wang, E., Wang, R., DeParasis, J., Loughrin, J. H., Gan, S., & Wagner, G. J. (2001). Suppression of a P450 hydroxylase gene in plant trichome glands enhances natural-product-based aphid resistance. *Nature Biotechnology*, *19*, 371–374.

Wang, J., Chen, Z., Du, J., Sun, Y., & Liang, A. (2005). Novel insect resistance in *Brassica napus* developed by transformation of chitinase and scorpion toxin genes. *Plant Cell Reports*, *24*, 549–555.

Wolfenbarger, L. L. R., Naranjo, S. E., Lundgren, J. G., Bitzer, R. J., & Watrud, L. S. (2008). Bt crop effects on functional guilds of non-target arthropods: A meta-analysis. *PLoS ONE*, *3*, e2118.

Wu, F., Miller, J. D., & Casman, E. A. (2004). The economic impact of Bt corn resulting from mycotoxin reduction. *Journal of Toxicology – Toxin Reviews*, *23*, 397–424.

Wu, J., Luo, X., Wang, Z., Tian, Y., Liang, A., & Sun, Y. (2008). Transgenic cotton expressing synthesized scorpion insect toxin AaHIT gene confers enhanced resistance to cotton bollworm (*Heliothis armigera*) larvae. *Biotechnology Letters*, *30*, 547–554.

Wu, N., Sun, Q., Yao, B., Fan, Y., Rao, H., & Huang, M., et al. (2000). Insect-resistant transgenic poplar expressing *AaIT* gene. *Chinese Journal of Biotechnology*, *16*, 126–133.

Xu, D., Xue, Q., McElroy, D., Mawal, Y., Hilder, V. A., & Wu, R. (1996). Constitutive expression of a cowpea trypsin inhibitor gene, *CpTI*, in transgenic rice plants confers resistance to two major rice insect pests. *Molecular Breeding*, *2*, 167–173.

Yadav, B. C., Veluthambi, K., & Subramaniam, K. (2006). Host-generated double stranded RNA induces RNAi in plant-parasitic nematodes and protects the host from infection. *Molecular and Biochemical Parasitology*, *148*, 219–222.

Zhang, X., Candas, M., Griko, N. B., Taussig, R., & Bulla, L. A. (2006). A mechanism of cell death involving an adenylyl cyclase/PKA signaling pathway is induced by the Cry1Ab toxin of *Bacillus thuringiensis*. *Proceedings of the National Academy of Sciences of the United States of America*, *103*, 9897–9902.

Zhang, X., Liu, J., & Zhao., Q. (1999). Transfer of high lysine-rich gene into maize by microprojectile bombardment and detection of transgenic plants. *Journal of Agricultural Biotechnology*, *7*, 363–367.

Zhao, J. Z., Cao, J., Collins, H. L., Bates, S. L., Roush, R. T., & Earle, E. D., et al. (2005). Concurrent use of transgenic plants expressing a single and two *Bacillus thuringiensis* genes speeds insect adaptation to pyramided plants. *Proceedings of the National Academy of Sciences*, *102*, 8426–8430.

Zhou, X., Wheeler, M. M., Oi, F. M., & Scharf, M. E. (2008). RNA interference in the termite *Reticulitermes flavipes* through ingestion of double-stranded RNA. *Insect Biochemistry and Molecular Biology*, *38*, 805–815.

Zwart, M. P., Van Der Werf, W., Van Oers, M. M., Hemerik, L., Van Lent, J., & De Visser, J., et al. (2009). Mixed infections and the competitive fitness of faster acting genetically modified viruses. *Evolutionary Applications*, *2*, 209–221.

Section E

Biotechnology for improvement of yield and quality traits

Growth Control of Root Architecture

24

Christopher N. Topp Philip N. Benfey

Department of Biology and Institute for Genome Science and Policy, Center for Systems Biology, Duke University, Durham, North Carolina, USA

TABLE OF CONTENTS

Introduction to Root System Architecture

Root systems provide the foundation for shoot growth and reproductive success. Environmental forces above and below ground constantly shape the landscape from which a plant must extract water and nutrients. How a plant integrates its endogenous developmental program with environmental perceptions to deploy its root system is a central question of plant biology. This question also bears heavily on mankind's ability to develop crop varieties that can improve current yields and thus satisfy rapidly increasing world food demand. New challenges have arisen since the research-driven Green Revolution that fed hundreds of millions in decades past. Today's climate is hotter, rainfall is less predictable, and soils have deteriorated from decades of intensive farming and industrial waste — all coincident with a worldwide reduction of plant biomass (Glover et al., 2010; Guo et al., 2010; Tester and Langridge, 2010; Zhao and Running, 2010). Crop breeding aided by biotechnology will continue to play a major role in agricultural improvement. However, such efforts have traditionally focused on the visible, aboveground characters of plants, while root systems have been relatively ignored. We must extend our view into the soil if we are to confront such fundamental issues as water and nutrient limitations.

Root system architecture (RSA) refers to the spatial organization of the root system in its growing environment, which reflects its capability to extract resources (Lynch, 1995). There are three basic components to root architecture: the initiation of root axes that cause branching, the growth rate and path of each axis, and the expansion of the root surface area. These features combined can produce highly complex topologies. Although root development occurs through reiterative processes such as branching and hair formation, environmental conditions heavily influence the extent of growth along a particular root axis. Thus RSA is difficult to accurately predict or even comprehensively describe.

With these concepts in mind, we review root architecture through the various aspects that influence its growth. We examine the endogenous genetic control of root development, how it is influenced by internal and external cues to shape the overall growth of the root system, and how evolutionary factors may or may not constrain architecture. Finally, we consider technological approaches that are useful to study the genetic basis of root architecture from an agricultural standpoint.

Genetic and Developmental Aspects of Root Growth

The shape and physiological attributes of plants depend on the coordinated growth of their tissues and organs. Central

DOI: 10.1016/B978-0-12-381466-1.00024-9

to the development of complex plants are meristems, regions of actively dividing stem cells that generate and organize new tissues. There are two general types of meristems that form in the developing embryo: shoot apical meristems (SAMs) responsible for generating aboveground growth; and root apical meristems (RAMs) responsible for the root system. Based on the activity of their apical meristems, plants generate most of their growth as extensions from their tips. Mature plants can have numerous apical meristems, the vast majority of which develop post-embryonically through integration of genetic networks and environmental signals. Consequently, plants can tailor the growth of their root architecture to suit the prevailing environmental conditions.

Stereotypical organization of root tissues

Due to its relatively simple form and ease of experimental manipulation, *Arabidopsis thaliana* has proven to be a powerful model species to study the mechanisms that govern meristem function in plants. In the *Arabidopsis* embryo, a single meristem generates one root primordium, which becomes the primary root upon germination. Tissue development after this point is genetically controlled by formative series of cell divisions in the RAM (Dolan et al., 1993; Howell, 1998). The key transcription factors that orchestrate these cell divisions have been identified through genetic and morphological analyses (Vandenberg et al., 1995; DiLaurenzio et al., 1996). These studies revealed the critical importance of asymmetric cell divisions, which generate a specialized progenitor cell for each type of tissue. The identity of this initial cell is passed on to daughters during subsequent symmetrical divisions, so that most cells in a given tissue layer are clonally related.

Divisions in the root are further classified based on their orientation to the longitudinal axis. Periclinal divisions are parallel to the longitudinal axis, whereas anticlinal divisions are usually orthogonal to the longitudinal axis. Thus, periclinal divisions can add to the number of concentric cell layers, whereas anticlinal divisions elongate the root by displacing the meristem downward. Asymmetric divisions of both peri- and anticlinal type contribute to new layers of tissue. For example, an anticlinal division invariantly followed by a periclinal division generates the cortical and endodermal cell lineages of *Arabidopsis* (Petricka et al., 2009). The combined activities of divisions in the meristem result in layers of tissues arranged around the root longitudinal axis. This radially symmetric organization is most easily understood by viewing a cross-section of the root. From the outside toward the center the basic tissues are the epidermis, the cortex, the endodermis, and the pericycle and vascular tissues that collectively comprise the stele. The numbers of cells in these layers are well defined in *Arabidopsis*, but can be highly variable in other species, particularly in the cereals that have numerous cortical cells and thicker roots (Esau, 1965).

The positions of neighboring plant cells are essentially fixed by their cell walls. Since cells are continually generated in the RAM, the age of a given cell increases as a function of its distance from the root tip, creating a temporal gradient. Just behind the meristem in what is known as the elongation zone, cells divide less often, and rapidly increase their volumes. Cell expansion not only lengthens the root, but also enables it to curve. Uneven expansion of cells on one side of the root causes the tip to bend away from the more quickly expanding side. Consequently, molecular tropic responses to gravity, touch, and so forth, converge on cell wall expansion (Swarup et al., 2005). Wall expansion is controlled in part by cortical microtubules that when arranged helically, can cause the root to twist around its long axis (Thitamadee et al., 2002; Buschmann et al.. 2009). Thus, the patterns and rates of cell expansion in the elongation zone heavily influence the topology of the growing root.

After the period of rapid expansion, further maturation occurs in the differentiation zone where cells establish their specialized physiological roles. This process often involves characteristic changes to cellular organization, some of which are initiated by evolutionarily well-conserved genetic programming. For example, the tracheal elements of xylem strands develop from procambial cells in a process of programmed cell death and cell wall restructuring (Esau, 1965; Groover and Jones, 1999). The development of xylem and other vascular elements is so tightly controlled that the patterns of these cells are nearly invariant within a species. As such, vascular morphologies are the historical gold standard for phylogenetic classifications (Esau, 1965). Patterning of tissues with core functions such as water transport must be consistent; therefore the influence of lineage on cell identity is strong.

Other important maturation processes in the root are significantly influenced by the environment. A defining physical feature of the transition from the elongation to the maturation zone is the extension of root hairs from epidermal cells. In this process, hair-forming cells are usually delineated from non-hair-forming cells by either positional information from neighbors or by asymmetric divisions (Gilroy and Jones, 2000; Muller and Schmidt, 2004). However, root hair outgrowth is also sensitive to nutrient scarcity. Under conditions of low iron, *Arabidopsis* root hairs can become bifurcated, and when phosphate is not available, root hairs may grow longer and generate from cells that are normally in non-hair-forming positions (Muller and Schmidt, 2004). These and other data suggest that root hair responses to nutrient availabilities are distinct, and in some cases are influential enough to override endogenous developmental programming (Schikora and Schmidt, 2001; Muller and Schmidt, 2004; Dinneny et al., 2008). Since root hairs can comprise a majority of root surface area, which has a direct correlation with water and nutritional intake ability, control of their differentiation is a key aspect of root system function (Gilroy and Jones, 2000; Forde and Lorenzo, 2001; Lopez-Bucio et al., 2003a).

Architectural possibilities

Active meristems are primarily maintained at the apices of shoots and roots. However, incipient meristems also typically form along the sides of shoots and roots (Esau, 1965). When activated, these laterally arranged meristems allow plants to shape the geometry of their architecture through secondary, tertiary, and higher order branching. Lateral SAMs are formed from stem cells as part of reiterative developmental units called phytomers, and their activity is largely controlled by systemic hormone gradients (Leyser, 2005). Lateral RAM control is more complicated because these meristems are not pre-formed. Instead they are generated when needed by a process of cell dedifferentiation and subsequent tissue reorganization that is initiated from pericycle cells flanking the xylem poles (Malamy and Benfey, 1997; De Smet et al., 2008; Moreno-Risueno et al., 2010). Thus, the possibilities of root branching are confined by the architecture of the vascular system. In *Arabidopsis*, the vasculature is arranged in a symmetrical diarch pattern, resulting in a bilaterally symmetric pattern of lateral roots, whereas the polyarch symmetry found in many cereals offers multiple angles around the root axis from which laterals may emerge (Esau, 1965; Malamy and Benfey, 1997). Furthermore, only specific pericycle cells are competent for initialization, conceivably limiting the possible branch density. Recent evidence in *Arabidopsis* indicates that oscillating genetic signals are responsible for patterning potential sites of lateral root formation (Moreno-Risueno et al., 2010). Whether this phenomenon applies to the more elaborated roots of crops is unknown, but these oscillations may be key parameters of branching potential.

Signaling

Coordinated cell differentiation is essential to generate the many specialized cell types that contribute to the physiology of a plant. However, plants do not benefit from a centrally organized nervous system that collects, interprets, and relays information throughout the body. Instead, plants coordinate a variety of short- and long-distance signals to amplify local information about their internal and environmental status. Through manipulation of the phytohormones auxin and cytokinin, classic *in vitro* experiments demonstrated the capacity of small pieces of differentiated tissue, such as leaves, stems, and roots, to regenerate whole plants (Steward et al., 1964). These results showed that plant cell fate is not absolutely constrained by developmental lineage, but is heavily influenced by extracellular information (Steward et al., 1964; Scheres, 2001). Production and distribution of hormones are used by plants to regulate growth as a result of the cross-talk among signaling pathways. Redistribution of hormones from their sources to sink tissues creates numerous overlapping local and systemic gradients that drive cell division and differentiation. For example, cytokinins can induce cell division in the presence of auxin (Scheres, 2001), and gibberellins induce cell wall expansion under certain conditions. Hormones can thus be considered as general systemic regulators, in that they enforce growth processes, but are not strictly interpreted by all cells in the same way (Busch and Benfey, 2010). The ultimate outcome of hormonal signaling depends on its integration with a multitude of other signals as well as endogenous genetic predisposition.

The movements of signals within plants are constrained by the physical connections among cells, tissues, and organs. Direct intercellular connections are made through membrane lined channels called plasmodesmata, in which small molecules such as mineral ions, sugars, and hormones are thought to be able to freely pass under general principles of chemical diffusion (Zambryski and Crawford, 2000; Jackson, 2001). Plasmodesmata can allow passage of larger molecules, although the precise mechanism by which this occurs is unknown. Hence, cells connected by a common intercellular space (the symplasm) can be considered to share a similar subset of regional information, which can be nuanced by the regulation of plasmodesmata. These connections between cells have a profound influence on plant development. For example, several transcription factors critical to root tissue patterning and cell differentiation are thought to move through plasmodesmata (Gallagher et al., 2004; Gallagher and Benfey, 2009; Busch and Benfey, 2010).

Cells are also connected by a common extracellular space, termed the apoplast, in which vast amounts of information and resources are trafficked. Some very small uncharged molecules, such as the hormone ethylene, are thought to pass uninhibited from the apoplast through the plasma membrane and into the symplasm. However, most molecules require recognition by receptor proteins in the plasma membrane before cellular uptake. The selectivity of uptake is apparent from the hundreds of putative extracellular receptors encoded in the *Arabidopsis* genome (Shiu and Bleecker, 2001). Well-characterized examples tie apoplastic transport to core aspects of plant development. Chiefly, cell-membrane-localized PIN transporters control the intercellular polar transport of auxin via the apoplast, and the Clavata receptor-ligand system regulates the shoot meristem through apoplast signaling (Rojo et al., 2002; Grieneisen et al., 2007).

Although signal transduction through the apoplast can be propagated over relatively long distances (Grieneisen et al., 2007), the capacity of the vasculature system for rapid, system-wide dissemination of molecules is far larger. The core functions of the vasculature are the transport of water and dissolved minerals from the root system through the xylem, and the distribution of hormones, amino acids, and photosynthates from source to sink tissues through the phloem. However, new functions of the vasculature in complex signaling are becoming increasingly apparent (Jorgensen et al., 1998; Lough and Lucas, 2006). One recent study demonstrated a second, physically distinct, phloem system in curcubit species seemingly devoted to signaling and alternative metabolite distribution (Zhang et al., 2010). Signaling through the vasculature is particularly important to maintain complementary growth of roots and shoots. For instance, shoot architecture is regulated by the transport of carotenoid-related hormones and other metabolites generated in the root, presumably in response to the soil environment (Van Norman et al., 2004; Gomez-Roldan et al., 2008; Umehara et al., 2008; Sieburth and Lee, 2010). Furthermore, microRNAs, sugars, and hormones are relayed between roots and shoots to coordinate nutrient homeostasis with growth (Forde and Lorenzo, 2001; Jiao et al., 2009).

Accurate predictions of growth and metabolism in new environmental and genetic scenarios are vital to a biotechnological approach to crop improvement (Nielsen et al., 1994; Lynch, et al., 1997; Hammer et al., 2004; Yin et al., 2004). It is therefore necessary to understand how particular signaling systems are integrated with one another and with intrinsic genetic developmental processes. These endeavors present an interesting challenge: system-wide information from a number of different types of molecules must be collected, but at the resolution of the individual cell types that interpret these signals.

Systems biology concept of cell identity

Rapid technological improvements have paced the advance of system-wide approaches to cell biology. The most obvious impacts are evident in the rise of "omics" fields (e.g., genomics, proteomics, metabolomics, ionomics), which seek to capture webs of information from various biological scales. The power of systems approaches have improved with the increasing resolution at which plants can be measured. Molecular analyses of the root have traditionally been conducted on relatively heterogeneous tissues such as whole roots or root sections, which muddle information from individual cell types. There are currently several robust approaches to dissect, purify, and analyze molecules from functional units as small as specific tissue types, cell types, and even individual cells (EmmertBuck et al., 1996; Levsky et al., 2002; Birnbaum et al., 2003; Heiman et al., 2008; Deal et al., 2010).

Cell-type-specific information can be further interrogated during tissue patterning and maturation processes, as well as in response to biotic and abiotic stresses (Brady et al., 2007; Dinneny et al., 2008; Sozzani et al., 2010). As a result, core elements that respond to multiple levels of information are being pinpointed. A cell-type-specific study in the *Arabidopsis* root established unique transcriptional responses to salt in the cortex, whereas iron stress induced the strongest responses in the stele (Dinneny et al., 2008). Metabolic profiling of *Arabidopsis* epidermal cells provided yet another type of tissue-specific physiological information (Ebert et al., 2010). Similar experiments are increasingly feasible in crop species. An atlas of cell-type-specific transcription indexed by developmental stage was recently completed in rice (Jiao et al., 2009), and cell-specific proteomic analysis of maize mutant tissue has begun to dissect the regulation of root formation at the cellular level (Muthreich et al., 2010). Computational and technological advances are paving the way for *in vivo* studies that measure the molecular dynamics of growth processes at fine scales (Lalonde et al., 2005; Mace et al., 2006). In these ways, the notion of cell identity is increasingly well defined by positional and temporal information, facilitating the identification of nodes of regulatory cross-talk. With this information predictive mathematical models of plant growth under various environmental conditions will become more accurate, allowing agricultural scientists to model and screen the presumed benefits of targets for genetic improvement *in silico* (Lynch et al., 1997; Hammer et al., 2004).

Plant–Environment Interactions

Plants not only sense and respond to their environment, but also actively manipulate their surroundings by chemical secretions and alliances with mutually beneficial organisms. The zone of influence around the root system is termed the rhizosphere, specifically the rhizosheath when referring to the physical contacts between plant cells, soil particles, and microorganisms (McCully, 1999). In the rhizosheath, root hairs are bound to soil particles, and microorganisms and soil solutes inundate the apoplast of epidermal and cortical cells (Darrah, 1993; Hinsinger et al., 2005; Clode et al., 2009). The micro-ecosystem around the root can have dramatic effects on plant nutrition and root architecture, as well as soil and aboveground ecophysiology (McCully, 1999; Hinsinger et al., 2005; Pineda et al., 2010). The following sections describe the dynamics of some of these interactions.

Environmental sensing and root exudation

Plant litter, animal decomposition, and microbial activity all contribute organic nutrients to the soil, but they create a highly heterogeneous distribution of resources for root foraging. Furthermore, plants are unable to use many organically derived nutrients directly, especially nitrogen and phosphorus, and must instead rely on their conversion to inorganic forms (Robinson, 1994). Even potentially available inorganic resources, particularly phosphorus and iron, are often insoluble and immobile in the soil, forcing roots to grow in the near vicinity of these nutrients. Acid soils, caused in part by the act of nutrient uptake, also constrain root foraging (Hinsinger et al., 2005). They can limit the growth of beneficial microbes, render nutrient ions unavailable, and produce toxic levels of metal ions, which can mitigate otherwise advantageous root architecture traits (Sponchiado et al., 1989; Kochian et al., 2004). To meet these challenges, plants have developed a number of strategies for exploiting organic resources through manipulation of the rhizosphere.

The root cap is a unique sensory organ at the very tip of the root that has key roles in plant–environment interactions. As the vanguard of root exploration, the root cap perceives an array of biotic and abiotic soil conditions to facilitate impending growth decisions. Root caps are organized in two nested functional compartments. The central columella houses statocyte cells, whose functions include gravity sensing. Cells of the surrounding domain comprise the lateral root cap, which, besides physically protecting the RAM, also exudes a battery of biochemical compounds into the rhizosphere (Dakora and Phillips, 2002; Arnaud et al., 2010). High molecular weight sugars secreted from the root caps of cereals comprise mucilage, a viscous hydrate that eases root passage and binds various components of the rhizosphere into a matrix (Kiesselbach, 1999; Arnaud et al., 2010). During typical root growth, border cells are released from the lateral root cap to the peripheral mucilaginous matrix. Border cells can remain viable for days, apparently with the purpose of secreting growth and defense compounds (Arnaud et al., 2010). Although some biochemical profiling has been conducted (Jiang et al., 2006; Ma et al., 2010), the specific functions of the root cap and its border cells are underexplored (Brigham et al., 1995).

Other parts of the root besides the cap also exude chemicals, in particular root hair epidermal cells. The general influence of plant roots on microbes in their vicinity was recognized as the "rhizosphere effect" over a century ago (Hiltner, 1903), but the scale and range of processes implied by this concept could hardly have been guessed at the time (Darrah, 1993; Forde and Lorenzo, 2001; Dakora and Phillips, 2002; Walker et al., 2003; Lugtenberg and Kamilova, 2009; Arnaud et al., 2010; Babalola, 2010). Root systems are commonly estimated to secrete 20% of their total fixed carbon into the soil, and under certain conditions have been measured to secrete as much as 50% (Hinsinger et al., 2005; Babalola, 2010). These amounts are considerable metabolic expenditures in competitive environments, suggesting that exudations are a core part of plant survival.

A major group of secreted chemicals are the organic acid metabolites citrate and malate (Forde and Lorenzo, 2001; Dakora and Phillips, 2002). Organic acids can solubilize several types of nutrient ions into forms usable by plants. A striking example of this process is the induction of proteoid roots as a systemic response to low phosphorus or low iron (Forde and Lorenzo, 2001). Proteoid roots form as dense clusters of lateral roots that secrete massive amounts of organic acids into the soil. In the white lupin response, 25-fold more C14-labeled carbon was exuded under phosphate-deficient conditions than under standard conditions (Johnson et al., 1996). Under similar phosphate deprivation, lupin roots were measured to produce 20-fold more acid phosphatases, a group of enzymes that aid phosphorus metabolism both within the root and in the rhizosphere (Gilbert et al., 1999; Dakora and Phillips, 2002). In cereal crops, a major response to iron deficiency is the exudation of phytosiderophores, a class of compounds that function to chelate iron and other micronutrients from the soil (Marschner et al., 1987; Dakora and Phillips, 2002). *Arabidopsis*, however, is not known to produce siderophores when iron stressed; instead it responds by altering root hair morphology (Muller and Schmidt, 2004). These examples highlight functional and evolutionary interplay between root architecture and root exudation.

Microbial interactions

Many functions of nutrient acquisition can also be achieved through biological collaboration. An estimated 80% of plant species form symbioses with mycorrhizal fungi (Wang and Qiu, 2006). Within these arrangements the fungus gains access to a rich carbon supply while the plant gains access to otherwise restricted organic nutrients, particularly phosphorus. Arbuscular associations are characterized by fungal penetration of plant cell walls to form functional arbuscules, whereas in ectomycorrhizal associations, the fungus insinuates the cortical apoplast to form a hartig net. In either case, there is a direct exchange of metabolites between host and symbiont. Legumes have evolved similar symbiotic associations with nitrogen-fixing rhizobial bacteria that allow the plant to exploit organic nitrogen sources. As detailed next, fungal and bacterial systems share several parallel mechanisms: plant-rhizosphere signaling to begin the symbiosis; followed by plant tissue reorganization instigated by calcium signaling, and finally, regulation of the symbiosis by the plant (Oldroyd et al., 2009; Ohkama-Ohtsu and Wasaki, 2010).

During rhizobium nodulation (RN), flavonoids are secreted from the host root system as chemoattractants. As a result, soil endemic rhizobia secrete nodulation (Nod) factors that allow access to the root by inducing tissue reorganization. As rhizobia penetrate into the cortex, they begin to establish symbiosomes, which are the functional compartments of nodules. However, a process involving auxin and cytokinin halts nodule maturation if the plant perceives that too little nitrogen is received in the exchange (Oldroyd et al., 2009). Nodule number is also autoregulated by a systemic signaling molecule sent from the shoot and recognized by the CLAVATA1-like HAR1 receptor in the root (Krusell et al., 2002; Nishimura et al., 2002). har1 knockout mutants therefore hypernodulate in the presence of rhizobia. If no rhizobia are present, primary root growth is abrogated in favor of excessive lateral branching, which is also a typical response to localized nitrogen resources (Drew and Saker, 1978). It has been suggested that control of rhizobial symbiosis was co-opted from an existing framework of root developmental regulation (Krusell et al., 2002). At minimum, systemic cross-talk exists between external nutrient perception and control of nodulation, because nodulation is inhibited under nitrogen-rich conditions (Forde and Lorenzo, 2001; Dakora and Phillips, 2002). Arbuscular mycorrhizal (AM) symbioses are similarly controlled by the plant depending on phosphorus availability (Oldroyd et al., 2009). The production and excretion of strigolactones, which prime AM hyphae for infection, are increased under phosphorus stress in some crop plants (Akiyama et al., 2005; Yoneyama et al., 2007). However, hyphal invasion of the root cortex is abrogated in the absence of appropriate phosphate transport through the plasma membrane, suggesting that plants monitor the quality of AM symbiosis (Oldroyd et al., 2009), possibly through the regulation of phosphate transporters that are triggered during infection (Paszkowski et al., 2002).

Overlap between root growth and symbiosis is unsurprising considering that fungal symbioses predate the evolutionary origins of roots by about 400 million years (Margulis and Schwartz, 1998; Landeweert et al., 2001). Fossilized evidence clearly shows that one of the earliest land plants, Rhynia, was rootless and engaged in widespread endomycorrhizal symbiosis. Modern day equivalents include Psilotum; rootless vascular plants that rely on fungal hyphae to supply root function (Margulis and Schwartz, 1998). There is some evidence to support the view that plants control mycorrhizal hyphae as metabolically low-cost, rapidly deployable extensions of the root system (Darrah, 1993; Tibbett, 2000). For one, fungal hyphae can greatly expand the functional surface area of root systems for water and nutrient absorption. Symbiotic fungal hyphae also appear to be extremely well suited to competitive root foraging processes that are important to the exploitation of localized nutrient patches (Tibbett, 2000). Mycorrhizae could also enable access to otherwise unavailable nutrient sources; for example, associations with saprotrophic fungi could extract resources directly from decaying organic matter.

Clearly, collaborative biological interactions play important roles in the growth and function of root systems. Besides the well-established examples of mycorrhizae and nodulation, multiple other contributions of soil or root-endemic bacteria to root function have been identified, such as production of plant hormones, mineral solubilization, protection against toxic metals, biotic defense and so forth (Cook et al., 1995; Malik et al., 1997; Elbeltagy et al., 2001; Ryu et al., 2003; Lugtenberg and Kamilova, 2009; Babalola, 2010; Pineda et al., 2010). These and likely undiscovered microbial capabilities suggest additional strategies to improve root function, for instance, through applied "biofertilizers" (Babalola, 2010) or bioengineering associative endophytes for specific agricultural conditions (Taghavi et al., 2009). Yet significant gaps exist in our understanding of how plants control and benefit from these alliances. Functional ecological approaches combined with molecular biology will greatly contribute to our understanding of the extent to which microbes aid in responses to competitive root interactions, pathogens, and water and nutrient scarcity (Guimil et al., 2005). Revised concepts of plant adaptive growth can then be accurately incorporated into predictive models of root architecture (Darrah, 1993; Lynch et al., 1997; Lynch, 2007).

Architectural responses to nutrient availability

Plant growth plasticity in response to nutrient availability has been termed "trophomorphogenesis" by Forde and Lorenzo (2001), and these responses largely determine root architecture in various environments. Classic experiments analyzing trophomorphogenesis revealed several general phenomena (Stryker et al., 1974; Drew, 1975; Drew and Saker, 1978; Jackson et al., 1990; Robinson, 1994; Hodge, 2004). First, root systems usually increase in size and lateral branch number when encountering a localized area of nutrients, or "patch." Secondly, the rate of nutrient uptake per unit of root is increased in patches. Thirdly, increased growth in the nutrient-rich region often compensates for reduced growth elsewhere. The third phenomenon results in root:shoot mass ratios similar to evenly well-supplied root systems, whereas nutrient-deficient plants commonly shift their mass ratio to favor the roots (Drew, 1975; Robinson, 1994). These generalities distinguish two systems that control trophomorphogenesis: one that functions in a regional response to external nutrients, and another that monitors systemic levels of nutrients (Gersani and Sachs, 1992; Forde and Lorenzo, 2001; Dakora and Phillips, 2002).

Molecular genetic analyses of nutrient transport and growth response have borne these concepts out. Nutrients enter plant cells via both high- and low-affinity plasma membrane transporters. Under plentiful nutrient conditions, low-affinity types generally predominate; however, under nutrient stress, high-affinity transporters (HATS) are induced (Malamy and Benfey, 1997; Glass et al., 2002; Paszkowski et al., 2002; Osmont et al., 2007). Depending on its phosphorylation status, the *Arabidopsis* transporter NRT1.1 (CHL1) changes its affinity for nitrate (NO_3; Ho et al., 2009), indicating a potential for rapid response to changing external nutrient concentrations. Nitrate influx through NRT1.1 activates a key transcriptional regulator, ANR1, which is required for lateral root outgrowth in response to external NO_3 patches (Zhang and Forde, 1998). Accordingly, mutations in NRT1.1 have a profound effect on nitrogen patch response (Remans et al., 2006). However, the NRT1.1 patch response can be separated from its nitrate transport function, demonstrating a dual function as a nutrient sensor (Remans et al., 2006; Ho et al., 2009). Remarkably, NRT1.1 is also capable of transporting auxin, prompting the proposal of a novel mechanism by which nutrient sensing could directly alter root architecture via hormonal control (Krouk et al., 2010). When extracellular nitrate is low, epidermal cell-localized NRT1.1 in lateral root primordia could facilitate auxin transport away from the root apex, inhibiting meristem activity. When extracellular nitrate is supplied in the vicinity of lateral root primordia, NRT1.1 may become occupied transporting nitrate, allowing auxin to flow to the root tip and stimulate growth (Krouk et al., 2010).

In comparison to the concentration dependent response of *Arabidopsis* lateral roots to nitrogen patches, the reorganization of root architecture under low phosphate conditions appears to involve a switch response. In uniformly low phosphorus, the root system is characterized by a short primary root and highly branched laterals (Lopez-Bucio et al., 2003b). Supplied with phosphate, the primary root maintains indeterminate growth with moderate lateral branching. However, contact with low-phosphate media induces a rapid and profound reorganization in which primary root growth ceases and lateral roots proliferate. This response is mediated by the expression of two multi-copper oxidase genes in the root cap, which are proposed to trigger a switch to meristem determinacy (Svistoonoff et al., 2007). The generality of this finding remains to be determined, however, as the primary axes of barley roots under similar conditions were not similarly inhibited (Drew and Saker, 1978).

If RSA was determined simply by responses to local nutrient conditions, root architecture under uniformly well-supplied nutrients would resemble patch response writ large. Instead, evidence suggests that local growth is conditioned by internal nutrient status (Drew, 1975; Drew and Saker, 1978; Forde and Lorenzo, 2001; Forde, 2002; Jiao et al., 2009). Internal nutrient homeostasis is maintained in part by feedback regulation of plasma membrane and vacuolar nutrient transporters. Either the concentrations of ions or of their metabolites (e.g., glutamate in nitrogen homeostasis) can be used as regulatory signals (Glass et al., 2002; Jiao et al., 2009). Nutrient status can be shared among local groups of cells through apoplastic and symplastic connections, but this information must ultimately be passed on to the vasculature. Therefore, the regulation of cargo through xylem and phloem cells should be the key link between local and systemic nutrient signaling (Lough and Lucas, 2006). Analysis of the *POPEYE* gene in *Arabidopsis* provides evidence for this view (Long et al., 2010). *POPEYE* modulates the influx of iron into the shoot, in part by regulating genes that control vacuolar iron transport in root cells that supply the vasculature. Under iron starvation, *POPEYE* also limits the amount of systemically transported iron, resulting in weak shoots. A co-regulated putative iron-binding domain protein, BRUTUS, acts to dampen this effect (Long et al., 2010). It is tempting to speculate that BRUTUS plays a reciprocal role by conveying systemic iron needs of the shoot to local regions of the root system, which would be an especially important function in a patchy environment where some roots are well supplied with iron and others were iron-limited.

Shoot and root system cross-talk also ensures that carbon metabolism in the shoot is synchronized with nutrient cycles in the root. Shoots use both light and sugar signals to regulate many types of nutrient transporters and thus coordinate root metabolism with photosynthesis (Lejay et al., 2003). Analysis of the *pho3* mutation provides functional evidence for photosynthate and mineral nutrient cross-talk. *pho3* was originally identified in a screen for reduced acid-phosphatase secretion in response to phosphate starvation, but was later identified as a mutation in the SUC2 transporter that unloads sugar from the phloem to the root system (Hammond and White, 2008). Sugar in this regard could act as a quantitative signal, since the production and excretion of phosphatases and organic acids place a high demand on photosynthate precursors (Johnson et al., 1996). Sugars applied exogenously to roots elicit substantially increased lateral branching in low phosphorus conditions (Karthikeyan et al., 2007). These data suggest that root architectural response to phosphate stress relies to a large degree on local sugar levels, rather than on systemic sugar transport *per se*. Coincidently, shoots regulate phosphate homeostasis in roots through direct transcriptional control. Under stress, shoots upregulate miR399, which travels to the roots through the phloem to suppress *PHO2* gene expression and facilitate phosphate accumulation in the shoot (Bari et al., 2006; Lin et al., 2008; Pant et al., 2008). Transpecific grafting experiments and phylogenetic analyses provide evidence that this form of phosphorus regulation is well conserved among plants (Bari et al., 2006; Lin et al., 2008; Pant et al., 2008). The phloem content of miR398 and miR395 is upregulated in *Brassica rapa* phloem under copper and sulfur stress, respectively, suggesting that the systemic transport of small RNAs may be generally applicable to nutrient regulation (Jiao et al., 2009).

Vascular plant species face a similar set of challenges in extracting resources from competitive and unpredictable soil environments, while successfully completing reproduction in an aboveground environment fraught with the same difficulties. Most plants seem to have evolved strategies that focus their metabolism on exploiting promising areas of the soil (Drew, 1975; Robinson, 1994; Hodge, 2004). A variety of tactics are employed, such as root proliferation, increased nutrient uptake, symbioses, and chemical secretions. The mechanisms involved in these processes are clear targets for biotechnological improvements to agriculture. However, the effective integration of scientific advances with agriculture will also require a view of the larger ecological and evolutionary pictures.

Crop Root Systems

Types of root systems

There are two main types of root systems in vascular plants, taproot and fibrous. The taproot system, common to gymnosperms and dicots such as *Arabidopsis*, is characterized by continuous growth and maturation of the embryonic primary root, which is facilitated by an indeterminate apical meristem. The stereotypical taproot extends deep into the soil and relies on lateral roots emanating from the central root axis for horizontal exploration. A relatively long and narrow geometry can be efficient at finding deep-water reserves in arid or drought prone environments (Gallardo et al., 1996), but it potentially limits the acquisition of nutrients that are most abundant in the topsoil (Lynch, 1995). Indeed, some taproot-forming crops, such as legumes (beans and pulses), use nodulation as an additional strategy for nutrient acquisition. It is intriguing to consider the evolutionary interplay between symbiotic associations and root architecture.

Fibrous root systems are characteristic of monocots, which include the cereal crops corn, rice, wheat, barley, sorghum, millet, oats, rye, teff, and so forth. They differ from taproot systems in two major ways: embryonic primary root growth is determinate due to cessation of apical meristem activity and in lieu of a prominent primary root, fibrous root systems can develop numerous shoot-borne/nodal/crown roots during post-embryonic growth. Each of these roots can form a main growth axis capable of several orders of lateral branching, resulting in superior coverage of the topsoil and wide variation in root system topology. For example, current maize cultivars typically have several dozen shoot-borne roots that can spread over several horizontal meters (Freeling and Walbot, 1994; Kiesselbach, 1999). In one classic example, a rye plant was estimated at 13 million root branches, covering 500 km of soil (Forbes and Watson, 1992). This example underscores the significant challenges of studying the architectural and functional characteristics of cereal crop root systems.

Embryonic and post-embryonic root systems

The branching capacity of fibrous root systems is evident early in the development of cereals. Whereas a single RAM is formed in the *Arabidopsis* embryo, cereal embryos can have several root primordia. These additional meristems are activated soon after germination, producing roots that play a significant role in soil exploration and physical support for the young plant. Excluding the primary root, embryonically derived roots collectively form the seminal root system. The numbers, relative positions, and growth rates of seminal roots are broadly characteristic of particular cereals (Weaver, 1926). However, considerable variation exists within species, even among inbred lines (Kiesselbach, 1999; Hochholdinger and Zimmermann, 2008). Growth plasticity of the seminal root system among genetically identical individuals suggests a significant environmental component to the development of embryonically patterned root architecture.

Examinations of maize and rice mutants at the seedling stage have led to several insights regarding the genetic programming of seminal and nodal root systems. For example, a mutation in Rootless Concerning Crown And Seminal Roots (RTCS) prevents the formation of both seminal and nodal root primordia in maize plants (Hetz et al., 1996; Taramino et al., 2007). RTCS is a lateral organ boundaries domain (LBD) gene, and part of a conserved group of plant-specific transcription factors that are emerging as major regulators of root architecture (Husbands et al., 2007). Additional evidence comes from rice mutants deficient in the RTCS ortholog, Adventitious Rootless1/Crown Rootless1 (ARL1/CRL1; Inukai et al., 2005; Liu et al., 2005). Rice typically forms several embryonic crown roots from the base of the coleoptile prior to the emergence of post-embryonic crown roots, but arl1/crl1 mutants are absent of all crown roots or primordia. Thus, the LBD genes in cereals provide a major regulatory link between embryonic and post-embryonic root systems. However, mutations in related *Arabidopsis* LOB-Domain genes, LBD16 and LBD29, affect only post-embryonic lateral root formation, apparently through auxin response factors (ARFs; Coudert et al., 2010). Although ARF-binding elements are present in the promoters of maize RTCS and rice ARL1/CRL1, lateral root formation in the respective mutants is unaffected (Hetz et al., 1996; Inukai et al., 2005; Liu et al., 2005). Auxin does induce lateral root outgrowth from the seminal and early crown roots in maize, but through the LRT1 gene (Hochholdinger and Feix, 1998). Remarkably, auxin-insensitive lrt1 mutant seedlings can still form lateral roots when supplied with high phosphate or a mycorrhizal symbiont (Paszkowski and Boller, 2002; Hochholdinger et al., 2004), suggesting that breakdowns in conserved growth mechanisms can be compensated for by environmental response pathways.

Developmental compensation is further exemplified by maize rtcs mutant plants, whose root system consists of only the embryonic primary root and its lateral extensions, similar to a taproot. Remarkably, rtcs plants are fully fertile under field conditions (Hetz et al., 1996). Considering that the maize embryonic root system typically supports the plant for only a matter of days (Freeling and Walbot, 1994; Kiesselbach, 1999), it is evident that a massive remobilization of resources occurs in rtcs plants, perhaps as a result of prolonged indeterminacy of their root apical meristems. Systems genetic approaches currently being employed in crops and *Arabidopsis* (Dinneny et al., 2008; Hoecker et al., 2008; Jiao et al., 2009) will be instrumental in identifying the molecular factors involved in repartitioning growth under developmental and environmental constraints.

Evolutionary strategies and trade-offs

The partitioning of carbon, nitrogen, and other resources into vegetative and reproductive growth determines the ultimate success of the plant. New structures such as root axes and flowers require an investment in resources whose returns depend on how appropriate the structure is for the prevailing environmental conditions. Due to a finite metabolic budget and environmental unpredictability, resource allocation decisions can establish trade-offs that limit certain types of growth in favor of others (Lynch, 1995, 2007; Hodge, 2004). Life history strategies of annual and perennial growth are often juxtaposed in terms of their trade-offs. Annual plant growth is seasonally determinate, culminating in senescence of root and shoot tissues to allocate maximum resources to developing seeds. In contrast, perennials maintain persistent belowground growth, liquidating resources from aging tissues as needed. Perennials are generally viewed as placing less emphasis on seed set, instead focusing their resources on environmental stability (Thomas et al., 2000; DeHaan et al., 2005). Because the seed is the primary focus of human oil and food production, our current grain crops are predominately annuals.

However, the long-standing assumption that a rigid evolutionary trade-off exists between yield and longevity is under increased scrutiny, mainly in response to worldwide agricultural challenges (Thomas et al., 2000; DeHaan et al., 2005; Glover et al., 2010). Proponents argue that perennial crops offer many advantages over annual systems, especially when viewed over the long term and from an ecological perspective. One major advantage of a perennial growth habit is the establishment of a permanent root system. Whereas many modern annual crops rely on high-input agricultural practice to re-establish root systems every season, perennials experience a precipitous drop in resource demands of the root system after the first year. A pre-existing root system at the start of the growth cycle allows perennials to capture more carbon by establishing their shoot systems earlier and senescing them later (DeHaan et al., 2005). To support this growth, perennial root systems may be more adept at collecting water and nutrient resources, especially in marginal soils and under low- or no-input agricultural conditions. Furthermore, the selective pressures facing perennials (k-selection, max density) are likely to have fostered evolutionary adaptations to intense intraspecific competition, disease, long-term drought, and heat (DeHaan et al., 2005; Glover et al., 2010).

Thus, the exploitation of perennial crops has strong potential to address many of the concerns of modern agriculture. Yet annuality and perenniality are not necessarily opposing evolutionary strategies. Instead, a confluence of environmental conditions and genetic traits may determine the growth habits of the root and shoot systems (Thomas et al., 2000). Local resources can significantly influence the choice between annual or biennial strategies, reinforcing the notion that life history trade-offs can be environment specific (DeHaan et al., 2005; Johnson, 2007). Many annual crops including maize, rice, wheat, sorghum, and soybean have wild and semi-domesticated perennial relatives that provide sources of novel genetic variation for plant breeders (Thomas et al., 2000). The quantitative basis of annuality and perenniality in crop species can be examined in hybrid populations through quantitative trait loci (QTL)s and association mapping, as discussed in the following section. These strategies could be used to dissect the genetic and environmental control of root meristem determinacy by comparing annual and perennial root architectures in recombinant inbred or introgression lines.

Approaches to Study Root Architecture

Quantitative analysis

Root system architecture has been manipulated since the beginning of agriculture, for instance, during selection for drought or density-tolerant crops. Nonetheless, little attention has been paid to the underlying genetic factors. Agronomic traits often have complex inheritances because they are controlled by many genes and are heavily influenced by epistatic genetic effects, by environmental interactions (GxE), and by ontogeny. Such traits are quantitative, in that their phenotypes are expressed in a continuous range throughout the population (Lynch and Walsh, 1998). Most RSA traits of agricultural interest, such as root length, angle, surface area, thoroughness of soil exploration, and so forth are quantitative traits, and thus are difficult to specifically manipulate. While traditional breeding has consistently improved yields over the years, it is highly labor-intensive and responds slowly to the type of rapidly changing environmental conditions we face today (Glover et al., 2010). Advanced breeding methods rely on the identification of QTLs, which are regions of the genome that contribute to phenotypic variance for the trait of interest. Numerous QTLs for specific root traits under a variety of biotic and abiotic stresses have been identified and bred for in recent years (de Dorlodot et al., 2007; Collins et al., 2008). Molecular markers associated with QTL are often used in marker-assisted selection (MAS) to rapidly screen for promising plant lines (Collins et al., 2008). However, the translation from RSA QTL to improved yield is not straightforward, and can hinge on the accurate identification and genetic background of the QTL. While polymorphisms in specific genes, regulatory elements, and non-coding DNA are the molecular bases of QTLs, quantitative loci are often broadly defined and encompass hundreds of candidate sequences, making the causative agents difficult to identify.

QTL studies that incorporate multiple traits, or one trait in multiple environments, are promising ways to home in on functional bases for root architecture traits (MacMillan et al., 2006a,b). In a combined greenhouse and field study of an interspecific cross between wild and cultivated lettuce, large-effect QTLs for rooting depth and efficient deep-soil water extraction were co-localized (Johnson et al., 2000). However, the precise molecular basis for this correlation is not known. Deep roots and deep-water uptake capacity may be pleiotropic effects, or may be correlated by tight genetic linkage of polymorphisms in separate genes (Chen and Lubberstedt, 2010). Discerning between these two scenarios is of particular importance when co-localized QTLs for key agronomic traits exhibit negative correlations, as do late senescence and early maturity under drought stress in sorghum (Crasta et al., 1999; Chen and Lubberstedt, 2010). In this sense pleiotropy would suggest a mechanistic trade-off, whereas simple linkage could be broken by recombination. Negative correlations have similarly been identified or suggested for other RSA traits (Sponchiado et al., 1989; Beebe et al., 2006). These and other studies identify the dichotomy of deep-water extraction and topsoil nutrient foraging as an especially important area to examine trade-off relationships for RSA traits.

Improvements in QTL resolution and underlying gene identification will be crucial to a comprehensive understanding of the functional mechanisms of root architecture. However, lack of accurate phenotyping methods, lack of mathematical descriptions, and strong G×E interactions have made the quantitative genetic bases of RSA traits notoriously difficult to pinpoint (Lynch, 1995; de Dorlodot et al., 2007; Collins et al., 2008). In the remaining sections, we survey how modern technological approaches aim to meet these challenges.

High-throughput sequencing

Traditional breeding methods are greatly bolstered by widespread genome sequencing and bioinformatic technologies in several ways. The candidate gene approach uses *a priori* molecular, developmental, and evolutionary information to identify genes potentially involved in agricultural traits. A brute-force candidate method was used with the plant-specific NAC transcription factor family that had been previously linked to stress response (Jeong et al., 2010). This study demonstrated that root-specific expression of OsNAC10 dramatically improved the yield of field-grown rice under drought stress, which was correlated with a 25% increase in root diameter. Functional and phylogenetic information can also be used to identify gene candidates within an existing QTL, greatly reducing the labor involved in fine-mapping. A major aluminum tolerance gene in elite *Sorghum* lines was identified in this way (Sasaki et al., 2004; Magalhaes et al., 2007), and homologs in several other cereals have subsequently been identified as targets for breeding. Similar methods were used to clone sodium and freezing tolerance genes (Collins et al., 2008). QTLs may also be mapped using high-throughput sequencing (Severin et al., 2010). For example, transcriptional profiling in combination with metaQTL analysis identified polymorphisms underlying several root architecture QTLs in rice (Norton et al., 2008).

Dense markers generated through genome sequencing have also facilitated increasingly powerful genome-wide association (GWA) studies. Association mapping uses evidence of historical recombination events to accurately identify QTL polymorphisms. This technique is especially powerful to study populations of high genetic diversity, including natural populations (Clark et al., 2007; Nordborg and Weigel, 2008; Yu et al., 2008). Analyses of genetically diverse native landraces and wild germplasms are revealing vast amounts of natural variation that were lost during domestication and breeding bottlenecks, but could potentially be reintroduced into elite crops (Johnson et al., 2000; de Dorlodot et al., 2007; Weber et al., 2008, 2009). The Nested Association Mapping (NAM) population was designed to combine both association and traditional linkage mapping to take advantage of the substantial genetic diversity present in current maize cultivars (Yu et al., 2008). The power of the NAM population was recently leveraged in field experiments to phenotype over one million plants across a range of environments, thus revealing the fundamental genetic bases of flowering time in maize (Buckler et al., 2009).

Phenomics

Despite profound advances in sequencing technologies, our ability to discover the underlying genetic mechanisms of complex traits is limited by available phenotyping methods. The field of phenomics strives toward high-throughput, high-resolution, dynamic, accurate, and affordable plant measurements. In field studies, groups of measurements such as water transpiration, soil moisture, temperature flux, and light capture can be taken from individual plants using automatic imaging platforms (Berger et al., 2010). However, similar high-throughput methods have yet to be established for root systems grown in the field at the resolution necessary to accurately measure functional RSA traits.

Alternatively, new methods to analyze root systems in laboratory and greenhouse settings have made feasible high-resolution phenotyping in controlled environments. Non-invasive technologies allow the observation of dynamic aspects of root growth *in situ* (de Dorlodot et al., 2007). Automated growth and imaging systems using soil-filled pots or hydroponic chambers can measure shoot and root traits of hundreds of individuals per day, facilitating complex trait dissection and the identification of the underlying genes (Berger et al., 2010; Tester and Langridge, 2010). Recently, the simultaneous monitoring of gene expression and lateral branching of *Arabidopsis* roots grown on agar plates demonstrated that oscillating transcriptional networks pattern the formation of lateral root branch sites (Moreno-Risueno et al., 2010). High-throughput, automated image analysis of similarly grown *Arabidopsis* described the growth response plasticity of roots to gravity as conditioned by a matrix of nutrient and developmental factors (Brooks et al., 2010). To fully capture the three-dimensional topology of monocot roots, rice plants were grown in a clear gel-substrate and imaged in 360 degrees at high resolution with a standard digital camera. The resulting images were combined in an analysis pipeline that automatically extracted sixteen RSA traits and accurately separated the root systems of diverse cultivars by genotype (Iyer-Pascuzzi et al., 2010).

Several methods exist for capturing images of root systems grown in solid substrates. With various precision, these data can be tomographically reconstructed into three-dimensional models that display root system topologies (Gregory et al., 2003; Zhu et al., 2006a; Fang et al., 2009; Jahnke et al., 2009; Dhondt et al., 2010; Tracy et al., 2010). Laser line-scanning was used to image gel-grown roots and generate three-dimensional reconstructions of rice and soybean root architectural responses to phosphorus availability over several days (Fang et al., 2009). These experiments followed an earlier study using a standard camera to image excavated roots (Zhu et al., 2006b). X-ray tomography has also been used to image root systems in natural soils, but at relatively low resolution (Gregory et al., 2003). Recent improvements to the X-ray system now allow high-resolution imaging of root systems (Dhondt et al., 2010; Tracy et al., 2010), including structural soil features such as mineral grains (Tracy et al., 2010). However, imaging times are prohibitive for high-throughput analysis of large samples, and repeated imaging was reported to inhibit normal growth (Dhondt et al., 2010), suggesting that the method is ill-suited for dynamic growth studies. In contrast, positron emission tomography (PET) offers extremely dynamic systems information, including carbon metabolism, root exudation, and hormone signaling (Thorpe et al., 2007; Kiser et al., 2008). Although PET is limited by throughput, resolution, and the involvement of radioisotopes, it can be combined with other technologies for a comprehensive view of RSA (Jahnke et al., 2009).

Thus, a continued focus on improving phenomics technologies promises to connect spatial and temporal aspects of RSA with underlying genetic and metabolic aspects (de Dorlodot et al., 2007; Tester and Langridge, 2010). Advances in machine vision and computing are beginning to yield automatic evaluation of traits and digital root system reconstructions at high-throughput (Iyer-Pascuzzi et al., 2010). These types of data should aid in the development of geometric and other mathematical descriptions of complex root system topologies, particularly in response to environmental cues (Nielsen et al., 1994; Lynch et al., 1997).

Concluding Remarks

Here, we illuminated root system research at many levels to understand the fundamentals of how plants interact with their environment and how this knowledge can improve crops to meet global agricultural challenges. New strategies are paramount as world food demand increases, agricultural conditions worsen, and advanced breeding lines reach maximum potential (Graybosch and Peterson, 2010). A blending of many disciplines from mathematics, to molecular and cellular biology, to agronomy, will be required to realize the concept of "breeding by design," envisioned to bring about the next generation of crops (Peleman and van der Voort, 2003). These efforts will be bolstered by accurate functional descriptions of RSA and an increased focus on introgressing useful traits gleaned from wild and semi-domesticated germplasms. Cell-type-specific genetic and metabolic information will inform biotechnological efforts to engineer crops that will thrive despite current and impending environmental adversities. As the foundation of plants, roots must also be the foundation of our agricultural prospects.

References

Akiyama, K., Matsuzaki, K., & Hayashi, H. (2005). Plant sesquiterpenes induce hyphal branching in arbuscular mycorrhizal fungi. *Nature*, *435*, 824–827.

Arnaud, C., Bonnot, C., Desnos, T., & Nussaume, L. (2010). The root cap at the forefront. *Comptes Rendus Biologies*, *333*, 335–343.

Babalola O. O. (2010). Beneficial bacteria of agricultural importance. *Biotechnology Letters* *32*(11), 170–175.

Bari, R., Datt Pant, B., Stitt, M., & Scheible, W. R. (2006). PHO2, microRNA399, and PHR1 define a phosphate-signaling pathway in plants. *Plant Physiology*, *141*, 988–999.

Beebe, S. E., Rojas-Pierce, M., Yan, X. L., Blair, M. W., Pedraza, F., & Munoz, F., et al. (2006). Quantitative trait loci for root architecture traits correlated with phosphorus acquisition in common bean. *Crop Science*, *46*, 413–423.

Berger, B., Parent, B., & Tester, M. (2010). High-throughput shoot imaging to study drought responses. *Journal of Experimental Botany*, *61*, 3519–3528.

Birnbaum, K., Shasha, D. E., Wang, J. Y., Jung, J. W., Lambert, G. M., & Galbraith, D. W., et al. (2003). A gene expression map of the *Arabidopsis* root. *Science*, *302*, 1956–1960.

Brady, S. M., Orlando, D. A., Lee, J. Y., Wang, J. Y., Koch, J., & Dinneny, J. R., et al. (2007).

A high-resolution root spatiotemporal map reveals dominant expression patterns. *Science*, *318*, 801–806.

Brigham, L. A., Woo, H. H., Nicoll, S. M., & Hawes, M. C. (1995). Differential expression of proteins and messenger-Rnas from border cells and root-tips of pea. *Plant Physiology*, *109*, 457–463.

Brooks, T. L., Miller, N. D., & Spalding, E. P. (2010). Plasticity of *Arabidopsis* root gravitropism throughout a multidimensional condition space quantified by automated image analysis. *Plant Physiology*, *152*, 206–216.

Buckler, E. S., et al. (2009). The genetic architecture of maize flowering time. *Science*, *325*, 714–718.

Busch, W., & Benfey, P. N. (2010). Information processing without brains — the power of intercellular regulators in plants. *Development*, *137*, 1215–1226.

Buschmann, H., Hauptmann, M., Niessing, D., Lloyd, C. W., & Schaffner, A. R. (2009). Helical growth of the *Arabidopsis* mutant tortifolia2 does not depend on cell division patterns but involves handed twisting of isolated cells. *Plant Cell*, *21*, 2090–2106.

Chen, Y. S., & Lubberstedt, T. (2010). Molecular basis of trait correlations. *Trends in Plant Science*, *15*, 454–461.

Clark, R. M., et al. (2007). Common sequence polymorphisms shaping genetic diversity in *Arabidopsis thaliana*. *Science*, *317*, 338–342.

Clode, P. L., Kilburn, M. R., Jones, D. L., Stockdale, E. A., Cliff, J. B., 3rd, & Herrmann, A. M., et al. (2009). *In situ* mapping of nutrient uptake in the rhizosphere using nanoscale secondary ion mass spectrometry. *Plant Physiology*, *151*, 1751–1757.

Collins, N. C., Tardieu, F., & Tuberosa, R. (2008). Quantitative trait loci and crop performance under abiotic stress: Where do we stand? *Plant Physiology*, *147*, 469–486.

Cook, R. J., Thomashow, L. S., Weller, D. M., Fujimoto, D., Mazzola, M., & Bangera, G., et al. (1995). Molecular mechanisms of defense by rhizobacteria against root disease. *Proceedings of the National Academy of Sciences of the United States of America*, *92*, 4197–4201.

Coudert, Y., Perin, C., Courtois, B., Khong, N. G., & Gantet, P. (2010). Genetic control of root development in rice, the model cereal. *Trends in Plant Science*, *15*, 219–226.

Crasta, O. R., Xu, W. W., Rosenow, D. T., Mullet, J., & Nguyen, H. T. (1999). Mapping of post-flowering drought resistance traits in grain sorghum: Association between QTLs influencing premature senescence and maturity. *Molecular and General Genetics*, *262*, 579–588.

Dakora, F. D., & Phillips, D. A. (2002). Root exudates as mediators of mineral acquisition in low-nutrient environments. *Plant and Soil*, *245*, 35–47.

Darrah, P. R. (1993). The rhizosphere and plant nutrition — a quantitative approach. *Plant and Soil*, *156*, 1–20.

de Dorlodot, S., Forster, B., Pages, L., Price, A., Tuberosa, R., & Draye, X. (2007). Root system architecture: Opportunities and constraints for genetic improvement of crops. *Trends in Plant Science*, *12*, 474–481.

De Smet, I., et al. (2008). Receptor-like kinase ACR4 restricts formative cell divisions in the *Arabidopsis* root. *Science*, *322*, 594–597.

Deal, R. B., Henikoff, J. G., & Henikoff, S. (2010). Genome-wide kinetics of nucleosome turnover determined by metabolic labeling of histones. *Science*, *328*, 1161–1164.

DeHaan, L. R., Van Tassel, D. L., & Cox, T. S. (2005). Perennial grain crops: A synthesis of ecology and plant breeding. *Renewable Agriculture and Food Systems*, *20*, 5–14.

Dhondt, S., Vanhaeren, H., Van Loo, D., Cnudde, V., & Inze, D. (2010). Plant structure visualization by high-resolution X-ray computed tomography. *Trends in Plant Science*, *15*, 419–422.

DiLaurenzio, L., WysockaDiller, J., Malamy, J. E., Pysh, L., Helariutta, Y., & Freshour, G., et al. (1996). The SCARECROW gene regulates an asymmetric cell division that is essential for generating the radial organization of the *Arabidopsis* root. *Cell*, *86*, 423–433.

Dinneny, J. R., Long, T. A., Wang, J. Y., Jung, J. W., Mace, D., & Pointer, S., et al. (2008). Cell identity mediates the response of *Arabidopsis* roots to abiotic stress. *Science*, *320*, 942–945.

Dolan, L., Janmaat, K., Willemsen, V., Linstead, P., Poethig, S., & Roberts, K., et al. (1993). Cellular-organization of the *Arabidopsis thaliana* root. *Development*, *119*, 71–84.

Drew, M. C. (1975). Comparison of effects of a localized supply of phosphate, nitrate, ammonium and potassium on growth of seminal root system, and shoot, in Barley. *New Phytologist*, *75*, 479–490.

Drew, M. C., & Saker, L. R. (1978). Nutrient supply and growth of seminal root-system in barley. 3. Compensatory increases in growth of lateral roots, and in rates of phosphate uptake, in response to a localized supply of phosphate. *Journal of Experimental Botany*, *29*, 435–451.

Ebert, B., Zoller, D., Erban, A., Fehrle, I., Hartmann, J., & Niehl, A., et al. (2010). Metabolic profiling of *Arabidopsis thaliana* epidermal cells. *Journal of Experimental Botany*, *61*, 1321–1335.

Elbeltagy, A., Nishioka, K., Sato, T., Suzuki, H., Ye, B., & Hamada, T., et al. (2001). Endophytic colonization and in planta nitrogen fixation by a Herbaspirillum sp isolated from wild rice species. *Applied and Environmental Microbiology*, *67*, 5285–5293.

EmmertBuck, M. R., Bonner, R. F., Smith, P. D., Chuaqui, R. F., Zhuang, Z. P., & Goldstein, S. R., et al. (1996). Laser capture microdissection. *Science*, *274*, 998–1001.

Esau, K. (1965). *Plant anatomy*. New York: Wiley.

Fang, S., Yan, X., & Liao, H. (2009). 3D reconstruction and dynamic modeling of root architecture *in situ* and its application to crop phosphorus research. *Plant Journal*, *60*, 1096–1108.

Forbes, J. C., & Watson, R. D. (1992). *Plants in agriculture*. Cambridge England; New York, NY, USA: Cambridge University Press.

Forde, B., & Lorenzo, H. (2001). The nutritional control of root development. *Plant and Soil*, *232*, 51–68.

Forde, B. G. (2002). The role of long-distance signalling in plant responses to nitrate and other nutrients. *Journal of Experimental Botany*, *53*, 39–43.

Freeling, M., & Walbot, V. (1994). *The Maize handbook*. New York: Springer-Verlag.

Gallagher, K. L., & Benfey, P. N. (2009). Both the conserved GRAS domain and nuclear localization are required for SHORT-ROOT *movement*. *Plant Journal*, *57*, 785–797.

Gallagher, K. L., Paquette, A. J., Nakajima, K., & Benfey, P. N. (2004). Mechanisms regulating SHORT-ROOT intercellular movement. *Current Biology*, *14*, 1847–1851.

Gallardo, M., Jackson, L. E., & Thompson, R. B. (1996). Shoot and root physiological responses to localized zones of soil moisture in cultivated and wild lettuce (*Lactuca* spp). *Plant Cell and Environment*, *19*, 1169–1178.

Gersani, M., & Sachs, T. (1992). Development correlations between roots in heterogeneous environments. *Plant Cell and Environment*, *15*, 463–469.

Gilbert, G. A., Knight, J. D., Vance, C. P., & Allan, D. L. (1999). Acid phosphatase activity in phosphorus-deficient white lupin roots. *Plant Cell and Environment*, *22*, 801–810.

Gilroy, S., & Jones, D. L. (2000). Through form to function: Root hair development and nutrient uptake. *Trends in Plant Science*, *5*, 56–60.

Glass, A. D. M., et al. (2002). The regulation of nitrate and ammonium transport systems in plants. *Journal of Experimental Botany*, *53*, 855–864.

Glover, J. D., et al. (2010). Increased food and ecosystem security via perennial grains. *Science*, *328*, 1638–1639.

Gomez-Roldan, V., et al. (2008). Strigolactone inhibition of shoot branching. *Nature*, *455* 189-U122

Graybosch, R. A., & Peterson, C. J. (2010). Genetic improvement in winter wheat yields in the great plains of North America, 1959–2008. *Crop Science*, *50*, 1882–1890.

Gregory, P. J., Hutchison, D. J., Read, D. B., Jenneson, P. M., Gilboy, W. B., & Morton, E. J. (2003). Non-invasive imaging of roots with high resolution X-ray micro-tomography. *Plant and Soil*, *255*, 351–359.

Grieneisen, V. A., Xu, J., Maree, A. F. M., Hogeweg, P., & Scheres, B. (2007). Auxin transport is sufficient to generate a maximum and gradient guiding root growth. *Nature*, *449*, 1008–1013.

Groover, A., & Jones, A. M. (1999). Tracheary element differentiation uses a novel mechanism coordinating programmed cell death and secondary cell wall synthesis. *Plant Physiology*, *119*, 375–384.

Guimil, S., et al. (2005). Comparative transcriptomics of rice reveals an ancient pattern of response to microbial colonization. *Proceedings of the National Academy of Sciences of the United States of America*, *102*, 8066–8070.

Guo, J. H., Liu, X. J., Zhang, Y., Shen, J. L., Han, W. X., & Zhang, W. F., et al. (2010).

Significant acidification in major Chinese croplands. *Science*, *327*, 1008–1010.

Hammer, G. L., Sinclair, T. R., Chapman, S. C., & van Oosterom, E. (2004). On systems thinking, systems biology, and the *in silico* plant. *Plant Physiology*, *134*, 909–911.

Hammond, J. P., & White, P. J. (2008). Sucrose transport in the phloem: Integrating root responses to phosphorus starvation. *Journal of Experimental Botany*, *59*, 93–109.

Heiman, M., et al. (2008). A translational profiling approach for the molecular characterization of CNS cell types. *Cell*, *135*, 738–748.

Hetz, W., Hochholdinger, F., Schwall, M., & Feix, G. (1996). Isolation and characterization of rtcs, a maize mutant deficient in the formation of nodal roots. *Plant Journal*, *10*, 845–857.

Hiltner, L. (1903). Neue untersuchungen über die wurzel knöllchen der leguminosen und deren erreger. n.p.

Hinsinger, P., Gobran, G. R., Gregory, P. J., & Wenzel, W. W. (2005). Rhizosphere geometry and heterogeneity arising from root-mediated physical and chemical processes. *New Phytologist*, *168*, 293–303.

Ho, C. H., Lin, S. H., Hu, H. C., & Tsay, Y. F. (2009). CHL1 functions as a nitrate sensor in plants. *Cell*, *138*, 1184–1194.

Hochholdinger, F., & Feix, G. (1998). Early post-embryonic root formation is specifically affected in the maize mutant lrt1. *Plant Journal*, *16*, 247–255.

Hochholdinger, F., Woll, K., Sauer, M., & Dembinsky, D. (2004). Genetic dissection of root formation in maize (*Zea mays*) reveals root-type specific developmental programmes. *Annals of Botany*, *93*, 359–368.

Hochholdinger, F., & Zimmermann, R. (2008). Conserved and diverse mechanisms in root development. *Current Opinion in Plant Biology*, *11*, 70–74.

Hodge, A. (2004). The plastic plant: Root responses to heterogeneous supplies of nutrients. *New Phytologist*, *162*, 9–24.

Hoecker, N., Keller, B., Muthreich, N., Chollet, D., Descombes, P., & Piepho, H. P., et al. (2008). Comparison of maize (*Zea mays* L.) F-1-hybrid and parental inbred line primary root transcriptomes suggests organ-specific patterns of nonadditive gene expression and conserved expression trends. *Genetics*, *179*, 1275–1283.

Howell, S. H. (1998). *Molecular genetics of plant development*. Cambridge, UK; New York: Cambridge University Press.

Husbands, A., Bell, E. M., Shuai, B., Smith, H. M. S., & Springer, P. S. (2007). LATERAL ORGAN BOUNDARIES defines a new family of DNA-binding transcription factors and can interact with specific bHLH proteins. *Nucleic Acids Research*, *35*, 6663–6671.

Inukai, Y., Sakamoto, T., Ueguchi-Tanaka, M., Shibata, Y., Gomi, K., & Umemura, I., et al. (2005). Crown rootless1, which is essential for crown root formation in rice, is a target of an AUXIN RESPONSE FACTOR in auxin signaling. *Plant Cell*, *17*, 1387–1396.

Iyer-Pascuzzi, A. S., Symonova, O., Mileyko, Y., Hao, Y. L., Belcher, H., & Harer, J., et al. (2010). Imaging and analysis platform for automatic phenotyping and trait ranking of plant root systems. *Plant Physiology*, *152*, 1148–1157.

Jackson, D. (2001). The long and the short of it: Signaling development through plasmodesmata. *Plant Cell*, *13*, 2569–2572.

Jackson, R. B., Manwaring, J. H., & Caldwell, M. M. (1990). Rapid physiological adjustment of roots to localized soil enrichment. *Nature*, *344*, 58–60.

Jahnke, S., et al. (2009). Combined MRI-PET dissects dynamic changes in plant structures and functions. *Plant Journal*, *59*, 634–644.

Jeong, J. S., Kim, Y. S., Baek, K. H., Jung, H., Ha, S. H., & Do Choi, Y., et al. (2010). Root-specific expression of OsNAC10 improves drought tolerance and grain yield in rice under field drought conditions. *Plant Physiology*, *153*, 185–197.

Jiang, K., Zhang, S. B., Lee, S., Tsai, G., Kim, K., & Huang, H. Y., et al. (2006). Transcription profile analyses identify genes and pathways central to root cap functions in maize. *Plant Molecular Biology*, *60*, 343–363.

Jiao, Y. L., et al. (2009). A transcriptome atlas of rice cell types uncovers cellular, functional and developmental hierarchies. *Nature Genetics*, *41*, 258–263.

Johnson, J. F., Allan, D. L., Vance, C. P., & Weiblen, G. (1996). Root carbon dioxide fixation by phosphorus-deficient Lupinus albus — Contribution to organic acid exudation by proteoid roots. *Plant Physiology*, *112*, 19–30.

Johnson, M. T. J. (2007). Genotype-by-environment interactions leads to variable selection on life-history strategy in common evening primrose (*Oenothera biennis*). *Journal of Evolutionary Biology*, *20*, 190–200.

Johnson, W. C., Jackson, L. E., Ochoa, O., van Wijk, R., Peleman, J., & St Clair, D. A., et al. (2000). Lettuce, a shallow-rooted crop, and *Lactuca serriola*, its wild progenitor, differ at QTL determining root architecture and deep soil water exploitation. *Theoretical and Applied Genetics*, *101*, 1066–1073.

Jorgensen, R. A., Atkinson, R. G., Forster, R. L., & Lucas, W. J. (1998). An RNA-based information superhighway in plants. *Science*, *279*, 1486–1487.

Karthikeyan, A. S., Varadarajan, D. K., Jain, A., Held, M. A., Carpita, N. C., & Raghothama, K. G. (2007). Phosphate starvation responses are mediated by sugar signaling in *Arabidopsis*. *Planta*, *225*, 907–918.

Kiesselbach, T. A. (1999). *The structure and reproduction of corn*. Cold Spring Harbor, N.Y.: Cold Spring Harbor Laboratory Press.

Kiser, M. R., Reid, C. D., Crowell, A. S., Phillips, R. P., & Howell, C. R. (2008). Exploring the transport of plant metabolites using positron emitting radiotracers. *Hfsp Journal*, *2*, 189–204.

Kochian, L. V., Hoekenga, O. A., & Pineros, M. A. (2004). How do crop plants tolerate acid soils? Mechanisms of aluminum tolerance and phosphorous efficiency. *Annual Review of Plant Biology*, *55*, 459–493.

Krouk, G., et al. (2010). Nitrate-regulated auxin transport by NRT1.1 defines a mechanism for nutrient sensing in plants. *Developmental Cell*, *18*, 927–937.

Krusell, L., et al. (2002). Shoot control of root development and nodulation is mediated by a receptor-like kinase. *Nature*, *420*, 422–426.

Lalonde, S., Ehrhardt, D. W., & Frommer, W. B. (2005). Shining light on signaling and metabolic networks by genetically encoded biosensors. *Current Opinion in Plant Biology*, *8*, 574–581.

Landeweert, R., Hoffland, E., Finlay, R. D., Kuyper, T. W., & van Breemen, N. (2001). Linking plants to rocks: Ectomycorrhizal fungi mobilize nutrients from minerals. *Trends in Ecology & Evolution*, *16*, 248–254.

Lejay, L., Gansel, X., Cerezo, M., Tillard, P., Muller, C., & Krapp, A., et al. (2003). Regulation of root ion transporters by photosynthesis: Functional importance and relation with hexokinase. *Plant Cell*, *15*, 2218–2232.

Levsky, J. M., Shenoy, S. M., Pezo, R. C., & Singer, R. H. (2002). Single-cell gene expression profiling. *Science*, *297*, 836–840.

Leyser, O. (2005). The fall and rise of apical dominance — Commentary. *Current Opinion in Genetics & Development*, *15*, 468–471.

Lin, S. I., Chiang, S. F., Lin, W. Y., Chen, J. W., Tseng, C. Y., & Wu, P. C., et al. (2008). Regulatory network of microRNA399 and PHO2 by systemic signaling. *Plant Physiology*, *147*, 732–746.

Liu, H. J., Wang, S. F., Yu, X. B., Yu, J., He, X. W., & Zhang, S. L., et al. (2005). ARL1, a LOB-domain protein required for adventitious root formation in rice. *Plant Journal*, *43*, 47–56.

Long, T. A., Tsukagoshi, H., Busch, W., Lahner, B., Salt, D. E., & Benfey, P. N. (2010). The bHLH transcription factor POPEYE regulates response to iron deficiency in *Arabidopsis* roots. *Plant Cell*, *22*, 2219–2236.

Lopez-Bucio, J., Cruz-Ramirez, A., & Herrera-Estrella, L. (2003a). The role of nutrient availability in regulating root architecture. *Current Opinion in Plant Biology*, *6*, 280–287.

Lopez-Bucio, J., Ramirez, V., Nieto, F., O'Connor, A., Herrera-Estrella, L. (2003b). Engineering enhanced nutrient uptake in transgenic plants. *Plant Biotechnology 2002 and Beyond*, 179–182.

Lough, T. J., & Lucas, W. J. (2006). Integrative plant biology: Role of phloem long-distance macromolecular trafficking. *Annual Review of Plant Biology*, *57*, 203–232, 619.

Lugtenberg, B., & Kamilova, F. (2009). Plant-growth-promoting rhizobacteria. *Annual Review of Microbiology*, *63*, 541–556.

Lynch, J. (1995). Root architecture and plant productivity. *Plant Physiology*, *109*, 7–13.

Lynch, J. P. (2007). Rhizoeconomics: The roots of shoot growth limitations. *Horttechnology*, *42*, 1107–1109.

Lynch, J. P., Nielsen, K. L., Davis, R. D., & Jablokow, A. G. (1997). SimRoot: Modelling and visualization of root systems. *Plant and Soil*, *188*, 139–151.

Lynch, M., & Walsh, B. (1998). *Genetics and analysis of quantitative traits*. Sunderland, Mass: Sinauer.

Ma, W., Muthreich, N., Liao, C. S., Franz-Wachtel, M., Schutz, W., & Zhang, F. S., et al. (2010).

The mucilage proteome of maize (*Zea mays* L.) primary roots. *Journal of Proteome Research*, *9*, 2968–2976.

Mace, D. L., Lee, J. Y., Twigg, R. W., Colinas, J., Benfey, P. N., & Ohler, U. (2006). Quantification of transcription factor expression from *Arabidopsis* images. *Bioinformatics*, *22*, E323–E331.

MacMillan, K., Emrich, K., Piepho, H. P., Mullins, C. E., & Price, A. H. (2006). Assessing the importance of genotype × environment interaction for root traits in rice using a mapping population II: Conventional QTL analysis. *Theoretical and Applied Genetics*, *113*, 953–964.

MacMillan, K., Emrich, K., Piepho, H. P., Mullins, C. E., & Price, A. H. (2006). Assessing the importance of genotype × environment interaction for root traits in rice using a mapping population. I: A soil-filled box screen. *Theoretical and Applied Genetics*, *113*, 977–986.

Magalhaes, J. V., et al. (2007). A gene in the multidrug and toxic compound extrusion (MATE) family confers aluminum tolerance in sorghum. *Nature Genetics*, *39*, 1156–1161.

Malamy, J. E., & Benfey, P. N. (1997). Organization and cell differentiation in lateral roots of *Arabidopsis thaliana*. *Development*, *124*, 33–44.

Malik, K. A., Bilal, R., Mehnaz, S., Rasul, G., Mirza, M. S., & Ali, S. (1997). Association of nitrogen-fixing, plant-growth-promoting rhizobacteria (PGPR) with kallar grass and rice. *Plant and Soil*, *194*, 37–44.

Margulis, L., & Schwartz, K. V. (1998). *Five kingdoms: An illustrated guide to the phyla of life on earth*. New York: W.H. Freeman.

Marschner, H., Romheld, V., & Kissel, M. (1987). Localization of phytosiderophore release and of iron uptake along intact barley roots. *Physiologia Plantarum*, *71*, 157–162.

McCully, M. E. (1999). Roots in soil: Unearthing the complexities of roots and their rhizospheres. *Annual Review of Plant Physiology and Plant Molecular Biology*, *50* 695.

Moreno-Risueno, M. A., Van Norman, J. M., Moreno, A., Zhang, J., Ahnert, S. E., & Benfey, P. N. (2010). Oscillating gene expression determines competence for periodic *Arabidopsis* root branching. *Science*, *329*, 1306–1311.

Muller, M., & Schmidt, W. (2004). Environmentally induced plasticity of root hair development in *Arabidopsis*. *Plant Physiology*, *134*, 409–419.

Muthreich, N., Schutzenmeister, A., Schutz, W., Madlung, J., Krug, K., & Nordheim, A., et al. (2010). Regulation of the maize (*Zea mays* L.) embryo proteome by RTCS which controls seminal root initiation. *European Journal of Cell Biology*, *89*, 242–249.

Nielsen, K. L., Lynch, J. P., Jablokow, A. G., & Curtis, P. S. (1994). Carbon cost of root systems — an architectural approach. *Plant and Soil*, *165*, 161–169.

Nishimura, R., et al. (2002). HAR1 mediates systemic regulation of symbiotic organ development. *Nature*, *420*, 426–429.

Nordborg, M., & Weigel, D. (2008). Next-generation genetics in plants. *Nature*, *456*, 720–723.

Norton, G. J., Aitkenhead, M. J., Khowaja, F. S., Whalley, W. R., & Price, A. H. (2008). A bioinformatic and transcriptomic approach to identifying positional candidate genes without fine mapping: An example using rice root-growth QTLs. *Genomics*, *92*, 344–352.

Ohkama-Ohtsu, N., & Wasaki, J. (2010). Recent progress in plant nutrition research: Cross-talk between nutrients, plant physiology and soil microorganisms. *Plant and Cell Physiology*, *51*, 1255–1264.

Oldroyd, G. E., Harrison, M. J., & Paszkowski, U. (2009). Reprogramming plant cells for endosymbiosis. *Science*, *324*, 753–754.

Osmont, K. S., Sibout, R., & Hardtke, C. S. (2007). Hidden branches: Developments in root system architecture. *Annual Review of Plant Biology*, *58*, 93–113.

Pant, B. D., Buhtz, A., Kehr, J., & Scheible, W. R. (2008). MicroRNA399 is a long-distance signal for the regulation of plant phosphate homeostasis. *Plant Journal*, *53*, 731–738.

Paszkowski, U., & Boller, T. (2002). The growth defect of lrt1, a maize mutant lacking lateral roots, can be complemented by symbiotic fungi or high phosphate nutrition. *Planta*, *214*, 584–590.

Paszkowski, U., Kroken, S., Roux, C., & Briggs, S. P. (2002). Rice phosphate transporters include an evolutionarily divergent gene specifically activated in arbuscular mycorrhizal symbiosis. *Proceedings of the National Academy of Sciences of the United States of America*, *99*, 13324–13329.

Peleman, J. D., & van der Voort, J. R. (2003). Breeding by design. *Trends in Plant Science*, *8*, 330–334.

Petricka, J. J., Van Norman, J. M., Benfey, P. N. (2009). Symmetry breaking in plants: Molecular mechanisms regulating asymmetric cell divisions in *Arabidopsis*. *Cold Spring Harbor Perspectives in Biology*, *1*(5), a000497.

Pineda, A., Zheng, S. J., van Loon, J. J., Pieterse, C. M., Dicke, M. (2010). Helping plants to deal with insects: The role of beneficial soil-borne microbes. *Trends in Plant Science*, *15*(9), 507–514.

Remans, T., Nacry, P., Pervent, M., Filleur, S., Diatloff, E., & Mounier, E., et al. (2006). The *Arabidopsis* NRT1.1 transporter participates in the signaling pathway triggering root colonization of nitrate-rich patches. *Proceedings of the National Academy of Sciences of the United States of America*, *103*, 19206–19211.

Robinson, D. (1994). The responses of plants to nonuniform supplies of nutrients. *New Phytologist*, *127*, 635–674.

Rojo, E., Sharma, V. K., Kovaleva, V., Raikhel, N. V., & Fletcher, J. C. (2002). CLV3 is localized to the extracellular space, where it activates the *Arabidopsis* CLAVATA stem cell signaling pathway. *Plant Cell*, *14*, 969–977.

Ryu, C. M., Farag, M. A., Hu, C. H., Reddy, M. S., Wei, H. X., & Pare, P. W., et al. (2003). Bacterial volatiles promote growth in *Arabidopsis*. *Proceedings of the National Academy of Sciences of the United States of America*, *100*, 4927–4932.

Sasaki, T., Yamamoto, Y., Ezaki, B., Katsuhara, M., Ahn, S. J., & Ryan, P. R., et al. (2004). A wheat gene encoding an aluminum-activated malate transporter. *Plant Journal*, *37*, 645–653.

Scheres, B. (2001). Plant cell identity. The role of position and lineage. *Plant Physiology*, *125*, 112–114.

Schikora, A., & Schmidt, W. (2001). Iron stress-induced changes in root epidermal cell fate are regulated independently from physiological responses to low iron availability. *Plant Physiology*, *125*, 1679–1687.

Severin, A. J., et al. (2010). An integrative approach to genomic introgression mapping. *Plant Physiology*, *154*, 3–12.

Shiu, S. H., & Bleecker, A. B. (2001). Receptor-like kinases from *Arabidopsis* form a monophyletic gene family related to animal receptor kinases. *Proceedings of the National Academy of Sciences of the United States of America*, *98*, 10763–10768.

Sieburth, L. E., & Lee, D. K. (2010). BYPASS1: How a tiny mutant tells a big story about root-to-shoot signaling. *Journal of Integrative Plant Biology*, *52*, 77–85.

Sozzani, R., Cui, H., Moreno-Risueno, M. A., Busch, W., Van Norman, J. M., & Vernoux, T., et al. (2010). Spatiotemporal regulation of cell-cycle genes by SHORTROOT links patterning and growth. *Nature*, *466*, 128–149.

Sponchiado, B. N., White, J. W., Castillo, J. A., & Jones, P. G. (1989). Root-growth of 4 common bean cultivars in relation to drought tolerance in environments with contrasting soil types. *Experimental Agriculture*, *25*, 249–257.

Steward, F. C., Mapes, M. O., Kent, A. E., & Holsten, R. D. (1964). Growth and development of cultured plant cells. *Science*, *143*, 20–27.

Stryker, R. B., Gilliam, J. W., & Jackson, W. A. (1974). Nonuniform phosphorus distribution in root zone of corn — growth and phosphorus uptake. *Soil Science Society of America Journal*, *38*, 334–340.

Svistoonoff, S., Creff, A., Reymond, M., Sigoillot-Claude, C., Ricaud, L., & Blanchet, A., et al. (2007). Root tip contact with low-phosphate media reprograms plant root architecture. *Nature Genetics*, *39*, 792–796.

Swarup, R., Kramer, E. M., Perry, P., Knox, K., Leyser, H. M. O., & Haseloff, J., et al. (2005). Root gravitropism requires lateral root cap and epidermal cells for transport and response to a mobile auxin signal. *Nature Cell Biology*, *7*, 1057–1065.

Taghavi, S., Garafola, C., Monchy, S., Newman, L., Hoffman, A., & Weyens, N., et al. (2009). Genome survey and characterization of endophytic bacteria exhibiting a beneficial effect on growth and development of poplar trees. *Applied and environmental microbiology*, *75*, 748–757.

Taramino, G., Sauer, M., Stauffer, J. L., Multani, D., Niu, X. M., & Sakai, H., et al. (2007). The maize (*Zea mays* L.) RTCS gene encodes a LOB domain protein that is a key regulator of embryonic seminal and post-embryonic shoot-borne root initiation. *Plant Journal*, *50*, 649–659.

Tester, M., & Langridge, P. (2010). Breeding technologies to increase crop production in a changing world. *Science*, *327*, 818–822.

Thitamadee, S., Tuchihara, K., & Hashimoto, T. (2002). Microtubule basis for left-handed helical growth in *Arabidopsis*. *Nature*, *417*, 193–196.

Thomas, H., Thomas, H. M., & Ougham, H. (2000). Annuality, perenniality and cell death. *Journal of Experimental Botany*, *51*, 1781–1788.

Thorpe, M. R., Ferrieri, A. P., Herth, M. M., & Ferrieri, R. A. (2007). 11C-imaging: Methyl jasmonate moves in both phloem and xylem, promotes transport of jasmonate, and of photoassimilate even after proton transport is decoupled. *Planta*, *226*, 541–551.

Tibbett, M. (2000). Roots, foraging and the exploitation of soil nutrient patches: The role of mycorrhizal symbiosis. *Functional Ecology*, *14*, 397–399.

Tracy, S. R., Roberts, J. A., Black, C. R., McNeill, A., Davidson, R., & Mooney, S. J. (2010). The X-factor: Visualizing undisturbed root architecture in soils using X-ray computed tomography. *Journal of Experimental Botany*, *61*, 311–313.

Umehara, M., et al. (2008). Inhibition of shoot branching by new terpenoid plant hormones. *Nature*, *455* 195-U129.

Van Norman, J. M., Frederick, R. L., & Sieburth, L. E. (2004). BYPASS1 negatively regulates a root-derived signal that controls plant architecture. *Current Biology*, *14*, 1739–1746.

Vandenberg, C., Willemsen, V., Hage, W., Weisbeek, P., & Scheres, B. (1995). Cell fate in the *Arabidopsis* root-meristem determined by directional signaling. *Nature*, *378*, 62–65.

Walker, T. S., Bais, H. P., Grotewold, E., & Vivanco, J. M. (2003). Root exudation and rhizosphere biology. *Plant Physiology*, *132*, 44–51.

Wang, B., & Qiu, Y. L. (2006). Phylogenetic distribution and evolution of mycorrhizas in land plants. *Mycorrhiza*, *16*, 299–363.

Weaver, J. E. (1926). *Root development of field crops*. New York: McGraw-Hill Book Co.

Weber, A. L., Briggs, W. H., Rucker, J., Baltazar, B. M., Sanchez-Gonzalez, J. D., & Feng, P., et al. (2008). The genetic architecture of complex traits in teosinte (*Zea mays* ssp *parviglumis*): New evidence from association mapping. *Genetics*, *180*, 1221–1232.

Weber, A. L., Zhao, Q., McMullen, M. D., & Doebley, J. F. (2009). Using association mapping in teosinte to investigate the function of maize selection-candidate genes. *Plos One*, *4*, e8227.

Yin, X., Struik, P. C., & Kropff, M. J. (2004). Role of crop physiology in predicting gene-to-phenotype relationships. *Trends in Plant Science*, *9*, 426–432.

Yoneyama, K., Xie, X., Kusumoto, D., Sekimoto, H., Sugimoto, Y., & Takeuchi, Y. (2007). Nitrogen deficiency as well as phosphorus deficiency in sorghum promotes the production and exudation of 5-deoxystrigol, the host recognition signal for arbuscular mycorrhizal fungi and root parasites. *Planta*, *227*, 125–132.

Yu, J. M., Holland, J. B., McMullen, M. D., & Buckler, E. S. (2008). Genetic design and statistical power of nested association mapping in maize. *Genetics*, *178*, 539–551.

Zambryski, P., & Crawford, K. (2000). Plasmodesmata: Gatekeepers for cell-to-cell transport of developmental signals in plants. *Annual review of cell and developmental biology*, *16*, 393–421.

Zhang, B., Tolstikov, V., Turnbull, C., Hicks, L. M., & Fiehn, O. (2010). Divergent metabolome and proteome suggest functional independence of dual phloem transport systems in cucurbits. *Proceedings of the National Academy of Sciences of the United States of America*, *107*, 13532–13537.

Zhang, H. M., & Forde, B. G. (1998). An *Arabidopsis* MADS box gene that controls nutrient-induced changes in root architecture. *Science*, *279*, 407–409.

Zhao, M., & Running, S. W. (2010). Drought-induced reduction in global terrestrial net primary production from 2000 through 2009. *Science*, *329*, 940–943.

Zhu, J. M., Mickelson, S. M., Kaeppler, S. M., & Lynch, J. P. (2006). Detection of quantitative trait loci for seminal root traits in maize (*Zea mays* L.) seedlings grown under differential phosphorus levels. *Theoretical and Applied Genetics*, *113*, 1–10.

Zhu, T., Fang, S., Li, Z., Liu, Y., Liao, H., & Yan, X. (2006). Quantitative analysis of 3-dimensional root architecture based on image reconstruction and its application to research on phosphorus uptake in soybean. *Chinese Science Bulletin*, *51*, 11p.

Control of flowering

25

Alon Samach

The R. H. Smith Institute for Plant Sciences and Genetics in Agriculture, the Hebrew University of Jerusalem, Israel

TABLE OF CONTENTS

Introduction

A plant's perspective

The vast amounts of diverse plant species currently on our planet are still here because they have succeeded in reproducing over many generations, surviving many environmental and human-caused catastrophes. The timing in which different plant species attempt to sexually reproduce is influenced by the need for sufficient resources to support reproduction as well as the vulnerability of the exposed plant reproductive organs to unfavorable environmental conditions. Thus, many species have developed a sensitive mechanism both to avoid flowering when they lack the sufficient resources, and also to avoid flowering during certain seasons, or better yet, to flower at a specific, best time of the year. This mechanism requires an ability to sense and react to internal changes as well as seasonal and daily changes in the environment, including temperature, light quality/intensity, and daylength. Three major decisions regarding timing of flowering are (Figure 25.1): the length of the juvenile phase ("juvenility"), a period lasting from germination until plants allow themselves to enter the reproductive cycle; when to flower during the year ("seasonality"); and when to time the next cycle of reproduction (in polycarpic plants), that is, the "reproductive cycle". There are quite a few comprehensive current reviews on plant reproduction (Michaels, 2009; Albani and Coupland, 2010; Amasino, 2010; Crevillen and Dean, 2010).

A farmer's perspective

The product of most agricultural crops is flowers or fruit, thus the need for farmers to get more product seems to fit well with the need of plants to reproduce. Conflicts arise when humans move plants that are adjusted to a specific environment into a new environment. How will plants know when to flower when facing a different annual cycle of environmental changes? For example, how will a plant that is geared to respond to changes in photoperiod respond if it is moved to the equator?

A different point of conflict is that plants are happy to produce fruit at a certain time of year, while humans feel the need to produce a certain commodity in different seasons—consumers "need" their favorite vegetable/fruit all year round. A new generation of seedlings uses the juvenile phase to establish themselves and are in no hurry to enter the reproductive cycle; if plant breeders would like to improve fruiting traits of these seedlings, they need great patience and the ability to convince others why they are progressing so slowly. Lastly, in polycarpic plants like fruit trees, a high quality and quantity of fruit every

DOI: 10.1016/B978-0-12-381466-1.00025-0

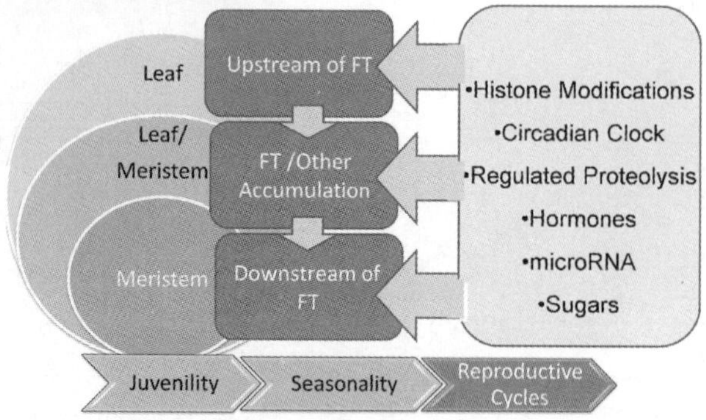

Figure 25.1 Diagram describing three decisions (as described in the section Developmental Decisions on Timing of Flowering) regarding time of flowering in plants (juvenility, seasonal flowering, and reproductive cycles). The place where FLOWERING LOCUS T (FT) is made (leaves) and the place where it acts (meristem) are shown. The proteins (described in the section Proteins Controlling Flowering Time) that regulate *FT* expression, can work instead of FT (SPLs), and genes that are downstream of FT, Processes (described in the section Processes Affecting Flowering Time Proteins), such as histone modifications or microRNA regulation, that affect the levels of the different genes or proteins involved in flowering control.

year is desirable, but many plants are just as happy and able to make what humans consider "low quality" fruits once every two years.

Reaching human goals by overcoming or circumventing plant mechanisms is a slow process that likely began with the start of agriculture. Introducing agricultural techniques such as "forcing" to partially overcome juvenility, artificial cold treatment to convince a plant it had gone through winter, or fruit thinning to intervene in fruiting cycles are some examples. Humans also bred or selected for new varieties with preferable flowering traits likely caused by modified alleles of genes (originating from wild species or mutations). Today, knowing more about the processes can help in devising more precise agricultural techniques. Our knowledge of genes that regulate these processes can serve in marker-assisted breeding and in the use of transgenic manipulations (Flachowsky et al., 2009; Jung and Muller, 2009).

Proteins Controlling Flowering Time

In this section we will concentrate on what is known in *Arabidopsis* regarding the proteins involved in setting the switch within the meristem, from formation of leaf primordia to formation of flower primordia. Similar roles of similar proteins were identified in other species such as rice, maize, and tomato, and we will mention a few examples. These proteins act at four sites: leaf; phloem; meristem; and the initial flower primordium formed by it. The proteins that decide the identity and quantity of organs within the flower, known as flower-organ identity proteins (Liu et al., 2009a), will not be discussed in this section.

Florigen and FLOWERING LOCUS T (FT)

Plants are subject to annual cycles in climate including temperature, light quality, light intensity, humidity, precipitation, and other environmental events; yet the timing and intensity of those phenomena are subject to change and are influenced by human impact. Changes in daylength (photoperiod) are a reliable annual clock, unlikely to be disrupted by human or other impact. Annual amplitude in daylength cycle is influenced by where organisms reside on Earth. In regions further from the equator, many plants use these predicted changes in photoperiod to correctly time the transition to flowering as well as other developmental processes. Some plants respond to shortening of daylength while others respond to increasing daylengths. The switch occurs in the apical meristem which starts making flowers instead of leaves. Scientists noted this photoperiodic affect on flowering time (Tournois, 1914; Klebs, 1918; Garner and Allard, 1920), and the need for leaves to perceive daylength suggested the existence of a moving signal termed "Florigen" formed in leaves and reaching the meristem (Knott, 1934; Chailakhyan, 1936). By grafting between species with opposite daylength requirements, it was shown that Florigen could be a common mobile flowering chemical (Zeevaart, 1962).

Molecular genetic studies in *Arabidopsis* (Redei, 1962; Koornneef et al., 1991; Kobayashi et al., 1999; Kardailsky et al., 1999; Michaels et al., 2005; Yamaguchi et al., 2005; Jang et al., 2009) as well as rice (Kojima et al., 2002; Ishikawa et al., 2005), tomato (Lifschitz et al., 2006), and wheat (Yan et al., 2006) provided the identity of one, and perhaps the only, component of Florigen (Turck et al., 2008), which is a small (23 kDa) protein encoded by *FLOWERING LOCUS T* (*FT*) and *TWIN SISTER OF FT* (*TSF*) in *Arabidopsis*, *SINGLE FLOWER TRUSS* (*SFT*) in tomato, *HEADING DATE 3* (*HD3A*) in rice, and *VRN3* in wheat. Here we will use the name FT to describe proteins of different species that are similar in structure and function to the protein encoded by *FT* of *Arabidopsis*. Lack of FT causes late flowering, and high levels of FT are correlated with early flowering. Many environmental cues regulate flowering time through regulating FT-encoding gene expression. Under conditions that promote flowering in *Arabidopsis*, *FT* is expressed in the phloem of leaves (Takada and Goto, 2003), and after translation, FT is loaded into the phloem and moves toward the meristem (Corbesier et al., 2007; Jaeger and Wigge, 2007; Lin et al., 2007; Mathieu et al., 2007; Tamaki et al., 2007). Once reaching the vegetative meristem, FT was shown to trigger a change in its identity, turning it into an inflorescence meristem producing flower rather than leaf primordia.

Based on present knowledge, it is likely that FT-like proteins are formed in leaves of all plant species and once they reach the meristem they can potentially trigger the flowering response in any plant species (Figure 25.1). The timing of FT formation, together with internal signals that intervene in steps that occur after FT formation, will define the flowering behavior of each plant. Polycarpic plants require a mechanism preventing all meristems from reacting to the FT signal, once it is formed. This would allow certain meristems to remain vegetative, and to act as a source of a new round of meristems that will be induced in the next flowering cycle.

There is no current evidence that FT-like proteins are DNA binding proteins, so they likely require other proteins to regulate changes in gene expression. One such candidate is the bZIP transcription factor FLOWERING LOCUS D (FD), expressed in meristems and young leaf primordia (anlagen), which interacts with FT-like proteins and in its absence, flowering is delayed (Abe et al., 2005; Wigge et al., 2005). The fact that a mutation in *FD* does not mimic the severity in delay of flowering shown by loss-of-function of both FT and TSF (Wang et al., 2009a), and that this mutation does not completely suppress early flowering caused by high expression of FT (Teper-Bamnolker and Samach, 2005) suggests that, in the absence of FD, paralogs or proteins of a different nature help FT turn on the flowering switch. Two proteins that negate FT function in what seems to be direct protein–protein interaction are FWA and 14-3-3. FWA is not usually expressed outside of the seed, but when it is, it causes a delay in flowering (Soppe et al., 2000; Ikeda et al., 2007). 14-3-3 proteins were shown to interact with FT-like proteins (Pnueli et al., 2001), and a role for such a protein in delaying flowering by interfering with FT entry into the nucleus was recently shown in rice (Purwestri et al., 2009).

TERMINAL FLOWER1 (*TFL1*) encodes a protein similar to FT and TSF yet with an opposite role. *TFL1* is expressed in the vegetative and inflorescence meristem, and in its absence plants are early flowering and the inflorescence meristem is rapidly transformed into a flower, causing a terminal flower phenotype (Shannon and Meeks-Wagner, 1991). The transformation of the inflorescence meristem into flower primordia is accompanied by the appearance in the meristem of flower-primodia-specific genes, such as *APETALA1* and *LEAFY* (Liljegren et al., 1999). TFL1 over-expression causes late flowering (Ratcliffe et al., 1998; Kobayashi et al., 1999). The interaction between these two proteins (FT and TFL1) with negating affects has been nicely studied in tomato where the TFL1 ortholog is encoded by the *SELF PRUNING* (*SP*) gene (Pnueli et al., 1998; Krieger et al., 2010).

Transcription factors regulating *FT*

FLOWERING LOCUS C (FLC) and MADS AFFECTING FLOWERING (MAF) proteins

FLC is a MADS-box transcription factor expressed in both leaves and meristem, that represses the transition to flowering (Michaels and Amasino, 1999; Sheldon et al., 1999) through repressing *FT* and *TSF* expression, and additionally repressing expression of the FT partner *FD*, and one of their downstream targets *SOC1* (see the next section). There is evidence for the presence of FLC on regulatory elements of these genes (Hepworth et al., 2002; Michaels et al., 2005; Yamaguchi et al., 2005; Helliwell et al., 2006; Searle et al., 2006). In *Arabidopsis*, there are several additional genes encoding proteins similar to FLC, named together they become the *MADS AFFECTING FLOWERING* (*MAF*) genes. Some of them have been shown to repress flowering, likely through the same mechanism (Alexandre and Hennig, 2008). Among them, FLOWERING LOCUS M (FLM, also known as MAF1 and AGL27) was better characterized (Ratcliffe et al., 2001, 2003; Scortecci et al., 2001, 2003; Werner et al., 2005; Balasubramanian et al., 2006). FLM loss-of-function causes early flowering, especially in short days. Genetic analysis suggests that FLM works together with an additional flower repressor (see the next section), SHORT VEGETATIVE PHASE (SVP) as well as with SCHLAFMÜTZE (SMZ; Scortecci et al., 2003; Mathieu et al., 2009).

SHORT VEGETATIVE PHASE (SVP)

This is an additional MADS-box transcription factor expressed in both leaves and meristem (Hartmann et al., 2000; Jang et al., 2009) that represses the transition to flowering through repressing *FT*, *TSF*, and *SOC1* expression (Lee et al., 2007a; Li et al., 2008; Jang et al., 2009). Its levels in the apical meristem decrease together with the increase in the FT target, *SOC1* levels (Jang et al., 2009). Less similar in sequence to FLC compared to members of the *MAF* clade, there is both genetic and biochemical evidence that repression of these genes is achieved by interaction of SVP and a member of the MAF clade (Scortecci et al., 2003; Fujiwara et al., 2008). In other words, in the absence of SVP, the MAF proteins are less capable of delaying flowering. There is quite a lot of evidence suggesting that the FLC clade of MADS-box repressors act through repression of *FT* transcription through CARG box binding sites in the promoter and first intron (Searle et al., 2006; Li et al., 2008; Adrian et al., 2010).

CONSTANS (CO)

This directly activates *FT* expression (Samach et al., 2000) in response to daylength in the phloem of leaves. Expression of CONSTANS is affected by the circadian clock (Suarez-Lopez et al., 2001), whereas protein stability is affected by light quality (Valverde et al., 2004). CONSTANS has a CCT domain found in many other proteins in plants. CO does not contain a known DNA binding domain, and it has recently been proposed that it might interact with DNA through interaction with proteins that comprise the trimeric CCAAT binding factor also known as Nuclear Factor Y (NF-Y) complex (Ben-Naim et al., 2006). Since the CCT domain shares similarity to the DNA binding domain of one of the components, HAP2/NFYA (Wenkel et al., 2006; Distelfeld et al., 2009a), it is possible that CCT domain proteins like CO replace that component in the complex (Wenkel et al., 2006). Indeed members of the NF-Y complex play a role in photoperiodic flowering, and without members of this complex, the ability of CO to cause early flowering is attenuated (Cai et al., 2007; Chen et al., 2007; Kumimoto et al., 2008, 2010; Tiwari et al., 2010). On the other hand it was recently made clear that CO can activate *FT* promoter without regions containing a CCAAT domain, and a new unique cis-element that CO uses was identified in the FT promoter (Tiwari et al., 2010).

APETALA2-like flowering time repressors

Combined loss-of-function of four members of this group, TARGET OF EAT1,2 (TOE1 and TOE2), SCHLAFMÜTZE

(SMZ), and SCHNARCHZAPFEN causes early flowering under short days (Mathieu et al., 2009). Over-expression of these genes causes late flowering (Schmid et al., 2003). This group of transcription factors is regulated by miR172 (see the next section). SMZ was shown to repress *FT* expression and this ability was dependent on the presence of the MAF protein FLM (Mathieu et al., 2009). Over-expression of SMZ could significantly reduce the ability of CO to upregulate *FT*. It is clear that aside from regulating *FT* expression, this protein regulates other meristem identity genes because it causes a change in inflorescence identity when expressed exclusively in the meristem (where FT transcription does not need to be regulated), and causes slight suppression of *35S:FT* early flowering (Mathieu et al., 2009). It is likely that similar to FLC and FLM, it targets genes like *SOC1*.

TEMPRANILLO (TEM)

TEM1 and TEM2 are two RAV transcription factors that contain two DNA-binding domains: an AP2/ERF and a B3 DNA-binding domain. Loss-of-function of both genes causes early flowering and abnormally high levels of *FT*, whereas over-expression of these genes causes late flowering and low levels of *FT*. When a plant contains both high levels of TEM1 and of CO, there seems to be a competition between them rather than an epistasis relationship (Castillejo and Pelaz, 2008). TEM1 seems to regulate *FT* by binding the 5′UTR of the gene.

Proteins parallel or downstream of FT

SOC1 and FRUITFULL

In the SAM, FT associates with the b-ZIP transcription factor FD to mediate the floral transition and specify flower meristems by activating expression of MADS-box genes such as *SUPPRESSOR OF OVEREXPRESSION OF CONSTANS1* (*SOC1*) and *FRUITFULL* (*FUL*; Abe et al., 2005; Wigge et al., 2005; Teper-Bamnolker and Samach, 2005; Yoo et al., 2005; Lee and Lee, 2010). Expression of both genes in the meristem occurs within 24h from introducing receptive (20 days or older) plants to an inductive photoperiod, and these are currently the earliest known changes in gene expression in the meristem (Menzel et al., 1996; Samach et al., 2000; Wang et al., 2009a). A mutation in SOC1 causes late flowering (Samach et al., 2000; Lee et al., 2000) and in FUL causes slight late flowering (Ferrandiz et al., 2000; Teper-Bamnolker and Samach, 2005; Wang et al., 2009a;). A plant mutated in both genes is severely late flowering (Melzer et al., 2008; Wang et al., 2009a) suggesting that they act in parallel to promote flowering. Expressing *FUL* in the meristem can partially compensate for the loss of FT (Wang et al., 2009a). It seems that in the absence of SOC1, loss of *FUL* in the apical meristem (using the FD promoter) is sufficient to delay flowering. Reversion to vegetative growth after the floral transition suggests that the floral transition decision in the double mutant is unstable (Melzer et al., 2008). Combining a knockout of both FT and TSF (p35S:amiR-FT/TSF) the over-expression of miR156 in the apex (pFD:mir156), which (see the next section) should repress both SOC1 and FUL expression, caused a severe delay compared to either construct. This might suggest that FT and TSF activate genes that are not suppressed by pFD:mir156. SOC1 seems to have a dependence on an additional MADS-box transcription factor, AGL24 (Michaels et al., 2003). AGL24 is itself a flowering promoter, the mutant being late flowering and over-expression causing early flowering. Its expression increases in the inflorescence meristem with the flowering transition and it is also expressed in flower primordia (Yu et al., 2002). Although *AGL24* is induced by vernalization, similar to *SOC1*, it is not repressed by FLC (Michaels et al., 2003). Both proteins were shown to activate each other's expression (Liu et al., 2008a) and together with SVP to downregulate expression of *SEPALLATA3* (SEP3, see next section; Liu et al., 2009b), a gene upregulated by FT (Teper-Bamnolker and Samach, 2005). SOC1s entry into the nucleus was shown to be dependent on AGL24 (Lee et al., 2008), and it was claimed that they both are needed to cause expression of LEAFY (see next section).

APETALA1 and LEAFY

The MADS-BOX transcription factor *APETALA1* (*AP1*) transcript accumulates in flower primordia produced by the inflorescence meristem. A mutation in both *AP1* and its close paralog *CAULIFLOWER* (*CAL*) causes the inflorescence meristem to make more and more inflorescence meristems instead of flower primordia, forming a cauliflower-like structure (Kempin et al., 1995). The gene encoding AP1 is considered in many instances in the literature as a direct and major target of FT, due to the *in vitro* binding assays of FD (Wigge et al., 2005). Still, non-flowering time phenotype of its loss-of-function both in wt and plants that over-express FT (Kardailsky et al., 1999) suggest it is not what FT uses to promote a change in flowering time. In addition, *AP1* comes up 48h after both *FUL* and *SOC1* in the transition to long days (Wang et al., 2009a), and unlike these other FT targets it is expressed in flower primordia and not the inflorescence meristem that forms these primordia. This protein is required to establish the fate of the flower primordia, and in addition, to coordinately organize the expression of the flower organ identity genes. AP1 was recently shown to repress expression of flowering promoters that are required to establish the inflorescence meristem, such as *SOC1* and AGL24 (Liu et al., 2007), *FD*, and its putative paralog *FDP*, as well as SQUAMOSA-PROMOTER BINDING PROTEIN LIKE 9, 15 (see the next section; Kaufmann et al., 2010). At the same time it also represses expression of flowering repressor SVP (Liu et al., 2007), repressors from the APETALA2 family, *TEM1,2*, and *TFL1* (Kaufmann et al., 2010). It is interesting that AP1 downregulates three genes (*SVP*, *SOC1*, and *AGL24*) that downregulate *SEP3*, and that AP1 and SEP3 work together in a complex to induce expression of many class B genes (Kaufmann et al., 2010).

LEAFY (LFY), a plant-specific transcription factor, is also required to establish the fate of the flower primordia. It is expressed in all primordia, yet expression is much higher in flower primordia. In its absence, primordia arising from the inflorescence meristem have shoot-like phenotypes, and when over-expressed, inflorescence meristems become terminal flowers. LFY directly activates the expression of AP1 (Wagner et al., 1999), and in its role to establish the fate of the flower primordia, AP1 activates *LFY* (Ferrandiz et al., 2000; Kaufmann et al., 2010; Weigel et al., 1992), so there seems to be a mutual backing of each other's expression. In an *ap1cal* background, the knockout of LFY causes the inflorescence meristem to make "flowers" containing leaf-like structures (Bowman et al., 1993). In the same background, the knockout of FUL causes the compound inflorescence meristem to make shoot-like structures (Ferrandiz et al., 2000). Over-expression of LEAFY can cause flower formation in *ful ap1 cal* (Ferrandiz et al., 2000) and in other non-flowering backgrounds described later, suggesting that LFY alone can establish the flowering program within a primordia if introduced at high enough levels. Many transcription factors have been shown to regulate *LFY* expression besides AP1. This includes MADS-box proteins such as SOC1, SPL3, and MYBs (Blazquez et al., 2006).

SEPALLATA3 (SEP3)

SEP3 is expressed in leaves and in the upper portion of late stage 2 flower primordia (Smyth et al., 1990; Mandel and Yanofsky, 1998). *SEP3* is upregulated by FT in leaves (Teper-Bamnolker and Samach, 2005). It does not switch on earlier in the inflorescence meristem because it is repressed by SOC1, SVP, and AGL24 (Liu et al., 2009b). The triple mutant *soc1 svp agl24* has flower defects that can be attributed to miss-expression of *SEP3* in the meristem and can be partially overcome by adding a mutation in *SEP3* (Liu et al., 2009b). This repression involves recruitment of histone modification enzymes HDAC and LHP1/TFL2 (see the next section) through SAP18 (Liu et al., 2009b). The accumulation of AP1 in stage 1 flowers enables the downregulation of *SEP3* repressors in the flower primordia. Once it is accumulated in flower primordia, SEP3 seems to mediate in multimeric MADS-box complexes (Immink et al., 2009). It functions together with AP1 to regulate many flower organ identity genes as well as repression of *SOC1* (Pelaz et al., 2001; Kaufmann et al., 2009; Liu et al., 2009b; Kaufmann et al., 2010).

SQUAMOSA-PROMOTER BINDING PROTEIN LIKE (SPL)

Transcription factors belonging to the SPL family play a conserved central role in the floral transition. This has not been so obvious by studying single loss-of-function mutations, likely due to redundancy. Ten out of the 16 genes encoding this family are regulated by microRNA156 (see the section miR156). Over-expression of miR156, supposedly affecting the 10 genes, causes late flowering and loss of apical dominance (Schwab et al., 2005; Wang et al., 2009a). This would suggest that as a group these genes are positive regulators of flowering. This is also evident in early flowering of transgenic plants that overexpress an miR156-resistant version of *SPL3* (Gandikota et al., 2007; Wang et al., 2009a). Expressing this construct in the meristem and early primordia causes early flowering in the *ft-10* mutant background. This suggests that SPL3 can activate flowering time genes downstream of FT in the apical meristem (Wang et al., 2009a). SPL3 seems to act as a direct transcriptional activator of *FUL* (Wang et al., 2009a). The ability of miR-resistant *SPL3* to cause early flowering is significantly reduced in the absence of FUL or FD (Wang et al., 2009a). Over-expression of miR156 is less late flowering in plants that also over-express *FUL* or *FD*. Still, SPL3 targets other genes such as flower-primordia markers *AP1* and *LFY*. SPL9 seems to induce expression of *SOC1*, *AGL42*, and the microRNA 172 (Wu et al., 2009; see the next section). FD and SPLs are two types of transcription factors that seem to work partially in parallel to induce expression of flowering time genes. Indeed the loss of FD together with over-expression of miR156 causes an additive severe delay in flowering time (Wang et al., 2009a).

PENNYWISE and PENNYFOOLISH

Pennywise (PNY) and Pennyfoolish (PNF) belong to the BEL1-like (BELL) subgroup of Three-Amino-Acid-Loop-Extension (TALE) plant homeobox proteins (Hamant and Pautot, 2010). BELL and KNOX proteins (the second subgroup of TALE homeobox proteins) seem to regulate transcription as heterodimers (Bellaoui et al., 2001; Smith et al., 2002). PNY and PNF seem to be required for at least part of FT function. Like FD, they are expressed in the shoot apical meristem. A *pny pnf* plant does not flower under long days (Smith et al., 2004; Kanrar et al., 2008). It seems that the double mutant does respond to photoperiodic induction, by increasing shoot apical meristem size and the appearance of *SOC1* in the meristem, yet meristems do not reach the size of a wt inflorescence meristem and do not produce flower primordia (lack of evocation, as described by the authors). Together with the lack of flower primordia, *AP1*, *LFY*, and *CAL* transcripts do not appear. As mentioned earlier, a triple mutant of *ap1 lfy cal* produces "flowers" containing leaf-like organs (Bowman et al., 1993), so that *pny pnf* mutants lack something else. The ability of FT over-expression to cause early flowering is severely perturbed in a *pny pnf* background (Kanrar et al., 2008). The abnormally high levels of *AP1* associated with 35S:FT plants (Teper-Bamnolker and Samach, 2005) are not present in the *pny; pnf* background, yet more immediate targets of FT were not tested (Kanrar et al., 2008). Interestingly, over-expression of *LFY* can cause flowering of the *pny pnf* non-flowering mutant (Kanrar et al., 2006) as it caused flowering of the *ful ap1 cal* triple non-flowering mutant (Ferrandiz et al., 2000). Although the *pny pnf* mutant continues to form leaf primordia after photoperiodic induction, these primordia are different from normal leaf primordia because they can take on the identity of a flower if high levels of LFY are introduced to them.

An additional BELL protein, ATH1, is required to maintain high levels of *FLC* expression, so that in its absence plants flower earlier (Proveniers et al., 2007). Ectopic expression of the BELL protein BLH3 together with its suggested KNOX partner, SHOOT MERISTEMLESS (STM), caused very early flowering (Cole et al., 2006).

Processes Affecting Flowering Time Proteins

Histone modifications

DNA is not bare within the nucleus; it is wound around a histone protein core (together called a nucleosome). The histone core is an octamer built from four different types of histones. Access to DNA by transcription factors is influenced by localized covalent modifications in specific histones. The involvement of histone modifications in gene transcription has been well documented in different organisms. Many plant components in this process have initially been revealed through cloning of *Arabidopsis* genes that, when mutated, cause a flowering time phenotype. The reason for this is mostly because the chromatin surrounding the gene encoding the flowering repressor FLC undergoes environmentally induced changes in modifications that close down its transcription. Any mutations that ruin the initial modifications or the environment-dependent modifications will therefore lead to a change in flowering time. Once these genes were revealed, they were shown to be involved in chromatin modifications in other genes involved in flowering time as well as other processes (Veley and Michaels, 2008).

Modifications associated with active genes

Highly transcribed genes normally wrap a nucleosome in which the N-terminal tails of histones 3 and 4 are acetylated. Deacetylation in the N-terminal tails of histones 3 and 4 can cause a reduction in gene transcription. In *Arabidopsis*, FVE partly mediates this deacetylation and its absence causes even higher levels of expression of the flowering repressor *FLC*. An additional histone deacetylase (HDAC), named HDA19, was shown to be linked to the repressors of SEP3 transcription, SOC1, SVP, and AGL24 (mentioned previously) via the structural SAP18 protein (Liu et al., 2009b).

A gene that is actively transcribed is likely to have its 5′ region wrapped around a nucleosome that contains a histone 3 that is trimethylated on Lysine 4 (H3K4me3). This happens when RNA polymerase II associates with two protein complexes, PAF1 (RNA Polymerase II Associated Factor 1) and COMPASS (Complex Proteins Associated with Set1) containing an H3K4 methyltransferase known as SET1, that catalyzes H3K4 trimethylation. The FRIGIDA (FRI) protein is required in the enrichment of a COMPASS-like complex at *FLC* chromatin (Jiang et al., 2009). Thus, *Arabidopsis* accessions that have an active FRI allele have higher expression of *FLC* and obtain a winter-annual growth habit. Mutations in components of the PAF1 complex cause less H3K4me3 around certain genes, which is correlated with reduced transcription. Since a major target of this modification is *FLC*, these mutations cause accumulation of less *FLC* transcript and, as a result, are early flowering. Mutations in H3K4 demethylases (FLD, LDL1, LDL2) increase H3K4me3 and *FLC* expression. H3K36me2 and H3K36me3 modifications are also associated with highly transcribed genes, including *FLC*. The balance between H3K4- and H3K36-specific histone methyl transferases versus H3K4- and H3K36-specific histone demethylases play an important part in regulation of gene expression, as recently shown for FLC (Ko et al., 2010).

Another mark associated with transcriptional activity is monoubiquitination of histone 2B (H2Bub1). In yeast, this modification seems to be a prerequisite for H3K4me3, and in *Arabidopsis* it seems to promote the latter modification.

Histone 2A can be replaced by a variant named histone 2A.Z, and this is performed by the conserved complex called SWR1c. The deposition of histone 2A.Z in chromatin of the *FLC* gene is associated with high *FLC* expression (Martin-Trillo et al., 2006; Choi et al., 2007; Deal et al., 2007). It was recently shown that the presence of this variant in chromatin allows plants to react to changes in ambient temperatures (Kumar and Wigge, 2010). Depletion of H2A.Z from many sites, including the *FT* promoter, was observed at higher temperatures. Higher temperatures cause early flowering through increase in *FT* expression (Balasubramanian et al., 2006). Unlike *FLC*, with *FT* the absence of this variant is correlated with higher transcription. Since FLC represses *FT* expression, the increase in *FT* expression may be due to depleted H2A.Z in *FLC* and other *MAF* loci.

Modifications associated with inactive genes

Silencing short-interfering RNAs (siRNAs) that recognize sequence in the 3′ region of the *FLC* gene seems to recruit chromatin modifying components capable of increasing H3K9me2, a modification associated with gene repression (Swiezewski et al., 2007). There is also evidence for RNA-mediated chromatin silencing mediated by FCA and FPA (Baurle et al., 2007; Veley and Michaels, 2008). Methylation of arginine in tails of histones 3 and 4, for example symmetric methylation of arginine 3 on histone 4 (H4R3sme2), has been linked to gene silencing.

While H3K4me3 is a mark for high gene expression, trimethylation of Lysine 27 on histone 3 (H3K27me3) is a mark for reduced expression. In *Arabidopsis*, this modification was found in chromatin surrounding the regulatory regions of 4400 genes (Zhang et al., 2007a). This modification is performed by a conserved protein complex known as Polycomb repressive complex 2 (PRC2). LIKE HETEROCHROMATIN 1 (LHP1/TFL2) binds this modification and causes transcriptional repression (Zhang et al., 2007b; Turck et al., 2007). Different regions of the same gene can wrap around histones that have opposing modifications (H3K4me3 versus H3K27me3), and their relative levels seem to modulate gene expression, as shown for *FLC*. Expression of *FT* and other *MAF* genes is also repressed by H3K27me3 modifications (Jiang et al., 2008). Recently, histone modifications on chromatin surrounding the FT gene have been documented (Adrian et al., 2010).

Gibberellin

Many plant hormones are associated with flowering (Davis, 2009), and here, for lack of space, I will concentrate on gibberellins (GAs). GAs are involved in flowering of many annuals (King et al., 2006) and perennials, in some plants they promote flowering while in other plants they prevent flowering. A good deal, if not all of GA function is to degrade DELLA transcription factors (Cheng et al., 2004). That is why plants containing non-degradable versions of DELLA proteins are insensitive to GAs. Other hormones seem to also affect DELLA proteins (Achard et al., 2007) so that DELLAs are at the summit of hormone cross-talk.

In *Arabidopsis* plants grown under short days (no external stimuli induces flowering under these conditions), there is a three-fold accumulation of sucrose and GA_4 in the meristem together or perhaps even prior to the induction of flowering (Eriksson et al., 2006). GA biosynthesis mutants or mutants that are insensitive to GA do not flower under short days (Reeves and Coupland, 2001). In published flowering time models, *SOC1* appears as the major target of GA with regard to flowering time since expression of *SOC1* is influenced by GA, and plants that either over-express SOC1 or contain a mutation in *SOC1* have a reduced response to GA (Moon et al., 2003). Still, the facts seem to be that the rise in *SOC1* expression after GA application takes quite a while and requires an additional protein (AGL24). Also, in the absence of SOC1, GA still strongly affects flowering time (Liu et al., 2008a), and flowering time of plants that over-express *SOC1* is still delayed by a GA biosynthesis inhibitor (Blazquez et al., 2002) or by a mutation in a GA biosynthesis gene (Moon et al., 2003). How DELLA proteins interact with flowering time genes is still an open question. As described in the next section, DELLAs downregulate miR159 levels, and miR159 might have a role in flowering time that is not yet clear.

MicroRNAs

MicroRNAs (miRNA) are a subgroup of siRNAs made by different organisms including plants (Voinnet, 2009). An miRNA gene includes a transcribed inverted repeat, so that the RNA forms an imperfect stem-loop foldback structure. Its original form (pri-miRNA) is edited by DICER LIKE-1 into a 20–24 bp dsRNA structure. This structure is methylated and goes into the cytoplasm where it associates with the RISC complex. This complex will now use a specific region of the micro RNA as a template for recognizing mRNAs with complementary sequence within the cell. The complex will either "slice" these mRNAs, leading to their degradation, or inhibit their translation. The ARGONAUTE1 (AGO) protein is the main "slicer" in *Arabidopsis*.

Plants reduce the action of a certain microRNA family using a mechanism recently identified and termed "target mimicry" (Franco-Zorrilla et al., 2007). Instead of a normal miRNA, they include a 3-nucleotide insertion at the site where cleavage should have taken place. The fake miRNA is sequestered in the RISC complex causing a reduction in the normal activity of the specific miRNA family.

This mechanism can be artificially introduced into plants to study how plants behave without a certain family of miRNAs (Todesco et al., 2010). This type of phenotype was difficult to assess using TDNA knockout procedures since they belong to families, and the genes are relatively small so that chances of getting a TDNA insertion in them is low. The other option was to express a target gene with a ruined microRNA target site (miR-resistant), yet in this case we are only fixing one target and expressing it in abnormal tissues at abnormal levels. Using target mimicry, a clear developmental shoot phenotype was observed in 14 miRNA families out of the 73 families found in *Arabidopsis* (Todesco et al., 2010). Among these, several are involved in flowering time (Poethig, 2009).

miR156

In *Arabidopsis*, miR156 levels decrease with age, while expression of the genes it represses, encoding SPL transcription factors that promote flowering (see previous section), increase with age (Wang et al., 2009a). Over-expression of miR156 causes late flowering under both short and long days in *Arabidopsis* (Schwab et al., 2006; Wu and Poethig, 2006; Wang et al., 2009a) as well as rice (Xie et al., 2006) and maize (Chuck et al., 2007). Target mimicry of miR156 causes early flowering under both daylengths (Wang et al., 2009a), although the rate of leaf production is much slower (Todesco et al., 2010). miR156 levels decrease with age while miR172 levels increase with age (Wu et al., 2009). miR156 levels affect miR172 levels: over-expression of miR156 reduces miR172 and target mimicry of miR156 increases miR172 (Wu et al., 2009). The effect of miR156 on miR172 is likely through SPL9 and SPL10, which promote transcription of miR172 (Wu et al., 2009). Over-expressing an miR-resistant *SPL3* gene causes early flowering under both daylengths (see previous section).

miR159

Over-expression of miR159 causes late flowering under short days in *Arabidopsis* (Achard et al., 2004). Plants were not late under long photoperiods. This suggests that long photoperiods directly overcome the delay caused by miR159 or circumvent this problem. Targets of miR159 are a subgroup of MYB transcription factors, some associated with GA signal transduction (GAMYB). DELLA downregulates miR159 levels (Achard et al., 2004) so that GA dependent DELLA degradation increases *miR159* levels and thus increases the degradation of its target MYBs, such as *MYB33* in *Arabidopsis*. Interestingly, while GA promotes flowering under SDs in *Arabidopsis*, it increases the levels of miR159, which was shown here to delay flowering under SDs. No flowering time phenotype was attributed to target mimicry of miR159 (Todesco et al., 2010), suggesting that the role of this miRNA in flowering time is still obscure.

miR167

Target mimicry of the miR167 group causes late flowering under long days (Todesco et al., 2010). Its known targets, *ARF6* and *ARF8*, were *not* shown to be involved in flowering time. The double mutant has miss-expressed KNOX-encoding genes, and many of its phenotypes can be mimicked by over-expression of STM (Tabata et al., 2010). Recently, new targets have been identified for this and other miRNAs (Alves et al., 2009), which might explain this discrepancy.

miR169

Mimicry of the miR169 group causes plants with smaller leaves (Todesco et al., 2010). This microRNA targets HAP2/NFYA encoding genes (Combier et al., 2006), which are part of the trimeric CCAAT binding factor complex (Nuclear Factor Y). These genes seem to have a role in flowering time, perhaps in the ability of CO to activate FT (see previous section). Recently this microRNA was found to be abundant in phloem sap in *Brassica* (Buhtz et al., 2010), suggesting that it might travel and cause changes in flowering time.

miR172

Over-expressing miR172 caused early flowering (Chen, 2004; Jung et al., 2007). Target mimicry of the miR172 group causes late flowering under long days (Todesco et al., 2010). Without miR172, its targets, APETALA2-like transcription factors, are expected to be highly expressed. These proteins were shown to act as flowering time repressors when over-expressed (Schmid et al., 2003; Mathieu et al., 2009), as they work through transcriptional regulation of *FT* and *TSF*.

The circadian clock

The circadian clock generates near 24h oscillations in gene expression that persist under constant environmental conditions (Mas and Yanovsky, 2009; Harmer, 2009). In *Arabidopsis*, about 25% of protein coding genes oscillate, and in addition oscillations in introns, intragenic regions, and non-coding RNAs were recently detected (Hazen et al., 2009). In *Arabidopsis*, and possibly in any other plant that changes in photoperiod to affect flowering time, the circadian clock sets the expression of CO (Suarez-Lopez et al., 2001), which eventually activates expression of FT (Samach et al., 2000). Any mutation in genes that have a role in the input, the central oscillator, or the output of the clock may affect the expression pattern of CO, and this will likely change flowering time (Samach and Coupland, 2000). One example is the CYCLING DOF FACTORS (CDFs) shown to repress CO transcription (Imaizumi et al., 2005; Fornara et al., 2009). These proteins are degraded by controlled proteolysis, and the use of controlled proteolysis seems to be central to clock function (Yakir et al., 2009) and CO stability. In addition, microRNAs such as miR167 (Hazen et al., 2009) also display a circadian rhythm, but whether this affects flowering time is still unknown. We will not discuss the current model of the proteins that are thought to comprise the *Arabidopsis* oscillator (Mas and Yanovsky, 2009) in this section. Interestingly, chromatin modifications (previously discussed) play an important role in the regulation of these oscillators

Regulated proteolysis

Regulated proteolysis is required for plant hormone signal transduction (Santner and Estelle, 2010), circadian clock function (Mas and Yanovsky, 2009), photoreceptor function (Henriques et al., 2009), and for the stability of CO (Valverde et al., 2004; Liu et al., 2008b), among many other processes. Thus, any problem with this process is likely to affect flowering time. It is clear from research in many other plants that photoperiod is influenced by light quality (Zeevaart, 1962), and the molecular basis of this could well be through protein stability of the output from the clock as well as the clock itself, which relies on regulated proteolysis to function correctly (Henriques et al., 2009).

Sugars

Signals in plants can use the phloem to travel (Turgeon and Wolf, 2009). Sucrose is the main sugar traveling in the phloem in *Arabidopsis* and many other plants. Its levels in the phloem sap increase during photoperiodic floral induction (Corbesier et al., 1998). Also under non-inductive short days there is a three-fold accumulation of sucrose in the meristem together with or perhaps even prior to the induction of flowering (Eriksson et al., 2006).

The uploading of sucrose into the companion cells of the sieve elements requires non-symplastic movement through plant membranes via membrane bound sucrose transporters (SUTs or SUCs). Some of these transporters are tissue specific and their loss-of-function can have different effects including changes in flowering time (Sauer, 2007). In tobacco, reduction of *NtSUT1* expression reduces loading of sucrose to the phloem and causes delayed flowering (Burkle et al., 1998), whereas reducing levels of the potato sucrose transporter *SUT4* caused an earlier increase in sucrose levels in the meristem, leading to earlier flowering (Chincinska et al., 2008). In *Arabidopsis*, a mutation in the *AtSUC9* gene encoding a sucrose transporter caused early flowering under short days without affecting flowering time under long days (Sivitz et al., 2007).

Sucrose application in growth media can reduce flowering time of wild-type plants grown under suboptimal photosynthetic conditions, and recover the late flowering phenotypes of mutations in genes that act upstream to *FT* (Roldan et al., 1999). Sucrose could not rescue mutations in *FT* (Roldan et al., 1999), so it seems to require FT for its function, or perhaps FT requires sucrose for its function. On the other hand, high amounts of sucrose application in growth media can delay flowering through a reduction in *FT* expression (Ohto et al., 2001). Recently sucrose was also shown to affect circadian clock function (Knight et al., 2008).

Trehalose 6-phosphate (T6P) is a carbohydrate that seems to be made by many different plants yet at very low abundance, thus not serving as a carbon source. T6P levels seem to correlate with sugar availability, so in other words, T6P level can serve as an indicator of the metabolic state of the plant. T6P has been shown to regulate sugar utilization and starch metabolism and interacts with other signaling pathways, including those mediated by sugars and plant hormones (Ramon and Rolland, 2007; Smeekens et al., 2010). T6P has been recently shown to pass on the signal by inhibiting the activity of SNF1-Related Protein Kinase 1 (SnRK1), which is part of the family of calcium-independent Ser/Thr protein kinases (Zhang et al., 2009). SNF1 kinases are conserved in eukaryotic cells and seem to have a universal role in mediating the reaction to a state of energy limitation and lack of carbon source. In such situations, the reaction mediated by SNF1s is activation of catabolic pathways and switching off biosynthesis (proteins, carbohydrates, and lipids), cell growth, and cell proliferation (Hardie, 2007). In *Arabidopsis*, SnRK1 regulates close to 1000 genes partially through post-translational modification of a subgroup of bZIP transcription factors (Baena-Gonzalez et al., 2007).

Genetic evidence has recently connected the T6P/SnRK1 relationship to the transition to flowering in *Arabidopsis*. Double knockout of two genes encoding SnRK1, *KIN10*, and *KIN11* caused, among other things, a phenotype of bolting without flower formation. Over-expression of *KIN10* caused late flowering under long days (Baena-Gonzalez, 2007). Trehalose-6-phosphate Synthase 1 is required for T6P biosynthesis, and a loss-of-function mutation in this gene in *Arabidopsis* is embryo lethal (Eastmond et al., 2002; van Dijken et al., 2004). Interestingly, plants containing weak alleles are very late flowering (Gomez et al., 2010), suggesting that this unique sugar is somehow required for flower induction. The fact that low levels of T6P and high levels of SnRK1 are associated with low sugar levels and that they cause the same late flowering phenotype suggest an interesting connection worth further investigation.

Developmental Decisions on Timing of Flowering

Juvenility

Juvenility describes the phase of growth following germination from seed during which flowering does not occur and the bud meristem is not "competent" to respond to seasonal environmental inductive cues, and hence remains vegetative. This could take days or years depending on the species. Juvenility in most trees is a significant impediment in breeding programs. For example, apple seedlings usually remain vegetative for 7–9 years after seed germination (Flachowsky et al., 2009), while olive seedlings remain juvenile for 4–5 years (Moreno-Alias et al., 2010). This can be reduced to a certain degree by growing the seedling tree under "forcing" conditions, which accelerates its growth until reaching a minimum of main stem nodes (Aldwinckle, 1975; Santos-Antunes et al., 2005). Still, it has been estimated that creating a new apple cultivar takes 15–20 years and costs approximately €400,000 (Flachowsky et al., 2009). Wild-type *Arabidopsis* can respond to photoperiodic induction two weeks after germination (Corbesier et al., 1996), so it does *not* seem like the perfect model system to study juvenility. Still, even in its short vegetative stage *Arabidopsis* goes through a phase change that can be studied using the powerful tools of molecular genetics (Poethig, 2003). The fact that there is conservation among distant species might suggest a common theme among plants. Indeed, both miR172 and miR156 were identified in apple (Gleave et al., 2008).

The last decade provided quite a lot of evidence that one can shorten the juvenile phase of trees by transgenic manipulation of internal genes or ectopic expression of foreign genes (Flachowsky et al., 2009). Over-expression of the *LEAFY* gene from *Arabidopsis* was shown to reduce juvenility in aspen (Weigel and Nilsson, 1995), poplar (Rottmann et al., 2000), and citrus (Pena et al., 2001). Over-expression of *FT* in citrus, poplar, pear, and apple (Endo et al., 2005; Bohlenius et al., 2006; Hsu et al., 2006; Matsuda et al., 2009; Kotoda et al., 2010) or downregulation of a gene encoding a TFL1-like protein in apple (Kotoda et al., 2006) significantly reduced the juvenile phase. Introducing a construct over-expressing a *FUL* ortholog gene from silver birch caused very early flowering of apple trees (Flachowsky et al., 2007). There are ongoing attempts to over-express FT in apple using inducible promoters and making use of its ability to move in the phloem, so that transgenic root-stocks can be used to overcome juvenility of one-year-seedlings through grafting (Flachowsky et al., 2009).

Seasonality

Photoperiod

An increase in daylength causes a gradual accumulation of the CONSTANS protein in leaves (Putterill et al., 1995; Suarez-Lopez et al., 2001; Valverde et al., 2004). *Arabidopsis* plants flower much earlier, and after producing fewer leaves in long days (16/8 hours of light/dark cycles) compared to short days (8/16). This is mainly because under long photoperiods high levels of the CONSTANS (CO) transcription factor accumulate and can directly activate *FT* and *TSF* expression in the vascular bundles of the leaves, leading to movement of these proteins via the phloem to the SAM. The meeting between FT and FD proteins is assumed to occur in the shoot apical meristem.

In rice and *Pharbitis nil*, flowering is promoted by short days, and indeed FT-encoding genes accumulate in short days (Hayama et al., 2003, 2007). In rice, a gene encoding a CO-like protein acts to delay flowering, which might be the reason why FT-like genes accumulate under short days (Yano et al., 2000, 2001).

In passion fruit (*Passiflora edulis*) after a relatively short juvenile period (a few months), the shoot apical meristem produces new leaf primordia that also contain a subtending meristem that produces both a flower and a tendril (Nave et al., 2010). As long as environmental conditions and the state of the vine allow growth of the apical meristem, flower primordia will be present in every node. Flower development depends on long photoperiods so that photoperiods below 11 h will cause the flower to stop developing after formation of sepals. So seasonality in passion fruit is not caused by controlling flower induction, but by controlling flower development. In passion fruit, similar to other photoperiodic plants, photoperiod is perceived by the leaves and a positive signal can move through graft junctions. So passion fruit and *Arabidopsis* share the existence of a mobile signal originating from the leaf under inductive photoperiods, yet in *Arabidopsis* it controls flower induction, while in passion fruit it controls flower development. The nature of the signal in passion fruit is unknown.

Vernalization

To avoid flowering before the end of winter, several annuals will not flower before experiencing a certain long period of cold temperatures. The mechanism is relatively well understood at the molecular level in *Arabidopsis* (Amasino, 2010; Crevillen and Dean, 2010). Explaining this mechanism is simplified when concentrating on FLC, the transcriptional repressor of *FT*, and what we already learned regarding histone modifications in the *FLC* chromatin. Still, it is quite clear that other *MAF* genes are involved, and that they do not only repress *FT*.

The expression of *FLC* is downregulated by cold temperature through stable changes in chromatin structure caused by specific changes in histone modification at the *FLC* locus from modifications that cause high expression, such as H3K4me3, to those that cause repression of transcription, such as H3K27me3. One trigger for this change seems to be the *VERNALIZATION INSENSITIVE 3* (*VIN3*) gene (Sung and Amasino, 2004), whose expression increases in response to prolonged low-temperature exposure. Its product, a plant homeodomain finger (PHD Finger) containing protein, interacts with two additional similar proteins (Greb et al., 2007) and the PRC2 complex described in the next section, in proximity to FLC. This complex causes H3K27me3 enrichment (De Lucia et al., 2008). Histone modifications surrounding *FLC* in response to vernalization do not mirror changes in all genes. Those that are activated show opposite modifications (Schonrock et al., 2006; Adrian et al., 2010). Without dividing cells vernalization does not occur (Wellensiek, 1962), and this is likely because the memory of winter, characterized by continued H3K27me3 modifications surrounding *FLC*, is only maintained in dividing cells (Finnegan and Dennis, 2007). Natural variation in the need for vernalization to cause flowering was found to be caused by inactive alleles of *FRIGIDA* (Johanson et al., 2000) (described previously) and *FLC*. In *Arabidopsis* ecotypes respond to vernalization, which relieves *FLC* transcriptional repression in both the leaf (*FT*) and meristem (*SOC1* and *FD*) allowing photoperiod-dependent production of systemic signals in the leaves and conferring competence on the meristem to respond to these signals (Searle et al., 2006).

In the perennial *Arabidopsis alpina*, the memory of winter is only temporary, so that expression of the gene encoding the FLC ortholog PERPETUAL FLOWERING 1 (PEP1) goes down in winter and up again in warm temperatures. This seems to provide a mechanism to avoid all meristems from becoming reproductive, thus allowing this species to be polycarpic (Wang et al., 2009b).

In both wheat (*Triticum aestivum* L.) and barley (*Hordeum vulgare* L.) there are cultivars with a vernalization requirement and cultivars with no such requirement. This provided a means to identify genes responsible for natural variation in the vernalization response (Distelfeld et al., 2009). Here, the FT wheat ortholog is called *VRN3* (Yan et al., 2006), and it seems to act together with an FD-like protein (Li and Dubcovsky, 2008). One of their targets is a gene encoding a protein similar to FUL and AP1, called *VRN1* in wheat and *BM5* in barley (Yan et al., 2003). In response to vernalization, the FUL-encoding gene is upregulated, and in some cultivars this gene is upregulated without vernalization, which seems to be enough for vernalization not to occur. Repressors of flowering include a CCT-domain protein (a domain found in the activator of flowering CO) called VRN2 (Yan et al., 2004; Trevaskis et al., 2006) and a gene encoding an SVP-like protein called *TaVRT-2* (Kane et al., 2005), both downregulated by vernalization. What seems different between *Arabidopsis* and wheat is that so far no connection to chromatin modifications has been made, and that the FUL-like gene seems to be upregulated by vernalization independently and before the FT-like gene.

Olives (*Olea europa* L) need to be exposed to cold temperatures before they will initiate flowering (Hartmann and Porlingis, 1953, 1957; Hartmann and Whisler, 1975). For some cultivars, exposure to 12°C was sufficient to induce flowering. Thus, similar to *Arabidopsis*, wheat, and barley, olives respond to a vernalization treatment. Flowering (anthesis) occurs in spring (~April in Israel, mid-May in Italy/Spain; Orlandi et al., 2005). Using a microscope, the first clear sight of an inflorescence meristem is in early February (Andreini et al., 2008; Kitsaki et al., 2010). The first sign of sepal formation occurs two months before full bloom (Hartmann, 1951). Before that there are many reports of changes in the meristem shape, RNA accumulation, production of additional primordia, and accumulation of different hormones (Andreini et al., 2008), yet none of these "signs" clearly distinguish an inflorescence producing bud from a vegetative bud. Genes that might be involved in flowering time have not been characterized in olive.

Warm ambient temperatures

While *Arabidopsis* plants grown under short photoperiods flower late, raising temperatures to 25–30°C in the same photoperiod cause a significant reduction in time and number of leaves until flowering (Balasubramanian et al., 2006; Paltiel et al., 2006). This early flowering is dependent on the *FT* gene (Balasubramanian et al., 2006), and FLM seems to be involved in this process. An *Arabidopsis* accession from Niederzenz (Nd) has a natural deletion of this gene and this accession flowers relatively early under short days, and increasing temperatures cause a reduced early flowering response (Werner et al., 2005; Balasubramanian et al., 2006).

SVP was also associated with flowering dependent on ambient temperatures, when comparing long day plants under 16 or 23°C (Lee et al., 2007a). In the Col-0 ecotype, 16°C causes delayed flowering and in the *svp* early flowering mutant it does not. Late flowering is likely due to high levels of *FLC* (Lee et al., 2007a) or other MAF-like proteins that require SVP to cause late flowering (see previous section). In plants that have high *FLC* levels at 23°C, SVP functions as a perfect flowering repressor under this temperature as well (Lee et al., 2007b). It is likely that SVP is a partner to proteins affected by changes in ambient temperatures, thus its presence is required to observe temperature-dependent changes in flowering time.

Apples (*Malus* × *domestica* Borkh.) flower (anthesis) in the spring and the meristem that made these flowers went through a transition 11 months beforehand, in the beginning of summer. The first structural evidence of change in meristem fate is the transition of the apical meristem from a flattened to a domed shape, with an intermediate stage in which the meristem apex is broadened (Kotoda et al., 2000; Foster et al., 2003). These microscopic morphological changes during flower initiation occur, depending on cultivar, bud type, climatic conditions, and tree factors, 3–6 weeks after full bloom of the previous cycle. Thus, unlike olives, apples do not require exposure to cold temperatures to form flower primordia. Unlike olives, apples are deciduous and enter dormancy before winter (with inflorescences present) and require chilling for bud break.

Several studies on gene expression in apple have provided a very nice picture on the expression of flowering time genes during the transition to flowering (Sung et al., 1999; Kotoda et al., 2000, 2010; Wada et al., 2002; Kotoda and Wada, 2005; Hattasch et al., 2008). When looking at expression within the apical bud of a fruit-bearing shoot (induction late June, formation of flower buds mid-July), expression of one FT-encoding gene increases during June and peaks in mid-July, and expression of an AP1-encoding gene starts rising a few weeks later (Kotoda et al., 2010). SOC1 and FUL-encoding genes were transcribed in mid-July (Hattasch et al., 2008). Additional genes that showed expected expression patterns were those encoding for TFL1 and LFY-like proteins. Transgenic apples that have high levels of the apple *FT* gene are early flowering (skipping juvenility) and express high levels of the FUL, SOC1, and AP1 encoding genes (Kotoda et al., 2010). This basically shows that the transition to flowering in apple shares many components with *Arabidopsis*. What may be different is the site in which *FT* is transcribed, and whether or not FT needs to travel in apple, or is made where it acts.

Reproductive cycles and alternate bearing

Within the first six months after giving birth, breastfeeding new mothers have a severely reduced chance of becoming pregnant (Howie and McNeily, 1982). In cucumbers, a seed containing cucumber fruit inhibits fruit set of the following flowers on the branch. In olives, a tree that produces many fruit in a certain year will likely produce very little fruit in the following year. These are diverse examples of reproductive cycles that occur in nature that are not due to environmental input, but to an internal rhythm, which clearly makes sense. Reproduction requires many resources and for an organism to produce offspring in more than one cycle it needs to spread its reproduction cycles to enable sufficient time for recuperation. It is unknown why alternate bearing is more pronounced is some species.

In curcubits, for example cucumber and cantaloupe (Bangerth, 1989; Valantin et al., 1999), the first fruits that set normally inhibit subsequent flower maturation and fruit set. This effect is pronounced in seed containing fruit, and there is genetic variation in the intensity of the response.

With perennial plants that reach anthesis at a specific time of year (normally spring), in many cases a high crop load will reduce the number of flowers appearing in the following year. Thus, a biannual program of reproduction occurs, also known as alternate bearing (Goldschmidt and Monselise, 1972; Jonkers, 1979; Monselise and Goldschmidt, 1982; Lavee, 2007).

The size of fruit on a tree with a heavy load is normally smaller and prices are normally reduced due to high supply and reduced quality. The "off" year is financially damaging because there is very little yield to sell. An entire fruit-growing region may enter this cycle causing a phenomenon of huge economical importance. This is a great financial burden on the growers and on the chain of production and sales. Thus, this rhythm that plants keep to protect themselves negates the interests of growers, and requires a solution that will satisfy both sides.

In alternate bearing fruit crops, high fruit load has two major effects: a reduction of vegetative growth (and as a result, a reduction in new formed buds); and a reduced chance of a formed bud to go through a transition to flowering. The timing of the transition to flowering is very different among fruit trees, even if anthesis occurs at the same time; for example, while apples and olives reach anthesis in spring, in apples, inflorescences differentiate 10 months before anthesis, whereas in olives this happens only a couple of months before anthesis. The ability of fruit to repress growth and induction occurs in both species.

A year with low yield can occur due to two different events, one internal and one external. The internal event is the high crop load from the previous year, and this is a mechanism that is tree specific and in several examples, branch specific. Meaning, in olive, one can remove all flowers from a branch of a tree that is full of flowers, and as a result, the next year, that branch will be the only branch with high amounts of flowers. An external event is some environmental or management catastrophe that caused the loss of all fruit in the orchard, or in a growing area. This will be an "off" year and many genotypes of fruit trees will respond to this by flowering profusely in the following year, creating a cycle of bearing that is regional. This cycle can be stopped or subdued by farmer intervention if flowers or young fruit are removed early during their development (Dennis, 2000; Tromp, 2000) or by an additional catastrophe that can cause two "off" years in a row. Non-alternate-bearing genotypes seem to have the ability to avoid profuse flowering after an "off" year and by doing so, they avoid entering into this cycle. They can be just as sensitive as alternate-bearing genotypes to catastrophes such as weather conditions that cause all flowers to abscise, or conditions that reduce pollination or fertilization, or conditions that enhance fruit drop early in development.

How does fruit load affect flower induction in nearby buds? Fruits are important for creating the inhibitor signal since removal of fruit early on in development can allow high induction (Fernandez-Escobar et al., 1992; Bangerth, 2000; Al-Shdiefat and Qrunfleh, 2006; Dag et al., 2010). In some species, there is evidence that the presence of seed is necessary for inhibiting induction. Killing an embryo in olive seemed to have reduced the inhibiting signal (Stutte and Martin, 1986). The "Spencer Seedless" apple cultivar has apetalous parthenocarpy and can be either seedless or with seeds if pollinated. Non-pollinated parthenocarpic fruit cause no reduction in return bloom, while pollinated seed-containing fruit reduced flowering in shoots of certain length (Chan and Cain, 1967; Neilsen and Dennis, 2000). In longer shoots seed did not have an inhibitory affect. It could be that the influence of seeds in this case is indirect. Diffusates from seeded and non-seeded fruit were compared for GA activity (Hoad, 1978) and higher levels of GA activity were identified in diffusates from seeded fruit. Still, in citrus, there are seedless cultivars that have a strong alternate bearing tendency. Many have shown changes in hormone and sugar content in diffusates coming out of the pedicels of fruit (Marino and Greene, 1981; Bangerth, 1989). On the other hand, transport in the phloem is supposed to be in the opposite direction, from stem to fruit (Oparka and Cruz, 2000; Turgeon and Wolf, 2009). Whether there is a signal coming out of fruit (Bangerth, 2009), or whether some other source organ, like the leaf, can perceive fruit load and send out a signal is still an unknown. It is clear that hormone and carbohydrate levels are different in a branch with heavy fruit load compared to a branch without fruit (Li et al., 2003; Baktir et al., 2004), yet which hormone/carbohydrate, if any, is responsible for lack of flower induction is still unknown. It is also unknown in what way the signal(s) interact(s) with the mechanism of flower induction.

Summary

Man has been attempting to manipulate time of flowering of plants for agricultural purposes for quite a long time. Our current understanding of the mechanisms plant use provides us with better tools to intervene in these processes (Figure 25.1). Our initial knowledge came mostly from the model plant *Arabidopsis*. Our current knowledge of mechanisms in other plants provides many examples in which different species have been using the same set of proteins to create a unique timing mechanism that answers their specific needs. It has taken plants a very long time to reach the perfect control of these processes. It still remains a challenge to intervene in this process, for the good of mankind, without causing damage to the plant, and long-term damage to agriculture.

Acknowledgment

Research in the AS lab is supported by German Israeli Project Cooperation (DIP project H3.1) the Israeli Science Foundation (grant no. 1464/07) and the Chief Scientist of the Israeli Ministry of Agriculture and Rural Development.

References

Abe, M., Kobayashi, Y., Yamamoto, S., Daimon, Y., Yamaguchi, A., & Ikeda, Y., et al. (2005). FD, a bZIP protein mediating signals from the floral pathway integrator FT at the shoot apex. *Science*, *309*, 1052–1056.

Achard, P., Baghour, M., Chapple, A., Hedden, P., Van Der Straeten, D., & Genschik, P., et al. (2007). The plant stress hormone ethylene controls floral transition via DELLA-dependent regulation of floral meristem-identity genes. *Proceedings of the National Academy of Sciences United States of America*, *104*, 6484–6489.

Achard, P., Herr, A., Baulcombe, D. C., & Harberd, N. P. (2004). Modulation of floral development by a gibberellin-regulated microRNA. *Development*, *131*, 3357–3365.

Adrian, J., Farrona, S., Reimer, J. J., Albani, M. C., Coupland, G., & Turck, F. (2010). cis-regulatory elements and chromatin state coordinately control temporal and spatial expression of FLOWERING LOCUS T in *Arabidopsis*. *Plant Cell*, *22*, 1425–1440.

Al-Shdiefat, S. M., & Qrunfleh, M. M. (2006). Effect of fruit thinning on endogenous plant hormones and bearing of the olive (*Olea europaea* L.) nabali muhasa. *Jordan Journal of Agricultural Sciences*, *2*, 348–360.

Albani, M. C., & Coupland, G. (2010). Comparative analysis of flowering in annual and perennial plants. Current topics in developmental biology. In M. C. P. (2010). Timmermans (Ed.), *Plant Development* (Vol. 91, pp. 323–348). Academic Press.

Aldwinckle, H. S. (1975). Flowering of apple seedlings 16–20 months after germination. *Horticultural. Science*, *10*, 124–126.

Alexandre, C. M., & Hennig, L. (2008). FLC or not FLC: The other side of vernalization. *Journal of Experimental Botany*, *59*, 1127–1135.

Alves, L., Jr., Niemeier, S., Hauenschild, A., Rehmsmeier, M., & Merkle, T. (2009). Comprehensive prediction of novel microRNA targets in *Arabidopsis thaliana*. *Nucleic Acids Research*, *37*, 4010–4021.

Amasino, R. (2010). Seasonal and developmental timing of flowering. *The Plant Journal*, *61*, 1001–1013.

Andreini, L., Bartolini, S., Guivarc'h, A., Chriqui, D., & Vitagliano, C. (2008). Histological and immunohistochemical studies on flower induction in the olive tree (*Olea europaea* L.). *Plant Biology*, *10*, 588–595.

Baena-Gonzalez, E., Rolland, F., Thevelein, J. M., & Sheen, J. (2007). A central integrator of transcription networks in plant stress and energy signalling. *Nature*, *448*, 938–942.

Baktir, I., Ulger, S., Kaynak, L., & Himelrick, D. G. (2004). Relationship of seasonal changes in endogenous plant hormones and alternate bearing of olive trees. *HortScience: A Publication of the American Society for Horticultural Science*, *39*, 987–990.

Balasubramanian, S., Sureshkumar, S., Lempe, J., & Weigel, D. (2006). Potent induction of *Arabidopsis thaliana* flowering by elevated growth temperature. *PLoS Genetics*, *2*, e106.

Bangerth, F. (1989). Dominace among fruit/sinks and the search for a correlative signal. *Physiologia Plantarum*, *76*, 608–614.

Bangerth, F. (2000). Abscission and thinning of young fruit and their regulation by plant hormones and bioregulators. In *Plant growth regulation*, (Vol. 31, pp. 43–59). Netherlands: Springer.

Bangerth, K. F. (2009). Floral induction in mature, perennial angiosperm fruit trees: Similarities and discrepancies with annual/biennial plants and the involvement of plant hormones. *Scientia Horticulturae*, *122*, 153–163.

Baurle, I., Smith, L., Baulcombe, D. C., & Dean, C. (2007). Widespread role for the flowering-time regulators FCA and FPA in RNA-mediated chromatin silencing. *Science*, *318*, 109–112.

Bellaoui, M., Pidkowich, M. S., Samach, A., Kushalappa, K., Kohalmi, S. E., & Modrusan, Z., et al. (2001). The *Arabidopsis* BELL1 and KNOX TALE homeodomain proteins interact through a domain conserved between plants and animals. *Plant Cell*, *13*, 2455–2470.

Ben-Naim, O., Eshed, R., Parnis, A., Teper-Bamnolker, P., Shalit, A., & Coupland, G., et al. (2006). The CCAAT binding factor can mediate between CONSTANS-like proteins and DNA. *Plant Journal*, *46*, 462–476.

Blazquez, M. A., Ferrandiz, C., Madueno, F., & Parcy, F. (2006). How floral meristems are built. *Plant Molecular Biology*, *60*, 855–870.

Blazquez, M. A., Trenor, M., & Weigel, D. (2002). Independent control of gibberellin biosynthesis and flowering time by the circadian clock in *Arabidopsis*. *Plant Physiology*, *130*, 1770–1775.

Bohlenius, H., Huang, T., Charbonnel-Campaa, L., Brunner, A. M., Jansson, S., & Strauss, S. H., et al. (2006). CO/FT regulatory module controls timing of flowering and seasonal growth cessation in trees. *Science*, *312*, 1040–1043.

Bowman, J. L., Alvarez, J., Weigel, D., Meyerowitz, E. M., & Smyth, D. R. (1993). Control of flower development in *Arabidopsis thaliana* by APETALA1 and interacting genes. *Development*, *119*, 721–743.

Buhtz, A., Pieritz, J., Springer, F., Kehr, J. (2010). Phloem small RNAs, nutrient stress responses, and systemic mobility. *BMC Plant Biology 10*, 64.

Burkle, L., Hibberd, J. M., Quick, W. P., Kuhn, C., Hirner, B., & Frommer, W. B. (1998). The H^+-sucrose cotransporter NtSUT1 is essential for sugar export from tobacco leaves. *Plant Physiology*, *118*, 59–68.

Cai, X., Ballif, J., Endo, S., Davis, E., Liang, M., & Chen, D., et al. (2007). A putative CCAAT-binding transcription factor is a regulator of flowering timing in *Arabidopsis*. *Plant Physiology*, *145*, 98–105.

Castillejo, C., & Pelaz, S. (2008). The balance between CONSTANS and TEMPRANILLO activities determines FT expression to trigger flowering. *Current Biology*, *18*, 1338–1343.

Chailakhyan, M. K. (1936). New facts in support of the hormonal theory of plant development. C. R. *(Dokl.) Academy of Sciences USSR*, *13*, 79–83.

Chan, B. G., & Cain, J. C. (1967). The effect of seed formation on subsequent flowering in apple. *Proceedings of the American Society for Horticultural Science*, *91*, 63–67.

Chen, K.-Y., Cong, B., Wing, R., Vrebalov, J., & Tanksley, S. D. (2007). Changes in regulation of a transcription factor lead to autogamy in cultivated tomatoes. *Science*, *318*, 643–645.

Chen, X. (2004). A MicroRNA as a translational repressor of APETALA2 in *Arabidopsis* flower development. *Science*, *303*, 2022–2025.

Cheng, H., Qin, L., Lee, S., Fu, X., Richards, D. E., & Cao, D., et al. (2004). Gibberellin regulates *Arabidopsis* floral development via suppression of DELLA protein function. *Development*, *131*, 1055–1064.

Chincinska, I. A., Liesche, J., Krugel, U., Michalska, J., Geigenberger, P., & Grimm, B., et al. (2008). Sucrose transporter StSUT4 from potato affects flowering, tuberization, and shade avoidance response. *Plant Physiology*, *146*, 515–528.

Choi, K., Park, C., Lee, J., Oh, M., Noh, B., & Lee, I. (2007). *Arabidopsis* homologs of components of the SWR1 complex regulate flowering and plant development. *Development*, *134*, 1931–1941.

Chuck, G., Cigan, A. M., Saeteurn, K., & Hake, S. (2007). The heterochronic maize mutant Corngrass1 results from overexpression of a tandem microRNA. *Nature Genetics*, *39*, 544–549.

Cole, M., Nolte, C., & Werr, W. (2006). Nuclear import of the transcription factor SHOOT MERISTEMLESS depends on heterodimerization with BLH proteins expressed in discrete sub-domains of the shoot apical meristem of *Arabidopsis thaliana*. *Nucleic Acids Research*, *34*, 1281–1292.

Combier, J. P., Frugier, F., de Billy, F., Boualem, A., El-Yahyaoui, F., & Moreau, S., et al. (2006). MtHAP2-1 is a key transcriptional regulator of symbiotic nodule development regulated by microRNA169 in *Medicago truncatula*. *Genes & Development* 20.

Corbesier, L., Gadisseur, I., Silvestre, G., Jacqmard, A., & Bernier, G. (1996). Design in *Arabidopsis thaliana* of a synchronous system of floral induction by one long day. *Plant Journal*, *9*, 947–952.

Corbesier, L., Lejeune, P., & Bernier, G. (1998). The role of carbohydrates in the induction of flowering in *Arabidopsis thaliana*: Comparison between the wild type and a starchless mutant. *Planta*, *206*, 131–137.

Corbesier, L., Vincent, C., Jang, S., Fornara, F., Fan, Q., & Searle, I., et al. (2007). FT protein movement contributes to long-distance signaling in floral induction of *Arabidopsis*. *Science*, *316*, 1030–1033.

Crevillen, P., & Dean, C. (2010). Regulation of the floral repressor gene FLC: The complexity of transcription in a chromatin context. *Current Opinion in Plant Biology*.

Dag, A., Bustan, A., Avni, A., Tzipori, I., Lavee, S., & Riov, J. (2010). Timing of fruit removal affects concurrent vegetative growth and subsequent return bloom and yield in olive (*Olea europaea* L.). *Scientia Horticulturae*, *123*, 469–472.

Davis, S. J. (2009). Integrating hormones into the floral-transition pathway of *Arabidopsis thaliana*. *Plant, Cell & Environment*, *32*, 1201–1210.

De Lucia, F., Crevillen, P., Jones, A. M., Greb, T., & Dean, C. (2008). A PHD-polycomb repressive complex 2 triggers the epigenetic silencing of FLC during vernalization. *Proceedings of the National Academy of Sciences of the United States of America*, *105*, 16831–16836.

Deal, R. B., Topp, C. N., McKinney, E. C., & Meagher, R. B. (2007). Repression of flowering in *Arabidopsis* requires activation of FLOWERING LOCUS C expression by the histone variant H2A.Z. *Plant Cell*, *19*, 74–83.

Dennis, F. G. J. (2000). The history of fruit thinning. In *Plant Growth Regulation*, (Vol. 31, pp. 1–16). Netherlands: Springer.

Distelfeld, A., Li, C., & Dubcovsky, J. (2009b). Regulation of flowering in temperate cereals. *Current Opinion in Plant Biology*, *12*, 178–184.

Distelfeld, A., Tranquilli, G., Li, C., Yan, L., & Dubcovsky, J. (2009a). Genetic and molecular characterization of the VRN2 loci in tetraploid wheat. *Plant Physiology*, *149*, 245–257.

Eastmond, P. J., van Dijken, A. J., Spielman, M., Kerr, A., Tissier, A. F., & Dickinson, H. G., et al. (2002). Trehalose-6-phosphate synthase 1, which catalyses the first step in trehalose synthesis, is essential for *Arabidopsis* embryo maturation. *Plant Journal*, *29*, 225–235.

Endo, T., Shimada, T., Fujii, H., Kobayashi, Y., Araki, T., & Omura, M. (2005). Ectopic expression of an FT homolog from citrus confers an early flowering phenotype on trifoliate orange (*Poncirus trifoliata* L. Raf.). *Transgenic Research*, *14*, 703–712.

Eriksson, S., Bohlenius, H., Moritz, T., & Nilsson, O. (2006). GA4 is the active gibberellin in the regulation of LEAFY transcription and *Arabidopsis* floral initiation. *Plant Cell*, *18*, 2172–2181.

Fernandez-Escobar, R., Benlloch, M., Navarro, C., & Martin, G. C. (1992). The time of floral induction in the olive. *Journal of the American Society for Horticultural Science*, *117*, 304–307.

Ferrandiz, C., Gu, Q., Martienssen, R., & Yanofsky, M. F. (2000). Redundant regulation of meristem identity and plant architecture by FRUITFULL, APETALA1 and CAULIFLOWER. *Development*, *127*, 725–734.

Finnegan, E. J., & Dennis, E. S. (2007). Vernalization-induced trimethylation of histone H3 lysine 27 at FLC is not maintained in mitotically quiescent cells. *Current Biology*, *17*, 1978–1983.

Flachowsky, H., Hanke, M., Peil, A., Strauss, S. H., & Fladung, M. (2009). A review on transgenic approaches to accelerate breeding of woody plants. *Plant Breeding*, *128*, 217–226.

Flachowsky, H., Peil, A., Sopanen, T., Elo, A., & Hanke, V. (2007). Overexpression of BpMADS4 from silver birch (*Betula pendula* Roth.) induces early-flowering in apple (*Malus domestica Borkh.*). *Plant Breeding*, *126*, 137–145.

Fornara, F., Panigrahi, K. C., Gissot, L., Sauerbrunn, N., Ruhl, M., & Jarillo, J. A., et al. (2009). *Arabidopsis* DOF transcription factors act redundantly to reduce CONSTANS expression and are essential for a photoperiodic flowering response. *Developmental Cell*, *17*, 75–86.

Foster, T., Johnston, R., & Seleznyova, A. (2003). A morphological and quantitative characterization of early floral development in apple (*Malus x domestica Borkh.*). *Annals of Botany*, *92*, 199–206.

Franco-Zorrilla, J. M., Valli, A., Todesco, M., Mateos, I., Puga, M. I., & Rubio-Somoza, I., et al. (2007). Target mimicry provides a new mechanism for regulation of microRNA activity. *Nature Genetics*, *39*, 1033–1037.

Fujiwara, S., Oda, A., Yoshida, R., Niinuma, K., Miyata, K., & Tomozoe, Y., et al. (2008). Circadian clock proteins LHY and CCA1 regulate SVP protein accumulation to control flowering in *Arabidopsis*. *Plant Cell*, *20*, 2960–2971.

Gandikota, M., Birkenbihl, R. P., Hohmann, S., Cardon, G. H., Saedler, H., & Huijser, P. (2007). The miRNA156/157 recognition element in the 3' UTR of the *Arabidopsis* SBP box gene SPL3 prevents early flowering by translational inhibition in seedlings. *The Plant Journal*, *49*, 683–693.

Garner, W. W., & Allard, H. A. (1920). Effects of the relative length of night and day and other factors of the environment on growth and reproduction in plants. *Journal of Agricultural Research*, *18*, 553–606.

Gleave, A., Ampomah-Dwamena, C., Berthold, S., Dejnoprat, S., Karunairetnam, S., & Nain, B., et al. (2008). Identification and characterisation of primary microRNAs from apple (*Malus domestica* cv. Royal Gala) expressed sequence tags. *Tree Genetics & Genomes*, *4*, 343–358.

Goldschmidt, E. E., & Monselise, S. P. (1972). Hormonal control of flowering in *Citrus* and some other woody perennials. In D. J. Carr (Ed.), *Plant growth substances* (pp. 758–766). Berlin/Heidelberg/New York: Springer-Verlag.

Gomez, L. D., Gilday, A., Feil, R., Lunn, J. E., & Graham, I. A. (2010). AtTPS1-mediated trehalose 6-phosphate synthesis is essential for embryogenic and vegetative growth and responsiveness to ABA in germinating seeds and stomatal guard cells. *The Plant Journal*, *64*, 1–13.

Greb, T., Mylne, J. S., Crevillen, P., Geraldo, N., An, H., & Gendall, A. R., et al. (2007). The PHD finger protein VRN5 functions in the epigenetic silencing of *Arabidopsis* FLC. *Current Biology*, *17*, 73–78.

Hamant, O., & Pautot, V. (2010). Plant development: A TALE story. *Comptes Rendus Biologies D veloppement v g tatif des plantes, Georges Pelletier, Jean-Fran ois Morot-Gaudy*, *333*, 371–381.

Hardie, D. G. (2007). AMP-activated/SNF1 protein kinases: Conserved guardians of cellular energy. *Nature Reviews Molecular Cell Biology*, *8*, 774–785.

Harmer, S. L. (2009). The circadian system in higher plants. *Annual Review of Plant Biology, 60*, 357–377.

Hartmann, H., & Porlingis, I. (1953). Effect of winter chilling on fruitfulness and vegetative growth in the olive. *Proceedings of the National Academy of Sciences United States of America, 62*, 184–190.

Hartmann, H., & Porlingis, I. (1957). Effect of different amounts of winter chilling on fruitfulness of several olive varieties. *Botanical Gazette, 119*, 102–104.

Hartmann, H. T. (1951). Time of floral differentiation of the olive in California. *Botanical Gazette, 112*, 323–327.

Hartmann, H. T., & Whisler, J. E. (1975). Flower production in olive as influenced by various chilling temperature regimes. *Journal of the American Society for Horticultural Science, 100*, 670–674.

Hartmann, U., Hohmann, S., Nettesheim, K., Wisman, E., Saedler, H., & Huijser, P. (2000). Molecular cloning of SVP: A negative regulator of the floral transition in *Arabidopsis*. *Plant Journal, 21*, 351–360.

Hattasch, C., Flachowsky, H., Kapturska, D., & Hanke, M. -V. (2008). Isolation of flowering genes and seasonal changes in their transcript levels related to flower induction and initiation in apple (*Malus domestica*). *Tree Physiology, 28*, 1459–1466.

Hayama, R., Agashe, B., Luley, E., King, R., & Coupland, G. (2007). A circadian rhythm set by dusk determines the expression of FT homologs and the short-day photoperiodic flowering response in Pharbitis. *Plant Cell, 19*, 2988–3000.

Hayama, R., Yokoi, S., Tamaki, S., Yano, M., & Shimamoto, K. (2003). Adaptation of photoperiodic control pathways produces short-day flowering in rice. *Nature, 422*, 719–722.

Hazen, S. P., Naef, F., Quisel, T., Gendron, J. M., Chen, H., & Ecker, J. R., et al. (2009). Exploring the transcriptional landscape of plant circadian rhythms using genome tiling arrays. *Genome Biology, 10*, R17.

Helliwell, C. A., Wood, C. C., Robertson, M., James Peacock, W., & Dennis, E. S. (2006). The *Arabidopsis* FLC protein interacts directly *in vivo* with SOC1 and FT chromatin and is part of a high-molecular-weight protein complex. *The Plant Journal, 46*, 183–192.

Henriques, R., Jang, I. C., & Chua, N. H. (2009). Regulated proteolysis in light-related signaling pathways. *Current Opinion in Plant Biology, 12*, 49–56.

Hepworth, S. R., Valverde, F., Ravenscroft, D., Mouradov, A., & Coupland, G. (2002). Antagonistic regulation of flowering-time gene SOC1 by CONSTANS and FLC via separate promoter motifs. *The EMBO Journal, 21*, 4327–4337.

Hoad, G. V. (1978). The role of seed derived hormones in the control of flowering in apple. *Acta Horticulturae, 80*, 93–103.

Howie, P. W., & McNeily, A. S. (1982). Effect of breast-feeding patterns on human birth intervals. *Journal of Reproduction & Fertility, 65*, 545–557.

Hsu, C. Y., Liu, Y., Luthe, D. S., & Yuceer, C. (2006). Poplar FT2 shortens the juvenile phase and promotes seasonal flowering. *Plant Cell, 18*, 1846–1861.

Ikeda, Y., Kobayashi, Y., Yamaguchi, A., Abe, M., & Araki, T. (2007). Molecular basis of late-flowering phenotype caused by dominant epi-alleles of the FWA locus in *Arabidopsis*. *Plant and Cell Physiology, 48*, 205–220.

Imaizumi, T., Schultz, T. F., Harmon, F. G., Ho, L. A., & Kay, S. A. (2005). FKF1 F-box protein mediates cyclic degradation of a repressor of CONSTANS in *Arabidopsis*. *Science, 309*, 293–297.

Immink, R. G., Tonaco, I. A., de Folter, S., Shchennikova, A., van Dijk, A. D., & Busscher-Lange, J., et al. (2009). SEPALLATA3: The "glue" for MADS box transcription factor complex formation. *Genome Biology, 10*, R24.

Ishikawa, R., Tamaki, S., Yokoi, S., Inagaki, N., Shinomura, T., & Takano, M., et al. (2005). Suppression of the floral activator Hd3a is the principal cause of the night break effect in rice. *Plant Cell, 17*, 3326–3336.

Jaeger, K. E., & Wigge, P. A. (2007). FT protein acts as a long-range signal in *Arabidopsis*. *Current Biology, 17*, 1050–1054.

Jang, S., Torti, S., & Coupland, G. (2009). Genetic and spatial interactions between FT, TSF and SVP during the early stages of floral induction in Arabidopsis. *Plant Journal, 60*, 614–625.

Jiang, D., Gu, X., & He, Y. (2009). Establishment of the winter-annual growth habit via FRIGIDA-mediated histone methylation at FLOWERING LOCUS C in *Arabidopsis*. *Plant Cell, 21*, 1733–1746.

Jiang, D., Wang, Y., & He, Y. (2008). Repression of FLOWERING LOCUS C and FLOWERING LOCUS T by the *Arabidopsis* Polycomb repressive complex 2 components. *PLoS One, 3*, e3404.

Johanson, U., West, J., Lister, C., Michaels, S., Amasino, R., & Dean, C. (2000). Molecular analysis of FRIGIDA, a major determinant of natural variation in *Arabidopsis* flowering time. *Science, 290*, 344–347.

Jonkers, H. (1979). Biennial bearing in apple and pear: A literature survey. *Scientia Horticulturae, 11*, 303–317.

Jung, C., & Muller, A. E. (2009). Flowering time control and applications in plant breeding. *Trends in Plant Science, 14*, 563–573.

Jung, J. H., Seo, Y. H., Seo, P. J., Reyes, J. L., Yun, J., & Chua, N. H., et al. (2007). The GIGANTEA-regulated microRNA172 mediates photoperiodic flowering independent of CONSTANS in *Arabidopsis*. *Plant Cell, 19*, 2736–2748.

Kane, N. A., Danyluk, J., Tardif, G., Ouellet, F., Laliberte, J. F., & Limin, A. E., et al. (2005). TaVRT-2, a member of the StMADS-11 clade of flowering repressors, is regulated by vernalization and photoperiod in wheat. *Plant Physiology, 138*, 2354–2363.

Kanrar, S., Bhattacharya, M., Arthur, B., Courtier, J., & Smith, H. M. (2008). Regulatory networks that function to specify flower meristems require the function of homeobox genes PENNYWISE and POUND-FOOLISH in *Arabidopsis*. *Plant Journal, 54*, 924–937.

Kanrar, S., Onguka, O., & Smith, H. M. (2006). *Arabidopsis* inflorescence architecture requires the activities of KNOX-BELL homeodomain heterodimers. *Planta, 224*, 1163–1173.

Kardailsky, I., Shukla, V. K., Ahn, J. H., Dagenais, N., Christensen, S. K., & Nguyen, J. T., et al. (1999). Activation tagging of the floral inducer FT. *Science, 286*, 1962–1965.

Kaufmann, K., Muino, J. M., Jauregui, R., Airoldi, C. A., Smaczniak, C., & Krajewski, P., et al. (2009). Target genes of the MADS transcription factor SEPALLATA3: Integration of developmental and hormonal pathways in the *Arabidopsis* flower. *PLoS Biology, 7*, e1000090.

Kaufmann, K., Wellmer, F., Muino, J. M., Ferrier, T., Wuest, S. E., & Kumar, V., et al. (2010). Orchestration of floral initiation by APETALA1. *Science, 328*, 85–89.

Kempin, S. A., Savidge, B., & Yanofsky, M. F. (1995). Molecular basis of the cauliflower phenotype in *Arabidopsis*. *Science, 267*, 522–525.

King, R. W., Moritz, T., Evans, L. T., Martin, J., Andersen, C. H., & Blundell, C., et al. (2006). Regulation of flowering in the long-day grass *Lolium temulentum* by gibberellins and the FLOWERING LOCUS T gene. *Plant Physiology, 141*, 498–507.

Kitsaki, C. K., Andreadis, E., & Bouranis, D. L. (2010). Developmental events in differentiating floral buds of four olive (*Olea europaea* L.) cultivars during late winter to early spring. *Flora, 205*, 599–607.

Klebs, G. (1918). Uber die Blutentbildung bei Sempervivum. *Flora (Jena), 128*, 111–112.

Knight, H., Thomson, A. J. W., & McWatters, H. G. (2008). SENSITIVE TO FREEZING6 Integrates cellular and environmental inputs to the plant circadian Clock. *Plant Physiology, 148*, 293–303.

Knott, J. E. (1934). Effect of localized photoperiod on spinach. *Proceedings of the Society for Horticultural Science, 31*, 152–154.

Ko, J. H., Mitina, I., Tamada, Y., Hyun, Y., Choi, Y., & Amasino, R. M., et al. (2010). Growth habit determination by the balance of histone methylation activities in *Arabidopsis*. *Embo Journal, 29*, 3208–3215.

Kobayashi, Y., Kaya, H., Goto, K., Iwabuchi, M., & Araki, T. (1999). A pair of related genes with antagonistic roles in mediating flowering signals. *Science, 286*, 1960–1962.

Kojima, S., Takahashi, Y., Kobayashi, Y., Monna, L., Sasaki, T., & Araki, T., et al. (2002). Hd3a, a rice ortholog of the *Arabidopsis* FT Gene, promotes transition to flowering downstream of Hd1 under short-day conditions. *Plant and Cell Physiology, 43*, 1096–1105.

Koornneef, M., Hanhart, C. J., & van der Veen, J. H. (1991). A genetic and physiological analysis of late flowering mutants in *Arabidopsis thaliana*. *Molecular & General Genetics, 229*, 57–66.

Kotoda, N., Hayashi, H., Suzuki, M., Igarashi, M., Hatsuyama, Y., & Kidou, S. -i., et al. (2010). Molecular characterization of FLOWERING LOCUS T-like genes of apple (*Malus domestica Borkh.*). *Plant and Cell Physiology, 51*, 561–575.

Kotoda, N., Iwanami, H., Takahashi, S., & Abe, K. (2006). Antisense expression of MdTFL1, a TFL1-like gene, reduces the juvenile phase in apple. *Journal of the American Society Horticultural Science, 131*, 74–81.

Kotoda, N., & Wada, M. (2005). MdTFL1, a TFL1-like gene of apple, retards the transition from the vegetative to reproductive phase in transgenic *Arabidopsis*. *Plant Science, 168*, 95–104.

Kotoda, N., Wada, M., Komori, S., Kidou, S. -i., Abe, K., & Masuda, T., et al. (2000). Expression pattern of homologues of floral meristem identity genes LFY and AP1 during flower development in apple. *Journal of the American Society for Horticultural Science, 125*, 398–403.

Krieger, U., Lippman, Z. B., & Zamir, D. (2010). The flowering gene SINGLE FLOWER TRUSS drives heterosis for yield in tomato. *Nature Genetics, 42*, 459–463.

Kumar, S. V., & Wigge, P. A. (2010). H2A.Z-containing nucleosomes mediate the thermosensory response in *Arabidopsis*. *Cell, 140*, 136–147.

Kumimoto, R. W., Adam, L., Hymus, G. J., Repetti, P. P., Reuber, T. L., & Marion, C. M., et al. (2008). The nuclear factor Y subunits NF-YB2 and NF-YB3 play additive roles in the promotion of flowering by inductive long-day photoperiods in *Arabidopsis*. *Planta, 228*, 709–723.

Kumimoto, R. W., Zhang, Y., Siefers, N., & Holt, B. F. (2010). NF-YC3, NF-YC4 and NF-YC9 are required for CONSTANS-mediated, photoperiod-dependent flowering in *Arabidopsis thaliana*. *The Plant Journal, 63*, 379–391.

Lavee, S. (2007). Biennial bearing in olive (*Olea europea*). *Annales Ser Hist Nat, 17*, 101–112.

Lee, H., Suh, S. S., Park, E., Cho, E., Ahn, J. H., & Kim, S. G., et al. (2000). The AGAMOUS-LIKE 20 MADS domain protein integrates floral inductive pathways in *Arabidopsis*. *Genes & Development, 14*, 2366–2376.

Lee, J., & Lee, I. (2010). Regulation and function of SOC1, a flowering pathway integrator. *Journal of Experimental Botany, 61*, 2247–2254.

Lee, J., Oh, M., Park, H., & Lee, I. (2008). SOC1 translocated to the nucleus by interaction with AGL24 directly regulates leafy. *Plant Journal, 55*, 832–843.

Lee, J. H., Park, S. H., Lee, J. S., & Ahn, J. H. (2007b). A conserved role of SHORT VEGETATIVE PHASE (SVP) in controlling flowering time of Brassica plants. *Biochimica et Biophysica Acta, 1769*, 455–461.

Lee, J. H., Yoo, S. J., Park, S. H., Hwang, I., Lee, J. S., & Ahn, J. H. (2007b). Role of SVP in the control of flowering time by ambient temperature in *Arabidopsis*. *Genes & Development, 21*, 397–402.

Li, C., & Dubcovsky, J. (2008). Wheat FT protein regulates VRN1 transcription through interactions with FDL2. *Plant Journal, 55*, 543–554.

Li, C. Y., Weiss, D., & Goldschmidt, E. E. (2003). Girdling affects carbohydrate-related gene expression in leaves, bark and roots of alternate-bearing citrus trees. *Annals of Botany (London), 92*, 137–143.

Li, D., Liu, C., Shen, L., Wu, Y., Chen, H., & Robertson, M., et al. (2008). A repressor complex governs the integration of flowering signals in *Arabidopsis*. *Developmental Cell, 15*, 110–120.

Lifschitz, E., Eviatar, T., Rozman, A., Shalit, A., Goldshmidt, A., & Amsellem, Z., et al. (2006). The tomato FT ortholog triggers systemic signals that regulate growth and flowering and substitute for diverse environmental stimuli. *Proceedings of the National Academy of Sciences of the United States of America, 103*, 6398–6403.

Liljegren, S. J., Gustafson-Brown, C., Pinyopich, A., Ditta, G. S., & Yanofsky, M. F. (1999). Interactions among APETALA1, LEAFY, and TERMINAL FLOWER1 specify meristem fate. *Plant Cell, 11*, 1007–1018.

Lin, M. -K., Belanger, H., Lee, Y. -J., Varkonyi-Gasic, E., Taoka, K. -I., & Miura, E., et al. (2007). FLOWERING LOCUS T protein may act as the long-distance florigenic signal in the cucurbits. *Plant Cell, 19*, 1488–1506.

Liu, C., Chen, H., Er, H. L., Soo, H. M., Kumar, P. P., & Han, J. -H., et al. (2008a). Direct interaction of AGL24 and SOC1 integrates flowering signals in *Arabidopsis*. *Development, 135*, 1481–1491.

Liu, C., Thong, Z., & Yu, H. (2009a). Coming into bloom: The specification of floral meristems. *Development, 136*, 3379–3391.

Liu, C., Xi, W., Shen, L., Tan, C., & Yu, H. (2009b). Regulation of floral patterning by flowering time genes. *Developmental Cell, 16*, 711–722.

Liu, C., Zhou, J., Bracha-Drori, K., Yalovsky, S., Ito, T., & Yu, H. (2007). Specification of *Arabidopsis* floral meristem identity by repression of flowering time genes. *Development, 134*, 1901–1910.

Liu, L. -J., Zhang, Y. -C., Li, Q. -H., Sang, Y., Mao, J., & Lian, H. -L., et al. (2008b). COP1-mediated ubiquitination of CONSTANS is implicated in cryptochrome regulation of flowering in *Arabidopsis*. *Plant Cell, 20*, 292–306.

Mandel, M. A., & Yanofsky, M. F. (1998). The *Arabidopsis* AGL9 MADS box gene is expressed in young flower primordia. *Sexual Plant Reproduction, 11*, 22–28.

Marino, F., & Greene, D. W. (1981). Involvement of gibberellins in the biennial bearing of "Early Mcintosh" apples. *Journal of the American Society for Horticultural Science, 106*, 593–596.

Martin-Trillo, M., Lazaro, A., Poethig, R. S., Gomez-Mena, C., Pineiro, M. A., & Martinez-Zapater, J. M., et al. (2006). EARLY IN SHORT DAYS 1 (ESD1) encodes ACTIN-RELATED PROTEIN 6 (AtARP6), a putative component of chromatin remodelling complexes that positively regulates FLC accumulation in *Arabidopsis*. *Development, 133*, 1241–1252.

Mas, P., & Yanovsky, M. J. (2009). Time for circadian rhythms: Plants get synchronized. *Current Opinion in Plant Biology, 12*, 574–579.

Mathieu, J., Warthmann, N., Kuttner, F., & Schmid, M. (2007). Export of FT protein from phloem companion cells is sufficient for floral induction in *Arabidopsis*. *Current Biology, 17*, 1055–1060.

Mathieu, J., Yant, L. J., Murdter, F., Kuttner, F., & Schmid, M. (2009). Repression of flowering by the miR172 target SMZ. *PLoS Biology, 7*, e1000148.

Matsuda, N., Ikeda, K., Kurosaka, M., Takashina, T., Isuzugawa, K., & Endo, T., et al. (2009). Early flowering phenotype in transgenic pears (*Pyrus communis* l.) expressing the cift gene. *Journal of the Japanese Society for Horticultural Science, 78*, 410–416.

Melzer, S., Lens, F., Gennen, J., Vanneste, S., Rohde, A., & Beeckman, T. (2008). Flowering-time genes modulate meristem determinacy and growth form in *Arabidopsis thaliana*. *Nature Genetics, 40*, 1489–1492.

Menzel, G., Apel, K., & Melzer, S. (1996). Identification of two MADS box genes that are expressed in the apical meristem of the long-day plant *Sinapis alba* in transition to flowering. *Plant Journal, 9*, 399–408.

Michaels, S. D. (2009). Flowering time regulation produces much fruit. *Current Opinion in Plant Biology, 12*, 75–80.

Michaels, S. D., & Amasino, R. M. (1999). FLOWERING LOCUS C encodes a novel MADS domain protein that acts as a repressor of flowering. *Plant Cell, 11*, 949–956.

Michaels, S. D., Ditta, G., Gustafson-Brown, C., Pelaz, S., Yanofsky, M., & Amasino, R. M. (2003). AGL24 acts as a promoter of flowering in *Arabidopsis* and is positively regulated by vernalization. *Plant Journal, 33*, 867–874.

Michaels, S. D., Himelblau, E., Kim, S. Y., Schomburg, F. M., & Amasino, R. M. (2005). Integration of flowering signals in winter-annual *Arabidopsis*. *Plant Physiology, 137*, 149–156.

Monselise, S. P., & Goldschmidt, E. E. (1982). Alternate bearing in fruit trees. *Horticultural Reviews, 4*, 128–173.

Moon, J., Suh, S. S., Lee, H., Choi, K. R., Hong, C. B., & Paek, N. C., et al. (2003). The SOC1 MADS-box gene integrates vernalization and gibberellin signals for flowering in *Arabidopsis*. *Plant Journal, 35*, 613–623.

Moreno-Alias, I., Rapoport, H. F., Leon, L., & de la Rosa, R. (2010). Olive seedling first-flowering position and management. *Scientia Horticulturae, 124*, 74–77.

Nave, N., Katz, E., Chayut, N., Gazit, S. Samach, A. (2010). Flower development in the passion fruit *Passiflora edulis* requires a photoperiod-induced systemic graft-transmissible signal. *Plant Cell Environment* DOI: 10.1111/j.1365-3040.2010.02206.x.

Neilsen, J. C., & Dennis, F. G. J. (2000). Effects of seed number, fruit removal, bourse shoot length and crop density on flowering in "Spencer Seedless" apple. *Acta Horticulturae, 527*, 137–146.

Ohto, M., Onai, K., Furukawa, Y., Aoki, E., Araki, T., & Nakamura, K. (2001). Effects of sugar on vegetative development and floral transition in *Arabidopsis*. *Plant Physiology, 127*, 252–261.

Oparka, K. J., & Cruz, S. S. (2000). THE GREAT ESCAPE: Phloem transport and unloading of macromolecules. *Annual Review of Plant Physiology and Plant Molecular Biology*, *51*, 323–347.

Orlandi, F., Vazquez, L. M., Ruga, L., Bonofiglio, T., Fornaciari, M., & Garcia-Mozo, H., et al. (2005). Bioclimatic requirements for olive flowering in two Mediterranean regions located at the same latitude (Andalucia, Spain and Sicily, Italy). *Annals of Agricultural and Environmental Medicine*, *12*, 47–52.

Paltiel, J., Amin, R., Gover, A., Ori, N., & Samach, A. (2006). Novel roles for GIGANTEA revealed under environmental conditions that modify its expression in *Arabidopsis* and *Medicago truncatula*. *Planta*, *224*, 1255–1268.

Pelaz, S., Gustafson-Brown, C., Kohalmi, S. E., Crosby, W. L., & Yanofsky, M. F. (2001). APETALA1 and SEPALLATA3 interact to promote flower development. *Plant Journal*, *26*, 385–394.

Pena, L., Martin-Trillo, M., Juarez, J., Pina, J. A., Navarro, L., & Martinez-Zapater, J. M. (2001). Constitutive expression of *Arabidopsis* LEAFY or APETALA1 genes in citrus reduces their generation time. *Nature Biotechnology*, *19*, 263–267.

Pnueli, L., Carmel-Goren, L., Hareven, D., Gutfinger, T., Alvarez, J., & Ganal, M., et al. (1998). The SELF-PRUNING gene of tomato regulates vegetative to reproductive switching of sympodial meristems and is the ortholog of CEN and TFL1. *Development*, *125*, 1979–1989.

Pnueli, L., Gutfinger, T., Hareven, D., Ben-Naim, O., Ron, N., & Adir, N., et al. (2001). Tomato SP-interacting proteins define a conserved signaling system that regulates shoot architecture and flowering. *Plant Cell*, *13*, 2687–2702.

Poethig, R. S. (2003). Phase change and the regulation of developmental timing in plants. *Science*, *301*, 334–336.

Poethig, R. S. (2009). Small RNAs and developmental timing in plants. *Current Opinion in Genetics & Development*, *19*, 374–378.

Proveniers, M., Rutjens, B., Brand, M., & Smeekens, S. (2007). The *Arabidopsis* TALE homeobox gene ATH1 controls floral competency through positive regulation of FLC. *The Plant Journal*, *52*, 899–913.

Purwestri, Y. A., Ogaki, Y., Tamaki, S., Tsuji, H., & Shimamoto, K. (2009). The 14-3-3 protein GF14c acts as a negative regulator of flowering in rice by interacting with the florigen Hd3a. *Plant and Cell Physiology*, *50*, 429–438.

Putterill, J., Robson, F., Lee, K., Simon, R., & Coupland, G. (1995). The CONSTANS gene of *Arabidopsis* promotes flowering and encodes a protein showing similarities to zinc finger transcription factors. *Cell*, *80*, 847–857.

Ramon, M., & Rolland, F. (2007). Plant development: Introducing trehalose metabolism. *Trends in Plant Science*, *12*, 185–188.

Ratcliffe, O., Kumimoto, R. W., Wong, B. J., & Riechmann, J. L. (2003). Analysis of *Arabidopsis* MADS AFFECTING FLOWERING gene family: MAF2 prevents vernalization by short periods of cold. *The Plant Cell*, *15*, 1159–1169.

Ratcliffe, O. J., Amaya, I., Vincent, C. A., Rothstein, S., Carpenter, R., & Coen, E. S., et al. (1998). A common mechanism controls the life cycle and architecture of plants. *Development*, *125*, 1609–1615.

Ratcliffe, O. J., Nadzan, G. C., Reuber, T. L., & Riechmann, J. L. (2001). Regulation of flowering in *Arabidopsis* by an FLC homologue. *Plant Physiology*, *126*, 122–132.

Redei, G. P. (1962). Supervital mutants of *Arabidopsis*. *Genetics*, *47*, 443–460.

Reeves, P. H., & Coupland, G. (2001). Analysis of flowering time control in *Arabidopsis* by comparison of double and triple mutants. *Plant Physiology*, *126*, 1085–1091.

Roldan, M., Gomez-Mena, C., Ruiz-Garcia, L., Salinas, J., & Martinez-Zapater, J. M. (1999). Sucrose availability on the aerial part of the plant promotes morphogenesis and flowering of *Arabidopsis* in the dark. *Plant Journal*, *20*, 581–590.

Rottmann, W. H., Meilan, R., Sheppard, L. A., Brunner, A. M., Skinner, J. S., & Ma, C., et al. (2000). Diverse effects of overexpression of LEAFY and PTLF, a poplar (*Populus*) homolog of LEAFY/FLORICAULA, in transgenic poplar and *Arabidopsis*. *Plant Journal*, *22*, 235–245.

Samach, A., & Coupland, G. (2000). Time measurement and the control of flowering in plants. *BioEssays*, *22*, 38–47.

Samach, A., Onouchi, H., Gold, S. E., Ditta, G. S., Schwarz-Sommer, Z., & Yanofsky, M. F., et al. (2000). Distinct roles of CONSTANS target genes in reproductive development of *Arabidopsis*. *Science*, *288*, 1613–1616.

Santner, A., & Estelle, M. (2010). The ubiquitin-proteasome system regulates plant hormone signaling. *The Plant Journal*, *61*, 1029–1040.

Santos-Antunes, F., Leon, L., de la Rosa, R., Alvarado, J., Mohedo, A., & Trujillo, I., et al. (2005). The length of the juvenile period in olive as influenced by vigor of the seedlings and precocity of the parents. *HortScience*, *40*, 1213–1215.

Sauer, N. (2007). Molecular physiology of higher plant sucrose transporters. *FEBS Letters*, *581*, 2309–2317.

Schmid, M., Uhlenhaut, N. H., Godard, F., Demar, M., Bressan, R., & Weigel, D., et al. (2003). Dissection of floral induction pathways using global expression analysis. *Development*, *130*, 6001–6012.

Schonrock, N., Bouveret, R., Leroy, O., Borghi, L., Kohler, C., & Gruissem, W., et al. (2006). Polycomb-group proteins repress the floral activator AGL19 in the FLC-independent vernalization pathway. *Genes & Development*, *20*, 1667–1678.

Schwab, R., Ossowski, S., Riester, M., Warthmann, N., & Weigel, D. (2006). Highly specific gene silencing by artificial MicroRNAs in *Arabidopsis*. *Plant Cell*, *18*, 1121–1133.

Schwab, R., Palatnik, J. F., Riester, M., Schommer, C., Schmid, M., & Weigel, D. (2005). Specific effects of MicroRNAs on the plant transcriptome. *Developmental Cell*, *8*, 517–527.

Scortecci, K., Michaels, S. D., & Amasino, R. M. (2003). Genetic interactions between FLM and other flowering-time genes in *Arabidopsis thaliana*. *Plant Molecular Biology*, *52*, 915–922.

Scortecci, K. C., Michaels, S. D., & Amasino, R. M. (2001). Identification of a MADS-box gene, FLOWERING LOCUS M, that represses flowering. *Plant Journal*, *26*, 229–236.

Searle, I., He, Y., Turck, F., Vincent, C., Fornara, F., & Krober, S., et al. (2006). The transcription factor FLC confers a flowering response to vernalization by repressing meristem competence and systemic signaling in *Arabidopsis*. *Genes & Development*, *20*, 898–912.

Shannon, S., & Meeks-Wagner, D. R. (1991). A mutation in the *Arabidopsis* tfl1 gene affects inflorescence meristem development. *Plant Cell*, *3*, 877–892.

Sheldon, C. C., Burn, J. E., Perez, P. P., Metzger, J., Edwards, J. A., & Peacock, W. J., et al. (1999). The FLF MADS box gene: A repressor of flowering in *Arabidopsis* regulated by vernalization and methylation. *Plant Cell*, *11*, 445–458.

Sivitz, A. B., Reinders, A., Johnson, M. E., Krentz, A. D., Grof, C. P., & Perroux, J. M., et al. (2007). *Arabidopsis* sucrose transporter AtSUC9. High-affinity transport activity, intragenic control of expression, and early flowering mutant phenotype. *Plant Physiology*, *143*, 188–198.

Smeekens, S., Ma, J., Hanson, J., & Rolland, F. (2010). Sugar signals and molecular networks controlling plant growth. *Current Opinion in Plant Biology*, *13*, 273–278.

Smith, H. M. S., Boschke, I., & Hake, S. (2002). Selective interaction of plant homeodomain proteins mediates high DNA-binding affinity. *Proceedings of the National Academy of Sciences of the United States of America*, *99*, 9579–9584.

Smith, H. M. S., Campbell, B. C., & Hake, S. (2004). Competence to respond to floral inductive signals requires the homeobox genes PENNYWISE and POUND-FOOLISH. *Current Biology*, *14*, 812–817.

Smyth, D. R., Bowman, J. L., & Meyerowitz, E. M. (1990). Early flower development in *Arabidopsis*. *Plant Cell*, *2*, 755–767.

Soppe, W. J., Jacobsen, S. E., Alonso-Blanco, C., Jackson, J. P., Kakutani, T., & Koornneef, M., et al. (2000). The late flowering phenotype of fwa mutants is caused by gain-of- function epigenetic alleles of a homeodomain gene. *Molecular Cell*, *6*, 791–802.

Stutte, G. W., & Martin, G. C. (1986). Effect of killing the seed on return bloom of olive. *Scientia Horticulturae*, *29*, 107–111.

Suarez-Lopez, P., Wheatley, K., Robson, F., Onouchi, H., Valverde, F., & Coupland, G. (2001). CONSTANS mediates between the circadian clock and the control of flowering in *Arabidopsis*. *Nature*, *410*, 1116–11120.

Sung, S., & Amasino, R. M. (2004). Vernalization in *Arabidopsis thaliana* is mediated by the PHD finger protein VIN3. *Nature*, *427*, 159–164.

Sung, S. -K., Yu, G. -H., & An, G. (1999). Characterization of MdMADS2, a member of the SQUAMOSA subfamily of genes, in apple. *Plant Physiology, 120*, 969–978.

Swiezewski, S., Crevillen, P., Liu, F., Ecker, J. R., Jerzmanowski, A., & Dean, C. (2007). Small RNA-mediated chromatin silencing directed to the 3' region of the *Arabidopsis* gene encoding the developmental regulator, FLC. *Proceedings of the National Academy of Sciences of the United States of America, 104*, 3633–3638.

Tabata, R., Ikezaki, M., Fujibe, T., Aida, M., Tian, C. E., & Ueno, Y., et al. (2010). *Arabidopsis* auxin response factor6 and 8 regulate jasmonic acid biosynthesis and floral organ development via repression of class 1 KNOX genes. *Plant and Cell Physiology, 51*, 164–175.

Takada, S., & Goto, K. (2003). TERMINAL FLOWER2, an *Arabidopsis* homolog of HETEROCHROMATIN PROTEIN1, counteracts the activation of FLOWERING LOCUS T by CONSTANS in the vascular tissues of leaves to regulate flowering time. *Plant Cell, 15*, 2856–2865.

Tamaki, S., Matsuo, S., Wong, H. L., Yokoi, S., & Shimamoto, K. (2007). Hd3a protein is a mobile flowering signal in rice. *Science, 316*, 1033–1036.

Teper-Bamnolker, P., & Samach, A. (2005). The flowering integrator FT regulates SEPALLATA3 and FRUITFULL accumulation in *Arabidopsis* leaves. *Plant Cell, 17*, 2661–2675.

Tiwari, S. B., Shen, Y., Chang, H. C., Hou, Y., Harris, A., & Ma, S. F., et al. (2010). The flowering time regulator CONSTANS is recruited to the FLOWERING LOCUS T promoter via a unique cis-element. *The New Phytologist, 187*, 57–66.

Todesco, M., Rubio-Somoza, I., Paz-Ares, J., & Weigel, D. (2010). A collection of target mimics for comprehensive analysis of microRNA function in *Arabidopsis thaliana*. *PLoS Genetics, 6*, e1001031.

Tournois, J. (1914). Etudes sur la sexualite du houblon. *Annals des Sciences Naturelles, 19*, 49–191.

Trevaskis, B., Hemming, M. N., Peacock, W. J., & Dennis, E. S. (2006). HvVRN2 responds to daylength, whereas HvVRN1 is regulated by vernalization and developmental status. *Plant Physiology, 140*, 1397–1405.

Tromp, J. (2000). Flower-bud formation in pome fruits as affected by fruit thinning. In *Plant Growth Regulation* (Vol. 31, pp. 27–34). Netherlands: Springer.

Turck, F., Fornara, F., & Coupland, G. (2008). Regulation and identity of florigen: FLOWERING LOCUS T moves center stage. *Annual Review of Plant Biology, 59*, 573–594.

Turck, F., Roudier, F., Farrona, S., Martin-Magniette, M. L., Guillaume, E., & Buisine, N., et al. (2007). *Arabidopsis* TFL2/LHP1 specifically associates with genes marked by trimethylation of histone H3 lysine 27. *PLoS Genetics, 3*, e86.

Turgeon, R., & Wolf, S. (2009). Phloem transport: Cellular pathways and molecular trafficking. *Annual Review of Plant Biology, 60*, 207–221.

Valantin, M., Gary, C., Vassiere, B. E. & Frossard, J. S. (1999). Effect of fruit load on partitioning of dry matter and Energy in cantaloupe (*Cucumis melo* L.). *Annals of Botany., 84*(2), 173–181.

Valverde, F., Mouradov, A., Soppe, W., Ravenscroft, D., Samach, A., & Coupland, G. (2004). Photoreceptor regulation of CONSTANS protein in photoperiodic flowering. *Science, 303*, 1003–1006.

van Dijken, A. J., Schluepmann, H., & Smeekens, S. C. (2004). *Arabidopsis* trehalose-6-phosphate synthase 1 is essential for normal vegetative growth and transition to flowering. *Plant Physiology, 135*, 969–977.

Veley, K. M., & Michaels, S. D. (2008). Functional redundancy and new roles for genes of the autonomous floral-promotion pathway. *Plant Physiology, 147*, 682–695.

Voinnet, O. (2009). Origin, biogenesis, and activity of plant MicroRNAs. *Cell, 136*, 669–687.

Wada, M., Cao, Q. F., Kotoda, N., Soejima, J., & Masuda, T. (2002). Apple has two orthologues of FLORICAULA/LEAFY involved in flowering. *Plant Molecular Biology, 49*, 567–577.

Wagner, D., Sablowski, R. W., & Meyerowitz, E. M. (1999). Transcriptional activation of APETALA1 by LEAFY. *Science, 285*, 582–584.

Wang, J. W., Czech, B., & Weigel, D. (2009a). miR156-regulated SPL transcription factors define an endogenous flowering pathway in *Arabidopsis thaliana*. *Cell, 138*, 738–749.

Wang, R., Farrona, S., Vincent, C., Joecker, A., Schoof, H., & Turck, F., et al. (2009b). PEP1 regulates perennial flowering in *Arabis alpina*. *Nature, 459*, 423–427.

Weigel, D., Alvarez, J., Smyth, D. R., Yanofsky, M. F., & Meyerowitz, E. M. (1992). LEAFY controls floral meristem identity in *Arabidopsis*. *Cell, 69*, 843–859.

Weigel, D., & Nilsson, O. (1995). A developmental switch sufficient for flower initiation in diverse plants. *Nature, 377*, 495–500.

Wellensiek, S. J. (1962). Dividing cells as the locus for vernalization. *Nature, 195*, 307–308.

Wenkel, S., Turck, F., Singer, K., Gissot, L., Le Gourrierec, J., & Samach, A., et al. (2006). CONSTANS and the CCAAT box binding complex share a functionally important domain and interact to regulate flowering of *Arabidopsis*. *Plant Cell, 18*, 2971–2984.

Werner, J. D., Borevitz, J. O., Warthmann, N., Trainer, G. T., Ecker, J. R., & Chory, J., et al. (2005). Quantitative trait locus mapping and DNA array hybridization identify an FLM deletion as a cause for natural flowering-time variation. *Proceedings of the National Academy of Sciences of the United States of America, 102*, 2460–2465.

Wigge, P. A., Kim, M. C., Jaeger, K. E., Busch, W., Schmid, M., & Lohmann, J. U., et al. (2005). Integration of spatial and temporal information during floral induction in *Arabidopsis*. *Science, 309*, 1056–1059.

Wu, G., Park, M. Y., Conway, S. R., Wang, J. W., Weigel, D., & Poethig, R. S. (2009). The sequential action of miR156 and miR172 regulates developmental timing in *Arabidopsis*. *Cell, 138*, 750–759.

Wu, G., & Poethig, R. S. (2006). Temporal regulation of shoot development in *Arabidopsis thaliana* by miR156 and its target SPL3. *Development, 133*, 3539–3547.

Xie, K., Wu, C., & Xiong, L. (2006). Genomic organization, differential expression, and interaction of SQUAMOSA promoter-binding-like transcription factors and microRNA156 in rice. *Plant Physiology, 142*, 280–293.

Yakir, E., Hilman, D., Kron, I., Hassidim, M., Melamed-Book, N., & Green, R. M. (2009). Posttranslational regulation of CIRCADIAN CLOCK ASSOCIATED1 in the circadian oscillator of *Arabidopsis*. *Plant Physiology, 150*, 844–857.

Yamaguchi, A., Kobayashi, Y., Goto, K., Abe, M., & Araki, T. (2005). TWIN SISTER OF FT (TSF) acts as a floral pathway integrator redundantly with FT. *Plant and Cell Physiology, 46*, 1175–1189.

Yan, L., Fu, D., Li, C., Blechl, A., Tranquilli, G., & Bonafede, M., et al. (2006). The wheat and barley vernalization gene VRN3 is an orthologue of FT. *Proceedings of the National Academy of Sciences of the United States of America, 103*, 19581–19586.

Yan, L., Loukoianov, A., Blechl, A., Tranquilli, G., Ramakrishna, W., & SanMiguel, P., et al. (2004). The wheat VRN2 gene is a flowering repressor down-regulated by vernalization. *Science, 303*, 1640–1644.

Yan, L., Loukoianov, A., Tranquilli, G., Helguera, M., Fahima, T., & Dubcovsky, J. (2003). Positional cloning of the wheat vernalization gene VRN1. *Proceedings of the National Academy of Sciences of the United States of America, 100*, 6263–6268.

Yano, M., Katayose, Y., Ashikari, M., Yamanouchi, U., Monna, L., & Fuse, T., et al. (2000). Hd1, a major photoperiod sensitivity quantitative trait locus in rice, is closely related to the *Arabidopsis* flowering time gene CONSTANS. *Plant Cell, 12*, 2473–2484.

Yano, M., Kojima, S., Takahashi, Y., Lin, H., & Sasaki, T. (2001). Genetic control of flowering time in rice, a short-day plant. *Plant Physiology, 127*, 1425–1429.

Yoo, S. K., Chung, K. S., Kim, J., Lee, J. H., Hong, S. M., & Yoo, S. J., et al. (2005). CONSTANS activates SUPPRESSOR OF OVEREXPRESSION OF CONSTANS 1 through FLOWERING LOCUS T to promote flowering in *Arabidopsis*. *Plant Physiology, 139*, 770–778.

Yu, H., Xu, Y., Tan, E. L., & Kumar, P. P. (2002). AGAMOUS-LIKE 24, a dosage-dependent mediator of the flowering signals. *Proceedings of the National Academy of Sciences of the United States of America, 99*, 16336–16341.

Zeevaart, J. A. (1962). Physiology of flower formation. *Science, 137*, 723–731.

Zhang, X., Clarenz, O., Cokus, S., Bernatavichute, Y. V., Pellegrini, M., & Goodrich, J., et al. (2007). Whole-genome analysis of histone H3 Lysine 27 Trimethylation in *Arabidopsis*. *PLoS Biology, 5*, e129.

Zhang, X., Germann, S., Blus, B. J., Khorasanizadeh, S., Gaudin, V., & Jacobsen, S. E. (2007). The *Arabidopsis* LHP1 protein colocalizes with histone H3 Lys27 trimethylation. *Nature Structural & Molecular Biology*, *14*, 869–871.

Zhang, Y., Primavesi, L. F., Jhurreea, D., Andralojc, P. J., Mitchell, R. A., & Powers, S. J., et al. (2009). Inhibition of SNF1-related protein kinase1 activity and regulation of metabolic pathways by trehalose-6-phosphate. *Plant Physiology*, *149*, 1860–1871.

Fruit development and ripening: a molecular perspective

26

Avtar K. Handa[1] Martín-Ernesto Tiznado-Hernández[1,2]
Autar K. Mattoo[3]

[1]*Purdue University, West Lafayette, Indiana,* [2]*Centro de Investigación en Alimentación y Desarrollo, Sonora, México,* [3]*USDA-ARS, Sustainable Agricultural Systems Laboratory, Beltsville Agricultural Research Center, Beltsville, Maryland*

TABLE OF CONTENTS

Fruit Classification

Botanically, a fruit is derived either from a flower or an inflorescence. Fruits developed from one flower are divisible into two broad categories referred to as simple or aggregate, whereas the inflorescence-derived fruit are considered multiple fruit (Spjut, 1994). However, edible fruits also derived partially or fully from a flower are referred to as accessory fruit. Simple fruit are further divided into fleshy, dry, and dehiscent, and dry and indehiscent categories with each representing several types (Stuppy and Kesseler, 2008). Fleshy fruits include drupes, berries, pome, pepo, and hesperidium. A drupe has one seed, also known as a pit, which arises from hardened endocarp or inner ovary walls (coconut, cherries, olive, peach, avocado), whereas a berry contains many seeds with a thin skin representing the outer casing of the ovary. The common edible berry fruits include blackcurrant, redcurrant, gooseberry, guava, pepper, pomegranate, kiwifruit, grape, cranberry, and blueberry. Pomes, such as apple and pear, have a fleshy hypanthium that primarily originates from an inferior ovary that lies below the attachment of other floral parts. Pepo fruits also form from an inferior ovary, but have a thick leathery rind (pumpkin, squash, melons, watermelon, cucumber, and zucchini). Hesperidium fruits, which include most citrus fruit, have a leathery rind embedded with glands producing volatile compounds.

Dry and dehiscent fruit are subdivided into several types including capsule, silique, follicle, legume, loment, and schizocarp, which open up on maturation to shed seeds. Capsule pods are formed from two or more carpels and, depending on species, shed seeds in a number of ways (nigella, poppy, lily, orchid, willow, cotton, and jimson weed). Silique fruit, whose length is generally at least twice its width, are formed after fusion of two carpels (*Brassicaceae* family). Follicle, a dry unilocular many seeded fruit, results from a single carpel (larkspur, magnolia, banksia, peony, and milkweed). Also, a simple carpel develops into a legume fruit that normally dehisces (alfalfa, clover, peas, beans, lentils, lupins, mesquite, carob, soy, and peanuts). A loment fruit is a modified legume with narrowing between the seed segment and is dehiscent at maturity. A schizocarp fruit is a product of multiple carpels that generally split up into one-seeded mericarps (these include fruits from the genders *Malva*, *Malvastrum*, and *Sida*).

Dry and indehiscent fruit retain their seeds upon maturation and form the third major category of simple fruit, which includes caryopsis, nuts, achene, cypsela, samara, and utricle. The caryopsis fruit, also known as grain, includes major agronomical crops such as rice, wheat, and corn. These indehiscent fruits arise from a single carpel (monocarpelate) with pericarp fused with the seed coat. Nuts develop from the pistil with the inferior ovary. These indehiscent fruit normally contain one seed with the ovary walls becoming hard (woody or stony) at maturity. An achene fruit develops from one

DOI: 10.1016/B978-0-12-381466-1.00026-2

carpel and contains a single seed that nearly fills the pericarp without adhering to it, for example, buttercup, buckwheat, and cannabis. Cypsela is an achene-like fruit derived from an inferior ovary and contains only one locule and one seed per fruit (dandelion and *Helianthus*). Utricles, often compared with a one-seeded achene, comprise a small, bladder-like, thin-walled fruit with pericarp that is loose and fragile, which are rarely seen (e.g., in duckweed). A samara fruit, which develops from the ovary wall, belongs to a simple dry fruit and indehiscent type. Its shape as a flattened wing of fibrous and papery tissue helps the wind to carry the seed farther away from the parent tree (elms, maples, and ashes).

Aggregate fruits are derived from multiple carpels of the same flower where each of the unjoined pistil converts to a fruitlet. Etaerio, a collection of these fruitlets, develops into an aggregate fruit, which is further subdivided into achene, follicle, drupelet, or berry types. Although technically strawberry is an aggregate fruit, it consists of many achenes embedded in a fleshy edible receptacle. Raspberry is derived from separate ovaries within a single flower and consists of many individual small fruitlets attached to a common receptacle. Other examples of aggregate fruits include blackberry and boysenberry. Unlike aggregate fruits, the multiple fruits are derived from an inflorescence (a cluster of flowers). Fruit produced from each flower matures into a single mass. Examples are the pineapple, fig, mulberry, osage-orange, and breadfruit. A number of fruits in which some or the entire edible portion is not derived from ovary/ovaries are classified as accessory fruits. Generally, when inferior ovaries, and the receptacle or hypanthium, become part of the fruit (pseudocarp), they are known as false or spurious fruit, including examples such as cashew fruits, apples, and pears.

Each facet of the development, maturation, and ripening of diverse classes of these fruits is understood to different degrees, but presenting a detailed overview on these is clearly outside the scope of this chapter. Instead, we attempt to summarize some of the recent studies that have led to an increased understanding of the development and ripening of fleshy fruit at a molecular level. The solanaceous tomato has provided the most information in this regard and is therefore the prime model system described and discussed here and, wherever appropriate, references to studies on other fruit species are made in the text.

Fruit Development

Fruit development involves three basic phases. The first phase is the development of the ovary and the initiation of the cell division, together called a fruit set. In the second phase, cell division is the predominant feature. During the third phase, fruit increases in size mainly by cell expansion. Once the fruit cells have fully expanded and the fruit matured, the ripening process ensues (Gillaspy et al., 1993; Ezura and Hiwasa-Tanase, 2010). Each phase of fruit development and ripening involves specific gene activity as revealed by transcriptomic analyses (Carrari et al., 2006; Mounet et al., 2009; Srivastava et al., 2010). As expected many genes identified include transcription factors known to control large alterations of fruit development and ripening, like the *Cnr* locus (Manning et al., 2006) and several MADS-box genes expressed during the early stages of tomato fruit ripening (Busi et al., 2003; Giovannoni, 2007). Studies geared toward our understanding of genetic regulation of fruit ripening are growing exponentially (Giovannoni, 2007; Fatima et al., 2008; Bapat et al., 2010).

Fruit shape, size, and mass

The transformation from their wild characteristics that made fruits inedible to non-toxic, attractive and edible fruits for human consumption involved alterations through generations of domestication. Most modern fruit varieties are usually bigger compared to their wild-type as a result of the domestication process (Smartt and Simmonds, 1995). For instance, the wild-type *Solanum lycopersicum* var. *cerasiforme* bears fruit that has two locules and weighs only a few grams, whereas some of the modern domesticated tomato varieties contain many more locules and could weigh as much as 1 kg (Tanksley, 2004). In recent years, quantitative trait loci (QTL) mapping and the development of molecular marker technology have made it possible to locate the genomic regions and genes responsible for several desirable fruit phenotypic attributes (van der Knaap et al., 2002; Tanksley, 2004). These studies led to the identification of genetic loci as well as genes regulating fruit size, weight, and shape in several species including tomato. About 30 QTL affecting the tomato fruit size, yield, and shape have been identified. However, less than 10 loci account for most changes associated with tomato domestication (Grandillo et al., 1999). The same is the case with other species. Moreover, several loci first identified in other fruits have also been identified in tomato (Frary et al., 2000; Tanksley, 2004; Zygier et al., 2005; Brewer et al., 2007; Cong et al., 2008; Zhang et al., 2008; Bertin et al., 2009), apple (Liebhard et al., 2003), and peach (Yamamoto et al., 2001). The orthologous loci of tomato QTL regulating fruit shape, mass, and size are also present in green pepper (*Capsicum annuum*) and eggplant (*Solanum melongena*), supporting the hypothesis that fruit domestication proceeded via mutations of a small number of genes with conserved function (Tanksley, 2004; Frary et al., 2004).

Mutation in *ovate*, *sun*, and *fs8.1* loci cause elongated tomato fruit by modulation of early stages of carpel development (Gonzalo and van der Knaap, 2008). All elongated tomato shape varieties carry mutation in at least one of these three loci. The tomato variety Howard German carries the *sun* mutation, the Sausage in the *ovate*, and the Rio Grande in the *fs8.1* locus (Gonzalo and van der Knaap, 2008). The *sun* and *ovate* loci play major roles in controlling the ratio of fruit height over width, which determines the fruit shape index. A nonsense mutation in the *OVATE* gene is responsible for the development of a pear-shaped fruit instead of an oval-shaped in tomato (Table 26.1). All tested pear-shaped tomato varieties share the same nonsense recessive mutation in *ovate* (Ku et al., 1999; Liu et al., 2002). The *Ovate* gene encodes a transcription repressor regulating *AtGA20ox1*, a gibberellic acid (GA) biosynthesis enzyme in *Arabidopsis*. The mutation in this gene causes the deficiency in gibberellin biosynthesis resulting in a reduction in cell elongation (Wang et al., 2007).

Table 26.1 Functional characterization of genes regulating fruit size, shape, mass, ripening, and composition in tomato

Gene	Transgene	Phenotype	Reference
Fruit shape, size, mass			
fw 2.2 Negative regulator of cell division	Over-expression	Reduced fruit size and weight	Frary et al. (2000)
SUN IQ67 domain-containing gene	Over-expression	Fruit shape after anthesis	Xiao et al. (2008)
CNR1 Maize fw2.2 like gene	Over-expression	Reduced plant and organ size	Guo et al. (2010)
	Downregulation	Increased growth, biomass, ear length, kernel per row	Guo et al. (2010)
OVATE Negative regulatory transcription factor	Over-expression	Fruit shape pear to round	Liu et al. (2002)
Ethylene regulated ripening			
ACC synthase	Antisense	Impaired ethylene production and fruit ripening, extended fruit shelf life	Oeller (1991)
ACC oxidase	Under-expression	Prolonged fruit shelf life	Xiong et al. (2005)
ETR1 Ethylene receptor	Antisense	Delayed abscission, shorter internode length, no effect on pigmentation and fruit softening	Whitelaw et al. (2002)
SAMdc SAM decarboxylase	Over-expression	Increased spermidine and spermine, lycopene, juice quality, and vine life	Mehta et al. (2002)
SAM hydrolase	Over-expression	Delayed ripening, softening, extended shelf life	Good et al. (1994)
SfIAP Inhibitor of apoptotic from *Spodoptera frugiperda*	Over-expression	Delayed ripening	Li et al. (2010)
Ripening mutations			
MADS-RIN transcription factor	Over-expression Under-expression	Complements *RIN* mutation Regulates fruit ripening	Vrebalov et al. (2002)
Green ripe	Over-expression	Complement *Gr* mutation	Barry and Giovannoni (2006)
Green flesh homolog of rice STAYGREEN protein	Over-expression	Override *Gf* phenotype	Barry et al. (2008)
CNR, Colorless non-ripening SPB-box transcription factor	VIGS	Inhibits fruit ripening	Manning et al. (2006)
DDB1 UV-damaged DNA binding protein 1	Over-expression	Complemented *high pigment 1* mutation	Azari et al. (2010a,b)
det1 High-pigment-2	Over-expression	Elevated carotenoid and flavonoid accumulation	Davuluri et al. (2004)
Fruit texture			
PG Polygalacturonase	Over-expression	Increased pectin solubilization	Giovannoni et al. (1989)
	Antisense	Slightly firmer fruit, increased juice viscosity, reduced pectin solubilization	Schuch et al. (1991)
PG β-subunit	Antisense	Enhanced fruit softening, decreased middle lamella cohesion, reduced tissue integrity	Watson et al. (1994); Chun and Huber (1998)
PME Pectin methylesterase	Antisense	Increased degree of methoxylation, molecular size of pecti, higher solids	Tieman et al. (1992)

(Continued)

Table 26.1 (Continued)

Gene	Transgene	Phenotype	Reference
		Reduced tissue integrity at late ripening, decreased bound calcium, no effect on softening	Tieman and Handa (1994)
		Increased juice and serum viscosity, paste consistency	Thakur et al. (1996a,b)
PL Pectate lyase	Antisense	Increased strawberry fruit firmness	Jiménez-Bermúdez (2002)
TBG1 β-galactosidase1	Antisense	No effect of cell wall GalA loss and fruit softening	Carey et al. (2001)
TBG3 β-galactosidase3	Antisense	Reduction in TBG1 and TBG4 transcripts, higher GalA in cell wall, reduced fruit deterioration	de Silva and Verhoeyen (1998)
TBG4 β-galactosidase4	Antisense	Reduced fruit softening, higher GalA in cell wall	Smith et al. (2002)
Exp Expansin	Antisense	Increased firmness, reduced polymer size	Brummell et al. (1999b)
Cel Cellulase	Antisense	No reduction in xyloglucan depolymerization	Harpster et al. (2002a)
Endo-1,4-β –glucanase	Over-expression	No increase in pepper's xyloglucan depolymerization	Harpster et al. (2002b)
EXP and PG	Both antisense	Increased fruit firmness, modified juice rheology	Kalamaki et al. (2003)
PG and PME	Both antisense	Modified juice rheology	Errington et al. (1998)

The *sun* gene encodes a member of the IQ67 protein family containing the calmodulin binding domain (Xiao et al., 2008). This protein plays a role in expansion of the tissue that likely changes the fruit shape by affecting the pattern along the apical basal axis (Xiao et al., 2008; Table 26.1). An unusual 24.7kb gene duplication event mediated by the long terminal repeat retrotransposon rider caused the genomic context resulting in the formation of this locus with increased *sun* expression relative to that of the ancestral copy. Although the mode of action of the *sun* gene is yet to be uncovered, altered expression of many genes involved in phytohormone biosynthesis or signaling, organ identity, and patterning of tomato fruit is associated with this mutation (Xiao et al., 2009).

The *fs8.1* gene is known to exhibit pleiotropic effects, including increased number of flowers and altered carpel length during pre-anthesis, consequently increasing the fruit size (Ku et al., 2000). A mutation in *fs8.1* initiates changes in floral/carpel development that are responsible for increased fruit size (Ku et al., 2000). The *fs8.1* locus is allelic of the *bell8.1* (van der Knaap and Tanksley, 2003). The *bell8.1* locus, along with *bell2.1* and *bell2.2*, was identified as responsible for a bell pepper shape of fruits of tomato cultivar Yellow Stuffer (van der Knaap and Tanksley, 2003). The locus *bell2.1* increases the locule number, whereas the locus *bell2.2* has been suggested to enhance the fruit weight. Tomato fruit with Yellow Stuffer allele of *fs8.1* exhibit convex locule walls with extended bumpy shape, characteristic of bell peppers (van der Knaap and Tanksley, 2003).

Crosses between small-fruited wild tomatoes and their large-fruited cultivated counterparts led to the identification of *fw1.1*, *fw2.2*, *fw3.1*, and *fw4.1* loci that play roles in fruit size and shape in tomato (Grandillo et al., 1999). With a possible exception of *fw3.1*, these loci increase fruit mass by enlarging fruit by as much as 30% (Tanksley, 2004). The QTL *fw2.2* seems to have the strongest effect on the fruit size (Table 26.1). The *fw2.2* locus encodes a membrane-spanning protein that is expressed at low levels only during the cell division phase of fruit development (Frary et al., 2000; Cong et al., 2002; Tanksley, 2004). It has been suggested that *fw2.2* acts as a negative regulator of cell division and affects fruit size by regulating either transcriptional activity (Cong et al., 2002; Nesbitt and Tanksley, 2002) or photoassimilate availability (Baldet et al., 2006). The *fw2.2* protein has also been suggested to regulate the cell cycle controlled signal transduction pathway by physically interacting with the regulatory β-subunit of a CKII kinase at or near the plasma membrane (Cong and Tanksley, 2006). *Pafw2.2*, a gene similar to *Slfw2.2*, has been cloned from avocado and its higher expression is associated with smaller fruit phenotype in this fruit (Dahan et al., 2010).

The *fasciated* locus (*lcn11.1*) at the end of the long arm of chromosome 11 and the *locule number* locus (*lcn2.1*) on chromosome 2 are associated with a higher number of flower carpels (Barrero and Tanksley, 2004; Barrero et al., 2006). A mutation in *fasciated* locus seems to be associated with enlarging the fruit size during tomato domestication (Barrero and Tanksley, 2004). The *fasciated* gene is expressed prior to the earliest stage of flower organogenesis and thus can alter the floral meristem to form more carpels (locules) and greater fruit size. The exact nature of the gene(s) at these loci is yet to be characterized. The putative tomato homologs of *Arabidopsis* genes known to be involved in floral meristem development did not map to the *lcn11* locus (Barrero et al., 2006). However, putative tomato homologs of *Arabidopsis WUSCHEL* (*WUS*, a homeodomain TAF associated with reduction in floral organs) and *WIGGUM* (*WIG*, a farnesyl-transferase subunit protein associated with increased sepals and petals) mapped to *lcn2.1*, with *WUS* showing the highest association. Impaired expression of *WEE1* kinase (Solly; *WEE1*) by its antisense gene caused reduction in fruit and plant sizes that correlated with lower DNA-ploidy of cells (Gonzalez et al., 2007). These latter studies implied that *WEE1* kinase plays a role in regulating the time of the G phase; its reduction would lengthen the G phase causing a higher degree of endoreduplication. The resulting hypertrophy of the nucleus would then cause cell enlargement.

Fruit Ripening

Fruit ripening is a complex process regulated by both genetic and epigenetic factors. It includes changes occurring in physiologically mature green fruit through the early stages of senescence and is responsible for characteristic aesthetic and/or food quality attributes of fruit (Brady, 1987). During the ripening phenomenon, a fruit undergoes a dramatic shift in gene expression that results in many desirable changes including texture and firmness, sugar accumulation, reduction in organic acids, alterations in pigments leading to development of characteristic color, and production of volatiles responsible for flavor and aroma (Brady, 1987; Giovannoni, 2004; Fatima et al., 2008; Bapat et al., 2010). These transitions make fruit desirable for human consumption, but overripening of a fruit results in large losses of fruit crops (Negi and Handa, 2008). Significant progress has been made in understanding genetic factors that regulate the fruit ripening process, and a number of genes that play a role in fruit development and ripening have been cloned and functionally characterized (Table 26.1; Alba et al., 2005; Seymour et al., 2008; Bapat et al., 2010).

Ripening mutations

The availability of a number of tomato mutants altered in their ripening attributes, and pigment quality and intensity, have greatly facilitated the identification and cloning of the genes that play a central role in the fruit ripening process (Table 26.1). These mutants include *RIPENING INHIBITOR* (*rin*), *COLOURLESS NON-RIPENING* (*cnr*), *NON RIPENING* (*nor*), *NEVER RIPE* (*NR*), GREEN RIPE (*Gr*)/NEVER-RIPE 2 (*Nr-2*), *ALCOBACA* (*ALC*), *HIGH PIGMENT* (*hp*), *DARK Green* (*dg*), *ATROVIOLACEA* (*atv*), *INTENSE PIGMENT* (*IP*), and the *DELAYED FRUIT DETERIORATION* (*DFD*; Kendrick et al., 1994; White, 2002; Barry and Giovannoni, 2006; Levin et al., 2006; Saladié et al., 2007; Lavi et al., 2009). Since these mutations primarily affect the fruit ripening process with little effect on growth and development of the parent plant, they became powerful tools to elucidate and understand the genetic basis of fruit development and ripening. Although the physiological attributes of many of these mutants — including climacteric versus non-climacteric behavior, accumulation of pigments, alteration of shelf life, changes in fruit texture and aroma — had been known for some time (Tigchelaar et al., 1978), it was recombinant DNA technology that made it possible to identify and characterize genes responsible for the observed phenotypes (Figure 26.1). The molecular characterization of these mutations has provided new insight into the fruit ripening process and made tools available to intervene in this agriculturally important process.

Chromosomal walking identified two MADS-box genes in tandem (*SlMADS-RIN* and *SlMADS-MC*) associated with *rin* locus (Vrebalov et al., 2002). The MADS-box genes represent a family of transcription factors that regulate diverse developmental processes in flowering plants, particularly the molecular architecture during flower morphogenesis. The ectopic expression of *SlMADS-RIN* complemented mutation in the *rin* fruits and its downregulation by the antisense gene imparted non-ripening phenotype similar to *rin* mutation (Vrebalov et al., 2002). The other MADS-box gene *SlMADS-MC* affected sepal development and inflorescence determinancy (Vrebalov et al., 2002). The normal *RIN* protein forms a stable homodimer that binds to MADS domain-specific DNA sites, whereas *in vivo* it bound to the CArG-box, a highly conserved DNA motif that binds to MADS-box proteins of *SlACS2* (Ito et al., 2008). Although the *rin* mutation truncated the *RIN* protein by 27 amino acid residues at the C-terminus, it still exhibited a DNA binding activity. Nonetheless, it could not activate transcription and ethylene production (Ito et al., 2008). These results provided an explanation as to why *rin* tomato fruit was not affected in ethylene perception and responds to exogenous ethylene treatment. A fruit-specific ortholog of *SlMADS-RIN*, *FvMADS9*, was also identified in strawberry (Vrebalov et al., 2002).

Identification of tomato *rin* mutation gene as a member of the MADS-box gene led to isolation of other MADS-box genes, including *TAG1*, *TAGL2*, *TAGL11*, *TAGL12*, *TAGL1*, *TDR6*, and *TDR4* from tomato (Vrebalov et al., 2002; Busi et al., 2003). Functional roles of some of these MADS-box paralogs were evaluated using over-expression or downregulation approaches. The constitutive expression of *TAG1* resulted in the conversion of first whorl sepals into mature pericarpic tissue with ripening-like cell wall metabolism, thus supporting its role in fruit development. The downregulation of *TAG1* by its antisense caused homeotic conversion of the third whorl stamens into petaloid organs (Vrebalov et al., 2009). The tomato fruits with reduced *TAGL1* expression exhibited a yellow-orange fruit phenotype, and reduced carotenoid levels and decreased ethylene production and thin pericarp, typical symptoms of impaired ripening due to lack of availability of sufficient ethylene (Vrebalov, 2009). The ectopic expression of *TAGL1* caused expansion of sepals and accumulation of lycopene. Together these results provided evidence that tomato *TAGL1* (*AGAMOUS-LIKE1*), an ortholog of

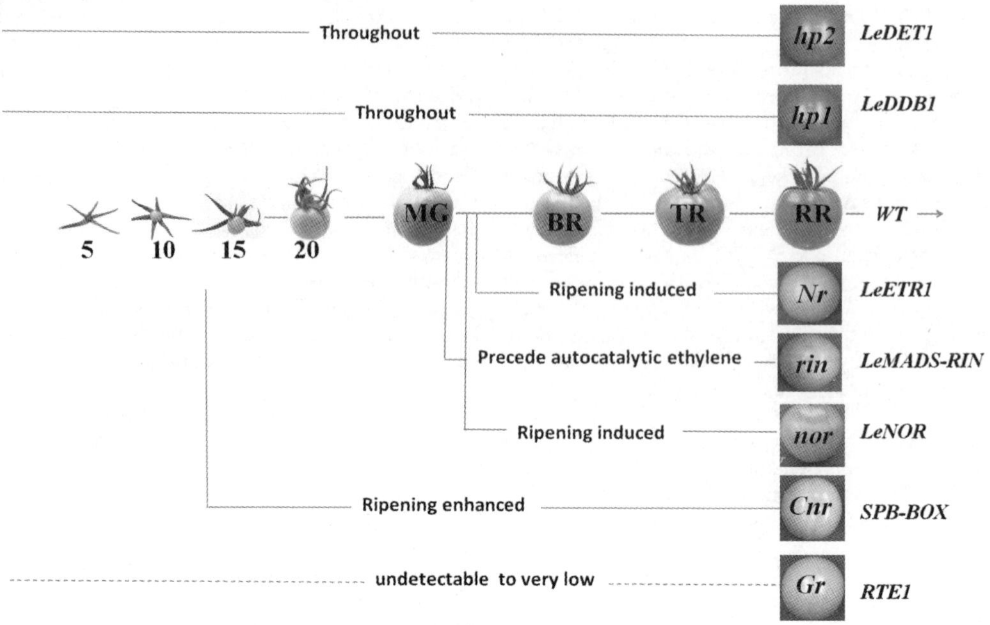

Figure 26.1 • Temporal and developmental regulation of ripening and composition-altering genes in the wild-type fruits • Shown are the developmental (young to fully mature) and ripening (BR to RR) stages of the wild-type (WT) fruits. Also shown are the typical phenotypes of various altered ripening or composition mutants of tomato at a physiological stage comparable to BR + 10 days WT fruits. The lines to the left of each mutant fruit represent the stages when the WT fruit exhibit expression of ripening and the composition-altering genes. The broken line indicates that *RTE1* expresses at very low level throughout development and ripening of the WT fruit. 5, 10, 15, and 20 represent days after anthesis and MG, BR, TR, and RR represent mature green, breaker, turning, and red ripe, respectively. Abbreviations: *hp1*, high pigment 1 (*DDB1, UV-damaged DNA binding*); *hp2,* high pigment 2 (*SlDET1, deetiolated 1*); *Nr, never ripe* (*SlETR1, ethylene response 1*); *rin, ripening inhibitor* (*SlRIN-MADS, RIN-MADS* box-transcription factor); *nor, non-ripening* (*nor, nonripening*); *Cnr, Colorless non ripening* (*SlSPB-Cnr,* squamosa promoter binding protein); and *Gr, green ripe* (*RTE1, reversion to ethylene sensitivity 1).* Except for the *Gr* mutation that activated expression of *RTE1 in Gr* mutant, all other shown mutations resulted in loss of functional expression of their respective genes. (Pictures of the various mutant fruits courtesy of Dr. J. Giovannoni.) Please see color plate section at the end of the book.

SHATTERPROOF (*SHP*) MADS-box genes of *Arabidopsis thaliana)* regulates fruit ripening (Pnueli et al., 1994).

The *TAGL2* is a tomato homolog of the *Arabidopsis SEPALLATA* (*SEP*), which has been shown to interact with four MADS-box proteins (Busi et al., 2003). The transgenic plants with downregulated *TAGL2* (*TM29*) exhibited several phenotypes including production of aberrant flowers, morphogenetic alterations in the organs of the inner three whorls, green instead of yellow petals, infertile stamens, and parthenocarpic fruit set. These attributes collectively suggested this gene to be a negative regulator of fruit development (Ampomah-Dwamena et al., 2002). The MADS-box gene *TD4* is likely a tomato ortholog of the *Arabidopsis FRUITFULL* (*FUL*) MADS-box gene that plays a role in silique development. The constitutive expression of *FUL* prevented the formation of the dehiscence zone in *Arabidopsis* (Ferrándiz et al., 2000). The expression of *TD4* increases during fruit development and is not affected in the *rin* and *nor* genetic backgrounds; however, *Cnr* mutation eliminates fruit-development-related increase in *TD4* transcripts (Seymour et al., 2002). The *TD4* gene may play a role in cell adhesion in tomato fruit as the expression of its apple (*Malus domestica*) ortholog, *MdMADS2.1*, was significantly associated with flesh firmness in apple fruit (Cevik et al., 2010).

The *COLOURLESS NON-RIPENING* (*Cnr*) mutation, a dominant single gene mutation that results in a non-ripening fruit phenotype, was observed first in the commercial population. The *Cnr* fruits had reduced ethylene production, impaired softening, severely reduced cell-to-cell adhesion, and a yellow skin lacking carotenoids in the pericarp tissue (Thompson et al., 1999; Fraser et al., 2001). The ripe *Cnr* fruit has a mealy texture and the pericarp contains approximately 50% more intercellular spaces than the wild-type fruit (Orfila et al., 2001). The *Cnr* locus was mapped and its positional cloning showed that it contained an SBP-box (SQUAMOSA promoter binding protein-like) gene (Manning et al., 2006). An epiallele of *SlSPL-CNR* is responsible for a large degree of methylation in the 286 bp region located 2.4 kb upstream from the first ATG of the *SlSPL-CNR* gene in the *Cnr* mutant (Manning et al., 2006). The high degree methylation was observed only in fruit showing the *Cnr* phenotype (Manning et al., 2006). It has been proposed that the

SlSPL-CNR protein interacts with sequence motifs in the promoters of MADS-box genes of the SQUAMOSA family and modulates expression of a MADS-box transcription factor such as *TDR4*. As discussed previously, *Cnr* mutation reduces transcription of a MADS-box gene *TDR4*. This gene family has also been found associated with the development of dry fruiting structures during maize kernel development (Wang et al., 2005). These results suggest that SBP-box genes play roles in modulating ripening in both dry and fleshy fruits.

The MADS-box genes are also present in other fruit crops and regulate fruit development and ripening. Among the MADS-box genes reported in apples, six (*MdMADS5–9* and *MdMADS11*) are similar to *APETALA1* (*AP1*) and one (*MdMADS10*) to the *AGAMOUS* (*AG*) group. These are expressed during early fruit development (Yao et al., 1999). The insertion mutations in Rae Ime, Spencer Seedless, and Wellington Bloomless apple mutants were identified as homologs of *pistillata*, a MADS-box gene involved in identity determination of petals and stamen in *Arabidopsis* (Weigel and Meyerowitz, 1994; Yao et al., 2001). The functional characterization of one of the peach MADS-box genes, carried out in transgenic tomato, suggested that *PpPLENA* plays a role in fruit formation (Tadiello et al., 2009). Ectopic expression of an *AGAMOUS/SHATTERPROOF* homolog from grape, *Vvmads1*, altered the morphology in the outer two floral whorls in tobacco (Boss et al., 2001). Other MADS-box genes isolated from grapes, *VvMADS2* and *VvMADS4*, show relatedness to the *SEP*, *VvMADS3* to *AGL6* and *AGL13*, and *VvMADS5* to *AGL11* (Boss et al., 2002). Several *MADS*-box genes, isolated from banana (Liu et al., 2009; Elitzur et al., 2010), had differential temporal expression with respect to climacteric ethylene production and exogenous treatment with 1-MCP, suggesting a role in fruit ripening. However, *MaMADS2* and *MaMADS1*, the two genes that showed low similarity to *SlRIN*, did not complement the tomato *rin* mutation (Elitzur et al., 2010). MADS-box genes are expressed in other fruits including hot pepper (Sung et al., 2001), and hazelnut (Rigola et al., 1998), but their role in fruit development and ripening remains to be investigated.

The gene encoding the *never ripe* (*SlNr*) mutation was the first gene to be identified from a ripening impaired tomato mutant (Wilkinson et al., 1995). The *Nr* fruit does not undergo normal fruit ripening and externally exhibits orange color with marginal softening only after several months (Rick and Butler, 1956). The *Nr* seedlings show lack of characteristic triple response to exogenous ethylene, flowers exhibit impaired senescence and abscission, and petioles display an epinastic response, all suggesting that *Nr* mutation causes ethylene insensitivity (Lanahan et al., 1994). Molecular cloning and characterization of gene mutated in *Nr* mutant identified a homolog of the *Arabidopsis* ethylene receptor *ETR1*, but the *Nr* gene was missing the response regulator domain found in the *Arabidopsis ETR1* and related prokaryotic two-component signal transducers (Wilkinson et al., 1995). Variation in *Nr* gene expression regulated ethylene response in ripening tomato fruit and a single amino acid change in the sensor domain eliminated ethylene response (Wilkinson et al., 1995). The *ETR1* ethylene receptor interacts with three different tomato *Constitutive Triple Response* (*CTR*) genes, designated as *SlCTR1*, *SlCTR3*, and *SlCTR4*, suggesting that this receptor plays a role in downstream signaling of ethylene through Ser/Thr type kinases (Zhong et al., 2008).

Although *Green-ripe* (*Gr*) and the *Never-ripe 2* (*Nr-2*) mutations arose independently, their close physical proximity within the genome and molecular cloning showed them to be allelic (Barry et al., 2005). Both mutants were identified based on the retention of chlorophyll in mature fruit and impaired fruit softening and lycopene accumulation. The mutant fruits were insensitive to ethylene and subdued ethylene responses were observed during floral senescence, abscission, and root elongation, but the ethylene-regulated inhibition of hypocotyl elongation and petiole epinasty remained intact (Barry et al., 2005). The ethylene biosynthetic pathway was not impaired, but the expression of the ethylene-regulated gene *E4* was not restored by exogenous application of ethylene. The positional cloning of *GR/NR-2* locus showed an identical 334bp deletion in *Gr/Gr* and *Nr-2/Nr-2* genotypes, confirming the allelic nature of mutations (Barry and Giovannoni, 2006). The deletion was in the 5′-flanking region and ectopic expression of *Gr* in tomato recreated the *Gr* mutant phenotype, except that the whole transgenic plants did not exhibit ethylene insensitivity (Barry and Giovannoni, 2006). The molecular evaluation of the *Gr/Nr-2* gene-encoded protein indicates it to be an evolutionary conserved protein with several *trans*-membrane domains, but its functional role is not yet fully understood (Barry and Giovannoni, 2006).

The *green-flesh* (*gf*) and *chlorophyll retainer* (*cl*) are two mutations in tomato and pepper, respectively, which alter fruit color to brown. The impaired chlorophyll degradation with concomitant accumulation of carotenoids is responsible for the observed brown color of fruits from these two mutants. The positional cloning led to the characterization of a point mutation at the *gf* locus in tomato, which causes an amino acid substitution in an invariant residue of a tomato homolog of the STAY-GREEN (SGR) protein of rice (Barry et al., 2008). A similar mutation resulting in an amino acid substitution at an invariant residue in a pepper homolog of SGR was shown to be responsible for the chlorophyll retainer (*cl*) phenotype in pepper (Barry et al., 2008). Expression of both *GF* and *CL* are highly induced at the onset of fruit ripening and are coincident to the ripening-associated chlorophyll decline. However, the way the SGR subfamily associates with the chlorophyll degradation process is not yet elucidated.

Nutritional mutations

Since fruit is a good source of dietary vitamins, minerals, and many phytonutrients that promote good nutrition, there has been considerable interest in boosting their nutrient levels through genetics, breeding, or molecular engineering (Mehta et al., 2002; Razdan and Mattoo, 2007). Several single-point mutations and quantitative trait loci affecting levels of phytonutrients have been identified and characterized in tomato (Table 26.1). Among these mutations, the recessive high pigment (*hp-1*, *hp-1w*, *hp-2*, and *hp-2j*) and dark green (*dg*) mutations show higher sensitivity to light than wild-type, and accumulate higher levels of the health-promoting metabolites, such as carotenoids (mainly lycopene) and flavonoids in the red ripe fruit (Levin et al., 2006). The *hp-1* is a spontaneous mutant, reported first in 1917 at the Campbell Soup Company farms (Riverton, NJ; Reynard, 1956), while the *hp-2* mutant was discovered in the Italian San Marzano variety in 1975 (Soressi, 1975). Seedlings homozygous in the mutated loci also show higher levels of anthocyanin and shorter hypocotyls compared to the corresponding isogenic lines. It was initially thought that the *hp* mutations were lesions in the genes encoding enzymes in the carotenoid biosynthetic pathway, but mapping showed that they are mutations in two genes regulating light signal transduction responsible for photomorphogenesis (Levin et al., 2006). Whereas the mutation *hp*-1 was allelic to *hp-1w, the hp-2* was allelic to *hp-2j* and *dg*. The role of light signaling machinery in regulating fruit phytonutrient contents has been reviewed (Azari et al., 2010a).

Mapping of a tomato EST sequence, which was highly homologous to the human and *A. thaliana* UV-damaged *DNA binding protein 1* (*DDB1*), to the *hp-1* region in chromosome 2 suggested that *hp-1* and *hp-1(w)* mutant phenotypes are associated with a mutation in *DDB1*. This was confirmed by sequencing the entire coding region of the *DDB1* gene in *hp-1*and and *hp-1(w)* mutants and their near-isogenic normal plants. A single A^{931}-to-T^{931} base transversion in the *hp-1* mutant and G^{2392}-to-A^{2392} transversion in the *hp-1(w)* mutant, respectively, were found in the coding sequence of the *DDB1* gene (Lieberman et al., 2004). These mutations resulted in the substitution of an Asp by Tyr at residue 311 in *hp-1* and Glu by Lys at position 798 in *hp-1(w)* proteins, respectively. Further, these are two highly conserved residues in the DDB1 protein both in plants and animals. An unambiguous demonstration that *DBB1* gene mutation is responsible for the phenotypic characteristics of *hp-1* mutants was obtained by complementing the *hp-1* tomato mutant by overexpression of a non-mutated tomato *DBB1* gene under the control of CaMV35S promoter (Azari et al., 2010a).

A high degree of similarity between CT151, an RFLP marker that mapped to the region containing the *hp-2* mutation, and the *Arabidopsis DEETIOLATED1* (*DET1*) gene suggested that the *hp-2* mutant phenotype resulted from a mutation in the tomato homolog of *DET1* (Mustilli et al., 1999). Isolation and sequencing of the tomato *DET1* homolog confirmed this hypothesis. *DET1* is implicated in photomorphogenic regulation since a mutation in this gene results in the deetiolated phenotype in *Arabidopsis* (Pepper et al., 1994). DET1 is a nuclear protein present in a 350 kDa complex that also contains DDB1 (Schroeder et al., 2002). Since DET1 binds to hypoacetylated amino-terminal tails of the core histone H2B, its involvement in chromatin remodeling around photoregulated genes has been proposed (Benvenuto et al., 2002; Schäfer and Bowler, 2002). The *hp-2* and *hp-2j* phenotypes were shown to result from two independent and different mutations of the *DET1* homolog. In the *hp-2* mutant, a transversion of A to T introduced an alternative splicing to the tenth intron causing a deletion of three amino acids from the nuclear localization signal, which likely leads to mislocalization of the DET1 protein (Mustilli et al., 1999). The other mutation in *hp-2j* resulted from a C to T transition, which gave rise to the substitution of a serine for a conserved proline residue in the C-terminal domain. Although post-transcriptional silencing of *DET1* expression in tomato resulted in its hyper-responsiveness to light, which was often lethal at the early stages of plant development, it increased carotenoids in the transgenic fruits (Davuluri et al., 2004). The transgenic repression of the tomato *DET1/HP2* gene using a fruit-specific promoter allowed elevated carotenoid and flavonoid accumulation in fruit without the deleterious effects such as dwarfism and reduced yield of such manipulations in non-fruit tissues (Davuluri et al., 2005).

Mapping of the *dg* mutant found this gene to be allelic to the *hp-2* gene but with a mutation in a different part of the gene sequence. The *dg* mutation was due to a single A to T transversion leading to the substitution of an Ile by Asn in the N-terminal domain of the protein in contrast to the mutation in the two alleles of *hp-2*, which are located toward the C-terminal domain of the protein (Levin et al., 2003). The role of the light signal transduction in regulating the levels of carotenoids and other antioxidants has also been shown by ectopic expression of positive (*CONSTITUTIVELY PHOTOMORPHOGENIC 1* [*COP1*]-like) and negative (tomato *ELONGATED HYPOCOTYL 5* [*HY5*]) light signaling regulators and *CRY2* photoreceptor (Liu et al., 2004; Giliberto et al., 2005).

High pigment mutants were also selected using ethyl methanesulfonate (EMS) and fast neutron bombardment mutagenesis. One of these mutants designated as *high pigment 3* (*hp3*) accumulated higher levels of carotenoids and chlorophyll in leaves and green fruit, with 30% more carotenoids found in ripe fruits compared to the parental wild-type fruit (Galpaz et al., 2008). The *hp3* mutation was localized in a gene encoding the enzyme zeaxanthin epoxidase that converts zeaxanthin to violaxanthin. This mutation lowered the ABA levels and provided evidence, in addition to that obtained from two other ABA-deficient mutants *flacca* and *sitiens*, that ABA plays a role in pigment biosynthesis and accumulation in tomato fruit (Galpaz et al., 2008).

Other mutants include *Atroviolacea* (*atv*; Kendrick et al., 1994) and the *INTENSE PIGMENT* (*IP*; Levin et al., 2006) that have altered responses to light. The *atv* gene has been proposed to encode a repressor of phytochrome signal amplification based on a loss-of-function effect. The *IP* mutant was a result of a cross between *S. chmielewskii* and the cultivated tomato *S. lycopersicum*. This mutant produces smaller and dark-red mature fruits as well as greener leaves than the nearly isogenic tomato lines. The red ripe *IP* fruit had higher amounts of total soluble solids, different carotenoids, and more chloroplasts in outer mesocarp, inner mesocarp, and jelly tissues than the parental wild-type fruits (Lavi et al., 2009).

Shelf life mutations

In addition to the ripening impaired tomato mutants discussed above, several other mutants are known to extend fruit shelf life. These include *ALCOBACA (alc)* and *"Delayed Fruit Deterioration"* (*DFD*) mutations. The *alc* fruit has reduced ethylene production, respiration, carotenoids, and polygalacturonase (PG) activity, and maintains firmer texture during storage. The *alc* mutation, in combination with high pigment mutations, has provided an effective way to develop commercial tomato varieties with a longer fruit shelf life and a saturated red color (Kuzemenskii, 2007). The *alc* fruit has about a three-fold higher putrescine level and altered wax profiles during fruit development, which are likely effects of this mutation (Kosma et al., 2010).

The *DFD* fruit also has a longer shelf life despite the exhibition of a normal climacteric peak of carbon dioxide and a burst of ethylene production during ripening (Saladié et al., 2007). However, the mutant fruit shows a very low transpirational water loss and substantially elevated cellular turgor during postharvest storage. Since biochemical characterization revealed several changes in cuticle chemical composition, it was proposed that the altered cuticles in *DFD* fruits affect the softening of intact tomato fruit by providing both physical support and regulating water loss (Saladié et al., 2007). However, the nature of mutations in both *alc* and *DFD* cultivars has yet to be determined.

Ethylene and Fruit Ripening

Ethylene, a gaseous plant hormone, plays a critical role in many physiological processes during plant development, including a massive metabolic shift during the ripening of climacteric fruits (Lin et al., 2009; Bapat et al., 2010). Based on respiration patterns, fruits are divided into two broad categories: climacteric and non-climacteric. The climacteric fruits exhibit a surge in the rate of respiration at the onset of ripening that is associated with increased biosynthesis of ethylene. The role of respiratory surge in climacteric fruits at the onset of ripening is still a mystery, but it has been proposed to be related to coordination and synchronization of the ripening process in these fruits. Both climacteric and non-climacteric fruits show similar changes in various metabolic processes leading to pigment alterations, sugar accumulation, textural change, fruit softening, volatiles production, and enhanced susceptibility to pathogens. Also, the climacteric and non-climacteric distinction of fruits is not determined by phylogeny, since fruits from closely related species exhibit this distinction (Giovannoni, 2004). Although ethylene plays a significant role in the development of ripening attributes in climacteric fruit, it is not required for the metabolic shifts seen in non-climacteric fruits such as grape, citrus, and strawberry (Oeller et al., 1991; Giovannoni, 2004; Bapat et al., 2010). These observations have raised an interesting debate on ethylene-dependent and -independent processes during fruit ripening. Significant information has begun to emerge on ethylene regulation of the ripening process in climacteric fruits (Pnueli et al., 1994; Cara and Giovannoni, 2008; Bapat et al., 2010). Notably, the control of the onset of ripening in non-climacteric fruit is poorly understood, and suggestions of a control by sugar sensing and hormones such as abscisic acid (ABA) have been put forth (Gambetta et al., 2010). Tomato fruit has been used extensively as a model system to understand ripening in climacteric fruit and strawberry as a model to investigate ripening in non-climacteric fruits (Zhang et al., 2010). Other climacteric and non-climacteric fruits are also being investigated to study the fruit ripening process (Bapat et al., 2010).

Ethylene biosynthesis

The biochemical pathway of ethylene synthesis in plants starts from methionine (Lieberman et al., 1966), which is then S-adenosylated to form S-adenosylmethionine (SAM) by SAM synthetase (Fluhr and Mattoo, 1996). SAM is subsequently metabolized to 1-aminocyclopropane-1-carboxylic acid (ACC; Adams and Yang, 1979) and 5′-methylthioadenosine (MTA) by ACC synthase (ACS). ACC, the first committed intermediate in the ethylene biosynthesis pathway, is converted to ethylene by ACC oxidase (ACO). MTA, the other product of ACS reaction, enters the methionine cycle. Genes encoding ACS and ACO were first identified from zucchini (Sato and Theologis, 1989) and tomato fruit (Hamilton et al., 1990), respectively. Although several genes involved in methionine metabolism show altered expression during plant development and in response to the environment, mostly activities of ACS and in some cases ACO are generally considered the rate-limiting steps in ethylene production in fruit (Fluhr and Mattoo, 1996; Mattoo and Handa, 2004; Cara and Giovannoni, 2008).

ACS and *ACO* belong to a family of genes that have been characterized and from tomato, melon, apple, banana, pear, kiwifruit, peach, and persimmon (Cara and Giovannoni, 2008; Lin et al., 2009; Bapat et al., 2010). The members in each gene family are differentially expressed during development and in response to environmental cues (Cara and Giovannoni, 2008; Lin et al., 2009). Nine *ACS* and five *ACO* genes were identified in tomato. Among them, four *ACS* (*SlACS1A*, *SlACS2*, *SlACS4*, and *SlACS6*) and three *ACO* (*SlACO1*, *SlACO3*, and *SlACO4*) members are differentially expressed during tomato fruit development (Cara and Giovannoni, 2008; Lin et al., 2009). *SlACS6* is predominantly expressed in the green fruit and is associated with ethylene production during the early stages of fruit development (Figure 26.2). *SlACS1A* is also expressed in green fruit but at lower levels than *SlACS6*. *SlACS1A* and *SlACS4* are induced alongside ripening transition and proposed to be responsible for the induction of *SlACS2*, a gene implicated in the autocatalytic production of ethylene during the robust ripening process (Figure 26.2). The autocatalytic ethylene production may exert negative feedback inhibition on early ethylene production by reducing the expression of *SlACS1A* and *SlACS6* (Barry et al., 2000). The RIN MADS-box transcription factor regulates the expression of *SlACS1A* and *SlACS4* during ripening (Vrebalov et al., 2002). Among the *ACO*s, *SlACO1* and *SlACO4* are expressed throughout fruit development, but their expression increases dramatically at the onset of ripening and remains elevated thereafter. The ripening-associated induction of these *ACO*s is ethylene dependent, since their increased expression is impaired by treatment with 1-methylcyclopropene (1-MCP), an ethylene perception inhibitor (Mattoo and Suttle, 1991; Abeles et al., 1992). *SlACO3* is expressed during fruit maturation and is transiently induced at the breaker stage of fruit ripening (Cara and Giovannoni, 2008; Bapat et al., 2010).

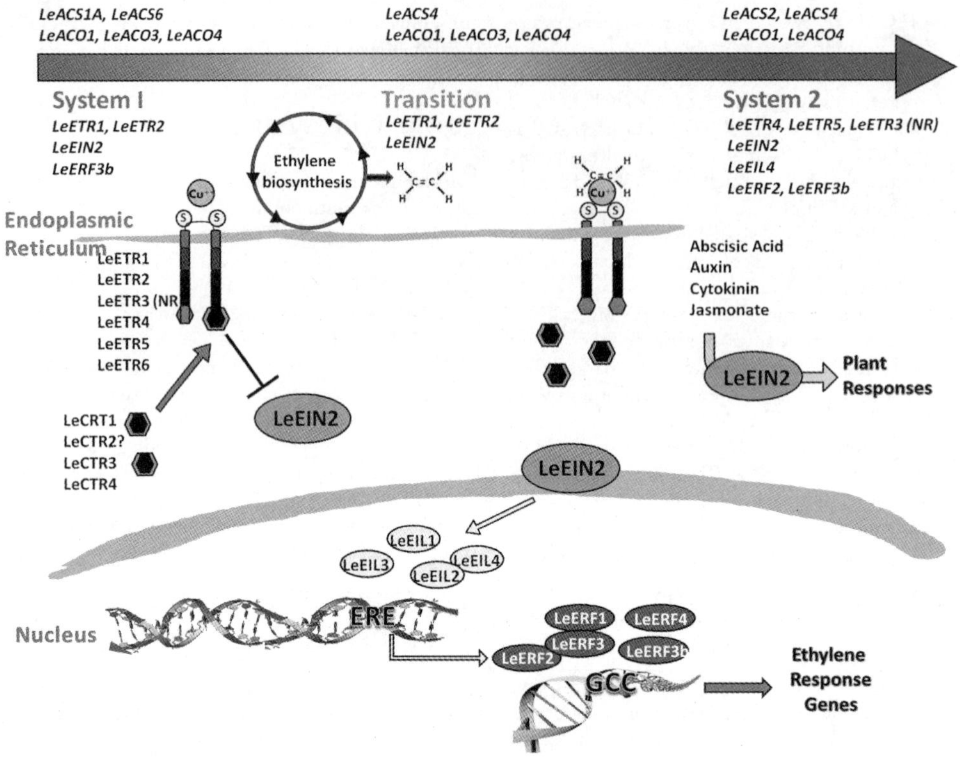

Figure 26.2 • Regulation of ethylene production and perception during tomato fruit ripening • Top panel: The participation of various *ACC synthase* (*ACS*), *ACC oxidase* (*ACO*), and *ethylene receptor* (*ETR*) genes in the System 1 (lower and auto-inhibitory amounts of ethylene), System 2 (autocatalytic ethylene production), and ripening-transition ethylene production and perception during fruit ripening. Middle panel: ETRs that are localized in the endoplasmic reticulum form disulfide bridges (yellow circles) and their dimers interact with copper ion (green circle) to bind ethylene. With the exception of NR, ETRs have four different domains depicted in different colors (green, sensor; gray, GAF; black, histidine kinase; and red, response regulator or receiver domains, respectively). In the absence of ethylene, constitutive triple response (CTR) protein binds to an ETR. This complex interacts with an ETHYLENE INSENSITIVE2 (EIN2) protein to negatively regulate ethylene response. Ethylene binding to ETR inhibits its interaction with CTR resulting in releasing inhibition on EIN2 activity. Other phytohormones may also use EIN2 to transduce their response. Lower Panel: Free EIN2 moves to nuclear membranes where it activates transacting protein (ETHYLENE INSENSITIVE3) EIN3 and EIN3-like proteins (EIL) by an unknown mechanism. This facilitates binding of EIN3 and EILs to ethylene response elements (ERE) present in the promoter regions of *ethylene response factors* (ERFs). This cascade activates expression *ERFs*, which in turn enables expression of ethylene responsive genes. Please see color plate section at the end of the book.

Ethylene perception and signal transduction

During the last two decades, significant understanding of how plants perceive and respond to ethylene has emerged (Klee, 2004; Cara and Giovannoni, 2008; Lin et al., 2009). Most of the initial information about the molecular components of ethylene perception and action was obtained using *A. thaliana* mutants altered in seedling triple response to exogenous ethylene treatment. A series of elegant experiments including the epistatic behavior of these mutations led to the development of a linear model for ethylene action in plants. Figure 26.2 is a model showing the current understanding of the various interactions of the molecular components in the biosynthesis and perception of ethylene during tomato fruit ripening. Molecular mechanisms regulating ethylene perception and signaling have been elegantly reviewed (Cara and Giovannoni, 2008; Lin et al., 2009), therefore we briefly discuss only the salient features in this chapter.

Ethylene is perceived by receptors encoded by a multigene family (*ETRs*). Five *ETRs* (*ETR1*, *ETR2*, *ERS1*, *ERS2*, and *EIN4*) genes have been characterized in *Arabidopsis*. ETRs show homology to bacterial two-component regulators and structurally they are separable into three domains designated as sensor, kinase, and receiver domains (Klee, 2004; Cara and Giovannoni, 2008; Lin et al., 2009). The sensor domain present at the N-terminal of ETR contains three or more hydrophobic *trans*-membrane stretches that bind ethylene and participate in their localization to the endoplasmic reticulum (and possibly to the Golgi apparatus; Chen et al., 2002). All of the known dominant *ETR1* mutations are located in this domain. The kinase domain follows the GAF motif (a ubiquitous motif found in sensory proteins of both prokaryotic and eukaryotic cells) domain and is present in the middle of ETRs. It shows sequence homology to His kinases and contains up to five subdomains including one with the His. The receiver domain is present at the C-terminus of these polypeptides and shares sequence identity with the output portion of bacterial two-component systems. Based on peptide structural attributes, the ETRs have been classified into two subfamilies (Lin et al., 2009). The *Arabidopsis* ETR subfamily 1 includes *ETR1* and *ERS1*, characterized by the presence of three N-terminal transmembrane domains and a conserved histidine kinase domain. *ETR2*, *ERS2*, and *EIN4* represent the subfamily 2 and are characterized by the presence of an additional transmembrane domain and lack of one or more of the catalytic subdomains, including the His. The subfamily 2 ETRs are thought to function as Ser/Thr kinases (Moussatche and Klee, 2004). The expression of an altered *ETR1* gene that lacks the active histidine kinase rescued *Arabidopsis* plants harboring double mutations in both *ers1*and *etr1*, indicating that the histidine kinase activity may not be required for the ethylene signal transmission (Wang et al., 2003). The tomato ETR family contains six members (*SlETR1*; *SlETR2*; *SlETR3*, also called *NR*; *SlETR4*; *SlETR5*; and *SlETR6*). Although they show significant divergence in primary sequence, all bind to ethylene (Zhang et al., 2010). *SlETR1*, *SlETR2*, and *NR* represent the three members of the tomato *ETR* subfamily 1 with only NR lacking the receiver domain. *SlETR4*, *SlETR5*, and *SlETR6* resemble *Arabidopsis* subfamily 2 as they have an extra amino terminal membrane-spanning domain but are missing one or more conserved HK domains (Klee and Tieman, 2002). Ethylene seems to enhance degradation of *SlETR4* or *SlETR6* proteins, likely by the 26S proteasome-dependent pathway (Kevany et al., 2007). Tomato *ETR1* is constitutively expressed in vegetative and reproductive tissues (Zhou et al., 1996).

The functional ethylene receptors are disulfide-linked dimers and need copper ions to bind ethylene (Figure 26.2; Rodriguez et al., 1999). All dominant ethylene insensitive *Arabidopsis* mutations have a single amino acid substitution in ETR1; the loss-of-function mutations in *ETR1*, *ETR2*, *EIN4*, and *ERS2* have normal phenotypes (Hua and Meyerowitz, 1998). However, quadruple mutants harboring mutations in *ETR1*, *ETR2*, *EIN4*, and *ERS2* are constitutively ethylene responsive, providing evidence that ethylene receptors are functionally redundant and negatively regulate ethylene responses. In addition to the formation of a homodimer, heteromers of the ethylene receptors are known, which may mediate ethylene signaling in *Arabidopsis* (Gao et al., 2008; Schaller et al., 2008). The GAF domain that follows the sensor domain in all of the ETRs has been implicated in their dimerization, since ETR1 truncated in the GAF domain did not dimerize when expressed in yeast (Gao et al., 2008). The role of GAF motif in ethylene signal transduction remains to be investigated.

Ethylene receptors negatively regulate ethylene response through a *constitutive triple response1* (*CRT1*) gene that acts downstream of ETR (Kieber et al., 1993). *CTR1* encodes a Raf-like ser/thr kinase that physically interacts with ETR (Clark et al., 1998). *Arabidopsis* contains a single *CTR*, but four homologs — *SlCTR1*, *SlCTR2*, *SlCTR3*, and *SlCTR4* — have been isolated from tomato. All tomato *CTR* homologs, with the exception of *SlCTR2*, are capable of complementing *ctr1* mutations in *Arabidopsis* (Adams-Phillips et al., 2004; Lin et al., 2008). *SlCTR2* shows higher similarity with *Arabidopsis* Enhanced Disease Resistance 1 (*EDR1*) than with *CTR1*, suggesting that it may have a different function than a general role in ethylene action (Tang et al., 2005). The four members of tomato *CTR* are differentially expressed during plant development, but *CTR1* is predominantly expressed in abscission zones, flower senescence, and fruit ripening (Adams-Phillips et al., 2004).

ETHYLENE INSENSITIVE2 (*EIN2*) transduces ethylene response through downstream transcription factors such as ETHYLENE INSENSITIVE3 (EIN3) and other EIN3-like proteins (EILs) that activate ethylene regulated transcriptional (MAP kinase) cascade (Roman et al., 1995). In the absence of ethylene, ETR binds to CTR1, which then interacts with EIN2 and inactivates the downstream signals of ethylene response. Ethylene binding to ETR eliminates CTR-regulated suppression of ethylene signaling pathway. How EIN2 works at the molecular level to relay ethylene signal to EIN3, EILs, and ETHYLENE RESPONSE FACTORs (ERFs), transcription factors that regulate ethylene response, is not yet understood (Klee, 2004). Notably, the N-terminal domain of EIN2 shares sequence homology with the NRAMP family of metal ion carriers, and loss-of-function mutations in this domain result in ethylene insensitivity in *Arabidopsis* (Alonso et al., 1999). The C-terminal domain of EIN2 interacts with two F-box proteins, ETP1 and ETP2 (EIN2 TARGETING PROTEIN). ETP1 and EPT2 have been suggested to regulate proteasome-mediated degradation (Qaio et al., 2009). *EIN2* expression is not regulated by ethylene and is constant throughout fruit development (Zhu et al., 2006). *EIN2* has also been suggested to participate in signal transduction pathways of other phytohormones including ABA, auxin, cytokinin, and jasmonate, and thus may regulate cross-talk between multiple hormone signaling pathways (Cara and Giovannoni, 2008).

EIN3 and EILs comprise a family of DNA binding proteins that bind to ethylene response elements (ERE) and regulate expression of ethylene sensitive genes (Solano et al., 1998). *ETHYLENE-RESPONSE-FACTOR1* (*ERF1*), which contains the tomato E4-element in its promoter, is an immediate target for EIN3 and is necessary and sufficient for *ERF1* expression (Deikman et al., 1998). *ERF1* acts downstream of *EIN3* and its constitutive expression activates a variety of ethylene response genes and phenotypes in *Arabidopsis* (Solano et al., 1998). Four EILs (*SlEIL1*, *SlEIL2*, *SlEIL3*, and *SlEIL4*) are present in tomato fruit (Tieman et al., 2001; Yokotani et al., 2009). Neither ripening nor ethylene induces *SlEIL1*, *SlEIL2*, and *SlEIL3*, but expression of *SlEIL4* is ripening inducible (Tieman et al., 2001; Yokotani et al., 2009). Antisense RNA-based inhibition studies suggested that *SlEIL1*, *SlEIL2*, and *SlEIL3* were functionally redundant (Tieman et al., 2001). However, ectopic expression of *SlEIL1* in *Nr* was not able to restore expression of ripening-regulated genes, suggesting independent roles for specific EILs (Chen et al., 2004).

A complex post-transcriptional regulation seems to govern the cellular levels of *EIN3* and EIL proteins in response to ethylene (Yoo et al., 2009). Two F-box proteins (EBF1/2), which are Skp–Cullin–F-box (SCF) E3 ligases, target EIN3 and EILs for 26S proteasome degradation (Guo and Ecker, 2003) and MAPKK9–MAPK3/6 cascade that phosphorylates EIN3 at specific sites influences their stability (Yoo et al., 2009).

Ethylene response factors (ERFs), also called ethylene-responsive element binding proteins (EREBPs), are plant transcriptional regulators that mediate ethylene response likely via binding to a conserved motif AGCCGCC (GCC-box) located in the promoter region of ethylene-regulated genes. Characterization of *E4* and *E8* (ethylene regulated genes) promoters resulted in the identification of two E4-UpEREBP and E4/E8BP factors that exhibited DNA binding activity and were required for the expression of these genes (Deikman et al., 1998; Guo and Ecker, 2003). E4-UpEREBP was present in unripe fruit and its activity was reduced in response to ethylene treatment suggesting it to be a transcriptional repressor (Guo and Ecker, 2003). Activity of E4/E8BP increases during fruit ripening and is implicated as a positive regulator of ethylene response (Deikman et al., 1998).

In tomato, five ERFs (SlERF1, SlERF2, SlERF3, SlERF4, and SlERF3b) show binding capability to GCC-box elements (Montgomery et al., 1993). These genes exhibit differential expression patterns during plant development and fruit ripening. *SlERF2* (Sl-ERF2) and *SlERF3b* transcripts accumulate during fruit ripening. The promoter of *SlERF2* contains five putative EREs and its expression is similar to E8, in that both are induced during ripening and absent in the non-ripening tomato mutants *NR*, *rin*, and *nor* (Montgomery et al., 1993). The *SlERF3b* transcript accumulates in green fruits with a sharp decline at the onset of ripening. Its levels also increased in fruits that produce low levels of ethylene, such as *NR* fruit or transgenic fruit with impaired ACC expression. Unlike the regulation by RIN-MADS, CLEAR NON-RIPENING, TAGL1, and LeHB-1 genes act as positive regulators of ripening phenomenon. APETALA2/ERF, SlAP2a, a ripening induced gene, acts as a negative regulator of fruit ripening (Chung et al., 2010). Repression of *SlAP2a* by targeted RNAi transgene resulted in fruit that produced higher levels of ethylene, ripened early, and had altered carotenoid accumulation (Chung et al., 2010).

Genetic intervention in ethylene biosynthesis and perception

The emerging understanding of ethylene biosynthesis and perception has allowed both pharmacological and genetic approaches to reduce undesirable effects of ethylene in fruit. A number of metabolic inhibitors are available that alter production or perception of ethylene. Among them, 1-methylcyclopropene, which inhibits ethylene perception, is now widely used to reduce postharvest losses of fruit and vegetable crops (Martínez-Romero et al., 2007). 1-MCP is active at low concentrations and has 10 times more affinity for the ethylene receptor than ethylene. Because it is a non-toxic compound to humans, its use is expanding rapidly as a safe chemical to enhance postharvest shelf life of ethylene-sensitive crops. Nonetheless, 1-MCP is a volatile compound and attempts have been made to develop a non-volatile and water-soluble cyclopropene derivative as an ethylene antagonist. Importantly, a water-soluble sodium salt of 3-cyclopropyl-1-enyl-propanoic acid was synthesized and found to effectively delay fruit ripening and increase shelf life of fruit and cut flowers (Huberman et al., 2009).

As an alternative to synthetic chemical use to inhibit fruit ripening, molecular engineering has provided novel and genetic toolkits to overcome the pro-ripening and -senescence effects of ethylene and enhance postharvest shelf life and manage disease in horticultural crops. Downregulation of ACC synthase expression by its antisense gene in tomato resulted in significant increase in fruit shelf life, a study that established ethylene as a ripening hormone (Oeller et al., 1991). Since then, ACS expression has been modified in several fruits including bitter apples (Dandekar et al., 2004), melon (Lin and Do, 2008), and tomato (Gao et al., 2008), whereas ACO expression was modified in tomato (Hamilton et al., 1990), melon (Flores et al., 2001), apple (Dandekar et al., 2004), and papaya (López-Gómez et al., 2009). Fruit from these transgenic lines are altered in ripening behavior, particularly ripening-associated pigment alterations, fruit softening, and volatiles production. Transgenic pineapple with reduced amounts of ACC synthase, which decreased ethylene production, was accompanied by a delay in the flowering time (Trusov and Botella, 2006). Transgenic seedlings of pears expressing antisense gene for an ACC oxidase produced 15 to 71% ethylene compared to parental wild-type; however, effects on fruit quality were not reported (Gao et al., 2007). Reduction in ethylene production was also obtained by expressing a bacterial gene encoding ACC deaminase (Klee et al., 1991) or bacteriophage T3-encoded SAM hydrolase (Good et al., 1994) in tomato, which resulted in altered fruit ripening (Klee, 1993). Bacteriophage T3-SAM hydrolase was also introduced in avocado under a fruit-ripening-specific cellulase promoter, but the effects on fruit ripening were not evaluated (Litz et al., 2007).

Manipulation of the ethylene receptors to impact fruits and other plant parts was achieved by up- and downregulation of various ETRs under the control of constitutive or ripening induced promoters. Downregulation of *SlETR4* and *SlETR6* receptors using an antisense RNA approach under the CaMV35S promoter resulted in severe epinasty, enhanced flower senescence, and accelerated fruit ripening in tomato (Tieman et al., 2000; Kevany et al., 2007, 2008). Ethylene-associated developmental effects were not seen in transgenic lines that had reduced expression of *SlNR*, *SlETR1*, *SlETR2*, or *SlETR5*, suggesting that *SlETR4* or *SlETR6* plays a more important role in mediating ethylene response (Tieman et al., 2000; Kevany et al., 2007). Transgenic fruits with reduced *NR* expression showed higher accumulation of *SlETR4*. However, its over-expression resulted in tomato lines with lowered *SlETR4* gene expression and ethylene-resistant phenotype, implicating that in spite of structural differences, these receptors could functionally complement each other and are therefore functionally redundant (Martínez-Romero et al., 2007; Tieman et al., 2000). Downregulation of *SlETR4* using *tmf7*, a promoter active during early fruit development, led to early fruit ripening but without an effect on fruit size, yield, and flavor-related chemical composition (Kevany et al., 2008). Impaired *SlETR1* expression by its antisense did not affect fruit ripening but delayed abscission and reduced plant size in tomato (Whitelaw et al., 2002). Studies with transgenic tomato expressing *Arabidopsis etr1-1* under an inducible promoter showed that the transgenic plants were sensitive to ethylene in the absence of the inducer but insensitive in the presence of inducers. Furthermore, the level of ethylene sensitivity had a linear relationship with the amount of inducer used (Gallie, 2010).

Fruit Texture

Fruit texture, especially softening, during ripening is an essential attribute that makes fruits desirable for human consumption. However, it is also associated with reduced postharvest shelf life and increased susceptibility to microbial infection. Because of the economic significance of fruit textural changes, there is a significant interest in understanding the biochemical mechanisms regulating this process. Ripening-associated fruit softening usually involves modifications to the polysaccharide and protein components of the primary cell wall and middle lamella, resulting in a weakening of the structure (Brummell, 2006; Vicente et al., 2007; Negi and Handa, 2008). However, the process of fruit textural changes is very complex and many protein families along with water relations and free radicals contribute to it. Developmental regulation of cell wall stabilizing and depolymerizing proteins as well as protein glycosylation have been demonstrated during fruit ripening (Brummell, 2006; Vicente et al., 2007; Negi and Handa, 2008). These include polygalacturonase (PG), pectin methylesterases (PMEs), β-galactosidase, β-mannanase, xyloglucan xyloglucosyltransferase/endohydrolase (XET), pectate lyase (PL), endo-β-1,4-glucanases (EGases, cellulases), expansins, α-Man, and β-Hex (Table 26.1).

Cell wall depolymerizing enzymes

Since middle lamella is made primarily of pectin, it was hypothesized that the pectin depolymerizing enzymes regulate fruit softening by controlling cell separation. This hypothesis gained additional support when a correlation between reduced softening and impaired expression of *SlPG2* was found in ripening-impaired tomato mutants (Tigchelaar et al., 1978; Biggs and Handa, 1989; DellaPenna et al., 1989). However, ectopic expression of *SlPG2*, a ripening induced enzyme, failed to enhance softening of *rin* tomato, suggesting that PG played only a limited role in this process (Giovannoni et al., 1989). Antisense inhibition of *SlPG2* led to the development of "Flavr Savr" tomato, the first commercially marketed genetically modified edible crop (Kramer and Redenbaugh, 1994). Unfortunately, Flavr Savr did not meet the market expectation of the extended shelf life of fruit (Kramer and Redenbaugh, 1994). The antisense inhibition of *SlPG2*, however, resulted in increased viscosity of processed tomato juice (Thakur et al., 1997). Antisensing of *FaPG1* was found to reduce ripening-associated fruit softening in strawberry (Quesada et al., 2009). These transgenic fruit were about 95% reduced in *FaPG1* expression and had higher amounts of covalently bound pectins. As the expression of *FaPG2* was not impaired by the *FaPG1* antisense, the fruit PG activity was only slightly reduced.

PMEs exist as multiple isozymes during fruit ripening, but their role in fruit texture is not yet clear (Harriman et al., 1991; Tieman and Handa, 1994; Gaffe et al., 1994; Phan et al., 2007). Transgenic tomato fruits expressing *SlPME3* and with over 95% reduction in PME transcripts, protein, and enzymatic activity, were unaffected in fruit softening (Tieman and Handa, 1994). However, processing attributes of low PME fruit were greatly enhanced: the transgenic fruits had higher juice viscosity and increased total soluble solids (Tieman et al., 1992; Thakur et al., 1996a,b). The negative effect was on fruit integrity, which was compromised upon extended storage of fully ripe fruits. Silencing of a ubiquitously expressed PME isozyme, *SlPMEU1*, did not impart any detectable phenotype to leaf tissue, but enhanced fruit softening even though the transgenic fruits had only a 25% reduction in PME activity (Phan et al., 2007; Gaffe et al., 1997).

Loss of galactose and/or arabinose from cell walls during early fruit ripening (Gross and Sams, 1984) led to the suggestion that β-galactosidase plays an important role in fruit textural changes. At least seven β-galactosidase genes (*SlTBG1* to *7*) are expressed during tomato fruit development (Smith and Gross, 2000). The homology-dependent silencing of *SlTBG1*that resulted in about 90% reduction in its transcripts in tomato fruit did not affect the total exo-galactanase activity, cell wall galactose content, or fruit softening (Carey et al., 2001). Antisense RNA suppression of *SlTBG3* caused about 75% reduction in *TBG1* and *TBG4* transcript levels and extractable exo-galactanase enzymatic activity and higher retention of the cell wall galactose, but still had little effect on fruit softening (de Silva and Verhoeyen, 1998). Transgenic fruits with impaired expression of *SlTGB4* showed about 90% reduction in extractable exo-galactanase activity, free galactose levels in mature green fruit, and 40% firmer fruit as compared to the control (Smith et al., 2002). The enzyme endo-β-mannanase (β-Man) hydrolyzes mannose in hemicellulose polymers to mannobiose and mannotriose. Transgenic tomato fruits with reduced β-Man were developed by expressing an antisense RNA or by gene-specific hairpin RNAi. These transgenic fruits exhibited reduced β-Man activity, but a clear correlation between fruit firmness and β-Man activity was not obtained (Wang et al., 2009).

It is suggested that xyloglucan xyloglucosyltransferase/endohydrolase (XTHs: EC 2.4.1.207 and/or EC 3.2.1.151) plays a dual role in cell wall chemistry by integrating newly secreted xyloglucan chains into an existing wall-bound xyloglucan, and restructuring existing cell wall material by catalyzing transglucosylation between previously wall-bound xyloglucan molecules. Its role in fruit development and texture was examined by ectopic expression of *SlXTH*, a tomato homolog of *Nicotiana tabacum NtXET-1* gene, under the CaMV35S promoter. The transgenic fruit had more than a four-fold increase in XET activity and reduced in xyloglucan depolymerization and fruit softening, suggesting a role in maintaining the structural integrity of cell walls (Miedes et al., 2010). Multiple genes for the pectate lyase

(PL), an enzyme that hydrolyzes the unesterified galacturonosyl linkages by a β-elimination reaction, have been identified in many fruit species. Their expression increases during fruit ripening, and understanding their role in pectinolysis and fruit texture changes has only recently begun. The antisense inhibition of *PL* gene expression in strawberry led to extended postharvest shelf life, reduction in pectin solubility, decreased depolymerization of bound polyuronides, and loss of cell–cell adhesion in the transgenic fruits (Jiménez-Bermúdez et al., 2002; Santiago-Doménech et al., 2008). A family of endo-β-1,4-glucanase (EGases, cellulases) genes is present in many fruit, some of them are expressed in an ethylene-dependent manner and inducible during fruit ripening (Brummell et al., 1999a). Repression of *SlCEL1* and *SlCEL2*, the two ripening-related Endo-β-1,4-glucanases (EGases), led to reduced abscission of pedicel and fruit, respectively, but had little effect on tomato fruit softening (Lashbrook et al., 1998). Similar results were obtained in strawberry where downregulation of *Cel1* and *Cel2* did not influence fruit softening but resulted in slightly higher abundance of the larger hemicellulosic polymers (Mercado et al., 2010).

Expansins

Expansins induce extension in isolated plant cell walls and are therefore implicated in fruit textural changes (Rose et al., 1997). Expansins belong to a multi-gene family with different members exhibiting differential expression during fruit development and ripening (Choi et al., 2006). Antisense RNA inhibition of *SlExp1*, a ripening-specific expansin, resulted in firmer fruits that exhibited reduced polyuronide depolymerization but were not impaired in the breakdown of structurally important hemicelluloses, a major component of fruit softening. The transgenic fruit with constitutive expression of *SlExp1* had an opposite phenotype, showing enhanced fruit softening correlated with the precocious and extensive depolymerization of structural hemicelluloses (Brummell et al., 1999). However, polyuronide depolymerization was not affected. It was proposed (Brummell et al., 1999) that *Exp1* regulates polyuronide depolymerization late in ripening by direct modulation of relaxation of the cell walls, likely by controlling access of a pectinase to its substrate. The depolymerization of hemicellulose occurs independent of or requires only very small amounts of Exp1 protein. Simultaneous downregulation of *Exp1* and PG resulted in fruit that retained firmer texture and maintained cellular integrity for a longer period compared to parental wild-type fruit (Powell et al., 2003).

Protein glycosylation

Protein glycosylation may play an important role in regulating fruit softening and control shelf life. The role of protein glycosylation in fruit ripening was first demonstrated by treating tomato pericarp disc with tunicamycin, which inhibits the UDP-HexNAc:polyprenol-P HexNAc-1-P family of enzymes and blocks the synthesis of all N-linked glycoproteins (N-glycans). Tunicamycine treatment impairs their ripening (Handa et al., 1985). Later, it was reported that injecting Man3(Xyl)GlcNAc(Fuc)GlcNAc and Man3GlcNAc into fruit enhanced fruit ripening (Priem and Gross, 1992). Genetic proof of this concept was only recently provided when N-glycan processing enzymes were suppressed and the resulting transgenic tomato fruit found to have enhanced fruit shelf life (Meli et al., 2010). These authors have identified two ripening-specific N-glycoprotein modifying enzymes, α-Man and β-Hex, and showed that their suppression reduced the rate of ripening-associated softening leading to improved fruit shelf life. Over-expression of *α-Man* or *β-Hex* genes caused excessive softening of the transgenic fruit. These studies provided a novel way to alter fruit ripening and extend their shelf life.

Future Perspectives

Fruit development and ripening have occupied a major stage in plant biology and horticulture. Economic losses due to short shelf life of produce have intensified research in this important basic and applied field. Therefore, it is not surprising that the first genetically engineered marketed crop was a fruit, tomato (or you could call it a vegetable). It has been known for a long time that ethylene plays a central role in fruit ripening, and only recently the recognition has also dawned that other molecules and/or processes likely interact and impact the fruit physiology and desirable attributes of edible fruits. Molecular dissection of fruit development and ripening processes has revealed a map of complex interactions that regulate fruit quality and shelf life. Thus, studies with mutants impacted in ripening, nutritional attributes, or photomorphogenesis have shown the importance of not only ethylene biosynthesis and perception, but also a connection with light responsive elements in achieving the desired changes. In addition to ethylene, other phytohormones including auxin, cytokinin, gibberellins, ABA, jasmonates, and brassinosteroids play significant roles in fruit development and ripening (Srivastava and Handa, 2005; Seymour et al., 2008). Recently, polyamines have emerged as significant growth regulators of both fruit quality and shelf life (Mehta et al., 2002; Nambeesan et al., 2008, 2010). The tools of breeding and molecular genetics are providing significant insight into specific genetic determinants critical in improving the shelf life and response of fruits to abiotic and biotic signals. While research is ongoing to meet such major challenges, it is becoming clear that added efforts are needed to unravel the role of epigenetic factors in these processes (Martienssen and Colot, 2001; Seymour et al., 2008). The role of protein glycosylation, particularly of those enzymes that modify N-glycosylation of enzymes, in altering shelf life and fruit texture has been demonstrated. Thus, added challenges include identifying the hierarchy of regulators or a specific pattern of events that control desirable attributes and then use genetic intervention to modify critical and beneficial processes without any negative impact to the plant, ecosystem in which it is grown, or to the consumer. There is a definite need to develop and field test transgenic plants that are particularly suited to grow well in ecologically friendly, sustainable agricultural systems that have minimal reliance on chemical input and synergistically (positively) influence plant metabolism (Neelam et al., 2008; Mattoo and Teasdale, 2010).

References

Abeles, F., Morgan, P., & Saltveit, M. (1992). *Ethylene in plant biology*. San Diego: Academic Press.

Adams, D. O., & Yang, S. F. (1979). Ethylene biosynthesis: Identification of 1-aminocyclopropane-1-carboxylic acid as an intermediate in the conversion of methionine to ethylene. *Proceedings of the National Academy of Sciences of the United States of America*, *76*, 170–174.

Adams-Phillips, L., Barry, C., Kannanz, P., Leclercq, J., Bouzayen, M., & Giovannoni, J. (2004). Evidence that CTR1-mediated ethylene signal transduction in tomato is encoded by a multigene family whose members display distinct regulatory features. *Plant Molecular Biology*, *54*, 387–404.

Alba, R., Payton, P., Fei, Z., McQuinn, R., Debbie, P., & Martin, G., et al. (2005). Transcriptome and selected fruit metabolite analysis reveal multiple points of ethylene regulatory control during tomato fruit development. *Plant Cell*, *17*, 2954–2965.

Alonso, J. M., Hirayama, T., Roman, G., Nourizadeh, S., & Ecker, J. R. (1999). EIN2, a bifunctional transducer of ethylene and stress responses in *Arabidopsis*. *Science*, *284*, 2148–2152.

Ampomah-Dwamena, C., Morris, B., Sutherland, P., Veit, B., & Yao, J. (2002). Down-regulation of TM29, a tomato SEPALLATA homolog, causes parthenocarpic fruit development and floral reversion. *Plant Physiology*, *130*, 605–617.

Azari, R., Reuveni, M., Evenor, D., Nahon, S., Shlomo, H., & Chen, L., et al. (2010). Overexpression of UV-DAMAGED DNA BINDING PROTEIN 1 links plant development and phytonutrient accumulation in high pigment-1 tomato. *Journal of Experimental Botany*, *61*, 3627–3637.

Azari, R., Tadmor, Y., Meir, A., Reuveni, M., Evenor, D., & Nahon, S., et al. (2010). Light signaling genes and their manipulation towards modulation of phytonutrient content in tomato fruits. *Biotechnology Advances*, *1*, 108–118.

Baldet, P., Hernould, M., Laporte, F., Mounet, F., Just, D., & Mouras, A., et al. (2006). The expression of cell proliferation-related genes in early developing flowers is affected by a fruit load reduction in tomato plants. *Journal of Experimental Botany*, *57*, 961–970.

Bapat, V. A., Trivedi, P. K., Ghosh, A., Sane, V. A, Ganapathi, T. R., & Nath, P. (2010). Ripening of fleshy fruit: Molecular insight and the role of ethylene. *Biotechnology Advances*, *28*, 94–107.

Barrero, L. S., Cong, B., Wu, F., & Tanksley, S. D. (2006). Developmental characterization of the *fasciated* locus and mapping of *Arabidopsis* candidate genes involved in the control of floral meristem size and carpel number in tomato. *Genome*, *49*, 991–1006.

Barrero, L. S., & Tanksley, S. D. (2004). Evaluating the genetic basis of multiple-locule fruit in a broad cross section of tomato cultivars. *Theoretical and Applied Genetics*, *109*, 669–679.

Barry, C. S., & Giovannoni, J. (2006). Ripening inhibition in the tomato *Green-ripe* mutant results from ectopic expression of a novel protein which disrupts ethylene signal transduction. *Proceedings of the National Academy of Sciences of the United States of America*, *103*, 7923–7928.

Barry, C. S., Llop-Tous, I., & Grierson, D. (2000). The regulation of 1-aminocyclopropane-1-carboxylic acid synthase gene expression during the transition from system-1 to system-2 ethylene synthesis in tomato. *Plant Physiology*, *123*, 979–986.

Barry, C. S., McQuinn, R. P., Chung, M. Y., Besuden, A., & Giovannoni, J. J. (2008). Amino acid substitutions in homologs of the STAY-GREEN protein are responsible for the green-flesh and chlorophyll retainer mutations of tomato and pepper. *Plant Physiology*, *147*, 179–187.

Barry, C. S., McQuinn, R. P., Thompson, A. J., Seymour, G. B., Grierson, D., & Giovannoni, J. J. (2005). Ethylene insensitivity conferred by the Green-ripe and Never-ripe 2 ripening mutants of Tomato. *Plant Physiology*, *138*, 267–275.

Benvenuto, G., Formiggini, F., Laflamme, P., Malakhov, M., & Bowler, C. (2002). The photomorphogenesis regulator DET1 binds the amino-terminal tail of histone H2B in a nucleosome context. *Current Biology*, *12*, 1529–1534.

Bertin, N., Causse, M., Brunel, B., Tricon, D., & Génard, M. (2009). Identification of growth processes involved in QTLs for tomato fruit size and composition. *Journal of Experimental Botany*, *60*, 237–248.

Biggs, M. S., & Handa, A. K. (1989). Temporal regulation of PG gene expression in fruits from normal, mutant and heterozygous tomato genotypes. *Plant Physiology*, *89*, 117–125.

Boss, P. K., Sensi, E., Hua, C., Davies, C., & Thomas, M. R. (2002). Cloning and characterization of grapevine (*Vitis vinifera* L.) MADS-box genes expressed during inflorescence and berry development. *Plant Science*, *162*, 887–895.

Boss, P. K., Vivier, M., Matsumoto, S., Dry, I. B., & Thomas, M. R. (2001). A cDNA from grapevine (*Vitis vinifera* L.), which shows homology to AGAMOUS and SHATTERPROOF, is not only expressed in flowers but also throughout berry development. *Plant Molecular Biology*, *45*, 541–553.

Brady, C. J. (1987). Fruit ripening. *Annual Review of Plant Physiology*, *38*, 155–178.

Brewer, M. T., Moyseenko, J. B., Monforte, A. J., & van der Knaap, E. (2007). Morphological variation in tomato: A comprehensive study of quantitative trait loci controlling fruit shape and development. *Journal of Experimental Botany*, *58*, 1339–1349.

Brummell, D. A. (2006). Cell wall disassembly in ripening fruit. *Functional Plant Biology*, *33*, 103–119.

Brummell, D. A., Hall, B. D., & Bennett, A. B. (1999). Antisense suppression of tomato endo-beta-1,4-glucanase Cel2 mRNA accumulation increases the force required to break fruit abscission zones but does not affect fruit softening. *Plant Molecular Biology*, *40*, 615–622.

Brummell, D. A., Harpster, M. H., Civello, P. M., Bennett, A. B., & Dunsmuir, P. (1999). Modification of expansin protein abundance in tomato fruit alters softening and cell wall polymer metabolism during ripening. *Plant Cell*, *11*, 2203–2216.

Brummell, D. A., Harpster, M. H., & Dunsmuir, P. (1999). Differential expression of expansin gene family members during growth and ripening of tomato fruit. *Plant Molecular Biology*, *39*, 161–169.

Busi, M. V., Bustamante, C., D'Angelo, C., Hidalgo-Cuevas, M., Boggio, S. B., & Valle, E. M., et al. (2003). MADS-box genes expressed during tomato seed and fruit development. *Plant Molecular Biology*, *52*, 801–815.

Cara, B., & Giovannoni, J. J. (2008). Molecular biology of ethylene during tomato fruit development and maturation. *Plant Science*, *175*, 106–113.

Carey, A. T., Smith, D. L., Harrison, E., Bird, C. R., Gross, K. C., & Seymour, G. B., et al. (2001). Down-regulation of a ripening-related beta-galactosidase gene (TBG1) in transgenic tomato fruits. *Journal of Experimental Botany*, *52*, 663–668.

Carrari, F., Baxter, C., Usadel, B., Urbanczyk-Wochniak, E., Zanor, M. -I., & Nunes-Nesi, A., et al. (2006). Integrated analysis of metabolite and transcript levels reveals the metabolic shifts that underlie tomato fruit development and highlight regulatory aspects of metabolic network behavior. *Plant Physiology*, *142*, 1380–1396.

Cevik, V., Ryder, C. D., Popovich, A., Manning, K., King, G. J., & Seymour, G. B. (2010). A FRUITFULL-like gene is associated with genetic variation for fruit flesh firmness in apple (*Malus domestica Borkh.*). *Tree Genetics Genomes*, *6*, 271–279.

Chen, G., Alexander, L., & Grierson, D. (2004). Constitutive expression of EIL-like transcription factor partially restores ripening in the ethyleneinsensitive Nr tomato mutant. *Journal of Experimental Botany*, *55*, 1491–1497.

Chen, Y. F., Randlett, M. D., Findell, J. L., & Schaller, G. E. (2002). Localization of the ethylene receptor ETR1 to the endoplasmic reticulum of *Arabidopsis*. *Journal of Biological Chemistry*, *277*, 19861–19866.

Choi, D., Cho, H. T., & Lee, Y. (2006). Expansins: Expanding importance in plant growth and development. *Physiologia Plantarum*, *126*, 511–518.

Chun, J. P., & Huber, D. J. (1998). Polygalacturonase-mediated solubilization and depolymerization of pectic polymers in tomato fruit cell walls—regulation by pH and ionic conditions. *Plant Physiology*, *117*, 1293–1299.

Chung, M.-Y., Vrebalov, J., Alba, R., Lee, J., McQuinn, R., Chung, J.-D., et al. (2010). A tomato (*Solanum lycopersicum*) *APETALA2/ERF* gene, *SlAP2a*, is a negative regulator of fruit ripening. *Plant J.*, *64*(6), 936–947.

Clark, KL, Larsen, PB, Wang, X, & Chang, C. (1998). Association of the *Arabidopsis* CTR1 Raf-like kinase with the ETR1 and ERS ethylene receptors. *Proceedings of the National Academy of Sciences of the United States of America*, *95*, 5401–5406.

Cong, B., Barrero, L. S., & Tanksley, S. D. (2008). Regulatory change in YABBY-like transcription factor led to evolution of extreme fruit size during tomato domestication. *Nature Genetics*, *40*, 800–804.

Cong, B., Liu, J., & Tanksley, S. D. (2002). Natural alleles at a tomato fruit size quantitative trait locus differ by heterochronic regulatory mutations. *Proceedings of the National Academy of Sciences of the United States of America*, *99*, 13606–13611.

Cong, B., & Tanksley, S. D. (2006). FW2.2 and cell cycle control in developing tomato fruit: A possible example of gene co-option in the evolution of a novel organ. *Plant Molecular Biology*, *62*, 867–880.

Dahan, Y., Rosenfeld, R., Zadiranov, V., & Irihimovitch, V. (2010). A proposed conserved role for an avocado fw2.2-like gene as a negative regulator of fruit cell division. *Planta*, *232*, 663–676.

Dandekar, A. M., Teo, G., Defilippi, B. G., Uratsu, S. L., Passey, A. J., & Kader, A. A., et al. (2004). Effect of down-regulation of ethylene biosynthesis on fruit flavor complex in apple fruit. *Transgenic Research*, *13*, 373–384.

Davuluri, G. R., van Tuinen, A., Fraser, P. D., Manfredonia, A., Newman, R., & Burgess, D., et al. (2005). Fruit specific RNAi-mediated suppression of DET1 enhances carotenoid and flavonoid content in tomatoes. *Nature Biotechnology*, *23*, 890–895.

Davuluri, G. R., van Tuinen, A., Mustilli, A. C., Manfredonia, A., Newman, R., & Burgess, D., et al. (2004). Manipulation of DET1 expression in tomato results in photomorphogenic phenotypes caused by post-transcriptional gene silencing. *Plant Journal*, *40*, 344–354.

de Silva, J., & Verhoeyen, M. E. (1998). Production and characterization of antisense-exogalactanase tomatoes. In H. A. Kuiper (Ed.), *Report of the demonstration programme on food safety evaluation of genetically modified foods as a basis for market introduction* (pp. 99–106). The Hague: The Netherlands Ministry of Economic Affairs.

Deikman, J., Xu, R. L., Kneissl, M. L., Ciardi, J. A., Kim, K. N., & Pelah, D. (1998). Separation of cis elements responsive to ethylene, fruit development, and ripening in the 50-flanking region of the ripening-related E8 gene. *Plant Molecular Biology*, *37*, 1001–1011.

DellaPenna, D., Lincoln, J. E., Fischer, R. L., & Bennett, A. B. (1989). Transcriptional analysis of PG and other ripening associated genes in Rutgers, *rin*, *nor*, and *Nr* tomato fruit. *Plant Physiology*, *90*, 1372–1377.

Elitzur, T., Vrebalov, J., Giovannoni, J. J., Goldschmidt, E. E., & Friedman, H. (2010). The regulation of MADS-box gene expression during ripening of banana and their regulatory interaction with ethylene. *Journal of Experimental Botany*, *61*, 1523–1535.

Errington, N., Tucker, G. A., & Mitchell, J. R. (1998). Effect of genetic down-regulation of PG and pectin esterase activity on rheology and composition of tomato juice. *Journal of the Science of Food and Agriculture*, *76*, 515–519.

Ezura, H., & Hiwasa-Tanase, K. (2010). Fruit Development. In E. C. Pua & M. R. Davey (Eds.), *Plant developmental biology – Biotechnological perspectives* (Vol. 1, pp. 301–318). Berlin Heidelberg: Springer-Verlag.

Fatima, T., Rivera-Domínguez, M., Troncoso-Rojas, R., Tiznado-Hernández, M. E., Handa, A. K., & Mattoo, A. K. (2008). Tomato. In C. Kole & T. C. Hall (Eds.), *Compendium of transgenic crop plants: Transgenic vegetable crops* (pp. 1–46). Oxford, UK: Blackwell Publishing.

Ferrándiz, C., Liljegren, S. J., & Yanofsky, M. F. (2000). Negative regulation of the *SHATTERPROOF* genes by FRUITFULL during *Arabidopsis* fruit development. *Science*, *289*, 436–438.

Flores, F. B., Martínez-Madrid, M. C., Sánchez-Hidalgo, F. J., & Romojaro, F. (2001). Differential rind and pulp ripening of transgenic antisense ACC oxidase melon. *Plant Physiology and Biochemistry*, *39*, 37–43.

Fluhr, R., & Mattoo, A. K. (1996). Ethylene – Biosynthesis and perception. In B. V. (1996). Conger (Ed.), *Critical reviews in plant sciences* (Vol. 15, pp. 479–523). Boca Raton, Florida: CRC Press, Inc..

Frary, A., Fulton, T. M., Zamir, D., & Tanksley, S. D. (2004). Advanced backcross QTL analysis of a *Lycopersicon esculentum* × *L. pennellii* cross and identification of possible orthologs in the *Solanaceae*. *Theoretical and Applied Genetics*, *108*, 485–496.

Frary, A., Nesbitt, T. C., Frary, A., Grandillo, S., van der Knaap, E., & Cong, B., et al. (2000). Cloning and transgenic expression of *fw2.2*: A quantitative trait locus key to the evolution of tomato fruit size. *Science*, *289*, 85–88.

Fraser, P. D., Bramley, P., & Seymour, G. B. (2001). Effect of the *Cnr* mutation on carotenoid formation during tomato fruit ripening. *Phytochemistry*, *58*, 75–79.

Gaffe, J., Tieman, D. M., & Handa, A. K. (1994). Pectin methyesterase isoforms in tomato (*Lycopersicon esculentum*) tissues: Effects of expression of a pectinmethylesterase antisense gene. *Plant Physiology*, *105*, 199–203.

Gaffe, J., Tiznado, M. E., & Handa, A. K. (1997). Characterization and functional expression of a ubiquitously expressed tomato pectin methylesterase. *Plant Physiology*, *114*, 1547–1556.

Gallie, D. R. (2010). Regulated ethylene insensitivity through the inducible expression of the *Arabidopsis etr1-1* mutant ethylene receptor in tomato. *Plant Physiology*, *152*, 1928–1939.

Galpaz, N., Wang, Q., Menda, N., Zamir, D., & Hirschberg, J. (2008). Abscisic acid deficiency in the tomato mutant high-pigment 3 leading to increased plastid number and higher fruit lycopene content. *Plant Journal*, *53*, 717–730.

Gambetta, G. A., Matthews, M. A., Shaghasi, T. H., McElrone, A. J., & Castellarin, S. D. (2010). Sugar and abscisic acid signaling orthologs are activated at the onset of ripening in grape. *Planta*, *232*, 219–234.

Gao, H., Zhu, H., Shao, Y., Chen, A., Lu, C., & Zhu, B., et al. (2008). Lycopene accumulation affects the biosynthesis of some carotenoid-related volatiles independent of ethylene in tomato. *International Journal of Plant Biology*, *50*, 991–996.

Gao, M., Matsuta, N., Murayama, H., Toyomasu, T., Mitsuhashi, W., & Dandekar, A. M., et al. (2007). Gene expression and ethylene production in transgenic pear (*Pyrus communis* cv. 'La France') with sense or antisense cDNA encoding ACC oxidase. *Plant Science*, *173*, 32–42.

Giliberto, L., Perrotta, G., Pallara, P., Weller, J. L., Fraser, P. D., & Bramley, P. M., et al. (2005). Manipulation of the blue light photoreceptor cryptochrome 2 in tomato affects vegetative development, flowering time, and fruit antioxidant content. *Plant Physiology*, *137*, 199–208.

Gillaspy, G., Ben-David, H., & Gruissem, W. (1993). Fruits: A developmental perspective. *Plant Cell*, *5*, 1439–1451.

Giovannoni, J. (2007). Fruit ripening mutants yields insights into ripening control. *Current Opinion in Plant Biology*, *10*, 283–289.

Giovannoni, J. J. (2004). Genetic regulation of fruit development and ripening. *Plant Cell*, *16*, S170–S180.

Giovannoni, J. J., DellaPenna, D., Bennett, A. B., & Fischer, R. L. (1989). Expression of a chimeric polygalacturonase gene in transgenic *rin* (ripening inhibitor) tomato fruit results in polyuronide degradation but not fruit softening. *Plant Cell*, *1*, 53–63.

Gonzalez, N., Gevaudant, F., Hernould, M., Chevalier, C., & Mouras, A. (2007). The cell cycle-associated protein kinase WEE1 regulates cell size in relation to endoreduplication in developing tomato fruit. *Plant Journal*, *54*, 642–655.

Gonzalo, M. J., & van der Knaap, E. (2008). A comparative analysis into the genetic bases of morphology in tomato varieties exhibiting elongated fruit shape. *Theoretical and Applied Genetics*, *116*, 647–656.

Good, X., Kellogg, J. A., Wagoner, W., Langhoff, D., Matsumura, W., & Bestwick, R. K. (1994). Reduced ethylene synthesis by transgenic tomatoes expressing S-adenosylmethionine hydrolase. *Plant Molecular Biology*, *26*, 781–790.

Grandillo, S., Ku, H. -M., & Tanksley, S. D. (1999). Identifying loci responsible for natural variation in fruit size and shape in tomato. *Theoretical and Applied Genetics*, *99*, 978–987.

Gross, K. C., & Sams, C. E. (1984). Changes in cell wall neutral sugar composition during fruit ripening: A species survey. *Phytochemistry*, *23*, 2457–2461.

Guo, H., & Ecker, J. (2003). Plant responses to ethylene gas are mediated by SCFEBF1/EBF2-dependent proteolysis of EIN3 transcription factor. *Cell*, *115*, 667–677.

Guo, M., Rupe, M. A., Dieter, J. A., Zou, J., Spielbauer, D., & Duncan, K. E., et al. (2010). *Cell Number Regulator1* affects plant and organ size in maize: Implications for crop yield enhancement and heterosis. *Plant Cell*, *22*, 1057–1073.

Hamilton, A., Lycett, G., & Grierson, D. (1990). Antisense gene that inhibits synthesis of the hormone ethylene in transgenic plants. *Nature*, *346*, 284–287.

Handa, A. K., Singh, N. K., & Biggs, M. S. (1985). Effect of tunicamycin on *in vitro* ripening of tomato pericarp tissue. *Physiologia Plantarum*, *63*, 417–424.

Harpster, M. H., Brummell, D. A., & Dunsmuir, P. (2002). Suppression of a ripening-related endo-1,4-β-glucanase in transgenic pepper fruit does not prevent depolymerization of cell wall polysaccharides during ripening. *Plant Molecular Biology*, *50*, 345–355.

Harpster, M. H., Dawson, D. M., Nevins, D. J., Dunsmuir, P., & Brummell, D. A. (2002). Constitutive overexpression of a ripening-related pepper endo-1,4-β-glucanase in transgenic tomato fruit does not increase xyloglucan depolymerization or fruit softening. *Plant Molecular Biology*, *50*, 357–369.

Harriman, R. W., Tieman, D. M., & Handa, A. K. (1991). Molecular cloning of tomato pectin methylesterase gene and its expression in Rutgers, *rin*, *nor* and *Nr* tomato mutants. *Plant Physiology*, *97*, 80–87.

Hua, J., & Meyerowitz, E. (1998). Ethylene responses are negatively regulated by a receptor gene family in *Arabidopsis thaliana*. *Cell*, *94*, 261–271.

Huberman, M., Riov, J., Goldschmidt, E. E., Apelbaum, A., & Goren, R. (2009). The antagonizing effect of a water soluble cyclopropene derivative 3-(cyclopropyl-1-enyl-propanoic acid) sodium salt on ethylene-induced action in field crops. Thirty-seventh PGRSA Annual Meeting in Portland, Oregon, August 7–12, 2010.

Ito, Y., Kitagawa, M., Ihashi, N., Yabe, K., Kimbara, J., & Yasuda, J., et al. (2008). DNA-binding specificity, transcriptional activation potential, and the rin mutation effect for the tomato fruit-ripening regulator RIN. *Plant Journal*, *55*, 212–223.

Jiménez-Bermúdez, S., Redondo-Nevado, J., Muñoz-Blanco, J., Caballero, J. L., López-Aranda, J. M., & Valpuesta, V., et al. (2002). Manipulation of fruit softening by antisense expression of a pectate lyase gene. *Plant Physiology*, *128*, 751–759.

Kalamaki, M. S., Harpster, M. H., Palys, J. M., Labavitch, J. M., Reid, D. S., & Brummell, D. A. (2003). Simultaneous transgenic suppression of LePG and LeExp1 influences rheological properties of juice and concentrates from a processing tomato variety. *Journal of Agricultural and Food Chemistry*, *51*, 7456–7464.

Kendrick, R. E., Kerckhoffs, L. H. J., Pundsnes, A. S., Van Tuinen, A., Koornneef, M., & Nagatani, A., et al. (1994). Photomorphogenic mutants of tomato. *Euphytica*, *79*, 227–234.

Kevany, B. M., Taylor, M. G., & Klee, H. J. (2008). Fruit-specific suppression of the ethylene receptor *LeETR4* results in early-ripening tomato fruit. *Plant Biotechnology Journal*, *6*, 295–300.

Kevany, B. M., Tieman, D. M., Taylor, M. G., Cin, V. D., & Klee, H. J. (2007). Ethylene receptor degradation controls the timing of ripening in tomato fruit. *Plant Journal*, *51*, 458–467.

Kieber, J. J., Rothenberg, M., Roman, G., Feldman, K. A., & Ecker, J. R. (1993). CTR1, a negative regulator of the ethylene response pathway in *Arabidopsis*, encodes a member of the Raf family of protein kinases. *Cell*, *72*, 427–441.

Klee, H., & Tieman, D. (2002). The tomato ethylene receptor gene family: Form and function. *Physiologia Plantarum*, *115*, 336–341.

Klee, H. J. (1993). Ripening physiology of fruit from transgenic tomato (*Lycopersicon esculentum*) plants with reduced ethylene synthesis. *Plant Physiology*, *102*, 911–916.

Klee, H. J., Hayford, M. B., Kretzmer, K. A., Barry, G. F., & Kishore, G. M. (1991). Control of ethylene synthesis by expression of a bacterial enzyme in transgenic tomato plants. *Plant Cell*, *3*, 1187–1193.

Klee, H. J. (2004). Ethylene signal transduction. Moving beyond *Arabidopsis*. *Plant Physiology*, *135*, 660–667.

Kosma, D. K., Parsons, E. P., Isaacson, T., Lü, S., Rose, J. K., & Jenks, M. A. (2010). Fruit cuticle lipid composition during development in tomato ripening mutants. *Physiologia Plantarum*, *139*, 107–117.

Kramer, M. G., & Redenbaugh, K. (1994). Commercialization of a tomato with an antisense polygalacturonase gene: The FLAVR SAVR™ tomato story. *Euphytica*, *79*, 293–297.

Ku, H. -M., Doganlar, S., Chen, K. Y., & Tanksley, S. D. (1999). The genetic basis of pear shaped tomato fruit. *Theoretical and Applied Genetics*, *99*, 844–850.

Ku, H. -M., Grandillo, S., & Tanksley, S. D. (2000). *fs8.1*, a major QTL, sets the pattern of tomato carpel shape well before anthesis. *Theoretical and Applied Genetics*, *101*, 873–878.

Kuzemenskii, A. V. (2007). Effects of interaction of alc (alcobaca) keeping-life gene with elevated fruit pigmentation genes. *Cytology and Genetics*, *41*, 357–364.

Lanahan, M. B., Yen, H. C., Giovannoni, J. J., & Klee, H. J. (1994). The *never ripe* mutation blocks ethylene perception in tomato. *Plant Cell*, *6*, 521–530.

Lashbrook, C. C., Giovannoni, J. J., Hall, B. D., & Bennett, A. B. (1998). Transgenic analysis of tomato endo-ß-1,4-glucanase gene function. Role of cel1 in floral abscission. *Plant Journal*, *13*, 303–310.

Lavi, N., Tadmor, Y., Meir, A., Bechar, A., Oren-Shamir, M., & Ovadia, R., et al. (2009). Characterization of the INTENSE PIGMENT tomato genotype emphasizing targeted fruit metabolites and chloroplast biogenesis. *Journal of Agricultural and Food Chemistry*, *57*, 4818–4826.

Levin, I., De Vos, C. H. R., Tadmor, Y., Bovy, A., Lieberman, M., & Oren-Shamir, M., et al. (2006). High pigment tomato mutants—more than just lycopene (a review). *Israel Journal of Plant Sciences*, *54*, 179–190.

Levin, I., Frankel, P., Gilboa, N., Tanny, S., & Lalazar, A. (2003). The tomato dark green mutation is a novel allele of the tomato homolog of the DEETIOLATED1 gene. *Theoretical and Applied Genetics*, *106*, 454–460.

Li, W., Kabbage, M., & Dickman, M. B. (2010). Transgenic expression of an insect inhibitor of apoptosis gene, Sf-IAP, confers abiotic and biotic stress tolerance and delays tomato fruit ripening. *Physiological and Molecular Plant Pathology*, *74*, 363–375.

Lieberman, M., Kunishi, A., Mapson, L. W., & Wardale, D. A. (1966). Stimulation of ethylene production in apple tissue slices by methionine. *Plant Physiology*, *41*, 376–382.

Lieberman, M., Segev, O., Gilboa, N., Lalazar, A., & Levin, I. (2004). The tomato homolog of the gene encoding UV-damaged DNA binding protein 1 (DDB1) underlined as the gene that causes the high pigment-1 mutant phenotype. *Theoretical and Applied Genetics*, *108*, 1574–1581.

Liebhard, R., Kellerhals, M., Pfammatter, W., Jertmini, M., & Gessler, C. (2003). Mapping quantitative physiological traits in apple (*Malus × domestica Borkh*). *Plant Molecular Biology*, *52*, 511–526.

Lin, S. -H., & Do, Y. -Y. (2008). Studies on delaying fruit ripening by transforming sense ACC synthase gene into *Momordica charantia* L. *Journal of Taiwan Society for Horticultural Science*, *54*, 115–125.

Lin, Z., Alexander, L., Hackett, R., & Grierson, D. (2008). LeCTR2, a CTR1-like protein kinase from tomato, plays a role in ethylene signaling, development and defense. *Plant Journal*, *54*, 1083–1093.

Lin, Z., Zhong, S., & Grierson, D. (2009). Recent advances in ethylene research. *Journal of Experimental Botany*, *60*, 3311–3336.

Litz, R. E., Raharjo, S. H. T., Efendi, D., Witjaksono, M., & Gómez-Lim, M. A. (2007). Plant recovery following transformation of avocado with anti-fungal protein and SAM hydrolase genes. *Acta Horticulturae*, *738*, 447–450.

Liu, J., Van Eck, J., Cong, B., & Tanksley, S. D. (2002). A new class of regulatory genes underlying the cause of pear-shaped tomato fruit. *Proceedings of the National Academy of Sciences of the United States of America*, *99*, 13302–13306.

Liu, J., Xu, B., Hu, L., Li, M., Su, W., & Wu, J., et al. (2009). Involvement of a banana MADS-box transcription factor gene in ethylene-induced fruit ripening. *Plant Cell Reports*, *28*, 103–111.

Liu, Y., Roof, S., Ye, Z., Barry, C., van Tuinen, A., & Vrebalov, J., et al. (2004). Manipulation of light signal transduction as a means of modifying fruit nutritional quality in tomato. *Proceedings of the National*

Academy of Sciences of the United States of America, 26, 9897–9902.

López-Gómez, R., Cabrera-Ponce, J. L., Saucedo-Arias, L. J., Carreto-Montoya, L., Villanueva-Arce, R., & Díaz-Perez, J. C., et al. (2009). Ripening in papaya fruit is altered by ACC oxidase cosuppression. *Transgenic Research, 1*, 89–97.

Manning, R., Tor, K., Poole, M., Hong, M., Thompson, Y., & King, A., et al. (2006). A naturally occurring epigenetic mutation in a gene encoding an SPB-box transcription factor inhibits tomato fruit ripening. *Nature Genetics, 38*, 949–952.

Martienssen, R. A., & Colot, V. (2001). DNA methylation and epigenetic inheritance in plants and filamentous fungi. *Science, 293*, 1070–1074.

Martínez-Romero, D., Bailén, G., Serrano, M., Guillén, F., Valverde, J. M., & Zapata, P., et al. (2007). Tools to maintain postharvest fruit and vegetable quality through the inhibition of ethylene action: A review. *Critical Reviews in Food Science and Nutrition, 47*, 543–560.

Mattoo, A. K., & Suttle, J. C. (Eds.). (1991). *The plant hormone ethylene*. Boca Raton, Florida, USA: CRC Press Inc..

Mattoo, A. K., & Handa, A. K. (2004). Ethylene signaling in plant cell death. In L. Nooden (Ed.), *Plant cell death processes* (pp. 125–142). New York: Academic Press.

Mattoo, A. K., & Teasdale, J. R. (2010). Ecological and genetic systems underlying sustainable horticulture. *Horticultural Reviews, 37*, 331–362.

Mehta, R. A., Cassol, T., Li, N., Ali, N., Handa, A. K., & Mattoo, A. K. (2002). Engineered polyamine accumulation in tomato enhances phytonutrient content, juice quality and vine life. *Nature Biotechnology, 20*, 613–618.

Meli, V. S., Ghosh, S., Prabha, T. N., Chakraborty, N., Chakraborty, S., & Datta, A. (2010). Enhancement of fruit shelf life by suppressing N-glycan processing enzymes. *Proceedings of the National Academy of Sciences of the United States of America, 107*, 2413–2418.

Mercado, J. A., Trainotti, L., Jiménez-Bermúdez, L., Santiago-Doménech, N., Posé, S., & Donolli, R., et al. (2010). Evaluation of the role of the endo-β-(1,4)-glucanase gene *FaEG3* in strawberry fruit softening. *Postharvest Biology and Technology, 55*, 8–14.

Miedes, E., Herbers, K., Sonnewald, U., & Lorences, E. P. (2010). Overexpression of a cell wall enzyme reduces xyloglucan depolymerization and softening of transgenic tomato fruits. *Journal of Agricultural and Food Chemistry, 58*, 5708–5713.

Montgomery, J., Goldman, S., Deikman, J., Margossian, L., & Fischer, R. L. (1993). Identification of an ethylene-responsive region in the promoter of a fruit ripening gene. *Proceedings of the National Academy of Sciences of the United States of America, 90*, 5939–5943.

Mounet, F., Moing, A., Garcia, V., Petit, J., Maucourt, M., & Deborde, C., et al. (2009). Gene and metabolite regulatory network analysis of early developing fruit tissues highlights new candidate genes for the control of tomato fruit composition and development. *Plant Physiology, 149*, 1505–1528.

Moussatche, P., & Klee, H. (2004). Autophosphorylation activity of the *Arabidopsis* ethylene receptor multigene family. *Journal of Biological Chemistry, 279*, 48734–48741.

Mustilli, A. C., Fenzi, F., Ciliento, R., Alfano, F., & Bowler, C. (1999). Phenotype of the tomato *high pigment-2* mutant is caused by a mutation in the tomato homolog of DEETIOLATED1. *Plant Cell, 11*, 145–157.

Nambeesan, S., Datsenka, T., Ferruzzi, M. G., Malladi, A., Mattoo, A. K., & Handa, A. K. (2010). Overexpression of yeast spermidine synthase impacts ripening, senescence and decay symptoms in tomato. *Plant Journal, 63*, 836–847.

Nambeesan, S., Handa, A. K., & Mattoo, A. K. (2008). Polyamines and regulation of ripening and senescence. In G. Paliyath, D. P. Murr, A. K. Handa & S. Lurie (Eds.), *Postharvest biology and technology of fruits, vegetables and flowers* (pp. 319–340). Oxford, UK: Blackwell Publishers.

Neelam, A., Cassol, T., Mehta, R. A., Abdul-Baki, A. A., Sobolev, A., & Goyal, R. K., et al. (2008). A field-grown transgenic tomato line expressing higher levels of polyamines reveals legume cover crop mulch-specific perturbations in fruit phenotype at the levels of metabolite profiles, gene expression and agronomic characteristics. *Journal of Experimental Botany, 59*, 2337–2346.

Negi, P. S., & Handa, A. K. (2008). Structural deterioration of the produce—The breakdown of cell wall components. In G. Paliyath, D. P. Murr, A. K. Handa & S. Lurie (Eds.), *Postharvest biology and technology of fruits, vegetables and flowers* (pp. 162–194). Ames, IA: Wiley-Blackwell.

Nesbitt, C. T., & Tanksley, S. D. (2002). Comparative sequencing in the genus Lycopersicon: Implications for the evolution of fruit size in the domestication of cultivated tomatoes. *Genetics, 162*, 365–379.

Oeller, P. W., Wong, L. M., Taylor, L. P., Pike, D. A., & Theologis, A. (1991). Reversible inhibition of tomato fruit senescence by antisense 1-aminocyclopropane-1-carboxylate synthase. *Science, 254*, 427–439.

Orfila, C., Seymour, G. B., Willats, W. G. T., Huxham, I. M., Jarvis, M. C., & Dover, C. J., et al. (2001). Altered middle lamella homogalacturonan and disrupted deposition of (1- > 5)-α-L-arabinan in the pericarp of *Cnr*, a ripening mutant of tomato. *Plant Physiology, 126*, 210–221.

Pepper, A., Delaney, T., Washburn, T., Poole, D., & Chory, J. (1994). DET1, a negative regulator of light-mediated development and gene expression in *Arabidopsis*, encodes a novel nuclear-localized protein. *Cell, 78*, 109–116.

Phan, T. D., Bo, W., West, G., Lycett, G. W., & Tucker, G. A. (2007). Silencing of the major salt-dependent isoform of pectinesterase in tomato alters fruit softening. *Plant Physiology, 144*, 1960–1967.

Pnueli, L., Hareven, D., Rounsley, S. D., Yanofsky, M. F., & Lifschitz, E. (1994). Isolation of the tomato AGAMOUS gene TAG1 and analysis of its homeotic role in transgenic plants. *Plant Cell, 6*, 163–173.

Powell, A. L. T., Kalamaki, M. S., Kurien, P. A., Gurrieri, S., & Bennett, A. B. (2003). Simultaneous transgenic suppression of LePG and LeExp1 influences fruit texture and juice viscosity in a fresh market tomato variety. *Journal of Agricultural and Food Chemistry, 51*, 7450–7455.

Priem, B., & Gross, K. (1992). Mannosyl and Xylosyl-containing glycans promote tomato (*Lycopersicon esculentum Mill.*) fruit ripening. *Plant Physiology, 98*, 399–401.

Qiao, H., Chang, K. N., Yazaki, J., & Ecker, J. R. (2009). Interplay between ethylene, ETP1/ETP2 F-box proteins, and degradation of EIN2 triggers ethylene responses in *Arabidopsis*. *Genes and Development, 23*, 512–521.

Quesada, M. A., Blanco-Portales, R., Posé, S., García-Gago, J. A., Jiménez-Bermúdez, S., & Muñoz-Serrano, A., et al. (2009). Antisense down-regulation of the *FaPG1* gene reveals an unexpected central role for polygalacturonase in strawberry fruit softening. *Plant Physiology, 150*, 1022–1032.

Razdan, M. K., & Mattoo, A. K. (2007). *Genetic improvement of solanaceous crops: Volume 2: Tomato*. (p. 637) Enfield: Science Publishers, Inc.

Reynard, G. B. (1956). Origin of Webb special (Black Queen) in tomato. *Report of the Tomato Genetics Cooperative, 6*, 22.

Rick, C. M., & Butler, L. (1956). Phytogenetics of the tomato. *Advances in Genetics, 8*, 267–382.

Rigola, D., Pe, M. E., Fabrizio, C., Me, G., & Sari Gorla, M. (1998). CaMADS1, a MADS box gene expressed in the carpel of hazelnut. *Plant Molecular Biology, 38*, 1147–1160.

Rodriguez, F. I., Esch, J. J., Hall, A. E., Binder, B. M., Schaller, G. E., & Bleecker, A. B. (1999). A copper cofactor for the ethylene receptor ETR1 from *Arabidopsis*. *Science, 283*, 996–998.

Roman, G., Lubarsky, B., Kieber, J. J., Rothenberg, M., & Ecker, J. R. (1995). Genetic analysis of ethylene signal transduction in *Arabidopsis thaliana*: Five novel mutant loci integrated into a stress response pathway. *Genetics, 139*, 1393–1409.

Rose, J. K. C., Lee, H. H., & Bennett, A. B. (1997). Expression of a divergent expansin gene is fruit-specific and ripening-regulated. *Proceedings of the National Academy of Sciences of the United States of America, 94*, 5955–5960.

Saladié, M., Matas, A. J., Isaacson, T., Jenks, M. A., Goodwin, S. M., & Niklas, K. J., et al. (2007). A reevaluation of the key factors that influence tomato fruit softening and integrity. *Plant Physiology, 144*, 1012–1028.

Santiago-Doménech, N., Jiménez-Bemúdez, S., Matas, A. J., Rose, J. K. C., Muñoz-Blanco, J., & Mercado, J. A., et al. (2008). Antisense inhibition of a pectate lyase gene supports a role for pectin depolymerization in strawberry fruit softening. *Journal of Experimental Botany, 59*, 2769–2779.

Sato, T., & Theologis, A. (1989). Cloning the mRNA encoding 1-aminocyclopropane-1-carboxylate synthase, the key enzyme for ethylene biosynthesis in plants. *Proceedings of the National Academy of Sciences of the United States of America, 86*, 6621–6625.

Schäfer, E., & Bowler, C. (2002). Phytochrome-mediated photoperception and signal transduction in higher plants. *EMBO Reports, 3*, 1042–1048.

Schaller, G. E., Kieber, J. J., & Shiu, S. H. (2008). *Two-component signaling elements and histidyl-aspartyl phosphorelays. The Arabidopsis book*. Rockville, MD: American Society of Plant Biologists. (Online: <http://www.aspb.org/publications/arabidopsis/>)

Schroeder, D. F., Gahrtz, M., Maxwell, B. B., Cook, R. K., Kan, J. M., & Alonso, J. M., et al. (2002). De-etiolated 1 and damaged DNA binding protein 1 interact to regulate *Arabidopsis* photomorphogenesis. *Current Biology, 12*, 1462–1472.

Schuch, W., Kanczler, J, Robertson, D., Hobson, G., Tucker, G., & Grierson, D., et al. (1991). Fruit quality characteristics of transgenic tomato fruit with altered polygalacturonase activity. *HortScience, 26*, 1517–1520.

Seymour, G., Poole, M., Manning, K., & King, G. J (2008). Genetics and epigenetics of fruit development and ripening. *Current Opinion in Plant Biology, 11*, 58–63.

Seymour, G. B., Manning, K., Eriksson, E. M., Popovich, A. H., & King, G. J. (2002). Genetic identification and genomic organization of factors affecting fruit texture. *Journal of Experimental Botany, 53*, 2065–2071.

Smartt, J., & Simmonds, N. W. (1995). *Evolution of crop plants* (2nd ed.). (531 pp.) Essex, UK: Longman Scientific & Technical.

Smith, D. L., Abbott, J. A., & Gross, K. C. (2002). Down-regulation of tomato ß-galactosidase 4 results in decreased fruit softening. *Plant Physiology, 129*, 1755–1762.

Smith, D. L., & Gross, K. C. (2000). A family of a least seven β-galactosidase genes is expressed during tomato fruit development. *Plant Physiology, 123*, 1173–1183.

Solano, R., Stepanova, A., Chao, Q., & Ecker, J. R. (1998). Nuclear events in ethylene signaling: A transcriptional cascade mediated by ETHYLENE-INSENSITIVE3 and ETHYLENE-RESPONSE-FACTOR1. *Genes and Development, 12*, 3703–3714.

Soressi, G. P. (1975). New spontaneous or chemically-induced fruit ripening tomato mutants. *Report of the Tomato Genetics Cooperative, 25*, 21–22.

Spjut, R. W. (1994). A systematic treatment of fruit types. *Memoirs of the New York Botanical Garden, 70*, 1–182.

Srivastava, A., Gupta, A. K., Datsenka, T., Mattoo, A. K., & Handa, A. K. (2010). Maturity and ripening-stage specific modulation of tomato (*Solanum lycopersicum*) fruit transcriptome. *GM Crops, 1*, 1–13.

Srivastava, A., & Handa, A. K. (2005). Hormonal regulation of fruit development: A molecular perspective. *Journal of Plant Growth Regulation, 24*, 67–82.

Stuppy, W., & Kesseler, R. (2008). *Fruit. Edible, inedible, incredible*. (pp. 264) London: Papadakis.

Sung, S. K., Moon, Y. H., Chung, J. E., Lee, S. Y., Park, H. G., & An, G. (2001). Characterization of MADS box genes from hot pepper. *Molecules Cells, 11*, 352–359.

Tadiello, A., Pavanello, A., Zanin, D., Caporali, E., Colombo, L., & Rotino, G. L., et al. (2009). A *PLENA*-like gene of peach is involved in carpel formation and subsequent transformation into a fleshy fruit. *Journal of Experimental Botany, 60*, 651–661.

Tang, D., Christiansen, K. M., & Innes, R. W. (2005). Regulation of plant disease resistance, stress responses, cell death, and ethylene signaling in *Arabidopsis* by the EDR1 protein kinase. *Plant Physiology, 138*, 1018–1026.

Tanksley, S. D. (2004). The genetic, developmental, and molecular bases of fruit size and shape variation in tomato. *Plant Cell, 16*, S181–S189.

Thakur, B. R., Singh, R. K., & Handa, A. K. (1996). Effects of an antisense pectin methylesterase gene on the chemistry of pectins in tomato (*Lycopersicon esculentum*) fruit juice. *Journal of the Science of Food and Agriculture, 44*, 628–630.

Thakur, B. R., Singh, R. K., & Handa, A. K. (1997). Chemistry and uses of pectin—A review. *Critical Reviews in Food Science and Nutrition, 37*, 47–73.

Thakur, B. R., Singh, R. K., Tieman, D. M., & Handa, A. K. (1996). Quality of processed tomato products from transgenic tomato fruits with reduced levels of pectin methylesterase activity. *Journal of Food Science, 61*, 245–248.

Thompson, A., Barry, C., Jarvis, M., Vrebalov, J., Giovannoni, J., & Grierson, D., et al. (1999). Molecular and genetic characterization of a novel pleiotropic tomato-ripening mutant. *Plant Physiology, 120*, 383–389.

Tieman, D. M., Ciardi, J. A., Taylor, M. G., & Klee, H. J. (2001). Members of the tomato LeEIL (EIN3-like) gene family are functionally redundant and regulate ethylene responses throughout plant development. *Plant Journal, 26*, 47–58.

Tieman, D. M., & Handa, A. K. (1994). Reduction in pectin methylesterase activity modifies tissue integrity and cation levels in ripening tomato (*Lycopersicon esculentum* Mill.) fruits. *Plant Physiology, 106*, 429–436.

Tieman, D. M., Harriman, R. W., Ramamohan, G., & Handa, A. K. (1992). An antisense pectin methylesterase gene alters pectin chemistry and soluble solids in tomato fruit. *Plant Cell, 4*, 667–679.

Tieman, D. M., Taylor, M. G., Ciardi, J. A., & Klee, H. J. (2000). The tomato ethylene receptors NR and LeETR4 are negative regulators of ethylene response and exhibit functional compensation within a multigene family. *Proceedings of the National Academy of Sciences of the United States of America, 97*, 5663–5668.

Tigchelaar, E. C., McGlasson, W. B., & Buescher, R. W. (1978). Genetic regulation of tomato fruit ripening. *HortScience, 13*, 508–513.

Trusov, Y., & Botella, J. R. (2006). Delayed flowering in pineapples (*Ananas comosus* (L.) Merr.) caused by co-suppression of the AcACS2 gene. *Acta Horticulturae, 702*, 29–36.

van der Knaap, E., Lippman, Z. B., & Tanksley, S. D. (2002). Extremely elongated tomato fruit controlled by four quantitative trait loci with epistatic interactions. *Theoretical and Applied Genetics, 104*, 241–247.

van der Knaap, E., & Tanksley, S. D. (2003). The making of a bell pepper-shaped tomato fruit: Identification of loci controlling fruit morphology in Yellow Stuffer tomato. *Theoretical and Applied Genetics, 107*, 139–147.

Vicente, A. R., Saladié, M., Rose, J. K. C., & Labavitch, J. M. (2007). The linkage between cell wall metabolism and fruit softening: Looking to the future. *Journal of the Science of Food and Agriculture, 87*, 1435–1448.

Vrebalov, J., Pan, I., Matas Arroyo, A. J., McQuinn, R., Chung, M., & Poole, M., et al. (2009). Fleshy fruit expansion and ripening are regulated by the tomato SHATTERPROOF gene TAGL1. *Plant Cell, 21*, 3041–3062.

Vrebalov, J., Ruezinsky, D., Padmanabhan, V., White, R., Medrano, D., & Drake, R., et al. (2002). A MADS-box gene necessary for fruit ripening at the tomato *ripening-inhibitor* (*rin*) locus. *Science, 296*, 343–346.

Wang, A., Li, J., Zhang, B., Xu, X., & Bewley, J. D. (2009). Expression and location of endo-β-mannanase during the ripening of tomato fruit, and the relationship between its activity and softening. *Journal of Plant Physiology, 166*, 1672–1684.

Wang, H., Nussbaum-Wagler, T., Li, B., Zhao, Q., Vigouroux, Y., & Faller, M., et al. (2005). The origin of the naked grains of maize. *Nature, 436*, 714–719.

Wang, S., Chang, Y., Guo, J., & Chen, J. G. (2007). Arabidopsis Ovate Family Protein 1 is a transcriptional repressor that suppresses cell elongation. *Plant Journal, 50*, 858–872.

Wang, W., Hall, A., O'Malley, R., & Bleecker, A. B. (2003). Canonical histidine kinase activity of the transmitter domain of the ETR1 ethylene receptor from *Arabidopsis* is not required for signal transmission. *Proceedings of the National Academy of Sciences of the United States of America, 100*, 352–357.

Watson, C. F., Zheng, L. S., & DellaPenna, D. (1994). Reduction of tomato polygalacturonase beta-subunit expression affects pectin solubilization and degradation during fuit ripening. *Plant Cell, 6*, 1623–1634.

Weigel, D., & Meyerowitz, E. M. (1994). The ABCs of floral homeotic genes. *Cell, 78*, 203–209.

White, P. J. (2002). Recent advances in fruit development and ripening: An overview. *Journal of Experimental Botany, 53*, 1995–2000.

Whitelaw, C. A., Lyssenko, N. N., Chen, L., Zhou, D., Mattoo, A. K., & Tucker, M. L. (2002). Delayed abscission and shorter internodes correlate with a reduction in the ethylene receptor *LeETR1* transcript in transgenic tomato. *Plant Physiology, 128*, 978–987.

Wilkinson, J. Q., Lanahan, M., Yen, H., Giovannoni, J., & Klee, H. (1995). An ethylene-inducible component of signal transduction encoded by *Never-ripe*. *Science*, *270*, 1807–1809.

Xiao, H., Jiang, N., Schaffner, E., Stockinger, E. J., & van der Knaap, E. (2008). A retrotransposon-mediated gene duplication underlies morphological variation of tomato fruit. *Science*, *319*, 1527–1530.

Xiao, H., Radovich, C., Welty, N., Hsu, J., Li, D., & Meulia, T., et al. (2009). Integration of tomato reproductive developmental landmarks and expression profiles, and the effect of SUN on fruit shape. *BMC Plant Biology*, *9*, 49.

Xiong, A. S., Yao, Q. H., Peng, R. H., Li, X., Han, P. L., & Fan, H. Q. (2005). Different effects on ACC oxidase gene silencing triggered by RNA interference in transgenic tomato. *Plant Cell Reports*, *23*, 639–646.

Yamamoto, T., Shimada, T., Imai, T., Yaegaki, H., Haji, T., & Matsuta, N., et al. (2001). Characterization of morphological traits based on a genetic linkage map of peach. *Breeding Science*, *51*, 271–278.

Yao, J., Dong, Y., & Morris, B. (2001). Parthenocarpic apple fruit production conferred by transposon insertion mutations in a MADSbox transcription factor. *Proceedings of the National Academy of Sciences of the United States of America*, *98*, 1306–1311.

Yao, J. -L., Dong, Y. H., Kvarnheden, A., & Morris, B. A. M. (1999). Seven MADS-box genes in apple are expressed in different parts of the fruit. *Journal of the American Society for Horticultural Science*, *124*, 8–13.

Yokotani, N., Nakano, R., Imanishi, S., Nagata, M., Inaba, A., & Kubo, Y. (2009). Ripening-associated ethylene biosynthesis in tomato fruit is autocatalytically and developmentally regulated. *Journal of Experimental Botany*, *60*, 3433–3442.

Yoo, S. -D., Cho, Y., & Sheen, J. (2009). Emerging connections in the ethylene signalling network. *Trends in Plant Science*, *14*, 270–279.

Zhang, S. G., Jiang, Z. R., Xing, Y. P., & Li, C. L. (2008). Genetic analysis of fruit traits based on a double haploid population from a hybrid of pepper (*Capsicum annuum* L.). *Acta Horticulturae Sinica*, *35*, 515–520.

Zhang, Z., Huber, D. J., & Rao, J. (2010). Short-term hypoxic hypobaria transiently decreases internal ethylene levels and increases sensitivity of tomato fruit to subsequent 1-methylcyclopropene treatments. *Postharvest Biology and Technology*, *56*, 131–137.

Zhong, S. L., Lin, Z. F., & Grierson, D. (2008). Tomato ethylene receptor-CTR interactions, visualization of NEVER-RIPE interactions with multiple CTRs at the endoplasmic reticulum. *Journal of Experimental Botany*, *59*, 965–972.

Zhou, D., Kalaitzis, P., Mattoo, A. K., & Tucker, M. L. (1996). The mRNA for an ETR1 homologue in tomato is constitutively expressed in vegetative and reproductive tissues. *Plant Molecular Biology*, *30*, 1331–1338.

Zhu, H., Zhu, B., Shao, Y., Wang, X., Lin, X., & Xie, Y., et al. (2006). Tomato fruit development and ripening are altered by the silencing of LeEIN2 gene. *Journal of Integrative Plant Biology*, *48*, 1478–1485.

Zygier, S., Chaim, A. B., Efrati, A., Kaluzky, G., Borovsky, Y., & Paran, I. (2005). QTLs mapping for fruit size and shape in chromosomes 2 and 4 in pepper and a comparison of the pepper QTL map with that of tomato. *Theoretical and Applied Genetics*, *111*, 437–445.

Potential application of biotechnology to maintain fresh produce postharvest quality and reduce losses during storage

27

Amnon Lers

Department of Postharvest Science of Fresh Produce, Agricultural Research Organization, the Volcani Center, Bet-Dagan, Israel

TABLE OF CONTENTS

Introduction

A sufficient supply of food is anticipated to become a very serious problem for humanity in the near future (Hubert et al., 2010; Anon., 2010). Already today major parts of the world population suffer from hunger. A great deal of effort is invested in basic and applied research to develop new approaches and strategies to increase the supply of agriculture-based fresh produce. Many of these efforts are directed toward improving biomass/yield production, as well as improving plants' adaptation to environmental stress (Long and Ort, 2010). Whereas efforts to increase yield are likely to improve the fresh produce food supply, a complementary effort must be implemented to decrease the extremely heavy losses in both quality and quantity of the agricultural fresh produce already produced that occurs during the postharvest period (Kader, 2005).

During the time between harvest and consumption, fresh produce undergoes changes in different aspects that lead to a major reduction in overall quality. These changes can result in food waste and major economic damage. Quality losses, besides affecting the edibility of the fresh produce, may have a major impact on the nutritional, taste, or caloric value, or the acceptability of the produce to the consumer due to changes in the appearance of fruits, vegetables, or flowers. Worldwide losses of plant-origin fresh (non-processed) produce are estimated to be around 33% of that harvested (Kantor et al., 1997; Kader, 2005; Kitinoja and Cantwell, 2010). The degree of loss varies greatly and is dependent on the commodity and location of the produce. Accumulating losses between production and retail sites are estimated to be in the range of 2 to 24% in developed countries and 5 to 50% in developing countries. Further losses at retail and consumer sites make these losses even more dramatic, with higher percentages in the developed countries. Especially in the developing countries losses are due to insufficient professional knowledge, lack of resources, and industrial infrastructure including inaccessibility of advanced applied technologies required for storage of

DOI: 10.1016/B978-0-12-381466-1.00027-4

the fresh produce under optimal conditions. Still, losses of both quality and quantity are substantial even in developed countries where advanced postharvest professional knowledge and resources are available.

An important and, in many cases, crucial factor determining postharvest qualities of fresh produce such as fruit, flower, tuber, bulb, or leaf, and the ability to maintain these postharvest qualities, is the biology of the parent plant from which the produce is derived or harvested. Progress of different biochemical, physiological, and pathological processes that take place during postharvest life and storage of the fresh produce can eventually lead to accumulated losses with much of the fresh produce going to waste or, in the best case, being transferred to industrial use. The degree of postharvest losses varies depending on the commodity but also on the season and production site, which may indirectly affect the biology of the fresh produce and its behavior during postharvest life.

Genetically different plants have different physiological, biochemical, and molecular characteristics. These differences affect their postharvest behavior and the ability to be adequately stored to maintain their quality until consumed. Two examples of physiological characteristics that have large implications for postharvest are respiration rate and cold sensitivity. Commodities derived from plants or organs that have higher rates of respiration tend to have a more limited postharvest life and are perishable when compared with those that have low rates of respiration. Respiration plays a major role in fresh produce postharvest life, because it reflects the metabolic activity of the tissue. Cold sensitivity of fruits or vegetables limits our ability to apply low-temperature storage, which is one of the best and most efficient methods available for keeping fresh produce in long-term storage while maintaining its quality. The inherent cold sensitivity of plants is genetically encoded and dictates the lowest possible temperature for use without development of chilling injuries.

Traditionally, classical breeding has been used for domestication of crop plants, but novel biotechnology-based breeding offers important advantages in terms of both focusing on single traits or genes and the timescale required for progress in the development of new varieties with improved relevant agricultural properties. Detailed knowledge is now available regarding the relationship between the biology of the fresh produce and its postharvest characteristics. This knowledge has great potential to be used in biotechnology-based breeding to improve specific postharvest-related traits. Some postharvest-related physiological, or biochemical processes, such as respiration, might be biologically too complex to be modified using biotechnology. However, in some other cases, such as cold sensitivity, strategies for biotechnological manipulation for the benefit of postharvest quality can already be implemented, since proof of concepts in model plants have already been demonstrated.

In this chapter the biological processes most relevant to postharvest quality are discussed. Furthermore, some potential gene targets that could be used for biotechnologically-based improvements of postharvest quality are identified.

Although no full coverage of all, or even most, possibilities is claimed, we do claim that the time is right to successfully implement our insights on the molecular biology of plants and our knowledge of postharvest biology to employ biotechnology for the development of crop plants with better postharvest-related qualities.

Some of the major biological processes relevant to the way a given fresh produce will behave during its postharvest life, storage, and shelf life condition are discussed. It is very important to determine in advance, for a given fruit, which of these biological processes is most relevant for its postharvest behavior and the way quality is affected. Focus on the most relevant physiological aspects may result in high impact on postharvest qualities. In this chapter we include examples of genes that were found to be involved in postharvest-relevant processes. Some of these genes were identified in model plants such as *Arabidopsis*, and their relevance to the postharvest of crop plants must be examined. Still, these genes constitute great potential, and the current challenge is to forward and transform this knowledge for its application in the postharvest agricultural industry. This may result in very big increases in the availability of fresh food produce worldwide.

Ethylene Biosynthesis or Perception and Its Relation to Postharvest Quality of Fresh Produce

Ethylene has a central role in the regulation of different biological processes associated with the postharvest life of many different crops including fruits and vegetables, leafy vegetables, flowers, tubers, and bulbs. The progress of these processes during postharvest life of the fresh produce determines its quality and the level of losses. Postharvest-related processes in which ethylene action is well documented include fruit ripening, senescence, abscission, softening and texture changes, development of physiological disorders, taste and aroma development, nutritional value, and fruit–pathogen interactions (Saltveit, 1999). The biology and molecular biology of fruit ripening and the role of ethylene in this process and its implications to postharvest issues, as well as biotechnological approaches to control postharvest problems by manipulation of ethylene biosynthesis or perception, were subject to several reviews (Stearns and Glick, 2003; Martinez-Romero et al., 2007; Cara and Giovannoni, 2008; Lin et al., 2009; Little et al., 2009; Matas et al., 2009; Bapat et al., 2010). In this chapter we discuss additional implications of manipulating ethylene function to improve postharvest qualities. The reader is also referred to Chapter 26.

Many of the manipulations of ethylene biosynthesis were achieved by inhibition of the two central biosynthesis-related genes, 1-aminocyclopropane-1-carboxylate oxidase (ACC) synthase (ACS) or ACC oxidase (ACO). Hormone perception was manipulated in most cases by introduction of a mutated form of ethylene receptor genes found to confer insensitivity when introduced into heterologous plant systems (Wilkinson et al., 1997). Manipulating function or regulation of such ethylene-related genes was successful in many different plants, resulting in marked physiological consequences to plant development and response to the environment (Stearns and Glick, 2003; Little et al., 2009). Although ethylene action in many postharvest-related contexts has negative implications and generally its inhibition has a postharvest advantage, in some cases ethylene action is required for maintaining or improving postharvest qualities. For example, downregulation of ethylene synthesis or perception aimed at extending the shelf life of climacteric fruits often results in lower production of aroma compounds (Kader, 2008). Excessive inhibition of ethylene likely results in non-attractive fresh produce. One of the main challenges in biotechnological applications for modifying ethylene biosynthesis or perception is the use of appropriate regulatory elements that will lead to the desired and appropriate effect in a given plant system for optimal postharvest benefit. Recently regulated ethylene insensitivity was obtained through the use of inducible expression of the *Arabidopsis etr1-1* mutant ethylene receptor in tomato (Gallie, 2010). This study demonstrated the feasibility of this approach, which could be applied for the benefit of the postharvest quality of fresh produce in which ethylene action is required, but the ability to decrease it at some specific point during postharvest storage may be greatly beneficial.

Senescence in Postharvest of Leafy Vegetables and Flowers

Background

Postharvest senescence contributes to the loss of quality of flowers, foliage, and vegetables. Visible and undesired yellowing of the tissue, as well as degradation of macromolecules during storage, may reduce the nutritive value of the fresh produce. Another undesirable result of senescence is an enhanced vulnerability to pathogens which, in the case of stored fresh produce, can lead to high losses.

Basic research of the senescence process in plants has resulted in significant and interesting insights, although it is far from finding a full understanding of the process and its regulatory mechanisms. Based on this knowledge, various approaches for successful retardation of senescence in a transgenic model or crop plants by manipulation of the relevant genes were applied (Lim et al., 2007; Guo and Gan, 2007). This knowledge can be further used for development and application of biotechnological strategies designed to control postharvest senescence in crop plants for practical applications.

It is widely accepted that the genetic regulatory system involved in senescence induction and progress is highly complex. The process possibly involves a network of senescence-promoting pathways responsible for activation of senescence-associated genes (SAGs) with the involvement of both plant internal signals and environmental/stress factors (Buchanan-Wollaston et al., 2005; Lim et al., 2007).

The accelerated senescence observed many times under postharvest storage conditions is possibly related to the acceleration of natural senescence under environmental stress conditions (Lers, 2007). Storage of fresh produce is frequently associated with abiotic stress such as darkness, extreme temperatures, wounding, or drought. It is likely that the same molecular mechanisms and genes associated with senescence induction in the attached organ during environmental stress are also involved in postharvest-induced senescence. The examples included are of studies in which genetic manipulations were shown to affect artificial senescence of detached plant, leaf, or flower relevant to postharvest senescence.

Senescence regulatory genes

The complexity of the senescence regulatory mechanism is reflected by the relatively large number of regulatory genes, encoding mainly for transcriptional factors suggested to be associated with senescence (Buchanan-Wollaston et al., 2005; Guo and Gan, 2005; Balazadeh et al., 2008). Manipulation of some of these regulatory genes has been found to also affect senescence induced under conditions relevant for postharvest. In *Arabidopsis*, AtNAP, an NAC family transcription factor, was found to have an important role in the regulation of natural leaf senescence. Although AtNAP is not readily induced by darkness, dark-induced senescence of detached leaves was delayed in the *atnap* null mutant, suggesting that AtNAP may function in dark-induced senescence downstream of the dark-responsive signaling pathway (Guo and Gan, 2006). In rice a nuclear-localized CCCH-type zinc finger protein "*Oryza sativa* delay of the onset of senescence" (OsDOS) was found to be involved in the regulation of leaf senescence (Kong et al., 2006). RNAi knockdown of OsDOS caused accelerated leaf senescence, whereas its over-expression resulted in a marked delay of the process, suggesting that it acts as a negative regulator of leaf senescence. Manipulation of the *Arabidopsis* mitogen-activated protein kinase cascade, MKK9-MPK6, was shown to have a role in leaf senescence (Zhou et al., 2009). When either MKK9 or MPK6 is knocked out, leaf senescence is delayed. This signal transduction cascade might be involved in mediating the action of ethylene on leaf senescence.

Possible involvement of post-transcriptional regulatory factors was demonstrated in *Arabidopsis* plants suppressed in expression of the deoxyhypusine synthase *(DHS)* gene. Delay of postharvest leaf senescence was observed in these plants (Wang et al., 2003). DHS is involved in the conversion of inactive eukaryotic translation initiation factor-5A (eIF-5A) to an activated form, and is thought to facilitate translation. These observations were interpreted as indicating that DHS plays a central role in the initiation of senescence.

Protein degradation mechanisms had been recognized to have important regulatory functions. The *Arabidopsis ore9* and *dls1* mutations affected the ubiquitin-dependent protein degradation pathways and revealed the consequences of both age-dependent and artificially induced senescence (Woo et al., 2001; Yoshida et al., 2002).

When manipulation of gene expression is performed, it is highly desirable to have the changes confined to a specific time and tissue to avoid pleiotropic or side effects to plant development. This can be achieved by the activation of the transgene by an appropriate specific promoter that regulates gene expression at the right time and tissue. Such an approach was applied by using a maize homeobox gene *Knotted1 (kn1)*, known to affect senescence, under the control of the poplar wound-inducible promoter *win3.12*. The resulting transgenic tobacco plants were morphologically normal; however, artificial senescence was delayed by at least two weeks compared with wild-type leaves (Luo et al., 2006). Thus, expression of the *kn1* gene driven by the wound-inducible promoter *win3.12* is potentially useful to delay senescence of fresh produce after harvest. A similar approach was utilized for suppression of the cysteine protease, aleurain, demonstrated to delay floret senescence in broccoli (Eason et al., 2005). Regulation of the antisense construct was achieved by the senescence-inducible promoter of the *Asparagus officinalis* asparagine synthetase (*AS*) gene (Winichayakul et al., 2004). In some of the resulting transgenic broccoli lines, postharvest floret senescence was delayed compared to wild-type plants during floret storage at 20°C (Eason et al., 2005).

Senescence-associated hormone biosynthesis or perception

Plant senescence is regulated by a coordinated genetic program mediated, in part, by changes in different plant hormones including ethylene, jasmonic acid, abscisic acid, and cytokinin (Schippers et al., 2007). In recent years much progress has been made in our understanding of the molecular mechanisms involved in hormone biosynthesis or perception, which is potentially useful for modulation of the senescence process. Delayed senescence can be achieved by either suppressing senescence-promoting hormone action, such as ethylene, by inhibition of their synthesis or perception or overproducing senescence-inhibiting hormones such as cytokinin, or elevating sensitivity of their perception.

Cytokinin has been long recognized to have a senescence retarding effect in different plants (Richmond and Lang, 1957). Accordingly strategies for modulating its activity were subjected to a great deal of research when trying to develop biotechnological approaches for delaying senescence. Whereas senescence is accompanied by a decline in leaf cytokinin content, exogenous application of cytokinins or an increase in their endogenous concentration delays senescence. Modification of cytokinin biosynthesis using genetic manipulation in transgenic plants, and especially during the senescence phase, can significantly delay senescence. The *Agrobacterium tumefaciens IPT* gene, encoding for isopentenyl transferase (IPT), a key enzyme in cytokinin biosynthesis, was recognized to be a good target for manipulation. The use of the senescence-specific *SAG12* promoter for activation of the *IPT* gene resulted in an efficient autoregulatory cytokinin production system (Gan and Amasino, 1995, 1997). At the onset of leaf senescence, the senescence-specific promoter activates the expression of *IPT*, resulting in increased cytokinin levels, which in turn prevent the leaf from senescing. The inhibition of leaf senescence renders the senescence-specific promoter inactive to prevent cytokinins from accumulating to very high levels, since overproduction of cytokinins may interfere with other aspects of plant development. Because cytokinin production is targeted to senescing leaves, overproduction of cytokinins before senescence should be avoided. The initial use of this autoregulated system in tobacco plants was followed successfully in other plants as well, demonstrating its wide application (Guo and Gan, 2007). Transformation of lettuce with the *SAG12-IPT* gene resulted in a significant delay of developmental and postharvest leaf senescence in mature heads of transgenic lettuce (McCabe et al., 2001). Apart from retardation of leaf senescence, mature plants exhibited normal morphology with no significant differences in head diameter or fresh weight of leaves and roots. Also in broccoli, the *PSAG12-IPT* construct was used successfully for retarding postharvest yellowing (Chen et al., 2001). The ability to use the *PSAG12-IPT* system for retardation of flower senescence was also demonstrated in *Petunia* in which overproduction of cytokinins in transformed flowers resulted with a delay in corolla senescence and a decrease in sensitivity to ethylene (Chang et al., 2003).

In addition to the use of the *SAG12* promoter for feedback control and specific expression of the *IPT* gene, the use of a cold-inducible promoter, *cor15a* from *Arabidopsis*, was successfully applied for *Petunia* and chrysanthemum (Khodakovskaya et al., 2005). Excised leaves from the transgenic plants remained green and healthy during prolonged dark storage after an initial exposure to a brief cold-induction period that resulted in accumulation of cytokinin before storage.

Insights regarding cytokinin perception open up the possibility of manipulating perception of the hormone as well. A cytokinin signal is known to be perceived by histidine kinase receptors. The *Arabidopsis* cytokinin receptor AHK3 was found to play a major role in controlling cytokinin-mediated leaf longevity through a specific phosphorylation of a response regulator, ARR2 (Kim et al., 2006). A gain-of-function *Arabidopsis* mutant was identified that exhibits delayed leaf senescence because of a missense mutation in *AHK3*. Consistently, transgenic over-expression of wild-type *ARR2* led to delayed senescence of leaves (Kim et al., 2006). In addition, the activity of extracellular invertase was found to be an essential component of cytokinin-mediated delay of senescence (Lara et al., 2004). Expression of the invertase gene under control of the senescence-induced *SAG12* promoter results in a delay of senescence.

Ethylene has long been recognized to have a senescence-promoting effect in many different plant tissues (Burg, 1968; Grbic and Bleecker, 1995). Knowledge of both the pathway of ethylene biosynthesis and the mechanism of its perception enabled the development of molecular approaches for nullifying ethylene action by interfering in either its synthesis or its perception (Stearns and Glick, 2003; Lin et al., 2009). Both inhibition of ethylene biosynthesis and perception was used in biotechnology-based applications for retardation of flower senescence and the relevant literature was recently reviewed (Czarny et al., 2006; Tanaka et al., 2005).

The observations that the mutant ethylene receptor exerts a dominant ethylene-insensitive phenotype were utilized for modulating ethylene sensitivity in different plants. For example, heterologous expression of *Arabidopsis ERS1* resulted in delayed senescence in coriander in which accelerated postharvest senescence otherwise occurs (Wang and Kumar, 2004). In broccoli, a mutant *ERS* receptor gene was obtained through site-directed mutagenesis to manipulate ethylene sensitivity following genetic transformation (Chen et al., 2004). Transgenic lines showing delayed senescence in leaves and floral heads were obtained, although in most lines the yellowing was only delayed by one to two days.

As previously mentioned, for cytokinin biosynthesis specific temporal and spatial regulation of the manipulated gene is generally required to eliminate any undesired consequences to plant development. This approach was used in inhibition of ethylene production in broccoli, mainly during the senescence process (Gapper et al., 2005). The senescence-associated *AS* promoter (Winichayakul et al., 2004) was used to drive the expression of an antisense *ACO* cDNA from broccoli to reduce ethylene production following harvest. Physiological analyses revealed that transgenic broccoli lines displayed delayed senescence in detached leaves and heads.

Marker genes are an essential component of plant transformation systems but are considered undesirable in genetically modified (GM) products. One way to produce marker-free transgenic plants is by co-transformation. This involves the stable integration of at least two different T-DNA sequences into the plant genome, which is sometimes achieved with high efficiency. If the two T-DNA sequences integrate into the plant genome at physically unlinked loci, it is possible to segregate out the T-DNA with the selectable marker from the T-DNA carrying the gene in the following generation using sexual crossing. This marker-free approach was applied successfully for a generation of transgenic broccoli with increased shelf life due to manipulation of the *ACS* or *ACO* genes involved in ethylene biosynthesis (Higgins et al., 2006). The green fluorescent protein gene (*GFP*) marker used for initial identification of co-transformants was later segregated out. This study demonstrated that the shelf life of broccoli may be increased by at least 2 days at 20°C by reducing postharvest ethylene production following inhibition of the ethylene biosynthesis-related genes.

Manipulation of auxin responses was shown in *Arabidopsis* to affect dark-induced senescence of detached *Arabidopsis* leaves (Ellis et al., 2005). Retardation of the process was observed in mutant plants lacking the auxin response factors (ARF) associated with responses to the plant hormone. Still, cytokinin maintained its senescence-retarding activity, suggesting that manipulation of both auxin and cytokinin responses may have an additive effect for retardation of senescence.

Oxidative stress involvement in senescence

The involvement of oxidative stress and plant tissue antioxidative status in senescence initiation and progress has been suggested based on physiological studies (Zimmermann and Zentgraf, 2005; Rosenvasser et al., 2006). Thus, biotechnological manipulations aimed at improving the leaf antioxidative potential during postharvest storage may result in retardation of senescence. Accordingly, deficiency in ascorbic acid in the *Arabidopsis* mutant vitamin C-1 (*vtc1*) resulted in accelerated dark-induced senescence (Barth et al., 2004). Thus, approaches aimed at elevating vitamin C may result in retardation of postharvest senescence. Additional support for the involvement of oxidative stress in senescence is the observation that the process is inhibited in transgenic tobacco *ndhF* plastid mutant in which the gene is knocked out (Zapata et al., 2005). The Ndh complex was proposed to trigger senescence by increasing reactive oxygen species (ROS). An additional component relevant to the involvement of oxidative stress was described in *Arabidopsis* as the nitric oxide synthase1 (NOS1), which was found to be targeted to mitochondria and thought to protect it against oxidative damage and dark-induced senescence (Guo and Crawford, 2005). Dark-induced senescence of detached leaves and intact plants progressed more rapidly in *NOS1* mutant compared with the wild-type, suggesting the possibility that elevation of NOS1 may result in senescence retardation. This study supports the notion that nitric oxide (NO) acts as a negative regulator of leaf senescence and, therefore, its manipulation using biotechnology may have the potential for retarding postharvest senescence in different plant tissues.

Chlorophyll degradation

An important visual aspect of senescence is yellowing due to chlorophyll degradation, which is only one aspect of senescence and in most cases its inhibition will not result in significantly improved postharvest quality. However, in some cases, such as citrus or melon peel/skin, the change in color may be significant in determining postharvest quality. Recent advances were made in the identification of catabolic enzymes responsible for chlorophyll degradation (Pruzinska et al., 2005; Hortensteiner, 2009; Schelbert et al., 2009). This information opens the way to use chlorophyll catabolic enzyme encoding genes for manipulation of chlorophyll content when it has a postharvest significance.

The ability to retard senescence by manipulating different genes involved in different senescence subprocesses was demonstrated in broccoli (Chen et al., 2008). Using biotechnology to manipulate more than one gene simultaneously in one plant may have a more pronounced effect on the delay of postharvest senescence due to the effect of different aspects of the senescence process, and will have an additive or even synergistic effect.

Abscission of Fruits, Flowers, and Leaves During Postharvest

Background

The occurrence of abscission has a large agricultural implication both at the level of crop plant growth and yield production, as well as during postharvest storage of fresh produce. In different agricultural commodities it is necessary to either inhibit or induce abscission. The premature induction of abscission in the flowers of trees and very young fruits as a result of extreme temperature (Kondo and Takahashi, 1989; Zhao et al., 2005) has a significant negative implication on yield. Similarly, acceleration of abscission in different fresh produce during postharvest storage results in a significant loss of quality; for example, abscission of cherry tomato from the bunch (Beno-Moualem et al., 2004), inflorescence or leaves in ornamentals (van Doorn and Stead; 1997; Michaeli et al., 1999; van Doorn, 2002), or basil leaves during low-temperature storage. To overcome the problem, crops are treated after harvest with plant growth regulators to delay abscission (van Doorn and Woltering, 1991). Dehiscence is a process responsible for the dispersal of mature seeds from the silique. Accumulating evidence suggests that the biochemical and molecular events involved in this process share similarities with those occurring during flower or leaf abscission (Roberts et al., 2002). In some crops (soybean, canola) delay of the seed dispersal form of the dry pods is required to decrease losses. In canola, for example, under certain conditions up to 50% of the yield can be lost due to seed dispersal (Moffat, 2000).

In some agricultural commodities the acceleration of abscission is required at the appropriate time to allow improved mechanization and labor saving; for example, defoliation of cotton leaves before boll harvest (Faircloth et al., 2004) or weakening of the abscission zone (AZ) of fruit tree flowers or young fruitlets for thinning the fruit load (Link, 2000). In citrus, intensive research is aimed at developing methods for weakening the AZ tissue to allow mechanized harvest (Burns, 2002), which is also important for olive harvest.

Currently two main strategies are used for modifying the abscission process in crop plants. In the genetic breeding approach screening of new lines with desirable related traits is performed, which is a relatively slow and unfocused process. The second approach is development and use of chemicals or growth regulators to artificially manipulate the process (Whitney et al., 2000), such as in cotton or citrus.

Abscission is a natural phase of plant development in which leaves, flowers, or fruits separate from the mother plant. It is generally similar for different organs and among different plants (Bleecker and Patterson, 1997; Roberts et al., 2002; Lewis et al., 2006). The AZ has a crucial role in the process; it includes a few layers of differentiated cells localized between the organ to be detached (fruit, flower, or leaf) and the plant body. The AZ becomes competent enough to respond to ethylene and auxin, which are known to induce or inhibit abscission, respectively. The central role of ethylene is demonstrated by its ability to induce premature abscission (Taylor and Whitelaw, 2001). In many cases the abscission process is induced by ethylene, whereas the rate and degree of abscission depends upon the endogenous balance between auxin and ethylene levels in the tissue (Taylor and Whitelaw, 2001; Patterson, 2001; Roberts et al., 2002). Actual organ separation is mediated by activation of hydrolytic enzymes within the AZ that hydrolyze the middle lamella (Jarvis et al., 2003). Based on their expression pattern and the consequences of their manipulation in transgenic plants, genes encoding for polygalacturonases and β-1,4-glucanases (cellulases) were thought to have a central role in executing cell separation (Roberts and Gonzalez-Carranza, 2009).

In general, three different stages of the abscission process could be targeted for biotechnological manipulation by focusing on different genes that were found to be involved. The following section discusses some examples of the possible target genes that have been identified to be associated with these different stages of abscission.

Development of the dedicated AZ tissue

Development of a functional AZ is a prerequisite for abscission. Thus, genes involved in the proper differentiation and development of this tissue are very relevant for biotechnology-based approaches aimed at inhibiting abscission. Among the known regulatory genes associated with AZ development is the tomato *JOINTLESS* identified as a MADS-box gene controlling tomato flower AZ development (Mao et al., 2000). The presence of a functional *JOINTLESS* is essential for AZ development. The *jointless* mutation has great agronomic value and is widely used in the tomato processing industry where lack of AZs on the fruits' pedicels aids mechanical harvesting and prevents physical wounding during transportation. The *BLADE-ON-PETIOLE* (*bop*) genes are essential in *Arabidopsis* for AZ formation and histological analysis; petal break strength measurements of *bop1 bop2* flowers show no differentiation in the floral AZs (McKim et al., 2008).

Regulatory genes involved in abscission control or mediating hormonal signal transduction

Recent studies indicated that before the AZ can become abscission competent various subprocesses should take place in which different regulatory proteins are involved; for example, molecular mechanisms that lead to increased tissue sensitivity to ethylene in response to IAA deficiency in the AZs. These mechanisms are as yet unknown. The following are a few examples for abscission-associated genes identified in *Arabidopsis* in which regulatory function was shown to be required for normal progress of the process.

Mutation in the *hawaiian skirt* F-Box gene was found to delay flower petal abscission and is thought to be necessary for cell separation to proceed at the normal rate (Gonzalez-Carranza et al., 2007). An ADP-ribosylation factor GTPase-activating protein, NEVERSHED (NEV), that regulates membrane trafficking is essential for floral organ shedding in *Arabidopsis* (Liljegren et al., 2009). In a screen for mutations that restore organ separation in *nev* flowers, a leucine-rich repeat receptor-like kinase, EVERSHED (EVR), was identified to function as an inhibitor of abscission (Leslie et al., 2010). The leucine-rich repeat receptor-like kinase SOMATIC EMBRYOGENESIS RECEPTOR-LIKE KINASE1 (SERK1) was reported to act as a negative regulator of flower abscission in *Arabidopsis*. Mutations in SERK1 rescue abscission without functional NEV (Lewis et al., 2010). Floral organ abscission was shown to be regulated by the INFLORESCENCE DEFICIENT IN ABSCISSION (IDA) regulatory protein. *ida* mutants fail to abscise floral organs, and plants over-expressing IDA display earlier abscission (Butenko et al., 2003). HAESA is a plasma membrane serine/threonine protein kinase, and inhibition of expression of its encoding gene was shown to result in delayed abscission of floral organs (Jinn et al., 2000). Over-expression of the *Arabidopsis* ZINC FINGER PROTEIN2 transcription factor encoding gene, *AtZFP2*, resulted in enhanced retention of floral organs, indicating its regulatory role (Cai and Lashbrook, 2008). A similar effect for inhibition of floral organ abscission was demonstrated by over-expression of the *Arabidopsis* transcription factor AtDOF4.7, which was shown to interact with the ZINC FINGER PROTEIN2 (Lewis et al., 2010). AtDOF4.7 was thought to participate in the regulation of genes encoding for cell wall hydrolysis enzymes.

All of the previously mentioned regulatory functions involving the abscission of floral organs were identified in *Arabidopsis*. However, further research is needed to examine the presence and function of related proteins in the abscission process of relevant crop plants. Once their existence and function is confirmed, they might be used for biotechnologically based genetic manipulations that affect abscission.

Genes involved in actual execution of cell separation in the later stage of abscission

Even if AZ is developed and its functional maturation is achieved, successful completion of abscission is dependent on the functionality of the enzymatic system involved in the actual cell separation process. Functional studies of enzymes/genes involved in this very late stage of abscission demonstrated the ability to affect its progress by manipulation of the expression of the relevant genes.

In *Arabidopsis* inhibition of expression of the genes encoding for ADPG2 and QRT2 polygalacturonases resulted in retardation of floral organ abscission (Ogawa et al., 2009). In tomato, inhibition of expression of the polygalacturonase encoding *TAPG1* gene or that of the endo-1,4-β-glucanase *Cel2* resulted in a marked delay of abscission (Brummell et al., 1999; Jiang et al., 2008). On the other hand, over-expression in transgenic apple of a polygalacturonase gene led to premature leaf shedding due to reduced cell adhesion in leaf abscission zones (Atkinson et al., 2002). The suppression of another hydrolytic enzyme, the LX ribonuclease, which is not involved in cell wall metabolism, resulted in marked delay of abscission in tomato (Lers et al., 2006). It was hypothesized that LX is involved in the programmed cell death process (PCD) required for the normal progression of abscission, which occurs at the late stage of the process.

Ethylene and abscission

Due to the central role of ethylene in the induction of abscission in many, but not all, plants, the manipulation of either ethylene biosynthesis or perception may be targeted for modulating abscission (Patterson and Bleecker, 2004; Binder and Patterson, 2009). As previously mentioned, ethylene biosynthesis or perception were successfully biotechnologically manipulated in many different plants (Stearns and Glick, 2003; Little et al., 2009). Although not directed originally for manipulation of abscission, in some studies an effect was observed on abscission as well when ethylene synthesis or perception was manipulated by targeting the relevant genes. In antisense *ACO* melon, abscission was delayed and fruit did not drop, even at a very late stage of development, whereas external ethylene application led to activation of abscission (Ayub et al., 1996). In a few reports an inhibitory effect on abscission was observed following the introduction of a mutated form of the ethylene receptor into transgenic plants, or when its expression level was reduced (Ecker, 1995; Wilkinson et al., 1997; Bleecker and Patterson, 1997; Whitelaw et al., 2002; Yang et al., 2008).

Regulated manipulation of abscission

Care must be taken to avoid negative secondary effects when applying biotechnology to manipulate gene expression to affect abscission by limiting the effect to the AZ tissue at the appropriate time. For this, an appropriate promoter could be utilized to regulate expression for targeting the AZ tissue. Few promoters were described in the literature to be active in the AZ tissue. These include in *Arabidopsis* the promoters of the genes encoding for polygalacturonases (Ogawa et al., 2009; Gonzalez-Carranza et al., 2002), promoters of genes encoding for the ACC oxidases (Tsuchisaka and Theologis, 2004), and the *PECTATE LYASE-LIKE* (*PLL*) gene promoter (Sun and van Nocker, 2010). In tomato the promoters of the polygalacturonase genes *TAPG1* and *TAPG4* (Hong et al., 2000) were shown to be active in AZs. Examples are available for promoters that exhibit AZ-specific expression in heterologous plant systems such as the promoter of the *Arabidopsis BFN1* nuclease gene, which is active in tomato leaf and fruit AZs as well (Farage-Barhom et al., 2008), or the promoter of the bean abscission-related cellulase gene, *BAC*, which was found to be expressed in AZs of *Arabidopsis* and tomato as well as in bean (Koehler et al., 1996; Tucker et al., 2002; Patterson and Bleecker, 2004). Thus, such promoters could be useful for the development of general biotechnological tools for the manipulation of abscission in different agriculturally important crop plants. Still, care must be taken as in most cases these promoters will not be absolutely specific for AZ tissue and are likely to be active in other tissues or times during development or in response to changing environmental conditions.

Reducing Postharvest Chilling Sensitivity

Background

Low-temperature storage is widely used because it is considered as one of the best storage methods to delay different physiological, biochemical, and pathological processes associated with the loss of postharvest quality (McGlasson et al., 1979). Low temperature slows down the cell metabolism rate and delays senescence and ripening processes that are very important for postharvest storage. Low temperature also inhibits the rate of growth and level of pathogenicity of many postharvest pathogens. However, the ability to use low-temperature storage is limited by the specific inherited cold sensitivity of the fresh produce. Exposure to temperatures below the cold tolerability threshold leads to various chilling-induced physiological disorders termed chilling injuries (Lyons, 1973; Sevillano et al., 2009). These result in quality loss and accelerated deterioration due to secondary pathogen-mediated decay of the fresh produce. The symptoms of chilling injuries in fruits or vegetables or other types of fresh produce are diverse and develop as a result of the effects of low temperatures on the normal progress of different biological processes. Sometimes chilling injuries are evident as developmental or metabolic disorders such as incomplete or unbalanced ripening and lack of normal development of some aspects of normal ripening such as reduced development of flavor and aroma (Forney, 2001) or development of a mealiness/wooliness texture due to an unbalanced softening process (Brummell et al., 2004). Frequently physiological chilling injury symptoms are apparent by ultrastructural abnormalities. On the microscopic level swelling and disorganization of cell organelles can be observed, whereas on the macroscopic level chilling injuries can be manifested as skin depression or pitting, abnormal skin coloration, internal or surface browning (scald), water-soaked tissue, and lower resistance to mechanical injury and pathogen attack (Kratsch and Wise, 2000; Sevillano et al., 2009). The different manifestations of postharvest chilling injuries observed in different plants and different organs indicate that the mechanisms involved are likely to be diverse.

Low-temperature or chilling sensitivity is closely linked to the environmental adaptation of a given plant. Plants originating from temperate or cold climates are genetically more adapted to cope with chilling temperatures compared to tropical and subtropical plants (Sung et al., 2003). Fruits and vegetables originating from low-temperature-sensitive plants have high sensitivity to low temperature during storage, resulting in physiological disorders that can have an enormous impact on quality. Plants that are better adapted to low temperature have evolved systems for efficiently sensing low-temperature stress and biological mechanisms for protection against low-temperature stress. A great deal of progress has been achieved in recent years to understand these low-temperature-response mechanisms including signal transduction pathways mediating cold stress and specific pathways and genes involved in higher resistance to chilling temperature (Iba, 2002; Penfield, 2008; Hua, 2009; Janska et al., 2010). It is clear today that both cold temperature sensing and signaling and even more resistance mechanisms are very complex and differ as a function of the specific crop. Also, there are various types of symptoms that develop as a result of chilling injury that add to the complexity of the problem. Different postharvest technologies were developed over the years for inhibiting or delaying the development of chilling injuries (Sevillano et al., 2009). These include pre-storage exposure of fresh produce to different physical treatments such as high temperature, conditioning at moderate low to intermediate low temperatures, and different storage conditions such as changes of humidity and use of a controlled or modified atmosphere (Sevillano et al., 2009). It is possible that these treatments/conditions induce genes and biological mechanisms involved in chilling tolerance.

Careful consideration should be given when biotechnological strategies are planned for the development of transgenic crops in which sensitivity to chilling temperatures will be reduced. Targeting genes involved in the biological processes associated with the primary events of the development of chilling injuries or activation of basic, wide-range resistance mechanisms should get high priority. Two possible major targets for genetic manipulation to improve postharvest quality during low-temperature storage are cell membrane structure and oxidative stress response mechanisms. The use of genetic engineering techniques to improve the adaptability of plants to temperature stress was already shown to be possible in different plant model systems (Iba, 2002).

Membrane structure and chilling sensitivity

The cell membrane is considered to be the site of initial events leading to chilling injuries that occur because of membrane structure and permeability (Lyons, 1973; Nishida and Murata, 1996). Low temperature results in damage from biophysical changes in the membrane properties because of the lipid phase transition and chemical damage to the membrane lipids from ROS activity (Marangoni et al., 1996; Kratsch and Wise, 2000; Wolfe, 1978; Sharom et al., 1994). Consequently, the properties and functionality of the membrane are damaged. Membrane lipids undergo enzymatic and non-enzymatic lipid peroxidation and degradation. The resulting defects could be ion leakage and cellular decompartmentalization and reduced or loss of functionality of different membrane proteins such as enzymes or transporters (Wolfe, 1978; Marangoni et al., 1996). In general, the chemical structure of lipids will be highly significant in determining low-temperature sensitivity of the cell membrane with the degree of saturation having high significance (Nishida and Murata, 1996; Kratsch and Wise, 2000). Adaptation to low temperatures was found frequently to involve an increase in the level of unsaturated fatty acids also resulting in increased membrane fluidity during postharvest cold storage (Zhang and Tian, 2009). The central role of the membrane in the development of chilling injuries in fresh produce during cold storage was demonstrated in peach fruit (Wongsheree et al., 2009) and basil leaf (Zhang et al., 2010).

Over the years many studies demonstrated the ability to modify and reduce plant sensitivity to cold temperatures by manipulating the expression of single genes involved in lipid metabolism. In many of these studies model plant systems were used and attenuation of cold sensitivity was examined in the leaves of plants growing in the light. The relevance of these approaches and the specific genes used for manipulation will have to be examined for low-temperature sensitivity during postharvest storage of fresh produce, which in many cases is a non-leafy organ such as a fruit or a flower.

The following are a few examples of specific genetic manipulation that resulted in a successful change in low-temperature sensitivity. The level of fatty acid unsaturation and the degree of chilling sensitivity of tobacco were successfully manipulated by transformation with different glycerol-3-phosphate acyltransferase (GAPT) encoding genes (Murata et al., 1992). Different genes encode for different isoforms of the enzyme, each having different preferences for the acyl-ACP substrate, thus leading to a membrane with different fatty acid composition (Murata et al., 1992). Over-expression in tomato of *LeGPAT* increased total activity of LeGPAT and *cis*-unsaturated fatty acids in the membranes (Sui et al., 2007). Chilling treatment induced less ion leakage from the transgenic plants compared to wild-type. The enzyme encoded by the cyanobacterium *Delta 9 desaturase* gene has a broad specificity and introduces a *cis*-double bond at the Delta 9 position of both 16 and 18 carbon saturated fatty acids linked in many kinds of membrane lipids. The *Delta 9* gene was introduced into tobacco, and in the resulting transgenic plants a highly reduced level of saturated fatty acid content was observed in most membrane lipids, and the plants exhibited a significant increase in chilling resistance (Ishizaki-Nishizawa et al., 1996). Chilling tolerance in *Arabidopsis* was reported to involve ALA1, a putative aminophospholipid translocase (Gomes et al., 2000). Downregulation of *ALA1* in *Arabidopsis* results in cold-sensitive plants that are much smaller than the wild-type. Transformation of tobacco with a gene for the thermophilic acyl-lipid desaturase was found to enhance chilling tolerance. Lipid content and the extent of fatty acid unsaturation significantly increased in leaves, and the chilling tolerance of those plants also increased (Orlova et al., 2003). Enhanced cold tolerance was demonstrated in transgenic tobacco expressing a chloroplast omega-3 fatty acid desaturase gene (*FAD7*) under the control of a cold-inducible promoter (Khodakovskaya et al., 2006). This approach of utilization of a cold-responsive regulatory element might be adopted for postharvest applications when precise temperature conditions during storage are known or could be modulated in a controlled way.

Oxidative stress and chilling sensitivity or tolerance

Low-temperature stress increases the level of ROS in the cells, which can lead to several different and major destructive side effects if not neutralized (Suzuki and Mittler, 2006). The production of these toxic species is accelerated when homeostasis is unbalanced and their elevated level can result in major uncontrolled damage to different cellular components (Scandalios, 1993; Prasad, 1996). In plants a major cause for low-temperature-induced ROS generation is inefficient photosynthesis, which takes place in the light (Foyer et al., 1994; Wise, 1995). However, cold storage of fresh produce in the dark frequently results in elevation of toxic ROS, which leads to different damage within the cells, including lipid peroxidation leading to membrane disintegration, protein oxidation leading to enzymatic activity inhibition, and finally cell death (Sevillano et al., 2009). For example, postharvest chilling was found to induce oxidative stress in tomato fruit (Malacrida et al., 2006) and enhanced leaf abscission via increased oxidative processes (Michaeli et al., 2001).

The involvement of oxidative stress in different environmental biotic and abiotic stresses was demonstrated in different plants. Manipulation of different genes involved in antioxidative function in plants was reported to affect the response of these transgenic plants to environmental stress (Perl et al., 1993; Dat et al., 2001; Kim et al., 2010). These observations demonstrate the potential of using biotechnology for affecting the ability of crop plants to grow under extreme conditions (Mittler and Blumwald, 2010). An association for low-temperature stress was also found to exist between the degree of sensitivity or resistance and the level of different components of the plant antioxidative systems (Lukatkin, 2002; Sevillano et al., 2009). Direct evidence linking the expressions of specific genes encoding for antioxidative enzymes and the degree of chilling sensitivity supplies a basis for the possibility of targeting these or similar genes for modulating chilling sensitivity during postharvest storage of fresh produce. High sensitivity to cold storage limits storage, as is the case in basil (Aharoni et al., 2010).

Although the antioxidative system is composed of a few different enzymes and protective molecules with antioxidative activity, the ability to affect sensitivity to cold temperature by modulation of a single component or gene was demonstrated. Inhibition of a catalase gene increases susceptibility to oxidative stress and chilling injury in transgenic tomato plants (Kerdnaimongkol and Woodson, 1999). Overproduction of maize or potato superoxic dismutase (SOD) and tomato cytosolic ascorbate peroxidase increased tolerance to chilling (van Breusegem et al., 1999; Wang et al., 2005; Waterer et al., 2010). Along with the manipulation of genes encoding for enzymes known to be part of the enzymatic system responsible for detoxification of ROS, there are other components for which manipulation may contribute to elimination of chilling-induced oxidative stress. Galactinol and raffinose constitute a novel function to protect plants from oxidative damage. *Arabidopsis* plants that had increased levels of galactinol and raffinose in the leaves due to over-expression of the relevant biosynthetic genes had an increased tolerance to chilling stress and other stresses. Genetic manipulation, which indirectly affects the levels of chilling-induced ROS, may also result in enhanced resistance to postharvest chilling stress. Such an indirect effect was observed in Charentais cantaloupe melons in which ethylene biosynthesis was inhibited due to suppression of the *ACO* gene (Ben-Amor et al., 1999). In contrast to wild-type fruit, antisense *ACO* melons did not develop the characteristic chilling injury of pitting and browning of the rind when stored at low temperature or upon re-warming. The tolerance to chilling was associated with reduced membrane deterioration and higher capacity of the fruit to remove active oxygen species.

Regulation of low-temperature responses

When planning to use biotechnology for manipulation of genes for improving major aspects of plant development or in response to the environment, it could be generally more efficient to target a regulatory gene that has a more central role in a given process compared to a gene that is involved in a very specific and limited function (Century et al., 2008). Thus, targeting a regulatory gene that controls a wide aspect of the plant low-temperature response can be very efficient for improving fresh produce resistance to chilling temperature. In recent years new and detailed information was accumulated regarding the networks involved in cold temperature perception and signal transduction, and later activation and regulation of the expression of sets of genes involved with the designated resistance mechanisms (Sung et al., 2003; Yamaguchi-Shinozaki and Shinozaki, 2006; Chinnusamy et al., 2007; Zhu et al., 2007; Hua, 2009). The reader is referred to Chapter 19 of this book in which regulation of the cold response in plants and how the cold signal cross-talks with other internal and external cues is discussed.

The plant hormone ABA, known to mediate the signal transduction of different abiotic stresses, has a central role in increasing plant cold tolerance (S.Y. Kim, 2007). A great deal of information is available regarding the cold-tolerance-related signaling networks ABA activates. Specific genes were identified that control major pathways involved in ABA-regulated cold tolerance. Manipulation of the expression of these genes can be targeted for biotechnologically improving plant cold tolerance (Xue-Xuan et al., 2010).

The *CBF* (*DREB*) genes act as major transcription regulators of *Arabidopsis* response to cold stress, which are conserved in other plants as well (Gilmour et al., 1998; Jaglo-Ottosen et al., 1998; Owens et al., 2002; Dubouzet et al., 2003; van Buskirk and Thomashow, 2006; Pino et al., 2008). CBF activates multiple components of the cold acclimation response (Gilmour et al., 2000). Plant tolerance to low temperatures can be improved by over-expression of *CBF/DREB* genes, which in turn results in enhancement of pathways associated with resistance to low-temperature stress (Sung et al., 2003; J. Kim, 2007; Sevillano et al., 2009). Heterologous expression of the *Arabidopsis CBF/DREB* genes was shown to improve tolerance to cold stress in crop plants such as grape and potato (Pino et al., 2008; Jin et al., 2009). Since the CBF/DREB system controls a wide array of genes affecting different subprocesses, the constitutive over-expression of these genes can result in deleterious effects to plant growth or development. This can be limited by the use of a regulated promoter (Kasuga et al., 1999). In tomato fruit the *LeCBF1* expression level was reported to positively correlate with cold tolerance and showed a high correlation to chilling injury index with ethylene and cold regulating its expression (Zhao et al., 2009a,b). This observation suggests CBF as a potential target for manipulation to improve fresh produce tolerance to low-temperature postharvest storage. Along with the CBF/DREB regulatory system, there are other regulatory genes involved with the induction of plants' tolerance to low temperatures. Whereas CDF/DREB is considered to have a quick response to the cold stress signal, rice MYBS3 was suggested to be part of a slower responsive regulatory system (Su et al., 2010). Gain- and loss-of-function analyses indicated that MYBS3 is sufficient and necessary for enhancing cold tolerance (Su et al., 2010).

Much insight is available from recent studies regarding the regulatory components and signal transduction involved in cold sensing in plants that mediate protection responses to minimize damage. Much of this knowledge originates from studies with plant model systems, some of which are temperate zone plants that have adapted to low-temperature environments. It will be necessary to examine if the same or similar regulatory circuits and proteins function in high-temperature-adapted plants before their manipulation could be used for developing plants better adapted to low-temperature storage conditions. Another challenge is to examine if the cold responses of fruit are similar to those that occur in leafy tissue, since much of the available information originates from leaves in model systems.

Molecules with protective functions during cold stress

During abiotic stress the biosynthesis and accumulation of different molecules thought to have protective functions in the cells is induced. These molecules are thought to mediate their protective function by their interaction with, or stabilizing of, different cellular components such as membrane elements or proteins/enzymes whose structure or function are sensitive and can be damaged as a result of the low temperature. For some of these protective molecules the biosynthetic pathways were described and genes encoding specific key enzymes in the pathways were identified and subjected to molecular manipulations to modulate biosynthesis (Bhatnagar-Mathur et al., 2008). In cases where these protective molecules contribute to plant chilling tolerance, key biosynthetic or regulatory genes could serve as potential targets for biotechnological manipulations for improving cold tolerance for fresh produce during postharvest storage.

Glycinebetaine is an effective protectant against abiotic stress in plants including chilling and freezing (Chen and Murata, 2008). Much progress was made in recent years toward improving plant tolerance to abiotic stress by genetic engineering of glycinebetaine biosynthesis (Khan et al., 2009). Overproduction in tomato of glycinebetaine, otherwise not synthesized in tomato, resulted in plants more tolerant to chilling (Park et al., 2004). This was obtained by over-expression of the *codA* gene, which encodes choline oxidase catalyzing the conversion of choline to glycinebetaine.

Heat shock pre-treatment in some plants has been found to induce induction of chilling tolerance and was demonstrated also to have a beneficial effect as a pre-storage treatment for some fruits or vegetables (Porat et al., 2000; Paull and Chen, 2000; Fallik, 2004). Heat-shock proteins (HSPs) act to stabilize proteins in the cell under temperature stress conditions by their activity as molecular chaperones. The HSPs support protein folding, translocation, assembly, and degradation in the cells during optimal and normal growth conditions. Their function is also vital under stress conditions such as extreme temperatures (Waters et al., 1996; Wang et al., 2004).

Dehydrins, a group of late embryogenesis abundant (LEA) proteins, are characterized by the presence of a lysine-rich amino acid motif. LEA proteins are proposed to act as chaperones in the cells and function to protect proteins and membranes under stress conditions leading to dehydration (Kosova et al., 2007). Induction of the level of LEA proteins was associated with improved tolerance to low temperatures (Close, 1997). The over-expression of multiple dehydrin genes was demonstrated to enhance tolerance to freezing stress in *Arabidopsis* (Puhakainen et al., 2004).

Polyamines (PAs) such as tetramine spermine (Spm), the-triamine spermidine (Spd), and diamine putrescine (Put) are considered components of plants' defense mechanisms against different types of abiotic stresses (Groppa and Benavides, 2008; Alcazar et al., 2010;). Results from different studies demonstrated an increase in PA levels following postharvest exposure of fresh produce to chilling stress, mainly in chilling-injury-tolerant fruits (Kramer and Wang, 1989; McDonald, 1989). Manipulation of genes involved in the biosynthesis of PAs can be used to alter PA levels. Over-expression of the apple spermidine synthase gene in pear confers multiple abiotic stress tolerance by altering the amount of PA (Wen et al., 2008). No reports are available yet regarding the chilling tolerance effects by using this approach. Interestingly, over-expression of yeast spermidine synthase was recently shown to increase tomato fruit shelf life, probably by reducing postharvest senescence and decay (Nambeesan et al., 2010).

It is not clear yet to what extent the mechanisms and genes found to be involved in tolerance against low-temperature stress during growth are also involved in tolerance to the stress imposed on sensitive, fresh produce when dark stored at low temperatures that induce chilling injuries. Both membrane structure characteristics and the antioxidative system seem to play a major role in both of these low-temperature stress conditions. Furthermore, manipulation of these biological processes may benefit other aspects such as plant development and the ability of crop plants to cope with different abiotic stress conditions such as drought, salinity, or high temperatures.

Affecting Postharvest Texture and Appearance Qualities

Background

The quality of texture of fresh fruits and vegetables is very important both as a sensory property that directly influences consumer attraction and eating quality. Description of sensory evaluation or experience can be described in different terms that relate to different texture characteristics such as firmness, softening, crispness, mealiness, crunchiness, grittiness, and other properties contributing to mouth feel (Harker et al., 1997; Toivonen and Brummell, 2008). Frequently firmness or softening are measured and used as the primary criteria for evaluation of postharvest quality. Although texture characteristics are an inherit quality of the plant and its relevant genetics, the conditions of postharvest storage and shelf life can dramatically affect texture qualities.

Softening and cell wall hydrolysis

The softening of fruits is an integral part of fruit ripening and is thought to be dependent to a large extent on cell wall metabolism. Many studies demonstrated association between changes in cell wall structure and cell wall hydrolases, including molecular studies performed to examine how manipulation of different cell wall hydrolase genes affect softening during ripening and postharvest storage (Brummee, 2006; Vicente et al., 2007; Goulao and Oliveira, 2008; Li et al., 2010).

Most studies that examined the consequences of manipulating the cell wall metabolism were performed to examine the effect on texture or softening of fruit tissues. A recent study demonstrated that manipulation of cell wall hydrolases can be important to the postharvest qualities of leafy vegetables. Transgenic lettuce plants were generated in which the production of the cell-wall-modifying enzyme xyloglucan endotransglucosylase/hydrolase (XTH) was downregulated (Wagstaff et al., 2010). In transgenic plants the leaf area and fresh weight were decreased, whereas leaf strength was increased and membrane permeability was reduced toward the end of shelf life. Overall an extended shelf life of transgenic lines was observed relative to the non-transgenic control plants, which illustrated the potential for manipulation of cell-wall-related genes for improving postharvest quality of leafy crops.

Softening and turgor

Softening of fruits during ripening has been generally considered to result from cell wall disassembly due to activation of cell wall hydrolases, as previously mentioned. Whereas some studies demonstrated a link between expression of specific genes associated with cell wall metabolism and advancement of softening, it is clear that the relationship between the two is not simple and depends upon the specific plant system examined. Few studies suggested that turgor pressure is an important factor in determining changes in the resistance of the fruit tissue to pressure, thus contributing to softening during postharvest storage (Vicente et al., 2007). Thus, water loss that is generally a major problem during storage seems to also be relevant to softening in some cases. In apples it was found that cultivars with little softening during storage had low rates of reduction in turgor, indicating that it closely relates to postharvest apple fruit softening (Iwanami et al., 2008). The association of a single gene, *Delayed Fruit Deterioration*, (*DFD*) to both softening and water loss was demonstrated in tomato (Saladie et al., 2007). Cell wall metabolism was not altered in DFD fruits, whereas water loss was inhibited leading to higher cellular turgor compared to wild-type. Based on observed changes in cuticle composition and structure in fruits of DFD, it was suggested that cuticle has an influence on intact tomato fruit firmness and ripening physiology. This function makes it a potential biotechnological target to prolong fruit quality (Saladie et al., 2007). An additional interesting link between water status and softening was proposed to be in strawberry based on the association found between expression of fruit-specific aquaporin genes and changes in firmness during ripening (Alleva et al., 2010). These studies showed that along with the suggested role of cell wall metabolism, water balance in fresh produce may have an important contribution to changes observed in firmness.

Tissue lignifications

In most cases softening is associated with increased storage periods of fresh produce, but in some fruits or vegetables under some storage conditions an increase in tissue firmness is observed resulting in flesh lignification. Lignification of fruits and vegetables during postharvest storage leads to loss of quality and commercial losses. Accumulation of lignin affects digestibility as well as marked reduction in texture quality, which in turn results in loss of consumer attraction. Lignin accumulation during postharvest storage was reported for various fruits and vegetables such as apple (Valentines et al., 2005), asparagus (Hennion et al., 1992), carrot (Bolin, 1992), and celery (Vina and Chaves, 2003). Although the problem of lignin accumulation is more limited compared to softening, it is still an important postharvest problem.

Lignins are phenolic compounds that are important cell wall components. Increases in firmness and lignin content are associated with increased activities of a few enzymes involved in lignin biosynthesis including phenylalanineammonia lyase (PAL), 4-coumarate:CoA ligase (4CL), cinnamyl alcohol dehydrogenase (CAD), Omethyltransferase (OMT), and peroxidase (POD; Li et al., 2010). The role of these enzymes in fruit ripening has been investigated recently, including functional characterization of these key phenylpropanoid biosynthetic enzymes/genes during the fruit ripening processes (Singh et al., 2010).

Molecular manipulations for reducing lignin level, mostly in tree plants, were reported to be successful. These studies focused on inhibiting the expression of the genes encoding three of the key enzymes in the phenylpropanoid biosynthetic pathways such as *4CL* in which suppression led to up to a 45% reduction of lignin (Hu et al., 1999; Voelker et al., 2010). In plants in which *CAD* gene expression was inhibited, although absolute level was not markedly changed, lignin characteristics were altered in both composition and structure (Halpin et al., 1994). Alteration in lignin characteristics and reduction in their level was also obtained in plants inhibited for *OMT* gene expression (Dwivedi et al., 1994; Rastogi and Dwivedi, 2008;). Thus, it is possible that by manipulation of genes encoding the above lignin-biosynthesis related key enzymes, it would be possible to reduce the problem of lignin accumulation in certain fresh fruits or vegetables during postharvest storage.

Implications for Plant and Agricultural Biotechnology

The use of biotechnology for future improvement of postharvest quality and reduction of losses could potentially have a profound effect on agriculture and the ability to supply the growing demand for food. Many efforts are being invested into developing new and more efficient postharvest storage technologies, treatments, and strategies to further improve the ability to keep fresh produce quality. However, the ability to further extend the available postharvest technologies seems to be limited. Currently it seems that there are two main approaches that may yield the greatest benefit and hold much promise for further improving worldwide postharvest qualities of fresh produce. The first, and more simple and immediate, is to efficiently distribute the already available postharvest professional knowledge to sites around the world where such knowledge is limiting. The other approach is to utilize biotechnology or biotechnology-assisted classical breeding for the development of new crop varieties that will be improved in postharvest-related traits, thus becoming more adapted to postharvest storage.

In this chapter several highly relevant examples for postharvest storage of fresh produce were discussed. Studies of gene manipulation in transgenic plants resulting in new phenotypes relevant to postharvest issues were presented. The basic research conducted so far had identified potential targets for genetic manipulation, some of which could be used as markers in genetic breeding programs to identify progenies better adapted for postharvest storage. Along with the target genes, appropriate regulatory elements of gene expression should be used to activate the manipulated genes at the right time and the right tissue. This is crucial to obtain a specific effect during postharvest storage without any undesired consequences to normal plant development or response to environmental stress. To obtain such specificity of gene expression it might be possible to take advantage of the fact that in many cases the postharvest fresh produce is exposed to specific and defined conditions during storage. These, for example, can include certain CO_2/O_2 concentrations very different from that of air, darkness, or low temperatures. If gene expression regulatory elements that respond to these postharvest-specific conditions are used, it might be possible to limit the manipulation of the expression of the targeted genes only to the storage period. In such cases the biotechnological manipulation will be silent during the development or growth of the genetically modified plant and decrease the chances of undesired negative side effects.

Additional molecular and biotechnological studies should be performed in crop plants to fully apply biotechnological tools for improving postharvest quality. However, this knowledge allows us to use biotechnology to improve postharvest-related qualities of fresh produce.

References

Aharoni, N., Kenigsbuch, D., Chalupowicz, D., Faura-Mlinski, M., Aharon, Z., Maurer, D., et al. (2010). Reducing chilling injury and decay in stored sweet basil. *Israel Journal of Plant Sciences*, *58*, 167–168.

Alcazar, R., Altabella, T., Marco, F., Bortolotti, C., Reymond, M., & Koncz, C., et al. (2010). Polyamines: Molecules with regulatory functions in plant abiotic stress tolerance. *Planta*, *231*, 1237–1249.

Alleva, K., Marquez, M., Villarreal, N., Mut, P., Bustamante, C., & Bellati, J., et al. (2010). Cloning, functional characterization, and co-expression studies of a novel aquaporin (FaPIP2;1) of strawberry fruit. *Journal of Experimental Botany*, *61*, 3935–3945.

Anonymous, How to feed a hungry world. *Nature*, *466*, 531–532.

Atkinson, R. G., Schroder, R., Hallett, I. C., Cohen, D., & MacRae, E. A. (2002). Overexpression of polygalacturonase in transgenic apple trees leads to a range of novel phenotypes involving changes in cell adhesion. *Plant Physiology*, *129*, 122–133.

Ayub, R., Guis, M., BenAmor, M., Gillot, L., Roustan, J. P., & Latche, A., et al. (1996). Expression of *ACC oxidase* antisense gene inhibits ripening of cantaloupe melon fruits. *Nature Biotechnology*, *14*, 862–866.

Balazadeh, S., Riano-Pachon, D. M., & Mueller-Roeber, B. (2008). Transcription factors regulating leaf senescence in *Arabidopsis thaliana*. *Plant Biology*, *10*, 63–75.

Bapat, V. A., Trivedi, P. K., Ghosh, A., Sane, V. A., Ganapathi, T. R., & Nath, P. (2010). Ripening of fleshy fruit: Molecular insight and the role of ethylene. *Biotechnology Advances*, *28*, 94–107.

Barth, C., Moeder, W., Klessig, D. F., & Conklin, P. L. (2004). The timing of senescence and response to pathogens is altered in the ascorbate-deficient *Arabidopsis* mutant vitamin c-1. *Plant Physiology*, *134*, 1784–1792.

Ben-Amor, M., Flores, B., Latche, A., Bouzayen, M., Pech, J. C., & Romojaro, F. (1999). Inhibition of ethylene biosynthesis by antisense *ACC oxidase* RNA prevents chilling injury in Charentais cantaloupe melons. *Plant Cell and Environment*, *22*, 1579–1586.

Beno-Moualem, D., Gusev, L., Dvir, O., Pesis, E., Meir, S., & Lichter, A. (2004). The effects of ethylene, methyljasmonate and 1-MCP on abscission of cherry tomatoes from the bunch and expression of endo-1,4-beta-glucanases. *Plant Science*, *167*, 499–507.

Bhatnagar-Mathur, P., Vadez, V., & Sharma, K. K. (2008). Transgenic approaches for abiotic stress tolerance in plants: Retrospect and prospects. *Plant Cell Reports*, *27*, 411–424.

Binder, B. M., & Patterson, S. E. (2009). Ethylene-dependent and -independent regulation of abscission. *Stewart Postharvest Review*, 5(1), 1–10.

Bleecker, A. B., & Patterson, S. E. (1997). Last exit: Senescence, abscission, and meristem arrest in *Arabidopsis*. *Plant Cell*, *9*, 1169–1179.

Bolin, H. R. (1992). Retardation of surface lignification on fresh peeled carrots. *Journal of Food Processing and Preservation*, *16*, 99–104.

Brummell, D. A. (2006). Cell wall disassembly in ripening fruit. *Functional Plant Biology*, *33*, 103–119.

Brummell, D. A., Dal Cin, V., Lurie, S., Crisosto, C. H., & Labavitch, J. M. (2004). Cell wall metabolism during the development of chilling injury in cold-stored peach fruit: Association of mealiness with arrested disassembly of cell wall pectins. *Journal of Experimental Botany*, *55*, 2041–2052.

Brummell, D. A., Hall, B. D., & Bennett, A. B. (1999). Antisense suppression of tomato endo-1,4-beta-glucanase *Cel2* mRNA accumulation increases the force required to break fruit abscission zones but does not affect fruit softening. *Plant Molecular Biology*, *40*, 615–622.

Buchanan-Wollaston, V., Page, T., Harrison, E., Breeze, E., Lim, P. O., & Nam, H. G., et al. (2005). Comparative transcriptome analysis reveals significant differences in gene expression and signalling pathways between developmental and dark/starvation-induced senescence in *Arabidopsis*. *Plant Journal*, *42*, 567–585.

Burg, S. P. (1968). Ethylene plant senescence and abscission. *Plant Physiology*, *43*, 1503–1509.

Burns, J. K. (2002). Using molecular biology tools to identify abscission materials for citrus. *HortTechnology*, *37*, 459–464.

Butenko, M. A., Patterson, S. E., Grini, P. E., Stenvik, G. E., Amundsen, S. S., & Mandal, A., et al. (2003). INFLORESCENCE DEFICIENT IN ABSCISSION controls floral organ abscission in *Arabidopsis* and identifies a novel family of putative ligands in plants. *Plant Cell*, *15*, 2296–2307.

Cai, S. Q., & Lashbrook, C. C. (2008). Stamen abscission zone transcriptome profiling reveals new candidates for abscission control: Enhanced retention of floral organs in transgenic plants overexpressing *Arabidopsis* ZINC FINGER PROTEIN2. *Plant Physiology*, *146*, 1305–1321.

Cara, B., & Giovannoni, J. J. (2008). Molecular biology of ethylene during tomato fruit development and maturation. *Plant Science*, *175*, 106–113.

Century, K., Reuber, T. L., & Ratcliffe, O. J. (2008). Regulating the regulators: The future prospects for transcription-factor-based agricultural biotechnology products. *Plant Physiology*, *147*, 20–29.

Chang, H. S., Jones, M. L., Banowetz, G. M., & Clark, D. G. (2003). Overproduction of cytokinins in petunia flowers transformed with *P-SAG12-IPT* delays corolla senescence and decreases sensitivity to ethylene. *Plant Physiology*, *132*, 2174–2183.

Chen, L. F. O., Huang, J. Y., Wang, Y. H., Chen, Y. T., & Shaw, J. F. (2004). Ethylene insensitive and post-harvest yellowing retardation in mutant ethylene response sensor (boers) gene transformed broccoli (*Brassica olercea* var. italica). *Molecular Breeding*, *14*, 199–213.

Chen, L. F. O., Hwang, J. Y., Charng, Y. Y., Sun, C. W., & Yang, S. F. (2001). Transformation of broccoli (*Brassica oleracea* var. italica) with isopentenyltransferase gene via *Agrobacterium tumefaciens* for post-harvest yellowing retardation. *Molecular Breeding*, *7*, 243–257.

Chen, T. H. H., & Murata, N. (2008). Glycinebetaine: An effective protectant against abiotic stress in plants. *Trends in Plant Science*, *13*, 499–505.

Chen, Y. T., Chen, L. F. O., & Shaw, J. F. (2008). Senescence-associated genes in harvested broccoli florets. *Plant Science*, *175*, 137–144.

Chinnusamy, V., Zhu, J., & Zhu, J. K. (2007). Cold stress regulation of gene expression in plants. *Trends in Plant Science*, *12*, 444–451.

Close, T. J. (1997). Dehydrins: A commonality in the response of plants to dehydration and low temperature. *Physiologia Plantarum*, *100*, 291–296.

Czarny, J. C., Grichko, V. P., & Glick, B. R. (2006). Genetic modulation of ethylene biosynthesis and signaling in plants. *Biotechnology Advances*, *24*, 410–419.

Dat, J. F., Inze, D., & Van Breusegem, F. (2001). Catalase-deficient tobacco plants: Tools for *in planta* studies on the role of hydrogen peroxide. *Redox Report*, *6*, 37–42.

Dubouzet, J. G., Sakuma, Y., Ito, Y., Kasuga, M., Dubouzet, E. G., & Miura, S., et al. (2003). OsDREB genes in rice, *Oryza sativa* L., encode transcription activators that function in drought-, high-salt- and cold-responsive gene expression. *Plant Journal*, *33*, 751–763.

Dwivedi, U. N., Campbell, W. H., Yu, J., Datla, R. S. S., Bugos, R. C., & Chiang, V. L., et al. (1994). Modification of lignin biosynthesis in transgenic nicotiana through expression of an antisense *o-methyltransferase* gene from *populus*. *Plant Molecular Biology*, *26*, 61–71.

Eason, J. R., Ryan, D. J., Watson, L. M., Hedderley, D., Christey, M. C., & Braun, R. H., et al. (2005). Suppression of the cysteine protease, aleurain, delays floret senescence in *Brassica oleracea*. *Plant Molecular Biology*, *57*, 645–657.

Ecker, J. R. (1995). The ethylene signal-transduction pathway in plants. *Science*, *268*, 667–675.

Ellis, C. M., Nagpal, P., Young, J. C., Hagen, G., Guilfoyle, T. J., & Reed, J. W. (2005). AUXIN RESPONSE FACTOR1 and AUXIN RESPONSE FACTOR2 regulate senescence and floral organ abscission in *Arabidopsis thaliana*. *Development*, *132*, 4563–4574.

Faircloth, J. C., Edmisten, K. L., Wells, R., & Stewart, A. M. (2004). Timing defoliation applications for maximum yields and optimum quality in cotton containing a fruiting gap. *Crop Science*, *44*, 158–164.

Fallik, E. (2004). Prestorage hot water treatments (immersion, rinsing and brushing). *Postharvest Biology and Technology*, *32*, 125–134.

Farage-Barhom, S., Burd, S., Sonego, L., Perl-Treves, R., & Lers, A. (2008). Expression analysis of the *BFN1* nuclease gene promoter during senescence, abscission, and programmed cell death-related processes. *Journal of Experimental Botany*, *59*, 3247–3258.

Forney, C. F. (2001). Horticultural and other factors affecting aroma volatile composition of small fruit. *HortTechnology*, *11*, 529–538.

Foyer, C. H., Lelandais, M., & Kunert, K. J. (1994). Photooxidative stress in plants. *Physiologia Plantarum*, *92*, 696–717.

Gallie, D. R. (2010). Regulated ethylene insensitivity through the inducible expression of the *Arabidopsis* etr1-1 mutant ethylene receptor in tomato. *Plant Physiology*, *152*, 1928–1939.

Gan, S. S., & Amasino, R. M. (1995). Inhibition of leaf senescence by autoregulated production of cytokinin. *Science*, *270*, 1986–1988.

Gan, S. S., & Amasino, R. M. (1997). Making sense of senescence—Molecular genetic regulation and manipulation of leaf senescence. *Plant Physiology*, *113*, 313–319.

Gapper, N. E., Coupe, S. A., McKenzie, M. J., Scott, R. W., Christey, M. C., & Lill, R. E., et al. (2005). Senescence-associated down-regulation of 1-aminocyclopropane-1-carboxylate (ACC) oxidase delays harvest-induced senescence in broccoli. *Functional Plant Biology*, *32*, 891–901.

Gilmour, S. J., Sebolt, A. M., Salazar, M. P., Everard, J. D., & Thomashow, M. F. (2000). Overexpression of the *Arabidopsis CBF3* transcriptional activator mimics multiple biochemical changes associated with cold acclimation. *Plant Physiology, 124*, 1854–1865.

Gilmour, S. J., Zarka, D. G., Stockinger, E. J., Salazar, M. P., Houghton, J. M., & Thomashow, M. F. (1998). Low temperature regulation of the *Arabidopsis* CBF family of AP2 transcriptional activators as an early step in cold-induced *COR* gene expression. *Plant Journal, 16*, 433–442.

Gomes, E., Jakobsen, M. K., Axelsen, K. B., Geisler, M., & Palmgren, M. G. (2000). Chilling tolerance in *Arabidopsis* involves ALA1, a member of a new family of putative aminophospholipid translocases. *Plant Cell, 12*, 2441–2453.

Gonzalez-Carranza, Z. H., Rompa, U., Peters, J. L., Bhatt, A. M., Wagstaff, C., & Stead, A. D., et al. (2007). *HAWAIIAN SKIRT*: An F-box gene that regulates organ fusion and growth in *Arabidopsis*. *Plant Physiology, 144*, 1370–1382.

Gonzalez-Carranza, Z. H., Whitelaw, C. A., Swarup, R., & Roberts, J. A. (2002). Temporal and spatial expression of a polygalacturonase during leaf and flower abscission in oilseed rape and *Arabidopsis*. *Plant Physiology, 128*, 534–543.

Goulao, L. F., & Oliveira, C. M. (2008). Cell wall modifications during fruit ripening: When a fruit is not the fruit. *Trends in Food Science and Technology, 19*, 4–25.

Grbic, V., & Bleecker, A. B. (1995). Ethylene regulates the timing of leaf senescence in *Arabidopsis*. *Plant Journal, 8*, 595–602.

Groppa, M. D., & Benavides, M. P. (2008). Polyamines and abiotic stress: Recent advances. *Amino Acids, 34*, 35–45.

Guo, F. Q., & Crawford, N. M. (2005). *Arabidopsis* nitric oxide synthase1 is targeted to mitochondria and protects against oxidative damage and dark-induced senescence. *Plant Cell, 17*, 3436–3450.

Guo, Y., & Gan, S. (2007). Genetic manipulation of leaf senescence. In S. Gan (Ed.), *Senescence processes in plants* (pp. 304–322). Oxford, UK: Blackwell Publishing.

Guo, Y. F., & Gan, S. S. (2005). Leaf senescence: Signals, execution, and regulation. *Current Topics in Developmental Biology, 71*, 83.

Guo, Y. F., & Gan, S. S. (2006). AtNAP, a NAC family transcription factor, has an important role in leaf senescence. *Plant Journal, 46*, 601–612.

Halpin, C., Knight, M. E., Foxon, G. A., Campbell, M. M., Boudet, A. M., & Boon, J. J., et al. (1994). Manipulation of lignin quality by down-regulation of cinnamyl alcohol-dehydrogenase. *Plant Journal, 6*, 339–350.

Harker, F. R., Stec, M. G. H., Hallett, I. C., & Bennett, C. L. (1997). Texture of parenchymatous plant tissue: A comparison between tensile and other instrumental and sensory measurements of tissue strength and juiciness. *Postharvest Biology and Technology, 11*, 63–72.

Hennion, S., Little, C. H. A., & Hartmann, C. (1992). Activities of enzymes involved in lignification during the postharvest storage of etiolated asparagus spears. *Physiologia Plantarum, 86*, 474–478.

Higgins, J. D., Newbury, H. J., Barbara, D. J., Muthumeenakshi, S., & Puddephat, I. J. (2006). The production of marker-free genetically engineered broccoli with sense and antisense ACC synthase 1 and ACC oxidases 1 and 2 to extend shelf-life. *Molecular Breeding, 17*, 7–20.

Hong, S. B., Sexton, R., & Tucker, M. L. (2000). Analysis of gene promoters for two tomato polygalacturonases expressed in abscission zones and the stigma. *Plant Physiology, 123*, 869–881.

Hortensteiner, S. (2009). Stay-green regulates chlorophyll and chlorophyll-binding protein degradation during senescence. *Trends in Plant Science, 14*, 155–162.

Hu, W. J., Harding, S. A., Lung, J., Popko, J. L., Ralph, J., & Stokke, D. D., et al. (1999). Repression of lignin biosynthesis promotes cellulose accumulation and growth in transgenic trees. *Nature Biotechnology, 17*, 808–812.

Hua, J. (2009). From freezing to scorching, transcriptional responses to temperature variations in plants. *Current Opinion in Plant Biology, 12*, 568–573.

Hubert, B., Rosegrant, M., van Boekel, M., & Ortiz, R. (2010). The future of food: Scenarios for 2050. *Crop Science, 50*, S33–S50.

Iba, K. (2002). Acclimative response to temperature stress in higher plants: Approaches of gene engineering for temperature tolerance. *Annual Review of Plant Biology, 53*, 225–245.

Ishizaki-Nishizawa, O., Fujii, T., Azuma, M., Sekiguchi, K., Murata, N., & Ohtani, T., et al. (1996). Low-temperature resistance of higher plants is significantly enhanced by a nonspecific cyanobacterial desaturase. *Nature Biotechnology, 14*, 1003–1006.

Iwanami, H., Moriya, S., Kotoda, N., & Abe, K. (2008). Turgor closely relates to postharvest fruit softening and can be a useful index to select a parent for producing cultivars with good storage potential in apple. *HortTechnology, 43*, 1377–1381.

Jaglo-Ottosen, K. R., Gilmour, S. J., Zarka, D. G., Schabenberger, O., & Thomashow, M. F. (1998). *Arabidopsis* CBF1 overexpression induces COR genes and enhances freezing tolerance. *Science, 280*, 104–106.

Janska, A., Marsik, P., Zelenkova, S., & Ovesna, J. (2010). Cold stress and acclimation—what is important for metabolic adjustment? *Plant Biology, 12*, 395–405.

Jarvis, M. C., Briggs, S. P. H., & Knox, J. P. (2003). Intercellular adhesion and cell separation in plants. *Plant Cell and Environmental, 26*, 977–989.

Jiang, C. Z., Lu, F., Imsabai, W., Meir, S., & Reid, M. S. (2008). Silencing polygalacturonase expression inhibits tomato petiole abscission. *Journal of Experimental Botany, 59*, 973–979.

Jin, W. M., Dong, J., Hu, Y. L., Lin, Z. P., Xu, X. F., & Han, Z. H. (2009). Improved cold-resistant performance in transgenic grape (*vitis vinifera* L.) overexpressing cold-inducible transcription factors *AtDREB1b*. *HortTechnology, 44*, 35–39.

Jinn, T. L., Stone, J. M., & Walker, J. C. (2000). HAESA, an *Arabidopsis* leucine-rich repeat receptor kinase, controls floral organ abscission. *Genes and Development, 14*, 108–117.

Kader, A. A. (2005). Increasing food availability by reducing postharvest losses of fresh produce. *Acta Horticulturae, 682*, 2169–2175.

Kader, A. A. (2008). Flavor quality of fruits and vegetables. *Journal of the Science of Food and Agriculture, 88*, 1863–1868.

Kantor, L. S., Lipton, K., Manchester, A., & Oliveira, V. (1997). *Estimating and addressing America's food losses*. Economic Research Service, U.S. Department of Agriculture. <http://www.ers.usda.gov/Publications/FoodReview/Jan1997/Jan97a.pdf/>

Kasuga, M., Liu, Q., Miura, S., Yamaguchi-Shinozaki, K., & Shinozaki, K. (1999). Improving plant drought, salt, and freezing tolerance by gene transfer of a single stress-inducible transcription factor. *Nature Biotechnology, 17*, 287–291.

Kerdnaimongkol, K., & Woodson, W. R. (1999). Inhibition of catalase by antisense RNA increases susceptibility to oxidative stress and chilling injury in transgenic tomato plants. *Journal of the American Society for Horticultural Science, 124*, 330–336.

Khan, M. S., Yu, X., Kikuchi, A., Asahina, M., & Watanabe, K. N. (2009). Genetic engineering of glycine betaine biosynthesis to enhance abiotic stress tolerance in plants. *Plant Biotechnology, 26*, 125–134.

Khodakovskaya, M., Li, Y., Li, J. S., Vankova, R., Malbeck, J., & McAvoy, R. (2005). Effects of cor15a-IPT gene expression on leaf senescence in transgenic Petunia x hybrida and Dendranthema x grandiflorum. *Journal of Experimental Botany, 56*, 1165–1175.

Khodakovskaya, M., McAvoy, R., Peters, J., Wu, H., & Li, Y. (2006). Enhanced cold tolerance in transgenic tobacco expressing a chloroplast *omega-3 fatty acid desaturase* gene under the control of a cold-inducible promoter. *Planta, 223*, 1090–1100.

Kim, H. J., Ryu, H., Hong, S. H., Woo, H. R., Lim, P. O., & Lee, I. C., et al. (2006). Cytokinin-mediated control of leaf longevity by AHK3 through phosphorylation of ARR2 in *Arabidopsis*. *Proceedings of the National Academy of Sciences of the United States of America, 103*, 814–819.

Kim, J. (2007). Perception, transduction, and networks in cold signaling. *Journal of Plant Biology, 50*, 139–147.

Kim, M. D., Kim, Y. H., Kwon, S. Y., Yun, D. J., Kwak, S. S., & Lee, H. S. (2010). Enhanced tolerance to methyl viologen-induced oxidative stress and high temperature in transgenic potato plants overexpressing the *CuZnSOD, APX* and *NDPK2* genes. *Physiologia Plantarum, 140*, 153–162.

Kim, S. Y. (2007). Recent advances in ABA signaling. *Journal of Plant Biology, 50*, 117–121.

Kitinoja, L., & Cantwell, M. (2010). *WFLO grant final report—Identification of appropriate postharvest technologies for improving market access and incomes for small horticultural farmers in Sub-Saharan Africa and South Asia*. (p. 316.) World Food Logistics Organization.

Koehler, S. M., Matters, G. L., Nath, P., Kemmerer, E. C., & Tucker, M. L. (1996). The gene promoter for a bean abscission cellulase is ethylene-induced in transgenic tomato and shows high sequence conservation with a soybean abscission cellulase. *Plant Molecular Biology*, *31*, 595–606.

Kondo, S., & Takahashi, Y. (1989). Relation between early drop of apple fruit and ethylene evolution under high night-temperature conditions. *Journal of the Japanese Society for Horticultural Science*, *58*, 1–8.

Kong, Z. S., Li, M. N., Yang, W. Q., Xu, W. Y., & Xue, Y. B. (2006). A novel nuclear-localized CCCH-type zinc finger protein, OsDOS, is involved in delaying leaf senescence in rice. *Plant Physiology*, *141*, 1376–1388.

Kosova, K., Vitamvas, P., & Prasil, I. T. (2007). The role of dehydrins in plant response to cold. *Biologia Plantarum*, *51*, 601–617.

Kramer, G. F., & Wang, C. Y. (1989). Correlation of reduced chilling injury with increased spermine and spermidine levels in zucchini squash. *Physiologia Plantarum*, *76*, 479–484.

Kratsch, H. A., & Wise, R. R. (2000). The ultrastructure of chilling stress. *Plant Cell and Environment*, *23*, 337–350.

Lara, M. E. B., Garcia, M. C. G., Fatima, T., Ehness, R., Lee, T. K., & Proels, R., et al. (2004). Extracellular invertase is an essential component of cytokinin-mediated delay of senescence. *Plant Cell*, *16*, 1276–1287.

Lers, A. (2007). Environmental regulation of leaf senescence. In S. Gan (Ed.), *Senescence processes in plants* (pp. 108–144). Oxford, UK: Blackwell Publishing.

Lers, A., Sonego, L., Green, P. J., & Burd, S. (2006). Suppression of LX ribonuclease in tomato results in a delay of leaf senescence and abscission. *Plant Physiology*, *142*, 710–721.

Leslie, M. E., Lewis, M. W., Youn, J. Y., Daniels, M. J., & Liljegren, S. J. (2010). The EVERSHED receptor-like kinase modulates floral organ shedding in *Arabidopsis*. *Development*, *137*, 467–476.

Lewis, M. W., Leslie, M. E., Fulcher, E. H., Darnielle, L., Healy, P. N., & Youn, J. Y., et al. (2010). The SERK1 receptor-like kinase regulates organ separation in *Arabidopsis* flowers. *Plant Journal*, *62*, 817–828.

Lewis, M. W., Leslie, M. E., & Liljegren, S. J. (2006). Plant separation: 50 ways to leave your mother. *Current Opinion in Plant Biology*, *9*, 59–65.

Li, X., Xu, C. J., Korban, S. S., & Chen, K. S. (2010). Regulatory mechanisms of textural changes in ripening fruits. *CRC Critical Reviews in Plant Sciences*, *29*, 222–243.

Liljegren, S. J., Leslie, M. E., Darnielle, L., Lewis, M. W., Taylor, S. M., & Luo, R. B., et al. (2009). Regulation of membrane trafficking and organ separation by the NEVERSHED ARF-GAP protein. *Development*, *136*, 1909–1918.

Lim, P. O., Kim, H. J., & Nam, H. G. (2007). Leaf senescence. *Annual Review of Plant Biology*, *58*, 115–138.

Lin, Z. F., Zhong, S. L., & Grierson, D. (2009). Recent advances in ethylene research. *Journal of Experimental Botany*, *60*, 3311–3336.

Link, H. (2000). Significance of flower and fruit thinning on fruit quality. *Plant Growth Regulation*, *31*, 17–26.

Little, H. A., Grumet, R., & Hancock, J. F. (2009). Modified ethylene signaling as an example of engineering for complex traits: Secondary effects and implications for environmental risk assessment. *HortTechnology*, *44*, 94–101.

Long, S. P., & Ort, D. R. (2010). More than taking the heat: Crops and global change. *Current Opinion in Plant Biology*, *13*, 241–248.

Lukatkin, A. S. (2002). Contribution of oxidative stress to the development of cold-induced damage to leaves of chilling-sensitive plants: 2. The activity of antioxidant enzymes during plant chilling. *Russian Journal of Plant Physiology*, *49*, 782–788.

Luo, K. M., Deng, W., Xiao, Y. H., Zheng, X. L., Li, Y., & Pei, Y. (2006). Leaf senescence is delayed in tobacco plants expressing the maize knotted1 gene under the control of a wound-inducible promoter. *Plant Cell Reports*, *25*, 1246–1254.

Lyons, J. M. (1973). Chilling injury in plants. *Annual Review of Plant Physiology and Plant Molecular Biology*, *24*, 445–466.

Malacrida, C., Valle, E. M., & Boggio, S. B. (2006). Postharvest chilling induces oxidative stress response in the dwarf tomato cultivar Micro-Tom. *Physiologia Plantarum*, *127*, 10–18.

Mao, L., Begum, D., Chuang, H. W., Budiman, M. A., Szymkowiak, E. J., & Irish, E. E., et al. (2000). *JOINTLESS* is a MADS-box gene controlling tomato flower abscission zone development. *Nature*, *406*, 910–913.

Marangoni, A. G., Palma, T., & Stanley, D. W. (1996). Membrane effects in postharvest physiology. *Postharvest Biology and Technology*, *7*, 193–217.

Martinez-Romero, D., Bailen, G., Serrano, M., Guillen, F., Valverde, J. M., & Zapata, P., et al. (2007). Tools to maintain postharvest fruit and vegetable quality through the inhibition of ethylene action: A review. *Critical Reviews in Food Science and Nutrition*, *47*, 543–560.

Matas, A. J., Gapper, N. E., Chung, M. Y., Giovannoni, J. J., & Rose, J. K. C. (2009). Biology and genetic engineering of fruit maturation for enhanced quality and shelf-life. *Current Opinion in Biotechnology*, *20*, 197–203.

McCabe, M. S., Garratt, L. C., Schepers, F., Jordi, W., Stoopen, G. M., & Davelaar, E., et al. (2001). Effects of *P-SAG12-IPT* gene expression on development and senescence in transgenic lettuce. *Plant Physiology*, *127*, 505–516.

McDonald, R. E. (1989). Temperature-conditioning affects polyamines of lemon fruits stored at chilling temperatures. *HortTechnology*, *24*, 475–477.

McGlasson, W. B., Scott, K. J., & Mendoza, D. B., Jr (1979). The refrigerated storage of tropical and subtropical products. *International Journal of Refrigeration*, *2*, 199–206.

McKim, S. M., Stenvik, G. E., Butenko, M. A., Kristiansen, W., Cho, S. K., & Hepworth, S. R., et al. (2008). The *BLADE-ON-PETIOLE* genes are essential for abscission zone formation in *Arabidopsis*. *Development*, *135*, 1537–1546.

Michaeli, R., Philosoph-Hadas, S., Riov, J., & Meir, S. (1999). Chilling-induced leaf abscission of *Ixora coccinea* plants. I. Induction by oxidative stress via increased sensitivity to ethylene. *Physiologia Plantarum*, *107*, 166–173.

Michaeli, R., Philosoph-Hadas, S., Riov, J., Shahak, Y., Ratner, K., & Meir, S. (2001). Chilling-induced leaf abscission of *Ixora coccinea* plants. III. Enhancement by high light via increased oxidative processes. *Physiologia Plantarum*, *113*, 338–345.

Mittler, R., & Blumwald, E. (2010). Genetic engineering for modern agriculture: Challenges and perspectives. *Annual Review of Plant Biology*, *61*, 443–462. (Annual Reviews, Palo Alto)

Moffat, A. S. (2000). Plant research—Can genetically modified crops go "greener"? *Science*, *290*, 253–254.

Murata, N., Ishizakinishizawa, Q., Higashi, S., Hayashi, H., Tasaka, Y., & Nishida, I. (1992). Genetically engineered alteration in the chilling sensitivity of plants. *Nature*, *356*, 710–713.

Nambeesan, S., Datsenka, T., Ferruzzi, M. G., Malladi, A., Mattoo, A. K., & Handa, A. K. (2010). Overexpression of yeast *spermidine synthase* impacts ripening, senescence and decay symptoms in tomato. *Plant Journal*, *63*, 836–847.

Nishida, I., & Murata, N. (1996). Chilling sensitivity in plants and cyanobacteria: The crucial contribution of membrane lipids. *Annual Review of Plant Physiology and Plant Molecular Biology*, *47*, 541–568.

Ogawa, M., Kay, P., Wilson, S., & Swain, S. M. (2009). ARABIDOPSIS DEHISCENCE ZONE POLYGALACTURONASE1 (ADPG1), ADPG2, and QUARTET2. Are polygalacturonases required for cell separation during reproductive development in *Arabidopsis*. *Plant Cell*, *21*, 216–233.

Orlova, I. V., Serebriiskaya, T. S., Popov, V., Merkulova, N., Nosov, A. M., & Trunova, T. I., et al. (2003). Transformation of tobacco with a gene for the thermophilic acyl-lipid desaturase enhances the chilling tolerance of plants. *Plant and Cell Physiology*, *44*, 447–450.

Owens, C. L., Thomashow, M. F., Hancock, J. F., & Iezzoni, A. F. (2002). CBF1 orthologs in sour cherry and strawberry and the heterologous expression of *CBF1* in strawberry. *Journal of the American Society for Horticultural Science*, *127*, 489–494.

Park, E. J., Jeknic, Z., Sakamoto, A., DeNoma, J., Yuwansiri, R., & Murata, N., et al. (2004). Genetic engineering of glycinebetaine synthesis in tomato protects seeds, plants, and flowers from chilling damage. *Plant Journal*, *40*, 474–487.

Patterson, S. E. (2001). Cutting loose. Abscission and dehiscence in *Arabidopsis*. *Plant Physiology*, *126*, 494–500.

Patterson, S. E., & Bleecker, A. B. (2004). Ethylene-dependent and -independent processes associated with floral organ abscission in *Arabidopsis*. *Plant Physiology*, *134*, 194–203.

Paull, R. E., & Chen, N. J. (2000). Heat treatment and fruit ripening. *Postharvest Biology and Technology*, *21*, 21–37.

Penfield, S. (2008). Temperature perception and signal transduction in plants. *New Phytologist*, *179*, 615–628.

Perl, A., Perltreves, R., Galili, S., Aviv, D., Shalgi, E., & Malkin, S., et al. (1993). Enhanced oxidative-stress defense in transgenic potato expressing tomato cu,zn superoxide dismutases. *Theoretical and Applied Genetics*, *85*, 568–576.

Pino, M. T., Skinner, J. S., Jeknic, Z., Hayes, P. M., Soeldner, A. H., & Thomashow, M. F., et al. (2008). Ectopic *AtCBF1* over-expression enhances freezing tolerance and induces cold acclimation-associated physiological modifications in potato. *Plant Cell and Environment*, *31*, 393–406.

Porat, R., Pavoncello, D., Peretz, J., Ben-Yehoshua, S., & Lurie, S. (2000). Effects of various heat treatments on the induction of cold tolerance and on the postharvest qualities of "Star Ruby" grapefruit. *Postharvest Biology and Technology*, *18*, 159–165.

Prasad, T. K. (1996). Mechanisms of chilling-induced oxidative stress injury and tolerance in developing maize seedlings: Changes in antioxidant system, oxidation of proteins and lipids, and protease activities. *Plant Journal*, *10*, 1017–1026.

Pruzinska, A., Tanner, G., Aubry, S., Anders, I., Moser, S., & Muller, T., et al. (2005). Chlorophyll breakdown in senescent *Arabidopsis* leaves. Characterization of chlorophyll catabolites and of chlorophyll catabolic enzymes involved in the degreening reaction. *Plant Physiology*, *139*, 52–63.

Puhakainen, T., Hess, M. W., Makela, P., Svensson, J., Heino, P., & Palva, E. T. (2004). Overexpression of multiple dehydrin genes enhances tolerance to freezing stress in *Arabidopsis*. *Plant Molecular Biology*, *54*, 743–753.

Rastogi, S., & Dwivedi, U. N. (2008). Manipulation of lignin in plants with special reference to O-methyltransferase. *Plant Science*, *174*, 264–277.

Richmond, A. E., & Lang, A. (1957). Effect of kinetin on protein content and survival of detached *Xanthium* leaves. *Science*, *125*, 650–651.

Roberts, J. A., Elliott, K. A., & Gonzalez-Carranza, Z. H. (2002). Abscission, dehiscence, and other cell separation processes. *Annals Reviews Plant Biology*, *53*, 131–158.

Roberts, J. A., & Gonzalez-Carranza, Z. H. (2009). Pectinase functions in abscission. *Stewart Postharvest Review*, *1*, 2.

Rosenvasser, S., Mayak, S., & Friedman, H. (2006). Increase in reactive oxygen species (ROS) and in senescence-associated gene transcript (SAG) levels during dark-induced senescence of Pelargonium cuttings, and the effect of gibberellic acid. *Plant Science*, *170*, 873–879.

Saladie, M., Matas, A. J., Isaacson, T., Jenks, M. A., Goodwin, S. M., & Niklas, K. J., et al. (2007). A reevaluation of the key factors that influence tomato fruit softening and integrity. *Plant Physiology*, *144*, 1012–1028.

Saltveit, M. E. (1999). Effect of ethylene on quality of fresh fruits and vegetables. *Postharvest Biology and Technology*, *15*, 279–292.

Scandalios, J. G. (1993). Oxygen stress and superoxide dismutases. *Plant Physiology*, *101*, 7–12.

Schelbert, S., Aubry, S., Burla, B., Agne, B., Kessler, F., & Krupinska, K., et al. (2009). Pheophytin pheophorbide hydrolase (pheophytinase) is involved in chlorophyll breakdown during leaf senescence in *Arabidopsis*. *Plant Cell*, *21*, 767–785.

Schippers, J. H. M., Jing, H. C., Hille, J., & Dijkwel, P. (2007). Developmental and hormonal control of leaf senescence. In S. Gan (Ed.), *Senescence processes in plants* (pp. 145–170). Oxford, UK: Blackwell Publishing.

Sevillano, L., Sanchez-Ballesta, M. T., Romojaro, F., & Flores, F. B. (2009). Physiological, hormonal and molecular mechanisms regulating chilling injury in horticultural species. Postharvest technologies applied to reduce its impact. *Journal of the Science of Food and Agriculture*, *89*, 555–573.

Sharom, M., Willemot, C., & Thompson, J. E. (1994). Chilling injury induces lipid phase-changes in membranes of tomato fruit. *Plant Physiology*, *105*, 305–308.

Singh, R., Rastogi, S., & Dwivedi, U. N. (2010). Phenylpropanoid metabolism in ripening fruits. *Comprehensive Reviews in Food Science and Food Safety*, *9*, 398–416.

Stearns, J. C., & Glick, B. R. (2003). Transgenic plants with altered ethylene biosynthesis or perception. *Biotechnology Advances*, *21*, 193–210.

Su, C. F., Wang, Y. C., Hsieh, T. H., Lu, C. A., Tseng, T. H., & Yu, S. M. (2010). A novel MYBS3-dependent pathway confers cold tolerance in rice. *Plant Physiology*, *153*, 145–158.

Sui, N., Li, M., Zhao, S. J., Li, F., Liang, H., & Meng, Q. W. (2007). Overexpression of glycerol-3-phosphate acyltransferase gene improves chilling tolerance in tomato. *Planta*, *226*, 1097–1108.

Sun, L. X., & van Nocker, S. (2010). Analysis of promoter activity of members of the *PECTATE LYASE-LIKE (PLL)* gene family in cell separation in *Arabidopsis*. *BMC Plant Biology*, *10*, 13.

Sung, D. Y., Kaplan, F., Lee, K. J., & Guy, C. L. (2003). Acquired tolerance to temperature extremes. *Trends in Plant Science*, *8*, 179–187.

Suzuki, N., & Mittler, R. (2006). Reactive oxygen species and temperature stresses: A delicate balance between signaling and destruction. *Physiologia Plantarum*, *126*, 45–51.

Tanaka, Y., Katsumoto, Y., Brugliera, F., & Mason, J. (2005). Genetic engineering in floriculture. *Plant Cell Tissue and Organ Culture*, *80*, 1–24.

Taylor, J. E., & Whitelaw, C. A. (2001). Signals in abscission. *New Phytologist*, *151*, 323–339.

Toivonen, P. M. A., & Brummell, D. A. (2008). Biochemical bases of appearance and texture changes in fresh-cut fruit and vegetables. *Postharvest Biology and Technology*, *48*, 1–14.

Tsuchisaka, A., & Theologis, A. (2004). Unique and overlapping expression patterns among the arabidopsis 1-amino-cyclopropane-1-carboxylate synthase gene family members. *Plant Physiology*, *136*, 2982–3000.

Tucker, M. L., Whitelaw, C. A., Lyssenko, N. N., & Nath, P. (2002). Functional analysis of regulatory elements in the gene promoter for an abscission-specific cellulase from bean and isolation, expression, and binding affinity of three TGA-Type basic leucine zipper transcription factors. *Plant Physiology*, *130*, 1487–1496.

Valentines, M. C., Vilaplana, R., Torres, R., Usall, J., & Larrigaudiere, C. (2005). Specific roles of enzymatic browning and lignification in apple disease resistance. *Postharvest Biology and Technology*, *36*, 227–234.

van Breusegem, F., Slooten, L., Stassart, J. M., Botterman, J., Moens, T., & Van Montagu, M., et al. (1999). Effects of overproduction of tobacco MnSOD in maize chloroplasts on foliar tolerance to cold and oxidative stress. *Journal of Experimental Botany*, *50*, 71–78.

van Buskirk, H. A., & Thomashow, M. F. (2006). *Arabidopsis* transcription factors regulating cold acclimation. *Physiologia Plantarum*, *126*, 72–80.

van Doorn, W. G. (2002). Effect of ethylene on flower abscission: A survey. *Annals of Botany*, *89*, 689–693.

van Doorn, W. G., & Stead, A. D. (1997). Abscission of flowers and floral parts. *Journal of Experimental Botany*, *48*, 821–837.

van Doorn, W. G., & Woltering, E. J. (1991). Developments in the use of growth regulators for the maintenance of postharvest quality in cut flowers and potted plants. *Acta Horticulturae*, *298*, 195–204.

Vicente, A. R., Saladie, M., Rose, J. K. C., & Labavitch, J. M. (2007). The linkage between cell wall metabolism and fruit softening: Looking to the future. *Journal of the Science of Food and Agriculture*, *87*, 1435–1448.

Vina, S. Z., & Chaves, A. R. (2003). Texture changes in fresh cut celery during refrigerated storage. *Journal of the Science of Food and Agriculture*, *83*, 1308–1314.

Voelker, S. L., Lachenbruch, B., Meinzer, F. C., Jourdes, M., Ki, C., & Patten, A. M., et al. (2010). Antisense down-regulation of *4CL* expression alters lignification, tree growth, and saccharification potential of field-grown poplar. *Plant Physiology*, *154*, 874–886.

Wagstaff, C., Clarkson, G. J. J., Zhang, F. Z., Rothwell, S. D., Fry, S. C., & Taylor, G., et al. (2010). Modification of cell wall properties in lettuce improves shelf life. *Journal of Experimental Botany*, *61*, 1239–1248.

Wang, T. W., Lu, L., Zhang, C. G., Taylor, C., & Thompson, J. E. (2003). Pleiotropic effects of suppressing deoxyhypusine synthase expression in *Arabidopsis thaliana*. *Plant Molecular Biology*, *52*, 1223–1235.

Wang, W. X., Vinocur, B., Shoseyov, O., & Altman, A. (2004). Role of plant heat-shock proteins and molecular chaperones in the abiotic stress response. *Trends in Plant Science*, *9*, 244–252.

Wang, Y., & Kumar, P. P. (2004). Heterologous expression of *Arabidopsis* ERS1 causes delayed senescence in coriander. *Plant Cell Reports*, *22*, 678–683.

Wang, Y. J., Wisniewski, M., Meilan, R., Cui, M. G., Webb, R., & Fuchigami, L. (2005). Overexpression of cytosolic ascorbate peroxidase in tomato confers tolerance to chilling and salt stress. *Journal of the American Society for Horticultural Science*, *130*, 167–173.

Waterer, D., Benning, N. T., Wu, G. H., Luo, X. M., Liu, X. J., & Gusta, M., et al. (2010). Evaluation of abiotic stress tolerance of genetically modified potatoes (*Solanum tuberosum* cv. Desiree). *Molecular Breeding*, *25*, 527–540.

Waters, E. R., Lee, G. J., & Vierling, E. (1996). Evolution, structure and function of the small heat shock proteins in plants. *Journal of Experimental Botany*, *47*, 325–338.

Wen, X. P., Pang, X. M., Matsuda, N., Kita, M., Inoue, H., & Hao, Y. J., et al. (2008). Over-expression of the apple *spermidine synthase* gene in pear confers multiple abiotic stress tolerance by altering polyamine titers. *Transgenic Research*, *17*, 251–263.

Whitelaw, C. A., Lyssenko, N. N., Chen, L. W., Zhou, D. B., Mattoo, A. K., & Tucker, M. L. (2002). Delayed abscission and shorter internodes correlate with a reduction in the ethylene receptor *LeETR1* transcript in transgenic tomato. *Plant Physiology*, *128*, 978–987.

Whitney, J. D., Hartmond, U., Kender, W. J., Burns, J. K., & Salyani, M. (2000). Orange removal with trunk shakers and abscission chemicals. *Applied Engineering in Agriculture*, *16*, 367–371.

Wilkinson, J. Q., Lanahan, M. B., Clark, D. G., Bleecker, A. B., Chang, C., & Meyerowitz, E. M., et al. (1997). A dominant mutant receptor from *Arabidopsis* confers ethylene insensitivity in heterologous plants. *Nature Biotechnology*, *15*, 444–447.

Winichayakul, S., Moyle, R. L., Ryan, D. J., Farnden, K. J. F., Davies, K. M., & Coupe, S. A. (2004). Distinct cis-elements in the Asparagus officinalis asparagine synthetase promoter respond to carbohydrate and senescence signals. *Functional Plant Biology*, *31*, 573–582.

Wise, R. R. (1995). Chilling-enhanced photooxidation—the production, action and study of reactive oxygen species produced during chilling in the light. *Photosynthesis Research*, *45*, 79–97.

Wolfe, J. (1978). Chilling injury in plants — The role of membrane lipid fluidity. *Plant, Cell and Environmental*, *1*, 241–247.

Wongsheree, T., Ketsa, S., & van Doorn, W. G. (2009). The relationship between chilling injury and membrane damage in lemon basil (*Ocimum* x *citriodourum*) leaves. *Postharvest Biology and Technology*, *51*, 91–96.

Woo, H. R., Chung, K. M., Park, J. H., Oh, S. A., Ahn, T., & Hong, S. H., et al. (2001). ORE9, an F-box protein that regulates leaf senescence in *Arabidopsis*. *Plant Cell*, *13*, 1779–1790.

Xue-Xuan, X., Hong-Bo, S., Yuan-Yuan, M., Gang, X., Jun-Na, S., & Dong-Gang, G., et al. (2010). Biotechnological implications from abscisic acid (ABA) roles in cold stress and leaf senescence as an important signal for improving plant sustainable survival under abiotic-stressed conditions. *Critical Reviews in Biotechnology*, *30*, 222–230.

Yamaguchi-Shinozaki, K., & Shinozaki, K. (2006). Transcriptional regulatory networks in cellular responses and tolerance to dehydration and cold stresses. *Annual Review of Plant Biology*, *57*, 781–803.

Yang, T. F., Gonzalez-Carranza, Z. H., Maunders, M. J., & Roberts, J. A. (2008). Ethylene and the regulation of senescence processes in transgenic *Nicotiana sylvestris* plants. *Annals of Botany*, *101*, 301–310.

Yoshida, S., Ito, M., Callis, J., Nishida, I., & Watanabe, A. (2002). A delayed leaf senescence mutant is defective in arginyl-tRNA: Protein arginyltransferase, a component of the N-end rule pathway in *Arabidopsis*. *Plant Journal*, *32*, 129–137.

Zapata, J. M., Guera, A., Esteban-Carrasco, A., Martin, M., & Sabater, B. (2005). Chloroplasts regulate leaf senescence: Delayed senescence in transgenic ndhF-defective tobacco. *Cell Death and Differentiation*, *12*, 1277–1284.

Zhang, C. F., Ding, Z. S., Xu, X. B., Wang, Q., Qin, G. Z., & Tian, S. P. (2010). Crucial roles of membrane stability and its related proteins in the tolerance of peach fruit to chilling injury. *Amino Acids*, *39*, 181–194.

Zhang, C. F., & Tian, S. P. (2009). Crucial contribution of membrane lipids' unsaturation to acquisition of chilling-tolerance in peach fruit stored at 0 degrees C. *Food Chemistry*, *115*, 405–411.

Zhao, D., Reddy, K. R., Kakani, V. G., Koti, S., & Gao, W. (2005). Physiological causes of cotton fruit abscission under conditions of high temperature and enhanced ultraviolet-B radiation. *Physiologia Plantarum*, *124*, 189–199.

Zhao, D. Y., Shen, L., Fan, B., Liu, K. L., Yu, M. M., & Zheng, Y., et al. (2009a). Physiological and genetic properties of tomato fruits from 2 cultivars differing in chilling tolerance at cold storage. *Journal of Food Science*, *74*, C348–C352.

Zhao, D. Y., Shen, L., Fan, B., Yu, M. M., Zheng, Y., & Lv, S. N., et al. (2009b). Ethylene and cold participate in the regulation of LeCBF1 gene expression in postharvest tomato fruits. *FEBS Letters*, *583*, 3329–3334.

Zhou, C. J., Cai, Z. H., Guo, Y. F., & Gan, S. S. (2009). An *Arabidopsis* mitogen-activated protein kinase cascade, MKK9-MPK6, plays a role in leaf senescence. *Plant Physiology*, *150*, 167–177.

Zhu, J., Dong, C. -H., & Zhu, J. -K. (2007). Interplay between cold-responsive gene regulation, metabolism and RNA processing during plant cold acclimation. *Current Opinion in Plant Biology*, *10*, 290–295.

Zimmermann, P., & Zentgraf, U. (2005). The correlation between oxidative stress and leaf senescence during plant development. *Cellular and Molecular Biology Letters*, *10*, 515–534.

Engineering the biosynthesis of low molecular weight metabolites for quality traits (essential nutrients, health-promoting phytochemicals, volatiles, and aroma compounds)

28

Fumihiko Sato[1] Kenji Matsui[2]

[1]*Graduate School of Biostudies, Kyoto University, Sakyo, Kyoto, Japan,* [2]*Faculty of Agriculture and Graduate School of Medicine, Yamaguchi University, Yamaguchi, Japan*

TABLE OF CONTENTS

General Introduction

Plants provide staple food and feed for humans and livestock, and are also an important nutritional resource for essential nutrients, such as amino acids, fatty acids, vitamins, and minerals (Hounsome et al., 2008), as well as health-promoting phytochemicals (also called nutraceuticals or functional foods) and aromas that can enhance our quality of life (Figure 28.1). Now that the Green Revolution based on plant breeding has dramatically improved the yield of crops over the past several decades, we need methods to improve the quality traits of crops, such as their production of essential nutrients, health-promoting phytochemicals, and life-enriching chemicals, along with further improvements in yields without the excess use of fossil fuels.

While quality traits based on low molecular weight metabolites are difficult to evaluate during breeding, rapid advances in biochemistry and molecular biology, especially genetic engineering and analytical development, have made these quality improvements possible. Some conceptual strategies to enhance the formation of desired secondary metabolites through genetic modification are outlined in Figure 28.2A–E. In general, metabolic flux is tightly regulated, and there is little room for modification (Figure 28.2A). Accumulating evidence indicates that over-expression of the enzyme(s) for a rate-limiting step, especially enzymes in primary metabolism that are insensitive to feedback regulation, can increase the production of desired metabolites such as amino acids (Figure 28.2B). Another approach is the introduction of a new branch pathway to produce novel compounds that are not normally produced by the host plants, such as very long chain polyunsaturated fatty acids (VLC-PUFAs; Figure 28.2C). A more attractive approach is the use of a transcription factor (TF) that controls an entire enzymatic pathway in metabolism. In this case, even if we do not fully understand the details of a

DOI: 10.1016/B978-0-12-381466-1.00028-6

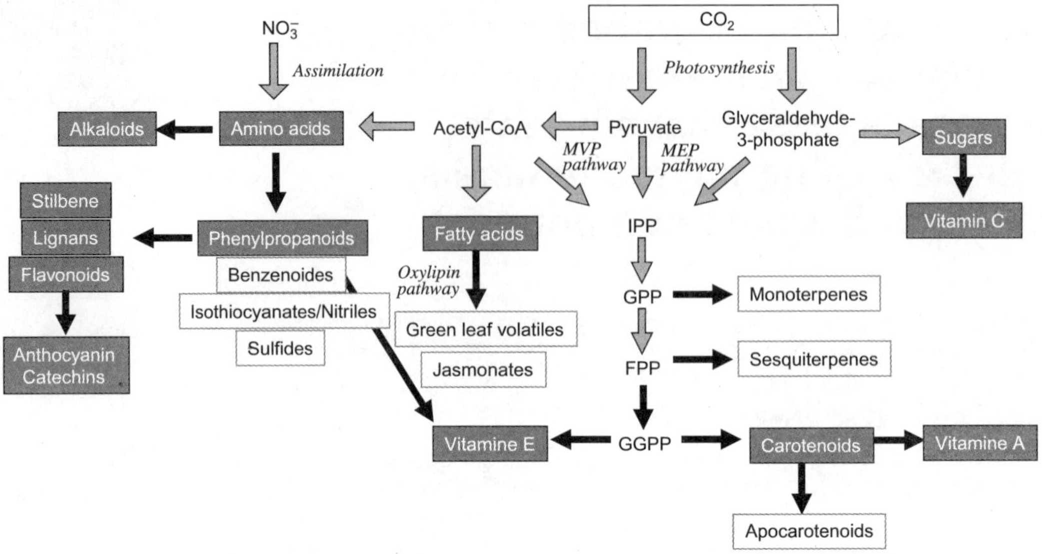

Figure 28.1 • An overview of plant secondary metabolism • The reactions in primary metabolism are shown with gray arrows, and those in secondary metabolism are shown with black arrows. The non-volatile metabolites that are important for the agronomic traits of plants are shown in white with gray boxes, while volatile metabolites are shown in black with white boxes. All of the secondary metabolites are formed from primary metabolites, and thus an unconsidered attempt to enhance the formation of secondary metabolites might disturb the primary metabolism, which could negatively affect important traits of crop plants.

biosynthetic pathway, the introduction of just one TF gene into a crop plant can promote the entire pathway and increase the accumulation of a metabolite (Figure 28.2D). Finally, there is a growing body of evidence that the supply of a primary metabolite precursor can limit the production of secondary metabolites. Thus, the enhanced flow of a precursor may be essential for the production of desired metabolites, although this should be achieved without disturbing the primary metabolism (Figure 28.2E).

Primary metabolites as essential nutrients are being more intensively studied based on these concepts, and their engineering has often been reviewed (Zhu et al., 2007; Mayer et al., 2008; Newell-McGloughlin, 2008). Thus, in this chapter we first outline the metabolic engineering of essential nutrients, such as essential amino acids, lipids, vitamins, and minerals, and then summarize the recent advances in the modification of a more specific secondary metabolism for the improvement of quality traits.

Lessons from Essential Nutrients

While plants provide staple food and feed, no crop is nutritionally perfect. For example, cereal grains, such as rice and maize, are poor sources of lysine, vitamin A, folic acid, iron, zinc, and selenium, which are essential for normal growth and metabolism, whereas legumes lack methionine and cysteine (Molvig et al. 1997). Since about one-third of the world's population (mostly in sub-Saharan Africa and Southeast Asia) depends on cereals as their source of nutrition (FAO, 2006), it is important to improve the essential nutrients in cereals to reduce malnutrition in these areas (Zhu et al., 2007; Mayer et al., 2008; Newell-McGloughlin, 2008).

While the modification of storage protein profiles would be useful for improving desirable amino acid profiles; for example, elimination of 22 kDa zeins with RNAi to increase lysine content in maize (Segal et al., 2003), the deregulation of rate-limiting-step enzymes is often used to improve quality traits, as outlined in the next section.

Essential amino acids

Essential amino acid biosynthesis is strictly regulated by some key enzymes, such as dihydrodipicolinate synthase (DHDPS) in lysine (Lys) biosynthesis, and anthranilate synthase in tryptophan (Trp) synthesis (Figure 28.3A,B). Thus, introduction of a mutant enzyme gene without feedback regulation is critical to improve the content of essential amino acids.

The first successful increases in Lys content in the seeds of canola and soybean plants were achieved by the introduction of Lys-feedback-insensitive bacterial DHDPS and aspartokinase (AK) enzymes encoded by the *Corynebacterium dapA* gene and a mutant *Escherichia coli lysC* gene (Falco et al., 1995). The expression of *Corynebacterium* DHDPS resulted in more than a 100-fold increase in the accumulation of free Lys in the seeds of canola, and the total seed Lys content approximately doubled. The expression of *Corynebacterium* DHDPS plus Lys-insensitive *E. coli* AK in soybean transformants similarly increased the free Lys level by several hundred-fold and the total seed Lys content by as much as five-fold. However, genes should be linked to a plastid transit peptide

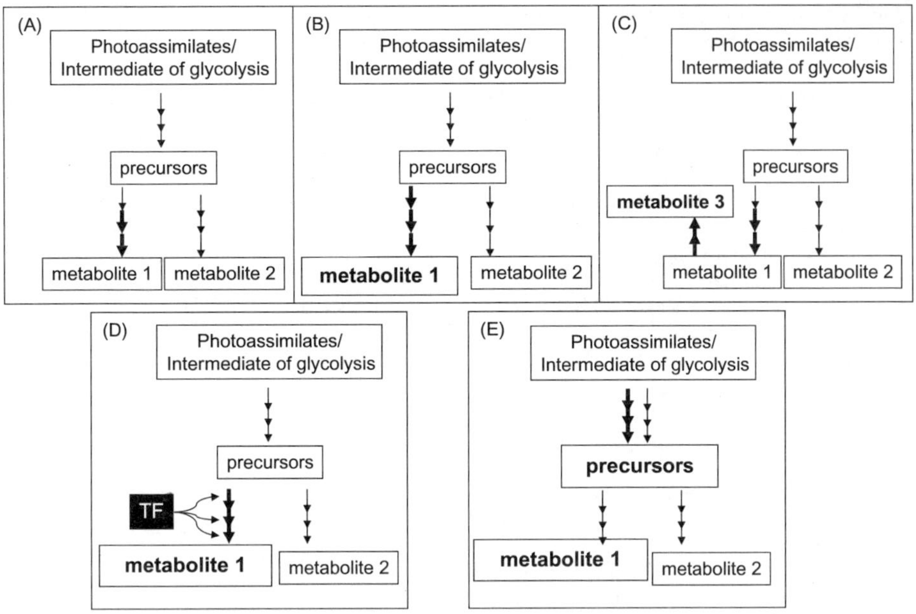

Figure 28.2 • Conceptual drawing of strategies for enhancing the formation of desired secondary metabolites through genetic modification • The existing pathway is generally tightly regulated (A). Over-expression of enzyme(s) in the rate-limiting step with insensitive feedback regulation increases the production of desired metabolites (B). Introduction of a new branch pathway produces novel compounds that are not normally produced by the host plants (C). Ectopic expression of a transcription factor (TF) that controls an entire enzymatic pathway in metabolism may increase the accumulation of a metabolite (D). Enhanced supply of a precursor primary metabolite(s) would be essential for the production of desired metabolites (E). Please refer to the text for more details.

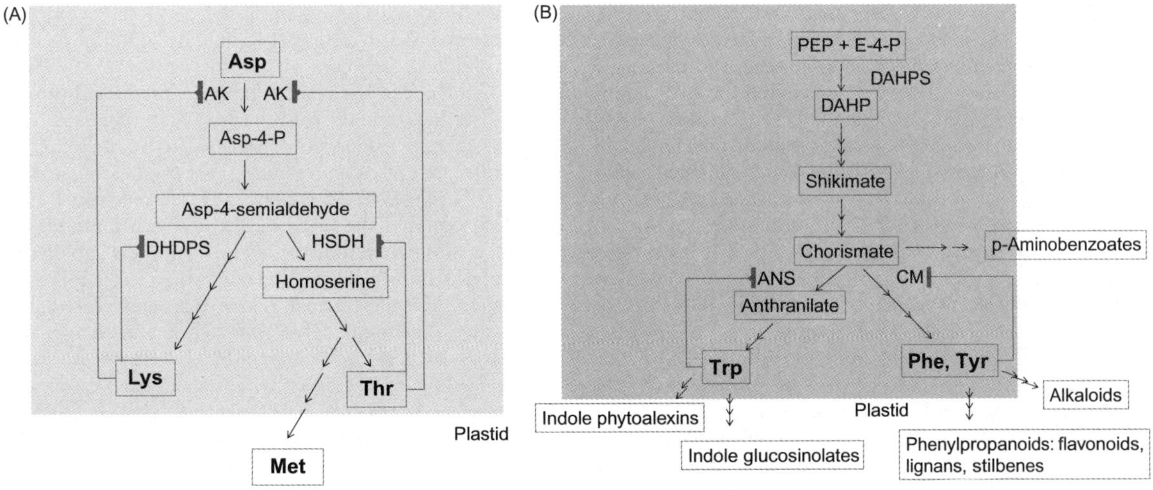

Figure 28.3 • Amino acid biosynthetic pathways and feedback regulation • Simplified pathways for lysine (Lys; A) and aromatic amino acids (B) such as tryptophan (Trp), phenylalanine (Phe), and tyrosine (Tyr) are shown. Red bars show the feedback inhibition of enzymes in the rate-limiting step by products such as Lys and Trp. A green background indicates that these reactions occur in plastids. Abbreviations: Asp, aspartate; AK, aspartate kinase; Asp-4-P, aspartate-4-phosphate; DHDPS, dihydrodipicolinate synthase; HSDH, homoserine dehydrogenase; Met, methionine; Thr, threonine; PEP, phosphoenol pyruvate; E-4-P, erythrose-4-phosphate; DAHP, 3-deoxy-D-arabino-heptulosonate-7-phosphate; DAHPS, DAHP synthase; ANS, anthranilate synthase; and CM, chorismate mutase. Please see color plate section at the end of the book.

and expressed from a seed-specific promoter in transgenic canola and soybean seeds, since these biosyntheses occur in plastids in seeds.

The first successful engineering of Lys biosynthesis also illustrated enhanced catabolism: the accumulation of α-amino adipic acid in canola and saccharopine in soybean. This indicates that engineering can influence catabolism through the modification of biosynthesis (Zhu and Galili, 2003). Transgenic *Arabidopsis* expressing the feedback-insensitive DHDPS in a knockout mutant in the Lys catabolism pathway showed a synergistic 80-fold increase in the seed-free Lys level compared to a single trait line; a single DHDPS overexpresser and a Lys catabolism mutant contained, respectively, only 12- and 5-fold higher levels of seed-free Lys than wild-type plants.

A feedback-insensitive α-subunit of rice anthranilate synthase is another successful application of the accumulation of tryptophan (Trp; Wakasa et al., 2006). The amount of free Trp in transgenic rice seeds was increased by about two orders of magnitude compared with that in wild-type, and the total Trp content was also increased. However, it is possible that the marked overproduction of Trp might reduce spikelet fertility, yield, and seed germination ability, although agronomic traits varied among lines. Interestingly, no substantial changes in the amounts of other phenolic compounds were detected, except for about a two-fold increase in indole acetic acid in the transgenic seeds.

Fatty acids

Fatty acids are another target for nutritional improvement because some of them are critical in metabolism, cardiovascular health, inflammatory responses, and blood pressure regulation. Many of the enzymes involved in fatty acid biosynthesis and catabolism have been characterized, and transgenic approaches have been attempted to modify the oil content and composition in plants (Qi et al., 2004; Truksa et al., 2006; Clemente and Cahoon, 2009). Modification of the content of unsaturated fatty acids (e.g., linoleic acid and α-linolenic acid) and the production of VLC-PUFAs (e.g., arachidonic acid, ARA; eicosapentaenoic acid, EPA; and docosahexaenoic acid, DHA), which are usually obtained from fish oils, have been intensively studied (Truksa et al., 2006). One of the most promising applications of metabolic engineering in the area of nutritional fatty acids is the production of two VLC-PUFAs, AA and EPA, in a higher plant using genes in the Ω3/6 Δ^8-desaturation biosynthetic pathway for the formation of C20 polyunsaturated fatty acids (Qi et al., 2004).

The sequential introduction of genes encoding a Δ^9-specific elongating activity from *Isochrysis galbana*, a Δ^8-desaturase from *Euglena gracilis*, and a Δ^5-desaturase from *Mortierella alpina* reconstituted a bacterial pathway that is not present in plants (Clemente and Cahoon, 2009). The constitutive expression of three transgenes in *Arabidopsis* produced EPA and ARA in vegetative tissues at 3.0 and 6.6% of total fatty acids. *I. galbana* C18-Δ^9-elongating activity may bypass rate-limiting steps present in the conventional Δ^6-desaturase/elongase pathways. No deleterious effect on plant growth was detected with the elevated accumulation of VLC-PUFAs, whereas some growth defect was reported for the modification of natural unsaturated fatty acids to increase the oleic and stearic acid contents (Clemente and Cahoon, 2009).

In addition to the introduction of novel genes from other organisms, such as microbes, to improve the lipid composition, gene knockdown is commonly used to modify the oil composition, especially to improve the oxidative stability and nutritional quality through the reduction of unsaturated fatty acids. The silencing or knocking out of Δ^{12} desaturase (FAD2) and prevention of the formation of Δ^9 desaturase successfully increased the contents of oleic (18:1) and stearic (18:0) acids (read Clemente and Cahoon, 2009, for details on soybean oil metabolic engineering).

Vitamins

Plants are rich resources of vitamins. While the biosynthesis of vitamins A and E has been most extensively studied, extensive work has also been done on other vitamins such as vitamin C and folate (Van Eenennaam et al., 2003; Zhu et al., 2007; Mayer et al., 2008; Newell-McGloughlin, 2008). These biosynthetic pathways have been studied in both higher plants and microorganisms, and some microbial genes are used to generate vitamin overproducing plants such as "golden rice" by bypassing the limiting step for plant enzymes.

Vitamin A

The carotenoid biosynthesis pathway provides not only an important photosynthetic pigment, but also a precursor for vitamin A (Figure 28.4). Humans can synthesize vitamin A from β-carotene (known as pro-vitamin A). Vitamin A deficiency is still a serious problem in the developing world, and contributes to the deaths of two million children annually, and to blindness in many of the survivors (WHO Report, 1995). Golden rice has been created to produce β-carotene in rice. The biosynthetic pathway for β-carotene was reconstituted in rice endosperm using a daffodil (*Narcissus pseudonarcissus*) phytoene synthase (*psy*) and carotene desaturase (*ctri*) from *Erwinia uredovora*. More recently, Paine et al. (2005) replaced the rate-limiting *psy* with more efficient maize *psy* through systematic testing. The new "golden rice II" variety had up to 23-fold (maximum 37 μg/g) more total carotenoids than the original golden rice, and preferentially accumulated β-carotene. This approach has been applied to other crops such as "golden potato" (Diretto et al. 2007), "orange cauliflower," carrots with enhanced β-carotene in the taproot and tomatoes have been established. In potato, Diretto et al. (2007) reconstituted the β-carotene biosynthetic pathway from geranylgeranyl diphosphate with three genes, encoding phytoene synthase (CrtB), phytoene desaturase (CrtI), and lycopene β-cyclase (CrtY), from *Erwinia*, under tuber-specific or constitutive promoter control. Interestingly, constitutive expression of the CrtY and/or CrtI genes was harmful for the establishment of transgenic plants, whereas the expression of all three genes under a tuber-specific promoter resulted in tubers with a deep yellow ("golden2") phenotype without any adverse leaf phenotypes.

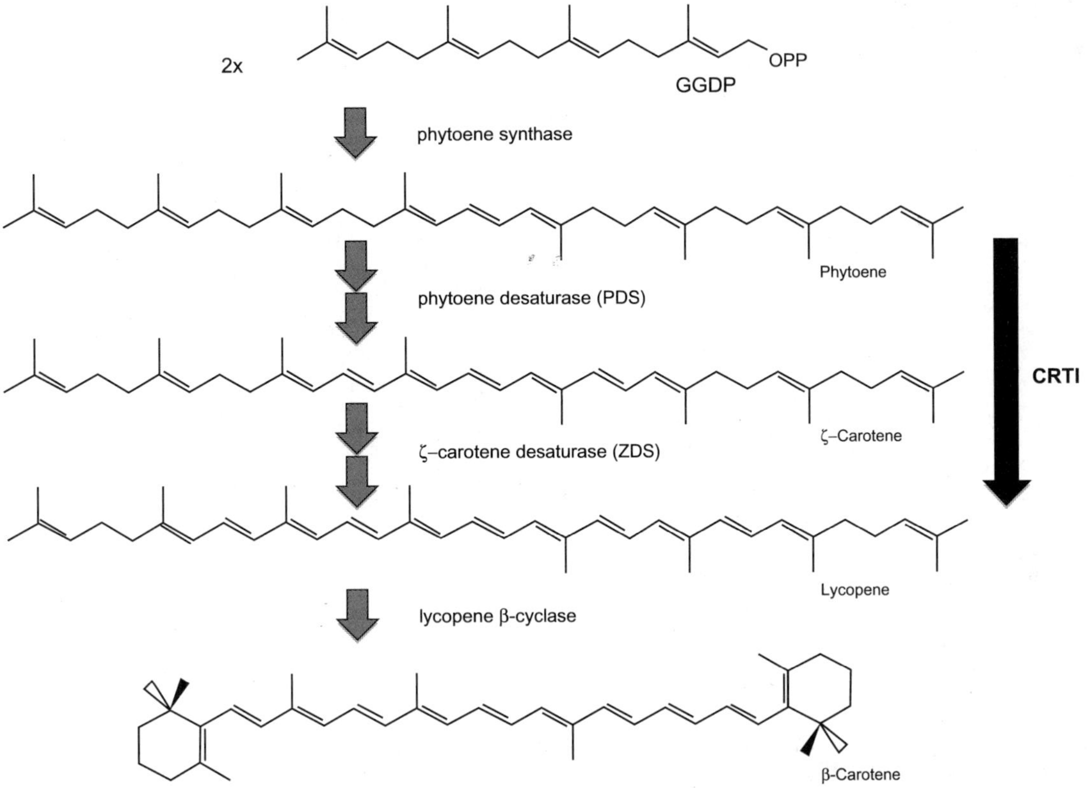

Figure 28.4 • Carotenoid biosynthetic pathway and engineering with a microbial gene (*CRTI*) • Bacterial CRTI is a bifunctional enzyme that catalyzes both phytoene desaturase and ζ-carotene desaturase reactions.

The carotenoid contents in tubers were increased approximately 20-fold (to 114 μg/g dry weight) and β-carotene was increased 3600-fold (to 47 μg/g dry weight). Based on the conversion rate of β-carotene to retinol (6:1), 250 g (fresh weight) of "golden" potatoes should provide 50% of the recommended daily allowance of Vitamin A.

Attempts have been made to produce derivatives of β-carotene, such as astaxanthin, in transgenic plants, since astaxanthin contributes to the red pigmentation of fish and crustacean and is important for human health. The introduction of *CrtO* from the alga *Haematococcus pluvialis*, encoding β-carotene ketolase, under regulation of the tomato phytoene desaturase gene promoter with the transit peptide of phytoene desaturase resulted in the successful accumulation of (3*S*, 3′*S*) astaxanthin and other ketocarotenoids in plants (Mann et al., 2000).

Vitamin E

Vitamin E is actually a group of eight hydrophobic compounds (known as vitamers) that prevent the oxidation and polymerization of unsaturated fatty acids; vitamin E deficiency leads to lipid peroxidation, neurological dysfunction, and infertility (Brigelius-Flohe and Traber, 1999). Plant oils typically contain the most potent α-tocopherol as a minor component, and high levels of its biosynthetic precursor γ-tocopherol. Since many biosynthetic genes including homogentisic acid dioxygenase (HPPD), homogentisic acid prenyltransferase (HPT, VTE2 in *Arabidopsis*), 2-methyl-6-phytylbenzoquinol (MPBQ) methyltransferase (VTE3 in *Arabidopsis*), tocopherol cyclase (VTE1 in *Arabidopsis*), and γ-tocopherol methyltransferase (γ-TMT, VTE4 in *Arabidopsis*) have been isolated, the quantity and quality of tocopherols has been intensively engineered after the pioneering work by Shintani and Della-Penna (1998; Van Eenennaam, 2003; Figure 28.5A). In particular, γ-tocopherol methyltransferase is crucial for the improvement of tocopherol activity; for example, the simultaneous expression of MPBQ methyltransferase and γ-TMT in soybean significantly increased the total vitamin E activity (five-fold greater than that of wild-type plants) through an eight-fold increase in the level of α-tocopherol (10% of total vitamin E in wild-type to over 95% in transgenic seeds).

Folate

Folate has received considerable attention, since its deficiency induces neural tube defects during early pregnancy (Laurence et al., 1981). However, the biosynthesis of folates

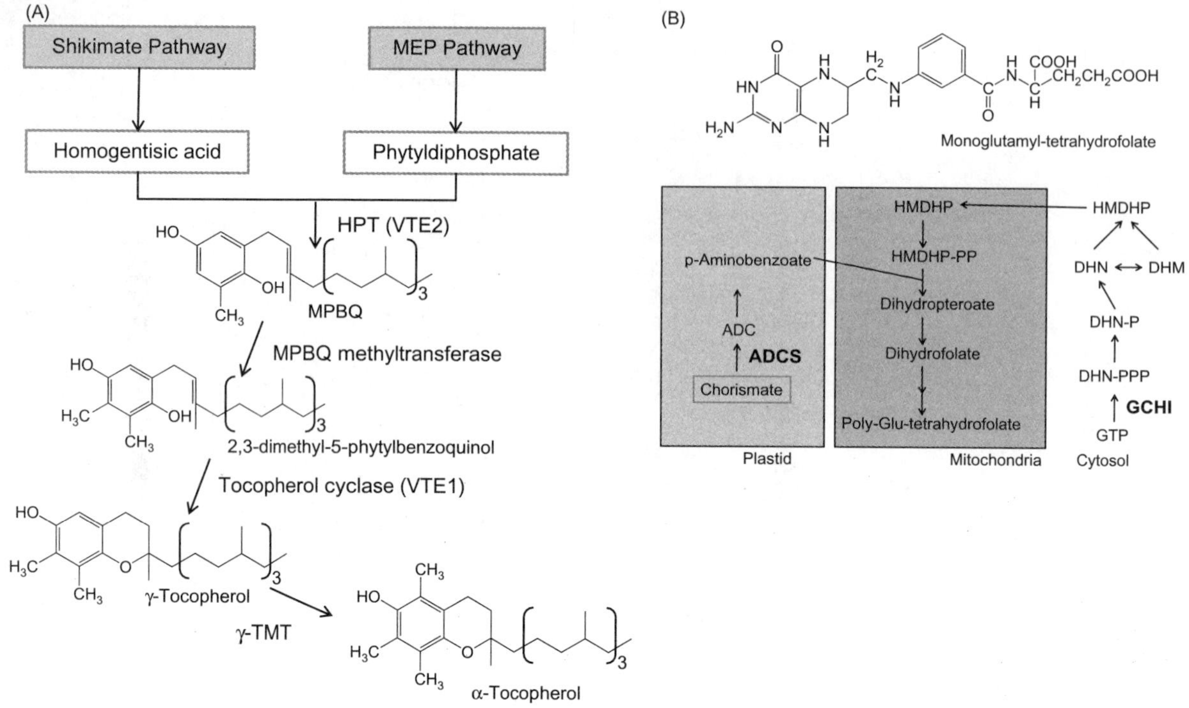

Figure 28.5 • Biosynthetic pathways of vitamin E (A) and folate (B) • Vitamin E (tocopherols and tocotrienols) is a group of eight hydrophobic compounds. Vitamin E biosynthesis involves homogentisic acid dioxygenase (HPPD), homogentisic acid prenyltransferase (HPT, VTE2 in *Arabidopsis*), 2-methyl-6- phytylbenzoquinol (MPBQ) methyltransferase (VTE3 in *Arabidopsis*), tocopherol cyclase (VTE1 in *Arabidopsis*), and γ-tocopherol methyltransferase (γ-TMT, VTE4 in *Arabidopsis*). (B) Folate is synthesized from three moieties of pteridine, *p*-aminobenzoate (PABA), and glutamate. Pteridines synthesized in the cytosol and PABA synthesized in the plastids are transported to the mitochondria to form dihydropteroate and conjugated to glutamate. Abbreviations: ADC, aminodeoxychorismate; ADCS, ADC synthase; DHM, dihydromonopterin; DHN, dihydronopterin; GCHI, GTP cyclohydrolase I; and HMDHP, hydroxymethyldihydropterin.

is complicated, since not only are they synthesized from pteridine, *p*-aminobenzoate (PABA), and glutamate, but because it also requires three subcellular compartments (Figure 28.5B). Pteridines synthesized in the cytosol and PABA synthesized in the plastids are transported to the mitochondria to form dihydropteroate and conjugated to glutamate. The initial attempts to increase folate levels through the over-expression of GTP cyclohydrolase I (GCHI) in the pteridine branch of the pathway led to partial success with an overall two-fold increase in the folate level due to the limitation of the free PABA pool in the plastids, whereas the amount of pteridine increased 140-fold (Diaz de la Garza et al., 2004; Hossain et al., 2004). Further over-expression of aminodeoxychorismate synthase (ADCS), a key enzyme in the PABA branch, with GCHI increased the levels of both precursors, and the folate content in tomato fruits increased up to 20 times that in the wild type (Diaz de la Garza et al., 2007). The resulting fruit (100 g) contains the full adult recommended daily allowance (400 μg) of folate.

The attempt by Diaz de la Garza et al. (2007) to maximize the production of folate through the expression of both ADCS and GCHI from strong fruit-specific promoters gave high levels of expression but also elevated the levels of pathway intermediates. Free pteridines (0.36 mg per portion) that accumulated in transgenic tomato would represent <10% of the pteridines that humans excrete in the urine per day, and more studies are needed to assess the safety of pteridine accumulation and to control the accumulation of intermediates.

Vitamin C

Ascorbate (vitamin C) is an essential metabolic substrate, electron donor enzyme co-factor, and antioxidant, and its absence leads to scurvy (Peterkofsky, 1991). At least three separate metabolic pathways converge on ascorbate in plants, which means that several strategies can be used to enhance its synthesis. The most successful experiments include over-expression of the rat L-gulonolactone oxidase gene in tobacco, which resulted in a seven-fold increase in the ascorbate pool (Jain and Nessler, 2000), and the expression of myo-inositol oxidase in *Arabidopsis*, which resulted in a three-fold enhancement (Lorence et al., 2004). Encouraging results have also been achieved by reducing the rate at which ascorbate is recycled, which can be up to 40% per day in some plants (Chen et al., 2003).

Improvement of the bioavailability of minerals through metabolic engineering

Minerals, such as Zn, I, and Fe are also essential for health. In particular, Fe is critical for blood cell formation (Panel on Micronutrients, 2001). Some metal-binding proteins are used to improve the uptake/bioavailability of metals, for example, ferritin for Fe (Goto et al., 1999). In addition to such metal-binding proteins, transporters, reductase (e.g., Fe(III) reductases, which convert Fe(III) to Fe(II) in many plants except grasses and cereals), and some low molecular weight metal-chelating compounds in grasses and cereals, called phytosiderophores, are important for the uptake of metals into plants. Phytosiderophores are critical for the growth of plants under alkaline salt conditions, but they also improve the nutritional quality of plants. To date, the barley *naat-A* and *naat-B* genes, which encode nicotianamine aminotransferases in phytosiderophore biosynthesis, have been successfully expressed in rice to increase iron uptake and enhance tolerance to alkaline soils (Takahashi et al., 2001). On the other hand, nicotianamine has also been shown to be a key component for mineral nutrition in plants. Thus, transgenic rice plants that over-express nicotianamine synthase (OsNAS3) have been produced through activation tagging with cauliflower mosaic virus 35S enhancer elements or through expression of the barley NAS gene under an endosperm-specific promoter. These transgenic rice seeds have been shown to contain increased nicotianamine and elevated amounts of Fe, Zn, and Cu (Usuda et al., 2008; Lee et al., 2009). A feeding experiment in anemic mice with engineered rice seeds showed that Fe was more bioavailable, which led to normal levels of hemoglobin (Lee et al., 2009). Interestingly, nicotianamine has also been shown to have an antihypertensive effect in humans.

On the other hand, plants also contain anti-nutritional chemicals, such as phytic acid (also known as phytate), which reduce the bioavailability of minerals. Phytate not only chelates minerals, but also reduces the bioavailability of phosphate, since its complex is difficult to digest and absorb. Whereas the addition or expression of phytase, which is used for phytate digestion, from a microbe in cereal feed has often been used to promote the digestion of phytate, modification of a key synthetic gene for phytate such as IP synthase, should help to increase the bioavailability of iron and phosphate (Raboy, 2002)

Multigene transfer for improved food quality

Based on the accumulation of information regarding metabolism, an interesting challenge is to increase the levels of three vitamins in a single crop, such as corn. The enhancement of multiple aspects of the nutritional value of crops through the modification of multiple genes should be a new trend for future crop breeding. Naqvi et al. (2009) created elite inbred South African transgenic corn plants in which the levels of three vitamins were increased specifically in the endosperm through the simultaneous modification of three separate metabolic pathways. In their combinatorial transformation strategy, they randomly introduced multiple transgenes by bombardment and selected plants that contained and expressed all of the input genes, and recapitulated the entire pathway. The transgenic kernels contained 169-fold of the normal amount of β-carotene, six-fold of the normal amount of ascorbate, and double the normal amount of folate. These high vitamin contents were reported to be stable through to the T3 homozygous generation, but a limited substrate supply often hinders the improvement of multiple quality traits, as exemplified by competition for the substrate in phenylpropanoid engineering (see the section Resveratrol).

General Strategy for the Engineering of Secondary Metabolites with Nutritional Value

As illustrated by our experience with the enhancement of primary metabolites, metabolic engineering is a powerful tool for improving the quantity and quality of a metabolic profile, even though primary metabolism is highly tuned, for example, through feedback regulation, and difficult to modify. It is evident that progress in the engineering of primary metabolism is strongly stimulated by knowledge regarding microbial cells. On the other hand, specific plant metabolites are more complicated and the characterization of secondary metabolism lags due to the low abundance of biosynthetic enzymes and a lack of experimental materials such as chemicals (substrates). However, recent progress in molecular biology and instrumental developments, such as genome sequencing, transcriptome analysis using a microarray, and highly sensitive instrumental analysis such as with LC-MS, TOF-MS, and LC-NMR, enables the comprehensive analysis of metabolites and genes, and lets us integrate them into systems biology. There are now several good reviews on the progress regarding the identification of biosynthetic genes and metabolic engineering (Verpoorte and Memelink, 2002; Hashimoto and Yamada, 2003; Sato et al., 2007a,b). We highlight some breakthrough topics in the following sections.

Identification of biosynthetic genes

Since microbial information cannot be used to identify biosynthetic genes in secondary metabolism in plants, such identification is largely dependent on the preparation of high-metabolite-producing materials (Hashimoto and Yamada, 2003). For example, the isolation of a metabolite-producing organ such as a root culture of tropane-alkaloid producing plants, or even specific cells such as oil-gland cells in herbs, has been shown to be very useful for isolating biosynthetic genes (Lange et al., 2000; Hashimoto and Yamada, 2003). Additionally, mutant cells and/or plant species with differences in metabolite productivity, and/or some chemical-treated materials (e.g., elicitors or methyljasmonate), would be useful for the identification of gene(s) (Verpoorte and Memelink, 2002; Ziegler et al., 2005; Takemura et al., 2010) when these chemicals activate target metabolism. Alternatively, when a general transcriptional factor in the target metabolism is available, transgenic cells with enhanced gene expression might be a suitable material; for example, *Arabidopsis thaliana* that over-expressed the PAP1 gene (an MYB transcriptional factor) was used to identify novel genes in flavonoid biosynthesis (Tohge et al., 2005).

In either case, the molecular characterization of transcripts by EST sequencing (Lange et al., 2000), microarray (Ziegler et al., 2005; Tohge et al., 2005), reverse-transcription PCR (Takemura et al., 2010), or the cDNA-amplified fragment length polymorphism (AFLP) method (Goossens et al., 2003) is commonly used to identify the desired biosynthetic genes in combination with a target metabolite analysis. Further advances in genomic sequencing methods, as well as proteomics and metabolomics, can stimulate the identification of genes. Especially, a comparative genomic approach based on genome sequences would be a powerful method, for identifying a gene, when metabolism, is restricted to certain plant species (Ziegler et al., 2005; Takabayashi et al., 2009).

Once a candidate enzyme in biosynthesis is identified by a combination of bioinformatic analysis, a recombinant protein can be produced in either *E. coli*, yeast, or insect cells, and its enzyme activity can be determined. On the other hand, gene function might be directly evaluated by a reverse-genetics approach in either homologous or heterologous host plant cells. The latter approach, especially heterologous expression, should be able to overcome the limitation of substrate in an enzyme assay, as well as the difficulties of transformation in some plant species such as medicinal plants. Protein engineering of biosynthetic enzymes is another resource for the future engineering of compounds involved in metabolism (Runguphan and O'Connor, 2009; Morishige et al., 2010). Molecular engineering of protein structure based on the crystal structure or random mutagenesis has enabled the creation of enzyme mutants with broadened substrate specificities or enzyme reactivities (Runguphan and O'Connor, 2009; Morishige et al., 2010).

Identification of transcription factors and engineering through integrated "omics"

As mentioned previously, the identification of a transcription factor is very useful for the isolation of biosynthetic genes and for metabolic engineering (see Figure 28.2); that is, the introduction of transcriptional regulator-R and C1 of maize-induced anthocyanin biosynthesis in a heterologous system (Lloyd et al., 1992). However, it is more difficult to isolate a transcription factor than a biosynthetic gene since isolation of the latter is based on the characterization of the promoter sequence of biosynthetic genes. Thus, many researchers have tried to overcome this difficulty with novel approaches. Whereas the activation-tagging method (Borevitz et al., 2000) has shown high potential for the isolation of candidate gene(s) in signal transduction, its application is still limited due to the complexity of the transcriptional regulation system (i.e., biosynthetic genes in secondary metabolism are controlled by multiple transcriptional regulators; Kato et al., 2007). In this sense, RNAi of a transcription factor could be a more effective approach than activation tagging for isolating candidate genes for a general regulator (Kato et al., 2007), since knockdown of a single transcriptional regulator in a complex affects whole transcriptional activity. However, an increased opportunity to isolate a transcriptional regulator does not necessarily mean the isolation of a potential general regulator of an entire biosynthetic pathway, since transcriptional regulation is complicated, even in secondary metabolism (Kato et al., 2007). Thus, we need a more comprehensive approach to isolate whole sets of transcriptional regulators to understand the metabolic network and control it.

Recently, Yokota-Hirai et al. (2007) reported a more sophisticated approach to understanding plant metabolism as an integrated system for the engineering of compounds involved in secondary metabolism. They used the "omics" approach and integrated transcriptome and metabolome data into a single dataset to identify the regulatory network of poorly described metabolic pathways of glucosinolates (GSLs). GSLs are metabolites produced by the family Brassicaceae (i.e., broccoli and cabbage), that have received considerable attention due to their anticarcinogenic, antioxidative, and antimicrobial activity. They assumed that if a set of genes co-expressed under a given experimental regimen would be involved in the same or related metabolic pathway, then candidate genes involved in the regulation or synthesis steps of a particular metabolic pathway could be comprehensively identified, or at least predicted with some confidence. Furthermore, they compared a co-expression profile for a specific condition (i.e., sulfur-deficiency stress) with a "condition-independent" profile derived from public datasets, along with metabolic profiling, and increased the reliability and feasibility of predicting a gene function. These combined transcriptome co-expression analyses and analyses of knockout mutant and ectopic expression of the gene identified *Myb28* as a positive regulator for the basal-level production of aliphatic GSLs.

Modulation of organelle development

While recent advances in the characterization of transcriptional factors have provided a powerful approach to increase metabolite biosynthesis, the control of plastid development would be another option for modifying the metabolite content, since several metabolites are produced and accumulated in plastids (see Figures 28.1, 28.3, and 28.5; Davuluri et al., 2005; Enfissi et al., 2010). While both carotenoids and flavonoids are highly beneficial for human health, the simultaneous improvement of both is difficult since these metabolites are generated through independent pathways. While the manipulation of light signal transduction components or photoreceptors would facilitate the enhancement of these biosynthetic pathways simultaneously, such manipulation has been accompanied by a reduction of yield and plant vigor. Interestingly, Davuluri et al. (2005) reported the successful increase of both carotenoid and flavonoid contents in tomato fruit by suppressing an endogenous photomorphogenesis regulatory gene, *DET1*, using fruit-specific promoters combined with RNA interference (RNAi) technology without the development of detrimental agronomic traits. Recent metabolomic and transcriptomic analyses further revealed that high levels of carotenoids are more likely to be related to increased plastid volume per cell, whereas the increased transcription of key biosynthetic genes is a likely mechanism for producing elevated phenolic-based metabolites, such as flavonoids (Enfissi et al., 2010), which highlights how important it can be to control organelle biogenesis.

Quality Improvement of Plants as Functional or Medicinal Food

A functional or medicinal food is any food that is claimed to have a health-promoting or disease-preventing property beyond the basic function of supplying nutrients (Wikipedia). The market for functional foods is rapidly growing. Many functional foods contain plant secondary metabolites derived from a phenylpropanoid pathway such as resveratrol (stilbene), anthocyanins, catechin (polyphenol), sesamin (lignan), and so on. The phenylpropanoid pathway has been extensively investigated and many biosynthetic genes and transcriptional factors have been isolated. We list here some metabolites in the phenylpropanoid pathway as examples of quality trait improvement in functional foods (Figure 28.6).

Resveratrol

Resveratrol (3,5,4′-trihydroxy-trans-stilbene) is a non-flavonoid phenolic compound produced by a limited number of plant species. Grape (*Vitis* sp.), peanut (*Arachis hypogaea*), and several berry varieties are the major dietary sources of resveratrol. The health benefits of resveratrol have been intensively studied, and its ability to prevent cardiovascular diseases and cancers, and to promote longevity in several animal systems has recently been reported (Baur and Sinclair, 2006). The pathway and enzymes in resveratrol biosynthesis have been well characterized and considerable work has been performed on metabolic engineering (Figure 28.6; see Hall and Yu, 2008). The resveratrol pathway is branched from 4-coumaroyl-CoA, which is generated through the common phenylpropanoid pathway from phenylalanine by phenylalanine ammonia lyase (PAL), cinnamic acid 4-hydroxylase (C4H), and 4-coumarate:CoA ligase (4CL). Stilbene synthase (STS), a type III polyketide synthase, catalyzes the condensation of resveratrol from one molecule of 4-coumaroyl-CoA and three molecules of malonyl-CoA.

Resveratrol has long been known to be a defense compound (phytoalexin) in plant species, and the accumulation of resveratrol in tobacco by the expression of grape STS increased its resistance to the tobacco fungal pathogen *Botrytis cinerea* (Hain et al., 1993). Interestingly, resveratrol engineering not only increased disease resistance in transgenic plants, but also extended the postharvest shelf life of the fruit. Most of the resveratrol produced in transgenic plants is conjugated with a glucose moiety. When other endogenous flavonoid compounds were analyzed, in some cases the overexpression of STS led to reduced levels of ferrulic acid, rutin,

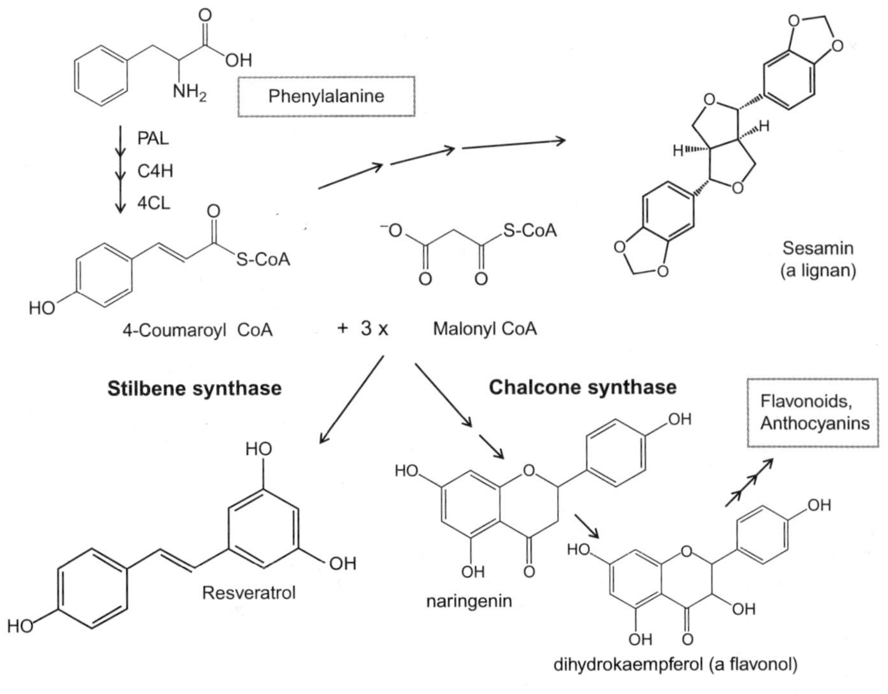

Figure 28.6 • Some phytochemicals in the phenylpropanoid pathway • Abbreviations: PAL, phenylalanine ammonia lyase; C4H, cinnamate 4-hydroxylase; and 4CL, coumarate; CoA ligase.

quercetin, and naringenin, suggesting that under a strong promoter, STS might redirect the metabolic flux and alter the total flavonoid profile of the transgenic plant (Nicoletti et al., 2007).

Anthocyanins and flavonoids

Anthocyanins and flavonoids are popular flower pigments and their biosynthesis has been studied intensively at both enzymological and molecular-regulation levels (Tanaka et al., 2008). Genes for biosynthetic enzymes and transcriptional regulators have been isolated and used for metabolic engineering. Blue carnation and rose are successful examples of metabolic engineering in ornamental plants, and similar engineering has also been attempted in vegetables and fruits since anthocyanins help protect against certain cancers, cardiovascular disease, and age-related degenerative diseases, while the levels achieved through dietary consumption may be inadequate to confer optimal benefits (http://www.fruitsandveggiesmatter.gov).

Since quantity improvement is essential, enhancement of the overall biosynthetic activity of anthocyanin has been attempted using transcription factors after the pioneering work by Lloyd et al. (1992) using maize R and C1. When maize transcription factor genes *LC* and *C1* were expressed in tomato, the expression of both genes sufficiently upregulated the flavonoid pathway and fruit accumulated high levels of the flavonol kaempferol, but anthocyanins could not be detected in ripe LC/C1 fruit (Bovy et al., 2002). When AtMYB12, which was originally identified as a flavonol-specific transcriptional activator in *A. thaliana*, was expressed in a tissue-specific manner in tomato, it activated the caffeoyl quinic acid biosynthetic pathway in addition to the flavonol biosynthetic pathway (Luo et al., 2008). These results indicated that transcription factors may have different specificities for target genes in different plants, although the basic regulatory systems would be common among plant species. Butelli et al. (2008) clearly indicated that two transcription factors selected from snapdragon were functional in tomato and increased anthocyanin contents comparable to those in blackberries and blueberries. Furthermore, they proved that transgenic tomato fruits with increased anthocyanins have enhanced hydrophilic antioxidant capacity and extended the life span of cancer-susceptible Trp53$^{-/-}$ mice in a pilot test.

Catechins and proanthocyanidins

Proanthocyanidins (PAs, also known as condensed tannins) are plant-derived flavonoid polymers with a wide range of benefits for human health (Xie et al., 2006). They are present in many foods and drinks, such as barley, tea, and wine, and are strong antioxidants. PAs from various plant species block the growth of human cancer cells *in vitro* and protect cells against UV-light-induced damage. Many commercial PA products are widely used as dietary supplements. PAs are formed by the polymerization of monomeric flavan-3-ols; that is, catechins and epicatechins and their derivatives, which also have potential health-promoting activity against Alzheimer's disease, oral and stomach cancer, and cardiovascular disease. The presence of PAs is also important in forage crops to protect ruminant animals from potentially lethal pasture bloat associated with the consumption of forage, such as alfalfa (*Medicago sativa*). However, much of the forage consumed by domestic animals lacks PAs. Since pathway engineering of catechin biosynthesis is still difficult, Xie et al. (2006) used an MYB family transcription factor for anthocyanidin synthesis and anthocyanidin reductase to convert anthocyanidin into (epi)-flavan-3-ol. Transgenes successfully induced the accumulation of epicatechin and gallocatechin monomers, and a series of dimers and oligomers consisting primarily of epicatechin units at the level of PAs to confer bloat-reduction in forage species. The expression of anthocyanidin reductase in anthocyanin-containing leaves of the forage legume *M. truncatula* resulted in the production of a specific subset of PA oligomers (Xie et al., 2006).

Sesamins

Lignans are a large class of secondary metabolites in plants that have numerous biological effects in mammals, including antitumor and antioxidant activities. Some plant lignans, such as sesamin, can be converted by intestinal microbiota to the mammalian lignans (i.e., enterodiol and enterolactone), which may have protective effects against hormone-related diseases such as breast cancer (Liu et al., 2006). Sesamin, the most abundant lignan in sesame seeds (*Sesamum* plants), is produced by the cytochrome P450 enzyme CYP81Q1 from the precursor lignan pinoresinol. Recently, the CYP81Q1 gene was isolated (Ono et al., 2006), and some attempts have been made to increase the sesamin content in plants. When a model plant, *Forsythia*, was transformed with pinoresinol/lariciresinol reductase RNAi construct to inhibit the conversion of pinoresinol to matairesinol and further co-transformed with CYP81Q1 over-expression construct, transgenic cells produced sesamin and accumulated pinoresinol glucoside (Kim et al., 2009). These data suggested that the metabolic engineering of lignan production may be possible.

Beloved Metabolites: Plant Volatiles

Most biological organisms use volatile compounds to find food resources, to avoid potentially toxic food materials, or to avoid detrimental conditions. In the animal kingdom, primitive lampreys first acquired an olfactory system to find carrion for food (Watson, 2000). After a long period of evolution, higher animals developed the ability to use volatile compounds to find mating partners, detect enemies, and establish social communications, as well as to find preferable or distasteful foods. Volatile compounds, when perceived by humans, are called flavor, aroma, odor, or scent. They reach the olfactory receptors located at the top of the nasal cavity either directly through the nose during breathing or indirectly through the mouth during mastication. In the latter case, the volatile compounds reach the receptors in a retronasal manner, and in this case the sensation is called flavor, which is a cooperative sensation that includes both gustatory and olfactory components. Humans have 347 olfactory receptors, and can sense more than 7000 flavor volatiles (Goff and Klee, 2006). This can be explained by many-to-many interactions between the receptors and volatiles. Additionally, other sensory inputs, such as the color or texture of foods, and even the past experience of an individual, can affect the olfactory sensation (Taylor and Roberts, 2004). After these inputs are integrated together in the brain, the overall sensation is established. Some volatile compounds have physiological, psychological, and pharmacological effects on humans. Plants are the most important sources of volatile secondary metabolites. Thousands of plant volatiles have been described (Pichersky et al., 2006). Some are common among plants, but others are unique to a peculiar genus or species, and thus there is a wide variety of biosynthetic pathways that essentially branch out from the following major biosynthetic pathways.

Biochemistry of plant volatile secondary metabolites

The biochemistry of plant volatile secondary metabolites has been largely clarified through extensive studies (Figure 28.1). Aromatic amino acids, such as phenylalanine and tyrosine, are the precursors of benzenoids and phenylpropanoids (Figure 28.7). The deamination of amino acids by ammonia lyases to give precursors (cinnamic acid or 4-coumaric acid) for phenylpropanoids lies at the branch point between primary and secondary metabolism (Vogt, 2010). Benzenoids are also formed from the same amino acid substrates, but through a decarboxylation reaction to yield phenylethylamine from phenylalanine (Sakai et al., 2007). Glucosinolates are a group of plant secondary metabolites that consist of anionic thioglucosides with a diverse array of carbon skeletons (Yan and Chen, 2007). They are formed predominantly in cruciferous plants. Tissue damage facilitates their hydrolysis into various nitrogen-containing molecules such as isothiocyanate, nitrile, and so on. Glucosinolates are also biosynthesized from amino acids, such as methionine and tryptophan.

Terpenoids (or isoprenoids) are compounds consisting of five-carbon units with widely diversified structures. Monoterpenoids (C10) and sesquiterpenoids (C15) are generally volatile. The five-carbon units are derived from isopentenyl pyrophosphate (IPP) that is formed either in plastids from pyruvate and glyceraldehyde 3-phosphate (methylerythritol phosphate: MEP pathway) or in cytosol from acetyl-CoA (mevalonic acid: MVA pathway; Arimura et al., 2009). In general, monoterpenoids are formed through the MEP pathway, while sesquiterpenoids are formed through the MVA pathway, although this is not entirely strict because of cross-talk between the pathways. A cumulative chain elongation reaction yields higher terpenoids such as phytosterols (C30) and carotenoids (C40). The cleavage of carotenoids by a carotenoid cleavage oxygenase yields apo-carotenoids (Walter et al., 2010). Some of them are also volatiles and important constituents of the aroma of various fruits.

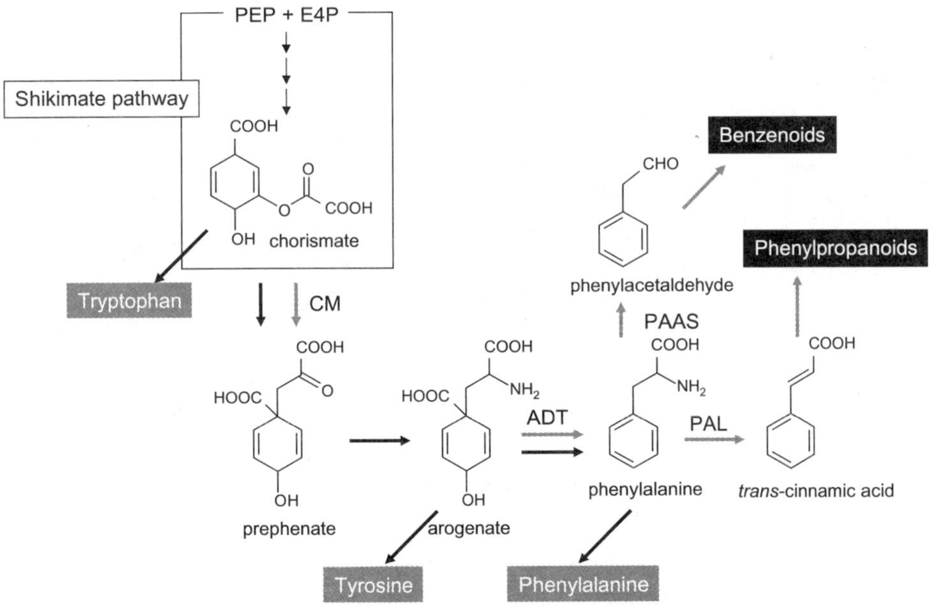

Figure 28.7 • Phenylalanine is formed specifically for benzenoids/phenylpropanoids in *P. hybrida* petals • For most of the biosynthetic steps downstream of chorismate, there is a special set of enzymes dedicated to benzenoid/phenylpropanoid formation (red arrows). This specialization offers a new strategy to modify volatile secondary metabolism. Abbreviations: CM, chorismate mutase; ADT, arogenate dehydratase; PAL, phenylalanine ammonia lyase; and PAAS, phenylacetaldehyde synthase. Please see color plate section at the end of the book.

Fatty acids are also precursors of volatile secondary metabolites in plants. Lipoxygenases add oxygen to polyunsaturated fatty acids such as linoleic and linolenic acids to form their hydroperoxides. They are further converted into a diverse array of phytooxylipins (Mosblech et al., 2009). Some of them, such as green leaf volatiles and some jasmonates, are volatiles.

Flavor compounds in fruits

The reproductive tissues of plants form volatile metabolites as infochemicals to disperse seeds and attract pollinators. With fruits such as tomatoes, these volatiles are somehow related to their nutritional qualities because they are formed from nutritionally valuable constituents, such as amino acids, fatty acids, or carotenoids (Goff and Klee, 2006). This implies that fruit flavors contribute to the positive perception of food and that we are intrinsically attracted to fruit flavors and develop an appetite after perceiving favorable flavor compounds in fruits. Consequently, flavor properties are criteria that consumers use to choose fruits, vegetables, and processed foods made thereof.

Crop plants that are currently available have been established through extensive domestication. Breeding programs generally target yield, fruit size, or tolerance against stresses, and volatile properties have received less attention. As a result, current crop plants are somewhat "flavorless" compared to their original, wild cultivars (Gutterson, 1993). One introgression line made from modern tomato (*Solanum lycopersicon*) and wild tomato (*Solanum pennellii*) smelled like cucumber (Matsui et al., 2007). It was revealed that *S. pennellii* contains active 9/13-hydroperoxide lyase (HPL), which converts 9-hydroperoxide of linolenic acid into (*Z,Z*)-3,6-nonadienal, which has a cucumber-like flavor. The modern tomato, *S. lycopersicon*, has a corresponding gene; however, it has one different amino acid, and as a result it encodes an inactive enzyme. Introgression of the chromosome region spanning the 9/13-HPL gene from *S. pennellii* into *S. lycopersicon* restores the biosynthetic pathway to form carbon 9 volatiles in the modern tomato. This can be achieved because *S. lycopersicon* fruits have high lipoxygenase (LOX) activity to provide a substrate, that is, 9-hydroperoxide of linolenic acid, for the 9/13-HPL derived from *S. pennellii*. In contrast, when the 9/13-HPL gene derived from cucumber was introduced into *S. lycopersicon* under the control of the CaMV35S promoter, there was little change in the flavor profile even though expression of the introduced gene was evident (Matsui. et al. 2001). This indicates that tomato 9/13-HPL and 9-LOX are compatible with each other, while cucumber 9/13-HPL is incompatible with tomato 9-LOX. Even though the molecular mechanism of this compatibility has not been elucidated, this implies that the biosynthetic pathway for volatiles is closely regulated in plants.

Tomato fruits vigorously synthesize carotenoids like lycopene during maturation. Since lycopene is formed through the terpenoid pathway, it is possible to intercept the intermediate for monoterpene synthesis. Linalool synthase from *Clarkia breweri* was introduced into tomato under the control of the tomato *E8* promoter, which is active in maturing tomato fruits (Lewinsohn et al., 2001). With this enzyme, a monoterpene alcohol, linalool, is formed from geranyl diphosphate (GPP) in a single step. Introduction of the synthase enhanced the amount of linalool in the transgenic tomato fruits; however, the amount was insufficient to change the overall flavor properties. In fact, there was little change in the carotenoid content, which indicated that redirection of the metabolic flow from lycopene synthesis to linalool synthesis was not so obvious.

Introduction of the lemon basil geraniol synthase gene under the control of the tomato polygalacturonase promoter into tomato resulted in an enhanced monoterpene content in tomato fruits (Davidovich-Rihanati et al., 2007). Due to the rather broad product specificity of geraniol synthase, a wide range of monoterpenes, such as geraniol, rose oxide, citronellol, and geranial, which are essentially absent from control tomato fruits, were found. In a sensory evaluation, more than 90% of the panelists distinguished the smell of the transgenic tomato fruits from that of the control fruits, and more than 60% of the panelists preferred the transgenic tomato fruits. This success might be due to the nature of the promoter used to drive geraniol synthase, but the threshold values of flavor compounds must also be taken into account. For example, the odor threshold value of linalool is 6 ppb, while that of rose oxide is 0.5 ppb (http://www.leffingwell.com/odorthre.htm). Thus, a small increase in the amount of rose oxide in the geraniol synthase-over-expressing tomato might strongly affect its organoleptic properties. Therefore, a clear understanding of the composition of flavor compounds that considers the organoleptic properties of the designed bouquet of volatiles is needed when designing transgenic fruits with novel flavor properties.

In the geraniol synthase-over-expressing tomato, the carotenoid contents were markedly reduced (Davidovich-Rihanati et al., 2007). Since the carotenoid content is an important trait of tomato fruits, this creates an apparent dilemma. To overcome this dilemma, an *in silico* simulation on the impact of the modification of metabolic pathways is needed, based on accumulating information regarding metabolomics (Iijima and Aoki, 2009).

Edible seeds such as grains and legumes also contain volatile compounds. Generally, the amounts of volatile secondary metabolites formed in seeds are not so high, partly because they do not want to attract organisms. However, even though the amounts of volatiles are not very high, some pests, such as the granary weevil, can respond to a wide range of cereal volatiles (Germinara et al., 2008), which results in a serious problem worldwide. For humans, GLVs derived from fatty acids, for example, usually have negative effects on the flavor properties of foods made from the seeds. For soymilk, *n*-hexanal and *n*-hexan-1-ol formed predominantly from soybean seed lipoxygenase-2 give an off flavor (Furuta et al., 1996). (*E*)-2-Nonenal formed from linoleic acid via lipoxygenase reaction also gives an off flavor for beer, which is derived from barley grains (Hirota et al., 2006). In both cases, breeding studies have established soybean seeds and barley grains that have no lipoxygenase activity (Furuta et al., 1996; Hirota et al., 2006). However, the established varieties are not always suitable for cultivation, and thus a molecular biotechnological approach may be an alternative way to solve this problem.

Scent/aroma of flowers

Terpenoids and benzenoids are the major volatiles formed in flower organs. While they are not directly related to any nutritional value, flowers provide rewards (honey or nectar) to pollinators, and thus insects and birds are attracted to flowers. In return, plants can achieve more efficient pollination. In this way, flowers should advertise themselves with their scent as well as their shape and color, and humans have loved such inherent properties of flowers since ancient times. Improvements in the quality of flower scents may lead to new products with higher value, and may be welcomed by consumers. Even consumers who dislike GM crops tend to tolerate transgenic ornamental flowers because they are not used for food.

The first attempt to modify the scent of flowers was made with petunia and carnation by introducing *Clarkia* linalool synthase under the control of 35S promoter (Lücker et al., 2001; Lavy et al., 2002). This introduction was successful and a high production of linalool was observed with transgenic petunia and carnation; however, the resulting linalool was largely converted into non-volatile glucoside in petunia, and oxidized to volatile linalyl oxides in carnation. In both cases, the organoleptic properties of the transgenic flowers toward humans were indistinguishable from those of the respective wild-types. Again, this implies that rational design based on the metabolic flow of the target plant tissues and the organoleptic properties of expected volatile compositions are important prerequisites for success.

The introduction of three terpene synthases (i.e., γ-terpinene synthase, limonene synthase, and β-pinene synthase) into tobacco, which produces low levels of terpenoids in flowers, resulted in the accumulation of γ-terpinene, limonene, and β-pinene in the flowers (Lücker et al., 2004). Other monoterpenoids that are not usually detected in wild-type tobacco plants were also detected, probably because of the broad product specificities of the introduced enzymes. In this case, a sensory evaluation indicated that the transgenic tobacco flowers had a more intense and flowery scent than the wild-type flowers (El Tamer et al., 2003). In the transgenic tobacco, the level of β-caryophyllene (a sesquiterpene formed from farnesyl pyrophosphate, FPP), which was the most abundant terpenoid in the wild-type tobacco, was about three-fold lower than that in the control plants, which again implies that a metabolic shift took place through competition for intermediates. This result was unexpected because monoterpenes are formed from the plastidic MEP pathway, whereas sesquiterpenes are formed from the cytosolic MVA pathway. This indicated that the two compartmentalized pathways share some intermediates, and the modification of one affects the other.

An alternative to the time-consuming introduction of several genes into a plant may be enhancement of the biochemical and transcriptional bottlenecks in a metabolic pathway for the formation of volatile compounds. As an amino acid, phenylalanine is a primary metabolite. Secondary metabolites, such as benzenoids and phenylpropanoids, are formed from phenylalanine. Thus, it has been recognized that the benzenoid/phenylpropanoid pathway should "steal" phenylalanine from the primary metabolic pathway. However, in petals of *Petunia hybrida* there is a pathway to form phenylalanine that is dedicated to secondary metabolites (Colquhoun et al., 2010). The expression of a chorismate mutase in *P. hybrida*, *PhCM1*, correlated with emission of its floral scent. Prephenate that is formed by this chorismate mutase is converted into arogenate, and then further converted into phenylalanine with arogenate dehydratase1, the expression of which is again positively correlated with phenylalanine levels and phenylpropanoid/benzenoid volatiles in the petals (Maeda et al., 2010). Taken together, these results indicate that there is a specified pathway for the formation of floral scents in *P. hybrida* from unexpected upstream steps, at least from upstream of phenylalanine synthesis. Fortification of the upstream steps to provide a sufficient amount of a primary substrate, such as phenylalanine for phenylpropanoid/benzenoid formation, might be a useful approach to improve the scent of ornamental flowers.

ODORANT1 (*ODO1*) is an MYB transcription factor that regulates volatile benzenoid biosynthesis (Verdonk et al., 2005). Silencing of *ODO1* reduced the expression of most genes involved in benzenoid production from phenylalanine in flowers of *P. hybrida*, and as a result reduced volatile benzenoid production in the flower. The expression of *ODO1* in petals of *P. hybrida* flowers shows a diurnal cycle, which correlates with the emission of scent from the flower. Suppression of *ODO1* had little effect on the color of the petals, which is mainly the result of anthocyanins formed from phenylalanine. Therefore, it was expected that *ODO1* was the master switch that controlled volatile benzenoid production. An *Arabidopsis* MYB transcription factor (Pap1) regulates anthocyanin pigmentation. When *Arabidopsis Pap1* was introduced into *P. hybrida* under regulation of the 35S promoter, both the pigmentation and volatile benzenoid/phenylpropanoid production were enhanced (Zvi et al., 2008), which indicates that gross modification of a biosynthetic pathway through the introduction of a transcription factor may represent a new strategy for the metabolic engineering of flower properties. Even though the transcription factors that control volatile secondary metabolism are still being identified, transcription factors appear to be useful tools for plant metabolic engineering (Grotewold, 2008).

Accordingly, we are now elucidating the regulatory machinery required for floral scent formation at the biochemical and transcriptional levels. This knowledge will likely be useful for the design of flowers with improved scent.

Volatile organic chemicals in vegetative organs of plants

In vegetative tissues of plants, volatile metabolites are formed to cope with stresses mainly caused by herbivores and pathogens (Arimura et al., 2009). Herbs usually accumulate volatile compounds in glandular trichomes distributed on the leaf surface. Upon herbivore attack, the trichomes are disrupted, and high concentrations of volatile compounds, which repel herbivores, are released (Schilmiller et al., 2008). After being damaged by an herbivore, plant leaves quickly form GLVs, which repel herbivores or pathogens that attempt to invade the plant tissues through the wound site (Matsui, 2006). Isothiocyanates that are formed after wounding in cruciferous plants also have anti-herbivore effects (Zhang et al., 2006). Accordingly, the rational engineering of these herbal volatiles should improve the ability of crops to tolerate biotic stresses. In addition, these herbal volatiles often refresh and relieve humans, probably because these odors are associated with the absence of harmful insects or dangerous microbes. Some herbal volatiles are called "phytoncides" because of their potent bactericidal activities, and Asians love "forest bathing" to inhale phytoncides.

The volatile compounds emitted from vegetative tissues function not only to provide the previously mentioned direct defense effects, but also to exert indirect defense effects (Arimura et al., 2009; Dicke, 2009). Plants emit a subset of volatile compounds after they are attacked by herbivores. This bouquet is an attractive "smell" to carnivorous or symbiotic animals. Through this higher trophic interaction, plants receive a benefit because the carnivorous or symbiotic animals impair the action of herbivores. Accordingly, plant volatile secondary metabolites are useful weapons for plants to improve their fitness in a stressful ecosystem. Thus, the rational design of "supersmelly" plants could enhance their tolerance to biotic stresses either directly or indirectly and could be useful in agriculture.

In the oxylipin pathway in plants, fatty acid hydroperoxides are the common intermediates for each branch, and allene oxide synthase and HPL are the most important enzymes that divert the metabolic flow at the branch point to jasmonates and GLVs, respectively (Matsui, 2006; Arimura et al., 2009). When the expression of HPL was enhanced by the introduction of the corresponding gene under regulation of the 35S promoter in *Arabidopsis*, the GLV levels in intact transgenic plants were still equivalent to those in the wild-type plants (Shiojiri et al., 2006). However, when the transgenic plants were infested by cabbage white butterfly (*Pieris rapae*) or infected by gray mold (*B. cinerea*), higher amounts were emitted from the transgenic plants. This is because the amount of GLV is directly regulated by the supply of its substrate, and not by the level of HPL activity. As a result, the transgenic plants recruited a higher number of parasitic wasps (*Cotesia glomerata*) that attacked *P. rapae* larvae and also showed higher tolerance against gray mold disease. This is because of the positive effects of GLVs in the direct and indirect defense responses. This was the first attempt to enhance GLV formation to improve plant traits in the context of integrated pest management.

Strawberry linalool/nerolidol synthase (FaNES) was targeted to the mitochondria of *Arabidopsis* (Kappers et al., 2005). Mitochondria are the sites of ubiquinone biosynthesis, and therefore possess an FPP synthase that provides FPP (C15 compound) that can be used as a substrate for FaNES to form a C15 volatile, nerolidol. The amount of nerolidol formed in transgenic plants was 20–30 times higher than that found in transgenic plants that were made by introducing FaNES with a plastid-targeted signal sequence. The growth of transgenic plants was almost the same as that of wild-type plants, which suggested that mitochondria in the transgenic plants still produced enough FPP for both ubiquinone and heme A, as well as for the new end product, nerolidol. The transgenic plants recruited a greater number of carnivorous predatory mites, which suggested that the transgenic plants were more fit for survival if they were to be attacked by herbivores.

The addition of a new branch to an existing pathway in a subcellular compartment is a promising strategy in metabolic engineering because, in some cases, there may be an abundance of metabolic flow in a compartment so that there would be little effect if a small portion of the intermediate is provided to the new branch. This concept was applied to sesquiterpene formation in tobacco plants by redirecting the MEP pathway in plastids to form patchoulol through the introduction of FPP synthase and patchoulol synthase in plastids (Wu et al., 2006). The amounts of the sesquiterpenes patchoulol and amorpha-4,11-diene increased more than 1000-fold and the transgenic tobacco became less preferable than the wild type to tobacco hornworm larvae. However, some of the transgenic lines formed extremely high levels of sesquiterpenes, and growth reduction and leaf chlorosis were evident, indicating that the introduced enzymes consumed too much of the carbon source for sesquiterpene synthesis.

Maize roots emit β-caryophyllene into their rhizosphere in response to feeding by larvae of the *Diabrotica virgifera virgifera* (Rasmann et al., 2005). An entomopathogenic nematode was strongly attracted by the volatile sesquiterpene, and this then efficiently infected the larvae. Even though this below-ground tritrophic system is found in most maize lines and in their closest wild relative, teosinte, expression of the β-caryophyllene synthase gene (TPS23) in most North American maize lines is reduced, which results in a loss of β-caryophyllene production. Due to this low level of β-caryophyllene, North American maize lines are more susceptible to this herbivore. The β-caryophyllene synthase gene from oregano was introduced into a North American maize line. The transgenic maize was able to form β-caryophyllene, and, as a result, it suffered less root damage in a field test because of the efficient recruitment of entomopathogenic nematodes (Degenhardt et al., 2009). Breeding programs usually have a definitive objective, such as fruit size, yield, and pathogen tolerance. During breeding of the North American maize line, the breeders did not pay much attention to the ability to form β-caryophyllene from roots, and malfunctional β-caryophyllene synthase was propagated. If this background information can be clarified, a transgenic strategy to "restore" a forgotten trait should be feasible. Such restoration was also successful with tomato 9/13-HPL, as previously mentioned, even though the physiological significance of the formation of a C9-volatile in wild tomatoes has not been elucidated.

Perspectives

As previously mentioned, plants are rich resources for health-promoting phytochemicals. While the biosynthetic pathways of many important metabolites have been characterized, little is known about other biosynthetic pathways, such as those of capsaicin in hot pepper, gingerol in ginger, and apocarotenoid in saffron (*Cocus sativus*). However, recent advances in secondary metabolism have also been very rapid, and the effects of this progress have been amazing, as shown in onion.

Onion (*Allium cepa*), garlic (*A. sativum*), and many of the *Allium* species are important vegetables due to their flavors. Intensive studies have revealed the relevant biosynthetic pathways (Jones et al., 2004). *Allium* species synthesize a unique set of secondary sulfur metabolites derived from Cys; for example, the S-alk(en)yl-L-Cys sulfoxides, including S-2-propenyl-L-cysteine sulfoxide (alliin; 2-PRENCSO) and trans-S-1-propenyl-L-cysteine sulfoxide (isoalliin; 1-PRENCSO). When the tissues of any *Allium* species are disrupted, these amino acid derivatives are cleaved by the enzyme alliinase into the corresponding sulfenic acids, and the resulting volatile sulfur compounds give the characteristic flavor and bioactivity of the species. In garlic (*A. sativum*), tissue disruption produces di-2-propenyl thiosulfinate (allicin) from 2-PRENCSO, which then liberates di-2-propenyl disulfide, the dominant volatile component. In onion (*A. cepa*), however, propanthial S-oxide (lachrymatory factor, LF), 1-propenyl methane thiosulfinate, and di-propyl disulfide are produced from 1-PRENCSO upon tissue disruption (Figure 28.8).

It had been considered that lachrymatory factor is formed spontaneously in onion, but after a careful enzymological analysis, Imai et al. (2002) identified lachrymatory factor synthase (LFS), which converts 1-propenyl sulfenic acid to LF. Later, Eady et al. (2008) transformed onion with RNAi vector for the lachrymatory factor synthase gene, and then significantly reduced the levels of tear-inducing lachrymatory factor after wounding. They also confirmed that more 1-propenyl sulfenic acid was converted into di-1-propenyl thiosulfinate and markedly increased the production of a zwiebelane isomer and other volatile sulfur compounds, di-1-propenyl disulfide and 2-mercapto-3,4-dimethyl-2,3-dihydrothiophene, which had previously been reported in trace amounts or had not been detected in onion. This successful engineering is expected to increase the health and flavor attributes of the onion.

This work in the onion clearly indicates the importance of traditional biochemical analysis, and the power of the integration of molecular biology and metabolomics. Many other metabolites are awaiting a similar investigation.

However, we should also note that additional progress is needed in the metabolic engineering of low molecular weight compounds for their application, since we do not know the exact mechanisms of the physiological functions of many phytochemicals for health promotion. A more integrated analysis of the physiological mechanisms of phytochemicals and their interaction with other components is needed, as with medicinal plants. In addition, a more detailed investigation of the consequences of metabolic engineering will be needed for the development of functional food, since metabolic engineering can considerably modify chemical profiles. The effects of metabolic engineering on plant growth and yield should also be fine-tuned, since modification of the flow of primary metabolites for an engineered metabolism might impair plant growth. While there are several obstacles for the application of these techniques to the development of functional foods,

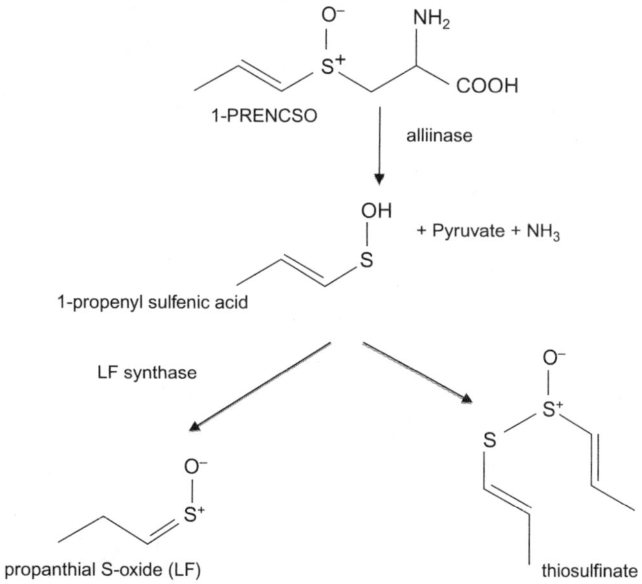

Figure 28.8 • Onion flavor formation and an engineering target • When onion cells (*Allium cepa*) are disrupted, trans-S-1-propenyl-L-cysteine sulfoxide (isoalliin; 1-PRENCSO) is cleaved by the enzyme alliinase into 1-propenyl sulfenic acids, and propanthial S-oxide (lachrymatory factor, LF], 1-propenyl methane thiosulfinate, and di-propyl disulfide are produced. Based on the discovery of LF synthase, the formation of a more volatile flavor compound has been attempted.

engineered materials should certainly be useful for the industrial preparation of these chemicals.

Conclusion

The underlying science of low molecular weight metabolites, including essential nutrients, phytochemicals, and volatiles, has been recognized fairly recently. Since low molecular weight metabolites have not been targets for breeders, who have provided many crops with advantageous agronomic traits such as high yield or low susceptibility to biotic and abiotic stresses, these small compounds have been overlooked or even deteriorated in the pursuit of other agronomic traits. Now, however, we recognize that low molecular weight metabolites are important components of crop plants for both consumers and producers. This is why many researchers are now trying to modify the composition and/or the amounts of these metabolites through molecular biological techniques. Even though several important achievements have been reported so far, not every effort to improve crop plants through the modification of low molecular weight compounds, especially secondary metabolites, has been successful. Since secondary metabolites are formed when their physiological functions are needed, plants fine-tune their formation. In fact, the formation of these compounds is often limited to a specific developmental stage or a specific organ, tissue, or even cell. The subcellular compartments where their biosynthesis occurs are also tightly controlled; otherwise, they would disturb the primary metabolism. Plant metabolisms are highly organized and are controlled by innate regulatory and homeostatic mechanisms. Therefore, the molecular design of a transgenic plant that shows an improved trait through a modified ability to form secondary metabolites should be based on the molecular mechanisms of secondary metabolite formation in plants, which are now being extensively studied. A kinetic modeling approach has been successfully used to predict the metabolic flux of benzenoids in petunia flowers (Colon et al., 2010). This approach predicted that phenylacetaldehyde synthase plays a critical role in determining the flux to form benzenoids. Even though this prediction still requires experimental validation, this approach may provide valuable clues for modifying plant volatiles through genetic modification. A recent omics-based approach that considers both the transcriptome and metabolome may also be useful for obtaining a basic understanding of plant secondary metabolism (Yokota-Hirai et al., 2007). Research on plant systems through this omics-based approach may provide a deeper understanding of the network structure of plant secondary metabolism, and, based on such information, it may be possible to design crop and ornamental plants with improved quality traits that involve low molecular weight compounds.

Another important point to consider in attempting to improve the traits of plants through the modification of secondary metabolites is how organisms, such as humans, insects, or pathogens, perceive these chemicals. For example, the threshold value of a volatile compound can vary greatly. Furthermore, in humans, olfactory sensation caused by volatile compounds is quite complicated and other factors, such as texture and color, must also be taken into account (Taylor and Roberts, 2004). Therefore, every attempt to improve the flavor quality of foods must be evaluated by an organoleptic assessment. In contrast, insects usually show inherent behavioral responses to volatile compounds. However, it has also been shown that insects can learn odors that might be beneficial to them (De Boer et al., 2005). The ecosystem that surrounds a plant consists of a network of multitrophic interactions (Dicke, 2009), and thus an improvement in one trophic system might have an adverse effect on another trophic system. Since metabolic engineering is a powerful tool for modifying the quality traits of plant metabolites, practical agricultural and nutritional evaluations will be essential for the application of this approach.

Acknowledgments

This work was supported in part by a Grant-in-Aid from the Ministry of Education, Culture, Sports, Science, and Technology of Japan (21248013 to F.S.) and by Grants-in-Aid for Scientific Research (18580105 and 19101009 to K.M.) from the Japan Society for the Promotion of Science.

References

Arimura, G., Matsui, K., & Takabayashi, J. (2009). Chemical and molecular ecology of herbivore-induced plant volatiles: Proximate factors and their ultimate functions. *Plant and Cell Physiology*, *50*, 911–923.

Baur, J. A., & Sinclair, D. A. (2006). Therapeutic potential of resveratrol: The *in vivo* evidence. *Nature Reviews Drug Discovery*, *5*, 493–506.

Borevitz, J. O., Xia, Y. J., Blount, J., Dixon, R. A., & Lamb, C. (2000). Activation tagging identifies a conserved MYB regulator of phenylpropanoid biosynthesis. *Plant Cell*, *12*, 2383–2393.

Bovy, A., de Vos, R., Kemper, M., Schijlen, E., Pertejo, M. A., & Muir, S., et al. (2002). High-flavonol tomatoes resulting from the heterologous expression of the maize transcription factor genes LC and C1. *Plant Cell*, *14*, 2509–2526.

Brigelius-Flohe, R., & Traber, M. G. (1999). Vitamin E: Function and metabolism. *The FASEB Journal*, *13*, 1145–1155.

Butelli, E., Titta, L., Giorgio, M., Mock, H. -P., Matros, A., & Peterek, S., et al. (2008). Enrichment of tomato fruit with health-promoting anthocyanins by expression of select transcription factors. *Nature Biotechnology*, *26*, 1301–1308.

Chen, Z., Young, T. E., Ling, J., Chang, S-C., & Gallie, D. R. (2003). Increasing vitamin C content of plants through enhanced ascorbate recycling. *Proceedings of the National Academy of Sciences of the United States of America*, *100*, 3525–3530.

Clemente, T. E., & Cahoon, E. B. (2009). Soybean oil: Genetic approaches for modification of functionality and total content. *Plant Physiology*, *151*, 1030–1040.

Colon, A. M., Sengupta, N., Rhodes, D., Dudareva, N., & Morgan, J. (2010). A kinetic model describes metabolic response to perturbations and distribution of flux control in the benzenoid network of Petunia hybrida. *Plant Journal*, *62*, 64–76.

Colquhoun, T. A., Schimmel, B. C. J., Kim, J. Y., Reinhardt, D., Cline, K., & Clark, D. G. (2010). A petunia chorismate mutase specialized for the production of floral volatiles. *Plant Journal, 61*, 145–155.

Davidovich-Rihanati, R., Sitrit, Y., Tadmor, Y., Iijima, Y., Bilenko, N., & Bar, E., et al. (2007). Enrichment of tomato flavor by diversion of the early plastidial terpenoid pathway. *Nature Biotechnology, 25*, 899–901.

Davuluri, G. D., van Tuinen, A., Fraser, P. D., Manfredonia, A., Newman, R., & Durgess, D., et al. (2005). Fruit-specific RNAi-mediated suppression of *DET1* enhances carotenoid and flavonoid content in tomatoes. *Nature Biotechnology, 23*, 890–895.

De Boer, J. G., Snoeren, T. A. L., & Dicke, M. (2005). Predatory mites learn to discriminate between plant volatiles induced by prey and nonprey herbivores. *Animal Behaviour, 69*, 869–879.

Degenhardt, J., Hiltpold, I., Köller, T. G., Frey, M., Gierl, A., & Gershenzon, J., et al. (2009). Restoring a maize root signal that attracts insect-killing nematodes to control a major pest. *Proceedings of the National Academy of Sciences of the United States of America, 106*, 13213–13218.

Diaz de la Garza, R., Quinlivan, E. P., Klaus, S. M., Basset, G. J., Gregory, J. F., III, & Hanson, A. D. (2004). Folate biofortification in tomatoes by engineering the pteridine branch of folate synthesis. *Proceedings of the National Academy of Sciences of the United States of America, 101*, 13720–13725.

Diaz de la Garza, R. I., Gregory, J. F., III, & Hanson, A. D. (2007). Folate biofortification of tomato fruit. *Proceedings of the National Academy of Sciences of the United States of America, 104*, 4218–4222.

Dicke, M. (2009). Behavioural and community ecology of plants that cry for help. *Plant Cell and Environment, 32*, 654–665.

Diretto, G., Al-Babili, S., Tavazza, R., Papacchioli, V., Beyer, P., & Giuliano, G. (2007). Metabolic engineering of potato carotenoid content through tuber specific overexpression of a bacterial mini-pathway. *PLoS ONE, 2*, e350.

Eady, C. C., Kamoi, T., Kato, M., Porter, N. G., Davis, S., & Shaw, M., et al. (2008). Silencing onion lachrymatory factor synthase causes a significant change in the sulfur secondary metabolite profile. *Plant Physiology, 147*, 2096–2106.

El Tamer, M. K., Smeets, M., Holthuysen, N., Lücker, J., Tang, A., & Roozen, J., et al. (2003). The influence of monoterpene synthase transformation on the odour of tobacco. *Journal of Biotechnology, 106*, 15–21.

Enfissi, E. M. A., Barneche, F., Ahmed, I., Lichtle, C., Gerrish, C., & MaQuinn, R. P., et al. (2010). Integrative transcript and metabolite analysis of nutritionally enhanced DE-ETIOLATED1 downregulated tomato fruit. *Plant Cell, 22*, 1190–1215.

Falco, S. C., Guida, T., Locke, M., Mauvais, J., Sanders, C., & Ward, R. T., et al. (1995). Transgenic canola and soybean seeds with increased lysine. *Nature Biotechnology, 13*, 577–582.

FAO, *The state of food insecurity in the World 2006*. FAO.

Furuta, S., Nishida, Y., Hajika, M., Igita, K., & Suda, I. (1996). DETBA value and hexanal production with the combination of unsaturated fatty acids and extracts prepared from soybean seeds lacking two or three lipoxygenase isozymes, *Journal of Agricultural and Food Chemistry, 44*, 236–239.

Germinara, G. S., Cristofaro, A. D., & Rotundo, G. (2008). Behavioral responses of adult *Sitophilus granaries* to individual cereal volatiles. *Journal of Chemical Ecology, 34*, 523–529.

Goff, S. A., & Klee, H. J. (2006). Plant volatile compounds: Sensory cues for health and nutritional value? *Science, 311*, 815–819.

Goossens, A., Häkkinen, S. T., Laakso, I., Seppänen-Laakso, T., Biondi, S., & De Sutter, V., et al. (2003). A functional genomics approach toward the understanding of secondary metabolism in plant cells. *Proceedings of the National Academy of Sciences of the United States of America, 100*, 8595–8600.

Goto, F., Yoshihara, T., Shigemoto, N., Toki, S., & Takaiwa, F. (1999). Iron fortification of rice seeds by the soybean ferritin gene. *Nature Biotechnology, 17*, 282–286.

Grotewold, E. (2008). Transcription factors for predictive plant metabolic engineering: Are we there yet? *Current Opinion in Biotechnology, 19*, 138–144.

Gutterson, N. C. (1993). Molecular breeding for color, flavor and fragrance. *Scientia Horticulturae, 55*, 141–160.

Hain, R., Reif, H-J., Krause, E., Langebartels, R., Kindl, H., & Vornam, B., et al. (1993). Disease resistance results from foreign phytoalexin expression in a novel plant. *Nature, 361*, 153–156.

Hall, C., & Yu, O. (2008). Potential for metabolic engineering of resveratrol biosynthesis. *Trends in Biotechnology, 26*, 77–81.

Hashimoto, T., & Yamada, Y. (2003). New genes in alkaloid metabolism and transport. *Current Opinion in Biotechnology, 14*, 163–168.

Hirota, N., Kuroda, H., Takoi, K., Kaneko, T., Kaneda, H., & Yoshida, I., et al. (2006). Brewing performance of malted lipoxygenase-1 null barley and effect on the flavor stability of beer. *Cereal Chemistry, 83*, 250–254.

Hossain, T., Rosenberg, I., Selhub, J., Kishore, G., Beachy, R., & Schubert, K. (2004). Enhancement of folates in plants through metabolic engineering. *Proceedings of the National Academy of Sciences of the United States of America, 101*, 5158–5163.

Hounsome, N., Hounsome, B., Tomos, D., & Edwards-Jones, G. (2008). Plant metabolites and nutritional quality of vegetables. *Journal of Food Science, 73*, R48–R65.

http://www.fruitsandveggiesmatter.gov/>

Iijima, Y., & Aoki, K. (2009). Application of metabolomics to improve tomato fruit productivity and quality. *Journal of the Japanese Society for Horticultural Science, 78*, 14–22.

Imai, S., Tsuge, N., Tomotake, M., Nagatome, Y., Sawada, H., & Nagata, T., et al. (2002). Plant biochemistry: An onion enzyme that makes the eyes water. *Nature, 419*, 685.

Jain, A. K., & Nessler, C. L. (2000). Metabolic engineering of an alternative pathway for ascorbic acid biosynthesis in plants. *Molecular Breeding, 6*, 73–78.

Jones, M. G., Hughes, J., Tregova, A., Milne, J., Tomsett, A. B., & Collin, H. A. (2004). Biosynthesis of the flavour precursors of onion and garlic. *Journal of Experimental Botany, 55*, 1903–1918.

Kappers, I. F., Aharoni, A., van Herpen, T. W. J. M., Luckerhoff, L. L. P., Dicke, M., & Bouwmeester, H. J. (2005). Genetic engineering of terpenoid metabolism attract bodyguards to *Arabidopsis*. *Nature, 309*, 2070–2072.

Kato, N., Dubouzet, E., Kokabu, Y., Yoshida, S., Dubouzet, J., & Yazaki, K., et al. (2007). Identification of a WRKY protein as a transcriptional regulator of benzylisoquinoline alkaloid biosynthesis in *Coptis japonica*. *Plant and Cell Physiology, 48*, 8–18.

Kim, H. J., Ono, E., Morimoto, K., Yamagaki, T., Okazawa, A., & Kobayashi, A., et al. (2009). Metabolic engineering of lignan biosynthesis in *Forsythia* cell culture. *Plant and Cell Physiology, 50*, 2200–2209.

Lange, B. M., Wildung, M. R., Stauber, E. J., Sanchez, C., Pouchnik, D., & Croteau, R. (2000). Probing essential oil biosynthesis and secretion by functional evaluation of expressed sequence tags from mint glandular trichomes. *Proceedings of the National Academy of Sciences of the United States of America, 97*, 2934–2939.

Laurence, K. M., James, N., Miller, M. H., Tennants, G. B., & Campbell, H. (1981). Double-blind randomized controlled trial of folate treatment before conception to prevent recurrence of neural-tube defects. *British Medical Journal, 282*, 1509–1511.

Lavy, M., Zuker, A., Lewinsohn, E., Larkov, O., Ravid, U., & Vainstein, A., et al. (2002). Linalool and linalool oxide production in transgenic carnation flowers expressing the *Clarkia breweri* linalool synthase gene. *Molecular Breeding, 9*, 103–110.

Lee, S., Jeon, U. S., Lee, S. J., Kim, Y. K., Persson, D. P., & Husted, S., et al. (2009). Iron fortification of rice seeds through activation of the nicotianamine synthase gene. *Proceedings of the National Academy of Sciences of the United States of America, 106*, 22014–22019.

Lewinsohn, E., Schalechet, F., Wilkinson, J., Matsui, K., Tadmor, Y., & Nam, K. -H., et al. (2001). Enhanced levels of the aroma and flavor compound S-linalool by metabolic engineering of the terpenoid pathway in tomato fruits. *Plant Physiology, 127*, 1256–1265.

Liu, Z., Saarinen, N. M., & Thompson, L. U. (2006). Sesamin is one of the major precursors of mammalian lignans in sesame seed (*Sesamum indicum*) as observed *in vitro* and in rats. *The Journal of Nutrition, 136*, 906–912.

Lloyd, A. M., Walbot, V., & Davis, R. W. (1992). *Arabidopsis* and *Nicotiana* anthocyanin production activated by maize regulator-r and regulator-C1. *Science, 258*, 1773–1775.

Lorence, A., Chevone, B. I., Mendes, P., & Nessler, C. L. (2004). Myo-inositol oxygenase offers a possible entry point into plant ascorbate biosynthesis. *Plant Physiology*, *134*, 1200–1205.

Luo, J., Butelli, E., Hill, L., Parr, A., Niggeweg, R., & Bailey, P., et al. (2008). AtMYB12 regulates caffeoyl quinic acid and flavonol synthesis in tomato: Expression in fruit results in very high levels of both types of polyphenol. *Plant Journal*, *56*, 316–326.

Lücker, J., Bouwmeester, H. J., Schwab, W., Blaas, J., van der Plas, L. H. W., & Verhoeven, H. A. (2001). Expression of *Clarkia* S-linalool synthase in transgenic petunia plants results in the accumulation of S-linalyl β-D-glucopyranoside. *Plant Journal*, *27*, 315–324.

Lücker, J., Schwab, W., van Hautum, B., Blaas, J., van der Plas, L. H. W., & Bouwmeester, H. J., et al. (2004). Increased and altered fragrance of tobacco plants after metabolic engineering using three monoterpene synthases from lemon. *Plant Physiology*, *134*, 510–519.

Maeda, H., Shasany, A. K., Schnepp, J., Orlova, I., Taguchi, G., & Cooper, B. R., et al. (2010). RNAi suppression of arogenate dehydratase1 reveals that phenylalanine is synthesized predominantly via the arogenate pathway in petunia petals. *Plant Cell*, *22*, 832–849.

Mann, V., Harker, M., Pecker, I., & Hirschberg, J. (2000). Metabolic engineering of astaxanthin production in tobacco flowers. *Nature Biotechnology*, *18*, 888–892.

Matsui, K. (2006). Green leaf volatiles: Hydroperoxide lyase pathway of oxylipin metabolism. *Current Opinion in Plant Biology*, *9*, 274–280.

Matsui, K., Ishii, M., Sasaki, M., Rabinowitch, H. D., & Ben-Oliel, G. (2007). Identification of an allele attributable to formation of cucumber-like flavor in wild tomato species (*Solanum pennellii*) that was inactivated during domestication. *Journal of Agricultural and Food Chemistry*, *55*, 4080–4086.

Matsui, K., Fukutomi, S., Wilkinson, J., Hiatt, B., Knauf, V., & Kajiwara, T. (2001). Effect of overexpression of fatty acid 9-hydroperoxide lyase in tomatoes (*Lycopersicon esculentum* Mill.). *Journal of Agricultural and Food Chemistry*, *49*, 5418–5424.

Mayer, J. E., Pfeiffer, W. H., & Beyer, P. (2008). Biofortified crops to alleviate micronutrient malnutrition. *Current Opinion in Plant Biology*, *11*, 166–170.

Molvig, L., Tabe, L. M., Eggum, B. O., Moore, A. E., Craig, S., & Spencer, D., et al. (1997). Enhanced methionine levels and increased nutritive values of seeds of transgenic lupins (*Lupinus angustifolius* L.) expressing a sunflower seed abumin gene. *Proceedings of the National Academy of Sciences of the United States of America*, *94*, 8393–8398.

Morishige, T., Tamakoshi, M., Takemura, T., & Sato, F. (2010). Molecular characterization of O-methyltransferases involved in isoquinoline alkaloid biosynthesis in *Coptis japonica*. *Proceedings of the Japan Academy, Series B*, *86*, 757–768.

Mosblech, A., Feussner, I., & Heilmann, I. (2009). Oxylipins: Structurally diverse metabolites from fatty acid oxidation. *Plant Physiology and Biochemistry*, *47*, 511–517.

Naqvi, S., Zhu, C. F., Farre, G., Ramessar, L, Bassie, L., & Breitenbach, J., et al. (2009). Transgenic multivitamin corn through biofortification of endosperm with three vitamins representing three distinct metabolic pathways. *Proceedings of the National Academy of Sciences of the United States of America*, *106*, 7762–7767.

Newell-McGloughlin, M. (2008). Nutritionally improved agricultural crops. *Plant Physiology*, *147*, 939–953.

Nicoletti, I., De Rossi, A., Giovinazzo, G., & Corradini, D. (2007). Identification and quantification of stilbenes in fruits of transgenic tomato plants (*Lycopersicon esculentum* Mill.) by reversed phase HPLC with photodiode array and mass spectrometry detection. *Journal of Agricultural and Food Chemistry*, *55*, 3304–3311.

Ono, E., Nakai, M., Fukui, Y., Tomimori, N., Fukuchi-Mizutani, M., & Saito, M., et al. (2006). Formation of two methylenedioxy bridges by a *Sesamum* CYP81Q protein yielding a furofuran lignan, (+)-sesamin. *Proceedings of the National Academy of Sciences of the United States of America*, *103*, 10116–10121.

Paine, J. A., Shipton, C. A., Chaggar, S., Howells, R. M., Kennedy, J. J., & Vernon, G., et al. (2005). Improving the nutritional value of Golden Rice through increased provitamin A content. *Nature Biotechnology*, *23*, 482–487.

Panel on Micronutrients, Subcommittees on Upper Reference Levels of Nutrients and of Interpretation and Use of Dietary Reference Intakes, and the Standing Committee on the Scientific Evaluation of Dietary Reference Intakes (2001). Dietary reference intakes for vitamin A, vitamin K, arsenic, boron, chromium, copper, iodine, iron, manganese, molybdenum, nickel, silicon, vanadium, and zinc. The National Acad. Press. <http://books.nap.edu/catalog.php?record_id = 10026/>.

Peterkofsky, B. (1991). Ascorbate requirement for hydroxylation and secretion of procollagen: Relationship to inhibition of collagen synthesis in scurvy. *The American Journal of Clinical Nutrition*, *54*, 1135S–1140S.

Pichersky, E., Noel, J. P., & Dudareva, N. (2006). Biosynthesis of plant volatiles: Nature's diversity and ingenuity. *Science*, *311*, 808–811.

Qi, B., Fraser, T., Mugford, S., Dobson, G., Sayanova, O., & Butler, J., et al. (2004). Production of very long chain polyunsaturated omega-3 and omega-6 fatty acids in plants. *Nature Biotechnology*, *22*, 739–745.

Raboy, V. (2002). Progress in breeding low phytate crops. *The Journal of Nutrition*, *132*, 503S–505S.

Rasmann, S., Köllner, T. G., Degenhardt, J., Hiltpold, I., Toepfer, S., & Kuhlmann, U., et al. (2005). Recruitment of entomopathogenic nematodes by insect-damaged maize roots. *Nature*, *434*, 732–737.

Runguphan, W., & O'Connor, S. E. (2009). Metabolic reprogramming of periwinkle plant culture. *Nature Chemical Biology*, *5*, 151–153.

Sakai, M., Hirata, H., Sayama, H., Sekiguchi, K., Itano, H., & Asai, T., et al. (2007). Production of 2-phenylethanol in roses as the dominant floral scent compound from L-phenylalanine by two key enzymes, a PLP-dependent decarboxylase and a phenylacetaldehyde reductase. *Bioscience, Biotechnology, and Biochemistry*, *71*, 2408–2419.

Sato, F., Inai, K., & Hashimoto, T. (2007a). Metabolic engineering in alkaloid biosynthesis: Case studies in tyrosine- and putrescine-derived alkaloids. In R. Verpoorte, A. W. Alfermann & T. S. Johnson (Eds.), *Applications of Plant Metabolic Engineering* (pp. 145–173). Dordrecht, Netherlands: Springer.

Sato, F., Inui, T., & Takemura, T (2007b). Metabolic engineering in isoquinoline alkaloid biosynthesis. *Current Pharmaceutical Biotechnology*, *8*(4), 211–218.

Schilmiller, A. L., Last, R. L., & Pichersky, E. (2008). Harnessing plant trichome biochemistry for the production of useful compounds: *Plant Journal*, *54*, 702–711.

Segal, G., Song, R., & Messing, J. (2003). A new opaque variant of maize by a single dominant RNA interference-inducing transgene. *Genetics*, *165*, 387–397.

Shintani, D., & Della-Penna, D. (1998). Elevating the vitamin E content of plants through metabolic engineering. *Science*, *282*, 2098–2100.

Shiojiri, K., Kishimoto, K., Ozawa, R., Kugimiya, S., Urashimo, S., & Arimura, G., et al. (2006). Changing green leaf volatile biosynthesis in plants: An approach for improving plant resistance against both herbivores and pathogens. *Proceedings of the National Academy of Sciences of the United States of America*, *103*, 16672–16676.

Takabayashi, A., Ishikawa, N., Obayashi, T., Ishida, S., Obokata, J., & Endo, T., et al. (2009). Three novel subunits of *Arabidopsis* chloroplastic NAD(P)H dehydrogenase identified by bioinformatic and reverse genetic approaches. *Plant Journal*, *57*, 207–219.

Takahashi, M., Nakanishi, H., Kawasaki, S., Nishizawa, N. K., & Mori, S. (2001). Enhanced tolerance of rice to low iron availability in alkaline soils using barley nicotianamine aminotransferase genes. *Nature Biotechnology*, *19*, 466–469.

Takemura, T., Ikezawa, N., Iwasa, K., & Sato, F. (2010). Metabolic diversification of benzylisoquinoline alkaloid biosynthesis through the introduction of a branch pathway in *Eschscholzia californica*. *Plant and Cell Physiology*, *51*, 949–959.

Tanaka, Y., Sasaki, N., & Ohmiya, A. (2008). Biosynthesis of plant pigments: Anthocyanins, betalains and carotenoids. *Plant Journal*, *54*, 733–749.

Taylor, A. J., & Roberts, D. D. (2004). *Flavor perception*. Oxford, UK: Blackwell.

Tohge, T., Nishiyama, Y., Hirai, M. Y., Yano, M., Nakajima, J., & Awazuhara, M., et al.

(2005). Functional genomics by integrated analysis of metabolome and transcriptome of *Arabidopsis* plants over-expressing an MYB transcription factor. *Plant Journal, 42*, 218–235.

Truksa, M., Wu, G., Vrinte, P., & Qiu, X. (2006). Metabolic engineering of plants to produce very long-chain polyunsaturated fatty acids. *Transgenic Research, 15*, 131–137.

Usuda, K., Wada, Y., Ishimaru, Y., Kobayashi, T., Takahashi, M., & Nakanishi, H., et al. (2008). Genetically engineered rice containing larger amounts of nicotianamine to enhance the antihypertensive effect. *Plant Biotechnology Journal, 7*, 87–95.

Van Eenennaam, A. L., Lincoln, K., Durrett, T. P., Valentin, H. E., Shewmaker, C. K., & Thorne, G. M., et al. (2003). Engineering vitamin E content: From *Arabidopsis* mutant to soy oil. *Plant Cell, 15*, 3007–3019.

Verdonk, J. C., Haring, M. A., van Tunen, A. J., & Schuurink, R. C. (2005). *ODORANT1* regulates fragrance biosynthesis in petunia flowers. *Plant Cell, 17*, 1612–1624.

Verpoorte, R., & Memelink, J. (2002). Engineering secondary metabolite production in plants. *Current Opinion in Biotechnology, 13*, 181–187.

Vogt, T. (2010). Phenylpropanoid biosynthesis. *Molecular Plant, 3*, 2–20.

Wakasa, K., Hasegawa, H., Nemoto, H., Matsuda, F., Miyazawa, H., & Tozawa, Y., et al. (2006). High-level tryptophan accumulation in seeds of transgenic rice and its limited effects on agronomic traits and seed metabolite profile. *Journal of Experimental Botany, 57*, 3069–3078.

Walter, M. H., Floss, D. S., & Strack, D. (2010). Apocarotenoids: Hormones, mycorrhizal metabolites and aroma volatiles. *Planta, 232*, 1–27.

Watson, L. (2000). *Jacobson's organ and the remarkable nature of smell*. New York: W. W. Norton & Company.

WHO report (1995). Global prevalence of Vitamin A deficiency; <http://www.who.int/nutrition/publications/vad_global_prevalence/en/index.html/>

Wu, S., Schalk, M., Clark, A., Miles, R. B., Coates, R., & Chappell, J. (2006). Redirection of cytosolic or plastidic isoprenoid precursors elevates terpene production in plants. *Nature Biotechnology, 24*, 1441–1447.

Xie, D-Y., Sharma, S. B., Wright, E., Wang, Z-Y., & Dixon, R. A. (2006). Metabolic engineering of proanthocyanidins through co-expression of anthocyanidin reductase and the PAP1 MYB transcription factor. *Plant Journal, 45*, 895–907.

Yan, X., & Chen, S. (2007). Regulation of plant glucosinolate metabolism. *Planta, 226*, 1343–1352.

Yokota-Hirai, M., Sugiyama, K., Sawada, Y., Tohge, T., Obayashi, T., & Suzuki, A., et al. (2007). Omics-based identification of *Arabidopsis* Myb transcription factors regulating aliphatic glucosinolate biosynthesis. *Proceedings of the National Academy of Sciences of the United States of America, 104*, 6478–6483.

Zhang, Z., Ober, J. A., & Kliebenstein, D. J. (2006). The gene controlling the quantitative trait locus *EPITHIOSPECIFIER MODIFIER1* alters glucosinolate hydrolysis and insect resistance in *Arabidopsis*. *Plant Cell, 18*, 1524–1536.

Zhu, C., Naqvi, S., Gomez-Galera, S., Pelacho, A. M., Capell, T., & Christou, P. (2007). Transgenic strategies for the nutritional enhancement of plants. *Trends in Plant Science, 12*, 548–555.

Zhu, X., & Galili, G. (2003). Increased lysine synthesis coupled with a knockout of its catabolism synergistically boosts lysine content and also trans-regulates the metabolism of other amino acids in *Arabidopsis* seeds. *Plant Cell, 15*, 845–853.

Ziegler, J., Diaz-Chavez, M. L., Kramell, R., Ammer, C., & Kutchan, T. M. (2005). Comparative macroarray analysis of morphine-containing *Papaver somniferum* and eight morphine-free *Papaver* species identifies an O-methyltransferase involved in benzylisoquinoline biosynthesis. *Planta, 222*, 458–471.

Zvi, M. M. B., Negre-Zakharov, F., Masci, T., Ovadis, M., Shklarman, E., & Ben-Meir, H., et al. (2008). Interlinking showy traits: Co-engineering of scent and colour biosynthesis in flowers. *Plant Biotechnology Journal, 6*, 403–415.

Section F

Plants as factories for industrial products, pharmaceuticals, biomaterials, and bioenergy

Vaccines, antibodies, and pharmaceutical proteins

29

Yuri Y. Gleba Anatoli Giritch

Nomad Bioscience GmbH, Halle, Germany

Table of Contents

Introduction

Plants are rapidly becoming accepted manufacturing hosts for biopharmaceuticals. Until recently, investors and large pharmaceutical companies have avoided plants as manufacturing hosts to avoid double (product *and* process) development risk, but such an attitude is changing. Table 29.1 provides a list of plant-derived pharmaceutical proteins that have been approved by regulatory agencies or are in active clinical trials. There are currently three pharmaceutical proteins approved for market manufacturing: anti-caries antibody (developed by Planet Biotechnology, USA); Newcastle virus subunit vaccine (Dow AgroSciences, USA); and glucocerebrosidase (Protalix Biotherapeutics, Israel); the last product is the most significant both in terms of potential market and because it is the first injectable biopharmaceutical made in plants. Protalix, an Israeli company specializing in the use of cultured plant cells for manufacturing biopharmaceuticals, has successfully completed the Phase III clinical trial with its biosimilar version of glucocerebrosidase, and has permission to use it on a compassionate basis (approval by the Food and Drug Administration, FDA, is expected shortly). Protalix has also entered into a joint venture deal with Pfizer, the world's largest pharmaceutical company, thus ensuring the best commercialization effort for the product. All previously mentioned biopharmaceuticals have been developed on the basis of stable nuclear genetic transformation, and their successful approval significantly lowers both real and perceived risks associated with plants as new manufacturing hosts, paving the way to the new, more efficient plant-based manufacturing processes. Among those novel processes that have made the most significant progress in recent years are, in addition to stable nuclear transformation, plastid transformation as well as various transient expression technologies.

In this regard, it is important to mention that most of the currently announced clinical trials are based on the biopharmaceuticals produced using transient expression systems. These include recombinant antigens and immunoglobulins produced using deconstructed viral vectors (Icon Genetics/Bayer, Germany; Mapp Pharmaceuticals, USA; Fraunhofer Institute, USA) or a non-viral transient system (Medicago, Canada). Such popularity of the transient systems reflects their speed and yields, but it also illustrates their scalability and acceptance by the regulatory agencies.

The newly found confidence of investors and large companies is also in line with the previously mentioned information. For example, Protalix has made a successful reverse merger and is listed on Nasdaq with a cap of $500 million; SemBioSys and Medicago, two Canadian companies, have both made an initial public offering (IPO) on the Canadian stock exchange; in 2008 alone, Biolex, a U.S.-based company, has raised $60 million in private investment for clinical

DOI: 10.1016/B978-0-12-381466-1.00029-8

development of interferon alpha, whereas Medicago was partially (49% of shares) acquired by Philip Morris International and subsequently entered into a number of important business deals. Acquisition of Icon Genetics by Bayer has been successful to the acquirer and resulted in an active development of plant-made individualized vaccines by Bayer. Recently, the U.S. governmental agency DARPA awarded grants worth over $100 million to Fraunhofer Institute/USA, G-Con, KBP, and Medicago, thus recognizing plant-based transient expression as the technology of choice for rapid large-scale manufacturing processes of vaccines and antibodies. Manufacturing platforms have also matured, with a number of companies and contract manufacturing organizations (CMOs) possessing cGMP-compliant manufacturing facilities (KBP, Protalix, Medicago/PMI, Biolex, Icon/Bayer, Fraunhofer/Germany, Fraunhofer Institute/USA). Thus, the overall perceived and real risks of using plants as manufacturing hosts have been greatly reduced. Equally important, at least some companies have developed new processes that are superior over existing manufacturing alternatives in many respects.

The area of plant-made pharmaceuticals is an actively developing field and its progress has recently been the subject of a number of reviews (Gleba et al., 2004, 2005, 2007a,b; Hellwig et al., 2004; Canizares et al., 2005a,b; Santi et al., 2006; Bock, 2007; McCormick and Palmer, 2008; Lico et al., 2008; Yusibov and Rabindran, 2008; Mett et al., 2008; Rybicki, 2009; Daniell et al., 2009b; Smith et al., 2009; Karg and Kallio, 2009; Ko et al., 2009; Plasson et al., 2009; Sainsbury et al., 2009; De Muynck et al., 2010). This chapter provides a general overview of the plant-based manufacturing platform as a novel technical process and a new business opportunity, and discusses the strengths and weaknesses of plants and different plant expression systems, as compared to the established manufacturing hosts and expression platforms.

Expression Technologies: Nuclear Transformation

Transgenic plants obtained through nuclear genetic transformation are historically the first production hosts. Nuclear transformation is a very well established expression technology; however, most applied work has been done with just one constitutive promoter — the 35S promoter of the cauliflower mosaic virus. This promoter provides expression in most plant tissues, albeit at low levels of milligrams to tens of milligrams per kilogram of fresh tissue weight. These levels in some cases can be significantly improved by co-expression of a silencing suppressor such as Hc-Pro or P19 (e.g., Marillonnet et al., 2004; Kalantidis et al., 2006). The best results have, nevertheless, been obtained using tissue-specific promoters such as promoters of seed proteins, as well as regulatory elements of proteins highly expressed in leaves. Examples of high expression in nuclear transformants are listed in Table 29.2; and it is clearly evident that the best

Table 29.1 Plant-made pharmaceutical proteins approved or currently in clinical trials

Biopharmaceutical	Approval	Company	Expression system, host
CaroRx™ anti-caries Mab	EU (medical device)	Planet Biotechnology	Transgenic, tobacco
Newcastle virus subunit vaccine	U.S.	DowAgro	Transgenic, tobacco
Taliglucerase™ glucocerebrosidase	U.S.	Protalix	Transgenic, carrot cells
Interferon α	Phase II	Biolex	Transgenic, duckweed
Insulin	Phase II	SemBioSys	Transgenic, safflower
H5N1 Influenza virus VLP vaccine	Phase I	Medicago	Transient, *Nicotiana*
H1N1 Influenza virus subunit vaccine	Phase I	Fraunhofer, U.S.	Transient *Nicotiana*
Non-Hodgkin's lymphoma individualized vaccine	Phase I	Icon Genetics/Bayer	Transient, *Nicotiana*
Anti-Ebola Mab	Phase I (2010)	Mapp Biopharmaceuticals	Transient, *Nicotiana*
Anti-*Clostridium difficile* Mab	Phase I (2010)	Mapp Biopharmaceuticals	Transient, *Nicotiana*
noroVAXX™ VLP vaccine	Phase I (2010)	VAXX	Transient, *Nicotiana*
Anit-HIV Mab	Phase I (2010)	Pharma-Planta	Transgenic, maize
Acetylcholinesterase	Phase I (2010)	Protalix	Transgenic, carrot cells
Lysozyme[a]	—	Ventria Bioscience	Transgenic, rice
Lacotrrin[a]	—	Ventria Bioscience	Transgenic, rice

[a]Although not in clinical trials *per se*, these two products have undergone extensive toxicology and activity characterization.

Table 29.2 Very high expression levels of recombinant proteins produced in plants

Protein	Expression cassette	Yield, mg/g (%tsp)	Reference
Nuclear transformants			
Phytase	35S	(14.4%)	Verwoerd et al. (1995)
ScFV	Arc5-I	(36.5%)	De Jaeger et al. (2002)
β-Glucuronidase	rbcS1	(10%)	Outchkourov et al. (2003)
Lactoferrin	Gt1	5.0	Nandi et al. (2005)
Lysozyme	Gt1	5.0 (45%)	Huang et al. (2002)
Lysozyme	Puroindoline, Gt1	5.2–9.2	Hennegan et al. (2005)
Cellulase 1	Globulin 1	(16%)	Hood et al. (2007)
Cellobiohydrolase 1	Globulin 1	(16%)	Hood et al. (2007)
Transplastomic plants			
Tetanus toxin C fragment		(25%)	Tregoning et al. (2003)
Canine parvovirus CTB-2L21		(31%)	Molina et al. (2005)
Bacillus anthracis pagA		(4.5–14%)	Koya et al. (2005)
Interferon α 2b		(8–21%)	Arlen et al. (2007)
Human papillomavirus L1		(20–26%)	Fernandez-San Millan et al. (2008)
HIV-1 p24 antigen		0.45 (4.5%)	McCabe et al. (2008)
Yersinia pestis V-F1 antigen		(14.8%)	Arlen et al. (2008)
Lysin PlyGBS		(>70%)	Oey et al. (2009a)
Lysin Pal		2.0 (30%)	Oey et al. (2009b)
Lysin Cpl-1		0.5 (10%)	Oey et al. (2009b)
Insulin-like growth factors		(32%)	Daniell et al. (2009a)
Transient systems			
α-Amylase	TMV	(5% tsp)	Kumagai et al. (2000)
Non-Hodgkin's lymphoma ScFv	TMV	0.5	McCormick et al. (2003)
GFP, DsRed	TMV	5.0	Marillonnet et al. (2004)
GFP	TMV	2.5–4.0	Marillonnet et al. (2005)
Human somatotropin	TMV	1.0	Gils et al. (2005)
IgG1, Mab A5	TMV/PVX	0.5	Giritch et al. (2006)
TMV CP-Protein A	TMV	2–3	Werner et al. (2006)
Yersinia pestis F1, V, F1-V	TMV	1–2	Santi et al. (2006a)
Hepatitis B core antigen	TMV	2.4	Huang et al. (2006)
GFP	TMV	5.5	Lindbo (2007a,b)
At adenosine kinase	TMV	5.0	Lindbo (2007a)
Tuberculosis ESAT6 antigen	TMV	0.8	Dorokhov et al. (2007)

(*Continued*)

Table 29.2 (Continued)

Protein	Expression cassette	Yield, mg/g (%tsp)	Reference
HPV antigen L1	35S	0.5	Maclean et al. (2007)
Plasmodium PyMSP4/5 antigen	TMV	2.0	Webster et al. (2009)
Norwalk virus CP	TMV	0.8	Santi et al. (2008)
GFP	35S	1.6	Sainsbury and Lomonossoff (2008)
IgG HT-2G12	35S	0.3	Sainsbury and Lomonossoff (2008)
GFP	CMV	0.5	Fujiki et al. (2008)
GFP	AMV/CMV	0.4	Green et al. (2009)
Hepatitis B core antigen	BeYDV	0.8	Huang et al. (2009a)
Norwalk CP antigen	BeYDV	0.3	Huang et al. (2009a)
Griffithsin	TMV	1.0	O'Keefe et al. (2009)
Murine IgG C5-1	Plastocyanin	0.4–1.5	Vezina et al. (2009)
Human IgG	BeYDV	0.5	Huang et al. (2009b)
G-CSF, GM-CSF	TMV	0.3–0.5	Zvereva et al. (2009)
Aprotinin	TMV	0.3–1.0	Vancanneyt et al. (2009)
Non-Hodgkin's lymphoma twenty different IgG	TMV/PVX	0.5–4.8	Bendandi et al. (2010)
Mini-insulin	TMV	5.0	Nomad Bioscience data
Human interferon α	TMV	5.0	Nomad Bioscience data

Abbreviations: 35S, 35S promoter of cauliflower mosaic virus; arc5-I, arcelin promoter of *Phaseolus vulgaris*; rbcS1, promoter of RUBISCO small subunit of *Chrysanthemum*; Gt1, glutenin promoter of *Oryza sativa*; puroindoline, purindoline B promoter of *Triticum aestivum*; globulin 1, globulin 1 promoter from *Zea mays*; TMV, tobacco mosaic virus; PVX, potato virus X; CMV, cucumber mosaic virus; AMV, alfalfa mosaic virus ; BeYDV, bean yellow dwarf virus; plastocyanin, *Medicago sativa* plastocyanin gene promoter.

results have thus far been obtained using expression cassettes with seed-specific promoters.

Seeds have proven to be versatile vehicles for recombinant proteins of all types, including short peptides and complex heterooligomeric proteins such as antibodies. The downstream processing of seeds is greatly assisted by the stability of recombinant proteins in seeds during storage, thus allowing decoupling of processing from the growth and harvest cycle. In many cases, the purification is also simplified because seeds, compared to leaves (another main production organ), do not contain reactive small molecules (chlorophylls, phenolics) that complicate primary processing of the biomass. In terms of net protein level per biomass, transgenic rice plants expressing over 9 g/kg of seed (Hennegan et al., 2005) hold the productivity record. However, it should be mentioned that, since seed yields per hectare are far below respective yields of green biomass (standard yields would be 8–20 tons per hectare for seed versus 100–300 ton per hectare for leaves), the highest productivity on a per hectare basis is achieved with plants expressing recombinant proteins in leaves such as transplastomics in which foreign genes have been inserted in the DNA of plastids or transiently transfected plants. A number of companies are exploring seed-produced recombinant proteins as a manufacturing platform, including SemBioSys (Canada), Ventria (USA), and Plantechno (Italy). Among the production hosts used are safflower, rice, and tobacco; in the past, maize and barley have also been used. A recent general review of seed-based expression systems has been published by Boothe et al. (2010).

There are several publications that describe high levels of recombinant protein also obtained in leaf tissues of nuclear transgenic plants, starting with an early paper of Verwoerd et al. (1995) who described significant levels of expression (10% of total soluble protein, TSP) of phytase in leaves of transgenic tobacco using a 35S CaMV promoter. Later, superior results were obtained with a leaf-specific promoter and terminator signals such as those of the ribulose-1,4-bisphosphate carboxylase-oxygenase (RUBISCO) small subunit (Outchkourov et al., 2003). A special case is the manufacturing platform of Biolex (USA); the company is using duckweed, an aquatic plant, as a production host.

Expression takes place in green photosynthesizing tissues and expression levels are high compared to those in terrestrial plants (Cox, 2006).

Last but not least, numerous experiments have been directed toward the expression of recombinant proteins in cultured plant cells and cultured organs. We did not find exceptionally high expression levels in the literature, so this area seems to require further exploration. Plant cell fermentation is a very promising manufacturing platform and two out of the three approved products (Newcastle virus subunit vaccine of Dow AgroSciences and glucocerebrosidase of Protalix) are manufactured in plant cell fermenters.

Nuclear transformation allows for expression of proteins in all cellular compartments, thus it offers possibilities for various post-translational modifications of the expressed proteins including glycosylation. Nuclear transformants are often subject to silencing, therefore, the development requires testing multiple transformation events over a number of generations, thus slowing down the research and development. Expression of proteins in seeds seems to have reached a certain level of maturity, however, in our opinion future efforts should concentrate on secondary crops (e.g., safflower developed by SemBioSys) as biopharmaceutical production hosts, rather than relying on major crops and risk the segregation problems.

Another interesting application of nuclear transformation was demonstrated by Inge Broer and colleagues (Huhns et al., 2008, 2009). They developed stable transgenic lines of tobacco and potato expressing the cyanophycin synthase gene from cyanobacterium *Thermosynechococcus elongatus*, which resulted in the accumulation of cyanobacterial storage material cyanophycin, a copolymer of aspartic acid and arginine. Plastid targeting of expressed enzyme allowed a cyanophycin accumulation of about 6.8% of dry weight in tobacco leaves (Huhns et al., 2008) and 7.5% of dry weight in potato tubers (Huhns et al., 2009).

Expression Technologies: Plastid Transformation

In the plastid transformation process, foreign genes are incorporated into the chloroplast genome by homologous recombination, eliminating variation of expression typically common for nuclear transformants. Moreover, gene silencing does not seem to occur in transplastomic plants. The expression levels are facilitated by the fact that there are over 10,000 copies of transgene in each transformed cell. For many practically important proteins, very high levels of expression (up to 70% of TSP or up to 5g/kg of fresh biomass, see also Table 29.2) have been achieved using this technology. In addition, unlike seed harvest, leaf biomass can be harvested several times per year, and for crops such as tobacco or alfalfa, annual biomass yields of up to 300 tons per hectare per year are possible. Since the plastid genes are inherited (in most crop species) strictly maternally, the ecological risk of transgene escape through pollen to other crops is minimal. Although most work has been done with transplastomic tobaccos, plastid transformation can be performed on crops such as lettuce, which eventually would allow using an unprocessed or semi-purified biomass as edible medicinal material.

The gene expression in plastids (most often in chloroplasts) is similar to expression in bacterial cells with the same limitations: the expressed protein will start with methionine; and plastid-produced proteins are not glycosylated. This second limitation especially narrows down the usefulness of the platform, as most biopharmaceuticals used today are glycosylated proteins. We are, however, left with several important classes of polypeptides and proteins that can be produced in plastids, the most obvious cases are vaccines (especially bacterial subunit vaccines and adjuvants), and antimicrobial and antiviral polypeptides and proteins, as well as some cytokines and growth hormones.

The first chloroplast-derived antigen that proved to be immunologically active in animal studies was a subunit vaccine against *Clostridium tetani*, the causal agent of tetanus (Tregoning et al., 2003). In the original work, the TetC fragment of tetanus toxin was expressed in transplastomic plants at levels of 10–25% of TSP. The material was immunologically active and protected mice from the *Clostridium* challenge. There are a number of publications describing high expression levels in chloroplasts of immunologically active antigens and adjuvants from bacteria and protozoans, including *Bacillus anthracis* (Watson et al., 2004; Koya et al., 2005), *Borrelia burgdorferi* (Glenz et al., 2006), *Vibrio cholerae* (Davoodi-Semiromi et al., 2009, 2010), *Yersinia pestis* (Arlen et al., 2008), and *Entamoeba histolytica* (Chebolu and Daniell, 2007). Some cases reflecting very high expression levels of functional antigens are listed in Table 29.2. Several papers also describe expression of viral antigens in a form of subunit or virus-like particles (Molina et al., 2005; Fernandez-San Millan et al., 2008).

Another fascinating opportunity is reflected in the work of R. Bock and colleagues who developed transplastomic plants expressing high levels of antibacterial proteins such as bacteriophage lysins that are toxic to bacteria and thus cannot be produced easily by microbial fermentation (Oey et al., 2009a,b). Lysins act through hydrolysis of the bacterial walls, and since plastids do not have one, expression in plants can reach very high (over 70% of TSP) levels without obvious toxic effects. Specific lysins such as Pal and Cpl-1 effective against *Streptococcus pneumoniae* and plyGBS active against pathogenic group A and B streptococci have been successfully produced and exhibited full enzymatic activity.

It is a pity that the only company created to commercialize the transplastomic platform, Chlorogen (USA), was founded before it had a serious product candidate; we hope that in view of the current successes, there will be new business attempts.

Expression Technologies: Transient Expression Systems

Plant virus vectors have been developed as an alternative to other expression methods (stably transformed transgenic or transplastomic plants; reviewed in Gleba et al., 2004, 2007a) because of some anticipated advantages, such as expression speed and yield, reduced cost/duration of research and development, very high throughput, and so forth. There has been impressive technical progress in the past few years in the development of production processes and industrial plant hosts, based on different approaches, including the virus vector-based approach. Most progress has been achieved with RNA viruses, and the most advanced vectors have been built using just a handful of plant virus species such as tobacco mosaic virus (TMV), but also including potato virus X (PVX), alfalfa mosaic virus (AMV), cowpea mosaic virus (CPMV), and so forth.

"Full virus" vectors

Historically, the first approach was to design a vector capable of infecting a plant in the same way as a wild-type virus, but that has additionally been engineered to carry and express a heterologous sequence coding for a gene of interest (this strategy is also known as a "full virus" vector strategy). The most extensive program resulting in a series of commercial vectors used in various laboratories worldwide has been executed at Large Scale Biology Corp., Vacaville, California (LSBC; formerly Biosource Genetics), together with associates from several universities. LSBC filed for bankruptcy in 2006, but most of its assets, including intellectual property, are owned and are being further developed by Kentucky Bioprocessing in Owensboro, Kentucky. That effort focused on TMV as the vector backbone, and the resultant technology and viral vectors are known as Geneware® technology, and are documented in numerous publications (Donson et al., 1991; Kumagai et al., 1993, 1995, 2000; Turpen et al., 1995; Shivprasad et al., 1999; Turpen, 1999; McCormick et al., 1999, 2003, 2008; Rabindran and Dawson, 2001; Lindbo, 2007a; O'Keefe et al., 2009). In several cases, high yields (10% and up to 20% of TSP) have indeed been observed (Table 29.2).

LSBC has developed a concept of an individualized vaccine for non-Hodgkin's lymphoma that in essence is a single-chain antibody individually tailored to induce immune response against cancerous B-cells (McCormick et al., 1999, 2003, 2008). The company has performed a successful Phase I clinical trial, but economic problems shortly thereafter led to the discontinuation of product development. LSBC and others have demonstrated that a number of proteins can be well expressed using Geneware® technology. Among the products described in the literature are α-galactosidase (Turpen, 1999), α-amylase (Kumagai et al., 2000), and, recently, griffithsin (O'Keefe et al., 2009). LSBC's manufacturing capabilities are currently owned and operated by Kentucky Bioprocessing, a contract manufacturing organization. Currently, there are several proteins in late preclinical development that are expected to be made using full virus technology. Probably the most advanced case is that of griffithsin, an antiviral protein that can be expressed in *Nicotiana benthamiana* plants at high yield (over 1 g/kg) and in a functional form by using Geneware® vectors (O'Keefe et al., 2009). Plant viral vectors have also been successfully used to present different epitopes on its surface through expression of fusions between a coat protein gene and a gene encoding short polypeptide (usually immunogenic epitope).

Magnifection

The "second generation" expression systems such as "magnifection" (derivative from "magnum infection") integrate elements of viral machinery, such as RNA/DNA amplification, and cell-to-cell movement, along with non-viral processes such as replicon formation via *Agrobacterium*-mediated delivery of T-DNA encoding viral vector or via activation from a plant chromosomal DNA resulting in intracellular generation of replicons from a chromosomally encoded pro-replicon or pro-virus. The most advanced transient version of this approach is the technology developed by Icon Genetics, Halle, Germany, and trademarked as magnICON® technology and vectors. The magnICON® vectors are being used by dozens of business and research laboratories worldwide, and the technology, as well as its applications, have been covered in a number of publications (Gleba et al., 2004, 2005, 2007a; Marillonnet et al., 2004, 2005; Gils et al., 2005; Giritch et al., 2006; Werner et al., 2006; Santi et al., 2006a, 2008; Huang et al., 2006; Vancanneyt et al., 2009; Bendandi et al., 2010).

The magnifection, an indication that two infection agents are being used such as a bacterium and a virus process, is a simple and indefinitely scalable protocol for heterologous protein expression in plants, which relies on transient amplification of viral vectors delivered to multiple areas of a plant body (systemic delivery) by *Agrobacterium* as DNA precursors. Such a process is an infiltration of whole mature plants with a diluted suspension of agrobacteria carrying T-DNAs encoding viral RNA replicons. The scale-up (industrial) version requires simple apparatuses for high-throughput *Agrobacterium* delivery to whole plants. The magnifection platform effectively addresses most of the major shortcomings of earlier plant-based technologies by providing an overall best combination of the following features: highest expression levels approaching biological limits; highest relative yields; ability to express complex proteins; low up- and downstream costs; very fast and low R&D costs; and low biosafety concerns. The process has been brought to a QA/GMP compliance level at Bayer's new pilot manufacturing facility in Halle, Germany, and a pilot manufacturing with a throughput capacity of 0.6 ton/day based on Bayer's technology was started at Kentucky Bioprocessing LLC in 2009. There are several magnICON®-based products, mainly pharmaceutical proteins that are being developed by Bayer and its licensees; Bayer's lead product candidate, individualized vaccines for non-Hodgkin's lymphoma produced in a cGMP-compliant manufacturing facility in Halle, entered clinical trials in January 2010 (Table 29.1).

Depending on the vector used (the host organism and the initial density of bacteria), this much more efficient and synchronous agro-mediated local transfection leads in 4 to 10 days to a dramatic synthesis of proteins, providing yields as high as 5 g/kg of fresh leaf biomass. The viral vectors effectively shut down normal biosynthetic processes in cells. As a result, in 10–14 days post-infection, the most abundant plant leaf protein, RUBISCO, all but disappears, and the amount of viral vector-encoded protein reaches as much as 80% of the TSP in a leaf. These reported data clearly demonstrated that TMV- and PVX-based vectors provide yields approaching biological limits of the plant-leaf-based production. Infiltration of plants/detached leaves with bacteria has been done in many different ways; the simplest one is vacuum infiltration done by immersing whole aerial parts of plants in a bacterial suspension and applying a weak vacuum. Further development and automation of these protocols could increase the throughput of the process and reduce the amount of required manpower.

Table 29.2 provides yield data for selected recombinant proteins of practical interest based on publications of numerous groups as well as our unpublished results. It is obvious that many proteins, including multiple biopharmaceuticals (interferons, insulins, growth hormones, colony-stimulating factors, numerous antigens, selected full IgG antibodies, etc.), can be expressed at very high yields of 1–5 g/kg. Most proteins expressed to date using magnifection have been properly processed and folded, and were functional. This is true for high molecular weight proteins (up to 100 kDa), proteins requiring specific cleavage of signal or transit peptides, proteins requiring extensive disulfide bond formation, proteins requiring complex processing (e.g., hemagglutinin of Influenza virus), and proteins forming virus-like particles (such as hepatitis B core protein, Norwalk virus capsid protein, papilloma virus L1 protein).

Derisking the new manufacturing process

There were several real and/or perceived risk factors in connection with the new process that have been carefully considered and addressed when developing the process and later, preparing the business plan. These include quality of plant-made proteins, quality of proteins made by transient process, effects of plant-specific protein modifications, risk of public acceptance of plant-made pharmaceuticals made in plants and, last but not least, a potential new regulatory burden imposed by regulatory agencies on plant-based production. Extensive studies at Icon/Bayer and other companies demonstrated that the plant-based manufacturing in general and the transient expression in particular, provides proteins of the quality and purity required by the pharmaceutical industry. Our manufacturing plant host, *N. benthamiana*, has been modified to be free of plant-specific sugars added to recombinant proteins, thus removing a potential risk of undesired host-specific immunogenicity of plant-derived proteins. Unlike other plant-based manufacturing processes, because of the very high protein yields provided by the magnICON® process, this technology can, and is expected to be, commercially run in a contained greenhouse facility, thus avoiding any real or political risks in connection with the release of genetically modified organisms into the environment. The interaction of Icon/Bayer and other companies (LSBC, Protalix, Biolex, SemBioSys, etc.) with the FDA and other regulatory agencies did not result in any new essential regulations imposed on plant-based transgenic or transient technology. It has been proven that the process of making pharmaceutical proteins in plants can be run under proper quality control, and where needed, in compliance with cGMP regulations. Existing functioning pilot manufacturing facilities have also proven that the process can be scaled up to intended industrial levels without new unknown regulation risk.

Plant-Made Pharmaceuticals: A Unique Selling Proposition?

Recombinant proteins can be manufactured in different living cells and organisms, and historically, for pharmaceutical proteins, bacteria, yeasts, and animal cells have been the main industrial manufacturing hosts. It is not so easy to come up with a general reason as to why plants should be better industrial hosts. For many years, proponents of the idea of making pharmaceutical proteins in plants have claimed the low cost of this manufacturing process as the main argument. The experimental data provided did not support this claim; however, because the expression levels of proteins achieved in most early experiments were very low (0.1–1% TSP), thus magnifying the purification costs that already accounted for more than 50% of *manufacturing costs* (direct "costs of goods"). As a general rule, the purification costs are inversely proportional to the initial concentration of the active ingredient in a primary extract. To circumvent this obvious problem, many researchers proposed the use of unpurified or partially purified plant extracts as medicines, applied orally. There are probably some medical needs that can be addressed by such an approach, but generally speaking the high purification level of a biopharmaceutical is a prerequisite for the great majority of potential applications. We are therefore left with the costs of downstream processing as the major bottleneck that needs to be addressed before the plant-based manufacturing can be a really appealing cost-reduction proposition. Some of the research addressing the issue of the purification costs is discussed next. In conclusion, our best message to an interested party is that plant-based manufacturing is 20–25% cheaper in terms of running costs compared to the animal-cell-based process.

A much more plant-made pharmaceuitical (PMP) friendly conclusion can be made about the *capital costs* associated with green plants as production hosts. We estimate that, compared to the standard animal-cell-based manufacturing process, the plant-based production will be at least 40% cheaper in capital expenses, and it will provide much more flexibility to the manufacturer. In the case of Herceptin® (trastuzumab) and other antibodies, the only industrial production method used is the expression of antibodies in transgenic animal cells such as CHO cells, in combination with affinity capture-based purification. Development of such a production facility is connected with very high capital and other upfront costs. According to various sources, for example Bernstein Research, capital costs for a mammalian cell facility with a capacity of 500 kg/year are over $300 million, and additional costs of running it for the initial three years prior to entering the manufacturing phase will add another $120 million, also required upfront (http://www.ftc.gov/bc/workshops/hcbio/docs/fob/rgal.pdf). Based on existing prototypes such as Icon and KBP facilities, we have made estimates of capital costs and costs of goods associated with our new integrated process. Our calculations show that a facility with the capacity of 1 ton of antibody per year would cost less than $50 million and operation costs for the first 3 years would amount to $45–60 million. Since this new manufacturing platform has been essentially derisked, such a low financial entry barrier represents a very attractive opportunity for medium-size companies in developed countries, as well as excellent opportunities for manufacturers in developing countries.

Other important advantages presented by plants include the speed of manufacturing, speed of research and development, and the inherent flexibility of the transient expression systems. So far, only the speed of the system has been recognized, for example, by DARPA, which concluded that plant-based transient systems are overall the best manufacturing technologies when dealing with bioterrorism and pandemics. We expect that in the near future, the unique flexibility and linear scalability of these technologies will win over many companies.

Many active PMP players have chosen biosimilars (follow-on versions of already approved biopharmaceuticals) as their first products; the main rationale is to avoid the product development risk. The biosimilars market has grown fast and many developing countries are ready to accept the new process in return for lower capital costs and lower costs of goods (COGs) or improved clinical performance. Many important patents will expire around 2010–2017 and as a result many highly successful antibodies (including Herceptin®, Avastin®, and Rituxan®) and other proteins (with market worth of over $50 billion) will lose protection. The biosimilars market has a number of big and medium-size players (Teva, Novartis, Stada, Shanta Biotech, etc.), and several pharmaceutical companies (most recently Merck) announced their interest in biosimilars, thus providing various exit scenarios for plant biotech investors. The attractive niche in the biosimilars market is the so-called "biobetters," or biosimilars which, through a relatively simple modification of their structure such as modification of glycans, show better therapeutic properties (see the following section).

Yet another approach taken by several companies is to develop new products for which plant-based expression provides a unique technical opportunity. Examples of such products would be medicinal foods developed by Ventria (high volumes of production, low cost of products that do not require purification, etc.) or products requiring rapid manufacturing such as vaccines and antibodies needed in the case of pandemics and bioterrorism and individualized medicines (speed, flexibility, and low cost manufacturing).

Plant-Based Manufacturing, Post-Translational Modifications, and Plant-Specific Sugars

With a few exceptions (insulin, serum albumin, apoliporotein A), proteins made by the human body and used as therapeutics are glycosylated. Glycans are sometimes not essential and do not influence the clinical performance; for example, in the case of some interferons and colony-stimulating factors. It appears that for many subunit and virus-like particle-based vaccines, glycosylation does not seriously lower immunogenicity. In a few other cases, such as erythropoietins, a complex and specific glycosylation is critical for product performance. For most other proteins that require glycosylation, such as antibodies, the issues can be reliably resolved through the use of the glycoengineered plants mentioned next.

Plants are able to assemble complex proteins, and in cellular compartments such as Golgi, different proteins of human origin are properly processed, cleaved, folded, assembled, and post-translationally modified. In particular, plants provide a good host for heterooligomers such as human immunoglobulins. The only essential difference in post-translational modification by a plant cell is glycosylation. Although plants add main sugar scaffolds to the protein, the resultant glycoforms are different from those made by a human body. Currently, *Physcomitrella* moss as well as *Lemna* (duckweed) and *N. benthamiana* plants lacking xylose and fucose, produced via RNAi technology, are available, whereas *N. benthamiana* xyl-/fuc-plants obtained via mutagenesis at Bayer Crop Science will be available soon.

New developments in plant glycoengineering provide unique opportunities both in the area of new-generation antibody-based therapeutics, but especially in the area of so-called biosimilars that, with this technology, could be developed as proprietary biobetter analogs superior to the currently licensed anticancer antibodies. Glycoengineered production hosts include plants that, unlike industrial CHO cells, do not add fucose and galactose to the glycans attached to the recombinant proteins (Figure 29.1). Current studies demonstrated that the effector functions (antibody-dependent cell cytotoxicity, ADCC) and correspondingly the clinical efficacy of existing licensed therapeutic antibodies such as Avastin® (bevacizumab), Rituxan® (retuximab), or Herceptin® (trastuzumab) are strongly inhibited due to non-specific IgG competing for binding of the therapeutics to the receptor on natural killer cells, which leads to the requirement for significantly higher amounts of drug. This problem is effectively remedied by using a production host that does not add a fucose to the glycan structure. The combined market for the three most successful antineoplastic antibodies (Avastin®, Rituxan®, or Herceptin®) in 2007 was $6.8 billion, and the peak sales are projected to reach $21.5 billion by 2016. By 2017, all three will run out of patent protection in both the United States and the EU. Development of non-fucosylated (and perhaps non-galactosylated) analogs that show superior therapeutic properties compared to the existing therapeutics represents a compelling business case.

Such a development, because it is based on similar, already registered therapeutic product(s), but takes into account the latest scientific understanding of the deficiencies of the licensed therapies, would decrease an inherent risk of failure in (late) clinical trials, which is historically above 85% for cancer biopharmaceuticals. Coupled with lower capital costs and costs of goods of plant-based production, as well as anticipated higher acceptance of a plant-based platform (in view of the successful Phase III clinical trials of plant-made biosimilar glucocerebrosidase by Protalix), such a project could raise a PMP company to the position of a serious player in the area of second-generation biobetter anticancer therapies. Essentially

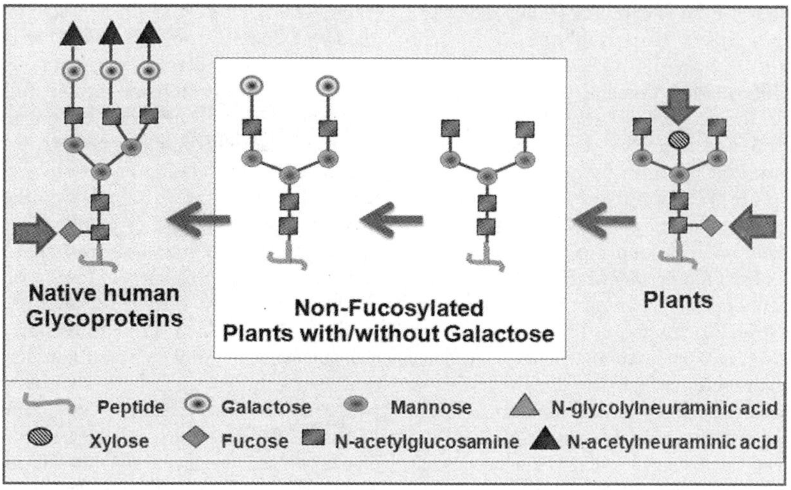

Figure 29.1 • Plant glycan engineering • Glycan structures of human and plant secreted proteins. Red arrows indicate human-specific (α-1,6-linked fucose) and plant-specific (α-1,3- linked fucose and β-1,2-linked xylose) sugars. Magenta arrows indicate glycoengineering of plants through RNAi silencing of specific plant genes, mutagenesis, and the addition of new glycosyl-transferases. Please see color plate section at the end of the book.

the same approach could also be applied to novel antibody-based therapeutics under development. Technically, the best studied effect is expected for antineoplastic antibodies and antibody-based molecules, but glycoengineered plants could be a source of molecules with new therapeutic properties also in other indication areas.

Production of non-fucosylated antibodies has been recently achieved in animal host lines in which the endogenous fucosyltransferase gene is knocked out. Several clinical trials are underway at present, with much of the attention devoted to existing commercial therapeutics such as Rituxan® (rituximab) and Herceptin® (trastuzumab), which have been improved through removal of fucose (biosimilars now termed biobetters).

A number of research groups have engineered plants to perform a designer glycosylation. As a first step, galactose residues have been added to protein glycans via expression of proper galactosyl-transferases (Palacpac et al., 1999; Bakker et al., 2001). As a second step, plants lacking plant-specific sugars such as xylose and α-1,3-linked fucose have been created using RNAi technology (Cox et al., 2006; Schahs et al., 2007; Strasser et al., 2008) and, recently, by chemical mutagenesis (Bayer Crop Science). The final missing step, the addition of N-acetylneuraminic acid, has also been demonstrated in transient assays (Decker and Reski, 2008; Castilho et al., 2010). Such designer glycosylation does not have to be done as a stable genetic transformation. Recent work from Vezina et al. (2009) demonstrated that transient co-expression of the genes encoding the protein of interest together with the RNAs suppressing fucosyl- and xylosystransferases provides for the appropriately modified recombinant proteins.

For those in need of plants that have a fully controlled and defined glycosylation pattern, the platform developed by Greenovation, Germany, relying on manufacturing in cells of a moss, *Physcomitrella patens*, could be the ultimate solution (Gorr and Altmann, 2006; Schuster et al., 2007; Nechansky et al., 2007; Weise et al., 2007). The glycans added to proteins in this system are defined as those of glycoengineered yeasts.

Plant-Based Manufacturing and Downstream Issues

Usually, standard "low-cost arguments" of the proponents of plant-based manufacturing are met with skepticism. A typical investor's answer is that even if the plant-based production (upstream part of the manufacturing process) is 10 times cheaper, the overall process is unlikely to be 2 times cheaper because of high costs of the purification (downstream part), and even this saving would be difficult to achieve with an entirely new process. Commercial downstream processes employed today are extremely complex and costly, accounting for as much as 50–80% of total manufacturing costs (Roque et al., 2004). The most significant part of the downstream costs is costs of immunoadsorbents and other affinity adsorbents. For example, the cost of Protein A-based immunoadsorbent, a standard reagent used for industrial purification of monoclonal antibodies, is so high that the packed columns have to be reused up to 50 times, resulting in significant new costs due to cleaning and re-validation of the columns. An ideal alternative has to be a single-use reagent that is cheap enough to provide sufficient economic advantages.

A number of research groups have tried to develop downstream protocols that are simpler and cheaper, most of them addressing the cost of affinity ligands such as immunoadsorbents. Several potential solutions described in the literature propose fusing Protein A (or other affinity ligands) to elements capable of polymerizing or to structures of high molecular weight: bacterial S-layer proteins (Vollenkle et al., 2004; Nomellini et al., 2007); bacteriophages (Kushwaha et al., 1994); oleosins recognizing oil bodies (see the following information); cellulose-binding domains (Cao et al., 2007; Hussack et al., 2010); and so forth. The process based on oleosin fusions (to Protein A domains) has been advanced to the stage of a commercial platform by SemBioSys (www.sembiosys.com). The process is robust and utilizes principles and industrial flotation centrifugation machinery used in the separation of lipids from aqueous emulsions such as milk or seed extracts. Since oleosin and oleosin-Protein A fusion move together with oil bodies, a simple, inexpensive, and highly selective purification process step is feasible.

Plant tobamoviruses provide an extremely cheap source of protein biopolymers of discrete nanoparticle sizes than can be manufactured rapidly and under very simple conditions. In our studies, we have found that a functional fragment of Protein A (133 aa) can be displayed on the surface of TMV as a carboxy-terminal fusion to the Coat Protein (CP) via a 15 aa linker. The macromolecular nature of these nanoparticles allowed the design of a simple laboratory protocol for purification of monoclonal antibodies (Mab) with 50% recovery yield and >90% product purity. The extremely dense packing of Protein A on the nanoparticles (over 2000 copies per TMV particle) results in an immunoadsorbent with a binding capacity of up to 2g Mab per g. This characteristic, combined with the high level of expression of the nanoparticles (up to 5g adsorbent per kilogram of leaf biomass), provides a very inexpensive, self-assembling matrix that could meet criteria for a single-use industrial immunoadsorbent for antibody purification (Werner et al., 2006). Our studies also showed that such nanoparticles can be engineered or conjugated chemically to add a second functionality, for example, a magnetic property, thus allowing for an efficient separation of the nanoparticles in complex solutions.

There are a number of attempts to develop other purification principles published recently; the most interesting one is based on the use of elastin-like polypeptides that reversibly change solubility at different temperatures, thus allowing for a simple non-chromatographic purification of proteins fused to or having an affinity for elastins (Floss et al., 2008, 2009, 2010; Conley et al., 2009a,b). Expressed as fusions to proteins of interest such as single chain antibodies or antigens, they provide for a higher expression level and for an easy purification step (Floss et al., 2008, 2009, 2010; Conrad et al., 2011). There is a need for subsequent separation from the elastin, which can be engineered by the addition of a cleavage site; the alternative is using elastin fusions with affinity ligands such as Protein A domains. This technology principle has been "borrowed" by plant scientists, and it could be a very interesting solution for a number of industrial processes and products.

A more plant-specific approach has been advocated by M.D. Ludevid and colleagues (Kogan et al., 2001; Torrent et al., 2009a,b; Alvarez et al., 2010). Their platform utilizes the natural ability of plants to accumulate large amounts of storage proteins in plant seeds by storing them as "plant bodies." Plant body sequestration is controlled by a proline-rich N-terminal domain of a seed storage protein; such a domain can be added as a fusion to many proteins of interest. This technology has been exemplified using a number of commercially interesting proteins and is being currently developed by a Spanish company called Era Biotech.

Another possibility, at least for a limited number of cases, is the use of plant biomass without full purification. Such an approach can be exemplified by the products developed by Ventria — lactoferrin and lysozyme. Both proteins and products of their degradation in gut are extremely powerful antibacterials with a known safety and efficacy. They can also be used to replace antibiotics or other antibacterials, for example, in animal feed and infant formula. Since the products are expressed at very high levels (5g/kg of seed), the transgenic rice seed could be eventually used as a basis for medicinal foods or animal feed admixtures and for controlling bacterial gut flora, as well as for other purposes. These product candidates, however, require a different regulatory approval path from the one used to manufacture biopharmaceuticals. During the early years of plant biotechnology, a great deal of excitement was generated by the idea of edible vaccines. Although the idea is still appealing, because of the lack of regulatory rules about such a product, neither investors nor companies supported such projects.

Plant-Based Expression Systems: Advantages and Limitations

Because of the inherent speed of the recently introduced transient systems and because of the latest successful attempts to engineer plants with a designer glycosylation, green plants represent a very attractive production host when compared to other manufacturing hosts. In addition, recent clinical trials have also clearly demonstrated that plant-based platforms can meet criteria set by the regulatory agencies and in this respect, plant-based processes *per se* do not represent an added process risk any more. Unlike other production organisms used for the manufacturing of biopharmaceuticals, plants and plant cells do not contain pathogens capable of infecting humans. Materials used to grow plants, transgenic plants, and to cultivate agrobacteria (in the case of transient expression) are or can be made animal-material-free. In this respect, plant-derived recombinant proteins are inherently safer than materials generated using animal cells.

A general comparison of plants and other living cells as manufacturing hosts is represented in Figure 29.2. The conclusions from this "strengths-weaknesses-opportunities-threats" (SWOT) analysis are different from those made just a few years ago (Fischer et al., 2004) and show plants as overall the most attractive manufacturing platform for biopharmaceuticals. Production processes based on bacteria, yeasts, and animal cells are current industrial standards, and plant based-processes will not displace them any time soon, for a variety of reasons. We expect that the first successes will be led by small- and medium-size companies (attracted primarily by lower capital costs and higher development speed), and that the overall best chances for adoption of manufacturing based on plants will be in developing countries that so far have not invested in traditional manufacturing infrastructures and are more open toward new generation processes that promise lower manufacturing costs. Such an adoption would be greatly facilitated by the development of less expensive purification processes, thus we continue to face numerous technology challenges.

The plant-based manufacturing platform represents a number of different manufacturing processes, each with certain advantages and disadvantages. A general summary of those is given in the following sections.

Nuclear transformation

An important advantage of the nuclear transformation platform is its versatility regarding expression in various subcellular compartments, tissues, and organs, including seeds and cultured cells. Expression in seeds was regarded by many as the main advantage, because it allows inexpensive storage of the harvested material in a stable form, thus allowing the disconnection of the seasonal plant cultivation (upstream) from the continuous processing (downstream) phases of the manufacturing process. Also, the manufacturing based on fermentation of cultured plant cells is currently utilizing nuclear transformants only, although there are attempts to develop protocols based on transient expression. The main problem of nuclear transformants is generally low expression levels, with important exceptions representing expression in seeds using seed-specific promoters and to a lesser extent in leaves. The other problem is the long time needed to develop a transgenic production host (in part because of potential silencing of the transgene of interest in subsequent progenies). The great majority of studies conducted thus far have relied on the 35S CaMV promoter, which is known to provide for constitutive expression in most plant tissues, but at fairly low (milligrams to tens of milligrams per kilogram) levels. It is not surprising that the best results have been obtained with other promoters, and we hope that in the future, researchers will use more imagination when developing expression cassettes for nuclear transformants.

Plastid transformation

The main advantages of the transplastomic plants are very high expression levels of many pharmaceutically relevant proteins and the ability of plastids to fold and assemble fully functional proteins of various classes. Transplastomics are usually stable and not subject to silencing, and because of the maternal inheritance of plastids in most crops, ecological control of the transplastomics is easier than with nuclear transformants. Among the main disadvantages is the lack of glycosylation of proteins made in plastids, the long time needed to generate a production host, and to a lesser extent, the confinement of the (high) expression to photosynthesizing tissues such as leaves. The proteins expressed in plastids are

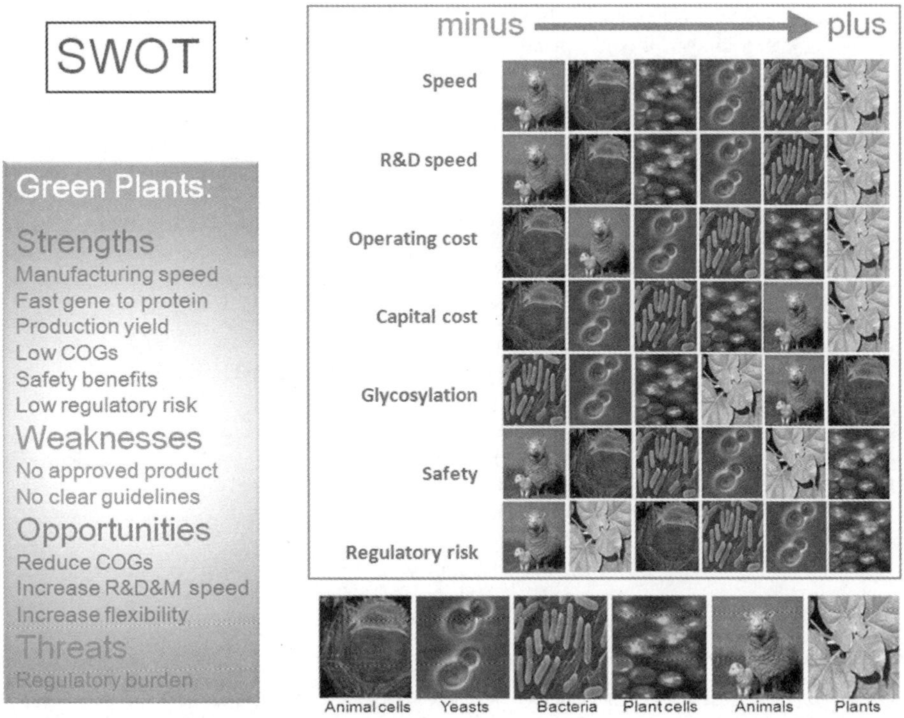

Figure 29.2 • Strengths-weaknesses-opportunities-threats (SWOT) analysis of production hosts for manufacturing biopharmaceuticals. Due to the recent advancements of the transient expression platforms and glycosylation engineering, plants are overall the most attractive manufacturing host. Please see color plate section at the end of the book.

devoid of glycans, and since most pharmaceutically relevant proteins do currently require glycosylation, plastid transformation is regarded as a valid platform for only a few biopharmaceutical classes such as vaccines and antimicrobial proteins and polypeptides, and perhaps some cytokines. The classical host organism for plastid manipulation is tobacco, and most studies aimed at production of pharmaceutical proteins have been made using this organism. Another suitable organism could be an edible leafy plant such as spinach.

Transient expression

Advantages of the transient systems are development and manufacturing speed, along with very high expression levels. Existing technologies allow the manufacture of all classes of recombinant proteins, including antibodies, proteins with designer glycosylation, and so forth. The main limitations stem from the need to deliver the genetically engineered *Agrobacterium* into the intercellular space of plants (the throughput of the current "vacuum-infiltration apparatuses" is 1 ton of green biomass per day but the process, in principle, is easily scalable) and to provide for physical containment of the bacteria. However, because of the very high expression levels provided by transient systems, the whole process can be contained in greenhouses retrofitted for the containment of plants treated with *Agrobacterium*. The manufacturing organs are leaves, and in some cases, other aerial parts of plant. The transient processes have been developed around two plant hosts: tobacco and its wild relative *N. benthamiana*. In principle, the technology works with a number of edible plants such as beets and spinach.

Conclusions and Outlook

Although major risk factors associated with the plant-based manufacturing of biopharmaceuticals have been effectively eliminated and a number of earlier anticipated advantages of PMP have indeed been demonstrated, this research and development area still faces a number of challenges. The main one is the simple fact that the green plants are just one of many manufacturing hosts available to the pharmaceutical industry and as such they do not bring unique qualitative features that the industry urgently needs. Plants are "late comers": the major companies have already invested tens of billions of dollars in microbial- and animal-cell-based manufacturing platforms. These platforms are mature (i.e., they are accepted by regulatory agencies, robust, and provide high expression levels), and the related capital investments have been to a great extent amortized. The advantages of plants are more difficult to realize in the context of the current regulations that have been "morphed" around earlier developed platforms; some of the theoretically most appealing opportunities such as use of semi-processed plant biomass as edible medicine, are not easily compatible with the current regulations, or they appear too extravagant and risky to companies and investors capable of financing such projects. Last, but not least, the cost of goods parameter has never been the main factor driving pharmaceutical companies' strategies.

One is tempted to conclude that the most optimistic scenario for PMP would be its initial adoption in developing countries such as the BRIC (Brazil-Russia-India-China) countries. For the biopharmaceutical players in the developing world, initial capital cost barriers and cost-of-goods barriers are real problems. In addition, they have not yet invested so much in animal-cell-based manufacturing — they are themselves late comers as well. They probably also have more latitude in asking regulatory bodies in their countries to look at novel processes and novel product opportunities, and stand to benefit directly from lower costs of clinical trials in their regions and faster time to market provided by some plant-based platforms. Using biosimilar (biobetter) molecules as initial targets, these companies would also be able to lower their overall risk by minimizing the product failure risk, whereas using the parameters qualifying their product as "better" to avoid the market price erosion inherent in generic competition. For all of this to happen, many remaining technical hurdles still have to be overcome, meaning that researchers active in this area will find themselves useful for many years to come. Among the main unresolved issues affecting the overall manufacturing cost of biopharmaceuticals are the costs of protein purification, especially the cost of affinity chromatography. Although new inexpensive purification processes would benefit both plant- and other cell-based manufacturing, the added low cost of manufacturing in green plants should provide an additional incentive to invest in such a new integrated (upstream plus downstream) process.

Acknowledgments

The Authors wish to thank their colleagues at Nomad Bioscience (Halle, Germany) and Icon Genetics (Halle, Germany) for valuable discussions.

References

Alvarez, M. L., Topal, E., Martin, F., & Cardineau, G. A. (2010). Higher accumulation of F1-V fusion recombinant protein in plants after induction of protein body formation. *Plant Molecular Biology*, *72*, 75–89.

Arlen, P. A., Falconer, R., Cherukumilli, S., Cole, A., Cole, A. M., & Oishi, K. K., et al. (2007). Field production and functional evaluation of chloroplast-derived interferon-alpha2b. *Plant Biotechnology Journal*, *5*, 511–525.

Arlen, P. A., Singleton, M., Adamovicz, J. J., Ding, Y., Davoodi-Semiromi, A., & Daniell, H. (2008). Effective plague vaccination via oral delivery of plant cells expressing F1-V antigens in chloroplasts. *Infection Control and Hospital Epidemiology*, *76*, 3640–3650.

Bakker, H., Bardor, M., Molthoff, J. W., Gomord, V., Elbers, I., & Stevens, L. H., et al. (2001). Galactose-extended glycans of antibodies produced by transgenic plants. *Proceedings of the National Academy of*

Sciences of the United States of America, 98, 2899–2904.

Bendandi, M., Marillonnet, S., Kandzia, R., Thieme, F., Nickstadt, A., Herz, S., et al. (2010). Rapid, high-yield production in plants of individualized idiotype vaccines for non-Hodgkin's lymphoma. *Annals of Oncology, 21*(12), 2420–2427.

Bock, R. (2007). Plastid biotechnology: prospects for herbicide and insect resistance, metabolic engineering and molecular farming. *Current Opinion in Biotechnology, 18*, 100–106.

Boothe, J., Nykiforuk, C., Shen, Y., Zaplachinski, S., Szarka, S., & Kuhlman, P., et al. (2010). Seed-based expression systems for plant molecular farming. *Plant Biotechnology Journal, 8*, 588–606.

Canizares, M. C., Lomonossoff, G. P., & Nicholson, L. (2005). Development of cowpea mosaic virus-based vectors for the production of vaccines in plants. *Expert Review of Vaccines, 4*, 687–697.

Canizares, M. C., Nicholson, L., & Lomonossoff, G. P. (2005). Use of viral vectors for vaccine production in plants. *Immunology and Cell Biology, 83*, 263–270.

Cao, Y., Zhang, Q., Wang, C., Zhu, Y., & Bai, G. (2007). Preparation of novel immunomagnetic cellulose microspheres via cellulose binding domain-protein A linkage and its use for the isolation of interferon alpha-2b. *Journal of Chromatography A, 1149*, 228–235.

Castilho, A., Strasser, R., Stadlmann, J., Grass, J., Jez, J., & Gattinger, P., et al. (2010). *In planta* protein sialylation through overexpression of the respective mammalian pathway. *Journal of Biological Chemistry, 285*, 15923–15930.

Chebolu, S., & Daniell, H. (2007). Stable expression of Gal/GalNAc lectin of Entamoeba histolytica in transgenic chloroplasts and immunogenicity in mice towards vaccine development for amoebiasis. *Plant Biotechnology Journal, 5*, 230–239.

Conley, A. J., Joensuu, J. J., Jevnikar, A. M., Menassa, R., & Brandle, J. E. (2009a). Optimization of elastin-like polypeptide fusions for expression and purification of recombinant proteins in plants. *Biotechnology and Bioengineering, 103*, 562–573.

Conley, A. J., Joensuu, J. J., Menassa, R., & Brandle, J. E. (2009b). Induction of protein body formation in plant leaves by elastin-like polypeptide fusions. *BMC Biology, 7*, 48.

Conrad, U., Plagmann, I., Malchow, S., Sack, M., Floss, D. M., Kruglov, A. A., et al. (2011). ELPylated anti-human TNF therapeutic single-domain antibodies for prevention of lethal septic shock. *Plant Biotechnology Journal, 9*(1), 22–31.

Cox, K. M., Sterling, J. D., Regan, J. T., Gasdaska, J. R., Frantz, K. K., & Peele, C. G., et al. (2006). Glycan optimization of a human monoclonal antibody in the aquatic plant Lemna minor. *Nature Biotechnology, 24*, 1591–1597.

Daniell, H., Ruiz, G., Denes, B., Sandberg, L., & Langridge, W. (2009). Optimization of codon composition and regulatory elements for expression of human insulin like growth factor-1 in transgenic chloroplasts and evaluation of structural identity and function. *BMC Biotechnology, 9*, 33.

Daniell, H., Singh, N. D., Mason, H., & Streatfield, S. J. (2009). Plant-made vaccine antigens and biopharmaceuticals. *Trends in Plant Science, 14*, 669–679.

Davoodi-Semiromi, A., Samson, N., & Daniell, H. (2009). The green vaccine: A global strategy to combat infectious and autoimmune diseases. *Human Vaccines, 5*, 488–493.

Davoodi-Semiromi, A., Schreiber, M., Nalapalli, S., Verma, D., Singh, N. D., & Banks, R. K., et al. (2010). Chloroplast-derived vaccine antigens confer dual immunity against cholera and malaria by oral or injectable delivery. *Plant Biotechnology Journal, 8*, 223–242.

De Jaeger, G., Scheffer, S., Jacobs, A., Zambre, M., Zobell, O., & Goossens, A., et al. (2002). Boosting heterologous protein production in transgenic dicotyledonous seeds using Phaseolus vulgaris regulatory sequences. *Nature Biotechnology, 20*, 1265–1268.

De Muynck, B., Navarre, C., & Boutry, M. (2010). Production of antibodies in plants: Status after twenty years. *Plant Biotechnology Journal, 8*, 529–563.

Decker, E. L., & Reski, R. (2008). Current achievements in the production of complex biopharmaceuticals with moss bioreactors. *Bioprocess and Biosystems Engineering, 31*, 3–9.

Donson, J., Kearney, C. M., Hilf, M. E., & Dawson, W. O. (1991). Systemic expression of a bacterial gene by a tobacco mosaic virus-based vector. *Proceedings of the National Academy of Sciences of the United States of America, 88*, 7204–7208.

Dorokhov, Y. L., Sheveleva, A. A., Frolova, O. Y., Komarova, T. V., Zvereva, A. S., & Ivanov, P. A., et al. (2007). Superexpression of tuberculosis antigens in plant leaves. *Tuberculosis (Edinb), 87*, 218–224.

Fernandez-San Millan, A., Ortigosa, S. M., Hervas-Stubbs, S., Corral-Martinez, P., Segui-Simarro, J. M., & Gaetan, J., et al. (2008). Human papillomavirus L1 protein expressed in tobacco chloroplasts self-assembles into virus-like particles that are highly immunogenic. *Plant Biotechnology Journal, 6*, 427–441.

Fischer, R., Stoger, E., Schillberg, S., Christou, P., & Twyman, R. M. (2004). Plant-based production of biopharmaceuticals. *Current Opinion in Plant Biology, 7*, 152–158.

Floss, D. M., Sack, M., Arcalis, E., Stadlmann, J., Quendler, H., & Rademacher, T., et al. (2009). Influence of elastin-like peptide fusions on the quantity and quality of a tobacco-derived human immunodeficiency virus-neutralizing antibody. *Plant Biotechnology Journal, 7*, 899–913.

Floss, D. M., Sack, M., Stadlmann, J., Rademacher, T., Scheller, J., & Stoger, E., et al. (2008). Biochemical and functional characterization of anti-HIV antibody-ELP fusion proteins from transgenic plants. *Plant Biotechnology Journal, 6*, 379–391.

Floss, D. M., Schallau, K., Rose-John, S., Conrad, U., & Scheller, J. (2010). Elastin-like polypeptides revolutionize recombinant protein expression and their biomedical application. *Trends in Biotechnology, 28*, 37–45.

Fujiki, M., Kaczmarczyk, J. F., Yusibov, V., & Rabindran, S. (2008). Development of a new cucumber mosaic virus-based plant expression vector with truncated 3a movement protein. *Virology, 381*, 136–142.

Gils, M., Kandzia, R., Marillonnet, S., Klimyuk, V., & Gleba, Y. (2005). High-yield production of authentic human growth hormone using a plant virus-based expression system. *Plant Biotechnology Journal, 3*, 613–620.

Giritch, A., Marillonnet, S., Engler, C., van Eldik, G., Botterman, J., & Klimyuk, V., et al. (2006). Rapid high-yield expression of full-size IgG antibodies in plants coinfected with noncompeting viral vectors. *Proceedings of the National Academy of Sciences of the United States of America, 103*, 14701–14707.

Gleba, Y., Klimyuk, V., & Marillonnet, S. (2005). Magnifection — a new platform for expressing recombinant vaccines in plants. *Vaccine, 23*, 2042–2048.

Gleba, Y., Klimyuk, V., & Marillonnet, S. (2007). Viral vectors for the expression of proteins in plants. *Current Opinion in Biotechnology, 18*, 134–141.

Gleba, Y., Marillonnet, S., & Klimyuk, V. (2004). Engineering viral expression vectors for plants: The "full virus" and the "deconstructed virus" strategies. *Current Opinion in Plant Biology, 7*, 182–188.

Gleba, Y., Marillonnet, S., & Klimyuk, V. (2007). Plant virus vectors (gene expression systems) (3rd ed.). In M. H. V. van Regenmortel & B. W. J. Mahy (Eds.), *Encyclopedia of virology* (Vol. 4, pp. 229–237). San Diego: Academic Press.

Glenz, K., Bouchon, B., Stehle, T., Wallich, R., Simon, M. M., & Warzecha, H. (2006). Production of a recombinant bacterial lipoprotein in higher plant chloroplasts. *Nature Biotechnology, 24*, 76–77.

Gorr, G., & Altmann, F. (2006). Glycosylation of recombinant proteins in plants. In O. Kayser & W. Quax (Eds.), *Medicinal Plant Biotechnology* (Vol. 2, pp. 345–374). Wiley-VCH.

Green, B. J., Fujiki, M., Mett, V., Kaczmarczyk, J., Shamloul, M., & Musiychuk, K., et al. (2009). Transient protein expression in three *Pisum sativum* (green pea) varieties. *Biotechnology Journal, 4*, 230–237.

Hellwig, S., Drossard, J., Twyman, R. M., & Fischer, R. (2004). Plant cell cultures for the production of recombinant proteins. *Nature Biotechnology, 22*, 1415–1422.

Hennegan, K., Yang, D., Nguyen, D., Wu, L., Goding, J., & Huang, J., et al. (2005). Improvement of human lysozyme expression in transgenic rice grain by combining wheat (*Triticum aestivum*) puroindoline b and rice (*Oryza sativa*) Gt1 promoters and signal peptides. *Transgenic Research, 14*, 583–592.

Hood, E. E., Love, R., Lane, J., Bray, J., Clough, R., & Pappu, K., et al. (2007). Subcellular targeting is a key condition for high-level accumulation of cellulase protein in transgenic maize seed. *Plant Biotechnology Journal, 5*, 709–719.

Huang, J., Nandi, S., Wu, L., Yalda, D., Bartley, G., & Rodriguez, R., et al. (2002). Expression of natural antimicrobial human lysozyme in rice grains. *Molecular Breeding, 10*, 83–94.

Huang, Z., Chen, Q., Hjelm, B., Arntzen, C., & Mason, H. (2009a). A DNA replicon system for rapid high-level production of virus-like particles in plants. *Biotechnology and Bioengineering, 103*, 706–714.

Huang, Z., Phoolcharoen, W., Lai, H., Piensook, K., Cardineau, G., Zeitlin, L., et al. (2009b). High-level rapid production of full-size monoclonal antibodies in plants by a single-vector DNA replicon system. *Biotechnology and Bioengineering, 106*(1), 9–17.

Huang, Z., Santi, L., LePore, K., Kilbourne, J., Arntzen, C. J., & Mason, H. S. (2006). Rapid, high-level production of hepatitis B core antigen in plant leaf and its immunogenicity in mice. *Vaccine, 24*, 2506–2513.

Huhns, M., Neumann, K., Hausmann, T., Klemke, F., Lockau, W., & Kahmann, U., et al. (2009). Tuber-specific cphA expression to enhance cyanophycin production in potatoes. *Plant Biotechnology Journal, 7*, 883–898.

Huhns, M., Neumann, K., Hausmann, T., Ziegler, K., Klemke, F., & Kahmann, U., et al. (2008). Plastid targeting strategies for cyanophycin synthetase to achieve high-level polymer accumulation in *Nicotiana tabacum*. *Plant Biotechnology Journal, 6*, 321–336.

Hussack, G., Grohs, B. M., Almquist, K. C., McLean, M. D., Ghosh, R., & Hall, J. C. (2010). Purification of plant-derived antibodies through direct immobilization of affinity ligands on cellulose. *Journal of Agricultural and Food Chemistry, 58*, 3451–3459.

Kalantidis, K., Tsagris, M., & Tabler, M. (2006). Spontaneous short-range silencing of a GFP transgene in *Nicotiana benthamiana* is possibly mediated by small quantities of siRNA that do not trigger systemic silencing. *Plant Journal, 45*, 1006–1016.

Karg, S. R., & Kallio, P. T. (2009). The production of biopharmaceuticals in plant systems. *Biotechnology Advances, 27*, 879–884.

Ko, K., Brodzik, R., & Steplewski, Z. (2009). Production of antibodies in plants: Approaches and perspectives. *Current Topics in Microbiology and Immunology, 332*, 55–78.

Kogan, M. J., Dalcol, I., Gorostiza, P., Lopez-Iglesias, C., Pons, M., & Sanz, F., et al. (2001). Self-assembly of the amphipathic helix (VHLPPP)8. A mechanism for zein protein body formation. *Journal of Molecular Biology, 312*, 907–913.

Koya, V., Moayeri, M., Leppla, S. H., & Daniell, H. (2005). Plant-based vaccine: Mice immunized with chloroplast-derived anthrax protective antigen survive anthrax lethal toxin challenge. *Infection Control and Hospital Epidemiology, 73*, 8266–8274.

Kumagai, M. H., Donson, J., della-Cioppa, G., & Grill, L. K. (2000). Rapid, high-level expression of glycosylated rice alpha-amylase in transfected plants by an RNA viral vector. *Gene, 245*, 169–174.

Kumagai, M. H., Donson, J., della-Cioppa, G., Harvey, D., Hanley, K., & Grill, L. K. (1995). Cytoplasmic inhibition of carotenoid biosynthesis with virus-derived RNA. *Proceedings of the National Academy of Sciences of the United States of America, 92*, 1679–1683.

Kumagai, M. H., Turpen, T. H., Weinzettl, N., della-Cioppa, G., Turpen, A. M., & Donson, J., et al. (1993). Rapid, high-level expression of biologically active alpha-trichosanthin in transfected plants by an RNA viral vector. *Proceedings of the National Academy of Sciences of the United States of America, 90*, 427–430.

Kushwaha, A., Chowdhury, P. S., Arora, K., Abrol, S., & Chaudhary, V. K. (1994). Construction and characterization of M13 bacteriophages displaying functional IgG-binding domains of staphylococcal protein A. *Gene, 151*, 45–51.

Lico, C., Chen, Q., & Santi, L. (2008). Viral vectors for production of recombinant proteins in plants. *Journal of Cellular Physiology, 216*, 366–377.

Lindbo, J. A. (2007). High-efficiency protein expression in plants from agroinfection-compatible Tobacco mosaic virus expression vectors. *BMC Biotechnology, 7*, 52.

Lindbo, J. A. (2007). TRBO: A high-efficiency tobacco mosaic virus RNA-based overexpression vector. *Plant Physiology, 145*, 1232–1240.

Maclean, J., Koekemoer, M., Olivier, A. J., Stewart, D., Hitzeroth, II, & Rademacher, T., et al. (2007). Optimization of human papillomavirus type 16 (HPV-16) L1 expression in plants: Comparison of the suitability of different HPV-16 L1 gene variants and different cell-compartment localization. *Journal of General Virology, 88*, 1460–1469.

Marillonnet, S., Giritch, A., Gils, M., Kandzia, R., Klimyuk, V., & Gleba, Y. (2004). *In planta* engineering of viral RNA replicons: Efficient assembly by recombination of DNA modules delivered by *Agrobacterium*. *Proceedings of the National Academy of Sciences of the United States of America, 101*, 6852–6857.

Marillonnet, S., Thoeringer, C., Kandzia, R., Klimyuk, V., & Gleba, Y. (2005). Systemic *Agrobacterium tumefaciens*-mediated transfection of viral replicons for efficient transient expression in plants. *Nature Biotechnology, 23*, 718–723.

McCabe, M. S., Klaas, M., Gonzalez-Rabade, N., Poage, M., Badillo-Corona, J. A., & Zhou, F., et al. (2008). Plastid transformation of high-biomass tobacco variety Maryland Mammoth for production of human immunodeficiency virus type 1 (HIV-1) p24 antigen. *Plant Biotechnology Journal, 6*, 914–929.

McCormick, A. A., & Palmer, K. E. (2008). Genetically engineered Tobacco mosaic virus as nanoparticle vaccines. *Expert Review of Vaccines, 7*, 33–41.

McCormick, A. A., Kumagai, M. H., Hanley, K., Turpen, T. H., Hakim, I., & Grill, L. K., et al. (1999). Rapid production of specific vaccines for lymphoma by expression of the tumor-derived single-chain Fv epitopes in tobacco plants. *Proceedings of the National Academy of Sciences of the United States of America, 96*, 703–708.

McCormick, A. A., Reddy, S., Reinl, S. J., Cameron, T. I., Czerwinkski, D. K., & Vojdani, F., et al. (2008). Plant-produced idiotype vaccines for the treatment of non-Hodgkin's lymphoma: Safety and immunogenicity in a phase I clinical study. *Proceedings of the National Academy of Sciences of the United States of America, 105*, 10131–10136.

McCormick, A. A., Reinl, S. J., Cameron, T. I., Vojdani, F., Fronefield, M., & Levy, R., et al. (2003). Individualized human scFv vaccines produced in plants: Humoral anti-idiotype responses in vaccinated mice confirm relevance to the tumor Ig. *Journal of Immunological Methods, 278*, 95–104.

Mett, V., Farrance, C. E., Green, B. J., & Yusibov, V. (2008). Plants as biofactories. *Biologicals, 36*, 354–358.

Molina, A., Veramendi, J., & Hervas-Stubbs, S. (2005). Induction of neutralizing antibodies by a tobacco chloroplast-derived vaccine based on a B cell epitope from canine parvovirus. *Virology, 342*, 266–275.

Nandi, S., Yalda, D., Lu, S., Nikolov, Z., Misaki, R., & Fujiyama, K., et al. (2005). Process development and economic evaluation of recombinant human lactoferrin expressed in rice grain. *Transgenic Research, 14*, 237–249.

Nechansky, A., Schuster, M., Jost, W., Siegl, P., Wiederkum, S., & Gorr, G., et al. (2007). Compensation of endogenous IgG mediated inhibition of antibody-dependent cellular cytotoxicity by glyco-engineering of therapeutic antibodies. *Molecular Immunology, 44*, 1815–1817.

Nomellini, J. F., Duncan, G., Dorocicz, I. R., & Smit, J. (2007). S-layer-mediated display of the immunoglobulin G-binding domain of streptococcal protein G on the surface of *Caulobacter crescentus*: Development of an immunoactive reagent. *Applied and Environmental Microbiology, 73*, 3245–3253.

Oey, M., Lohse, M., Kreikemeyer, B., & Bock, R. (2009). Exhaustion of the chloroplast protein synthesis capacity by massive expression of a highly stable protein antibiotic. *Plant Journal, 57*, 436–445.

Oey, M., Lohse, M., Scharff, L. B., Kreikemeyer, B., & Bock, R. (2009). Plastid production of protein antibiotics against pneumonia via a new strategy for high-level expression of antimicrobial proteins. *Proceedings of the National Academy of Sciences of the United States of America, 106*, 6579–6584.

O'Keefe, B. R., Vojdani, F., Buffa, V., Shattock, R. J., Montefiori, D. C., & Bakke, J., et al. (2009). Scaleable manufacture of HIV-1 entry inhibitor griffithsin and validation of its safety and efficacy as a topical microbicide component. *Proceedings of the National Academy of Sciences of the United States of America, 106*, 6099–6104.

Outchkourov, N. S., Peters, J., de Jong, J., Rademakers, W., & Jongsma, M. A. (2003). The promoter-terminator of chrysanthemum rbcS1 directs very high expression levels in plants. *Planta*, *216*, 1003–1012.

Palacpac, N. Q., Yoshida, S., Sakai, H., Kimura, Y., Fujiyama, K., & Yoshida, T., et al. (1999). Stable expression of human beta1, 4-galactosyltransferase in plant cells modifies N-linked glycosylation patterns. *Proceedings of the National Academy of Sciences of the United States of America*, *96*, 4692–4697.

Plasson, C., Michel, R., Lienard, D., Saint-Jore-Dupas, C., Sourrouille, C., & de March, G. G., et al. (2009). Production of recombinant proteins in suspension-cultured plant cells. *Methods in Molecular Biology*, *483*, 145–161.

Rabindran, S., & Dawson, W. O. (2001). Assessment of recombinants that arise from the use of a TMV-based transient expression vector. *Virology*, *284*, 182–189.

Roque, A. C., Lowe, C. R., & Taipa, M. A. (2004). Antibodies and genetically engineered related molecules: Production and purification. *Biotechnology Progress*, *20*, 639–654.

Rybicki, E. P. (2009). Plant-produced vaccines: Promise and reality. *Drug Discovery Today*, *14*, 16–24.

Sainsbury, F., & Lomonossoff, G. P. (2008). Extremely high-level and rapid transient protein production in plants without the use of viral replication. *Plant Physiology*, *148*, 1212–1218.

Sainsbury, F., Liu, L., & Lomonossoff, G. P. (2009). Cowpea mosaic virus-based systems for the expression of antigens and antibodies in plants. *Methods in Molecular Biology*, *483*, 25–39.

Santi, L., Batchelor, L., Huang, Z., Hjelm, B., Kilbourne, J., & Arntzen, C. J., et al. (2008). An efficient plant viral expression system generating orally immunogenic Norwalk virus-like particles. *Vaccine*, *26*, 1846–1854.

Santi, L., Giritch, A., Roy, C. J., Marillonnet, S., Klimyuk, V., & Gleba, Y., et al. (2006). Protection conferred by recombinant Yersinia pestis antigens produced by a rapid and highly scalable plant expression system. *Proceedings of the National Academy of Sciences of the United States of America*, *103*, 861–866.

Santi, L., Huang, Z., & Mason, H. (2006). Virus-like particles production in green plants. *Methods*, *40*, 66–76.

Schahs, M., Strasser, R., Stadlmann, J., Kunert, R., Rademacher, T., & Steinkellner, H. (2007). Production of a monoclonal antibody in plants with a humanized N-glycosylation pattern. *Plant Biotechnology Journal*, *5*, 657–663.

Schuster, M., Jost, W., Mudde, G. C., Wiederkum, S., Schwager, C., & Janzek, E., et al. (2007). *In vivo* glyco-engineered antibody with improved lytic potential produced by an innovative non-mammalian expression system. *Biotechnology Journal*, *2*, 700–708.

Shivprasad, S., Pogue, G. P., Lewandowski, D. J., Hidalgo, J., Donson, J., & Grill, L. K., et al. (1999). Heterologous sequences greatly affect foreign gene expression in tobacco mosaic virus-based vectors. *Virology*, *255*, 312–323.

Smith, M. L., Fitzmaurice, W. P., Turpen, T. H., & Palmer, K. E. (2009). Display of peptides on the surface of tobacco mosaic virus particles. *Current Topics in Microbiology and Immunology*, *332*, 13–31.

Strasser, R., Stadlmann, J., Schahs, M., Stiegler, G., Quendler, H., & Mach, L., et al. (2008). Generation of glyco-engineered *Nicotiana benthamiana* for the production of monoclonal antibodies with a homogeneous human-like N-glycan structure. *Plant Biotechnology Journal*, *6*, 392–402.

Torrent, M., Llompart, B., Lasserre-Ramassamy, S., Llop-Tous, I., Bastida, M., & Marzabal, P., et al. (2009). Eukaryotic protein production in designed storage organelles. *BMC Biology*, *7*, 5.

Torrent, M., Llop-Tous, I., & Ludevid, M. D. (2009). Protein body induction: A new tool to produce and recover recombinant proteins in plants. *Methods in Molecular Biology*, *483*, 193–208.

Tregoning, J. S., Nixon, P., Kuroda, H., Svab, Z., Clare, S., & Bowe, F., et al. (2003). Expression of tetanus toxin Fragment C in tobacco chloroplasts. *Nucleic Acids Research*, *31*, 1174–1179.

Turpen, T. H. (1999). Tobacco mosaic virus and the virescence of biotechnology. *Philosophical Transactions of the Royal Society of London Series B, Biological Sciences*, *354*, 665–673.

Turpen, T. H., Reinl, S. J., Charoenvit, Y., Hoffman, S. L., Fallarme, V., & Grill, L. K. (1995). Malarial epitopes expressed on the surface of recombinant tobacco mosaic virus. *Biotechnology (NY)*, *13*, 53–57.

Vancanneyt, G., Dubald, M., Schroder, W., Peters, J., & Botterman, J. (2009). A case study for plant-made pharmaceuticals comparing different plant expression and production systems. *Methods in Molecular Biology*, *483*, 209–221.

Verwoerd, T. C., van Paridon, P. A., van Ooyen, A. J., van Lent, J. W., Hoekema, A., & Pen, J. (1995). Stable accumulation of Aspergillus niger phytase in transgenic tobacco leaves. *Plant Physiology*, *109*, 1199–1205.

Vezina, L. P., Faye, L., Lerouge, P., D'Aoust, M. A., Marquet-Blouin, E., & Burel, C., et al. (2009). Transient co-expression for fast and high-yield production of antibodies with human-like N-glycans in plants. *Plant Biotechnology Journal*, *7*, 442–455.

Vollenkle, C., Weigert, S., Ilk, N., Egelseer, E., Weber, V., & Loth, F., et al. (2004). Construction of a functional S-layer fusion protein comprising an immunoglobulin G-binding domain for development of specific adsorbents for extracorporeal blood purification. *Applied and Environmental Microbiology*, *70*, 1514–1521.

Watson, J., Koya, V., Leppla, S. H., & Daniell, H. (2004). Expression of Bacillus anthracis protective antigen in transgenic chloroplasts of tobacco, a non-food/feed crop. *Vaccine*, *22*, 4374–4384.

Webster, D. E., Wang, L., Mulcair, M., Ma, C., Santi, L., & Mason, H. S., et al. (2009). Production and characterization of an orally immunogenic Plasmodium antigen in plants using a virus-based expression system. *Plant Biotechnology Journal*, *7*, 846–855.

Weise, A., Altmann, F., Rodriguez-Franco, M., Sjoberg, E. R., Baumer, W., & Launhardt, H., et al. (2007). High-level expression of secreted complex glycosylated recombinant human erythropoietin in the Physcomitrella Delta-fuc-t Delta-xyl-t mutant. *Plant Biotechnology Journal*, *5*, 389–401.

Werner, S., Marillonnet, S., Hause, G., Klimyuk, V., & Gleba, Y. (2006). Immunoabsorbent nanoparticles based on a tobamovirus displaying protein A. *Proceedings of the National Academy of Sciences of the United States of America*, *103*, 17678–17683.

Yusibov, V., & Rabindran, S. (2008). Recent progress in the development of plant derived vaccines. *Expert Review of Vaccines*, *7*, 1173–1183.

Zvereva, A. S., Petrovskaya, L. E., Rodina, A. V., Frolova, O. Y., Ivanov, P. A., & Shingarova, L. N., et al. (2009). Production of biologically active human myelocytokines in plants. *Biochemistry (Mosc)*, *74*, 1187–1194.

Plants as factories for bioplastics and other novel biomaterials

30

Jan B. van Beilen Yves Poirier

Département de Biologie Moléculaire Végétale, Université de Lausanne, Lausanne, Switzerland

TABLE OF CONTENTS

Introduction

Well into the nineteenth century, biomass was a main source of materials and energy in most countries. The rapid development of petrochemistry in the twentieth century led to a shift from biomass to fossil carbon (coal, oil, gas) as the major source of fuels as well as of building blocks for the generation of a wide variety of chemicals and materials. The worldwide share of biomass in energy production is around 10% and biomass still plays a significant role in some OECD countries, such as Sweden and Austria. However, fossil carbon is currently the dominant source of both energy and the materials on which our society depend.

Climate change and resource constraints now force us to reconsider the use of plant biomass as a source of materials and fuels. However, it is well known that prior to the nineteenth century, overuse of biomass led to deforestation, soil degeneration, and desertification. In the early twenty-first century, the world fed, clothed, and sheltered almost 7 billion people instead of the 1 billion population at the start of the nineteenth century, while increased wealth, changed habits, and excessive consumption drastically increased the per capita use of energy and materials. At the same time, increasing droughts and changes in weather patterns linked to climate change have reversed the decades-long trend of increased global plant growth, according to a new analysis of NASA satellite data (Zhao and Running, 2010); ocean acidification threatens to destroy marine ecosystems providing food, feed, and materials (Fabry et al., 2008); global biodiversity is collapsing (Ehrlich and Pringle, 2008); and human appropriation of net primary production (fraction of the net amount of solar energy converted to plant organic matter through photosynthesis, used or lost by human activity, HANPP) may have already reached 50% (Erb et al., 2009). The inescapable conclusion is that biomass is a limited resource that should be put to the best (and most efficient) use possible.

Since mankind now depends on fossil fuels for 80–90% of its energy demands and the present level of biomass use is already critically high, biomass cannot provide more than a few percent of present fuel consumption, especially if the greenhouse gas balance and other unwanted side effects (soil degradation, water use, eutrophication, loss of nature, loss of jobs) are taken into account. It is evident that bioenergy will not be a major solution to our energy and climate change predicament; energy (carriers) for transport and heating will have to come from other, non-biological, sources.

The story is different for carbon-based materials, because most major polymers and chemicals depend on a source of carbon. If fossil fuels are no longer available, affordable, or their use politically acceptable, the chemical industry must become based on heterogeneous biomass sources (agricultural and forestry waste) or well-defined biomass fractions (sugars, starch, cellulose, vegetable oils, etc.). The first option is complicated by the necessity to maintain soil carbon content and fertility, while the latter competes with the food market as it

DOI: 10.1016/B978-0-12-381466-1.00030-4

is based on existing crops. A completely different scenario is based on the assumption that there will be sufficient oil available for chemicals and polymers if it is no longer squandered as fuel. Ultimately, the question is one of deciding between scenarios, and life cycle analysis is essential in deciding what works best.

Several recent reviews have focused on the specific products (chemicals, polymers, bioactives) that are produced in plants (van Beilen, 2008; Sharma and Sharma, 2009; Bornke and Broer, 2010). This chapter focuses on the biopolymers. The first section covers major biopolymers based on naturally produced plant polymers that are applied widely in industry. The second section discusses efforts to produce new polymers in transgenic plants, discussing their promise and potential shortcomings. The multitude of minor plant and algal biopolymers is not discussed here.

Major Natural Plant Biopolymers

The following sections discuss natural plant biopolymers that are used on a large scale in industry for material application. Major plant polymers that are mainly used for energy (e.g., lignin, hemicellulose), in food applications (e.g., inulin, pectin), or in their original state (e.g., suberin in cork) are not discussed.

Starch

Starch is a ubiquitous and abundant biological energy storage polymer that is used on a very large scale in feed, food, fuel, and materials. Eighty percent of the calories in our diet come from starch; it is the second major agricultural commodity after cellulose, and it is the least expensive processed food commodity at around \$0.5/kg. Non-food applications of starch already amounted to 3.6 million tons per year or about 13% of the total starch market in the EU15 (van Beilen and Poirier, 2007a). Accordingly, biosynthesis, genetics, properties, and applications of starch are well characterized.

Starch can be converted to a thermoplastic, and currently about 50,000 tons/year of starch is converted to plastic materials by a range of small and large companies worldwide (van Beilen and Poirier, 2007a). Most of this bioplastic is marketed as biodegradable, and it is used for packaging films and foams and for disposables such as cups and plates, plant pots, and bags. The growth potential of this market is high, with studies referring to future market sizes in the range of 1,000,000 tons/year (Crank et al., 2004). However, polymers made primarily of starch have relatively poor physical properties that can be improved by the addition of various plasticizers and copolymers, but there are still major limitations. Starch-based polymers typically have low resistance to stress, are sensitive to moisture, and the glucoside links start to break at temperatures higher than 150°C. Fortunately, the multiple −OH groups of the starch molecule are susceptible to substitution reactions, for example, with acetate, producing a more hydrophobic polymer retaining its tensile properties in aqueous environments, or with silane, which improves dispersion in polyethylene. Thermoplastic starch can be blended with polycaprolactone or coated with a water barrier (Bastioli, 2005) to produce more water-resistant materials. Blending with fibers such as cellulose, or fillers such as lignin and clay, yields nano- or biocomposites, which may have markedly improved mechanical properties, better thermal resistance, reduced water sensitivity, and better post-processing aging (Averous and Halley, 2009).

Because the native starches of many crops do not have ideal properties for specific food and non-food applications, efforts to create transgenic crops with altered starches or starch metabolism are ongoing (Jobling, 2004). Yet, as the authors of a recent review stated: "despite this remarkable progress and the obvious economic importance, very little has been achieved in terms of adding value to starch or increasing starch yield, particularly in cereal crops" (Keeling and Myers, 2010). Many plants have been bred and engineered for altered starch composition, resulting in high- and low-amylose starches, amylose-free amylopectin (*waxy* crops), long chain amylopectin starch, highly branched amylopectin, and phosphorylated amylopectin. Although most of the research on the *in planta* modification of starch was carried out with food applications in mind, some of these modified starches have new and potentially useful properties helping their applications as thermoplastic starches (Averous and Halley, 2009). However, the altered starches also still need further chemical or physical treatments for optimal functionality.

Two major reasons for a lack of progress are the great number of (starch biosynthesis) enzyme variants that can be generated, and the lack of understanding of the relationship between a given starch structure and its physiochemical properties. To test the properties of starch, significant quantities are necessary, which is costly for large numbers of genotypes (Zeeman et al., 2010). Furthermore, there are potential deleterious consequences of starch modifications on the plant's physiology. For example, starch granule structure is likely to be affected by the extensive modifications, which in turn will affect the amount of starch produced by the plant. Seed germination and other agronomic properties of the crop may be affected. On another level, starch mutants produced by genetic engineering were deemed too problematic due to a lack of public acceptance and regulatory hurdles, especially in Europe. For example, only after waiting for 13 years, the European Commission finally approved AMFLORA of BASF, a transgenic potato producing a pure amylopectin starch. Thus, companies such as AVEBE have gone to great lengths to reproduce the desired mutations in the amylopectin potato by classical breeding, resulting in the non-genetically-modified amylopectin potato ELIANE (www.avebe.com). At least in Europe, other transgenic plants face similar hurdles.

Cellulose

In its unmodified form, cellulose is the dominant component of many major agricultural and forestry commodities, such as timber, paper, board, cotton, linen, and many other fiber products. Purified cellulose is produced mainly from the bleached wood pulp resulting from the sulfite or Kraft process, which removes most of the associated hemicellulose, lignin, pectin, and other compounds. Cellulose as isolated cannot be used for plastics, because it is crystalline, has a stiff rod-like conformation, and the individual chains are too tightly connected by hydrogen bonds. Unmodified, cellulose fibers are used as a reinforcing agent in biocomposites, improving tensile and flexural modus (Averous and Halley, 2009).

Various chemical modifications or treatments have enabled a wide range of major applications for cellulose (Mooney, 2009). Nitrocellulose combined with camphor and other components yielded the first thermoplastic, known as Parkesine and Celluloid, which was used as an ivory replacement. Because of its flammability and easy decomposition, only minor applications such as guitar picks and table tennis balls survived. Cellophane is made of restructured cellulose produced by dissolving it in base and carbon disulfide (the solution is known as viscose), extruding through a slit into dilute sulfuric acid and sodium sulfate, washing, and then adding glycerol. The resulting film is still widely used for food packaging. Extruding through a spinneret (small hole) produces rayon fibers. Lyocell or Tencel is a similar material produced using a more environmentally friendly process, in which the wood pulp is dissolved in N-methylmorpholine N-oxide, which is recovered after spinning the fibers. However, the process still consumes solvents and is energy intensive. Cellulose-acetate, resulting from a treatment of purified cellulose with glacial acetic acid, is used in frames for eyeglasses, adhesives, photography film, and many other applications amounting to about 750,000 tons/year. Other derivatives are carboxymethyl cellulose (CMC), methyl cellulose (MC), hydroxypropyl cellulose (HPC), and hydroxyethyl cellulose.

Most of the bulk applications of cellulose have not survived the advent of oil-based alternatives, often because of the high cost of purified cellulose and the chemical or physical modification processes. The high cost of cellulose is due to the processing methods; it is difficult and costs a lot of energy to separate cellulose fibers from lignin, hemicellulose, and other compounds present in the woodchips used. A century of research and industrial developments has brought improvements, but has not solved these issues.

Rubber

Natural rubber (hereafter simply referred to as rubber, in contrast to synthetic rubber) is a remarkable polymer consisting of isoprene units linked together in a 1,4-*cis* configuration. Although rubber is produced in over 2500 plant species, commercial rubber is currently almost exclusively harvested from one biological source, the Brazilian Para rubber tree (*Hevea brasiliensis*). Nearly 90% of all rubber is harvested in Southeast Asia, and the major producer countries include Indonesia, Thailand, and Malaysia. Rubber is used at nearly 10^7 tons/year, or about 1.5 times more than the production of synthetic rubber, despite its labor-intensive production, by manually tapping the latex from the rubber trees. This occurs because of the superior properties of natural rubber: efficient heat dispersion, superior resilience, elasticity, resealing after puncture, abrasion and impact resistance, and malleability at cold temperatures, properties that are a function of its enantiopure structure, high molecular weight, and additional — although rather ill-defined — components present in the latex, such as proteins, lipids, carbohydrates, and minerals (Cataldo, 2000; Cornish, 2001a). Thus, natural rubber is almost irreplaceable in many applications, such as heavy-duty tires for trucks, buses, and airplanes, as well as in latex products for medical applications (Cornish, 2001a). Synthetic rubbers (e.g., styrene-butadiene or acrylonitrile-butadiene copolymers) are also produced at a scale of several million tons per year, but cannot match the price–performance ratio of natural rubber. Whereas some plants can produce other types of polyisoprenes, none of them have material properties or value similar to *Hevea* rubber; for example, gutta-percha or balata is a flexible but inelastic material from *Palaquium gutta* and *Manilkara bidentata*, respectively, made of isoprene units linked together in a 1,4-*trans* configuration (Polhamus, 1962). The inertness of gutta-percha or balata to biodegradation made them useful as an impermeable coating for undersea cables, and gutta-percha is still used in dentistry as a filling material. Chicle is a polyisoprene from *Achras sapota* containing both *cis* and *trans* bonds in a 1:2 ratio (Polhamus, 1962), which is still used in some chewing gums.

Because rubber is an essential material largely produced in one region of the world, it is considered as a strategic commodity. Despite this fact, the biosynthesis of rubber still is only partly characterized (Cornish, 2001b). Rubber synthesis starts with cytosolic acetyl-CoA, which is converted via 3-hydroxy-glutaryl-CoA to isopentenyl diphosphate (IPP). Farnesyl diphosphate is the likely physiological initiator molecule for rubber synthesis (Stubbe et al., 2005). The rubber polymerase, or *cis*-prenyltransferase, is thought to be embedded in the membrane monolayer surrounding the rubber granules, and adds isoprenyl units from IPP to form the rubber polymer. A major open question is the nature of the protein (or protein complex) required for the synthesis of high molecular weight rubber. A cloned *cis*-prenyltransferase from *H. brasiliensis* expressed in *Escherichia coli* was found to catalyze the formation of only short chain *cis*-polyisoprenes (Ko et al., 2003), while long chain rubber polymer was only synthesized when the purified *cis*-prenyltransferase was combined with washed rubber particles (i.e., the rubber granules with associated proteins and lipids; Asawatreratanakul et al., 2003). Thus, it is not clear whether the rubber particles contain the sought after long chain rubber polymerase, or whether another protein or co-factor present on the rubber granules aids the expressed *cis*-prenyltransferase in producing long chain rubber. Sequence comparisons of dominant proteins attached to rubber particles in different plant species producing rubber do not provide clear answers, as between species these proteins differ strongly in number, size, expression level, and predicted function (Cornish, 2001a; van Beilen and Poirier, 2007b).

The production of rubber from *H. brasiliensis* is threatened by the pathogen *Microcyclus ulei*, the causative agent of South American leaf blight (SALB; Lieberei, 2007; Le Guen et al., 2007), and attempts to create resistant trees by breeding have not been very successful until now. SALB, a pathogen endemic to the Amazon basin, has made large-scale *Hevea* rubber production in South America almost impossible. The rubber tree used in commercial production in Southeast Asia has a very narrow genetic basis, coming essentially from a few seeds collected at one location in Brazil by Dr. Henry Wickham in 1876 (Polhamus, 1962). Rubber trees in Southeast Asian plantations are thus sensitive to SALB and would be devastated by the establishment of the pathogen (Davis, 1997). Currently, the only barrier to prevent the spread of SALB to Asia is a strict quarantine. It is thus essential to further understand the biology of SALB on both the pathogen and host response sides to devise adequate defense strategies, which may, in the long term, involve the development of a genetically modified rubber tree. Efficient transformation of calli and regeneration of plants was shown to be possible (Blanc et al., 2006). However, the narrow genetic base, prolonged breeding cycles and juvenile period, and highly heterozygous nature of *H. brasiliensis* make breeding complex, time-consuming, and labor-intensive (Lieberei, 2007).

World production of rubber is also influenced by other factors, such as changes in weather patterns or socioeconomic aspects. Climate change already affects producer countries in Southeast Asia; production in 2009 was lowered by 5% due to severe droughts on the Indochina peninsula, and heavy rains during the tapping season in Indonesia. Socioeconomic changes increase labor costs in producer countries, especially in Malaysia, and fewer workers are available for tapping, while competition for land by oil palm, food, nature reserves, and other uses is increasing. Furthermore, trade restrictions, including bilateral agreements with consumer countries in the region (e.g., China) and the possible establishment of a producer cartel, may limit the amount of natural rubber available on the free market.

In view of the problems related to rubber from *H. brasiliensis*, the strategic importance of natural rubber and the uncertainties of its future supply, the development of alternative sources of natural rubber was identified as a clear priority (Mooibroek and Cornish, 2000). Large research programs were established during WWII that aimed at identifying plants that could be used as alternative crops for large-scale rubber production. While plants such as Goldenrod (*Solidago leavenworthii*), lettuce, or sunflower produce rubber having unfavorable characteristics, such as very low yield or low molecular weight of the polymer, which makes them difficult subjects for the development of rubber producing crops, two main plants have been selected as promising for rubber production — the Russian dandelion (*Taraxacum koksaghyz*) and guayule (*Parthenium argentatum* Gray). Both plants produce high amounts of high molecular weight rubber, an essential determinant of rubber quality (van Beilen and Poirier, 2007b).

Russian dandelion accumulates rubber in the roots and leaves and is considered a potentially attractive annual crop for temperate regions. Although Russian dandelion has laticifers like the rubber tree, rubber must be harvested by homogenizing the roots. Rubber yields of 150–500 kg/ha/year have been reported (van Beilen and Poirier, 2007b), and dandelion rubber was produced in the USSR and Germany during WWII to make tires. However, limited progress has been achieved since then, as research projects were stopped when *Hevea* rubber became widely available again after WWII. Turning Russian dandelion into a viable crop for rubber production would require an increase of vigor and more favorable agronomic properties, such as larger roots that are easier to harvest and increased rubber accumulation. In addition, the low yield per hectare, labor-intensive cultivation, crosses and seed contamination with other dandelions, and weed potential should be tackled (van Beilen and Poirier, 2007c).

Guayule is a shrub native to Mexico and the Southwestern United States that can also be grown in southern Europe and northern Africa. One advantage of guayule rubber is that its latex contains less proteins, both in number of species and total amount, than latex from *H. brasiliensis*, and these proteins do not cross-react with immunoglobulins reactive against *H. brasiliensis* latex proteins (Siler et al., 1996). This makes allergenic reactions to guayule rubber by consumers sensitized to *Hevea* rubber unlikely. Indeed, the Yulex Corporation (www.yulex.com) is currently marketing guayule latex as a source of hypoallergenic rubber, particularly useful for medical applications. In guayule, rubber latex accumulates intracellularly in the bark parenchymal cells. Thus, the latex cannot be harvested by tapping, as with the rubber tree, but the guayule shoot material must be harvested and thoroughly disrupted to release the rubber particles (Cornish et al., 2006). Thus, while rubber harvest can be mechanized for guayule, extraction is technically more complicated than for *H. brasiliensis*. Productivity of rubber from guayule up to 2000 kg/ha/year has been reported, while up to 3000 kg/ha/year can be produced from the Brazilian rubber tree. While high-value applications, such as hypoallergenic latex for medical application, at present justify the higher cost of guayule rubber, large-scale use of guayule rubber as an alternative to *Hevea* rubber in lower value items, such as tires, will require improvement in the level of productivity of the varieties used and agronomic and harvesting practices, as well as optimization of the extraction method (van Beilen and Poirier, 2007c).

Modern molecular approaches are required to better understand and modulate rubber biosynthesis in both guayule and Russian dandelion. Although large gaps are present in our knowledge of the rubber polymerase, there is generally a good understanding of the enzymes involved in the synthesis of IPP and farnesyl diphosphate and this knowledge could be used, via transgenesis, to over-express or downregulate key genes involved in rubber biosynthesis and accumulation. This type of functional approach is easier to apply to Russian dandelion with its short life cycle, and because detectable rubber phenotypes can be obtained after six months. In contrast, for *H. brasiliensis*, several years of growth are needed before clear rubber phenotypes can be determined, whereas for guayule one to two years is required. Furthermore, to make rubber from either Russian dandelion or guayule competitive for large-scale, low-value applications, it is essential to consider the plants in the context of a biorefinery, where all components, including waste material, would be used. Co-products from both guayule and Russian dandelion could include biogas or bioethanol from the cellulosic waste, or various specialty chemicals from secondary metabolites. Side products from Russian dandelion processing could also include inulin, a polymer of fructose, which is the major storage sugar of dandelions (25–40% of dry root weight), which could be used directly in non-food applications, used to make the building block 5-hydroxymethylfurfural, used in polymer synthesis, or fermented to bioethanol (van Beilen and Poirier, 2007b). Public research programs in the United States (PENRA; http://oardc.osu.edu/penra/) and Europe (EU-PEARLS; http://www.eu-pearls.eu) are now aimed at improving Russian dandelion and guayule for rubber production.

Proteins

Various materials (bioplastic and fibers) can be derived from natural plant proteins obtained as a co-product of starch, vegetable oil, or biofuel production (Mooney, 2009). Zein (the major corn protein; Shukla and Cheryan, 2001; Lawton, 2002), soy protein (Mohanty et al., 2005), and wheat gluten (Woerdeman et al., 2004; Pallos et al., 2006) have been converted to plastics by cross-linking with formaldehyde, glutaraldehyde, or other chemicals. In 1950 about 2700 tons/year of zein plastics for coatings, and 2200 tons/year of Vicara, a zein-based fiber, were produced. The current cost of zein is about $40€/kg, but as a by-product from ethanol production, costs could be as low as $4/kg (Lawton, 2002).

Soy protein-based bioplastics were used in the 1930s by Henry T. Ford to construct car parts. Soon after, however, petroleum-based plastics took over, in part because of poor properties of the protein-based plastics and fibers: the soy-based materials were (and still are) plagued by susceptibility to microbial degradation, water permeability, and a persistent bad smell (formaldehyde, other residual chemicals), as well as a relatively high price (Mohanty et al., 2005). Progress has been made in processing by extrusion (Verbeek and van den Berg, 2010) and the processing of wheat gluten. Nevertheless, "current gluten materials are still outperformed by their synthetic polymer counterparts" (Lagrain et al., 2010).

As gluten, zein, and soy protein are by-products derived from food and feed crops, optimization for material applications by genetic engineering is not a likely option. Also, in view of the complex structure of proteins, the effort required to make significantly better protein bioplastics may be out of proportion to the benefits. Natural limits to the properties are given by the peptide bond linking the amino acids, and the reactivity of pendant groups: proteins are sensitive to hydrolysis by strong acids and bases, oxidation, heat denaturation, and decomposition. Protein engineering can significantly change properties, but the fundamental limits cannot be changed. Thus, progress is incremental, and mainly in the field of processing technology, biocomposites, plasticizers, and formulations.

The amount of protein co-products from large-scale biofuel production potentially greatly exceeds the amount required for the food and feed markets, which could help the development of a protein-based bioplastics industry (Sanders et al., 2007). If 10% of transportation fuels will be bio-based (providing that issues such as indirect land-use change do not interfere), up to 100 million tons of protein side product could become available (Scott et al., 2007). Even if a large fraction is absorbed as feed, replacing protein crops such as soybeans, a significant amount of protein would still be available for the production of bioplastics.

Novel Polymers Produced in Transgenic Plants

A role for transgenic crops in the production of biopolymers?

Three basic production strategies can be distinguished for renewable biomaterials from plants (Figure 30.1). The first is the direct use of the plant biomass (wood, cork) or biomass components (natural rubber, starch, cellulose) in biomaterials. The second option is the conversion of plant carbon storage products (lipids, fatty acids, or sugars) into new compounds by microbial biotechnology (also called white biotechnology) involving fermentation or by chemical polymerization methods. For example, PHA, biocellulose, xanthan, silk, and polythioesters can be produced by recombinant or wild-type microorganisms in fermentation processes using carbohydrates from plants (Thakor et al., 2005), while polyesters such as polylactic acid (PLA), polycaprolactone, polytrimethylene terephthalate (e.g., Sorona® by Dupont), and polybutylene succinate (e.g., Bionolle® by Showa) are produced using chemical polymerization of substrates that are, at least in part, renewable and generated by fermentation of plant sugars (Mecking, 2004). The third option is the production of novel or modified biopolymers directly in transgenic plants. As the production of biopolymers from white and green biotechnology compete on the market, advantages and disadvantages of these two technologies must be compared.

Life cycle assessments suggest that biopolymers produced in plants may bring considerable savings in energy use and greenhouse gas emissions, compared to petrochemical polymers or biopolymers produced by fermentation, provided that the product concentration in the plant is high and the remaining biomass is used to generate energy (Kurdikar et al., 2001). This occurs because polymers produced in plants are direct products of photosynthesis-driven plant metabolism, while in white biotechnology, the polymers (or polymer building blocks) are indirectly produced from a carbon source synthesized and extracted from crop plants, such as sugars or lipids. Thus, the direct synthesis of biopolymers in plants is theoretically more energy efficient and requires fewer steps. Furthermore, agricultural crops offer the possibility of producing biopolymers on a larger scale than that possible by microbial biotechnology, an important factor for biopolymers used in commodity products that are typically produced above the million tons per year range. Under the condition that processing is not appreciably more difficult, direct production in plants obviously involves fewer steps and could be cost-effective compared to microbial biotechnology (van Beilen and Poirier, 2007a). However, several factors affect the competitive advantage of green versus white biotechnology. The first is that the level of control over key factors affecting biopolymers properties, such as molecular weight, polydispersity, and monomer composition, is likely to be lower in transgenic plants compared to bacterial fermentations or chemical synthesis, which involve closed systems with extensive control

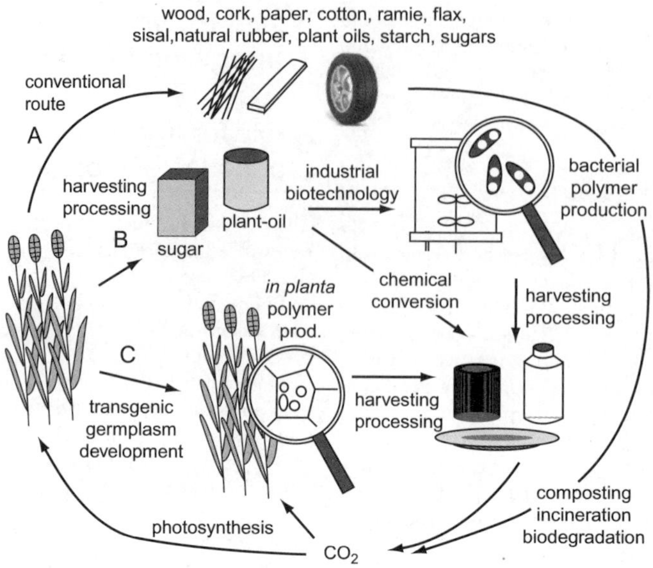

Figure 30.1 • Production cycles • Route A. Direct production of materials from plants. Route B. Conversion of plant products by industrial biotechnology or chemical methods to polymers. Route C. Production of new chemical building blocks or polymers in transgenic plants. In all cases, the cycle is closed by composting, incineration, or biodegradation. Reproduced with permission from van Beilen and Poirier (2008).

over the production conditions. Plant metabolism can be difficult to manipulate without affecting plant health and yield, while external environmental conditions can change fluxes through metabolic pathways. The synthesis of polymers in plants also brings up the issue of seasonal harvesting and processing, and consequently, the stability of the polymer in tissues between harvest and extraction. Fermentation and chemical synthesis are more flexible, with both continuous production throughout the year or peaks of synthesis depending on the possibility of demand. A partial remedy is the production of polymers in plant organs that can be easily stored over long time periods and where polymers are likely to be stable, such as in dry seeds. Alternatively, crops that can be harvested throughout the year, such as sugarcane, may be used.

The requirement of transgenesis for the synthesis of novel biopolymers in plants is also a significant factor affecting its competitiveness. The process of developing a transgenic crop is more complicated and costly than is generally realized, in part because of regulatory issues and the lack of public acceptance (Devine, 2005). The overall costs of discovery, R&D, breeding, production, admission, and other regulatory matters associated with bringing a transgenic crop into the field may run into several tens of millions of dollars (Devine, 2005). Moreover, the required time for development of new registered crop varieties may take 8 to 12 years, making it very difficult to respond quickly to shifts in market demands using genetically modified crops relative to white biotechnology (Wenzel, 2006). Thus, only a few novel products or characteristics can be developed to marketable crop varieties, which should bring a high return on investment, either through the high value of the product or because of its high market volume. Thus, while transgenic plants with increased resistance to pests or herbicides have been a commercial success, with annual growth rates of 12% and a total area planted of almost 115 million hectares in 23 countries (Khan and Liu, 2009), few plants engineered for new products or product properties have reached the commercial stage, or promise to reach that stage in the near future. Promising transgenic crop products include altered starch, polyunsaturated fatty acids, new sugars, and organic acids (van Beilen, 2008).

Which biopolymers should be targeted for production in transgenic crops?

Only a few biopolymers (natural rubber stands out in this respect) are now competitive with their petrochemical equivalents based on price and performance. Polymers that: (1) can be produced more efficiently in plants than in microbes; (2) can be extracted easily from plant biomass; and (3) are needed on a large scale at low cost are worthwhile targets, in view of the strengths and weaknesses of plant biotechnology compared to white biotechnology.

The advantages of direct biopolymer synthesis in plants must be weighed against the cost of polymer extraction. Because extraction and processing typically is a very large contributor to the costs of biopolymer, the question of whether specific polymers can be isolated easily from plants must be answered early. Natural rubber is harvested as a liquid (latex) that leaks from cuts in the bark of rubber trees. This is labor intensive (70% of production costs are labor costs), but little technology is needed (chemical additives, cleaning, drying). Starch granules are easily separated from seeds and tubers by technical processes. In contrast, the most abundant biopolymer cellulose is very difficult to separate from the matrix containing hemicellulose, lignin, proteins, and other polysaccharides. Initially a widely used base for plastics, cellulose lost out to petrochemical plastics. Other abundant polymers such as lignin suffer from the same drawback.

Because of the scale requirement (the cost of developing a transgenic crop), the target biopolymer should be able to compete with commodity petrochemical plastics used in consumer products (e.g., polyethylene and polypropylene) based on price and properties. The current capacity of bioplastics production plants is estimated at 400,000 tons/year worldwide (www.european-bioplastics.org) or about 0.15% of annual worldwide production of petroleum-based plastics (260 million tons/year). A target of 10% of bioplastics or 26 million tons per year of bioplastics could be met by synthesis in crop plants. For example, assuming synthesis of a biopolymer at a level of 10% dry weight in leaves of *Miscanthus*, which would yield 2–3 tons of biopolymer/hectare, would require 10 million hectares of such a crop, or 0.2% of worldwide agricultural land.

Which crops should be targeted?

The choice of the crop is crucial to successfully establish biopolymer synthesis in plants. Transgenic crops are well accepted in North America, but this is not the case in Europe and other regions of the world. In the case of biomaterials, a clear-cut choice for industrial crops or non-food crops lacking close relatives in the cultivation area seems prudent, as this clearly reduces real and perceived risks to consumers and the food supply. Most industrially processed crops (sugar beet, corn, potato varieties, and fiber crops) belong to the same species as food varieties, and out-crossing, weedy variants, and mix-ups are inevitable. New crops developed as biomass or energy crops, such as *Miscanthus* and switchgrass, could also be used to produce biopolymers in the context of a future biorefinery infrastructure, as they are high yielding, not used in food, and sterile variants can be used. Tobacco, a non-food crop native to South America; is another promising alternative with many potential advantages: it lacks close relatives in Europe and North America; does not persist in northern climates; seed production can be prevented by cytoplasmic male sterility and by regular harvesting of biomass before flowering; greenhouses suffice for seed production; and plastid transformation is very well established. Yields of leaf biomass can be amazingly high at up to 14 tons/hectare if not grown for smoking, which compares reasonably well with sugar beet at 18 tons/hectare (average yield in Europe), but is lower than sugarcane or *Miscanthus* at 20–25 tons/hectare and 20–30 tons/hectare, respectively (van Beilen et al., 2007).

Fibrous proteins

Proteins differ from other polymers in that they are composed of 20 different natural monomers (amino acids), the order and composition of which is specified by the DNA code and can be controlled at will. This implies that the properties of the resulting materials can be changed and, to some extent, predicted. With further progress in our understanding of the structure–function relationships, proteins with the desired properties can be designed and produced in appropriate hosts (Sanford and Kumar, 2005). The proposed applications include tissue engineering scaffolds, drug carriers, hydrogels, coatings, glues, elastomers, and fibers.

The so-called fibrous proteins (silk, elastin, collagen, etc.) typically contain short blocks of repeated amino acids. They have received a good share of attention as the original materials already have quite favorable properties, including unique strength-to-weight, elastic, or adhesive properties. Well-known fibrous proteins are silk, collagen, elastin, keratin, and mussel adhesive proteins. Silks can be loosely defined as externally spun fibrous protein secretion. The best known silk is the one produced from the mulberry silkworm cocoons and used for clothing. Many other insects produce silks, but spider silk has received special attention due to its exceptional mechanical properties. For example, silks of the spider dragline, the main structural web silk, and the main spider's lifeline are stronger than steel, when compared on a weight basis, and have a similar strength but are more elastic than Kevlar® (Hinman et al., 2000). Synthetic silk genes have been expressed in transgenic tobacco, potato, and *Arabidopsis thaliana*, which were cultivated in greenhouses as well as in field trials (Menassa et al., 2004; Scheller and Conrad, 2005). Targeting to the endoplasmic reticulum (ER) in tobacco and potato leaf cells yielded 2% of total soluble protein (TSP; Scheller et al., 2001). Expression in leaf apoplasts of *A. thaliana* yielded 8.5% TSP, while targeting to seed endoplasmic reticulum yielded 18% TSP (Yang et al., 2005), close to the level of 10 and 30% TSP reported in *E. coli* and *Pichia pastoris*, respectively (Fahnestock and Bedzyk, 1997; Fahnestock and Irwin, 1997). However, clone instability due to homologous recombination of repetitive sequences, gene size limitations, a distinct codon usage, and the formation of inclusion bodies are general problems limiting high-level and stable expression of silk-like proteins in microorganisms.

Production of a silk-elastin fusion protein in tobacco and potato plants was achieved through the targeting of the protein to the ER using the KDEL ER retention signal (Scheller and Conrad, 2004). Laboratory-scale production of 80 mg of pure protein per kilogram of tobacco leaves was obtained by heating a simple buffer-extract to 95°C for 1 hour, causing aggregation of most proteins, while leaving the silk protein in the supernatant. However, this protein preparation lacks the structural properties of silks, which are conferred by assembly of the proteins during the process of spinning in specialized organs in the spider (Huang et al., 2007).

Floss and co-workers have developed elastin-like polypeptides (ELPs) that are biocompatible and exhibit a thermally responsive reversible phase transition, which makes ELPs attractive for various applications in drug delivery and tissue repair or engineering (Floss et al., 2010a). ELPs also aid in the purification of recombinant proteins as fusion proteins (ELPylated proteins), which also exhibit the reversible phase transition property. A number of plant-based expression systems have been evaluated for the production of such proteins, which may help in using plants as bioreactors to synthesize both biopharmaceuticals and industrial proteins (Floss et al., 2010b). Elastin-like polypeptides also appear to allow a wide range of expression levels in plants, helping to overcome the typically low expression levels by depositing concentrated heterologous protein in stable protein bodies (Conley et al., 2009).

Expression of collagen, of a synthetic protein made from repeats of a motif found in elastin, and of a chimeric protein made of silk and elastin domains, have also been reported in tobacco or potato (Guda et al., 2000; Ruggiero et al., 2000; Scheller et al., 2004). A recent study reported the synthesis of a biofunctional procollagen at 2% of soluble protein in transgenic tobacco (Stein et al., 2009).

Improvements in fibrous protein synthesis in plants may require several approaches including optimization of the amino acid and tRNA pools for the few amino acids that are over-represented in these proteins (such as glycine and alanine in spider silk), co-expression of several fibrous proteins as found in natural silk, and improved targeting to subcellular compartments and tissues optimal for protein synthesis and storage. Beyond these strategies to optimize expression, effective synthesis of a fibrous protein does not ensure production of a good fiber. The properties of silk fibers depend to a large extent on the correct assembly by spinning of the different types of proteins. In the spider or silkworm glands, various types of silk proteins are spun together via an elaborate liquid spinning process, which has not been reproduced in the lab (Huang et al., 2007). Recombinant spider silk obtained from mammalian cells has been spun into filaments showing similar toughness to dragline silk, but with a lower tenacity (Lazaris et al., 2002). Thus, while processing of fibrous proteins is facilitated by the fact that proteins such as spider fibroins are extremely heat-stable and acid-soluble (Scheller et al., 2001; Scheller et al., 2004), advances in micro-spinning and other processing technologies are essential to make good fibers from fibrous proteins.

Assuming that expression and processing limitations can be solved, heterologous expression in plants would enable production on a much larger scale than is possible in bioreactors. However, the question should be asked if any of the fibrous proteins has a (potential) market size that would justify the significant cost and time required for the development of a transgenic crop germplasm. Plant-based fibrous protein production would be better than bacterial fermentation only if the fibrous protein is to be used on a commodity scale, for example, as a technical or textile fiber. In this scenario, production costs in plants have been estimated at 10–50% of the cost for production in bioreactors (Scheller et al., 2001).

Cyanophycin

Cyanophycin is a nitrogen-storage polymer composed of a polyaspartate backbone with arginine side chains attached with their α-amino group to the β-carboxy group of each aspartate (Figure 30.2). It is synthesized by the enzyme cyanophycin synthase and deposited as granules in cyanobacteria as well as several photosynthetic and non-photosynthetic bacteria (Berg et al., 2000; Oppermann-Sanio and Steinbüchel, 2002). While arginine is the most common side chain attached to the polyaspartate backbone, other amino acids have also been found in cyanophycin produced in heterologous hosts, such as lysine, ornithine, or citrulline (Steinle et al., 2009).

Although cyanophycin is not suitable for material applications, it is a useful source of polyaspartate, which can be used

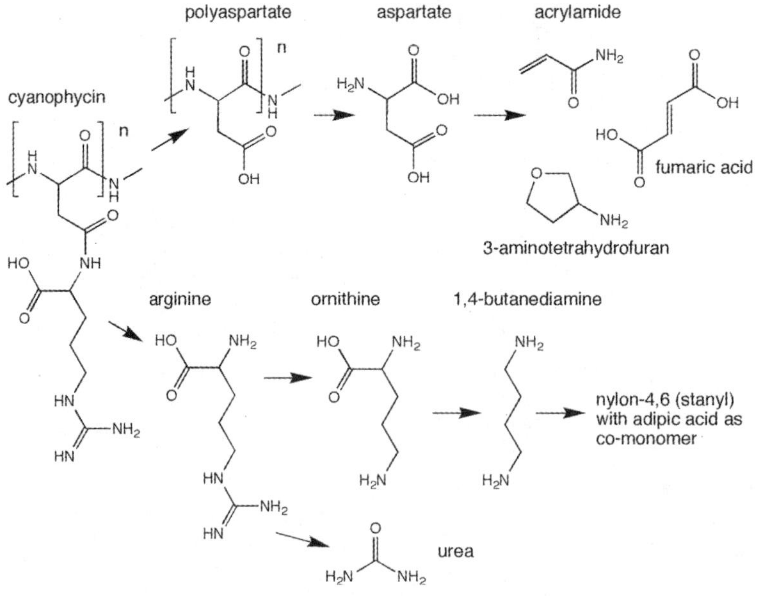

Figure 30.2 • Structure of cyanophycin and potential products derived from cyanophycin • Reproduced with permission from van Beilen and Poirier (2008).

as a super-adsorbent or antiscalant (potential market $450 million per year; Tsao et al., 1999), replacing the chemically synthesized compound (Oppermann-Sanio and Steinbüchel). Cyanophycin could also be used as a valuable source of dipeptide and of amino acids in the nutritional and medical fields; for example, arginine has numerous physiological roles in many cardiovascular, gastrointestinal, and immune disorders (Sallam and Steinbuchel, 2010). Aspartate and other amino acids found in cyanophycin could also serve as a starting point for the synthesis of a range of chemicals that can be obtained by reductions, dehydrations, polymerization, decarboxylation, and deamination reactions (Werpy and Pedersen, 2004; Scott et al., 2007). As an example, reduction of aspartic acid would produce 3-aminotetrahydrofuran and 2-amino-1,4-butanediol, close analogs of high volume chemicals currently used in industry, while arginine could be converted to 1,4-butanediamine used to synthesize nylon-4,6 (Figure 30.2).

Expression of cyanophycin synthase in plants has been pioneered by Broer and co-workers (Neumann et al., 2005). Transgenic tobacco and potato contain cyanophycin up to 1.1% dry weight through expression of the cyanophycin synthase in the cytoplasm of leaf cells (Neumann et al., 2005), with some deleterious effects such as changes in leaf morphology and decreased growth. The adverse effects were mitigated by translocation of cyanophycin synthesis to plastids, which brought an increase in cyanophycin content up to 9.4% of dry weight in tobacco leaves (Hühns et al., 2008). Minimal effects on growth and morphology were also observed when expression of cyanophycin synthase was restricted to potato tubers. Here, up to 7.5% of dry weight was accomplished (Hühns et al., 2009). For comparison, recombinant *E. coli*, *Saccharomyces cerevisiae*, and *P. pastoris* could produce cyanophycin at 39%, 15%, and 23% of the cell dry weight, respectively (Hai et al., 2006; Steinle et al., 2009, 2010). Economically viable levels in plants may require optimization of the pathways involved in supplying aspartic acid and arginine, as well as engineering the cyanophycin synthase for maximal activity in the plant cell environment.

Polyhydroxyalkanoate

Polyhydroxyalkanoates (PHAs) are biological polyesters produced by a wide variety of bacteria as carbon and energy storage compounds. PHAs accumulate as osmotically inert intracellular inclusions. Almost all naturally occurring PHAs consists of 3-hydroxy fatty acids with a chain length of 4 to 16 carbons polymerized by a PHA synthase using *R*-3-hydroxyacyl-CoA intermediates. Under special circumstances or with specific feed compounds, bacterial PHAs may also contain 4-, 5-, and 6-hydroxy fatty acids or monomers substituted with various groups such as aromatic rings, methyl- or hydroxy groups, ether-, and double-bonds (Steinbüchel and Valentin, 1995). PHA can accumulate to high levels by bacterial fermentation, typically between 50–80% cell dry weight, depending on the type of PHA, with some reports of up to 90% cell dry weight.

Depending on the composition and resulting properties, PHAs have many and wide-ranging potential applications. Due to their impermeability to water and air, PHAs are considered very suitable for consumer products such as bottles, films, and fibers. The simplest PHA, poly-3-hydroxybutyrate (PHB) is a relatively hard and brittle material (Lenz and Marchessault, 2005). Whereas the presence of C5-monomers to produce poly-3-hydroxybutyrate-*co*-3-hydroxyvalerate (P(HB-co-HV)) improves the properties of the material to some extent, inclusion of a small proportion of longer monomers (C6 and longer) has resulted in materials that are much easier to process, and are similar to polypropylene (Noda et al., 2005). PHAs consisting of higher molecular weight monomers (C6–C16, referred to as medium chain length PHA; mclPHA) typically are rubber-like materials with an amorphous soft-sticky consistence. These polymers still have to find large-scale applications, which may include the production of enantiopure *R*-3-hydroxy-carboxylic acids (Witholt and Kessler, 1999). The range of properties and applications may also be greatly extended by chemical modification of the polymer after extraction (Hazer and Steinbüchel, 2007). In general, PHAs are considered for material that ends up in the environment, such as flowerpots used in planting, foils, bags, fishing lines and nets (which should decompose if lost), material used in biomedical applications, and for other disposable items such as bottles, cups, plates, and cutlery that can be composted but not recycled.

Considering the potential of using PHAs as substitutes for petrochemical plastics in numerous applications, PHAs are good candidates as large-scale commodity plastics that could be produced in transgenic plants in the million ton range, setting PHA apart from fibrous proteins and cyanophycin (Poirier et al., 1995). The proof of concept was demonstrated in 1992 for the synthesis of PHB in transgenic *Arabidopsis* expressing the biosynthetic enzymes from *Ralstonia eutropha* in the cytoplasm (Poirier et al., 1992). Since then, synthesis of various types of PHA in transgenic plants has been reported in more than 10 different plant species (Table 30.1; reviewed in van Beilen, 2008; Poirier and Brumbley, 2010). Most experience has been gathered for the synthesis of PHB. Whereas synthesis of PHB from the cytoplasmic pool of acetyl-CoA led to only low (≤0.3% dwt) levels of polymer accumulation, synthesis of PHB in the plastid led to the highest level of accumulation, reaching 40% dry weight in *Arabidopsis* leaves (Figure 30.3; Bohmert et al., 2000; Nawrath et al., 1994). However, expression of such high levels of PHB led to severe growth retardation. Collectively, data from the synthesis of PHB in the plastid leaves of various plants indicate that accumulation of PHB in leaves at 5–8% dry weight is possible without any apparent growth defect, but higher levels typically lead to chlorosis and progressively more growth retardation. Interestingly, accumulation of 8% PHB in the leucoplasts of seeds of oilseed rape was not associated with any growth retardation or defect in germination, indicating that seed accumulation of PHA may be a better alternative than leaves (Houmiel et al., 1999). PHB accumulation in the range of 1–2% dry weight using acetyl-CoA produced in the peroxisomes has been reported in both corn suspension cells and sugarcane leaves (Hahn et al., 1999; Tilbrook et al., 2011).

Since PHB is a rather stiff and brittle plastic with limited application, synthesis of PHA co-polymers with better

Table 30.1 Polyhydroxyalkanoates produced in transgenic plants

Plant species	Subcellular compartment	Tissue	PHA produced	PHA yield (% dwt)
Arabidopsis thaliana	Plastid	Shoot	P(3HB)	14–40
	Plastid	Shoot	P(3HB-3HV)	1.6
	Cytoplasm	Shoot	P(3HB)	0.1
	Cytoplasm	Whole plant	P(3HB-co-3HV)	0.6
	Peroxisome	Whole plant	mclPHA	0.4
	Peroxisome	Whole plant	scl-mclPHA	0.04
	Peroxisome	Whole seedling	P(3HB)	1.8
Alfalfa	Plastid	Shoot	P(3HB)	0.2
Corn	Plastid	Shoot	P(3HB)	6
	Peroxisome	Cell suspension	P(3HB)	2
Cotton	Cytoplasm	Vascular bundles	P(3HB)	0.3
	Plastid	Vascular bundles	P(3HB)	0.05
Potato	Plastid	Shoot	P(3HB)	0.02
	Cytoplasm	Cell line	mclPHA	1
	Plastid	Shoot	mclPHA	0.03
Oilseed rape	Cytoplasm	Shoot	P(3HB)	0.1
	Plastid	Seed	P(3HB)	8
	Plastid	Seed	P(3HB-co-3HV)	2.3
Tobacco	Cytoplasm	Shoot	P(3HB)	0.01
	Plastid	Shoot	P(3HB)	<1.7
	Plastid	Leaves	mclPHA	0.005
Sugar beet	Plastid	Hairy roots	P(3HB)	5
Sugarcane	Plastids	Leaves	P(3HB)	1.8
Flax	Plastids	Stem	P(3HB)	0.005

Reproduced with permission from van Beilen and Poirier (2008).

properties and broader applications is an important goal. In this perspective, synthesis of the PHA copolymer P(HB-co-HV) copolymer was demonstrated in *Arabidopsis* leaves at 1.6% dry weight, and seeds of oilseed rape at 2.3% dry weight. This was achieved by a rather complex genetic engineering feat, implicating the expression of four bacterial genes for the supply and copolymerization of 3-hydroxybutyryl-CoA and 3-hydroxyvaleryl-CoA in the plastid (Figure 30.3; Slater et al., 1999). Interestingly, there was an inverse relationship between the amount of PHA and the proportion of the hydroxyvalerate monomer, indicating a bottleneck in providing 3-hydroxyvaleryl-CoA to the PHA-synthase. Other types of PHA copolymers have also been synthesized in low amounts in plants, including mclPHA consisting of C6–C16 monomers, produced to a level of 0.4–1% through the polymerization of 3-hydroxyacyl-CoA intermediates from the ß-oxidation cycle of fatty acids (Figure 30.3; Mittendorf et al., 1998, 1999; Poirier, 1999; Arai et al., 2002; Matsumoto et al., 2006).

Accumulation of PHA in plants at a level of 15% dry weight is considered a minimal threshold level for commercial viability. Considering the current experience, it would seem that reaching this level, although feasible, will still require considerable research efforts. It will be essential to gain a better understanding of how the synthesis of the precursors is regulated and how it could be more efficiently channeled (e.g., acetyl-CoA, propionyl-CoA, 3-hydroxyacyl-ACP) toward PHA without affecting plant growth. Very little has been done to understand how carbon flux is affected through the introduction of a new carbon sink such as PHA, and it is likely that the tools of genomics, metabolomics, and flux analysis would provide very valuable information in this context; the same way these technologies have been instrumental in optimizing synthesis of new compounds through fermentation.

As with several other biopolymers, commercially viable PHA production in plants is not only determined by the costs and technological possibilities of production in plants, but also depends on the competing production methods. PHA production through bacterial fermentation is now a commercial reality with the PHA copolymer Mirel™ produced by Telles (http://www.mirelplastics.com/), as well as the P(HB-co-HV) co-polymer by Tianan in China (http://www.tiananenmat.com). The PHA accumulation level in bacteria reaches 50–85% dry weight, which is probably unattainable in plants. Such high level PHA accumulation leads to easier recovery of the polymer than is likely to be possible in plants. Although several methods for PHA extraction in plants are found in the (largely prophetic) patent literature, actual experimentation on PHA extraction from plants is very limited and clearly deserves more attention (Poirier and Gruys, 2001). An outstanding feature of bacterial PHAs is the enormous flexibility in properties as a function of the nature and ratio of the different monomers. Bacterial fermentation allows the synthesis of a wide spectrum of PHA with various physical properties (by choice of polymerase, host, feedstock, and conditions), with some PHAs suitable for low-value commodity applications and others for specific high-value niche market application, such as medical implants (Philip et al., 2007). Reaching similar levels of control and producing a wide spectrum of PHA types is unrealistic in the context of transgenic crops. It is thus clear that synthesis of PHA in crops must be limited to one or perhaps two types of PHAs that would be used for large-scale and low-cost bulk applications, such as a substitute for plastics used in consumer products, while bacterial PHA would be left for higher value, lower volume applications.

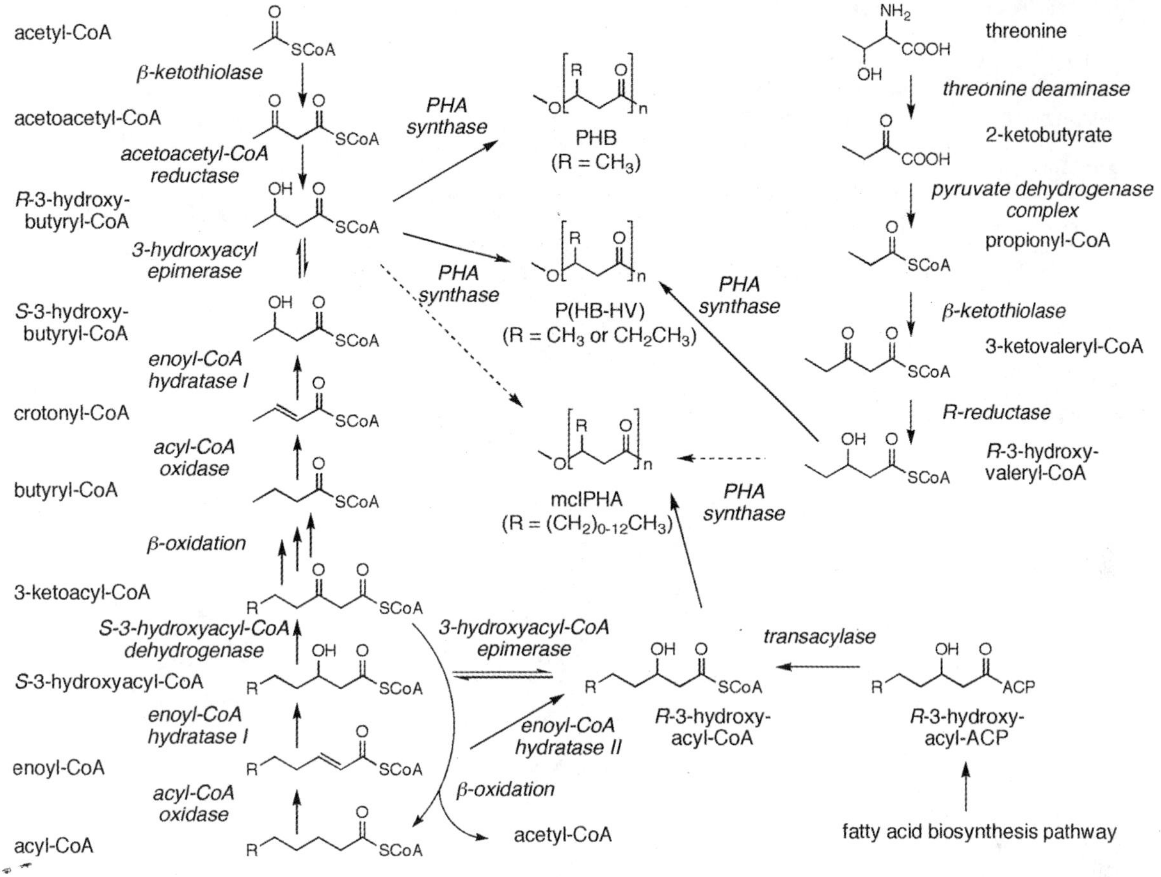

Figure 30.3 • Metabolic pathways leading to the production of PHB, P(HB-co-HV), and mclPHA in plants • The pathway for PHB synthesis from acetyl-CoA (top left) has been implemented in the cytosol, plastid, and peroxisome. The pathway for P(HB-co-HV) synthesis in the plastid was created by combining the PHB biosynthesis pathway with a pathway generating 3-propionyl-CoA via a threonine deaminase and pyruvate decarboxylase (top right). Synthesis of P(HB-co-HV) in the cytosol involved an unidentified source of propionyl-CoA or 3-hydroxyvaleryl-CoA. Other PHA copolymers, such as mclPHA, have been mainly synthesized in the peroxisome using the 3-hydroxyacyl-CoA intermediates generated in the ß-oxidation cycle (bottom left). Synthesis of mclPHA from the conversion of *R*-3-hydroxyacyl-ACP to *R*-3-hydroxyacyl-CoA is achieved via a bacterial 3-hydroxyacyl-CoA-ACP transacylase (lower right corner). Reproduced with permission from van Beilen and Poirier (2008).

The cost of synthesis of PHA in plants versus bacteria still remains somewhat speculative. Previously, PHAs produced by microbial fermentation were considered too expensive and lacking in environmental benefits (Gerngross, 1999). However, the rapid development of biogas technology to utilize waste biomass for the production of process heat and electricity makes PHA production by fermentation more energy and CO_2 efficient (Kim and Dale, 2005). According to some sources, PHAs could be produced by fermentation for approximately \$2/kg, although the current price for Mirel is currently \$5/kg (Ramseier, 2010). Thus, the cost of PHA produced by bacterial fermentation is substantially higher than the cost of polypropylene and polyethylene at around \$1/kg (Philip et al., 2007). Competitive PHA production in plants must assume that PHA is only one of several valuable products obtained from the crop. For example, strategies such as developing *Miscanthus*, sugar beet, or sugarcane as crops for biofuel, commodity chemicals, and PHA would considerably enhance the economic benefit of PHA in crops (Snell and Peoples, 2009).

Conclusion and Prospects

Despite the large potential that plants may have as providers of biomaterials, progress in unlocking this potential and bringing it into commercial applications has been slow, with the exception of rubber, which has been used on a large scale for over a century. Various factors have contributed to this. First, it is clear that competing with petroleum-derived plastics is a challenging task. Biomass is a much more complicated material to handle than crude oil. Oil can be cheaply transported over large distances by simply pushing it through a pipeline, and it can be distilled into fractions based on vapor pressure, with heavier fractions amenable to cracking or other physical or chemical process steps. In contrast, biomass is bulky and more expensive to transport. It contains many different compounds in variable amounts, of different polarity, molecular weight, solubility, macromolecular structure, and water content. Whereas some plant biopolymers are rather easy to purify, such as starch, fibrous proteins, and cyanophycin, others are much more problematic, such as cellulose and polyhydroxyalkanoates. Nevertheless, this comparison has its limits, since the finite supply of petroleum and the carbon footprint associated with its use creates a necessity to explore all possible means to find adequate substitutes. Whether direct production of polymers *in planta* or microbiological synthesis via plant-derived carbon will, in the end, be used in commercial biopolymers, depends on numerous factors: quality, quantity, and cost of the biopolymers; their potential value-added properties; the development of efficient and economical methods in biorefineries, including success in economical conversion of wood and straw into simple sugars; and the capacity or willingness of the industry to replace petroleum-derived polymers with either identical polymers made from (at least partially) biological sources, such as polyethylene made from bioethanol (http://ethanolproducer.com/article.jsp?article_id = 7022), or with new substitutes, such as PHA.

References

Arai, Y., Nakashita, H., Suzuki, Y., Kobayashi, Y., Shimizu, T., & Yasuda, M., et al. (2002). Synthesis of a novel class of polyhydroxyalkanoates in *Arabidopsis* peroxisomes, and their use in monitoring short-chain-length intermediates of β-oxidation. *Plant and Cell Physiology*, *43*, 555–562.

Asawatreratanakul, K., Zhang, Y. W., Wititsuwannakul, D., Wititsuwannakul, R., Takahashi, S., & Rattanapittayaporn, A., et al. (2003). Molecular cloning, expression and characterization of cDNA encoding cis-prenyltransferases from *Hevea brasiliensis* — A key factor participating in natural rubber biosynthesis. *European Journal of Biochemistry*, *270*, 4671–4680.

Averous, L., & Halley, P. J. (2009). Biocomposites based on plasticized starch. *Biofuels Bioproducts Biorefining*, *3*, 329–343.

Bastioli, C. (Ed.). (2005). Handbook of Biodegradable Polymers. Shawbury: RAPRA Technology Limited.

Berg, H., Ziegler, K., Piotukh, K., Baier, K., Lockau, W., & Volkmer-Engert, R. (2000). Biosynthesis of the cyanobacterial reserve polymer multi-L-arginyl-poly-L-aspartic acid (cyanophycin) — Mechanism of the cyanophycin synthetase reaction studied with synthetic primers. *European Journal of Biochemistry*, *267*, 5561–5570.

Blanc, G., Baptiste, C., Oliver, G., Martin, F., & Montoro, P. (2006). Efficient *Agrobacterium tumefaciens*-mediated transformation of embryogenic calli and regeneration of *Hevea brasiliensis* Muell Arg. plants. *Plant Cell Reports*, *24*, 724–733.

Bohmert, K., Balbo, I., Kopka, J., Mittendorf, V., Nawrath, C., & Poirier, Y., et al. (2000). Transgenic *Arabidopsis* plants can accumulate polyhydroxybutyrate to up to 4% of their fresh weight. *Planta*, *211*, 841–845.

Bornke, F., & Broer, I. (2010). Tailoring plant metabolism for the production of novel polymers and platform chemicals. *Current Opinion in Plant Biology*, *13*, 354–362.

Cataldo, F. (2000). Guayule rubber: A new possible world scenario for the production of natural rubber. *Progress in Rubber and Plastics Technology*, *16*, 31–59.

Conley, A. J., Joensuu, J. J., Menassa, R., & Brandle, J. E. (2009). Induction of protein body formation in plant leaves by elastin-like polypeptide fusions. *BMC Biology*, *7*.

Cornish, K. (2001). Similarities and differences in rubber biochemistry among plant species. *Phytochemistry*, *57*, 1123–1134.

Cornish, K. (2001). Biochemistry of natural rubber, a vital raw material, emphasizing biosynthetic rate, molecular weight and compartmentalization, in evolutionarily divergent plant species. *Natural Product Reports*, *18*, 182–189.

Cornish, K., McCoy, R. G., Martin, J. A., Williams, J., & Nocera, A. (2006). *Biopolymer extraction from plant materials*. 14 pp.

Crank, M., Patel, M., Marscheider-Weidemann, F., Schleich, J., Hüsing, B., & Angerer, G. (2004). PRO-BIP. Techno-economic feasibility of large-scale production of bio-based polymers in Europe. European Commission's Institute for Prospective Technological Studies, Utrecht, Karlsruhe.

Davis, W. (1997). The rubber industry's biological nightmare. *Fortune Magazine*, *4*, 86–95.

Devine, M. D. (2005). Why are there not more herbicide-tolerant crops? *Pest Management Science*, *61*, 312–317.

Ehrlich, P. R., & Pringle, R. M. (2008). Where does biodiversity go from here? A grim business-as-usual forecast and a hopeful portfolio of partial solutions. *Proceedings of the National Academy of Sciences of the United States of America*, *105*, 11579–11586.

Erb, K. H., Krausmann, F., Gaube, V., Gingrich, S., Bondeau, A., & Fischer-Kowalski, M., et al. (2009). Analyzing the global human appropriation of net primary production — processes, trajectories, implications. An introduction. *Ecological Economics*, *69*, 250–259.

Fabry, V. J., Seibel, B. A., Feely, R. A., & Orr, J. C. (2008). Impacts of ocean acidification on marine fauna and ecosystem processes. *ICES Journal of Marine Science*, *65*, 414–432.

Fahnestock, S. R., & Bedzyk, L. A. (1997). Production of synthetic spider dragline silk protein in *Pichia pastoris*. *Applied Microbiology and Biotechnology*, *47*, 33–39.

Fahnestock, S. R., & Irwin, S. L. (1997). Synthetic spider dragline silk proteins and their production in *Escherichia coli*. *Applied Microbiology and Biotechnology*, *47*, 23–32.

Floss, D. M., Mockey, M., Zanello, G., Brosson, D., Diogon, M., & Frutos, R., et al. (2010). Expression and immunogenicity of the mycobacterial Ag85B/ESAT-6 antigens produced in transgenic plants by elastin-like peptide fusion strategy. *Journal of Biomedicine & Biotechnology*, *2010*, 274346. 10.1155/2010/274346

Floss, D. M., Schallau, K., Rose-John, S., Conrad, U., & Scheller, J. (2010). Elastin-like polypeptides revolutionize recombinant protein expression and their biomedical application. *Trends in Biotechnology*, *28*, 37–45.

Gerngross, T. U. (1999). Can biotechnology move us toward a sustainable society? *Nature Biotechnology*, *17*, 541–544.

Guda, C., Lee, S. B., & Daniell, H. (2000). Stable expression of a biodegradable protein-based polymer in tobacco chloroplasts. *Plant Cell Reports*, *19*, 257–262.

Hahn, J. J., Eschenlauer, A. C., Sleytr, U. B., Somers, D. A., & Srienc, F. (1999). Peroxisomes as sites for synthesis of polyhydroxyalkanoates in transgenic plants. *Biotechnology Progress*, *15*, 1053–1057.

Hai, T., Frey, K. M., & Steinbüchel, A. (2006). Engineered cyanophycin synthetase (CphA) from *Nostoc ellipsosporum* confers enhanced CphA activity and cyanophycin accumulation to *Escherichia coli*. *Applied and Environmental Microbiology*, *72*, 7652–7660.

Hazer, B., & Steinbüchel, A. (2007). Increased diversification of polyhydroxyalkanoates by modification reactions for industrial and medical applications. *Applied Microbiology and Biotechnology*, *74*, 1–12.

Hinman, M. B., Jones, J. A., & Lewis, R. V. (2000). Synthetic spider silk: A modular fiber. *Trends in Biotechnology*, *18*, 374–379.

Houmiel, K. L., Slater, S., Broyles, D., Casagrande, L., Colburn, S., & Gonzalez, K., et al. (1999). Poly(β-hydroxybutyrate) production in oilseed leukoplasts of *Brassica napus*. *Planta*, *209*, 547–550.

Huang, J., Foo, C. W. P., & Kaplan, D. L. (2007). Biosynthesis and applications of silk-like and collagen-like proteins. *Polymer Reviews*, *47*, 29–62.

Hühns, M., Neumann, K., Hausmann, T., Klemke, F., Lockau, W., & Kahmann, U., et al. (2009). Tuber-specific cphA expression to enhance cyanophycin production in potatoes. *Plant Biotechnology Journal*, *7*, 883–898.

Hühns, M., Neumann, K., Hausmann, T., Ziegler, K., Klemke, F., & Kahmann, U., et al. (2008). Plastid targeting strategies for cyanophycin synthetase to achieve high-level polymer accumulation in *Nicotiana tabacum*. *Plant Biotechnology Journal*, *6*, 321–336.

Jobling, S. (2004). Improving starch for food and industrial applications. *Current Opinion in Plant Biology*, *7*, 210–218.

Keeling, P. L., & Myers, A. M. (2010). Biochemistry and genetics of starch synthesis. *Annual Review of Food Science and Technology*, *Vol 1*, 271–303.

Khan, E. U., & Liu, J. H. (2009). Plant biotechnological approaches for the production and commercialization of transgenic crops. *Biotechnology & Biotechnological Equipment*, *23*, 1281–1288.

Kim, S., & Dale, B. E. (2005). Life cycle assessment study of biopolymers (polyhydroxyalkanoates) derived from no-tilled corn. *The International Journal of Life Cycle Assessment*, *10*, 200–210.

Ko, J. H., Chow, K. S., & Han, K. H. (2003). Transcriptome analysis reveals novel features of the molecular events occurring in the laticifers of *Hevea brasiliensis* (Para rubber tree). *Plant Molecular Biology*, *53*, 479–492.

Kurdikar, D., Fournet, L., Slater, S. C., Paster, M., Gruys, K. J., & Gerngross, T. U., et al. (2001). Greenhouse gas profile of a plastic material derived from a genetically modified plant. *Journal of Industrial Ecology*, *4*, 107–122.

Lagrain, B., Goderis, B., Brijs, K., & Delcour, J. A. (2010). Molecular basis of processing wheat gluten toward biobased materials. *Biomacromolecules*, *11*, 533–541.

Lawton, J. W. (2002). Zein: A history of processing and use. *Cereal Chemistry*, *79*, 1–18.

Lazaris, A., Arcidiacono, S., Huang, Y., Zhou, J. -F., Duguay, F., & Chretien, N., et al. (2002). Spider silk fibers spun from soluble recombinant silk produced in mammalian cells. *Science*, *295*, 472–476.

Le Guen, V., Garcia, D., Mattos, C. R. R., Doare, F., Lespinasse, D., & Seguin, M. (2007). Bypassing of a polygenic Microcyclus ulei resistance in rubber tree, analyzed by QTL detection. *The New Phytologist*, *173*, 335–345.

Lenz, R. W., & Marchessault, R. H. (2005). Bacterial polyesters: Biosynthesis, biodegradable plastics and biotechnology. *Biomacromolecules*, *6*, 1–8.

Lieberei, R. (2007). South American leaf blight of the rubber tree (*Hevea* spp.): New steps in plant domestication using physiological features and molecular markers. *Annals of Botany*, *100*, 1125–1142.

Matsumoto, K., Arai, Y., Nagao, R., Murata, T., Takase, K., & Nakashita, H., et al. (2006). Synthesis of short-chain-length/medium-chain-length polyhydroxyalkanoate (PHA) copolymers in peroxisome of the transgenic *Arabidopsis thaliana* harboring the PHA synthase gene from *Pseudomonas* sp. 61-3. *Journal of Polymers and the Environment*, *14*, 369–374.

Mecking, S. (2004). Nature or petrochemistry? — biologically degradable materials. *Angewandte Chemie*, *43*, 1078–1085.

Menassa, R., Hong, Z., Karatzas, C. N., Lazaris, A., Richman, A., & Brandle, J. (2004). Spider dragline silk proteins in transgenic tobacco leaves: Accumulation and field production. *Plant Biotechnology Journal*, *2*, 431–438.

Mittendorf, V., Bongcam, V., Allenbach, L., Coullerez, G., Martini, N., & Poirier, Y. (1999). Polyhydroxyalkanoate synthesis in transgenic plants as a new tool to study carbon flow through β-oxidation. *Plant Journal*, *20*, 45–55.

Mittendorf, V., Robertson, E. J., Leech, R. M., Kruger, N., Steinbüchel, A., & Poirier, Y. (1998). Synthesis of medium-chain-length polyhydroxyalkanoates in *Arabidopsis thaliana* using intermediates of peroxisomal fatty acid β-oxidation. *Proceedings of the National Academy of Sciences of the United States of America*, *95*, 13397–13402.

Mohanty, A. K., Liu, W., Tummala, P., Drzal, L. T., Misra, M., & Narayan, R. (2005). Soy protein-based plastics, blends, and composites. In: Mohanty, A. K., Misra, M. and Drzal, L. T., (Eds.), *Natural Fibers, Biopolymers, and Biocomposites* (pp. 699–725).

Mooibroek, H., & Cornish, K. (2000). Alternative sources of natural rubber. *Applied Microbiology and Biotechnology*, *53*, 355–365.

Mooney, B. P. (2009). The second green revolution? Production of plant-based biodegradable plastics. *The Biochemical Journal*, *418*, 219–232.

Nawrath, C., Poirier, Y., & Somerville, C. (1994). Targeting of the polyhydroxybutyrate biosynthetic-pathway to the plastids of *Arabidopsis thaliana* results in high-levels of polymer accumulation. *Proceedings of the National Academy of Sciences of the United States of America*, *91*, 12760–12764.

Neumann, K., Stephan, D. P., Ziegler, K., Huhns, M., Broer, I., & Lockau, W., et al. (2005). Production of cyanophycin, a suitable source for the biodegradable polymer polyaspartate, in transgenic plants. *Plant Biotechnology Journal*, *3*, 249–258.

Noda, I., Green, P. R., Satkowski, M. M., & Schechtman, L. A. (2005). Preparation and properties of a novel class of polyhydroxyalkanoate copolymers. *Biomacromolecules*, *6*, 580–586.

Oppermann-Sanio, F. B., & Steinbüchel, A. (2002). Occurrence, functions and biosynthesis of polyamides in microorganisms and biotechnological production. *Naturwissenschaften*, *89*, 11–22.

Pallos, F. M., Robertson, G. H., Pavlath, A. E., & Orts, W. J. (2006). Thermoformed wheat gluten biopolymers. *Journal of Agricultural and Food Chemistry*, *54*, 349–352.

Philip, S., Keshavarz, T., & Roy, I. (2007). Polyhydroxyalkanoates: Biodegradable polymers with a range of applications. *Journal of Chemical Technology and Biotechnology*, *82*, 233–247.

Poirier, Y. (1999). Production of new polymeric compounds in plants. *Current Opinion in Biotechnology*, *10*, 181–185.

Poirier, Y., & Brumbley, S. M. (2010). Metabolic engineering of plants for the synthesis of polyhydroxyalkanoates. In: Chen, G.-Q. (eds.). *Plastics from Bacteria: Natural functions and applications* (pp. 187–210).

Poirier, Y., Dennis, D. E., Klomparens, K., & Somerville, C. (1992). Polyhydroxybutyrate, a biodegradable thermoplastic, produced in transgenic plants. *Science*, *256*, 520–523.

Poirier, Y., & Gruys, K. J. (2001). Production of PHAs in transgenic plants. In: Y. Doi & A. Steinbüchel (Eds.), *Biopolyesters* (pp. 401–435). Weinheim: Wiley-VCH.

Poirier, Y., Nawrath, C., & Somerville, C. (1995). Production of polyhydroxyalkanoates, a family of biodegradable plastics and elastomers, in bacteria and plants. *Biotechnology Advances*, *13*, 142–150.

Polhamus, L. G. (1962). *Rubber. Botany, production, and utilization*. London: Leonard Hill Limited.

Ramseier, T. (2010). Biopolymers from North America. International Symposium on Biopolymers, Stuttgart, Germany.

Ruggiero, F., Exposito, J. Y., Bournat, P., Gruber, V., Perret, S., & Comte, J., et al. (2000). Triple helix assembly and processing of human collagen produced in transgenic tobacco plants. *FEBS Letters*, *469*, 132–136.

Sallam, A., & Steinbuchel, A. (2010). Dipeptides in nutrition and therapy: Cyanophycin-derived dipeptides as natural alternatives and their biotechnological production. *Applied Microbiology and Biotechnology*, *87*, 815–828.

Sanders, J., Scott, E., Weusthuis, R., & Mooibroek, H. (2007). Bio-refinery as the bio-inspired process to bulk chemicals. *Macromolecular Bioscience*, *7*, 105–117.

Sanford, K., & Kumar, M. (2005). New proteins in a materials world. *Current Opinion in Biotechnology*, *16*, 416–421.

Scheller, J., & Conrad, U. (2004). Production of spider silk proteins in transgenic tobacco and potato. In: R. Fischer & S. Schillberg (Eds.), *Molecular Farming* (pp. 171–181). Weinheim: Wiley-VCH.

Scheller, J., & Conrad, U. (2005). Plant-based material, protein and biodegradable plastic. *Current Opinion in Plant Biology*, *8*, 188–196.

Scheller, J., Gührs, K. H., Grosse, F., & Conrad, U. (2001). Production of spider silk proteins in tobacco and potato. *Nature Biotechnology*, *19*, 573–577.

Scheller, J., Henggeler, D., Viviani, A., & Conrad, U. (2004). Purification of spider silk-elastin from transgenic plants and application for human chondrocyte proliferation. *Transgenic Research*, *13*, 51–57.

Scott, E., Peter, F., & Sanders, J. (2007). Biomass in the manufacture of industrial products — the use of proteins and amino acids. *Applied Microbiology and Biotechnology*, *75*, 751–762.

Sharma, A. K., & Sharma, M. K. (2009). Plants as bioreactors: Recent developments and emerging opportunities. *Biotechnology Advances*, *27*, 811–832.

Shukla, R., & Cheryan, M. (2001). Zein: The industrial protein from corn. *Industrial Crops and Products*, *13*, 171–192.

Siler, D. J., Cornish, K., & Hamilton, R. G. (1996). Absence of cross-reactivity of IgE antibodies from subjects allergic to *Hevea brasiliensis* latex with a new source of natural rubber latex from guayule (*Parthenium argentatum*). *The Journal of Allergy and Clinical Immunology*, *98*, 895–902.

Slater, S., Mitsky, T. A., Houmiel, K. L., Hao, M., Reiser, S. E., & Taylor, N. B., et al. (1999). Metabolic engineering of *Arabidopsis* and *Brassica* for poly(3-hydroxybutyrate-co-3-hydroxyvalerate) copolymer production. *Nature Biotechnology*, *17*, 1011–1016.

Snell, K. D., & Peoples, O. P. (2009). PHA bioplastic: A value-added coproduct for biomass biorefineries. *Biofuels Bioproducts Biorefining*, *3*, 456–467.

Stein, H., Wilensky, M., Tsafrir, Y., Rosenthal, M., Amir, R., & Avraham, T., et al. (2009). Production of bioactive, post-translationally modified, heterotrimeric, human recombinant Type-I collagen in transgenic tobacco. *Biomacromolecules*, *10*, 2640–2645.

Steinbüchel, A., & Valentin, H. E. (1995). Diversity of bacterial polyhydroxyalkanoic acids. *FEMS Microbiology Letters*, *128*, 219–228.

Steinle, A., Bergander, K., & Steinbuchel, A. (2009). Metabolic engineering of *Saccharomyces cerevisiae* for production of novel cyanophycins with an extended range of constituent amino acids. *Applied and Environmental Microbiology*, *75*, 3437–3446.

Steinle, A., Witthoff, S., Krause, J. P., & Steinbuchel, A. (2010). Establishment of cyanophycin biosynthesis in *Pichia pastoris* and optimization by use of engineered cyanophycin synthetases. *Applied and Environmental Microbiology*, *76*, 1062–1070.

Stubbe, J., Tian, J. M., He, A. M., Sinskey, A. J., Lawrence, A. G., & Liu, P. H. (2005). Nontemplate-dependent polymerization processes: Polyhydroxyalkanoate synthases as a paradigms. *Annual Review of Biochemistry*, *74*, 433–480.

Thakor, N., Luetke-Eversloh, T., & Steinbuechel, A. (2005). Application of the BPEC pathway for large-scale biotechnological production of poly(3-mercaptopropionate) by recombinant *Escherichia coli*, including a novel *in situ* isolation method. *Applied and Environmental Microbiology*, *71*, 835–841.

Tilbrook, K., Gebbie, L., Schenk, P.M., Poirier, Y., & Brumbley, S.M. (2011). Peroxisomal polyhydroxyalkanoate biosynthesis is a promising strategy for bioplastic production in high biomass crops. *Plant Biotechnology Journal*, in press.

Tsao, G. T., Cao, N. J., Du, J., & Gong, C. S. (1999). Production of multifunctional organic acids from renewable resources. *Advances in Biochemical Engineering/Biotechnology*, *65*, 243–280.

van Beilen, J. B. (2008). Transgenic plant factories for the production of biopolymers and platform chemicals. *Biofuels Bioproducts Biorefining*, *2*, 215–228.

van Beilen, J. B., Möller, R., Toonen, M., Salentijn, E., & Clayton, D. (2007). Industrial crop platforms for the production of chemicals and biopolymers. In: D. Bowles (Ed.), Outputs of the EPOBIO project. Berks: CNAP, University of York.

van Beilen, J. B., & Poirier, Y. (2007a). Prospects for biopolymer production in plants. *Advances in Biochemical Engineering/Biotechnology*, *107*, 133–151.

van Beilen, J. B., & Poirier, Y. (2007b). Guayule and Russian dandelion as alternative sources of natural rubber. *Critical Reviews in Biotechnology*, *27*, 217–231.

van Beilen, J. B., & Poirier, Y. (2007c). Establishment of new crops for the production of natural rubber. *Trends in Biotechnology*, *25*, 522–529.

van Beilen, J. B., & Poirier, Y. (2008). Production of renewable polymers from crop plants. *Plant Journal*, *54*, 684–701.

Verbeek, C. J. R., & van den Berg, L. E. (2010). Extrusion processing and properties of protein-based thermoplastics. *Macromolecular Materials and Engineering*, *295*, 10–21.

Wenzel, G. (2006). Molecular plant breeding: achievements in green biotechnology and future perspectives. *Applied Microbiology and Biotechnology*, *70*, 642–650.

Werpy, T., & Pedersen, G., eds. (2004). Top value added chemicals from biomass, Volume 1 – Results of screening for potential candidates from sugars and synthesis gas. PNNL, NREL, EERE (report 8674), Oak Ridge.

Witholt, B., & Kessler, B. (1999). Perspectives of medium chain length poly(hydroxyalkanoates), a versatile set of bacterial bioplastics. *Current Opinion in Biotechnology*, *10*, 279–285.

Woerdeman, D. L., Veraverbeke, W. S., Parnas, R. S., Johnson, D., Delcour, J. A., & Verpoest, I., et al. (2004). Designing new materials from wheat protein. *Biomacromolecules*, *5*, 1262–1269.

Yang, J., Barr, L. A., Fahnestock, S. R., & Liu, Z. -B. (2005). High yield recombinant silk-like protein production in transgenic plants through protein targeting. *Transgenic Research*, *14*, 313–324.

Zeeman, S. C., Kossmann, J., & Smith, A. M. (2010). Starch: Its metabolism, evolution, and biotechnological modification in plants. *Annual Review of Plant Biology*, *61*, 209–234.

Zhao, M., & Running, S. W. (2010). Drought-induced reduction in global terrestrial net primary production from 2000 through 2009. *Science*, *329*, 940–943.

31

Bioenergy from plants and plant residues

Blake A. Simmons

Joint BioEnergy Institute, Emeryville, California

TABLE OF CONTENTS

Introduction

The development of efficient routes of bioenergy production, defined here as the outputs of biofuels (liquid transportation fuels) and biopower (electrons produced by the consumption of biomass), that are scalable, cost-competitive, and sustainable, and can potentially displace a significant amount of petroleum and coal within the global energy sector has re-emerged as a topic of significant global interest, effort, and investment over the past decade. This renewed interest is attributed to two major drivers: (1) concerns over energy security and the reliance on finite fossil fuel supplies; and (2) the significant environmental risks associated with unabated carbon emissions of these fossil fuels. In 2009, the United States consumed 94.6 quads (one quad = 10^{15} BTU = 1.055 exajoules [EJ]) of energy (see Figure 31.1), with the transportation and electricity sector accounting for 26.98 and 38.19 quads, respectively, with biomass contributing 3.88 quads. Globally, it is calculated that more than half of the current energy consumption of Earth is currently met by the consumption of liquid fossil fuels, and in 2009 over 20% of global carbon emissions were generated by the transportation sector alone (Energy Information Administration, 2009).

Recent estimates by the International Energy Agency (IEA) calculate that global demand for transport will increase by 45% by 2030, placing even further strains on a global fossil fuel production and distribution system that has reached its maximum in terms of production, and is a source of unmitigated carbon emissions. These constraints underscore the need for the realization and rapid commercialization of scalable and cost-effective means of generating low-carbon fuels and energy to meet the growing global demand in a sustainable and environmentally responsible fashion that promotes global energy security and environmental sustainability.

In light of these concerns, a significant number of governments worldwide are committed to displacing fossil fuels with renewable, low-carbon fuels produced from plant biomass grown on a sustainable basis. For example, the U.S. federal government has set a target of displacing 36 billion gallons of current U.S. petroleum consumption within the transportation sector by 2022 under the Renewable Fuel Standard (RFS2) legislation. With a production cap of 15 billion gallons per year placed on corn ethanol, this leaves a gap of 21 billion gallons of renewable liquid fuels per year that must be met by other sources and conversion technologies. Moreover, with total fossil fuel consumption within the transportation sector currently running at levels of ~200 billion gallons per year in the United States alone, the realization of advanced biofuels require the development of a significant commercial infrastructure capable of producing about this level of non-starch biofuels per year over a very short time frame. China, New Zealand, the European Union, United Kingdom, Brazil, Canada, and Australia are examples of others that have also established similar renewable bioenergy production targets, although most of the proposed deployment schedules have either been postponed or revised from the originally stated scale of production due to the global recession that started in 2008.

There are numerous non-food biomass feedstocks that are considered appropriate, viable, and potentially sustainable for biofuel and biopower production worldwide. The current primary metrics used in determining their commercial viability

DOI: 10.1016/B978-0-12-381466-1.00031-6

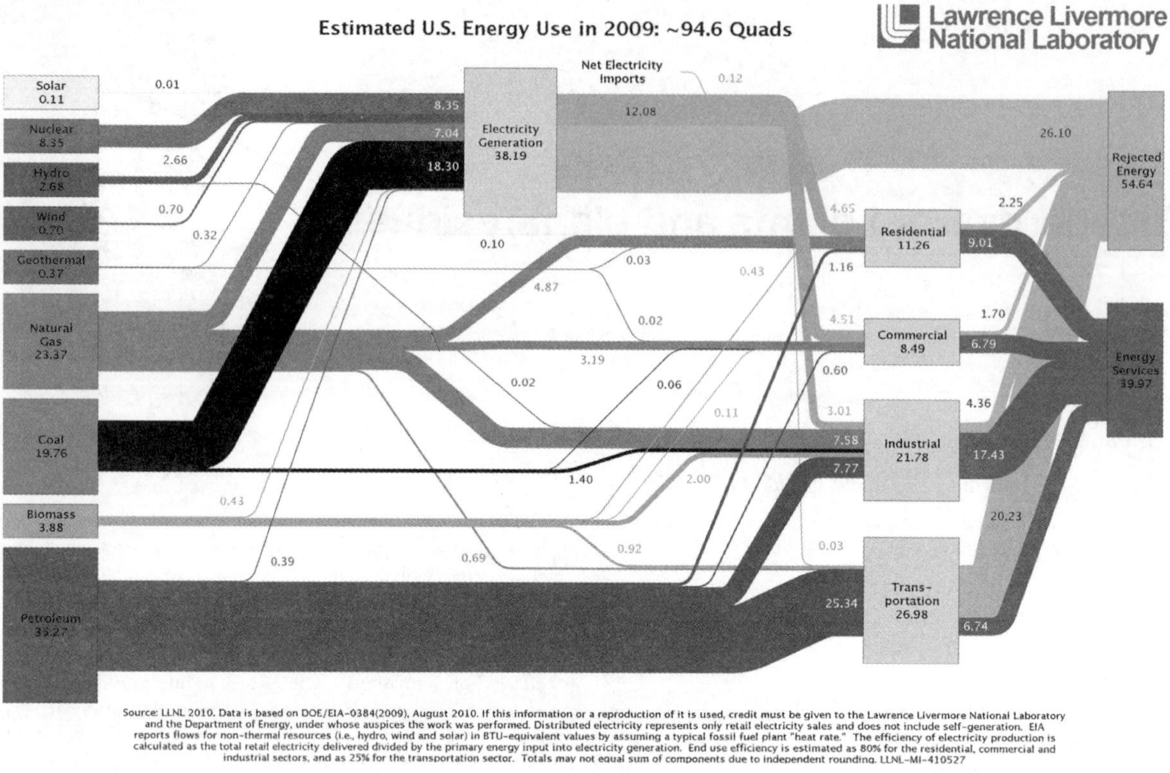

Figure 31.1 • Estimated energy use in the United States by source and sector • From Lawrence Livermore National Laboratory.

are cost, energy density, and availability. These include oilseeds (e.g., rapeseed, soy, palm oil), agricultural residues (e.g., corn stover, wheat straw, rice straw, sugarcane bagasse), woody biomass (e.g., loblolly pine, eucalyptus, hybrid poplar), and dedicated energy crops (e.g., energy cane, sorghum, switchgrass, *Miscanthus*). The selection of the most appropriate feedstock for a particular conversion technology is influenced by regional influences such as land availability, land ownership, rainfall, disposition of water rights, soil type, and use of fertilizers. A summary of the major current and future feedstocks by type and region is shown in Table 31.1, and the projections for global biofuel production are presented in Figure 31.2.

The assessments of the potential impact and deployment of bioenergy, and the land required to support it, vary widely. In one report, it was calculated that the available global agricultural energy is 25.9–33.5 EJ/year with the production of food crops accounting for 25.6 EJ/year, leaving little room for expanded biofuel production (Johansson et al., 2010). This report found that the primary agricultural resources available for bioenergy production are residues and bioorganic waste, and that the conversion of these resources could replace one-fourth of current global fossil fuel consumption (Johansson et al., 2010). In contrast, another report calculated that by 2050, the bioenergy potential available of land not needed for the production of feed and food could produce between 215–1272 EJ/year, with the majority of these resources located in sub-Saharan Africa, the Caribbean, and South America (Smeets et al., 2007). Some of the most critical (and controversial) issues around the sustainability and carbon benefits of increased bioenergy production are associated with how to measure and attribute direct and indirect land use change (Miller et al., 2009; Cherubini and Ulgiati, 2009). All energy sources have direct and indirect effects that must be accounted for from a life cycle perspective, but to date most of the attention has been focused on the evaluation of biofuels alone (Kim et al., 2009). Although this topic has sparked significant conversation and debate worldwide, there remains significant uncertainty in the modeling tools developed to date that can accurately assess and attribute these effects and, more important, be validated against known data sets (Mullins et al., 2011).

As with all terrestrial systems of production, the ultimate potential for bioenergy depends to a great extent on the land available for production and the realization of sustainable methods of growing and harvesting this plant biomass resource responsibly. Currently, the amount of land devoted to growing energy crops for biomass fuels is calculated to represent only 0.19% of the world's total land area and only 0.5–1.7% of the available agricultural land. Although there

Table 31.1 Summary of global feedstock availability by type and region

	Sugar, starch, and oil crops					Cellulosic feedstocks		
	Sugarcane	Corn	Wheat	Palm	Soybeans	Bagasse	Agricultural residue	Other
Argentina	(✓)	(✓)	✓		✓	✓	✓	✓
Brazil	✓	✓			✓	✓	✓	✓
Canada		✓	✓				✓	✓
China	(✓)	(✓)	(✓)		(✓)	✓	✓	✓
Colombia	(✓)			(✓)		✓		✓
India	✓					✓		✓
Mexico	✓	✓				✓	✓	✓
Caribbean Basin (CBI)	✓			✓		✓		✓
Share of world production (excluding U.S.)	73%	55%	27%	3%	81%	–	–	–

Source: World Biofuels Production Potential, Office of Policy Analysis, U.S. Department of Energy, 2008.

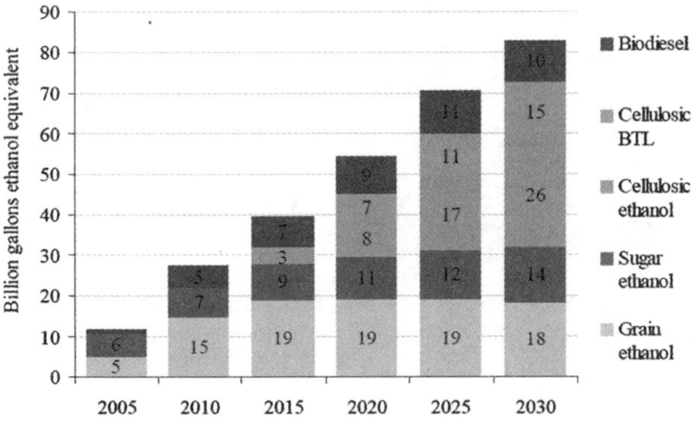

Figure 31.2 • Projected levels of global biofuel production by type and conversion technology • From *World Biofuels Production Potential*, Office of Policy Analysis, U.S. Department of Energy (2008).

remains significant concerns around the sustainability of non-food feedstocks grown for biofuel and biopower production (Cherubini and Stroemman, 2010), the "food vs. fuel" debate (Knocke and Vogt, 2009), and other related issues such as centralized versus distributed modes of bioenergy production (Serio et al., 2003), there is little doubt that plant biomass has a significant role to play in the renewable energy sector, especially in biopower and biofuel production if it is realized using rational, responsible, and reasonable development and deployment schedules (Larson, 2006). This chapter breaks down the ongoing plant biomass activities into two main categories based on the type of conversion technology employed: (1) biochemical conversion; and (2) thermochemical conversion.

Biochemical Conversion

The biochemical conversion route to biofuels has been extensively studied for centuries, primarily owing to the well-established route of converting starches and grains into ethanol that we benefit from for a wide range of applications, including human consumption. So-called "first-generation" biofuels are defined as those derived from starches (e.g., corn) and vegetable oils (e.g., soybean oil; Rude and Schirmer, 2009). These biofuels have rapidly achieved significant production levels, especially in Brazil and the United States, and have proven that biofuels are an effective means to displace fossil fuels. These first-generation biofuels have also generated significant concerns around sustainability, environmental impact, and direct and indirect land use change. This section will not present any details of these "first-generation" biofuels, but will instead focus on the much more complex and challenging pathway of converting non-food lignocellulosic biomass into a wide range of biofuels and co-products (Plass and Reimelt, 2007). A general overview of the biochemical conversion pathway for lignocellulosic biomass into biofuels is shown in Figure 31.3.

The plant cell walls in biomass — primarily comprised of cellulose, hemicellulose, and lignin — are more complex and heterogeneous than starches and therefore require more energy as an input into the biofuel conversion process (Himmel et al., 2007). The polysaccharides present in biomass that account for between 50–75% of the dry weight of biomass are the primary targets for biofuel conversion. The 5-C (e.g., xylose) and 6-C (e.g., glucose) sugars that are the monomeric units of hemicellulose and cellulose, respectively, are ideal carbon sources for several microorganisms that produce biofuels through fermentation, such as *Saccharomyces*

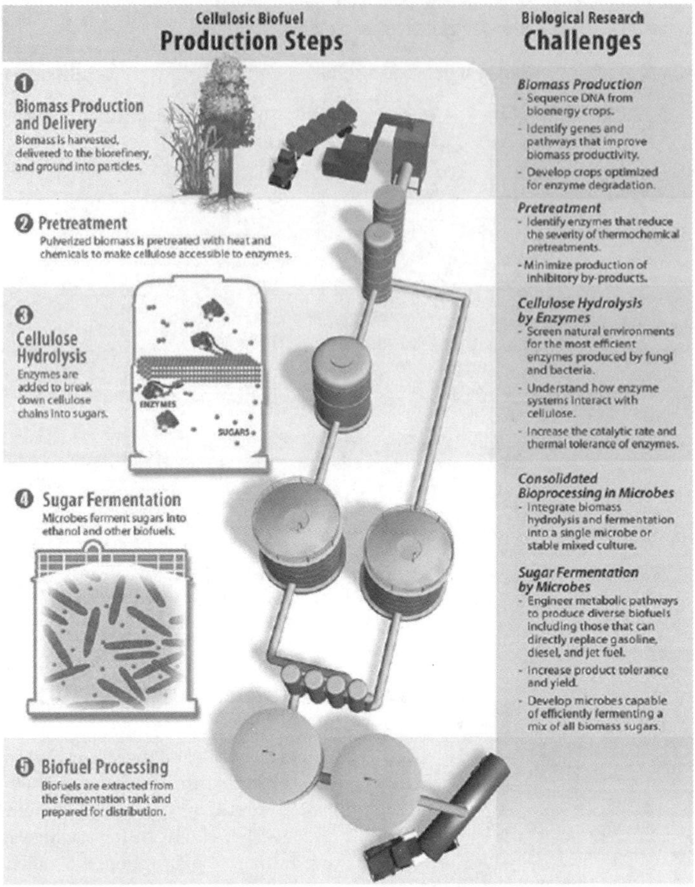

Figure 31.3 • Schematic depiction of the major unit operations and associated challenges for the efficient conversion of plant biomass into biofuels, such as ethanol • From U.S. Department of Energy, genomicsgtl.energy.gov/centers/brcbrochure.pdf.

cerevisiae (yeast; Argueso et al., 2009). One of the great barriers to the commercialization of second-generation biofuels is the lack of viable technologies that can efficiently and cheaply liberate these monomeric sugars from lignocellulose (Kumar et al., 2009a). In addition to developing new and improved methods of breaking down biomass into fermentable sugars, there has been significant interest in genetically engineering biomass feedstocks that are easier to break down and/or fractionate into desired output streams (Vanhohlme et al., 2010; Lorenz et al., 2010). This section will provide a brief overview of the biochemical conversion pathway and highlight some of the challenges and opportunities for each unit operation.

Comminution

The first step in the conversion of biomass is to reduce its particle size to a manageable range, starting with the initial harvesting that breaks down the whole plant by chopping and/or cutting. It should be noted that the development of specialized equipment for the harvesting of biomass feedstocks has received significant attention recently, but no commercial harvesters designed specifically for proposed dedicated energy crops such as switchgrass and *Miscanthus* remain, although several appear to be near the early stages of commercialization (Sokhansanj et al., 2009). Further particle size reduction, or comminution, can be achieved using several approaches such as milling, grinding, and chipping. Particle size reduction increases the active surface area of the biomass and is known to increase downstream process efficiency in terms of sugar yields, but there is an inverse relationship between particle size and the energy required for the operation. The energy requirements for comminution are influenced by factors including the initial and final particle sizes desired, moisture content, biomass type, and feed rate (Miao et al., 2011). Chipping typically results in biomass particles between 10–30 mm in size, while grinding and milling result in small biomass particles between 0.2–2.0 mm in size. If the

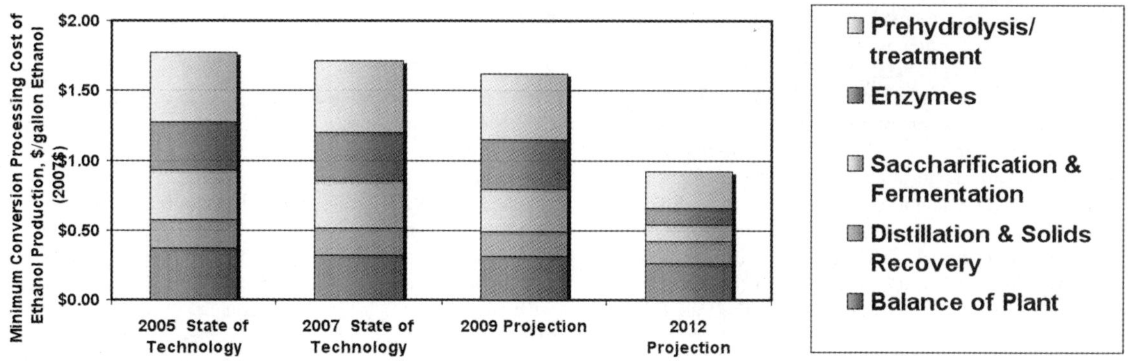

Figure 31.4 • Projected cost of a gallon of cellulosic ethanol based on biochemical conversion of corn stover that equates to a Minimum Ethanol Selling Price of $1.49 in 2012 • From U.S. Department of Energy, Office of Biomass Program.

targeted particle size is between 3–6 mm, the energy required is ~30 kWh/tonne biomass. Some of the more established milling techniques that can efficiently generate very small particles are rotary knife and hammer mills (0.2–2 mm). It has been reported that the composition of the biomass can be influenced by particle size (Bridgeman et al., 2006).

Pre-treatment

Pre-treatment is one of the critical steps in the biochemical conversion pathway, and is generally considered as one of the most costly unit operations within any proposed lignocellulosic biorefinery (see Figure 31.4). The primary goal of the pre-treatment process is to increase the accessibility of the polysaccharides to hydrolytic enzymes and, if possible, reduce the lignin content and decrease cellulose crystallinity. The majority of the biomass pre-treatment techniques studied have been derived from those used in the pulp and paper industry. These include the following list.

Dilute acid pre-treatment: The use of dilute acids to pre-treat biomass is one of the most mature technologies available. The dilute acids, typically sulfuric acid, can effectively solubilize the majority of the hemicellulose present in biomass and generate a material that is more readily hydrolyzed than untreated biomass (Guo et al., 2008; Foston and Ragauskas, 2010). Some of the disadvantages of this approach are the formation of microbial inhibitors from sugar degradation at higher temperatures, limited impact on lignin except at high temperatures (Selig et al., 2007), and retention of cellulose crystallinity. There have been recent examples where the use of concentrated acids has been shown to be effective as a pre-treatment alternative.

Liquid hot water pre-treatment (hydrothermolysis): Solvolysis by hot liquid water at 200–230°C for 15 min at 350–400 psig results in conversion of all hemicellulose to monomeric sugars, as well as the removal of 4–22% of the cellulose and 35–60% of the lignin (Bonn et al., 1983; Bobleter et al., 1983; Burtscher et al., 1987). The remaining solids have improved enzymatic digestibility relative to untreated biomass. Extensive size reduction of biomass is not required as hot water pre-treatment causes lignocellulosics to break apart, but softwoods are less susceptible to the pre-treatment and the effects on lignin are poorly understood.

Ammonia fiber expansion (AFEX) pre-treatment: This process yields very high rates of enzymatic hydrolysis with low enzyme loadings, and is best suited for agricultural residues (Balan et al., 2009). This technique can use gaseous or liquid ammonia. For the gas phase pre-treatment, gaseous ammonia is introduced into the pre-treatment reactors at specific temperatures and pressures and then released and the ammonia recovered (Lau et al., 2010). For the liquid phase pre-treatment, typically a 5–15% ammonia solution at 160–180°C is passed through the biomass, and the aqueous ammonia reacts, depolymerizes, and solubilizes the lignin, thus improving enzymatic hydrolysis (Gupta and Lee, 2010).

Ionic liquid (IL) pretreatment: A relatively new biomass pre-treatment technique utilizes room temperature molten salts as solvents for the dissolution and pre-treatment of lignocellulosic biomass. Several ionic liquids show substantial cellulose solubility; for example, 3-methyl-N-butylpyridinium chloride [C4ampy]Cl and 1-n-butyl-3-methylimidazolium chloride [C4mim]Cl have reported cellulose solubilities of 39 and 25%, respectively (Dadi et al., 2007). Regenerated cellulose can be precipitated with antisolvents such as water or alcohols with significantly reduced crystallinity and lignin content (Cetinkol et al., 2010). Ionic liquid pre-treatment of switchgrass has been shown to be superior to dilute acid pre-treatment in terms of sugar yield and enzyme kinetics (Li, et al., 2010). Although IL pre-treatment is promising, the issues around cost and recycling of the IL must be fully addressed before this technology becomes truly competitive with other established pre-treatment technologies (Simmons et al., 2010).

Organosolv pre-treatment: Lignocellulosics can be fractionated by hot aqueous ethanol, typically with sulfuric or Lewis acids acting as catalysts (Zhu and Pan, 2010). The resulting liquor contains substantial amounts of solubilized lignin, and the pulp contains the residual lignin and most of the carbohydrates that were initially present in the plant biomass are separated. Whereas lignin is recovered from the aqueous ethanol solvent by dilution with water, subsequent washing of the pulp yields a material suitable for subsequent enzymatic hydrolysis and sugar production at comparatively high yields relative to untreated biomass (Brosse et al., 2010; Mesa et al., 2010; Park et al., 2010).

Saccharification

As with pre-treatment, the enzymes required for hydrolyzing the polysaccharides into monomeric sugars are one of the most significant costs at a biorefinery (see Figure 31.4; Joergensen et al., 2007). Typically, there are several enzymes required to break down polysaccharides into monomeric sugars, commonly referred to as "enzyme cocktails." For cellulose, the enzyme cocktail is comprised of 5–10 different enzymes that possess three general activities — endocellulase, exocellulase, and β-glucosidase — whereas for hemicellulose, which is much more complex in terms of the functional groups present, a typical enzyme cocktail contains between 14 to 20 different enzymes. In some anaerobic microorganisms, these saccharolytic enzymes are immobilized within a scaffold, known as a cellulosome (Hemme et al., 2010; Nataf et al., 2010; Sakka et al., 2010; Tamaru et al., 2010; Vodovnik and Logar, 2010), that improves overall sugar yields from biomass and localizes sugar production at or near the cellular membrane. In addition, the catalytic domains of the enzymes are sometimes linked with a carbohydrate binding module that, as its name implies, binds to the polysaccharide and increases the time and specificity of the enzyme-substrate interaction (Abbott et al., 2009; Michel et al., 2009; Nakahira et al., 2009; Wilson et al., 2009; Su et al., 2010). There are known synergies between the enzymes present (Wen et al., 2010), and there are non-catalytic proteins that "boost" enzyme performance dramatically. Thus, the development of optimal enzyme cocktails for a wide range of substrates requires a comprehensive understanding of the interactions and chemistries involved.

Most of the commercial cellulases and hemicellulases are derived from fungal hosts, such as *Trichoderma reesei* or *T. viride*. These enzymatic "cocktails" perform well in specific operating environments and on certain substrates, but they are not ideally suited for industrial applications such as high temperature and extremes of pH. The current approach taken in the biofuels community typically develops a tailored enzyme cocktail for each pairing of feedstock and pre-treatment, which is far from an optimal process. One of the key components in realizing improved enzymes relevant to lignocellulose deconstruction is the establishment of a fundamental and comprehensive library of the specific activities, operational performance limits, and kinetics of the reactions that take place (Wilson et al., 2009). While the majority of the industrial enzymes known to date have been derived from bacteria and fungi, Archaea, which represent the third domain of life, are increasingly finding applications in biotechnology (Datta et al., 2010). This is in no small part due to the fact that most of the archaeal species have been identified from extreme environments such as geothermal and deep-sea volcanic sites (80–121°C), polar regions (−20°C), acidic solfatara fields (pH <4), alkaline springs (pH >8), and hypersaline lakes (2–5 M NaCl), that were thought to be unable to support life. Recent studies have indicated that these environments are heavily populated with archaeal microorganisms with very specialized metabolisms. These discoveries, coupled with the advances in genome sequencing and bioinformatics, offer an opportunity to harvest genes compatible with the integration of process steps associated with cellulosic biomass deconstruction (Wu et al., 2009; Hemme et al., 2010; Hess et al., 2011).

This is especially true in the case of integration of pre-treatment and hydrolysis at elevated temperatures – a critical link that must be established and realized on a commercial scale. Several β-glucosidases have been detected in *Sulfolobales* species, an archaeaon that grows optimally at high temperatures (>70°C) and low pH (pH 2–3) and offers the possibility of a simultaneous enzymatic hydrolysis and pre-treatment of cellulosic biomass. Also, thermostable endoglucanases, which degrade β-1,4 linkages of β-glucans and cellulose, have been identified in *Pyrococcus furiosus*, *P. horikoshii*, and *Sulfolobus solfataricus* (Blumer-Schuette et al., 2008; Chou et al., 2008). The recombinant endoglucanase from *P. furiosus* is active at 100°C and hydrolyzes β-1,4, but not β-1,3, glycosidic linkages, whereas the endoglucanase from *P. horikoshii* has been shown to be active even toward crystalline cellulose (Kim and Ishikawa, 2010). These archaeal enzymes are promising candidates to optimize bioconversion of recalcitrant cellulose to fermentable sugars by running the process of saccharification at lower pH and higher temperatures and serve as model systems for the bioinformatics analysis of the process. The industrial process benefits of using enzymes from extremophiles include reduced risk of contamination, improved mass transfer rates, lower viscosity, and higher solubility of substrates. There have also been recent efforts in expressing these enzymes *in planta* to improve sugar yields and decrease the amount of exogenous enzyme required (Nakahira et al., 2009; Patell et al., 2010).

Fuel synthesis

The basic central metabolic pathway at the heart of all biochemical biofuel production is the conversion of sugars through fermentation. There are several variants of the overall biochemical conversion process shown in Figure 31.3, and some of the most promising are those that involve metabolic engineering and/or synthetic biology to generate pathways that efficiently produce advanced biofuels from sugars that are closer in energy density and water miscibility to their petroleum counterparts when compared to ethanol. The ability to simultaneously ferment 5-C and 6-C sugars is also highly desirable from an operational standpoint, as long as the overall biofuel titers either remain constant or improve.

Another goal is the development of an organism that secretes hydrolytic enzymes and ferments the liberated sugars into an advanced biofuel (aka consolidated bioprocessing; Stanley and Hahn-Haegerdal, 2010; Linger et al., 2010; Hyeon et al., 2011; Anasontzis et al., 2011).

Researchers have engineered and classified the modular biological components that perform specific input/output functions in order to understand and manipulate the genetic circuits present, and in the case of biofuels the targeted output is a specific molecule suitable for use as a blendstock for gasoline, diesel, and jet fuels (Peralta-Yahya and Keasling, 2010). This effort relies upon the generation of a "Parts Registry" that establishes and standardizes well-known metabolic components as a genetic toolbox that can then be incorporated into industrially relevant and genetically tractable microorganisms, such as *S. cerevisiae* and *Escherichia coli*. Once a basal level of output is achieved in these organisms, the system is then optimized further to attain higher production levels of the targeted compounds, as well as further enhancements with other parts derived from other metabolic pathways to produce a commercially viable biofuel technology, such as fatty acid esters (Steen et al., 2010).

A great deal of attention over several decades has been paid to the production of cellulosic ethanol (Gray, 2007), but recently there has been a growing effort in the realization of "drop-in" biofuels such as butanol (Duerre, 2007), isobutanol (Taylor et al., 2010), isopentanol, farnesene, and fatty acid esters (Steen et al., 2010) that intrinsically possess higher energy densities than ethanol and would be more compatible in terms of water miscibility with the existing petroleum infrastructure that stores and distributes liquid fuels (Peralta-Yahya and Keasling, 2010).

Thermochemical Conversion

There are three basic types of thermochemical conversion technologies for the processing of biomass: gasification; pyrolysis; and combustion (Kumar et al., 2009b). The combustion of biomass is the simplest and most mature of these in terms of processing biomass into a desired singular output — heat — and the overall efficiencies are low in comparison to the other two. In comparison to biochemical conversion technologies, the thermochemical conversion technologies have the following advantages: most can use a wide range of biomass feedstocks with little or no impact on efficiency; in comparison to ethanol, the fuels typically generated are more compatible with the existing petroleum infrastructure; and the gaseous volatile organic product streams can be converted into a variety of fuels and chemicals using known catalytic processes (Wang et al., 2008). The primary disadvantages are that the thermochemical approaches are typically more costly than the biochemical routes and require extensive post-processing to remove unwanted contaminants, high energy demands due to high process temperatures, and the instability of some of the products formed (e.g., pyrolysis oil). The thermochemical routes also enable the realization of multiple processing options for biofuel and biopower generation (Kumar et al., 2008), such as an integrated gasifier combined cycle (IGCC) and consolidated heat and power generation at biorefineries. There have also been recent developments around hybrid thermochemical–biochemical approaches, such as the fermentation of syngas (Munasinghe and Khanal, 2010), although this section will focus on gasification and pyrolysis.

Pyrolysis

Pyrolysis is a technology that can trace its origins back as far as ancient Egypt, where pyrolysis of wood was used to produce tars for boats and embalming agents. Wood pyrolysis became a very large industry in the 1800s to produce charcoal, with a heating value equal to coal (~24 MJ/kg) but no sulfur content, which helped supply power for the industrial revolution. Biomass pyrolysis is classically defined as the chemical decomposition of lignocellulose by rapid heating that typically occurs in the absence of oxygen, or at least reduced levels of oxygen relative to those required for complete combustion (Demirbas, 2009). The pyrolysis process does not directly involve chemical reactions with oxygen or with any other chemical components and the biomass.

The biomass pyrolysis process typically results in the generation of a wide range of products — including gases, solids, and liquids — and the stoichiometric balance achieved between these three products, and the composition of each, is determined by the processing conditions used (Mohan et al., 2006). For instance, high yields of the liquid product tend to favor short residence times, fast heating rates, and moderate processing temperatures. The solid pyrolysis product is often referred to as "bio-char," which can be used as a soil amendment or as a feedstock for gasification (Demirbas, 2006). In general, the liquid pyrolysis product is referred to as "pyrolysis oil" or "bio-oil" (Balat, 2011). Bio-oil from biomass pyrolysis consists of a mixture of more than 200 compounds including aliphatic, aromatic, and naphthenic hydrocarbons as well as a variety of oxygenated compounds including furans, ethers, ketones, and aldehydes (Balat, 2011). Bio-oil has a higher energy density than the initial biomass feedstock, can be readily stored and transported although there are some stability and storage issues that must be addressed, and can be used either as a renewable liquid transportation fuel and/or for the production of chemical products. These bio-oil product streams can theoretically be fed into a conventional refinery stream and, through hydrotreating and hydrocracking (French et al., 2010), produce fuels equivalent and/or chemically identical to conventional petroleum products (Xu et al., 2010).

The basic process steps for biomass pyrolysis include: (1) heat transfer from a heat source that heats the biomass; (2) initiation of the pyrolysis reaction in the biomass due to the increased temperature; (3) product formation in the vapor phase, resulting in heat transfer between the hot vapor products and the unreacted biomass; (4) condensation of products from the vapor phase; and (5) autocatalytic reactions between the products (Babu et al., 2008). The kinetics of the pyrolysis reaction are dictated by the reaction conditions and reactor configurations (heating rate, feed rate, particle size, etc.). There are three proposed reaction mechanisms for the pyrolysis of biomass: (1) single-step global models; (2) single-stage multi-reaction models; and (3) two-stage semi-global models.

There are several different modes of biomass pyrolysis that are being actively developed by academics and industry. A short synopsis of each is presented in the following list.

Conventional pyrolysis: This is a comparatively low temperature, long residence time technology that has existed for thousands of years that has been optimized for the production of charcoal, or bio-char, from biomass, traditionally from woody biomass feedstocks (Demirbas and Arin, 2002). In a typical conventional pyrolysis process, the biomass is slowly heated to ~600°C, with residence times between 5 and 30 minutes, and the molecules in the vapor product continue to react as they are removed from the system. The bio-char product has been proposed as a significant means of greenhouse gas reduction through soil amendments that could result in long-term carbon sequestration. A recent report carried out a life cycle assessment to determine the potential climate change impacts and economics of bio-char pyrolysis systems and that the net greenhouse gas emissions for corn stover is − 864 kg CO_2e/tonne (Roberts et al., 2010).

Fast pyrolysis: This is a high temperature process that involves rapid heating of the biomass to generate vapors, aerosols, and solids. There are four characteristic steps that define fast pyrolysis: (1) high heating and heat transfer rates (~500–1000°C/s); (2) carefully controlled pyrolysis reaction temperature typically between 425 and 600°C; (3) short residence times for the vapor phase (0.5–5 s); and (4) bio-oil is formed through condensing the vapor and aerosol product streams (Yanik et al., 2007). Typical reaction stoichiometries for the fast pyrolysis process are 10–20 wt% gases, 15–25 wt% solids, and 60–75 wt% bio-oil. The bio-oil is produced through rapid heating and rapid quenching and the high reaction rates reduce char formation and favor the production of either gas or liquid products.

Catalytic fast pyrolysis: One variant of the fast pyrolysis process is catalytic fast pyrolysis — a technique that combines rapid heating and the formation of volatile organics, gases, and char — with zeolite catalysts to transform the organic pyrolysis products (Carlson et al., 2011). The volatile organics generated are delivered to a zeolite catalyst bed and are converted into aromatics, carbon monoxide, carbon dioxide, water, and coke (Carlson et al., 2008, 2010). During the zeolite catalysis step, the organics undergo a series of reactions: dehydration; decarbonylation; decarboxylation; isomerization; oligomerization; and dehydrogenation reactions (Huber, 2010). Although this technology offers several advantages relative to other pyrolysis technologies, challenges are associated with reaction selectivity and specificity, generation of coke as an undesired co-product, and the stability and cost of the catalysts. Three parameters are used to maximize the production of desired products in this process: (1) rapid heating rates; (2) identification of the optimal catalyst; and (3) high catalyst-to-feed ratios.

Thermal plasma pyrolysis: In this variant of pyrolysis, the biomass particles are exposed to very rapid heating in a highly reactive plasma (e.g., argon/hydrogen) arc discharge (Babu, 2008). The products from the thermal plasma pyrolysis are gases and solids with no liquid output. During the rapid heating stage, there is an explosive release of volatile organics followed by gasification of the homogeneous phase (Zhao et al., 2001). The volatile organics are converted into low molecular weight hydrocarbons (e.g., acetylene, methane) and hydrogen.

Gasification

There are four basic unit operations in biomass gasification: (1) the pre-processing of biomass; (2) biomass gasification and generation of syngas; (3) syngas clean-up and reforming; and (4) gas utilization. Gasification of lignocellulosic biomass provides an extremely flexible way to produce different liquid fuels, primarily through Fischer-Tropsch Synthesis (FTS) or mixed alcohol synthesis, of the generated syngas. Syngas is comprised of CO, CO_2, H_2O, H_2, and a mix of impurities. The FTS technology is a relatively mature technology and has been applied to syngas generated from coal gasification where the syngas is converted into usable liquid fuels through the water–gas shift reaction coupled with CO hydrogenation (Lappas and Heracleous, 2011).

There are several types of gasifiers that have been studied, developed, and/or commercialized (Wang et al., 2008; Balat et al., 2009). The general classifications divide the different types based on flow and the type of bed utilized — the bed can be fluidized or fixed (updraft or downdraft; Melgar et al., 2009). Gasification occurs at high temperatures (600–1000°C) and uses gasification agents (e.g., oxidizing agents) such as steam, nitrogen, air, oxygen, carbon dioxide, or a combination of these. The combination of high temperature and gasification agents breaks down the biomass components into smaller molecular weight fragments and ultimately into gases such as light hydrocarbons, methane, hydrogen, and carbon monoxide (Hanaoka et al., 2008). The efficiency of syngas production is directly impacted by the biomass flow rate and type, the air and steam flow rate, and the temperature profile of the gasifier (Babu and Sheth, 2007).

The conversion of syngas using FTS enables the production of a wide variety of fuels with targeted properties. FTS also affords the production of multiple products simultaneously, and it is inherently more flexible than other conversion technologies. As FTS is an exothermic process, it should be possible to recover some of the heat generated to assist in drying the biomass with a regenerative heat exchanger. The most significant challenges in using FTS for plant biomass are similar to those associated with the conversion of coal, with the exception that contaminants generated by upstream process steps need to be removed before being delivered to the FTS catalyst bed. Additionally, FTS typically requires massive centralization to make the process cost-competitive and maximize efficiency. The most serious challenges associated with the FTS process are the cost of cleaning up and reforming the tars that can develop and accumulate over time (Carpenter et al., 2008). Theses tars can create multiple problems including fouling of particulate filters, clogging fuel line filters in vehicles, and coke formation in the FTS catalyst bed (Nimlos et al., 2009). It has been reported that tar formation can be minimized or avoided by using entrained-flow gasifiers that operate at high temperatures (Cui and Turn, 2009).

Concluding Remarks

There is a pressing need for the development of scalable and competitive technologies that can efficiently convert renewable lignocellulosic feedstocks into second-generation (and beyond) biofuels. There are multiple pathways being researched and developed across the globe that are based on biochemical, thermochemical, and hybrid approaches. For the biochemical conversion platform, there is a clear need for horizontally integrated systems of technologies that can efficiently handle a wide range of feedstocks into targeted advanced biofuels. This includes the development of feedstocks tailored for energy production, the development of more efficient enzymes and pre-treatments, and the development of robust microorganisms that can attain high titers of these advanced biofuels. To reach maturity and commercial viability, a more fundamental understanding of the entire system of plants, pre-treatments, enzymes, and enzyme complexes responsible for biomass deconstruction must be realized and deployed.

Despite the advances in genomics, proteomics, transcriptomics, and metabolomics, currently there are an insufficient number of microbial genomes and synthetic biology transformation systems. Further advances in biomass pre-treatment, enzyme structure–function relationships, enzyme–enzyme interaction, systems biology, and bioinformatics will create the toolbox required to generate these next-generation biorefineries. Substantial progress has also been made in thermochemical conversion technologies, but there remain several obstacles for pyrolysis and gasification that must be addressed before they are truly cost-competitive and scalable. For pyrolysis, central issues are bio-oil stability and the efficiency of catalytic upgrading of this material into fuels and chemicals. For gasification, the issues are those associated with the requirement for centralized conversion facilities that are significant in scale and catalyst development.

Regardless of the particulars of the proposed conversion pathway, progress that is made in the controlled laboratory environment must also be tested and validated by extensive pilot scale trials to generate the appropriate material and energy balances required to demonstrate the commercial viability of any new biofuel production technology. In conclusion, it is certain that plant biomass will play a central role in the development of low-carbon, renewable energy supplies capable of displacing fossil fuels, but the extent and time required to achieve significant levels of production will be determined by public policy, sustainability metrics, available capital funds, and sustained public and private funding of research and development activities.

Acknowledgment

This work conducted by the joint BioEnergy Institute was supported by the Office of Science, Office of Biological and Environmental Research, of the U.S. Department of Energy under contract No. DE-AC02-05CH 11231.

References

Abbott, D. W., et al. (2009). Analysis of the structural and functional diversity of plant cell wall specific family 6 carbohydrate binding modules. *Biochemistry*, *48*(43), 10395–10404.

Anasontzis, G. E., et al. (2011). Homologous overexpression of xylanase in *Fusarium oxysporum* increases ethanol productivity during consolidated bioprocessing (CBP) of lignocellulosics. *Journal of Biotechnology*, *152*(1–2), 16–23.

Argueso, J. L., et al. (2009). Genome structure of a *Saccharomyces cerevisiae* strain widely used in bioethanol production. *Genome Research*, *19*(12), 2258–2270.

Babu, B. V., & Sheth, P. N. (2007). Modeling and simulation of reduction zone of downdraft biomass gasifier: Effect of air to fuel ratio. *Journal of Future Engineering and Technology*, *2*(3), 35–40.

Babu, B. V. (2008). Biomass pyrolysis: A state-of-the-art review. *Biofuels, Bioproducts and Biorefining*, *2*(5), 393–414.

Balan, V., et al. (2009). Lignocellulosic biomass pretreatment using AFEX. *Methods in Molecular Biology (Totowa, NJ, U.S.)*, *581*, 61–77. (Biofuels).

Balat, M. (2011). An overview of the properties and applications of biomass pyrolysis oils. *Energy Sources, Part A*, *33*(7), 674–689.

Balat, M., et al. (2009). Main routes for the thermo-conversion of biomass into fuels and chemicals. Part 2: Gasification systems. *Energy Conversion Management*, *50*(12), 3158–3168.

Blumer-Schuette, S. E., et al. (2008). Extremely thermophilic microorganisms for biomass conversion: Status and prospects. *Current Opinion in Biotechnology*, *19*(3), 210–217.

Bobleter, O., Bonn, G., & Concin, R. (1983). Hydrothermolysis of biomass—production of raw material for alcohol fermentation and other motor fuels. *Alternative Energy Sources*, *3*(3), 323–332.

Bonn, G., Concin, R., & Bobleter, O. (1983). Hydrothermolysis—a new process for the utilization of biomass. *Wood Science and Technology*, *17*(3), 195–202.

Bridgeman, T. G., et al. (2006). Influence of particle size on the analytical and chemical properties of two energy crops. *Fuel*, *86*(1–2), 60–72.

Brosse, N., et al. (2010). Dilute sulphuric acid and ethanol organosolv pretreatment of *Miscanthus x Giganteus Cellulose Chemistry and Technology*, *44*(1–3), 71–78.

Burtscher, E., et al. (1987). Chromatographic analysis of biomass reaction products produced by hydrothermolysis of poplar wood. *Journal of Chromatography*, *390*(2), 401–412.

Carlson, T. R., et al. (2010). Catalytic fast pyrolysis of glucose with HZSM-5: The combined homogeneous and heterogeneous reactions. *Journal of Catalysis*, *270*(1), 110–124.

Carlson, T. R., et al. (2011). Production of green aromatics and olefins by catalytic fast pyrolysis of wood sawdust. *Energy Environmental Science*, *4*(1), 145–161.

Carlson, T. R., Vispute, T. P., & Huber, G. W. (2008). Green gasoline by catalytic fast pyrolysis of solid biomass derived compounds. *Chem Sus Chem*, *1*(5), 397–400.

Carpenter, D., et al. (2008). Temperature and feedstock effects on tar formation during pilot-scale biomass gasification. *Preprint Symposium American Chemical Society, Division of Fuel Chemistry*, *53*(1), 425–426.

Cetinkol, O. P., et al. (2010). Understanding the impact of ionic liquid pretreatment on eucalyptus. *Biofuels*, *1*(1), 33–46.

Cherubini, F., & Stroemman, A. H. (2010). Production of biofuels and biochemicals from lignocellulosic biomass: Estimation of maximum theoretical yields and efficiencies using matrix algebra. *Energy & Fuels*, *24*(4), 2657–2666.

Cherubini, F., & Ulgiati, S. (2009). Crop residues as raw materials for biorefinery systems—A LCA case study. *Applied Energy*, *87*(1), 47–57.

Chou, C.-J., et al. (2008). Hydrogenesis in hyperthermophilic microorganisms: Implications for biofuels. *Metabolic Engineering*, *10*(6), 394–404.

Cui, H., & Turn, S. Q. (2009). Tar quantification from a fluidized bed biomass gasification – gas chromatographic and gravimetric methods. *Preprint Symposium American Chemical Society, Division of Fuel Chemistry*, *54*(1), 198–199.

Dadi, A. P., Schall, C. A., & Varanasi, S. (2007). Mitigation of cellulose recalcitrance to enzymatic hydrolysis by ionic liquid pretreatment. *Applied Biochemistry and Biotechnology*, *137-140*, 407–421.

Datta, S., et al. (2010). Ionic liquid tolerant hyperthermophilic cellulases for biomass pretreatment and hydrolysis. *Greem Chemistry*, *12*(2), 338–345.

Demirbas, A., & Arin, G. (2002). An overview of biomass pyrolysis. *Energy Sources*, *24*(5), 471–482.

Demirbas, A. (2006). Production and characterization of bio-chars from biomass via pyrolysis. *Energy Sources, Part A*, *28*(5), 413–422.

Demirbas, A. (2009). Pyrolysis of biomass for fuels and chemicals. *Energy Sources, Part A*, *31*(12), 1028–1037.

Duerre, P. (2007). Biobutanol: An attractive biofuel. *Biotechnology Journal*, *2*(12), 1525–1534.

Energy Information Administration (2009) http://www.eia.gov/FTPRoot/environment/2009_c02_analysis.pdf.

Foston, M., & Ragauskas, A. J. (2010). Changes in lignocellulosic supramolecular and ultrastructure during dilute acid pretreatment of *Populus* and switchgrass. *Biomass Bioenergy*, *34*(12), 1885–1895.

French, R. J., Hrdlicka, J., & Baldwin, R. (2010). Mild hydrotreating of biomass pyrolysis oils to produce a suitable refinery feedstock. *Environ. Program Sustainable Energy*, *29*(2), 142–150.

Gray, K. A. (2007). Cellulosic ethanol – state of the technology. *International Sugar Journal*, *109*(1299), 145–146. 148, 150–151

Guo, G.-L., et al. (2008). Characterization of dilute acid pretreatment of silvergrass for ethanol production. *Bioresource Technology*, *99*(14), 6046–6053.

Gupta, R., & Lee, Y. Y. (2010). Investigation of biomass degradation mechanism in pretreatment of switchgrass by aqueous ammonia and sodium hydroxide. *Bioresource Technology*, *101*(21), 8185–8191.

Hanaoka, T., et al. (2008). Bench-scale production of hydrocarbon liquid fuel from woody biomass via gasification. *Journal of the Japan Institute Energy*, *87*(9), 737–743.

Hemme, C. L., et al. (2010). Sequencing of multiple clostridial genomes related to biomass conversion and biofuel production. *Journal of Bacteriology*, *192*(24), 6494–6496.

Hess, M., et al. (2011). Metagenomic discovery of biomass-degrading genes and genomes from cow rumen. *Science (Washington, DC, U.S.)*, *331*(6016), 463–467.

Himmel, M. E., et al. (2007). Biomass recalcitrance: Engineering plants and enzymes for biofuels production. *Science (Washington, DC, U.S.)*, *315*(5813), 804–807.

Huber, G. W. (2010). Green gasoline and petrochemicals from lignocellulosic biomass by pyrolysis based technologies. *Preprint Symposium American Chemical Society, Division of Fuel Chemistry*, *55*(1), 228–229.

Hyeon, J. E., et al. (2011). Production of minicellulosomes for the enhanced hydrolysis of cellulosic substrates by recombinant *Corynebacterium glutamicum*. *Enzyme and Microbial Technology*, *48*(4–5), 371–377.

Joergensen, H., Kristensen, J. B., & Felby, C. (2007). Enzymatic conversion of lignocellulose into fermentable sugars: Challenges and opportunities. *Biofuels, Bioproducts and Biorefin*, *1*(2), 119–134.

Johansson, K., et al. (2010). Agriculture as provider of both food and fuel. *Ambio*, *39*(2), 91–99.

Kim, H., Kim, S., & Dale, B. E. (2009). Biofuels, land use change, and greenhouse gas emissions: Some unexplored variables. *Environmental Science and Technology*, *43*(3), 961–967.

Kim, H.-W., & Ishikawa, K. (2010). Complete saccharification of cellulose at high temperature using endocellulase and beta-glucosidase from *Pyrococcus* sp. *Journal of Microbiology and Biotechnology*, *20*(5), 889–892.

Knocke, C., & Vogt, J. (2009). Biofuels – challenges & chances: How biofuel development can benefit from advanced process technology. *Engineering Life Science*, *9*(2), 96–99.

Kumar, A., Jones, D. D., & Hanna, M. A. (2009). Thermochemical biomass gasification: A review of the current status of the technology. *Energies (Basel, Switz.)*, *2*(3), 556–581.

Kumar, A., Flynn, P., & Sokhansanj, S. (2008). Biopower generation from mountain pine infested wood in Canada: An economical opportunity for greenhouse gas mitigation. *Renewable Energy*, *33*(6), 1354–1363.

Kumar, S., et al. (2009). Recent advances in production of bioethanol from lignocellulosic biomass. *Chemical Engineering and Technology*, *32*(4), 517–526.

Lappas, A., & Heracleous, E. (2011). Production of biofuels via Fischer-Tropsch synthesis: Biomass-to-liquids. *Woodhead Publishing Series in Energy*, *15*, 493–529. (Handbook of Biofuels Production).

Larson, E. D. (2006). A review of life-cycle analysis studies on liquid biofuel systems for the transport sector. *Energy Sustainable Development*, *10*(2), 109–126.

Lau, M. J., et al. (2010). Ammonia fiber expansion (AFEX) pretreatment, enzymatic hydrolysis, and fermentation on empty palm fruit bunch fiber (EPFBF) for cellulosic ethanol production. *Applied Biochemistry and Biotechnology*, *162*(7), 1847–1857.

Li, C., et al. (2010). Comparison of dilute acid and ionic liquid pretreatment of switchgrass: Biomass recalcitrance, delignification and enzymatic saccharification. *Bioresource Technology*, *101*(13), 4900–4906.

Linger, J. G., Adney, W. S., & Darzins, A. (2010). Heterologous expression and extracellular secretion of cellulolytic enzymes by *Zymomonas mobilis*. *Applied and Environment Microbiology*, *76*(19), 6360–6369.

Lorenz, A. J., et al. (2010). Genetic analysis of cell wall traits relevant to cellulosic ethanol production in maize (*Zea mays* L.). *Crop Science*, *50*(3), 842–852.

Melgar, A., Perez, J., & Horrillo, A. (2009). Biomass gasification process in a downdraft fixed bed gasifier: A real time diagnosis model based on gas composition analysis. *Revista Facultad de Ingeniería Universidad de Antioquia*, *49*, 9–18.

Mesa, L., et al. (2010). An approach to optimization of enzymatic hydrolysis from sugarcane bagasse based on organosolv pretreatment. *Journal of Chemical Technology and Biotechnology*, *85*(8), 1092–1098.

Miao, Z., et al. (2011). Energy requirement for comminution of biomass in relation to particle physical properties. *Industrial Crops and Products*, *33*(2), 504–513.

Michel, G., et al. (2009). The family 6 carbohydrate-binding modules have coevolved with their appended catalytic modules toward similar substrate specificity. *Glycobiology*, *19*(6), 615–623.

Miller, C. A. (2009). *Environmental impacts of biofuel production and use*. Abstract of Papers, 238th ACS National Meeting, Washington, DC, United States, August 16-20, 2009, p. FUEL-285.

Mohan, D., Pittman, C. U., Jr., & Steele, P. H. (2006). Pyrolysis of wood/biomass for bio-oil: A critical review. *Energy & Fuels*, *20*(3), 848–889.

Mullins, K. A., Griffin, W. M., & Matthews, H. S. (2011). Policy implications of uncertainty in modeled life-cycle greenhouse gas emissions of biofuels. *Environmental Science and Technology*, *45*(1), 132–138.

Munasinghe, P. C., & Khanal, S. K. (2010). Biomass-derived syngas fermentation into biofuels: Opportunities and challenges. *Bioresource Technology*, *101*(13), 5013–5022.

Nakahira, Y., et al. (2009). *The transgenic plants with chloroplasts expressing thermostable cellulase for enzymic saccharification of cell wall components*. (Japan). Application: JP. p. 18pp.

Nataf, Y., et al. (2010). *Clostridium thermocellum* cellulosomal genes are regulated by extracytoplasmic polysaccharides via alternative sigma factors. *Proceedings of the National Academy of Sciences of the United States of America*, *107*(43), 18646–18651. (S18646/1-S18646/4).

Nimlos, M. R., et al. (2009). Global mechanisms of tar formation during gasification of biomass. *Preprint Symposium American Chemical Society, Division of Fuel Chemistry*, *54*(1), 372–373.

Park, N. -H., et al. (2010). Organosolv pretreatment with various catalysts for enhancing enzymatic hydrolysis of pitch pine (*Pinus rigida*) *Bioresource Technology*, *101*(18), 7046–7053.

Patell, V. M., et al. (2010). *Methods for generating transgenic plants expressing cellulases from diatoms for production of bioethanol*. (Avesthagen Limited, India). Application: IN IN. p. 14pp.

Peralta-Yahya, P. P., & Keasling, J. D. (2010). Advanced biofuel production in microbes. *Biotechnol Journal*, *5*(2), 147–162.

Plass, L., & Reimelt, S. (2007). Second generation biofuels. *Hydrocarbon Engineering*, *12*(6), 71–74.

Roberts, K. G., et al. (2010). Life cycle assessment of biochar systems: Estimating the energetic, economic, and climate change potential. *Environmental Science and Technology*, *44*(2), 827–833.

Rude, M. A., & Schirmer, A. (2009). New microbial fuels: A biotech perspective. *Current Opinion in Microbiology*, *12*(3), 274–281.

Sakka, M., et al. (2010). Analysis of a *Clostridium josui* cellulase gene cluster containing the man5A gene and characterization of recombinant Man5A. *Bioscience Biotechnology and Biochemistry*, *74*(10), 2077–2082.

Selig, M. J., et al. (2007). Deposition of lignin droplets produced during dilute acid pretreatment of maize stems retards enzymatic hydrolysis of cellulose. *Biotechnology Progress*, *23*(6), 1333–1339.

Serio, M. A., Kroo, E., & Wojtowicz, M. A. (2003). Biomass pyrolysis for distributed energy generation. *Preprint Symposium American Chemical Society, Division of Fuel Chemistry*, *48*(2), 584–589.

Simmons, B. A., et al. (2010). Ionic liquid pretreatment. *Chemical Engineering Progress*, *106*(3), 50–55.

Smeets, E. M. W., et al. (2007). A bottom-up assessment and review of global bio-energy potentials to 2050. *Progress in Energy and Combustion Science*, *33*(1), 56–106.

Sokhansanj, S., et al. (2009). Large-scale production, harvest and logistics of switchgrass (*Panicum virgatum* L.) – current technology and envisioning a mature technology. *Biofuels, Bioproducts and Biorefining*, *3*(2), 124–141.

Stanley, G., & Hahn-Haegerdal, B. (2010). Fuel ethanol production from lignocellulosic raw materials using recombinant yeasts. *Biomass and Biofuels*, 261–291.

Steen, E. J., et al. (2010). Microbial production of fatty-acid-derived fuels and chemicals from plant biomass. *Nature (London, U.K.)*, *463*(7280), 559–562.

Su, X., et al. (2010). Mutational insights into the roles of amino acid residues in ligand binding for two closely related family 16 carbohydrate binding modules. *Journal of Biological Chemistry*, *285*(45), 34665–34676.

Tamaru, Y., et al. (2010). Comparison of the mesophilic cellulosome-producing *Clostridium cellulovorans* genome with other cellulosome-related clostridial genomes. *Microbial Biotechnology*, *4*(1), 64–73.

Taylor, J. D., Jenni, M. M., & Peters, M. W. (2010). Dehydration of fermented isobutanol for the production of renewable chemicals and fuels. *Topics in Catalysis*, *53*(15–18), 1224–1230.

Vanholme, R., et al. (2010). Engineering traditional monolignols out of lignin by concomitant up-regulation of F5H1 and down-regulation of COMT in *Arabidopsis*. *The Plant Journal: For Cell and Molecular Biology*, *64*(6), 885–897.

Vodovnik, M., & Logar, R. M. (2010). Cellulosomes: Promising supramolecular machines of anaerobic cellulolytic microorganisms. *Acta Chimica Slovenica*, *57*(4), 767–774.

Wang, L., et al. (2008). Contemporary issues in thermal gasification of biomass and its application to electricity and fuel production. *Biomass Bioenergy*, *32*(7), 573–581.

Wen, F., Sun, J., & Zhao, H. (2010). Yeast surface display of trifunctional minicellulosomes for simultaneous saccharification and fermentation of cellulose to ethanol. *Applied and Environment Microbiology*, *76*(4), 1251–1260.

Wilson, D. B. (2009). Cellulases and biofuels. *Current Opinion in Biotechnology*, *20*(3), 295–299.

Wu, D., et al. (2009). A phylogeny-driven genomic encyclopaedia of Bacteria and Archaea. *Nature (London, U.K.)*, *462*(7276), 1056–1060.

Xu, J. -M., et al. (2010). Refining and application of fractions of biomass-pyrolysis oil. *Linchan Huaxue Yu Gongye*, *30*(2), 1–5.

Yanik, J., et al. (2007). Fast pyrolysis of agricultural wastes: Characterization of pyrolysis products. *Fuel Processing Technology*, *88*(10), 942–947.

Zhao, Z., et al. (2001). Biomass pyrolysis in an argon/hydrogen plasma reactor. *Engineering in Life Sciences*, *1*(5), 197–199.

Zhu, J. Y., & Pan, X. J. (2010). Woody biomass pretreatment for cellulosic ethanol production: Technology and energy consumption evaluation. *Bioresource Technology*, *101*(13), 4992–5002.

Section G

Commercial, legal, sociological, and public aspects of agricultural plant biotechnologies

Containing and mitigating transgene flow from crops to weeds, to wild species, and to crops

32

Jonathan Gressel
Plant Sciences, Weizmann Institute of Science, Rehovot, Israel

TABLE OF CONTENTS

Introduction: Does Transgene Flow Matter?

Genes have been moving among species for time immemorial, most often between closely related species where there are few genetic barriers (vertical gene flow). More rarely, genes flow between genera within a family where barriers exist and the vast majority of the rare offspring are sterile, but a few produce pollen and a gene can eventually establish (diagonal gene flow). Far rarer, especially in eukaryotic organisms, genes can move from one unrelated species to another, often asexually carried by a vector such as a virus or a plasmid (horizontal gene flow). In eukaryotic organisms horizontal gene flow is so rare that it is considered to occur more in evolutionary than in human time. The genetic barriers to all types of gene flow seem to be lower when an organism is under stress, resulting in promiscuous intergeneric crosses. This issue has been ignored in most discussions of gene flow.

Because gene flow is a matter of nature, there are some that state that we should no more worry about transgene flow than we worry about normal gene flow. This cavalier argument ignores the fact that some transgenes confer traits that are not found in the related species and could not have easily arrived there had there not been a transgene. Thus, one must ask more specific questions: What trait does the transgene confer? Will the trait confer an advantage on the recipient? If the trait confers a disadvantage, are there enough other normal offspring that the gene movement will have an inconsequential effect? In the case of crop-to-crop transgene flow, we can ask whether the disadvantage conferred is economic; for example, is the transgenic trait a pharmaceutical trait that would render a recipient crop unfit to enter the food chain? Indeed, can the transgene even get to a potential recipient; distance may make the heart grow fonder, but it is an impediment to procreation. There is much literature on the distances transgenic pollen can flow, usually using male sterile recipients at various distances in fields where there is no wild-type of the crop. The longer

DOI: 10.1016/B978-0-12-381466-1.00032-8

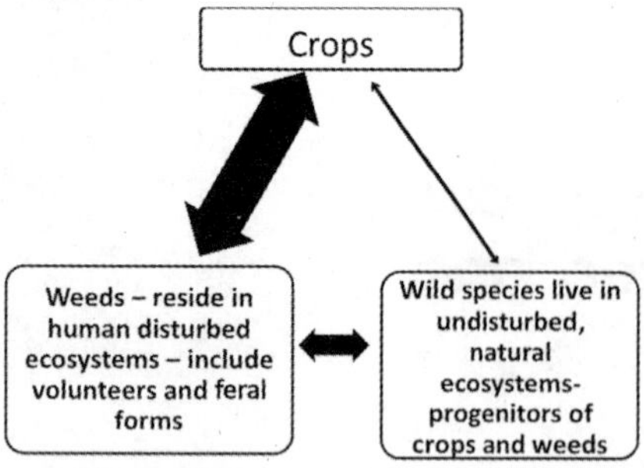

Figure 32.1 • The present relationship between related crops, weeds, and wild species • The thickness of arrows denotes the average relative likelihood of gene flow between them — there can be differences depending on biology and geography. When crops were domesticated by the sparse human population living in the wild thousands of years ago, the arrows would have been at near equal width. Based on concepts developed in Warwick and Stewart (2005) and this author's interpretation.

pollen flies, whether by wind or by insect, the weaker it gets. When it has to compete with indigenous healthy pollen, it is at a disadvantage, but when it is the only compatible pollen it has a much greater likelihood of success. There is thus a strong dependence on where the potential recipient may reside.

Transgene flow: To what ecosystem?

There is considerable confusion in the literature about gene flow due to the use of exceedingly loose terminology, and articles abound about transgene flow from crops to "wild" species, when actually the gene flow was to weeds (Figure 32.1). This is good for obtaining headlines in the popular press and publishing in prestigious journals and books (Ellstrand, 2003; Hooftman et al., 2008). A close perusal of the articles revealed that natural ecosystems have rarely been affected (a rare case is with oilseed rape and other *Brassica* species in Europe; Allainguillaume et al., 2009), as the gene flow was to agricultural or ruderal weeds (Hooftman et al., 2008; Zapiola et al., 2008). Many transgenic ruderal populations found are most often due to seeds falling from conveyances bringing the crop to market, or from port to processor (von der Lippe and Kowarik, 2007). Some populations are first-generation "volunteers"; more rarely do they persist and cross with indigenous weeds or establish themselves as feral (partially de-domesticated) weeds. There are consequences from such occurrences, as feral populations can become weedier by selection or by hybridization with related weeds, and they do have a selective advantage, as herbicides are often used to control ruderal weeds where they live on the field borders and roadsides. The feral transgenic glyphosate- and glufosinate-resistant oilseed rape populations (some containing both transgenes due to hybridization) have extreme advantages when those herbicides are used, and can later move back into the fields as weeds that can reintroduce the transgenes into other varieties of the crop. These volunteer cum feral ruderal weeds are at a loss, and are lost when 2,4-D or atrazine are used on the roadsides. The misuse of the word "wild" for such transgenic populations implies that they are in natural ecosystems, conferring a problem on conservationists. Such populations do present a problem, but it is not to the protectors of natural ecosystems; it is to farmers and road authorities, despite the headlines implying consequences to natural biodiversity.

It is thus necessary to carefully distinguish between a crop and its weedy relative, and a crop and its wild relative. The ecosystems in which each resides are different (Figure 32.1; Warwick and Stewart, 2005). The weeds are evolutionarily adapted to live in agroecosystems and ruderal ecosystems where humans cultivate or otherwise "disturb" the ecosystems. Wild species live in pristine natural ecosystems largely unimpacted by humans. The vast majority of wild species cannot cope with the conditions of cultivated fields (agroecosystems), cultivated field margins, or roadsides (ruderal ecosystems), and weeds cannot compete with wild species in the wild. Weeds, crops, and wild species each have a syndrome of traits that differentiate between them. Wild species obviously were the progenitors of crops, as well as weeds, in the distant past. A recent example is a *Amaranthus rudis* (*A. tuberculatus*), which literally emerged from the swamp and became a major weed, presumably by crossing with a weedy *Amaranthus* species and changing its wild syndrome toward weediness (Trucco et al., 2009).

Why is it so important to distinguish between crop, wild, and weedy ecosystems and their inhabitants? The movement of a given transgenic trait will have different implications depending on the recipient. A transgene for an oral vaccine or an industrial feedstock moving into another variety of a crop will clearly be undesirable, but would have little consequence in a wild relative. If it were to move into a related weed, the question would arise as to whether the weed could be a reservoir from which the trait could cross back into other varieties of the related crop. A transgene conferring resistance to a disease or a pest might be good for a wild species if the related species is not already resistant to it. If the insect or disease exerted a frequent, strong pressure on the wild or weedy relative, the selective advantage conferred by the trait might affect ecological balances, but if the infestation is only sporadic, the advantage may hardly affect such balances. Should a transgene for herbicide resistance move to a wild relative, it would be of no advantage, as herbicides are not used in the wild, but it would be very beneficial to weeds, thus making weed control far more problematic.

It is not surprising that most cases of reported gene flow are of movement of herbicide resistance for two reasons: most of the transgenic crops are those with herbicide resistance transgenes, and herbicide resistance is the easiest to find in the field due to treatment of fields and ruderal areas with herbicides, decimating all but the resistant individuals.

Much of our information on the dire implications of herbicide resistance movement from crop to weed does not come from a transgenic crop; it derives from a mutant acetolactate synthase gene used in rice to confer resistance to imidazolinone herbicides (Clearfield™ rice), and more recently in transgenic glufosinate-resistant rice (Song et al., 2011). Its release was much heralded as the herbicides excellently controlled weeds, and especially the previously intractable weedy rice (often called red or feral rice). As it is the same species as rice it is naturally resistant and susceptible to the same herbicides as rice, meaning that there could be no selective control before this mutant resistance was used. Weedy rice had been partially controlled by hand-transplanting rice from nurseries, a procedure being supplanted by direct seeding. The ability to control the weedy rice was quickly lost as the resistance gene introgressed by crossing and backcrossing into the weedy rice, rendering the technology useless (Gressel and Valverde, 2009b). This demonstrates the need for mechanisms to prevent gene flow from crops to their related weeds. In such cases, the planting distances that prevent crop-to-crop gene movement are not applicable; the weed lives within and next to the crop. Many other crops have feral, conspecific weeds living in their midst, including sorghum, sunflowers, sugar beets, carrots, and radish, allowing similar "vertical" gene flow. Other crops have closely related weeds partially sharing a genome (wheat, oilseed rape, sorghum) or similar enough genomes to cross and have some partially fertile offspring to allow "diagonal" gene flow. Other crops only have wild or weedy relatives near their centers of origin (e.g., maize, potatoes). In the case of soybeans, it is questioned whether there are any remaining wild populations at the center of origin in China, or only weedy soybean in agroecosystems that seem quite crop-like (Lu, 2005).

Because crop–wild gene flow seem to be inconsequential, and there are some (but hardly all) transgenic crops where crop–crop and crop–weed gene flow is an issue, the focus of this chapter will deal only with these consequential issues. It will do so not by decrying the issues, but by constructively dealing with methods to inherently prevent such crop–crop and crop–weed gene flow from happening (transgene containment), or if it eventually happens, to prevent the transgene from establishing and spreading within the population (transgene mitigation).

Thresholds matter

There is a desire by some that there should be a zero threshold for transgenic material in non-transgenic material. Such proponents expose their political inclinations by using terminology last used by the Third Reich. They demand "genetic purity," eschewing "genetic contamination," and so forth. It is the task of toxicologists together with regulators to set the threshold levels of allowable admixtures, and they usually do so at order(s) of magnitude above known danger levels. Thus, most milk is pasteurized with some live bacteria remaining and not sterilized. Wheat ground into bread flour has (surprisingly high) thresholds for weed seeds, dirt, rodent hairs, stones, and fecal material. We live with such thresholds. Analytical methods such as PCR easily detect transgenic DNA from infinitesimal amounts of admixture to wild-type, for example, DNA from a speck of maize leaf from a neighboring field in a container of soybeans. Is this dangerous? It seems to be considered as such, as immense resources have been expended in training and setting up laboratories worldwide to make such DNA determinations, while little effort is made to assay for carcinogenic mycotoxins at above regulatory thresholds in the same commodities. Are all those who support this ultrasensitive DNA testing for transgenes performing a service to humanity?

Gene containment and/or mitigation is often necessary

From the previous section it is obvious that that there are many cases where gene flow is of little consequence, yet there are some cases where it is undesirable. Clearly any technology used to physically, genetically, or transgenically contain or mitigate transgene flow cannot be 100% perfect. Logically there should be toxicologically and scientifically determined thresholds, not those determined by the sensitivity of PCR. If the reader is looking for absolute zero levels of transgene introgression, there is no reason to read on.

The goal of gene containment is to keep the transgene of choice within the crop and thus prevent it from introgressing into related weeds, crop varieties, and wild species. It will soon be apparent that all containers "leak"; the transgene will eventually get out, and a second level of protection is needed to prevent the transgene from establishing and then spreading. The transgene will spread if it has a selective advantage, and the task of mitigation is to suppressively maintain the transgene at frequencies beneath the regulatory thresholds where they do not belong.

Dealing with transgene flow is the sole subject of a recent book by Oliver and Li (2011), as well as an excellent review by Warwick et al. (2009), which is referred to those wanting more in-depth coverage of the topic.

Methods of Containment[1]

It has often been suggested that choosing the traits that genetically exist within a crop can be used to prevent gene flow, such as choosing varieties with a flowering time that does not overlap with that of the related weed/wild species. Unfortunately there are usually tails of one and leading edges of the timing of the other that overlap. Besides, that may limit when a crop may be cultivated. Before experiments were performed, it was suggested that gene flow would not be an issue in rice, as it is a cleistogamous species; the anthers incestuously pollinate the pistil before the flower even opens. When good marker genes appeared it became apparent that a few percent of the seeds derived from cross-pollination. Besides, breeders have been selecting for non-cleistogamous rice strains for use to breed hybrid rice varieties. Cleistogamy has also been proposed for oilseed rape, but at best it is only able to "significantly reduce pollen mediated gene flow" (Gruber and Hüsken, 2011), which is hardly enough.

Several transgenic molecular mechanisms that have been suggested for containing genes and especially transgenes within the crop (i.e., to prevent inflow from and/or outflow to related species) are described in the following section. As will be presented, each has its benefits and limitations.

[1] The text in the following sections of this chapter is condensed and updated from Sections 4.8–4.9 of Chapter 4, "Biosafety Considerations with Further Domesticated Crops" in *Genetic Glass Ceiling: Transgenics for Crop Biodiversity* (Gressel, 2008) with permission of the publisher.

Containment by targeting genes to a cytoplasmic genome

The most widely discussed containment possibility is to integrate the transgene of choice in the plastid or mitochondrial genomes (Maliga, 2004). The opportunity for gene outflow is limited due to the predominantly maternal inheritance of these genomes in many, but far from all, species. This presently arduous technology of transforming genes into chloroplasts, which so far is limited to a few crops, does not preclude the wild or weedy relative from pollinating the crop, giving rise to the same F_1 hybrid that would have been obtained if the crop had been pollinated by the weedy/wild species. Instead it bears the plastomic or mitochondrial trait. Then, if the wild or weedy species acts as the recurrent pollen parent, the plastid or mitochondrial trait can be fixed by backcrossing the wild/weedy relative. Even though the hybrids into the crop and into the wild-type may be the same, the likelihood of a crop–wild hybrid of any of the major crops discussed surviving in the wild is minimal but finite, especially over long periods. Thus, problems ensue when the crop and its relative inhabit the same ecosystem in close proximity.

The claim of strict maternal inheritance of plastome-encoded traits in many species (Daniell, 2002; Maliga, 2004; Ruf et al., 2007) has not been substantiated, and has even later been repudiated by some of its initial proponents (Maliga, 2004; Svab and Maliga, 2007). Experiments initially performed by proponents to demonstrate that the frequency is below a low number were not large enough, for example, less than the frequency of typical mutations. Tobacco (Avni and Edelman, 1991) and other species (Darmency, 1994) have between a one in one thousand and one in ten thousand frequency of pollen transfer of plastid inherited traits in the laboratory. Tobacco seems to be the species of choice for chloroplast production of pharmaceuticals (e.g., Lentz et al., 2010), so this rate of gene flow would be far too high.

Pollen transmission of plastome-encoded traits can best be detected using large samples, together with nuclear as well as plastomic selectable genetic markers. Such a large-scale field experiment was set out using a *Setaria italica* (foxtail millet) with chloroplast-inherited atrazine resistance, with the resistant plants also bearing a nuclear dominant red leaf-base marker (Wang et al., 2004). These were allowed to naturally cross in the field with male sterile herbicide susceptible lines, and the results of cross-pollination were easy to distinguish due to the nuclear markers. Chloroplast-inherited resistance was pollen-transmitted at a frequency of 3 in 10,000 hybrids among the >78,000 offspring (Wang et al., 2004). More recently it was shown that chloroplasts from oilseed rape (*B. napus*) easily integrate into and establish in weedy *B. rapa* without the oilseed rape nuclear chromosomes (Haider et al., 2009).

The probability of transgene movement via plastomic gene flow is orders of magnitude greater than by spontaneous dominant nuclear genome mutations. This renders chloroplast transformation unacceptable for preventing transgene outflow to related weeds, unless stacked with additional mechanisms. The technology may be sufficient to preclude substantial transgene flow to neighboring varieties of the same crop, especially when coupled with standard separation distances.

There is a possibility of selectively turning on plastid transgenes, discussed in a later section, which will greatly enhance the biosafety of plastomic transgenes; a gene that is not being expressed cannot be making a product dangerous.

Male sterility

Male sterility has been tested as a specific mechanism to contain gene outflow in various crops; the transgene of choice is transformed into the male sterile (female) inbred to form a hybrid, and is much discussed as a containment method (Moon et al., 2010). A great deal of male sterility is cytoplasmic and inherited on the chondriome (mitochondrial genome). This will be difficult to perform as chondriome engineering is yet unknown. The ability of cytoplasmic male sterility to preclude sorghum transgene flow through pollen (using non-transgenic pollinators) to decrease the risk of viable pollen flow was tested under field conditions and four male fertile individuals from a population of a thousand contained the marker (Pedersen et al., 2003). This severe restriction of gene flow through pollen would be helpful, but a trait that increases fitness of related weeds would establish, albeit slowly, because male sterility slows gene flow, but transgenes would quickly spread once they appeared in sexually crossing material. Such male sterility only prevents the outflow of genes. The male sterile variety could be pollinated by a related weed, and the same F_1 forms as would have with pollen flow.

A simple fail-safe mechanism may be possible with nuclear sterility in hybrid crops. If a dominant transgene of choice is placed in the male sterile line in close linkage with the male sterility gene, there will be little possibility of introgression into wild or weedy relatives in crop-production areas. Care will have to be taken in the seed production areas when the male sterile line is restored. Such areas must be kept free of related weeds, a typical precaution in seed production that was generally practiced well before the advent of transgenics.

Instead of using indigenous genes, if one is engineering male sterility into the crop by one of the newer technologies for nuclear male sterility (Williams, 1995), chromosome position-specific transformation using sequence homology is chosen. This has recently been done in eggplant, and could prevent pollination of non-transgenic eggplant varieties (Cao et al., 2010).

Utilizing male sterility stacked with transplastomic traits considerably lowers the risk of plastome gene outflow within a field (but not gene influx from related strains or species). This would vastly reduce the risk of transgene flow, except in the small isolated areas required for line maintenance, and might be considered sufficient where there are highly stringent requirements for preventing gene outflow to other varieties (e.g., to organically cultivated ones) or where pharmaceutical or industrial traits are engineered into a species. Plastome-encoded transgenes for non-selectable traits (e.g., for pharmaceutical production) have been transformed into the chloroplasts together with a trait such as tentoxin or atrazine resistance as a selectable plastome marker (Darmency, 1994).

When they are coupled with the recently reported transplastomic engineered male sterility (Ruiz and Daniell, 2005) to further reduce out-crossing risk, plastome transformation can possibly meet the initial expectations. This stacked technology will still not overcome having a wild or weedy species as a recurrent pollen parent with selection for the plastid traits, and this is a risk that must be further evaluated.

Many other aspects of engineering sterility have been recently reviewed at length (Liu and Singer, 2011), especially RNA silencing technologies that have been used for purposes other than engineering sterility for prevention of gene flow.

Rendering crops asexual

Some vegetatively propagated crops such as potato and elite tissue culture propagated forestry material also flower. In forestry, this is especially problematic, as the long-term implications of gene movement may take longer than human lifetimes to measure. The introgression of traits from these species to wild populations has been discussed (Gressel, 2002). Some landscaping trees such as decorative plantings of olives create an urban problem of allergies from pollen (Chapter 23 in Gressel, 2008). Such ornamental trees could be vegetatively propagated if there was a way to prevent allergy-causing pollen clouds and messy fruits. There is also the possibility that preventing allocation of resources to sex will increase the growth of the vegetative tissue, which could be advantageous in many ornamentals and in forestry, that is, whereever vegetative propagation is possible. Thus, a transgenic mitigation (TM) trait that prevents pollen formation or fruit set could be coupled to herbicide resistance or other primary traits. An ideal gene for doing this is *barnase* under the T29 tapetum-specific promoter (Mariani et al., 1990). The ribonuclease is only produced in the tapetum tissue in the flower and prevents pollen formation with no other ill effects. There is a good chance that the shelf life of many flower species (e.g., roses and carnations) could be enhanced as well by preventing pollen production; fertilization starts the process of floral degeneration and fruit set.

A flower-specific promoter from poplar coupled to a cytotoxin gene caused flower ablation (Skinner et al., 2003), requiring vegetative propagation of the trees. Another way to achieve total sterility is to transform *barnase* under a gamete-specific promoter that kills both male and female gametes (Kobayashi et al., 2006). If one has an important crop in which transgenics are exceedingly worthwhile, yet the risks of gene flow are too great, one could envisage using a pollen sterility system coupled with flower drop, as described earlier. The crop could then be propagated by artificial seed, such as artificially encased somatic embryos produced in mechanized tissue culture systems.

Had the glyphosate herbicide-resistant creeping bentgrass (*Agrostis stolonifera*) that spread 100 km by wind from one site in Oregon (Zapiola et al., 2008) been rendered male sterile, or totally sterile by these techniques, and cultivated by stolons (as is often done), then the problem would not have occurred. The public would have benefited by not having hay fever from its allergenic pollen.

Genetic use restriction technologies: Alias "terminators"

Other molecular approaches suggested for crop transgene containment include seed sterility utilizing genetic use restriction technologies (GURT), or the "terminator genes," as they are referred to in the popular press (Oliver et al., 1998; 2004; Crouch, 1998; Oliver and Hake, 2011). In these proposed systems, the transgenes of choice are inserted behind an inducible promoter that causes the inactivation or physical excision of the genes of choice in the flowers. The inducer was to be turned on in the seed stage, for example, before sale to the farmer. The plants would grow normally after induction, expressing the transgene throughout vegetative growth, but neither the seed nor the pollen carry the transgenic trait to further generations. In the excision version of GURT technologies, the pollen and seed are viable (in contrast to press reports), but they do not carry the proprietary or patented transgene. It does not prevent farmers from saving seed (as intimated in the press), but the saved seed does not bear the transgenic trait. The GURT systems were originally developed to protect intellectual property, not to prevent gene flow. There has not been enough documented experimentation with these systems to know whether the transgenes would be excised from all offspring after induction, that is, whether the induction is incomplete. This is immaterial for the first proposed use of the technology, which requires purchase of seed every generation; for this purpose 90% effectiveness is sufficient. Near 100% efficiency is required to prevent transgene flow, and the leakage rate is currently unknown. Theoretically, if the inducible controlling ("terminating") element of the transgene is silenced by mutation, then expression would occur, which is a potential defect in principle and possibly in practice. The frequency of loss of such controlling elements is yet unclear, as there have been no large-scale field trials to test this.

About 0.5% of the crop area sown is planted with seed for future planting, that is, seed that is not "terminator" induced. This area would have to be contained by other means to prevent transgene flow, and if the transgenes do escape, there is no way to "terminate" them once out.

Chemically induced promoters for containment

If a transgene encoding the desired trait is placed behind a strong chemically induced promoter, the desired trait will be expressed when the chemical inducer is used. Such a promoter system was patented for use with a glyphosate-resistant EPSP synthase gene (Jepson et al., 1998). The chemical inducer can be treated together with glyphosate, as glyphosate kills slowly, and inducers supply products within hours. The herbicide can be used the following season without the inducer as a treatment just before planting the rotational crop, or in a naturally resistant crop to control the transgenic volunteer weed. If the crop introgressed the gene into a weed, the weed could be controlled by the herbicide (without inducer).

If the herbicide-resistant gene was to introgress into a wild species that does not inhabit agroecosystems where herbicides are not used, it would be of little value and would probably have enough of a fitness penalty not to establish. Similarly, other transgenes of choice would have no benefit (selective advantage) to wild or weedy relatives if the inducible promoter does not turn on the transgenes.

A system that turns on transgenes such as this may be preferable to one that turns them off, such as GURT. Theoretically, if the GURT gene is silenced, there is a possibility for introgression. If the inducer gene is silenced, then those individuals possessing the mutant are killed and the germline is non-transferable. Still, there is the remote possibility of an inducible promoter mutating to become a constitutive promoter.

There is no inexpensive foolproof inducible promoter system presently available for plants. The copper and alcohol inducible promoters that have been developed do not work in the field, there is enough copper in most soils to trip the former, and environmental conditions are often sufficient to cause enough alcohol to be naturally present in a crop to trigger the latter (Zuo and Chua, 2000). There is also an estrogen inducible promoter that can be used (Curtis et al., 2005), but the inducer is expensive and is a hormone. The auxins of the plant hormones can also be used with a new inducible promoter (Cazzonelli et al., 2005), but the auxins may have side effects on the plants. Antibiotic inducible promoters are known, but widespread use of antibiotics in the field would not be allowed. For certain resistance traits, it might be advisable to use pathogen- (Zhu et al., 2004) or wound-inducible (Breitler et al., 2004) promoters, but these will not be fail-safe mechanisms, as the same disease or insect that induces the gene in the crop will do so in the related species.

A synthetic modification of a bacterial riboswitch activated by theophylline is an effective tightly regulated promoter that can be used to turn on plastid-targeted transgenes (Verhounig et al., 2010). Much remains to be elucidated about the system. What plant secondary metabolites other than theophylline will activate the riboswitch? One would not want to discover that a stressed plant turns on secondary metabolite biosynthesis, which in turn activates the riboswitch. Still, it is an interesting concept.

Recoverable block of function

Various technologies have been developed using the *barnase/barstar* gene system. *Barnase* encodes a potent ribonuclease, which when expressed, kills a cell because it chews up the whole protein manufacturing system. The action of *barnase* is held in check by *barstar*, a strong repressor gene that prevents *barnase* action. It is only when *barstar* is not present that *barnase* can exert its lethal effect. In the recoverable block of function strategy to prevent transgene flow, *barnase* is inserted in a large synthetic intron inserted in the midst of the gene of choice in such a way that transcription of the two genes is in opposite directions (Kuvshinov et al., 2001, 2004). Both genes are thus genetically linked, in what the developers refer to as a "blocking construct"; when one is inherited, so is the other. This is in the same manner as proposed for TM (see the section Mitigating Transgene Flow) and is a derivative of this technology. To prevent this lethality, they propose a "recovering construct" containing *barstar*, but under an inducible promoter. They demonstrate the efficacy of this with a heat shock promoter. Both constructs are inserted (randomly) on different chromosomes. In crosses with a wild/weedy species, or another variety, all F_1 progeny of the parents of homozygous progeny will live if *barstar* is induced, all will die if not. Heat shock promoters are unlikely to be efficacious in most agroecosystems, as there are enough hot days in most areas to trigger the promoter. They also developed the same system using a tetracycline antibiotic inducible promoter for *barstar* (Kuvshinov et al., 2004).

If the crop is a heterozygous hybrid containing *barnase* and *barstar*, then half of the progeny will die and half will have uninduced *barstar*. This would not be an effective fail-safe if it were not for the inducibility of *barstar*. The transgene cannot be expressed in the progeny containing *barnase* and *barstar* without the promoter action. The recoverable block of function system seems like a complicated way to use an inducible promoter, especially when a simple one has already been proposed (see the section Chemically Induced Promoters for Containment).

Repressible seed-lethal technologies

An impractical technology has been proposed to use a "repressible seed-lethal system" (Schernthaner et al., 2003). The seed-lethal trait and its repressor must be simultaneously inserted at the same locus on homologous chromosomes in the hybrid the farmer sows to prevent recombination (crossing over). Such site-specific transformation technologies are not yet workable in plants. The hemizygote transgenic seed-lethal parent of the hybrid cannot reproduce by itself, as its seeds are not viable. If the hybrid could be made, then half the progeny would not carry the seed-lethal trait (or the trait of interest linked to it) and they would have to be culled, which would not be easy without a marker gene. A containment technology should leave no viable volunteers with the transgene, but this complex technology would kill only a quarter of the hybrid progeny, half would be like the hybrid parents, and a quarter would contain just the repressor. Thus, the repressor can cross from the volunteers to related weeds as can the trait of choice linked with the lethal trait, and viable hybrid weeds could form. The death of a quarter of the seeds in all future weed generations is inconsequential to most weedy and wild species that copiously produce seed, as long as the transgenic trait provides some selective advantage.

Trans-splicing to prevent movement

A system has been proposed and partly demonstrated (Sun et al., 2001) that was designed for the generation of transgenic hybrids, where only part of the segregating F_2 generation would bear the transgenic trait. Enzyme splicing in *trans* was demonstrated using the *DnaE* intein, which reconstituted to a functional DnaE protein. The gene for herbicide-resistant *ALS*II was fused in frame to *DnaE* intein segments capable of promoting protein splicing in *trans* and was expressed as two unlinked fragments. Co-transformation with the two plasmids led to production of a functional enzyme by protein splicing in *trans*, which then conferred herbicide resistance (Sun et al., 2001). If each plasmid integrates into a different chromosome, introgressing into a readily crossing weed will render 25% of the weeds resistant, which is hardly a fail-safe. If one of the genes is on a nuclear chromosome and the other in the plastome, the rate of introgression will be half that of a whole gene being on the plastome. The rate of introgression will be near zero if one half of the genes are on the plastome and the other half on the chondriome, but chondriome engineering is still closer to science fiction than reality.

A genetic chaperon to prevent promiscuous transgene flow from wheat to its wild and weedy relatives

Some *Aegilops* species are closely related to wheat and interbreed with it, especially at the center of origin. Some species such as *Aegilops cylindrica* have spread with wheat and have become uncontrollable weeds in wheat. There is ample evidence that inserting transgenes into allopolyploid crop chromosomes that are homoeologous (similar to but not homologous) to related weed/wild species does not preclude transfer (Weissmann et al., 2005). Still, wheat needs transgenes for herbicide resistance to control resistant weeds, newly evolved diseases, and quality traits. Thus a specific method was conceived for wheat, based on the presence of a specific gene in wheat, and the possibility of gene insertion in specific chromosomal locations (Weissmann et al., 2008). Wheat bears the gene *Ph*1 that specifically prevents the promiscuous pairing of homoeologous chromosomes, preventing recombination within the three genomes of wheat (in the absence of homologous chromosomes) and with related species (Sears, 1976). Reduction in homoeologous pairing may also be due to the rapid elimination of many sequences from homoeologous chromosomes of progenitor species after allopolyploidization (Ozkan and Feldman, 2001), which increases the differences between the homoeologous chromosomes.

Using targeted introgression (also called, confusingly to geneticists, homologous recombination; Puchta, 2002; Shaked et al., 2005), it should be possible to insert the transgenes of choice in close proximity to *Ph*1 on chromosome 5BL (Weissmann et al., 2008). Thus, the transgene of choice will remain genetically linked with the ever-watchful *Ph*1, and will segregate with it in the hybrid and backcrosses, thus not introgressing into the chromosomes of wild/weedy relatives (Weissmann et al., 2008). During backcrosses with the wild/weedy species, the excess wheat chromosomes are selectively eliminated due to lagging during anaphase. Chromosome 5BL with the *Ph*1 gene and the linked transgene will be retained in a small proportion of offspring only as long as the transgene confers a selective advantage, for example, when the transgene is for disease or herbicide resistance and the particular disease or herbicide is present (Shaked et al., 2005). In seasons where other herbicides are used or the disease is not active, the selective disadvantage will eliminate 5BL and its linked transgene. Such a solution, especially if coupled with other solutions, such as mitigating genes (see the section Mitigating Transgene Flow), could considerably lower the risk of gene flow between wheat and its relatives. *Ph*1 was recently localized to a 2.5 megabase region of chromosome 5BL containing a structure consisting of a segment of subtelomeric heterochromatin that inserted into a cluster of *cdc*2-related genes — genes that affect chromosome condensation (Griffiths et al., 2006). The correlation of the presence of this structure with Ph1 activity makes the structure a candidate for the *Ph*1 locus (Griffiths et al., 2006). When *Ph*1 is fully isolated and described, it might be possible to use it directly as a mitigator gene, no longer necessitating targeted insertion, randomly inserting it linked to the gene of choice.

An alternative way to curtail the movement of a transgene from wheat into wild relatives is by inserting the transgene in tandem with a suicide gene on any chromosome arm other than 5BL, and inserting a gene encoding a suppressor of the suicide gene product on chromosome arm 5BL adjacent to *Ph*1 (Weissmann et al., 2008). The linkage between the suicide-suppressor gene and *Ph*1 on 5BL will prevent the transfer of the suppressor to a wild chromosome and, consequently, the establishment of this gene in the wild population. The suicide gene can encode any heterologous protein that is toxic to plants possessing it (Weissmann et al., 2008). *Barnase* is such a gene (see section Male Sterility), and another suicide gene encodes a ribosome-inhibitor protein (RIP) that destroys ribosomes (Jach et al., 1995). The suicide-suppressor gene can be any gene that encodes a heterologous protein that inactivates either the suicide gene or the toxic protein encoded by the suicide gene, such as a *barstar*. Any backcross progeny having the tandem transgene-suicide gene, but without the suppressor of the suicide gene, will die. Chromosome arm 5BL will be eliminated during the continuous backcrossing to the wild parent because it cannot recombine with homoeologous material, due to the presence of *Ph*1 (Weissmann et al., 2008).

This concept might be less efficient at preventing gene flow if the recipient wild/weedy species contains a gene that suppresses the *Ph*1 gene. There is evidence that several diploid relatives of wheat — *Ae. speltoides*, *Amplyopyrum muticum* (Riley et al., 1966), and several diploid and tetraploid *Agropyron* species — have such a suppressor (Chen et al., 1993). Thus, this might necessitate ascertaining that transgenic wheat will not transfer genes to the indigenous wild/weedy diploid relatives in each locality where transgenic wheat is to be cultivated to ascertain the value of this chaperoning system. Clearly either mechanism using the *Ph*1 chaperon can be effective in preventing gene flow from wheat to *Ae. cylindrica* in North America where it stably introgresses genes from wheat (Weissmann et al., 2008).

Transiently transgenic crops

Various pathogens that are not transmissible via meiosis have been tested as single-generation vectors of transgenes such that the transgenes cannot spread sexually. Endophytic fungi and bacteria that normally grow inside plant tissues have been used to carry useful genes into plants (e.g., Bt genes) causing a yield reduction, probably due to an overload of the endophytes. One could also introduce the transgene of choice into a disarmed (non-disease producing) strain of a pathogen such as *Clavibacter* that also dwarfs the infected crop (Carlson et al., 1992).

Systemic RNA viruses that are not carried through meiosis into reproductive cells have also been used as single-generation vectors of transgenes. *Arabidopsis* plants infected with Tobacco Etch virus (Whitham et al., 1999) and cucurbits artificially infected with an attenuated Zucchini Yellow Mosaic potyvirus (Shiboleth et al., 2001) each carry the *bar* gene that renders the plants resistant to the herbicide glufosinate. The attenuated Zucchini Yellow Mosaic potyvirus was also used to produce an antitumor protein (Arazi et al., 2002). A wheat streak mosaic virus carrying an NTPII antibiotic resistance gene was used to infect various grain leaves (Choi et al., 2000), but the plants were antibiotic resistant. The virus carrying the genes was expressed in wheat roots following leaf infection, although not in all tissues (Choi et al., 2000). The technology has been used to simultaneously introduce three chimeric genes into a tobacco, using potyvirus Potato virus A (Kelloniemi et al., 2008).

Hopefully, such infection procedures could be used to introduce useful genes into crop seeds, perhaps using pressure infiltration systems (Marillonet et al., 2005). An alternative to seed inoculation is to infect standing forage crops by modified commercial sandblasting equipment (Chapter 8 in Gressel, 2008). So far this has been performed in the field with gold particles instead of sand (Sikorskaite et al., 2010). Considerable technological obstacles to efficient infection will have to be worked out. There are safety issues about the mode of disarming the virus that must be considered. There must be a total lack of gene introgression from the virus to the plant chromosome, and near total non-transmission of the virus through ovules or pollen. Still, there are many crops, especially those with related, introgressing weeds (e.g., sorghum, barley, rice, sunflowers), where such a technology could be very worthwhile in safely solving problems without the fear of gene flow.

Mitigating Transgene Flow

All of the above containment mechanisms that can be sexually propagated are leaky, and once gene transfer occurs, the new bearer of the transgene could disperse throughout the population if the transgene confers just a small fitness advantage. Thus it is necessary to also utilize technologies that will prevent the establishment and/or spread of transgenes in the population, such as mitigation strategies.

When a transgene confers even a small fitness disadvantage, the progeny should only be able to exist as a very small proportion of the population. Therefore, it should be possible to mitigate progeny establishment by lowering the fitness of transgene recipients below the fitness of competitors so that the reproduction of volunteer or hybrid offspring will be severely inhibited. The concept of TM was proposed where mitigation genes are linked or fused to the desired primary transgene (Gressel, 1999).

This TM approach is based on the premises that: (1) tandem constructs act as tightly linked genes, and their segregation from each other is exceedingly rare; (2) the gain-of-function dominant or semi-dominant TM traits chosen are neutral or favorable to crops, but deleterious to volunteer weed progeny and their hybrids due to a negative selection pressure; and (3) individuals bearing even mildly harmful TM traits will be kept at very low frequencies in volunteer/hybrid populations. The strong competition with their own wild-type or with other species should mostly eliminate even marginally unfit individuals and prevent them from persisting in the field at anything above a very low frequency (Gressel, 1999).

Thus, it was predicted that the primary gene of agronomic advantage engineered into a crop will not persist in future generations if it is flanked by one or more TM gene(s), such as genes encoding dwarfing, strong apical dominance to prevent tillering (in grains) or multi-heading (in crops such as sunflowers), determinate growth, non-bolting (premature flowering) genes, uniform seed ripening, non-shattering, anti-secondary dormancy, and so forth. When such a TM gene or genes is in a tandem construct with the transgene of choice, the overall effect would be deleterious to the volunteer progeny and to hybrids. Indeed, a TM gene such as anti-seed shattering will lower the number of initial volunteers (Gressel, 1999, 2002).

Many but not all of the mitigation traits described next are specifically useful for mitigating against transgene establishment in wild or weedy relatives of a crop, and may not always be useful in preventing establishment of transgenes in other varieties of the crop. Still, a mechanism for specifically mitigating against transgenes encoding pharmaceuticals in maize from surviving as volunteers or establishing in other varieties is described, as are mitigation methods of controlling volunteer transgenic rice in other rice varieties.

Demonstration of transgenic mitigation

Tobacco and oilseed rape were used as model crops to test the TM concept: a tandem construct was made containing an *ahas*^{R} (acetohydroxy acid synthase) gene for herbicide resistance as the primary desirable gene of choice, and the dwarfing Δ*gai* (gibberellic-acid-insensitive) mutant gene as a TM mitigator (Al-Ahmad et al., 2004). Dwarfing would be disadvantageous to the rare weeds introgressing the TM construct, as they could no longer compete with other crops or with fellow weeds, but it is desirable in many crops, preventing lodging and producing more grain yield at the expense of stem tissue. The dwarf- and herbicide-resistant TM heterozygous hybrid plants were more productive than the wild-type when cultivated alone (without herbicide). They formed many more flowers and seeds than the wild-type when cultivated by themselves, which is an indication of a higher harvest index. Conversely, the TM transgenic hybrids were weak competitors and highly unfit when co-cultivated with the wild-type in experiments simulating ecological competition. The inability to achieve flowering on the transgenically mitigated tobacco plants in the competitive situation resulted in zero reproductive fitness of the transgenically mitigated plants grown in an equal mixture (Al-Ahmad et al., 2004), or at different ratios

(Al-Ahmad et al., 2005) in a replacement series with the wild-type at a spacing representative of weed spacing in the field. The TM oilseed rape could neither compete well with its wild-type (Al-Ahmad et al., 2006) nor could hybrids with a related weed compete with either parent (Al-Ahmad and Gressel, 2006) in greenhouse experiments. Similarly, in a field experiment where the TM oilseed rape hybrids competed with wild-type in a wheat field, establishment was well mitigated, which was not the case when the desirable transgene was not in tandem with a mitigating gene (Rose et al., 2009).

From the previous data it is clear that TM should be advantageous to a crop growing alone and disastrous to a crop–weed hybrid living in a competitive environment. If a rare pollen grain bearing tandem transgenic traits bypasses containment, it must compete with multitudes of wild-type pollen to produce a hybrid. Its rare progeny must then compete with fitter wild-type cohorts during self-thinning when hundreds of plants are thinned to a single plant that establishes, replacing a parent. Even a small degree of unfitness encoded in the TM construct would bring about the elimination of the vast majority of progeny in all future generations, as long as the primary gene provides no selective advantage that counterbalances the unfitness of the linked TM gene. The rare hybrid offspring from escaped pollen bearing transgenic mitigator genes would not pose a dire threat, especially to wild species outside fields, as the amount of pollen reaching the pristine wild environment would only be a minuscule fraction of the pollen from the wild-type.

In cases where the greatest biosafety is required, it is probably wisest to flank the gene of choice with two mitigator genes, such as dwarfing and non-shattering, so that seeds from the few surviving dwarf plants are harvested and discarded. The use of two flanking mitigator genes will mathematically compound (the yet unknown) infrequency of mutation to loss-of-function of TM. Large-scale field studies will be needed with crop/weed pairs to continue to evaluate the positive implications of TM to risk mitigation.

Will transgenic mitigation traits adversely affect wild relatives of the crop? Models that suggest that mitigation is deleterious

One model claims that "demographic swamping" by crop transgenes would cause "migrational meltdown" of wild species related to the crop, especially if the introgressed genes confer unfitness (Haygood et al., 2003). Untested modifications of the untested model arrive at similar conclusions (Meirmans et al., 2009). This proposition that recurrent gene flow from crops, even unfit gene flow, could affect wild relatives deserves some discussion, as it flies in the face of Darwinian concepts of survival of the fittest (Gressel, 2005a). If the model was correct, it would reduce the utility of TM.

Data with conventional crops already belie the possibility that recurrent gene flow from transgenic crops with less fit genes will cause wild populations to shrink. Major domesticated crops are not fit to compete in wild ecosystems, so their normal genes confer a modicum of unfitness. Crop × wild hybrids continually form at a low frequency, yet there is no published evidence that demographic swamping has occurred from recurrent gene flow from conventional crops to wild species in natural ecosystems, or to suggest concrete situations where the model may apply. No data or examples are supplied to support their model simulations. Indeed, considerable evidence has appeared that many crops exist near their wild or weedy progenitors, without causing the extinction of the progenitors, despite continuous gene flow of crop genes that are unfit naturally in the wild (Gressel, 2005b). As an example, some grass biotypes have lived for >2000 years on Roman mine tailings in Wales, evolving heavy metal resistance, and are unfit compared to the wild-type on normal soil. Their pollen has continuously blown to their sensitive cohorts centimeters away without the modeled swamping (Bradshaw, 1982). Maize and teosinte grow in close proximity, and F_1 hybrids form, but the teosinte has not been "swamped" by massive maize cultivation in close proximity (Doebley, 1992), as would be predicted by such models. At worst, hybrid swarms often appear at the boundary between crop and wild species, but they remain contained.

Three problematic issues seem to invalidate such models for the vast majority of conceivable crop/wild species systems, and they are discussed in the following list:

1. To get the level of swamping discussed (Gressel, 2005a; Meirmans et al., 2009), the wild relative and the crop would have to live in the same ecosystem, which by definition is not the case. Pollen flow from crops decreases exponentially with distance, usually to a low asymptote due to wind currents or insects, not fully obeying simple physics. There would be far more wild pollen produced in the wild ecosystems than pollen inflow from agroecosystems, so hybridization events in the wild from crop pollen will be rare. Indeed, even assuming an enormous 10% hybridization with the wild each generation from crop pollen, according to their model it will take about 20 generations of recurrent pollination for the unfit crop allele to become fixed in half the population, and 50 generations to asymptotically reach 80% of the population. This model is also contradicted by data from a replacement series with a transgenic crop bearing an unfitness gene competitively intermingled with the wild-type (Al-Ahmad et al., 2005). When nine times more unfit individuals swamped fit individuals, the result was much less than 90% unfit progeny (Al-Ahmad et al., 2005).
2. The models assume that the crop and the theoretical wild species growing in its midst will have synchronous flowering, no self-fertilization, and no genetic or other barriers to cross-fertilization negating the concept of speciation. It is exceedingly rare for crop pollen to fertilize another species such as a wild relative without any genetic barriers. No cases were discussed where this might happen. Even conspecific wild sunflowers have genomic deterrents to introgression (reviewed in Stewart et al., 2003). Weedy sunflowers, growing in or near domestic sunflowers, have often introgressed advantageous crop genes and have not been wiped out by unfit crop genes. Truly wild sunflowers growing on a native prairie are far less affected by crops than weedy sunflowers.

3. The model (Haygood et al., 2003) assumes animal-type replacement rates where just a few progeny per mating is typical, allowing lower fitness to indeed become fixed. Most wild relatives of crops produce copious amounts of seed to replace parents. Self-thinning, except by sperm during fertilization, is far less an important factor in animals than plants due to the low progeny number. Most animals are "perennial" perpetuating unfitness for several breeding cycles. Most wild relatives of crops produce a single generation in their adult life.

The conclusion that "the most striking implication of this model is the possibility of thresholds and hysteresis, such that a small increase in (unfit gene) immigration can lead to fixation of a disfavored crop allele...." (Haygood et al., 2003) contradicts classic and contemporary field data showing that only near-neutral genes exist in pockets of the evolutionary landscape of plants. Blatantly unfit plant genes only exist at frequencies below the mutation frequency to the unfitness trait.

Such models seem invalid in referring to transgene establishment in the wild, but may have some validity for a few weeds (not wild species) related to crops in agroecosystems or ruderal ecosystems. When flowering weeds are at a low density in an agricultural ecosystem (and perhaps close by in ruderal systems) the model might be predictive, but would it be bad for a weed to go extinct? Weeds are (inadvertently) man-made domesticated species (Warwick et al., 2005). Shouldn't people also have the right to eliminate them? The nature of weeds is such that they do not become extinct, as much as the farmer would desire. It is far more likely that such evolutionarily threatened weeds would evolve exclusionary mechanisms that would block evolutionarily threatening gene flow. For example, they could evolve a shift from out-crossing to predominant self-fertilization that would protect them from crop pollen bearing unfit genes.

Traits that can be Used in Tandem Transgenic Mitigation Constructs

Mitigation can be achieved by suppressing weedy traits in hybrids or by rendering the hybrids hypersensitive to chemicals.

Morphological traits and genes for mitigation

Not all TM traits have known sequenced genes that are immediately available for insertion in tandem constructs, but it is assumed that many of the needed TM genes will soon be found due to the spate of whole genome sequencing and reliable annotation.

Secondary dormancy

Unfortunately, *Arabidopsis*, the typical source of genes, has already been sufficiently domesticated so that it is unlike related weeds; the laboratory strains no longer have strong secondary dormancy (van der Schaar et al., 1997). A mutant that is insensitive to abscisic acid and lacks secondary dormancy was found in a wild, undomesticated *Arabidopsis* strain (Steber et al., 1998). Such a gene might be useful.

Seed shattering

Physiologically, one way to avoid premature seed shattering is to have uniform ripening. Early maturing seeds of oilseed rape on indeterminate, continuously flowering varieties typically shatter. Determinacy, with its single uniform flush of flowering, is one method to prevent shattering, but this often shortens the season and reduces yield. The hormonology of the abscission zone controls whether shattering will occur, and it is possible that if cytokinins are overproduced then shattering will be delayed. The cytokinin pathway is well documented and there are genes that could be put in constructs for cytokinin overproduction (Kusaba et al., 1998).

Various genes have been isolated from *Arabidopsis* that prevent seed shatter by delaying valve opening on the silique (Liljegren et al., 2000, 2004). Some of these genes may be useful for closely related oilseed rape, and one was tested in a related mustard (Ostergaard et al., 2006). None of these genes may necessarily be appropriate for grass crops where the stem breaking mode of shattering is quite different from *Arabidopsis* silique or legume pod opening. Grasses have multiple pathways for seed shattering, even in the same species (Li and Gill, 2006). As shattering is typically dominant, and the non-shattering of domesticated species is recessive, our ancestors had to select for a number of recessive traits to obtain what we now cultivate. In wheat, *Aegilops* complex shattering can be due to (Feldman, 2001):

- Spike disarticulation (breakage): Rachis is brittle at the base, and the whole spike shatters as a unit. This occurs only in *Aegilops* species at present.
- Barrel type spikelet disarticulation in some wheat strains: Shattering occurs at the lower side of the junction between the rachis and the spikelet base.
- Wedge-shaped spikelet disarticulation: The rachis breaks at the upper side of the attachment with the spikelet base.

The first cultivated wheats were selected for non-shattering at these loci. When wheat de-domesticated and became feral, as it had in Tibet (Sun et al., 1998), different accessions had different shattering types.

Rice has a cloned shattering gene *qSH-1* (Konishi et al., 2005), which has a high degree of homology with barley gene *JuBel2* that does not seem to control shattering on that species (Li et al., 2006). Maize shattering genes are not orthologous to any of these (Li et al., 2006). Because multiple genetic pathways seem to control shattering, it is unlikely that there will be a "one gene fits all" single anti-shattering mitigation gene for all grain crops. Antisense or RNAi-based anti-shattering mitigators should be based on the shattering mechanism in the related weed, not the crop.

Dwarfing

Many of the genes controlling dwarfism seem to have an unknown function. Still many other genes are known that control height.

Preventing the biosyntheses of gibberellins reduces height, which is the basis of many chemical dwarfing agents used commercially on wheat. The three enzymes and genes controlling steps in gibberellin biosyntheses have been cloned (Helliwell et al., 1998; Lange, 1998; Yamaguchi, 2008). Additionally, a defective GA receptor gene confers gibberellin insensitivity when transformed into crops (Δ*GAI*) by competing with the native receptor, inducing dwarfing (Peng et al., 1999). Interestingly, despite dwarfing oilseed rape at the early rosette stage, plants transformed with the Δ*GAI* gene do have flowers on long stalks (Al-Ahmad et al., 2006). This is despite a requirement for gibberellin activity to bolt. Presumably either less ΔGAI is produced in the stalk, or more native receptor or a different gibberellin receptor is produced when it is time to flower. Gibberellins GA_1 and GA_4 are specifically responsible for flower stalk bolting in radish (Nishijima et al., 1998). This might suggest that oilseed rape has different receptors in the stalk controlling bolting. Known and cloned genes coding for the monooxygenases and dioxygenases are responsible for these later steps of GA biosynthesis (Kusaba et al., 1998).

Another anti-bolting gene has recently been reported in Chinese cabbage, *BrpFLC*, which encodes a MADS-domain transcription factor (Li et al., 2005). Its level was higher in varieties that required longer vernalization, suggesting that its transgenic modulation could be utilized to prevent premature bolting in climatic conditions where this is a problem. It is yet unclear which genes this transcription factor controls.

Brassinosteroids also control stem elongation in many plant species, and their absence results in dwarf plants. A 22 d-hydroxylase cytochrome P_{450} controls a series of these steps in brassinosteroid biosynthesis, and plants lacking the enzyme are dwarfed (Azpiroz et al., 1998). Plants are also dwarfed when they produce normal levels of these growth regulators but are mutated in the *bri1* gene coding for the receptor (Noguchi et al., 1999). Additionally, suppressive over-expression of a sterol C24-methyl transferase in the pathway also causes dwarfing (Schaller et al., 1998).

Shade avoidance

Shade avoidance is controlled by various forms of the pigment phytochrome that interact and detect whether a plant is being shaded (Devlin et al., 1998). It is advantageous for a crop plant to grow taller when shaded by a weed, but not when shaded by cohorts, as less grain is produced on the taller stalks. The engineering of suppressive over-expression constructs of one of these phytochromes led to plants that did not elongate in response to shading (Robson et al., 1996) .

Chemical mitigation of transgene flow

Mitigation can also be achieved chemically by using TM genes that are either activated by chemicals, or by conferring supersusceptibility to chemicals such as herbicides used in crop rotations.

Activatable genes for susceptibility to chemicals

A bacterial P_{450} that activates an experimental sulfonylurea pro-herbicide has been used under a tapetum-specific promoter to prevent pollen formation (O'Keefe et al., 1994). The same gene could be used under an inducible promoter that can be generally expressed, which would allow the use of the pro-herbicide to cull crop–weed hybrids as well as volunteer crop weeds. Other such pro-herbicides with exogenous activating genes could also be envisaged for use as mitigating fail-safes, should the primary transgene escape.

Hypersensitivity to herbicides as transgenic mitigation

A TM construct with a gene conferring glyphosate resistance and a TM gene conferring herbicide hypersensitivity to the herbicide bentazon, which is used in rice. It was then possible to selectively cull transgenic volunteers and hybrids in rotations when the wild-type crops are cultivated, because the wild-type crop naturally degrades bentazon (Lin et al., 2008).

A multi-chemical approach was then devised for monoculture crops such as rice and sorghum, proposing a rotation of chemicals and transgenic varieties (Gressel and Valverde, 2009a). It could provide cost-effective and sustainable control of weedy relatives of crops while preventing transgene establishment in the weedy relatives, as well as confounding the evolution of herbicide resistance in other weeds. The concept requires rotating different metabolically herbicide-resistant crop varieties with other varieties or lines of the crop that are resistant to other herbicides, but are hypersensitive to the first herbicide, such that the herbicide used with the second variety will preferentially kill hybrids with the first variety. For example, one variety has a gene conferring metabolic resistance to glyphosate in tandem with a TM antisense or RNAi gene suppressing a glufosinate resistance transgene. The first variety would be used in crop rotations in conjunction with the herbicide glyphosate. The second variety has a gene conferring metabolic resistance to glufosinate in tandem with a TM antisense or RNAi gene suppressing a glyphosate resistance transgene. The second variety would be used in crop rotations in conjunction with the herbicide glufosinate and would selectively control volunteers and weed × crop hybrids from the first variety. Using glyphosate with the first variety will selectively control volunteers and weed × crop hybrids from the second variety (Gressel and Valverde, 2009a).

Similar constructs could be made to ensure long-term sustainability with other crops having pernicious interbreeding weeds. These could be made with metabolic resistance/hypersensitivities to other herbicides as well, using the morphological mitigation genes such as anti-shattering and dwarfing. An important gene to further ensure long-term sustainability will be an anti-secondary dormancy gene, when it is found (Gressel and Al-Ahmad, 2011). It will guarantee that hybrids with weedy rice will germinate the following season, because they will be controlled by the reciprocal herbicide in rotation.

Special cases where transgenic mitigation is needed

There are some special cases where other genes can be envisaged to confer an unfitness upon volunteers or on feral forms coming from the crop. The TM genes are typically still neutral or positive to the crop but give unfit offspring.

Mitigation for biennial and annual "root" crops

Mitigating genes should easily prevent both the premature and volunteer flowering in sugar beets, carrots, onions, celery, and radishes as well as in other biennial or two-phase crops where the vegetative material is marketed, and premature flowering (bolting) is detrimental. This could easily be affected by preventing gibberellic acid biosynthesis (Yamaguchi, 2008) in a TM construct. Suppression of kaurene oxidase, a key enzyme in gibberellic acid biosynthesis, would require the use of gibberellic acid to "force" flowering for seed production. There should be a concomitant biosafety requirement that seed production areas are far removed from areas where weedy or other feral or wild relatives grow to prevent pollen transfer. If, despite all isolation distances, a TM construct or a mutant in a seed production area introgresses with a wild species, the progeny will be biennial or be too delayed, such that the transgenic hybrid would be non-competitive with cohorts that reproduce in a single year and do not need to overwinter.

Delaying of bolting and flowering by using other transgenes was achieved using an antisense engineered fragment of the *GIGANTEA* gene into radish, encoding a protein that is part of the photoperiod recognition system. Bolting was considerably delayed, but seed production came about without reversal mechanisms by waiting (Curtis et al., 2005). *Flowering Locus* C gene also controls bolting in Chinese cabbage with higher transcript levels resulting in the less bolting (Li et al., 2005).

Other transgenes can be considered for mitigating the risks of introgression with root crops, such as genes promoting partitioning to roots, which would be advantageous to cultivated root crops, but detrimental to feral forms.

Transgenically mitigated genes for crop-produced pharmaceuticals and industrial products

Pharmaceuticals and industrial proteins, especially enzymes and antibodies, can be inexpensively produced in plants without the need for animal tissue culture cells grown in a medium of expensive serum albumin that is all too easily contaminated with pathogenic mycoplasms, prions, and viruses. Whereas there are compelling economic and biosafety considerations propelling the production of pharmaceuticals in crops, there are also reasons to exclude the pharmaceutical and industrial transgenes from introgressing into other varieties of the crop, related species, or to remain in viable volunteers in the field (e.g., Breyer et al., 2009).

The containment of pharmaceutical transgenes has been physical as evidenced by human error that reportedly allowed temporary volunteer escape of "Prodigene" maize containing such genes. The biological containment strategies previously described may be preferable to a dependence on physical containment by humans, and the TM strategies should work as well, especially as a secondary fail-safe.

Pharmaceutical transgenes in maize are expressed in embryo tissues, and a potential tandem mitigating gene could be any RNAi type suppression of genes that affect the endosperm, such as the various "shrunken seed" loci, especially those where sugar transformation to starch is inhibited (Chourey et al., 1998). Such shrunken seeds, with their high sugar content, are harder to store than normal maize, and are extremely unfit in the field and unlikely to overwinter and produce volunteers. Hybrids with other varieties would have shrunken seeds that would be culled during seed cleaning. Their volunteers would be unfit and also could not overwinter as volunteers in the field. Should such pollen fertilize a few seeds in adjacent fields, they will be shrunken and will be sorted out during processing. Thus, such a technology would mitigate against both outflow and influx of pollen.

Mitigation in species used for phytoremediation

Plants have been used to correct human error over the ages. A few species are capable of revegetating Roman lead and zinc mine tailing in Wales (Smith and Bradshaw, 1979). Of these, plants that can withstand toxic wastes after they have been taken up are of interest for phytoremediation. Two types of multi-cut species are usually considered for phytoremediation, with the cut material burnt to extract the heavy metals or to oxidize the organic wastes: herbaceous species such as *B. juncea* and *Spartina* spp. (cord grasses), which are most efficient at dealing with surface wastes and trees such as *Populus* spp., for dealing with deeper wastes (Pilon-Smits and Pilon, 2002)

Heavy metal tolerance has been brought into *B. juncea* (Indian mustard) from slow-growing *Thlaspi* by protoplast fusion (along with many other genes; Dushenkov et al., 2002). It was better yet to transgenically transfer genes leading to enhanced glutathione content (Bennett et al., 2003) to make the necessary phytochelatins that complex the heavy metals. A single cropping of *B. juncea* does not clean up a toxic site. Many growth cycles are required, with multiple harvests and natural reseeding. *B. juncea*, even more than its close relative *B. napus* (oilseed rape), is not fully domesticated, and the multiple cycles of cropping would allow the possibility of selecting for ferality. Thus, mitigation seems necessary to prevent volunteers from becoming feral. One gene that might specifically fulfill the need for a mitigator gene is over-expression of a cytokinin oxidase (Bilyeu et al., 2001), which reduces cytokinin levels. This in turn led to phenotypes with far reduced shoot systems (unfitness to compete) but with faster growing, more extensive root systems (Werner et al., 2003), all the better for extracting toxic wastes. Genes that delay or prevent flowering may also be useful with the *Brassica* species, allowing multiple cuts of larger vegetative plants and preventing gene flow.

Concluding Remarks

Systems exist that can theoretically preclude the movement to and establishment of crop transgenes in related crops, weeds, or weedy relatives, whether by containing gene flow or by preventing the establishment of hybrid plants in the field by mitigation. Thus, if a risk is discerned, it should not preclude developing transgenic crops, it should stimulate the imagination to devise and test appropriate systems to deal with the potential problems.

References

Al-Ahmad, H., & Gressel, J. (2006). Mitigation using a tandem construct containing a selectively unfit gene precludes establishment of *Brassica napus* transgenes in hybrids and backcrosses with weedy *Brassica rapa*. *Plant Biotechnology Journal*, *4*, 23–33.

Al-Ahmad, H. I., Galili, S., & Gressel, J. (2004). Tandem constructs mitigate risks of transgene flow from crops: Tobacco as a model. *Molecular Ecology*, *13*, 687–710.

Al-Ahmad, H., Dwyer, J., Moloney, M. M., & Gressel, J. (2006). Mitigation of establishment of *Brassica napus* transgenes in volunteers using a tandem construct containing a selectively unfit gene. *Plant Biotechnology Journal*, *4*, 7–21.

Al-Ahmad, H., Galili, S., & Gressel, J. (2005). Poor competitive fitness of transgenically mitigated tobacco in competition with the wild type in a replacement series. *Planta*, *272*, 372–385.

Allainguillaume, J., Harwood, T., Ford, C. S., Cuccato, G., Norris, C., & Allender, C. J., et al. (2009). Rapeseed cytoplasm gives advantage in wild relatives and complicates genetically modified crop biocontainment. *New Phytologist*, *183*, 1201–1211.

Arazi, T., Huang, P. L., Huang, P. L., Zhang, L., Shiboleth, Y. M., & Gal-On, A., et al. (2002). Production of antiviral and antitumor proteins MAP30 and GAP31 in cucurbits using the plant virus vector ZYMV-AGII. *Biochemical and Biophysical Research Communications*, *292*, 441–448.

Avni, A., & Edelman, M. (1991). Direct selection for paternal inheritance of chloroplasts in sexual progeny of *Nicotiana*. *Molecular & General Genetics*, *225*, 273–277.

Azpiroz, R., Wu, Y., LoCascio, J. C., & Feldmann, K. A. (1998). An *Arabidopsis* brassinosteroid-dependent mutant is blocked in cell elongation. *Plant Cell*, *10*, 219–230.

Bennett, L. E., Burkhead, J. L., Hale, K. L., Terry, N., Pilon, M., & Pilon-Smits, E. A. H. (2003). Analysis of transgenic Indian mustard plants for phytoremediation of metal-contaminated mine tailings. *Journal of Environmental Quality*, *32*, 432–440.

Bilyeu, K. D., Cole, J. L., Laskey, J. G., Riekhof, W. R., Esparza, T. J., & Kramer, M. D., et al. (2001). Molecular and biochemical characterization of a cytokinin oxidase from maize. *Plant Physiology*, *125*, 378–386.

Bradshaw, A. D. (1982). *Evolution of heavy metal resistance—an analogy for herbicide resistance?* (pp. 293–307) New York: Wiley.

Breitler, J. C., Vassal, J. M., Catala, M. D., Meynard, D., Marfa, V., & Mele, E., et al. (2004). Bt rice harbouring *cry* genes controlled by a constitutive or wound-inducible promoter: Protection and transgene expression under Mediterranean field conditions. *Plant Biotechnology Journal*, *2*, 417–430.

Breyer, D., Goossens, M., Herman, P., & Sneyers, M. (2009). Biosafety considerations associated with molecular farming in genetically modified plants. *Journal of Medicinal Plants Research*, *3*, 825–838.

Cao, B. H., Huang, Z. Y., Chen, G. J., & Lei, J. J. (2010). Restoring pollen fertility in transgenic male-sterile eggplant by Cre/lox p-mediated site-specific recombination system. *Genetics and Molecular Biology*, *33*, 298–307.

Carlson, P. S., Fahey, J. W., & Flynn, J. L. (1992). Modified plant containing a bacterial inoculant. U.S. Patent, 5,157,207.

Cazzonelli, C. I., McCallum, E. J., Lee, R., & Botella, J. R. (2005). Characterization of a strong, constitutive mung bean (*Vigna radiata* L.) promoter with a complex mode of regulation in planta. *Transgenic Research*, *14*, 941–967.

Chen, Q., Jahier, J., & Cauderon, Y. (1993). The B-chromosome system of inner Mongolian *Agropyron gaertn*. 3. Cytogenetical evidence for B-A pairing at metaphase-I. *Hereditas*, *119*, 53–58.

Choi, I.-R., Stander, D. C., Morris, T. J., & French, R. (2000). A plant virus vector for systemic expression of foreign genes in cereals. *Plant Journal*, *23*, 547–555.

Chourey, P. S., Taliercio, E. W., Carlson, S. J., & Ruan, Y.-L. (1998). Genetic evidence that the two isozymes of sucrose synthase present in developing maize endosperm are critical, one for cell wall integrity and the other for starch biosynthesis. *Molecular & General Genetics*, *259*, 88–96.

Crouch, M. L. (1998). How the terminator terminates: An explanation for the non-scientist of a remarkable patent for killing second generation seeds of crop plants < http://www.bio.indiana.edu/people/terminator/html/ >. The Edmonds Institute, Edmond WA, USA.

Curtis, I. S., Hanada, A., Yamaguchi, S., & Kamiya, Y. (2005). Modification of plant architecture through the expression of GA 2-oxidase under the control of an estrogen inducible promoter in *Arabidopsis thaliana* L. *Planta*, *222*, 957–967.

Daniell, H. (2002). Molecular strategies for gene containment in transgenic crops. *Nature Biotechnology*, *20*, 581–586.

Darmency, H. (1994). Genetics of herbicide resistance in weeds and crops. In S. B. Powles & J. A. M. Holtum (Eds.), *Herbicide resistance in plants: Biology and biochemistry* (pp. 263–298). Boca-Raton: Lewis.

Devlin, P. F., Patel, S. R., & Whitelam, G. C. (1998). Phytochrome E influences internode elongation and flowering time in *Arabidopsis*. *Plant Cell*, *10*, 1479–1487.

Doebley, J. (1992). Molecular systematics and crop evolution. In P. S. Soltis, D. E. Soltis & J. J. Doyle (Eds.), *Molecular systematics of plants* (pp. 202–222). New York: Chapman and Hall.

Dushenkov, S., Skarzhinskaya, M., Glimelius, K., Gleba, D., & Raskin, I. (2002). Bioengineering of a phytoremediation plant by means of somatic hybridization. *International Journal of Phytoremediation*, *4*, 117–126.

Ellstrand, N. C. (2003). *Dangerous liaisons – when cultivated plants mate with their wild relatives*. Baltimore MD: Johns Hopkins University Press.

Feldman, M. (2001). The origin of cultivated wheat. In A. Bonjean & W. Angus (Eds.), *The world wheat book* (pp. 3–56). Paris: Lavousier Tech & Doc.

Gressel, J. (1999). Tandem constructs: Preventing the rise of superweeds. *Trends in Biotechnology*, *17*, 361–366.

Gressel, J. (2002). *Molecular biology of weed control*. London: Taylor & Francis.

Gressel, J. (2005a). Problems in qualifying and quantifying assumptions in plant protection models: Resultant simulations can be mistaken by a factor of million. *Crop Protection*, *24*, 1007–1015.

Gressel, J. (2005b). *Crop ferality and volunteerism*. Boca Raton: CRC Press.

Gressel, J. (2008). *Genetic glass ceilings: Transgenics for crop biodiversity*. Baltimore: Johns Hopkins University Press.

Gressel, J., & Al-Ahmad, H. (2011). Transgenic mitigation of gene flow. In M. J. Oliver & Y. Li (Eds.), *Gene containment*. Ames: Wiley.

Gressel, J., & Valverde, B. E. (2009a). A strategy to provide long-term control of weedy rice while mitigating herbicide resistance transgene flow, and its potential use for other crops with related weeds. *Pest Management Science*, *65*, 723–731.

Gressel, J., & Valverde, B. E. (2009b). The other, ignored HIV – Highly invasive vegetation. *Food Security*, *1*, 463–478.

Griffiths, S., Sharp, R., Foote, T. N., Bertin, I., Wanous, M., & Reader, S., et al. (2006). Molecular characterization of *Ph1* as a major chromosome pairing locus in polyploid wheat. *Nature*, *439*, 749–752.

Gruber, S., & Hüsken, A. (2011). Control of cleistogamy and seed dormancy for biological gene containment in oilseed rape (*Brassica napus* L). In M. J. Oliver & Y. Li (Eds.), *Gene containment*. Ames: Wiley.

Haider, N., Allainguillaume, J., & Wilkinson, M. J. (2009). Spontaneous capture of oilseed rape (*Brassica napus*) chloroplasts by wild *B. rapa*: Implications for the use of chloroplast transformation for biocontainment. *Current Genetics*, *55*, 139–150.

Haygood, R., Ives, A. R., & Andow, D. A. (2003). Consequences of recurrent gene flow from crops to wild relatives. *Proceedings of the Royal Society of London Series B*, *270*, 1879–1886.

Helliwell, C. A., Sheldon, C. C., Olive, A. M., Walker, R. R., Zeevaart, J. A. D., & Peacock, W. J., et al. (1998). Cloning of the *Arabidopsis ent*-kaurene oxidase gene *GA3*. *Proceedings of the National Academy of Science, USA*, *95*, 9019–9024.

Hooftman, D. A. P., Gerard, J., Oostermeijer, B., Marquard, E., & den Nijs, H. C. M. (2008). Modelling the consequences of crop-wild relative gene flow: A sensitivity analysis of the effects of outcrossing rates and hybrid vigour breakdown in *Lactuca*. *Journal of Applied Ecology*, *45*, 1094–1103.

Jach, G., Gornhardt, B., Mundy, J., Logemann, J., Pinsdorf, P., & Leah, R., et al. (1995). Enhanced quantitative resistance against fungal disease by combinatorial expression of different barley antifungal proteins in transgenic tobacco. *Plant Journal*, *8*, 97–109.

Jepson, I., Martinez, A., & Sweetman, J. P. (1998). Chemical-inducible gene expression systems for plants—a review. *Pesticide Science*, *54*, 360–367.

Kelloniemi, J., Makinen, K., & Valkonen, J. P. T. (2008). Three heterologous proteins simultaneously expressed from a chimeric potyvirus: Infectivity, stability and the correlation of genome and virion lengths. *Virus Research*, *135*, 282–291.

Kobayashi, K., Munemura, I., Hinata, K., & Yamamura, S. (2006). Bisexual sterility conferred by the differential expression of Barnase and Barstar: A simple and efficient method of transgene containment. *Plant Cell Reports*, *25*, 1347–1354.

Konishi, S., Lin, S. Y., Ebana, K., Fukuta, Y., Izawa, T., & Sasaki, T., et al. (2005). Molecular cloning of a major QTL, *QSH-1*, controlling seed shattering habit in rice. *Plant and Cell Physiology*, *46*, S198-S198.

Kusaba, S., Kano-Murakami, Y., Matsuoka, M., Tamaoki, M., Sakamoto, T., & Yamaguchi, I., et al. (1998). Alteration of hormone levels in transgenic tobacco plants overexpressing the rice homeobox gene *OSH1*. *Plant Physiology*, *116*, 471–476.

Kuvshinov, V., Koivu, K., Kanerva, A., & Pehu, E. (2001). Molecular control of transgene escape from genetically modified plants. *Plant Science*, *160*, 517–522.

Kuvshinov, V., Anissimov, A., & Yahya, B. M. (2004). *Barnase* gene inserted in the intron of GUS—a model for controlling transgene flow in host plants. *Plant Science*, *167*, 173–182.

Lange, T. (1998). Molecular biology of gibberellin synthesis. *Planta*, *204*, 409–419.

Lentz, E. M., Segretin, M. E., Morgenfeld, M. M., Wirth, S. A., Santos, M. J. D., & Mozgovoj, M. V., et al. (2010). High expression level of a foot and mouth disease virus epitope in tobacco transplastomic plants. *Planta*, *231*, 387–395.

Li, W., & Gill, B. S. (2006). Multiple genetic pathways for seed shattering in the grasses. *Functional and Integrative Genomics*, *6*, 300–330.

Li, Z., Zhao, L., Cui, C., Kai, G., Zhang, L., & Sun, X., et al. (2005). Molecular cloning and characterization of an anti-bolting related gene (*BrpFLC*) from *Brassica rapa* ssp. *Pekinensis*. *Plant Science*, *168*, 407–413.

Li, C., Zhou, A., & Sang, T. (2006). Rice domestication by reduced shattering. *Science*, *311*, 1936–2193.

Liljegren, S. J., Ditta, G. S., Eshed, Y., Savidge, B., Bowman, J. L., & Yanofsky, M. F. (2000). *SHATTERPROOF MADS*-box genes control seed dispersal in *Arabidopsis*. *Nature*, *404*, 766–770.

Liljegren, S. J., Roeder, A. H. K., Kempin, S. A., Gremski, K., Ostergaard, L., & Guimil, S., et al. (2004). Control of fruit patterning in *Arabidopsis* by *INDEHISCENT*. *Cell*, *116*, 843–853.

Lin, C., Jun, F., Xu, X., Te, Z., Cheng, J., & Tu, J., et al. (2008). A built-in strategy for containment of transgenic plants: Creation of selectively terminable transgenic rice. *PLoS ONE*, *3*(3), e1818.

Liu, Z., & Singer, S. D. (2011). Tapping RNA silencing for transgene containment through the engineering of sterility in plants. In M. J. Oliver & Y. Li (Eds.), *Gene containment*. Ames: Wiley.

Lu, B. R. (2005). Multi-directional gene flow among wild, weedy and cultivated soybeans. In J. Gressel (Ed.), *Crop ferality and volunteerism*. Boca Raton: CRC Press.

Maliga, P. (2004). Plastid transformation in higher plants. *Annual Review of Plant Biology*, *55*, 289–313.

Mariani, C., Debeuckeleer, M., Truettner, J., Leemans, J., & Goldberg, R. B. (1990). Induction of male sterility in plants by a chimeric ribonuclease gene. *Nature*, *347*, 737–741.

Marillonet, S., Thoeringer, C., Kandzia, R., Klimyuk, V., & Gleba, Y. (2005). Systemic *Agrobacterium tumefaciens*-mediated transfection of viral replicons for efficient transient expression in plants. *Nature Biotechnology*, *25*, 718–723.

Meirmans, P. G., Bousquet, J., & Isabel, N. (2009). A metapopulation model for the introgression from genetically modified plants into their wild relatives. *Evolutionary Applications*, *2*, 160–171.

Moon, H. S., Li, Y., & Stewart, C. N. (2010). Keeping the genie in the bottle: Transgene biocontainment by excision in pollen. *Trends in Biotechnology*, *28*, 3–8.

Nishijima, T., Katsura, N., Koshioka, M., Yamazaki, H., Nakayama, M., & Yamane, H., et al. (1998). Effects of gibberellins and gibberellin-biosynthesis inhibitors on stem elongation and flowering of *Raphanus sativus* L. *Journal of the Japanese Society for Horticultural Science*, *67*, 325–330.

Noguchi, T., Fujioka, S., Choe, S., Takatsuto, S., Yoshida, S., & Yuan, H., et al. (1999). Brassinosteroid-insensitive dwarf mutants of *Arabidopsis* accumulate brassinosteroids. *Plant Physiology*, *121*, 743–752.

O'Keefe, D. P., Tepperman, J. M., Dean, C., Leto, K. J., Erbes, D. L., & Odell, J. T. (1994). Plant expression of a bacterial cytochrome-P450 that catalyzes activation of a sulfonylurea pro-herbicide. *Plant Physiology*, *105*, 473–482.

Oliver, M. J., & Hake, K. (2011). Seed based strategies for gene containment. In M. J. Oliver & Y. Li (Eds.), *Gene containment*. Ames: Wiley.

Oliver, M. J., & Li, Y. (2011). *Gene containment*. Ames: Wiley.

Oliver, M. J., Quisenberry, J. E., Trolinder, N. L. G., & Keim, D. L. (1998). Control of plant gene expression. US Patent 5,723,765.

Oliver, M. J., Luo, H., Kausch, A., & Collins, H. (2004). Seed-based strategies for transgene containment. In *Eighth international symposium on the biosafety of genetically modified organisms*, INRA, Montpellier, pp. 154–161.

Ostergaard, L., Kempin, S. A., Bies, D., Klee, H. J., & Yanofsky, M. F. (2006). Pod shatter resistant *Brassica* fruit produced by ectopic expression of the *FRUITFULL* gene. *Plant Biotechnology Journal*, *4*, 45–51.

Ozkan, H., & Feldman, M. (2001). Genotypic variation in tetraploid wheat affecting homoeologous pairing in hybrids with *Aegilops peregrina*. *Genome*, *44*, 1000–1006.

Pedersen, J. F., Marx, D. B., & Funnell, D. L. (2003). Use of A(3) cytoplasm to reduce risk of gene flow through sorghum pollen. *Crop Science*, *43*, 1506–1509.

Peng, J., Richards, D. E., Hartley, N. M., Murphy, G. P., Devos, K. M., & Flintham, J. J., et al. (1999). "Green revolution" genes encode mutant gibberellin response modulators. *Nature*, *400*, 25.

Pilon-Smits, E., & Pilon, M. (2002). Phytoremediation of metals using transgenic plants. *Critical Reviews in Plant Sciences*, *21*, 439–456.

Puchta, H. (2002). Gene replacement by homologous recombination in plants. *Plant Molecular Biology*, *48*, 173–182.

Riley, R., Campbell, C. S., Young, R. M., & Belfield, A. M. (1966). Control of meiotic chromosome pairing by the chromosomes of homoeologous group 5 of *Triticum aestivum*. *Nature*, *212*, 1475–1477.

Robson, P. R. H., McCormac, A. C., Irvine, A. S., & Smith, H. (1996). Genetic engineering of harvest index in tobacco through overexpression of a phytochrome gene. *Nature Biotechnology*, *14*, 995–998.

Rose, C. W., Millwood, R. J., Moon, H. S., Rao, M. R., Halfhill, M. D., & Raymer, P. L. (2009). Genetic load and transgenic mitigating genes in transgenic *Brassica rapa* (field mustard) × *Brassica napus* (oilseed rape) hybrid populations. *BMC Biotechnology*, *9*, 93.

Ruf, S., Karcher, D., & Bock, R. (2007). Determining the transgene containment level provided by chloroplast transformation. *Proceedings of the National Academy of Sciences USA*, *104*, 6998–7002.

Ruiz, O. N., & Daniell, H. (2005). Engineering cytoplasmic male sterility via the chloroplast genome by expression of beta-ketothiolase. *Plant Physiology, 138*, 1232–1246.

Schaller, H., Bouvier-Navé, P., & Benveniste, P. (1998). Overexpression of an *Arabidopsis* cDNA encoding a sterol-C24-methyltransferase in tobacco modifies the ratio of 24-methyl cholesterol to sitosterol and is associated with growth reduction. *Plant Physiology, 18*, 461–469.

Schernthaner, J. P., Fabijanski, S. F., Arnison, P. G., Racicot, M., & Robert, L. S. (2003). Control of seed germination in transgenic plants based on the segregation of a two-component genetic system. *Proceedings of the National Academy of Sciences of the United States of America, 100*, 6855–6859.

Sears, E. R. (1976). Genetic control of chromosome pairing in wheat. *Annual Review of Genetics, 10*, 31–51.

Shaked, H., Melamed-Bessudo, C., & Levy, A. A. (2005). High-frequency gene targeting in *Arabidopsis* plants expressing the yeast RAD54 gene. *Proceedings of the National Academy of Science of the United States of America, 102*, 12265–12269.

Shiboleth, Y. M., Arazi, T., Wang, Y., & Gal-On, A. (2001). A new approach for weed control in a cucurbit field employing an attenuated potyvirus-vector for herbicide resistance. *Journal of Biotechnology, 92*, 37–46.

Sikorskaite, S., Vuorinen, A. L., Rajamaki, M. L., Nieminen, A., Gaba, V., & Valkonen, J. P. T. (2010). HandyGun: An improved custom-designed, non-vacuum gene gun suitable for virus inoculation. *Journal of Virological Methods, 165*, 320–324.

Skinner, J. S., Meilan, R., Ma, C. P., & Strauss, S. H. (2003). The *Populus PTD* promoter imparts floral-predominant expression and enables high levels of floral-organ ablation in *Populus, Nicotiana* and *Arabidopsis*. *Molecular Breeding, 12*, 119–132.

Smith, R. A. H., & Bradshaw, A. D. (1979). Use of metal tolerant plant populations for the reclamation of metalliferous wastes. *Journal of Applied Ecology, 16*, 595–603.

Song, X., Wang, Z., & Qiang, S. (2011). Agronomic performance of F1, F2 and F3 hybirds between weedy rice and transgenic glufosinate-resistant rice. *Pest Management Science, 67*, 921–931.

Steber, C. M., Cooney, S. E., & McCourt, P. (1998). Isolation of the GA-response mutant sly1 as a suppressor of ABI1-1 in *Arabidopsis thaliana*. *Genetics, 149*, 509–521.

Stewart, C. N., Halfhill, M. D., & Warwick, S. I. (2003). Transgene introgression from genetically modified crops to their wild relatives. *Nature Reviews Genetics, 4*, 806–817.

Sun, Q. X., Ni, Z. F., Liu, Z. Y., Gao, J. W., & Huang, T. C. (1998). Genetic relationships and diversity among Tibetan wheat, common wheat, and European spelt wheat revealed by rapid markers. *Euphytica, 99*, 205–211.

Sun, L., Ghosh, I., Paulus, H., & Xu, M. Q. (2001). Protein trans-splicing to produce herbicide-resistant acetolactate synthase. *Applied and Environmental Microbiology, 67*, 1025–1029.

Svab, Z., & Maliga, P. (2007). Exceptional transmission of plastids and mitochondria from the transplastomic pollen parent and its impact on transgene containment. *Proceedings of The National Academy of Sciences of the United States of America, 104*, 7003–7008.

Trucco, F., Tatum, T., Rayburn, A. L., & Tranel, P. J. (2009). Out of the swamp: Unidirectional hybridization with weedy species may explain the prevalence of *Amaranthus tuberculatus* as a weed. *New Phytologist, 184*, 819–827.

van der Schaar, W., Blanco, C. L. A., Kloosterziel, K. M., Jansen, R. C., Van Ooijen, J. W., & Koornneef, M. (1997). QTL analysis of seed dormancy in *Arabidopsis* using recombinant inbred lines and MQM mapping. *Heredity, 79*, 190–200.

Verhounig, A., Karcher, D., & Bock, R. (2010). Inducible gene expression from the plastid genome by a synthetic riboswitch. *Proceedings of the National Academy of Sciences of the United States of America, 107*, 6204–6209.

von der Lippe, M., & Kowarik, I. (2007). Crop seed spillage along roads: A factor of uncertainty in the containment of GMO. *Ecography, 30*, 483–490.

Wang, T., Li, Y., Shi, Y., Reboud, X., Darmency, H., & Gressel, J. (2004). Low frequency transmission of a plastid encoded trait in *Setaria italica*. *Theoretical and Applied Genetics, 108*, 315–320.

Warwick, S. I., & Stewart, C. N., Jr. (2005). Crops come from wild plants: How domestication, transgenes, and linkage together shape ferality. In J. Gressel (Ed.), *Crop ferality and volunteerism*. Boca Raton: CRC Press.

Warwick, S. I., Beckie, H. J., & Hall, L. M. (2009). Gene flow, invasiveness, and ecological impact of genetically modified crops. *Annals of the New York Academy of Sciences, 1168*, 72–99.

Weissmann, S., Feldman, M., & Gressel, J. (2005). Sporadic inter-generic DNA introgression from wheat into wild *Aegilops* spp. *Molecular Biology and Evolution, 22*, 2055–2062.

Weissmann, S., Feldman, M., & Gressel, J. (2008). Hypothesis: Transgene establishment in wild relatives of wheat can be prevented by utilizing the *Ph1* gene as a *senso stricto* chaperon to prevent homoeologous recombination. *Plant Science, 175*, 410–414.

Werner, T., Motyka, V., Laucou, V., Smets, R., Van Onckelen, H., & Schmulling, T. (2003). Cytokinin-deficient transgenic *Arabidopsis* plants show multiple developmental alterations indicating opposite functions of cytokinins in the regulation of shoot and root meristem activity. *Plant Cell, 15*, 2532–2550.

Whitham, S. A., Yamamoto, M. L., & Carrington, J. C. (1999). Selectable viruses and altered susceptibility mutants in *Arabidopsis thaliana*. *Proceedings of the National Academy of Sciences of the United States of America, 96*, 772–777.

Williams, M. E. (1995). Genetic engineering for pollination control. *Trends in Biotechnology, 13*, 344–349.

Yamaguchi, S. (2008). Gibberellin metabolism and its regulation. *Annual Review of Plant Biology, 59*, 225–251.

Zapiola, M. L., Campbell, C. K., Butler, M. D., & Mallory-Smith, C. A. (2008). Escape and establishment of transgenic glyphosate-resistant creeping bentgrass *Agrostis stolonifera* in Oregon, USA: A 4-year study. *Journal of Applied Ecology, 45*, 486–494.

Zhu, Y. J., Agbayani, R., Jackson, M. C., Tang, C. S., & Moore, P. H. (2004). Expression of the grapevine stilbene synthase gene VST1 in papaya provides increased resistance against diseases caused by *Phytophthora palmivora*. *Planta, 220*, 241–250.

Zuo, J. R., & Chua, N. H. (2000). Chemical-inducible systems for regulated expression of plant genes. *Current Opinion in Biotechnology, 11*, 146–151.

Intellectual property rights of biotechnologically improved plants

33

Antoine Harfouche[1] Richard Meilan[2] Kannan Grant[3]
Vincent K. Shier[4]

[1]*Department for Innovation in Biological, Agro-food and Forest Systems, University of Tuscia, Viterbo, Italy,*
[2]*Department of Forestry and Natural Resources, Purdue University, West Lafayette, Indiana,*
[3]*Office of Technology Commercialization, University of Alabama Huntsville, Huntsville, Alabama,*
[4]*Oblon, Spivak, McClelland, Maier, & Neustadt, LLP, Alexandria, Virginia*

TABLE OF CONTENTS

Introduction: Capitalizing on Research and Development in Agricultural Biotechnology with Intellectual Property Protection

Agricultural biotechnology is a fast-expanding industry in many countries of the world that will continue to offer remarkable economic, environmental, and social opportunities in the years ahead. Since its introduction about 15 years ago, plant biotechnology has achieved very important milestones in increasing global crop productivity to improve food, feed, and fiber security, and in reducing the environmental footprint of agriculture.

With the world turning to agricultural technologies for solutions, the twenty-first century will witness a major development of genetically improved bioenergy crops that help to mitigate climate change and contribute to the cost-effective production of biofuels (Harfouche et al., 2010). The sequencing of genomes from several plant species and advances made in plant genetics and other aspects of agricultural sciences have led to technological breakthroughs that will provide the building blocks for what are likely to be major industries, with profound implications for agriculture. Not only will humankind benefit from these innovative technologies, but the efforts that spawned them are contributing to the existing body of scientific knowledge and the development of global biotechnology industries (Singh et al., 2009).

Sustaining this plant biotechnology revolution requires long-term commitment to both public and private sector research and development (R&D). In the agricultural sector, R&D is unique among industries in at least two aspects: the truly global reach of a majority of agricultural R&D; and the historical success of what has been largely a public enterprise. In relation to other industries, research and innovation in agriculture are far more geographically dispersed (Boettiger et al., 2004). Private enterprises contribute roughly one-third of global agricultural R&D expenditures, whereas public research institutions make up the other two-thirds, which is evenly split between developed and developing countries (Pardey and Beintema, 2001).

Through the ages, innovations in agriculture have greatly benefited humanity. It is no small irony, then, that controversy surrounds agriculture and the intellectual property (IP) laws that were enacted for the very purpose of fostering invention (Bent, 2006). As the pace of scientific discovery in plant biotechnology has accelerated over the past few decades, the use of IP rights to protect these inventions has increased tremendously. These advances can only be achieved through resource-intensive R&D. Capitalizing on R&D investments using IP protection is just one part of the research-to-marketplace continuum. Protecting IP rights is becoming increasingly more important in public, private, and non-profit sectors, both as a means of fueling innovation and providing incentives for researchers to produce new and improved agricultural technologies.

DOI: 10.1016/B978-0-12-381466-1.00033-X

In addition, IP protection in agriculture is crucial in driving economic growth and competitiveness. Innovative biotechnologists are expected to make major contributions as the world emerges from the current period of economic uncertainty (Adams et al., 2009). The IP protection differences between countries remain a contentious issue. Countries with strong innovation systems (technology producers) often favor strong IP protection because it stimulates innovation by granting innovators temporary market exclusivity for their new, non-obvious, and useful inventions. Countries with less-developed innovation systems (technology consumers) often prefer weaker patent protection, because it enables them to access inventions developed elsewhere without having to forfeit licensing or manufacturing fees that may be beyond their capacity (or desire) to pay (Friedman, 2009).

Although investments in R&D and agricultural innovations have been fundamental to long-term economic growth worldwide, global resource allocation has been uneven. While some developing nations are approaching the scientific capacities of developed countries, others are regaining ground lost over the past decade or so, but, unfortunately, many others are either stalled or slipping behind (Wright et al., 2007). Substantial shifts in the scientific basis for a great deal of biological research have created new and promising opportunities for innovation in agricultural biotechnologies. Moreover, the institutional, regulatory regimes, especially those that affect agricultural R&D IP, are also undergoing rapid change and providing enhanced incentives for applications of biotechnology in agriculture. However, the nature of these developments is raising real concerns about the extent to which agriculture in developing countries will be able to partake in the benefits offered by these advancements (Wright et al., 2007).

In this chapter, we review the major types of IP in agricultural biotechnology, each having different rights and requirements for their protection, including: Plant breeder's rights; plant variety protection (PVP); plant patents; utility patents; gene patenting; material transfer agreements (MTAs); trademarks; trade secrets; know-how; and geographical designations of origin. We will then provide an overview of IP-related technology transfer (TT). One of the goals of this chapter is to shed light on the freedom-to-operate (FTO) and risk-management issues that are critical to scientists. Afterward, a case study on how to conduct an FTO analysis will be presented. A discussion of the most important concerns raised by developing countries related to IP protection and access will conclude the chapter.

Intellectual Property Protection of Biotechnologically Improved Plants

Plant biotechnologists around the world are in hot pursuit of new innovations in agriculture; however, one of the main features of plant biotechnology is its increasingly proprietary nature. Biotechnologically related plant products and processes can be copied much easier than in other fields of technology, and seeds from improved varieties can be harvested and planted. Likewise, with the advent of bioinformatics, gene(s) of interest can be cloned and incorporated into plants according to established and readily accessible protocols. This means that protection and enforcement of IP rights is becoming increasingly difficult.

Since 1989, over 10 multilateral agreements, guided by external forces, have been put into place to strengthen IP protection. These include: (1) Union International pour la Protection des Obtentions Végétales, known more simply as UPOV; (2) agreement on Trade-related IP Rights (TRIPS); and (3) International Treaty on Plant Genetic Resources (ITPGR). All are directly related to agricultural biotechnology inventions and are discussed in this chapter. Moreover, newly developed plant varieties can be protected by different types of regulations depending on the legal jurisdiction, the nature of the plant, the means of reproduction (sexual or vegetative), and other characteristics. The most common IP rights in agricultural biotechnology are plant patents, PVP certificates, and utility patents, as described in the following section. Trademarks and geographical designations, know-how, trade secrets, and contractual agreements, if available, are other forms of IP protection that may be relevant in plant biotechnology, but are only briefly discussed.

International intellectual property protection agreements

Union International pour la Protection des Obtentions Végétales

Because the IP protection for plants varies widely between countries, the UPOV came into existence as a global agreement setting a minimum standard for the protection of plant varieties (Union International pour la Protection des Obtentions Végétale, 1991). Member states that have signed the UPOV Convention must guarantee that these standards are met within their own legislation. UPOV was first brought into law in 1961 and revised in 1972, 1978, and 1991. Currently two versions of UPOV, 1978 and 1991, are in force. Despite the fact that these versions are similar in that they provide protection for a plant variety that is distinct from existing, known varieties and all that is uniform, stable, and novel, there are several important changes in the 1991 Act. For example, the definition of propagating material has been tightened and provisions relating to farmers' rights (e.g., permitting a farmer to harvest and sow seed for personal use) have been defined (Sechley and Schroeder, 2002). Even if plant-variety protection meeting the standard established under UPOV is accepted in many countries, such protection has not been uniformly adopted, and countries with strong histories of farmers' privilege have yet to comply with the convention (Cullet, 1999).

UPOV acknowledges that the generation of new plant varieties is desirable, and that the farmer should be allowed to continue the traditional farming practices of reserving part of the harvested seed for planting to produce a new crop (Sterckx, 2010). While UPOV of 1961 preserved the breeder's right to use the seed of a new plant variety to generate additional new varieties, the 1978 version enshrined the farmer's privilege to keep part of his harvest for subsequent sowing.

Significant amendments have been added to the 1991 Act. These pertain to the number of plant genera and species that was increased from a selected list of plants in the UPOV of 1978 to all plants. The UPOV of 1991 extends the minimum period of protection for most plants from 15 to 20 years, and for trees and vines the minimum is 25 years. The 1991 Act also removed farmers' privilege to sell or exchange seeds with other farmers for propagating purposes by protecting only the production and reproduction of a plant variety. Yet, the application of this provision is optional for each member state of UPOV and a member state can provide an exemption in its laws to permit farmers' privilege, if desired (Article 15, 1991). Another important amendment in the 1991 Act includes the ability to cover essentially derived varieties — comprising the properties of the protected variety along with only a minor change (Article 14, 1991). Essentially derived varieties may be obtained by the selection of a natural or induced mutant, a somaclonal variant, or a variant individual of the initial variety obtained by backcrossing or transformation by genetic engineering. The insertion of a gene into a protected plant variety by means of genetic engineering technologies might not be sufficient to surpass the essentially derived criteria unless a significant plant phenotype was caused by the introduced gene (Article 14(5)(c), 1991).

As of October 2009, 68 countries had become members of the UPOV Convention, with 45 countries under the terms of the 1991 Act and 22 countries under the 1978 Act; only Belgium is still bound by the older 1961/1972 Act. The United States (U.S.) became a party to the 1978 UPOV in 1981. With the exception of Cyprus, Greece, Luxembourg, and Malta, all member states of the European Union (EU) are members of UPOV. The Norwegian government made an unusual move by turning down a proposed law enabling its membership of UPOV in 1991. A thorough debate is continuing on the effect of PVP and associated MTAs on sharing and developing new germplasm (Price, 1999). Furthermore, as individual countries enact their national PVP laws, more differences may arise. For example, a PVP act recently enacted by the Indian government provides lower levels of IP protection for plant breeders than the UPOV of 1978.

Trade-related Intellectual Property Rights

The TRIPS agreement came into effect on January 1, 1995. To date, it is the most comprehensive multilateral agreement on IP (Agreement on Trade-related Aspects, 1996). The protection and enforcement of IP rights under TRIPS are intended to contribute to the promotion of technological innovations and to the transfer and dissemination of technology, to the mutual benefit of its producers and users, in a manner conducive to social and economic welfare, and to balance rights and obligations (Article 7, 1996; Article 8, 1996). The areas of IP covered by TRIPS include: copyright; trademarks; geographical indications; trade secrets; and patents.

With respect to each of the main areas of IP that TRIPS covers, the minimum standard of protection that should be provided by each member is defined. Each of the main elements of protection is defined, namely the subject matter to be protected, the rights to be conferred and permissible exceptions to those rights, and the minimum duration of protection. Member countries that have signed this agreement are free to determine the appropriate method(s) for implementing the provisions of the Agreement within their own legislation. Under TRIPS, any inventions, whether products or processes, in all fields of technology, are patentable, provided that they are new, involve an inventive step, and are suitable for industrial application (Article 27(1), 1996). Although this statement seems consistent with the requirements for obtaining a patent in most developed countries, patenting biotechnological inventions is often controversial.

With respect to plant biotechnology, Article 27(2) of TRIPS permits several grounds for exclusions, which are optional and vary from country to country, in issuing a patent related to essentially biological processes (consisting entirely of natural phenomena such as crossing or selection) for the production of plants (Article 27(2), 1996). Following adoption of the TRIPS Agreement, the question of *sui generis* (one of a kind, unique) IP rights protection for plant varieties has become a matter of great importance. TRIPS states that member countries must protect plant varieties either by patents, by an effective *sui generis* system, or by a combination thereof (Article 27(3)(b), 1996). Beyond the protections afforded or possibly afforded by treaty, common law has evolved to offer greater options and flexibility when pursuing IP rights in plant biotechnology. For example, in 2001, the U.S. judicial system recognized that dual protection by way of utility patents and plant variant protection is permissible for plant materials. Specifically, the U.S. Supreme Court held that "newly developed plant breeds fall within terms of 35 U.S.C. § 101, and neither Plant Patent Act, 35 U.S.C. § § 161–164, nor PVP Act, 7 U.S.C. § 2321 et seq., limits scope of subject matter patentable under Section 101" (decision in *J.E.M. Ag Supply, Inc. v. Pioneer Hi-Bred International, Inc.* (2001).

International Treaty on Plant Genetic Resources

This treaty originated from, and ultimately replaced, the International Undertaking on Plant Genetic Resources (IU), a previous, voluntary agreement that is not legally binding and was adopted in 1983 by the Food and Agriculture Organization (FAO) of the United Nations. The IU was the first international tool to ensure that plant genetic resources of economic and social interest, particularly for agriculture, would be explored, preserved, evaluated, and made available for plant breeding and research purposes. Accordingly, ITPGR provides a general framework for conservation and sustainable use of plant genetic resources.

The Treaty was adopted by consensus during the 31st session of the FAO conference on November 3, 2001, and entered into force on June 29, 2004, 90 days after the ratification, acceptance, approval, or accession of the 40th country. The Treaty currently has 120 contracting parties and ensures that the plant genetic resources that serve as the raw material used by plant breeders to develop new crop varieties remain publicly available. In short, ITPGR promotes the conservation and utilization of plant genetic resources for food and agriculture. The fundamental purpose of the Treaty is to preserve farmers' privilege and develop a Multilateral System of Access and Benefit Sharing (MLS). The MLS was created to ease access to genetic resources of major food and feed crop species and to share, in a fair way, the benefits arising from the exploitation of these materials, in accordance with multilaterally agreed terms and conditions. For example, genetic materials from the member countries are accessed by members of the Treaty free of charge, or, when a fee is charged, it shall not exceed the minimal cost involved. The wording in the Treaty recognizes implicitly that farmers may have unlimited rights under national law. In addition, the Treaty's preamble indicates that nothing in it shall be interpreted as implying in any way a change in the rights and obligations of the contracting parties under other international agreements.

However, there is a vigorous debate as to whether the Treaty will remain subordinate to TRIPS and UPOV. The U.S. has not yet taken any further action on the Treaty since it was submitted to its Senate for advice and ratification in 2007. The Senate Foreign Relations Committee held a Treaty hearing in 2009 with testimony in support of ratification, but further action is still pending. Even though proponents of the Treaty assert that the U.S. might provide needed leadership and resources to the Treaty, it abstained from signing this Treaty partially because of the lack of clarity in the IP provisions. An American presence could help to foster more trust between the contracting parties and possibly provide incentive for becoming a signatory to the Treaty, especially for countries like Japan and China. However, without the participation by the U.S., this Treaty is not likely to have a major international effect.

Types of intellectual property protection in plant biotechnology

Plant variety protection

In most fields of technology, the favored mode of legal protection for inventions is patents (a time-limited exclusionary right). In agriculture, an alternative form of IP rights called plant variety rights, also known as plant breeders' rights (PBRs) outside of the United States, was developed. Importantly, most of the existing PVP systems are based on one of the versions of the UPOV Convention. PBRs are used to protect new varieties of plants by giving exclusive commercial rights for 20 years (25 years for trees or vines) to market a new variety or its reproductive material. Additionally, PBRs prevent others from securing PVP or utility patent rights. The subject matter of PBRs is a plant variety. To be eligible for protection by way of a PVP certificate, a particular variety must be novel (new and distinct from other varieties), genetically uniform, and stable through successive generations. This protection thwarts anyone from growing or selling the variety without the owner's permission. However, exceptions may be made for research use or the sowing of saved seeds. Of particular concern is that the requirements for evaluation in each country, and the need for local legal representation, can render broad international application for PBRs costly and time-intensive.

The European Patent Convention (EPC), to which a large majority of EU member states are a contracting party, states that plant varieties are excluded from patent protection (Article 52(b), 1973). The exception that plant varieties as such cannot be protected by a patent has been confirmed in an EU biotechnology directive (European Directive 98/44/CE, 1998). Most of the countries in the EU have adopted legislation based on one of the versions of the UPOV Convention. There are two plausible explanations for this exclusion. First, no consensus on this controversial issue has been achieved. Secondly, in parallel with those national systems, a Community PVP system has been created to offer PVP for the whole territory of the EU. On the basis of one application, a Community plant variety right can be obtained, which is valid in all member states. This Community system, based on the UPOV of 1991, has thus far been quite successful.

As of 2003, more than 2500 applications for Community plant variety rights were received by the Community Plant Variety Office, and more than 8500 Community plant variety rights are in force. The demarcation line between the scope of a patent and the PVP system will be discussed in the utility patent section.

In the U.S., the legislative situation relating to the protection of plants and plant varieties is quite different from the EU. In the U.S. breeders and inventors of new plant varieties can choose between three types of IP rights to protect their inventions. Although the U.S. is a UPOV member, its PVP legislation does not cover asexually reproduced plants (except tubers). PVP certificates are administered by the U.S. Department of Agriculture (USDA) under the legal authority of the PVP Act (PVPA) of 1970, which has been amended six times and is based on the UPOV 1991 Convention. It provides protection for any sexually or tuber-propagated plant variety if the UPOV requirements for protection are met. The U.S. PVP Office (PVPO) considers a variety to be new and, therefore, eligible for PVP certification, only if propagated or harvested material of the variety has not been sold, or otherwise made available to other persons, for the purpose of exploiting the variety for more than one year in the United States or four years outside of the United States. It is very important that after a PVP certificate has been issued owners should be diligent about watching for infringements (Pardee, 2007). The PVP certificate offers the certificate holder the right to sell or market the variety, condition or stock the variety, offer the variety for sale or for reproduction, and to import or export the variety.

U.S. plant patents

Legislation governing IP rights continues to evolve in the agricultural sector. The U.S. Plant Patent Act (PPA) of 1930 was a major twentieth-century innovation relating to germplasm. Plant patents, which are a unique form of IP protection offered in the U.S., are different from utility patents. Numbers of American and foreign plant patents granted at the U.S. Patent and Trademark Office (USPTO) are listed in Table 33.1. The right to a plant patent stems from Title 35 of the U.S. Code U.S.C. § 161. *Patents for plants*. The PPA provides protection for asexually reproduced, novel plant varieties, not consisting of tuber-propagated plants or plants found

Table 33.1 Numbers of U.S. plant patents granted at the USPTO and assigned by type of inventing organization, 1977–2009.[a]

Organization	U.S. Patent and Trademark Office
Private sector	
U.S. corporation	7084
Foreign corporation	5554
U.S. individual	2328
Foreign individual	1523
Subtotal	16,489
Public sector	
U.S. government	21
Foreign government	62
Subtotal	83
Total	16,572

[a]Data are taken from the USPTO Web site (http://www.uspto.gov/) as of July 6, 2010.

in an uncultivated state. This protection was introduced primarily to benefit the horticulture industry by encouraging plant breeding and enhancement of plant genetic diversity.

One of the conditions the U.S. placed on becoming a UPOV signatory was that it could continue to provide protection for asexually reproduced plants under its existing plant patent legislation. The term of a plant patent is for 20 years from the earliest effective U.S. filing date (35 U.S.C. § 154). Plant patents protect the inventor's right to exclude others from asexually reproducing the plant, and from using, offering for sale, or selling the plant so reproduced, or any of its parts, throughout the U.S., or from importing the plant so reproduced, or any parts thereof, into the U.S. (35 U.S.C. § 163).

In contrast with utility patents, where deposit of biological material into a public germplasm repository may be required by the USPTO to comply with description and enablement requirements, it is not necessary to deposit the claimed plant to secure a plant patent. Another key difference between utility and plant patents is that the latter can have only a single claim tied to the plant variety specified in the application.

Recently, the USPTO has started rejecting plant patent applications with a UPOV-based certificate that had been granted before filing for a plant patent application, if the certificate had been filed more than one year before filing the plant patent application. Rejection on these grounds is in direct opposition to that decided *In re LeGrice* (1962), but argued by the USPTO as consistent with *Ex parte Thomson* (1992). A one-year grace period is granted to the applicant. Thus, a public disclosure of a new plant variety or a granted patent of the new variety in a foreign jurisdiction does not preclude patentability, provided that the disclosure or foreign patent occurred less than 12 months prior to the plant patent application being lodged in the U.S. If publication occurred in any other country prior to the 12-month grace period, the plant patent application will be refused (*Ex parte Thomson*, 1992). This has led to strong pressure from industry, resulting in a review of the USPTO position. A proposal to extend the grace period to 10 years on the basis that plants from foreign jurisdictions are often held in quarantine by the USDA for a period exceeding 12 months, effectively nullifying patentability, is still under discussion.

The PPA was amended effective October 27, 1998, to extend the exclusive right to plant parts obtained from varieties protected by plant patents that are issued on or after the effective date. Interestingly, in the U.S. a dual protection is allowed; a plant and utility patent (*infra*) can both be obtained to protect the same plant, provided that the requirements for both types of patents are fulfilled.

Utility patents applied to plant biotechnology

Another type of IP protection available for plant biotechnology is the utility patent. Any plant inventions that meet the statutory requirements of novelty, utility, and non-obviousness are eligible for utility patent protection. On April 10, 1790, the first U.S. Patent Act was passed into law. It specified that U.S. utility patent shall provide protection for any new and useful manufacture, or composition of matter, or any new and useful improvement thereof (35 U.S.C. § 101). Initially, the 1980 Supreme Court decision in *Diamond v. Chakrabarty* (1980) held that utility patents could be granted for inventions involving living organisms. The 2001 landmark decision in the case of *J.E.M. AG Supply, Inc. v. Pioneer Hi-Bred International, Inc.*, (2001) further established that such protection is valid also for plants. This case also validated that plants are a composition of matter, as ruled earlier by the U.S. Patent Board of Appeal.

Patents have been issued in the U.S. for plant inbreds; hybrids; plant parts (e.g., seeds, pollen, fruits, and flowers); biotechnology methods; genes; and many other innovative plant-derived products. Although it is the most expensive and complex process, utility patents provide the strongest protection for biotechnologically improved materials. It may protect DNA sequences and processes used to produce transgenic plants, as well as the plants themselves, and uses thereof for a period of time (usually 20 years from the earliest effective filing date or, for much older patents, 17 years from granting). In addition, it protects plant breeders' rights to control the use, sale, import, and reproduction of protected plants. With a utility patent, one can prevent others from using a protected plant for subsequent breeding. Besides, it may be useful where an invention is not limited to a particular variety, or where method claims are desired. There is a complex interplay between plant and utility patents. Any asexually reproduced plant may alternatively be protected under 35 U.S.C. § 101, as the PPA is not an exclusive form of protection that conflicts with the granting of utility patents for plants. *Ex parte Hibberd* (1985) was affirmed by the Supreme Court in *J.E.M. Ag Supply, Inc. v. Pioneer Hi-Bred International, Inc.* (2001). Inventions claimed under 35 U.S.C. § 101 may include the same asexually reproduced plant that is claimed under 35 U.S.C. § 161, as well as plant materials and processes involved in producing the plant materials.

In conclusion, there is a broad range of ways to protect plant-related innovations. It is critical to look at the intended use of the plant(s) and/or plant material(s) to devise a multi-tiered approach to protecting IP. Conceptually, one cannot obtain a PVP certificate and a plant patent for the same plant, but it is possible to simultaneously pursue a utility patent together with either. Alternatively, one can obtain a PVP for a sexually produced variety and then make an asexually reproduced variant that would be eligible for plant patent protection.

Advances in genetic engineering have generated a completely new set of concerns related to patent law. Although many countries have taken an inclusive approach, allowing various higher life forms (HLFs) to be patented, the Canadian Supreme Court has expressed an opposing opinion. In the landmark *Harvard College v. Canada* decision (2002), it denied patent protection to Harvard College for what was known as an Oncomouse — a genetically modified mouse (and for all similarly modified non-human mammals), which was deemed valuable for cancer research. Patents on the genetic alteration process were granted, excluding the end result, the genetically modified mouse itself. The Court argued that HLFs do not fall within the definition of invention and do not qualify as composition of matter. The USPTO was the first to allow the patent for the Oncomouse to Harvard College, followed by the European Patent Office (EPO).

Pursuant to the ruling in the *Harvard College* case, plants are HLFs, and consequently, are not patentable. However, this opinion is open to interpretation. For example, one can argue that even if a herbicide-tolerant transgenic plant is not technically covered by patent protection, any use of the transgenic plant would implicitly include use of the underlying process and genes. Consequently, use of the plant itself would be prohibited, effectively giving patent protection to the herbicide-tolerant plant.

A second decision was made a few years later, in *Monsanto Canada, Inc. v. Schmeiser* (2004) in which Monsanto was the owner of a patent containing claims to a chimeric gene, a Glyphosate-resistant plant cell, and to genetic engineering methods for producing a herbicide-resistant plant, but without claims to a plant, knowing that this was not allowed by the Canadian Patent Office. The case involved a canola farmer, Schmeiser, who infringed on Monsanto's patent by using canola plants that contained a patented gene. This time, the Court ruled in favor of patent protection.

In Europe, based on the 1999 formalization of the EU Biotechnology Directive (European Directive 98/44/CE, 1998), the rules of the EPC were amended to comply with the provisions outlined in the Directive. Accordingly, EPO has accepted claims to plants and animals if the technical feasibility is not confined to a particular plant or animal variety. Shortly afterward, the EPO's Enlarged Board of Appeal ruled that a claim that might encompass plant varieties is not excluded from patentability under Article 53(b) (1999) if the claim does not specify certain plant varieties. As stated in Article 8 of the European Directive involving the protection of biotechnological inventions and a decision of the Enlarged Boards of Appeal of the EU in the Novartis case (Fleck and Baldock, 2003), germplasm can be included in the scope of a patent claim so long as protection is not limited to a single variety (Le Buanec, 2004). However, some countries (e.g., France and Germany) have incorporated breeder exemption clauses in their patent laws (Smith, 2008). As the effective level of IP protection available to plant breeders is also influenced by the use of new technologies and changes in breeding practices (Le Buanec, 2004, 2005; Donnenwirth et al., 2004; McConnell, 2004; Heckenberger et al., 2005; Lence et al., 2005; Smolders, 2005; Kock et al., 2006; Smith et al., 2008), further complexity arose. For example, transgenic approaches resulted in the rapid development of new distinct varieties from existing ones (Fleck and Baldock, 2003; Smolders, 2005). Thus, under PVP laws enacted in accordance with the 1978 revision of UPOV, it is possible to pirate existing varieties, for example, through their conversion to herbicide- or insect-resistant versions (Smith, 2008).

Some of the decision's critics could argue that the Court's decision not to permit patents for HLFs was based on public morality and an outcry rather than a technical interpretation of the Patent Act. Thus, by further amending the patent law, patent offices around the world should allow the issuance of patents that claim not only the process, but also the plant as the product; otherwise, we run the risk of opening the door to parties to circumvent patent protection. The extension of HLFs to plants makes perfect sense from both scientific and commercial perspectives.

Gene patenting

Plant biotechnology or genetic engineering allows breeders to modify plants in an effort to meet the demands of a fast-growing global community. Many plant biotechnology companies and universities are behind a wave of patent claims on hundreds of genes that confer tolerance to herbicides and biotic and abiotic stresses; genes that improve wood properties for the biofuel and paper and pulp industries; and genes that have important applications in phytoremediation, nutrition, pest resistance, and plant-based pharmaceuticals. The patents cover the development of new crops capable of resisting attack by insects and diseases, and withstanding salt, drought, high and low temperatures, or other environmental changes brought on by global climate change. Worldwide, patent offices have documented thousands of patent applications for genes that confer these characteristics. They have been filed in multiple countries and several of the patents cover many genes across a range of plant species.

However, one of the issues relating to the patenting of genetic inventions revolves around the question of whether a DNA sequence can be considered a discovery or an invention. In Europe, the rules of the EPC (Official Journal of the EPC, 1999) were recently amended to comply with the EU Biotechnology Directive. These amendments state that biological material isolated from its natural environment or produced by means of a technical process, even if it occurred previously in nature, is patentable (Rule 23(c) EPC, 1999). In addition to the need for the sequence to be novel and non-obvious, the process of isolating and characterizing it must also be patentable in order to patent the sequence.

The USPTO takes the position that genes are chemical compounds, albeit complex ones, and, as such, they qualify for potential patenting as compositions of matter. Whereas a naturally occurring product, as it exists in nature, cannot be patented, the USPTO has allowed patents on naturally occurring products that have been purified, isolated, or otherwise altered. For this reason, gene patents have been issued in the U.S. for decades (Harfouche et al., 2010). Additionally, the USPTO produced a set of guidelines arguing that a gene sequence is patentable if it is novel, isolated, its molecular structure is known, and if it has substantive and credible utility (Utility Guidelines, 2001; Written Description Guidelines, 2001). This position is consistent with the European rule amendments; however, on March 29, 2010, the U.S. District Court for the Southern District of New York delivered a serious blow to human-gene patent holders. The opinion, the first round in the closely followed *Association for Molecular Pathology v. United States Patent and Trademark Office* (widely known as the *ACLU v. Myriad Genetics*) case, invalidates several patents with claims covering genes associated with breast cancer. The result surprised many in the biotechnology industry because the broad ruling, if upheld on appeal, would eliminate future gene patenting, as well as effectively make existing gene patents invalid. Weakening patent protection for human genes could lead to a market change for future R&D involving genes not only related to human health, but also to agricultural biotechnology.

The complex nature of gene patents and enabling technologies involved in the development of transgenic plants presents other concerns related to IP rights. These issues are coming to the fore more frequently in developing countries, where no possibilities for gene patenting exist and no PVP laws are in place. For example, India's experience with insect-resistant cotton, transformed with a *Bacillus thuringiensis* (*Bt*) gene, was first developed by a joint venture between Mahyco, India's largest seed company, and Monsanto. Despite the fact that these varieties were hybrids and, thus, providing at least some protection against farmers saving seed, there were widespread reports of black markets for seeds and clandestine plant breeding to incorporate the *Bt* gene (Jayaraman, 2004).

Another example is herbicide-tolerant soybean, which is the most widely grown transgenic crop in the world. It has become very popular in Argentina (Qaim and Traxler, 2005), where although genes may be patented, Monsanto was unable to make an application deadline, so its Roundup Ready® (RR) varieties were simply protected by PVP. But much of the RR soybean in Argentina is grown from seed that is saved or informally traded. In Paraguay and Brazil, RR soybean has gained great popularity based on the sale of black-market seed. Because only a small fraction of the RR soybean seed used in these countries is from authorized sources, Monsanto is beginning to pursue legal action for patent infringement against shipments of soybeans from South America that arrive in the ports of European countries, where the technology is patented. The solution currently under discussion is to charge farmers a royalty based on grain sold at harvest, to be collected by grain dealers, processors, or farmer associations. One of the major points of disagreement here is farmers' insistence on their legal right to save seed (Tripp et al., 2007).

Hence, in developing countries, controlling the use of genes and access to transgenic varieties will require the implementation of restrictive IP protection, such as strong laws related to patents and plant breeder's rights, and the vigorous enforcement of these laws, as well as biosafety regulations.

Material transfer agreements

While the defense of R&D related to biotechnologically improved plant products has traditionally been based on patents, PVP, and other means of IP protection, academic research is conducted in an open environment, with relatively unrestricted access to fundamental scientific knowledge and products, often with few formal arrangements. However, under the patent laws, there is no explicit exemption for academic use, prompting administrators at publicly funded research institutions to execute MTAs granting permission to use patented materials in laboratory research and in breeding programs. Although this action allows these institutions to capitalize on their R&D investments, it constitutes another method for controlling access to research results. MTAs are a preferred mechanism for the transfer of tangible property (e.g., plant genetic resources, plasmids containing genes of interest, and genetic transformation vectors) in countries where enforcement of IP rights is not appropriate (Kowalski et al., 2002).

Material Transfer Agreements are the most common legal arrangement by which centers of the Consultative Group on International Agricultural Research (CGIAR) system obtain permission to utilize proprietary biotechnologies (Cohen et al., 1998). If access to the materials is not otherwise restricted, this protection may be effective in preserving the provider's rights to the germplasm. Effective January 2007, germplasm exchange for all parties in TRIPS, including the whole CGIAR system, is now covered by a standardized MTA, known as the Uniform Biological MTA (UBMTA).

An MTA is a type of two-party legal agreement. A typical MTA would stipulate the following: (1) materials are to be used for research purposes only; (2) a separate agreement is required for commercial use; and (3) materials may not be transferred to any third party without prior written approval from the provider. However, these terms can go beyond those established for patents. For example, they may restrict the recipient's rights to modify, improve, re-sell, or commercialize the transferred material. But, plant materials created by the recipient that contain or incorporate the transferred materials are considered modifications. Ownership of such modifications is determined according to appropriate inventorship laws using terms that are consistent with the UBMTA. They can also take the form of a first right of refusal to negotiate a non-exclusive license to utilize products that were improved using the original material provided. Licensing structure, fees, and royalties on products resulting from material provided in the agreement may be discussed in *bona fide* between the provider and recipient. In transgenic plant product development, such modifications occur frequently. A gene conferring a desirable trait could be provided by one source and deployed in an elite genotype (obtained through traditional breeding) derived from another source, and perhaps using enabling technologies developed by a third or fourth source.

There is also considerable concern over the possible role of MTAs in raising research costs. Others suggest that the contractual structure provided by MTAs may reduce transaction costs and facilitate exchange (Mowery and Ziedonis, 2007). In their work, Mowery and Ziedonis (2007) undertook a preliminary analysis of the role of MTAs in the biomedical research enterprise at the University of Michigan, a significant patenter and licensor of biomedical IP. After examining the relationship among invention disclosures, patents, licenses, and the existence of an MTA, their analysis suggested that the increased assertion of property rights by universities through MTAs does not appear to impede the commercialization of the products of university research.

Trademarks, trade secrets, know-how, and geographical designations

Protection of IP rights also includes trade secrets, know-how, trademarks, and geographical designations. Trademarks are valuable for plant biotechnology R&D because they identify plant biotechnology products and distinguish them from those manufactured or sold by others. Registration grants the exclusive right to use them for product and/or services. Trademarks are awarded by national governments, and the protection is valid only in countries in which they are issued. A significant part of the value of agricultural biotechnology variety often lies not in its agronomic characteristics *per se*, but rather in its identification and reputation, not to mention that of its producer, among consumers. A plant biotechnology product also can be given a brand name that is protected as IP by registering it as a trademark with the appropriate patent office. For example, Bollgard® cotton contains a protein encoded by a gene from *Bt* that protects cotton plants from specific lepidopteran insect pests. Bollgard® is a registered trademark of the Monsanto Company. Other trademarks, such as Roundup Ready® or Liberty Link® refer to specific traits imparted by a transgene and may be of some value in marketing to farmers, although the identities of such agronomic traits receive little attention from final consumers (Graff et al., 2004).

Trade secrets are simply described as a form of IP protection that assists in maintaining confidentiality by imposing penalties if information held as secret is used improperly. For a certain period of time, a former employee who goes to work for a competitor is enjoined to not reveal secret information. Although no application is needed to protect trade secrets, two conditions need to be met: (1) the trade secret must have a commercial value; and (2) an effort must be made to keep it secret. In agriculture, even an F_1 hybrid may be considered a trade secret. Importantly, trade secrets are also compatible with and complementary to patents used to protect transgenic plants. The glue holding a patent and trade secret together, and jointly translating research ideas into application, is often the technical know-how, which is the grist of trade secrets. Know-how is another key type of IP produced by an R&D program. This often occurs when research institutions do not enjoy the benefit of patent protection. Know-how can be in the form of an aggregation of processes, procedures, protocols, a secret formula, or a combination of these. Access to know-how is granted by patent licensing or under MTAs.

The identification of a product by the consumer has an important bearing on its value. Thus, a registered trademark or geographical designation under which a plant variety or genetic trait is sold can provide a significant means of protecting the genetic improvements embodied in the product (Boettiger et al., 2004). Geographical designations of origin give the right to prevent unauthorized parties from using misleading or false geographical indications for their plant varieties. Geographical indicators identify a product as originating from the specific territory, region, or locality, where a given quality, reputation, or other characteristics may be attributable to geographic origin. In agricultural biotechnology, geographical designations can also play a positive and complementary role in protecting traditionally bred lines that contain genes or alleles that impart agronomic traits that may be desirable in a certain area.

Freedom-to-Operate in Agricultural Biotechnology: The Road from a Research Idea to Commercialization of a Biotechnologically Improved Plant Product

Developing a plant biotechnology product is often both capital- and time-intensive, and access to IP may present an insurmountable obstacle for developers. Ownership of IP in agricultural biotechnology is highly fragmented (Figure 33.1) and has many layers of complexity. This fact leads to situations where no single public- or private-sector organization is able to provide a complete set of IP rights to guarantee FTO in a particular jurisdiction using a certain technology. Plant products improved through genetic engineering may encapsulate several components, proprietary technological processes, and know-how, each of which may be protected under IP/tangible property (TP) law. Therefore, it is imperative to conduct an FTO analysis that identifies these rights by dissecting the product's IP and TP. Although it is a simple and straightforward concept, FTO means that for a given product or service, at a given point in time, with respect to a given market or geography, no IP from any third party is infringed. Translating this concept into a productive strategy for companies and for public-sector institutions alike is not always straightforward (Krattiger, 2007).

The advantages of incorporating FTO considerations into plant biotechnology product development strategies are manifold. For public-sector organizations, they may include benefits such as greater prestige, the ability to bring about cultural change, or the forging of strong partnerships with providers of IP and technology (Krattiger, 2007). On the other hand, an integrated FTO that relies not only on technical and legal skills, but also on strategic business development planning, becomes a tactical component in a private organization's risk-management strategy.

Scientific research advancements are frequently nurtured by a highly synergistic environment, as was evidently the case for deducing the mechanism by which RNA interference (RNAi) mediates gene suppression (Chi-Ham et al., 2010). Because of its potential commercial application, an increasing number of patent applications are seeking exclusive rights

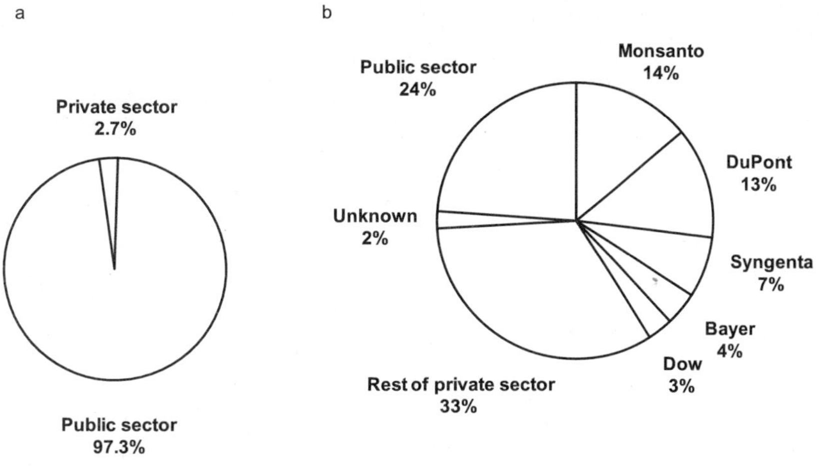

Source: Graff et al. 2003. Reproduced with permission of the copyright owner (Nature Publishing Group). The original figure includes additional data on public-sector agricultural biotechnology patents.

Figure 33.1 • Distribution of assignment for U.S. patents granted from 1982 to 2001. • (a) All U.S. patents. (b) U.S. agricultural biotechnology patents (Graff et al., 2003).

to RNAi-based discoveries. However, when claiming patent rights, patent law favors clear and well-defined invention boundaries. Failure to design around existing IP or to license technologies claimed in patents may trigger legal repercussions, which may prevent product commercialization (Chi-Ham et al., 2010). Thus, due diligence when evaluating potential legal risks associated with IP is an important step in the commercialization of products, and requires both technical and legal expertise (Fenton et al., 2007). Assessing the FTO to determine whether a product under development infringes third-party IP rights is a vital step in strategic risk-management. Because of the prevalence of national jurisdictions over patents, together with IP laws, which can vary between countries, it is crucial to carry out an FTO analysis for each country in which a product will be developed and commercialized. Such a process necessarily involves engaging IP attorneys in each jurisdiction where product marketing is desired.

It is essential that an FTO analysis be planned and executed before starting a new project. A schematic representation of this process is shown in Figure 33.2. A simplified listing of IP used to develop a plant biotechnology product is presented in Table 33.2. Concerns about FTO remain throughout the product development process, and beyond. A set of 10 strategies for managing potential IP infringement are proposed by Krattiger et al. (2007). In practice, typically two or more of these strategies will be adopted, with the specific mix of strategies varying by project. Which strategies will be appropriate depend on, for example, how advanced the product is, the type of organization that develops the product, and relevant market dynamics. Not all of the listed strategies are feasible for public-sector institutions.

Another concern lies in the efficiencies of accessing IP, which appears to be hampering valuable R&D related to horticulture and the development of ornamental crop varieties, as well as forest trees. Horticultural and ornamental markets are fragmented and provide little incentive, discouraging large, multi-national corporations with strong IP positions and FTO from pursuing product development in these sectors. Because of its weak IP position and in order to secure FTO, the public sector often joins forces with industry to support horticultural, floriculture, and forest projects. This can be mutually advantageous, as has been the case with cross-licensing deals struck by leaders in the private sector. As a result of FTO issues in the public and non-profit sectors, the Public Intellectual Property Resource for Agriculture (PIPRA), a coalition of American universities and foundations that are committed to this strategy, was formed (Graff et al., 2004). PIPRA attempts to mobilize collaborative support from a wide range of public-sector institutions worldwide. In addition to this broad base of institutional support, PIPRA's molecular biology labs and its network of *pro bono* IP attorneys coordinate the scientific and legal resources needed to support agricultural advances in developing countries (Boettiger and Bennett, 2007). Implicit in PIPRA's mission is finding ways to provide access to technologies, such as PIPRA's plant transformation vector, which is comprised of numerous patented components (owned by PIPRA members).

There is also concern that the private sector may be blocking access to IP in developing countries. According to Pardey et al. (2003), this concern is exaggerated and largely misdirected. They showed that international and national agricultural research centers currently have far greater FTO in agricultural research on food crops for the developing world than is commonly perceived. A plausible example of this is the public-private partnership between the African Agricultural Technology Foundation (AATF) that is developing drought-tolerant varieties of maize for Africa. The partnership, known as Water-Efficient Maize for Africa (WEMA), was formed in response to a growing call by African farmers, political leaders, and scientists to address

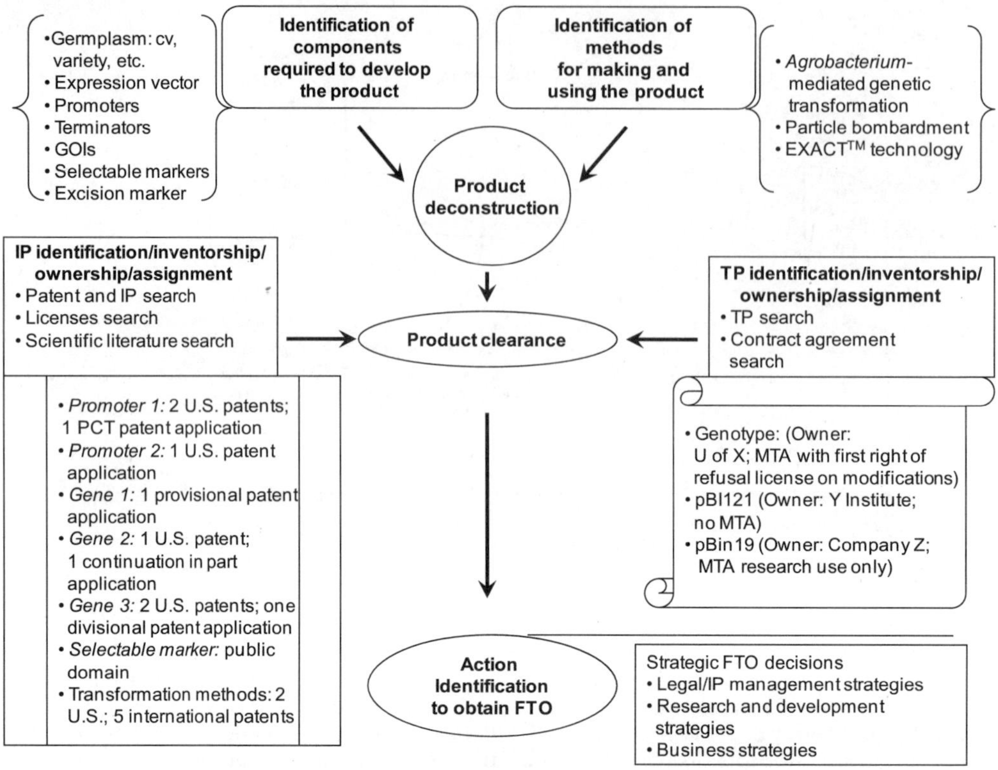

Figure 33.2 • Schematic representation of the steps needed for FTO analysis of a transgenic plant product. • The freedom to operate (FTO) analysis has three main steps, beginning with product deconstruction, leading to product clearance (PC), and concluding with FTO decisions. The product deconstruction starts in the very early stages of FTO analysis and involves the identification of all components, processes, enabling technologies, and methods required to develop the product. It is done by intensive discussions with researchers and review of the scientific literature to identify key players involved in the development of related technologies. PC analyzes the intellectual property (IP) and tangible property (TP) landscape of the product using extensive patent and literature searches, allowing the identification of the potential patents, MTAs, and license agreements related to the proposed product. In addition, it involves studying the patent literature, issued patents, and patent applications published or publicly available via patent offices such as the United States Patent and Trademark Office (USPTO), the World Intellectual Property Organization (WIPO), The European Patent Office (EPO), Japan Patent Office (JPO), etc. PC can also determine if the patent is still in force. An issued utility patent remains active for 20 years from the date of filing. Claims in an expired patent are in the public domain, and the proposed product is free to operate based on those features. PC also provides information on patent inventor, assignee, and licensee(s). Finally, the various components and methods required for making and using the product and their IP/TP issues are used to make strategic FTO decisions that help reduce the risk of IP litigation. An example of PC output is presented in Table 33.2.

the devastating effects of drought on farmers and their families. Marker-assisted breeding and biotechnology will be used to develop African maize varieties, with the long-term goal of making drought-tolerant, royalty-free maize available to small-scale African farmers. The AATF will work with the International Maize and Wheat Improvement Center (CIMMYT) and Monsanto. CIMMYT will provide conventionally developed, drought-tolerant, high-yielding maize varieties that are adapted to conditions in Africa, and expertise in conventional breeding and testing for drought tolerance. Monsanto will provide proprietary germplasm, advanced breeding tools, and other needed expertise. Additionally, Monsanto and BASF will provide drought-tolerance genes that they developed collaboratively. The national agricultural research systems, farmers' groups, and seed companies who are participating in this project will contribute their expertise with breeding and regulatory issues, and will be responsible for country-specific implementation, including project governance, testing, germplasm evaluation, and seed production and distribution. The Bill and Melinda Gates and Howard G. Buffet Foundations are funding this partnership.

Technology Transfer as a Means to Facilitate the Development of Biotechnology-Based Agriculture

Academic institutions excel at research, but often need collaborations with industry to turn science into commercial

Table 33.2 Intellectual property used to assemble a plant genetic transformation vector: An example used to develop transgenic turfgrass[a].

Components	No. of patents	No. of assignees
Rice actin promoter	1 U.S.	1 (Cornell Research Foundation)
Octopine synthase (ocs) and mannopine synthase (mas) super promoter	1 U.S.	2 (Biotechnology Research and Development Corporation and Purdue Research Foundation)
Four copies of an ABA-responsive complex sequence (4ABRC) promoter	1 U.S.	2 (Cornell Research Foundation and Washington University)
Cauliflower mosaic virus (CaMV) 35S terminator	None found	None found
A gene that confers resistance to the herbicide bialaphos (bar)	2 (1 U.S. and 1 EP)	1 (Bayer Crop Sciences)
Neomycin phosphotransferase type II	2 U.S.	2 (Monsanto & Rockefeller Foundation)
Arabidopsis thaliana white-brown complex homolog 19 gene (*Atwbc19*) gene for antibiotic resistance	2 U.S. and 1 WO	1 (University of Tennessee Research Foundation)
Duplicated CaMV 35S enhancer sequences	5 U.S., 1 CA, 1 Brazil	2 for U.S. patents (University of British Columbia and Monsanto)
Cotton leaf curl virus (CLCuV) promoter	1 U.S. (family patents: 1 AU, 3 CN, 1 U.S., 1 WO)	1 (Institute of Genetics, Chinese Academy of Science)
PureIntro Technology™	4 (2 U.S., 1 EP, 1 AU)	1 (Japan Tobacco)
PureIntro Turfgrass (*Agrobacterium*-mediated transformation of turfgrass)	6 (1 WO, 2 U.S., 2 EP)	1 (Rutgers)
Agrobacterium tumefaciens transformation of stolons	1 U.S. and 1 WO	1 (The Samuel Roberts Nobel Foundation)

[a]Data are taken from the CAMBIA's patent lens (http://www.patentlens.net/) and the USPTO (http://www.uspto.gov/) Web sites.

products. Universities in many countries, as well as research institutions, are under increasing pressure from government agencies to partner with industry to be eligible for public funding. This is being done both to leverage public money and to propel new inventions from discovery to the commercial development stage.

The U.S. has a long history of generating inventions with commercial value and, thus, is a good example of successful TT (Rosenberg and Nelson, 1994; Mowery and Sampat, 2001; Geiger, 2004). Countries in the EU are now following the same pattern. In the U.S., the establishment of TT offices (TTOs) and the biotechnology industry are often linked to the passage of the Bayh-Dole Act in 1980, which, for the first time, allowed universities to take ownership of patent(s) on discoveries made as a result of federal funding (University and Small Business Patent Procedures Act, 1980). Prior to this Act, IP ownership remained with the federal government, and, as a consequence, relatively few inventions were licensed. Thus, the Bayh-Dole Act signaled U.S. government support for commercialization of research products. There were two motivations behind this Act: (1) patents resulting from federally funded research were unexploited due to uncertainty regarding their ownership (Eisenberg, 1996; Mowery et al., 2004; Berman, 2008) and (2) a belief that universities are a source of innovation that can be used to strengthen the U.S. economy and maintain its preeminence (Brooks, 1993; Stevens, 2004; Berman, 2008).

There is now a growing consensus that although the Bayh-Dole Act has been beneficial, improvements are needed. At most universities, TTOs will be in the best position to protect and transfer the IP through commercialization, a service that is increasingly appreciated by university inventors. However, it is in those cases where there is a significant difference of opinion between the TTO and the inventor regarding the process of commercialization, or where the TTO is not capable of providing the needed services, that difficulties occur. Thus, in discussing modifications to the current Bayh-Dole regime, a system where the inventor holds the IP could resolve most of these difficulties (Kenny and Patton, 2009). But, if the Act is eliminated, many new technologies with the potential to save lives, benefit society, and stimulate the economy are less likely to make it out of the laboratory. It would also put the universities at risk for liabilities surrounding the product in the marketplace.

Further, some university TTOs have reputations for being difficult to deal with, while others can play an important role in the knowledge translation and commercialization processes by promoting public-private collaboration. In fact, many biotechnology companies originated as university start-ups, and many prominent biotechnology clusters are located near universities. Although TT can also occur through licensing to a third-party, a new paradigm was recently proposed by Renault et al. (2007). Instead of measuring the success of TT by counting numbers of patents and licensing deals, they suggest measuring knowledge flows (e.g., training of students, building of knowledge infrastructure, and supporting a creative milieu) between universities and their localities. This approach should reveal a more accurate picture of the full impact that universities have on their regions (Renault et al., 2007). Recently, interviews with professionals at Canadian TTOs have revealed that, at their best, TTOs support the social and academic missions of their institutions by facilitating knowledge mobilization and research relationships with other sectors, including industry; however, this does not always produce obvious or traditional commercial outputs. Thus, the existing metrics used to measure the success of TTOs do not capture this reality and, as such, realignment is needed (Bubela and Caulfield, 2010).

In plant biotechnology, TT is especially important given the close relationship in many countries between research centers and the agricultural industry. The fact that the variety of inventions in plant biotechnology from universities is far greater than those of any private-sector firm, might indeed offer a platform of technologies adequate to facilitate the development of new plant varieties through genetic engineering.

Agricultural biotechnology has become a priority area for TT in developing countries, where production of food, feed, and fiber is of vital concern. Crops grown for subsistence in developing countries and the technologies that are used to develop such crops are clearly of little commercial interest to the private sector. Thus, promoting TT of plant biotechnologies to develop such crops for humanitarian purposes is of paramount importance. Developing countries frequently lack the required IP management capacity and resources to perform product clearance (PC) analyses and evaluations that facilitate the legitimate import, use, and/or export of technologically advanced products (Tripp et al., 2007). For maximum impact, researchers in developing countries must strengthen their IP rights knowledge and management capacity. Also, international regulations and IP rights need to be tailored to promote discovery-oriented research to help meet the needs of developing nations.

Finally, in many parts of the world, IP management and commercialization of university research products still occurs in the absence of legislative frameworks, but in the developing countries TT is beginning to occur. For example, the Philippines recently launched a new law to encourage TT. The TT Act of 2009, largely inspired by the Bayh-Dole Act, enables R&D institutions to be the owner of IP rights arising from the results of government-funded research.

Conclusion and Future Needs

Protecting IP in agricultural biotechnology helps accelerate economic and social development. While the private sector primarily focuses on classical cash crops, where markets are large, the public sector (e.g., universities, national research institutions, and non-profit organizations) is often left to develop biotechnologies for small, specialty crops that are important to developing countries. However, there are IP concerns in developing countries and a need for greater protection and access.

Intellectual property rights are sometimes viewed as barriers to access to innovations in health and agriculture. However, it is not IP *per se* that raises barriers, but rather how IP is used and managed, particularly by public-sector institutions (Lei et al., 2009). The literature offers almost no direct evidence of scientists' views of the trade-offs associated with protecting research tools. A recent survey conducted by Lei et al. (2009) showed that scientists attributed problems of delayed or blocked access to needed research tools to MTAs. Another major challenge for management of public-sector IP is the fragmentation of technology ownership and the need for many technology components to provide FTO in transgenic crops (Graff et al., 2003). Recently, a model of public-sector collaboration has been suggested that could directly address this issue by including data sharing and patent pooling — the equivalent of an IP clearinghouse (Graff and Zilberman, 2001; Atkinson et al., 2003). Graff et al. (2003) believed that using this model, IP patented by the public sector might provide a platform of technologies sufficient to enable the development of new transgenic varieties and cultivars. This approach may promote sharing access to enabling technologies (gene delivery methods, marker-gene excision techniques, selectable markers, promoters, etc.) that play a key role in early transgenic plant product development. Besides, it will solve the problem caused by a technology owner trying to delay or hinder a competitor's development program. Furthermore, considerable research effort is spent to design around existing enabling technology patents, and to search for ways to invent bargaining tools.

These issues help to explain why plant biologists have been pushing for open-source biology (Biological Innovation for Open Society, BiOS; Jefferson, 2007) and the creation of institutional collaborations to facilitate access to crucial, enabling technologies (PIPRA; Atkinson et al., 2003). An open-access patent database could be used to collect IP data from several national patent offices and, thus, improve the ability to interpret and filter IP with overlapping and competing rights (Patently transparent, 2006).

Increasing access to fundamental innovations in agricultural biotechnology for developing countries will require capacity building, active engagement, and wide participation. These factors must also be coupled with improved institutional IP management and stewardship capabilities. In addition, developing countries urgently need to implement IP legislation modeled after what has been created by more advanced economies, albeit some may be of little relevance, given their needs and stages of development (Huete-Perez, 2003). Strengthening of IP protection and access necessitates a strong collaboration between the public and private sectors, with the public sector pursuing IP protection as vigorously as the private one. The public sectors in these developing countries need to become generators of IP. Other sound strategies that could be adopted include: worldwide harmonization of IP rights; identification of areas of common interest and opportunity between the private and public sectors; and an increase in R&D investments.

Recently, clear evidence of accelerating IP protection in developing countries was provided by China, where patent applications for agricultural biotechnology innovations are on the rise. China is now fifth among countries that signed the Patent Cooperation Treaty (PCT), an agreement by which patents can be simultaneously lodged in different countries. Although guiding policies still need to be improved, this example demonstrates an interest in competing internationally.

Liberalized TT and increased foreign investment are ways in which developing countries will benefit from the harmonization of IP management and enforcement. Ultimately, harmonization will improve the world's capacity to innovate, so that the whole will be greater than the sum from its parts. The World Intellectual Property Office's (WIPO) proposed Substantive Patent Law Treaty (SPLT) is an exploratory step in the right direction (Bera, 2009).

One variant of the harmonization of plant IP legislation was the formation of Europe's Community Plant Variety Office (Koo et al., 2004). Another example is the recent EU political breakthrough on an enhanced patent system that will lead to a single EU patent, and establish a new patent court. This move has been seen as a very strong signal from the Council that the EU is committed to achieving a single market for patents. A new U.S. IP enforcement strategy was even recently released. Among other things, it is an attempt to increase transparency, and to encourage public participation in policy-making for IP enforcement and international negotiations.

Last, but not least, a number of improvements are needed for patent laws and licensing agreements. Broad patent claims or claims on gene sequences without a true invention should not be allowed. In addition, patent filings should be delayed until there is an actual reduction to practice of the claimed biotechnology. A provisional patent application procedure could also be used to promote earlier disclosure of novel scientific knowledge (Bentwich, 2010). On the other hand, TTOs need to become more transparent in their licensing agreements and other activities related to managing publicly funded research outputs (Pressman, 2006). Such transparency should involve reporting not only the value and number of licenses, but also whether the licenses are exclusive or non-exclusive, whether research rights have been retained for academic and not-for-profit institutions, and provisions that enable widespread access (e.g., for researchers in developing countries). In effect, this amounts to reporting on whether they are adhering to broadly recognized best practices for licensing university research (AUTM, 2007). Making agricultural biotechnologies available to the poor via a broad, humanitarian license system will have huge impacts and improve an untold number of lives.

Acknowledgments

AH is supported by the Brain Gain Program (*Rientro dei cervelli*) of the Italian Ministry of Education, University and Research (MIUR). We would like to thank Dr. J. Derek Mason for his valuable contributions during the editing of this chapter.

References

Article 52(b) EPC. *In* Convention on the Grant of European Patents of October 5, 1973 (European Patent Convention) (10th ed. 2000), European Patent Office, Germany, ISBN 3-89605-038-9.

Article 53(b) EPC. *In* Convention on the Grant of European Patents of October 5, 1973 (European Patent Convention) (10th ed. 2000), European Patent Office, Germany, ISBN 3-89605-038-9.

Adams, D. J., Beniston, L. J., & Childs, P. R. N. (2009). Promoting creativity and innovation in biotechnology. *Trends in Biotechnology*, *27*, 445–447.

Agreement on Trade Related Aspects of Intellectual Property Rights (TRIPS) World Intellectual Property Office (WIPO) Publication No. 223(E) (1996). ISBN 92-805-0640-4.

Article 14. Scope of the Breeder's Right. *In* Union International pour la Protection des Obtentions Végétales (UPOV) International Convention for the Protection of New Varieties of Plants of December 2, 1961, as revised in Geneva on November 10, 1972, on October 23, 1978, and on March 19, 1991. UPOV Publication No. 221(E), ISBN 92-805-0332-4.

Article 14(5)(c). *In* Union International pour la Protection des Obtentions Végétales (UPOV) International Convention for the Protection of New Varieties of Plants of December 2, 1961, as revised in Geneva on November 10, 1972, on October 23, 1978, and on March 19, 1991. UPOV Publication No. 221(E), ISBN 92-805-0332-4.

Article 15. Exceptions to the breeder's right. *In* Union International pour la Protection des Obtentions Végétales (UPOV) International Convention for the Protection of New Varieties of Plants of December 2, 1961, as revised in Geneva on November 10, 1972, on October 23, 1978, and on March 19, 1991. UPOV Publication No. 221(E), ISBN 92-805-0332-4.

Article 27(1). *In* Agreement on Trade Related Aspects of Intellectual Property Rights (TRIPS), World Intellectual Property Office (WIPO) Publication No. 223(E) (1996). ISBN 92-805-0640-4.

Article 27(2). *In* Agreement on Trade Related Aspects of Intellectual Property Rights (TRIPS), World Intellectual Property Office (WIPO) Publication No. 223(E) (1996). ISBN 92-805-0640-4.

Article 27(3)(b). *In* Agreement on Trade Related Aspects of Intellectual Property Rights (TRIPS), World Intellectual Property Office (WIPO) Publication No. 223(E) (1996). ISBN 92-805-0640-4.

Article 7. *In* Agreement on Trade Related Aspects of Intellectual Property Rights (TRIPS), World Intellectual Property Office (WIPO) Publication No. 223(E) (1996). ISBN 92-805-0640-4.

Article 8. *In* Agreement on Trade Related Aspects of Intellectual Property Rights (TRIPS), World Intellectual Property Office (WIPO) Publication No. 223(E) (1996). ISBN 92-805-0640-4.

Atkinson, R., Beachy, R., Conway, G., Cordova, F., Fox, M., & Holbrook, K., et al. (2003). Public sector collaboration for agricultural management. *Science*, *301*, 174–175.

AUTM. (2007). In the Public Interest: Nine Points to Consider in Licensing University Technology (<http://www.autm.net/Nine_Points_to_Consider.htm/>).

Bent, S. A. (2006). *History of patents and their use in Europe and the United States of America. Patent Protection of Plant-Related Innovations: Facts and issues*. Copenhagen: International Seed Federation International Seminar.

Bentwich, M. (2010). Changing the rules of the game: Addressing the conflict between free access to scientific discovery and intellectual property rights. *Nature Biotechnology*, *28*, 137–140.

Bera, R. K. (2009). Harmonization of patent laws. *Current Science*, *96*, 457–458.

Berman, E. P. (2008). Why did universities start patenting? Institution-building and the road to the Bayh-Dole Act. *Social Studies of Science*, *38*, 835–871.

Boettiger, S., & Bennett, A. (2007). PIPRA: A resource for collaborative intellectual property management in agriculture. *Journal of Intellectual Property Rights*, *12*, 86–91.

Boettiger, S., Graff, G., Pardey, P. G., van Dusen, E., & Wright, B. D. (2004). Intellectual property rights for plant biotechnology: International aspects. In P. Christou & H. Klee (Eds.), *Handbook of plant biotechnology* (Vol. 2, pp. 1089–1113). Chichester: John Wiley & Sons, Ltd.

Brooks, H. (1993). Research universities and the social contract for science. In L. M. Branscomb (Ed.), *Empowering technology: Implementing a U.S. Strategy* (pp. 202–234). Boston: MIT Press.

Bubela, T. M., & Caulfield, T. (2010). Role and reality: Technology transfer at Canadian universities. *Trends in Biotechnology*, *28* (9), 447–451.

Chi-Ham, C. L., Clark, K. L., & Bennett, A. (2010). The intellectual property landscape for gene suppression technologies in plants. *Nature Biotechnology*, *28*, 32–36.

Cohen, J. I., Falconi, C., Komen, J., & Blakeney, M. (1998). *Proprietary biotechnology inputs and international agricultural research*. ISNAR Briefing Paper No. 39. The Hague: International Service for National Agricultural Research. <www.isnar.cgiar.org/publications/briefing/BP39.htm/>.

Cullet, P. (1999). Revision of the TRIPS agreement concerning the protection of plant varieties – lessons from India concerning the development of a *sui generis* system. *Journal of World Intellectual Property*, *2*, 617–656.

Diamond v. Chakrabarty 447 U.S. 303, 206 USPQ 193 (1980).

Donnenwirth, J., Grace, J., & Smith, S. (2004). Intellectual property rights, patents, plant variety protection and contracts: A perspective from the private sector. *IP Strategy Today*, *9*, 19–34.

Eisenberg, R. S. (1996). Public research and private development: Patents and technology transfer in government-sponsored research. *Virginia Law Review*, *82*, 1663–1727.

European Directive 98/44/CE on the protection of biotechnological inventions of July 6th, 1998, (1998) OJEC L213/13.

Ex parte Hibberd, (1985) 227 USPQ 443 (USPTO Board of Patent Appeals and Interferences.

Ex parte Thomson, (1992) 24 USPQ2d 1618 (USPTO Board of Patent Appeals and Interferences.

Fenton, G. M., Chi-Ham, C., & Boettiger, S. (2007). Freedom to operate: The law firm's approach and role. In A. Krattiger, R. T. Mahoney, L. Nelsen, J. A. Thomson, A. B. Bennett, K. Satyanarayana, G. D. Graff, C. Fernandez & S. P. Kowalski (Eds.), *Intellectual property management in health and agricultural innovation: A handbook of best practices* (pp. 1363–1383). Oxford: MIHR. and PIPRA, Davis. Available online at <www.ipHandbook.org/>.

Fleck, B., & Baldock, C. (2003). Intellectual property protection for plant-related inventions in Europe. *Nature Biotechnology*, *4*, 834–838.

Friedman, Y. (2009). Intellectual property and biotechnology innovation: To protect or not protect? *Journal of Commercial Biotechnology*, *15*, 285–286.

Geiger, R. (2004). *Knowledge and money: Universities and the paradox of the marketplace*. Stanford: Stanford University Press.

Graff, G., & Zilberman, D. (2001). Towards an intellectual property clearinghouse for agricultural biotechnology. *IP Strategy Today*, *3*, 1–13.

Graff, G. D., Cullen, S. E., Bradford, K. J., Zilberman, D., & Bennett, A. B. (2003). The public-private structure of intellectual property ownership in agricultural biotechnology. *Nature Biotechnology*, *21*, 989–995.

Graff, G. D., Wright, B. D., Bennett, A. B., & Zilberman, D. (2004). Access to intellectual property is a major obstacle to developing transgenic horticultural crops. *California Agriculture*, *58*, 120–126.

Harfouche, A., Grant, K., Selig, M., Tsai, D., & Meilan, R. (2010). Protecting innovation: Genomics-based intellectual property for the development of feedstock for second-generation biofuels. *Recent Patents on DNA & Gene Sequences 4* (2), 94–105.

Harvard College v. Canada (Commissioner of Patents) 4 S.C.R. 45, 2002 Supreme Court of Canada 76 (2002).

Heckenberger, M., Bohn, M., Klein, D., & Melchinger, A. E. (2005). Identification of essentially derived varieties obtained from biparental crosses of homozygous lines: II. Morphological distances and heterosis in comparison with simple sequence repeat and amplified fragment length polymorphism data in maize. *Crop Science*, *45*, 1132–1140.

Huete-Perez, J. A. (2003). Agricultural IP and the public sector. *Science*, *302*, 781–783.

In re LeGrice, 301 F.2d 929, 133 USPQ 365 (U.S. Court of Customs and Patent Appeals 1962).

J.E.M. Ag Supply, Inc. v. Pioneer Hi-Bred International, Inc., 60 USPQ2d 1865, (2001).

Jayaraman, K. (2004). Illegal seeds overtake India's cotton fields. *Nature Biotechnology*, *22*, 1333–1334.

Jefferson, R. (2007). Science as social enterprise: The CAMBIA BiOS Initiative. *Innovations: Technology, Governance, Globalization*, *l*, 13–44.

Kenny, M., & Patton, D. (2009). Reconsidering the Bayh-Dole Act and the Current University Invention Ownership Model. *Research Policy*, *38*, 1407–1422.

Kock, M. A., Porzig, S., & Willnegger, E. (2006). The legal protection of plant-biotechnological inventions and plant varieties in light of the EC biopatent directive. *International Review of Intellectual Property and Competition Law*, *37*, 135–244.

Koo, B., Nottenberg, C., & Pardey, P. G. (2004). Plants and intellectual property: An international appraisal. *Science*, *306*, 1295–1297.

Kowalski, S. P., Ebora, R., Kryder, R. D., & Potter, R. (2002). Transgenic crops, biotechnology and ownership rights: What scientists need to know. *The Plant journal*, *31*, 407–421.

Krattiger, A. (2007). Freedom to operate, public sector research and product-development partnerships: Strategies and risk-management options. In A. Krattiger, R. T. Mahoney, L. Nelsen, J. A. Thomson, A. B. Bennett, K. Satyanarayana, G. D. Graff, C. Fernandez & S. P. Kowalski (Eds.), *Intellectual property management in health and agricultural innovation: A handbook of best practices* (pp. 1317–1327). Oxford: MIHR. and PIPRA, Davis. Available online at <www.ipHandbook.org/>.

Le Buanec, B. (2004). Protection of plant-related innovations: Evolution and current discussion. *IP Strategy Today*, *9*, 1–18.

Le Buanec, B. (2005). Plant genetic resources and freedom to operate. *Euphytica*, *146*, 1–8.

Lei, Z., Juneja, R., & Wright, B. D. (2009). Patents versus patenting: Implications of intellectual property protection for biological research. *Nature Biotechnology*, *27*, 36–40.

Lence, S. H., Hayes, D. J., McCunn, A., Smith, S., & Niebur, W. S. (2005). Welfare impacts of intellectual property protection in the seed industry. *American Journal of Agricultural Economics*, *87*, 951–968.

McConnell, R. (2004). Developing genetic resources for the future: The long look back. *Bio-Science Law Review*, *7*, 57–60.

Monsanto Canada, Inc. v. Schmeiser 1 S.C.R. 902, 2004 Supreme Court of Canada 34 (2004).

Mowery, D. C., Nelson, R. R., Sampat, B. N., & Ziedonis, A. A. (2004). *Ivory Tower and Industrial Innovation. University-Industry Technology before and after the Bayh-Dole Act*. Stanford: Stanford University Press.

Mowery, D. C., & Sampat, B. N. (2001). Patenting and licensing university inventions: Lessons from the history of the Research Corporation. *Industrial and Corporate Change*, *10*, 317–355.

Mowery, D. C., & Ziedonis, A. A. (2007). Academic patents and material transfer agreements: Substitutes or complements? *Journal of Technology Transfer*, *32*, 157–172.

Official Journal of the European Patent Office (1999) 573.

Pardee, W. D. (2007). Protecting new plant varieties through PVP: Practical suggestions from a plant breeder for plant breeders. In A. Krattiger, R. T. Mahoney, L. Nelsen, J. A. Thomson, A. B. Bennett, K. Satyanarayana, G. D. Graff, C. Fernandez & S. P. Kowalski (Eds.), *Intellectual property management in health and agricultural innovation: A handbook of best practices* (pp. 981–987). Oxford: MIHR. and PIPRA, Davis. Available online at <www.ipHandbook.org/>.

Pardey, P. G., & Beintema, N. M. (2001). *Slow magic: Agricultural R&D a century after Mendel*. Washington, DC: Food Policy Report, International Food Policy Research Institute.

Pardey, P. G., Wright, B. D., Nottenburg, C., Binenbaum, E., and Zambrano, P. (2003). Intellectual property and developing countries: Freedom to operate in agricultural biotechnology. *Biotechnology and Genetic Resource Policies*, Brief 3, pp. 1–6. International Food Policy Research Institute, Washington, DC.

Patently transparent, *Nature Biotechnology*, *24*, 474.
Pressman, L. (2006). The licensing of DNA patents by US academic institutions: An empirical survey. *Nature Biotechnology*, *24*, 31–39.
Price, S. C. (1999). Public and private plant breeding. *Nature Biotechnology*, *17*, 938.
Qaim, M., & Traxler, G. (2005). Roundup Ready soybeans in Argentina: Farm level and aggregate welfare effects. *Agricultural Economics*, *32*, 73–86.
Renault, C. S., Cope, J., Dix, M., & Hersey, K. (2007). A new technology transfer paradigm: How state universities can collaborate with industry. *AUTM*, *14*, 13–26.
Rosenberg, N., & Nelson, R. R. (1994). American university and technical advance in industry. *Research Policy*, *23*, 323–348.
Rule 23(c) EPC. *In* Convention on the Grant of European Patents of October 5, 1973 and revised on December 9, 2004 (European Patent Convention) (10th ed.), European Patent Office, Germany. ISBN 3-89605-038-9
Sechley, K. A., & Schroeder, H. (2002). Intellectual property protection of plant biotechnology inventions. *Trends in Biotechnology*, *20*, 456–461.
Singh, A., Hallihosur, S., & Ranga, A. (2009). Changing landscape in biotechnology patenting. *World Patent Information*, *31*, 219–225.
Smith, J. S. C., Hussain, T., Jones, E. S., Graham, G., Podlich, D., & Wall, S., et al. (2008). Use of doubled haploids in maize breeding: Implications for intellectual property protection and genetic diversity in hybrid crops. *Molecular Breeding*, *22*, 51–59.
Smith, S. (2008). Intellectual property protection for plant varieties in the 21st century. *Crop Science*, *48*, 1277–1290.
Smolders, W. (2005). Plant genetic resources for food and agriculture: Facilitated access or utility plants on plant varieties. *IP Strategy Today*, *13*, 1–17.
Sterckx, S. (2010). Is the non-patentability of "Essentially Biological Processes" under threat? *Journal of World Intellectual Property*, *13*, 1–23.
Stevens, A. J. (2004). The enactment of Bayh–Dole. *Journal of Technology Transfer*, *29*, 93–99.
Tripp, R., Louwaars, N., & Eaton, D. (2007). Plant variety protection in developing countries. A report from the field. *Food Policy*, *32*, 354–371.
Union International pour la Protection des Obtentions Végétales (UPOV) International Convention for the Protection of New Varieties of Plants of December 2, 1961, as revised in Geneva on November 10, 1972, on October 23, 1978, and on March 19, 1991. UPOV Publication No. 221(E). ISBN 92-805-0332-4.
University and small business patent procedures Act, 35U.S.C. § 203 (1980).
Utility Guidelines. (2001) 66 Federal Registry 1093.
Wright, B. D., Pardey, P. G., Nottenburg, C., and Bonwoo, K. (2007). Agricultural innovation: Investments and incentives, In R. Evenson & P. Pingali (Eds.), *Handbook of agricultural economics*, (Vol. III, pp. 2533–2603). Amesterdam: Elsevier. (Chapter 48)
Written Description Guidelines. (2001) 66 Federal Registry 1099.

Regulatory issues of biotechnologically improved plants

34

Elizabeth E. Hood[1] Deborah Vicuna Requesens[1] Kellye A. Eversole[2]

[1]*Arkansas Biosciences Institute, Arkansas State University, Jonesboro, Arkansas*
[2]*Eversole Associates, Bethesda, Maryland*

TABLE OF CONTENTS

Introduction

Plant biotechnology can be defined in many ways, but it is most often considered the genetic engineering of plants through the use of recombinant DNA (rDNA; Sharma et al., 2002; Lemaux, 2008; Akhond and Machray, 2009). For the technology to work properly, many factors must be considered including which plant is being altered and for what purpose, which molecular factors will best express the gene when and where it is needed, and what germplasm can be used to enhance gene expression of the input or output trait of interest. Each of these issues is addressed in other chapters in this volume. Here, we focus on regulatory requirements for genetically engineered (GE) plants.

Modern agriculture faces the grand challenge of meeting the increasing demand for food, feed, and non-food products in an environmentally sustainable manner while maintaining quality and yield, although pest and disease stress is rising and extreme weather patterns are surfacing, putting even greater demands on the system. To meet these demands for the growing human population, plant breeders must have access to a complete toolbox, the wherewithal to exploit all available technologies, and the opportunity to commercialize results. During the last 25 years, biotechnology has allowed the production of genetically improved varieties to begin to address some of these challenges. Although field trials of GE traits have been conducted for 130 plant species, only 4 crops (soybean, maize, cotton, and rapeseed or canola) account for the overwhelming majority of all acreage planted with GE crops. Over the past 20 years, there has been a dramatic reduction in the number of firms conducting GE field trials, partially due to the increasing costs of development but also because of high regulatory costs (Arundel and Sawaya, 2009). Most crop developers encounter significant barriers to commercialization as they cannot afford the high cost of regulatory approval for GE crops, in addition to development and marketing costs. To be commercialized and grown without restrictions, GE crops must go through an extensive food, feed, and environmental safety assessment process to prove that they are "substantially equivalent" to comparable non-GE crops (OECD, 1993; ILSI, 2004, 2008). Achieving substantial equivalence requires a prohibitive amount of data, some of which may duplicate existing data. Estimates of the cost of meeting regulatory requirements for GE crops range from \$20–30 million (McElroy, 2003) to over \$100 million (The Economist, 2009). According to the Phillips McDougall consultancy, the global seed market in 2008 was worth \$26 billion, with \$9 billion coming from sales of corn seed alone (Kaskey and Ligi, 2010). In comparison, the Context Network (http://www.contextnet.com) estimated that the 2008 value of the global seed market for all of the different vegetables produced was approximately \$4 billion. Thus, specialty crops represent a significant percentage of total agricultural receipts, and although several GE specialty crops have received initial regulatory approval, only a few have been marketed (Miller and Bradford, 2010). Miller and Bradford (2010) reported that GE research is underway on 46 species of specialty crops, but the additional regulatory costs per insertion event is "out of proportion" to the additional market value for these crops. Unfortunately, the prohibitive cost of achieving regulatory approval has limited the commercialization of GE crops to a few multi-national biotechnology companies and a

DOI: 10.1016/B978-0-12-381466-1.00034-1

few traits on a few crops that have a market value high enough to recover the cost of deregulation and development.

The process of genetic engineering was developed to improve traits for agronomic performance — disease, insect, and herbicide resistance. These traits have led to a significant reduction in the environmental impact associated with insecticide and herbicide use and greenhouse gas emissions at the same time that yields were enhanced (James, 2009, 2010; Lemaux, 2009; Brookes and Barfoot, 2010; Carpenter, 2010). Genetic engineering also can be used to alter metabolism to enhance nutrition or to make small molecule pharmaceuticals. Protein production in plants through genetic engineering can be utilized for pharmaceutical proteins, vaccines, and industrial enzymes (ILSI, 2004; Howard and Hood, 2005; Chapotin and Wolt, 2007; Akhond and Machray, 2009). These output traits are human-pathogen-free and inexpensive to produce. Many interesting input and output traits have been developed by academic labs, and although research is interesting, impact is better. Nevertheless, most results from genetic engineering have not been commercialized, thus the full potential impact of this research has not been realized.

Commercializing an Agricultural Biotechnology Product

Moving from a concept to a marketed product is a complex process regardless of the nature of the product. When the product is derived from a scientific application, the complexity goes up many-fold. Scientific experimentation often leads to interesting discoveries that are predicted to have a useful outcome. However, to achieve this outcome, the idea should be protected so that the business/developer can have a chance to develop a strong business foundation. Even though an idea is interesting and novel, it may not be marketable. Market considerations must be addressed to understand if the idea has merit in the business world (i.e., scientific, health-related, industrial, or agricultural) or if it will be acceptable to consumers. If the research being conducted is to fill a specific need, such as building resistance in wheat to Hessian fly infestations, a market pull probably drove the research and hence a market is likely to develop because wheat is a large volume crop and Hessian fly is a major pest. However, if one is exploring powdery mildew resistant okra, the same investment of time and money is required to develop the trait and push it to market. In contrast to the wheat market, the okra seed market is small, perhaps not over $2 million annually. Thus the return on investment for these plants must be carefully considered before investing in trait-improvement technology.

Regardless of the type of crop developed, the process of commercialization requires significant effort in several areas:

- Scientific capability: Is it possible to develop the crop/trait combination to yield the desired result?
- Intellectual property: Does the developer have the freedom to use the transformation method, and can the developer establish and maintain intellectual property on the invention or crop?
- Market realities: Is there a market for the crop?
- Regulations: Can the developer afford the costs associated with the various governmental regulatory requirements and can these be achieved?

In this chapter, we are focusing on the last step in this process, that is, regulatory requirements, as they relate specifically to GE crops. For scientists, particularly public sector and small company scientists, this is the most daunting challenge partially because a simple checklist of tasks does not exist, and even when the developer begins to see the light at the end of the tunnel, the regulators can request even more data and testing. Some of the regulatory barriers that crop developers face in commercializing GE crops include the:

- Complex regulatory framework
- Lack of a road map that delineates how to navigate through the regulatory framework to achieve the freedom to commercialize (e.g., in the United States, achieving non-regulated status from the U.S. Department of Agriculture's Animal Plant Health Inspection Service, APHIS)
- Costs associated with meeting the safety standard of "substantial equivalence"
- Challenges and costs related to collecting extensive, relevant, and comprehensive data sets
- Lack of standardization for data collection and analysis.

Generally, regulations cover the environmental release of plants produced through genetic engineering. In the United States, the U.S. Department of Agriculture (USDA) regulates these GE products, protecting agriculture from potential plant pests including weeds; the Environmental Protection Agency (EPA) protects human health and the environment from adverse effects; and the Food and Drug Administration (FDA) ensures safe food and feed. Each agency must be satisfied that the product the scientist is developing is safe before the product is allowed in the environment and to be in food or feed products. Thus, the opportunity and promise of genetic engineering has many hoops through which to jump before it can be commercialized.

The goals of a regulatory system should be and often are flexibility, science-based rigor, stakeholder-responsiveness, nationally and internationally coordinated, and risk-proportional. However, one of the difficulties in establishing public policies is that they typically are the culmination of inputs from many often competing sectors and stakeholders (e.g., governmental representatives, elected representatives, citizens, general media, courts, industry, consumers, and scientists), with different levels of awareness and knowledge, and with different goals and objectives. The resulting legislation and implementing regulations that emerge are often complex and occasionally have contradictory policies with many checkpoints. Public opinion, often uninformed, has been a significant factor in the development of regulations and in the commercialization of GE crops (Fedoroff and Brown, 2004; Fedoroff et al. 2010). Several activists and non-representative groups have taken advantage of public doubts and lack of confidence, spreading fear of the potential threats of GE crops to the environment and human health. To address these issues, regulatory agencies require rigid testing of GE crops and regulatory decisions are always taken on a case-by-case basis (Trewavas and Leaver, 2001). In the EU, in an effort to impose public opinion, especially the opinions of activist anti-GE crop groups on the marketing of GE crops, politicians play a major role in GE-crop approvals. In particular, after a GE crop has gone through a rigorous risk and safety assessment, the European Commission (EC) and the individual Member States must individually approve market entry (Rommens, 2009). Layered on top of these regulatory considerations are the market drivers that either embrace or reject biotechnology-derived crops and the specific politics of those markets (Kalaitzandonakes and Bijman, 2003).

The Regulatory Framework

The focus in this chapter is primarily on the United States regulatory system, because the first steps toward regulating GE crops were taken in the United States in the 1970s. Three regulatory agencies require different and sometimes duplicate data despite operating under a "Coordinated Framework" (see the section The U.S. Coordinated Framework). The majority of the events that are marketed today were approved first in the United States, and almost half of the global acreage devoted to GE crops is planted there (James, 2009). Further, many of the concepts that are operating today in all of the regulatory frameworks for GE crops are the result first of the United States system, and secondly of the responses to that system, particularly in Europe. To understand where we are now, it is first necessary to gain an appreciation for how we arrived at this point.

The various United States and international regulatory requirements for GE crops are the result of concerns expressed in the mid-1970s about the uncertainties of the new technologies that involved the use of recombinant DNA in research. As molecular biologists began to use restriction enzymes to study animal viruses, concerns arose about laboratory practices, potential hybrid organisms, and the possibility for infecting laboratory workers with deadly viruses (Singer and Soll, 1973). A National Academy of Sciences committee recommended that the U.S. National Institutes of Health (NIH) form an advisory committee on rDNA to evaluate the potential biological and ecological hazards. In 1974, the NIH responded by establishing the Recombinant DNA Advisory Committee (RAC) to assist in the development of guidelines or procedures that would minimize the risks. The first draft guidelines were presented in December 1975 and have been revised several times. Agriculture was not the first target, and the guidelines did affect agricultural research by banning experiments to increase the pathogenicity of plant pathogens, requiring certain containment practices, and by prohibiting the deliberate release into the environment of an organism containing rDNA (Tolin and Vidaver, 1989). Waiver authority was granted in a subsequent revision to the guidelines under which the NIH developed a process for determining the conditions for permitting the deliberate release into the environment (Tolin and Vidaver, 1989). This "waiver authority" was used for the first transgenic plants and serves as the precursor to current U.S. regulations governing GE crops.

The first transgenic plants were produced in 1982, announced at the Miami Winter Symposium in January 1983, and later described in detail in several journal articles (Herrera-Estrella et al., 1983; Fraley et al., 1983a,b; Schell et al. 1983; Barton et al., 1983). The first environmental release of a genetically modified organism for agricultural applications was an ice-minus bacterium (*Pseudomonas syringae*), which prohibited the formation of ice crystals. With NIH approval in mid-1983, it was tested first on potatoes and several years later gained EPA approval for use on strawberries (McHughen and Smyth, 2008).

Protests and court cases emerged immediately questioning the authority of the NIH to approve this field test. With respect to the court cases, claimants alleged that the NIH had not prepared a specific Environmental Impact Statement (EIS) as required under the National Environmental Policy Act (42 U.S.C. §4321 et seq. 1969) for all major Federal actions significantly affecting the quality of the human environment. In response to the challenges and the continuing rapid technology developments, the White House Office of Science and Technology Policy (OSTP) established the Biotechnology Science Coordinating Committee (BSCC) in 1984 with representatives from 18 agencies including the NIH, the USDA, the EPA, the FDA, and the U.S. National Science Foundation (Tolin and Vidaver, 1989; McHughen and Smyth, 2008). After two years of effort, the Coordinated Framework for Regulation of Biotechnology was published in the U.S. *Federal Register* and included policy statements from the agencies involved, as well as a proposed USDA regulation governing the introduction of organisms and products developed through genetic engineering (plant pests; Kingsbury, 1986).

Several important concepts were included in the Coordinated Framework. First, there was agreement that genetically engineered organisms are not inherently risky and thus regulations should focus on the products and not the process used for their generation. Secondly, rDNA modified organisms should be evaluated on a case-by-case basis for potential risk prior to applications in agriculture and the environment, and not every case will require a review as some instances may be excluded. Thirdly, despite considerable debate about whether new laws were needed, the BSCC determined that existing statutes and agencies were adequate for regulating biotechnology. Fourthly, the USDA, the EPA, and the FDA would share responsibility for regulating biotechnology under their own statutes (Kingsbury, 1986; OSTP, 1986; Belson, 2000; McHughen and Smyth, 2008).

Perhaps the most significant aspect of the U.S. framework was the decision to base regulations and safety considerations on "product" and without regard to the technology used to develop the crop. In and of itself, this theoretically did not target GE crops; rather, it targeted the products that had not been in the environment and not previously consumed. The assumption is that food and feed crops existing before the advent of biotechnology are safe (i.e., generally recognized as safe, GRAS), and thus the safety questions regarding GE crops focus on whether these new products have impacts different from non-GE crops that had been in the environment before — specifically, determining whether the GE crop is substantially equivalent to the comparable non-GE crop.

The Organization for Economic Development and Cooperation (OECD) began to focus on genetically engineered plants beginning in 1983 with discussions that culminated with the publication of the OECD publication "Recombinant DNA Safety Considerations," referred to as the "OECD Blue Book" and occurring at the same time as the development of the U.S. Coordinated Framework (OECD, 1986). Importantly, the OECD Blue Book included key risk assessment concepts such as "case-by-case" and "step-by-step" approaches (Bergmans, 2006). The "step-by-step" approach refers to the process of progressively decreasing physical containment of genetically modified organisms (i.e., from the lab, to contained research environments, small-scale field testing, and subsequently large field testing) during which safety assessments and performance data can be assessed (Bergmans, 2006). Subsequent to the publication of the OECD Blue Book, an effort was undertaken to collect a body of knowledge from traditional crop breeding practices that could serve as a baseline for assessments (Bergmans, 2006). The OECD followed with an update in 1992 (OECD, 1992) and the release in 1993 of another document that detailed the concepts and principles for evaluating the safety of foods derived from biotechnology (OECD, 1993). The basic premise underlying the OECD recommendations was that regulations should be based on an assessment of the product (and not the technology), and the OECD concluded that no scientific basis exists to require specific legislation to regulate the use of rDNA organisms (Tzotzos et al., 2009).

In contrast to the U.S. framework, regulations in Europe are focused first and foremost on the kind of technology used to develop the crop without regard to whether it may be safer than a conventionally bred crop. This "process-based" system is dependent upon the breeding technique used to create the crop, and a genetically modified organism (GMO) is one that is genetically modified in a way that does not occur naturally by crossing or natural recombination (European Commission, 2001; Devos et al., 2010). Whether this definition is an appropriate one is beyond the scope of this chapter. Suffice it to say that all crops have been genetically modified in one way or another through natural selection, crossing, natural recombination, induced mutations, or genetic engineering (NAS, 2004). The point of the EC regulations is that if genetic engineering is used, it is thus genetically modified. The operational framework of the European Union (EU) is principally defined by Directive 2001/18/EC and a number of specific regulations (Tzotzos, 2009; Devos et al., 2010).

Several of these so-called "process-based" regulatory frameworks incorporate the "Precautionary Principle," including the EU regulations. With the adoption of Directive 2001/18/EC, the EU replaced previous directives and explicitly adopted the precautionary principle as a guide for assessment. Among other things, risk assessment criteria were broadened to include direct, indirect, immediate, delayed, and cumulative long-term adverse effects. Further, post-market environmental monitoring became mandatory (Devos et al., 2010). As applied in the EU, the precautionary principle applies where scientific evidence is insufficient, inconclusive, or uncertain, and preliminary evaluation indicates that there are reasonable grounds for concern (European Commission, 2003; Tzotzos, et al., 2009). In the case of some advocacy groups, the precautionary principle has been interpreted to require absolute certainty of no possible harm. Under the precautionary principle, the burden of proof to demonstrate no possible harm is placed directly on the developer of the GE crop. GE crop opponents, on the other hand, do not have to substantiate their claims of harm; rather, they can merely raise the possibility of future harm and stall the marketing of GE crops (Tzotzos et al., 2009). Despite the fact that no consensus exists on the exact interpretation of the precautionary principle and how it should be applied, most recent international environmental agreements have incorporated the principle, including the Cartagena Protocol on Biosafety that regulates international trade of GE crops (McAllister, 2005). Another regulatory concept that is becoming increasingly challenging is the demand for absolutely no GE presence in non-GE agricultural products. This concept of zero tolerance, driven by fear of GE crops, is essentially unachievable using current farming practices (Kershen and McHughen, 2005; Gaines et al., 2007). Adventitious presence refers to the presence of "other" matter in a seed lot that occurs through non-intended means, possibly by cross-pollination, volunteer regrowth, or mechanical mixing from machinery. The comingling of foreign materials with a crop can be among conventional farm produce, organic produce, and genetically engineered products. Because deregulated GE crops have been extensively tested, the issue is not related to food or environmental safety, but rather to economic concerns (Kershen and McHughen, 2005). Standards for seed production exist with allowances for adventitious presence. For example, certified wheat seed allows up to 0.1% adventitious presence to remain certified as "pure" (Gaines et al., 2007). This level of purity is required in the seed industry and can be achieved through conventional agricultural practices with diligent stewardship. Establishment of an accepted level of adventitious presence of GE material in non-GE harvested products would drive the establishment of cultural practices that would achieve such levels. Currently in the United States, the onus of stewardship is on the farmer that wishes to avoid having the adventitious GE material in their product (Endres, 2005). Coexistence would be highly possible if regulations on the amount of acceptable "contamination" were established.

Although the theories underpinning product- and process-based regulatory frameworks are fundamentally different, the data and information required for risk assessment are generally comparable and follow the basic principles of risk assessment (Bergmans, 2006; Johnson et al., 2007; Tzotzos et al., 2009; Raybould et al., 2010). In the United States, the basis for formal regulatory risk assessment began with the publication in 1983 of the "Red Book" by the National Research Council (National Research Council, 1983; Nickson, 2008). The two essential elements that form the basis of regulations are outlined in the Red Book — risk assessment and risk management. Risk assessment is the process of using factual information to define the effects of exposure to hazardous materials and situations; whereas risk management is the process of weighing policy alternatives and selecting the most appropriate regulatory action (National Research Council, 1983). The management of risk or the determination of what is an acceptable risk level is not a scientific decision as it balances the interest of society with the results of a risk assessment (National Research Council, 1983; Johnson et al., 2007). It is important to keep this distinction in mind as one prepares dossiers for regulatory agencies.

A model was published recently for determining risk of proteins in transgenic plants and is based not only on the protein, but also on the likelihood of exposure to that protein (Howard and Donnelly, 2004). The benefits to society of the products generated through genetic engineering have largely been ignored in the emotional fights over the technology. The focus in this chapter is on the process of risk assessment as required by various regulatory agencies. The regulations that have been implemented by the various agencies for managing the potential risks to human health and the environment associated with GE crops all rely on these principles: risk assessment should be on a case-by-case basis, analysis should be "stepwise" (i.e., performed on diminishing levels of physical containment), and risk should be measured on the effects of the product in relation to a non-GE comparator crop.

Comparative risk assessment is a fundamental principle of GE safety assessments with end point measurements made relative to comparator plants and growth practices (Wolt et al. 2010), and the basic premise is that there are no biologically meaningful differences between a GE phenotype and the non-GE comparator. Phenotypic changes that may arise from the molecular change must be identified (Nickson, 2008). To do this efficiently, the parameters of the assessment are determined and the risks of greatest relevance are defined, that is, a problem formulation (Raybould, 2006; Wolt et al., 2010). To develop meaningful case-specific attributes of the GE crop under conditions of release and to define the scope of the assessment, all available and existing knowledge from product development, field trials, and compositional analyses; previous regulatory decisions; and peer reviewed literature should be applied (Wolt et al., 2010). By effectively using existing data, crop developers do not submit a deluge of new data of negligible utility in characterizing the risks; they can reduce the costs of regulatory studies and avoid opportunity costs from delays in the regulatory review (Raybould, 2006, 2010; Romeis et al., 2009; Raybould et al., 2010).

The assessment for a particular GE crop is based on very specific hypotheses regarding the amount of risk to particular, clearly defined assessment end points, that is, a particular environmental element that is measurable and is considered important to society (e.g., wildlife; Johnson et al., 2007; Raybould et al., 2010). To perform the analysis, risk is characterized and scientific knowledge is acquired by tests that attempt to falsify the hypotheses (Raybould, 2006; Raybould et al., 2010). GE crop risk assessments are most effective when the hypothesis is a finding of no harm, the assessment end points are clear, and adverse effects to the end points are defined, as opposed to merely comparing the GE crop with a non-transgenic comparator (Raybould et al., 2010; Raybould, 2010).

The U.S. Coordinated Framework

The majority of commercialized GE crops have achieved regulatory clearance in the United States under the Coordinated Framework. As discussed earlier, the United States concurred with the OECD recommendation that new legislation was not necessary for GE crops. The United States opted instead to regulate GE crops on the basis of existing statutes administered by three different regulatory agencies. Thus, GE crops and products derived from them are under the jurisdiction of the USDA, the EPA, and the FDA. Each of these agencies was established to fulfill specific missions, and it is within the scope of these missions that regulations governing GE crops and their products were implemented. The USDA has jurisdiction over plant pests, plants, and veterinary products; the EPA has jurisdiction over planting and food/feed uses of pesticidal plants, new uses of existing pesticides, and novel microorganisms; and the FDA regulates food and food additives, feed, and veterinary products (McHughen and Smyth, 2008; Lemaux, 2009; Tzotzos et al., 2009). Which agencies have jurisdiction over a given GE crop is dependent upon the nature of the crop and its intended uses.

USDA–APHIS

APHIS is the primary regulatory arm of the USDA that deals with plants and crops. Protecting United States agriculture, food, and natural resources is a particular focus of APHIS, and it is in this context that it regulates GE crops (Nestmann et al., 2002). APHIS is primarily involved with regulating specific GE organisms capable of posing risks to plant or animal health. It has an important role in biotechnology and the modification of living organisms through genetic engineering and the use of recombinant DNA technology.

The U.S. Plant Protection Act (Title IV Public Law 106-224, 114 Statute 438, 7 U.S.C. §7701–7772) requires the USDA to prevent the introduction and dissemination of plant pests into the United States and provides the specific authority to prohibit or restrict the importation, exportation, and interstate movement of plants, plant products, certain biological control organisms, noxious weeds, and plant pests. As discussed previously, the U.S. guidelines for rDNA prohibits the deliberate release into the environment of an organism containing rDNA; however, it provides for the authority to waive this prohibition. APHIS regulations under 7 CFR §339.1 stipulate that any organism that has been altered or produced through genetic engineering is considered a regulated article if the donor organism, recipient organism, or vector or vector agent meets the definition of a plant pest, or is an unclassified organism and/or an organism whose classification is unknown, or any product that contains such an organism, or any other organism or product altered or produced through genetic engineering that the Administrator determines is a plant pest or has reason to believe is a plant pest (Cowan, 2010). APHIS defines plant pests broadly to include any organisms that can damage or cause injury to plants or plant parts. Since many plant pathogens are used as vectors or promoters in genetic engineering, they are considered plant pests. APHIS regulation 7 CFR Part 339.6 permits an applicant to petition APHIS to evaluate submitted data to determine that a particular regulated article does not present a plant pest risk and should no longer be regulated (McHughen and Smyth, 2008; Cowan, 2010).

As a regulated article, before a GE crop can be field tested or released into the environment in any way, the developer must obtain an authorization through either the: (1) permit process; or (2) notification process. Movement and planting restrictions are imposed by permits, whereas notification can be used in lieu of a permit when the plant is not considered a noxious weed (Cowan, 2010). Notifications are used primarily for field trial approval, as well as importation and transport of certain GE plants within the United States. To be eligible for a notification, specific criteria as well as designated performance and characteristics standards must be met. For example, a GE plant cannot be considered a noxious weed under USDA regulations or a weed in the area of release if these criteria are met: the function of the genetic material must be known and its expression must not cause plant disease, the modification should not contain human or animal pathogenic sequence, and the GE crop cannot express a plant-made pharmaceutical or a plant-made industrial product (Belson, 2000). Permits are used when the GE plant does not meet the requirements for notification. As with the notifications, developers must provide APHIS scientists with details about the nature of the GE organism. The main consideration for field trials is biosafety or the possibility of unintended release and spread of a potential plant pest to the environment.

Once a developer has sufficient evidence that its GE organism does not pose more risks than the equivalent non-GE organism, the developer may petition APHIS for a determination of non-regulated status to introduce the GE organism into U.S. agriculture and commerce without oversight from APHIS (i.e., to become "deregulated"). Petitions must contain, among other information, details about plant genetics, the nature and origin of the genetic material, field test reports, and effects on other plants. Petitions are reviewed by APHIS with a focus on environmental impact, including the potential for cross-pollination, the effect on wildlife, and the potential for the GE plant to become a weed or plant pest (Belson, 2000; Cowan, 2010). During the process of evaluation, APHIS, at a minimum, prepares an Environmental Assessment (EA) and a "Determination of Non-Regulated Status" as well as considering comments from the public.

A list of Petitions of Non-regulated Status Granted or Pending by APHIS can be found at http://www.aphis.usda.gov/biotechnology/not_reg.html.

EPA

The main role of the EPA is to ensure that all Americans are protected from significant risks to human health and to the environment where they live, learn, and work. EPA regulations contribute to making American communities and ecosystems diverse, sustainable, and economically productive (Nestmann et al., 2002). The regulatory activities of the EPA are governed by the authorities and requirements of specific environmental laws. These laws direct the EPA to regulate many kinds of "pollution," including pesticides, with a focus on how they affect different aspects of the environment. "Environment" is generally defined in these statutes as including human health, plant and animal wildlife, water quality, and air quality (National Research Council, 2009). There is a general recognition by the courts, the scientific community, and the regulatory community that the EPA should consider other factors along with the science when it makes decisions about risk management (EPA, 2005).

Under the Federal Insecticide, Fungicide, and Rodenticide Act (FIFRA; (7 U.S.C. §135 et seq., 1972), before a pesticide, including a GE pesticide, can be registered for use and commercialized, the EPA must determine that the pesticide is able to perform its intended function without unreasonable adverse effects on human health and the environment (FIFRA §3). Plant incorporated protectants (PIPs), such as the Bt toxin in corn, are considered pesticides under FIFRA because they are introduced into plants with the intention of preventing, destroying, repelling, or mitigating a pest (FIFRA §2). In 2001, the EPA published a rule that established the regulatory framework for PIPs and included: (1) an exemption from FIFRA requirements for those PIPs that are derived entirely through conventional breeding between sexually compatible plants; and (2) an exemption from the requirement of the Federal Food, Drug, and Cosmetic Act (FFDCA; 21 U.S.C. § 301 et seq.) for the establishment of a residue tolerance level for pesticides (McHughen and Smyth, 2008).

In contrast to the APHIS process of "deregulation," a developer of a crop that contains a pesticidal property must submit an application to the EPA to register a PIP. This involves regulating GE plants with obvious herbicide-tolerance, designed to be sprayed with a pesticidal chemical, as well as GE virus-resistant plants that are not to be sprayed with new pesticidal substances or have them synthesized internally.

The EPA evaluates each submission to deregulate a GE plant on a case-by-case basis. It requires data related to the product characterization that are similar to the data required by other agencies regarding the description of the GE plant and the details of its modification, such as an organic source of the pesticidal gene construct (promoter, terminator, enhancer, marker genes, and other inserted DNA). Molecular techniques should be utilized to demonstrate genetic stability and integrity, assuring number and location of insertion of foreign DNA. The dossier should include information on the biology of the plant and the anti-nutritional substances produced by the plant, among others.

Additionally, the EPA requires data on mammalian toxicity, the effects on non-target organisms, environmental metabolism, and an insect resistance management program for plants specifically expressing Bt products. Since the EPA focuses on the pesticidal properties, developers should include a full biochemical and amino acid description of the pesticidal protein. Since most PIPs are proteins, the issue of mammalian allergenicity is highly important. Acute digestibility assays and amino acid sequence homology are necessary to demonstrate the absence of allergenicity concerns. The EPA is also concerned with gene flow issues as they may result in expanded exposure to novel pesticidal substances, as well as the environmental fate of PIP substances and their effect on non-target organisms (McHughen and Smyth, 2008).

In addition to the requirement for a PIP to be registered before use unless it is exempted (as previously noted), FIFRA Section 7 requires that the producers of pesticides register the facilities where the pesticide is produced and report the amount produced. Comparable to APHIS requirements, pre-commercial regulation is through a system of notifications for small-scale field tests, or experimental use permits for larger field trials. In an effort to delay the onset of resistance to pesticides, the EPA considers management strategies after commercialization. Farmers cultivating GE crops with PIPs are required to follow the appropriate insect resistance management (IRM) practices required and approved by the EPA that may include refuge areas.

The Pesticide Registration Improvement Act of 2003 (PRIA), contained in P.L. 108–199 (HR 2673), created a registration service fee system for pesticide applications and establishment of tolerances under FIFRA and FFDCA. PRIA was reauthorized in 2007 via the Pesticide Registration Improvement Renewal Act (PRIA 2). PRIA 2 requires the payment of registration fees to be made at the time of application to the agency, and an application that does not include the payment is required to be rejected. Under PRIA 2, the category or type of application and the type of action (e.g., PIP with a new food use) determines the registration service fee and the decision review period. For fiscal years 2011–2012, an application that requests an experimental use permit, the registration of an active ingredient, and the establishment or waiver of tolerances with review by an independent science advisory panel requires a fee of more than $390,000 (*Federal Register* 75–154, pp. 48672–48683). Registration fees can be reduced by up to 75% for small businesses. Federal and State agencies are exempted from the fee and minor use applications associated with the Inter-Regional Project 4 (IR-4) program for minor uses may be exempt as well if it is determined to be in the public interest. Once registered by the EPA, all registrants are also required to pay an annual maintenance fee to the EPA pursuant to FIFRA Section 4.

FDA

The FDA is an agency within the U.S. Department of Health and Human Services. It primarily ensures the safety, effectiveness, and security of food, animal feed additives, and human and animal drugs under the regulatory authority of the FFDCA (21 U.S.C. § 301 et seq.) and the Public Health Service Act (42 U.S.C. § 201 et seq; Cowan, 2010). They also regulate tobacco products and help the public obtain accurate, science-based information for their use of medicines and foods to improve their health (www.fda.org; Nestmann et al., 2002). Under the FFDCA, food and feed manufacturers must ensure the safety and proper labeling of their products.

In 1992, the FDA issued a new policy statement (*Federal Register*, May 29, 1992, 57 FR 22984) indicating that it would regulate foods and feeds regardless of their method of breeding, and it would not require special labeling of GE foods unless the composition of the food differed significantly from its conventional comparator (FDA, 1992; Belson, 2000). The FDA regulates the presence of allergens and any changes in composition or levels of nutritional and anti-nutritional substances. This means that any food with nearly identical composition to the regular version is not considered "adulterated," therefore it is not regulated by the FDA. The method for producing the new food or feed is not a consideration for this statement, which means bioengineered foods composed of the same or nearly the same substances as their available counterpart are not considered adulterated (Belson, 2000; Smyth and McHughen, 2008; Cowan, 2010).

Nonetheless, the FDA recommended that developers consult with the agency about bioengineered foods under development; the FDA review is considered to be "voluntary." Since 1992, all developers of genetically modified foods have routinely consulted the FDA. A list of the FDA consultations is available on the FDA Web site http://www.fda.gov/Food/Biotechnology/Submissions/default.htm.

The FDA is divided into five centers that focus on different regulatory targets such as food or drugs. The Center for Food Safety and Nutrition (CFSAN) regulates food and color additives and has been the primary regulatory focus for GE crops (Belson, 2000). CFSAN and the Center for Veterinary Medicine evaluate new foods and feed composition, including those from GE plants. This agency follows the requirements of the FFDCA with respect to human foods and animal feeds derived from new plant varieties, and provides guidance to industry on scientific and regulatory issues related to these foods. The FDA regulations are applied to all foods derived from all new plant varieties, including varieties that are developed using rDNA technology (e.g., bioengineered foods).

The FDA describes the procedures as useful for developers who intend to commercialize a bioengineered food. The agency is interested in relevant safety, nutritional, or other regulatory issues regarding the bioengineered food. Developers should submit a summary of the scientific and regulatory assessment of the food. The dossier must include information regarding the genetic stability of the plant, nutritional and compositional analyses, and allergenicity and toxicology studies, as well as a complete assessment of the introduced gene products. In addition to these compositional analyses, the FDA might require information regarding the expected dietary exposure to any risk groups, such as children or immunosuppressed patients, or particular religious or ethnic groups that might be overly exposed to the new food.

It is important to remark that up until now, the FDA has not identified any new food or feed produced by rDNA technology that has unexpected changes in composition or in the presence of allergens and other toxicants when compared to the same naturally occurring type of food (NAS, 2004).

After careful review, the FDA issues a "memo" indicating the characteristics of the new food and summarizing the possible safety concerns. They do not formally approve the new food or feed to be safe *per se*. Instead, they indicate it is not materially different from the unmodified version with respect to composition or safety as it is "as safe as its non-modified counterparts" (McHughen and Smyth, 2008).

Perspectives

Specialty crops regulatory assistance: A new paradigm

The regulatory system has evolved into a system in which regulations are approached on a case-by-case basis. Each new crop/gene combination requires a separate data package and petition to the USDA as a starting point, then to either the FDA and/or the EPA depending on the crop (food or feed) and the gene (plant incorporated protectant), respectively. The current climate in regulatory affairs assumes that the genetically modified crop is *not* substantially equivalent to the non-modified crop and that there is a greater risk associated with the newly formed crop, in contrast to the Coordinated Framework recommendations in 1986. Baseline studies are not available for most crops other than the major commodity crops, so comparative data must be generated for each new species moving through the regulatory system. The costs associated with each of these petitions for non-regulated status are staggering — in the tens of millions of dollars (McElroy, 2003) and possibly even more (The Economist, 2009). For academics or small crop developers, the possibility of the product of interest making it to market is very slim since each non-commodity crop such as peas, squash, okra, or tomato, has a small market and very little chance of recovering the money invested in deregulation.

A new paradigm is needed for data collection and coordination to simplify and standardize the petition process for federal agencies — fulfilling data needs for crop assessment while ensuring safety of the product. The barrier to regulatory approval for non-commodity or specialty crops was recognized by the USDA in the early 2000s. The Specialty Crops Regulatory Initiative (SCRI) was formed as the outcome of the workshop Public Research and the Regulatory Review of Small Market (Specialty) Biotechnology Derived Crops, which was held in November 2004 at the USDA APHIS headquarters in Riverdale, Maryland. This workshop was organized and initially funded through cooperation among several government agencies (http://www.csrees.usda.gov/nea/biotech/in_focus/biotechnology_if_smallmarket.html). A working group was established to continue discussing and revising models and plans for structuring an organization that would benefit the specialty crop community. It was and is essential to engage the stakeholders — specialty crop developers in the public and private sector — who have participated in this organization. A steering committee of volunteers was formed from the working group and an executive committee identified to continue working toward developing the SCRI concept. Follow-up workshops were held and initial funding to launch the SCRI concept was provided by the USDA Chief Economist and the USDA Office of the Secretary. Because this was a voluntary association, the steering committee used funds provided by the USDA to identify a consultant to help build the SCRI structure. One was engaged in 2006 to work with the group.

Because the SCRI acronym was used by a number of entities not associated with the Specialty Crops Regulatory Initiative, the steering committee approved changing the name of the group to Specialty Crops Regulatory Assistance (SCRA). The SCRA mission is to facilitate regulatory clearance for biotechnology-derived specialty crops, including assistance in the scientific and business categories as monetary and personnel resources permit. To accomplish the mission of this public sector assistance organization, private organizations must be designed and built to support the goals.

A recent study commissioned by the Specialty Crop Regulatory Assistance initiative (www.specialtycropassistance.org) indicated that there are 46 different crops and a multitude of unique biotechnology-derived traits in the development pipeline in the United States alone. Despite this robust pipeline, small crop developers have nicknamed the deregulation process the "Valley of Death" in the development of GE specialty crops because deregulation may be achieved only after a significant number of studies have been conducted. These may include molecular characterization of the genetic modification, compositional assessments, protein production and characterization, studies of impact on non-target organisms, and toxicology studies. These studies are expensive and time-consuming, and for developers of small market crops, prohibitive. Some studies could be expedited significantly if there was an established concept of "the normal range of values" for each crop characteristic that could be targeted.

Standardization

To facilitate the curation of data and develop extremely robust databases, standard formats for entering basic agronomic and compositional information about a GE crop would be extremely useful. Currently, there are no standard formats for reporting or describing the results of field and laboratory analyses of GE crops. Throughout the field of human genomics, there is a growing interest in developing standards for reporting information. For example, to ensure that data are captured electronically in a standard format, a Genomic Standards Consortium was launched and it proposed the minimum information about a genome sequence (MIGS) specification (Field et al. 2008). The FDA, in partnership with IBM, created JANUS (http://crix.nci.nih.gov/projects/janus/), a standards-based clinical data repository that forms the basis of a clinical trial review application. The objectives for JANUS include more efficient submissions-based, custom datasets for analysis; less data redundancy; less ambiguity in information; easier peer review; and the overall use of common standards. It would be useful to both the crop developer and the regulatory agencies if comparable standards were established to support the development of regulatory dossiers.

Conclusions

A basic road atlas from which a specific regulatory plan can be derived should be created. Certain data are required in every petition, and standardization of the reporting of that information will greatly simplify the process. Experiments still need to be conducted for each GE event, and how to collect the data and conduct the experiments can be a part of a customized regulatory plan based on the general regulatory atlas prepared by an independent group such as SCRA. With more products from GE plants on the market without incident, the social fears surrounding genetic engineering can be allayed.

In the end, the ultimate goal of genetic engineering of plants is to benefit society through improvements in agricultural practices, plant health, and plant productivity. Additional benefits can be reaped through improved nutrition and products derived from renewable plant resources. In all regulations of transgenic (GE) plants, the GE plant is compared to its non-GE counterpart. The question is whether or not this is the correct comparator or if the improvement in agronomic practices and productivity outweigh any potential drawbacks of the plants produced through GE technology. Agriculture itself has had an enormous impact on land use and biodiversity (Edwards and Hilbeck, 2001). The balance between positive and negative impacts of GE plants in the environment and on human health should be considered in this broader agronomic context to define the full costs and benefits of agricultural systems (Pretty, 2008). The goal should be to make the regulations reflect the measured rather than the perceived risks to allow the benefits of this new technology to be realized.

References

Akhond, M. A. Y., & Machray, G. C. (2009). Biotech crops: Technologies, achievements and prospects. *Euphytica*, *166*, 47–59.

Arundel, A., & Sawaya, D. (2009). Biotechnologies in agriculture and related natural resources to 2015. *OECD Journal: General Papers*, 7–111.

Barton, K. A., Binns, A. N., Matzke, A. J. M., & Chilton, M. -D. (1983). Regeneration of intact tobacco plants containing full length copies of genetically engineered T-DNA, and transmission of T-DNA to R1 progeny. *Cell*, *32*, 1033–1043.

Belson, N. A. (2000). US Regulation of agricultural biotechnology: An overview. *AgBioForum*, *3*, 268–280.

Bergmans, H. (2006). Basic framework for risk assessment for transgenic plants developed by the OECD: 20 years after the OECD "Blue Book". *Environmental Biosafety Research*, *5*, 213–219.

Brookes, G., & Barfoot, P. (2010). Global impact of biotech crops: Environmental effects, 1996–2008. *AgBioForum*, *13*, 76–94.

Carpenter, J. E. (2010). Peer-reviewed surveys indicate positive impact of commercialized GM crops. *Nature Biotechnology*, *28*, 319–321.

Chapotin, S. M., & Wolt, J. D. (2007). Genetically modified crops for the bioeconomy: Meeting public and regulatory expectations. *Transgenic Research*, *16*, 675–688.

Cowan, T. (2010). *Agricultural biotechnology: Background and recent issues. (Development, N. R. a. R.)*. Washington, DC: Congressional Research Service.

Devos, Y., Lheureux, K., & Schiemann, J. (2010). Regulatory oversight and safety assessment of plants with novel traits. In F. Kempken & C. Jung (Eds.), *Genetic modification of plants* (pp. 553–574). Berlin Heidelberg: Springer.

Economist. (2009). The parable of the sower. *The Economist*, *November 19*, 71–73.

Edwards, P. J., & Hilbeck, A. (2001). Biodiversity of agroecosystems: Past, present and uncertain future. In J. Nosberger, H. H. Geiger & P. C. Struik (Eds.), *Crop science: Progress and prospects* (pp. 213–229). London, UK: CABI Publishing.

Endres, A. (2005). Revising seed purity laws to account for the adventitious presence of genetically modified varieties: A first step towards coexistence. *Food Law and Policy*, *1*, 131–139.

EPA. (2005). *An examination of EPA risk assessment principles and practices. (Risk Assessment Task Force, O. o. t. S. A.)*. Washington, DC: Environmental Protection Agency.

European Commission. (2001). Directive 2001/18/EC of the European parliament and of the council of 12 March 2001 on the deliberate release into the environment of genetically modified organisms and repealing Council Directive 90/220/EEC. (EC, O. J.,). pp. 1–39.

European Commission. (2003). *Communication from the commission on the precautionary principle*. European Commission.

FDA. (1992). Statement of policy: Foods derived from new plant varieties. pp. 984–923 005, US Federal Register.

Fedoroff, N. V., Battisti, D. S., Beachy, R. N., Cooper, P. J. M., Fischhoff, D. A., Hodges, C. N., et al. (2010). Radically rethinking agriculture for the 21st century. *Science*, *327*, 833–834.

Fedoroff, N. V., & Brown, N. M. (2004). *Mendel in the ktichen: A scientist's view of genetically modified foods*. Washington, DC: Joseph Henry Press.

Field, D., Garrity, G., Gray, T., Morrison, N., Selengut, J., Sterk, P., et al. (2008). The minimum information about a genome sequence (MIGS) specification. *Nature Biotechnology*, *26*, 541–547.

Fraley, R. T., Rogers, S. B., & Horsch, R. B. (1983). Use of a chimeric gene to confer antibiotic resistance to plant cells. In K. Downey (Ed.), *Advances in gene technology: Molecular genetics of plants and animals: Proceedings of the Miami Winter Symposium* (pp. 211–221). Miami: Academic Press.

Fraley, R. T., Rogers, S. G., Horsch, R. B., Sanders, P. R., Flick, J. S., Adams, S. P., et al. (1983). Expression of bacterial genes in plant cells. *Proceedings of the National Academy of Sciences of the United States of America*, *80*, 4803–4807.

Gaines, T., Preston, C., Byrne, P., Henry, W. B., & Westra, P. (2007). Adventitious presence of herbicide resistant wheat in certified and farm-saved seed lots. *Crop Science*, *47*, 751–754.

Herrera-Estrella, L., Depicker, A., Van Montagu, M., & Schell, J. (1983). Expression of chimaeric genes transferred into plant cells using a Ti-plasmid-derived vector. *Nature*, *303*, 209–213.

Howard, J. A., & Donnelly, K. C. (2004). A quantitative safety assessment model for transgenic protein products produced in agricultural crops. *Journal of Agricultural and Environmental Ethics*, *17*, 545–558.

Howard, J. A., & Hood, E. (2005). Bioindustrial and biopharmaceutical products produced in plants. *Advances in Agronomy*, *85*, 91–124.

ILSI. (2004). Nutritional and safety assessments of foods and feeds nutritionally improved through biotechnology: An executive summary a task force report by the international life sciences institute Washington, D.C. *Comprehensive Reviews in Food Science and Food Safety*, *3*, 35–104.

ILSI. (2008). Nutritional and safety assessments of foods and feeds nutritionally improved through biotechnology: Case studies. *Comprehensive Reviews in Food Science and Food Safety*, *7*, 50–113.

James, C. (2009). *Global status of commercialized biotech/GM crops*. Ithaca, NY: ISAAA.

James, C. (2010). A global overview of biotech (GM) crops: Adoption, impact and future prospects. *GM Crops*, *1*, 8–12.

Johnson, K. L., Raybould, A. F., Hudson, M. D., & Poppy, G. M. (2007). How does scientific risk assessment of GM crops fit within the wider risk analysis? *Trends in Plant Science*, *12*, 1–5.

Kalaitzandonakes, N., & Bijman, J. (2003). Who is driving biotechnology acceptance? *Nature Biotechnology*, *21*, 366–369.

Kaskey, J., & Ligi, A. (2010). Monsanto, DuPont race to win $2.7 billion drought-corn market. *Bloomberg Businessweek*.

Kershen, D., & McHughen, A. (2005). *Adventitious presence: Inadvertent co-mingling and co-existence among farming methods in CAST commentary*. Ames, Iowa: Council for Agricultural Science and Technology.

Kingsbury, D. T. (1986). The regulatory "Coordinated Framework" for biotechnology. *Bio/Technology*, *4*, 1071–1073.

Lemaux, P. G. (2008). Genetically engineered plants and foods: A scientist's analysis of the issues (Part I). *Annual Review of Plant Biology*, *59*, 771–812.

Lemaux, P. G. (2009). Genetically engineered plants and foods: A scientist's analysis of the issues (Part II). *Annual Review of Plant Biology*, *2009*, 511–559.

McAllister, L. K. (2005). Judging GMOs: Judicial application of the precautionary principle in Brazil. *Ecology Law Quarterly*, *32*, 149.

McElroy, D. (2003). Sustaining agbiotechnology through lean times. *Nature Biotechnology*, *21*, 996–1002.

McHughen, A., & Smyth, S. (2008). US regulatory system for genetically modified organisms (GMO), rDNA or transgenic crop cultivars. *Plant Biotechnology Journal*, *6*, 2–12.

Miller, J. K., & Bradford, K. J. (2010). The regulatory bottleneck for biotech specialty crops. *Nature Biotechnology*, *28*, 1012–1014.

NAS. (2004). *Safety of genetically engineered foods: Approaches to assessing unintended health effects*. Washington: National Academies Press.

National Research Council. (1983). *Risk assessment in the federal government: Managing the process*. Washington, DC: National Academies Press.

National Research Council. (2009). *Science and decisions: Advancing risk assessment. (EPA, C. o. I. R. A. A. U. b. t. U.)*. Washington, DC: National Academies Press.

Nestmann, E., Copeland, T., & Hlywka, J. (2002). The regulatory and science-based safety evaluation of genetically modified food crops; A USA perspective. In K. T. Atherton (Ed.), *Genetically modified crops, assessing safety* (pp. 1–44). London, UK: Taylor and Francis.

Nickson, T. E. (2008). Planning environmental risk assessment for genetically modified crops: Problem formulation for stress-tolerant crops. *Plant Physiology*, *147*, 494–502.

OECD. (1986). *Recombinant DNA safety considerations*. Paris, France: Organization for Economic Co-operation and Development.

OECD. (1992). *Safety considerations for biotechnology*. Paris, France: Organization for Economic Co-operation and Development.

OECD. (1993). *Safety evaluation of foods derived by modern biotechnology: Concepts and principles*. Paris, France: Organization for Economic Co-operation and Development.

OSTP. (1986). Coordinated framework for regulation of biotechnology: Announcement of policy and notice for public comment. *Federal Register*, *51*, 302–323. 393

Pretty, J. (2008). Agricultural sustainability: Concepts, principles and evidence. *Philosophical Transactions of the Royal Society B*, *363*, 447–466.

Raybould, A. (2006). Problem formulation and hypothesis testing for environmental risk assessments of genetically modified crops. *Environmental Biosafety Research*, *5*, 119–125.

Raybould, A. (2010). Reducing uncertainty in regulatory decision-making for transgenic crops: More ecological research or shrewder environmental risk assessment? *GM Crops*, *1*, 1–7.

Raybould, A., Tuttle, A., Shore, S., & Stone, T. (2010). Environmental risk assessments for transgenic crops producing output trait enzymes. *Transgenic Research*, *19*, 595–609.

Romeis, J., Lawo, N. C., & Raybould, A. (2009). Making effective use of existing data for case-by-case risk assessments of genetically engineered crops. *Journal of Applied Entomology*, *133*, 571–583.

Rommens, C. M. (2009). Barriers and paths to market for genetically engineered crops. *Plant Biotechnology Journal*, *8*, 101–111.

Schell, J., Van Montagu, M., Holsters, M., Zambryski, P. H. J., Inze, D., Herrera-Estrella, L., et al. (1983). Ti plasmids as experimental gene vectors for plants. In K. Downey (Ed.), *Advances in gene technology: Molecular genetics of plants and animals: Proceedings of the Miami Winter Symp* (pp. 191–209). Miami: Academic Press.

Sharma, H. C., Crouch, J. H., Sharma, K. K., Seetharama, N., & Hash, C. T. (2002). Applications of biotechnology for crop improvement: Prospects and constraints. *Plant Science*, *163*, 381–395.

Singer, M. F., & Soll, D. (1973). Guidelines for DNA hybrid molecules. *Science*, *181*, 1114.

Smyth, S., & McHughen, A. (2008). Regulating innovative crop technologies in Canada: The case of regulating genetically modified crops. *Plant Biotechnology Journal*, *6*, 213–225.

Tolin, S. A., & Vidaver, A. K. (1989). Guidelines and regulations for research with genetically modified organisms: A view from academe. *Annual Review of Phytopathology*, *27*, 551–581.

Trewavas, A., & Leaver, C. (2001). Is opposition to GM crops science or politics? An investigation into the arguments that GM crops pose a particular threat to the environment. *EMBO Reports*, *2*, 445–459.

Tzotzos, G. T., Head, G. P., & Hull, R. (2009). *Genetically modified plants: Assessing safety and managing risks*. San Diego: Academic Press.

Wolt, J., Keese, P., Raybould, A., Fitzpatrick, J., Burachik, M., Gray, A., et al. (2010). Problem formulation in the environmental risk assessment for genetically modified plants. *Transgenic Research*, *19*, 425–436.

Prospects for increased food production and poverty alleviation: What plant biotechnology can practically deliver and what it cannot

35

Martina Newell McGloughlin

Director, University of California Systemwide Biotechnology Research and Education Program (UCBREP), University of California Davis, California

TABLE OF CONTENTS

Introduction

During the coming decades, food and agricultural production systems will need to be significantly enhanced to respond to a number of remarkable changes, such as a growing world population; increasing international competition; globalization; shifts to increased meat consumption in developing countries and rising consumer demands for improved food quality, safety, health enhancement, and convenience. New and innovative techniques will be required to ensure an ample supply of healthy food by improving the efficiency of the global agriculture sector. Innovation is essential for sustaining and enhancing agricultural productivity. This involves new, science-based products and processes that contribute reliable methods for improving quality, productivity, and environmental sustainability. Biotechnology has introduced a new dimension to such innovation, offering efficient and cost-effective means to produce a diverse array of novel, value-added products and tools. It has the potential to improve qualitative and quantitative aspects of food, feed, fiber, and biofuel production; reduce the dependency of agriculture on chemicals and fossil fuels; diminish over-cultivation and erosion; and lower the cost of raw materials, all in an environmentally sustainable manner. Commercialization of the first generation of products of recombinant DNA technology was another facet in a long history of human intervention in nature for agricultural and food production purposes. As such, the same parameters of risk-based assessment should apply. Commercialization of products must be undertaken within a regulatory framework that ensures adequate protection of the consumer, the environment, and alternate production systems, while not stymieing innovation.

In a world whose population is increasing disproportionally in disadvantaged regions, it is hard to envisage feeding and sustaining these numbers in a livable environment without the use of biotechnology. From 1800 onward, more food was simply produced by plowing up virgin land and forest. The land area used for farming increased about five-fold up to the middle of the twentieth century in step with population increases. The Green Revolution put a brake on this expansion, increasing yields three-fold with limited need for further expansion. Since 1950, the proportion of the land devoted to farming has barely increased, even though the world population doubled over the same period. We currently use at least half the available good quality soil for agriculture, with the remainder under tropical forests. Couple this with the ever-diminishing, non-renewable resources and the compounding effects of climate change on the limitation of land usage and this leads to an obvious dilemma. Unless we can pull off a second Green Revolution, increasing yield but limiting it to land currently used for farming, there will be further deterioration of natural habitats and biodiversity that may threaten more than our lifestyles.

During the global food crisis in 2008, which was erroneously laid disproportionately on the shoulders of biofuel production and most especially grain ethanol, the Gates Foundation announced $306 million in grants to boost agricultural yields in the developing world, with nearly $165 million to replenish depleted soils in Africa. These efforts are not without controversy: critics charge that western philanthropists are violating African "food sovereignty" and promoting American at the expense of peasant farmers knowledgeable about local practices. But local practices have yielded scarcity. A farmer in India grows three to four times as much food on the same amount of land as a farmer in Africa, and a farmer in China, roughly seven times as much.

The Food and Agricultural Organization (FAO) reports that global demand for food could easily double between 2000

DOI: 10.1016/B978-0-12-381466-1.00035-3

and 2050, with a two-and-a-half- to three-fold increase in the poorest countries (FAO/WHO, 2004). They found that biotechnology and genetic engineering of crops hold great promise for agriculture in developing countries. The report noted that more than 70% of the world's poor still live in rural areas and depend directly on agriculture for their survival. The World Health Organization (WHO) estimates that 800 million people worldwide suffer from malnutrition. It is difficult to imagine a promising alternative to biotechnology and industrial agriculture that will sustain such numbers without catastrophic consequences. As far back as 2004 the Economic and Social Council of the United Nations (ECOSOC) noted that most developing countries are unlikely to meet the Millennium Development Goals without a clear political commitment to making science and technology among the top priorities in their development agenda (Economic and Social Council Resolution, 2004). FAO members called for strengthening efforts in maximizing the benefits and minimizing the potential adverse consequences of biotechnology, through the development of a multidisciplinary, cross-sectoral program in conjunction with the committee on Agriculture and the council. In response, the Biotechnology Applications in Food and Agriculture, Forestry, and Fisheries Priority Area for Interdisciplinary Action (Biotech-PAIA) was established and an Interdepartmental Working Group on Biotechnology was set up to oversee its planning and implementation. Prior to that the U.S. National Academy of Sciences, joined by six other academies from around the world (Royal Society of London, Third World Academy of Sciences, and national academies of Brazil, China, India, and Mexico) issued a report in 2000 declaring that biotechnology should be used to increase the production of main food staples, improve the efficiency of production, reduce the environmental impact of agriculture, and provide access to food for small-scale farmers. Without question agricultural research of all forms holds an important key to meeting these needs, as the FAO noted biotechnology can accelerate conventional breeding programs and may offer solutions where conventional methods fail. That is a positive for growers, consumers, and clearly contributes to environmental sustainability.

Progress to Date

Modifications of crop plants can be organized into two broad-based, non-mutually exclusive categories: those that benefit the producer; and those that benefit the consumer. Modifications that protect the crop from either biotic or abiotic stress (biotic stress is damage by predators such as insects and nematodes, and disease agents such as viruses, fungi, bacteria, and weeds; abiotic stress comes in the form of drought, cold, heat, and poor soils) or increase total crop yield benefit the producer, and are called "input traits." The majority of modified crops in commercial use fit in this group. Scientists have just begun to tap the large potential of biotechnology to produce varieties of plants that confer a wide spectrum of advantages to consumers. These varieties are modified with "output traits." Developing and commercializing plants with these improved traits involves overcoming a variety of technical, regulatory, and perception challenges inherent in perceived and real challenges of complex modifications. Both the panoply of traditional plant breeding tools and modern biotechnology-based techniques are required to produce plants with the desired quality traits. In addition to the older gene transfer technology where mostly single genes were modified, newer techniques such as the use of RNA interference to manipulate endogenous genes, and especially the use of transcription factors to modulate whole suites of genes and metabolic networks will become increasingly important tools in the effort to introduce valuable traits. The latter approach is already a major focus in multigenic and quantitative traits, such as developing stress tolerance crops and modifying paths for improving nutritional characteristics.

Since the first biotech crop was commercialized in 1996, genetically modified (GM) crops have grown commercially by 15.4 million farmers in 29 countries on 366 million acres (James, 2011). More than half of the 63 countries engaged in biotech research, development, and production are developing countries. Whereas North America still leads with U.S. acreage accounting for about 45% of the total acreage worldwide, 19 of the 29 countries are developing countries, and of the 15.4 million farmers that grew these crops, a full 14.4 million (90%) are resource-poor LDC farmers. The most recent countries joined this group in 2009 including South America (Uruguay, Paraguay, and Bolivia) and Africa (Egypt and Burkina Faso). In 2010, three countries planted approved biotech crops for the first time and Germany resumed planting. Pakistan planted Bt cotton, as did Myanmar, and notably Sweden (the first Scandinavian country to plant a biotech crop) planted "Amflora," a potato with high amylase starch for industrial applications. Germany also resumed adoption of biotech crops by planting Amflora for a net gain of four countries in 2010 (James, 2011). The first generation of such crops focused largely on input agronomic traits; the next generation will focus more on value-added output traits. In the next decade some studies estimate the global value of biotech crops will increase nearly five-fold to $210 billion.

Agricultural biotechnology has helped farmers around the world boost their productivity and grow crops in more ecologically healthy fields while allowing much more efficient use of resources. This technology allows reduced tillage, which cuts down on greenhouse gas emissions, water runoff, machinery use, and soil erosion. Meanwhile, the benefits experienced by larger scale farmers in both industrialized nations and lesser developed countries are already considerable. Research by Brookes and Barfoot (2005, 2008) shows that in the first 11 years of GM crop cultivation, global net farm income increased by $33.8 billion — the environmental footprint associated with pesticide use was reduced by 15.4%, and there was a reduction in carbon dioxide emissions in 2006 equivalent to taking nearly 6.6 million cars off the road for a year. More recent assessments indicate a 35% reduction in the footprint for corn and 25% in the footprint for cotton.

An earlier study by researchers at Denmark's National Environmental Research Institute (NERI) monitored fields of conventional and Glyphosate-tolerant sugar beet. They found that the GM plots supported more plant species and insects than the conventional plots, thus providing more food for birds and other types of wildlife. Use of transgenic crops increased biodiversity compared to traditional herbicide treatments (Elmegaard and Pedersen, 2001). Proper measurements in the UK indicate that no-till (directly compared with plowed organic fields on the same farm and using the same farmer) uses only one-third the fossil fuel, uses land much more efficiently, reduces nitrate (and pesticide) runoff by at least half, and increases soil carbon that is lost when plowed. In addition, bird territories are orders of magnitude higher; soil erosion almost vanishes; soil invertebrates such as earthworms soar in numbers, as do predatory arthropods to keep pests down. Organic fields in the UK see a three-fold rise in weeds on conversion that necessitates use of the plow (Trewavas, 2001).

Therefore reduced-till agriculture means healthier soil with reduced erosion and far less carbon dioxide release. Soil carbon sequestration will be an important part of any international strategy to mitigate the increase in atmospheric CO_2 concentrations. By adopting more sustainable management practices, agriculture may play a large part in enhancing soil carbon sequestration across the globe. One way is by reducing the amount of conventional tillage; after long-term tillage soil carbon stocks are depleted. In general, cultivation is not a sustainable practice. It is energy intensive and exposes soil to wind and water erosion. It allows rain to compact the soil, increasing the oxygen content and allowing organic matter to oxidize away. In turn, lower organic matter in the soil allows more compaction and more nutrient loss. Additionally, in warmer and drier climates evaporative water loss may be reduced as residue remains on the soil surface creating a wetter and cooler soil microclimate.

The 2008 Brookes and Barfoot study indicated that, with the adoption of Br cotton and maize, pesticide use fell by over 286 million kg (−7.8%; which is equivalent to about 40% of the annual volume of pesticide active ingredient applied to arable crops in the EU). Less spraying means fewer tractor passes, contributing to lower carbon dioxide emissions. Insect-resistant Bt maize also has a collateral effect — less insect damage results in much less infection by fungal molds, which reduces mycotoxins that are known health risks that may result in the development of liver cancer in humans and animals. Bt corn resulted in a 90% reduction in mycotoxin fungal fumonisins. In addition to the obvious health benefits, the total U.S. economic benefit is estimated to be approximately $23 million annually (Wu, 2006). The only "natural" way to control those fungi is the use of copper sulfate, which has one of the highest toxic hazard ratings of acceptable pesticides and selects for antibiotic-resistant bacteria in the soil.

A 2005 paper from the Royal Society suggested that intensive high-yield farming on less land is better for wildlife than "wildlife friendly," less efficient farming (Green et al., 2005). The paper provided convincing evidence that without yield increase land use will double by 2050, and this effect will be especially significant in developing countries like China and India where, without greater productivity, they will need four times the land area to support their expanding populations. They show that in Latin America where increased productivity was achieved there was a significant decrease in deforestation; those producers with greatest yield increase had lower land use.

While North America remains the epicenter for cutting edge GM research, other regions, such as China, are emerging as contenders on the global stage. Agricultural science is now China's fastest growing research field, with China's share of global publications in agricultural science growing from 1.5% in 1999 to 5% in 2008 (James, 2011). China's early experience with Bt cotton demonstrated the direct and indirect benefits of its investment in plant biotechnology research and product development. In 2002, Bt cotton was grown in 2.1 million hectares by around five million farmers. At that time the average Bt cotton farmer had reduced pesticide sprayings for the Asian bollworm from 20 to 6 times per year. Pesticide applications were reduced by 59–80% compared to conventional cotton (assessed in 3 years of use) and Bt cotton farmers produced a kilogram of cotton for 28% less cost than the farmers using non-Bt varieties. Net revenues increased by $357–549 ha compared to conventional cotton (assessed in 3 years of use; Huang et al., 2002). Ultimately, however, it is the social benefits from reducing exposure to insecticides and saving lives that is the real payoff.

The demand for productivity-enhancing technologies by farmers and for cost savings by consumers, the rate of increase in research investments, and success with Bt cotton suggest that products from China's research program will one day become widespread inside China. Indeed, China is emerging as one of the trendsetters in the adoption of novel traits as the country is setting the pace for new approvals for Bt rice and phytase maize that were accepted on November 27, 2009 (James, 2011). Rice is the principal staple for much of the world and maize is the largest animal feed source. Bt rice has the potential to increase yields up to 8%, decrease pesticide use by 80% (17 kg/ha), and generate $4 billion in benefits annually. The phytase approval is a major step forward in approvals as it is the first since the Flavr Savr tomato focusing on a "quality" trait. However, it is far more than this both literally and figuratively since this single trait addresses several issues from nutritional to environmental, as will be expanded on later.

The first GM crop to be released for commercial cultivation in India was Bt cotton, which was developed by the Maharashtra Seed Company (Mahyco) in partnership with Monsanto. The approval, given in 2002, came after several years of field trials following the biosafety procedures laid down by the government. Three cotton hybrids were granted permission for field sowing in six states for three years. For the first season, farmer demand for Bt cotton seed was very high; it is estimated that 44,500 hectares of certified Bt cotton were planted by nearly 55,000 farmers. However, the initial events thrived in regions that resembled the area in which they were originally developed, but did not perform as well in growing regions with disparate climate challenges. It was not until the trait was introgressed into locally adapted varieties that Bt cotton thrived in all growing regions. Between 2005 and 2006 the biggest impact of this approach was realized. From 3 Bt cotton hybrids in 2002 to 62 in 2006 the rapid deployment of Bt cotton hybrids based on different agro-climatic conditions resulted in decreased insecticide sprays by 39%, and increased yield by 31%, resulting in increased profit per hectare of 88% or $250. Over this period of rapid deployment the average cotton yields increased from 308 kg/ha to 450 kg/ha of lint (of this increase 50% could be attributed to Bt technology). Also over this period raw cotton exports rose from 0.9 million bales in 2005 to 4.7 million in 2006 and 5.9 by 2007 (Huang et al., 2002). By 2009 5.6 million resource-poor farmers in India planted 8.4 million hectares of Bt cotton, equivalent to 87% of the 9.6 million hectare national cotton crop. The increase from 50,000 hectares when Bt cotton was first commercialized in 2002 to 8.4 million hectares in 2009 represents an unprecedented 168-fold increase in 8 years. Between 2002 and 2008, Bt cotton generated economic benefits for farmers valued at $5.1 billion, halved insecticide requirements, contributed to the doubling of yield, and transformed India from a cotton importer to a major exporter. Choudhary contends that the deployment of Bt cotton over the last eight years has resulted in India becoming the number one exporter of cotton globally, as well as the second largest cotton producer in the world (Choudhary and Gaur, 2010).

However, despite the success of Bt cotton, the expected successful commercialization of Bt eggplant never materialized as an effective opposition managed to scupper its approval. Bt eggplant, or brinjal as it is referred to in India, was found to be effective against fruit and shoot borer (FSB), with 98% insect mortality in shoots and 100% in fruits compared to less than 30% mortality in non-Bt counterparts. The multi-location research trials confirmed that Bt brinjal required, on average, 77% less insecticides than non-Bt counterparts for control of FSB, and 42% less for the control of all insect pests of brinjal. The benefits of Bt brinjal translate to an average increase of 116% in marketable fruits over conventional hybrids, and a 166% increase over popular open-pollinated varieties (Choudhary and Gaur, 2008). Furthermore, the significant decrease in insecticide usage reduced the farmers' exposure to insecticides and resulted in a substantial decline in pesticide residues on brinjal fruits. Scientists have estimated that Bt brinjal will deliver farmers a net economic benefit ranging from $330 to 397 per acre with national benefits to India exceeding $400 million per year. However, in February 2010, the environmental minister announced a moratorium citing that "There is no overriding food security argument for Bt brinjal. Our objective is to restore public confidence and trust in Bt brinjal" (Bricknel, 2010). This statement clearly demonstrated that the decision was not based on scientific analysis or risk assessment but rather on unsupported and ill-defined concerns which should never form the basis of regulatory oversight.

A number of other multi-institutional projects have also been launched in India, including the development of transgenics for resistance to geminiviruses in cotton, mung bean, and tomato; resistance to rice tungro disease; development of a nutritionally enhanced potato with a balanced amino acid composition; and development of molecular methods for heterosis breeding. Other transgenic crops that are awaiting approval for commercial cultivation include transgenic herbicide-tolerant mustard hybrids and nutritionally enhanced potato varieties. Despite the resounding success of Bt cotton, and given the experience with Bt brinjal, it is difficult to be optimistic about the prospects for commercialization of food crops.

A somewhat similar but even more insidious situation was experienced by Egypt. In 2008, Egypt became the first country in the Arab world (and only one of three in Africa) to commercialize biotech crops by planting 700 hectares of Bt yellow maize. The variety commercialized, Ajeeb-YG, is a cross between MON 810 and an Egyptian maize variety with resistance to three corn borer pests. The reason that the amount was so low can in part be attributed to the fact that France's President Sarkozy had intimated that Egypt's deal with France may be in jeopardy if they pursued growing their approved Bt maize.

The Next Generation

The vast majority of products approved to date are in the area of agronomic traits, most specifically biotic stress. The principal focus in the immediate future will remain on agronomic traits, especially in the area of pest control, but with an increasing interest in abiotic stress tolerance, which is gaining prominence as external pressures from climate change to land use change are rising.

On the biotic stress tolerance side the focus is expanding to multitiered control systems. This, in theory, serves a double advantage: primarily expanding the effectiveness of the broad based resistance events, but also allowing more effective management of the resistance trait since there is less selective pressure when genes are stacked. SmartStax, an eight-trait event developed through collaboration between Monsanto and Dow, takes advantage of multiple modes of insect protection and herbicide tolerance against above- and below-ground insects and provides broad herbicide tolerance, including Yieldgard VT Triple (Monsanto), Herculex Xtra (Dow), RoundUp Ready 2 (Monsanto), and Liberty Link (Dow). SmartStax is currently available for corn, but cotton, soybean, and specialty crop variations are to be released. It is estimated that this should require only 5% refuge acres as opposed to the 20% required of older technologies to mitigate against pest tolerance (Monsanto News Release, 2010).

On the second area of agronomic traits, namely abiotic stress, there is a meta-issue that overlays much of the individual efforts — climate change. This poses a real challenge in terms of available agricultural land and fresh water use. Apart from the obvious effects, the decline of crop yields, ocean acidification, poor nutrition and abiotic stress, population displacement, and threatened ecosystems are effects underlined by the Stern Report (Stern, 2006) as potential consequences of climate change. In addition there are also broader, more systemic effects of drought beyond food insecurity, such as decreased household income, the loss of assets due to slaughter of livestock, health threats due to the lack of water for hygiene and household uses, environmental degradation, and less sustainable land management .

These effects should be considered in the light of growing population levels.

To feed the overall population, the world will have to double its rate of agricultural production over the next 25 years, despite having already quadrupled it in the last 50 years. Severe drought accounts for half of the world's food emergencies annually (World Bank, 2007). In 2003, the World Food Program spent $565 million in response to drought in Sub-Saharan Africa (SSA). In this context solutions must be developed to adapt crops to existing but also evolving conditions, such as marginal soils or harsher conditions (cold, heat, drought, and salinity). The agriculture sector is both a contributor and provider of potential solutions to this phenomenon. It impacts two of the principal components of climate change: greenhouse gases and water. Agriculture is a major source of the former emissions. Practices such as deforestation, cattle feedlots, and use of fertilizer currently account for about 25% of greenhouse gas emissions. When broken down — this amounts to 14% of total carbon dioxide emission, 48% of methane, and 52% of all nitrous oxide emissions (Stern, 2006). In addition this sector uses a significant amount of available fresh water; approximately 70% of water currently consumed by humans is used in agriculture, and this is likely to increase as temperatures rise. The impact between resource-poor LDCs and developed countries is likely to be asymmetrical.

Given the potential impacts of climate change on the range and extent of agricultural productivity and the impact of agricultural practices on global warming, techniques should play a substantial part in mitigating against climate change. This is especially relevant in emerging countries where producers and consumers are subjected more to the vagaries of climate fluctuations than in the west, where there is a greater ability to respond to the effects and manage resources. Green biotechnology offers a set of tools that can help producers limit greenhouse gas emissions as well as adapting their agricultural techniques to shifting climates. The three major contributions of green biotechnology to the mitigation of the impact of climate change are greenhouse gas reduction, crop adaptation (environmental stress, changing niches), and protection and yield increase in less desirable and marginal soils.

On the first of these issues, greenhouse gas reduction in addition to carbon dioxide agriculture contributes to two of the other major gases; indeed, one of them is nitrous oxide, which has a global warming potential of about 300 times that of carbon dioxide. In addition, nitrous oxides stay in the atmosphere for a considerable period. Nitrous oxide is produced through bacterial degradation of applied nitrogen fertilizer. In addition, fertilizer can contribute to eutrophication at ground level so its reduction is desirable on several levels. However, nitrogen is essential for crop production, because it is quantitatively the most essential nutrient for plants and a major factor limiting crop productivity. One of the critical steps limiting the efficient use of nitrogen is the ability of plants to acquire it from applied fertilizer. Therefore, the development of crop plants that absorb and use nitrogen more efficiently can serve both the plant and the environment. Arcadia Biosciences of Davis, California, developed nitrogen-efficient crops by introducing a barley AlaAT (alanine aminotransferase) into both rice and canola. Arcadia's Nitrogen Use Efficiency (NUE) technology produces plants with yields that are equivalent to conventional varieties, but require significantly less nitrogen fertilizer because the AlaAT gene allows more efficient use. Compared with controls, transgenic plants also demonstrated significant changes in key metabolites and total nitrogen content, confirming increased nitrogen uptake efficiency. This technology has the potential to reduce the amount of nitrogen fertilizer that is lost by farmers every year from leaching into the air, soil, and waterways. In addition to environmental pressures, nitrogen costs can represent a significant portion of a farmer's input costs and can significantly impact farmer profitability. Farmers spend $60 billion annually for 150 million tons of fertilizer. This technology has been licensed to DuPont for maize and to Monsanto for application in canola.

The second area where green technology can help in a changing climate is crop adaptation to environmental stress and changing niches. Under stress plants will divert energy into survival instead of producing biomass and reproduction, so addressing this impact should have substantial effect on yield. In addition, improved stress tolerance allows an expanded growing season (especially earlier planting) and further reduces yield variability and grower financial risk. The most critical of these stresses is water. One of the most effective methods of addressing water limitation problems (irrigation) unfortunately is also one of the major causes of arable land degradation. It is estimated that 24.7 million acres of farmland worldwide is lost each year due to salinity build-up resulting from over irrigation. Crops are now limited by salinity on 40% of the world's irrigated land (25% in the United States). To address the extreme end of irrigation impact Eduardo Blumwald at University of California Davis used AtNHX1, the most abundant vacuolar Na^+/H^+ antiporter in *Arabidopsis thaliana*, which mediates the transport of Na^+ and K^+ into the vacuole. By over-expressing this vacuolar Na^+/H^+ antiporter, transgenic tomatoes were able to grow, flower, and produce fruit in the presence of 200 mM sodium chloride (Sottosanto et al., 2007). Arcadia Biosciences introduced this gene into economically important crops.

Even at a more moderate level of impact it is estimated that about 70–80 million acres in the United States suffer yield losses due to moderate water stress. The most critical time for water stress is near pollination and flowering, where yields with or without irrigation can vary by up to 100%. This effect is clearly demonstrable in dry land production where yields can be cut in half in the absence of irrigation. At this time about 15% of U.S. maize acres are irrigated. Given the negative effective and cost of irrigation, it is estimated that 20 million acres in the United States would benefit from a drought tolerance gene that gives a 10% yield increase. It would also allow shifting of high-value crops into production on more marginal land.

One of the first commercialized products to have included a "yield gene" is Monsanto's second generation Roundup Ready 2 Yield® soybeans, which include the Glyphosate-tolerant trait and was developed using extensive gene mapping to identify specific DNA regions that segregated with yield increase. It is a perfect example of the power of combining recombinant DNA technology with genomics tools. Monsanto claims that following four years of field trials across six states in the United States showed 7–11% higher yields, compared to the first generation of Roundup Ready soybeans. At the National Technical Biosafety Committee (CTNBio) meeting in Brazil in August 2010, the committee approved the Bt-enhanced version of this product for planting in Brazil (Monsanto News Release, 2010).

As noted, transcription factors are some of the most versatile tools employed in developing stress-tolerant plants. One of the most versatile classes of transcription factors regarding environmental response is the dehydration-responsive element binding (DREB) protein transcription factors, which are involved in the biotic stress signaling pathway. They can also activate as many as 12 resistant functional genes relying on DRE members of *cis*-regulation under adverse conditions (e.g., rd29, cor15, and rd17) and cause proline content to rise enabling plants to improve in many traits such as drought, freezing, and salt tolerance. It has been possible to engineer stress tolerance in transgenic plants by manipulating the expression of DREBs (Agarwal et al., 2006). One DREB isolated from *Arabidopsis* has improved drought tolerance by increasing productivity by at least two-fold during severe water stress. In Monsanto field trials using this approach, maize yields have increased under water stress by up to 30%.

Other approaches include modification of individual genes involved in stress response and cell signaling. For example, drought-tolerant canola engineered to reduce the levels of poly(ADP-ribose) polymerase (PARP), a key stress-related protein in many organisms, shows relative yield increases of up to +44% compared to control varieties. A subset of the transcription factor's homeodomain leucine zipper proteins (HDZip) plays a role in regulating adaptation responses including developmental adjustment to environmental cues such as water stress in plants (Deng et al., 2006). One of these effectors is abscisic acid (ABA), an important plant regulator controlling many environmental responses including stomata movement, which is modulated by the DREB elements. Some work is being done on modifying HDZip directly and others are working indirectly such as downregulating farnesyltransferase, a signaling system in the production of abscisic acid and stomata control that results in stomata closure and water retention.

Eduardo Blumwald is also working on modifying basic acid to enhance the tolerance of plants to water deficiency by delaying the drought-induced leaf senescence and abscission during the stress episode. Using tobacco plants expressing an isopentenyltransferase (IPT) gene under the control of a stress- and maturation-induced promoter (PSARK), they showed that delayed drought-induced leaf senescence resulted in remarkable drought-tolerant phenotypes, as well as minimal yield loss when plants were watered with only 30% of the water used under controlled conditions (Rivero et al., 2009). This is now being introduced into rice among other crops. This work is done in conjunction with Arcadia Biosciences. In addition Bayer CropScience, Pioneer Hi-Bred, BASF, and Dow among others, are conducting research on maize, cotton, canola, and rice to develop a new generation of stress-tolerant, high-performance crop varieties. Clearly stress-tolerant traits are of paramount importance in LDCs, especially in SSA and Asia. Major efforts are already underway on this front. The partnership, known as Water Efficient Maize for Africa (WEMA), was formed in response to a growing call by African farmers, leaders, and scientists to address the devastating effects of drought on small-scale farmers. Frequent drought leads to crop failure, hunger, and poverty. Climate change can only aggravate this situation (African Agricultural Technology Association, 2007).

On the other end of the spectrum of climate change impact is flooding due to changing rain patterns and rising sea levels. This is already a major cause of rice crop loss. It is estimated that 4 million tons of rice, which is sufficient to feed 30 million people, are lost every year because of flooding. Rice is not grown in flooded fields by necessity but to control weeds; however, most rice varieties die after more than three days of complete submergence. FR13A is one rice variety that researchers know can tolerate flooding for longer periods, but conventional breeding failed to create an event that was acceptable to farmers. The Ronald laboratory at University of California Davis cloned the submergence tolerance (Sub1) locus from this resistance variety using a map-based cloning approach. The Sub1 locus encodes three putative transcription regulators, one of which increases dramatically in response to oxygen deprivation in Sub1 seedlings; whereas Sub1C levels decrease. Transgenic lines over-expressing the Sub1A-1 gene have been introgressed into a submergence intolerant line and display enhanced submergence tolerance (Ki-Hong, 2010).

There is also some research in the final abiotic stress focus area, namely expansion of crops into, and increased yield from, less desirable and marginal soils. For example, a gene that produces citric acid in roots can protect plants from soils contaminated with aluminum as it binds to the contaminant preventing uptake by the root system (Lopez-Bucio et al., 2000). Genes such as these allow crops to be cultivated in hostile soils and temperatures, increasing geographic range while reducing potential impact on fragile ecosystems.

All of this is exciting and very relevant, but research in abiotic stress tolerance is still an input trait. The first generation of biotechnology crops focused largely on those input agronomic traits, and the next generation will focus more on value-added output traits. This will include identifying and isolating genes and metabolites that will make the enhancement of valuable traits possible, with some of the later compounds produced in mass quantities for niche markets. Two of the more promising markets are improved nutrition including nutraceuticals, or so-called "functional foods," and plants developed as bioreactors (production factories) for the commercial level production of valuable proteins and compounds, a field known as plant molecular farming (Newell-McGloughlin, 2005). Functional foods are defined as any modified food or food ingredient that may provide a health benefit beyond the traditional nutrients it contains. Scientific evidence is accumulating to support the role of phytochemicals and functional foods in the prevention and treatment of disease. Epidemiological research has shown a positive association with dietary intake of food components. Developing plants with improved quality traits involves overcoming a variety of technical challenges inherent to metabolic engineering programs. Both traditional plant breeding and biotechnology techniques are needed to produce plants carrying the desired quality traits. Examples include improvement of nutritional quality at the macro- (protein, carbohydrates, lipids, fiber) and micro-level (vitamins, minerals, phytochemicals), and amelioration of anti-nutrients, allergens, and toxins (Newell-McGloughlin, 2008). Examples include pro-vitamin A enhanced rice, improved protein content maize and sweet potato, and increased vital minerals in rice and maize.

In addition to functional foods, this area has the potential to address both nutritional and environmental impact. A good example of this is the addition of transgenic phytase enzymes to crops to reduce the need to add phosphate to feed (James, 2011). Most of the phosphate is added to counteract the non-bioavailability of phosphorus in phytic acid and the sequestering effect of phytic acid on uptake of divalent mineral ions such as iron, calcium, and zinc. Unfortunately, excess phosphate is excreted causing major environmental impact through eutrophication and fish kills in regions with intense pig and poultry farming (Zhang et al., 2000; Lucca et al., 2002). In addition, in humans such mineral deficiencies due to phytate binding are estimated to afflict 2–3 billion people, primarily in the developing world. Several studies have shown that *Aspergillus*-derived phytases can be produced in large quantities in a range of plants including cereals, with clear-cut positive effects on phytate degradation and phosphate and mineral bioavailability in animal-feeding trials (Zhang et al., 2000; Lucca et al., 2002). It is thus conceivable that genetic engineering of staples for increased phytase expression could potentially improve iron and zinc bioavailability, alleviating the need for supplementation in all monogastrics and consequent reduction in polluting run-off in non-ruminant animals. As noted earlier, China leads the way in approval of this "output" trait in maize as the first country to approve commercialization in November 2009 (James, 2011). Continuing improvements in molecular and genomic technologies are contributing to the acceleration of such product development. One estimate states that foods used for functional purposes made up 10% of the $503 billion total U.S. retail food market.

In addition to being a source of nutrition, plants have been a valuable wellspring of therapeutics for centuries. During the past decade, however, intensive research has focused on expanding this source through rDNA biotechnology and essentially using plants and animals as living factories for the commercial production of vaccines, therapeutics, and other valuable products such as industrial enzymes and biosynthetic feedstocks. More pressingly, with the increasing costs in economic and environmental terms of our dependency on fossil fuels, biotechnology offers innovative means to improve plant material for biomass conversion and enzymes to do the converting.

Barriers to Introduction

Most of the crops approved to date appear to support the notion that the deregulation process is prohibitive for any but well-financed companies whose focus is primarily the large commodity crops that were previously discussed. Worldwide there is clear asymmetry and lack of consensus in regulatory systems. This discourages research on anything but the most mundane of crops and traits, and is a real disincentive to creative research. For all intents and purposes there is just one trait from a public institution that has successfully traversed the regulatory minefields and been translated into a commercially viable commodity — the viral coat protein protection system initially developed for the papaya ringspot virus (PRSV) pandemic in Hawaii. Papaya is a major tropical fruit crop in the Asian region; however, production in many Asian countries is set back by the prevalence of the PRSV disease, as well as postharvest losses. The PRSV-resistant papaya, based on RNAi suppression of the coat protein expression, literally saved the $17 million economy in Hawaii and is of significant importance in Taiwan and other southeast Asian countries. Coat-protein-based resistance is a demonstration of what is known as post-transcriptional gene silencing (PTGS). RNA interference (RNAi) in animals and basal eukaryotes, quelling in fungi, and PTGS in plants are examples of a broad family of phenomena collectively called RNA silencing. This system has now been applied to many species. A five-year effort to combat plum pox virus disease through PTGS resistance paid off. In 1990, USDA/Agricultural Research Service (ARS) scientists began their efforts with a PRSV coat protein gene obtained from Dennis Gonsalves (1998). This gene shows 70% homology to the plum pox gene and has been used to control other viruses similarly related to papaya ringspot. However, irrespective of the mechanism, it is important that resistance based on a single gene is managed well and alternate control mechanisms are introduced to reduce pressure on the development of viral resistance. Other approaches include expression of the RNA replicating enzymes of the virus, expression of satellite RNA, replicating RNA molecules that are molecular parasites of the virus, or the use of protease inhibitors to interfere with processing of the viral proteins.

Rather interestingly, it has been reported anecdotally that organic papaya growers now surround their plots with the transgenic rainbow variety as the PTGS system proves to be a most effective method to reduce the viral reservoir, thus protecting susceptible varieties through a mechanism similar to herd immunity in mammalian systems. A recent paper by Hutchison et al. (Potrykus, 1999) demonstrated a variation of this halo effect for Bt maize. As noted, to mitigate against the evolution of resistance to the cry protein growers are required to maintain a 20% refugia of non-Bt maize. Despite predictions that this single gene protection may select for the development of resistance in corn borer larvae on maize or bollworm in cotton, it has proven to be remarkably resilient. Some resistance has arisen, most specifically in the latter, where a mutation in the cadherin receptor has led to resistance, but it is much lower than might be expected given the extent of usage of the Bt phenotype. Mutations providing small decreases in susceptibility to Bt proteins are relatively common, but those conferring sufficient resistance to enable survival on some types of Bt corn are exceedingly rare. Hutchison demonstrated that this has led to cumulative benefits over 14 years of between $3.2 and $3.6 billion with between $1.9 and $2.4 billion of this total accruing to non-Bt maize growers. They postulate that these results affirm theoretical predictions of pest population suppression, and highlight economic incentives for growers to maintain non-Bt maize refugia for sustainable insect resistance management (Potrykus, 1999). Whereas initially these refugia were required to be in specified plots, it has been determined that mixing Bt and non-Bt maize grain during planting may be an equally effective management strategy. Again, enforcement of these requirements had been relatively easy in the developed world, but to do so in some regions may prove more challenging.

While translation of biotech research into value-added products for producers and consumers is a challenge in the

United States, it is exponentially more difficult in LDCs. A problem facing Africa in particular is the lack of a dynamic private sector to take technologies to the farmer. It has also been estimated that regulatory costs might exceed the costs of research and experimentation needed to develop a given GM crop, which is a major problem in releasing such crops to the market. A way to reduce the costs of generating food and environmental safety data is to develop regional "centers of excellence" with complementary facilities where food safety testing can be done reliably and regulatory costs could be reduced. The economic gains from using GM crop technology in SSA are potentially large according to the World Bank Group. The results suggest that welfare gains are potentially very large, especially from golden rice and nutritionally enhanced GM wheat, and that those benefits are diminished only slightly by the presence of the EU's current ban on imports of GM foods.

The authors of this chapter used the global economy-wide computable general equilibrium model known as GTAP. They specifically noted that if SSA countries impose bans on GM crop imports in deference to EU market demand for non-GM products, the domestic consumer loss net of that protectionism boost to SSA farmers would be more than the small gain derived from greater market access to the EU.

Problems cited for the slow passage of GM crops from experimental, to trial, to the commercial stage include the lack of capacity to negotiate licenses to use genes and research techniques patented by others, especially for crops with export potential. In addition, there are difficulties in meeting regulatory requirements and a lack of effective public commercialization modalities and working extension networks. Biosafety regulation and intellectual property rights (IPR) still have to be enforced in many countries for an effective and safe use of genetically engineered crops, especially if their production is meant for the export market.

As noted, intellectual property constraints continue to be a significant barrier in some regions and for some technologies. At the research level this is usually not of major import in developing regions, but at the commercialization stage this has the potential to be a significant barrier to initial adoption and most specifically enforcement. In the United States, Technology Use Agreements (TUAs) have been used to prohibit replanting patented seed by farmers. So far, there is no indication that Fine Seeds International will be selling seed under a TUA. Furthermore, the vast majority of farmers in Egypt have very small farms (less than 2 acres), making enforcement of a TUA more difficult than in the United States where farm land holdings are typically quite large and accessible. In addition, it is unclear whether a TUA would be upheld by an Egyptian court. Until more information is available, this criticism is unfounded. Without a TUA farmers are able to save and use seed for their own consumption.

In summary, Egyptian plant variety protection law does not put farmers at the mercy of private industry commercializers, either multi-national or domestic companies. Besides, farmers must ultimately decide whether or not to spend their money on this new variety.

Scientific, civic, and religious opinion leaders from all over the world have expressed support for the value of this technology. Florence Wambugu of Kenya states that the great potential of biotechnology to increase agriculture in Africa lies in its "packaged technology in the seed," which ensures technology benefits without changing local cultural practices. For example, over 120 million children worldwide are deficient in vitamin A. The Potrykus (1999) group has engineered rice to accumulate provitamin A (beta carotene). Incorporation of this trait into rice cultivars and widespread distribution of this packaged technology in the seed could prevent 1 to 2 million deaths each year. She observes that in the past many foreign donors funded high-input projects, which have not been sustainable because they have failed to address social and economic issues such as changes in cultural practice (Wambugu, 1999). In concurrence with this, Ismail Serageldin, former Chairman of the Consultative Group on International Agricultural Research (CGIAR) noted that, *a priori*, biotechnology could contribute to food security by helping to promote sustainable agriculture centered on smallholder farmers in developing countries.

American consumer attitudes also tend to be relatively positive about agricultural biotechnology. In a 2010 IFIC survey consumers were determined to be largely familiar with the term "biotechnology" (International Food Information Council, 2010): More than two-thirds of surveyed consumers (69%) have read or heard at least "a little" about the concept. Half (51%) of surveyed consumers say they are favorable toward farmers using biotechnology to grow more crops that would help meet food demand. In addition, significantly more consumers this year (28% vs. 23% in 2008) believe foods produced through biotechnology are available in the supermarket today, although this figure is still relatively low. Certain benefits of biotechnology are found to resonate better with consumers than others. These tend to be consumer-facing qualities such as improved health or better taste. For example, the majority of consumers say they are somewhat or very likely to purchase foods produced through biotechnology to provide more healthful fats like Omega-3s (76%), to avoid trans fat (74%), and to make foods taste better/fresher (67%). This is consistent from 2008. Additionally, more than three-quarters (77%) of surveyed consumers say they would be likely to purchase foods produced through biotechnology for their ability to reduce pesticide use, and 73% of those consumers said they would be likely to purchase bread, crackers, cookies, cereal, or pasta made with flour from wheat that had been modified by biotechnology to use less land, water, and/or pesticides. Of the 18% who would like to see additional information on the FDA label, only 3% mentioned anything about biotechnology.

But what of the context in which these crops are grown? Can all cropping systems coexist in harmony? According to Brookes and Barfoot (2004, 2008) it is important to determine the relative importance of different crop production systems based on planted area, production, and economic value to the region in question. The issue is what, if any, are the economic consequences of the adventitious presence of material from one crop system within another based on the notion that farmers should be able to freely cultivate the crops of their choice using the production system that works best in any given context (GM, conventional, or organic). It is never a food or environmental safety issue; instead it is a production and marketing matter. The heart of the issue is assessing the likelihood of the adventitious presence of material from one production system affecting another and the potential impact. This requires consistency when dealing with the adventitious presence of any unwanted material including, but most definitely not limited to, biotechnology-derived material. Adventitious presence is simply the unintended incidence of something other than the desired crop such as small quantities of weed seeds, seeds from other crops, dirt, insects, or foreign material (e.g., stones). It is unrealistic to expect 100% purity for any crops, or products derived from them, so thresholds that are consistent across all materials should be set and should not discriminate (e.g., thresholds for the adventitious presence of biotech material should be the same as those applied to thresholds for other unwanted material and *vice versa*). All measures should be proportionate, non-discriminatory, and science-based.

The issue of economic liability provisions that compensate growers for the adventitious presence of biotech material is often raised. Historically, the worldwide market has adequately addressed economic liability issues relating to the adventitious presence of unwanted material in any agricultural crop. For example, for certified seed the onus is on the producers who require isolation from undesired pollination for the purity of their product to ensure such purity. This is not their neighbor's problem. By extension the onus is on growers of any specialty crops to take action to protect the purity of their crops, since these are self-imposed standards for and by that market. Growers who have chosen a more stringent standard than that established in EU legislation should not expect their neighbors to bear the special management costs of meeting that self-imposed standard; to do so would reverse fundamental freedoms of economic activity and would establish a dangerous precedent. To allow specialty operators to formulate unrealistic standards for GM in their own produce would impose impossibly high standards on neighbors and would effectively impose a ban on the choice of other producers. Such growers are usually rewarded by higher prices and niche markets for taking such actions. Their neighbors enjoy no such advantage.

Existing legislation in North America and the EU is more than adequate to protect all grower and consumer interests, but if new regulations were considered to address economic liability provisions for any negative economic consequences of the adventitious presence of unwanted material, the same principle should apply to all farmers regardless of their chosen production methods. On equity grounds, biotech growers should have equal access to compensation for the adventitious presence of material from conventional or organic crops (such as fungal contamination) as conventional and organic producers have from biotech growers. No one sector should be able to unfairly prohibit another — access and choice work both ways. All coexistence measures should be based on legal, practical, and scientific realities and not on commercial or niche marketing objectives. Where unintended presence has occurred such as the presence of minute levels of Bayer CropScience regulated material LLRICE 604 found in Clearfield 131 (CL131) rice seed in 2007 and Mycogen's event "32" in maize in 2008, the agencies cooperated and determined that these events did not prove any risk as they carried similar constructs to those already having achieved non-regulated status.

According to Brookes and Barfoot (2004, 2008), biotech crops coexist successfully with conventional and organic crops in North America (where biotech crops account for the majority of acreage of important arable crops like soybeans, cotton, and maize), Spain, and more recently in the Czech Republic. The market has developed practical, proportionate, and workable coexistence measures without new regulations or any government intervention. Where isolated instances of the adventitious presence of biotech material have been found in conventional or organic crops, they have usually been caused by inadequate implementation of good coexistence practices (e.g., inefficient segregation of crops in storage and transport, and non-use of tested, certified seed). Under civil liability (i.e., tort damages) and for intellectual property infringement (except for the unauthorized StarLink), there have been no lawsuits brought by any parties for adventitious presence. Every case brought by a seed company for infringement has involved a claim that the farmer charged with infringement was an intentional infringer (i.e., adventitious presence was not the issue). To date, each of these cases was upheld by the courts. Indeed, all except one notable exception in North America has conceded to this claim.

Virtually all EU member states have transcribed EU Directive 2001/18 and implement EU regulations on traceability and labeling. Within the EU, provision has been made for a *de minimis* threshold for the unavoidable presence of GM organizations, but no actual threshold has been set. Therefore the default state of the 0.9% on labeling and traceability is the one enforced. In the United States, organic products cannot be (legally) downgraded or the producer decertified by unintentional presence when all required measures and best practices are adhered to, and no producer has been so impacted to date.

Going forward there are four major stanchions to the furtherance of coexistence, and all of them are incumbent on cooperation.

1. Monitoring: Verify the models and predictions about cost, isolation standards, and generally learn how the farming community copes with the requirements for keeping the product streams separated.
2. Dialog: Strategy development takes place in a dialog between the scientific and technical community and all relevant stakeholders (Denmark).
3. Stewardship: Stewardship programs should take into account the interests of both GM and non-GM farmers. Existing product stewardship programs for non-GM crops in farming should be a starting point for developing stewardship schemes for GM crops.
4. Research: The scientific community should be encouraged to fill the knowledge gaps that have been identified. Projects are needed to validate models and guidelines, including long-term studies. Building up mechanistic, probabilistic, and predictive models of gene flow as well as methods for restricting gene flow by eliminating the fertility of pollen or seeds (apomixis, cytoplasmic male sterility, plastid transformation, Genetic Use Restriction Technology (GURT), etc.) is necessary.

The World Trade Organization ruled in 2006 that a six-year European ban on genetically engineered crops violates international trade. The three-person panel issued its decision ruling in favor of the three countries — United States, Canada, and Argentina — on a large majority of the 25 crops under dispute in the case and issued mixed rulings on a few crops. The panel also ruled in favor of challenging national bans on specific biotech crops issued by Austria, France, Germany, Greece, Italy, and Luxembourg. The EU had argued that it did not have a moratorium but that it just took more time to weigh the possible risks to health and the environment posed by genetically engineered foods. It said it needed to take a "precautionary" approach to regulation, which is different from what it called Washington's "laissez-faire" stance.

The trade organization panel appears not to have challenged Europe's regulatory process for biotech crops. Rather, it said Europe failed to follow its own procedures, resulting in undue delay of decisions. Interestingly, one of the most comprehensive assessments on the technology was conducted by EU scientists. An EU Commission Report (2001) that summarized biosafety research of 400 scientific teams from all original 15 EU countries conducted over 15 years stated that research on biotechnology-derived plants and -derived products so far developed and marketed, following usual risk assessment procedures, had not shown any new risks to human health or the environment beyond the usual uncertainties of conventional plant breeding. Indeed, the use of more precise technology and the greater regulatory scrutiny probably make these plants and products even safer than conventional plants and foods. If there are unforeseen environmental effects — none have appeared as yet — these should be rapidly detected by existing monitoring systems. This analysis was repeated in a 2008 EU Joint Research Centre (JRC) Report commissioned by Members of the European Parliament conducted by world experts including the European Food Safety Authority (EFSA) and WHO (European Union Joint Research Council, 2008). Their report concluded that there is no evidence that GM foods have any harmful effects; a declaration signed by over 3500 scientists including 25 Nobel Laureates reiterates this position.

While biotech research and development in Europe slowed significantly following the EU's 1998 *de facto* moratorium on approvals and individual countries such as France and Germany bowing to internal minority pressure, Europe's stance on biotech crops cannot prevent biotech adoption in the rest of the world and, as noted, Asia is forging ahead. According to a study by Runge and Ryan (2005), as the EU becomes increasingly isolated, it will discourage its young scientists and technicians from pursuing European careers. They opine that if, on the other hand, the EU engages biotech in an orderly regulatory framework harmonized with the rest of the world, then it will encourage a more rapid international diffusion of the technology. More nations will join the top tiers of commercial production, and emerging nations will continue to expand the sector.

The challenges going forward are: (1) technical, as we strive to modify intricate metabolic pathways and networks as opposed to single genes, the scientific challenges are not trivial; (2) intellectual property restrictions that may limit translation of public research if not managed judiciously; (3) liability concerns over abuse or misuse of constructs; (4) prohibitive and asymmetric biosafety regimes; and finally, public acceptance. The last is often predicated on how much of the former challenges are perceived, and how positions are presented by the opposing factions. Indeed, the actual commercialization of biotech products may have little to do with technical limitations and more to do with these external constraints, primarily the process of regulatory approval. The flagship of improved nutritional varieties, such as beta carotene enhanced rice commonly referred to as golden rice, despite being under consideration since the late 1990s and subject to a barrage of risk assessments is unlikely to be approved until 2012 at the earliest. Ingo Potrykus, the developer, says an unreasonable amount of testing has been required without scientific justification. In a recent *Nature* article (Potrykus, 2010) he lays the blame largely on the regulatory process, which he considers excessive, observing that unjustified and impractical legal requirements are stopping genetically engineered crops from saving millions from starvation and malnutrition.

In the final analysis resources are finite, true sustainability can come only from an enlightened philosophy that promotes the development of resource-enhancing technologies. The only sure way to protect the planet's resources is not to settle into the complacency of maintaining the status quo, but to engage in continual, constructive change based on scientific knowledge. Biotechnology is not a panacea for all the world's ills; it is one of many sets of tools and technologies that will be required to ensure such sustainable development.

References

African Agricultural Technology Association. (2007). Combining Breeding and Biotechnology to Develop Water Efficient Maize for Africa (WEMA) < http://www.aatf-africa.org/userfiles/Wema-Concept-Note.pdf >.

Agarwal, P. K., Agarwal, P., Reddy, M. K., & Sopory, S. K. (2006). Role of DREB transcription factors in abiotic and biotic stress tolerance in plants. *Plant Cell Reports*, *25*(12), 1263–1274.

Bricknel, T. (2010). India delays Bt brinjal decision. Fruitnet.com <http://www.fruitnet.com/content.aspx?cid = 5731&rid = 2> accessed August 8, 2010.

Brookes, G., & Barfoot, P. (2004). Co-existence of GM and non-GM arable crops: The non-GM and organic context in the EU. PG Economics Ltd <http://www.pgeconomics.co.uk>.

Brookes, G., & Barfoot, P. (2005). GM crops: The global economic and environmental impact — the first nine years 1996–2004. *AgBioForum*, *8*(2&3), 187–196. (Available at: <http://www.agbioforum.missouri.edu/v8n23/v8n23a15-brookes.htm>).

Brookes, G., & Barfoot, P. (2008). Co-existence of GM and non-GM arable crops: The non-GM and organic context in the EU. PG Economics Ltd.

Choudhary, B., & Gaur, K. (2008). The development and regulation of Bt brinjal in India (Eggplant/Aubergine), ISAAA Brief No. 38. Ithaca, NY: ISAAA.

Choudhary, B., & Gaur, K. (2010). *Bt Cotton in India: A country profile. ISAAA Series of Biotech Crop profiles*. Ithaca, NY: ISAAA.

Deng, X., Phillips, J., Bräutigam, A., Engström, P., Johannesson, H., & Ouwerkerk, P. B., et al. (2006). A homeodomain leucine zipper gene from *Craterostigma plantagineum* regulates abscisic acid responsive gene expression and physiological responses. *Plant Molecular Biology*, *61*(3), 469–489.

Economic and Social Council Resolution. (2004). 68. Science and Technology for Development (E/2004/31).

Elmegaard, N., & Bruus Pedersen, M. (2001). *Flora and fauna in roundup tolerant fodder beet fields*. National environmental research institute. 40 pp. Technical Report No. 349.

EU Commission Report. (2001). EC-sponsored research into the safety of Genetically Modified Organisms. Fifth Framework Programme — External Advisory Groups "GMO research in perspective" Report of a workshop held by External Advisory Groups of the "Quality of Life and Management of Living Resources" Programme <http://europa.eu.int/comm/research/quality-of-life/gmo/index.html http://europa.eu.int/comm/research/fp5/eag-gmo.html>.

European Union Joint Research Council. (2008). Scientific and technical contribution to the development of an overall health strategy in the area of GMOs <http://ec.europa.eu/dgs/jrc/downloads/jrc_20080910_gmo_study_en.pdf>

FAO/WHO. (2004). The State Of Food and Agriculture 2003–2004, Agricultural Biotechnology Meeting the needs of the poor? Food and agriculture organization of the United Nations Rome, 2004. <http://www.fao.org/documents/show_cdr.asp?url_file = /docrep/006/Y5160E/Y5160E00.HTM> (accessed March 23, 2011).

Gonsalves, D. (1998). Control of papaya ringspot virus in papaya: A case study. *Annual Review of Phytopathology*, *36*, 415–437.

Green, R. E., Cornell, S. J., Scharlemann, J. P. W., & Balmford, A. (2005). Farming and the fate of wild nature. *Science*, *307*, 550–555.

Huang, J., Rozelle, S., Pray, C., & Wang, Q. (2002). Plant biotechnology in China. *Science*, *295*, 674–677.

International Food Information Council. (2010). 2010 "Consumer Perceptions of Food Technology". Survey conducted by Cogent Research of Cambridge, Massachusetts for the International Food Information Council, Washington, DC. <http://www.foodinsight.org/Content/3843/Final_Executive%20Summary%20Food%20Tech%20Report_Website%20version_7-7-10.pdf>.

James, C. (2011). Global status of commercialized biotech/GM crops: 2010. ISAAA briefs 42-2010. Ithaca, NY: ISAAA. <http://www.isaaa.org/resources/publications/briefs/42/executivesummary/default.asp> (accessed March 23, 2011).

Ki-Hong, J., Seo, Y. -S., Walia, H., Cao, P., Fukao, T., Canlas, P. E., et al. (2010). The submergence tolerance regulator Sub1A mediates stress-responsive expression of AP2/ERF transcription factors. *Plant Physiology*, *152*, 1674–1692.

Lopez-Bucio, J., Nieto-Jacobo, M. F., Ramirez-Rodriguez, V. V., & Herrera-Estrella, L. (2000). Organic acid metabolism in plants: From adaptive physiology to transgenic varieties for cultivation in extreme soils. *Plant Science*, *160*(1), 1–13.

Lucca, P., Hurrell, R., & Potrykus, I. (2002). Fighting iron deficiency anemia with iron-rich rice. *Journal of the American College of Nutrition*, *21*, 184S–190S.

Monsanto News Release. (2010). Insect-protected Roundup Ready 2 yield soybeans achieve important milestone toward commercialization in Brazil <http://monsanto.mediaroom.com/index.php?s = 43&item = 875>.

Newell-McGloughlin, M. (2005). "Functional Foods" and biopharmaceuticals: The next generation of the GM revolution (pp. 163–178). In: *Let them eat precaution*. American Enterprise Institute.

Newell-McGloughlin, M. (2008). Nutritionally improved agricultural crops. Editor's choice series on the next generation of biotech crops. *Plant Physiology*, *147*, 939–953. < http://www.plantphysiol.org/cgi/content/full/147/3/939>.

Potrykus, I. (1999). Vitamin-A and iron-enriched rices may hold key to combating blindness and malnutrition: A biotechnology advance. *Nature Biotechnology*, *17*, 37.

Potrykus, I. (2010). Regulation must be revolutionized. *Nature*, *466*(7306), 561.

Rivero, R. M., Shulaev, V., & Blumwald, E. (2009). Cytokinin-dependent photorespiration and the protection of photosynthesis during water deficit. *Plant Physiology*, *150*(3), 1530–1540.

Runge, C. F., & Ryan, B. (2005). The global diffusion of plant biotechnology: International adoption and research in 2004. <http://www.apec.umn.edu/faculty/frunge/globalbiotech04.pdf>.

Sottosanto, J. B., Saranga, Y., & Blumwald, E. (2007). Impact of AtNHX1, a vacuolar Na^+/H^+ antiporter, upon gene expression during short and long term salt stress in *Arabidopsis thaliana*. *BMC Plant Biology*, *7*, 18.

Stern, N. (2006). Review on the economics of climate change. HM Treasury.

Trewavas, A. (2001). Urban myths of organic farming. *Nature*, *410*, 409–410.

Wambugu, F. (1999). Why Africa needs agricultural biotech. *Nature*, *400*, 15–16.

World Bank. (2007). World Development Report 2008 Agriculture for Development, Overview (No. ISBN 978-0-8213-7297-5).

Wu, F. (2006). Mycotoxin reduction in Bt corn: Potential economic, health, and regulatory impacts. *Transgenic Research*, *15*(3), 277–289.

Zhang, Z. B., Kornegay, E. T., Radcliffe, J. S., Wilson, J. H., & Veit, H. P. (2000). Comparison of phytase from genetically engineered *Aspergillus* and canola in weanling pig diets. *Journal of Animal Science*, *78*, 2868–2878.

Crop biotechnology in developing countries

36

Hugo De Groote
International Maize and Wheat Improvement Centre (CIMMYT), Nairobi, Kenya

TABLE OF CONTENTS

Introduction

Since its first commercial release in 1996, transgenic crops have been adopted at a rate never before seen in the developed world (James, 2010). The current first-generation genetically modified (GM) varieties mainly use two traits, herbicide tolerance (HT) and insect resistance (Bt), in four major crops: soybeans, maize, cotton, and canola. The production of these crops is used, for the most part, for feed and fiber. A set of second-generation GM crops is currently under development, with the aim to improve the nutritional status and other qualities of major food crops (Qaim, 2009). Golden rice, for example, has been biofortified with high levels of provitamin A (Zimmermann and Qaim, 2004).

GM crops were developed by and for the developed world, in particular Europe and North America. The basic research was often done at public research institutes, but the applications and crop varieties are almost uniquely developed by private companies in response to the needs of commercial farmers in the developed world (Pingali and Traxler, 2002). Still, the area in GM crops in the developing world is large and increasing, and the number of farmers growing GM crops is actually much higher in developing, rather than in developed countries. These varieties were not developed in response to the needs of poor smallholders, and currently these varieties are mainly for feed and fiber, not food crops.

In the United States, where most GM crops have been developed, they are not seen as essentially different from other crops and follow the same regulations. But the use of GM technology in food is still controversial, and many countries, notably in Europe and Japan, have not embraced it. The principal objections concern possible harm to human health, damage to the environment, and unease about the "unnatural" status of the technology (Nuffield Council on Bioethics, 1999).

DOI: 10.1016/B978-0-12-381466-1.00036-5

Consumers in Europe also perceive the technology as having limited benefits. Although there is little scientific evidence to support consumers' preference for non-GM food (FAO, 2004; ICSU, 2004), this preference is clear and well documented (Lusk et al., 2005). GM technology does not generate many benefits for countries that already have large agricultural surpluses (Demont et al., 2004). This combination has led Europe to adopt the precautionary principle. While this principle is not easily defined, it does put GM crops in a special category, where applications are dealt with on a case-by-case basis. The regulatory system based on this principle also puts a heavy burden of proof on the research institute or company applying for the commercial use of GM technology.

Many developing countries have now adopted biosafety laws and regulatory systems on the same principle, even though their conditions are very different (Paarlberg, 2008). Unlike Europe, developing countries could benefit strongly from GM crops. First, both the yield-increasing and cost-reducing effects of the first-generation and the nutritional improvement of the second-generation GM crops could bring large benefits to large numbers of farmers and consumers. Secondly, consumers in developing countries do not necessarily share the concerns of their European counterparts. The regulatory institutions in many developing countries are new and, apart from Latin America and South Africa, have only recently started to approve GM crops. It is, therefore, important for regulators and policy makers to bring scientific evidence to the debate, based not only on objective risk assessment and cost benefit analysis, but also on the subjective opinion of consumers, their gatekeepers, and the policy makers themselves. This evidence is particularly important in the developing debate on the desirability of labeling.

To contribute to the debate, an overview of the evidence so far is presented in this chapter, based on the literature on the use of GM crops in developing countries, their benefits and costs, the risks, the perceptions of consumers, and the regulatory process.

Agriculture and Food in Developing Countries: The Needs

Feeding a growing world population

The world today faces the daunting challenge of feeding a still rapidly growing population at a time when the limits of conventional technologies are being felt. At the beginning of the nineteenth century, the world had fewer than 1 billion people, and it was relatively easy to increase food production over the next 100 years to feed another 0.6 billion (James, 2008). This was mainly achieved by increasing the area of land under cultivation, in particular in the Americas, Eastern Europe, and Australia (James, 2010). At the beginning of the twentieth century, the world population stood at 1.6 billion. To feed the increasing population during this century, an increase in global food production years was achieved mainly by increasing crop productivity through the Green Revolution (Evenson and Gollin, 2003).

Agricultural statistics from the 1960s and 1970s show a phenomenal growth of cereal crop productivity in the developing world (Pingali and Heisey, 1999). The Green Revolution, the initial phase of this growth, resulted from an increase in land productivity which occurred in areas of increasing land scarcity and areas with high land values. Significant investments in research and infrastructure development, especially irrigation, were the strategic components of this increased productivity. In a second phase, particularly in Asia, productivity growth has been sustained through increased input use and, more recently, through more efficient use of inputs.

Despite the success of the Green Revolution in keeping food production abreast of population growth, there are over 800 million people in the world today who are chronically undernourished, and 180 million children who are severely underweight for their age (Conway, 2000). At the same time, the work population continues to increase rapidly. At the beginning of the twenty-first century, the population stood at 6.1 billion and is expected to rise to 9.2 billion by 2050. Moreover, dietary patterns change when people become more affluent. Projections by the Food and Agriculture Organization (FAO) and the International Food Policy Research Institute (IFPRI) on the future supply and demand of food necessary to keep pace with population growth and changing dietary habits predict large increases in global demand for food (Rosegrant et al., 2001; FAO, 2009). Cereal production for food and feed needs to increase by 40% by 2050 (FAO, 2009). The global demand for food is expected to double again in 50 years (James, 2010).

Undernutrition and poverty

While various econometric models predict that market supply will meet market demand by 2020, there could still be nearly one billion people who lie outside the market and are chronically undernourished (Conway, 2000). Moreover, 400 million women of childbearing age suffer from iron deficiency, and the resultant anemia leads to infant and maternal mortality. Undernutrition is estimated to cause more than one in three child deaths globally (Caulfield et al., 2004; Black et al., 2008). Lack of access to nutritious food is a major cause of child mortality (Caulfield et al., 2004; Black et al., 2008). In rural areas, household diets are dominated by staples such as cereals and tubers, most of it produced at the homestead, while the consumption of other foods that would improve dietary quality, such as legumes, vegetables, fruits, and animal sourced foods, is limited by availability and price.

Subsistence farmers also depend on storing their crops on-farm to bridge seasons. High temperatures and high levels of humidity in many developing countries, combined with poor storage and drying facilities, lead to contamination by aflatoxins and other mycotoxins. An estimated 4.5 billion people in developing countries are chronically exposed to large amounts of aflatoxins, a well-known cause of liver cancer, that also compromise immunity and interfere with the metabolism of protein and micronutrients (Williams et al., 2004).

The situation is particularly difficult in Africa. Despite the Millennium Development Goal to halve the proportion of people who suffer from hunger by 2015, Africa is the only region where both the proportion and the number of underweight children are increasing (Rosegrant et al., 2001; de Onis et al., 2004). In East Africa particularly, food production per capita decreased from the mid-1970s to the early 1990s with no improvement since then (FAOSTAT, 2010).

Technology

Despite the need to feed an increasing population, the world stock of arable land is exhausted. Fossil fuels, the basis for fertilizer, are about to be depleted, and water levels and supply are decreasing in many areas of the world (The World Bank, 2008). The effects of global warming need to be addressed urgently, and environmental degradation brought under control to feed future generations in a sustainable way.

The classic technologies and conventional breeding that have helped to sustain the world population in the past, in particular with improved varieties, fertilizer, tillage, irrigation, and pesticides, clearly have their limitations. Indicators show a decrease in the growth rate of productivity of two of the three primary cereals: rice and wheat (Pingali and Heisey, 1999). The maintenance of soil fertility is a major problem, particularly in Africa (Sanchez et al., 1997).

Reducing tillage is one of the promising options, but this technology needs cost-efficient weed control. Water levels have been decreasing, especially in Asia, and the expansion of irrigation for staple crops is unlikely. Finally, the use of pesticides is problematic in developing countries, where low education levels, lack of equipment and training, and poor regulatory systems can lead to the high exposure of farmers and farm workers to toxic pesticides. The resultant health effects of pesticide use have been documented in many countries (Pingali et al., 1994; Houndekon et al., 2006).

Biotechnology is not a panacea to solve the world's food problem. It does, however, offer a range of opportunities and methods that can help lift the limitations of conventional technologies. Moreover, it dramatically expands the range of possibilities for increasing the quality of food staples, still the major source of food for the world's poor.

Current State of GM Crops

Geographic distribution

GM technology has been developed since the early 1980s, and the first commercial GM varieties were planted in 1996. Area and production statistics, as documented annually by the International Service for the Acquisition of Agri-biotech Applications (ISAAA), show their immediate popularity with farmers (James, 2010): once the crops are approved, their adoption is fast. Approval, however, has been slow, and only a small number of countries so far have commercial production of GM crops.

By 2009, after only 13 years, 134 million hectares had been planted as GM crops (James, 2010). No other agricultural technology has ever spread as quickly and on such a massive scale. Still, GM crops are only grown by 25 countries, with only 15 serious producers who cultivate more than 100,000 hectares each. Most of the GM area (95%) is situated in just six countries: the United States, Brazil, Argentina, India, Canada, and China (Table 36.1). Of the ten major producers, each with more than 0.8 million hectares, eight are developing countries; the most important are Brazil and Argentina with more than 20 million hectares each, India and China with more than 8 million hectares each, and Paraguay and South Africa with more than 2 million hectares each.

In the developing world, most GM crops are grown in South America (eight countries) and Asia (four countries). In Africa, only South Africa is a major producer, while Egypt and Burkina Faso have only recently started. South Africa is the only country in the world (up to 2009) that grew a GM crop specifically for food: white Bt maize. The map of countries growing GM crops (Figure 36.1) clearly shows that most of the Americas and large areas in Asia are covered, but large areas in Europe and Africa are left blank. The major areas in developing countries that do not grow GM crops, or at least not on a substantial scale, are in Europe and Japan. Whereas five European countries grow GM crops, only Spain grows them on more than 100,000 hectares. The limited acreage in Europe is due to strong consumer objections (Gaskell et al., 2000) driven by a lack of perceived benefit (Gaskell et al., 2004).

Crops, traits, and farmers

In 2009, more than half of the GM area worldwide was planted in herbicide-tolerant soybean (52%), followed by maize (31%), cotton (12%), and canola (5%). Other GM crops are papaya, sugar beet, tomato, squash, and sweet pepper, but these are only grown by three countries and in small quantities. Within the three major crops, a large proportion is now grown using GM varieties: more than three-quarters of soybean (77%), one-half of the cotton (49%), and one-quarter of the maize (26%).

From the first commercialization of GM crops, herbicide tolerance (HT) has consistently been the dominant trait. In the four main crops (soybean, maize, cotton, and canola), HT varieties covered 62% of the area worldwide in 2009. The stacked double and triple traits covered 21%, whereas varieties with only insect resistance covered 15%. Most countries in the developing world only produce industrial crops (cotton) and feed crops (soybean and maize). Only South Africa produces a specific GM staple food crop (white Bt maize).

Worldwide, 14 million farmers grow GM crops. The large majority (90%) are small and resource-poor farmers in developing countries, who grow mainly Bt cotton (7 million in China and 5.6 million in India). About one million farmers are large farmers from industrial countries, such as the United States and Canada, and developing countries, such as Argentina and Brazil. HT technology is less appreciated in developing countries, where labor costs are low and smallholders usually rely on manual weeding.

Future and trends

The trends indicate an increased use of stacked traits, an increased interest in food crops, and an increasing number of

Table 36.1 Countries growing different GM crops

Country	Area (million ha)	Soybean	Maize	Cotton	Other crops
United States*	64.0	x	x	x	Canola, squash, papaya, alfalfa, sugar beet
Brazil	21.4	x	x	x	
Argentina	21.3	x	x	x	
India	8.4			x	
Canada	8.4	x	x		Canola, sugar beet
China	8.2			x	Papaya, tomato, sweet pepper
Paraguay	2.2	x			
South Africa	2.1	x	x	x	
Uruguay	0.8	x	x		
Bolivia	0.8	x			
Philippines	0.5		x		
Australia	0.2	x			Canola
Burkina Faso	0.1			x	
Spain	0.1		x		
Mexico	0.1	x		x	
Chile	< 0.1	x	x		Canola
Colombia	< 0.1			x	
Honduras	< 0.1		x		
Czech Republic	< 0.1		x		
Portugal	< 0.1		x		
Costa Rica	< 0.1	x		x	
Egypt	< 0.1		x		
Slovakia	< 0.1		x		

Source: James, 2010.

private–public partnerships (PPPs) to develop GM food crop. Increasingly, seed companies develop GM crop varieties with stacked traits. In the United States, 85% of the maize area and 75% of the cotton area are now sown in varieties with stacked traits. This poses particular problems for the regulatory systems in developing countries, given the low level of expertise. The HT trait is not of major interest to subsistence and small-scale farmers.

The second important trend is the increasing importance of food crops. Until 2009, only South Africa was producing a specific GM food crop. In that same year, however, China approved the release and commercial production of Bt rice and India approved the use of Bt eggplant. The approval of Bt rice in China is of major importance to the developing world, because it was the first staple food crop in a large developing country and the first GM rice variety. This experience can lead the way for other GM food crops, particularly golden rice. The release of Bt eggplant in India was extremely important because it was the first vegetable GM crop outside of the United States and China, and could have been very influential for this increasingly important sector for small-scale farmers in developing countries. However, India recalled the release of Bt eggplant in 2010 under pressure from consumer organizations and non-governmental organizations (NGOs).

For the developing world, PPPs are increasingly important, because the development and the release of GM crops is very expensive. Most developing countries do not have the scientific capacity to develop GM crops, and the resources for biotechnology in the international agricultural research institutes are limited. The International Maize and

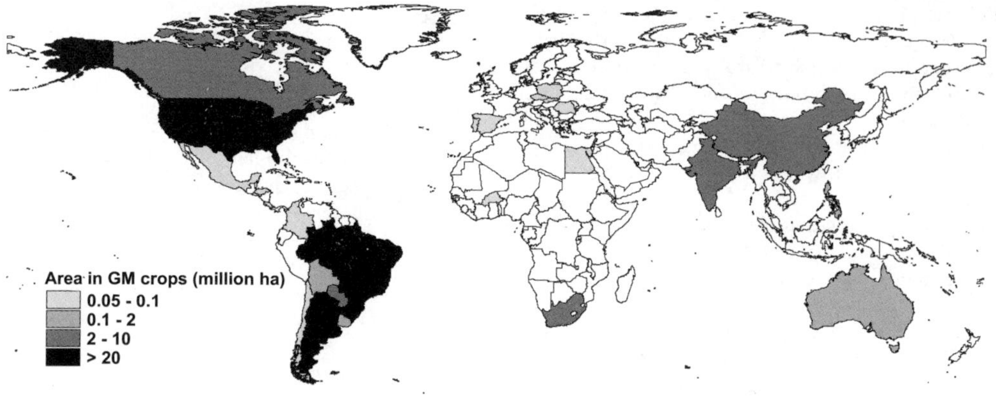

Figure 36.1 • Countries growing GM crops • From ISAAA (James, 2010). Please see color plate section at the end of the book.

Wheat Improvement Centre (CIMMYT) and the Kenya Agricultural Research Institute (KARI) have tried to develop Bt maize varieties for Africa with Bt events from the public sector, in particular events developed by CIMMYT and the University of Ottawa (Mugo et al., 2011). These public sector events did not, however, provide sufficient control of some stem borer species, so CIMMYT and KARI are now collaborating with private companies to use their Bt events (De Groote et al., 2011b). CIMMYT is working with the private sector as well to develop drought-tolerant and nitrogen-efficient maize varieties for developing countries in different PPP projects.

Economic Impact of Transgenic Crops in Developing Countries

Main effects of current GM crops

At the farm level, the economic impact of the current transgenic crops, based on the Bt and HT genes, is realized through yield increase, cost reduction, or both. The Bt gene, inserted in cotton or maize varieties, causes the plant to produce a protein that is toxic to certain insects, in particular lepidopteran pests such as the maize stem borers and the cotton boll weevil (FAO, 2004). Several Bt genes are highly effective in controlling these pests but, by themselves, do not increase crop yield; they only reduce the losses due to better pest control, or make pest control cheaper (Zilberman et al., 2007). Moreover, Bt genes do not control all insect pests; therefore, they do not eliminate all crop loss due to insects or all pesticide use.

In developed countries, farmers usually have access to the means and the know-how to control most pests with pesticides. Bt varieties, therefore, do not generally increase crop yields, but they will provide a cheaper alternative to pesticides, and reduce their applications and cost. Insects usually thrive in the warmer and more humid climates typical of many developing countries, leading to higher pest pressure. Moreover, resource-poor farmers often face serious constraints in using pesticides, in particular having access, know-how, equipment, and cash or credit, which often leads to the suboptimal use of pesticides (Qaim and Zilberman, 2003; Qaim, 2009; Smale et al., 2009). In such cases, Bt varieties will reduce crop loss through better control of the target pests, leading to higher yields. For those farmers who have used pesticides before, switching to Bt varieties will reduce their production costs.

HT crops work on another principle: they are tolerant to the broad spectrum herbicide Glyphosate (FAO, 2004). Therefore, the use of HT technology can reduce production costs through the substitution of Glyphosate for an array of more expensive, and usually more toxic, herbicides. HT would only increase yields in cases where weeds were not properly controlled before. An extra advantage of the improved weed control from HT technology is the option of reducing tillage, because weed control is one of the main objectives of tilling. Unlike Bt crops, HT crops are mainly grown in developing countries and by large-scale farmers in Latin America, while farmers in developing countries usually rely on manual weeding (James, 2010).

Apart from estimating the farm level benefits, the distribution of these benefits is also important, in particular to small-scale versus large-scale farmers. Moreover, the total economic benefit of new crop varieties is not limited to increased yields or reduced costs at the farm level: price effects and impact on the rest of the economy also need to be considered. An increase of production levels or a decrease in production costs generally reduces the market price, benefiting the consumer. If consumer demand is inelastic, as is often the case with basic agricultural products, most benefits will go to the consumer. Increased production also increases trade and employment in other sectors.

Empirical evidence of farm level benefits

Bt cotton

The economic impact of Bt cotton has been the subject of a large number of studies, based on farmer surveys, farm records, or trials. A recent IFPRI review of the economic impact of GM crops in developing countries identified 15 studies of Bt cotton in South Africa, 11 in China, 16 in India, and 5 in Latin America (Smale et al., 2009). In South Africa, the analyses comparing economic returns per hectare between smallholders and large-scale producers found that smallholders in South Africa are major beneficiaries of Bt cotton (Smale et al., 2009). In China, the synthesis of many studies is that Bt cotton leads to sustained, widespread, and positive effects on farm profits.

Studies from India were not as clear: Bt cotton is grown in all agroecological zones and yields are highly variable, authors use different methods, and there is a difference between types of Bt varieties (Smale et al., 2009). In another review of studies on the impact of GM crops, estimated yield effects of Bt cotton in developing countries varied from 9% in Mexico to 33% in Argentina and 37% in India (Qaim, 2009). In Argentina, conventional cotton farmers underutilize chemical insecticides and insect pests are not effectively controlled (Qaim and de Janvry, 2003). In India, however, insecticide use in conventional cotton is much higher. This suggests that factors other than insecticide quantity influence damage control in conventional cotton and, thus, the yield effects of Bt technology. These factors include insecticide quality, insecticide resistance, and the correct choice of products and timing of sprays. The same review found increases in gross margins from $23/ha in Argentina to $470/ha in China, as compared to $58/ha in the United States.

Bt maize

The effects of Bt maize are similar to those observed in Bt cotton, albeit generally at a lower magnitude (Qaim, 2009). Except for Spain, where the percentage reduction in insecticide use is large, the more important result of the use of Bt maize is an increase in effective yields. Mean yield effects are higher in developing countries such as Argentina (9%), South Africa (11%), and the Philippines (34%) because pest pressure is more severe there than in temperate climates. The average yield in South Africa is mostly from yellow Bt hybrids produced by large commercial farms. On-farm trials carried out with smallholder farmers and white Bt maize hybrids found average yield gains of 32% (Gouse et al., 2006).

Estimates of gross margin increases from using Bt maize in developing countries ranged from $20/ha in Argentina to $42 in South Africa and $53 in Argentina. In developed countries, estimates ranged from only $12/ha in the United States to $70/ha in Spain (Qaim, 2009).

In Africa only Kenya, apart from South Africa, has been testing Bt maize. Here, crop losses due to stem borers in maize are estimated at 13.5% or 0.4 million tons, valued at $80 million (De Groote et al., 2011a). Benefits of Bt maize for the country are calculated at $208 million over 25 years (66% of which is consumer surplus) as compared to a cost of $5.7 million.

Some preliminary evidence from field trials is available for maize and other Bt crops. In Bt rice in China, high insecticide reductions, but relatively small yield effects were observed (Huang et al., 2005), while in Bt eggplants in India, significant insecticide as well as yield effects were observed (Krishna and Qaim, 2008a).

HT crops

The use of HT in developing countries has not resulted in increased yields, but (as in North America) it has resulted in a reduction in herbicide costs (Qaim and Traxler, 2005). As in the United States, Glyphosate applications have increased, but the number of applications for other herbicides has decreased. This increase was caused by the increase in reduced tillage — farming techniques that limit soil disturbance to increase water retention and to reduce erosion. Reduced tillage methods usually rely on herbicides, in combination with HT varieties, to control weeds.

Because HT crops were developed and commercialized by private companies, a technology fee is charged on seeds, which varies among crops as well as countries. In North America, the fee costs for HT crops were similar or sometimes higher than the average cost reduction, so that gross margin effects were small or sometimes even negative (Qaim, 2009). The main reasons for farmers in such situations to continue using HT technologies were easier weed control and savings in management time.

In South American countries, however, the agronomic advantages are similar but the fees charged are lower because HT technology is not patented there, and many soybean farmers use farm-saved GM seeds (Qaim, 2009). This has led to larger average gross margin effects of HT crops, especially HT soybeans (Trigo and Cap, 2006). The average gross margin gains through HT soybean adoption in Argentina, for example, were estimated at $23/ha (Qaim and Traxler, 2005).

Effect of GM crops on poverty and inequality

So far, HT crops have not been widely adopted in the small-farm sector. Smallholders often weed manually, so HT crops are inappropriate, unless labor shortages or weeds that are difficult to control justify conversion to chemical practices (Qaim, 2009).

Bt crops, on the other hand, are very popular with small-scale farmers, especially Bt cotton, which is produced by many small-scale farmers in China (Huang et al., 2002), India (Smale et al., 2009), and South Africa (Gouse et al., 2004).

In South Africa, many smallholders now grow Bt white maize as their staple food, with substantial increases in yield and gross margin (Gouse et al., 2006). Several studies show that Bt technology advantages for small-scale farmers are of a similar magnitude as those of larger scale producers and, in some cases, the advantages can be even greater (Qaim, 2009).

Combined effects on farmer income

The total increase of farmer income worldwide from the first 13 years of commercialization of GM crops (1996 to 2008) is estimated at $52 billion (Brookes and Barfoot, 2010). About half of this gain was generated in developing countries. The income increase in the last year of the study (2008) was estimated at $9.4 billion, with slightly more than half going to farmers in developing countries (Figure 36.2).

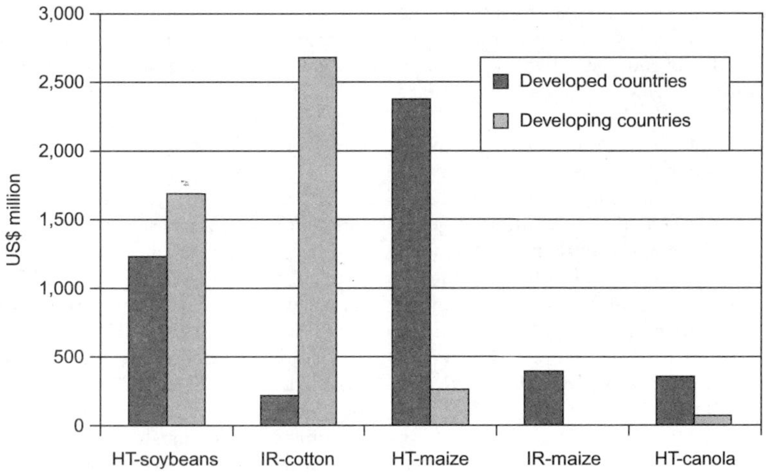

Figure 36.2 • Increases of global farm level income from GM technologies • From Brookes and Barfoot (2010). Please see color plate section at the end of the book.

HT soybeans were the most important crop, with increased revenue estimated at \$2.9 billion, which was almost equally divided between developed and developing countries. Most of this increased revenue from HT soybeans, however, went to large-scale farmers. The benefits of insect resistant (IR) cotton (largely Bt cotton) are of the same magnitude (\$2.9 billion), but, in contrast to HT soybeans, their benefits go almost uniquely to developing countries and to many smallholders. The benefits of GM maize are again of the same magnitude, but most of it is from IR maize (\$2.6 billion), and most of it goes to developed countries.

Macro level impacts

When a new technology increases production and the welfare of producers, the supply curve shifts and leads to a decrease in food prices that affects consumers' welfare (Alston et al., 1998). The combined effects are often analyzed with the economic surplus method, which is a partial equilibrium method. In the United States, the economic surplus gain from Bt cotton, for example, was estimated at 164%, of which 37% went to farmers, 18% to consumers, and 45% to the innovating companies. Economic surplus in China was estimated at a similar \$140 million, but only 1.5% went to the innovating company and the rest to farmers, whereas consumers did not receive any benefit because prices were fixed by the government (Pray et al., 2001). GM crops can bring sizable welfare gains, but the distributional effects depend on intellectual property rights (IPRs) and institutional conditions (Qaim, 2009).

Global effects from new technologies include their effects on other sectors of the economy as well as their effects, through trade, on markets in other countries and regions. Increased production can increase the demand for labor at harvest, increasing income and demand for other products, whereas reduced prices on international markets are beneficial to consumers worldwide. These global effects can be captured by estimating general equilibrium models. Global welfare gains of Bt cotton, as estimated by different studies, range from \$0.7 to \$1.8 billion per year, while GM rice (Bt, HT, and drought tolerance) could bring gains of \$2.1–2.5 billion per year (Qaim, 2009) .

Health Impact

Safety concerns

The power of the new discoveries in genetics raises concerns in many societies as to the ethics and safety of their use and the risks they may pose to human health, biodiversity, and the environment (ICSU, 2004). The introduction of GM crops has become highly controversial in the UK and some other parts of the world. The principal objections concern possible harm to human health, damage to the environment, and unease about the unnatural status of the technology (Nuffield Council on Bioethics, 1999).

Many reviews by national and international science organizations, and reviews of these reviews, synthesizing the scientific knowledge on GM crops on human health show a wide consensus among the scientific community that currently available GM foods are safe to eat (FAO, 2004; ICSU, 2004). The most important review, by the International Council for Science, concludes that food safety assessments by national regulatory agencies in several countries have deemed currently available GM foods to be as safe to eat as their conventional counterparts and suitable for human consumption (ICSU, 2004). This view is shared by several intergovernmental agencies, including the FAO/WHO Codex Alimentarius Commission on food safety which has 162 member countries, the European Commission, and the Organization for Economic Cooperation and Development (OECD). Further, there is no evidence of any ill effects from the consumption of foods containing GM ingredients. Since GM crops were first cultivated commercially in 1995, many millions of meals have been made with GM ingredients and consumed by people in a number of countries with no demonstrated adverse effects (ICSU, 2004).

Still, the reviews do point out that, although currently available GM foods are considered safe to eat, this does not guarantee that no risks will be encountered as more foods are developed with novel characteristics, and different genes are stacked into one crop. An ongoing evaluation of emerging products is, therefore, required to ensure that new foods coming to market are safe for consumers. They recommend that food safety evaluation must be undertaken on a case-by-case basis, and that the extent of the risk evaluation should be proportionate to the possible risks involved with particular foods (FAO, 2004; ICSU, 2004).

Nutritional benefits of biofortification

Undernutrition remains a persistent problem in developing countries, especially in rural areas where the poor largely depend on staples and have limited access to a diverse diet. Malnutrition is still widespread, with both protein-energy deficiencies and micronutrient deficiencies, in particular vitamin A, zinc, and iron.

Biofortification, or improvement of the nutritional quality of food crops using conventional plant breeding or genetic engineering (White and Broadley, 2005; Nestel et al., 2006), is one strategy to improve access to nutritious food. Because staple foods are eaten in large quantities every day by the malnourished poor, delivery of biofortified foods can rely on existing consumer behavior (Bouis, 1999). Studies have shown that biofortification is a cost-effective way to reduce micronutrient deficiencies (Meenakshi et al., 2010), and it is more sustainable than nutrient supplementation (Bouis, 1999). Therefore, it is a viable complementary strategy to food fortification, which relies on the central processing of food and whose products may be less accessible, particularly to the rural poor (Bouis, 1999; Horton, 2006).

While both conventional breeding and biotechnology can be used for biofortification, conventional methods require sufficient variation in the genetic pools to justify the effort. Maize and sweet potatoes have sufficient variability in provitamin A content, but variability for iron in maize or provitamin A in rice, for example, are not sufficient for conventional breeding. Transgenic engineering, on the other hand, can draw from a much wider pool of genes, can increase micronutrients to much higher levels (as the example of golden rice shows; Unnevehr et al., 2007), and can progress much more quickly.

So far, no GM biofortified crops have been released, but two conventionally biofortified crops have been released: quality protein maize (QPM) and orange fleshed sweet potatoes (OFSP). Both have shown to increase the nutritional status of children and are indicative of the potential impact of GM biofortification.

Although the term biofortification most commonly refers to the process of increasing micronutrient levels, more recently it has been used in a larger sense to include other nutritional enhancements, such as the increased bioavailability of existing nutrients, improved oil profiles, or improved amino acid profiles resulting in improved protein quality (Unnevehr et al., 2007). Under this definition, the first biofortified crop was *opaque-2* (*o2*) maize, based on a natural mutation that nearly doubles the lysine and tryptophan content of maize grain (Mertz et al., 1964). Further breeding, using conventional methods, led to the development of agronomically competitive maize varieties with improved protein quality, collectively called quality protein maize (National Research Council, 1988; Vasal, 2000). QPM varieties are currently being developed and disseminated, particularly in sub-Saharan Africa (Krivanek et al., 2007).

A study in Ethiopia, where randomized households were provided with QPM seed or seed of a conventional variety for home production and consumption, found that children consuming conventional maize progressively faltered in their growth during the hungry season, whereas children consuming QPM did not change significantly in height-for-age and had a marginal increase in weight-for-age (Akalu et al., 2010). A meta-analysis of available randomized, controlled studies in target communities indicated that consumption of QPM instead of conventional maize leads to a 12% increase in the rate of growth in weight and a 9% increase in the rate of growth in height in infants and young children with mild to moderate undernutrition from populations in which maize is the major staple food (Gunaratna et al., 2010).

The second biofortified crop released was OFSP, which shows a high potential to reduce vitamin A deficiencies. In a study in Mozambique, combining agricultural and nutritional interventions to disseminate and promote OFSP (using a quasi-experimental design with a control group), vitamin A intakes were much higher in intervention children than in control children, leading to a significant increase of serum retinol concentrations (Low et al., 2007).

Nutritional impact of GM biofortification

Building on the experience of conventional biofortification, investments in biofortification research are likely to result in high returns due to the high malnutrition rates in developing countries and their high costs to human welfare and productivity, the low cost of breeding biofortified crops, and the relative ease of disseminating them to large groups of people (Bouis, 1999).

So far, several GM biofortified crops have been developed, but none has been released. The potential impact of these and future biofortified varieties can, however, be estimated from trials and *ex ante* impact assessments.

Rice is the most important food crop in the world, so golden rice, which has received the most attention, has the potential to dramatically reduce vitamin A (VA) deficiency in Asia. A first study, based on actual consumption patterns in the Cebu region of the Philippines, concluded that VA enriched rice is likely to deliver substantial amounts of VA, although it is unlikely to fulfill all requirements since rice is not the major staple of this country (Dawe et al., 2002). Using national data for the Philippines, developing a link between VA intake and disease, and quantifying the disease burden using the concept of disease adjusted life years (DALYs), high rates of return for the development and dissemination of golden rice were predicted (Zimmermann and Qaim, 2004). Given broad public support, golden rice could more than halve the disease burden of VA deficiency in India (Stein et al., 2008a)

Similarly, through biofortification of rice and wheat in India, the disease burden associated with iron deficiency (estimated at 4.0 million DALYs per year) could be reduced by 19–58% (Stein et al., 2008b) and the burden from zinc deficiency (estimated at 2.8 million DALYs per year) by 20–51% (Stein et al., 2007). The cost of saving one DALY through improved health from increased zinc intake through biofortification in India is estimated at $0.73–$7.31, which is lower than that of most other micronutrient interventions (Stein et al., 2007).

An *ex ante* impact assessment of the biofortification of globally important staple food crops with provitamin A, iron, and zinc for 12 countries in Africa, Asia, and Latin America, again using a modification of the DALY framework, indicated that the intervention can make a significant impact on the burden of micronutrient deficiencies in the developing world in a highly cost-effective manner (Meenakshi et al., 2010).

Reduced exposure to toxins, pesticides, and anti-nutrients

Evidence shows that GM crops have additional benefits for human health by reducing exposure to mycotoxins and pesticides, and have the potential to reduce anti-nutrients. Insect-resistant food crops suffer less insect damage in storage and are, therefore, less likely to be infected by fungi. Bt maize kernels, for example, accumulate fewer mycotoxins from opportunistic fungi than maize kernels from insect susceptible varieties. Mycotoxins are harmful, and often lethal, to humans and livestock (Miller et al., 2006).

Many studies have documented the reduced pesticide use in Bt cotton and Bt maize (Smale et al., 2009). Data from South Africa suggest that the number of accidental pesticide poisoning cases there has declined since the introduction of Bt crops (Bennett et al., 2003).

Finally, food crops can be engineered to reduce the effect of anti-nutrients and improve digestion. No such varieties have been released for human consumption, but China approved GM phytase maize for animal feed in 2009. Pork production with phytase maize will be more efficient because pigs can more easily digest phosphorus, coincidentally enhancing growth and reducing pollution due to the lower phosphate animal waste (James, 2010).

The Environment

Several reviews conclude that the environmental impact of GM crops may be either positive or negative depending on how and where they are used (FAO, 2004; ICSU, 2004). Possible negative environmental effects include gene transfer to wild relatives of conventional crops, weediness, and trait effects on non-target species (FAO, 2004). Positive effects on the environment include reduced pesticide use (insecticides and herbicides) and reduced tillage from Bt and HT crops (Qaim, 2009). These risks might affect developing countries differently where farmers have low levels of education, and regulatory and monitoring agencies have limited expertise and resources.

Currently available evidence suggests that genes can move from GM crops into landraces and related wild species, generally at low frequency and only in areas where compatible wild relatives are found (ICSU, 2004). In theory, GM traits like pest resistance could provide a fitness advantage (FAO, 2004), but evidence suggests that they do not significantly increase the fitness of the plants in semi-natural habitats. Hybridization between GM crops and wild relatives are likely to transfer those genes that are advantageous in agricultural environments, but they will not prosper in the wild (GM Science Review Panel, 2003).Traits that make them desirable as crops often make them less fit to survive in the wild (FAO, 2004). So far, no evidence of any deleterious environmental effects has occurred from the trait and species combinations currently available (ICSU, 2004).

Evidence suggests that genes can move from GM crops into landraces (ICSU, 2004). This is important for developing countries where, despite the availability of improved varieties, many farmers still grow a wide range of local varieties. These varieties have particular characteristics desirable to farmers, in particular, hardiness and storability, or to consumers, such as taste, texture, and color. If GM varieties of open pollinated crops with dominant traits (e.g., Bt maize) are grown in the neighborhood of local varieties, these traits are likely to cross over. Farmers often select desirable plants in the field for seed production, and since they skip plants with insect damage, those plants with the Bt gene will have a higher probability of being selected. Since maize pollen is heavy and does not travel far, buffer zones could, in theory, stop this flow, albeit a cost (Tumusiime et al., 2010). However, small-scale farmers do not have the required land, and regulatory agencies are unlikely to enforce a mandatory buffer. The release of GM crops in centers of origin is, therefore, one possible strategy (FAO, 2004) currently used in Mexico. Another strategy would be to engineer GM crops to limit their reproduction, but these so-called terminator genes were heavily contested and several seed companies have now agreed not to use them (Stein, 2005).

By using GM crops, farmers have been able to substantially reduce pesticide applications (ICSU, 2004). Globally, by 2004, the reduced use of pesticides in cotton was estimated at 14%, with several country studies showing reductions of 40–60% (ICSU, 2004). This is of particular benefit to the millions of small Bt cotton producers in developing countries. While HT technology decreased the use of more toxic herbicides, the total use of herbicides did not decrease. The technology does, however, reduce tillage and, therefore, erosion (Qaim and Traxler, 2005). While these effects are similar in developing and developed countries, they are not of major interest to smallholders.

In crop-associated biodiversity, the number of beneficial insects has increased (ICSU, 2004), but the effect on crop biodiversity is not clear. If farmers choose a limited number of popular GM varieties, diversity could decrease, but if the traits become available in a wide range of varieties, such as Bt maize in South Africa, this does not have to be the case.

GM technology can make crops more suitable to new areas, for example, drought-tolerant maize for dry lands, and rice tolerant to salinity (ICSU, 2004). At the same time, expanding crop acreage into marginal dry land areas or into mangroves can threaten delicate ecosystems. Clearly, a case-by-case environmental impact assessment of specific applications in specific agroecosystems is needed. The scientific community generally agrees that these assessments, as well as ecological monitoring of the environment for unintended effects after the release of GM crops, are necessary (FAO, 2004). To realize this, however, the capacity of environmental impact assessment and monitoring in developing countries needs to be increased.

Consumer Acceptance of GM Food

Regional differences

Although GM crops might be technologically superior and of great benefit to the farmer, they cannot have an impact if consumers are not willing to accept them. Despite the available scientific evidence, presented in this chapter, that GM food is safe to eat and GM crops have not demonstrated any negative effect on the environment, a large number of studies show that consumers in developed countries consistently show a preference for non-GM food over GM food (Lusk et al., 2005; Costa-Font et al., 2008). Only a few studies have quantified consumer preferences in developing countries, with one review (Smale et al., 2009), but these studies clearly indicate a more positive attitude than in the developed countries.

The preference for non-GM food is particularly strong in Europe, where the latest survey of public perceptions of biotechnology reveals widespread opposition to GM food in much of Europe, although public attitudes to medical and environmental applications remain positive (Gaskell et al., 2000). In most European countries, consumers find benefits associated with GM food insufficient to overcome their associated perceived risk, and they are willing to pay a premium for GM-free food (Costa-Font et al., 2008). Worldwide, consumers associate, on the whole, more negative than positive attributes to agrobiotechnology, and individual behaviors are driven by perceptions of beliefs about risks (Costa-Font et al., 2008). A meta-analysis of 25 studies reporting 57 valuations for GM food concluded that consumers on average place a 23–43% premium on non-GM products, and that premium is higher for European consumers than for consumers from North America and Asia (Lusk et al., 2005). Only one of these studies, however, was in a developing country (China) where consumers actually showed a preference for GM food.

Studies on consumer preferences in developing countries are few, but the available evidence indicates a low awareness of GM food and a more positive attitude. A recent review covering the 1996–2007 period identified 33 studies on consumers' attitudes toward GM food. Only 12 of them, however, tried to quantify consumers' willingness to pay (WTP) for GM food, and all of those were in Asia (Smale et al., 2009). The studies in China and India all found positive consumer attitudes toward GM food. The studies in China indicated that consumers there were actually willing to pay a premium for GM food. In Beijing, for example, consumers were willing to pay a premium of 38% for GM rice (Li et al., 2002). The high level of acceptance of biotech products in China can be ascribed in part to government policies and also to cultural and political history (Smale et al., 2009).

Similarly, Indian consumers (in New Delhi and Patna) were willing to pay a small premium for biotech chapattis (Anand et al., 2007) even if no information was provided, and this premium increased slightly with "producer friendly" information. A second Indian study (in Ahmedabad) found very low awareness among consumers, but still 70% were willing to buy GM food at the same price as conventional food (Deodhar et al., 2007). These results were confirmed by a later study in India, showing that 60% of urban consumers would purchase Bt vegetables at current conventional vegetable prices (Krishna and Qaim, 2008b).

In Africa, three studies on consumer acceptance of GM food are available. Among urban maize consumers in Kenya, only 38% were aware of GM crops, but two-thirds would buy GM maize at the same price as conventional maize (Kimenju and De Groote, 2008). Many of these consumers were, however, concerned about the loss of biodiversity and the associated impact on non-target insects. A second study in urban centers of northern Nigeria, on the other hand, found 90% of the respondents were aware of GM products, but two-thirds disapproved of the use of GM cowpeas (Kushwaha et al., 2008). The third study, in Kenya, shows that only a small proportion of rural consumers were aware of GM crops, but a large majority would be willing to buy GM maize meal at the same price as conventional maize (those unaware were provided with neutral information; Kimenju et al., 2011). A survey study among the gatekeepers of the food industry found that most respondents were largely non-committal on the use of GM products in their companies, preferring to decide on a case-by-case basis (Bett et al., 2010).

Factors influencing acceptance

Since consumer acceptance has been such a driving force in the regulatory process, a good understanding of the factors that influence consumers' perceptions and WTP is important. Consumers' attitudes and behaviors are complex and derive from subjective knowledge, information, and values. Important factors that were found to play a role, apart from country or region of origin, were awareness and information, risk and benefit perceptions, and the type of product. In quantifying WTP, the methods used in consumer surveys created major differences.

In general, most consumers reluctant about GM foods are typically those who are relatively more risk conscious and who exhibit attitudes favoring sluggish technology innovation in the food sector (Costa-Font et al., 2008). Consumer characteristics, the food studied, and the research method used were the most important factors in estimating WTP for non-GM food (Lusk et al., 2005). GM meat products were found least acceptable and GM oil most acceptable, whereas GM products that provided tangible benefits such as increased nutrition significantly decreased the premiums for non-GM food. Methodologically, non-hypothetical premiums were lower than hypothetical premiums, whereas conducting the valuation task in person generated lower premiums for non-GM food as compared to when the valuation was elicited over the phone or by mail.

In developing countries, the impact of information on the preferences expressed by consumers was found to be crucial, irrespective of the region of study, and the attitudes of consumers change significantly as they absorb new information, particularly negative messages (Smale et al., 2009). In Nigeria, those respondents who were most concerned about the ethics of genetic transformation were also more likely to disapprove of such products (Kushwaha et al., 2008). In Kenya, perceptions of health risk and ethical and equity concerns had a negative influence on the likelihood of purchasing GM maize meal, whereas trust in government to ensure food quality had a positive influence on WTP (Kimenju and De Groote, 2008). All studies in developing countries reported so far used hypothetical methods, which are likely to raise the estimated WTP, but they were also mostly conducted in person.

Regulatory Systems

Importance of regulatory systems

When scientists develop technologies based on living organisms, which are the essence of biotechnology, the outcome of their applications cannot be predicted with 100% accuracy, because the application of biotechnology always carries certain risks. Regulatory systems are, therefore, needed to assess those risks, in particular with respect to food safety and the environment. These impacts need to be quantified where possible, and the benefits of proposed technologies compared to their benefits.

Apart from the Codex Alimentarius, which governs food safety standards, there are no international standards for health and environmental assessment, and each country has to set up its own regulatory system. So when a company has developed a new GM product, it needs to apply for a license for that product in each of the different countries where it wants to deploy the product. The regulation of biotech products can be expensive and time-consuming, depending on the standards of risk assessment maintained. Moreover, this can delay access to the technology by producers and consumers, or it can even deprive target populations of the potential benefits altogether (Paarlberg, 2008).

Currently, the heavy regulatory systems arguably pose the most important single constraint to the adoption of biotech crops in most developing countries (James, 2010). Proper economic analysis of these systems is, therefore, needed (Qaim, 2009). Regional harmonization needs to be explored to reduce the burden of regulation to make the benefits of biotechnology available to the poor, while maintaining high health and environmental safety standards.

Regional differences

Despite large potential benefits, the use of GM varieties remains controversial, largely driven by negative perceptions from Western consumers, in particular, Europeans. In the United States, scientific studies have concluded that GM crops do not pose any more risk than their conventional counterparts, and they have basically been deregulated (Paarlberg, 2000). In Europe, the potential benefits of biotechnology are limited (Demont et al., 2004), but consumers are worried about the safety of their food quality and of their food system and are well organized, and trade barriers offer protection for local farmers. As a result, a regulatory system has been built upon the precautionary principle (McMahon, 2003), which requires that new technologies demonstrate no harm (ICSU, 2004). In practice, this has put high burdens of proof on the applicants, and as a result the production of GM crops is all but banned from the EU.

In developing countries, however, food crop yields are often stagnating, and the need for and potential gains from biotech crops are much more important. Moreover, given the food security situation there, consumers in developing countries are likely more concerned about sufficient food than about perceived risks. This is supported by most studies of consumer acceptance in developing countries (Qaim, 2009; Smale et al., 2009).

Still, governments and the public have been alarmed, mostly by concerns from Western consumers (Paarlberg, 2002). In Africa, the political elite has strong cultural ties with Europe, where many have been trained, go on vacation, and receive medical treatment, and from where they receive important news and information services. Therefore, with the help of donors and some international agencies, most African countries have, or are in the process of, developing regulatory frameworks and systems based on the European example (Paarlberg, 2008).

It can be argued that these regulatory systems, designed almost 15 years ago to answer the initial needs of wealthy industrial countries dealing with a new technology and with access to significant resources for regulation, are not appropriate for the current GM crops in poor developing countries that do not have the resources or the expertise required (James, 2010).

Economics of regulation

The regulatory process of having a GM product approved in one country, and so far each country approves them separately, can be very expensive. Estimates vary from $1 million (James, 2010) to $6–15 million for a new Bt or HT maize technology in one country (Kalaitzandonakes et al., 2007).

Delays in forgone benefits are also an important indirect cost of regulation (Qaim, 2009). The cost of a two-year delay in the approval of Bt cotton in India led to estimated aggregated losses to farmers of more than $100 million (Pray et al., 2005). In Brazil, the approval of HT soybean took from 1998 to 2006, due to a cumbersome process and legal challenges. The cost of this delay to the economy has been estimated at $4.5 billion, most of which ($3.10 billion) was borne by the farmers (Anderson et al., 2008).

The cost, complexity, and uncertainty of the regulation of GM crops make regulatory requirements a barrier for public research institutes, poor countries, and small companies (ICSU, 2004), and can impede their commercialization of GM technologies in small countries and for minor crops (Qaim, 2009), limiting the choices for generating public good (ICSU, 2004).

Biological systems cannot deliver certainty; zero risk is an unattainable standard (ICSU, 2004). Probabilistic risk assessment models, however, can and should be used, similar to models used in pesticide regulation (Qaim, 2009). These models should balance objective, scientific risk assessment with subjective, but clearly stated and transparent safety rule decision making.

The way forward

Within the scientific community, there is broad agreement that regulatory systems need to be science-based and transparent, and must involve community participation (ICSU, 2004). Further, safety assessments should be undertaken on a case-by-case basis, using the best available techniques. International standards and regional harmonization would reduce the cost and make GM technologies more available to the people who would benefit the most from them: farmers and consumers in developing countries.

For food, there is the Codex Alimentarius, an intergovernmental forum that seeks to achieve international agreement in standards of food safety, including GM food. A similar forum is needed to facilitate international agreement on standards for assessing the environmental impact of gene technology (ICSU, 2004). The Cartagena Protocol of the Convention on Biological Diversity (CBD) provides for one aspect, but a broader forum is needed to enable the development of internationally agreed standards for comprehensive environmental impact assessments of the risks and benefits of new genetics in agriculture.

The data sought by regulators are similar, but the interpretation in risk assessment and management differs among countries and regions, particularly in dealing with areas of uncertainty. The development of probabilistic models would help in making the discussion more scientific.

With the accumulated knowledge of the last fourteen years, it is now possible to design appropriate regulatory systems that are responsible, rigorous, and yet not onerous, requiring only modest resources that are within the means of most developing countries (James, 2010).

Conclusions

GM crops have now been successfully deployed in both developing and developed countries for the last 15 years. With much of the basic research in genetic engineering provided by the public sector, most GM crops were developed by private companies in developed countries for the needs of farmers there. However, current economic benefits to farmers are equally divided between developed and developing countries. Moreover, a majority of the farmers using GM crops are small-scale farmers in developing countries who have benefited from technologies that fit their needs, particularly from Bt crops and especially from Bt cotton.

Biotech crops have also been shown to benefit the environment, especially in reducing pesticide applications and tillage, and to increase biodiversity in non-target mechanisms. Environmental risks such as the development of insect resistance and gene flow are clearly possible, but so far have not had any major effect. The crossover of biotech genes into local varieties, while not yet demonstrated, seems likely to happen and would be hard to contain.

Further applications of GM technology to address the specific needs of poor farmers and consumers in developing countries look promising. The experience with conventionally biofortified crops and *ex ante* impact assessment indicates that GM biofortified crops have a high potential to reduce micronutrient deficiencies and improve nutrition among poor consumers. Several other GM technologies are under development, many under PPPs, to address the needs of poor farmers. These include crops resistant to abiotic stresses, such as drought, low fertility, and salinity. These crops can increase yields, food security, and income, as well as reduce the variability of those outcomes.

However, several factors hinder the use of GM crops in developing countries and prevent them from reaching their full potential. They cause costly delays in foregone benefits to the poor. First, the specific needs of developing countries are addressed by public research institutes in national and international systems, but their resources and capacity are limited and, therefore, need to be expanded. Secondly, awareness of biotechnology and knowledge of the benefits and risks of GM crops and food is limited. Consumers in developing countries, both urban and rural, need better information and need to be engaged in the ongoing debate. A better understanding of their attitudes and perceptions of GM food is, therefore, needed. Thirdly, the regulatory system is cumbersome and expensive. Evaluations need to be objective and scientific, and based on objective risk assessment. Moreover, clear and transparent safety rule decision mechanisms are needed that should include the opinions of farmers and consumers. The cost of these evaluations and the regulatory process should be balanced with the perceived risks and the potential benefits.

Finally, GM crops are not risk-free, as with any biological intervention. Proper evaluation on a case-by-case basis is needed for every combination of trait, crop, and agroecological system. Apart from regulation, continuous monitoring of these crops in the environment will be needed. Similarly, the use of GM food in the food systems and its effect on health needs to be monitored.

Acknowledgments

This chapter was written during a sabbatical at Cornell University. I would like to thank the faculty and staff of the Charles H. Dyson School of Applied Economics and Management for hosting me, the Frosty Hill Foundation for their financial support, and the Mann Library for their assistance. I also would like to thank Kathleen Sinclair for editing the manuscript.

References

Akalu, G., Taffesse, S., Gunaratna, N. S., & De Groote, H. (2010). The effectiveness of quality protein maize in improving the nutritional status of young children in the ethiopian highlands. *Food and Nutrition Bulletin*, *31*, 418–430.

Alston, J. M., Norton, G. W., & Pardey, P. G. (1998). *Science under scarcity: Principles and practices for agricultural research evaluation and priority setting*. Wallingford, U.K: CAB International (CABI).

Anand, A., Mittelhammer, R. C., & McCluskey, J. J. (2007). Consumer response to information and second-generation genetically modified food in India. *Journal of Agricultural and Food Industrial Organization*, *5*, 1–18.

Anderson, K., Valenzuela, E., & Jackson, L. A. (2008). Recent and prospective adoption of genetically modified cotton: A global computable general equilibrium analysis of economic impacts. *Economic Development and Cultural Change*, *56*, 265–296.

Bennett, R., Morse, S., & Ismael, Y. (2003). Bt cotton, pesticides, labour and health: A case study of smallholder farmers in the Makhathini Flats, Republic of South Africa. *Outlook on Agriculture*, *32*, 123–128.

Bett, C., Ouma, J. O., & De Groote, H. (2010). Perspectives of gatekeepers in the Kenyan food industry towards genetically modified food. *Food Policy*, *35*, 332–340.

Black, R. E., Allen, L. H., Bhutta, Z. A., Caulfield, L. E., de Onis, M., Ezzati, M., et al. (2008). Maternal and child undernutrition: Global and regional exposures and health consequences. *Lancet*, *371*, 243–260.

Bouis, H. E. (1999). Economics of enhanced micronutrient density in food staples. *Field Crops Research*, *60*, 165–173.

Brookes, G., & Barfoot, P. (2010). *GM crops: Global socio-economic and environmental impacts 1996–2008*. Dorchester, UK: PG Economics Ltd.

Caulfield, L. E., de Onis, M., Blössner, M., & Black, R. E. (2004). Undernutrition as an underlying cause of child deaths associated with diarrhea, pneumonia, malaria, and measles. *American Journal of Clinical Nutrition*, *80*, 193–198.

Conway, G. (2000). Genetically modified crops: Risks and promise. *Conservation Ecology*, *4* (1), 2. [online] URL: < http://www.consecol.org/vol4/iss1/art2/ >

Costa-Font, M., Gil, J. M., & Traill, W. B. (2008). Consumer acceptance, valuation of and attitudes towards genetically modified food: Review and implications for food policy. *Food Policy*, *33*, 99–111.

Dawe, D., Robertson, R., & Unnevehr, L. (2002). Golden rice: What role could it play in alleviation of vitamin A deficiency? *Food Policy*, *27*, 541–560.

De Groote, H., Overholt, W. A., Ouma, J. O. & Wanyama, J. (2011a). Assessing the potential economic impact of *Bacillus thuringiensis* (Bt) maize in Kenya. *African Journal of Biotechnology*, *10*, 4741–4751.

De Groote, H., Hall, M. D., Spielman, D. J., Mugo, S. N., Andam, K., & Munyua, B. G. et al. (2011b). Options for pro-poor maize seed market segmentation in Kenya. *African Journal of Biotechnology*, *10*, 4699–4712.

de Onis, M., Blossner, M., Borghi, E., Frongillo, E. A., & Morris, R. (2004). Estimates of global prevalence of childhood underweight in 1990 and 2015. *Journal of the American Medical Association*, *291*, 2600–2606.

Demont, M., Wesseler, J., & Tollens, E. (2004). Biodiversity versus transgenic sugar beets: The one Euro question. *European Review of Agricultural Economics*, *31*, 1–18.

Deodhar, S. Y., Ganesh, S., & Chern, W. S. (2007). *Emerging markets for GM foods: An Indian perspective on consumer understanding and willingness to pay*. Ahmedabad, India: Indian Institute of Management. IIMA Working Paper 2007-06-08

Evenson, R. E., & Gollin, D. (2003). Assessing the impact of the green revolution. *Science*, *300*, 758–762.

FAO. (2004). *The state of food and agriculture 2003–2004. Agricultural Biotechnology: Meeting the needs of the poor*. Rome: Food and Agriculture Organization of the United Nations.

FAO. (2009). How to Feed the World in 2050 [Online]. Available by FAO < http://www.fao.org/fileadmin/templates/wsfs/docs/expert_paper/How_to_Feed_the_World_in_2050.pdf > (verified March 1, 2011).

FAOSTAT. (2010). Core production data (online). http://faostat.fao.org/site/339/default.aspx (verified May 1, 2010).

Gaskell, G., Wolfgang, W., Nicole, K., Helge, T., Juergen, H., & Julie, B. (2004). GM foods and the misperception of risk perception. *Risk Analysis*, *24*, 185–194.

Gaskell, G., Allum, N., Bauer, M., Durant, J., Allansdottir, A., & Bonfadelli, H., et al. (2000). Biotechnology and the European public. *Nature Biotechnology*, *18*, 935–938.

GM Science Review Panel. (2003). GM Science Review, First Report. An open review of the science relevant to GM crops and food based on the interests and concerns of the public, London.

Gouse, M., Pray, C., & Schimmelpfennig, D. (2004). The distribution of benefits from Bt cotton adoption in South Africa. *AgBioForum*, *7*, 187–194.

Gouse, M., Pray, C., Schimmelpfennig, D., & Kirsten, J. (2006). Three seasons of subsistence insect-resistant maize in South Africa: Have smallholders benefited?. *AgBioForum*, *9*, 15–22.

Gunaratna, N. S., De Groote, H., Nestel, P., Pixley, K. V., & McCabe, G. P. (2010). A meta-analysis of community-level studies on quality protein maize. *Food Policy*, *35*, 202–210.

Horton, S. (2006). The economics of food fortification. *Journal of Nutrition*, *136*, 1068–1071.

Houndekon, V., De Groote, H., & Lomer, C. (2006). Health costs and externalities of pesticide use in the sahel. *Outlook on Agriculture*, *35*, 25–31.

Huang, J., Rozelle, S., Pray, C., & Wang, Q. (2002). Plant Biotechnology in China. *Science*, *25*, 674–676.

Huang, J., Hu, R., Rozelle, S., & Pray, C. (2005). Insect-resistant GM rice in farmers' fields: Assessing productivity and health effects in China. *Science*, *308*, 688–690.

ICSU. (2004). *New Genetics, Food and Agriculture: Scientific Discoveries – Societal Dilemmas*. International Council for Science.

James, C. (2008). *Global Status of Commercialized Biotech/GM Crops: 2007*. Ithaca NY, New York: International Service for the Acquisition of Agri-Biotech Applications (ISAAA). ISAAA Briefs No. 37-2007.

James, C. (2010). *Global Status of Commercialized Biotech/GM Crops: 2009. The first fourteen years, 1996 to 2009*. Ithaca NY, New York: International Service for the Acquisition of Agri-Biotech Applications (ISAAA). ISAAA Briefs No. 41-2009.

Kalaitzandonakes, N., Alston, J. M., & Bradford, K. J. (2007). Compliance costs for regulatory approval of new biotech crops. *Nature Biotechnology*, *25*, 509–511.

Kimenju, S. C., & De Groote, H. (2008). Consumers' willingness to pay for genetically modified food in Kenya. *Agricultural Economics*, *38*, 35–46.

Kimenju, S. C., Bett, C., & De Groote, H. (2011). Attitudes and perceptions of maize consumers and food industry stakeholders towards Genetically Modified Food in Kenya. *African Journal of Biotechnology*, *10*, 4767–4776.

Krishna, V. V., & Qaim, M. (2008a). Potential impacts of Bt eggplant on economic surplus and farmers' health in India. *38*, 167–180.

Krishna, V. V., & Qaim, M. (2008b). Consumer attitudes toward GM food and pesticide residues in India. *Applied Economic Perspectives and Policy*, *30*, 233–251.

Krivanek, A. F., De Groote, H., Gunaratna, N. S., Diallo, A. O., & Friesen, D. (2007). Breeding and disseminating quality protein maize (QPM) for Africa. *African Journal of Biotechnology*, *6*, 312–324.

Kushwaha, S., Musa, A. S., Lowenberg-DeBoer, J., & Fulton, J. (2008). Consumer acceptance of genetically modified (GM) — Cowpeas in sub-Saharan Africa. *Journal of International Food & Agribusiness Marketing*, *20*, 7–23.

Li, Q., Curtis, K. R., McCluskey, J. J., & Wahl, T. I. (2002). Consumer attitudes toward genetically modified foods in Beijing, China. *AgiBioForum*, *5*, 145–152.

Low, J. W., Arimond, M., Osman, N., Cunguara, B., Zano, F., & Tschirley, D. (2007). A food-based approach introducing orange-fleshed sweet potatoes increased vitamin A intake and serum retinol concentrations in young children in rural

Mozambique. *Journal of Nutrition*, *137*, 1320–1327.

Lusk, J. L., Jamal, M., Kurlander, L., Roucan, M., & Taulman, L. (2005). A meta analysis of genetically modified food valuation studies. *Journal of Agricultural and Resource Economics*, *30*, 28–44.

McMahon, J. A. (2003). Food safety and the precautionary principle. *EuroChoices*, *41*, 42–46.

Meenakshi, J. V., Johnson, N. L., Manyong, V. M., DeGroote, H., Javelosa, J., Yanggen, D. R., et al. (2010). How cost-effective is biofortification in combating micronutrient malnutrition? An *ex ante* assessment. *World Development*, *38*, 64–75.

Mertz, E. T., Bates, L. S., & Nelson, O. E. (1964). Mutant gene that changes protein composition and increases lysine content of maize endosperm. *Science*, *145*, 279–280.

Miller, H. A., Conko, G., & Kershen, D. L. (2006). Why spurning food biotech has become a liability. *Nature Biotechnology*, *24*, 1075–1077.

Mugo, S. N., Mwimali, M. G., Taracha, C. O., Songa, J. M., Gichuki, S. T., & Tende, R., et al. (2011). Testing public Bt maize events for control of stem borers in the first confined field trials in Kenya. *African Journal of Biotechnology*, *10*, 4713–4718.

National Research Council. (1988). *Quality-protein maize*. Washington, D.C., USA: National Academy Press.

Nestel, P., Bouis, H. E., Meenakshi, J. V., & Pfeiffer, W. (2006). Biofortification of staple food crops. *Journal of Nutrition*, *136*, 1064–1067.

Nuffield Council on Bioethics. (1999). *Genetically modified crops: The ethical and social issues*. London, UK: Nuffield Council on Bioethics.

Paarlberg, R. (2008). *Starved for science: How biotechnology is being kept out of Africa*. Cambridge, MA: Harvard University Press.

Paarlberg, R. L. (2000). *Governing the GM crop revolution policy choices for developing countries*. Washington, D. C: International Food Policy Research Institute. Food, Agriculture, and the Environment. Discussion Paper 33.

Paarlberg, R. L. (2002). The real threat to GM crops in poor countries: Consumer and policy resistance to GM foods in rich countries. *Food Policy*, *27*, 247–250.

Pingali, P. L., & Heisey, P. W. (1999). *Cereal Crop Productivity in Developing Countries*. Mexico D.F.: CIMMYT. CIMMYT Economics Paper 99-03.

Pingali, P. L., & Traxler, G. (2002). Changing locus of agricultural research: Will the poor benefit from biotechnology and privatization trends? *Food Policy*, *27*, 223–238.

Pingali, P. L., Marquez, C. B., & Palis, G. F. (1994). Pesticides and Philippine rice farmer health: A medical and economic analysis. *American Journal of Agricultural Economics*, *76*, 587–592.

Pray, C. E., Bengali, P., & Ramaswami, B. (2005). The cost of biosafety regulations: The Indian experience. *Quarterly Journal of International Agriculture*, *44*, 267–289.

Pray, C. E., Ma, D., Huang, J., & Qiao, F. (2001). Impact of Bt cotton in China. *World Development*, *29*, 813–825.

Qaim, M. (2009). The economics of genetically modified crops. *Annual Review of Resource Economics*, *1*, 665–694.

Qaim, M., & de Janvry, A. (2003). Genetically modified crops, corporate pricing strategies, and farmers' adoption: The case of Bt cotton in Argentina. *American Journal of Agricultural Economics*, *85*, 814–828.

Qaim, M., & Zilberman, D. (2003). Yield effects of genetically modified crops in developing countries. *Science*, *299*, 900–902.

Qaim, M., & Traxler, G. (2005). Roundup ready soybeans in Argentina: Farm level and aggregate welfare effects. *Agricultural Economics*, *32*, 73–86.

Rosegrant, M. W., Paisner, M. S., Meijer, S., & Witcover, J. (2001). *2020 Global food outlook trends, alternatives, and choices*. Washington, D.C: International Food Policy Research Institute (IFPRI).

Sanchez, P. A., Shepherd, K. D., Soule, M. J., Place, F. M., Mokwunye, A. U., & Buresh, R. J., et al. (1997). Soil fertility replinishment in Africa: An investment in natural resource capital. In R. J. Buresh (Ed.), *Replenishing soil fertility in Africa* (pp. 1–46). Madison WI: Soil Science Society of America.

Smale, M., Zambrano, P., Gruerè, G., Falck-Zepeda, J., Matuschke, I., Horna, D., et al. (2009). Measuring the Economic Impacts of Transgenic Crops in Developing Agriculture during the First Decade. Approaches, Findings, and Future Directions. Food Policy Review 10 IFPRI, Washington DC.

Stein, A. J., Sachdev, H. P. S., & Qaim, M. (2008a). Genetic engineering for the poor: Golden rice and public health in India. *World Development*, *36*, 144–158.

Stein, A. J., Nestel, P., Meenakshi, J. V., Qaim, M., Sachdev, H. P. S., & Bhutta, Z. A. (2007). Plant breeding to control zinc deficiency in India: How cost-effective is biofortification? *Public Health Nutrition*, *10*, 492–501.

Stein, A. J., Meenakshi, J. V., Qaim, M., Nestel, P., Sachdev, H. P. S., & Bhutta, Z. A. (2008b). Potential impacts of iron biofortification in India. *Social Science & Medicine*, *66*, 1797–1808.

Stein, H. (2005). Intellectual property and genetically modified seeds: The United States, trade, and the developing world. *North Western Journal of Technology and Intellectual Property*, *3*, 160–178.

The World Bank, *World Development Report 2008*. Washington DC: Agriculture for Development The World Bank.

Trigo, E. J., & Cap, E. J. (2006). *Ten years of genetically modified crops in Argentine agriculture*. Buenos Aires: Argent. Counc. Inf. Dev. Biotechnol.

Tumusiime, E., De Groote, H., Vitale, J., & Adam, B. (2010). The cost of coexistence between Bt maize and open pollinated maize varieties in lowland coastal Kenya. *AgBioForum*, *13*, 208–221.

Unnevehr, L., Pray, C., & Paarlberg, R. (2007). Addressing micronutrient deficiencies: Alternative interventions and technologies. *AgBioForum*, *10*, 124–134.

Vasal, S. K. (2000). The quality protein maize story. *Food and Nutrition Bulletin*, *21*, 445–450.

White, P. J., & Broadley, M. R. (2005). Biofortifying crops with essential mineral elements. *Trends in Plant Science*, *10*, 586–593.

Williams, J. H., Phillips, T. D., Jolly, P. E., Stiles, J. K., Jolly, C. M., & Aggarwal, D. (2004). Human aflatoxicosis in developing countries: A review of toxicology, exposure, potential health consequences, and interventions. *The American Journal of Clinical Nutrition*, *80*, 1106–1122.

Zilberman, D., Ameden, H., & Qaim, M. (2007). The impact of agricultural biotechnology on yields, risks, and biodiversity in low-income countries. *Journal of Development Studies*, *43*, 63–78.

Zimmermann, R., & Qaim, M. (2004). Potential health benefits of golden rice: A Philippine case study. *Food Policy*, *29*, 147–168.

Index

A

B

C

D

E

F

G

N

O

P

Q

R

S

T

U

V

W

X

Y

Z

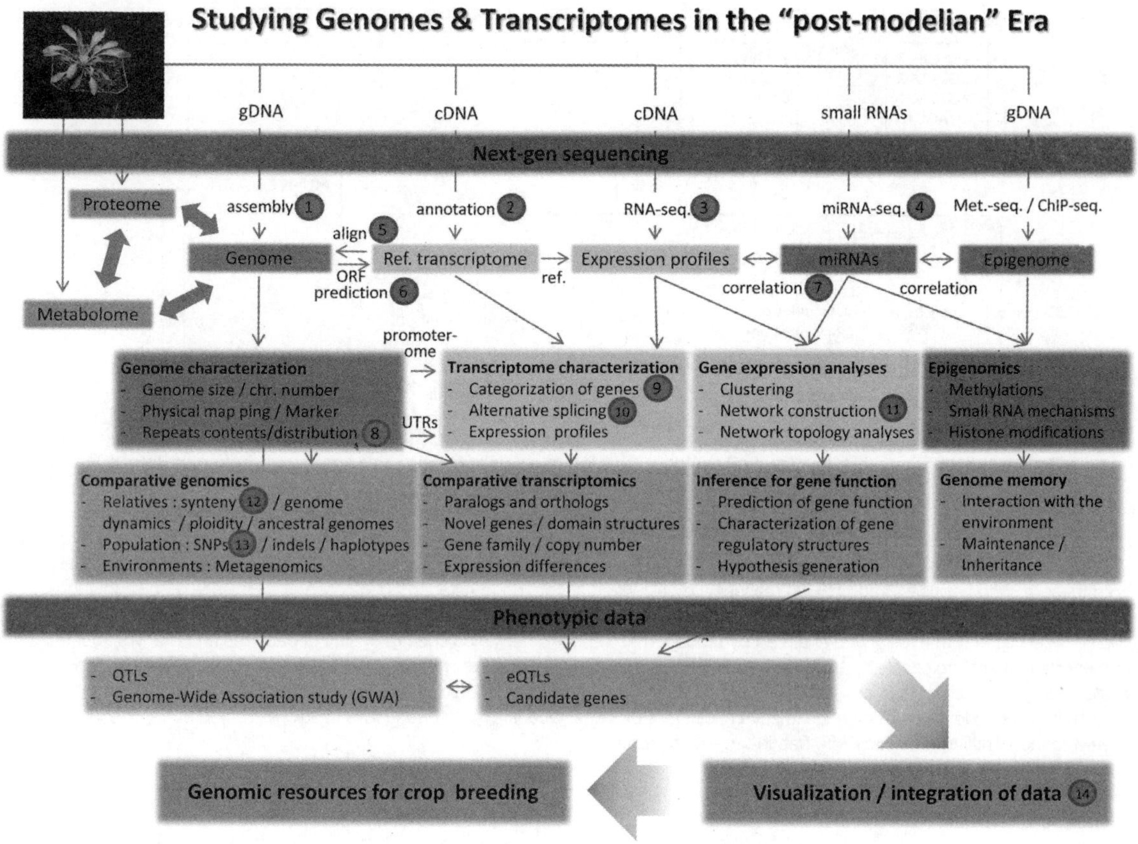
Studying Genomes & Transcriptomes in the "post-modelian" Era
gDNA
cDNA
cDNA
small RNAs
gDNA
Next-gen sequencing
Proteome
assembly 1
annotation 2
RNA-seq. 3
miRNA-seq. 4
Met.-seq. / ChIP-seq.
align 5
Genome
Ref. transcriptome
Expression profiles
miRNAs
Epigenome
ORF prediction 6
ref.
correlation 7
correlation
Metabolome
promoter-ome
Genome characterization
- Genome size / chr. number
- Physical map ping / Marker
- Repeats contents/distribution 8
UTRs
Transcriptome characterization
- Categorization of genes 9
- Alternative splicing 10
- Expression profiles
Gene expression analyses
- Clustering
- Network construction 11
- Network topology analyses
Epigenomics
- Methylations
- Small RNA mechanisms
- Histone modifications
Comparative genomics
- Relatives : synteny 12 / genome dynamics / ploidity / ancestral genomes
- Population : SNPs 13 / indels / haplotypes
- Environments : Metagenomics
Comparative transcriptomics
- Paralogs and orthologs
- Novel genes / domain structures
- Gene family / copy number
- Expression differences
Inference for gene function
- Prediction of gene function
- Characterization of gene regulatory structures
- Hypothesis generation
Genome memory
- Interaction with the environment
- Maintenance / Inheritance
Phenotypic data
- QTLs
- Genome-Wide Association study (GWA)
- eQTLs
- Candidate genes
Genomic resources for crop breeding
Visualization / integration of data 14

Figure 2.1 • Studying genomes and transcriptomes in the "post-modelian" era.

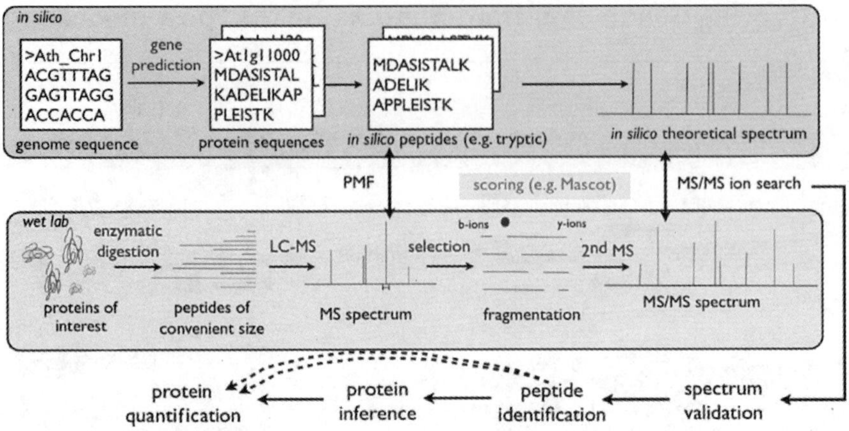

Figure 4.1 • Typical proteomics workflow for protein identification and quantification in complex sample mixtures. • Two parallel procedures constitute a typical proteomics workflow. Protein identification from measured spectra in the wet lab (in blue) directly relay on genomic sequences processed *in silico* (in green). *In silico*, protein sequence databases are generated from available genomic sequences. In this process, a gene prediction algorithm is used to predict open reading frames from which protein sequences are subsequently extracted. Protein sequences are then digested *in silico* into small polypeptides (peptides) following the same enzymatic rules as used in the wet lab. From these *in silico* peptides, fragment ion masses are calculated according to established fragmentation rules (e.g., b- and y-ions for collision-induced dissociation activation) resulting in the theoretical MS/MS spectrum. In the wet lab, protein extracts are enzymatically digested (in most cases with trypsin) and analyzed by tandem mass spectrometry. MS spectra of eluting peptides and MS/MS spectra obtained after fragmentation of selected peptides are recorded. For protein identification, measured MS and MS/MS spectra are compared with *in silico* spectra by using specific algorithms such as Mascot or Sequest. Spectra matching results in peptide and subsequent protein identification if the peptide is not ambiguous to several different proteins. Quantitative information is usually extracted at peptide level and then propagated to the identified proteins. PMF: Peptide mass fingerprint.

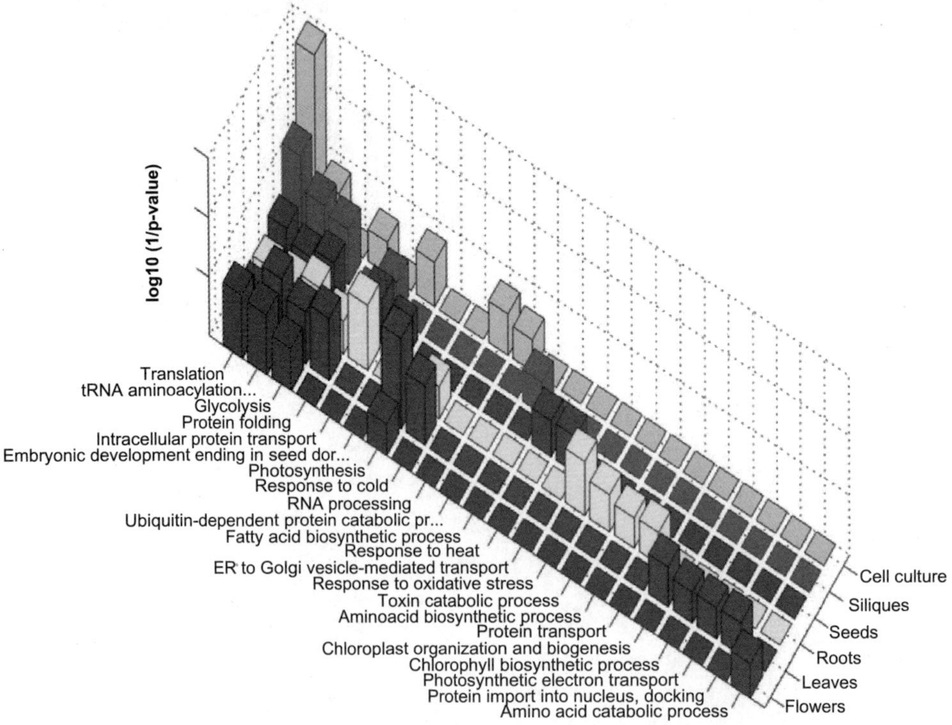

Figure 4.2 • Protein landscape of identified proteins in *Arabidopsis* tissues. • Functional classification of proteins identified in a large-scale proteome study of various *Arabidopsis* tissues. Functional classification is based on TAIR GO categories from the aspect "biological process." Figure and legend are adapted from Baerenfaller et al. (2008).

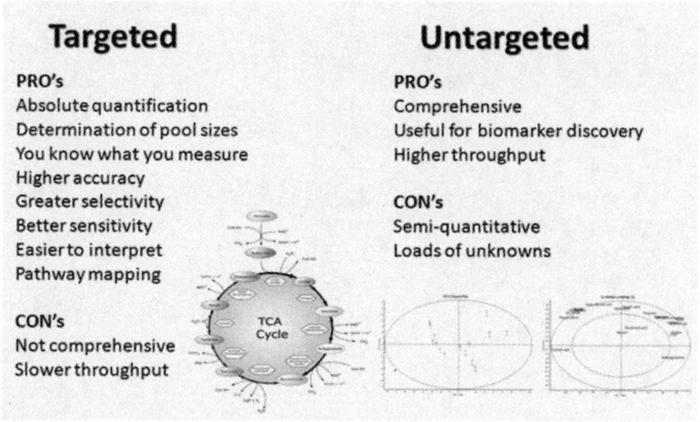

Figure 5.2 • Schematics of the advantages and disadvantages of targeted versus untargeted metabolite analysis.

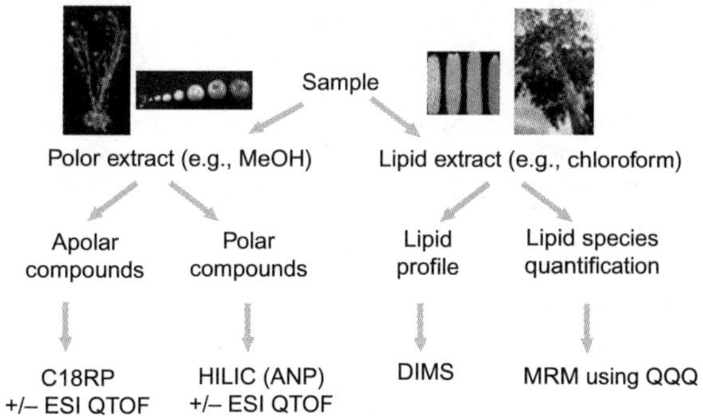

Figure 5.3 • Schematic of the potential methodologies required for untargeted and highly comprehensive metabolite analyses.

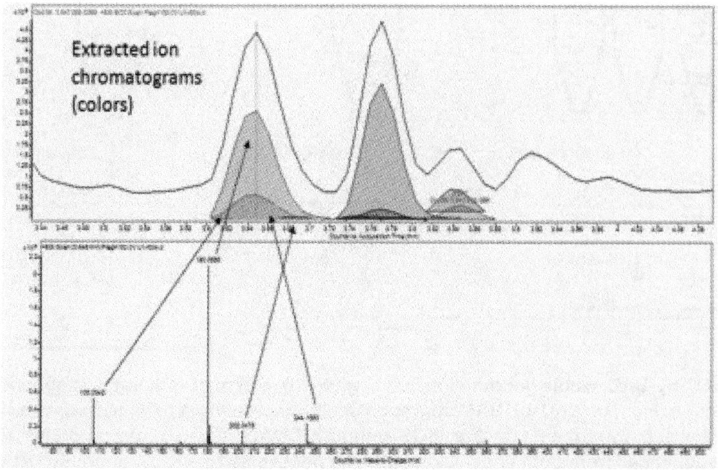

Figure 5.4 • Example of deconvolution. • Upper panel shows the total ion chromatogram (in black) and the deconvoluted extracted ion chromatograms of each compound detected (in color). The lower panel shows the respective accurate mass peaks for the deconvoluted extracted ion chromatograms for the leftmost peak in the upper panel only.

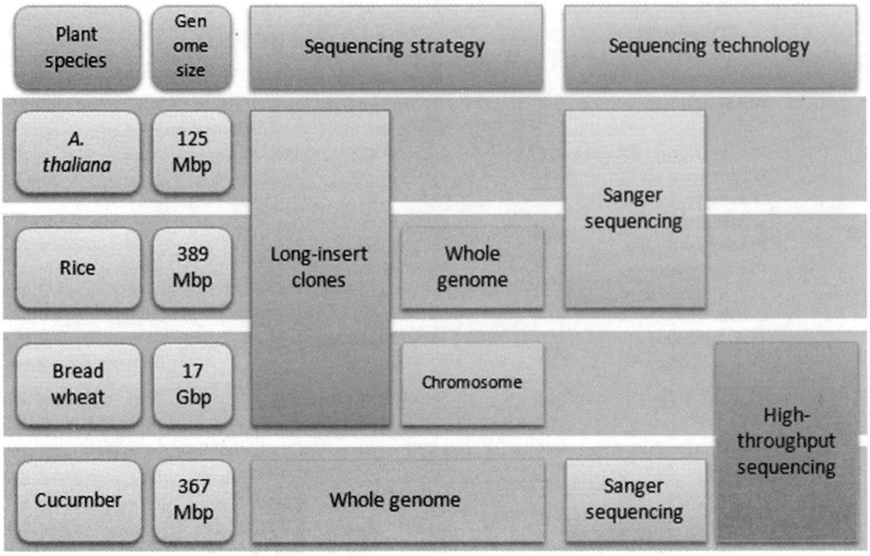

Figure 6.1 • Strategies used for plant genome sequencing. • Depending on the genome size, different strategies have been deployed in plants, using long-insert clones such as bacterial artificial chromosomes (BAC) and shotgun sequencing of the whole genome or of specific chromosomes. Sanger sequencing technology, high-throughput sequencing (or next-generation sequencing), and combinations of both technologies were utilized.

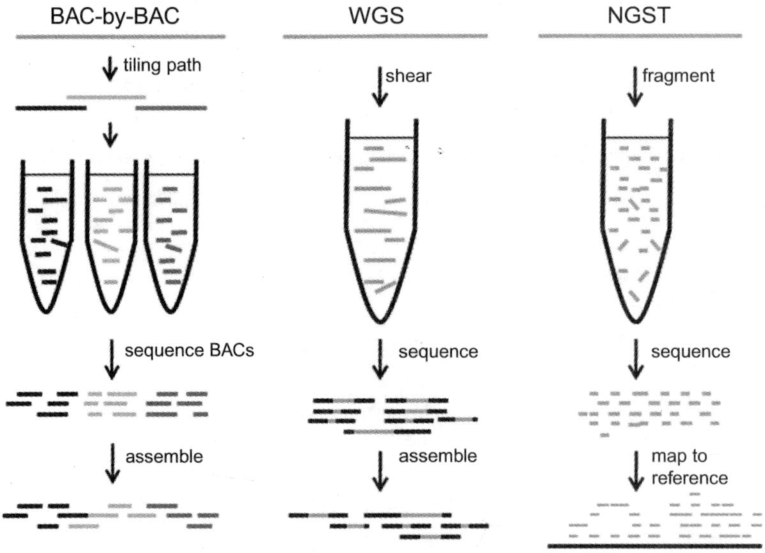

Figure 6.2 • Principles of BAC-by-BAC, whole genome sequencing (WGS), and next-generation sequencing technologies (NGST) approaches of genome sequencing. The BAC-by-BAC approach starts by creating a tiling path of overlapping BACs followed by shotgun sequencing and assembly of the BACs. For WGS, genomic DNA is sheared into random fragments that are cloned into a plasmid vector and sequenced from both ends. Cloning is not necessary for NGST: genomic DNA is fragmented and sequenced, and sequence reads are eventually mapped to a reference genome. •

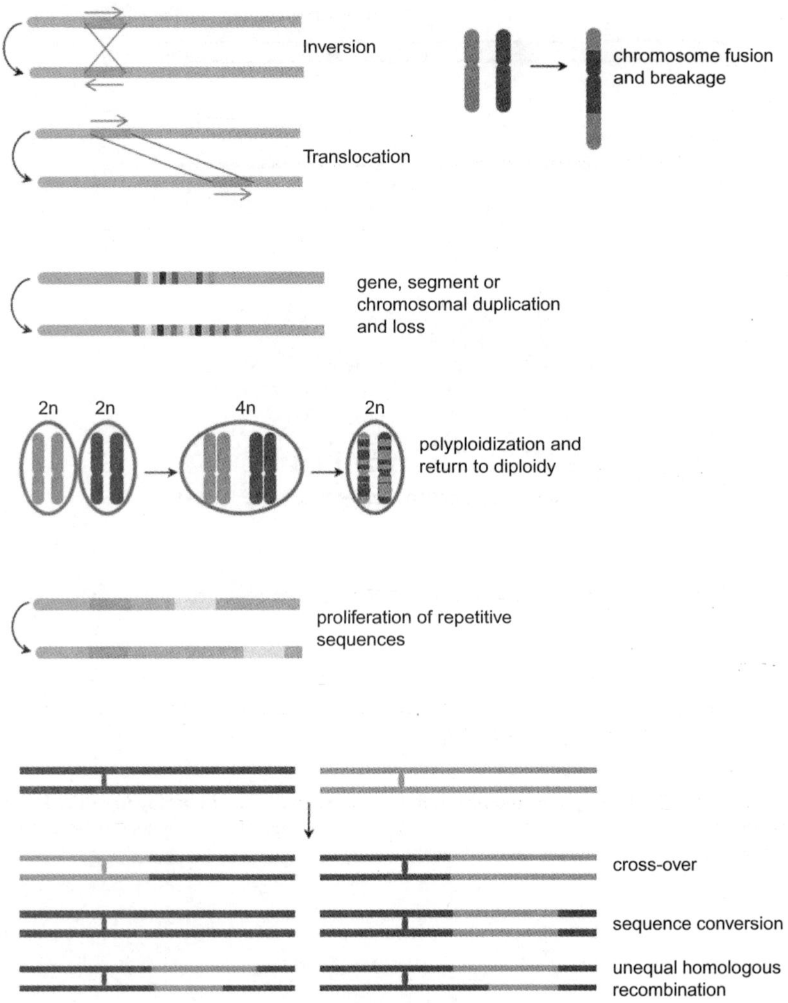

Figure 6.3 • Chromosomal rearrangement processes leading to disruption of synteny. • Disruptions of synteny commonly observed involve translocation, inversion, duplication, and loss of DNA segments. A change of chromosome number between species might be due to chromosome fusion and breakage. A common phenomenon in plant evolution involves an increase of ploidy level followed by gradual loss of DNA and return to diploidy. Some rearrangements involve proliferation of repetitive elements such as transposons. Sequence conversion, unequal homologous recombination, and illegitimate recombination might also modify the synteny between homologous chromosomes.

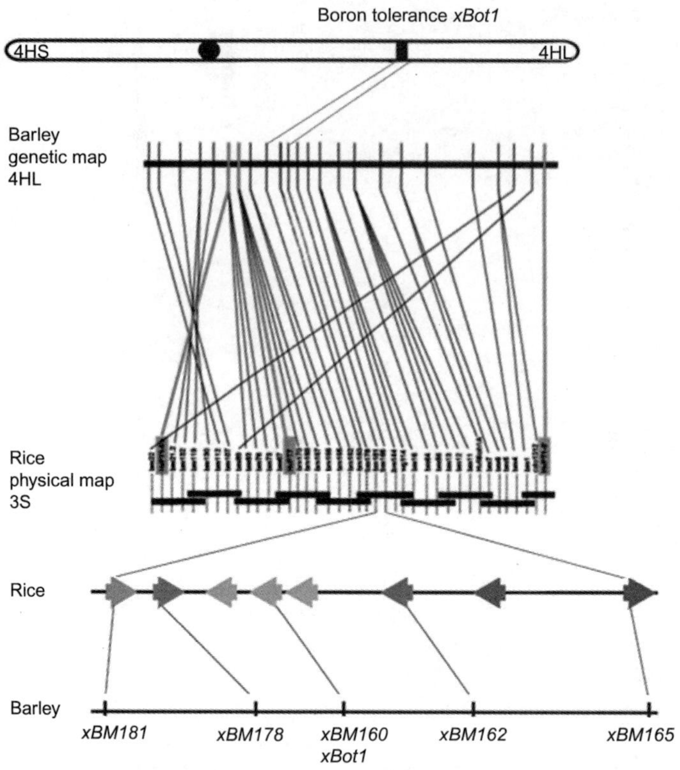

Figure 6.4 • Positional cloning of QTL: Example of the boron tolerance • *Bot1* locus in barley (modified from Sutton et al., 2007). • Synteny between barley chromosome 4HL and rice chromosome 3S enabled the identification of new molecular markers in barley using rice sequences that were collinear to the boron tolerance locus *xBot1*.

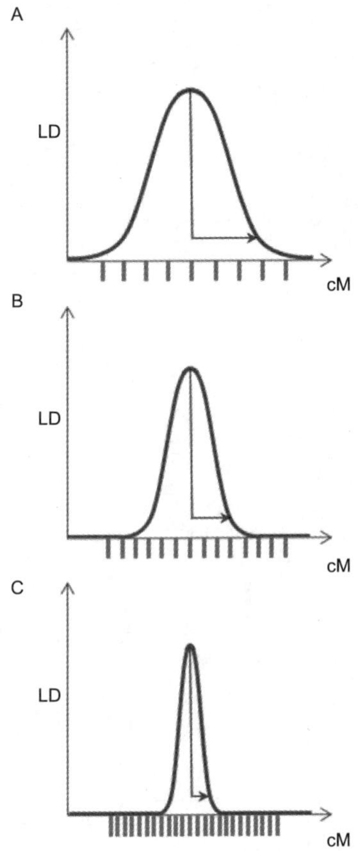

Figure 6.5 • Decay of linkage disequilibrium (LD) with genetic distance along a chromosome. • Red arrows represent the LD extent in centimorgan (cM) in different species (ex: A: wheat and C: maize) and germplasm collections (ex: in barley A: cultivars, B: landraces, C: wild). Green bars represent the density of markers necessary to detect a peak of LD.

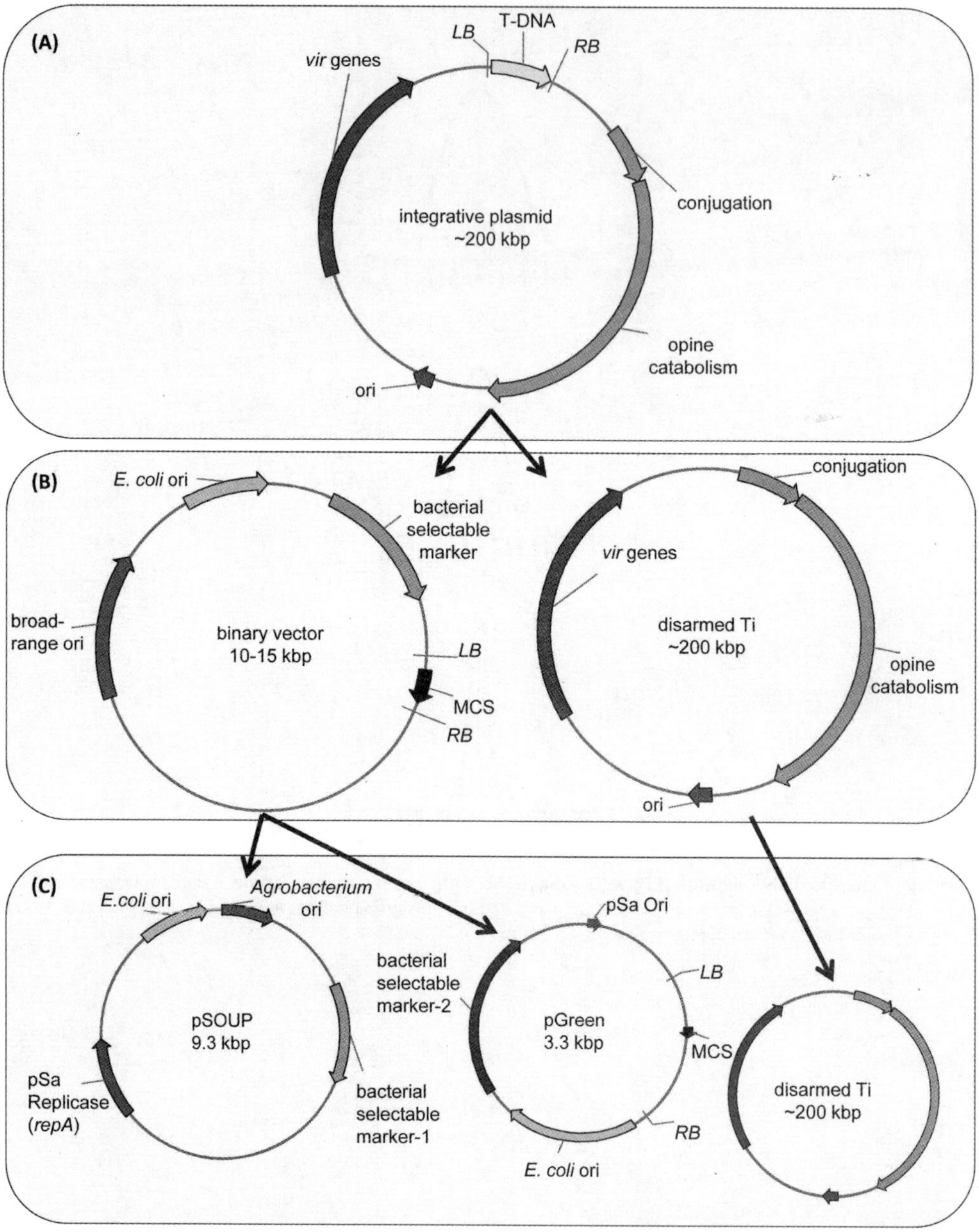

Figure 7.1 • Structure of three types of *Agrobacterium* vector systems. • (A) Integrative Ti plasmid: the native T-DNA region was replaced by recombinant T-DNA and is present in *cis* to the *vir* region. Integrative Ti plasmids also typically carry opine catabolism and conjugation genes. (B) Binary system: composed of a disarmed Ti plasmid (which is essentially similar to an integrative Ti plasmid, only without a T-DNA region) and a small, autonomous binary vector. The binary plasmid is engineered to carry an MCS in which genes of interest can be cloned to produce a T-DNA, an *E. coli* ori, and a bacterial selectable marker. Binary plasmids are significantly smaller than Ti plasmids and are easier to manipulate in *E. coli* cells. A broad-range ori facilitates the use of binary vectors in various *Agrobacterium* strains. (C) Split binary system: composed of a disarmed Ti plasmid and a split binary vector, in which the minimized pGreen vector carries the T-DNA region and the pSoup vector is engineered to support the replication and maintenance of pGreen in *Agrobacterium* cells (the *repA* gene on pSoup is required to support the function of pSa ori on pGreen). Abbreviations: LB, left border; RB, right border; ori, origin of replication.

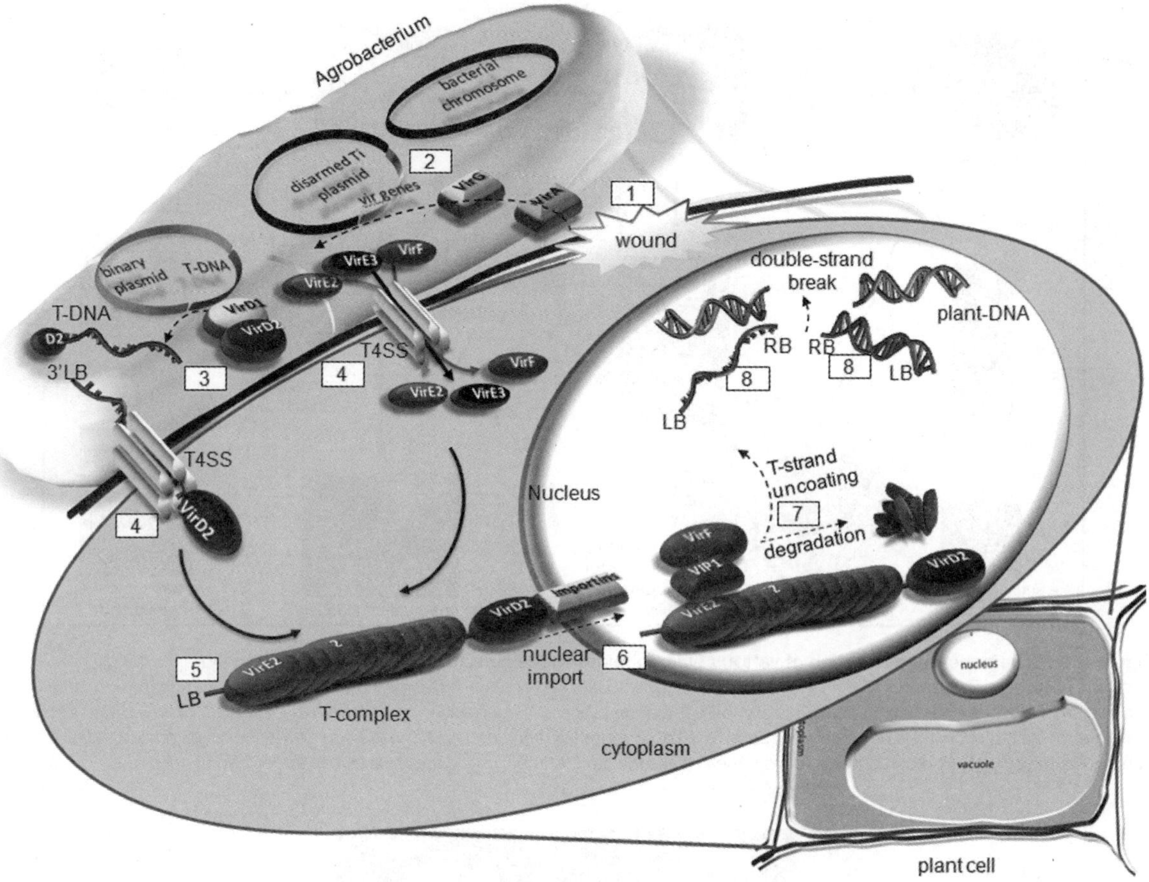

Figure 7.2 • Model for *Agrobacterium*-mediated plant genetic transformation. • A typical engineered *Agrobacterium* cell carries, in addition to its own natural chromosome, a disarmed Ti plasmid and a binary plasmid with a recombinant T-DNA region. We divide the transformation process into eight general steps: (1) sensing of wounded plant cells by the VirA/VirG signal-transduction system; (2) activation of the virulence (*vir*) region and production of virulence proteins; (3) production of immature T-complex; (4) export of immature T-complex and Vir proteins via the T4SS; (5) formation of mature T-complex and its cytoplasmic transport; (6) T-complex nuclear import; (7) uncoating of the T-complex and intranuclear transport; (8) T-DNA integration as a single- or double-stranded intermediate, presumably into genomic DSBs.

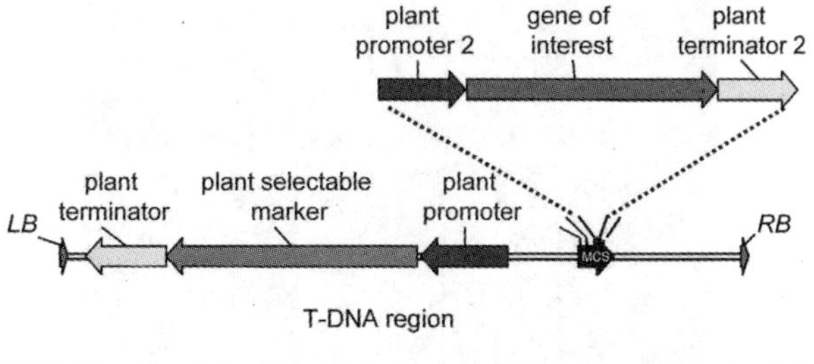

Figure 7.3 • General structure of T-DNA region. • Current practice calls for placing the plant selectable marker expression cassette close to the T-DNA's left border (LB) and constructing a versatile MCS next to the T-DNA's right border (RB). Independent plant expression cassette(s) can next be cloned into the MCS. This structure ensures that a high proportion of the selected plants will carry both the selection marker and the gene(s) of interest.

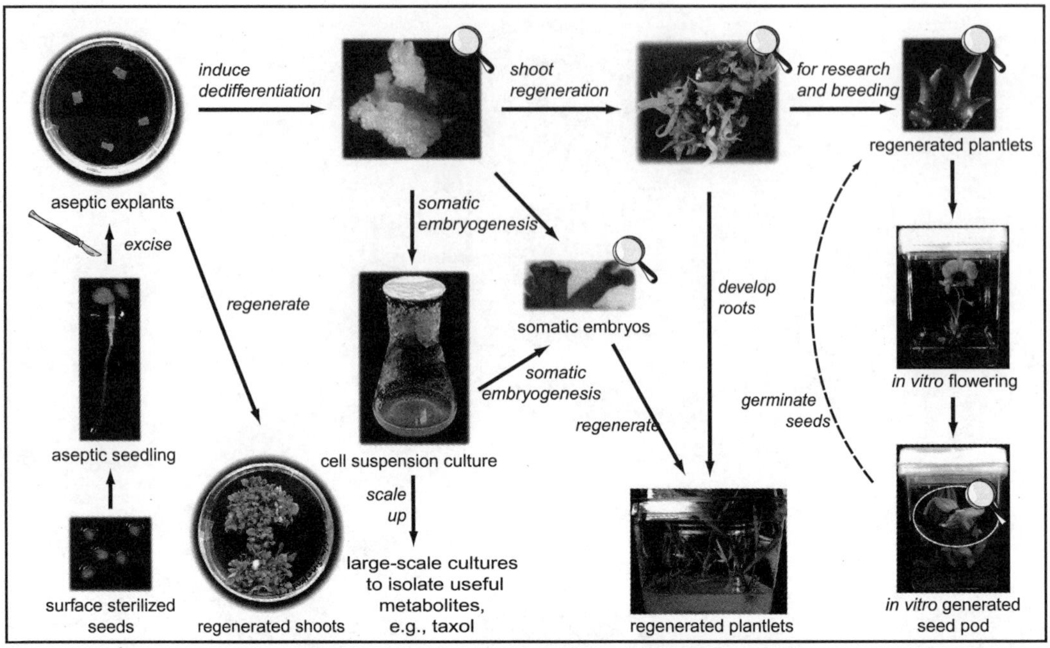

Figure 9.1 • Schematic representation of various types of plant tissue cultures. • Photographs corresponding to the different stages of culture (from the left) leading to plant regeneration from excised leaf segments or establishment of callus and cell suspension cultures with somatic embryogenesis are arranged sequentially to illustrate the process. Regeneration of switchgrass plantlets from callus and orchid plant regeneration via protocorm-like bodies as well as *in vitro* flower induction and seed pod formation in *Dendrobium* sp. are illustrated on the right half of the collage. Photographs with a magnifying lens indicate that they are close-up views of a larger field of culture.

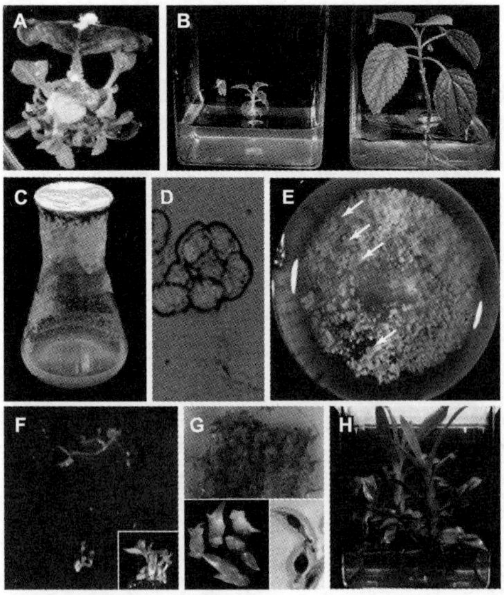

Figure 9.2 • Examples of various types of tissue culture. • (A, B) Adventitious shoot regeneration from excised leaves of *Paulownia fortunei*. (B) Well-developed individual rooted plantlets from the leaf cultures in (A). (C) Cell suspension culture of *Brassica napus* ssp. *oleifera* in liquid medium. (D) Cultured cell clusters from the suspension. (E) Somatic embryogenesis in cell suspension culture. (F) Tobacco anther cultures with germinated haploid embryos. (G) Orchid propagation starting from seeds (inset) and multiplication of protocorm-like bodies resulting from the seed germination *in vitro*. (H) Well-established orchid plantlets arising from the cultures in (G). This method is routinely used for large-scale propagation of various ornamental orchid species.

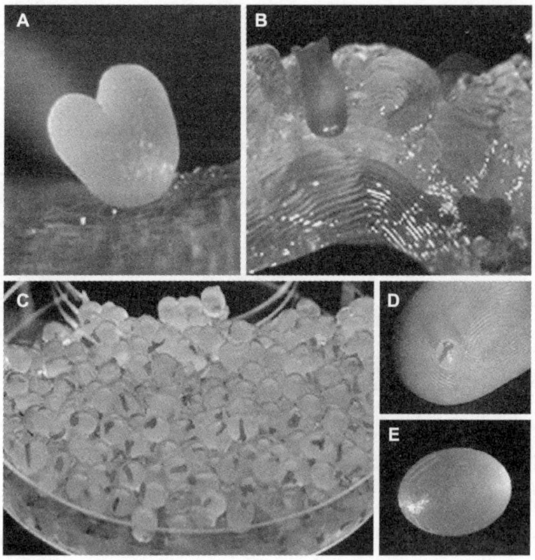

Figure 9.3 • (A) Heart-shaped somatic embryo of *Brassica napus* ssp. *oleifera*. (B) Secondary somatic embryogenesis from the hypocotyls of *B. napus* ssp. *oleifera*. (C, D, E) Artificial seeds prepared by embedding somatic embryos of *B. napus* ssp. *oleifera* in calcium alginate.

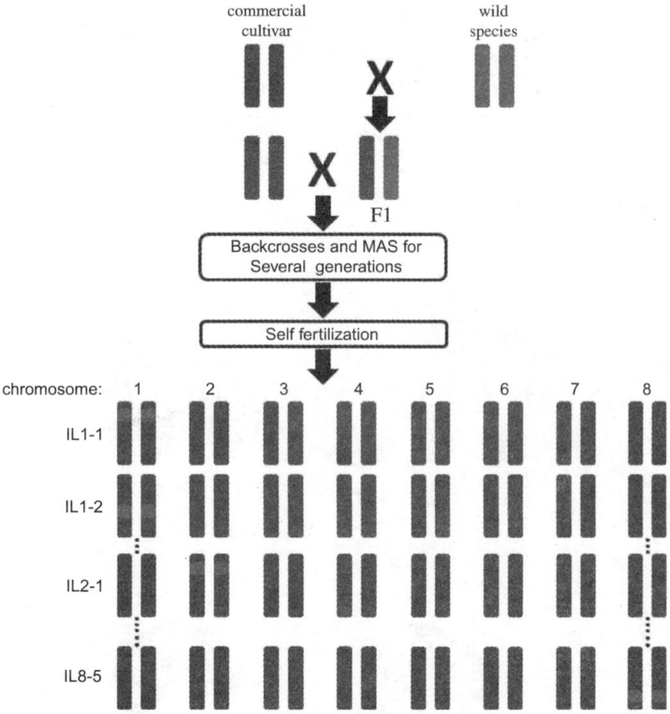

Figure 11.2 • Introgression lines (ILs) library • A wild-type parent (red genome) is crossed with the commercial cultivar (blue genome) to generate the F1 hybrid, which is backcrossed for several generations with the commercial cultivar. The progenies are then selfed for several generations. This procedure results in a series of plants, each of which are carrying a known, different, homozygous segment of the wild species genome. These chromosomal segments are identified by genotyping the library with genome-wide DNA markers distinguishing between the wild and commercial parental strains. The whole IL library covers all of the wild genome. For example, IL1-1 consists of the entire commercial cultivar genome with the exception of a single segment donated by the wild species located at the north of chromosome 1, whereas IL8-5 contains the wild segment on the south of chromosome 8.

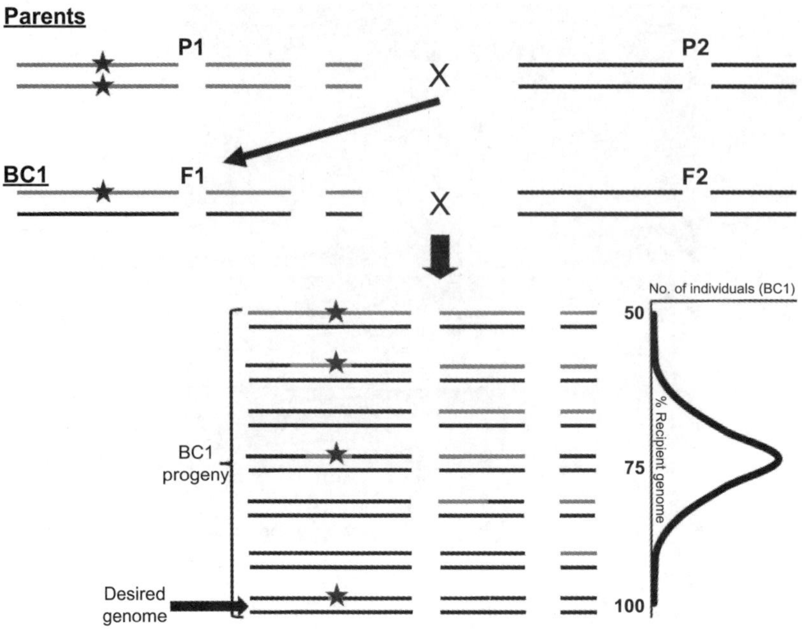

Figure 11.3 • Gene introgression • Carried out by crossing a donor parent (P1; represented by a red line) and a recipient parent (P2; represented by a black line) followed by repeated backcrosses (BC) of the F1 with the recipient parent to recover the recipient genome. The target gene (star) is selected for in each of the backcross generations. The average content of the recipient genome (black and red continuous line) in BC1 progeny is 75%. However, the content varies among the progeny from 50 to 100% and is characterized by a normal distribution curve (right graph). Recovery of the recipient genome can be enhanced by using DNA markers distributed across the entire genome and distinguishing the two parents. Thus, by the proper use of DNA markers one can significantly decrease the number of required back crosses.

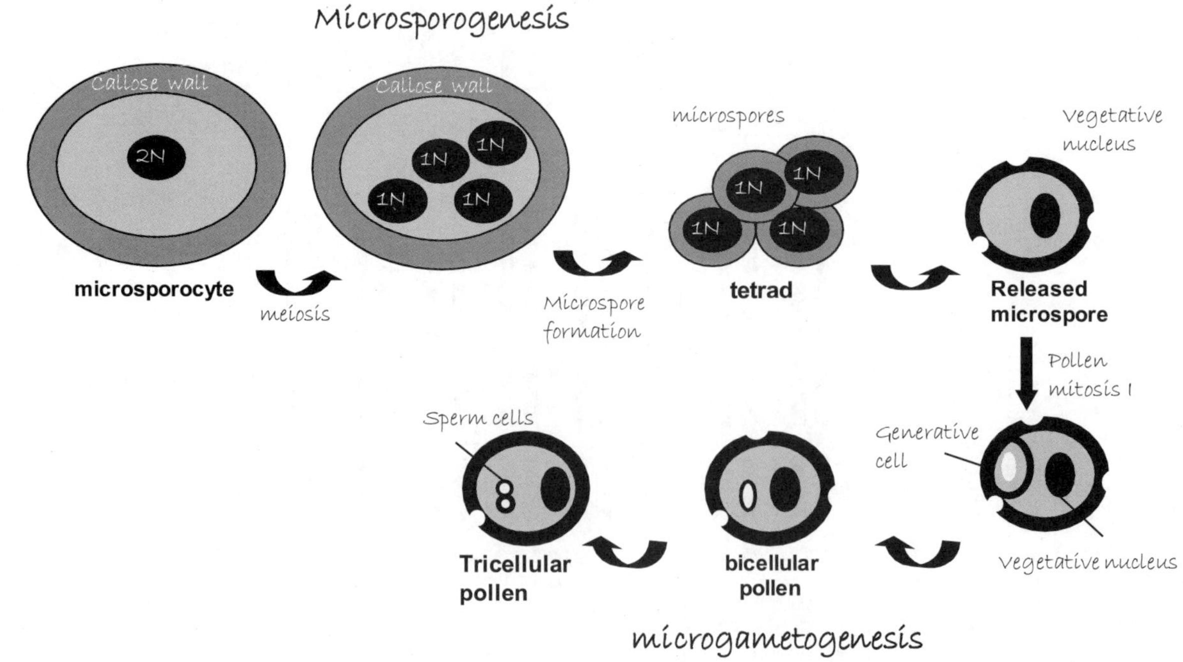

Figure 12.1 • The process of microgametogenesis •

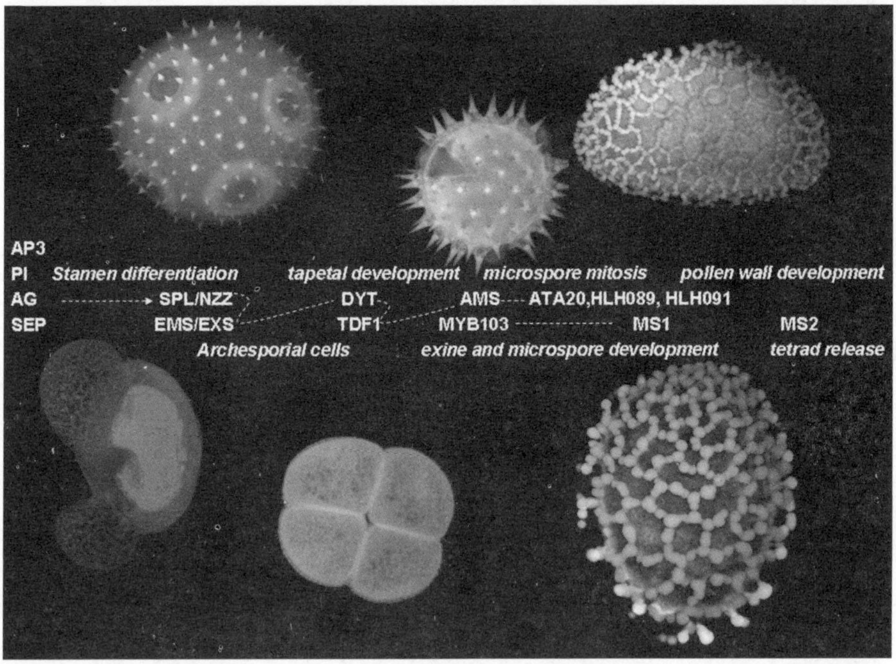

Figure 12.2 • Laser scanning confocal microscopy reveals the extensive diversity and remarkable beauty of plant pollen • Mutant analysis in *Arabidopsis* has allowed the identification of several genetic components regulating the development of these structures, shown here in their predicted functional order relative to the developmental stages at which they act. Mutation at many of these loci results in a male-sterile plant phenotype.

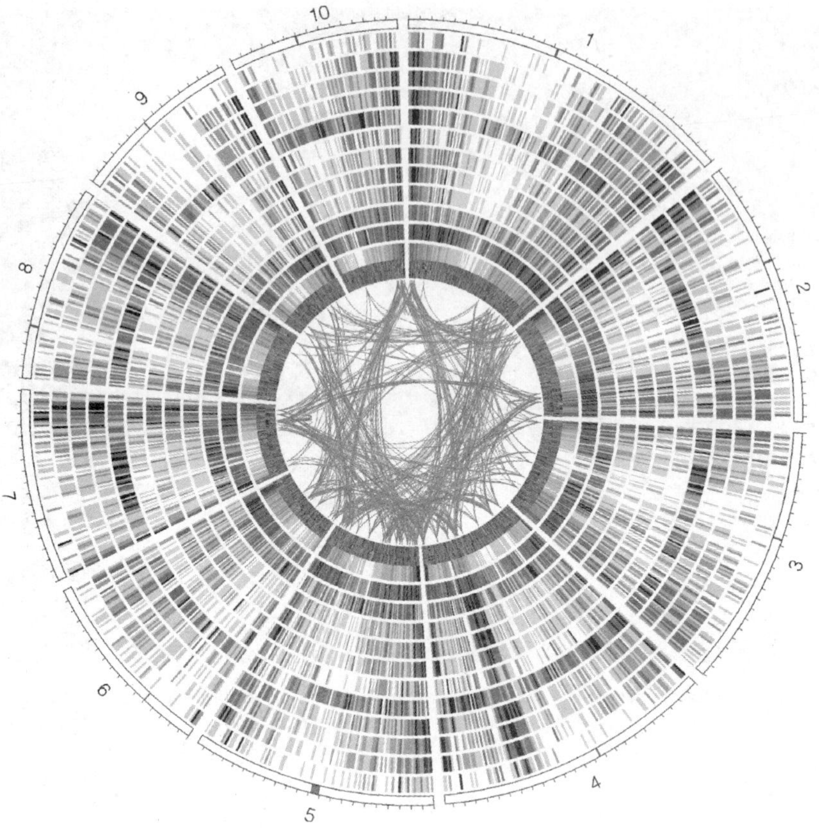

Figure 14.1 • The ten chromosomes of *Zea mays* B73, encircling layers of epigenomic information (red and green tracks), specific retrotransposable elements (blue track), genes (gray track), haplotype SNPs between the B73 and MO17 maize lines (multicolor double-track), and miRNAs (red and green links in the center). All tracks are heat maps of frequencies, from white to the darkest shade of the perspective color, apart from the SNP data — see details in the text. Visualization was created with Circos and in-house PERL scripts.

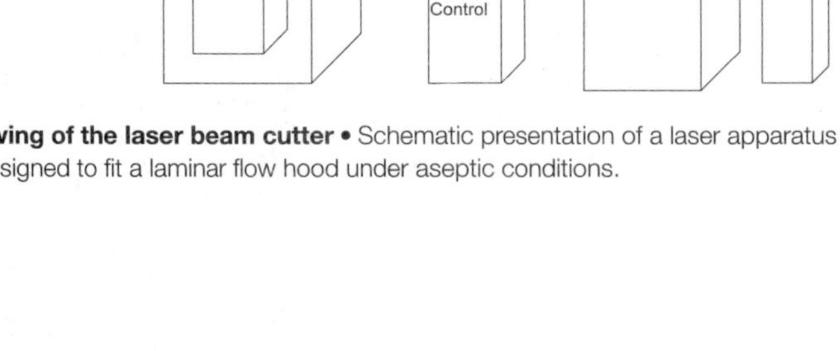

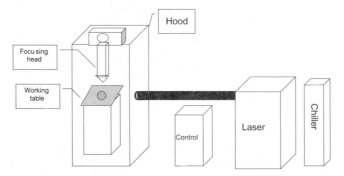

Figure 15.1 • Schematic drawing of the laser beam cutter • Schematic presentation of a laser apparatus for mechanical cutting of meristems. The apparatus is designed to fit a laminar flow hood under aseptic conditions.

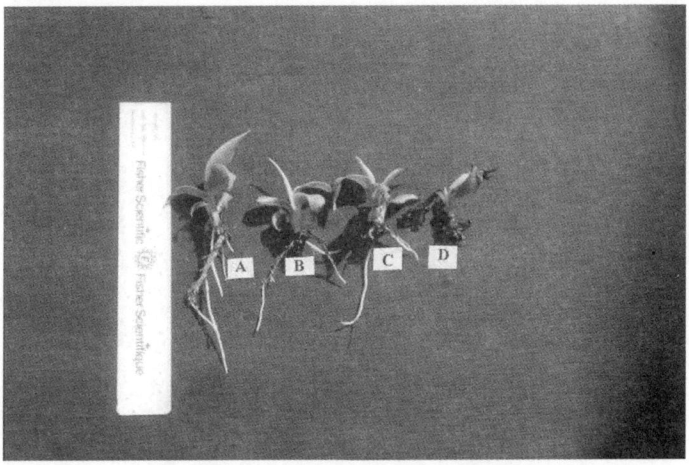

Figure 15.2 • Effect of bacterial strains and *Fusarium oxysporum* on banana plants • (A) control, (B) cocktail combination of bacterial strains and *G. intraradices*, (C) *Glomus intraradices*, and (D) *F. oxysporum*.

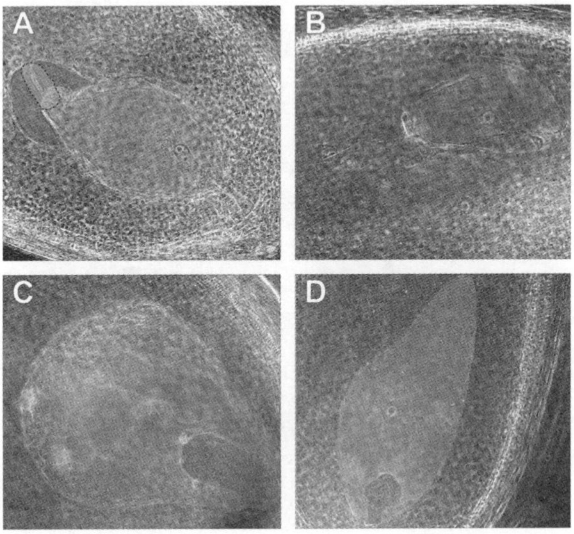

Figure 16.3 • Cleared *Pennisetum* sexual and apomictic ovaries • Ovaries were cleared and mounted using methyl salicylate and photographed using differential interference contrast (DIC) optics. Features of interest are highlighted with color. (A) Sexual *Pennisetum glaucum* ovary with embryo sac consisting of antipodal cells (gray); antipodal cell on a different plane of focus (orange), central cell (yellow) with unfused polar nuclei (blue), and reduced egg and synergid cells (pink). (B) Apomictic *Pennisetum squamulatum* ovary with the formation of multiple aposporous embryo sacs. Central cells (green) with a single 2N polar nucleus (blue) are shown. (C) Developing endosperm (red) and embryo (purple) in an ovary from a BC_8 *P. glaucum* introgression line. (D) Apomictic *P. squamulatum* ovary with parthenogenetic precocious embryo development (purple) in the embryo sac with a uninucleate (blue) central cell (green).

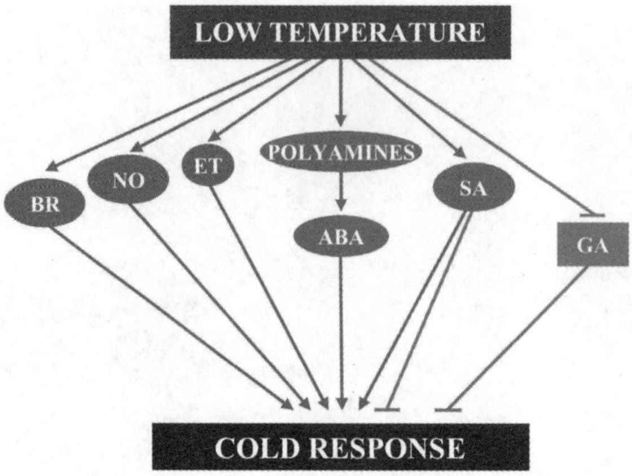

Figure 19.1 • Schematic representation showing the role of plant hormones in low-temperature response. Brassinosteroids (BR), nitric oxide (NO), ethylene (ET), ABA, and polyamines act as positive regulators. Gibberellins (GA) act as negative regulators. Salicylic acid (SA) can act as a positive (cereals, banana) or negative (*Arabidopsis*) regulator. Arrowheads and end lines indicate positive and negative regulation, respectively.

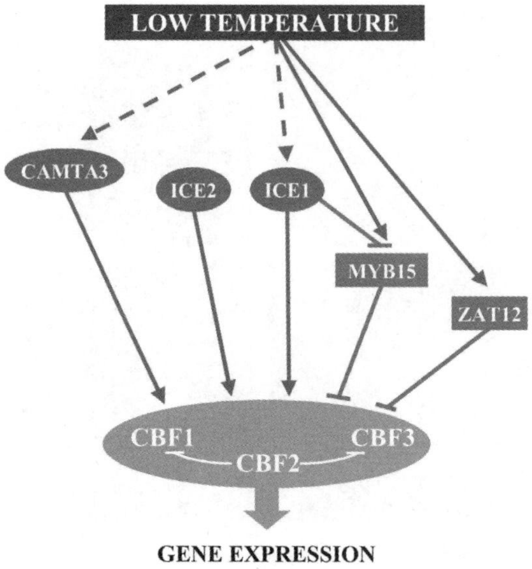

Figure 19.2 • CBFs constitute central intermediates in cold signaling to induce gene expression. ICE1, ICE2, and CAMTA3 positively regulate *CBF* expression. MYB15 and ZAT12 act as negative regulators of *CBFs*. CBF2 negatively regulates the expression of *CBF1* and *CBF3*. ICE1 inhibits *MYB15* expression. Broken arrow means activation by unknown mechanisms. Arrowheads and end lines indicate positive and negative regulation, respectively.

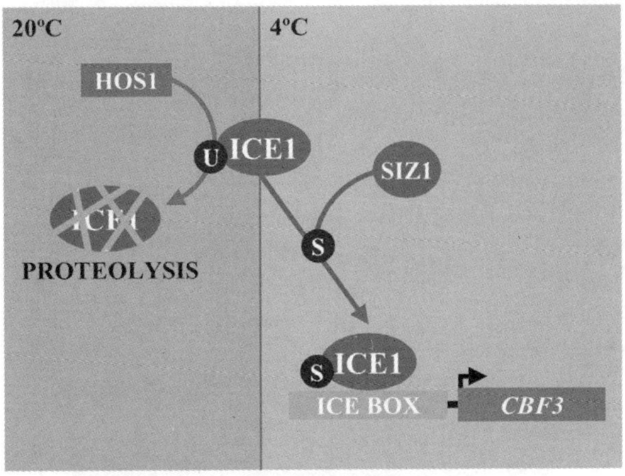

Figure 19.3 • Hypothetical model for post-translational regulation of *CBF3* expression. Ubiquitin (U) and SUMO (S). From Medina et al. (2010).

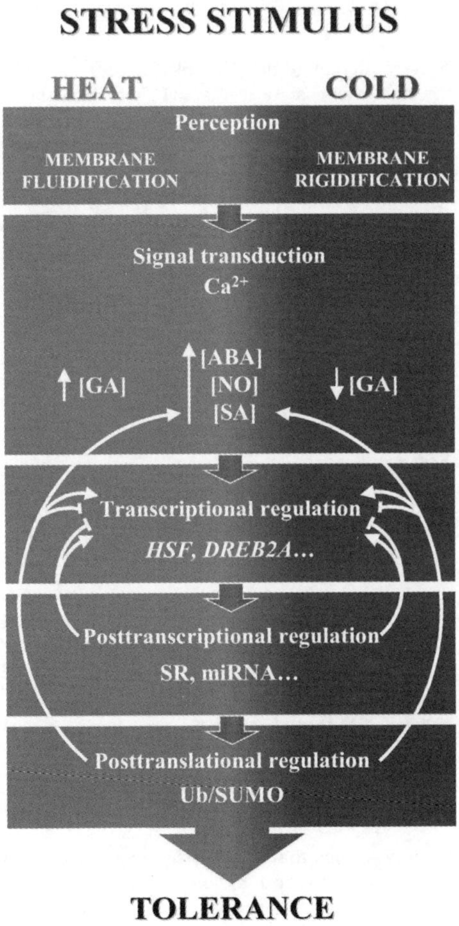

Figure 19.4 • Proposed model for integration of low- and high-temperature signaling pathways. Changes in membrane fluidity are thought to be one of the first steps in the perception of temperature variation. Calcium acts as a second messenger in both low- and high-temperature responses, transducing the signal generated. On the other hand, cold and heat stresses cause an increase in ABA, NO, and SA levels that mediate changes in gene expression and are required for plant tolerance. Gibberellins (GA) are also involved in extreme temperature responses. Finally, plant responses to both low and high temperatures are regulated at the transcriptional, post-transcriptional and post-translational levels. Arrowheads and end lines indicate positive and negative regulation, respectively.

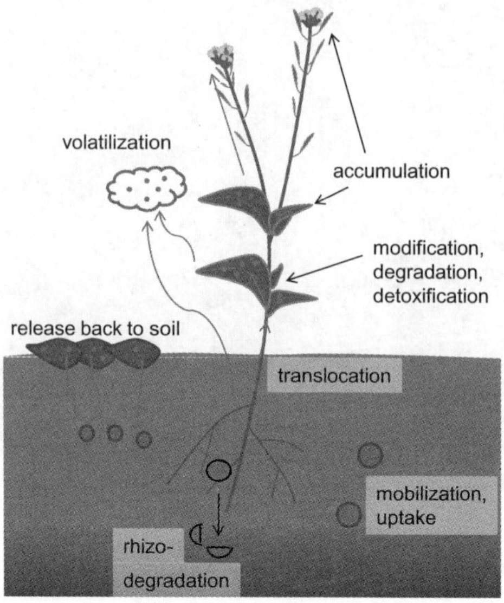

Figure 20.1 • Overview of plant–pollutant interactions and the possible fates of the pollutant (represented as circles) • Degradation of certain organic pollutants in the rhizosphere can be facilitated by root-released compounds. Some root-released compounds can also mobilize pollutants from soil particles and affect uptake. Pollutants can be taken up into the root symplast and translocated via the xylem (apoplast) to the shoot, in the transpiration stream. In the shoot the pollutant can be taken up into the shoot symplast. There it may be further modified (assimilated, degraded, side groups attached, conjugate/chelator attached) and either sequestered in vacuole or cell wall, or volatilized. From leaves pollutants may be remobilized via the phloem to reproductive tissues. Pollutants (or their downstream products) may also be returned to the soil after leaf drop.

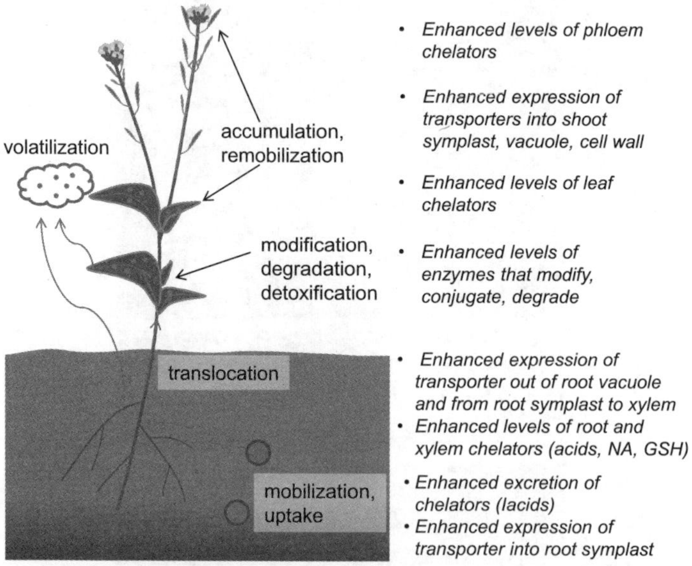

Figure 20.2 • Overview of biotechnological approaches that may enhance various rate-limiting steps in phytoremediation • Excreted compounds may facilitate mobilization, and enhanced expression of transporters in the root cell membrane may facilitate import into the root symplast. Enhanced expression of exporters out of the root vacuole and out of the root symplast into the xylem may facilitate translocation to the shoot. Increased levels of root and xylem chelators (acids, GSH) may enhance plant tolerance to the pollutant and pollutant mobilization in the xylem. Uptake into the leaf symplast may be enhanced by increased expression of transporters in the mesophyll cell membrane. Inside leaf cells, enhanced levels of enzymes that modify, conjugate, or degrade pollutants can facilitate tolerance, degradation, sequestration, or volatilization. Tolerance and sequestration are also enhanced by higher levels of leaf chelators or transporter proteins that export pollutants out of the cytosol and into the vacuole or cell wall. Enhanced levels of phloem chelators may facilitate remobilization to reproductive tissues.

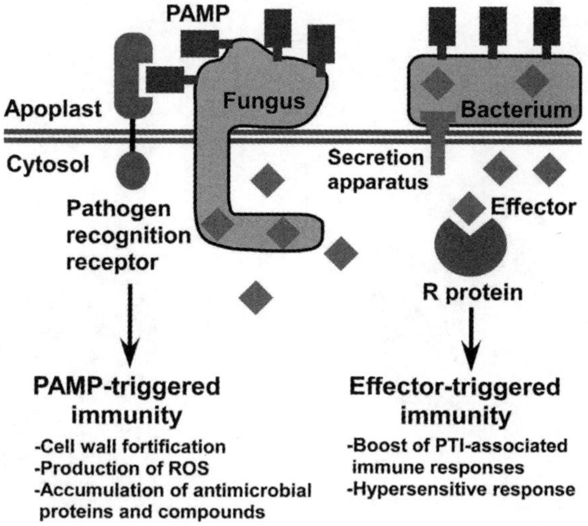

Figure 21.1 • A model illustrating the two main branches of plant immunity • Plant cells detect invasion by a fungal or bacterial microbe through pathogen recognition receptors (PRRs). Most of the PRRs are embedded in the plasma membrane and recognize conserved pathogen determinants, designated as pathogen- or microbe-associated molecular pattern (PAMPs or MAMPs). This recognition event activates a network of signaling cascades resulting in the onset of a first line of immune responses, collectively referred to as PAMP-triggered immunity (PTI). PTI includes fortification of the cell wall, production of reactive oxygen species (ROS), and accumulation of antimicrobial proteins and compounds. A second line of immunity is represented by plant resistance (R) proteins, mostly cytoplasmic, that specifically recognize a pathogen effector protein and mount very efficient immune responses, referred to as effector-triggered immunity (ETI). ETI boosts PTI-associated immune response, often culminating in a localized hypersensitive response (HR) cell death, and inhibits spread of the pathogen to neighboring healthy tissues.

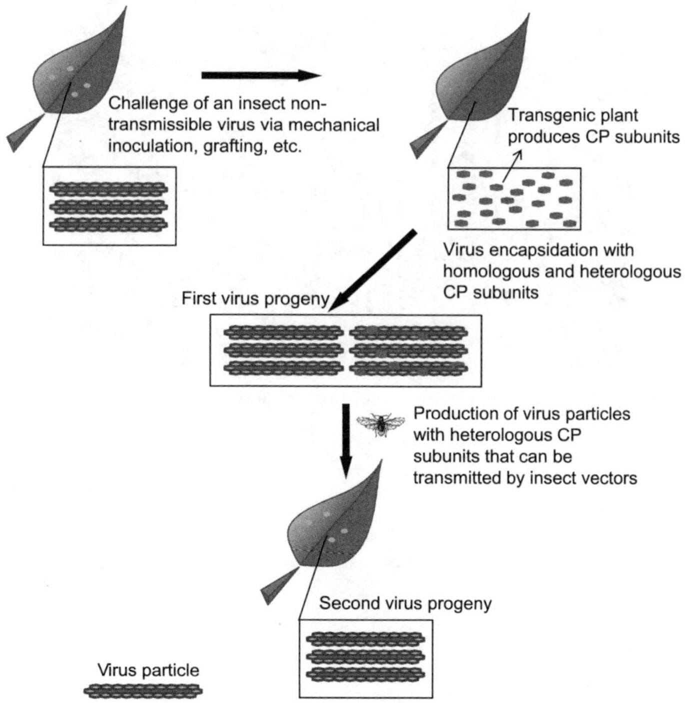

Figure 22.1 • Schematic illustration of heteroencapsidation • A challenge virus, which is non-transmittable by insect vectors, can be transmitted from infected plant to a transgenic plant by mechanical inoculation or grafting. In the transgenic plant, the progeny of the challenge virus can be encapsidated with homologous CP subunits as well as with those heterologous CP subunits encoded by the transgene. Those virus particles with the heterologous CP subunits can be acquired by insect vectors and further transmitted.

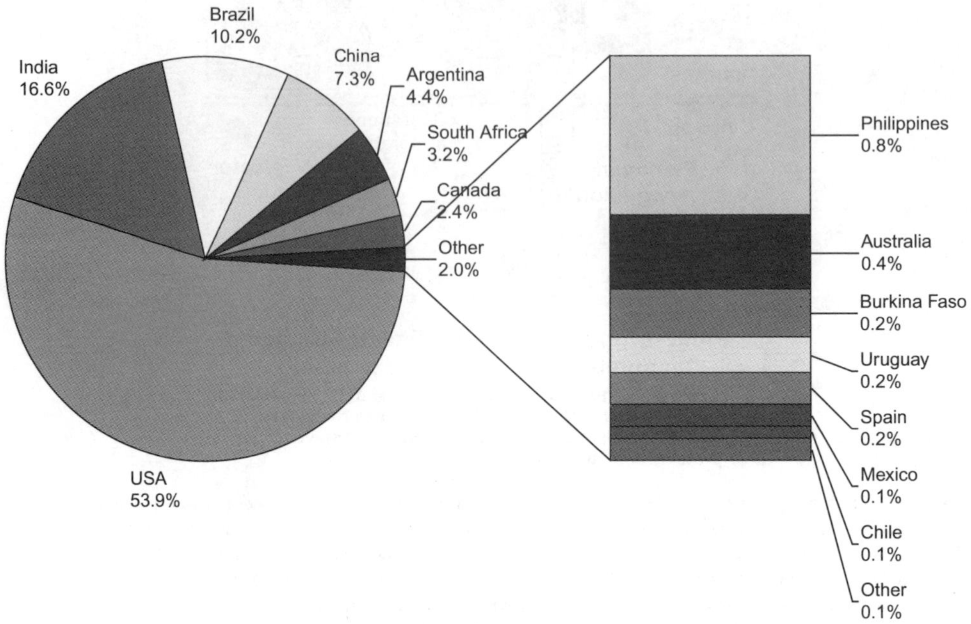

Figure 23.2 • Global distribution of Bt Crops in 2009. (Adapted from James 2009.)

Figure 23.3 • Unsprayed, non-transgenic cotton (left) and unsprayed Bt (Cry 1Ac) cotton (right) prior to harvest. (Photo courtesy of C. Mares.)

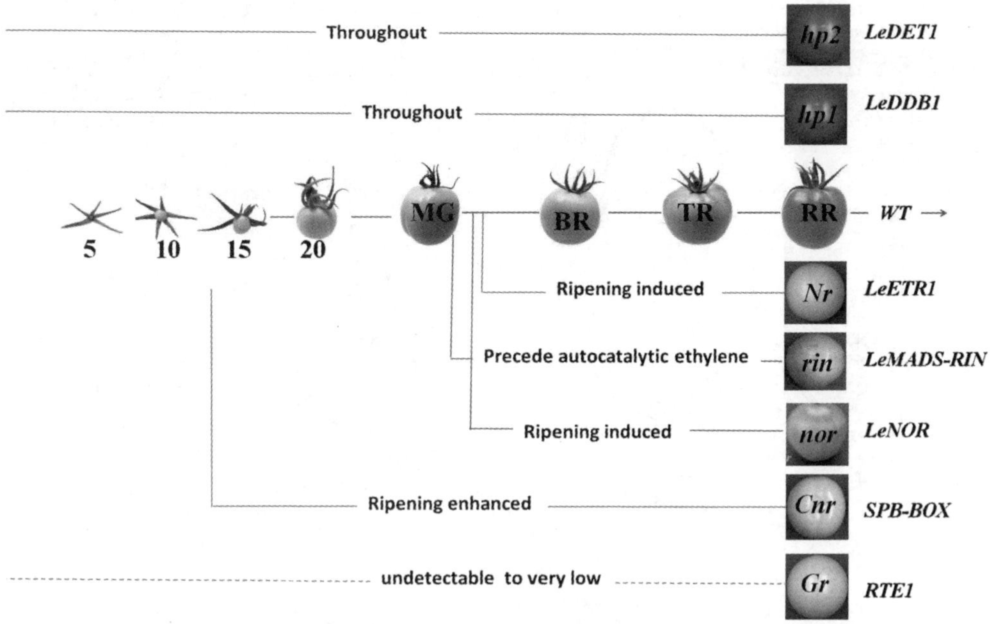

Figure 26.1 • Temporal and developmental regulation of ripening and composition-altering genes in the wild-type fruits • Shown are the developmental (young to fully mature) and ripening stages (BR to RR) of the wild-type (WT) fruits. Also shown are the typical phenotypes of various altered ripening or composition of tomato mutants at a physiological stage comparable to BR + 10 days WT fruits. The lines to the left of each mutant fruit represent the stages of WT fruit exhibiting expression (or lack of expression of RTE1) of ripening and the composition-altering genes in WT fruits. 5, 10, 15, and 20 represent days after anthesis and MG, BR, TU, and RR represent mature green, breaker, turning, and red ripe, respectively. Abbreviations: *hp1,* high pigment 1 (*DDB1, UV-damaged DNA binding*); *hp2,* high pigment 2 (*SlDET1, deetiolated 1*); *Nr, never ripe* (*SlETR1, ethylene response 1)*; *rin*, *ripening inhibitor* (*SlRIN-MADS, RIN-MADS* box-transcription factor); *nor*, *non-ripening* (*nor, nonripening*); *Cnr*, *Colorless non ripening (SlSPB-Cnr,* squamosa promoter binding protein); and *Gr*, *green ripe* (*RTE1, reversion to ethylene sensitivity 1).* Except for the Gr mutation that activated expression of *RTE1 in Gr* mutant, all other shown mutations resulted in loss of functional expression of their respective genes. (Pictures of the various mutant fruits courtesy of Dr. J. Giovannoni.)

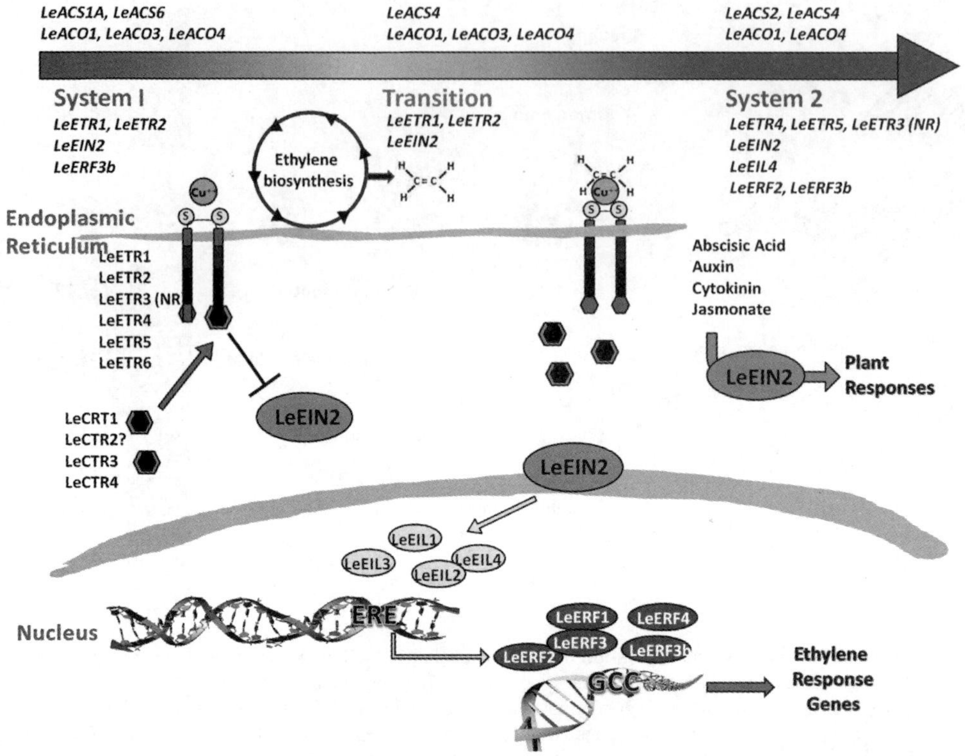

Figure 26.2 • Regulation of ethylene production and perception during tomato fruit ripening • Top panel: The participation of various ACC synthase (*ACS*), ACC oxidase (*ACO*), and ethylene receptor (*ETR*) genes in the System 1 (lower and auto-inhibitory amounts of ethylene), System 2 (autocatalytic ethylene production), and ripening-transition ethylene production and perception during fruit ripening. Middle panel: ETRs that are localized in the endoplasmic reticulum form disulfide bridges (yellow circles) and dimers interact with copper ion (green circle) to bind ethylene. With the exception of NR, ETRs have four different domains depicted in different colors (green, sensor; gray, GAF; black, histidine kinase; and red, response regulator or receiver domains, respectively. In the absence of ethylene, constitutive triple response (CTR) protein binds to an ETR. This complex interacts with an ETHYLENE INSENSITIVE2 (EIN2) protein to negatively regulate ethylene response. Ethylene binding to ETR inhibits its interaction with CTR resulting in releasing inhibition on EIN2 activity. Other phytohormones may also use EIN2 to transduce their response. Lower Panel: Free EIN2 moves to nuclear membranes where it activates transacting protein (ETHYLENE INSENSITIVE3) EIN3 and EIN3-like proteins (EIL) by an unknown mechanism. This facilitates binding of EIN3 and EILs to ethylene response elements (ERE) present in the promoter regions of ethylene response factors (ERFs). This cascade activates expression ERFs, which in turn enables expression of ethylene responsive genes.

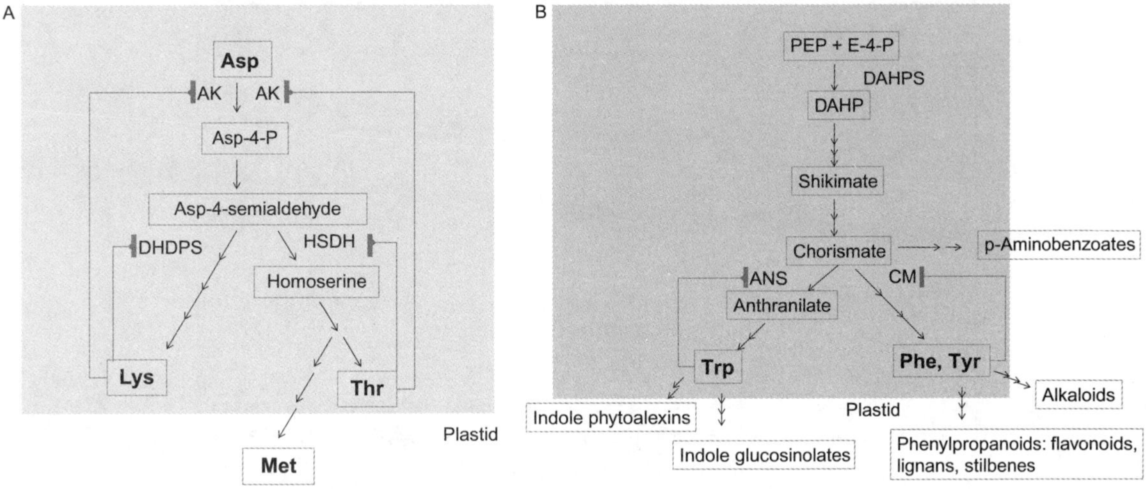

Figure 28.3 • Amino acid biosynthetic pathways and feedback regulation • Simplified pathways for lysine (Lys; A) and aromatic amino acids (B) such as tryptophan (Trp), phenylalanine (Phe), and tyrosine (Tyr) are shown. Red bars show the feedback inhibition of enzymes in the rate-limiting step by products such as Lys and Trp. A green background indicates that these reactions occur in plastids. Abbreviations: Asp, aspartate; AK, aspartate kinase; Asp-4-P, aspartate-4-phosphate; DHDPS, dihydrodipicolinate synthase; HSDH, homoserine dehydrogenase; Met, methionine; Thr, threonine; PEP, phosphoenol pyruvate; E-4-P, erythrose-4-phosphate; DAHP, 3-deoxy-D-arabino-heptulosonate-7-phosphate; DAHPS, DAHP synthase; ANS, anthranilate synthase; and CM, chorismate mutase.

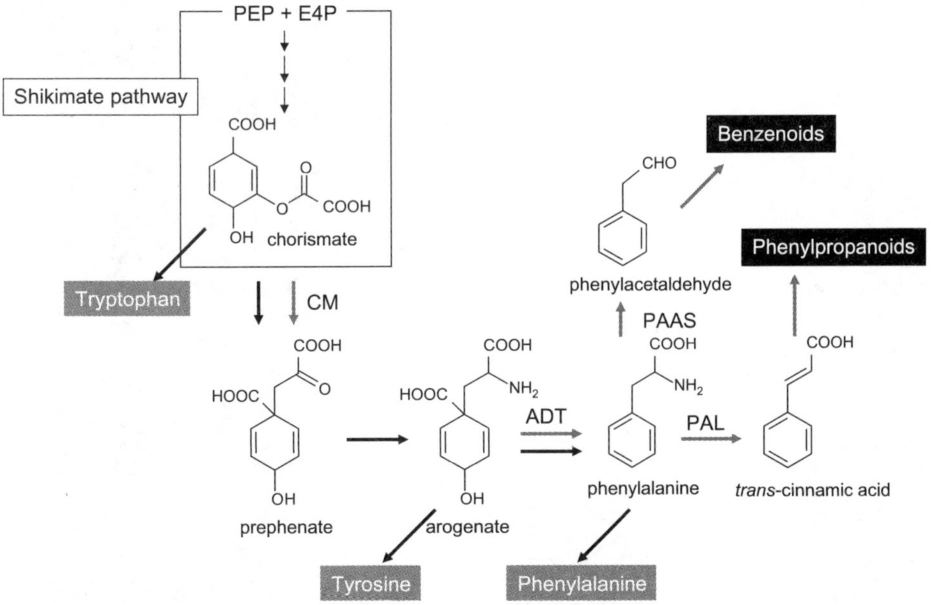

Figure 28.7 • Phenylalanine is formed specifically for benzenoids/phenylpropanoids in *P. hybrida* petals • For most of the biosynthetic steps downstream of chorismate, there is a special set of enzymes dedicated to benzenoid/phenylpropanoid formation (red arrows). This specialization offers a new strategy to modify volatile secondary metabolism. Abbreviations: CM, chorismate mutase; ADT, arogenate dehydratase; PAL, phenylalanine ammonia lyase; and PAAS, phenylacetaldehyde synthase.

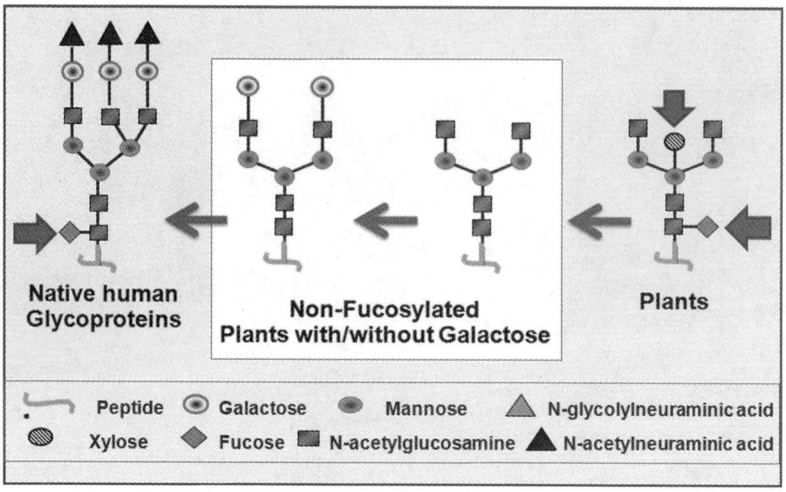

Figure 29.1 • Plant glycan engineering • Glycan structures of human and plant secreted proteins. Red arrows indicate human-specific (α-1,6-linked fucose) and plant-specific (α-1,3- linked fucose and β-1,2-linked xylose) sugars. Pink arrows indicate glycoengineering of plants through RNAi silencing of specific plant genes, mutagenesis, and the addition of new glycosyl-transferases.

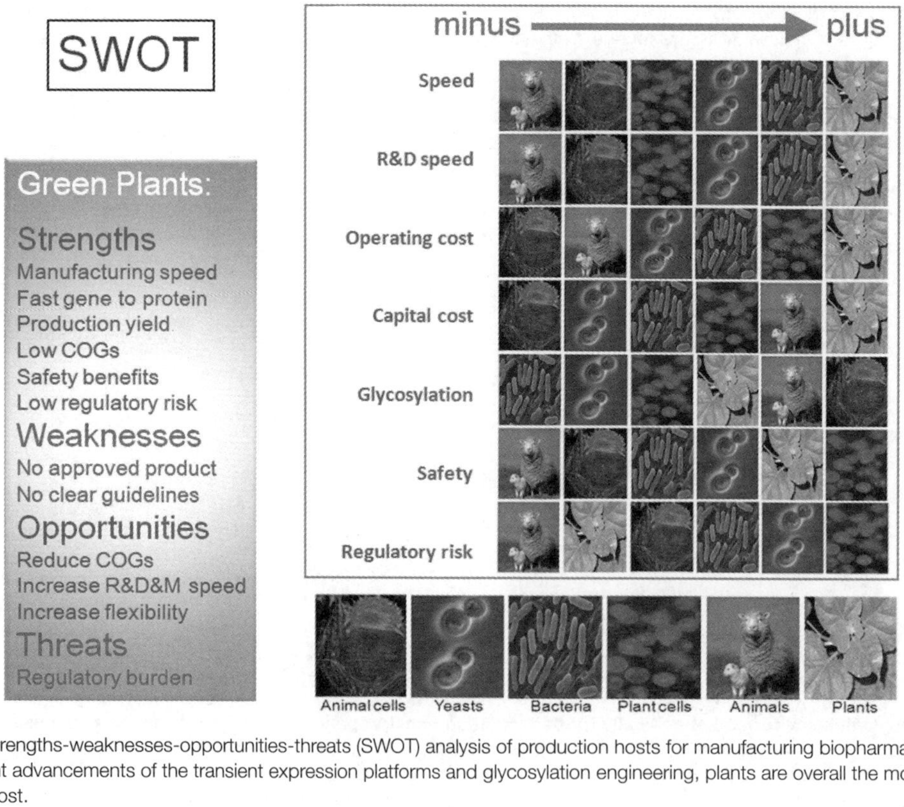

Figure 29.2 • Strengths-weaknesses-opportunities-threats (SWOT) analysis of production hosts for manufacturing biopharmaceuticals. Due to the recent advancements of the transient expression platforms and glycosylation engineering, plants are overall the most attractive manufacturing host.

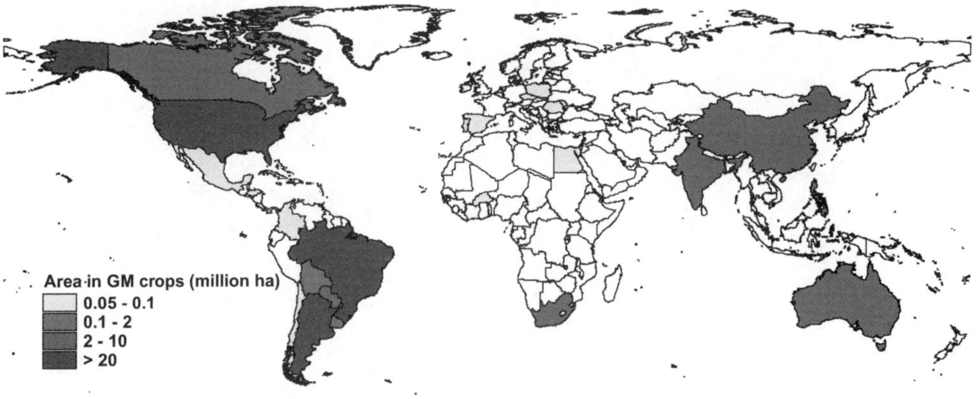

Figure 36.1 • Countries growing GM crops • From ISAAA (James, 2010).

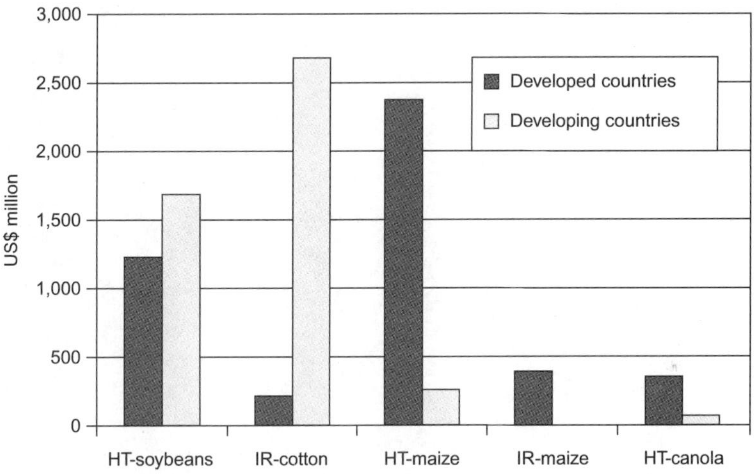

Figure 36.2 • Increases of global farm level income from GM technologies • From Brookes and Barfoot (2010).